CW00371457

eps hep 97 Jerusalem

Springer
Berlin
Heidelberg
New York
Barcelona
Hong Kong
London
Milan
Paris
Singapore
Tokyo

Daniel Lellouch Giora Mikenberg
Eliezer Rabinovici (Eds.)

International Europhysics Conference on High Energy Physics

Proceedings of the
International Europhysics Conference
on High Energy Physics
Held at Jerusalem, Israel, 19–25 August 1997

Daniel Lellouch
Giora Mikenberg
Department of Particle Physics
The Weizmann Institute of Science
76100 Rehovot, Israel

Eliezer Rabinovici
The Racah Institute of Physics
Hebrew University
91904 Jerusalem, Israel

Library of Congress Cataloging-in-Publication Data

International Europhysics Conference on High Energy Physics (1997: Jerusalem, Israel)
International Europhysics Conference on High Energy Physics: proceedings of the International Europhysics Conference on High Energy Physics held at Jerusalem, Israel, 19–25 August 1997 / Daniel Lellouch, Giora Mikenberg, Eliezer Rabinovici (eds.).
p. cm. Includes bibliographical references.
ISBN 3-540-64970-0 (hardcover: alk. paper)
1. Particles (Nuclear physics) - Congresses. I. Lellouch, Daniel, 1957- II. Mikenberg, Giora, 1947-
III. Rabinovici, Eliezer, 1946- IV. Title.
QC793.I5544 1997 539.7'2–ddc21 98-37586 CIP

ISBN 3-540-64970-0 Springer-Verlag Berlin Heidelberg New York

Typesetting: Camera-ready by authors/editors
Cover design: *design & production* GmbH, Heidelberg

SPIN 10676471 55/3144 - 5 4 3 2 1 0 - Printed on acid-free paper

Foreword

The 1997 International Europhysics Conference on High Energy Physics was held at the campus of the Hebrew University of Jerusalem and at the Jerusalem Renaissance Hotel, from August 19th to August 25th, 1997. This was the first time that the European Physical Society had its High Energy Physics Conference outside the boundary of Europe. A total of 550 physicists participated in the conference with a total of 250 presentations in the parallel sessions and 26 presentations in the plenary sessions.

The Board of the of the High Energy and Particle Physics division (HEPP) of the EPS acted as the Scientific Organizing Committee. The Board acknowledges the help of the International Advisory Committee as well as that of the Local Organizing Committee.

The conference was co-organized by the Hebrew University of Jerusalem and by the Weizmann Institute of Science, with important help by physicists from the Israeli Institute of Technology (Technion) and the Tel Aviv University.

A large number of new experimental results were presented, ranging from very precise measurements of the weak parameters (Z and W couplings and masses) to new results in High Energy QCD from HERA, as well as evidence for the existence of Glueballs. Heavy Quark properties (Top and Bottom) led to a large variety of new results. Some of the highlights of the conference were the new results from Super-Kamiokande, showing a clear hint for the possibility of neutrino oscillations, the first observation of a two-neutrino decay of the K and the excess of high Q^2 events seen at HERA. These subjects were the main issue of many discussions.

On the theoretical side CP violation and heavy quark systems and outstanding issues in QCD were described. The status of research on the possible extensions of the Standard Model was reported. The new dialog between String Theory and Super-Symmetric Gauge Theories was shown to lead to insights in non-perturbative aspects of Low Energy Field Theories. Experimental and theoretical results in cosmology were reviewed.

As it is traditional at this conference series, the plenary sessions were preceeded by the presentation of the 1997 EPS prize in Particle Physics by the EPS President, Denis Weare. The prize was awarded to Robert Brout, Francois Englert and Peter Higgs (not present), for work centered on the idea of spontaneous symmetry breakdown. The conference also coincided with the 100 year celebration of the discovery of the electron, which was featured in a public lecture by Jack Steinberger.

Finally, many thanks are due to the technical personnel from the Hebrew University and the Weizmann Institute. Their effort is strongly appreciated by the Local and International Organizing Committees.

Jerusalem, July 1998

Local Organizing Committee:

Daniel Lellouch (*Scientific Secretary*)
Giora Mikenberg (*Co-Chairman*)
Eliezer Rabinovici (*Chairman*)

INTERNATIONAL ADVISORY COMMITTEE

W. Bardeen, *FERMILAB*
H. Chen, *Beijing*
E. Fernandez, *Barcelona*
J. Haissinkski, *Saclay*
H. Harari, *Weizmann*
C. Jarlskog, *Lund*
G. Jarlskog, **Chairman**, *Lund*
C.H. Llewellyn-Smith, *CERN*
L. Maiani, *INFN*
G. van Middelkoop, *NIKHEF*
D. Miller, *West Lafayette*
J. Peoples, *FERMILAB*
F.M. Renard, *Montpellier*
B. Richter, *SLAC*
D. Saxon, *Glasgow*
A. Skrinsky, *Novosibirsk*
R. Sosnowski, *Warsaw*
H. Sugawara, *KEK*
M.A. Virasoro, *ICTP*
B.H. Wiik, *DESY*

INTERNATIONAL ORGANIZING COMMITTEE

J.J. Aubert, *Marseille*
M. Baldo-Ceolin, *Padua*
F. Barreiro, *Madrid*
Ch. Berger, *Aachen*
F. Berends, *Leiden*
A. Bialas, *Krakow*
M. Davier, *Orsay*
L. Evans, *CERN*
E. Fernandez, *Barcelona*
M. Greco, *Rome and Frascati*
G. Jarlskog, *Lund*
P. Kalmus, *QMW*
W. Kummer, **Chairman**, *Vienna*
J. Lemonne, *Brussels*
G. Mikenberg, *Weizmann*
J. Niederle, *Prague*
R. Orava, *Helsinki*
J. Pisut, *Bratislava*
E. Rabinovici, *Jerusalem*
G. Ross, *Oxford*
J. Wess, *Munchen*
P. Zerwas, *DESY*

LOCAL ORGANIZING COMMITTEE

H. Abramowicz, *Tel Aviv*
G. Alexander, *Tel Aviv*
A. Davidson, *Ben Gurion*
A. Dar, *Technion*
S. Elitzur, *Jerusalem*
M. El-Fiki, *NIS, Cairo*
Y. Frishman, *Weizmann*
A. Giveon, *Jerusalem*
E. Gross, *Weizmann*
J. Goldberg, *Technion*
U. Karshon, *Weizmann*
M. Karliner, *Tel Aviv*
D. Lellouch, **Sc. Secretary** *Weizmann*
G. Mikenberg, **Co-Chairman** *Weizmann*
Y. Nir, *Weizmann*
E. Rabinovici, **Chairman** *Jerusalem*
P. Singer, *Technion*
S. Solomon, *Jerusalem*
S. Tarem, *Technion*
S. Yankielowicz, *Tel Aviv*

SPONSORS

The Hebrew University of Jerusalem
Technion, Israel Institute of Technology
The Tel-Aviv University
The Weizmann Insitute of Science
The European Union, DG-XII
The Israeli Academy of Science
The City of Jerusalem
The Jerusalem Fund
The Israeli Ministry of Foreign Affairs
The Israeli Ministry of Science
The Goldschleger Conference Foundation
The Einstein Center of Theoretical Physics
The N. Cummings Chair of Particle Physics
The Israeli HEP Committee

Parallel Sessions and convenors

1. Hadron Spectroscopy:
   Pavel Chliapnikov, *IHEP Serpukhov*; Stephan Narison, *Montpellier.*
2. Soft interactions: Martin Erdmann, *Heidelberg*; Leonid Frankfurt, *Tel-Aviv.*
3. Structure functions and low x physics (also 313):
   Rudiger Voss, *CERN*; Paul Kooijman, *NIKHEF*; Wilfried Buchmuller, *DESY.*
4. Hard processes and perturbative QCD:
   Steve Geer, *Fermilab*; Erwin Mirkes, *Karlsruhe.*
5. Production and decays of heavy flavours:
   Shlomit Tarem, *Technion*; Ikaros Bigi, *Notre Dame.*
6. High energy nuclear interactions:
   John Sullivan, *LANL*; Kari J. Eskola, *CERN.*
7. Precision tests of the Standard Model:
   Joachim Mnich, *CERN*; Ramon Miquel, *Universitat de Barcelona*; Bernd Kniehl, *MPI Munich*
8. W-boson physics: Rasmus Moller, *Niels Bohr Inst.*; Wim Beenakker, *Leiden.*
9. Quark masses and mixing, CP violation and rare phenomena:
   Ludwig Tauscher, *Basel*; Michael Gronau, *Technion.*
10. Neutrino physics:
    Franz v.Feilitzsch, *Munich*; Leslie L. Camilleri, *CERN*; Rabindra Mohapatra, *Maryland.*
11. Physics beyond the Standard Model:
    Carlos Wagner, *CERN* Antonio Masiero, *Perugia and SISSA.*
12. Phenomenology of GUTS, Supergravity and Superstring developments:
    Costas Kounnas, *CERN and ENS Paris*; Carlos Munoz, *Madrid.*
13. Search for new particles at present colliders (also 313):
    Eilam Gross, *Weizmann*; Lee Sawyer, *Louisiana Tech University.*
14. Astroparticle physics and cosmology:
    Pierre Bareyre, *Saclay and College de France*; Edward W. Kolb,*Fermilab.*
15. Lattice calculations and non-perturbative methods:
    Rainer Sommer, *DESY Zeuthen & CERN*; Michael Teper, *Oxford.*
16. New developments in field theory and string theory:
    Constantin Bachas, *Ecole Polytechnique*; Amit Giveon, *Jerusalem.*
17. New detectors and experimental techniques:
    Walter Van Doninck, *Brussels*; Michele Livan, *Pavia.*
18. Physics at future machines: Fabiola Gianotti, *CERN*; Michael Dittmar, *Zurich*; Jonathan Dorfan, *SLAC*; Abdelhak Djouadi, *Montpellier.*

Table of Contents

PLENARY SESSIONS

PARALLEL SESSIONS

Part I. Hadron Spectroscopy

Part II. Soft Interactions

Part III. Structure Functions and Low-x Physics

Part IV. Hard Processes and Perturbative QCD

Part V. Production and Decays of Heavy Flavours

Part VI. High Energy Nuclear Interactions and Heavy Ion Collisions

Part VII. Precision Tests of the Standard Model

Part VIII. W-Boson Physics

Part IX. Quark Masses and Mixing, *CP* Violation and Rare Phenomena

Part X. Neutrino Physics

Part XI. Physics Beyond the Standard Model

Part XII. Phenomenology of GUTs, Supergravity and Superstrings

Part XV. Lattice Calculations and Non-pertubative Methods

Part XVI. New Developments in Field Theory and String Theories

Part XVII. New Detectors and Experimental Techniques

Part XVIII. Physics at Future Machines

Part XIX. High Q^2 Events at HERA

List of Participants

S. ABDULLIN (ITEP Moscow, Russia)
abdullin@mail.cern.ch
H. ABRAMOWICZ (Tel Aviv, Israel)

D. ACOSTA (Ohio State U., USA)
acosta@mps.ohio-state.edu
R. ALEKSAN (CEA Saclay, France)
aleksan@hep.saclay.cea.fr
A. ALEKSEJEVS (NRC Salaspils, Latvia)
nrc@mailbox.neonet.lv
G. ALEMANNI (Lausanne, Switzerland)
gilberto.alemanni@ipn.unil.ch
M. ALTMANN (TU Munich, Germany)
altmann@e15.physik.tu-muenchen.de
B. ARMBRUSTER (Karlsruhe, Germany)
armbruster@ik1.fzk.de
R. ARNOLD (Strasbourg, France)
roger.arnold@ires.in2p3.fr
V. ARTEMOV (JINR Dubna, Russia)
artemov@nu.jinr.ru
S. ASAI (Tokyo, Japan)
Shoji.Asai@cern.ch
C. BACHAS (Polytechnique, France)
bachas@cpt.polytechnique.fr
J. BAEHR (DESY Zeuthen, Germany)
baehr@ifh.de
P. BAMBADE (LAL Orsay, France)
bambade@lal.in2p3.fr
S. BANERJEE (Tata Inst., India)
sunanda.banerjee@cern.ch
M. BANNER (LPNHE Paris 6/7, France)
banner@lpnax1.in2p3.fr
Z. BAR YAM (Massachusetts Darmouth, USA)
zbaryam@umassd.edu
S. BAR-SHALOM (Riverside, USA)
shaouly@phyun0.ucr.edu
P. BAREYRE (CEA Saclay, France)
bareyre@hep.saclay.cea.fr
S. BARKANOVA (NRC Salaspils, Latvia)
brzs@lanet.lv
E. BARRELET (LPNHE Paris 6/7, France)
barrelet@in2p3.fr

J. BARTELSKI *(Warsaw, Poland)*
janbart@fuw.edu.pl
U. BASSLER *(LPNHE Paris 6/7, France)*
bassler@mail.desy.de
G. BAUM *(Bielefeld, Germany)*
Baum@Physik.uni-bielefeld.de
U. BECKER *(MIT, USA)*
becker@mitlns.mit.edu
W. BEENAKKER *(Leiden, Netherlands)*
wimb@lorentz.leidenuniv.nl
F. BEHNER *(ETH Zurich, Switzerland)*
Frank.Behner@cern.ch
G. BELLA *(Tel Aviv, Israel)*
Bella@lep1.tau.ac.il
S. BELLUCCI *(Frascati, Italy)*
bellucci@lnf.infn.it
C. BEMPORAD *(Pisa, Italy)*
bemporad@axpia.pi.infn.it
M. BENEKE *(CERN, Switzerland)*
martin.beneke@cern.ch
Y. BENHAMOU *(IPN Lyon, France)*
ybenham@lyohp5.in2p3.fr
M. BENKEBIL *(LAL Orsay, France)*
benkebil@lal.in2p3.fr
S. BENTVELSEN *(CERN, Switzerland)*
stan.bentvelsen@cern.ch
S. BERGMANN *(Weizmann Inst., Israel)*

K. BERKHAN *(DESY Zeuthen, Germany)*
berkhan@ifh.de
B. BERTUCCI *(Perugia, Italy)*
Bruna.Bertucci@cern.ch
M. BESANCON *(CEA Saclay, France)*
BESANCON@FRCPN11.IN2P3.FR
D. BESSON *(U. of Kansas, USA)*
dzb@kuhep4.phsx.ukans.edu
P. BILLOIR *(LPNHE Paris 6/7, France)*
billoir@in2p3.fr
P. BINETRUY *(LPTHE Orsay, France)*
binetruy@qcd.th.u-psud.fr
D. BISELLO *(Padova, Italy)*
bisello@mxsld2.pd.infn.it
V. BLOBEL *(Hamburg, Germany)*
blobel@mail.desy.de

D. BLOCH *(Strasbourg, France)*
BLOCH@IN2P3.FR

A. BLONDEL *(Polytechnique, France)*
alain.blondel@cern.ch

J. BLOUW *(NIKHEF, Netherlands)*
johanb@nikhef.nl

J. BLUMLEIN *(DESY Zeuthen, Germany)*
blumlein@ifh.de

A. BODEK *(Rochester, USA)*
bodek@urhep.pas.rochester.edu

V. BOLDEA *(Bucharest, Romania)*
boldea@roifa.ifa.ro

M. BONESINI *(Milano, Italy)*
bonesini@mi.infn.it

J. BONN *(Mainz, Germany)*
Jochen.Bonn@uni-mainz.de

G. BONNEAUD *(Polytechnique, France)*
bonneaud.polhp6.in2p3.fr

E. BOOS *(MSU Moscow, Russia)*
boos@ifh.de

F. BOTELLA *(IFIC Valencia, Spain)*
botella@evalvx.ific.uv.es

J. BOUCHEZ *(CEA Saclay, France)*
BOUCHEZ@HEP.saclay.cea.fr

A. BRESKIN *(Weizmann Inst., Israel)*
fnbresk@weizmann.weizmann.ac.il

D. BROADHURST *(Open U. London, UK)*
D.Broadhurst@open.ac.uk

J. BROSE *(TU Dresden, Germany)*
jbrose@slac.stanford.edu

W. BUCHMULLER *(DESY, Germany)*
buchmuwi@vxdesy.desy.de

E. BUCKLEY-GEER *(FNAL, USA)*
buckley@fnal.gov

L. BUGGE *(Oslo, Norway)*
lars.bugge@fys.uio.no

C. BURGARD *(CERN, Switzerland)*
christoph.burgard@cern.ch

G. BUSCHHORN *(MPI Munich, Germany)*
gwb@mppmu.mpg.de

D. BUSKULIC *(LAPP Annecy, France)*
buskulic@lapp.in2p3.fr

J. BUTLER *(FNAL, USA)*
butler@fnal.gov

J. BUTLER (*Boston U., USA*)
jmbutler@bu.edu
M. CAFFO (*Bologna, Italy*)
caffo@bo.infn.it
L. CAMILLERI (*CERN, Switzerland*)
leslie.camilleri@cern.ch
T. CAMPORESI (*CERN, Switzerland*)
Tiziano.Camporesi@cern.ch
V. CANALE (*Roma 2, Italy*)
canale@vxcern.cern.ch
T. CARLI (*MPI Munich, Germany*)
h01rtc@rec06.desy.de
N. CARTIGLIA (*Nevis Labs, USA*)
nicolo@nevis1.nevis.columbia.edu
J. CARVALHO (*Coimbra, Portugal*)
jcarlos@piranha.fis.uc.pt
A. CASAS (*CSIC Madrid, Spain*)
casas@cc.csic.es
R. CASHMORE (*Oxford, UK*)
r.cashmore@physics.oxford.ac.uk
D. CASPER (*CERN, Switzerland*)
david.casper@cern.ch
A. CECCUCCI (*CERN, Switzerland*)
augusto.ceccucci@cern.ch
P. CENCI (*Perugia, Italy*)
cenci@pg.infn.it
L. CHAUSSARD (*IPN Lyon, France*)
chaussad@in2p3.fr
M. CHEMARIN (*IPN Lyon, France*)
chemarin@cern.ch
M. CHEN (*Mons-Hainaut, Belgium*)
maocun.chen@cern.ch
A. CHERLIN (*Weizmann Inst., Israel*)
cherlin@ceres.weizmann.ac.il
M. CHEVALIER (*CEA Saclay, France*)
CHEVALIER@HEP.SACLAY.CEA.FR
P. CHLIAPNIKOV (*CERN, Switzerland*)
chliap@vxcern.cern.ch
D. CHRISMAN (*CERN, Switzerland*)
chrisman@cern.ch
J. CHYLA (*Prague, Czech R.*)
chyla@fzu.cz
J. CIBOROWSKI (*Warsaw, Poland*)
cib@fuw.edu.pl

A. CLARK (U. Geneve, Switzerland)
Allan.Clark@cern.ch
W. CLELAND (Pittsburg, USA)
cleland@vms.cis.pitt.edu
P. COLAS (CEA Saclay, France)
COLAS@HEP.saclay.cea.fr
P. COLLINS (CERN, Switzerland)
paula.collins@cern.ch
J. CONWAY (Rutgers U., USA)
conway@physics.rutgers.edu
C. CORMACK (Liverpool, UK)
ccormack@mail.desy.de
J. CORTES (Zaragoza, Spain)
cortes@dftuz.unizar.es
F. COSSUTTI (CEA Saclay, France)
fabio.cossutti@cern.ch
P. COYLE (CPPM Marseille, France)
paschal@cppm.in2p3.fr
F. CSIKOR (Budapest, Hungary)
csikor@hal9000.elte.hu
M. DANILOV (ITEP Moscow, Russia)
danilov@hera-b.desy.de
M. DAOUDI (SLAC, USA)
daoudi@SLAC.Stanford.EDU
M. DAVID (CEA Saclay, France)
dvd@frcpn11.in2p3.fr
A. DAVIES (Glasgow, UK)
A.Davies@physics.gla.ac.uk
G. DAVIES (Imperial Coll., UK)
g.j.davies@ic.ac.uk
L. DE BARBARO (Nothwestern, USA)
lucyna@miranda.fnal.gov
P. DE BARBARO (Rochester, USA)
barbaro@fnald.fnal.gov
W. DE BOER (Karlsruhe, Germany)
wim.de.boer@cern.ch
N. DE BOTTON (CEA Saclay, France)
botton@hep.saclay.cea.fr
A. DE MIN (Padova, Italy)
demin@padova.infn.it
F. DEL AGUILA (Granada, Spain)
faguila@goliat.ugr.es
R. DEVENISH (Oxford, UK)
r.devenish@ph.ox.ac.uk

A. DI GIACOMO (Pisa, Italy)
digiacomo@pi.infn.it
E. DI SALVO (Genova, Italy)
disalvo@ge.infn.it
G. DIAMBRINI PALAZZI (Roma 1, Italy)
diambrini@roma1.infn.it
M. DIAZ (IFIC Valencia, Spain)
mad@flamenco.ific.uv.es
S. DITA (Bucharest, Romania)
sdita@roifa.ifa.ro
S. DITTMAIER (CERN, Switzerland)
Stefan.Dittmaier@cern.ch
M. DITTMAR (ETH Zurich, Switzerland)
michael.dittmar@cern.ch
F. DITTUS (CERN, Switzerland)
Fido.Dittus@cern.ch
Y. DOKCHITZER (Milano, Italy)
yuri@mi.infn.it
J. DOLBEAU (CDF Paris, France)
dolbeau@cdf.in2p3.fr
J. DORFAN (SLAC, USA)
jonathan@slac.stanford.edu
V. DOROFEEV (IHEP Protvino, Russia)
dorofeev@mx.ihep.su
S. DUBNICKA (Bratislava, Slovakia)
fyzidubn@nic.savba.sk
A. DUBNICKOVA (Bratislava, Slovakia)
dubnickova@fmph.uniba.sk
E. DUCHOVNI (Weizmann Inst., Israel)
fhducho1@weizmann.weizmann.ac.il
G. DUCKECK (LMU Munchen, Germany)
Guenter.Duckeck@Physik.Uni-Muenchen.DE
L. DUFLOT (LAL Orsay, France)
duflot@lal.in2p3.fr
R. EHRLICH (Cornell U., USA)
rde4@cornell.edu
R. EICHLER (ETH Zurich, Switzerland)
eichler@mail.desy.de
Y. EISENBERG (Weizmann Inst., Israel)
yehuda@mail.desy.de
S. ELITZUR (Jerusalem, Israel)
elitzur@vms.huji.ac.il
J. ELLIS (CERN, Switzerland)
johne@mail.cern.ch

E. ELSEN *(DESY, Germany)*
elsen@mail.desy.de

A. ENGLER *(Carnegie Mellon, USA)*
engler@defoe.phys.cmu.edu

M. ERDMANN *(Heidelberg, Germany)*
erdmann@physi.uni-eidelberg.de

A. ERWIN *(U. of Wisconsin, USA)*
erwin@wishpa.physics.wisc.edu

D. ESPRIU *(Barcelona, Spain)*
espriu@ecm.ub.es

E. ETZION *(U. of Wisconsin, USA)*
Erez@SLAC.Stanford.EDU

F. FABBRI *(Frascati, Italy)*
fabbri@lnf.infn.it

M. FAESSLER *(LMU Munchen, Germany)*
martin.faessler@cern.ch

A. FAHR *(Hamburg, Germany)*
fahr@desy.de

A. FALK *(Johns Hopkins, USA)*
falk@jhu.edu

D. FASSOULIOTIS *(NCSR Demokritos, Greece)*
fassoul@vxcern.cern.ch

L. FAVART *(LAL Orsay, France)*
favart@lal.in2p3.fr

M. FEINDT *(Karlsruhe, Germany)*
michael.feindt@cern.ch

M. FELCINI *(ETH Zurich, Switzerland)*
marta.felcini@cern.ch

R. FELST *(DESY, Germany)*
felstro@mail.desy.de

E. FERNANDEZ *(UA Barcelona, Spain)*
fernandez@ifae.es

A. FERRANDO *(CIEMAT Madrid, Spain)*
ferrando@ae.ciemat.es

I. FERRANTE *(Pisa, Italy)*
ferrante@galileo.pi.infn.it

A. FERRER *(IFIC Valencia, Spain)*
ferrer@evalvx.ific.uv.es

F. FERUGLIO *(Padova, Italy)*
feruglio@padova.infn.it

P. FILIP *(Bratislava, Slovakia)*
filip@savba.sk

I. FLECK *(CERN, Switzerland)*
ivor@hpopb1.cern.ch

E. FOCARDI (Firenze, Italy)
focardi@fi.infn.it
R. FOLMAN (Weizmann Inst., Israel)
folman@mail.cern.ch
B. FOSTER (Bristol, UK)
bf@siva.bris.ac.uk
Z. FRAENKEL (Weizmann Inst., Israel)
fnfrankl@physics.weizmann.ac.il
J. FRANK (BNL, USA)
frank@bnlk04.phy.bnl.gov
Y. FRISHMAN (Weizmann Inst., Israel)
fnfrishm@wicc.weizmann.ac.il
M. FUNK (DESY, Germany)
maf@hermes.desy.de
J. FUSTER (IFIC Valencia, Spain)
fuster@evalo1.ific.uv.es
E. GALLAS (U of Texas, USA)
eggs@fnal.gov
B. GARY (Riverside, USA)
bill.gary@cern.ch
M. GAY DUCATI (Porto Alegre, BRAZIL)
gay@if.ufrgs.br
S. GEER (FNAL, USA)
sgeer@fnal.gov
D. GERDES (Johns Hopkins, USA)
gerdes@jhu.edu
Y. GIRAUD-HERAUD (CDF Paris, France)
ygh@cdf.in2p3.fr
M. GIRONE (Imperial Coll., UK)
maria.girone@cern.ch
C. GIUNTI (Torino, Italy)
GIUNTI@TO.INFN.IT
A. GIVEON (Jerusalem, Israel)
giveon@vms.huji.ac.il
C. GLASMAN (DESY, Germany)
claudia@mail.desy.de
F. GLUCK (Budapest, Hungary)
gluck@rmkthe.rmki.kfki.hu
A. GNAENSKI (Weizmann Inst., Israel)
gnaenski@ceres.weizmanm.ac.il
A. GO (CEA Saclay, France)
apollo@hep.saclay.cea.fr
L. GODENKO (Solomon U. Kiev, Ukraine)
mashkevich@gluk.apc.org

J. GOLDBERG (Technion, Israel)
Jacques.Goldberg@cern.ch
T. GOLDBRUNNER (TU Munich, Germany)
goldbrunner@e15.physik.tu-muenchen.de
L. GONZALEZ MESTRES (CDF Paris, France)
lgonzalz@vxcern.cern.ch
E. GOTSMAN (Tel Aviv, Israel)
gotsman@post.tau.ac.il
P. GOTTLICHER (DESY, Germany)
goettlicher@vxdesy.desy.de
J. GOURBER (CERN, Switzerland)
jean-pierre.gourber@cern.ch
A. GOUSSIOU (Stony Brook, USA)
goussiou@fnal.gov
J. GOVAERTS (Louvain-la-Neuve, Belgium)
govaerts@fynu.ucl.ac.be
H. GRABOSCH (DESY Zeuthen, Germany)
grabosch@ifh.de
P. GRAFSTROM (CERN, Switzerland)
per.grafstrom@cern.ch
P. GRANNIS (Stony Brook, USA)
pgrannis@sunysb.edu
E. GRAZIANO (ISS Roma, Italy)
graziani@vaxsan.iss.infn.it
M. GRONAU (Technion, Israel)
gronau@physics.technion.ac.il
E. GROSS (Weizmann Inst., Israel)
eilam.gross@cern.ch
C. GROSSE KNETTER (Bielefeld, Germany)
knetter@physik.uni-bielefeld.de
G. GRUNBERG (Polytechnique, France)
grunberg@pth.polytechnique.fr
M. GRUNEWALD (CERN, Switzerland)
Martin.Grunewald@cern.ch
D. GUETTA (Weizmann Inst., Israel)
A. GURTU (Tata Inst., India)
Gurtu@tifrvax.tifr.res.in
E. GURVITCH (Tel Aviv, Israel)
gurvich@alzt.tau.ac.il
S. HAHN (Wuppertal, Germany)
hahns@vxcern.cern.ch
G. HANSON (Indiana U., USA)
gail.hanson@cern.ch

H. HARARI (Weizmann Inst., Israel)
ftharari@weizmann.weizmann.ac.il
J. HART (RAL, UK)
J.C.Hart@rl.ac.uk
Y. HASEGAWA (CERN, Switzerland)
Yoji.Hasegawa@cern.ch
M. HAUSCHILD (CERN, Switzerland)
Michael.Hauschild@cern.ch
P. HERNANDEZ (CERN, Switzerland)
pilar@thwgs.cern.ch
R. HEUER (CERN, Switzerland)
rolf.heuer@cern.ch
K. HILLER (DESY Zeuthen, Germany)
hiller@ifh.de
D. HOCHMAN (Weizmann Inst., Israel)
fhhochmn@wicc.weizmann.ac.il
K. HOEPFNER (Technion, Israel)
hoepfner@phep1.technion.ac.il
K. HOFFMAN (Purdue, USA)
kara@fnal.gov
R. HOMER (Birmingham, UK)
rjh@axprl1.rl.ac.uk
J. HOREJSI (Prague, Czech R.)
jiri.horejsi@mff.cuni.cz
D. HORN (Tel Aviv, Israel)

J. HOSEK (NPI Rez, Czech R.)
hosek@ujf.cas.cz
G. HOU (Nat. Taiwan U., Taiwan)
wshou@phys.ntu.edu.tw
K. HUITU (Helsinki, Finland)
katri.huitu@helsinki.fi
G. IACOBUCCI (Bologna, Italy)
IACOBUCCI@vxdesy.desy.de
V. ILYIN (MSU Moscow, Russia)
ilyin@theory.npi.msu.su
A. JACHOLKOWSKA (LAL Orsay, France)
jachol@aloha.cern.ch
A. JACHOLKOWSKI (CERN, Switzerland)
Adam.Jacholkowski@cern.ch
D. JACKSON (RAL, UK)
djackson@slac.stanford.edu
P. JANOT (CERN, Switzerland)
patrick.janot@cern.ch

C. JARLSKOG (Lund, Sweden)
cecilia.jarlskog@matfys.lth.se
G. JARLSKOG (Lund, Sweden)
goran@hplund1.cern.ch
R. JESIK (Indiana U., USA)
jesik@fnal.gov
M. JIMACK (Birmingham, UK)
martin.jimack@cern.ch
R. JONES (Lancaster, UK)
Roger.Jones@cern.ch
C. JOSEPH (Lausanne, Switzerland)
Claude.Joseph@ipn.unil.ch
R. JUAREZ WYSOZKA (ESFM Mexico City, Mexico)
rebeca@esfm.ipn.mx
V. KADYCHEVSKI (JINR Dubna, Russia)
kadyshev@jinr.dubna.su
A. KAGAN (U. of Cincinnati, USA)
kagan@physunc.phy.uc.edu
A. KALLONIATIS (Erlangen, Germany)
ack@theorie3.physik.uni-erlangen.de
D. KARLEN (Carleton, Canada)
Dean.Karlen@cern.ch
M. KARLINER (Tel Aviv, Israel)

U. KARSHON (Weizmann Inst., Israel)
fhkarsho@ultra4.weizmann.ac.il
M. KASEMANN (DESY, Germany)
Matthias.Kasemann@desy.de
H. KASHA (Yale, USA)
henry.kasha@yale.edu
G. KERNEL (Ljubljana, Slovenia)
gabrijel.kernel@ijs.si
T. KETEL (VU Amsterdam, Netherlands)
tjeerd@nat.vu.nl
V. KHOVANSKY (ITEP Moscow, Russia)
Valerii.Khovanski@cern.ch
M. KIENZLE (U. Geneve, Switzerland)
maria.kienzle@cern.ch
D. KISIELEWSKA (Krakow, Poland)
KIS@vsk02.if.edu.pl
A. KLIER (Weizmann Inst., Israel)
fhamit@wicc.weizmann.ac.il
B. KLIMA (FNAL, USA)
Klima@fnal.gov

P. KLUIT (NIKHEF, Netherlands)
kluit@vxcern.cern.ch
B. KNIEHL (MPI Munich, Germany)
kniehl@vms.mppmu.mpg.de
E. KOLB (FNAL, USA)
rocky@fnal.gov
J. KONIGSBERG (U. of Florida, USA)
konigsberg@fnald.fnal.gov
P. KOOIJMAN (NIKHEF, Netherlands)
h84@nikhef.nl
G. KORCHEMSKY (LPTHE Orsay, France)
korchems@qcd.th.u-psud.fr
V. KOROTKOV (DESY Zeuthen, Germany)
korotkov@ifh.de
C. KORTHALS ALTES (Marseille, France)
altes@cpt.univ-mrs.fr
P. KOSTKA (DESY Zeuthen, Germany)
kostka@ifh.de
A. KOTWAL (Nevis Labs, USA)
kotwal@fnal.gov
C. KOUNNAS (CERN, Switzerland)
kounnas@nxth04.cern.ch
P. KRIEGER (CERN, Switzerland)
krieger@hpopb1.cern.ch
P. KRIZAN (Ljubljana, Slovenia)
peter.krizan@ijs.si
H. KROHA (MPI Munich, Germany)
kroha@mppmu.mpg.de
M. KUGLER (Weizmann Inst., Israel)
fhkugler@wicc.weizmann.ac.il
M. KUHLEN (MPI Munich, Germany)
kuhlen@desy.de
D. KUHN (Innsbruck, Austria)
dietmar.kuhn@uibk.ac.at
W. KUMMER (TU Wien, Austria)
wkummer@tph.tuwien.ac.at
V. KUNDRAT (Prague, Czech R.)
kundrat@fzu.cz
D. KUTASOV (Weizmann Inst., Israel)
fnkutaso@weizmann.weizmann.ac.il
V. KUVSHINOV (Minsk, Belarus)
kuvshino@dragon.bas-net.by
R. KVATADZE (IPN Lyon, France)
kvatadze@vxcern.cern.ch

L. LABARGA (*UA Madrid, Spain*)
labarga@hepdc1.ft.uam.es
M. LAMANNA (*Trieste, Italy*)
Massimo.Lamanna@cern.ch
K. LANG (*U of Texas, USA*)
lang@hep.utexas.edu
A. LANGE (*Siegen, Germany*)
lange@alwa01.physik.uni-siegen.de
D. LANSKE (*Aachen, Germany*)
d.lanske@physik.rwth-aachen.de
F. LAPLANCHE (*LAL Orsay, France*)
laplanch@lal.in2p3.fr
J. LAUBER (*Univ. College, UK*)
jal@hep.ucl.ac.uk
J. LAYSSAC (*Montpellier, France*)
layssac@lpm.univ-montp2.fr
I. LAZZIZZERA (*U. Tento, Italy*)
lazi@abacus.science.unitn.it
J. LE GOFF (*CEA Saclay, France*)
Jean-Marc.Le.Goff@cern.ch
D. LELLOUCH (*Weizmann Inst., Israel*)
fhlellou@wicc.weizmann.ac.il
M. LELTCHOUK (*Nevis Labs, USA*)
leltchouk@nevis.columbia.edu
V. LEMAITRE (*DESY, Germany*)
lemaitre@mail.desy.de
J. LEMONNE (*VU Brussel, Belgium*)
lemonne@hep.iihe.ac.be
A. LEVY (*Tel Aviv, Israel*)
levy@alzt.tau.ac.il
G. LIN (*Chiao Tung U., Taiwan*)
glin@beauty.phys.nctu.edu.tw
L. LIN (*Chung Hsing U., Taiwan*)
llin@phys.nchu.edu.tw
A. LINDNER (*Hamburg, Germany*)
lindner@mail.desy.de
I. LIPPI (*Padova, Italy*)
Lippi@cern.ch
M. LIVAN (*Pavia, Italy*)
livan@pv.infn.it
C. LLEWELLYN SMITH (*CERN, Switzerland*)
cllewell@mail.cern.ch
B. LOEHR (*DESY, Germany*)
loehr@desy.de

W. LOHMANN (DESY Zeuthen, Germany)
wlo@ifh.de
T. LOHSE (HU Berlin, Germany)
lohse@ifh.de
M. LOSTY (Carleton, Canada)
losty@crpp.carleton.ca
A. LOUNIS (Strasbourg, France)
abl@crnhp3.in2p3.fr
D. LOWENSTEIN (BNL, USA)
lowenstein@bnldag.ags.bnl.gov
W. LUCHA (Wien, Austria)
v2032dac@helios.edvz.univie.ac.at
L. LUDOVICI (CERN, Switzerland)
lucio.ludovici@cern.ch
R. MADDEN (Ben-Gurion, Israel)
madden@bgumail.bgu.ac.il
A. MALAKHOV (JINR Dubna, Russia)
malakhov@lhe.jinr.ru
V. MANDELZWEIG (Jerusalem, Israel)
victor@vms.huji.ac.il
F. MANDL (Wien, Austria)
mandl@hephy.oeaw.ac.at
R. MANKEL (HU Berlin, Germany)
mankel@ifh.de
N. MANKOC-BORSTNIK (Ljubljana, Slovenia)
norma.mankoc@ijs.si
B. MARANGELLI (Bari, Italy)
Meo.Marangelli@ba.infn.it
C. MARIOTTI (CERN, Switzerland)
chiara.mariotti@cern.ch
M. MARKYTAN (Wien, Austria)
markytan@qhepu1.oeaw.ac.at
E. MARSCHALKOWSKI (Mainz, Germany)
marschal@julia.physik.uni-mainz.de
K. MARTENS (ICRR Tokyo, Japan)
kai@icrr.u-tokyo.ac.jp
C. MARTIN (IPN Orsay, France)
martinc@ipno.in2p3.fr
M. MARTINEZ (IFAE Barcelona, Spain)
martinez@dux.ifae.es
H. MARTYN (Aachen, Germany)
martyn@mail.desy.de
A. MASAIKE (Kyoto, Japan)
masaike@pn.scphys.kyoto-u.ac.jp

M. MASERA *(Torino, Italy)*
masera@to.infn.it
S. MASHKEVICH *(ITP Kiev, Ukraine)*
mash@ap3.gluk.apc.org
V. MASHKEVICH *(Solomon U. Kiev, Ukraine)*
mashkevich@gluk.apc.org
A. MASIERO *(SISSA Trieste, Italy)*
Masiero@sissa.it
G. MATTHIAE *(Roma 2, Italy)*

P. MATTIG *(Weizmann Inst., Israel)*
peter.mattig@cern.ch
B. MAYER *(CEA Saclay, France)*
mayer@phnx7.saclay.cea.fr
R. MCCARTHY *(Stony Brook, USA)*
mccarthy@sbhep.physics.sunysb.edu
J. MCDONALD *(Helsinki, Finland)*
mcdonald@outo.helsinki.fi
A. MEHTA *(RAL, UK)*
mehta@mail.desy.de
A. MENEGUZZO *(Padova, Italy)*
Anna.Meneguzzo@pd.infn.it
C. MERONI *(Milano, Italy)*
chiara.meroni@mi.infn.it
W. METZGER *(Nijmegen, Netherlands)*
wes@hef.kun.nl
G. MIKENBERG *(Weizmann Inst., Israel)*
giora.mikenberg@cern.ch
D. MILLER *(Purdue, USA)*
miller@purdd.physics.purdue.edu
D. MILLER *(RAL, UK)*
dmiller@hephp1.rl.ac.uk
A. MILOV *(Weizmann Inst., Israel)*
milov@ceres.weizmanm.ac.il
P. MINKOWSKI *(Bern, Switzerland)*
Mink@butp.unibe.ch
E. MIRKES *(Karlsruhe, Germany)*
mirkes@ttpux5.physik.uni-karlsruhe.de
G. MITSELMAKHER *(U. of Florida, USA)*
mitselmakher@phys.ufl.edu
J. MNICH *(CERN, Switzerland)*
joachim.mnich@cern.ch
R. MOLLER *(Niels Bohr Inst., Denmark)*
moller@nbi.dk

E. MONNIER (Chicago, USA)
monnier@hep.uchicago.edu
F. MUHEIM (U. Geneve, Switzerland)
franz.muheim@cern.ch
A. MULLER (CEA Saclay, France)
muller@hep.saclay.cea.fr
D. MULLER (SLAC, USA)
muller@slac.stanford.edu
K. MULLER (DESY, Germany)
kmueller@mail.desy.de
C. MUNOZ (KAIST Taeion, Korea)
cmunoz@chep6.kaist.ac.kr
W. MURRAY (RAL, UK)
w.murray@rl.ac.uk
K. NAGAI (Weizmann Inst., Israel)
fhnagai@wicc.weizmann.ac.il
M. NAKAHATA (ICRR Tokyo, Japan)
nakahata@icrr.u-tokyo.ac.jp
S. NARISON (Montpellier, France)
narison@lpm.univ-montp2.fr
J. NASSALSKI (Warsaw, Poland)
Jan.Nassalski@fuw.edu.pl
U. NAUENBERG (U. of Colorado, USA)
uriel@colohe.colorado.edu
Y. NE'EMAN (Tel Aviv, Israel)
matilda@cc.nov.tau.ac.il
H. NEAL (CERN, Switzerland)
homer.neal@cern.ch
M. NEUBERT (CERN, Switzerland)
matthias.neubert@cern.ch
N. NEUMEISTER (Wien, Austria)
Norbert.Neumeister@cern.ch
B. NICOLESCU (IPN Orsay, France)
nicolesc@ipno.in2p3.fr
U. NIERSTE (DESY, Germany)
nierste@mail.desy.de
Y. NIR (Weizmann Inst., Israel)
ftnir@weizmann.weizmann.ac.il
R. NISIUS (CERN, Switzerland)
Richard.Nisius@cern.ch
L. NODULMAN (Argonne, USA)
LJN@FNAL.GOV
A. NORTON (CERN, Switzerland)
alan.norton@cern.ch

S. NOWAK (*DESY Zeuthen, Germany*)
nowaks@ifh.de
W. OCHS (*MPI Munich, Germany*)
wwo@mppmu.mpg.de
I. ODA (*Edogawa U., Japan*)
ioda@edogawa-u.ac.jp
T. OEST (*DESY, Germany*)
Thorsten.Oest@desy.de
F. OHLSSON-MALEK (*IPN Lyon, France*)
fmalek@ipnl.in2p3.fr
M. OLSSON (*U. of Wisconsin, USA*)
olsson@pheno.physics.wisc.edu
G. ORGANTINI (*INFN Roma, Italy*)
organtini@roma1.infn.it
I. OTS (*Tartu, Estonia*)
OTS@park.TARTU.EE
B. OVRUT (*U. of Penn., USA*)
ovrut@ovrut.hep.upenn.edu
C. PADILLA (*DESY, Germany*)
cristobal.padilla@desy.de
H. PAES (*Heidelberg, Germany*)
Heinrich.Paes@mpi-hd.mpg.de
F. PAIGE (*BNL, USA*)
paige@bnl.gov
C. PAJARES (*S. de Compostela, Spain*)
pajares@gaes.usc.es
Y. PAN (*U. of Wisconsin, USA*)
pan@wisconsin.cern.ch
G. PARRINI (*Firenze, Italy*)
parrini@fi.infn.it
J. PARSONS (*Nevis Labs, USA*)
parsons@nevis.nevis.columbia.edu
S. PATRICELLI (*Napoli, Italy*)
PATRICELLI@NAPOLI.INFN.IT
T. PATZAK (*CDF Paris, France*)
Patzak@cdf.in2p3.fr
C. PAUS (*CERN, Switzerland*)
christoph.paus@cern.ch
D. PEDRINI (*Milano, Italy*)
Daniele.Pedrini@mi.infn.it
Y. PEI (*CERN, Switzerland*)
Yi-Jin.Pei@cern.ch
O. PELC (*Jerusalem, Israel*)
oskar@shum.cc.huji.ac.il

J. PENARROCHA *(IFIC Valencia, Spain)*
PENARROC@EVALVX.IFIC.UV.ES
O. PERDEREAU *(LAL Orsay, France)*
perdereau@lalcls.in2p3.fr
P. PEREZ *(CEA Saclay, France)*
p.perez@hep.saclay.cea.fr
P. PERRODO *(LAPP Annecy, France)*
perrodo@lapp.in2p3.fr
R. PETRONZIO *(Roma 2, Italy)*
petronzio@roma2.infn.it
G. PIERAZZINI *(Pisa, Italy)*
pierazzini@pisa.infn.it
K. PIOTRZKOWSKI *(DESY, Germany)*
piotrzkowski@desy.de
T. PIRAN *(Jerusalem, Israel)*
tsvi@vms.huji.ac.il
R. PITTAU *(PSI Villigen, Switzerland)*
PITTAU@PSW218.PSI.CH
D. PLANE *(CERN, Switzerland)*
david.plane@cern.ch
F. PLASIL *(ORNL, USA)*
plasilf@ornl.gov
G. POELZ *(Hamburg, Germany)*
poelz@mail.desy.de
L. POGGIOLI *(LPNHE Paris 6/7, France)*
Luc.Poggioli@cern.ch
G. POLIVKA *(Basel, Switzerland)*
Guido.Polivka@cern.ch
R. PREPOST *(U. of Wisconsin, USA)*
prepost@wishep.physics.wisc.edu
M. PRIMAVERA *(Lecce, Italy)*
primavera@le.infn.it
A. PRINIAS *(Imperial Coll., UK)*
prinias@mail.desy.de
A. PROSKURYAKOV *(DESY, Germany)*
proskur@vxdesy.desy.de
G. QUAST *(Mainz, Germany)*
quastg@cern.ch
M. QUIROS *(CSIC Madrid, Spain)*
quiros@pinar1.csic.es
E. RABINOVICI *(Jerusalem, Israel)*
eliezer@vms.huji.ac.il
S. RABY *(Ohio State U., USA)*
raby@mps.ohio-state.edu

G. RAFFELT (MPI Munich, Germany)
raffelt@mppmu.mpg.de
S. RATTI (Pavia, Italy)
ratti@pv.infn.it
I. RAVINOVICH (Weizmann Inst., Israel)

D. REEDER (U. of Wisconsin, USA)
reeder@wishep.physics.wisc.edu
A. RIBON (FNAL, USA)
RIBON@FNALD.FNAL.GOV
J. RIDKY (CERN, Switzerland)
Jan.Ridky@cern.ch
M. RINGNER (Lund, Sweden)
markus@thep.lu.se
S. ROBINS (Technion, Israel)
robins@phep6.technion.ac.il
S. ROCK (SLAC, USA)
SER@SLAC.STANFORD.EDU
C. RODA (Pisa, Italy)
chiara.roda@cern.ch
B. ROE (U. of Michigan, USA)
Byron.Roe@umich.edu
D. ROEHRICH (, Germany)
roehrich@ikf.uni-frankfurt.de
J. ROLDAN GARCIA (DESY, Germany)
roldan@desy.de
A. ROMERO (Torino, Italy)
romero@to.infn.it
M. RONEY (Victoria, Canada)
mroney@uvic.ca
M. ROOS (U. of Helsinki, Finland)
mroos@phcu.helsinki.fi
L. ROSENSON (MIT, USA)
rosenson@mitlns.mit.edu
B. ROSENSTEIN (Chia Tung U., Taiwan)
baruch@phys.nthu.edu.tw
A. ROSSI (Bologna, Italy)
rossi@bo.infn.it
C. ROYON (CEA Saclay, France)
royon@hep.saclay.cea.fr
Y. ROZEN (Technion, Israel)

A. RUIZ (Santander, Spain)
ruiz@gae1.ifca.unican.es

H. RYSECK *(DESY, Germany)*
Hans-Eckhard.Ryseck@desy.de
H. SAARIKKO *(U. of Helsinki, Finland)*
heimo.saarikko@helsinki.fi
A. SADOFF *(Cornell U., USA)*
AJS@LNS62.LNS.CORNELL.EDU
M. SADZIKOWSKI *(Krakow, Poland)*
ufsadzik@thp3.if.uj.edu.pl
M. SAMUEL *(Oklahoma State U, USA)*
physmas@mvs.ucc.okstate.edu
M. SARAKINOS *(CERN, Switzerland)*
miltiadis.sarakinos@cern.ch
G. SARKISSIAN *(Jerusalem, Israel)*
GOR@vms.huji.ac.il
E. SARKISYAN *(Tel Aviv, Israel)*
edward@lep1.tau.ac.il
L. SAWYER *(Louisiana State U., USA)*
sawyer@phys.latech.edu
A. SCHAFFER *(LAL Orsay, France)*
R.D.Schaffer@cern.ch
H. SCHREIBER *(DESY Zeuthen, Germany)*
schreibe@ifh.de
K. SCHUBERT *(TU Dresden, Germany)*
schubert@physik.tu-dresden.de
M. SCHUMACHER *(Bonn, Germany)*
Markus.Schumacher@cern.ch
P. SCHWEMLING *(LPNHE Paris 6/7, France)*
schwemli@lpnhp1.in2p3.fr
B. SEGEV *(Harvard, USA)*
bsegev@cfa.harvard.edu
L. SEHGAL *(Aachen, Germany)*
sehgal@physik.rwth-aachen.de
K. SELIVANOV *(ITEP Moscow, Russia)*
selivano@heron.itep.ru
S. SEMENOV *(ITEP Moscow, Russia)*
ssemenov@aix0.itep.ru
M. SENE *(CDF Paris, France)*
monique.sene@cdf.in2p3.fr
A. SHAPIRA *(Weizmann Inst., Israel)*

A. SHARMA *(Darmstadt, Germany)*
Archana.Sharma@cern.ch
E. SHEFER *(Weizmann Inst., Israel)*
Fnefrat.Weizmann@wis.weizmann.ac.il

V. SHOUTKO (CERN, Switzerland)
shoutko@vxcern.cern.ch
N. SHUMEIKO (Minsk, Belarus)
shum@hep.by
A. SILENKO (Minsk, Belarus)
silenko@inp.belpak.minsk.by
V. SIMAK (Prague, Czech R.)
simak@fzu.cz
P. SINGER (Technion, Israel)
phr26ps@physics.technion.ac.il
A. SINGOVSKI (CERN, Switzerland)
Alexander.Singovski@cern.ch
Y. SIROIS (Polytechnique, France)
sirois@in2p3.fr
A. SISSAKIAN (JINR Dubna, Russia)
sisakian@jinr.dubna.su
A. SKUJA (U. of Maryland, USA)
skuja@umdhep.umd.edu
A. SMITH (U of Minnesota, USA)
smith@mnhep.hep.umn.edu
V. SOERGEL (MPI Munich, Germany)
soergel@mppmu.mpg.de
J. SOFFER (Marseille, France)
soffer@cpt.univ-mrs.fr
S. SOLOMON (Jerusalem, Israel)

S. SOMALWAR (Rutgers U., USA)
sunil@ruthep.rutgers.edu
R. SOMMER (DESY Zeuthen, Germany)
sommer@ifh.de
C. SOMMERFIELD (Yale, USA)
charles.sommerfield@yale.edu
A. SONI (BNL, USA)
soni@penguin.phy.bnl.gov
J. SONNENSCHEIN (Tel Aviv, Israel)
cobi@ccsg.tau.ac.il
P. SOUDER (Syracuse, USA)
souder@suhep.phy.syr.edu
S. SPANIER (CERN, Switzerland)
stefan.spanier@cern.ch
M. SPIRA (CERN, Switzerland)
spira@cern.ch
A. STAIANO (Torino, Italy)
staiano@to.infn.it

M. STEINHAUSER (MPI Munich, Germany)
Matthias.Steinhauser@mppmu.mpg.de
H. STENZEL (MPI Munich, Germany)
stenzel@mppmu.mpg.de
M. STRAUSS (U. of Oklahoma, USA)
strauss@mail.nhn.ou.edu
M. STROVINK (LBL, USA)
strovink@lbl.gov
R. STUART (U. of Michigan, USA)
stuartr@umich.edu
J. SULLIVAN (Los Alamos, USA)
sullivan@lanl.gov
P. SUTTON (Lund, Sweden)
peter@thep.lu.se
B. SVETITSKY (Tel Aviv, Israel)
bqs@julian.tau.ac.il
M. SZCZEKOWSKI (Warsaw, Poland)
marek.szczekowski@cern.ch
J. SZWED (Krakow, Poland)
szwed@if.uj.edu.pl
S. TAREM (Technion, Israel)
s.tarem@cern.ch
G. TARTARELLI (CERN, Switzerland)
Giuseppe.Tartarelli@cern.ch
S. TATUR (Warsaw, Poland)
st@camk.edu.pl
L. TAUSCHER (CERN, Switzerland)
ludwig.tauscher@cern.ch
M. TEPER (Oxford, UK)
teper@thphys.ox.ac.uk
N. TESCH (DESY, Germany)
tesch@x4u2.desy.de
G. THOMPSON (U. of London, UK)
G.Thompson@qmw.ac.uk
S. THURNER (TU Wien, Austria)
thurner@ds1.kph.tuwien.ac.at
H. TICHO (UCSD, USA)
hticho@ucsd.edu
L. TIKHONOVA (MSU Moscow, Russia)
larisa@monet.npi.msu.su
I. TOMALIN (CERN, Switzerland)
ian.tomalin@cern.ch
K. TOTH (Budapest, Hungary)
ktoth@rmki.kfki.hu

M. TRAN *(Lausanne, Switzerland)*
Minh-Tam.Tran@ipn.unil.ch
T. TREFZGER *(LMU Munchen, Germany)*
thomas.trefzger@cern.ch
G. TRISTRAM *(CDF Paris, France)*
tristram@cdf.in2p3.fr
U. TRITTMANN *(Weizmann Inst., Israel)*
trittman@wicc.weizmann.ac.il
D. TSABAR *(Jerusalem, Israel)*

I. TSAKOV *(Sofia, Bulgaria)*
itsak@inrne.acad.bg
J. TUOMINIEMI *(Helsinki, Finland)*
tuominiemi@phcu.helsinki.fi
G. UNAL *(LAL Orsay, France)*
unal@in2p3.fr
V. VALUEV *(LAPP Annecy, France)*
valuev@axnd02.cern.ch
R. VAN DANTZIG *(NIKHEF, Netherlands)*
rvd@nikhef.nl
W. VAN DONINCK *(IIHE Brussels, Belgium)*
VANDONINCK@hep.iihe.ac.be
R. VAN KOOTEN *(Indiana U., USA)*
rickv@paoli.physics.indiana.edu
G. VENEZIANO *(CERN, Switzerland)*
gabriele.veneziano@cern.ch
W. VENUS *(RAL, UK)*
wilbur.venus@cern.ch
C. VERZEGNASSI *(Lecce, Italy)*
Claudio.Verzegnassi@trieste.infn.it
A. VICINI *(DESY Zeuthen, Germany)*
vicini@ifh.de
L. VITALE *(Trieste, Italy)*
vitale@trieste.infn.it
S. VLACHOS *(Basel, Switzerland)*
sotirios.vlachos@cern.ch
H. VOGEL *(Carnegie Mellon, USA)*
vogel@cmphys.phys.cmu.edu
F. VON FEILITZSCH *(TU Munich, Germany)*
feilitzsch@e15.physik.tu-muenchen.de
I. VOROBIEV *(CERN, Switzerland)*
igor.vorobiev@cern.ch
R. VOSS *(CERN, Switzerland)*
Rudiger.Voss@cern.ch

A. WAGNER (DESY, Germany)
albrecht.wagner@desy.de
H. WAHLEN (Wuppertal, Germany)
wahlen@wpcl1.physik.uni-wuppertal.de
Y. WANG (Stanford , USA)
yfwang@hep.Stanford.EDU
D. WARD (Cambridge, UK)
drw1@hep.phy.cam.ac.uk
P. WARD (Cambridge, UK)
cpw1@hep.phy.cam.ac.uk
A. WEBER (Aachen, Germany)
Alfons.Weber@cern.ch
F. WEBER (Harvard, USA)
weber@sunom1.cern.ch
H. WEERTS (Michigan State U., USA)
weerts@pa.msu.edu
G. WEIGLEIN (Karlsruhe, Germany)
georg@itpaxp2.physik.uni-karlsruhe.de
S. WEINZIERL (CEA Saclay , France)
stefanw@spht.saclay.cea.fr
N. WEISS (Tel Aviv, Israel)

H. WELLISCH (CERN, Switzerland)
Hans-Peter.Wellisch@cern.ch
K. WICK (Hamburg, Germany)
wick@maren.desy.de
B. WIIK (DESY, Germany)
bjoern.Wiik@desy.de
G. WILKINSON (CERN, Switzerland)
guy.wilkinson@cern.ch
E. WITTMANN (Heidelberg, Germany)
wittmann@mpi-hd.mpd.ge
G. WOLF (MPI Munich, Germany)
wolf@vms.mppmu.mpg.de
K. WOLLER (DESY, Germany)
woller@orion.desy.de
J. WOMERSLEY (FNAL, USA)
womersley@fnal.gov
S. WONG (Wuppertal, Germany)
wong@theorie.physik.uni-wuppertal.de
G. WORMSER (LAL Orsay, France)
wormser@lal.in2p3.fr
B. WOSIEK (Krakow, Poland)
wosiek@chopin.ifj.edu.pl

J. WOSIEK *(Krakow, Poland)*
wosiek@thp4.if.uj.edu.pl
A. WRIGHT *(RAL, UK)*
awright@aloha.cern.ch
X. WU *(U. Geneve, Switzerland)*
xin.wu@cern.ch
J. WUDKA *(Riverside, USA)*
jose.wudka@ucr.edu
A. YAGIL *(FNAL, USA)*
yagil@fnal.gov
G. YEKUTIELI *(Weizmann Inst., Israel)*
fhyekut3@weizmann.weizmann.ac.il
J. YU *(FNAL, USA)*
yu@fnal.gov
M. YVERT *(LAPP Annecy, France)*
yvert@lapp.in2p3.fr
O. ZAIMIDOROGA *(JINR Dubna, Russia)*
oleg@ljap5.jinr.dubna.su
I. ZAKOUT *(Betlehem University, NPA)*
zakout@rorqual.cc.metu.edu.tr
A. ZALEWSKA *(Krakow, Poland)*
zalewska@vxcern.cern.ch
A. ZARNECKI *(DESY, Germany)*

P. ZARUBIN *(JINR Dubna, Russia)*
zarubin@lhe.jinr.ru
D. ZAVRTANIK *(Ljubljana, Slovenia)*
danilo.zavrtanik@ses-ng.si
D. ZER-ZION *(CERN, Switzerland)*
daniel.zer-zion@cern.ch
G. ZOUPANOS *(HU Berlin, Germany)*
George.Zoupanos@cern.ch
C. ZUCCHELLI *(Stockholm, Sweden)*
gcz@physto.se

PLENARY SESSIONS

Award Ceremony: Spontaneous Symmetry Breaking in Gauge Theories – A Historical Survey

R. Brout (rbrout@ulb.ac.be) and F. Englert (fenglert@ulb.ac.be)

Service de Physique Théorique, Université Libre de Bruxelles

Abstract. The personal and scientific history of the discovery of spontaneous symmetry breaking in gauge theories is outlined and its scientific content is reviewed.

1 History[1]

Our discovery of spontaneous symmetry breaking in gauge theory is intimately linked to the history of our collaboration. Evoking this period of our life, I shall survey the scientific history of spontaneous symmetry breaking. A more detailed historical review can be found in the talk presented by Veltman at the International Symposium on Electron and Photon Interactions at High Energies in 1973 at Bonn.

I came to Cornell University in 1959 as a Research Associate to Robert Brout who was Professor there. My background was mainly in solid state physics and many body problems, and Robert Brout was already well known for his work in these fields. Our first contact was unexpectedly warm. It started right when he came to take me at the airport with his near century old Buick and took me to a drink which lasted up to the middle of the night. When we left, we knew that we would become friends.

At Cornell, we realized that, in our approach to physics, we were different. Robert had an amazing easiness in translating abstract concepts into tangible intuitive images, whereas I, with my latin oriented education, had on the contrary always a tendency to express images in terms of formal structures. But this difference turned into a fruitful complementarity because we quickly learned to understand the functioning of each others mind. Still today, talking physics, each of us gets somehow frustrated not to be able to terminate a sentence, as the other does it for him, and is apparently very happy to do so.

Playing together on many-body problems, we got involved in the study of phase transitions and particularly in ferromagnetism. We understood the importance of the spin wave excitations for the description of the ferromagnetic phase. The order parameter, the magnetization, is the manifestation of spontaneously broken rotation invariance and the spin waves are collective modes whose energy goes to zero when the wavelength goes to infinity.

[1] Presented by F. Englert

These are the massless Nambu-Goldstone bosons [1, 2, 3] associated to the spontaneously broken symmetry. Their dynamics essentially determine the magnetization curve. When the range of the forces between spins is extended to all spins instead of, as is usually considered, limited to near neighbors, the spin waves become effectively massive and the magnetization curve encodes the emergence of such a gap in the spectrum. This was the first time we realized that the Nambu-Goldstone bosons, which signal in general the spontaneous breakdown of a global symmetry, cannot survive in presence of long range forces.

Our study of phase transition led us also to the analysis of superconductivity. We were extremely impressed by Nambu's formulation of the BCS theory [4]. This paper and the related papers of Nambu and Jona-Lasinio [2] on spontaneously broken chiral invariance brought to light, in full field theoretic terms, the emergence of the massless Nambu-Goldstone boson. They also pointed out the existence of a massive scalar bound state in the channel orthogonal to the massless mode. The significance of this massive scalar is more transparent in the context of the Goldstone scalar field model when the symmetry breaking is driven by the scalar field potential itself [3]: it describes the response to the order parameter. In the particular case of the abovementioned ferromagnetic transition, this response is the longitudinal susceptibility. These beautiful papers were certainly an element which later drove us into field theory.

In fall 1961, I was scheduled to return to Belgium. By that time our collaboration and our friendship had become deeply rooted. I received an offer of a professorship at Cornell but I was missing Europe very much. I decided not to accept it and to return to Belgium. Robert and his wife Martine had a similar attraction for the Old Continent; Robert got a Guggenheim fellowship and they joined me in Belgium. After a few months, the social life there and our personal relations decided Robert to resign from his professorship at Cornell University and to settle permanently at Brussels University.

We then resumed in Belgium our analysis of broken symmetry. We knew from our study of ferromagnetism that long range forces give mass to the spin waves and we were aware, from Anderson's analysis of superconductivity [5], of the fact that the massless mode of neutral superconductors, which is also a Nambu-Goldstone mode, disappears in charged superconductors in favor of the usual massive plasma oscillations resulting from the long range coulomb interactions in metals. Comforted by these facts, we decided to confront, in relativistic field theory, the long range forces of Yang-Mills gauge fields with the Nambu-Goldstone bosons of a broken symmetry.

The latter arose from the breaking of a *global* symmetry and Yang-Mills theory extends the symmetry to a *local* one [6]. Although the problem in this case is more subtle because of gauge invariance, the emergence of the Nambu-Goldstone massless boson is very similar. We indeed found that there were well defined gauges in which the broken symmetry induces such modes. But, as we expected, the long range forces of the Yang Mills fields were conflicting with those of the massless Nambu Goldstone fields. The conflict is resolved by the generation of a mass reducing long range forces to short range ones. In addition, gauge invariance requires the Nambu-Goldstone mode to combine with the Yang Mills excitations. In this way, the gauge fields acquire a gauge invariant mass!

This work was finalized in 1964. We obtained the mass formula for gauge fields, abelian and non abelian, where symmetry breaking arises from non vanishing expectation values of scalar fields [7]. These play the role of "order parameters". We also, guided by Nambu's work on superconductivity, obtained through Ward identities a mass formula for a dynamical symmetry breaking by fermion condensate [7]. Thus the scalar fields could be either fundamental or constitute a phenomenological description of a condensate. Only future experimental and theoretical development can tell. Their massive excitations is a property of global symmetry breaking and are to be identified with the Goldstone (or with the Nambu) massive scalar bosons describing the response to the "order parameters"; they are not altered by the introduction of local symmetry. On the contrary, the Nambu-Goldstone bosons which arise in group space in directions orthogonal to the massive scalars, and also exist whether the scalar fields are fundamental or composite, are "eaten up" by the gauge fields which lay in these directions. These acquire mass. This fate of the Nambu-Goldstone bosons is the characteristic feature of the symmetry breaking mechanism in a local gauge symmetry.

I shall not dwell on subsequent developments but briefly recall the most relevant steps. First there is the work of Higgs who obtained essentially the same results in a somewhat different way [8]. He showed in simple field theoretic terms that the Nambu-Goldstone boson was unobservable as such but provided the required longitudinal polarization for the gauge fields to get mass [9]. This fact is deeply related to the unitarity of the scheme and is less explicit in our approach. On the other hand, our formulation of the problem, using covariant gauges and Ward identities, puts into evidence that power counting for Feynman graphs was consistent with renormalizability. This is why we were led in 1966 to suggest that the theory of vector mesons with mass generated by the symmetry breaking mechanism was renormalizable [10]. But the full proof of renormalizability was much more involved and in fact required a detailed analysis of the consistency of the power counting in covariant gauges and of the unitarity of the theory. This was worked out by Veltman and 't Hooft and the proof was essentially completed in 1971 [11]. It rendered the electroweak theory [12, 13], hitherto the most impressive application of the mechanism discovered by Higgs and ourselves, a truly consistent and predictive scheme whose experimental verification confirmed the validity of the symmetry breaking mechanism.

This ends the story of the genesis of the unification scheme which relates short and long range forces in gauge field theory, indicating a path to the search for more general laws of nature. For us, it was only the beginning of our lasting collaboration and of our lasting friendship.

2 Scientific Review [1]

The acquisition of mass by gauge vector mesons results from the mutual coupling of two fields each of which, in other circumstances, had vanishing mass. These are, on one hand, the zero mass excitations which result from the spontaneous breakdown of symmetry (SBS) and, on the other, the zero mass vector field which is necessitated when this same symmetry is promoted from global to local, in which case it is called a gauge symmetry. In relativistic theory these are called respectively the Nambu-Goldstone (NG) and Yang-Mills (YM) fields, although they have had a long prior history, the first in statistical mechanics and solid state physics and the second, of course, dating from Maxwell. I shall first explain their masslessness in conceptual terms and then indicate how their coupling gives rise to the vector mass.

We begin with the NG field which arises in consequence of SBS. But firstly what is SBS? An amusing image has been given by Abdus Salam. Consider an ensemble of dinner companions seated round the table set with plates between which is placed a spoon. When the first guest chooses a spoon, to his right *or* to his left, all others must follow suit. This is SBS.

The above example is a case of discrete SBS. A spoon on the right (left) of a plate is represented by a spin which is on a lattice site that is polarized up (down). Interaction favors that neighboring spins be parallel. So the ground state is all spins up (or down). An angel who fixes the polarization of some spin then determines the polarization of all. Clearly it costs a finite amount of energy to turn one spin against the other so that this model presents an excitation spectrum which has a gap.

Contrast this to SBS for the case of continuous symmetry. For simplicity take the group $U(1)$. Then the spin has two components (S_x, S_y) such that $S_x^2 + S_y^2 = 1$. Once more the ferromagnetic interaction favors neighbors to be parallel. Imagine a state where all sit at an angle θ with respect to the x-axis in group space. Clearly it costs no energy to rotate them all through an angle $\Delta\theta$, since they all remain parallel. The ground state is thus degenerate with respect to θ. Now divide the system in two and rotate the two halves against each other. Only the spins that "rub" against each other require an energy to make such a configuration. Call this one node's worth of energy. If one divides in thirds, it will cost two nodes' worths of energy, and so on. Since translational symmetry of the lattice requires that the excitations be classed by wave number, k, we see that the energy grows with k, and furthermore $\omega(k = 0) = 0$. So $k = 0$ is the terminal point of a spectrum which starts at zero frequency. This is the NG excitation.

---

[1] Presented by R. Brout

It is noteworthy that this process of the monotonic increase of energy with the number of nodes stops once k reaches the inverse range of the force between spins, for then it cost no more energy to make more nodes. Thus as the range tends to infinity, the spectrum develops a gap, that is a mass, i.e. $\lim_{k\to 0}\omega(k)$ is finite. This is a precursor of what happens in general when long range forces are present.

I now explain why it is natural that gauge fields are also massless. Take once more the simple case of spins. Global symmetry is the invariance of the energy when all spins are rotated "en masse". Local symmetry is realized when different portions can be rotated differently (in group space) at no cost of energy. This is possible only if there is a "messenger" which transmits from portion to portion the information that such local rotations indeed do not cost any energy and have no physical effect. In technical terms, this messenger is called a connection or a gauge field. It transforms under local rotations exactly in a way to compensate for the energy that would otherwise follow from relative rotations of neighboring spins. It is this beautiful idea which governs all the presently known interactions of nature.

¿From the above one understands that it is natural that the gauge field has zero mass. Indeed under global transformations a gauge field is not required to ensure invariance. So it should not manifest itself. But a global transformation corresponds to one whose wave vector k_μ vanishes: it should cost no energy ($k_0 = 0$) to make a gauge field excitation which is everywhere the same ($\overline{k} = 0$). In relativity the condition $k_\mu = 0$ becomes the invariant statement $k_\mu k^\mu = 0$ (or $k_0^2 - \overline{k}^2 = 0$) which is the statement of masslessness.

It is to be expected that a dramatic situation arises when these two kinds of zero mass excitations are put together in the context of a local symmetry. What happens is that they combine into one massy vector field. The gauge field, of itself, due to relativistic constraints has two degrees of freedom. These are encoded in the polarization transverse to the direction of propagation. Massive vector fields have a longitudinal polarization as well and this is induced by the coupling to the NG field. To see how this mechanism works it is convenient to express things in terms of Feynman graphs.

In the case of no SBS, gauge fields propagate in vacuum by taking into account the dielectric constant of the vacuum. This is represented by loop insertion in the gauge field propagator. For matter represented by a scalar field, a single loop insertion is drawn as follows

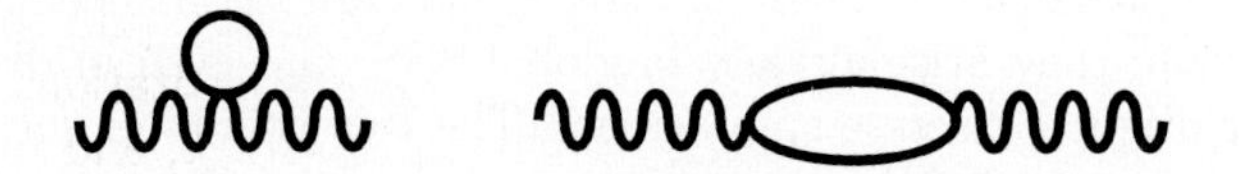

These loops insertions in the YM propagator, represented by wavy lines, conserve the transverse character of the gauge field and keep it massless.

Their effect is to change the value of the coupling constant of the gauge field to matter.

In the case of SBS, the finite expectation value of the scalar field causes additional graphs to arise which are found by cutting the loop, generating the so-called tadpole graphs. One must then addend to the above the graphs

In these graphs the wavy lines still represent the gauge field and the solid line tadpoles are the expectation values of the scalar fields which play the role of order parameters. The dashed line is the propagator of a NG excitation. These arise in directions orthogonal in group space to the order parameters. The latter graphs show how the NG field gets absorbed into the gauge field, the net effect being to give to the latter a mass proportional to the order parameter and to increase the number of degrees of freedom of the gauge field from two to three. Although this "order parameter" is here gauge dependent, there are Ward identities ensuring the gauge invariance of the mass arising from these graphs. These two elements are of utmost importance since the gauge invariance ensures that the divergences of the graphs remain under control, indicating that the theory could be renormalizable, and the new longitudinal degree of freedom renders the perturbation series unitary. It is the combination of these two elements that Veltman and 't Hooft used in their masterful works to prove that the theory is indeed renormalizable, thereby really setting the standard model on a sound basis.

As just mentioned, the appearance of the massless NG bosons is guaranteed by the Ward identities, and as such does not rely on perturbation theory. They therefore also appear if SBS is realized dynamically through a fermion condensate, as in the BCS theory of superconductivity or in the Nambu Jona-Lasinio theory of broken global chiral symmetry. In presence of a local symmetry, they would then still generate a mass for the gauge vector mesons. In that case, the scalar fields would be phenomenological rather than fundamental objects but the mechanism would remain essentially the same. Whether fundamental or not, the scalar fields describing the order parameters, have massive quanta. These massive scalars are not a specific feature of the mechanism: they arise already in global SBS, and even in discrete ones such as our original discrete spin system. The free energy of such a system presents as a function of the magnetization, below the Curie point, the double dip shape typical of the Landau-Ginsburg potential V represented in the figure below. This potential is the same as the one driving global SBS in the Goldstone scalar field model. The distance of the dip to the origin and the curvature at this point are respectively the expectation value and the mass squared of the Goldstone massive scalar boson. The latter, or more precisely its inverse mass squared, measures the longitudinal susceptibility. This is the response of a field parallel to the order parameter and appears in any second order phase transition.

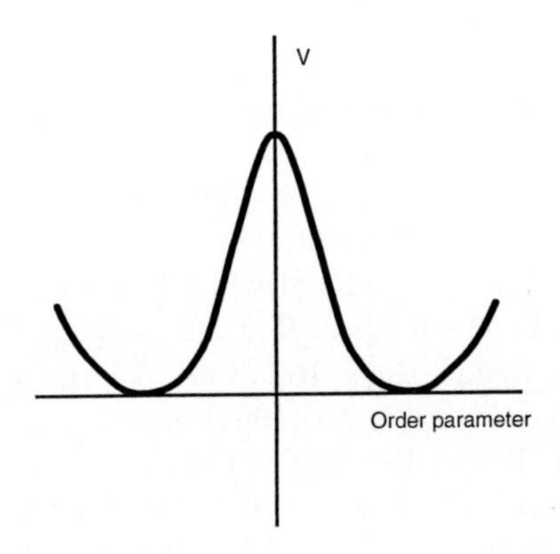

I conclude by restating the main results with some emphasis on their phenomenological implications.

Massive gauge vector mesons are an inevitable consequence of SBS, independently of the dynamical mechanism which causes the breaking, scalar fields[1], bound state condensate... Observationally in the electroweak sector they occur as narrow resonances because their coupling to the continuum is small. This latter is an observational fact: the electric charge is small and thus governs the scale of all gauge field couplings.

Massive scalar occur in channels orthogonal to the NG channels. They appear in any global SBS and are thus not a specific feature of mass generation for gauge fields; their physics is accordingly much more sensitive to dynamical assumptions. They too would appear as resonances whether or not they be manifestations of an elementary scalar field or as composites due to a more elaborate mechanism. Whether or not these resonances are swamped into the continuum depends on the parameters of the theory and we do not control these in the same way as we control coupling to the gauge vectors. Thus observation of both mass and width in these channels will deliver to us precious indications of the mechanism at work.

[1] Within the past few years, an interesting development has occurred principally due to the mathematician A. Connes who has applied techniques of non commutative geometry to construct the standard model. In this, the key point is that the scalar field plays the role of a gauge connection in the "motion" of a fermion (whose mass is generated by spontaneously broken chiral symmetry) during its zitterbewegung.

In all cases SBS in gauge theories is characterized by NG bosons which are "eaten up" by the YM fields, giving them longitudinal polarization and mass. It is this phenomenon which allows for consistent renormalizable theories of massive gauge vector mesons.

References

[1] Y. Nambu, Phys.Rev.Lett. **4** (1960) 380.
[2] Y. Nambu and G. Jona-Lasinio, Phys. Rev. **122** (1961) 345; Phys. Rev. **124** 1961 246.
[3] J. Goldstone, Il Nuovo Cimento **19** (1961) 154.
[4] Y. Nambu, Phys. Rev. **117** (1960) 648.
[5] P.W. Anderson, Phys. Rev. **112** (1958) 1900.
[6] C.N. Yang and R.L. Mills, Phys. Rev. **96** (1954) 191.
[7] F. Englert and R. Brout, Phys. Rev. Lett. **13** (1964) 321.
[8] P.W. Higgs, Phys. Rev. Lett. **13** (1964) 508.
[9] P.W. Higgs, Phys. Rev. **145** (1966) 1156.
[10] F. Englert, R.Brout and M. Thiry, Il Nuovo Cimento **43A** (1966) 244; see also the Proceedings of the 1997 Solvay Conference, *Fundamental Problems in Elementary Particle Physics*, Interscience Publishers J. Wiley ans Sons, p 18.
[11] G. 't Hooft, Nucl. Phys. **B35** (1971) 167.
[12] S.L. Glashow, Nucl. Phys. **22** (1961) 579.
[13] S. Weinberg, Phys. Rev. Lett. **19** (1967) 1264.

QCD at Hadron Colliders

Harry Weerts (weerts@pa.msu.edu)

Michigan State University, East Lansing MI, USA

Abstract. This paper reviews the status of Quantum Chromo Dynamics, when compared to measurements at hadron colliders. The emphasis is on a confrontation with results from the Tevatron collider experiments, varying from inclusive cross sections to more exclusive or differential measurements.

1 Introduction

All physics processes at a hadron collider involve the strong interaction and are described within the framework of quantum chromodynamics (QCD). To interpret a measurement in this environment relies heavily on QCD being verified in other more clean circumstances, for example e^+e^- or $lepton - hadron$ scattering. From the point of view of confronting QCD based predictions at a hadron collider machine, we take the view that this environment is rich in QCD, although one might be tempted to think that it borders on chaos.

Within the framework of QCD the interaction between hadrons is seen as the hard interaction between two massless, pointlike partons from the scattering hadrons. The parton-parton interaction is described by a hard parton cross section $\hat{\sigma}$. Because there are several parton species, with varying momenta in the hadron, the theoretical prediction for any measured cross section(σ^{meas})has the following form in perturbative QCD:

$$\sigma^{meas} \propto \sum_{ij} \int_{x_1} \int_{x_2} \phi_i(x_1, \mu_f) \hat{\sigma}_{ij}(\mu_r) \phi_j(x_2, \mu_f) \tag{1}$$

where $\phi_{i,j}$ is the parton momentum distribution (PDF) for parton types i and j in the hadrons, μ_f is the factorization scale at which it is evaluated, $\hat{\sigma}_{ij}(\mu_r)$ is the cross section for an interaction between parton i and j, μ_r is the renormalization scale at which the strong coupling constant (α_s) is evaluated and x_1, x_2 are the momentum fractions of the partons in the hadrons. The PDF 's are derived mainly from lepton hadron (lh) scattering cross section (=structure function) measurements (σ_{lh}^{meas}) using the same underlying QCD theoretical framework ($\phi_i = \sigma_{lh}^{meas}/\hat{\sigma}_{lh}$) and they are only defined within this framework. Since they come from experiment there are uncertainties associated with them, but these uncertainties are not well defined at this time. It should be noted that without the high precision lepton hadron experiments, especially at HERA, it would be practically impossible to study QCD quantitatively at a hadron collider.

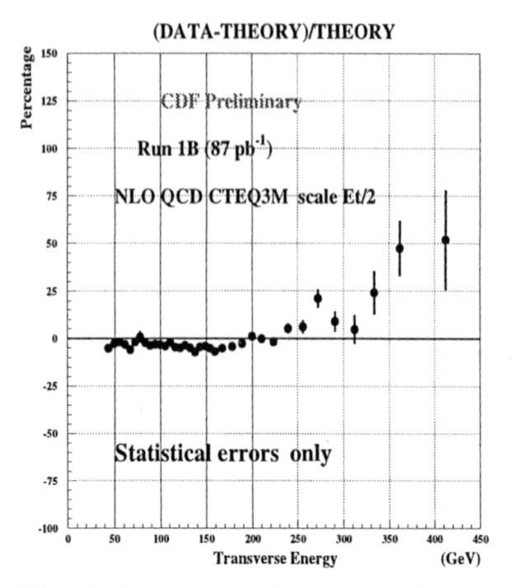

Fig. 1. Inclusive jet cross section from the CDF experiment, $0.1 \leq| \eta |\leq 0.7$

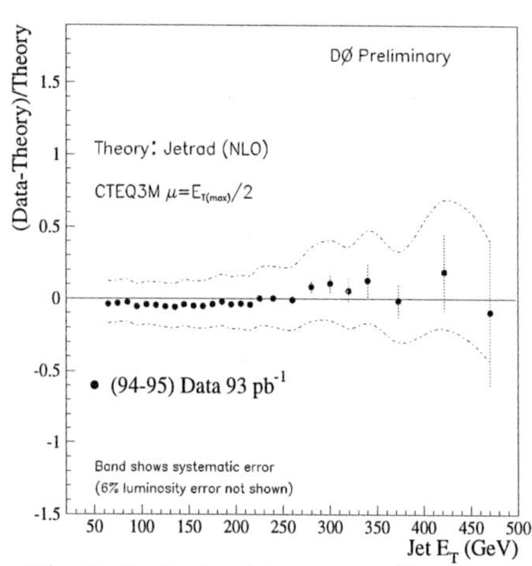

Fig. 2. Inclusive jet cross section from the DØ experiment, $| \eta |\leq .0.5$

In the following we will concentrate on results from the Tevatron $p\bar{p}$ collider. The theoretical predictions will mainly be based on next-to-leading order (NLO) perturbative QCD. Most experimental measurements presented here are made with quantities that were corrected back to the particle level and are directly compared to the NLO parton level predictions, without any corrections for parton showering and/or fragmentation. Since this is an experimental summary the paper is organized by final states, instead of underlying physics. In many cases the underlying physics is similar, but at this time, the experimental results seem to be requiring additional features in the theoretical predictions, so they dictate the order. Strictly speaking NLO QCD predictions are only valid for inclusive cross section predictions, so in each final state we will start with the inclusive cross section.

2 Jet final states

Jets observed in hadron-hadron collisions are the clearest manifestation that the interaction between two hadrons is described by the scattering of two partons and results in final state partons, which are experimentally observed as a localized deposition of energy in the form of jets. The production of jets is copious and the inclusive jet cross section ($d^2\sigma/dE_T d\eta$) is typically measured by counting all jets in a bin of transverse energy (E_T) and pseudorapidity ($\eta = -\ln(\tan(\theta/2))$), where θ is the polar angle with respect to the beam. This cross section has been measured in the central rapidity region ($| \eta |\leq 0.7$) at $\sqrt{s} = 1800$ GeV[1, 2], by the CDF and DØ experiments. Over the E_T range 50-450 GeV the cross section drops by seven orders of magnitude. The high E_T region probes the smallest distance scales accessible by experiments today and is therefore the most sensitive probe of substructure in quarks.

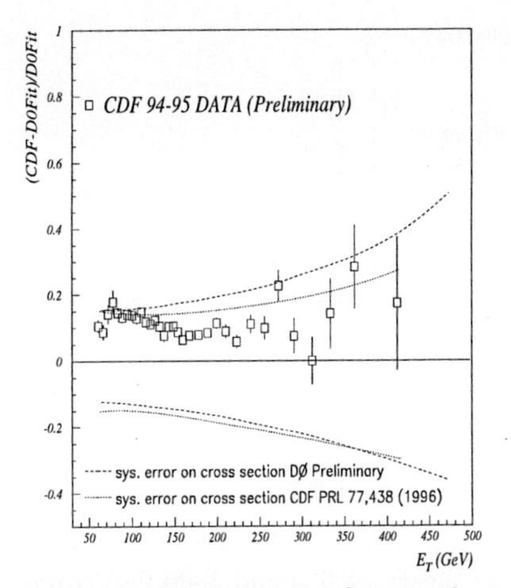

Fig. 3. Comparison of experimental CDF and DØ cross sections in $0.1 \leq |\eta| \leq 0.7$

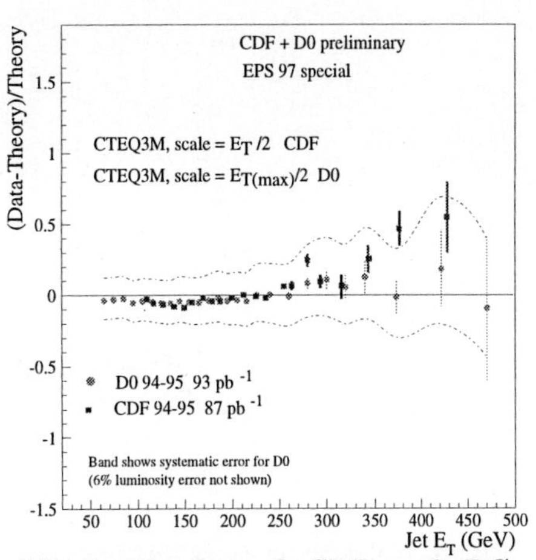

Fig. 4. Overlay of CDF and DØ (DATA-THEORY) / THEORY graphs.

To facilitate a comparison with QCD based theoretical predictions, Figs. 1 and 2 show the quantity (DATA-THEORY) / THEORY for this measurement by both experiments. In both cases jets are defined by a fixed cone algorithm with a cone size $\Delta R = \sqrt{\Delta^2\eta + \Delta^2\phi} = 0.7$ in pseudorapidity and azimuthal coordinates, according to the Snowmass definition [4]. The same algorithm is used in the NLO ($O(\alpha_s^3)$) theory prediction, because a jet at the parton level can consist of two partons. To obtain better matching between experimentally defined jets, with experiment specific merging/splitting criteria, and parton jets the distance between two partons belonging to one jet is required to be less than $R_{sep} \times 0.7$ with $R_{sep} = 1.3$ for DØ and 2.0 for CDF. The rapidity regions covered are slightly different: CDF covers $0.1 \leq |\eta| \leq 0.7$ and DØ covers $|\eta| \leq 0.5$. The theory predictions used are both $O(\alpha_s^3)$, but based on the EKS program [5] with $\mu_f, \mu_r = E_T^{jet}/2$ for CDF and on JETRAD [6] with $\mu_f, \mu_r = E_T^{max}/2$ for DØ . Here E_T^{max} is the largest transverse energy observed among all jets in an event. Both predictions use the CTEQ3M [7] parton distributions. The DØ data agree with the NLO QCD predictions rather well, especially with the systematic error taken into account. The CDF data agree very well at medium transverse energies, but the data seem to show an excess at large E_T. There is no apparent excess in the DØ data.

Questions are: are the theoretical predictions the same and do the experimental data agree? The main difference in the prediction would be due to a difference in the choice of scale . The variations of the theory predictions, due to uncertainties in the parameters used, has been studied extensively [8]. It was found that the different choice of scale results in roughly a 2% normalization shift above E_T =300GeV and an additional 5% change in shape

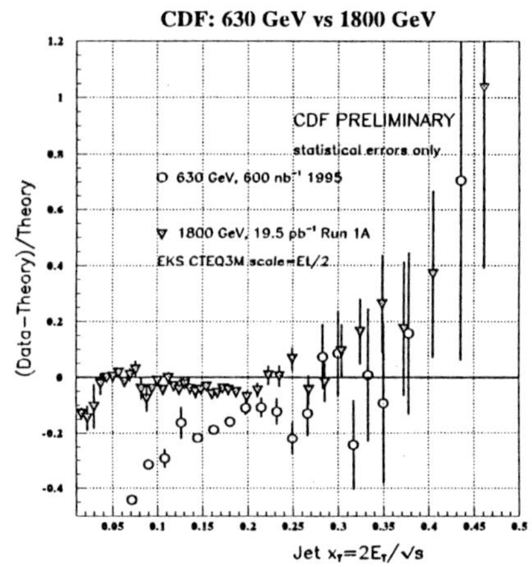

Fig. 5. Inclusive jet cross sections compared to theory at 630 and 1800 GeV from the CDF experiment as a function of x_T.

as a function of E_T between 50 and 300GeV. This cannot explain the difference between the experiments. To check the experimental results, DØ has repeated the measurement for $0.1 \leq \mid \eta \mid \leq 0.7$ where the systematic errors are not as well understood as in the $\mid \eta \mid \leq 0.5$ region. A functional form was fit to these data points and called $D0fit$. Figure 3 shows the quantity $(CDF - D0fit)/D0fit$, where CDF are the values of the cross section measured by CDF and systematic error bands are shown for both experiments. The experimental data measured in the same η region agree very well within errors. In Fig. 4 the data from the Fig. 1 have been put on the DØ graph and both results are shown in the same graph. From this figure it is evident that the two results are in very good agreement, even if only the DØ systematic error is taken into account. There is no need to invoke new physics yet and combining the data might result in a better and more clear conclusion.

The CDF experiment has also measured the same inclusive jet cross section at 630 GeV $p\bar{p}$ collisions. Fig. 5 displays the (DATA-THEORY) / THEORY quantity but now plotted as a function of $x_T = 2E_T/\sqrt{s}$ for $\sqrt{s}$=1800 and 630 GeV. The 630 GeV data show a strong E_T dependence and the shape is in obvious disagreement with the NLO prediction. At this time this is not understood. Previous results at the same center of mass energy by the UA2 experiment [3], are in very good agreement with leading order $(O(\alpha_s^2))$ QCD predictions. The DØ results for this energy are eagerly awaited.

Predictions of the inclusive jet cross section assume that the partonic hard scattering cross sections ($\hat{\sigma}_{ij}$ in Eq. 1) are correctly given by QCD. The angular distribution of the two final state jets in the center of mass system (cms) of the two initial partons is dominated by t-channel vector gluon exchange. This results in the characteristic Rutherford type angular distribution for spin=1 exchange: $dN/d\cos\theta^* \propto (1 - \cos\theta^*)^{-2}$, where θ^* is the angle between the incoming and outgoing partons. The shape of this distribution with its pole at $\cos\theta^* = 1$ is not very well suited for a detailed comparison

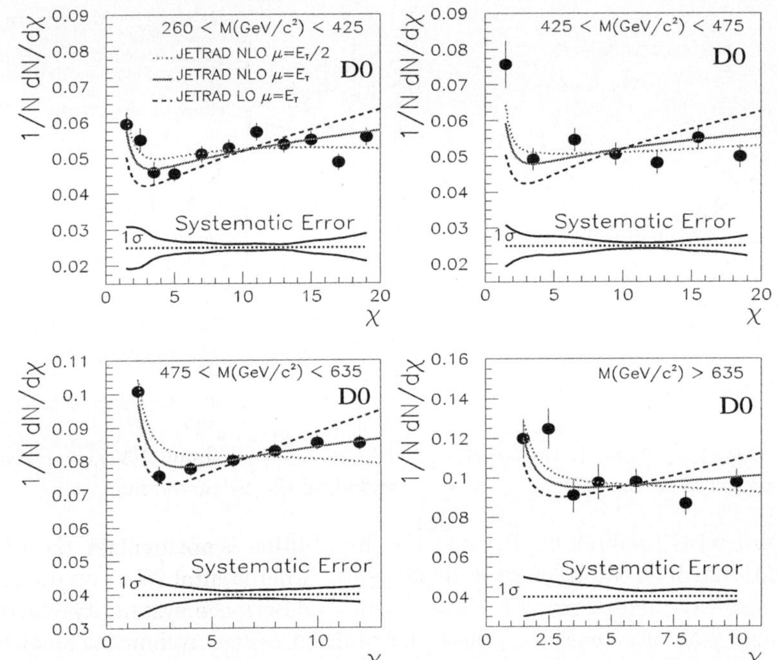

Fig. 6. The normalized dijet angular distribution $N^{-1}dN/d\chi$ for different regions of M_{jj}.

between theory and experiment. For that reason the variable χ, defined in Eq. 3 is used. This variable transforms a $(1 - \cos\theta^*)^{-2}$ distribution into a flat distribution. The relationships between the variables used to describe the dijet system are given by following equations.

$$\eta^* = \frac{\eta_1 - \eta_2}{2} \qquad \cos\theta^* = \tanh\eta^* \qquad \eta_{boost} = \frac{\eta_1 + \eta_2}{2} \tag{2}$$

$$M_{jj} = 2E_T^1 E_T^2(\cosh 2\eta^* - \cos(\phi_1 - \phi_2)) \qquad \chi = e^{2|\eta^*|} = \frac{(1+\cos\theta^*)}{(1-\cos\theta^*)} \tag{3}$$

Here all the quantities with a "*" are in the *cms* system and the indices 1, 2 refer to the final state jets. The inclusive cross section describing the final state is $d^3\sigma/dM_{jj}d\eta^*d\eta_{boost}$. Integrating over large fractions of the dijet invariant mass M_{jj} and η_{boost} space results in the normalized distribution $N^{-1}dN/d\chi$. This is typically referred to as the dijet angular distribution and its shape is practically independent of parton distributions, because all contributing graphs are dominated by one gluon exchange and have the same angular distribution. Experimentally the pseudorapidities (η_1, η_2) of the two leading E_T jets are used. Figure 6 shows the angular distribution as measured by the DØ experiment in different regions of M_{jj}. The data are compared

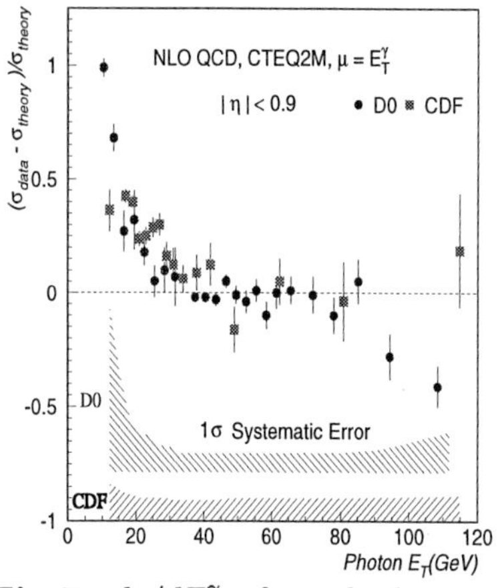

Fig. 7. $d\sigma/dE_T^\gamma$, from both experiments compared to theory.

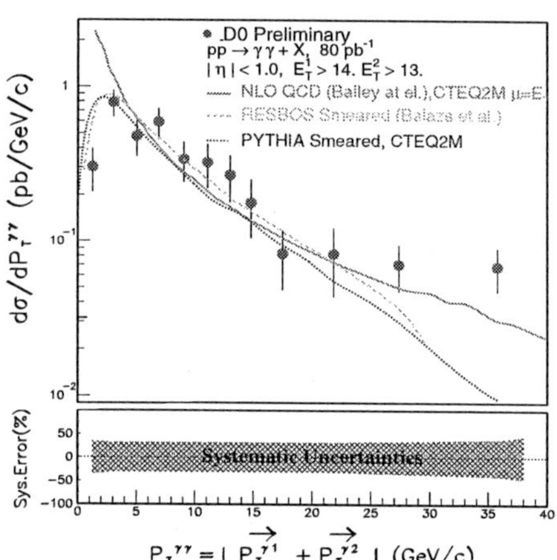

Fig. 8. $d\sigma/p_T^{\gamma\gamma}$ from DØ and corresponding theory predictions.

to LO and NLO predictions. Because the distribution is normalized, the LO prediction hardly exhibits a scale dependence, whereas the NLO one does. In this case LO predictions are not sufficient to describe the data. The CDF experiment has also measured these distribution over a somewhat smaller range in χ and those data are described well by LO and/or NLO predictions. In the highest M_{jj} bin these data can be used set a limit on quark compositeness. If a contact interaction with a uniform angular distribution and strength proportional to $(1/\Lambda)^2$, is added to the QCD Lagrangian, the DØ data require $\Lambda > 2.0$ TeV for all contact interaction scales [1, 2].

3 Photon production

The production of photons in hadron-hadron collisions has been pursued for about 20 years. The two basic processes contributing are $q\bar{q} \to \gamma g$ and $qg \to \gamma q$. Because of the qg initial state this process has always been considered as the best way to probe the gluon density. Practically, it has not been possible to achieve this in a quantitative way either because of experimental difficulties (π^0 backgrounds) and/or uncertainties in the theoretical predictions. Experimentally only the isolated photon cross section can be measured, because photons inside jets can not be resolved. Theoretical predictions attempt to take the experimental isolation cuts into account. Both Tevatron experiments have measured the isolated, inclusive photon cross section for $|\eta| < 0.9$. Fig. 7 shows the comparison of both data sets with a NLO QCD prediction in the familiar form (DATA-THEORY) / THEORY as a function of the transverse energy of the photon (E_T^γ). The bulk of the experimental data are in very good agreement and both data sets show a clear excess of events for $E_T^\gamma < 50$ GeV compared to the prediction. The significance is greatest in the

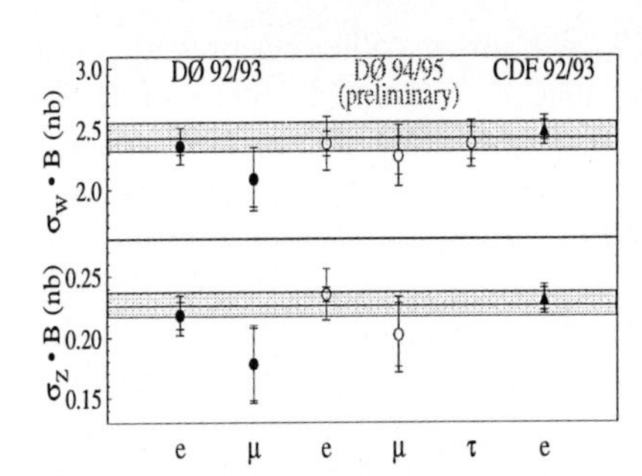

Fig. 9. W, Z cross sections from both experiments compared to theory.

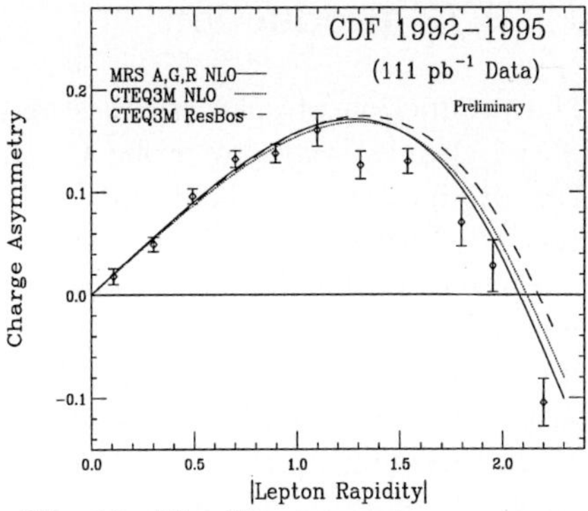

Fig. 10. The W asymmetry measurement from CDF and theory predictions.

CDF data, because of the smaller systematic error. In this case the CTEQ2M [7] PDF was used. In the meantime more precise HERA data have resulted in the CTEQ4M [7] distribution, which reduces the discrepancy around $E_T^\gamma = 15$ GeV by about 15%. This is in the right direction but not sufficient to explain the difference between data and theory. In a comprehensive study by the CTEQ group [9] it has been shown that this behavior is observed in all direct photon experiments, performed at different energies and in fixed target or collider mode. The NLO prediction is believed to lack a sufficient amount of additional gluon radiation. It has been shown [10] that adding parton showers to the NLO partons improves the data-theory comparison significantly. Also adding an ad hoc "k_T" to the initial state partons (equivalent to more radiation) improves the comparison. Especially the very precise data from the E706 [11] experiment, which studied γ and π^0 production in 800 GeV pBe collisions, exhibit the clearest need for additional radiation. They clearly require an average $k_T \approx 1.5$ GeV to even begin to agree with perturbative QCD predictions. Because of these problems the impact of direct photon data in determining PDF 's has been less then orginally expected.

Diphoton final states allow an independent and sensitive test of the need for additional radiation in the NLO predictions. Fig. 8 shows $d\sigma/dp_T^{\gamma\gamma}$ where $p_T^{\gamma\gamma}$ is the transverse momentum of the two photon final state. In lowest order $p_T^{\gamma\gamma} = 0$, but higher order processes cause it to increase. The NLO prediction shown in the figure clearly keeps rising for $p_T^{\gamma\gamma} \to 0$. Only the analytical predictions including resummed higher order contributions (RESBOS) [14] or predictions based on parton showers (PYTHIA) agree with the data. Unfortunately the RESBOS calculation is not available for single photon final states. Direct photon production is an area where theoretical work is needed and it is obvious that strict NLO predictions are not good enough to describe the rather precise experimental data.

4 W/Z production

The production of colorless W's and Z's in hadron collisions provides one of the cleanest ways to probe QCD, because it is a special case of Drell-Yan production at a fixed and large mass. As in most other processes the first comparison to be done with theory is the inclusive cross section. The prediction in this case is to $O(\alpha_s^2)$ (NNLO) using the CTEQ2M PDF . The experiments measure crossing section $\times$ branching ratio ($\sigma^{W,Z} \cdot B$) for e, μ and τ final states. Because the final state leptons are only detected in instrumented regions of the detectors, the data are corrected for acceptance with simulations that either resum higher order contributions or use parton shower generators. These measurements from both DØ and CDF and the QCD predictions are shown in Fig. 9 for all final states. The experimental results are in excellent agreement with the $O(\alpha_s^2)$ prediction and this is an important confirmation of the assumptions underlying the theoretical QCD predictions. The uncertainty in the prediction is dominated by the error in the PDF .

We now turn to the more differential cross sections and start with the measurement of the W charge lepton asymmetry measured by CDF. The charge asymmetry is defined as: $A(y_L) = [d\sigma^+/dy_L - d\sigma^-/dy_L]/[d\sigma^+/dy_L + d\sigma^-/dy_L]$. Here y_L is the rapidity of the final state lepton and its charge is indicated by the index on the cross section. Considering the basic process for W production: $ud \to W$, the asymmetry would be zero if the momentum distributions for u and d quarks in the (anti)proton were identical. The asymmetry measures the difference or the ratio of the two distributions and the lepton rapidity corresponds to certain momentum fractions x of the quarks. The x-range probed by this measurement is 0.01 to 0.04. Fig. 10 shows the CDF data for an integrated luminosity of 111 pb^{-1}, which is a factor of 6 more data than a previous measurement. When comparing to NLO predictions using PDF 's including the previous asymmetry results, it is clear that these old PDF 's need to be updated. A prediction based on a resummed calculation, using RESBOS, is also shown, but in this case the difference between resummed and NLO is rather small. This result is a beautiful example of how a single precise measurement can very accurately determine an aspect of parton distributions, in this case the ratio d/u.

Production of W's and Z's to first order is a $q\bar{q}$ annihilation process involving no other constituents. In this picture the bosons are produced with no transverse momentum (p_T), but can be boosted along the beam direction. However any initial or final state radiation will produce a finite p_T. Therefore the measurement of this quantity constitutes an important test of QCD predictions. Only Z production will be considered because the boson transverse momentum is more accurately measured in this case. Fig. 11 shows the experimental result $d\sigma/p_T^Z$, not corrected for acceptances and resolution smearing, obtained by DØ . The peak at small p_T^Z with a rapid fall off towards higher values is evident. The distribution is not well described by NLO

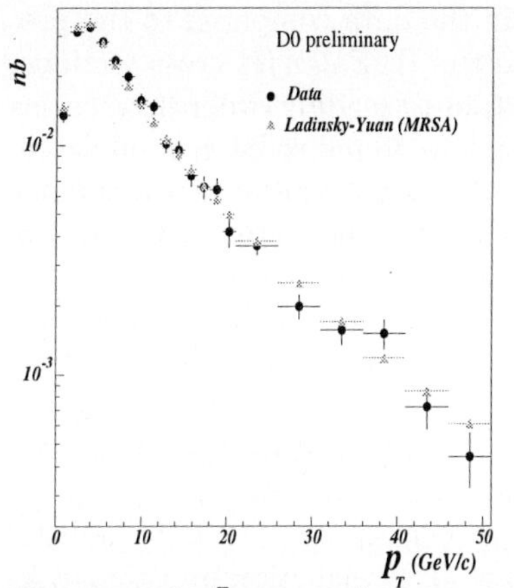

Fig. 11. $d\sigma/dp_T^Z$ from DØ with prediction from ref.[12].

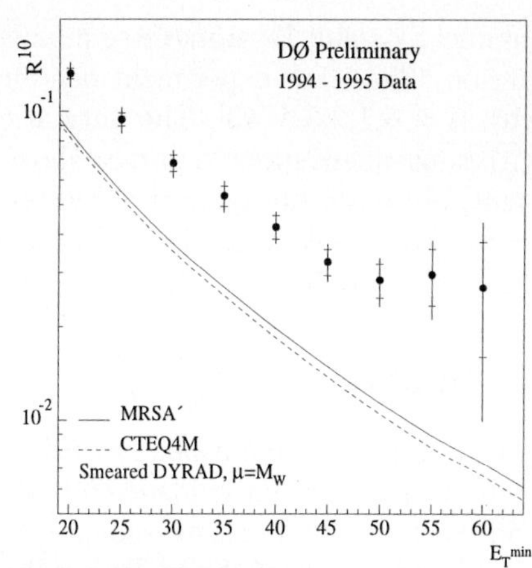

Fig. 12. R^{10} as a function of E_T^{min} as measured by DØ and the NLO prediction.

predictions and especially the turnover at small transverse momenta requires an approach where additional radiation is taken into account. This is done in a resummation approach, which is valid in the small p_T^Z region and which is matched to a perturbative $O(\alpha_s^2)$ calculation at high p_T^Z. This has been implemented [12] in a manner where several parameters are introduced that have to be derived from data. These parameters are considered to be universal i.e. process independent. Using fixed target, low energy Drell-Yan data these parameters were fit and then used to predict $d\sigma/p_T^Z$ at Tevatron energies. Fig. 11 shows this prediction, including detector acceptances and resolutions, and it agrees very well with the data. A fit χ^2 fit gives $\chi^2/dof = 24.4/20$.

The measurement of the ratio of $W + 1jet$ to $W + 0jet$ is in principle proportional to α_s and the desire to measure α_s motivated the original measurement of this quantity. The availability of $O(\alpha_s^2)$ calculations in the form of the program DYRAD [6] was an additional incentive, because it predicts these cross sections with small scale dependence. DØ has measured $R^{10} = W + 1j/W + 0j$ with the final state $W \to e\nu$. Jets are defined with a fixed cone size of 0.7 and are required to have a transverse energy $> E_T^{min}$. Fig. 12 shows the experimental result for a variety of E_T^{min} values. The theoretical parton level prediction for two PDF 's is shown as well. Contrary to the inclusive W cross section, any acceptance and kinematic cuts as well as experimental resolutions are now applied in the NLO calculation, after clustering partons into jets using the same algorithm as in the data. There is an apparent disagreement between data and theory, which at this time is not understood and is under investigation. A more consistent treatment of the resolution, especially in missing transverse momentum, reduces the discrepancy between data and theory somewhat, but it remains several standard deviations. It is also observed that there is an excess of events in the $W + 1j$

channel at small W transverse momentum in the data compared to the prediction. The CDF experiment has measured the $W, Z + njet$ cross sections, with n = 0,1,2,3,4 [13]. The data are compared to leading order predictions with subsequent HERWIG parton showering and are found to be in good agreement, although the level of agreement depends on the choice of scale. Once these data are compared to NLO predictions they may shine some light on the "R^{10}" puzzle.

5 Summary

Several topics have not been mentioned in this paper: b-quark and J/ψ cross sections, searches for evidence of BFKL signatures, double parton scattering and color coherence measurements in jet and W final states.

The precision of the hadron collider data require NLO QCD predictions and the agreement between data end theory is good as far as inclusive quantities are concerned. The theoretical predictions have greatly improved, through more precise parton distributions. At the moment the vast amount of rather precise hadron collider data are pointing to shortcomings in the theoretical predictions, some of which have been addressed by resumming higher order contributions. In the necessary interplay between theory and experiment it seems for now that experiment is pushing the limits of what theory can predict. The Tevatron data have stimulated a lot of activity in the area of QCD phenomenology and a lot of progress has been made, but there is still a lot of room for improvements and refinements with the goal to achieve more reliable QCD predictions in the future.

References

[1] E.Buckley-Geer, proceedings of this conference

[2] R.McCarthy, proceedings of this conference

[3] J.Alitti *et al.*, Phys. Lett. B257, 232 (1991)

[4] Proc. of the 1990 Summer Study on HEP, July 1990, Snowmass, CO, World Scientific, p.134-136 (Snowmass Jet Accord).

[5] S.Ellis,Z.Kunzst and D.Soper, Phys. Rev Lett. 64, 2121(1990).

[6] W.Giele, E.Glover, D.Kosower, Nucl. Phys. B403, 633 (1993).

[7] H.L.Lai *et al.*,Phys Rev D55, 1280 (1997) and references therein.

[8] B.Abbott *et al.*, hep-ph/9801285.

[9] J.Huston *et al.*, Phys. Rev D51, 51 (1995)

[10] H.Baer and M.Reno, Phys. Rev. D54, 2017 (1996)

[11] L.Apanasevich *et al.*,((E706 collaboration),hep-ex/971017.

[12] G.Ladinsky and C.P.Yuan, Phys. Rev. D50, 4239 (1994) and references therein.

[13] F.Abe *et al.*, Phys. Rev. Lett. 79,4760 (1997)

[14] C.Balazs and C.P.Yuan, Phys. Rev D56, 5558 (1997)

Soft Interactions and Diffractive Phenomena

Ralph Eichler (eichler@particle.phys.ethz.ch)

Institute for Particle Physics, ETH-Zürich, CH 8093 Zurich

Abstract. Recent results on hard diffraction at HERA and the Tevatron are presented. Charged particle multiplicities in diffraction and differences in multiplicity in quark and gluon jets measured at LEP are discussed. Spin effects in the fragmentation of leading quarks show some interesting features.

1 Total and Elastic Cross Section

The energy dependence on the center of mass energy $\sqrt{s}$ of all hadron-hadron total cross sections is described by

$$\sigma_{tot}(s) = A_{IP} s^{\alpha_{IP}(0)-1} + A_{IR} s^{\alpha_{IR}(0)-1} \ .$$

with universal exponents $\alpha_{IP}(0)-1$ and $\alpha_{IR}(0)-1$ and process dependent constants A_{IP} and A_{IR} [1]. Regge theory, which relates poles in a t-channel scattering amplitude to energy behaviour in the s-channel gives a link to Regge trajectories $\alpha(t)$. The trajectories have been determined in hadron scattering and describe also surprisingly well $\gamma\gamma$ and γp scattering. The first term in the total cross section formula dominates the high energy behaviour and the corresponding Pomeron trajectory $\alpha_{IP}(t)$ is unique in Regge theory with largest intercept and vacuum quantum numbers: $\alpha_{IP}(t) = 1.08 + 0.25\ GeV^{-2}\ t$. Next leading reggeons have approximately degenerate trajectories and carry the quantum numbers of ρ, ω, a_2, f_2 mesons. They are combined in an effective trajectory $\alpha_{IR}(t) = 0.55 + 0.9\ GeV^{-2}\ t$.

Figure 1a shows a measurement by the L3 collaboration [2] of the total $\gamma\gamma$ cross section where the universal exponents have been assumed and coefficients $A_{IP} = 173 \pm 7\ nb/GeV^{-2}$ and $A_{IR} = 519 \pm 125\ nb/GeV^{-2}$ were fitted. Figure 1b presents the γp cross section as a function of the γp energy $W_{\gamma p}$ at photon virtualities $Q^2 \sim 0$. The elastic vector meson production $\gamma p \rightarrow Vp$ with $V = \rho, \omega, \Phi, J/\Psi$ shows also a power law behaviour. The elastic cross section is related to the total cross section via the optical theorem

$$\frac{d\sigma^{elastic}}{dt}|_{t=0}(\gamma p \rightarrow Vp) \propto \sigma_{tot}^2 \propto (W^2)^{2(\alpha_{IP}(0)-1)}.$$

Figure 1b suggests that the growth $(W^2)^{2\lambda}$ is steeper for the heavier vector mesons, namely $\lambda = 0.22$ for J/Ψ-production [3] compared to the value for σ_{tot}, namely $\lambda = \alpha_{IP}(0) - 1 = 0.08$. For elastic ρ-production the exponent

λ grows from 0.08 at low Q^2 to 0.19±0.07 at $Q^2 = 20\ GeV^2$ [4]. Apparently λ is a function of the hardness of the scale (quark mass, Q^2) involved in the scattering.

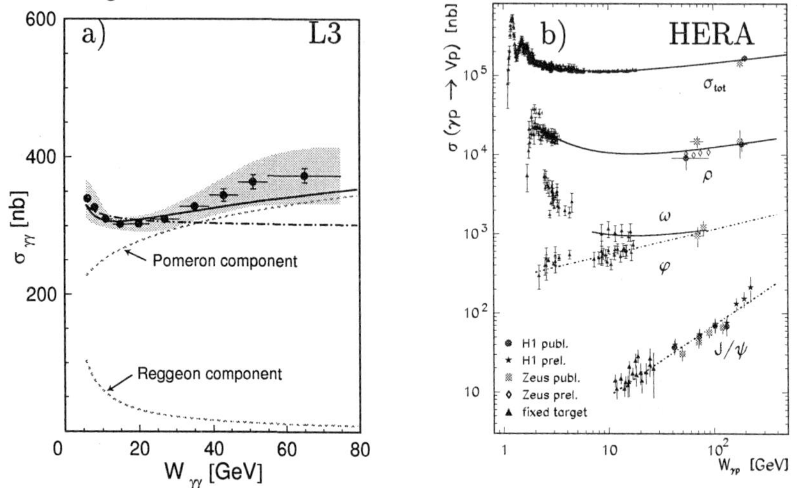

Fig. 1. a) total $\gamma\gamma$ cross section versus $\gamma\gamma$ energy $W_{\gamma\gamma}$. The shaded band is the systematic error, solid line a fit of the sum of Pomeron and Reggeon component. b) quasireal total and exclusive vector meson γp cross section with photon virtuality $Q^2 \sim 0$ vs γp center of mass energy $W_{\gamma p}$. The lines are a fit to the data.

2 Inelastic Diffraction

Diffractive scattering has been observed in $h-h$ interactions a long time ago. A typical signature is a forward peaked beam particle which remains intact or is excited to a small mass M_Y and a rapidity gap between it and the rest of the final state X (Figure 2a).

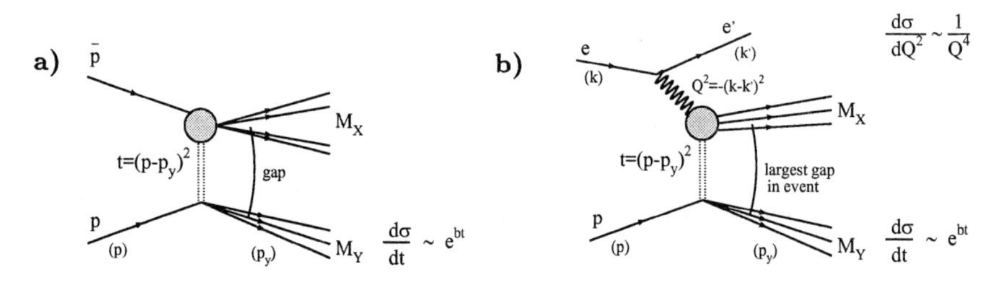

Fig. 2. a) Diffractive scattering in $p\bar{p}$ interactions. Hard diffraction is observed in reactions, where the system X consists of jets. b) diffractive scattering in deep inelastic ep interactions. Without detection of the leading proton experimental cuts constrain $t < 1\ GeV^2$ and the mass $M_Y < 1.6\ GeV$ (H1), $< 4\ GeV$ (ZEUS).

Diffractive scattering has also been observed in deep inelastic scattering (DIS) at HERA with a fraction of roughly 10% of total DIS. The kinematics is defined through the measurement of the scattered positron, the mass M_X of the system X via the hadronic final state and the observation of a gap between X and Y, fig. 2b. The standard DIS variables are $q = k - k'$, $Q^2 = -q^2$, $W^2 = (p+q)^2$, $x_{Bj} = q^2/2P \cdot q$ and the additionally measurable variables

$$x_{IP} \equiv \xi = \frac{M_X^2 + Q^2 - t}{W^2 + Q^2 - m_p^2}, \; \beta = \frac{Q^2}{2q \cdot (p - p_Y)} = \frac{x_{Bj}}{\xi} = \frac{Q^2}{M_X^2 + Q^2}.$$

The variable $t = (p - p_Y)^2$ is known only if the scattered proton is detected in the leading proton spectrometer, otherwise data from the forward detectors only limit t and M_Y (see caption of figure 2). The intuitive meaning of $\xi = x_{IP}$ and β can be best understood in the infinite momentum frame of the proton. Consider the Feynman diagrams of Figure 3. The virtual photon γ^* probes partons in the diffractively exchanged object which carries a momentum fraction x_{IP} of the proton. These partons carry a fraction β of this object [5]. In the following it will be shown, that this diffractively exchanged object can be viewed as in the formula of σ_{tot} as a sum of a Pomeron and a Reggeon term.

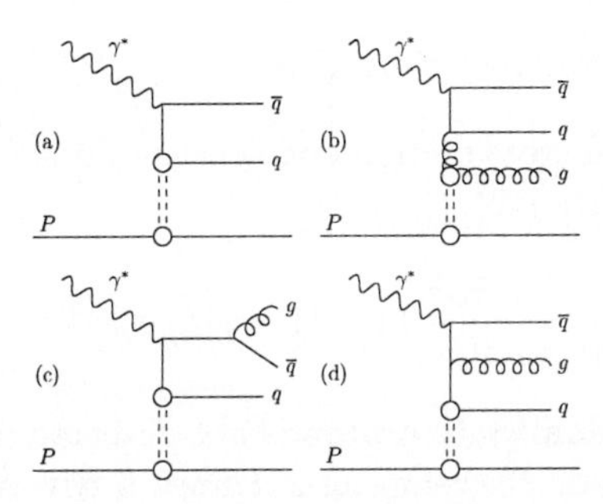

Fig. 3. Lowest order Feynman diagrams of inelastic diffraction in DIS. The dotted double line indicates the diffractively exchanged object in the infinite momentum frame of the proton.

An equivalent picture of diffraction is seen in the proton rest frame [6, 7, 8, 9, 10]. The Fock states of the photon $|\gamma>=|q\bar{q}>, |q\bar{q}g>, ...$ etc are formed long before the target (Ioffe length $\ell \sim 1/x$ ~several 100 fm) and the quark pairs interact softly through multiple gluon exchange with the target. Despite hard scattering at large Q^2 the target is only hit gently. The rapidity gap formation is then a long time scale process after the quark pair has traversed the proton and the diffractively produced final states X are expected to be sensitive to the topology and colour structure of the partonic fluctuations of the virtual photon.

2.1 Results from Leading Proton Spectrometer at HERA

Figure 4 shows preliminary results from the ZEUS leading proton spectrometer [11]. The device measures the energy E_p' of the scattered proton and its transverse momentum p_t. The kinematical variables are defined $x_L = 1 - \xi = E'_p/E_p$ and $t = -p_t^2/x_L - m_p^2(1-x_L)^2/x_L$. A clear diffractive peak at $x_L = 1 - \xi > 0.97$ is observed and the b-parameter ($d\sigma/dt \propto e^{bt}$) was fitted and found to be between 4 and 10 depending on x_L [11].

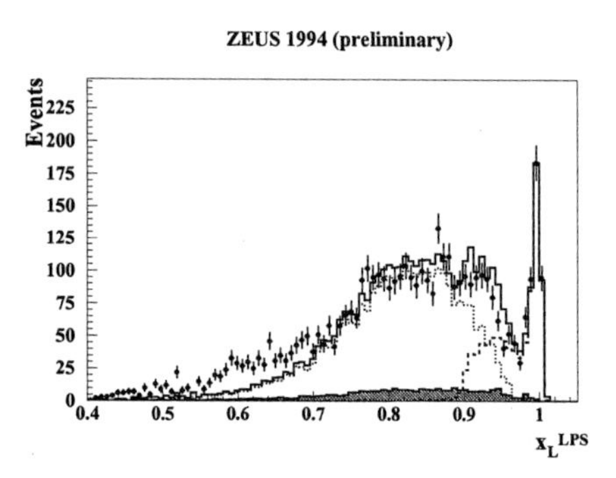

Fig. 4. Preliminary results from the ZEUS leading proton spectrometer.

2.2 Diffractive Structure Function

The fivefold differential cross section defines the diffractive structure function $F_2^{D(5)}$ with five variables [12]

$$\frac{d^5\sigma_{ep\to eXY}}{d\beta dQ^2 d\xi dt dM_Y} = \frac{4\pi\alpha^2}{\beta Q^4}\left(1+(1-y)^2\right)F_2^{D(5)}(\beta,\xi,Q^2,t,M_Y).$$

To gain statistics the final state proton is not detected and the variables M_Y and t are not measured. Therefore one defines a new structure function by integrating over the variables t and M_Y. (H1: $|t| < 1$ GeV^2 and $M_Y < 1.6$ GeV, ZEUS: $|t| < 1$ GeV^2 and $M_Y < 4.0$ GeV)

$$\frac{d^3\sigma_{ep\to eXY}}{d\beta dQ^2 d\xi} = \frac{4\pi\alpha^2}{\beta Q^4}\left(1+(1-y)^2\right)F_2^{D(3)}(\beta,\xi,Q^2).$$

H1 made an Ansatz for $F_2^{D(3)}$ as a sum of Pomeron exchange and Reggeon (Meson) exchange [13]. Each term factorises into a Pomeron/Meson flux dependent function of ξ, t and a structure function dependent on β, Q^2:

$$F_2^{D(3)}(\beta,\xi,Q^2) = \int dt\left[\frac{e^{B_{IP}t}}{\xi^{\alpha_{IP}(t)-1}}F_2^{IP}(\beta,Q^2,t) + C_M\frac{e^{B_{IR}t}}{\xi^{2\alpha_{IR}(t)-1}}F_2^{M}(\beta,Q^2,t)\right]$$

plus a possible interference term (since the f_2-meson and the IP have the same C-parity). In the above expression the integral is over $|t| <1$ GeV^2.

The measurements of H1 were fitted in 47 bins in $4.5\ GeV^2 \leq Q^2 \leq 75\ GeV^2$, $0.04 \leq \beta \leq 0.9$ with free parameters $\alpha_{IP}(0), \alpha_{IR}(0), C_M$. A very good fit confirms the Ansatz and two such examples are shown in Figure 5.

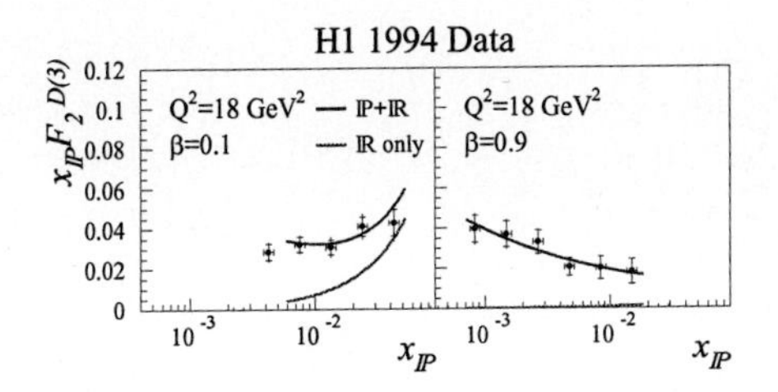

Fig. 5. $x_{IP}F_2^{D(3)}(\beta, Q^2)$ as a function of x_{IP} for two bins in Q^2 and β.

Contributions of the meson term is noticable at $\xi = x_{IP} > 0.01$ and small β (large M_X). As a result H1 gets [13]

$$\alpha_{IP}(0) = 1.203 \pm 0.02(stat) \pm 0.013(syst)^{+0.03}_{-0.035}(model)$$

$$\alpha_{IR}(0) = 0.50 \pm 0.11(stat) \pm 0.11(syst) \pm 0.10(model).$$

An analysis of the ZEUS collaboration [14] with a fit at fixed Q^2

$$\frac{d\sigma^D_{ep\to eXY}}{dM_X} \sim (W^2)^{2\overline{\alpha_{IP}}-2}$$

where the non-diffractive background has been parametrised and subtracted first yields a similar result for $\alpha_{IP}(0)$ (see Figure 6).

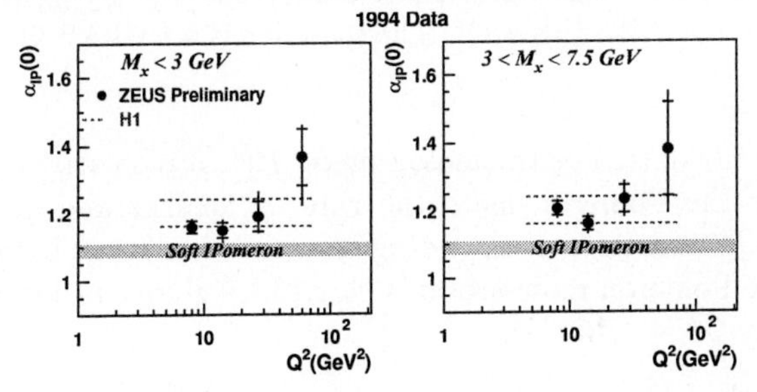

Fig. 6. $\alpha_{IP}(0)$ vs Q^2 in two bins of M_X. The band limited by the dotted lines indicate the range in $\alpha_{IP}(0)$ from the H1 analysis. The shaded band is the value found in soft processes like σ_{tot}.

The results of the two experiments are consistent with each other and the intercept in hard diffraction is larger than $\alpha_{IP}(0) = 1.08$ determined from σ_{tot}.

The diffractive structure function shows scaling violation as depicted in Figure 7. The structure function rises with Q^2 up to very high $\beta = 0.65$ values [13, 16]. The H1-Ansatz for $F_2^{D(3)}$ has been taken and the two structure functions F_2^{IP} and F_2^M parametrised at a fixed $Q_0^2 = 3GeV^2$ through parton distributions. These parton distributions are evolved according to Dokshitzer-Gribov-Lipatov-Altarelli-Parisi (DGLAP) [15] equation to any other Q^2 and

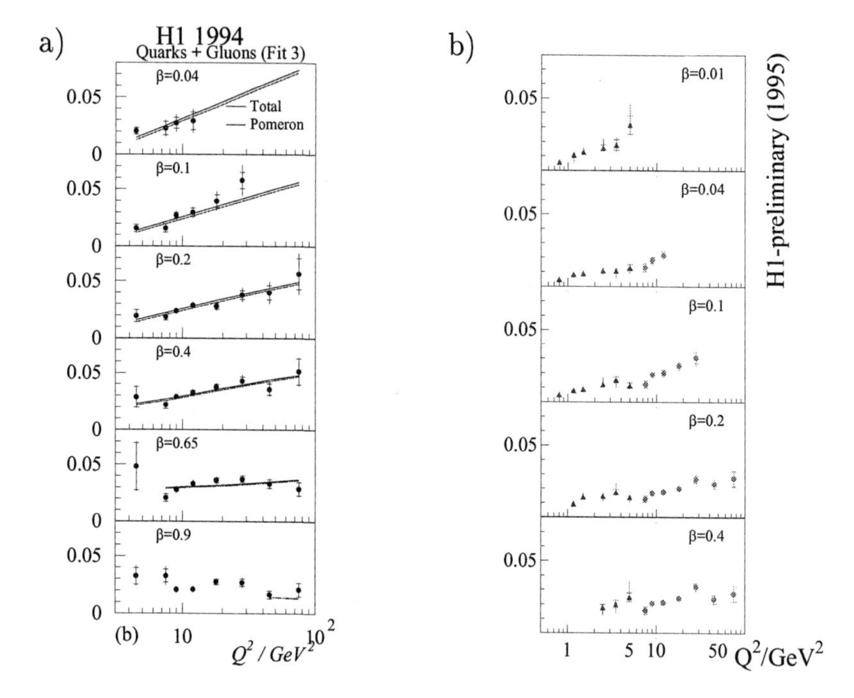

Fig. 7. Scaling violation of the diffractive structure function $\xi F_2^D(\xi, \beta, Q^2)$ at fixed $\xi = 0.003$ (a) and $\xi = 0.005$ (b) vs Q^2. The solid line is a DGLAP fit.

compared with the data. For the meson piece, F_2^M, a pion structure function was taken. For the Pomeron piece the ratio of quarks and gluons at the starting scale Q_0^2 and its corresponding shapes were fitted with the result that 90% of the Pomeron momentum is carried by gluons at $Q^2 = 4.5\ GeV^2$ and still 80% at $Q^2 = 75\ GeV^2$.

2.3 Hard Diffraction at the Tevatron

Hard diffraction has also been observed in $p\bar{p}$ collisions at the Tevatron where the hardness scale is given by the required jets in the final state or the production of a heavy gauge boson like W. Three processes are considered:

1. $p + \bar{p} \rightarrow p + gap + (X \rightarrow jet1 + jet2 + X')$. The kinematics has been defined in Figure 2a. Without the measurement of the final state proton the kinematical variable ξ cannot be measured. Its distribution is taken from MC model calculations (POMPYT) with a factorised Pomeron flux $f_{IP/p}(\xi, t) = \frac{K}{\xi^{2\alpha_{IP}(t)-1}} F^2(t)$ and a hard Pomeron structure function $\beta G(\beta) = 6\beta(1-\beta)$ [5]. In this model ξ is restricted through the final state topology to $0.005 \leq \xi \leq 0.015$. The ratio of diffractive dijets to all dijets with $E_T > 20 GeV$ is $R_{JJ} = 0.75 \pm 0.05(stat) \pm 0.09(syst)\%$ (CDF [17]), and $0.67 \pm 0.05(stat)\%$

(D0). This ratio is sensitive to the relative ratio of quarks and gluons in the Pomeron.

2. $p + \bar{p} \rightarrow p + (W \rightarrow e\nu + X')$. The gluon content in the Pomeron does not contribute much (suppressed by α_s). The result from CDF [18] for the ratio of diffractive to non-diffractive W-production $R_W = 1.15 \pm 0.55(stat) \pm 0.20(syst)\%$.

3. $p + \bar{p} \rightarrow jet1 + gap + jet2$. Rapidity gaps of size $\Delta\eta$ between jets from fluctuations in fragmentation are exponentially suppressed. Figure 8 shows data from D0 [19] where the fraction of events with a large gap as a function of the jet E_T (fig. 8a) and as a function of $\Delta\eta$ (fig. 8b) are plotted.

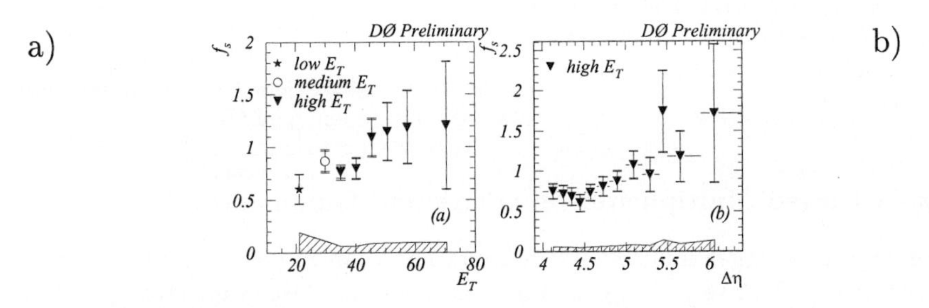

Fig. 8. Fraction $f_s[\%]$ of events with a rapidity gap between jets vs jet E_T (a) and vs the gap size $\Delta\eta$ (b).

It has become clear during the year, that not all 3 types of observations can be explained by this simple factorised POMPYT model. It remains to be seen if factorisation could be restored as in the case of HERA.

3 Fragmentation

3.1 Charged Multiplicity in Diffraction at HERA

Fragmentation gives the link between cross section on the parton level and observed particles in the detector. The charged particle multiplicity grows with the CM energy, with $\sqrt{s}$ in e^+e^- annihilation and with W in lepton-proton scattering. It has been shown that the multiplicity is independent of Q^2 [20]. The question arises if the diffractive system X defined in Figure 2b fragment as an independent object, i.e. is the multiplicity in diffraction governed by M_X or still by W?

In Figure 9 we compare the particle density $\frac{1}{N}\frac{dn}{dy}$ at the plateau [21] as a function of M_X and compare it to inclusive (non-diff) DIS at $W = M_X$. The multiplicity in diffractive events is about 0.6 units higher. We conclude that hadronisation in diffraction cannot be described by the fragmentation of a single colour string as implemented in the Lund model. The dominant diagram according to [7] is the one in Figure 3c where the system X is rather a colour octet string.

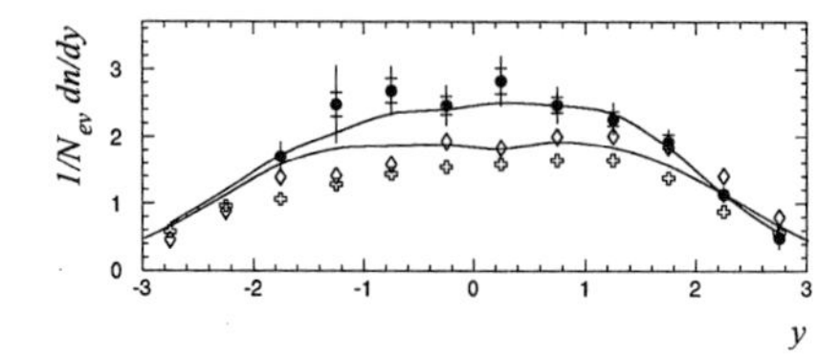

Fig. 9. Particle density $1/Ndn/dy$ as a function of rapidity for $M_X = 22.5\ GeV$ (full symbols, H1), for $W = 19\ GeV$ (open crosses, EMC μp) and for $W = 23.4\ GeV$ (open circle, E665 μD). The full curve is the RAPGAP simulation for γp processes and the dashed line from JETSET (e^+e^-).

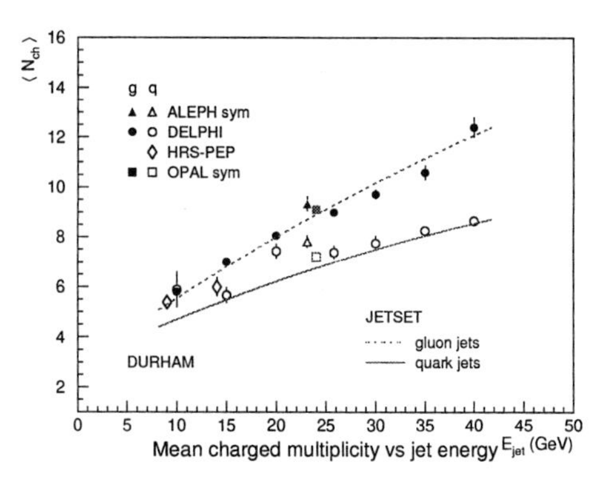

Fig. 10. Mean charged multiplicity vs jet energy for gluon jets (full symbols) and quark jets (open symbols).

3.2 Charged Multiplicities in Quark and Gluon Jets

One expects a larger multiplicity in gluon jets than in quark jets because of the stronger effective $g \to gg$ than $q \to qg$ coupling. Gluon initiated jets are selected at LEP in a 3-jet $q\bar{q}g$ topology, where the jets with identified heavy quarks are taken as quark initiated jets. The effect of a larger multiplicity of the remaining gluon jets (75-90% purity) is clearly seen in Figure 10 where results from all four LEP experiments are shown [23].

3.3 Leading Particles in Fragmentation

Leading particles carry quantum numbers such as flavour and spin of the primary produced partons. This is demonstrated for flavours in a result of SLD (Figure 11) and for spin in LEP results.

The polarised initial state in $e^+e^- \to q\bar{q}$ at SLC allows the distinction of a quark initiated jet as opposed to an antiquark jet. Consider jets from light flavours u, d, s and let $N(q \to h)$ be the number of hadrons of type h from quark jets. The production rate R and the difference D are defined

$$R_h^q = \frac{1}{2N_{evts}} \frac{d}{dx_p} \left[N(q \to h) + N(\bar{q} \to \bar{h})\right], \qquad D_h = \frac{R_h^q - R_{\bar{h}}^q}{R_h^q + R_{\bar{h}}^q}.$$

Figure 11 plots D_h vs x_p [25]. One sees clear evidence for production of leading baryons at high x_p. Also K^- and $\overline{K^*}$ are dominant over their antiparticles at high x_p.

Quarks from $Z^0 \to q\bar{q}$ are polarised. It has been known since several years, that this polarisation is transferred to the first rank baryons $\Lambda, \Lambda_c, \Lambda_B$. LEP measured a Λ-polarisation of $p_\Lambda = -32.9 \pm 7.6\%$(OPAL [22]) and $= -32 \pm 7\%$ (ALEPH). This is explained if the s-quark carries the spin of the Λ particle.

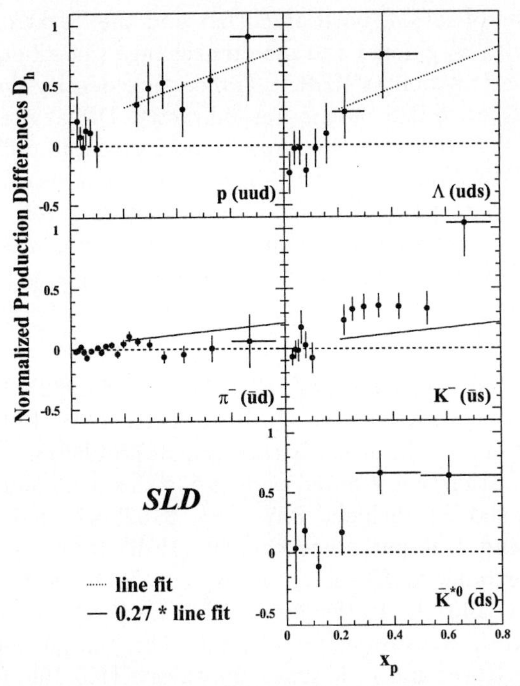

Fig. 11. Normalized production differences D_h between particles in quark jets and particles in antiquark jets vs x_p. The dotted lines represent the results of a linear fit to the baryon data for $x_p > 0.2$, and the solid lines represent this fit scaled by a dilution factor 0.27 valid for pions.

In $J^P = 1^-$ vector mesons any spin alignment must arise from hadronisation. The spin density matrix ρ_{ij} can be measured through angular dependence of the decay mesons with respect to the spin quantisation axis $W(\cos\theta_H) = \frac{3}{4}\left[(1-\rho_{00}) + (3\rho_{00}-1)\cos^2\theta_H\right]$.

In a statistical model [24] $\rho_{00} = \frac{1}{2}(1 - P/V)$ with P/V ratio of pseudoscalar to vector mesons in fragmentation. In this model $\rho_{00} \leq 0.5$ and $\rho_{00} = 0$ for $P/V = 1$, $\rho_{00} = 1/3$ for $P/V = 1/3$.

Results from CLEO, HRS, TPC-2γ, ALEPH, DELPHI, OPAL give $\rho_{00} \sim 1/3$ for ρ, D^*, B^*-mesons, $\rho_{00} > 1/3$ for D^*, Φ, K^*-mesons at large fractional momentum, and even $\rho_{00} = 0.66 \pm 0.11$ for K^* at $x_p > 0.7$ [26]. A value of $\rho_{00} > 0.5$ is an interesting observation and is not explained in any fragmentation model.

4 Summary

Multi-dimensional distributions in diffraction at the Tevatron and the precision data of HERA help the understanding of diffractive phenomena. The semi-inclusive cross sections rise with energy proportional $(W^2)^\lambda$ where λ grows with the hard scale involved. Inelastic diffraction using only Pomeron exchange breaks factorisation, both at HERA and the Tevatron. A sum of Pomeron exchange (80% gluons) and meson exchange (quark-dominated exchange) restores factorisation at HERA. The particle density dn/dy at fixed M_X is larger in diffractive DIS than in non-diffractive DIS at the corresponding $W_{\gamma p}$ which is related to a more complex colour string in diffraction than in non-diffractive DIS. Fragmentation properties of gluon jets are well measured at LEP and show a larger charged multiplicity in gluon initiated jets than in quark jets. Large spin alignment of leading vector mesons at Z^0 is observed, which is an interesting observation and not explained so far.

References

[1] A. Donnachie and P.V. Landshoff, Phys.Lett.B296(1992)227
[2] L3-Collaboration, M. Acciarri et al., Phys. Lett. B408 (1997) 450
[3] H1-Collaboration, S. Aid et al., Nucl.Phys. B472(1996)3
[4] ZEUS-Collaboration, contributed paper#639 this conference
[5] G. Ingelmann and P.E. Schlein, Phys.Lett. B152(1985)256
[6] J.D. Bjorken and J. Kogut, Phys.Rev. D8 (1973) 1341
[7] W. Buchmüller and A. Hebecker, Phys.Lett. B355(1995)573; A.Edin, G. Ingelmann and J. Rathsmann, Phys.Lett. B366(1996)371; W. Buchmüller, M.F. McDermott and A. Hebecker, hep-ph/9607290
[8] S. Brodsky, P. Hoyer and L. Magnea, Phys.Rev. D55(1997)5585
[9] N.N. Nikolaev and B.G. Zakharov, Z.Phys. C49(1991) 607; Z.Phys.C64(1994)631
[10] D.E. Soper, hep-ph/9707384
[11] ZEUS-Collaboration, contributed paper#644 this conference
[12] A. Berera and D.E. Soper, Phys.Rev. D50(1994) 4328; Phys.Rev. D53(1996) 6162; G. Veneziano, L. Trentadue Phys.Lett. B323(1994)201
[13] H1-Collaboration, C. Adloff et al., DESY 97-158
[14] Zeus-Collaboration, contributed paper#638 this conference
[15] Yu.L. Dokshitzer, Sov.Phys. JETP 46 (1977) 641; V.N. Gribov and L.N. Lipatov , Sov. J. Nucl. Phys. 15 (1972) 438,675; G. Altarelli and G. Parisi, Nucl.Phys. B126 (1977) 298
[16] H1-Collaboration, contributed paper#377 this conference
[17] CDF-Collaboration, contributed paper LP97, Hamburg (1997)
[18] CDF-Collaboration, F. Abe et al., Phys.Rev.Lett. 78(1997)2698
[19] D0-Collaboration, contributed paper#800 this conference
[20] H1-Collaboration, C. Adloff et al., Z.Phys. C72(1996)573
[21] H1-Collaboration, contributed paper#250 this conference
[22] Opal-Collaboration, K. Ackerstaff et al., CERN-PPE/97-104
[23] LEP experiments, see contributed paper #683 this conference
[24] J.F. Donoghue, Phys.Rev. D19(1979)2806
[25] SLD-Collaboration, contributed paper#287 this conference
[26] OPAL-Collaboration, K. Ackerstaff et al., CERN-PPE/97-094,

Structure Functions

Robin Devenish (r.devenish@physics.ox.ac.uk)

Physics Dept., Oxford University, UK

Abstract. Recent measurements of unpolarised and polarised nucleon structure functions and F_2^γ are reviewed. The implications for QCD and the gluon momentum distribution are discussed. The status of the understanding of $\sigma_{tot}^{\gamma^* p}$ in the transition region between real photoproduction and deep-inelastic scattering is summarised briefly.

1 Introduction

This talk covers three areas: unpolarised deep-inelastic scattering (DIS) data, parton distributions and associated phenomenology (Sec. 2); nucleon spin structure (Sec. 3) and the status of F_2^γ measurements (Sec. 4). New measurements from the Tevatron relevant for parton determination such as W decay asymmetries, Drell-Yan asymmetries, direct γ and inclusive jet cross-sections are covered by Weerts [1]. Diffractive DIS and the diffractive structure function are covered by Eichler [2], recent measurements of α_S by Ward [3] and the status of DIS measurements at very large Q^2 from HERA are summarised by Elsen [4].

2 Unpolarised Deep Inelastic Scattering

The kinematic variables describing DIS are $Q^2 = -(k - k')^2$, $x = Q^2/(2p.q)$, $y = (p.q)/(p.k)$, where $q = k - k'$ and k, k', p are the 4-momenta of the initial and final lepton and target nucleon respectively. At fixed s, where $s = (k+p)^2$, and ignoring masses the variables are related by $Q^2 = sxy$. The expression for the double differential neutral-current DIS cross-section is

$$\frac{d^2\sigma(l^\pm N)}{dxdQ^2} = \frac{2\pi\alpha^2}{Q^4 x}\left[Y_+ F_2(x,Q^2) - y^2 F_L(x,Q^2) \mp Y_- xF_3(x,Q^2)\right], \qquad (1)$$

where $Y_\pm = 1 \pm (1-y)^2$ and F_i $(i = 2, 3, L)$ are the nucleon structure functions. For Q^2 values much below that of the Z^0 mass squared, the parity violating structure function xF_3 is negligible. F_L is a significant contribution only at large y. At HERA both F_3 and F_L are treated as calculated corrections and the F_2 data quoted is that corresponding to γ^* exchange only. The kinematic coverage of recent fixed target and HERA collider experiments and some other details are given in Table 1.

The vast bulk of nucleon structure function data is for F_2 and here the overall situation is rather pleasing. The fixed target programme is complete

Table 1. Summary of recent structure function experiments. All the data referred in this table are available from the Durham HEPDATA database, at http://durpdg.dur.ac.uk/HEPDATA on the world wide web.

Beam(s)	Targets	Experiment	Q^2 (GeV^2)	x	R	Status
e^-	p,d,A	SLAC	$0.6 - 30$	$0.06 - 0.9$	yes	complete
μ	p,d,A	BCDMS	$7 - 260$	$0.06 - 0.8$	yes	complete
μ	p,d,A	NMC	$0.5 - 75$	$0.0045 - 0.6$	yes	complete
μ	p,d,A	E665	$0.2 - 75$	$8 \cdot 10^{-4} - 0.6$	no	compete
$\nu, \bar{\nu}$	Fe	CCFR	$1. - 500.$	$0.015 - 0.65$	yes	complete
$e^{\pm}$, p	-	H1	$0.35 - 5000$	$6 \cdot 10^{-6} - 0.32$	estimate	running
$e^{\pm}$, p	-	ZEUS	$0.16 - 5000$	$3 \cdot 10^{-6} - 0.5$	no	running

with the publication in the last 18 months of the final data from NMC [5] and E665 [6] to add to the older data from SLAC and BCDMS that still play an important role in global fits to determine parton distribution functions (PDFs). The first high statistics data from the 1994 HERA run were published by H1 [7] and ZEUS [8] last year. The F_2 data now covering 4 decades in Q^2 and 5 decades in x are summarised in Fig. 1. The data from the fixed target and HERA collider experiments are consistent with each other in shape and normalisation and show the pattern of scaling violations expected from perturbative QCD (pQCD). The systematic errors for the fixed target experiments are typically less than 5% and those for H1 and ZEUS around 5% for $Q^2 < 100\,\text{GeV}^2$, above this value the errors become statistics dominated. Fairly recently CCFR published an update of their high statistics $F_2^{\nu Fe}$ and $xF_3^{\nu Fe}$ data [9], following an improved determination of energy calibrations. The CCFR data and a determination of α_S are described in more detail by De Barbaro [10].

H1 has submitted preliminary values of F_2 from more recent HERA runs. The data are shown in Fig. 2: on the left for $1 < Q^2 < 100\,\text{GeV}^2$ (the region covered by the improved H1 rear detector) is from $5.4\,\text{pb}^{-1}$ taken in 1995/96 [11]; on the right for $150 < Q^2 < 5000\,\text{GeV}^2$ is from $22\,\text{pb}^{-1}$ accumulated over the period 1995-97 [12]. Also shown in Fig. 2 is a NLO QCD fit to H1 data with $Q^2 < 120\,\text{GeV}^2$, which is evolved to cover the region of the higher Q^2 data. All the new data are well described by the QCD curves and the characteristic steep rise of F_2 as x decreases is seen up to the largest Q^2 values.

2.1 The low Q^2 transition region

One of the surprises of the HERA F_2 data is the low scale from which NLO QCD evolution seems to work. H1 and ZEUS have now measured the cross-sections and hence F_2 from the safely DIS at $Q^2 \sim 6\,\text{GeV}^2$ through the

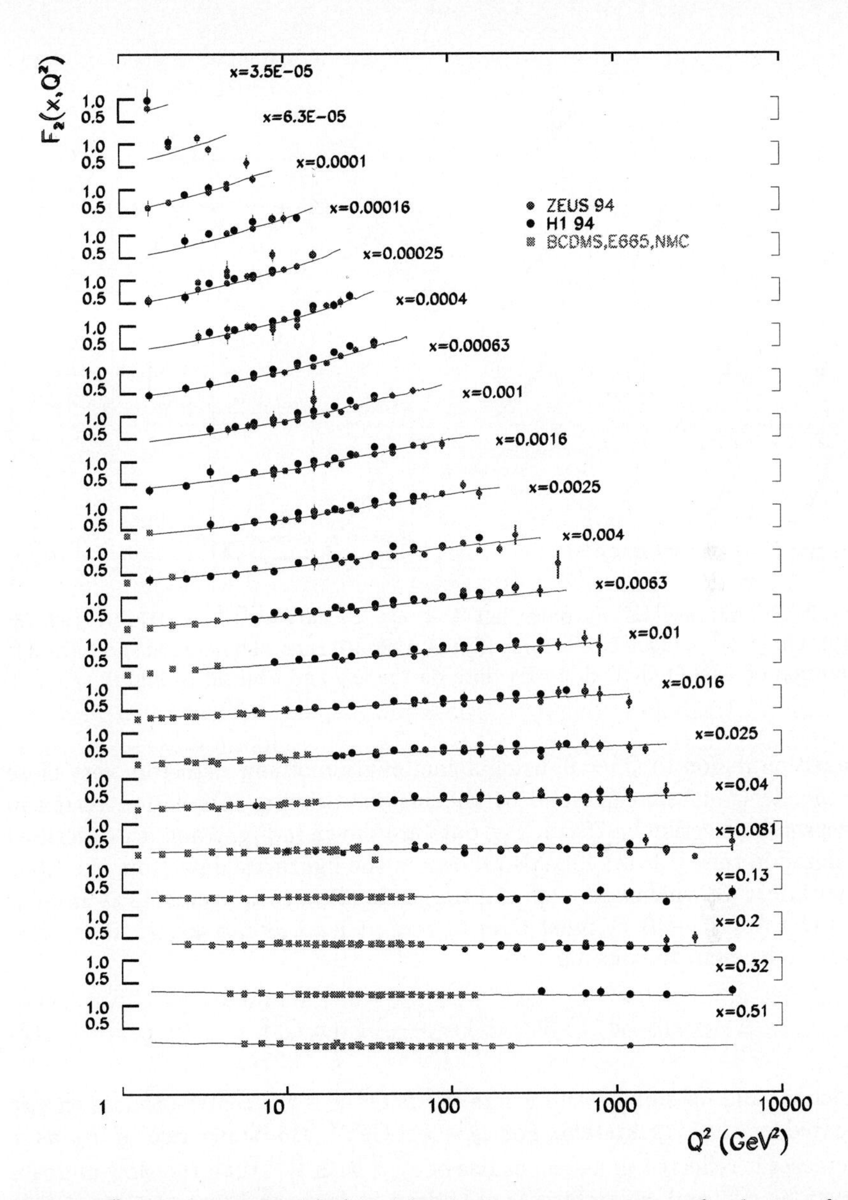

Fig. 1. F_2^p data from HERA(94) and fixed target experiments at fixed x as a function of Q^2. The curves shown are the NLO DGLAP QCD fit used to smooth the data during unfolding.

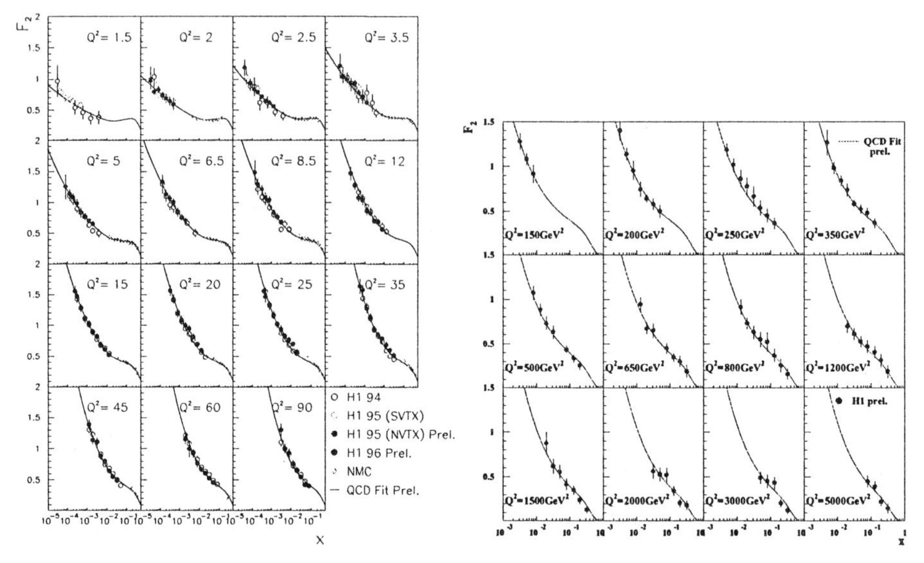

Fig. 2. Preliminary H1 F_2 data: left $1 < Q^2 < 100\,\mathrm{GeV}^2$ from HERA 1995/6; right $150 < Q^2 < 5000\,\mathrm{GeV}^2$ from HERA 1995-97 runs. The curves show the Q^2 evolution of a NLO QCD fit to H1 data on the left and evolved to higher Q^2.

transition region to $Q^2 = 0$, using a combination of new detectors very close to the electron beam line and by shifting the primary interaction vertex in the proton direction by 70 cm. The data are shown in Fig. 3 and are described in detail in refs. [13, 14, 15]. Also shown in the figure are data from the E665 experiment [6] which had a special trigger to allow measurements at small x and Q^2. As $Q^2 \to 0$ F_2 must tend to zero at least as fast as Q^2, it is often more convenient to consider

$$\sigma_{tot}^{\gamma^* p}(W^2, Q^2) \approx \frac{4\pi^2\alpha}{Q^2} F_2(x, Q^2) \tag{2}$$

which is valid for small x and where $W^2 \approx Q^2/x$ is the centre-of-mass energy squared of the $\gamma^* p$ system. For $Q^2 > 1\,\mathrm{GeV}^2$, the steep rise of F_2 as x decreases is reflected in a steeper rise of $\sigma_{tot}^{\gamma^* p}$ with W^2 than the slow increase shown by $\sigma_{tot}^{\gamma p}$ and characteristic of hadron-hadron total cross-sections.

Two very different approaches, both proposed before the HERA measurements, may be taken as paradigms. Glück, Reya and Vogt (GRV) [16] have long advocated a very low starting scale as part of their approach to generate PDFs 'dynamically' using NLO QCD. Predictions from their most recent parameterisation [17] are shown as the black solid line in Fig. 3, starting in the $Q^2 = 0.4\,\mathrm{GeV}^2$ bin and upwards. The data are in reasonable agreement with the theory down as far as the $Q^2 = 0.92\,\mathrm{GeV}^2$ bin. The other approach, that of Donnachie and Landshoff (DL) [18], is an extension of Regge parameterisations that describe hadron-hadron and real photoproduction data well.

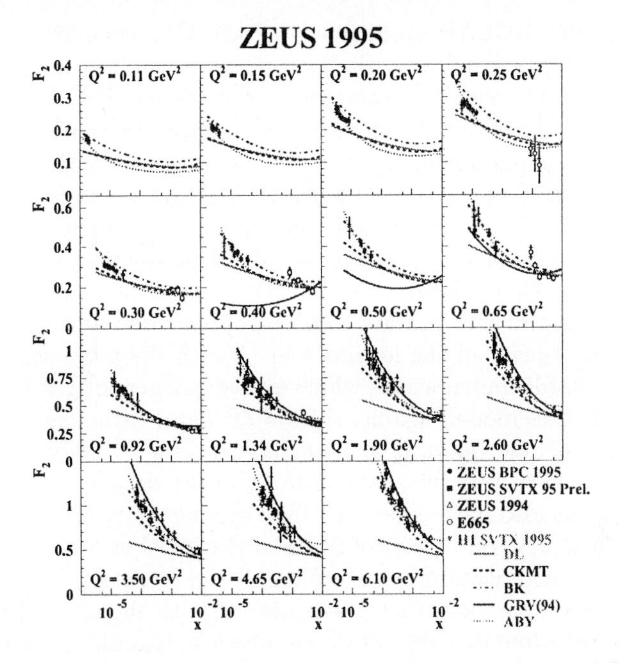

Fig. 3. F_2 data from E665, H1 and ZEUS at very small values of Q^2. The curves are described in the text.

The form that DL use to describe $\sigma_{tot}^{\gamma^* p}$ is

$$\sigma_{DL} = A(Q^2)(W^2)^{\alpha_P - 1} + B(Q^2)(W^2)^{\alpha_R - 1}, \tag{3}$$

where α_P and α_R are the intercepts of the Pomeron and Reggeon trajectories respectively with values $\alpha_P = 1.08$, $\alpha_R = 0.05$, determined from hadron-hadron data. The DL model gives the trend of the energy dependence of the very low Q^2 $\sigma_{tot}^{\gamma^* p}$ HERA data, up to $Q^2 \sim 0.4\,\text{GeV}^2$, though the normalisation of the model is a bit on the low side. The DL curves in Fig. 3 are the solid grey lines. In [19] the ZEUS collaboration has investigated the transition region. From NLO QCD fits with starting scales of $Q_0^2 = 0.4,\ 0.8,\ 1.2\,\text{GeV}^2$ it is found that only the latter two give acceptable descriptions of the data. For the limit $Q^2 \to 0$ a DL form is used. From these two approaches the transition to pQCD occurs in the Q^2 range $0.8 - 1.2\,\text{GeV}^2$.

The advent of accurate data from HERA has prompted many groups to try to model the behaviour of $\sigma_{tot}^{\gamma^* p}$ throughout the transition region. Very briefly: the model of Capella et al (CKMT) [20] uses a DL form at low Q^2 but allows the Regge intercepts α_P, α_R to become Q^2 dependent, for $Q^2 > 2\,\mathrm{GeV}^2$ DGLAP evolution gives the Q^2 dependence; Abramowicz et al (ALLM) [21] follow a similar approach but use a QCD inspired parameterisation at large Q^2; Badelek & Kwiecinski (BK) [22] take $F_2 = F_2^{VMD} + Q^2 F_2^{QCD}/(Q^2 + Q_0^2)$ where F_2^{VMD} is given by strict ρ, ω, ϕ VMD and the QCD scale parameter Q_0^2 is chosen to be $1.2\,\mathrm{GeV}^2$; Schildknecht and Spiesberger (ScSp) [23] revive the idea of GVMD to fit data for $0 < x < 0.05$ and $0 < Q^2 < 350\,\mathrm{GeV}^2$; Kerley and Shaw [24] modify the idea of long-lived hadronic fluctuations of the photon to include jet production; Gotsman, Levin & Maor [25] also follow this approach but have an additional hard QCD term; finally Adel, Barreiro & Yndurain (ABY) [26] have developed a model with an input x dependence of the form $a + bx^{-\lambda}$ with the two terms representing 'soft' and 'hard' contributions which evolve independetly with Q^2. Fig. 3 shows some of these models against the low Q^2 data. Although most of them give a reasonable description of the trends in x and Q^2, only the ScSp and ABY models (which were fit to the data) give the details correctly. In fact these two models also have defects as they are not able to describe the low energy $\sigma_{tot}^{\gamma p}$ data [27]. Very recently Abramowicz and Levy [28] have updated the ALLM parameterisation by including all the recent HERA data in the fit and result gives a satisfactory description of both the Q^2 and W^2 dependence. However, while this represents an advance, it is still true to say that more work needs to be done before the low Q^2, low x region is completely understood. More details of many of these models are given in the review by Badelek and Kwiecinski [29].

2.2 QCD and parton distributions

The striking rise of F_2 as x decreases was at first thought, at least by some, to be evidence for the singular behaviour of the gluon density $xg \sim x^{-\lambda}$ with $\lambda \sim 0.3 - 0.5$ proposed by Balitsky et al [30] (BFKL) and a breakdown of 'conventional' pQCD as embodied in the DGLAP equations. By the time of the EPS HEP95 conference in Brussels [31] the pendulum had swung the other way, largely through the work of Ball & Forte [32] on 'double asymptotoc scaling' (DAS) and the success of GRV(94) [17] in describing the data. In both cases the rise in F_2 is generated through the DGLAP kernels with a non-singular input. It is clear from Figs. 1, 2 that NLO DGLAP Q^2 evolution can describe the F_2 data from $Q^2 \approx 1.5\,\mathrm{GeV}^2$ to the highest values of $5000\,\mathrm{GeV}^2$. The two global fitting teams in their most recent determinations of the PDFs (CTEQ4 [33] and MRS(R) [34]), which include the HERA 1994 data, now use starting scales of around $1\,\mathrm{GeV}^2$. The quality of the fits is good with χ^2/ndf in the range $1.06 - 1.33$ and it is found that the gluon distribution is now non-singular in x at the input scale with the quark sea still mildly singular. Both CTEQ and MRS give PDFs for $\alpha_S(M_Z^2)$ in the range $0.113 - 0.120$ as there is some indication that the more recent determinations [3, 35] give a somewhat larger value than the 'DIS value' of 0.113

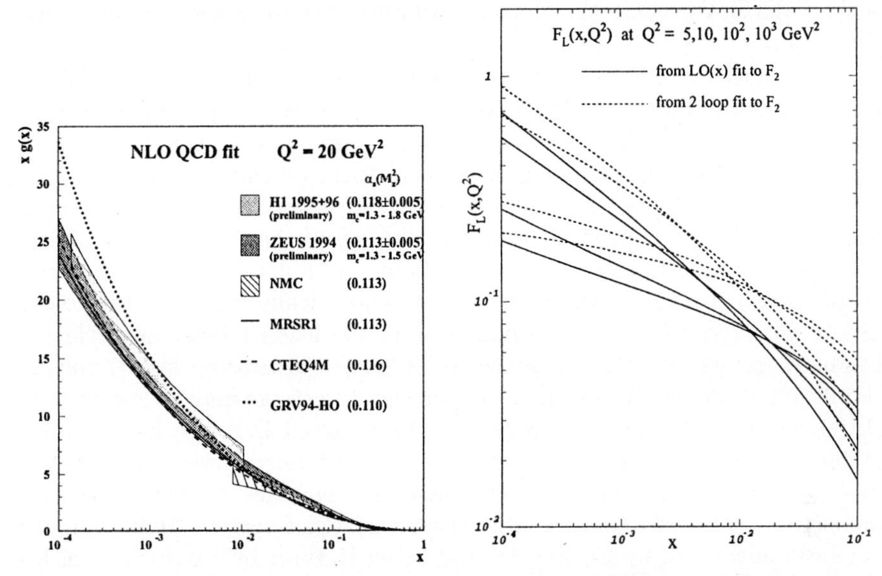

Fig. 4. Left: the gluon momentum density from H1 and ZEUS, together with an earlier result from NMC. The error bands are from the experimental systematic errors. The curves are from some recent global fits. Right: predictions for F_L at four values of Q^2 from Thorne showing the result of a standard two-loop QCD calculation and his LORSC procedure (labelled LO(x)).

determined from a fit to BCDMS and SLAC data [36]. The extension of accurate measurements to low x provided by the HERA(94) data has led to a big improvement in the knowledge of the gluon density. At low x the gluon drives the scaling violations through $\dfrac{dF_2}{d\ln Q^2} \sim \alpha_S P_{gg} g(x, Q^2)$. Apart from the global fits already mentioned both ZEUS and H1 have performed NLO QCD fits to extract $xg(x, Q^2)$. The advantage that the experimental teams have is that they can include a full treatment of systematic errors. Since HERA data does not extend to large x fixed target DIS data has to be included to fix the parameters of the valence quark distributions. ZEUS uses its 1994 HERA data and a fixed $\alpha_S = 0.113$, H1 fits HERA 1995/96 data and $\alpha_S = 0.118$ More details are given by Prinias [37]. The resulting gluon distributions are shown in Fig. 4(left) together with that from the NMC experiment and some curves from global fits. The total error is about 10% at the lowest x values. All determinations agree within the error bands except for GRV(94) which was not fit to the HERA(94) data and which does not describe the recent HERA data in detail. Part of the discrepancy comes from the lower value of α_S used, but it is also known that the very low starting scale of 0.3 GeV^2 makes the gluon distribution rise too steeply at moderate Q^2 values.

Despite the manifest success of DGLAP evolution in describing F_2 data, the argument about low x QCD continues. If DGLAP is the full story then why are the large $\ln(1/x)$ terms suppressed? A number of authors [38] have investigated the need for including the $\ln(1/x)$ terms ('resummation') but come to different conclusions. The most complete approach is that of Kwiecinski, Martin and Stasto [39] which combines the BFKL and DGLAP equations and gives a reasonable representation of the low x data. Another approach to BFKL which is quite successful phenomenologically is that of the colour dipole [40]. Apart from the resummation of the leading twist log terms, it has been argued recently that higher twist (power corrections in Q^2) may be significant at low x [41] and that shadowing corrections may be larger than BFKL effects in the kinematic region of HERA data [42]. It may be that some of diferences in outcome can be traced to different renormalisation schemes. A way to avoid such difficulties is to formulate the problem in terms of *physical* quantities, such as F_2 and F_L, rather than parton densities. This approach has been advocated by Catani [43] and taken furthest by Thorne [44] in his Leading Order Renormalisation Scheme Consistent (LORSC) framework. Although only at leading order he gets slightly better fits to the low x data than the conventional DGLAP global fits. What is crucially needed to sort out these various ideas are measurements of another observable as the different schemes can all fit F_2 but then differ for the other. This is demonstrated in Fig. 4(right) for F_L.

2.3 F_L and F_2^c

All fixed target experiments, except E665, have provided measurements of F_L. The measurement requires collecting data at high y for at least two centre-of-mass energies. The most recent measurements are from SLAC/E140X [45], NMC [5, 46] and CCFR [47]. At the smallest x value of these data, $4 \cdot 10^{-3}$ from NMC, F_L is possibly rising, but the errors are rather large. The x range and precision of the F_L data are both insufficient for them to discriminate between low x models. To date HERA has run essentially at a fixed centre-of-mass enegry of 300 GeV thus precluding a direct measurement of F_L. In the meantime H1 has used NLO QCD and their high statistics data to make an estimate of F_L [48]. The essence of the idea is to determine F_2 for $y < 0.35$ (where the contribution of F_L to the cross-section is negligible) by a NLO QCD fit. The fit is then extrapolated to larger y and used to subtract F_2 from the measured cross-section. At this conference the results for F_L were updated by preliminary data from the HERA 1996 run [11], giving F_L at $y = 0.68$ and 0.82. The extrapolation is the most uncertain part of the analysis. H1 has checked that using other models for the extrapolation gives the same value for F_L to within a few percent, but it has been argued that the error could be larger [49]. The H1 estimate for F_L is compatible with pQCD calculations using recent global PDFs.

The calculation of the NLO coefficient functions for massive quarks by Laenen et al [50] gave an impetus for the question of how massive quarks should be included in NLO global fits. The GRV(94) fit and the fits by H1 and ZEUS include charm only by the boson-gluon fusion (BGF) process. It has been argued that this cannot be correct well above threshold when the charm mass becomes negligible, charm should then be treated as any other light quark. This interesting subject will not be pursued here as it can be followed in refs [51], rather the status of measurements of F_2^c will be reviewed briefly. At HERA charm can contribute up to 30% of the cross-section, so it is important to understand both how to describe it theoretically and to measure it directly. The methods used by H1 and ZEUS to tag charm are by D^*, D^0 two body decays and by the $D^* - D^0$ mass difference. Statistics are limited by the small combined $D^* \to K\pi\pi$ branching ratio of only 2.6%. A major source of systematic error is the extrapolation of the measured D^* production cross-section to the full phase space in rapidity and p_T. All these matters are covered in more detail by Prinias [37]. The HERA results for F_2^c are shown in Prinias Fig. 5 together with a NLO calculation including the uncertainty in the charm quark mass. The results are encouraging and their precision will improve through higher luminosity and the use of microvertex detectors (installed in H1, planned for ZEUS). In another contribution [52], also covered by Prinias, H1 have used tagged DIS charm events from HERA(95) data to make a direct determination of the gluon density at four x values between $0.7 \cdot 10^{-3}$ and $0.5 \cdot 10^{-1}$ (Prinias Fig. 6).

The material in section 2 is covered in greater detail in a recent review by Cooper-Sarkar, Devenish and De Roeck [53].

3 Nucleon Spin Structure

The challenge of polarised DIS is to understand the dynamical distribution of spin amongst the nucleon's constituents, summarised by the relation $\frac{1}{2} = \frac{1}{2}\Delta\Sigma + \Delta g + \langle L_z \rangle$ where $\Delta\Sigma$, Δg are the contributions of the quarks and gluons respectively and $\langle L_z \rangle$ is the contribution from parton orbital angular momentum. The primary measurements are the spin asymmetries for nucleon spin parallel and perpendicular to the longitudinally polarised lepton spin. They are related to the polarised structure functions g_1, g_2 by kinematic factors. Only g_1 has a simple interpretation in terms of polarised PDFs, namely

$$g_1(x) = \frac{1}{2} \sum_f e_f^2 (q_f^{\uparrow}(x) - q_f^{\downarrow}(x)) = \frac{1}{2} \sum_f e_f^2 \Delta q_f(x) \tag{4}$$

where the sum is over quark and antiquarks with flavour f and $q_f^{\uparrow}$, $q_f^{\downarrow}$ are the quark distribution functions with spins parallel and antiparallel to the nucleon spin. Full details of the formalism and QCD evolution equations may be found in ref. [54]. The observed asymmetries are reduced by the beam and target polarisations and the target dilution factor. Polarisations are usually greater than 50%, but the dilution factor is generally quite small for solid or liquid targets, typically 0.13 for butanol and 0.3 for ^{3}He. The HERMES experiment at DESY uses a polarised internal gas jet target in the HERA-e beam and thus achieves a dilution factor of 1. Apart from HERMES, the latest round of experiments from SLAC and CERN is almost complete, details are given in Table 2. From the table it can be seen that the measurements of polarised DIS do not reach very small values of x, the largest range is that of the SMC experiment.

Table 2. Summary of recent polarised structure function experiments.

Lab	Beam	Targets	Experiment	x	Status
SLAC	e 29(GeV)	^{3}He, NH_3, ND_3	E142/3	$0.03 - 0.8$	complete
SLAC	e 48	^{3}He, NH_3, LiD	E154/5	$0.014 - 0.7$	analysis
CERN	μ 190	D- H- butanol, NH_3	SMC	$0.003 - 0.7$	complete
DESY	e 27	H, ^{3}He	HERMES	$0.023 - 0.6$	running

Most of the data is for g_1 and there is nice agreement between the different experiments as can be seen from Fig. 5(left). There is a small amount of data for g_2 from the SLAC experiments, g_2 is small and consistent both with zero and the expectation of the twist-2 calculation. More details on the individual experiments are to be found in the contributions of Souder [55] (E154/5), Le Goff [56] (SMC) and Blouw [57] (HERMES). New preliminary data on the proton asymmetry comes from the HERMES collaboration [57] and the SLAC E155 experiment [58], both offer the prospect of reduced statistical errors as can be seen for E155 from Fig. 5(right).

Apart from more accurate data, the big advance this year has been the extensive use of NLO QCD fitting. Apart from the intrinsic interest in testing QCD, the NLO fit also gives the best extrapolation of the data to a common Q^2 for the evaluation of sum rum integrals $\Gamma_1^i(Q^2) = \int_0^1 g_1^i(x, Q^2)dx$ and the evaluation of separate parton components. The first NLO fits were performed in 1995/6 [59], this year the experimental groups SLAC/E154 [55] and SMC [56] and the theoretical teams of Altarelli et al (ABFR) [60] and Leader et al (LSS) [61] have published such analyses. There are considerable differences of detail in the approaches taken by the different groups, perhaps the most important is the choice of factorisation scheme, LSS use $\overline{\text{MS}}$ and all other groups follow the Adler-Bardeen scheme to give a scale independent first moment for $\Delta\Sigma$, $\Delta\Sigma_{AB} = \Delta q_0 + n_f \frac{\alpha_S}{2\pi} \Delta g$. All groups assume a

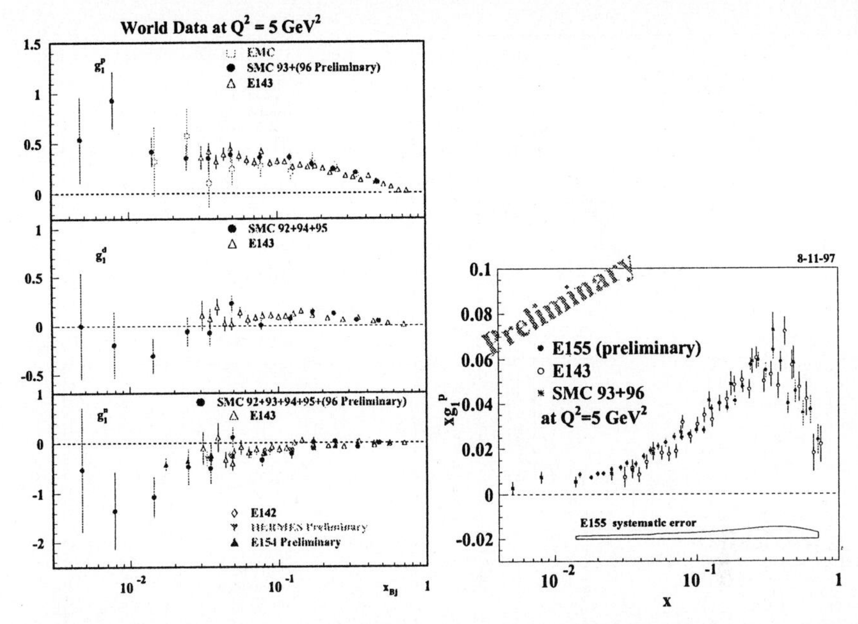

Fig. 5. Left: a compilation by the SMC collaboration of data on g_1 for protons, neutrons and deuterons from SMC, SLAC and HERMES polarised DIS experiments. Right: Preliminary data for xg_1^p from the SLAC E155 polarised DIS experiment, compared to earlier data.

non-singular x dependence for the input distributions at Q_0^2 and the partonic constraint $|\Delta q_{NS}| < q_{NS}$, where NS refers to the non-singlet contribution. The quality of the fits is good and one finds that the non-singlet valence quark distributions are quite well determined. The results for the quark singlet and gluon distributions are less good as there are no data for $x < 3 \cdot 10^{-3}$. These features are shown in Fig. 6 from the ABFR fits, the two left hand plots show the quality of the fit to data (fit B) and the two right plots show $\Delta\Sigma$ (upper) and Δg (lower) for a variety of different assumptions about the low x behaviour (see [60] for details).

In addition to extrapolation in Q^2, to evaluate Γ_1^i the data must also be extrapolated in x. There is no problem as $x \to 1$, but there is still considerable uncertainty as $x \to 0$. This is of course a reflection of both the lack of data and the range of possible behaviours for the singlet distributions at small x. SMC has investigated this point in some detail [56] for the evaluation of Γ_1^p. Generally Γ_1^p is measured to about 10% and Γ_1^n to about 20%. For the parton components, the quark integral is known to about 10% but the gluon integral only to 40% (ABFR) and more like 100% error from the experimenters fits. All agree that the gluon contribution is positive.

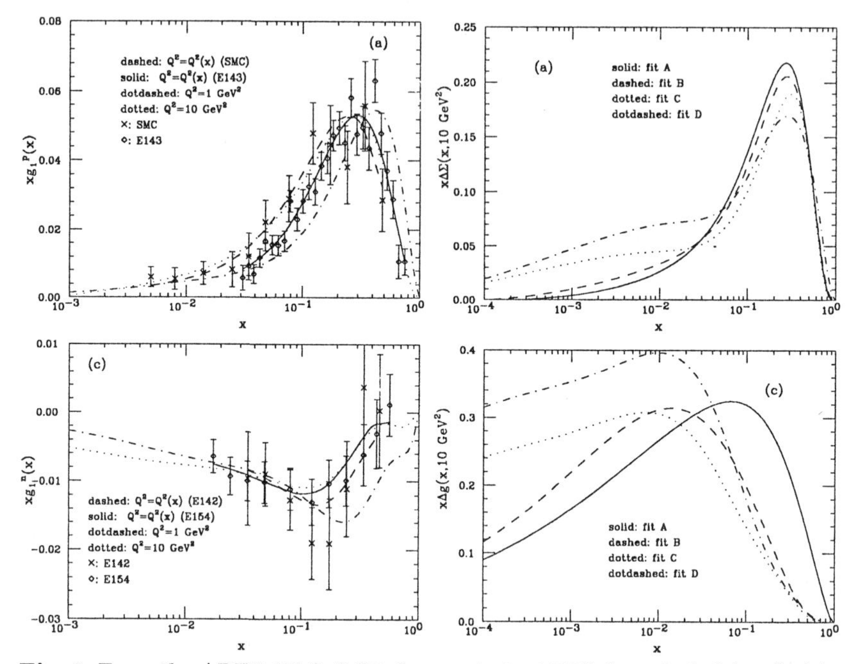

Fig. 6. From the ABFR NLO QCD fit to polarised DIS data. Left (a) xg_1^p, (c) xg_1^n; right (a) $x\Delta\Sigma$, (c) $x\Delta g$. Curves A, B, C, D refer to fits with different assumptions for low x behaviour.

What does this mean for the sum rules? The fundamental Bjorken sum rule $\Gamma_1^p - \Gamma_1^n = \frac{C_1^{NS}(Q^2)}{6}\left|\frac{g_A}{g_V}\right|$, where C_1^{NS} is a QCD coefficient known to order α_S^3, is found to be reasonably well satisfied, at about the 10% level, by all groups. The theoretically less well found Ellis-Jaffe sum rules for Γ_1^p, Γ_1^n separately are violated at the 2σ level. The overall situation is summarised in Fig. 7.

For inclusive measurements, this situation will not improve until there is data at smaller values of x, from RHIC or a polarised HERA collider. Another way to learn about individual parton distributions is through the measurement of semi-inclusive asymmetries. This type of measurement has been pioneered by the SMC whose latest results are reported by Baum [62]. HERMES has also presented some preliminary semi-inclusive results to this conference [57]. Asymmetries for identified particles such as positive or negative hadrons have been measured and from these the valence contributions Δu_V, Δd_V and the sea quark $\Delta \overline{q}$ (with some additional assumptions) determined. In the future such techniques applied to charmed particles will help to pin down the gluon contribution as well.

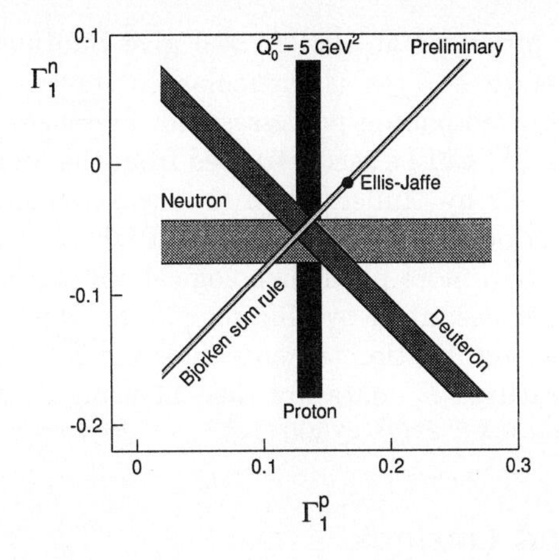

Fig. 7. Summary of the status of experimental determination of the integrals Γ_1^p, Γ_1^d, Γ_1^d and the Bjorken and Ellis-Jaffe sum rules by the SMC collaboration.

4 F_2^γ

The photon structure functions both for the two-lepton final states and the hadronic final state, F_2^γ, have been measured from two-photon interactions at LEP. The details of the measurements by ALEPH, DELPHI and OPAL are covered in the mini-review talk by Nisius [63]. In such measurements, only one scattered $e^\pm$ is detected (to give the Q^2 of the event), the other giving the 'target' photon is lost down the beam pipe. Q^2 is measured almost directly from the tagged lepton, but x has to be deduced from the measured final state particles. The Monte Carlo modelling of the physics and detectors is thus very important and some significant improvements have been made in these areas recently [64]. The increase of the beam energies in LEP2 operations has increased both the phase space and the statistics for two-photon physics.

Apart from the interest in determining the partonic content of the photon, the QCD evolution equations involve an inhomogeneous term which can be calculated from $\gamma \to q\bar{q}$ splitting. The data for F_2^γ are well described by NLO QCD fits and the larger lever arm in Q^2 allows one to see for the first time the logarithmic increase of F_2^γ with Q^2, as demonstrated in Nisius [63] Fig. 3. Because of the limited reach in small x at LEP, it has not been possible to determine if F_2^γ rises steeply as x decreases. For the same reason the gluon component of the NLO fits is not well determined.

Photoproduction processes at HERA also give information on photon structure. The process $\gamma p \to j_1 j_2 X$ is particularly attractive as it is sensitive to both direct and resolved photon processes and the kinematic variables x_γ and p_t^2 (equivalent to Q^2) can be reconstructed from the final state jets. Details are given in the talk by Muller [65] of the measurement of this process by H1, and the extraction of an effective photon PDF $\overline{f_\gamma} = f_{q/\gamma} + \frac{9}{4} f_{g/\gamma}$ for $0.2 < x_\gamma < 0.7$. Finally a more phenomenological approach to the problem of the small x region is described by Gurvich [66] in which F_2^p data at low x is used with Gribov factorisation to generate low x F_2^γ 'data'. The generated and directly measured F_2^γ data are then fit using leading order QCD evolution over the range $4.3 < Q^2 < 390\,\mathrm{GeV}^2$.

5 Summary and Outlook

Generally the measurements of structure functions are in good shape and the data are well described by NLO QCD. For F_2 at low x more work is needed to understand fully the implications for QCD and accurate data for an other observable such as F_L or F_2^c are essential. The understanding of both polarised structure functions and F_2^γ at small x is hampered by lack of data. Information on the nucleon gluon density is being provided by the use of charm tagging and this will improve.

For the future we can look forward to the completion of the two-photon programme at LEP2 and the large increase in luminosity promised by the HERA upgrade. The COMPASS experiment at CERN, polarised scattering at RHIC and maybe a fully polarised HERA hold out the promise of finally unravelling the mysteries of nucleon spin.

The measurement and analysis of DIS and structure functions are still challenging our understanding of hadronic structure and QCD 30 years after the discovery of scaling.

Acknowledgements

I thank my colleagues in H1 and ZEUS for providing information and insights on the latest HERA results. I thank A. Brüll, J. Le Goff and E. Rondio for instructions on spin physics, likewise D. Miller, R. Nisius and S. Söldner-Rembold on two-photon physics. I thank the organisers of the Conference, particularly D. Lellouch, and IT staff at Oxford and Jerusalem for much practical help.

References

[1] H. Weerts, Plenary talk PL1, these proceedings.
[2] R. Eichler, Plenary talk PL2, these proceedings.
[3] D. Ward, Plenary talk PL15, these proceedings.

[4] E. Elsen, Plenary talk PL26, these proceedings.
[5] NMC, M. Arneodo et al, Nucl. Phys. B483 (1997), 3.
[6] E665, M. R. Adams et al, Phys. Rev. D54 (1997), 3006
[7] H1, S. Aid et al, Nucl. Phys. B470 (1996), 3.
[8] ZEUS, M. Derrick et al, Z. Phys. C72 (1996), 399.
[9] CCFR, W. G. Seligman et al, Phys. Rev. Lett. 79 (1997) 1213.
[10] L. De Barbaro, Invited talk 316, these proceedings.
[11] H1, Submitted paper 260 to this conference.
[12] U. Bassler, Invited talk 31302, these proceedings.
[13] H1, C. Adloff et al, Nucl. Phys. B497 (1997) 3.
[14] ZEUS, J. Breitweg et al, Phys. Lett. B407 (1997) 43.
[15] ZEUS, Submitted paper 646 to this conference.
[16] M. Glück, E. Reya & A. Vogt, Z. Phys. C48 (1990) 471; C53 (1992) 127.
[17] M. Glück, E. Reya & A. Vogt, Z. Phys. C67 (1995) 433.
[18] A. Donnachie & P. V. Landshoff, Phys. Lett. B296 (1992) 227; Z. Phys. C61 (1994) 139.
[19] ZEUS, Submitted paper 647 to this conference.
[20] A. Capella et al, Phys. Lett. B337 (1994) 358.
[21] H. Abramowicz et al, Phys. Lett. B269 (1991) 465.
[22] B. Badelek & J. Kwiecinski, Phys. Lett. B295 (1992) 263.
[23] D. Schildknecht & H. Spiesberger, preprint BI-TP-97-25
[24] G. Kerley & G. Shaw, preprint MC-TH-97-12
[25] E. Gotsman, E. M. Levin & U. Maor, hep-ph/9708275.
[26] K. Adel, F. Barreiro & F. J. Yndurain, Nucl. Phys. B495 (1997) 221.
[27] V. Shekelyan, invited talk at the 1997 International Symposium on Lepton and Photon Interactions, Hamburg July 1997.
[28] H. Abramowicz & A. Levy, DESY 97-251 hep-ph/9712415.
[29] B. Badelek & J. Kwiecinski, Rev. Mod. Phys. 68 (1996) 445.
[30] E. A. Kuraev, L. N. Lipatov & V. Fadin, Soviet Phys. JETP 45 (1977) 199; Ya. Balitsky & L. N. Lipatov, Soviet J. Nucl. Phys. 28 (1978) 822.
[31] F. Eisele, Proc. of EPS HEP95 Conference Brussels Aug. 1995 eds J. Lemonne, C. Vander Velde & F. Verbeure, World Scientific p709.
[32] R. D. Ball & S. Forte, Phys. Lett. B335 (1994) 77; Phys. Lett. B336 (1994) 77; Acta Physica Polonica B26 (1995) 2097.
[33] CTEQ4, H. Lai et al, Phys. Rev. D55 (1997) 1280.
[34] MRS(R), A. D. Martin, R. G. Roberts & W. J. Stirling, Phys. Lett. B387 (1996) 419.
[35] S. Bethke, Nucl. Phys. Proc. Supp. 54A (1997) 314.
[36] M. Virchaux & A. Milsztajn, Phys. Lett. B274 (1992) 221.
[37] A. Prinias, Invited talk 315, these proceedings.
[38] For example: J. Blümlein & A. Vogt, submitted paper 751 to this conference; J. R. Forshaw et al, Phys. Lett. B356 (1995) 79; I. Bojak & M. Ernst, Phys. Lett. B397 (1997) 296.

[39] J. Kwiecinski, A. D. Martin & A. M. Stasto, Phys. Rev. D56 (1997) 3991.
[40] C. Royon, Invited talk 313, these proceedings.
[41] J. Bartels & C. Bontus, Proc. DIS97 Chicago, `http://www.hep.anl.gov/dis97/`.
[42] E. M. Levin, Invited talk at the Conference on Perspectives in Hadronic Physics Trieste May 1997 `hep-ph/9706448`; M. Gay Ducati, Invited talk 318, these proceedings
[43] S. Catani, Z. Phys. C75 (1997) 665.
[44] R. S. Thorne, Phys. Lett. B392 (1997) 463; `hep-ph/9710541`.
[45] L. H. Tao et al, Z. Phys. C70 (1996) 387.
[46] A. J. Milsztajn, Proc. DIS96 Rome Eds G. D'Agostini & A. Nigro, World Scientific 1997 p213.
[47] U. K. Yang, J. Phys. G22 (1996) 775.
[48] H1, C. Adloff et al, Phys. Lett. B393 (1997) 452.
[49] R. S. Thorne, `hep-ph/9708302`.
[50] E.Laenen et al, Nucl. Phys. B291 (1992) 325; B392 (1993) 162,229; S. Riemersma et al, Phys. Lett. B347 (1995) 143.
[51] See the discussion at DIS97 Chicago and H. L. Lai & W. K. Tung, Z. Phys. C74 (1997) 463; A. D. Martin et al `hep-ph/9612449`; M. Buza et al `hep-ph/9612398`; R. G. Roberts & R. S. Thorne `hep-ph/9711213`.
[52] H1, submitted paper 275 to this conference.
[53] A. M. Cooper-Sarkar, R. Devenish & A. De Roeck, `hep-ph/9712301`, to be published in IJMPA.
[54] M. Anselmino, A. Efremov & E. Leader, Phys. Rep. 261 (1995) 1, erratum 281 (1997) 399.
[55] E154/4, P. Souder, Invited talk 301, these proceedings; K. Abe et al, Phys. Rev. Lett. 79 (1997) 26.
[56] SMC, J. Le Goff, Invited talk 303, these proceedings; D. Adams et al, Phys. Rev. D56 (1997) 5330.
[57] HERMES, J. Blouw, Invited talk 302, these proceedings.
[58] E155, S. E. Rock, private communication.
[59] M. Glück et al, Phys. Rev. D53 (1996) 4775; T. Gehrmann & W. J. Stirling, Phys. Rev. D53 (1996) 6100.
[60] G. Altarelli et al, Nucl. Phys. B496 (1997) 337.
[61] E. Leader, A. V. Sidorov & D. B. Stamenov, tt hep-ph/9708335.
[62] SMC, G. Baum, Invited talk 304, these proceedings.
[63] R. Nisius, Invited talk 306, these proceedings.
[64] D. Miller, Summary talk at Photon97 Amsterdam May 1997, `hep-ex/9708002`.
[65] H1, K. Muller, Invited talk 307, these proceedings.
[66] E. Gurvich, Invited talk 308, these proceedings.

QCD, Theoretical Issues

Yuri L. Dokshitzer (yuri@mi.infn.it)

INFN, sezione di Milano, Italy, and
St. Petersburg Nuclear Physics Institute, Russia

Abstract. Today's QCD problems, prospects and achievements are reviewed.

1 Introduction: QCD in 13 puzzles

Exploring the gap between small- and large-distance-dominated phenomena remains the challenge for quantum chromodynamics. QCD is the strangest construction in the history of modern physics. Practically nobody doubts today that it **is** the microscopic theory of strong interactions; at the same time we are practically as vague in addressing the basic questions of this Field Theory as we were 25 years ago.

Puzzle of

Objects: quarks/gluons are the Truth *but* hadrons are the Reality.
Rules: sacred *but* absent. The QCD Lagrangian is a beautiful construction. At the microscopic level, *bal tosif, bal tigra* [1]. No initiative is allowed, as in a FSU "administrative economy". When it comes to the macroscopic hadron world, on the contrary, no rules seem to exist. As in a market economy, what matters is not the quality of the product, but whether you manage to fool your customer into buying it.
Responsibility: gluons are essential *but* quarks dominate. The non-Abelian gluon self-interaction is crucial for asymptotic freedom and thus for causing an infrared instability. However, the hadron world as we know it remains persistently quark-driven.
Scales: finite interaction radius *but* $m_{gluon} \equiv 0$. No field-theoretical mechanism is known that would protect a theory possessing a strictly massless gluon field from developing long-range forces.
Binding, Relativity, Multiplicity: we deal with strongly interacting practically massless particles, hence one could expect many-body ($N \gg 1$) relativistic ($1-v \ll 1$) dynamics, *but* we see additive quarks, surprising successes of the non-relativistic valence quark model.
Goals & Means: We aim at describing a colourless world in terms of propagation and interaction of coloured quarks and gluons. Strictly speaking, we do not know how to do it consistently, that is how to properly define gluon degrees of freedom, even within perturbation theory. The problem of "Gribov copies" is there and remains unsolved.

Perturbation: $\alpha_s(Q^2)$ is small *but* often all orders are essential. Depending on the problem under consideration, the true perturbative expansion parameter is often enhanced by power(s) of $\log Q^2$.
Evolution: Quantum Field Theory *but* classical probabilities.
Freedom: quarks imprisoned *but* fly away for about 10^3 fm these days.
Confinement: Inevitable *but* elusive. Certain global characteristics of multi-hadron production in hard processes, such as inclusive energy spectra of light hadrons (pions, all charged) in jets, multiplicity hadron flows in the inter-jet regions, show no sign of hadronisation effects.
Coupling: "Strong interactions" *but* $\left\langle\frac{\alpha_s}{\pi}\right\rangle_{\text{infrared.}} \sim 0.2$.

"Confinement? What's the problem? Hasn't it been solved long ago by W. [the name may vary]?" — would be a mean theoretician's response. As a reflection of this attitude, the authors of experimental papers dealing with hadroproduction increasingly often tend to equate QCD with Monte Carlo hadron generators. Not only does time cure: it also tends to put out of focus, for the sake of psychological protection, difficult problems that have resisted head-on theoretical attacks for far too long. We may appreciate the former property of time; we'd better be alarmed by the latter.

Vladimir Gribov, a brilliant physicist and a strong character, had enough curiosity, motivation and stamina to pursue a non-stop 20-year study of QCD confinement, till that night of August 13, 1997, when he passed away.

The founder and the leader of the renouned "Leningrad school of theoretical physics", Gribov belongs to the generation of physicists, now almost extinct, who did not take the Quantum Field Theory (QFT) for granted. Thus he was able to raise heretic questions, putting under scrutiny the very basics of the QFT as we (think we) know it. Let me list for you some of them:

- Are the three approaches — secondary quantisation, Feynman diagrams, functional integrals — really equivalent?
- Does classical topology play any rôle in QFT?
- Can Euclidean rotation be justified and thus a statistical system substituted for QFT in the case of infrared unstable dynamics?
- How should an ultraviolet renormalization programme be carried out in a theory where the physical states do not resemble the fundamental fields the Lagrangian is made of?
- Whether the Dirac picture of the vacuum, with negative-energy levels being occupied and the positive-energy ones being kept empty, applies to quarks?
- How to bind together massless particles?
- Shouldn't the electro-weak symmetry breaking scale and the QCD scale be related with one another by a pion which is an electro-weak point-like Goldstone and, at the same time, a quark-antiquark bound state from the QCD point of view?

In general, the confinement problem is a problem of understanding and describing the physical states of a Field Theory with (unbroken) non-Abelian gauge symmetry. There can be many confinements, that is many solutions to the non-Abelian instability. The core observation made by V.N. Gribov was that *the* confinement, that is confinement in the world we live and experiment in, is largely determined by the fact that practically massless quarks are present in the theory.

Gribov's ideas remain to be discovered, understood and developed.

2 Small x and Confinement (Heron vs. Pomeron)

A message from DIS 97, Chicago:

A rare thing is more damaging to empirical science than an improper name. In perturbative QCD we may discuss the BFKL *approximation*, the BFKL *equation*, BFKL *dynamics* but should restrain from talking about a "BFKL or Hard Pomeron". There are two reasons for that. First of all, the Pomeron (Gell-Mann's name for the Gribov vacuum pole or vacuum singularity) is a reserved word... More importantly, the very term "Hard Pomeron" is nonsensical.

- **Pomeron:** a leading high-energy contribution to elastic hadron scattering amplitudes driven by the leading singularity in the t-channel (complex) angular momentum ω.
- **Hard:** determined by small distances, and therefore perturbatively controllable.

These two words simply don't merge, since the position and the nature of the leading singularity in ω is entirely off the books of perturbative QCD.

2.1 BFKL and confinement

"I am but mad north-north-west:
when the wind is southerly I know
a hawk from a handsaw."
Hamlet, Prince of Denmark

The fact that the behaviour of DIS structure functions at fixed Q^2 and $x \to 0$ is an entirely *non*-perturbative phenomenon, was recently demonstrated, beyond any reasonable doubt, by Camici & Ciafaloni and by A. Mueller.

In Mueller's paper [2] an estimate is given of the range of x beyond which the perturbative treatment (operator product expansion, OPE) breaks down due to diffusion into the small-transverse-momentum (strong interaction) region. Numerically,

$$Y = \ln x^{-1} \ll (2\alpha_s(Q_0))^{-3},$$

with Q_0 the *minimal* of the two scales, to which high energy $s \propto 1/x$ is applied. This leaves no hope for *predicting* the x-behaviour of the initial parton distributions at virtualities as low as 1–2 GeV, and, therefore, the x-behaviour of the DIS structure functions.

Camici and Ciafaloni have shown in [3], in a very nice physically transparent way, that the position and the nature of the leading complex-angular-momentum singularity (Pomeron) depend crucially on the behaviour of the QCD coupling α_s in the infrared region. Read: have non-perturbative origin.

This does not mean that the BFKL analysis and results are nice but irrelevant. We have to come to terms with the fact that the BFKL-predicted increase of cross sections with energy is of little relevance for DIS structure functions. However, in special circumstances it remains a perfectly sound and unquestionable QCD prediction which should be seen experimentally.

Instead of the "BFKL Pomeron" we may say the BFKL **HER**on, a temporary **H**igh-**E**nergy **R**egime of increasing *small* interaction cross sections of *small* hadronic objects.

2.2 Next-to-leading BFKL dynamics

The next-to-leading analysis of the BFKL problem is about to be completed. The necessary ingredients have been computed, in recent years, in a series of highly technical works by Camici, Ciafaloni, Fadin, Kotsky, Lipatov, Quartarolo. What remains to be done is to deduce how the BFKL equation is modified beyond leading logarithms, whether the Regge-Gribov factorisation is respected, how the coupling runs in the BFKL kernel, how much down the energy-growth exponent (the "BFKL intercept") goes. These questions are not easy to answer, even though all the next-to-leading corrections, to this and that, are known. To put it simple, a severe "ideological" problem remains of how to *formulate* the improved answer. For example, the very notion of the BFKL exponent λ is elusive: how to quantify "*asymptotic behaviour*" of a *non-asymptotic* amplitude? (The true asymptotia, as we know, is the soft = non-perturbative Pomeron.)

Let us present the behaviour of the forward amplitude (total cross section) for the scattering of two objects with small sizes $p_t^{-1} \ll 1\text{fm}$ at invariant energy $\hat{s}$ as

$$\text{Heron}(\hat{s}, p_t^2; 0) \propto \left(\frac{\hat{s}}{p_t^2}\right)^{\lambda(\alpha_s(p_t^2))} .$$

(The third argument of H is the momentum transfer along the BFKL-ladder, $t = 0$ for the forward scattering amplitude.)

A *preliminary* result for the Heron "intercept" 1997 is [4]

$$\lambda^{1997}(\alpha_s) = \lambda^{1976}(\alpha_s) \cdot \left(1 - 3.4 N_c \frac{\alpha_s}{\pi} - 0.15 n_f \frac{\alpha_s}{\pi} + \mathcal{O}(\alpha_s^2)\right) .$$

Numerically,

$$\lambda = 0.5 \ \rightarrow \ \lambda = 0.2 \qquad \text{for } \alpha_s = 0.15 \ \ (?)$$

If we close our eyes on the inapplicability of the BFKL Heron to DIS structure functions and translate the next-to-leading BFKL correction into the DIS anomalous dimension, weird results may follow. As was demonstrated by Blümlein & Vogt, the gluon splitting yields negative probability, structure functions (a gluon-driven F_L in the first place) start decreasing at small x, and the like [5]. Not that the BFKL-corrected anomalous dimension adopted by B&V was free of criticism [6]. What seems to me a more important shortfall is asking a wrong question. Let DIS SF rest in peace!

The crisis is over. Back to work.

2.3 Two large-and-equal scales

In ep-scattering we look at forward jet production (Mueller-Navelet jets) with jet transverse momentum $p_t^2 \simeq Q^2$. This is a two-large-scale problem, and the corresponding cross section should increase with the boson–jet pair energy as

$$d\sigma^{\mathrm{MN}} \propto \mathrm{Heron}(\hat{s}, Q^2; 0), \qquad \hat{s} = Q^2 x_{jet}/x_{Bjorken} \gg Q^2 \ .$$

In pp-scattering we may discuss at least two options. The first is *inclusive* production of a high-invariant-mass pair of jets with large (and comparable) transverse momenta, $s \gg \hat{s} = (p_1 + p_2)^2 \gg p_t^2$. In this case we expect

$$d\sigma^{\mathrm{incl.}} \propto \mathrm{Heron}(\hat{s}, p_t^2; 0) \, .$$

If we impose a condition of having a *rapidity gap* between the jets, then for rapidities large enough so as to eliminate the Sudakov-suppressed one-gluon exchange, we should have

$$d\sigma^{\mathrm{gap}} \propto \left(\mathrm{Heron}(\hat{s}, p_t^2; -p_t^2)\right)^2 .$$

(Recall, the last argument of the Heron amplitude is the momentum transfer; here $t \approx -p_t^2$.) The latter "double-diffractive" process is a unitarity-shadow of the former (total, inclusive) one.

The Heron may also be searched for in $\gamma\gamma$ collisions in conditions similar to pp, as well as in double-hard double-diffractive processes like the "dream process" $\gamma\gamma \to J/\psi + J/\psi$,

$$d\sigma^{\mathrm{frwrd.,excl.}}_{\gamma\gamma\to J/\psi J/\psi} \propto \left(\mathrm{Heron}(\hat{s}, m_c^2; 0)\right)^2 \ .$$

Two comments are due concerning experimental BFKL searches, one optimistic, another rather pessimistic.

Mueller-Navelet experiment and "BFKL at hadron level". ZEUS exercises some caution before claiming that the BFKL-predicted increase of MN-jet cross section is observed experimentally [7]. This caution is well grounded but may be slightly exaggerated. As I understand, it stems from the observation that the standard cascade-type generators (Lepto, Herwig) exhibit an alarmingly large mismatch between the parton- and hadron-level cross sections.

However, this does not necessarily imply that the lacking hadron-level BFKL prediction is really necessary. The very idea of jet finding algorithms was to keep the correspondence between the parton and hadron ensembles. The fact that the MC generators fail to preserve this correspondence may be simply due to their inability to properly estimate the parton level cross section: they can produce the MN-events only as improbable fluctuations.

Indeed, a starting point for BFKL in the MN-experiment is the α_s^2 QCD matrix element. For small values of Bjorken x, production of a forward jet with $p_\perp^2 \sim Q^2$ is accompanied by two more quark jets with transverse momenta of the same order, coming from the Boson-Gluon-Fusion box. None of the standard MC generators based on the logarithmic DGLAP evolution picture ever pretended to embody such high-order configurations with 3 partons at the same hardness scale (corresponding to the two-loop correction to the coefficient function).

Therefore, MC models are likely to be responsible for the mismatch between parton- and hadron-level results, not the physics, which is the BFKL physics in this case.

Azimuthal de-correlation as a sign of the BFKL dynamics. Accompanying gluon radiation in the production of two jets with large p_t has a double-logarithmic nature. It consists of gluons with transverse momenta distributed logarithmically from small $k_{ti}^2 \ll p_t^2$ up to $k_{ti}^2 \sim p_t^2$ at a given rapidity (energy log) and uniformly in rapidity (angular log). In the inclusive two-jet cross section the double-logs disappear (real and virtual gluons with $k_{ti}^2 \ll p_t^2$ cancel). The result can be expressed in terms of (an imaginary part of) the single-logarithmic forward amplitude known as BFKL.

Studying the $\hat{s}$ dependence of the two-jet cross section, $\hat{s} = x_1 x_2 s$ (keeping the x_1, x_2 fixed as to factor out the initial parton distributions) is a straight road to verifying the BFKL dynamics. However, the cancellation of small-$k_\perp$ gluons mentioned above is not present in less inclusive quantities such as the distribution in the total transverse momentum of the two triggered jets, or in the azimuthal angle between the jets, in particular in the back-to-back region, $\Delta\phi = \pi - \phi_{12} \ll 1$. Sudakov form factor suppression broadens the $\Delta\phi$–distribution. Moreover, this broadening (azimuthal de-correlation) increases with $\hat{s}$ roughly as

$$\frac{d\sigma}{d\phi} \propto (\Delta\phi)^{-1+N_c \frac{\alpha_s}{\pi} \ln \frac{\hat{s}}{p_\perp^2}} .$$

The $\hat{s}$–dependent part of the de-correlation is due to the dynamical suppression of soft gluon emission at "large angles", that is within the broad rapidity interval between the jets. Since in two-jet production at large $\hat{s}$ one-gluon exchange dominates, this coherent radiation is determined by the colour charge of the t-channel exchange, i.e. by that of the gluon. Hence, the N_c factor in the radiation intensity. It is related with QCD Reggeization of the *gluon* and has little to do with the vacuum-exchange dynamics, that is with the BFKL phenomenon.

There is a hope to see the BFKL-motivated energy dependence of the first *moments* of the ϕ_{12}–distribution, in which the back-to-back region is suppressed [8,9]. This is not impossible. However, the necessity of hunting down single-log effects in the presence of double-logs in the differential distribution makes one feel suspicious about this option of visualising the BFKL dynamics.

2.4 0– and 1–hard-scale diffractive (quasi-elastic) processes

The following HERA chart illustrates the transition from 0– to 1–hard-scale processes:

$\sigma_{\mathrm{tot}}(\gamma p)$	Pomeron
$\sigma_{\mathrm{D}}(\gamma p)$	[Pomeron]^2
$\sigma_{\mathrm{tot}}(\gamma^* p)$	$G(x,Q^2)$
$\sigma_{\mathrm{D}}(\gamma^* p)$	$G^2(x,Q^2)$
$\sigma(\gamma^{(*)}p \to J/\psi + p^*)$	$[G(x,m_c^2)]^2 \to [G(x,Q^2)]^2$
$\sigma(\gamma^{(*)}p \to 2\mathrm{jets}(k_t) + p^*)$	$[G(x,k_t^2)]^2 \to [G(x,Q^2)]^2$

In the second block the rôle of a hard scale may be played by the initial photon virtuality Q^2, a heavy quark mass, or large *transverse momenta* inside a diffractively produced system. (Notice that a large *invariant mass* of the latter does not qualify.)

In the two last cases γp and $\gamma^* p$ (DIS) stand on an equal footing. With increasing photon virtuality, Q^2 may take over from the quark mass or the jet transverse momentum as the hardness argument determining the scale of the gluon distribution.

Here there is no room for the Heron amplitude which applies to the scattering of **two** small objects:
The energy dependence of total (inclusive) cross sections are, for the

scattering of
- Large on Large: **Pomeron**,
- Small on Large: **Gluon** (in the proton),
- Small on Small: **Heron**.

The energy dependence of diffractive cross sections are, for the

$$\text{scattering of} \begin{cases} \text{Large on Large:} & (\textbf{Pomeron})^2, \\ \text{Small on Large:} & (\textbf{Gluon})^2, \\ \text{Small on Small:} & (\textbf{Heron})^2. \end{cases}$$

There is an interesting option to ensure two small scales, and thus to approach BFKL dynamics, namely by studying diffractive processes, e.g. $\gamma p \to \rho + p^*$ at sufficiently large momentum transfer t. Forshaw and Ryskin have recently verified that a finite momentum transfer $|t|$ suppresses large-distance-directed diffusion and thus keeps the BFKL gluon system under perturbative control. For example, one can expect the two-gluon exchange to turn into a fully-fleshed Heron in the $s/t \gg 1$ limit of (double-) diffraction processes with $|t| >$ few GeV^2,

$\gamma p \to \gamma^{(**)} + p^{(*)}$	$2g \to [\text{Heron}(\hat{s}, 0; t)]^2$

where $\gamma^{(**)} = \gamma,\ \rho,\ J/\psi, Z^0$, whatever.

3 Coherence and the Nucleus as Colorometer

Our field has emerged as a result of the digression: natural philosophy $\to$ physics $\to$ quantum physics $\to$ elementary particle physics. The older generation participated in the next step, elementary particle physics $\to$ high energy physics. In the past 20 years we have witnessed the final split

$$\text{high energy physics} \to \begin{cases} \text{soft physics} \\ \text{hard physics} \end{cases}.$$

It is about time to restore the integrity of the subject. *Coherence* is the key-word for re-integrating *soft* and *hard* physics into *high energy physics.*

You can smell new **quasi-classics** in the air when you hear discussions of high gluon densities (hot spots), disoriented chiral condensates, percolating strings, quark-gluon plasmas, the effective colour field of a nucleus, etc. What remains to be done is to condense this smell into a marketable fragrance.

High-energy hadron-scattering phenomena may be more *classical* than we use to think they are. To illustrate the point let us look into Scattering phenomenon and High Energies.

3.1 Scattering

V.N. Gribov produced an argument in favour of Scattering in the Quantum Field Theory framework being an inch closer to Classical Scattering than, paradoxically, that in Quantum Mechanics. He addressed the question "Why does the total cross section rise in the first place?"

Total cross sections of classical potential scattering are typically infinite. For example, the tail of the Yukava potential, $V(r) = \mu \exp(-\mu r)/r$, produces small-angle scattering which makes $\sigma_{tot} = \infty$.

What keeps σ_{tot} finite in quantum mechanics, as we know, is the impossibility to separate a go-straight wave from a diffracted one in the case of too small a scattering angle, $p_\perp = p\theta < 1/\rho$.

QFT, however, gives us such an option by introducing *inelastic diffraction.* The final state being different from the initial one, the QM argument breaks down, and large impact parameters start contributing to the total cross section. The latter increases with energy due to the decrease of $t_{\min}$ with s.

3.2 High Energies

The hadron as a QFT object is a coherent sum of various configurations. The quantum portrait of a projectile, its field fluctuation, stays frozen in the course of interaction at high energies, so that each fluctuation scatters independently [10]. The total interaction cross section, for example, emerges as a classical (incoherent) sum of cross sections of different configurations inside the projectile hadron h, each of which interacts with its private σ. Introducing the corresponding distribution [11], we may write

$$\sigma_{\text{tot}}^h = \langle \sigma \rangle_h \equiv \int d\sigma\, \sigma \cdot p_h(\sigma)\,.$$

The issue of coherence comes onto the stage when we turn from the total cross sections to more subtle phenomena, **diffraction** being rightfully the first among many.

Let us repeat, an incident proton is a coherent sum of different field fluctuations. Talking hadrons, they are $p \to \pi^+ n \to p$, $p \to K^+ \Lambda \to p$, etc. Switching to the quark language, we may talk about three valence quarks at various relative distances, with a correspondingly suppressed or enhanced gluon field between them, that is to picture the proton as being virtually squeezed or swollen. As was noticed by Feinberg and Pomeranchuk, if all the configurations inside the projectile interacted *identically* with the target, such interaction would not induce *inelastic diffraction.* Indeed, in this case the interaction Hamiltonian preserves the coherence of the initial state which, therefore, would scatter elastically but would not break apart producing small-mass diffractive states. In other words, inelastic diffraction is sensitive to the *σ-distribution* [12]. The dispersion of the latter can be directly measured in diffraction on nuclei

$$\left.\frac{\sigma(hA \to h^* A)}{\sigma(hA \to hA)}\right|_{t=0} = \frac{\langle \sigma^2 \rangle_h}{\langle \sigma \rangle_h^2} - 1\,.$$

In general, high energy scattering off nuclear targets provides indispensable tools for studying the internal structure of hadrons.

Proton-penetrator. The A-dependence of inelastic diffraction gives a classical example of the use of the nucleus as a *colorometer*. If an incident hadron (p, π) *always* interacted strongly with the target in central collisions, inelastic diffraction would be peripheral and therefore its cross section would grow as $A^{1/3}$ (total absorption being an example of an "identical interaction"). Experimentally, the A-dependence is much faster, $\sigma_D \sim A^{0.8}$ or so. This shows that there are *small* configurations inside the proton that penetrate the nucleus. It is not easy to visualise such a configuration ("penetrator") using the hadronic fluctuation picture. However, it is readily understood in the quark language: tightly placed valence quarks with abnormally small gluon field between them, due to quasi-local compensation of the colour charge.

The interaction of such fluctuations with the target stays under the jurisdiction of pQCD. Theoretically, there is a nice tool for describing the interaction of a simpler projectile, that is a tight colourless $q\bar{q}$ pair:

$$\sigma(q\bar{q}, A) = \frac{\pi^2}{3}\, b^2\, \alpha_s(b^{-2})\, xG_A(x, b^{-2})\,.$$

It relates the interaction cross section of a quark and an antiquark at small relative transverse distance b (inside a real/virtual photon or a meson) with the gluon distribution in the target [13]. Its application to the A- and s-dependence of shadowing, to colour transparency phenomena, and many other intriguing subjects marked the beginning of QCD "Geometric colour optics", the name of the game invented by Frankfurt and Strikman.

Proton-perpetrator. Thus, diffraction and transparency phenomena can be employed to study squeezed, penetrating hadrons. Their opposite members — swollen hadrons — can also be visualised in scattering on nuclei.

Contrary to the penetrator, the proton-perpetrator should willingly interact with the target. In such configurations "strings" are pulled, colour (or, pion) fields are stronger than typical, the vacuum is virtually broken. So, one can expect an enhanced yield of antiquarks pre-prepared to be released when coherence of the projectile is destroyed. Perpetrators can take responsibility for a bunch of puzzling phenomena observed in pA and ion-ion (AB) collisions, such as baryon stopping, the increase of the Λ/p ratio (Kharzeev) and an amazingly large relative yield of *antibaryons*.

To enhance the rôle of non-typical configurations one should look at the *tails* of various distributions. In particular, by looking into events with *very large* E_t in ion-ion collisions one may see unusual percolated nuclei consisting of merged nucleons, instead of the much too familiar nucleon gas. The quark-gluon plasma state is usually thought to be formed in the collision, which provides a melting pot for individual nucleons. A large E_t-yield is considered to be a sign of the phase transition into such a state, in the course of the collision. Alternatively, very large E_t may be looked upon as a *precondition* for the collision, rather than the result of it: to observe a larger than typical transverse energy yield, we catch the colliding nuclei in rare, confinement-perpetrating, virtually melted configurations a'la the desired plasma state.

In such configurations the yield of Drell-Yan lepton pairs should be higher (extra pions, or antiquarks, around); the reduced yield of J/ψ (heavy traffic) is to be expected.

3.3 Praising J/ψ — a boarder-guard

Heavy onia remain a pseudo-perturbative tool for probing strong interaction dynamics, and a source of surprises at the same time. One of the striking puzzles is offered by the comparison of *medium-dependent* transverse momentum broadening of Drell-Yan pairs, J/ψ and Υ observed in pA collisions: $\ell^+\ell^- : J/\psi : \Upsilon \simeq 1 : 2 : 8$. Future explanation of this puzzle should shade light on the production mechanism of vector $Q\bar{Q}$ onia.

The cross section for J/ψ production in pp collisions is a long-standing problem: the pQCD treatment based on the gluon-gluon fusion into J/ψ plus the final-state gluon, $gg \to J/\psi + g$, is way off.

To avoid an extra α_s-cost the so-called "octet mechanism" was suggested based on producing a colour-*octet* $c\bar{c}$ which then gets rid of colour by radiating off "a soft, non-PT gluon", for free. This quite popular picture, however, misses one essential point: According to the renouned Low-Barnet-Kroll theorem, soft radiation does not change the state of the radiating system. In other words, soft radiation does not carry away quantum numbers. Colour is no exception: paradoxical though it may sound, soft gluons cannot blanch the octet $c\bar{c}$ into a white J/ψ.

Physically, soft gluons are radiated *coherently* by the colour charges involved, and enter the stage, formally speaking, long after the J/ψ formation time. Technically, a soft gluon, that is one with the characteristic logarithmic energy spectrum $d\omega/\omega$, sits, colour-wise, in an *antisymmetric* f_{abc} configuration with its two partners. However, to form a C-odd J/ψ state one needs a *symmetric* d_{abc} colour coupling instead. This is the reason why in the decay channel, $J/\psi \to ggg$, none of the gluons is allowed to be soft: in the small energy domain one has the $\omega d\omega$ distribution rather than the classical $d\omega/\omega$.

There is no doubt that J/ψ is not an entirely small-distance object, not a pure non-relativistic Coulomb $c\bar{c}$ system. Therefore the idea of introducing into the game new non-perturbative matrix elements, in other words, large-distance configurations in the J/ψ field wave function (Braaten et al) is well grounded. What makes me feel uneasy, however, is an accompanying (hopefully, unnecessary) chant of the free-of-charge soft-gluon blanching of the system, which sounds too much like a free lunch.

4 QCD Checks 1997

Let me start by citing George Sterman at the DIS conference in Chicago last spring:

"The study of quantum chromodynamics and the investigation of hadronic scattering are the most challenging problems in quantum field theory that are currently accessible in the laboratory."

4.1 QCD in danger

A decade after the JADE collaboration, which pioneered the e^+e^- jet studies at PETRA, was dissolved, impressive new JADE results have appeared. Reanalysis of the old data was necessitated by the new jet measure (broadening), new jet finder (Durham) and new wisdom (confinement $1/Q$ effects in jet shape observables, see below).

Have a look at few citations from "A study of Event Shapes and Determination of α_s using data of e^+e^- Annihilation at $\sqrt{s}$=22 to 44 GeV", and you will agree that the paper could have been rightfully submitted to a journal on Archaeology.

> "The retrieval of the data eleven years after shutdown of the experiment turned out to be cumbersome and finally incomplete."
> "... missing data sets of about 250 events around 22 GeV and about 450 events around 44 GeV. In addition, the original files containing information about the luminosity of different running periods could not be retrieved..."

Even the Acknowledgement section sounds alarming:

> "We thank the DESY computer centre for copying old IBM format tapes to modern data storage devices before the shutdown of the DESY-IBM. We also acknowledge the effort... to search for files and tapes..."

Hadron physics is forever, because today's devotion to high energies is *temporary.* High energies let us watch the vacuum being excited and think about it for some (Lorentz-dilatated) time. Once the vacuum structure has been understood (the most important and the most difficult step still to make), hadron physics will turn back to small and medium energies.

Anyone willing to revisit celebrated ISR data? Forget it! What makes the old experiments dead as a doornail?

1. the data is gone;
2. the tapes are kept somewhere, but cannot be read out (the recording media have changed);
3. the data can be retrieved but nobody remembers how the information storage was organised: the students who knew are no longer around.

To make you shiver, just imagine a similar fate awaiting the LEP-1 data! We will never have anything comparable with the Z-hadron-physics-factory in the future. No one has enough imagination to foresee what sort of questions theoreticians will fancy in ten years from now. It will be a major catastrophe for the field if a well organised representative set of pre-processed LEP-1 data is not available in the future, for theoreticians to probe new ideas against the wealth of hadronic LEP-1 information. A special task force should be established at CERN to carry out the project.

4.2 Basic stuff

George Sterman again:

> "Perturbative methods have been found to be surprisingly, sometimes *amazingly*, flexible, when the *right questions* are asked."

I took the liberty to emphasise the two ingredients crucial for the discussion that follows, which are: "amazing" and "right questions".

LEP and SLAC e^+e^- experiments have reached a high level of sophistication. These days they talk about identified particles in perfectly identified (heavy quark, light quark, gluon) jets. Theoretical expectations of gluon jets being softer and broader are verified, even an elusive C_A/C_F ratio is now well in place. Extracting gluon jets from three-jet events is a tricky business which involves choosing a proper event-geometry-dependent hardness scale to describe the gluon subjet (see, in particular, [14]); e^+e^- annihilation events with gluon jet recoiling against heavy $Q\bar{Q}$ flying into the opposite hemisphere, though rare, provide a bias-free environment for studying glue [15].

The HERA experiments are catching up, the global properties of the struck quark jet (inclusive energy spectra and the scaling violation pattern, KNO multiplicity fluctuations, etc.) converging with those of quark jets seen in e^+e^- [16,17]. What makes the HERA jet studies even more exciting, is an ability to scan through the moderate (and small) Q^2 range in order to shed some light onto the transition between hard and soft phenomena.

A comparative study of the yield of different hadron species in quark and gluon jets is underway. Gluon jets are reportedly richer with baryons (as was expected from the times of hadronic Υ decays). The yield of η and η' mesons in gluon jets is also under focus (L3, ALEPH). The numbers are still uncertain. For example, the relative excess of Λ hyperons was reported from almost none [18] up to 40% [19]. There is no doubt that in the near future the situation will be clarified[1].

From the theoretical side, one warning is due. According to the present-day wisdom, the production of accompanying hadrons with *relatively small* momenta in jets is always "*gluonic*": it is driven by multiple radiation of soft gluons off the primary hard quark/gluon parton which determining the nature

[1] I remember hearing during the conference of a 100% Λ–excess, but failed to find any documented evidence.

of the jet. The gluon radiation being *universal*, so should be the relative abundance of different hadron species in the "sea".

From this point of view, the difference in the yield of hadrons should be there only in the leading-parton fragmentation region: hadron fragmentation of the valence quark can differ from that of a gluon. So, the crucial question is: whether the differences between the quark- and gluon-initiated jets is concentrated near the tip of the jet. If it is not, that would be evidence for (unexpected) gluon density effects in the dynamics of hadronisation.

4.3 Subtleties

The effects of QCD coherence in hadron flows **in** and **in-between** jets are well established experimentally. Gluon coherence inside jets leads to the so-called "hump-backed" plateau in one-particle inclusive energy spectra. A quantitative theoretical prediction known as the "MLLA-LPHD prediction" was derived in 1984. It has survived the LEP-1 scrutiny; more recently, it has been confirmed by a detailed CDF analysis [20]; these days it is seen also at HERA[16].

This QCD prediction has two ingredients.

MLLA stands for the "modified leading log approximation" of pQCD and represents, in a certain sense, the resummed next-to-leading-order approximation. This step is necessary for deriving *asymptotically correct* predictions concerning multiple particle production in jets. This means that the MLLA parton-level predictions become exact in the $W^2 \to \infty$ limit.

LPHD (local parton-hadron duality) is a hypothesis rather than a solid QCD prediction. It was based on the idea of soft confinement, motivated by the analysis of the space-time picture of parton multiplication, and stated that observable spectra of hadrons should be mathematically similar to the calculated spectrum of partons (the bulk of which are relatively soft gluons).

Experiment does respect LPHD [21].

What makes the story really surprising is that the perturbative QCD spectrum is mirrored by that of the pions (which constitute 90% of all charged hadrons produced in jets), even at momenta below 1 GeV! Moreover, the ratios of particle flows in the inter-jet regions in various multi-jet configurations, which reveal the so-called "string" or "drag" effects also respect the parameter-free pQCD predictions based on the coherent soft gluon radiation picture. These observables are dominated, at present energies, by junky pions in the 100–300 MeV momentum range!

Is there any sense in applying the quark-gluon language at such low scales? What this tells us is that the production of hadrons is driven by the strength of the underlying colour fields, the perturbative gluon radiation probability being a mere tool for quantifying the latter.

There is no need to experimentally verify the MLLA, that is to check quantum mechanics for ca 200 SF/Z^0 (LEP-1). However, it is not difficult to imagine *a world* without "LPHD". The QCD string is a nice image for encoding the structure of the basic underlying hadronisation pattern (Feynman hot-dog). What the experimental verification of LPHD tells us, is that hadronisation is an amazingly soft phenomenon. As far as the global characteristics of final states are concerned, such as inclusive energy and angular distributions of particle flows, there is no visible re-shuffling of particle momenta when the transformation from coloured quarks and gluons to blanched hadrons occurs. This means that the QCD string is not a dynamical object, in a sense: it does not *pull.* It could, if there were no light quarks around.

4.4 Some nasty theoretical remarks

Disproving QCD is no longer in fashion. We are now at the stage of trying to understand QCD and learn to apply it to a broader range of phenomena. To this end we should ask "proper questions" and use proper tools to avoid confusion.

It is not experimentors' fault that the present-day theory is not smart enough to extract valuable information from a given experimental observation. Therefore I had better restrain myself from listing "improper questions" (IQs). But let me mention one famous case: the long-standing problem of the (W+1-jet) to (W+0-jet) Tevatron ratio may belong to the IQ-club. The ratio (W+1-jet)/(W+all) would be a safer quantity to check against the fixed-order QCD prediction.

Two examples of the necessity of "proper tools".

There seems to be a problem with electro-production of large-p_t jets. However, in the region $p_\perp^2 \gg Q^2$ HERA becomes a "Tevatron" with a virtual photon substituted for one of the protons. Hence, a proper tool here would be merging the parton structure of the proton with that of the virtual photon. (A general plea: you don't have a proper Monte-Carlo $\neq$ QCD fails!)

Another example is the E706 finding of very broad distributions over the total transverse momentum of $\pi^0\pi^0$, $\pi^0\gamma$ and $\gamma\gamma$ pairs at Fermilab. The observed phenomenon is similar to that known for almost 20 years in the Drell-Yan pair production. The proper tool here would be all-order double logarithmic Sudakov form factor effects as a substitute for the claimed large intrinsic transverse momentum, $\langle k_t \rangle > 1$ GeV [22].

5 ICE Observables and Confinement

The much debated problem of power uncertainties in the perturbative expansion (the concept of renormalons pioneered by G. 't Hooft and A. Mueller) has mutated, all of a sudden, into brave attempts to *quantify* power-behaving contributions to various Infrared and Collinear safe (ICS) observables. This field was initiated and is being pursued by Korchemsky and Sterman, Akhoury and Zakharov, Beneke and Braun, Nason and Seymour, Shifman, Vainstein and Uraltsev, Marchesini and Webber, and many others.

The notion of ICS pQCD predictions ascends to Sterman and Weinberg. ICS are the observables that do not contain logarithms of collinear and/or infrared cutoff μ in pQCD calculations, and therefore have a finite $\mu \to 0$ limit. The contribution of small momentum scales to such quantities should therefore be proportional to $(\mu^2/Q^2)^p$ with $p > 0$, modulo logarithms.

Simply by examining Feynman diagrams, one can find the powers p for different observables. This information is already useful: it tells us how (in)sensitive to confinement physics a given observable is. For example, one can compare the performance of different jet finding algorithms in this respect to see that hadronisation corrections to jet rates defined with use of the Durham jet-finder are expected to be smaller ($1/Q^2$) than those for the JADE algorithms ($1/Q$)[23]. More ambitious a programme aims at the *magnitudes* of power-suppressed contributions to hard cross sections/jet observables.

In the last two or three years first steps have been made towards a joint technology for triggering and quantifying non-perturbative effects in "Euclid-translatable" cross sections (vacuum condensates) and, at the same time, in the essentially Minkowskian characteristics of hadronic final states. In the systematic approach known as the "Wise Dispersive Method" (WDM) [24] new dimension-full parameters $A_{2p.q}$ emerge that normalise genuine non-perturbative contributions to dimensionless ICS observables V,

$$\delta^{(\mathrm{NP})}V = \sum_{q=0}^{q_m} \rho_q^{(V)} \left(\ln \frac{Q^2}{\mu^2}\right)^q \cdot \frac{A_{2p,q}}{(Q^2)^p} + \dots$$

Perturbative analysis (PT) provides the observable-dependent factors $\rho_q^{(V)}$ and the leading power p. The non-perturbative (NP) parameters A are expressed in terms of log-moments of the "effective coupling modification"

$$A_{2p,q} = \frac{C_F}{2\pi}\int_0^\infty \frac{dm^2}{m^2}(m^2)^p \left(\ln\frac{m^2}{\mu^2}\right)^{q_m-q} \delta\alpha_{\mathrm{eff}}(m^2)\,,$$

where p is half-integer or $q > 0$. For different observables, $q_m = 0, 1, 2$. (If $q_m > 0$, the contributions combine so as to produce an answer that does not depend on the arbitrary parameter μ.)

The moments of $\delta\alpha_{\mathrm{eff}}(m^2)$ converge at m^2 of the order of Λ^2_{QCD}. The function α_{eff} is related to the standard QCD coupling via the dispersive integral,

$$\alpha_s(k^2) = \int_0^\infty \frac{dm^2\, k^2}{(m^2+k^2)^2}\,\alpha_{\mathrm{eff}}(m^2)\,; \quad \alpha_{\mathrm{eff}}(m^2) = \frac{\sin(\pi D)}{\pi D}\alpha_s(m^2)\,,\ D \equiv \frac{m^2\, d}{dm^2}\,.$$

It is thought to be a universal function that characterises, in an effective way, the strength of the QCD interaction all the way down to small momentum scales. Given this universality, it becomes possible to *predict* the ratios of the Q^{-2p} contributions to observables belonging to the same class p.

Those who believe that a school of little fish can be mistaken for a baby-whale, are aware of mounting evidence in favour of the notion of an infrared-finite QCD coupling[2]. Phenomena range from simple estimates of hadron interaction cross sections in the Low-Nussinov two-gluon model of the Pomeron, all the way up to a detailed sophisticated analysis of meson properties in the framework of the "relativised" potential model of Godfrey and Isgur [25].

Of primary interest are ICS jet-shape observables many of which exhibit the $1/Q$ leading power corrections, $p = \frac{1}{2}$. These include the thrust T, the so called C-parameter, invariant jet masses, the jet broadening B ($\ln Q$–enhanced). The energy-energy correlation function in e^+e^- annihilation, EEC(χ), also contains the $1/Q$ confinement contribution (away from the back-to-back region, $\chi \neq \pi$).

A crucial question is that of marrying PT and NP contributions. At the PT level, only the first few orders of the α_s expansion are available, for most observables up to the next-to-leading α_s^2 order. This fact is not too disappointing: a full knowledge of the PT expansion would be of little help anyway. Indeed, the series diverge factorially, so that an intrinsic uncertainty of the sum of PT terms is at the level of that very same power Q^{-2p} (infrared renormalon ambiguity).

The price offered in [26] for resolving this ambiguity was the introduction of a matching infrared scale μ_I, above which the coupling is well matched by its famous logarithmic PT expression. The genuine NP $1/Q$ effects can then be expressed in terms of the effective magnitude of the coupling in the infrared region, $\bar{\alpha}_0$,

$$\bar{\alpha}_0(\mu_I) \equiv \frac{1}{\mu_I} \int_0^{\mu_I} dk \, \alpha_s(k^2) \, .$$

Experimental analyses carried out by ALEPH, DELPHI, H1, JADE and OPAL have demonstrated consistency between power terms in the *mean values* of $1-T$, M^2 and, to a lesser extent, B. They pointed at the value for $\bar{\alpha}_0(2\text{GeV})$ in the ball-park of 0.5–0.54[3].

It is clear that the data at *smaller* Q^2 are more sensitive to power effects. Hence, a potential advantage of HERA and the necessity of revisiting JADE and TASSO data. "Parasitic" radiative LEP-1 events $e^+e^- \to Z^0 + \gamma$ also provide a nice opportunity for studying jet shapes at reduced hardness scales. The L3 collaboration has it all, but, for the time being, has conservatively restricted itself to comparison with the MC models only [27].

The most exciting result has emerged from the recent ALEPH study of the thrust *distribution* in the energy range from 14 up to 180 GeV [28]. The

[2] A certain irony is necessary since little can be rigourously proved in the game.

[3] OPAL came up with a smaller value, the reason being an implementation of a strongly reduced scale of the PT contribution, as substitute for part of the power effect. An experimental verification of the renormalon phenomenon, if you wish.

quality of the two-parameter fit of the form [26]

$$\frac{d\sigma}{dT}(T) = \left(\frac{d\sigma}{dT}\right)_{\mathrm{PT}} (T - A/Q) \,,$$

proves to be *better* than that of the fits that incorporate MC-generated hadronisation effects! The expression in the right-hand-side is the all-order resummed perturbative spectrum *shifted* by A/Q, that very same NP correction term that appears in the mean, $\langle 1-T \rangle$. The ALEPH result is

$$\bar{\alpha}_0 = 0.529 \pm 0.002 \pm 0.0034 \,, \quad \text{with } \alpha_{\overline{\mathrm{MS}}}(M_Z^2) = 0.1194 \pm 0.0003 \pm 0.0035 \,.$$

A free fit to the power in the form

$$\langle 1-T \rangle = c_1 \alpha_s(Q^2) + c_2 \alpha_s^2(Q^2) + \frac{\mathrm{const}}{Q^P} \,,$$

with c_1 and c_2 the known PT coefficients, yielded $P = 0.98 \pm 0.19$, in accord with the theoretical expectation.

It would be premature, however, to celebrate the success of the PT-motivated approach to the NP-physics outlined above. Experimental studies do not yet incorporate the latest theoretical findings. First is the so-called Milan factor, the two-loop renormalisation effect that multiplies the $1/Q$ power terms by the factor 1.8 (for three active quark flavours) [29]. The good news is that this factor is universal. In spite of this, its inclusion may affect the fits.

On top of it, in the studies of the *jet broadening* (H1, JADE) the wrong relation $\rho_1^{(B)} = \rho_0^{(1-T)}$, instead of the correct $\rho_1^{(B)} = \frac{1}{2}\rho_0^{(1-T)}$, was being used (stemming from an unfortunate misprint in the theoretical paper). Moreover, an improved PT description of the B-distribution is now available [30] and should be implemented.

These reservations do not undermine the main amazing finding that a pure perturbative analysis is capable of predicting the power of the Q-dependence and the magnitude of genuine confinement effects in hard observables in general, and in jet shapes in particular.

With the notion of an infrared-finite coupling gaining grounds, we shall be able to speculate about the characteristic QCD parameter,

$$\frac{\alpha_s}{\pi} \simeq 0.16 \,, \qquad \text{versus} \quad \frac{\alpha_{\mathrm{crit}}}{\pi} = \frac{1}{C_F}\left(1 - \sqrt{\frac{2}{3}}\right) \approx 0.137 \,,$$

being sufficiently small to allow the application of perturbative language, at least semi-quantitatively, down to small momentum scales. At the same time, it appears to be sufficiently large to activate the Gribov super-critical light-quark confinement mechanism [31].

6 Conclusions

6.1 Theory

QCD is an infrared-unstable theory. Physical states are Swedish miles[4] away from the fundamental objects making up the QCD Lagrangian. In such circumstances we had better be sceptical and put under Cartesian scrutiny our field-theoretical concepts and tools. In particular, we expect quark and gluon Green's functions to have weird analytic properties as they ought to describe *decaying* objects, in a rather unprecedented way. This fact makes the concept of "Euclidean rotation" far from secure and, in principle, undermines the familiar statistical mechanics substitute for Minkowskian field theory (read: lattice).

To understand the structure of the QCD vacuum and that of hadrons in the real world, we have to address the general problem of binding massless particles. The Gribov super-critical confinement remains, at present, the only dynamical mechanism proposed for that.

The 13 puzzles will stay with us for a while longer.

6.2 Phenomenology

The quantitative theory of hadrons, which theoreticians ought to be looking for, gets more and more restricted by the findings of our experimenting colleagues. The news is, that the small-distance (pQCD) approach, using quarks and gluons, works too often too well. Hadronisation effects, when viewed *globally* seem to behave surprisingly amicably: they either stay *invisible* (inclusive energy and angular hadron spectra) or can be quantified (power effects).

A pQCD-motivated technology for triggering and quantifying genuine non-perturbative (confinement) effects is under construction. These effects show up as power-behaving contributions to Infrared/Collinear-Safe observables, and jet shapes in particular. From within perturbation theory the leading powers can be detected, and the *relative* magnitude of power terms predicted. The absolute values of new dimensional parameters, which we find phenomenologically these days, can be related to the shape of the effective interaction strength (effective QCD coupling) in the infrared region, $\langle \alpha_s/\pi \rangle \sim 0.2$.

6.3 Experiment ("what am I doing this for?")

The epoch of basic QCD tests is over. Today's quest is to understand hadron structure via hadron interactions. The major goal is to study the interface between small and large distances.

[4] 1 Swedish mile = 10 km

The best laboratory for that is "almost-photo-production" at HERA, that is the interaction of small-virtuality photons, $0 < Q^2 < 4\,\mathrm{GeV}^2$, or so, with protons and, hopefully, nuclei.

Diffraction phenomena also target smaller-than-typical hadronic states (e.g., t- and Q^2–dependence of vector meson photo/electro-production).

Studying Drell-Yan pairs, J/ψ and Υ in pp ($p\bar{p}$), pA and AB interactions remains a top priority. In addition to the total production cross sections and the p_t–distributions, various correlation experiments are extremely informative (a famous example being the NA 50 study of the J/ψ yield as a function of accompanying hadronic activity, E_T, in ion-ion collisions).

Jet studies looking for differences in hadron abundances between quark- and gluon-initiated jets should be pursued. Similarity of the yield of hadrons of different species in q and g jets in the "sea" region, if confirmed, would be of major importance for understanding hadronisation dynamics.

Differential jet rates and internal jet-substructure of jets in hard interactions also provide a handle on the soft-hard interface, when one increases jet resolution (by decreasing y_{cut}). To look for genuine confinement effects in such observables, special care should be taken to preserve, as much as possible, the correspondence between parton and hadron ensembles at the perturbative level. To this end the recently proposed modified Durham jet finder, the so-called Cambridge jet algorithm, will have to be used.

6.4 Overall

QCD is on the move, and the pace is good.

Acknowledgements

I am grateful to Marek Karliner for the opportunity to visit Tel Aviv University where this talk was prepared. I enjoyed the hospitality of and invaluable on-line help from my dear friends Halina Abramowicz and Aharon Levy. I would also like to thank Lenya Frankfurt for teaching me high energy nuclear physics.

References

[1] "nothing to add, nothing to subtract"; Torah (about Torah).
[2] A.H. Mueller, Phys.Lett. B396 (1997) 251.
[3] G. Camici and M. Ciafaloni, Phys.Lett. B395 (1997) 118.
[4] ibid. B412 (1997) 396.
[5] J. Blümlein and A. Vogt, hep-ph/9712546.
[6] M. Ciafaloni, private communication.
[7] ZEUS contribution 659.
[8] V. Del Duca and C.R. Schmidt, Nucl.Phys.Proc.Suppl. 39BC (1995) 137.
[9] D0 contribution 087.
[10] E.L. Feinberg and I.Y. Pomeranchuk, Suppl.Nuovov.Cim. III (1956) 652.
[11] M.L. Good and W.D. Walker, Phys.Rev. 120 (1960) 1857.
[12] H. Miettinen and J. Pumplin, Phys.Rev.Lett. 42 (1979) 204.
[13] L. Frankfurt, A. Radyushkin and M. Strikman, Phys.Rev. D55 (1997) 98, and references therein.
[14] DELPHI contribution 545.
[15] OPAL contribution 181.
[16] H1 contribution 251
[17] ZEUS contribution 662.
[18] L3 contribution 506.
[19] OPAL contribution 192.
[20] CDF contribution 562.
[21] review: V.A. Khoze and W. Ochs, Int.J.Mod.Phys. A12 (1997) 2949.
[22] E706 contribution 256.
[23] B.R. Webber, talk at the Workshop on Deep Inelastic Scattering and QCD, Paris, April 1995; hep-ph/9510283.
[24] Yu.L. Dokshitzer, G. Marchesini and B.R. Webber, Nucl.Phys. B469 (1996) 93.
[25] A.C. Mattingley and P.M. Stevenson, Phys. Rev. D49 (1994) 437, and references therein.
[26] Yu.L. Dokshitzer and B.R. Webber, Phys.Lett. B352 (1995) 451; ibid. B404 (1997) 321.
[27] L3 contribution 499.
[28] ALEPH contribution 610.
[29] Yu.L. Dokshitzer, A. Lucenti, G. Marchesini and G.P. Salam, hep-ph/9707532.
[30] ibid. hep-ph/9801324.
[31] V.N. Gribov, Physica Scripta T15 (1987) 164; Lund preprint LU 91–7 (1991).

CP Violation in K and B Systems

C. Jarlskog

Division of Mathematical Physics, LHT, Lund University
email: cecilia.jarlskog@matfys.lth.se

Abstract. Since the last EPS and Rochester Conferences on particle physics no new experimental results have been presented in the field of CP violation. This field is very much in a state of data taking, data analysis and preparing for future experiments. My talks will, therefore, be rather general. I will discuss the following items:

- Historical remarks
- The quark mixing matrix
- The Parameter ϵ'
- "Forked tracks of a very striking character"
- The decay $K^+ \to \pi^+ \nu\bar{\nu}$ and other rare decays of kaons
- T reversal
- Beauty
- What about the future?

1 Historical remarks

CP violation was discovered in 1964 [1] in the decay of $K_L \to \pi^+\pi^-$. It immediately became a very hot topic in particle physics. One looked for experimental signatures of CP "everywhere", in strong interactions, in electromagnetic interactions as well as in weak interactions. Several theoretical reasons for the existence of CP violation were proposed. In some of the more fanciful proposals the CP violation was supposed to be caused by

- a long-range cosmic force
- an octet of strongly interacting vector bosons
- new particles called the a-particles or chimerons
- mirror fermions
- leptoquarks
- gravitational dipole, etc.

This was the situation in 1960's. However, no experimental signatures of CP violation were seen anywhere except in a few decay modes of K_L. By the summer of 1974 the experiments had become rather accurate. After several years, with confusing results concerning the equality or non-equality of the CP parameters η_{+-} and η_{00}, the conclusion, in 1974, was that these two parameters are equal, within the experimental errors. Wolfenstein's superweak model [2] became the standard theory of CP violation. This model says that magnitude of CP violation is very small, second order in weak interactions. Hence, the only place it had a chance to manifest itself was in $K^0 - \bar{K}^0$

system because of several fortunate circumstances. The main reason being that in the absence of weak interactions, these two mesons are degenerate. A degenerate system can be extremely sensitive to even tiny perturbations. The other reasons were, non-conservation of strangeness by weak interactions that causes the mixing of the two neutral kaons as well as the fact that long-lived kaon and the short-lived one have very different lifetimes and can be "separated", etc.

Somehow, in 1974 the chapter on CP violation seemed essentially closed. Some physicists participating in the 1974 International Conference on High Energy Physics, that took place in London, were considering the Conference as a kind of "funeral" of CP violation. How could anything exciting happen in that field?

Fortunately, the above pessimistic state of affairs didn't last long. Already late in 1974 the hidden charm was discovered. The mesons D^0 and $\bar{D}^0$ provided a kind of copy of the neutral kaons and stimulated new interest in CP violation. A few years later the B mesons were discovered providing an even richer landscape. Another reason for the great interest in CP violation had to do with the so-called marriage of particle physics and cosmology. Understanding the baryon asymmetry of the universe is considered to be one of the major issues in cosmology. It requires understanding CP violation.

Actually, since the discovery of parity violation in electron scattering off quarks, in 1978, where the effect was found to be in agreement with the predictions of the standard electroweak model, this theory has been the main framework for discussing CP violation. There is also much interest in CP violation in the framework of supersymmetric theories. At this Conference the topic is reviewed by Binetruy [3].

Perhaps the most important of all is the fact that there is a lot of experimental activity in the field of CP violation. In 1970's who could ever imagine that before the end of this century several huge detectors (BABAR, at SLAC, BELLE at KEK, HERA-B at DESY, KLOE at the phi-factory DAΦNE) would be constructed, their main purpose being to look for and study CP violation?

As one of the hotest topics in particle physics CP violation is much discussed at the international conferences on high energy physics (see, for example [4]). Therefore, I will be rather brief here.

2 The quark mixing matrix

The standard explanation of CP violation is that it originates from the quark mass matrices, for the up-type and down-type quarks. It is telling us that these mass matrices are not relatively real. For the case of three families this is equivalent to the statement that the commutator of the two quark mass matrices is non-singular. Of course the quark mass matrices have their origin in the interactions of the Higgs sector with the quark fields. So, CP violation is due to the as yet unobserved Higgs sector of the standard model. All this is very exciting but it can't possibly be the final answer. In the standard model all the elements of the quark mass matrices are free parameters. In the real life we know that there is a great deal of hierarchy in these mass matrices, as is demonstrated in the structure of the quark mixing matrix and the pattern of masses. The quark mass matrices for the up-type and down-type quarks are very much "aligned" [5]. Where do these structures come from? This makes us believe that there must be a more fundamental theory that dictates the observed structures.

The quark mixing matrix, often referred to as the CKM matrix,

$$\begin{pmatrix} V_{ud} & V_{us} & V_{ub} \\ V_{cd} & V_{cs} & V_{cb} \\ V_{td} & V_{ts} & V_{tb} \end{pmatrix}$$

can be parametrized by four real parameters. These are often taken to be the original Kobayashi-Maskawa parameters [6] θ_1, θ_2 ,θ_3 ,δ or the Wolfenstein parameters λ, A, ρ, η,

$$\begin{pmatrix} 1-\lambda^2/2 & \lambda & A\lambda^3(\rho - i\eta) \\ -\lambda & 1-\lambda^2/2 & A\lambda^2 \\ A\lambda^3(1-\rho-i\eta) & -A\lambda^2 & 1 \end{pmatrix}$$

up to corrections of order λ^4. Other parametrizations of this matrix are discussed in the Review of Particle Physics [7].

The quark mixing matrix being a unitary matrix, the scalar product of any of its two columns i and j, in the form $\mathbf{c}_i \bullet \mathbf{c}_j{}^\star$ is zero for $i \neq j$. This defines three triangles. Similarly, there are three triangles coming from the scalar products of the rows. All the six triangles have the same area, J/2. The quantity J is the invariant measure of CP violation in the standard model. If CP were conserved all the six triangles would collapse to lines. Conversely, if the area of any of the triangles is non-zero CP must be violated. The three angles of the most triangular-looking of the six triangles are denoted by α, β, γ. For example, in the Wolfenstein parametrization one has

$$\rho - i\eta = \sqrt{\rho^2 + \eta^2}\; exp(-i\gamma) \tag{1}$$

$$1 - \rho - i\eta = \sqrt{(1-\rho)^2 + \eta^2}\; exp(-i\beta) \tag{2}$$

and the third angle is, evidently given by $\alpha + \beta + \gamma = \pi$. The area of the unitarity triangle, in this parmetrization is given by $J/2 \approx A^2\lambda^6\eta/2$.

Actually, one could just as well forget about all parametrizations and just stick to say the absolute values of the four elements in the left-upper corner of the quark mixing matrix, i.e., $|V_{ud}|$, $|V_{us}|$, $|V_{cd}|$, $|V_{cs}|$. Measuring these four quantities is, in principle, sufficient for reconstructing all the six unitarity triangles and learning all there is to know about CP violation in the standard three family model, up to a two-fold ambiguity having to do with the overall orientation of the triangles, i.e., the sign of J. For a detailed discussion and references see [8]. In the real life, however, the precision needed is beyond our reach.

The present status of the quark mixing matrix can be found in the latest issue of the "Review of Particle Physics". In the future one expects a great deal of improvement on the determination of the sides and the angles of the unitarity triangle discussed above, as was discussed at this Conference by several authors, see for example the talk by Schubert [9].

3 The parameter ϵ'

The only evidence for CP violation, since its discovery in 1964, is that [7]

$$\begin{aligned} \Gamma(K_L \to \pi^+\pi^-) &\neq 0, \qquad BR \approx 0.2\%) \\ \Gamma(K_L \to \pi^0\pi^0) &\neq 0, \qquad BR \approx 0.1\%) \\ \Gamma(K_L \to \pi^- l^+ \nu) &\neq \Gamma(K_L \to \pi^+ l^- \nu), \qquad l = e, \mu \\ \Gamma(K_L \to \pi\pi\gamma) &\neq 0 \end{aligned}$$

All the above can be accounted for by assumming that the long-lived kaon is not a pure CP eigenstate but a linear combination

$$|K_L \approx |CP = -1 > +\epsilon\ |CP = +1 >$$

where ϵ is a small parameter, its magnitude being $(2.28 \pm 0.02)10^{-3}$. The fact that a physical state is not pure, but is a linear combination of two opposite CP states is called indirect CP violation. For many years one has been in search of direct CP violation, i.e., for a CP violating transition between two CP pure states. In the K-system, a signal of direct CP violation would be to find that the parameter ϵ' is non-zero, where

$$Re(\epsilon'/\epsilon) = \frac{1}{6}\{1 - \frac{\Gamma_L{}^{00}/\Gamma_S{}^{00}}{\Gamma_L{}^{+-}/\Gamma_S{}^{+-}}\}$$

where $\Gamma_X{}^{ij}$ stands for $\Gamma(K_X \to \pi^i\pi^j)$, X=L,S and i, j denote the electric charge of the pion. The present value of the above ratio is $(2.3 \pm 0.7) \times 10^{-3}$ from NA31 Collaboration at CERN [10] and $(0.74 \pm 0.60) \times 10^{-3}$ from E731 Collaboration at Fermilab. [11].

Several new experiments on ϵ' are now in progress. These are:

-E832 which is an experiment at Fermilab. It has taken data since October 1996. The experiment has two K_L beams and uses a regenerator to produce K_S. It measures the modes $\pi^+\pi^-$ and $\pi^0\pi^0$ simultaneously.

-NA48 experiment at CERN. It has also two beams and has just gone into operation this summer (1997).

-KLOE, at the Φ factory DAΦNE, in Rome will start taking data in 1998. This factory is expected to produce about 5000 Φ's per second and $\Phi \rightarrow K_L K_S$ has a large branching ratio (34%). This experiment promises to measure the quantity ϵ'/ϵ to an accuracy of about 10^{-4}.

Hopefully, in a couple of years we should know much more about ϵ'/ϵ.

On the theoretical side, the situation is not at all so clear. It has been realized that the prediction of ϵ'/ϵ is very difficult. Because of strong interactions, there are several effective operators that contribute to this quantity. These operators are denoted by Q_i, i=1,2, ... Two of these operators (Q_6 and Q_8) give the largest contribution but, unfortunately, with opposite signs. Thus there are a lot of cancellations. Recent compilations by Ciuchini [12] and by Buras and Fleischer [13] make this point very clear. For example, Buras and Fleischer quote a large range for ϵ'/ϵ, from -50 to +16 in the units of 10^{-4}. A non-zero value of ϵ'/ϵ would show that there is direct CP violation and that by itself is a very important piece of knowledge. Indeed, it is the experiment that is going to put constraints on the theory.

4 "Forked tracks of a very striking character"

This year marks the 50th anniversary of a monumental paper, written by George Dixen Rochester and Clifford Charles Butler. The paper was published in Nature, Dec. 20, 1947. What Rochester and Butler report on is two cloud chamber photographs containing "forked tracks of a very striking character". Their interpretation of one of the photographs is that it shows a neutral particle decaying inte two charged particles. They estimate the mass of the decaying particle to be $\approx 1000 m_e$ and its lifetime to be 5×10^{-8}. Now, suppose that Rochester and Butler had had a kind of time machine and could leap forward in time by 50 years and look into our beloved tables in "Review of Particle Physics", produced by the Particle Data Group. They would have found a unique solution to their second photograph: the CP violating decay $K_L \rightarrow \pi^+\pi^-$. Of course, one can not claim that Rochester and Butler discovered CP violation. Nonetheless their paper is great.

5 The decay $K^+ \rightarrow \pi^+ \nu\bar{\nu}$ and other rare decays of kaons

Now, at this Conference, we had another great event presented to us by J. Frank [14] from the E787 Collaboration at Brookhaven. In this experiment kaons are stopped in plastic scintillating fibers and the decay chain $\pi \rightarrow \mu \rightarrow e$ is observed. The experiment has found one candidate for the decay $K^+ \rightarrow \pi^+\nu\bar{\nu}$. The point made by Frank was that the expected background is much smaller than one. The estimated branching ratio [14], based on the one observed event is

$$BR(K^+ \to \pi^+ \nu\bar{\nu} = 4.2^{+9.7}_{-3.5}) \times 10^{-10}\ (67\%CL)$$

The theoretical analysis of the decay $K^+ \to \pi^+ \nu\bar{\nu}$ is rather clean. In the standard model the underlying process is $\bar{s} \to \bar{d}\nu\bar{\nu}$ which is caused by penguin diagrams, $\bar{s} \to \bar{d}Z$ involving the top quark as well as box diagrams. The branching ratio is estimated to be $0.6 - 1.5$ in the units of 10^{-10}. From the one event one obtains $0.006 < |V_{td}| < 0.06$. The future of the experiment looks promising and determination of $|V_{td}|$ to an accuracy of about 30% should be possible [14].

Theorists most beloved decay in this field is the CP violating process $K_L \to \pi^0 \nu\bar{\nu}$ as its theoretical analysis is very clean indeed. The branching ratio for this decay is proportional to the square of the area of the unitarity triangle ($\sim J^2$) and is estimated to be roughly 3×10^{-11} [13]. From the above two decays on would be able to determine rather accurately both the quantity $|V_{td}|$ and the height of the unitarity triangle.

Rare decays of the neutral kaons are being studied at Fermilab, by the KTeV Collaboration. The results obtained are very impressive indeed and could be further improved in the future [15].

I should also mention that at Fermilab., an experiment called E871 or HyperCP is looking for CP violation in the decays of hyperons, by comparing the decay parameters of hyperons with those of antihyperons [16].

6 T reversal

The T symmetry is referred to as time reversal which is unfortunate since no experiment has ever succeeded in reversing the direction of time. Motion reversal would have been a better name, as was actually suggested by Lüders. A direct test of time reversal (read motion reversal) has been done by the CPLEAR Collaboration [17] where the rates of $K^0 \to \bar{K^0}$ and its motion reversed image $\bar{K^0} \to K^0$ are measured and compared. Defining

$$A_T = \frac{R(\bar{K^0} \to K^0) - R(K^0 \to \bar{K^0})}{R(\bar{K^0} \to K^0) + R(K^0 \to \bar{K^0})}$$

the CPLEAR experiment quotes the preliminary result [17]

$$A_T = (6.3 \pm 2.1 \pm 1.8) \times 10^{-3}$$

where the first (second) error is statistical (systematical). A summary of results obtained from CPLEAR experiment can be found in the contribution by Zavrtanik [17].

7 Beauty

The year 1997 marks the 20th anniversary of the discovery of beauty [18]. The systems B_d^0 and B_s^0 are again copies of K^0 and $\bar{B}_d^0$ and $\bar{B}_s^0$ are copies of $\bar{K}^0$. The analysis presented for the neutral kaons can be repeated for these systems. There is a long-lived B_d and a short-lived one, etc.

At this Conference a review of the measurements of Δm_d and Δm_s was presented [19]. The first (second) quantity denotes the mass difference between the long-lived and the short-lived neutral B-mesons containing a d-quark (s-quark) or its antiquark. Δm_d is now known to a precision of a few percents, $\Delta m_d = (0.472 \pm 0.018)ps^{-1}$. For further information see Ref. [19].

A substantial progress in measuring B decays was reported by the CLEO Collaboration [20]. I shall not discuss B-decays, as they are reviewed by Neubert [21] at this conference. A somewhat disturbing fact, however, is that the upper limit for the branching ratio of $B \to \pi^+\pi^-$ is now as low as 15×10^{-6} [20]. As this decay is one of the important candidates for determining the angle α of the unitarity triangle one would have liked its branching ratio to be as large as possible. The quoted upper limit falls in the range predicted for the branching ratio, in the standard model, but is now on the low side.

In principle, it is easy to look for CP violation in the neutral B system. One can compare the rates of CP conjugate transitions

$$\begin{aligned} B &\to f \\ \bar{B} &\to \bar{f} \\ B &\to \bar{f} \\ \bar{B} &\to f \end{aligned}$$

Life is somewhat simpler if $f = \bar{f}$. In addition, the mixing of B and $\bar{B}$ must be taken into account. In other words, starting with say a pure B, it will evolve into a linear combination of B and $\bar{B}$. The CP asymmetry, for a self conjugate final state is given by

$$a_{CP} = \frac{\Gamma(B \to f) - \Gamma(\bar{B} \to f)}{\Gamma(B \to f) + \Gamma(\bar{B} \to f)}$$

The gold-plated case is expected with $f = J/\psi K_S$. The reason being that its branching ratio is relatively speaking large (a few times 10^{-4}) and the theoretical prediction of the effect is clean as the above decay is described by tree diagrams. The measurement of the asymmetry of this decay will determine the angle β of the unitarity triangle. Other decays, such as $B \to \Phi K_S$ and $B \to D^+D^-$ that also give the angle β are less clean. The former is described by penguin diagrams and the latter receives both tree and penguin contributions.

Determination of the angles of the unitarity triangle, from B decays is reviewed in some detail Ref. [7]. I shall not repeat the theoretical analysis here.

The future of CP violation in the beauty sector looks very promising indeed.

In 1998 CLEOIII at Cornell should provide valuable new knowledge on B decays. Also the experiment HERA-B at DESY is to start. In this experiment the proton beam from HERA will hit an internal target. B mesons will be produced and studied.

In 1999 the experiments BABAR at SLAC and BELLE at KEK should start taking data. These detectors will observe $B, \bar{B}$ produced in asymmetric electron-positron colliders dedicated to producing beauty mesons. In addition the CDF and DØ experiments at Fermilab are expected to study B decays.

Early next century we will hopefully learn even more about B decays and CP violation from the BTeV experiment at FNAL and the three particle physics experiments (ATLAS, CMS and LHC-B) at the Large Hadron Collider (LHC) now being built at CERN.

"And God said" let $detC \neq 0$ and there was CP violation, where C is the commutator of the quark mass matrices [8] and

$$
\begin{aligned}
detC &= -2J \times F \times F' \\
F &= (m_u^2 - m_c^2)(m_c^2 - m_t^2)(m_t^2 - m_u^2) \\
F &= (m_d^2 - m_s^2)(m_s^2 - m_b^2)(m_b^2 - m_d^2)
\end{aligned}
$$

J being twice the area of the unitarity triangle, but we don't understand the reason.

8 What about the future?

Before looking into future let us have a look into the past. The year of this Conference, 1997, is a fantastic year. It is:

• the 100th anniversary of the discovery of the electron by Thomson the Father. He presented his results at a meeting of the Royal Society on 30 April 1897. He demonstrated the existence of the electron as a particle. Thomson the Son showed that it is a wave.

• the 50th anniversary of the discovery of the pion (see Nature 160 (1947))

• the 50th anniversary of the monumental paper on "forked tracks of a very striking character" that I mentioned before

• the 40th anniversary of the experimental discovery of parity violation (see Phys Rev. Jan. 1957)

• the 20th anniversary of the discovery of beauty (see Phys. Rev. Lett. Aug. 1977)

There can be no doubt that in the contemporary particle physics the most important concept is SYMMETRY. Discrete symmetries are not considered to be as important as say nonabelian local symmetries, because the latter give rise to forces. So the answer to the question "why are the discrete symmetries, P, C and T violated?" is often "why shouldn't they be?".

Historically, the discrete symmetries, P, C and T were considered to be sacred. Thus the fall of parity came as a "super surprise" to the physics community. I believe that most young people nowadays would have difficulty in understanding why it created so much excitement. After all, the young people these days are brought up on left-handed doublets and right-handed singlets of the standard model. The standard model, of course, assumes violation of parity without explaining it.

The experimental papers on parity violation were published in a January 1957 issue of Physical Review. In the September of the same year, i.e., almost exactly 40 years ago, a conference on nuclear structure took place in Rehovot, here in Israel [22]. Because of the discovery of parity violation the scientific program of the conference was modified to discuss the new revolution in physics. There were many very distinguished participants, among them Wolfgang Pauli and one of the discoverers of parity violation, Madam C. S. Wu. The President of Israel, D. Ben-Gurion was also excited about parity violation and wanted to have a discussion with Madam Wu. He tried to contact her. Not having found her, President Ben-Gurion left Madam Wu an interesting note asking her opinion on a book he had read on "polarity". The note ends with the question "Does it make sense from the point of view of physics?". Now, I ask you: when did a president, or a prime minister or a head of state ask your opinion on a physical matter?

In the proceedings of the Rehovot Conference, there are many interesting articles, including a short contribution by Wolfgang Pauli, that even now, after 40 years is very much worth reading.

It is generally believed that there must be new sources of CP violation in nature. The reason is that the standard model does not provide enough CP violation to describe the baryon asymmetry of the universe. Actually, this is one of the very few empirical signs that we have at present for the existence of a theory beyond the standard model. The solar and atmospheric neutrino problems are other such signs. However, the neutrino problems do not necessarily imply a major revision of the standard model. Massive neutrinos can be accommodated in the standard model with little effort. But the CP problem will require a major revision. Thus it is not unreasonable to expect surprising results from the future CP violation experiments. These might turn out to provide a bridge (tunnel?) to the continent of New Physics which we believe lies out there somewhere. However, we have to be patient. Let me close by quoting what Madam Wu said at the 1957 Rehovoth Conference:

"But Rome wasn't built in a day,
and this is just the very beginning"

9 Acknowledgements

In preparing this talk I have had useful discussions and email correspondences with many colleagues. They have kindly informed of the present status of their experiments or theories. I wish to thank Juliet Lee-Franzini, Bruce Winstein, Sheldon Stone, Stew Smith, Ken Peach, Giles Barr, Philip Harris, Noulis Pavlopoulos, Fumihiko Takasaki, Tatsuya Nakada, Michael Gronau, Amarjit Soni, Ed Thorndike, Andreas Schwarz, Lawrence Gibbons, Marco Ciuchini, Walther Schmidt-Parzefall, Klaus Schubert, David Miller, as well as the speakers of the PA9 session.

References

[1] J. H. Christenson, J. W. Cronin, V. L. Fitch and R. Turlay, Phys. Rev. Lett. 13 (1964) 138

[2] L. Wolfenstein, Phys. Rev. Lett. 13 (1964) 569

[3] P. Binetruy, these Proceedings; see also
Y. Nir, talk given at the 1997 Lepton Photon Conference, Hamburg, to be published

[4] R. Aleksan, Proc. of the International Europhysics Conference on High Energy Physics (Brussels, 1995), Eds. J. Lemonne et al. (World Scientific 1996) p. 743
J.-M. Gerard, *ibid* p. 813
L. K. Gibbons, Proc. of the conf. ICHEP'96 (Warsaw), Eds. Z. Ajduk and A. K. Wroblewski (World Scientific 1997) p. 183
A. J. Buras, *ibid* p. 243

[5] P. H. Frampton and C. Jarlskog, Phys. Lett. 154 B (1985) 421

[6] M. Kobayashi and T. Maskawa, Prog. Theor. Phys. 49 (1973) 652

[7] Particle Data Group, Phys. Rev. D54 (1996) 1

[8] C. Jarlskog *in* CP Violation, Ed. C. Jarlskog (World Scientific, 1989) p. 3

[9] K. Schubert, these Proceedings

[10] The NA31 Collaboration, G. Barr et al., Phys. Lett. B317 (1993) 233

[11] The E731 Collaboration, L. K. Gibbons, Phys. Rev. Lett. 70 (1993) 1203

[12] M. Ciuchini, talk presented at the 4th KEK Conference on "Flavour Physics", preprint CERN-TH/97-2

[13] A. J. Buras and R. Fleischer, preprint TUM-HEP-275/97 (hep-ph/9704376)

[14] J. Frank, These Proceedings

[15] Letter of intent at Fermilab., E. Cheu et al. (June 2, 1997)

[16] A. Chan et al. "Search for CP violation in the decays of $\Xi^-/\bar{\Xi}^+$ and $\Lambda/\bar{\Lambda}$ hyperons"
[17] D. Zavrtanik, These Proceedings
[18] S. W. Herb et al., Phys. Rev. Lett. 39 (1977) 252
[19] J. Jimack, These Proceedings
[20] D. Miller, These Proceedings
[21] M. Neubert, These Proceedings
[22] See the Proc. of the Rehovoth Conf. on Nuclear Structure, Ed. H. Lipkin (North-Holland, 1958)

Tau and Charm Physics

Dave Besson[1] (dzb@kuhep4.phsx.ukans.edu)

University of Kansas, Lawrence KS 66045 USA

Abstract. A selection of results in charm and tau physics are reviewed.

1 Charm

Charm production, and the study of the decays of particles containing charm, is now well-established at virtually all of the world's major accelerator facilities. The experimental environments, however, vary considerably from one site to the next. Excluding contributions from B's, there are much different visible production cross-sections for open (D^{*+}) and hidden (J/ψ) charm at the various facilities, as briefly summarized in the Table below:

	e^+e^- (10 GeV)	ep (HERA)	$p\bar{p}$
$\frac{\sigma^{vis}(D^{*+}\to D^0(K^-\pi^+)\pi^+)}{\sigma^{vis}(J/\psi\to l^+l^-)}$	50	1	0.04

Contributions to this conference in this field included a wide variety of subjects. Due to space limitations, it will be impossible to be encyclopedic; for more details on contributions to this conference, the interested reader is referred to the appropriate Web pages.

1.1 Study of Charm Hadron Decays

The study of charm meson decays typically begins with the long-standing pattern of charm hadron lifetimes. In the most naive picture, in which the only accessible decay diagram is the simple external W-emission spectator diagram (W_{ext}), the lifetime of all the charm particles should be identical and calculable from the mass of the charm quark. The fact that there are large discrepancies between the various charm lifetimes is an indication that other diagrams contribute, as well. The pattern of lifetimes, as well as the primary diagrams expected to contribute to charm decay, is presented below:

Particle	τ (ps)[1]	Main Diagrams
D^+	1.057±0.015	$\|W_{ext} - W_{int}\|^2$, $W_{annihilation}$
D_s	0.467±0.017	W_{ext}, $W_{annihilation}$
D^0	0.415±0.004	W_{ext}, $W_{exchange}$
Λ_c	0.206 ± 0.012	$W_{ext}, W_{int}, W_{exchange}$
Ξ_c^+	$0.35^{+0.07}_{-0.04}$	W_{ext}, W_{int}
Ξ_c^0	$0.098^{+0.023}_{-0.015}$	$W_{ext}, W_{int}, W_{exchange}$
Ω_c^0	0.064 ± 0.020	W_{ext}, W_{int}

The 'standard' explanation for this pattern, at least in the case of charged and neutral D-mesons, is that destructive interference between the internal and external W-emission diagrams which can contribute to the same final state (e.g., $D^+ \rightarrow \overline{K^0}\pi^+$) dilates the D^+ lifetime[2]. Nonetheless, it is likely that other diagrams, such as annihilation and exchange, also modify the D-meson lifetimes – the fact that the D_s and D^0 lifetimes deviate (whereas in the naive quark model, their lifetimes would be identical) is fairly compelling evidence that additional non-external sepctator effects must be present. Note that, for baryon decays, the presence of a third quark can 'lift' the color/helicity suppression.

Among the more interesting possible diagrams (from the standpoint of connection to the underlying physics) is the annihilation diagram. In principle, this rate is straightforward to calculate: for the decay of a (spin-zero) pseudoscalar, the annihilation of the $c\bar{s}$ system can be written in terms of q^μ (in this case, the only four-vector in the hadronic current) as: $< 0|J^\mu|D_s >= iV_{cs}f_{D_s}q^\mu$. The term f_{D_s} is the D_s decay constant, which depends on the wave function overlap of the c and $\bar{s}$ quarks $|\Psi(0)|^2/m$ at the origin and therefore gives fundamental information on the quark-antiquark interaction. The momentum transfer $q^\mu \propto (1 - \frac{m_l^2}{m_{D_s}^2})$. In long form, the total decay rate can be written as:

$$\Gamma(D_s \rightarrow \mu\nu_\mu) = G_F^2 m_D f_{D_s}^2 m_\mu^2 |V_{cs}|^2 (1 - m_\mu^2/m_{D_s}^2),$$

where the mass term m_μ^2 reflects helicity suppression at the light quark vertex. Consequently, for annihilation into leptonic modes, e.g., $\Gamma_{D_s\rightarrow\tau\nu_\tau} > \Gamma_{D_s\rightarrow\mu\nu_\mu} > \Gamma_{D_s\rightarrow e\nu_e}$; the $\tau\nu$ mode is larger than the $\mu\nu$ mode by approximately a factor of 10:1. In the case where the annihilation proceeds into an hadronic final state, it may be the case that gluon emission in the hadronic case may mitigate the helicity suppression, so that D_s annhiliation into hadronic final states may be enhanced relative to the leptonic final states. Note that not only is the D_s decay constant a fundamental quantity of interest in and of itself, but present lattice gauge calculations claim to be able to bootstrap from measurements of the D_s decay constant to determine the B-decay constant f_B with ~5% precision. The B-decay constant, of course, is one of the parameters that is used to constrain the geometry of the CKM triangle.

Annihilation Contributions to the D_s width: $D_s \rightarrow l\nu_l$ There are new results on D_s decay to leptons from both DELPHI[3] and L3[4] at LEP, as well as CLEO at CESR[5]. The techniques are similar - in all cases, the missing four-vector in the event is identified with the energy/momentum carried off by the neutrino(s), and is used to reconstruct the $D_s \rightarrow l\nu_l$ decay. In the case of LEP, these higher-energy experiments can take advantage of the higher collimation of the events to give better resolution on the direction of the missing momentum; they can therefore accept the poorer resolution on the missing energy and take advantage of the higher branching fraction for $D_s \to \tau\nu_\tau$; $\tau \to (e/\mu)\nu_\tau\nu_{(e/\mu)}$. In the case of CESR, the high statistics sample allows one to concentrate on the $D_s \to \mu\nu_\mu$ decay mode.

In both cases, the relatively poor resolution on the D_s itself can be compensated for by combining the candidate D_s with photons in the event in a search for D_s^*; the resolution on the $D_s\gamma$ mass less the D_s mass (i.e., the mass difference) is considerably better than the D_s mass alone.

Performing such an exercise, L3[4] obtains $D_s \to \tau\nu_\tau = 0.074 \pm 0.028 \pm 0.016 \pm 0.018$, where the last error reflects the uncertainty in the normalizing mode (i.e., the uncertainty in $\mathcal{B}(\phi\pi)$). This branching fraction corresponds to a D_s decay constant $f_{D_s} = 309 \pm 58 \pm 33 \pm 38(\text{norm})$ MeV. DELPHI[3] performs a similar analysis, obtaining $D_s \to \tau\nu_\tau = 0.085 \pm 0.042 \pm 0.026$, or $f_{D_s} = 330 \pm 95$ MeV.

CLEO[5] has redone their original analysis of $D_s \to \mu\nu$, with an enlarged data sample, an improved analysis technique which more fully exploits the kinematics of the $D_s \to \mu\nu_\mu$ decay, and taking advantage of an improved understanding of leptonic fake rates. Their original result, based on a signal of 39 ± 8 events, after adjustment for the most recently tabulated $D_s \to \phi\pi$ absolute branching fraction, as well as CLEO's most recent leptonic fake rates, corresponds to a D_s decay constant: $f_{D_s} = 278 \pm 35 \pm 30 \pm 24$ MeV. Their new result has approximately four times the signal size (resulting from improvements in a factor of two in both data sample, as well as efficiency) and yields: $f_{D_s} = 275 \pm 14 \pm 25 \pm 24$ MeV.

The CLEO measurement now dominates the world average, which now corresponds to: $\mathcal{B}(D_s \to \mu\nu) = 0.48 \pm 0.07 \pm 0.12$ (the $D_s \to \tau\nu_\tau$ measurements from LEP have been 'adjusted' to their equivalent $D_s \to \mu\nu_\mu$ branching fractions knowing the expected helicity suppression of $\frac{D_s \to \mu\nu}{D_s \to \tau\nu_\tau}$). This $D_s \to \mu\nu_\mu$ branching fraction corresponds to an implied D_s decay constant: $f_{D_s} = 250 \pm 20 \pm 31$, which agrees with the expectation for the D_s decay constant as the decay constant value derived from factorization measurements in B-decay.[1] Note that the biggest error on the D_s decay constant at this point is the uncertainty in the absolute scale, as determined by the normalization mode: $D_s \to \phi\pi$. Note also that lattice gauge calculations by the MILC[6] collaboration have yielded predictions for both the absolute D_s decay constant (predicting $f_{D_s} = 199 \pm 8^{+40+10}_{-11-0}$), and a rather precise

[1] In factorization, one can relate $B \to DD_s$ to $B \to Dl\nu_l$, provided the $l\nu_l$ system is evaluated at a value of q^2 corresponding to the mass of the D_s. I.e., we assume both processes are essentially $B \to DW^+$ and we liken $W \to l\nu_l$ to $W \to c\bar{s}$ at the same q^2, with the exception that the production of D_s requires fusion of the $c\bar{s}$ system into a hadron (the inverse of D_s decay). The rate for $B \to DD_s$ is then related to the rate for $B \to Dl\nu_l$ at this value of q^2 by the likelihood for $c\bar{s}$ wavefunction overlap, or f_{D_s}.

calculation for the ratio of the B-decay constant to the D_s decay constant: $\frac{f_B}{f_{D_s}} = 0.76 \pm 0.03^{+0.07+0.02}_{-0.04-0.01}$.

Gluon Emission Contributions to Heavy Meson Decays: Having explored D_s annihilation in the leptonic sector, it is natural to consider whether evidence for D_s annihilation can be found in the hadronic sector. Perhaps the simplest decay to search for such annihilation would be: $D_s \rightarrow W^+ \rightarrow u\bar{d} \rightarrow (\rho/\omega)\pi$. Unfortunately, $\rho\pi$ is hampered by the large combinatoric background attendant to the three pion final state; $\omega\pi$ is, in principle, prohibited from proceeding through the simple decay chain outlined above by simple quantum numbers (this decay constitutes a second-class current).

Similar to reconstruction of the D_s annihilation into the leptonic mode, the search for this decay by CLEO[7] Collaboration similarly relies on reconstruction of $D_s^* \rightarrow D_s\gamma$; with $D_s \rightarrow \omega\pi$. As before, the signal is observed as a peak in the $D_s^* - D_s$ mass difference plot. Since the ω is observed via $\omega \rightarrow \pi^+\pi^-\pi^0$, it is convenient to normalize this branching fraction to: $D_s \rightarrow \eta\pi$; $\eta \rightarrow \pi^+\pi^-\pi^0$. CLEO observes a statistically significant signal in this search, corresponding to $\mathcal{B}(\frac{D_s \rightarrow \omega\pi^+}{D_s \rightarrow \eta\pi^+}) = 0.16 \pm 0.04 \pm 0.03$. Since this mode is prohibited from directly coupling to the D_s by the second-class current argument cited earlier, one is therefore prompted to consider whether there are final state interaction (FSI) effects which are feeding the $\omega\pi$ final state. However, the fact that the $\omega\pi$ system has uncommon quantum numbers ($J^P = 0^-$; $I^G = 1^+$) argues against a large resonant FSI contribution, although non-resonant rescattering of $D_s \rightarrow K^*K$, e.g., is certainly allowed.

Of particular interest is the possibility that this mode is proceeding through $c\bar{s} \rightarrow gggW^+ \rightarrow \omega\pi^+$; with the gluonic emission lifting the helicity suppression that one would otherwise naively expect to be operative for direct $D_s \rightarrow W^+ \rightarrow u\bar{d}$. In fact, there are suggestions that multigluonic emission may be present in other charm decays. In particular, the anomalously large η' rate in two-body decays relative to η may indicate the presence of such couplings. The pattern of large $\frac{D_s \rightarrow \eta' X}{D_s \rightarrow \eta X}$ is evidenced by:

- $\frac{\Gamma(D_s \rightarrow \eta\pi)}{\Gamma(D_s \rightarrow \phi\pi)} = 0.48 \pm 0.03 \pm 0.04$[8]
- $\frac{\Gamma(D_s \rightarrow \eta'\pi)}{\Gamma(D_s \rightarrow \phi\pi)} = 1.03 \pm 0.06 \pm 0.07$[8]
- $\frac{\Gamma(D_s \rightarrow \eta\rho)}{\Gamma(D_s \rightarrow \phi\pi)} = 2.98 \pm 0.20 \pm 0.39$[8]
- $\frac{\Gamma(D_s \rightarrow \eta'\rho)}{\Gamma(D_s \rightarrow \phi\pi)} = 2.78 \pm 0.28 \pm 0.30$[8]: 4-7 $\times$ all models!

The larger mass of the η', of course, would naively be expected to produce ratios of branching ratios which were consistently less than unity, contrary to observation (particularly in the case of $\frac{\eta'\rho}{\eta\rho}$). One is led to the conjecture that, in the event that the digluonic fraction of the η' wave function is large, there may be direct external gluon lines connecting the D_s to the η' through: $D_s \rightarrow W^+(gg) \rightarrow \pi\eta'$. Related experimental information which bears on this result comes from the fact that: $\frac{\psi\to\gamma\eta'}{\psi\to\gamma\eta} = \frac{(4.31\pm0.3)\times10^{-3}}{(8.6\pm0.8)\times10^{-4}} \sim 5$[1], and that the two-photon production of η vs. η' implies a larger conventional $q\bar{q}$ component of the η, consistent with the conjecture given above.

As a check of the possibility of enhancements in hadronic decays, it is natural to consider the magnitude of η' production in semileptonic decays. In this case, which excludes the possibility of η' enhancements through external gluon lines, we find: $\frac{D_s\to\eta' l\nu_l}{D_s\to\phi l\nu_l}=0.44\pm0.13$[1], while $\frac{D_s\to\eta l\nu_l}{D_s\to\phi l\nu_l}=1.27\pm0.19$[1]; in principle, these measurements can be considered as measurements of the $s\bar{s}$ admixture of the η and η' wave functions. Again, factorization can be used to 'bootstrap' from the semileptonic rate for $D_s \to \eta' l\nu_l$ to obtain a prediction for the absolute rate for $D_s \to \eta'\rho$, for the case where $M(l\nu_l) \sim M_\rho$. In long form, we expect: $\Gamma(D_s \to \eta^{(')}\rho) = 6\pi^2 a_1^2 f_\rho^2 |V_{ud}|^2 \frac{d\Gamma}{dq^2}(D_s \to \eta^{(')} l\nu)|_{q^2=m_\rho^2}$.

The expectation from factorization for the case of the η is consistent with observation: $\frac{\Gamma(D_s\to\eta\rho)}{\Gamma(D_s\to\eta l\nu_l)} = 4.3 \pm 1.1$; while factorization predicts 2.9 for this ratio of rates. For the case of the η', consistency is allowed by the large errors, although the large value experimentally obtained for the absolute $\eta'\rho$ rate is conspicuous: $\frac{\Gamma(D_s\to\eta'\rho)}{\Gamma(D_s\to\eta' l\nu_l)} = 14.8 \pm 5.8$; again the naive factorization prediction is 2.9. This model may also have implications for B-decay, where CLEO has recently observed 'large' (i.e., of order 10^{-4}) branching fractions for two-body decays of the type: $B \to \omega h$, $B \to \eta' h$[9]. It may be that such two-body decays are enhanced by the same sorts of gluonic diagrams which may be feeding $D_s \to \omega\pi$ and $D_s \to \eta'\pi$.

1.2 Charmed Hadron Spectroscopy

The spectroscopy of charmed hadrons is very familiar; the lowest-lying mesons can be catalogued according to their values of $J(PC)$ and also according to the orbital angular momentum between the heavy and the light quark (or di-quark, in the case of baryons – in this case, the diquark obviously has integral spin and we tabulate the quantum numbers J(P)):

Pseudoscalar	Vector	Orb. Excited
η_b (0-+)	Υ (1- -)	χ_b (L=1)
D (0-+)	D^* (1- -)	$D^{**}(\equiv D_J)$ (L=1)
D_s (0-+)	D_s^* (1- -)	$D_s^{**}(\equiv D_{s,J})$ (L=1)†
Σ_c (3/2)-	Σ_c^* (3/2)-††	$\Lambda_c^*(2593)$ (L=1, (1/2)-)

†L=1 states have parity $P = -1^\ell P_q P_{\bar{q}}$=+1)
††Note that the Λ_c has no mesonic analog, as $S_{diquark}$=0

For each of the above states, there are, in principle, radial excitations which are expected, as well (the heavy-light analogs of the ψ', ψ'', Υ', Υ'', etc.). DELPHI has, at this conference, reported the first observation of such a radially excited mesonic state, the $D^{*'}$[10]. As expected from the pattern observed in charmonium and bottomonium, one would search for such a particle through its decay to $D^*\pi\pi$. Based on experience in the quarkonium sector, it would be reasonable to expect a mass difference which is somewhat larger than the $\psi' - \psi$ mass difference, owing to the larger relativistic corrections in the case of the light-heavy $D^{*'}$. DELPHI observe this signal in the mass difference plot $\Delta M(D^*\pi^+\pi^-) - M(D^*)$, and claim a measured $D^{*'}$ mass of $M = 2637 \pm 2 \pm 7$ MeV. This value is approximately within 10 MeV of the mass predicted originally by Godfrey & Isgur[11]. The width of this state is observed to be smaller than the experimental resolution, consistent with the suppression of two orders of α_s as expected in the general formalism of dipion transitions. The production rate is measured relative to the inclusive decays of orbitally excited D_J^{*0} into the $D^{*+}\pi^-$ final state and determined to be: $(\sigma \cdot B) \sim (0.5 \pm 0.2)(\Sigma(D_J^{(*)0} \to D^{*+}\pi^-))$.

There have also been substantial improvements in our cataloging of excited charmed baryons. Considering that the first orbitally excited charmed mesons (the Λ_c^*) were only observed four years ago by ARGUS[12], the progress since has been fairly remarkable. Baryonic states are classified according to their quantum numbers. Realizing that these quantum numbers reflect, among other things, the symmetry of the light diquark spin wavefunction (with the antisymmetric state corresponding to ↑↓ and the symmetric state to ↑↑), we can categorize baryons as follows:

quark content	1/2+ (Mixed Anti)	1/2+ (Mixed sym)	3/2+ (Symm)	L=1 1/2-	L=1 3/2-
uds	$\Lambda(1115)$	$\Sigma_0(1193)$	$\Sigma^{*0}(1385)$	$\Lambda(1405)$	$\Lambda(1520)$
udc	$\Lambda_c(2285)$	$\Sigma_c^+(2455)$	$\Sigma_c^{*+}(2520)$	$\Lambda_c(2593)$	$\Lambda_c(2625)$
qsc	$\Xi_c(2465)$	$\Xi_c'(2572)$	$\Xi_c^*(2645)$	(not seen)	$\Xi_{c,1}(2815)$

In more detail, the charmed strange baryons can be written as follows (in order of increasing mass), as categorized by the angular momenta of the light quarks:

- Ξ_c : c↑(s↑q↓) (L=0, J=S=1/2)
- Ξ_c' : c↑(s↓q↓) (L=0, J=S=1/2) (sq diquark in a *symmetric* wave function with respect to interchange)
- Ξ_c^*: c↑(s↑q↑) (L=0, J=S=3/2) (Dominant decay mode expected $\to \Xi_c\pi$, analogous to $D^* \to D\pi$)
- Ξ_c^{**}: c↑(s↑q↓) (L=1, S=1/2)

CLEO has first observations of the $\Xi_c^{'+}$, the $\Xi_c^{'0}$[2] and the Ξ_c^{*+} in this sector[13]. The $\Xi_c^{'+}$ is the decuplet partner of the Ξ_c^+, in the same way that the Ξ' is the decuplet partner of the Ξ, for which the diquark is in a spin 1 state ($\Xi' : \Xi = \Sigma : \Lambda$). The Ξ_c^{*+} is the L=1 partner of the Ξ, with the diquark in a spin 0 state, but with one unit of orbital angular momentum

[2] A previous measurement of the Ξ_c' by the WA89 Collaboration three years ago has not yet been published.

of the diquark relative to the 'heavy' central quark (i.e., $\Xi^* : \Xi = \Lambda_c^{**} : \Lambda_c$). We can consider the question of what the mass splitting is expected to be for the Ξ_c'-Ξ_c; qualitatively, it can be expected to be smaller than the corresponding mass difference between the Ξ' and the Ξ of 209 MeV ($\Delta M = M(\Xi'(1530) - \Xi(1321))$, due to the smaller relativistic corrections for the case of charm. By contrast, CLEO measures $M(\Xi_c^{+'} - \Xi_c^+) = 107.8 \pm 1.7 \pm 2.5$ MeV and, for the neutral state: $M(\Xi_c^{0'} - \Xi_c^0) = 107.0 \pm 1.4 \pm 2.5$ MeV.

Additionally, CLEO observes a particle decaying into $\Xi_c^+\pi^+\pi^-$, with a mass difference $\Delta[M(\Xi_c^+\pi^+\pi^-) - M(\Xi_c)]$ of $349.4 \pm 0.7 \pm 1.0$ MeV. In analogy to the orbitally excited Λ_c^{**} doublet similarly decaying to $\Lambda_c\pi^+\pi^-$ (the $\Lambda_c(2593)$ and the $\Lambda_c(2630)$, corresponding to the J=1/2- and J=3/2- states), this particle is interpreted as the J=3/2 orbital excitation of the Ξ_c.

1.3 Color-octet vs. color-singlet charm production

Although space does not allow a full discussion here, it should be mentioned that, among the areas of study to emerge within the last 3-4 years is that of charm production (both open, in the form of $D^{(*)}$ and also 'hidden' in the form of charmonium) from gluon fragmentation, particularly in the low-x regime. ZEUS[14] and H1[15] at HERA and CDF at the Tevatron have extended the original work of CDF[16] which first observed unexpectedly large J/ψ production at high p_T relative to the beam axis. The expectation at that time was that charmonium production could be adequately handled in the context of the color-singlet model, in which, e.g., at the Tevatron, J/ψ production resulted from the chain: p$\bar{\text{p}}\rightarrow gg$(color singlet)$\rightarrow \chi_c \rightarrow \psi$+X; production at high p_T was dominated by single partons (i.e., color octets). The CDF result, indicating excessive high p_T production, prompted the development of models which allowed for both 'color-octet' as well as 'color-singlet' contributions to the observed charmonium rate. In this approach, charmonium production is handled by non-relativistic perturbative QCD at short-distances; lattice gauge results were used for long-distance terms. In this formalism, c$\bar{\text{c}}$ pairs are originally produced as color-octets, and evolve into color-singlets through soft gluon emission. Results from HERA and the Tevatron, contributed to this conference, are generally in support of the color-octet model; contributions on inclusive charm production at low-x from LEP have also helped to elucidate the role of charm production from gluons.

2 Tau Results

The tau lepton is approximately of the same mass as the charm quark; this similarity in mass would naively lead to the expectation of similarity in lifetime; in fact, the D^0 and the τ lifetimes are within approximately 35% of each other. The crucial difference between the two lies, of course, in the near absence of strong corrections in studies of tau decays. The tau, therefore, affords a cleaner study of weak processes than for charm.

The very large (and clean) data samples accumulated recently by LEP and CLEO have afforded considerable advances in our understanding of rare tau decays and the resonant substructure in multi-prong tau decays, as summarized in the table below (relevant Branching fractions are taken from reference [1]):

Mode	1990	1997
$\tau \to \pi^-\pi^0\nu_\tau$	$\tau \to \rho\nu_\tau$	$\frac{\tau\to\rho'(1450)\nu_\tau}{\tau\to\rho\nu_\tau} \sim 0.1;\ \Delta\phi = \pi$
$\tau \to \pi\pi\pi\nu_\tau$	$\tau \to a_1\nu_\tau;\ a_1 \to \rho\pi$	$\tau \to a_1' \to \rho\pi?$
$\tau \to K^0_S\pi^-\nu_\tau$	$\tau \to K^{*-}\nu_\tau$	$\tau \to K^{*-}\nu_\tau + \tau \to K^{*-'}\nu_\tau$
$\tau^- \to 3h\pi^0\nu_\tau$	$\tau \to \omega\pi\nu_\tau$	
		$\tau \to \eta K^-\nu_\tau;\ \eta \to \pi^-\pi^+\pi^0$
$\tau \to 5h\nu_\tau$	(0.064±0.025)%	(0.075±0.007)%
$\tau \to 3h\pi^0\pi^0\nu_\tau$	–	$\tau \to \omega^0\pi^-\pi^0\nu_\tau \sim 80\%$
		$\tau \to \eta^0\pi^-\pi^0\nu_\tau \sim 10\%$
$\tau \to 3h3\pi^0\nu_\tau$	–	$\tau \to f_1(1285)(\to \eta\pi\pi)\pi^-\nu_\tau$
		$\tau \to \omega^0\pi^-\pi^0\pi^0\nu_\tau \sim 10\%$
$\tau \to 7h^-$	-	$< 2.4 \times 10^{-6}$

Experimental studies of charged kaon production in tau decay is considered in several contributions to this conference[17, 19, 18]. The theoretical aspect of this question has been treated by Finkemeir & Mirkes[20]. The general prescription of Finkemeir & Mirkes is to factorize the matrix element into the product of a leptonic current×hadronic current (the tau decay into the W followed by the hadronization of the W); i.e. $M.E. = \frac{G}{\sqrt{2}}(cos\theta_c M_\mu J^\mu + sin\theta_c M_\mu J^\mu)$ For the leptonic current, a standard V-A form is assumed: $M_\mu = \overline{u}(l', s')\gamma_\mu(g_V - g_A\gamma_5)u(l, s)$. For the hadronic current, one must take into account creation of the strange system at the W-decay vertex. In the case of $\tau \to K^-\nu_\tau$, there is coupling only to the axial vector portion of the current, with: $< K(q)|A^\mu(0)|0 >= i\sqrt{2}f_K q^\mu$. Including radiative corrections leads to the prediction $\mathcal{B}(K\nu_\tau) = 0.723 \pm 0.006\%$, for the one-prong decay of the tau, in excellent agreement with the experimental measurement of: $\mathcal{B}(K\nu_\tau) = 0.692 \pm 0.028\%$[1].

Applying this formalism to the newly measured three-prong $\tau \to KX$ states, one must take into the fact that, for the K_1, there are, in principle four form factors which can be written for the most general form of the matrix element, with both Vector and Axial vector contributions. These form factors are, not unexpectedly, written as Breit-Wigner forms, with the decay rates somewhat sensitive to the width and masses of these Breit-Wigner forms. There is an obvious caveat here, insofar as the a_1 is known to have a larger width in tau decay compared to hadroproduction. Similarly, the predictions for $\tau \to K_1\nu_\tau$ are dependent on the assumed K_1 parameters - if the K_1 is somewhat wider than the presently tabulated PDG values, the predicted rate comes into considerably better agreement with the data.

Production of inclusive charged kaons in three-prong tau decay is expected to be dominated by the K_1 resonances, in analogy to the well-studied $\tau \rightarrow a_1\nu_\tau$; with $a_1 \rightarrow \pi^+ pi^- \pi^+$. In fact, in the non-strange sector, the three pion decays, in principle, would be expected to decay through the J(P)=1+ current into the triplet a_1 ($^3P_1$) and the singlet b_1 ($^1P_1$). The decay $\tau \rightarrow b_1$ constitutes a second-class current and is therefore prohibited. In the strange sector, the states analogous to the a_1 and b_1 are the states K_A and K_B. However, due to SU(3) symmetry breaking, the second-class current prohibition does not hold here, and, in principle, both the K_A and the K_B are allowed. In practice, what is observed are the mass eigenstates $K_1(1270)$ and $K_1(1430)$, which are mixtures of K_A and K_B.

In recent years the large data samples accumulated at CLEO and LEP have allowed much-improved measurements of inclusive decays of tau leptons to charged kaons, complementing similar measurements of inclusive decays of tau leptons to neutral kaons. We now consider specifically the decays of $\tau^- \rightarrow K^- h^+ \pi^-(\pi^0)\nu_\tau$ and $\tau^- \rightarrow K^- K^+ \pi^-(\pi^0)\nu_\tau$ relative to $\tau^- \rightarrow \pi^-\pi^+\pi^-(\pi^0)\nu_\tau$, where $h^\pm$ can be either a charged pion or kaon.

To find the number of events with kaons, the three experiments with recent measurements (ALEPH, CLEO, and OPAL, preliminary) use specific ionization (dE/dx) information from their central tracking chambers. All three experiments have comparable separation capabilities of pions and kaons ($\sim 2\sigma$); because of the modest separation, a good understanding of the dE/dx calibration is crucial. For each track, one calculates the deviation of the measured energy loss relative to that expected for true kaons or pions.

The number of kaon and pion tracks in the sample of selected 1vs3 events is found statistically by fitting the dE/dx ionization distribution for charged tracks in the three-prong hemisphere to the sum of the pion shape plus the kaon shape. For OPAL, these shapes are determined from Monte Carlo, for ALEPH, the shapes are obtained from data (e.g., the pion shape is taken from a sample of $\tau \rightarrow \omega\pi\nu_\tau$ events which have a topology similar to that expected for three-prong tau decays to kaons). For CLEO, the kaon and pion shapes are derived from data, using large samples of charged kaons and pions tagged by the very clean decay chain: $D^{*+} \rightarrow D^0\pi^+$; $D^0 \rightarrow K^-\pi^+$. The fitted area under the kaon curve directly gives the number of kaons and the area under the pion curve gives the number of pions.

Table 1 summarizes the results. The weighted average of the experimental results is found to be somewhat different than the theoretical predictions, although, as has been mentioned before, this discrepancy can be ameliorated to large measure, by using a different K_1 width as input to the theoretical calculation.

Branching fractions	CLEO97 %	ALEPH97[18]	OPAL97[19]
$Br(\tau \to K\pi\pi\nu_\tau)$	$0.317 \pm 0.027 \pm 0.072$	$0.214 \pm 0.037 \pm 0.029$	
$Br(\tau \to K\pi\pi\pi^0\nu_\tau))$	$0.165 \pm 0.045 \pm 0.047$	$0.061 \pm 0.029 \pm 0.015$	
$Br(\tau \to KK\pi\nu_\tau)$	$0.155 \pm 0.017 \pm 0.027$	$0.168 \pm 0.022 \pm 0.020$	
$Br(\tau \to K\pi\pi\nu_\tau)n(\pi^0))$			$0.358^{+0.089+0.046}_{-0.084-0.065}$
$Br(\tau \to KK\pi\pi^0\nu_\tau)$	$0.070 \pm 0.37 \pm 0.14$	$0.075 \pm 0.029 \pm 0.015$	
$Br(\tau \to KK\pi\nu_\tau)n(\pi^0))$			$0.036^{+0.047+0.024}_{-0.042-0.022}$

Table 1. Results

3 Outlook

Tau and charm studies should continue vigorously for the next decade as the B-factory era dawns. With anticipated integrated luminosities of order 10^2 fb^{-1}, corresponding to 10^8 $\tau\tau$ and $c\bar{c}$ events, increasingly sensitive tests of the Standard Model will be afforded, including searches for CP-violation in tau and charm decays and mixing in $D\bar{D}$ events. Within the next year, results from the Fermilab fixed target experiments FOCUS and SELEX should substantially augment the world's data on charm, particularly charmed baryons (SELEX). However, as of this writing, there is no obvious site for the long-proposed tau-charm factory, which is probably the only experimental facility where systematic errors can be made comparable to statistical errors for high-precision studies.

Acknowledgments: My thanks go to all of the individuals who contributed to studies of particle physics in the last year, and especially the conference organizers.

References

[1] R.M. Barnett *et al.*, Particle Data Group, Phys. Rev. **D54**, 19 (1996)

[2] M. Bauer *et al.*, Z. Phys. **C34**, 103 (1987).

[3] "Measurement of the branching fraction $D_s^+ \to \tau^+\nu_\tau$", (The DELPHI Collaboration), submitted to this conference.

[4] "Measurement of $D_s \to \tau^-\overline{\nu_\tau}$ and a new limit for $B^- \to \tau^-\overline{\nu_\tau}$", The L3 Collaboration, submitted to this conference.

[5] "Improved measurement of the pseudoscalar decay constant f_{D_s}", (The CLEO Collaboration), CLNS-97-1526

[6] "Heavy-light decay constants - MILC results with the Wilson action", C. Bernard, Nucl. Phys. Proc. Suppl. **60A**, 106, 1998

[7] "Observation of the Decay $D_s \to \omega\pi$", (The CLEO Collaboration), submitted to this conference.

[8] "Measurement of the Branching Ratios for the Decays of $d_s^+ \to \eta\pi^+$, $\eta'\pi^+$, $\eta\rho^+$, and $\eta'\rho^+$", (The CLEO Collaboration), CLNS-97-1515

[9] "Observation of Inclusive $B^o \to \eta' X_s$ Decays" (The CLEO Collaboration), "Study of Charmless Hadronic Decays of B-Mesons With Final States Including η or η' Mesons" (The CLEO Collaboration), "Observation of Exclusive Rare Charmless B Decays" (The CLEO Collaboration), and "Search for Charmless Hadronic Decays of B-Mesons to Final States Including ω or ϕ Mesons" (The CLEO Collaboration), submitted to this conference.

[10] "First evidence of a radially excited $D^{*'}$", The DELPHI Collaboration, submitted to this conference.

[11] S. Godfrey and N. Isgur, Phys. Rev. **D32**, 189 (1985).

[12] "Observation of a new charmed baryon", (The ARGUS Collaboration), Phys. Lett. **B317**, 227, (1993).

[13] "Observation of Two Narrow States Decaying into $\Xi_c^+\gamma$ and $\Xi_c^0\gamma$" (The CLEO Collaboration), and "Search for excited charmed baryons decaying into $\Xi_c n(\pi)$" (The CLEO Collaboration), submitted to this conference.

[14] "Measurement of Inelastic J-psi photoproduction at HERA" (The ZEUS Collaboration), "D^* meson production in DIS at HERA with the ZEUS detector" (The ZEUS Collaboration), and "Photoproduction of D^* in ep Collisions at HERA" (The ZEUS Collaboration), submitted to this conference.

[15] "Elastic Production of J/ψ Mesons in Photoproduction and at High Q^2 at HERA" (The H1 Collaboration), "D^* Meson Production in $e-p$ Collisions at HERA" (The H1 Collaboration), "$D^{*\pm}$ Meson Production in Deep–Inelastic Diffractive Interactions at HERA" (The H1 Collaboration) and: Photo-production of ψ(2S) Mesons at HERA (The H1 Collaboration), submitted to this conference.

[16] "J/ψ and ψ(2S) production in p$\bar{\text{p}}$ collisions at $\sqrt{s}$=1.8 TeV" (The CDF Collaboration), Phys. Rev. Lett., **79**, 572 (1997), and references cited therein.

[17] "Meaurement of the branching fractions $\mathcal{B}(\tau \to K^-h^+h^-(\pi^0)\nu_\tau)/\mathcal{B}(\tau \to \pi^-\pi^+\pi^-(\pi^0)\nu_\tau)$" (The CLEO Collaboration), submitted to this conference.

[18] "Three-prong tau decays with charged kaons" (The ALEPH Collaboration), CERN PPE/97-069

[19] "Determination of Tau Branching Ratios to Three-Prong Final States with Charged Kaons", (The OPAL Collaboration), OPAL Physics Note PN-304 (1997)

[20] "Theoretical Aspects of $\tau \to K\pi\pi\nu_\tau$ decays and the K_1 widths", E. Mirkes, submitted to this conference.

B Physics

Michael Feindt (feindt@cern.ch.)

Institut für Experimentelle Kernphysik, Universität Karlsruhe, Germany

Abstract. A review over recent experimental progress in the physics of the fifth quark is given.

1 Introduction

The value of the mass of the fifth, the beauty-quark, around 5 GeV leads to a special role of the b-hadrons. The most heavy quark, the top quark, is too heavy to build hadrons. This is because it can decay by "weak" interaction into a real W-boson and a b-quark. This decay occurs much faster than the typical time needed to bind with an antiquark into a meson by the strong interaction. Thus, hadrons containing a b-quark are the heaviest hadrons. On the other hand, the b-quark mass is much larger than the typical scale of the strong interaction, Λ_{QCD}, responsible for the binding of quarks into hadrons. This is the reason for the success of Heavy Quark Effective Theory. A B-meson consisting of a heavy b-quark and a light antiquark thus resembles lots of properties of the hydrogen atom.

This review gives a short (due to space limitations) summary of the state of the field. More extensive recent summaries of LEP results are found in [1] and [2], on B-decays in [3]. Note also the companion theoretical talk at this conference [4]. The transparencies of this talk can be found in [5].

At this conference the first experimental verification of the running of the b-quark mass has been presented [6, 7]. The relative 3-jet rate in $b\bar{b}$ events is compared to the relative 3-jet rate in light quark events. A three jet topology originates from gluon radiation off a quark line. This radiation is suppressed for heavy quarks. Comparing the measured ratio to new next-to-leading-order (NLO) QCD calculations [8] the running $\overline{MS}$ mass has been determined to $m_b(M_Z) = 2.67 \pm 0.25(stat.) \pm 0.34(frag.) \pm 0.27(theo.)GeV/c^2$, to be compared to $m(b(M_\Upsilon/2) = 4.16 \pm 0.18GeV/c^2$ from Upsilon spectroscopy. There is clear evidence for a running, see Fig.1a.

2 b quark production

2.1 b production at the Tevatron

Both CDF and D0 find b production cross sections about 2 times as large as NLO QCD predictions [9]. The shape of the transverse momentum spectrum

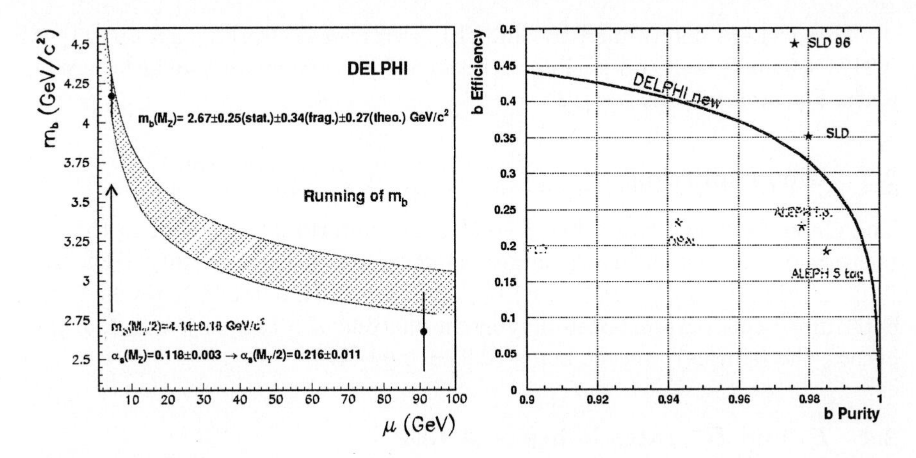

Fig. 1. left: b-quark mass as function of scale; right: b tagging effiency vs. purity for different experiments

as well as the dependence on the c.m.s. energy is roughly described, not however the rapidity distribution: the discrepancy gets larger in the forward region.

2.2 b production in Z decays

The ratio R_b of Z decays into $b\bar{b}$-quark relative to all hadronic (i.e. $q\bar{q}$) decays is interesting due to the large mass of the b-quark, since an enhanced R_b could indicate a Higgs-like Yukawa coupling to masses.

There was lots of excitement because with increasing precision the world average actually showed a discrepancy, with more than 3σ in 1995. At last year's Warsaw Conference ALEPH [10] presented an analysis with very small errors ($R_b = 0.2161 \pm 0.0009 \pm 0.0011$), in perfect agreement with the SM prediction of 0.2158. This was less dependent on charm background and had strongly reduced hemisphere correlations.

This year also other collaborations presented new analyses. Especially DELPHI [eps419] has undertaken a major effort to reprocess all their data with a strongly improved pattern recognition and track fitting procedure, leading to a much cleaner reconstruction especially in dense jets. Also the b-tagging algorithm was optimised by including the z-measurements of the silicon vertex detector, vertexing, and additional variables like invariant mass and track rapidities. The b-tagging performances of the different detectors are summarized in fig 1b.. Note the SLD 96 point which is due to a new vertex detector with twice as good resolution. Unfortunately, SLD is lacking statistics compared to the LEP Collaborations.

The average determined by the LEP electroweak working group [11] is 0.2171 ± 0.0009, well compatible with the SM prediction with an accuracy of 0.4%. For more details see [12].

2.3 Gluon splitting

The gluon splitting probability $f \to b\bar{b}$ is an important ingredient in the R_b measurement, constituting the largest single systematic uncertainty. DELPHI ($0.21 \pm 0.11 \pm 0.09$)[13] and ALEPH ($0.257 \pm 0.040 \pm 0.087$)%)[eps606] have determined this parameter employing an analysis of b-tagged jets in four jet events, the (simple) average being $(0.24 \pm 0.09)\%$.

2.4 B_s and B^+ rates in b-jets at LEP

The classical method of determining the primary B_s rate f_s consists of an comparison of the integrated $B\bar{B}$ mixing χ in Z decays with the expectation $\chi = f_d\chi_d + f_s\chi_s$, taking the measured x_d (see below) and assuming a fast B_s mixing frequency leading to $\chi_s = 0.5$. The baryon contribution is estimated from lepton-Λ_c and lepton-Ξ correlations. The results of the LEP mixing working group [14] are $f_{Baryon} = 0.1062^{+0.0373}_{-0.0273}$, $f_d = f_u = 0.3954^{+0.0156}_{-0.0203}$ and $f_s = 0.1031^{+0.0158}_{-0.0153}$. DELPHI [eps451] has performed a search for a charged fragmentation kaon accompanying a primary B_s (including the excited states B_s^*, B_s^{**}) at high rapidity, separating out background contributions from B^+ accompanied by a K^-. The primary B_s rate f_S' has been determined to $(12.0 \pm 1.4 \pm 2.5)\%$. This value [15] is smaller than that of the contributed paper, due to a different, probably more solid assumption about the validity of the model. With an estimated B_s^{**} rate of $(27 \pm 6)\%$ this corresponds to a rate $f_{B_s} = (8.8 \pm 1.0 \pm 1.8 \pm 1.0)\%$ of weakly decaying B_s mesons, where the last error is due to the B_s^{**} rate.

In the same paper [eps451] an analysis of the rate of charged versus neutral weak B-hadrons is presented: $B(\bar{b} \to X_b^0) = (57.8 \pm 0.5 \pm 1.0)\%$, $B(\bar{b} \to X_b^+) = (42.2 \pm 0.5 \pm 1.0)\%$. Making an assumption about the small contribution of charged Ξ_b and Ω_b production ($(1.0 \pm 0.6)\%$) leads to $B(\bar{b} \to B^+) = (41.2 \pm 1.3)\%$.

ALEPH has also presented a new measurement of the b-baryon rate of $(12.1 \pm 0.9 \pm 3.1)\%$ [eps597].

3 Spectroscopy

The mesons B^0, B^+ and B_s are clearly established and their masses measured. The vector meson B^* is seen in its decays into $B\gamma$ and Be^+e^- [eps450]. The existence of the $L = 1$ orbitally excited B^{**} mesons is established, but there is not yet a clear decomposition into the expected 2 narrow and 2 broad states. DELPHI has preliminary evidence for two narrow B_s^{**} states decaying into BK and some evidence for a radial excitation in $B\pi^+\pi^-$, which need confirmation. The latter analysis triggered a similar analysis in the charm sector, where a narrow resonance in $D^*\pi^+\pi^-$ has been found [eps452]. No $L > 1$ B-mesons are known. Also searches for the beautiful charmed B_c-meson have been hitherto unsuccessful.

In the baryon sector, the Λ_b is clearly established now by CDF [16], the mass being measured to $5621 \pm 4 \pm 3$ MeV. There are Σ_b and Σ_b^* candidates seen by DELPHI, which need confirmation. The existence of the Ξ_b is proven, but there is no mass measurement available. No other b-baryon states are known: no Ξ_b', Ξ_b^*, Ω_b or Ω_b^*, also no orbital or radial excitation.

For more extensive summaries on B-spectroscopy see e.g.[17, 18].

4 Lifetimes

Again there have been many new lifetime measurement contributions, which are averaged by a b-lifetime working group [19] taking into account correlated systematics etc. The results are:

$$\begin{aligned}
\tau_{average} &= 1.554 \pm 0.013 ps \\
\tau(B^0) &= 1.57 \pm 0.04 ps \\
\tau(B^+) &= 1.67 \pm 0.04 ps \\
\tau(B^+)/\tau(B^0) &= 1.07 \pm 0.04 \\
\tau(B_s) &= 1.54 \pm 0.06 ps \\
\tau(b - baryon) &= 1.22 \pm 0.05 ps
\end{aligned}$$

Thus the qualitative picture remains intact: Charged B mesons live slightly longer, the B^0 and B_s lifetimes are roughly the same, and the Λ_b has a much shorter lifetime than the mesonic states. The Λ_b lifetime is correlated with a small semileptonic branching ratio, see below. The origin of the low b-baryon lifetime is not yet clarified.

5 Mixing

Second order weak interactions lead to particle-antiparticle oscillations between B_0 and $\bar{B}_0$ and B_s and $\bar{B}_s$. They are described by a mass difference of the CP-eigenstates constructed by the sum and difference of the original wave functions. In the Standard Model, the mass differences are related to the Kobayashi Maskawa matrix elements V_{td} and V_{ts}, respectively. To measure the time dependence of the mixing, one needs to know the b-flavour at production and decay time to define whether a mixing occured or not, as well as the decay length and energy to reconstruct the proper decay time. Many different methods have been developed for this purpose. Fig. 4 gives

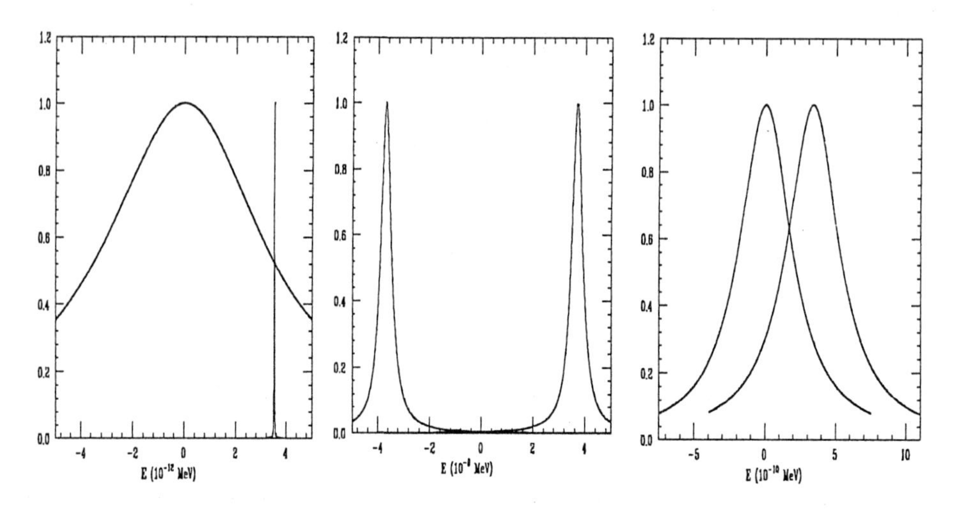

Fig. 2. Mass excitation curves of the $K^0 - \bar{K}^0$ ($\Gamma_s > \Delta m \approx \Gamma_l$, $\Delta\Gamma$ large), $B_s - \overline{B_s}$ ($\Delta m > \Gamma$, $\Delta\Gamma$ small), and $B^0 - \bar{B}^0$ ($\Delta m \approx \Gamma$, $\Delta\Gamma$ small) systems. Curves are due to E. Golowich, Moriond 1995

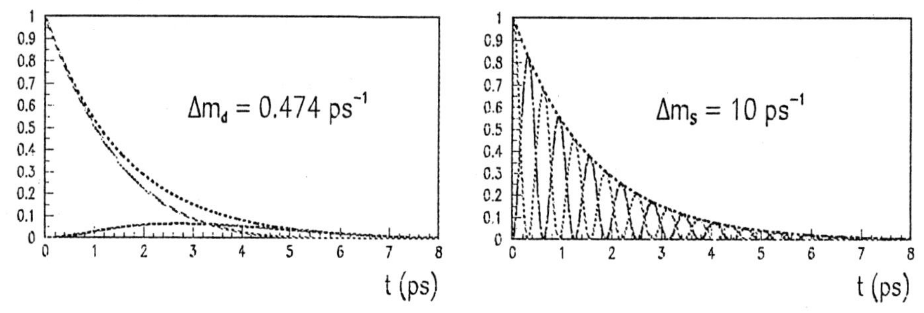

Fig. 3. Expected oscillation patterns for B_d with $\Delta m = 0.474 ps^{-1}$ and B_s with $\Delta m = 10 ps^{-1}$.

an overview of the available results for the B^0 mass difference Δm_d, which is proportional to the oscillation frequency: The preliminary average derived by the LEP oscillation working group is $0.472 \pm 0.018\, ps^{-1}$.

B_s mixing proceeds much faster than B^0 mixing, and the time evolution has not yet been resolved. Only a lower limit could be derived: $\Delta m_s > 10.2\, ps^{-1}$ at 95% c.l. Especially noteworthy at this conference is the new ALEPH measurement using an inclusive lepton ansatz [eps612]. The large sensitivity came somewhat surprising, since it was common belief that more exclusive methods with better resolution (but less statistics) are superior.

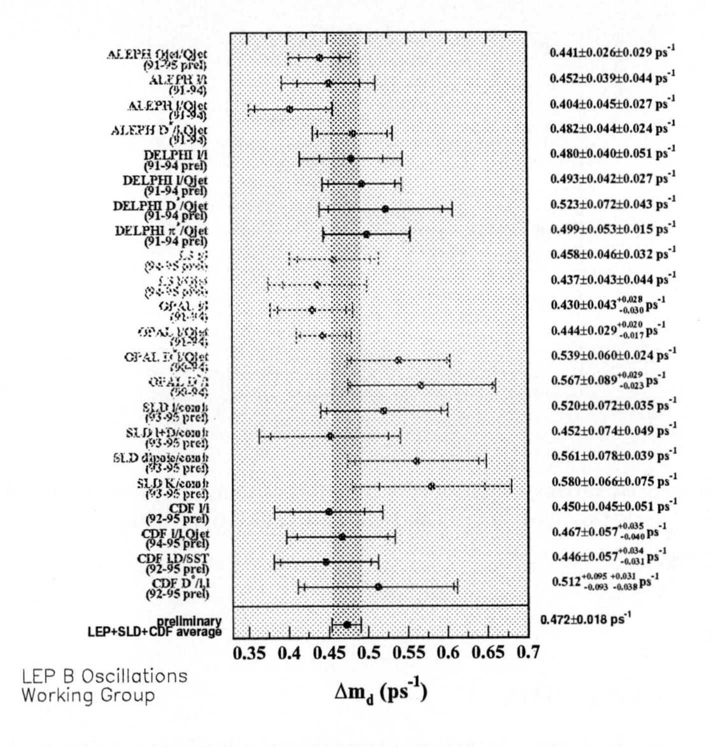

Fig. 4. Measurements of Δm_d

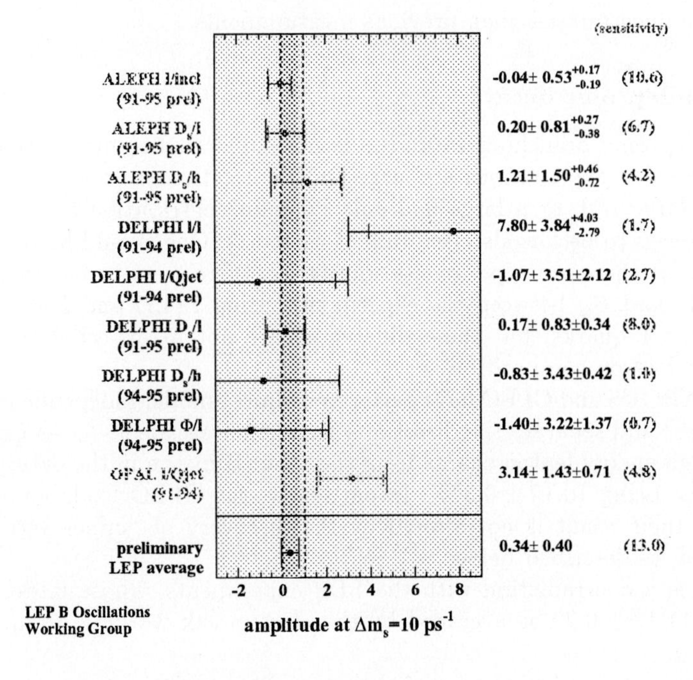

Fig. 5. Sensitivity on Δm_s

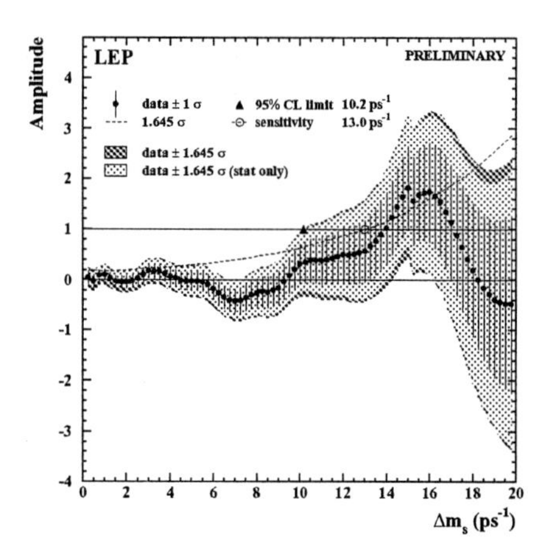

Fig. 6. Combined B_s mixing amplitude as function of Δm_s. Physical values are 0 (i.e. no mixing with this frequency) and 1 (data compatible with mixing of the given frequency).

6 Decays

6.1 Inclusive properties

A new analysis of the mean charged multiplicity in b-hadron decays produced at the Z by DELPHI [eps850] of $n(B) = 4.96 \pm 0.03 \pm 0.05$ has much smaller systematic uncertainties than previous measurements.

6.2 Semileptonic decays

The semileptonic branching ratio B_{sl} is measured to be smaller than expected. Possible explanations are large QCD corrections to $b \to c\bar{c}s$ (which implies a large N_c), or a large hadronic Penguin contribution $b \to sg$. N_c however seems to be low also (see talk of Neubert). One should however not overlook that there are experimental mysteries: there seem to be differences both in N_c and B_{sl} between the experiments using $\Upsilon(4S)$ and Z decays to produce the b-quarks, but these differences are not as expected from the different b-hadron composition.

Both ARGUS and CLEO have performed almost model-independent analyses. They could separate the direct $b \to l$ and the cascade $b \to c \to l$ decays using a high energy lepton charge from the other B meson in the event, their mean value being $10.19 \pm 0.37\%$. In particular, the CLEO Collaboration is sure that their result is not altered by the discovery of “upper vertex” D production, as described below.

There is a contradiction with the LEP experiments, whose latest value is $B_{sl} = 11.12 \pm 0.20$ as averaged by the electroweak working group. It is interesting to have a closer look at this average. For the LEP HF-EW working group B_{sl} is an auxiliary quantity in the complete electroweak heavy flavour fit. The early measurements essentially measured $R_b \times B_{sl}$, i.e. there are large correlation coefficients between R_b and B_{sl}. With the current very precise values of R_b, the old 1991 measurements get huge weights (and are on the high side) for the average B_{sl} determination. This result however crucially depends on the correlation matrix elements, including the estimates of the systematics correlation. Furthermore, it might be allowed to doubt that the systematics was as well under control in 1991 as now. It should also be remarked that the upper vertex D production is not yet included in the Monte Carlo models used for acceptance and background calculations, and there is no serious study yet about the possible influence.

The value at LEP is not necessarily the same as at CLEO, due to the different B-hadron contribution. The Λ_b has been measured to have a lower semileptonic branching ratio (see below), but this should lead to a smaller value at LEP than in $\Upsilon(4S)$ decays.

6.3 Λ_b semileptonic branching ratio

OPAL [eps153] measured the ratio $R_{\Lambda l} = B(\Lambda_b \to \Lambda l^- X)/B(\Lambda_b \to \Lambda X) = (7.0 \pm 1.2 \pm 0.7)\%$, and ALEPH [eps597] the corresponding ratio $R_{pl} = (7.8 \pm 1.2 \pm 1.4)\%$ (i.e. Λ replaced by proton), both of which can be assumed to be very similar to $B(\Lambda_b \to lX)$. Both are significanty smaller than the average semileptonic branching ratio. Given the apparent smaller lifetime of the Λ_b, this is consistent with the hypothesis of a constant semileptonic width for all B-hadrons.

6.4 Wrong sign charm

The determination of the average number of charm and anticharm quarks per b-hadron decay is called charm counting. In most of the cases exactly one charm quark is produced from the b-quark by W-emission. Only a small number of cases without c-quark is expected, either from $b \to u$ transitions or due to loop (Penguin) processes. A second (anti-) charm quark is produced when the W decays into $s\bar{c}$. Up to recently it was thought that these two quarks always end up in a single $\bar{D}_s$ meson. Through the measurements $B(B \to DX)/B(B \to \bar{D}X) = 0.100 \pm 0.026 \pm 0.016$ (CLEO) [eps383], $B(B^{0,-} \to D^0\bar{D}^0X, D^0D^-X, \bar{D}^0D^+X) = 12.8 \pm 2.7 \pm 2.6$ (ALEPH) [20] and $B(B^{0,-} \to D^{*+}D^{*-}X) = 1.0 \pm 0.2 \pm 0.3$ (DELPHI) [21] we know that this is not true. CLEO has performed this measurement by analysing angular correlations of D-mesons with high momentum leptons. CLEO also has observed four exclusive double charm decay modes and has placed limits on three others [eps337]. The observed large rates including D^* mesons suggest that the wrong sign D mesons have a very soft spectrum in the B cms. CLEO also has searched for resonances (especially the $J^P = 1^+ D_{s1}^+(2536)$) in the upper vertex D^*K spectra, but didn't find any enhancement[eps384]. From their upper limit $B(B \to D_{s1}^+ X) < 0.95\%$ at $95\% c.l.$ one can deduce that the axial vector coupling constant $f_{D_{s1}^+}$ is at least a factor 2.5 lower than that of the pseudocalar D_s^+.

6.5 Charm Counting

Classical charm counting experiments consist in measuring the rates of the weakly decaying D-hadrons in selected b-events. The published values differ quite a bit: CLEO: $N_c = 111.9 \pm 1.8 \pm 2.3 \pm 3.3\%$ [22], ALEPH: $N_c = 123.0 \pm 3.6 \pm 3.8 \pm 5.3\%$ [23], OPAL: $N_c = 106.1 \pm 4.5 \pm 6.0 \pm 3.7\%$ [24], where the last error is due to D branching ratios, largely correlated between the experiments. OPAL measures comparatively small D^0 and D^+ rates. A main difference between the experiments however are assumptions made about the unmeasured Ξ_c contribution, which is set to 0 in the case of OPAL, whereas ALEPH estimates it to be $6.3 \pm 2.1\%$. Accepting this last estimate and including also DELPHI's measurement of D^0 and D^+ rates [27] the averaged result is $N_c = (120.2 \pm 4.0(stat + syst) \pm 5.3(BR))\%$.

Two alternative methods to determine the fraction of b-decays into 0,1 and 2 charmed hadrons have been suggested by the DELPHI Collaboration [25, 26]: An analysis of the hemisphere b-tagging probability distribution in terms of Monte Carlo expectations of the three components delivers the result $B(b \to 0c) = 4.4 \pm 2.5\%$, $B(b \to 2c) = 16.3 \pm 4.6\%$, and $N_c = 116.3 \pm 4.5\%$. In another ansatz correlations of identified charged kaons with inclusively reconstructed D mesons are analysed. A fit of the transverse momentum spectra of same sign and opposite sign K pairs results in $B(b \to 2c) = 17.0 \pm 3.5 \pm 3.2\%$ and $B(b \to \bar{D}D_s X)/B(b \to 2c) = 0.84 \pm 0.16 \pm 0.09$. Large rates of $b \to sg$, as proposed by Kagan[28], would show up as extra source of charged kaons, especially visible at high momentum in the B c.m.s.. This is not seen in DELPHI, and an upper limit $B(b \to sg) < 5\%$ at 95% c.l. is derived. The SLD Collaboration [29] however finds a small excess in the kaon p_T spectrum, when they demand that the tracks form a good single vertex (to enhance b-decays without secondary charm decay), but they did not present a numerical analysis yet.

In their wrong sign charm paper CLEO [eps383] also derives the numbers $B(b \to sg) = 0.2 \pm 4.0\% (< 6.8\%$ at 90%c.l.), $B(b \to c\bar{c}s) = 21.9 \pm 3.6\%$ and $n_c = 120.4 \pm 3.7\%$.

Although there still are some discrepancies at the 2 sigma level, and there were controversial discussions on many of the analyses involved, it seems that there is not a serious N_c problem any more. Combining all the numbers leads to $N_c = 117.6 \pm 2.3\%$.

6.6 B^+ branching fractions

Although many inclusive branching ratios have been measured at ARGUS and CLEO, most of them are B-inclusive and do not distinguish between B^+, B^0, $\bar{B^0}$, and B^-. DELPHI [eps473] has presented a feasibility study of a method to enrich inclusively B^+ mesons and to measure π^+, π^-, K^+, K^-, e^+, e^-, and μ^+ and μ^- rates as function of the momentum in the B-meson c.m.s.. Also the rates of different D-hadron species in B^+ decays are largely unknown and can be addressed with this method. Furthermore the method can be modified to a flavour-specific (b-$\bar{b}$) study.

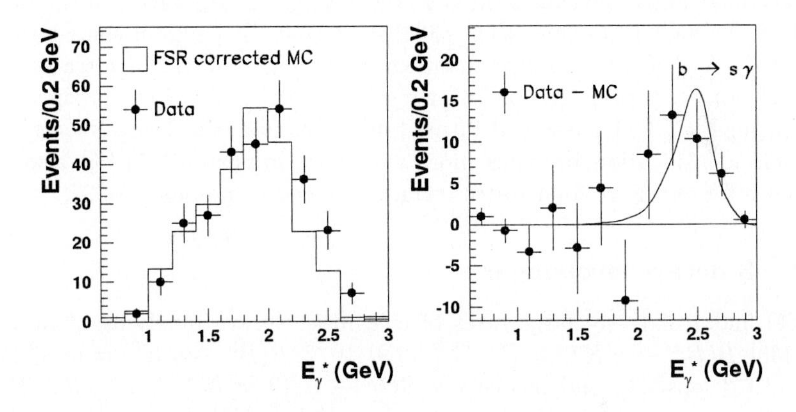

Fig. 7. ALEPH signal of $B \to s\gamma$

6.7 V_{cb}

There is not much news on V_{cb}. Mean values from different reactions are [3] $B \to D^*l\nu$: $(38.7 \pm 3.1) \cdot 10^{-3}$, $B \to Dl\nu$: $(39.4 \pm 5) \cdot 10^{-3}$, $\Upsilon(4S)$ inclusive: $(38.7 \pm 2.1) \cdot 10^{-3}$, Z^0 inclusive: $(40.6 \pm 2.1) \cdot 10^{-3}$. The last two correlated values can be combined into $(39.9 \pm 2.2) \cdot 10^{-3}$, leading to an overall average value of $(39.5 \pm 1.7) \cdot 10^{-3}$. A limiting factor in exclusive and semiexclusive analyses is the bad knowledge of D^{**} and nonresonant $D\pi$ production.

6.8 V_{ub}

Three measurements of V_{ub} are available: CLEO's lepton endpoint spectrum $(3.1 \pm 0.8) \cdot 10^{-3}$, CLEO's exclusive $B \to \pi/\rho l\nu$ value $3.3 \pm 0.3^{+0.3}_{-0.4} \pm 0.7) \cdot 10^{-3}$ and ALEPH's neural network analysis [30] with $(4.3 \pm 0.6 \pm 0.6) \cdot 10^{-3}$. All of them are strongly model dependent, however in different ways. The good agreement between the numbers is thus comforting.

6.9 $b \to s\gamma$

The electomagnetic penguin $b \to s\gamma$ has now also been observed by the ALEPH Collaboration at a rate of $B(b \to s\gamma) = (3.29 \pm 0.71 \pm 0.68) \cdot 10^{-4}$. Averaged with the 1994 CLEO result this corresponds to a new mean value of $B(b \to s\gamma) = (2.578 \pm 0.57) \cdot 10^{-4}$. New next to leading order calculations [31] are $(3.28 \pm 0.31) \cdot 10^{-4}$ and $(3.48 \pm 0.33) \cdot 10^{-4}$, slightly larger than the measured value. One has to wait for an updated CLEO number with smaller errors.

6.10 Hadronic penguins

CLEO, ALEPH and DELPHI have observed a number of still very small signals on exclusive charmless final states with branching ratios in the order of 10^{-5}. A new CLEO analysis [eps334] show that the penguin contributions (e.g. $B \to K\pi$) might be larger than expected compared to $b \to u$ transitions (like $B \to \pi\pi$). Especially this latter result is worrisome for the prospects of measuring the CKM-phase γ at future b-factories from the decay $B \to \pi^+\pi^-$. Particle identification becomes more and more important! CLEO also has evidence for exclusive final states including ω and ϕ mesons [eps335].

6.11 B decays involving η'

CLEO finds relatively large rates of charmless decays involving η' mesons [eps333]: $B(B^\pm \to \eta' K^\pm) = (7.1^{+2.5}_{-2.1} \pm 0.9) \cdot 10^{-5}$, $B(B^0 \to \eta' K^0) = (5.3^{+2.8}_{-2.2} \pm 1.2) \cdot 10^{-5}$ [eps333], and inclusively $B(B \to \eta' X) = (6.2^{\pm}1.6 \pm 1.1) \cdot 10^{-4}$ (with $2.0 < p(\eta') < 2.7GeV$) [eps332]. One might speculate whether there is a larger than expected $c\bar{c}$ or glueball component in the η' wave function. In this respect also another CLEO analysis is of interest: The measurement of the electromagnetic form factors of the π^0,η and η', as pioneered by TPC/2γ [33] and CELLO [34] , now is measured up to such high Q^2 that a very remarkable qualitative statement can be made [eps703]: The Q^2 dependence of the η' form factor cannot be described simultaneously at low and high Q^2 with the same formalism as the π^0 and η mesons. This might be another clue that there is something more than just light quarks inside the η'.

CLEO finds much smaller branching ratios for decays incolving η mesons, as predicted by Lipkin as interference effect between creating the η and η' by their $u\bar{u}, d\bar{d}$ component and their $s\bar{s}$ component.

7 Other puzzles and open questions, further studies

7.1 Interference

CLEO [eps339] has measured the interference sign between colour suppressed and colour allowed amplitudes (which are closely connected to internal and

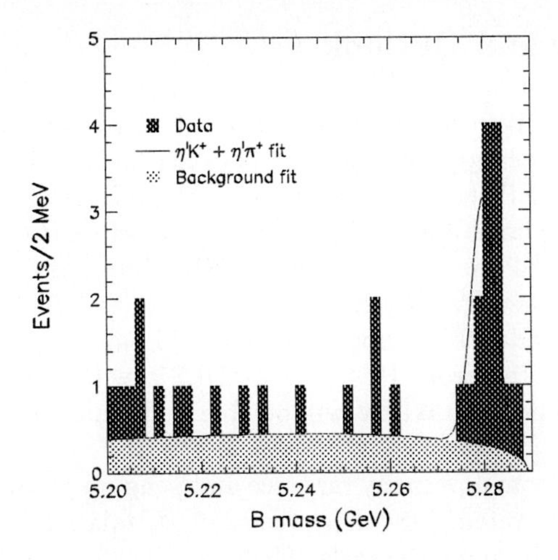

Fig. 8. CLEO signal of $B \to \eta' K$

external spectator diagrams) to be positive and consistent with equal in the six final states $D\pi, D\rho, Da_1$ and $D^*\pi, D^*\rho$ and D^*a_1. From this they would expect an up to 15% larger B^0- than B^+-lifetime, in contradiction to experiment. This must mean that this interference pattern is not typical for the whole set of hadronic B-decays.

7.2 B^+, B^0 production rates in $\Upsilon(4S)$ decays

It is worrying to observe that the B^+ to B^0 production ratio in $\Upsilon(4S)$ decays still is known only very badly: $f_{+-}/f_{00} = 1.21 \pm 0.12 \pm 0.17$ (CLEO alone), or $f_{+-}/f_{00} = 1.074 \pm 0.129$ (CLEO, with world average lifetimes). Most analyses assume that the ratio is 1, but due to phase space effects a non-zero value is not excluded. If the charged and neutral lifetime are really different, and the production ratio is not 1, there could be quite some surprises in CLEO-LEP comparisons. However, a larger semileptonic branching ratio at LEP than at the $\Upsilon(4S)$ is not achievable with current input data.

7.3 Search for CP-violation

Both OPAL[eps162] and DELPHI[eps449] have searched for CP-violating effects and established limits on $Re(\epsilon_b)$.

7.4 CKM matrix fits

A combination of B-mixing results, V_{ub}/V_{cb} and CP-violation parameters from K^0 decay shows clear evidence of a non-trivial unitarity triangle. With some input from theory two angles of the triangle can already be determined with quite good precision [1]: $sin2\alpha = -0.10 \pm 0.40$ and $sin2\beta = 0.68 \pm 0.10$.

8 Summary and Outlook

There is a bright future for b-physics in front of us - it will dominate the experimental high energy physics scene between 2000 and the LHC startup in 2007 or so. At LEP the final analyses with optimised algorithms are being prepared. SLC will hopefully get higher statistics before it will be shut down. CLEO has upgraded the detector, and CESR's luminosity will continue to improve. In 1999 the b-factory detectors BABAR and BELLE will start to take data. CDF and D0 get upgraded and will take data at much higher luminosity at the Tevatron, HERA-B at DESY will enter the scene, and finally LHCB and perhaps BTEV will be able to do many analyses with huge precision.

In a few years from now many rare tree and penguin decays will be known, the B_c will be discovered, time-dependence of B_s mixing detected, and CP-violation observed in many channels. Then we will probably laugh about the few 2 sigma discrepancies that we have to deal with now.

9 Acknowledgements

Many thanks to the ALEPH, CDF, CLEO, DELPHI, D0, L3, OPAL and SLD Collaborations, the LEP Electroweak Heavy Flavour, B-lifetime and B-oscillation working groups, and in particular P. Antilogus, P. Drell, L. Gibbons, T. Hessing, C. Kreuter, J. Kroll, C. Mariotti, P. Maettig, K. Moenig, M. Neubert, P. Roudeau, E. Thorndike, I. Tomalin and S. Willocq for their help in preparing the review.

References

[1] P. Roudeau, LAL97-96, to appear in Nuovo Cimento
[2] O. Schneider, CERN-PPE/97-143
[3] P. Drell, CLNS 97/1521
[4] M. Neubert, talk 19, these proceedings and CERN-TH-98-2
[5] M. Feindt,http://wwwinfo.cern.ch/~feindt
[6] DELPHI Coll., CERN-PPE-97-141
[7] J. Fuster, talk 514, these proceedings
[8] G. Rodrigo, Nucl. Phys. Proc. Suppl. 54A (1997) 60;
G. Rodrigo, M. Bilenky, A. Santamaria, Phys. Rev. Lett. 79 (1997) 193;
W. Bernreuther, A. Brandenburg, P. Uwer, Phys. Rev. Lett. 79 (1997) 189;
P. Nason, O. Olcari, Phys. lett. B407 (1997) 57
[9] R. Jesik, talk 506, these proceedings
[10] ALEPH Coll., R. Barate et al., Phys.Lett.B401 (1997) 150, Phys.Lett.B401 (1997) 163

[11] The LEP Experiments and the LEP Electroweak Working Group, CERN-PPE-97-154
[12] C. Mariotti, talk 707, these proceedings
[13] DELPHI Coll., P. Abreu et al., Phys.Lett. B405 (1997) 202
[14] LEP Oscillation Working Group, http://www.cern.ch/LEPBOSC/
[15] C. Weiser, Ph.D. thesis, University of Karlsruhe, in print
[16] CDF Coll., F. Abe et al.,Phys. Rev. D 55 (1997) 1142
[17] V. Canale, talk 103, these proceedings
[18] M. Feindt, Procs. Hadron 95, Manchester, UK, World Scientific, p.241
M. Feindt, Acta Phys. Pol. B 28 (1997) 789
[19] LEP Lifetime Working Group, http://wwwinfo.cern.ch/~claires/lepblife.html
[20] ALEPH Coll., contributed paper to ICHEP'96 pa05-060, Warsaw 1996
[21] DELPHI Coll., contributed paper to ICHEP'96 pa01-108, Warsaw 1996
[22] CLEO Coll., L. Gibbons et al., CLNS 96/1454, hep-ex/97030006
[23] ALEPH Coll., D. Buskulic et al., Z. Phys. C 69 (1996) 585
[24] OPAL Coll., G. Alexander et al., Z. Phys. C 72 (1996) 1
[25] P. Kluit, talk 906 at this conference
[26] DELPHI Coll., CERN-EP-98-007
[27] DELPHI Coll., contributed paper to ICHEP'96 pa01-058
[28] A. Kagan, talk 512 at this conference
[29] D. Jackson, talk 511 at this conference
[30] ALEPH Coll., contributed paper to ICHEP'96 pa05-59, Warsaw 1996
[31] K.G. Chertyrkin, M. Misiak and M. Munz, Phys. Lett. B400 (1997) 206;
A.J.Buras, A. Kwiatkowski and N. Pott, TUM-HEP-287/97
[32] D. Miller, talk 510 at this conference
[33] TPC/2γ Coll., H. Aihara et al., Phys. Rev. Lett. 64 (1990) 172
[34] CELLO Coll., H.-J. Behrend et al., Z. Phys. C49 (1991) 401

Glueballs and Other Non-$\bar{q}q$ Mesons

Martin Faessler (martin.faessler@cern.ch)

University of Munich, Germany

Abstract. Experimental evidence has become strong for an exotic, non-$\bar{q}q$ resonance at 1400 MeV with $J^{PC} = 1^{-+}$. Among $0^{-+}, 1^{+-}$ and 2^{++} mesons, the best candidates for non-$\bar{q}q$ states are still weak. The extraordinary and difficult sector of isoscalar 0^{++} (f_0) mesons approaches experimental consistency; theoretical explanations of the observed states need the glueball component.

1 Introduction

The main task of meson spectroscopy today is to verify or falsify the existence of non-$\bar{q}q$ mesons predicted by Quantumchromodynamics (QCD).

Non-$\bar{q}q$ is a short name for mesons whose constituents are not just a quark and an antiquark. The constituents could be, for instance, two gluons, forming a glueball. It is not obvious that gluons, even if they carry colour and attract other gluons, will form bound states [1]. The massless gauge gluons of QCD have to get dressed in a similar way as current quarks in order to behave like massive constituents of a hadron. The notion of a massive constituent gluon could indeed be exotic. However, the fact that lattice QCD finds glueballs can be considered as an existence proof for constituent gluons.

The three lowest glueball states, found in the quenched approximation of lattice QCD are: a scalar, i.e. $J^{PC} = 0^{++}$ ground state at around 1600 MeV; a pseudoscalar and a tensor excited state, both 600 MeV above the g.s. [2]. (J=Spin, P= intrinsic parity, C= C- parity). The nagging question, not answered in the quenched approximation, is whether these states are distinct from, and additional to the SU(3)-flavour singlet states of $\bar{q}q$ mesons.

The main difficulty in searching for glueballs and other non-$\bar{q}q$ mesons is due to the fact that even some of the lowest orbital excitations of ordinary $\bar{q}q$ mesons have not been safely identified. In particular, this applies to the sector of scalar (0^{++}) isoscalar (Isospin $I = 0$) mesons, called f_0.

This talk concentrates on experimental evidence for non-$\bar{q}q$ mesons and on consistency of observed signals rather than on interpretation or theoretical work. It consists of 3 sections:

- The following section is exclusively devoted to the best candidate for a meson which has a quantum number combination not possible for $\bar{q}q$.
- The next section deals with potentially supernumerary or else suspicious pseudoscalar, axial and tensor mesons.
- The final section is about the very complex scalar isoscalar sector which houses the ground state glueball.

2 Mesons with non-$\bar{q}q$ quantum numbers

The most direct proof for non-$\bar{q}q$ mesons would be to establish the existence of a meson with quantum numbers or a combination of quantum numbers excluded for a $\bar{q}q$ meson. To find one single meson of this kind would be sufficient. Rather than discussing numerous, often weak, experimental hints for such mesons, I shall focus onto the by now best candidate.

The quantum number combination $J^{PC} = 1^{-+}$ is not possible for a $\bar{q}q$ system. It could, for instance, be carried by a hybrid meson - containing a constituent gluon in addition to a $\bar{q}q$ pair. There is rising evidence for an isovector (Isospin $I = 1$) meson with $J^{PC} = 1^{-+}$.

The first claim of such a meson has been made by the GAMS Collaboration. In a partial wave analysis of the reaction $\pi^- p \to (\eta\pi^0)n$, they found an $(\eta\pi)$ P-wave resonance at a mass 1405 ±20 MeV with a width of 180 ±20 MeV, called $\hat{\rho}(1405)$ [3]. S, P and D in connection with 'waves' refers to orbital angular momentum of 0, 1 and 2. It will also be used as subscript, e.g. $(\eta\pi)_P$. The $(\eta\pi^0)_P$ wave has $J^{PC} = 1^{-+}$ and $I^G = 1^-$.

Later, again $(\eta\pi)_P$ intensities were found, but at different mass and with different width [4, 5]. The VES Collaboration at Serpukhov [4] studied the reactions (1) $\pi^- N \to (\eta\pi^-)N$ and (2) $\pi^- N \to (\eta'\pi^-)N$ on a nucleon N of a Beryllium target. The dominant partial wave in $(\eta\pi^-)$ is D^+ where the $^+$ stands for positive exchange naturality. It is due to the $a_2(1320)$ meson, a safe $\bar{q}q$ meson. The exotic P- wave signal is significant but relatively weak. The phase difference between D and P waves varies rapidly at the a_2 but there is no apparent phase motion due to the exotic P wave. The partial intensities for reaction (2) showed a D wave due to $a_2(1320) \to \eta'\pi^-$ and an exotic $(\eta'\pi^-)_P$ wave activity, here the largest partial wave. However the latter was seen at a different mass and width than in $\eta\pi^-$. The mean masses are $\approx$ 1300 MeV for $\eta\pi^-$ and 1600 MeV for $\eta'\pi^-$, the widths 300-400 MeV.

Almost certainly, there is exotic $(\eta\pi^-)_P$ and $(\eta'\pi^-)_P$ activity but is it resonant, a meson, or a non-resonant interaction? To qualify for resonant, the phase of the partial amplitude has to move in a typical way as a function of the invariant mass. Two recent experiments, indeed, report evidence for resonant phase motion of the $(\eta\pi^-)_P$ wave. E852 at Brookhaven, using the MultiParticle Spectrometer (MPS), has investigated the reaction $\pi^- p \to (\eta\pi^-)p$ with 18 GeV/c incoming π^- [6]. Again, the D+ wave dominates; the P+ intensity is weaker by an order of magnitude, but significant, at a mass of 1370 $\pm 20 \pm 50/30$ MeV, with a width of $385 \pm 40 \pm 70/100$ MeV. The relative phase motions are precisely as expected for resonant behaviour.

Even more evidence comes from the Crystal Barrel experiment at CERN which studied $\bar{p}p$ annihilations at LEAR until end of 1996. In an analysis [7] of $\bar{p}p \to \eta\pi^0\pi^0$ the need for an exotic $(\eta\pi^0)_P$ amplitude was noted, however it was hiding behind a complex interference pattern of other amplitudes. The masss and width could not be well determined. Recently, more convincing evidence came from the analysis of $\bar{p}d \to \eta\pi^0\pi^- p_{spectator}$ [8]. This channel is particularly sensitive to the $(\eta\pi)_P$ wave because only two other intermediate states, $\rho\eta$ and $a_2\pi$, are allowed from the initial $\bar{p}n$ state $^3S_1$, owing to parity, G-parity and angular momentum conservation. For the other possible $\bar{p}n$ initial state, $^1P_1$, also the intermediate state $a_0\pi$ is allowed, but its contribution turns out to be negligible. The intermediate states are seen in the Dalitzplot, fig. 1a. The diagonal band is due to the ρ meson decaying into $\pi^0\pi^-$. Due to interference with the $(\eta\pi)_P$ amplitude, the a_2 appears as horizontal and vertical steps of intensity at $M^2 = 1.74\text{GeV}^2$, left side (lhs) of the ρ, and as an accumulation where the two a_2 bands cross, rhs of the ρ. The simple structure of the Dalitz plot allows a very precise measurement of the $(\eta\pi)_P$ amplitude. The presence of this exotic wave can be seen with the naked eye after inspecting the individual partial intensities and interference terms (fig. 1b). All imaginable alternatives have been tried, for instance, an effective range ansatz which allows the fit to choose between resonant and nonresonant behaviour. The fit definitely needs the exotic $(\eta\pi)_P$ activity and wants a Breit Wigner amplitude, i.e. resonant behaviour. The mass is $M = (1400 \pm 20_{stat} \pm 10_{sys})\text{MeV}$, the width $\Gamma = (310 \pm 50_{stat} \pm 40_{sys})\text{MeV}$ and the intensity is strong, about 50% of the a_2 intensity.

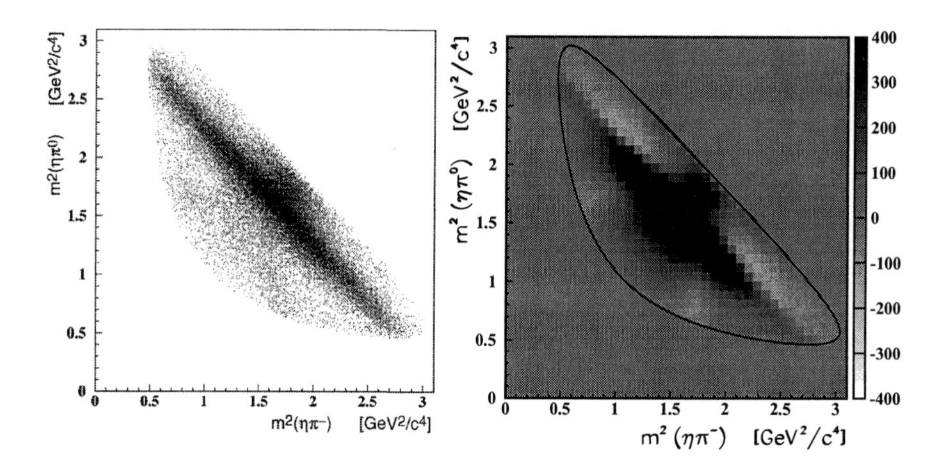

Fig. 1. a) Measured Dalitz plot for $\bar{p}d \to \eta\pi^0\pi^- p_{spect}$, 53k events. b) Contributions of the exotic 1^{-+} amplitude, all interference terms included.

In conclusion of section 2: The evidence for an exotic $J^{PC} = 1^{-+}$ $(\eta\pi)_P$ amplitude is almost beyond doubt. Resonant or not? At present, the experimental indications favour resonant behaviour. However, the fact that the $(\eta'\pi)_P$ intensity seen by VES peaks at a different mass, is uncomfortable, although it does not exclude an $(\eta\pi)_P$ resonance. The $f_1\pi$ threshold could play a role which should be subject of further investigation.

3 Pseudoscalar, axial and tensor non-$\bar{q}q$ mesons

The best non-$\bar{q}q$ candidates, i.e. the states claimed to be supernumerary, in the sector of pseudoscalar, axial and tensor mesons are in all three cases rather narrow. How strong is the evidence for their existence?

Pseudoscalar mesons ($J^{PC} = 0^{-+}$): The E/ι, now called $\eta(1440)$ was first seen in the H_2 bubble chamber at CERN [10], in $\bar{p}p \to \bar{K}K3\pi$. After it had been observed in J/ψ radiative decays at SPEAR in 1980 [11] it became the world best glueball candidate, for good reasonings, for a number of years.

Its existence is beyond doubt. Many experiments have seen it, in several decay modes. However, it was shown [12, 13] that it could be, together with $\eta(1295)$ a reasonable candidate for the first radial excitations of η and η'. But the case became more complicated by the fact that three experiments, MarkIII, DM2 [15, 14] (J/ψ radiative decays), and OBELIX [16] ($\bar{p}$ annihilation) have reported evidence for **two** narrow neighbouring states at the place of $\eta(1440)$. In this case, one of them would clearly not fit into the nonet of first-radially excited pseudoscalars. The lower state with a mass of about 1410 MeV apparently prefers to decay to $a_0\pi \to (\bar{K}K)\pi$, $(\eta\pi)\pi$, or to $(K\pi)_S\pi$, the upper state with a mass of 1460 to 1490 MeV to $K^*\bar{K}$.

The question is whether by using a Flatté parametrization instead of Breit Wigner functions, the two peaks could not be explained by a single resonance at 1410 MeV. Since the $K^*\bar{K}$ threshold is around 1.38 Gev, this decay channel would lead to a higher apparent mass.

In this context, a very recent analysis of J/ψ radiative decays to $\gamma(\eta\pi\pi)$, $\gamma(\bar{K}K\pi)$, $\gamma K^*\bar{K}^*$, $\gamma\rho\rho$, $\gamma\omega\omega$ and $\gamma\phi\phi$ should be mentioned [17]. The authors conclude that there is a narrow and a broad 0^{-+} resonance, both being mixtures of a glueball and a strangeonium ($\bar{s}s$). The broad component is found to decay nearly flavour blind, like an SU(3) singlet or a glueball.

The situation in the sector of **axial** vector mesons ($J^{PC} = 1^{+-}$) is similar to that of pseudoscalars: In addition to two resonances beyond doubt, $f_1(1285)$ and $f_1(1420)$, both seen in central production [18], and $f_1(1420)$, recently also seen in $\bar{p}p$ annihilation by OBELIX [16], there are indications of a nearby second state, $f_1(1510)$. The latter has been reported as $\bar{K}K\pi$ resonance by two low-statistics experiments [19] (HBC at CERN and LASS at SLAC) which studied $K^-p \to K_S K^\pm \pi^\mp \Lambda$. Further, weak support came from a study of $\pi^-p \to K_S K^\pm \pi^\mp n$ at the MPS at BNL [20]. But the LEPTON-F experiment at Serpukhov, investigating, with higher statistics than LASS, $K^-p \to K^+K^-\pi Y$, has seen $f_1(1420)$, but no $f_1(1520)$ [21].

Assuming the existence of two neighbouring axial vector mesons $f_1(1420)$ and $f_1(1510)$, in addition to the $f_1(1285)$, the narrow $f_1(1420)$ has been discussed as prime candidate for an axial non-$\bar{q}q$ meson, e.g. a mesonic molecule. However, the experimental evidence for $f_1(1510)$ has been questioned in the past years [22, 23] and in reference [23] $f_1(1285)$ and $f_1(1420)$ are considered to be members of the lowest axial vector $\bar{q}q$ nonet. Obviously, the K^-p experiments need to be repeated with higher statistics and good mass resolution.

Experimentally established **tensors** ($J^{PC} = 2^{++}$) are $f_2(1270)$ and $f_2'(1525)$, good candidates for the lowest tensor $\bar{q}q$ mesons. Further established tensors are: $f_2(1430)$, $f_2(1520)$, $f_2(1810)$, $f_2(2010)$, $f_2(2150)$, $f_2(2300)$, $f_2(2340)$, considered as potential non-$\bar{q}q$- candidates [9]. Some of the latter are probably radial excitations 2^3P_2, 3^3P_2, some perhaps molecules like $f_2(1520)$ close to $\omega\omega, \rho\rho$ threshold.

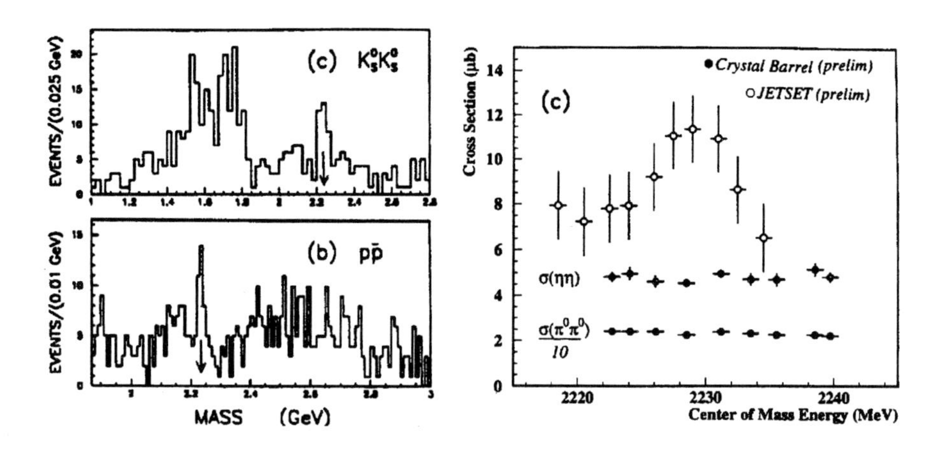

Fig. 2. Invariant mass distribution for a) $J/\psi \to \gamma K_S K_S$, b) $J/\psi \to \gamma \bar{p}p$ [25]. c) Partial cross section as function of cm energy for $\bar{p}p \to \eta\eta$ and $\to \pi^0\pi^0$ from JETSET [26] and Crystal Barrel [29].

Although it may not be the strongest case among the tensors, $\xi(2230)$ is the most popular 2^{++} glueball candidate these days. It is remarkably narrow, was seen at SPEAR [24] (Mark III), recently also by BES, Bejing [25], in various decay modes and by Jetset LEAR [26] in $\eta\eta$, see fig.2. Two contributions at HEP97, from e^+e^- experiments, are relevant to $\xi(2230)$ [27, 28]. CLEO [27] gives an upper limit for the coupling to 2γ of 1.3 eV. Using an average product branching ratio from MarkIII and BES

$$BR(J/\psi \to \gamma\xi(2230) \to \gamma K_S K_S) = (2.2 \pm 0.6) \cdot 10^{-5}$$

and a ξ width of 19 MeV, this translates into a 'stickiness' $S_x > 82$ –stickiness is a measure of the glueball nature, $S_x = 1$ per definition for a $\bar{q}q$-mesons; a large S_x signifies small coupling to $\bar{q}q$ and hence small constituent quark or large constituent gluon content of the meson.

L3 [28] have reported the observation of a signal at 2230 MeV in 3-jet events at this conference. The peak is very prominent, the large width of about 150 MeV experimental. The signal is apparently associated with the gluon jet. The identity with the very narrow $\xi(2230)$ (width smaller than 20 MeV) has yet to be shown.

The significance of the narrow $\xi(2230)$ in any individual decay channel where it had been seen is only 2-3 σ. The narrow peak seen by JETSET at LEAR [26] in $\bar{p}p \to \eta\eta$ is shown in fig. 2c. During the last weeks of LEAR operation Crystal Barrel has made a high statistics scan around $\sqrt{s} = 2230$MeV. The black dots in fig. 2c show the event rate as a function of cm energy for $\bar{p}p \to \eta\eta$ and $\to \pi^0\pi^0$ [29]. There is no sign for any structure. Combined with the BES results for $BR(J/\psi \to \gamma\xi(2230) \to \gamma\bar{p}p)$ and $BR(.. \to \gamma\eta\eta)$, the following limits are obtained: $BR(J/\psi \to \gamma\xi) \geq 10^{-2}$, $BR(\xi(2230) \to \bar{p}p) \leq 1.4 \cdot 10^{-3}$, and $BR(\xi(2230) \to \eta\eta) \leq 27 \cdot 10^{-3}$ [30].

Concluding section 3: All pseudoscalar, axial and tensor non- $\bar{q}q$ candidates need more experimental evidence to be established.

4 Light scalar isoscalar (f_0) mesons

The f_0 mesons are by far the most complex and most interesting sector of meson spectroscopy. They have the quantum numbers of the vacuum - which explains much of their specific properties, and the ground state glueball ought to be one of them. The PDG review [9] lists five established mesons:
$f_0(400-1200)$, a broad structure, seen in $(\pi\pi)_S$ scattering and production.
The narrow $f_0(980)$ seen in $(\pi\pi)_S$ and $(\bar{K}K)_S$.
The $f_0(1370)$ seen in (4π), $(\pi\pi)_S$, $(\eta\eta)_S$, $(\bar{K}K)_S$ and $(\gamma\gamma)$.
The fairly narrow $f_0(1500)$ seen in $(\pi\pi)_S$, $(\eta\eta)_S$, $(\eta\eta')_S$, $(\bar{K}K)_S$ and (4π).
The fairly narrow $f_J(1720)$ seen in $(\bar{K}K)_S$, $(\eta\eta)_S$ and $(\pi\pi)_S$.

For quite some time, ' G(1590)' was the best scalar (f_0) glueball candidate. It was discovered as $\eta\eta$, $\eta\eta'$ and 4π resonance in $\pi^- p$ interactions, by the GAMS Collaboration, by means of 2000 or 4000 lead glass crystals. Fig. 3a shows the scalar $(\eta\eta)_S$ wave intensity from a partial wave analysis of $\pi^- p \to \eta\eta n$ [31]. GAMS has claimed two resonances from this, an $\epsilon(1250)$ and the G(1590).

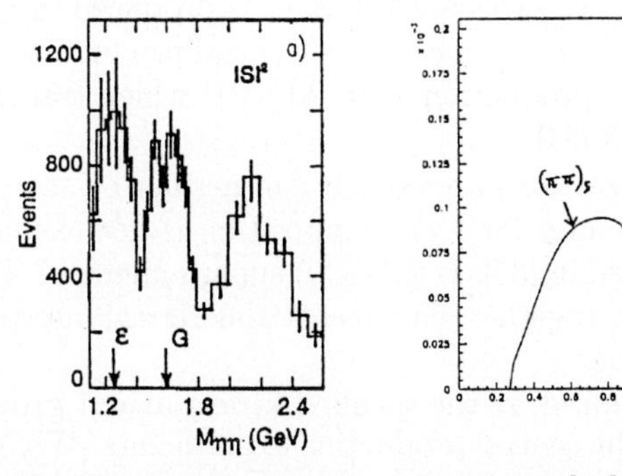

Fig. 3. a) Partial $(\eta\eta)_S$ intensity from GAMS [31]. b) Partial $(\eta\eta)_S$ and $(\pi\pi)_S$ intensity from Crystal Barrel [32, 33] as a function of invariant mass.

G(1590) has been superseded by the $f_0(1500)$ of Crystal Barrel, observed in $\bar{p}p$ annihilation into $3\pi^0, \eta\eta\pi^0, \bar{K}K\pi^0$ and $5\pi^0$, by means of 1380 CsI crystals, in a very-high statistics experiment. The partial $(\eta\eta)_S$ intensity obtained by Crystal Barrel[32] has a similar structure as that of GAMS (fig. 3b). The two peaks and the minimum are almost at the same places where GAMS observes them. This agreement may be fortuitous. Assuming, the same two overlapping resonances are produced in $\pi^- p$ and $\bar{p}p$, but with different relative production strengths, the peaks could be at different positions. But there is not doubt here, that the same two resonances are observed.

The masses of the two resonances reported by GAMS and CB disagree, since CB uses the K matrix formalism, appropriate for overlapping resonances with identical quantum numbers, whereas GAMS has used Breit-Wigner functions. The disagreement in branching ratios of $f_0(1500)$/ G(1590) are also likely due to the different analyses and to different phase space.

Crystal Barrel has discovered the $f_0(1500)$ in a Dalitz plot with 750 k events for $\bar{p}p \to 3\pi^0$ [33]. The scalar $(\pi\pi)_S$ partial intensity for a single combination of the 3 pions extracted from the Dalitz plot analysis of [33] is superimposed in fig. 3b. Above the $(\eta\eta)$ threshold, again two peaks are seen at nearly the same locations as '$\epsilon(1250)$' and 'G(1590)'; their modern interpretation is in terms of $f_0(1370)$ and $f_0(1500)$.

But the $(\pi\pi)_S$ wave is more complicated than $(\eta\eta)_S$. Below the $(\eta\eta)$ threshold there is a broad maximum, associated with the σ meson; it is part of $f_0(400-1200)$; in addition there is a sharp dip due to $f_0(980)$.

If one compares this picture with the $(\pi\pi)_P$ wave intensity for the same mass range which is is dominated by a single Breit-Wigner resonances, the $\rho(770)$, one realizes how complicated the scalar sector is. (And one may pray for having a simpler scalar sector of the electroweak interaction, i.e. for elementary Higgs particles rather than a new strong interaction.)

The $f_0(1500)$ has been considered to be the best candidate for the ground state glueball because 1) of its unusually narrow width of 120 MeV – for the basic $\bar{q}q$ meson states f_0 and f_0', with expected masses of 1090 and 1360 MeV, widths of 850 and 600 MeV are expected[12]. 2) It is produced in gluon-rich reactions: $\bar{p}p$ annihilation, J/ψ radiative decays, central production. 3) There is now an upper limit for $\gamma\gamma$ production from ALEPH which translates into a stickiness of larger than 13 [34].

However, its branching fractions are odd. It can neither be strangeonium nor a glueball, but a dominantly $\bar{u}u + \bar{d}d$ composition is not excluded. The decay properties are discussed in [35]. Is $f_0(1500)$ supernumerary? This question will be discussed below, together with recent theoretical interpretations of the scalar isoscalar mesons.

4) Its production is enhanced by the socalled **kinematical glueball filter**, discovered recently by the central production experiments (WA76,91,102) at CERN [36]. Consider a pp interaction where the centrally produced system of mesons called X emerges in addition to a fast and a slow proton:

$$pp \to p_{slow} X p_{fast}$$

Part of central production at SPS energies is due to double Pomeron exchange, since long considered an ideal process to produce glueballs. They found triggering on events where both protons are scattered to the same side, i.e. when the two exchanged Pomerons are emitted in about the same direction, that the known glueball candidates are favoured relative to known $\bar{q}q$ mesons. Later on the selection criterion was made more quantitative by cutting on $\Delta p_T = \sqrt{(\mathbf{p}_{T1}^2 - \mathbf{p}_{T2}^2)}$ where $\mathbf{p}_{T1}$ and $\mathbf{p}_{T2}$ are the transverse momentum vectors of the exchanged particles. An example of this cut is shown in fig. 4 where X = 4π and the invariant 4π mass is plotted for two Δp_T cuts. The $f_1(1285)$, a known $\bar{q}q$ meson, becomes more visible at larger Δp_T. The glueball candidate $f_0(1500)$ and an other presumed glueball candidate - $f_2(1900)$, are more pronounced at smaller Δp_T.

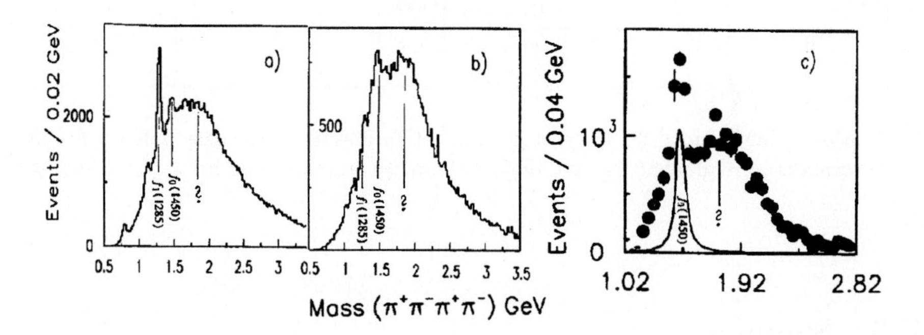

Fig. 4. a) Invariant (4π) mass spectrum from WA102 [36] for a) $\Delta p_T \geq 500$ MeV/c and b) $\Delta p_T \leq 200$ MeV/c. c) Partial $(4\pi)_S$ intensity from WA91 [37].

How does this glueball filter work? At present, this is a mystery. A possible explanation could be that resonances X with low spin are favoured at low Δp_T. Intuitively, Δp_T limits the spin transferred to the centrally produced system: $l \leq \Delta p_T \cdot R/\hbar = 1$, for Δp_T= 200 MeV/c and $R = 1fm$, the range of interaction. An argument against this explanation would be that also a tensor state $f_2(1900)$ is favoured at low Δp_T. However, it is possible that the broader peak at 1900MeV is due to a scalar intensity, seen in the partial wave analysis of WA91 [37] (fig. 4c).

A number of serious attempts have been made recently to explain the very complex pattern of scalar isoscalar f_0 mesons. They are summarized in table 1. The lines correspond to the observed physical f_0 mesons in the mass range below $2m_{proton}$. The columns describe the composition of these states in terms of pure $\bar{q}q$ or glueball states, before these basic states mix through their coupling to common decay channels. Despite of major disagreement in all the details, the glueball component is needed in all attempts who account for the $f_0(1500)$. In the latest attempts, the glueball component is distributed over at least two physical states.

PDG96 [9]	Anisovich.. [38]	Close.. [39]	Narison[40]	Törnqvist[41]	Weingarten[42]
$f_0(400-1200)$	47% gg 41% $\bar{n}n,^3P_0$		50% gg 50% $\bar{n}n$	$\bar{n}n, \bar{s}s$ $\simeq SU(3)_1$	
$f_0(980)$	$\bar{s}s, 1^3P_0$	KK molecule	50% gg 50% $\bar{s}s$	$\bar{s}s + KK$	
$f_0(1370)$	38% gg 62% $\bar{n}n,^3P_0$	25% gg 73% $\bar{n}n$	50% gg 50% $\bar{s}s$	$\bar{n}n, \bar{s}s$ $\simeq SU(3)_8$	18% gg 75% $\bar{n}n$
$f_0(1500)$	64% $\bar{n}n, 2^3P_0$ 35% gg	37% gg 37% $\bar{s}s$ 18% $\bar{n}n$	50% gg 50% $\bar{s}s$ ****		82% $\bar{s}s$ 12% $\bar{n}n$
$f_J(1710)$	99% $\bar{s}s, 2^3P_0$	36% gg 57% $\bar{s}s$	$\bar{s}s, 2^3P_0$		77% gg

Table 1. Suggestions for the composition of f_0 mesons with mass below 1.8 GeV, in terms of pure glueball gg, and of $\bar{q}q = \bar{s}s$ or $\bar{n}n$ states, where $\bar{n}n = (\bar{u}u + \bar{d}d)/\sqrt{2}$.

5 Summary

Here we are, after many years in search of the QCD-predicted non-$\bar{q}q$ mesons. Evidence has accumulated for the existence of an exotic 1^{-+} resonance. In the sector of pseusocalar, axial and tensor mesons, the best candidates for supernumerary states are still weak and need more experimental work. The complex sector of light scalar, f_0, mesons is basically clarified, on the experimental side, up to 1.6 GeV mass, thanks to modern high statistics experiments at LEAR and in high energy hadron beams. Theoretical attempts to explain the observed f_0 states all need a glueball component, but are still diverging in other details. As long as the fundamental questions raised by QCD are not answered, meson spectroscopy will remain an active field of particle physics, with new, improved experiments like COMPASS at CERN.

References

[1] H.Fritzsch and M.Gell-Mann, Proc. XVI Int. Conf. on High Energy Phys., Chicago 1972, Vol. II p. 135.
[2] M. Teper, Contribution #768, EPS Jerusalem HEP97.
[3] D. Alde et al.(GAMS), PLB 205 (1988) 397.
[4] G.M. Beladidze et al.(VES), PLB 313 (1993) 274.

[5] H. Aoyagi et al.(KEK), PLB 314 (1993) 246.
[6] D.R. Thompson et al.(E852), PRL 79 (1997) 1630.
[7] Crystal Barrel Coll, C.Amsler et al., PLB 333 (1994) 277.
[8] Crystal Barrel Coll., K. Hüttmann, Diploma Thesis, Univ. Munich 1997.
[9] Particle Data Group, R.M.Barnett et al., PRD 54 Part I (1996) 1.
[10] P. Baillon et al.(HBC), Nouvo Cim. 50A (1967) 393.
[11] D.L.Scharre et al.(MARKIII), PLB 97 (1980) 329; C.Edwards et al. (CBAL), PRL 49 (1982) 259 and PRL 50 (1983) 219.
[12] S.Godfrey and N. Isgur, PRD 32 (1985) 189.
[13] I. Cohen et al., PRL 48 (1982) 1074.
[14] J.E.Augustin et al.(DM2), PR D 46 (1992) 1951.
[15] Z. Bai et al.(MARKIII), PRL 65 (1990) 2507.
[16] A. Bertin et al.(OBELIX), PLB 400 (1997) 226.
[17] D.V.Bugg and B.S.Zou, PLB 397 (1997) 295.
[18] T. Armstrong et al.(WA76), Z. Phys.C56 (1992) 29.
[19] Ph. Gavillet et al., Z. Phys.C 16 (1982) 119; D. Aston et al.(LASS), PLB 201 (1988) 573.
[20] A. Birman et al.(MPS), PRL 61 (1988) 1557.
[21] S.I. Bityukov et al.(LEPTON-F), Sov. J. Nucl. Phys. 39 (1984) 735.
[22] M. Faessler in Proc. of the NATO ASI on Hadron Spectroscopy and the Confinement Problem, 1995 London and Swansea, UK, ed. D.V.Bugg, Plenum Press, New York 1996, p.1.
[23] F.E.Close and A. Kirk, Preprint RAL-97-029, BHAM-HEP/97-03.
[24] R.M.Baltrussitis et al.(MARKIII), PRL 56 (1986) 107.
[25] Y.Zhu (BES), 28th Int. Conf. on HEP (ICHEP96), Warsaw.
[26] J.Ritter et al.(JETSET), 56 A (1997) 291.
[27] CLEO Coll., Contribution ♯389, EPS Jerusalem HEP97.
[28] L3 Coll., Contribution ♯507, EPS Jerusalem HEP97.
[29] Crystal Barrel, preliminary data.
[30] K.K. Seth, Moriond 1997, to be published.
[31] M. Boutemeur, Proc. of the NATO ASI as ref. [22], p. 131.
[32] (Crystal Barrel Coll.) C.Amsler et al., PLB 353 (1995) 571.
[33] (Crystal Barrel Coll.) V.V.Anisovich et al., PLB 323 (1994) 233; Amsler et al., PLB 342 (1995) 433.
[34] ALEPH Coll., Contribution ♯626, EPS Jerusalem HEP97.
[35] S.Spanier (Crystal Barrel), Contr. #183, EPS Jerusalem HEP97.
[36] D.Barberis et al.(WA102), PLB 397 (1997) 339; F.E.Close and A. Kirk, PLB 397 (1997) 333.
[37] F.Antinori et al. (WA91), PLB 353 (1995) 589.
[38] V.V.Anisovich et al., PLB 395 (1997) 123.
[39] F.E. Close et al., PRD 55 (1997) 5749.
[40] S.Narison, Univ. Montpellier preprint P96/37, hep-ph/9612448.
[41] N.A.Törnqvist, Z.Phys. C 68 (1995) 647.
[42] D.Weingarten, Proc. Lattice'96, NPB (Proc.Suppl.) 53 (1997) 232.

Astroparticle Physics

Georg G. Raffelt (raffelt@mppmu.mpg.de)

Max-Planck-Institut für Physik (Werner-Heisenberg-Institut)
Föhringer Ring 6, 80805 München, Germany

Abstract. Recent developments of those areas of astro-particle physics are discussed that were represented at the HEP97 conference. In particular, the current status of direct and indirect dark-matter searches and of TeV neutrino and γ-ray astronomy will be reviewed.

1 Introduction

Astro-particle physics is such a wide field that it is certainly impossible to review its current status in a single lecture. To make a sensible selection it seemed most appropriate to cover those areas which were represented in the parallel sessions of this conference, i.e. mostly experimental topics in the areas of dark-matter detection and of neutrino and γ-ray astronomy. One of the most cherished dark-matter candidates is the lightest supersymmetric particle so that accelerator searches for supersymmetry are of immediate cosmological importance, yet I consider this topic to lie outside of my assignment. Likewise the laboratory searches for neutrino masses and oscillations are of direct astrophysical and cosmological significance, yet they exceed the boundaries of my task. Finally, I will not cover the very exciting recent developments in MeV to GeV neutrino astronomy (solar, supernova, and atmospheric neutrinos) because they are reviewed by another speaker [1].

2 Dark Matter Searches

2.1 Dark Stars (MACHOs)

The existence of huge amounts of dark matter in the universe is now established beyond any reasonable doubt, but its physical nature remains an unresolved mystery [2, 3]. A number of well-known arguments negate the possibility of a purely baryonic universe, but also point to significant amounts of nonluminous baryons. If some of them are in the galactic halo one most naturally expects them to be in the form of Massive Astrophysical Compact Halo Objects (MACHOs)—small and thus dim stars (brown dwarfs, M-dwarfs) or stellar remnants (white dwarfs, neutron stars, black holes). Stellar remnants and M-dwarfs are virtually excluded [4], which leaves us with brown dwarfs, i.e. normal stars with a mass below $0.08\,M_\odot$ (solar masses) so that they are too small to ignite hydrogen.

Paczyński proposed in 1986 to search for dim stars by the "microlensing" technique [5]. A distant star brightens with a characteristic lightcurve if a gravitational deflector passes near the line of sight. Gravitational lensing produces two images, but if their angular separation is too small the only observable effect is the apparent brightening of the source. A convenient sample of target stars is provided by the Large Magellanic Cloud (LMC), a satellite galaxy of the Milky Way. The LMC has enough bright stars and it is far enough away and far enough above the galactic plane that one intersects a good fraction of the galactic halo. If MACHOs comprise the halo, the lensing probability ("optical depth" for microlensing) is about 10^{-6} so that one has to monitor $\sim 10^6$ stars in the LMC. The duration of the brightness excursion depends on the lens mass; for $1\,M_\odot$ it is typically 3 months, for $10^{-2}\,M_\odot$ it is 9 days, for $10^{-4}\,M_\odot$ it is 1 day, and for $10^{-6}\,M_\odot$ it is 2 hours.

The microlensing search was taken up by the MACHO and the EROS Collaborations, both reporting candidates toward the LMC since 1993 [6]. Moreover, the galactic bulge has been used as another target where many more events occur through microlensing by ordinary disk stars. While these

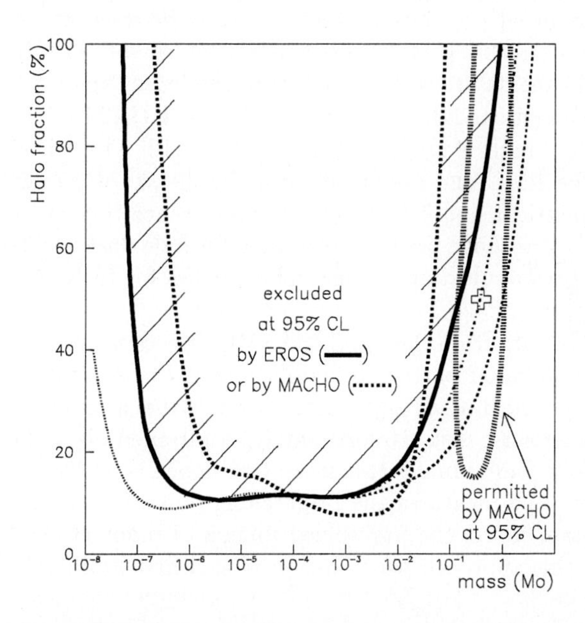

Fig. 1. Exclusion diagram at 95% C.L. for the halo fraction and mass of the assumed MACHOs [7]. Their masses were taken to be fixed and a standard model for the galactic halo was used. The dotted line on the left is the EROS limit when blending and finite size effects are ignored. The dot-dashed and dotted lines on the right are the EROS limits when 1 or 2 of their events are attributed to MACHOs. The cross is centered on the 95% C.L. permitted range of the MACHO Collaboration [8].

observations are not sensitive to halo dark matter, they allow one to develop a good understanding of the microlensing technique and are an interesting method to study the galaxy and its stellar content. Within the past few years microlensing has established itself as a completely new approach to astronomy, with at least half a dozen collaborations pursuing observations of various target regions. They also produce a huge database of intrinsically variable stars, which is an invaluable project in its own right.

Far from clarifying the status of dim stars as a galactic dark matter contribution, the microlensing results toward the LMC with about a dozen events [7, 8] are quite confusing. For a standard spherical halo the absence of short-duration events excludes a large range of MACHO masses as a dominant component (Fig. 1). On the other hand, the observed events indicate a halo fraction between about 10% and 100% of MACHOs with masses around $0.4\,M_\odot$ (Fig. 1). This is characteristic of white dwarfs, but a galactic halo consisting primarily of white dwarfs is highly implausible and almost excluded [4]. Attributing the events to brown dwarfs ($M \lesssim 0.08\,M_\odot$) requires a very nonstandard density and/or velocity distribution. Other explanations include an unexpectedly large contribution from LMC stars, a thick galactic disk, an unrecognized population of normal stars between us and the LMC, and other speculations [9]. It is quite unclear which sort of objects the microlensing experiments are seeing and where the lenses are.

Meanwhile a first candidate has appeared in both the MACHO and EROS data toward the Small Magellanic Cloud (SMC) [10] which is slightly more distant than the LMC and about 20° away in the sky. One event does not carry much statistical significance, but its appearance is consistent with the LMC data if they are interpreted as evidence for halo dark matter. However, this interpretation would imply a few solar masses for the SMC lens due to the large duration.

Besides more data from the LMC and SMC directions, other lines of sight would be invaluable. Of particular significance is the Andromeda galaxy as a target because the line of sight cuts through the halo almost vertically relative to the galactic disk. Unfortunately, Andromeda is so far away that one cannot resolve individual target stars. One depends on the "pixel lensing" technique where one measures the brightening of a single pixel of the CCD camera; one pixel covers the unresolved images of many stars. At least two groups pursue this approach which has produced first limits [11].

2.2 Axions

While the microlensing searches seem to indicate that some fraction of the galactic halo may consist of MACHOs, perhaps even in the form of primordial black holes [12], particle dark matter *aficionados* should not get disheartened—weakly interacting particles are still the best motivated option for the cold dark matter which apparently dominates the universe.

One of two well motivated possibilities are axions which appear as Nambu-Goldstone bosons of the spontaneously broken Peccei-Quinn symmetry which is motivated as a solution of the CP problem of strong interactions [13]. Apart from numerical parameters of order unity, these models are characterized by a single unknown quantity, the Peccei-Quinn scale f_a or the axion mass $m_a = 0.62\,\mathrm{eV}\,(10^7\,\mathrm{GeV}/f_a)$. In the early universe axions form nonthermally as highly occupied and thus quasi-classical low-momentum oscillations of the axion field. If axions are the dark matter, a broad class of early-universe scenarios predicts m_a to lie in the range $1\,\mu\mathrm{eV}$ to $1\,\mathrm{meV}$ [14].

In a magnetic field axions convert into photons by the Primakoff process because they have a two-photon coupling [15]. A frequency of 1 GHz corresponds to $4\,\mu\mathrm{eV}$; a search experiment for galactic axions consists of a high-Q microwave resonator placed in a strong magnetic field. At low temperature one looks for the appearance of microwave power beyond thermal and amplifier noise. Two pilot experiments [16, 17] could not reach realistic axion models, but two ongoing experiments with much larger cavity volumes have the requisite sensitivity (Fig. 2). In its current setup, the Livermore experiment [18] uses conventional microwave amplifiers while the Kyoto experiment [19] employs a completely novel detection technique based on the excitation of a beam of Rydberg atoms which passes through the cavity.

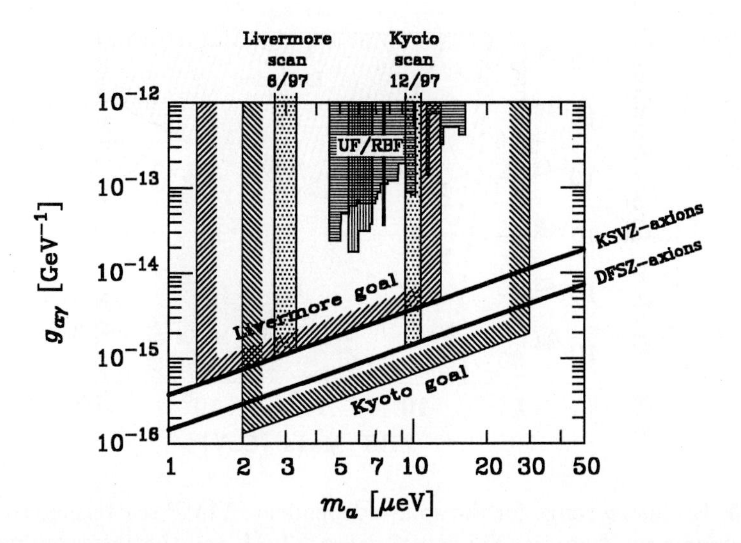

Fig. 2. Current limits on galactic dark matter axions from the University of Florida (UF) [16] and the Rochester-Brookhaven-Fermilab (RBF) [17] experiments and search goals of the Livermore [18] and Kyoto [19] experiments. It was assumed that the local galactic axion density is $300\,\mathrm{MeV\,cm^{-3}}$. The axion-photon coupling is given by $\mathcal{L}_{\mathrm{int}} = g_{a\gamma}\mathbf{E}\cdot\mathbf{B}\,a$. The relationship between $g_{a\gamma}$ and m_a for the popular DFSZ and KSVZ models is indicated.

2.3 Weakly Interacting Massive Particles (WIMPs)

The other favored class of particle dark matter candidates are WIMPs, notably the lightest supersymmetric particles in the form of neutralinos [20]. Direct searches rely on WIMP-nucleus scattering, for example in Ge or NaI crystals [21]. The expected counting rate is of order 1 event $\mathrm{kg}^{-1}\,\mathrm{day}^{-1}$ and thus extremely small. To beat cosmic-ray and radioactive backgrounds one must go deeply underground and use ultra pure materials. The recoils for 10–100 GeV WIMP masses are of order 10 keV. Such small energy depositions can be measured by electronic, bolometric, and scintillation techniques. The number of experimental projects is too large to even list them here [22].

The current limits already dig into the supersymmetric parameter space (Fig. 3). The DAMA/NaI experiment has actually reported a WIMP signature [30] which would point to neutralinos just below their previous exclusion range [31]. The significance of this result is very low, and tentative signals are bound to appear just below the previous exclusion range. Still, the good news is that this detection *could* be true in the sense that one has reached the sensitivity necessary to find supersymmetric dark matter. In the near future the large-scale cryogenic detectors CRESST [28] and CDMS [29] will explore a vast space of WIMP-nucleon cross-sections (Fig. 3).

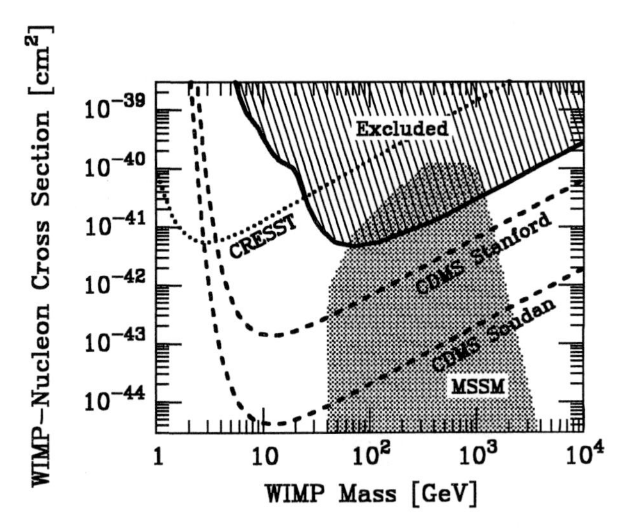

Fig. 3. Exclusion range for the spin-independent WIMP scattering cross section per nucleon from the NaI experiments [23, 24] and the germanium detectors [25]. Also shown is the range of expected counting rates for dark-matter neutralinos in the minimal supersymmetric standard model (MSSM) without universal scalar mass unification [26, 27]. The search goals for the upcoming large-scale cryogenic experiments CRESST [28] and CDMS [29] are also shown, where CDMS is located at a shallow site at Stanford, but will improve its sensitivity after the planned move to a deep site in the Soudan mine.

3 Neutrino Astronomy

Indirect methods to search for WIMPs rely on their annihilation in the galactic halo or in the center of the Sun or Earth where WIMPs can be trapped [20]. The search for GeV–TeV neutrinos from the Sun or Earth in the Kamiokande, Baksan, and MACRO detectors [32] already touch the parameter range relevant for neutralino dark matter [33]. Neutrino telescopes are thus competitive with direct dark-matter searches, where it depends on details of the supersymmetric models which approach has a better chance of finding neutralinos. Roughly, a water or ice Cherenkov detector requires a km^3 volume to be competitive with the CDMS-Soudan search goal.

After the sad demise of the deep-sea DUMAND project a km^3 neutrino telescope has again come within realistic reach after the breathtaking progress in the development of the AMANDA ice Cherenkov detector at the south pole [34]. The lake Baikal water Cherenkov detector [37] is another operational neutrino telescope, but probably it cannot reach the km^3 size. The prospects of the deep-sea projects NESTOR [35] and ANTARES [36] in the Mediterranean depend on the outcome of their current "demonstrator" phase. Either way, after the explosive development of solar and atmospheric neutrino observatories (MeV–GeV energies), high-energy neutrino astronomy is set to become a reality in the very near future.

Besides the search for dark matter, neutrino astronomy addresses another old and enigmatic astrophysical problem, the origin of cosmic rays which engulf the Earth with energies up to $\sim 10^{20}\,\mathrm{eV}$. They consist of protons and nuclei which must be accelerated somewhere in the universe. Whenever they run into stuff they produce pions and thus neutrinos and photons in roughly equal proportions ("cosmic beam dumps"). Because the universe is opaque to photons with energies exceeding a few 10 TeV due to pair production on the cosmic microwave background, and because charged particles are deflected by magnetic fields, high-energy neutrino astronomy offers a unique observational window to the universe, and especially a chance to identify the sites of cosmic-ray acceleration [38].

4 TeV γ-Ray Astronomy

Perhaps the most attractive sites for the cosmic-ray acceleration are active galactic nuclei (AGN) which are likely powered by accreting black holes. These objects tend to eject huge jets in opposite directions; for an estimate of the expected neutrino flux see Ref. [39]. If this is indeed the case one would equally expect high-energy γ-rays from these sources. Remarkably, for the past few years TeV γ-rays have indeed been observed [40, 41, 42] from the two nearby ($\sim$ 300 million light-years) AGNs Markarian 421 and 501 which have jets pointing toward Earth.

Until recently γ-ray astronomy reached only up to $\sim 20\,\text{GeV}$ because the low fluxes at higher energies require forbiddingly large satellites. The observational break-through in the TeV range arose from Imaging Air Cherenkov Telescopes (IACTs) on the ground [43]. A high-energy γ-ray hits the upper atmosphere at an altitude of $\sim 16\,\text{km}$ and produces an electromagnetic shower which in turn produces Cherenkov light. With a relatively crude telescope one can thus take an image of the shower. The axes of the cigar-shaped shower images of many γ-rays intersect in one point which corresponds to the location of the source in the sky and thus allows one to discriminate against the much larger but isotropic flux of hadronic cosmic rays. A number of galactic sources are now routinely observed, notably the Crab nebula, which is seen at energies up to $50\,\text{TeV}$ and serves as a "calibration" source. There remains an unexplored spectral range between about 20 and $300\,\text{GeV}$ which requires much larger IACTs than are currently available.

The Markarians are the first extragalactic sources in the TeV γ-sky. Their behavior is quite tantalizing in that they are hugely variable on sub-hour time scales (Fig. 4). Moreover, Mrk 501 essentially "switched on" from a state of low activity with about 0.1 of the Crab flux in 1995, about 0.3 in 1996, to

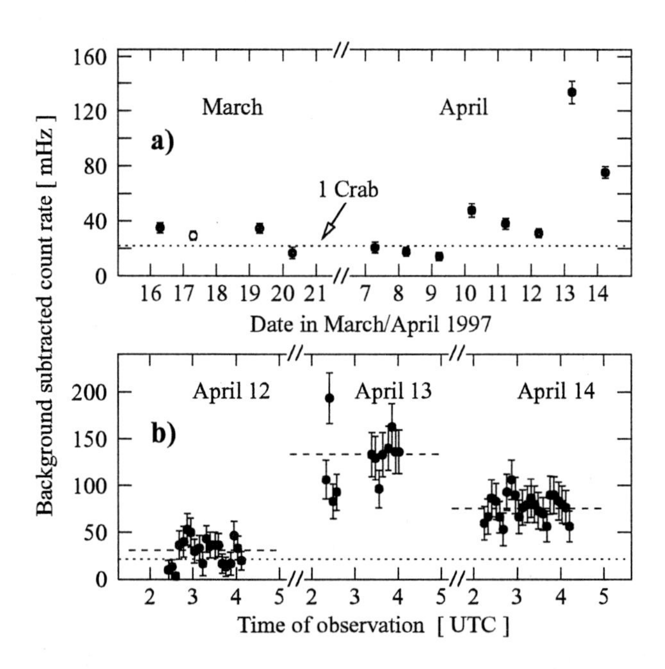

Fig. 4. Detection rate of the active galaxy Mrk 501 in the HEGRA stereoscopic system of Imaging Air Cherenkov Telescopes on a night by night basis, **a)** for the whole data set and **b)** for the last 3 nights in 5 min. intervals [42]. The dashed lines indicate the average per night, the dotted line shows the Crab detection rate as a reference. Errors are statistical only.

about 10 times the Crab since February 1997. If the high-energy photons are produced by a proton beam or, say, photon upscattering by accelerated electrons is by no means obvious. One will need neutrino telescopes to decide this question and thus to understand these intruiging objects.

5 Summary

Experiments to identify the physical nature of the galactic dark matter have recently made great strides. The microlensing searches for MACHOs have observed around a dozen candidates toward the Large Magellanic Cloud, but an interpretation of these events is quite puzzling in that their best-fit mass puts them into the white-dwarf category which is highly implausible. This year a first candidate toward the Small Magellanic Cloud has been reported, with an even larger apparent mass. From virtually any cosmological perspective cold dark matter remains the favored hypothesis for the dominant mass fraction of the universe. Full-scale searches for the most favored particle candidates (axions and WIMPs) are in progress. One can hunt WIMPs also by searching for their annihilation products in the form of high-energy neutrinos from the Sun or the center of the Earth. With a km^3 water or ice Cherenkov detector one could cover a significant fraction of the parameter space for supersymmetric dark matter. The impressive progress of the AMANDA south pole detector and the appearance of new deep-sea projects (ANTARES and NESTOR) bode well for this approach—a km^3 detector could be up and running before the LHC. High-energy neutrino astronomy has other intruiging objectives, notably searching for the sites of cosmic-ray acceleration. The breathtaking recent observations of TeV γ-rays from the two nearby active galaxies Markarian 421 and 501 have zoomed this physics target into sharper focus. Whether or not active galactic nuclei accelerate protons is a question that can be answered only by neutrino astronomy.

Acknowledgments

This work was supported, in part, by the Deutsche Forschungsgemeinschaft under grant No. SFB 375.

References

[1] M. Nakahata, PL11, these Proceedings.
[2] V. Trimble, Annu. Rev. Astron. Astrophys. 25 (1987) 425. S. Tremaine, Physics Today 45:2 (1992) 28.
[3] E.W. Kolb and M.S. Turner, *The Early Universe* (Addison-Wesley, Redwood City, 1990). G. Börner, *The Early Universe*, 2nd edition (Springer, Berlin, 1992).

[4] B.J. Carr, Comments Astrophys. 14 (1990) 257; Annu. Rev. Astron. Astrophys. 32 (1994) 531. B.J. Carr and M. Sakellariadou, Fermilab-Pub-97-299-A, Astrophys. J. (submitted 1997). D.J. Hegyi and K.A. Olive, Astrophys. J. 303 (1986) 56.
[5] B. Paczyński, Astrophys. J. **304** (1986) 1.
[6] C. Alcock et al. (MACHO Collab.), Nature 365 (1993) 621; Phys. Rev. Lett. 74 (1995) 2867. E. Aubourg et al. (EROS Collab.), Nature 365 (1993) 623; Astron. Astrophys. 301 (1995) 1. R. Ansari et al. (EROS Collab.), Astron. Astrophys. 314 (1996) 94.
[7] C. Renault et al. (EROS Collab.), Astron. Astrophys. 324 (1997) L69.
[8] C. Alcock et al. (MACHO Collab.), Astrophys. J. 471 (1996) 774; Astrophys. J. 486 (1997) 697.
[9] K. Sahu, Nature 370 (1994) 275. H.S. Zhao, astro-ph/9606166, 9703097. C. Alcock et al. (MACHO Collab.), Astrophys. J. 490 (1997) L59. A. Gould, astro-ph/9709263. N. Evans et al., astro-ph/9711224. E. Gates et al., astro-ph/9711110. D. Zaritsky and D. Lin, astro-ph/9709055.
[10] C. Alcock et al. (MACHO Collab.), Astrophys. J. 491 (1997) L11. N. Palanque-Delabrouille (EROS Collab.), astro-ph/9710194.
[11] A.P.S. Crotts and A.B. Tomaney, Astrophys. J. 473 (1996) L87. R. Ansari et al. (AGAPE Collab.), Astron. Astrophys. 324 (1997) 843.
[12] J. Yokoyama, Astron. Astrophys. 318 (1997) 673. K. Jedamzik, Phys. Rev. D 55 (1997) R5871. J.C. Niemeyer and K. Jedamzik, astro-ph/9709072.
[13] R.D. Peccei and H.R. Quinn, Phys. Rev. Lett. 38 (1977) 1440; Phys. Rev. D 16 (1977) 1791. S. Weinberg, Phys. Rev. Lett. 40 (1978) 223. F. Wilczek, Phys. Rev. Lett. 40 (1978) 279. J.E. Kim, Phys. Rept. 150 (1987) 1. H.-Y. Cheng, Phys. Rept. 158 (1988) 1.
[14] R.L. Davis, Phys. Lett. B 180 (1986) 225. R.A. Battye and E.P.S. Shellard, Phys. Rev. Lett. 73 (1994) 2954; (E) ibid. 76 (1996) 2203. D. Harari and P. Sikivie, Phys. Lett. B 195 (1987) 361. C. Hagmann and P. Sikivie, Nucl. Phys. B 363 (1991) 247.
[15] P. Sikivie, Phys. Rev. Lett. 51 (1983) 1415; Phys. Rev. D 32 (1985) 2988.
[16] C. Hagmann, P. Sikivie, N.S. Sullivan, D.B. Tanner, Phys. Rev. D 42 (1990) 1297.
[17] W.U. Wuensch et al., Phys. Rev. D 40 (1989) 3153.
[18] C. Hagmann et al., Nucl. Phys. Proc. Suppl. 51B (1996) 209.
[19] I. Ogawa, S. Matsuki and K. Yamamoto, Phys. Rev. D 53 (1996) R1740.
[20] G. Jungman, M. Kamionkowski and K. Griest, Phys. Rept. 267 (1996) 195.
[21] M.W. Goodman and E. Witten, Phys. Rev. D 31 (1985) 3059.
[22] J. Primack, D. Seckel and B. Sadoulet, Annu. Rev. Nucl. Part. Sci. 38 (1988) 751. P.F. Smith and J.D. Lewin, Phys. Rept. 187 (1990) 203. D.O. Caldwell, Mod. Phys. Lett. A 5 (1990) 1543; Proc. TAUP97, to be published. N.E. Booth, B. Cabrera and E. Fiorini, Annu. Rev. Nucl. Part. Sci. 46 (1996) 471. A. Watson, Science 275 (1997) 1736.

[23] P.F. Smith et al., Phys. Lett. B 379 (1996) 299. J.J. Quenby, Astropart. Phys. 5 (1996) 249.
[24] R. Bernabei et al., Phys. Lett. B 389 (1996) 757.
[25] D. Reusser et al., Phys. Lett. B 255 (1991) 143. E. Garcia et al., Phys. Rev. D 51 (1995) 1458. M. Beck et al., Phys. Lett. B 336 (1994) 141.
[26] P. Gondolo, private communication (1997).
[27] L. Bergström and P. Gondolo, Astropart. Phys. 5 (1996) 263. A. Bottino, F. Donato, G. Mignola and S. Scopel, Phys. Lett. B 402 (1997) 113. J. Edsjö and P. Gondolo, Phys. Rev. D 56 (1997) 1879.
[28] M. Sisti et al., in: Proc. 7th Int. Workshop on Low Temperature Detectors (LTD-7), 27 July–2 August 1997, Munich, Germany, ed. by S. Cooper (Max-Planck-Institut für Physik, Munich, 1997).
[29] S.W. Nam et al., in: Proc. 7th Int. Workshop on Low Temperature Detectors (LTD-7), 27 July - 2 August 1997, Munich, Germany, ed. by S. Cooper (Max-Planck-Institut für Physik, Munich, 1997).
[30] R. Bernabei et al., astro-ph/9710290, to be published in Proc. TAUP97.
[31] A. Bottino, F. Donato, N. Fornengo and S. Scopel, astro-ph/9709292 and 9710295.
[32] M. Mori et al. (Kamiokande Collab.), Phys. Rev. D 48 (1993) 5505. M.M. Boliev et al. (Baksan telescope), in: Proc. TAUP95, ed. by A. Morales, Nucl. Phys. B (Proc. Suppl.) 48 (1996) 83. M. Ambrosio et al. (MACRO Collab.), Preprint INFN-AE-97-23.
[33] M. Kamionkowski et al., Phys. Rev. Lett. 26 (1995) 5174. V. Berezinskii, et al., Astropart. Phys. 5 (1996) 333. L. Bergström, J. Edsjö and P. Gondolo, Phys. Rev. D 55 (1997) 1765.
[34] L. Berström, astro-ph/9612122, Proc. Identification of Dark Matter, Sheffield, UK, Sept. 1997. F. Halzen, astro-ph/9707289.
[35] E.G. Anassontzis et al. (NESTOR Collab.) Preprint DFF-283-7-1997, Jul 1997. Proc. 18th International Symposium on Lepton-Photon Interactions (LP 97), Hamburg, Germany, 28 July–1 Aug. 1997.
[36] ANTARES Collaboration, Proposal, astro-ph/9707136.
[37] V.A. Balkanov (Baikal Collab.), astro-ph/9705017, Proc. XXXII Rencontres de Moriond, Les Arcs, France, Jan. 18–25, 1997.
[38] T.K. Gaisser, astro-ph/9707283, Proc. OECD Megascience Forum Workshop, Taormina, Italy, May 22–23, 1997. F. Halzen, astro-ph/9605014, Proc. Venitian Neutrino Conference, Feb. 1996.
[39] F. Halzen and E. Zas, Astrophys. J. 488 (1997) 669.
[40] M. Punch et al., Nature 358 (1992) 477. A.D. Kerrick et al., Astrophys. J. 438 (1995) L59. D. Petry et al., Astron. Astrophys. 311 (1996) L13.
[41] J. Quinn et al., Astrophys. J. 456 (1996) L83.
[42] F. Aharonian et al. (HEGRA Collab.), Astron. Astrophys. 327 (1997) L5. M. Catanese et al., astro-ph/9707179.
[43] J.V. Jelley and T.C. Weekes, Sky & Telescope, Sept. 1995, pg.20.

Supersymmetric Extensions of the Standard Model

Pierre Binétruy (binetruy@qcd.th.u-psud.fr)

LPTHE, Université Paris-Sud, France

Abstract. The status of the Minimal Supersymmetric Model (MSSM) is reviewed before some recent developments on coupling unification are described. The different generalisations of the MSSM are then discussed with respect to their theoretical interest as well as new experimental results: R-parity violation, non-minimal models, supersymmetry-breaking sector, quark and lepton masses and mixings.

1 The MSSM

Let us first consider the stripped version of the Minimal Supersymmetric Model (MSSM), that is the model with

- minimum field content: the gauge fields of $SU(3) \times SU(2) \times U(1)$, quarks, leptons, two Higgs doublets H_1 and H_2 and their superpartners,
- minimum coupling content: in the supersymmetric sector, gauge couplings, Yukawa couplings and the mu-term which couples the two Higgs doublets in the superpotential ($W \ni \mu H_1 H_2$: μ is the only supersymmetric dimensionful coupling); in the soft supersymmetry-breaking sector, scalar and gaugino mass terms, trilinear A-terms and the $B\mu$ term which couples the two Higgs doublets in the potential ($\mathcal{L} \ni B\mu H_1 H_2 + h.c.$)

In this section, we will consider the MSSM to be a genuinely low energy model and disregard the constraints from coupling unification (see section 2).

In view of the many successes of the Standard Model under precision tests at LEP, it might come as a surprise to realize that the MSSM is surviving all these tests almost just as well. The buzzword here is custodial $SU(2)$ symmetry which is a global symmetry under which the $SU(2)_L$ gauge fields transform as a triplet and which ensures that the parameter ρ is 1. Deviation from 1 is associated with the breaking of this symmetry: for example since (t, b) is a doublet of custodial $SU(2)$, the top mass being much larger than the bottom mass is the major source of breaking. The precision tests from LEP have taught us that the breaking of custodial $SU(2)$ in the Standard Model is in good agreement with the one observed in the data.

If one considers radiative corrections to the propagators (oblique corrections), the most significant contributions come from the third generation, through the left-handed squark mass difference $m^2_{\tilde{t}_L} - m^2_{\tilde{b}_L} = m^2_{\tilde{t}} - m^2_{\tilde{b}} +$

$M_W^2 \cos 2\beta$ and the trilinear A_t term [1] and they are usually down by some powers of the squark masses (thus decoupling for large squark masses).

Thus if the squarks of the third generation are not too large, one may obtain predictions for $\sin^2\theta_W$ or M_W which are sensibly different from their Standard Model values. Hence the interest of settling the discrepancy between LEP and SLD about the precise value of $\sin^2\theta_W$, or of obtaining more precision on the W mass in order to start discreminating between the Standard Model and the MSSM. However if the squarks of the third generation are heavy enough (the squarks of the first two generations and the sleptons being within their experimental limits), the MSSM is difficult to distinguish from the Standard Model.

As has been stressed many times, vertex corrections may provide some of the key tests of low energy supersymmetry through the processes $Z \to b\bar{b}$ and $b \to s\gamma$. They involve mainly the charged Higgs $H^\pm$, the chargino, the stop and the pseudoscalar A; of particular relevance [2] is the Higgsino-stop-bottom supersymmetric coupling $g(m_t/M_W)(\tan\beta)^{-1}\,\bar{b}_L\tilde{t}_R\tilde{h}^-$.

In general, the charged Higgs loop has an opposite effect to the stop-chargino loop (for example [3] in the supersymmetric limit $\mathrm{Br}(b \to s\gamma) = 0$). Thus, the heavier the superpartners (squarks et charginos) are, the heavier the charged Higgs must be in order to reproduce the successes of the Standard Model [1]. This is why a key experimental result is the limit on chargino masses: $m_{\chi^\pm} > 90$ GeV [4]

With heavy superpartners, the charged Higgs mass (hence M_A) needs to be even larger in the small $\tan\beta$ region in order to counterbalance the large Higgs-top-bottom coupling. For example, the experimental value of R_b tends to forbid the region low $\tan\beta$ and low or moderate M_A. In the case of large $\tan\beta$, R_b allows a light M_A but $b \to s\gamma$ requires a cancellation between large chargino-stop contributions and large charged Higgs contribution, thus leading to some amount of fine-tuning. The limits obtained have been analysed in a series of works [5, 6]. Also constraints on $\tan\beta$ are obtained from: the decay $b \to c\tau\bar{\nu}$ [7] ($\tan\beta < 42M_{H^\pm}/M_W$), the condition for electroweak radiative breaking ($1 < \tan\beta < m_t/m_b$) and a constraint on the finiteness of λ_t up to 10^{16} GeV [8] ($1.5 < \tan\beta < 65$).

To conclude, the MSSM with heavy enough squarks of the third generation, chargino and stop masses beyond the experimental limit and heavy enough pseudoscalar A^0 (that is heavy enough charged Higgs) looks very similar to the Standard Model in electroweak precision tests.

[1] This translates into a lower bound on M_A since, to a good approximation, $M_{H^\pm}^2 = M_A^2 + M_W^2$.

2 Unification of Couplings

2.1 Grand Unified Theories (GUT)

If the gauge symmetry of the Standard Model $SU(3) \times SU(2) \times U(1)$ is embedded in a simple gauge group G broken at a scale M_U, then at this scale all gauge couplings are equal to the GUT coupling α_U:

$$\alpha_1(M_U) = \alpha_2(M_U) = \alpha_s(M_U) = \alpha_U(M_U) \ . \tag{1}$$

One can start with the values of $\sin^2\theta_W$ and α_{em} at M_Z and deduce $\alpha_s(M_Z) = 0.130 \pm 0.001$, $M_U = 3 \times 10^{16}$ GeV and $\alpha_U(M_U)^{-1} = 23$. The central value for $\alpha_s(M_Z)$ lies somewhat higher than the most recent determination [9] $\alpha_s(M_Z) = 0.119 \pm 0.005$ but the error quoted is from the input parameters only. One must also include uncertainties coming from the low-scale thresholds – masses of the superpartners, Higgs bosons, top quark – or high-scale thresholds due to the mass splittings between superheavy particles as well as possible effects of non-renormalizable operators. One may estimate the overall uncertainty to about 10%, which leads to an agreement between the predicted value for $\alpha_s(M_Z)$ and the data.

One may also wonder whether there is unification of the Yukawa couplings. There might actually be an indication coming from the third family. It has been stressed many times that the top mass lies close to the quasi-infrared fixed point value for the corresponding Yukawa coupling λ_t. Such a value is reached if the top Yukawa coupling is large at M_U. For low and moderate values of $\tan\beta$ one may neglect the bottom Yukawa coupling and one obtains [10]:

$$m_t = 196 \text{ GeV } \left[1 + 2(\alpha_s(M_Z) - 0.12)\right] \frac{\tan\beta}{\sqrt{1+\tan^2\beta}}, \tag{2}$$

which singles out regions of low $\tan\beta$ ($\tan\beta < 2$). Large values of $\tan\beta$ (say 40) are also allowed if one restores the bottom coupling: they tend to correspond to a top-bottom-tau unification.

In any case, the large values of $\lambda_t(M_{GUT})$ which seem to be favored by the infrared fixed point solution may indeed be an indication of bottom-tau unification: the condition $\lambda_b(M_U) = \lambda_\tau(M_U)$ yields too large a bottom mass unless the top Yukawa coupling (which influences the evolution of λ_b) is large. Thus, contrary to the first two families, it seems that the mass spectrum of the third family is in possible agreement with the ideas of Yukawa unification.

Let me take this opportunity to stress the predominant role played by the third family in unified supersymmetric models. Indeed, because the first and second generations of quarks and leptons are weakly coupled to the Higgs *it is the third family which is directly related to the hierarchy problem.* Consider for example the tree level mass relation:

$$\frac{1}{2}M_Z^2 = \frac{m_1^2 - m_2^2\tan^2\beta}{\tan^2\beta - 1} - |\mu|^2, \tag{3}$$

where m_1 and m_2 are the soft mass terms of the two Higgs doublets. In the scenario of radiative breaking it is through the large top coupling to H_2 that m_2^2 turns negative at low energy.

This has been stressed by Cohen, Kaplan and Nelson who, in their *effective supersymmetry* scenario [11], point out that a theory with stop and sbottom but no squark or slepton fields corresponding to the (massless) quarks and leptons of the first two generations generates no quadratic divergences. Such a theory can be viewed as an effective low energy theory where very massive (through supersymmetry breaking) squarks and sleptons of the first two families are integrated out. This illustrates the fact that the hierarchy problem is solved just by keeping the third generation squarks not too heavy.

2.2 Minimal supergravity model

In order to further constrain the MSSM, one often assumes universality of the soft terms at the scale of unification M_U. If one assumes that at M_U all gaugino fields have a mass $M_{1/2}$, one obtains the so-called *constrained* MSSM model where the following (fairly robust) relation holds at all scale μ:

$$\frac{M_1(\mu)}{\alpha_1(\mu)} = \frac{M_2(\mu)}{\alpha_2(\mu)} = \frac{M_3(\mu)}{\alpha_s(\mu)}. \tag{4}$$

If one further assumes that all scalar masses are equal to m_0 and all trilinear terms to A_0 at M_U, one obtains the *minimal supergravity* model whose hypotheses are less well motivated. This model has a minimal number of new parameters $(m_0, m_{1/2}, \tan\beta, A_0, \mathrm{sign}(\mu))$ compared with the Standard Model, which is the reason for its popularity.

For small $\tan\beta$ (say smaller than 2) the model is compatible with the infrared fixed point solution and thus, as we saw, with bottom-tau unification. Given the present limits on superpartners, there is some fine-tuning involved with the radiative breaking of $SU(2) \times U(1)$. One way to see this is to express (3) in terms of the original parameters m_0 and $M_{1/2}$; for example, for $\tan\beta = 1.65$, it reads:

$$M_Z^2 = 5.9 m_0^2 + 0.1 A_0^2 - 0.3 A_0 M_{1/2} + 15 M_{1/2}^2 - 2|\mu|^2, \tag{5}$$

which involves cancellation between large numbers. This fine tuning is usually described more quantitatively in terms of the variable $\Delta = \max(a/M_Z^2)(\partial M_Z^2/\partial a)$ expressing the *global* sensitivity of M_Z with respect to the variations of an input parameter a [12]. Fine-tunings of order 10^{-1} to 10^{-2} have recently [13] been advocated for the low $\tan\beta$ minimal supergravity model. It should be stressed however that: i) including one-loop corrections tends to decrease the amount of fine-tuning [2][14] ii) any discussion of fine tuning contains assumptions about how the fundamental parameters are distributed [16] iii)

[2] Not including them may actually make the definition of $\tan\beta$ strongly renormalisation scale-dependent [15].

the variable Δ is quite sensitive to correlations among parameters, such as appear in any given scenario of supersymmetry breaking [3].

This low $\tan\beta$ scenario predicts a light Higgs of mass smaller than 105 GeV, which should therefore be seen or excluded at LEP running at 200 GeV. It has also been advocated [17] as a possible scenario for baryogenesis at the electroweak scale if the stop is light enough.

For large $\tan\beta$ (of order m_t/m_b), unification of Yukawa couplings imposes rather large values of m_0 and $M_{1/2}$ [18]. If one wants to avoid the fine tuning associated with the cancellation necessary between these large scales in order to obtain satisfactory radiative breaking, one is led to relax the universality of scalar masses [19, 20]. This scenario yields a light Higgs boson heavier than 120 GeV.

Finally, an intermediate $\tan\beta$ scenario is not compatible with Yukawa unification but may have some welcome features: less severe fine tuning and less sensitivity to non-universality of scalar masses. The light Higgs lies in the intermediate region 100-115 GeV.

2.3 String models

In the case of strings, there are some important modifications to the unification scenario:

- Unification of gauge interactions is the sign of an underlying structure (the string) but not necessarily of a simple group of unification. Since the unification also involves gravitation, the unification scale M_U (string scale M_S) is close to the Planck scale M_{Pl}. Unification of gauge couplings takes a generalized form:

$$k_1\alpha_1(M_U) = k_2\alpha_2(M_U) = k_3\alpha_s(M_U), \tag{6}$$

  where the k_i are numbers (integers in the case of non-abelian symmetries) called Kac-Moody levels and fixed for each string model considered.
- Massive string modes induce the presence of new threshold effects at the string scale.

¿From this point of view, string theories face a serious problem of scales. Indeed, since these theories have a unique fundamental scale, all scales are related to one another. In particular, for the weakly coupled string theory,

$$M_S = M_{Pl} g_S. \tag{7}$$

Identifying M_S and the string coupling g_S with M_U and $g_U(M_U)$ as found above from gauge coupling unification, one finds that $M_U/g_U(M_U) \sim 5.10^{16}$ GeV, which misses M_{Pl} by a factor 20!

[3] For example, a linear correlation between μ and $M_{1/2}$ decreases Δ significantly [13].

Several ways out of this problem have been devised [4]. One may invoke large string threshold effects but these turn out to be usually small [22] apart from some field-dependent contributions (in which case it is necessary to impose large vacuum expectation values for these fields). Intermediate thresholds may also be present since realistic string models tend to have a large number of additional matter (and gauge) fields: they modify the evolution of the gauge couplings and could thus increase the value of M_U. One may also think of the extra (six) dimensions which introduce towers of Kaluza-Klein states but the gravitational coupling is affected likewise. As we will see momentarily this turns out not to be the case in a recent proposal by Hořava and Witten [23], based on the strongly coupled string.

Let us first try to be more quantitative [24]. The heterotic string, which among the weakly coupled string theories is the best candidate for a theory of all known interactions, is characterized by a fundamental length scale $l_s = \sqrt{\alpha'}$, the value of the dilaton field $< \phi >$ which fixes the string coupling $g_s = e^{2<\phi>}$, and the volume V_6 of the 6-dimensional compact manifold ($V_6 \sim M_U^{-6}$). One can express Newton's constant as well as the unified coupling α_U in terms of them:

$$G_N = \frac{\alpha'^4 e^{2<\phi>}}{64\pi V_6}, \quad \alpha_U = \frac{\alpha'^3 e^{2<\phi>}}{16\pi V_6}. \tag{8}$$

The whole approach makes sense if $e^{2<\phi>} < 1$, that is $G_N > \alpha_U^{4/3}/M_U^2$, which is too large. There are several ways out, among which assuming that $M_U > 3 \times 10^{16}$ GeV as discussed above. A further possibility is to suppose that $e^{2<\phi>} > 1$, *i.e. that the string is strongly coupled* [25]. With the recent advances in string duality, one can hope to go somewhat further along this road. The strong coupling regime is believed to be described by the so-called M-theory whose effective field theory would be as follows: 11-dimensional supergravity coupled to gauge fields which live on two 10-dimensional hyperplanes [23]. The radius R_{11} of the (compact) eleventh dimension which interpolates between the two hyperplanes introduces a new scale in the game which modifies the expressions:

$$G_N = \frac{\kappa^2}{16\pi^2 V_6 R_{11}}, \quad \alpha_U = \frac{(4\pi\kappa^2)^{2/3}}{2V_6}. \tag{9}$$

where κ is the eleven-dimensional supergravity coupling: $\kappa = (M_{Pl}^{(11)})^{-9/2}$. Putting numbers one finds that $R_{11}^{-1} \sim M_U/4$ and $M_{Pl}^{(11)} \sim 2M_U$. Thus, even before reaching the unification scale, the gravitational sector turns 5-dimensional (whereas matter which originates from the 10-dimensional gauge fields remains 4-dimensional). And the theory becomes 11-dimensional just

[4] Note however that gauge coupling unification only determines through the renormalisation group equations the logarithm of scales and the above discrepancy represents a miss in logarithmic scale by only 15% [21].

above the unification scale, thus avoiding the singular behavior at the Planck scale. This scenario should lead to a very different cosmology compared with the standard case [26].

3 Enlarging the MSSM

For the sake of print space I only summarize the content of this section.

3.1 More couplings: *R*-parity violating terms

The MSSM has the minimal number of couplings and therefore assumes R-parity which prevents the most dangerous violations of baryon and lepton number that are allowed by the sole requirement of supersymmetry and gauge symmetry. The arguments in favor of R-parity are, besides the proton stability: a possible origin from R-symmetries present at the supergravity level; and most of all dark matter for which the lightest supersymmetric particle (LSP) represents the best candidate. An important body of work has been devoted to extract from data the limits on R-parity-violating couplings [27]. In doing so, it is often assumed that a single coupling dominates but the most stringent bounds apply to products of couplings.

A huge interest for R-parity violation has been spurned by the high-Q^2 events observed at HERA. This has been reviewed by J. Ellis at this conference and I refer to his talk for details [28].

¿From a theoretical point of view, it is important to understand the structure of R-parity violating couplings. Efforts have been undertaken in two directions: using grand unified symmetries, it is possible to understand why only one kind of couplings (baryon-violating or lepton-violating) should be present [29]; family symmetries naturally lead to hierarchies among R-parity violating couplings [30].

3.2 More fields

One often has to introduce more fields to solve some of the problems of the MSSM. For example, μ is the only dimensionful parameter of the low energy supersymmetric theory. If the underlying theory has a typical scale of 10^{16} or 10^{19} GeV, why is μ so much smaller? This is the well-known μ problem which has received several solutions. One consists in introducing a gauge singlet S and allowing the renormalisable coupling $\lambda S H_1 H_2$: $\mu = \lambda < S >$. This gives the minimal extension to the MSSM and is sometimes called (M+1)SSM. The phenomenology of this model has been studied recently in detail [31]: the existing experimental constraints impose the singlet sector to decouple from the other fields and, surprisingly, the model looks very much like the MSSM with extra constraints.

Other extensions which have been discussed recently [32] include models with extra $U(1)$ (the S field just discussed could be charged under this new symmetry, thus avoiding domain wall problems associated with a discrete symmetry present in the (M+1)SSM). Such models or other extensions could very well find their origin in superstring constructions [33].

4 Supersymmetry breaking

Let us first discuss some phenomenological issues relevant for supersymmetry breaking:

- scalar masses and flavor-changing neutral currents (FCNC)
  In all generality, it is supersymmetry breaking which generates the non-diagonal squark and slepton mass matrices, whereas gauge symmetry breaking generates through the Yukawa couplings the non-diagonal quark and lepton mass matrices. Thus the diagonalisation matrices $\tilde{V}^{q,l}$ of the former are in principle independent of the diagonalisation matrices $V^{q,l}$ of the latter. If one takes for example the quark-squark-gaugino coupling, this induces a rotation matrix between quark and squark states $W \equiv V^q \tilde{V}^{q\dagger}$ (with L, R indices suppressed). Such a matrix appears in the box diagram contribution to $\Delta m_K / m_K$ involving two internal gluinos and two internal squarks $\tilde{u}_i$, $\tilde{u}_j$:
$$\frac{\Delta m_K}{m_K} \sim \alpha_s^2 \sum_{ij} W^\dagger_{di} W_{is} W^\dagger_{dj} W_{js} f \left(\frac{m^2_{\tilde{u}_i}}{m^2_{\tilde{g}}}, \frac{m^2_{\tilde{u}_i}}{m^2_{\tilde{g}}} \right) \tag{10}$$
  which puts some strong constraints on the rotation matrix W or on the squark masses $m_{\tilde{u}_i}$, and thus on supersymmetry breaking.
- the $\mu/B\mu$ problem
  The soft term $B\mu$ which connects the two Higgs doublets in the scalar potential is necessary to obtain both $< H_1 > \neq 0$ and $< H_2 > \neq 0$. Indeed $\sin 2\beta$ which vanishes for vanishing $< H_1 >$ or $< H_2 >$ reads at tree level:
$$\sin 2\beta = \frac{2B\mu}{2\mu^2 + m_1^2 + m_2^2} \tag{11}$$
  This not only shows that $B\mu$ cannot be nonzero but also that it should not be too large or too small compared to the μ parameter, otherwise the breaking of $SU(2) \times U(1)$ may not be possible.
- trilinear A-terms and charge and color breaking minima
  Trilinear A-terms give essentially negative contributions to the scalar potential, and thus potentially dangerous minima [34].

The history of supersymmetry breaking has gone through several phases. It was soon realized that breaking at tree level leads to some problems: F-term breaking yields vanishing supertrace and thus, on average, squarks and sleptons in the same mass range as quarks and leptons [35]; D-term breaking yields color breaking minima or involves at low energy an anomalous $U(1)$ symmetry. This is why in the early 1980's people resorted to models where supersymmetry breaking appears in the observable sector through loops: it is broken in a hidden sector and messenger fields ensure the coupling between observable and hidden sectors. Early attempts at implementing such a mechanism at low energy encountered many problems: charge or color breaking minima, light scalars due to the breaking of a global R-symmetry,...; a large number of new fields were needed in order to solve them.

Gravity-mediated models

In the light of the increasing complexity of the low energy models, one turned to gravity for the messenger interaction. Then the supertrace is of the order of the gravitino mass, whose nonzero value is the sign of supersymmetry breaking. As a bonus, it was thought that "gravity is flavor-blind", all soft terms are universal at the Planck scale.This is a wrong statement in most explicit models.

Let us take for example a class of superstring models which has been studied in some generality by Brignole, Ibanez and Munoz [37]. The fields responsible for supersymmetry breaking that they consider are the dilaton S (introduced earlier as a ten-dimensional scalar field e^{ϕ}) and the moduli fields T_i which describe the volume and shape of the compact manifold. In a dilaton-dominated scenario ($< F_S > \neq 0$, $< F_{T_i} >= 0$), soft terms are universal (up to string loop effects [38]) because the dilaton has universal couplings. In the simplest case, $m_0 = |M_{1/2}|/\sqrt{3} = |A_0| = B/2 = \mu = m_{3/2}$ at M_U, which obviously shows correlations between the soft parameters. Renormalisation group equations then fix the low energy spectrum. Indeed, the constraints are too strong and the spectrum is not compatible with electroweak radiative breaking. In a mixed dilaton-moduli scenario ($< F_S > \neq 0$, $< F_{T_i} >= 0$), gaugino masses are still universal but one loses universality for the other soft terms (flavor-independent gluino loop contributions may help fulfiling FCNC constraints). In most cases, scalar masses are smaller than gaugino masses at M_U, which leads to $m_{\tilde{l}} < m_{\tilde{q}} \sim M_{\tilde{g}}$ at M_Z.

This type of analysis does not assume a specific mechanism for supersymmetry breaking in the hidden sector: it is its strength. It is important however to be able to produce an explicit example, as an existence proof, as well as to check the hypotheses made. Let us take for instance gaugino condensation in the hidden sector of superstring models, an ever popular model [39]. Specifically the model I will consider [40] incorporates invariance under modular transformations, a Green-Schwarz term for anomaly cancellation, string threshold corrections, and stabilization of the dilaton field through non-perturbative corrections to the Kähler potential [41]. Although non-perturbative corrections are invoked to stabilize the dilaton, the ground state remains in the perturbative string regime and does not depend much on the exact forms of these corrections. One finds at the minimum that $T_i = 1$ and $F_{T_i} = 0$: one is thus in a dilaton-dominated scenario but some of the previous conclusions are evaded because there are extra scalar fields, those describing the gaugino condensate effective degrees of freedom.

Gauge-mediated models

The potential problems of gravity-messenger models with FCNC together with new developments in dynamical supersymmetry breaking [42] have led a certain number of authors to reconsider models with "low energy" messengers [43]. These are the gauge-mediated models who rely on standard gauge interactions as the mediator: since "standard gauge interactions are flavor-blind", soft masses are universal.

One can discuss the nature of messenger fields on general grounds. These fields are charged under $SU(3) \times SU(2) \times U(1)$ and since their masses are larger than the electroweak scale they must appear in vectorlike pairs. One can parametrize this through a (renormalizable) coupling to a singlet field S, fundamental or composite: $W \ni \lambda S M \bar{M}$ with $\lambda < S > \gg M_W, M_Z$. Since $M, \bar{M}$ must feel supersymmetry breaking, $< F_S > \neq 0$. In order not to spoil perturbative gauge coupling unification, the messengers must appear in complete representations of $SU(5)$.

Gaugino masses appear at one loop whereas scalar masses appear at two loops[5]:

$$M_a \sim \frac{\alpha_a}{4\pi} \frac{< F_S >}{< S >} \tag{12}$$

$$\tilde{m}^2 = 2 \left(\frac{< F_S >}{< S >} \right)^2 \left[\sum_a C_i^a \left(\frac{\alpha_a}{4\pi} \right)^2 \right], \tag{13}$$

where the index a refers to the gauge group, $C_i^3 = 4/3, C_i^2 = 3/4, C_i^1 = 5Y_i^2/12$. As advertized, squark masses are family independent because the $SU(3) \times SU(2) \times U(1)$ quantum numbers are so. This is welcome to suppress FCNC.

Since squarks and sleptons must be heavy enough, the scale $\Lambda \equiv < F_S > / < S >$ must be larger than 30 TeV. Requiring zero vevs for the messengers yields $< F_S > < \lambda < S >^2$. Thus the messenger mass $\lambda < S >$ is larger than Λ.

In the minimal model, the gaugino masses obey (4) and typically $m_{\tilde{q}}^2 \sim 8M_3^2/3$, $m_{\tilde{l}_L}^2 \sim 3M_2^2/2$, $m_{\tilde{l}_R}^2 \sim 5M_1^2/6$. Hence there are strong correlations in the supersymmetric spectrum [44]. But there are possible modifications: presence of several singlet fields, non-zero D-term for $U(1)_Y$, direct couplings of messengers to the observable sector [45, 46].

One of the reasons for the success of gauge-mediated model is that they yield a phenomenology quite different from the standard gravity-mediated approach. This has generated a lot of activity recently. The most notable

[5] A-terms vanish at the messenger scale.

differences come from the fact that, quite often, the gravitino is the lightest supersymmetric particle (LSP): for example $< F_S >= (10^5 \text{ GeV})^2$ gives $m_{3/2} =< F_S > /M_{Pl} \sim 10$ eV. It is then important to determine which is the next lightest (NLSP). Since typically the NLSP decays into gravitino plus photon, the signatures will involve missing transverse energy and one or several photons [4, 47].

On the theoretical side, gauge-mediated models face several problems. One concerns the radiative breaking of $SU(2) \times U(1)$. In standard supergravity models, as discussed above, it is the coupling to the top and large $\ln M_U/M_Z$ which are responsible for this breaking. In gauge-mediated models, it is the large stop mass:

$$m_2^2(m_{\tilde{t}}) \sim m_2^2(\Lambda) - \frac{3}{8\pi^2} h_t^2 (m_{\tilde{t}_L}^2 + m_{\tilde{t}_R}^2) \ln \Lambda / m_{\tilde{t}} \qquad (14)$$

Indeed this mechanism works so well that it involves some undesirable fine-tuning [48].

The worst problem faced by gauge-mediated models is however the $\mu/B\mu$ problem: since all soft supersymmetry-breaking terms scale as Λ, it is very difficult to avoid the relation:

$$B\mu \sim \mu\Lambda \qquad (15)$$

Given the value of Λ, this is obviously incompatible with a relation such as (11). There has been attempts to decouple the origin of μ and of $B\mu$ [49, 45]. For instance, one may introduce a new singlet S' with coupling $\lambda' S' H_1 H_2$: $\mu = \lambda' < S' >$. But it is then extremely difficult to decouple S and S'. And, to my knowledge, there is no completely satisfactory solution to this $\mu/B\mu$ problem.

An interesting new development is the elaboration of gauge-mediated models without messsengers. These models take more advantage of the new developments in dynamical symmetry breaking (the corresponding DSB sector was often a black box in the previous models): the role of the messengers is played by effective degrees of freedom of the DSB sector. This implies that the gauge symmetry of the Standard Model is a subgroup of the flavor symmetry group of the DSB sector. This group is therefore rather large and there are many effective messengers. Unlss they are heavy, this spoils the perturbative unification of gauge couplings. How to make effective messengers heavy? Remember that the messenger mass is $\lambda < S >$ whereas $< F_S > / < S >$ is fixed by supersymmetry breaking; since $< S >$ itself is not fixed, the idea is to require $< S > \gg < F_S >^{1/2}$. Explicit realizations involve non-renormalisable terms [50] or inverted hierarchy [51].

Other scenarios

Let me finish this section by discussing other supersymmetry breaking scenarios recently proposed:

- anomalous $U(1)$ [52]
  Many superstring models have a local $U(1)_X$ symmetry which is anomalous: anomaly is cancelled by the Green-Schwarz mechanism. Because of this anomaly cancellation mechanism, fields of both hidden and observable sector have non-zero charge under $U(1)_X$. Hence it is this abelian gauge interaction that serves as a mediator between hidden and observable sector. It is broken at a superheavy scale: one or two orders of magnitude below the string scale. Such a scenario has thus properties in commmon with both gravity and gauge-mediated models.
- M-theory
  Interesting attempts have been made at using the specifics of effective M-theory to devise new scenarios of supersymmetry-breaking [53]. In particular, the boundary conditions on the two 10-dimensional hyperplanes may break supersymmetry.

5 The problem of mass

Universality of the soft terms is not the only way to avoid unwanted FCNC. One can alternatively:

• make the squarks of the first two families heavy.

Since FCNC constraints are only crucial for the first two families, whereas the fine tuning associated with the hierarchy problem concerns mostly the third family, one can keep the third family sleptons relatively light whereas making the first two family sleptons very heavy [54]. Often in this approach a crucial role is played by an anomalous $U(1)$ symmetry of the type discussed above. One should however pay a careful attention in this context to charge or color breaking minima [55].

Such superheavy sfermions would be difficult to find directly at the future colliders. It has been noted [56] that they could be discovered through precision tests of the physics involving the lighter superpartners: they lead to hard [6] supersymetry breaking corrections, such as a difference between a gauge and a gaugino coupling.

• use family symmetries to orient the squark mass matrices along the quark mass matrices.

If $V^{\tilde{q}} = V^q$, then $W = 1$ in (10), thus leading to a vanishing contribution to $\Delta m_K / m_K$. Of course in practical cases, the alignment is only approximate and the contribution is nonzero but small [57].

This relates two problems: the structure of squark mass matrices which appear at the scale Λ_S of supersymmetry-breaking and the problem of quark and lepton masses and mixings associated with the flavor dynamics scale Λ_F. If $\Lambda_S \ll \Lambda_F$, one may expect a particle spectrum which is approximately flavor-free (universality); if $\Lambda_S \gg \Lambda_F$, the two problems are connected: this is

[6] Obviously since supersymmetry is assumed to be broken softly in order not to generate quadratic divergences these hard corrections are finite and calculable.

the *supersymmetric flavor problem.* Horizontal (or family) symmetries should account for such mass hierarchies.

Horizontal symmetries

How such symmetries would work has been explained by C. Froggatt and H. Nielsen almost some 20 years ago [58], when they proposed an illustrative example which remains the prototype of such models. They assume the existence of a symmetry which requires some quark and lepton masses to be zero and generate a finite mass at some order in a symmetry breaking interaction. This line of research has been pursued by many authors since [59]-[65]. Let me illustrate it on an example.

Consider an abelian gauge symmetry $U(1)_X$ which forbids any renormalisable coupling of the Yukawa type except the top quark coupling. Hence the Yukawa matrix Λ_U which appears in the superpotential through the term $\Lambda^U_{ij} Q_i \bar{u}_j H_2$ has the form:

$$\Lambda_U = \begin{pmatrix} 0\,0\,0 \\ 0\,0\,0 \\ 0\,0\,1 \end{pmatrix} \tag{16}$$

where 1 in the last entry means a matrix element of order one. The presence of such a non-zero entry means that the charges under $U(1)_X$ obey the relation:

$$X_{Q_3} + X_{U_3^c} + X_{H_u} = 0 \tag{17}$$

whereas similar combinations for the other field are non-vanishing and prevent the presence of a non-zero entry elsewhere in the matrix Λ_U.

We assume that this symmetry is spontaneously broken through the vacuum expectation value of a field θ of charge X_θ normalized to -1: $< \theta > \neq 0$. The presence of non-renormalizable terms of the form $Q_i U_j^c H_u (\theta/M)^{n_{ij}}$ induces in the effective theory below the scale of $U(1)_X$ breaking an effective Yukawa matrix of the form:

$$\Lambda_U = \begin{pmatrix} \lambda^{n_{11}} & \lambda^{n_{12}} & \lambda^{n_{13}} \\ \lambda^{n_{21}} & \lambda^{n_{22}} & \lambda^{n_{23}} \\ \lambda^{n_{31}} & \lambda^{n_{32}} & 1 \end{pmatrix} \tag{18}$$

where $\lambda =< \theta > /M$ and

$$n_{ij} = X_{Q_i} + X_{U_j^c} + X_{H_u} \tag{19}$$

$(n_{33} = 0)$.

It might seem on this example that, by choosing the charges of the different fields, one may accomodate any observed pattern of masses. There are however constraints on the symmetry: in particular those coming from the cancellation of anomalies. The observed pattern of masses is incompatible with a nonanomalous family symmetry: the $U(1)_X$ symmetry must be anomalous [59, 63].

We have already discussed one instance where one finds a seemingly anomalous symmetry: in some superstring models, there is a $U(1)$ symmetry whose anomaly is compensated by the 4-dimensional version [66] of the Green-Schwarz mechanism [67]. This is possible through the couplings of the gauge fields to a dilaton-axion-dilatino supermultiplet. Moreover, it has been shown [68] that the value of the Weinberg angle may be related to the pattern of anomaly coefficients. Indeed, if the horizontal abelian symmetry is precisely the anomalous $U(1)_X$, observed hierarchies of fermion masses are compatible with the standard value of $\sin^2\theta_W$ at gauge coupling unification.

We already stressed that a theory of mass should also discuss sfermion masses. It turns out that sfermion masses are also constrained by the symmetry $U(1)_X$ [60, 57].

Non-abelian horizontal symmetries have also been considered. They allow for a quasi-degeneracy among the sfermions of the first two families if the three families transform as $2+1$ under the flavor symmetry group. Discrete [69] as well as continuous [70] symmetries have been considered.

Finally, it might be that the sector responsible for flavor symmetry breaking has some connection with the sector where supersymmetry is dynamically broken, much in the way the gauge mediator sector connects the supersymmetry breaking sector with the observable sector in gauge-mediated models [71]. This leads to some new ways of relating the two scales Λ_F and Λ_S.

Composite fermions

Until recently, the supersymmetric approach to physics beyond the Standard Model inherently assumed that the fields of the Standard Model are fundamental. Now that one understands better strongly coupled supersymmetric theories [42], one may relax this assumption and consider composite fermions. Obviously, since the scale of the relevant dynamics lies above the electroweak scale, one needs to consider nearly massless composite fermions. This requires that the underlying theory possesses a confining phase where chiral symmetry is not completely broken [72],[73].

Such an endeavour has been undertaken recently by several groups [74]-[76]. The toy models proposed have some interesting properties. For example, the approximate symmetry is typically a product of $U(1)$'s, which are associated with the conservation of each type of subconstituent. Hierarchies of masses arise from hierarchies of interactions depending on the number and nature of the constituents. Also such a picture might give a clue for the number of families, a question usually not addressed by model builders.

Such a possibility may lead to models very different from the ones that we consider today. It remains to be seen then whether there is more to supersymmetry than to allow us to tackle the difficult nonperturbative issues.

Supersymmetric CP problem

Connected with the flavor problem is the supersymmetric CP problem. In a supersymmetric model, there are new sources of CP violation besides the CKM matrix: new phases for the parameters μ, A, $B\mu$, $M_{\tilde{g}}$ as well as in the off-diagonal terms of the squark mass matrices. The most significant effects are on the electric dipole moment of the neutron and on the parameter ϵ_K in the neutral K system. Existing experimental limits impose severe constraints on the supersymmetric models. General model-independent predictions are hard to make but obviously the supersymmetric CP problem must be discussed in connection with the flavor problem, for which one can identify several distinct classes of models: CP violation can then be discussed within each class of models [77].

Acknowledgments

I wish to acknowledge the hospitality of the CERN Theory Division where this talk was finally prepared and partly written. I thank my collaborators, in particular E. Dudas, M.K. Gaillard, P. Ramond and C. Savoy, and wish to thank W. deBoer, S. Pokorski and F. Richard for valuable discussions.

References

[1] R. Barbieri, M. Frigeni, F. Giuliani and H. Haber, Nucl. Phys. B341 (1990) 309; D. García and J. Solà, Mod. Phys. Lett. A9 (1994) 211; P. Chankowski, A. Dabelstein, W. Hollik, W. Mösle, S. Pokorski and J. Rosiek, Nucl. Phys. B417 (1994) 101.
[2] M. Boulware and D. Finnell, Phys. Rev. D44 (1991) 2054.
[3] R. Barbieri and G.F. Giudice, Phys. Lett. B309 (1993) 86.
[4] P. Janot, these Proceedings.
[5] G. Altarelli, R. Barbieri and F. Caravaglios, Phys. Lett. B314 (1993) 357; J.D. Wells, C. Kolda and G.L. Kane, Phys. Lett. B338 (1994) 219; D. Garcia, R. Jimenez and J. Sola, Phys. Lett. B347 (1995) 321; P.H. Chankowski and S. Pokorski, Phys. Lett. B356 (1995) 307, Phys. Lett. B366 (1996) 188, Nucl.Phys. B475 (1996); E. Ma and D. Ng, Phys. Rev. D53 (1996) 255; G.L. Kane and J.D. Wells, Phys. Rev. Lett. 76 (1996) 869; J. Ellis, J.L. Lopez and D.V. Nanopoulos, Phys. Lett. B372 (1996) 95, B397 (1997) 88; A. Brignole, F. Feruglio and F. Zwirner, Z. Phys. C71 (1996) 679; M. Drees et al. Phys. Rev. D54 (1996) 5598.
[6] S. Bertolini, F. Borzumati, A. Masiero and G. Ridolfi, Nucl. Phys. B353 (1991) 591; R. Barbieri and G.F. Giudice, Phys. Lett. B309 (1993) 86; N. Osimo, Nucl. Phys. B404 (1993) 20; R. Garisto and J.N. Ng, Phys. Lett. B315 (1993) 372; Y. Okad, Phys. Lett. B315 (1993) 119; M. Diaz, Phys. Lett. B322 (1994) 207; F. Borzumati, Zeit. Phys. C63 (1994) 395; J.L. Lopez, D.V. Nanopoulos, G.T. Park and A. Zichichi, Phys. Rev. D49 (1994) 355; R. Arnowitt and P. Nath, Phys. Lett. B336 (1994) 395; V. Barger, M.S. Berger, P. Ohmann and R.J.N. Phillips, Phys. Rev. D51 (1995) 2438; S. Bertolini and F.. Vissani, Zeit. Phys.

C67 (1995) 513; F. Borzumati, M. Olechowski and S. Pokorski, Phys. Lett. B349 (1995) 311; B. de Carlos and J.A. Casas, Phys. Lett. B349 (1995) 300; H. Baer and M. Brhlik, Phys. Rev. D55 (1997) 3201; J.L. Hewett and J.D. Wells, Phys. Rev. D55 (1997) 5549; T. Blažek and S. Raby, hep-ph/9712257.

[7] Y. Grossman, H.E. Haber and Y. Nir, Phys. Lett. B357 (1995) 630; J.A. Coarasa, R.A. Jimenez, J. Sola, Phys. Lett. B406 (1997) 337.

[8] B. Schrempp and M. Wimmer hep-ph/9606386.

[9] S. Catani, talk presented at the Lepton-Photon 97 conference, Hamburg.

[10] M. Carena, S. Pokorski and C. Wagner, Nucl. Phys. B406 (1993) 59.

[11] A. Cohen, D. Kaplan and A. Nelson, Phys. Lett. B388 (1996) 588.

[12] J. Ellis, K. Enqvist, D.V. Nanopoulos and F. Zwirner, Nucl. Phys. B276 (1986) 14; R. Barbieri and G. Giudice, Nucl. Phys. B306 (1988) 63.

[13] P.H. Chankowski and S. Pokorski, hep-ph/9702431, to appear in *Perspectives on Supersymmetry*, ed. G. Kane, World Scientific; P.H. Chankowski, J. Ellis and S. Pokorski, hep-ph/9712234.

[14] M. Olechowski and S. Pokorski, Nucl. Phys. B404 (1993) 590; B. de Carlos and J.A. Casas, Phys. Lett. B309 (1993) 320.

[15] G. Gamberini, G. Ridolfi and F. Zwirner, Nucl. Phys. B331 (1990) 331.

[16] G.W. Anderson and D.J. Castaño, Phys. Lett. B347 (1995) 300, Phys. Rev. D52 (1995) 1693 and D53 (1996) 2403; G.W. Anderson, D.J. Castaño and A. Riotto, Phys. Rev. D55 (1997) 2950.

[17] M. Carena, M. Quiros and C.E.M. Wagner, hep-ph/9710401.

[18] W. deBoer, these Proceedings.

[19] M. Olechowski and S. Pokorski, Phys. Lett. B 344 (1995) 201.

[20] S. Dimopoulos and G. Giudice, Phys. Lett. B357 (1995) 573.

[21] P. Binétruy and P. Langacker, unpublished.

[22] K. Dienes, Phys. Rep. 287 (1997) 447 and references therein; A. Faraggi, hep-ph/9707311.

[23] P. Hořava and E. Witten, Nucl. Phys. B460 (1996) 506, B475 (1996) 96.

[24] E. Witten, Nucl. Phys. B471 (1996) 135.

[25] M. Dine and N. Seiberg, Phys. Rev. Lett. 55 (1985) 366; V. Kaplunovsky, Phys. Rev. Lett. 55 (1985) 1036.

[26] B. Ovrut, these Proceedings.

[27] H. Dreiner, hep-ph/9707435, to appear in *Perspectives on Supersymmetry*, ed. G. Kane, World Scientific.

[28] J. Ellis, these Proceedings.

[29] G. Giudice and R. Rattazzi, Phys. Lett. B406 (1997) 321; R. Barbieri, A. Strumia and Z. Berezhiani, Phys. Lett. B407 (1997) 250.

[30] T. Banks, Y. Grossman, E. Nardi and Y. Nir, Phys. Rev. D52 (1995) 5319; P. Binétruy, E. Dudas, S. Lavignac and C. Savoy, hep-ph/9711517 (to be published in Physics Letters).

[31] U. Ellwanger, M. Rausch de Traubenberg and C.A. Savoy, Nucl. Phys. B492 (1997) 21.

[32] M. Cvetic and P Langacker, Phys. Rev. D54 (1996) 3570, Mod. Phys. Lett. A11 (1996) 1247; M. Cvetic, D.A. Demir, J.R. Espinosa, L. Everett and P. Langacker, Phys. Rev. D56 (1997) 2861; G. Cleaver, M. Cvetic, J.R. Espinosa, L. Everett and P. Langacker, hep-ph/9705391.

[33] A. Faraggi, hep-ph/9707311.

[34] J.M. Frère, D.R.T. Jones and S. Raby, Nucl. Phys. B222 (1983) 11; J.A. Casas, A. Lleyda and C. Muñóz, Nucl. Phys. B471 (1996) 3; A.Kusenko, P. Langacker and G. Segré, Phys. Rev. D54 (1996) 5824; for more complete references see the review by A. Casas, hep-ph/9707475.

[35] P. Fayet, Phys. Lett. 84B (1979) 416; S. Dimopoulos and H. Georgi, Nucl. Phys. B193 (1981) 150.

[36] P. Fayet, Phys. Lett. 69B (1977) 489.

[37] A. Brignole, L. Ibanez and C. Muñoz, Nucl. Phys. B422 (1994) 125, (E) B436 (1995) 747 and hep-ph/9707209; C. Muñoz, these Proceedings.

[38] J. Louis and Y. Nir, Nucl. Phys. B447 (1995) 18.

[39] See also the contributions of P. Minkowski and of P. Hernandez to these Proceedings.

[40] P. Binétruy, M.K. Gaillard and Y.Y. Wu, Nucl. Phys. B481 (1996) 109, B493 (1997) 27, Phys. Lett. B412 (1997) 288; J.A. Casas, Phys. Lett B384 (1996) 103.

[41] T. Banks and M. Dine, Phys. Rev. D50 (1994) 7454.

[42] N. Seiberg, Nucl. Phys. B431 (1995) 129.

[43] M. Dine and A. Nelson, Phys. Rev. D48 (1993) 1277; M.Dine, A. Nelson, and Y. Shirman, Phys. Rev. D51 (1995) 1362; M. Dine, A. Nelson, Y. Nir and Y. Shirman, Phys. Rev. D53 (1996) 2658.

[44] S. Dimopoulos, S. Thomas and J.D. Wells, Nucl. Phys. B488 (1997) 39.

[45] M. Dine, Y. Nir and Y. Shirman, Phys. Rev. D55 (1997) 1501.

[46] S. Raby, these Proceedings.

[47] M. Chemarin, these Proceedings.

[48] P. Ciafaloni and A. Strumia, Nucl. Phys. B494 (1997) 41.

[49] G. Dvali, G.F. Giudice and A. Pomarol, Nucl. Phys. B478 (1996) 31.

[50] E. Poppitz and S. Trivedi, Phys. Rev. D55 (1997) 5508 and hep-ph/9707439, to appear in Proceedings of *SUSY97*, Philadelphia, 1997; N. Arkani-Hamed, J. March-Russell and H. Murayama, Nucl. Phys. B509 (1998) 3.

[51] H. Murayama, Phys. Rev. Lett. 79 (1997) 18; S. Dimopoulos, G. Dvali, R. Rattazzi and G.F. Giudice, hep-ph/9705307.

[52] P. Binétruy and E. Dudas, Phys. Lett. B389 (1996) 503; G. Dvali and A. Pomarol, Phys. Rev. Lett. 77 (1996) 3728.

[53] P. Hořava, Phys. Rev. D54 (1996) 7561; E. Dudas and C. Grojean, hep-th/9704177; I. Antoniadis and M. Quiros, Nucl. Phys. B505 (1997) 109, hep-th/9707208; H.P. Nilles, M. Olechowski and M. Yamaguchi, hep-th/9707143; M. Quiros, these Proceedings.

[54] S. Dimopoulos and G. Giudice, 1995; A. Pomarol and Tommasini, 1996; A.G. Cohen, D.B. Kaplan and A.E. Nelson, Phys. Lett. B388 (1996) 588; G. Dvali and A. Pomarol, Phys. Rev. Lett. 77 (1996) 3728; R.N. Mohapatra and A. Riotto, Phys. Rev. D55 (1997) 1138, Phys. Rev. D55 (1997) 4262; Zhang, hep-ph/9702333; A.E. Nelson and D. Wright, Phys. Rev. D56 (1997) 1598.

[55] N. Arkani-Hamed and H. Murayama, Phys. Rev. D56 (1997) 6733.

[56] H-C. Cheng, J. Feng and N. Polonski, Phys. Rev. D56 (1997) 6875 and hep-ph/9706476; L. Randall, E. Katz and S. Su, hep-ph/9706478; M.M. Nojiri, D. Pierce and Y. Yamada, hep-ph/9707244.

[57] Y. Nir and N. Seiberg, Phys. Lett. B 309 (1993) 337.

[58] C.D. Froggatt and H.B. Nielsen, Nucl. Phys. B147 (1979) 277 and B164 (1979) 114.

[59] J. Bijnens and C. Wetterich, Nucl. Phys. B283 (1987) 237.

[60] M. Leurer, Y. Nir and N. Seiberg, Nucl. Phys. B398 (1993) 319, B420 (1994) 468.

[61] L. Ibáñez and G.G. Ross, Phys. Lett. B332 (1994) 100.

[62] V. Jain and R. Shrock, Phys. Lett. B352 (1995) 83.

[63] P. Binétruy and P. Ramond, Phys. Lett. B350 (1995) 49; P. Binétruy, S. Lavignac and P. Ramond, Nucl. Phys. B477 (1996) 353; J. Elwood, N. Irges and P. Ramond, hep-ph/9705270.

[64] Y. Nir, Phys. Lett. B354 (1995) 107.

[65] E. Dudas, S. Pokorski and C.A. Savoy, Phys. Lett. B356 (1995) 45; E. Dudas, S. Pokorski and C.A. Savoy, Phys. Lett. B369 (1996) 255; E. Dudas, C. Grojean, S. Pokorski and C.A. Savoy, Nucl. Phys. B481 (1996) 85.

[66] M. Dine, N. Seiberg and E. Witten, Nucl. Phys. B289 (1987) 317.

[67] M. Green and J. Schwarz, Phys. Lett. B149 (1984) 117.

[68] L. Ibáñez, Phys. Lett. B303 (1993) 55.

[69] D. Kaplan and M. Schmalz, Phys. Rev. D49 (1994) 3741; L. Hall and H. Murayama, Phys. Rev. Lett. 75 (1995) 3985; A. Pomarol and D. Tommasini, Nucl. Phys. B466 (1996) 3; C. Carone, L. Hall and H. Murayama, Phys. Rev. D53 (1996) 6282; P. Frampton and O. Kong, Phys. Rev. D53 (1995) 2293; K.C. Chou and Y.L. Wu, Phys. Rev. D53 (1996) 3492.

[70] L. Hall and L. Randall, Phys. Rev. Lett. 65 (1990) 2939; M. Dine, A. Kagan and R. Leigh, Phys. Rev. D48 (1993) 4269; P. Pouliot and N. Seiberg, Phys. Lett. B318 (1993) 169; R. Barbieri, G. Dvali, and L. Hall, Phys. Lett. B377 (1996) 76; R. Barbieri, L. Hall, S. Raby and A. Romanino, Nucl. Phys. B493 (1997) 3; R. Barbieri, L. Hall, and A. Romanino, Phys. Rev. D56 (1997) 7183; K.S. Babu and S.M. Barr, Phys. Lett. B387 (1996) 3728; A. Rašin, hep-ph/9705210.

[71] N. Arkani-Hamed, C. Carone, L. Hall and H. Murayama, Phys. Rev. D54 (1996) 7032; C. Carone, L. Hall and T. Moroi, hep-ph/9705383.

[72] G. 't Hooft, 1979 Cargèse Lectures, published in Recent Developments in Gauge Theories, Plenum Press 1980.
[73] C. Csáki, M. Schmaltz and W. Skiba, Phys. Rev. Lett. 78 (1997) 799, Phys. Rev. D55 (1997) 7840.
[74] M. Strassler, Phys. Lett. 376B (1996) 119; A. Nelson and M. Strassler, hep-ph/9607362.
[75] M.A. Luty and R.N. Mohapatra, Phys. Lett. B396 (1997) 161; N. Arkani-Hamed, M.A. Luty and J. Terning, hep-ph/9712389.
[76] D. Kaplan, F. Lepeintre and M. Schmaltz, Phys. Rev. D56 (1997) 7193.
[77] Y. Grossman, Y. Nir and R. Rattazzi, hep-ph/9701231 and references therein.

New Directions in Quantum Field Theory

David Kutasov (kutasov@yukawa.uchicago.edu)

Weizmann Institute, Israel

Abstract. I briefly review some of the new insights into the dynamics of supersymmetric gauge theories in different dimensions that were obtained recently by realizing them as low energy theories on branes.

1 Introduction

In the last few years important progress was made in the study of the strongly coupled dynamics of supersymmetric gauge theories. New understanding of the constraints due to supersymmetry (which leads to holomorphicity of some quantities), the importance of solitonic objects and electric-magnetic duality led to many exact results on the vacuum structure of these theories (see *e.g.* [1, 2, 3, 4, 5] for reviews).

Many of the qualitative new phenomena that were discovered came as a surprise and even in hindsight have not found a satisfactory conceptual explanation in the framework of gauge theory. A few example are:

- The low energy coupling matrix of $N = 2$ SUSY gauge theory in $3 + 1$ dimensions has been shown by Seiberg and Witten [6] to be the period matrix of a two dimensional Riemann surface, whose physical role seemed rather mysterious.
- Seiberg [7] has shown that *different* $N = 1$ SUSY gauge theories in $3 + 1$ dimensions sometimes have the *same* low energy behavior. Following his work numerous examples of this phenomenon were found in other $N = 1$ SUSY theories, but the underlying reason for its existence seemed mysterious.
- Intriligator and Seiberg [8] have found examples of different $N = 4$ SUSY gauge theories in $2 + 1$ dimensions with the same low energy behavior. Again, the underlying mechanism for this duality was not clear in the context of gauge theory.

In the last year, the results mentioned above (and many others that were not mentioned) were significantly clarified by embedding the gauge theories in question in string theory. String theory of course has vastly more degrees of freedom than field theory but most of these are irrelevant at low energies. The conventional view in the past was that to study low energy gauge dynamics, it is sufficient to retain the light (*i.e.* gauge theory) degrees of freedom and therefore string theory can not shed light on issues having to do with strong coupling at long distances.

Recent work suggests that while most of the degrees of freedom of string theory are indeed irrelevant for understanding low energy dynamics, there is a sector of the theory that is significantly larger than the gauge theory in question that should be kept to understand the low energy structure. This sector involves degrees of freedom living on branes and describing their internal fluctuations and embedding in spacetime.

In this note I will briefly review these developments, focusing on our own work [9, 10] on $N = 1$ gauge theories in $3 + 1$ dimensions. A much more detailed account on the interplay of brane dynamics and supersymmetric gauge theories can be found in [11], which also includes an extensive list of references to the rather rich recent literature on the subject.

2 Branes and the Nahm Construction of Monopoles

Starting from the fact that a fundamental string can end on a D-brane (which is one way of defining D-branes [12]) and using U-duality, one can deduce that a Dp-brane can end on a $D(p+2)$-brane or on an $NS5$-brane. The ground state of such a configuration preserves 1/4 of the 32 supercharges of type II string theory (or M-theory).

Similarly, the fact that fundamental strings can be suspended between Dp-branes [12] implies that Dp-branes can be suspended between $D(p+2)$-branes and/or $NS5$-branes. These simple observations led recently to very useful embeddings of gauge dynamics in string theory.

Consider as an example a configuration of two parallel $D3$-branes stretched in (x^0, x^1, x^2, x^3). At low energies $E << M_p$ the dynamics on the worldvolume of the threebranes is that of $N = 4$ SYM in $3+1$ dimensions with gauge group[1] $SU(2)$, and coupling $g^2_{SYM} = g_s$ [13, 14, 15]. One can decouple the $N = 4$ gauge theory from gravity and massive string modes by sending the IIB string tension $l_s^{-2} \to \infty$ keeping the string coupling g_s (and thus g_{SYM}) fixed.

The separation of the two threebranes in the transverse $(x^4, \cdots, x^9)$ directions parametrizes the six dimensional Coulomb branch of the $N = 4$ SUSY $SU(2)$ gauge theory. Fundamental IIB strings stretched between the two threebranes correspond to the charged gauge bosons $W^{\pm}$; their mass is proportional to the separation (or Higgs v.e.v.).

D-strings stretched between the $D3$-branes correspond to monopoles in the broken $SU(2)$ gauge theory. This can be seen *e.g.* by performing a strong-weak duality transformation $g_s \to 1/g_s$, a symmetry of type IIB string theory which acts as electric-magnetic duality on the threebrane worldvolume gauge theory, and exchanges the fundamental and D-strings [14, 15].

[1] Actually, the gauge group is $U(2)$, but none of the fields we will consider are charged under the diagonal $U(1) \subset U(2)$, so we can ignore it.

Thus, to describe the moduli space of k monopoles in $SU(2)$ gauge theory, $\mathcal{M}_k$, one is instructed to study the space of configurations of k parallel D-strings stretched between the two $D3$-branes [16]. When viewed from the point of view of the D-strings this space can be thought of as the moduli space of vacua of the non-abelian $1+1$ dimensional $U(k)$ gauge theory on the k stretched D-strings.

Taking the two $D3$-branes to be separated in the x^6 direction, the theory on the D-string lives in the (1+1) dimensions (x^0, x^6). The spatial dimension, x^6, is confined to a finite line segment (say) $-x \leq x_6 \leq x$; hence the theory reduces at low energies to Supersymmetric Quantum Mechanics (SQM). Of course, SQM does not have a moduli space of vacua, but there is an approximate Born-Oppenheimer notion of a space of vacua, which arises after integrating out all the fast modes of the system. The low energy dynamics is described by a sigma model on the moduli space $\mathcal{M}_k$.

The theory on the D-strings has eight supercharges and the following matter content. The $U(k)$ gauge field A_0 and five adjoint scalars $(X^4, X^5, X^7, X^8, X^9)$ have Dirichlet boundary conditions at $x^6 = \pm x$ (the locations of the two threebranes). The remaining component of the D-string worldvolume gauge field A_6 and the three adjoint scalars (X^1, X^2, X^3) have (formally) Neumann boundary conditions.

To study the dynamics on the worldvolume of the D-string we can set to zero all the fields which satisfy Dirichlet boundary conditions, and the gauge field A_6 (by a gauge choice). From the Lagrangian for X^1, X^2, X^3

$$\mathcal{L} \sim \mathrm{Tr}\Big(\sum_{I=1}^{3} \partial_s X^I \partial_s X^I - \sum_{I,J} [X^I, X^J]^2\Big) \tag{1}$$

(where we have denoted x^6 by s) it is clear that ground states satisfy

$$\partial_s X^I + \frac{1}{2}\epsilon^{IJK}[X^J, X^K] = 0 \tag{2}$$

The boundary conditions of the fields X^I at the edges of the interval $s = \pm x$ are interesting. For well separated D-strings the matrices X^I are diagonal

$$X^I = \mathrm{diag}(x_1^I, \cdots, x_k^I) \tag{3}$$

Such configurations describe k infinitely separated monopoles. The k monopole system is described by non-commuting matrix solutions of (2) satisfying the boundary conditions

$$X^I \sim \frac{T^I}{s - x} \tag{4}$$

as $s \rightarrow x$. The $k \times k$ matrices T^I satisfy

$$[T^I, T^J] = \epsilon_{IJK} T^K \tag{5}$$

and form a k dimensional representation of $SU(2)$.

One can think of (4) as describing a bound state whose size R diverges as $s \to x$ as

$$R^2 = X^I X^I = \frac{T^I T^I}{(s-x)^2} = \frac{(k-1)(k+1)}{4(s-x)^2} \tag{6}$$

$R \simeq k/2(s-x)$. This is consistent with the fact that from the threebrane perspective the D-strings look like magnetic charges which curve the threebrane according to $s - x \simeq k/|\mathbf{X}|$ (see [11] for details).

Equations (2, 4, 5) define the Nahm construction of the moduli space of k $SU(2)$ monopoles [17]. The brane realization provides a new perspective and in particular a physical rationale for Nahm's description of monopoles. It also makes it easy to describe generalizations, *e.g.* to the case of monopoles in (broken) $SU(N_c)$ gauge theory, and clarifies the origin and scope of the "reciprocity" discussed in [18].

3 Branes and Three Dimensional N=4 SUSY Gauge Theory

Additional information about the k monopole moduli space $\mathcal{M}_k$ can be obtained by mapping the configuration of D-strings stretched between $D3$-branes discussed above into other brane configurations, using U-duality. As an example, starting with the configuration of the previous section, compactify (x^4, x^5), T-dualize $R_i \to l_s^2/R_i$ $(i = 4, 5)$, and decompactify the new (x^4, x^5). Under T-duality the $D3/D1$-branes turn into $D5/D3$-branes. It is convenient to employ a further S-duality, $g_s \to 1/g_s$, which turns the $D5$-branes into $NS5$-branes and leaves the $D3$-branes invariant.

The resulting configuration of k $D3$-branes stretched between two parallel $NS5$-branes still describes the moduli space of k monopoles in broken $SU(2)$ gauge theory. As before, it is useful to study the same space from the point of view of the threebrane dynamics. The fact that the threebranes are finite in x^6 implies that their low energy dynamics is $2+1$ dimensional, governed by an $N = 4$ SUSY gauge theory with gauge group $U(k)$ [19]. The moduli space $\mathcal{M}_k$ is the space of vacua of this theory. Thus we learn that two apriori different spaces – the moduli space of vacua of an $N = 4$ SUSY gauge theory in $2+1$ dimensions, and the moduli space of monopoles $\mathcal{M}_k$ – coincide [19].

The construction of $N = 4$ SYM theories in $2+1$ dimensions in terms of $D3$-branes stretched between $NS5$-branes can be significantly generalized and used to study other aspects of these theories [19]. For example, to study $U(N_c)$ $N = 4$ SUSY gauge theory with N_f "flavors" of fundamental hypermultiplets, $Q, \tilde{Q}$, one considers N_c threebranes stretched between two $NS5$-branes in the presence of N_f $D5$-branes.

All the branes are taken to stretch in (x^0, x^1, x^2), the $1+2$ dimensional "spacetime" of the gauge theory. The $NS5$-branes are further extended in (x^4, x^5) and located at different values of x^6, say $x^6 = 0, L_6$ and the same (x^7, x^8, x^9). The $D3$-branes are stretched in x^6 between the two $NS5$-branes;

the $D5$-branes are extended in (x^7, x^8, x^9) and located between the two $NS5$-branes in x^6. It is easy to check that this configuration preserves eight supercharges.

The $U(N_c)$ vector multiplet arises from ground states of 3-3 strings with both ends on the threebranes. The N_f flavors of hypermultiplets arise from 3–5 strings, stretched between the $D3$ and $D5$-branes. The global R-symmetry of $N = 4$ SYM in $2+1$ dimensions, $SU(2)_R \times SU(2)_X$ is realized in the brane construction as part of the unbroken subgroup of the Lorentz group. $SU(2)_R$ is the $SO(3)$ symmetry acting on (x^7, x^8, x^9), while $SU(2)_X$ is the $SO(3)$ acting on (x^3, x^4, x^5). The unbroken supercharges which are spinors of the full $Spin(1,9)$ clearly transform as doublets under the two $SU(2)$'s, as appropriate for R-symmetries.

One interesting feature of this description of the $N = 4$ SYM theory is the geometric realization of the Higgs branch of the theory. In gauge theory it is well known that the above theory has, for $N_f \geq 2N_c$ a Higgs moduli space of vacua at generic points of which the gauge group is completely broken, and the low energy spectrum consists of $2(N_f N_c - N_c^2)$ complex scalar fields (and the fermions required for $N = 4$ SUSY), which can be parametrized by the gauge invariant mesons $M = \tilde{Q}Q$, and baryons $B = Q^{N_c}$, $\tilde{B} = \tilde{Q}^{N_c}$ (subject to various compositeness constraints). There are also mixed Higgs-Coulomb branches at which part of the gauge symmetry remains unbroken.

In the brane description the transition to a Higgs branch is described by the process of breaking $D3$-branes on $D5$-branes. When a $D3$-brane and a $D5$-brane coincide in (x^3, x^4, x^5) the former can break on the latter with each of the two resulting threebranes free to move independently along the $D5$-brane.

Taking this process into account one can describe the Higgs branch of the moduli space of vacua by moving threebranes stretched between $NS5$-branes in (x^3, x^4, x^5) until they encounter $D5$-branes, and allowing the threebranes to break on the $D5$-branes. In gauge theory this corresponds to going to points in the Coulomb branch from which a Higgs branch emanates and entering it.

A threebrane stretched between an $NS5$-brane and a $D5$-brane has no massless fields living on it. Therefore, to have a Higgs branch, at least two $D5$-branes must coincide in (x^3, x^4, x^5) (they are still, generically, at different values of x^6). One can then break a $D3$-brane into segments some of which connect $NS5$ and $D5$-branes and some connnect different $D5$-branes. The latter are free to move in (x^7, x^8, x^9) (and by SUSY have a fourth massless scalar, A_6, living on them) and hence give rise to massless moduli.

Furthermore, it is clear that in the process of breaking we lose vector multiplets and gain hypermultiplets in precisely the way familiar from the Higgs phenomenon in gauge theory. By studying the geometrical deformations of the above brane configuration one finds the known phase structure of $N = 4$ SYM.

The brane description of SYM can also be used to learn about properties of branes in string theory by comparing them to known gauge theory physics. An example is the "s-rule" of [19]: to reproduce the correct moduli space of vacua of $N = 4$ SYM using branes one is forced to postulate that a configuration in which an $NS5$-brane and a $D5$-brane are connected by more than one $D3$-brane is not supersymmetric. This phenomenological rule of brane dynamics is not well understood from "first principles".

One of the main results of [19] is an explanation of Intriliagtor and Seiberg's duality [8] mentioned in the introduction. If one performs an S-duality transformation $S : g_s \to 1/g_s$ on the type IIB string vacuum including the branes described above, $NS5$-branes are exchanged with $D5$-branes and $D3$-branes are invariant.

Vectormultiplets corresponding to $D3$-branes between $NS5$-branes are exchanged with hypermultiplets corresponding to $D3$-branes between $D5$-branes; therefore, the gauge group changes under S. The Coulomb branch of the model is exchanged with the Higgs branch. All this is in agreement with [8].

The string construction furthermore generalizes the results of [8] to an *exact* equivalence between the theories and leads to an efficient tool for generalizing these results to large classes of new theories (see *e.g.* [19, 11] for more details).

4 Four Dimensional N = 2 Models

To study four dimensional gauge theories one can use brane configurations similar to those of [19]. Specifically, [9], one should consider configurations of $D4$-branes stretched between $NS5$-branes in the presence of $D6$-branes in type IIA string theory. The use of such configurations led recently (among other things) to a much better understanding of the physical role of the Riemann surface of [6] which is used to describe the Coulomb branch of $N = 2$ SYM in $3 + 1$ dimensions, and to a much better understanding of Seiberg's duality in $N = 1$ SYM. In this section we will briefly describe this progress; details and many additional results obtained using branes appear in [11].

$N = 2$ SYM with gauge group $SU(N_c)$ and no matter can be described as the worldvolume theory on N_c $D4$-branes with worldvolume $(x^0, x^1, x^2, x^3, x^6)$ stretched in x^6 between two $NS5$-branes with worldvolume $(x^0, x^1, \cdots, x^5)$. The gauge coupling is

$$\frac{1}{g_{SYM}^2} = \frac{L_6}{g_s l_s} = \frac{L_6}{R_{10}} \tag{7}$$

where L_6 is the distance between the $NS5$-branes and R_{10} is the radius of the eleventh dimension that appears at finite coupling in IIA string theory (or M-theory). To study gauge theory dynamics one should take the limit

$L_6, R_{10} \to 0$ with fixed g_{SYM} (7), which decouples gravity and massive string modes.

The classical picture of a $D4$-brane ending on $NS5$-branes receives important corrections for finite R_{10}/L_6 [20]. First, the end of the fourbrane looks like a charge in the $NS5$-brane worldvolume theory. To preserve SUSY the fivebrane must curve in response to it. Denoting

$$\begin{aligned} s &= x^6 + ix^{10} \\ v &= x^4 + ix^5 \end{aligned} \tag{8}$$

one finds that the classical $NS5$-brane at $s = $ const stretched in v is (asymptotically) curved by q_L fourbranes ending on it from the left at $v = a_1, \cdots, a_{q_L}$, and q_R fourbranes ending on if from the right at $v = b_1, \cdots, b_{q_R}$ according to:

$$s = R_{10} \sum_{i=1}^{q_L} \log(v - a_i) - R_{10} \sum_{i=1}^{q_R} \log(v - b_i) \tag{9}$$

Second, the $D4$-brane is reinterpreted for finite R_{10} as an M-theory fivebrane ($M5$-brane) wrapped around the x^{10} circle. Thus, in M-theory on $R^{1,9} \times S^1$ the configuration of N_c $D4$-branes stretched between two $NS5$-branes corresponds to a single $M5$-brane with a convoluted worldvolume, $R^{1,3} \times \Sigma$, where $R^{1,3}$ corresponds to the spacetime shared by all the branes (labeled by (x^0, x^1, x^2, x^3)) and Σ is a Riemann surface embedded in the four dimensional space $Q \simeq R^3 \times S^1$ labeled by (s, v).

Vacua of $N = 2$ SYM correspond to surfaces Σ which preserve SUSY having the right global structure and asymptotic boundary conditions (9). That means that they are described by the holomorphic equation

$$t^2 + B(v)t + 1 = 0 \tag{10}$$

where we have set the QCD scale to one, $t = \exp(-s/R_{10})$ and

$$B(v) = v^{N_c} + u_2 v^{N_c - 2} + \cdots + u_{N_c} \tag{11}$$

For fixed t, (10) has N_c roots v_i corresponding to the N_c "fourbranes" whose position is determined by the values of the moduli $\{u_i\}$. Note that while classically there should only be such solutions for t between the $NS5$-branes, because of the bending (9) there are in fact N_c solutions for v for any $t \neq 0$.

Similarly, for all v there are two solutions for t, which for large v behave like

$$t_\pm \simeq v^{\pm N_c} \tag{12}$$

in agreement with the general structure expected from (9).

Interestingly, (10) is identical to the Seiberg-Witten curve for pure $SU(N_c)$ SYM theory [6, 21, 22]. Its period matrix describes the low energy $U(1)^{N_c - 1}$ coupling matrix and, by $N = 2$ SUSY, also the metric on the moduli space of vacua.

We see that in brane theory the Riemann surface Σ becomes physical and low energy SYM theory arises by compactifucation on Σ of a six dimensional theory – the $M5$-brane worldvolume theory. It should be noted that a similar description of Seiberg-Witten theory first appeared in [23]. It can also be used for studying the spectrum of BPS states which correspond to membranes stretched between different parts of the $M5$-brane.

5 Four Dimensional N = 1 Models

$N = 1$ SYM is obtained by changing the orientation of one of the $NS5$-branes [9]. An example is $N = 1$ SQCD, a theory with gauge group $G_e = SU(N_c)$ and N_f flavors of quarks Q_i, $\tilde{Q}^i$ $i = 1, \cdots, N_f$.

It is obtained by stretching N_c $D4$-branes with worldvolume $(x^0, x^1, x^2, x^3, x^6)$ between an $NS5$-brane stretched in $(x^0, x^1, x^2, x^3, x^4, x^5)$ and a rotated $NS5$-brane, that one may denote by $NS5'$-brane, stretched in $(x^0, x^1, x^2, x^3, x^8, x^9)$, in the presence of N_f $D6$-branes stretched in $(x^0, x^1, x^2, x^3, x^7, x^8, x^9)$. For concreteness, we will take the $NS5$-brane to be to the left of the $NS5'$-brane and the $D6$-branes (in x^6). It is easy to see that this configuration preserves $N = 1$ SUSY in the $3+1$ dimensions (x^0, x^1, x^2, x^3) shared by all the branes.

One can repeat the analysis described for the $N = 2$ case in this example [24, 25, 26] to study the space of vacua of the theory. This has been discussed by other speakers at this conference.

Alternatively, one may ask whether the brane picture sheds any light on Seiberg's duality [7], which in this case is the claim that the low energy behavior of the above theory (for $N_f > N_c$) is the same as that of a different theory with gauge group $G_m = SU(N_f - N_c)$, N_f "magnetic quarks" q^i, $\tilde{q}_i$, and a singlet meson M^i_j which couples to the magnetic quarks via the cubic superpotential

$$W = M^i_j \tilde{q}_i q^j \tag{13}$$

The singlet mesons M should be thought of as representing the "electric mesons" $\tilde{Q}^i Q_j$ in the magnetic theory.

In the brane description, one finds [9] that the electric and magnetic theories provide different parametrizations of the same quantum moduli space of vacua and therefore Seiberg's duality is manifest!

To see that, it is convenient to proceed in two steps. Consider first the classical limit $R_{10} \to 0$ (and therefore (7) $g_{SYM} \to 0$). Seiberg's duality [7] is a quantum symmetry, but it has classical consequences in situations where the gauge symmetry is completely broken and there is no strong infrared dynamics. In such situations Seiberg's duality reduces to a classical equivalence of Higgs branches and their deformations.

To exhibit this relation using branes start, for example, with the electric theory with gauge group $SU(N_c)$. As we learned above, higgsing corresponds to breaking $D4$-branes on $D6$-branes. It is convenient before entering the Higgs branch to move the $NS5$-brane to the right through the N_f D sixbranes.

This is a smooth process which does not influence the low energy physics on the fourbranes [19]. When the $NS5$-brane passes through the N_f $D6$-branes, it generates N_f $D4$-branes connecting it to the $D6$-branes; as before, these new $D4$-branes are rigid.

We now enter the Higgs phase by reconnecting the N_c original fourbranes to N_c of the N_f new fourbranes created previously; we then further reconnect the resulting fourbranes in the most general way consistent with the rules described above. The resulting moduli space is $2N_fN_c - N_c^2$ complex dimensional, in agreement with the gauge theory analysis[2]. Note that, generically, there are now $N_f - N_c$ $D4$-branes attached to the $NS5$-branes, and N_c $D4$-branes connected to the $NS5'$-branes(the other ends of all these fourbranes lie on different $D6$-branes).

Once we are in the Higgs phase, we can freely move the $NS5$-brane relative to the $NS5'$-brane, and in particular the two branes can pass each other in the x^6 direction without ever meeting in space. This can be achieved by taking the NS fivebrane around the $NS5'$-brane in the x^7 direction.

In the gauge theory this relative motion can be described by weakly gauging the $U(1)$ symmetry corresponding to baryon number and turning on a Fayet-Iliopoulos D-term for that $U(1)$. At a generic point in the Higgs branch of the electric theory, turning on such a D-term is a completely smooth procedure; this is particularly clear from the brane description, where in the absence of $D4$-branes connecting the $NS5$-brane to the $NS5'$-brane, the relative displacement of the two in the x^7 direction can be varied freely.

After exchanging the $NS5$ and $NS5'$-branes, the brane configuration we find can be interpreted as describing the Higgs phase of *another* gauge theory. To find out what that theory is, we approach the root of the Higgs branch by aligning the $N_f - N_c$ $D4$-branes emanating from the $NS5$-brane with the $NS5'$-brane, and the N_c $D4$-branes emanating from the $NS5'$-brane with $D4$-branes stretched between $D6$-branes.

We then reconnect the $D4$-branes to obtain a configuration consisting of $N_f - N_c$ $D4$-branes connecting the $NS5'$-brane to the $NS5$-brane which is to its left; the $NS5'$-brane is further connected by N_f $D4$-branes to the N_f $D6$-branes which are to its right. This is the brane description of the magnetic SQCD with gauge group $SU(N_f - N_c)$ described above. The magnetic mesons M^i_j correspond to $4-4$ strings describing fluctuations of the $D4$-branes connecting the $NS5'$-brane to the $D6$-branes. These fourbranes can slide along the common directions to the two branes, (x^8, x^9), and the magnetic mesons are the light degrees of freedom describing these motions.

To summarize, we have shown that the moduli space of vacua of the electric SQCD theory with (completely broken) gauge group $SU(N_c)$ and N_f flavors of quarks Q^i, $\tilde{Q}_i$, and the moduli space of vacua of the magnetic

---

[2] Actually, the gauge theory moduli space is $2N_fN_c - N_c^2 + 1$ dimensional; the extra complex dimension corresponds to the baryonic branch and does not appear to be geometrically realized in the brane construction. It will be ignored below.

SQCD model with (broken) gauge group $SU(N_f - N_c)$, can be thought of as providing different descriptions of a single moduli space of supersymmetric brane configurations. One can smoothly interpolate between them by varying the "QCD scale" Λ (or equivalently the microscopic gauge coupling), keeping the FI D-term fixed but non-zero.

Since the only role of Λ in the low energy theory is to normalize the operators [27], theories with different values of Λ are equivalent. The electric and magnetic theories will thus share all features, such as the structure of the chiral ring (which can be thought of as the ring of functions on moduli space), that are independent of the interpolation parameter Λ.

The above smooth interpolation relies on the fact that the gauge symmetry is completely broken, due to the presence of the FI D-term. As mentioned above, it is not surprising that duality appears classically in this situation since there is no strong infrared gauge dynamics.

The next step is to analyze what happens as the gauge symmetry is restored when the D-term goes to zero and we approach the origin of moduli space. Classically, we find a disagreement. In the electric theory, it is easy to see that nothing special happens when the gauge symmetry is restored. New massless degrees of freedom appear, but there are no new branches of the moduli space that one gains access to.

In the magnetic theory the situation is different. When we set the FI D-term to zero, a large moduli space of previously inaccessible vacua becomes available. While the electric theory has a $2N_f N_c - N_c^2$ dimensional smooth moduli space, the classical magnetic theory experiences a jump in the dimension of its moduli space from $2N_f N_c - N_c^2$ for non-vanishing FI D-term, to N_f^2 when the D-term is zero.

These correspond to giving vanishing expectation values to the magnetic quarks q, $\tilde{q}$ and turning on arbitrary expectation values to the singlet mesons M. Clearly, the $SU(N_f - N_c)$ gauge symmetry is restored, and to understand what really happens we must study the quantum dynamics.

The quantum corrections to the moduli space are described in brane language in [10], where it is found that different $D4$-branes ending on the same $NS5$-brane sometimes attract or repel each other. These forces between branes are not well understood from first principles (just like the s-rule mentioned above); we refer the reader to [10] for a more detailed discussion.

Taking them into account one finds that the quantum effects eliminate the discontinuous jump in the magnetic moduli space as the FI D-term is tuned to zero. Quantum mechanically one does not have access to the full N_f^2 dimensional classical moduli space, but to a $2N_f N_c - N_c^2$ dimensional subspace thereof – precisely the subspace that connects smoothly to the electric moduli space via the brane construction!

6 Discussion

The point of view on gauge theory suggested by the brane constructions described above is potentially important beyond the particular applications discussed in the literature so far.

Many of the results of the last few years seem to suggest the need for a new description of gauge theory, in which gauge invariance and the formulation in terms of interacting gauge bosons and quarks, are less fundamental.

The embedding of gauge theories in brane dynamics provides an attractive scenario to achieve that. In the brane description, gauge theory is seen to arise as an effective collective theory that is useful in some region in the moduli space of vacua. Different descriptions are useful in different regions of moduli space, but the underlying degrees of freedom are always the same – branes in string theory.

Relations between different gauge theories such as Seiberg's duality, and even relations between gauge theories in different dimensions with different amounts of SUSY, which are hidden in the conventional description, become manifest in brane theory. In fact all the phenomena discovered in gauge theory in the last few years have been exhibited in string theory using branes.

Thus brane dynamics seems to, at the very least, provide a uniform and powerful geometrical description of an astonishingly diverse set of gauge theory phenomena and point to hidden relations between them. It is possible that the recent success actually implies that formulating gauge theory using brane degrees of freedom will allow to go beyond the holomorphic, vacuum sector, and solve for the mass spectrum and other features of strongly coupled gauge theories.

To proceed it is necessary to understand better the theory on the $M5$-brane in different limits and for different compactifications. There are tantalizing analogies between the brane constructions of SYM theories and non-critical superstrings, and one possible avenue for progress is to construct the QCD string, based on a better understanding of the fivebrane theory in the large N_c limit.

Other directions which are relatively unexplored at this time are chiral models, in particular models that break SUSY dynamically (see *e.g.* [28]), and theories with explicitly broken SUSY (see *e.g.* [25]).

Acknowledgements

I thank the organizers for the invitation to speak. It is a pleasure to thank my collaborators S. Elitzur, A. Giveon, E. Rabinovici and A. Schwimmer for numerous discussions. This work is supported in part by a DOE OJI grant.

References

[1] N. Seiberg, hep-th/9408013; hep-th/9506077.
[2] K. Intriligator, N. Seiberg, hep-th/9509066.
[3] A. Bilal, hep-th/9601007.
[4] W. Lerche, hep-th/9611190.
[5] M. Shifman, hep-th/9704114, Prog. Part. Nucl. Phys. **39** (1997) 1.
[6] N. Seiberg and E. Witten, hep-th/9407087, Nucl. Phys. **B426** (1994) 19; hep-th/9408099, Nucl. Phys. **B431** (1994) 484.
[7] N. Seiberg, hep-th/9411149, Nucl. Phys. **B435** (1995) 129.
[8] K. Intriligator, N. Seiberg, hep-th/9607207, Phys. Lett. **387B** (1996) 513.
[9] S. Elitzur, A. Giveon and D. Kutasov, hep-th/9702014, Phys. Lett. **400B** (1997) 269.
[10] S. Elitzur, A. Giveon, D. Kutasov, E. Rabinovici and A. Schwimmer, hep-th/9704104.
[11] A. Giveon and D. Kutasov, to appear.
[12] J. Polchinski, hep-th/9611050, 1996 TASI lectures.
[13] E. Witten, hep-th/9510135, Nucl. Phys. **B460** (1996) 335.
[14] A. Tseytlin, hep-th/9602064, Nucl. Phys. **B469** (1996) 51.
[15] M. Green and M. Gutperle, hep-th/9602077, Phys. Lett. **377B** (1996) 28.
[16] D. Diaconescu, hep-th/9608163.
[17] W. Nahm, Phys. Lett. **90B** (1980) 413.
[18] E. Corrigan and P. Goddard, Ann. Phys. **154** (1984) 253.
[19] A. Hanany and E. Witten, hep-th/9611230, Nucl. Phys. **B492** (1997) 152.
[20] E. Witten, hep-th/9703166, Nucl. Phys. **B500** (1997) 3.
[21] P. Argyres and A. Faraggi, hep-th/9411057, Phys. Rev. Lett. **74** (1995) 3931.
[22] A. Klemm, W. Lerche, S. Theisen and S. Yankielowicz, hep-th/9411048, Phys. Lett. **344B** (1995) 169.
[23] A. Klemm, W. Lerche, P. Mayr, C.Vafa and N. Warner, hep-th/9604034, Nucl. Phys. **B477** (1996) 746.
[24] K. Hori, H. Ooguri and Y. Oz, hep-th/9706082.
[25] E. Witten, hep-th/9706109.
[26] A. Brandhuber, N. Itzhaki, V. Kaplunovsky, J. Sonnenschein and S. Yankielowicz, hep-th/9706127.
[27] D. Kutasov, A. Schwimmer and N. Seiberg, hep-th/9510222, Nucl. Phys. **B459** (1996) 455.
[28] J. Lykken, E. Poppitz and S. Trivedi, hep-th/9708134.

Neutrino Masses and Oscillations

Masayuki Nakahata (nakahata@icrr.u-tokyo.ac.jp)

Institute for Cosmic Ray Research, University of Tokyo, Japan

Abstract. New experimental results on neutrino masses and oscillations are reviewed. Notable result is the atmospheric neutrino data from Super-Kamiokande, which shows a strong distortion in angular distributions. The result gives a hint of neutrino oscillations. Other subjects are neutrino mass experiments, accelerator-based neutrino oscillation experiments, solar neutrinos and future neutrino experiments.

1 Introduction

Currently unresolved issues about neutrinos are masses and oscillations. If neutrinos have masses, it gives a strong clue for physics beyond the standard model. Furthermore, neutrinos may have played an important role in astrophysics through dark matter.

Direct searches for neutrino masses are performed by Tritium beta decay experiments for $\overline{\nu}_e$, $\pi^+ \rightarrow \mu^+\nu_\mu$ decay for ν_μ, and τ decay measurements for ν_τ. Double beta decay gives an information of Majorana neutrino masses.

Although neutrino oscillation experiments are sensitive to only mass differences of neutrino species, much smaller mass ranges can be investigated than direct search experiments. A probability of neutrino oscillations is given by the following formula:

$$P(\nu_\alpha \rightarrow \nu_\beta) = sin^2 2\theta sin^2(1.27\frac{L}{E}\Delta m^2), \tag{1}$$

where θ is the mixing angle of neutrinos, L is the distance from a neutrino source to a detector in unit of kilometer, E is the energy of neutrinos in unit of GeV, and Δm^2 is the difference of mass squares in unit of eV^2. As seen in the equation, one needs longer distance and smaller energy of neutrinos for investigating small Δm^2 . The neutrino oscillation experiments with a baseline of sun-to-earth or earth-size, namely solar neutrino and atmospheric neutrino experiments, are best suited for searching for small Δm^2 . To search for small mixing angle, one needs high statistics of neutrino events, which are available at accelerator-based experiments. The MSW mechanism[1] of solar neutrino oscillations has given a tool to investigate small mixing angle for $\Delta m^2 = 10^{-7} - 10^{-4}$ eV^2 range.

In this report, recent results on neutrino masses and oscillation experiments are described.

2 Neutrino Mass Experiments

A finite mass of $\overline{\nu}_e$ should bend down the Tritium beta decay spectrum in the vicinity of the end-point. Results of Tritium experiments are summarized in Table 1. As seen in the table, all experiments give negative value of m_ν^2. The

Experiment	m_ν^2	m_ν limit
INS Tokyo 91	$-65 \pm 85 \pm 65\text{eV}^2$	13.1 eV
LANL91	$-147 \pm 68 \pm 41\text{eV}^2$	9.3 eV
Zürich 92	$-24 \pm 48 \pm 61\text{eV}^2$	11.7 eV
China 93	$-31 \pm 75 \pm 48\text{eV}^2$	12.4 eV
Livermore 95	$-130 \pm 20 \pm 15\text{eV}^2$	7 eV
Mainz 94	$-22 \pm 17 \pm 14\text{eV}^2$	5.6 eV
Troitsk 94	$-22 \pm 4.8\text{eV}^2$	4.35 eV
Troitsk 94+96	$-1 \pm 6.3\text{eV}^2$	3.5 eV
PDG (96)	$-27 \pm 20\text{eV}^2$	

Table 1. Results of Tritium neutrino mass experiments.[2] The fitted m_ν^2 and 95 % C.L. limits are shown. The results of "Troitsk 94+96" are obtained accounting for the anomaly.

first five experiments in Table 1 have used magnetic spectrometers. Recent experiments, TROITSK[3] and MAINZ[4], use integral electrostatic analyzers, which have achieved an energy resolution of ~6 eV. The negative value of m_ν^2 is due to an anomalous excess of the event rate in the vicinity of the end point. TROITSK fitted the observed spectrum assuming an additional spike-like local enhancement near the end-point[3] and obtained m_ν^2 of -1±6.3 eV^2.

In this conference, MAINZ experiment has presented the status of 1997 data taking.[4] The main source of the background in the previous data taking was back-scattered electrons from the source. The background was reduced by putting the T_2 source further upstream of spectrometer and transporting electrons through new guiding magnets. The reduction of the background enable us to increase the source intensity. Thus, the 1997 data was taken between June 23 and August 18 with an increased intensity by a factor of 6, whereas the background level is reduced by 40 % compared with the 1995 data. The resolution of m^2 is expected to be ~6 eV^2 only using the last 70 eV range below the end-point. Results of 1997 data of MAINZ will be published soon.

ALEPH and DELPHI presented tau neutrino mass limits in this conference.[5],[6] ALEPH analyzed $5\pi^{\pm}(\pi^0)\nu_\tau$ and $3\pi^{\pm}\nu_\tau$ decay modes.[5] The two-dimensional

3 Atmospheric Neutrinos

Atmospheric neutrinos are the decay products of hadronic showers produced by primary cosmic ray interactions in the atmosphere. In recent years, the double-ratio of neutrino events classified by lepton types, $R \equiv (\mu/e)_{DATA}/(\mu/e)_{MC}$, has been studied to approximate the atmospheric neutrino flavor ratio ν_μ/ν_e and to cancel uncertainties in the neutrino flux and cross sections. The measurements of R in underground experiments are summarized in Table 3.

Experiment	exposure (kt·yr)	events	R
NUSEX	0.74	50	$0.96\ ^{+0.32}_{-0.28}$
Frejus	2.0	200	$1.00 \pm 0.15 \pm 0.08$
Kamiokande sub-GeV	7.7	482	$0.60\ ^{+0.06}_{-0.05} \pm 0.05$
Kamiokande multi-GeV	6/8.2	233	$0.57\ ^{+0.08}_{-0.07} \pm 0.07$
IMB	7.7	610	$0.54 \pm 0.05 \pm 0.11$
Soudan-II	2.83	331	$0.61 \pm 0.14\ ^{+0.05}_{-0.07}$
Super-K sub-GeV	20.1	1453	$0.64 \pm 0.04 \pm 0.05$
Super-K multi-GeV	20.1	444	$0.60 \pm 0.05 \pm 0.05$

Table 3. Measurements of R in atmospheric neutrino experiments.[9]

The results of NUSEX and Frejus gave R close to one. Results of Kamiokande, IMB, Soudan-II and Super-Kamiokande gave significantly small R value. Especially, the high statistical data of Super-Kamiokande re-confirmed the atmospheric neutrino anomaly.

The Super-Kamiokande is a huge water Cherenkov detector located in Kamioka mine in Japan. The detector consists of inner and outer detectors. In the inner-detector, 11,146 photomultiplier tubes(PMT's), each 20 inch in diameter, are uniformly placed facing inward on a 0.707 m grid on the entire surface with dimensions 33.8 m in diameter by 36.2 m high, which contains 32,000 metric tons of water. The total photocathod surface area of all PMT's is 40 % of the surface of the inner-detector. A 4π solid-angle outer-detector surrounds the inner-detector. The outer-detector is also a water Cherenkov counter with 1,885 sets consisting of a wavelength shifter plate and an 8 inch PMT. The thickness of the outer-detector is 2 meters. The outer-detector is designed to reduce external gamma-rays from surrounding rocks and tag cosmic-ray muons. The fiducial volume of the detector is defined to be more than 2 m from the detector wall, which amounts to 22 ktons of water. The data taking of Super-Kamiokande was started in April 1996. Data of 20.1

kt·yr was accumulated between May 1996 and June 1997. The atmospheric neutrino events are classified into fully contained (FC) events and partially contained (PC) events. The FC events are selected by requiring no hits in outer-detector. The PC events are selected essentially by requiring single outer-detector hit-cluster close to an exit point of the particles observed in the inner-detector. The vertex positions of PC events are required to be in the fiducial volume. In total, 2591 FC events and 156 PC events were observed in the 20.1 kt·yr data sample. The FC and PC events are sub-divided into sub-GeV and multi-GeV events. The selection of sub-GeV events in the FC sample is that (1) visible energy (E_{vis}) is less than 1.33 GeV, (2) momentum is greater than 100 MeV/c for e-like events and 200 MeV/c for μ-like, and (3) maximum pulse height of a single PMT is less than 200 p.e., which rejects events with a particle stopping very close to the detector wall. The multi-GeV events are required to be $E_{vis} > 1.33$ GeV. The number of sub-GeV and multi-GeV events are 1986 and 605 in the FC sample. Then, the events are analyzed through Cherenkov ring number counting program and particle identification program. The misidentification probability of the μ/e separation is estimated to be 1% and 2% in the sub-GeV and multi-GeV samples, respectively. The observed number of single ring e-like and μ-like events are 718 (149) and 735 (139) in sub (multi)-GeV data sample. The all PC events are assumed to be μ-like events. The observed events are compared with a Monte Carlo simulation which uses the flux calculation by Honda et al.[10] and a neutrino interaction program based on accelerator neutrino experiments.[11] Nuclear effects, Pauli principle and a mass difference between μ and e are also taken into account in the Monte Carlo simulation. The results of R obtained by using single ring events are shown in Table 3 together with other experiments. The obtained R is significantly smaller than unity and it confirms the results of Kamiokande and IMB results. The zenith angle distributions of observed e-like and μ-like events are shown in Figure 1. As seen in the figure, the upward-going μ-like events are significantly smaller than the MC expectations. It is seen both in the sub-GeV and multi-GeV samples. The distortion of the zenith angle distribution strongly indicates anomaly in the atmospheric neutrinos, because the systematic error of up/down asymmetry of the detector is negligibly small and the shape of the angular distribution cancels various uncertainties in neutrino cross sections, nuclear effects and etc. Possible solution of the anomaly is neutrino oscillations. Using the zenith angle information together with energy information of events, an allowed region of neutrino oscillation parameters is obtained as Fig.2. The overlapped allowed regions of Super-Kamiokande and Kamiokande is around $\Delta m^2 = 5 \times 10^{-5}$ eV^2 with large mixing angle. The dotted histograms in Fig.1 show expected zenith angle distributions assuming neutrino oscillations of $\nu_\mu \leftrightarrow \nu_\tau$ with parameters of $\Delta m^2 = 5 \times 10^{-5}$ eV^2 and $\sin^2 2\theta = 1.0$. The neutrino oscillation hypothesis reproduces the observed zenith angle distributions quite well.

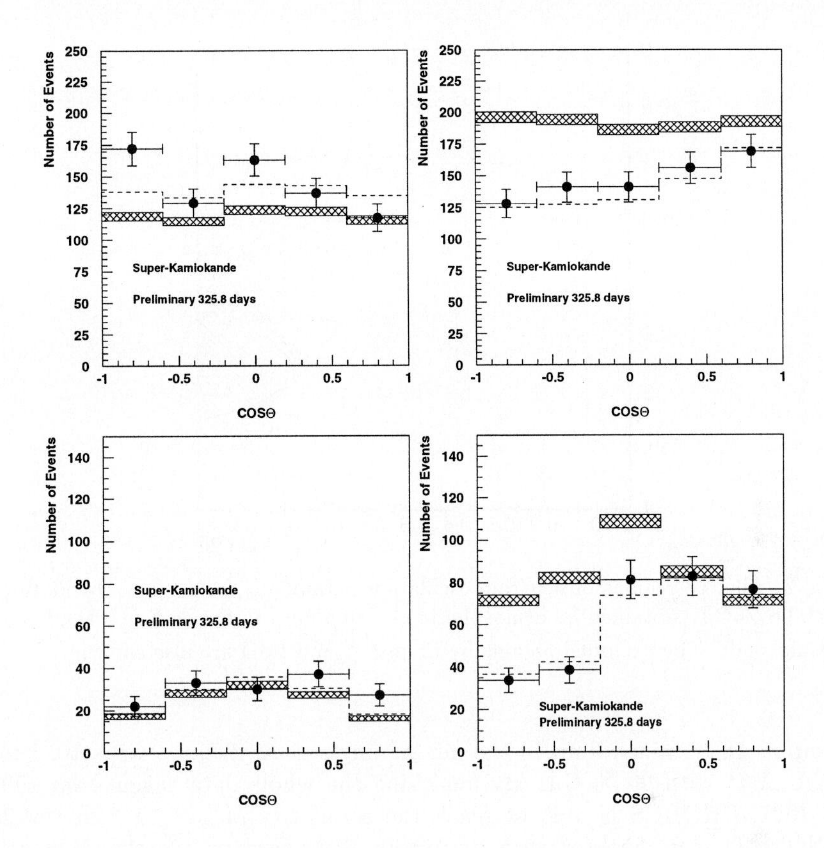

Fig. 1. Zenith angle distributions of e-like and μ-like atmospheric neutrino data in Super-Kamiokande. Upper and lower figures are for sub-GeV and multi-GeV data samples, respectively. $cos(\theta)$=1 corresponds to downward-going direction. The hatched bands show MC simulation and dotted histograms show the ones assuming $\nu_\mu \leftrightarrow \nu_\tau$ oscillations with $\Delta m^2 = 5 \times 10^{-5}$ eV^2 and $\sin^2 2\theta = 1.0$.

4 Accelerator neutrino oscillation experiments

The detectors located in the CERN wide band neutrino beamline, CHORUS and NOMAD, had presented their preliminary results in this conference.[14],[15] The salient feature of the CHORUS detector is the 0.8 tons nuclear emulsion target, which enable us to identify τ neutrinos unambiguously. CHORUS searched for $\nu_\mu \to \nu_\tau$ oscillations using charged current (CC) interactions of ν_τ. 68 % of the τ decay in the nuclear emulsion shows a kink topology through the decay channels of $\tau^- \to \mu^- \nu_\tau \overline{\nu}_\mu$ (1μ) and $\tau^- \to h^-$+neutrals (0μ). 73 % (32 %) of the data taken in 1994 were analyzed for $1\mu(0\mu)$ decay modes and no kink candidate which have more than 250 MeV/c transverse momentum was found. Estimated background is 0.15 events in the analyzed

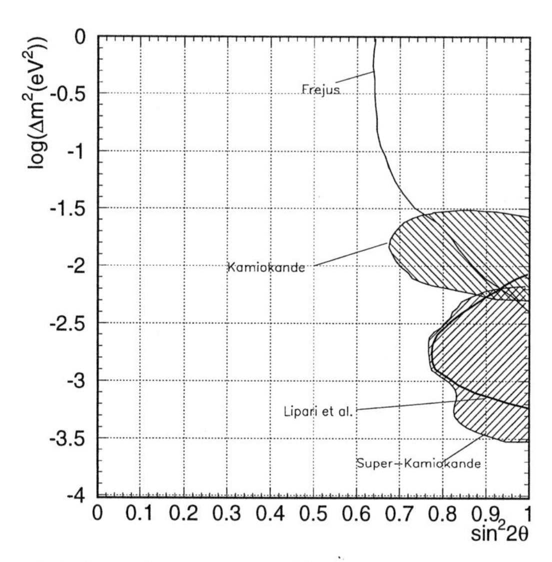

Fig. 2. Allowed regions of neutrino oscillation parameters for $\nu_\mu \leftrightarrow \nu_\tau$ oscillations with 90 % C.L. obtained by atmospheric neutrino data in Super-Kamiokande and Kamiokande. The excluded regions by Frejus[12] and IMB are also shown.

sample. An obtained limit in the mixing angle is $sin^2 2\theta_{\mu\tau} < 4.5 \times 10^{-3}$ for large Δm^2 with 90 % C.L. By analyzing the whole data taken from 1994 to 1997, CHORUS is able to reach the sensitivity of 2×10^{-4} in $\sin^2 2\theta$. NOMAD is a detector with extremely good electron identification and charge separation. NOMAD searched for $\nu_\mu \rightarrow \nu_\tau$ oscillations using decay modes of $\tau^- \rightarrow \mu^- \nu_\tau \overline{\nu}_\mu$, $\tau^- \rightarrow e^- \nu_\tau \overline{\nu}_e$, $\tau^- \rightarrow \pi^-(K^-)\nu_\tau$, $\tau^- \rightarrow \rho^- \nu_\tau$ and $\tau^- \rightarrow \pi^-\pi^-\pi^+(n\pi^0)\nu_\tau$. No candidate was found in the 1995 data and an upper limit of $sin^2 2\theta_{\mu\tau} < 3.4 \times 10^{-3}$ (90 % C.L.) was obtained. $\nu_\mu \rightarrow \nu_e$ oscillations were also analyzed in NOMAD using the energy dependence of the ratio of charged current events of ν_e and ν_μ ($R_{e\mu}(E_\nu)$). The observed $R_{e\mu}(E_\nu)$ distribution is consistent with the estimated ν_e contamination of the beam. The result excluded the region of $\Delta m^2 > 10$ eV^2 in the LSND allowed region.

LSND group had published a search for $\overline{\nu}_\mu \rightarrow \overline{\nu}_e$ oscillations using a $\overline{\nu}_\mu$ beam decay at rest (DAR) as an evidence of neutrino oscillations.[16] The number of selected events with signatures of $\overline{\nu}_e$ appearance is 22 events in the energy range between 36 MeV and 60 MeV, whereas the estimated background is 4.6 events. Recently, LSND analyzed also $\nu_\mu \rightarrow \nu_e$ oscillations using ν_μ from π^+ decay in flight (DIF) and an excess of the ν_e events was observed.[17] The neutrino oscillation probabilities given by LSND are $(3.1\pm1.2\pm0.5)\times10^{-3}$ and $(2.6\pm1.0\pm0.5)\times10^{-3}$ for $\overline{\nu}_\mu \to \overline{\nu}_e$ and $\nu_\mu \to \nu_e$ analyses. KARMEN had analyzed both oscillation modes using the data till 1995.[18] Although KARMEN had observed no evidence of the neutrino oscillations, the sensitivity of the detector was not good enough to discuss whole LSND allowed oscillation parameter regions because of cosmic-ray background. In 1996, KARMEN built an additional active veto layer and reduced the cosmic-ray background by a factor of ~40. KARMEN will cover the whole LSND allowed region in 2 - 3 years.

5 Solar neutrinos

Thirty years have passed since Davis started the pioneering solar neutrino experiment. The second generation solar neutrino experiments, Kamikande, GALLEX and SAGE, confirmed the deficit of the solar neutrinos which was first presented by Davis as "solar neutrino problem" (SNP). Now, the third generation experiments, Super-Kamiokande, SNO and BOREXINO are started or under construction. The event rate of first and second generation experiments were <1 neutrino events per day. The rate was increased to several tens of events per day in the third generation experiments.

Results of currently running solar neutrino experiments are shown in Table-4.

Experiment	method	flux	Data/SSM(BP95)
^{37}Cl	ν_e^{37}Cl	$2.54\pm0.14\pm0.14$ SNU	0.27 ± 0.02
GALLEX	ν_e^{71}Ga	$69.7\pm6.7^{+3.9}_{-4.5}$ SNU	0.51 ± 0.06
SAGE	ν_e^{71}Ga	73^{+10}_{-11} SNU	$0.53^{+0.07}_{-0.08}$
Kamiokande	νe scat.	$(2.80\pm0.19\pm0.33)\times10^6$ /cm^2/sec	0.42 ± 0.06
Super-K.	νe scat.	$(2.44\pm0.06^{+0.25}_{-0.09})\times10^6$ /cm^2/sec	$0.37^{+0.04}_{-0.02}$

Table 4. Results of running solar neutrino experiments.[19]

The comparisons between the observed fluxes and the expectations from the SSM[20] are also shown in Table-4. The flux ratio, data/SSM, is small in ν_e^{37}Cl experiment and almost half in Gallium experiments. The νe scattering experiments for ^{8}B neutrinos give data/SSM =~0.4. A detailed study of the relative difference in the flux ratio indicates that the astrophysical solutions have difficulty in explaining SNP.[21] An analysis on neutrino oscillations shows that the possible solutions of SNP are a small mixing solution ($\Delta m^2 \sim 0.5 \times 10^{-5}$ eV^2 and $\sin^2 2\theta \sim 6 \times 10^{-3}$) or a large mixing solution($\Delta m^2 \sim 1 \times 10^{-5}$ eV^2 and $\sin^2 2\theta \sim 0.6$).[21]

Super-Kamiokande(SK) has observed 4395 solar neutrino events during 306 days of live time with an energy threshold of 6.5 MeV, . The high statistics data of SK enable us to discuss not only the absolute flux value but also short time variations of the flux, such as day/night effect, and shape of the energy spectrum. Those checks purely depend on properties of neutrinos and are free from any ambiguities in SSM.

The obtained flux difference between day-time and night-time was

$$\frac{Day - Night}{Day + Night} = -0.017 \pm 0.026 \pm 0.017. \tag{2}$$

The night-time was further sub-divided into five time bins and flux of each bin was compared. No significant variation of flux was observed in the day/night analysis. The result already excludes lower half of the large mixing solution. The analysis of the energy spectrum is in progress. The largest contribution of the systematic errors in the energy spectrum analysis is the uncertainty of the absolute energy scale of the detector. The SK is using electron LINAC for the energy calibration. The current estimate of the energy scale uncertainty is $^{+4.1}_{-1.5}$ %. By lowering the uncertainty down to 0.5−1.0 %, SK data enable us to discuss small mixing solution only using the shape of the energy spectrum. Further systematic data taking of LINAC will be able achieve such small uncertainty. Several future solar neutrino experiments are under construction or being proposed. GNO, the upgrade of GALLEX, plans to use 100 tons of Gallium and achieve ~4% accuracy in flux.[22] ICARUS, which is under construction in Grand Sasso, will detect ^{8}B neutrino by ν_e^{40}Ar interactions.[23] The HELLAZ is a proposed experiment, which measures the energy spectra of pp and ^{7}Be neutrinos with νe scattering.[24] SK, SNO, BOREXINO, and the future solar neutrino experiments will solve the SNP within several years without relying on solar models.

6 Long baseline experiments

The atmospheric neutrino data of several underground experiments indicate neutrino oscillations of $\Delta m^2 = \sim 5 \times 10^{-3}$eV^2 with a large mixing angle. To check the oscillations with terrestrial experiments, L(km)/E(GeV) of ~100 is needed. Such experiments can be done by long baseline experiments using reactor neutrinos for $\overline{\nu}_e$ disappearance and using accelerator neutrinos for $\nu_\mu \leftrightarrow \nu_X$ oscillations. The summary and status of those experiments are shown in Table-5 and -6.

Experiment	Reactor	Distance	Detector	Event rate	Status
CHOOZ	2 × 4.2 GWth	1.0 km	5t Gd loaded scinitillator	25 ev./day	running
Palo Verde	3 × 3.6 GWth	0.75 km	12t Gd loaded scinitillator	50 ev./day	start in fall 1997
Kam-LAND	many reactors	150 km	1000t liquid scintillator	~2 ev./day	data taking 2000~

Table 5. Reactor long baseline experiments.[25]

	K2K[26]	MINOS[27]	Europe
Baseline	250 km	730 km	730 km
Accelerator	KEK-PS (12 GeV)	FNAL M.I. (120 GeV)	CERN SPS (450 GeV)
Far detector	Super-K.	MINOS	ICARUS, NOE (Gran Sasso)
Event rate	400 CC	140,000 (CC+NC)	1600/1 ICARUS
Status	Construction	R&D	Construction (ICARUS)
Data taking	1999~	~2001	> 2000

Table 6. Accelerator long baseline experiments.

7 Summary

It is very interesting time for neutrino oscillation physics now. Several underground experiments suggest neutrino oscillations in atmospheric neutrinos and in solar neutrinos. Future experiments should check neutrino oscillations with improved sensitivities, and also neutrino masses themselves should be investigated. A summary of suggested oscillation parameter regions are shown in Fig.3 together with sensitivities of future neutrino oscillation experiments.

References

[1] L. Wolfenstein,
Phys. Rev. D17(1978)2369; S. P. Mikheyev and A. Yu. Smirnov, Nuovo cimento 9C(1986)17; H. A. Bethe, Phys. Rev. Lett. 56(1986)1305.

[2] References are in order: H. Kawakami a et al., Phys. Lett. B256(1991)105; R. G. H. Robertson et al., Phys. Rev. Lett. 67(1991)957; E. Holzschuh et al., Phys. Lett. B287(1992)381; Sun et al., CJNP 15(1993)261; W. Stoeffl et al., Phys. Rev. Lett. 75(1995)3237; J. Bonn, Proceedings of the 17th International Conference on Neutrino Physics and Astrophysics, ed. by K. Enqvist et al., World Scientific(1996)259; A. I. Belesev et al., Phys. Lett. B350(1995)263; V. M. Lobashev, Proceedings of the 17th International Conference on Neutrino Physics and As-

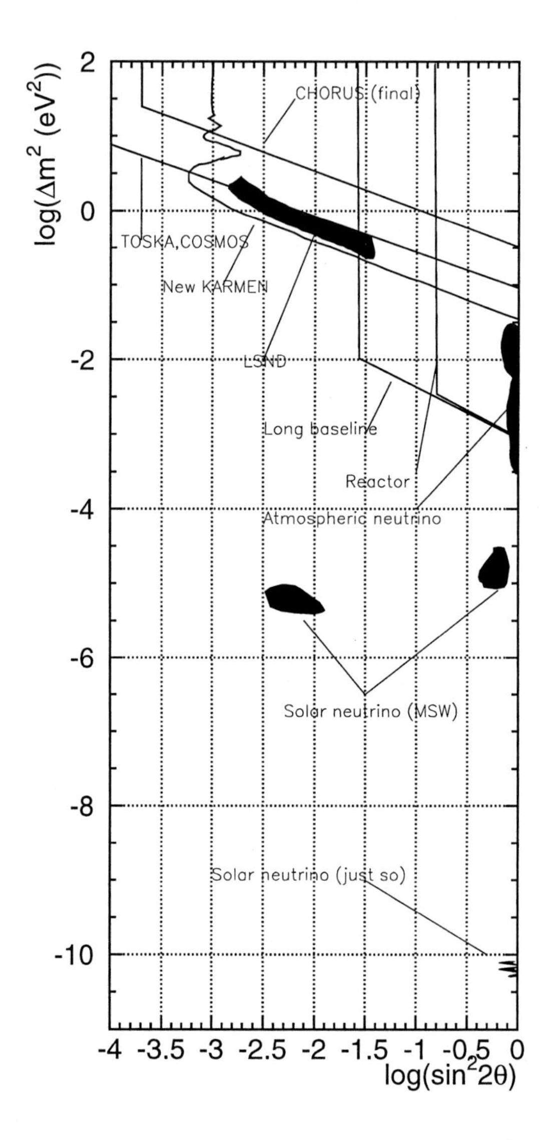

Fig. 3. Suggested oscillation parameter regions shown in filled areas. Sensitivities of future neutrino oscillation experiments are also shown.

trophysics, ed. by K. Enqvist et al., World Scientific(1996)264; Particle Data Group, Phys. Rev. D54(1996)280.

[3] V. M. Lobashev, Proceedings of the 17th International Conference on Neutrino Physics and Astrophysics, ed. by K. Enqvist et al., World Scientific(1996)264;

[4] Presentation by J. Bonn in PA10, 1004.

[5] Presentation by M. Girone in PA10, 1003.

[6] Presentation by L. Bugge in PA10, 1002.

[7] Presentation by H. Paes in PA10, 1018.

[8] Presentation by F. Laplanche in PA10, 1017.

[9] References are in order: M. Aglietta et al., Eourophys. Lett. 8(1989)611; Ch.Berger et al., Phys. Lett. B245(1990)305; K. S. Hirata et al., Phys. Lett. B205(1988)416; K. S. Hirata et al., Phys. Lett. B280(1992)146; R. Becker-Szendy et al., Phys. Rev. D46(1992)3720; W. W. M Allison et al., Phys. Lett. B391(1997)491; Presentation by K. Martens in PA10, 1013.

[10] M. Honda et al., Phys. Rev. D52(1995)4985.

[11] M. Nakahata et al., J. Phys. Soc. Jpn. 55(1986)3786.

[12] C. Berger et al., Phys. Lett. B245(1990)305.

[13] P. Lipari et al., Phys. Rev. Lett. 74(1995)4384.

[14] Presentation by L. Ludovici in PA10, 1009.

[15] Presentation by V. Valuev in PA10, 1008.

[16] C. Athanassopoulos et al., Phys. Rev. C54(1996)2685.

[17] LSND collaboration, nucl-ex/9706006.

[18] Presentation by B. Armbruster in PA10, 1007.

[19] K. Lande Proceedings of the 17th International Conference on Neutrino Physics and Astrophysics, ed. by K. Enqvist et al., World Scientific(1996)25; W. Hampel et al., Phys. Lett. B357(1996)384; Results of SAGE I+II+III, private communication with V. N. Gavrin; Y. Fukuda et al., Phys. Rev. Lett. 77(1996)1683; Presentation by K. Martens in PA10, 1013.

[20] J. N. Bahcall and M. H. Pinsonneault, Rev. Mod. Phys. 67(1995)781.

[21] N. Hata and P. Langacker, LASSNS-AST 97/29, hep-ph/9705339.

[22] E. Bellotti et al., Proposal of GNO, February 1996.

[23] F. Arneodo et al., Nucl. Phys. B (Proc. Suppl.) 54B(1997)95.

[24] Presentation by T. Patzak in PA10, 1016.

[25] Presentations by C. Bemporad in PA10, 1010; by Y. Wang in PA10, 1011; private communication with A. Suzuki. The initial results from CHOOZ is submitted to Phys. Lett. B after the conference. A preprint is available in hep-ex/9711002.

[26] K. Nishikawa, Report No. INS-Rep-924, 1992 (unpublished).

[27] Presentation by K. Lang in PA10, 1012.

High Energy Nuclear Collisions

F. Plasil (plasil@mail.phy.ornl.gov)

Physics Division, Oak Ridge National Laboratory
Oak Ridge, Tennessee 37831, U.S.A.

Abstract. The field of nucleus-nucleus collisions at "ultrarelativistic" energies is surveyed with emphasis on photon and lepton measurements. Prospects for future experiments at the Relativistic Heavy Ion Collider and the Large Hadron Collider are discussed.

1 Introduction

The foundations of studies of nucleus-nucleus collisions at relativistic energies were laid in the mid-seventies at the Bevalac facility located at the Lawrence Berkeley National Laboratory. An extensive and very productive program was carried out, making use of beams of highly-ionized heavy nuclei (heavy ions) with energies of 1-2 GeV/nucleon. Insights gained from these studies, together with a number of theoretical developments, led to the somewhat intuitive postulation that at certain sufficiently high combinations of temperature and baryon density, which may be achieved in nucleus-nucleus collisions at sufficiently high energies, quarks and gluons may become deconfined within a volume of nuclear dimensions, leading to the formation of the quark-gluon plasma (QGP). It is believed that the entire universe existed in the QGP state a few microseconds after the Big Bang of creation, before free quarks and gluons hadronized to protons and neutrons. The intriguing possibility of creating this state of matter in the laboratory has driven the construction of several experimental facilities and has given rise to an extensive and vibrant field of research at the intersection of nuclear and particle physics.

This presentation covers three broad topics: a brief introduction to the field of nucleus-nucleus collisions at relativistic energies; a discussion of several topics illustrating what we have learned after more than a decade of fixed target experiments; and an indication of what the future may bring at the Relativistic Heavy Ion Collider (RHIC) under construction at the Brookhaven National Laboratory (BNL) and at the Large Hadron Collider (LHC) planned at CERN.

Present experimental results from nucleus-nucleus collisions at very high energies (colloquially referred to as "ultrarelativistic" energies) have been obtained at BNL's AGS and at CERN's SPS. Both facilities have provided us with beams of heavy ions ranging from relatively light nuclei, such as Be, C, and O, to very heavy ions, such as Au and Pb. The highest energies available are 15 GeV/nucleon at the AGS and 200 GeV/nucleon at the SPS. Since it is not possible in the available time to cover systematically all available experimental results, I have chosen to highlight data obtained from measurements of photons and leptons. These so-called "penetrating probes" have mean free paths that are much larger than the size of a system of nuclear dimensions and are very likely to escape the reaction zone without interacting with the surrounding hot dense medium. Thus, they are likely to provide us with direct information on the conditions that existed at the time of their production. All available measurements making use of these penetrating probes (measurements of photons, electrons, and muons) have been carried out at the SPS and they constitute the most intriguing results obtained thus far. Extensive direct measurements of hadrons have been made both at the AGS and at the SPS. However, hadron measurement results that I present here for illustration purposes were also all obtained at the SPS. This choice was made in part due to the fact that no AGS data have been presented in parallel sessions at this conference and because the points that need to be made do not depend, to any great extent, on the difference in the energies available at the AGS and at the SPS.

Most recent QCD calculations provide us with strong theoretical indications that the phase transition to the QGP is likely to exist. This transition should manifest itself in sudden changes in system properties such as the specific heat, the energy density, and the magnitude of the quark mass scale. From state-of-the-art studies, it is concluded that the critical temperature, T_c, associated with the transition is in the 160- to 200-MeV range at an energy density of 2.5 ± 0.8 GeV/fm^3. For example, QCD calculations on an $8^3 \times 4$ lattice with two dynamical quark flavors [1] exhibit a sharp discontinuity in color mobility (related to deconfinement) as a function of increasing inverse lattice coupling constant (related to the temperature of a thermalized system). At very nearly the same value of the inverse coupling constant, the quark mass is found to drop steeply. This suggests that the quarks (and, hence, the hadrons) loose their mass at the critical temperature. This process is called chiral symmetry restoration, and within the framework of these calculations, it is predicted to manifest itself simultaneously with deconfinement.

The evolution of a head-on, central, collision between two large nuclei such as Pb at the SPS energy of 158 GeV/nucleon is likely to proceed as follows: first, if the target and projectile are not transparent to each other, there will be a rapid buildup of baryon density consisting of target and projectile nucleons. This process is often referred to as "stopping." If the energy density attained at this stage is sufficiently high, partonic matter may be formed. There are two crucial questions at these early reaction stages. If partonic matter is formed, does it achieve thermal equilibrium as would be expected if the QGP is formed? Is chiral symmetry simultaneously restored? In any event, whether the system consists of the QGP or of hot dense hadronic matter, it is known that as the reaction evolves, the initial "fireball" expands both in the longitudinal and in the transverse direction accompanied by cooling.

Hadronization takes place leading, in turn, to chemical freeze-out and, finally, to kinetic freeze-out, i.e., to a complete decoupling of the constituents. Thus, the degree of stopping of the colliding nuclei and the energy density attained in the early phase of the collision relate directly to the probability of QGP formation and play a crucial role in the subsequent evolution of the reaction.

2 Global event characteristics

Proton rapidity distributions, which relate to the degree of stopping, are shown in Fig. 1 for 200-GeV/nucleon S+S and 158-GeV/nucleon Pb+Pb collisions [2]. In the Pb case the proton rapidity distribution peaks at midrapidity, indicating a high degree of stopping. In contrast, in the case of the smaller S+S system, the degree of stopping is lower, as is evidenced by only a partial shift of the original target and projectile protons toward midrapidity. Thus, the highest energy density attained in the early phase of the reaction is likely to be significantly larger in the Pb+Pb case than in the S+S case.

Traditionally, estimates of attained energy densities are obtained from measured transverse energy rapidity (or pseudorapidity) distributions. There is, however, a conceptual problem with this process, since it is not clear to what stage of the reaction the extracted values refer. The buildup of energy density results from the dynamics of the collision (sharp increase in baryon density per unit volume) as well as from particle formation. Yet, only educated guesses are available for the formation time (e.g., 1 fm/c), and it remains virtually a free parameter. It is even likely that the expansion process is initiated before particle formation is terminated. The most frequently used method of estimating the attained energy density is attributed to Bjorken [3]. It assumes longitudinal growth and free hydrodynamic flow in a baryon-free region. Strictly speaking, given the assumptions of the Bjorken approach, it should be applied only to cases where there is a broad (baryon-free) plateau in the $dE_T/d\eta$ distribution, and not to distributions with large midrapidity baryon admixtures as is the case at SPS energies (see Fig. 1). Consequently, applying the Bjorken ansatz to our systems results in overestimates of attained energy densities. This reservation notwithstanding, Bjorken energy densities are found to be in the range of 2.5 to 3 GeV/fm^3 for the reactions considered here. As was pointed out earlier, theoretical estimates of the energy density required for the phase transition to the QGP to take place are also in the 2- to 3-GeV/fm^3 range, with a broad band of uncertainty around this value. Thus, if the required energy densities have not actually been attained at SPS energies, we are probably not very far from this goal.

3 Measurements of single photons

Directly radiated single photons, because of their low interaction probability, are likely to reflect the thermal properties of hot and dense matter, whether

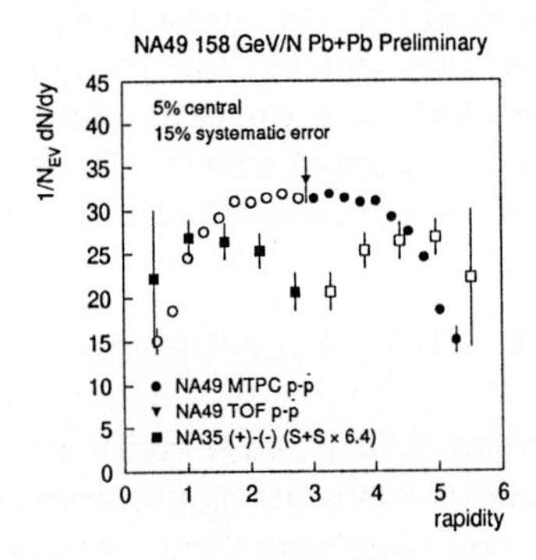

Fig. 1. Rapidity distribution of participant protons in central Pb+Pb and S+S collisions.

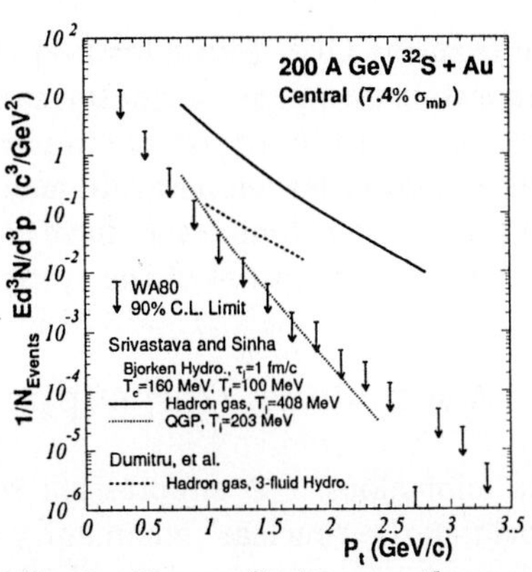

Fig. 2. Upper limits on the excess photon yield for collisions of ^{32}S+Au at 200 GeV/nucleon. For theoretical curves see the text.

it be a hadron gas, the QGP, or a mixture of both. Expected contributions to the observed photon yield from the QGP very probably include $q - \overline{q}$ annihilation and the QCD equivalent of the Compton process in which a quark interacts with a gluon to produce a photon. However, a hot hadron gas is expected to radiate photons with emission rates that are similar to those that are expected from the QGP [4]. The dominant production mechanisms in a hot hadron medium are expected to be $\pi^+\pi^- \rightarrow \rho\gamma$ and $\pi\rho \rightarrow \pi\gamma$. Given a fixed energy density produced in a given collision, a higher temperature is attained in a thermalized system that has a lower number of degrees of freedom. In the case of the QGP, considering spin/polarization, isospin and color, the number of degrees of freedom is 12 for the quarks and 16 for the gluons. For a hadron gas the estimated number of degrees of freedom depends on specific assumptions regarding its constituents. However, even when all plausible hadron resonances are taken into account, the number of degrees of freedom is lower, and hence the predicted temperature and photon yield are higher in a hadron-gas scenario than in a QGP scenario.

The measurement of direct photons is difficult since they are embedded in a very large combinatorial background from the decay of π^0 mesons. The WA80 Collaboration [5] has deduced direct photon yields for ^{32}S+Au at 200 GeV/nucleon on a statistical basis, as a function of p_T, by comparing the total photon yield to that which can be attributed to all long-lived decays. The results, which are sensitive at the 5% level of inclusive photons, are consistent (within 1σ) with the absence of photon excess in both central and peripheral collisions. The deduced upper limits at the 90% confidence level are shown as a function of p_T for central collisions in Fig. 2. Also shown in Fig. 2 are calculated thermal photon production yields [6] expected from a hot hadron gas with only a limited number of degrees of freedom (solid curve) and from the QGP (dotted curve). The dashed curve gives results from more elaborate hadron-gas calculations of Dumitru et al [7]. The upper limits on direct photon production are important in that they rule out the possibility that a high initial temperature may have been attained in the early phase of the collision. Sollfrank et al. have concluded on the basis of reference [5] that the initial temperature of the system could not have exceeded 250 MeV [8].

4 Anomalous *J/psi* suppression in Pb+Pb collisions

The anomalous J/ψ suppression recently observed in Pb+Pb collisions is providing the "nuclear" community with perhaps the most intriguing results. J/ψ suppression was proposed by Matsui and Satz as an unambiguous signature of the QGP in 1986 [9], before the start of the experimental program. The predicted suppression of charmonium states in the QGP is due to the Debye screening of the $c\bar{c}$ binding potential by the freely-moving color charges. Early in the collision process, $c\bar{c}$ pairs are created at small separations, and if they are to lead to the production of a bound vector meson state, they must separate to its appropriate asymptotic size. If the radius of the vector meson of interest is larger than the screening distance, the quark and antiquark may "lose contact," leading to open charm $(D, \overline{D})$ production. Since the screening distance is small in the presence of many colored objects, the conditions for suppression of the J/ψ and the ψ' in the QGP are very good. The NA38 collaboration set out to determine the predicted J/ψ suppression by measuring muon pairs and reported results consistent with the theoretical predictions as early as 1987. There were, however, several developments which cast doubt on the QGP interpretation of the measurements. First, similar effects were found in p-A reactions. Second, several alternative theoretical interpretations were proposed. These include pre-resonance absorption to the $c\bar{c}g$ state and the breaking-up effects of "comovers" in a confined hadronic medium. Before lead beams became available at CERN, extensive J/ψ measurements (relative to the Drell-Yan, D-Y, continuum resulting from the annihilation of light quarks) were made by the NA38/50/51 collaborations ranging from p-p and p-A collisions to oxygen- and sulfur-induced A-A reactions. The results are shown in Fig. 3. It is seen that the J/ψ to D-Y ratios follow closely a straight line when plotted as a function of a geometrical parameter L, which represents the average length of the $c\bar{c}$ path through nuclear matter. (L is related to the impact parameter and, hence, to E_T.) The common view, prior to the availability in 1996 of the lead-beam data, was that all data can be accounted for by absorption and that there was no need to postulate a QGP scenario.

The recent NA 50 J/ψ results from Pb+Pb reactions at 158 GeV/nucleon are also shown in Fig. 3 [10]. It is seen that the lead data deviate from the earlier systematic trend. Furthermore, there appear to be two "steps" in the data, indicating possible threshold effects. Thus, the J/ψ suppression appears to be much stronger than expected from standard absorption scenarios. Fur-

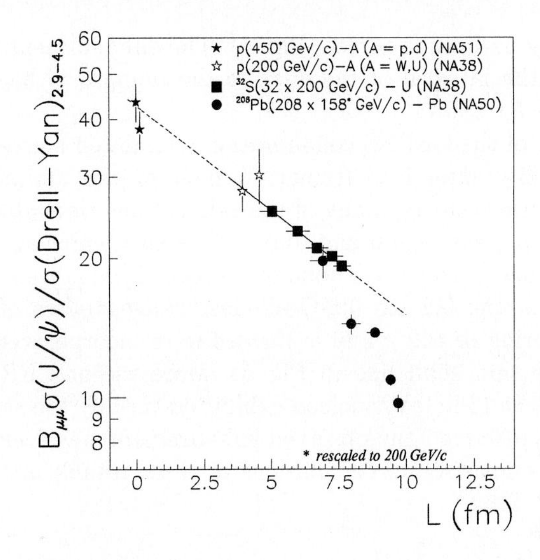

Fig. 3. Ratios of J/ψ to Drell-Yan cross sections versus the average length L of the path in nuclear matter [10].

thermore, the suppression depends strongly on E_T and, hence, on reaction centrality and on energy density. Although these results are still the subject of differing interpretations, many hold the belief that they provide us with strong evidence for collective parton behavior in nucleus-nucleus collisions. To explore these interesting findings further, it would be desirable to have available results on the p_T dependence of the suppression as well as studies indicating the onset of the anomalous suppression, possibly via the investigation of peripheral collisions. Studies making use of inverse kinematics which would minimize absorption in confined matter would also be of interest.

5 Low-mass electron pair production

Also of great interest are the results of the CERES (NA45) collaboration on vector meson production as deduced from measurements of low-mass electron pairs [11]. These data indicate that the chiral symmetry restoration transition may have manifested itself in certain reactions. The electron pairs are measured in the mass range from about 50 MeV to 1.2 GeV by means of "hadron-blind" tracking, using two RICH detectors. Studies of the p-Au reaction at 450 GeV indicate that the inclusive invariant mass spectrum can be fully accounted for by electron pairs stemming from known hadronic decays. In contrast, in the case of the S+Au reaction at 200 GeV/nucleon, there is a large enhancement of the electron pair yield over the yield that can be accounted for by hadronic sources (Fig. 4). The enhancement factor, defined as the ratio of the integral of the data to the integral of hadronic sources, was found to be 5.0 ± 0.7.

The findings of the CERES collaboration stimulated a great deal of theoretical activity. By taking into account in-medium pion annihilation, bremsstrahlung, and other effects, many of the calculations were able to reproduce the data in the ρ-mass region and above. One such calculation is shown in Fig. 5 (dotted curve). However, none of the calculations were able to reproduce the data in the 0.2- to 0.6-GeV mass range, unless effects of chiral restoration lowering of the ρ and ω masses were incorporated in the calculations [12] (see thin solid line in Fig. 5). More recent CERES data from Pb-Au collisions at 158 GeV/nucleon exhibit essentially the same features as the S+Au data, although the observed enhancement was found to be somewhat lower. A comparison of calculations to these results was shown at this conference by P. Filip.

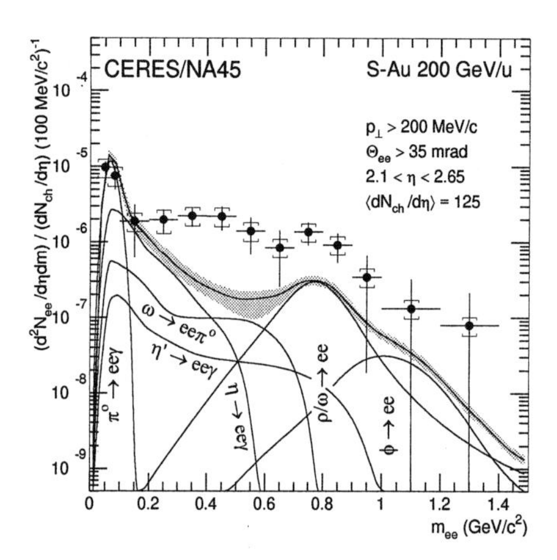

Fig. 4. Dilepton invariant mass spectrum from S+Au at 200 GeV/ nucleon. Data are shown together with known hadronic decay contributions.

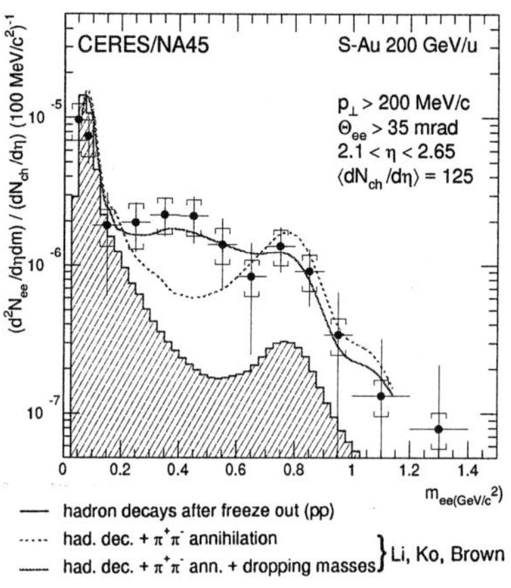

Fig. 5. Same as Fig. 4 but with comparisons to theoretical calculations (see text). Contributions from hadron sources are given by the shaded area.

6 Expansion, hadronization, and freeze-out

The interesting data on strangeness enhancement covered at the conference will not be presented here due to the space limitations. However, measurements of hyperons and antihyperons also provide us with valuable information on primordial yield ratios resulting from partonic coalescence essentially free of reequilibration in the subsequent hadronic expansion phase. The measurements were made by the WA85 collaboration for S+W at 200 GeV/nucleon [13], and an analysis was performed [14] in which various yield ratios describe curves of allowed Hagedorn model solutions in a plane of temperature, T, and baryochemical potential, μ_B. The curves were found to intersect at about $T = 185$ MeV and $\mu_B = 0.24$ GeV. These values can be viewed as an indication of the hadronization phase-transition point at SPS energies.

Transverse mass spectra of all produced particles (with the exception of pions at low m_T values) are reasonably well reproduced by exponential functions from which inverse slope parameters T can be extracted. Values of T are shown in Fig. 6 as a function of particle mass ranging from pions and kaons to deuterons, for particles produced in central Pb+Pb collisions [2]. The inverse slope parameters for the heaviest particles far exceed the Hagedorn limit, and the increase with mass is not consistent with a simple hadronic fireball model in which all T values should be equal for primary emitted hadrons. It is possible that the large T values may be explained by a large transverse flow within the framework of a hydrodynamical model.

A picture that emerges from the above findings is that observed hadrons may not all stem from a common global chemical equilibrium. Species with a large chemical relaxation time, such as hyperons, "freeze-in" at high values of temperature and baryon density. On the other hand, pions and kaons with short relaxation times follow the expansion, staying in equilibrium until their turn comes to freeze-in at a lower temperature and baryon density. Thus, sequential freeze-in stages provide us with markers of successive expansion stages.

Within model assumptions, it is possible to extract from the m_T spectra values of the temperature and of the transverse flow velocity, $\beta_\perp$, at thermal freeze-out [15]. Results are shown for negative hadrons (mostly pions) and for deuterons in Fig. 7. It is seen that many combinations of T and $\beta_\perp$ fit the data equally well. Fortunately, the dependence of the two-pion Bose-Einstein correlation on the average transverse momentum of the pion pair constrains the values of $\beta_\perp$ [16]. This result is also shown in Fig. 7. The conclusion is that at final kinetic freeze-out the temperature of the system is in the 115- to 125-MeV range and that it is expanding in the transverse direction with a velocity of 0.5 to 0.6 c. In addition, from other considerations, it is concluded that at this stage the system has a lifetime of about 8 fm/c and that it has undergone a substantial longitudinal expansion.

7 The future at RHIC and at the LHC

If the probability of QGP formation is marginal at the SPS, there is no question that the required energy densities can be achieved at the new colliders. Simple estimates indicate energy densities relative to the SPS that are a

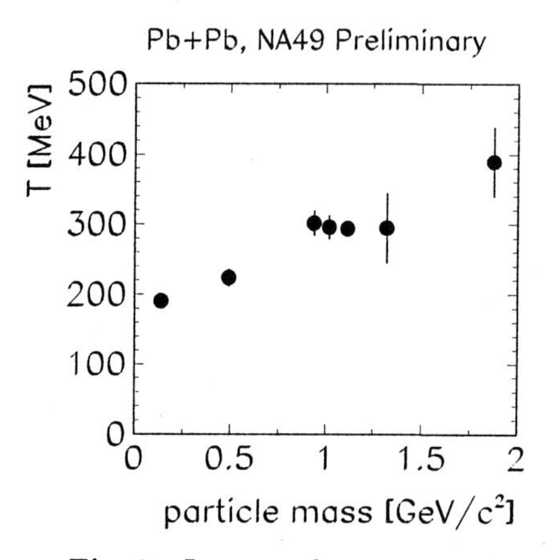

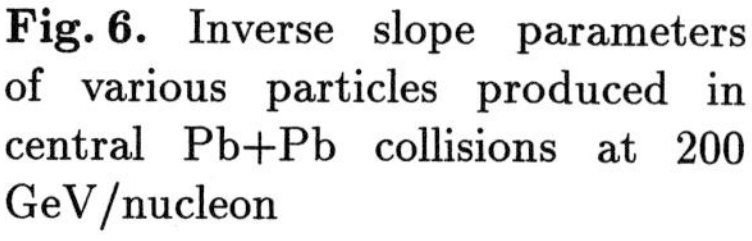

Fig. 6. Inverse slope parameters of various particles produced in central Pb+Pb collisions at 200 GeV/nucleon

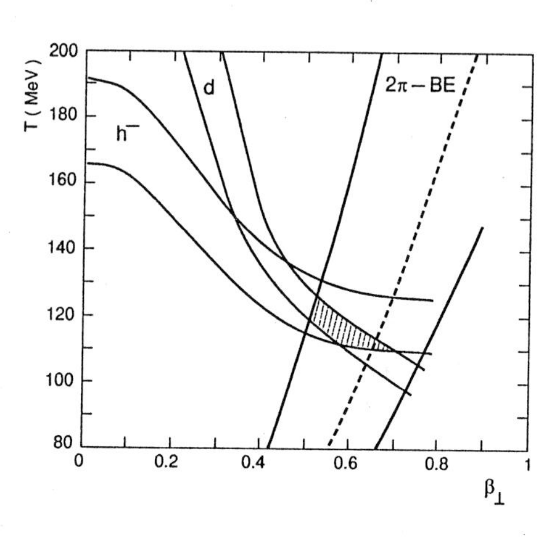

Fig. 7. Temperature vs transverse velocity. Curves depict constraints on the variables from m_T spectra and from particle correlations

factor of 2 higher at RHIC and a factor of 7 higher at the LHC. Correspondingly, the reaction volumes are expected to be seven times larger at RHIC and twenty times larger at the LHC. All of these factors significantly enhance the QGP discovery potential.

RHIC will be a dedicated heavy-ion machine and is expected to start up in 1999. It will be able to deliver Au nuclei with energies ranging from 30 to 100 GeV/nucleon (at a luminosity of 2.10^{26} $\mathrm{cm^{-2}\ s^{-1}}$) and protons at 250 GeV (luminosity 10^{32} $\mathrm{cm^{-2}\ s^{-1}}$). Four intersection regions will house two large experiments (PHENIX and STAR) and two smaller experiments (PHOBOS and BRAMS). PHENIX will concentrate on measurements of penetrating electromagnetic probes (photons, electrons, and muons), although it will also have a hadron-measurement capability. The midrapidity region will be covered by two electron-photon-hadron spectrometer arms in an open axial-field magnet. There will also be two back-to-back muon spectrometers covering large rapidities. STAR is dedicated to hadronic probes. Its centerpiece is a large-acceptance TPC in a solenoid magnet, and emphasis will be on event-by-event measurements of a large number of observables. PHOBOS is essentially a "table-top" experiment consisting on many silicon telescopes and a time-of-flight array. The primary capability of the experiment will be low p_T measurements of hadrons. BRAHMS will measure inclusive hadrons in two movable spectrometer arms. It will be able to make measurements at very large rapidities. In combination, the four experiments will be able to address all conceivable, predicted (as well as unexpected) signatures of the QGP.

At the LHC it is expected that heavy-ion operation will take place during six weeks per year. The expected start-up is in 2005. The energy of lead ions is expected to be 2.8 TeV/nucleon at a luminosity of 2.10^{27} cm^{-2} s^{-1}. Only one dedicated heavy-ion detector, ALICE, is planned, although other detectors (e.g., CMS, see R. Katadze, these proceedings) are planning to develop heavy-ion measurement capabilities. ALICE is a large-scale version of STAR, housed in the L3 magnet, with additional forward–backward muon measurement capability. It is also expected to concentrate on event-by-event measurements. Electrons and hadrons will be measured via tracking in a weak field. A high-resolution EM calorimeter will be used to measure photons.

8 Acknowledgements

I am indebted to the following individuals for help in the preparation of the presentation and/or the manuscript:
T.C. Awes, S.J. Ball, C. Gerschel, T.J. Hallman, B.M. Johnson, I. Ravinovich, D. Röhrich, Y. Tserruya, J. Schukraft, P.W. Stankus, and R. Stock.

Oak Ridge National Laboratory is managed by Lockheed Martin Energy Systems Corp. under contract DE-AC05-96OR22464 with the U.S. Department of Enegy.

References

[1] E. Laerman, Nucl. Phys. A610 (1996) 1c
[2] S. V. Afanasiev et al. (NA49 Collaboration), Nucl. Phys. A 610 (1996) 188c; D. Rörich (NA49 Collaboration), these proceedings
[3] J. D. Bjorken, Phys. Rev. D 27 (1983) 140
[4] J. Kapusta, P. Lichard and D. Seibert, Phys. Rev. D 44 (1991) 2774
[5] R. Albrecht et al., Phys. Rev. Lett. 76 (1996) 3506
[6] D. K. Srivastava and B. Sinha, Phys. Rev. Lett. 73 (1994) 2421
[7] A. Dumitru et al., Phys. Rev. C 51 (1995) 2166
[8] J. Solfrank et al., Phys. Rev. C 55 (1997) 392
[9] T. Matsui and H. Satz, Phys. Lett. B 178 (1986) 416
[10] M. C. Abreu et al. (NA50 Collaboration), Nucl. Phys. A610 (1996) 404c
[11] G. Agakichiev et al.(NA45 Collaboration), Nucl. Phys. A610 (1996) 317c
[12] G. Q. Li et al., Phys. Rev. Lett. 75 (1995) 4007; C. M. Ko et al., Nucl. Phys. A 610 (1996) 342c
[13] S. Abatzis et al. (WA85 Collaboration), Nucl. Phys. A 566 (1994) 225c
[14] J. Cleymans and H. Satz, Z. Phys. C 57 (1993) 135; D. Röhrich (NA49 Collaboration), these proceedings
[15] B. Kaempfer, preprint FZR-149, hep-ph/9612336
[16] U. Heinz, Nucl. Phys. A 610 (1996) 264

Recent Developments in Detectors

R.J. Cashmore (r.cashmore1@physics.ox.ac.uk)

Department of Physics, University of Oxford, U.K.

Abstract. Many of the recent developments have been driven by the new collider experiments, particularly those at the LHC. Muon detectors, hadron calorimetry are briefly reviewed with the major attention being devoted to progress in tracking systems and particle identification.

1 Introduction

The last year has brought the now-steady progress in detector developments and many of these developments were reported in the parallel session. Inevitably this summary can only deal with a subset of those developments and it is equally inevitable that the treatment is comparatively superficial so that the inquisitive physicist is best referred directly to the primary authors. Much of the work has been driven by preparations for the LHC experiments, although this is not exclusively the case. The LHC programme required major developments and still requires major developments, but some areas are now ready for construction to begin. Developments from ATLAS [1], CMS [2], ALICE [3] and LHC-B [4] will be prominent but there are major activities associated with CDF [5], DØ [6], HERA -B [7], BaBar [8], Belle [9] and CLEO III [10], as well as primary detector research. From this list it is already clear that accelerator based activities are the prime topic and unfortunately non-accelerator detector developments will not be given the attention they deserve. The review will deal with

(i) Muon Detection Systems
(ii) Calorimetry
(iii) Tracking systems
(iv) Particle Identification.

(i) and (ii) will only be dealt with briefly, major attention being devoted to (iii) and (iv), the particle identification being particularly driven by the needs of B physics experiments.

2 Muon Detection Systems

The enormous systems which are required at the LHC – $\sim$ 10000 m^2 of each chamber type and 100000's of channels of read out – are a major challenge. First to produce chambers of the required specifications and then to complete a system which can retain this performance.

2.1 Individual Systems

In ATLAS there are four types of chamber – Monitored Drift Tubes (MDT), Cathode Strip Chambers (CSC), Resistive Plate Chambers (RPC) and Thin Gap Chambers (TGC) – while in CMS there are 3 types – Drift Tubes (DT), RPC and CSC. Each have their specific roles: Precision measurements (MDT, DT, CSC) and Triggers (RPC, TGC, CSC). In the two experiments the specifications are different but detectors have now been developed which achieve these specifications as shown in Table 1.

Chamber	Spatial resolution	Timing resolution
MDT	$\sim 80\ \mu m$ [1]	
DT	$150 - 200\ \mu m$ [2]	
CSC	$\sim 60\ \mu m$	~ 5 ns [2]
TGC		< 25 ns [1]
RPC		~ 2 ns [1, 2]

Table 1. Performance of muon detectors

There does remain work to be done on aging and in the case of RPC's a continued study of the requirement or not for linseed oil, although current results [2] (see Fig 1) appear to give optimism.

2.2 System Design

This is the next major step dealing with long term stability, electronics, DAQ services etc, including the issue of alignment and this cannot be underestimated. Full size systems are required for the latter and the ATLAS collaboration have shown that with a laser system (RASNIK) and cosmic rays the position of the chambers can be monitored with the necessary accuracy as indicated in Table 2.

Inner Chamber	Laser system	Cosmic rays
displacement (μm)	measured displacement (μm)	
$+255 \pm 5$	254 ± 2	237 ± 53
$+510 \pm 8$	512 ± 2	520 ± 53
-1785 ± 30	-1769 ± 2	-1776 ± 58

Table 2. Calibration of chamber displacement in the ATLAS test system.

2.3 Summary

The required specifications, both for individual chambers and systems, can be achieved which was not obvious initially. Construction is about to begin for the ATLAS and CMS detector systems.

3 Calorimetry

Already construction has been approved for both the ATLAS Liquid Argon and Tile calorimeters [1] while the CMS Technical Design Report [2] for the Hadron calorimeter – copper and scintillator – is about to be considered. It is in the Lead Tungstate crystal calorimeter for CMS [2], but also ALICE [3], that further R&D was and still is necessary. However, substantial progress has been made here too. The main attraction of lead tungstate is its density leading to a very compact calorimeter and a good energy resolution, claimed for CMS [2], of

$$\frac{\sigma}{E} = \frac{2\%}{\sqrt{E}} \oplus 0.5\%$$

Achieving the 0.5% is a real challenge. The major outstanding issue is the radiation hardness. In order to achieve this good performance the induced absorption length λ should satisfy, $\lambda \geq 2\times$ crystal length ie an induced absorption coefficient $\lesssim 2.5\ \mathrm{m}^{-1}$. It has long been appreciated that doping with Niobium is necessary to give satisfactory performance up to the radiation levels expected at the LHC (ie $\lesssim 10$ MRad). What came as a surprise was that the crystals exhibited a rapid change with small radiation doses as shown in Fig 2 from which the crystal recovered and which in turn could be reproduced. This clearly represents a major problem to achieving a constant term of $\sim 0.5\%$ as the immediate history of the crystal would need to be known. This short term damage is now known to be dependent on the stoechiometry of the crystals and thus careful control of the initial composition (or perhaps by annealing the crystals) can reduce this problem, see Fig 3. CMS have also demonstrated that it is possibly to track the crystal light output with an LED system, Fig 4, so that the radiation history of the crystals can be incorporated into any energy measurements.

Progress is being made, but it is equally clear that further R&D is still required. However, it is possible to be optimistic that eventually a good detector (for CMS and the LHC) meeting reliably the performance specifications will be obtained.

4 Tracking Systems

The generic tracking detector for collider experiments consists, in general, of an inner and outer system, as indicated in Table 3. The inner system is invariably some form of silicon detector while there is more diversity in the outer system.

	ATLAS	CMS	CDF	DØ
Outer	TRT straw tubes	MSGC	Wire chamber	Scintillating fibres
Inner	Si microstrips and pixels	Si microstrips and pixels	Si microstrips	Si microstrips

Table 3. Tracking detectors at Hadron Colliders

Similar layouts exist for the LEP experiments, HERA experiments, the SLD, and the B factory experiments. The SLD is somewhat different in having a pixel CCD system. Of course the major problem at the LHC is once again radiation damage and particularly the survival of those detectors close to the intersection point (which experience $\sim 10^{14}$ charged hadrons per cm^2 per year and a similar neutron fluence). This issue has driven most of the recent detector development.

4.1 Outer Tracking Systems

ATLAS TRT Straw Tubes This detector is not only a tracking detector but also a transition radiation detector. Aging studies have shown no degradation in performance up to deposited charges of 8 C/cm on the anode wires and 18 C/cm on the cathodes, equivalent to between 10-20 years of expected LHC running, and a satisfactory gas has been identified. The performance at LHC counting rates is shown in Fig 5, the efficiency of 60% still leading to $\gtrsim 20$ hits/track while retaining an accuracy of $\sim 170\ \mu$m which meets specifications. ATLAS is thus confident that it will work at a luminosity of 10^{34} cm^{-2}s^{-1}.

Microstrip Gas Chambers MSGC The basic idea of the MSGC is indicated in Fig 6. However, at this time last year severe doubts were aired over the operation of such chambers. At high gain (10^4) there was a tendency for breakdown which destroyed the electrode structure (observed by HERA-B where such gains are necessary). In the last year, since 1996, there has been substantial progress. Two types of solution have been found to attack this problem. The first is advanced passivation with a polyimide layer along the edge of the cathode. Along with the lower gain required by CMS ($\sim 10^3$) there is reasonable hope of a working chamber. The second approach is to combine a GEM with the MSGC [11] as shown in Fig 7. In this concept there are two regions of modest amplification. Results are shown in Fig 8 which demonstrate the operation of such a device and indicates that high gains can be achieved with comparatively low MSGC cathode voltages. This gives some relief to the HERA-B collaboration.

Questions still remain with respect to long term stability but there is now the promise of working chambers that will satisfy the rigorous demands of HERA-B and CMS.

It is worth noting that there has been an explosion of ideas based on this type of concept – Micromegas, Double GEM, Double GEM + PCB's, Microdot chambers – to name but a few. These have in turn been combined into a variety of devices – visible light imaging photomultipliers and even TPC readout. Furthermore, it is clear from this that there are enormous numbers of applications for the developments made for particle physics detectors.

Scintillating Fibre Trackers The DØ experiment will implement a major fibre tracking system for the next stage of Tevatron running. This is unique in that it will use VLPC's for photon detection. These are devices operating in the 6-12 K temperature range with quantum efficiencies up to 80%. The fibres are arranged in 8 doublets to that a reasonable tracking efficiency per layer ($\sim$ 99.5%) can be obtained despite single fibre efficiency of $\sim$ 85%. Results from test beams have confirmed the fibre MC simulations.

DØ is not the only experiment where fibres are contemplated. Some are successfully in use in the CHORUS experiment [12] while other systems are considered for HERA-B and LHC-B.

4.2 Inner Tracking Systems

Silicon devices now dominate tracking systems close to the intersection point in collider experiments. This is the result of the developments of the last 10-15 years and still more are required for the LHC. However, before proceeding to this, it is worth drawing attention to two other developments.

1. DELPHI has a combined microstrip and pixel system now operating [13]; and
2. SLD has a major new CCD pixel detector consisting of over 300 million pixels of 20 μm $\times$ 20 μm granularity, due to steady improvements in CCD manufacture and design [14].

Of course the benefit of pixels is that 3-d tracking is possible, with a dramatic improvement in b-tagging performance, and the CCD type of detector is of great potential for any future linear collider.

Silicon Radiation Damage and Operation The silicon devices considered for the LHC (and elsewhere) are general either p or n+ implants on an n substrate, the detectors being $\sim$ 300 μm thick. After radiation damage type inversion of the bulk material occurs and the major issue in whether these detectors can then be operated efficiently and safely for another 10 years. In Fig 9 results from ATLAS demonstrates that n+ on n detectors can be operated with a depletion voltage of $\sim 250 - 300\,V$ after a dose of $\sim 2 \times 10^{14}$ p/cm^2. Using p on n detectors is more ambitious (and almost certainly cheaper) requiring higher voltages, but CMS simulations, Fig 10, seem to indicate that operations for $\sim$ 10 years at the LHC should be feasible providing that the silicon detector environment (ie temperature) is carefully maintained.

The issue of type inversion, together with many other properties, has been studied by the ROSE (RD48) collaboration [15]. They have demonstrated that careful control of the initial resistivity implies that the radiation damage properties and subsequent operation can in turn be controlled. This is demonstrated in Fig 11 where type inversion occurs at different neutron fluences indicating that an optimum design can be made limiting the range of operating voltages. This work is of enormous value for future experiments, but given the long time required for construction of the LHC tracking detectors it may have limited impact on ATLAS and CMS.

It now appears that it should be possible to operate silicon microstrip detectors at radii $> 20-30$ cm while inside this pixels are obligatory, requiring further development of the electronics for these detectors.

Possible Alternatives to Silicon in High Radiation Environments

GaAs [16]: For a long time the resistance of this material to neutrons indicated great promise. Unfortunately, and rather surprisingly, charged particles produce dramatic damage and it is no longer viable in the LHC environment. However because of the higher Z of the material it may well find applications, amongst other things, in X-ray imaging.

Diamond [17]: A major study of diamond as a detector has been made by the RD42 collaboration. Diamond has many attractive properties, (a) mechanically strong, (b) excellent thermal conductivity, (c) low Z, low dielectric constant (and hence low noise), (d) high resistivity (and hence small leakage current), (e) radiation hard and (f) of simple construction – to be balanced by two main disadvantages – a) smaller (than silicon) deposited charge and b) a limited carrier mean free path.

Most efforts have concentrated on improving the charge collection distance (ccd) with the result that in the last 8 years this quantity has risen from $\sim 1\mu$m to $\sim 200\ \mu$m today, resulting in a charge collection of $\sim$ 7000 e as indicated in Fig 12. Furthermore a CVD microstrip detector has been operated with fast electronics giving a $S/N \sim 7$ so that a useful detector (particularly a pixel detector) is almost in sight. The radiation hardness up to $\sim 2 \times 10^{15}\pi/\text{cm}^2$ is good, no degradation being observed as indicated in Fig 13. Results from neutron irradiation at fluences of $> 10^{15}/\text{cm}^2$ are eagerly awaited.

Continuing R&D is clearly needed, together with a more detailed understanding of the material, but one can hope for pixel devices in the not too distant future.

4.3 Electronics for tracking detectors on high radiation environments

After $\sim$ 10 years of R&D on radiation hard electronics, a reliable working system does not yet exist. This is of particular concern for ATLAS and CMS as these experiments are rapidly approaching the point where construction must begin. The R&D will continue both to give a reliable system and hopefully increase the number of suppliers (with, presumably, cost benefits). Optimistically the job is nearly complete, but effort cannot be reduced until an affordable radiation hard system is delivered.

Most attention has been concentrated on the electronics close to collision points, but it is now being appreciated that, although the radiation levels are much reduced (by factors of $\sim 10^3 - 10^4$) it must still be essential to dentify radiation tolerant electronics if the experiments are to have lifetimes $\sim$ 10 years at the LHC.

5 Particle Identification

The interest and activity in particle identification has been driven by 3 major activities.

1. B physics at e^+e^- B factories in the Belle, BaBar and CLEO III detectors
2. Particle compositions in heavy ion collisions at the LHC in the ALICE experiment; and
3. B physics in "fixed target" configuration in the HERA-B and LHC-B experiments.

In all cases, except for the Pestov counters being developed for ALICE, the identification is based on Cherenkov radiation either in the threshold mode or in variants of the ring imaging technique.

5.1 Cherenkov detectors at B factory experiments

In general it is necessary to identify particles up to momentum of $\sim$ 4.5 GeV mainly to ensure well identified decays B $\rightarrow \pi\pi$, so essential for CP violation studies. Different approaches have been followed in all cases.

Belle: Here the threshold technique is used by varying the refractive index of aerogel $n = 1.01$ to $n = 1.03$ through the appropriate solid angle. 960 separate modules are used to give sufficient granularity. In each case the light is detected with FM-PMT's within the 1.5 T magnetic field. Test beam results at KEK have demonstrated sufficient photoelectrons (around 20) for all refractive indices so that Belle should begin operation with a good particle identification system.

BaBar and CLEO III

The principle of operation of these two detectors is summarised in Fig 14. In each case a high refractive index radiator (quartz for BaBar and LiF for CLEO III) is the source of Cherenkov radiation.

In BaBar the light is trapped by total internal reflection (preserving the Cherenkov angle), transferred through a water ($n_{\text{water}} \cong n_{\text{quartz}}$) filled stand off box at one end and the light eventually detected by an array of photomultiplier tubes. This has the advantage of placing the light collection outside the active volume, although this brings added engineering challenges. One particular virtue is that the number of Cherenkov photons increases at large polar angles, ie in regions where particles are of highest momentum. Tests at CERN have confirmed the light yield and detection method so that the simulated K/π separations of Fig 15 can be achieved.

In CLEO III proximity focussing is used to produce a ring of photons which are detected using a CH_4 and TEA mixture as absorber and the subsequent photoelectrons in turn detected with a cathode pad chamber. What was an advantage for BaBar is an impediment to CLEO III ie the total internal reflection in the radiator. However, CLEO III have introduced an ingenious solution – a sawtooth surface to the radiator which allows a greater number of photons to emerge. Cosmic ray tests have shown that for a $\beta = 1$ particle 13 photoelectrons are detected and the Cherenkov angle measured with a precision of 14 mrad. This translates into a 3.3σ K/π separation at 2.8 GeV/c. (At CLEO III momenta are not as great as the CESR ring is of symmetric design).

5.2 ALICE particle identification

Over a comparatively small angle the particle composition is determined in heavy ion collisions. At high momentum (3σ separation for π/K at momenta < 3.4 GeV/c, K/p at momenta < 6.5 GeV/c) a ring imaging system is again used, but in this case a liquid C_6F_{14} radiator is proposed with the photons detected using a CsI coated PCB with the subsequent photoelectrons identified on the pad cathodes of a MWPC. The great challenge has been to develop large CsI coated photocathodes which can be reliably reproduced with high quantum efficiency (> 0.2) and which can be preserved over a long period of time. Fig 16 demonstrates the success in this activity, but it does require careful attention to conditions eg the PC's are kept under Argon flow. Tests have now been made of a large prototype and rings observed with ~ 18 photoelectrons.

Particle identification at lower momenta in ALICE is achieved using a TOF system. It now appears that, with careful attention to surface processing, reliable Pestov counters can be constructed which have a timing resolution of ~ 50 ps (even better can be achieved, but at the expense of higher voltage operation). Coupled with a start signal, this means that TOF's can be measured, over a large solid angle, with resolution of $\sim$ 100 ps. This is sufficient to give good $\pi/K/p$ separation up to momenta where the RICH system becomes relevant.

5.3 *B* Physics in "fixed-target" mode experiments

In these cases, HERA-B and LHC-B, the momenta of particles are substantially higher and it is important to achieve π/K separation up to momenta of $\sim$100-150 GeV/c. There is no substitute for the ring imaging technique, but once again the challenge is in the photon detection. At HERA-B (due to start operating in 1998/99) multianode (16 and 4 anode) PMT's outside the acceptances of the spectrometer are the chosen method. From tests it appears that using a C_4F_{10} radiator $\sim$ 34 photons will be detected. Given the time scales LHC-B can study new photon detectors. In this case a Hybrid PhotoDiode (HPD) which has 61 pixels of dimensions 2×2 mm, has been studied [18] and single photoelectron detection demonstrated. The final arrangement will have many such HPD's, but already in test beams rings have been observed from aerogel and C_4F_{10} radiators in 7 HPD's. Thus there will be various possibilities available for the final choice.

5.4 Summary

As indicated above, there is an immense variation in the specific particle identification techniques. They all appear to work well and this should give some reassurance when the precision physics of low branching ratio channels is studied in detail.

6 Summary

It is an impossible task to review fairly this enormous field and I must apologise to all those working in areas which I did not cover, particularly in the non-accelerator experiments. Triggering and DAQ are areas also uncovered in this review, but are absolutely vital to the subject.

If the LHC experiments appeared to feature over strongly, it is because such stringent challenges have been set there from which all applications of particle detectors will benefit. There still remains much to be done for the LHC and it will continue to assume this central importance for a while yet. However, the variety of particle physics ensures that there are many other interesting and exciting technical developments.

7 Acknowledgements

It is a pleasure to thank all those people who willingly provided me with information both before and at the conference. Special thanks must go to the parallel session speakers and organisers and particularly to my scientific secretary.

8 Figures

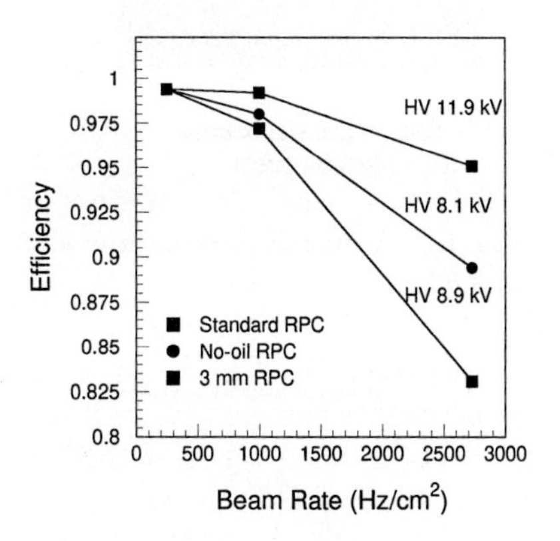

Fig. 1. Efficiency of RPC's.

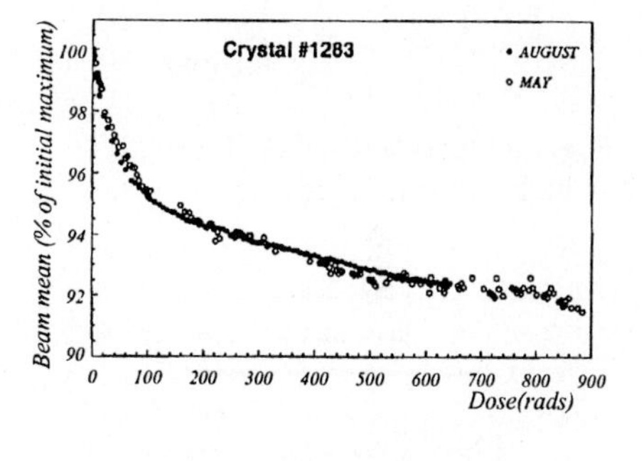

Fig. 2. Low dose radiation damage in $PbWO_4$.

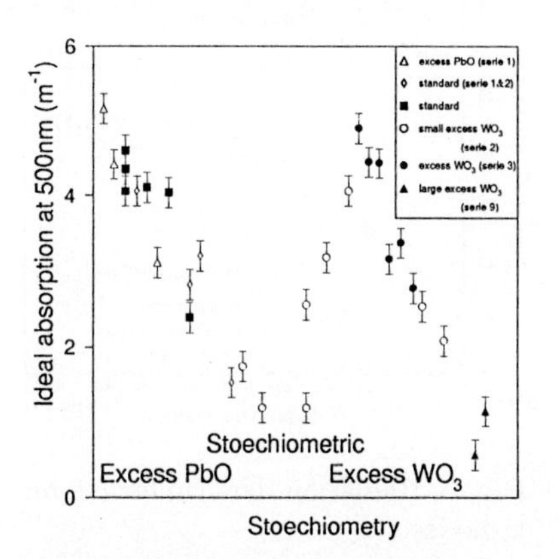

Fig. 3. Absorption as a function of the stoechiometry of $PbWO_4$.

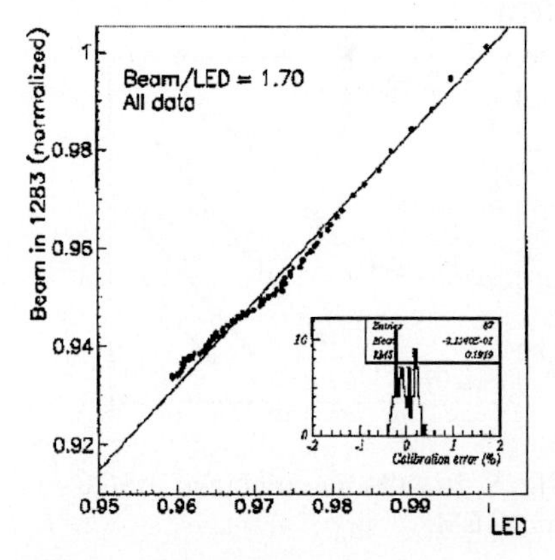

Fig. 4. Correlation of crystal light output with LED output.

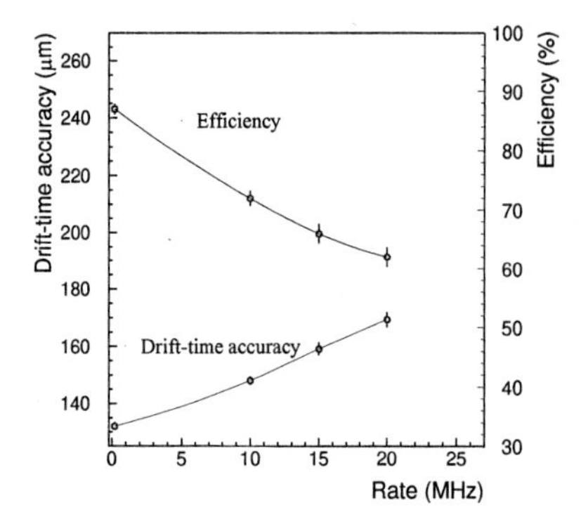

Fig. 5. Operation of ATLAS TRT at high counting rates.

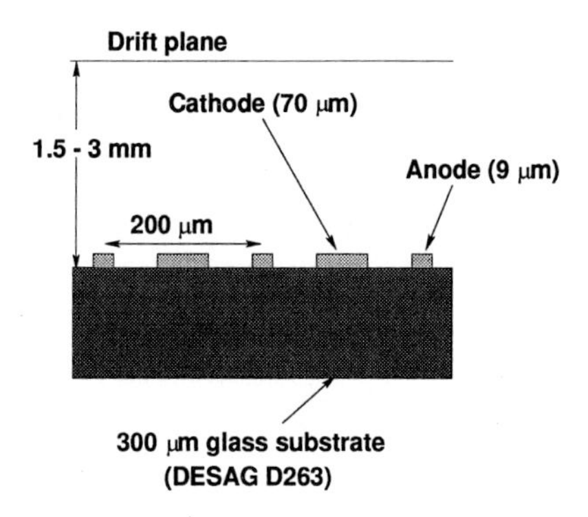

Fig. 6. Principle structure of MSGC.

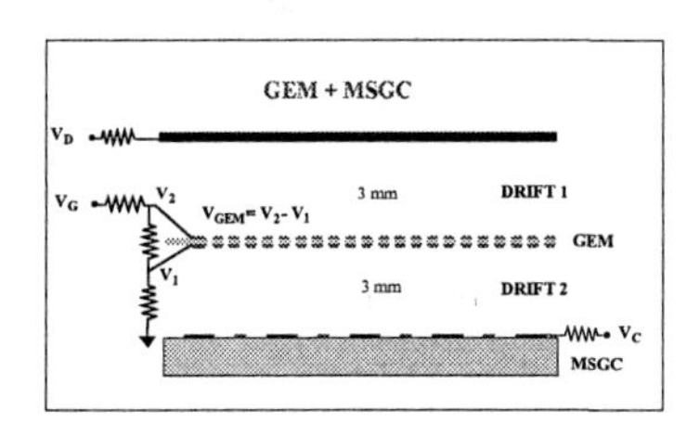

Fig. 7. Combining MSGC and GEM.

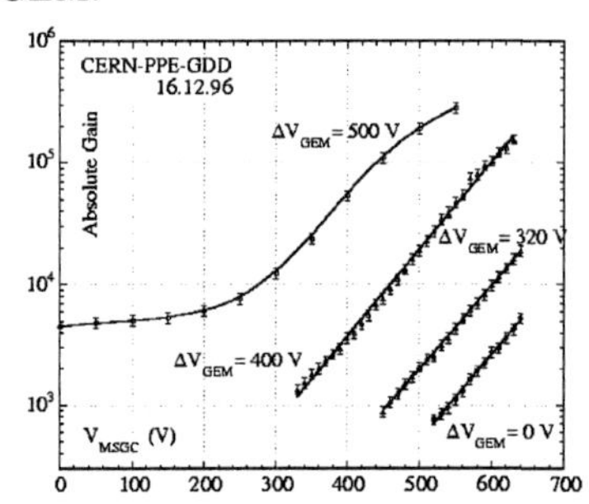

Fig. 8. Results for combined MSGC and GEM.

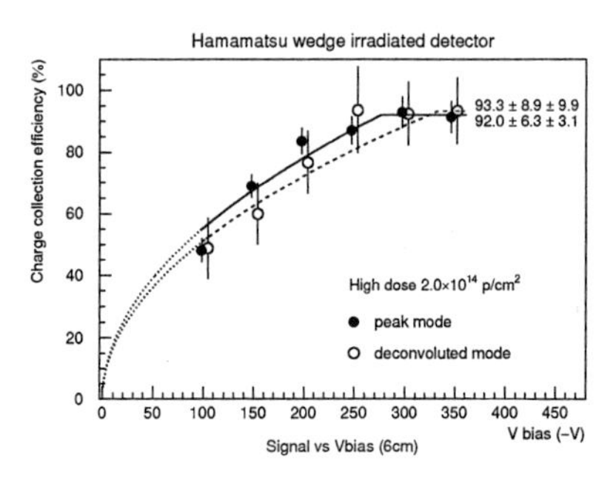

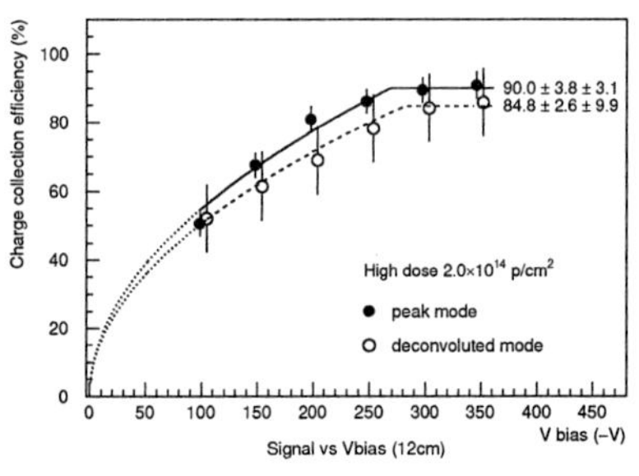

Fig. 9. Radiation damage for n+ on n diodes.

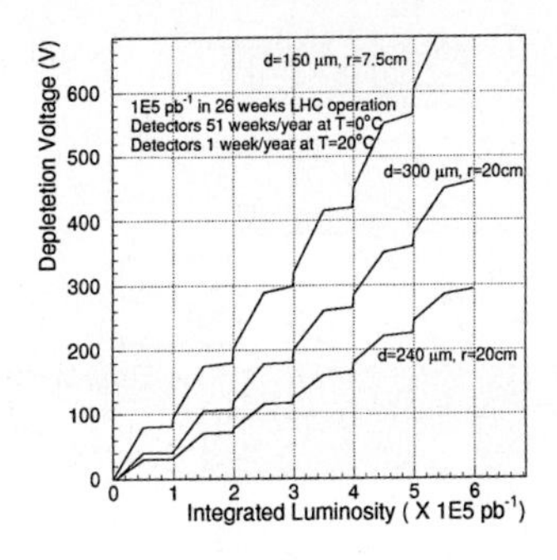

Fig. 10. Simulated performance of p on n diodes.

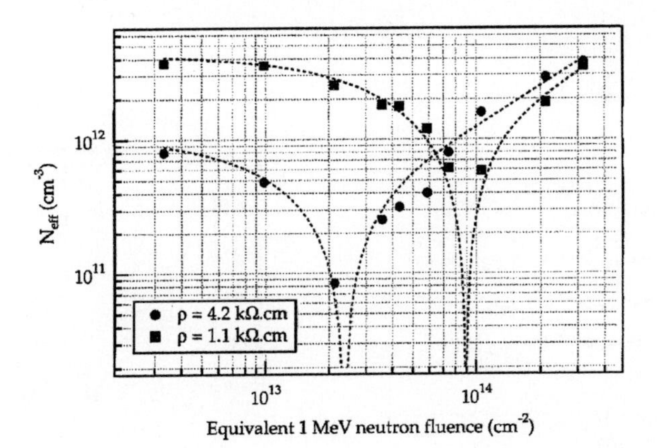

Fig. 11. Dependence of the effective number of carriers in silicon on initial resistivity.

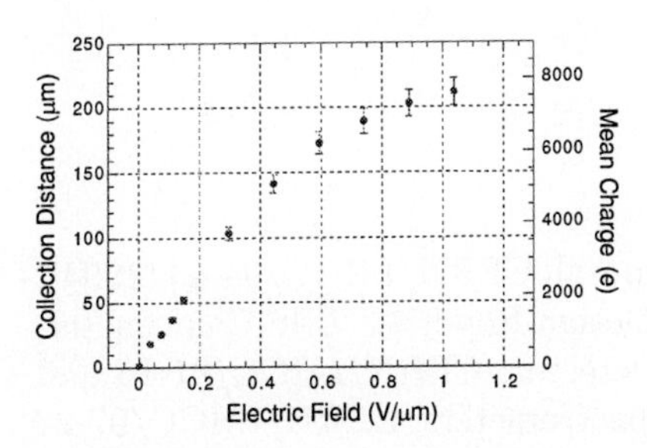

Fig. 12. Charge collection distance for diamond.

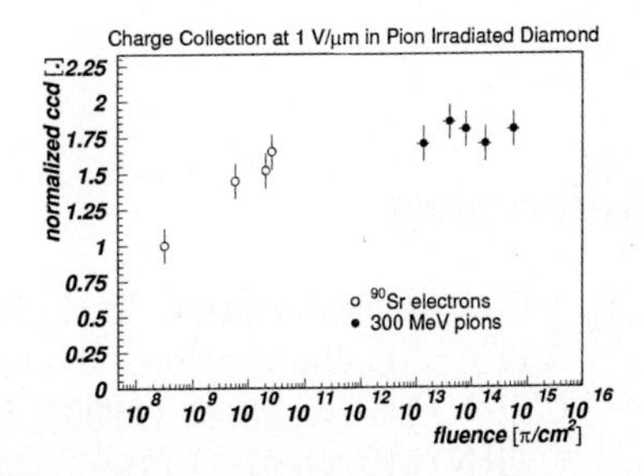

Fig. 13. Radiation hardness of diamond detectors.

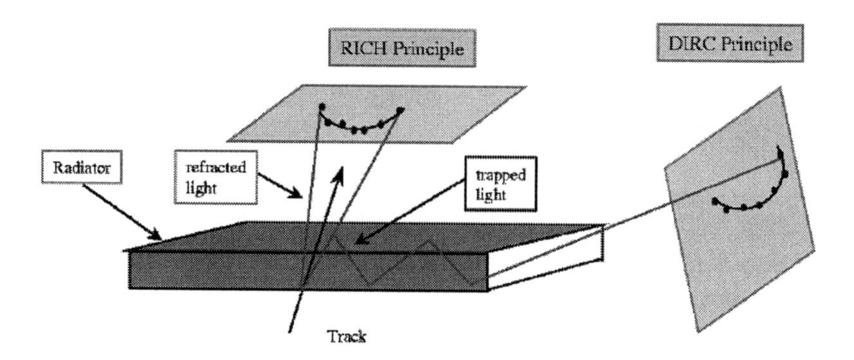

Fig. 14. Principle of RICH and DIRC.

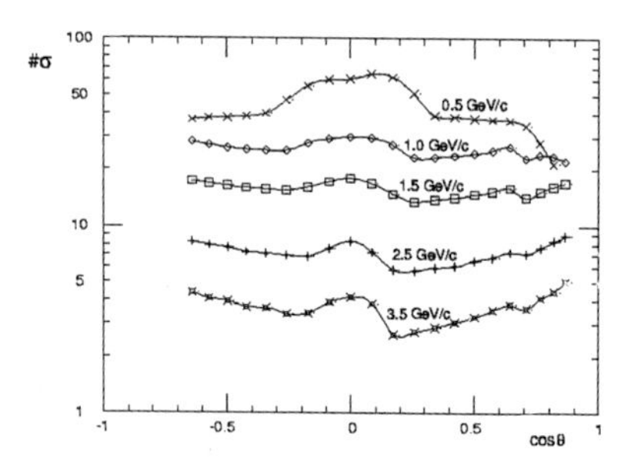

Fig. 15. K/π separation in BaBar.

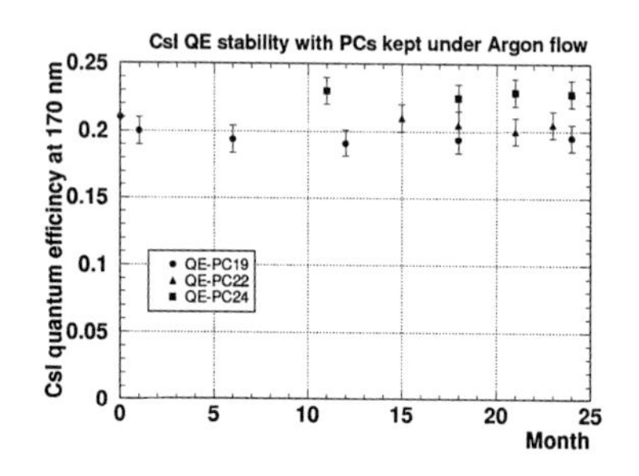

Fig. 16. Long term stability of the quantum efficiency of CsI coated PCB's.

References

[1] ATLAS Collaboration, Technical Proposal, CERN/LHCC/94-43 (1994). ATLAS Collaboration, Technical Design Reports: LAr-Calorimeter, CERN/LHCC/96-41 (1996); Inner Detector, CERN/LHCC/97-16 and CERN/LHCC/97-17 (1997); Muon Spectrometers, CERN/LHCC/97-22 (1997).

[2] CMS Collaboration, Technical Proposal CERN/LHCC/94-38 (1994). CMS Collaboration, Technical Design Reports: Hadron Calorimeter, CERN/LHCC/97-31 (1997); Muon Detector, CERN/LHCC/97-32 (1997); Electromagnetic Calorimeter, CERN/LHCC/97-33 (1997).

[3] ALICE Collaboration, Technical Proposal CERN/LHCC/95-71 (1995).

[4] LHC-B Collaboration, Letter of Intent, CERN/LHCC/95-5, (1995).
[5] CDF Collaboration, CDF II Technical Design Report, FERMILAB-Pub-96/390-E (1996).
[6] DØ Collaboration, *The DØ Upgrade: The Detector and its Physics* (1996).
[7] HERA-B Collaboration, Technical Design Report, DESY-PRC 95/01 (1995).
A. Spengler, Nucl. Instr. Methods **A384** (1996) 106.
[8] BaBar Collaboration, Technical Design Report (1996).
[9] Belle Collaboration, Technical Design Report, KEK-Report 95-1 (1995); Progress Reports 96-1 (1996) and 97-1 (1997).
[10] CLEO Collaboration, *The CLEO III Detector: Design and Physics Goals* (1994).
[11] A. Sharma, CERN-OPEN-97-032, subm. to HEP97, Jerusalem (1997); R. Bouclier, CERN-PPE-97-32 and 97-73 (1997).
[12] CHORUS Collaboration, P. Annis et al., CERN-PPE-97-100 (1997), subm. to Nucl. Instr. Methods A.
[13] DELPHI Collaboration, DELPHI 97-121 CONF 103, subm. to HEP97, Jerusalem (1997);
P.Chochula et al., *The DELPHI Silicon Tracker at LEP2*, subm. to Nucl. Instr. Methods A.
[14] K. Abe et al., *Design and Performance of the SLD Vertex Detector*, SLAC-PUB-7385 (1997).
[15] ROSE Collaboration, R&D Proposal, CERN/LHCC/96-23 (1996).
[16] RD8 Collaboration, R.L. Bates et al., Nucl. Instr. Methods **A395** (1997) 54.
[17] RD42 Collaboration, W. Adam et al., subm. to Proceeding of VERTEX 97, Mangaratiba (1997).
[18] E. Albrecht et al., *First observation of Cherenkov Ring Images using Hybrid Photon Detectors*, Imperial College, London preprint IC/HEP/97-16 (1997), to be subm. to Nucl. Instr. Methods A.

Tests of the Standard Model: W Mass and WWZ Couplings

David Ward (drw1@cam.ac.uk)

Cavendish Laboratory, University of Cambridge, U.K.

Abstract. Recent tests of the electroweak Standard Model are reviewed, covering the precise measurements of Z decays at LEP I and SLC and measurements of fermion pair production at higher energies at LEP II. Special emphasis is given to new results on W physics from LEP and FNAL.

1 Precision measurements of the Z boson

1.1 Lineshape and leptonic forward-backward asymmetries

The accurate measurement of the Z mass is essential for precise tests of the Standard Model. This measurement is dominated by the energy scans performed at LEP in 1993 and 1995, in which approximately 40 pb^{-1} of data were recorded at energies ± 1.8 GeV away from the Z peak. Important progress has been made in the last year in understanding the LEP beam energy for these scans, and final results became available shortly before the conference. The estimated errors on the centre-of-mass energy (in MeV) are:

	peak−2	peak	peak+2
1993	3.5	6.7	3.0
1994		3.7	
1995	1.8	5.4	1.7

The path is now clear for the LEP experiments to complete their analyses of the data. The principal changes to the results compared to those presented in 1996 [1, 2] are (see [3] for more details):

- The new beam energy values have already been incorporated by ALEPH and DELPHI; OPAL and L3 have chosen not to update their results at this stage, but an appropriate correction to the Z mass and width values has been applied.
- Some data from the 1995 run have been added by OPAL and L3, and there have been significant changes to the $\tau^+\tau^-$ measurements from ALEPH.
- L3 have included the 1995 data in their τ polarization measurements [4].
- A preliminary measurement of A_{LR} from the SLD 1996 data is available.

Overall the changes in the combined results since last year are quite small. The basic combined LEP measurements are [3]:

Without lepton universality		Assuming lepton universality	
M_Z /GeV	91.1867 ± 0.0020	M_Z /GeV	91.1867 ± 0.0020
Γ_Z /GeV	2.4948 ± 0.0025	Γ_Z /GeV	2.4948 ± 0.0025
σ_h^0 /nb	41.486 ± 0.053	σ_h^0 /nb	41.486 ± 0.053
R_e	20.757 ± 0.056		
R_μ	20.783 ± 0.037	R_ℓ	20.775 ± 0.027
R_τ	20.823 ± 0.050		
$A_{FB}^{0,e}$	0.0160 ± 0.0024		
$A_{FB}^{0,\mu}$	0.0163 ± 0.0014	$A_{FB}^{0,\ell}$	0.0171 ± 0.0010
$A_{FB}^{0,\tau}$	0.0192 ± 0.0018		
χ^2/dof=21/27		χ^2/dof=23/31	

The data from the four experiments are in excellent agreement, as shown by the values of χ^2/dof.

From these measurements, it is possible to infer a value for the invisible width of the Z:

$$\Gamma_{inv} = 500.1 \pm 1.8 \text{ MeV}$$

which can be converted into a measurement of the number of light neutrino species assuming they have Standard Model couplings:

$$N_\nu = 2.993 \pm 0.011$$

A limit on the possible additional invisible width arising from new physics can also be inferred: $\Delta\Gamma_{inv} < 2.8$ MeV at 95% c.l.

The results from the Z leptonic lineshape and asymmetries can be combined with measurements of the τ polarization and its asymmetry (which are sensitive to the electron and τ couplings separately) to perform the best tests of lepton universality. In fig. 1 we show the vector and axial couplings for each lepton species, as inferred from the LEP data. The measurements are clearly consistent with universality of the Z leptonic couplings, with a precision of ~0.2% for g_A and 5-10% for g_V, and also with the Standard Model expectation, indicated by the shaded area with uncertainties arising from varying the Higgs mass between 60 and 1000 GeV and the t-quark mass in the range 175.6 ± 5.5 GeV. The measurement of A_{LR} from SLD is also shown. Under the assumption of lepton universality, the following values for the couplings are obtained:

	LEP	LEP+SLD
$g_{V\ell}$	-0.03681 ± 0.00085	-0.03793 ± 0.00058
$g_{A\ell}$	-0.50112 ± 0.00032	-0.50103 ± 0.00031

1.2 Heavy flavour electroweak measurements

The main changes in the past year, reviewed in [5, 6], are:

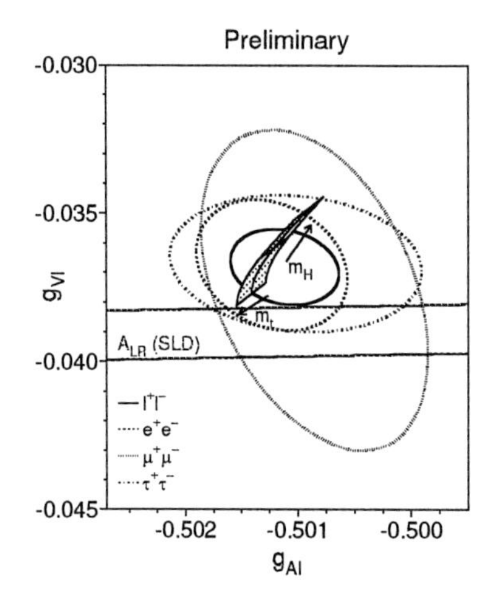

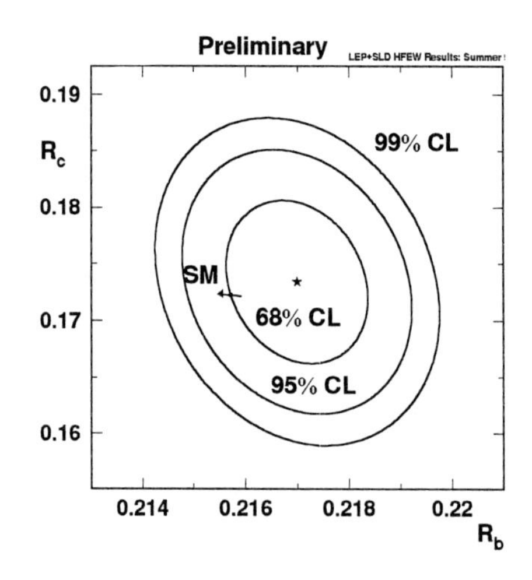

Fig. 1. 68% probability contours of vector and axial vector couplings of the Z to leptons.

Fig. 2. Measurements of R_{b} and R_{c}, compared with the Standard Model.

- Measurements of $R_{\mathrm{b}} = \Gamma_{\mathrm{b\bar{b}}}/\Gamma_{\mathrm{had}}$ from ALEPH [7], OPAL [8] and SLD have been finalised for publication in the last year, and new preliminary measurements from DELPHI [9], L3 [10] and SLD [11] have led to significantly improved precision.
- New determinations of R_{c} from OPAL [12], ALEPH [13] and SLD [14] (the latter exploiting a double vertex tag), have led to continued improvement in the precision of this measurement.
- There are new measurements of the forward backward asymmetry $A_{\mathrm{FB}}^{\mathrm{b}}$ from L3 [15] and OPAL [16], but an ALEPH result has been withdrawn, so the overall precision of the measurement is unchanged.
- The SLD measurements of the polarized asymmetries [17] have led to a much improved determination of $\mathcal{A}_{\mathrm{c}}$.

The combined LEP/SLD heavy flavour measurements may be summarized as follows:

R_{b}	0.2170 ± 0.0009
R_{c}	0.1734 ± 0.0048
$A_{\mathrm{FB}}^{0,\mathrm{b}}$	0.0984 ± 0.0024
$A_{\mathrm{FB}}^{0,\mathrm{c}}$	0.0741 ± 0.0048
$\mathcal{A}_{\mathrm{b}}$	0.900 ± 0.050
$\mathcal{A}_{\mathrm{c}}$	0.650 ± 0.058

In fig. 2 we compare the measured values of R_{b} and R_{c} with the Standard Model expectations. The apparent disagreement which excited much interest in previous years has evaporated.

The various measurements of polarizations and asymmetries at LEP and SLD can be interpreted, in the context of the Standard Model, as measurements of the effective electroweak mixing parameter $\sin^2\theta_{\mathrm{eff}}^{\mathrm{lept}}$. In fig. 3 we show a comparison of the various determinations of $\sin^2\theta_{\mathrm{eff}}^{\mathrm{lept}}$. The values are not incompatible with a common value (χ^2/dof=12.6/6), though it should be noted that the two most precise determinations (from A_{LR} and the forward-backward b-quark asymmetry) show the largest discrepancies from the mean. In fig. 4 we show the measured values of $\sin^2\theta_{\mathrm{eff}}^{\mathrm{lept}}$ and the leptonic width Γ_ℓ, which are in good agreement with the Standard Model expectation. The star indicates the prediction if only photonic radiative corrections are applied, with the arrow showing the non-negligible uncertainty induced by the running of the electromagnetic coupling. The data clearly demonstrate the need for electroweak radiative corrections, and their sensitivity to the Higgs mass m_{H} is also evident.

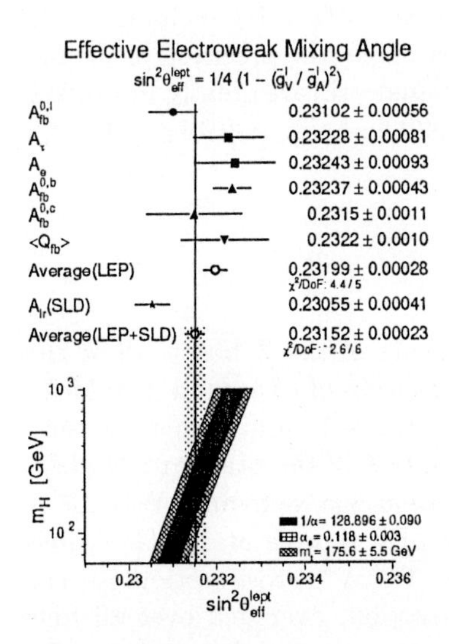

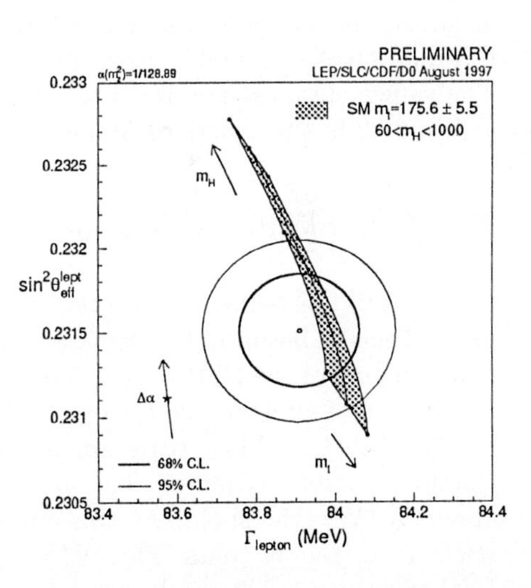

Fig. 3. Determinations of $\sin^2\theta_{\mathrm{eff}}^{\mathrm{lept}}$ from asymmetry and τ polarization measurements. The Standard Model expectation as a function of m_{H} is shown.

Fig. 4. Combined measurements of $\sin^2\theta_{\mathrm{eff}}^{\mathrm{lept}}$ and Γ_ℓ, compared with the Standard Model expectation with and without electroweak radiative corrections.

2 Fermion pair production at LEP II

Fermion-pair production at energies well above the Z resonance is characterized by a tendency for *radiative return* to the Z through the emission of one or more photons from the initial state. The main physics interest, and the greatest sensitivity to new physics, lies in the events with only a small amount of initial state radiation. *Non-radiative* events are selected by imposing a cut on the effective c.m. energy of the fermion-pair, $\sqrt{s'}$, which can be reconstructed from the event kinematics. Measurements now exist of fermion pair cross-sections and forward-backward asymmetries up to 183 GeV (see [18] for a compilation and detailed references).

The measurements are all in excellent agreement with the Standard Model. They may be interpreted in various ways – either to constrain parameters of the Standard Model, or to place limits on new physics (for more details, see [18]). For example, the hadronic cross-section at LEP II can be used to constrain Z$-\gamma$ interference. The data taken on the Z peak constrain this interference only weakly, so it is normally fixed to its Standard Model expectation. A more model independent interpretation of the data can be performed using the S-matrix formalism [19], allowing the parameter $j_{\rm had}^{\rm tot}$ which parametrizes γ-W interference in the hadronic cross-section to be free. By including data above and below the Z, where interference is sizeable, the precision of the determination of $j_{\rm had}^{\rm tot}$, and hence of $M_{\rm Z}$ in this framework, are greatly improved. The values obtained are $M_{\rm Z} = 91.1882 \pm 0.0029$ and $j_{\rm had}^{\rm tot} = 0.14 \pm 0.12$ (c.f. $j_{\rm had}^{\rm tot} = 0.22$ in the Standard Model).

3 Mass of the W boson

As we shall see below, the precise measurements of the Z boson allow the mass of the W boson to be predicted with a precision of of around ± 40 MeV. A major goal of the LEP II and Tevatron programs is to match this precision by direct measurement, so as to provide a new test of the Standard Model.

At LEP II, $\rm W^+W^-$ pairs are produced either via s-channel $\rm W/\gamma$ or t-channel neutrino exchange. The first runs at LEP II were at 161 GeV, just above $\rm W^+W^-$ threshold. At this energy, the $\rm W^+W^-$ cross-section is very sensitive to the W mass. The $\rm W^+W^-$ cross-section, averaged over all four LEP experiments [20], is shown in fig. 5. From the cross-section at 161 GeV, the W mass is obtained as $M_{\rm W} = 80.40 \pm 0.22$ GeV, where the error is predominantly statistical.

At higher energies, the cross-section is much less sensitive to the W mass, and the better technique is to reconstruct the W mass directly from the invariant mass of its decay products. The final states $\rm W^+W^- \to q\bar{q}q\bar{q}$ and $\rm W^+W^- \to q\bar{q}\ell\nu_\ell$ can both be used for this purpose. Kinematic fit techniques, imposing energy and momentum conservation and equality of the two W masses in each event, are used to improve the mass resolution. The results obtained after averaging the measurements from each LEP experiment [21, 22], are:

Channel	M_W /GeV
$W^+W^- \to q\bar{q}q\bar{q}$	80.62±0.26
$W^+W^- \to q\bar{q}\ell\nu_\ell$	80.46±0.24
Combined (172 GeV)	80.53±0.18
LEP 161 and 172 GeV	80.48±0.14

In this direct reconstruction approach, the $q\bar{q}q\bar{q}$ final state is potentially more problematical than $q\bar{q}\ell\nu_\ell$ because it can be affected by hadronic final state interaction effects. These can arise because the two Ws typically decay so close together that they are within the range of the strong interaction. It has been suggested that *colour reconnection* effects [23] could bias the reconstructed W mass in the $q\bar{q}q\bar{q}$ channel by several hundred MeV. However, the models which predict the largest effect [24] also predict other observable effects, such as a $\sim$ 10% reduction in the hadron multiplicity in the $q\bar{q}q\bar{q}$ case. Data already exist [25] comparing the hadronic W decay multiplicity in the $q\bar{q}q\bar{q}$ and $q\bar{q}\ell\nu$ final states, yielding a ratio of 1.04±0.03. At first sight this appears inconsistent with the most extreme colour reconnection model, though caution is needed because the models used to correct the data do not include the colour reconnection effect. Another possible problem could result from Bose-Einstein correlations between pions from different Ws. Data from LEP so far [26], with large errors, show no evidence for such correlations, consistent with the most recent theoretical investigations [27].

At the Tevatron, W bosons are produced singly from $q\bar{q}'$ fusion. The leptonic W decays $W \to \ell\nu$ ($\ell = e/\mu$) are used to defeat QCD background, with the neutrino inferred from missing momentum. The value of M_W may be extracted from the distribution of transverse mass of $\ell\nu$, as shown in fig. 6. The current results from CDF and D0 are 80.375±0.120 GeV and 80.44±0.11 GeV respectively [28, 29, 30]. The combined W mass measurement from hadron collider experiments (including UA2) is 80.41 ± 0.09 GeV.

The measurements of M_W from hadron colliders and from LEP II are therefore in excellent agreement. The combined "World Average" is:

$$M_W = 80.43 \pm 0.08 \text{ GeV}$$

At present this average is dominated by the Tevatron measurements. However, if LEP delivers say 50 pb^{-1} of data per experiment in 1997, the LEP error can be expected to reduce to around 0.08 GeV. By the end of LEP II and after the Tevatron upgrade, both LEP and Fermilab expect to be able to reach a precision on M_W around 0.03–0.04 GeV.

4 Global standard model fits

Combined fits of the Standard Model have been performed to the measurements of Z decays and the W mass outlined above. Additional measurements

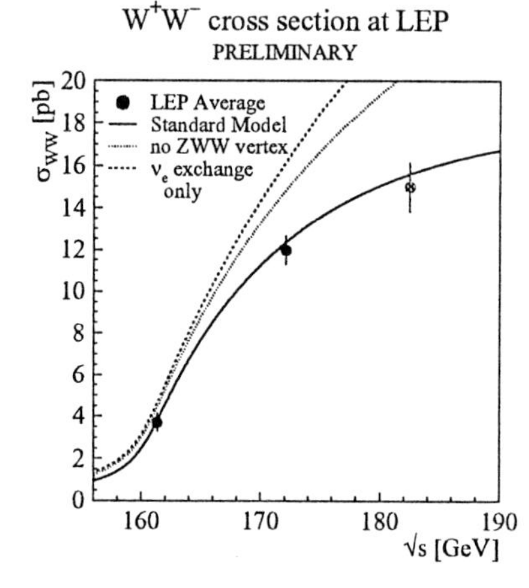

Fig. 5. Measurements of the $e^+e^- \to W^+W^-$ cross-section at LEP II.

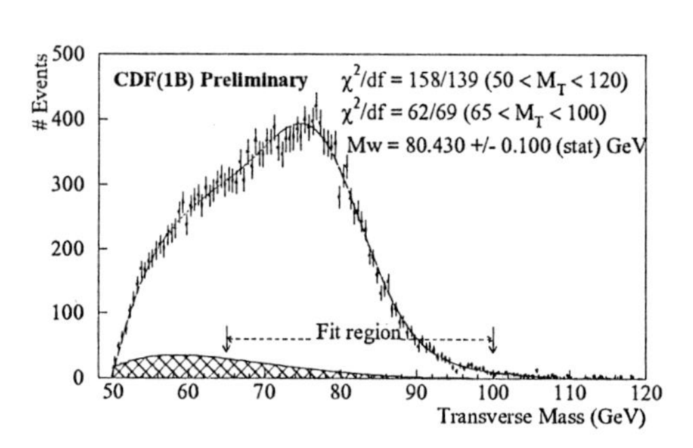

Fig. 6. Transverse mass distribution in CDF, from which the W mass can be extracted.

can also be included: the top quark mass $m_{\mathrm{t}} = 175.6 \pm 5.5$ GeV [31], the value of $1 - M_{\mathrm{W}}^2/M_{\mathrm{Z}}^2 = 0.2254 \pm 0.0037$ from νN scattering (which includes a new result from CCFR [32]) and the value of the electromagnetic coupling, $1/\alpha(M_{\mathrm{Z}}) = 128.894 \pm 0.090$ [33], which carries an error because of the need to run it to scale M_{Z}. In the fits, m_{H}, $\alpha_{\mathrm{s}}(M_{\mathrm{Z}})$ and optionally m_{t} and M_{W} are treated as free parameters.

Three fits have been performed, with the results shown below:

i) A fit to LEP I and LEP II data only.

ii) A fit to all data except the direct measurements of M_{W} and m_{t}. This permits a direct comparison between the direct and indirect determinations of these masses, as shown in fig. 7. The two sets of measurements are seen to be consistent.

iii) A fit to all data. As in the previous two fits, the χ^2/dof value is excellent, showing that the data are globally compatible with the Standard Model. In fig. 8 we show the input measurements and the *pulls* for this fit, i.e. the difference between fitted and measured values divided by the error. The distribution of pulls is satisfactory, with only one measurement ($\sin^2\theta_{\mathrm{eff}}^{\mathrm{lept}}$ from A_{LR}) more than two standard deviations from the Standard Model.

	i) LEP (inc. M_W)	*ii)* All but M_W, m_t	*iii)* All data
m_t / GeV	158^{+14}_{-11}	157^{+10}_{-9}	173.1 ± 5.4
m_H / GeV	83^{+168}_{-49}	41^{+64}_{-21}	115^{+116}_{-66}
$\log m_H$	$1.92^{+0.48}_{-0.39}$	$1.62^{+0.41}_{-0.31}$	$2.06^{+0.30}_{-0.37}$
$\alpha_s(M_Z)$	0.121 ± 0.003	0.120 ± 0.003	0.120 ± 0.003
χ^2/dof	8/9	14/12	17/15
M_W / GeV	80.298 ± 0.043	80.329 ± 0.041	80.375 ± 0.030

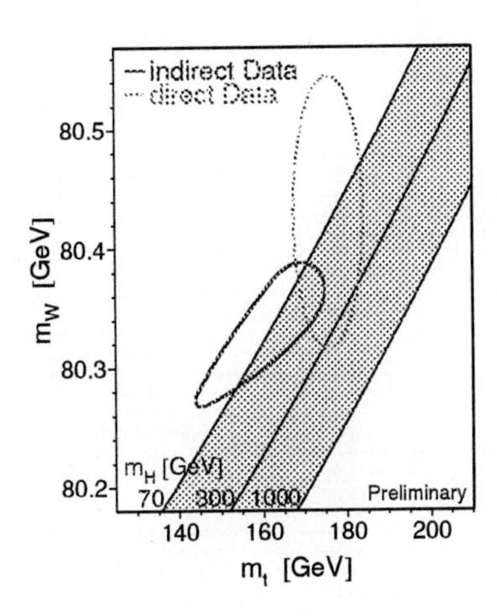

Fig. 7. Direct and indirect measurements of M_W and m_t, compared with the Standard Model for various values of m_H (68% probability contours).

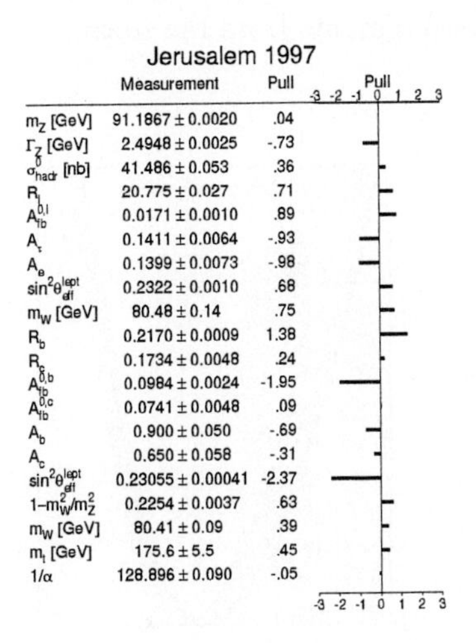

Fig. 8. Measured quantities and their pulls in the global Standard Model fit.

The results of the fits provide an indirect estimate of the mass of the Higgs boson, m_H. The most precise estimate comes from the fit to all data, yielding $m_H = 115^{+116}_{-66}$ GeV. The two most discrepant measurements in the Standard Model fit tend to pull m_H in opposite directions, A_{LR} favouring a low m_H, and A^b_{FB} preferring a high value. The dependence on m_H of the difference between χ^2 and its minimum value is shown in fig. 9. The band indicates an estimate of the theoretical uncertainties resulting from uncomputed higher order terms. Taking this into account, an upper limit on m_H may be placed:

$$m_H < 420 \text{ GeV} \quad (95\% \text{ c.l.})$$

In deriving this limit, the lower mass limit derived from direct searches, $m_H > 77$ GeV [34, 35], has not been taken into account.

The value obtained for $\alpha_s(M_Z) = 0.120 \pm 0.003$ is one of the most accurate measurements, even after including a theoretical systematic error of about ± 0.002. This measurement is compared with other recent determinations [36] in fig. 10. The measurements display a good level of consistency, which is a pleasing improvement on the situation a couple of years ago. A reasonable World Average value is

$$\alpha_s(M_Z) = 0.119 \pm 0.004$$

where the error has been estimated very simply as the r.m.s. deviation of the measurements from the mean.

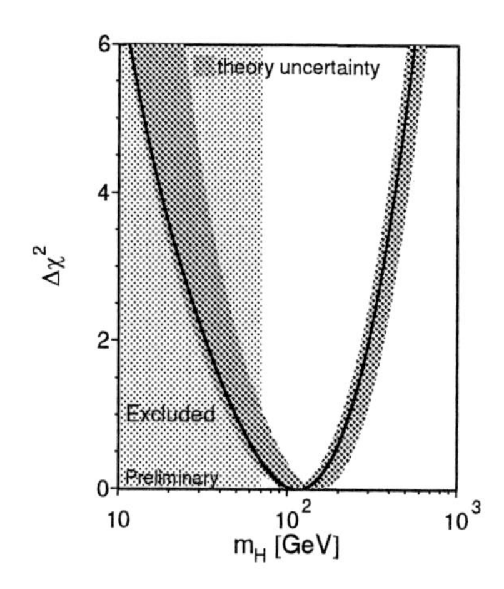

Fig. 9. Dependence of χ^2 of the global electroweak fit on m_H.

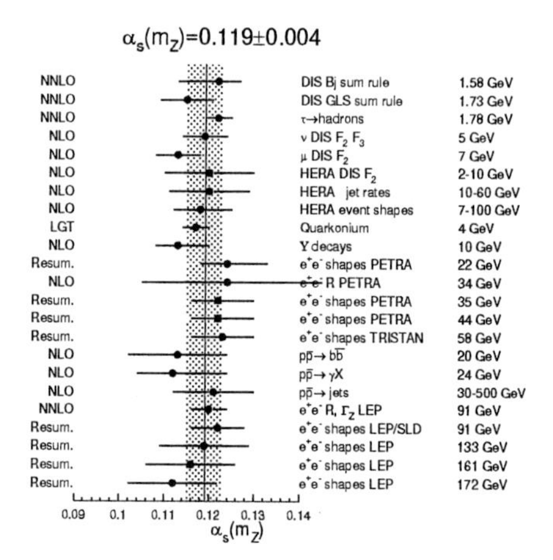

Fig. 10. A compilation of recent measurements of α_s.

5 Width and branching ratios of the W boson

5.1 W Width

The most precise (but indirect) estimate of the W width is based on the observed W and Z cross-sections at the Tevatron, using the relation

$$\frac{\sigma_W \cdot \mathrm{BR}(W \to \ell\nu)}{\sigma_Z \cdot \mathrm{BR}(Z \to \ell\ell)} = \frac{\sigma_W \cdot \Gamma(W \to \ell\nu) \cdot \Gamma_Z}{\sigma_Z \cdot \Gamma(Z \to \ell\ell) \cdot \Gamma_W}$$

Taking the cross-sections from QCD, the Z data from LEP, and $\mathrm{BR}(W \to \ell\nu)$ from the Standard Model, the combined CDF/D0 data yield $\Gamma_W = 2.06 \pm$

0.06 GeV, to be compared with the Standard Model expectation of 2.077 ± 0.014 GeV. A more direct measurement can be made from a detailed study of the tail of the transverse mass distribution (fig. 6). The latest result from CDF [29] is $\Gamma_W = 2.11 \pm 0.17 \pm 0.09$ GeV. First results from LEP [21], based on direct observation of the W lineshape, have started to appear:

$$\Gamma_W = 1.74^{+0.88}_{-0.78}(\text{stat.}) \pm 0.25(\text{syst.}) \text{ (L3)}$$
$$\Gamma_W = 1.30^{+0.62}_{-0.55}(\text{stat.}) \pm 0.18(\text{syst.}) \text{ (OPAL)}$$

At present the errors are not competitive, but an interesting measurement should be possible by the end of the 1997 run.

5.2 W branching ratios and V_{cs}

The observation of W^+W^- production at LEP permits a direct determination of the W branching ratios. The combined results from LEP are [37]:

W decay channel	Branching Ratio (%)
$e\nu$	10.8 ± 1.3
$\mu\nu$	9.2 ± 1.1
$\tau\nu$	12.7 ± 1.7
$\ell\nu$	10.9 ± 0.6
Hadrons	67.2 ± 1.7

The leptonic results are consistent with lepton universality, though not yet competitive with results from other processes, such as τ decays. The hadronic branching ratio can be related to elements of the CKM matrix:

$$\frac{B_h}{1 - B_h} = \sum_{i=u,c;\ j=d,s,b} |V_{ij}|^2 \left(1 + \frac{\alpha_s}{\pi}\right)$$

Amongst these CKM elements, V_{cs} is by far the least well measured (V_{cs}=1.01$\pm$0.18 from D meson decays). One can therefore take the other elements from the PDG world averages, and infer a value $|V_{cs}| = 0.96 \pm 0.08$.

Direct measurements of W $\to$ c using charm tagging are also appearing [38], which yield a more direct determination of V_{cs}. The values to date are:

$$V_{cs} = 1.13 \pm 0.43(\text{stat.}) \pm 0.03(\text{syst.}) \text{ (ALEPH)}$$
$$V_{cs} = 0.87^{+0.26}_{-0.22}(\text{stat.}) \pm 0.11(\text{syst.}) \text{ (DELPHI)}$$

Although V_{cs} can be constrained much more strongly by the unitarity of the CKM matrix, these direct measurements will ultimately provide an interesting check.

6 Triple gauge couplings

The WWZ and WWγ couplings are predicted by the Standard Model, and can be tested by the LEP and Tevatron experiments. The effective Lagrangian used to parametrize any anomalous couplings involves 2×7 parameters to describe most general Lorentz invariant WWV (V=Z,γ) vertices. By assuming C, P, and electromagnetic gauge invariance, this can be reduced to a more practicable set of five parameters: λ_γ, λ_Z (=0 in Standard Model) and κ_γ, κ_Z, g_1^Z (=1 in Standard Model); $g_1^\gamma = 1$ results from electromagnetic gauge invariance. These parameters may be related to the static moments of the W:

Charge	$Q_W = e g_1^\gamma$
Magnetic dipole moment	$\mu_W = (e/2m_W)(g_1^\gamma + \kappa_\gamma + \lambda_\gamma)$
Electric Quadrupole moment	$q_W = -(e/m_W^2)(\kappa_\gamma - \lambda_\gamma)$

At LEP II, anomalous values for these couplings generally increase the W^+W^- cross-section. We see from fig. 5 that the measured cross-sections clearly require the existence of both WWZ and WWγ couplings. Anomalous couplings also influence the production angle of the W^- and affect the helicity states, and hence the decay angles, of the $W^\pm$. The $W^+W^- \to q\bar{q}\ell\nu$ final states are particularly sensitive, because the lepton charge allows an unambiguous assignment of the W charges. Futher information can be obtained from "single W production" (i.e. $q\bar{q}e\nu$ final states with only a single on-shell W) and $\nu\bar{\nu}\gamma$ final states, which are particularly sensitive to the WWγ vertex.

The precise measurements of the Z already constrain possible anomalous couplings. For this reason, the LEP experiments have focussed on the following combinations of anomalous couplings, which do not affect gauge boson propagators at tree level, and are therefore not already indirectly constrained:

$$\begin{aligned} \Delta\kappa_\gamma - \Delta g_1^Z \cos^2\theta_W &= \alpha_{B\phi} \\ \Delta g_1^Z \cos^2\theta_W &= \alpha_{W\phi} \\ \lambda_Z = \lambda_\gamma &= \alpha_W \end{aligned}$$

with the constraint $\Delta\kappa_Z = \Delta g_1^Z - \Delta\kappa_\gamma \tan^2\theta_W$. All these couplings should be zero according to the Standard Model. Combined results from the four experiments have been obtained by adding likelihood curves from each experiment, taking both cross-section and angular information into account, as illustrated in fig. 11. The results are [39, 40]:

	95% c.l. limits
$\alpha_{W\phi} = 0.02 \pm^{0.16}_{0.15}$	$-0.28 < \alpha < 0.33$
$\alpha_W = 0.15 \pm^{0.27}_{0.27}$	$-0.37 < \alpha < 0.68$
$\alpha_{B\phi} = 0.45 \pm^{0.56}_{0.67}$	$-0.81 < \alpha < 1.50$

Thus no discrepancy from the Standard Model is observed.

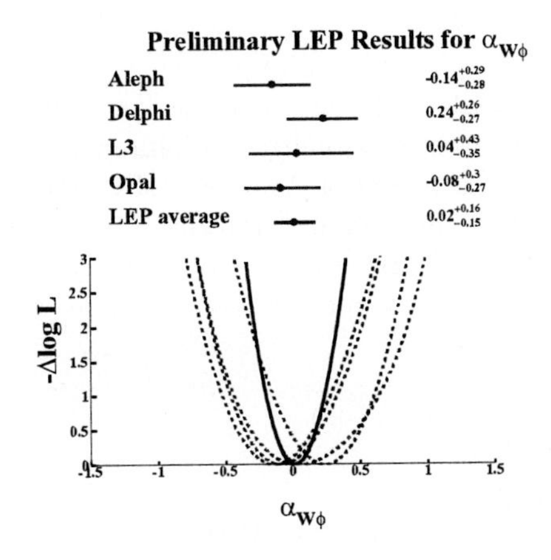

Fig. 11. Likelihood curves for the combined LEP measurement of $\alpha_{W\phi}$.

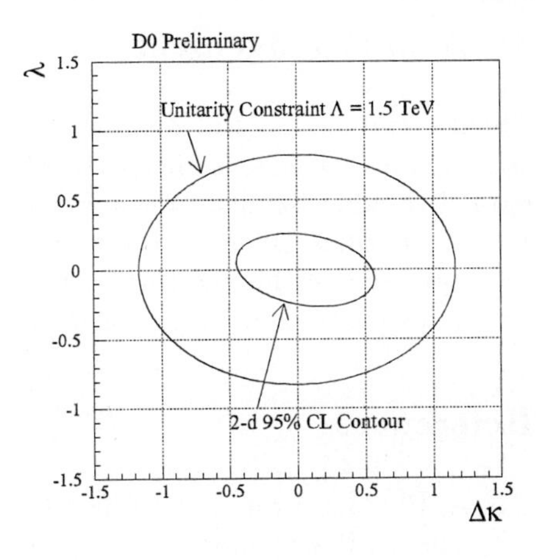

Fig. 12. Limits on the couplings λ and $\Delta\kappa$ from the combined analysis of D0.

In $\overline{\text{p}}$p experiments, information may be gleaned in two ways. Observation of the rate of Wγ production, especially for high p_T photons, is sensitive to the WWγ coupling (and thus complementary to LEP II, which is sensitive to WWZ as well). The results are:

	$\lambda_\gamma = 0$	$\Delta\kappa_\gamma = 0$
D0 [30, 41]	$-0.93 < \Delta\kappa_\gamma < 0.94$	$-0.31 < \lambda_\gamma < 0.29$
CDF [29]	$-1.8 < \Delta\kappa_\gamma < 2.0$	$-0.7 < \lambda_\gamma < 0.6$

In addition a few instances of WW or WZ pair production are observed in $\ell\nu\ell\nu$ and $\ell\nu jj$ final states. The must stringent limits are obtained by D0 [30, 41] in a combined fit to all channels, assuming equal WWZ and WWγ couplings: $-0.33 < \Delta\kappa < 0.45$ and $-0.20 < \lambda < 0.20$, as shown in fig. 12. Where comparison is possible, at present the Tevatron limits are typically better than the LEP II limits by a factor 2. This situation should start to change by the end of the 1997 LEP run.

7 Summary

In summary, the electroweak sector of the Standard Model continues to stand up to all tests. The precise electroweak measurements in Z decays at LEP I are coming to an end, and only modest improvements can be expected. The distinctive contribution from polarized beam measurements at SLC is set to continue. The results on W physics from LEP II and the Tevatron are starting to appear, and over the next few years a factor ~ 20 more data is anticipated at both machines, to pursue tests of the Standard Model.

8 Acknowledgements

Thanks are due to the members of the LEP/SLD Electroweak Working Group, who have done much of the work of professionally combining the data. I should also like to thank the many members of ALEPH, DELPHI, L3, OPAL, SLD, D0 and CDF who have given me help and information, and the organisers of this meeting in Jerusalem for inviting me to speak at such an interesting and well run conference.

References

[1] A. Blondel, Proc. ICHEP'96 (Warsaw), p.205.
[2] The LEP Collaborations and the LEP Electroweak Working Group, CERN-PPE/96-183.
[3] G. Quast, talk #709; these proceedings.
[4] L3 Collab., contributed paper #481.
[5] C. Mariotti, talk #707; these proceedings.
[6] E. Etzion, talk #708; these proceedings.
[7] ALEPH Collab., Phys. Lett. **B401** (1997) 150, 163.
[8] OPAL Collab., Z. Phys. **C74** (1997) 1.
[9] DELPHI Collab., contributed paper #419.
[10] L3 Collab., contributed paper #489.
[11] SLD Collab., contributed paper #118.
[12] OPAL Collab., CERN-PPE/97-093; Z. Phys. **C**, to be published.
[13] ALEPH Collab., contributed paper #623.
[14] SLD Collab., contributed paper #120.
[15] L3 Collab., contributed paper #490.
[16] OPAL Collab., Z. Phys. **C75** (1997) 385.
[17] SLD Collab., contributed papers #122,#123,#124,#125.
[18] C.P. Ward, talk #712; these proceedings.
[19] A. Leike, T. Riemann, J. Rose, Phys. Lett. **B273** (1991) 513
T. Riemann Phys. Lett. **B293** (1992) 451
S. Kirsch, T. Riemann, Comput. Phys. Commun. **88** (1995) 89.
[20] ALEPH Collab., Phys. Lett. **B401** (1997) 347;
DELPHI Collab., Phys. Lett. **B397** (1997) 158;
L3 Collab., Phys. Lett. **B398** (1997) 223;
OPAL Collab., Phys. Lett. **B389** (1997) 416.
[21] ALEPH Collab., contributed paper #600;
DELPHI Collab., contributed paper #347;
L3 Collab., CERN-PPE/97-98, Phys. Lett. **B**, to be published;
OPAL Collab., CERN-PPE/97-116, Z. Phys. **C**, to be published.
[22] D. Fassouliotis, talk #806; these proceedings.
[23] Physics at LEP2, CERN 96-01 vol. 1, pp. 141-205.
[24] J. Ellis and K. Geiger, hep-ph/9703348.

[25] DELPHI Collab., contributed papers #307, #667;
L3 Collab., contributed paper #814;
OPAL Collab., in [21].
[26] DELPHI Collab., Phys. Lett. **B401** (1997) 181;
DELPHI Collab., contributed paper #307;
ALEPH Collab., contributed paper #590.
[27] V. Kartvelishvili, R. Kvatadze, R.Møller, contributed paper #233.
[28] CDF Collab., Phys. Rev. **78** (1997) 4536;
D0 Collab., Fermilab-Pub-97/1136-E, Phys. Rev. Lett. , to be published.
[29] P. de Barbaro, talk #803; these proceedings.
[30] A. Kotwal, talk #804; these proceedings.
[31] A. Yagil, talk #16; these proceedings.
[32] J. Yu, talk #701; these proceedings.
[33] S. Eidelmann, F. Jegerlehner, Z. Phys. **C67** (1995) 585.
[34] W.J. Murray, talk #1307; these proceedings.
[35] P. Janot, talk #17; these proceedings.
[36] Based on: S. Bethke, Proc. QCD'97 Montpellier, hep-ex/9710030;
Updated with results from contributed papers #176, #177, #498, #544, #629.
[37] Ch. Burgard, talk #805; these proceedings.
[38] DELPHI Collab., contributed paper #535;
ALEPH Collab., contributed paper #667.
[39] A. Weber, talk #807; these proceedings.
[40] ALEPH Collab., contributed papers #599,#601;
DELPHI Collab., contributed paper #295;
L3 Collab., contributed papers #513,#514;
OPAL Collab., contributed papers #170,#171.
[41] D0 Collab., contributed paper #106;

High Q^2 Events at HERA

Eckhard Elsen (elsen@mail.desy.de)

Deutsches Elektronen Synchrotron, Hamburg, Germany

Abstract. The H1 and ZEUS collaborations have reported on the observation of an excess of events at very large Q^2 in deep-inelastic ep collisions at HERA. New data from the 97 data-taking up to the summer are included in this paper. In addition, events with a high p_t lepton, hadronic activity and large missing transverse momentum are observed. Their rate is compared to the expectations for $W^{\pm}$ production and background processes.

1 Introduction

The ep collider HERA allows the proton to be probed at distances down to $\simeq 10^{-16}$ cm via t-channel exchange of a highly virtual gauge boson [1, 2, 3], and to search for s-channel formation of new particles that couple to lepton-parton pairs with masses up to the kinematic limit of $\sqrt{s} \simeq 300\,\mathrm{GeV}$ [4, 5]. Earlier this year [6, 7] both collider experiments, H1 and ZEUS, reported on a measurement of deep-inelastic scattering at high Q^2 using all e^+p data recorded from 1994 to 1996 with respective luminosities of $14.2\,\mathrm{pb}^{-1}$ and $20.1\,\mathrm{pb}^{-1}$ at beam energies of 27.5 GeV and 820 GeV for positrons and protons. The data indicated an appreciable excess of events over the expectation from the DIS Standard Model. This paper gives an update of these analyses, which includes the original data and comprises $23.7\,\mathrm{pb}^{-1}$ and $33.5\,\mathrm{pb}^{-1}$ for H1 and ZEUS, respectively [8]. It extends the previous charged current analysis by a complementary study of a sub-sample with moderate Q^2: some events of the large missing transverse momentum selection yield isolated leptons with extraordinarily large p_T^{ℓ} which, a priori, can only be caused by decays of real $W^{\pm}$.

2 Experimental Considerations

In both experiments [9, 10] the calorimeter serves as the primary means of reconstructing the final state. H1 uses liquid argon with lead (steel) absorber as an electromagnetic (hadronic) calorimeter, which is placed inside the large solenoid of 1.2 T. The granularity is optimized to provide approximately uniform segmentation in laboratory pseudorapidity [1] and azimuthal angle ϕ. This arrangement provides particularly good energy resolution for electrons (table 1). ZEUS employs a uranium/scintillator calorimeter optimized for roughly equal response to electrons and pions.

[1] z direction is defined as the direction of the incident proton.

Table 1. Compilation of calorimeter resolutions. σ describes the energy resolution and Δ the uncertainty in the absolute energy scale.

	H1	ZEUS
elm. $\sigma(E)/E$	$12\%\sqrt{E/\mathrm{GeV}} \oplus 1\%$	$18\%\sqrt{E/\mathrm{GeV}} \oplus 2\%$
hadr. $\sigma(E)/E$	$50\%\sqrt{E/\mathrm{GeV}} \oplus 4\%$	$35\%\sqrt{E/\mathrm{GeV}} \oplus 3\%$
elm. $\Delta E/E$	3%	3%
hadr. $\Delta E/E$	4%	3%
σ_{θ_e}	2-5 mrad	5 mrad

The kinematics of lepton-proton scattering at fixed collider beam energies is described by two independent variables. Several Lorentz invariant quantities can be calculated of which the most common are Q^2, x and y (eq. (1) through (5)). Here k and P refer to the 4-vectors of the incoming e^+ and p beams and k' to the 4-vector of the scattered positron. Q^2 describes the virtuality of the exchanged boson, while, in the parton picture, y and M are related to the scattering angle θ^* in the CM system and to the mass, respectively, of the positron-quark interaction.

$$Q^2 = -q^2 = (k - k')^2 \tag{1}$$

$$y = \frac{Pq}{Pk} = \frac{1}{2}(1 + \cos\theta^*) \tag{2}$$

$$x = \frac{Q^2}{2Pq} \tag{3}$$

$$M = \sqrt{sx} \tag{4}$$

$$Q^2 = \frac{P_T^2}{(1-y)} \tag{5}$$

Experimentally, convenient approximations are based on using equation (5) or derivations thereof, where the transverse momentum P_T and the angle related y are measured from either the scattered lepton, the momenta of the scattered hadrons (X) or both. Of the four variables available in neutral current (NC) events H1 obtains the best resolutions when using the angle and energy of the scattered positron (e-method) while ZEUS takes advantage of reducing the energy scale uncertainties by measuring the angle of the scattered lepton and an effective hadron angle (double angle method).

3 Hard Scale

The selections described subsequently refer to measurements of large p_T-processes. Associating each calorimeter cluster with a momentum vector $\mathbf{p}_{\mathrm{cl}}$ ($E_{\mathrm{cl}} = |\mathbf{p}_{\mathrm{cl}}|$) from the origin, both the scalar (S) and the vector (V) sums of the transverse momentum components can be used to distinguish different physics processes.

Table 2. Kinematics of the H1 events with leptons and large transverse momentum imbalance. The calculation of the global event quantities includes the measurement of the μ-momenta in the H1 tracking system. When calculating $M_T^{\ell\nu}$, the transverse $\ell\nu$-mass, the missing transverse momentum is assumed to be carried by a single ν. MUON-1 has been published before [5].

	ELECTRON	MUON-1	MUON-2	MUON-3	MUON-4
charge$^\ell$	neg.(5σ)	pos.(5σ)	pos.(5σ)	neg.(5σ)	neg.(2.7σ)
p_T^ℓ [GeV]	$37.6^{+1.3}_{-1.3}$	$23.4^{+7.5}_{-5.5}$	$27.7^{+6.4}_{-4.4}$	$39.4^{+9.4}_{-6.4}$	$81.5^{+47.1}_{-21.9}$
θ^ℓ [*deg.*]	27.3 ± 0.2	46.2 ± 1.3	28.9 ± 0.1	35.2 ± 0.1	28.5 ± 0.1
p_T^X [GeV]	8.0 ± 0.8	42.2 ± 3.8	67.5 ± 5.4	27.5 ± 2.7	59.8 ± 5.9
Event global					
$P_{T,\mathrm{miss}}$ [GeV]	30.6 ± 1.5	$18.9^{+6.6}_{-8.3}$	$43.5^{+5.4}_{-6.4}$	$43.3^{+8.0}_{-5.3}$	$29.2^{+44.1}_{-13.4}$
δ [GeV]	10.4 ± 0.7	$18.9^{+3.9}_{-3.2}$	$17.0^{+1.9}_{-1.5}$	$27.7^{+3.4}_{-2.6}$	$43.6^{+12.2}_{-6.1}$
$M_T^{\ell\nu}$ [GeV]	67.7 ± 2.7	$3.0^{+1.5}_{-0.9}$	$22.6^{+5.0}_{-3.6}$	$77.8^{+18.0}_{-12.2}$	$93.2^{+100.0}_{-46.5}$

3.1 Large Transverse Momentum Imbalance

A transverse momentum requirement $V > 25\,\mathrm{GeV}$, a priori, preferentially selects charged current (CC) events, in which the ν remains undetected. H1 has performed such a search and found 336 candidate events after applying the standard cosmic and halo filtering [2]. The surprising observation among this sample is a yield of five events with a high-p_t isolated track. The isolation criterion has been based on separation in rapidity-azimuth both from the nearest charged tracks and from a calorimeter jet, separately. All five tracks are well compatible with a lepton: four leptons are identified as muons and one event as an electron. The parameters of the events are listed in table 2.

Production of real $W^\pm$ in $ep \to eW^\pm X$ naturally leads to events with large p_t^ℓ when the W decays leptonically. H1 expects to see 0.41 ± 0.07 (1.34 ± 0.20) in the $\mu^\pm$ and $e^\pm-$ channels respectively, as estimated from a full simulation [11] of the underlying processes. Note, however, that such processes preferentially lead to $M_T^{\ell\nu} \sim m_W$ (and small p_T^X), which is not really evidenced by all events of table 2. Other background processes, such as photoproduction of large E_T-jets or of heavy quarks providing an unobserved ν seem to contribute only little. Another background stems from $\gamma\gamma \to \ell\ell$ ($\ell = \mu$ or e), when one of the leptons remains unobserved in the beam pipe. This process has been studied and cannot contribute to largely imbalanced events (once the μ-momentum is taken into consideration).

For a similar selection [12] ZEUS find no $\mu^\pm$- and 2 e^+-events, when the expected $W^\pm$ signal and background is estimated to be 0.48 and 4. The two events can be understood as a high-Q^2 di-muon event $e^+p \to e^+\mu\mu X$ and an acoplanar e^++jet event.

Table 3. The observed number of CC events as a function of a lower Q^2-cutoff $Q^2_{\rm min}$. The expectations are based on the SM DIS model. The syst. errors of ZEUS have been separated into an uncertainty due to the energy scale and due to parton distributions (PDF).

	H1		ZEUS			
$Q^2_{\rm min}$ [GeV^2]	seen	expected	seen	expected	E scale	PDF
1000			**455**	419	$\pm$ 13	$\pm$ 33
2500	**61**	56.30 ± 9.40	**192**	178	$\pm$ 13	$\pm$ 17
5000	**42**	34.70 ± 6.90	**63**	58.5	$\pm$ 9.0	$\pm$ 7.3
10000	**14**	8.33 ± 3.10	**15**	9.4	$\pm$ 2.5	$\pm$ 1.6
15000	**6**	2.92 ± 1.44	**5**	2.0	$^{+0.81}_{-0.54}$	$\pm$ 0.4
20000	**3**	1.21 ± 0.64	**1**	0.46	$^{+0.28}_{-0.16}$	$\pm$ 0.10
30000			**1**	0.034	$^{+0.037}_{-0.016}$	$\pm$ 0.008

3.2 Charged Current Events at High Q^2

Concentrating on the high Q^2 regime a CC selection is accomplished by requiring an even more significant transverse momentum imbalance. The detailed cuts of H1 are: $V > 50\,\text{GeV}$ and $V/S > 0.5$ and $y < 0.9$ to largely eliminate the background from photoproduction. H1 and ZEUS obtain the events of table 3 as a function of Q^2 for their respective selections. Both experiments observe slightly larger event yields for $Q^2 > 10000\,\text{GeV}^2$ than expected. ZEUS finds an event with $Q^2 = 34000\,\text{GeV}^2$. The expectation is based on the Standard Model of DIS. The uncertainty in this prediction is related to the knowledge of the parton distribution functions (PDF) (see below). The largest uncertainty arises from the incomplete knowledge of the absolute energy scale (table 1).

3.3 Neutral Currents

H1 has provided new data at this conference [13] on the measurement of the NC DIS cross sections up to $5000\,\text{GeV}^2$, which are shown in their reduced form

$$\sigma(e^+p) = \frac{xQ^4}{2\pi\alpha^2}\frac{1}{1+(1-y)^2}\frac{\mathrm{d}^2\sigma(e^+p \to e^+X)}{\mathrm{d}x\mathrm{d}Q^2} \tag{6}$$

in fig. 1. Despite the fact that measurements at smaller Q^2 had to be extrapolated over two orders of magnitude the canonical parameterisation of the parton distributions evolved through the DGLAP mechanism is seen to hold up to the largest values of Q^2. It is also seen that for very large x and Q^2 the effect of the structure function F_3 is seen as an apparent decrease of $\sigma(e^+p)$. Not more than $\mathcal{O}(5\%)$ is left as an uncertainty on the cross section and hence the parton evolution at high Q^2 (see also refs. [6, 7]).

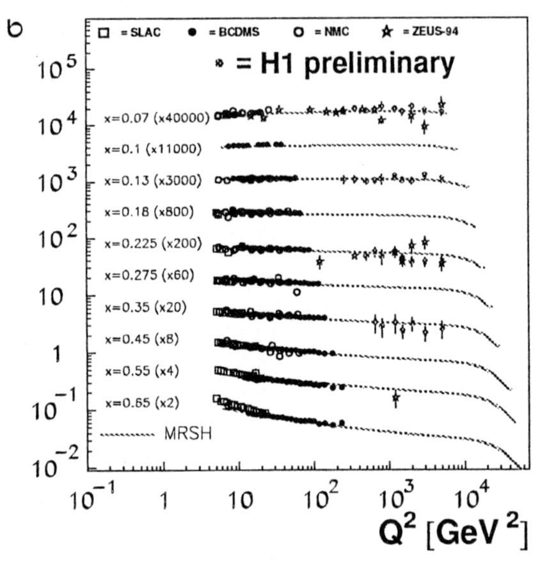

Fig. 1. The reduced cross section of NC scattering (eq. 6) including the preliminary data of ref. [13].

3.4 Neutral Current Events at high Q^2

H1 and ZEUS have updated [13, 14] their selection of NC events at large Q^2, where, in addition to the positron requirement sufficient balance in $\delta = \sum E(1 - \cos\theta)$ as a safeguard against catastrophic initial state radiation has been demanded. The respective numbers of events are 724 (714 ± 69) for $Q^2 > 2500\,\mathrm{GeV}^2$ and 326 (328 ± 15) for $Q^2 > 5000\,\mathrm{GeV}^2$ for H1 and ZEUS, where the numbers in parentheses refer to the expectation.

The comparison of the observed Q^2 dependence is seen in fig. 2. At $Q^2 > 15000\,\mathrm{GeV}^2$ the data start to exceed the SM expectation. Again, ZEUS finds events with *very* large Q^2. These measurements continue to indicate the trend reported on in ref. [6, 7] of exceeding the SM model expectation with probabilities hovering around 1%.

The individual events are shown in the M-y plane in fig. 3 for both experiments. H1 had reported a major contribution to the excess at $M \sim 200\,\mathrm{GeV}$, for which ZEUS sees no particular enhancement. The quantities have been calculated in different manners and are hence susceptible to different systematics, notably with respect to initial state radiation. Initial state radiation leads to an underestimate of the true value of M for the e-method (H1) and to an overestimate in the double angle method (ZEUS). Nevertheless, the consistency of the measurement has been verified [15] experimentally by comparing the results of the two methods of calculation on an event basis.

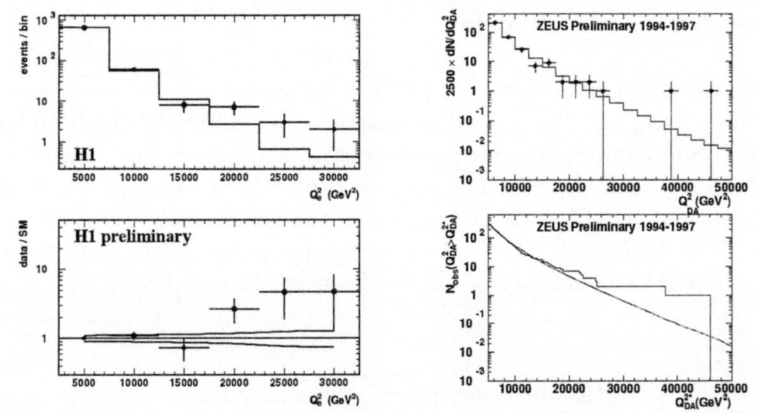

Fig. 2. Q^2 dependence of NC scattering; left: H1, right ZEUS. The ZEUS numbers are given for $Q^2 > Q^2_{DA}$ (integr. distribution). The solid lines refer to the DIS SM model.

As a result it seems unlikely that the enhancement observed in both experiments could be made to agree on a *single* value of M. (ZEUS have confirmed in a separate study on MC events that a large deviation in M due to initial state radiation is indeed very unlikely). On the other hand, if all the residual uncertainty on the energy scale is allowed for, the differences become less striking.

It remains to be noted that H1 continues to see the biggest enhancement around $M \sim 200\,\text{GeV}$, while ZEUS sees several events at extremely large x and y.

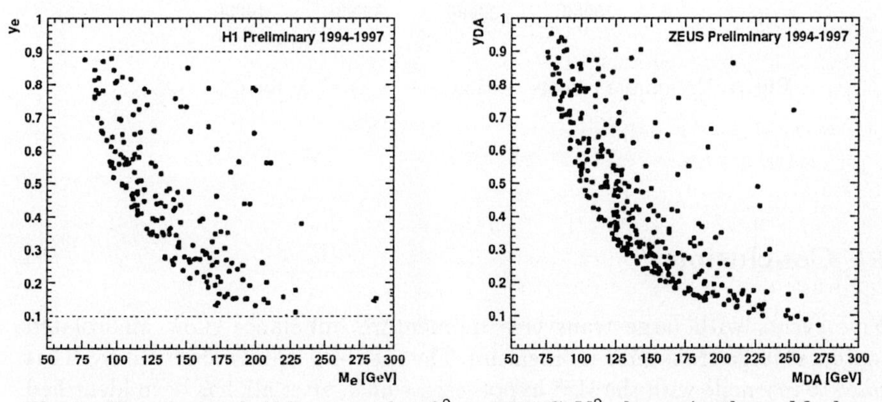

Fig. 3. The observed NC events for $Q^2 > 5000\,\text{GeV}^2$ shown in the y-M plane. Each experiment uses its preferred method to compute M (see text).

4 The NC Cross Section

The observed number of events have been converted into a cross section $\sigma(Q^2 > Q^2_{\mathrm{min}})$ (fig. 4). The procedure assumes the validity of the Standard Model of DIS to compute migration corrections, efficiencies, and to extrapolate into the region of large y which is either explicitly excluded ($y < 0.9$, H1) or implicitly restricted by the acceptance of the calorimeter. ZEUS records events to $Q^2 > 40000\,\mathrm{GeV}^2$. (A corresponding upper limit on the cross section in bins, where H1 sees no events, has not been included in fig. 4).

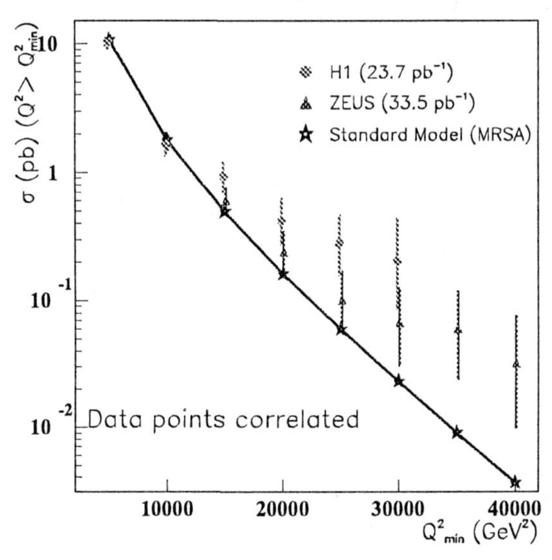

Fig. 4. Preliminary cross section of $e^+p \to e^+X$ for $Q^2 > Q^2_{min}$.

5 Conclusion

Five events with large transverse momentum imbalance show an isolated lepton at large transverse momentum. The rate of μ-events observed by H1 is hard to reconcile with the $W^\pm$ hypothesis, which, after all, has been identified as the most likely production mechanism within the Standard Model.

H1 and ZEUS continue to find an excess of events in deep-inelastic scattering. In NC processes for $Q^2 > 15000\,\mathrm{GeV}^2$ H1 sees 18 events when 8.0 ± 1.2 events are expected. For $Q^2 > 35000\,\mathrm{GeV}^2$ ZEUS sees 2 events with 0.24 ± 0.02

expected. These measurements are in agreement with the observations reported earlier using a sub-sample of the data but they do not improve the statistical significance appreciably. The excess seen in CC processes is based on a small number of events.

Some $10\,\mathrm{pb}^{-1}$ will be added to the data sample by each experiment from the remainder of the 1997 data-taking. 1998 makes possible the study of e^-p data, with a significantly different high energy behaviour. If the observation survives these stringent statistical tests we will be faced with accepting one of the few remaining theoretical explanations [16] beyond the Standard Model which are left open.

References

[1] ZEUS Collaboration, M. Derrick et al., *Z. Phys.* C **67** (1995), 81; H1 Collaboration, S. Aid et al., *Phys. Lett.* B **353** (1995), 578; ZEUS Collaboration, M. Derrick et al., *Z. Phys.* C **72** (1996), 47.
[2] H1 Collaboration, S. Aid et al., *Z. Phys.* C **67** (1995), 565.
[3] H1 Collaboration, S. Aid et al., *Phys. Lett.* B **379** (1996), 319.
[4] H1 Collaboration, I. Abt et al., *Nucl. Phys.* B **396** (1993), 3; ZEUS Collaboration, M. Derrick et al., *Phys. Lett.* B **306** (1993), 173; H1 Collaboration, T. Ahmed et al., *Z. Phys.* C **64** (1994), 545; H1 Collaboration, S. Aid et al., *Phys. Lett.* B **369** (1996), 173; ZEUS Collaboration, M. Derrick et al., *Z. Phys.* C **73** (1997), 613.
[5] H1 Collaboration, I. Abt et al., *Z. Phys.* C **71** (1996), 211
[6] H1 Collaboration, C. Adloff, et al., *Z. Phys.* C **74** (1997), 191.
[7] ZEUS Collaboration, J. Breitweg, et al., *Z. Phys.* C **74** (1997), 207.
[8] B. Straub, Plenary Talk, *Proceedings of the 1997 Lepton Photon Symposium*, Hamburg 1997.
[9] H1 Collaboration, I. Abt et al., *Nucl. Instrum. Methods* A **386** (1997), 310; H1 Collaboration, I. Abt et al., *Nucl. Instrum. Methods* A **386** (1997), 348
[10] ZEUS Collaboration, M. Derrick, et al., *Phys. Lett.* B **293** (1992), 465;ZEUS Collaboration, M. Derrick, et al., *Z. Phys.* C **63** (1994), 391;ZEUS Collaboration, The ZEUS Detector, Status Report 1993, DESY 1993.
[11] U. Baur, J.A.M. Vermaseren and D. Zeppenfeld, *Nucl. Phys.* B **375** (1992), 3.
[12] M. Kuze, ZEUS Collaboration, private communication.
[13] U. Bassler, Parallel Session Talk at this conference.
[14] D. Acosta, Parallel Session Talk at this conference.
[15] U. Bassler and G. Bernardi, *Z. Phys.* C **76** (1997), 223.
[16] J. Ellis, Parallel Session Talk at this conference.

Searches for New Particles at Present Colliders

Patrick Janot (Patrick.Janot@cern.ch)

CERN, Geneva, Switzerland

Abstract. Most recent results of searches for new particles at LEP, at HERA and at the Tevatron are reviewed, including discoveries, hints for new physics, and a few 95% confidence level exclusion limits on Higgs bosons and supersymmetric particles.

1 Introduction

The results of new particle searches presented here were obtained by the experiments installed at the three highest energy colliders currently in operation: LEP at CERN, an e^+e^- collider with a centre-of-mass energy upgraded in steps from 130 to 183 GeV between 1995 and 1997 (results from ALEPH, DELPHI, L3 and OPAL); HERA at DESY, an asymmetric machine where 27 GeV positrons collide on 800 GeV protons, corresponding to a centre-of-mass energy of 300 GeV (results from H1 and ZEUS); and the Tevatron at Fermilab, a $p\bar{p}$ collider with a centre-of-mass energy of 1.8 TeV (results from CDF and D0). Substantial luminosity improvements were achieved both at HERA and the Tevatron, with about 30 and 110 pb^{-1} analysed respectively, to be compared to the 3 and 20 pb^{-1} available at the time of the 1995 EPS conference in Brussels [1]. In addition, the second phase of LEP (LEP 2) started its operation in 1995 with centre-of-mass energies well above the Z mass. Altogether, a total of 35 pb^{-1} was accumulated in the past two years by each of the four LEP experiments (by the time of the conference): 6 pb^{-1} at 130-136 GeV in 1995, 22 pb^{-1} in 1996 at 161-172 GeV, and 7 pb^{-1} at 183 GeV in the first two weeks of operation in 1997. Since then, in 1997, a total of 60 pb^{-1} at 183 GeV, and an additional 7 pb^{-1} at 130-136 GeV, were recorded.

An impressive quantity of new experimental results was therefore reported at this conference. It is impossible to do justice to the vast amount of material which has been presented in the various parallel sessions, and only a few selected topics are covered in this talk. First, the most important discoveries are described, followed by a review of the possible hints for new physics which have been reported in the past two years, the subsequent checks and their present status. Finally, the most recent results of the searches for Higgs bosons and supersymmetric particles are listed. In the following, all limits reported are given at the 95% confidence level, and all discoveries at the 5σ level.

2 Discoveries

Unfortunately, no discoveries were reported this year... We all look forward to the next EPS conference.

3 Hints for New Physics

A few deviations with respect to the standard model predictions were observed and reported in the past two years. They can be due to either statistical fluctuations or systematic effects, but may also be the first hints for new physics. The four most popular "effects" are: the HERA high Q^2 events, the CDF ee$\gamma\gamma\not{E}_T$ event, the ALEPH four-jet events, and the CDF High E_T excess. For each, the facts are briefly reminded, the analysis of recent data is presented, and the checks made by the other experiments, and/or at other colliders, are reported. My personal conclusions are also given.

3.1 The HERA High Q^2 events

The experimental status of the HERA High Q^2 events has been reported elsewhere [2]. When presented for the first time [3, 4], an excess of neutral current events at high x and Q^2 was found in the data taken up to 1996 with both the H1 and the ZEUS detectors. With 14.2 and 20.1 pb^{-1} analysed, H1 and ZEUS found seven and four events with 0.95 and 0.91 expected, respectively, in a high x/high Q^2 window. The H1 excess was found in addition to be clustered around an electron-quark invariant mass of 200 GeV/c^2, with possible leptoquark or squark (with R-parity violation) interpretations. However, the H1 and ZEUS windows, chosen to maximize the significance of a possible effect (this procedure was also followed on Monte Carlo gedanken experiments to compute unbiased probabilities), were not only different, but disjointed, therefore rendering questionable the compatibility of the two observations. Still, the cumulative Q^2 spectrum seemed to exhibit a quite significant excess at high values when combining H1 and ZEUS data, as if a new electron-quark contact interaction had taken place. The significance of the excess was reinforced by charged current events. (See Table 1.)

An additional integrated luminosity of 9.5 and 13.4 pb^{-1} was recorded by H1 and ZEUS until June 1997, and can be used to check the reproducibility of the effect (without, of course, including the former data where the effect was observed!). In the window defined by the H1 94-96 data, H1 found one additional event in 1997, while ZEUS saw altogether three events, making a total of four events observed with 3.6 expected. Similarly, in the window defined by the ZEUS 94-96 data, ZEUS found one additional event in 1997, while H1 saw altogether one event, for a total of two events observed with 1.4 expected. When the two windows are merged, six events are therefore observed with 5.0 expected. The combined cumulative 1997 Q^2 spectra (charged and neutral currents) are given in Table 2.

The natural conclusion of these new numbers, based on unbiased data samples and obtained with the same selection as that used in 1996, is that the former excesses are not confirmed, thus favouring the hypothesis of an unlikely fluctuation (either in 1996, or in 1997).

Several independent checks were anyway performed at LEP and at the Tevatron. If a new electron-quark contact interaction were responsible for the 1996 HERA excesses, it could be seen as deviations in the total $e^+e^- \to q\bar{q}$ (LEP)

Table 1. Numbers of neutral (NC) and charged (CC) current events observed (Obs.) and expected (Exp.) above a given Q^2 value in H1 and ZEUS 1994-96 data

$Q^2 > \ldots$ (GeV^2)	10000	15000	20000	25000	30000
Obs./Exp. (NC)		24/13.4	10/ 4.1	6/ 1.5	4/ 0.6
Obs./Exp. (CC)	19/10.6	8/ 2.9	5/ 1.0		

Table 2. Numbers of neutral (NC) and charged (CC) current events observed (Obs.) and expected (Exp.) above a given Q^2 value in H1 and ZEUS 1997 data

$Q^2 > \ldots$ (GeV^2)	10000	15000	20000	25000	30000
Obs./Exp. (NC)		12/ 9.5	4/ 3.0	1/ 1.0	0/ 0.3
Obs./Exp. (CC)	9/ 7.1	3/ 2.0	0/ 0.7		

or $q\bar{q} \rightarrow e^+e^-$ (Tevatron) cross-sections. No such deviation was observed, either in the combined LEP $q\bar{q}$ cross-section [5] or in the CDF di-electron mass spectrum [6], allowing stringent limits to be set on the relevant compositeness scales [7]. These limits are reported in Fig. 1, together with the cumulative HERA Q^2 spectrum and its standard model prediction. It can be seen that the contact interaction interpretation could not explain the shape of the low part of this spectrum , and is unambiguously excluded by the LEP/CDF constraints for the high part.

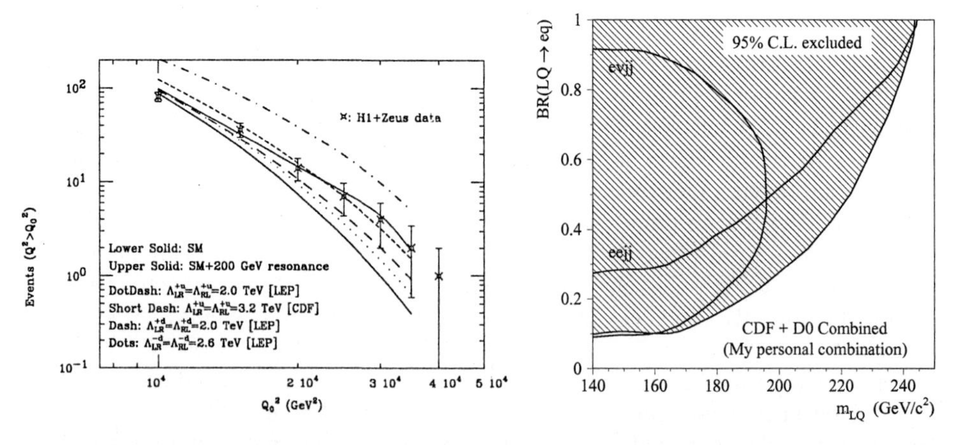

Fig. 1. Left: Comparison of the HERA neutral current cumulative Q^2 spectrum with compositeness (as excluded at 95% C.L. by LEP and CDF) and with a 200 GeV/c^2 resonance. Right: Region excluded (from a combination of CDF and D0) for the mass of a first generation leptoquark as a function of its branching ratio into eq.

Still, a 200 GeV/c^2 resonance (see Fig. 1), such as a first generation leptoquark, could accommodate the HERA observations. Such a resonance would then be copiously pair-produced at the Tevatron *via* gluon s-channel exchange. The two particles thus produced would decay either in eq, with a large branching fraction, or in ν_eq, with a smaller but sizeable branching fraction, to explain both neutral and charged current HERA excesses. The two relevant final states (eeqq and e$\nu_e$qq) were looked for by the CDF and D0 experiments [8, 9], by searching for events with two jets with large E_T, two electrons with large E_T or only one such electron accompanied with large $\not{E}_T$. No significant excess was observed (a total of 5 events was observed with 10.6 expected, with a signal selection efficiency of 20-30%). The resulting limit on the leptoquark mass is shown in Fig. 1b as a function of its branching fraction into eq, when the two searches, with both CDF and D0 data, are combined. (This is my personal combination.) An hypothetical 200 GeV/c^2 leptoquark is excluded for any value of this branching fraction in excess of 25%. This excludes the leptoquark interpretation of the 1996 HERA excess of high Q^2 events.

Another possible interpretation of this excess could be a R-parity violating squark s-channel production, through a $\lambda' L_i Q_j D_k^c$ coupling. The only viable candidate for HERA is [10]: $e^+d_R \rightarrow \tilde{c}_L \rightarrow e^+d_R$, competing with a R-parity conserving decay $\tilde{c}_L \rightarrow c_L\chi^0$ to evade the leptoquark negative search at the Tevatron, followed by a R-parity violating neutralino decay $\chi^0 \rightarrow q\bar{q}'\ell^\pm, \nu$. A preliminary search for such objects, which would be pair-produced at the Tevatron, has been carried out by CDF [8], with a final state containing two like-sign electrons, multijets with high E_T and no significant $\not{E}_T$. No events were found, with a signal selection efficiency of 15%. This would exclude a 200 GeV/c^2 c-squark if its R-parity conserving decay were 100%. Further investigation is needed for smaller branching fraction values.

To summarize, the contact interaction and the leptoquark interpretation of the high Q^2 events seen in HERA in 1994-96 are ruled out by LEP and Tevatron negative searches. The only remaining viable explanation, a c-squark with R-parity violating decay, in not in particularly good shape. Since, in addition, the earlier reported excess was not confirmed by more recent data, its most likely interpretation is a statistical fluctuation. This will undoubtedly be sorted out with further statistics.

3.2 The CDF ee$\gamma\gamma\not{E}_T$ event

In 1995, a spectacular ee$\gamma\gamma$ event accompanied by large $\not{E}_T$ was found in the first 20 pb^{-1} of the CDF data. No other such events were found since then, either by CDF or by D0. Although never published (this event is still thoroughly studied by the CDF collaboration to understand possible experimental flaws), theorists have been speculating a lot about its origin in the past two years. There are indeed many possible interpretations, of which the most likely is certainly the standard model: more than 0.01 WW$\gamma\gamma$ events were expected to be seen in the D0 and CDF data [11]. The fact that, in this particular event, the two W's decayed — which they have to do! — into eν_e is anecdotic and should not be taken into consideration when computing the relevant probabilities.

Supersymmetry, involving radiative neutralino decays, could also explain the occurrence of this event, either *via* $\chi_2^0 \rightarrow \chi_1^0\gamma$ in which case the particular event kinematics requires non-universal gaugino SUSY breaking terms, or *via* $\chi_1^0 \rightarrow \tilde{G}\gamma$, as in models with gauge mediated supersymmetry breaking (GMSB). In both cases, the lightest neutralino χ_1^0 or the massless gravitino $\tilde{G}$ would escape undetected and be responsible for the observed $\not{E}_T$. Two processes could give rise to an $ee\gamma\gamma\not{E}_T$ event: a chargino pair $\chi^+\chi^-$ production with subsequent chargino decay into $W\chi^0$, each W decaying into $e\nu_e$, and selectron pair $\tilde{e}^+\tilde{e}^-$ production, with a subsequent selectron decay into $e\chi^0$. In the first case, the production cross-section would be such that, considering all decay channels of the W's, many $\gamma\gamma\not{E}_T + X$ events would be expected to be found in the CDF and D0 data.

Inclusive searches were therefore carried out by the two experiments: events with two large E_T photons were selected and their $\not{E}_T$ distribution scrutinized [12]. No apparent excess was found in either experiment (see Fig. 2). Given the constraints from the kinematics of the CDF event, D0 was able to exclude its chargino interpretation.

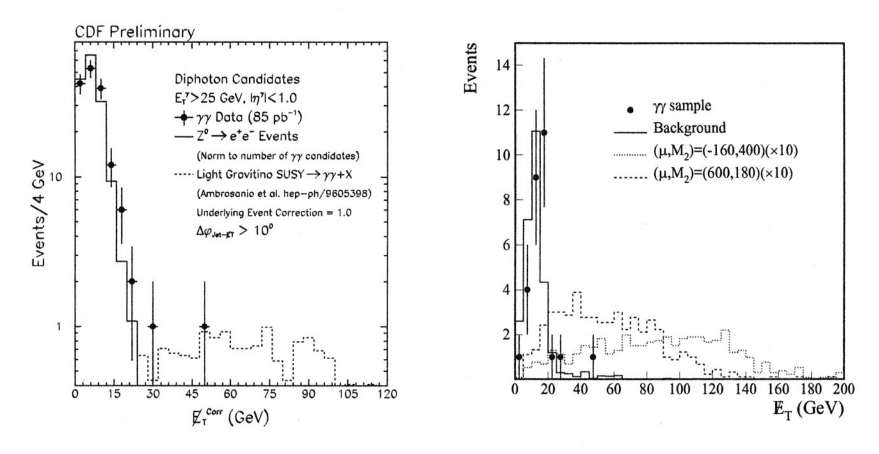

Fig. 2. Inclusive $\not{E}_T$ distribution in CDF (left) and D0 (right) for $\gamma\gamma + X$ events and comparison to the predictions from various supersymmetric scenarii.

The selectron interpretation was further studied at LEP, by looking for $\chi^0\chi^0$ or $\chi^0\tilde{G}$ production *via* a selectron exchange in the t-channel, giving rise to events with one photon or two acoplanar photons in the final state, accompanied with large missing energy. The events found in both final state topologies were in close agreement with the prediction from the standard model process $e^+e^- \rightarrow \nu\bar{\nu}\gamma(\gamma)$ for which the missing mass is expected to cluster around the Z mass [13]. For instance, in the acoplanar photon final state, 14 events were observed by the four LEP experiments combined with 18 expected, of which seven with a missing mass above 100 GeV/c^2(with 6.6 expected) and none below 80 GeV/c^2, while many would have been produced there according to the prediction of GMSB models. This agreement with the standard model allows a large fraction of the $(m_{\chi^0}, m_{\tilde{e}_R})$ plane to be excluded by LEP. In Fig. 3, it is compared to the region defined by the kinematics of the CDF event in its selectron interpretation. Most of this region will be tested by the end of LEP 2.

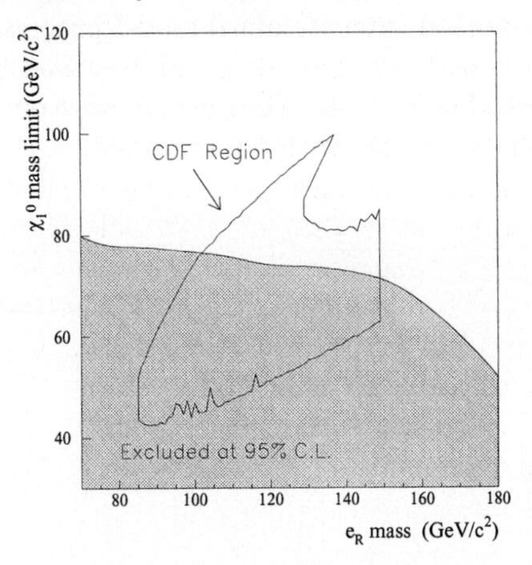

Fig. 3. Region of the $(m_{\chi^0}, m_{\tilde{e}_R})$ plane exluded [13] by the LEP acoplanar photon searches, compared to the region allowed by the kinematics of the CDF $ee\gamma\gamma\not{E}_T$ event in its selectron interpretation.

To conclude, while waiting for an official statement from the CDF collaboration about possible experimental flaws in this spectacular event, its most likely interpretation remains standard, with a probability in excess of 1%. Other exotic interpretations are or are being excluded.

3.3 The ALEPH Four-Jet events

With the data recorded by ALEPH in November 1995 at centre-of-mass energies of 130/136 GeV, 16 events were selected in the four-jet topology with 7.3 expected, and were found to present an unexpected accumulation in the dijet mass sum spectrum around 105 GeV/c^2 [14]. Nine events were selected with a dijet mass sum in a mass interval corresponding to $\pm 2\sigma$ of the detector resolution, for the jet pairing with the smallest mass difference, while 0.7 were expected from standard model processes. (See Fig. 4a.) These events were characterized in addition by unusual parton dynamics and jet charges, and by a large dijet charge separation. The other three LEP Collaborations, DELPHI, L3 and OPAL, developed selection procedures very similar to that of ALEPH. Their measured rate during the 1995 run at 130–136 GeV was also high, although compatible within statistics with expectations: a total of seven events with a dijet mass sum around 105 GeV/c^2 was reported with 2.9 events expected [15]. No significant deviation was observed at the higher centre-of-mass energies, including the most recent 183 GeV data, by either experiment.

Given that no systematic bias was found, either in the analysis or in the events themselves, and that it was shown that all experiments have similar performance to detect this kind of events, the origin of the effect at 130 and 136 GeV, new physics or statistical fluctuation, remained open. A new short run at these centre-of-mass energies was thus scheduled at LEP to settle the problem in October 1997 (thus after the conference). A total of eight events was observed in ALEPH, with 9.2 expected from standard model processes. In the dijet mass sum window around 105 GeV/c^2, the one event observed is in agreement with the 0.9 events predicted. (See Fig. 4b.) Four events were observed in the window by the other three Collaborations, with 3.0 expected.

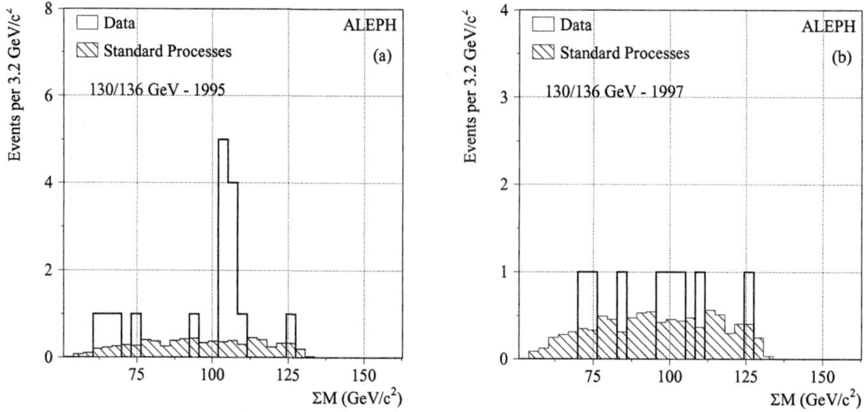

Fig. 4. Distributions of the dijet mass sum for the jet pair combination with the smallest dijet mass difference in the ALEPH data taken at centre-of-mass energies of 130 – 136 GeV (a) in 1995, and (b) in 1997.

No other explanation was found than a statistical fluctuation.

3.4 The CDF High E_T events

In 1996, with the first 20 pb^{-1} recorded by CDF, the inclusive jet cross-section was found to be [16] in good agreement with the QCD predictions for a total transverse energy below 200 GeV, but presented a significant excess above 200 GeV. The additional 90 pb^{-1} were subsequently analysed in the same way and, as shown in Fig. 5a, confirmed the observation. (See also Ref. [17].) The possible interpretations of this effect are the following: *(i)* the modeling of the theory is incorrect; *(ii)* the measurement is affected by large systematic uncertainties; and *(iii)* the quarks are composite, in which case the compositeness scale would amount to 1.8 TeV when fitted to explain the CDF data.

The jet inclusive cross-section was also measured by D0, and a good agreement with QCD was found over the whole E_T range [17]. Had the same modeling of the theory been used in the D0 and CDF analyses, this would favour the hypothesis *(ii)*, but it turned out that it was not the case. To remove this modeling dependence, D0 repeated their analysis using CDF cuts, and compared the CDF measurement to their fitted cross-section, thus used as "theory" here (with no error in the modeling!). A global agreement was found over the whole E_T range, apart from an overall systematic shift in normalization of 10%. (See Fig. 5b.) This shift was found to be well inside the estimated systematic uncertainties. This observation strongly favours *(i)* over *(iii)*.

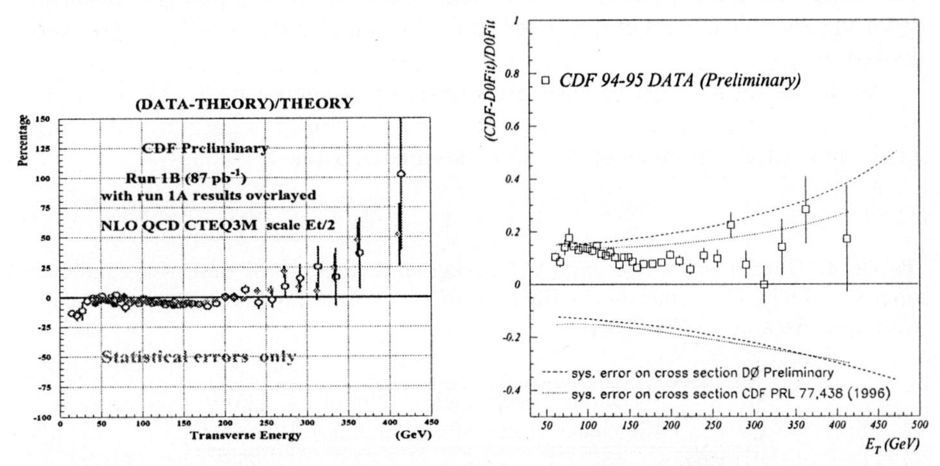

Fig. 5. Inclusive distribution of the total transverse energy in CDF. Left: Normalized difference with the the theoretical expectation (NLO QCD, EKS implementation of CTEQ3M). Right: Normalized difference to the D0 fitted distribution.

Compositeness was nevertheless further tested by CDF and D0 by studying the dijet angular distributions: dijet events were selected, and boosted into their rest frame to build the $\cos\theta^*$ distribution. This distribution presents the advantage, unlike the total transverse energy, of being insensitive to the QCD modeling or to the gluon/quark content of the proton, while it is very sensitive to new interactions due to a quark substructure. No deviations with respect to QCD were found in these angular distributions, in particular in the large E_T dijet events. This agreement allowed a lower limit on the quark compositeness scale to be set at 2 TeV by D0 (and slightly below by CDF).

This excludes the compositeness interpretation of the CDF high E_T excess, which is thus most likely due to an incorrect description of more standard processes and/or of their simulation.

4 A few 95% C.L. Exclusion Limits

A number of searches were carried out and their negative results reported at this conference. As already mentioned, the appropriate credit could not be given to all of them in this talk, but the reader can find the relevant missing information in the write-up of the parallel talks [18]. Here, only results from Higgs boson and supersymmetry searches are briefly accounted for.

4.1 Higgs boson searches

The standard model Higgs boson Since its startup in 1989, LEP has the monopoly of the searches for the standard model Higgs boson, and will keep it until the end of LEP 2 in the year 2000. Searches were also performed at the Tevatron [19] in the $p\bar{p} \to WH, ZH$ channels, but their sensitivity is still limited to a cross-section about two orders of magnitude above the standard model cross-section. At LEP, the standard model Higgs boson would be produced *via* the Higgs-strahlung process $e^+e^- \to HZ$. All final state topologies resulting from the possible decays of the Higgs ($b\bar{b}$, $\tau^+\tau^-$) and the Z ($\ell^+\ell^-$, $\nu\bar{\nu}$, $q\bar{q}$) were looked for [20].

With the data recorded with a centre-of-mass energy up to 172 GeV, no significant excess of events was observed above the expected background by any of the four LEP experiments, as can be seen from Table 3.

Table 3. Overall efficiency, expected background, number of observed candidate events, and 95% C.L. limit on the Higgs boson mass, from the four LEP experiments, with their data up to 172 GeV.

	Overall Efficiency	Expected Background	Nb. of Candidates	Limit (GeV/c^2)
ALEPH	29%	0.9	0	70.7
DELPHI	29%	4.3	2	66.2
L3	40%	38.2	33	69.5
OPAL	31%	4.1	2	69.4

The highest individual limit on the standard model Higgs boson was at that time 70.7 GeV/c^2, updated to 74 GeV/c^2 with the first 7 pb^{-1} recorded at 183 GeV. The four individual limits obtained with the 172 GeV data were combined [21] into a new lower limit of 77.5 GeV/c^2. Since then (and after the conference), the whole 183 GeV data sample has been analysed, bringing the highest individual limit [22] to 88.6 GeV/c^2. (For comparison, the combined LEP limit on m_H after seven years of LEP 1 operation was 65.6 GeV/c^2 [23].)

The neutral Higgs bosons of the MSSM In the MSSM, the Higgs sector consists of five physical states, namely three neutral bosons — two CP-even h and H, and one CP-odd A — and a pair of charged bosons $H^\pm$. In this section, only searches for h and A are considered (the heavier CP-even Higgs boson H is too heavy to be produced at LEP). The MSSM Higgs sector can be described, at tree-level, by only two parameters, chosen here to be $\tan\beta$, the ratio of the vacuum expectation values of the two Higgs doublets, and the mass m_h, because the experimental results are essentially independent of any radiative correction when expressed in the (m_h, $\tan\beta$) plane. The neutral Higgs bosons can be produced at LEP *via* two complementary processes, the standard model-like Higgs-strahlung process $e^+e^- \to hZ$ with a cross-section proportional to $\sin^2(\beta - \alpha)$, and the associated pair-production $e^+e^- \to hA$ with a cross-section proportional to $\cos^2(\beta - \alpha)$. (Here α is the the mixing angle in the CP-even sector.)

The Higgs-strahlung process has been searched for by the four LEP Collaborations in the topologies described in the previous section and the results were reinterpreted in the MSSM framework by multiplying the number of signal events expected by a reduction factor $\sin^2(\beta - \alpha)$. Specific "MSSM Higgs decays", such as the invisible decays $h \to \chi^0\chi^0$ or $h \to AA$ were also looked for, with efficiencies at least as good as for the standard analyses. (For instance, limits of 69.6 and 71.2 GeV/c^2 were set on m_h by L3 and ALEPH in the invisible decay case, assuming $\sin^2(\beta - \alpha) = 1$.) Only the configuration in which $h \to AA$ with $m_A < 2m_b$ was not efficiently covered, but the part of this configuration not excluded by LEP 1 analyses also corresponds to a region where $\tan\beta \lesssim 1$ which is treated in the following section.

The pair production process has also been searched for in the $\tau^+\tau^- q\bar{q}$ and $b\bar{b}b\bar{b}$ final states [24]. As can be seen from Table 4, no significant excess of events was observed in these topologies either. Also indicated in this Table is the limit on m_h and m_A when $\cos^2(\beta - \alpha) \sim 1$ (large $\tan\beta$ case). The highest individual limit was at that time 62.5 GeV/c^2, updated to 64.5 GeV/c^2 with the first 7 pb^{-1} recorded at 183 GeV. It increased since then to 73 GeV/c^2 (ALEPH, DELPHI) with the whole 183 GeV data sample. (For comparison, this limit was 45 GeV/c^2 after seven years of LEP 1 operation.)

Table 4. Overall efficiency, expected background, number of observed candidate events, and 95% C.L. limit on the Higgs boson mass, from the four LEP experiments, with their data up to 172 GeV.

	Overall Efficiency	Expected Background	Nb. of Candidates	Limit (GeV/c^2)
ALEPH	54%	0.8	0	62.5
DELPHI	36%	2.4	0	59.5
L3	??%	?	?	58.4
OPAL	38%	7.4	8	56.1

As an illustration, the domains excluded in the $(m_h, \tan\beta)$ by the ALEPH experiment are shown in Fig. 6, for a specific choice of the soft SUSY breaking MSSM parameters (so-called maximal mixing) responsible for the largest possible radiative corrections on m_h. All values of m_h below 62.5 GeV/c^2 are excluded, almost independently of the parameters of the model, and in particular of $\tan\beta$. All values of m_A below 62.5 GeV/c^2 are also excluded if $\tan\beta \geq 1$, but no absolute limit on m_A could be obtained if this constraint on $\tan\beta$ is relaxed. In addition, a systematic scan of the MSSM parameter space (squark

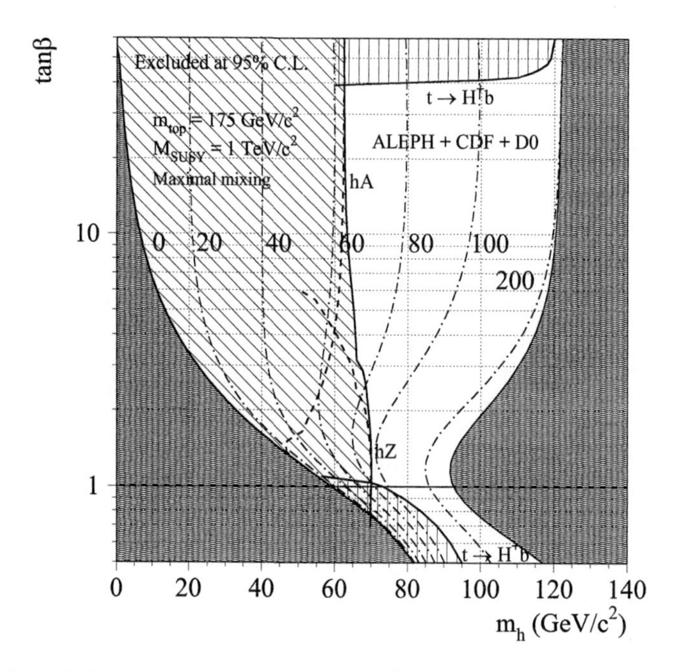

Fig. 6. Excluded domains in the (m_{h}, $\tan\beta$) plane by the ALEPH direct searches for $\mathrm{e^+e^-} \rightarrow \mathrm{hZ}$ and hA at centre-of-mass energies up to 172 GeV, and by the indirect charged Higgs boson CDF (at low $\tan\beta$) and D0 (at large $\tan\beta$) searches in top decays. The dot-dashed lines are equal-m_{A} lines.

masses and mixing, Higgs mixing, ...) was done by OPAL and revealed a few fine-tuned and exotic unexcluded sets of parameters with A or h masses smaller than 62.5 GeV/c^2.

The charged Higgs bosons Pair produced charged Higgs bosons, followed by decays into $\tau\nu_\tau$ and/or $\mathrm{c\bar{s}}$, can also be searched for at LEP 2 [24]. Lower limits on $m_{\mathrm{H}^\pm}$ as high as 52 (ALEPH), 52 (OPAL) and 54.5 GeV/c^2 (DELPHI) were set independently of $\tan\beta$. However, in the MSSM, the charged Higgs bosons are expected to be heavier than the W, their mass being given at tree level by $m^2_{\mathrm{H}^\pm} = m^2_{\mathrm{W}} + m^2_{\mathrm{A}}$.

Heavier charged Higgs bosons can only be looked for in the decay products of the top quarks produced at the Tevatron. Indeed, if the charged Higgs boson were lighter than the top quark, the two decay channels $\mathrm{t} \rightarrow \mathrm{W^+b}$ and $\mathrm{t} \rightarrow \mathrm{H^+b}$ would coexist. The decay into $\mathrm{H^+b}$ would even dominate typically for $\tan\beta$ in excess of 100 (already excluded by $\mathrm{b} \rightarrow \tau\bar{\nu}_\tau\mathrm{c}$ searches at LEP) or $\tan\beta$ below 1, for $m_{\mathrm{H}^\pm} = 100$ GeV/c^2. Since, for these low values of $\tan\beta$, the charged Higgs boson decays predominantly into $\mathrm{c\bar{s}}$, the vast majority of $\mathrm{t\bar{t}}$ events produced at the Tevatron would have turned into a six-jet topology, and not to the lepton + jets and dilepton events which lead to the top quark discovery.

Since the expected $t\bar{t}$ production cross-section can theoretically not be larger than 5.6 pb [25], the existence of such leptonic events allows a lower limit to be set on the top branching ratio into W^+b, and therefore, an upper limit on its branching fraction into H^+b. This limit, as obtained by both CDF and D0 [26], is about 25% and can be turned into excluded regions in the $(m_{H^\pm}, \tan\beta)$ plane. For instance, for $m_{H^\pm} = 100$ GeV/c^2, all $\tan\beta$ values below 1.1 are excluded by the CDF search. Similarly, $\tan\beta$ values in excess of 40 are also excluded by the D0 search for this mass.

In the MSSM, as mentioned above, the charged Higgs boson mass is directly related to m_A, and therefore to m_h and $\tan\beta$. The corresponding regions in the $(m_h, \tan\beta)$ plane, as excluded by CDF at low $\tan\beta$ and D0 at large $\tan\beta$, are indicated in Fig. 6. The low $\tan\beta$ part covers entirely the domain not considered by LEP 2 searches where $m_A < 2m_b$ and, as a bonus, covers a good fraction of the unexcluded fine-tuned points revealed by the OPAL scan. When ALEPH and CDF excluded domains are combined, any value of m_A below 62.5 GeV/c^2 is unambiguously excluded, independently of $\tan\beta$. With the recently analysed 183 GeV data, this limit is increased to 73 GeV/c^2.

4.2 Supersymmetry

In the MSSM, with minimal particle content, there are two Higgs doublets and hence two charginos $\chi_1^\pm$, $\chi_2^\pm$ and four neutralinos χ_1^0, χ_2^0, χ_3^0, χ_4^0. The gaugino sector is completed with eight gluinos $\tilde{g}_i$. The scalar sector is made of left/right sleptons, sneutrinos and squarks, all supersymmetric partners of the usual matter left-handed and right-handed fermions.

The parameters needed to specify this minimal model are still numerous. They consist (in addition to the Higgs sector parameters, m_A, $\tan\beta$ and μ) of a set of soft SUSY breaking masses m_i and trilinear couplings A_i for all scalar doublets and singlets, and a set of soft SUSY breaking masses M_i for the three gauginos. Additional simplifying assumptions inspired by supergravity (SUGRA) are often made: a universal scalar mass m_0 and a universal trilinear coupling A at the GUT scale are assumed for all sleptons, sneutrinos and squarks, and a universal gaugino mass $m_{1/2}$ for all gauginos, while the low energy parameters are determined with the renormalization group equations. As a result, the gluinos are expected to be much heavier (typically 3.5 and 7 times) than the lightest chargino and neutralino. It is also expected that right sleptons are lighter than left sleptons and sneutrinos, in turn much lighter than squarks. The section therefore emphasizes the chargino, neutralino and slepton searches (performed at LEP) rather than gluino and squark searches (performed at the Tevatron).

In the conventional scenario, R-parity is conserved, with the consequences that SUSY particles are pair produced, all of them decay into the lightest SUSY particle (LSP) and this LSP is stable. From cosmological arguments, the LSP has to be colourless and electrically neutral, which implies that it interacts weakly with matter, hence the missing energy signature of supersymmetry. The usual choice for the LSP is the lightest neutralino, with the sneutrino and the gravitino (see Section 3.2) as possible alternatives. Apart from the last paragraph, only this conventional scenario is reviewed here.

Charginos and neutralinos Pair production of charginos and neutralinos at LEP proceeds *via* s-channel Z/γ exchange, and by t-channel sneutrino (for $\chi^+\chi^-$) and selectron (for $\chi_i^0\chi_j^0$) exchange. The latter becomes important relative to the s-channel diagram when the sneutrino and the selectron are light, *i.e.*, for small values of m_0. The first case discussed here is the large m_0 configuration. Charginos and neutralinos decay into the LSP, accompanied by a (virtual or not) W or Z boson, which in turns decays hadronically or leptonically. The mass of this visible system is therefore directly related to the mass difference ΔM between the original chargino/neutralino and the LSP. Many selections were developed to account for up to four ΔM regimes (large, medium, small, very small), and the three possible final state topologies arising from the W and Z decays: hadrons with missing energy, acoplanar lepton pair with missing energy, and mixed final state (hadrons and leptons) with missing energy. (DELPHI also systematically searched for topologies arising in the "unnatural" very small ΔM regime, such as stable heavy charged particles, decays in flight, and ISR tags, turning into an absolute limit on the chargino mass of 56.8 GeV$/c^2$, in the large m_0 configuration) [27].

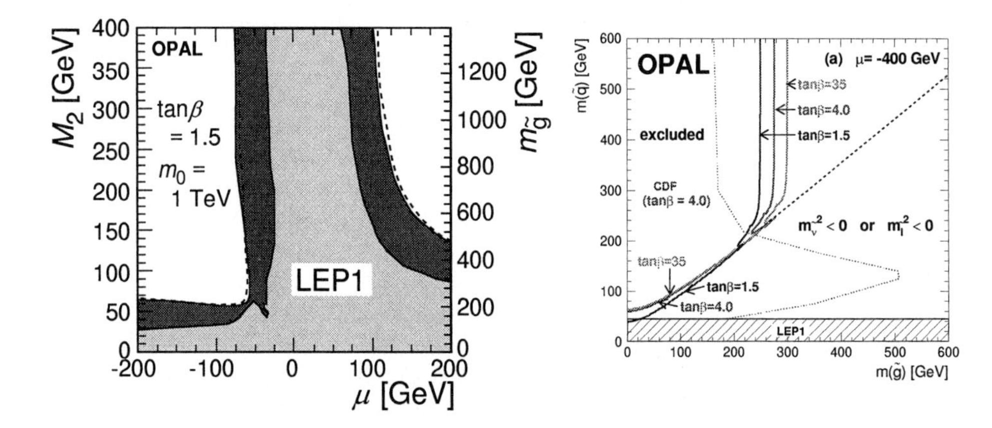

Fig. 7. Left: Excluded domains in the (μ, M_2) plane by the OPAL searches for charginos and neutralinos at centre-of-mass energies up to 172 GeV, for $\tan\beta\ = 1.5$ and in the large m_0 configuration. The right axis indicates the value of the gluino mass within SUGRA, and the dashed lines the kinematic limit. Right: comparison of this gluino mass limit in the ($m_{\tilde{g}}$, $m_{\tilde{q}}$) plane with that obtained by direct searches by CDF, for specific sets of parameters.

Altogether, a total of 29 events was observed by the four LEP experiments (for a centre-of mass energy up to 172 GeV) with 32.4 events expected from standard processes. This led to model-independent limits on the production cross-sections, which are usually interpreted within SUGRA as excluded regions in the (M_2, μ) plane. (See Fig. 7a.) The bottom line is that the kinematical limit for chargino production is reached and even exceeded (thanks to the neutralino searches) almost everywhere, independently of $\tan\beta$. With the recent 183 GeV data, this lower limit is as high as 91 GeV/c^2, but decreases slowly for small ΔM, which occurs only for very large M_2 values. In SUGRA, this limit turns automatically into a limit on the universal gaugino mass parameter $m_{1/2}$ and therefore on the gluino mass. As can be seen from Fig. 7b, this limit is better than what can be achieved at the Tevatron with their present statistics [28].

Sleptons The previous result applies only in the large m_0 configuration. For smaller m_0, the t-channel exchange of light sneutrino interferes destructively with the s-channel diagram for chargino production, and the limits obtained on chargino masses are therefore much weaker. In contrast, the interference is normally constructive with the slepton t-channel exchange for neutralino production, but the invisible neutralino decay channel $\chi^0 \to \tilde{\nu}\nu$ becomes dominant. Chargino and neutralino searches are therefore inoperative for small m_0 values.

However, sleptons are light in this configuration and can be directly searched for. At LEP, they are pair-produced, and decay to a normal lepton and a LSP. The signature is therefore an acoplanar lepton pair with missing energy. Altogether, the four LEP experiments found 29 such events with 38.1 events expected from standard processes [29]. As can be seen from Fig. 8, combined LEP limits of 80 and 70 GeV/c^2 can be set on the right-selectron and smuon masses, for large ΔM values.

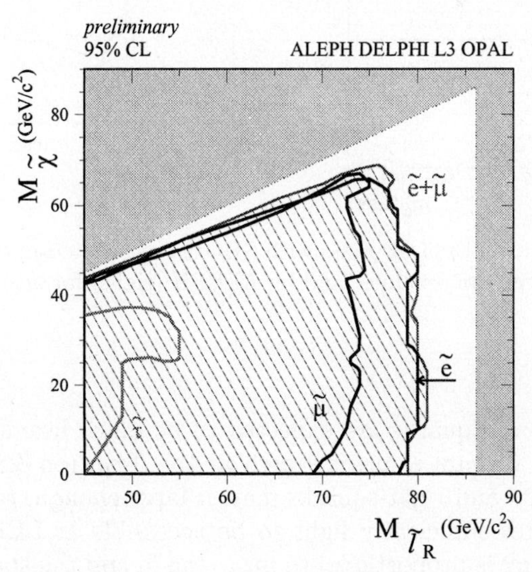

Fig. 8. Excluded domains in the $(m_{\tilde{\ell}_R}, m_{\chi^0})$ plane by a combination of LEP searches for selectron, smuon and staus, at centre-of-mass energies up to 172 GeV.

Limit on the LSP mass Within SUGRA, the two above searches can be combined into an absolute lower limit on the LSP mass, in the following way. As already mentioned, the chargino and neutralino production cross-sections are large for large m_0 values, thus allowing a relevant limit on $m_{1/2}$ to be set for any given value of $\tan\beta$, constraining in turn the lightest neutralino mass. For small m_0 values, the slepton searches allow limits to be set on the slepton masses, which can be expressed in terms of m_0 and $m_{1/2}$. Consequently, this translates also to a limit on $m_{1/2}$ and therefore on the LSP mass. For intermediate m_0 values (60 to 80 GeV/c^2), the constraints from sleptons are weak, the chargino production cross-section is still small, and the neutralino final states mostly invisible. It is therefore to be expected that the weakest limit on the LSP mass be obtained here.

Constraints on m_{χ^0} were obtained by L3 and ALEPH [30] with a systematic scan of the MSSM parameters (while OPAL [31] had done a less general scan) as a function of m_0 for all $\tan\beta$ values, as displayed in Fig. 9a. The envelope of these curves (see Fig. 9b) gives the limit on m_{χ^0} irrespective of $\tan\beta$. For large m_0, it amounts to $\sim$ 25 GeV/c^2 for all LEP experiments. The weakest ALEPH point, obtained for m_0 = 68.4 GeV/c^2 and $\tan\beta$ = 2.4, leads to the best absolute limit on the LSP mass today, 14 GeV/c^2.

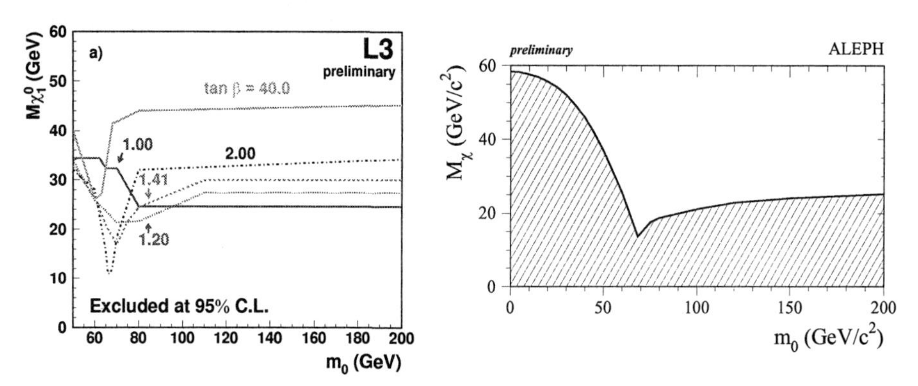

Fig. 9. Absolute limit on the LSP mass with SUGRA as a function of m_0, as obtained by L3 (left) for various $\tan\beta$ values, and by ALEPH (right) for any $\tan\beta$.

Stop and sbottom Squarks are expected to be much heavier than sleptons in SUGRA, which favours a direct search at the Tevatron [28]. However, the mixing between left- and right-squarks may, if large enough, render one of the two mass eigenstates sufficiently light to be accessible at LEP. In particular, since the stop mixing is proportional to $m_{\mathrm{top}}/\tan\beta$, and the sbottom mixing to $m_{\mathrm{b}}\tan\beta$, the lighter stop or the lighter sbottom may meet this requirement.

Stops and sbottoms would be pair-produced at LEP and followed by decays into cχ^0 (or b$\ell\tilde{\nu}$ if kinematically allowed) for the stop, and to bχ^0 for the sbottom (decay into cχ^+ is excluded in the LEP search region by the current limits on the chargino mass). The relevant final state topology is therefore an acoplanar pair of jets (with or without leptons) with missing energy. Two events were observed by the four LEP experiments with 5.7 events expected from standard processes [29]. Excluded domains are displayed in Fig. 10, together with the result of the former D0 stop search [32].

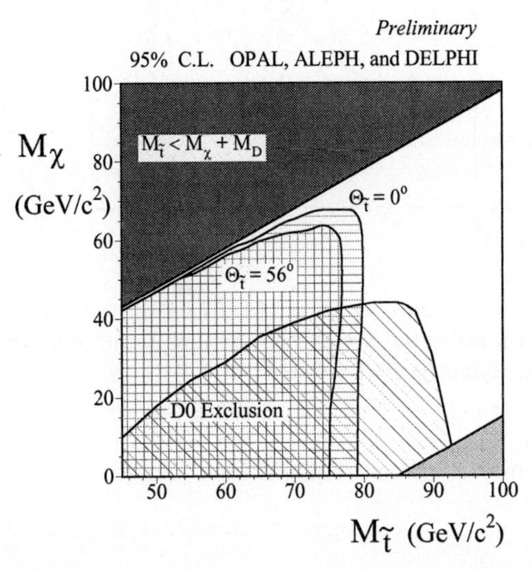

Fig. 10. Excluded domains in the $(m_{\tilde{t}}, m_{\chi^0})$ plane by a combination of LEP searches for $\tilde{t} \to c\chi^0$ at centre-of-mass energies up to 172 GeV. Also indicated is the domain excluded by the former D0 stop search.

The light gluino window In most of the analyses presented above, it was assumed that gluinos are too heavy to affect the decay pattern of the SUSY particles considered. Indeed, stringent limits for gluinos have been obtained both at the Sp$\bar{\text{p}}$S and the Tevatron (direct searches) and at LEP (indirect searches, within SUGRA). In 1996, however, a small window for very light gluinos (below 1.5 GeV/c^2 and between 3 and 5 GeV/c^2) remained officially unexcluded by any of the existing searches.

Such light gluinos would modify the usual phenomenology of QCD. For instance, they would affect the topology of four-jet events *via* $g \to \tilde{g}\tilde{g}$ splitting. A full four-jet analysis, recently carried out at LEP 1 by ALEPH [33] allowed a simultaneous measurement of α_s and n_f, the number of active flavours (which ought to be 5 in normal QCD, but 8 with an additional massless gluino). The result of that measurement is displayed in Fig. 11, turning into a limit on the gluino mass of 6.4 GeV/c^2.

It has been argued that the errors claimed by ALEPH were underestimated by a large factor and that the limit would then be invalidated. At this conference, a combination of this four-jet analysis with enlarged errors and the α_s running (from the measurements of R_γ, R_τ, and R_Z) was presented [34]. As a result, the light gluino window was shown to be excluded at more than 99.9% C.L.

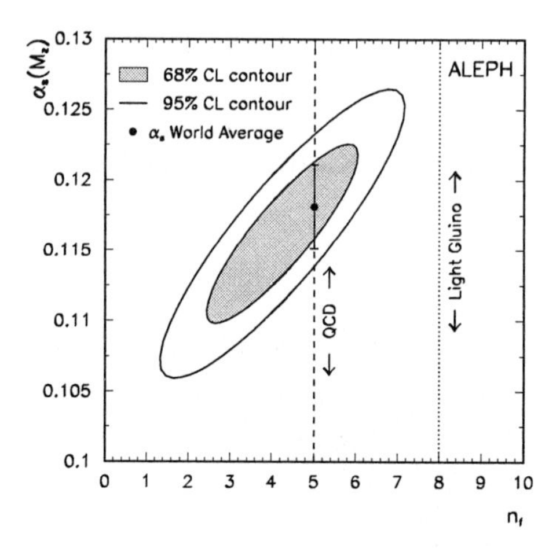

Fig. 11. Measurement of α_s and n_f from the ALEPH study of the four-jet event topology at LEP 1. Also indicated are the predictions form QCD ($n_f = 5$) and from an additional massless gluino ($n_f = 8$).

The light gluino window is therefore definitively closed.

5 Conclusion

To summarize this talk, no discovery was presented, all the recent exciting hints for unexpected new physics were shown to be slowly dying, while limits on expected new physics are becoming more and more stringent. Let us hope, nevertheless, that the higher energies/luminosities foreseen in the very near future will bring us some surprises!

Acknowledgements

It is a real pleasure to thank the organizers and the secretariat of the '97 E.P.S. conference for their hospitality and for having provided us with such an inspiring atmosphere in Jerusalem.
I am indebted to the LEP Higgs and Susy Working Groups, to François Richard (DELPHI), Joachim Mnich and Sylvie Rosier-Lees (L3), Rick Van Kooten and Peter Mättig (OPAL), Eckhard Elsen (H1), Patrizia Azzi and Kara Hoffman (CDF), for their precious cooperation during the preparation of this talk.
I am also grateful to Laurent Duflot, Jean-François Grivaz and Nikos Konstantinidis for their critical reading of the manuscript.

References

[1] J.-F. Grivaz, *"New particle searches"*, Talk given at the E.P.S. conference in Brussels, July 1995, published in the proceedings.

[2] U. Bassler, *"Search for new physics at high Q^2 and high p_T"*, these proceedings and references therein;
E. Elsen, *"High Q^2 events"*, these proceedings and references therein;
B. Straub, *"New results from neutral and charged current scattering at high Q^2 from H1 and ZEUS"*, talk given at the 23rd Lepton-Photon Symposium, Hamburg, July 1997.

[3] E. Perez, *"Observation of events at very high Q^2 with the H1 detector at HERA"*, Proc. of the XXXIInd Rencontres de Moriond, March 1997.

[4] B. Straub, *"Search for a deviation from the standard model at high x and Q^2 with the ZEUS detector at HERA"*, Proc. of the XXXIInd Rencontres de Moriond, March 1997.

[5] M. Besancon, *"Search for new physics at LEP: excited leptons, compositeness and Z'"*, these proceedings, and references therein.

[6] E. Gallas, *"Searches for compositeness at the Tevatron"*, these proceedings, and references therein.

[7] G. Altarelli, J. Ellis, G.F. Giudice, S. Lola and M.L. Mangano, *Nucl. Phys.* **B506** (1997) 3;
The figure presented in this talk is a straight update with HERA 1997 data, and was privately provided to me by M. Mangano.

[8] X. Wu, *"CDF limits on leptoquarks and R-parity violation SUSY from CDF"*, these proceedings, and references therein.

[9] B. Klima. *"Search for leptoquarks at D0"*, these proceedings, and references therein.

[10] J. Ellis, *"Pursuing interpretations of the HERA high Q^2 events"*;
P. Binetruy, *"Supersymmetric Extensions of the Standard Model"*;
these proceedings, and references therein.

[11] P. Azzi, *"Search for new phenomena with the CDF detector"*, Proc. of the XXXIInd Rencontres de Moriond, March 1997.

[12] J. Womersley, *"Search for new phenomena in photonic final states"*, these proceedings and references therein.

[13] M. Chemarin, *"Search for anomalous photonic events at LEP"*, these proceedings and references therein.

[14] ALEPH Coll., *Z. Phys.* **C71** (1996) 179.

[15] The LEP Working Group on Four Jets, Internal Note, ALEPH 97-056, DELPHI 97-57, L3 # 2090, OPAL TN 486 (June 1997), and references therein.

[16] CDF Coll., *Phys. Rev. Lett.* **77** (1996) 438.

[17] H.J. Weerts, *"QCD at Hadron Colliders"*, these proceedings and references therein.

[18] I. Fleck, *"Search for R parity violation and leptoquarks signature at LEP"*;
A. Zarnecki, *"Search for new particles at ZEUS"*;
D. Casper, *"Search for new physics at LEP: heavy leptons, ..."*;
these proceedings.

[19] K. Hoffman, *"Search for high mass states and dijet searches at the Tevatron"*, these proceedings and references therein.

[20] W. Murray, *"Search for the Standard Model Higgs boson at LEP"*, these proceedings and references therein.
[21] The LEP working group for Higgs boson searches, CERN/LEPC 97-11 (1997), and references therein.
[22] P. Dornan, talk given for the ALEPH Coll. at the LEPC meeting (Nov. 11th, 1997).
[23] For a detailed review, see for instance:
P. Janot, *"Searching for Higgs bosons at LEP 1 and LEP 2"*, to appear in Perspectives on Higgs Physics II, World Scientific, ed. G.L. Kane.
[24] Y. Pan, *"Search for Higgs bosons beyond the Standard Model at LEP"*, these proceedings and references therein.
[25] S. Catani et al. , *Phys. Lett.* **B378** (1996) 328;
E. Berger and H. Contopanagos, *Phys. Rev.* **D54** (1996) 3085.
[26] J. Konigsberg, *"Top results from CDF"*;
M. Strovink, *"Top results from D0"*;
A. Yagil, *"Top Physics"*;
these proceedings and references therein.
[27] M. Felcini, *"Search for SUSY signature at LEP: charginos, neutralinos, and interpretations"*, these proceedings and references therein.
[28] J. Conway, *"SUSY searches at the Tevatron"*, these proceedings and references therein.
[29] S. Asai, *"Searches for sfermions at LEP"*, these proceedings and references therein.
[30] L. Duflot, *"Implications of Aleph SUSY Searches for the MSSM"*, these proceedings and references therein.
[31] H. Neal, *"Improved Constraints on the Neutralino Cold Dark Matter Candidate from the Search for Chargino and Neutralino Production with OPAL at* $170 \div 172$ *Gev"*, these proceedings and references therein.
[32] D0 Coll., *Phys. Rev. Lett.* **76** (1996) 2222.
[33] ALEPH Coll., CERN-PPE/97-02, to be published in *Z. Phys.* **C**.
[34] F. Csikor, *"Closing the light gluino window"*, these proceedings and references therein.

Status of the LHC

J.-P. Gourber (jean-pierre.gourber@cern.ch)

CERN, European Organization for Nuclear Research, Geneva, Switzerland

Abstract. Since the approval of the Large Hadron Collider (LHC) by the CERN Council in December 1994, considerable progress has been made in the assessment of the beam parameters and the refining of the design of the machine components and experimental areas. Thanks to the strong support from a number of countries outside the Member States, the machine will be constructed in one single stage with first physics in 2005. The first large calls for tenders are being launched. The status of the project and the future plans are presented.

1 Introduction

The Large Hadron Collider (LHC) was originally approved by the CERN Council in December 1994 as a two-stage machine subject to a final review in 1997. Thanks to the strong support of a number of countries outside the CERN Member States, it is now fully approved for construction in one stage with the first physics run in 2005.

Considerable progress has been made in the last two years [1, 2] to better assess the beam parameters and to refine the design of the machine components and experimental areas. The project is now fully in its implementation phase and the first large-scale contracts have been launched.

Three experiments ATLAS, CMS and ALICE are now approved and a fourth one LHC-B optimized for B-physics is at the stage of preparation of a technical proposal. The allocation of the other five long straight sections for machine utilities remains unchanged (Fig. 1).

The only changes in the parameter list (Table 1) concern the beam separation which has been increased from 180 mm to 194 mm and the dipole field which has been further reduced from 8.4 to 8.3 T for the same nominal beam energy of 7 TeV thanks to a better optimization of the junctions between the magnets.

2 Optics

The optics studies have been pursued with the aim of getting a more robust and flexible lattice [3]. In the new version-5 optics one quadrupole has been added to the outer triplet (Fig. 2) and the last quadrupole of the regular arc is equipped with a trim quadrupole and used together with the dispersion suppressor quadrupoles to adjust the beam parameters in the interaction

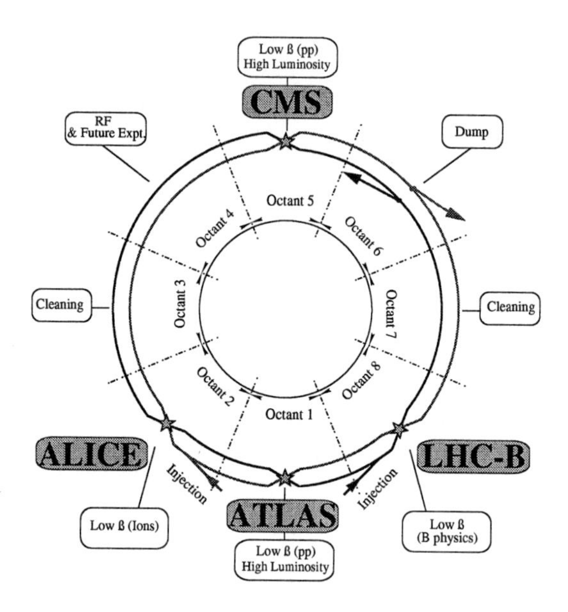

Fig. 1. LHC layout

region. In this way, the Q split $|Q_x - Q_y|$ can be varied from 0 to 8 and the optical functions of the interaction regions are much more favourable. The high-luminosity insertions for ATLAS and CMS (Fig. 2) have a reduced $\beta_{\max}$ for the same $\beta^* = 0.5$ m and can be detuned at injection to $\beta^* = 20$ m with $\beta_{\max} = 160$ m instead of 300 m in the previous version. The ALICE insertion can be tuned from $\beta^* = 0.5$ m to $\beta^* = 250$ m. This allows ALICE to get the maximum luminosity of 2×10^{27} $\text{cm}^{-2}\,\text{s}^{-1}$ for normal operation with ${}^{208}\text{Pb}^{82+} - {}^{208}\,\text{Pb}^{82+}$ collisions and not more than 10^{30} $\text{cm}^{-2}\,\text{s}^{-1}$ with p–p collisions when the other experiments are at full luminosity.

A lot of effort has also been devoted to the optimization of the geometrical and dynamic apertures [4]. The latter is defined as the limit expressed in number of σ (r.m.s. beam radius) beyond which the movement of the particles becomes chaotic and the particles are lost (Fig. 3).The dynamic aperture is mainly critical at injection where the beam is large and it is directly related to the field errors of the superconducting magnets. These errors include systematic and random components which in turn result from many contributions: coil geometry, iron yoke saturation, persistent currents in the cable, power supply ripple, etc. All these sources of error have been carefully investigated from the measurements of magnet prototypes and from the results obtained in series production with the HERA and RHIC machines. The particles are generally tracked up to 10^5 or 10^6 turns using powerful computing hardware and software and then the results are extrapolated to 10^7 turns which represents the actual injection time. The last results show that the dynamic aperture is larger than the collimating setting of $6\,\sigma$ which is needed

Table 1. Main machine parameters

Parameters		p–p	$^{208}Pb^{82+}$	
Beam energy	TeV/charge	7.0	7	
Centre-of-mass energy	TeV	14	1148	
Injection energy	GeV/charge	450	450	
Dipole field	T	8.3	8.3	
Coil aperture	mm	56	56	
Distance between apertures	mm	194	194	
Bunch spacing	ns	25	124.75	
Number of bunches per ring		2835	608	
Normalized transverse emittance	μm	3.75	1.5	
r.m.s. bunch length	m	0.075	0.075	
Beta values at IP	m	0.5	0.5	
Full crossing angle	μrad	200	200	
Particles per bunch		1×10^{11}	6.3×10^{7}	9.4×10^{7}
Intensity per beam	mA	530	5.2	7.8
Initial luminosity	cm^{-2} s^{-1}	1×10^{34}	0.85×10^{27}	1.95×10^{27}
Luminosity lifetime	h	10	10	6.7

to guarantee the beam lifetimes. This requires, however, a very tight control of the magnet manufacturing errors.

3 Injector complex

Most of the links of the LHC injection chain exist (Fig. 4). However, these machines have to be upgraded and some equipment must be added in order to provide the required bunch spacing and the high phase space density [5]. The PS booster must be upgraded from 1 to 1.4 GeV to alleviate space charge problems and new RF cavities ($h = 1$, 7 kV) must be added. The PS must be equipped with new 40 and 80 MHz RF systems which are being developed in collaboration with TRIUMF. The antiproton decelerator LEAR will be converted into an ion accumulator (LEIR).

The SPS requires four new 400 MHz superconducting cavities for bunch compression and an upgrade of the existing 200 MHz system. In order to preserve a low emittance during acceleration, the dominant impedances which have been identified with the vacuum ports and the septa must be severely reduced. Beams with the nominal parameters are expected in 1999.

The new SPS–LHC transfer lines totalize 2×2.5 km of beam lines and 2.7 km of magnets which are being built by BINP (Novosibirsk). A new ejection system is also required in the SPS for injection into the anticlockwise LHC beam.

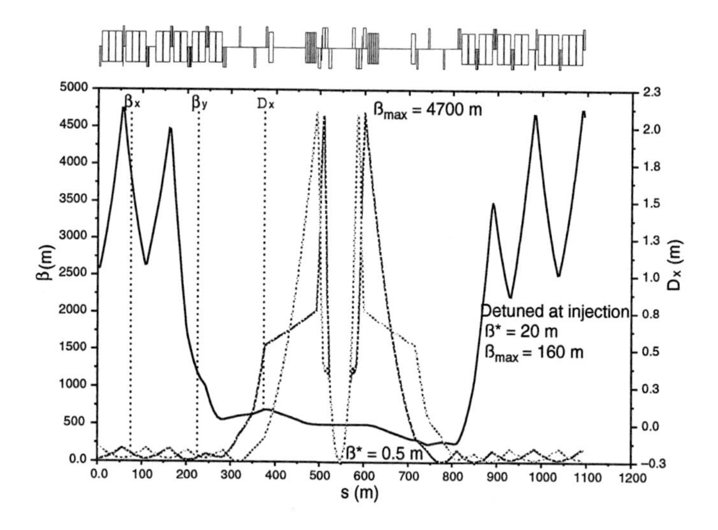

Fig. 2. Optics for ATLAS or CMS

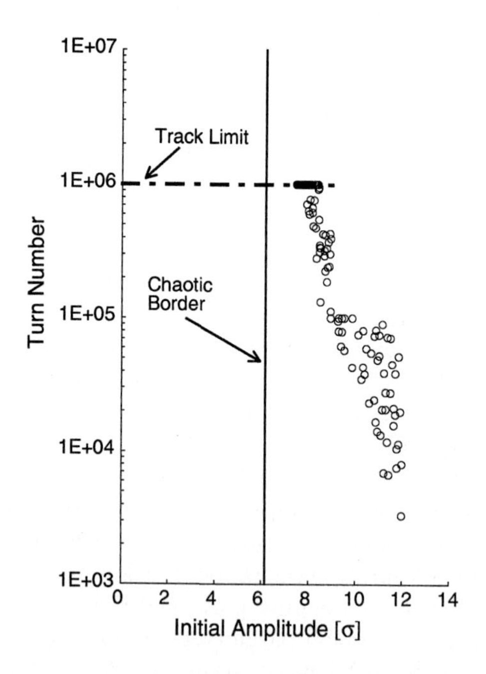

Fig. 3. Particle survival plot

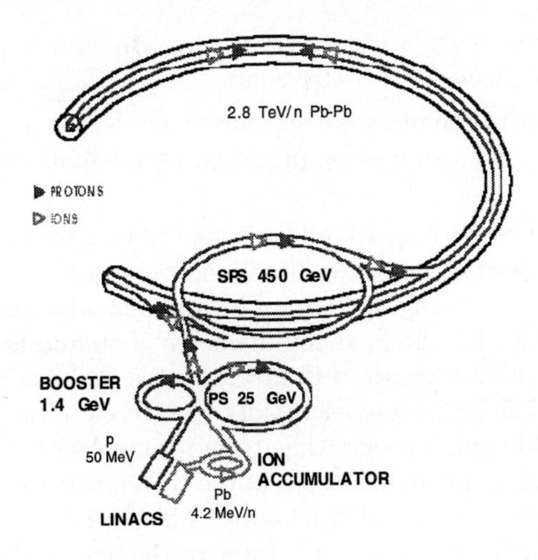

Fig. 4. The LHC injector complex

4 Magnets

A considerable amount of development of work has been done on the different types of the 8000 magnets which are needed in the LHC [6] and which include the 1232 two-in-one superconducting dipoles.

The design of the dipoles and the manufacturing procedures have been further optimized in order to improve their performance and to facilitate series production.

The measurements of the minimum quench energy on the superconducting cables made in collaboration with BNL and LBL have shown the importance of a sufficient helium penetration inside the cable. This has led in particular to a small decrease of the keystone angle. A great deal of attention has also been given to the cable insulation (now all polyimide) and to the interstrand resistance which shall stay between 10 and 30 $\mu\Omega$ in order to avoid interstrand eddy currents yet maintaining sufficient current sharing. So far, the best results have been obtained with SnAg ('Staybrite') coated cables. More than 20 t of cables have been produced to the required characteristics by the four potential European suppliers. The call for tenders for the series production was sent out in May 1997 for an adjudication of contracts foreseen in December 1997.

The cross-section of the dipole is shown in Fig. 5. The beam separation has been increased from 180 to 194 mm to reduce the prestress of the coils during collaring and to simplify the design of the magnets. A coil design with six blocks is now preferred to the five-block design of the yellow book [1] since it gives a better compensation of the field distortion due to persistent currents at injection, a slight increase of the margin with respect to the short sample limit, and allows for a better control of fabrication errors.

The manufacturing changes are systematically tested on 1-m-long single-aperture models which can now be produced at a rate of more than one per month.

The transfer of technology to industry via the long prototype programme has continued. Following the seven 10 m long prototypes of the first generation which had an inner coil diameter of 50 mm and which all have exceeded 9 T, six prototypes with the final coil diameter of 56 mm are in fabrication. The first one still with a length of 10 m was tested in June 1997. After a few quenches around 7.5 T the magnet rapidly trained close to the short sample limit (Fig. 6). Field quality measurements performed before and after this initial plateau showed a global movement of the coils inside the collars resulting from a defect in the coil collaring mandrel. This stresses the importance of the tooling and quality control at all stages of the magnet manufacturing.

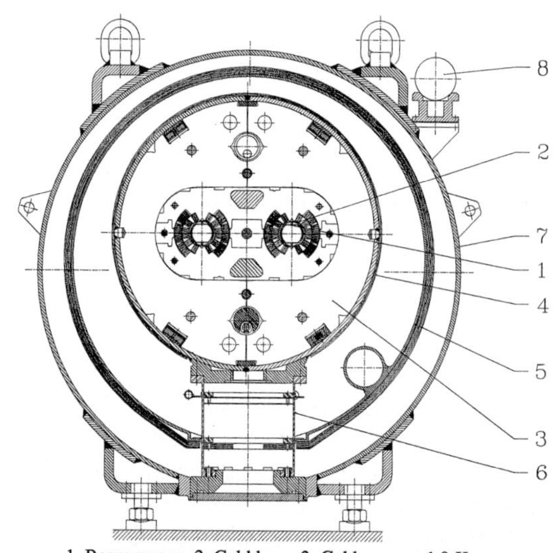

1. Beam screen, 2. Cold bore, 3. Cold mass at 1.9 K,
4. Radiative insulation, 5. Thermal shield (55 to 75 K),
6. Support post, 7. Vacuum vessel, 8. Alignment target

Fig. 5. Cross section of the dipole magnet and cryostat

The development of the 3.1 m long two-in-one quadrupoles for the arcs has been continued in collaboration with CEA and CNRS in France. Two prototypes have already been successfully tested with the first quench above the nominal gradient. Two more prototypes with the final coil diameter of 56 mm are in fabrication. The tests will start before the end of 1998.

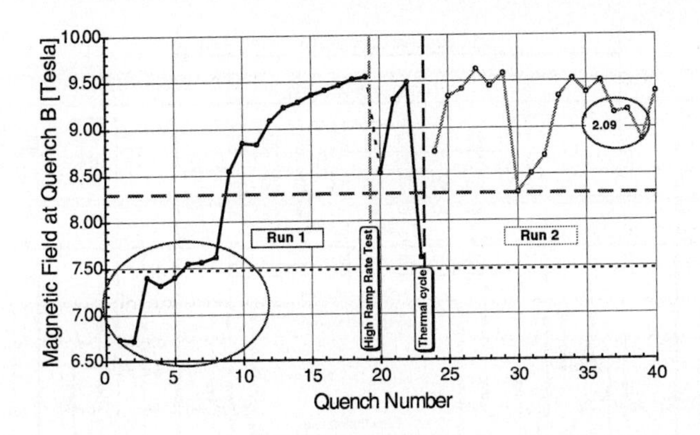

Fig. 6. Training quenches at 1.8 K of dipole MBL1N1

CERN is collaborating with FNAL and KEK to finalize the design of the 70 mm aperture low-beta quadrupoles with a view to fabricating the full series. A short model has already been successfully tested in industry [7].

Prototypes of most of the corrector magnets which are attached to the dipoles or to the quadrupoles have already been tested.

5 Cryogenics

The design of the cryogenic system has also undergone a number of improvements. The distribution of the cryogenic fluids along the tunnel is now done (Fig. 7) via a separate cryoline connected to the magnet string every two quadrupoles (1 cell length = 106.9 m). The subcooled helium line at 2.2 K has been suppressed as well as the large 120 g/s heat exchanger 1.9 K, 16 mbar/2.2 K, 1.3 bar which was needed near each cryoplant. The latter is replaced by small and more conventional 5 g/s heat exchangers (HX on Fig. 7) located in each jumper connection cryoline quadrupole. This new design [8] facilitates the installation and maintenance of the magnet and cryogenic systems and leads to a substantial reduction in the number of components and cost.

The helium cooling capacity required for the LHC will be provided by eight cryoplants of 18 kW equivalent capacity at 4.5 K. Four of them are those currently used for LEP; the capacity will be upgraded from 12 to 18 kW. The original symmetrical layout with the plants grouped in pairs at the even points has been altered by locating one of the new plants at Point 1.8 where a large shaft and surface facilities already exist. This will avoid any major civil engineering at Point 2 and will reduce the temperature increase of the magnets at Point 1 due to the tunnel slope. The calls for tenders for the upgrading of the LEP cryoplants and for the four new cryoplants will be

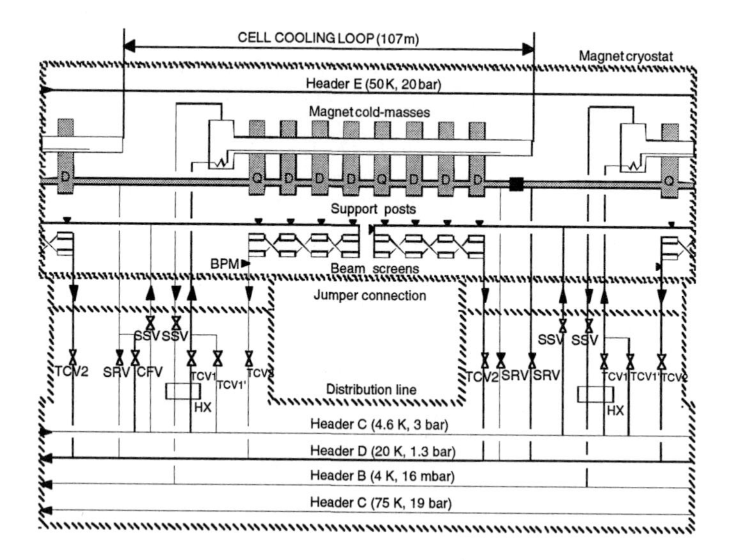

Fig. 7. Simplified distribution scheme

issued in autumn 1997. It is expected that the first new cryoplant will be operational at Point 1.8 as from early 2000 for testing cryogenic equipment.

The pumping capacity of 120 g/s at 16 mbar which is needed near each cryoplant will be provided by connecting in series and parallel several cold compressors. In order to assess their performance, three prototypes of the same characteristics (flow of 18 g/s at 16 mbar and compression rate of 3) have been ordered from three different firms, two in Europe and one in Japan. Two of them have already been successfully tested (Fig. 8) and the third one will be delivered this autumn.

6 Vacuum

For cryogenic efficiency, the synchrotron radiation emitted by the protons (about 4 kW per ring at 7 TeV) must be absorbed by a liner inserted in the magnet cold bore and cooled at an intermediate temperature between 4.5 and 20 K. This liner is a thin tube of very low permeability stainless steel (high-manganese steel) coated with copper on the inside. It is pierced with many slots to allow cryopumping of the desorbed molecules on the cold bore. These slots are randomly disposed along the tube to avoid resonance modes. All these engineering aspects are now mastered and prototypes have been made and successfully tested.

Quite recently a new problem, linked to the presence of electrons, has been revealed [9]. The photoelectrons emitted in the median plane at the impact of the synchrotron light are accelerated towards the proton beam and deposit their energy on the other side of the liner. There, they also create

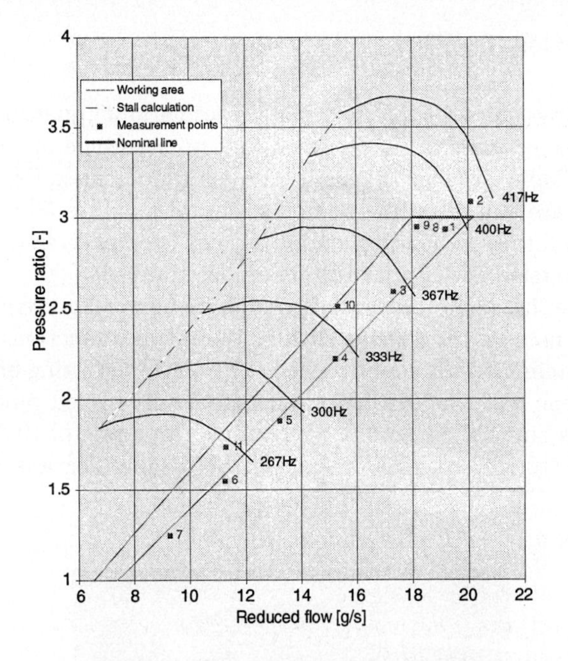

Fig. 8. Performances of the cold compressor CCU2

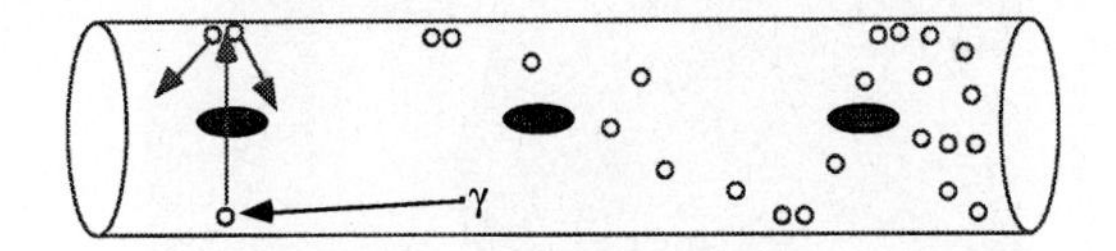

Fig. 9. Beam-induced electrons

secondary electrons which are accelerated backwards by the following bunch of protons (Fig. 9). Calculations show that the secondary electron yield σ_{max} must be smaller than 1.4 in order to avoid a multipacting effect with an exponential growth of electron density and the destruction of the proton beam. The phenomenon is less critical in the dipole regions since the primary electrons are forced to spiral around the field lines and are reabsorbed in the wall near their emission point. In the 18 kW equivalent capacity at 4.5 K of the cryoplants, 1.7 kW has been allowed for the increase of heat load on the liner. Studies and experiments in the laboratory and with synchrotron light at CERN in EPA and at BINP are going on to select the best coating and treatment of the inner surface of the liner.

7 String test facility

A string of three dipoles and one straight section unit (quadrupole with its attached correctors, beam-position monitor and cryogenic module) simulating one LHC half-cell (Fig. 10) has been operational since December 1994. The magnets are still all from the first generation with a 50 mm inner coil diameter but all the surrounding equipment has been progressively upgraded to the latest version [10]. So far the string has totalled 1500 hours of operation with more than 30 forced quenches from its nominal current or above. It has validated many critical items such as the magnet junctions, the quench protection system, and the cooling scheme. Recently the string has undergone a forced lifetime test where it has been repetitively cycled more than 2100 times up to its nominal current at the nominal ramp rate of 10 A/s, simulating more than 10 years of LHC operation. No magnet degradation has been observed.

Fig. 10. String test facility

8 Civil engineering works

The LHC makes use of the existing LEP infrastructure. Nevertheless a large investment in civil engineering works equivalent to half that of LEP is needed to house the two large detectors ATLAS and CMS, for the two transfer tunnels from the SPS, and for the surface infrastructure needed for assembly halls and cryogenic equipment.

The ATLAS detector requires two huge underground caverns of a total volume of about 70 000 m^3. The CMS detector requires about the same volume of cavern as well as considerable surface infrastructure for the assembly and testing of the 14 000 t solenoid. Calls for tenders for the three civil engineering packages (Point 1, Point 5 and the remaining work) have been launched. Adjudication of the contracts is foreseen in November 1997 allowing site preparation to start at the beginning of 1998.

9 Planning

It is planned to run LEP until the end of 1999 although the LHC planning has been adapted in order to allow a further year of LEP running if the physics prospects justifies it and if appropriate funding can be found. The civil engineering work is planned such that the sites will be delivered in July 2002 for ATLAS and in July 2003 for CMS. This difference of one year is due to the fact that the geological conditions are less favourable for CMS, but is acceptable as the magnet will be assembled and tested at the surface before installation.

An injection test in the octant from Point 8 to Point 7 is foreseen in 2003 and the first physics run is expected in the second half of 2005.

10 Conclusions

The Large Hadron Collider is the first large CERN project in which there is a participation of external institutes in the machine construction. It is now well engaged in its implementation phase, with a total of about 700 MCHF of contracts to be adjudicated by the end of 1997.

11 Acknowledgements

The achievements reported in this paper owe very much to the enthusiastic work of many teams from CERN and other laboratories both in the Member States and in Canada, India, Japan, Russia, and the USA.

References

[1] The LHC Study Group, The Large Hadron Collider, Conceptual Design, CERN/AC/95–05 (LHC).

[2] L.R. Evans, LHC Status and Plans, LHC Proj. Rep. 101, Invited Paper at the 1997 Part. Accel. Conf., Vancouver, 12–16 May 1997.

[3] A. Faus-Golfe et al., A more Robust and Flexible Lattice for LHC, LHC Proj. Rep. 107, Paper presented at the 1997 Part. Accel. Conf., Vancouver, 12–16 May 1997.
[4] M. Boege et al., Overview of the LHC Dynamic Aperture Studies, LHC Proj. Rep. 106.
[5] K. Schindl, Conversion of the PS Complex as LHC Proton Pre-Injector, Paper presented at the 1997 Part. Accel. Conf., Vancouver, 12–16 May 1997.
[6] R. Perin, State of the LHC Main Magnets, LHC Proj. Rep. 108, Paper presented at the 1997 Part. Accel. Conf., Vancouver, 12–16 May 1997.
[7] R. Ostojic et al., Quench Performance and Field Quality Measurements of the first LHC Low-Beta Quadrupole Model, LHC Proj Rep 130, Paper presented at the 1997 Part. Accel. Conf., Vancouver, 12–16 May 1997.
[8] M. Chorowski et al., A Simplified Cryogenic Distribution Scheme for the Large Hadron Collider, LHC Proj. Rep. 143, Paper presented at CEC-ICMC 1997, Portland, USA, 29 July–1 August 1997.
[9] O. Gröbner, Beam Induced Multipacting, LHC Proj. Rep. 127, Paper presented at the 1997 Part. Accel. Conf., Vancouver, 12–16 May 1997.
[10] J. Casas-Cubillos et al., Experiments and Cycling at the LHC prototype Half Cell, LHC Proj. Rep. 110, Paper presented at the 1997 Part. Accel. Conf., Vancouver, 12–16 May 1997.

Heavy Quark Effective Theory and Weak Matrix Elements

Matthias Neubert (Matthias.Neubert@cern.ch)

Theory Division, CERN, CH-1211 Geneva 23, Switzerland

Abstract. Recent developments in the theory of weak decays of heavy flavours are reviewed. Applications to exclusive semileptonic B decays, the semileptonic branching ratio and charm counting, beauty lifetimes, and hadronic B decays are discussed.

1 Introduction and theoretical concepts

To discuss even the most significant recent theoretical developments in heavy-flavour physics in a single talk is a difficult task. Fortunately, this field is blooming and will continue to be of great importance in view of several new experimental facilities (B factories) to start operating in the near future. Below, I will review the latest theoretical developments in this field and discuss the most important phenomenological applications. They concern semileptonic B decays and the measurements of the CKM parameters $|V_{cb}|$ and $|V_{ub}|$, the semileptonic branching ratio and charm yield in inclusive B decays, the lifetimes of beauty mesons and baryons, and hadronic B decays, including the rare decays into two light mesons. I will start with an introduction to the main theoretical concepts used in the analysis of these processes.

The properties of hadrons containing a single heavy quark Q are characterized by the large separation of two length scales: the Compton wave length $1/m_Q$ of the heavy quark is much smaller than the typical size $1/\Lambda_{\rm QCD}$ of hadronic bound states in QCD (see the left plot in Fig. 1). In the limit $m_Q \to \infty$, the configuration of the light degrees of freedom in the hadron becomes independent of the spin and flavour of the heavy quark. In that limit there is a global SU($2n_h$) spin–flavour symmetry of the strong interactions, where n_h is the number of heavy-quark flavours [1]. This symmetry helps in understanding the spectroscopy and decays of heavy hadrons from first principles. It does not allow us to solve QCD, but to parametrize the strong-interaction effects of heavy-quark systems by a minimal number of reduced matrix elements, thus giving rise to nontrivial relations between observables. In particular, all form factors for the weak $\bar{B} \to D^{(*)}\ell\,\bar{\nu}$ transitions are proportional to a universal function $\xi(w)$, where $w = v_B \cdot v_{D^{(*)}}$ is the product of the meson velocities. At the zero-recoil point $w = 1$, corresponding to $v_B = v_{D^{(*)}}$, this function is normalized to unity: $\xi(1) = 1$. The symmetry-breaking corrections to the heavy-quark limit can be organized in an expansion in powers of the small parameters $\alpha_s(m_Q)$ and $\Lambda_{\rm QCD}/m_Q$. A convenient way to do this is provided by the heavy-quark effective theory (HQET), whose purpose is to separate the short- and long-distance physics associated with the two length scales, making all dependence on the large mass scale m_Q explicit. This allows us to derive scaling laws relating different observables to each other. The philosophy behind the HQET is illustrated in the right plot in Fig. 1.

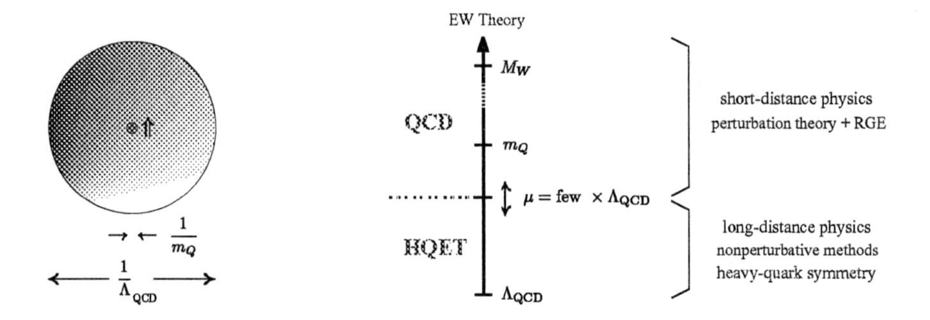

Fig. 1. Length scales of a heavy hadron, and construction of the HQET

The effective Lagrangian of the HQET is

$$\mathcal{L}_{\rm eff} = \bar{h}\, iv\!\cdot\! Dh + \frac{1}{2m_Q}\,\bar{h}\,(iD)^2 h + \frac{C(m_Q)}{4m_Q}\,\bar{h}\,\sigma_{\mu\nu}G^{\mu\nu}h + \dots ,$$

where v is the velocity of the hadron containing the heavy quark, and h is the heavy-quark field. The Lagrangian incorporates the spin–flavour symmetry to leading order in $1/m_Q$. The only nontrivial short-distance coefficient has recently been calculated to next-to-leading and even next-to-next-to-leading order [2]–[4]. The result is

$$C(m_Q) = \left[\alpha_s(m_Q)\right]^{9/25}\Big\{1 + 0.672\alpha_s(m_Q) + (1.33 \pm 0.04)\alpha_s^2(m_Q) + \dots\Big\}.$$

The long-distance physics is encoded in the hadronic matrix elements of HQET operators, e.g.

$$\mu_\pi^2(H) = -\langle H|\bar{h}\,(iD)^2 h|H\rangle\,, \qquad \mu_G^2(H) = \frac{C(m_Q)}{2}\,\langle H|\bar{h}\,\sigma_{\mu\nu}G^{\mu\nu}h|H\rangle\,,$$

$$\bar{\Lambda}(H) = M_H - m_Q + \frac{\mu_G^2(H) - \mu_\pi^2(H)}{2M_H} + \dots ,$$

which are referred to as the "kinetic energy", the "chromomagnetic interaction", and the "binding energy", respectively. They are important parameters entering many applications of the heavy-quark expansion, the calculation of inclusive decay rates and spectra in particular. Certain combinations of these parameters can be extracted from spectroscopy:

$$\mu_G^2(B) = \frac{3}{4}\,(M_{B^*}^2 - M_B^2) \approx 0.36\,\text{GeV}^2\,, \qquad \mu_G^2(\Lambda_b) = 0\,,$$

$$\mu_\pi^2(\Lambda_b) - \mu_\pi^2(B) = -\frac{M_B M_D}{2}\left(\frac{M_{\Lambda_b} - M_{\Lambda_c}}{M_B - M_D} - \frac{3}{4}\,\frac{M_{B^*} - M_{D^*}}{M_B - M_D} - \frac{1}{4}\right) \approx 0\,,$$

$$\bar\Lambda(\Lambda_b) - \bar\Lambda(B) \approx 0.31\,\text{GeV}\,.$$

The individual parameters μ_π^2 and $\bar\Lambda$ are scheme dependent, however, because of renormalon ambiguities in their definition. Once a scheme is chosen, they can be calculated nonperturbatively, or extracted from moments of inclusive decay spectra. Table 1 shows a collection of some experimental determinations of these parameters in the on-shell scheme.

Table 1. Determinations of the parameters $\bar\Lambda$ and μ_π^2 from inclusive decay spectra

Reference	Method	$\bar\Lambda(B)$ [GeV]	$\mu_\pi^2(B)$ [GeV^2]
Falk et al. [5]	Hadron Spectrum	≈ 0.45	≈ 0.1
Gremm et al. [6]	Lepton Spectrum	0.39 ± 0.11	0.19 ± 0.10
Chernyak [7]	$(\bar B \to X\,\ell\,\bar\nu)$	0.28 ± 0.04	0.14 ± 0.03
Gremm, Stewart [8]		0.33 ± 0.11	0.17 ± 0.10
Li, Yu [9]	Photon Spectrum $(\bar B \to X_s\gamma)$	$0.65^{+0.42}_{-0.30}$	$0.71^{+1.16}_{-0.70}$

Using the operator product expansion, any inclusive decay rate of a beauty hadron can be expanded as

$$\Gamma(H) = \frac{G_F^2 m_b^5}{192\pi^3}\left(1 - \frac{\mu_\pi^2(H)}{2m_b^2}\right)\left\{c_3 + c_5\,\frac{\mu_G^2(H)}{m_b^2} + \sum_n c_6^{(n)}\,\frac{\langle O_n\rangle_H}{m_b^3} + \dots\right\},$$

where $\langle O_n\rangle_H$ are the matrix elements of local four-quark operators, which parametrize nonspectator effects in these decays, and c_i are calculable short-distance coefficients, which depend on CKM parameters, the ratios of quark masses, and the renormalization scheme. The free quark decay emerges as the leading term in a systematic $1/m_b$ expansion, with bound-state corrections suppressed by two inverse powers of the heavy-quark mass. Note that ratios of inclusive decay rates are independent of μ_π^2 and the common factor m_b^5, as well as of most CKM parameters. The application of the operator product expansion to the calculation of inclusive decay rates relies on the assumption of quark–hadron duality. Strictly speaking, the theoretical description of such processes is thus not entirely from first principles.

2 Exclusive semileptonic decays

The most important applications of the HQET concern the description of exclusive semileptonic decays based on the quark transition $b \to c\,\ell\,\bar{\nu}$. This is where the theory is well tested, and theoretical uncertainties are best understood. The most important result is a precision determination of the CKM parameter $|V_{cb}|$ [10].

2.1 Determination of $|V_{cb}|$ from $\bar{B} \to D^{(*)}\ell\,\bar{\nu}$ decays

The differential semileptonic decay rates as a function of the kinematical variable $w = v_B \cdot v_{D^{(*)}}$ are given by

$$\begin{aligned}\frac{\mathrm{d}\Gamma(\bar{B} \to D^*\ell\,\bar{\nu})}{\mathrm{d}w} &= \frac{G_F^2 M_B^5}{48\pi^3}\, r_*^3 (1-r_*)^2 \sqrt{w^2-1}\,(w+1)^2 \\ &\quad \times \left[1 + \frac{4w}{w+1}\,\frac{1-2wr_*+r_*^2}{(1-r_*)^2}\right] |V_{cb}|^2\, \mathcal{F}^2(w)\,, \\ \frac{\mathrm{d}\Gamma(\bar{B} \to D\,\ell\,\bar{\nu})}{\mathrm{d}w} &= \frac{G_F^2 M_B^5}{48\pi^3}\, r^3 (1+r)^2\, (w^2-1)^{3/2}\, |V_{cb}|^2\, \mathcal{G}^2(w)\,,\end{aligned}$$

where $r_{(*)} = M_{D^{(*)}}/M_B$. In the heavy-quark limit, the form factors $\mathcal{F}(w)$ and $\mathcal{G}(w)$ coincide with the universal function $\xi(w)$ and are thus normalized to unity at $w = 1$. Much effort has gone into calculating the symmetry-breaking corrections to this limit, with the result that [11]

$$\begin{aligned}\mathcal{F}(1) &= 1 + c_A(\alpha_s) + 0 + \delta_{1/m^2} + \ldots = 0.924 \pm 0.027\,, \\ \mathcal{G}(1) &= 1 + c_V(\alpha_s) + \delta'_{1/m} + \delta'_{1/m^2} + \ldots = 1.00 \pm 0.07\,,\end{aligned}$$

where the short-distance coefficients c_A and c_V are known to two-loop order [12, 13]. These numbers include the leading-logarithmic QED corrections. The absence of first-order power corrections to $\mathcal{F}(1)$ is a consequence of Luke's theorem [14]. The theoretical errors quoted above include the perturbative uncertainty and the uncertainty in the calculation (and truncation) of power corrections, added in quadrature. If instead the errors are added linearly, the result for $\mathcal{F}(1)$ changes to 0.924 ± 0.041. This value has recently been confirmed in a different regularization scheme, in which the separation between short- and long-distance contributions is achieved by means of a hard momentum cutoff [15].

A value for $|V_{cb}|$ can be obtained by extrapolating experimental data for the differential decay rates to the zero-recoil point, using theoretical constraints on the shape of the form factors. Model-independent bounds on the physical $\bar{B} \to D^{(*)}\ell\,\bar{\nu}$ form factors can be derived using analyticity properties of QCD correlators, unitarity and dispersion relations [16, 17]. Combining these methods with the approximate heavy-quark symmetry, very powerful one-parameter functions can be derived, which approximate the physical form factors in the semileptonic region with an accuracy of better than 2% [18]. For instance, the function $\mathcal{G}(w)$ can be parametrized as

$$\frac{\mathcal{G}(w)}{\mathcal{G}(1)} \approx 1 - 8\rho_1^2 z(w) + (51.\rho_1^2 - 10.)z^2(w) - (252.\rho_1^2 - 84.)z^3(w)\,,$$

where $z(w) = (\sqrt{w+1} - \sqrt{2})/(\sqrt{w+1} + \sqrt{2})$, and ρ_1^2 is the (negative) slope of the form factor at zero recoil. A similar parametrization can be given for the function $\mathcal{F}(w)$. At present, these constraints are only partially included in the analyses of experimental data.

The world-average results of such analyses are [19, 20]

$$|V_{cb}|\,\mathcal{F}(1) = (35.2 \pm 2.6) \times 10^{-3}\,,$$
$$|V_{cb}|\,\mathcal{G}(1) = (38.6 \pm 4.1) \times 10^{-3}\,.$$

A good fraction of the present errors reflect the uncertainty in the extrapolation to zero recoil, which could be avoided by implementing the dispersive constraints mentioned above. When combined with the theoretical predictions for the normalization of the form factors at zero recoil, the data yield the accurate value

$$|V_{cb}| = (38.2 \pm 2.3_{\rm exp} \pm 1.2_{\rm th}) \times 10^{-3}\,,$$

which is in good agreement with an independent determination from inclusive B decays [1, 19].

2.2 Tests of heavy-quark symmetry

In general, the decays $\bar{B} \to D\,\ell\,\bar{\nu}$ and $\bar{B} \to D^*\ell\,\bar{\nu}$ are described by four independent form factors: $\mathcal{G}(w)$ for the former process, and $h_{A1}(w)$, $R_1(w)$, $R_2(w)$ – a combination of which defines the function $\mathcal{F}(w)$ – for the latter one. In the heavy-quark limit, $\mathcal{G}(w)$, $\mathcal{F}(w)$ and $h_{A1}(w)$ become equal to the function $\xi(w)$, whereas $R_1(w)$ and $R_2(w)$ approach unity. The universality of the function $\xi(w)$ can be tested by measuring the ratio $\mathcal{G}(w)/\mathcal{F}(w)$ as a function of w. The ALEPH data for this ratio [21] are shown in Fig. 2; a similar measurement has also been reported by CLEO [22]. Within errors, the data are compatible with a universal form factor. At large recoil, where the experimental errors are smallest, this provides a test of heavy-quark symmetry at the level of 10–15%.

A more refined analysis of symmetry-breaking corrections has been done by measuring the ratios R_1 and R_2 close to zero recoil. There the HQET predicts that [1]

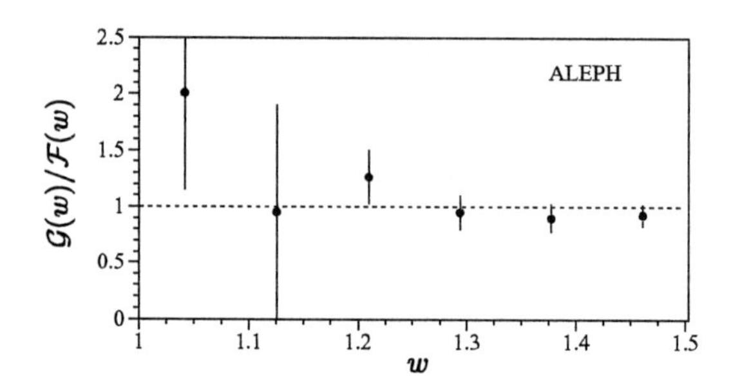

Fig. 2. Ratio of the two form factors $\mathcal{G}(w)$ and $\mathcal{F}(w)$, from Ref. [21]

$$R_1 \approx 1 + \frac{4\alpha_s(m_c)}{3\pi} + \frac{\bar{\Lambda}}{2m_c} = 1.3 \pm 0.1\,,$$

$$R_2 \approx 1 - \frac{\bar{\Lambda}}{2m_c} = 0.8 \pm 0.2\,,$$

$$\rho_{A_1}^2 - \rho_{\mathcal{F}}^2 = \frac{R_1 - 1}{6} + \frac{1 - R_2}{3(1 - r_*)} = 0.2 \pm 0.1\,.$$

The CLEO data for these three quantities are $R_1 = 1.24 \pm 0.29$, $R_2 = 0.72 \pm 0.19$, and $\rho_{A_1}^2 - \rho_{\mathcal{F}}^2 = 0.20 \pm 0.19$ [23], in good agreement with the theoretical predictions. With a little more precision, the data would start to test the pattern of symmetry-breaking effects in the heavy-quark expansion.

2.3 Total exclusive semileptonic rates

Summing up the average experimental results $\mathrm{B}(\bar{B} \to D\,\ell\,\bar{\nu}) = (1.95 \pm 0.27)\%$, $\mathrm{B}(\bar{B} \to D^*\ell\,\bar{\nu}) = (5.05 \pm 0.25)\%$, $\mathrm{B}(\bar{B} \to D^{(*)}\pi\,\ell\,\bar{\nu}) = (2.30 \pm 0.44)\%$, and adding $\mathrm{B}(\bar{B} \to X_u\,\ell\,\bar{\nu}) = (0.15 \pm 0.10)\%$ for the inclusive branching ratio for semileptonic decays into charmless final states, one gets a total of $(9.45 \pm 0.58)\%$ [19], which is not far below the average value for the inclusive semileptonic branching ratio, $\mathrm{B}(\bar{B} \to X\,\ell\,\bar{\nu}) = (10.19 \pm 0.37)\%$, measured at the $\Upsilon(4s)$ resonance (see below). Thus, there is little room for extra contributions.

A solid theoretical understanding of B decays into p-wave charm meson resonances would be important in order to address the question of whether there are additional contributions not accounted for in the above sum, and also to understand the main source of background in the determination of $|V_{cb}|$ from $\bar{B} \to D^*\ell\,\bar{\nu}$ decays. The description of these processes in the context of the HQET involves, at leading order, two new universal functions: $\tau_{3/2}(w)$ for the decays into the narrow states (D_1, D_2^*), and $\tau_{1/2}(w)$ for those into the broad states (D_0^*, D_1^*). In the heavy-quark limit, the differential (in w) decay rates vanish at zero recoil, since in that limit the p-wave states are orthogonal to the ground state. There is a sizable $1/m_c$ correction to the $\bar{B} \to D_1\,\ell\,\bar{\nu}$ decay rate at zero recoil, which can be calculated in a model-independent way in terms of known charm meson masses [24]. It is important, since the kinematical region is restricted close to zero recoil ($1 < w < 1.3$). Detailed theoretical predictions for the semileptonic and hadronic decay rates into p-wave charm states, which incorporate the constraints imposed by heavy-quark symmetry, can be found in Refs. [24, 25].

3 Semileptonic $b \to u$ decays and $|V_{ub}|$

Two exclusive semileptonic B decay modes into charmless hadrons have been observed by CLEO; the corresponding branching ratios are [26]

$$\mathrm{B}(\bar{B} \to \pi\,\ell\,\bar{\nu}) = (1.8 \pm 0.5_{\mathrm{exp}} \pm 0.2_{\mathrm{model}}) \times 10^{-4}\,,$$

$$\mathrm{B}(\bar{B} \to \rho\,\ell\,\bar{\nu}) = (2.5^{+0.6}_{-0.8\mathrm{exp}} \pm 0.5_{\mathrm{model}}) \times 10^{-4}\,.$$

That the theoretical description of these processes involves heavy-to-light form factors implies a certain amount of model dependence, since heavy-quark symmetry does not help to fix their normalization in a precise way. To some extent, a discrimination between models can be obtained by requiring a simultaneous fit of both exclusive channels. From a χ^2-weighted average of models, CLEO obtains [26]

$$|V_{ub}| = (3.3 \pm 0.4_{\mathrm{exp}} \pm 0.7_{\mathrm{model}}) \times 10^{-3}\,.$$

Even with the present, very limited statistics of the measurements, the theoretical uncertainties are the limiting factor in the determination of $|V_{ub}|$. In the future, the model dependence can and will be reduced mainly by combining different theoretical approaches, in particular: lattice calculations, which are restricted to the region of large q^2 [27]; light-cone QCD sum rules, including $O(\alpha_s)$ and higher-twist corrections [28, 29]; analyticity and unitarity constraints [30, 31]; dispersion relations [32]. Although the optimal strategy is not yet clear at present, I believe it will be possible to reach the level of 15% theoretical uncertainty.

The traditional way to extract $|V_{ub}|$ from inclusive $\bar{B} \to X\,\ell\,\bar{\nu}$ decays has been to look at the endpoint region of the charged-lepton energy spectrum, where there is a tiny window not accessible to $\bar{B} \to X_c\,\ell\,\bar{\nu}$ decays. However, this method involves a large extrapolation and is plagued by uncontrolled (and often underestimated) theoretical uncertainties. A reanalysis of the available experimental data (using the ISGW2 model) gives $|V_{ub}| = (3.7 \pm 0.6_{\mathrm{exp}}) \times 10^{-3}$ [33], in agreement with the value quoted above. A better discrimination between $b \to u$ and $b \to c$ transitions should use vertex information combined with a cut on the invariant mass M_h (or energy E_h) of the hadronic final state [34]–[40]. Parton model calculations (with Fermi motion included) indicate that about 90% of all $\bar{B} \to X_u\,\ell\,\bar{\nu}$ decays have $M_h < M_D$, as shown in Fig. 3. Ideally, this cut would thus provide for a very efficient discriminator. In practise, there will be some leakage so that presumably one will be forced to require $M_h < M_{\rm max}$ with some threshold $M_{\rm max} < M_D$. The task for theorists is to calculate the fraction of events with hadronic mass $M_h < M_{\rm max}$,

$$\Phi(M_{\rm max}) = \frac{1}{\Gamma} \int\limits_0^{M_{\rm max}} {\rm d}M_h \, \frac{{\rm d}\Gamma(\bar{B} \to X_u \, \ell \, \bar{\nu})}{{\rm d}M_h} \, .$$

To calculate this fraction requires an ansatz for the "shape function", which describes the Fermi motion of the b quark inside the B meson [41, 42]. The first three moments of this function are determined in terms of known HQET matrix elements. Still, some theoretical uncertainty remains, mainly associated with the values of the b-quark mass and the kinetic energy μ_π^2, as well as unknown $O(\alpha_s^2)$ corrections. As an example, Fig. 3 shows the dependence of $\Phi(M_{\rm max})$ on the value of m_b. It turns out that the resulting theoretical uncertainty strongly depends on the value of the threshold $M_{\rm max}$. First estimates yield $\delta|V_{ub}|/|V_{ub}| \approx 10\%$ for $M_{\rm max} = M_D$, and $\delta|V_{ub}|/|V_{ub}| \approx 20\%$ for $M_{\rm max} = 1.5\,{\rm GeV}$ [38, 39]. This new method is challenging both for theorists and for experimenters, but it is superior to the endpoint method. I believe that ultimately a theoretical accuracy of 10% can be reached.

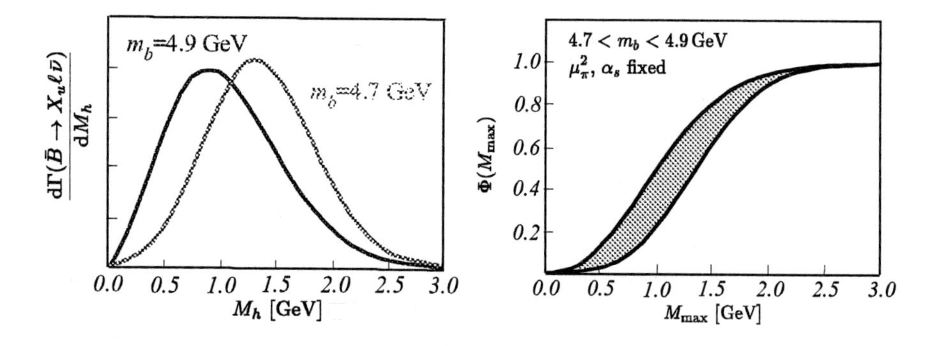

Fig. 3. Hadronic mass distribution (left) and fraction of $\bar{B} \to X_u \, \ell \, \bar{\nu}$ decays with invariant hadronic mass below $M_{\rm max}$ (right), from Ref. [38]

4 Semileptonic branching ratio and charm multiplicity

For many years, the apparent discrepancy between (some) measurements of the semileptonic branching ratio of B mesons and theoretical predictions for this quantity has given rise to controverse dispute and speculations about deviations from the Standard Model. Another important input to this discussion has been the charm multiplicity n_c, i.e. the average number of charm (or anticharm) quarks in the hadronic final state of a B decay. From the theoretical point of view,

$$\mathrm{B_{SL}} = \frac{\Gamma(\bar{B} \to X\, e\, \bar{\nu})}{\sum_\ell \Gamma(\bar{B} \to X\, \ell\, \bar{\nu}) + \Gamma_{\mathrm{had}} + \Gamma_{\mathrm{rare}}},$$

$$n_c = 1 + \mathrm{B}(\bar{B} \to X_{c\bar{c}}) - \mathrm{B}(\bar{B} \to \text{no charm})$$

are governed by the same partial inclusive decay rates. The theoretical predictions for these two quantities depend mainly on two parameters: the quark-mass ratio m_c/m_b and the renormalization scale μ [43, 44]. The latter dependence reflects our ignorance about higher-order QCD corrections to the decay rates. The results obtained by allowing reasonable ranges for these parameters are represented by the dark-shaded area in Fig. 4. The two data points show the average experimental results obtained from experiments operating at the $\Upsilon(4s)$ resonance: $\mathrm{B_{SL}} = (10.19 \pm 0.37)\%$ and $n_c = 1.12 \pm 0.05$, and at the Z resonance: $\mathrm{B_{SL}} = (11.12 \pm 0.20)\%$ and $n_c = 1.20 \pm 0.07$ [19, 45]. At this conference, it has been emphasized that a dedicated reanalysis of the LEP data for the semileptonic branching ratio is necessary, because some sources of systematic errors had previously been underestimated [45]. To account for this, I have doubled the corresponding error bar on $\mathrm{B_{SL}}$ in Fig. 4. If we ignore the LEP point for the moment, it appears that the theoretical predictions for both $\mathrm{B_{SL}}$ and n_c lie significantly higher than the experimental results.

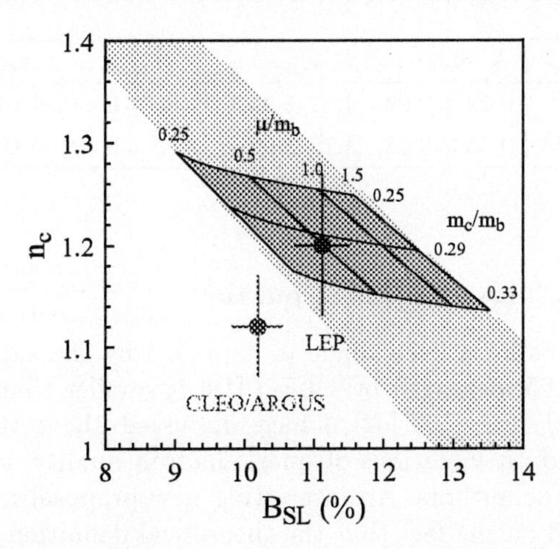

Fig. 4. Theory [44] versus experiment for the semileptonic branching ratio and charm multiplicity

At this conference, several new ingredients have been presented that shed some light on this problem. First, there are some nontrivial tests of the theory. Table 2 shows details of the theoretical calculation in the form of the dominant partial decay rates normalized to $\mathrm{B}(\bar{B} \to X_c \ell \bar{\nu})$, for $m_c/m_b = 0.29 \pm 0.03$, using the results of Refs. [43, 44, 46]. (Note that the values for $r_{c\bar{u}d'}$ are larger than the value 4.0 ± 0.4 often used in the literature.) It has been argued that the assumption of local quark–hadron duality, which underlies the theoretical treatment of inclusive decay rates, may fail for the decays $\bar{B} \to X_{c\bar{c}s'}$, because there is only little kinetic energy released to the final state. To test this hypothesis, one can eliminate the corresponding partial decay rate, in which case one obtains a linear relation between n_c and $\mathrm{B_{SL}}$ [47]. The result is shown as the light band in Fig. 4. The fact that, within this band, the original prediction (i.e. the dark-shaded area) is closest to the data indicates that there is no problem with quark–hadron duality. Another important check is provided by an experimental determination of the ratio $r_{c\bar{u}d'}$ using flavour-specific measurements of charm branching ratios [48]. The result is [49]–[51]

$$r_{c\bar{u}d'} = \frac{\mathrm{B}(\bar{B} \to \text{open } c) - \mathrm{B}(\bar{B} \to \text{open } \bar{c})}{\mathrm{B_{SL}}} - (2 + r_{c\tau\bar{\nu}} - r_{u\bar{c}s'}) = 4.1 \pm 0.7 \,,$$

in good agreement with the theoretical predictions given in Table 2. The theory input $r_{c\tau\bar{\nu}} - r_{u\bar{c}s'} = 0.19 \pm 0.03$ in this extraction has a very small uncertainty. In summary, it appears that the heavy-quark expansion works well for both of the hadronic decay rates.

Table 2. Predictions for ratios of partial inclusive decay rates

	$r_{c\tau\bar{\nu}}$	$r_{c\bar{u}d'}$	$r_{c\bar{c}s'}$	$r_{\text{no charm}}$
$\mu = m_b$	0.22 ∓ 0.03	4.21 ± 0.01	1.89 ∓ 0.44	0.14 ± 0.04
$\mu = m_b/2$	0.23 ∓ 0.03	4.75 ± 0.02	2.20 ∓ 0.49	0.19 ± 0.04

4.1 Is there a "missing charm puzzle"?

Several suggestions have been made to explain why the experimental value of $n_c = 1.12 \pm 0.05$ measured by CLEO [19] is smaller than the theoretical prediction $n_c = 1.20 \pm 0.06$ [44]. I have discussed above that the problem cannot be blamed on violations of quark–hadron duality, as was originally speculated by some authors. An interesting new proposal made by Dunietz et al. [49] is based on the fact that the theoretical definition of n_c refers to a fully inclusive quantity counting the number of charm and anticharm quarks per B decay, irrespective of whether they end up as "open" or "hidden" charm. If there were a sizable branching ratio for decays into hadronic final states containing undetected $(c\bar{c})$ pairs,

$$\mathrm{B}(\bar{B} \to (c\bar{c})_{\text{undetected}} + X) \equiv b_{(c\bar{c})} \,,$$

then the experimentally observed value of n_c would be lower than the theoretical one: $n_c^{\rm obs} = n_c^{\rm th} - 2b_{(c\bar{c})}$. Note that the conventional charmonium states J/ψ, ψ', χ_{c1}, χ_{c2}, η_c are included in the charm counting and thus do not contribute to $b_{(c\bar{c})}$. However, it has been speculated that there may exist an enhanced production of exotic $(c\bar{c}g)$ hybrids, which decay into noncharmed light mesons and could yield a value of $b_{(c\bar{c})}$ of about 10% [49, 52].

Alternatively, enhanced flavour-changing neutral current (FCNC) processes such as $b \to sg$ could, simultaneously, lower the predictions for $\mathrm{B_{SL}}$ and n_c by a factor of $[1+\mathrm{B(new\ FCNC)}]^{-1}$, thus providing for a "new physics explanation" of the missing charm puzzle [53, 54].

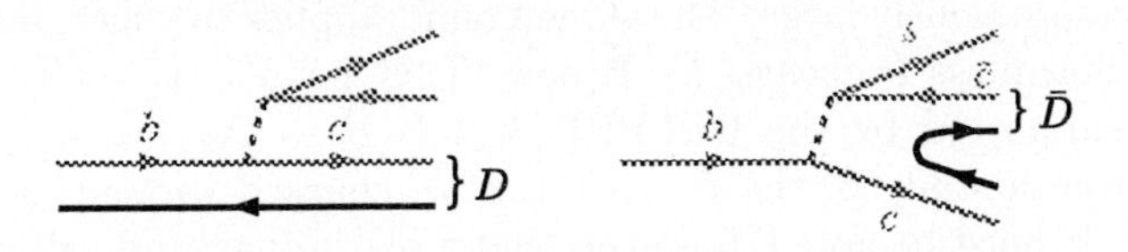

Fig. 5. Mechanisms for D and $\bar{D}$ production in $\bar{B}$-meson decays

At least a partial answer to the question of whether there is any support for such speculations is provided by new, flavour-specific measurements of charm production. Using charm-particle–lepton correlations in tagged B decays, it is possible to measure the relative rate of the two mechanisms shown in Fig. 5. Whereas it was previously assumed that the first mechanism was responsible for all D-meson production, recent measurements indicate a rather large "wrong-charm" yield [48]:

$$\frac{\mathrm{B}(\bar{B} \to \bar{D}\,X)}{\mathrm{B}(\bar{B} \to D\,X)} = 0.100 \pm 0.026 \pm 0.016\,,$$

corresponding to a "wrong-charm" branching ratio $\mathrm{B}(\bar{B} \to \bar{D}\,X) = (7.9 \pm 2.2)\%$. This observation is supported by measurements of both exclusive [55] and inclusive [56, 57] production of $D\bar{D}$ meson pairs. This effect was not included in Monte Carlo B-decay generators and may be partly responsible for the discrepancy in the CLEO and LEP measurements of $\mathrm{B_{SL}}$ and n_c [49]. When combined with previously measured decay rates, the new CLEO result implies [48]

$$\mathrm{B}(\bar{B} \to X_{c\bar{c}}) - b_{(c\bar{c})} = (21.9 \pm 3.7)\%\,,$$

which is close to the theoretical prediction $\mathrm{B}(\bar{B} \to X_{c\bar{c}}) = (22 \pm 6)\%$ [44], indicating that there is not much room for decays into undetected $(c\bar{c})$ pairs, represented by $b_{(c\bar{c})}$. This measurement is supported by a DELPHI result on the production of two open charm particles [58]. Furthermore, by measuring a double ratio of flavour-specific rates, CLEO obtains a bound on charmless modes that is largely independent of detection efficiencies and charm branching ratios. The result is $\mathrm{B}(\bar{B} \to \text{no open charm}) = (3.2 \pm 4.0)\% < 9.6\%$ (90%

CL), which after subtraction of the known charmonium contributions implies [48]

$$\mathrm{B}(\bar{B} \to \text{no charm}) + b_{(c\bar{c})} = (0.2 \pm 4.1)\% < 6.8\% \quad (90\%\ \mathrm{CL}) .$$

Given that the Standard Model prediction for the charmless rate is $\mathrm{B}(\bar{B} \to \text{no charm}) = (1.6 \pm 0.8)\%$ [46], it appears that there is little room for either hidden $(c\bar{c})$ production or new physics contributions. Again, this conclusion is supported by a DELPHI result obtained using impact parameter measurements: $\mathrm{B}(\bar{B} \to \text{no open charm}) = (4.5 \pm 2.5)\% < 8.4\%$ (90% CL) [58].

Combining these two measurements, I conclude that $b_{(c\bar{c})} < 5\%$ (90% CL) cannot be anomalously large. The same bound applies to other, nonstandard sources of charmless B decays, i.e. B(new FCNC) $< 5\%$ (90% CL). This conclusion is supported by the DELPHI limit $\mathrm{B}(\bar{B} \to X_{sg}) < 5\%$ (95% CL) obtained from a study of the $p_\perp$ spectrum of charged kaons produced in B decays [58]. It must be noted, however, that a preliminary indication of a kaon excess at large $p_\perp$, as expected from enhanced $b \to sg$ transitions [50], has been reported by SLD at this conference [59]. Even though a definite conclusion can therefore not been drawn before these measurements become final, at present there is no compelling experimental evidence of any nonstandard physics in inclusive B decays.

The two CLEO measurements quoted above can be combined to give a new determination of the charm multiplicity, in which the unknown quantity $b_{(c\bar{c})}$ cancels out. The result

$$n_c = 1 + \left[\mathrm{B}(\bar{B} \to X_{c\bar{c}}) - b_{(c\bar{c})}\right] - \left[\mathrm{B}(\bar{B} \to \text{no charm}) + b_{(c\bar{c})}\right] = 1.22 \pm 0.06$$

is significantly higher than (though consistent with) the value 1.12 ± 0.05 obtained using the conventional method of charm counting, and in excellent agreement with the theoretical prediction. I believe that, ultimately, this new way of measuring n_c will be less affected by systematic uncertainties than the traditional one. It seem that the "missing charm puzzle" is about to disappear.

5 Beauty lifetime ratios

The current world-average experimental results for the lifetime ratios of different beauty hadrons are [60]

$$\frac{\tau(B^-)}{\tau(B^0)} = 1.06 \pm 0.04 , \quad \frac{\tau(B_s)}{\tau(B_d)} = 0.98 \pm 0.07 , \quad \frac{\tau(\Lambda_b)}{\tau(B^0)} = 0.78 \pm 0.04 .$$

Theory predicts that $|\tau(B_s)/\tau(B_d) - 1| < 1\%$ [61], and it will be very difficult to push the experimental accuracy to a level where one would become sensitive to sub-1% effects. The theoretical predictions for the other two ratios have been analysed to third order in the heavy-quark expansion. Unfortunately, these predictions depend on some yet unknown "bag parameters" B_1, B_2, ε_1, ε_2, $\widetilde{B}$, r parametrizing the hadronic matrix elements of local four-quark operators. In terms of these parameters, the results are [44]

$$\frac{\tau(B^-)}{\tau(B^0)} = 1 + 16\pi^2 \frac{f_B^2 M_B}{m_b^3} \left[k_1 B_1 + k_2 B_2 + k_3\varepsilon_1 + k_4\varepsilon_2\right],$$

$$\frac{\tau(\Lambda_b)}{\tau(B^0)} = 0.98 + 16\pi^2 \frac{f_B^2 M_B}{m_b^3} \left[p_1 B_1 + p_2 B_2 + p_3\varepsilon_1 + p_4\varepsilon_2 + (p_5 + p_6\widetilde{B})r\right],$$

where k_i and p_i are short-distance coefficients, whose values depend on the ratio m_c/m_b and on the renormalization scale. The large-N_c counting rules of QCD imply that $B_i = O(1)$ and $\varepsilon_i = O(1/N_c)$. The factorization approximation for the meson matrix elements suggests that $B_i \approx 1$ and $\varepsilon_i \approx 0$ [62]. Similarly, the constituent quark model suggests that $\widetilde{B} \approx 1$ and $r \approx |\psi_{qq}^{\Lambda_b}(0)|^2/|\psi_q^B(0)|^2$. The parameter r is the most uncertain one entering the predictions for the lifetime ratios. Existing theoretical estimates for this parameter range from 0.1 to 2. Some recent estimates can be found in Refs. [63, 64].

Without a reliable field-theoretical calculation of the bag parameters, no lifetime "predictions" can be obtained. However, one can see which ranges of the lifetime ratios can be covered using sensible values for the bag parameters [65]. To this end, I scan the following parameter space: $B_i, \widetilde{B} \in [2/3, 4/3]$, $\varepsilon_i \in [-1/3, 1/3]$, $r \in [0.25, 2.5]$, $m_c/m_b = 0.29 \pm 0.03$, $f_B = (200 \pm 20)$ MeV. The resulting distributions for the lifetime ratios are shown in Fig. 6. The most important observation of this exercise is that it is possible to reproduce both lifetime ratios, within their experimental uncertainties, using the same set of input parameters. The theoretically allowed range for the ratio $\tau(B^-)/\tau(B^0)$ is centered around the experimental value; however, the width of the allowed region is so large that no accurate "prediction" of this ratio could have been made. Any value between 0.8 and 1.3 could be easily accommodated by theory. The predictions for the ratio $\tau(\Lambda_b)/\tau(B^0)$, on the other hand, center around a value of 0.95, and the spread of the results is much narrower. Only a small tail extends into the region preferred by experiment. Requiring that the theoretical results be inside the 2σ ellipse around the central experimental values, I find that the value of the parameter ε_1 must be close to zero, $\varepsilon_1 \approx -(0.1 \pm 0.1)$, in agreement with the expectation based on factorization. In addition, this requirement maps out the region of parameter space for ε_2 and r shown in the right lower plot of Fig. 6. Thus, the data indicate a large (with respect to most model predictions) value of r and a negative value of ε_2. Ultimately, it will be important to perform reliable field-theoretical calculations of the hadronic parameters, for instance using lattice gauge theory or QCD sum rules, to see whether indeed the observed beauty lifetime ratios can be accounted for by the heavy-quark expansion (for a recent sum-rule estimate of the parameters B_i and ε_i, see Ref. [66]).

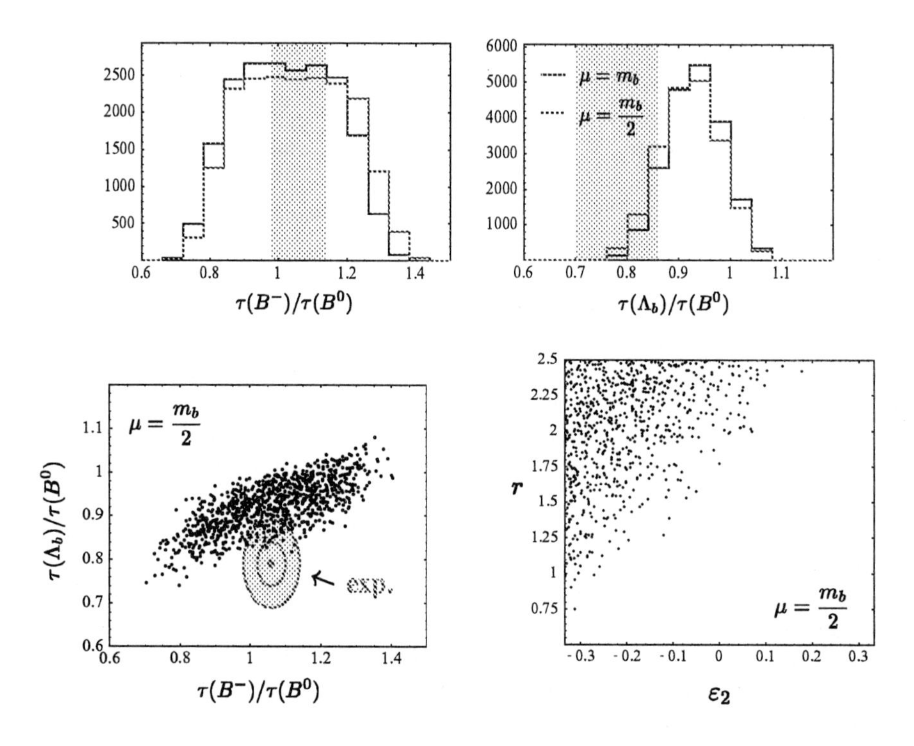

Fig. 6. Upper plots: Theoretical distributions for beauty lifetime ratios. The two curves in each figure correspond to different choices of the renormalization scale. The shaded areas show the experimental values with their 2σ error bands. Lower left: Two-dimensional distribution of lifetime ratios, together with the 1σ and 2σ contours around the central experimental values. Lower right: Distribution of the hadronic parameters ε_2 and r for simulations with results inside the 2σ ellipse.

This is particularly important for the parameter r, large values of which are not excluded a priori. Using a variant of the quark-model relation derived in Ref. [63], combined with a preliminary DELPHI measurement of the mass-splitting between the Σ_b^* and Σ_b baryons [67], I find

$$r \approx \frac{4}{3}\,\frac{M_{\Sigma_b^*}^2 - M_{\Sigma_b}^2}{M_{B^*}^2 - M_B^2} = 1.8 \pm 0.5\,.$$

With such a large value, it is possible to explain the short Λ_b lifetime without invoking violations of local quark–hadron duality.

Although at present there is thus no convincing evidence that the low value of the Λ_b lifetime could not be accommodated in the context of the heavy-quark expansion, it remains a possibility that in the future such a discrepancy may emerge, for instance if lattice calculations would show that $r \ll 1$. Then, one would have to blame violations of local quark–hadron duality to be responsible for the failure of the heavy-quark expansion for inclusive, nonleptonic B-decay rates. Recently, several authors have studied in QCD-inspired models how quark–hadron duality may be violated. Based on a simple model for the difference of two heavy–light current correlators in the chiral and large-N_c limits, Shifman has argued that deviations from local duality are due to the asymptotic nature of the operator product expansion (OPE) [68]. More recently, Blok et al. have studied QCD_2 (i.e. QCD in $1+1$ space–time dimensions) at next-to-leading order in $1/N_c$, where finite resonance widths provide for a dynamical smearing of correlation functions [69]. They find that, in the Euclidean region, deviations from local duality are indeed due to the divergence of the OPE, and are exponentially suppressed with Q^2. In the physical (Minkowskian) region, on the other hand, they find that two mass scales, Λ_1 and Λ_2, are dynamically generated. For $\Lambda_1^2 < q^2 < \Lambda_2^2$, there are strong oscillations of correlation functions. Global duality works, but local duality may be strongly violated. For $q^2 > \Lambda_2$, the oscillations become exponentially damped, and local duality is restored. They conjecture that a similar pattern may hold for real, four-dimensional QCD. Similar results have also been obtained by Grinstein and Lebed [70], and by Chibisov et al. studying the instanton vacuum model [71]. Quite different conclusions have been reached by Colangelo et al., who have considered a more "realistic" version of Shifman's model, finding that global duality holds, but local duality is spoiled by $1/Q$ effects not present in the OPE [72]. This study gives some support for the conjecture by Altarelli et al. that there may be non-OPE terms of order $\bar{\Lambda}/m_Q$ increasing the nonleptonic decay rates of heavy hadrons [73]. Such terms could easily explain the short Λ_b lifetime. However, I stress that, although the model of Ref. [72] is interesting, it must not be taken as an existence proof of non-OPE power corrections. Nature may be careful enough to avoid power-like deviations from local duality.

6 Hadronic B decays

The strong interaction effects in hadronic decays are much more difficult to understand, even at a qualitative level, than in leptonic or semileptonic decays. The problem is that multiple gluon exchanges can redistribute the quarks in the final state of a hadronic decay. As a consequence, some phenomenological assumptions are unavoidable is trying to understand these processes.

6.1 Energetic two-body decays

It has been argued that energetic two-body decays with a large energy release are easier to understand because of the colour-transparency phenomenon: a pair of fast-moving quarks in a colour-singlet state, which is produced in a local interaction, effectively decouples from long-wavelength gluons [74]. This intuitive argument suggests that nonleptonic amplitudes factorize to a good approximation. The main strong-interaction effects are then of a short-distance nature and simply renormalize the operators in the effective weak Hamiltonian [75]. For the case of $b \to c\bar{u}d$ transitions, for instance,

$$H_{\text{eff}} = \frac{G_F}{\sqrt{2}}\, V_{cb} V_{ud}^* \Big\{ c_1(\mu)\, (\bar{d}u)(\bar{c}b) + c_2(\mu)\, (\bar{c}u)(\bar{d}b) \Big\} + \dots ,$$

where $(\bar{d}u) = \bar{d}\gamma^\mu(1-\gamma_5)u$ etc. are left-handed, colour-singlet quark currents, and $c_1(m_b) \approx 1.1$ and $c_2(m_b) \approx -0.3$ are Wilson coefficients taking into account the short-distance corrections arising from the exchange of hard gluons. The effects of soft gluons remain in the hadronic matrix elements of the local four-quark operators. In general, a reliable field-theoretical calculation of these matrix elements is the obstacle to a quantitative theory of hadronic weak decays.

Using Fierz identities, the four-quark operators in the effective Hamiltonian may be rewritten in various forms. It is particularly convenient to rearrange them in such a way that the flavour quantum numbers of one of the quark currents match those of one of the hadrons in the final state of the considered decay process. As an example, consider the decays $\bar{B} \to D\pi$. Omitting common factors, the various amplitudes can be written as

$$\begin{aligned}
A_{\bar{B}^0 \to D^+\pi^-} &= \Big(c_1 + \frac{c_2}{N_c}\Big) \langle D^+\pi^- | (\bar{d}u)(\bar{c}b) | \bar{B}^0 \rangle + 2c_2\, \langle D^+\pi^- | (\bar{d}t_a u)(\bar{c}t_a b) | \bar{B}^0 \rangle \\
&\equiv a_1\, \langle \pi^- | (\bar{d}u) | 0 \rangle\, \langle D^+ | (\bar{c}b) | \bar{B}^0 \rangle , \\
A_{\bar{B}^0 \to D^0\pi^0} &= \Big(c_2 + \frac{c_1}{N_c}\Big) \langle D^0\pi^0 | (\bar{c}u)(\bar{d}b) | \bar{B}^0 \rangle + 2c_1\, \langle D^0\pi^0 | (\bar{c}t_a u)(\bar{d}t_a b) | \bar{B}^0 \rangle \\
&\equiv a_2\, \langle D^0 | (\bar{c}u) | 0 \rangle\, \langle \pi^0 | (\bar{d}b) | \bar{B}^0 \rangle ,
\end{aligned}$$

where t_a are the SU(3) colour matrices. The two classes of decays shown above are referred to as class-1 and class-2, respectively. The factorized matrix elements in the last steps are known in terms of the meson decay constants f_π and f_D, and the transition form factors for the decays $\bar{B} \to D$ and $\bar{B} \to \pi$. Most of these quantities are accessible experimentally. Of course, the above matrix elements also contain other, nonfactorizable contributions. They are absorbed into the definition of the hadronic parameters a_1 and a_2, which in general are process dependent. Recently, some progress in the understanding of these parameters has been made, leading to the "generalized factorization hypothesis" that [76]–[78]

$$a_1 \approx c_1(m_b)\,, \qquad a_2 \approx c_2(m_b) + \zeta c_1(m_b)\,,$$

where ζ is a process-independent hadronic parameter (for energetic two-body decays only!), which accounts for the dominant nonfactorizable contributions to the decay amplitudes. To derive these results, one combines the $1/N_c$ expansion with an argument inspired by colour transparency [78].

Table 3. Experimental tests of the generalized factorization hypothesis

	a_1	a_2/a_1
$\bar{B} \to D^{(*)}\pi$	1.08 ± 0.06	0.21 ± 0.07
$\bar{B} \to D^{(*)}\rho$	1.07 ∓ 0.07	0.23 ± 0.14
$\bar{B} \to \psi^{(\prime)}K^{(*)}$		$\lvert a_2\rvert = 0.21 \pm 0.04$
$\bar{B} \to D^{(*)}\bar{D}_s^{(*)}$	1.10 ± 0.18	

Some experimental tests of the generalized factorization hypothesis are shown in Table 3, where I quote the values of a_1 and a_2 extracted from the analysis of different classes of decay modes, using data reported by CLEO [79]–[81]. Within the present experimental errors, there is indeed no evidence for any process dependence of the hadronic parameters. The generalized factorization prescription provides a simultaneous description of all measured Cabibbo-allowed two-body decays with a single parameter $\zeta = 0.45 \pm 0.05$ extracted from the data on class-2 decays [78]. So far, this theoretical framework is fully supported by the data.

The factorization hypothesis can be used to obtain rather precise values for the decay constants of the D_s and D_s^* mesons, by comparing the theoretical predictions for various ratios of $\bar{B}^0$ decay rates,

$$\frac{\Gamma(D^+D_s^-)}{\Gamma(D^+\pi^-)} = 1.01\left(\frac{f_{D_s}}{f_\pi}\right)^2 , \qquad \frac{\Gamma(D^{*+}D_s^-)}{\Gamma(D^{*+}\pi^-)} = 0.72\left(\frac{f_{D_s}}{f_\pi}\right)^2 ,$$

$$\frac{\Gamma(D^+D_s^{*-})}{\Gamma(D^+\rho^-)} = 0.74\left(\frac{f_{D_s^*}}{f_\rho}\right)^2 , \qquad \frac{\Gamma(D^{*+}D_s^{*-})}{\Gamma(D^{*+}\rho^-)} = 1.68\left(\frac{f_{D_s^*}}{f_\rho}\right)^2 ,$$

with data. These predictions are rather clean for the following reasons: first, all decays involve class-1 transitions, so that deviations from factorization are probably very small; secondly, the parameter a_1 cancels in the ratios; thirdly, the two processes in each ratio have a similar kinematics, so that the corresponding decay rates are sensitive to the same form factors, however evaluated at different q^2 values. Finally, some of the experimental systematic errors cancel in the ratios (however, I do not assume this in quoting errors below). Combining these predictions with the average experimental branching ratios [79] yields

$$f_{D_s} = (234 \pm 25)\ \text{MeV} , \qquad f_{D_s^*} = (271 \pm 33)\ \text{MeV} .$$

The result for f_{D_s} is in good agreement with the value $f_{D_s} = 250 \pm 37$ MeV extracted from the leptonic decay $D_s \to \mu^+\nu$ [82]. The ratio of decay constants, $f_{D_s^*}/f_{D_s} = 1.16 \pm 0.19$, which cannot be determined from leptonic decays, is in good agreement with theoretical expectations [83, 84]. Finally, I note that, assuming SU(3)-breaking effects of order 10–20%, the established value of f_{D_s} implies $f_D \gtrsim 200$ MeV, which is larger than most theoretical predictions.

6.2 *B* Decays into two light mesons

Since last year, CLEO has observed (or put strict upper limits on) a number of rare B decay modes, many of which have strongly suppressed tree amplitudes and are thus dominated by loop processes. The relevant quark diagrams for the transitions $b \to s(d)\bar{q}q$ are shown in Fig. 7, where I also give the powers of the Wolfenstein parameter $\lambda \approx 0.22$ associated with these diagrams. Because the penguin digram shown on the right is a loop process, it is sensitive to new heavy particles and thus potentially probes physics beyond the Standard Model.

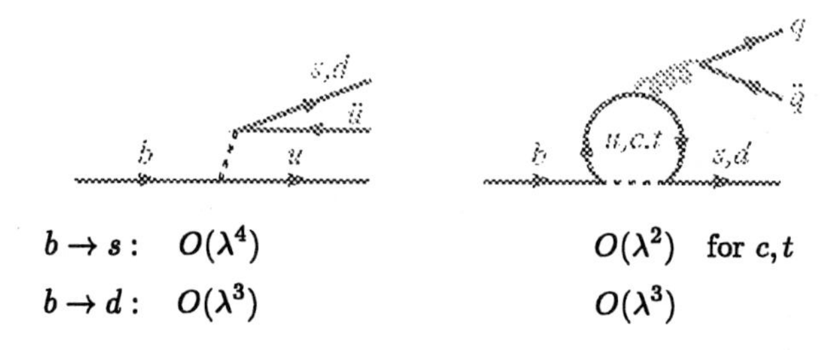

Fig. 7. Tree and penguin diagrams for rare B decays into light hadrons

Some of the experimental results that have caused a lot of excitement are (all in units of 10^{-5} and averaged over CP-conjugate modes; upper limits at 90% CL) [81, 85, 86]:

$$\mathrm{B}(B^0 \to K^+\pi^-) = 1.5^{+0.5}_{-0.4} \pm 0.1 \pm 0.1\,,$$

$$\mathrm{B}(B^+ \to K^0\pi^+) = 2.3^{+1.1}_{-1.0} \pm 0.3 \pm 0.2\,,$$

$$\mathrm{B}(B^0 \to \pi^+\pi^-) = 0.7 \pm 0.4 < 1.5\,,$$

$$\mathrm{B}(B^+ \to \pi^+\pi^0) = 1.0^{+0.6}_{-0.5} < 2.0\,,$$

$$\mathrm{B}(B^+ \to K^+\eta') = 7.1^{+2.5}_{-2.1} \pm 0.9\,,$$

$$\mathrm{B}(B \to \eta' + X_s) = 62 \pm 16 \pm 13 \quad (2.0 < p_{\eta'}\ [\mathrm{GeV}] < 2.7)\,.$$

In Table 4, I show the results of an analysis of these results in terms of SU(3)-invariant amplitudes representing diagrams having tree topology $T^{(\prime)}$, colour-suppressed tree topology $C^{(\prime)}$, penguin topology $P^{(\prime)}$, and SU(3)-singlet penguin topology $S^{(\prime)}$, where primed (unprimed) quantities refer to $b \to s(d)$

transitions [87]. The results of this analysis establish the presence of a significant SU(3)-singlet contribution S' active in the decays with an η' meson in the final state. They also indicate a rather large "penguin pollution" $|P/T| \approx 0.3 \pm 0.1$ in the decays $B \to \pi\pi$. This has important (and optimistic) implications for CP-violation studies using these decay modes [88].

Table 4. Amplitude analysis for some rare B decays into two light mesons. The dominant amplitudes are highlighted.

Decay mode	SU(3) invariant amplitudes	(in units of 10^{-6})
$B^0 \to K^+\pi^-$ $B^+ \to K^0\pi^+$	$-(T' + P')$ P'	$\|P'\|^2 \approx 16 \pm 4$ $\Rightarrow \|P\|^2 \approx 0.8 \pm 0.4$
$B^0 \to \pi^+\pi^-$ $B^+ \to \pi^+\pi^0$	$-(T + P)$ $-\frac{1}{\sqrt{2}}(T + C)$	$\|T\|^2 \approx 8 \pm 4$ $\Rightarrow \|T'\|^2 \approx 0.4 \pm 0.2$
$B^+ \to \eta' K^+$	$\frac{1}{\sqrt{6}}(T' + C' + 3P' + 4S')$	$\|S'\|^2 \approx 18 \pm 3$ no interference $\|S'\|^2 \approx 5 \pm 2$ constr. interference

Several authors have discussed the question whether the penguin amplitudes in these transitions are anomalously large. There seems to be some agreement that a standard analysis, using next-to-leading order Wilson coefficients combined with factorized matrix elements, can accommodate the experimental results for the exclusive rare decay rates [89, 90]. However, it has been stressed that significant long-distance contributions to the charm penguin are required to fit the data [91, 92]. This is not too surprising, since the energy release in B decays is such that the $c\bar{c}$ pair in the charm penguin is not far from its mass shell, and hence this penguin is really more a long-distance than a short-distance process [93]. This is illustrated in Fig. 8. In the case of $B \to K\pi$ decays, the box may represent final-state rescattering processes such as $B \to D_s\bar{D} \to K\pi$ [94]–[97]. If the final state contains an η' meson, the box may represent an anomaly-mediated coupling of the η' to glue, or an intrinsic charm component in the η' wave function. At least to some extent these are different words for the same physics.

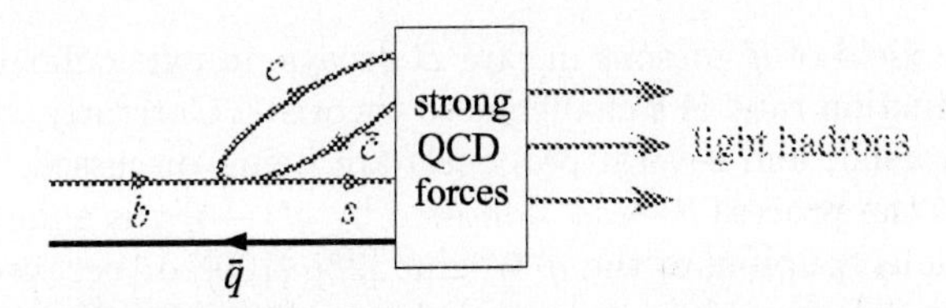

Fig. 8. Long-distance effects in "charming penguins"

In a way, the charm penguin is all there is in $b \to s$ FCNC processes [93], because the unitarity of the CKM matrix implies that

$$\sum_{q=u,c,t} V_{qs}^* V_{qb} P_q \approx V_{cs}^* V_{cb} (P_c - P_t) \, .$$

The top penguin P_t simply provides the GIM cutoff for large loop momenta, whereas the up penguin P_u is strongly CKM suppressed. The construction of the effective weak Hamiltonian is used to separate the short- and long-distance contributions to the charm penguin, in a way that is illustrated in Fig. 9. The short-distance contributions from large loop momenta are contained in the local operators $Q_{3,\dots,6}$ and Q_8, while long-distance effects from small momenta are contained in the matrix elements of the current–current operators $Q_{1,2}$. From the physical argument presented above, one expects sizable long-distance contributions, since the energy release in B decays is in the region of charm resonances. Therefore, purely short-distance estimates of the charm penguin may be very much misleading.

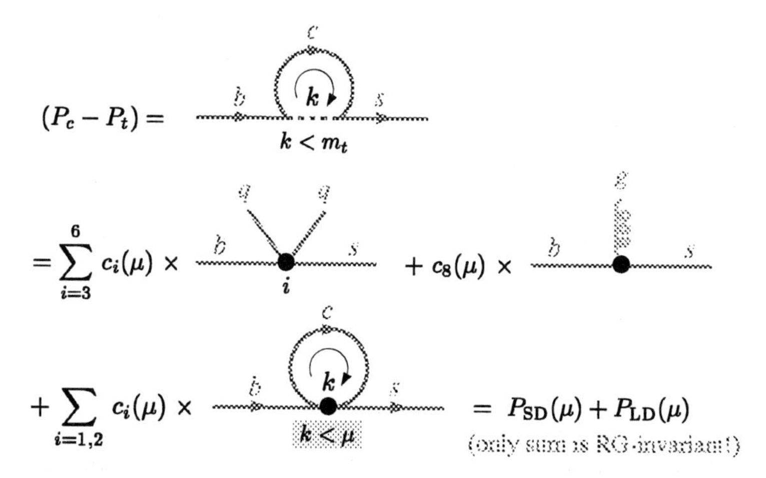

Fig. 9. Operator product expansion for the charm penguin

6.3 Explanations of the η' modes

To explain the yield of η' mesons in rare B decays, in particular the large inclusive η' production rate, is a challenge to theorists. Currently, the situation is still controversial, and several proposals are being discussed. It has been suggested that the process $b \to sg^*$ followed by $g^* \to \eta' g$ is enhanced, either by the anomalous coupling of the η' to glue [98]–[100], or because the initial $b \to sg$ rate is enhanced by some new physics [101]. Because of its three-body nature, this mechanism would mainly be responsible for multi-particle inclusive processes containing η' mesons, but give a minor contribution to the two-body modes $B \to \eta' K^{(*)}$. It has also been argued that an enhanced η' production could result from a significant intrinsic charm component in the η' wave function, either through the colour-singlet mechanism $b \to (c\bar{c})_1\, s$ followed by $(c\bar{c})_1 \to \eta'$ [102, 103], or through the colour-octet mechanism $b \to (c\bar{c})_8\, s$ followed by $(c\bar{c})_8 \to \eta' X$ [104], which is strongly favoured by the structure of the effective weak Hamiltonian. The relevance of the intrinsic charm mechanism depends on the size of a "decay constant" $f_{\eta'}^{(c)}$, estimates of which range from 6 MeV to 180 MeV. A clarification of this issue is needed (see, e.g., the discussion in Refs. [89, 105]). Yet other possibilities to explain the large η' yield have been explored in Refs. [106, 107].

Table 5. Model predictions for ratios of rare B decay rates

	Ali, Greub [89]	Lipkin [109]	Cheng, Tseng [110]	Datta et al. [111]	Halperin et al. [102]
$\dfrac{B^+ \to \eta' K^{*+}}{B^+ \to \eta' K^+}$	0.01–0.2		0.20–0.24	0.36–0.38	≈ 2
$\dfrac{B^+ \to \eta K^+}{B^+ \to \eta' K^+}$	0.04–0.14	0–0.02	0.03–0.08	0.02–0.09	
$\dfrac{B^+ \to \eta K^{*+}}{B^+ \to \eta' K^{*+}}$	0.6–2.9	0.5–8.2	0.26–0.48	1.0–1.4	

Future measurements of other decay channels, as well as of the distribution of the invariant hadronic mass $M(X_s)$ in the decay $B \to \eta' X_s$, will help to clarify the situation. Preliminary CLEO data indicate that relatively large values of $M(X_s)$ are preferred [81, 108]. Another good discriminator between models is provided by the ratios of the various $B \to \eta^{(\prime)} K^{(*)}$ decay rates. A collection of theoretical predictions for such ratios is shown in Table 5. In summary, the rare decays of B mesons into light mesons provide unique opportunities to extract many yet unknown hadronic matrix elements. At present, there are no convincing arguments that these processes could not be understood in a conventional Standard Model framework. In particular, because of the complexity of the dynamics of the η' meson, I believe it is premature to deduce evidence for new physics from the yield of η' production in B decays.

7 Summary and outlook

Owing to the combined efforts of experimenters and theorists, there has been significant progress in heavy-quark physics in the past few years. The exclusive and inclusive semileptonic decays mediated by the transition $b \to c\,\ell\,\bar{\nu}$ are understood from first principles, using the heavy-quark expansion. Heavy-quark effective theory (HQET) works well in describing these processes, and starts being tested at the level of symmetry-breaking corrections. This, by itself, is a remarkable development, which puts the determination of $|V_{cb}| = 0.038 \pm 0.003$ on firm theoretical grounds. New ideas are being developed for getting a precise ($\sim$ 10%) determination of $|V_{ub}|$, combining exclusive and inclusive methods. The error in the current value, $|V_{ub}| = (3.3 \pm 0.8) \times 10^{-3}$, is still dominated by the theoretical uncertainty.

New flavour-specific measurements of semi-inclusive charm yields indicate that the "problem" of the charm deficit and low semileptonic branching ratio is disappearing. The theoretical predictions for $\mathrm{B}(\bar{B} \to X_{c\bar{c}})$ and $\mathrm{B}(\bar{B} \to \text{no charm})$ are confirmed by the data. There is no compelling evidence for violations of quark–hadron duality, nor for new physics or exotic decay modes in B decays. The low value of the Λ_b lifetime remains a surprise, but can be accommodated in a small corner of parameter space. Lattice calculations of baryon matrix elements of four-quark operators would help to clarify the situation. However, this may also be the first sign of departures from local quark–hadron duality.

Hadronic two-body decays of B mesons with a large energy release can be understood in terms of a generalized factorization hypothesis, which includes the leading nonfactorizable corrections. Finally, new data on B decays into two light mesons open a window to study the details of nonperturbative dynamics of QCD, including long-distance penguins and other exotica. So far, the Standard Model can account for the data, but we should be prepared to find surprises as measurements and theory become more precise. For now, it remains to stress that the observed penguin effects are large and thus raise hopes for having large CP asymmetries in these decays. Hence, there is more *B*eauty to come in a bright future.

References

[1] For a review and a comprehensive list of references before 1996, see: M. Neubert, Phys. Rep. **245**, 259 (1994); Int. J. Mod. Phys. A **11**, 4173 (1996).
[2] G. Amorós, M. Beneke and M. Neubert, Phys. Lett. B **401**, 81 (1997).
[3] A. Czarnecki and A.G. Grozin, Phys. Lett. B **405**, 142 (1997).
[4] A.G. Grozin and M. Neubert, Nucl. Phys. B **508**, 311 (1997).
[5] A.F. Falk, M. Luke and M.J. Savage, Phys. Rev. D **53**, 6316 (1996).
[6] M. Gremm et al. Phys. Rev. Lett. **77**, 20 (1996).
[7] V. Chernyak, Nucl. Phys. B **457**, 96 (1995); Phys. Lett. B **387**, 173 (1996).
[8] M. Gremm and I. Stewart, Phys. Rev. D **55**, 1226 (1997).
[9] H. Li and H.-L. Yu, Phys. Rev. D **55**, 2833 (1997).
[10] M. Neubert, Phys. Lett. B **264**, 455 (1991); **338**, 84 (1994).
[11] The theoretical error in the prediction for $\mathcal{F}(1)$ has been reanalysed in *The* BABAR *Physics Books*, Chapter 8, to appear as a SLAC Report.

[12] A. Czarnecki, Phys. Rev. Lett. **76**, 4124 (1996); A. Czarnecki and K. Melnikov, Nucl. Phys. B **505**, 65 (1997).

[13] J. Franzkowski and J.B. Tausk, Preprint MZ-TH/97-37 [hep-ph/9712205].

[14] M.E. Luke, Phys. Lett. B **252**, 447 (1990).

[15] N. Uraltsev, Nucl. Phys. B **491**, 303 (1997); A. Czarnecki, K. Melnikov and N. Uraltsev, Preprint TPI-MINN-97-19 [hep-ph/9706311].

[16] C.G. Boyd, B. Grinstein and R.F. Lebed, Phys. Lett. B **353**, 306 (1995); Nucl. Phys. B **461**, 493 (1996); Phys. Rev. D **56**, 6895 (1997).

[17] I. Caprini and M. Neubert, Phys. Lett. B **380**, 376 (1996).

[18] I. Caprini, L. Lellouch and M. Neubert, Preprint CERN-TH/97-91 [hep-ph/9712417].

[19] P. Drell, Preprint CLNS-97-1521 [hep-ex/9711020], to appear in the Proceedings of the 18th International Symposium on Lepton–Photon Interactions, Hamburg, Germany, July 1997.

[20] L.K. Gibbons, Proceedings of the 28th International Conference on High Energy Physics, Warsaw, Poland, July 1996, edited by Z. Ajduk and A.K. Wroblewski (World Scientific, Singapore, 1997), p. 183.

[21] ALEPH Coll. (D. Buskulic et al.), Phys. Lett. B **395**, 373 (1997).

[22] CLEO Coll. (M. Athanas et al.), Phys. Rev. Lett. **79**, 2208 (1997).

[23] CLEO Coll. (J.E. Duboscq et al.), Phys. Rev. Lett. **76**, 3898 (1996); CLEO Coll. (A. Anastassov et al.), CLEO CONF 96-8, contributed paper to the 28th International Conference on High Energy Physics, Warsaw, Poland, July 1996.

[24] A.K. Leibovich et al., Phys. Rev. Lett. **78**, 3995 (1997); Phys. Rev. D **57**, 308 (1998).

[25] M. Neubert, Preprint CERN-TH/97-240 [hep-ph/9709327], to appear in Phys. Lett. B.

[26] CLEO Coll. (J.P. Alexander et al.), Phys. Rev. Lett. **77**, 5000 (1996).

[27] UKQCD Coll. (J.M. Flynn et al.), Nucl. Phys. B **461**, 327 (1996); for a review, see: J.M. Flynn and C.T. Sachrajda, Preprint SHEP-97-20 [hep-lat/9710057], to appear in *Heavy Flavours*, Second Edition, edited by A.J. Buras and M. Linder (World Scientific, Singapore).

[28] A. Khodjamirian et al., Phys. Lett. B **410**, 275 (1997).

[29] P. Ball and V.M. Braun, Phys. Rev. D **55**, 5561 (1997); E. Bagan, P. Ball and V.M. Braun, Preprint NORDITA-97-59 [hep-ph/9709243].

[30] L. Lellouch, Nucl. Phys. B **479**, 353 (1996).

[31] C.G. Boyd, B. Grinstein and R.F. Lebed, Phys. Rev. Lett. **74**, 4603 (1995); C.G. Boyd and M.J. Savage, Phys. Rev. D **56**, 303 (1997).

[32] G. Burdman and J. Kambor, Phys. Rev. D **55**, 2817 (1997).

[33] L.K. Gibbons, to appear in the Proceedings of the 7th International Symposium on Heavy Flavour Physics, Santa Barbara, California, July 1997.

[34] V. Barger, C.S. Kim and R.J.N. Phillips, Phys. Lett. B **251**, 629 (1990).

[35] J. Dai, Phys. Lett. B **333**, 212 (1994).
[36] A.O. Bouzas and D. Zappala, Phys. Lett. B **333**, 215 (1994).
[37] C. Greub and S.-J. Rey, Phys. Rev. D **56**, 4250 (1997).
[38] R.D. Dikeman and N.G. Uraltsev, Preprint TPI-MINN-97-06-T [hep-ph/9703437]; I. Bigi, R.D. Dikeman and N. Uraltsev, Preprint TPI-MINN-97-21-T [hep-ph/9706520].
[39] A.F. Falk, Z. Ligeti and M.B. Wise, Phys. Lett. B **406**, 225 (1997).
[40] C. Jin, Preprint DO-TH-97-27 [hep-ph/9801230].
[41] M. Neubert, Phys. Rev. D **49**, 3392 and 4623 (1994).
[42] I.I. Bigi et al., Int. J. Mod. Phys. A **9**, 2467 (1994).
[43] E. Bagan et al., Nucl. Phys. B **432**, 3 (1994); Phys. Lett. B **342**, 362 (1995) [E: **374**, 363 (1996)]; E. Bagan et al., Phys. Lett. B **351**, 546 (1995).
[44] M. Neubert and C.T. Sachrajda, Nucl. Phys. B **483**, 339 (1997).
[45] M. Feindt, these Proceedings.
[46] A. Lenz, U. Nierste and G. Ostermaier, Phys. Rev. D **56**, 7228 (1997).
[47] G. Buchalla, I. Dunietz and H. Yamamoto, Phys. Lett. B **364**, 188 (1995).
[48] CLEO Coll. (T.E. Coan et al.), Preprint CLNS-97-1516 [hep-ex/9710028].
[49] I. Dunietz et al., Eur. Phys. J. C **1**, 211 (1998).
[50] A.L. Kagan and J. Rathsman, Preprint [hep-ph/9701300]; A.L. Kagan, these Proceedings.
[51] T. Browder, to appear in the Proceedings of the 7th International Symposium on Heavy Flavor Physics, Santa Barbara, California, July 1997.
[52] F.E. Close et al., Preprint RAL-97-036 [hep-ph/9708265].
[53] A. Kagan, Phys. Rev. D **51**, 6196 (1995).
[54] L. Roszkowski and M. Shifman, Phys. Rev. D **53**, 404 (1996).
[55] CLEO Coll., CLEO CONF 97-26, contributed paper eps97-337 to this Conference.
[56] ALEPH Coll., contributed paper pa05-060 to the 28th International Conference on High Energy Physics, Warsaw, Poland, July 1996.
[57] DELPHI Coll., contributed paper pa01-108 to the 28th International Conference on High Energy Physics, Warsaw, Poland, July 1996.
[58] DELPHI Coll., DELPHI 97-80 CONF 66, contributed paper eps97-448 to this Conference; P. Kluit, these Proceedings.
[59] M. Douadi, these Proceedings.
[60] T. Junk, to appear in the Proceedings of the 2nd International Conference on B Physics and CP Violation, Honolulu, Hawaii, March 1997.
[61] M. Beneke, G. Buchalla and I. Dunietz, Phys. Rev. D **54**, 4419 (1996).
[62] B. Blok et al., Phys. Rev. D **49**, 3356 (1994) [E: **50**, 3572 (1994)]; I.I. Bigi et al., in: B Decays, edited by S. Stone, Second Edition (World Scientific, Singapore, 1994), p. 134.
[63] J.L. Rosner, Phys. Lett. B **379**, 267 (1996).

[64] P. Colangelo and F. De Fazio, Phys. Lett. B **387**, 371 (1996).

[65] M. Neubert, Preprint CERN-TH/97-148 [hep-ph/9707217], to appear in the Proceedings of the 2nd International Conference on B Physics and CP Violation, Honolulu, Hawaii, March 1997.

[66] M.S. Baek, J. Lee, C. Liu and H.S. Song, Preprint SNUTP-97-064 [hep-ph/9709386].

[67] DELPHI Coll. (P. Abreu et al.), DELPHI 95-107 PHYS 542, contributed paper to the 28th International Conference on High Energy Physics, Warsaw, Poland, July 1996.

[68] M. Shifman, in *Particles, Strings and Cosmology*, Proceedings of the Joint Meeting of the International Symposium on Particles, Strings and Cosmology and the 19th Johns Hopkins Workshop on Current Problems in Particle Theory, edited by J. Bagger et al. (World Scientific, Singapore, 1996), p. 69.

[69] B. Blok, M. Shifman and D.-X. Zhang, Preprint TPI-MINN-13-97-T [hep-ph/9709333].

[70] B. Grinstein and R.F. Lebed, Preprint UCSD/PTH 97-20 [hep-ph/9708396].

[71] B. Chibisov et al., Int. J. Mod. Phys. A **12**, 2075 (1997).

[72] P. Colangelo, C.A. Dominguez and G. Nardulli, Phys. Lett. B **409**, 417 (1997).

[73] G. Altarelli et al., Phys. Lett. B **382**, 409 (1996).

[74] J.D. Bjorken, Nucl. Phys. B (Proc. Suppl.) **11**, 325 (1989).

[75] For a review, see: G. Buchalla, A.J. Buras and M.E. Lautenbacher, Rev. Mod. Phys. **68**, 1125 (1996).

[76] H.-Y. Cheng, Phys. Lett. B **335**, 428 (1994); H.-Y. Cheng and B. Tseng, Preprint IP-ASTP-04-97 [hep-ph/9708211].

[77] J.M. Soares, Phys. Rev. D **51**, 3518 (1995).

[78] M. Neubert and B. Stech, Preprint CERN-TH/97-99 [hep-ph/9705292], to appear in *Heavy Flavours*, Second Edition, edited by A.J. Buras and M. Lindner (World Scientific, Singapore); M. Neubert, Preprint CERN-TH/97-169 [hep-ph/9707368], to appear in the Proceedings of the High-Energy Physics Euroconference on Quantum Chromodynamics (QCD 97), Montpellier, France, July 1997.

[79] T.E. Browder, K. Honscheid and D. Pedrini, Ann. Rev. Nucl. Part. Sci. **46**, 395 (1996).

[80] J.L Rodriguez, to appear in the Proceedings of the 2nd International Conference on B Physics and CP Violation, Honolulu, Hawaii, March 1997.

[81] D. Miller, these Proceedings.

[82] D. Besson, these Proceedings.

[83] S. Narison, Phys. Lett. B **322**, 247 (1994).

[84] T. Huang and C.W. Luo, Phys. Rev. D **53**, 5042 (1996).

[85] CLEO Coll. (R. Godang et al.), Preprint CLNS-97-1522 [hep-ex/9711010].
[86] CLEO Coll., CLEO CONF 97-22, contributed paper eps97-333 to this Conference; CLEO CONF 97-13 (1997).
[87] A.S. Dighe, M. Gronau and J.L. Rosner, Phys. Rev. Lett. **79**, 4333 (1997).
[88] For a review, see: R. Fleischer, Int. J. Mod. Phys. A **12**, 2459 (1997).
[89] A. Ali and C. Greub, Preprint DESY-97-126 [hep-ph/9707251]; A. Ali et al., Preprint DESY-97-235 [hep-ph/9712372].
[90] N.G. Deshpande, B. Dutta and S. Oh, Preprint OITS-641 [hep-ph/9710354].
[91] A.J. Buras and R. Fleischer, Phys. Lett. B **341**, 379 (1995).
[92] M. Ciuchini et al., Nucl. Phys. B **501**, 271 (1997); M. Ciuchini et al., Preprint CERN-TH/97-188 [hep-ph/9708222].
[93] J.D. Bjorken, Preprint SLAC-PUB-7521 [hep-ph/9706524], to appear in the Proceedings of the 2nd International Conference on B Physics and CP Violation, Honolulu, Hawaii, March 1997.
[94] A.J. Buras, R. Fleischer and T. Mannel, Preprint CERN-TH/97-307 [hep-ph/9711262].
[95] M. Neubert, Preprint CERN-TH/97-342 [hep-ph/9712224].
[96] A.F. Falk et al., Preprint JHU-TIPAC-97018 [hep-ph/9712225].
[97] D. Atwood and A. Soni, Preprint [hep-ph/9712287]; these Proceedings.
[98] D. Atwood and A. Soni, Phys. Lett. B **405**, 150 (1997).
[99] W.-S. Hou and B. Tseng, Preprint HEPPH-9705304 [hep-ph/9705304]; X.-G. He, W.-S. Hou and C.S. Huang, Preprint [hep-ph/9712478].
[100] H. Fritzsch, Preprint CERN-TH-97-200 [hep-ph/9708348].
[101] A.L. Kagan and A.A. Petrov, Preprint UCHEP-27 [hep-ph/9707354].
[102] I. Halperin and A. Zhitnitsky, Phys. Rev. D **56**, 7247 (1997); Preprint HEPPH-9705251 [hep-ph/9705251].
[103] E.V. Shuryak and A.R. Zhitnitsky, Preprint NI-97033-NQF [hep-ph/9706316].
[104] F. Yuan and K.-T. Chao, Phys. Rev. D **56**, 2495 (1997).
[105] A.A. Petrov, Preprint JHU-TIPAC-97016 [hep-ph/9712497].
[106] M.R. Ahmady, E. Kou and A. Sugamoto, Preprint RIKEN-AF-NP-274 [hep-ph/9710509].
[107] D. Du, C.S. Kim and Y. Yang, Preprint BIHEP-TH-97-15 [hep-ph/9711428]; D. Du and M. Yang, Preprint BIHEP-TH-97-17 [hep-ph/9711272].
[108] CLEO Coll., CLEO CONF 97-13, contributed paper to this Conference.
[109] H.J. Lipkin, Preprint ANL-HEP-CP-97-45A [hep-ph/9708253].
[110] H.-Y. Cheng and B. Tseng, Preprint IP-ASTP-03-97 [hep-ph/9707316].
[111] A. Datta, X.G. He and S. Pakvasa, Preprint UH-511-864-97 [hep-ph/9707259].

Lattice Gauge Theories

Roberto Petronzio (petronzio@roma2.infn.it)

University of Roma Tor Vergata, Rome, Italy

Abstract. This talk is a selected and qualitative summary of the main results on lattice gauge theories simulations presented at this conference and at the topical lattice meeting that was held in Edinburgh last July.

Most of the activities of the lattice community are concentrated in getting accurate results for QCD but there are interesting new developments in the non perturbative definition of chiral gauge theories and in the implementation of supersymmetry on the lattice.

1 QCD: the quest for the continuum limit

In a lattice QCD simulation one generates by a Monte Carlo procedure important samples, according to the functional measure, of gauge field configurations on a space-time grid with N^d points (usually $d = 4$) forming a hypercube with a physical volume of $L^d = (Na)^d$ where "a" is the lattice spacing in physical units. The calculation of the quantum averages of field operators is simply obtained by arithmetically averaging the values of the operators over the sample of gauge configurations generated by Monte Carlo. The integration over fermions that cannot be done numerically because of their Grassmanian nature (i.e. they are anticommuting variables) is instead duable and done analytically for each fixed gauge field configuration. According to the Wick theorem this results in a product of fermion propagators, inverse of the Dirac operator with the lattice covariant derivative, that are functions of the external fixed gauge field and in a modification of the gauge fields functional integral due to the fermion determinant. The numerical inversion of the Dirac operator and the inclusion of the determinant are the most time consuming operations.

The accuracy of lattice simulations ultimately depends upon how close they are to the "continuum limit", i.e. the limit of zero lattice spacing.

At fixed physical volume, the value of the lattice spacing is trivially related to the magnitude of the lattice. The "distance" from the continuum limit depends crucially upon how big can be N that in turn depends upon:
i) the speed of the forthcoming generation of computers
ii) the way numerical simulations scale with the number of lattice points.
Figure 1 gives an overview of the existing or forthcoming machines, of their

Next generation computers for LQCD

machine	peak speed [Gflops]	memory [Gbytes]	installation
CP - PACS (Tsukuba)	614	128	1996 (full machine)
QCDSP (Columbia)	819	33	1997 (400 Gflops)
APEmille (INFN)	1081 @ 66 MHz	16 - 64	1998 (135 Gflops)

Fig. 1. The existing or forthcoming machines: their peak speed and memory.

peak speed and memory [1]. One should bear in mind that the highest efficiencies of typical codes of lattice simulations are of the order of 70%. The values of figure 1 should be compared with the curves in figure 2: they give, as a function of the number of lattice points for each lattice direction, the number of days required for three types of simulations (QCD, QQCD and D^{-1}) in a calculation involving 100 inversions of the Dirac operators, for a single point on the lattice, on a gauge sample of 100 independent (i.e. separated by a suitable number of Monte Carlo "sweeps") gauge configurations [2]. The three curves refer to the time spent for the inversion of the Dirac operator (D^{-1}), for a quenched calculation (QQCD), neglecting the presence of dynamical fermion loops, and for a full QCD calculation (QCD). In the last case the inclusion of the fermion determinant representing the effect of dynamical fermion loops requires naively speaking the inversion of the Dirac operator for *every* point on the lattice and leads to the highest computing demand. From the figure one can see that a 100 Gflops computer can perform quenched simulations up to N of order 70 and unquenched up to N of order 20: this explains why the majority of simulations is still done for the quenched case while unquenched results are mostly qualitative.

In any case the "continuum limit" can only be reached by extrapolating results obtained at finite lattice spacing: besides improving computers technology and numerical algorithms the second fundamental strategy for accurate calculations is to make the extrapolation to the continuum the smoothest possible, by reducing the size of the lattice artifacts, i.e. of the corrections to the continuum theory due to finite lattice spacing.

The lattice artifacts are of order :

– a^2 for a pure gauge theory.

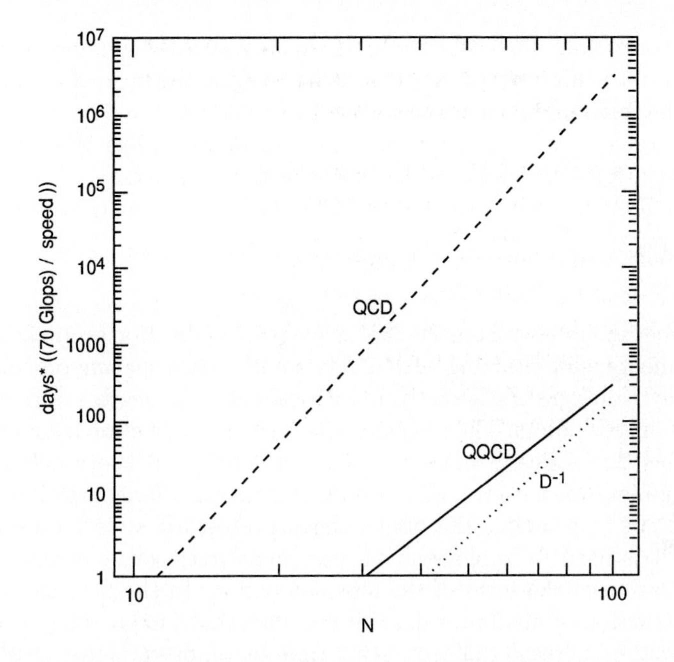

Fig. 2. The number of days required with a sustained speed of 70 Gflops for a typical (see text) simulation in full QCD (QCD), quenched QCD (QQCD) and for 100 inversions of the Dirac operator(D^{-1}) as a function of the number of lattice points per side.

- a for Wilson fermions [3]:
  - in this lattice fermion formulation, chiral symmetry is explicitely broken (for N_f favours) from the global symmetry group $SU(N_f)_L \times SU(N_f)_R$ to its vector part $SU(N_f)_V$
- a^2 for staggered fermions[4] :
  - in this formulation an axial chiral symmetry survives the lattice regularization: $SU(N_f)_L \times SU(N_f)_R$ is broken to $U(1)_L \times U(1)_R$ but the flavour symmetry is lost.

In the continuum limit both formulations recover the symmetries broken by the regularization. In spite of the nominal faster convergence for the staggered fermion formulation, the order a^2 corrections for staggered are rather severe and most simulations are done for Wilson fermions. In the latter case lattice artefacts errors for current cutoff values (i.e. a^{-1}) of about 2.5 GeV are of order:

- Λa i.e. of order 6%
- $m_{quark}a$ that for a charm quark can be as large as 50%
- pa where p are typical external momenta that cannot be smaller than $2\pi/N$

The continuum limit of renormalised quantities is unique but not the approach to the limit that in general depends upon the form of the discretised action: the freedom in the choice of the lattice action can be used to improve the convergence to the continuum limit also in cases, like Wilson fermions, where the original formulation had large lattice artefacts.

There are two main strategies in order to improve the continuum limit:

– the Wilson approach [5] aiming to construct a "perfect action" along the renormalised trajectory.

The method follows from the real space renormalisation group approach to critical phenomena. Starting with a very small lattice spacing one moves to a larger one by integrating, exactly, in an ideal case, the degrees of freedom of a subset of lattice points. The resulting effective action lives on a coarser lattice but retains the lattice artefacts related to the original theory defined with a higher momentum cutoff. The method, if iterated a large (infinite) number of times, has in principle the bonus of **removing all cutoff effects** but in practice it is hard to implement without sometimes severe errors related to the truncation of the form of the effective action. In the best cases it needs the introduction of medium range interactions that , expanding at least over an hypercube, increase the computing time by an extra factor $O(2^4)$.

– the Symanzik approach [6] aiming to eliminate cutoff effects **up to a given order in the lattice spacing**

This method aims to a more modeste result which however may be totally adequate for practical purposes, it is easier to implement than the previous one and requires a negligible computing overhead. It is based on the introduction of suitable counterterms in the action and in the definition of the operators whose coefficients must be determined in a non perturbative way by requiring the fulfillment, up to corrections of order a^2, of identities valid in the continuum limit. The choice of the continuum quantities used to match the improvement coefficients is somewhat arbitrary and "bad choices" may leave still with large order a^2 lattice artefacts.

Over the last years there has been considerable progress in both approaches. Concerning the Wilson approach, **classically** perfect actions have been constructed free from cutoff effects to all orders in the lattice spacing **at tree level** [7].

To be more specific, the renormalisation group transformation leading from a coarse to a fine lattice is identified by its "kernel" T(U,V), a function of the "coarse" lattice variables V and of the "fine" lattice variables U. The effective action on the coarse lattice is given by:

$$e^{-\beta' S'(V)} = \int \mathcal{D}U e^{-\beta(S(U)+T(U,V))},$$

The classic theory is reached when quantum fluctuations in the functional integral are completely suppressed, i.e. in the β going to infinity limit. From the point of view of a statistical system in that limit the theory sits on the critical surface: finding the classically perfect action is equivalent to find the fixed point of the renormalisation transformation on that surface. In that limit the equation above reduces to a saddle point equation:

$$S^{FP}(V) = \min_U (S^{FP}(U) + T(U,V)),$$

This equation holds for an arbitrary choice of the coarse variables V and it can be solved for the unknown parameters of the classically perfect action.

This method has been successfully applied to the $O(3)$ non linea sigma model [8] where it leads, contrary to what happens for a standard action, to a very precocious scaling of the mass gap. In QCD at finite temperature it leads to the scaling laws of a free gas at a temperatures, above the critical one, lower than with the standard Wilson action [9]. An interesting appplication of the saddle point equation is in providing, once the parameters of the perfect action are fixed, an interpolation from course to finer lattices that preserves the classical topological content of a given gauge configuration [10].

It was speculated that the classically prefect actions would keep their "perfection" including one loop corrections: unfortunately, this conjecture has found counterexamples and does not hold anymore [11] .

Concerning the Symanzik approach for the Wilson fermions the key observation is that at energies E $\ll 1/a$ the lattice theory is equivalent to an effective continuum theory.

$$S_{eff} = S_{QCD} + \int d^4x[aL_1(x) + a^2 L_2(x) + ...],$$

$$L_1 = c_1 \bar{\psi} \sigma_{\mu\nu} F_{\mu\nu} \psi + c_2 \bar{\psi} D_\mu D_\mu \bar{\psi}$$

c_1 and c_2 can be made vanish for "on shell quantities", i.e. those where one can reduce the basis of possible counterterms by the equations of motion, by replacing [13]:

$$S^{lattice}_{Wilson} \rightarrow S^{lattice}_{Wilson} + a^5 \sum_x c_{sw} \bar{\psi} \frac{i}{4} \sigma_{\mu\nu} F_{\mu\nu} \psi,$$

and tuning c_{sw} to make a reference physical quantity free from lattice artefacts up to $O(a^2)$ corrections. Operator expectation values need additional and specific improvements through mixing with lower dimensional operators and multiplication by appropriate renormalization factors. As an example, consider the improved axial current J^{axial}_μ [14] :

$$J^{axial}_\mu(ren) = Z_A(m)(J^{axial}_\mu(lat) + c_A \partial_\mu P_5(lat));$$

where P_5 is the axial density and $Z_A(m)$ is a renormalisation factor which can be expanded around the massless limit as:

$$Z_A(m) = Z_A(0)(1 + b_A ma)$$

These "mass lattice artefacts" are crucial in correlation functions involving heavy quarks. They also appear in the relation between the bare and the renormalised mass on the lattice:

$$m_{ren}(a) = m_{bare}(a) Z_m(0)(1 + b_m ma)$$

Improvement coefficients have been evaluated:

- at one loop in perturbation theory and are available the expressions for c_{sw}, c_A, $Z_i(0)$ and b_i for i= V, A, P, S [15]
- in tadpole improved or mean field approximation, i.e. by resumming to all orders in perturbation theory a set of universal contributions (the tadpole diagrams) taken as the relevant ones [16]
- non perturbatively, with the Schrödinger functional technique [12] that allows to evaluate quark matrix elements on a finite volume in a guage invariant way and by preserving the renormalisation properties of infinite volume QCD: the results are avalable for c_{sw}, c_A, b_V, $Z_V(0)$, $Z_A(0)$ [17] and b_m , b_A - b_P [18]
- non perturbatively, by evaluating quark matrix elements in a fixed gauge and a momentum subtraction scheme: using the renormalisation coefficients calculated with this method, one can obtain the correct chiral behaviour for the operators entering the hadronic matrix elements of the ΔS=2 weak Hamiltonian [19].

The non perturbative determination of the dependence of c_{sw} upon the lattice coupling constant deviates substantially from the one loop and tadpole improved estimates at current values (around one) of the lattice coupling constant. Figures 3-5 show some of the recent progress in the Symanzik improvement programme. Fig.3 shows the relation between the divergence of the axial current and the pseudoscalar density (i.e. the axial Ward identity) after removing the $O(a)$ lattice artefacts. Fig.4 shows how earlier is in the improved case the scaling of the vector meson mass as a function of the lattice spacing compared with the unimproved one. Fig.5 shows the chiral behaviour of the ΔS=2 four fermion operator after a non perturbative renormalisation.

When the lattice artefacts are mild the extrapolation to the continuum limit can be performed with controllable systematic errors. Fig. 6 shows the continuum non perturbative evolution of the QCD running coupling constant in the quenched case. The resulting determination of Λ_{QCD} (**in the quenched case**) is quite accurate:

$$\Lambda_{QCD}^{MS-bar} = 251 \pm 21 MeV$$

Such an error on Λ_{QCD} leads to an accuracy better than 1% at LEP energies. Similar non perturbative, continuum estimates are under way for the evolution of the running mass.

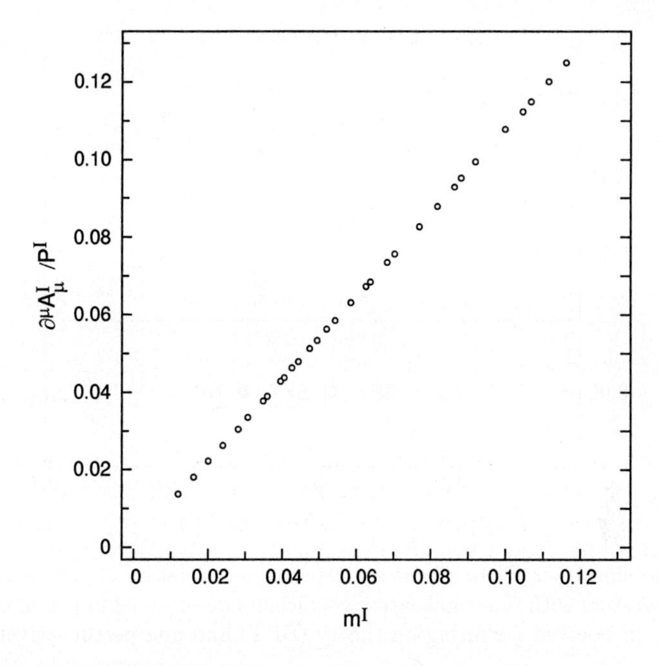

Fig. 3. The ratio, in the improved case, of the divergence of the axial current over the pseudoscalar density as a function of the quark mass.

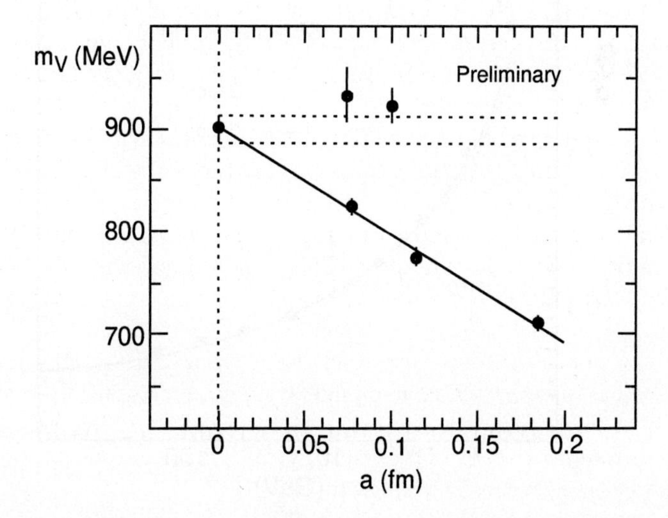

Fig. 4. The flat dependence of mass of the vector meson as a function of the lattice spacing in the improved case is compared with the significant dependence of the unimproved case.

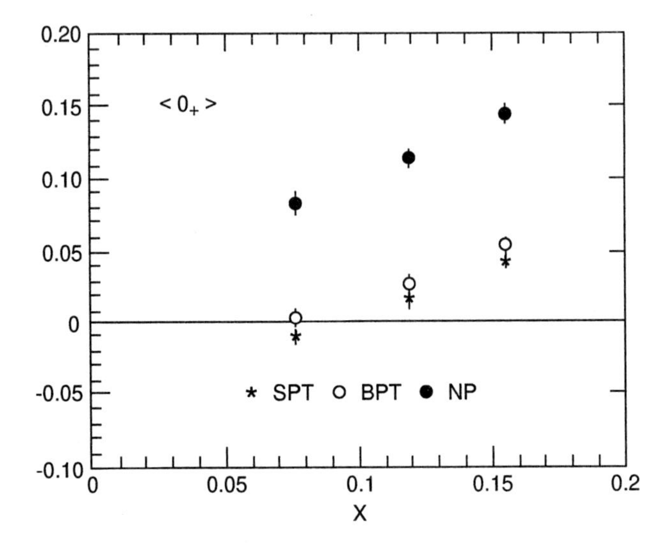

Fig. 5. The chiral behaviour of the appropriate combination of lattice ΔS=2 four fermion operators with renormalization coefficients determined in perturbation theory (SPT), in boosted perturbation theory (BPT) and non perturbatively (NP).

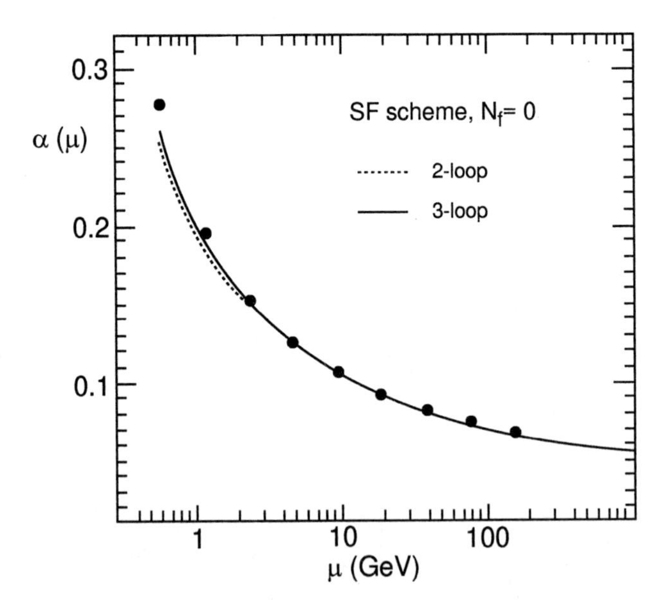

Fig. 6. The non perturbative **continuum** evolution of the running coupling constant in the Schrodinger functional scheme.

This section can be summarized by saying that the approach to the continuum limit is now under better control: results for "simple" quantities like the light hadron spectrum, ward identities and chiral behaviour, running coupling and running masses are rather reliable. More difficult problems like the $\Delta I = 1/2$ rule, the Isgur wise function, the normalization and the running of hadronic structure functions and the full QCD running coupling constant can be approached with a better control of systematic errors.

2 The unquenched case

In this field the main activity and the main progress concerns the development of fast algorithm to perform dynamical fermion simulations. There are basically two competing algorithm: the hybrid Monte Carlo [20] and the multiboson algorithm [21].

In any case the computing demand for an unqhenched simulation can easily become two orders of magnitude more expensive than the corresponding quenched simulation and this explains why unquenched simulations are currently done on smaller lattices, with fewer statistics and for a limited range of parameters. Nevertheless some expected and distinctive features of dynamical fermion loops have been seen and in particular the flavourdynamics of the deconfining finite temperature phase transition [22] and the splitting of the η' from the standard octet of pseudoscalar mesons. This last feature has been seen also within the "bermion" approach, a shortcut to unquenched estimates where dynamical fermions are replaced by fields with a Dirac action but Bose statistics (which therefore can be sampled by Monte Carlo) [25]. This is equivalent to compute the effects of **negative** flavour numbers. In this scheme, according to the Witten Veneziano formula, the η', whose mass is proportional to the flavour numbers, should be **lighter** than the pion, as it is found and shown in figure 7 [23].

Some apparent discrepancies of quenched calculations like the derivative of vector meson masses with respect to the square of pseudoscalar masses, at the moment roughly twenty percent below the experimental number are expected to be reconciliated by future accurate full QCD simulations to which the improvement programme is being extended [24].

3 Chiral gauge theories

The problem with the constyruction of chiral gauge theories on the lattice was stated by Nielsen and Ninomiya with a theorem[26] : a theory with local, bilinear fermion interactions and chiral invariance has **doublers** which make the theory vector like. Indeed doublers can be avoided by breaking chiral invariance (a' la Wilson) but, in chiral gauge theories, like the $SU(2)$ sector of the standard model, this breaks also gauge invariance. There are two classes of proposed solutions.

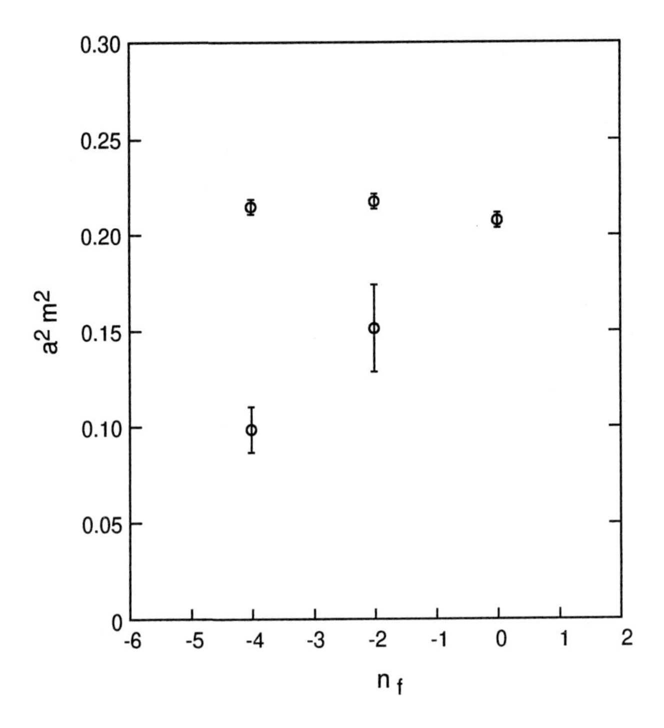

Fig. 7. The squared mass of the pseudoscalar octet (upper points) is compared with the flavour singlet pseudoscalar in the approach to unquenching from negative flavour numbers.

To the first belongs the "Rome approach" [27], where one accepts an explicit breaking of gauge invariance that however is recovered by imposing, by a non perturbative tuning of the coefficients of appropriate counterterms, the BRST identities of the continuum, chiral invariant, theory. This approach is theoretically solid, but the non perturbative fine tuning of the counterterms has so far discouraged the lattice community.

In the second class of solutions one tries to evade the N-N theorem by generating non local interactions from the integration of **extra degrees of freedom**. To this class belong three proposals.

The domain wall [28] approach introduces a five dimensional space with a modification of the Dirac mass term and free boundary conditions on the four dimensional spaces boundaries where chiral modes get trapped in the limit where the fith dimension goes to infinity. In this method it is difficult to estimate the systematic error of keeping the fifth dimension finite: simulations in two dimensional models have shown that the approach to the infinite dimension limit may depend upon the topological sector of the theory [29]. Although this method may fail for chiral gauge theories [30], it may offer an interesting approach to **vector** like models. Indeed, an interesting application of the method to QCD has been presented at this conference by A. Soni.

In the "overlap" approach [31] the infinite fifth dimension limit is done analytically, but the result of the integration over the chiral modes is expressed by the overlap of the ground state wavefunctions of the two four dimensional "hamiltonian" defined at the boundaries of the fifth infinite dimension. The computing demand of this approach is similar to a full QCD calculation.

Finally, in the two cutoffs approach [32] fermions live on a finer space time grid than gluons: they are introduced in a Wilson formulation that breaks gauge invariance with lattice artefacts of the order of the finer lattice spacing. At the scale of gauge interactions that are defined on a coarser lattice, the breaking should be suppressed by the ratio of gluon to fermion lattice spacings. It is a strong statement of this approach that loop effects do not compensate such a suppression by linear divergences porportional to the inverse of the fermion lattice spacing. Only two dimensional models have been successfully explored so far with this method where the computing demand is governed by the smaller lattice spacing and the accuracy by the larger.

All proposed solutions are very highly computer demanding and have been tried mostly in two dimensional models; some of them need approximations to be implemented and a systematic investigation of their validity in the real four dimensional world will require some time.

4 Lattice Susy (SYM)

An emerging field of activity is lattice Super Yang Mills: its formulation is rather old [33] but its implementation poses some major problems. According to the original paper, SYM is realized in the chiral limit of a theory of gluons and massive Majorana fermions. It is difficult to define such a limit in the full theory where the fermion loops lift the mass of the would be Goldstone boson associated with the anomalous current, but fermion loops are essential to the realisation of supersymmetry. So far, only "quenched" spectrum calculations and exploratory studies of unquenching have been made. In the quenched calculations, there are interesting relations in **broken** SYM [34] that can be tested as discussed in the talk by P. Hernandez at this conference. In addition the authors claim that the chiral limit can be defined from a quenched calculation, where massless pions do exist also with fermion in the adjoint representation, very much like in ordinary QCD. In the unquenched calculations [35], there is a problem associated with the sign of the "square root" of the fermion determinant asssociated with Majorana fermions that remains still unsolved.

5 Summary

There are important omissions in this talk that I want to mention: these include in QCD the calculation of semileptonic form factors, of heavy meson decay constants, of the "B" parameter of CP violating amplitudes, of quark masses, of the spectrum of hybrids and glueballs. Beyond QCD considerable progress has been made in the simulations of random surfaces and in understanding their relation with 2-d quantum gravity.

Summarising, quenched QCD is close to the continuum limit, a non perturbative determination of the renormalisation coefficients of two and four fermion operators is becoming a standard, some discrepancies of the quenched approximation with experimental data have been identified and important checks of specific features of full QCD have been performed. Outside QCD, the various approaches to the formulation of chiral gauge theories on the lattice, after a succesfful testing period in two dimensinal models, will be tested in a near future in the real four dimensional world. A byproduct of these approaches may be some new ways for approaching the chiral limit of Wilson type fermions in vectorlike theories like QCD. Finally, exploratory studies of super Yang Mills have started.

Acknowledgments: I am grateful to M.Lüscher and K.Jansen for many interesting discussions.

References

[1] M. Lüscher, Talk given at the 18th International Symposium on Lepton-Photon Interactions, Hamburg, 28 July – 1 August.
[2] K.Jansen private communication.
[3] K.G.Wilson, Phys. Rev. D10 (1974) 2445
[4] For a recent discussion of generalized staggered fermion actions, see A. Peikert et al, HEP-LAT/9709157
[5] K.G.Wilson, Rev. Mod. Phys.47 (1975) 773, ibidem 55 (1983) 583
[6] M.Lüscher, P.Weisz, Comm.Math.Phys 97 (1972) 3543, B.Sheikholeslami, R.Wohlert, Nucl. Phys. B259 (1985) 572
[7] For a review, see F. Niedermayer, Nucl.Phys.Proc.Suppl. 53 (1997) 56
[8] P. Hasenfratz and F. Niedermayer, Nucl. Phys. B414 (1994) 785
[9] F. Karsch et al, Nucl.Phys.Proc.Suppl. 53 (1997) 413-416, A. Papa, Nucl.Phys. B478 (1996) 335
[10] T.A. DeGrand et al, Nucl.Phys.Proc.Suppl. 53 (1997) 945
[11] P.Hasenfratz, F.Niedermayer, HEP-LAT/9706002
[12] M.Lüscher et al, Nucl. Phys. B384 (1992) 168
[13] M.Lüscher, S. Sint, R. Sommer, P. Weisz, H. Wittig, U. Wolff, Nucl.Phys.Proc.Suppl. 53 (1997) 905
[14] M.Lüscher, S.Sint, R.Sommer, H.Wittig, Nucl.Phys. B491 (1997) 344
[15] S.Sint, P.Weisz, Nucl.Phys. B502 (1997) 251 and ref. therein

[16] G.Parisi, Proc. XX ICHEP, Madison, AIP (1981) (L.Durand, L.G.Pondrom eds.), G.P.Lepage and P.B.Mackenzie, Phys.Rev. D48 (1993) 2250

[17] M.Lüscher, S.Sint, R.Sommer, P.Weisz, U.Wolff, Nucl.Phys. B491 (1997) 323

[18] G. M. de Divitiis, R. Petronzio, HEP-LAT/9710071

[19] A. Donini et al, Nucl.Phys.Proc.Suppl. 47 (1996) 489

[20] S.Duane, A.D.Kennedy, B.J.Pendleton and D.Roweth, Phys. Lett. B195 (1987) 216

[21] M. Lüscher, Nucl. Phys. B418 (1994) 637, For a review, see K. Jansen, Nucl. Phys. B (Proc. Suppl.) 53 (1997) 127

[22] JLQCD coll., HEP-LAt /9710048, S.Aoki et al, HEP-LAT/9710031

[23] R. Frezzotti, M. Masetti, R. Petronzio, Nucl.Phys. B480 (1996) 381

[24] K. Jansen and R. Sommer,CERN preprint, CERN-TH/97-239,HEP-LAT/9709022

[25] G. M.de Divitiis et al, Nucl.Phys. B455 (1995) 274

[26] H.B.Nielsen, M.Ninomiya, Nucl.Phys. B193 (1981) 173

[27] A.Borrelli et al, Nucl.Phys. B333 (1990) 335. For a review, see M.Testa, HEP-LAT/9707007

[28] D.B.Kaplan, Phys.Lett. B288 (1992) 342,Y.Shamir, Nucl.Phys. B406 (1993) 90

[29] P.M.Vranas, HEP-LAT//9705023

[30] M.F.L. Golterman, K. Jansen, D.N. Petcher and J.C Vink, Phys. Rev.D49 (1994) 1606

[31] R.Narayanan, H.Neuberger, Nucl. Phys. B443 (1995) 305

[32] P.Hernandez, R.Sundrum, Nucl.Phys. B472 (1996) 334

[33] G.Curci and G.Veneziano, Nucl.Phys. B292 (1987) 555

[34] A. Donini et al, HEP-LAT/9708006, G.Koutsoumbas, I.Montvay, Phys.Lett. B398 (1997) 130

[35] I. Montvay, HEP-LAT/9709080

Physics at Future Colliders

John Ellis (John.Ellis@cern.ch)

Theoretical Physics Division, CERN, CH 1211 Geneva 23

Abstract. After a brief review of the Big Issues in particle physics, we discuss the contributions to resolving that could be made by various planned and proposed future colliders. These include future runs of LEP and the Fermilab Tevatron collider, B factories, RHIC, the LHC, a linear e^+e^- collider, an $e-p$ collider in the LEP/LHC tunnel, $\mu^+\mu^-$ collider and a future larger hadron collider (FLHC). The Higgs boson and supersymmetry are used as benchmarks for assessing their capabilities. The LHC has great capacities for precision measurements as well as exploration, but also shortcomings where the complementary strengths of a linear e^+e^- collider would be invaluable. It is not too soon to study seriously possible subsequent colliders.

1 The Big Issues

The discoveries for which future colliders will probably be remembered are not those which are anticipated. Nevertheless, we cannot avoid comparing their capabilities to address our present prejudices as to what big issues they should resolve. These include, first and foremost, the **Problem of Mass**: is there an elementary Higgs boson, or is it replaced by some composite technicolour scenario, and is any Higgs boson accompanied by a protective bodyguard of supersymmetric particles? As we were reminded at this meeting [1], precision electroweak data persist in preferring a relatively light Higgs boson with mass

$$m_H = 115^{+116}_{-66} \text{ GeV} \tag{1}$$

which is difficult to reconcile with calculable composite scenarios. However, we should be warned that the range (1) is compatible with the validity of the Standard Model all the way to the Planck scale [2], as seen in Fig. 1, with no new physics to stabilize the electroweak coupling or keep the Standard Model couplings finite. Nevertheless, the range (1) is highly consistent [3] with the prediction of the minimal supersymmetric extension of the Standard Model [4], so we use in this talk supersymmetry [5] as one of our benchmarks for future colliders. The **Problem of Flavour** includes the questions why there are just six quarks and six leptons, what is the origin of their mass ratios and the generalized Cabibbo mixing angles, and what is the origin of CP violation? The Standard Model predicts the presence of CP violation, but we do not yet gave any quantitative tests of the Kobayashi-Maskawa mechanism [6]: detailed studies may reveal its inadequacy. Finally, the **Problem of Unification** raises the possibility of neutrino masses and proton decay, that are not addressed by colliders. However, GUTs also predict many relations between couplings, such as $\sin^2\theta_W$ [7], and masses, such as m_b/m_τ [8], that can be tested at colliders, such as detailed predictions for the spectroscopy of sparticles - if there are any!

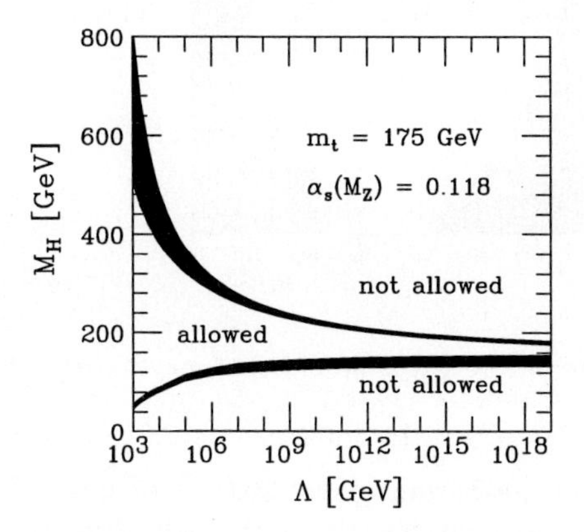

Fig. 1. The range of M_H allowed if the Standard Model remains valid, unmodified, up to an energy scale Λ [2].

2 Catalogue of Future Colliders

Table 1 lists future high-energy colliders, both approved and under discussion, together with some of their key parameters and their planned start-up dates [9]. To these should be added LEP 2000, which is the proposal to extend the scheduled running of LEP through the year 2000, for which approval and extra funding is now being sought from the CERN Member States. By using some of the LHC cryogenic facilities, this and the 1999 run of LEP could be at a higher energy, possibly the 200 GeV foreseen in the original LEP design. This would extend the LEP reach for the Standard Model Higgs boson through the LEP/LHC transition region to above 100 GeV. The principal gain for Higgs physics of the extra year's run would be to be sure of overcoming the Z background around 90 GeV. Running LEP at 200 GeV would also extend significantly the LEP coverage of MSSM Higgs bosons [10], passing a definitive verdict on the region $\tan\beta \lesssim 2$ favoured in many theoretical models, as seen in Fig. 2, and would also provide closure on supersymmetric interpretations of the CDF $e^+e^-\gamma\gamma p_T$ event [11].

Also in the category of future runs of present accelerators is the TeV 2000 programme [12]. This actually comprises two runs: Run II, which is approved to run from 1999 to about 2002, and is slated to accumulate $4fb^{-1}$ of data, and Run III, which is proposed to start a couple of years later and continue

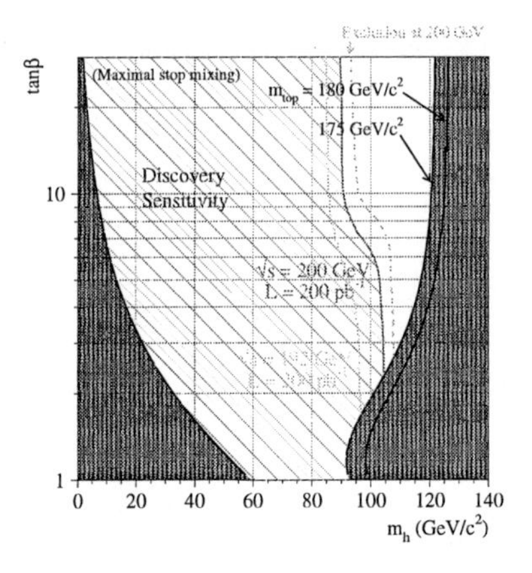

Fig. 2. The sensitivity to MSSM Higgs bosons that may be achieved at LEP 2 [10].

until the LHC kicks in, gathering about $20fb^{-1}$ of data. Top of the TeV 2000 physics agenda is top physics [12], with such objectives as decreasing the uncertainty in m_t to $\sim$ 2 GeV in Run II and perhaps 1 GeV in Run III, measuring $\sigma_{\bar{t}t}$ with an accuracy of 7(5) % (which would test "topcolour" models), and determining Γ_t to 20(12) %. TeV 2000 also offers new prospects in W physics: $\delta m_W \simeq 40(20)$ MeV, $\delta\Gamma \simeq 15$ MeV and measuring triple-gauge couplings at the 10 % level. The TeV 2000 programme also has a nice chance of finding the Higgs boson of the Standard Model, with a reach extending above 100 GeV in Run II and perhaps to 125 GeV in Run III [12]. Beyond this Standard Model physics, TeV 2000 can search for gluinos up to 400 GeV or so, and charginos up to about 220 GeV, depending on the details of the model [12].

The next new colliders to start taking data will presumably be the B factories at SLAC and KEK, with their experiments BaBar and Belle [13], respectively. Their tasks will be to understand CP violation, possibly within the Standard Model by measuring or overconstraining the unitarity triangle [6]

$$V_{ud}V_{ub}^* + V_{cd}V_{cb}^* + V_{td}V_{tb}^* = 0 \tag{2}$$

or (we can but hope!) finding new physics beyond it, as well as to improve limits on (and measurements of) rare B decays. This will be a crowded field, with HERA-B [14] and the Tevatron Run II starting in 1999, as well as BaBar and Belle. The field will become even more crowded during the next decade, with BTeV [15] perhaps starting around 2003, as well as CMS [16, 17], ATLAS [18] and LHC-B [19] in 2005. To these should be added CLEO, which will continue to provide useful complementary information on B physics, as well as SLD, which also has a chance to measure the B_s mixing parameter x_s.

Another collider due to start taking data in 1999 is RHIC, which expects to reach nuclear energy densities $\sim$ 6 GeV/fm^3. As we heard here from Plasil [20], this will have two large experiments: STAR [21] which will measure general event charactersitics and statistical signatures, and PHENIX [22] which will concentrate more on hard probes such as $\ell^+\ell^-$ pairs. RHIC will also have two smaller experiments: PHOBOS [23] and BRAHMS [20]. In the longer run, the LHC also offers relativistic heavy-ion collisions at an nuclear energy density which is model-dependent, but expected to be considerably higher than at RHIC. There will be a dedicated heavy-ion experiment at the LHC, ALICE [24], which aims at both statistical signatures and $\ell^+\ell^-$ pairs. There is also interest in heavy-ion physics from CMS [16, 26], which may be able to observe jet quenching in events with a large$-p_T Z^0$ or γ trigger, and ATLAS would also have interesting capabilities for heavy-ion physics.

3 LHC Physics

The primary task of the LHC, approved in late 1994 and scheduled for first beams in 2005 [25], is to explore the 1 TeV energy range. The two major "discovery" experiments ATLAS [18] and CMS [16] were approved in early 1996, and construction of some of their detectors has begun. ALICE [24] was approved in early 1997, and the dedicated CP-violation experiment LHC-B [19] has passed successfully through the preliminary stages of approval and is expected to receive final approval soon.

Top of the physics agenda for the LHC is the elucidation of the origin of particle masses, i.e., the mechanism of spontaneous electroweak symmetry breakdown. Within the Standard Model, this means looking for the Higgs boson, whose mass is currently estimated in (1). The branching ratios and production cross sections for the Higgs at the LHC are well understood, including first-order QCD corrections, as reviewed here by Spira [27] and shown in Fig. 3.

However, work is still needed on the QCD corrections to some backgrounds, such as $\gamma\gamma$. Favoured search signatures [18, 16] include $H \to \gamma\gamma$ for 100 GeV $\lesssim m_H \lesssim$ 140 GeV, $H \to 4\ell^\pm$ for 130 GeV $\lesssim m_H \lesssim$ 700 GeV and $H \to W^+W^-, Z^0Z^0 \to \ell^+\ell^- jj, \bar{\nu}\nu\ell^+\ell^-, \ell\nu jj$ for $m_H \gtrsim$ 500 GeV. As seen in Fig. 4, one delicate mass region is $m_H \lesssim$ 120 GeV, where the requirements on electromagnetic calorimeters are particularly stringent [28]. Recent studies indicate that the channel $W + (H \to \bar{b}b)$ may play a useful rôle in this mass region [29]. Another delicate mass region is $m_H \sim$ 170 GeV, where the branching ratio for the preferred $H \to ZZ^* \to 4\ell^\pm$ signal is reduced, as seen in Fig. 4, because the $H \to W^+W^-$ channel opens up. The possibility of isolating the $H \to W^+W^- \to (\ell^+\nu)(\ell^-\bar{\nu})$ decay mode in this mass region has recently been re-examined [30, 29]. By making suitable cuts on the charged leptons ($|\eta| < 2.4, p_T >$ 25 GeV, $p_{T_2} >$ 10 GeV, $m_{\ell^+\ell^-} >$ 10 GeV, $|m_{\ell^+\ell^-} - m_Z| >$ 5 GeV) and vetoing events with jets (no $p_{T_j} >$ 20

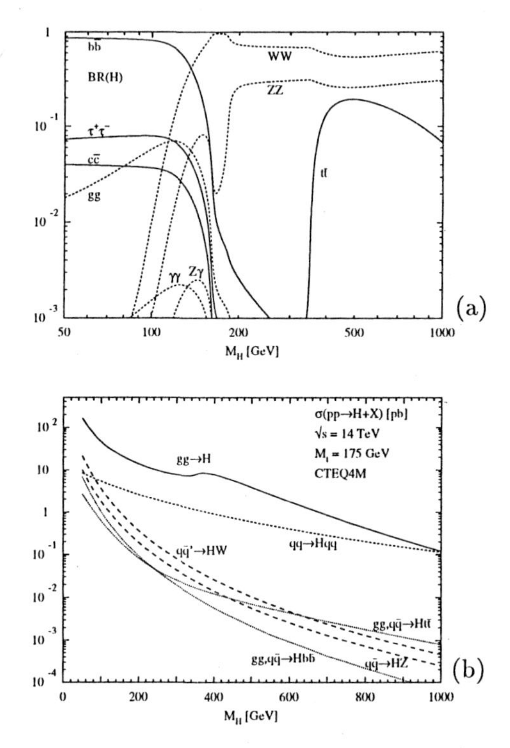

Fig. 3. (a) Branching ratios and (b) cross sections for Higgs production at the LHC, including radiative corrections [28].

GeV in $|\eta| < 3$), as well as cuts on the polar and transverse opening angles of the $\ell^+\ell^-$, it was found possible to display an excess above the continuum W^+W^- background, which is clearest when $m_H \sim 170$ GeV, as seen in Fig. 5. However, this technique does not produce a well-defined resonance peak.

It is hard to find any theorist who thinks that the single Higgs boson of the Standard Model is the whole of the story, and many would plump for the minimal supersymmetric extension of the Standard Model (MSSM). The mass of the lightest neutral Higgs boson h in the MSSM is restricted to $m_h \lesssim 120$ GeV [4], which is, as already mentioned, quite consistent [3] with the range (1) indicated by precision electroweak measurements [1]. The MSSM Higgs branching ratios and production cross sections at the LHC have also been studied intensively, including leading-order QCD corrections [27]. It has been known for some time that, although there are extensive regions of the MSSM Higgs parameter space where one or more of the MSSM Higgs bosons may be detected [31], there is a region around $m_A \sim 150$ to 200 GeV and $\tan\beta \sim 5$ to 10 which is difficult to cover. As seen in Fig. 6, after several years of running the LHC at the design luminosity, even this region may be covered by combining data from ATLAS and CMS [31]. However, one would

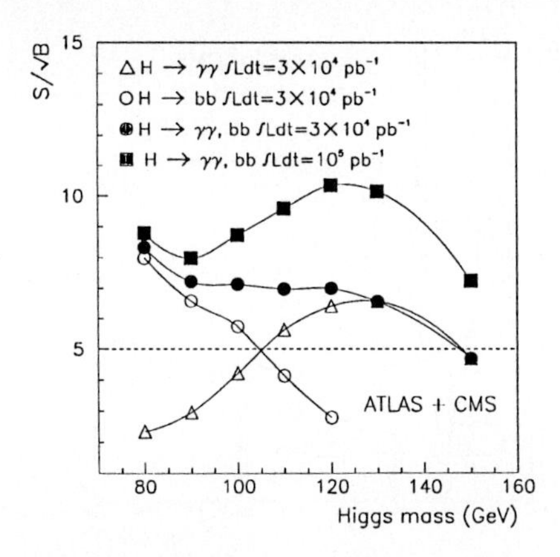

Fig. 4. Estimated significance of light-mass Higgs detection at the LHC: note the "delicate" regions $m_H \lesssim 120$ GeV and ~ 170 GeV [30].

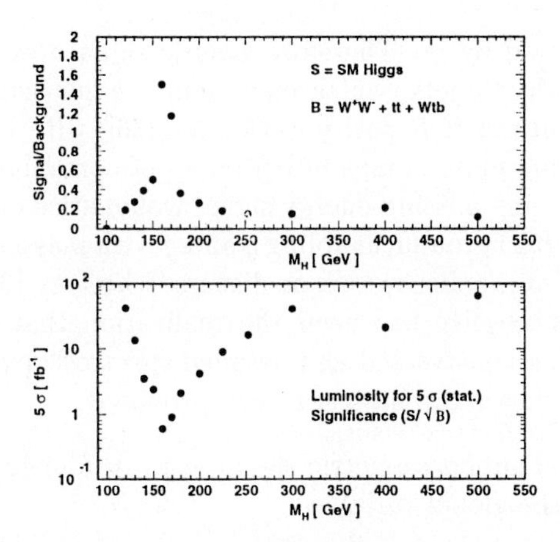

Fig. 5. Possible improvement in the sensitivity to $m_H \sim 170$ GeV at the LHC, using cuts optimized for $H \rightarrow W^+W^-$ decay [31].

be more comfortable if there were more coverage, and in particular with more help from LEP by covering a larger area of the $(m_A, \tan\beta)$ plane by running at $E_{cm} = 200$ GeV [10].

We can still hope that supersymmetry may be discovered before the start-up of the LHC [32], but the LHC has unprecedented mass reach in the search for supersymmetric particles [33]. Cross sections for producing the strongly-interacting squarks $\tilde{q}$ and gluons $\tilde{g}$ have been calculated including leading-order QCD corrections [34]. Since these are expected to be among the heaviest sparticles, if R parity is conserved one expects their generic decays to involve complicated cascades such as $\tilde{g} \rightarrow \tilde{b}\bar{b}, \tilde{b} \rightarrow \chi_2 b, \chi_2 \rightarrow \chi_1 \ell^+ \ell^-$ where

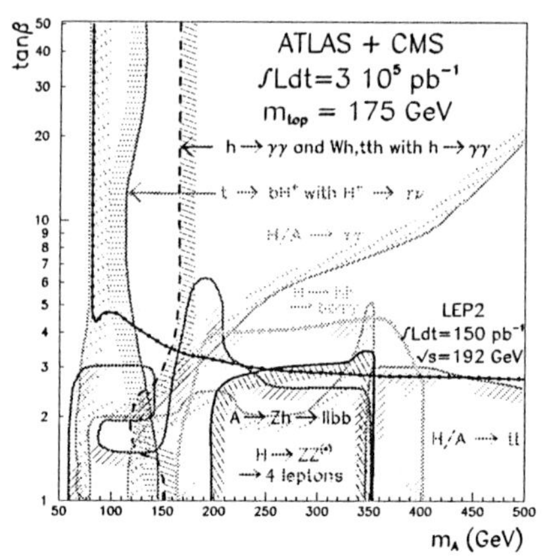

Fig. 6. Detectability of MSSM Higgs bosons at the LHC. In most of the plane, more than one experimental signature is visible [32].

neutralinos are denoted by χ_i. Therefore generic signatures are missing energy, leptons and hadronic jets (which may include b quarks) [35]. These are also interesting signatures if R parity is violated [36], with the added possibility of reconstructing mass bumps in lepton + jet combinations. As is well known, the R-conserving missing-energy signal would stick out clearly above the Standard Model background, enabling $\tilde{q}$ and $\tilde{g}$ with masses between about 300 GeV and 2 TeV to be discovered, as discussed here by [33].

The main recent novelty has been the realization that the $\tilde{q}$ or $\tilde{g}$ decay cascades may be reconstructed and detailed spectroscopic measurements made [37] The following is the basic strategy proposed.

1. Identify a general supersymmetric signal, e.g., in four-jet + missing energy events via the global variable

$$m_{eff} = p_{T_{j_1}} + p_{T_{j_2}} + p_{T_{j_3}} + p_{T_{j_4}} + \not{p}_t \tag{3}$$

   as seen in Fig. 7, whose mean is found from the Monte Carlo studies to be around $2m_{(\tilde{q} \text{ or } \tilde{g})}$.
2. Reconstruct decay chains starting from the end, e.g., in the above case via $\chi_2 \to \chi_1(\ell^+\ell^-)$ which should exhibit a characteristic edge in the spectrum at $m_{\ell^+\ell^-} = m_{\chi_2} - m_{\chi_1}$, then adding (for example) b and $\bar{b}$ jets to reconstruct $m_{\tilde{b}}$ and $m_{\tilde{g}}$.
3. Finally, make a global fit to MSSM parameters within an assumed standard parametrization such as that suggested by supergravity, namely $m_{1/2}$ (a common gaugino mass), m_0 (a common scalar mass), A (a common trilinear soft supersymmetry-breaking parameter), $\tan\beta$ (the ratio of Higgs vacuum expectation values), and the sign of the Higgs mixing parameter μ [5].

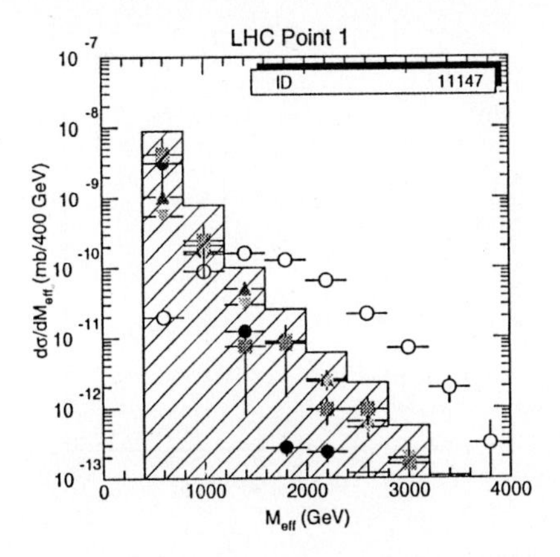

Fig. 7. Significance of m_{eff} (3) signal at the LHC, compared to Standard Model backgrounds [38].

This strategy has been applied within the framework of a study commissioned by the LHC experiments committee of five particular points in this parameter space [38]. The values of representative sparticle masses for these parameter choices are shown in Table 2. A typical $(\ell^+\ell^-)$ spectrum is shown in Fig. 8, where we see a sharp edge that would enable $m_{\chi_2} - m_{\chi_1}$, to be measured with a precision of 100 MeV [33, 37]. Events close to this edge can be used to reconstruct $\tilde{q} \to q\chi_2$ decays, such as $\tilde{b} \to b\chi_2$ and $\tilde{g} \to \tilde{b}\bar{b}$ decays [33, 37], as seen in Fig. 9. For generic other parameter choices, as seen in Fig. 10, one may reconstruct $h \to \bar{b}b$ decays in the cascade [33, 35]. Another possibility discussed here [33] is that $\chi_2 \to \chi_1 + (\tau^+\tau^-)$ decays dominate at large $\tan\beta$, in which case one may observe an excess in the $M_{\tau^+\tau^-}$ distribution. Table 3 shows the MSSM particles that may be discovered at each of the five points in parameter space that have been explored in detail [38]. We see that a sizeable fraction of the spectrum may be accessible at the LHC.

Moreover, precision determinations of the supergravity model parameters are possible if one combines the different measurements of endpoints, masses, products of cross sections and branching ratios, etc. [33, 37]. For example, it has been estimated that one could attain

$$\Delta(m_{\chi_2} - m_{\chi_1}) = \pm \left\{ \begin{array}{c} 50\ \text{MeV} \\ 1\ \text{GeV} \end{array} \right. @\ \text{point} \left\{ \begin{array}{c} 3 \\ 4 \end{array} \right. \tag{4}$$

$$\left. \begin{array}{r} \Delta m_{\tilde{b}_1} = \pm 1.5 \Delta m_{\chi_1} \pm 3\ \text{GeV} \\ \Delta(m_{\tilde{g}} - m_{\tilde{b}_1}) = \pm 2\ \text{GeV} \end{array} \right\} @\ \text{point 3} \tag{5}$$

A global fit at point 5 yielded

$$\Delta m_0 = \pm 5(\pm 3)\ \text{GeV}, \quad \Delta m_{1/2} = \pm 8(\pm 4)\ \text{GeV}\ , \quad \Delta \tan\beta = \pm 0.11(\pm 0.02) \tag{6}$$

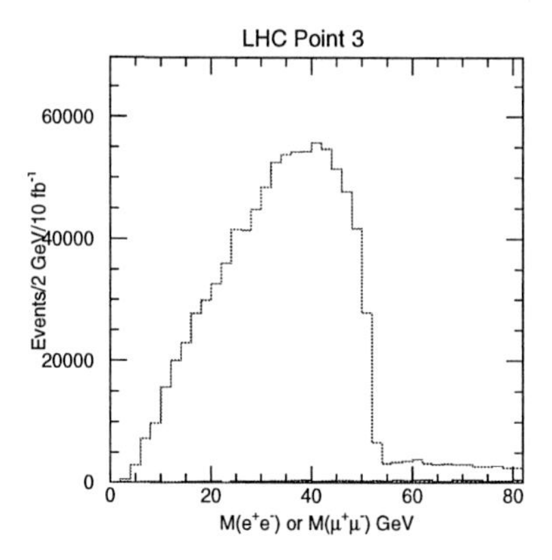

Fig. 8. Typical $\ell^+\ell^-$ spectrum from $\chi_2 \to \chi_1, \ell^+\ell^-$ decay in cascade decays of $\tilde{q}/\tilde{g}$ at the LHC [38].

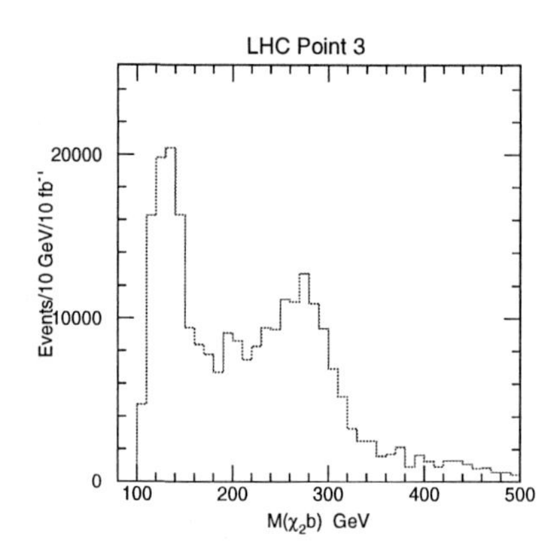

Fig. 9. Typical $\tilde{b} \to \chi_2 b$ decay signature at the LHC [38].

where the different errors refer to different stages in the sophistication of the analysis, and other examples are shown in [33].

These are not the only precision aspects of physics at the LHC. Parton distributions are the limiting uncertainties in present experiments, cf the large-E_T jet cross section at FNAL [39] and the large-x data at HERA [40]. Previously, it had been thought difficult to determine the pp collision luminosity with a precision better than 5%. A new approach [41, 42] is to bypass this step, and measure directly the parton-parton luminosity functions via the rapidity distributions of the $W^{\pm}, Z^0$, which could fix the products $q(x_1)\bar{q}(x_2)$ for $10^{-4} \gtrsim x \gtrsim 0.1$ with an accuracy of $\pm 1/2\%$, cf the LEP luminosity error of $\pm 1/4\%$ from Bhabha scattering. The problem in exploiting this is primarily

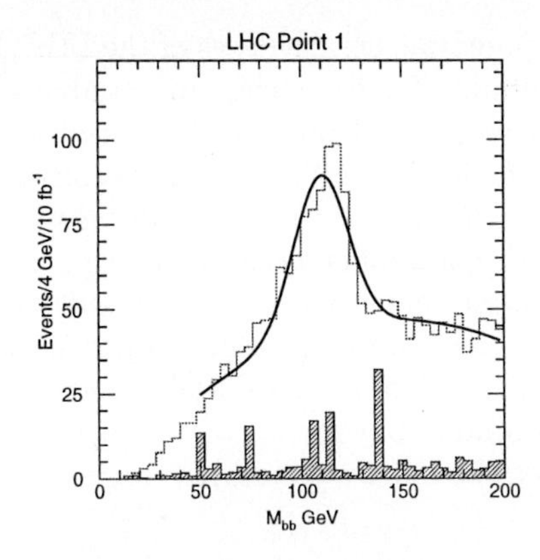

Fig. 10. Typical $h \to \bar{b}b$ decay signal from cascade $\tilde{q}/\tilde{g}$ decays at the LHC [38].

theoretical: can one relate the cross section for $W^{\pm}$ or Z^0 production to other production cross sections, e.g., for W^+W^- pairs or the Higgs, with comparable accuracy? This would require a significant advance in the state of the art of higher-order QCD calculations. So far, one-loop corrections to jet cross sections are known [43], including those due to strongly-interacting sparticle loops [44]. These amount to several percent in some subprocess cross sections, and may have observable structures at to the supersymmetric threshold $M_{jj} = 2m_{\tilde{q}}$ or $m_{\tilde{q}} + m_{\tilde{g}}$ or $2m_{\tilde{g}}$.

Beyond the approved (or almost approved) parts of the LHC programme, there are two further experimental initiatives now circulating. One is for a full-acceptance ($|\eta| < 11$) detector called FELIX [45] to operate at moderate luminosity ($\mathcal{L}_{pp} \gtrsim 2 \times 10^{32}$ cm$^{-2}s^{-1}$) with a physics agenda centred on QCD. This would include novel hard phenomena, such as small-x physics, had diffraction, rapidity-gap physics, the BFKL pomeron and minijets, as well as soft QCD effects at large impact parameters and in central collisions. The proposed programme also includes hard and soft QCD in pA collisions and $\gamma\gamma$ physics in pp and AA collisions, as well as the search for an astroparticle connection to "anomalies" in cosmic-ray collisions, possibly via disordered chiral condensates. FELIX would require bringing the LHC beams into collision in a new interaction point, and has yet to leap the hurdle of assembling a large involved community of experimentalists.

TOTEM [46] is a modest "classical" proposal to measure the total pp cross section ($\Delta\sigma \sim 1$ mb), elastic scattering in the range 5×10^{-4} GeV$^2 \gtrsim |t| \gtrsim 10$ GeV^2 and diffractive production of systems weighing up to 3 TeV. It could very likely be placed at the same interaction point as one of the (almost) approved experiments.

To conclude this review of the LHC physics programme, and set the stage for proposed subsequent accelerators it is appropriate to conclude this section by reviewing some potential sweakenesses of the LHC, as regards probing the MSSM. It is not suitable for producing and studying the heavier charginos $\chi_2^{\pm}$ and neutralinos $\chi_{3,4}$, nor sleptons if they weigh $\gtrsim 400$ GeV. It may have difficulty studying the heavier MSSM Higgs bosons $H, A, H^{\pm}$, and even the lightest Higgs boson h if it has non-standard decay modes. It is not well suited for measuring squark mass differences, since, e.g., it cannot distinguish $\tilde{u}, \tilde{d}$ and $\tilde{s}$. Thus it is questionable whether it provides enough cross-checks on the validity of the MSSM and test simplified parametrizations such as those based on supergravity with a universal scalar mass m_0. There will be plenty of scope for further studies of supersymmetry, even if it discovered earlier at the LHC (or LEP or the FNAL collider).

4 e^+e^- Linear Collider Physics

These machines offer a very clean experimental environment and egalitarian production of new weakly-interacting particles. Moreover, polarizing the beam is easy and can yield interesting physics signatures, and $e\gamma, \gamma\gamma$ and e^-e^- collisions can also be arranged quite easily. Thus linear colliders have many features complementary to those of the LHC [47].

It is likely that the first linear collider would have an initial centre-of-mass energy of a few hundred GeV. Some of the cross sections for important physics processes [48] are shown in Fig. 11 [49]. In the absence (so far) of any clear indication of a threshold in this energy range for new physics, we start by reviewing the bread-and-butter Standard Model physics agenda of such a machine. Pride of place goes to top physics. Detailed measurements of $\sigma_{\bar{t}t}$ and momentum spectra around the threshold at $E_{cm} \sim 350$ GeV, shown in Fig. 12, should enable the error in the top quark mass to be reduced to $\sim$ 120 MeV, and it should be possible to measure Γ_t with an error around 10%. It will also be possible to search very cleanly for non-standard top decays and measure static parameters of the t quark, such as $g^Z_{A,V}$, μ_t, its Higgs coupling [50] and its electric dipole moment. Turning to $W^{\pm}$ and Z^0 physics, precision on the triple-gauge couplings can be improved over the LHC down to the 10^{-3} level. Moreover, if one is able to run the collider with high luminosity at the Z^0 peak, one can quickly obtain a precise measurement of $\sin^2\theta_W$ ($\pm$ 0.0001), and running at the W^+W^- threshold could enable the error on m_W to be reduced to 15 MeV.

The range of Higgs masses preferred (1) by the precision electroweak measurements gives hope that the Higgs boson may lie within the kinematic reach of a first linear collider, via the reactions $e^+e^- \to Z + H$ and $e^+e^- \to H\bar{\nu}\nu$. Moreover, it is easy to detect a Higgs boson in the mass ranges that are "delicate" at the LHC, namely $m_Z \gtrsim m_H \gtrsim 120$ GeV and $m_H \sim$ 170 GeV [49, 51]. Even if the Higgs discovery is made elsewhere, a linear collider could tell us much more about its couplings and branching ratios: $g_{ZZH}, B(\bar{b}b), B(WW^*), B(\tau^+\tau^-)$ and $B(\bar{c}c + gg)$, as seen in Fig. 13, enabling

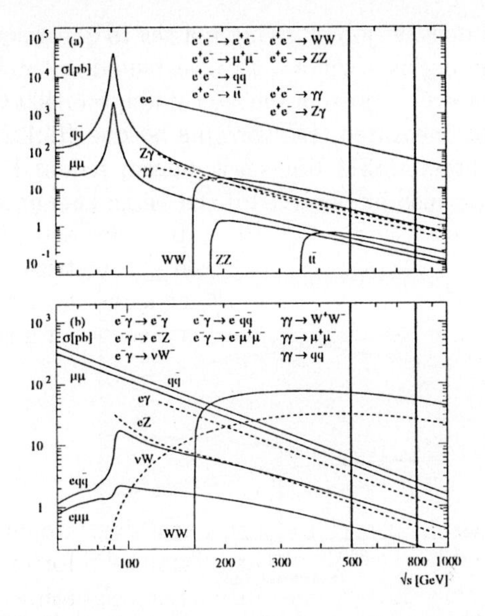

Fig. 11. Important cross sections at a linear collider [50].

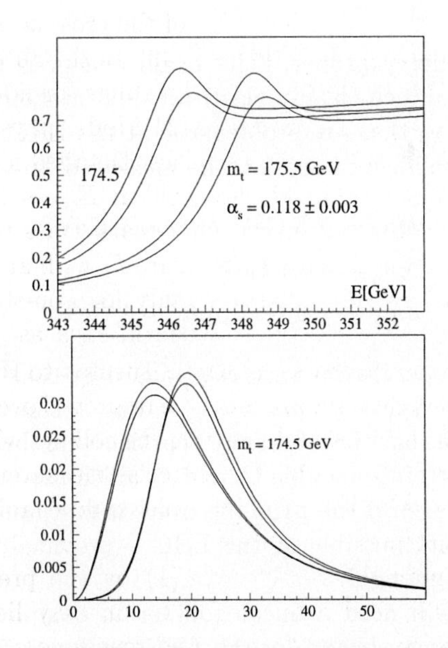

Fig. 12. (a) Cross section and (b) kinematic measurements of $e^+e^- \to \bar{t}t$ at a linear collider [50].

us to verify that it does its job of giving masses to ghe gauge bosons, quarks and leptons, and giving us a window on possible non-minimal Higgs models. One can also measure $\Gamma(\gamma\gamma)$ using the $\gamma\gamma$ collider modes and the spin-parity of the Higgs can be measured [49]. Turning to the MSSM, production and detection of the lightest MSSM Higgs h is guaranteed, and the heavier Higgs bosons $H^{\pm}, H, A$ can also be observed if the beam energy is high enough.

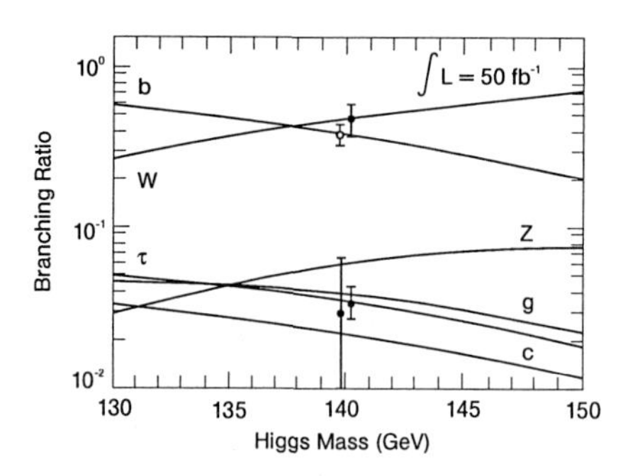

Fig. 13. Accuracy with which Higgs decay branching ratios may be measured at a linear collider [50].

As for supersymmetry proper, if its beam energy is above threshold, a linear collider will produce cleanly electroweakly-interacting sparticles such as the $\tilde{\ell}^{\pm}, \tilde{\nu}, \chi^{\pm}$ and χ_i that are problematic at the LHC [49, 52], as seen in Fig. 14. Moreover, sparticle masses can be measured accurately:

$$\begin{aligned} &\delta m_{\tilde{\mu}} \sim 1.8 \text{ GeV}, \delta m_{\tilde{\nu}} \sim 5 \text{ GeV}, \\ &\delta m_{\chi^{\pm}} \sim 0.1 \text{ GeV} \\ &\delta m_{\chi} \sim 0.6 \text{ GeV}, \delta m_{\tilde{t}} \sim 4 \text{ GeV} \end{aligned} \tag{7}$$

enabling one to test supergravity mass relations, over-constrain model parameters and check universality (is $m_{\tilde{\ell}} = m_{\tilde{\mu}} = m_{\tilde{\tau}}$, for example?). Moreover, the couplings and spin-parities of manuy sparticles can be measured. Thus a linear collider will certainly be able to add significantly to our knowledge of supersymmetry [49], even if the LHC discovers it first, and despite the large range of measurements possible at the LHC – provided the linear collider beam energy is large enough!

In my view, we will need a linear collider to complete our exploration of the TeV energy range, begun by the LHC, to pin down the mechanism of electroweak symmetry breaking, and to complement the LHC programme with more precision measurements. To have physics reach comparable to the LHC, the collider energy should be able to reach $E_{cm} \sim 2$ TeV, and it would be desirable to be able to operate back in the LEP energy range and the Z peak and the W^+W^- threshold. Thus the machine should be

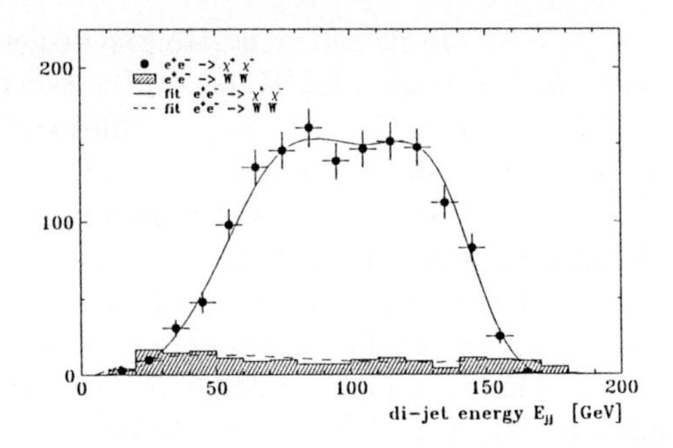

Fig. 14. Possible measurement of $e^+e^- \rightarrow \chi^+\chi^-, \chi^\pm \rightarrow jj\chi$ at a linear collider [50].

flexible, with an initial E_{cm} in the few hundred GeV and the possibility of subsequent upgrades. Unfortunately, we do not yet know exactly where to start, in the absence of clear information on new physics thresholds. A final personal comment is that I hope very much that the linear collider community can converge on a single project. Can the world afford two such colliders? In the recent past, our political masters have decided to support just one hadron collider.

5 Beyond the Standard Colliders

Even though construction of the LHC will not be completed for another 8 years, and no linear collider proposal has even been submitted, it is already time to think what might come next, since we need to maintain a long-term vision of the future of accelerator-based particle physics, and the $R \times D$ lead time for any new accelerator project is necessarily very long.

One possible future option which as been kept in mind since the inception of the LHC project has been an ep collider in the LEP tunnel, using an LHC beam and an $e^\pm$ beam circulating in a rearranged LEP ring [53]. The latest design envisages beam energies of 7 TeV and 67 GeV, yielding collisions at a centre-of-mass energy of 1.37 TeV with a luminosity $\mathcal{L}_{ep} \sim 10^{32} cm^{-2} s^{-1}$. The physics interest of this ep option will be easier to judge after the physics potential of the HERA collider (and in particular the interpretation of the current large-Q^2 anomaly [40]) has been further explored: certainly it would be great for producting leptoquarks or R-violating squarks up to masses around 1 TeV.

However, this option is presumably not a complete future for CERN, let alone the world high-energy physics community. More complete possibilities for the future are $\mu^+\mu^-$ colliders [54], which may be the best way to collide leptons at $E_{cm} = 4$ TeV or more, and a possible next-generation pp collider (variously named the Eloisatron, RLHC or VLHC, called here a 'Future Larger Hadronic Collider' or FLHC) with $E_{cm} = 100$ to 200 TeV [55]. We now discuss each of these possibilities in turn.

Many technical issues need to be resolved before a multi-TeV $\mu^+\mu^-$ collider can be proposed: the accumulation of the $\mu^\pm$, their cooling, shielding the detectors and the surrounding populace from their decay radiation, etc. These should be addressed by a smaller-scale demonstrator project, much as the SLC demonstrates the linear-collider principle. A very exciting possibility for this demonstration is a Higgs factory [56], exploiting the non-universality of the HL^+L^- couplings, which implies that $\sigma(\mu^+\mu^- \to H)/\sigma(e^+e^- \to H) \sim$ 40,000, and the possibly reduced energy spread in $\mu^+\mu^-$ collisions, which may be as small as 0.01%. Neglecting the energy spread, the $\mu^+\mu^-$ cross section in the neighbourhood of the H peak is given by

$$\sigma_H(s) = \frac{4\pi\,\Gamma(H \to \mu^+\mu^-)\,\Gamma(H \to X)}{(s - m_H^2)^2 + m_H^2 \Gamma_H^2} \tag{8}$$

where the natural width of a 100 GeV Higgs is expected to be about 3 MeV, whereas $\Delta\sqrt{s}$ may be as low as 10 MeV. Typical line shapes for Standard Model and MSSM Higgs bosons in this mass range are shown in Fig. 15. Such a Higgs factory would be able to measure Higgs decay branching ratios into channels such as $\bar{b}b, \tau^+\tau^-$, WW^* and ZZ^*. It could also draw a clear distinction between the Standard Model H and the MSSM h, could (at higher energies) separate the H and A of the MSSM, and also make detailed studies of their properties. Other possible applications of the narrow $\mu^+\mu^-$ spread in E_{cm} include the measurement of m_H with a precision $\sim$ 45 MeV, and improved precision in the values of m_t and m_W [54].

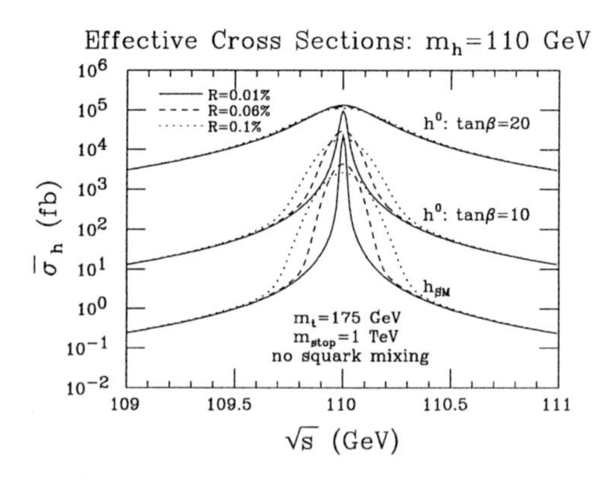

Fig. 15. Cross sections for Higgs production at a $\mu^+\mu^-$ collider run as a "Higgs factory" [55,57].

As for a possible FLHC, this is clearly the tool of choice for exploring the 10 TeV energy range, as is shown by one example in Fig. 16 [57]. At present, we do not know what physics may lie there – the messenger sector of a gauge-mediated scenario for supersymmetry breaking [5]? a Z' ? a fifth dimension? We will never know unless we go and look.

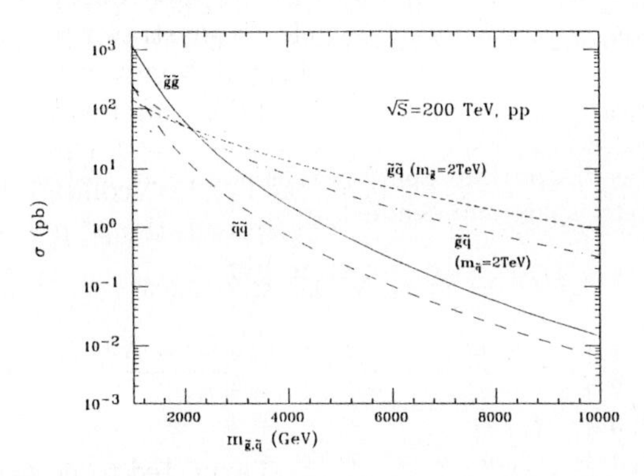

Fig. 16. Cross sections for $\tilde{q}$ and $\tilde{g}$ production at a FLHC [58].

Table 1. Table of future collider parameters

Collider	Particles	E_{cm} (GeV)	Luminosity($cm^{-2}s^{-1}$)	Starting Date
PEP II	e^+e^-	10	3×10^{33}	1999
KEK-B	e^+e^-	10	10^{33}	1999
RHIC	$Au - Au$	200A	2×10^{26}	1999
LHC	pp	14×10^3	10^{34}	2005
LHC	$Pb - Pb$	1.15×10^6	10^{27}	2005
LHC	ep	1300	10^{32}	?
LC	e^+e^-	500/2000	$5 \times 10^{33}/10^{34}$	2008 ?
FMC	$\mu^+\mu^-$	$m_H/4000$	$2 \times 10^{33}/10^{35}$	?
FLHC	pp	$1/2 \times 10^5$	10^{34}	?

Table 2. Test points for supersymmetry studies at the LHC (masses in GeV)

	m_0	$m_{1/2}$	A_0	$\tan\beta$	$m_{\tilde{g}}$	$m_{\tilde{u}_R}$	$m_{\chi^\pm}$	$m_{\tilde{e}_R}$	m_h
1	400	400	0	2	1004	925	325	430	111
2	400	400	0	10	1008	933	321	431	125
3	200	100	0	2	298	313	96	207	68
4	800	200	0	10	582	910	147	805	117
5	100	300	300	2.1	767	664	232	157	104

Table 3. The LHC as "Bevatrino":Sparticles detectable at selected points in supersymmetric parameter space and denoted by +

	h	H/A	χ_2^0	χ_3^0	χ_1^-	$\chi_1^\pm$	$\chi_2^\pm$	$\tilde{q}$	$\tilde{b}$	$\tilde{t}$	$\tilde{g}$	$\tilde{\ell}$
1	+		+					+	+	+	+	
2	+		+					+	+	+	+	
3	+	+	+			+		+	+		+	
4	+		+	+	+	+	+	+			+	
5	+		+					+	+	+	+	+

References

[1] D.R. Ward, Plenary Talk 15 at this conference, hep-ph/9711515.

[2] T. Hambye and K. Riesselmann, hep-ph/9708416.

[3] J. Ellis, G.L. Fogli and E. Lisi, Physics Letters B389 (1996), 321-326, and references therein.

[4] M. Carena, M. Quiros and C.E.M. Wagner, Nuclear Physics B461 (1996) 407-436;
H.E. Haber, R. Hempfling and A.H. Hoang, Zeit. für Physik C75 (1997) 539-554.

[5] P. Binetruy, Plenary Talk 9 at this conference.

[6] C. Jarlskog, Plenary Talk 5 at this conference,
M. Feindt, Plenary Talk 7 at this conference.

[7] H. Georgi, H. Quinn and S. Weinberg, Physical Review Letters 33 (1974), 451-454.

[8] M.S. Chanowitz, J. Ellis and M.K. Gaillard, Nuclear Physics B128 (1977), 506.

[9] P. Mättig, Parallel session Talk 1801 at this conference.

[10] P. Janot, Plenary Talk 17 at this conference.

[11] S. Park, in *Proceedings of the 10th Topical Workshop on Proton-Antiproton Collider Physics*, Fermilab, 1995, edited by R.Raja and J. Yoh (AIP, New York, 1995), p. 62.

[12] J.M. Butler, Parallel Session Talk 1802 at this conference;
D. Gerdes, Contributed Paper 238.
[13] K.Schubert, Parallel Session Talk 1819 at this conference;
BaBar collaboration, Contributed Paper 005.
[14] P. Krizan, Parallel Session Talk 1818 at this conference.
[15] J.N. Butler, Parallel Session Talk 1820 at this conference.
[16] CMS collaboration, G.L. Bayatian et al., Technical Proposal, CERN/LHCC/94-38, LHCC/P1 (1994).
[17] CMS collaboration, Contributed Paper 542.
[18] ATLAS collaboration, W.W. Armstrong et al., Technical Proposal, CERN/LHCC/94-43, LHCC/P2 (1994).
[19] S. Semenov, Parallel Session Talk 1821 at this conference.
[20] F. Plasil, Plenary Talk 12 at this conference.
[21] STAR collaboration, J.W. Harris et al., Nuclear Physics A566 (1994), 277c-286c.
[22] PHENIX collaboration, J.C. Gregory et al., BNL-PROPOSAL-R2 (1992).
[23] B. Wosiek, Parallel Session talk 604 at this conference.
[24] ALICE collaboration, S. Boelè et al., Technical Proposal, CERN/LHCC 95-71, LHCC/P3 (1995) and Addendum CERN/LHCC 96-32 (1996).
[25] J.-C. Gourber, Plenary Talk 18 at this conference.
[26] R. Kvatadze, Parallel Session Talk 606 at this conference.
[27] M. Spira, Parallel Session Talk 1805 at this conference, hep-ph/9711394.
[28] R.J. Cashmore, Plenary Talk 13 at this conference.
[29] L. Poggioli, Parallel Session Talk 1806 at this conference.
[30] M. Dittmar and H. Dreiner, Physical Review D55 (1997), 167-172.
[31] J. Tuominiemi, Parallel Session Talk 1807 at this conference.
[32] M. Dittmar, Parallel Session Talk 1812 at this conference.
[33] F. Paige, Parallel Session Talk 1814 at this conference.
[34] W. Beenakker, R. Hopker, M. Spira and P. Zerwas, Nuclear Physics B492 (1997), 51-103.
[35] S. Abdullin, Parallel Session Talk 1813 at this conference;
D. Denegri, A. Kharchilava, W. Majerotto and L. Rurua, Contributed Paper 540;
D. Denegri, W. Majerotto and L. Rurua, hep-ph/9711357.
[36] H. Baer, C.-H. Chen and X. Tata, Physical Review D55 (1997) 1466-1470.
[37] I. Hinchliffe, F.E. Paige, M.D. Shapiro, J. Soderqvist and W. Yao, Physical Review D55 (1997), 5520-5540.
[38] LHCC supersymmetry workshop, `http://www.cern.ch/Committees/LHCC/SUSY96.html`.
[39] CDF collaboration, F. Abe et al., Physical Review Letters 77 (1996) 438-443.
[40] E. Elsen, Plenary Talk 26 at this conference.

[41] F. Behner, Parallel Session Talk 1803 at this conference.
[42] M. Dittmar, F. Pauss and Zürcher, Physical Review D56 (1997), 7284-7290.
[43] R.K. Ellis and J.C. Sexton, Nuclear Physics B269 (1986), 445.
[44] J. Ellis and D.A. Ross, hep-ph/9708312.
[45] FELIX collaboration, E. Lippmaa et al., CERN/LHCC 97-45, LHCC/I10 (1997).
[46] TOTEM collaboration, W. Kienzle et al., CERN/LHCC 97-49, LHCC/I11 (1997).
[47] B. Wiik, Plenary Talk 14 at this conference.
[48] M. Martinez, Parallel Session Talk 1804 at this conference;
E. Boos, Parallel Session Talk 1809 at this conference;
V. Ilyin Parallel session Talk 1811 at this conference.
[49] ECFA/DESY LC Physics Working Group, E. Accomando et al., DESY-97-100 (1997), Contributed Paper 673;
see also H. Murayama and M.E. Peskin, Annual Reviews of Nuclear and Particle Science 46 (1997), 533.
[50] S. Bar-Shalom, Parallel session Talk 1816 at this conference, hep-ph/9710355.
[51] H.-J. Schreiber, Parallel Session Talk 1808 at this conference, hep-ph/9711468;
G.-L. Lin, Parallel session Talk 1817 at this conference;
G. Montagna et al. Contributed Paper 281.
[52] H. Martyn, Parallel Session Talk 1815 at this conference.
[53] E. Keil, *LHC e–p Option*, LHC Project Report 93 (1997).
[54] The $\mu^+\mu^-$ Collider collaboration, *Muon Muon Collider: A Feasibility Study*, BNL-52503, FERMILAB-CONF-96/092, LBNL-38946 (1996).
[55] See, for example, D. Denisov and S. Keller, Summary of the Very Large Hadron Collider Physics and Detector Subgroup, in *Proc. New Directions for High-Energy Physics*, Snowmass 1996, p.277;
G. Anderson et al., Summary of the Very Large Hadron Collider Physics and Detector Workshop, FERMILAB-CONF-97/318-T.
[56] V. Barger, M. Berger, J. Gunion and T. Han, Physical Review Letters 78 (1997), 3991-3994 and Physical Review D56 (1997), 1714-1722.
[57] M. Mangano, private communication (1997).

New Results on Cosmology

Tsvi Piran (tsvi@nikki.fiz.huji.ac.il)

The Hebrew University, Jerusalem, Israel 91904

Abstract. Cosmology is characterized during the last few years by rapid and exciting developments. Most of these developments are observational. New astronomical tools turn cosmology into a higher precision science at a rapid pace. Numerous preliminary results demonstrate the power of these new tools. The trend of high precision cosmological measurements is certain to continue for the next few years with the introduction of new CMBR detectors. These should be able to nail down most cosmological parameters to within a few percents. I briefly review some of the recent developments and then I focus on what seems to me the greatest success of the last year - the discovery of GRB (gamma-ray burst) afterglow - which confirmed their cosmological origin and verified the GRB fireball model.

1 Introduction

Observational cosmology is blooming during the last few years. New techniques, larger surveys, more sensitive and more precise detectors and large automated searches are introduced at an impressive rate. These are just beginning to bear the first fruits now. While we have not reached yet high precision in the determination of most cosmological parameters it is clear that this goal is within reach soon. The scope of this review is too short to summarize the wealth of new observational techniques. I will not try to do that. Instead I will give a brief and subjective list of what seems to me the most exciting developments and provide a somewhat subjective "best guess" table of cosmological parameters.

Then I turn to what seems to me the most exciting development during the last year - the discovery of Gamma-ray burst (GRB) afterglow. This discovery have proven the cosmological origin of GRBs and has demonstrated that these are the most (electromagnetically) luminous objects in the Universe. It brought us close to the resolution of this mystery, which has been puzzling astrophysicists for the last thirty years.

2 Cosmological Parameters

Observational cosmology focuses on the quest to find several cosmological parameters. These can be divided to two groups. The first and most fundamental group includes the parameters that appear in Friedmann's Equation:

- H_0 - The Hubble parameter which measures the expansion rate of the Universe. It is related to the age of the Universe as $t_0 = C(\Omega, \Lambda)H_0^{-1}$.

- $\Omega = \rho/\rho_0$ - The ratio of the density to the critical density $\rho_c = 3H_0^2/8\pi G$.
- Λ - The cosmological constant (measured in units of the critical density).
- t_0 - The age of the oldest observed objects in the Universe.

The parameters H_0, Ω, and Λ determine the overall evolution of the Universe. Ω is also called the "closure parameter". It is generally believed that (for $\Lambda = 0$) the Universe is spatially closed, open or flat if $\Omega > 1$, $\Omega < 1$ or $\Omega = 1$. It should be stressed that Ω (or $\Omega + \Lambda$) measures only the local curvature of the Universe. It does not determine the overall topology. Closed spaces with local negative curvature (which correspond to $\Omega < 1$) have been recently discovered [1]. Thus the Universe might be closed even if $\Omega < 1$! The name "closure parameter" is misleading.

The forth parameter in this group is t_0, the age of the oldest observed objects in the Universe. This should be compared with the age of the Universe calculated from the first three parameters: H_0, Ω, and Λ. An independent measurement of t_0 yields the first and most basic consistency check. The Universe must be older than the oldest stars. Currently, the observed age is larger than the calculated one for most cosmological models. However, the observations are not precise enough to worry about the validity of the Big Bang yet. Furthermore, new observations tend to close this gap. The recent Hipparcos measurement of Cepheid distances [3] are larger by 10% than previous estimates. It lowers H_0 (and thus increases t_0) by 10%. At the same time it lowers the age of Globular Clusters by 20%.

In addition to Ω, H_0 and Λ other parameters determine the growth of structure in the Universe. These parameters are:

- Ω_b - The baryonic density (in terms ρ_c).
- Ω_c and Ω_h - The density of cold, Ω_c, and hot, Ω_h, dark matter.
- n - The index of the primordial fluctuation spectrum $P(k) \propto k^n$.

Ω_c and Ω_h are related of course to the question of what is the the constitution of the dark matter, which is of great interest and particularly so to a particle physics. However, quite generally cosmology is insensitive to the specific nature of the dark matter. It is only sensitive to the question whether the dark matter particle should be considered as hot (like a neutrino) or cold (like a WIMP). Similarly, n is related to the nature of the inflation, or more generally, to the process of formation of primordial fluctuation. Once more Cosmology is sensitive only to these two parameters.

Different observations aim at measuring various combinations of these parameters. The progress is rapid and impressive. To me the advances concerning the Hubble constant are particularly encouraging. As a student I grew up was watching the H_0 controversy: is H_0 100 or is it 50? I was almost lead to believe that Homo Sapiens are not allowed to know the age of the Universe to better than a factor of two! It is a relieve to see that different techniques converge today to a single value of H_0 around 65 Mpc/km/sec [2].

I will not attempt to summarize all the recent results here. Instead I provide a single best guess table of cosmological parameters. I refer the reader to some longer and more extensive reviews [4, 5] for a more critical description of the observations and their results.

Table 1. *Cosmological Parameters*

Parameter	"best guess"	Range
H_0	65 Mpc/km/sec	60-70 Mpc/km/sec
Ω	0.4	0.2-1
Λ	0	< 0.7
t_0	12 Gyr	10-14 Gyr
Ω_b	.04	0.005-0.2
Ω_c	0.3	0.15-1
Ω_h	0.1	0 - 1
n	1.1	0.9-1.3

3 Gamma-Ray Bursts

GRBs are short and intense bursts of $\sim$ 100keV-1MeV photons that appear from random places in the sky. They last a few seconds and disappear, leaving (until recently) no detectable trace of their appearance. GRBs were discovered accidentally in the sixties by the Vela satellites - defense satellites sent to monitor nuclear explosions in space. A wonderful by-product of this effort was the discovery of GRBs, which was was announced in 1973 [6]. Since then GRBs have been one of the greatest astronomical puzzles. Recent observation and theoretical progress bring us almost to the resolution of this mystery.

The BATSE detector on the COMPTON-GRO (Gamma-Ray Observatory) was launched in the spring of 1991. It has revolutionized GRB observations and our ideas on their nature. BATSE's observations suggested that GRBs are extra-galactic at cosmological distances [7]. A cosmological GRB emits a total energy of 10^{51}ergs (as inferred from the observed flux and the implied distance of a cosmological source).

3.1 Compactness, Relativistic Motion and the Fireball Model

Standard considerations suggest that the temporal variability observed in GRBs implies that the sources are compact with a size, $R_i < c\delta T \approx 3000$km. The observed spectrum contains a large fraction of the high energy γ-ray photons. These photons could interact with lower energy photons and produce electron-positron pairs via $\gamma\gamma \rightarrow e^+e^-$. The average optical depth for this process is $\sim 10^{15}(E/10^{51}\text{ergs})(\delta T/10 \text{ msec})^{-2}$ [8]. However, the observed non-thermal spectrum indicates with certainty that the source must be optically thin.

This problem, the compactness problem, can be resolved if the emitting region is moving towards us with a relativistic velocity characterized by a Lorentz factor, $\gamma \gg 1$. In this case we detect blue-shifted photons whose energy at the source is lower by a factor γ. Fewer photons have sufficient energy to produce pairs. Additionally, relativistic effects allow the radius from which the radiation is emitted to be larger than the previous estimate by a factor of γ^2. The resulting optical depth is lower by a factor $\gamma^{(4+2\alpha)}$ (where $\alpha \sim 2$ is the spectral index). The minimal Lorentz factors must satisfy $\gamma > 10^{15/(4+2\alpha)} \approx 10^2$.

The potential of relativistic motion to resolve the compactness problem was realized in the eighties by Goodman [9], Paczyński [10] and Krolik & Pier [11]. While Krolik & Pier [11] considered a kinematical solution, Goodman [9] and Paczyński [10] considered a dynamical solution in which the relativistic motion results naturally when a large amount of energy is released within a small volume. They have shown that this would result in a relativistic explosion - a fireball. Goodman [9] and Paczyński [10] considered pure radiation fireballs. Shemi & Piran [12] have shown latter that if the fireball contains a small amount of baryonic mass, M, it will reach an asymptotic Lorentz factor $\gamma = E/Mc^2$. At this stage all the initial energy of the fireball will be converted to the kinetic energy of the baryons.

A generic scheme for a cosmological GRB model has emerged in the last few years - the fireball model (see [8] for a detailed review). The observed γ-rays are emitted when ultra-relativistic particles are slowed down and their kinetic energy is converted to radiation. The kinetic energy is converted via shocks to "thermal" energy and then to radiation. Both the low energy spectrum of GRBs and the high energy spectrum of the afterglow provide indirect evidence for relativistic shocks in GRBs [13] and in the afterglow [14]. Slowing down can take place in two ways via (i) External shocks, which are due to interaction with an external medium like the ISM [15], or via (ii) Internal shocks that arise due to shocks within the flow when fast moving particles catch up with slower ones [16, 17].

External shocks are practically inevitable if the fireball is surrounded by some external medium, such as the ISM. External shocks were considered, therefore, as the canonical model. However, Sari & Piran [18] have recently shown that external shocks cannot produce the complicated highly variable temporal structure observed in most GRBs and consequently Internal shocks are essential. Internal shocks would take place if the "inner engine" produces a highly irregular wind. Internal shocks would then take place when faster shells catch up with slower ones. Numerical simulations of internal shocks show that, unlike external shocks, they can reproduce the temporal structure observed in GRBs [19].

3.2 The Internal-External Model

Internal shocks can convert only a fraction of the total energy to radiation [20, 19]. A few months before the discovery of the afterglow by BeppoSAX Sari & Piran [18] have pointed out that after the GRB has been produced via internal shocks the flow has enough energy and it will interact via an external shock with the surrounding medium. This shock will produce an afterglow - a signal that will follow the GRB. The idea of an afterglow in other wavelengths was suggested earlier [21, 22, 23] as an extrapolation of the GRB, which was believe to be produce by the external shock. In this case the afterglow would have been a direct continuation of the GRB activity and its properties would have scaled directly to the properties of the GRB.

According to internal-external model (internal shocks for the GRB and external shocks for the afterglow) different mechanisms produce the GRB and the afterglow. Therefore the afterglow should not be scaled to the GRB. This behavior was observed in the recent afterglow observations.

3.3 GRB Afterglow

GRB observations were revolutionized on February 28 1997 with discovery of an X-ray counterpart to GRB970228 by the Italian-Dutch satellite BeppoSAX [24]. The accurate position determined by BeppoSAX enabled the identification of an optical afterglow [25] - a decaying point source surrounded by a red nebulae. Following observations with HST [26] revealed that the nebula is roughly circular with a diameter of 0".8. The nebula's intensity does not vary, while the point source decays with a power law index ≈ -1.2 [27].

Afterglow was also detected from GRB970508. This γ-ray burst lasted for $\sim$ 15sec, with a γ-ray fluence of $\sim 3 \times 10^{-6}$ergs/cm^{-2}. Variable emission in X-rays, optical [28] and radio [29] followed the γ-rays. The spectrum of the optical transient revealed a set of absorption lines with a redshift $z = 0.835$ [30]. A second absorption line system with $z = 0.767$ is also seen. This sets the cosmological redshift of this bursts to $z \geq 0.835$. The lack of Lyman alpha absorption lines sets an upper limit of $z < 2.1$. The optical light curves show a clear peak at around 2 days after the burst. After that it shows a continuous power law decay $\propto t^{-1.18}$ [31]. Radio emission was observed first one week after the burst [29]. This emission showed intensive oscillations which were interpreted as scintillations [32]. The subsequent disappearance of these oscillations after about three weeks enables Frail and Kulkarni [29] to estimate the size of the fireball at this stage to be $\sim 10^{17}$cm This was supported by the indication that the radio emission was initially optically thick [29], which yields a similar estimate to the size [33]. These size estimates show clearly a relativistic expansion supporting the fireball model.

These observations confirm the general fireball picture. The observed X-ray, optical and radio afterglow were predicted by this model [21, 22, 23, 18]. In particular the observations are in agreement as stressed before with the "internal-external" model according to which the GRB is produced by internal shocks within the flow while the afterglow is produce by external shocks with the ISM. Afterglow observations agree qualitatively with synchrotron cooling from a slowing down relativistic shell [14, 33, 34, 35, 36]. In this model the shell expands, collects external matter and slows down. The shock front accelerated the electrons to a power law distribution and these electrons cool via synchrotron (or Inverse Compton) emission. The Lorentz factor of the shell decreases gradually and this leads to a decrease in the typical synchrotron frequency.

3.4 The "Inner Engine"

The fireball model is based on an "inner engine" that supplies the relativistic baryons. This "engine" is well hidden from direct observations. Currently it is impossible to determine directly what is it. Unfortunately, the discovery of afterglow does not shed a new light on this issue. For a long time the only direct clues that existed on the nature of the "inner engine" were the rate - once per 10^6years per galaxy [37] - and the energy output - $\sim 10^{52}$ergs. Even these limits are uncertainty due to the possibility of beaming. A beaming angle, θ, leads to an increase of order $4\pi/\theta^2$ in the rate and a corresponding decrease in the total energy involved.

The recent realization that energy conversion process is via internal shock rather than via external shocks provides additional information. The inner engine must be capable of producing irregular flow. It must be variable on a short time scale (this time scale is seen in the variability of the bursts) and it must be active for up to a few hundred seconds (this determines the observed duration of the bursts). The engine must be compact ($\sim 10^7$cm) to produce the observed variability and it must operate for a few hundred seconds (million times longer than its light crossing time) to produce a few hundred seconds signal. These demands rule out all explosive models and require some sort of a flaring mechanism.

3.5 Neutron Star Mergers

Binary neutron star mergers (NS^2Ms) are, in my mind the best candidates for the "inner engines" [38]. These mergers take place because of the decay of the binary orbit due to gravitational radiation emission. Binary pulsar observations suggest that NS^2Ms take place at a rate of $\approx 10^{-6}$ events per year per galaxy [39, 40], in amazing agreement with the GRB event rate [37]. NS^2Ms result, most likely, in rotating black holes [41]. The process releases $\approx 5\times10^{53}$ ergs [42]. Most of this energy escapes as neutrinos and gravitational radiation. A small fraction of this energy suffices to power a GRB. We see that the observed rate of NS^2Ms is similar to the observed rate of GRBs and that in principle they can produced the required energy. This is not a lot - but this is more than can be said, at present, about any other GRB model.

3.6 Implications of GRBs to Particle Physics

GRBs have several important implications to other branches of astrophysics and in particular to high energy astrophysics and to particle physics. This is not surprising in view of the unique character of the fireball model that involves relativistic motion of a significant amount of matter. I turn now to some of these implications.

It is likely that in addition to γ-rays other particles, denoted here x, are emitted in GRBs. Let $f_{x-\gamma}$ be the ratio of energy emitted in x particles relative to γ-rays[1]. These particles will appear as a burst accompanying the GRB. The fluence of a typical GRB observed by BATSE, F_γ, is 10^{-7}ergs/cm^2, We expect, therefore, accompanying bursts with a fluence of:

$$F_{x\ |burst} = 0.001 \frac{\text{particles}}{\text{cm}^2} f_{x-\gamma} \left(\frac{F_\gamma}{10^{-7}\text{ergs/cm}^2}\right) \left(\frac{E_x}{\text{GeV}}\right)^{-1}, \quad (1)$$

where E_x is the typical energy of an x particles. This burst will be spread in time and delayed relative to the GRB if the particles do not move at the speed of light. Relativistic time delay will be significant (larger than 10 seconds) if the particles are not massless and their Lorentz factor is smaller than 10^8! Similarly a deflection angle of 10^{-8} will cause a significant time delay.

In addition to the prompt burst we expect a continuous background of x particles. With one 10^{51}ergs GRB per 10^6 years per galaxy we expect $\sim 10^4$ events per galaxy in a Hubble time (provided of course that the event rate remains a constant). This corresponds to a background flux of

$$F_{x\ |bg} = 3 \cdot 10^{-8} \frac{\text{particles}}{\text{cm}^2\text{sec}} f_{x-\gamma} \left(\frac{E_\gamma}{10^{51}\text{ergs}}\right) \left(\frac{R}{10^{-6}\text{years/galaxy}}\right) \left(\frac{E_x}{\text{GeV}}\right)^{-1}. \quad (2)$$

For any specific particle that could be produced one should calculate the ratio $f_{x-\gamma}$ and then compare the expected fluxes with fluxes from other sources and with the capabilities of current and future detectors.

One should distinguish between two types of predictions: (i) Predictions of the generic fireball model which include low energy cosmic rays [12], UCHERs [43, 44] and high energy neutrinos [45]. (ii) Predictions of a specific GRB model and in particular of the NS^{2}M model, which include low energy neutrino bursts [42] and gravitational waves.

[1] I assume in the following that the γ-rays from the GRB and the x particles have the same angular distribution. This is a reasonable assumption if both are produced by the fireball's shocks. It might not be the case if the x particles are produced by the "inner engine". A modification that takes care of this correction is trivial

Cosmic Rays Already in 1990, Shemi & Piran [12] pointed out that fireball model is closely related to Cosmic Rays. A "standard" fireball model involves the acceleration of $\sim 10^{-7} M_\odot$ of baryons to a typical energy of 100GeV per baryon. Protons that leak out of the fireball become low energy cosmic rays. However, a comparison of the GRB rate (one per 10^6 years per galaxy) with the observed flux of low energy cosmic rays, suggests that even if $f_{CR-\gamma} \approx 1$ this will amount only to 1% to 10% of the observed cosmic ray flux at these energies. Cosmic rays are believed to be produced by SNRs. Since supernovae are ten thousand times more frequent than GRBs, unless GRBs are much more efficient in producing Cosmic Rays in some specific energy range their contribution will be swamped by the SNR contribution.

UCHERs - Ultra High Energy Cosmic Rays Waxman [43] and Vietri [44] have shown that the observed flux of UCHERs (above 10^{19}eV) is consistent with the idea that these are produced by the fireball shocks provided that $f_{UCHERs-\gamma} \approx 1$. The fireball's relativistic shocks might be capable of accelerating particles to such high energies. Waxman [46] has shown that the spectrum of UCHERs is consistent with the spectrum expected from Fermi acceleration within those shocks.

High Energy Neutrinos Waxman and Bahcall [45] suggested that collisions between protons and photons within the relativistic fireball shocks produce pions. These pions produce high energy neutrinos with $E_\nu \sim 10^{14}$eV and $f_{\text{he. }\nu-\gamma} > 0.1$. The corresponding flux is comparable to the flux of atmospheric neutrinos. But these neutrinos will be correlated with the positions of strong GRBs. This signal might be detected in future km^2 size neutrino detectors.

Gravitational Waves If GRBs are associated with NS^{2}Ms then they will be associated with gravitational waves and with low energy neutrinos. The spiraling in phase of a NS^{2}M produces a clean chirping gravitational radiation signal. This signal is the prime target of LIGO and VIRGO, the two large interferometers that are build now in the USA [47] and in Europe. The observational scheme of these detectors is heavily dependent on digging deeply into the noise. A coincidence with a GRB could enhance greatly the statistical significance of detection of a gravitational radiation signal [48]. It will also verify at the same time this model.

Low Energy Neutrinos Most of the energy released in a NS^{2}M will be released as low energy ($\sim 5-10$MeV) neutrinos [42]. The total energy is quite large $\sim$ a few $\times 10^{53}$ergs, leading to $f_{\text{low energy }\nu-\gamma} \approx 100$. However, this neutrino signal will be quite similar to a supernova neutrino signal - which can be detected at present only if it is galactic. As Supernovae are ten thousand times more frequent then GRBs, these NS^2M neutrinos constitute an insignificant contribution to the background at this energy range.

4 Concluding Remarks

After thirty years we are finally beginning to understand the nature of GRBs. The discovery of the afterglow has demonstrated that we are on the right track, at least as far as the γ-ray producing regions are concerned. This by itself have some fascinating implications on accompanying UCHER and high energy neutrino signals. However, we are still uncertain what are the engines that power the whole phenomenon. My personal impression is that binary neutron mergers are the best candidates. This model has one specific prediction - a correlation between GRBs and gravitational radiation signals. This would confirm or rule out this model next decade when the next generation of gravitational radiation detectors will begin to operate.

I thank E. Cohen, J. Katz, S. Kobayashi, R. Narayan, and R. Sari for helpful discussions. This work was supported by the US-Israel BSF grant 95-328 and by NASA grant NAG5-3516

References

[1] DeWitt, B., S., 1997, in *The Proceedings of the VIII Marcel Grossmann Meeting*, Ed. Piran, T., in press.
[2] Press, W., in *Unsolved Problems in Astrophysics* Eds. Bahcall, J. N., and Ostriker, J. P., Princeton University Press.
[3] Madore, B. F., & Freedman, W. L., 1998, Ap. J. Lett., **492**, 110.
[4] Freedman, W., 1997, in *The Proceedings of the XVIII Texas Symposium*, Eds. D. Schramm, J. Friedman and A. Oliento, in press.
[5] Dekel, A., Burstein, D., & White, S.D.M. 1997, in *Critical Dialogues in Cosmology - Proceedings of the Princeton 250th Anniversary Conference, June 1996*, Ed. N. Turok (World Scientific), p. 175.
[6] Klebesadel, R. W., Strong, I. B., & Olson, R. A. 1973, Ap. J. Lett., 182, L85.
[7] Meegan, C. A., *et. al.,* 1992, Nature, **355**, 134.
[8] Piran, T., 1996, in *Unsolved Problems in Astrophysics* Eds. Bahcall, J. N., and Ostriker, J. P., Princeton University Press, 343.
[9] Goodman, J. 1986, Ap. J. Lett., 308, L47.
[10] Paczyński, B. 1986, Ap. J. Lett., 308, L51.
[11] Krolik, J. H., & Pier, E. A. 1991, Ap. J. , 373, 277.
[12] Shemi, A., & Piran, T. 1990, Ap. J. Lett., 365, L55.
[13] Cohen, E. *et. al.,* 1997, Ap. J., **488**, 330.
[14] Wijers, A. M. J., Rees, M. J., & Mészáros, P., 1997, MNRAS, **288**. L5.
[15] Mészaros, P., & Rees, M. J. 1992, MNRAS, 258, 41p.

[16] Narayan, R., Paczyński, B., & Piran, T. 1992, Ap. J. Lett., 395, L83.
[17] Rees, M. J., & Mészáros, P. 1994, Ap. J. Lett., 430, L93.
[18] Sari, R., & Piran, T., 1997, Ap. J., **485**, 270.
[19] Kobayashi, S., Piran, T., & Sari, R., 1997, Ap. J., **490**, 92.
[20] Mochkovitch, R., Maitia, V., & Marques, R. 1995, in *Proceeding of 29th ESLAB Symposium*, eds. Bennett, K. & Winkler, C., 531.
[21] Paczyński, B. and Rhodas, J., 1993, Ap. J. Lett., **418**, L5.
[22] Katz, J. I., 1994, Ap. J. , 422, 248.
[23] Mészáros, P., & Rees, M. J. 1997, Ap. J., **476**, 232.
[24] Costa, E., *et. al.,* 1997, Nature, **387**, 783.
[25] van Paradijs, J., *et. al.,* 1997, Nature, **386**, 686.
[26] Sahu, K., *et. al.,* 1997, Nature **387**, 476.
[27] Galma, T., J., *et. al.,* 1997, Nature, **387**, 497.
[28] Bond, H. E., 1997, IAU circ. 6665.
[29] Frail, D., A., *et. al.,* 1997, Nature, **389**, 261.
[30] Metzger, M., R., *et. al.,* 1997, Nature, **387**, 878.
[31] Sokolov, V. V., *et. al.,* 1997, in *The 4th Huntsville Meeting.*
[32] Goodman, J., 1997, New Astronomy, **2**, 449.
[33] Katz, J. I., & Piran, T., 1997, Ap. J., **490**, 772.
[34] Waxman, E., 1997, Ap. J. Lett. **485**, L9.
[35] Vietri, M., 1997, Ap. J. Lett, 478. L9.
[36] Mészáros, P., Rees, M. J., & Wijers, A. M. J., 1997, Astro-ph9709273.
[37] Piran, T. 1992, Ap. J. Lett., 389, L45.
[38] Eichler, D., Livio, M., Piran, T., & Schramm, D. N. 1989, Nature, 340, 126.
[39] Narayan, R., Piran, T., & Shemi, A. 1991, Ap. J. Lett., 379, L1.
[40] Phinney, E. S. 1991, Ap. J. Lett., 380, L17.
[41] Davies, M. B., Benz, W., Piran, T., & Thielemann, F. K. 1994, ApJ, 431, 742.
[42] Clark, J. P. A., & Eardley, D. 1977, ApJ, 215, 311.
[43] Waxman, E., 1995 Ap. J. Lett., **452**, 1.
[44] Vietri, M., 1995, Ap. J., **453**, 883.
[45] Waxman, E., & Bahcall, J. N., 1997, Phys. Rev. Lett., **78**, 2292.
[46] Waxman, E., 1995 Phys. Rev. Lett., **75**, 386.
[47] Abramovichi, A., *et. al.,* 1992, Science, **256**, 325.
[48] Kochaneck C. & Piran, T., 1993, Ap. J. Lett., **417**, L17.

Summary of Experimental Results

Albrecht Wagner (wagnera@desy.de)

DESY and University of Hamburg, D 22603 Hamburg

Abstract. The highlights of the experimental results presented at the International Europhysics Conference on High Energy Physics in Jerusalem are summarised. They include neutrino masses and oscillations, tests of the electroweak and strong interactions, and the search for signs of new physics.

1 Introduction

Many beautiful and new results were presented at this conference. Facing the usual and impossible fate of a summary speaker to review 17 plenary talks and 250 talks in parallel sessions, accompanied by more than 800 contributed papers, I have decided to present my personal selection of highlights. These come under four headings: Neutrino Masses and Oscillations, Tests of the Electroweak Interaction, Tests of the Strong Interaction, and New Physics. This summary is therefore far from being complete.

The new results would not have been possible without the outstanding performance of the accelerators at which the experiments were performed and of the experiments themselves. LEP, in spite of a major fire, started data taking at 182 GeV shortly before the conference, yielding already first results from this previously unexplored energy region. HERA, taking an early start this year, has accumulated until now more data than in any of its previous years of operation. These data were in part already analysed in order to see if the excess of events at very large Q^2, reported earlier this year, could be confirmed.

2 Neutrino Masses and Oscillations

From LEP we know the number of light neutrino species, N_ν, with a precision of 0.3%. Assuming the Standard Model value for $\Gamma_{\nu\nu}/\Gamma_{ll}$, the measurement of Γ_{inv}/Γ_{ll} yields a value of [1]

$$N_\nu = 2.993 \pm 0.011.$$

In spite of this impressive accuracy the third neutrino species, ν_τ, remains so far unobserved.

While in the Standard Model $m_\nu = 0$, a very light neutrino mass cannot be excluded. It would be an indication for physics beyond the Standard

Model. The sea-saw mechanism relates the neutrino mass to the mass of the corresponding lepton

$$m_\nu \sim m_l^2/M \ll m_l$$

where M is the scale of new physics. Massive neutrinos, even of low mass, would have major cosmological implications, e.g. in the interpretation of the origin of dark matter.

Direct searches for neutrino masses have led to the following limits [2]:

$$m(\nu_e) < 3.5\,eV,$$

$$m(\nu_\mu) < 170\,keV(95\%\,c.l.),$$

$$m(\nu_\tau) < 18.2\,MeV(95\%\,c.l.).$$

At present new measurements are in progress, attempting a better limit on $m(\nu_e)$ and to eliminate the problem of negative m_ν^2, found in all previous measurements.

The only method to probe neutrino masses in the sub-eV region is the study of neutrino oscillations which occur if neutrinos have a finite mass. Three pieces of evidence exist today hinting at a possible neutrino mass, a mass difference between neutrino species and a finite mixing angle, which could either be vacuum oscillations or mass induced: i) The solar neutrino deficit, ii) the anomaly in the ratio of muons to electrons from atmospheric neutrinos, and iii) the excess of $\overline{\nu}_e$ events observed by the experiment LSND [3]. Other searches for oscillations performed at accelerators - (CHORUS, NOMAD) [4] and KARMEN [5]- and reactors have so far produced no evidence.

In a system of two neutrino generations the oscillation probability P is given by

$$P(\nu_\alpha \rightarrow \nu_\beta) = sin^2 2\theta_{\alpha\beta}\, sin^2(1.27(L/E)\Delta m^2_{\alpha\beta})$$

where $sin^2 2\theta_{\alpha\beta}$ is the two generation mixing angle, $\Delta m_{\alpha\beta}$ the mass difference (in eV), L the baseline (in km) of the measurement and E the neutrino energy in (GeV).

For many years the observed flux of solar neutrinos disagreed with the predictions by the solar model. Table 1 shows the rates measured by the experiments Homestake, SAGE, GALLEX, Kamiokande and Super-Kamiokande (SK).

In recent years the statistical significance of the experiments has increased and different methods of calibration were used to further reduce the systematic errors. The deficit is observed using experimental techniques which differ by more than a factor of 10 in their energy threshold and which therefore are sensitive to neutrinos from different processes. In conclusion, also the recent high statistics rate measurements confirm a substantial deficit in the flux of solar neutrinos. This deficit could be a hint for neutrino oscillations but could also, with a small probability, be due to the solar model and the

Experiment	units	meas. rate	exp. rate	meas/exp	Ref.
Homestake	SNU	2.54±0.21	9.3±1.3	0.27± 0.02	[6]
SAGE	SNU	73.0±10.4	137.0±8.0	0.53± 0.08	[7]
GALLEX	SNU	69.7±7.9	137.0±8.0	0.51± 0.06	[8]
Kamiokande	$cm^{-2}s^{-1}$	(2.80±0.38) × 10^6	(6.62±1.00) × 10^6	0.42±0.06	[9]
Super-K.	$cm^{-2}s^{-1}$	$(2.44^{+0.26}_{-0.11}) \times 10^6$	(6.62±1.00) × 10^6	$0.37\ ^{+0.04}_{-0.02}$	[10]

Table 1. The measured and expected solar neutrino flux. 'SNU' stands for 'solar neutrino units'.

resulting neutrino fluxes, which have however recently been cross checked by helioseismology-inspired calculations.

The Super-Kamiokande collaboration has exploited the time resolution of its detector to study possible day-night differences in the neutrino flux and reported first results [2]. These differences would occur if neutrino oscillations were at the origin of the solar neutrino deficit. Based on 306 days of data taking Super-Kamiokande reported

$$\frac{(D-N)}{(D+N)} = -0.017 \pm 0.026(stat) \pm 0.017(syst)$$

where D and N stand for daytime and night-time rates. Within the experimental accuracy no significant effect is observed. This may change with additional data and an improved systematic error. Today, the interpretation of the solar neutrino deficit in terms of neutrino oscillations is not yet conclusive.

Atmospheric neutrinos, another non-accelerator neutrino source, originate from the interaction of cosmic rays in the atmosphere, in which $\pi^{\pm}$ are produced which subsequently decay via $\mu^{\pm}$, yielding in the end ν_e, ν_μ, $\overline{\nu}_e$, $\overline{\nu}_\mu$. These neutrinos, after having traversed some fraction of the earth, interact in or near the underground detector and are observed as μ-like or e-like events. They are classified according to their energy and direction (downward or upward going neutrinos). For low energy neutrinos ($<$ 10 GeV) one expects the ratio $(\nu_\mu + \overline{\nu}_\mu)/(\nu_e + \overline{\nu}_e)$ to be close to 2, as the decays of the $\pi^{\pm}$ and the daughter $\mu^{\pm}$ yield two $\nu_\mu$and one ν_e. Kamiokande observed that this ratio was significantly smaller than expected, but the overall experimental evidence was inconclusive.

New results have been presented this year by the Super-Kamiokande and Soudan 2 underground experiments. They are expressed in terms of a double ratio R,

$$R = (N_\mu/N_e)_{data}/(N_\mu/N_e)_{MC}$$

where MC stands for the expected ratio. By taking the double ratio, a number of uncertainties cancel. The new results are shown in table 2 and establish that R is significantly smaller than expected. The results will improve further as both experiments continue data taking.

Experiment	$p_l(MeV/c)$	Exposure	R	Ref.
Soudan 2	(e) >150 (μ) >100	2.83	$0.61 \pm 0.14^{+0.05}_{-0.07}$	[11]
Super-K	(e) 100-1330 (μ) 200-1400	20.1	$0.64 \pm 0.04 \pm 0.05$	[2]
Super-K	(e) >1330 (μ) >1400	20.1	$0.60 \pm 0.05 \pm 0.05$	[2]

Table 2. The double ratio R as measured by the experiments Soudan 2 and Super-Kamiokande. p_l denotes the momentum range of the observed charged leptons, the exposure is given in kiloton years and represents a measure for the number of events.

For high energy neutrinos the charged lepton follows the original neutrino direction. The zenith angle ($cos\theta$) of the neutrino and its flight length L can therefore be measured. A deviation from the expected zenith angle distribution could be an indication for neutrino oscillation. The zenith angle distribution as measured by Super-Kamiokande for different sub-samples of events shows that there are fewer μ-like events in the upward direction (corresponding to longer L) whereas the number of downward-going neutrinos agrees with the expectation. Super-Kamiokande has performed a fit in terms of a ν_μ- ν_τ oscillation to the data which yields $\Delta m^2 \sim 5 \cdot 10^{-3}$ and $sin^2 2\theta \sim 1$.

In conclusion, the new data confirm the anomaly in the flux of atmospheric neutrinos and could be due to a neutrino oscillation.

No new results from oscillation searches using neutrinos from reactors were reported this year. Recent accelerator experiments are performed using low-energy (LSND, KARMEN) and high-energy beams (CHORUS, NOMAD). So far, only one of these experiments, LSND, has reported evidence for neutrino oscillation in the $\overline{\nu}_\mu \rightarrow \overline{\nu}_e$ decay-at-rest channel. The KARMEN experiment at present can neither confirm nor disprove the result of LSND. The CERN experiments have recently reported first preliminary results from a search for $\nu_\mu \rightarrow \nu_e$ oscillations. They do not find any evidence for oscillations.

The present status of the search for neutrino oscillations is not conclusive. While the LSND rates, the solar neutrino rate and both rate and angular distribution of atmospheric neutrinos hint at neutrino oscillations, no unique solution has emerged yet. Several mixing scenarios have been proposed, for example the one presented in the summary talk at the last EPS conference

[12]. The conclusion of this analysis is that all existing disappearance data are compatible with a large mixing angle and $\Delta m^2 \approx 10^{-2} eV^2$. A clarification of the question whether neutrinos oscillate will most likely require the next generation of experiments, e.g. long base line experiments.

The total mass present in the universe is more than ten times greater than the luminous mass. Neutrinos with finite mass could be one explanation for the origin of this dark matter. At present a number of astrophysical dark matter searches are being performed [13]. The experiments Macho and Eros looked for an apparent brightening of stars. This would occur, due to gravitational micro-lensing, if a dark matter object, e.g. a brown dwarf, would cross the light path from the star to the observer. Both, Macho and Eros, have observed in total more than 10 such micro-lensing candidates in the Large and Small Magellanian Cloud. This rate indicates however that the missing galactic dark matter can probably not be explained by brown dwarfs.

Other dark matter candidates are axions and weakly interacting massive particles (WIMPs). So far there exists no experimental evidence for these particles, but the detectors have recently been significantly improved and reach now the critical sensitivity.

3 Tests of the Electroweak Interaction

The Standard Model of electroweak interactions has in recent years been tested with an unprecedented accuracy. The newest results were reviewed at this conference [1]. The Z^0 boson mass, a key parameter of the Standard Model, was obtained from a lineshape fit to the data from LEP operating at centre-of-mass energies around 90 GeV

$$M_{Z^0} = 91186.7 \pm 2.0\, MeV.$$

In this fit the γZ interference term is fixed to its Standard Model value. Using also data from LEP2 an S-matrix fit was recently performed without this constraint which yields within errors the same value. The improved error results largely from a better LEP energy calibration ($\Delta E = \pm 1.5\, MeV$). The measurements of the angular asymmetries A_{FB} and A_{LR} of leptonic Z-decays and of the τ polarisation agree with the expectations of the SM, providing a test of lepton universality at the level of 0.2 % for the axial-vector coupling g_A and of 5-10 % for the vector coupling g_V. Combing the data from LEP and SLC and assuming lepton universality the following values for the coupling constants are found:

$$g_{Vl} = -0.03793 \pm 0.00058,$$

$$g_{Al} = -0.50103 \pm 0.00031.$$

Two years ago the largest deviation from the Standard Model predictions was observed in the ratios of the b- and c-quark partial widths of the $Z^0$to the total hadronic partial width, R_c and R_b [14]. Already one year later this discrepancy started to go away [15]. At this conference the results from new high statistics analyses were reported which use improved tagging methods and correct both for gluon splitting into heavy quarks and for hemisphere correlations in the analysis. The new measurements agree well with the Standard Model predictions.

The effective electroweak mixing angle $sin^2\theta_{eff}^{lept} = \frac{1}{4}(1-(g_V/g_A)^2)$ can be determined using the forward-backward asymmetry of leptons and quarks as well as the left-right asymmetry measured with polarised electrons at SLC. The overall average is

$$sin^2\theta_{eff}^{lept} = 0.23152 \pm 0.00023.$$

with a $\chi^2/dof = 12.6/6$. This relatively large value for χ^2 is mainly due to two measurements, $A_{FB}^b(LEP)$ and $A_{LR}(SLD)$ which differ by 3.1 σ. As this is the only larger deviation from the predictions of the Standard Model it should not be taken as a sign for some new physics.

In 1996 LEP ran for the first time at energies above the threshold for W-pair production. Only days before the conference LEP has reached for the first time an energy of more than 180 GeV. The production of fermion pairs at energies well above the Z resonance proceeds frequently through the so-called 'radiative return' to the Z^0 via the emission of a photon in the initital state. Non-radiative events are selected by a cut on the effective c.m. energy of the fermion pair, $\sqrt{s'}$. The measured cross sections and forward-backward asymmetries agree very well with the Standard Model up to energies of 183 GeV.

The production of W^+W^- pairs proceeds either via the γ/Z exchange in the s-channel or a neutrino exchange in the t-channel. The W^+W^- cross-section has been measured at three energies: at 161 GeV, very close to the threshold, at 172 GeV and at 182 GeV. The measured energy dependence of the cross-section proves that all three production mechanisms and their interferences, as incorporated in the Standard Model, are needed to describe the data.

The W-mass is determined at threshold from the cross section which is very sensitive to the mass value, and at higher energies directly from the invariant mass of the decay products. After averaging the mass measurements from the four LEP experiments at 161 and 172 GeV, one obtains $M_W = 80.48 \pm 0.14\, GeV$.

The W-production at the Tevatron proceeds through $q\overline{q}$ fusion. The W-mass is obtained from a fit to the transverse mass of the $l\nu$ system. Combining the results of UA2, CDF, and D0 one obtains $M_W = 80.41 \pm 0.09\, GeV$. The mass measurements from LEP and the hadron collider are in excellent agreement and the combined average is

$$M_W = 80.43 \pm 0.08\, GeV.$$

The results from the precision measurements of Z decays from LEP and SLC and of the W mass from LEP and Tevatron have been used as inputs for a global Standard Model fit. Also included in the fit were a combined measurement from CDF and D0 [16] of

$$M_t = 175.6 \pm 5.5\, GeV,$$

a value of $(1 - M_W^2/M_Z^2)$ from νN scattering [17], and a value for $(1/\ \alpha(M_Z))$ [18]. The χ^2/dof of the fit is excellent, (17/15), underlining that all data are very well compatible with the Standard Model. Only one measurement, $sin^2\theta_{eff}^{lept}$, deviates by −2.4 standard deviations.

The fit can be used to estimate also the mass of the Higgs boson, $m_H = 115^{+116}_{-66}$ GeV, which, expressed as upper limit, corresponds to

$$m_H < 420\, GeV (95\%\, c.l.).$$

In this limit the results from the direct searches, which yield $m_H > 77.1\, GeV$ $(95\%\, c.l.)$ [19], are not taken into account.

A wealth of information has been presented at this conference [20] on the production and decays of heavy quarks. The analyses are based in the case of CLEO on an integrated luminosity of 3.1 fb^{-1}, in the case of LEP and SLD on improved efficiencies and purities of the flavour identification. Of special interest is a new preliminary result from LEP on B_s oscillations, yielding $\Delta m_s > 10.2 ps^{-1}$ at 95% c.l. This limit provides a constraint on the CKM matrix element ratio $\mid V_{ts}/V_{td} \mid > 3.8$ at 95% c.l.

No new results on CP violation were reported this year. Two major experiments, KTeV at Fermilab and NA48 at CERN, are taking data, aiming at a significant improvement of the precision in the measurement of the CP violation in K decays. In a few years the first measurements of the CP violation in B-meson decays is expected.

Rare K-decays are being studied as they offer the possibility to probe the physics beyond the Standard Model. At this conference the first observation of the decay $K^+ \to \pi^+ \nu\bar{\nu}$ was reported [21]. In a sample of $1.49 \cdot 10^{12}$ stopped K^+ mesons one event with a clean signature of this decay was observed, whereas the expected background was estimated to be 0.08 events. Taking this event at face value, one can calculate a branching ratio $BR(K^+ \to \pi^+ \nu\bar{\nu}) = 4.2^{+9.7}_{-3.5} \cdot 10^{-10}$ and $0.006 <\mid V_{td} \mid < 0.6$. More data are needed however to establish this decay firmly.

4 Tests of the Strong Interaction

The inclusive jet cross section as a function of E_T^{jet} as measured at the Tevatron by CDF and D0 in the central rapidity region agrees with the next-to-leading order QCD calculation over 7 orders of magnitude, representing one of the most stringent tests of the validity of QCD [22]. However, for values of E_T^{jet} around 400 GeV the CDF data from runs 1A and 1B lie consistently above the QCD prediction while the data of D0 show no deviation. This observation has caused considerable excitation and was even interpreted as possible indication for quark substructure. A modification of the structure functions, assuming an increased gluon content at high x, was shown to describe the data better. However, a recent experiment on direct photon production [23], which is sensitive to the gluon distribution at high x, favours the parton distribution without enhanced gluon content. D0 has repeated its analysis using the fiducial cuts of CDF and compared the CDF data with the QCD fit to their own data. They find that data and fit agree within the systematic error of D0. It seems therefore that NLO QCD describes the inclusive jet cross section data well up to highest energies and that the disagreement between CDF and D0 is not significant.

When comparing the measured cross section for the b-quark production in $p\bar{p}$ collision as a function of p_T^b with NLO perturbative QCD predictions one finds an agreement in the overall shape of the distribution, but a discrepancy in the overall scale, the data lying a factor of 2.1 ± 0.2 above the theory. The origin of this discrepancy is not understood but could lie in the assumptions about the fragmentation and in the fact that the running of m_b is not taken into account in the calculations.

The DELPHI collaboration has reported at this conference the first evidence for the energy dependence of the b-mass as expected in NLO QCD [20]. The value of $m_b(M_Z)$ has been derived from the measured ratio of three-jet rates in b-quark and light quark events, as these rates have sizeable quark mass corrections. The result is $m_b(M_Z) = 2.67 \pm 0.50$ GeV, which can be compared with the value of $m_b(M_\Upsilon/2) = 4.16 \pm 0.18$ GeV. The observed change in the b mass value corresponds to nearly 3 σ [24].

Nucleon structure functions are a vital ingredient to many QCD predictions. As they cannot (yet) be calculated ab initio, a precise measurement over a wide kinematic range is very important. The unpolarised and polarised structure functions are measured in deep-inelastic scattering (DIS) experiments using fixed targets and most recently the HERA collider [25]. The double differential neutral current DIS cross section is

$$\frac{d^2\sigma(l^\pm N)}{dxdQ^2} = \frac{2\pi\alpha^2}{Q^4 x}[Y_+ F_2(x,Q^2) - y^2 F_L(x,Q^2) \mp Y_- xF_3(x,Q^2)]$$

where $Y_\pm = 1 \pm (1-y)^2$ is a kinematic factor and F_2, F_3 and F_L are the nucleon structure functions, while x, Q^2, y are the kinematic variables used to describe DIS. The kinematic coverage of the recent HERA data extends in x down to nearly 10^{-6} and in Q^2 from 10^{-1} to 10^4. As shown in [25] the new data are well described by NLO QCD fits and the steep rise in F_2 as x decreases persists up to the largest values of Q^2. The extension of accurate measurements down to low x has led to a much better knowledge of the gluon density in the proton. H1 and ZEUS have extracted $xg(x,Q^2)$ by performing a NLO QCD fit to their own as well as fixed target data taken at large x, including a full treatment of systematic errors. They find a good agreement between the experiments and also with the global fits. At HERA a significant contribution to the cross section comes from charm, through photon-gluon fusion, which can be tagged by the D^*, D^0 mesons and their characteristic decays. H1 has used these tagged charm events for a direct determination of the gluon density, finding good agreement with results from the QCD fits.

Using new detectors close to the beam line and by shifting the interaction vertex in beam direction the experiments H1 and ZEUS were recently able to measure F_2 down to $Q^2 = 0.11\,GeV^2$. The data can be compared with two theoretical predictions, GRV [26] and DL [27], and are found to agree with GRV down to $Q^2 = 0.9\,GeV^2$. The DL model fails at large Q^2 but agrees with the shape of the x-dependence of F_2 for $Q^2 < 0.9\,GeV^2$, while lying slightly below the data points. The new precision data from HERA have stimulated the development of additional models, all of which describe the data more or less well. However, it is not yet clear why the NLO QCD evolution works at such low scales.

Deep inelastic scattering experiments with polarised beams and targets probe the spin structure of the nucleon through the measurement of the polarised structure functions. The distribution of the spin among the constituents of the nucleon is given by the relation $\frac{1}{2} = \frac{1}{2}\Delta\Sigma + \Delta g + L_z$, where $\Delta\Sigma$, Δg and L_z are the contributions to the spin from the quarks, the gluons and the parton orbital angular momenta. New data from inclusive measurements were reported at this conference by the experiments E154/5 at SLAC, SMC at CERN and HERMES at DESY [25]. The SMC data do no longer suggest a rise of the polarised structure function g_1^p for $x < 10^{-2}$. For the evaluation of the sum rule integrals NLO QCD fits were used to evolve all data to a common Q^2 and to extrapolate to $x \to 0$. Also the new analyses confirm the previous observation that only approx. 30 % of the nucleon spin is carried by the quarks, that the Bjorken sum rule is satisfied whereas the Ellis-Jaffe sum rules are violated at the level of 2σ. A further improvement in the understanding of the origin of the spin will come from semi-inclusive analyses which are in progress.

The recent results on soft interactions and diffraction were summarised in [28]. Considerable progress was made in the understanding of diffraction in terms of QCD. At HERA detailed studies were made about the internal structure of the colour-neutral object exchanged in diffractive processes, the pomeron. These studies show that 80-90% of the pomeron momentum is carried by gluons. Hard diffraction has also been studied at the Tevatron.

Fragmentation studies have been performed in many experiments yielding consistent results. The mean charged multiplicity n for gluon and quark jets was measured at LEP in 3-jet events, yielding $R = n(g)/n(q) \approx 1.5$, as expected due to the stronger effective coupling of gluons.

A recent re-analysis of the data taken by the CCFR collaboration in deep inelastic neutrino scattering ($\nu_\mu Fe \to \mu X$), using an in-situ beam calibration of the detector, has resulted in a higher value of α_s than quoted previously: $\alpha_s = 0.119 \pm 0.005$. Due to this new measurement the value for α_s as obtained in DIS agrees now very well with the world average [29]:

$$\alpha_s = 0.119 \pm 0.005,$$

where the error is now mainly limited by theoretical uncertainties.

High energy nucleus-nucleus collisions carry the potential to lead to the formation of a quark-gluon plasma (QGP), as theoretically predicted. In recent years many experiments have been performed searching for QGP signatures [30]. J/Ψ suppression has been proposed as an unambiguous signature by [31], and was recently observed in $Pb + Pb$ collisions at CERN [32]. At this conference new results were reported from $Pb + Pb$ collisions at 158 GeV/nucleon [33]. In this experiment a fit is performed to the $\mu^+\mu^-$ spectra including contributions from J/Ψ, Ψ' and Drell-Yan. When plotting $B_{\mu\mu}\sigma(J\Psi)/\sigma(Drell - Yan)$ as function of L, the geometric path length in the nuclear medium, one observes a strong J/Ψ suppression above a certain L, indicating possible threshold effects. This effect is interpreted as strong evidence for a collective parton behaviour in nucleus-nucleus collisions.

5 New Physics?

A wide spectrum of searches for new particles and for physics beyond the Standard Model has been performed, unfortunately without success. The most recent results have been reviewed in [19]. The greatest excitement during this year came from an excess of events observed in deep inelastic scattering at the HERA collider at a centre-of-mass energy of $\sqrt{s} = 300$ GeV and at Q^2 exceeding M_Z^2, a kinematic region previously unexplored [34, 35]. This observation has stimulated a large number of papers on the interpretation of the effect [36] in terms of signals of new physics, like leptoquarks or supersymmetric particles. The published results were recently updated, including the data taken during the first half of 1997 [37].

The selection of DIS events at very high Q^2 proceeds through very few cuts only. The selected events are strikingly clean and can neither be missed nor mis-interpreted, the remaining background is far less than 1 event. Comparing the distributions of the reconstructed events in the kinematic variables x, y, Q^2 with the Standard Model expectations one observes in general a good agreement, especially at low Q^2. However, at high Q^2, typically above 10000 GeV^2, an excess of events above expectation is observed in both experiments, as shown in table 3. The table gives also the Standard Model predictions. Both collaborations have estimated the error on the Standard Model prediction due to uncertainties in the knowledge of the structure functions to be about 6.5%, and the total error on the prediction to be around 10%. The excess reported previously remains with the new data, but did not become more significant.

	$\int Ldt\ [pb^{-1}]$		N ($Q^2 > 15000\ GeV^2$)
H1	14.2 + 9.5 = 23.7	observed	12 + 6 = 18
		expected	4.7 + 3.3 = 8.0 ± 1.2
ZEUS	20.1 +13.5 = 33.5	observed	12 + 6 = 18
		expected	8.7 + 5.8 = 14.5 ± 1.1

Table 3. Number of observed and expected events from H1 and ZEUS for $Q^2 > 15000\ GeV^2$. The first number corresponds to the published data, taken until the end of 1996, the second to the new data taken until June 1997.

The distribution of the data in x has been studied by both experiments to see if the observed deviation is clustered around some particular value of x or an equivalent mass $M = \sqrt{xs}$. This analysis has shown a certain clustering of events in H1 in a mass range around 200 GeV (or x around 0.45), whereas the excess in ZEUS occurs at larger values of x: For $x > 0.36$ and $y > 0.4$ H1 observes (7 + 1) = 8 events in the (old + new) data, while ZEUS for $x > 0.55$ and $y > 0.25$ finds (4 + 1) = 5 events. As the largest deviation occurs in both experiments in different kinematical regions it is unlikely that the effect is caused by one single narrow resonance.

In conclusion it is clear that at present a statistical fluctuation as origin of the effect cannot be excluded. Therefore HERA has embarked on a concentrated effort to increase the integrated luminosity in the hope to shed more light on this question within the next few years.

6 Conclusion

Many new and exciting results have been presented at this conference. The Standard Model wins again: its key parameters are measured with an unprecedented accuracy of mostly better than 0.1 %. This precision allows significant tests of the electroweak radiative corrections: the mass of the top as derived from a global fit agrees well with the new direct measurement of m_t. From the same fit an upper limit on the Standard Model Higgs mass has been derived. Previous indications for deviations from the Standard Model have disappeared.

Nevertheless, some new results might be signs of a physics beyond the Standard Model. The new results on the anomaly in the flux of atmospheric neutrinos as measured by Super-Kamiokande can be interpreted as sign of a finite, yet very small neutrino mass. An excess of events at very large Q^2 has been observed at HERA. The statistical significance of the effect however is not big enough to establish a deviation from the Standard Model predictions. Experimenters have hunted in vain so far for first signs of a Higgs or a supersymmetric particle.

In summary, experiments of unprecedented precision have confirmed a theory which for many reasons cannot be considered as the 'theory of everything'. In order to know how to proceed from here new experimental observations are needed, those hoped for and those unexpected. They might come from LEP, HERA, the Tevatron, from the cosmic accelerators or from one of the facilities under construction. It might also be that we will have to wait for the start of the LHC. Exciting times lie ahead of us.

Acknowledgement

I would like to thank the organisers of the EPS Conference, especially the Chairmen E. Rabinovici and G. Mikenberg, for their hospitality in a town of special flair, the scientific secretary of the conference, D. Lellouch, and my scientific secretary, S. Robins, for their untiring help and patience and the many speakers for their assistance in preparing this talk. I would also like to acknowledge the help of S. Petzold in the preparation of this paper.

References

[1] D. Ward, plenary talk PL 15, these proceedings
[2] M. Nakahata, plenary talk PL 11, these proceedings
[3] C. Athanassopoulos et al., Phys. Rev. C54, 2685 (1996); nucl-ex/9706006
[4] L. Ludovici, talk contributed to the parallel session PA 10,
V. Valuev, talk contributed to the parallel session PA 10.
For a review of preliminary results from CHORUS and NOMAD see: A. Rubbia, in Proceedings of the 18th International Symposium on Lepton and Photon Interactions, Hamburg 1997, Editors A. DeRoeck and A. Wagner, to be published by World Scientific.
[5] B. Armbruster et al., talk contributed to the parallel session PA 10
[6] B.T. Cleveland et al., Nucl. Phys. (Proc. Suppl.) B38 (1995) 47
[7] J.N. Abdurashitov et al., Phys. Lett. B328 (1994) 234
[8] W. Hampel et al., Phys. Lett. B388 (1996) 384
[9] K.S. Hirata et al., Phys. Rev. D44 (1991) 2241
[10] T.K. Gaisser et al., Phys. Rev. D54 (1996) 5578
[11] W.W.M. Allison et al., Phys. Lett B391 (1997) 491
[12] D.H. Perkins, Proceedings of the International Europhysics Conference on High Energy Physics, Brussels 1995, World Scientific, Editors J. Lemonne et al. and
P.F. Harrison et al., Phys. Lett. B349 (1995) 137
[13] G. Raffelt, plenary talk PL 8, these proceedings
[14] A. Olchevski, Proceedings of the International Europhysics Conference on High Energy Physics, Brussels 1995, World Scientific, Editors J. Lemonne et al.

[15] A. Blondel, Proceedings of the 28th International Conference on High Energy Physics, Warsow 1996, World Scientific, Editors Z. Ajduk and A.K. Wroblewski
[16] A. Yagil, plenary talk PL 16, these proceedings
[17] J. Yu, talk contributed to the parallel session, these proceedings
[18] S. Eidelmann, F. Jegerlehner, Z. Phys. C67 (1995) 585
[19] P. Janot, plenary talk PL 17, these proceedings
[20] M. Feindt, plenary talk PL 7, these proceedings
[21] C. Jarlskog, plenary talk PL 5, these proceedings
[22] H.J. Weerts, plenary talk PL 1, these proceedings
[23] H. Schellman, in Proceedings of the 18th International Symposium on Lepton and Photon Interactions, Hamburg 1997, Editors A. DeRoeck and A. Wagner, to be published by World Scientific.
[24] P. Abreu et al., CERN-PPE/97-141 (1997)
[25] R. Devenish, plenary talk PL 3, these proceedings
[26] M. Glueck, E. Reya, A. Vogt, Z. Phys. C48 (1990) 471; C53 (1992) 127; C67 (1995) 433
[27] A. Donnachie, P.V. Landshoff, Phys. Lett. B296 (1992) 227; Z. Phys. C61 (1994) 139
[28] R. Eichler, plenary talk PL 2, these proceedings
[29] S. Catani, in Proceedings of the 18th International Symposium on Lepton and Photon Interactions, Hamburg 1997, Editors A. DeRoeck and A. Wagner, to be published by World Scientific.
[30] F. Plasil, plenary talk PL12, these proceedings
[31] T. Matsui and H. Satz, Phys. Lett. B178 (1986) 416
[32] M.C. Abreu et al. (NA50 Collaboration), Nucl. Phys. A610 (1996) 404
[33] F. Ohlsson-Malek, talk contributed to the parallel session, these proceedings
[34] H1 Collaboration, C. Adloff et al., Z. Phys C74 (1997) 191.
[35] ZEUS Collaboration, J. Breitweg et al., Z. Phys C74 (1997) 207
[36] For a recent summary see: J. Ellis, CERN-TH-97-356 (1997)
[37] E. Elsen, plenary talk PL 26, these proceedings

New Directions in Particle Theory*

Gabriele Veneziano, (venezia@nxth04.cern.ch)

Theoretical Physics Division, CERN, CH 1211 Geneva 23

Abstract. A personal selection of theoretical sopics, and *not* a summary, is presented.

1 Particle theory 1967-1997

Particle theory, the subject of this talk, has evolved in an incredible way during the thirty or so years of my own scientific life. It has grown in complexity, diversity, and mathematical sophistication, much beyond what I could ever have expected. Progress has been enormous, as exemplified by the overwelming success of the Standard Model. Areas of particle theory have a growing (excessive?) role in interpreting, suggesting (biasing?) particle-physics experiments. Other areas are daring beyond imagination, look somewhat (too?) pretentious, and, to some of us, somewhat (too?) remote from the real world.

A corollary of what precedes is that the field is hard to follow in its entirety. The biased sample of subjects I will discuss below should give an idea of what I mean. Fortunately, the organizers have limited to 45' the length of my talk (further squeezed by a marriage ...) and to 10 pages its write-up. In order to meet these space–time constraints I will apply three theoretical "cuts": familiarity, novelty, interdisciplinarity.

Can we see a common denominator in the trends characterizing modern particle theory? I claim that there is at least one: *the idea and the use of effective theories*. Let me explain.

In spite of its astonishing successes, most theorists no longer regard Quantum Field Theory (QFT), our present tool to check particle data against theory, as an internally consistent, final framework, mostly because of its ultraviolet (UV) divergences. Although we have learned how to live with renormalization, infinities are conceptually highly unsatisfactory and, from a more practical viewpoint, limit predictivity. All this without mentioning the fact that we do not know how to quantize General Relativity, the classical theory of gravitational phenomena. My own attitude on this is summarized as:

---

* Dedicated to the memory of Vladimir Gribov, who inspired much of my early research, and whose unsurpassed passion for physics is still so alive in my memory.

**The final theory must be finite,
but a finite theory is not necessarily the final one.**

As shown by S. Weinberg in his recent books on QFT [1], the familiar renormalizable quantum field theories emerge anyway as the only possible *low-energy effective theories* that correctly incorporate the constraints of special relativity and quantum mechanics. Gauge theories are needed to describe spin 1 particles, general-covariant theories for including spin up to 2. The relevant questions in particle theory then become:

- Which is the effective theory that describes a particular class of HEP phenomena?
- Which are the symmetries that determine its dynamics?
- Which is the structure (degeneracy, symmetry, etc.) of its ground state?
- Which are its relevant/irrelevant degrees of freedom?
- Can we integrate out the latter and derive the effective dynamics of the former?
- Can we solve that effective dynamics and make contact with the data?

My presentation will illustrate what precedes through specific examples taken from gauge theories of non-gravitational phenomena (Section 2), from theories of gravity/cosmology (Section 3), and from the most recent attempts to a Grand-Synthesis (Section 4).

2 Gauge theories of non-gravitational phenomena

A renormalizable chiral gauge theory, known as the Standard Model (SM), appears to describe well all non-gravitational phenomena. Nonetheless, the SM is still facing difficult, unsolved theoretical problems, such as the large-distance behaviour of QCD, or the nature of the EW phase transition at finite temperature. The way the theoretical effort is being directed these days is shown in Fig. 1, in which the vertical axis denotes the amount of supersymmetry in the gauge theory under study. Two arrows order points on that axis. The "relevance" arrow goes upwards, from $N = 4$ to $N = 0$, the non-supersymmetric case, while the arrow of "tractability" goes in the opposite direction: formal theorists love supersymmetry because it makes theoretical analysis much easier. Unfortunately, even if there are indications that Nature too likes SUSY, this must be largely broken, and the relevance to the real world of understanding SUSY dynamics is still to be proved. My presentation in this section will follow the direction of increasing (decreasing) tractability (relevance).

2.1 $N = 0$

I picked up five subjects, according to the criteria I explained and will briefly comment on them in turn.

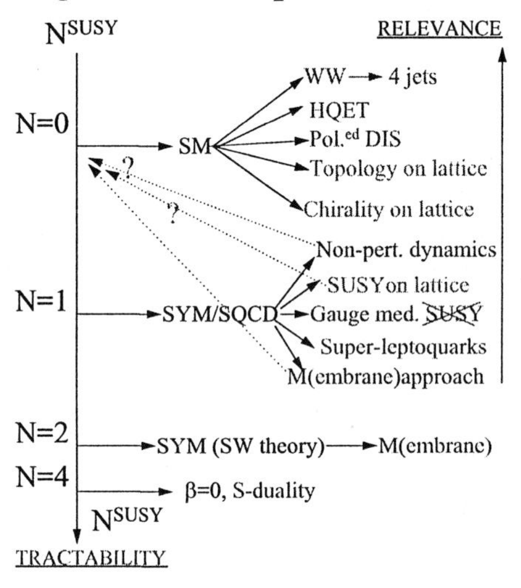

- $WW \to 4$ jets.
  This is an example of a theoretically interesting problem with a nice practical application, the reconstruction of M_W, by measuring the four jets of the hadronic decay of a pair of W's produced near threshold. Looks trivial. Why is it not?
  The problem (called "colour reconnection" or "exogamy") has to do [2] with the impossibility (in principle) to attribute unambiguously each one of the jets to a well-defined parent (W). The theoretical analysis is non-perturbative and thus not obvious. What I find amusing is that, while the no-reconnection diagrams should lead to two ordinary, back-to-back, single-W jets, contributions with colour exchange should give two *boosted* jets. Furthermore, the latter process resembles a 25 year old model for the Pomeron, actually a possible starting point of Gribov's Reggeon field theory. Will we learn something about the Pomeron from W physics and the other way around? That would be amusing!
- Heavy-quark effective theory HQET
  The theory of heavy-quark systems is a beautiful example of the use of effective theories in the sense explained in the introduction. It can be

compared with the effective theory of very light quarks, i.e. with good old chiral effective Lagrangians/perturbation theory (still an active field, incidentally). The basic point of the HQET is similar to the use of the operator product expansion (OPE) in deep inelastic lepton–hadron scattering. Through OPE, after using asymptotic freedom, the effective dynamics is reduced to evaluating the matrix elements of several composite operators. If the same matrix elements appear in different processes, testable relations can be derived. For lack of time I will refer you to the excellent account given by M. Neubert [3].

– Polarized DIS

OPE is, of course, the basic tool in this game. It allows the expression of the first moment of the polarized structure function of the nucleon in terms of matrix elements of various axial-vector currents. Non-singlet axial currents show no surprises (e.g. the Bjorken sum rule is well satisfied). However, flavour-singlet matrix elements appear to be suppressed (by a factor of 0.5 or so) relative to the naive (OZI–EJ) expectation. Before claiming a "crisis", we should remember that, while the Bj sum rule is a consequence of QCD, the Ellis–Jaffe sum rule involves extra assumptions similar to those that would predict a light η' particle.

The origin of this suppression may lie in the universal screening of *topological* charge in QCD, something analogous to the screening of *electric* charge in QED. Both are purely "t-channel" effects and thus "target"-independent. The idea can be tested experimentally, by comparing various semi-inclusive polarized DIS spectra in the target fragmentation region. Although the experiment looks tough, it is being considered [4] in the eventuality that HERA decides to go for polarization.

– Topology on the lattice

For a long time the belief has been that topological charge fluctuations in the QCD vacuum have to do with the large η' mass (relative to the non-singlet pseudoscalar mesons). As just stated, topological charge screening could be the mechanism by which (axial-vector) flavour-singlet matrix elements are suppressed (relative to their naïve relation to the non-singlet). This explains the interest of the lattice community in getting hold of the QCD topological charge. The report by Di Giacomo [5] in a parallel session shows that continuous progress is being made on this important but difficult issue.

– Chirality on the lattice

Formulating chiral gauge theories (e.g. the Standard Model!) on the lattice is a notoriously difficult problem. Some interesting ideas have appeared though, e.g.,

(i) by introducing a fifth dimension chiral fermions can be put chiral fermions on two hyperplanes (ordinary 4-D space–times) a distance d_5 apart. The problem is that, to recover the desired chiral theory, d_5 should be sent to ∞.

(ii) Another method consists of using separate lattices, with different lattice spacing, for fermions and for bosons. It is still early to assess the value of this interesting idea.
(iii) Quite apart from the problem of chiral gauge theories, the domain-wall idea described in (i) can be used to better study the chiral limit of vector-like theories such as QCD (see A. Soni's report [6]).

2.2 $N = 1$

For lack of time I will confine myself to just three of the topics shown in Fig. 1.

- SUSY dynamics the old way
  Following my own guidelines I will be very brief here; just recall that there have been two periods of intense activity on this subject even before the new wave described in the next subsection.
  The first period (roughly 1981–85, see [7] for a review) saw the combined use of Witten-index arguments, of effective Lagrangians, and of supersymmetric instanton calculations, to provide a consistent and detailed understanding of the ground state(s) and low-energy excitations of several SUSY gauge theories. These include the so-called super Yang–Mills (SYM) case (just gluons and gluinos), the supersymmetric version of QCD (so-called SQCD) with different numbers of "colours" and "flavours" and different quark–squark mass matrices, and certain chiral theories (such as the supersymmetric Georgi–Glashow $SU(5)$) exhibiting dynamical supersymmetry breaking.
  After a gap (probably a "superstring revolution" effect) the second period (roughly 1990–96, and mainly due to N. Seiberg and his collaborators) led to a more rigorous understanding of the consequences of holomorphy and to the use of strong–weak coupling duality. Considerable progress was thus made on SQCD at $N_f/N_c > 1$ or for other gauge groups (e.g. $Sp(n)$ and $O(n)$). Of particular importance was the discovery, by N. Seiberg [8], of new (electric–magnetic-type) dualities between the infrared limit of pairs of distinct SQCD theories. New possible phases of $N = 1$ theories have thus been uncovered and extended to various dimensions.
- SUSY dynamics the new way
  This new approach to SUSY dynamics uses string (M)-theory as a tool, not as a basic, more fundamental theory. This is analogous to the use of string-loop techniques made in the past in order to simplify the book-keeping/calculation of Feynman diagrams. Let me try to give the flavour of the idea.
  The older part of this audience is certainly accustomed to thinking of $SU(2)_I$ (I for isospin) as rotations in some "internal" three-dimensional space. In the M (membrane) approach to gauge dynamics, there are also

extra internal coordinates, but the game is different: some extended objects, called membranes, occupy ordinary space–time as well as (parts of) the internal space. There are gauge and matter fields propagating and interacting on these high-dimensional membranes. A given multi-membrane configuration thus defines a given gauge theory. For instance, the number and relative orientation of the membranes in the extra dimensions define the rank of the gauge group, the amount of supersymmetry present, the number of matter fields, the free parameters of the corresponding field theory (masses, couplings,...), and so on. Since all the membranes contain, as a subspace, ordinary space–time, we may look at gauge dynamics in that subspace (our own world) by new powerful techniques. So far, the construction has led to vector-like gauge theories with at least $N = 1$ SUSY. It will be interesting to see whether, in a similar spirit, non-supersymmetric theories and/or chiral gauge theories can be constructed.

Space–time limitations force me to refer you to Kutasov's talk [9] for a summary of the results so far obtained in this area: I will only mention that Seiberg's duality for various massless SQCD theories is geometrically reinterpreted in M-QCD and that the same is true for the condensates characterizing the ground state of massive SQCD.

– SUSY on the lattice

R. Petronzio and P. Hernandez [10] reported on new interesting attempts to study SUSY dynamics on the lattice. Experts feel that it will soon be possible to study at least the $N = 1$ SYM case using this new technique. This would be very interesting for at least two reasons:

(i) In lattice QCD the numerical results are eventually compared with actual data as a way to test QCD. In the supersymmetric case there are no data to compare with, but there are instead the theoretical expectations found by the analytic methods described in the previous paragraghs. Here the lattice approach actually replaces experiments as a way to check theoretical expectations!

(ii) Any lattice regularization explicitly breaks supersymmetry (after all it breaks translational invariance). SUSY can only be recovered in the continuum limit, after some parameters are suitably tuned. I see this as an advantage rather than a disadvantage. Indeed, by varying these lattice parameters, one can switch on and off SUSY-breaking terms. For instance, in SYM theory the supersymmetric limit corresponds to a vanishing gluino mass, while the opposite limit, a very heavy gluino, should give back the already much studied pure-glue theory (cf. glueball mass calculations). It would be interesting to see also how switching on a gluino mass affects the results, e.g., how the gluino condensate of SYM goes over to the gluon condensate of YM as the gluino mass is increased beyond the confinement scale.

2.3 $N > 1$

So much has already been said on the celebrated Seiberg–Witten solution [11] of $N = 2$ SYM theory (and generalizations) that I will only add the statement that the new (M)-approach can also be used to give some very nice geometrical interpretation to the SW surface: this emerges particularly nicely in the 11-dimensional (M-theory) embedding of string theory.

I should also mention very recent work by Witten [12], who makes a first attempt to use M(embrane) techniques to understand confinement and the appearance of string-like flux tubes in ordinary (non-supersymmetric) QCD.

3 Gravity/Cosmology

This field was reviewed very nicely in the plenary talks by T. Piran and G. Raffelt [13]. Here I will limit myself to a few topics.

3.1 $\mathbf{B}_{cosmic}$, $\mathbf{B}^{Y}_{cosmic}$

The origin of cosmic magnetic fields (of about μG intensity) on galactic or supergalactic scales has long been a mystery. Even if a cosmic-dynamo mechanism can be invoked as an amplifier of tiny primordial fields, a minimal amount of "seeds" is needed and even that is very hard to achieve. Various unconventional mechanisms have been proposed, including some that violate gauge invariance in the early Universe. In my opinion, the most appealing proposal is the one by Ratra [14], in which a scalar field coupled to $F^2_{\mu\nu}$ amplifies electromagnetic vacuum fluctuations as it evolves during the early Universe. Quite amazingly, this mechanism finds a natural realization in the so-called pre-big bang scenario of string cosmology, which I will briefly mention below.

Quite apart from the question of their origin it is important to understand the impact of cosmic magnetic fields on various cosmological processes such as primordial nucleosynthesis and, possibly, baryogenesis. The success of standard nucleosynthesis implies upper bounds on the amount of energy stored in magnetic fields (cf. nucleosynthesis bounds on additional light neutrinos). On the other hand, a primordial "electro-magnetic" field in the $U(1)_Y$ subgroup of the SM, with a non-vanishing $\mathbf{E} \cdot \mathbf{B}$, can lead to B- and L-number violation through the Adler–Bell–Jackiw anomaly and thus, potentially, to a new mechanism for baryogenesis. Work is in progress [15] to fully explore this possibility.

3.2 Topological defects vs. Doppler

An interesting development has recently taken place [16] on distinguishing competing models of large-scale structure formation on the basis of their predictions for the so-called Doppler peaks (peaks in the multipole expansions of $\Delta T/T$ correlations around $l = 200$). Recent numerical calculations have shown that the Doppler-peak structure predicted by topological-defect models (in particular cosmic strings) is close to being excluded by present data, which instead favour standard-inflation's predictions. New data will soon be available to further confirm (or contradict) this indication.

3.3 Reheating

It is well known that during inflation the Universe cools by an enormous factor, simply as a consequence of the enormous red-shift. It is believed that the Universe gets reheated by the decay of the inflaton as it relaxes to its present value. If this decay proceeds very quickly all the potential energy stored in the inflaton at the end of inflation is converted into heat/temperature. If, instead, the decay proceeds slowly, the final temperature is further red-shifted and may end up being insufficient to ignite processes such as primordial nucleo-synthesis. New theories [17] of reheating, known as pre-heating or resonant decay, take advantage of an amusing coherence effect that enhances the decay rate (as in a stimulated emission of radiation in a LASER) and can thus produce higher final temperatures. This field is now being actively investigated by several groups.

3.4 Black holes and D-branes

Much theoretical activity is still being devoted to the connection between black holes and the D-branes we have already encountered in the previous section. Let me recall you that, after some pioneering work by J. Bekenstein, attributing to a black hole an entropy proportional to the area of its horizon $A = 4\pi R^2 = 4\pi(2GM)^2$, Hawking's quantum analysis of black-hole radiation and temperature led to the famous Bekenstein–Hawking (BH) formula for the entropy of a Schwarzschild (black hole) reads (with $c = 1$):

$$S = \frac{1}{4}\frac{A}{G\hbar} \Rightarrow T^{-1} = dS/dM = (8\pi GM)/\hbar \,. \tag{3.1}$$

In quantum statistical mechanics entropy is associated with the (logarithm of the) number of microscopically distinct states sharing the same macroscopic properties. So far, it had not been possible to construct an explicit model of a quantum mechanical black hole in which the above identification did work. Since a few years the situation has changed: certain D-brane configurations have all the characteristics to be identified with (some type of) black holes, for which there exist equivalents of the BH entropy formula. At the same time the number of D-brane states can be evaluated quite precisely. The connection between entropy and density of states turns out to work perfectly well. After this original discovery several other desirable properties of quantum black holes have been rederived by using string-theory [18] techniques on D-branes, and so far no contradiction has been found. This leads to the hope that the information-loss paradox of black-hole physics can eventually be resolved in a complete quantum theory of gravity such as string (or M-) theory.

A related interesting question concerns the ultimate fate of black holes as they keep evaporating. If we trust Hawking's results down to arbitrarily small black-hole masses, we end up in a sort of naked singularity. String theory appears to provide a solution also to that kind of puzzle [19]. I will explain this by using what is known as the Beckenstein bound (BB) on entropy [20] (not a rigorous theorem, yet a bound respected by all known physical systems). The BB tells us that the entropy S of a system of mass M and size d satisfies, apart from factors $O(1)$, the bound:

$$S \leq S_{BB} \equiv \frac{2\pi \cdot d \cdot M \cdot c}{\hbar} \,. \tag{3.2}$$

Black holes saturate this bound, with $d \sim GM$, thus with $S \sim M^2$. Elementary strings have an entropy proportional to M (i.e. to their length) rather than to M^2. Thus, for large enough M, they easily fulfill the bound. On the other hand, at sufficiently small M, there is a potential danger vis a vis the BB. The danger is avoided by interpreting Eq. (3.2) as a *lower* bound on d, $d > l_s$, where l_s is the fundamental scale of string theory. The conclusion is that, in string theory, black holes whose Schwarzschild radius is smaller than l_s do not exist [21] since they are not inside their Schwarzschild radius. Thus, as a black hole evaporates down to a mass M_{Pl}/g, where g is the string coupling constant, it makes a transition to a non-collapsed perturbative string state, which eventually decays into radiation and smartly avoids the singularity problem. Precisely at $M = M_{Pl}/g$, the Hawking temperature and entropy of the black hole coincide with those of an elementary string state [19], and the Hawking temperature coincides (up to a numerical factor) with the Hagedorn temperature of string theory.

3.5 Before Bereshit

"Bereshit" is, of course, the first (Hebrew) word in the Bible. This is actually my favourite subject of research these days, a new cosmological scenario, called pre-big bang, in which the origin of the Universe and its inflationary phase goes much farther back in time than the big bang itself. A characteristic is that, during pre-big bang inflation, the coupling constant grows, which can lead to large enough seeds for the cosmic magnetic fields discussed above. Another consequence of this scenario is the prediction of a characteristic stochastic background of gravitational waves, which should surround us today like the CMB does. It would be too long to go into further detail here and thus I rather refer you to a very complete home page, now available on the subject [22].

4 A grand synthesis?

4.1 String unification and M-theory

This subject, reviewed by P. Binétruy [23] and it represents a large fraction of what is coming out of the more formal side of theoretical physics these days.

The main point about M-theory is that ... it should exist! What it actually is, is a different story, clouded in mystery (hence the M?). A possible answer could be what is described in the next subsection. A few things, however, are believed to be firmly established about M-theory [24].

- It is defined in a world with $10 + 1$ space–time dimensions just as good old 11-dimensional supergravity and, indeed,
- The low-energy limit of M-theory *is* 11-dimensional supergravity (11 dimensions is believed to be the largest number in which supergravity can be consistently defined).
- Different compactifications from 11 to 10 dimensions (with a small 11th dimension) lead to all five known consistent 10-dimensional superstring theories at weak couplings. The coupling, which in string theory is controlled by the value of a scalar field, the dilaton, is related, in M-theory, to the size of the 11th dimension with small (large) coupling corresponding to small (large) size. Thus the strong-coupling limit of all superstring theories should be related to a truly 11-dimensional theory.
- Theories with chiral fermions can emerge from non-trivial compactifications [25]. For instance, $E_8 \otimes E_8$ heterotic-string theory corresponds to the 11th dimension being a segment, with the two E_8 living at its ends (i.e. at the two fixed extreme values of x_{11}).

M-theory has been used to further improve the analysis of supersymmetric gauge theories by D-brane techniques. It turns out that going from 10 to 11 dimensions, i.e. going from string to M-theory, helps incorporating the non-perturbative quantum phenomena that characterize the ground states of these theories.

Another physically interesting consequence of M-theory is that it may help solving a long-standing problem with string unification of all forces. In short, without M-theory ideas, string unification of gravitational and gauge interactions occurs at a scale which is close, but definitely above, the one at which supersymmetric unification of gauge interactions seems to occur ($\sim 10^{16}$ GeV). As first pointed out by Witten [26], and as discussed by M. Quiros at this conference [27], such a conclusion can be avoided in the strong-coupling situation described by M-theory.

4.2 M(atrix) theory

There are some educated guesses about what M-theory might actually be. The most popular suggestion [28] is that it is related to the large-N limit of some Matrix model (hence the funny M(atrix)) in two-dimensions. This is still, however, a matter of speculation.

4.3 Reflections on high energy theory's 1997 vintage

I find the new directions in particle theory **exciting, fascinating, and ... frightening**. Sometimes I wonder whether the "hard" theorists are solving the puzzles of Nature or just some self-posed problems. Is the field physics- or mathematics-driven?

To be sure the problem is not just with theorists' choices: if anything, it has to do with the enormous success they had in the recent past (see the EPS prize attributed to Brout, Englert and Higgs). The Standard Model works so amazingly well that the real crisis is a ... lack of healthy crises. Previous hints of new physics seem to have faded away once more in Jerusalem (as they did in Warsaw a year ago). If neutrino masses/oscillations are true facts, this will be refreshing news. Otherwise, we have no major problems, just some puzzles such as: Who ordered the SM? It is not particularly elegant, with its repeated complex representations of a non-simple gauge group, with its funny-looking CKM matrix and its arbitrary Higgs sector, but it works.

Theorists have tried to make it look nicer by supersymmetrizing it. Did they succeed? My own opinion (I am sorry!) is that they did not. The MSSM is not nicer than the SM. It does solve one problem (stabilizing the Higgs mass), ... creating 10 others (most notably a flavour-conservation problem). Furthermore there is no real hint of SUSY, except for its better performance in Grand Unification or the indirect indications favouring a light Higgs particle. In any case the approval of the LHC (if not an upgraded LEP before) promises to give definite answers to the questions of the nature of the EW-breaking mechanism and of SUSY itself.

However, also, particle experiments/theory is not confined to accelerator physics. Look at the sky! Astroparticle physics, cosmology, gravity have no shortage of crises: they remind me of strong interactions when I started my career ... Both phenomenological and theoretical puzzles abund:

- The Standard Cosmological Model (unlike the SM) is in deep trouble
- Its successor, inflation, is only a paradigm looking for a theoretical framework
- Dark matter, antimatter, a small cosmological constant may be needed/there.
- High-energy cosmic rays, cosmic magnetic fields are mysterious
- Black holes and gravitational waves should exist but are hard to "see"
- Quantum gravity/cosmology still needs a lot of work
- Black holes still pose conceptual puzzles
- The big bang singularity is still a big theoretical headache
- etc.

So, to conclude in the same spirit as in my Warsaw talk last year, my plea is that we should get rid of the historical barriers that still divide accelerator and non-accelerator experiments, gravitational and non-gravitational theories, join our efforts, our human and financial resources, because the chapter on *FUNDAMENTAL PHENOMENA* in the physics textbooks is still far from being written.

References

[1] S. Weinberg, *The Quantum Theory of Fields I and II* (Cambridge University Press, 1995).
[2] D. Ward, Session PL 15.
[3] M. Neubert, Session PL 19.
[4] D. de Florian, G. M. Shore and G. Veneziano, hep-ph/9711353, and references therein.
[5] A. Di Giacomo, Session PA 15.
[6] A. Soni, Session PA 15.
[7] D. Amati et al., Physics Reports **162** (1988) 169.
[8] See, e.g., N. Seiberg, Proceedings of *Particles Strings and Cosmology*, Syracuse, N.Y. 1994 (K. C. Wali, editor).
[9] D. Kutasov, Session PL 10.
[10] R. Petronzio, Session PL 20;
P. Hernandez, Session PA 15.
[11] N. Seiberg and E. Witten, Nucl. Phys. **B426** (1994) 19; **B431** (1994) 484.
[12] E. Witten, hep-th/9706109.
[13] G. Raffelt, Session PL 8; T. Piran, Session PL 22.
[14] B. Ratra, Astrophys. J. Lett. **391** (1992) L1.
[15] M. Giovannini and M. E. Shaposhnikov, hep-ph/9710234.
[16] N. Turok , Ue-Li Pen and U. Seljak, astro-ph/9706250.
[17] L. Kofman, A. Linde and A.A. Starobinsky, Phys. Rev. Lett. **76** (1996) 1011, and references therein.
[18] See e.g. C. Vafa and A. Strominger, hep-th/9601029;
J. Maldacena, hep-th/9607235, and references therein.
[19] M. J. Bowick L. Smolin and L.C.R. Wijewardhane, Gen. Rel. Grav. **19** (1987) 113;
G. Veneziano, talk given at the NATO Advanced Research Workshop: "Hot Hadronic Matter: Theory and Experiment", Divonne-les-Bains (June 1994).
[20] J. Bekenstein, *Phys. Rev.* **D23** (1991) 287.
[21] G. Veneziano, Proceedings of the 1989 Erice Summer School, *The Challenging Questions* (A. Zichichi, editor), p. 199.
[22] http://www.to.infn.it/teorici/gasperini/.
[23] P. Binétruy, Session PL
[24] E. Witten, Nucl. Phys. **B471** (1996) 135.
[25] P. Horava and E. Witten, Nucl. Phys. **B460** (1996) 506.

[26] E. Witten, hep-th/9602070.
[27] M. Quiros, Session PA16.
[28] See e.g. R. Dijkgraaf, E. Verlinde and H. Verlinde, hep-th/9703030, and references therein.

PARALLEL SESSIONS

Part I

Hadron Spectroscopy

Crystal Barrel Results on Two-Body Decays of the Scalar Glueball

Stefan Spanier (stefan.spanier@cern.ch)

University of Zürich, Switzerland

Abstract. The Crystal Barrel Collaboration observes scalar meson resonances in $\bar{p}p$ annihilation. Based on the measurements and partial wave analyses these are candidates for the $^3P_0$ groundstate nonet. The supernummerary $f_0(1500)$ resonance is identified as a scalar groundstate glueball. Important information for its characterization comes from the decay pattern into pseudoscalar and scalar mesons. Data on kaonic decays in the mass region up to 1700 MeV are now avaible at Crystal Barrel. New analysis results are presented.

1 The Crystal Barrel Detector

The Crystal Barrel detector [1] was located at the Low Energy Antiproton Ring (LEAR) at CERN. Antiprotons with a momentum of 200 MeV/c were stopped in a liquid hydrogen target placed in the center of the apparatus. Here the proton-antiproton annihilation occured. Measurement of charged tracks was performed with two cylindrical proportional wire chambers, which could be replaced by a silicon micro strip detector, and a jet drift chamber with 23 sensitive wire layers. A barrel-shaped calorimeter consisting of 1380 CsI(Tl) crystals with photodiode readout and coverig 94% of the solid angle 4π detected photons from the decay of neutral mesons like π^0 and η with a precision in the energy of $2.5\%/\sqrt[4]{E}$. The assembly was embedded in a solenoid providing a homogeneous magnetic field of 1.5 T parallel to the incident antiproton beam.

2 Three Pseudoscalar Final States

Suitable channels to explore the scalar mesons are the three neutral pseudoscalar meson final states. The reaction proceeds in two steps: the scalar resonance is produced together with a recoil particle and in the second step decays into two pseudoscalar mesons. Scalar mesons decaying into two pseudoscalar mesons are dominantely produced from $^1S_0$ initial state of the $\bar{p}p$ -system in liquid hydrogen. Centrifugal barriers hindering the reaction are absent. A large sample of events triggered to include only neutral particles which decay into photons was accumulated. An amount of 712,000 $\pi^0\pi^0\pi^0$, 198,000 $\pi^0\eta\eta$ and 977 $\pi^0\eta\eta'$ events could be reconstructed. To extract the resonance content in these annihilation channels a partial wave analysis was performed using the K-matrix formalism and describing the final states simultaneously [2]. In this analysis model the decay from a proton-antiproton initial state into the three final state particles proceeds successively via intermediate two-body states with a certain spin parity J^{PC}. If the intermediate states are resonant they are parametrized by a mass pole and couplings to the two-body channels. A common feature of the all neutral final states mentioned is the isoscalar scalar resonance $f_0(1500)$. It could be described with a common mass and couplings to $\pi\pi$, $\eta\eta$ and $\eta\eta'$.

The Crystal Barrel collaboration observes the $f_0(1500)$ also in the $5\pi^0$ final state in $\bar{p}p$ annihilation at rest [3]. The selection of this 10 photon final state made use of the all-neutral data sample. The $4\pi^0$ invariant mass shows after subtraction of the $\eta \to 3\pi^0$ events a peak at a mass of 1450 MeV. This structure strongly deviates from phase space distribution. The partial wave analysis explains it as $f_0(1500)$ decaying to $\sigma\sigma$ (σ is a name for the low energy part of the $\pi\pi$ S-wave) with $\sigma \to \pi^0\pi^0$ and either produced together with the $3\pi^0$ resonance $\pi(1300)$ or produced together with $f_0(1370)$ and the low mass tail of a resonance above 1700 MeV in the same S-wave. The σ is interpreted as a glue-rich object.

The $f_0(1500)$ is observed in the decay into K_LK_L by studying the final state $K_LK_L\pi^0$ of annihilation at rest [4]. The π^0 is fully reconstructed, one K_L is missing and one K_L undergoes a hadronic interaction in the CsI(TL) crystals with an average probability of 54%. This is sufficient information to reconstruct the kinematics of an event and to perform a partial wave analysis on the Dalitz plot. The plot is shown in fig. 1 a). The resonance features in the $K\pi$ system are the $K^*(892)$ and the $K^*(1430)$. In the $\bar{K}K$ subsystem isospin I=0 and I=1 is possible. Therefore $f_2(1270)$ and $a_2(1320)$ are seen together. The $f_2'(1525)$ adds to the $I = 0$ $\bar{K}K$ D-wave. A strong contribution of the $\bar{K}K$ S-wave is found. At least two poles in a 1×1 K-matrix are needed in the S-wave to arrive at a satisfactory description of the data. These belong to the I=0 resonances $f_0(1370)$ and $f_0(1500)$. The result stays ambiguous since one also expects the presence of the $a_0(1450)$ resonance which was observed in its $\pi^0\eta$ decay [5]. Any contribution of the $a_0(1450)$ between 0% and 15% does only affect the $\bar{K}K$ S-wave but not the quality of the fit.

To resolve the isospin ambiguity the final state $K_LK^\pm\pi^\mp$ of $\bar{p}p$ annihilation at rest was selected [6]. In the reaction $\bar{p}p \to K_LK^\pm\pi^\mp$ only the $I = 1$ $\bar{K}K$ resonances are produced. By applying isospin symmetry one can calculate their contributions to the $K_LK_L\pi^0$ channel. The $K_LK^\pm\pi^\mp$ final state was reconstructed by requiring a missing K_L and two charged particles, which are identified by dE/dx [6]. An amount of 11,373 events went into the Dalitz plot displayed in fig. 1 b). The branching ratio for the proton antiproton annihilation at rest into this final state is found compatible with earlier bubble chamber determinations on less statistics. The average is: $BR(\bar{p}p \to K_LK^\pm\pi^\mp) = 2.74\pm0.10)\cdot 10^{-3}$ [6]. The total fraction of background from other annihilation channels is below 2%. The partial wave analysis revealed necessity for the introduction of the $I = 1$ resonance $a_0(1450)$. Mass and width were determined as $m = (1480{\pm}30)$ MeV and $\Gamma = (265{\pm}15)$ MeV, respectively. The comparison with the annihilation channel $\pi^0\pi^0\eta$ yields the relative ratio $B(a_0 \rightarrow \bar{K}K)/B(a_0 \rightarrow \pi\eta) = 0.88 \pm 0.23$ which agrees well with the prediction from SU(3) flavour symmetry assuming that this object is member of the scalar nonet. Having determined its contribution the $f_0(1500)$ decay to $\bar{K}K$ can be fixed. The relation is shown in fig. 2. The branching ratio for $f_0(1500)$ from this combined analysis is: $B(\bar{p}p \rightarrow \pi f_0; f_0 \rightarrow \bar{K}K) = (4.52 \pm 0.36) \cdot 10^{-4}$.

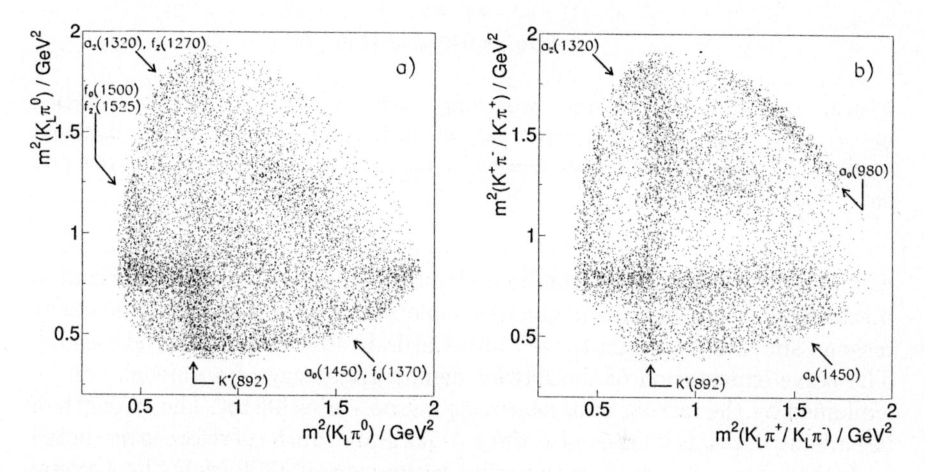

Fig. 1. The Dalitz plots for the annihilation reactions a) $\bar{p}p \rightarrow K_L K_L \pi^0$ and b) $\bar{p}p \rightarrow K_L K^{\pm} \pi^{m} p$ in liquid hydrogen.

3 Summary

The phase space corrected couplings of $f_0(1500)$ are:
$\pi\pi : \eta\eta : \eta\eta' : \bar{K}K = 1 : 0.25 \pm 0.11 : 0.35 \pm 0.15 : 0.24 \pm 0.09$. Due to these couplings and the strong $\sigma\sigma$ decay the $f_0(1500)$ appears as ω-like member of the scalar meson nonet. Other candidates for a nonet in this mass range are the $I = 0$ $f_0(1370)$, $I = \frac{1}{2}$ $K^*(1430)$ and $I = 1$ $a_0(1450)$. The mass of $m = 1505 \pm 9$ MeV would fit but the $I = 0$ nonet position is also occupied by the $f_0(1370)$. The width $\Gamma = 111 \pm 12$ MeV of $f_0(1500)$ is relatively small in comparison to the other scalar mesons having Γ greater than 250 MeV. Hence, it appears supernummerary. The $f_0(1500)$ matches the mass range of lattice gauge calculations for the scalar groundstate glueball [7]. The strong

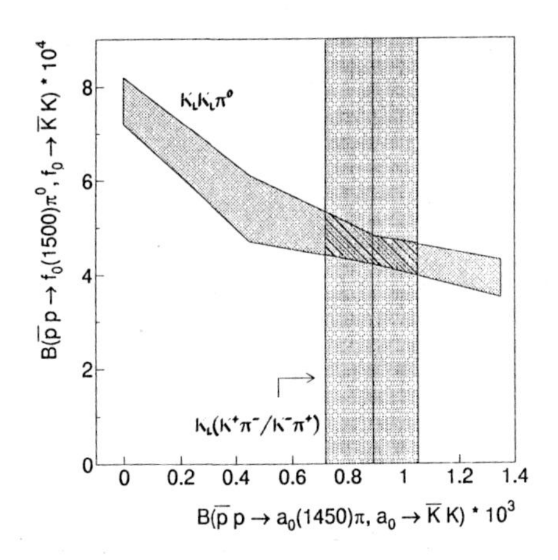

Fig. 2. Determination of the branching ratio of $f_0(1500)$ in the reaction $\bar{p}p \to K_L K_L \pi^0$ [4] (light hatched area), where it is correlated with the $a_0(1450)$ production. The $a_0(1450)$ contribution is fixed in the analysis of the reaction $\bar{p}p \to K_L K^{\pm} \pi^{m} p$ [6].

coupling to $\eta\eta$, $\eta\eta'$ and $\sigma\sigma$ can be understood in terms of the decolorization mechanism: The constituent gluons couple to the glue-content of the decay mesons and are color-neutralized afterwards by the exchange of gluons [8]. The naive expectation of the flavour democratic decay of a glueball can be explained by the mixing with nearby $\bar{q}q$ meson states [9][10]. The strength of the mixing depends on the mass of the ϕ-like state which therefore is predicted at higher mass. To explore the mass region above 1700 MeV the Crystal Barrel Collaboration is analyzing annihilation channels at higher momenta of the incoming antiprotons.

References

[1] E.Aker et al. (CBAR), NIM A321 (1992), 69
[2] C.Amsler et al. (CBAR), Phys. Lett. B355 (1995), 425
[3] A. Abele et al. (CBAR), Phys. Lett. B380 (1996), 453
[4] A. Abele et al. (CBAR), Phys. Lett. B385 (1996), 425
[5] C.Amsler et al. (CBAR), Phys. Lett. B333 (1994), 277
[6] A. Abele et al. (CBAR), submitted to Phys. Rev. D
[7] G.S.Bali et al., Phys. Lett. B309 (1993), 378
J.C.Sexton et al., Phys. Rev. Lett. 75 (1995), 4563
[8] S.S.Gershtein et al., Z. Phys. C24 (1984), 305
[9] C.Amsler and F.E.Close, Phys. Rev. D53 (1996), 295
[10] F.E.Close, G.R.Farrar, Z.Li, Phys. Rev. D55 (1997), 5749

Study of the Glueball Candidates $f_0(1500)$ and $f_J(2220)$ at LEP

Alison Wright (Alison.Wright@cern.ch)

Rutherford Appleton Laboratory, UK

Abstract. Data taken with the ALEPH detector at LEP have been used to study $\gamma\gamma$ production of the glueball candidate $f_0(1500)$ via its decay to $\pi^+\pi^-$. No events were observed in excess of background and an upper limit to the $\gamma\gamma$ width of the $f_0(1500)$ has been calculated.

The first observation of $f_J(2220)$ production in hadronic Z decays in the L3 detector is also presented. A signal in the $K_s^0 K_s^0$ channel is observed in the least energetic jet of three-jet events.

1 Introduction

Lattice QCD has predicted the lightest scalar glueball (0^{++}) to be in the 1.5 GeV region [1], while the tensor glueball (2^{++}) may have a mass of around 2.2 GeV . Experimental observations have revealed two particular glueball candidates to match these predictions: the Crystal Barrel collaboration has observed the $f_0(1500)$ resonance in the gluon–rich environment of $p\bar{p}$ interactions at rest [2]; the BES collaboration has observed $f_J(2220)$ [formerly $\xi(2230)$] production in radiative J/ψ decays [3]. If J=2 for the $f_J(2220)$ resonance it would be consistent with being a glueball, and the CLEO collaboration recently published a stringent limit on its two–photon width [4].

2 $f_0(1500)$ in $\gamma\gamma$ events

If the $f_0(1500)$ were a pure glue state, it would have negligible coupling to photons and so no signal for the process $\gamma\gamma \to f_0(1500)$ would be seen. However, various models suggest mixing of the bare glueball with nearby $q\bar{q}$ nonet states [5]. In particular, the $f_0(1500)$ could be mostly glue but with some $q\bar{q}$ admixture, enhancing coupling to photons. A measurement of the two-photon width, $\Gamma_{\gamma\gamma}$, of the $f_0(1500)$ would allow the determination of the quark content of the state and thus of a mixing angle between $q\bar{q}$ and the glueball in this scenario.

The process $\gamma\gamma \to f_0(1500) \to \pi^+\pi^-$ has been studied, using 160.9pb^{-1}of Z^0–peak data taken from 1990 to 1995 with the ALEPH detector [6] at the e^+e^- collider, LEP. The $\pi^+\pi^-$ invariant mass spectrum obtained is shown in figure 1. The background is dominated at low invariant mass by mis–identified, very low energy muons from the $\gamma\gamma \to \mu^+\mu^-$ process.The clear peak in the spectrum above 1 GeV can be identified with the known tensor resonance $f_2(1270)$. Around the 2 GeV region, there is no evidence of any signal that could be associated with the $f_J(2220)$ resonance.

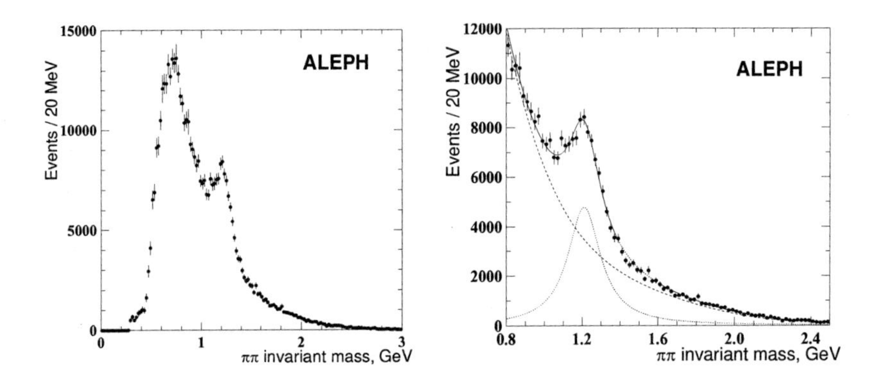

Fig. 1. The invariant mass distribution for two-pion final states: data (left) and the fit to data (right) with the Breit-Wigner for the $f_2(1270)$ (dotted line), the polynomial for the background (dashed line) and the combination of these functions (solid line) are shown.

The experimental resolution of the masses of the $f_2(1270)$ and the $f_0(1500)$ resonances is of order 10 MeV . The background spectrum is fitted with a 5th order polynomial. The mass region 0.8 to 2.5 GeV is used in the fit, but with a window from 1.38 to 1.62 GeV excluded ($2 \times \Gamma_{tot}$ of the $f_0(1500)$). The fit to the data is also shown in figure 1: the Breit-Wigner shape for the $f_2(1270)$, the polynomial for background processes and the combination of both these are indicated. The χ^2/d.o.f. for the fit is 1.1.

The background to the $f_0(1500)$ signal is defined by extrapolation of the fitted background spectrum through the excluded mass window of 1.38 to 1.62 GeV . A binned likelihood fit is performed in this window to determine the upper limit on the number of events $\gamma\gamma \to f_0(1500) \to \pi^+\pi^-$. The expected signal is taken to be a Breit-Wigner resonance of mass 1503 MeV and width 120 MeV (although this shape could be distorted by interference effects). The maximum likelihood is seen for the number of signal events equal to -905. The area under the graph was integrated above zero to obtain the 95% C.L. limit on the number of signal events: it is found to be 128 events.

The limit on the number of signal events is used to calculate the upper limit on $\Gamma_{\gamma\gamma}(f_0(1500))$. The acceptance and selection efficiency is 22%, while the trigger efficiency in the mass window is $(64 \pm 3)\%$. The branching ratio $\mathcal{BR}(f_0(1500) \to \pi^+\pi^-)$, taking all interference effects into account, has recently been calculated to be 0.24±0.05 [7]. Thus, the upper limit on the two-photon width of the $f_0(1500)$ is $\Gamma_{\gamma\gamma} < 0.17$keV ,95%C.L. .

This limit can be used to calculate the *stickiness* [8] of the resonance, by comparison of the two photon width with the width for $f_0(1500)$ production in radiative J/ψ decay. The lower limit on the stickiness of the $f_0(1500)$ is found to be 13, higher than would be expected for a $q\bar{q}$ state (~ 1).

3 $f_J(2220)$ in hadronic Z decays

If the $f_J(2220)$ is the tensor glueball state, its production in gluon fragmentation might be expected to be enhanced with respect to its production in quark fragmentation. From 130pb^{-1}of data collected by the L3 detector [9] at LEP (again at $\sqrt{s} \approx 91$GeV), two-jet and three-jet events are selected using the LUCLUS jet algorithm [10] (with d_{join}=5GeV).

Oppositely-charged track pairs from secondary vertices are selected and each track assigned the charged pion mass. The combined mass of the pair must then fall within a 50MeV window around the nominal K_s^0 mass. Comparison with Monte Carlo simulation indicates a purity of ~80% by this method. $K_s^0 K_s^0$ pairs are combined within jets, and their invariant mass spectrum is shown in figure 2 for two-jet and three-jet samples. There is little evidence of the $f_J(2220)$ in two-jet events, but a clear peak at 2.2GeV is seen for events containing a gluon jet. Also in figure 2 is a comparison of the observed peak with Monte Carlo simulation, where the $f_J(2220)$ has been included as a meson of mass 2230MeV and width 25MeV .

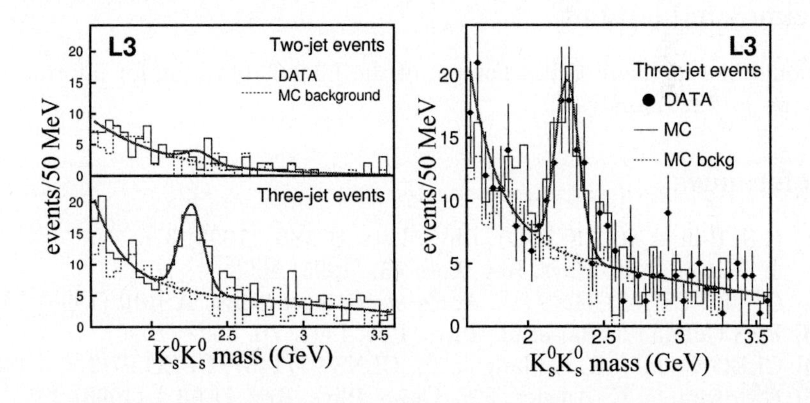

Fig. 2. Invariant distributions for $K_s^0 K_s^0$ pairs in two-jet and three-jet events (left), and comparison of the mass peak with simulation of $f_J(2220)$ production (right).

In three-jet events the lowest energy jet is most likely (72%) to be a gluonic jet. The $K_s^0 K_s^0$ mass spectrum for each of the three jets after energy-ordering is shown in figure 3 – the $f_J(2220)$ peak is certainly most prominant in the least energetic jet, *ie.* the jet most likely to have developed from a radiated gluon.

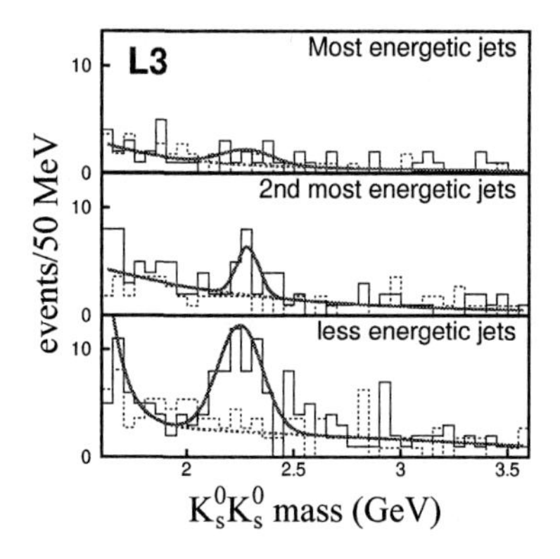

Fig. 3. Invariant distributions for $K^0_s K^0_s$ pairs in energy-ordered jets.

4 Conclusion

Production of the $f_0(1500)$ resonance in $\gamma\gamma$ collisions at LEP1 has been studied, and the upper limit on its two–photon width has been calculated as $\Gamma_{\gamma\gamma} < 0.17\text{keV}$, at 95% C.L. . The lower limit on the stickiness of the resonance suggests it may not be a simple $q\bar{q}$ state. Production of the $f_J(2220)$ resonance in hadronic Z^0 decays has been observed to be much enhanced in gluonic jets, consistent with the glueball hypothesis for this resonance.

Acknowledgement

I would like to thank Gilles Forconi of the L3 collaboration for information on the $f_J(2220)$ analysis.

References

[1] G.S. Bali *et al.*(UKQCD), Phys. Lett. B **389**, (1993) 378;
J. Sexton *et al.*, Phys. Rev. Lett **75**, (1995) 4563;
[2] Crystal Barrel Collab., A. Abele *et al.*, Nucl. Phys. **A 609** (1996) 562.
[3] BES Collab., J. Bai *et al.*, Phys. Rev. Lett. **76**, 3502 (1996).
[4] CLEO Collab., R. Godang et al., CLNS 97/1467, CLEO 97-3.
[5] For example, C. Amsler, F.E. Close, Phys. Rev. D **53** 1 (1996) 295.
[6] ALEPH Collab., D. Buskulic *et al.*, Nucl. Instr. Meth **A 360** (1995) 481.
[7] D.Bugg, B. Zou, private communication.
[8] M. Chanowitz, Proc. VI Int. Workshop on $\gamma\gamma$ Coll. (1984) Lake Tahoe.
[9] L3 Collab., B.Adriani *et al.*, Phys.Rep. **236** (1993) 1.
[10] T.Sjöstrand, Comp. Phys. Comm. **28** (1983) 229.

LEP Results on Beauty and Charm Spectroscopy

Vincenzo Canale (canale@axtov1.roma2.infn.it)

University of Roma "Tor Vergata" and INFN Roma 2, Italy

Abstract. We report on the most recent results on heavy flavour spectroscopy at LEP from the 1991-1995 running period at the Z pole. $Q\bar{Q}$, charm and beauty sectors are revised, several results are still preliminary.

1 Introduction

In this report some of the last results on the heavy flavour spectroscopy at LEP are presented. Heavy flavour physics at LEP profits of several favourable conditions. During the running period 1991-1995 at the Z pole about $\sim$ $880,000 b\bar{b}$ and $\sim 270,000$ $c\bar{c}$ pairs have been produced for each of the four experiments (ALEPH, DELPHI, L3 and OPAL). All spectrocopy states are accessible in the hadronization processes $c \rightarrow C$ and $b \rightarrow B$. Hard fragmentation and long lifetimes give rise to tipical decay length of few millimeter for the heavy flavour states.

To exploit such conditions experiments at LEP are equipped with silicon track detectors [1] which allow very high resolution tracking and vertexing. Tipical experimental resolutions on decay lenght are $\sigma_L \sim 300 \mu m$. Powerfull b-tagging algorithm, using impact parameters, have been developped which allow to isolate samples of very high b-purity ($\sim 95\%$) with sufficiently high efficiency ($\sim 50\%$).

In most of the analysis the heavy flavour signature is obtained exploiting the presence of known resonances within golden decay channels (e.g. $J/\psi \rightarrow l^+l^-$, $D^* \rightarrow D^0\pi$ etc...). Then the particle identification plays a key role in combinatorial background reduction and also in differentiating the different heavy flavour states (e.g. exploiting the enhancement of branching ratios like $D_s, B_s \rightarrow K^{\pm} + X$ and $\Lambda_b \rightarrow p + X$).

2 Results on $Q\bar{Q}$ States

The measurement of the production rate of $Q\bar{Q}$ states is very interesting to test the relative contributions of the different models. In particular the measurements at the TEVATRON collider [2] gave evidence of prompt quarkonium production at a much higher rate than predicted by the first order color singlet models [3]. Production via color octet model [4] could explain this difference.

At LEP the inclusive production of J/ψ and ψ' has been measured, the results are shown in table 1. These results are not directly sensitive singlet-octet models because the main source of J/ψ is B decyas. The ratio $R^b_{\chi c1} = \frac{BR(b \to \chi_{c1})}{BR(b \to J/\psi)}$ is sensitive to both singlet and octet production. The L3 collaboration measured $R^b_{\chi c1} = 0.75 \pm 0.19 \pm 0.15$ [7] while color singlet models would predict ~ 0.2. Direct evidence of prompt J/ψ decays has also been proven, the results are reported in table 2 in agreement with the theoretical predictions which also include the octet production mechanism.

Search for Υ production has been performed at LEP. Analysis are performed in the leptonic channels and include all the possible contributions from the $\Upsilon(1s), \Upsilon(2s), \Upsilon(3s)$ states. Results are shown in table 2, there is still a discrepancy between OPAL and the other collaborations.

	$BR(Z \to J/\psi + X)$	$BR(Z \to \psi' + X)$
ALEPH [5]	$(3.9 \pm 0.2 \pm 0.3)10^{-3}$	-
DELPHI [6]	$(3.7 \pm 0.4 \pm 0.2)10^{-3}$	$(1.5 \pm 0.7 \pm 0.3)10^{-3}$
L3 [7]	$(3.40 \pm 0.23 \pm 0.27)10^{-3}$	$(1.6 \pm 0.5 \pm 0.3)10^{-3}$
OPAL [8]	$(3.9 \pm 0.2 \pm 0.3)10^{-3}$	$(1.6 \pm 0.3 \pm 0.2)10^{-3}$

Table 1. Results on J/ψ and ψ' production

	$BR(Z \to "prompt"\ J\psi\ +\ X) \times 10^4$	$BR(Z \to \Upsilon + X) \times 10^5$
ALEPH	$3.00 \pm 0.78 \pm 0.30_{syst.} \pm 0.15_{theo.}$[9]	$< 7.3\ (90\%CL)$ [12]
DELPHI	$< 7.6\ (90\%CL)$ [10]	-
L3	-	$< 7.6\ (90\%CL)$ [13]
OPAL	$1.9 \pm 0.7 \pm 0.5_{syst.} \pm 0.4_{theo.}$[11]	$10 \pm 4 \pm 2$ [14]

Table 2. Results on "prompt" J/ψ and Υ production

3 Results on Charm

A lot of new results have been obtained in the charm sector. The production rates of the lowest energy states $D^0, D^\pm, D_s$ and Λ_c has been revised by ALEPH [15], from which the determination of the fractional decay width of the Z into $c\bar{c}$ was obtained $R_c = 0.1756 \pm 0.0048(stat.) \pm 0.0085(syst.) \pm 0.0068(B.R.)$. This result is in good agreement with the Standard Model predictions [16] and the accuracy is similar to other determinations [17] .

The production of the excited state D^* has also been extensively studied. New results have been otained by ALEPH [15] and OPAL [18]. The study of the energy spectrum of the D^* combined with vertexing and b-tagging algorithm allows to disentangle the different contributions from b and c-quark hadronization and from gluon splitting in $c\bar{c}$ pairs. The results are summarised in table 3. The ALEPH collaboration [15] has also reported the evidence for D_s^* production.

OPAL observed [19] the production of orbitally excited charm mesons in Z decays. Through the chain $D^{**0} \to D^{*+}\pi^-$ they measured the production probability of the two P-wave states $D_1^0(2420)$ and $D_2^{*0}(2460)$. The evidence for D_{s1}^+ production in Z decays was also reported.

	ALEPH [15]	OPAL [18] [19]
$f(b \to D^*)$	-	$0.173 \pm 0.016 \pm 0.012$
$f(c \to D^*)$	-	$0.222 \pm 0.014 \pm 0.014$
$\frac{R_b.f(b\to D^*)}{R_c.f(c\to D^*)}$	1.15 ± 0.06	1.28 ± 0.11
$< X_{D^*}^{c\bar{c}} >$	$0.489 \pm 0.005 \pm 0.006$	$0.515 \pm 0.002 \pm 0.009$
$< n_{Z\to D^*} >$	$0.195 \pm 0.003 \pm 0.003$	$0.185 \pm 0.004 \pm 0.009$
$< n_{g\to c\bar{c}} >$	$0.047 \pm 0.010 \pm 0.011$	-
$f(c \to D_s^*)$	$0.075 \pm 0.022 \pm 0.005 \pm 0.018_{BR}$	-
$f(b \to D_s^*)$	$0.102 \pm 0.031 \pm 0.005 \pm 0.025_{BR}$	-
$f(c \to D_1^0)$	-	$0.021 \pm 0.007 \pm 0.003$
$f(c \to D_2^{*0})$	-	$0.052 \pm 0.022 \pm 0.013$
$f(b \to D_1^0)$	-	$0.050 \pm 0.014 \pm 0.006$
$f(b \to D_2^{*0})$	-	$0.047 \pm 0.024 \pm 0.013$
$f(c \to D_{s1}^+)$	-	$0.016 \pm 0.004 \pm 0.003$

Table 3. Results on production of excited charm states

4 Results on Beauty

Several new measurement have been performed on the beauty production at LEP. The gluon splitting into $b\bar{b}$ pairs has been studied. Events with four jets topology were selected and b-tagging algorithms were applied to the single jets. The probability $g_{b\bar{b}}$ for secondary production of a beauty quark pair from a gluon per hadronic Z decay was then derived. The results are

$g_{b\bar{b}} = (2.6 \pm 0.4 \pm 0.9)\ 10^{-3}$ for ALEPH [20] and $g_{b\bar{b}} = (2.1 \pm 1.1 \pm 0.9)\ 10^{-3}$ for DELPHI [21].

An interesting result has been reported by DELPHI [22] on the production of B_s mesons at the hadronization level. Exploiting the powerfull discrimination of the RICH [23] to identify $K^{\pm}$ accompanying B hadrons, it is possible to correlate the charge of the kaon and the strangeness content of the B-hadron. The probability $f(b \rightarrow "B_s") = 0.144 \pm 0.017 \pm 0.030$ was measured. This result is almost unaffected by the uncertainties on the production and decay of $B_s^{**} \rightarrow B_{u,d}^{*} K$ which is the major systematic error when the strange B-hadrons are tagged at the decay level.

For excited states no update have been presented on B^{**} and B-baryons. OPAL presented a new measurement [27] on $B^* \rightarrow B\gamma$ production which is in agreement with other determinations. These results are summarized in table 4. DELPHI reported [28] the evidence for the Dalitz decay $B^* \rightarrow Be^+e^-$. The preliminary result $\frac{\Gamma(Be^+e^-)}{\Gamma(B\gamma)} = (4.7 \pm 1.1 \pm 1.0)\ 10^{-3}$ is in good agreement with the theoretical QED predictions.

	$\Delta M\ (MeV)$	$\frac{\sigma_{B^*}}{\sigma_B}$
ALEPH [24]	$45.3 \pm 0.4 \pm 0.9$	$0.77 \pm 0.03 \pm 0.07$
DELPHI [25]	$45.5 \pm 0.3 \pm 0.8$	$0.72 \pm 0.03 \pm 0.06$
L3 [26]	46.3 ± 1.9	$0.76 \pm 0.08 \pm 0.06$
OPAL [27]	$46.2 \pm 0.3 \pm 0.8$	$0.76 \pm 0.04 \pm 0.08$

Table 4. Results on B^* production

The B_c meson has not yet been observed experimentally. Decays containing J/ψ should be enhanced. The dominant production at LEP occurs in $b\bar{b}$ events through hard gluon radiation and successive $g \rightarrow c\bar{c}$ splitting, and it is expected to be small, moreover the consequent softer spectrum of B_c reduces the power of usual B experimental cuts (momentum, rapidity, etc...). Experiments have set limits studying the decay channels in $J/\psi\pi$,$J/\psi 3\pi$ and $J/\psi l\nu$. The results are reported in table 5. ALEPH [29] has one serious candidate in the semileptonic channel with an estimated mass $m_{B_c} = (5.96^{+0.25}_{-0.19})GeV$ and a decay time $t = (1.77 \pm 0.17)ps$.

5 Conclusion

The running of LEP at the Z pole has provided a lot of usefull data for the spectroscopy of heavy flavours. The favourable physical conditions at LEP allowed complementary and/or peculiar results with respect to the experiments

	"$J/\psi\ \pi$"	"$J/\psi\ l\nu$"	"$J/\psi\ 3\pi$"
ALEPH [29]	2.5 10^{-5}	3.6 10^{-5}	-
DELPHI [30]	10.5 10^{-5}	5.8 10^{-5}	17.5 10^{-5}
OPAL [31]	7.1 10^{-5}	4.9 10^{-5}	33.0 10^{-5}

Table 5. Summary of B_c production. The results correspond to limits at 90% C.L. on $BR(Z \to B_c) \times BR(B_c \to \text{"}channel\text{"})$.

at the $\Upsilon(4s)$. In the near future the inclusion of all the available statistics (1995 data are often not yet included in the results), and the refinement of anlysis techniques (e.g vertexing, b-tagging etc...) would produce new interesting results. ALEPH, DELPHI and OPAL have or plan to reprocess their data with optimized tracking alogorithm and therefore large improvements are expected in a variety of analysis.

I would like to thank the organizers of the conference and my colleagues from the ALPEH, DELPHI, OPAL and L3 collaborations for providing me the material and for very helpful discussions.

References

[1] D.Creanza et al., Conf. Front. Detec. for Front. Phys., Isola d'Elba ,Italy May 25-31 1997
V.Chabaud et al., Nucl. Instr. and Meth. A 368(1996)314
M.Acciarri et al., Nucl. Instr. and Meth. A 289(1994)300
P.Allport et al., Nucl. Instr. and Meth. A 346(1994)476

[2] CDF Collaboration, F.Abe et al., Phys. Rev. Lett. 69(1992)3704
CDF Collaboration, F.Abe et al., Phys. Rev. Lett. 79(1997)572

[3] E.L. Berger, D.Jones, Phys. Rev. D 23(1981)1521
E.L. Berger, D.Jones, Phys. Lett. B 121(1983)61
R.Baier, R.Ruckl, Zeit. Phys. C 19(1983)251

[4] G.T.Bodwin et al., Phys. Rev. D 46(1992)3703
G.T.Bodwin et al., Phys. Rev. D 51(1995)1125
E.Braaten et al., Ann.Rev.Nucl.Part. Sci. 46(1996)197

[5] ALEPH Collaboration, D.Buskulic et al., Phys. Lett. B 295(1992)396

[6] DELPHI Collaboration, P.Abreu et al., Phys. Lett. B 341(1994)109

[7] L3 Collaboration, M.Acciarri et al., Phys. Lett. B 407(1997)351

[8] OPAL Collaboration, G.Alexander et al., Zeit. Phys. C 70(1996)197

[9] ALEPH Collaboration, EPS HEP 97, Paper 624

[10] DELPHI Collaboration, P.Abreu et al., Zeit. Phys. C 69(1996)575

[11] OPAL Collaboration, G.Alexander et al., Phys. Lett. B 384(1996)343

[12] ALEPH Collaboration, ICHEP 96, PA05-066

[13] L3 Collaboration, CERN-PPE / 97-78
[14] OPAL Collaboration, G.Alexander et al., CERN-PPE / 95-181
[15] ALEPH Collaboration, EPS HEP 97, Paper 623
[16] M. A. Samuel, Phys. Lett. B 397(1997)241
[17] DELPHI Collaboration, P.Abreu et al., Zeit. Phys. C 59(1993)533
OPAL Collaboration, G.Alexander et al., Zeit. Phys. C 72(1996)1
ALEPH Collaboration, ICHEP 96, PA10-016
[18] OPAL Collaboration, G.Alexander et al., CERN-PPE / 97-093
[19] OPAL Collaboration, G.Alexander et al., CERN-PPE / 97-035
[20] ALEPH Collaboration, EPS HEP 97, Paper 606
[21] DELPHI Collaboration, P.Abreu et al., Phys. Lett. B 405(1997)202
[22] DELPHI Collaboration, EPS HEP 97, Paper 451
[23] DELPHI Collaboration, Nucl. Instr. and Meth. A 378(1996)57
[24] ALEPH Collaboration, D.Buskulic et al., Zeit. Phys. C 69(1996)393
[25] DELPHI Collaboration, P.Abreu et al., Zeit. Phys. C 68(1995)353
[26] L3 Collaboration, M.Acciarri et al., Phys. Lett. B 345(1995)589
[27] OPAL Collaboration, G.Alexander et al., Zeit. Phys. C 74(1997)413
[28] DELPHI Collaboration, EPS HEP 97, Paper 450
[29] ALEPH Colaboration, D.Buskulic et al., Phys. Lett. B 402(1997)213
[30] DELPHI Collaboration, P.Abreu et al., Phys. Lett. B 398(1997)207
[31] OPAL Physics Note PN297, (submitted to the XVIII International Symposium on Lepton-Photon Interactions. Hamburg 1997)

CLEO Results on Hadron Spectroscopy

Alexander Smith (smith@mnhep.hep.umn.edu)

University of Minnesota, Minneapolis, MN

Abstract. I present the results of three recent analyses from the CLEO Collaboration: 1) a search for two-photon production of the glueball candidate $f_J(2220)$, 2) a search for two-body radiative $\Upsilon(1S)$ decays and study of a 3.8σ enhancement observed in the mass range 1.2-1.4 GeV/c^2, and 3) a study of the dipion transitions $\Upsilon(2S) \rightarrow \Upsilon(1S)\pi\pi$.

1 Search for Two-photon Production of the Glueball Candidate $f_J(2220)$

There are several glueball candidates in the mass range 1-2.5 GeV/c^2, where QCD predicts the existence of gg bound states. In general, such gg states are expected to mix with neighboring $q\overline{q}$ states. Therefore, we define a glueball to be a particle which is composed predominantly of gluons. The $f_J(2220)$, also known as $\xi(2230)$, is a glueball candidate due to its narrow width[1, 2], its observation in glue-rich environments[1, 2, 3, 4, 5], and its mass, which is near the QCD predictions for a tensor glueball[6, 7].

We report on a search for two-photon production of $f_J(2220)$ and set an upper limit on the product of the two-photon partial width and branching fraction to K_sK_s. This measurement is an improvement over a previous upper limit by ARGUS[8], using the K^+K^- decay channel.

The gluon content can be measured relative to the quark content of the meson by comparing the rate of two-photon production, where the coupling is only to the electric charge of the quarks, with the rate in radiative J/ψ decays, where the particle is produced from gluons and therefore only couples to the color charge. We define a quantity called the "stickiness" [9], S_X, which is a measure of this ratio:

$$S_X \sim \frac{|<X|gg>|^2}{|<X|\gamma\gamma>|^2} \sim N_L\left(\frac{m_X}{k_{J/\psi\gamma X}}\right)^{2L+1}\frac{\Gamma(J/\psi \rightarrow \gamma X)}{\Gamma(X \rightarrow \gamma\gamma)} \quad . \qquad (1)$$

We searched for the $f_J(2220)$ in two-photon events detected in the CLEO II detector[10], reconstructing the decay to K_sK_s, where the K_s subsequently decays to $\pi^+\pi^-$:

$$\gamma\gamma \rightarrow f_J(2220) \rightarrow K_{s,1}K_{s,2}$$

$$K_{s,2} \rightarrow (\pi^+\pi^-)_2$$

$$K_{s,1} \rightarrow (\pi^+\pi^-)_1$$

Two-photon events were selected from $3.0\,\mathrm{fb}^{-1}$ of e^+e^- annihilation data collected at center-of-mass energy of $\sim 10.6\,\mathrm{GeV}$. By requiring $|\Sigma \mathbf{p}_T| < 0.2\,\mathrm{GeV/c}$ and total observed energy $< 6\,\mathrm{GeV}$, we selected a two-photon event topology, where at least one of the electrons scattered through small angles and was not detected. Events were required to have four good tracks and zero net charge. Two pairs of tracks of opposite charge were required to form vertices corresponding to K_s decays, where the two vertices are separated by more than 5 mm. Finally, the invariant mass of the $\pi^+\pi^-$ pairs was required to be within a $10\,\mathrm{MeV/c}^2$ circle in $m(\pi^+\pi^-)_1 m(\pi^+\pi^-)_2$-space. The result is a very clean signal with less than 5% background, as is shown in Fig. 1. The K_sK_s invariant mass distribution was formed from this sample. As is shown in Fig. 1, there are four events within the signal region delineated by the vertical arrows. This signal region was determined with a Monte Carlo simulation which used the $f_J(2220)$ mass and width from the Mark III and BES measurements, and a GEANT simulation of the CLEO II detector. Using the standard technique for a Poisson distribution[11], we extract an upper limit of 4.9 events at the 95% C.L. Using this limit and the Monte Carlo width and branching fraction, the expression

$$\Gamma_{\gamma\gamma}^{data}\mathcal{B}_{K_SK_s}^{data} = \frac{n^{data}}{n^{MC}}\frac{L^{MC}}{L^{data}}\Gamma_{\gamma\gamma}^{MC}\mathcal{B}_{K_SK_s}^{MC}, \tag{2}$$

gives an upper limit on $\Gamma_{\gamma\gamma}\mathcal{B}_{K_SK_s}$ of 1.3 eV at the 95% C.L.

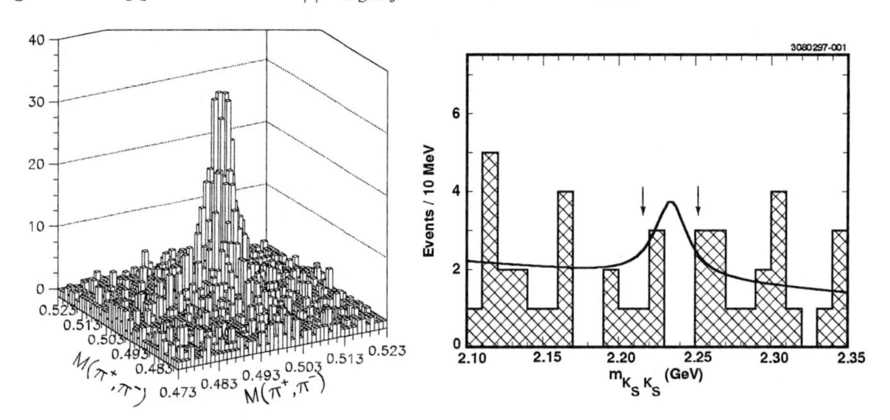

Fig. 1. Distribution of $m(\pi^+\pi^-)_1$ vs $m(\pi^+\pi^-)_2$ in data. Events outside a circle of $10\ \mathrm{MeV/c}^2$ about the point $(m(K_s), m(K_s))$ are rejected (left). The K_sK_s mass distribution near the $f_J(2220)$ resonance (right). The vertical lines indicate the signal region, which contains four events. The curve is the result of a fit to the background and a signal corresponding to the 95% C.L. upper limit of 4.9 events.

A lower limit on the stickiness is obtained by combining this measurement with the Mark III[1] and BES[2] measurements of the width times branching fraction in radiative J/ψ decay, as in Eq. 1. The factor $\mathcal{B}_{K_sK_s}$ drops out, since it appears in both the numerator and denominator in Eq. 1. This results in a lower limit on the stickiness of 82 at the 95% C.L.

This lower bound for stickiness is much larger than one would expect for a $q\bar{q}$ resonance. Most explanations that give a small two-photon width, such as the meson being a radial or angular excitation, also give a small radiative J/ψ decay branching fraction. A cancelation resulting in a small two-photon width is possible, but very unlikely, since it demands specific values of the singlet-octet mixing and the ratio of matrix elements.

2 Search for Two-body Radiative $\Upsilon(1S)$ Decays

Several modes of radiative and hadronic $\Upsilon(1S)$ decays to multiparticle final states have previously been observed. While two-body or *resonant* radiative decays of the J/ψ have been observed at the 10^{-4} level[11], no corresponding two-body decays of the $\Upsilon(1S)$ have been observed. Such decays may provide a window to exotic states, such as Higgs, non-interacting particles[12], glueballs, and axions. As discussed in Section 1, branching fractions for radiative two-body heavy quarkonium decays can be combined with two-photon production widths to study the quark vs gluon content of a glueball candidate.

The search was performed by looking at the decay $\Upsilon(1S) \rightarrow \gamma\pi^+\pi^-$. Such events were selected from a data sample of $78.9\,\mathrm{pb}^{-1}$ collected on the $\Upsilon(1S)$ resonance. Events were required to have two charged tracks of zero net charge, neither of which was identified as a lepton. At least one photon of $E > 4\,\mathrm{GeV}$ was required to be detected in the central region of the CsI calorimeter. Energy and momentum conservation was required in the event.

One expects a large non-$\Upsilon(1S)$ continuum background, dominated by $\gamma\rho$ and $\gamma\gamma$ production. The $\pi^+\pi^-$ invariant mass spectrum for this background was obtained from a data sample of $500.4\,\mathrm{pb}^{-1}$ collected at the $\Upsilon(4S)$, which was corrected for the differences in energy, efficiency, and luminosity. A fit to this data was used to obtain the background $\pi^+\pi^-$ invariant mass spectrum, shown in Fig. 2.

In the $\Upsilon(1S)$ sample, the $\pi^+\pi^-$ invariant mass spectrum, shown in Fig. 2, contains a significant enhancement over the expected continuum distribution (dotted line) in the mass region 1.2-1.4 $\mathrm{GeV/c}^2$. When this distribution is fitted with a Breit-Wigner signal line shape and the continuum distribution as background, one finds 43 ± 11.4 events in a peak of width $\Gamma = 163\pm73\,\mathrm{MeV/c}^2$ centered at a mass of $1.28 \pm 0.03\,\mathrm{GeV/c}^2$. The photon polar angle distributions were studied in the mass regions above, below, and on the enhancement in both continuum and $\Upsilon(1S)$ data. While the statistics of the sample were limited, data from the $\Upsilon(1S)$ resonance in this region seem to obey a different distribution, in support of the hypothesis that the enhancement is from resonant $\Upsilon(1S)$ decay rather than continuum. The mass and width of the enhancement is consistent with that of the $f_2(1270)$. *If we assume this* 3.8σ

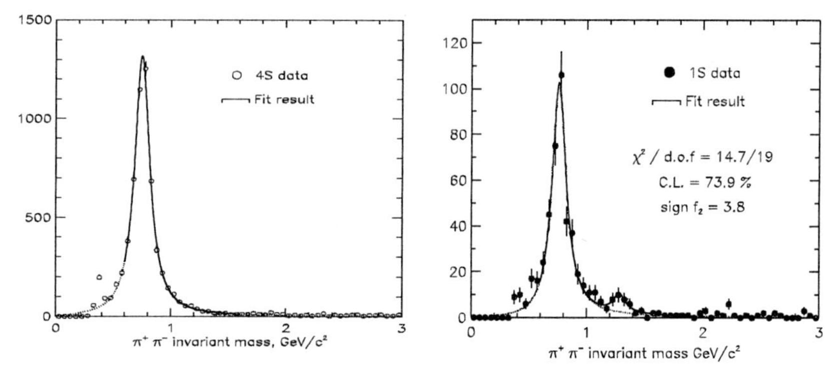

Fig. 2. Fits to the continuum background data (left) and to the $\Upsilon(1S)$ data (right). A significant excess above the continuum is observed in the mass region 1.2-1.4 GeV/c^2.

excess is due entirely to $\gamma f_2(1270)$ production, this would correspond to a branching fraction of $\mathcal{B}(\Upsilon(1S) \to \gamma f_2(1270)) = (0.85 \pm 0.23) \times 10^{-4}$.

3 Study of the Dipion Transitions $\Upsilon(2S) \to \Upsilon(1S)\pi\pi$

The transitions $\Upsilon(2S) \to \Upsilon(1S)\pi\pi$ provide a good testing ground for non-perturbative QCD theoretical models and approximations. Such decays of heavy quarkonium ($b\bar{b}$ and $c\bar{c}$) are typically treated as a factorizable product of a short range transition $q\overline{q}' \to q\overline{q}$ with gluon emission, and a subsequent long range hadronization of gluons to form the final state pions. The observable quantities examined are the branching fraction, dipion invariant mass spectra, and the angular distributions.

We considered both charged pion, $\Upsilon(2S) \to \Upsilon(1S)\pi^+\pi^-$, and neutral pion, $\Upsilon(2S) \to \Upsilon(1S)\pi^0\pi^0$ decay channels. In the charged pion channel, both exclusive decay modes, where the $\Upsilon(1S)$ was required to decay to e^+e^- or $\mu^+\mu^-$, and inclusive modes, where the $\Upsilon(1S)$ could decay to anything, were studied. For the neutral pion mode, only exclusive decays were studied.

This analysis is based on the world's largest sample of approximately $500,000$ events collected on the $\Upsilon(2S)$ resonance (73.5 pb^{-1} on resonance, 5.2 pb^{-1} off), by the CLEO II detector.

For the $\Upsilon(2S) \to \Upsilon(1S)\pi^+\pi^-$ analyses, at least two tracks with momentum less than 0.5 GeV/c were required. For the exclusive charged pion analysis the additional requirement of two lepton tracks with $p > 3.5$ GeV/c and $\cos\theta_{\pi\pi} < 0.9$ was imposed. From these samples distributions of missing mass, defined as

$$M_{miss} \equiv \sqrt{(M_{\Upsilon(2S)}) - E_{\pi\pi}^2 - \mathbf{p}_{\pi\pi}^2}\,, \tag{3}$$

were measured. We expect these distributions to be peaked at the mass of the $\Upsilon(1S)$. In the exclusive channels, large signals of 956 ± 31 and 1130 ± 34 events were found, in the e^+e^- and $\mu^+\mu^-$ channels, respectively, with essentially no background, as shown in Fig. 3. From this signal the branching fraction was determined to be $\mathcal{B}(\Upsilon(2S) \to \Upsilon(1S)\pi^+\pi^-) = 0.189\pm0.004\,(\text{stat})\pm0.009\,(\text{sys})$.

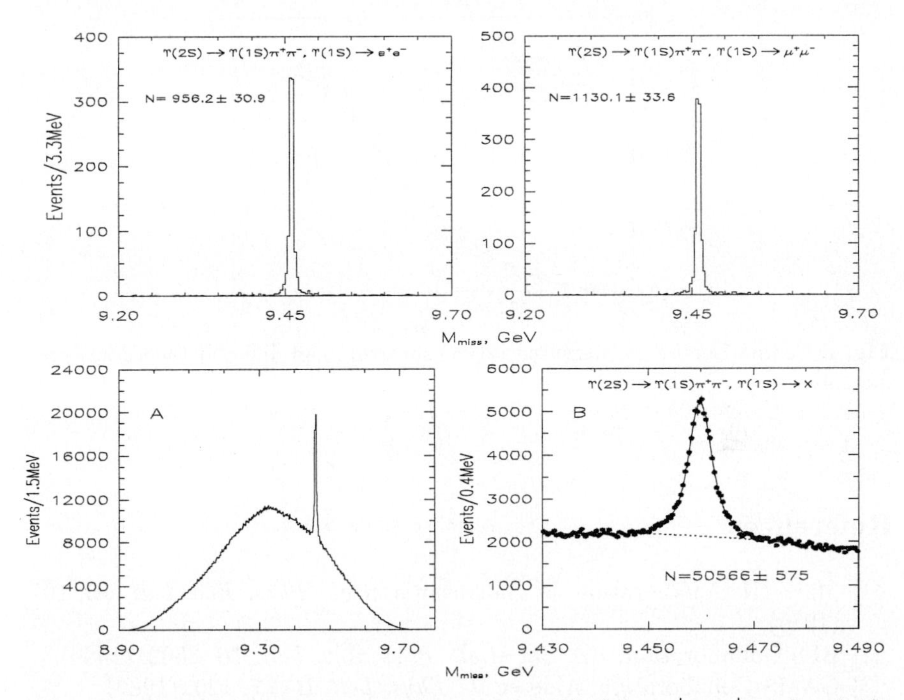

Fig. 3. Measured missing mass distributions for $\Upsilon(2S) \to \pi^+\pi^-e^+e^-$ (top left), $\Upsilon(2S) \to \pi^+\pi^-\mu^+\mu^-$ (top right), and $\Upsilon(2S) \to \pi^+\pi^-X$ (bottom two) final states.

The missing mass distribution for $\Upsilon(2S) \to \pi^+\pi^-X$, shown in Fig. 3, has a large, well-defined peak at the $\Upsilon(1S)$ mass on a much larger background. A fit found $50,566\pm575$ events in the peak. After correcting for efficiency and luminosity, the branching fraction was found to be $\mathcal{B}(\Upsilon(2S) \to \Upsilon(1S)\pi^+\pi^-) = 0.196 \pm 0.003\,(\text{stat}) \pm 0.012\,(\text{sys})$, consistent with the exclusive analysis.

The decays of $\Upsilon(2S) \to \pi^0\pi^0\ell^+\ell^-$ were selected by requiring at least two lepton tracks with $p > 3.5\,\text{GeV/c}$ and four or five showers in the calorimeter that satisfy a single-photon hypothesis and are in the energy range $50\,\text{MeV} < E_\gamma < 430\,\text{MeV}$. The missing mass distributions have peaks containing 133.2 ± 11.5 and 142.5 ± 11.9 events in the e^+e^- and $\mu^+\mu^-$ channels, respectively, with negligible backgrounds. From this signal the branching fraction was measured to be $\mathcal{B}(\Upsilon(2S) \to \Upsilon(1S)\pi^0\pi^0) = 0.092 \pm 0.006\,(\text{stat}) \pm 0.007\,(\text{sys})$.

The theoretical predictions[13, 14, 15] of the $m(\pi\pi)$ spectra are in excellent agreement with our measurements, as shown for the acceptance corrected $\pi^+\pi^-$ data in Fig. 3. The angular distributions were also measured and found to be in good agreement with the theoretical prediction of mostly s-wave and a small d-wave contribution.

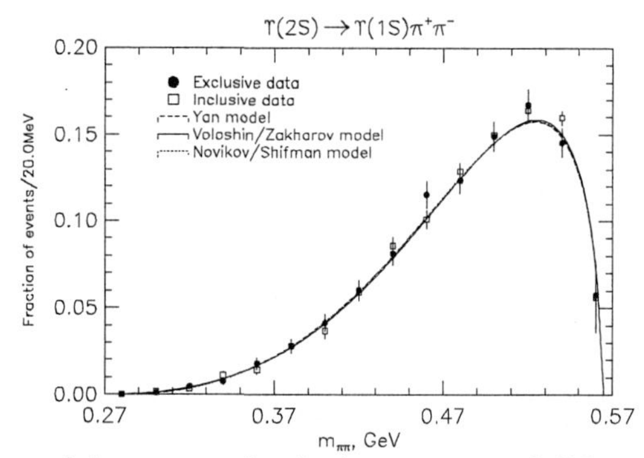

Fig. 4. Comparison of the measured $m(\pi\pi)$ spectrum and different theoretical predictions.

References

[1] Mark III Collaboration, R. Baltrusaitis *et al., Phys. Rev. Lett.* **56**, 107 (1986).

[2] BES Collaboration, J.Z. Bai *et al., Phys. Rev. Lett.* **76**, 3502, (1996).

[3] GAMS Collaboration, Alde *et al., Phys. Lett.* **B177**, 120, (1986).

[4] LASS Collaboration, D. Aston *et al., Phys. Lett.* **B215**, 199, (1988).

[5] MSS Collaboration, B.V. Bolonkin *et al., Nucl. Phys.* **B309**, 426, (1988).

[6] C. Michael, "Glueballs and Hybrid Mesons" hep-ph/96052443 (1996).

[7] C. Morningstar and M. Peardon, *Nucl. Phys. Proc. Suppl.* **53**, 917 (1997), hep-lat/9608050.

[8] ARGUS Collaboration, H. Albrecht *et al., Z. Phys. C* **48**, 183, (1990).

[9] M. Chanowitz "Resonances in Photon-Photon Scattering" Proceedings of the VIth International Workshop on Photon-Photon Collisions (1984).

[10] CLEO Collaboration, Y. Kubota *et al., Nucl. Inst. & Meth.* **A320**, 66, (1992).

[11] Particle Data Group, *Phys. Rev. D* **54** (1996).

[12] P. Fayet, J. Kaplan, *Phys. Lett.* **B269** 213(1991).

[13] T.-M. Yan, *Phys. Rev. D* **22** 1652 (1980).

[14] M. Voloshin, V. Zakharov, *Phys. Rev. Lett.* **45** 688 (1980).

[15] V. Novikov, M. Shifman, *Z. Phys. C* **8** 43 (1981).

Light Quark Photoproduction at Fermilab

Sergio P. Ratti (for the E687 Collaboration) ⋆ (ratti@pv.infn.it)

University of Pavia and I.N.F.N. - Sezione di Pavia

1 Quasi-exclusive $\pi^+\pi^-$ States

This analysis is based on approximately 900,000 quasi-exclusive $\pi^+\pi^-$ pairs recorded with the E687 spectrometer during the 1990/91 fixed target runs [1] at high energy ($7.5 < \sqrt{s} < 17.0$). Exclusive final states are selected based on the total number of charged tracks seen in the spectrometer and by requiring no visible energy in our electromagnetic calorimeters. Most of the data ($\approx$ 98%) has been taken with a Beryllium target (a small part of the date were taken with Al and Pb targets to prove that our samples are indeed diffractive by relating the highest t-slope to the size of the nucleus).

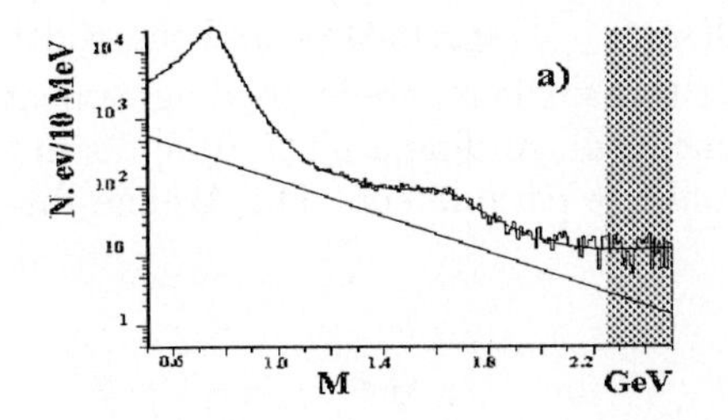

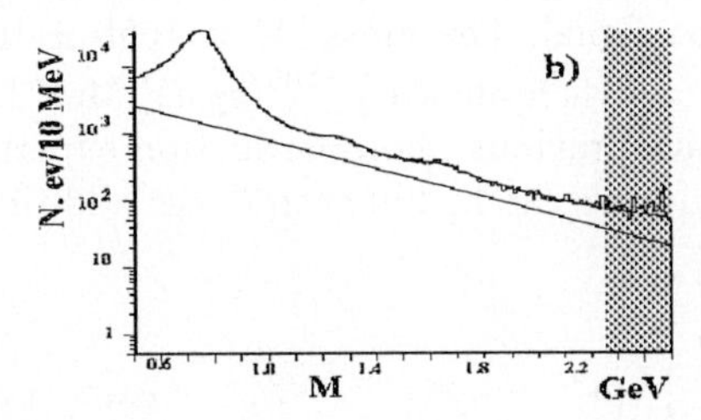

Fig. 1. $M_{2\pi}$ spectra (10 MeV bin) fitted with multiple Breit-Wigners, interfering Drell amplitude and incoherent background (straight lines): a- small t; b) all events; 162 d.o.f., $\chi^2 = 161$ case a), $\chi^2 = 147$ case b). Shaded areas excluded from fit.

Our data is well consistent with an s-channel helicity conserving (schc) production mechanism[2]a, characterized by a $sin^2(\theta)$ angular distribution in the Gottfried-Jackson frame, and with a photoproduction cross section independent of $\sqrt{s}$ (given by the reconstructed longitudinal momentum P_z of the $\pi^+\pi^-$ pair), as observed by earlier experiments[2]b. The acceptance corrected $M_{2\pi}$ spectra, shown for small t ($t < 0.0625$ $(GeV/c)^2$ (fig. 1a) and for all events (fig. 1b), are fitted with spin 1 relativistic Breit-Wigner amplitudes A_{BW}, plus a "Drell" type[3] amplitude $A_{Drell} \approx F_{nr}\, e^{i\phi_{nr}}(M_r^2 -$

⋆ co-authors are: P.L.Frabetti-Bologna; H.W.K.Cheung, J.P.Cumalat, C.Dallapiccola, J.F.Ginkel, W.E.Johns, M.S.Nehring-Colorado; J.N.Butler, S.Cihangir, I.Gaines, P.H.Garbincius, L.Garren, S.A.Gourlay, D.J.Harding, P.Kasper, A.Kreymer, P.Lebrun, S.Shukla, M.Vittone-Fermilab; S.Bianco, F.L.Fabbri, S.Sarwar, A.Zallo-Frascati; R.Culbertson, R.W.Gardner, R.Greene, J.Wiss-Illinois; G.Alimonti, G.Bellini, M.Boschini, D.Brambilla, B.Caccianiga, L.Cinquini, M.DiCorato, M.Giammarchi, M.Gullotta, P.Inzani, F.Leveraro, S.Malvezzi, D.Menasce, E.Meroni, L.Moroni, D.Pedrini, L.Perasso, F.Prelz, A.Sala, S.Sala, D.Torretta-Milano; D.Buchholz, D.Claes, B.Gobbi, B.O'Reilly-Northwestern; J.M.Bishop, N.M.Cason, C.J.Kennedy, G.N.Kim,T.F.Lin, D.L.Puseljic, R.C.Rutchi, W.D.Shephard, J.A.Swiatek, Z.Y.Wu-Notre Dame; V.Arena, G.Boca, G.Bonomi, C.Castoldi, G.Gianini, M.Merlo, C.Riccardi, L.Viola, P.Vitulo-Pavia; A.Lopez-Puerto-Rico; G.P.Grim, V.S.Paolone, P.M.Yager-Davis; J.R.Wilson-South Carolina; P.D.Sheldon-Vanderbilt; F.Davenport-North Carolina; K.Danyo, T.Handler-Tennessee; B.G.Cheon, J.S.Kang, K.Y.Kim, K.B.Lee-Korea.

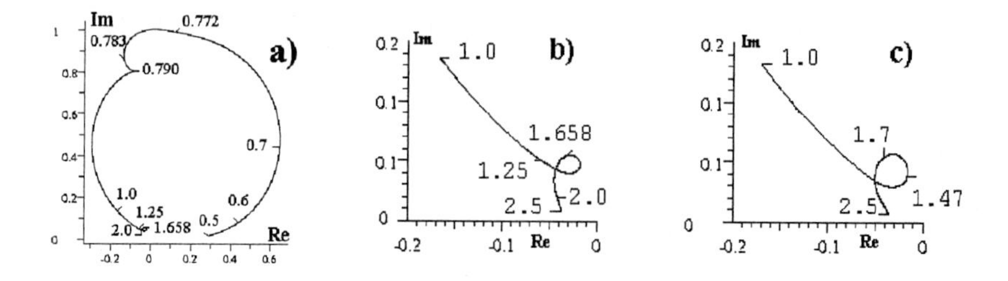

Fig. 2. a- Argand diagrams for the p-wave amplitude assuming single $\rho(1650)$; b- enlargement of $M_{2\pi} \geq 1.0$ GeV region; c- region $M_{2\pi} \geq 1.0$ GeV, assuming 2 ρ', at $M = 1450$ and $M = 1700$MeV.

$M^2)/[(M_r^2 - M^2) + iM_r\Gamma_r(M)]$ where $F_{nr} = [q(m)/M_{\pi\pi}] * e^{\frac{-q^2(m)}{2k_0^2}}$ and q the momentum transfer in the center of mass of the 2π pair[1]. To these we add the $\omega - \rho$ interference term. In fig. 1b the $f_2(1270)$ has been added showing that the schc mechanism works well at our energies. In both distributions there is no evidence for an $f_0(980)$ scalar meson. The fits to fig.s 1 assume that only one ρ' meson in the high mass region contributes to the signal. The mass $[M = 1656 \pm 10(stat)\,^{+200}_{-20}(syst)]$ MeV and the width $\Gamma = [344 \pm 36(stat)\,^{+100}_{-150}(syst)]$ MeV for this single ρ', are in good agreement with previous photoproduction experiments[4]. Adding a $\rho'(1700)$[5] doesn't improve the fit but introduces additional free parameters. The Argand dia-

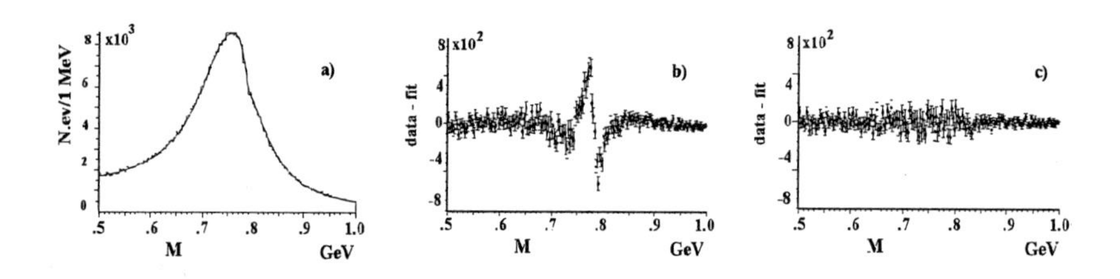

Fig. 3. a- $M_{2\pi}$ distribution around the $\rho-\omega$ pole mass; b- residual (fit-data)- mixing amplitude set to 0; c- residuals (fit-data) with interfering amplitude included.

grams for the p-waves, shown in fig.s 2, summarizes our analysis; to first order, the Drell amplitude pushes the entire diagram towards real values for the total amplitude. If one requires two distinct ρ', with masses around 1450 and 1700 MeV, (diagram of fig. 2c) the two resonances add up coherently to simulate a single resonance, with a mass around 1660 MeV. The latter solution (diagram of fig. 2b) is preferred for its simplicity. In the low mass region the

[1] In our kinematical regime, "t" is given by P_t; t-distributions provide a consistency check on the exclusive, diffractive character of the reaction.

$\rho-\omega$ mixing process is investigated with an unprecedented statistical sample that provides the most precise measurement of the mixing parameter. Fig.s 3 show the mass spectrum (1 MeV bin) in the region 0.5-1 GeV and the residual with and without mixing (fig.s 3). The $\rho(770)$ mass is a free parameter in the fit; the $\omega(783)$ mass is fixed at the nominal value[5]. The mixing amplitude is not strongly correlated to the exact shape of the Drell form factor. Assuming the ratio of the exclusive photoproduction cross sections for $\rho(770)$ and $\omega(783)$ to be 13.6, as predicted by broken SU(3)[6], the "2π" branching ratio of the ω relative to the ρ is: $\frac{\Gamma(\omega\to\pi^+\pi^-)}{\Gamma(\rho\to\pi^+\pi^-)} = [1.95 \pm 0.02(stat) \pm 0.015(syst)]$.

2 Observation of the $f_0(980)$ scalar meson

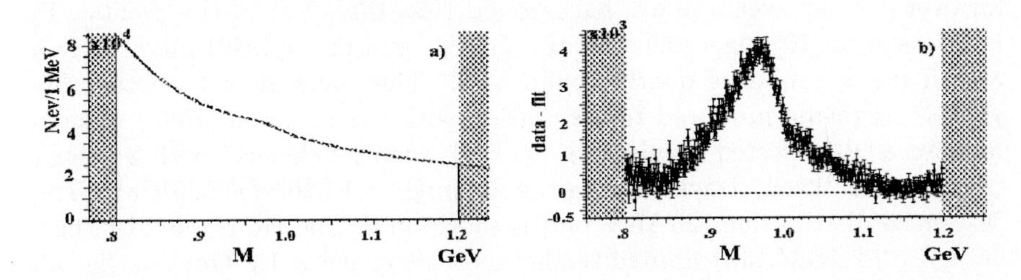

Fig. 4. a- $M_{2\pi}$ spectrum fitted with one Breit-Wigner; b- residuals (data-fit) for $M_{2\pi}$ if the $f_0(980)$ amplitude is set to zero.

For this analysis a special Light Quark Skim (LQS) has been used, selecting about 140 million triggers with $n_{ch} = 2, 3, 4, 5, 6$ as well as two independent sub-selections: a- the first, aiming at an inclusive search of the $f_0(980)$ selected $\sim$ 33 million events with $0.8 < M(\pi^+\pi^-) < 1.2$ GeV); b- a second, to search for the $f_0(980)$ in the channel $K^+K^-\pi^+\pi^-$ selected 100,510 events. A signal around 970 MeV emerges in fig.4a, after the tail of the $\rho(770)$, in spite of the huge background. The effect is present when $n_{ch} > 2$; it persists if no e.m. energy is detected; it is not enhanced by p_t cuts. For the $n_{ch} = 2$ sample, the signal is strongly suppressed; no narrow $f_0(980)$ is detected in fig.s1. To show the signal we describe the shape of the 2π mass spectrum as a "non resonant" continuum and a parametrisation af the $\rho(770)$ tail as: $d\sigma/dm_{2\pi} \propto N_{NR}e^{-\alpha}$ where $\alpha = \sum_{i=1}^{5} C_i*(m_{2\pi}-1.0)^i$ and add to it an s-wave Breit-Wigner function $A_{f_0} = e^{i\delta_{f_0}} m_{f_0}\Gamma_{f_0}(m)/[(m_{f_0}^2 - m_{2\pi}^2) - im_{f_0}\Gamma_{f_0}(m)]$. This single Breit-Wigner fit clearly identifies the accumulation of events as due to the $f_0(980)$. Fig. 4b shows the residuals in absence of the $f_0(980)$ amplitude. No improvement on the mass and width can be achieved however, without introducing other broader amplitudes at play. As shown in fig. 4b, the signal is asymmetric, a clear indication of interference effects. At least a coupled Breit-Wigner of the "Flatté" type[7] is needed. As already observed[8], this relatively narrow structure could interfere with broad s-wave resonances in the nearby mass region[9]. Nonetheless we can't account

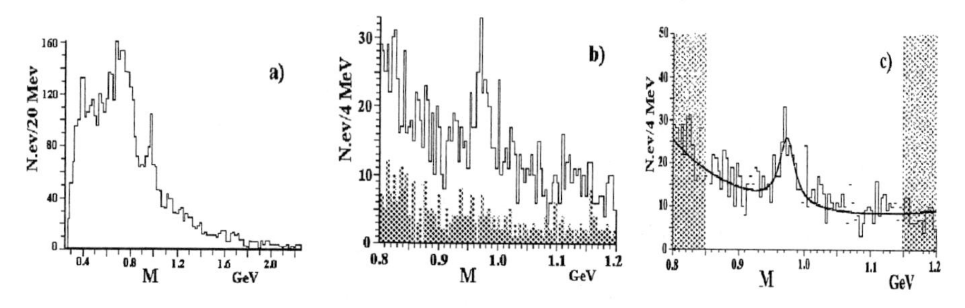

Fig. 5. a- $M(2\pi)$ spectrum (20 MeV bin) associated to a $\phi(1020)$; b- same (4 MeV bin), for $0.8 < M(2\pi) < 1.2$ GeV (black histogram); dashed histogram is background; c- same as b, fitted by Breit-Wigner plus background (3 parameters).

for over 4×10^5 events above background (less than 1 % of the events). F. E. Close et al.[10] suggested that the $f_0(980)$ and the $a_0(980)$ might play a role in the dynamics of quarks confinement. Thus we searched whether the $f_0(980)$ might be produced in association with the narrow $\phi(1020)$ di-kaon resonance. We selected events with $n_{ch} = 4$, energy balanced, with 2 "pion" tracks and 2 "kaon" tracks having a mass $|m_{2K} - 1.020| < 0.005$ GeV. The $M_{2\pi}$ mass distribution (20 MeV bin) is shown in fig. 5a and repeated in fig.s 5b and c (4 MeV bin) limited to the mass range $0.8 - 1.2$ GeV. In fig. 5b the dashed histogram is the background ($|m_{2K} - 1.045| < 0.005$ GeV). We find a $\sim 5\sigma$ signal (167 ± 47 events over background), a $\chi^2 = 73.45$ for 68 d.o.f., $m_{f_0(980)} = 973 \pm 2.5$ MeV, $\Gamma_{f_0(980)} = 28 \pm 10$ MeV and an unnormalized amplitude of 0.674 ± 0.20 (fit shown in fig. 5c). The significance of the signal is clearly enhanced by the presence of the ϕ meson in the event so that the signal seems to be favoured by the presence of hidden strangeness. The exclusive $\phi(1020)f_0(980)$ enhancement does not seem to come from a narrow higher mass resonance. This is the first observation of quasi-exclusive $\phi(1020)f_0(980)$ associated photoproduction.

References

[1] a- P.L. Frabetti et al. N.I.M. in Phys. Res. **A320**,(1992)519; b- ibidem **A329**,(1993)62;

[2] a- K.Gottfried and J.D.Jackson, N. Cim. **33**,(1964)309; b- P.Callahan et al., Fermilab Preprint FERMILAB-PUB-84/36-E;

[3] S.D.Drell and J.S.Trefil, Phys. Rev. Lett. **16**,(1966)552;

[4] A.Donnachie and A.B.Clegg, Zeit. Phys. **C51**,(1991)689;

[5] Particle Data Group, R.M.Barnett et al. Phys. Rev. **D54**, (1996)1;

[6] R.J.Oakes and J.J.Sakurai, Phys. Rev. Lett. **19**,(1967) 1266;

[7] T.A.Armstrong *et al.*, Zeit. Phys. **C51**,(1991)351;

[8] D.Morgan and M.R.Pennington, Zeit. Phys. **C48**,(1990)623;

[9] M.Harada *et al.*, Phys.Rev. **D54**,(1996),1991 and ref. therein;

[10] F.E.Close *et al.*, Phys. Lett. **B319**,(1993)291;

Study of $\eta\pi^-$ Production by Pions in Coulomb Field

Valeri Dorofeev[1], VES collaboration (dorofeev@mx.ihep.su)

Institute for High Energy Physics, Russia

Abstract. Coulomb production of $\eta\pi^-$-system at $M_{\eta\pi^-} < 1.18\,GeV$ in the reaction $\pi^- + Be \rightarrow \eta\pi^- + Be$ was observed. The coupling constant was extracted from the measured cross section and compared to that of radiative decay $\eta \rightarrow \pi^+\pi^-\gamma$.

1 Introduction

Coulomb production of $\eta\pi^-$-system (1) by negatively charged pions with momentum $37\,GeV$ was studied by VES collaboration at Serpukhov PS.

$$\pi^- + Z \rightarrow \eta\pi^- + Z \qquad (1)$$

The subject of this report are the features of subprocess (2) with $\eta\pi^-$-system mass $M_{\eta\pi^-} < 1.18\,GeV$.

$$\pi^- + \gamma^* \rightarrow \pi^- + \eta \qquad (2)$$

The amplitude of reaction (2) is:

$$M_{\eta\pi\pi\gamma} = -i\epsilon_{\mu\nu\rho\sigma} A^\mu p^\nu_{\pi^+} p^\rho_{\pi^-} p^\sigma_\eta \times F_{\eta\pi\pi\gamma}(s,t,u) \ . \qquad (3)$$

where A^μ - polarization vector, p - hadron four momentum, $F_{\eta\pi\pi\gamma}(s,t,u)$ - coupling constant, depending on kinematical invariants of reaction (2).

Reaction of Coulomb production is interesting object for exploration from the point of using chiral langrangian approach for study of hadron properties. Four vertex anomaly dominates in the chiral limit. Reaction (2) can be connected to the decay $\eta \rightarrow \pi^+\pi^-\gamma$ by crossing symmtery. The pole of ρ meson dominates in the $\pi^+\pi^-$ pair form factor and the correction to the chiral limit is significant.

Cross section of (1) is:

$$\frac{d\sigma}{dsdq^2} = \frac{Z^2\alpha}{\pi}\frac{|q^2 - q^2_{min}|}{q^4}\frac{1}{s - m^2_\pi} \times \frac{p^3_\eta p_\pi}{48\pi} F^2_{\eta\pi\pi\gamma} G^2(q^2) \ . \qquad (4)$$

where s - $\eta\pi^-$ invariant mass squared, q^2 - momentum transfer from π^- to $\eta\pi^-$ squared, Z, $G(q^2)$ - nucleus charge and form factor, p_η, p_π - η and beam CMS momentum respectively of reaction (1). The main background to reaction (1) is diffractive production of $\eta\pi^-$-system. Sharp q^2 dependence is the key point for selection of the process (1).

Diffractive production of $\eta\pi^-$ were previously carefully studied by VES collaboration [1] and BNL E852 experiment [2]. Both found relatively large production rate of $\eta\pi^-$ in the P-wave state. Intensity of this wave looks like a broad bump with maximum near $\sim 1.4\,GeV$.

2 Selection of $\pi^- + A \rightarrow \pi^- + \eta(\pi^+ + \pi^- + \pi^0) + A$.

Reaction:

$$\pi^- + A \to 2\pi^- + \pi^+ + \pi^0 + A \quad (5)$$

was selected from the experimental sample of $\sim 6 \times 10^7$ reconstructed events. Three track events with total charge equal to -1 and 2-3 reconstructed γ's were filtered. The vertex had to coinside with geometrical size and position of the target. Tracks with the ratio of energy deposited in calorimeter to momentum measured by spectrometer less than 0.8 were rejected as tracks of electrons. The energy of additional γ was required to be less than $0.7\,GeV$. For energy conservation only events with the total energy of final state particles $36\,GeV < E_{tot} < 39\,GeV$ were selected.

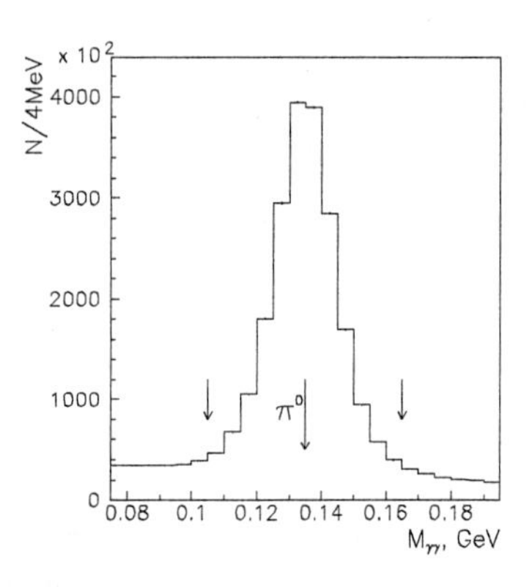

Fig. 1. Invariant mass of $\gamma\gamma$.

Fig. 1 shows the $\gamma\gamma$ invariant mass spectrum for selected events, where clear peak corresponding to π^0 can be seen. Events of $2\pi^-\pi^+\pi^0$ production were required to have $105\,MeV < M_{\gamma\gamma} < 165\,MeV$. Clear evidence of η meson production is seen in the $\pi^+\pi^-\pi^0$ invariant mass spectrum (fig. 2). Signal filtering procedure was applied to make distributions describing $\eta\pi^-$ production and decay. The number of η mesons in each bin was determined by the fit of $\pi^+\pi^-\pi^0$ invariant mass spectrum corresponding to this bin with Gaussian function and linear background. The mass and the width were fixed. We tried the usuall procedure of background subtraction. The results were identical and the difference was accounted in the errors.

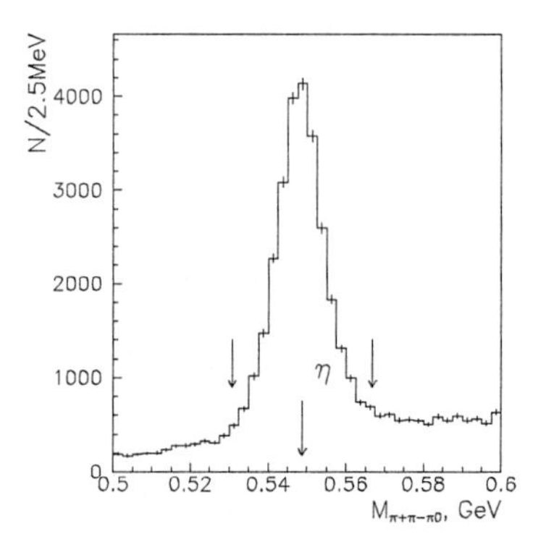

Fig. 2. Invariant mass of $\pi^+\pi^-\pi^0$

3 Selection of $\eta\pi^-$-system produced in the Coulomb field.

Copious production of $a_2(1320)$-meson is seen in $\eta\pi^-$ invariant mass spectrum in fig. 3. The region of $\eta\pi^-$ mass less than $1.18\,GeV$ was selected to study other than D-wave states of $\eta\pi^-$-system. Fig 5. shows the momentum transfer squared from the beam to the $\eta\pi^-$-system $t' = |t - t_{min}|$, with t_{min} to be the minimum possible value.

The sharp peak at low t' corresponds to Coulomb production. Relatively flat

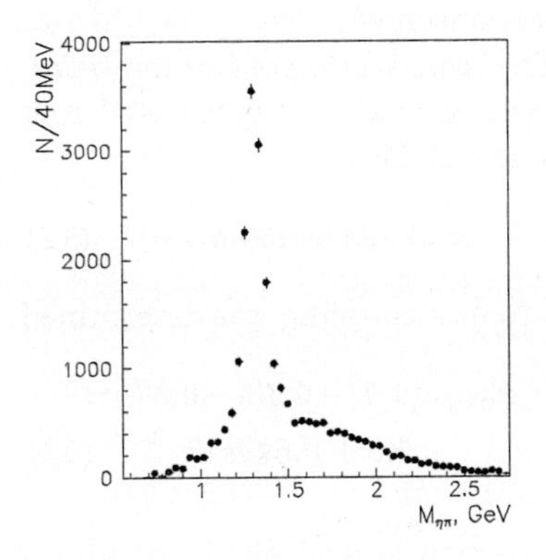

Fig. 3. Invariant mass of $\eta\pi^-$.

part of the distribution corresponds to the diffraction of π^- into $\eta\pi^-$. It is convenient to use the basis of states with definite naturality to describe the production of certain states. Naturality equals to $\eta = P(-1)^J$ in the low t' limit, where P - is parity and J - spin of the exchanged particle. States with positive naturality are dominantly produced in diffraction processes. Distribution of Treiman-Young angle is shown in fig. 4 for $t' > 0.01\,GeV^2$, where Coulomb production is negligidable. This distribution is well described by $\sin^2\phi$, corresponding to the diffractive production of waves with $M^\eta = 1^+$, where M is magnetic quantum number. Cross section of such processes is described by

$$\frac{d\sigma}{dt'} \sim t' e^{-Bt'} \quad . \tag{6}$$

It is important to mark that diffraction cross section tends to zero with t' decreasing and Coulomb production dominates at low $t' < 0.01$. It follows from above consideration that to estimate the Coulomb production rate it is sufficient to fit t'-distribution with sum of Coulomb term and (6).

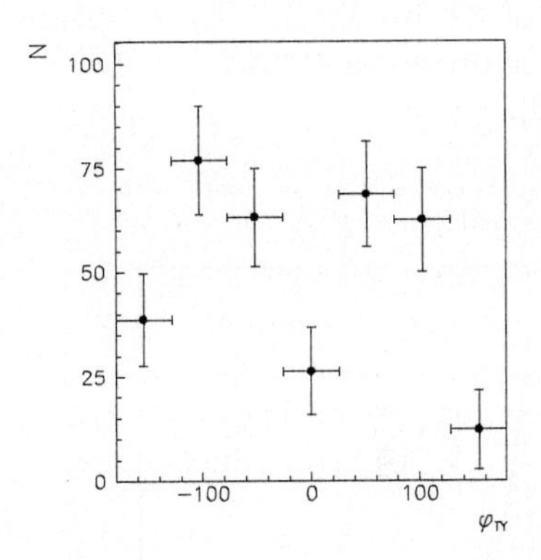

Fig. 4. Treiman-Young angle for $t' > 0.01\,GeV^2$.

4 Analysis of Coulomb produced low mass $\eta\pi^-$-system.

Coulomb production of P-wave $\eta\pi^-$ state dominates at low masses. The t'-distribution was fitted with coherent sum of (4) convoluted with experimental resolution function and (6) with constant phase. The phase turned out to be 90°. The number of Coulomb produced P-wave $\eta\pi^-$ state events was found by the fit with incoherent sum:

$$N = 182 \pm 29 \quad . \tag{7}$$

Background from a_2 Coulomb production was estimated using $a_2 \to \pi\gamma$ decay width [5]:

$$\sigma = 35 \pm 7\,nb \quad . \tag{8}$$

We determined the cross section of Coulomb production of $\eta\pi^-$-system with mass $M_{\eta\pi^-} < 1.18\,GeV$ and $t' < 0.09\,GeV^2$ to be:

$$\sigma = 188 \pm 43\,nb\ . \tag{9}$$

The cross section was normalized to that of 3π production by π^--beam with momentum $40\,GeV$ on Be target [3].

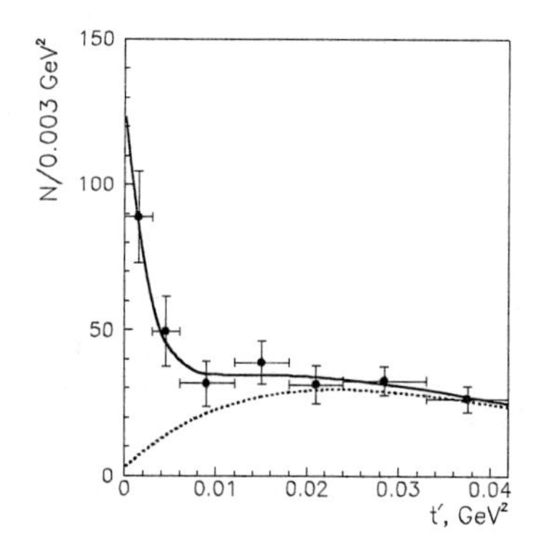

Fig. 5. Momentum transfer squared for $M_{\eta\pi^-} < 1.18\,GeV$. Background contribution is shown in a dashed line.

The value of $F_{\eta\pi\pi\gamma}(s,t,u)$ at the average kinematical point with $s = m^2_{\eta\pi} = 1.1\,GeV^2$, $t = m^2_{\pi\pi} = -0.38\ GeV^2$, $u = m^2_{\eta\pi_{in}} = -0.37\,GeV^2$ was determined from the cross-section according (4):

$$F_{\eta\pi\pi\gamma}(1.1, -0.38, -0.37) = 6.2 \pm 0.6\,GeV^{-3}\ . \tag{10}$$

From $\eta \to \pi^+\pi^-\gamma$ decay [4]

$$F_{\eta\pi\pi\gamma}(0.085, 0.167, 0.085) = 11.1 \pm 0.5\,GeV^{-3}\ . \tag{11}$$

It is evident the substantial dependence of $F_{\eta\pi\pi\gamma}$ on kinematical variables.

5 Conclusions.

- Coulomb production of P-wave $\eta\pi^-$-system with $M_{\eta\pi^-} < 1.18\,GeV$ was observed;
- The cross section of Coulomb production with $t' < 0.09\,GeV^2$ was found to be:

$$\sigma = 188 \pm 43\,nb\ . \tag{12}$$

- γ to $\eta\pi\pi$ coupling was determined

$$F_{\eta\pi\pi\gamma}(1.1, -0.38, -0.37) = 6.2 \pm 0.6\,GeV^{-3}\ . \tag{13}$$

References

[1] G.M. Beladidze et al, Physics Letters 313B (1993), 276-282
[2] D.R. Tompson et al, Physical Review Letters 79 (1997), 1630-1633
[3] G. Bellini et al, Nuclear Physics B199 (1982), 1-26
[4] Review of particle Physics, Physical Review D54 (1996), 325-326
[5] S. Cihangir et al, Physics Letters 117B (1982), 119-122

First Evidence for a Charm Radial Excitation $D^{*'}$

Daniel Bloch (Daniel.Bloch@in2p3.fr)

Institut de Recherches Subatomiques, Strasbourg, France

Abstract. Using D^{*+} mesons exclusively reconstructed in the DELPHI experiment at LEP, an excess of 67 ± 15(stat.) events is observed with a $D^{*+}\pi^+\pi^-$ invariant mass of 2637 ± 2(stat.) ± 7(syst.) MeV/c^2. This signal is compatible with the expected decay of a radially excited $D^{*'}$ ($J^P = 1^-$) meson.

1 Introduction

For mesons containing heavy and light quarks ($Q\bar{q}$) and in the limit where the heavy quark mass is much larger than the typical QCD scale ($m_Q \gg \Lambda_{QCD}$), the spin s_Q of the heavy quark and the total (spin+orbital) angular momentum $j_q = s_q + L$ of the light component are separately conserved. Heavy quark symmetry [1], together with the knowledge of lower mass mesons, allow to predict the mass and decay widths of heavy mesons of total spin $J = s_Q + j_q$ [2]. Previous estimates can also be found in ref. [3, 4, 5].

In the charm sector, the lowest $L = 0$ states of D and D^* mesons are well identified. Only the narrow orbital excitations ($L = 1$, $j_q = 3/2$) of charmed mesons have been clearly observed so far [6]. Broad states are also expected [3, 4, 5]. Together with these orbital excitations, radial excitations of heavy mesons are also foreseen [3]: based on a QCD inspired relativistic quark model, D' ($J^P = 0^-$) and $D^{*'}$ ($J^P = 1^-$) are expected with a mass of 2.58 GeV/c^2 and 2.64 GeV/c^2 , respectively. The uncertainty on the mass values is of the order of 10-25 MeV/c^2. The dominant decay modes are $D\pi\pi$ and $D^*\pi\pi$, respectively, but decays into $D^{(*)}\rho$ or through an intermediate orbital excitation are not excluded (see Table 1).

In the B sector, orbitally excited beauty mesons present a similar scheme. A narrow resonance compatible with the expectation of a $B^{(*)'}$ decaying into $B^{(*)}\pi^+\pi^-$ has been reported recently by the DELPHI experiment [7]. The purpose of the present study is to search for a narrow $D^{*+}\pi^+\pi^-$ resonance [1]. The followed strategy will be first to look for the known D_1^0 and D_2^{*0} decaying into $D^{*+}\pi^-$ in order to control the $(D^0\pi_*^+\pi^-)$ vertex finding procedure and the invariant mass estimate. Then $D^{*+}\pi^+\pi^-$ final states will be considered, fitting a similar $(D^0\pi^+\pi^-)$ vertex (see also reference [8] for more details).

[1] Throughout this note charge-conjugate states are implicitly included.

	J^P j_q	Mass (MeV/c^2)	Γ Width (MeV/c^2)
D_0^*	0^+ 1/2	~ 2360	≥ 170
D_1^*	1^+ 1/2	~ 2430	≥ 250
D_1^+	1^+ 3/2	2427 ± 5	28 ± 8
D_1^0	1^+ 3/2	2422.2 ± 1.8	$18.9^{+4.6}_{-3.5}$
D_2^{*+}	2^+ 3/2	2459 ± 4	25^{+8}_{-7}
D_2^{*0}	2^+ 3/2	2458.9 ± 2.0	23 ± 5

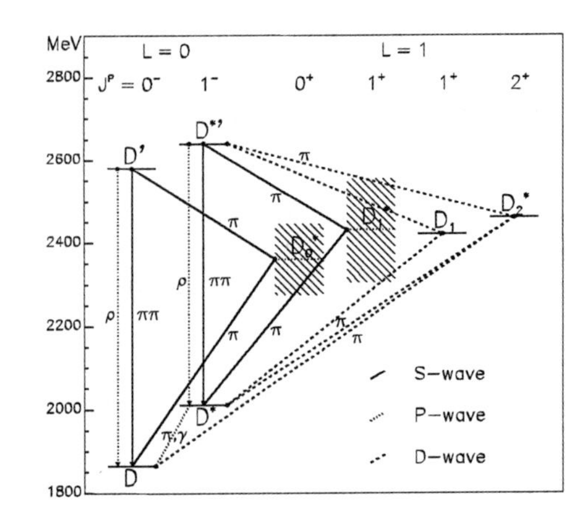

Table 1. Spectroscopy of non-strange D mesons. The broad $L = 1$ states ($j_q = 1/2$) are estimated according to [5]. The narrow $j_q = 3/2$ states are measured [6].

2 Selection of D^{*+} and orbitally excited states

The DELPHI detector has been described in detail elsewhere [9]. Using standard selection criteria, a total of 3.4 million hadronic events was obtained from the 1992-1995 data at centre-of-mass energies close to the Z mass.

The D^{*+} candidates were selected, with an energy $E(D^{*+})/E_{\rm beam} > 0.25$, using the decay channels $D^{*+} \to D^0\pi_*^+$ followed by $D^0 \to K^-\pi^+$ or $D^0 \to K^-\pi^+\pi^+\pi^-$. Only kinematical selections were applied in the $K\pi$ channel whereas, in order to reduce the combinatorial background, kaon identification, impact parameter cuts and a positive decay length were also required in the $K3\pi$ channel. Finally, applying cuts around the nominal D^0 mass and the $(D^{*+}-D^0)$ mass difference, a total amount of 4660 ± 95(stat.) and 3250 ± 90(stat.) $D^{*\pm}$ were reconstructed in both channels, respectively, with a signal-to-noise ratio larger than two.

In the $D^{*+}\pi^-$ final state, all pion candidates of charge opposite to the D^{*+} and momentum greater than 1.5 GeV/c were considered. Those pion candidates compatible with a kaon according to the combined RICH and dE/dx identification were rejected.

Then a $D^0\pi_*^+\pi^-$ vertex was fitted. In the $D^0 \to K^-\pi^+$ channel, the additional pion π^- was selected if its impact parameter with respect to the common $D^0\pi_*^+\pi^-$ vertex was less than 50 μm. In the $D^0 \to K^-\pi^+\pi^+\pi^-$ channel, the χ^2 probability of the $D^0\pi_*^+\pi^-$ vertex had to be larger than 0.001. These different vertex conditions were justified because the D^0 decay vertex was better defined in the case of the $K^-\pi^+\pi^+\pi^-$ channel.

In order to reduce further the combinatorial background, data samples enriched in $b\bar{b}$ or $c\bar{c}$ events were selected by using the D^0 apparent proper time $t(D^0)$, the decay length significance $\Delta L/\sigma_L$ between the secondary $D^0\pi_*^+\pi^-$ vertex and the primary vertex, and by using a B-tagging algorithm [9].

The $(D^{*+}\pi^-)$ invariant mass was computed as: $M(D^{*+}\pi^-) = M(D^0\pi_*^+\pi^-) - M(D^0\pi_*^+) + m_{D^*}$ with a resolution σ of about 6 MeV/c^2 , according to the simulation.

This invariant mass distribution is presented on Figure 1 (left). An excess of $(D^{*+}\pi^-)$ pairs is observed between 2.4 and 2.5 GeV/c^2 , but not in the $(D^{*+}\pi^+)$ sample. The $M(D^{*+}\pi^-)$ distribution was fitted as the sum of two contributions: a function of the form $\alpha(M(D^*\pi) - m_{D^*} - m_\pi)^\beta \cdot \exp(-\gamma(M(D^*\pi) - m_{D^*} - m_\pi))$ for the background, and two Breit-Wigner functions describing both D_1^0 and D_2^{*0} resonances. The overall number and relative proportions of these states and their average masses were left free in the fit, whereas both full widths were fixed to their world average value [6] (see Table 1). A total number of 311 ± 56 D_1^0 and D_2^{*0} was observed, $(65 \pm 8)\%$ being associated to the D_1^0 resonance. The obtained D_1^0 mass was 2421 ± 2 MeV/c^2, in good agreement with the nominal value of 2422 ± 2 MeV/c^2. The fitted D_2^{*0} mass of 2475 ± 5 MeV/c^2 is larger than the nominal value of 2459 ± 2 MeV/c^2 [6]. All fitted errors are statistical only.

3 Search for a radially excited $D^{*'}$ meson

Repeating a similar procedure as in the previous section, all remaining opposite charge $\pi^+\pi^-$ pairs produced in the same direction as the D^{*+} candidates were used to fit a $D^0\pi^+\pi^-$ vertex.

In the $K\pi$ ($K3\pi$) channel, the additional pions were selected with a momentum larger than 0.6 (1.0) GeV/c . The reconstructed invariant mass was computed as previously:
$M(D^{*+}\pi^+\pi^-) = M(D^0\pi_*^+\pi^+\pi^-) - M(D^0\pi_*^+) + m_{D^*}$ with a resolution σ of about 6 MeV/c^2, according to the simulation. The invariant mass distribution of all $D^{*+}\pi^+\pi^-$ candidates is presented on Figure 1 (right). A narrow peak is observed whereas the wrong charge combinations $D^{*+}\pi^-\pi^-$ do not show any excess. A binned maximum likelihood fit was performed on this $M(D^{*+}\pi^+\pi^-)$ distribution by taking into account two contributions: a function of the form $\alpha(M(D^*\pi\pi) - m_{D^*} - m_{\pi\pi})^\beta \cdot \exp(-\gamma(M(D^*\pi\pi) - m_{D^*} - m_{\pi\pi}))$ for the background, with $m_{\pi\pi}$ set to 340 MeV/c^2, and a Gaussian function with free parameters to describe the narrow peak. The χ^2 per degree of freedom is 64/59. The fitted number of events is 67 ± 15, the fitted σ width is 7 ± 2 MeV/c^2, in agreement with the expected resolution, and the average mass is found to be 2637 ± 2 MeV/c^2. Only statistical errors are quoted.

This 4.5 standard deviations signal is likely to come from the expected $D^{*'}$ radial excitation (see Table 1) [3]. A systematic uncertainty of ± 7 MeV/c^2 on its mass can be inferred by comparing the previously fitted invariant mass of the D_1^0 and D_2^{*0} to their world average values.

No signal was found for the decay $D_2^{*+}(2459) \rightarrow D^{*+}\rho^0$ which is kinematically disfavoured [2].

According to the simulation, $D^{*'}$ candidates had an overall reconstruction and selection efficiency of about 5% (2%) in the $K\pi$ ($K3\pi$) decay mode of the D^0, respectively. The corresponding efficiencies were about 13% (5%) for the observed orbital resonances. The production fraction of $D^{*'}$ can thus be compared to D_1^0 and D_2^{*0} :

$$\frac{\mathrm{Br}(Z \to (D^{*'} \to D^{*+}\pi^+\pi^-)\ X)}{\sum_{J=1,2} \mathrm{Br}(Z \to (D_J^{(*)})^0 \to D^{*+}\pi^-)\ X)} = 0.5 \pm 0.2 \text{ (stat. data + stat. MC)}.$$

This is the first evidence for a radially excited $D^{*'}$ ($J^P = 1^-$) meson, as predicted by S. Godfrey and N. Isgur [3].

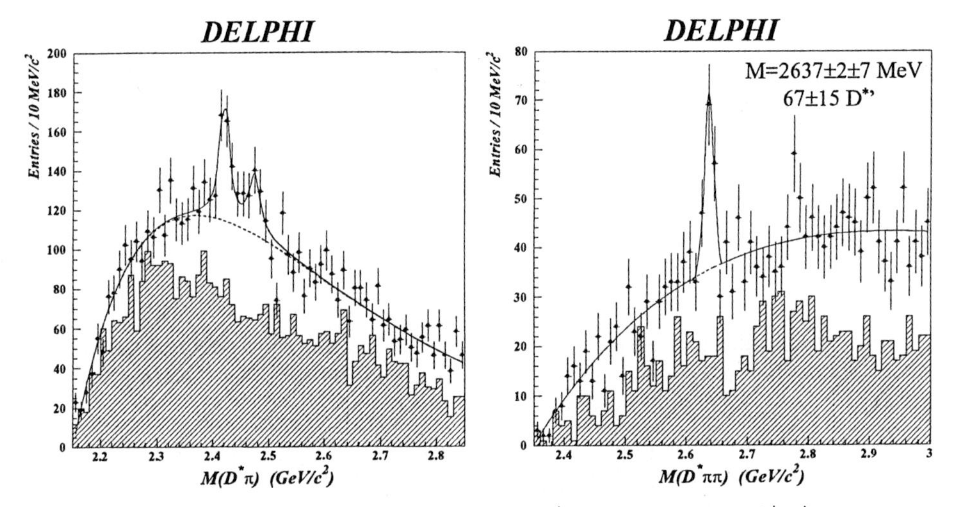

Fig. 1. Invariant mass distributions (left) $D^{*+}\pi^-$ and (right) $D^{*+}\pi^+\pi^-$. Wrong charge combinations $D^{*+}\pi^+$ and $D^{*+}\pi^-\pi^-$ are presented in hatched histograms. The mass computation and the fitted curves are explained in the text.

Acknowledgements

I warmly thank J.-P. Gerber, P. Juillot, M. Feindt, R. Strub and T. Spassov for their fruitful help and significant contributions.

References

[1] N. Isgur and M.B. Wise, Phys. Rev. **D66** (1991) 1130.
[2] E.J. Eichten, C.T. Hill and C. Quigg, Phys. Rev. Lett.**71** (1993) 4116.
[3] S. Godfrey and N. Isgur, Phys. Rev. **D32** (1985) 189.
[4] J. Rosner, Comments on Nucl. and. Part. Phys. **16** (1986) 109.
[5] S. Godfrey and R. Kokoski, Phys. Rev. **D43** (1991) 1679.
[6] Particle Data Group, Phys. Rev. **D54** (1996) 1.
[7] DELPHI collab., M. Feindt and O. Podobrin, ICHEP'96, pa01-021.
[8] DELPHI collab., D. Bloch et al., HEP'97 #452, DELPHI 97-102.
[9] DELPHI collab., P. Abreu et al., Nucl. Instr. & Meth. **A378** (1996) 57.

Properties of Light Particles Produced in Z^0 Decays

Volker Blobel (blobel@mail.desy.de)

University of Hamburg, Germany

Abstract. Properties of light particles in the decay of the Z^0 have been studied, the data are compared with Monte Carlo models. Spin alignment of leading particles with preference for helicity 0 is observed for several vector mesons. Measured characteristics of $f_0(980)$ production are consistent with its interpretation as a conventional $q\bar{q}$ scalar meson.

1 Vector meson spin alignment in hadronic Z^0 decays

The production and decay properties of vector mesons can be described in terms of a 3×3 spin density matrix, denoted $\rho_{\lambda\lambda'}$ in the helicity basis. For decay into two pseudoscalars the element ρ_{00} is determined from the decay angular distribution $W(\cos\vartheta) = (3/4)\left[(1-\rho_{00}) + (3\rho_{00}-1)\cos^2\vartheta\right]$ after φ integration $(-\pi \leq \varphi \leq +\pi)$. The distribution is isotropic for no spin alignment ($\rho_{00} = 1/3$), and is proportional to $\sin^2\vartheta$ for helicity ± 1 states and to $\cos^2\vartheta$ for helicity 0 states. Simple statistical models based on spin counting give $\rho_{00} = \frac{1}{2}(1-(P/V))$ in terms of P/V, the ratio of pseudoscalar to vector meson production. Any alignment of the vector meson spin must arise from the hadronisation phase; however spin aspects of particle production are essentially ignored in the commonly used Monte Carlo models of the hadronization. No spin alignment of vector mesons is expected in the simplest string model. The QCD cluster model of HERWIG has no mechanism to produce spin-aligned vector mesons. Neither model will generate non-zero values for the off-diagonal elements of the helicity density matrix.

Contributions to this conference are about ρ^0, K^{*0}, Φ mesons [1] by the DELPHI collaboration, on Φ, $D^{*\pm}$, B^* mesons [2] and on K^{*0} mesons [3] by the OPAL collaboration.

The DELPHI results in the region of scaled momentum $x_p \leq 0.3$ show no evidence for spin alignment of the K^{*0} and Φ; the results are consistent with $\rho_{00} = 1/3$ and with the ratio of $P/V = 1/3$ as expected from spin counting. Some indication for spin alignment however is found in the fragmentation region $x_p \geq 0.4$ of the ρ^0 and the K^{*0}. For the Φ no spin alignment is observed by DELPHI for $x_p \geq 0.4$, but for $x_p \geq 0.7$ with $\rho_{00} = 0.55 \pm 0.10$. The OPAL value $\rho_{00} = 0.54 \pm 0.06 \pm 0.05$ for $x_E > 0.7$ is almost identical. The OPAL results for the K^{*0} are show in figure 1 as a function of x_p. The figure shows for $x_p \geq 0.3$ a clear leading particle effect with $\rho_{00} > 1/3$. The D^* has $\rho_{00} > 1/3$ too, although the value is not als large as that for

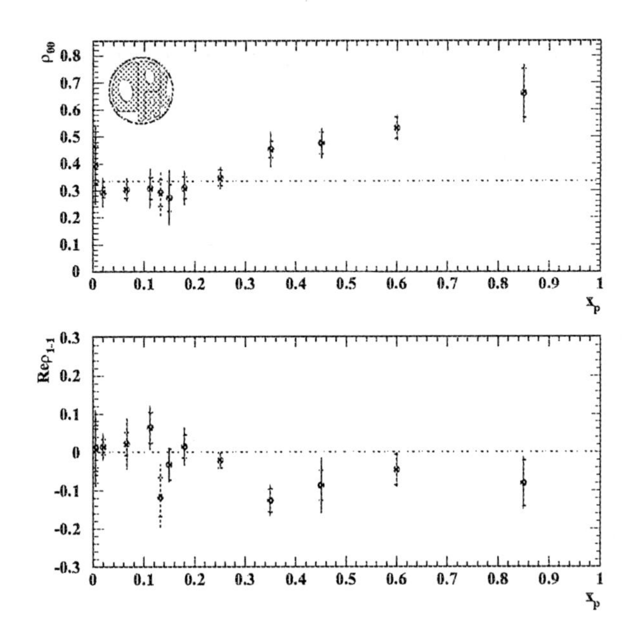

Fig. 1. Measured K^{*0} helicity density matrix elements, ρ_{00} and Re $\rho_{1,-1}$, as function of the scaled momentum, x_p. The error bars are statistical and systematic combined in quadrature, with the tick marks giving the statistical errors only.

the Φ. For the B^* the result $\rho_{00} = 0.36 \pm 0.06 \pm 0.07$ is consistent with *no spin alignment*, as expected, if the primary b quark hadronizes to produce vector and pseudoscalar mesons in the ratio 3:1. Approximately 20% of the B^* mesons may be decay products of the orbitally excited B^{**} states, so some alignment of the primary contribution is not ruled out.

In most cases deviations from isotropy are small, although the measurements show some evidence that spin plays a role in the production dynamics of vector mesons from parton hadronization at LEP energies. The values obtained for off-diagonal elements are in general small, and mostly consistent with zero. However OPAL observes in the large-momentum region non-zero (and negative) values of Re $\rho_{1,-1}$ in the case of Φ, D^* and especially K^{*0} (see figure 1). In the calculation of Anselmino et al. [4] negative values of Re $\rho_{1,-1}$ are generated for leading vector mesons by coherence in the production and fragmentation of the primary quark and antiquark from the Z^0 decay. Quantitatively the expectation for the ratio $\mathrm{Re}\,\rho_{1,-1}/(1-\rho_{00})$ is about -0.10 for mesons containing a primary quark. The OPAL result for the weighted average of the ratio is -0.19 ± 0.05 for $x_p > 0.3$. Other off-diagonal elements are found to be consistent with zero over the entire x_p range.

2 Polarization and Forward-Backward Asymmetry of Λ Baryons

Polarization and forward-backward asymmetry of Λ baryons in hadronic Z^0 decays has been studied by the OPAL collaboration [5]. For the s quarks from Z^0 decay a longitudinal polarization P_L of -0.91 is expected in the standard model. The polarization parameter P_L of the Λ can be determined from the decay distribution in the weak decay of Λ: $1/N\ dN/d\cos\vartheta^* = 1 + \alpha P_L \cos\vartheta^*$ (with the Λ decay parameter $\alpha = 0.642 \pm 0.013$). The OPAL result is $P_L = -32.9 \pm 7.6\%$ for $x_E > 0.3$. This is consistent with earlier ALEPH results. The results are reasonably well modelled using a simple quark model and the JETSET Monte Carlo, which has been tuned using LEP data. There is no significant evidence for any transverse polarization. A significant forward-backward asymmetry (0.083 ± 0.013 for $x_E > 0.3$), consistent with earlier ALEPH and DELPHI results, has been measured and can be described by a JETSET model.

3 Ξ production

Based on a sample of about 2500 Ξ^- and 2300 $\bar{\Xi}^+$ decays observed in the 92 to 95 data, the DELPHI collaboration [6] made direct measurements of the Ξ^- and $\bar{\Xi}^+$ masses, lifetimes and mass and lifetime differences. The inclusive production rates for Ξ^- plus $\bar{\Xi}^+$ in hadronic Z^0 decays and in $Z^0 \to b\bar{b}$ decays were found to be

$$\langle \Xi^- + \bar{\Xi}^+ \rangle_{q\bar{q}} = 0.0233 \pm 0.0007(\text{stat.}) \pm 0.0024(\text{syst.})$$
$$\langle \Xi^- + \bar{\Xi}^+ \rangle_{b\bar{b}} = 0.0183 \pm 0.0016(\text{stat.}) \pm 0.0030(\text{syst.})\ .$$

The predictions of JETSET are in numerical agreement with these measurements, while the predictions of HERWIG ($\langle \Xi^- + \bar{\Xi}^+ \rangle_{q\bar{q}} = 0.0730$ for HERWIG 5.9) are not.

4 Production of $f_0(980)$, $f_2(1270)$ and $\Phi(1020)$

Inclusive production of f_0, f_2 and Φ in Z^0 decay has been studied by the OPAL collaboration [7]. The $f_0(980)$ is well established as a scalar ($J^{PC} = 0^{++}$) state (with ${}^3P_0$?), but some properties (a small total and $\gamma\gamma$ width, a relatively large coupling to $K\bar{K}$) differ from the expectation for a conventional meson. A number of suggestions have been made as to the nature of the $f_0(980)$. It could be a "cryptoexotic" $qq\bar{q}\bar{q}$ state [Jaffe and Johnson], a $K\bar{K}$ molecule [Weinstein and Isgur] or a glueball, or the $f_0(980)$ plays the role of a novel "vacuum scalar" state (bound state of q and $\bar{q}$ with negative kinetic energy, interacting repulsively to give a state of positive energy) [Gribov]. Signature of a "vacuum scalar" state is a relatively larger yield in low multiplicity events and in events, where the f_0 is isolated in rapidity.

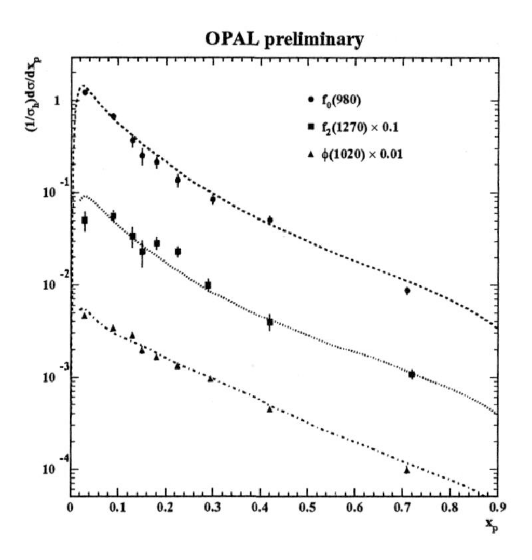

Fig. 2. Measured fragmentation functions for f_0, f_2 and Φ as a function of the scaled momentum x_p, together with curves from the JETSET 7.4 Monte Carlo generator, normalized for each particle to the total rate seen in the data. The errors are statistical only for the f_0 and the f_2, and include uncorrelated systematic errors for the Φ. Additional fully correlated systematic errors are 7.5 % for the f_0, 11 % for the f_2, and 2.8 % for the Φ.

The OPAL approach to analyse the $f_0(980)$ is to compare the production with the production of Φ (which is close in mass) and of $f_2(1270)$ (which is an established meson with $q\bar{q}$ in $^3P_2$ state), and a comparison with the LUND string model of hadronization, in which the f_0 is treated as a conventional meson. Measured fragmentation functions $(1/\sigma_\mathrm{h})\mathrm{d}\sigma/\mathrm{d}x_\mathrm{p}$ are shown in figure 2 along with curves from JETSET 7.4. Total inclusive rates (per hadronic Z^0 decay) are 0.160 ± 0.014 (for the $f_0(980)$) and 0.119 ± 0.016 (for the $f_2(1270)$). The values are consistent with f_0 and f_2 being $q\bar{q}$ mesons in the ^3P state composed mainly of $u\bar{u}$ and $d\bar{d}$. In addition OPAL has determined the ratio of data rate : JETSET 7.4 prediction versus the event charged-particle multiplity and the rapidity gap between the meson and the nearest charged particle. There are no significant differences in production characteristics between the three mesons. No evidence is observed for enhanced f_0 production at low multiplicities or at large rapidity gaps, thus no evidence to identify the f_0 with the vacuum scalar state proposed by Gribov. All measured characteristics of f_0 production in the Z^0 decay data of OPAL are consistent with its interpretation as a conventional scalar meson.

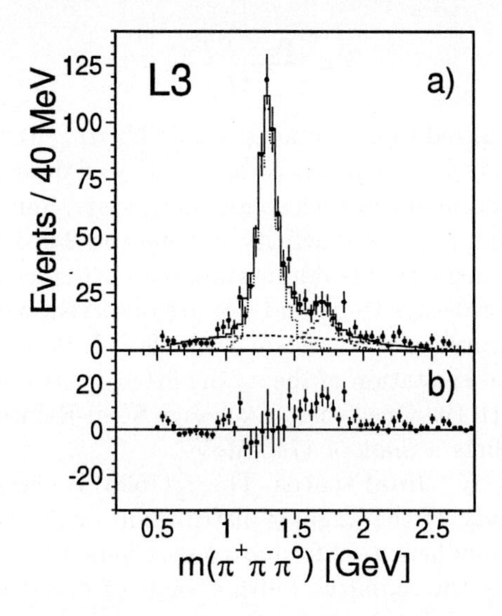

Fig. 3. (a) The $\pi^+\pi^-\pi^0$ mass spectrum. The solid line is a fit to the Monte Carlo sample for the a2 (dotted line), a Breit-Wigner distribution in the high-mass region (dotted-dashed line) and a background parametrized by a third order polynominal (dashed line). (b) The distribution with a_2 and background subtracted.

5 Light-quark Resonances in Two-Photon Collisions

The two-photon widths of several light-quark resonances formed in $\gamma\gamma$ collisions have been measured by the L3 collaboration [8].

$\pi^+\pi^-\gamma$ final state. For quasi-real photons ($Q^2 < 0.01$ GeV^2) the mass distribution is dominated by the $\eta'(958)$, with background from $a_2(1300)$ (one photon undetected). The result for the radiative width $\Gamma_{\gamma\gamma}(\eta') = 4.17 \pm 0.10$ (stat) ± 0.27 (syst) keV is the most precise value obtained in a single experiment. The η' transition form factor has been measured in the Q^2 ranges $0.01 \leq Q^2 < 0.9$ GeV^2 and (in singly tagged events) $1.5 \leq Q^2 \leq 10$ GeV^2. The data are well represented by a pole parametrization with $\Lambda = 0.900 \pm 0.046$ (stat) ± 0.022 (syst) GeV.

$\pi^+\pi^-\pi^0$ final state. The $\pi^+\pi^-\pi^0$ mass spectrum is shown in figure 3. The low-mass sample $1.0 < m(\pi^+\pi^-\pi^0) < 1.55$ GeV is dominated by the $J^{PC} = 2^{++}$ state $a_2(1300)$ with decay $a_2 \to \rho^\pm\pi^\mp$. The relative importance of helicity states 0 and 2 is determined from the $\cos\vartheta$ distribution with the result: $P(\text{helicity}\,2) = 0.92 \pm 0.05$ with a systematic uncertainty of 6 %.

The high-mass sample $1.55 < m(\pi^+\pi^-\pi^0) < 2.1$ GeV is studied under the hypothesis of a resonance (or resonances) decaying through $\rho\pi$ and $f_2\pi$. The Dalitz parameter Λ defined in the $\pi^+\pi^-\pi^0$ rest frame by

$$\Lambda = \left| \frac{\mathbf{p}_{\pi^+} \times \mathbf{p}_{\pi^-}}{Q_K^2} \right|^2$$

(π momenta normalized to maximum possible kinetic energy Q_K) is used for the spin-parity analysis. Λ depends on spin-parity of the state and the amplitude and phase parameters for each spin-helicity state. Spin-parity $J^P = 2^+$ is favored. Relative importance of helicity states 0 and 2 is determined from the $\cos\vartheta$ distribution. Helicity 2 is dominating with $P(\text{helicity}\,2) \approx 75\%$. Equal branching ratios for decays to $\rho\pi$ and $f_2\pi$ are observed, with the phase angle between the two amplitudes consistent with 180°. If the events are identified with the first radial excitation of the a_2 the fitted mass $1752 \pm 21 \pm 4$ MeV is consistent only with the One Gluon Exchange Semi-Relativistic Quark Mass model, which predicts a mass of 1740 MeV.

$K_S^0 K_S^0$ and K^+K^- final states. The $f_2^{'}(1525)$ is observed in the $K_S^0 K_S^0$ final state; the study of the angular distribution in the two-photon centre-of-mass system favors helicity-2 formation over helicity-0, as expected. From the measurement of the radiative width a value of $\vartheta = (29.4 \pm 1.4/1.6^{+1.4}_{-1.6})°$ is obtained for the mixing angle between the singlet and octet members of the tensor meson nonet, in agreement with the value of $\vartheta = (28 \pm 3)°$ found from the Gell-Man - Okubo mass formula; the result indicates a deviation fom ideal mixing (35.3°). Around 1300 MeV $f_2 - a2$ interference is observed, which is destructive in the $K_S^0 K_S^0$ final state and constructive in the K^+K^- final state, as theoretically predicted [Lipkin]. A 3σ enhancement is observed in the $K_S^0 K_S^0$ spectrum around 1750 MeV, which is probably due to the formation of a radially excited state of the $f_2^{'}(1525)$ tensor meson.

References

[1] DELPHI Collaboration, Measurement of the Spin Density Matrix for the ρ_0, $K^{*0}(892)$ and Φ Produced in Z^0 Decays, CERN-PPE/97-55, contributed paper # 235.
[2] OPAL Collaboration, Study of $\Phi(1020)$, $D^{*\pm}$ and B^* Spin Alignment in Hadronic Z^0 Decays, CERN-PPE/97-05, contributed paper # 190.
[3] OPAL Collaboration, Spin Alignment of Leading $K^{*0}(892)$ Mesons in Hadronic Z^0 Decays, contributed paper # 191.
[4] Anselmino et al., hep-ph/97004420
[5] OPAL Collaboration, Polarization and Forward-Backward Asymmetry of Λ Baryons in Hadronic Z^0 Decays, contributed paper # 197.
[6] DELPHI Collaboration, Masses, Lifetimes and Production Rates of Ξ^- and $\bar{\Xi}^+$ at LEP1, contributed paper #719.
[7] OPAL Collaboration, Production of $f_0(980)$, $f_2(1270)$ and $\Phi(1020)$ in Hadronic Z^0 Decays, contributed paper # 193.
[8] L3 Collaboration, Formation of Light-quark Resonances in Two-Photon Collisions at LEP, contributed paper # 474.

Meson Spectra and Couplings in a Relativistic Quark Model

I. Zakout

Department of Physics, Bethlehem University, Bethlehem P.O. Box 9, Palestinian National Authority

Abstract. The meson spectra and couplings of heavy quarkonia are studied in the relativistic quark model. The matrix elements of meson couplings are evaluated exactly in terms of energy plane wave amplitudes. The leptonic decay widths and pseudoscalar form factors are calculated. The electromagnetic decay rates are found in agreement with recent experiments after considering the one loop QCD radiative corrections. We find that the small-small component of the positive and negative energy solution in the canonical representation couples the P-state with the F-state. The exact calculation by taking the contribution of the F-state in the calculation of the electromagnetic decay rates $\Gamma(\mathbf{2}^{++} \to \gamma\gamma)$ raises the value of $R = \Gamma(\mathbf{2}^{++} \to \gamma\gamma)/\Gamma(\mathbf{0}^{++} \to \gamma\gamma)$ significantly.

I. INTRODUCTION

The meson spectra and couplings in the relativistic quark model have received much attention with the development of the experimental facilities. The decay rates of many processes are subjected to substantial relativistic corrections[1-2]. At present work, the meson couplings will be studied in the context of the quasipotential equation[3]. The outline is as follows. In sect. II, we present the projection of the wave function amplitude into energy plane waves and the normalization of the quasipotential equation. In sect. III, the pseudoscalar form factor and the leptonic decay width are given. The electromagnetic decay rates are considered thoroughly in the Mandelstam formalism. Finally, discussions and conclusions are presented in Sect. IV.

II. QUASIPOTENTIAL EQUATION

The quasipotential equation for the two-body bound state reads,

$$\mathcal{G}^{-1}(P,\mathbf{p})\phi(P,\mathbf{p}) = -\int \frac{d^3q}{(2\pi)^3}\mathcal{V}(\mathbf{p}-\mathbf{q})\phi(P,\mathbf{q}), \tag{1}$$

where $\mathcal{G}^{-1}(P,\mathbf{p}) = \mathcal{G}_a^{-1}(P,\mathbf{p})\mathcal{G}_b^{-1}(P,\mathbf{p})$ and $\mathcal{V}(\mathbf{p}-\mathbf{q})$ are the quark(-antiquark) propagator and the kernel interaction, respectively. The canonical wave function can be decomposed into energy plane waves as

$$\phi(\mathbf{p}) = \Lambda_a^+(\mathbf{p})\phi(\mathbf{p})\Lambda_b^-(-\mathbf{p}) + \Lambda_a^+(\mathbf{p})\phi(\mathbf{p})\Lambda_b^+(-\mathbf{p}) + \Lambda_a^-(\mathbf{p})\phi(\mathbf{p})\Lambda_b^-(-\mathbf{p}) + \Lambda_a^-(\mathbf{p})\phi(\mathbf{p})\Lambda_b^+(-\mathbf{p}) \tag{2}$$

The projection energy operator for the constituent quark reads $\Lambda_a^\pm = \frac{1}{2}\left[1 \pm \frac{\alpha\cdot\mathbf{p}+\beta m_a}{E_a}\right]$ where $E_a = \sqrt{\mathbf{p}^2+m_a^2}$ and $S_a = E_a + m_a$. The quasipotential equation propagator can be projected into energy channels. The BSLT propagator is adopted at present calculations. The propagator of the quasipotential equation in this representation reads

$$\mathcal{G}(\mathbf{p})\phi(\mathbf{p}) = \sum_{\pm\pm} \mathcal{G}_{\pm\pm}(\mathbf{p})\Lambda_a^{\pm}(\mathbf{p})\gamma_0\phi(\mathbf{p})\gamma_0\Lambda_b^{\pm}(-\mathbf{p}). \tag{3}$$

The form of the bound state equation reads[3]

$$\lambda\phi(\mathbf{p}) = -\mathcal{G}(M_{ab},\mathbf{p})\mathcal{V}(\mathbf{p},\mathbf{q})\phi(\mathbf{q}). \tag{4}$$

The bound state M_{ab}^i and its corresponding wave function $\phi(\mathbf{p}) = (\eta(\mathbf{p}), i\xi(\mathbf{p}), i\chi(\mathbf{p}), \zeta(\mathbf{p}))$ can be found by solving Eq.(4) as a standard eigenvalue problem with respect to the fictitious parameter λ and tunneling $\lambda_i \to 1$. The details are found in Ref. 3.

The normalization of Bethe-Salpeter wave function amplitude is essential to calculate the matrix elements of meson couplings and decay constants. The three-dimensional quasipotential normalization reads

$$\int \frac{d^3\mathbf{p}}{(2\pi)^3}\bar{\phi}(M_n,\mathbf{p})\frac{\partial}{\partial M_n}\left[\sum_{\pm\pm} -\mathcal{G}^{\pm\pm^{-1}}(M_n,\mathbf{p})\gamma_0\Lambda_a^{\pm}(\mathbf{p})\phi(M_n,\mathbf{p})\Lambda_b^{\pm}(-\mathbf{p})\gamma_0\right] = 1 \tag{5}$$

III. MESON COUPLINGS

The pseudoscalar form factor using the wave function amplitudes of the quasipotential equation reduces to

$$f_P = 2\sqrt{\frac{N_c}{M_n}}\int \frac{d^3\mathbf{p}}{(2\pi)^{3/2}}\left[\left(1-\frac{p^2}{S_aS_b}\right)\eta_{00}(\mathbf{p}) - \left(1-\frac{p^2}{S_aS_b}\right)\zeta_{00}(\mathbf{p})\right]. \tag{6}$$

The leptonic decay width $\Gamma_n(\mathbf{l}^+l^-)$ for the decay of a $J^{PC} = \mathbf{1}^{--}(q\bar{q})$ vector meson of mass M_n into a lepton pair through a single intermediate photon is given by

$$\Gamma_n(l^+l^-) = \frac{16\pi N_c\alpha^2 e_q^2}{3M_n^2}. \tag{7}$$

$$\left|\int \frac{d^3\mathbf{p}}{(2\pi)^{3/2}}\left[\left(1+\frac{1}{3}\frac{p^2}{S_aS_b}\right)\frac{\eta_{01}(|\mathbf{p}|)-\zeta_{01}(|\mathbf{p}|)}{\sqrt{4\pi}} - \left(\frac{2\sqrt{2}}{3}\frac{p^2}{S_aS_b}\right)\frac{\eta_{21}(|\mathbf{p}|)-\zeta_{21}(|\mathbf{p}|)}{\sqrt{4\pi}}\right]\right|^2 \tag{8}$$

where α is the fine structure constant and $N_c = 3$. The contribution of the D-state is found at the same order of the relativistic correction of the S-state for vector meson.

The T-matrix element for a meson with mass M decays into two photons with momentum and polarization of photons k_1, ε_1 and k_2, ε_2 in the Mandelstam formalisem can be written as

$$T = i\sqrt{N_c}(e_q)^2\alpha\int \frac{d^4p}{(2\pi)^4}\mathrm{Sp}\mathcal{G}_F(\frac{1}{2}P+p)\Gamma(\mathbf{p})\mathcal{G}_F(-\frac{1}{2}P+p)... \tag{9}$$

$$\left[\not{\varepsilon}_1\mathcal{G}_F(\frac{1}{2}P+p-k_1)\not{\varepsilon}_2 + \not{\varepsilon}_2\mathcal{G}_F(k_1-\frac{1}{2}P+p)\not{\varepsilon}_1\right], \tag{10}$$

where the vertex function is given by $\Gamma(\mathbf{p}) = -\int \frac{d^3q}{(2\pi)^{3/2}}\mathcal{V}(\mathbf{p}-\mathbf{q})[\sqrt{2M}\phi(\mathbf{p})]$.

The use of relativistic amplitude of the quasipotential equation and the off-shell treatment of the quarks allows one to avoid the use of the phenomenological "mock-meson" correction factors being increasingly uncertain for high angular momentum states. The electromagnetic decay rate width is written as

$$\Gamma(J^{PC}) = \frac{1}{2}\frac{1}{(2J+1)}\sum_{M_z}\sum_{\text{polar}}\int \delta(\mathbf{k}-k_0\hat{z})|T|^2 d\Omega_k \tag{11}$$

IV. RESULTS AND CONCLUSIONS

We have solved the quasipotential equation with BSLT propagator[3]. The mass spectra for an equal admixture vector-scalar confining potentials are found in the levels $[1^1S_0, 1^3S_1, 2^3S_1, 1^3D_1] = [2.983, 3.098, 3.691, 3.777]$ GeV and $[1^3S_1, 2^3S_1] = [9.461, 10.020]$ GeV for $c\bar{c}$ and $b\bar{b}$, respectively, for our fitting parameters.

The plane wave function amplitudes of the quasipotential equation are used to evaluate the matrix element of meson couplings. The pseudoscalar form factor f_P are found 510 ± 10 and 680 ± 10 MeV for $c\bar{c}$ and $b\bar{b}$, respectively. We found the leptonic decay widths with perturbative one loop QCD radiative correction $\Gamma(\bar{e}e) =$ 5.23 and 1.72 KeV for $\bar{c}c$ and $\bar{b}b$, respectively, in a good agreement with the available experimental values[5]. The transition amplitude from quark-antiquark state into two photons is formulated for off shell quarks in the Mandelstam formalism. The electromagnetic decay rates for the pseudoscalar meson are found $\Gamma^0_{\gamma\gamma}(\mathbf{0}^{-+}) = 9.40$ KeV and 431 eV for $c\bar{c}$ and $b\bar{b}$, respectively. They are found larger than those of the nonrelativistic ones[4]. They are acceptable in the light of the recent experimental data[1,5]. We have also considered the electromagnetic decay rates of $c\bar{c}$ for the P-states. The decay rate for scalar $c\bar{c}$ is found $\Gamma^{\text{Rad}}_{\gamma\gamma}(\mathbf{0}^{++}) = 7.22$ KeV. We considered the case of $\mathbf{2}^{++}$ state without any effect of the F-state. We also considered the case that small-small component couple the P-state with the F-state. They are found equal to $\Gamma^{Rad}_{\gamma\gamma}(\mathbf{2}^{++} \approx {}^3P_2) = 970$ eV and $\Gamma^{Rad}_{\gamma\gamma}(\mathbf{2}^{++} = {}^3P_2 + {}^3F_2) = 488$ eV for $c\bar{c}$. We have found that the ratio $R = \Gamma_{\gamma\gamma}(\mathbf{2}^{++})/\Gamma_{\gamma\gamma}(\mathbf{0}^{++})$ without the F-state gives the nonrelativistic limit $R^0 \approx 3.75$ [4] while those with the effect of the F-state gives larger value by factor 2 for $c\bar{c}$ states. Therefore, the small-small component has a significant contribution to $\mathbf{2}^{++}$ decay rates. After considering the radiative corrections and the coupling with the F-state due to the small-small component, we have found a large ratio $R^{Rad} \approx 12 - 14$ in a good agreement with recent results[1].

ACKNOWLEDGMENTS: I wish to thank Professor E. Rabinovici and Professor G. Mikenberg for their kind hospitality.

References

[1] H. W. Huang, C. F. Qiao and K. T. Chao, Phys. Rev. D **54**, 2123 (1996); S. N. Gupta, J. M. Johnson and W. W. Repko,*ibid.* **54**, 2075 (1996)

[2] Z. P. Li, F. E. Close and T. Barnes, Phys. Rev. D **43**, 2161 (1991); E. S. Ackleh and T. Barnes, *ibid.* **45**, 232 (1992). S. Godfrey and N. Isgure, Phys. Rev. D **32**, 189 (1985).

[3] I. Zakout, Phys. Rev C**54**, 2647 (1996); P. C. Tiemeijer and J. A. Tjon, *ibid.* **49**, 494 (1994); **48**, 896 (1993).
[4] R. Barbieri, R. Gatto, and R. Kögerler, Phys. Lett. **60 B**, 183 (1976).
[5] Data group, R. M. Barnett *et al.*, Phys. Rev. D **54**, 1, (1996).

Study of Nucleon Properties in Quark Potential Model Approach

A.Aleksejevs, S.Barkanova, T.Krasta, J.Tambergs

Nuclear Research Center of Latvian Academy of Sciences
Miera Str. 31, Salaspils, LV-2169, Latvia

The unambiguous evaluation of nucleon's electromagnetic characteristics - the rms radius of the proton $\sqrt{\langle r^2\rangle_p}$, neutron's mean square charge radius (MSCR) $\langle r^2\rangle_n$, neutron-electron scattering length b_{ne}, neutron electric and magnetic polarizabilities α_n and β_n - is still an open problem both from the experimental and theoretical points of view. The main aim of the present work was to calculate these parameters in the framework of non-relativistic quark model (NRQM) employing the approach of Nag et al. [1]. Our study of nucleon's properties in the framework of NRQM was based on the systematics of lightest baryons according to the requirements of $SU(6) \times O(3)$ symmetry [2].

The calculations of nucleon states and their wave functions were based on the solutions of three quark Schrödinger equation using qq interaction potential characterized by the confinement term $-V_0 + Ar_{ij}^2$, one-gluon exchange and electromagnetic Coulomb term $(k\alpha_s + \alpha Q_i Q_j)/r_{ij}$. The Schrödinger equation of such system one can present as follows:

$$\sum_{i=1}^{3}\left[-\frac{\hbar^2}{2m_i}\nabla_{r_i}^2\right] + \sum_{i,j>i}^{3}\left[-V_o + Ar_{ij}^2 + (k\alpha_s + \alpha Q_i Q_j)/r_{ij}\right]\Psi = E\Psi. \quad (1)$$

The qq interaction used in involves also the contact spin-spin $(\vec{S}_i, \vec{S}_j)$ interaction term $H_{ss}(i,f)$, which according to Fuch's theorem cannot be accounted for exactly in Eq.(1), and therefore is treated as perturbation (see ref.[1]):

$$H_{ss}(i,f) = -<\Psi_f\left|\sum_{i,j>i}^{3}(k\alpha_s + \alpha Q_i Q_j)\frac{C\pi}{2}\delta(r_{ij})f_{ij}^{ss}\right|\Psi_{i>}$$

Eq.(1) we solve using the hyperspherical harmonics (HH) method. The particular set of nucleon and their resonance states, which was used in our calculations, was obtained for corresponding unitary spin (F) and spin (S) multiplets $(F, 2S+1)$ and $SU(6)$ supermultiplets: $(56)^+_{L=0}, (70)^+_{L=0}, (70)^-_{L=0}$, using the HH expansion and taking into account only the minimal K value in each case. These states are listed in Table, where in the last column the calculated masses of nucleons and their resonances are given.

No.(p)	SU(6) supermultiplet structure	HH state (KL), symmetry type	J^P	Nucleon state, $M_{exp}[MeV]$	M_{calc} $[MeV]$
1	$\left\|56_0^+,(8,2)\right\rangle_N^{1/2}$	KL=00, $\|S>$	$1/2^+$	N(939)	939
	$\left\|56_0^+,(8,2)\right\rangle_N^{1/2}$	KL=00, $\|S>$	$1/2^+$	N(1440)	1538
2	$\left\|56_0^+,(10,4)\right\rangle_\Delta^{3/2}$	KL=00, $\|S>$	$3/2^+$	Δ(1232)	1232
	$\left\|56_0^+,(10,4)\right\rangle_\Delta^{3/2}$	KL=00, $\|S>$	$3/2^+$	Δ(1660)	1736
3	$\left\|70_0^+,(8,2)\right\rangle_N^{1/2}$	KL=20, $\|mS>$	$1/2^+$	N(1710)	1706
4	$\left\|70_0^+,(10,2)\right\rangle_\Delta^{1/2}$	KL=20, $\|mS>$	$1/2^+$	Δ(1750)	1746
5	$\left\|70_1^-,(8,2)\right\rangle_N^{3/2}$	KL=11, $\|mS>$	$3/2^-$	N(1520)	1535
6	$\left\|70_1^-,(8,2)\right\rangle_N^{1/2}$	KL=11, $\|mS>$	$1/2^-$	N(1535)	1485
7	$\left\|70_1^-,(10,2)\right\rangle_\Delta^{1/2}$	KL=11, $\|mS>$	$1/2^-$	N(1620)	1525
8	$\left\|70_1^-,(8,4)\right\rangle_N^{1/2}$	KL=11, $\|mS>$	$1/2^-$	N(1650)	1649
9	$\left\|70_1^-,(8,4)\right\rangle_N^{3/2}$	KL=11, $\|mS>$	$3/2^-$	N(1700)	1699
10	$\left\|10_1^-,(10,2)\right\rangle_\Delta^{3/2}$	KL=11, $\|mS>$	$3/2^-$	Δ(1700)	1575

The following values of qq interaction potential parameters have been adopted:

$$V_0 = 37.11[MeV],\ A = 47.0[MeV],\ \alpha_s = 1.238,\ C = 5.62 \tag{2}$$

The obtained wave functions of nucleon states have been used in order to calculate the nucleon's MSCR value:

$$\left\langle r^2\right\rangle_n = \sum_{i=1}^{3}\left\langle \Psi \left| \sum_i \widehat{e}_i \left(\overrightarrow{\mathbf{r}_i} - \overrightarrow{\mathbf{R}}\right)^2 \right| \Psi \right\rangle, \tag{3}$$

The electric (α_n) and magnetic (β_n) polarizabilities of the neutron we have calculated using expressions:

$$\alpha_n = 2\sum_{n\neq 0} \frac{\left|\left\langle \Psi_n \left| \widehat{d}_z \right| \Psi_0 \right\rangle\right|^2}{E_n - E_0}, \qquad \beta_n = 2\sum_{m\neq 0} \frac{\left|\left\langle \Psi_m \left| \widehat{m}_z \right| \Psi_0 \right\rangle\right|^2}{E_n - E_0},$$

where $\widehat{d}_z, \widehat{m}_z$ denote z-components of the neutron intrinsic electric and magnetic dipole operators.

The numerical values of nucleon parameters obtained using Eqs.(3),() are as follows:

$$\sqrt{\left\langle r^2\right\rangle_p} = 0.45\ [fm];\ \left\langle r^2\right\rangle_n = -0.105\ \left[fm^2\right];\ \alpha_n = 1.42{\cdot}10^{-3}[fm^3];\ \beta_n = 0.235{\cdot}10^{-3}[fm^3]. \tag{4}$$

Employing the following relationship between the neutron's MSCR $\left\langle r^2\right\rangle_n$ and neutron-electron scattering length b_{ne},

$$\langle r^2 \rangle_n = (3\hbar^2/(m_n e^2)) b_{ne} = 86.46 \cdot b_{ne} \ [fm^2] \,, \tag{5}$$

one obtains from our $\langle r^2 \rangle_n$ value (Eq.(4)) the value

$$b_{ne} = -1.21 \cdot 10^{-3} \ [fm] \,. \tag{6}$$

The quite reasonable description of the state energies was obtained, but evaluated proton charge radius of 0.45 fm is approximately twice smaller than the experimental value of ˜0.8 fm [3], but is very close to one from recent calculations [4] (0.46 fm). The obtained neutron parameters $\langle r^2 \rangle_n$ and b_{ne} values are in overall agreement with earlier experimental evaluations [5,6].

1 References

1. R. Nag, S. Sanyal, S. N. Mukherjee. Progr. Theor. Phys. 83, Nr.1 (1990) 51.
2. N.Isgur, G.Karl. Phys. Rev D 19, Nr.9 (1979) 2653.
3. D.J.Berkeland, E.A.Hinds, M.G.Boshier. Phys. Rev. Lett.75 (1995) 2470.
4. Z.Dziembowski, M.Fabre de la Ripelle, G.A.Miller. Phys.Rev.C 53,Nr.5(1996)R2039.
5. S. Kopecky,P.Riehs,J.A.Harvey,N.W.Hill. Phys.Rev.Lett.74,Nr.13 (1995)2427.
6. J. Schmiedmayer,P.Riehs,J.A.Harvey,H..W.Hill. Phys.Rev.Lett.66,Nr.8 (1991)1015

What Lattice Calculations Tell Us About the Glueball Spectrum

Michael Teper (teper@thphys.ox.ac.uk)

Theoretical Physics, University of Oxford, UK

Abstract. I review what lattice QCD simulations have to tell us about the glueball spectrum. We see that the various lattice calculations are in good agreement with each other. They predict that prior to mixing with nearby flavour singlet quarkonia the lightest glueball states are the scalar at 1.61 ± 0.15 GeV, the tensor at 2.26 ± 0.22 GeV, and the pseudoscalar at 2.19 ± 0.32 GeV.

1 Introduction

My topic here concerns the glueball spectrum. The physics question is: where, in the experimentally determined hadron spectrum, are the glueballs hiding? Ideally I should be telling you what happens when you simulate QCD with realistically light quarks. But it is going to be a few years yet before I can do that. What I will focus on here are lattice glueball calculations in the SU(3) gauge theory without quarks. We now know what are the lightest states in the continuum (rather than lattice) theory; and I will tell you what they are. The next step, if we want to make contact with the real world, is to introduce the physical GeV mass scale. This introduces uncertainties which I will try to estimate for you. The final step is to discuss possible mixing scenarios with nearby flavour singlet quarkonia. At this stage we can look at the experimental spectrum and pinpoint the experimental states most likely to have large glueball components. I will not have time to say much about the latter topics and refer you instead to ref [1] and ref [2] where you will also find a more complete set of references.

The states in the pure SU(3) gauge theory are glueballs by definition - we only have gluons in the theory. If you want hadrons with quarks then you can propagate quarks in this gluonic vacuum and tie such propagators together so that the object propagating has the appropriate hadronic quantum numbers. That is to say, you calculate hadron masses in the relativistic valence quark approximation (the 'quenched approximation') to QCD. The spectrum one obtains this way is a remarkably good approximation to the observed light hadron spectrum. This is not too surprising: one reason we were able to learn of the existence of quarks in the first place is because the low-lying hadrons are in fact well described by a simple valence quark picture.

Of course in this theory, with no vacuum quark loops, we don't have mixing between quarkonia and glueballs. There is however reason to believe that this mixing is weak in the real world – the Zweig (OZI) rule: hadron decays where the initial quarks all have to annihilate are strongly suppressed, e.g.in ϕ decays. Such a decay may be thought of as $quarks \rightarrow glue \rightarrow quarks$. Glueball mixing with quarks should therefore be $\sqrt{}$OZI suppressed. As should glueball decays into hadrons composed of quarks. The existence of such a suppression is supported by a recent lattice calculation ref [3].

The picture we have in mind is therefore as follows. The glueballs will only be mildly affected by the presence of light quarks. They will, of course, decay into pions etc. but their decay width will be relatively small; and there will be a correspondingly small mass shift. Only if there happens to be a flavour singlet quarkonium state close by in mass will things be very different, because of the mixing of these nearly degenerate states. In this context we expect 'close by' to mean within $\sim$ 100 MeV. So we view the glueballs in the pure SU(3) gauge theory as being the 'bare' glueballs of QCD which may mix with nearby quarkonia to produce the hadrons that are observed in experiments. All this is an assumption of course, albeit well motivated. If true it tells us that the glueballs, whether mixed with quarkonia or not, should lie close to the masses they have in the gauge theory. So we now turn to the calculation of those masses.

2 Glueballs in the SU(3) gauge theory

In lattice calculations (Euclidean) space-time is discretised to a hypercubic lattice, and the volume is made finite and (anti)periodic. So the first step is to be able to calculate masses reliably on such a lattice hypertorus. The second is to make sure the volume is large enough. Assuming this has been done we obtain the mass spectrum, $am_i(a)$, of the discretised theory in units of the lattice spacing a. What we actually want is the corresponding spectrum of the continuum theory, $a = 0$. To obtain this we proceed as follows. Theoretically we know that in this theory the leading lattice spacing corrections to dimensionless ratios of physical masses are $O(a^2)$ where a is the lattice spacing ref [4]. So for small enough a we can extrapolate our calculated mass ratios to $a = 0$ using

$$\frac{am_1(a)}{am_2(a)} \equiv \frac{m_1(a)}{m_2(a)} = \frac{m_1(a=0)}{m_2(a=0)} + c(am)^2 \tag{1}$$

where m may be chosen to be m_1 or m_2 or some other physical mass: the difference between these choices is clearly higher order in a^2 - which we are neglecting. In practice I shall use $am_2 = am = a\sqrt{\sigma}$, where $a^2\sigma$ is the confining string tension as calculated in lattice units, and $am_1 = am_G$ will be a glueball mass.

In Fig 1 I show how the calculated mass ratios, for the lightest scalar glueball, vary with $a^2\sigma$. As you see, the behaviour is linear - not surprising given the fact that $a^2\sigma$ is indeed small for the values plotted. One fits a straight line and obtains the continuum limit as the intercept at $a = 0$, i.e.

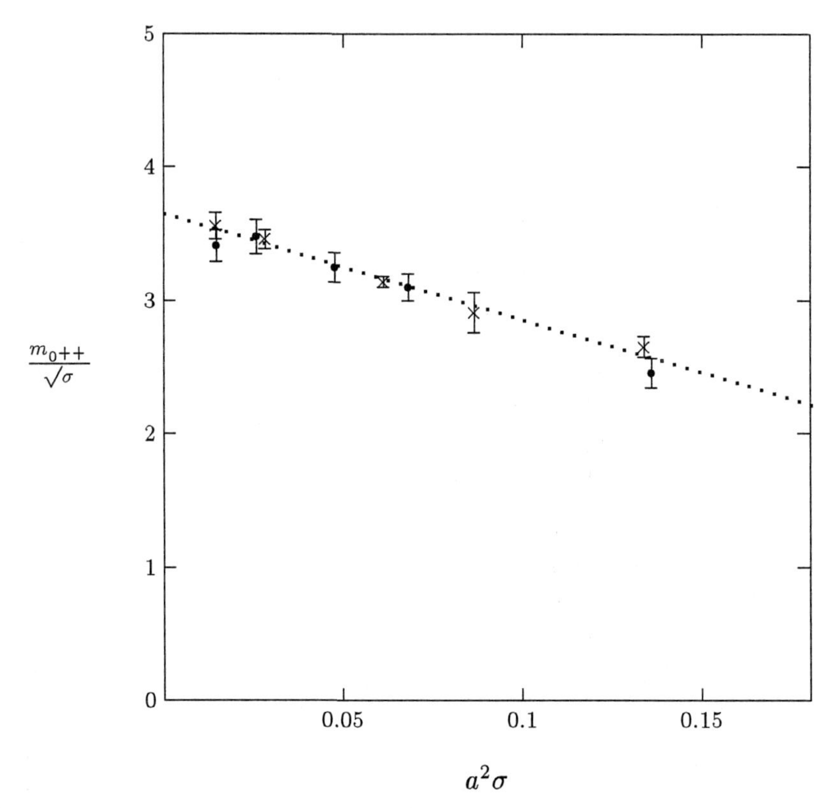

Fig. 1. The scalar glueball mass: the GF11 values ($\times$) and the rest ($\bullet$). The best linear extrapolation to the continuum limit is shown.

at $a^2\sigma = 0$. We obtain in this way the following lightest three masses in the *continuum* limit:

$$\frac{m_{0^{++}}}{\sqrt{\sigma}} = 3.65 \pm 0.11 \tag{2}$$

$$\frac{m_{2^{++}}}{\sqrt{\sigma}} = 5.15 \pm 0.21 \tag{3}$$

$$\frac{m_{0^{-+}}}{\sqrt{\sigma}} = 4.97 \pm 0.58 \tag{4}$$

Although we do not have continuum extrapolations for other masses, the UKQCD lattice results strongly suggest that glueballs with other J^{PC} are heavier ref [5].

The values of the glueball masses that we have used are from refs [5, 6, 7, 8] and for our sources of the string tension see refs [1, 2]. You may recall that a couple of years ago some publicity was given to an apparent discrepancy between the GF11 and UKQCD predictions for the 0^{++} glueball mass. However, as you can explicitly see in Fig 1, the various calculations are entirely in agreement with each other. The discrepancy was an illusion: it arose largely from different ways of setting the MeV scale. Such differences should be part of the final systematic error on the mass: as below.

3 Glueballs masses in GeV units

To introduce MeV units we need the string tension in these units. We can do this by calculating the mass of the ρ or nucleon or ... in the quenched approximation and extrapolate $m_\rho/\sqrt{\sigma}$ or $m_N/\sqrt{\sigma}$ or ... to the continuum limit as we did for the glueballs. We then set $m_\rho = 770$ MeV or $m_N = 930$ MeV or ... to obtain $a\sqrt{\sigma}$ in MeV units. Because the quenched spectrum differs slightly from the real world, these estimates differ slightly. This forms part of the error. One can estimate the error *ad nauseam*, as in refs [1, 2], and this leads to an estimate

$$\sqrt{\sigma} = 440 \pm 15 \pm 35 \text{ MeV} \tag{5}$$

where the first error is statistical and the second is systematic.

We can now use this value in eqns 2- 4 to express our glueball masses in GeV units. We obtain

$$m_{0^{++}} = 1.61 \pm 0.07 \pm 0.13 \text{ GeV} \tag{6}$$

$$m_{2^{++}} = 2.26 \pm 0.12 \pm 0.18 \text{ GeV} \tag{7}$$

$$m_{0^{-+}} = 2.19 \pm 0.26 \pm 0.18 \text{ GeV} \tag{8}$$

These, then, are our best lattice predictions for the lightest glueballs prior to any mixing with nearby quarkonium states. In the case of the 0^{++} the focus is naturally on the $f_0(1500)$ and any scalar lurking in the $f_J(1700)$. The tensor focus is naturally on the $f_2(1900)$ and the $G(2150)$. With the pseudoscalar things are murky: the lattice calculation has huge errors and is far from the obvious $\iota(1490)$ candidate; but in this case topology is important and that is a quantity that is sensitive to light quarks in the vacuum.

References

[1] F. E. Close and M. Teper, in preparation.
[2] M. Teper, Lectures at the Newton Institute, NATO-ASI School, June 1997 (hep-lat 9711011).
[3] J.Sexton et al., Phys Rev. Letters, 75 (1995) 4563.
[4] K. Symanzik, Nucl. Phys. B226 (1983) 187.
[5] G.Bali et al, Phys. Lett. B309 (1993) 378.
[6] H. Chen et al, Nucl. Phys. Proc. Suppl. B34 (1994) 357.
[7] C. Michael and M. Teper, Nucl. Phys. B314 (1989) 347.
[8] Ph.de Forcrand et al, Phys. Lett. B152 (1985) 107.

Gluonia in QCD with Massless Quarks

Stephan Narison,

Laboratoire de Physique Mathématique et Théorique,
UM2, Place Eugène Bataillon, 34095 Montpellier Cedex 05
E-mail: narison@lpm.univ-montp2.fr

Abstract. I briefly review [1] the estimate of the gluonia masses, decay and mixings in QCD with massless quarks from QCD spectral sum rules and some low-energy theorems. The data in the 0^{++} channel can be explained with some maximal gluonium-quarkonium mixing schemes, which then suggest a large violation of the OZI rule, similar to the case of the η' in the $U(1)_A$ sector.

1 Introduction

The possible existence of the gluon bound states (gluonia or glueballs) or/and of a gluon continuum is one of the main consequences of the non-perturbative aspects of QCD. In this talk, I summarize our recent results on these topics.

2 Gluonia masses from QCD spectral sum rules

We shall consider the lowest-dimension gauge-invariant currents J_G built from two gluon fields with the quantum numbers of the $J^{PC} = 0^{++}$, 2^{++} and 0^{-+} gluonia. The former two enter the QCD energy-momentum tensor $\theta_{\mu\nu}$, while the third one is the $U(1)_A$ axial anomaly. We shall also consider the three-gluon current asssociated to the 0^{++} gluonia. We shall work with the generic two-point correlator: $\psi_G(q^2) \equiv i \int d^4x\, e^{iqx}\ \langle 0|\mathcal{T} J_G(x)\,(J_G(0))^\dagger\,|0\rangle$, where its QCD expression can be parametrized by the usual perturbative terms plus the non-perturbative ones due to the vacuum condensates in the Wilson expansion [2]. In the massless quark limit $m_i = 0$, the dominant non-perturbative contribution is due to the dimension-four gluonic condensate $\langle \alpha_s G^2\rangle \simeq (0.07 \pm 0.01)$ GeV^4, recently estimated from the e^+e^- data and heavy-quark mass splittings [3]. We shall study the SVZ Laplace sum rules:

[1] For more details and complete references, see: S.N., hep-ph/9612457 (1996) [Nucl. Phys. B (in press)] and QCD 97 International Euroconference, Montpellier, hep-ph/9710281 (1997) [Nucl.Phys. B (Proc. Suppl.) (in press)].

[2] Higher dimension condensates, including the ones due instanton–anti-instanton ($D = 11$) operators, can be neglected at the sum rule optimization scale. UV renormalon and some eventual other effects induced by the resummation of the QCD series, and not included in the OPE, are estimated from the last known term of the truncated series, as is done in the extraction of α_s from τ decays.

[3] Throughout this paper, we shall use, for three active flavours, the value of the QCD scale $\Lambda = (375 \pm 125)$ MeV.

$$\mathcal{L}_G(\tau) = \int_{t_\leq}^{\infty} dt \exp(-t\tau) \frac{1}{\pi} \text{Im}\psi_G(t), \quad \text{and} \quad \mathcal{R}_G \equiv -\frac{d}{d\tau} \log \mathcal{L}_G, \tag{1}$$

where $t_\leq$ is the hadronic threshold. The latter, or its slight modification, is very useful, as it is equal to the resonance mass squared, in the simple duality ansatz parametrization of the spectral function: $\frac{1}{\pi}\ \text{Im}\psi_G(t) \simeq 2f_G^2 M_G^4 \delta(t - M_G^2)\delta(t - M_R^2)$ + "QCD continuum"$\Theta(t - t_c)$, where the decay constant f_G is analogous to $f_\pi = 93.3$ MeV; t_c is the continuum threshold which is, like the sum rule variable τ, an a priori arbitrary parameter. Our results in Table 1 satisfy the t_c and τ stability criteria, whilst the upper bounds have been obtained from the minimum of $\mathcal{R}_G$ combined with its positivity. Our results show the mass hierarchy $M_S < M_P \approx M_T$, which suggests that the scalar is the lightest gluonium state. However, the consistency of the different subtracted and unsubtracted sum rules in the scalar sector requires the existence of an additional lower mass and broad σ_B-meson coupled strongly both to gluons and to pairs of Goldstone bosons (a case similar to the η')[4]. The effect of the σ_B-meson can be missed in a one-resonance parametrization of the spectral function, and in the present lattice quenched approximation. The values of $\sqrt{t_c}$, which are about equal to the mass of the next radial excitations, indicate that the mass splitting between the ground state and the radial excitations is much smaller (30%) than for the ρ meson (about 70%), so that one can expect rich gluonia spectra in the vicinity of 2.0–2.2 GeV. We also conclude that:

- The $\zeta(2.2)$ is a good 2^{++} gluonium candidate because of its mass (see Table 1) [5] and small width in $\pi\pi$ (≤ 100 MeV). However, the associated value of t_c can suggest that the radial excitation state is also in the 2 GeV region, which should stimulate further experimental searches.
- The E/ι (1.44) or other particles in this region is too low for being the lowest pseudoscalar gluonium. One of these states is likely to be the first radial excitation of the η' because its coupling to the gluonic current is weaker than the one of the η' and of the gluonium (see Table 1).

3 Widths of the scalar gluonia

- The hadronic widths can be estimated from the vertex: $V(q^2) = \langle H_1 | \theta_\mu^\mu | H_2 \rangle$, where: $q = p_1 - p_2$, $H \equiv \pi, \eta_1, \sigma_B$, and B refers to the unmixed bare state. It satisfies the constraints $V(0) = M_H^2$ and, in the chiral limit, $V'(0) = 1$. Saturating it with the three resonances $H \equiv \sigma_B, \sigma'_B$ and G, one obtains the first and second NV sum rules (Goldberger–Treiman-like relation):

[4] An analogous large violation of the OZI rule is also necessary for explaining the proton spin crisis. These features are consequences of the $U(1)$ symmetry, which is not the case of the vector meson ϕ of the $SU(3)_F$ symmetry.

[5] The small quarkonium-gluonium (mass) mixing angle allows us to expect that the observed meson mass is about the same as the one in Table 1.

$$\frac{1}{4}\sum_{S\equiv\sigma_B,\sigma'_B,G} g_{SHH}\sqrt{2}f_S \simeq 2M_H^2, \qquad \frac{1}{4}\sum_{S\equiv\sigma_B,\sigma'_B,G} g_{S\pi\pi}\sqrt{2}f_S/M_S^2 = 1. \quad (2)$$

We identify the G with the $G(1.5 \sim 1.6)$ at GAMS (an almost pure gluonium candidate), and the σ_B and σ'_B (its radial excitation) with the broad resonance below 1 GeV and the $f_0(1.37)$. In this way, we obtain the predictions in Table 2, showing the presence of gluons inside the wave functions of the broad σ_B and σ'_B, which decay copiously into $\pi\pi$, signalling a large violation of the OZI rule in this channel. For the G meson, we deduce $g_{G\eta\eta} \simeq \sin\theta_P g_{G\eta\eta'}$ ($\theta_P = -18°$ being the pseudoscalar mixing angle), implying a ratio of widths of about 0.22 in agreement with the GAMS data $r \simeq 0.34 \pm 0.13$, but suggest that the Crystal Barrel particle having $r \approx 1$ is a mixing between this gluonium and other states. We also expect that the 4π decay of the $G(1.6)$ are mainly S-waves initiated from the decay of pairs of σ_B.

• The $\gamma\gamma$ widths of the σ_B, σ'_B and G can be obtained (Table 2) by identifying the Euler-Heisenberg box two gluons–two photons effective Lagrangian, with the scalar-$\gamma\gamma$ Lagrangian, while the $J/\psi \rightarrow \gamma S$ radiative decays can be estimated, using dispersion relation techniques, to this effective interaction.

4 "Mixing-ology" for the decay widths of scalar mesons

• We consider that the observed f_0 and σ result from the two-component mixing of the σ_B and $S_2 \equiv \frac{1}{\sqrt{2}}(\bar{u}u + \bar{d}d)$ unmixed bare states. Using the prediction on $\Gamma(\sigma_B \rightarrow \gamma\gamma)$, and the experimental width $\Gamma(f_0 \rightarrow \gamma\gamma) \approx 0.3$ keV, one obtains a maximal mixing angle: $\theta_S \approx (40 \sim 45)°$, which indicates that the f_0 and σ have a large amount of gluons in their wave functions [6]. Then, one can deduce (in units of GeV):

$$g_{f_0\pi^+\pi^-} \simeq (0.1 \sim 2.6), \quad g_{f_0K^+K^-} \simeq -(1.3 \sim 4.1), \tag{3}$$

which can give a simple explanation of the exceptional property of the f_0 (strong coupling to $\bar{K}K$: $\pi\pi$ and $\bar{K}K$ data, and $\phi \rightarrow \gamma f_0$ decay).

• The $f_0(1.37), f_0(1.50)$ can be well described by a 3×3 mixing scheme involving the $\sigma'_B(1.37), S_3(\bar{s}s)(1.47)$ and $G(1.5)$ gluonium, and contains a large amount of glue and $\bar{s}s$ components. The different widths are given in Table 3. The $f_J(1.71)$, if confirmed to be a 0^{++}, can be essentially composed by the radial excitation $S'_3(\bar{s}s)$ due to its main decay into $\bar{K}K$.

5 Conclusions and acknowledgements

I have shown that QCD spectral sum rules plus some QCD low-energy theorems can provide a plausible explanation of the complex gluonia spectra.

I thank the cern theory division for its hospitality.

[6] This situation is quite similar to the case of the η' in the pseudoscalar channel (mass given by its gluon component, but strong coupling to quarkonia).

Table 1. Unmixed gluonia masses and couplings from QSSR

J^{PC}	Name	Mass [GeV] Estimate	Upper bound	f_G [MeV]	$\sqrt{t_c}$ [GeV]
0^{++}	G	1.5 ± 0.2	2.16 ± 0.22	390 ± 145	$2.0 \sim 2.1$
	σ_B	1.00 (input)		1000	
	σ'_B	1.37 (input)		600	
	$3G$	3.1		62	
2^{++}	T	2.0 ± 0.1	2.7 ± 0.4	80 ± 14	2.2
0^{-+}	P	2.05 ± 0.19	2.34 ± 0.42	$8 \sim 17$	2.2
	E/ι	1.44 (input)		$7\ :\ J/\psi \to \gamma\iota$	

Table 2. Unmixed scalar gluonia and quarkonia decays

Name	Mass [GeV]	$\pi^+\pi^-$ [GeV]	K^+K^- [MeV]	$\eta\eta$ [MeV]	$\eta\eta'$ [MeV]	$(4\pi)_S$ [MeV]	$\gamma\gamma$ [keV]
heightGlue							
σ_B	$0.75 \sim 1.0$ (input)	$0.2 \sim 0.5$	$SU(3)$	$SU(3)$			$0.2 \sim 0.3$
σ'_B	1.37 (input)	$0.5 \sim 1.3$	$SU(3)$	$SU(3)$		$43 \sim 316$ (exp)	$0.7 \sim 1.0$
G	1.5	≈ 0	≈ 0	$1 \sim 2$	$5 \sim 10$	$60 \sim 138$	$0.2 \sim 1.8$
Quark							
S_2	1.	0.12	$SU(3)$	$SU(3)$			0.67
S'_2	$1.3 \approx \pi'$	0.30 ± 0.15	$SU(3)$	$SU(3)$			4 ± 2
S_3	1.47 ± 0.04		73 ± 27	15 ± 6			0.4 ± 0.04
S'_3	≈ 1.7		112 ± 50	$SU(3)$			1.1 ± 0.5

Table 3. Predicted decays of the observed scalar mesons

Name	$\pi^+\pi^-$ [MeV]	K^+K^- [MeV]	$\eta\eta$ [MeV]	$\eta\eta'$ [MeV]	$(4\pi^0)_S$ [MeV]	$\gamma\gamma$ [keV]
$f_0(0.98)$	$0.2 \sim 134$	Eq. (3)				≈ 0.3 (exp)
$\sigma(0.75 \sim 1)$	$300 \sim 700$	$SU(3)$	$SU(3)$			$0.2 \sim 0.5$
$f_0(1.37)$	$22 \sim 48$	≈ 0 (exp)	≤ 1	≤ 2.5	150	≤ 2.2
$f_0(1.5)$	25 (exp)	$3 \sim 12$	$1 \sim 2$	≤ 1	$68 \sim 105$ (exp)	≤ 1.6
$f_J(1.71)$	≈ 0	112 ± 50	$SU(3)$		≈ 0	1.1 ± 0.5

Determining Masses of First Excited States in the ^3P Channels of Charmonium by Means of QCD Sum Rules

E. Di Salvo (disalvo@ge.infn.it), M. Pallavicini and E. Robutti

Dipartimento di Fisica, Universita' di Genova, INFN, sez. Genova

Abstract. We apply a numerical method, based on QCD sum rules, and already used in $^1S_0$ and $^1P_1$ channels of Charmonium, for predicting masses of first excited states of the ^3P channels.

QCD sum rules[1, 2, 3, 4, 5] have been shown to be a very useful tool for predicting masses and decay widths of Charmonium states. In particular power moment sum rules [1, 2] at varying momentum[4], implemented with two numerical algorithms [5, 6], predict the mass of the first excited state and some partial decay widths of the ground state in a given channel of Charmonium. The method, which has been applied to the $^1S_0$ and $^1P_1$ channels[5, 7, 8], now is exploited for investigating the ^3P channels.

We consider the correlator of the quark current (or density) of a given channel Γ of Charmonium. We assume quark-hadron duality and correlator analyticity. Owing to the latter assumption we write a dispersion relation for the correlator. Moreover we take derivatives at any order of both sides of such a relation with respect to the square modulus of the overall four-momentum q of the $c-\bar{c}$ pair, thus getting the so-called power moment QCD sum rules[1, 2, 4]:

$$\Pi_n^\Gamma(Q^2) = \frac{1}{\pi}\int_0^\infty \frac{Im\Pi^\Gamma(s)}{(s+Q^2)^n}ds, \tag{1}$$

where $Q^2 = -q^2$. For spacelike four-momenta the l.h.s. of eq. (1) can be expanded according to a non-trivial Wilson operator product expansion (OPE) enriched by the contribution of the lowest-dimensional gluon condensate[3, 4].

As a consequence of duality the spectral function $Im\Pi^\Gamma(s)$, which appears at the r.h.s. of eq. (1), contains information on Charmonium masses; for the same reason this function may be parametrized by a θ-function, describing continuum states, and either a) one or b) two Dirac δ-functions. If we introduce parametrization a) into the r.h.s. of eq. (1), we obtain a prediction for the mass of the ground state, m_1, which depends on Q^2, n and on four parameters: the mass m_c^0 of the charmed quark, the threshold square energy s_0 of continuum, α_s and the gluon condensate, the last two quantities being known from preceding analyses. On the other hand, if we adopt parametrization b), we get an expression of the mass of the first excited state which depends on the same arguments and, moreover, on the mass of the ground state and on the inverse coupling constant g_1

of the ground state to the quark current (or density) in the channel considered. Since generally we know only the mass of the ground state, we have to elaborate a rather complex numerical method. As regards s_0, the same value is taken in calculating m_1 and m_2, in spite of the different parametrizations assumed for the spectral function; this becomes plausible if $Q^2_{(1)} > Q^2_{(2)}$ (where $Q^2_{(i)}$ is the Q^2-value involved in the calculation of m_i, $i = 1, 2$), since it can be shown[6] that s_0 - a fictitious threshold of continuum - is a decreasing function of Q^2. Indeed, it can be checked that this condition is systematically fulfilled in our analyses [6, 7, 8], including the present one.

We have elaborated two numerical algorithms based on the criteria of *analogy* among plots (e. g. $m_i(n)$) in different channels and of *stability* of the masses m_i at varying parameters. The first algorithm, which fixes n and Q^2 by requiring that m_i be as stable as possible with respect to n, may be used for determining m_c^0 and s_0. m_c^0 can be extracted from data of the vector channel, by imposing that $m_1 = m_{J/\psi}$ and $m_2 = m_{\psi'}$, g_1 being known from the electromagnetic decay width of the J/ψ resonance into e^+e^-[4]. In any other channel, exploiting this information, and imposing that m_1 equal the mass of the ground state, one gets s_0.

Still we need a second algorithm to determine g_1. To this end we study the plot of $\overline{D}_2$ vs g_1, where $\overline{D}_2$ is the average second derivative of m_2 with respect to n in correspondence of a very stable stationary point. In the vector channel this plot presents an oblique inflection in correspondence of the value of g_1 inferred from data. Also in any other channel of Charmonium examined till now the plot is endowed with at least an oblique inflection[5, 6, 7, 8], which for ^3P channels is unique, so that we assume it to provide the correct value of g_1; this in turn leads to determining m_2 through the first algorithm. Our analysis yields the following values for the masses of first excited states (in MeV): 4097^{+32}_{-44} ($^3P_0$), 4317^{+25}_{-35} ($^3P_1$), 4466^{+34}_{-21} ($^3P_2$). These masses are above the threshold of open charm decay, therefore the corresponding states have to be classsified as resonant ones, and no bound states besides the ground ones are present.

References

[1] V.A.Novikov et al., Phys.Rep. *41*, 1 (1978)

[2] M.A.Shifman et al., Phys.Lett. B *77*, 80 (1978)

[3] M.A.Shifman et al., Nucl.Phys. B *147*, 385, 448, 519 (1979)

[4] L.J.Reinders et al., Nucl.Phys. B *186*, 109 (1981)

[5] E.Di Salvo et al., Nucl.Phys. B *427*, 22 (1994)

[6] E.Di Salvo, Proc. Int. School of Physics "E. Fermi", Course CXXX, ed. A.DiGiacomo and D.Diakonov, IOS Press, Amsterdam 1996, pag 469.

[7] E. Di Salvo et al., Phys. Lett. B *387*, 395 (1996).

[8] E. Di Salvo et al., Nucl. Phys. B (Proc. Suppl.) 54A, 233 (1997).

Part II

Soft Interactions

The Hard Scale from the Forward Peak in Diffractive Leptoproduction of Vector Mesons

Errol Gotsman (gotsman@post.tau.ac.il)

School of Physics and Astronomy,University of Tel Aviv, Israel

Abstract. The measured forward slope in elastic and inelastic leptoproduction of vector mesons differ by a substantial amount.To investigate this phemomenon we construct a two radii model for the target proton, and estimate the effective parameters of the "hard" Pomeron obtained from a pQCD dipole model with eikonal shadowing corrections (SC). The SC reduce the intercept of the "hard" Pomeron, and generate an effective shrinkage of the forward peak. A diffractive dip, similar to that seen in pp and $p\bar{p}$ is predicted to occur at $| t | \approx 1\ Gev^2$, which effects the behaviour of the forward peak.

The behaviour of the forward slope in diffractive leptoproduction of vector mesons provides an ideal "laboratory" for exploring the long distance ("soft") and short distance ("hard" or perturbative) components of the forward diffractive slope. The fact that the forward slope has been measured over a wide range of the virtual photon momentum squared i.e. $0 < Q^2 < 20 GeV^2$ for different vector mesons, $\rho, \omega, \phi, J/\psi$, allows one to test the predictions of both the "soft" (nonperturbative) and "hard" (perturbative) approaches. Another attrractive feature of this process is that the t-channel quantum numbers are those of the vacuum, and hence at high energies we expect Pomeron exchange to dominate.

We would like to stress that the **slope** B(t) is defined as the **logarithmic derivative** of $\frac{d\sigma}{dt}$; and only in the special case when the fall-off of the differential cross section over a range of t is an exponential, is B(t) a constant. In general this is not so.

1 Soft characteristics

We expect the behaviour of the cross ection for the production of vector mesons to follow the D-L model [0].

$$\frac{d\sigma}{dt} \sim (\frac{s}{s_0})^{2\epsilon} F_1^2(t) F_{VM}^2(t) R(t) \tag{1}$$

where the Pomeron trajectory $\alpha_P(t) = 1 + \epsilon + \alpha'_P t$ with $\alpha'_P = 0.25\ GeV^{-2}$ and $\epsilon = 0.08$.

$F_1(t)$ denotes the Dirac form factor, $F_{VM}(t)$ the form factor of the produced vector meson, and R(t) the Regge shrinkage factor.

pp and $p\bar{p}$ elastic scattering display a diffractive dip of $\frac{d\sigma}{dt}$ at $|t| = 1.4$ GeV^2 at $\sqrt{s} = 23$GeV, which moves to lower $|t|$ values as $\sqrt{s}$ increases. Movement of the "diffractive dip" is associated with "screening corrections" which should also be present in diffractive photoproduction processes.

2 QCD or "hard" characteristics

In the short distance regime the Pomeron exchange can be represented by two gluon exchange [0] [0] where the predicted behaviour of the differential cross section is

$$\frac{d\sigma}{dt} \propto [F_N^{2G}(t)]^2$$

where $F_N^{2G}(t)$ denotes the gluon form factor of the nucleon. Eq.(2) suggests that the t dependence of the forward slope should be **universal** (i.e. independent of the mass of the produced vector meson). Also, within the LLA of pQCD we do not expect any shrinkage of the diffractive cone $\alpha'_{2G}(W) = 0$ (i.e. the "hard trajectory should be flat).

3 J/ψ production

The experimental data on the photoproduction (electroproduction) of the J/ψ vector meson is consistent with all our expectations of a "hard" process.
i) $\sigma_{J/\psi}(W) \propto W^{0.8}$ (while in a "soft process e.g. rho phoproduction we have $\sigma_\rho(W) \propto W^{0.22}$
ii) The forward slope of $\frac{d\sigma}{dt}(\gamma p \to J/\psi p)$ does not vary as a function of Q^2 and has a constant value of $B_{J/\psi} \approx 4\ GeV^{-2}$, and no shrinkage with W^2, compared to $B_\rho(Q^2 = 0) = 10\ GeV^2$ (with shrinkage).

Following [0] and [0] a general approach for including screening corrections (SC) for two gluon exchange in diffractive leptoproduction of vector mesons has been developed by [0]

The amplitude factorizes into three parts, the wave function of the initial off-shell photon $\Psi^{\gamma^*}(Q, r_\perp, z)$, the wave function of the final vector meson $\Psi^V(r_\perp = 0, z)$ and the dipole cross section in LLA is

$$\sigma(r_\perp, x) = \frac{2}{3}\pi^2 \alpha_s(\frac{4}{r_\perp}) r_\perp^2 [x G^{DGLAP}(\frac{4}{r_\perp}, x)]$$

The screened amplitude is

$$A(b; Q, x) = C \int \frac{d^2 r_\perp}{\pi} \int dz \Psi^{\gamma^*}(Q, r_\perp, z)\, 2\, [1 - e^{-\kappa(r_\perp, x; b)}] \Psi^V(r_\perp = 0, x)^* \tag{2}$$

where the amount of screening is determined by

$$\kappa(b; r_\perp, x) = \sigma(r_\perp, x) \Gamma(b) \tag{3}$$

The profile

$$\Gamma(b) \approx \frac{1}{\pi B'} e^{-\frac{b^2}{B'}} \tag{4}$$

with $B' = B + \frac{1}{a^2}$, and $a^2 = z(1-z)Q^2 + M_Q^2$ is the contribution to the diffraction slope due to the γ^* - 2 gluon-vector meson interaction. a^2 appears to be a new **hard** scale, where z denotes the fraction of the photon's energy carried by the quark. (More detailed information concerning the calculation can be found in [0]).

Data from Hera [0] find for J/ψ (i.e. "hard") production, that the process $\gamma^* + p \to J/\psi + p$ has a forward slope $B_{el} \approx 4.0 \pm 0.3 GeV^{-2}$, while $\gamma^* + p \to J/\psi + X$ has a forward slope $B_{in} \approx 1.6 \pm 0.3 GeV^{-2}$, on the other hand the integrated cross sections of both processes are the same. This suggests the possibility of there being two different radii in the target proton. A two radii model was constructed [0], and the amplitude calculated with the aid of generating functions. We took the wave function $\Psi_{J/\psi}(r_\perp = 0, z) = \delta(z - \frac{1}{2})\phi_{J/\psi}(0)$. Choosing values of the radii $R_1^2 = 6GeV^{-2}$ and $R_2^2 = 2GeV^{-2}$ reproduces the experimental J/ψ data. See

Screening corrections effect the slope $B_{J/\psi}(W, Q^2)$ (a "hard" process) and produce a shrinkage of the forward slope at $Q^2 = 0$, such that $\alpha'_{eff} \approx 0.08$ (about $\frac{1}{3}$ that expected from the "soft" Pomeron).

4 ρ production

For the case of rho meson production, we take the rho wave function to be $\Psi_\rho(r_\perp = 0, z) = z(1-z)\phi_\rho(0)$, and find that the "proton radii" chosen for the case J/ψ diffractive production do not reproduce the rho data. To be consistent with the experimental data for the diffractive rho slope [0], we require $R_1^2 = 10GeV^{-2}$ and $R_2^2 = 3GeV^{-2}$. The results are illustrated by dashed curves in ig. 1. This suggests that not **only** "hard" processes contribute to rho production i.e. that "soft" processes are not negligble in rho production for $Q^2 < 15\, GeV^2$, and only for values of $Q^2 > 20\, GeV^2$ (i.e. values of $a^2 > 5\, GeV^2$) can the long distance contribution be neglected.

5 Conclusions

The two radii model of the proton predicts that a diffractive dip should occur at $-t = 1.2\, GeV^2$ for W $\approx$ 90 GeV in J/ψ production (see ig. 2). If this dip is seen in the more accurate data for J/ψ production expected in the near future, it will be further evidence that screening corrections are improtant in "hard" processes as well.

In Fig.3, we display the experimental values of the forward diffractive slope of the different vector mesons $\rho, \omega, \phi, J/\psi$ as a function of $a^2 = (\frac{Q^2}{4} + m_{VM}^2)$. We also show the two radii model's prediction of the "hard" slope.

It appears that a^2 plays the role of a "hard" scale, which suggests that the "soft" contribution is negligible for $a^2 > 5.5\, GeV^2$ (in rho production only for values of $Q^2 > 20\, GeV^2$).

I would like to thank my friends Genya Levin and Uri Maor for an enjoyable collaboration, the fruits of which are presented here.

This research was supported by THE ISRAEL SCIENCE FOUNDATION founded by The Israel Academy of Sciences and Humanities.

References

[1] A. Donnachie and P.V. Landshoff, Nucl. Phys. B231 (1983), 189; Phys. Lett. B296 (1992) 227
[2] E.M. Levin and M.G. Ryskin, Sov. J. Nucl. Phys. 45 (1987) 150
[3] A.H. Mueller, Nucl. Phys. B335 (1990) 115
[4] A.H. Mueller and J. Qui, Nucl. Phys. B268 (1986) 427
[5] E. Gotsman, E.M. Levin and U. Maor, Phys. Lett. B353 (1995) 526
[6] E. Gotsman, E.M. Levin and U. Maor, Phys. Lett. B403 (1997) 120
[7] ZEUS Collab., M. Derrick et al., Phys. Lett. B305 (1996) 120; H1 Collab., S Aid et al., Nucl. Phys. B468 (1996) 3; E. Gallo, Rapporteur talk in Proceedings of 18th Int. Symp. on Lepton Photon Int. Hamburg July 1997 (hep-ex/9710013)

Vector Meson Production at HERA

Alexander Proskuryakov (proskur@mail.desy.de)

DESY, Germany and Moscow State University, Russia

Presented for the H1 and ZEUS collaborations

Abstract. Recent results on exclusive vector meson production at HERA are presented.

1 Introduction

Exclusive production of vector mesons, $\gamma^\star p \to V^0 p$, has been studied in a wide ranges of the photon-proton centre-of-mass energy W and the photon virtuality Q^2 in fixed target experiments [1] and at HERA [2, 3]. In the region of low Q^2 this reaction is well described in the framework of the Regge theory [4] and of the Vector Dominance Model (VDM) [5]. In such an approach, exclusive vector meson production is assumed to proceed at high energy through the exchange of a pomeron trajectory [6] with an effective intercept of $1+\epsilon = 1.08$ and a slope of $\alpha' = 0.25$ GeV^{-2}. The same approach fails to describe the recently measured energy dependence of the cross section at HERA for elastic J/ψ photoproduction [3]. The rapid rise with energy of the elastic J/ψ photoproduction cross section is consistent with perturbative QCD (pQCD) calculations [7], in which the hard scale is provided by the mass of the J/ψ.

The hard scale can also be introduced by the photon virtuality Q^2 for exclusive vector meson leptoproduction. This reaction was studied in the framework of perturbative and non-perturbative QCD [8, 9, 10]. The main prediction of the QCD models is a fast increase of the cross section with W, related to the Q^2 evolution of the gluon density of the proton, $xg(x, Q^2)$.

2 Vector meson photoproduction

The cross sections for photoproduction of ρ^0, ϕ, ω and J/ψ mesons measured at HERA are shown in Fig. 1 together with low energy data. A fit to the ZEUS data alone with a function of the type $\sigma \propto W^\beta$ gives $\beta = 0.16 \pm 0.06^{+0.11}_{-0.15}$. This weak energy dependence agrees with the expectation for the 'soft' pomeron exchange. In contrast, the rapid rise with energy of the elastic J/ψ photoproduction cross section is consistent with perturbative QCD (pQCD) calculations [7].

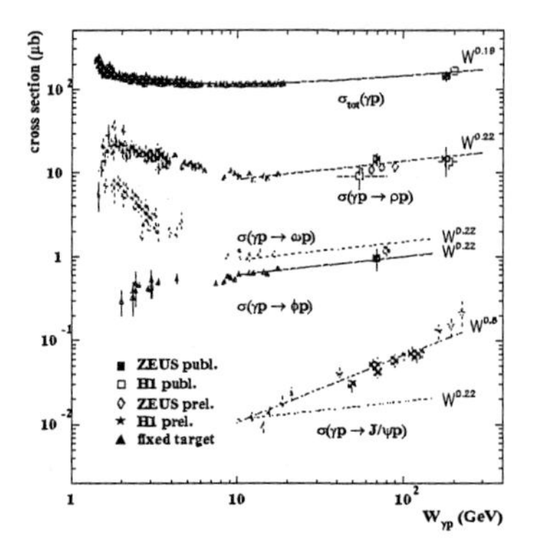

Fig. 1. Cross section for vector meson photoproduction as a functuon of W.

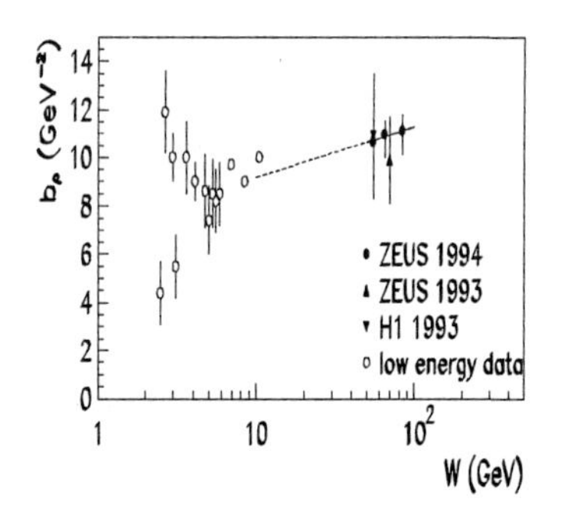

Fig. 2. The slope parameter b for the elastic ρ^0 photoproduction as a functuon of W.

The differential t distribution for exclusive vector meson production has the typical exponential shape (e^{bt}). The energy dependence of the slope parameter b has been measured using high statistics sample of reaction $ep \to ep\rho^0$. A fit of the form $b = b_0 + 2\alpha' lnW^2$ yields $\alpha' = 0.23 \pm 0.15^{+0.10}_{-0.07}$ GeV^{-2}.

3 Vector meson production at low and high Q^2

The W dependence of the exclusive ρ^0 production cross section at different values of Q^2 is shown in Fig. 3. The results of a fit to the ZEUS data with a function of the type $\sigma \propto W^\alpha$ are given in table. The W dependence of the cross section for the reaction $\gamma^\star p \to \rho^0 p$ is consistent with the soft pomeron approach at low values of Q^2. In the high Q^2 region where pQCD models predict steeper cross section dependence on W there is an indication that the rise of the cross section becomes steeper. It should be noted that the data still have rather large normalisation uncertainties, on the order of 20% for the ZEUS and NMC which are not shown in Fig. 3.

In contrast, the cross section for the exclusive J/ψ production rises rapidly with W already at low values of Q^2. The W dependence for the reaction $\gamma^\star p \to J/\psi p$ is displayed in Fig. 4. The HERA data alone exibit a strong W dependence of the cross section even at very low Q^2.

The differential $dN/d|t|$ distribution for the reaction $\gamma^\star p \to V^0 p$ was studied in wide range of photon virtuality. An exponential fit of the form e^{bt} in the kinematic range $0.25 < Q^2 < 0.85$ GeV^2, $50 < W < 90$ GeV and $|t| < 0.5$ GeV^2 gives $b = 9.2 \pm 0.3$(stat) $\pm$ 0.7(syst) GeV^{-2}. A similar fit for exclusive ρ^0 production in the kinematic range $5 < Q^2 < 30$ GeV^2 and $43 < W < 134$ GeV gives $b = 7.2 \pm 0.5$(stat) $^{+0.8}_{-0.6}$(syst) GeV^{-2}. The slope value b for the exclusive J/ψ production determined in the kinematic region $2 < Q^2 < 40$ GeV^2, $55 < W < 125$ GeV and $|t| < 1$ GeV^2 is

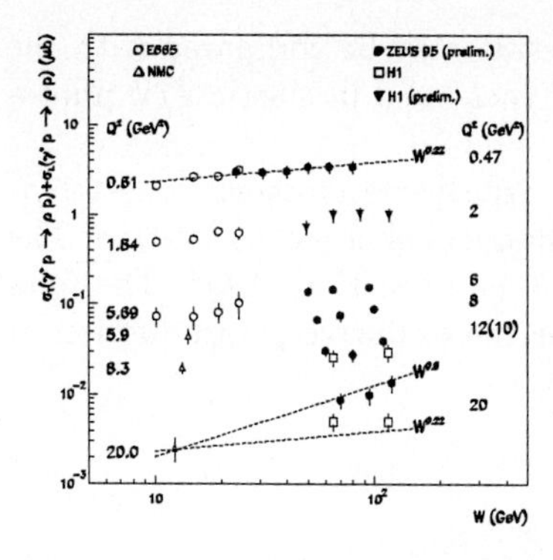

Fig. 3. Cross sections for exclusive ρ^0 production at different values of Q^2. The ZEUS measurements are displayed together with H1, E665 and NMC results.

$\sigma \propto W^{\alpha}$ (ZEUS 95 preliminary)

$< Q^2 >$ (GeV^2)	α
0.47	$0.18 \pm 0.05 \pm 0.13$
6.0	$0.18 \pm 0.16 \pm 0.05$
8.0	$0.47 \pm 0.16 \pm 0.09$
12.0	$0.38 \pm 0.21 \pm 0.13$
20.0	$0.77 \pm 0.19 \pm 0.19$

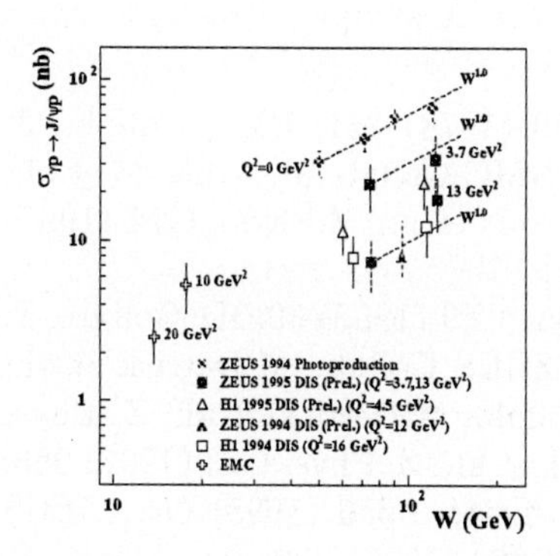

Fig. 4. Cross section for the reaction $\gamma^{\star}p \to J/\psi p$ as a function of W. The lines are are drawn to guide the eye and correspond to $\sigma \propto W^n$, with $n = 1$.

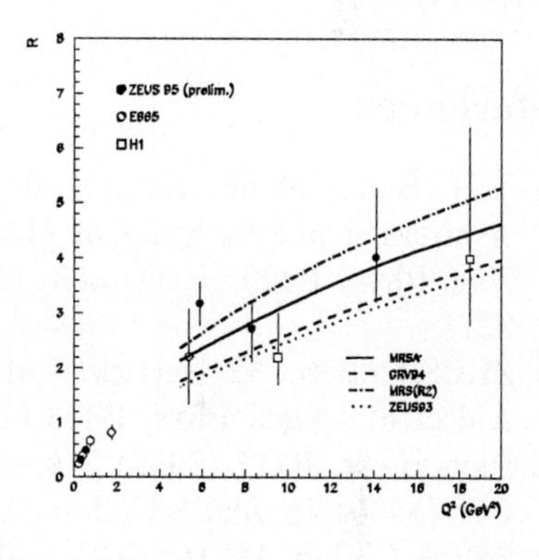

Fig. 5. The ratio of the longitudinal to transverse ρ^0 production cross section measured assuming s-channel helicity consevation. The lines represent the QCD based calculation for exclusive ρ^0 production at $W = 100$ GeV and $Q^2 > 5$ GeV^2.

4.5 ± 0.8(stat)± 1.0(syst) GeV^{-2}. This result may be compared with the value of $b = 4.6 \pm 0.4$(stat)$^{+0.4}_{-0.6}$(syst) GeV^{-2} measured in elastic J/ψ photoproduction [3].

The ratio $R = \sigma_L/\sigma_T$ of longitudinal to transverse cross sections for ρ^0 production can be determined, assuming s-channel helicity conservation. The measured values of R are displayed in Fig. 5 as a function of Q^2. The lines represent the QCD based calculation [10] for the exclusive ρ^0 production at $W = 100$ GeV.

4 Conclusions

Vector meson production has been measured at HERA in a wide range of Q^2. The photoproduction of light vector mesons exibits the characteristics of a soft diffractive process. There is an indication that in the high Q^2 region the W dependence of the ρ^0 production cross section becomes steeper. Elastic J/ψ production is consistent with perturbative QCD models already at low values of Q^2.

References

[1] T.H. Bauer et al., Rev. Mod. Phys. 50 (1978) 261; EMC Collab., J. Ashman et al., Z.Phys. C39 (1988)169; NMC Collab.,M. Arneodo et al., Nucl.Phys. B429 (1994) 503; M. R. Adams et al., Z.Phys., C74 (1997) 237.

[2] ZEUS Collab., M. Derrick et al., Z. Phys. C69 (1995) 39; H1 Collab., S. Aid et al., Nucl. Phys. B463 (1996) 3; ZEUS Collab., M.Derrick et al., Phys. Lett. B377 (1996) 259; ZEUS Collab., M.Derrick et al., Z. Phys. C73 (1996) 73; ZEUS Collab., M.Derrick et al., Z. Phys. C73 (1997) 253; ZEUS Collab.,M. Derrick et al., Phys. Lett., B356 (1995) 601; ZEUS Collab.,M. Derrick et al., Phys. Lett., B380 (1996) 220; H1 Collab., S. Aid et al., Nucl.Phys. B468 (1996) 3.

[3] ZEUS Collab., M.Derrick et al., Phys. Lett. B350 (1995) 120; H1 Collab., S. Aid et al., Nucl. Phys., B472 (1996) 3; ZEUS Collab., J.Breitweg et al., Z. Phys. C75 (1997) 215.

[4] P.D.B. Collins, Introduction to Regge Theory and High Energy Physics, Cambridge University Press (1977).

[5] J.J.Sakurai, Ann. Phys. 11 (1960) 1.

[6] A. Donnachie and P.V. Landshoff, Phys. Lett. B296 (1992) 227.

[7] M.G. Ryskin, Z. Phys. C57 (1993) 89; M.G. Ryskin, R.G. Roberts, A.D. Martin, E.M. Levin, RAL-TR-95-065.

[8] S. J. Brodsky et al., Phys. Rev. D50 (1994) 3134.

[9] W. Koepf et al., Proceedings of the Workshop 'Future Physics at HERA', Hamburg, Germany, 1995/96, Vol. 2, p. 674.

[10] A.D. Martin, M.G. Ryskin and T. Teubner, Phys. Rev. D55 (1997) 4329.

Production of Vector Meson Pairs in Two-Photon Reactions

Gabrijel Kernel (Gabrijel.Kernel@ijs.si)
presenting results of the ARGUS collaboration

University of Ljubljana and Institute J. Stefan, Ljubljana, Slovenia

Abstract. A summary is given of the present status of experimental results on two–photon production of pairs of vector mesons. The cross sections for production of all low–energy vector pairs were measured with the only exception of $\phi\phi$. The applicability of various phenomenological models is discussed.

1 Introduction

In the last decade, the two-photon production of pairs of vector mesons has been measured for all possible combinations of mesons ρ, ω, ϕ and K^* in the energy region of a few GeV above the thresholds (see ref.[1] and references cited there). The only exception is the combination $\phi\phi$ that has a so far undetectably small cross section. There exist several models [2, 3, 4, 5, 6] that describe various mechanisms for two–photon production of pairs of vector mesons. At this stage it was felt useful to summarize the existing experimental results and compare them to model predictions.

The talk is organized as follows. We first present a summary of measured cross sections that includes recent results on $\gamma\gamma \to K^*\overline{K^*}$ from ARGUS. Next we briefly discuss the partial wave analysis, and conclude with a comparison to existing model predictions.

2 Experimental Data

Fig. 1 shows a collection of available results of cross section measurements taken from ARGUS data [7, 8, 9]. The $\gamma\gamma \to K^{*+}K^{*-}$ and $\gamma\gamma \to K^{*0}\overline{K^{*0}}$ cross sections come from a recent analysis [9] that combines several reaction channels leading to the same $K^*\overline{K^*}$ states. The results are in excellent agreement with an older analysis [10] that involved only about $\frac{1}{3}$ of the data.

A feature common to all spectra is the peaking of cross sections at the vicinity of reaction thresholds. It is also seen that $\gamma\gamma \to \rho^0\rho^0$ dominates all other cross sections.

For the understanding of two-photon production of pairs of vector mesons it is important to know the partial wave composition of the cross section. Whenever the number of events was sufficient a partial wave analysis was performed based on the maximum likelihood method (see e.g. [7, 9]). This was the case for $\rho^0\rho^0$, $\rho^+\rho^-$ and $K^{*0}\overline{K^{*0}}$ where the wave $(J^P, J_z) = (2^+, 2)$ was found to dominate the cross section.

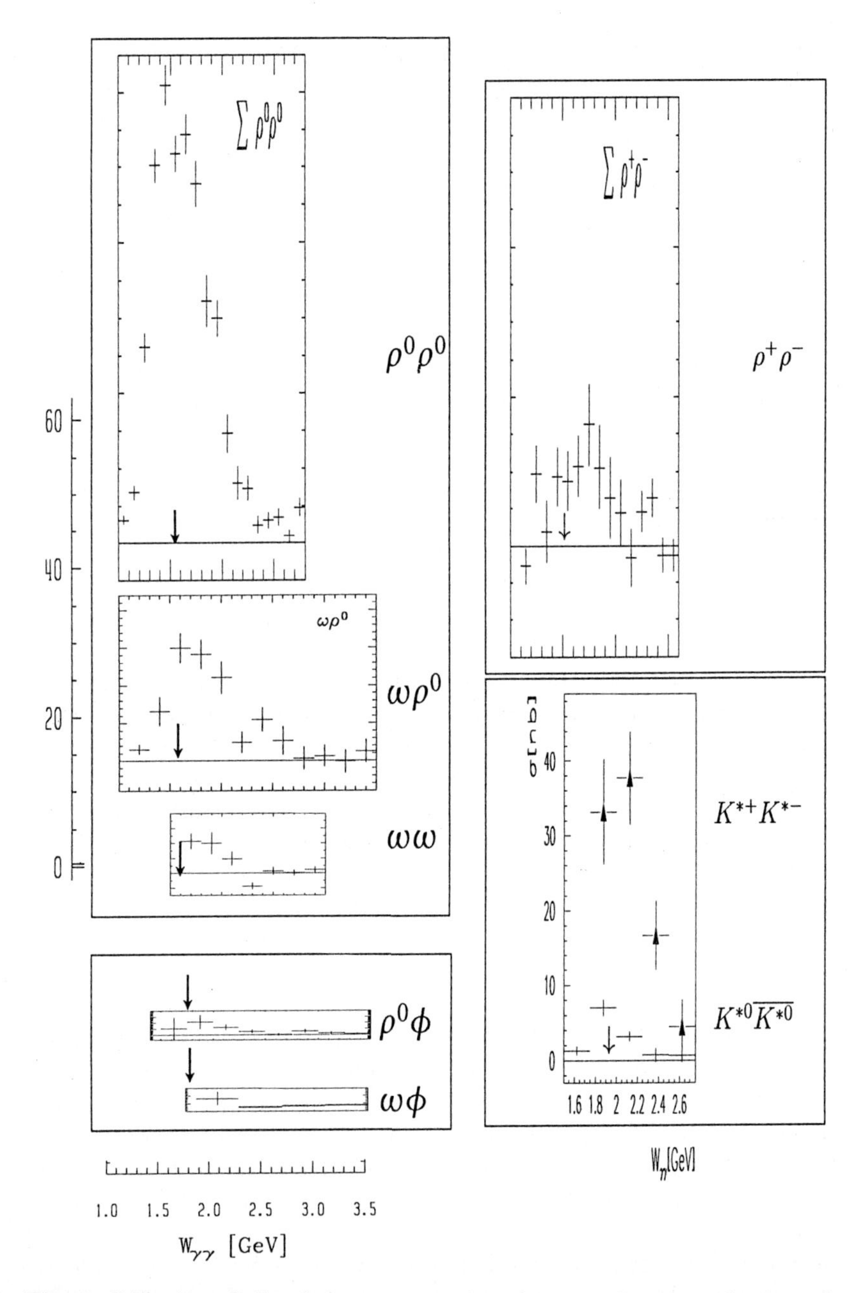

Fig. 1. Collection of all existing cross sections for two–photon production of pairs of vector mesons. Energy and cross section scales are adjusted so that they are equal for all spectra. Arrows indicate reaction thresholds.

We point out two main limitations of the partial wave analysis. Firstly, the final state is assumed to be composed of two isobars (ρ, ϕ, K^*, π, K etc.) and phase space distributed multiparticle states (fig. 2). Secondly, the analysis is usually limited to the partial waves of orbital angular momenta $L = 0, 1, 2$.

Fig. 2. A schematic drawing representing the isobar model with several particles in the final state. The final state partial waves are treated as being composed of two well defined isobars and phase space distributed parts.

The main origin of systematic errors in the partial wave analysis is a limited knowledge of acceptances for different partial waves. Typically, a relative error in acceptance of 1% may cause an error in cross section of about 10%. Therefore, it is important to have a good knowledge of detector performance. The analysis of the reaction $\gamma\gamma \to \pi^+\pi^-\pi^0$ was used as a test of this kind of systematics. It was found that the $J^P = 2^+$ nature of a_2 was excellently reproduced [11].

3 Interpretation of Experimental Results

Most models for two–photon production of vector meson pairs are based on *vector dominance*. The initial state creation of vector mesons is governed by charge factors for $\rho^0\rho^0 : \rho^0\omega : \omega\omega : \phi\rho^0 : \phi\omega : \phi\phi : \rho^+\rho^- : K^{*+}K^{*-} : K^{*0}\overline{K^{*0}}$ $= 1 : \frac{1}{9} : \frac{1}{81} : \frac{2}{9} : \frac{2}{81} : \frac{4}{81} : 0 : 0 : 0$. They are slightly modified due to flavour SU(3) symmetry breaking. As a first guess, one can see a qualitative tendency of integrated cross sections for neutral nonstrange final states to follow the above ratios, with the exception of processes involving ϕ mesons. On the other hand, there is a significant cross section for production of $\rho^+\rho^-$, $K^{*+}K^{*-}$ and $K^{*0}\overline{K^{*0}}$ that can only proceed through the interaction of initial vector mesons.

A simple VMD approach is to relate e.g. $\gamma\gamma \to \rho^0\rho^0$ to $\rho^0\rho^0$ scattering ($\rho^0\rho^0$ scattering being in turn related to high energy pp scattering). It gives a smooth energy variation of the cross section and cannot account for the enhancement at threshold.

Various Regge exchange mechanisms [3] seem to be able to reproduce to some extent two photon production of $\omega\omega$, $\rho^+\rho^-$, $K^{*0}\overline{K^{*0}}$ and $K^{*+}K^{*-}$ (including $J^P = 2^+$ dominance).

Resonance models like the model of Achasov et al. [2] use the MIT bag model to find two degenerate $qq\bar{q}\bar{q}$ states with $I = 2$ and $I = 0$. In VMD they interfere constructively in $\gamma\gamma \to \rho^0\rho^0$ and destructively in $\gamma\gamma \to \rho^+\rho^-$. The intermediate 2^+ states with OZI superallowed decays to ρ pairs may be long-lived enough to be observable if their mass is close or below the $\rho\rho$ threshold. In addition an $I = 1$ $qq\bar{q}\bar{q}$ state should show up in $\gamma\gamma \to \omega\rho$ and another in $\gamma\gamma \to \rho\phi$ channels. Experimentally there is little evidence for any of the latter two.

Alexander et al. [5] estimate the t-channel exchange contributions in a factorization model by extrapolating from high energies to the threshold region. As input they use experimental cross sections from vector meson production in elastic baryon–baryon scattering, and from vector meson photoproduction on baryons. It is not clear whether this model can be applied to the threshold region. Their results cannot account for the steep rise of the $\gamma\gamma \to \rho^0\rho^0$ cross section at threshold. Also, they are unable to reproduce the experimental data on $\gamma\gamma \to \rho^0\omega$ and $\gamma\gamma \to \omega\omega$.

In the perturbative QCD model of Brodsky et al. [6] the main contribution comes from the formation mechanism. The model reproduces the experimentally observed ratio 8 of integrated cross sections $\gamma\gamma \to K^{*+}K^{*-}$ and $\gamma\gamma \to K^{*0}\overline{K^{*0}}$, but fails to account for the large values of the cross sections.

In summary, the mechanism behind the reactions $\gamma\gamma \to V_1V_2$ is too complex to be understood by a single model. In particular, none of the models can explain satisfactorily the cross sections for $\gamma\gamma \to \rho^0\phi$ and $\gamma\gamma \to \omega\phi$.

The two–photon production of pairs of vector mesons appears to be a combination of several processes. Exceptionally it may happen that one of the processes prevails. Such may be the case with the presumably observed $qq\bar{q}\bar{q}$ states in $\gamma\gamma \to \rho\rho$. It is difficult to see how to prove or disprove the existence of such states. There is some hope that a careful analysis of high statistics experimental data may solve the problem.

References

[1] H. Albrecht et al. (ARGUS), Phys. Rep. 276 (1996), 223-405.

[2] N.N. Achasov, S.A. Devyanin and G.N. Shestakov, Phys. Lett. 108B (1982), 134-139.

[3] N.N. Achasov, V.A. Karnakov and G.N. Shestakov, Int. J. Mod. Phys. A5 (1990), 2705-2719.

[4] B.A. Li and K.F. Liu, Phys. Rev. Lett. 51 (1983) 1510-1513, K.F. Liu and B.A. Li, Phys. Rev. Lett. 58 (1987) 2288-2291.

[5] G. Alexander and U. Maor, Phys. Rev. D46 (1992) 2882-2890.

[6] S.J.Brodsky, G.Köpp and P.M.Zerwas, Phys.Rev.Lett.58(1987)443-446.

[7] H. Albrecht et al. (ARGUS), Z. Phys. C50 (1991) 1-10 and Phys. Lett. B267 (1991) 535-540, T. Živko, Doctoral thesis, University of Ljubljana, Slovenia, 1994.

[8] H. Albrecht et al. (ARGUS) Phys. Lett. B374 (1996) 265-270 and B332 (1994) 451-457, E. Križnič, Doctoral thesis, University of Ljubljana, Slovenia, 1993.

[9] G. Medin, Doctoral thesis, University of Ljubljana, Slovenia, 1997.

[10] H. Albrecht et al. (ARGUS) Phys. Lett. B198 (1987) 255-260 and B212 (1988) 528-532.

[11] H. Albrecht et al. (ARGUS) Z. Phys. C74 (1997) 469-477.

Measurements of the Quark and Gluon Fragmentation Functions and of Hadron Production at LEP

Fabio Cossutti (Fabio.Cossutti@cern.ch)

CEA-Saclay/DSM/DAPNIA/SPP

Abstract. The large amount of hadronic events collected at LEP during the run at the Z^0 peak allows a detailed study of the production of hadrons, both inclusively and on identified final states, from the primary $q\bar{q}$ pair produced in the hard process. The measurements can be used to test the predictions of perturbative QCD and of the models commonly used to describe the non perturbative part of the fragmentation. New results presented by the DELPHI Collaboration are here discussed.

1 Measurement of the quark and gluon fragmentation functions

The transverse (T), longitudinal (L) and asymmetric (A) components of the fragmentation function in the process $e^+e^- \to h + X$ are defined as [1, 2]:

$$F_P = \frac{1}{\sigma_{tot}} \sum_h \frac{d\sigma_P^h}{dx_p} \quad (P = T, L, A) \tag{1}$$

where σ_{tot} is the total cross section, x_p is beam momentum fraction carried by the hadron h and the sum runs on all the charged hadrons produced.

DELPHI has provided a measurement of the fragmentation functions [3], both on an inclusive sample and on a flavour tagged one, using a lifetime tag technique in order to separate heavy and light primary flavours.

The average charged particles multiplicities obtained integrating the fragmentation functions $< n >= \int_0^1 F_{T+L} dx_p$ are:

$$< n_{ch} >= 21.21 \pm 0.01 (stat.) \pm 0.20 (syst.)$$
$$< n_{uds} >= 20.35 \pm 0.01 (stat.) \pm 0.19 (syst.)$$
$$< n_b >= 23.47 \pm 0.07 (stat.) \pm 0.36 (syst.)$$

in good agreement with the previous measurements. The study of the charge asymmetric fragmentation function: $\tilde{F}_A = 1/\sigma_{tot}(d\sigma_A^{h+}/dx_p - d\sigma_A^{h-}/dx_p)$ shows a 2 σ discrepancy with respect to the QCD predictions:

	NLO,NNLO	LO	DELPHI
$\int_{0.1}^1 \tilde{F}_A dx_p =$	−0.016	−0.023	−0.028 ± 0.006
$\int_{0.1}^1 \tilde{F}_A \frac{x_p}{2} dx_p =$	−0.0020	−0.0027	−0.0036 ± 0.0008

which could be interpreted as an evidence for the importance of non perturbative effects.

Using NLO QCD prediction for $F_L(x_p, Q^2)$ [1, 4] the gluon fragmentation function $D_g(z, Q^2)$ has also been extracted from a fit to the measured F_L, F_T. The result shows a good agreement with the previous measurements in the x_p region above 0.1 (for low x_p non perturbative effects start to play a significant role).

The knowledge of the transverse and longitudinal component of the fragmentation function can be used to measure the strong coupling constant $\alpha_s(M_Z)$ using the QCD prediction at NLO [5]. A precise estimate of neutral particle contribution has been done using JETSET PS 7.4 and HERWIG 5.9. Taking into account also non perturbative "power" corrections and the renormalization scale uncertainties, $\alpha_s(M_Z)$ is found to be:

$$\alpha_s^{NLO+POW}(M_Z) = 0.101 \pm 0.002\ \mathit{(stat.)} \pm 0.013\ \mathit{(syst.)} \pm 0.007\ \mathit{(scale)}$$

2 Scaling violations in quark and gluon jets

The possibility to isolate samples of quark and gluon jets in hadronic events allows the study of the scaling violations of the fragmentation functions. DELPHI has presented a measurement of these scaling violating effects based on about 3000000 hadronic events collected in the period 1991-94[6].

Three jets events have been used; calling θ_1, θ_2 and θ_3 the interjets angles, either asymmetric topologies ($\theta_1, \theta_2 \in [135^0 \pm 35^0]$) or symmetric "Y" ones ($\theta_1, \theta_2 \in [(120^0, 130^0, 140^0, 150^0, 160^0) \pm 5^0]$) have been selected. The jet energy E has been computed assuming massless kinematics, and the jet scale has been given as the "hardness" scale $\kappa = E \sin\frac{\theta_{min}}{2}$ with θ_{min} the angle with respect to the closest jet. In this variable a particular angular range in the topology corresponds to a given jet scale range.

A lifetime tag technique has been used in order to separate quark and gluon jets: in events with evidence to come from a $b\bar{b}$ initial pair the jets not containing heavy hadrons decay products are isolated as gluon jets. Quark jets are selected from anti b-tagged events after subtraction of the gluon component previously measured.

Scaling violations are observed both in the quark and in the gluon sample (see fig. 1), and the violations are stronger in the gluon sample. There is a good agreement with the prediction of the DGLAP evolution equations [7]. The measurement of the ratio of the logarithmic derivatives of the fragmentation functions for gluons and quarks:

$$r_S(x_E) = \frac{S_g}{S_q} = \frac{\frac{d\ln D_g^H(x_E,s)}{d\ln s}}{\frac{d\ln D_q^H(x_E,s)}{d\ln s}} \tag{2}$$

allows an important test of QCD, being $\lim_{x_E\to 1} r_S(x_E) = C_A/C_F$ where C_A and C_F are the QCD Casimir factors. The data yields:

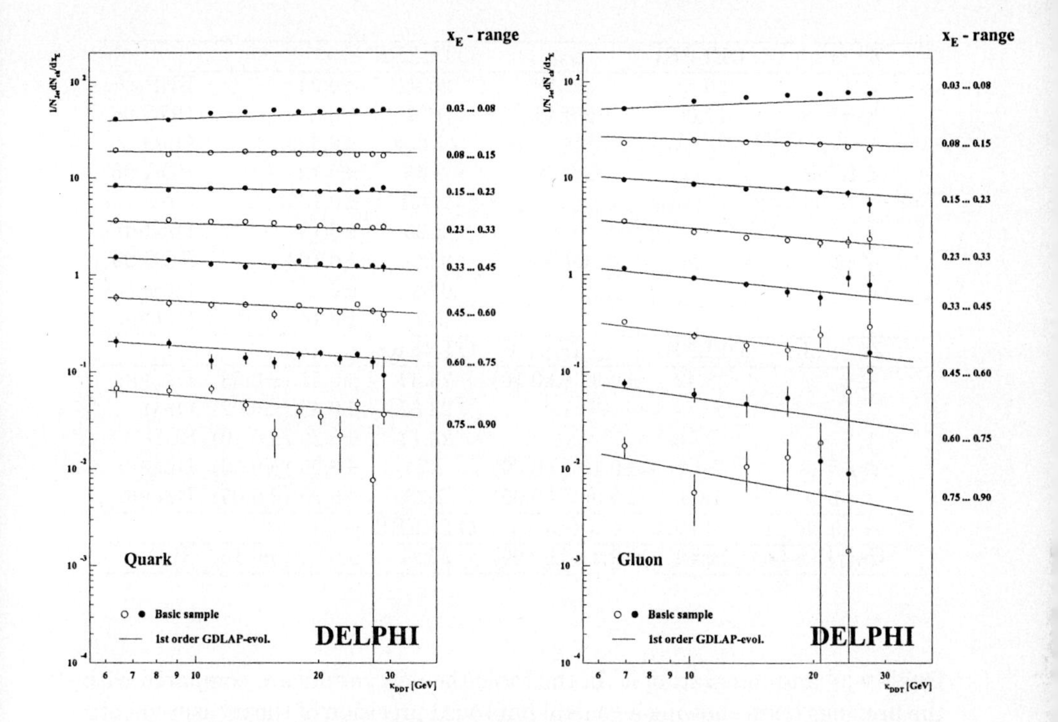

Fig. 1. Scale dependence of quark (left) and gluon (right) fragmentation functions.

$$C_A/C_F = S_g/S_q|_{x_E>0.5} = 2.7 \pm 0.7\,(stat.)$$

in good agreement with the prediction of QCD of 2.25.

3 Production of $\pi^\pm$, $K^\pm$, p and $\bar{p}$ in $Z^0 \to q\bar{q}$, $Z^0 \to b\bar{b}$, $Z^0 \to u\bar{u}, d\bar{d}, s\bar{s}$

The data sample collected by DELPHI in 1994 (about 1400000 hadronic events) has been used to measure the $\pi^\pm$, $K^\pm$ and p production on the full momentum spectrum ($0.7 \leq p \leq 45\,\mathrm{GeV}/c$) [8], exploiting the particle identification capability of the Barrel Ring Imaging Cherenkov detector. The production rates in flavour tagged samples have also been measured enriching samples in b and uds primary quarks with a lifetime tag technique.

The measured spectra show that in general JETSET PS 7.3, after the DELPHI tuning, has an higher capability to describe the hadron production than HERWIG 5.8, especially in the proton case.

The average particle multiplicities has been measured fitting the spectra with a distorted gaussian as predicted by MLLA in Local Parton Hadron

$Z^0 \to q\bar{q}$	DELPHI	$\epsilon_{tot.}$	OTHERS	$\epsilon_{tot.}$	
$< n_{ch} >$	20.43	±0.17	20.92	±0.24	LEP Av.
$< \pi^{\pm} >$	17.00	±0.43	17.1	±0.4	PDG'96
			17.052	±0.429	OPAL
$< \mathrm{K}^{\pm} >$	2.11	±0.08	2.39	±0.12	PDG'96
			2.421	±0.133	OPAL
			2.26	±0.18	DELPHI
$< p\bar{p} >$	1.04	±0.04	0.964	±0.102	PDG'96
			0.916	±0.111	OPAL
			1.07	±0.14	DELPHI
$Z^0 \to b\bar{b}$	DELPHI	$\epsilon_{tot.}(\epsilon_{stat.})$	OTHERS	$\epsilon_{tot.}(\epsilon_{stat.})$	
$< n_b >$	23.42	±0.35 (±0.10)	23.32	±0.51 (±0.08)	DELPHI
			23.62	±0.48 (±0.02)	OPAL
			23.14	±0.39 (±0.10)	SLD
$< \mathrm{K}^{\pm} >$	2.54	±0.16 (±0.09)	2.74	±0.50 (±0.10)	DELPHI
$< p\bar{p} >$	1.03	±0.06 (±0.06)	1.13	±0.26 (±0.05)	DELPHI
$Z^0 \to u\bar{u}, d\bar{d}, s\bar{s}$	DELPHI	$\epsilon_{tot.}(\epsilon_{stat.})$	OTHERS	$\epsilon_{tot.}(\epsilon_{stat.})$	
$< n_{uds} >$	19.44	±0.21 (±0.09)	20.21	±0.24 (±0.10)	SLD

Duality [9] and integrating it. In the table the new results are compared with the previous ones, showing a general improved precision of the measurements; *uds* sector results are completely new. A discrepancy in $< n_{uds} >$ with respect to the result presented in the first analysis can be observed, while $< n_{ch} >$ and $< n_b >$ are in good agreement.

Particularly interesting is the new value of $< \mathrm{K}^{\pm} >$, in better agreement with the average measured production of K^0, $< \mathrm{K}^0 + \overline{\mathrm{K}}^0 >= 2.01 \pm 0.04$, than the previous measurements.

References

[1] G. Altarelli et al., Nucl. Phys. B160 (1979) 301
[2] P. Nason, Br.R. Webber, Nucl. Phys. B421 (1994) 473
[3] DELPHI 97-69, contribution #234 to the EPS-HEP97 conference
[4] G. Altarelli et al., Phys. Rep. 81 N1 (1982) 1
[5] P.J. Rijken, W.L. van Neerven, Phys. Lett. B386 (1996) 422
[6] DELPHI 97-117, contribution #546 to the EPS-HEP97 conference
[7] V.N. Gribov, L.N. Lipatov, Sov. J. Nucl. Phys. 15:438 and 675 (1972); G. Altarelli, G. Parisi, Nucl. Phys. B126 (1977) 298; Y. Dokshitzer, Sov. Phys. JETP 46 (1977) 641
[8] DELPHI 97-110, contribution #541 to the EPS-HEP97 conference
[9] Y. Azimov, Y. Dokshitzer, V. Khoze and S. Troyan, Z. Phys. C31 (1986) 213

Properties of Quark and Gluon Jets - a LEP Mini Review

Roger Jones (Roger.Jones@cern.ch)

University of Lancaster, England

Abstract. Recent results on the structure of and differences between quark and gluon jets from the four LEP experiments are reviewed.

1 Introduction

Investigations of the differences in properties of quark and gluon jets have advanced considerably in recent years. The earliest technique to study the gluon jets was to select three-jet events; in typically 70% of these, the lowest energy jet was from a gluon. This method had the disadvantage that the quark and gluon jets studies were at different scales, which then required the use of observables insensitive to the scale, or else a model-dependent correction of the differences in scale. A great advance was the use of selected samples with the lower-energy jets symmetrically arranged about the highest energy jet. The two low energy jets are then at the same scale, in a similar track environment, and represent an equal mixture of quark and gluon jets. By tagging a quark in one of the jets, the other low energy jets may be anti-tagged as a pure sample of gluons. The quark and gluon properties at the same scale are obtained by algebraic decomposition. The quark tagging is usually done by standard b-tags; OPAL presented a new analysis where the quark tagging uses the higher collimation of the energy in quark jets to obtain a high-statistics gluon sample. This technique seems widely applicable.

Aside from the discrimination between quark and gluon jets, there is an arbitrary nature to our classification of jets at the particle level, and to their identification with original partons, Two analyses, one by ALEPH covered here, and one by OPAL described elsewhere, try to address this problem by reducing the jet-finder dependence.

This review can be classified into two parts: the first concerns the structure of the quark and gluon jets, particularly their splitting into subjets; and the second reports the identified particle composition in the jets.

2 Subjet structure

Both ALEPH[1] and DELPHI[2] have submitted papers continuing the investigation of the splitting of jets into subjets. ALEPH select events with a high scaled invariant mass cut $y = 0.1$ favouring 3-fold symmetric topologies and allows the sub-jet splitting to be probed well into the perturbative region; gluon jet samples are selected by finding b-tags in both of the other jets. The DELPHI analysis uses a lower y-cut, explicitly selecting symmetric topologies, and isolating the pure-gluon sample in the usual ways. Both investigate the mean number of sub-jets found in quark and gluon jets as the y-scale is decreased. Comparisons with Monte Carlo indicate that the high-y region is well described by the Monte Carlos (figure 1, left). Comparisons are made to LO+NLLA predictions; it is agreed that these describe the high-y gluon and quark distributions, but the experiments disagree as to whether the quark or the gluon predictions are more successful at lower y. As different calculations are used in the two cases, the disagreement may be theoretical not experimental. In neither experiment does the subjet multiplicity ratio $(n_{gluon} - 1)/(n_{quark} - 1)$ rise to naïve value of C_A/C_F. ALEPH have also studied the ratio of the corresponding dispersions and these also fall short of the naïve prediction of $\sqrt{C_A/C_F}$ for all y.

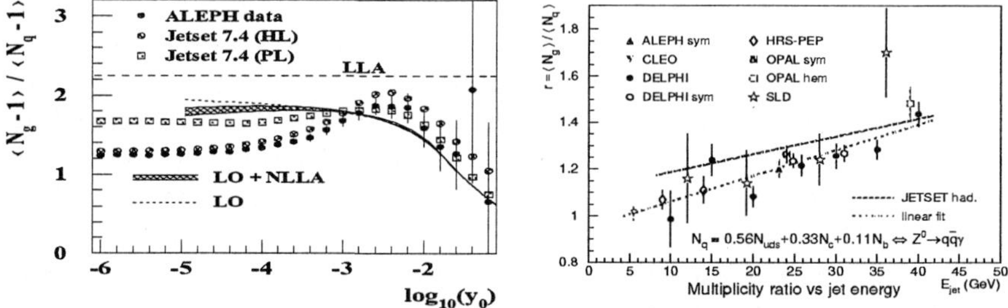

Fig. 1. Left, the ALEPH corrected ratio of subjet multiplicities in gluon and quark jets as a function of mass clustering variable y; HL and PL refer to hadron and parton level values. Right, the variation of quark and gluon jet particle multiplicities as a function of centre-of-mass energy.

In the limit of very low y, the subjet multiplicity becomes the particle multiplicity. DELPHI have submitted papers investigating the quark and gluon where jet particle multiplicities as a function of **energy**[3] and **scale**[4]. The energy analysis selects gluon jet samples of varying energy in symmetric 3-jet events using b-tagging, and quark jets in $q\bar{q}\gamma$ events. An approximately linear rise in the gluon-to-quark particle multiplicity ratio is found as a function of jet energy. This seems to be confirmed with the inclusion of data fromother experiments. The slope seems to disagree with that predicted by their JETSET simulation. The second analysis is performed in terms of effective k_T scales and is believed to reveal the more fundamental behaviour. For the quark jets, the scale is defined to be $\kappa_{DDT} = E_{jet} \sin\theta/2$, where θ is the angle to the nearest jet; for gluon jets, the scale is defined to be $\kappa_{dipole} = E_{jet}\sqrt{\sin\theta_1/2 \sin\theta_2/2}$, where θ_1 and θ_2 are the angles to the other two jets. The gluon jet multiplicity is seen to rise faster with scale than the quark jet; here, the agreement with JETSET predictions seems better.

3 Identified particle production

The LEP experiments have recently been studying the observed production of identified particles in quark and gluon jets. Several new papers were submitted on this topic, and several outstanding questions are addressed. The first concerns η production; L3 previously published the fragmentation function $1/\sigma d\sigma(\eta)/dx$[6] in 3-jet events, where $x = p(\eta)/p_{beam}$, and concluded that the measured rates exceeded their JETSET predictions in the third (mainly gluon) jet. Such an enhancement might be expected in various independent fragmentation models. ALEPH has presented a new paper with a similar analysis[5] in terms of $x = E(\eta)/E_{beam}$, and observe much better agreement with the JETSET predictions. The disagreement seems not to be in the data, but rather in the the Monte Carlo models, in particular the treatment of the deficit of 3-jet events overall in JETSET, rather than in the η production.

The next outstanding question concerns the relative rate of meson and baryon production in quark and gluon jets. Data from the $\Upsilon(1s) \to ggg$ decay and from continuum production at $\sim$ 10GeV suggest baryon production is enhanced in gluon jets over quark jets of the same energy; and that meson production is about the same in quark and gluon jets. DELPHI previously published rates for $K^{\pm}, K^0, \Lambda$ and p production in quark and gluon jets[7] (distinguished using symmetric event topologies), and also indicated, though with large errors, that the baryon rates are enhanced as expected. The relative rates were not well modelled by the Monte Carlos, particularly HERWIG.

OPAL have two analyses on K^0 and Λ production in quark and gluon jets[8], one using energy-ordering to discriminate between the two sorts of jets, and one using the novel energy-collimation quark-tagger mentioned earlier, combined with events with symmetric event topologies. L3 have also presented an analysis based on energy-ordered jets in 3-jet events[9]. As the particle multiplicity depends on the jet energy, the observable studied is usually the ratio $R = N_X/N_{ch}$, where N_X is the multiplicity for identified particle X, as this ratio is believed to be insensitive to the jet energy. A ratio of ratios R_g/R_q is then used to study gluon/quark jet differences. In the L3 case, the numbers quoted later are obtained from the main author. In addition, a new ALEPH analysis considers 2- and 3-jet events simultaneously, classifying them according to the two κ scale variables introduced above[10]. The quark and gluon jet multiplicities (either charged, K^0 or Λ) is fitted according to a MLLA prediction, differing only by a constant factor. Fits to this model give very consistent results for all particle species considered, and an effective ratio of ratios may be obtained.

The results for the ratio of ratios are given below. It is clear that OPAL and DELPHI see an enhanced production of Λ in gluon jets and no enhancement for K^0. The ALEPH result shows a smaller enhancement, but consider an average over very different topologies; the difference in the two OPAL results may also be topological in nature, requiring further investigation.

R_g/R_q	K^0	JETSET	Λ	JETSET
DELPHI	1.13 ± 0.16	0.98	1.40 ± 0.38	1.53
OPAL, E-order	1.10 ± 0.03	0.93	1.41 ± 0.06	1.27
OPAL collim.	1.05 ± 0.12	1.06	1.32 ± 0.21	1.56
L3	1.067 ± 0.023	—	1.25 ± 0.09	—
ALEPH f_g/f_q	1.04 ± 0.09	0.95	1.14 ± 0.08	1.07

Finally, OPAL have produced a conference paper in which the equivalent numbers are given for identified charged particles[11]. The results are in good agreement with those previously produced by DELPHI[7], but with much smaller numbers. The Monte Carlo predictions from the same model and version are quite different in the two experiments; experiment-specific modifications and tuning are clearly important. Note also that the proton production is much less enhanced in gluons that in the Λ case, and that the charged kaon production is somewhat surprisingly suppressed in the gluon jets, as opposed to a small enhancement in the K^0 case. The results from both experiments, along with the OPAL model predictions are given in the table below.

R	OPAL	DELPHI	JETSET 7.4	HERWIG 5.9
$\pi^\pm$	1.02 ± 0.01	—	1.01	1.05
$K^\pm$	0.95 ± 0.03	0.93 ± 0.04	0.90	0.73
K^0	1.10 ± 0.03	1.13 ± 0.16	0.93	0.73
$p(\overline{p})$	1.10 ± 0.04	1.12 ± 0.12	1.38	1.02
Λ	1.41 ± 0.06	1.40 ± 0.38	1.27	0.88

References

[1] The ALEPH Collaboration, HEP'97 Conference paper HEP97 #609
[2] The DELPHI Collaboration, HEP'97 Conference paper HEP97 #546
[3] The DELPHI Collaboration, Z. Phys. C70 (1996) 179
The DELPHI Collaboration, HEP'97 Conference paper HEP97 #683
[4] The DELPHI Collaboration, HEP'97 Conference paper HEP97 #546
[5] The ALEPH Collaboration, HEP'97 Conference paper HEP97 #598
[6] The L3 Collaboration, M. Acciarri *et al.*, Phys. Lett. B371 (1996) 126
[7] The DELPHI Collaboration, P. Abreu *et al.*, Z. Phys. C67 (1995) 543
[8] The OPAL Collaboration, Physics Note PN236;
The OPAL Collaboration, Physics Note PN289
[9] The L3 Collaboration, HEP'97 Conference paper HEP97 #506
[10] The ALEPH Collaboration, HEP'97 Conference paper HEP97 #627
[11] The OPAL Collaboration, HEP'97 Conference paper HEP97 #195

Interference Effects and Correlations Probing the Hadronization Process at LEP

Ulrich Becker (becker@mitlns.mit.edu)

Massachusetts Institute of Technology, Cambridge MA 02139

Abstract. Hadronization is studied starting with the clean diquarks system from Z decays. Color string effects are demonstrated by new observations of charge ordering. Analysis of the charged multiplicity in the ratio of the cumulative to factorial moments shows oscillations. They are not indicative of NNLLA as thought before. $\Lambda\bar{\Lambda}$ and $p\bar{p}$ correlate tightly, whereas $\Lambda\bar{p}$ and $\bar{\Lambda}p$ have a flatter rapidity distribution. Identical $\Lambda\Lambda$ pairs at $\Delta Q < 2$ GeV are suppressed, as expected from Fermi-Dirac statistics. The spin analysis is consistent but the source size small. Pions close in phase space show clear Bose-Einstein correlations.

Compared to hadron collisions the decays of Z's provide a clean quark-antiquark pair at the 50 GeV scale, free of projectile or target remnants. Perturbative QCD describes the subsequent quark-gluon cascade down to the ≈ 1.0 GeV scale [1, 2, 3], in principle properly accounting for interferences. In practice approximations are used like Double Leading Log, Modified Leading Log, Next to Leading Log, and Next to Next Leading Log(NNLLA). Descent to 100 MeV must proceed through a phase transition from the quark gluon plasma by freezing(crystallizing) out the colorless combinations forming hadrons. In this theoretically unknown region we encounter interferences and correlations which we will try to deduce from the experimental data, which are the decay products of these 'primordial' hadrons.

Hadronization is modelled by the Herwig program [4, 5], generating hadron clusters. The Jetset [6, 7] program reproduces the hadrons very well using many input parameters. Here the initial quark-antiquark pair stretches and breaks the color string, converting the 1 GeV/fm into hadrons.

1 Charge Ordering [8]

By the simple string picture breaking the color string produces a set of alternating charges, Figure 1a, diluted by π^0's (not shown). The Delphi Collaboration [8] assigns each charged hadron a rank, n_r, corresponding to the rapidity with respect to the event thrust axis,Figure 1b. The leading particle, $n_r = 1$, reflects the correct quark charge to $\approx 60\%$. Δy_{tag} denotes the rapidity difference to the second particle. The Compensation Charge, defined as:

$$\Delta\rho(n_r) = \frac{N_{opposite}(n_r) - N_{same}(n_r)}{N_{TaggedHemispheres}} \tag{1}$$

is given for $\Delta y_{\text{tag}} > 0.5$ in Figure 2a. The second particle typically has op-

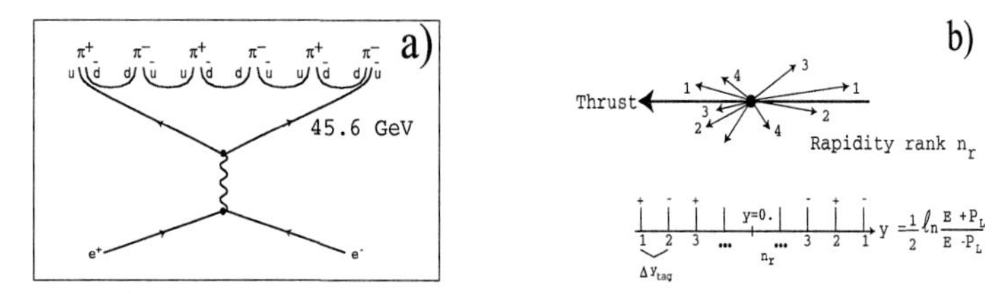

Fig. 1. a) String picture simplified for Z → charged pions b) Rapidity ranking

posite charge. Local charge compensation dominates in the same side hemisphere as expected. The distribution of the same events, with charge randomly assigned to each track, is very different and symmetric. Demanding rapidity gaps of $\Delta y > 0.5$ to the neighbors (Figure 2b) shows indications for alternating charges in the opposite hemisphere. To amplify the effect Figures 2(c,d) give the double ratio:

$$R(n_r) = \frac{\rho_{opposite}(n_r) - \rho_{same}(n_r)}{\rho_{opposite}(n_r)} \tag{2}$$

If the cut is sharpened to $\Delta y_{\text{tag}} \geq 1.$ as in Figure 2d, **charge ordering** is clearly observed, supporting the string picture of Figure 1a. The behavior is well described by Jetset7.3 and Herwig.

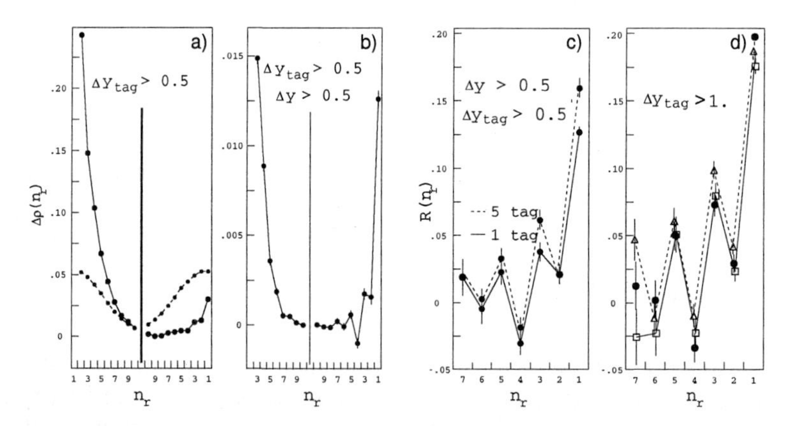

Fig. 2. Compensation charge (Eq. 1) a) Data (solid line) compared to random charge (broken line) b) Data sample with tracks spaced by $\Delta y > 0.5$ c)Ratio of compensation charges, R (Eq. 2), for data with $\Delta y_{\text{tag}} \geq 0.5$ and $\Delta y > 0.5$ d)R for $\Delta y_{\text{tag}} \geq 1.$ in comparison to Herwig(....) and Jetset(—-)

2 Charged Multiplicity [9]

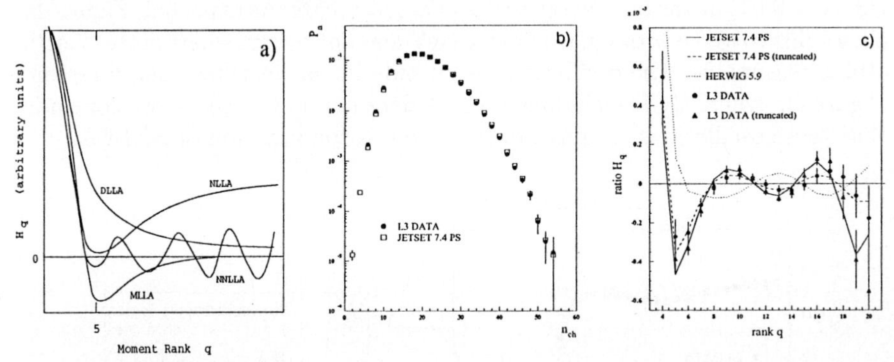

Fig. 3. a) QCD predictions for H_q b) Charged particle multiplicity of L3 c) H_q for the distribution of b)

The multiplicity distribution of charged particles from Z decays has been measured at LEP and SLC. Every distribution can be characterized by its moments. The normalized factorial moments are:

$$F_q = \frac{\sum_{n=q}^{\infty} n(n-1)....(n-q+1)P_n}{(\sum_{n=1}^{\infty} nP_n)^q} \tag{3}$$

and

$$K_q = F_q - (\sum_{m=0}^{q-1} \frac{(q-1)!}{m!(q-m-1)!} K_q - mF_m) \tag{4}$$

For a Poisson distribution all $F_q = 1$; if correlated $F_q < 1$. To project out the q-particle system the cumulative factorial moments K_q are used. Since both F_q and K_q are strongly rising it is customary to use the "double" ratio, $H_q = K_q/F_q$. QCD predictions for this sensitive ratio are given in Figure 3a. These assume that gluons and quarks translate directly to hadrons by the local parton-hadron duality(LPHD). LPHD is known to be good for single particle distributions but poor for resonances and heavy quarks. Striking are the large differences and the NNLLA oscillations as a function of rank q. These were observed in SLD data [11] and taken to be a confirmation of NNLLA.

L3 has carried out a new measurement and analysis [9] of the charged particle distribution, Figure 3b, to test LPHD-NNLLA at various energy scales selected by the Durham jet algorithm, $y_{ij} = (2\min(E_i^2, E_j^2))/(E_{vis}^2(1-cos\delta_{ij}))$ The charged tracks of 0.8 million hadronic decays were studied. Figure 3b attests that Jetset 7.4PS describes the data well.

The ratio H_q oscillates (Figure 3c). The data agree with NNLLA and SLD [11]. Jetset 7.4PS gives an accurate description, which is very surprising since it is not of order NNLLA! Probing into the subjet region, L3 analyzed the ratio of multiplicities from 3-jet (M_3) to 2-jet (M_2) events (separated by a cut $y_1 > 0.01$) at various energy scales given by y-cuts. As expected, Figure 4a shows this to agree with QCD above 1 GeV and not below, where Jetset 7.4PS still is reasonable. The oscillations occur only in the very low energy region, Figure 4b, where QCD including NNLLA does not apply. Hence, we conclude that these oscillations are fortuitous and not a confirmation of NNLLA.

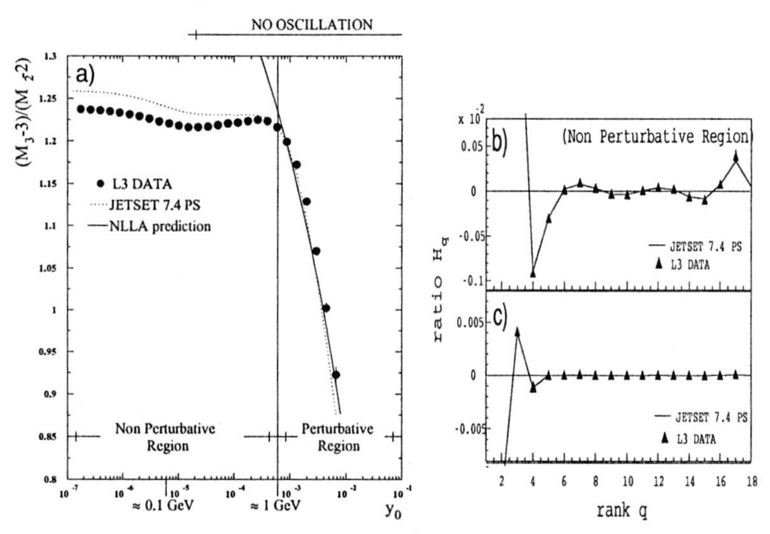

Fig. 4. a) Soft gluon production in 3-jet/2-jet events from NLLA (solid line) compared to Data and Jetset 7.4PS b) Oscillation are present at energy of 100 MeV ($y_{cut} \leq 6.0 \times 10^{-6}$) c) Oscillations are nearly gone at 200 MeV ($y_{cut} \leq 1.5 \times 10^{-6}$)

3 Correlations in Baryon Pairs [12]

The few baryons produced in Z decays provide a clean sample for the study of the transition from partons to hadrons. The hope is to follow the string interaction. Figure 5a displays the simplest string pictures for $\Lambda\bar{\Lambda}$ or $p\bar{p}$ baryon pairs produced closely in rapidity, i.e. $\Delta y \simeq 0.5$ at this energy. The wider quark loops of the 'Popcorn' graphs in Figures 5b and 5c should generate larger Δy. Delphi [12] has selected $\Lambda\bar{\Lambda}$ candidates by kinematics of V-type events. For protons dE/dx as well as the liquid and gaseous Rich Cerenkov counters were needed. After subtraction of background and corrections for solid angle, the spectra of Figure 5d were obtained. The observations are:

1. All nonidentical pairs are closely correlated with $\Delta y < 2$.
2. The $\Lambda\bar{p}$,$\bar{\Lambda}p$ are suppressed at low y suggesting 'Popcorn'-like production.
3. $\Lambda\bar{\Lambda}$,$p\bar{p}$ agree with Jetset 7.3 (f=0.5). For $\Lambda\bar{p}$ and $\bar{\Lambda}p$ a difficult choice of parameters is needed [12].

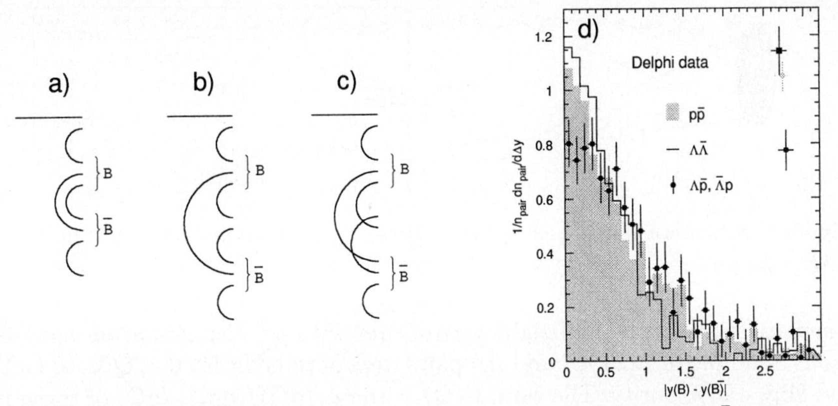

Fig. 5. a)b)c) Baryons in the string model d) Correlations in $B\bar{B}$ pairs

4. No long range correlations can be deduced from the data.

According to Figure 5c one naively expects to find more kaons between Λ's and protons. This is not seen. Also the rate of ΛKp is as frequent as $\Lambda\bar{K}$p. Delphi [12] concludes that 'decays smear the correspondence between rapidity and rank in the string.' Yet, within $\Delta y < 2$ from a Λ usually the correct sign kaon is found.

4 Fermi-Dirac Correlations in $\Lambda\Lambda$ and $\bar{\Lambda}\bar{\Lambda}$ Pairs [13]

Considering the Z decay fragmentation to be an extended source for $\Lambda\Lambda$ production, the Λ's from r_1 and r_2 may reach detectors at x_1 and x_2 either by the direct or crossed path in Figure 6a, which cannot be distinguished. In the plane wave approximation the two possibilities produce added a symmetric wavefunction, ϕ_s, or subtracted an antisymmetric wavefunction, ϕ_a:

$$\phi_{s,a}(x_1, x_2) = \frac{1}{2}\{e^{ip_1(x_1-r_1)}e^{ip_2(x_2-r_2)} \pm e^{ip_1(x_1-r_2)}e^{ip_2(x_2-r_1)}\} \qquad (5)$$

As the identical particles get very close, i.e. $Q = |p_1 - p_2| \to 0$, we see interference.

$$\text{destructive:} \quad |\phi_a|^2 = 1 - cos[(p_1 - p_2)(r_1 - r_2)] \quad \to 0\,,\ \text{as}\ \ Q \to 0 \quad (6)$$

$$\text{constructive:} \quad |\phi_s|^2 = 1 + cos[(p_1 - p_2)(r_1 - r_2)] \quad \to 2\,,\ \text{as}\ \ Q \to 0 \quad (7)$$

$\Lambda\Lambda$ can be in a s_0(odd) or s_1(even) spin state and the total wavefunction must be odd, hence:

$$\Psi^{tot}_{s=1} = \phi_a s_1 \quad \text{and} \quad \Psi^{tot}_{s=0} = \phi_s s_0 \qquad (8)$$

This offers a cross check, as the Aleph Collaboration [13] realized, since the the spin state can be deduced from the Λ decay asymmetry [14]. Using strict

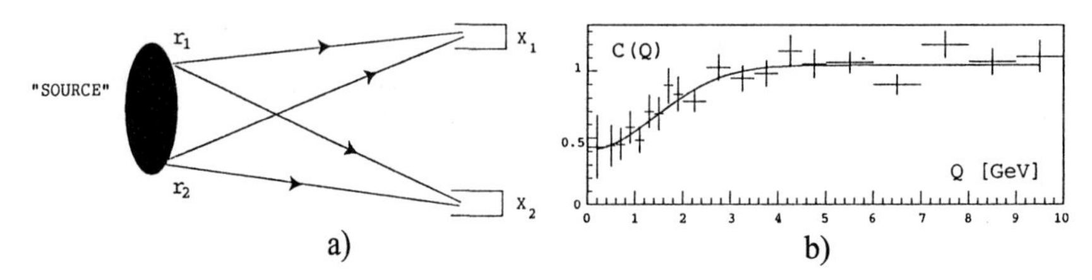

Fig. 6. a) Schematic illustration of Fermi-Dirac b) Correlation data and fit function C(Q)

kinematic constraints and tight vertex cuts with χ^2 discrimination against kaons a sample of 2223 $\Lambda\Lambda$ and $\bar{\Lambda}\bar{\Lambda}$ pairs were obtained with $0 < Q < 10$ GeV and 86% - 90% purity. The ratio $\mathrm{C(Q)} = (\mathrm{d}\sigma_{\Lambda\Lambda}/\mathrm{dQ})/(\mathrm{d}\sigma_{ref}/\mathrm{dQ})$ of these is given in Figure 6b. The reference sample was obtained from Jetset 7.4 $\Lambda\Lambda$ without Fermi-Dirac correlations. Destructive interference is observed at low Q, which was fitted to [14]:

$$C(Q) \propto (1 + \beta e^{-R^2 Q^2}) \tag{9}$$

with the result : $\beta = -0.61 \pm 0.08$ and $R = (0.10 \pm 0.02)$ fm. The source size parameter, $R = r_1 - r_2$, is surprisingly small.

The check by spin analysis is elegant, Figure 7. Plotted are the detector-corrected angular distributions, fitted with dN/dy $\propto 1 + 0.642$y for spin=0 and dN/dy $\propto 1 - 0.642$y for spin=1. Equal probability for the singlet and three triplet states yields a fraction of spin-1 $\epsilon = 3/4$. For $\Lambda\bar{\Lambda}$ pairs where no correlation is expected, this is indeed observed, Figures 7(c,d), also for $\Lambda\Lambda$ and $\bar{\Lambda}\bar{\Lambda}$ at high Q, Figure 6b. At low Q the spin=0 dominates, Figure 6a, because $\Psi^{tot}_{s=1}$ is suppressed by the destructive Φ_a indicating Fermi-Dirac behavior.

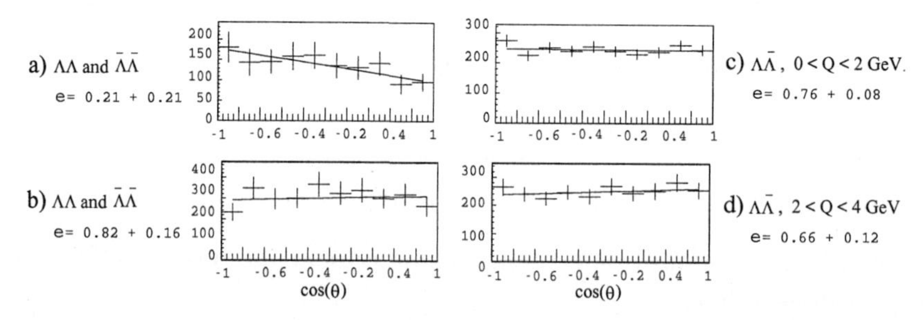

Fig. 7. Angular distributions of $\Lambda\Lambda$ with fits to $s = 0$ and $s = 1$ contributions.

5 Bose-Einstein Correlation Between Charged Pions [17]

A new measurement of like-sign particles and a correlation analysis has been reported by the L3 Collaboration [17]. With light cuts 1.5 million $Z \to q\bar{q}$ events were selected and analyzed for equal charge tracks, all taken to be pions. The distribution of unlike- and like-sign pairs is given in Figure 8a, where the Coulomb interaction was corrected for. The experimental correlation

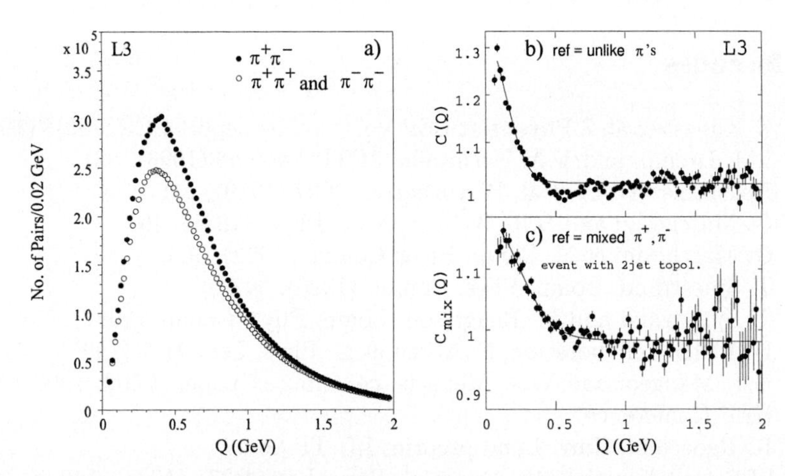

Fig. 8. a) pion pairs b) reference sample of unlike-sign pions c) reference sample from like-sign pions from opposite hemispheres

$C(Q) = [d\sigma_{like}(Q)/dQ]/[\,d\sigma_{ref}(Q)/dQ]$ with $Q^2 = (p_1 - p_2)^2 = M^2_{\pi\pi} - 4m^2_\pi$ is compared to the ansatz $C(Q) \simeq (1 + \lambda e^{-R^2Q^2})$, where R measures the size of a Gaussian source and λ measures correlations (1=completely incoherent with full overlap). To make C insensitive to purity(88% pions), efficiency, and acceptance effects, the ratio $C_d = N_{like}/N_{unlike}$ is formed for data and uncorrelated Monte Carlo events, so that $C(Q) = C_d/C_{MC}$. As usual the difficulty is to find a truly unbiased reference sample. Two extremes are represented by taking the unlike tracks, Figure 8b, or pairing tracks by reflecting the momentum vector through the origin for events with two jet topology, Figure 8c. Both were fitted and averaged, with the difference providing the systematic error. The result is:

$$R = 0.76 \pm 0.02(stat) \pm 0.15(syst) \tag{10}$$

and

$$\lambda = 0.25 \pm 0.01(stat) \pm 0.04(syst) \tag{11}$$

Note that if only pions < 4 fm from the origin contribute $\lambda_{max} < 0.4$. Comparing previous [18, 19, 20, 21] and present [22, 23, 24] data shows:

1. Bose-Einstein correlations are clearly present.
2. Comparing to PETRA/PEP data, there is no energy dependence.
3. The parameters in Jetset prior to 1994 were not appropriate.

In summary, much has been learned, yet the road to understanding will be long.

I gratefully acknowledge discussions with Drs. H.R. Hoorani, W.J Metzger, G. Wolf, P. Maettig, and W. Venus.

References

[1] Z. Kunzst et al, Z Physics at LEP Vol. 1 yellow report CERN 89-8 (1989)
[2] I.M. Dremin and V.A. Nechitailo, JETP Lett. 58 (1993), 881
[3] Dokshitzer Yu L. et al, Perturbative QCD (1989), 241
[4] G. Marchesini and B.R. Webber, Nucl. Phys. (1988), 461
[5] G. Marchesini et al, Comp. Phys. Comm. (1992), 465
[6] T. Sjoestrand, Comp. Phys. Comm. (1986), 347
[7] T. Sjoestrand and M. Bengtsson, Comp. Phys. Comm. (1987), 367
[8] DELPHI Collaboration, P. Abreu et al, Phys. Lett. B47 (1997), 174
[9] D.J. Mangeol and W.J. Metzger, contributed paper #502, 1997 EPS-HEP Conference
[10] R. Ugoccioni, Univ. Lund preprint LU TP 96-27
[11] SLD Collaboration, K. Abe et al, Phys. Lett. B371 (1996), 149
[12] DELPHI Collaboration, P. Abreu et al, CERN preprint CERN-PPE/97-27
[13] ALEPH Collaboration, D. Decamp et al, contributed paper #591, 1997 EPS-HEP Conference
[14] G. Goldhaber et al, Phys. Rev. 120 (1960), 300
[15] G. Alexander and H.J. Lipkin, Phys. Lett. B352 (1995), 162
[16] G. Goldhaber et al, Phys. Rev. Lett. 3 (1959), 181
[17] L3 Collaboration, M. Acciari et al, contributed paper#505, 1997 EPS-HEP Conference
[18] CLEO Collaboration, P. Avery et al, Phys. Rev. D32 (1985), 2294
[19] TPC Collaboration, H. Aihara et al, Phys. Rev. D31 (1985), 996
[20] TASSO Collaboration, M. Althoff et al, Z. Phys. C30 (1986), 355
[21] MARK II Collaboration, I. Juricic et al, Phys. Rev. D39 (1989), 1
[22] OPAL Collaboration, P.D. Acton et al, Phys. Lett. B267 (1991), 143
[23] ALEPH Collaboration, D. Decamp et al, Z. Phys. C54 (1992), 75
[24] DELPHI Collaboration, P. Abreu et al, Phys. Lett. B286 (1992), 201

Transverse and Longitudinal Bose-Einstein Correlations in e^+e^- Annihilation

Markus Ringnér (markus@thep.lu.se)

Lund University, Sweden

Abstract. We show how a difference in the correlation length longitudinally and transversely, with respect to the jet axis in e^+e^- annihilation, arises naturally in a model for Bose-Einstein correlations based on the Lund string model. The difference is more apparent in genuine three-particle correlations and they are therefore a good probe for the longitudinal stretching of the string field.

1 Introduction

Bosons obey Bose-Einstein statistics, which compared to an uncorrelated production leads to an enhancement of the production of identical bosons at small momentum separations. The enhancement is called the Bose-Einstein effect and it is very often parametrised in the form

$$R_2(q) = 1 + \lambda \exp(-Q^2 R^2) \qquad (1)$$

where Q is the relative four-momenta of a pair of bosons, $Q^2 = -q^2 = -(p_2 - p_1)^2$, and R and λ are two phenomenological parameters. The parameter R is often refered to as the radius of the boson emitting source. It is clear that the correlation function R_2 reflects the space–time region in which the particle production occurs but both the parametrisation in equation 1 and the interpretation of R are often given without very convincing arguments.

In reference [1] (an extension of reference [2] to multi-boson final states) a model for Bose-Einstein correlations based on a quantum-mechanical interpretation of the Lund string fragmentation model [3] is presented. In this work we will investigate some features of the model to see how the correlation lengths in the model arises and this will be used to show what the parameter R is sensitive to. We will in particular show that the model predicts, due to the properties of string fragmentation, a difference between the correlation length along the string and transverse to it. In practice this means that if we introduce the longitudinal and transverse components of the vector q (defined with respect to the thrust axis in e^+e^- annihilation) we obtain a noticable difference in the correlation distributions. This becomes even more apparent when we go to three-particle correlations because in this case one is even more sensitive to the longitudinal stretching of the string field.

2 Correlation lengths

The starting point of our Bose-Einstein model [1, 2] is an interpretation of the (non-normalised) Lund string area fragmentation probability for an n-particle state (see figure 1)

$$dP(p_1, p_2, \ldots, p_n) = \qquad (2)$$
$$\prod_1^n N dp_j \delta(p_j^2 - m_j^2) \delta(\sum p_j - P_{tot}) exp(-bA)$$

in accordance with a quantum mechanical transition probability containing the final state phase space multiplied with the square of a matrix element $\mathcal{M}$. In reference [2] and in more detail in reference [1] a possible matrix element is suggested in agreement with

(Schwinger) tunneling and the (Wilson) loop operators necessary to ensure gauge invariance. The matrix element is

$$\mathcal{M} = \exp(i\kappa - b/2)A \tag{3}$$

where the area A is interpreted in coordinate space, $\kappa \simeq 1$ GeV/fm is the string constant and $b \simeq 0.3$ GeV/fm is the decay constant.

The transverse momentum properties are in the Lund model taken into account by means of a Gaussian tunneling process. The produced ($q\bar{q}$)-pair in each vertex will in this way obtain $\pm\mathbf{k}_\perp$ and the hadron stemming from the combination of a $\bar{q}$ from one vertex and a q from the adjacent vertex obtains $\mathbf{p}_\perp = \mathbf{k}_{\perp \mathbf{j}+1} - \mathbf{k}_{\perp \mathbf{j}}$.

In case there are two or more identical bosons the matrix element should be symmetrised and in general we obtain the symmetrised production amplitude

$$\mathcal{M} = \sum_{\mathcal{P}} \mathcal{M}_{\mathcal{P}} \tag{4}$$

where the sum goes over all possible permutations of identical particles. Taking the square we get

$$|\mathcal{M}|^2 = \sum_{\mathcal{P}} |\mathcal{M}_{\mathcal{P}}|^2 \left(1 + \sum_{\mathcal{P}' \neq \mathcal{P}} \frac{2\mathrm{Re}(\mathcal{M}_{\mathcal{P}} \mathcal{M}^*_{\mathcal{P}'})}{|\mathcal{M}_{\mathcal{P}}|^2 + |\mathcal{M}_{\mathcal{P}'}|^2}\right) \tag{5}$$

The MC program JETSET [4] will provide the outer sum in equation 5 by the generation of many events but it is evident that the model predicts a quantum mechanical interference weight, $w_{\mathcal{P}}$, for each given final state characterised by the permutation $\mathcal{P}$:

$$w_{\mathcal{P}} = 1 + \sum_{\mathcal{P}' \neq \mathcal{P}} \frac{2\mathrm{Re}(\mathcal{M}_{\mathcal{P}} \mathcal{M}^*_{\mathcal{P}'})}{|\mathcal{M}_{\mathcal{P}}|^2 + |\mathcal{M}_{\mathcal{P}'}|^2} \tag{6}$$

In the Lund model we note in particular for the case exhibited in figure 1, with two identical bosons denoted 1 and 2 having a state I in between, that the decay area is different if the two identical particles are exchanged. It is evident that the interference between the two permutation matrices will contain the area difference, ΔA, and the resulting general weight formula will be

$$w_{\mathcal{P}} = 1 + \sum_{\mathcal{P}' \neq \mathcal{P}} \frac{\cos \frac{\Delta A}{2\kappa}}{\cosh\left(\frac{b\Delta A}{2} + \frac{\Delta(\sum \mathbf{k}^2_{\perp j})}{2\kappa}\right)} \tag{7}$$

where Δ stands for the difference between the configurations described by the permutations $\mathcal{P}$ and $\mathcal{P}'$ and the sum is taken over all the vertices. The calculation of the weight function

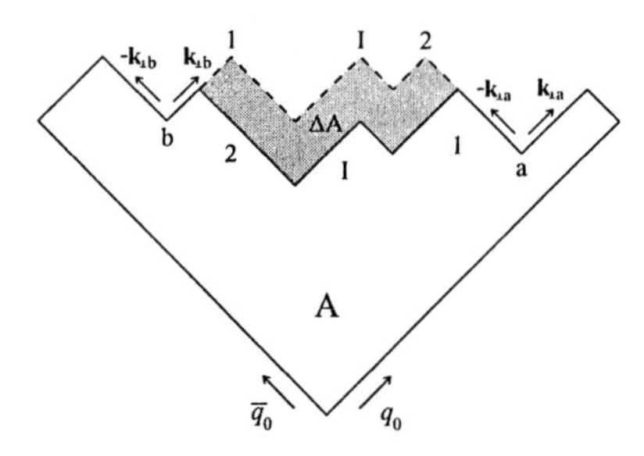

Fig. 1. *The decay of a Lund Model string spanning the space–time area A. The particles* 1 *and* 2 *are identical bosons and the particle(s) produced in between them is denoted by I. The two ways to produce the state are shown.*

for n identical bosons contains $n! - 1$ terms and it is therefore from a computational point of view of exponential-type. We have in reference [1] introduced approximate methods so that it becomes of power-type instead and we refer for details to this work.

We have seen that the transverse and longitudinal components of the particles momenta stem from different generation mechanisms. This is clearly manifested in the weight in equation 7 where they give different contributions. In the following we will therefore in some detail analyse the impact of this difference on the transverse and longitudinal correlation lengths, as implemented in the model.

In order to understand the properties of the weight in equation 7 we again consider the simple case in figure 1. The area difference of the two configurations depends upon the energy momentum vectors p_1, p_2 and p_I and can in a dimensionless and useful way be written as

$$\frac{\Delta A}{2\kappa} = \delta p \delta x_L \tag{8}$$

where $\delta p = p_2 - p_1$ and $\delta x_L = (\delta t; 0, 0, \delta z)$ is a reasonable estimate of the space-time difference, along the surface area, between the production points of the two identical bosons.

To preserve the transverse momenta of the particles in the state $(1, I, 2)$ it is necessary to change the generated $\mathbf{k}_\perp$ at the two internal vertices around the state I during the permutation, i.e. to change the Gaussian weights. Also in this case we may write a formula similar to equation 8 for the transverse momentum change:

$$\frac{\Delta(\sum \mathbf{k}_{\perp j}^2)}{2\kappa} = \delta \mathbf{p}_\perp \delta \mathbf{x}_\perp \tag{9}$$

where $\delta \mathbf{p}_\perp$ is the difference $\mathbf{p}_{\perp 2} - \mathbf{p}_{\perp 1}$ and $\delta \mathbf{x}_\perp = (\mathbf{k}_{\perp b} - (-\mathbf{k}_{\perp a}))/\kappa$. The two neighbour vertices to the state $(1, I, 2)$ $((2, I, 1))$ are denoted by a and b and $\mathbf{k}_{\perp b} + \mathbf{k}_{\perp a}$ corresponds to the state's transverse momentum exchange to the outside. Therefore $\delta \mathbf{x}_\perp$ constitutes a possible estimate of the transverse distance between the production points of the pair.

For the general case when the permutation $\mathcal{P}'$ is more than a two-particle exchange there are formulas similar to equations 8 and 9.

It is evident from the considerations leading to equations 8 and 9 that only particles with a finite longitudinal distance and small relative energy momenta will give significant contributions to the weights. We also note that we are in this way describing longitudinal correlation lengths along the color fields, inside which a given flavor combination is compensated. The corresponding transverse correlation length describes the tunneling (and in this model it provides a damping chaoticity).

The weight distribution we obtain is discussed in reference [1]. It is strongly centered around unity and we find negligible changes in the JETSET default observables (besides the correlation functions) by this extension of the Lund model.

3 Results

We have analysed two- and three-particle correlations in the Longitudinal Centre-of-Mass System (LCMS). For each pair (triplet) of particles the LCMS is the system in which the sum of the particles momentum components along the jet axis is zero. In the pair analysis we have used the kinematical variables

$$q_\perp = \sqrt{(p_{x2} - p_{x1})^2 + (p_{y2} - p_{y1})^2} \tag{10}$$
$$q_L = |p_{z2} - p_{z1}|$$

and in the triplet analysis we have used

$$q_\perp = \sqrt{q_{\perp 12}^2 + q_{\perp 13}^2 + q_{\perp 23}^2} \tag{11}$$
$$q_L = \sqrt{q_{L12}^2 + q_{L13}^2 + q_{L23}^2}$$

where the jet axis is along the z-axis.

We have taken the ratio of the two-particle probability density of pions, ρ_2, with and without BE weights applied as the two-particle correlation function, R_2

$$R_2(p_1, p_2) = \frac{\rho_{2w}(p_1, p_2)}{\rho_2(p_1, p_2)} \tag{12}$$

and the resulting distribution is shown in figure 2. It is clearly seen that it is not symmetric in q_L and $q_\perp$ and in particular that the correlation length , as measured by the inverse of the width of the correlation function, is longer in the longitudinal than in the transverse direction. This difference remains for reasonable changes of the width in the transverse momentum generation. Using all the final pion pairs in the analysis results in in a small decrease in the transverse correlation length and of course a large decrease in the height for $q_L \simeq q_\perp \simeq 0$, while the longitudinal correlation length is rather unaffected.

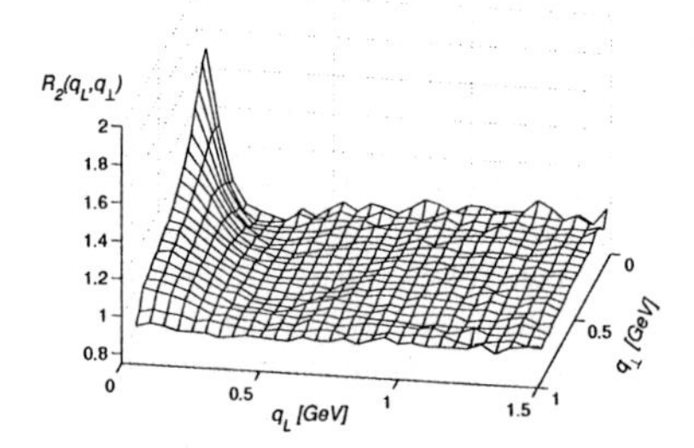

Fig. 2. The two-particle correlation function $R_2(q_L, q_\perp)$ for charged pions. The sample consistes of particles which are either initially produced or stemming from short-lived resonances.

The total three-particle correlation function is in analogy with equation 12

$$R_3''(p_1,p_2,p_3) = \frac{\rho_{3w}(p_1,p_2,p_3)}{\rho_3(p_1,p_2,p_3)} \qquad (13)$$

To get the genuine three-particle correlation function, R_3, the consequences of having two-particle correlations in the model have to be subtracted from R_3''. To this aim we have calculated the weight taking into account only configurations where pairs are exchanged, w'. In this way the three-particle correlations which only are a consequence of lower order correlations can be defined as

$$R_3'(p_1,p_2,p_3) = \frac{\rho_{3w'}(p_1,p_2,p_3)}{\rho_3(p_1,p_2,p_3)} \qquad (14)$$

The genuine three-particle correlation function, R_3, is then given by

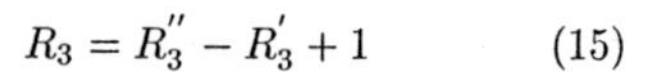

$$R_3 = R_3'' - R_3' + 1 \qquad (15)$$

This way of getting the genuine correlations is not possible in an experimental situation, where one has to find other ways to get a R_3' reference sample. We have suggested one possible option in reference [1]. The distribution R_3 is shown in figure 3. The effect of the higher order terms is to pull the triplets closer in longitudinal direction while the transverse direction is rather unaffected. This suggests that the higher order terms are more sensitive to the longitudinal stretching of the string field.

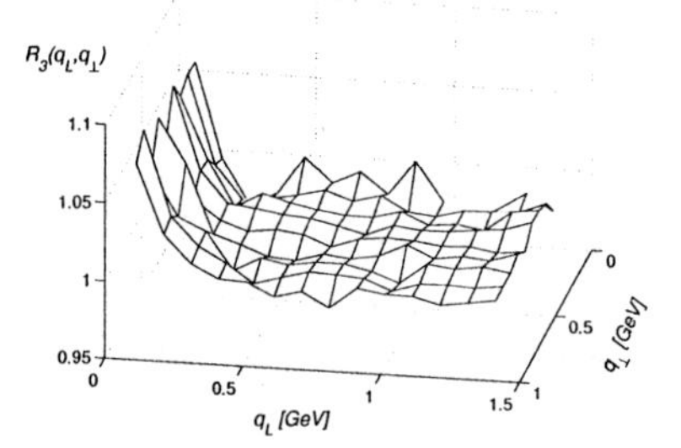

Fig. 3. The genuine three-particle correlation function $R_3(q_L, q_\perp)$ for all final state charged pions.

References

[1] B. Andersson and M. Ringnér,
LU TP 97-07 and hep-ph/9704383 (1997)
(accepted for publication in Nucl.Phys.B)

[2] B. Andersson and W. Hofmann,
Physics Letters B199 (1986), 364

[3] B. Andersson et al.,
Phys. Rep. 97 (1983), 31

[4] T. Sjöstrand,
Comp. Phys. Comm. 82 (1994), 74

Correcting Monte Carlo Generators for Bose–Einstein Effects

Jacek Wosiek (wosiek@thp4.if.uj.edu.pl)

Institute of Physics, Jagellonian University, Cracow

Abstract. An integral representation for weights, correcting Monte-Carlo generated events for the Bose Einstein effect, is derived. The saddle point approximation to these integrals results in a compact expression which sums effectively the original $n!$ terms with an accuracy better than 2% for $n > 7$ strongly correlated bosons.

The very nature of the Bose-Einstein correlations makes them rather difficult to include in the Monte Carlo event generators and consequently the problem became the area of intensive studies. See [1] for a comprehensive review of the generic phenomenon. Recently Białas and Krzywicki have proposed a simple recipe to include Bose-Einstein correlations in the event generating Monte Carlo programs [2]. The basic step in their procedure consists of calculating a positive weight for each generated event

$$W_n(p_1, ..., p_n) = \sum_{\mathcal{Q}_n} \prod_i^n A_{i\mathcal{Q}_n(i)} = perm(A), \tag{1}$$

where A is a "correlation matrix" which depends on particle momenta and on the choice of variables parameterizing the whole phenomenon. One commonly used form reads $A_{ik} = \exp\left(-(p_i - p_k)^2/2\sigma^2\right)$. The sum in Eq.(1) extends over all permutations, $\mathcal{Q}_n$, of n elements (1,2,...,n). Hence the numerical cost to calculate above weights grows like $n!$ and this limits practical applications [5],[6].

We have derived a simple $2n$-dimensional integral representation for a general permanent of a correlation matrix, $perm(A)$. This representation has two advantages. First, the integrals can be done analytically, in the saddle point approximation, providing relatively simple and accurate expression for the sum (1). Secondly, the integral representation admits a probabilistic interpretation. This indicates that implementation of the above corrections might be possible stochastically, and may advance the novel generation of MC programs.

Detailed derivation is published elsewhere [3], the key point consisting of the applications of the formalism of the free bosonic field theory to calculatulate the symmetrized sum (1). We have obtained the following integral representation

$$W_n = \int_{-\infty}^{\infty} \exp\left(-\frac{1}{2}\sum_j^n (x_i^2 + y_j^2)\right) \prod_\mu^n \frac{1}{2}((x \cdot e^{(\mu)})^2 + (y \cdot e^{(\mu)})^2) \prod_i^n \frac{dx_i dy_i}{2\pi}. \tag{2}$$

Vectors $\mathbf{e}^{(\mu)}$ depend on the momenta of generated particles and are given by $e_i^{(\mu)} = O_{\mu i}\sqrt{\lambda_i}$, where O diagonalizes the correlation matrix: $O^T A O = Diag(\lambda_1 \dots \lambda_n)$, and all integrals are along the real axis; see also Ref.[4] for the algebraic derivation of (2). An approximate expression for the wieghts has been derived in the saddle point approximation [3].

$$W_n = 2^{1-n}|s_0|\sqrt{\frac{\pi}{\prod_i'^{2n} \Lambda_i}} \exp\left(-S(s_0, s_0)\right), \tag{3}$$

where the "action" S denotes the logarithm of the inverse of the integrand in (1), and the product is taken over all, but one (equal to zero), eigenvalues of the 2n dimensional covariance matrix $S_{jk}^{(2)} = \frac{\partial^2 S(x,y)}{\partial z_j \partial z_k}|_{(s_0,s_0)}$, evaluated at the saddle point $(\mathbf{s}_0, \mathbf{s}_0)$, $z = x, y$.

Further improvement in the accuracy of the saddle point result can be easily achieved by calculating higher corrections. Including for example the two additional terms of the Taylor expansion around the saddle should not increase much the computational effort but would reduce the error of the approximation. As commented above, since the integrand in (1) is positve, there exists an interesting possibility of defining a new family of event generating MC programs which would have the corrections (1) included exactly on the stochastic basis.

To summarize, we have derived an exact integral representation for the Bose-Einstein weights, which until now have been computed by summing over $n!$ terms. Resulting $2n$-dimensional integral was calculated by the saddle point technique. Numerical comparison with Eq.(1) shows that the saddle point formula approximates the exact sum with error less than 2% for more than 7 correlated bosons. Possibile construction of the new event generators which accomodate Bose-Einstein correlations is emphacized.

This work is supported in part by the Polish Committee for Scientific Research under the grants no. 2P03B19609 and 2P03B04412.

References

[1] D. H. Boal, C-K. Gelbke and B. K. Jennings, Rev. Mod. Phys. **62** (1990) 553.
[2] A. Białas and A. Krzywicki, Phys. Lett. **B354** (1995) 134;
[3] J. Wosiek, Phys. Lett. **B399** (1997) 130.
[4] K. Zalewski, Acta. Phys. Polon. **B28** (1997) 1131.
[5] K. Fiałkowski and R. Wit, Acta Phys. Polon. **B28** (1997) 2039.
[6] L. Lönnblad and T. Sjöstrand, hep-ph/9711460.

Cross Section Measurements of Hadronic Final States in Two Photon Collisions at LEP

Jan A. Lauber, on behalf of the LEP collaborations (jal@hep.ucl.ac.uk)

University College London, Gower Street, London WC1E 6BT, U.K.

1 Introduction

Two-photon events occur in an e^+e^- collider when virtual photons emitted from both beams interact with each other. The photons participating in the interaction are described by their negative four-momentum transfer, Q^2. If the scattering angle θ of the corresponding electron is larger than about 30 mrad, it is detected or tagged in a typical LEP experiment. In this case the photon is highly virtual, and the event is labelled an $e\gamma$ collision. If, on the other hand, cuts are applied to ensure that no electron is detected, the event is anti-tagged. The corresponding photons are quasi-real. In this case the event is labelled a $\gamma\gamma$ collisions.

2 Hadron production in $e\gamma$ events

Monte Carlo models of DIS $e\gamma$ scattering must describe the pointlike structure as well as the soft hadronic structure of the photon. Comparisons with the observed experimental hadronic energy flow by OPAL [1] have shown that various models underestimate the high-p_T contribution, in particular at low x. In a LEP2 workshop [2] it was shown that the shape of the energy flow is very sensitive to the intrinsic transverse momentum of the quarks inside the struck photon. Changing this intrinsic k_T from a gaussian to a power-like distribution of the form $dk_T^2/(k_T^2 + k_0^2)$, $k_0 = 0.66$ GeV [3] has the effect of providing the extra transverse momentum, particularly for events with a large invariant mass and thus small x. Fig. 1 shows the transverse energy of the event perpendicular to the tag plane defined by the direction of the beam and the tagged electron. The tuned HERWIG model with the power-like k_T fits the data over the whole range. The aim of the improved Monte Carlo models is to be able to perform better unfoldings for detector effects which will reduce the systematic errors in measurements such as the photon structure function $F_2^\gamma(x, Q^2)$.

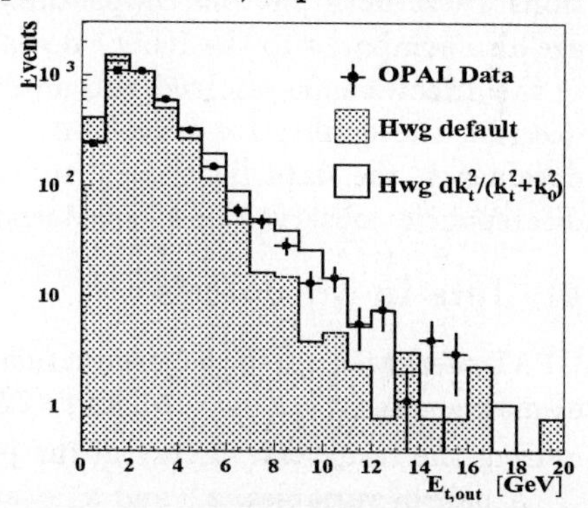

Fig. 1. Transverse energy out of the tag plane.

3 Hadron production in $\gamma\gamma$ events

The production of charged hadrons in the collision of quasi-real photons has been measured using several of the LEP detectors. It was found that the general features of the events are rather well reproduced by the PYTHIA and PHOJET generators, used in the L3 [4] and OPAL [5] analyses and TWOGAM in the case of DELPHI [6]. However, in some distributions, e.g. the normalised longitudinal momentum[4], discrepancies between the various generators and between the data and the Monte Carlo models as large as 30% occur. To compare production cross sections with other experiments and with NLO calculations, the visible invariant mass, W_{vis}, is unfolded to obtain the mass of the photon system, $W_{\gamma\gamma}$. Fig. 2 shows the differential cross section $d\sigma/dp_{\mathrm{T}}$ as a function of the transverse momentum of the hadrons measured by OPAL [5]. The data are compared with results from γp, πp and Kp scattering from the WA69 experiment which have been normalised to the $\gamma\gamma$ data. At $p_{\mathrm{T}} > 2$ GeV the $\gamma\gamma$ data lie significantly above the exponentially falling γp and hp data, evidence for the contribution form the direct $\gamma\gamma$ interactions where both photons couple directly to the quarks. The OPAL $\gamma\gamma$ data are also compared to the NLO calculations containing the three components of the direct, single-resolved and double-resolved processes. The calculations describe the qualitative behaviour, but at large p_{T}, where the direct term dominates, the data lie clearly above the NLO calculation, similar to the discrepancies observed in $e\gamma$ scattering data.

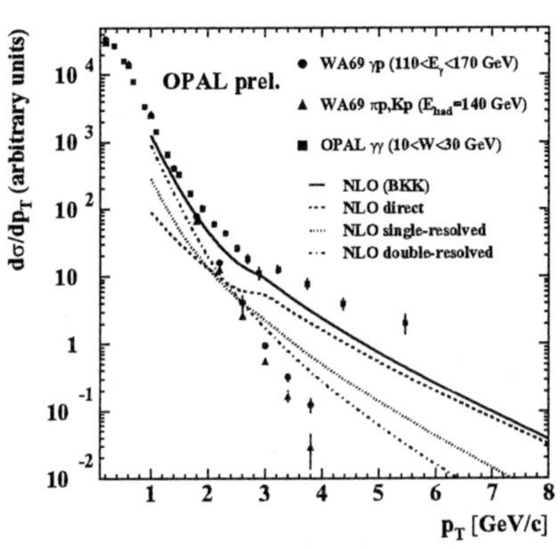

Fig. 2. The p_{T} distribution measured in $\gamma\gamma$ interactions is compared to the p_{T} distribution measured in γp and hp interactions in WA69.

4 Jets in $\gamma\gamma$ events

OPAL measured the di-jet production in anti-tagged $\gamma\gamma$ events [7] at centre-of-mass energies $\sqrt{s}_{ee} = 161$ and 172 GeV. The cone jet finder is used, requiring the transverse energy of the jets $E_T^{\mathrm{jet}} > 2$ GeV and $|\eta^{\mathrm{jet}}| < 2$.

A pair of variables, x_γ^+ and x_γ^-, can be defined which specify the fraction of the photon energy participating in the hard scattering:
$x_\gamma^\pm = \sum_{jets}(E \pm p_z)/\sum_{had}(E \pm p_z)$. Events with $x_\gamma^\pm > 0.8$ mainly stem from the direct process and the transverse energy flow measured relative to the direction of each jet are strongly collimated with no energy observed outside $\Delta\eta = 1$. Events with $x_\gamma^\pm < 0.8$ mainly stem from resolved events.

The transverse energy flow show significant tails out to large values of $\Delta\eta$, originating form the remnant jets of the resolved photons. Events with large contributions from direct and double-resolved processes show very different angular distributions, seen in Fig. 3. The inclusive di-jet production cross section, $d\sigma/dE_T^{\rm jet}$, unfolded for detector effects, is well described by NLO calculation [7].

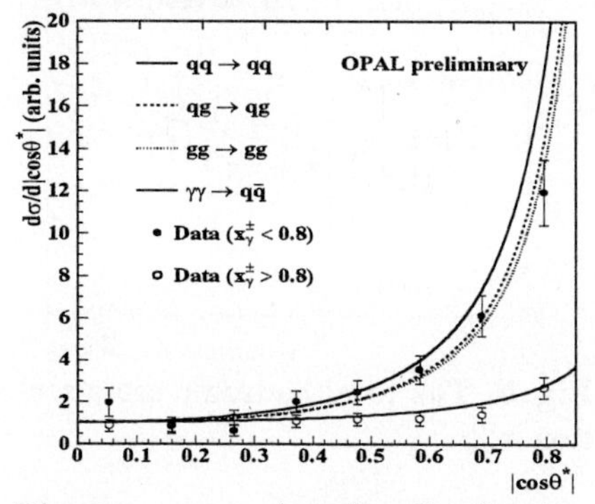

Fig. 3. Angular distribution of events with large direct and large double-resolved contributions.

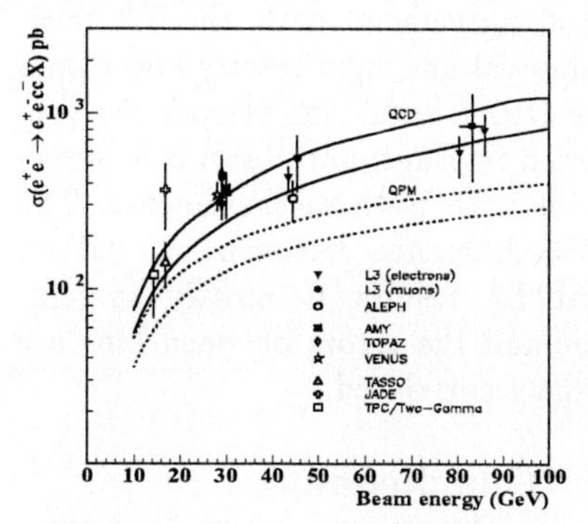

Fig. 4. Charm production cross section in $\gamma\gamma$ collisions

5 Charm production

The inclusive charm production in $\gamma\gamma$ collisions has been studied with the L3 detector at centre-of-mass energies of 91, 161 and 172 GeV. Charmed hadrons are identified by electrons and muons from semi-leptonic decays. About half of the selected events are expected to originate from other processes, such as τ decays. Fig 4 shows the total inclusive charm production cross section. The results are in agreement with the QCD predictions, within the theoretical uncertainties arising from variations of the charm mass. To obtain the total production cross section the differential cross section has to be extrapolated to $p_{\rm lept} = 0$ GeV. Due to the steeply falling momentum spectrum of the leptons this is a significant contribution to the systematic error of the measurement.

6 Total hadronic $\gamma\gamma$ cross section

The total hadronic cross section $\sigma_{\gamma\gamma}^{\rm tot}$ was measured by L3 [4] at centre-of-mass energies of 130–140 and 161 GeV, in the interval of the invariant mass $5 \le W_{\gamma\gamma} \le 75$ GeV and by OPAL [8] at 161 and 172 GeV in the interval $10 \le W_{\gamma\gamma} \le 110$ GeV (Fig 5).

The total cross section is compared to several theoretical models. Based on the Donnachie-Landshoff model, the assumption of a universal high energy behaviour of $\gamma\gamma$, γp and pp cross section was tested, which predicts a cross section of the form $\sigma = X_{I\!P} s^{\epsilon} + Y_{I\!R} s^{-\eta}$. The first term is due to the Pomeron exchange, the second term due to Reggeon exchange. This simple ansatz gives a reasonable description of the total $\gamma\gamma$ cross section. Schuler and Sjöstrand give a total cross section for the sum of the photon's direct, anomalous and VMD component. Also shown is the prediction of Engel and Ranft which is implemented in PHOJET. It is in good agreement with the L3 measurement and significantly lower than the OPAL data. The steeper rise predicted by Engel and Ranft is in agreement with both measurements. The 20% difference between the OPAL and L3 results is mostly systematic and the errors between bins are largely correlated.

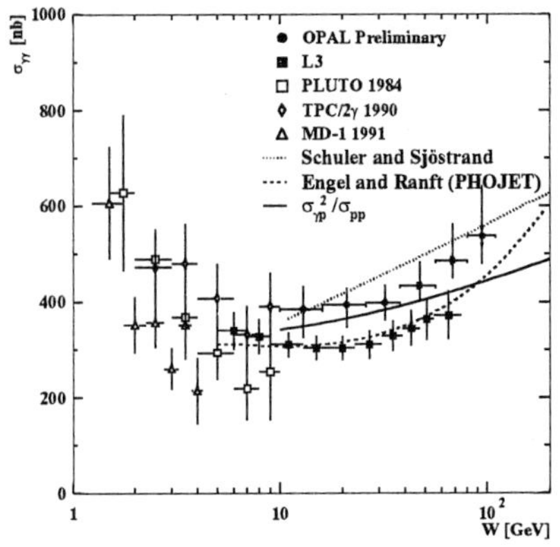

Fig. 5. The total hadronic cross section $\sigma_{\gamma\gamma}^{\text{tot}}$.

7 Conclusions

Although the description of the hadro- nic final state of $e\gamma$ events has been improved, a lot more theoretical work is needed to explain the origin of the observed discrepancies.

The differential cross section $d\sigma/dp_T$ in $\gamma\gamma$ events was found to have a much harder component than γp or hp scattering due to the presence of the direct photon-quark coupling. NLO calculations under-estimate this cross section at higher p_T. Are these discrepancies of the same origin as the ones seen in $e\gamma$ events?

New measurements of the total cross section show an expected rise at $W > 20$ GeV, but are plagued by large uncertainties from the contribution from diffractive processes outside the detector acceptance.

Acknowledgements

I would like to thank all LEP collaborations for their help in preparing this contribution, in particular M. Kienzle, M. Lehto, S. Söldner-Rembold and A. Zinchenko from the various two-photon working groups.

References

[1] OPAL Collaboration, *Z. Phys.* C**74** (1997) 33.
[2] S. Cartwright *et al.*, hep-ph/9708478
[3] J.A. Lauber, Photon'97, hep-ex/9707002
and ALEPH Collaboration, contrib. paper to EPS97, # 592
[4] L3 Collaboration, contrib. paper to EPS97,# 475, CERN-PPE/97-48
[5] OPAL Collaboration, contributed paper to EPS97, # 201
[6] DELPHI Collaboration, contributed paper to EPS97, # 422
[7] OPAL Collaboration, contributed paper to EPS97, # 200
[8] OPAL Collaboration, contributed paper to EPS97, # 202

A Simple Approach to Describe Hadron Production Rates in Inelastic pp and $p\bar{p}$ Collisions

Yi-Jin Pei (Yi-Jin.Pei@cern.ch)

I. Phys. Institut der RWTH, D-52074 Aachen, Germany

Abstract. We show that the hadron production rates in inelastic pp and p$\bar{\text{p}}$ collisions can be described by a simple approach used to describe data obtained in e^+e^- annihilation. Based on the idea of string fragmentation, the approach describes the production rates of mesons and baryons originating from fragmentation in terms of the spin, the binding energy of the particle, and a strangeness suppression factor. Apart from a normalization factor and the additional sea quark contribution in inelastic pp and p$\bar{\text{p}}$ collisions, pp, p$\bar{\text{p}}$ and e^+e^- data at various centre-of-mass energies are described simultaneously.

In Ref.[1] a simple approach based on the idea of string fragmentation to describe hadron production rates in e^+e^- annihilation is proposed. We consider that particle production proceeds in two stages, namely quark pair production in the colour string field and successive recombination. Quark pair production in the colour string field can be considered as a tunneling process. The probability of producing a $q\bar{q}$ pair is proportional to $\exp(-\pi m_q^2/\kappa)$, where m_q is the (constituent) quark mass, and κ the string constant. We assume that the probability of quarks recombining to a hadron with the mass M_h is proportional to $\exp(-E_{bind}/T)$, where T is the effective temperature in hadronization, and $E_{bind} = M_h - \sum_i m_{q_i}$ the hadron binding energy, which can be ascribed to the colour-magnetic hyperfine interaction.

Similar to an e^+e^- annihilation process, an inelastic pp (p$\bar{\text{p}}$) collision can also be considered to proceed in four steps: interaction between two initiator partons (valence quarks, gluons, sea quarks or antiquarks) with two beam remnants left behind, followed by a parton shower development from the initiator partons, and subsequently the transition from partons to hadrons. Finally, the unstable hadrons decay according to their branching ratios.

Taking into account the numbers of the primary quarks and diquarks in a collision, hadron production in inelastic pp and p$\bar{\text{p}}$ collisions can also be described by our approach discussed in Ref.[1]. The detailed description of the fit procedure can be found in Ref.[2].

The results of the fit to pp data at $\sqrt{s} = 27.4$–27.6 GeV, the most complete set of data on hadron production in inelastic pp collisions, are shown in Fig. 1 and listed in Table 2 in Ref.[2]. In the fit there are three free parameters: an overall normalization factor C, and two parameters for the fractions of the sea quarks. The results of the fit to pp and p$\bar{\text{p}}$ data at the other centre-of-mass energies are listed in Tables 1, 3, 4 and 5 in Ref.[2].

As can been seen from the fit results, the hadron production rates in inelastic pp and p$\bar{\text{p}}$ collisions are described well by our simple approach. Taking into account the different initial parton configurations and the additional sea quark contribution in inelastic pp and p$\bar{\rm p}$ collisions, pp, p$\bar{\rm p}$ and e^+e^- data on the production of light-flavoured hadrons at various centre-of-mass energies are described simultaneously (apart from a normalization factor, which reflects the rise of multiplicities with increasing energy). Moreover, as shown in Ref.[1], data on heavy flavour production in e^+e^- annihilation are also described by our approach. All this shows that our approach can provide a universal description of hadron production, irrespective of the type of the interaction and the initial parton configuration in various interactions.

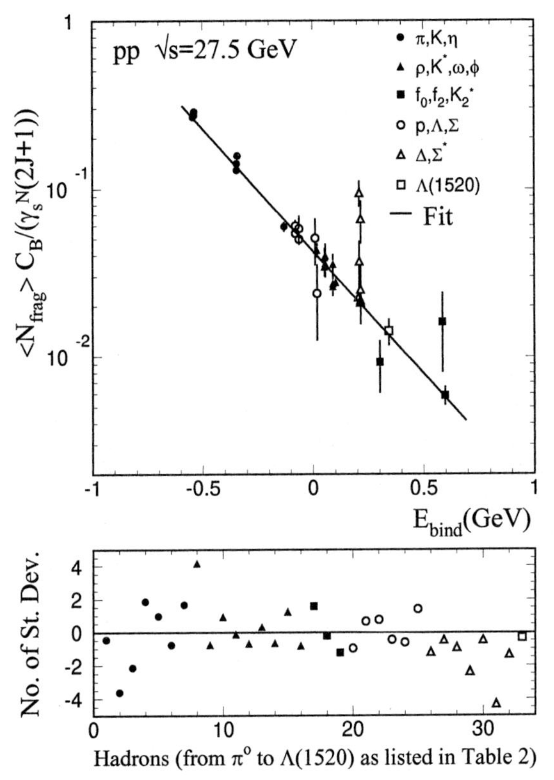

Fig. 1. The upper plot shows the average production rates of hadrons originating from fragmentation at $\sqrt{s}$ = 27.4–27.6 GeV in inelastic pp collisions (measured value × fraction of hadrons originating from fragmentation as determined by the fit), multiplied by the factor $C_B/[\gamma_s^{N_s}(2J+1)]$ (see Eq.(1) in Ref.[1]), as a function of the binding energy of hadrons. The fit results are shown as the line. The lower plot shows the difference between the fit results and the measured rates in terms of the number of standard deviations.

References

[1] Y.J. Pei, Z. Phys. **C72** (1996) 39.
[2] Y.J. Pei, hep-ph/9703243.

A New Measurement of the $\pi \rightarrow \mu\nu\gamma$ Decay

A.T.Meneguzzo (a) (meneguzzo@pd.infn.it), G. Bressi (b) , G. Carugno (a) , E. Conti (a) , S. Cerdonio (c) and D. Zanello (c)

(a) Università di Padova and I.N.F.N. Sez. di Padova-Italy, (b) I.N.F.N. Sez. di Pavia-Italy, (c) I.N.F.N. Sez. di Roma I, Università di Roma "La Sapienza"-Italy

Abstract. We measured the branching ratio of the radiative decay $\pi \rightarrow \mu\nu\gamma$. The energies of the gamma-ray and of the muon were both measured and the Dalitz-plot distribution of the decay was obtained. The data agree well with the theoretical prediction (QED internal bremsstrahlung) down to a γ-ray energy of 1 MeV. The discrepancy reported in a previous experiment is not confirmed.

Introduction

The experiment RAPID (RAdiative PIon Decay), run at the Paul Scherrer Institute (Villigen, CH), perfomed the first direct measurement of the decay

$$\pi^+ \rightarrow \mu^+\nu\gamma \tag{1}$$

Theory ([1],[2]) predict that Structure Dependent term is suppressed with respect to the QED Internal Bremsstrahlung. Neverthless in 1958 Castagnoli and Muchnik [3] ,who measured indirectly the decay (1) with the nuclear emulsion thecnique, suggested the possibility of a disagreement from QED at low γ-ray energy.

The goal of the present experiment is to identify both muon and gamma coming from reaction (1) and to measure their energies in order to obtain the branching ratio and the full two-dimensional distribution.

Experimental Layout and Data Taking

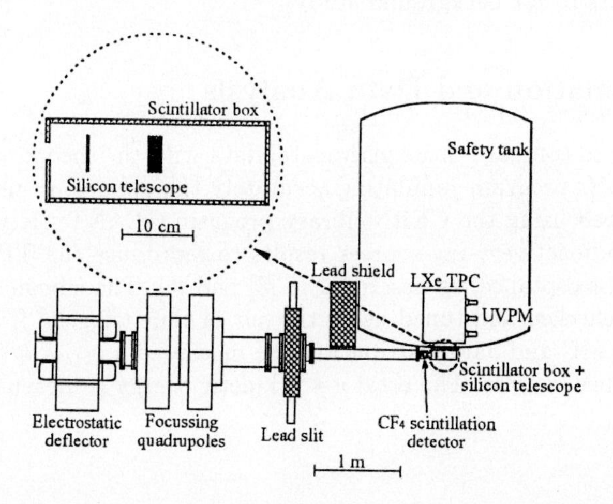

The incoming pion is first detected by a fast CF_4 gas scintillator counter (for good timing), then slowed down and brought to rest in a silicon pad telescope (ST). The decay γ is detected and measured in a 60 lt liquid xenon Time Projection Chamber (TPC). The ST measures the sum of the energies of the incoming pion and of the decay muon (E_{SILI}). The μ kinetic energy is obtained by subtracting the known energy of the pion. The silicon system is enclosed in a box of plastic scintillators to veto the positrons from the muon decays occurring within the trigger coincidence time.

The γ energy (E_{TPC}) is measured by means of the ionization charge produced in the active volume of the TPC [4] and collected by 12 anodes .The xenon scintillation light, used for triggering and timing, is detected by a system of 12 PM's (UVPM). The TPC response was accurately calibrated with radioactive sources [5]. The minimum detectable gamma energy, measured by means of the ionization charge, is 230 keV. The minimum energy seen by the UVPM used in the trigger is 170 keV. The energy resolution is parameterized as $\frac{\sigma(E_\gamma)}{E_\gamma} = \frac{7\%}{\sqrt{E_\gamma}} \oplus \frac{0.14MeV}{E_\gamma}$ where E_γ is in MeV and $\oplus$ indicates a quadratic sum. The acceptance of the TPC including geometry ,trigger efficiency and offline analysis losses is 5% at 500Kev and reach 11% above 1 Mev.

The π beam had large e^+ and π^+ contamination , $e^+/\pi^+ \approx 10, \mu^+/\pi^+ \approx 1$. Due to positrons electromagnetic showering, the TPC UVPM counted 4-5 kHz altough the TPC were shielded by 20 cm thick lead wall (the TPC counting rate was 50 Hz with beam off). Caused by the veto inefficiency the most serious source of background comes from accidental coincidences between a stopped pion and a background photon in the TPC.

We operated at $p = 50MeV/c$ and $\Delta p/p = 0.5\%$, with a pion rate of about $800\pi/s$. We worked with two kinds of triggers : a) prompt triggers which select either $\pi \to \mu\nu\gamma$ decays or accidental π-γ coincidences; b) delayed triggers, which select only accidental π-γ coincidences (i.e. stopping pions but with UVPM out of time). We stopped more than 207 millions pions for the study of decay (1) and collected .26 millions of trigger events a) and .25 millions of trigger events b) for background study.

MC Simulation and Data Analysis

In order to compare the experimental data with the theoretical QED predictions a MC program simulating accurately the apparatus and the trigger was developed using the CERN library program GEANT . It was tuned by means of radioactive γ-ray sources results to reproduce the TPC behaviour and trigger acceptance, as described in [5] and [6]. The silicon and the veto parts were checked and tuned using the out of time triggers; from minimum χ^2_{TOT} fit to MC and data distributions we measured a $p_{beam} = 50.04$ MeV/c with 0.03 MeV/c error. The total $\pi + \mu$ kinetic energy is measured with 0.20 MeV error .

In the MC the muon and gamma energies coming from (1) are generated according to QED ([1]) and ([2]) with an energy threshold for the γ-ray set to 100 keV, below the experimental threshold. The experimental trigger is reproduced with the appropriate thresholds for the TPC and the ST. Data selections ([6]) are applied to both the experimental data (prompt and delayed triggers) and the MC simulation data. Fig.1 shows the Dalitz-plot distribution obtained after this procedure for prompt-trigger, delayed-trigger and MC events. The shape of the irreducible background (caused by envi-

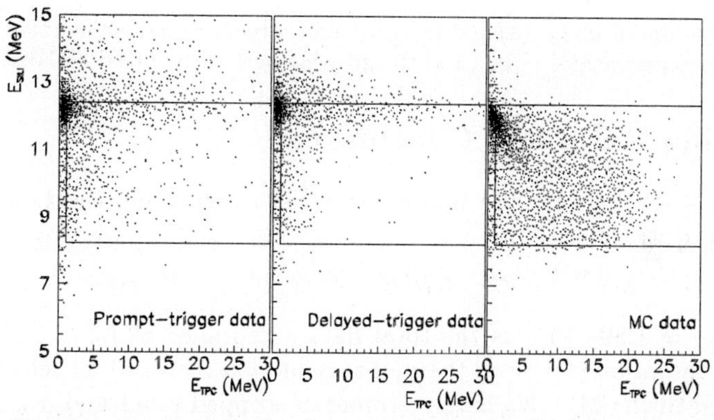

Fig. 1. Dalitz-plot distributions for the prompt-trigger, the delayed-trigger and the MC events after the cuts . The rectangular contours define the kinematical region where the fit has been performed.

ronmental gammas) is experimentally given by the delayed-trigger sample. The prompt-trigger data sample is therefore fitted to a linear combination of the delayed-trigger spectrum plus the QED spectrum given by the MC: [DATA]=α *[MC]+β *[BKG] where α and β are free parameters obtained by the χ^2 minimisation fit.

The fit with $\alpha = 0$ assesses our sensitivity to the QED term. The limits for E_{SILI} are $8.2 \leq E_{SILI} \leq 12.4$ MeV. They correspond to the kinematical limits for T_μ. The minimum E_{TPC} would be ≈ 250 keV, which is close to the MC estimate of our reconstruction threshold. However we measured that around 0.5 MeV the background spectrum has a huge peak and the total acceptance is steeply increasing so , being conservative we used for E_{TPC} the limits $1 \leq E_{TPC} \leq 30$ MeV. The fit yields $\alpha = (8.9 \pm 1.1) \cdot 10^{-3}$, $\beta = 0.99 \pm 0.02$ with χ^2/DoF = 95.6/101 (fit probability 63%). The fit with α fixed to 0 yields: $\beta = 1.06 \pm 0.02$ but $\chi^2/DoF = 159.7/102$ with a fit probability $\approx 2 \cdot 10^{-4}$.

From the analysis of the Dalitz plot distribution the QED term is needed and provides a good fit to our data. The gamma and muon spectra obtained by subtracting the background spectra, with weight β, from the prompt-trigger spectra are shown in fig.2 together with the fitted MC spectra

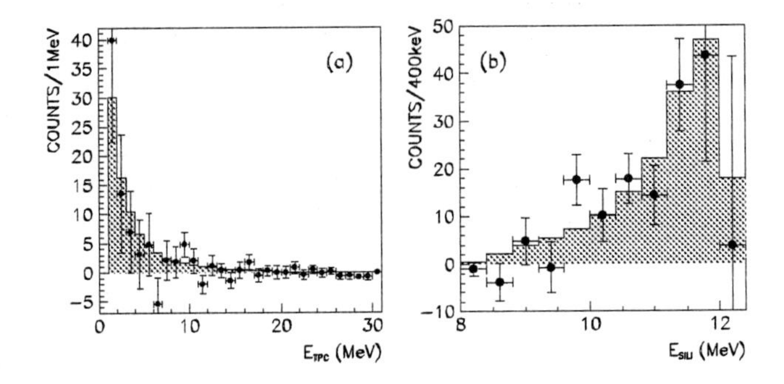

Fig. 2. Spectrum of E_{TPC} (a) and E_{SILI} (b) after background subtraction. The full dots are the experimental points and the grey histogram the fitted MC distribution.

Branching Ratio and Conclusions

The relation between the number of QED triggers and the experimental Branching Ratio is

$$\alpha N_{\gamma}^{MC} = N_{\pi} \cdot BR(E_{\gamma} > 1MeV)_{exp} \cdot F \cdot \epsilon_{MC} \tag{2}$$

where $\epsilon_{MC} = 1.89 \cdot 10^{-2}$ is the total data acceptance of $E_{\gamma} > 1MeV$ from the MC simulation (1) , F is the global reduction factor of all selection cuts normalized to the MC, N_{π} is the number of stopped pions and α is given by the fit described above. With our numbers ($F = 0.437$, $N_{\pi} = 207.4 \cdot 10^{6}$, $\alpha N_{\gamma}^{MC} = 335$) and with a total systematic error evaluated 4% ([6]) , the result is

$$BR(E_{\gamma} > 1MeV)_{exp} = 2.0 \cdot 10^{-4} \pm 12\%(stat) \pm 4\%(syst)$$

to be compared with the theoretical branching ratio $2.28 \cdot 10^{-4}$.

Since in this experiment we are triggering on the presence of a gamma-ray, our result shows unambiguously that reaction (1) is described correctly by QED , not only in the value of the branching ratio but also in the Dalitz-plot distribution. The discrepancy reported in [3] is not confirmed. We wish to point out that, in comparison with [3], this experiment measures both kinematical variables of decay (1) and covers a wider range of the same variables where the expected QED Branching Ratio is twice as large.

References

[1] S.G. Brown, S.A. Bludmann, Phys. Rev. **136** (1964), 1160
[2] D.E. Neville, Phys. Rev. **124** (1961), 2037
[3] C. Castagnoli, M. Muchnik, Phys. Rev. **112** (1958), 1779
[4] G. Carugno et al., Nucl. Instr. and Meth. **A376** (1996), 149
[5] G. Bressi et al., Nucl. Instr. and Meth. **A396** (1997), 67
[6] G.Bressi et al.,Preprint Univ. di Padova DFPD 97/EP/43,accepted for publ. in Nucl. Phys. B (oct 1997)

Part III

Structure Functions and Low-x Physics

Final Results on g_1^n and g_2^n from SLAC E154 and Interpretation via NLO QCD

Paul Souder, for the E154 Collaboration (souder@suhep.phy.syr.edu)

Syracuse University, Syracuse, NY

Abstract. Final results from SLAC Experiment E154 are presented. Data are presented for $g_1^n(x)$ and $g_2^n(x)$. The results are interpreted in the context of an NLO analysis..

1 Introduction

A number of technical advances have led to the ability to make precise measurements of the spin structure functions $g_1^n(x)$ and $g_2^n(x)$. In particular, electron beams with polarizations of ~80% [1], energies up to 50 GeV, and ^{3}He targets with polarizations >40% [2] were available for experiment E154 at SLAC. The E154 collaboration has recently published the final results for $g_1^n(x)$ [3], $g_2^n(x)$ [4], as well an NLO analysis [5] of our data together with other results on spin structure functions published prior to 1997.

2 Results

The results of the experiment are shown in Fig. 1. Total Systematic errors are given by the bands at the bottom of the figure. Previous data on g_1^n published before 1997 are also shown. All of the experimental results are consistent. However, a key feature of the new data is the inconsistency with the old assumption (Regge analysis)that as $x \to 0$, $g_1(x)$ is constant or decreasing. If we assume that

$$g_1^n(x) \sim Cx^{-\alpha} \tag{1}$$

we obtain $\alpha = 0.9 \pm 0.2$, totally inconsistent with the Regge analysis. Moreover, if we attempt to compute the first moment $\Gamma_1^n = \int_0^1 g_1^n(x)dx$ assuming the power behavior, we get an unexpectedly large result, even consistent with the integral diverging. In order to perform a more realistic low x extrapolation, we use an NLO analysis as described below. The results on $g_2^n(x)$ are given in Fig. 2. The data are consistent with zero, but with errors larger than the size of effects predicted by most models.

3 The NLO Analysis

In the simple quark-parton model, the polarized structure function g_1 may be expressed as

$$g_1(x) = \frac{1}{2}\sum e_q^2(\Delta q(x) + \Delta\overline{q}(x)) \tag{2}$$

where $\Delta q(x)$, and $\Delta\overline{q}(x)$ include all possible quarks u, d, and s. In QCD, the $\Delta q(x)$ become functions of Q^2 evolving according to the DGLAP equations and $g_1(x) \to g_1(x, Q^2)$ becomes a convolution of the $\Delta q(x, Q^2)$ that also involves the gluon polarized structure function $\Delta G(x, Q^2)$. The analysis follows the method of Gluck *et al.* [6], Ball, Forte, and Ridolfi [7], and Gehrmann and Stirling [8]. The procedure is to find functions $\Delta q(x, Q_0^2)\ldots$ at some fixed Q_0^2 value that together with the DGLAP equations fit all of the data. A remarkable feature of these analyses is that if Q_0^2 is chosen to be a low value, for example 0.34 $(\mathrm{Gev})^2$ as was chosen by ref. [6] and for the E154 analysis, very simple forms of the $\Delta q(x, Q_0^2)$ may be chosen. In particular, for our analysis, only four Δq's are used; $\Delta f_i = \Delta u_V = \Delta u - \Delta - \overline{u}$, Δd_V, $\Delta\overline{Q}$, and ΔG were used. Further, the x-dependence is assumed to be

$$\Delta f_i(x, Q_0^2) = A_i x^{\alpha_i} f_i(x, Q_0^2) \tag{3}$$

where the $f_1(x, Q_)^2)$ are the unpolarized structure functions. Thus all of the data are fit with just eight parameters, A_i and α_i for $i = 1, 4$.

The results of the fits, shown in Fig. 3 for the neutron data at low x, are excellent. Regge behavior is recovered so long as it is assumed at low Q^2 and QCD is used to evolve the prediction to the Q^2 of the experimental data. If, as was commonly done before 1996, Regge theory is assumed to apply at the Q^2 of the data, inconsistent values of Γ_1^n may be obtained, depending on the value of x where the extrapolation is started and whether ^{3}He or deuterium targets are used. With the present analysis, a reasonable value $\Gamma_1^n = -0.058 \pm 0.004(stat) \pm 0.007(syst) \pm 0.007(evol)$ at $Q^2 = 5\ (\mathrm{GeV})^2$ is obtained which differs from the prediction of Ellis and Jaffe. [9]

4 Implications and Future Work

The NLO fits also are consistent with the presence of a substantial contribution to g_1^n from the sea, namely gluons and antiquarks. However, this conclusion does not have overwhelming statistical significance and is also sensitive to the parameterization of the structure functions. Thus experiments directly measuring $\Delta\overline{Q}$ and ΔG are very important. Such experiments are planned at CERN and polarized RHIC. The fact that $g_1^n(x, Q^2)$ appears to be rising rapidly as $x \to 0$ could be tested at much lower x if a polarized ^{3}He beam were available at HERA. This effect is also highly dependent on Q^2 [5]. Finally, extensions to the kinematic range of g_1^n to higher x and lower Q^2 as well as precise measurements of g_2^n are being proposed at TJNAF.

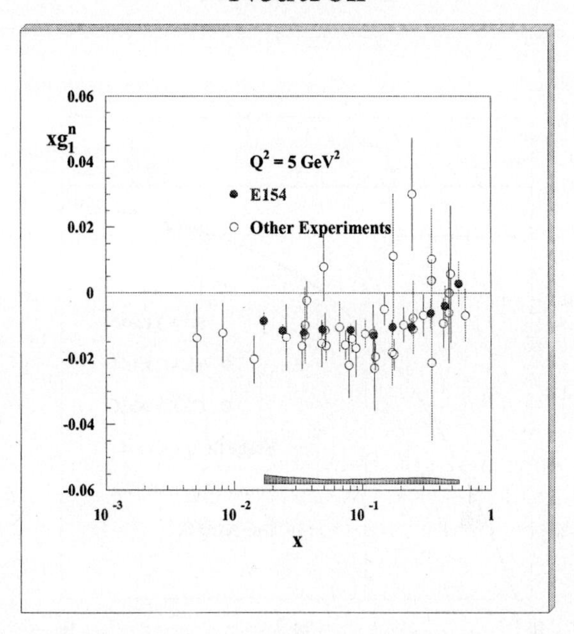

Fig. 1. Results for xg_1^n from SLAC Experiment E154 together with data from other experiments. All data is extrapolated to $Q^2 = 5\ (\text{GeV})^2$ assuming that g_1/F_1 is independent of Q^2.

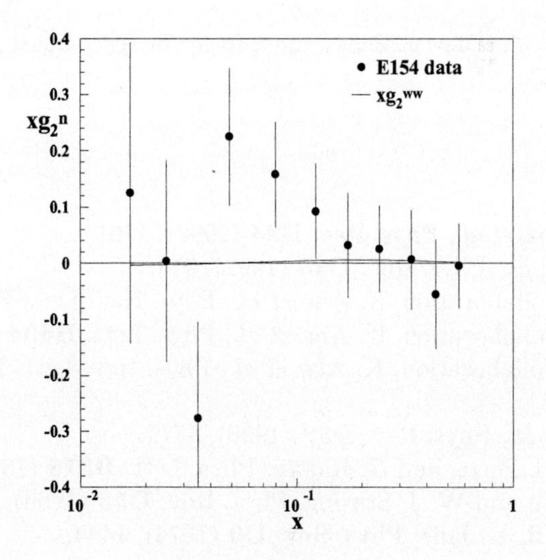

Fig. 2. Results for xg_2^n

Low x fits

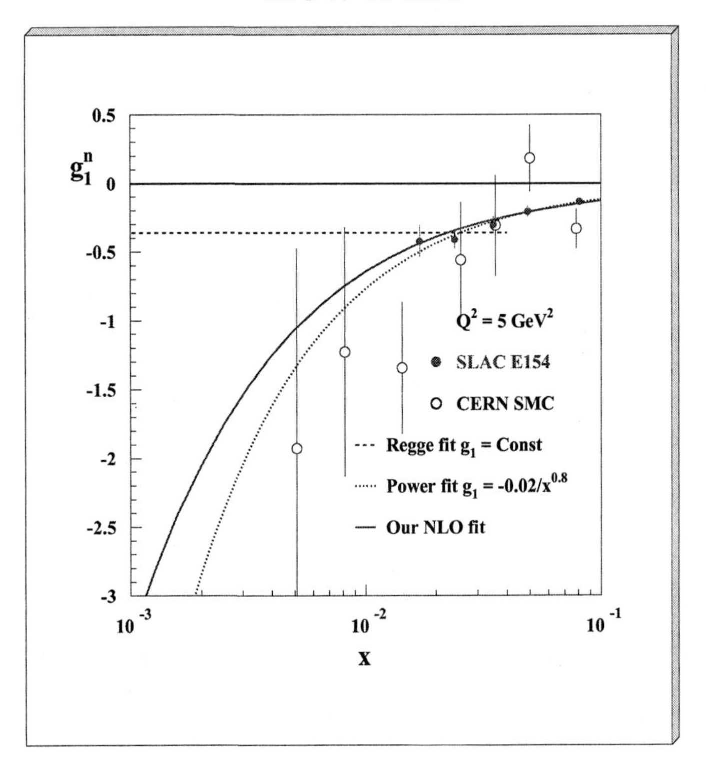

Fig. 3. Results for g_1^n at low x. Shown are a Regge fit (g_1^n =Const.), a power law fit, and the NLO fit.

References

[1] T. Maruyama *et al.*, Phys. Rev. **B34** (1992), 4261.
[2] T. Chupp *et al.*, Phys. Rev. **C45** (1992), 915.
[3] The E154 Collaboration, K. Abe *et al.*, Phys. Rev. Lett. **79** (1997), 26.
[4] The E154 Collaboration, K. Abe *et al.*, Phys. Lett. **B404** (1997), 377.
[5] The E154 Collaboration, K. Abe *et al.*, Phys. Rev. Lett. **B405** (1997), 180.
[6] M. Gluck *et al.*, Phys. Rev. **D53** (1996), 4775.
[7] R. D. Ball, S. Forte, and G. Ridolfi, Phys. Lett. **B378** (1996), 255.
[8] T. Gehrmann and W. J. Stirling, Phys. Rev. **D53** (1996), 6100.
[9] J. Ellis and R. L. Jaffe, Phys. Rev. **D9** (1974), 1444.

Inclusive and Semi-inclusive Results from Hermes

Johan Blouw[1], on behalf of the HERMES collaboration
(johanb@nikhef.nl)

NIKHEF, PO Box 41882, 1009 DB Amsterdam, The Netherlands

Abstract. The study of the spin structure of the nucleon is the principle object of the HERMES experiment. In 1995 the neutron structure function $g_1^n(x)$ was measured using a longitudinally polarized ^{3}He target. In addition, semi-inclusive pion asymmetries on ^{3}He are presented. In 1996 a longitudinally polarized Hydrogen target was employed, and first preliminary results of the inclusive proton asymmetry and hadron as well as pion asymmetries are presented.

1 The HERMES experiment

The HERMES experiment is aimed to measure the spin structure of the nucleon. The experiment uses the HERA facility at DESY in Hamburg, Germany. The HERMES spectrometer consists of tracking chambers, a dipole magnet for momentum determination, and four particle identification detectors; a Čerenkov detector, a Transition Radiation Detector, a Preshower detector and a Calorimeter.

The 27.5 GeV circulating positron beam of the HERA facility becomes naturally polarized through the Sokolov-Ternov effect in the transverse direction. Longitudinal beam polarization is required for the experiment and achieved by spin-rotators located around the experiment. A transverse and a longitudinal beam polarimeter at the experiment measure the beam polarization by Compton backscattering. Beam polarizations of typically 50 % to 60 % have been reached.

In the HERMES experiment, polarized positrons are scattered from a polarized gas target, internal to the beam. The gas is fed into a 40 cm long storage cell to increase the luminosity. Two types of polarized targets are employed. In 1995, a polarized ^{3}He target was used. The ^{3}He target makes use of metastability-exchange optical pumping [1]. Here a radio-frequency discharge populates the 2^3S_1 metastable state of a ^{3}He gas sample that is contained in the pumping cell. A 1083 nm laser is applied to induce transitions to the 2^3P_0 level and deplete selected hyperfine levels of the 2^3S_1 metastable population. The polarization is transferred to the ground-state via metastable exchange collisions. Typical polarization values of $\sim 50\%$ at a target areal density of $1.0 \cdot 10^{15}$ nucleons/cm^2 have been achieved.

Since 1996 a hydrogen/deuterium target [2] is used. Polarized hydrogen atoms from an Atomic Beam Source (ABS) are fed into the storage cell. The

ABS consists of a radio-frequency dissociator to break up the molecules, sets of permanent sextupole magnets to obtain high electron polarization via the Stern-Gerlach effect, and radio-frequency units to select the desired hyperfine states. Typical values of more than 90% nuclear polarization of hydrogen atoms have been achieved at target areal densities of $0.7 \cdot 10^{14}$atoms/cm^2.

2 Measurement of the spin-structure function of the neutron $g_1^n(x)$

Inclusive scattering of polarized positrons from polarized neutrons is characterized by two structure functions, $g_1^n(x)$ and $g_2^n(x)$. The asymmetry in polarized lepton neutron scattering is related to the cross section difference of longitudinally polarized positrons scattering from a longitudinally polarized target parallel and anti parallel to the beam.

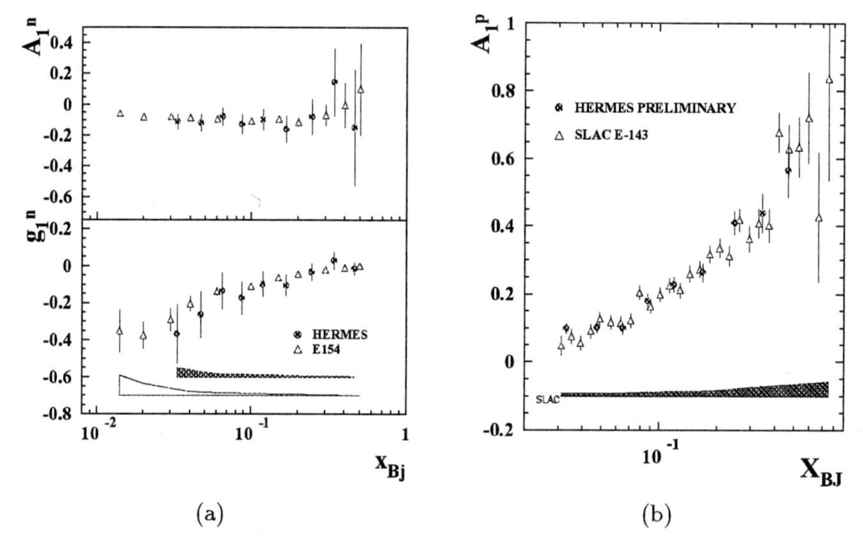

Fig. 1. Comparison of HERMES inclusive measurements on the neutron (Fig. a) and the proton (Fig. b) with data from SLAC E-154 and E-143 respectively.

In the Quark Parton Model (QPM) $g_1^n(x)$ is interpreted as the charge weighted sum of spin dependent quark momentum distributions. In this model $g_2^n(x) = 0$ (for the unpolarized cross section one has the Callan-Gross relation, $F_2(x) = 2xF_1(x)$). The Ellis-Jaffe sum $\Gamma^n = \int_0^1 dx g_1^n(x)$ is related to the fraction of the nucleon helicity carried by the quarks. Assuming $A_2^n(x,Q^2) = 0$ the counting rate asymmetry $A_1^n(x) = \frac{1}{D(y)} \frac{N^{\uparrow\downarrow} - N^{\uparrow\uparrow}}{N^{\uparrow\downarrow} + N^{\uparrow\downarrow}}$ is directly related to the spin structure function of the neutron by $g_1^n(x,Q^2) = \frac{F_2^n(x,Q^2)}{2x(1+R(x,Q^2))} A_1^n(x)$. Here, $R(x)$ is the longitudinal-transverse cross section ratio. The data are presented in Fig. 1(a). The HERMES data [3] compares well with SLAC's E-154 data [4]. The Ellis-Jaffe sum has been determined assuming that the asymmetry $A_1^n(x)$ is independent of the transferred four-momentum, and using a Regge extrapolation of the structure function to $x = 0$. A linear extrapolation of $A_1^n(x)$ from $x = 0.6$ to $x = 1$ is used. Calculated at the average of $Q^2 = 2.5$ GeV^2 of the measurement, the Ellis-Jaffe sum is determined to be $\int_0^1 dx g_1^n(x) = -0.037 \pm 0.013(\text{stat.}) \pm 0.008(\text{sys.})$ It compares well with SLAC's E-154 determination of the Ellis-Jaffe sum at $Q^2 = 5$ GeV^2; $\int_0^1 dx g_1^n(x) = -0.041 \pm 0.004(stat.) \pm 0.006(sys.)$.

Since 1996 the asymmetry on the proton is measured (see Fig. 1(b)). The data agrees well with the E-143 data [5] from SLAC. Note that in this preliminary result only one third of the now available data has been used.

3 Semi-inclusive results

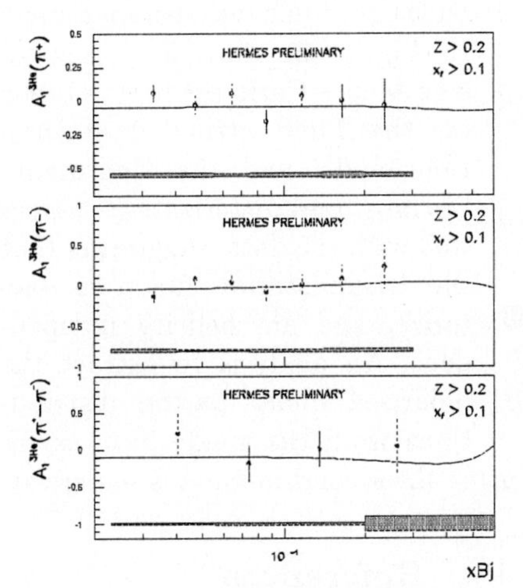

Fig. 2. *Pion asymmetries and the charge difference asymmetry on polarized* ^{3}*He. The solid line is a Monte-Carlo prediction.*

While the spin structure functions of the nucleon depend on the average contribution of the quark spin to the spin of the nucleon, one would like to specify this contribution for each quark flavor. One may determine the so-called flavor decomposition of the spin valence distributions using the charge difference asymmetries on the proton and the neutron. In the QPM the charge difference pion asymmetry is given by $A^p_{\pi^+-\pi^-} = \frac{4\Delta u_v(x) - \Delta d_v(x)}{4u_v(x) - d_v(x)}$. Here it is assumed that the quark scattering process and the fragmentation process factorize. Moreover, it is assumed that the fragmentation functions are helicity independent and obey isospin symmetry, and so divide out. A statistically favored method is to make use of single pion asymmetries instead, and determine the flavor decomposition from fitting the polarized quark distributions to the data using a Monte-Carlo. Assuming that the unfavored fragmentation functions for the light quarks and the strange quarks are the same, the single pion asymmetry for the proton in the QPM is given by $A^p_{\pi^+} = \frac{4\Delta u + \Delta \bar{d} + \eta(4\Delta\bar{u} + \Delta + \Delta s + \Delta\bar{s})}{4u + \bar{d} + \eta(4\bar{u} + d + s + \bar{s})}$, where η is the favored over unfavored fragmentation function. Similar expressions can be obtained for ^{3}He. The fragmentation functions are determined from analysis of unpolarized data. Furthermore, the Gehrmann-Stirling parameterizations [6] of the quark-spin distributions are used as input distributions. Finally one assumes that the leading hadron produced in the scattering process contains the struck quark, and hence comes from the current fragmentation region. Two cuts are applied to ensure that mainly hadrons from the current fragmentation region are used, namely $z = \frac{E_h}{\nu} > 0.2$ and $x_F = \frac{2p_{||}}{W} > 0.1$.

In figure 2 the preliminary pion asymmetries on ^{3}He are presented. The solid line in the figures represents a Monte-Carlo prediction using the Gehrmann-Stirling parameterization of the polarized parton distributions. In Fig. 3 we present the preliminary result of the hadron asymmetries on the proton. The HERMES data agrees well with the SMC data [7]. Again, only about 1/3 of the available data on the proton has been used. For the first time, also pion-asymmetries on the proton have been measured. In Fig. 4 the data is presented. The Monte-Carlo prediction based on the Lund string fragmentation model and the Gehrmann-Stirling parameterization agrees well with the data, suggesting that the fragmentation functions factorize, and are helicity independent. An analysis to extract the polarized quark parton distributions from the single pion asymmetries on the proton and the neutron using the second method is underway.

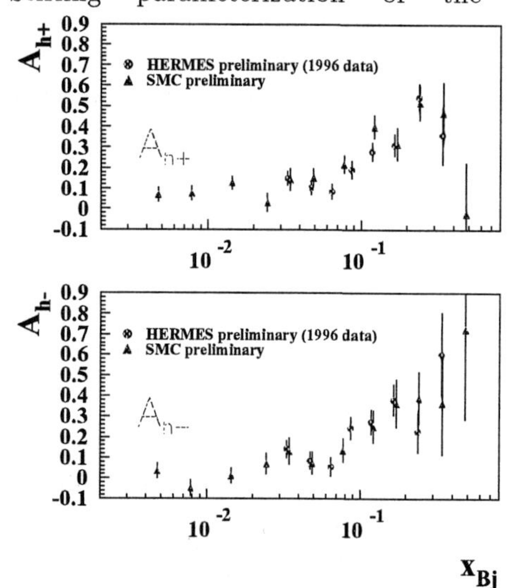

Fig. 3. *Preliminary hadron asymmetries. Solid circles:* HERMES *data, open triangles: SMC data.*

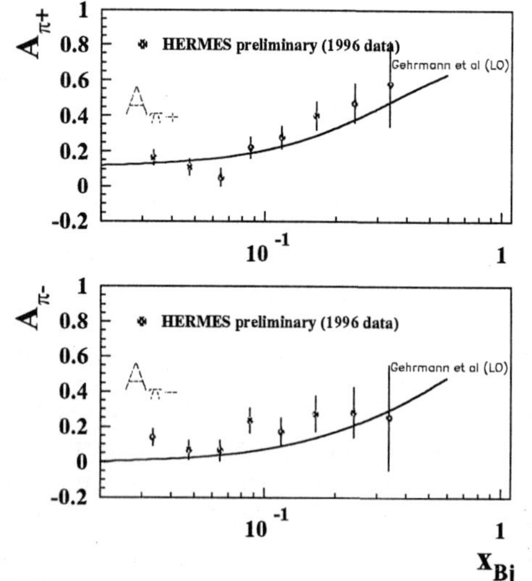

Fig. 4. *Single pion asymmetries on the proton.*

References

[1] K. Lee *et al.* *Nucl. Inst. and Meth.*, **A333**, 1993, p. 294.

[2] F. Stock *et al.* *Nucl. Instrum. Meth.*, A343, 1994, p. 334.

[3] K. Ackerstaff *et al. Phys. Lett.*, **B404**, 1997, p. 383.

[4] K. Abe *et al. Phys. Rev. Lett.*, **79**, 1997, p. 26.

[5] K. Abe *et al. Phys. Rev. Lett.*, **74**, (1995), p. 346.

[6] T. Gehrmann *et al. Phys. Rev.*, **D53**, 1996, p. 6100.

[7] SMC. *hep-ex/9711008.*

Recent Results on the Spin Structure Functions of the Nucleon from the SMC

Jean-Marc Le Goff, on behalf of the SMC collaboration (Jean-Marc.Le.Goff@cern.ch)

CEA Saclay, DAPNIA/SPhN, 91191 Gif-sur-Yvette, France

Abstract. The NA47/SMC experiment has measured in 1996 the spin structure function of the proton g_1^p using the 190 GeV/c polarized muon beam of the CERN SPS and a polarized Ammonia target. The statistics obtained in 1996 are equivalent to more than 3 times the statistics previously obtained by SMC in 1993 using a butanol polarized target. The results of 1996 are in agreement with the previous ones, however the trend for a rise of g_1^p at low x seen in the 1993 data is not confirmed in 1996. The value of the first moment depends on the approach used to describe the behavior of g_1^{p} at low x. We find that the Ellis-Jaffe sum rule is violated. With our published result for Γ_1^{d} we confirm the Bjorken sum rule with an accuracy of $\approx 15\%$.

We present a new measurement of the virtual photon proton asymmetry A_1^p by the Spin Muon Collaboration (SMC) in the kinematic range $0.0008 < x < 0.7$ and $0.2 < Q^2 < 100$ GeV^2. Using the data with $Q^2 > 1$ GeV^2 and $x > 0.003$ we determine the spin structure function g_1 of the proton.

The polarization of the CERN SPS muon beam was determined by measuring the cross section asymmetry for the scattering of polarized muons on polarized atomic electrons. For the average muon energy of 188 GeV, the polarization is $P_\mu = -0.77 \pm 0.03$. The energy dependence of the polarization is taken into account event by event.

The use of a solid polarized ammonia target during 1996 data taking, instead of butanol used in 1993, increased the dilution factor, which accounts for the fact that only a fraction of the target nucleons are polarizable, by $\approx 30\%$. The average polarization was $P_{\mathrm{p}} = \pm 0.89$ over the entire 1996 data taking and was known with a relative accuracy $\Delta P_{\mathrm{p}}/P_{\mathrm{p}} = 2.7\%$. Dedicated measurements of the nitrogen polarization P_{N}, with relative accuracy $\Delta P_{\mathrm{N}}/P_{\mathrm{N}}$ better than 10%, showed that P_{N} and P_{p} are related through the equal spin temperature relation [1]. In the analysis P_{N} could then be obtained from the measured P_{p}, the resulting average was $P_{\mathrm{N}} = \pm 0.14$.

The cross section asymmetry for parallel and antiparallel configurations of longitudinal beam and target polarizations $A_{\parallel}^{\mathrm{p}}$ and the spin-dependent structure function g_1^{p} are related to the virtual photon proton asymmetry A_1^{p} by [2]

$$A_{\parallel}^{\mathrm{p}} = D A_1^{\mathrm{p}}, \quad g_1^{\mathrm{p}} = \frac{F_2^{\mathrm{p}}}{2x(1+R)} A_1^{\mathrm{p}}, \qquad (1)$$

where the depolarization factor D depends on kinematic variables and on $R = \sigma_{\mathrm{L}}/\sigma_{\mathrm{T}}$. The contribution from A_2^{p} is neglected and we account for it in the systematic error.

In the shell model, ^{14}N is described as a spinless ^{12}C core with the valence proton and neutron being responsible for the nitrogen spin. A correction was computed in this framework, to account for the contribution to $A_{\parallel}^{\mathrm{p}}$ from scattering on ^{14}N. It amounts to less than 3% and results in a small systematic error [3].

The statistics obtained in 1996 are equivalent to more than 3 times the statistics previously obtained by SMC in 1993 [4] using a butanol polarized target. The two sets of measurements are in agreement, they are combined and the asymmetries are presented in Tab. 1 and Fig. 1. The comparison of SMC and EMC data to E143 data, obtained at lower Q^2, shows no evidence for a Q^2 dependence of A_1 .

x Range	$\langle x \rangle$	$\langle Q^2 \rangle$	A_1^p
0.003–0.006	0.005	1.3	0.017±0.018±0.003
0.006–0.010	0.008	2.1	0.047±0.016±0.004
0.010–0.020	0.014	3.6	0.035±0.014±0.003
0.020–0.030	0.025	5.7	0.058±0.018±0.005
0.030–0.040	0.035	7.8	0.067±0.022±0.005
0.040–0.060	0.049	10.4	0.115±0.019±0.008
0.060–0.100	0.077	14.9	0.176±0.019±0.013
0.100–0.150	0.122	21.3	0.267±0.025±0.018
0.150–0.200	0.173	27.8	0.318±0.035±0.021
0.200–0.300	0.242	35.6	0.400±0.036±0.028
0.300–0.400	0.342	45.9	0.568±0.058±0.042
0.400–0.700	0.480	58.0	0.658±0.079±0.055

Table 1. The virtual photon proton asymmetry A_1^p. The first error is statistical and the second is systematic.

We compute g_1^p for data with $Q^2 \geq 1$ GeV^2 using Eq. (1) and new parametrizations for F_2^p and R [5, 6] which include the latest experimental data. The use of these new parametrizations has a very small effect on g_1^p. The results for 1993 and 1996 data are shown in Fig. 2. They are statistically compatible and the combined results are also shown in the figure and listed in Tab. 2. However, the trend for a rise of g_1^p at low x seen in the 1993 data is not confirmed in 1996.

We evolve our data to $Q_0^2 = 10$ GeV^2 using a perturbative QCD analysis in NLO in the Adler-Bardeen scheme [8]. This analysis provides a fit $g_1^{\rm fit}(x, Q^2)$, illustrated in Fig. 3. By adding $g_1^{\rm fit}(x, Q_0^2) - g_1^{\rm fit}(x, Q^2)$ to g_1 at

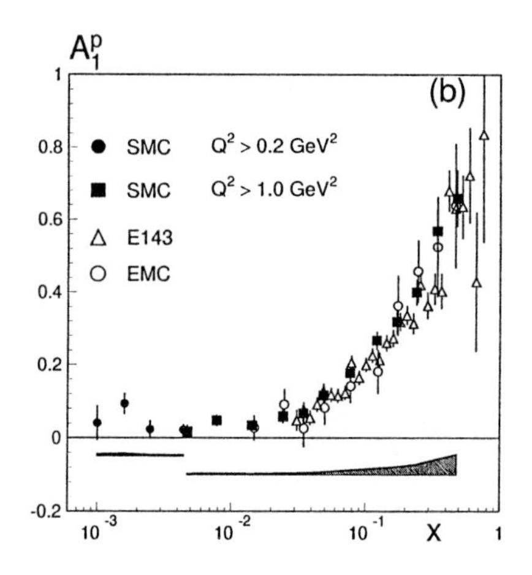

Fig. 1. A_1^p vs x from 1993 and 1996 SMC data combined is shown along with the measurements from EMC [2] and E143 [7] experiments. Statistical errors are shown as error bars while the shaded band below indicates the SMC systematic uncertainty.

$\langle x \rangle$	g_1^p	$g_1(Q_0^2$=10.GeV$^2)$
0.005	0.44±0.46±0.08	0.73±0.46±0.08±0.71
0.008	0.86±0.30±0.07	1.12±0.30±0.07±0.26
0.014	0.41±0.16±0.04	0.56±0.16±0.04±0.09
0.025	0.43±0.14±0.03	0.50±0.14±0.03±0.02
0.035	0.36±0.12±0.02	0.38±0.12±0.02±0.01
0.049	0.44±0.07±0.03	0.44±0.07±0.03±0.00
0.077	0.42±0.05±0.02	0.40±0.05±0.02±0.00
0.122	0.37±0.04±0.02	0.35±0.04±0.02±0.00
0.173	0.30±0.03±0.02	0.28±0.03±0.02±0.01
0.242	0.23±0.02±0.01	0.23±0.02±0.01±0.01
0.342	0.17±0.02±0.01	0.19±0.02±0.01±0.01
0.480	0.08±0.01±0.01	0.09±0.01±0.01±0.01

Table 2. The spin-dependent structure function $g_1^p(x)$ at the measured Q^2 and $g_1^p(x)$ evolved to $Q_0^2 = 10$ GeV^2. The first error is statistical and the second is systematic. In the last column, the third error indicates the uncertainty in the QCD evolution.

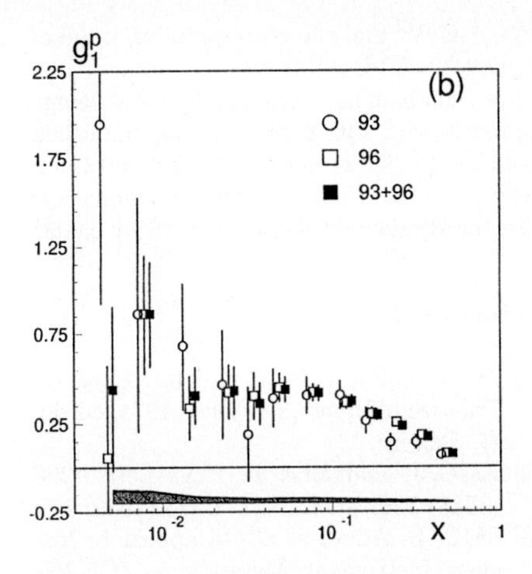

Fig. 2. SMC $g_1^{\rm p}$ values at measured Q^2 from 1993, 1996 and 1993+1996 combined data sets. In both figures error bars show the statistical uncertainty and the shaded band indicates the systematic uncertainty.

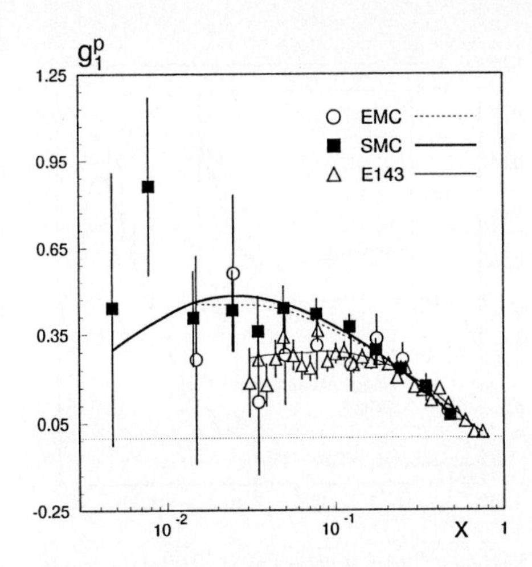

Fig. 3. Published data sets on $g_1^{\rm p}$ are shown. The curves represent the QCD fit at the measured Q^2 for each data set. Error bars represent the total error.

the measured x and Q^2, we obtain $g_1(x, Q_0^2)$, which is also given in Table 2.

In the measured range, $0.003 < x < 0.7$, at $Q_0^2 = 10$ GeV^2 the contribution to the first moment of the proton structure function is

$$\int_{0.003}^{0.7} g_1^{\rm p}(x, Q_0^2)dx = 0.139 \pm 0.006 \pm 0.008 \pm 0.006$$

where the first uncertainty is statistical, the second is systematic and the third is due to the uncertainty in the Q^2 evolution.

To estimate the contribution to the first moment from the unmeasured high x region $0.7 < x < 1.0$, we assume $A_1^p = 0.7 \pm 0.3$ which is consistent with the data and covers the upper bound $A_1 \leq 1$, and obtain a contribution of 0.0015 ± 0.0006. To estimate the contribution from the unmeasured low x region, as illustrated in Fig. 4, we either assume a Regge like behavior at low x, i.e. $g_1^{\rm p}$ = constant, which gives a contribution of 0.002 ± 0.002, or we integrate the QCD fit in this region and obtain -0.011 ± 0.011.

In the QCD approach we obtain at $Q_0^2 = 10$ GeV^2 $\Gamma_1^{\rm p} = 0.130 \pm 0.006 \pm 0.008 \pm 0.014$ where the first uncertainty is statistical, the second systematic and the third is due to the high and low x extrapolations and Q^2 evolution. In the Regge approach the last error appears smaller. The data do not allow us to exclude either approach so we keep the two results using the larger value for the third uncertainty. The Regge result is then $\Gamma_1^{\rm p} = 0.142 \pm 0.006 \pm 0.008 \pm 0.014$.

Assuming SU$(3)_f$ symmetry the flavor singlet axial charge a_o can be determined from Γ_1 using $a_3 = g_A/g_V$ and $a_8 = 3F - D$, we obtain $a_0 = 0.34 \pm 0.17$ and 0.22 ± 0.17 with the Regge and QCD approaches, respectively. The corresponding values for a_s are -0.08 ± 0.06 and -0.12 ± 0.06. Assuming $a_s = 0$, Ellis and Jaffe predicted $\Gamma_1^{\rm p} = 0.170 \pm 0.004$. Irrespec-

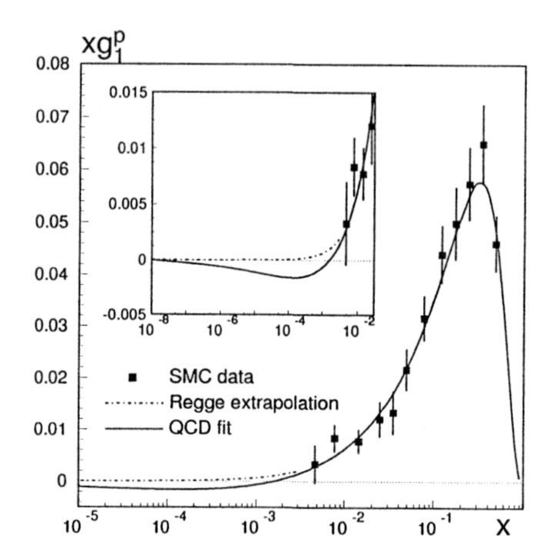

Fig. 4. xg_1^p as a function of x; SMC data points (squares) with the total error are shown together with the result of the QCD fit (continuous line), both at $Q^2 = 10$ GeV^2. For $x < 0.003$ the extrapolation assuming Regge behavior is indicated by the dot-dashed line. The inset is a close-up extending to lower x.

tive of the low x approach this sum rule is violated.

In the AB scheme the quark spin contribution to the nucleon spin, $\Delta\Sigma$, is independent of Q^2, which enables it to be interpreted as the intrinsic quark-spin content of the nucleon. In this scheme we have,

$$a_0(Q^2) = \Delta\Sigma - n_f \frac{\alpha_s(Q^2)}{2\pi} \Delta g(Q^2), \qquad (2)$$

where Δg is the gluon spin contribution to the nucleon spin. If we make the assumption $\Delta\Sigma = a_8$ corresponding to an unpolarized strange sea, our measurement of a_0 corresponds to $2 < \Delta g < 3$ at $Q^2 = 10$ GeV^2.

The QCD analysis done with all the published data along with the data presented here results in $\Delta g = 0.9 \pm 0.3(\text{exp}) \pm 1.0(\text{theory})$ at $Q^2 = 1$ GeV^2 and the corresponding value of Δg at $Q^2 = 10$ GeV^2 is 1.7.

If we combine our result for Γ_1^p using Regge approach with our corresponding published result for Γ_1^d [9], we obtain $\Gamma_1^p - \Gamma_1^n = 0.195 \pm 0.029$ at $Q^2 = 10$ Gev^2 which is compatible with the Bjorken prediction of 0.186 ± 0.003.

References

[1] M. Borghini, in Proc. 2nd Int. Conf. on Polarized Targets, Berkeley, 1971, ed. by G. Shapiro.
[2] EMC, J. Ashman et al., Phys. Lett. B206 (1988) 364; Nucl. Phys. B328 (1989) 1.
[3] SMC, B. Adeva et al., to appear in Nuclear Instrument Methods A. (CERN-PPE/97-66)
[4] SMC, D. Adams et al., Phys. Rev. **D56** (1997) 5330.
[5] NMC, M. Arneodo et al., Nucl. Phys. B483 (1997) 3.
[6] SMC, B. Adeva et al., Phys. Lett. B 412 (1997) 414
[7] E143 Collaboration, K. Abe et al., Phys. Rev. Lett. 74 (1995) 346.
[8] R. D. Ball, S. Forte, and G. Ridolfi, Phys. Lett. bf B378 (1996) 255.
[9] SMC, D. Adams et al., Phys. Lett. B396 (1997) 338.

Valence Quark Polarisations from Semi-inclusive Deep-Inelastic Muon Scattering

Günter Baum (baum@physik.uni-bielefeld.de)

Universität Bielefeld

Abstract. Results are presented from the analysis of all semi-inclusive data taken by the Spin Muon Collaboration (SMC). The measured spin asymmetries for charged hadrons are used to determine polarised valence and sea quark distributions and to evaluate their first and second moments.

1 Results

The Spin Muon Collaboration has reported previously [1] on a first ever measurement of polarised quark distributions from semi-inclusive spin asymmetries for positively and negatively charged hadrons. We present now results from all of our deep inelastic scattering data [2][3], taken with polarised muons of mostly 190 GeV on polarised protons and deuterons in the years 1992 to 1996. We detect semi-inclusive events by separating hadrons from electrons with the calorimeter [4] of the SMC apparatus. For the analysis hadrons with $z > 0.2$ are selected and a cut $Q^2 > 1$ GeV^2 is applied, covering a range in x of $0.003 < x < 0.7$ with an average $Q^2 = 10$ GeV^2. The four spin asymmetries shown in Fig.1 are analysed in a combined way, together with our inclusive

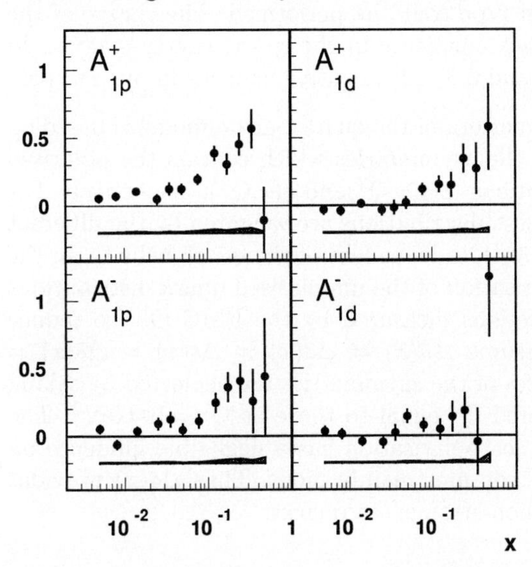

Fig.1. Semi-inclusive spin asymmetries for the proton and the deuteron as a function of x for positively and negatively charged hadrons. The error bars are statistical and the shaded areas represent the systematic uncertainties.

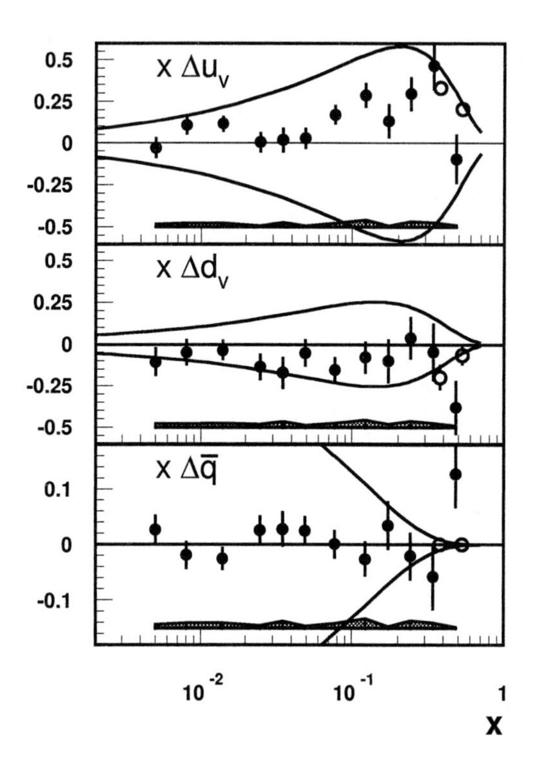

Fig.2. The polarised quark distributions $x\Delta u_v(x), x\Delta d_v(x)$ and $x\Delta\overline{q}(x)$ obtained with the assumption $\Delta\overline{u}(x) = \Delta\overline{d}(x)$. The open circles are obtained when the sea polarisation is set to zero while the closed circles are obtained without this assumption. Errors are as in Fig.1. The curves correspond to the upper and lower limits $\pm xq(x)$ from the unpolarised quark distribution [8] evaluated at Q^2 =10 GeV^2. In the bottom plot the curves are $\pm x(\overline{u}(x) + \overline{d}(x))/2$.

asymmetries for the proton [5][6] and for the deuteron [7] in order to improve on the significance of the results. For the semi-inclusive events no identification of the individual hadron type could be performed. The charge of the hadrons is known from the track curvature in the spectrometer magnet. In total there are $5 \cdot 10^6$ positive and $3.8 \cdot 10^6$ negative hadrons in our sample.

The analysis within the framework of the quark-parton model is based on the theoretical expressions for the asymmetries which contain the polarised and unpolarised quark distributions, $\Delta q(x,Q^2)$and $q(x,Q^2)$, respectively. For the semi-inclusive processes, these distributions are weighted by the different fragmentation functions, $D_q^h(z,Q^2)$, independent of the quark helicity. In the evaluation we use the parametrisation of the unpolarised quark distributions [8] and the fragmentation functions measured by the EMC [9]. To reduce the number of unknows we assume $\Delta\bar{u}(x) = \Delta\bar{d}(x) = \Delta s(x) = \Delta\bar{s}(x) = \Delta\bar{q}(x)$. Possible Q^2 dependences of the asymmetries are neglected by taking the asymmetries at the measured Q^2 equal to those at Q^2 =10 GeV^2. The assumption about the strange sea polarisation has a negligible influence on our results as our data sample contains mainly pions. Thus $\Delta\bar{q}(x)$ provides information mostly about the non-strange sea quarks.

In Fig.2 we show the polarised quark distributions, obtained separately for the valence up and down quarks, and for the non-strange sea quarks, as function of the momentum fraction carried by the struck quark, x. We observe that the up valence quarks have positive polarisation values over the whole x range ($\Delta u_v(x) > 0$), while the polarisation of the down valence quarks $\Delta d_v(x)$ is seen to be negative.

The first moments of the polarised quark distributions over the full x range are

$$\begin{aligned}\Delta u_v &= 0.77 \pm 0.10 \pm 0.08,\\ \Delta d_v &= -0.52 \pm 0.14 \pm 0.09,\\ \Delta \bar{q} &= 0.01 \pm 0.04 \pm 0.03.\end{aligned}$$

The statistical errors are reduced by almost a factor of two compared to our previous publication [1]. The contributions to the moments from the unmeasured low-x region are 0.04 ± 0.04 and -0.05 ± 0.05 for the u_v and d_v quarks, respectively. Details on this are given in our recent publication [2].

From our polarised quark distributions we determine for the first time the second moments:

$$\int_0^1 x\Delta u_v(x)\mathrm{d}x = 0.155 \pm 0.017 \pm 0.010 \quad \text{and}$$

$$\int_0^1 x\Delta d_v(x)\mathrm{d}x = -0.056 \pm 0.026 \pm 0.011\,,$$

where the first error is statistical and the second is systematic. Due to the additional factor x the low x extrapolations to the unmeasured region give negligible contributions while the large x contributions remain small. A calculation of the second moments in lattice QCD [10] predicts $\int_0^1 x\Delta u_v(x)\mathrm{d}x = 0.189 \pm 0.008$ and $\int_0^1 x\Delta d_v(x)\mathrm{d}x = -0.0455 \pm 0.0032$, which is in good agreement with our measured values.

References

[1] SMC, B. Adeva et al., Phys. Lett. B369 (1996) 93.
[2] SMC, B. Adeva et al., hep-ex/9711008, submitted to Phys. Lett. B.
[3] SMC; J. Pretz, Ph. D. thesis Mainz University (1997).
[4] O. C. Allkofer et al., Nucl. Instr. Methods 179 (1981) 445.
[5] SMC, B. Adeva et al., CERN-PPE-97-118, submitted to Phys. Lett. B.
[6] SMC, J. Le Goff, preceding contribution # 303.
[7] SMC, D. Adams et al., Phys. Lett. B396 (1997) 338.
[8] M. Glück et al., Z. Phys. C67 (1995) 433.
[9] EMC, A. Arneodo et al., Nucl. Phys. B321 (1989) 541.
[10] M.Göckeler et al., hep-ph/9708270 and DESY 97-117, accepted in Phys. Lett. B.

Quarks and Gluons in a Polarized Nucleon at RHIC

Jacques Soffer (soffer@cpt.univ-mrs.fr)

Centre de Physique Théorique, CNRS Marseille , France

Abstract. In view of the future operation of the RHIC facility at BNL, part of the time as a polarized proton-proton collider, we review some theoretical aspects of the physics programme which will be carried out by the two large detectors PHENIX and STAR, to improve our knowledge on quarks and gluons in a polarized nucleon.

Major progress have been made over the last ten years or so, in our understanding of the spin structure of the nucleon. This is due to a better determination of the polarized structure function $g_1^{p,n,d}(x, Q^2)$, from polarized deep inelastic scattering on different targets (hydrogen,deuterium,helium-3). However these fixed targets experiments [1], performed at CERN, DESY and SLAC, cover only a limited kinematic region, that is, $0.005 \leq x \leq 0.7$ with the corresponding average $< Q^2 >$ between $2GeV^2$ and $10GeV^2$. In spite of the constant progress realized in the accuracy of the data, they can still be described, non uniquely, in terms of several sets of polarized parton distributions. In particular, sea quark, antiquark and gluon distributions remain fairly ambiguous. The restricted Q^2 range accessible by the data makes also difficult, sensible tests of the Q^2 evolution, predicted by recent higher order QCD calculations. Moreover it is not possible to obtain a good flavor separation, to isolate the contribution of each quark to the nucleon spin.

Polarized hadronic collisions, another way to investigate this research field, have generated little progress due to the scarcity of the data in appropriate kinematic regions, and a low energy range, so far accessible to very few dedicated experiments.This situation will change drastically soon when the RHIC facility at BNL will start running, by 1999 or so, part of the time as a polarized *pp* collider. A vast spin programme will be undertaken by the two large detectors PHENIX and STAR, which will operate at RHIC and also by the *pp2pp* experiment, dedicated to *pp* elastic and total cross sections. Before we go on and explain very briefly what will be done, let us recall three key parameters, which will be crucial to answer some of the very challenging questions. The proton beam polarization will be maintained at the level of 70%, in both longitudinal and transverse directions, the center-of-mass energy $\sqrt{s}$ will be ranging between $100GeV$ and $500GeV$ and at its maximum value, the luminosity is expected to reach $2.10^{32}cm^{-2}sec^{-1}$.

The physical spin observables of interest are A_{LL}, the *double* helicity asymmetry and A_L the *single* helicity asymmetry. For a *pp* collision yielding a given final state (jet, γ, $W^{\pm}$, Z, etc...), with both initial beams longitudinally polarized, A_{LL} is defined as $(\sigma_{++} - \sigma_{+-})/(\sigma_{++} + \sigma_{+-})$. Here σ_{++} denotes the inclusive cross section when both protons have positive helicity and σ_{+-} when their helicities have opposite signs. Similarly A_L, which is a parity violating asymmetry, requieres only one beam polarized and is defined as $(\sigma_- - \sigma_+)/(\sigma_- + \sigma_+)$. The explicit expressions of A_{LL} and A_L in the QCD parton model, for the various processes we will consider, can be found in recent papers in the literature (see for example [2]).

The asymmetry A_{LL}^{γ} for direct photon production at high p_T^{γ} is strongly sensitive to sign and magnitude of the polarized gluon distribution $\Delta G(x, Q^2)$, because the cross section is dominated by the quark-gluon Compton subprocess $qG \to q\gamma$. Given the high event rate at $\sqrt{s} = 500GeV$ up to $p_L^{\gamma} = 50GeV/c$ or so, A_{LL}^{γ} will be determined with an error less than 5%, which therefore will allow to distinguish between the different possible $\Delta G(x, Q^2)$. This analysis can be done by studying A_{LL}^{γ} as a function of p_T^{γ}, but also as a function of the photon rapidity, for fixed p_T^{γ}. Single jet production is also dominated by gluon-quark elastic scattering in the medium p_T^{jet} range, say between $20GeV/c$ and $80GeV/c$. So this is, as well, a valuable process to pin down the gluon distribution by measuring A_{LL}^{jet} and a similar analysis to the direct photon case can be done. The very copious event rate could provide in fact a higher accuracy determination of $\Delta G(x, Q^2)$ for a broad Q^2 range, namely $10^2 \leq Q^2 \leq 10^4 GeV^2$.

Let us now consider the production of $W^{\pm}$ gauge bosons which is dominated by a Drell-Yan process. In the Standard Model, the W is a pure left-handed object so the single helicity asymmetry A_L has a simple expresssion [2]. The production of $W^{\pm}$ is known to raise with increasing energy, so the higher rates will be obtained at $\sqrt{s} = 500GeV$. At this energy, it can be shown that, by measuring A_L as a function of the W rapidity y, one can obtain a fairly clean separation both for quarks and antiquarks. More precisely for $y = -1$, one gets $\Delta\bar{d}(x, M_W^2)$ from W^+ and $\Delta\bar{u}(x, M_W^2)$ from W^- at $x = 0.06$. For $y = +1$, one gets $\Delta u(x, M_W^2)$ from W^+ and $\Delta d(x, M_W^2)$ from W^- at $x = 0.43$ and these determinations will be done at a level of a few percents error. The A_{LL} corresponding to a standard Drell-Yan lepton pair production l^+l^- of mass M is directly proportional to $\Delta\bar{q}(x, M^2)$ and this is an efficient way to pin down the antiquark distributions, but in this case the measurement must be done near the lowest available RHIC energy.

Finally, we emphasize the relevance of the measurement of parity violating asymmetries in single jet production to check the expected QCD electroweak interference effect. Any departure from this rather well known standard A_L, must provide a clean signature for *new physics*.

References

[1] For a short review see R.Devenish, these proceedings.
[2] J.Soffer and J.M.Virey,Preprint CPT-97/P.3495,to appear Nucl.Phys.B.

The Photon Structure Function Measurements from LEP

Richard Nisius, on behalf of the LEP Collaborations

PPE Division, CERN, CCH-1211 Genève 23, Switzerland
Richard.nisius@cern.ch

Abstract. The present knowledge of the structure of the photon based on measurements of photon structure functions is discussed. This review covers QED structure functions and the hadronic structure function F_2^γ.

1 Introduction

One of the most powerful tools to investigate the internal structure of quasi real photons is the measurement of photon structure functions in deep inelastic electron photon scattering at e^+e^- colliders. These measurements have by now a tradition of sixteen years since the first F_2^γ was obtained by PLUTO [1]. The LEP accelerator is a unique place for the measurements of photon structure functions until a high energy linear collider is realised. It is unique because of the large coverage in Q^2 owing to the various beam energies covered within the LEP programme, and due to the high luminosities delivered to the experiments. Structure function measurements are performed by all four LEP experiments, however concentrating on different aspects of the photon structure. The main idea is that by measuring the differential cross section

$$\frac{d^2\sigma_{e\gamma\rightarrow eX}}{dxdQ^2} = \frac{2\pi\alpha^2}{x\,Q^4}\left[\left(1+(1-y)^2\right)F_2^\gamma(x,Q^2) - y^2 F_{\mathrm{L}}^\gamma(x,Q^2)\right] \tag{1}$$

one obtains the photon structure function F_2^γ which is proportional to the parton content of the photon and therefore reveals the internal structure of the photon. Here Q^2 is the absolute value of the four momentum squared of the virtual photon, x and y are the usual dimensionless variables of deep inelastic scattering and α is the fine structure constant. In the region of small y studied ($y \ll 1$) the contribution of the term proportional to $F_{\mathrm{L}}^\gamma(x,Q^2)$ is small and it is usually neglected.

Because the energy of the quasi-real photon is not known, x has to be derived by measuring the invariant mass of the final state X, which consists of $\mu^+\mu^-$ pairs for $F_{\mathrm{k,QED}}^\gamma$, $k = 2, A, B$, and of hadrons created by a $\mathrm{q\bar{q}}$ pair in studies of F_2^γ. The invariant mass can be determined accurately in the case of $\mu^+\mu^-$ final states, and measurements of $F_{\mathrm{k,QED}}^\gamma$ are statistically limited. For hadronic final states the measurement of x is a source of significant uncertainties which makes measurements of F_2^γ mainly systematics limited.

2 The QED structure functions

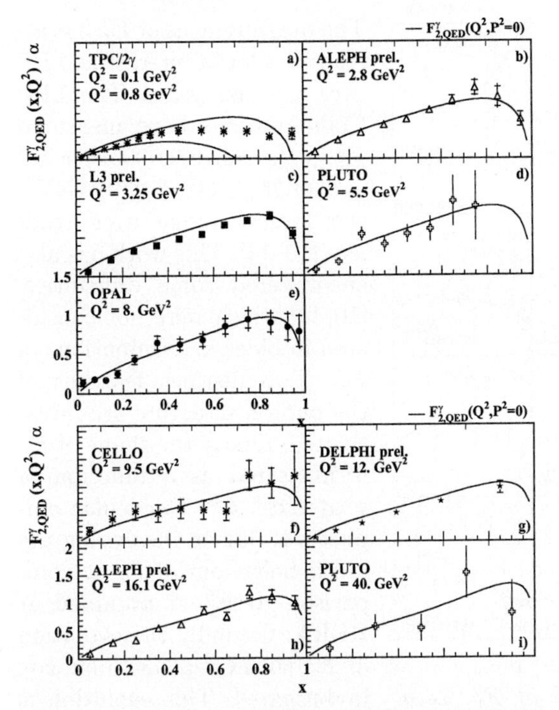

Fig. 1. *The measurements of $F^{\gamma}_{2,\mathrm{QED}}$ as a function of x for various Q^2 compared to QED assuming $P^2 = 0$. The Q^2 values for the predictions are taken from the publications. If the data were unfolded to a given Q^2 this value is taken, if only the average Q^2 of the sample is given, c), this value is taken, and if no value at all is quoted the Q^2 range as obtained from the information of the event selection is shown, a). The quoted errors for g) are statistical only.*

Several measurements of QED structure functions from LEP have been submitted to this conference. The $\mu^+\mu^-$ final state is such a clean environment that it allows for much more subtle measurements to be performed than in the case of hadronic final states. Therefore the investigation of QED structure functions serves not only as a test of QED but rather it is used to refine the experimentalists tools in a real but clean environment to investigate the possibilities of extracting similar information from the much more complex hadronic final states.

χ is defined as the azimuthal angle between the plane spanned by the muon pair and the plane spanned by the incoming quasi-real photon and the deep inelastically scattered electron. This angle has been used to extract structure functions of the photon, $F^{\gamma}_{\mathrm{A,QED}}$ and $F^{\gamma}_{\mathrm{B,QED}}$, which can not be assessed by the cross section measurement, which is dominated by the contribution of $F^{\gamma}_{2,\mathrm{QED}}$. Azimuthal correlations based on χ were investigated by ALEPH (prel.) [2], L3 (prel.) [3] and OPAL [4], and $F^{\gamma}_{2,\mathrm{QED}}$ has been obtained by all four experiments. Figure 1 shows the world summary of the $F^{\gamma}_{2,\mathrm{QED}}$ measurements [2, 3, 5–9]. All measurements are consistent with expectations from QED. The LEP data are so precise that it is possible to study the effect of the small virtuality P^2 of the quasi real photon [3, 6].

3 The hadronic structure function F_2^γ

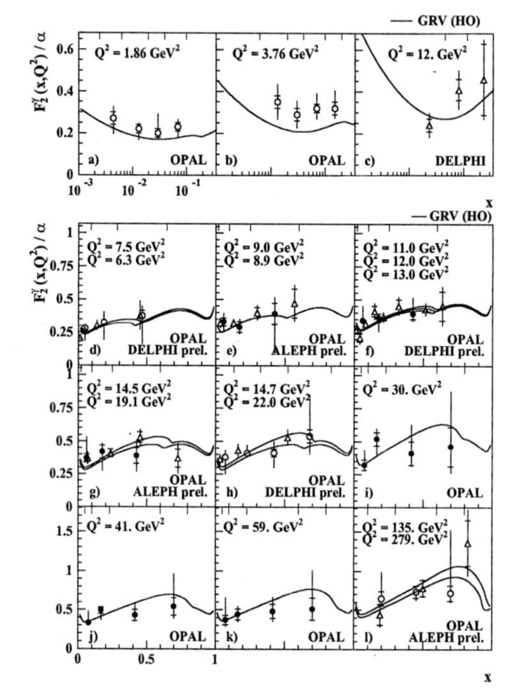

Fig. 2. *The measurements of F_2^γ as a function of x unfolded on a logarithmic x scale, a)–c), or on a linear x scale, d)–l), compared to the prediction of the GRV (HO) parametrisation. The OPAL data at 11* GeV^2 *(41* GeV^2*) are the combined data from 9 and 14.5* GeV^2 *(30 and 59* GeV^2*). The inner error bar is the statistical error and the outer the quadratic sum of statistical and systematic error. (ALEPH, DELPHI triangles, OPAL circles)*

The measurement of F_2^γ has attracted a lot of interest at LEP over the last years. The LEP Collaborations have measured F_2^γ in the range $0.0025 < x \lesssim 1$ and $(1.86 < Q^2 < 279)$ GeV^2, the largest range ever studied [10–14]. This work has also encountered some difficulties [10, 14] which were not considered in older determinations of F_2^γ. Two distinct features of the photon structure are investigated. Firstly the shape of F_2^γ is measured as a function of x at fixed Q^2. Particular emphasis is put on measuring the low-x behaviour of F_2^γ in comparison to F_2^{p} as obtained at HERA. Secondly the evolution of F_2^γ with Q^2 at medium x is investigated. This evolution is predicted by QCD to be logarithmic. In general F_2^γ is found to rise smoothly towards large x and there is some weak indication for a possible rise at low x for $Q^2 < 4$ GeV^2, as shown in Fig 2. This behaviour is satisfactorily described (except in Fig 2b) by several of the existing F_2^γ parametrisation, e.g. GRV [15,16] and SaS [17]. Experimentally there seems to emerge an inconsistency between the OPAL [12] and PLUTO [18] data on one hand and the TPC [19] data on the other hand at $Q^2 \approx 4$ GeV^2, see [12].

The evolution of F_2^γ with Q^2, Fig 3, has been studied using the large lever arm in Q^2, and also by comparing various ranges in x within one experiment [11]. Unfortunately the different experiments quote their results for different ranges in x which makes the comparison more difficult because the predictions for the various ranges in x start to be significantly different for $Q^2 > 100$ GeV^2, as can be seen in Fig 3. The measurements are consistent with each other and a clear rise of F_2^γ with Q^2 is observed. It is an interesting

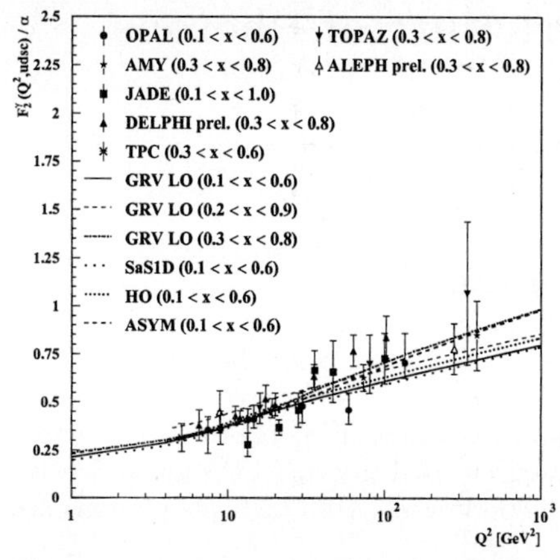

Fig. 3. *Summary of the measurements of F_2^γ at medium x without subtraction of the charm contribution compared to several theoretical predictions taking into account the various ranges in x.*

fact that this rise can be described reasonably well ($\mathcal{O}(15\%)$ accuracy) by the leading order asymptotic solution [20] for F_2^γ as predicted by perturbative QCD for $\alpha_s(M_z^2) = 0.128$ as detailed in [11].

4 Conclusion

New informations on the photon structure in so far unexplored Q^2 and x regions have been obtained based on structure function measurements at LEP. All observations are consistent with the QED and QCD predictions. The future LEP2 programme will allow to extend the study of the photon structure up to $Q^2 \approx 1000$ GeV^2.

Acknowledgement: I wish to thank the members of the LEP collaborations for their help during the preparation of this review.

References

[1] PLUTO Collab., C. Berger et al., Phys. Lett. **107B**, 168–172 (1981).
[2] ALEPH Collab., contributed paper to Photon 97, Egmond aan Zee.
[3] L3 Collab., contributed paper to this conference.
[4] OPAL Collab., K. Ackerstaff et al., Z. Phys. **C74**, 49–55 (1997).
[5] OPAL Collab., R. Akers et al., Z. Phys. **C60**, 593–600 (1993).
[6] DELPHI Collab., P. Abreu et al., Z. Phys. **C69**, 223–234 (1996).
[7] TPC/2γ Collab., M. P. Cain et al., Phys. Lett. **147B**, 232–236 (1984).
[8] CELLO Collab., H. J. Behrend et al., Phys. Lett. **126B**, 384–390 (1983).
[9] PLUTO Collab., C. Berger et al., Z. Phys. **C27**, 249–256 (1985).
[10] OPAL Collab., K. Ackerstaff et al., Z. Phys. **C74**, 33–48 (1997).
[11] OPAL Collab., K. Ackerstaff et al., Phys. Lett. **B411**, 387–401 (1997).
[12] OPAL Collab., K. Ackerstaff et al., Phys. Lett. **B412**, 225–234 (1997).
[13] DELPHI Collab., *Proceedings of Photon 97*, Egmond aan Zee.
[14] ALEPH Collab., contributed paper to this conference.
[15] M. Glück, E. Reya, and A. Vogt, Phys. Rev. **D45**, 3986–3994 (1992).
[16] M. Glück, E. Reya, and A. Vogt, Phys. Rev. **D46**, 1973–1979 (1992).
[17] G. A. Schuler and T. Sjöstrand, Z. Phys. **C68**, 607–623 (1995).
[18] PLUTO Collab., C. Berger et al., Phys. Lett. **142B**, 111–118 (1984).
[19] TPC/2γ Collab., H. Aihara et al., Z. Phys. **C34**, 1–13 (1987).
[20] E. Witten, Nucl. Phys. **B120**, 189–202 (1977).

Measurements of the Structure of Photons at HERA

Katharina Müller[1] (kmueller@mail.desy.de)

DESY, Notkestr. 85 D-22603 Hamburg, Germany

Abstract. Measurements of the structure of the photon with the H1 and ZEUS detector at HERA are presented. The double differential di-jet cross section is determined and an effective leading order parton distribution of the photon is extracted. A new method based on high p_T charged tracks is used to extract the gluon density of the photon. Cross sections of photoproduction events with high E_T photons are presented.

1 Introduction

The two experiments H1 and ZEUS at the positron-proton collider HERA (820 GeV proton, 27.5 GeV positrons) offer a unique possibility to study the structure of photons in a very large phase space. The study of hard processes in quasi-real photoproduction at HERA is in lowest order(LO) QCD described by the interaction of a parton from the proton with the incoming photon (direct process) or with a parton from the photon (resolved process). Both experiments analyse different final states, such as di-jet events, events with an isolated high p_T photon, so called prompt photon events, or high p_T charged tracks to investigate the structure of the photon.

2 Di-jet cross section and effective parton distribution

The double-differential di-jet cross section in photoproduction ($Q^2 < 4\,\mathrm{GeV}^2$) is measured with the H1 detector. Events were selected with two jets which were found using a cone algorithm with cone size R=0.7 and jet pseudorapidities in the range -0.5<η<2.5 in the laboratory system. The imbalance of the two highest E_T^{Jets} jets was required to be $\mid E_{T,1}^{Jets}$-$E_{T,2}^{Jets} \mid < 0.25\ E_T^{Jets}$. E_T^{Jets} is the average jet energy which had to be larger than 10 GeV. The fractional energy of the exchanged photon y is restricted to 0.2<y<0.83.

The measurement is shown in Figure 2a) as a function of the average transverse energy E_T^{Jets} in bins of x_γ^{Jets} the fraction of the photon momentum carried by the parton from the photon, as measured with the two jets: $x_\gamma^{Jets} = \Sigma_{jets} E_T^{Jet} e^{-\eta^{Jet}}/2E_\gamma$. The dominant systematical error comes from the uncertainty in the knowledge of the electromagnetic energy scale. Next-to-leading order QCD [1] calculations (GRV-HO [2] and GS96 [3] photon parton distributions) and leading order monte carlo simulations with parton showers (GRV-LO [2]) describe the data satisfactory except for the two highest x_γ^{Jets}

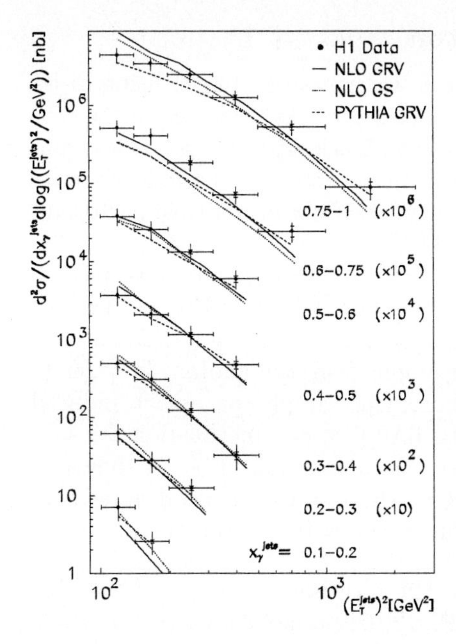

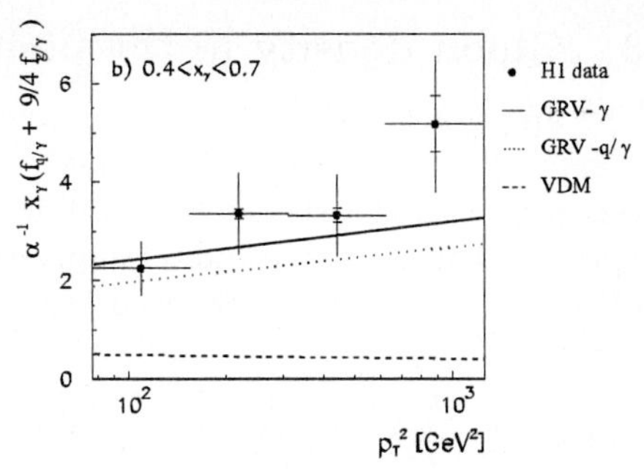

Fig. 1. a) Di-jet cross section as a function of the average jet energy squared $(E_T^{Jets})^2$ in bins of different fractional energy carried by the parton from the photon (x_γ^{Jets}). The data are compared to LO QCD simulation and two next to leading order calculations with different input structure functions. b) Leading order effective parton distribution in the photon as a function of the squared transverse momentum ${p_T}^2$ of the parton.

bins. Note that this high x_γ region was so far not accessible to measurements of the photon structure function by e^+e^-experiments.

Parton distributions of the photon at $x_\gamma \simeq 0.5$ are extracted using an approximation method [4]. The shape of the leading order matrix elements of the dominant partonic scattering processes qq→qq, qg→qg, and gg→gg are similar. Therefore the di-jet cross section is written as a product of one single effective subprocess M_{SES} and effective parton density functions f^γ_{eff} and f^p_{eff} which are a combination of quark and gluon densities.

$$f_{eff}(x,p_t^2) = \Sigma_q(q(x,p_t^2) + \bar{q}(x,p_t^2)) + 9/4 \cdot g(x,p_t^2)$$

p_T is the transverse momentum of the final state partons and is taken to be the scale of the process. With this ansatz f^γ_{eff} can be directly extracted from the data. The leading order effective parton distribution of the photon for $0.4 < x_\gamma < 0.7$ is shown in Figure 2b). The data rises with increasing p_T as is expected from the anomalous photon contribution [5]. The data are well reproduced by the GRV parametrization (full curve). This result extends the kinematic region of the measurement of the photon parton distributions to the scale of ${p_T}^2 = 1250$ GeV^2 with a precision comparable to e^+e^-measurements. The dotted line shows the GRV contribution from the quarks alone which amounts to 80% at large x_γ. A purely hadronic description of the photon as modelled in the vector meson dominance model (VDM) does not describe the data in shape and magnitude (dashed line).

3 Gluon density in the photon

The first gluon distribution of the photon was measured by H1 using di-jet events [6]. A new approach uses high p_T charged tracks measured in the central region of the H1 detector (at least one track with p_T >2.6 GeV and $|\eta|$<1). Compared to the direct approach its advantage is the independence of the energy scale and reduced sensitivity to effects of multiple parton interactions.

x_γ^{rec} is calculated as $x_\gamma^{rec} = \Sigma_N p_z e^{-\eta}/E_\gamma$ where the sum runs over all tracks with p_T >2 GeV. An unfolding procedure [7] is used to reconstruct x_γ^{true}. As in the di-jet analysis [6] the extracted x_γ distribution shows a large contribution of processes which include gluons from the photon. In order to extract the leading order gluon distribution in the photon, quark induced processes as modelled by PYTHIA with LAC1 [8] parametrisation are subtracted. The resulting gluon distribution as a function of x_γ is shown in Figure 2 at a scale of $<p_T{}^2>=38$ GeV^2. The result is in good agreement with the di-jet measuement but has reduced errors. It confirms the rise of the gluon density with decreasing x_γ.

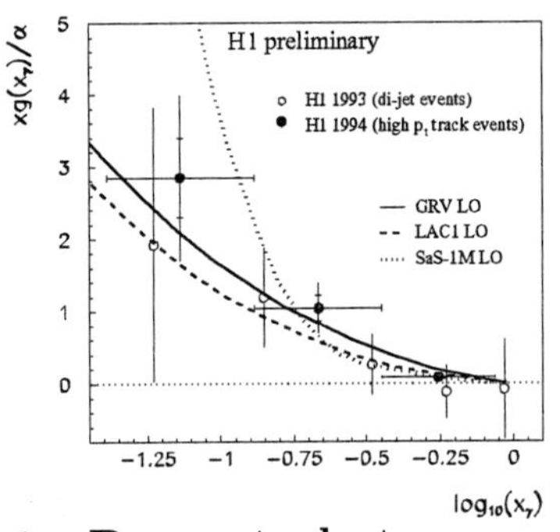

Fig. 2. Gluon distribution in the photon using high p_T charged particles (full circles). The measurement is compared to a previous measurement from H1 using di-jet events (open circles) which shows good agreement of the two methods. The full line shows the parametrisation of GRV, the dotted line SaS [9] and the dashed line LAC1.

4 Prompt photons

The production of prompt photons gives the opportunity to directly observe a particle from the hard subprocess, though with a substantially smaller cross section than di-jet production. Events are selected by both experiments with an isolated photon ($E_{T\gamma} > 5$ GeV) in a pseudorapidity range $-1.2(-0.7) < \eta_\gamma < 1.6(0.8)$ for H1 (ZEUS). Roughly 40% of the events have a jet measured in the detector which is needed for the determination of x_γ :$x_\gamma = (E_T^\gamma \cdot e^{-\eta^\gamma} + E_T^{Jet} \cdot e^{-\eta^{Jet}})/2E_\gamma$.

Isolation cuts and cuts on the cluster shape are applied to reduce background from hadronic jet events, where a neutral meson fakes a photon and from events with a photon radiated from a final state quarks. ZEUS estimates the remaining background from a fit to the distribution of the fractional energy deposited in the calorimeter cell with maximum energy. A photon deposits more than 75% of its energy in one cell, whereas π^0 or η tend to deposit their energy in several cells. A cross section is extracted by ZEUS for the sample with a jet($E_T^{Jet} > 5$ GeV, $-1.5 < \eta_{Jet} < 1.8$) for $x_\gamma > 0.8$ where the direct contribution is dominating: $\sigma_{\gamma p \to \gamma + Jet + X} = 15.3 \pm 3.8(stat) \pm 1.8(syst)$ pb which is in agreement with NLO calculations [10]: 14.05 pb (GS), 17.93 (GRV).

H1 estimates the background from Monte Carlo simulation to be of the order of 25%, it is subtracted binwise. The measured x_γ distribution is shown in Figure 3. The sample is dominated by the direct process. At x_γ below 0.9 where the contribution from the resolved process is substantial the data lie above the MC expectation which is an indication for a higher parton density in the photon than modelled in the LO-GRV parameterisation which was input for the Monte Carlo simulation. The cross section for the process $\gamma P \to \gamma + X$ ($E_{T\gamma} > 5$ GeV, $-1.2 < \eta_\gamma < 1.6$) was measured to be $\sigma = 104.8 \pm 5.9(stat) \pm 15.7(syst)$ pb. This is in broad agreement with QCD expectation, where a cross section in this E_T and η range of 84.1 pb is expected.

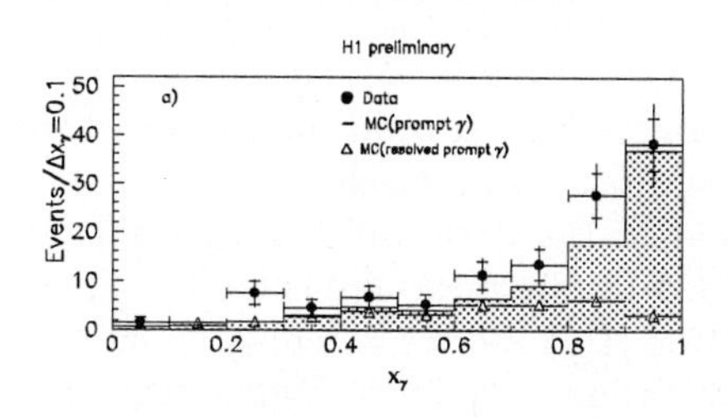

Fig. 3. x_γ distribution extracted from the event sample with a photon ($E_{T\gamma} > 5$ GeV, -1.2 < η_γ < 1.6) and a jet ($E_{T\,Jet} > 4$ GeV, -0.5 < η_{Jet} < 2.0). The full circles show uncorrected data after background subtraction. The shaded histogram is the QCD expectation for prompt photon production, the triangle symbols show the resolved photon part only.

5 Summary

Analyses of the structure of real photons were presented. The double differential di-jet cross section was presented and used to extract effective parton distributions with a precision comparable to e^+e^- data but at a higher scale. A logarithmic increase with $p_T{}^2$ is observed as predicted by perturbative QCD. A new method was presented to extract the gluon density in the photon using charged tracks. It is found to rise with decreasing x_γ. The prompt photon process was measured by both experiments and cross sections were given for the direct and the inclusive process.

References

[1] M. Klasen and G. Kramer, DESY-96-246, hep-ph@xxx.lsl.gov-9611450 (1996).
[2] M. Glück, E. Reya and A. Vogt, *Phys. Rev.* D **46**, 1973 (1992)
[3] L.E. Gordon, J.K. Storrow, ANL-HEP-PR-96-33, HEP-PH-9607370 (1996).
[4] B.L. Combridge and C.J. Maxwell, *Nucl. Phys.* B **239**, 429 (1984).
[5] E. Witten, *Nucl. Phys.* B **120**, 189 (1977).
[6] H1 Collab., T. Ahmed *et al*, *Nucl. Phys.* B **445**, 195 (1995).
[7] V. Blobel, DESY 84-119, proc. of CERN School of Computing, Aiguablava (Spain) CERN 1985.
[8] H. Abramowicz, K. Charchula, and A. Levy, *Phys. Lett.* B **269**, 1991 (458)
[9] A. Schuler, T. Sjostrand, *Phys. Lett.* B **376**, 1996 (193)
[10] L.E.Gordon, proc. Photon97, Egmond aan Zee, in prep. (hep-ph/9706355)

Parton Distributions in the Photon from $\gamma^*\gamma$ and γ^*p Scattering

H. Abramowicz, E. Gurvich and A. Levy

School of Physics and Astronomy, Raymond and Beverly Sackler Faculty of Exact Sciences, Tel–Aviv University, Tel–Aviv, Israel

Abstract. Leading order parton distributions in the photon are extracted from the existing F_2^γ measurements and the low-x proton structure function by assuming Gribov factorization to hold at low x. The resulting parton distributions in the photon are found to be consistent with the Frankfurt–Gurvich sum rule for the photon.

We present here new parameterization of leading order parton distributions in the photon. Known parameterizations all of which give similar values in the region where they were fitted to the data, differ appreciably in the low–x region which is not constrained by the measurements. The gluon distributions also differ in whole x region [1]. Here we attempt to fix this uncertainty assuming that Gribov factorization [2] holds at high energy (low x) for $\gamma^*\gamma$ scattering [3],

$$F_2^\gamma(x,Q^2) = F_2^p(x,Q^2)\frac{\sigma_{\gamma p}(W^2)}{\sigma_{pp}(W^2)} \quad . \tag{1}$$

Equation (1) gives a possibility to generate F_2^γ "data" from much more accurate F_2^p measurements at low x. We have used all available F_2^γ data together with the indirect "data" obtained through relation (1) using the data of the proton structure function [4] for $x \leq 0.01$. The results of global fit are shown in Fig.1

The presented parameterization, which gives a good fit to the data in the whole x region, can be confronted with the expectations of the Frankfurt–Gurvich sum rule [5]. This constraint was not used in the fit. The new parameterization agrees with the theoretical expectations within 5%.

References

[1] See e.g. H. Abramowicz et al., International journal of Modern Physics, A8, (1993)1005 and references therein.
[2] V. N. Gribov, L. Ya. Pomeranchuk, Phys. Rev. Lett., 8 (1962) 343.
[3] A. Levy, Physics Letters, B404 (1997)369.
[4] See e.g. H. Abramowicz, Rapporteur talk in *Proceedings of the XXVIII International Conference on High Energy Physics*, Warsaw, Poland, 1996, and references therein.
[5] L.L. Frankfurt, E.G. Gurvich, Physics Letters B386, (1996) 379.

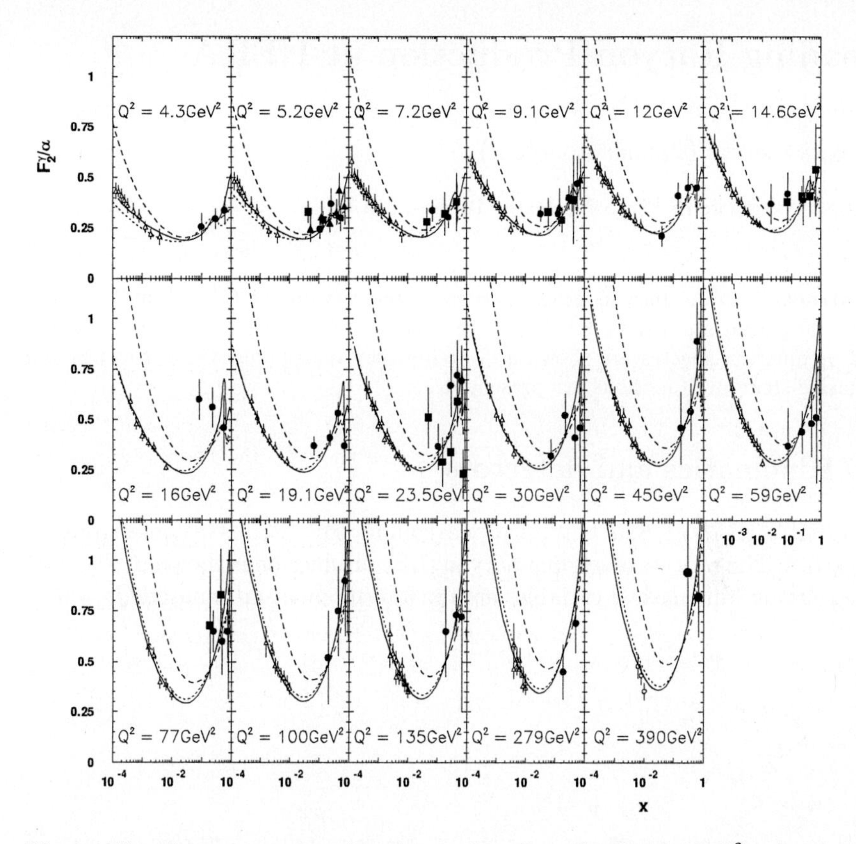

Fig. 1. The photon structure function as function of x for fixed Q^2 values as indicated in the figure. The full points are the direct measurements and the open triangles are those obtained from F_2^p through the Gribov factorization relation (1). The full line is the result of the present fit as described in the text, the dashed line is that of the GRV–LO parameterization, and the dotted line is that of SaS.

Leading Baryon Production at HERA

Amedeo Staiano (staiano@to.infn.it)

Istituto Nazionale di Fisica Nucleare, Torino-ITALY

Abstract. Leading baryon production in ep interactions at $Q^2 \sim 0$ and $Q^2 \geq 2$ GeV^2 has been studied at HERA by the H1 and ZEUS Collaborations. Energy and momentum spectra are shown and preliminary determinations of t-slopes and inclusive structure functions are presented.

1 Kinematics and detectors

At HERA (DESY) 820 GeV protons collide with 27.7 GeV electrons or positrons. The process of leading baryon (LB) production is shown in Fig. 1(A). The relevant kinematical variables are the four momentum squared $Q^2 = -q^2$

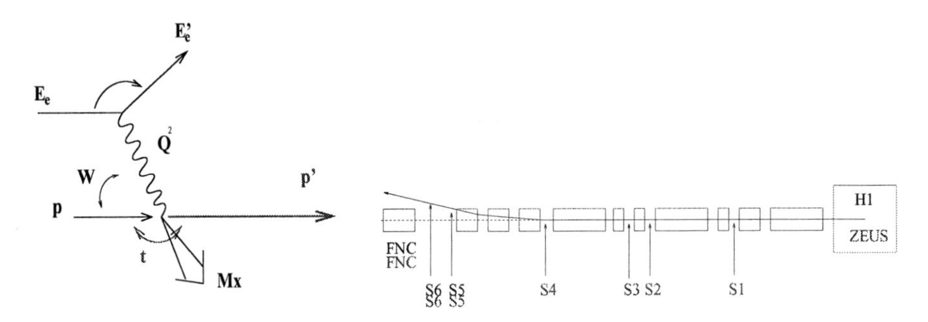

Fig. 1. A: Diagram for the reaction $ep \to NX$, B: H1 and ZEUS forward hadron detectors

exchanged at the lepton vertex , the four momenta p and p' of the incoming and outgoing proton, the c.m. energy in the γ-p system $W = (q+p)^2$ and x,y, the standard Bjorken variables. The kinematical variables associated to the proton vertex of Fig. 1(A) are $x_L = \frac{p'_z}{p_z}$, which represents the fraction of the proton beam momentum carried by the LB after the interaction, $t = (p-p')^2 \simeq -\left(\frac{m_p^2(1-x_L)^2}{x_L} + \frac{(p'_\perp)^2}{x_L}\right)$[1], which represents the four momentum squared transferred at the proton vertex, and $M_X^2 = W^2 \cdot (1-x_L)$, the invariant mass squared of the hadronic system when excluding the baryon.

[1] Here m_p denotes the proton mass and $p'_\perp$ is the transverse momentum of the scattered proton (neutron)

H1 and ZEUS are equipped to detect both forward produced neutrons and forward produced protons. The forward produced neutrons are detected by a calorimeter placed at 0° with respect to the incoming proton beam axis, and at a distance from the interaction point (IP) of 106 m for H1 and 107 m for ZEUS (see Fig.1(B)). The H1 *Forward Neutron Calorimeter* [1] is made of 75 hexagonal towers of 9.5 interaction lenghts (λ_I) thick consisting of layers of lead and 2 m long scintillating fibers. Two segmented planes of hodoscopes in front of the calorimeter are used to veto charged tracks. The ZEUS *Forward Neutron Calorimeter* [2] is a lead-scintillator sandwich calorimeter of 10 λ_I, with an energy resolution $\sigma(E_n) \sim 0.65\sqrt{E_n}$ and a geometrical acceptance larger than 25% for $\theta_{neutron} < 0.5$ mrad.

The proton tracking systems of both experiments consists of detectors installed in roman pots which can be moved close to the beam. The H1 system (*Forward Proton Spectrometer*) uses 1 mm ø scintillating fibers with two orientations, placed at 81 and 90 m from the IP. With a resolution of $\sim$ 0.1 mm in both projections, this detector has an acceptance > 50%, for $p_\perp <$ 200 MeV and $580 < E'_p < 740$ GeV.

The ZEUS *Leading Proton Spectrometer* [3] is made of 6 roman pot stations, located between 24 and 90 m from the IP, each containing 6 planes of silicon μstrip detectors with 3 orientations; the pitch is 115μm and and 81μm for 0° and $\pm 45^o$ orientations respectively. The resolution is $\frac{\Delta x_L}{x_L} \sim 3\%$ in the range $0.6 < x_L < 1.0$ for $0 < p_\perp^2 < 1$ GeV^2.

2 Properties of events with a leading baryon

The ratio of the number of events with a leading proton or a leading neutron in Deep Inelastic Scattering (DIS, $Q^2 > 4$ GeV^2 (ZEUS), $Q^2 > 5$ GeV^2 (H1)) to the total number of DIS events has been measured as a function of various kinematical variables by both experiments and found to be constant (see for example Fig. 2 for the H1 FNC results). A comparison of leading baryon energy spectra for different kinematic regimes is shown in Fig. 3. In Fig. 3(A) the neutron energy spectrum is shown (not corrected for acceptance), for DIS, photoproduction ($Q^2 \sim 0$), and *beam-gas* events, that is events in which the proton interacts with the residual molecules of air inside the beampipe. The three spectra show the same behaviour, indicating that the leading neutron production mechanism does not depend on Q^2 nor on the projectile. The independence of the leading proton production on Q^2 is also observed in the ZEUS data shown in Fig 3(B). ZEUS has also determined the size of the diffractive and non-diffractive contributions to leading baryon production [4]. An analysis based on an optimized cut on the pseudorapidity $\eta = -ln\left(tg\frac{\theta}{2}\right)$ [5], [2] shows that for values of $x_L < 0.9$, only ~6% of the events with a

[2] An event is accepted as diffractive with an efficiency > 99% if either the pseudorapidity of the most forward energy deposition in the central detector is less than

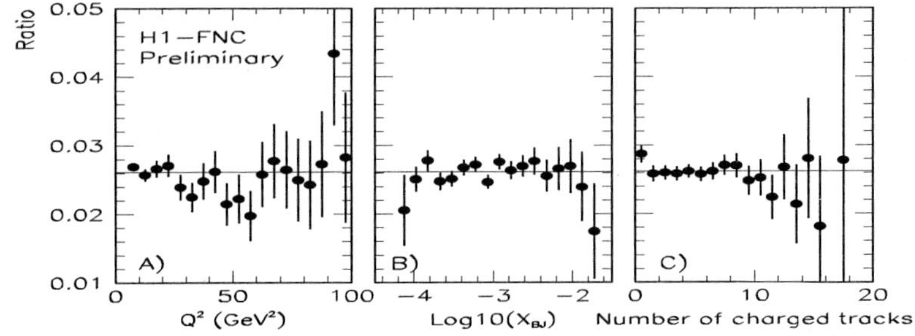

Fig. 2. H1 1996 Preliminary data: observed ratio of events with a neutron with $E_n \geq 100$ GeV relative to all DIS events, as a function of Q^2, x and the number of charged tracks.

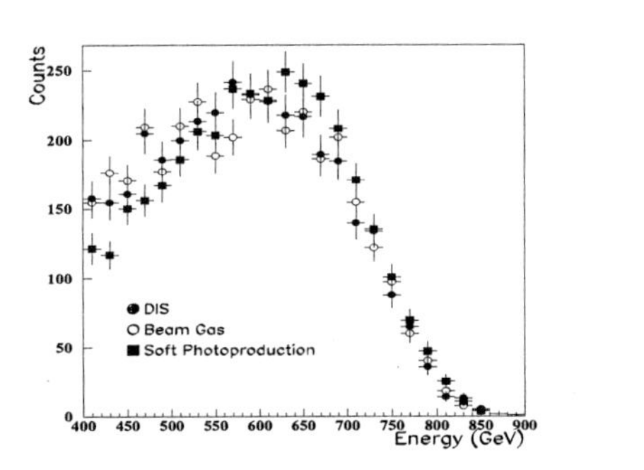

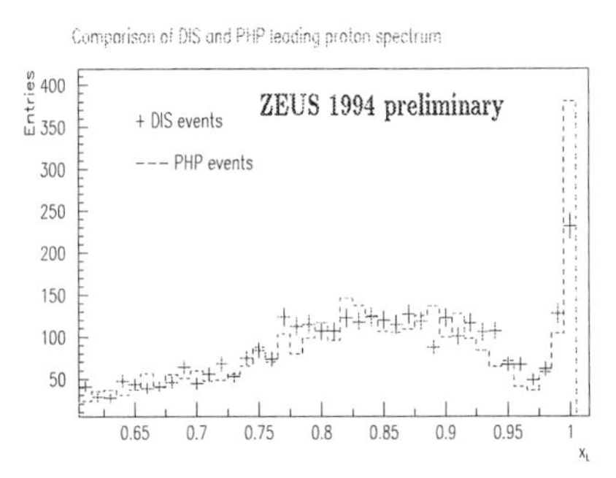

Fig. 3. ZEUS 1995 Preliminary data: A: Leading proton spectra in DIS and Photoproduction, B: Leading neutron spectra in DIS, photoproduction and beam-gas events.

leading baryon are produced diffractively. Moreover the amount of diffractive contribution is compatible for both leading proton and leading neutron events.

3 Measurement of the t-slope and F_2^{LP}

ZEUS has studied the t dependence of events with leading baryons as a function of x_L. In each x_L bin, the t distribution is well described by a function of the type $\frac{dN}{dt} \propto e^{bt}$. The fitted values of the t-slope b are shown in Fig. 4(A). The results for leading proton and leading neutron production are in good agreement suggesting a similar production mechanism for $x_L < 0.9$. The Q^2

1.8 or a pseudorapidity gap of at least 1.5 units is present between the forward edge of the detector and the most forward energy deposition between $2.5 < \eta < 4$.

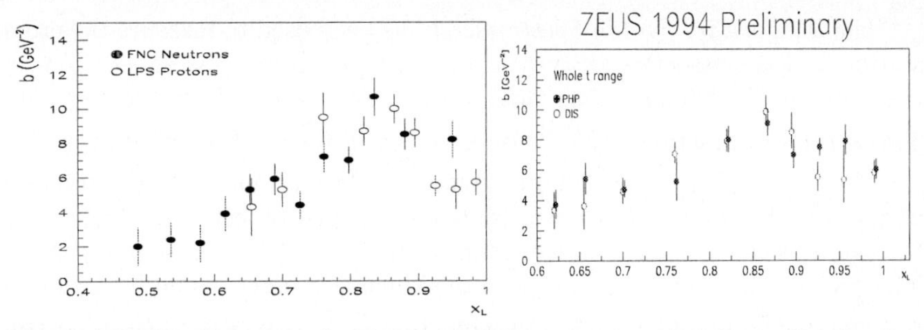

Fig. 4. ZEUS Preliminary: A: The slope of the neutron and proton t distributions in bins of x_L for DIS events. B: The slope of proton t distributions for DIS and photoproduction events. Errors are statistical only.

independence of the production mechanism can be checked by comparing the x_L dependence of b for DIS and photoproduction events, as shown in Fig. 4(B). The data have been compared with various models, but none can describe the full x_L spectrum [4]. *Pion-exchange*, in which the virtual photon scatters off a pion generated at the proton vertex via the process $p \to p\pi^0$ or $p \to n\pi^+$ is consistent with the data in the interval $0.7 < x_L < 0.9$, while diffractive models describe the leading proton data in the ranges $x_L < 0.7$ and $0.9 < x_L < 0.95$ for double dissociation and in the range $x_L > 0.95$ for single dissociation.

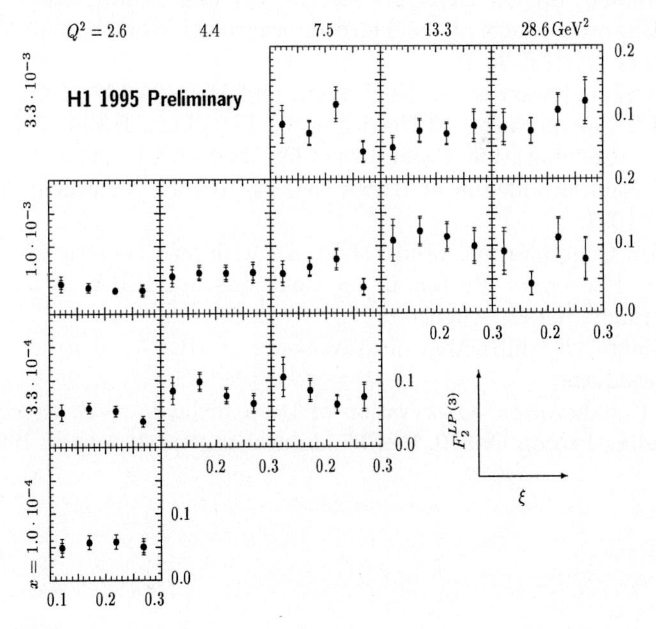

Fig. 5. $F_2^{LP(3)}(x, Q^2, \xi)$ in bins of x and Q^2 as function of ξ.

The H1 *Forward Proton Spectrometer* has been used to measure the cross section and the associated structure function for the process $ep \to e'p'X$ in the range $p_\perp < 200 MeV$, 580 GeV $\leq E'_p \leq$ 740 GeV and Q^2 >2 GeV^2 [6]. This structure function, $F_2^{LP(4)}$, is expressed in terms of 4 variables, t, Q^2, $\xi = \frac{q\cdot(p-p')}{q\cdot p}$, which in the infinite momentum frame represents the fraction of the proton momentum carried by the exchanged object, and $\beta = \frac{x}{\xi}$, the fraction of the exchanged object momentum carried by the struck parton. The exchanged object, depending on the kinematic range, could be a pion, a reggeon or a pomeron. The structure function $F_2^{LP(4)}$ reduces to $F_2^{LP(3)}$ when integrating over the unmeasured quantity t, as shown in Fig. 5. Following meson exchange models one can test a factorization hypothesis of the form $F_2^{LP(3)} = f_{\frac{\pi}{p}}(\xi) \cdot F_2^{\pi}(\beta, Q^2)$, by using a parametrization for the exchanged hadron structure function, with parameters extracted from a fit to the data. The measured value of the $\frac{\chi^2}{df}$ corresponds to a confidence level of 10% for the factorization hypothesis.

References

[1] Leading Neutron Production in Deep Inelastic Scattering at HERA, H1 collaboration, N-378, HEP97, Jerusalem, Israel, August 1997.

[2] Design and Test of a forward Neutron Calorimeter for the ZEUS Experiment, S.Bhadra et al., hep-ex/9701015, August 1997.

[3] A.Staiano, Silicon Detectors for the Leading Proton Spectrometer of ZEUS, proceedings of the Third International Workshop on Vertex Detectors, IUHEE-95-1;
K.O'Shaughnessy et al., Nucl. Instr. and Meth. **A342** (1994) 260;
ZEUS Collaboration; M.Derrick et al., Phys.Lett. **B384** (1996) 388.

[4] ZEUS Collaboration, Properties of Events with a Leading Proton in DIS and Photoproduction at HERA, N-644, HEP97, Jerusalem, Israel, August 1997.
ZEUS Collaboration, Study of Events with an Energetic Forward Neutron, Photon or Proton in ep Collisions at HERA, N-641, HEP97, Jerusalem, Israel, August 1997.

[5] A.Mehta, F_2 diffractive measurements at HERA, abstract 309, these proceedings.

[6] H1 Collaboration, Observation of Deep Inelastic *ep* Scattering with a Leading Proton, N-379, HEP97, Jerusalem, Israel, August 1997.

Rapidity Gaps and BFKL Tests at D

Anna Goussiou (goussiou@fnal.gov) (for the DØ Collaboration)

SUNY at Stony Brook, Stony Brook, NY 11794, USA

Abstract. Studies of hard color singlet exchange processes and BFKL tests in $p\bar{p}$ collisions are presented.

1 Hard Color Singlet Exchange

A signature for dijet production via hard color-singlet exchange is a rapidity gap (no particles in a region of rapidity) between the dijets. Hard color-singlet exchange has been observed at both the Tevatron [1] and HERA [2], at a rate of 1% and 10%, respectively. We present new measurements by the DØ Collaboration of the fraction of color-singlet exchange in dijet events as a function of dijet transverse energy (E_T), dijet pseudorapidity separation ($\Delta\eta$), and proton-antiproton center-of-mass energy ($\sqrt{s}$), which probe the color-singlet dynamics and its coupling to quarks and gluons. A color singlet that couples more strongly to quarks (gluons) is expected to produce a higher (lower) color-singlet fraction with increasing proportion of initial-state quark processes. The latter is achieved by decreasing $\sqrt{s}$ or increasing the dijet E_T or $\Delta\eta$ (i.e. increasing Bjorken x). The models for color singlet exchange that would result in production of rapidity gaps between jets are described in [3].

The color-singlet fraction calculated from models for the exchange of a hard color singlet (i.e. a two-gluon singlet or U(1) boson) includes the probability that the rapidity gap is not contaminated by particles from spectator interactions. This probability ($\sim$ 10%) is expected to be independent of the flavor of the initiating partons in the hard scattering and to have a weak logarithmic dependence on the proton-antiproton center-of-mass energy [4, 5].

Data from two center-of-mass proton-antiproton energies of $\sqrt{s} = 1800$ GeV and 630 GeV, taken during special low luminosity Tevatron running conditions, are used in this analysis. Jets are defined as E_T sums in a cone of radius $R = \sqrt{\Delta\eta^2 + \Delta\phi^2} = 0.7$ using calorimeter cell information. For the comparison of the 630 and 1800 GeV samples, two opposite-side jets with $E_T > 12$ GeV and $|\eta| > 1.9$ are required. For the measurement of the color-singlet fraction as a function of dijet E_T, three opposite-side dijet samples at 1800 GeV are used, with jet $E_T > 15$, 25 and 30 GeV, respectively, and jet $|\eta| > 1.9$. For the measurement of the color-singlet fraction as a function of dijet $\Delta\eta$, an opposite-side dijet sample at 1800 GeV is used, with jet $E_T > 30$ GeV and jet $|\eta| > 1.7$.

The particle multiplicity in the central rapidity region is approximated by the multiplicity, n_{cal}, of localized (0.1×0.1 in $\Delta\eta \times \Delta\phi$ space) transverse energy deposits above 200 MeV in the electromagnetic (EM) part of the calorimeter within $|\eta| < 1$, and by the track multiplicity in the central tracking chamber, n_{trk}. For the comparison of the 630 and 1800 data, the leading edge of each n_{cal} distribution is fitted using a single negative binomial distribution (NBD). The fraction of rapidity gap events is calculated from the excess of events over the fit in the first two bins ($n_{cal} = 0$ or 1) divided by the total number of entries. It is equal to $1.6 \pm 0.2\%$(stat.) at $\sqrt{s} = 630$ GeV and $0.6 \pm 0.1\%$(stat.) at $\sqrt{s} = 1800$ GeV. The ratio R of the rapidity gap fractions at 630 and 1800 GeV is equal to $2.6 \pm 0.6\%$(stat.). The color-singlet fraction as a function of E_T and $\Delta\eta$ is measured from the ($n_{cal} = n_{trk} = 0$) multiplicity bin, to avoid uncertainties in the color-exchange background subtraction. Figure 1(a) shows the color-singlet fraction at 1800 GeV binned as a function of the second leading jet E_T and plotted at the average dijet E_T for that bin. Figure 1(b) shows the color-singlet fraction at 1800 GeV as a function of dijet $\Delta\eta$ for the high E_T sample.

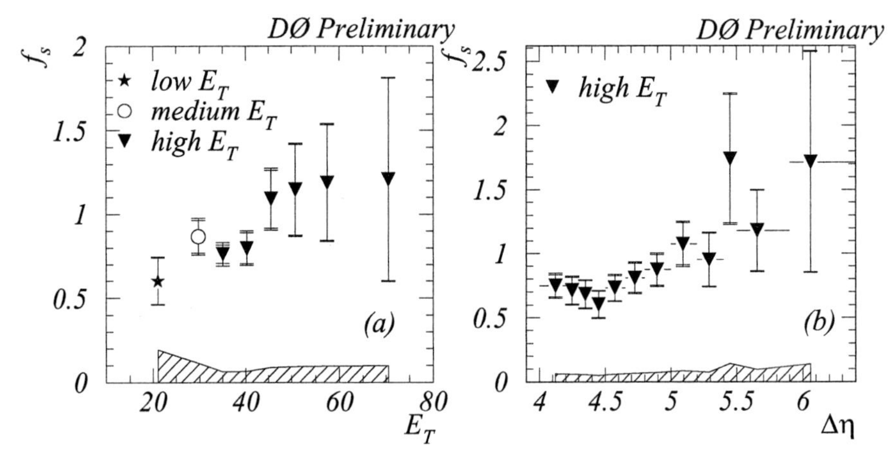

Fig. 1. The color-singlet fraction: (a) as a function of the second leading jet E_T plotted at the average dijet E_T for that bin; (b) as a function of $\Delta\eta$ between the leading dijets. Statistical (inner error bars) and statistical plus systematic errors (outer error bars) are shown. The error band at the bottom shows the normalization uncertainty (not included in the plotted systematic error).

The measured color-singlet fraction shows a slight rise as a function of dijet E_T and $\Delta\eta$, and the ratio of the color-singlet fraction at 630 and 1800 GeV is greater than one. Qualitatively, these results are consistent with a color-singlet fraction that rises with increasing initial-state quark content and thus appear to be inconsistent with current two-gluon models. However, these models may not properly take into account higher-order dynamical effects that must be understood in order to make direct comparison with the data.

2 Azimuthal Decorrelation in Dijet Production

An indirect signature for BFKL dynamics [6] in inclusive dijet production is the observation of azimuthal angle decorrelation with increasing $\Delta\eta$ between the two jets [7, 8]. New results from the 1994-95 Tevatron run with increased statistics, symmetric cuts on the two jet transverse energies $E_{T_{1,2}}$, and extended $\Delta\eta$ range are presented here.

In order to identify decorrelation in the azimuthal angle ϕ of the dijet system, we look for a decrease of the average of the $\cos(\pi - \Delta\phi)$ distribution (equal to unity in LO) with increasing $\Delta\eta$. Data from special low luminosity Tevatron runs at $\sqrt{s} = 1800$ are used. We require at least two jets with $E_T > 20$ GeV and order them in rapidity, i.e. $\eta_1 > \eta_i > \eta_2$, where η_1 and η_2 refer to the most forward and most backward jet, respectively. The absolute rapidity boost, defined as $|\bar{\eta}| = |(\eta_1 + \eta_2)/2|$, is required to be less than 0.5. We define $\Delta\eta = \eta_1 - \eta_2$ and $\Delta\phi = \phi_1 - \phi_2$. The $\cos(\pi - \Delta\phi)$ distribution is plotted in Fig. 2. Also shown in the plot are the theoretical predictions from the HERWIG parton shower Monte Carlo [9], from a calculation based on BFKL resummation in the Leading Logarithmic Approximation (LLA) [7], and from recent BFKL-based Monte Carlos [10]. The latter impose energy-momentum conservation and/or include the effect of the running of α_s; these characteristics are not present in LLA BFKL.

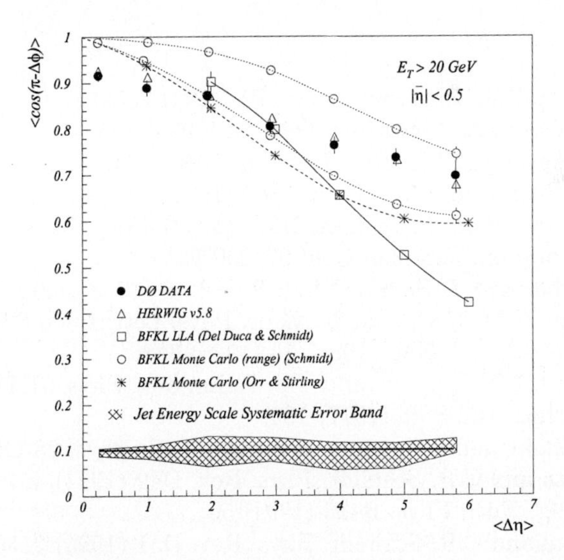

Fig. 2. The average value of $\cos(\pi - \Delta\phi)$ plotted vs. $\Delta\eta$, for the DØ data, HERWIG, the BFKL LLA calculation of Del Duca and Schmidt, and the BFKL Monte Carlos. The data errors are statistical and uncorrelated systematic added in quadrature. The shaded band represents the error due to the uncertainty in the jet energy scale. The theoretical errors are statistical only.

The data show a decorrelation effect in the azimuthal separation of the dijet system with increasing rapidity interval. The azimuthal decorrelation in HERWIG is similar to the one observed in the data. The LLA resummation calculation predicts stronger decorrelation effects than seen in the data. It would be interesting, however, to see the effect of the next-to-leading logarithmic terms in resummation, currently being calculated. The BFKL Monte Carlos, which include some higher order effects, improve the agreement between the resummation techniques and the data.

3 Conclusions

We have measured the fraction of color-singlet exchange in dijet events in $p\bar{p}$ collisions as a function of dijet transverse energy, dijet pseudorapidity separation, and proton-antiproton center-of-mass energy. The results are consistent with a color-singlet fraction that rises with increasing initial-state quark content. We have studied the azimuthal decorrelation in inclusive dijet production as a function of the dijet pseudorapidity separation. The results agree with predictions from a parton shower Monte Carlo program. The current analytical calculations based on LLA BFKL resummation are in disagreement with the data. BFKL-based Monte Carlos, including higher order effects, predict an azimuthal decorrelation closer to the data than the LLA calculations.

References

[1] DØ Collaboration, Phys. Rev. Lett. 72 (1994) 2332.
CDF Collaboration, Phys. Rev. Lett. 74 (1995) 855.
DØ Collaboration, Phys. Rev. Lett. 76 (1996) 734.

[2] ZEUS Collaboration, Phys. Lett. B369 (1996) 55.
ZEUS Collaboration, Phys. Lett. B315 (1993) 481.

[3] DØ Collaboration, Fermilab-Conf-97/250-E.

[4] R. S. Fletcher and T. Stelzer, Phys. Rev. D48 (1993) 5162.

[5] E. Gotsman, E.M. Levin and U. Maor, Phys. Lett. B309 (1993) 199.

[6] L.N. Lipatov, Sov. J. Nucl. Phys. 23 (1976) 338.
E.A. Kuraev, L.N. Lipatov, and V.S. Fadin, Sov. Phys. JETP 44 (1976) 443; Sov. Phys. JETP 45 (1977) 199.
Ya.Ya. Balitsky and L.N. Lipatov, Sov. J. Nucl. Phys. 28 (1978) 822.

[7] V. Del Duca and C.R. Schmidt, Phys. Rev. D49 (1994) 4510.
W.J. Stirling, Nucl. Phys. B423 (1994) 56.
V. Del Duca and C.R. Schmidt, Phys. Rev. D51 (1995) 2150.

[8] DØ Collaboration, Phys. Rev. Lett. 77 (1996) 595.

[9] G. Marchesini *et al.*, Comp. Phys. Comm. 67 (1992) 465.

[10] C.R. Schmidt, Phys. Rev. Lett. 78 (1997) 4531.
L.H. Orr and W.J. Stirling, Phys. Rev. D56 (1997) 5875.

Pomeron and Jet Events at HERA

Luis Labarga (labarga@hepdc1.ft.uam.es)

University Autónoma Madrid, Spain

Abstract. We study two and three jet events with a large rapidity gap at HERA. As a minimal model we take a scalar Pomeron ($I\!P$) with a pointlike coupling to quarks and gluons. The only degrees of freedom are the coupling constants of the Pomeron to the quarks, $g_{I\!Pqq}$, and the gluons, $g_{I\!Pgg}$. [1]

Recently the so-called 'large rapidity gap' (LRG) events observed in DIS *ep* interactions at HERA have caused new excitement about the Pomeron hypothesis. The outcoming events can be classified by the number of jets they show. This is similar to $e^-e^+ \to hadrons$. This number depends on the type of hard interaction between the Photon and the Pomeron. With our model, and because of its simplicity, we aim to have a useful tool to further understanding the Pomeron nature.

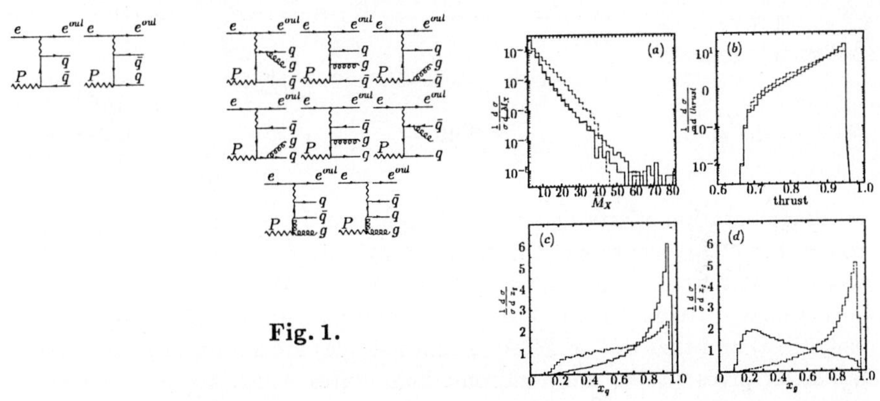

Fig. 1.

2-jet events are produced via the two diagrams shown in Fig. 1_{left} where the Pomeron couples to a $q\bar{q}$ pair. 3-jet events are produced in two ways. The first type of process occurs when the Pomeron couples to a $q\bar{q}$ pair and a gluon is emitted from one of the quark lines (first six diagrams of Fig. 1_{mid}). In the second process the Pomeron couples to a gluon pair, with one of the gluons splitting into a $q\bar{q}$ pair (last two diagrams of Fig. 1_{mid}). In all the cases the Pomeron is coupled to the proton via the Donnachie-Landshoff mechanism.

In Fig. $1_{right-a}$ we show $\frac{1}{\sigma}\frac{d\,\sigma}{d\,M_X}$ for the three reactions considered in the model. M_X is the invariant mass $\gamma^* I\!P$, being X the final state hadronic

[1] This talk is a summary of the work done in collaboration with J. Vermasseren, F. Barreiro and F.J. Ynduráin; preprint DESY 97-031, accepted by Phys. Lett. for publication. I refer the reader to that paper for the references used in this work

system. The three-jet processes through the $I\!\!Pgg$ coupling tend to give rise to larger masses than those coming from bremsstrahlung. In Fig. $1_{right-b}$ we show the thrust distribution for 3-parton final states in the $\gamma - I\!\!P$ centre of mass frame. Again, the tail in the thrust distribution associated with final states coming from a $I\!\!Pgg$ coupling is harder than that associated to bremsstrahlung from the $I\!\!Pq\bar{q}$ coupling, solid line. To deepen our insight into the roles played by gluons in the two different 3-parton processes we are considering, it is instructive to look at the parton fractional momenta, defined by $x_q = 2E_q/\sum E_{hadr}$ and $x_g = 2E_g/\sum E_{hadr}$ and computed in the rest system of X (Figures $1_{right-c}$ and $1_{right-d}$ respectively). A striking feature is that, while bremsstrahlung gluons are soft partons as expected, those coming from $I\!\!P$ splitting are much harder.

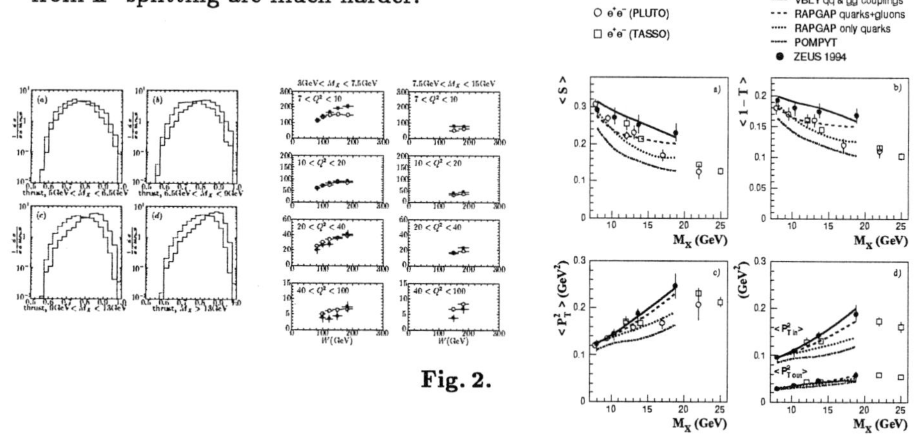

Fig. 2.

Fragmentation effects do not mask significantly the properties discussed above. Fig. 2_{right} shows the the thrust distributions in the hadronic centre of mass, after fragmentation, for various ranges in M_X. While at low masses both thrust distributions, i.e. the one stemming from the $I\!\!P \to q\bar{q}$ reaction (solid line) and that due to $I\!\!P \to gg$ (dashed line) are similar, at high masses the latter gives rise to multihadronic final states which are more spherical than the former.

Fig. 2_{mid} shows a comparison with the diffractive cross sections measured by the ZEUS Collaboration (solid dots), $d\sigma^{diff}(\gamma^* - p \to Xp)/dM_X)$, for various Q^2 and M_X intervals as a function of W, the photon proton invariant mass. $g_{I\!\!Pqq}$ is taken to be 0.6 and the relative fraction of gluons is taken 65% from a study of global event shapes in LRG DIS events done by the ZEUS Collaboration. For this large gluon fraction our model predicts important broadening effects in the hadronic final states. Fig. 2_{left} shows a comparison between the predictions from our model (VBLY in the figure) and ZEUS data for the mean sphericity, mean 1-thrust, mean squared transverse momenta w.r.t. the sphericity axis and its components in and out of the event plane, as a function of M_X. Our model describes the data fairly well.

I would like to thank the organizers for this superb conference.

QCD Dipole Description of Total and Diffractive Proton Structure Functions

Ch. Royon (royon@hep.saclay.cea.fr)

C.E.A./Saclay, DAPNIA-SPP, F-91191 Gif-sur-Yvette Cedex, France

Abstract. It is argued that the QCD dipole picture allows to build an unified theoretical description -based on BFKL dynamics- of the total and diffractive nucleon structure functions. This description is in qualitative agreement with the present collection of data obtained by the H1 collaboration.

1 BFKL dynamics in the framework of the QCD dipole model and F_2 fit

The quest for an unifying picture of total and diffractive structure functions based on a perturbative QCD framework is a challenge. The interest of using the QCD dipole approach [1] for deep-inelastic structure functions is to deal with an unified approach based on the BFKL resummation properties of perturbative QCD [2]. Indeed, starting from an unique non-perturbative input in terms of a primordial proton distribution of dipoles at low scale Q_0, it is possible to compute the theoretical predictions [3] for the (transverse and longitudinal) quark and the gluon distributions as functions of x and Q^2. In the same framework it is also possible to compute the two components of the dipole-model predictions for hard diffractive structure functions, namely the inelastic component and the quasi-elastic one [4].

To obtain the proton structure function F_2, we use the k_T factorisation theorem [5], valid at high energy (small x), to factorise the $(\gamma \, g(k) \rightarrow q \, \bar{q})$ cross section and the unintegrated gluon distribution of an onium state which contains the physics of the BFKL pomeron [1]. One can show that

$$F_2 = \mathcal{N} a^{1/2} e^{(\alpha_P - 1) \ln \frac{c}{x}} \frac{Q}{Q_0} e^{-\frac{a}{2} \ln^2 \frac{Q}{Q_0}} \tag{1}$$

where $\alpha_P - 1 = \frac{4 \bar{\alpha} N_C \ln 2}{\pi}$ and $a = \left(\frac{\bar{\alpha} N_c}{\pi} 7 \zeta(3) \ln \frac{c}{x} \right)^{-1}$. The detailed calculations can be found in [3]. It can be shown that the approximations are valid when $\ln Q/Q_0 / \ln(c/x) << 1$, that is in the kinematic domain of small x, and moderate Q/Q_0. The free parameters for the fit of the H1 data are $\mathcal{N}$, α_P, Q_0, and c. Finally, we get parameter free predictions for R, the ratio of the longitudinal and transverse structure functions, and the gluon density in the proton [3].

In order to test the accuracy of the F_2 parametrisation obtained in formula (1), a fit using the recently published data from the H1 experiment [6] has been performed [3]. We have only used the points with $Q^2 \leq 150GeV^2$ to remain in a reasonable domain of validity of the QCD dipole model. The result of the fit is given in detail in reference [3]. The χ^2 is 88.7 for 130 points, and the values of the parameters are $Q_0 = 0.522GeV$, $\mathcal{N} = 0.059$, and $c = 1.750$, while $\Delta_P = 0.282$. Commenting on the parameters, let us note that the effective coupling constant extracted using (3) from Δ_P is $\alpha = 0.11$, close to $\alpha(M_Z)$ used in the H1 QCD fit. It is an acceptable value for the small fixed value of the coupling constant required by the BFKL framework. The running of the coupling constant is not taken into account in the present BFKL scheme. This could explain the rather low value of the effective Δ_P which is expected to be reduced by the next to leading corrections.

2 Diffractive structure functions

The success of the dipole model applied to the proton structure function motivates its extension to the investigations to other inclusive processes, in particular to diffractive dissociation. We can distinguish two different components:
- the "elastic" term which represents the elastic scattering of the onium on the target proton;
- the "triple-pomeron" term which represents the sum of all dipole-dipole interactions (it is dominant at large masses of the excited system).
Let us describe in more detail each of the two components. The "triple-pomeron" term dominates at low β, where $\beta = x/x_P$, x_P is the proton momentum fraction carried by the "pomeron" [4]. This component, integrated over t, the momentum transfer, is factorisable in a part depending only on x_p (flux factor) and on a part depending only on β and Q^2 ("pomeron" structure function F) [4]:

$$F_2^{D(3),I} \simeq x_P^{-1-2\Delta_P} \left(\frac{2a}{\pi}\right)^3 \left(\frac{Q}{Q_0}\right) (\beta)^{-\Delta_P} \left(\frac{2a\,(\beta)}{\pi}\right)^{1/2} \exp\left\{-\frac{1}{2}a\,(\beta)\ln^2\left(\frac{Q}{Q_0}\right)\right\} \tag{2}$$

The important point to notice is that $a(x_p)$ is proportional of $ln 1/x_p$. The effective exponent (the slope of $ln F_2^D$ in $ln x_p$) is found to be dependent on x_p because of the term in $ln^3(x_p)$ coming from a, and is sizeably smaller than the BFKL exponent. This is why we can describe an apparently soft behaviour (a small exponent in x_p) with the BFKL equation, which predicts a hard behaviour (the exponent in x_p is close to 0.35). This is due to the fact that the effective exponent is smaller then the real one. It should be also noticed that the structure function F is directly proportional to the proton structure function.

The elastic component behaves quite differently [4]. The explicit formulae used in the prediction can be found in reference [8]. First it dominates at

Fig. 1. Prediction on F_2^D and comparison with the H1 measurement. The full line is the sum of the two components, the dashed line the inelastic one (which dominates at low β), and the dotted line the elastic one (which dominates at high β).

$\beta \sim 1$. It is also factorisable like the inelastic component, but with a different flux factor, which means that the sum of the two components will not be factorisable. This means that in this model, factorisation breaking is coming from the fact that we sum up two factorisable components with different flux factors. The β dependence is quite flat at large β, due to the interplay between the longitudinal and transverse components. The sum remains almost independent of β, whereas the ratio $R = F_L/F_T$ is strongly β dependent. Once more, a R measurement in diffractive processes will be an interesting way to distinguish the different models, as the dipole model predicts different β and Q^2 behaviours.

The sum of the two components shown in the figure describes quite well the H1 data [7]. There is no further fit of the data as we chose to take the different parameters (Q_0, α_P, c) from the F_2 fit. In this study, the normalisation is let free. The most striking point is that we describe quite well the factorisation breaking due to the resummation of the two components at low and large β [8]. The full line is the sum of the two components, the dashed line the inelastic one (which dominates at low β), and the dotted line the elastic one (which dominates at high β).

References

[1] A.H.Mueller and B.Patel, *Nucl. Phys.* **B425** (1994) 471., A.H.Mueller, *Nucl. Phys.* **B437** (1995) 107., A.H.Mueller, *Nucl. Phys.* **B415** (1994) 373, N.N.Nikolaev and B.G.Zakharov, *Zeit. für. Phys.* **C49** (1991) 607.

[2] V.S.Fadin, E.A.Kuraev, L.N.Lipatov *Phys. Lett.* **B60** (1975) 50; I.I.Balitsky and L.N.Lipatov, *Sov.J.Nucl.Phys.* **28** (1978) 822.

[3] H.Navelet, R.Peschanski, Ch.Royon, and S.Wallon, *Phys.Lett.*, **B385**, (1996) 357, H.Navelet, R.Peschanski, Ch.Royon, *Phys.Lett.*, **B366**, (1996) 329.

[4] A.Bialas, R.Peschanski, *Phys. Lett.* **B378** (1996) 302, A.Bialas, R.Peschanski, *Phys.Lett.* **B387** (1996) 405

[5] S.Catani, M.Ciafaloni and Hautmann, *Phys. Lett.* **B242** (1990) 97; *Nucl. Phys.* **B366** (1991) 135; J.C.Collins and R.K.Ellis, *Nucl. Phys.* **B360** (1991) 3; S.Catani and Hautmann, *Phys. Lett.* **B315** (1993) 157; *Nucl. Phys.* **B427** (1994) 475

[6] H1 coll., *Nucl.Phys.* **B470** (1996) 3

[7] H1 coll., "Inclusive measurement of Diffractive Deep-Inelastic ep Scattering", DESY 97-158, August 1997.

[8] A.Bialas, R.Peschanski, C.Royon, hep/ph97-12216, and references therein

F_2^c,F_L and Gluon Determination at HERA

Alexandros Prinias* on behalf of the H1 and ZEUS collaborations (prinias@mail.desy.de)

Imperial College of Science, Technology and Medicine, Blackett Laboratory, Prince Consort Road, London SW7 2BZ, United Kingdom

Abstract. A short overview of gluon and charm measurements from deep inelastic scattering data at HERA is presented.

1 Introduction

During the last five years HERA has provided the H1 and ZEUS Collaborations with high statistics deep inelastic scattering (DIS) data. Measurements of the proton structure function F_2 are presently extended over five orders of magnitude in Q^2 and Bjorken x, allowing for the first time an overlap with measurements from fixed target experiments at $Q^2 < 100$ GeV^2 and $x > 10^{-2}$. For the definitions of the DIS kinematic variables and a discussion of the latest measurements we refer to [1].

When plotted versus Q^2 at various values of x, F_2 exhibits strong scaling violations with decreasing x. In leading order (LO) Quantum Chromodynamics (QCD) these scaling violations are due to the presence of gluon radiation from the initial or the final quark line or gluon splitting into a $q\bar{q}$ pair. Whereas F_2 measures directly the quark content of the proton, the gluon is indirectly extracted by fits to the measured F_2 data assuming the Dokshitzer-Gribov-Lipatov-Altarelli-Parisi (DGLAP) model for the QCD evolution.

A large part of the DIS cross section, typically of $\sim 20\%$, in the low-x HERA regime is due to charm production. In LO perturbative QCD charm is produced in $c\bar{c}$ pairs *via* boson-gluon fusion (BGF). Recent next-to-leading order (NLO) theoretical calculations indicate that charm can be predicted, given the gluon and light flavour densities, and there is no need to be put in as an arbitrary density in the F_2 fits. Therefore, a measurement of the charm contribution $F_2^{c\bar{c}}$ to the inclusive proton structure function F_2 gives an excellent consistency check of the whole picture.

2 Gluon from NLO F_2 fits

F_2 was split into two parts, one coming from the light flavours and one coming from the charm

$$F_2 = F_2^{uds} + F_2^c \ . \tag{1}$$

The b quark contribution, heavily suppressed to typical values of $\sim 1\%$ by mass and charge effects, was neglected.

The three light flavours (or the valence and total sea) were parametrized at a low starting scale of $Q_0^2 = 1$ GeV^2 (H1) / 7 GeV^2 (ZEUS) as functions of x, e.g. (H1)

$$\begin{aligned} xg(x) &= A_g x^{B_g}(1-x)^{C_g} \\ xu_v(x) &= A_u x^{B_u}(1-x)^{C_u}(1 + D_u x + E_u\sqrt{x}) \\ xd_v(x) &= A_d x^{B_d}(1-x)^{C_d}(1 + D_d x + E_d\sqrt{x}) \\ xS(x) &= A_S x^{B_S}(1-x)^{C_S}(1 + D_S x + E_S\sqrt{x}) \end{aligned} \tag{2}$$

where A_i are normalisation factors, and the x^{B_i} and $(1-x)^{C_i}$ terms give the singular and the valence-type behaviour of the partons densities respectively. The s quark was assumed to be 20-25% of the total sea at this initial scale. The normalisations for the quarks were

---

* PPARC Post-doctoral Fellow

determined by the quark counting rules and for the gluon by the momentum sum rule. The total normalisation was left free. The value of $\alpha_s(M_Z^2) = 0.118(0.113)$ for H1(ZEUS) was given as an input to the fit.

The three light flavours were evolved to a higher scale using the next-to-leading order (NLO) splitting and coefficient functions in the $\overline{MS}$ factorisation and renormalisation scheme to give the F_2^{uds} part of eq. 1. F_2^c was calculated analytically at NLO using the procedure of [2], added to the light flavours and the resulting F_2 was fitted to the experimental data. H1 used their nominal vertex data from the 1994, 1995 and 1996 running periods and that part of the 1995 run where the interaction point was shifted allowing to access smaller scattering angles of the final state positron and, thus, smaller $Q^2 \simeq 1.5\text{GeV}^2$. ZEUS used their 1994 nominal and shifted vertex data as well as events where the initial positron has radiated a hard photon before interacting with the proton, allowing for a smaller c.m. energy and thus accessing $Q^2 \simeq 1.5\text{GeV}^2$. In addition to their own data, H1 and ZEUS used proton and deuteron F_2 data from NMC and BCDMS, to pin down the high x region.

The result of the H1 fit can be found in [1] where F_2 is plotted versus x in various bins of Q^2. The NLO fit gives an excellent description of the data. The ZEUS fit is shown in Fig. 1 where now F_2 is plotted versus $\log_{10} Q^2$ in various bins of x allowing to observe the transition from the region of approximate scaling at $x \sim 0.1$ to the region of strong scaling violations at $x \sim 0.0001$. The solid fit line, representing the result for eq. 1, gives again an excellent description of the data. The F_2^{uds} part is shown as a dashed line. It is clear that the charm contribution is very significant, especially at small x.

The result for the gluon density is shown in Fig. 2. At the scale of $Q^2 = 20$ GeV^2 the results of the two experiments, darker band for ZEUS and lighter for H1, are in good agreement. The bands represent the error estimations including contributions from the experimental statistical and systematic errors and the uncertainties on α_s and the charm mass. The hatched band at higher x is the result from NMC. On the same plot two results of global fits to DIS and related data (e.g. prompt photon data) from the MRS [3] and CTEQ [4] collaborations are shown (solid and dashed line respectively). The GRV [5] prediction (dotted line) overshoots the ZEUS and H1 fits, but it should be noted that this can be partially explained by the lower value of α_s used for this prediction which is compensated by a higher gluon density especially at low x. Generally, the gluon is known with a total error of ~10% at $x \sim 10^{-4}$.

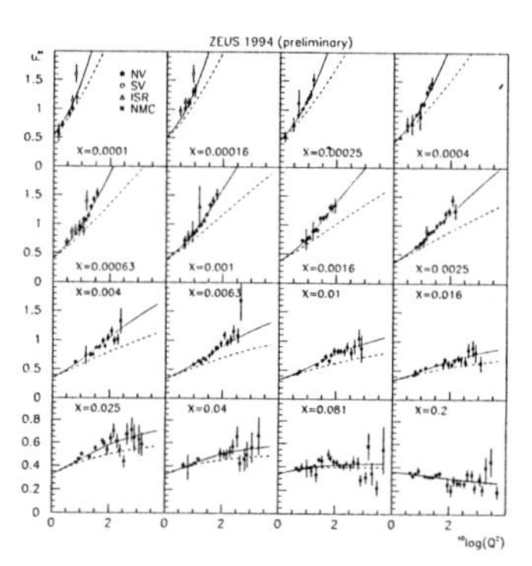

Fig. 1. The ZEUS NLO fit

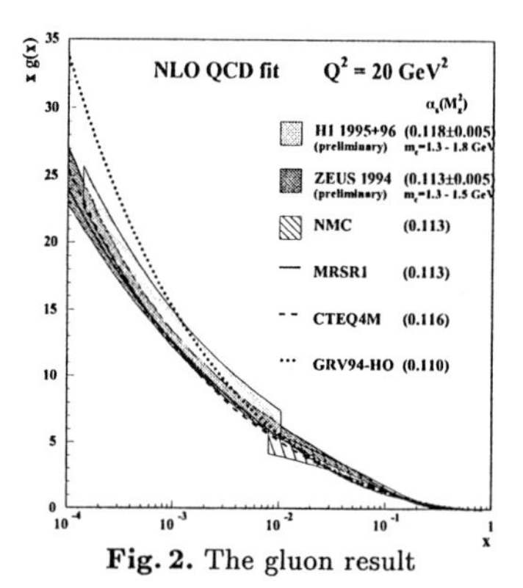

Fig. 2. The gluon result

3 Charm in DIS

In LO perturbative QCD charm is produced via the BGF process of Fig. 3. In its NLO

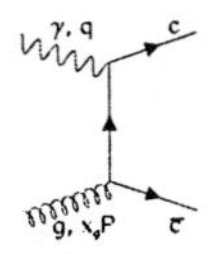

Fig. 3. The boson-gluon fusion process

extension, this approach was followed in the GRV analyses including the latest massive charm calculations [2] which actually predict a charm quark density given the gluon and light flavour densities. However, the inclusive F_2 behaviour can be equally well described by an arbitrary charm density $c(x, Q^2) = 0$ for $Q^2 \leq \mu_c^2$, evolved as a massless parton like the light flavours for $Q^2 > \mu_c^2$. The μ_c scale was chosen to fit the EMC data. It is however clear that the latter approach is not strictly correct. In the low x HERA regime BGF charm production is kinematically allowed even at very low Q^2 if the partonic centre of mass energy $\hat{s} = Q^2(1-z)/z \geq (2m_c)^2$, where $z = x/x_g$, x_g being the fraction of the proton momentum carried by the interacting gluon and m_c the charm mass. On the other hand, far from this energy threshold the charm is expected to behave like a massless parton. It is therefore very interesting to investigate both the onset of such a process and the transition between a heavy quark and a massless parton behaviour.

ZEUS and H1 have studied charm in DIS using D mesons in the final state. Current measurements are restricted in the region $1\ \mathrm{GeV}^2 \leq Q^2 \leq 600\ \mathrm{GeV}^2$ and $0.01 \leq y \leq 0.7$. The channels under investigation are $D^{*+} \to D^0\pi_s^+ \to (K^-\pi^+)\pi_s^+$ (+c.c.) and $D^0\pi_s^+ \to K^-\pi^+$ (+c.c.). It should be noted that the tracking detector acceptances limit the measurements in the region where $p_T(D) > 1.5$ GeV and $|\eta(D)| < 1.5$, where p_T is the transverse momentum with respect to the beam axis and $\eta = -\ln(\tan\frac{\theta}{2})$, the pseudorapidity of the D (θ being the polar angle w.r.t. the proton beam direction).

Preliminary results from the ZEUS analysis of the 1995 data on the cross sections for the production of D^* as functions of their p_T, η and the event Q^2 and hadronic invariant mass W for events in the foremen-tioned $Q^2, y, p_T(D^*), \eta(D^*)$ region are shown in Fig. 4. The cross sections are compared with the prediction of [2] (shaded band) as

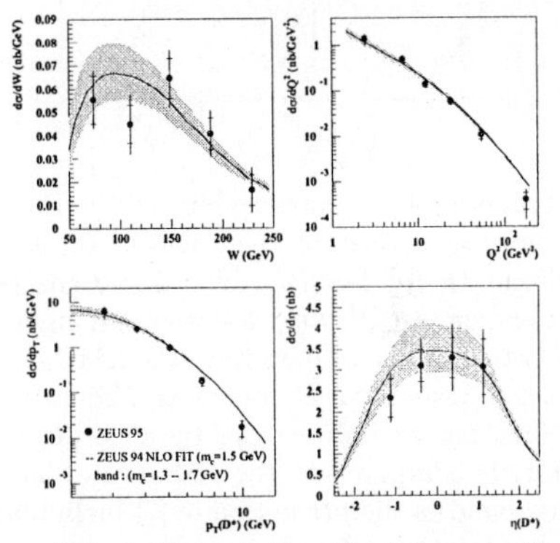

Fig. 4. Cross sections for D^* production

implemented in a 'Monte Carlo-like' program which allows charm to fragment to D^*. The Peterson fragmentation model with $\epsilon = 0.035$ was used. The input to this prediction is the gluon density and light flavours extracted from the ZEUS NLO fit to F_2 discussed in the previous section. The band width corresponds to the uncertainty on the charm mass. With the present experimental errors (quadratic sum of statistical and systematic) the theoretical predictions are in good agreement with the data in both shape and total normalisation.

By extrapolating the measured cross section to the whole $p_T(D^*), \eta(D^*)$ phase space and correcting for charm to D^* branching ratios the contribution $F_2^{c\bar{c}}$ of charm to the inclusive proton structure function F_2 is extracted. The result is shown in Fig. 5 where $F_2^{c\bar{c}}$ is plotted

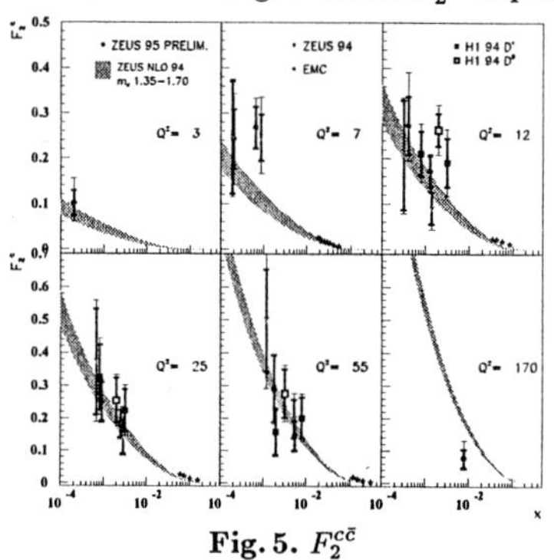

Fig. 5. $F_2^{c\bar{c}}$

as a function of x in various bins of Q^2. Data are shown from the 1994 analyses of H1 and ZEUS and the preliminary 1995 ZEUS analysis extending the Q^2 range to lower and higher Q^2. On the same plot data from the EMC collaboration are shown at higher x. The structure function at HERA rises by an order of magnitude compared to the EMC data. The shaded band is the prediction of [2] including the uncertainty on the charm mass. In general this prediction is in good agreement with the data of H1 and ZEUS.

In LO BGF the fraction of the proton momentum carried by the gluon, x_g, can be reconstructed from the virtual photon and one of the final state charm quarks, the momenta of which can be estimated from the scattered positron and the tagged D meson respectively. Although in higher orders the picture is not that simple, an estimator, x_g^{obs}, of the true x_g can always be reconstructed. A cross section for D production can then be measured in the restricted $p_T(D), \eta(D)$ region as a function of the x_g^{obs} observable quantity. Finally, using the Monte Carlo-like program of [2] to estimate the correlation between the observed and true x_g in NLO, the gluon momentum density $xg(x)$ can be extracted. The preliminary result of H1 from the 1995 data is shown in Fig. 6 as dots, compared to the H1 result

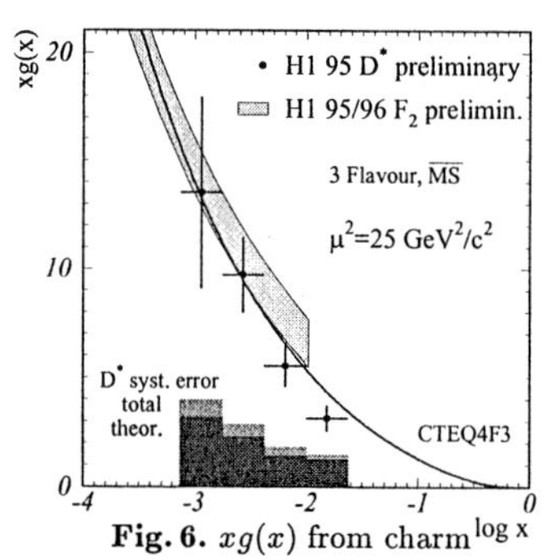

Fig. 6. $xg(x)$ from charm

for the gluon from their F_2 NLO fit discussed in the previous section, at an average scale of 25 GeV^2. The two results are in good agreement.

4 Summary and Conclusions

From NLO F_2 fits to the HERA data the gluon is constrained with an uncertainty of $\sim$ 10% at $x \sim 10^{-4}$. In these fits the charm part of the structure functions is calculated analytically from BGF in NLO QCD. These calculations are consistent with direct measurements of the charm contribution from c-tagged events responsible for up to 25% of the proton structure function in the HERA regime. Future higher statistics measurements will allow the HERA experiments to make precise investigations of the charm production mechanism in the threshold region.

References

[1] V. Lemaitre, these proceedings
[2] B. W. Harris and J. Smith, hep-ph/9706334 and references therein
[3] A. D. Martin *et al.*, Phys. Lett. B**387**(1996)419
[4] H. L. Wai and W. K. Tung, Z. Phys. C**74**(1997)463
[5] M. Glück *et al.*, Z. Phys. C**67**(1995)433

A New Determination of α_S from Neutrino–Nucleon Scattering in CCFR

Lucyna de Barbaro (lucyna@miranda.fnal.gov)

Northwestern University, Evanston, IL

Abstract. A new measurement of the strong coupling constant α_S from the ν-N deeply inelastic scattering experiment CCFR at Fermilab is presented. The combined fit to Q^2 evolution of nucleon structure functions F_2 and xF_3, measured in CCFR, yields the QCD scale parameter $\Lambda^{(4)}_{\overline{MS}}$ (NLO) $= 337 \pm 28(exp) \pm 13(HT)$ MeV. This corresponds to the value of α_S at $Q^2 = M_Z^2$ of $0.119 \pm 0.002(exp) \pm 0.001(HT) \pm 0.004(scale)$, one of the most precise measurement of this quantity.

Deep inelastic scattering of high energy neutrinos has long been used for testing of QCD and measurement of nucleon structure functions (SF) F_2 and xF_3. These functions not only constrain the valence, sea, and gluon parton densities of the nucleon, but, through their Q^2 evolution, allow for the extraction of the QCD scale parameter Λ. The $\alpha_S(\Lambda)$ determination from scaling violations in neutrino scattering has small theoretical uncertainty, since the electroweak radiative corrections are well understood and scale uncertainties are small.

We present an updated measurement of α_S, using CCFR neutrino data, taken during the 1985 and 1987 fixed-target runs at Fermilab, and originally analyzed in [1]. The beam for these runs was a mixed wide-band beam of neutrinos and antineutrinos. The interactions were observed in a target calorimeter, with hadronic energy resolution of $\sigma/E_{had} = 0.89/\sqrt{E_{had}}$, while outgoing muons momentum was measured in the toroidal spectrometer with resolution of $\sigma_p/p = 0.11$. In the earlier analysis, the muon and hadron energy calibrations were determined using a Monte Carlo technique in an attempt to reduce the dominant source of systematic error, the relative calibration between the muon and hadron energies. This paper presents α_S which resulted from a re-extraction of the SF and use of the calibrations directly determined from the test beam data. This changed the relative calibration by 2.1% and increased in the corresponding systematic error to 1.4%. Other changes included more complete radiative corrections and a new value of R_L[2]. In addition, the estimates of the experimental and theoretical systematic errors in the analysis were improved. The SF were corrected for the non-isoscalarity of the Fe target and the charm-production threshold (see [2] for more details).

For the QCD analysis of the SF, we performed a χ^2 fit which minimized the difference between a theoretical prediction and the measured values of F_2 and xF_3 in each (x, Q^2) bin. The theoretical prediction was obtained using NLO QCD evolution program[3]. Parton distributions were parmetrized

Table 1. Results of the global systematic fit to the CCFR data. The parton distributions at $Q_0^2 = 5\ GeV^2$ were parameterized by $xq^{NS}(x) = A_{NS}x^{\eta_1}(1-x)^{\eta_2}$, $xq^S(x) = xq^{NS}(x) + A_S(1-x)^{\eta_S}$, $xG(x) = A_G(1-x)^{\eta_G}$. The χ^2 of the fit was 158 / 164 degrees of freedom.

Parameter	Fit Results	Parameter	Fit Results
$\Lambda_{\overline{MS}}$	337 ± 28 MeV	A_G	2.22 ± 0.34
η_1	0.805 ± 0.009	η_G	4.65 ± 0.68
η_2	3.94 ± 0.03	A_S	1.47 ± 0.04
A_{NS}	8.60 ± 0.18	η_S	7.67 ± 0.13

as shown in Table 1. The statistical errors and the correlated systematic uncertainties were included in the χ^2 . The latter were determined from re-extraction of the SF with each systematic uncertainty changed by one standard error.

The values of the F_2 and xF_3 higher-twist (HT) corrections were taken to be one-half the values from Ref. [4], with a conservative systematic error[2]. Cuts of $Q^2 > 5\ GeV^2$ and the invariant mass-squared of the hadronic system $W^2 > 10\ GeV^2$ were applied to the data to include only the perturbative region, and an $x < 0.7$ cut included the x-bins where the resolution corrections were insensitive to Fermi motion. Q^2 range extended to 125 GeV^2.

From the fit, we obtained a value of $\Lambda_{\overline{MS}}$ in NLO QCD for 4 quark flavors of 337 ± 28(exp.)± 13(HT) MeV, which yielded $\alpha_S(M_Z^2) = 0.119 \pm 0.002$(exp.)$\pm 0.001(HT)\pm 0.004$(scale), where the error due to the renormalization and factorization scales comes from Ref. [5]. The fit also yielded a measurement of the gluon distribution (see Table 1) in the $0.04 < x < 0.70$ region. A fit to only the xF_3 data, which is not coupled to the gluon distribution, gave $\Lambda_{\overline{MS}} = 381 \pm 53$(exp.)$\pm 17$(HT) MeV, which is consistent with the result of the combined F_2 and xF_3 fit but has larger errors because effectively only half the data are used. When the systematic uncertainties were not allowed to vary in the F_2 and xF_3 fit and all effects of systematic uncertainties were added in quadrature, the value of $\Lambda_{\overline{MS}}$ was found to be 381 ± 23(stat.)± 58(syst.) MeV. Complete tables of the CCFR SF results can be obtained from the World Wide Web via the URL ftp://www.nevis.columbia.edu/pub/ccfr/seligman.

References

[1] P.Z. Quintas et al., Physical Review Letters 71 (1993), 1307
[2] W.G. Seligman et al., Physical Review Letters 79 (1997), 1213
[3] S.A. Devoto et al., Physical Review D27 (1983), 508
[4] M. Dasgupta and B.R. Webber, Physics Letters B382 (1996), 273
[5] M. Virchaux and A. Milsztajn, Physics Letters B274 (1992), 221

Part IV

Hard Processes and Perturbative QCD

QCD Connection Between Hadron and Jet Multiplicities

Wolfgang Ochs (wwo@mppmu.mpg.de)

Max-Planck-Institut für Physik, Föhringer Ring 6, D-80805 Munich, Germany

Abstract. The perturbative QCD provides a good overall description of both the jet and hadron multiplicities in the e^+e^- annihilation reaction. In this description the hadrons are considered as dual to partons at a small resolution parameter Q_0 characteristic of a hadronic scale of few hundred MeV.

1 Dual picture relating partons and hadrons

The production rates for jets and their distribution over the kinematic variables are described very well by the perturbative QCD and this success seems to be continued by recent measurements at TEVATRON and HERA. Such a success is at first sight astonishing as theory and experiment relate quite different objects: a 100 GeV jet seen in the experiment consists of a bundle of several dozens of hadronic particles whereas in the theoretical treatment this object is often represented by only one or two partons. Although an understanding at a fundamental level of the colour confinement process is not yet available there are models which explain how the initial partons first evolve perturbatively into a parton jet and then by nonperturbative processes into the observed hadrons. These nonperturbative processes look hopelessly complicated in view of the large variety of particle and resonance species.

Despite this complication it appears that after averaging over some degrees of freedom the perturbative description can actually be extended from jets to single hadrons. An example is the good description of the hadron energy spectrum in a jet which led to the notion of "local parton hadron duality" (LPHD) [1]. It is important to explore further the connection between hadronic and partonic final states to learn about the confinement process. Here we discuss a recent comparison [2] at the same but variable resolution scale, including the limit where jets are fully resolved into hadrons.

2 From jets to hadrons

The mean jet multiplicity in e^+e^- events can be described [2] by perturbative calculations as function of the ("Durham") resolution parameter Q_c in the full measured range down to $Q_c \sim 1$ GeV and smoothly connects to the multiplicity of hadrons in the limit of small $Q_c \to Q_0$; here Q_0 is a nonperturbative parameter of size 250 to 500 MeV depending on the approximation scheme. This is shown in Fig. 1 for the lower data set ("Jets") from LEP-1 (Q=91 GeV kept fixed); the upper data set ("Hadrons") refers to the hadronic multiplicity (taken as $\mathcal{N} = \frac{3}{2}\mathcal{N}_{ch}$) as function of total energy Q. The only free parameters are the QCD scale Λ and the hadronization scale Q_0 (here Λ=500 MeV and $\ln \frac{Q_0}{\Lambda}$=0.015). The curves represent the perturbative QCD

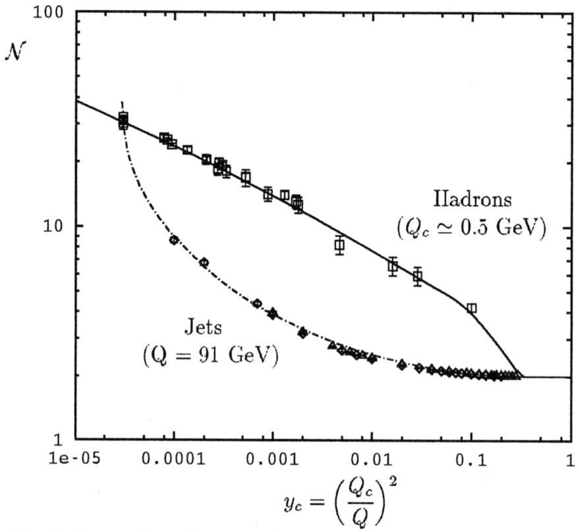

Fig. 1. Multiplicity of jets for varying resolution parameter Q_c and of hadrons for varying *cms* energy Q at fixed resolution parameter $Q_c = Q_0$ obtained from a fit to data in comparison with the perturbative calculation.

calculation: two coupled evolution equations for the multiplicities in quark and gluon jets which include the angular ordering condition for soft gluons and energy conservation are solved numerically with the threshold condition $\mathcal{N} = 2$ and then matched with the exact $O(\alpha_s)$ result. This improved accuracy of the solution yields a better description than previously in regions where the coupling becomes large, namely for hadrons near threshold and jets at high resolution. Remarkably, both the jet and hadron multiplicities can be described now with the same absolute normalization.

This analysis suggests that sufficiently inclusive properties of a jet of hadrons can be computed perturbatively from the jet of partons at all resolution scales down to the hadronization scale Q_0 of few hundred MeV and that the confinement process should be correspondingly soft.

References

[1] Ya.I. Azimov, Yu.L. Dokshitzer, V.A. Khoze and S.I. Troyan, Z. Phys. C27 (1985), 65 and C31 (1986), 213.

[2] S. Lupia and W. Ochs, Phys. Lett. B (in press) (hep-ph/9707393); further details in reports hep-ph 19709246 and hep-ph/9711255.

Towards the NLO Event Generator

Jiří Chýla (chyla@fzu.cz)

Institute of Physics, Prague

Abstract. A special choice of the definition of parton distribution functions which may provide a convenient framework for the construction of consistent NLO event generators is suggested.

There are numerous hard scattering processes which at the parton level are calculated to the NLO. Their consistent incorporation in currently popular Monte–Carlo event generators is, however, hampered by the fact that in all of them parton showers, which simulate the scale evolution of parton distribution functions (PDF), are basically of the LO only.

In analytical NLO QCD calculations of *inclusive quantities* hadron level cross–sections are given as convolutions of PDF $D_{i/h}(x, M, \mathrm{FS})$ and hard scattering cross–sections $\sigma_{ij}(x_1, x_2, \mathrm{FS}, M)$ (and eventually also parton fragmentation functions) evaluated in a particular factorization scheme (FS) and at a given factorization scale M. The PDF satisfy a system of coupled evolution equations involving the splitting functions $P_{ij}(x, \mathrm{FS})$

$$P_{ij}(z, \alpha_s(M), \mathrm{FS}) = P_{ij}^{(0)}(z) + \frac{\alpha_s(M)}{\pi} P_{ij}^{(1)}(z) + \cdots. \tag{1}$$

At the NLO only the first two terms in (1) are taken into account. While $P_{ij}^{(0)}(z)$ are unique, higher order splitting functions $P_{ij}^{(i)}(z), i \geq 1$ depend on the choice of FS and can as a result be essentially arbitrary. Turned around the set of functions $P_{ij}^{(k)}(z), k \leq n$ can conveniently serve as away of specifying this dependence at the n–th order. We can change the factorization scale M as well as the FS at will, but we must also correspondingly modify the NLO hard scattering cross–sections $\sigma_{ij}^{(1)}$ in the expansion

$$\sigma_{ij}(x_1, x_2, M, \mathrm{FS}) = \sigma_{ij}^{(0)}(x_1, x_2) + \frac{\alpha_s(M)}{\pi} \sigma_{ij}^{(1)}(x_1, x_2, M, \mathrm{FS}) + \cdots. \tag{2}$$

MC event generators differ from analytical calculations of inclusive quantities by the fact that they simulate *complete* events, not just the particles involved in the inclusive measurements. As, however, the conventional LO parton showers simulate merely the LO scale dependence of PDF, combining them with the NLO hard scattering cross–sections does not in general lead to a consistent NLO event generator.

There are two strategies how to proceed. One is to modify parton showers in such a way that they generate NLO effects in PDF. This is not easy as the NLO splitting functions do not correspond to elementary QCD vertices.

Another problem is that in the conventional $\overline{\text{MS}}$ FS [2] the NLO splitting functions $P_{GG}^{(1)}(x), P_{Gq}^{(1)}(x)$, $P_{qG}^{(1)}(x)$, and consequently also the full NLO splitting functions (1), become negative in part of the interval $0 < x < 1$. Although both problems have been overcome and in [1] a NLO event generator NLLJET has been constructed along this line, it has not gained wider use.

The purpose of this contribution it to suggest a different strategy to reach the same goal (the details will be described in [3]). Its essence is to exploit the freedom in the choice of the FS of PDF, which at the NLO is defined by the set of splitting functions $P_{ij}^{(1)}(x)$. If we cannot easily make NLO parton showers in the conventional $\overline{\text{MS}}$ FS then why not to choose a factorization scheme where $P_{ij}^{(1)}(z)$ vanish *by definition*? Within such a FS the NLO evolution equations for PDF, and consequently also the corresponding parton showers, will retain the simplicity of the LO ones and all the NLO corrections will be put into the NLO hard scattering cross–sections $\sigma_{ij}^{(1)}$. This FS, which I call "ZERO", is in a sense opposite to the familiar DIS, where on the contrary, the NLO hard scattering cross-sections $\sigma_{eq}^{(1)}$ are set to zero and all the NLO effects are shifted to the corresponding NLO splitting functions $P_{ij}^{(1)}$. To evaluate $\sigma_{ij}^{(1)}(\text{ZERO})$ requires finding the corresponding functions $A_{ij}^{(10)}(z)$ in the relation between the dressed $D_{i/h}(x, M, \text{FC})$ and the bare ones $D_{i/h}^{(0)}(x)$

$$D_{i/h}(x, M, \text{FC}) \equiv D_{i/h}^{(0)}(x) + \tag{3}$$

$$\alpha_s(M) \sum_j \int_x^1 \frac{\mathrm{d}y}{y} D_{j/h}^{(0)}(y) \left(\frac{A_{ij}^{(11)}(z)}{\epsilon} + A_{ij}^{(10)}(z) + \cdots \right),$$

where $z = x/y$ and the sum runs over all parton species. While all residues $A_{ij}^{(kl)}(z), l > 0$ are fixed by the factorization theorem, the finite terms $A_{ij}^{(k0)}(z)$ are arbitrary functions and determine the splitting functions $P_{ij}^{(k)}(z)$. At the NLO only $A_{ij}^{(10)}(z)$ enter and determine $P_{ij}^{(1)}(z)$ in (1). In [3] I discuss in detail this relation and then solve for such $A_{ij}^{(10)}(z)$ that yield vanishing $P_{ij}^{(1)}(z)$. The solution of the corresponding matrix equation is based on numerical evaluation of inverse Mellin transform. With $A_{ij}^{(10)}(z, \text{ZERO})$ at hand, the construction of $\sigma_{ij}(\text{ZERO})$ is in principle straightforward and will be described in future publication.

References

[1] K. Kato, T. Munehisa, Physical Review D36 (1986), 61
K. Kato, T. Munehisa, Physical Review D39 (1989), 156
[2] W. Furmanski, R. Petronzio: Zeitschrift für Physik C11 (1982), 293
[3] J. Chýla, in preparation

QCD Corrections to $e^+e^- \rightarrow 4$ Jets

Stefan Weinzierl, (stefanw@spht.saclay.cea.fr)

Service de Physique Théorique, Centre d'Etudes de Saclay, F-91191 Gif-sur-Yvette Cedex, France

Abstract. Four-jet production in e^+e^- annihilation is the lowest-order process in which the quark and gluon colour charges can be measured independently, and is thus sensitive to the presence of light coloured fermions such as gluinos. I report on the calculation of the one-loop corrections to $e^+e^- \rightarrow \bar{q}q\bar{Q}Q$ and $e^+e^- \rightarrow \bar{q}qgg$, along with a numerical implementation in a program for studying four-jet production at e^+e^- colliders. The incoporation of next-to-leading QCD corrections should reduce theoretical uncertainties.

QCD four-jet production in e^+e^- annihilation can be measured at LEP and can be studied in its own right. First of all, $e^+e^- \rightarrow 4$ jets is the lowest order process which contains the non-abelian three-gluon-vertex at tree level and thus allows for an measurement of the colour factors C_F, C_A and T_R of QCD. This in turn may be used to put exclusion limits on light gluinos. Furthermore the QCD process contributes as background to W-pair production, when both W's decay hadronically and to certain channels of the search of the Higgs boson like $e^+e^- \rightarrow Z^* \rightarrow ZH \rightarrow 4$ jets. In general leading-order calculations in QCD give a rough description of the process under consideration, but they suffer from large uncertainties. The arbitrary choice of the renormalization scale gives rise to an ambiguity, which is reduced only in an next-to-leading order calculation. Furthermore the internal structure of a jet and the sensitivity to the merging procedure of the jet algorithm are modelled only in an NLO analysis. Both uncertainties are related to the appearance of logarithms, ultraviolet in nature in the first case, infrared in the latter, which are calculated explicitly only in an NLO calculation.

One-loop amplitudes

The one-loop amplitudes for the first subprocess $e^+e^- \rightarrow q\bar{q}Q\bar{Q}$ were calculated in refs. [1, 2] and the amplitudes for the second subprocess $e^+e^- \rightarrow q\bar{q}gg$ in refs. [3, 4]. The calculation in [1, 3] used some non-conventional techniques in order to calculate the one-loop amplitudes efficiently. These include colour decomposition, where amplitudes are decomposed into simpler gauge-invariant partial amplitudes with definite colour structure, and the spinor helicity method, which consists in expressing all Lorentz four-vectors and Dirac spinors in terms of massless two-component Weyl-spinors. Their use divides the task into smaller, more manageable pieces. Also a decomposition inspired by supersymmetry proved to be useful, where the particles running around the loop are reexpressed in terms of supermultiplets. In a second step the cut technique and factorization in collinear limits are used to constrain the analytic form of the partial amplitudes. As an example I explain in more detail the cut technique [5], which is based on unitarity. To obtain the coefficients of the basic box, triangle or bubble integrals one considers the cuts in all possible channels. Each phase-space integral is rewritten with the help of the Cutkosky rules as the imaginary part of a loop amplitude. The power of this method lies within the fact, that on each side of the cut one has a full tree amplitude and not just a single Feynman diagram. This method allows one to reconstruct the one-loop amplitude up to terms without an imaginary part.

Phase-space integration

The second major part of a general purpose NLO program for four jets is coding the one-loop amplitudes and the five parton tree-level amplitudes in a numerical Monte Carlo program. Only the sum of the virtual corrections and the real emission part is infrared finite, whereas when taken separately, each part gives a divergent contribution. Several methods to handle this problem exist, such as the phase-space slicing method [8], the subtraction method [6] and the dipole formalism [7]. In this work we have chosen phase space slicing. It should be mentioned that there are two other numerical programs for $e^+e^- \to 4$ jets, MENLO PARC [9] and DEBRECEN [10], using the subtraction method and the dipole formalism, respectively.
Within the phase space slicing approach the $(n+1)$-parton phase space integration is split into a small region of order s_{min} containing all singularities, where the integration is done analytically and then combined with the n-parton phase space integration, and the remaining region, where the integration can be performed numerically. This procedure introduces a systematic error of order s_{min} and it has to be shown that the result does not depend on s_{min} within the desired accuracy. In order to improve the convergence of the Monte Carlo integration, the $(n+1)$-parton phase space is generated correlated with the n-parton phase space, allowing for the cancelation of infrared logarithms on a point-by-point basis.

References

[1] Z. Bern, L. Dixon, D.A. Kosower and S. Weinzierl, Nucl. Phys. B489, (1997), 3

[2] E.W.N. Glover and D.J. Miller, Phys. Lett. B396, (1997), 257

[3] Z.Bern, L.Dixon, D.A.Kosower, hep-ph/9708239

[4] J.M.Campbell, E.W.N.Glover and D.J.Miller, hep-ph/9706297

[5] Z. Bern, L. Dixon, D.C. Dunbar and D.A. Kosower, Nucl. Phys. B435, (1995), 39

[6] S. Frixione, Z. Kunszt and A. Signer, Nucl. Phys. B467, (1996), 399

[7] S. Catani and M.H. Seymour, Nucl. Phys. B485, (1997), 291

[8] W.T. Giele and E.W.N. Glover, Phys. Rev. D46, (1992), 1980

[9] L. Dixon and A. Signer, Phys. Rev. D56, (1997), 4031

[10] Z. Nagy and Z. Trócsányi, Phys. Rev. Lett. 79, (1997), 3604

Jet Production in DIS at NLO, Including Z and W Exchange

E. Mirkes[a], S. Willfahrt[a] and D. Zeppenfeld[b]

[a]Institut für Theoretische Teilchenphysik, Universität Karlsruhe, D-76128 Karlsruhe, Germany

[b]Department of Physics, University of Wisconsin, Madison, WI 53706, USA

Abstract. Next-to-leading order QCD predictions for 1-jet and 2-jet cross sections in deep inelastic scattering with complete neutral current (γ^* and Z) and charged current ($W^{\pm}$) exchange are presented.

Electroweak effects in DIS n-jet production are known to become important for $Q^2 \gtrsim 2500$ GeV^2. These effects can be studied with the fully differential $ep \to n$ jets event generator MEPJET [1, 2] which allows to analyze any 1- or 2-jet like observable in NLO in terms of parton 4-momenta. Numerical results for 1- and 2-jet cross sections in $e^{\pm}p$ scattering are shown in Fig. 1.

Jets are defined in a cone scheme (in the laboratory frame) with $R = 1$ and $p_T^{\rm lab}(j) > 5$ GeV. In addition, we require $0.04 < y < 1$, an energy cut of $E(e') > 10$ GeV on the scattered lepton (in NC scattering) and a cut on the pseudo-rapidity η of the scattered lepton and jets of $|\eta| < 3.5$.

The Q^2 distribution of the NC 2-jet cross section (including γ^* and Z exchange) is compared to the CC 2-jet cross section in Fig. 1a. The difference between the e^+p and e^-p results in the NC case is entirely due to the additional Z exchange, which increases (decreases) the 1-photon result in the e^-p (e^+p) case by more than a factor of two for very high Q^2. The Q^2-dependence of the 2-jet rate, defined as the ratio of the 2-jet to the 1-jet cross section, is shown in Fig. 1b for e^+p scattering in NC (solid) and CC (dashed) exchange.

The QCD corrections to the electroweak cross sections are investigated in Figs. 1c-f where the Q^2 dependence of the K-factors for NC and CC 1- and 2-jet cross sections are shown. While NC and CC effects differ markedly for e^+p and e^-p scattering, the QCD K-factors are largely independent of the initial state lepton. We find very similar K-factors for jets defined in the k_T scheme with $E_T^2 = 40$ GeV^2. More results can be found in [2].

References

[1] E. Mirkes and D. Zeppenfeld, *Phys. Lett.* **B380** (1996) 105, [hep-ph/9511448]; *Acta Phys. Pol.* **B27** (1996) 1393, [hep-ph/9604281].
[2] E. Mirkes, [hep-ph/9711224].
[3] M. Glück, E. Reya and A. Vogt, *Zeit. Phys.* **C67** (1995) 433.
[4] A.D. Martin, R.G. Roberts, W.J. Stirling, *Phys. Lett.* **B387** (1996) 419.

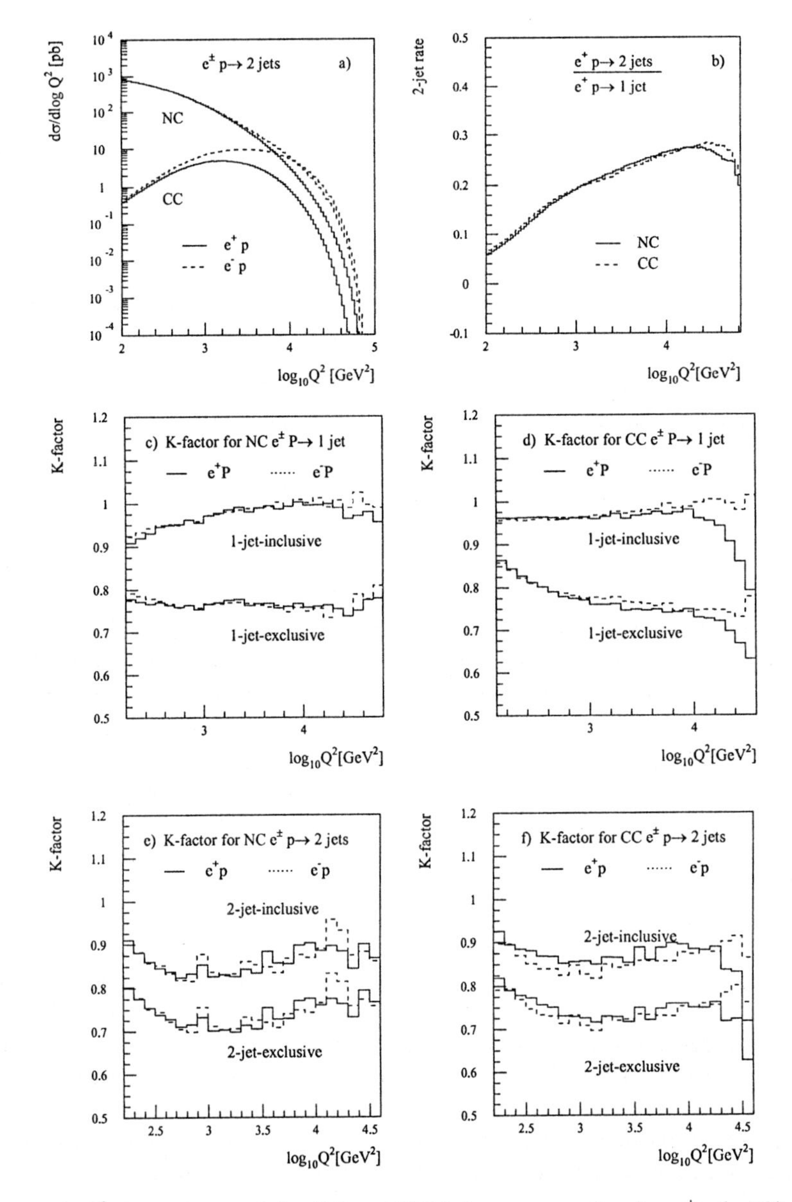

Fig. 1. a) Q^2 dependence of the CC and NC 2-jet cross section for e^+p (solid) and e^-p (dashed) scattering. b) 2-jet rate $\sigma(\text{2-jet})[Q^2]/\sigma(\text{1-jet})[Q^2]$ for NC (solid) and CC (dashed) exchange. The results in a,b are shown in LO with MRS Set (R1) [4] parton distributions and the two-loop formula for the strong coupling constant. c,d) Q^2 dependence of the K-factor for NC and CC 1-jet inclusive and exclusive cross sections in e^+p (solid) and e^-p (dashed) scattering. The (LO) NLO cross section, which enter the K-factor, are calculated with LO (NLO) GRV parton distributions [3] together with the 1-loop (2-loop) formula for the strong coupling constant. e,f) same as c,d) for 2-jet cross sections.

New Tests of QCD Using $Z^0 \to b\bar{b}g$ Events

David Muller (muller@slac.stanford.edu) For the **SLD** Collaboration

Stanford Linear Accelerator Center, Stanford, California, USA

Abstract. We present new studies of 3-jet final states from hadronic Z^0 decays recorded by the SLD experiment, in which the quark, antiquark, and gluon jets are identified. Our gluon energy spectrum, measured over the full kinematic range, is consistent with the predictions of QCD, and we derive a limit on an anomalous chromomagnetic bbg coupling. We measure the parity violation in Z^0 decays into $b\bar{b}g$ to be consistent with the predictions of electroweak theory and QCD, and perform new tests of T- and CP-conservation at the bbg vertex.

1 Introduction

Experimental studies of the structure of events containing three hadronic jets in e^+e^- annihilation have been limited by difficulties in identifying which jet is due to the quark, which to the antiquark and which to the gluon. Since the gluon is expected to be the lowest-energy jet in most events, the predictions of QCD have been tested using energy and angle distributions of energy-ordered jets, and this is sufficient to confirm the $q\bar{q}g$ origin of such events and to determine the gluon spin [1]. Tagging of the origin of any two of the three jets in such events would allow more complete and stringent tests of QCD predictions.

Here we present a study [2] of 3-jet final states in which two of the jets have been tagged as b or $\bar{b}$ jets using the long lifetime of the B-hadrons in the jets and the precision vertexing system of the SLD. The remaining jet is tagged as the gluon jet, and its energy spectrum is studied over its full kinematic range. Adding a tag of the charge of one of the $b/\bar{b}$ jets, and exploiting the high electron beam polarization of the SLC, we measure two angular asymmetries sensitive to parity violation in the Z^0 decay, and also construct new tests of T- and CP-conservation at the bbg vertex. The study of b-flavor events is especially useful as input to measurements of electroweak parameters such as R_b and A_b, and as a probe of new physics, which is expected in many cases to couple more strongly to heavier quarks.

2 The Gluon Energy Spectrum

Well contained hadronic events [3] in which exactly 3 jets are found using the JADE algorithm at $y_{cut} = 0.02$ are selected. The jet energies are calculated from the angles between them [3] and the jets are energy ordered such that

$E_1 > E_2 > E_3$. In each jet we count the number n_{sig} of 'significant' tracks, i.e. those with normalized transverse impact parameter with respect to the primary interaction point $d/\sigma_d > 3$. We require exactly two of the three jets to have $n_{sig} > 1$, and the remaining jet is tagged as the gluon jet. This yields 1533 events with an estimated purity of correctly tagged gluon jets of 91%. In 2.5% (12.5%) of these events, jet 1(2), the (second) highest energy jet, is tagged as the gluon jet, giving coverage over the full kinematic range.

The background from non-$b\bar{b}g$ events and events with an incorrect gluon tag is subtracted, and the resulting distribution of scaled gluon energy $z = 2E_g/\sqrt{s}$ is corrected for the effects of selection efficiency and resolution. The fully corrected spectrum is shown in fig. 1, and shows the expected falling behaviour with increasing z. The distribution is cut off at low z by the finite y_{cut} value used for jet finding. Also shown are the predictions of first and second order QCD. Both reproduce the general behaviour, but fail to describe the data in detail. The prediction of the JETSET [4] parton shower simulation is also shown and reproduces the data. Our data thus confirm the predictions of QCD, although higher order effects are clearly important in the intermediate gluon energy range, $0.2 < z < 0.4$.

The gluon energy spectrum is particularly sensitive to the presence of an anomalous chromomagnetic term in the QCD Lagrangian. A fit of the theoretical prediction [5] including an anomalous term parametrized by a relative coupling κ, yields a value of $\kappa = -0.03 \pm 0.06$ (*preliminary*), consistent with zero, and corresponding to limits on such contributions to the bbg coupling of $-0.15 < \kappa < 0.09$ at the 95% confidence level.

3 Parity Violation in 3-jet Z° Decays

We now consider two angles, the polar angle of the quark θ_q, and the angle between the quark-gluon and quark-electron beam planes $\chi = \cos^{-1}(\hat{p}_q \times \hat{p}_g) \cdot (\hat{p}_q \times \hat{p}_e)$. The cosine x of each of these angles is distributed as $1 + x^2 + 2A_P A_Z x$, where the Z^0 polarization $A_Z = (P_e - A_e)/(1 - P_e A_e)$ depends on the electron beam polarization P_e, and each A_P is predicted by QCD.

Three-jet events (Durham algorithm, $y_{cut} = 0.005$) are selected and energy ordered, and a topological vertex finder [6] is applied to the tracks in each jet. The 3420 events containing any vertex with mass above 1.5 GeV/c^2 are kept, having an estimated $b\bar{b}g$ purity of 87%. We calculate the momentum-weighted charge of each jet j, $Q_j = \Sigma_i q_i |\mathbf{p}_i \cdot \hat{p}_j|^{0.5}$, using the charge q_i and momentum $\mathbf{p}_i$ of each track i in the jet. In this case we assume that the highest-energy jet is not the gluon, and tag it as the b ($\bar{b}$) jet if $Q = Q_1 - Q_2 - Q_3$ is negative (positive). We then define the b-quark polar angle $\cos\theta_b = -\text{sign}(Q)(\hat{p}_e \cdot \hat{p}_1)$.

We construct the left-right-forward-backward asymmetry $\tilde{A}^b_{FB}$ in this polar angle, which is shown as a function of $|\cos\theta_b|$ in fig. 2a. A clear asymmetry is seen, which increases with $|\cos\theta_b|$ in the expected way. A fit yields an asym-

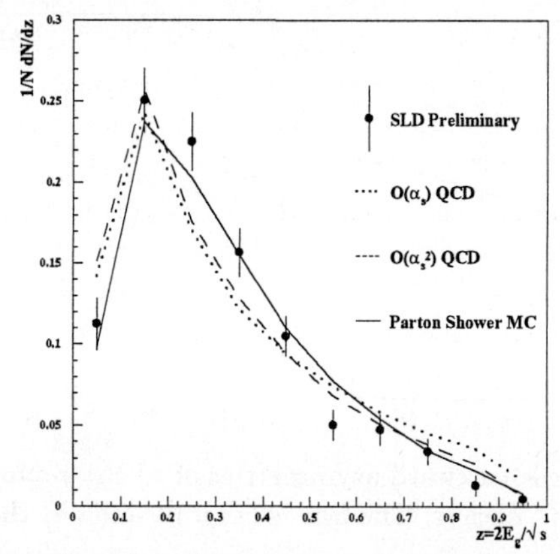

Fig. 1. The measured scaled gluon energy distribution compared with QCD.

metry parameter of $A_P = 0.99 \pm 0.09 \pm 0.07$ (*preliminary*), consistent with the QCD prediction of $A_P = 0.93 A_b = 0.87$.

We then tag one of the two lower energy jets as the gluon jet, using the impact parameters of their tracks. If jet 2 has $n_{sig} = 0$ and jet 3 has $n_{sig} > 0$, then jet 2 is tagged as the gluon jet; otherwise jet 3 is tagged as the gluon jet. In each event we construct the angle χ, and A^{χ}_{LRFB} is shown as a function of χ in fig. 2b. Here we expect only a small deviation from zero as indicated by the line on fig. 2b. Our measurement is consistent with the prediction, as well as with zero. A fit yields $A_\chi = 0.01 \pm 0.05(stat.)$ (*preliminary*), to be compared with an expectation of 0.047.

4 Symmetry tests in 3-jet Z° Decays

Using these fully tagged events, we can construct observables that are formally odd under time and/or CP reversal. For example, the energy-ordered triple product $\cos\omega^+ = \boldsymbol{\sigma}_Z \cdot (\hat{\boldsymbol{p}}_1 \times \hat{\boldsymbol{p}}_2)$, where $\boldsymbol{\sigma}_Z$ is the Z^0 polarization vector, is T_N-odd and CP-even. Since the true time reversed experiment is not performed, this quantity could have a nonzero $\tilde{A}_{FB}$, and we have previously set a limit [3] using events of all flavors. A calculation [7] predicts that $\tilde{A}^{\omega^+}_{FB}$ is largest for $b\bar{b}g$ events, but is only $\sim 10^{-5}$. The fully flavor-ordered triple product $\cos\omega^- = \boldsymbol{\sigma}_Z \cdot (\hat{\boldsymbol{p}}_q \times \hat{\boldsymbol{p}}_{\bar{q}})$ is both T_N-odd and CP-odd.

Our measured $\tilde{A}^{\omega^+}_{FB}$ and $\tilde{A}^{\omega^-}_{FB}$ are shown in fig. 3. They are consistent with zero and we set limits on possible T_N- and CP-violating asymmetries of $-0.056 < A^+_T < 0.051$ and $-0.056 < A^-_T < 0.051$, respectively.

5 Conclusion

In summary, we use the excellent vertexing capability of the SLD and the high electron beam polarization of the SLC to make several new tests of

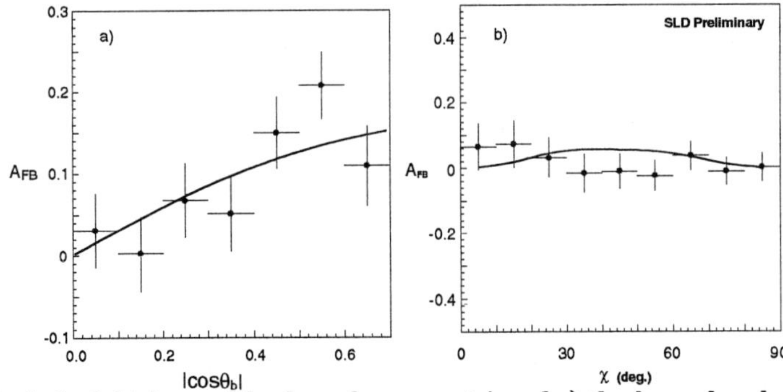

Fig. 2. Left-right-forward-backward asymmetries of a) the b-quark polar angle and b) the angle χ in 3-jet Z^0 decays. The line is (a) a fit and (b) the QCD prediction.

QCD, using 3-jet final states in which the quark, antiquark and gluon jets are identified. The gluon energy spectrum is measured over its full kinematic range; we confirm the prediction of QCD and set limits on anomalous chromomagnetic couplings. The parity violation in Z^0 decays to $b\bar{b}g$ is found to be consistent with the predictions of electroweak theory plus QCD, and new tests of T- and CP-conservation in strong interactions are performed.

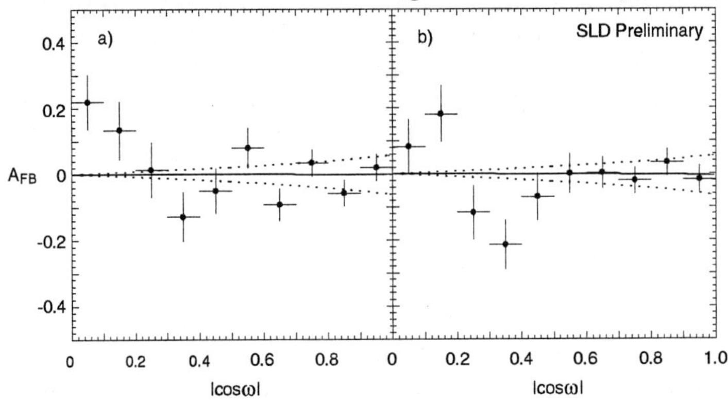

Fig. 3. LRFB asymmetries of the a) energy- and b) flavor-ordered triple product.

References

[1] See e.g. K. Abe, et al., Phys. Rev. **D55** (1997) 2533.
[2] For details, see papers **286, 288, 289**, contributed to this conference.
[3] K. Abe, et al., Phys. Rev. Lett. **75** (1996) 4173.
[4] T. Sjöstrand, Comp. Phys. Comm. **82** (1994) 74.
[5] T. Rizzo, Phys. Rev. **D50** (1994) 4478.
[6] D.J. Jackson, SLAC-PUB-7215 (1996), to appear in Nucl. Inst. Meth.
[7] A. Brandenburg, L. Dixon and Y. Shadmi, Phys. Rev. **D53** (1996) 1264.

QCD at LEPII with ALEPH

Hasko Stenzel[1], on behalf of the ALEPH collaboration

Max-Planck-Institut für Physik, Föhringer Ring 6, D-80805 München
stenzel@mppmu.mpg.de

Abstract. Hadronic events produced at LEP at centre-of-mass energies of 130, 136, 161 and 172 GeV have been studied and compared with QCD predictions. Distributions of event-shape observables and jet rates are presented and compared to the predictions of Monte Carlo models and analytic QCD calculations. From a fit of $\mathcal{O}(\alpha_s^2)$+NLLA QCD calculations to the differential two-jet rate, α_s has been determined at various energies. The mean charged particle multiplicities and the peak positions ξ^* in the $\xi = \ln(1/x_p)$ distribution have also been determined.

1 Introduction

In this paper preliminary analyses of hadronic events collected by the ALEPH detector at 161 and 172 GeV are presented. The measurements are based on integrated luminosities of 11.1 and 10.6 pb^{-1} at 161 and 172 GeV, respectively. In addition, analyses of hadronic events at 130 and 136 GeV, based on approximately 2.9 pb^{-1} at each energy and published in [1], are updated and extended. The primary goal is to investigate quantities for which the centre-of-mass energy dependence is well predicted by QCD.

A detailed description of the ALEPH detector is given in [2]. The measurements presented here are based on both charged particle measurements from the time projection chamber, inner tracking chamber, and vertex detector, as well as information on neutral particles from the electromagnetic and hadronic calorimeters [3]. At centre-of-mass energies higher than the Z resonance, there is a relatively high probability for initial state photon radiation (ISR), resulting in hadronic systems with an invariant mass $\sqrt{s'} \approx M_Z$. An additional source of background at $\sqrt{s} = 161$ and 172 GeV comes from production four-quark states through WW, ZZ and Zγ^*.

After all cuts, 182, 140, 292 and 254 events are selected at 130, 136, 161 and 172 GeV, respectively. Corrections for imperfections of the detector and for the residual effects of ISR are made by means of multiplicative factors, which are derived from the Monte Carlo model PYTHIA [4], and, for sytematic checks, from HERWIG [5]. Additional systematic uncertainties, mainly related to the simulation of the detector, are estimated by varying the cuts applied and repeating the analysis.

2 Inclusive charged particle distributions

Charged particle inclusive distributions were measured for the variables $\xi = -\ln x_p$, where $x_p = p/p_{\mathrm{beam}}$ and the rapidity $y = \frac{1}{2}\ln(E + p_{\|})/(E - p_{\|})$

with $p_{||}$ measured with respect to the thrust axis. By integrating the rapidity distribution, the mean multiplicity of charged particles can be determined. Results for 133, 161 and 172 GeV are given in Table 1. The multiplicities

$E_{\rm cm}$ (GeV)	$N_{\rm ch}$	stat. error	syst. error	ξ^*	stat. error	syst. error
133	23.99	0.47	0.36	3.968	0.048	0.050
161	26.70	0.58	0.52	4.085	0.071	0.050
172	26.32	0.66	0.53	4.064	0.062	0.071

Table 1. $N_{\rm ch}$ and ξ^* measured at $E_{\rm cm}$ = 133, 161 and 172 GeV.

measured here are shown in Fig.1 along with measurements from other experiments at various energies , and also with the predictions of the Monte Carlo models JETSET version 7.4 [4] and HERWIG. The peak position ξ^*

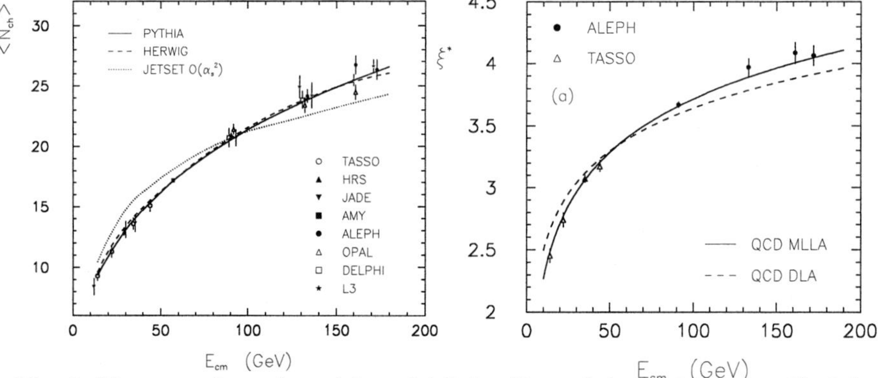

Fig. 1. The mean charged particle multiplicity $N_{\rm ch}$ and the peak position ξ^* of the distribution of $\xi = -\ln x_p$ as a function of the centre-of-mass energy.

of the inclusive distribution of ξ was determined by fitting a distorted Gaussian [6] to the central regions, defined by the width of the distribution at 60% of its maximum height; details of the analysis and error estimation procedure can be found in [1]. The values are given in Table 1 and shown in Fig.1, along with QCD predictions using the double logarithm approximation (DLA) and including higher order corrections (modified leading logarithm approximation – MLLA), and with the ξ^* values obtained with the same procedure from distributions at lower energies measured by the TASSO experiment [7].

3 Event shapes and jet rates

Distributions of event-shape variables are predicted to second order in QCD; some can also be resummed to all orders in α_s. The strong coupling constant may be determined from a fit of the theoretical prediction to the measured distribution. The primary objective is to observe the running of the coupling with centre-of-mass energy; this can be seen from the decrease of the mean values of event shape variables with $E_{\rm cm}$. The mean values corresponding to

$E_{\rm cm}$ (GeV)	$1-T$	M_H^2/s	B_W	C
133	0.0627 ± 0.0056	0.0515 ± 0.0042	0.0713 ± 0.0048	0.2305 ± 0.0157
161	0.0615 ± 0.0048	0.0487 ± 0.0035	0.0697 ± 0.0037	0.2305 ± 0.0145
172	0.0563 ± 0.0046	0.0454 ± 0.0043	0.0674 ± 0.0039	0.2147 ± 0.0149

Table 2. Mean values of event-shape variables measured at $E_{\rm cm} = 133$, 161 and 172 GeV. The error is the quadratic sum of statistical and systematic errors.

the event-shape variables thrust, heavy jet mass, wide jet broadening and C-parameter are given for each of the energies in Table 2.

Jet rates are defined by means of the Durham clustering algorithm [8] in the following way. For each pair of particles i and j in an event one computes $y_{ij} = \frac{2\min(E_i^2, E_j^2)(1-\cos\theta_{ij})}{E_{vis}^2}$. The pair of particles with the smallest value of y_{ij} is replaced by a pseudo-particle (cluster) with four-momentum taken to be the sum of the four momenta of particles, $p^\mu = p_i^\mu + p_j^\mu$. The clustering procedure is repeated until all y_{ij} values exceed a given threshold y_{cut}. The number of clusters remaining is defined to be the number of jets. Results are shown in Fig.2 along with the predictions of PYTHIA and HERWIG.

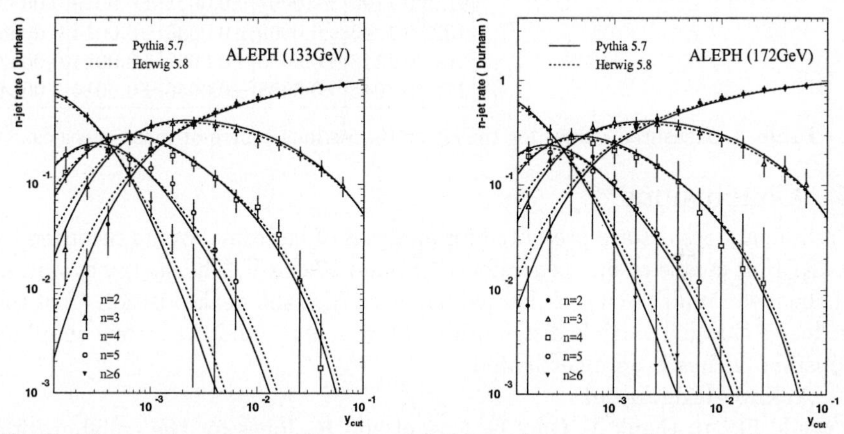

Fig. 2. Measured n-jet rates for $n = 2, 3, 4, 5$ and $n \geq 6$ and the predictions of Monte Carlo models, at centre-of-mass energies of 133 and 172 GeV.

4 Determination of α_s

QCD predicts that the value of the strong coupling constant, α_s, should fall by approximately 9% between 91 and 172 GeV. The analysis is constructed such that the systematic uncertainties are highly correlated between the measurements at the different energies, in order to observe the running of the coupling. The fitted variable is $-\ln y_3$, where y_3 is the value of cut on scaled invariant mass at which the event changes from being clustered into three jets to being clustered into two jets. The perturbative QCD prediction is based on the $O(\alpha_s^2)$ matrix element, improved by including resummed leading and next-to-leading logarithmic terms. The hadronisation corrections used in the analysis were taken from ARIADNE [9], which gave the best description of the data at 91.2 GeV. The corresponding uncertainty is estimated by using HERWIG or PYTHIA instead. To estimate the uncertainty due missing higher orders in the perturbative prediction, the two matching schemes R and $\ln R$ were used, and the renormalisation scale μ was varied in the range $-1 \leq \ln \mu^2/s \leq 1$. The resulting α_s measurements are given in Table 3 for the various energies, and are also shown in graphical form in Fig. 3, along with the fitted two-loop prediction of QCD.

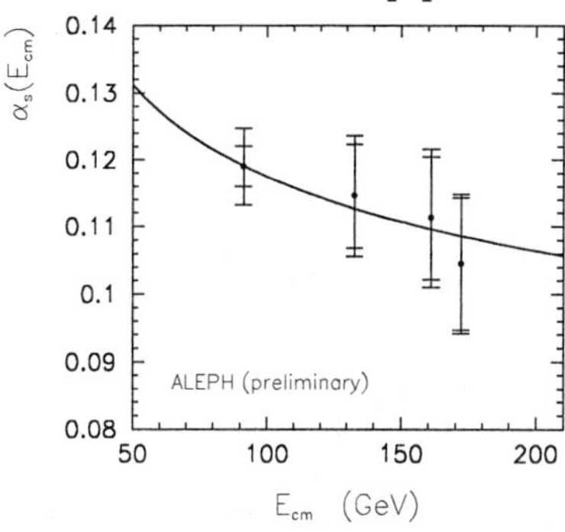

Fig. 3. The strong coupling constant α_s measured at 91.2, 133, 161 and 172 GeV using the distribution of the variable $-\ln y_3$. The outer error bars indicate the total error, the inner ones exclude the correlated theoretical error.

$\sqrt{s}$	α_s	stat.	syst.	hadron.	theory
91.2	0.1190	±0.0007	±0.0023	±0.0019	0.0049
133	0.1146	±0.0069	±0.0033	±0.0014	0.0046
161	0.1113	±0.0080	±0.0042	±0.0014	0.0047
172	0.1045	±0.0085	±0.0046	±0.0014	0.0034

Table 3. Measured values for the α_s at the various centre-of-mass energies.

5 Conclusions

Preliminary results are presented for analyses of hadronic events recorded by ALEPH at centre-of-mass energies of 161 and 172 GeV. The energy evolution of the mean multiplicity of charged particles N_{ch}, the peak position ξ^* of the inclusive charged particle distribution of $\xi = -\ln x_p$, and the strong coupling constant α_s has been investigated.

Acknowledgements
I would like to thank Y. Gao, G. Cowan and R. Jones for the collaboration which lead to the results presented in this talk.

References

[1] D. Buskulic et al., ALEPH Collab., Z. Phys. C73 (1997) 409.
[2] D. Decamp et al., ALEPH Collab., Nucl. Instr. Meth. A 294 (1990) 121.
[3] D. Buskulic et al., ALEPH Collab., Nucl. Instr. Meth. A360 (1995) 481.
[4] T. Sjöstrand, Computer Physics Commun. 82 (1994) 74.
[5] G. Marchesini et al., Comp. Phys. Comm. 67 (1992) 465.
[6] C.P. Fong and B.R. Webber, Phys. Lett. B229 (1989) 289.
[7] W. Braunschweig et al., TASSO Collab., Z. Phys. C47 (1990) 187.
[8] S. Catani et al., Phys. Lett. B269 (1991) 432;
[9] L. Lönnblad, Computer Physics Commun. 71 (1992) 15.

New Results on α_s from DELPHI

Siegfried Hahn (hahns@vxcern.cern.ch)

University of Wuppertal, 42119 Wuppertal, Germany

Abstract. Recent studies of the strong coupling α_s using the DELPHI detector at LEP [1, 2] are reviewed. For the analysis of the data at $\sqrt{s} = M_Z$ recorded in 1994, a large number of shape observables and their dependence on the event orientation are measured and compared to theoretical calculations in $\mathcal{O}(\alpha_s^2)$. The adjustment of the renormalization scale μ turns out to be of major importance. Only a poor description of the data can be achieved if the so-called physical scale $\mu^2 = Q^2$ is applied, whereas a combined fit of α_s and the renormalization scale value yields an excellent description of the high statistics data. For these scale values perfect consistency of the fit values of $\alpha_s(M_Z)$ is obtained for all event shape observables considered. The running of α_s is measured from the DELPHI data at $\sqrt{s} = 130$ GeV - 172 GeV. Here an analytical power ansatz for the hadronization corrections is compared with hadronization corrections obtained by using Monte Carlo generators.

1 Angular Dependence of IR- and Collinear-Safe Shape Observables and a Precise Determination of α_s

Within this analysis, the distributions of 18 different IR- and collinear-safe shape observables are determined as a function of the polar angle ϑ_T of the thrust axis with respect to the e^+e^- beam direction. Since the definition of thrust has a forward-backward ambiguity, we choose $\cos\vartheta_T \geq 0$. $\cos\vartheta_T$ is called the event orientation. The theoretical predictions in $\mathcal{O}(\alpha_s^2)$ are calculated with EVENT2 [3], which applies the matrix elements of the Leyden group [4]. Using this program, one can calculate the double differential cross section for any infrared- and collinear safe observable Y in e^+e^-annihilation in dependence on the event orientation:

$$\frac{1}{\sigma_{tot}}\frac{d^2\sigma(Y,\cos\vartheta_T)}{dY\,d\cos\vartheta_T} = \bar{\alpha_s}(\mu^2)A(Y,\cos\vartheta_T) + \bar{\alpha_s^2}(\mu^2)\left[B(Y,\cos\vartheta_T) + \left(2\pi\beta_0\ln(x_\mu) - 2\right)A(Y,\cos\vartheta_T)\right] \qquad (1)$$

where $\bar{\alpha_s} = \alpha_s/2\pi$ and $\beta_0 = (33-2n_f)/12\pi$. σ_{tot} is the one loop corrected cross section for the process $e^+e^-\rightarrow$ hadrons. The event orientation enters via ϑ_T, which denotes the polar angle of the thrust axis with respect to the e^+e^- beam direction. The renormalization scale factor x_μ is defined by $\mu^2 = x_\mu Q^2$ where $Q = M_Z$ is the center of mass energy. A and B denote

the $\mathcal{O}(\alpha_s)$ and $\mathcal{O}(\alpha_s^2)$ QCD coefficients, respectively. It should be noted that both coefficients explicitely depend on the event orientation. The definition of the observables studied within this analysis can be found in [1].
In $\mathcal{O}(\alpha_s^2)$, the running of the strong coupling α_s at the renormalization scale μ is given by

$$\alpha_s(\mu) = \frac{1}{\beta_0 \ln \frac{\mu^2}{\Lambda^2}} \left(1 - \frac{\beta_1}{\beta_0^2} \frac{\ln \ln \frac{\mu^2}{\Lambda^2}}{\ln \frac{\mu^2}{\Lambda^2}}\right) \tag{2}$$

where $\Lambda \equiv \Lambda^{(5)}_{\overline{\rm MS}}$ is the QCD scale parameter computed in the Modified Minimal Subtraction ($\overline{\rm MS}$) scheme for $n_f = 5$ flavors and $\beta_1 = (153 - 19n_f)/24\pi^2$.

The renormalization scale μ is a formally unphysical parameter and should not enter at all into an exact infinite order calculation. Within the context of a truncated finite order perturbative expansion for any particular process under consideration, the definition of μ depends on the renormalization scheme employed, and its value is in principle completely arbitrary. The traditional experimental approach is, to measure all observables at the same, fixed scale value, the so-called physical scale (PHS) $x_\mu = 1$ or equivalently $\mu^2 = Q^2$. Using this method, one finds in general quite large values for the $2^{\rm nd}$ order contibutions, in some cases the ratio of the $\mathcal{O}(\alpha_s^2)$ with respect to the $\mathcal{O}(\alpha_s)$ contribution r_{NLO} is almost $|r_{\rm NLO}| \simeq 1.0$, indicating a poor convergence behavior of the $\mathcal{O}(\alpha_s^2)$ predictions in the $\overline{\rm MS}$ scheme, which quite naturally results in a wide spread of the measured α_s values which indeed has been observed in a previous analysis in $\mathcal{O}(\alpha_s^2)$ QCD.
Within this analysis a combined fit on α_s and the scale parameter x_μ is applied, a method known as experimentally optimized scales (OPT). Here one finds in general much smaller contributions from the $\mathcal{O}(\alpha_s^2)$ term, we require $r_{NLO} \leq 0.3$.
In [1] $\alpha_s(M_Z^2)$ is studied as a function of the renormalization scale. It turns out that for most of the shape distributions the renormalization scale has to be fixed to a rather narrow range of values in order to be consistent with the data. For OPT, the $\mathcal{O}(\alpha_s^2)$ predictions including the event orientation yield an excellent description of the high statistics data. The $\chi^2/d.f.$ is for all observables considered $\chi^2/d.f. \simeq 1$, whereas for PHS the description is significantly worse for most of the observables, where one finds χ^2 values up to $\chi^2/d.f. \sim 20$. A detailed description of the results can be found in [1]. The $\alpha_s(M_Z^2)$ values determined from 15 different observables are shown in figure 1 for OPT in comparison with PHS. For the errors indicated, the uncertainty from the fit, the systematic experimental uncertainty and the hadronization uncertainty. have been added quadraticly. It should be noted that no additional uncertainty due to the variation of the renormalization scale has been added. For OPT the scatter among the different observables is significantly reduced. Here, the mean value is $\alpha_s(M_Z) = 0.1165 \pm 0.0031$, to be compared with $\alpha_s(M_Z) = 0.1237 \pm 0.0073$ for the PHS method.

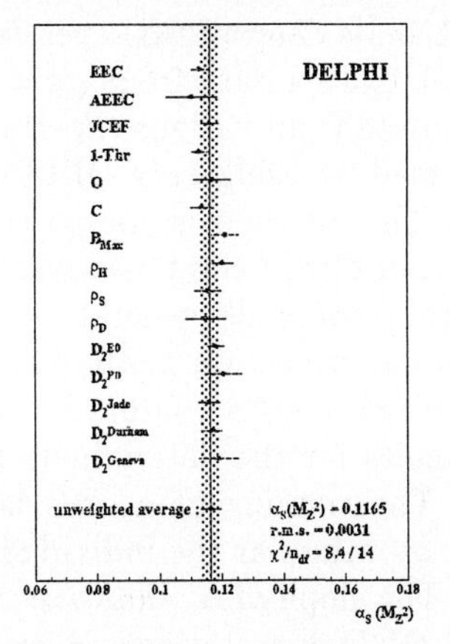

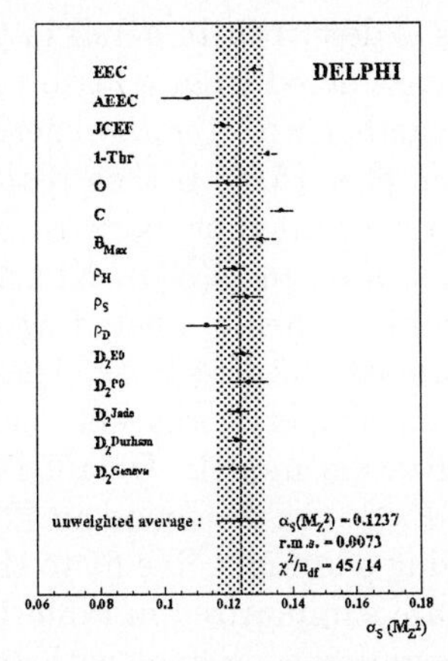

Fig. 1. Results of the $\alpha_s(M_Z^2)$ measurements from 15 event shape distributions. (a) fits using experimentally optimized scale values, (b) fixed scale fits : $x_\mu = 1$. For the errors on the $\alpha_s(M_Z^2)$ measurements, which are indicated by the horizontal lines, the uncertainty from the fit, the systematic experimental uncertainty and the hadronization uncertainty have been added quadraticly. Also shown are the unweighted averages.

The χ^2 value for the comparison of α_s from 15 observables, calculated by using the total uncertainty as mentioned above, yields $\chi^2 = 8.4$ for OPT and $\chi^2 = 45$ for PHS, i.e. the individual measurements are consistent with each other in the case of OPT, whereas they are clearly inconsistent for the PHS method. The rather small χ^2 value for OPT indicates that the individual observables are highly correlated. For the calculation of a weighted average the correlations have clearly to be taken into account. Since the exact correlation pattern is unknown and can hardly be worked out reliably, a method proposed in [5] is used in order to estimate the correlations between the observables. A weighted average based on this method yields $\alpha_s(M_Z) = 0.1164 \pm 0.0021$, which agrees well with the unweighted average given above.

2 Determination of α_s from DELPHI Data at $\sqrt{s} =$ 130 GeV - 172 GeV

For the QCD analysis from DELPHI data at $\sqrt{s} = 130$ Gev $= 172$ GeV the fragmentation model generator independent power correction ansatz from Y. L. Dokshitzer and B. R. Webber [6] (DW-model) have been applied to calculate the hadronization corrections for the mean values of event shape distributions. This ansatz has the advantage of making use of the data statistics over the whole kinematical range. Within the model, $\bar{\alpha_0}$ is a non-perturbative, free parameter, describing the contributions below an infrared matching scale μ_I.

The analysis is described in detail in [2]. $\bar{\alpha_0}$ is determined seperately for the observables considered from a fit to the DELPHI data from $\sqrt{s} = 91.2$ GeV - 172 GeV together with the low energy data from various experiments. The determination of $\alpha_s(M_Z^2)$ is then performed by using only DELPHI data at different center of mass energies E_{CM}. This method is compared with the determination of α_s in $\mathcal{O}(\alpha_s^2)$, NLLA, and $\mathcal{O}(\alpha_s^2)$ combined with NLLA in the $\ln R$ matching scheme, where fragmentation model generators are applied for the hadronization corrections. The power correction method for the mean values of event shapes works well and yields results compatible with those from hadronization models. Detailed results for the determination of α_s at the various E_{CM} can be found in [2]. The running of α_s has been determined applying a straight line fit to the α_s values at the individual energies. The results are summarized in table 1. The slope of α_s indicates running of the strong coupling consistent with the QCD expectations. However, in order to distinghuish from a QCD theory extended by light gluinos more data statistics is required.

Prediction	$d\alpha_s/dE_{CM}[10^{-3}GeV^{-1}]$
DW-model / $\mathcal{O}(\alpha_s^2)$	-0.16 ± 0.06
$\mathcal{O}(\alpha_s^2)$	-0.07 ± 0.07
NLLA	-0.20 ± 0.09
$\mathcal{O}(\alpha_s^2)$ + NLLA ($\ln R$ -scheme)	-0.14 ± 0.07
QCD expectation	-0.128
QCD + Gluinos	-0.095

Table 1. Running of α_s determined from DELPHI data at $\sqrt{s} = 130$ GeV - 172 GeV using the DW-model in comparison with measurements applying conventional hadronization correction methods. Also listed are the theoretical predictions for standard QCD and for QCD extended with light gluinos.

References

[1] J. Drees and S. Hahn, DELPHI 97-67 CONF 54, contributed to the EPS HEP conference in Jerusalem, 1997.
[2] J. Drees et al. , DELPHI 97-92 CONF 77, contributed to the EPS HEP conference in Jerusalem, 1997.
[3] M. Seymour, program EVENT2, URL http://suryal1.cern.ch/users/seymour/nlo see also S. Catani and M. Seymour, Nucl. Phys. B485 (1997) 291.
[4] E. B. Zijlstra and W. L. van Neerven, Nucl. Phys. B383 (1992) 525.
[5] M. Schmelling, Phys. Scripta 51 (1995) 676.
[6] Y. L. Dokshitzer and B. R. Webber, Phys. Lett. B352 (1995) 451.

QCD Studies and α_s Measurements at LEP1 and LEP2

Sunanda Banerjee (Sunanda.Banerjee@cern.ch)

L3 Collaboration &
Tata Institute of Fundamental Research, Bombay

Abstract. We present a study of the structure of hadronic events recorded by the L3 detector at LEP over a wide range of centre-of-mass energies. The distributions of event shape variables and the energy dependence of their mean values are well reproduced by QCD models. From a comparison of the data with resummed $\mathcal{O}(\alpha_s^2)$ QCD calculations, we determine the strong coupling constant at these energies. We find that the strong coupling constant decreases with increasing energy as expected in QCD.

1 Introduction

Multi-hadron production in e^+e^- interactions over a large range of centre of mass energies at LEP allows us to test the predictions of the theory of the strong interaction (QCD). Studies of events with isolated high energy photons in hadronic Z decays at LEP also offer an important probe of the short distance structure of QCD. The high energy photons are radiated early in the process either through initial state radiation or from quark bremsstrahlung. On the other hand, the development of the hadronic shower takes place over a larger time scale. So a study of the recoiling hadronic system in events with hard isolated photon radiation gives access to hadron production at reduced centre of mass energies.

We study here the characteristics of hadronic events from data collected by the L3 detector [1] in terms of the charged particle multiplicities, some infrared and collinear safe global event shape variables (thrust, heavy jet mass, total and wide jet broadenings), jet rates and mean jet multiplicities as a function of jet resolution parameters. The measured distributions are compared with event generators based on an improved leading log approximation (Parton Shower models including QCD coherence effects). The measured distributions of event shape variables at the different centre of mass energies have also been compared to the predictions of a second order QCD calculation with resummed leading and next-to-leading terms. This provides determinations of the strong coupling constant α_s at several centre of mass energies allowing a study the energy evolution of α_s.

2 Data

The data come from three sources : (a) data at Z peak; (b) high energy runs at LEP; (c) hadronic Z decays with hard isolated photons. The event selection procedure closely follows those in earlier studies [2]. Hadronic events are characterised by large visible energy and high multiplicity of final state particles and are selected using these characteristics.

The high energy data can be grouped to four centre of mass energies and are summarised in table 1. The high energy data have large contamination due to radiative return to Z and due to hadronic decays of W's. Events due to radiative returns to Z are rejected by demanding (a) low longitudinal imbalance $((E_{vis}/\sqrt{s}) > 2.0 (\mid E_{\parallel} \mid /E_{vis}) + 0.5)$; (b) no energetic photon seen in the detector ($E_\gamma < 30$ GeV). There is a large contribution from WW events at high centre of mass energies ($\sqrt{s} > 161$ GeV). W-pair events are first selected by (a) finding high multiplicity events; (b) forcing them into four jet configuration ($k_\perp$ algorithm); (c) selecting good four jet candidates by imposing cuts on most and least energetic jets; (d) using a cut on jet resolution parameter y^D_{34}. These W-pair events are then rejected resulting selection efficency of $\approx 90\%$ for the high energy hadron sample and a purity above 75%.

$\sqrt{s}$ (GeV)	$\int \mathcal{L}(\mathrm{pb}^{-1})$	# of Events
130–140	5.0	402
161	11.1	443
172	10.2	341
183	6.1	163

Table 1. Selected data sample at different centre of mass energies

For studies at $\sqrt{s} < m_Z$, events of the type $e^+e^- \to \gamma^*/Z \to$ hadrons + γ are selected from 142.4 pb^{-1} data collected between 1991 and 1995. The dominant background to these events are due to π^0 being misidentified as a photon. The background is reduced by (a) selecting neutral electromagnetic shower with $E > 5$ GeV; (b) demanding the photon candidate to be isolated locally (no other cluster with $E > 50$ MeV within $15°$) as well as from the nearest jet; (c) using a powerful π^0/γ discriminator based on neural network. The photon candidate is then removed from the event and the event is boosted back to the centre of mass frame of the remaining subsystem. This gives selection efficiency between 27.4% and 48.3% for high energy photon radiated events. The purity of the sample is between 68% and 90%.

Distributions at detector are level corrected bin-by-bin to particle level for remaining background, finite detector resolution, acceptance and initial and final state radiation. Charged particle multiplicity has been corrected using matrix unfolding method. Correction factors typically lie between 0.5 and 1.5 and detector and ISR corrections compensate each other.

3 Comparison with QCD Predictions

The measured distributions are compared with event generators based on an improved leading log approximation (Parton Shower models including QCD coherence effects). Four such Monte Carlo programs, ARIADNE 4.06 [3], COJETS 6.23 [4], HERWIG 5.8 [5] and JETSET 7.4 [6] have been used for these comparisons. All the models were tuned using Z peak data from L3. These programs differ in the variables used to define the Parton Shower evolution and also in the modelling of the hadronisation effects.

The jet rates as well as the global event shape distributions are well described by the Parton Shower models. For data at $\sqrt{s} < m_Z$, Monte Carlo events have been generated with the corrected $\sqrt{s}$ distribution and correct flavour composition but without photon radiation. Figure 1 shows the energy evolution of mean thrust and mean charged particle multiplicity compared with the different QCD models. One sees Parton Shower models with QCD coherence effects can describe the data.

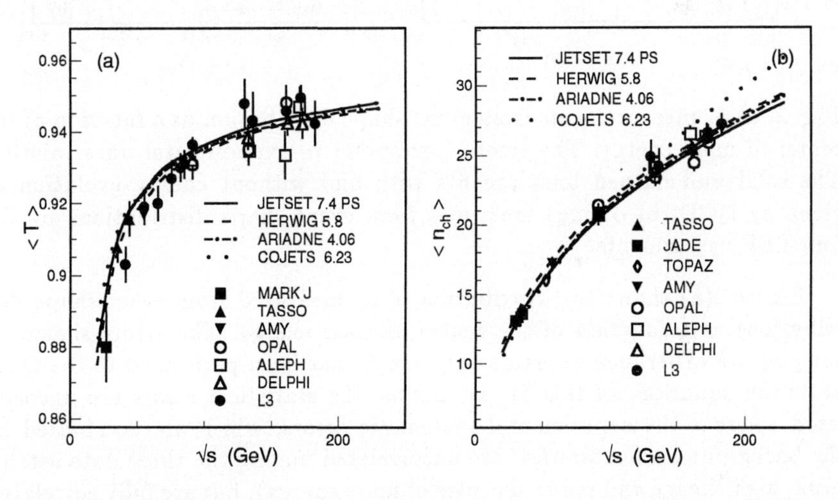

Fig. 1. Distributions of (a) mean thrust, $< T >$, and (b) mean charged particle multiplicity, $< n_{ch} >$, as a function of the centre of mass energy, compared to several QCD models.

The event shape distributions have also been compared to the analytical calculations where complete calculations to $\mathcal{O}({\alpha_s}^2)$ have been supplemented with resummed calculation of leading and next-to-leading log terms [7]. These combined calculations describe data over a wide kinematical range. Hadronisation effect has been folded in using Parton Shower Monte Carlo and a fit to the data yields a value of α_s at the corresponding centre of mass energy. Thrust, scaled heavy jet mass and the jet broadening variables at the 11 centre of mass energy points give reasonable fits. Systematic errors are estimated separately for each measurement of α_s (at each energy from each of these variables) by varying the hadronisation correction, renormalisation scale, matching algorithm. The measurements at each energy point are statistically correlated but have different systematic errors. So we take unweighted mean for the α_s value.

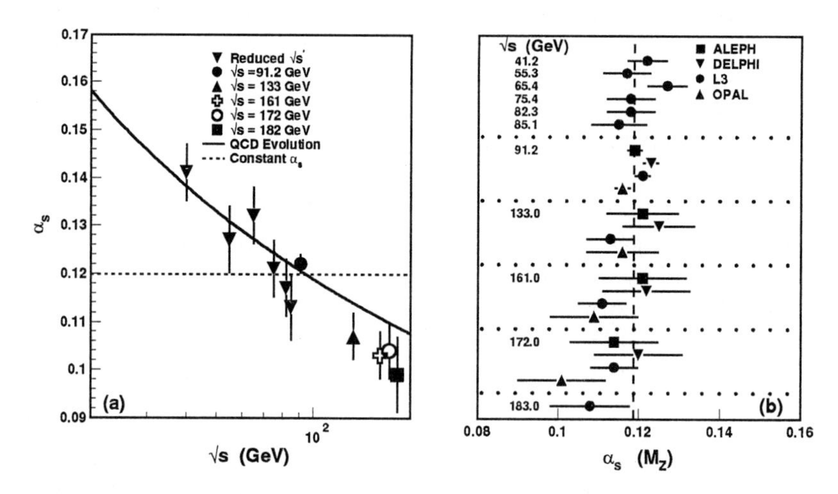

Fig. 2. a) α_s measurements from event shape distribution as a function of the center of mass energy. The errors correspond to experimental uncertainties. The solid and dashed lines are fits with and without energy evolution as given by QCD. b) $\alpha_s(m_Z)$ measured from event shape distributions by the four LEP experiments.

Figure 2(a) shows the distribution of α_s measured from event shape distributions as a function of the centre of mass energy. The errors shown on the plot are experimental errors only and fit has been performed to the QCD evolution equation. In this fit, we assume the statistical errors are uncorrelated whereas the experimental systematic errors, which are dominated by the background uncertainties, are uncorrelated among the three data sets (Z peak, high energy and reduced centre of mass energy), but are fully correlated between individual low energy or high energy measurements. The fit gives a χ^2 of 12.6 for ten degrees of freedom corresponding to a confidence level of 0.25 with a fitted value of α_s:

$$\alpha_s(m_Z) = 0.1205 \pm 0.0016\,(\text{expt}) \pm 0.0066\,(\text{theory})$$

On the other hand, a model with constant α_s gives a χ^2 of 47.5 which corresponds to a confidence level of 8×10^{-7}.

Figure 2(b) summarises α_s values from our measurements together with similar measurements from the other three LEP experiments [8], evaluated at the m_Z scale according to the QCD evolution equation. All the measurements are consistent among themselves and to a value

$$\alpha_s(m_Z) = 0.1188 \pm 0.0009\,(\text{expt})$$

obtained from a fit which ignores correlation among the experiments.

References

[1] L3 Collaboration, B. Adeva *et al.*, Nuclear Instrument & Methods **A 289** (1990) 35;
M. Chemarin *et al.*, Nuclear Instruments & Methods **A 349** (1994) 345;
M. Acciarri *et al.*, Nuclear Instruments & Methods **A 351** (1994) 300;
G. Basti *et al.*, Nuclear Instruments & Methods **A 374** (1996) 293;
A. Adam *et al.*, Nuclear Instruments & Methods **A 383** (1996) 342.

[2] L3 Collaboration, B. Adeva *et al.*, Physics Letters **284** (1992) 471;
L3 Collaboration, M. Acciarri *et al.*, Physics Letters **B371** (1996) 137;
L3 Collaboration, M. Acciarri *et al.*, Physics Letters **B404** (1997) 390;
L3 Collaboration, M. Acciarri *et al.*, CERN preprint CERN-PPE/97-74.

[3] U. Pettersson, "ARIADNE: A Monte Carlo for QCD Cascades in the Color Dipole Formulation", Lund Preprint, LU TP 88-5 (1988);
L. Lönnblad, "The Colour Dipole Cascade Model and the Ariadne Program", Lund Preprint, LU TP 91-11 (1991).

[4] R. Odorico, Nuclear Physics **B228** (1983) 381;
R. Odorico, Computer Physics Communications **32** (1984) 139, Erratum: **34** (1985) 43;
R. Mazzanti and R. Odorico, Nuclear Physics **B370** (1992) 23; Bologna preprint DFUB 92/1.

[5] G. Marchesini and B. Webber, Nuclear Physics **B310** (1988) 461;
I.G. Knowles, Nuclear Physics **B310** (1988) 571;
G. Marchesini *et al.*, Computer Physics Communications **67** (1992) 465.

[6] JETSET 7.4 Monte Carlo Program:
T. Sjöstrand, Computer Physics Communications **39** (1986) 347;
T. Sjöstrand and M. Bengtsson, Computer Physics Communications **43** (1987) 367.

[7] S. Catani *et al.*, Physics Letters **B263** (1991) 491;
S. Catani *et al.*, Physics Letters **B272** (1991) 360;
S. Catani *et al.*, Physics Letters **B295** (1992) 269.

[8] ALEPH Collaboration, H. Stenzel, contributed to this conference;
DELPHI Collaboration, S. Hahn, contributed to this conference;
OPAL Collaboration, D. Chrisman, contributed to this conference.

Recent QCD Results from OPAL

David Chrisman, on behalf of the OPAL collaboration (chrisman@cern.ch)

University of California, Riverside

Abstract. Here we present OPAL measurements which extend the LEP I program: a measurement of the inclusive quark-to-photon fragmentation function and an analysis of the multiplicity distributions of qluon and quark jets; as well as first preliminary results from analysis of LEP II data taken at a center-of-mass energy of $\sqrt{s} = 183$ GeV.

1 Quark-to-Photon Fragmentation Function

The properties of hadrons and charged leptons produced in e^+e^- collisions have been studied in great detail at different centre-of-mass (cms) energies. In the case of photons radiated off quarks, prompt photons, in hadronic e^+e^- collisions, much less information is available due to the difficulty of separating these photons from those produced in the decays of other particles. Both the shape and normalisation of the inclusive prompt photon energy spectrum in e^+e^- collisions are predicted, through the calculation of the quark-to-photon fragmentation function, by leading-order perturbative QCD [1, 2]. This asymptotic prediction has been parametrised in [3]. Non-perturbative effects can be included in the calculation through the vector-meson dominance ansatz as in [4, 5], where boundary terms missing in [3] were also accounted for.

Here we report on measurement of the inclusive prompt photon energy spectrum in hadronic Z^0 decays at LEP [6]. This method of studying the quark-to-photon fragmentation function was suggested in [5, 7]. To separate prompt photons from the photons from decays of other particles we use the following method. We selected clusters in the electromagnetic calorimeter not associated with charged tracks. A set of cuts were applied to reduce the background in the sample. The distribution of a variable characterising the transverse shape of the clusters in data was then fitted with a linear combination of the distributions for photons and for background to determine the fraction of prompt photons in the selected sample. The fully corrected energy spectrum of prompt photons in hadronic Z^0 decays is shown in Fig. 1 together with the theoretical predictions. To compare our result with leading-order theoretical predictions for the quark-to-photon fragmentation function we use the leading-order cross-section for prompt photon production in e^+e^- annihilation given by formula (12) in [7]. Our data are in agreement with these theoretical predictions. The experimental precision is not sufficient to discriminate between them.

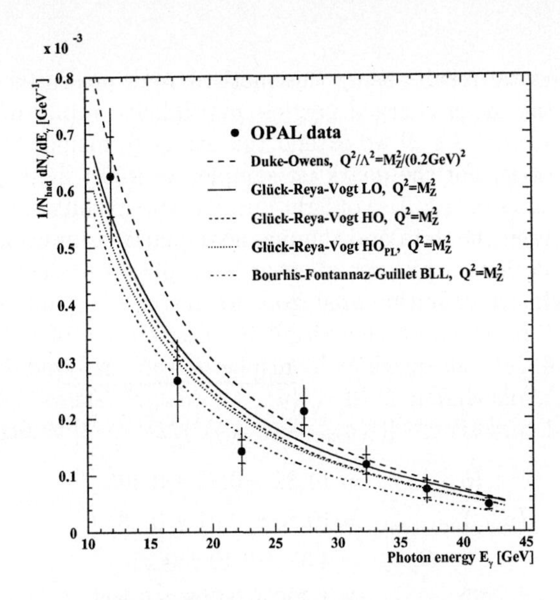

Fig. 1. The photon energy spectrum in hadronic Z^0 decays compared to various theoretical predictions: the Duke-Owens parametrisation [3], the Glück, Reya and Vogt predictions including leading-order (LO), higher-order (HO) and higher-order without the non-perturbative corrections (HO_{PL}) [4]. The Bourhis, Fontannaz and Guillet prediction shown include effects beyond leading logarithms (BLL) [5].

2 Multiplicity Distributions of Gluon and Quark Jets

In contrast to studies of quark jets, conclusive experimental studies of gluon jets have been rare. This is because the creation of a gluon jet pair, gg, from a color singlet point source – allowing an inclusive definition analogous to that for quark jets which arise uniquely from the point-like creation of quark-antiquark $q\overline{q}$ pairs – has been only rarely observed in nature.

In the analysis presented here (see Ref. [8] for details) we use a method proposed in [9] for LEP experiments to identify gluon jets using an inclusive definition similar to that used for analytic calculations. The method is based on rare events of the type $e^+e^- \to q\overline{q}\, g_{\text{incl.}}$ in which the q and $\overline{q}$ are identified quark (or antiquark) jets which appear in the same hemisphere of an event. The object $g_{\text{incl.}}$ taken to be the gluon jet, is defined by the sum of all particles observed in the hemisphere opposite to that containing the q and $\overline{q}$. In the limit that the q and $\overline{q}$ are collinear, the gluon jet $g_{\text{incl.}}$ is produced under the same conditions as gluon jets in gg events. The $g_{\text{incl.}}$ jets therefore correspond closely to single gluon jets in gg events, defined by dividing the gg events in half using the plane perpendicular to the principal event axis.

First experimental results using this method were presented in [10] and were limited to the mean charged particle multiplicity values of gluon and quark jets. In this analysis [8] we extend this study to include the full multiplicity distributions. For the quark jet sample, we select light quark (uds) event hemispheres, as in [10]. Use of light quark events results in a better correspondence between the data and the massless quark assumption employed for analytic calculations, while use of event hemispheres to define the quark jets yields an inclusive definition analogous to that of the gluon jets $\mathrm{g_{incl.}}$.

The corrected charged particle multiplicity distributions of 41.8 GeV $\mathrm{g_{incl.}}$ gluon jet and 45.6 GeV uds quark jet hemispheres were analyzed to determine their mean $\langle n_{\mathrm{ch}}\rangle$, dispersion $D \equiv \sqrt{\langle n^2_{\mathrm{ch.}}\rangle - \langle n_{\mathrm{ch.}}\rangle^2}$, skew $\gamma \equiv \langle(n_{\mathrm{ch.}} - \langle n_{\mathrm{ch.}}\rangle)^3\rangle/D^3$ and curtosis $c \equiv [(\langle(n_{\mathrm{ch.}} - \langle n_{\mathrm{ch.}}\rangle)^4\rangle/D^4) - 3]$ values:

$$
\begin{aligned}
\langle n_{\mathrm{ch.}}\rangle_{\mathrm{g\,incl.}} &= 14.32 \pm 0.23 \pm 0.40 \\
\langle n_{\mathrm{ch.}}\rangle_{\mathrm{uds\,hemis.}} &= 10.10 \pm 0.01 \pm 0.18 \\
D_{\mathrm{g\,incl.}} &= 4.37 \pm 0.19 \pm 0.26 \\
D_{\mathrm{uds\,hemis.}} &= 4.298 \pm 0.008 \pm 0.098 \\
\gamma_{\mathrm{g\,incl.}} &= 0.38 \pm 0.13 \pm 0.18 \\
\gamma_{\mathrm{uds\,hemis.}} &= 0.822 \pm 0.007 \pm 0.044 \\
c_{\mathrm{g\,incl.}} &= 0.18 \pm 0.34 \pm 0.30 \\
c_{\mathrm{uds\,hemis.}} &= 0.98 \pm 0.03 \pm 0.11 \quad ,
\end{aligned}
$$

where the first uncertainty is statistical and the second is systematic.

In addition, a factorial moment analysis of the gluon and quark jet multiplicity distributions was performed in order to test the predictions of QCD analytic calculations [11, 12] of those moments, details of which can be found in [8].

3 Preliminary Results at $\sqrt{s} = 183$ GeV

In July 1997 the LEP II program advanced to produce $\mathrm{e^+e^-}$ collisions at a cms energy of $\sqrt{s} \approx 183$ GeV. The initial study of this data outlined here (see Ref. [13] for details), based on an integrated luminosity of approximately 6.6 $\mathrm{pb^{-1}}$, focuses on general features of hadronic events in $\mathrm{e^+e^-}$ annihilations and is a continuation of our earlier publications at cms energies of $\sqrt{s} = 172$, 161, 130-136 GeV [14, 15, 16]. In [13] we examine the overall consistency of QCD at yet higher cms energies, including studies of event shape observables, jet production rates, and two topics we highlight here, charged particle multiplicity and charged particle momenta spectra.

The charged particle multiplicity distribution measured in [13] was compared with the PYTHIA, HERWIG and ARIADNE models and it was found that they described the data reasonably well. As we already observed at previous LEP II energies, the COJETS model predicts too many high multiplicity events and clearly disagrees with the data. We determine the mean value and the dispersion to be:

$$\langle n_{\mathrm{ch}} \rangle = 27.1 \pm 0.8 \pm 0.4$$
$$D = 9.0 \pm 0.6 \pm 0.7 \quad ,$$

where the first error is statistical and the second systematic.

The ξ_p distribution, $1/\sigma \cdot \mathrm{d}\sigma_{\mathrm{ch}}/\xi_p$, where $x_p = 2p/\sqrt{s}$, $\xi_p = \ln(1/x_p)$ and p is the measured track momentum, was also measured. The shape of the ξ_p distribution was also well described by PYTHIA, HERWIG and ARIADNE, in contrast to COJETS. The COJETS Monte Carlo predicts too many particles in the region of the peak and at large values of ξ_p, where low momentum particles contribute.

We fit a skewed Gaussian [15] to the region close to the peak of the ξ_p distribution and determine the position of the peak, ξ_0, to be

$$\xi_0 = 4.17 \pm 0.05 \pm 0.04 \quad ,$$

where the first error is statistical and the second systematic.

References

[1] E. Witten, Nucl. Phys. **B120** (1977) 189.
[2] C.H. Llewellyn Smith, Phys. Lett. **B79** (1978) 83.
[3] J.F. Owens, Reviews of Modern Physics **59** (1987) 465,
D.W. Duke, J.F. Owens, Phys. Rev. **D26** (1982) 1600.
[4] M. Glück, E. Reya, A. Vogt, Phys. Rev. **D48** (1993) 116.
[5] L. Bourhis, M. Fontannaz, J.Ph. Guillet, preprint **hep-ph 9704447**.
[6] OPAL Coll., *Measurement of the quark to photon fragmentation function through the inclusive production of prompt photons in hadronic* $\mathbf{Z^0}$ *decays*, preprint **CERN-PPE/97-086**.
[7] Z. Kunszt, Z. Trócsányi, Nucl. Phys. **B394** (1993) 139.
[8] OPAL Coll., *Multiplicity distributions of gluon and quark jets and tests of QCD analytic predictions*, preprint **CERN-PPE/97-105**.
[9] J. W. Gary, Phys. Rev. **D49** (1994) 4503.
[10] OPAL Coll., Phys. Lett. **B388** (1996) 659.
[11] E. D. Malaza and B. R. Webber, Phys. Lett. **B149** (1984) 501;
E. D. Malaza and B. R. Webber, Nucl. Phys. **B267** (1986) 702.
[12] I. M. Dremin and R. C. Hwa, Phys. Rev. **D49** (1994) 5805.
[13] OPAL Coll., *Initial studies of hadronic events at 183 GeV at LEP 2*, OPAL Physics Note **PN315** (1997).
[14] OPAL Coll., Z. Phys. **C72** (1996) 191
[15] OPAL Coll., Z. Phys. **C75** (1997) 193.
[16] OPAL Coll., *QCD studies with e+e- annihilation data at 172 GeV*, OPAL Physics Note **PN281** (1997).

Hard Processes as Tests of Perturbative QCD from D

Robert L. McCarthy (mccarthy@sbhep.physics.sunysb.edu)

State University of New York, Stony Brook, NY 11790, USA

Abstract. Results from the DØ experiment are presented on the jet inclusive cross section; the dijet mass distribution; the dijet angular distribution; the P_T distribution of W, Z, and diphotons; and the ratio of $(W + 1\text{jet})/(W + 0\text{jets})$ events. All results are consistent with the predictions of QCD except perhaps the production of jets in events with a W, which appears to be inconsistent with an NLO QCD prediction.

1 The DØ Experiment

The DØ experiment is specially suited to test the predictions of QCD because of the hermetic coverage and fine resolution of the DØ calorimeter for jets and missing E_T, combined with excellent coverage for electrons. The calorimeter has full coverage for hadronic measurements out to rapidities $|\eta| \leq 4.2$ with at least 7 interactions lengths of calorimetry over most of this region. Hence there are no long tails on the calorimeter energy measurement to cause confusion. DØ is in the final stages of determination of the calibration of the jet energy scale. Hence the errors due to the jet energy measurement are decreasing, and efforts are underway to include correlations in the goodness of fit to theory. Final results will appear shortly.

2 Inclusive Jet Production

DØ has measured the inclusive jet production cross section using 13 pb^{-1} of data from the 1992-93 run of the Fermilab Tevatron Collider and 93 pb^{-1} from the 1994-95 run. The measurements are compared to NLO QCD predictions from the JETRAD Monte Carlo written by Giele, Glover and Kosower[1]. Remarkably good agreement is seen in jet production with transverse energies ranging from 35 GeV to 450 GeV as the cross section falls by seven orders of magnitude. In Figure 1 we show the the difference between data and theory in ratio to the theory using the MRSA′ structure functions at a scale of the maximum E_T in the event divided by 2. Statistical errors are shown on the points. The systematic error (dominated by the uncertainty in the jet energy scale) is shown by the dashed band. Changes in the theoretical prediction due to variation of the choice of structure functions among modern sets can cause a change of up to 10% in the NLO prediction.

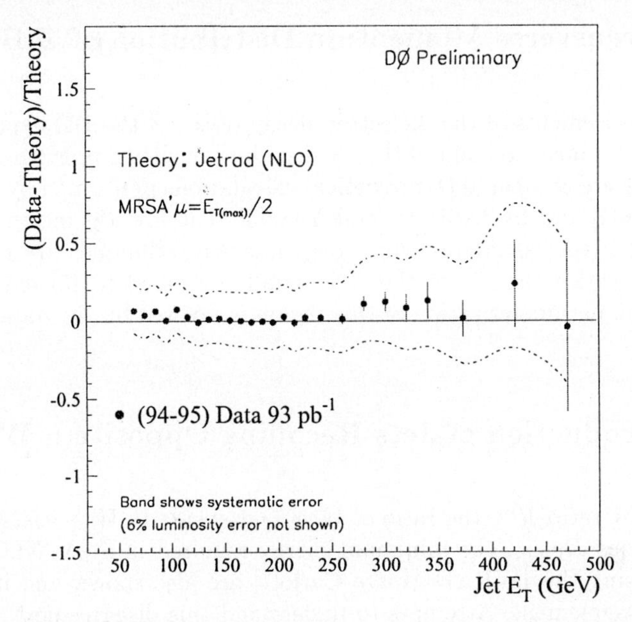

Fig. 1. The inclusive jet cross section in the region $|\eta| \leq 0.5$ is compared to the JETRAD NLO prediction[1].

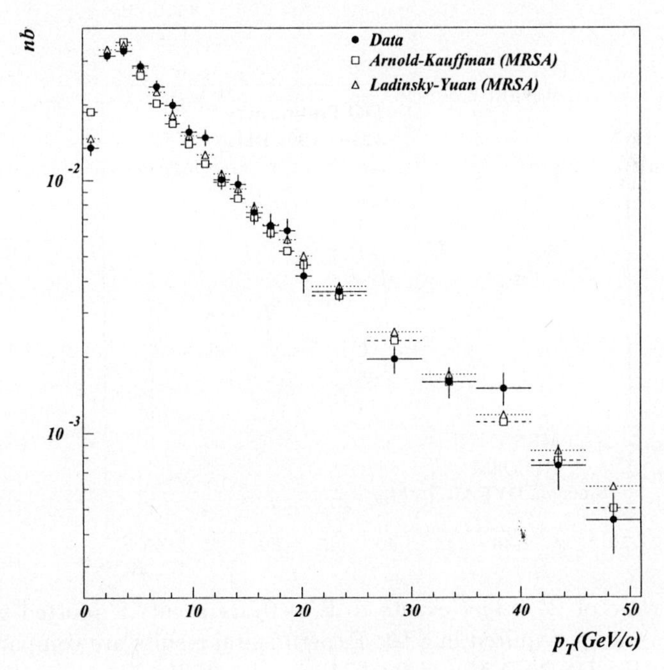

Fig. 2. The differential cross section $d\sigma/dp_T$ (nb/GeV) for Z production is compared to theoretical predictions [2] and [3].

3 The Transverse Momentum Distribution of Z Bosons

Through measurements of the dielectron decay of the Z the DØ experiment has carried out a measurement of the P_T distribution shown in Figure 2. The measurements are compared to theoretical calculations performed by Arnold and Kauffman[2] and by Ladinsky and Yuan[3]. The low P_T region (below 10 GeV/c) is of special interest since resummation techniques are required in the region where the P_T of the Z is small compared to its mass. The Ladinsky-Yuan parameterization of this region is favored by the data.

4 The Production of Jets Recoiling Opposite a W

In Figure 3 the ratio R^{10}, the ratio of $W + 1jet$ events to $W + 0jets$ events is plotted versus E_{Tmin}, the minimum energy required in a jet. NLO QCD predictions using the DYRAD Monte Carlo[1] are also shown and indicate significant disagreement. Attempts to understand this disagreement include investigation of the possibility that resummation techniques may be required in making the predictions, as is true for the p_T distribution of the Z and the W.

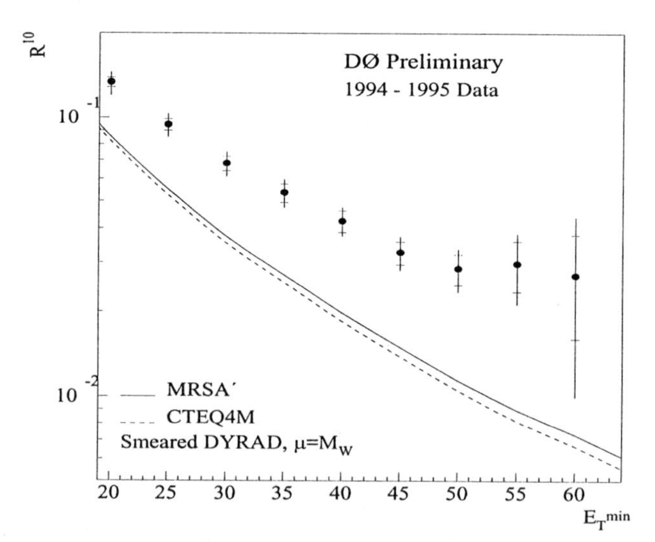

Fig. 3. The ratio of $W + 1jet$ events to $W + 0jets$ events is plotted versus the minimum energy required in a jet. Experimental results are compared to predictions of the DYRAD Monte Carlo[1].

References

[1] W. T. Giele, E. W. N. Glover, and D. A. Kosower, Nucl. Phys. **B403**, 633 (1993) and private communication.
[2] P. B. Arnold and R. Kaufman, Nucl. Phys. **B349**, 381 (1991).
[3] G. A. Ladinsky and C. P. Yuan, Phys. Rev. **D50**, 4239 (1994).

Hard Processes and Perturbative QCD Results from CDF

Elizabeth Buckley-Geer
for the CDF Collaboration (buckley@fnal.gov)

Fermi National Accelerator Laboratory

Abstract. We present results on the inclusive jet cross section at $\sqrt{s} = 1800$ GeV and 630 GeV, the two-jet cross section, multijet physics and the multijet differential cross section from the CDF experiment at the Fermilab Tevatron Collider.

1 The inclusive jet cross section at 1800 GeV and 630 GeV

The inclusive jet cross-section is obtained by measuring the number of events in a given bin of E_T normalized by the integrated luminosity and acceptance. The published CDF result based on a 19.5 pb^{-1} data sample showed an excess of events at high E_T [1]. The preliminary results from 87 pb^{-1} of data from Run IB are shown in Fig. 1(left) compared to NLO QCD predictions [2] using a renormalization scale $\mu = E_T/2$ and the CTEQ3M parton distribution functions (PDFs). The results are also compared to the previous data using the same PDF and scale. The two datasets are in good agreement. The systematic uncertainties are expected to be about the same size as the published result. Another way to test QCD is to measure the inclusive jet cross-section at two different center-of-mass energies. The scaling hypothesis predicts that if the cross-sections are written in a form that makes them dimensionless then they will be independent of $\sqrt{s}$. QCD predicts that there will be scaling violations due to the evolution of the PDFs with Q^2 and the running of α_s . The CDF experiment has recorded data at $\sqrt{s} = 546$ and 630 GeV in addition to 1800 GeV. In a previous measurement [3] using data at $\sqrt{s} = 546$ GeV, scaling was ruled out at the 95% C.L. but a disagreement with the NLO QCD predictions was observed in the low E_T region at the level of 1.5-2 σ. Fig. 1(right) shows the ratio of the scaled cross-sections plotted as a function of $x_T = 2E_T/\sqrt{s}$. The same disagreement that was observed at 546 GeV is observed in the low x_T region for the data at 630 GeV. The systematic uncertainties for the previous measurement at 546 GeV are shown, these are not expected to change significantly for 630 GeV.

2 Inclusive two-jet cross section

In this measurement, the two highest E_T jets are identified and one is required to be in the central ($0.1 \leq |\eta| \leq 0.7$) region. Because the central region

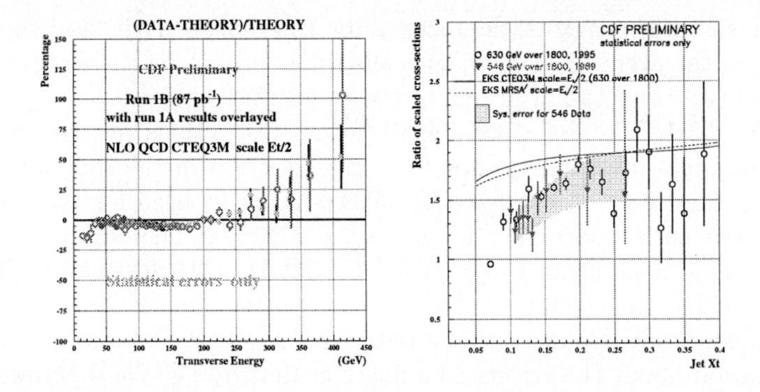

Fig. 1. Left: The preliminary Run 1B inclusive jet measurement is compared to the published CDF result and to NLO QCD predictions using the CTEQ3M PDF. Right: The ratio of scaled cross sections at $\sqrt{s} = 1800$ and 630 GeV.

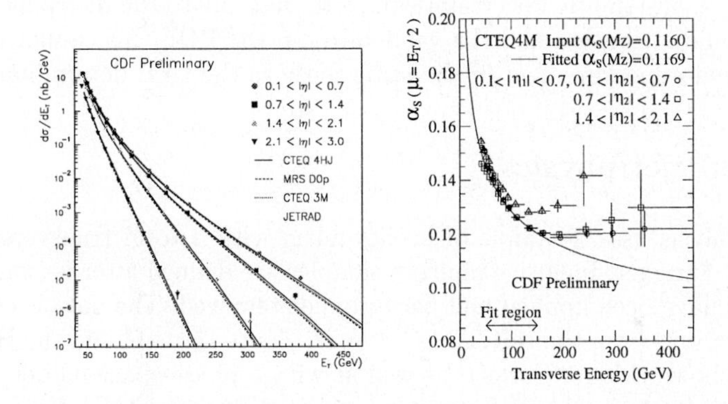

Fig. 2. Left: Comparison of data and NLO cross sections in the different rapidity bins. Right: Extracted α_s values as a function of jet E_T for CTEQ4M.

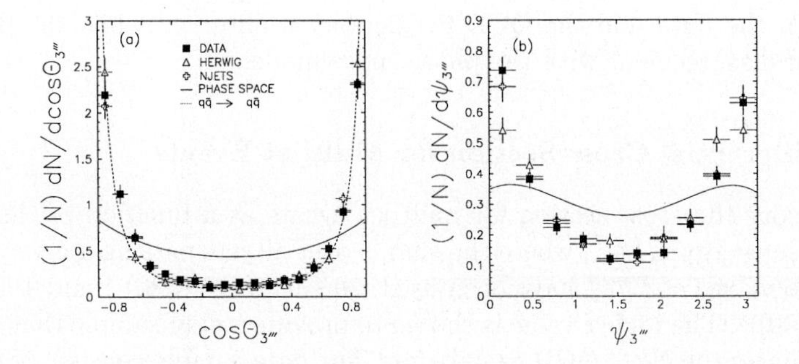

Fig. 3. Distributions of $cos\theta^*$ and ψ for events with ≥ 6 jets compared to theoretical predictions.

has the smallest energy scale uncertainty, the central jet is used to measure the E_T of the event. The other jet, called the "probe" jet, is required to have E_T >10 GeV and to fall in one of the η bins: 0.1≤ $|\eta|$ ≤0.7, 0.7≤ $|\eta|$ ≤1.4, 1.4≤ $|\eta|$ ≤2.1, 2.1≤ $|\eta|$ ≤3.0. There are no restrictions on the presence of additional jets. Fig. 2(left) shows the cross section in the individual η bins as a function of the central jet E_T. JETRAD [4] is used for the theoretical predictions with renormalization scale $\mu = E_T^{max}/2$. The data are compared to the predictions using three PDFs, CTEQ4HJ, MRSD0', and CTEQ3M. The statistical uncertainty is shown on the points; the systematic uncertainty is under study. Comparing the data with NLO QCD allows us to extract information about the strong coupling constant α_s [5]. We determine $\alpha_s(\mu)$ for each bin of E_T , η_1, η_2 using $\mu = E_T/2$. The fit region is $50 < E_T < 150$ GeV and 0.1≤ $|\eta_1|$ ≤0.7, 0.1≤ $|\eta_2|$ ≤0.7. The result of the fit for CTEQ4M is shown in Fig. 2(right). The running of α_s with E_T can be clearly seen. Evolving back to $\alpha_s(M_Z)$ yields $\alpha_s(M_Z) = 0.117 \pm 0.009$ (statistical + experimental systematic uncertainties). Note that due to the interplay between the gluon distribution and the value of α_s in the PDF this cannot really be considered a measurement in the same sense as the LEP determinations.

3 Multijet physics

This analysis uses a sample of events taken with a total transverse energy ($\sum E_T^{jet}$) trigger. Inclusive multijet samples are defined after jet energy corrections have been applied and backgrounds removed. The data is compared to predictions from the NJETS LO $2 \to N$ Monte Carlo[6] and the HERWIG parton shower Monte Carlo[7] as well as with a phase-space model.

In its rest frame an N-Jet system can be defined by $4N - 4$ independent variables [8]. The data contains 3, 4, 5 and 6 jet events, providing 56 variables which can be compared to theoretical predictions [9]. Both HERWIG and NJETS give reasonable descriptions of all 56 variables. Fig. 3 shows the angular distributions $\cos\theta_3$ and Ψ_3 for events with ≥ 6 jets. Good agreement between the data and the QCD predictions is observed while the data is clearly in disagreement with the phase space model.

3.1 Differential Cross Section for Multijet Events

We measure the cross section for multijet events as a function of the total transverse energy ΣE_T^{jet}, where the sum is over all jets passing a given E_T^{min} [10]. Two values of E_T^{min} have been used, 20 GeV ($E_T^{min}(20)$) and 100 GeV ($E_T^{min}(100)$). The higher value is chosen to provide a data sample that better approximates the NLO QCD calculation. The data sample consists of events with $\Sigma E_T^{jet} > 320$ GeV. The data have been corrected for the effects of detector resolution. The data are compared to predictions from HERWIG with CTEQ2L PDF's and renormalization scale $Q^2 = stu/2(s^2 + u^2 + t^2)$

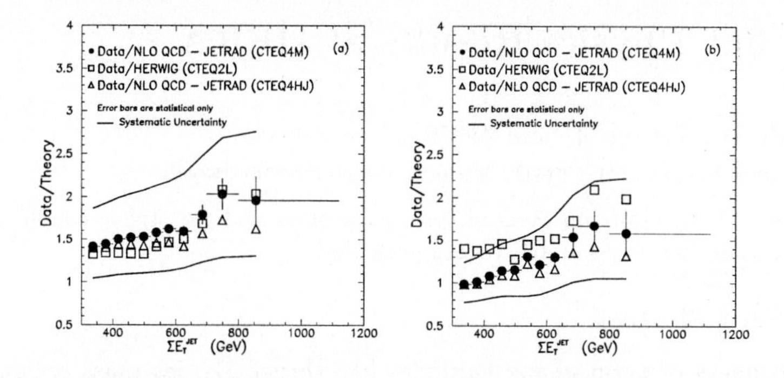

Fig. 4. Multijet differential cross section for (a) $E_T^{min}(20)$ and (b) $E_T^{min}(100)$ compared to theoretical predictions.

and to predictions from JETRAD with CTEQ4M PDF's, and renormalization scale $= 0.5\Sigma E_T^{jet}$. Figure 4 shows the CDF data compared to the predictions. The normalization for $E_T^{min}(20)$ is not well predicted by HERWIG or NLO QCD. For $E_T^{min}(20)$ 31% of the events have >3 jets which suggest that $\mathcal{O}(\alpha_s^4)$ corrections to the NLO $2 \to 2$ calculation may be important. There is much better agreement with NLO QCD for $E_T^{min}(100)$ but the agreement with HERWIG is still poor. This suggests that the NLO calculation can better describe the data once we are in a region where two-jet events dominate (95% have only two jets passing the threshold). Poor agreement is to be expected for HERWIG, because although it includes a parton shower, the underlying hard scattering cross section is only LO $2 \to 2$. Sensitivity to renormalization scale is also an indication of the influence of higher order terms. Changing the μ scale from $\Sigma E_T/2$ to $\Sigma E_T/4$ increases the predicted NLO cross section by 26% for $E_T^{min}(20)$ and only 7% for $E_T^{min}(100)$.

References

[1] F. Abe et al. (CDF Collaboration), *Phys. Rev. Lett.* **77**, 1996 (438).
[2] S. Ellis et al., *Phys. Rev. Lett.* **62**, 1989 (2188); **64**, 1990 (2121).
[3] F. Abe et al. (CDF Collaboration),*Phys. Rev. Lett.* **70**, 1993 (1376).
[4] W. Giele et al., *Nucl. Phys.* B **403**, 1993 (633).
[5] W. Giele et al., *Phys. Rev.* D **53**, 1996 (120).
[6] F.A. Berends, and H. Kuijf, *Nucl. Phys.* B **353**, 1991 (59).
[7] G. Marchesini and B. Webber,*Nucl. Phys.* B **310**, 1988 (461).
[8] S. Geer and T. Asakawa, *Phys. Rev.* D **53**, 1996 (4793).
[9] F. Abe et al. (CDF Collaboration),*Phys. Rev.* D **54**, 1996 (4221); F. Abe et al. (CDF Collaboration),*Phys. Rev.* D **56**, 1997 (2532);
[10] F. Abe et al., (CDF Collaboration), *Fermilab-PUB-97/093-E*, submitted to *Phys. Rev. Lett.*

New Jet Physics Results at HERA

Tancredi Carli for the H1 and ZEUS collaborations
Max-Planck-Institut für Physik, Munich; h01rtc@rec06.desy.de

Abstract. New results on event shapes, jet profiles and the strong coupling constant α_s from dijet rates at HERA are presented.

1 Event shapes

Measurements of event shape variables like thrust T_C, jet mass ρ_C and jet broadening B_C provide information about perturbative and non-perturbative aspects of QCD. They have been measured for momentum transfers Q between 7 and 100 GeV by the H1 collaboration [1]. They are defined in the current hemisphere of the Breit frame by[1]:

$$T_C := \max \frac{\sum_i |\mathbf{p}_i \cdot \mathbf{n}_T|}{\sum_i |\mathbf{p}_i|} \qquad B_C := \frac{\sum_i |\mathbf{p}_{Ti}|}{2\sum_i |\mathbf{p}_i|} \qquad \rho_C := \frac{M^2}{Q^2} = \frac{(\sum_i \mathbf{p}_i)^2}{Q^2} \tag{1}$$

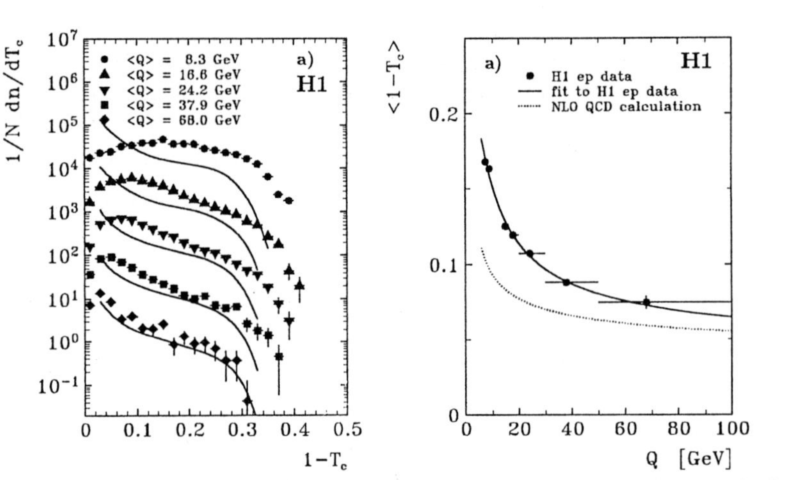

Fig. 1. Left: The differential thrust distribution compared to a NLO calculation. The spectra for $\langle Q \rangle = 8.3-68$ GeV are multiplied by factors of 10^n.Right: $\langle 1 - T_C \rangle$ as function of Q. The H1 data are compared to a NLO calculation and a fit inspired by power corrections.

As example the mean value and the differential thrust distribution of $1 - T_C$ is shown in Fig.1. A spherical (pencil-like) configuration corresponds to $1-T_C = 1/2$ (0). The energy flow along the event shape axis becomes more collimated as Q increases. The mean values of the event shape variables can be expressed by a perturbative part calculable to $\mathcal{O}(\alpha_s^2)$ of the strong coupling α_s using next-to-leading order (NLO) programs [2, 3] and a non-perturbative

---

[1] The summation of the four-vector $p_i = (E_i, \mathbf{p_i})$ extends over all objects (energy depositions, hadrons or partons). For the thrust calculation the unit vector $\mathbf{n_T}$ is chosen such that the normalized longitudinal momentum sum is maximal.

contribution suppressed by $\sim 1/Q$ which can be expressed as a function of only one free non-perturbative effective coupling parameter $\bar{\alpha_0}$ [4]. α_s and $\bar{\alpha_0}$ can be simultaneously fitted to the data corrected for detector effects without assuming any fragmentation model (see Fig.1 for $\langle 1 - T_C \rangle$). The power corrections are large at low Q, but become less important with increasing energy. For $\langle Q \rangle = 68$ GeV the perturbative contribution alone reproduces shape of the data for $(1 - T_C) > 0.03$. The fit results are summarized in Tab.1. The uncertainties on α_s and $\bar{\alpha_0}$ are due to the accuracy of the QCD calculation and to the choice of different possible scales. All investigated shape variables lead to consistent values for α_s and within 20% for $\bar{\alpha_0}$.

Observable	a_F	$\bar{\alpha_0}(\mu_I = 2\text{GeV})$	$\alpha_s(m_Z)$
$\langle 1 - T_C \rangle$	1	$0.497 \pm 0.005^{+0.070}_{-0.036}$	$0.123 \pm 0.002^{+0.007}_{-0.005}$
$\langle B_C \rangle$	2	$0.408 \pm 0.006^{+0.036}_{-0.022}$	$0.119 \pm 0.003^{+0.007}_{-0.004}$
$\langle \rho_C \rangle$	1/2	$0.519 \pm 0.009^{+0.025}_{-0.020}$	$0.130 \pm 0.003^{+0.007}_{-0.005}$

Table 1. Fit results of the event shape analysis.

2 Jet shapes

The internal structure of jets provides useful information on the transition of a parton to the complex aggregate of hadrons. The jet shape is defined as the average fraction of the transverse energy (E_T) of the jets inside an inner cone r concentric to the outer jet cone with radius R:

$$\Psi(r) = \frac{1}{N_{\text{jet}}} \sum_{\text{jets}} \frac{E_T(r)}{E_T(R)} \;, \quad E_T(r) = \int_0^r dr' dE_T(r')/dr' \tag{2}$$

where N_{jet} is the total number of selected jets. By definition $0 < \Psi < 1$ and $\Psi(R) = 1$. The steepness of the rise of Ψ describes the collimation of the jet. ZEUS has measured Ψ for jets with $\eta \in [-1, 2]$ selected with the CDF-cone algorithm [5] ($R = 1$) in the laboratory frame in a DIS sample with $Q^2 > 100$ GeV^2 and has compared them to jets in other reactions. In most cases the current jet recoiling from the scattered electron will be selected.

The jet shape Ψ corrected for detector effects for jets with $E_T > 37$ GeV is shown in Fig. 3. 90% of the jet E_T is contained in a cone with $r = 0.4$ around the jet axis. This is very similar to jets produced in e^+e^- collisions [7]. Jets in $p\bar{p}$ collisions from CDF (D0) [8, 9] are broader. This might be explained by a larger fraction of jets initiated by quarks in e^+e^- and ep collisions compared to $p\bar{p}$ collisions. Given the different considered reference frames one has, however, to be careful about strong conculsions. The shapes of jets produced in neutral (NC) or charged current (CC) reactions are very similar. H1 has reported on jet shapes measurements in a dijet sample for NC DIS in the kinematic region defined by: $10 < Q^2 < 100$ GeV^2, $y_{el} > 0.15$ and $E_{el} > 11$ GeV. Jets are defined using the CDF-cone algorithm ($R = 1$) requiring $E_T > 5$ GeV in the Breit frame and $-1 < \eta_{lab} < 2$ in the laboratory frame. As is shown in Fig.2 jets get narrower with increasing E_T. In the ZEUS measurement no significant η dependence is found. In the phase space region chosen by H1, jets get less collimated towards the proton remnant direction.

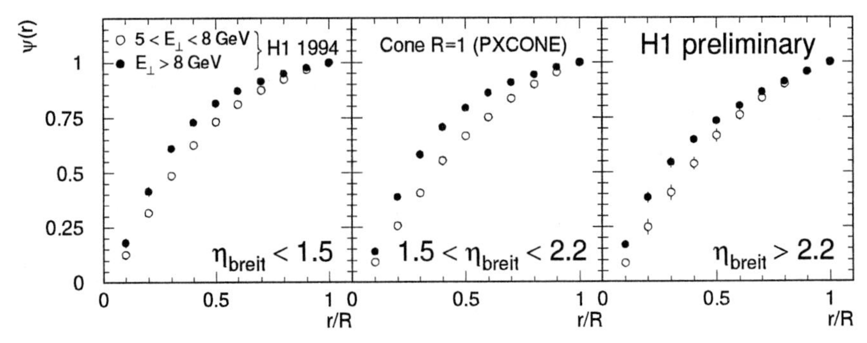

Fig. 2. The average jet profile for some η and E_T regions in the Breit frame in a dijet sample for NC DIS events with $10 < Q^2 < 100$ GeV^2. Syst. and stat. errors are added in quadrature. For large η the jets get further in the target region.

3 Determination of the strong coupling

The cross-section for events with two jets in the central part of the detector is proportional to α_s. Since at large x and Q^2 the parton densities are rather well known, it can be predicted by a NLO QCD calculation [2, 3]. The measurement of the ratio of dijet events to all events allows therefore $\alpha_s(Q^2)$ to be extracted. Jets are defined by the JADE algorithm [10] with a resolution parameter m_{ij}^2/W^2 where m_{ij} (W^2) is the invariant mass of a particle pair (total hadronic final state). Only the kinematic range of high x and Q^2 is considered to reduce the uncertainties from hadronisation and to avoid the region where higher order gluon emissions become important. Also the region towards the target is excluded. Inspired by the potential to measure the behaviour of α_s over a large span of Q^2, jet studies have focused on the determination of α_s right from the beginning of HERA data analysis [12, 13]. Since these early measurements used a NLO calculation [14] which was not fully applicable over the full phase space region [15], H1 [16] has now updated the α_s value obtained before[2]. The α_s values extracted at HERA are shown in Fig. 3. For increasing Q^2, $\alpha_s(Q^2)$ decreases as predicted by the renormalisation group equation. A fit to the data for $\alpha_s(m_Z^2)$ yields:

$$\text{H1 94/95}: \quad \alpha_s(m_z^2) = 0.115 \pm 0.003\,(\text{stat})^{+0.008}_{-0.011}\,(\text{sys}) + 0.006\,(\text{rec})$$

$$\text{ZEUS 94}: \quad \alpha_s(m_z^2) = 0.117 \pm 0.005(\text{stat})^{+0.004}_{-0.005}\,(\text{sys}_{\text{exp}}) \pm 0.007\,(\text{sys}_{\text{th}})$$

The dominant experimental errors stem from the uncertainty on the energy scale for hadrons, the model dependence of assigning sets to partons and the phenomenological description of the hadronisation process. Furthermore the renormalisation scale and the input parton density systematically influence the result. H1 has also given an error for the dependence on the recombination scheme (rec). The obtained value is compatible with the world average and has reached a competitive precision.

[2] The ZEUS result does not change when using the new NLO calculations [17].

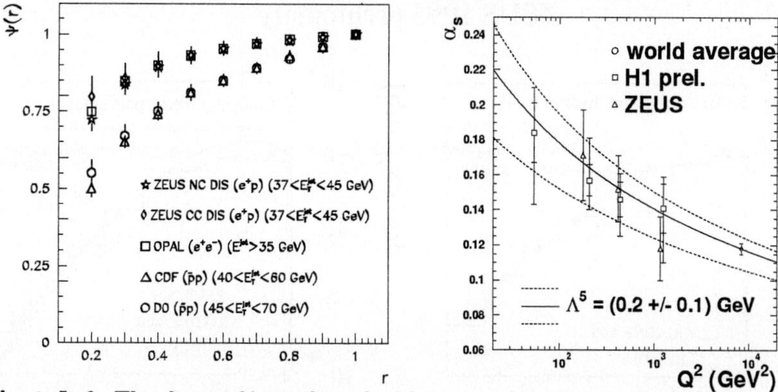

Fig. 3. Left: The shape of jets selected with a cone algorithm in the laboratory frame for DIS CC and NC events with $Q^2 > 100$ GeV^2, compared to jets in e^+e^- and $p\bar{p}$ events. Right: The strong coupling α_s in function of Q^2 as determined from the dijet rates. Shown are the HERA data together with the world average for $\alpha_s(M_Z^2)$ [11]. The lines give the expectation for $\Lambda^5 = 0.2 \pm 0.1$ GeV (for $n_f = 5$ flavours).

4 Forward jets

The DGLAP approximation for QCD evolution is exspected to fail in the region of small x. A possible signature of this incomplete aprroximation are an enhanced production of 'forward jets' [18]. They are characterized by E_T of the order of Q^2 and $x_{\text{jet}} = E_{\text{jet}}/E_p$ as large and x as small as kinematically possible[3]. The first requirement suppresses the strongly ordered DGLAP evolution. For x_{jet}/x large, the forward jet is separated by a large η interval from the struck quark such that the phase space of parton emission between the two is enlarged. For this kinematical configuration the $\alpha_s \ln x_{\text{jet}}/x$ terms are expected to become so large that their resummation should lead to a sizeable increase of the cross-section.

First measurements have been presented by H1 [20, 21]. A strong rise of the cross-section with decreasing x has been observed. The interpretation in terms of BFKL was however hindered due to large model dependences of hadronisation effects[22]. The ZEUS collaboration has reported on a new measurement of the forward jet cross-section [23] corrected for detector effects (see Fig. 4). The behaviour of the data is well described by the ARIADNE Monte Carlo [24] with an unsuppressed gluon radiation pattern while LEPTO [25] based on the $\mathcal{O}(\alpha_s)$ QCD matrix element and leading log parton showers cannot account for the steep rise towards small x. The NLO prediction from MEPJET [2] fails to describe the data at small x and agrees with an analytical calculation [19] (Born-BFKL) where only the quark box coupling to the photon in addition to the gluon leading to the forward jet is considered. If the full LO BFKL resummation is included, the cross-section rises towards low x, but overshoots the data. Since these analytical calculations do not include a jet algorithm, no firm conclusion can be drawn.

[3] E_{jet} (E_p) denotes the energy of the forward jet (proton).

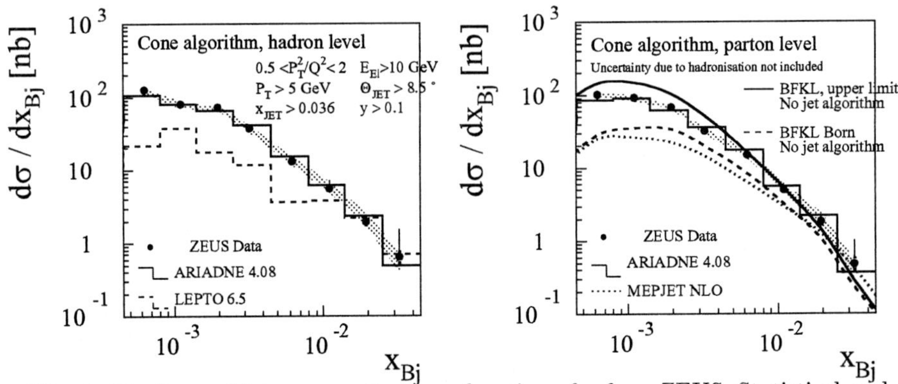

Fig. 4. The forward jet cross section as a function of x from ZEUS. Statistical and systematic errors are added in quadrature. The shaded bands give the uncertainty of the hadronic energy scale. Left: The data are corrected to the hadron level, and compared to the expectations from ARIADNE and LEPTO. Right: The data are corrected to the parton level. Systematic errors due to the hadronisation are not included. Also shown are the ARIADNE and the NLO predictions for jets [2], and the BFKL and Born graph calculations for parton cross sections [19].

References

[1] H1 Collab., C. Adloff et al., Phys.Lett. B406 (1997) 256.
[2] E. Mirkes and D. Zeppenfeld, Phys. Lett. B380 (1996) 205.
[3] S. Catani and M. Seymour, Nucl. Phys. B485 (1997) 291.
[4] B. Webber, Phys. Lett. 339 (1994) 148.
[5] CDF Collab., F. Abe et al., Phys. Rev. D45 (1992) 1148.
[6] ZEUS Collab., contrib. paper N-649 to HEP97, Jerusalem 1997.
[7] OPAL Collab., R. Akers et al., Z. Phys. C63 (1994) 197.
[8] CDF Collab., F. Abe et al., Phys. Rev. Lett. 70 (1993) 713.
[9] D0 Collab., S. Abachi et al., Phys. Lett. B357 (1995) 500.
[10] JADE Collab., W. Bartel et al., Z. Phys. C33 (1986) 23.
[11] Particle Data Group, R.M. Barnett et al., Phys. Rev. D54 (1996).
[12] H1 Collab., T. Ahmed et al., Phys. Lett. B346 (1995) 415.
[13] ZEUS Collab., M. Derrick et al., Phys. Lett. B363 (1995) 201.
[14] D. Graudenz, Phys. Comm. 92 (1995) 65.
[15] K. Rosenbauer, in: DIS 95, Rome, 1996, p. 444.
[16] H1 Collab., contrib. paper 246 to HEP97, Jerusalem 1997.
[17] T. Trefzger, in: DIS 95, Rome 1996, p. 434.
[18] A.H. Mueller, Nucl. Phys. B (Proc. Suppl.) 18C (1990) 125.
[19] J. Bartels et al., Phys. Lett. B384 (1996) 300.
[20] H1 Collab., S. Aid et al., Phys. Lett. B356 (1995) 118.
[21] H1 Collab., contrib. paper pa03-049 to ICHEP'96, Warsaw (1996).
[22] T. Carli, in: PIC 96, Mexico City (1996), DESY 97-10, Hamburg (1997).
[23] ZEUS Collab., contrib. paper N-659 to HEP97, Jerusalem 1997
[24] L. Lönnblad, Comp. Phys. Comm. 71 (1992) 15.
[25] G. Ingelman, A. Edin and J.Rathsman, Comp. Phys. Comm. 101 (1997) 108.

Jet Production at HERA

Claudia Glasman (claudia@mail.desy.de) representing the H1 and ZEUS Collaborations

DESY, Germany

Abstract. Studies on the structure of the photon are presented by means of the extraction of a leading order effective parton distribution in the photon and measurements of inclusive jet differential cross sections in photoproduction. Measurements of the internal structure of jets have been performed and are also presented as a function of the transverse energy and pseudorapidity of the jets. Dijet cross sections have been measured as a function of the dijet mass and centre-of-mass scattering angle.

The main source of jets at HERA[1] comes from collisions between protons and quasi-real photons ($Q^2 \approx 0$, where Q^2 is the virtuality of the photon) emitted by the positron beam (photoproduction). At lowest order QCD, two hard scattering processes contribute to jet production: the resolved process, in which a parton from the photon interacts with a parton from the proton, producing two jets in the final state; and the direct process, in which the photon interacts pointlike with a parton from the proton, also producing two jets in the final state.

The study of high-p_T jet photoproduction provides tests of QCD and allows to probe the structure of the photon. In perturbative QCD the cross section for jet production is given by

$$\frac{d^4\sigma}{dydx_\gamma dx_p d\cos\theta^*} \propto \frac{f_{\gamma/e}(y)}{y} \times$$

$$\sum_{ij} \frac{f_{i/\gamma}(x_\gamma, \hat{Q}^2)}{x_\gamma} \frac{f_{j/p}(x_p, \hat{Q}^2)}{x_p} |M_{ij}(\cos\theta^*)|^2 \quad (1)$$

[1] HERA provides collisions between positrons of energy $E_e = 27.5$ GeV and protons of energy $E_p = 820$ GeV.

where $f_{\gamma/e}(y)$ is the flux of photons from the positron approximated by the Weizsäcker-Williams formula; $y = E_\gamma/E_e$ is the inelasticity parameter; $f_{i/\gamma}(x_\gamma, \hat{Q}^2)$ ($f_{j/p}(x_p, \hat{Q}^2)$) are the parton densities in the photon[2] (proton) at a scale $\hat{Q}$; $x_\gamma = E_i/E_\gamma$ ($x_p = E_j/E_p$) is the fractional momentum of the incoming parton from the photon (proton); and $|M_{ij}(\cos\theta^*)|^2$ are the QCD matrix elements for the parton-parton scattering.

Two approaches to the study of the structure of the photon are presented here. One approach is to extract directly from the data an effective parton distribution in the photon. The second approach is to measure jet cross sections that can be calculated theoretically which then provide a testing ground for parametrisations of the photon structure function.

The method [1] to extract an effective parton distribution in the photon is based on the use of the leading order (LO) matrix elements for the subprocesses with gluon exchange, which give the dominant contribution to the jet cross section in resolved processes for the kinematic regime studied. Then, the quark and gluon densities are combined into an effective parton distribution:

$$\sum [f_{q/\gamma}(x_\gamma, p_T^2) + f_{\bar{q}/\gamma}(x_\gamma, p_T^2)] + \frac{9}{4} f_{g/\gamma}(x_\gamma, p_T^2) \quad (2)$$

A LO effective parton distribution in the photon was extracted by unfolding the double differential dijet cross section as a function of the average transverse energy of the

[2] The resolved and direct processes are included in $f_{i/\gamma}$.

two jets with highest transverse energy and of the fraction of the photon's energy (x_γ) participating in the production of the two highest-transverse-energy jets (see figure 1). The measurement was performed using a fixed cone algorithm [2] with radius $R = 0.7$ in the pseudorapidity[3] (η) – azimuth (φ) plane and in the kinematic range $0.2 < y < 0.83$ and $Q^2 < 4$ GeV^2. Figure 2 shows the extracted effective parton distribution as a function of the scale p_T (the transverse momentum of the parton) for two ranges of x_γ. The data exhibit an increase with the scale p_T which is compatible with the logarithmic increase predicted by QCD (i.e., the anomalous component of the photon structure function). The predictions using the GRV-LO [3] parametrisations of the parton densities in the photon and the pion are also shown in figure 2. The data disfavour a purely hadronic behaviour and are compatible with the prediction which includes the anomalous component.

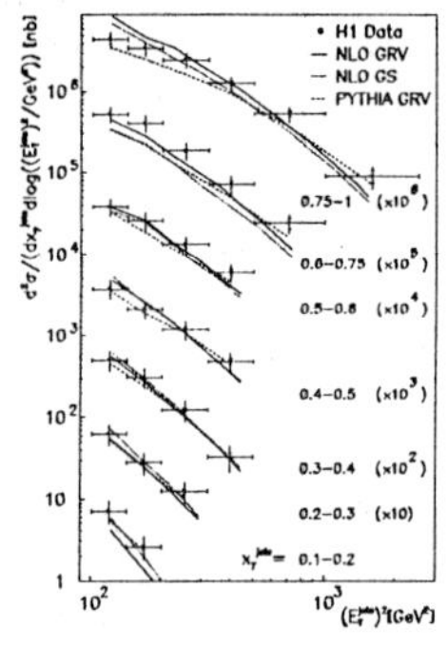

Fig. 1. Double differential dijet cross sections.

The second approach to the study of the photon structure function is to provide measurements of differential jet cross sections at high jet transverse energies, where the proton

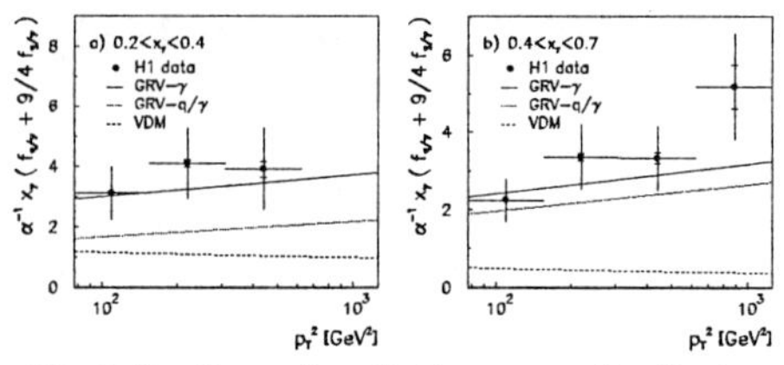

Fig. 2. Leading order effective parton distribution in the photon.

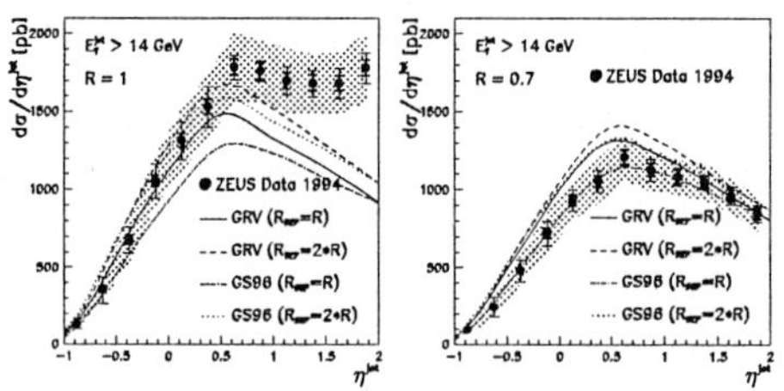

Fig. 3. Inclusive jet differential cross sections.

parton densities are constrained by other measurements, and, therefore, are sensitive to the photon structure function.

Figure 3 shows the measurements of the inclusive jet differential cross sections for jets searched with an iterative cone algorithm [2] as a function of the jet pseudorapidity (η^{jet}) for jets with transverse energy satisfying $E_T^{jet} > 14$ GeV. The measurements were performed in the kinematic region given by $0.2 < y < 0.85$ and $Q^2 \leq 4$ GeV^2, and for two cone radii: $R = 0.7$ and 1. The behaviour of the cross section as a function of η^{jet} in the region $\eta^{jet} > 1$ is very different for the two cone radii: it is flat for $R = 1$ whereas it decreases as η^{jet} increases for $R = 0.7$.

Next-to-leading order (NLO) QCD calculations [4] are compared to the measurements in figure 3 using two different parametrisations of the photon structure function: GRV-HO [5] and GS96 [6], and two values of[4] R_{SEP}. The

[3] The pseudorapidity is defined as $\eta = -\ln(\tan\frac{\theta}{2})$, where the polar angle θ is taken with respect to the proton beam direction.

[4] The parameter R_{SEP} is introduced into the NLO calculations in order to simulate the ex-

CTEQ4M [8] proton parton densities have been used in all cases. For forward jets with $R = 1$ an excess of the measurements with respect to the calculations is observed. This discrepancy is attributed to a possible contribution from non-perturbative effects (e.g., the so-called "underlying event"), which are not included in the theoretical calculations. This contribution is supposed to be reduced by decreasing the size of the cone since the transverse energy density inside the cone of the jet due to the underlying event is expected to be roughly proportional to the area covered by the cone. Good agreement between data and NLO calculations is observed for measurements performed using a cone radius of $R = 0.7$ for the entire η^{jet} range measured. The predictions using GRV-HO and GS96 show differences which are of the order of the largest systematic uncertainty of the measurements. Thus, these measurements exhibit a sensitivity to the parton densities in the photon and can be used in quantitative studies.

To study the internal structure of the jets, the jet shape $\psi(r)$ has been used. $\psi(r)$ is defined as the average fraction of the jet's transverse energy that lies inside an inner cone of radius r, concentric with the jet defining cone [7]:

$$\psi(r) = \frac{1}{N_{jets}} \sum_{jets} \frac{E_T(r)}{E_T(r = R)} \tag{3}$$

where $E_T(r)$ is the transverse energy within the inner cone and N_{jets} is the total number of jets in the sample. By definition, $\psi(r = R) = 1$. The jet shape is determined by fragmentation and gluon radiation. However, at sufficiently high E_T^{jet} the most important contribution is predicted to come from gluon emission off the primary parton.

perimental jet algorithm by adjusting the minimum distance in $\eta - \varphi$ at which two partons are no longer merged into a single jet [7].

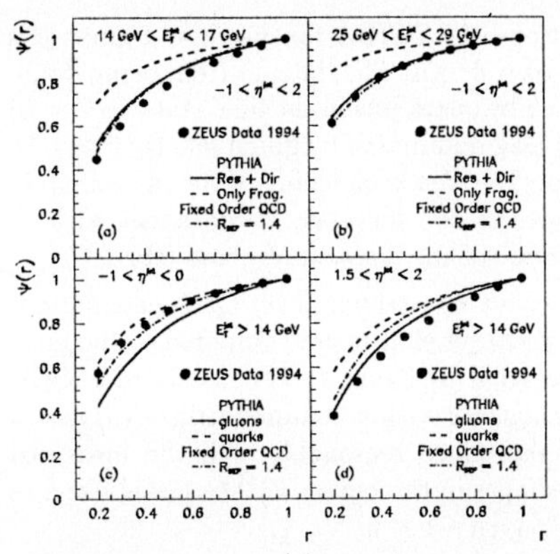

Fig. 4. Jet shapes: E_T^{jet} and η^{jet} dependence.

Figures 4a and 4b show the measured jet shapes $\psi(r)$ as a function of the inner cone radius r using a cone algorithm with radius $R = 1$ for jets with $-1 < \eta^{jet} < 2$ and in two regions of E_T^{jet}. As a jet becomes narrower, the value of $\psi(r)$ increases for a fixed value of r. It is observed that the jets become narrower as E_T^{jet} increases. For comparison, the predictions from leading-logarithm parton-shower Monte Carlo calculations as implemented in the PYTHIA generator for resolved plus direct processes are shown. The predictions reproduce reasonably well the data except in the lowest-E_T^{jet} region where small differences between data and the predictions are observed. PYTHIA including resolved plus direct processes but without initial and final state parton radiation predicts jet shapes which are too narrow in each region of E_T^{jet}. These comparisons show that parton radiation is the dominant mechanism responsible for the jet shape in the range of E_T^{jet} studied.

The η^{jet} dependence of the jet shape is presented in figures 4c and 4d. It is observed that the jets become broader as η^{jet} increases. Perturbative QCD predicts that gluon jets are broader than quark jets as a consequence of the fact that the gluon-gluon is larger than the quark-gluon coupling strength. The predictions of PYTHIA for quark and gluon jets are also shown. The data go from being dominated by quark jets in the final state ($\eta^{jet} < 0$) to being dominated by gluon jets ($\eta^{jet} > 1.5$). Therefore, the broadening of the measured jet shapes as η^{jet} increases is consistent with an increase of the fraction of gluon jets.

Lowest non-trivial-order QCD calculations [9] of the jet shapes are compared to the measurements in figure 4. The fixed-order QCD calculations with a common value of $R_{SEP} = 1.4$ reproduce reasonably well the measured jet shapes in the region $E_T^{jet} > 17$ GeV and in the region $-1 < \eta^{jet} < 1$.

The dijet mass (M^{JJ}) distribution provides a test of QCD and is sensitive to the presence of new particles or resonances that decay into two jets. The dijet cross section as a function of the scattering angle in the dijet centre-of-mass system ($\cos\theta^*$) reflects the underlying parton dynamics. New particles or resonances decaying into two jets may also be identified by deviations in the $|\cos\theta^*|$ distribution with respect to the predictions. $d\sigma/dM^{JJ}$ has been measured for $|\cos\theta^*| < 0.8$, and $d\sigma/d|\cos\theta^*|$ has been measured for $M^{JJ} > 47$ GeV. The results are presented in figure 5. The measured $d\sigma/d|\cos\theta^*|$ increases as $|\cos\theta^*|$ increases. The measured $d\sigma/dM^{JJ}$ exhibits a steep fall-off of 2 orders of magnitude in the M^{JJ} range considered. NLO QCD calculations [4] are compared to the measurements in figure 5. The CTEQ4M (GS96) parametrisations of the proton (photon) parton densities have been used. The prediction for $d\sigma/d|\cos\theta^*|$, which is normalised to the lowest-$|\cos\theta^*|$ data point, agrees in shape reasonably well with the measured distribution. The prediction for $d\sigma/dM^{JJ}$ describes the shape and magnitude of the measured distribution up to the highest M^{JJ} studied (~ 120 GeV).

The results on photoproduction of jets presented here constitute a step forward towards testing QCD and the extraction of the photon parton densities.

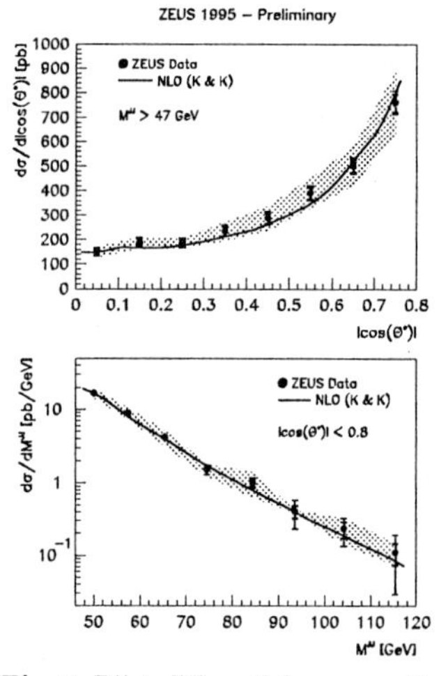

Fig. 5. Dijet differential cross sections.

References

[1] B.L. Combridge and C.J. Maxwell, *Nucl. Phys.* B **239**, 1984 (429).

[2] CDF Collab., F. Abe et al., *Phys. Rev.* D **45**, 1992 (1448); J. Huth et al., Proceedings of the 1990 DPF Summer Study on High Energy Physics, Snowmass, Colorado, edited by E.L. Berger (World Scientific, Singapore,1992) p. 134.

[3] M. Glück, E. Reya and A. Vogt, *Z. Phys.* C **53**, 1992 (127) and *Z. Phys.* C **53**, 1992 (651).

[4] M. Klasen, G. Kramer and S.G. Salesch, *Z. Phys.* C **68**, 1995 (113).

[5] M. Glück, E. Reya and A. Vogt, *Phys. Rev.* D **46**, 1992 (1973).

[6] L.E. Gordon and J.K. Storrow, *Nucl. Phys.* B **489**, 1997 (405).

[7] S.D. Ellis, Z. Kunszt and D.E. Soper, *Phys. Rev. Lett.* **69**, 1992 (3615).

[8] H.L. Lai et al., *Phys. Rev.* D **55**, 1997 (1280).

[9] M. Klasen and G. Kramer, *Phys. Rev.* D **56**, 1997 (2702).

Part V

Production and Decays of Heavy Flavours

Inclusive D^* and Inelastic J/ψ Production at HERA

Yehuda Eisenberg (fheisenb@rosinante.weizmann.ac.il)

Weizmann Institute of Science

Abstract. We report here the latest results of the H1 and ZEUS collaborations, on the photoproduction of open charm mesons (D^*) and inelastic J/ψ . The results are compared with several recent NLO pQCD calculations. We find that the energy dependence of both processes is close to that observed in hard reactions. For D^* photoproduction in a restricted kinematical region the cross sections agree better with NLO calculations based on the so called "massless charm scheme". The differential cross-sections show deviations from the above calculations, in particular in the forward psuedorapidity regions (proton direction). Clear indications for contributions from resolved γ reactions in charm photoproduction are observed. The inelasticity in inelastic J/ψ production is described well by the color singlet model.

1 Introduction

Heavy quark photoproduction can be used to probe pQCD calculations with a hard scale given by the heavy quark mass and the high transverse momentum of the produced parton. At leading order (LO) in QCD, two types of processes are responsible for the photoproduction of heavy quarks: the direct photon processes, where the photon participates as a point-like particle which interacts with a parton from the incoming proton, and the resolved photon processes, where the photon is a source of partons, one of which scatters off a parton from the proton. Charm quarks present in the parton distributions of the photon, as well as of the proton, lead to processes such as $cg \rightarrow cg$, which are called charm flavour excitation. In next-to-leading order (NLO) QCD only the sum of direct and resolved processes is unambiguously defined. Two types of NLO calculations using different approaches are available for comparison with measurements of charm photoproduction at HERA. The massive charm approach [1] assumes light quarks to be the only active flavours within the structure functions of the proton and the photon, while the massless charm approach [2, 3] also treats charm as an active flavour.

The data presented here correspond to an integrated luminosity of about 20 pb^{-1} taken by the H1 [4] and ZEUS collaborations [5] during 1994 to 1996, where a positron beam with energy E_e=27.5 GeV collided with a proton beam with energy E_p=820 GeV. Charm was tagged by the observations of J/ψ leptonic decays [6], and by the detection of $D^*(2010)$ mesons identified in the final state via their charged products detected in the central tracking detector. Cross sections were measured in the photoproduction range of photon virtualities $Q^2 < 4\,GeV^2$ ($Q^2_{median} \sim 5{\cdot}10^{-4}\,GeV^2$) within γp center-of-mass energy regions $115 < W_{\gamma p} < 280$ GeV for the ZEUS detector, $95 < W_{\gamma p} < 268$ GeV for the untagged H1 data, and $60 < W_{\gamma p} < 147$ GeV for the H1 tagged data. W was determined from the energy deposits in the calorimeter or by tagging the positron. The low Q^2 was ensured by the absence of an identified positron in the detector.

2 Open Charm D^* Photoproduction

D^* decays into $D^o\pi_s$ with D^o decaying into $(K\pi)$ or $(K\pi\pi\pi)$ states have been selected by means of the mass difference (ΔM) method [7] within the kinematic range $p_\perp^{D^*} > 3$ GeV and $-1.5 < \eta^{D^*} < 1.0$($p_\perp^{D^*} > 1.8$ GeV for H1 tagged data). Here $p_\perp^{D^*}$ is D^* transverse momentum and its pseudorapidity is $\eta^{D^*} = -ln(tan\frac{\theta}{2})$, where θ is the polar angle with respect to the proton beam direction. For the $d\sigma/dp_\perp^{D^*}$ measurement ZEUS extended the $p_\perp^{D^*}$ region down to 2 GeV and for the $d\sigma/d\eta^{D^*}$ measurement the η^{D^*} region was extended up to 1.5.

In the 1996 data, ZEUS observed 643$\pm$37 D^* mesons in the $(K\pi)\pi_s$ mode and 870 ± 86 in the $(K\pi\pi\pi)\pi_s$ mode. Integrated over the defined kinematic range, the preliminary results for the ZEUS cross section are $11.9\pm0.8(stat.)\pm 0.7(syst.)$ nb and $11.9 \pm 1.2(stat.)$ nb for the $(K\pi)\pi_s$ and $(K\pi\pi\pi)\pi_s$ decay modes respectively. The results are consistent with each other and in agreement with ZEUS 1994 published measurements[8] and with the preliminary 1995 results[9].

H1 has extrapolated the visible D^* cross section to the full phase space region, and converted the results [4] using the Weizsäcker-Williams Approximation [10] to $\sigma_{\gamma p \to c\bar{c}X}$. The resulting $\sigma_{\gamma p}$ as a function of $W_{\gamma p}$ is shown in fig.1, together with other HERA results and the low energy data. A steep rise of the cross section with $W_{\gamma p}$ is evident, and a fit of all the data to the form $W^{2\lambda}$ yields a value of $\lambda = 0.61 \pm 0.28$ which is the typical W dependence of hard processes.

ZEUS compared the differential cross sections $p_\perp^{D^*}$, η^{D^*} and W (Fig.2), in the above restricted kinematic range, with the NLO pQCD calculations in the massive charm[1] and massless charm[2] approaches. Both calculations use the Peterson fragmentation function[11] to describe the hadronization of charm into $D^{*\pm}$ mesons. The massless scheme is supposed to describe the data better for $p_\perp^{D^*} > m_c$ [12], where m_c is the mass of the charm quark.

ZEUS finds that the cross section calculated in the massless charm approach agrees better with the data whereas the massive charm one falls below the data by up to 50% depending on the charmed quark mass and renormalization scale (μ) values. With the improved 1996 data and the extended kinematic region in η^{D^*} there is a clear indication for a difference in shape and size between the data and the NLO calculations. The η^{D^*} distribution has

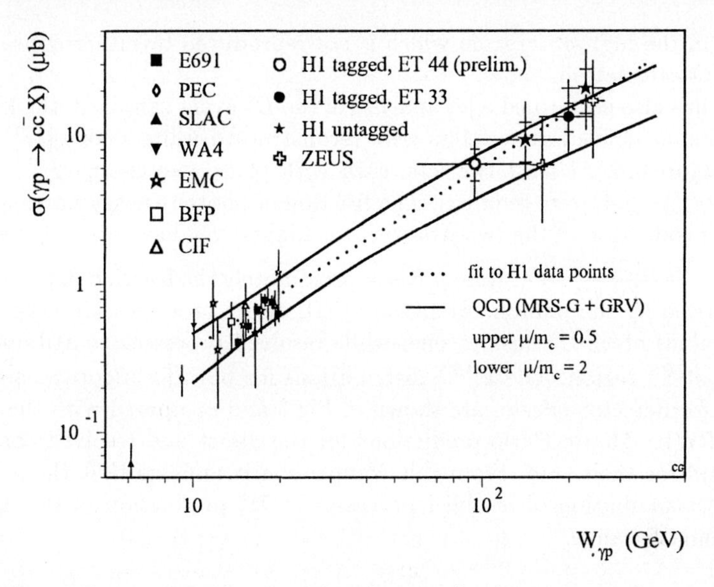

Fig. 1. Energy dependence of the photoproduction cross section $\gamma p \to c\bar{c}X$

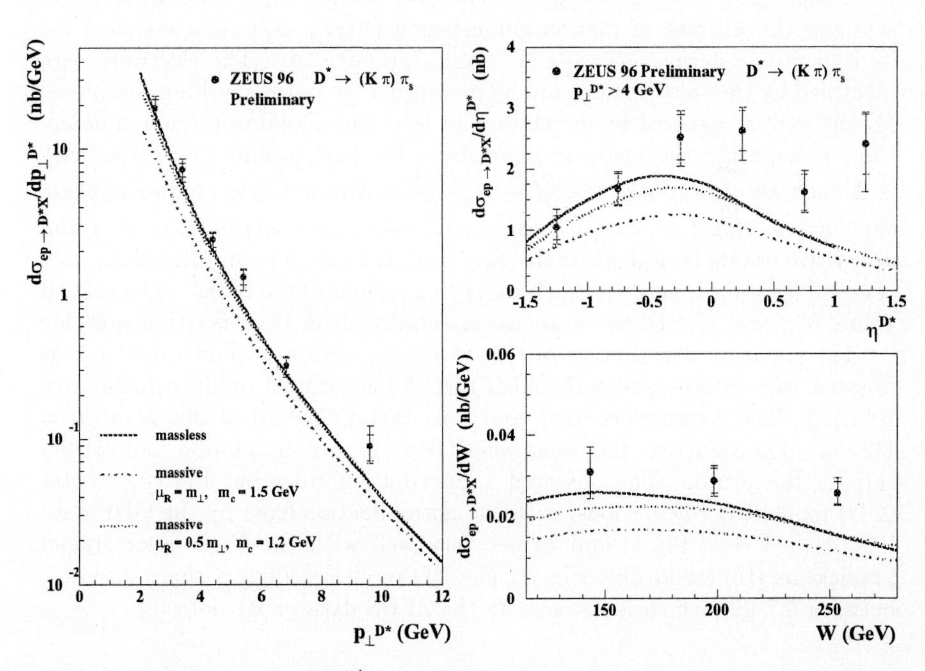

Fig. 2. Comparison of $d\sigma/dp_\perp^{D^*}$, $d\sigma/d\eta$, $d\sigma/dW$ with some NLO-QCD calculations

an excess in the high η^{D^*} region which is not reproduced by either approach in the NLO calculation.

ZEUS has also performed a jet analysis of the D^* event sample in the kinematic region as defined above. Jets were reconstructed using a cone [13] or a K_T [14] algorithm. At least two jets, each with transverse energy $E_T^{jet} > 4$ GeV and $\eta^{jet} < 2.4$, were required. The fraction of photon energy participating in the production of the two jets with the highest E_T has been calculated as $x_\gamma^{OBS} = \frac{\Sigma_{jets}(E_T^{jet} e^{-\eta^{jet}})}{2E_e y_j b}$. Here $y_j b$ is approximately the fraction of positron energy carried by the photon. At LO in QCD, direct processes are expected to yield values of x_γ^{OBS} close to one, while resolved processes contribute to the lower x_γ^{OBS} region. The x_γ^{OBS} distributions for both jet algorithms, not corrected for detector effects, are shown in Fig.3 and compared with the LO QCD HERWIG Monte Carlo predictions for the direct and resolved contributions and for their sum. From this comparison it appears that there is a significant contribution of resolved processes to D* production in the measured kinematic range.

3 J/ψ Inelastic Photoproduction

In measuring inelastic photoproduction cross section of J/ψ we measure essentially the process of photon-gluon fusion (PGF). J/ψ was searched for by its leptonic decays:$J/\psi \rightarrow e^+e^-$, and $J/\psi \rightarrow \mu^+\mu^-$. The electrons were identified by the calorimeters and by measuring $\frac{dE}{dx}$ in the tracking chambers. Muons were recognized by demanding at least one signal in the muon detectors. To separate the inelastic signal from the background J/ψ inelasticity is defined as : $z(J/\psi) = \frac{P_p \cdot P(J/\psi)}{P_p \cdot P_\gamma} = \frac{E(J/\psi)}{E_\gamma}$. Monte Carlo studies indicate that by selecting z = 0.4 - 0.9, one is supressing the contributions of proton diffractive events (leading to z strongly peaked in the forward direction), and resolved J/ψ photoproduction (leading to z regions of 0.0 to 0.4). The central values of z are selected to obtain an essentially clean PGF reaction sample.

The resulting distribution of the J/ψ cross section is shown in Fig.4, as function of z. Solid curve is the NLO pQCD color singlet model calculations, Ref. [15]. The parameters used were: $m_c = 1.4\,GeV$, $\Lambda = 300\,MeV$ $\mu = \sqrt{2} \cdot m_c$. The structure functions were GRV [16] for the photon and MRSG [17] for the proton. The observed z distribution rules out the color octet (CO) model LO calculations, with the normalization fixed by the CDF data [18],(dashed line, Fig.4) and agrees very well with the NLO color singlet predictions [15] (solid line, Fig.4). The CO model predicts a sharp rise at z between 0.7-0.9, which is not seen in the ZEUS data or H1 data [6].

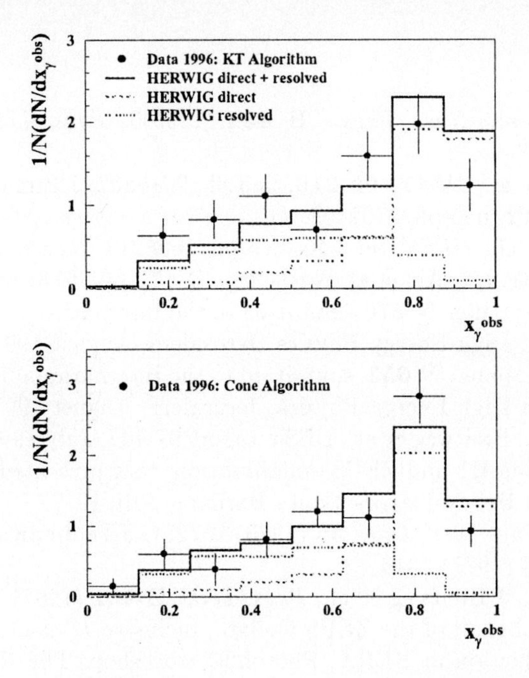

Fig. 3. Observed (preliminary, uncorrected) x_γ^{OBS} distribution for D^* events.

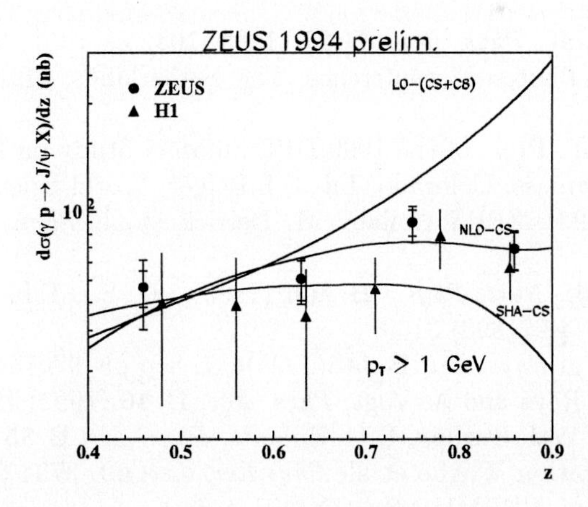

Fig. 4. J/ψ cross-section as function of the inelasticity z

References

[1] S.Frixione et al., *Nucl. Phys.* **B 454** (1995)3; *Phys. Lett.* **B 348** (1995)633.

[2] B.A.Kniehl et al., *DESY* **96-210**, hep-ph/9610267; J.Binnewies et al., *DESY* **97-012**, hep-ph/9702408.

[3] M.Cacciari et al., *DESY* **96-146**, hep-ph/9608213; *DESY* **97-029**.

[4] H1 Collaboration, S. Aid et al.,*Nucl. Phys.***B 472** (1996) 3; *Nucl. Phys.***B 472** (1996) 32; paper **N 276**,submitted to the International Europhysics Conference on High Energy Physics, Jerusalem, August 1997.

[5] ZEUS Collab., paper **N 653**, submitted to the International Europhysics Conference on High Energy Physics, Jerusalem, August 1997.

[6] ZEUS Collab., Breitweg et al., DESY report 97-147, July 1997; C. Grab, on behalf of the H1 and ZEUS collaboration, talk presented in the 7th Int. Symp. on Heavy Flavors, Santa Barbara, July 1997.

[7] S.Nussinov, *Phys. Rev. Lett.* **35** (1975) 1672; G.J.Feldman et al., *Phys. Rev. Lett.* **38** (1977) 1313.

[8] ZEUS Collab., J.Breitweg et al., *Phys. Lett.* **B 401** (1997) 192.

[9] L.Gladilin on behalf of the ZEUS Collab., Inclusive D^* and J/ψ inelastic photoproduction at HERA, Photon'97 workshop, The Netherlands, May 1997; C.Coldewey on behalf of the H1 and ZEUS Collab., Inclusive D^* photoproduction in ep collisions at HERA, Ringberg workshop, Germany, May 1997.

[10] C.F. Weizsäcker, *Z. Phys.***88** (1934) 612; E.J. Williams, *Phys. Rev.***45** (1934) 729.

[11] C.Peterson et al., *Phys. Rev.* **D 27** (1983) 105.

[12] J.R.Forshaw, Photon'97 conference, The Netherlands, May 1997 (hep-ph/9707238).

[13] J.E.Huth et al., Proc. of the 1990 DPF Summer Study on High Energy Physics, Snowmass, Colorado, Ed. E.L.Befger, World Scientific, Singapore 134 (1992); ZEUS Collab., M. Derrick et al, *Phys. Lett.***B 348** (1995) 665.

[14] S.Catani et al., *Nucl. Phys.* **B 406** (1993) 187; S.D.Ellis, D.E.Soper, *Phys. Rev.* **D 48** (1993) 3160.

[15] M.Krämer et al,*Phys. Lett.***B 348**(1995)657; hep-ph/9707449,July '97.

[16] M. Glück, E. Reya and A. Vogt, *Phys. Rev.* **D 46** (1992) 1973.

[17] A.D. Martin, W.J. Stirling, R.G. Roberts, *Phys. Lett.***B 354**(1995)155.

[18] CDF collaboration, F.Abe et al, *Phys.Rev. Lett.***69**, 3704 (1992); CDF coll., A.Sansoni, FERMILAB-CONF-95 /262-E.

Colour-Octet Contributions to J/Ψ Production via Fragmentation at HERA

Bernd A. Kniehl (kniehl@vms.mppmu.mpg.de)

Max-Planck-Institut für Physik, Föhringer Ring 6, 80805 Munich, Germany

Abstract. We study J/ψ photoproduction via fragmentation at next-to-leading order in the QCD-improved parton model, using the nonrelativistic factorization formalism proposed by Bodwin, Braaten, and Lepage. We predict that measurements of J/ψ photoproduction at DESY HERA should show a distinctive excess over the expectation based on the colour-singlet model at small values of the inelasticity variable z.

Since its discovery in 1974, the J/ψ meson has provided a useful laboratory for quantitatively testing quantum chromodynamics (QCD) and, in particular, the interplay of perturbative and nonperturbative phenomena. Recently, the cross section of J/ψ inclusive production measured at the Fermilab Tevatron turned out to be more than one order of magnitude in excess of what used to be the best theoretical prediction, based on the colour-singlet model (CSM). As a solution to this puzzle, Bodwin, Braaten, and Lepage [1] proposed the existence of so-called colour-octet processes to fill the gap. The idea is that $c\bar{c}$ pairs are produced at short distances in colour-octet states and subsequently evolve into physical (colour-singlet) charmonia by the nonperturbative emission of soft gluons. The underlying theoretical framework is provided by nonrelativistic QCD (NRQCD) endowed with a particular factorization theorem, which implies a separation of short-distance coefficients, which are amenable to perturbative QCD, from long-distance matrix elements, which must be extracted from experiment. This formalism involves a double expansion in the strong coupling constant α_s and the relative velocity v of the bound c quarks, and takes the complete structure of the charmonium Fock space into account.

In order to convincingly establish the phenomenological significance of the colour-octet mechanism, it is indispensable to identify it in other kinds of high-energy experiments as well. In this presentation, we briefly report on a next-to-leading-order (NLO) analysis [2] of inelastic J/ψ photoproduction via fragmentation at HERA. We study direct and resolved photoproduction of prompt J/ψ mesons and χ_{cJ} mesons ($J = 0, 1, 2$) that radiatively decay to $J/\psi + \gamma$, taking into account the formation of both colour-singlet and colour-octet $c\bar{c}$ states. The dominant J/ψ (χ_{cJ}) Fock states are $[\underline{1}, {}^3S_1]$ and $[\underline{8}, {}^3S_1]$ ($[\underline{1}, {}^3P_J]$ and $[\underline{8}, {}^3S_1]$), respectively. For comparison, we also consider γg fusion in the CSM.

From Fig. 1a, we observe that the fragmentation mechanism vastly dom-

variable $z = p_p \cdot p_{J/\psi}/p_p \cdot p_\gamma$. For a minimum-$p_T$ cut of 4 GeV (8 GeV), its contribution exceeds the one due to γg fusion for $z < 0.4$ (0.75), by factors of about 4 and 200 (20 and 700) at $z = 0.25$ and 0.05, respectively. The bulk of the fragmentation contribution is induced by resolved photons. In Fig. 1b, the ZEUS $J/\psi \to \mu^+\mu^-$ data [3] collected in the interval $0.3 < z < 0.8$ are compared with the respective fragmentation and γg-fusion predictions. We have divided the experimental data points by the estimated extrapolation factor 1.07 which was included in Ref. [3] to account for the unmeasured contribution from $0 < z < 0.3$. In fact, our combined analysis of fragmentation and γg fusion suggests that, at $p_T = 5$ GeV, this factor should be as large as 3.4. The corresponding value for $0.05 < z < 0.3$ is 2.6.

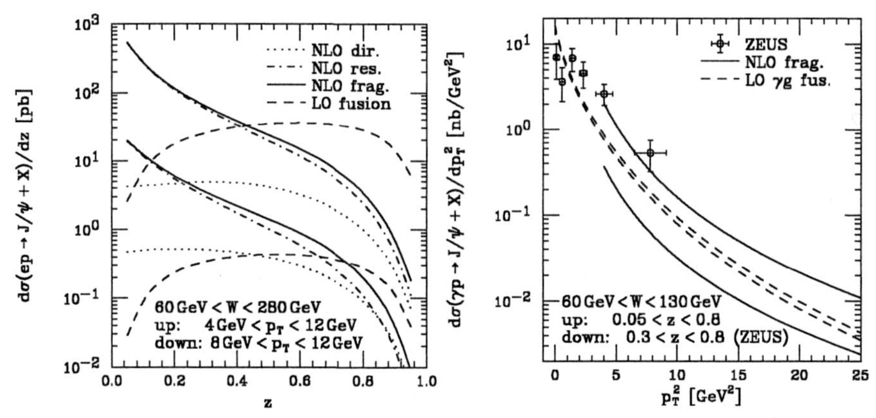

Fig. 1. (a) $d\sigma/dz$ and (b) $d\sigma/dp_T^2$ for inelastic J/ψ photoproduction at HERA.

In conclusion, the cross section of inelastic J/ψ photoproduction in ep collisions at low to intermediate z and large p_T, where fragmentation production is dominant, is very sensitive to the colour-octet matrix element $\langle 0|\mathcal{O}^{J/\psi}[\underline{8}, {}^3S_1]|0\rangle$. By contrast, γg fusion, which is only relevant in the upper z range and for $p_T \lesssim M_{J/\psi}$, probes $\langle 0|\mathcal{O}^{J/\psi}[\underline{8}, {}^1S_0]|0\rangle$ and $\langle 0|\mathcal{O}^{J/\psi}[\underline{8}, {}^3P_J]|0\rangle$. We propose to accordingly extend previous measurements of this cross section at HERA, in order to obtain an independent, nontrivial check of the Tevatron colour-octet charmonium puzzle.

References

[1] G.T. Bodwin, E. Braaten, and G.P. Lepage, Phys. Rev. D 51 (1995) 1125; 55 (1997) 5855 (E).
[2] B.A. Kniehl and G. Kramer, Phys. Rev. D 56 (1997) 5820; Phys. Lett. B 413 (1997) 416.
[3] ZEUS Collaboration, J. Breitweg et al., Contributed Paper No. pa02–047 to the 28th International Conference on High Energy Physics, Warsaw, Poland, 25–31 July 1996.

Quarkonium Polarization as a Test of Non-relativistic Effective Theory

M. Beneke (mbeneke@mail.cern.ch)

Theory Division, CERN, CH-1211 Geneva

Abstract. I compare current approaches to quarkonium production with regard to what they tell us about quarkonium polarization. Predictions for J/ψ polarization in hadron-hadron and photon-hadron collisions are summarized.

The production of charmonium involves physics at short and long distances. The short-distance part is given, uncontroversially, by the production cross section for a charm-anticharm-quark pair at small relative velocity v of the two quarks. The process through which the $c\bar{c}$ pair binds into a particular charmonium state is sensitive to long times $\tau \sim 1/(m_c v^2) \sim 1/(500\,\mathrm{MeV})$ and therefore it is non-perturbative. Various descriptions, with more or less contact to QCD, and relying on quite different physical pictures of this process, have been proposed, and used, over the years. The *colour-singlet model (CSM)* [1] assumes that only those $c\bar{c}$ pairs form J/ψ which are produced in a colour-singlet ${}^3S_1$ state already at short distances. The long-distance part is Coulomb-binding, accountable for by the wave-function at the origin. No gluons with energy less than $\mathcal{O}(m_c)$ in the J/ψ rest frame are emitted. The *colour-evaporation model (CEM)* [2] assumes that soft gluon emission from the $c\bar{c}$ pair is unsuppressed. The colour and spin quantum numbers of the $c\bar{c}$ pair at short distances are irrelevant. The long-distance physics is supposed to be described by a phenomenological parameter $f_{J/\psi}$, the fraction of 'open' $c\bar{c}$ pairs below threshold that bind into J/ψ. The *non-relativistic QCD (NRQCD) approach* [3] synthesizes elements of both approaches. J/ψ can be produced from $c\bar{c}$ pairs in any colour or angular momentum state at short distances but with probabilities that follow definite scaling rules [4] in v^2. Soft gluon emission does take place, but the interaction of soft gluons with the heavy quarks is determined by the NRQCD effective Lagrangian. Spin symmetry holds to leading order in v^2. There is a price to pay for the more detailed description of the long-distance part in NRQCD: It depends on (at least) four (rather than one) non-perturbative parameters, which have to be extracted from experiment. They are $\langle \mathcal{O}_1^{J/\psi}({}^3S_1)\rangle$, $\langle \mathcal{O}_8^{J/\psi}({}^3S_8)\rangle$, $\langle \mathcal{O}_8^{J/\psi}({}^1S_0)\rangle$, $\langle \mathcal{O}_8^{J/\psi}({}^3P_0)\rangle$, where the colour and angular momentum state indicated refers to the $c\bar{c}$ pair at short distances. The precise definition of these matrix elements is given in [3].

The following deals exclusively with polarization phenomena in J/ψ production. We discuss predictions for J/ψ production in hadron-hadron and photon-proton collisions, based on the CSM and the NRQCD approach. The prediction by the CEM is straightforward and universal: Because the model assumes that transitions ${}^3S_1 \leftrightarrow {}^1S_0$ are unsuppressed, we expect that J/ψ is always produced unpolarized. A polarization measurement has various discriminative powers. One can learn to what degree spin-flip transitions are suppressed and thereby check the basic assumption that distinguishes the CEM from the NRQCD approach. Since the octet production matrix elements of NRQCD (see above) lead to a polarization pattern different from the CSM, one can learn about the importance of colour-octet production mechanisms. In particular, production through a ${}^1S_0^{(8)}$ state yields unpolarized quarkonium. See [5, 6, 7] for the other production channels.

J/ψ production in fixed target hadron-hadron collisions. Polarization measurements exist for ψ and ψ' production in pion scattering fixed target experiments [8]. Both experiments observe an essentially flat angular distribution in the decay $\psi \to \mu^+\mu^-$ $(\psi = J/\psi, \psi')$,

$$\frac{d\sigma}{d\cos\theta} \propto 1 + \lambda\cos^2\theta,$$

where the angle θ is defined as the angle between the three-momentum vector of the positively charged muon and the beam axis in the rest frame of the quarkonium. The observed values for λ are 0.02 ± 0.14 for ψ', measured at $\sqrt{s} = 21.8\,\text{GeV}$ in the region $x_F > 0.25$ and 0.028 ± 0.04 for J/ψ measured at $\sqrt{s} = 15.3\,\text{GeV}$ in the region $x_F > 0$.

The colour-singlet contribution alone yields $\lambda \approx 0.25$ for the direct S-wave production cross section [9]. However, the total cross section is largely due to colour-octet production. The polarization in the colour-octet channels has been considered in [6] (see also [10]). If $\langle \mathcal{O}_8^{\psi'}({}^1S_0)\rangle$ and $\langle \mathcal{O}_8^{\psi'}({}^3P_0)\rangle$ are constrained to be positive, $0.15 < \lambda < 0.44$ is obtained for ψ' production at $\sqrt{s} = 21.8\,\text{GeV}$. The lower bound is obtained if production through a $c\bar{c}[{}^1S_0^{(8)}]$ intermediate state dominates. The analysis of J/ψ polarization is complicated by indirect J/ψ production through χ_c decays, which are not separated in the measurement above. In Fig. 1 the polar angle parameter λ is plotted as a function of r_L, the longitudinal polarization fraction of indirectly produced J/ψ (i.e. $r_L = 1/3$, if χ_c feed-down gives unpolarized J/ψ). r_L is difficult to obtain theoretically as χ_{c1} and perhaps even χ_{c2}

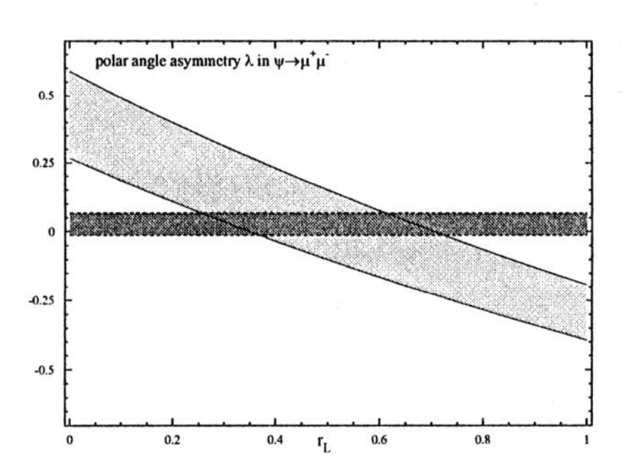

Fig. 1. Polar angle asymmetry for J/ψ production in pion-nucleus collisions at $\sqrt{s} = 15\,\text{GeV}$ as a function of the longitudinal polarization fraction of indirect J/ψ from radiative feed-down. The horizontal band shows the measurement of λ.

is dominantly produced through colour-octet states, whose polarization yield is described by too many phenomenological parameters to be predictable. The wide band in Fig. 1 is obtained by saturating the direct J/ψ production cross section by either $\langle \mathcal{O}_8^{J/\psi}({}^1S_0)\rangle$ (lower curve) or $\langle \mathcal{O}_8^{J/\psi}({}^3P_0)\rangle$ (upper curve). If the indirectly produced J/ψ are unpolarized, one would again have to assume that $\langle \mathcal{O}_8^{J/\psi}({}^1S_0)\rangle \gg \langle \mathcal{O}_8^{J/\psi}({}^3P_0)\rangle$ in order to reproduce the data (horizontal band in Fig. 1). A measurement of r_L could clarify the situation.

Since the total cross section is dominated by J/ψ production at small transverse momentum, non-factorizable final state interactions may be significant (though formally suppressed) and invalidate the predictions based on the CSM or NRQCD.

J/ψ polarization at the Tevatron. At transverse momentum $p_t \gg 2m_c$ J/ψ production in hadron-hadron collisions is now regarded as a gluon fragmentation process $g \to c\bar{c}[{}^3S_1^{(8)}] \to J/\psi + X$ [11]. Since the fragmenting gluon is nearly on-shell, this implies transversely polarized J/ψ as $p_t \to \infty$ [12] up to

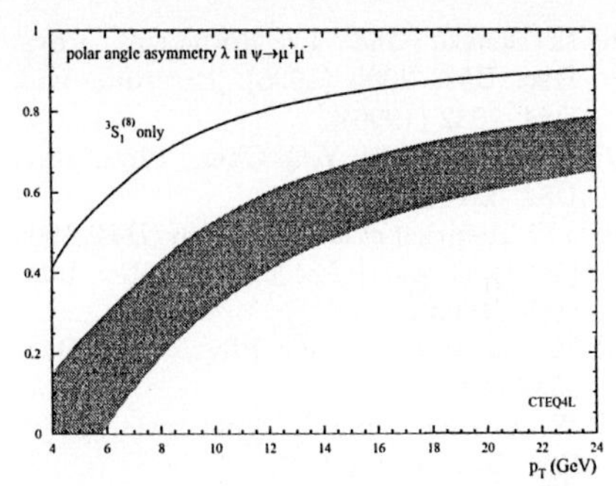

Fig. 2. λ as a function of p_t in $p+\bar{p} \to J/\psi + X$ at the Tevatron cms energy $\sqrt{s} = 1.8$ TeV. From [13].

spin-symmetry breaking corrections of order v^4 [5]. At finite p_t longitudinally polarized J/ψ can be produced, if a hard gluon is radiated in the fragmentation process [5] or the fragmentation approximation is relaxed [13, 14]. The non-fragmentation terms turn out to be particularly important. The predicted polar angle asymmetry λ is shown in Fig. 2. At $p_t \sim 5\,\mathrm{GeV}$ no trace of transverse polarization remains. As p_t increases, the angular distribution becomes rapidly more anisotropic. The observation of this pattern, even qualitatively, would already constitute strong support for the gluon fragmentation mechanism *and* the relevance of spin symmetry in quarkonium production. The polarization measurement will therefore rule out either the CEM or the applicability of NRQCD velocity power counting at the charmonium scale. Because of this, this measurement is probably the single most important one that can be done in the near future.

J/ψ **photo-production at HERA.** This production process deserves special attention in the context of NRQCD (and the CEM), since the colour-octet contributions to the energy distribution of inelastically produced J/ψ

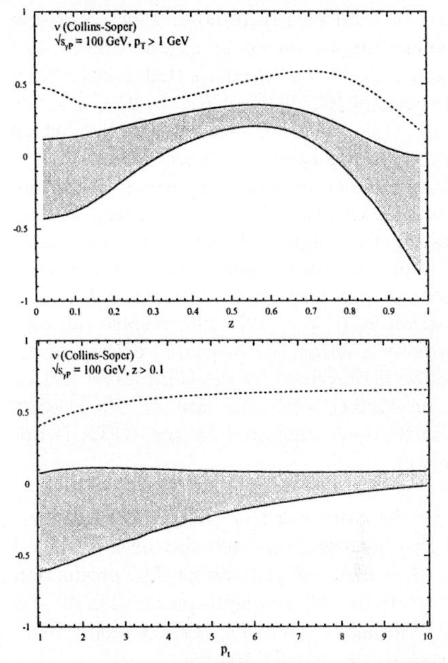

Fig. 3. Azimuthal angle parameter ν in the Collins-Soper frame for J/ψ energy and transverse momentum distributions in J/ψ photo-production at a typical HERA energy. From [17] where more details can be found.

seem to be too large close to the point $z = 1$ of maximal energy transfer [15]. Since the NRQCD velocity expansion is not valid in this endpoint region [16], one would like to infer the relevance of the colour-octet contribution from the data themselves. Polar and azimuthal J/ψ decay angular distributions may provide a clue to the answer to this problem, since the distributions are predicted to be quite different in the CSM and in the NRQCD approach [17]. [Recall that all angular distributions are isotropic in the J/ψ rest frame in the CEM.] The azimuthal dependence, characterized by two additional angular parameters μ and ν, is particularly instructive, as a function of both energy fraction z or transverse momentum as shown in Fig. 3. The shaded band reflects the variation that follows if either one of $\langle \mathcal{O}_8^{J/\psi}(^1S_0)\rangle$, $\langle \mathcal{O}_8^{J/\psi}(^3P_0)\rangle$ is set to zero, while the other saturates the sum which is constrained by the total production rate (in other production processes, primarily hadron-hadron collisions as discussed above). A measurement of angular distributions in inelastic J/ψ production comparable to the measurement of polarization in diffractive J/ψ production at HERA [18] could resolve the controversy whether the measured energy distribution is described by the CSM alone and/or is in conflict with the size of colour-octet contributions suggested by the NRQCD approach.

In the framework of NRQCD, predictions of J/ψ polarization have also been obtained for $B \to J/\psi + X$ [19], direct J/ψ production in Z^0 decay [20] and lepto-production of J/ψ [21]. Because of lack of space, the reader is referred to the original papers.

Acknowledgements. I would like to thank M. Krämer, I.Z. Rothstein and M. Vänttinen for their collaboration on this topic.

References

[1] For a review, see G.A. Schuler, CERN-TH-7170-94 [hep-ph/9403387], and references therein.
[2] H. Fritzsch, Phys. Lett. B67, 217 (1977); F. Halzen, Phys. Lett. B69, 105 (1977).
[3] G.T. Bodwin, E. Braaten and G.P. Lepage, Phys. Rev. D51, 1125 (1995) [Erratum: *ibid.* D55, 5853 (1997)].
[4] G.P. Lepage *et al.*, Phys. Rev. D46, 4052 (1992).
[5] M. Beneke and I.Z. Rothstein, Phys. Lett. B372, 157 (1996) [Erratum: *ibid.* B389, 789 (1996)].
[6] M. Beneke and I.Z. Rothstein, Phys. Rev. D54, 2005 (1996) [Erratum: *ibid.* D54, 7082 (1996)].
[7] E. Braaten and Y.Q. Chen, Phys. Rev. D54, 3216 (1996).
[8] J.G. Heinrich *et al.*, Phys. Rev. D44, 1909 (1991); C. Akerlof *et al.*, Phys. Rev. D48, 5067 (1993).
[9] M. Vänttinen *et al.*, Phys. Rev. D51, 3332 (1995).
[10] W.-K. Tang and M. Vänttinen, Phys. Rev. D54, 4349 (1996).
[11] E. Braaten and S. Fleming, Phys. Rev. Lett. 74, 3327 (1995).
[12] P. Cho and M.B. Wise, Phys. Lett. B346, 129 (1995).
[13] M. Beneke and M. Krämer, Phys. Rev. D55, 5269 (1997).
[14] A.K. Leibovich, Phys. Rev. D56, 4412 (1997).
[15] M. Cacciari and M. Krämer, Phys. Rev. Lett. 76, 4128 (1996); P. Ko, J. Lee and H.S. Song, Phys. Rev. D54, 4312 (1996).
[16] M. Beneke, I.Z. Rothstein and M.B. Wise, Phys. Lett. B408, 373 (1997).
[17] M. Beneke, M. Krämer and M. Vänttinen, [hep-ph/9709376].
[18] H1 Collaboration, S. Aid *et al.*, Nucl. Phys. B472, 3 (1996) and the update Paper pa02-085, submitted to the 28th International Conference on High Energy Physics, ICHEP'96, Warsaw, Poland, July 1996; ZEUS Collaboration, J. Breitweg *et al.*, Z. Phys. C75, 215 (1997).
[19] S. Fleming *et al.*, Phys. Rev. D55, 4098 (1997).
[20] S. Baek *et al.*, Phys. Rev. D55, 6839 (1997).
[21] S. Fleming and T. Mehen, JHU-TIPAC-96022 [hep-ph/9707365].

Subsidiary Sources of Heavy Quarks at LEP: Gluon Splitting and Onia

Paul Colas (Paul.Colas@cea.fr)

DAPNIA/SPP Saclay, France

Abstract. Recent progress in understanding other sources of heavy quarks at LEP than the mere decay of the Z boson into $b\bar{b}$ and $c\bar{c}$ is reviewed. The fraction of events with a gluon splitting into a $b\bar{b}$ pair is found to be $\bar{n}(g \to b\bar{b}) = (0.24 \pm 0.09)\%$. New inclusive charmonium production rates are given, as well as upper limits on the Υ production rate. Prompt charmonium production is discussed.

1 Introduction

The main source of heavy quarks at LEP (Q=c,b) is the decay $Z \to Q\bar{Q}$. About 17% of the hadronic decays are $c\bar{c}$, and about 21.5% are $b\bar{b}$. These two production fractions R_c and R_b, have been very accurately measured at LEP. After an excitement phase started in 1995 at the Brussel Conference, where raw averaging of the four LEP experiments showed deviations from the Standard Model prediction in these fractions, one or two years of careful studies of systematics lead to the 'nonsuit' conclusion that the agreement with the Standard Model is good. We now turn the focus to other sources of heavy quarks. These are the splitting of a gluon into a heavy quark pair and various production mechanisms of heavy quarkonia.

2 Gluon Splitting into $b\bar{b}$

The fraction of hadronic Z decay events with a gluon splitting into a $b\bar{b}$ pair, $\bar{n}(g \to b\bar{b})$, is the main systematics in some measurements of R_b (for instance, in ref. [1]). Though it is expected to be very small (0.18%) [2], a direct experimental measurement is of great interest. Such a study has been carried on by DELPHI [3] and ALEPH [4].

Both experiments first select four-jet events and require that the pair of jets with the smallest jet-jet angle (called jets number 1 and 2) are tagged as b. As confirmed by Monte-Carlo simulation, these two jets tend to be those from a gluon splitting. The other two jets (numbered 3 and 4) are the quark-antiquark pair originating from the Z decay. To further enrich their sample in gluon splitting events, the two collaborations use different topological and kinematic cuts. DELPHI rejects events where jets 1 and 2 are more energetic than jets 3 and 4 and they require that the rapidity of jet 1 (the most energetic of 1 and 2) with respect to the event thrust direction is less than 1.2. They also reject events where the angle α between the planes of jets (1,2) and (3,4) satisfies $\cos\alpha < 0.8$. The latter cut is known to distinguish $q\bar{q}gg$ from $q\bar{q}b\bar{b}$. In ALEPH, in the same spirit, the momenta of jets 3 and 4 are required to be in excess of 11 GeV/c while those of jets 1 and 2 must not exceed 27 GeV/c, and the angle between jets 1 and 2 must satisfy $0.2 < \cos\theta_{12} < 0.9$ and the angle between jets 3 and 4 must be such that $-0.9 < \cos\theta_{34} < 0.1$. The distribution of $\cos\theta_{34}$ (Fig.1) clearly shows the excess of $g \to b\bar{b}$ events (labelled as B in this figure) over $g \to c\bar{c}$ and other hadronic events (labelled C and Q) in this range. Systematic errors in this analysis arise from the simulation of the topology, the value of $\bar{n}(g \to c\bar{c})$ and Monte Carlo statistics. The DELPHI result is $\bar{n}(g \to b\bar{b}) = (0.21 \pm 0.11 \pm 0.09)\%$ and the ALEPH finds $\bar{n}(g \to b\bar{b}) = (0.26 \pm 0.04 \pm 0.09)\%$. Estimating conservatively the correlated part of the systematic error to be 0.05 we obtain the following weighted average:

$$\bar{n}(g \to b\bar{b}) = (0.24 \pm 0.09)\%.$$

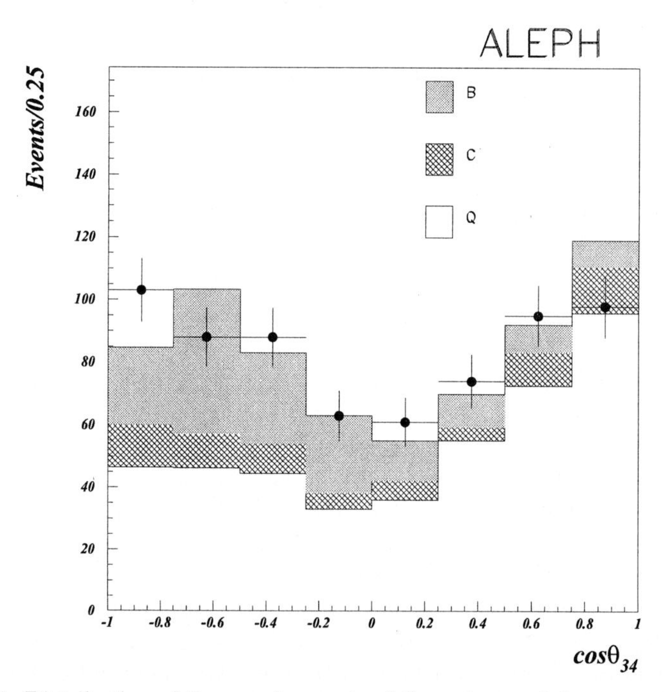

Fig. 1. Distribution of the opening angle of the pair remaining once the two closest jets have been removed, compared with the Monte Carlo prediction

3 Production of Heavy Quarkonia

3.1 Charmonium

The dominant, though trivial, source of charmonium is the decay of a b-hadron to J/ψ and its radial and orbital excitations. The branching fractions Z→ (c$\bar{c}$)X as measured by L3 [5] are given in table 1. The high resolution of the L3 calorimeter allows the the states decaying radiatively to J mesons to be separated. More interesting is the prompt charmonium production. A

Mode	Branching fraction
$Z \to$ JX	$(3.40 \pm 0.23 \pm 0.27) \times 10^{-3}$
$Z \to \psi'$X	$(1.6 \pm 0.5 \pm 0.3) \times 10^{-3}$
$Z \to \chi_{c1}$X	$(2.7 \pm 0.6 \pm 0.5) \times 10^{-3}$
$Z \to \chi_{c2}$X	$< 3.2 \times 10^{-3}$ at the 90%$C.L.$

Table 1. Inclusive branching fraction of the Z into various charmonium states, as measured by L3

prompt J/ψ signal can be found as a zero-lifetime component of the J/ψ signal. Though an order of magnitude less abundant than the B-decay source, it can be singled out thanks to the good decay length resolution provided by the vertex detectors of ALEPH, DELPHI and OPAL. One expects two contributions to this production. One is simply from fragmentation: in a c$\bar{c}$ event, one of the quarks of the decaying Z hadronizes by picking an anti-quark from a c$\bar{c}$ pair extracted from the vacuum. The other contribution is a gluon emitted by a q or a $\bar{q}$, followed by a splitting of this gluon into a c$\bar{c}$ pair which ends up into an onium. In the former process, there is an additional charmed quark in the event, leading to a less isolated charmonium. In a paper by OPAL [6], a cut was performed on the isolation of the J/ψ, introducing a large theoretical systematic error: $BR(Z \to prompt\, J/\psi) = (1.9 \pm 0.7 \pm 0.5 \pm 0.5(\text{theo})) \times 10^{-4}$. ALEPH carried out a similar analysis [7] without any isolation cut. They checked that the distribution of the isolation variable, chosen to be the energy in a cone of 30 deg. around the charmonium momentum direction, is consistent with a combination of contributions from both charm fragmentation and gluon→ J/ψ. The measured rate, $BR(\text{Z} \to \text{prompt J}/\psi\text{X}) = (3.00 \pm 0.78 \pm 0.30 \pm 0.17(\text{theo})) \times 10^{-4})$, also supports the presence of the two contributions, expected to be 0.8×10^{-4} for the charm fragmentation and 1.9×10^{-4} for the gluon splitting contribution in the colour octet model. The statistics is too low, however, to give a fully conclusive answer on the production mechanism.

3.2 Bottomium

Only the last two processes described in previous section (fragmentation and gluon splitting) can contribute to bottomium production.

After OPAL published a positive – though marginal – evidence for the production of Υ in Z decays, with an inclusive branching fraction of $(1.0 \pm 0.5) \times 10^{-4}$[8], this was not confirmed by any other experiment. ALEPH observed 2 events with an expected background of 1.2 ± 0.4, leading to a 95% C.L. upper limit of 7.3×10^{-5} on the inclusive branching ratio[9]. L3, at this conference, [10], published upper limits of respectively 5.5, 13.9 and 9.4 in units of 10^{-5} for the states 1S, 2S and 3S.

The expectations from b fragmentation and gluon splitting are respectively 1.6×10^{-5} and 4.1×10^{-5}, adding up to about 6×10^{-5}, at the border of the LEP sensitivity.

4 Conclusions

The knowledge of the gluon splitting into $b\bar{b}$ has reached enough accuracy not to be an issue anymore in the measurement of the fractional decay width of the Z into $b\bar{b}$. J/ψ production is now well measured, and in particular the prompt production and the limits on Υ production are consistent with a contribution from gluon splitting in the colour-octet model together with the fragmentation with a Heavy quark pair popping. An increase of statistics (a Z factory?) would be however necessary to really test the models.

References

[1] ALEPH Collab., Phys. Lett. B401 (1997) 163

[2] M. H. Seymour, Nucl. Phys. B436 (1995) 163

[3] DELPHI Collab., 'Measurement of the Multiplicity of Gluons Splitting to Bottom Quark Pairs in Hadronic Z^0 Decays', Contribution to this conference, number 716, CERN-PPE/97-39 (1997)

[4] ALEPH Collab., 'Measurement of the Gluon Splitting Rate into $b\bar{b}$ pairs in Hadronic Z Decays', Contribution to this conference, number 606 (1997)

[5] L3 Collab., 'Inclusive J, ψ' and χ_c Production in Hadronic Z Decays, CERN-PPE/97-44 (April 1997)

[6] OPAL Collab., Phys. Lett. B384 (1996) 343

[7] ALEPH Collab., 'Study of Prompt J/ψ Production in Hadronic Z Decays', Contribution to this conference, number 624 (1997)

[8] OPAL Collab., Phys. Lett. B370 (1996) 185

[9] ALEPH Collab., 'Inclusive Υ Production in Hadronic Z Decays', Contribution number PA05-066 to ICHEP96, Warsaw (1996)

[10] L3 Collab., 'Upsilon Production in Z Decays', CERN-PPE/97-78 (1997)

New Mass and Lifetime of the Ξ_c^+

Sergio P. Ratti (for the E687 Collaboration*) (ratti@pv.infn.it)

University of Pavia and I.N.F.N. - Sezione di Pavia

We report in this paper a new measurement of both mass and lifetime of the charm-strange baryon Ξ_c^+. Good measurements of the singly charmed baryon lifetimes help to discriminate among different theoretical models [1]. The data were collected in the Fermilab photoproduction experiment E687 during two run periods in 1990 and 1991 where we collected over 500 million hadronic triggers on tape. We published the mass and lifetime of the Ξ_c^+

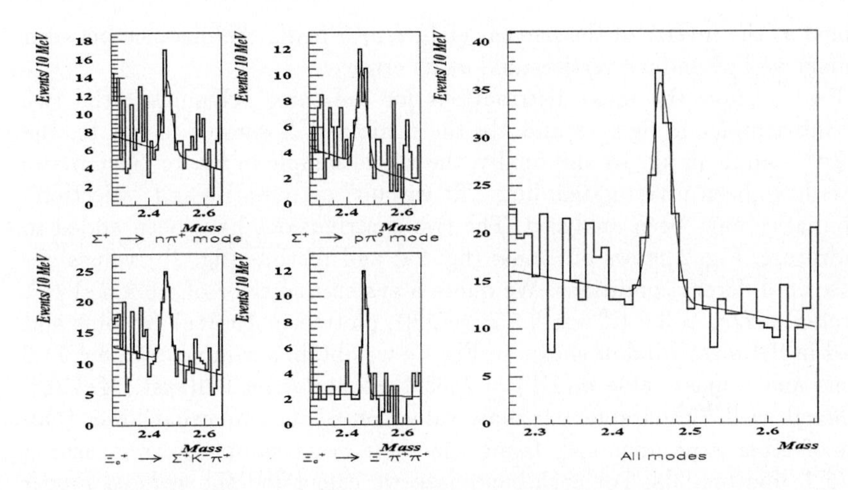

Fig. 1. Mass for the three Ξ_c^+ decay channels and for the total sample

baryon decaying into $\Xi^-\pi^+\pi^+$ with a sample of (29.7 ± 7.0) events [2]. About 45 events decaying into $\Sigma^+K^-\pi^+$ has been now reconstructed (the Σ^+ could decay both into $p\pi^o$ and $n\pi^+$) and added to the previous sample. The selection of the $\Xi^-\pi^+\pi^+$ sample has been described earlier [2]. Among usual selection criteria[2, 3], we isolate our signal using a *detachment parameter*

* co-authors are: P.L.Frabetti-Bologna; H.W.K.Cheung, J.P.Cumalat, C.Dallapiccola, J.F.Ginkel, W.E.Johns, M.S.Nehring-Colorado; J.N.Butler, S.Cihangir, I.Gaines, P.H.Garbincius, L.Garren, S.A.Gourlay, D.J.Harding, P.Kasper, A.Kreymer, P.Lebrun, S.Shukla, M.Vittone-Fermilab; S.Bianco, F.L.Fabbri, S.Sarwar, A.Zallo-Frascati; R.Culbertson, R.W.Gardner, R.Greene, J.Wiss-Illinois; G.Alimonti, G.Bellini, M.Boschini, D.Brambilla, B.Caccianiga, L.Cinquini, M.DiCorato, M.Giammarchi, M.Gullotta, P.Inzani, F.Leveraro, S.Malvezzi, D.Menasce, E.Meroni, L.Moroni, D.Pedrini, L.Perasso, F.Prelz, A.Sala, S.Sala, D.Torretta-Milano; D.Buchholz, D.Claes, B.Gobbi, B.O'Reilly-Northwestern; J.M.Bishop, N.M.Cason, C.J.Kennedy, G.N.Kim,T.F.Lin, D.L.Puseljic, R.C.Rutchi, W.D.Shephard, J.A.Swiatek, Z.Y.Wu-Notre Dame; V.Arena, G.Boca, G.Bonomi, C.Castoldi, G.Gianini, M.Merlo, C.Riccardi, L.Viola, P.Vitulo-Pavia; A.Lopez-Puerto-Rico; G.P.Grim, V.S.Paolone, P.M.Yager-Davis; J.R.Wilson-South Carolina; P.D.Sheldon-Vanderbilt; F.Davenport-North Carolina; K.Danyo, T.Handler-Tennessee; B.G.Cheon, J.S.Kang, K.Y.Kim, K.B.Lee-Korea.

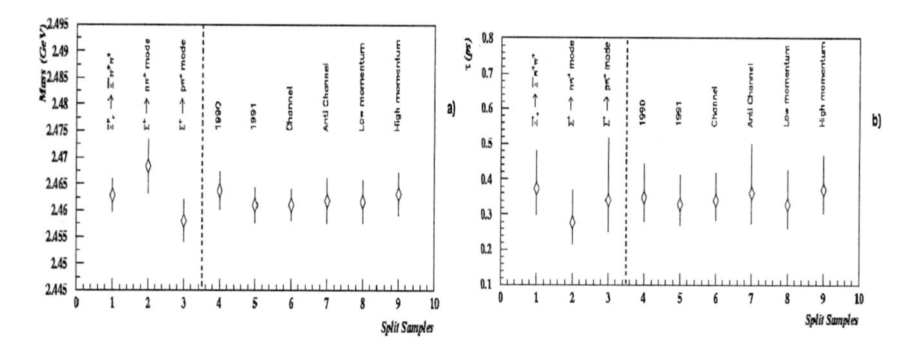

Fig. 2. Systematic studies: splitted samples; the 3 decays (left part); the 2 runs; channel-antichannel separation, low-high momentum separation (right part). a- mass values; b- lifetime values.

defined as the inverse of the percent error l/σ_l, l being the distance between primary and secondary vertices and σ_l its error.

Fig.s 1 show the mass distributions for the three "channels": the two Σ^+ sub-samples in fig.s 1a and 1b; the combined Σ sample in fig. 1c; the $\Xi^-2\pi^+$ sample in fig. 1d and finally, the whole sample in fig. 1e. Systematic errors have been investigated: both "fit variant" systematics and "selection" systematics have been analyzed. The two contributions have been added in quadrature. Fig. 2 shows our mass (fig. 2a) and lifetime (fig. 2b) values obtained in different conditions. We quote a systematic error of $\pm 0.9 MeV/c^2$. Selected at $l/\sigma_l > 3.6$ (it was 2.5 in ref.[2]), plotted in 10 MeV/c^2 bins and fitted in the mass window shown in Fig. 1e we obtain a sample of 72.8 ± 11.3 events and a mass value $m(\Xi_c^+) = 2463.4 \pm 2.6(stat.) \pm 0.9(syst.) MeV/c^2$. Although well comparable, our mass value comes out somewhat lower than that of other experiments[4]. Using the same event sample we remeasured the Ξ_c^+ lifetime [3]a. For each candidate we calculated the *reduced proper time t'*: $t' = (l - N\sigma_l)/(\beta\gamma c)$, where N=3.6 is the vertex detachment cut for the events in Fig. 1 and $\beta\gamma$ is the Lorentz boost. Montecarlo and data show that σ_l is indipendent of l. To calculate the lifetime τ and to estimate the background in the signal region, we used the maximum binned likelihood method[2, 3]. Our value for the lifetime is $\tau_{\Xi_c^+} = 0.35^{+0.06}_{-0.05} \pm 0.02$ ps. It is in excellent agreement with previous measurements[4] but has smaller errors.

References

[1] B.Blok and M.Shifman, in: *Tau-Charm Factory*, (Eds. J. and R. Kirkby, Ed. Frontiers, 1994) p. 847 and ref. therein;

[2] P.L. Frabetti et al.: Phys. Rev. Lett. **70**,(1993)1381;

[3] a- P.L. Frabetti et al.: Phys. Lett., **B328**,(1994)193; b- idem **B338**,(1994)106;

[4] R.M. Barnett et al. (Particle Data Group) Phys. Rev. **D54**,(1996)1.

LEP Recent Results on b-Baryon Decays

Dean Karlen (Dean.Karlen@cern.ch)

Carleton University, Ottawa, Canada

Abstract. An update of world averages of measurements of b-baryon lifetimes is presented. The lifetimes are significantly shorter than those of b-mesons which is posing a challenge to the theoretical understanding of b-hadron decays. New measurements of ratios of b-baryon branching ratios have been made by OPAL and ALEPH to further explore this puzzle. The results confirm that light quarks play a significant role in the decay of b-baryons.

1 Lifetimes of b-baryons

Figure 1 shows the evolution of the world average for published b-baryon lifetime measurements. The first round of measurements, published in 1992 and 1993, used Λ-lepton correlations to select b-baryon decays and gave an indication that b-baryons had a lifetime shorter than theoretical expectations. Later publications, which used nearly the complete LEP I data samples and included Λ_c-lepton correlations, confirmed this with much higher precision. In 1996 CDF published its first measurement of the average b-baryon lifetime, which falls in between the LEP average and theory. At the time of this conference ALEPH, DELPHI, and OPAL had several preliminary results on their full LEP I data set. Table 1 summarises these new measurements, shown in comparison with the previously published results.

Whereas the measurements using $\Lambda\ell^-$ correlations measure the average lifetime of an unknown mixture of b-baryons, those that use $\Lambda_c^+\ell^-$ correlations preferentially select events containing Λ_b^0 decays. The world average of all published and preliminary b-baryon lifetime measurements is, 1.22 ± 0.05 ps, assuming all b-baryons have the same lifetime. The average of just the $\Lambda_c^+\ell^-$ measurements, yields 1.23 ± 0.08 ps. [8] The ratio of b-baryon to B^0 meson lifetimes, 0.78 ± 0.04, is about three σ lower than the range of current theoretical expectations, $0.90 - 0.98$. [9] More precise or complementary measurements would be useful to confirm the discrepancy, and to help understand its origin. The next section describes such measurements.

2 Branching ratios of b-baryons

A measurement of the b-baryon semileptonic branching ratio would provide an independant determination of the hadronic partial width, whose value, according to the lifetime measurements, appears to be larger than theoretical expectations. The semileptonic branching ratio, however, is difficult to

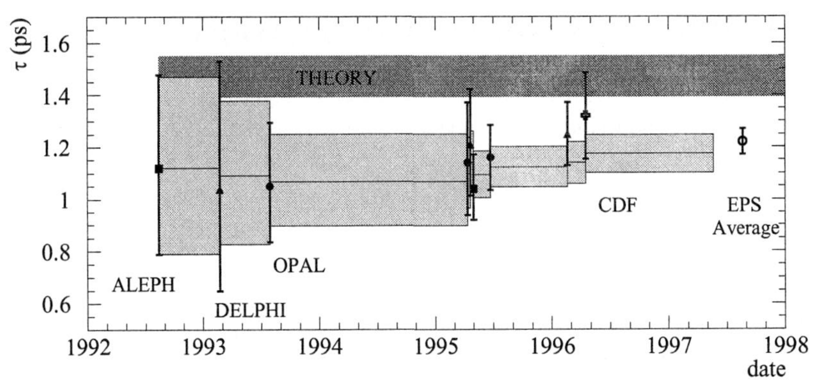

Fig. 1. The evolution of the average of published measurements of b-baryon lifetimes. The light band shows the 1σ uncertainty of the world average after each published measurement. The final data point (EPS) includes preliminary results available at the time of this conference.

Experiment	mode	method	dataset	lifetime (ps)	status
ALEPH	$\Lambda\ell^-$	i.p.	91-93	$1.05^{+0.12}_{-0.11} \pm 0.09$	pub. [1]
			91-95	$1.20 \pm 0.08 \pm 0.06$	prel. [2]
ALEPH	$\Lambda_c^+\ell^-$	d.l.	91-93	$1.02\ ^{+0.23}_{-0.18} \pm 0.06$	pub. [1]
			91-95	$1.18\ ^{+0.13}_{-0.12} \pm 0.03$	prel. [2]
ALEPH	$\Lambda\ell^+\ell^-$	d.l.	91-95	$1.30\ ^{+0.26}_{-0.21} \pm 0.04$	prel. [2]
DELPHI	$\mathrm{p}\ell^-$	d.l.	92	$1.27\ ^{+0.35}_{-0.29} \pm 0.09$	pub. [3]
			91-95	$1.30 \pm 0.13\ ^{+0.07}_{-0.16}$	prel. [4]
DELPHI	$\Lambda_c^+\ell^-$	d.l.	91-94	$1.19\ ^{+0.21}_{-0.18}\ ^{+0.07}_{-0.08}$	pub. [5]
			92-95	$1.11\ ^{+0.15}_{-0.13}\ ^{+0.07}_{-0.08}$	prel. [2]
DELPHI	$\Lambda\ell^+\ell^-$	d.l.	91-95	$1.36\ ^{+0.35}_{-0.28}\ ^{+0.09}_{-0.10}$	prel. [4]
OPAL	$\Lambda_c^+\ell^-$	d.l.	91-94	$1.14\ ^{+0.22}_{-0.19} \pm 0.07$	pub. [6]
			91-95	$1.34\ ^{+0.24}_{-0.22} \pm 0.06$	prel. [7]

Table 1. Summary of updates to previously published results and of new preliminary measurements. The decay mode used for selecting the b-baryon sample and the method (i.p. = impact parameter, d.l. = decay length) for extracting the lifetime are indicated.

measure at LEP, as no method to tag unbiased samples of b-baryons has been developed. Instead, OPAL and ALEPH have measured the following ratios of branching ratios,

$$R_{\Lambda\ell} = \frac{\mathrm{BR}(\Lambda_\mathrm{b} \to \Lambda\ell X)}{\mathrm{BR}(\Lambda_\mathrm{b} \to \Lambda X)} \qquad R_{\mathrm{p}\ell} = \frac{\mathrm{BR}(\Lambda_\mathrm{b} \to \mathrm{p}\ell X)}{\mathrm{BR}(\Lambda_\mathrm{b} \to \mathrm{p}X)}$$

in which the denominators refer to both hadronic and leptonic decays. If the light quarks did not participate in the decay of b-hadrons, these measurements would be equivalent to the semileptonic branching ratio of b-baryons and to b-mesons as well, since all b-hadrons would have the same semileptonic branching ratio. The extent to which $R_{\Lambda\ell}$, $R_{p\ell}$, and BR(b$\to \ell X$) differ depends on the strength of the amplitudes involving light quarks. A larger hadronic partial width for b-baryons tends to lower $R_{\Lambda\ell}$ and $R_{p\ell}$.

The greatest difficulty in making these measurements is in the determination of the denominators of the ratios and the two experiments have used quite different approaches. In the case of OPAL [10], $\Lambda_b \to \Lambda X$ events are counted by applying a high efficiency b-tag, selecting hard Λ's, and then the momentum of the second baryon in the jet (the *companion*, either a proton or Λ) is used to distinguish signal from background. Signal events have a hard spectrum for the Λ and a soft spectrum for the companion. In background events from baryonic decays of b-mesons, both baryons have hard spectra. For events in which both baryons come from fragmentation, the two baryons have soft spectra. A two dimensional fit to the observed distribution of baryon momenta is performed to extract the number of $\Lambda_b \to \Lambda X$ events. The reference spectra in the fit are from Monte Carlo simulation, but are adjusted using data from OPAL or CLEO. The projections of the fit are shown in figure 2.

The method used by ALEPH [11] to determine the ratio of branching ratios is more complex. The number of $\Lambda_b \to pX$ events is inferred by counting the number of b $\to pX$ events and correcting for the non b-baryon component by using the measured protonic branching ratios of b-mesons from CLEO, and estimates of the protonic branching ratios of the B_s^0 and b-baryons. The number of b $\to pX$ events is found by tagging the two hemispheres of events into 5 intervals of b purity. Double tagging is used to estimate 12 of the 15 tagging efficiencies (3 flavour categories {b,c,uds} $\times$ 5 purity intervals). The ionisation loss distributions for tracks in the opposite hemisphere are fit in 50 momentum bins for the 5 purity intervals. To distinguish fragmentation protons from those arising from b-hadron decay, further classification according to impact parameter and angle to thrust axis is used.

The results of the two measurements are,

$$\text{OPAL:} \quad R_{\Lambda\ell} = 7.0 \pm 1.2 \pm 0.7\ \%$$
$$\text{ALEPH:} \quad R_{p\ell} = 7.8 \pm 1.2 \pm 1.4\ \%$$

where R is the average of the results for electrons and muons. Both results are significantly less than the b semileptonic branching ratio [12],

$$\text{BR}^{4S}(\text{b} \to \ell X) = 10.49 \pm 0.46\ \%$$
$$\text{BR}^{Z}(\text{b} \to \ell X) = 11.12 \pm 0.20\ \%$$

thus confirming that the light quarks play a significant role in the decay of b-baryons, as suggested by the short b-baryon lifetime measurements.

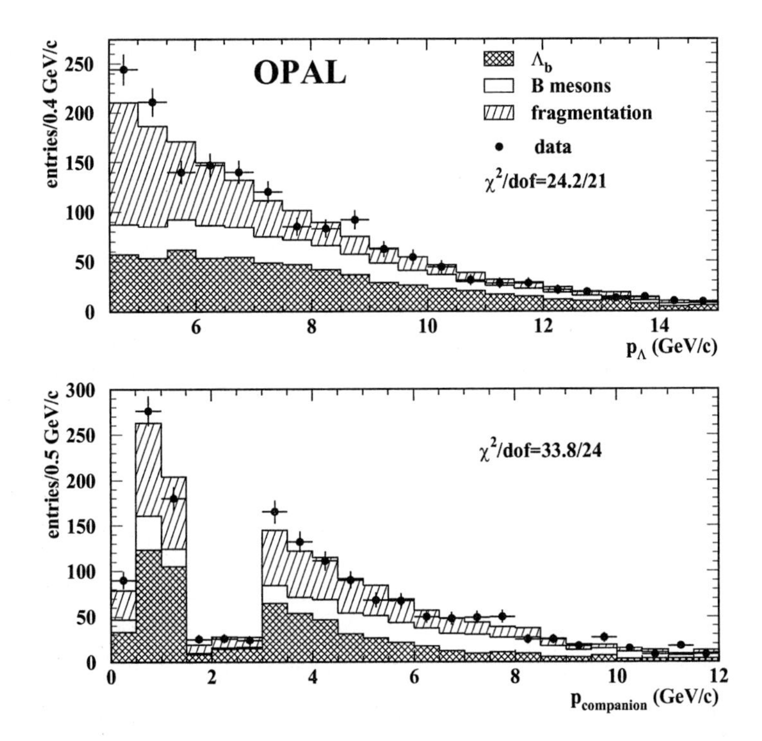

Fig. 2. The result of the two dimensional fit for the branching ratio ratio measurement by OPAL is shown by these projections. Proton candidates with momenta between 1.4 and 3.0 GeV/c are not included in the analysis because the particle identification, through ionisation loss, provides insufficient separation power to distinguish protons from pions.

References

[1] ALEPH collaboration, Phys. Lett. B357 (1995), 685.
[2] ALEPH collaboration, EPS contribution 604.
[3] DELPHI collaboration, Zeit. Phys. C68 (1995) 375.
[4] DELPHI collaboration, EPS contribution 454, DELPHI 97-104.
[5] DELPHI collaboration, Zeit. Phys. C71 (1996) 199.
[6] OPAL collaboration, Phys. Lett. B353 (1995) 402.
[7] OPAL collaboration, EPS contribution 156, OPAL PN305.
[8] LEP B Lifetime WG, `wwwcn.cern.ch/~claires/lepblife.html`
[9] I. Bigi, Nuovo Cim. 109A (1996) 713;
M. Neubert and C.T. Sachrajda, Nucl. Phys. B483 (1997) 339.
[10] OPAL collaboration, Zeit. Phys. C74 (1997) 423.
[11] ALEPH collaboration, EPS contribution 597.
[12] L. Chaussard, EPS presentation 508.

B Hadron Production and Inclusive b Decays at LEP

Lionel Chaussard (chaussad@in2p3.fr)

Institut de Physique Nucléaire, Université Lyon-I, Villeurbanne, France

Abstract. The production of beauty particles at LEP is reviewed. The results of inclusive studies are then presented, with an emphasis on semileptonic decays and studies of the number of charm in b decays.

1 B hadron production

Beauty quarks are produced at LEP through $e^+e^- \to Z^0 \to b\bar{b}$. These quarks can produce B^+, B_d^0, B_s^0 and beauty baryons through hadronization. In this section, the latest measurements of the production rates are reviewed.

From the studies of $\Lambda_c^+ - l^-$ and $\Xi^- - l^-$ correlations [1], the fraction of b quarks which hadronize into beauty baryons is estimated to be $(10.6^{+3.7}_{-2.7})\%$. The ALEPH Collaboration presents the results of a new analysis [2] based on the study of inclusive proton production in b tagged events : $f_{b-baryon} = (12.1 \pm 0.9(stat.) \pm 3.1(syst.))\%$. The average of the two values quoted above is $(11.3^{+2.5}_{-2.1})\%$.

From the studies of D_s-lepton correlations and of the mixing parameter $\chi = f_{B_d^0}\chi_d + f_{B_s^0}\chi_s$, the LEP B Oscillation Working Group [1] obtains the fraction of b quarks which hadronize into a B_s^0 meson : $f_{B_s^0} = (10.3^{+1.6}_{-1.5})\%$. The DELPHI Collaboration presents preliminary results of a new analysis [3] based on the study of the correlation between the charge of the produced quark (determined by a neural network) and the charge of the kaon which can be produced at the primary vertex when a B_s^0 is produced : $f_{B_s^0 + B_s^{**}} = (14.4 \pm 1.7(stat.) \pm 3.0(syst.))\%$. With some assumption on the fraction of charged beauty baryons (e.g. $f_{B_s^{**}}/(f_{B_s^0} + f_{B_s^{**}}) = 0.25 \pm 0.05$), the average value of $f_{B_s^0}$ is found to be $(10.4^{+1.4}_{-1.3})\%$.

Using the averages quoted above, assuming isospin symmetry ($f_{B^+} = f_{B_d^0}$) and the closure relation $\Sigma f_{b-species} = 1$, the fraction of beauty quarks hadronizing into a B^+ or a B_d^0 is then found to be $f_{B^+} = f_{B_d^0} = (39.15^{+1.2}_{-1.4})\%$.

The DELPHI Collaboration also presents the first measurements of the fraction of b quarks hadronizing into a charged or a neutral b hadron [3]. From a study of secondary vertex charge, DELPHI gives : $f_{H_B^0} = (57.8 \pm 0.5(stat.) \pm (1.0))\%$, and $f_{H_B^\pm} = (42.2 \pm 0.5(stat.) \pm (1.0))\%$. These results are consistent with the numbers quoted above, yet the systematic errors are very preliminary and some assumption on the fraction of charged b baryons is needed to extract usefull information from these measurements.

2 Charged multiplicity of weakly decaying B hadrons

New inclusive analysis have been developped by the LEP Collaborations to study b decays. A crucial parameter in such inclusive studies is the charged multiplicity of weakly decaying B hadrons (n_B): it is used (and needed) in particular to control the agreement between data and simulation.

The DELPHI Collaboration presents a new preliminary measurement of n_B, based on the study of positive and negative impact parameter distributions of tracks in b-jets [4]. DELPHI obtains $n_b = 4.96 \pm 0.03(stat.) \pm 0.05(syst.)$, which is much more precise than the older results of OPAL (obtained in 1994 with a systematical error of 0.49) and the previous results of DELPHI (obtained in 1995, with a systematical error of 0.38).

3 Semileptonic b decays

Among all the branching ratios of beauty particles, the semileptonic branching ratio is important for the estimation of the CKM matrix element V_{bc}, or to control the systematic in the measurements of the branching fraction of the Z^0 into $b\bar{b}$.

DELPHI is the only LEP Collaboration which gives an update of inclusive semileptonic branching ratios, from a re-analysis of 1994 data using a fit to momentum and transverse momentum distributions of leptons in b-tagged events containing one or two leptons [5]. The LEP averages provided by the LEP ElectroWeak Working Group [6] are $Br(b \to l) = (11.12 \pm 0.20)\%$ and $Br(b \to c \to l) = (8.04 \pm 0.33)\%$. The LEP average for $Br(b \to l)$, once noticed the presence of b-baryons and B_s^0 at LEP, is still surprisingly higher than the value obtained by CLEO at lower energy $((10.49 \pm 0.046)\%)$ [7].

The DELPHI and ALEPH Collaborations also presented some specific studies of these semileptonic decays. Using a neural network based on inclusive vertex charge, DELPHI obtains the first estimations of the semileptonic branching ratios of the B^+ meson [8]: $Br(B^+ \to l^+) = 0.136 \pm 0.008$ and $Br(B^+ \to l^-) = 0.060 \pm 0.006$, where the errors are statistical only. From the study of proton-lepton correlations in b-tagged events, ALEPH [2] gives $Br(b - baryon \to pl^- X)/Br(b - baryon \to pX) = (7.8 \pm 1.2(stat.)) \pm 1.4(syst.))\%$, which is compatible with an older OPAL result [9]. The semileptonic branching ratio of beauty baryons is then found to be lower than the semileptonic branching ratio of beauty mesons, which is consistent with the ratio of baryon/meson lifetimes.

It is then interesting to study the semi-exclusive modes of semileptonic b decays in order to compare the inclusive $b \to l$ result to the sum of its exclusive components. The DELPHI Collaboration presents a new preliminary measurement of the production of orbitally excited D^{**} in semileptonic b decays [10]. From a study of inclusive D^* production and a fit to the missing mass squared, DELPHI gives $f^{**} = 0.19 \pm 0.09$. Averaging this value with a result presented by ALEPH in 1996 [11], and taking into account D-lepton, D^*-lepton [12] and X_u-lepton [13] production, the exclusive results sum up to $(85 \pm 8)\%$ of the inclusive LEP value for $Br(b \to l)$, which shows that there is no serious discrepancy in this sector.

4 Charm in b decays

The b semileptonic branching ratio ($Br(b \to l)$) and the number of charm in b decays (n_c) should obviously be anti-correlated.

The number of charm in b decays can be estimated using $n_c = 1 + f(B \to D\overline{D}) + f(B \to \text{"}hidden\text{"}\ c\bar{c}) - f(B \to no\ c)$, where the *"hidden"* $c\bar{c}$ is the contribution of bound states (e.g. J/ψ). The DELPHI Collaboration gives a new preliminary result for the *"open charm"* term : $f(B \to D\overline{D}) = 0.166 \pm 0.036$. This result is obtained from an average of two results [14], one coming from the inclusive study of the distribution of a b-tagging probability variable, the other coming from the study of the transverse momentum distributions of kaons in b-tagged events containing two kaons. Using some assumptions or CLEO measurements for the *"hidden charm"* and *"no charm"* contributions [15], DELPHI obtains $n_c = 1.166 \pm 0.038$, which is compatible with the last CLEO result (1.10 ± 0.05) [16] or the results coming from charm counting at LEP (errors of 0.08) [17].

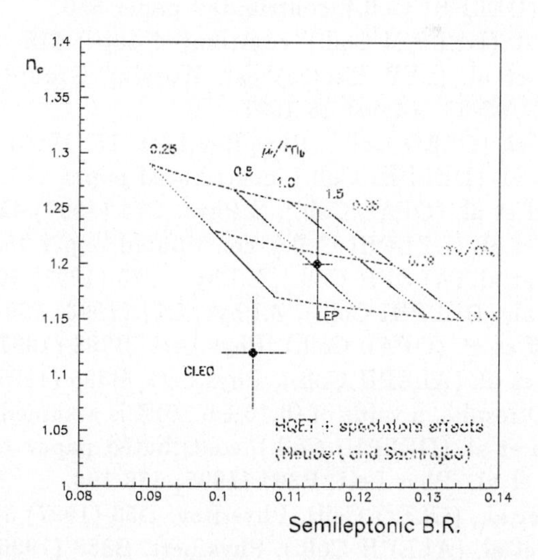

Fig. 1. Measurements of the number of charm in b decays vs the semileptonic branching ratio. The curves indicates theoretical expectations, the dots correspond to the LEP and CLEO data. The LEP point has been corrected by the ratio of meson/baryon lifetimes to account for baryon production at LEP.

The $Br(b \to l)$ and n_c results obtained at LEP and at lower energy (CLEO) may show some discrepancy. The values obtained at LEP are consistent with the HQET expectations [18] but the semileptonic branching ratio obtained at LEP is quite higher than the one obtained by CLEO (the LEP point has been corrected for the baryon production effect on Figure 1). The discrepancy between the two values of n_c is quite small, yet both $Br(b \to l)$ and n_c are found at high values at LEP. Some dedicated effort on this problem at LEP and by CLEO could clear the situation.

5 Acknowledgements

I thank my colleagues from the four LEP experiments who helped me to prepare this review. Special thanks to Clara Matteuzzi, for her help and advice, and to Michael Feindt, Peter Kluit, and Franco Simonetto for providing materials, references and fruitful discussions.

References

[1] V.Andreev et al. (LEP B Oscillations Working Group), LEPBOSC/97-02, August 18 1997.
[2] U.Becker (ALEPH Coll.), contributed paper 597.
[3] M.Feindt et al. (DELPHI Coll.), contributed paper 451.
[4] C.Mariotti (DELPHI Coll.), contributed paper 850.
[5] M.Calvi et al. (DELPHI Coll.), contributed paper 415.
[6] D.Abbaneo et al. (LEP ElectroWeak Working Group), Internal Note LEPEWWG/97-02, August 18 1997.
[7] B.Barish et al. (CLEO Coll.), Phys.Rev.Lett. 76 (1996) 1570-1574.
[8] M.Feindt et al. (DELPHI Coll.), contributed paper 473.
[9] K.Ackerstaff et al. (OPAL Coll.), Z.Phys. C74 (1997) 423-435.
[10] G.Borissov et al. (DELPHI Coll.), contributed paper 453.
[11] D.Buskulic et al. (ALEPH Coll.), Z.Phys. C73 (1997) 601-612.
[12] P.Abreu et al. (DELPHI Coll.), Z.Phys. C71 (1996) 539-554,
K.Ackerstaff et al. (OPAL Coll.), Phys.Lett. B395 (1997) 128-140,
D.Buskulic et al. (ALEPH Coll.), Phys.Lett. B395 (1997) 373-387.
[13] From CLEO results, a value of $(0.15 \pm 0.10)\%$ is assumed for $B \to X_u l\nu$.
[14] M.Battaglia et al. (DELPHI Coll.), contributed paper 448.
[15] G.Buchalla et al., Phys.Lett. B364 (1995) 188-194.
[16] L.Gibbons et al., (CLEO Coll), Phys.Rev. D56 (1997) 3783-3802.
[17] D.Busculic et al. (ALEPH Coll.), Phys.Lett. B388 (1996) 648-658,
G.Alexander et al. (OPAL Coll.), Z.Phys. C72 (1996) 1-16.
[18] M.Neubert et al., Nucl.Phys. B483 (1997) 339-370.

A Review of B Hadron Lifetimes

Franz Muheim (Franz.Muheim@cern.ch)

University of Geneva, Switzerland

Abstract. A review of B hadron lifetime measurements is presented. The new lifetime averages of B hadrons are:

$$\tau_{B^+} = 1.67 \pm 0.04 \text{ ps}, \qquad \tau_{B^0_d} = 1.57 \pm 0.04 \text{ ps},$$
$$\tau_{B^0_s} = 1.53 \pm 0.06 \text{ ps}, \qquad \tau_{b-baryon} = 1.22 \pm 0.05 \text{ ps},$$
$$\tau_b \text{ inclusive} = 1.554 \pm 0.013 \text{ ps}, \qquad \frac{\tau_{B^+}}{\tau_{B^0_d}} = 1.07 \pm 0.04.$$

1 Introduction and Methods

In the spectator model all heavy-quark hadrons have equal lifetimes. Non-spectator effects are due to Pauli Interference between internal and external spectator, weak annihilation, and W-exchange diagrams. These effects are calculated with Heavy Quark Expansion and scale like $1/m_Q^3$. In contrast to charmed hadrons, where $\tau_{D^+}/\tau_{D^0} = 2.55 \pm 0.04$, the lifetime differences for B hadrons are predicted to be small:

	Ref.[1]	Ref.[2]
$\tau_{B^+}/\tau_{B^0_d} \simeq$	$1 + 0.05(f_B/200\text{MeV})^2$	$1 + \mathcal{O}(1/m_b^3)$
$\tau_{B^0_s}/\tau_{B^0_d} \simeq$	$1 + \mathcal{O}(1\%)$	$(1.00 \pm 0.01) + \mathcal{O}(1/m_b^3)$
$\tau_{\Lambda_b}/\tau_{B^0_d} \simeq$	0.9	$(0.98 \pm 0.01) + \mathcal{O}(1/m_b^3)$

Thus one expects a hierarchy $\tau_{\Lambda_b} < \tau_{B^0_d} \approx \tau_{B^0_s} \leq \tau_{B^+}$. To test these predictions, precision measurements with errors on the percentage level are needed.

Measurements of the B hadron lifetimes are performed at LEP, SLD/SLC, and CDF/Fermilab. LEP experiments have collected up to 1.7 million B hadrons. SLD has 90k B hadrons, but has a smaller beam pipe and better beam spot constraints. CDF has a higher production rate (600 million B hadrons) but requires special B-hadron triggers. The mean decay length of a B hadron is 3 mm at LEP/SLD and 1.5 mm at CDF, respectively.

Two methods are being used to measure B hadron lifetimes. First, the impact parameter δ of lepton or hadron tracks stemming from B decays is positive on average ($\delta \sim c\tau_b \approx 300\ \mu$m at LEP/SLC). The lepton impact parameter was employed when measuring τ_b first. It has the advantage that it depends only weakly on the Lorentz boost of the B hadron. Second, the reconstruction of secondary vertices for fully or partially reconstructed B decays allows to measure the decay length L of B hadrons, directly. Since L equals $\frac{p}{m_B} c\tau_b$, the momentum p of the B hadron needs also to be measured to extract its proper time. Fully reconstructed hadronic B decays have good momentum resolution but lack in statistics. In semileptonic decays, the neutrino is missing, and the charmed hadron often is only partially reconstructed. The momentum is then estimated by using missing energy measurements or scaling of the measured $D^{(*)}$-lepton momentum.

The decay length method is also used in a more inclusive approach, called topological vertexing. The total charge at the secondary vertex allows to obtain two separate data samples which are enhanced in charged and neutral B mesons, respectively. In order to extract the lifetimes τ_{B^+} and $\tau_{B_d^0}$, the respective purities of the charged and neutral samples for B^+ and B_d^0 as well as the contaminations from B_s^0 mesons and B baryons have to be estimated from Monte Carlo simulations. The advantage of this method are the larger samples obtainable.

2 Measurements of B Hadron Lifetimes

A new measurement of the inclusive b-hadron lifetime τ_b is presented by L3[3]. In hadronic events, the best combination of all tracks to give three vertices is searched for. Beam spot and kinematic constraints are imposed. The lifetime τ_b is then extracted from a binned χ^2 fit to a) the impact parameter distribution of 2007k secondary vertex tracks, and b) the decay length L of 584k secondary vertices. Fit a) as compared to b) reduces the dependence of τ_b on the B fragmentation, i.e. the mean energy of B hadron $< \mathrm{x_E} >_\mathrm{b}$, from 30 to 9 fs while having similar statistical power. The two results are combined taking into account their different dependence on $< \mathrm{x_E} >_\mathrm{b}$ and correlations and yield $\tau_b = 1.556 \pm 0.010 \pm 0.017$ ps (Fig. 1). They also measure $< \mathrm{x_E} >_\mathrm{b} = 0.709 \pm 0.004$. The current LEP average is $< \mathrm{x_E} >_\mathrm{b} = 0.702 \pm 0.008$. L3 has also updated their lepton impact parameter measurement and ALEPH presents a new measurement using secondary vertices.

When averaging these results with other measurements from LEP, SLD, and CDF (Fig. 1 and Ref.[4]) one should keep in mind that the sample compositions are slightly different. By using leptons we have to take into account the semileptonic branching ratios of the different B hadrons which are themselves proportional to the lifetime. We expect τ_b to be 10 fs larger than by using hadrons, whereas we measure it τ_b to be smaller by 35 ± 20 fs. In most decay length analyses the largest systematic error is due to the B fragmentation. Thus an increase in the average of all $< \mathrm{x_E} >_\mathrm{b}$ measurements would decrease these τ_b values by as much as 20 fs. The total error on the new world average is 13 fs.

Most of the existing measurements of the exclusive B meson lifetimes τ_{B^+} and $\tau_{B_d^0}$ made by ALEPH, DELPHI, OPAL and CDF are based on reconstructing $D^{(*)}$ mesons in semileptonic B decays[4]. Simultaneously they measure the ratio of lifetimes $\tau_{B^+}/\tau_{B_d^0}$ where some systematic errors cancel. The update from CDF on these analyses now includes the full Run 1 data set[5]. Improvements are made by using more inclusive methods. DELPHI has

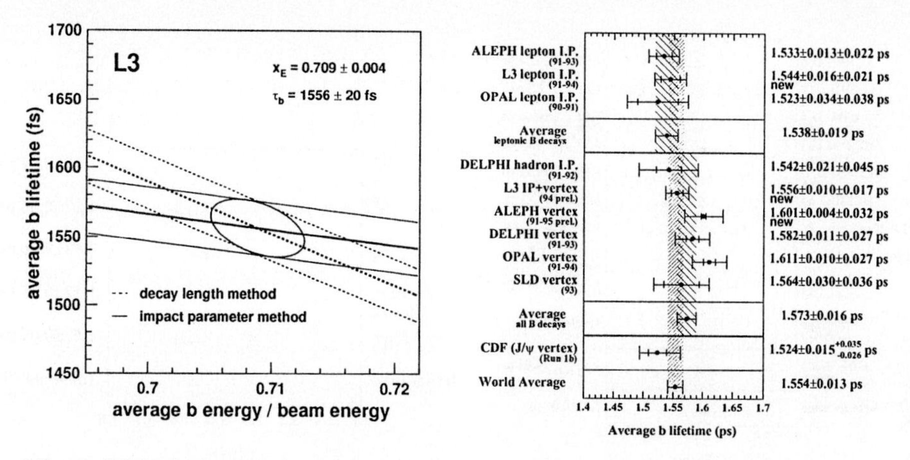

Fig. 1. L3 inclusive τ_b vs $< x_E >_b$; inclusive b-hadron lifetime measurements.

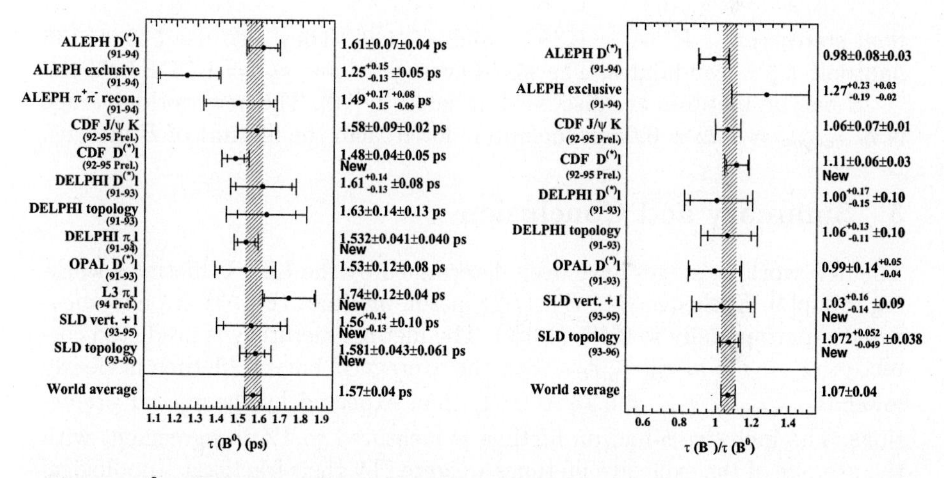

Fig. 2. B_d^0 lifetime and B^+/B_d^0 lifetime ratio measurements.

reconstructed 3523 ± 150 $\bar{B}_d^0 \to D^{*+}X\ell^-\bar{\nu}$, $D^{*+} \to D^0\pi_s^+$ candidates using $\pi_s^+\ell^-$ correlations while only inclusively reconstructing the D^0 candidate[6]. This method is also employed by a similar L3 analysis[7]. SLD uses topological vertexing and has included their 1996 data, presented in a separate contribution[8]. The statistical error on these B^+ and B_d^0 lifetime measurements is now ~40 fs, comparable to the systematic error (Fig. 2).

The best measurements of $\tau_{B_s^0}$, the lifetime of the B_s^0 meson, use the decay $B_s^0 \to D_s^-\ell^+\nu$ and reconstruct the D_s^- in many channels (ALEPH, DELPHI, OPAL and CDF, see Fig. 3 and Ref.[4]). OPAL has updated its $\tau_{B_s^0}$ measurement reconstructing 172 ± 28 $D_s^-\ell^+$ candidates in the following

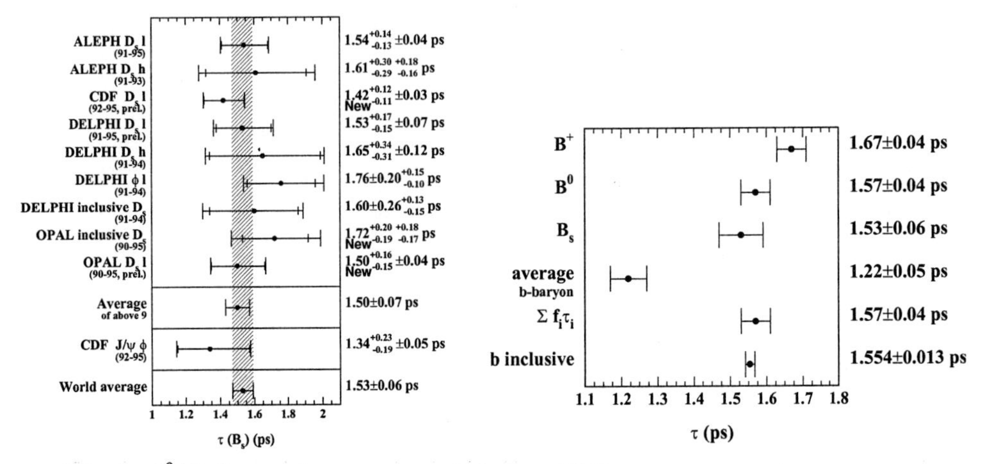

Fig. 3. B_s^0 lifetime and summary on B hadron lifetime measurements.

final states: $\phi\pi^-$, $K^{*0}K^-$, $K_s^0K^-$, and $\phi\ell^-\bar{\nu}$[9]. They also use 519 $\pm$ 136 hadronic D_s^-X candidates. The world average is now $\tau_{B_s^0} = 1.53 \pm 0.06$ ps.

B baryon lifetimes are discussed in detail in [10]. The new world average is $\tau_{b-baryon} = 1.22 \pm 0.05$ ps, definitely lower than the lifetime of B mesons.

3 Summary and Conclusion

The new world averages have been determined by the LEP B lifetimes working group[4]. Exclusive B^+, B_d^0, (B_s^0) meson lifetimes (Fig. 3) are now measured experimentally to 2.4% (4.0%). The lifetime hierarchy is predicted correctly: $\tau_{\Lambda_b} < \tau_{B_d^0} \approx \tau_{B_s^0} \leq \tau_{B^+}$. But the average B baryon lifetime is clearly lower $(\tau_{b-baryon}/\ \tau_{B_d^0} = 0.78 \pm 0.04)$ than expected by theoretical predictions. The inclusive b-hadron lifetime is measured to 1% in agreement with the average of the exclusive lifetimes weighted by their fractions. Topological vertexing methods are improving measurements.

Acknowledgements

I am grateful to the members of the LEP B lifetimes group for their support when preparing this review.

References

[1] I. Bigi *et al.*, in *B Decays*, ed. by S. Stone, World Scientific, (1994).
[2] M. Neubert *et al.*, Nucl. Phys. B483 (1997) 339-370.
[3] L3 Collab., M. Acciarri *et al.*, CERN-PPE/97-127, subm. to Phys. Lett.
[4] LEP B lifetimes working group; for references to measurements, see Web page http://wwwcn.cern.ch/~claires/lepblife.html.
[5] CDF Collab., http://www-cdf.fnal.gov/physics/new/bottom/cdf3786/-cdf3786.html
[6] DELPHI Collab. P. Abreu *et al.*, Zeit. Phys. C74 (1997) 19.
[7] L3 Collab., contrib. paper EPS/HEP 97-495.
[8] D. Jackson, in *these proceedings.*
[9] OPAL Collab., contrib. papers EPS/HEP 97-154 and 97-155.
[10] D. Karlen, in *these proceedings.*

Recent Results on B Decays from CLEO

David H. Miller[1] **(Miller@physics.purdue.edu)**

Purdue University, West Lafayette, IN, 47907 USA

Abstract. Using the CLEO detector[2] at the Cornell Electron Storage Ring (CESR), we have measured, or set limits on, a variety of hadronic decays of the B meson with branching ratios in the range of 10^{-3} to 10^{-5}. The data sample consists of 3.1 fb^{-1} on the $\Upsilon(4S)$ resonance which contains 3.3 million $B\bar{B}$ events and 10 million continuum events.

1 Introduction

Using 3.3 million $B\bar{B}$ events we have measured both exclusive and inclusive decays of the B meson. These include more precise measurements of known decays and new observations of rare decay modes. The most important new results are measurements of inclusive and exclusive upper vertex charm production and the observation of rare decays involving the etaprime, omega and phi mesons. This paper is a summary of the results presented at this conference and most, if not all, of the results will be refined and published prior to the publication of these proceedings. The reader is referred to the references on each section heading for the specific paper submitted to the conference.

2 $B \rightarrow D^{(*)}(n\pi)^-$ Decays[3]

Studies of the two body decays of the B meson provide important information on both the weak and hadronic interactions of heavy flavored mesons. In particular, tests of the factorization hypothesis and theoretical model predictions can be performed using precise measurements of the properties of these decays[4, 5, 6, 7, 8].

Determination of the magnitude and relative sign of the BSW parameters a_1 and a_2 allows for an estimate of the interference between internal and external spectator decay processes in charged B decays. To determine the parameters we use theoretical model predictions from Deandrea *et al.*(M1)[7], Neubert *et al.*(M2)[6]. Both models rely on the factorization approximation and employ HQET to determine heavy-to-heavy form-factors. They differ primarily in how they estimate the heavy-to-light form-factors. To allow for model comparison we have re-scaled the decay constants f_ρ and f_D in the M1 model to values used by M2. We also update the values of V_{cb} and the B lifetimes to current world averages. In addition, we use 1.0 for the B

production fraction f_{+-}/f_{00}, consistent with the the value used to compute the branching fractions. The tables below show our new values for a set of hadronic B decays and the results of a global fit to determine a_1 and a_2.

B Mode	Yield	Branching ratio %
$D^+\pi^-$	240	$0.250 \pm 0.022 \pm 0.024 \pm 0.024$
$D^{*+}\pi^-$	248	$0.234 \pm 0.021 \pm 0.022 \pm 0.010$
$D^0\pi^-$	1014	$0.473 \pm 0.025 \pm 0.033 \pm 0.014$
$D^{*0}\pi^-$	207	$0.392 \pm 0.046 \pm 0.033 \pm 0.028$
$D^+\rho^-$	251	$0.789 \pm 0.069 \pm 0.093 \pm 0.077$
$D^{*+}\rho^-$	250	$0.734 \pm 0.061 \pm 0.074 \pm 0.030$
$D^0\rho^-$	744	$0.920 \pm 0.080 \pm 0.072 \pm 0.028$
$D^{*0}\rho^-$	217	$1.277 \pm 0.129 \pm 0.113 \pm 0.090$
$D^+a_1^-$	136	$0.834 \pm 0.088 \pm 0.115 \pm 0.081$
$D^{*+}a_1^-$	160	$1.157 \pm 0.120 \pm 0.155 \pm 0.049$
$D^0a_1^-$	242	$0.887 \pm 0.104 \pm 0.107 \pm 0.028$
$D^{*0}a_1^-$	109	$1.597 \pm 0.250 \pm 0.209 \pm 0.115$

	$\|a_1\|$	a_2/a_1	χ^2/dof
M1	$0.953 \pm 0.026 \pm 0.042 \pm 0.020$	$0.232 \pm 0.037 \pm 0.022 \pm 0.018$	20.3/6
M2	$0.989 \pm 0.028 \pm 0.043 \pm 0.021$	$0.264 \pm 0.039 \pm 0.025 \pm 0.019$	9.1/6

3 Color suppressed decays[9]

In addition to the favoured hadronic decays we have improved the limits on a number of colour suppressed modes. The results are shown in the table below. These limits are consistent with the theoretical models and the values of a_1 and a_2 found in the analysis above.

Decay Mode	Branching Ratio (@90% C.L.)	Theoretical Predictions
$\bar{B}^0 \to D^0\pi^0$	$< 1.2\times10^{-4}$	0.7×10^{-4}
$\bar{B}^0 \to D^{*0}\pi^0$	$< 4.4\times10^{-4}$	1.0×10^{-4}
$\bar{B}^0 \to D^0\rho^0$	$< 3.9\times10^{-4}$	0.7×10^{-4}
$\bar{B}^0 \to D^{*0}\rho^0$	$< 5.6\times10^{-4}$	1.7×10^{-4}
$\bar{B}^0 \to D^0\eta$	$< 1.3\times10^{-4}$	0.5×10^{-4}
$\bar{B}^0 \to D^{*0}\eta$	$< 2.6\times10^{-4}$	0.6×10^{-4}
$\bar{B}^0 \to D^0\eta'$	$< 9.4\times10^{-4}$	
$\bar{B}^0 \to D^{*0}\eta'$	$< 19\times10^{-4}$	
$\bar{B}^0 \to D^0\omega$	$< 5.1\times10^{-4}$	0.7×10^{-4}
$\bar{B}^0 \to D^{*0}\omega$	$< 7.4\times10^{-4}$	1.7×10^{-4}

4 $\bar{B} \to D^{(*)}\bar{D}^{(*)}K^-$[10]

The dominant process for hadronic $\bar{B}$ decay is the $b \to c\bar{u}d$ process, which yields a single D meson and light mesons from the $\bar{u}d$ hadronization (the 'upper vertex'). The $b \to c\bar{c}s$ process is phase-space suppressed, with a rate expected to be 20-30% of the $b \to c\bar{u}d$ rate [11]. A $b \to c\bar{c}s$ decay often results in upper-vertex production of a $D_s^{(*)-}$. The D_s^- and D_s^{*-} are both below threshold for $\bar{D}\bar{K}$ decay, but if a light-quark pair is popped between the $\bar{c}$ and the s, we may then produce a $\bar{D}\bar{K}$ pair. These $\bar{D}$ mesons will be characterized by their wrong charm flavor relative to the $\bar{B}$, and their softer momentum spectrum. The measurement of upper vertex charm production is of particular interest because of the 'charm-counting' problem in B-decays [11]. where the inclusive charm yield from B decay and the inclusive semi-leptonic rate are somewhat inconsistent.

The simplest final states are the three-body decays of the form $D^{(*)}\bar{D}^{(*)}\bar{K}^{(*)}$and as shown in the table below we have observed four such decay modes to date.

$$\begin{aligned}
&\mathcal{B}(\bar{B}^0 \to D^{*+}\bar{D}^0K^-) = (0.45^{+0.25}_{-0.19} \pm 0.08)\% \\
&\mathcal{B}(B^- \to D^{*0}\bar{D}^0K^-) = (0.54^{+0.33}_{-0.24} \pm 0.12)\% \\
&\mathcal{B}(\bar{B}^0 \to D^{*+}\bar{D}^{*0}K^-) = (1.30^{+0.61}_{-0.47} \pm 0.27)\% \\
&\mathcal{B}(B^- \to D^{*0}\bar{D}^{*0}K^-) = (1.45^{+0.78}_{-0.58} \pm 0.36)\% \\
&\mathcal{B}(B^- \to D^0\bar{D}^0K^-) < (0.5)\% \quad (90\%CL) \\
&\mathcal{B}(B^- \to D^0\bar{D}^{*0}K^-) < (0.8)\% \quad (90\%CL) \\
&\mathcal{B}(B^- \to D^{*+}D^{*-}K^-) < (0.7)\% \quad (90\%CL)
\end{aligned}$$

We see a total $B \to D^*\bar{D}^{(*)0}K^-$ branching fraction of about 2% per B charge (B^- or $\bar{B}^0$). Applying upper-vertex isospin (i.e., equating $\bar{D}^{(*)}K^-$ with $D^{(*)-}\bar{K}^0$), this accounts for 4% per B charge. As yet unaccounted for are modes where the lower vertex is a D (not a D^*), for which we only have limits, and modes with a K^*, which we have not yet investigated.

5 Flavor-specific B decays to charm[12]

There has been a longstanding problem in heavy flavor physics of the measured B semileptonic decay branching fraction[13] being smaller than theoretical expectations. One possible explanation[14] is a larger-than-expected flavor-changing neutral current (FCNC) contribution, due to new physics. Another[11] is an enhanced rate for $b \to c\bar{c}s'$ (s' denotes the weak isospin partner of c). An argument against an enhanced $b \to c\bar{c}s'$ rate is that it would conflict with the measured branching fraction for $B \to \bar{D}X$ plus $B \to DX$. That measurement relies on a knowledge of $\mathcal{B}(D^0 \to K^-\pi^+)$, however, and if that is in error, the measurement of the branching fraction of B to charm or anticharm will be also. We address all three issues by measuring the yields of the flavor-specific inclusive B decay processes $B \to DX$, $B \to \bar{D}X$, and

$B \to \bar{D}X\ell^+\nu$ in a sample of $B\bar{B}$ events in which at least one B decays semileptonically. (Herein, "B" represents an average over B^0 and B^+, "D" a sum over D^0 and D^+, and "$\bar{D}$" a sum over $\bar{D}^0$ and D^- [15]. We use the term "upper vertex D" for a $\bar{D}$ produced from the charm quark from $W \to \bar{c}s$, and "lower vertex D" for a D produced from the charm quark from $b \to c$.) Experimentally we determine two ratios which are:

$$\frac{\Gamma(B \to DX)}{\Gamma(B \to \bar{D}X)} = 0.100 \pm 0.026 \pm 0.016,$$

$$\frac{\Gamma(B \to \bar{D}X)}{\Gamma(B \to all)} / \frac{\Gamma(B \to \bar{D}X\ell^+\nu)}{\Gamma(B \to X\ell^+\nu)} \equiv f_{all}/f_{SL} = 0.901 \pm 0.034 \pm 0.015.$$

One expects both f_{all} and f_{SL} to be close to 1.0. The first ratio will be less than 1.0 because of $b \to u$ transitions ($2|V_{ub}/V_{cb}|^2$, where the 2 is a phase space factor), lower vertex D_s (2%), bound $c\bar{c}$ states ($3.0 \pm 0.5\%$[16]), baryons ($6.5 \pm 1.5\%$[17]), and $b \to sg$ (to be extracted). The second ratio will be less than 1.0 because of $b \to u$ transitions ($3|V_{ub}/V_{cb}|^2$, enhanced by the 1.5 GeV/c lepton momentum requirement), and lower vertex D_s ($1.0 \pm 0.5\%$, suppressed by the lepton momentum requirement). This leads to

$$\begin{aligned} f_{all}/f_{SL} = 1.0 + |V_{ub}/V_{cb}|^2 - (0.010 \pm 0.005) - (0.030 \pm 0.005) \\ -(0.065 \pm 0.015) - \mathcal{B}(b \to sg) \end{aligned}$$

Here $b \to sg$ is symbolic for all FCNC processes. Using $|V_{ub}/V_{cb}|^2 = 0.008 \pm 0.003$, we obtain $\mathcal{B}(b \to sg) = (0.2 \pm 3.4 \pm 1.5 \pm 1.7)\%$. From these results and some previously measured branching fractions, we obtain $\mathcal{B}(b \to c\bar{c}s) = (21.9 \pm 3.7)\%$, $\mathcal{B}(b \to sg) < 6.8\%$ @ 90% c.l. We have also used this analysis to extract the branching ratio of $B \to K\pi$ and we obtain $\mathcal{B}(D^0 \to K^-\pi^+) = (3.69 \pm 0.20)\%$ in excellent agreement with the PDG value. The increase in the value of $\mathcal{B}(b \to c\bar{c}s)$ due to $B \to DX$ eliminates 40% of the discrepancy of the B semileptonic decay problem where the measured branching fraction is below theoretical expectations.

6 Rare decays[18]

Rare B decays are of great interest since they are sensitive to new heavy particles and open a possible window of discovery sensitive to high masses or unexpected interactions. In addition understanding CP violation in the B system is of great importance and a future goal of many experiments. It is clear that in order to disentangle particular effects that contribute to rare decays it is necessary to measure a large number of channels. We are engaged in a systemmatic study of rare decays and the results listed below are part of a continuing effort with data being added at the rate of a few inverse femtobarns per year. Of particular interest in the results below are the observations of decays into channels involving the etaprime, omega and phi as well as our older observation of $K\pi$ and $\pi\pi$.

6.1 Observation of inclusive $B \to \eta' X_s$ Decays

We have used a B reconstruction technique to isolate decays of the type $K\eta' + n\pi$ where n = 0, 1, 2, or 3. We have observed an inclusive etaprime signal in the region $2.7 > p > 2.0$ GeV/c of 39.0 ± 10.3 events. The possibility that the signal is due to charmless $b \to s$ processes is consistent with all the features of the data although other interpretations cannot be completely ruled out. Given the $b \to sg^*$ interpretation, the branching ratio is $\mathcal{B}(B \to \eta' X_s) = (6.2 \pm 1.6(stat) \pm 1.3(sys)) \times 10^{-4}$ for $2.0 < p < 2.7$ GeV/c.

$M(X_s)$ (GeV) Range	N (on)	N(off)	Yield
$0.4 < M(X) < 0.6$	4	0	4
$0.6 < M(X) < 1.2$	2.7 ± 2.1	0.6 ± 1.1	2.1 ± 2.9
$1.2 < M(X) < 1.8$	18.0 ± 4.9	6.6 ± 3.2	11.4 ± 7.6
$1.8 < M(X) < 2.5$	26.0 ± 6.4	-0.8 ± 2.3	26.8 ± 7.8

6.2 Exclusive rare two body decays

Decay Mode	Experiment	N_S	N_B	$\mathcal{B}$ (10^{-6})	Theory (10^{-6})
$\pi^+\pi^0$	CLEO II	11.3*	-	< 20	6–20
$\rho^+\pi^0$	CLEO II	8	5.5 ± 1.2	< 77	13–39
$\rho^0\pi^+$	CLEO II	4	2.3 ± 0.3	< 43	0–15
$\eta\pi^+$	CLEO II	0	2.2 ± 0.4	< 8	2–6
$\eta'\pi^+$	CLEO II	1.3*	-	< 45	8–17
$\omega\pi^+$	CLEO II	9.5*	-	11^{+6}_{-5}	0–15
$\pi^+\pi^-$	CLEO II	10.0*	-	< 15	8–26
$\pi^0\pi^0$	CLEO II	1.2*	-	< 20	0.2–0.6
$\rho^\pm\pi^\mp$	CLEO II	7	2.9 ± 0.7	< 88	23–81
$\rho^0\pi^0$	CLEO II	1	1.8 ± 0.6	< 24	0.7–3
$\eta\pi^0$	CLEO II	0*	-	< 11	2–4
$\eta\rho^0$	CLEO II	1	0.8 ± 0.2	< 84	0.1–8
$\eta'\pi^0$	CLEO II	0*	-	< 22	4–14
$\omega\rho^0$	CLEO II	0	1.2 ± 0.3	< 34	0.4

* Maximum likelihood fit - N_S does not include any background

Decay Mode	Experiment	N_S	N_B	$\mathcal{B}$ (10^{-6})	Theory (10^{-6})
$K^+\pi^0$	CLEO II	8.7*	-	< 16	5–10
$K^0\pi^+$	CLEO II	9.2*	-	23^{+11}_{-10}	6–13
$K^+\bar{K}^0$	CLEO II	0.6*	-	< 21	0.6–2.4
$K^{*+}\pi^0$	CLEO II	4	1.9 ± 0.7	< 99	3–9
$K^{*0}\pi^+$	CLEO II	2	1.0 ± 0.6	< 41	4–10
$K^+\rho^0$	CLEO II	1	3.8 ± 0.2	< 19	0.05–2
$K^0\rho^+$	CLEO II	0	0	< 48	0.1–0.4
$K^+\eta$	CLEO II	0	2.2 ± 0.4	< 8	0.3–5
$K^{*+}\eta$	CLEO II	0	0.5 ± 0.2	< 240	1–13
$K^+\eta'$	CLEO II	12.1*	-	71^{+27}_{-23}	10–41
$K^{*+}\eta'$	CLEO II	0	0.3 ± 0.1	< 290	0.2–9
$K^+\omega$	CLEO II	12.0*	-	15^{+8}_{-7}	1.5–5
$K^{*+}\omega$	CLEO II	0	0.9 ± 0.3	< 110	4–15
$K^+\phi$	CLEO II	0*	-	< 5.3	2–16
$K^{*+}\phi$	CLEO II	9.2*	-	13^{+7}_{-6}	2–31
$K^+\pi^-$	CLEO II	21.7*	-	15^{+5}_{-4}	9–17
$K^0\pi^0$	CLEO II	2.3*	-	< 40	2–7
K^+K^-	CLEO II	0.0*	-	< 4	
$K^{*+}\pi^-$	CLEO II	3	0.7 ± 0.2	< 72	6–19
$K^{*0}\pi^0$	CLEO II	0	1.1 ± 0.3	< 28	0.01–5
$K^+\rho^-$	CLEO II	2	2.0 ± 0.4	< 35	0.3–1.9
$K^0\rho^0$	CLEO II	0	0	< 39	0.05–0.4
$K^0\eta$	CLEO II	3.6*	-	< 75	0.07–2
$K^{*0}\eta$	CLEO II	1	1.3 ± 0.3	< 33	0.05–5
$K^0\eta'$	CLEO II	8.0*	-	53^{+30}_{-25}	9–33
$K^{*0}\eta'$	CLEO II	0	0.3 ± 0.1	< 99	3–4
$K^{*0}\omega$	CLEO II	1	1.7 ± 0.4	< 38	2–8
$K^0\phi$	CLEO II	2.2*	-	< 42	2–13
$K^{*0}\phi$	CLEO II	9.2*	-	13^{+7}_{-6}	2–31
$\phi\phi$	CLEO II	0	0	< 39	

* Maximum likelihood fit - N_S does not include any background

With the CLEO II detector we have observed eight rare exclusive two body hadronic decays of the B meson which are listed in the tables above. In addition upper limits close to, or even below, theoretical predictions have been measured.

References

[1] Representing the CLEO collaboration

[2] CLEO collaboration, Y. Kubota *et al.*, Nucl. Instr. and Meth. A320, 66 (1992). This contains a complete description of the CLEO II detector

[3] CLEO collaboration CONF 97-01

[4] M. Bauer, B. Stech and M. Wirbel, Z. Phys. C29, 639 (1985).

[5] M.J. Dugan, B. Grinstein, Phys. Lett. B225 583 (1990).

[6] M. Neubert, V. Rieckert, B. Stech and Q. P. Xu in *Heavy Flavours* edited by A. J. Buras and H. Lindner, World Scientific, Singapore (1992).

[7] A. Deandrea, N. Di Bartolomeo, R. Gatto and G. Nardulli, Phys. Lett. B 318, 549 (1993).

[8] M. Neubert, B. Stech, preprint hep-ph/9705292, to appear in: the Second Edition of Heavy Flavours, edited by A.J. Buras and M. Linder (World Scientific, Singapore)

[9] CLEO collaboration CLNS 97-19

[10] CLEO collaboration CONF 97-26

[11] G. Buchalla, I. Dunietz, and H. Yamamoto, "Hadronization of $b \to c\bar{c}s$," Phys. Lett. B 364, 188(1995).

[12] CLEO collaboration CLNS 97-23

[13] B. Barish *et al.* (CLEO), "Measurement of the B Semileptonic Branching Fraction with Lepton Tags," Phys. Rev. Lett. 76, 1570 (1996).

[14] I. Bigi, B. Blok, M.A. Shifman, and A. Vainshtein, Phys. Lett. B 323, 408 (1994).

[15] Throughout, charge conjugation of all equations is implied, i.e., $B \to \bar{D}X$ includes $\bar{B} \to D\bar{X}$.

[16] R. Balest *et al.* (CLEO), "Inclusive decays of B mesons to charmonium," Phys. Rev. D 52, 2661 (1995).

[17] G. Crawford *et al.* (CLEO),"Measurement of baryon production in B-meson decay," Phys. Rev. D 45, 725 (1992).

[18] CLEO collaboration CONF 97-22a, 97-23, 97-13

B^+ and B^0 Lifetime Measurement at SLD

David Jackson (djackson@slac.stanford.edu)

Rutherford Appleton Laboratory, Chilton, Didcot, Oxon, OX11 0QX, England

Abstract. The B^+ and B^0 lifetimes are measured from the 1993-5 data sample of 150,000 Z^0s plus 50,000 Z^0s collected in 1996 with an upgraded vertex detector.

1 *B* decay reconstruction

Topological vertex reconstruction [1] is applied to the charged tracks in each hadronic Z^0 event hemisphere. The vertices are reconstructed in 3D coordinate space by defining a vertex function $V(\mathbf{r})$ at each position $\mathbf{r}$. The helix parameters and errors for each track i are used to describe the 3D track trajectory as a Gaussian tube $f_i(\mathbf{r})$. $V(\mathbf{r})$ is defined as a function of the $f_i(\mathbf{r})$ such that it is large in regions of high track multiplicity. Tracks are associated with resolved maxima in $V(\mathbf{r})$ to form a set of topological vertices. The efficiency for reconstructing a secondary vertex in a b hemisphere is $\sim 50\%$ ($\sim 67\%$ for the upgraded vertex detector).

Further tracks may be associated with the B decay according to their trajectory relative to the axis drawn from the IP (interaction point) to the secondary vertex [1]. The invariant mass M of the B decay tracks and their total momentum transverse to the vertex axis P_T is determined. Requiring $M_{P_T} = (\sqrt{M^2 + P_T^2} + |P_T|) > 2$ GeV selects true b hemispheres with a purity of $\sim 98\%$. The total charge Q of the tracks associated with the B decay is used to define the charged ($|Q| = 1, 2, 3$) and neutral ($|Q| = 0$) samples. The decay length, measured from the IP to the vertex formed by the same tracks, is required to be greater than 1mm. A total of 20,783 such vertices are selected in the 1993-6 data. The results presented update previous SLD B lifetime measurements [2]. The sensitivity of the analysis has been improved by a factor of two and the 1996 data has been added.

2 Exclusive B^+ and B^0 lifetime measurement

The effective charge purity is enhanced by weighting vertices as a function of M_{P_T}. Lower mass vertices have lower analysing power since they are more likely to have an incomplete set of true B decay tracks and hence have poorer charge reconstruction. An initial state $b/\bar{b}$ tag (using the polarized forward-backward asymmetry and opposite hemisphere jet charge) is used to enhance the charged sample purity by giving a higher (lower) weight to the B^+ hypothesis if the vertex charge agrees (disagrees) with the $b/\bar{b}$ tag.

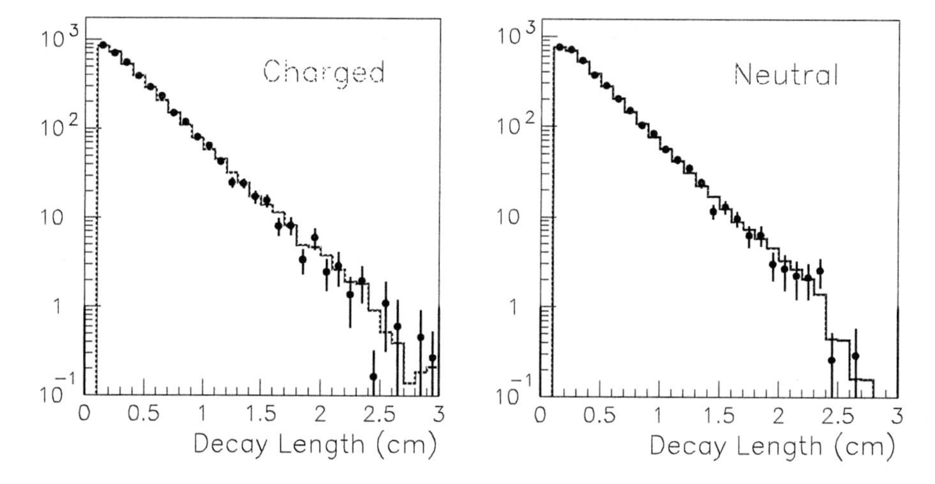

Fig. 1. Decay length distributions for data (points) and best fit MC (histogram).

The B^+ and B^0 lifetimes are extracted from a simultaneous binned χ^2 fit to the decay length distributions of the B vertices in the charged and neutral samples. These distributions are shown in fig. 1 for the full data sample and the best fit Monte Carlo. The statistical and systematic errors (detector, physics and fitting) are determined separately for the 1993-5 and 1996 measurements and then combined taking into account the correlated uncertainties. The combined 1993-5 and 1996 measurements yield the following result:

$$\tau_{B^+} = 1.698 \pm 0.040(\text{stat}) \pm 0.046(\text{syst})\ \text{ps},$$
$$\tau_{B^0} = 1.581 \pm 0.043(\text{stat}) \pm 0.061(\text{syst})\ \text{ps},$$
$$\frac{\tau_{B^+}}{\tau_{B^0}} = 1.072 \pm^{0.052}_{0.049} (\text{stat}) \pm 0.038(\text{syst}).$$

The uncertainty in the b fragmentation produces a significant systematic error on the exclusive B lifetimes but has little effect on the lifetime ratio. The largest systematic errors on the lifetime ratio are due to the uncertainties in the B baryon production fraction and the B_s lifetime. Full analysis details may be found in Ref [3]. These results are consistent with the expectation that the B^+ and B^0 lifetimes are almost equal and have a statistical accuracy among the best of the current world measurements.

References

[1] D. J. Jackson, Nucl. Inst. and Meth. **A388** (1997), 247.
[2] K. Abe *et al.*, Phys. Rev. Lett. **79**, 590 (1997).
[3] K. Abe *et al.*, SLAC-PUB-7635, contributed paper EPS-127, (1997)

Extraction of V_{ub} from the Hadronic Invariant Mass Spectrum in Semileptonic B Decays

Adam F. Falk (falk@jhu.edu)

The Johns Hopkins University, Baltimore, Maryland, USA

Abstract. I discuss the extraction of V_{ub} from the hadronic mass spectrum in the semileptonic decay $B \to X_u \ell \nu$. It is stressed that an accurate determination requires independent information about the hadronic parameter $\bar{\Lambda} = m_B - m_b$.

New strategies are needed to determine the CKM matrix element V_{ub} from the semileptonic decay $B \to X_u \ell \nu$. The current method is to look for charged leptons with energies beyond the kinematic endpoint for the background process $B \to X_c \ell \nu$. However, the shape of the spectrum in this endpoint region cannot be computed model independently, since the hard lepton energy cut destroys the convergence of the operator product expansion (OPE) on which the analysis is based. As a result, one is forced to use quark models to extrapolate the small observed region to the rest of phase space, which implies a large and uncontrolled theoretical uncertainty on V_{ub}.

Here I will summarize a proposal to measure V_{ub} by studying the spectrum $\mathrm{d}\Gamma/\mathrm{d}s_H$, where s_H is the mass of the hadronic state X_u. The quantity s_H can be inferred, inclusively, by measuring the energy and momentum of the charged lepton and reconstructing that of the neutrino. A complete discussion of this analysis can be found in Ref. [1], along with an extensive set of references. Another analysis, with similar conclusions, is given in Ref. [2].

In this measurement, the background $B \to X_c \ell \nu$ is rejected by imposing the cut $s_H < \Delta^2 \le m_D^2$. where Δ^2 is a "practical" experimental cut, which may be have to be less than m_D^2 because of details of the neutrino reconstruction algorithm. The virtue of this method is that, in contrast to the charged lepton spectrum, most of the decays pass the cut, and one only must extrapolate a small tail which "leaks" above Δ^2. While this "leakage" is still model dependent, the overall reliance on models is substantially reduced.

At tree level, the differential spectrum is $\mathrm{d}\Gamma/\mathrm{d}s_H = (G_F^2 m_b^3/192\pi^3) \times |V_{ub}|^2 y^2 (3-2y)\Theta(1-y)$, where $y = s_H/\bar{\Lambda} m_b$ is a scaling variable appropriate to these kinematics, and $\bar{\Lambda} = m_B - m_b$. The observable which must be computed is $\Gamma(\Delta^2) = \int_0^{\Delta^2} (\mathrm{d}\Gamma/\mathrm{d}s_H)\mathrm{d}s_H$. For $\bar{\Lambda} \sim 400\,\mathrm{MeV}$, $y = 1$ corresponds to $s_H \sim 2\,\mathrm{GeV}^2 < m_D^2$, so if $\Delta^2 \approx m_D^2$ then the tail excluded by the cut is due to physics beyond tree level or beyond the parton model.

There are two effects which can extend the spectrum $\mathrm{d}\Gamma/\mathrm{d}s_H$ into the region $s_H > \bar{\Lambda} m_b$. The first is the radiative correction from gluon bremmstrahlung. One finds an expansion of the form $\mathrm{d}\Gamma/\mathrm{d}s_H = (G_F^2 m_b^3/192\pi^3) \times$

$|V_{ub}|^2[(\alpha_s/\pi)\,X(s_H,\bar{\Lambda})+(\alpha_s/\pi)^2\beta_0\,Y(s_H,\bar{\Lambda})+\ldots]$, where $\beta_0=9$ is the first coefficient in the QCD beta function. The functions $X(s_H,\bar{\Lambda})$ and $Y(s_H,\bar{\Lambda})$ are presented in Ref. [1]. Here, we will simply note that both X and Y have a Sudakov-type singularity proportional to $\ln^2(y-1)$, for $y>1$. While this singularity is integrable, its position controls strongly the amount of "leakage" of the rate into the region $s_H>\Delta^2$, even for $\Delta^2\approx m_D^2$. This position, and hence the amount of "leakage", depends crucially on $\bar{\Lambda}$.

The second effect which can extend the spectrum above $y=1$ is the initial bound state motion of the b quark in the B meson. In the OPE, the moments of the b momentum distribution appear as nonperturbative hadronic matrix elements of the form $A_n^{\mu_1\cdots\mu_n}=\langle B|\,\bar{b}D^{\mu_1}\ldots D^{\mu_n}b\,|B\rangle$. For $\mathrm{d}\Gamma/\mathrm{d}s_H$, the OPE is effectively an expansion in powers of $1/(y-1)$. Since $y=2$ corresponds to $s_H=2\bar{\Lambda}m_b\approx m_D^2$, the OPE converges poorly in the region over which one must integrate. Instead of truncating it, the dominant terms in the expansion may be summed into a distribution $S(y)$, that is, $\mathrm{d}\Gamma/\mathrm{d}s_H=(G_F^2m_b^5/192\pi^3)|V_{ub}|^2S(y)$. We model $S(y)$ in the ACCMM model, where the B meson is treated as a nonrelativistic bound state of a b quark and a spectator antiquarkwith mass m_{sp} and momentum distribution given by the ansatz $\Phi(\mathbf{p})\propto\exp(-|\mathbf{p}|^2/p_F^2)$. In this model, the two parameters m_{sp} and p_F determine $\bar{\Lambda}$ and all of the moments A_n. The "leakage" of $s_H>\Delta^2$ depends on the form of $S(y)$, but it is shown in Ref. [1] that once $\bar{\Lambda}$ is fixed, the remaining model dependence is minimal. Thus it is crucial, once again, to determine $\bar{\Lambda}$ independently. Note that here a quark model is used only to determine a small correction to the observed integrated spectrum, *not* to infer the bulk of the rate from a small tail.

There exist strategies for extracting $\bar{\Lambda}$ from the inclusive decay $B\to X_c\ell\nu$ [3, 4, 5], and it is reasonable to hope for an eventual accuracy of 100 MeV or even better. The potential (theoretical) accuracy which could be obtained on V_{ub} depends both on $\bar{\Lambda}$ and on Δ. The analysis of Ref. [1] leads us to conclude that, for example, (*i*) if $\Delta^2\simeq m_D^2$ and $\bar{\Lambda}=400\pm100\,\mathrm{MeV}$, then $\delta|V_{ub}|\sim10\%$; (*ii*) if $\Delta^2\simeq1.5\,\mathrm{GeV}^2$ and $\bar{\Lambda}=400\pm100\,\mathrm{MeV}$, then $\delta|V_{ub}|\sim30\%$; (*iii*) if $\Delta^2\simeq1.5\,\mathrm{GeV}^2$ and $\bar{\Lambda}=200\pm100\,\mathrm{MeV}$, then $\delta|V_{ub}|\sim10\%$; and (*iv*) if $\Delta^2\simeq m_D^2$ and $\bar{\Lambda}=600\pm100\,\mathrm{MeV}$, then $\delta|V_{ub}|\sim50\%$. Clearly both an accurate value and a relatively *low* value of $\bar{\Lambda}$ are to be hoped for, as well as a large value of Δ^2. With luck, this method eventually could yield the most accurate value of $|V_{ub}|$ available.

References

[1] A.F. Falk, Z. Ligeti and M.B. Wise, Phys. Lett. B406 (1997) 225.
[2] R.D. Dikeman and N.G. Uraltsev, hep-ph/9703437.
[3] A.F. Falk, M. Luke and M.J. Savage, Phys. Rev. D53 (1996) 2491; 6316.
[4] A.F. Falk and M. Luke, hep-ph/9708327, Phys. Rev. D57, to appear.
[5] M. Gremm et al., Phys. Rev. Lett. 77 (1996) 20.

Determination of the b Quark Mass at the Z Mass Scale by DELPHI

J. Fuster (fuster@evalo1.ific.uv.es, Spain)

IFIC, Centre Mixte Unversitat de València – CSIC

Abstract. The relative ratio of the normalized three-jet cross section of b and uds quarks has been determined by the DELPHI collaboration in agreement with the QCD prediction including NLO radiative corrections with mass effects. In this study the running b quark mass at the M_Z scale has been measured to be

$$m_b(M_Z) = 2.67 \pm 0.25 \text{ (stat.)} \pm 0.34 \text{ (frag.)} \pm 0.27 \text{ (theo.) GeV}/c^2.$$

This value differs from that obtained using data extracted from the Υ-resonances by $m_b(M_Z) - m_b(M_\Upsilon/2) = -1.49 \pm 0.52$ GeV$/c^2$ and therefore provides the first experimental evidence of the running of the b quark mass. As a consequence of this study the flavour independence of α_s for b and light, uds, quarks is established within 1% accuracy.

1 Description of the analysis and results

The masses of quarks are fundamental quantities of the QCD lagrangian not predicted by the theory. The definition of the quark masses is however not unique. The perturbative pole mass, M_q, and the running mass, m_q, of the $\overline{MS}$ scheme are among the most currently used. Earlier calculations of the three-jet cross section in e^+e^- including mass terms already exist at LO [1, 2] and have been used to evaluate mass effects for the b-quark when testing the universality of α_s. They could not however be used to evaluate the mass of the b-quark, m_b, because these calculations are ambiguous in this parameter. Recently, expressions at NLO, for the multi-jet production rate in e^+e^- are available [3, 4, 5, 6] and, thus, they enable measuring m_b in case the flavour independence of α_s is assumed. For this purpose the calculation [3, 4, 5] of the ratio of the normalized three-jet cross sections between b-quarks and light uds-quarks has been performed

$$R_3^{b\ell}(y_c) = \frac{\Gamma_{3j}^{Z\to b\bar{b}g}(y_c)/\Gamma_{tot}^{Z\to b\bar{b}}}{\Gamma_{3j}^{Z\to \ell\bar{\ell}g}(y_c)/\Gamma_{tot}^{Z\to \ell\bar{\ell}}} \tag{1}$$

where $\Gamma_{3j}^{Z\to q\bar{q}g}$ and $\Gamma_{tot}^{Z\to q\bar{q}}$ are the differential three-jet and total cross-sections, respectively, for the b ($q = b$) and light ($q = \ell \equiv u, d, s$) quarks with respect to the jet-resolution parameter, y_c.

The experimental values of $R_3^{b\ell}$ have been measured by DELPHI using the DURHAM algorithm. The result obtained at $y_c = 0.02$ is

$$R_3^{b\ell}(0.02) = 0.971 \pm 0.005\ (\text{stat.}) \pm 0.007\ (\text{frag.}), \tag{2}$$

where the statistical and the fragmentation error have given.

Additional theoretical uncertainties enter when the measurement of R_3^{bl} is transformed into a determination of $m_b(M_Z)$ by means of Eq. (1). The result thus obtained for m_b at $\mu = M_Z$ is

$$m_b(M_Z) = 2.67 \pm 0.25\ (\text{stat.}) \pm 0.34\ (\text{frag.}) \pm 0.27\ (\text{theo.})\ \text{GeV/c}^2,$$

where the statistical, fragmentation and theoretical errors have been estimated. Evolving this result down to the b mass scale using $\alpha_s = 0.118 \pm 0.003$ would give $m_b(m_b) = 3.91 \pm 0.67$ GeV/c^2.

Adding all errors in quadrature the values of the running b mass at the M_Υ and M_Z scales can be compared. The measured difference between them is

$$m_b(M_Z) - m_b(M_\Upsilon/2) = -1.49 \pm 0.52\ \text{GeV/c}^2,$$

where the value of $m_b(M_\Upsilon/2) = 4.16 \pm 0.14$ GeV/c^2 has been considered as the non-weighted average of all the results at $\mu = M_\Upsilon/2$.

The observed change of the running b mass value from $M_\Upsilon/2$ to M_Z represents the first experimental evidence of the running property of any fermion mass. The result is in good agreement with the predicted QCD evolution and confirms that QCD radiative corrections including mass effects describe the data correctly from the $M_\Upsilon/2$ scale to the M_Z scale.

The result can also be interpreted as a test of the flavour independence of the strong coupling constant yielding to

$$\frac{\alpha_s^b}{\alpha_s^\ell} = 1.007 \pm 0.005\ (\text{stat.}) \pm 0.007\ (\text{frag.}) \pm 0.005\ (\text{theo.}),$$

which verifies the flavour independence of the strong coupling constant for b and light quarks.

References

[1] A. Ballestrero, E. Maina, S. Moretti, Phys. Lett. **B294** (1992) 425 and Nucl. Phys. **B415** (1994) 265.

[2] M. Bilenky, G. Rodrigo, A. Santamaría, Nucl. Phys. **B439** (1995) 505.

[3] G. Rodrigo, PhD Thesis, Universitat de València, 1996, hep-ph/9703359 and ISBN:84-370-2989-9.

[4] G. Rodrigo, M. Bilenky, A. Santamaría, Phys. Rev. Lett. **79** (1997) 193.

[5] W. Bernreuther, A. Brandenburg, P. Uwer, Phys. Rev. Lett. **79** (1997) 189.

[6] P. Nason and C. Oleari, Phys. Lett. **B407** (1997) 57.

Top Quark Results from D

Mark Strovink (strovink@lbl.gov) for the DØ Collaboration

University of California, Berkeley / E.O. Lawrence Berkeley National Laboratory

Abstract. This is a brief summary of DØ's top quark measurements, including $\sigma_{t\bar{t}}$ and m_t in the ℓ+jets and dilepton channels, $\sigma_{t\bar{t}}$ in the all jets channel, and the search for top disappearance via $t{\rightarrow}bH^+$, $H^+{\rightarrow}\tau\nu$ or $c\bar{s}$.

1 $\sigma(p\bar{p}{\rightarrow}t\bar{t}+X)$; m_t in the ℓ+jets channel

DØ recently published[1] its measurement of the top pair production cross section, $\sigma_{t\bar{t}}$=5.5±1.8 pb. It is dominated by the ℓ+jets channel, in which one of the W's from the decay $t{\rightarrow}bW$ decays to an isolated e or μ and the other W decays to a $q\bar{q}$ pair. Distinctive aspects of the analysis in this channel are the use of a logarithmic extrapolation in minimum jet multiplicity to estimate the main (W + ≥4 jets) background, and the use of stringent cuts on aplanarity ($\mathcal{A}$>0.65) and scalar transverse energy (H_T>180 GeV) to suppress it. Fig. 1 tabulates the main selection criteria and the signal and background for each channel. It includes a plot showing the excellent agreement of this result with theory.

Published[2] in the same journal issue is DØ's measurement in the ℓ+jets channel of the top quark mass, m_t=173.3±5.6±6.2 GeV/c^2. This analysis uses a multivariate discriminant $\mathcal{D}$ to separate top signal from background, based on input kinematic variables especially chosen to be only weakly correlated with the 2C fitted top mass $m_{\rm fit}$. The true top mass m_t is extracted from a likelihood (L) fit to events binned in both $m_{\rm fit}$ and $\mathcal{D}$. Fig. 1 shows the distribution of $m_{\rm fit}$ for both a top-enriched and a top-depleted sample, as well as L *vs.* m_t. Tabulated there are the fit parameters for two different definitions of $\mathcal{D}$, the systematic errors, and the result.

2 m_t in the dilepton channel

DØ's measurement of m_t in the dilepton channel has been submitted for publication[3]. Here, with both ν momenta unmeasured, the fit is −1C rather than +2C for ℓ+jets. If m_t is *assumed*, the system can be reconstructed via a quartic equation with 0, 2, or 4 real solutions, which usually exist for a wide range of m_t. More resolving power is gained by asking "if m_t had a certain value, how likely is it that the top decay products would appear in the detector as they did?" The factors[4] in this likelihood $\mathcal{L}(m_t)$ are: (A) $(1/\sigma_{t\bar{t}})(d\sigma_{t\bar{t}}/d\,\text{LIPS})$, (B) the lepton energy density dN/dE_ℓ in the top quark

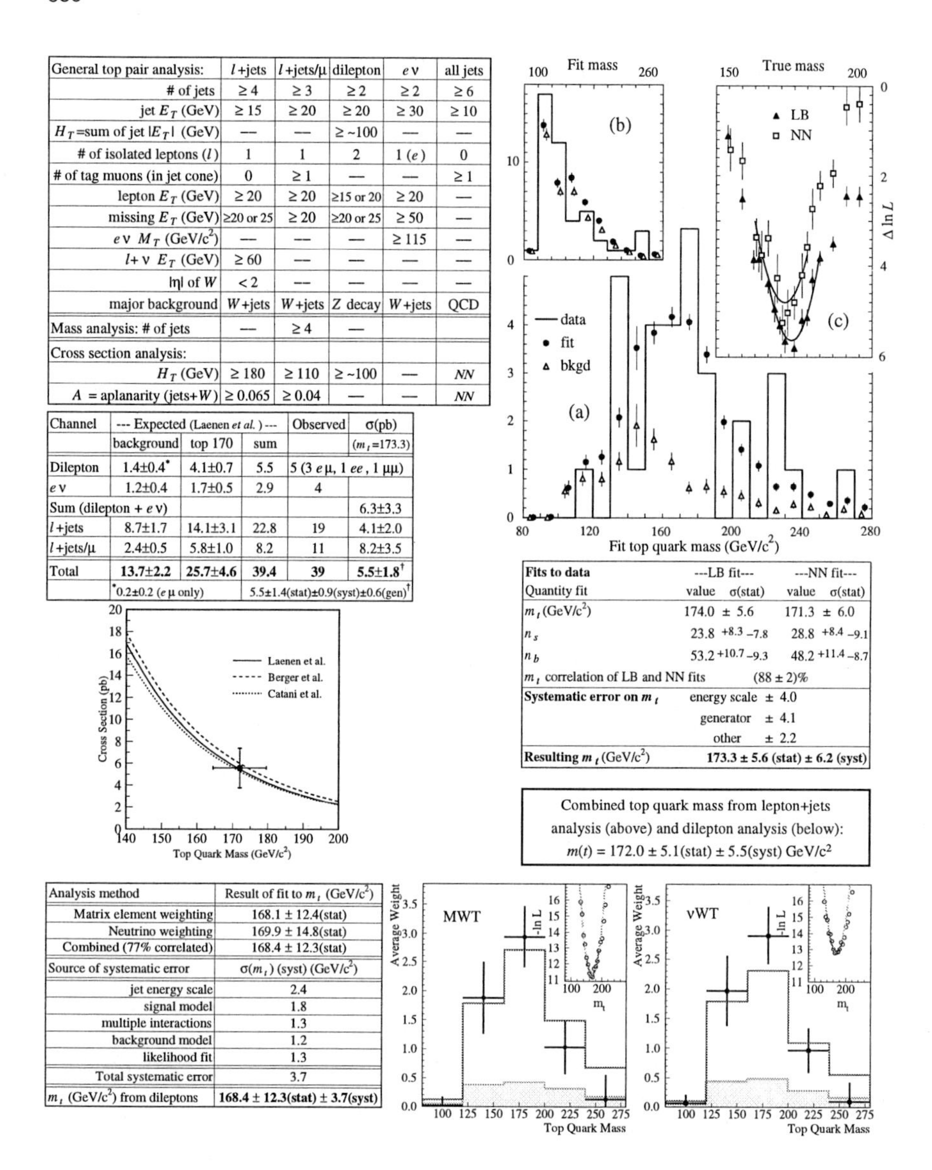

General top pair analysis:	l+jets	l+jets/μ	dilepton	$e\nu$	all jets
# of jets	≥ 4	≥ 3	≥ 2	≥ 2	≥ 6
jet E_T (GeV)	≥ 15	≥ 20	≥ 20	≥ 30	≥ 10
H_T=sum of jet $\vert E_T \vert$ (GeV)	—	—	≥ ~100	—	—
# of isolated leptons (l)	1	1	2	1 (e)	0
# of tag muons (in jet cone)	0	≥ 1	—	—	≥ 1
lepton E_T (GeV)	≥ 20	≥ 20	≥15 or 20	≥ 20	—
missing E_T (GeV)	≥20 or 25	≥ 20	≥20 or 25	≥ 50	—
$e\nu$ M_T (GeV/c^2)	—	—	—	≥ 115	—
l+ ν E_T (GeV)	≥ 60	—	—	—	—
$\vert\eta\vert$ of W	< 2	—	—	—	—
major background	W+jets	W+jets	Z decay	W+jets	QCD
Mass analysis: # of jets	—	≥ 4	—		
Cross section analysis:					
H_T (GeV)	≥ 180	≥ 110	≥ ~100	—	NN
A = aplanarity (jets+W)	≥ 0.065	≥ 0.04	—	—	NN

Channel	--- Expected (Laenen *et al.*) ---			Observed	σ(pb)
	background	top 170	sum		(m_t=173.3)
Dilepton	1.4±0.4*	4.1±0.7	5.5	5 (3 $e\mu$, 1 ee, 1 $\mu\mu$)	
$e\nu$	1.2±0.4	1.7±0.5	2.9	4	
Sum (dilepton + $e\nu$)					6.3±3.3
l+jets	8.7±1.7	14.1±3.1	22.8	19	4.1±2.0
l+jets/μ	2.4±0.5	5.8±1.0	8.2	11	8.2±3.5
Total	**13.7±2.2**	**25.7±4.6**	**39.4**	**39**	**5.5±1.8**†
	*0.2±0.2 ($e\mu$ only)		†5.5±1.4(stat)±0.9(syst)±0.6(gen)		

Fits to data	---LB fit---		---NN fit---	
Quantity fit	value	σ(stat)	value	σ(stat)
m_t (GeV/c^2)	174.0	± 5.6	171.3	± 6.0
n_s	23.8	$^{+8.3}_{-7.8}$	28.8	$^{+8.4}_{-9.1}$
n_b	53.2	$^{+10.7}_{-9.3}$	48.2	$^{+11.4}_{-8.7}$
m_t correlation of LB and NN fits	(88 ± 2)%			
Systematic error on m_t	energy scale ± 4.0			
	generator ± 4.1			
	other ± 2.2			
Resulting m_t (GeV/c^2)	**173.3 ± 5.6 (stat) ± 6.2 (syst)**			

Combined top quark mass from lepton+jets analysis (above) and dilepton analysis (below): $m(t)$ = 172.0 ± 5.1(stat) ± 5.5(syst) GeV/c^2

Analysis method	Result of fit to m_t (GeV/c^2)
Matrix element weighting	168.1 ± 12.4(stat)
Neutrino weighting	169.9 ± 14.8(stat)
Combined (77% correlated)	168.4 ± 12.3(stat)
Source of systematic error	σ(m_t) (syst) (GeV/c^2)
jet energy scale	2.4
signal model	1.8
multiple interactions	1.3
background model	1.2
likelihood fit	1.3
Total systematic error	3.7
m_t (GeV/c^2) from dileptons	**168.4 ± 12.3(stat) ± 3.7(syst)**

Fig. 1. Collage of DØ top quark results. Top left: main selection criteria by channel. Beneath: event statistics and $\sigma_{t\bar{t}}$, total and by channel; plot comparing measured $\sigma_{t\bar{t}}$ to theory. Top right: distributions in $m_{\rm fit}$ for (a) top enriched and (b) top depleted ℓ+jets samples; (c) likelihood *vs.* m_t. Beneath: fit parameters for two different discriminants; errors and resulting m_t. Bottom left: results and errors for m_t from dileptons. Bottom right: average weight in 5 m_t regions for both dilepton weights, with likelihood *vs.* m_t inset. Box: combined m_t.

rest frame, and (C) the Jacobian $|\partial\,\textsc{lips}/\partial\,\{o\}|$, where $\{o\}$ (LIPS) is the set of observed (Lorentz-Invariant Phase Space) variables.

We make two independent approximations to $\mathcal{L}(m_t)$. The *matrix element weight* (MWT) method ignores (C), includes (B), and approximates (A) using a product of proton pdf's with an empirical m_t dependent factor. The *neutrino phase space weight* (νWT) method ignores (A) and (B). It approximates (C) by predicting $\not{E}_T$ after fixing both ν rapidities to many different values. This is compared to the measured $\not{E}_T$ and a likelihood sum is incremented. To obtain the final weight, we sum over quartic solutions, jet assignments (including ISR and FSR), and many resolution-smeared versions of each event.

For both methods, a vector consisting of the fractional weight integrated over each of five m_t regions is stored for each event. To estimate the probability densities for signal and background in this vector space, we accumulate a Gaussian kernel for each event in the modeled sample. Plotted in Fig. 1 for each of the five m_t regions and both methods are the average weight for 6 data events, the best fit mixture, and the background. Inset is the likelihood *vs.* m_t. Tabulated also in Fig. 1 are the fit parameters for both methods, the systematic errors, and the result m_t=168.4±12.3±3.7 GeV/c^2. Combining this with DØ's ℓ+jets result, we obtain m_t=172.0±5.1±5.5 GeV/c^2.

3 $\sigma(p\bar{p}\to t\bar{t}+X)$ in the all jets channel

When both of the daughter W's decay into $q\bar{q}$, at least six jets are produced (we rank them with jet 1 highest in E_T). Compared to the huge background from QCD multijets, top events in this channel are harder, less planar, and more central, with stiffer nonleading jets. As inputs to the first of two neural networks (NN$_1$ and NN$_2$ with outputs $\mathcal{O}_1$ and $\mathcal{O}_2$), we use 2-3 kinematic variables for each property. These are H_T, $\sqrt{\hat{s}}$, E_T(jet 1)/H_T, $\mathcal{A}$, sphericity, centrality (=$H_T/\sum E$(jets)), rms η weighted by E_T, geometric mean η^2 and E_T of jets 5 and 6, H_T excluding jets 1 and 2, and E_T weighted no. of jets.

Events are required to have ≥1 non-isolated μ (tagging ≥1 jet as a b candidate). The inputs to NN$_2$ are $\mathcal{O}_1$; $p_T(\mu)$; a variable sensitive to the quality of a constrained fit to any top mass; and a Fisher discriminant sensitive to the jet width (considering signal to be quarks, background to be gluons). Both NN's are trained on HERWIG $t\bar{t}$ events as signal. The background model is non μ-tagged data to which a μ-tag-rate function $f(p_T(\mu), \text{jet } E_T, \text{detector } \eta)$ is applied. For all 14 NN inputs, observed (μ-tagged data) distributions agree with those of the model.

Fig. 2 exhibits the best fit for μ-tagged data of $\mathcal{O}_2$ to a sum of signal and background, with the background normalization and top cross section as free parameters. The preliminary result is $\sigma_{t\bar{t}}$=7.9±3.1±1.7 pb. The largest systematic uncertainties are in the background model (11%), $p_T(\mu)$ spectrum (7%), μ efficiency (7%), and μ-tag parametrization (7%), with nine smaller sources. Requiring $\mathcal{O}_2$>0.78, we obtain 44 events with an expected background of 25.3±7.3 and an expected top signal of 11.6±4.5. The observed excess corresponds to a Gaussian equivalent background fluctuation of ≈3σ.

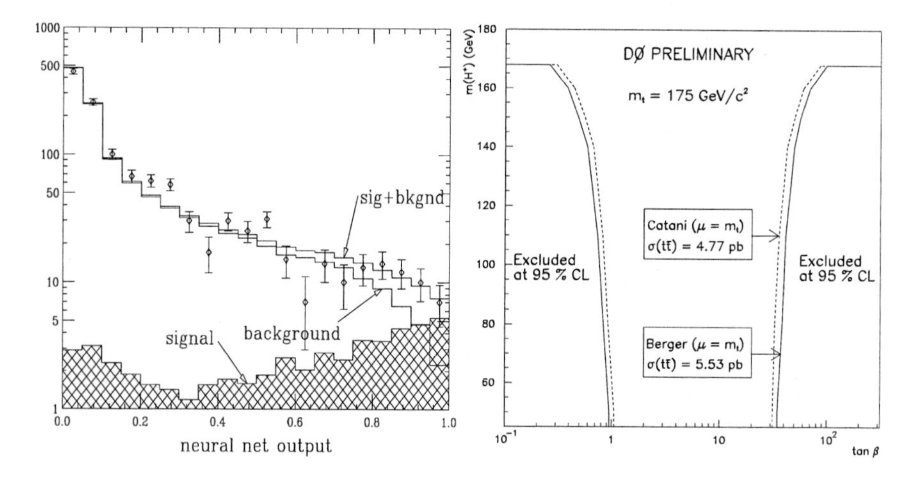

Fig. 2. Preliminary DØ results. Left: best fit for output of NN$_2$ to a mixture of all jets top signal and background. Right: regions in the M_{H+} *vs.* $\tan\beta$ plane excluded, each at 95% CL, by top disappearance search with the MSSM parameters shown.

4 Top disappearance via $t \to bH^+$, $H^+ \to \tau\nu$ or $c\bar{s}$?

If one or both of the produced $t\bar{t}$ were to decay to $H^{\pm}b$ rather than $W^{\pm}b$, the ℓ+jets analysis used for DØ's top cross section measurement would be less efficient, causing a shortfall in the measured cross section relative to the SM calculation. Within the MSSM, $t \to H^+b$ occurs primarily at low and high $\tan\beta$. The shortfall occurs both at low $\tan\beta$, where $H^+ \to c\bar{s}$, and at high $\tan\beta$, where $H^+ \to \tau\nu$, due mainly to a lack of energetic isolated leptons. It leads to exclusion regions at the $\tan\beta$ extremities of the M_{H+} *vs.* $\tan\beta$ plane, based on the relative likelihood *vs.* $\log\tan\beta$ of obtaining the observed number (30) of ℓ+jets events, for a given M_{H+}.

Fig. 2 shows this exclusion region for two values of calculated $\sigma_{t\bar{t}}$. Within the MSSM, taking m_t=175 GeV/c^2 and $\sigma_{t\bar{t}}$=5.53 pb, at 95% confidence the data require 0.96(0.26)< $\tan\beta$ for M_{H+}=50(168) GeV/c^2, or $\tan\beta$ <35(96) for M_{H+}=50(168) GeV/c^2. The dependence on m_t and on renormalization scale μ is modest over most of the M_{H+} *vs.* $\tan\beta$ plane.

References

[1] DØ Collaboration, S. Abachi *et al.*, Phys. Rev. Lett. **79** (1997), 1203.
[2] DØ Collaboration, S. Abachi *et al.*, Phys. Rev. Lett. **79** (1997), 1197.
[3] DØ Collaboration, B. Abbott *et al.*, submitted to Phys. Rev. Lett., hep-ex/970614 (1997).
[4] M. Strovink for the DØ Collaboration, HEP 93 (Proc. Europhysics Conf. on HEP, Marseille, J. Carr and M. Perrottet, eds.) (1994) 292.

Recent Tau Decay Results from OPAL

J. Michael Roney (mroney@uvic.ca)

University of Victoria, Canada

Abstract. **Four tau decay topics which have recently been explored by the OPAL experiment using the data collected at the Z^0 pole are summarized. The first is the muonic branching fraction; the second the study of the hadronic current and neutrino helicity in $\tau^- \to \pi^+\pi^-\pi^-\nu_\tau$ decays; the third is a set of measurements of strangeness content in multi-prong tau decays and the last reports a limit on the 7-prong branching ratio.**

1 The OPAL Detector and $e^+e^- \to \tau^+\tau^-$ Data Set

The $e^+e^- \to \tau^+\tau^-$ data presented here were collected mainly between 1991 and 1994 at centre-of-mass energies near the mass of the Z^0. They were recorded using the OPAL detector[1], which is a tracking magnetic spectrometer with excellent dE/dx resolution, surrounded by electromagnetic (ECAL) and hadronic (HCAL) calorimeters and outer muon chambers. The measurements reported here use data which are restricted to the barrel region of the OPAL detector.

At centre-of-mass energies near 91 GeV, the $e^+e^- \to \tau^+\tau^-$ events exhibit back-to-back low multiplicity jet-like topologies with substantial energy and transverse momentum carried off by the neutrinos in the tau decay. This signal can be cleanly discriminated from the major sources of background whilst maintaining a high efficiency[2]: a $e^+e^- \to \tau^+\tau^-$ selection efficiency of 93% within the fiducial acceptance and a background of less than 2% is achieved. The data sample used for these studies corresponds to 10^5 tau-pair events.

2 Precise Measurement of B.R.($\tau \to \mu\bar{\nu}_\mu\nu_\tau$)

The values of the charged current leptonic couplings are fundamental components of the Standard Model. Stringent tests of such aspects of the model as lepton unversality require measurements of the total and partial widths and masses of the electron, muon and tau lepton. OPAL has recently published the electronic branching ratio, (17.78±0.10±0.09)%[3], and lifetime, 289.2±1.7±1.2 fs[4], of the tau lepton. Presented here is a preliminary measurement of the muonic branching ratio of the tau using the 1991-1994 data set which also breaks the 1% precision barrier.

In order to achieve this high precision, the efficiency and purity of the decay selections were determined using data; problematic muon background from $e^+e^-\to\mu^+\mu^-$ and two-photon processes were kept low by cutting hard on very high and very low energy muons; and only very well modelled regions of the detector were employed in the analysis.

The efficiencies and purities were determined through the use of two selections which were designed to rely on independent detector components. The first selection employed only data from the external muon chambers and tracking detector whilst the second made use of the ECAL and HCAL and a 'mass' determined from the tracker and ECAL. The second selection was used to correct the momentum-dependent Monte Carlo efficiencies of the first. The corrections were within 2% of unity over the entire momentum range. This technique was shown to be valid at the sub-percent level using control samples of muons from two-photon processes and $e^+e^-\to\mu^+\mu^-$ events. The independence of the two samples was tested by showing that the efficiency of an analysis employing both selections was within approximately 1% of the product of the efficiencies of the two selections. The statistics of this 'data-based' determination of efficiency and purity are folded into the statistical error of the final result.

Uncertainties related to the background dominate the systematic error. These are evaluated by loosening the cuts in such a manner as to enhance the background by a factor of two. This exercise resulted in no significant change in the measured branching ratio, but the statistical uncertainty in the data-based corrections to the Monte Carlo estimates of the $\tau\to\pi(K)\nu_\tau$ and $\tau\to\rho\nu_\tau$ backgrounds are dominant contributions to the systematic uncertainty. Interestingly enough, another major systematic error arises from the modelling uncertainty of B^{**} production in the low multiplicity multihadron events which form part of the background in the tau-pair sample.

The OPAL preliminary result on the branching ratio of $\tau\to\mu\bar{\nu}_\mu\nu_\tau$ is (17.48±0.12±0.08)%. This is in good agreement with the previous world average and all other precise measurements. Combining this branching ratio with the electronic branching ratio yields a ratio of the muonic to electronic charged current couplings of 1.005±0.005 from the OPAL experiment alone, thus verifying charged-current e-μ universality at below the 1% level.

OPAL has also recently published details of an analysis of the radiative muonic decay of the tau ($\tau\to\mu\bar{\nu}_\mu\nu_\tau\gamma$)[5] in which good agreement with the QED-based Monte Carlo and earlier measurements is shown.

3 Hadronic Current and h_ν in $\tau^-\to\pi^+\pi^-\pi^-\nu_\tau$

The G-parity of the three pions produced in $\tau^-\to\pi^+\pi^-\pi^-\nu_\tau$ is -1 and PCAC suppresses pseudo-scalar currents, which leads to the a_1 ansatz for $\tau^-\to\pi^+\pi^-\pi^-\nu_\tau$ decays: $\tau\to a_1\nu_\tau$; $a_1\to\rho\pi$; $\rho\to\pi\pi$. Therefore, since its discovery, the tau lepton has been identified as an ideal lab for studies of the a_1 because of the clean nature of the electroweak initial state. In addition to the dominant $a_1\to\rho\pi$ channel, the $\pi\gamma$ mode has been 'seen' and the $K\bar{K}^*(892)$

mode 'possibly seen'[6]. Models are required to describe the a_1 and understanding the limitations these models is necessary for many measurements involving the a_1 , such as the tau polarization measurements at LEP. A detailed description of the OPAL $\tau \rightarrow a_1 \nu_\tau$ analyses have been published recently[7], therefore only the results are summarized here.

The two models studied here, (KS[8] and IMR[9]), both provide reasonable descriptions of the distributions of the 3π invariant mass squared (Q^2). However, both are imprecise in the modelling of the distributions of the $\pi^+\pi^-$ invariant mass squared distributions when examined as a function of Q^2. The mass and widths were determined to be $1.262 \pm 0.009 \pm 0.007$ GeV and $0.621 \pm 0.032 \pm 0.058$ GeV in the KS model and $1.210 \pm 0.007 \pm 0.017$ GeV and $0.457 \pm 0.015 \pm 0.017$ GeV in the IMR model.

The model independent hadronic structure function measurements only assumed that the current be pure axial-vector. Both KS and IMR models provide reasonable descriptions of these measured functions, although there is some suggestion that the $a_1 \rightarrow K\bar{K}^*$ treatment in the IMR model may be inadequate. The analysis also puts the first model independent limits on non-axial-vector contributions to be $< 26.1\%$ @ 95%CL and a model-dependent upper limit limit of 0.84% on any pseudo-scalar contribution. The sign of the neutrino helicity was also determined to be -1 in a model indendent analysis yielding a parity violating asymmetry parameter, $\gamma_A V$, measurement of $1.29 \pm 0.26 \pm 0.11$.

4 Strangeness Content and Multi-prong Tau Decays

Studies of strangeness content in tau decays provide input for issues in chiral perturbation theory and effective Lagrangians; $SU(3)_F$ and isospin symmetries; determinations of the s-quark mass; and the structure of the hadronic current. Very little data on the decay rates of $\tau^- \rightarrow \nu_\tau K^- \pi^+ \pi^- \geq 0\pi^0$, $\tau^- \rightarrow \nu_\tau K^- K^+ \pi^- \geq 0\pi^0$ and $\tau^- \rightarrow \nu_\tau K^0_S K^0_S \pi^- \geq 0\pi^0$ have been published and there exist some theoretical predictions of these rates[10]. Presented here are preliminary measurements of these branching ratios using the 1990-1995 OPAL tau data set.

The measurements of $\tau^- \rightarrow \nu_\tau K^- \pi^+ \pi^- \geq 0\pi^0$ and $\tau^- \rightarrow \nu_\tau K^- K^+ \pi^- \geq 0\pi^0$ branching ratios are based on a selection of three-prong tau decays which have had photon conversion and K^0_S candidates removed. The kaons are identified using the 159 dE/dx measurements in the 4 bar gas of the OPAL drift chamber tracker. The π-K separation exceeds two standard deviations up to track momenta of 40 GeV for isolated tracks and for those tracks in a multitrack environment which deposit ionisation closest to the anode plane. Particle identification was only applied to tracks above 3 GeV momentum with very high quality reconstruction: only tracks which were closest to the anode-plane and in the barrel region of the detector having high z-resolution measurements were used. The dE/dx stretch distribution of this class of tracks was determined from data using an ultra-high purity sample of 1400 π tracks from $\tau^- \to \pi^+\pi^-\pi^-\nu_\tau$ decays. These pions were selected on the basis of their charge being opposite to that of the parent tau lepton together with dE/dx identification of the like-charge tracks as pions.

The three-prong sample was separated into two samples: the first had tracks closest to the anode plane which had the same charge as the parent tau whilst the second sample had closest tracks with the opposite charge. An event-by-event maximum likelihood fit was performed on the dE/dx stretch parameter under a pion hypothesis in order to extract the fractions of electrons, pions and kaons. Fits were performed in different momentum bins and for the two samples. Under the rather good assumption that the highly suppressed $\tau^- \to \nu_\tau K^+\pi^+\pi^- \geq 0\pi^0$ branching ratio is 0, a simultaneous set of linear equations relates the kaon fractions in these two samples to the $\tau^- \to \nu_\tau K^-\pi^+\pi^- \geq 0\pi^0$ and $\tau^- \to \nu_\tau K^-K^+\pi^- \geq 0\pi^0$ branching ratios.

These data yield B.R.$(\tau^- \to \nu_\tau K^-\pi^+\pi^- \geq 0\pi^0) = (0.358\pm^{0.089}_{0.084}\pm^{0.046}_{0.065})\%$, which is in good agreement with the world average and various predictions; and B.R.$(\tau^- \to \nu_\tau K^-K^+\pi^- \geq 0\pi^0) = (0.036\pm^{0.047}_{0.042}\pm^{0.024}_{0.022})\%$ which agrees with measurements by DELCO and TPC/2γ (0.165±0.080)% [6] but is in poor agreement with a measurement reported by ALEPH (0.238±0.042)% [11] and is three standard deviations smaller than a calculation by Finkemeir *et al.*[10].

In an independent analysis[12], OPAL has also measured the $\tau^- \to \nu_\tau K^0_S K^0_S \pi^- \geq 0\pi^0$ branching ratio to be (0.054±0.030±0.015)%. This is consistent with published data and calcultations[10].

OPAL has also recently reported the search for evidence of 7-prong decays of the tau lepton[13] and has set a limit of 1.4×10^{-5} @ 90%CL on this rare process, improving the previous limit by a factor of ten.

References

[1] OPAL, N.I.M., A305 (1991) 275-319.
[2] OPAL, Z. Phys. C72 (1996) 365-376.
[3] OPAL, Phys. Lett. B 369 (1996) 163-172.
[4] OPAL, Phys. Lett. B374 (1996) 341-350.
[5] OPAL, Phys. Lett. B388 (1996) 437-449.
[6] PDG, Phys. Rev. D54, (1996) 1.
[7] OPAL, Z. Phys. C75 (1997) 593-605.
[8] J. H. Kühn and A. Santamaria, Z. Phys. C48 (1990) 445.
[9] N. Isgur, C. Morningstar, C. Reader, Phys. Rev. D39 (1989) 1357.
[10] M. Finkemeir *et al.*, Nucl.. Phys. B(Proc. Suppl) 55C (1997) 169; B. A. Li, Phys. Rev. D55 (1997) 1436.
[11] ALEPH, CERN-PPE/97-69
[12] S. Towers for OPAL, Nucl. Phys. B(Proc. Suppl.) 55C (1997) 137.
[13] OPAL, Phys. Lett. B404 (1997) 213-222.

D and B Decays to τ at LEP

Gian Claudio Zucchelli[1] (gcz@physto.se)

Stockholm University, Sweden

Abstract. Recent measurements of purely leptonic decay of D_s mesons from the DELPHI and the L3 experiments are presented, from which a new value of the decay constant f_{D_s} is derived. In the beauty sector, the DELPHI measurement of $BR(b \to \tau\nu_\tau X)$ and the L3 search for the purely leptonic decay $B^- \to \tau^-\bar{\nu}_\tau$ are reviewed.

1 Introduction

Decays of B- and D-mesons to τ leptons offer a benchmark to test the validity of the Standard Model. A measurement of those branching ratios allows one to constrain parameters in the MSSM model. In particular the measurement of the purely leptonic branching ratios are sensitive to the heavy mesons decay constants f_D and f_B which connect the absolute rate of various heavy-flavour transitions to CKM matrix elements.

2 Leptonic decay of charm mesons

The analysis of the leptonic decay $D_s \to \tau\nu_\tau$ is done by the DELPHI [1] and L3 [2] experiments using the decay chain $Z \to c\bar{c}$, $c \to D_s^{*-} \to \gamma D_s^-$, $D_s^- \to \tau^-\bar{\nu}_\tau$, $\tau \to \ell\nu_\ell\nu_\tau$. The signature of the decay is a combination of lepton, photon and missing energy in one hemisphere, and the resolution is improved constraining M_{D_s} in the kinematical fit. The DELPHI distribution of the mass difference between $D_s\gamma$ and D_s systems is shown in Figure 1 for the data with the estimated background superimposed. The measured value of branching ratio is $BR(D_s \to \tau\nu_\tau) = 8.5 \pm 4.2(stat) \pm 2.6(syst)\%$ which leads to $f_{D_s} = 330 \pm 95 MeV$. A similar analysis done by L3 gives $BR(D_s \to \tau\nu_\tau) = 7.4 \pm 2.8(stat) \pm 2.4(syst)\%$ which translates to $f_{D_s} = 309 \pm 77 MeV$. Using the previous results of WA75 [3], CLEO [4], BES [5], and E653 [6] the new world average is $f_{D_s} = 256 \pm 25 MeV$ which is in agreement with theoretical prediction [7] (Figure 2).

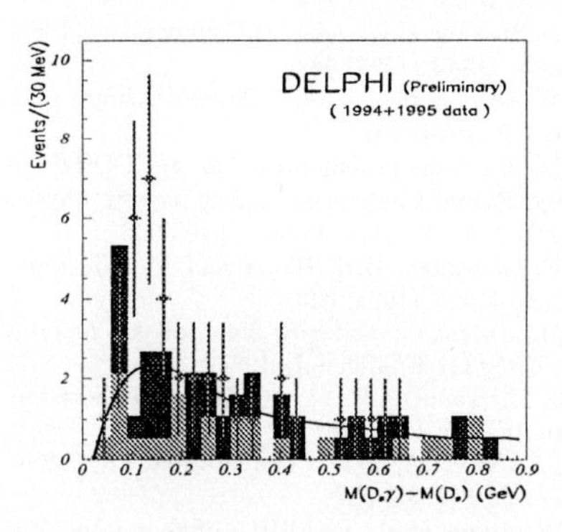

Fig. 1. *Distribution of invariant mass difference, $M(D_s\gamma) - M(D_s)$. The points with error bars correspond to the real data while the shaded histograms show the contribution of the simulated $b\bar{b}$ (clear) and $c\bar{c}$ (darker) events.*

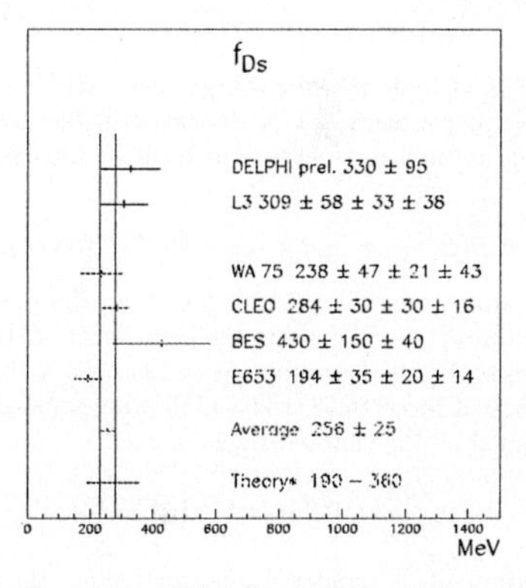

Fig. 2. *Summary of the f_{D_s} experimental measurements and comparison with the theoretical prediction*

3 Semileptonic decay of beauty hadrons

A measurement of the inclusive branching ratio of b to τ has been performed by the DELPHI experiment [8]. The analysis relies mostly on b-tagging and missing energy, and is done in the decay mode $\tau \to \nu h$ with h hadrons. The result $BR(b \to \tau\nu_\tau X) = 2.60 \pm 0.30(stat) \pm 0.65(syst)\%$ is consistent with the Standard Model prediction [9] of $2.30 \pm 0.25\%$ and with previous measurements at LEP experiments [10] [11] [12] (Figure 3). This measurement puts a constraint on the charged Higgs mass of $tan\beta/M_{H^\pm} < 0.50\, GeV^{-1}$ at 90% CL [13].

4 Search for leptonic decays of beauty mesons

A search for the exclusive decay chain $B^- \to \tau^-\bar{\nu}_\tau$, has been done by the L3 experiment [2]. The signal

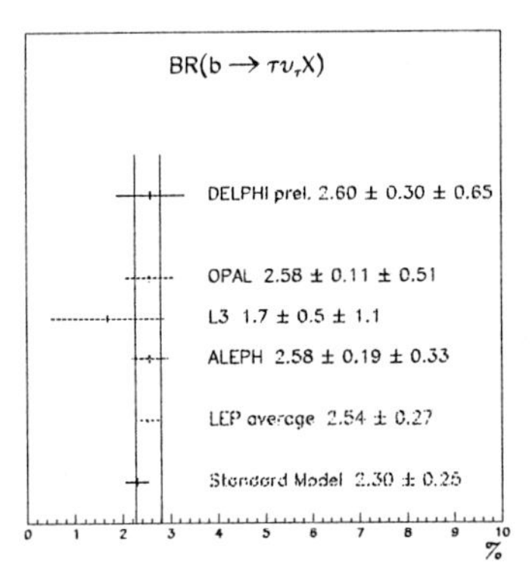

Fig. 3. *Summary of measurements for the branching ratio of $b \to \tau\nu_\tau X$ and comparison with the theoretical prediction*

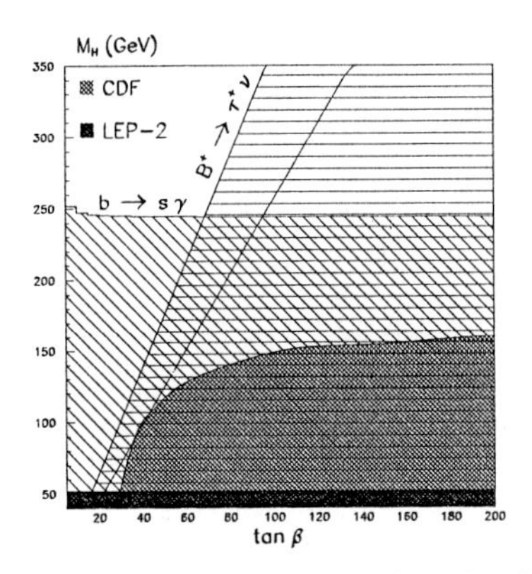

Fig. 4. *Present limits for $M_{H\pm}$ and $\tan\beta$ and the result from the search for $B \to \tau\nu_\tau$ decay, with (left line) and without (right line) B_c contamination*

consists of large missing energy, low particle multiplicity and a track not consistent with the primary vertex in one hemisphere. The limit is calculated to be

$$BR(B^- \to \tau^- \bar{\nu_\tau}) < 5.6 \times 10^{-4}\ (90\%\,CL).$$

This value is considerably better than the previous LEP limit [10] and close to the theoretical prediction of 6×10^{-5} [14]. Assuming $f_B = 190$ MeV and using $V_{ub} = 0.0033 \pm 0.0008$ [15] the following constraint is obtained on the charged Higgs mass:

$$\frac{tan\beta}{M_{H\pm}} < 0.38\,GeV^{-1}\ (90\%\,CL).$$

Inclusion of B_c contribution can reduce the limit down to 0.27 GeV^{-1} (90% CL) [14]. Figure 4 combines the results from the direct searches of $H^\pm$ (CDF collaboration [16] and DELPHI Collaboration [17]) from the measurements of the $b \to s\gamma$ decay (CLEO Collaboration) [18] and the present analysis on $B \to \tau\nu_\tau$.

5 Conclusions

In the D sector, the new measurements of the decay constant f_{D_s} are compatible with previous determinations and with the theoretical prediction. In the B sector, the new measurement of inclusive decay to τ confirm previous LEP branching ratios and the Standard Model prediction; the exclusive search puts a limit very close to the theoretical expectation and have a relevant impact on the extraction of constraints on new physics such as charged-Higgs contribution.

References

[1] F. Parodi *et al.*, DELPHI Collaboration, *Measurements of the branching fraction $D_s^+ \to \tau^+\nu_\tau$* contribution to this Conference, HEP-97, **#455**

[2] M. Acciarri *et al.*, L3 Collaboration, *Measurements of $D_s \to \tau^-\nu_\tau$ and a new limit for $B \to \tau^-\nu_\tau$*, CERN-PPE/96-198, contribution to this Conference, HEP-97, **#496**

[3] S. Aoki *et al.*, WA75 Collaboration, Prog. Theor. Phys. **89** (1993) 131

[4] D. Acosta *et al.*, CLEO Collaboration, Phys. Rev. **D67** (1994) 5690

[5] J.Z. Bai *et al.*, BES Collaboration, SLAC-PUB-7147, April 1996, submitted to Phys. Rev. D

[6] K. Kodama *et al.*, E653 Collaboration, Phys. Lett. **B382** (1996) 299

[7] J. Richman and P. Burchat, Rev. Mod. Phys. **67** (1995) 893

[8] G.C. Zucchelli *et al.*, DELPHI Collaboration, *Measurements of the branching ratio of $b \to \tau\nu_\tau X$* contribution to this Conference, HEP-97, **#459**

[9] A. Falk, Z. Ligeti, M. Neubert, Y. Nir, Phys. Lett. **B326** (1994) 145

[10] D. Buskulic *et al.*, ALEPH Collaboration, Phys. Lett. **B343** (1995) 444

[11] M. Acciarri *et al.*, L3 Collaboration, Phys. Lett. **B332** (1994) 201

[12] Opal Collaboration, presented at XXVIII International Conference in High Energy Physics July 1996, Warsaw, Poland, pa05-038

[13] Y. Grossman, H.E. Haber and Y. Nir, Phys. Lett. **B357** (1995) 630

[14] M.L. Mangano and S.R. Slabospitsky, preprint CERN-TH/97-150, July 1997

[15] J. Alexander *et al.*, CLEO Collaboration, ICHEP-96, PA05-081

[16] F. Abe *et al.*, CDF Collaboration, Phys. Rev. Lett. **79** (1997) 357

[17] W. Adam *et al.*, DELPHI Collaboration, Zeit. Phys. **C73** (1996) 1

[18] M.S. Alam *et al.*, CLEO Collaboration, Phys. Rev. Lett. **74** (1995) 2885

Tau and Charm Results from CLEO

Dave Besson[1] (dzb@kuhep4.phsx.ukans.edu)

University of Kansas, USA

Abstract. The CLEO-II detector has accumulated a sample of approximately 6 fb^{-1}, corresponding to approximately 20 Million total $e^+e^- \rightarrow$hadrons events. This large integrated luminosity comprises ~6 Million $c\bar{c}$ and ~6 Million $\tau\tau$ pair events. We herein focus on recent results in the charm and the tau sectors from CLEO - the observation of new excited charmed baryons and the study of inclusive charged kaon production in three-prong tau decays; other CLEO results can be found in other contributions to this conference, or from the CLEO homepage (http://www.lns.cornell.edu) With a silicon detector now in place, and a doubling of the event sample used for these analyses, CLEO should have many more tau and charm results for summer 1998, and beyond.

1 Detector Description

These data were collected with the CLEO-II detector[1] at the Cornell Electron Storage Ring. The CLEO II detector is a general purpose solenoidal magnet spectrometer and calorimeter. The detector was designed for efficient triggering and reconstruction of two-photon, tau-pair, and hadronic events. Measurements of charged particle momenta are made with three nested coaxial drift chambers consisting of 6, 10, and 51 layers, respectively. These chambers fill the volume from r=3 cm to r=1 m, with r the radial coordinate relative to the beam ($\hat{z}$) axis. This system is very efficient ($\epsilon \geq$98%) for detecting tracks that have transverse momenta (p_T) relative to the beam axis greater than 200 MeV/c, and that are contained within the good fiducial volume of the drift chamber ($|\cos\theta|$ <0.94, with θ defined as the polar angle relative to the beam axis).

2 Tau decays into charged kaons

Experimental studies of charged kaon production in tau decay is considered in several contributions to this conference[2, 4, 3]. The theoretical aspect of this question has been treated by Finkemeir & Mirkes[5]. The general prescription of Finkemeir & Mirkes is to factorize the matrix element into the product of a leptonic current×hadronic current (the tau decay into the W followed by the hadronization of the W); i.e. $M.E. = \frac{G}{\sqrt{2}}(cos\theta_c M_\mu J^\mu + sin\theta_c M_\mu J^\mu)$ For the leptonic current, a standard V-A form is assumed: $M_\mu = \overline{u}(l',s')\gamma_\mu(g_V - g_A\gamma_5)u(l,s)$. For the hadronic current, one must take into account creation of the strange system at the W-decay vertex. In the case of

$\tau \to K^- \nu_\tau$, there is coupling only to the axial vector portion of the current, with: $< K(q)|A^\mu(0)|0 >= i\sqrt{2} f_K q^\mu$. Including radiative corrections leads to the prediction $\mathcal{B}(K\nu_\tau) = 0.723 \pm 0.006\%$, for the one-prong decay of the tau, in excellent agreement with the experimental measurement of: $\mathcal{B}(K\nu_\tau) = 0.692 \pm 0.028\%$[6].

Applying this formalism to the newly measured three-prong $\tau \to KX$ states, one must take into the fact that, for the K_1, there are, in principle four form factors which can be written for the most general form of the matrix element, with both Vector and Axial vector contributions. These form factors are, not unexpectedly, written as Breit-Wigner forms, with the decay rates somewhat sensitive to the width and masses of these Breit-Wigner forms. There is an obvious caveat here, insofar as the a_1 is known to have a larger width in tau decay compared to hadroproduction. Similarly, the predictions for $\tau \to K_1 \nu_\tau$ are dependent on the assumed K_1 parameters - if the K_1 is somewhat wider than the presently tabulated PDG values, the predicted rate comes into considerably better agreement with the data.

Production of inclusive charged kaons in three-prong tau decay is expected to be dominated by the K_1 resonances, in analogy to the well-studied $\tau \to a_1 \nu_\tau$; with $a_1 \to \pi^+ pi^- \pi^+$. In fact, in the non-strange sector, the three pion decays, in principle, would be expected to decay through the J(P)=1+ current into the triplet a_1 ($^3P_1$) and the singlet b_1 ($^1P_1$). The decay $\tau \to b_1$ constitutes a second-class current and is therefore prohibited. In the strange sector, the states analogous to the a_1 and b_1 are the states K_A and K_B. However, due to SU(3) symmetry breaking, the second-class current prohibition does not hold here, and, in principle, both the K_A and the K_B are allowed. In practice, what is observed are the mass eigenstates $K_1(1270)$ and $K_1(1430)$, which are mixtures of K_A and K_B.

In recent years the large data samples accumulated at CLEO and LEP have allowed much-improved measurements of inclusive decays of tau leptons to charged kaons, complementing similar measurements of inclusive decays of tau leptons to neutral kaons. We now consider specifically the decays of $\tau^- \to K^- h^+ \pi^- (\pi^0) \nu_\tau$ and $\tau^- \to K^- K^+ \pi^- (\pi^0) \nu_\tau$ relative to $\tau^- \to \pi^- \pi^+ \pi^- (\pi^0) \nu_\tau$, where $h^\pm$ can be either a charged pion or kaon.

To find the number of events with kaons, the three experiments with recent measurements (ALEPH, CLEO, and OPAL, preliminary) use specific ionization (dE/dx) information from their central tracking chambers. All three experiments have comparable separation capabilities of pions and kaons ($\sim 2\sigma$); because of the modest separation, a good understanding of the dE/dx calibration is crucial. For each track, one calculates the deviation of the measured energy loss relative to that expected for true kaons or pions.

The number of kaon and pion tracks in the sample of selected 1vs3 events is found statistically by fitting the dE/dx ionization distribution for charged tracks in the three-prong hemisphere to the sum of the pion shape plus the kaon shape. For OPAL, these shapes are determined from Monte Carlo, for

ALEPH, the shapes are obtained from data (e.g., the pion shape is taken from a sample of $\tau \to \omega\pi\nu_\tau$ events which have a topology similar to that expected for three-prong tau decays to kaons). For CLEO, the kaon and pion shapes are derived from data, using large samples of charged kaons and pions tagged by the very clean decay chain: $D^{*+} \to D^0\pi^+$; $D^0 \to K^-\pi^+$. The fitted area under the kaon curve directly gives the number of kaons and the area under the pion curve gives the number of pions.

Table 1 summarizes the results. The weighted average of the experimental results is found to be somewhat different than the theoretical predictions, although, as has been mentioned before, this discrepancy can be ameliorated to large measure, by using a different K_1 width as input to the theoretical calculation.

Branching fractions	CLEO97 %	ALEPH97[3]	OPAL97[4]
$Br(\tau \to K\pi\pi\nu_\tau)$	$0.317 \pm 0.027 \pm 0.072$	$0.214 \pm 0.037 \pm 0.029$	
$Br(\tau \to K\pi\pi\pi^0\nu_\tau))$	$0.165 \pm 0.045 \pm 0.047$	$0.061 \pm 0.029 \pm 0.015$	
$Br(\tau \to KK\pi\nu_\tau)$	$0.155 \pm 0.017 \pm 0.027$	$0.168 \pm 0.022 \pm 0.020$	
$Br(\tau \to K\pi\pi\nu_\tau)n(\pi^0))$			$0.358^{+0.089+0.046}_{-0.084-0.065}$
$Br(\tau \to KK\pi\pi^0\nu_\tau)$	$0.070 \pm 0.37 \pm 0.14$	$0.075 \pm 0.029 \pm 0.015$	
$Br(\tau \to KK\pi\nu_\tau)n(\pi^0))$			$0.036^{+0.047+0.024}_{-0.042-0.022}$

Table 1. Results

3 Charm

There have also been substantial improvements in our cataloging of excited charmed baryons. Considering that the first orbitally excited charmed mesons (the Λ_c^*) were only observed four years ago by ARGUS[7], the progress since has been fairly remarkable. Baryonic states are classified according to their quantum numbers. Realizing that these quantum numbers reflect, among other things, the symmetry of the light diquark spin wavefunction (with the antisymmetric state corresponding to ↑↓ and the symmetric state to ↑↑), we can categorize baryons as follows:

quark content	1/2+ (Mixed Anti)	1/2+ (Mixed sym)	3/2+ (Symm)	L=1 1/2-	L=1 3/2-
uds	$\Lambda(1115)$	$\Sigma_0(1193)$	$\Sigma^{*0}(1385)$	$\Lambda(1405)$	$\Lambda(1520)$
udc	$\Lambda_c(2285)$	$\Sigma_c^+(2455)$	$\Sigma_c^{*+}(2520)$	$\Lambda_c(2593)$	$\Lambda_c(2625)$
qsc	$\Xi_c(2465)$	$\Xi_c'(2572)$	$\Xi_c^*(2645)$	(not seen)	$\Xi_{c,1}(2815)$

In more detail, the charmed strange baryons can be written as follows (in order of increasing mass), as categorized by the angular momenta of the light quarks:

- Ξ_c : c↑(s↑q↓) (L=0, J=S=1/2)
- Ξ_c' : c↑(s↓q↓) (L=0, J=S=1/2) (sq diquark in a *symmetric* wave function with respect to interchange)

- Ξ_c^*: c↑(s↑q↑) (L=0, J=S=3/2) (Dominant decay mode expected $\rightarrow \Xi_c\pi$, analogous to $D^* \rightarrow D\pi$)
- Ξ_c^{**}: c↑(s↑q↓) (L=1, S=1/2)

CLEO has first observations of the $\Xi_c^{'+}$, the $\Xi_c^{'01}$ and the Ξ_c^{*+} in this sector[8]. The $\Xi_c^{'+}$ is the decuplet partner of the Ξ_c^+, in the same way that the Ξ' is the decuplet partner of the Ξ, for which the diquark is in a spin 1 state ($\Xi' : \Xi = \Sigma : \Lambda$). The Ξ_c^{*+} is the L=1 partner of the Ξ, with the diquark in a spin 0 state, but with one unit of orbital angular momentum of the diquark relative to the 'heavy' central quark (i.e., $\Xi^* : \Xi = \Lambda_c^{**} : \Lambda_c$). We can consider the question of what the mass splitting is expected to be for the Ξ_c'-Ξ_c; qualitatively, it can be expected to be smaller than the corresponding mass difference between the Ξ' and the Ξ of 209 MeV ($\Delta M = M(\Xi'(1530) - \Xi(1321))$, due to the smaller relativistic corrections for the case of charm. By contrast, CLEO measures $M(\Xi_c^{+'} - \Xi_c^+) = 107.8 \pm 1.7 \pm 2.5$ MeV and, for the neutral state: $M(\Xi_c^{0'} - \Xi_c^0) = 107.0 \pm 1.4 \pm 2.5$ MeV.

Additionally, CLEO observes a particle decaying into $\Xi_c^+\pi^+\pi^-$, with a mass difference $\Delta[M(\Xi_c^+\pi^+\pi^-) - M(\Xi_c)]$ of $349.4 \pm 0.7 \pm 1.0$ MeV. In analogy to the orbitally excited Λ_c^{**} doublet similarly decaying to $\Lambda_c\pi^+\pi^-$ (the $\Lambda_c(2593)$ and the $\Lambda_c(2630)$, corresponding to the J=1/2- and J=3/2- states), this particle is interpreted as the J=3/2 orbital excitation of the Ξ_c.

References

[1] Y. Kubota *et al.* (CLEO II), Nucl. Instr. Meth. **A320**, 66 (1992).

[2] "Meaurement of the branching fractions $\mathcal{B}(\tau \rightarrow K^-h^+h^-(\pi^0)\nu_\tau)/\mathcal{B}(\tau \rightarrow \pi^-\pi^+\pi^-(\pi^0)\nu_\tau)$" (The CLEO Collaboration), submitted to this conference.

[3] "Three-prong tau decays with charged kaons" (The ALEPH Collaboration), CERN PPE/97-069

[4] "Determination of Tau Branching Ratios to Three-Prong Final States with Charged Kaons", (The OPAL Collaboration), OPAL Physics Note PN-304 (1997)

[5] "Theoretical Aspects of $\tau \rightarrow K\pi\pi\nu_\tau$ decays and the K_1 widths", E. Mirkes, submitted to this conference.

[6] R.M. Barnett *et al.*, Particle Data Group, Phys. Rev. **D54**, 19 (1996)

[7] "Observation of a new charmed baryon", (The ARGUS Collaboration), Phys. Lett. **B317**, 227, (1993).

[8] "Observation of Two Narrow States Decaying into $\Xi_c^+\gamma$ and $\Xi_c^0\gamma$" (The CLEO Collaboration), and "Search for excited charmed baryons decaying into $\Xi_c n(\pi)$" (The CLEO Collaboration), submitted to this conference.

---

[1] A previous measurement of the Ξ_c' by the WA89 Collaboration three years ago has not yet been published.

Part VI

High Energy Nuclear Interactions and Heavy Ion Collisions

Strange Baryon Production in Pb-Pb Collisions at 158A GeV/ c

A. Jacholkowski (Adam.Jacholkowski@cern.ch) for the WA97 collaboration

INFN Bari & CERN

E. Andersen[c], A. Andrighetto[k], F. Antinori[k], N. Armenise[b], J. Bán[g], D. Barberis[f], H. Beker[e], W. Beusch[e], J. Böhm[m], R.Caliandro[b], M. Campbell[e], E. Cantatore[e], N. Carrer[k], M.G. Catanesi[b], E. Chesi[e], M. Dameri[f], G. Darbo[f], J.P. Davies[d], A. Diaczek[l], D. Di Bari[b], S. Di Liberto[o], A. Di Mauro[b], D. Elia[b], D. Evans[d], K. Fanebust[c], R.A. Fini[b], J.C. Fontaine[i], J. Ftáčnik[g], W. Geist[r], B. Ghidini[b], G. Grella[p], M. Guida[p], E.H.M. Heijne[e], H. Helstrup[c], A.K. Holme[e], D. Huss[i], A. Jacholkowski[b], P. Jovanovic[d], A. Jusko[g], V.A. Kachanov[q], T. Kachelhoffer[r], J.B. Kinson[d], A. Kirk[d], W. Klempt[e], K. Knudson[e], I. Králik[e], J.C. Lassalle[e†], V. Lenti[b], J.A. Lien[j], R. Lietava[g], R.A. Loconsole[b], G. Løvhøiden[j], M. Lupták[g], I. Mácha[n], V. Mack[i], V. Manzari[b], P. Martinengo[e], M.A. Mazzoni[o], F. Meddi[o], A. Michalon[r], M.E. Michalon-Mentzer[r], P. Middelkamp[e], M. Morando[k], M.T. Muciaccia[b], E. Nappi[b], F. Navach[b], K. Norman[d], B. Osculati[f], B. Pastirčák[g], F. Pellegrini[k], K. Píška[m], F. Posa[b], E. Quercigh[e], R.A. Ricci[h], G. Romano[p], G. Rosa[p], L. Rossi[f], H. Rotscheidt[e], K. Šafařík[e], S. Saladino[b], C. Salvo[f], L. Šándor[e,g], T. Scognetti[b], G. Segato[k], M. Sené[l], R. Sené[l], P. Sennels[j], S. Simone[b], A. Singovski[q], B. Sopko[n], P. Staroba[m], J. Šťastný[m], T. Storås[j], S. Szafran[l], T.F. Thorsteinsen[c], G. Tomasicchio[b], J. Urbán[g], M. Vaníčková[m], G. Vassiliadis[a†], M. Venables[d], O. Villalobos Baillie[d], T. Virgili[p], A. Volte[l], C. Voltolini[r], M.F. Votruba[d] and P. Závada[m].

---

[a] Nuclear Physics Department, Athens University, Athens, Greece
[b] Dipartimento di Fisica dell'Università and Sezione INFN, Bari, Italy
[c] Fysisk Institutt, Universitetet i Bergen, Bergen, Norway
[d] University of Birmingham, Birmingham, UK
[e] CERN, European Laboratory for Particle Physics, Geneva, Switzerland
[f] Dipartimento di Fisica dell'Università and Sezione INFN, Genoa, Italy
[g] Institute of Experimental Physics, Košice, Slovakia
[h] INFN, Laboratori Nazionali di Legnaro, Legnaro, Italy
[i] GRPHE, Université de Haute Alsace, Mulhouse, France
[j] Fysisk institutt, Universitetet i Oslo, Oslo, Norway
[k] Dipartimento di Fisica dell'Università and Sezione INFN, Padua, Italy
[l] Collège de France and IN2P3, Paris, France
[m] Institute of Physics, Czech Academy of Sciences, Prague, Czech Republic
[n] Department of Physics, Technical University, Prague, Czech Republic
[o] Dipart. di Fisica dell'Università "La Sapienza" and Sezione INFN, Rome, Italy
[p] Dipartimento di Fisica dell'Università and Sezione INFN, Salerno, Italy
[q] Institute of High Energy Physics, Protvino, Russia
[r] Centre de Recherches Nucléaires, Strasbourg, France
[†] Deceased

Abstract. The first results on Λ, Ξ, and Ω particle production in p–Pb and Pb–Pb collisions obtained by the WA97 CERN experiment are presented.

1 Introduction

Strange particle production has been suggested long ago as a useful diagnostic tool in heavy ion reactions to study flavour equilibration [1], predicted to be much faster in a Quark–Gluon Plasma(QGP) than in a Hadron Gas [2, 3, 4]. In particular, multi-strange (anti-)baryons can be expected to probe the early phase of the collision [5], since hadronic production of such particles is suppressed due to high mass thresholds while in a QGP scenario they can be easily created due to a copious production of strange quark pairs from gluons.

An enhanced production of multistrange (anti-)baryons has already been observed in sulphur initiated reactions by the WA85 and WA94 collaborations [6, 7] and the WA97 experiment has extended this study to a Pb–Pb system.

2 The experimental setup

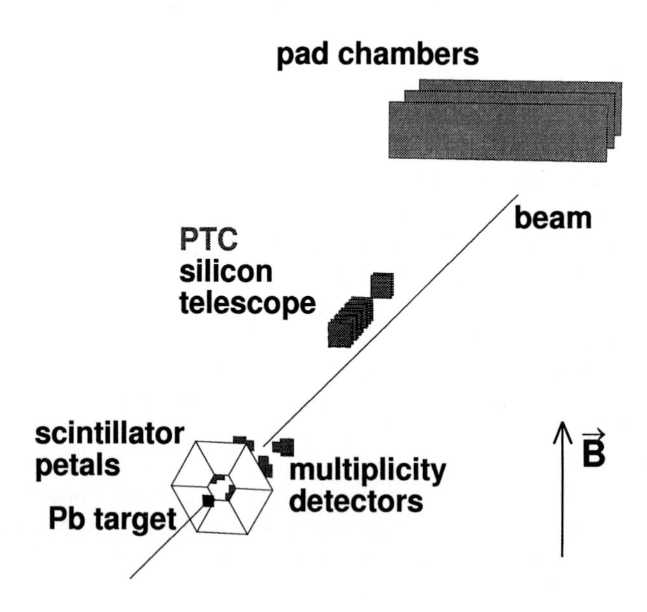

Fig. 1. The WA97 set-up in the Omega magnet.

The WA97 set-up, shown schematically in figure 1, is described in detail in ref. [9]. The target and the silicon telescope are placed inside the homogeneous region of the 1.8 T field of the CERN Omega spectrometer.

The tracks are detected using the silicon telescope as a Pixel Tracking Chamber (PTC). This makes use of 7 planes of the novel pixel detectors [10] with pixel dimensions of $75 \times 500\ \mu m^2$ (about 0.5×10^6 detecting elements) and of 10 silicon micro-strip planes with a pitch of 50 μm. The length of the first(compact) part of the telescope is 30 cm, the area of each plane is about 5×5 cm^2.

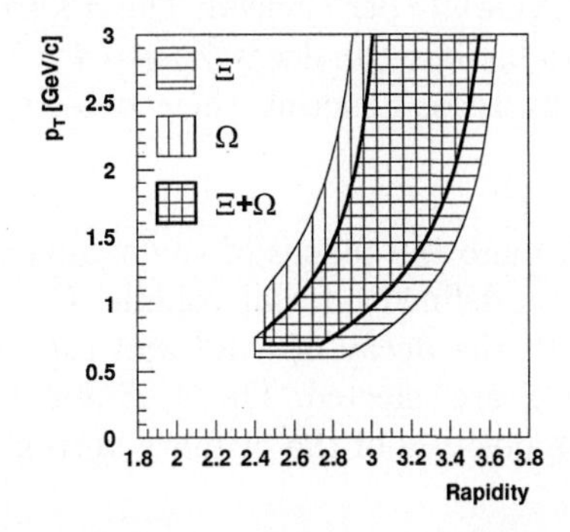

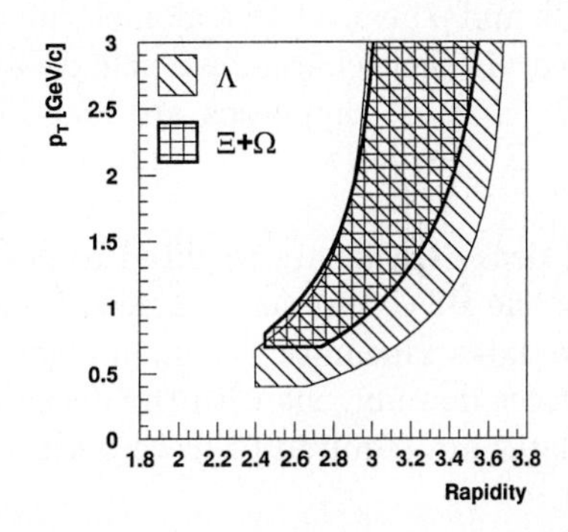

Fig. 2. The acceptance in transverse momentum and rapidity for Λ, Ξ and Ω and the overlap region for Pb–Pb collisions.

The telescope is placed slightly above the beam line and inclined (pointing to the target) in order to accept particles at central rapidity and medium transverse momentum. Figure 2 shows the acceptance windows for Λ, Ξ and Ω [1] and the region where the two acceptance windows overlap, for the Pb–Pb collisions. The acceptance windows (and their overlap) for Ω and Ξ coming from the p–Pb data are similar. The geometrical acceptance is the same for particles and antiparticles, since our apparatus is symmetric with respect to the magnetic field.

A set of scintillator petals provides a centrality trigger that selects about 30% of the geometric inelastic cross section. Two planes of silicon micro-strip detectors sample the charged particle multiplicity in the pseudorapidity interval $2 < \eta < 4$ and allow for more detailed off-line analysis of centrality dependences.

3 Data sample and selection criteria

The results presented here are based on the analysis of two data samples taken during the 1995 run:

- 42 M Pb–Pb events [2] taken with the PTC placed at a distance of 60 cm from the target with a trigger selecting central events,
- 120 M p–Pb events taken with the PTC 90 cm from the target with a trigger that required at least two tracks to pass through the compact part of the PTC.

Both Pb and p beams had a momentum of 158 GeV/c per nucleon. The Λ's are detected via their charged particle decay products in the decay $\Lambda \to \mathrm{p} + \pi^-$. The Ξ^- and Ω^- hyperons are identified by reconstructing their two-step decays:

$$\Xi^- \to \Lambda + \pi^- \qquad \Omega^- \to \Lambda + \mathrm{K}^-$$
$$\hookrightarrow \mathrm{p} + \pi^- \qquad \hookrightarrow \mathrm{p} + \pi^-$$

All decay tracks are required to pass through the planes of the compact part of the PTC and vertices need to be in a defined fiducial volume. Only Λ candidates kinematically unambigous with the decay of a K^0_s and the Ω candidates unambigous with the decay of a Ξ are selected. The Λ, Ξ and Ω candidates are required to trace back to the position of the primary vertex.

4 Corrections for acceptance and efficiency

A subsample of the Pb–Pb and all of the p–Pb data have been corrected for acceptance, detector and reconstruction efficiency. The correction procedure of the remaining data is under way.

A Monte Carlo simulation based on GEANT has been used. For each analysed Λ, Ξ and Ω a weight was calculated by generating hyperons with the measured transverse momentum, rapidity and position of the primary vertex. The pixel efficiencies corresponding to each read-out chip were taken into account. The background tracks and the electronic noise were simulated by embedding the hits of the Monte Carlo tracks into a real event which had a similar hit multiplicity in the PTC as the hyperon event.

These mixed events were processed in the same way as real events for pattern recognition, track fitting and signal selection. The resulting weight is the number of Monte Carlo hyperons generated divided by the number of Monte Carlo hyperons reconstructed.

All 19 Ω and 376 Ξ for the p–Pb and all 107 Ω for the Pb–Pb data were weighted using this procedure, while only subsamples of 253 Ξ and 475 Λ for the Pb–Pb data were corrected.

---

[2] This sample contains about 40% of the Pb–Pb events collected in 1995. Analysis of the remaining data is in progress.

5 Results

a) We have measured the ratio $(\Omega^- + \overline{\Omega}^+)/(\Xi^- + \overline{\Xi}^+)$ both for the p–Pb and Pb–Pb data in the overlap acceptance window. We find that this ratio is enhanced by a factor of 3.2 ± 1.1 when going from p–Pb to Pb–Pb.

b) The m_T distributions have been studied for Pb–Pb collisions for Λ and Ξ (present statistics did not allow us to do this study for Ω). We have used the parametrisation: $\frac{1}{m_\mathrm{t}^{3/2}} \frac{d\mathrm{N}}{dm_\mathrm{t}} = C \cdot \exp(-m_\mathrm{t}/T)$ and the maximum likelihood method in order to extract inverse slopes T and test a possible rapidity dependence of the cross sections. The inverse slopes obtained with this method, still very preliminary, can be seen in table 1. The WA97

Table 1. Inverse slopes (in MeV/c^2) obtained from the maximum likelihood fit for the 30% most central Pb–Pb events and for the S–S collisions (WA94 experiment).

Particle	WA97 (Pb–Pb)	WA94(S–S)
Λ^0	240 ± 21	213 ± 2
Ξ^-	270 ± 23	222 ± 10

slopes are slightly higher than those found by the WA94 experiment for S–S collisions [8]. This difference could result from a higher collective flow present in Pb–Pb collisions [11].

c) Cross sections and particle ratios are being calculated for Λ, Ξ^-, Ω^- and their antiparticles. We present here results on the Ξ^-/Λ ratio in Pb–Pb, which we compare to previous experiments, and on the Ω^-/Ξ^- ratio both from Pb–Pb and p–Pb collisions. The Ξ^-/Λ ratio has been calculated for $p_\mathrm{T} > 0.7$ GeV/c and one unit of rapidity for two intervals of the collision centrality, whose central values correspond to 180 and 340 participating nucleons respectively. The number of nucleons participating in the collision was estimated by means of the VENUS program from the measured multiplicities. The ratios are shown in fig. 3 together with the values previously obtained by our Collaboration in p–S, p–W, S–S and S–W collisions. For the comparison these values, measured at central rapidity unit for $p_\mathrm{T} > 1.2$ GeV/c, have been extrapolated down to $p_\mathrm{T} >$ 0.7 GeV/c which is the lower p_T limit for the Pb–Pb data. Such an extrapolation could be performed using the published slopes [8] of the m_T distributions. Forthcoming data from WA97, both for p–Be and p–Pb collisions, will allow us to compare directly this ratio in exactly the same p_T–rapidity window. There is clearly an enhancement of Ξ^- relative to Λ at central rapidity for larger colliding nuclei. The future increase of statistics should allow us to ascertain whether this ratio has a smooth

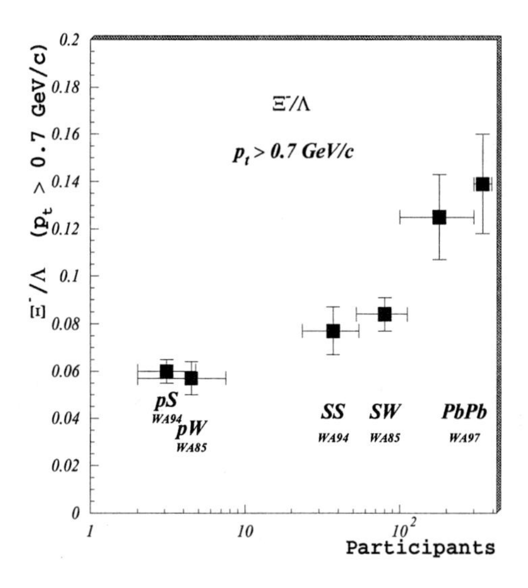

Fig. 3. Compilation of the Ξ^-/Λ ratio for the different colliding systems. Horizontal bars indicate approximate range of the number of participants for each experiment.

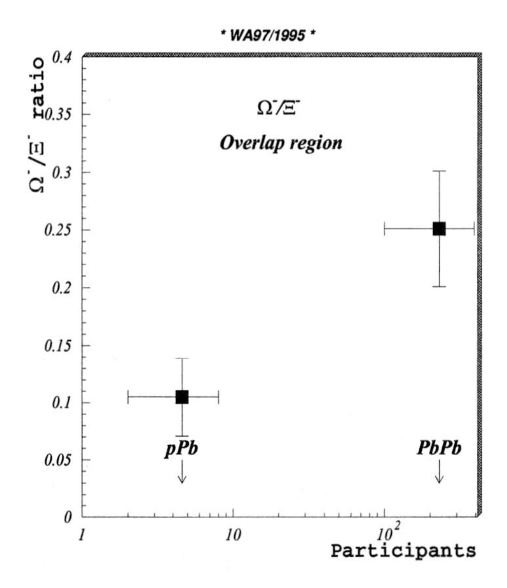

Fig. 4. Ω^-/Ξ^- ratio for the p–Pb and Pb–Pb data. Horizontal bars indicate approximate range of the number of participants.

or a step–like behaviour, when going from S–S to Pb–Pb. Fig.4 shows the Ω^-/Ξ^- ratio but only for data from our experiment. In this case Ω^- and Ξ^- are taken in the same kinematical window, *i.e.* the overlap window between Pb–Pb and p–Pb data. Again we can see an increase of a higher strangeness baryon production as compared to a lower strangeness baryon. This points to a pattern of strangeness enhancement increasing with strangeness content of a baryon.

6 Conclusions and outlook

The first results on fully corrected ratios of Ξ^- and Λ, detected by the WA97 experiment in Pb–Pb collisions have been presented. Also first results from the p–Pb run on Ω^- and Ξ^- are given.

The enhancement of the ratio $(\Omega^- + \overline{\Omega}^+)/(\Xi^- + \overline{\Xi}^+)$ is of 3.2 ± 1.1 when going from p–Pb to Pb– Pb collisions (it is 2.4 for the Ω^-/Ξ^- ratio). The Ξ^- hyperon production is in turn enhanced relative to Λ when going from p–S to Pb–Pb collisions by a factor of about 2 (the enhancement with respect to S–S is of 1.7).

The inverse slopes of transverse mass distributions for Λ and Ξ^- have been found larger in Pb–Pb than in S–S interactions.

The Pb–Pb data sample analysed till now corresponds to about one fourth of the statistics, of which all Ω, but only a fraction of Λ and Ξ have been passed through the correction procedure. For the p–Pb data all Ω and Ξ have been used. Further analysis is in progress which should enable us to complete these results and to improve their statistical precision.

References

[1] J. Rafelski and B. Müller, Phys. Rev. Lett. **48** (1982) 1066;
J. Rafelski and B. Müller, Phys. Rev. Lett. **56** (1986) 2334.
[2] P. Koch, B. Müller and J. Rafelski, Phys. Rep. **142** (1986) 167.
[3] U. Heinz, Nucl. Phys. **A566** (1994) 206c.
[4] H.C. Eggers and J. Rafelski, Int. Journ. of Mod. Phys. **A6** (1991) 1067.
[5] J. Rafelski, Phys. Lett. **B262** (1991) 333.
[6] WA85 Collaboration: S. Abatzis *et al.*, Nucl. Phys. **A590** (1995) 307c.
[7] WA94 Collaboration: S. Abatzis *et al.*, Nucl. Phys. **A590** (1995) 317c.
[8] WA94 Collaboration: M. Venables *et al.*, Proc. *Strangeness '96 Workshop*, Budapest Hungary, ed. T. Csörgo, P. Lévai and J. Zimányoı, APH N.S. Heavy Ion Physics **4** (1996) 91.
[9] WA97 Collaboration: G. Alexeev *et al.*, Nucl. Phys. **A590** (1995) 139c.
[10] E.H.M. Heijne *et al.*, Nucl. Instr. Meth. **A349** (1994) 138;
F. Antinori *et al.*, Nucl. Instr. Meth. **A360** (1995) 91.
[11] I.G. Bearden *et al.*, preprint CERN–PPE/96-163 , sub. to Phys. Rev. Lett.

A New Measurement of the J/ψ Production Rate in Lead-Lead Interactions at 158 GeV/c per Nucleon

Faïrouz Ohlsson-Malek[12,e] for the NA50 collaboration* (fmalek@isn.in2p3.fr)

* M.C. Abreu[6,a], B. Alessandro[11], C. Alexa[3], R. Arnaldi[11], J. Astruc[8], M. Atayan[13], C. Baglin[1], A. Baldit[2], M. Bedjidian[12], F. Bellaiche[12], S. Beolè[11], V. Boldea[3], P. Bordalo[6,b], A. Bussière[1], V. Capony[1], L. Casagrande[6], J. Castor[2], T. Chambon[2], B. Chaurand[9], I. Chevrot[2], B. Cheynis[12], E. Chiavassa[11], C. Cicalò[4], M.P. Comets[8], S. Constantinescu[3], J. Cruz[6], A. De Falco[4], N. De Marco[11], G. Dellacasa[11,c], A. Devaux[2], S. Dita[3], O. Drapier[12], B. Espagnon[2], J. Fargeix[2], S.N. Filippov[7], F. Fleuret[9], P. Force[2], M. Gallio[11], Y.K. Gavrilov[7], C. Gerschel[8], P. Giubellino[11], M.B. Golubeva[7], M. Gonin[9], A.A. Grigorian[13], J.Y. Grossiord[12], F.F. Guber[7], A. Guichard[12], H. Gulkanyan[13], R. Hakobyan[13], R. Haroutunian[12], M. Idzik[11,d], D. Jouan[8], T.L. Karavitcheva[7], L. Kluberg[9], A.B. Kurepin[7], Y. Le Bornec[8], C. Lourenço[5], M. Mac Cormick[8], P. Macciotta[4], A. Marzari-Chiesa[11], M. Masera[11], A. Masoni[4], S. Mehrabyan[13], S. Mourgues[2], A. Musso[11], F. Ohlsson-Malek[12,e], P. Petiau[9], A. Piccotti[11], J.R. Pizzi[12], W.L. Prado da Silva[11,f], G. Puddu[4], C. Quintans[6], C. Racca[10], L. Ramello[11,c], S. Ramos[6,b], P. Rato-Mendes[11], L. Riccati[11], A. Romana[9], S. Sartori[11], P. Saturnini[2], E. Scomparin[5,g], S. Serci[4], R. Shahoyan[6,h], S. Silva[6], C. Soave[11], P. Sonderegger[5,b], X. Tarrago[8], P. Temnikov[4], N.S. Topilskaya[7], G.L. Usai[4], C. Vale[6], E. Vercellin[11], N. Willis[8].

[1] LAPP, CNRS-IN2P3, Annecy-le-Vieux, France. [2] LPC, Univ. Blaise Pascal and CNRS-IN2P3, Aubière, France. [3] IFA, Bucharest, Romania. [4] Università di Cagliari/INFN, Cagliari, Italy. [5] CERN, Geneva, Switzerland. [6] LIP, Lisbon, Portugal. [7] INR, Moscow, Russia. [8] IPN, Univ. de Paris-Sud and CNRS-IN2P3, Orsay, France. [9] LPNHE, Ecole Polytechnique and CNRS-IN2P3, Palaiseau, France. [10] IReS, Univ. Louis Pasteur and CNRS-IN2P3, Strasbourg, France. [11] Università di Torino/INFN, Torino, Italy. [12] IPN, Univ. Claude Bernard and CNRS-IN2P3, Villeurbanne,France. [13] YerPhI, Yerevan, Armenia.

[a] also at FCUL, Universidade de Lisboa, Lisbon, Portugal. [b] also at IST, Universidade Técnica de Lisboa, Lisbon, Portugal. [c] Dipartimento di Scienze e Tecnologie Avanzate, II Facoltà di Scienze, Alessandria, Italy. [d] now at Faculty of Physics and Nuclear Techniques, University of Mining and Metallurgy, Cracow, Poland. [e] now at ISN, Univ. Joseph Fourier and CNRS-IN2P3, Grenoble, France. [f] now at UERJ, Rio de Janeiro, Brazil. [g] on leave of absence from Università di Torino/INFN, Torino, Italy. [h] on leave of absence of YerPhI, Yerevan, Armenia.

Abstract. We present preliminary results of a new measurement of J/ψ production in Pb-Pb interactions at 158 GeV/c. As a function of centrality, the J/ψ production rate shows peculiar discontinuities which, if confirmed, could imply a sudden change in the state of the reacting medium.

1 Introduction

Ultra-relativistic heavy ion collisions at the CERN-SPS are used to study the possible formation of quark gluon plasma (QGP). Among several predicted signatures of this phase transition, charmonia suppression is expected to occur at the early stages of the collision. This suppression is attributed to color screening in the QGP [1]. However, some "normal" suppression is also produced by absorption in nuclear matter [2, 3].

Measurements of J/ψ , ψ' and Drell-Yan cross-sections are performed by the NA38 and NA50 experiments. Charmonia and Drell-Yan are observed via their decay into muon pairs. Drell-Yan is used as a reference as it is insensitive to strong interactions.

In the NA38 experiment, the studied collisions were induced by incident protons, oxygen or sulfur ions at 200 GeV/c per nucleon on different targets. A significant suppression of the J/ψ resonance in central collisions [4, 5] in O-A_{target} and S-U collisions has already been observed. It has triggered a considerable amount of theoretical work intended to explain the experimental results.

More recently, the NA50 experiment, an upgraded version of the NA38 experiment, used a lead beam with an energy of 158 GeV/c per nucleon on a fixed lead target. Data were collected in 1995 and 1996. Results from the 1995 data show a strong suppression of charmonia [6]. It thus provides new constraints for any theoretical explanation.

We present the final results for the 1995 lead-lead data and compare them to the NA38 results. As for the 1996 lead-lead experiment, a new event selection procedure enables to increase significantly the statistics of the events produced in peripheral collisions. This new selection is explained. The latest results of charmonia suppression in lead-lead are presented and compared with the previous ones and with the NA38 results.

2 Experimental details

The NA50 detector [7] which is an upgraded version of the NA38 apparatus [4], see Fig 1, consists of a dimuon spectrometer based on an air-core toroidal magnet. A set of proportional muon chambers and scintillator hodoscopes provide muon pair tracking and trigger. A hadron absorber made of carbon and a BeO pre-absorber are placed upstream from the first muon chamber. This spectrometer covers a pseudo-rapidity interval $2.8 < \eta < 4.0$. A mass resolution of 3.1% is obtained for the J/ψ .

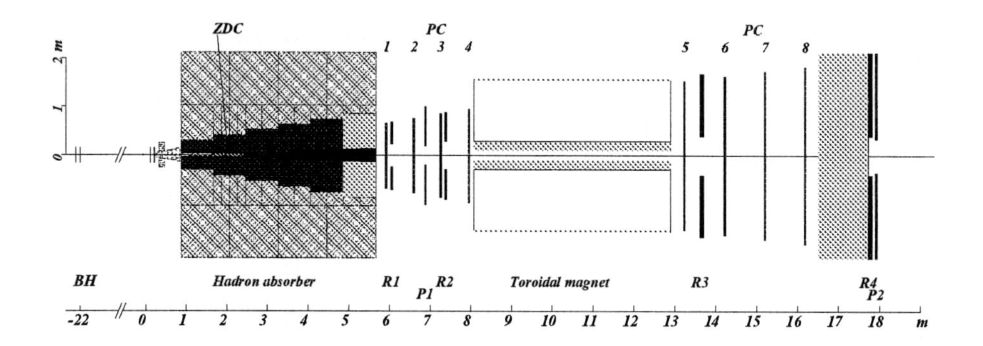

Fig. 1. Layout of the NA50 experiment.

The target is made of 7 lead subtargets in order to minimize secondary interactions. Its total thickness amounts to 30% of an interaction length (17.5% in 1995). A set of quartz blades allows the identification of the primary vertex. An electromagnetic calorimeter made of scintillating fibers embedded in lead measures the neutral transverse energy (E_T) produced in the collision. Its resolution is 5% for central Pb-Pb collisions. A "zero degree" calorimeter (ZDC) located along the beam axis and placed within the main muon absorber, downstream from the target, measures the energy carried out from the Pb-Pb reaction by the beam spectators. It is made of silica optical fibers embedded in a tantalum converter. Its energy resolution is 7% for 32.7 TeV incident Pb nuclei. The neutral transverse energy as well as the energy carried out by the beam spectators are used to estimate the centrality of the collision.

Events are selected for the final analysis if only one incident ion is detected, therefore avoiding pile-up effects. The tracking algorithm requires exactly two muon tracks in the acceptance of the spectrometer. The kinematical domain for the muon pair analysis is defined by a center-of-mass rapidity $0. < y_{cm} < 1.$ and a polar decay angle in the Collins-Soper reference frame $-0.5 < cos(\theta) < 0.5$.

3 Pb-Pb 1995 results and comparison with NA38 results

The invariant mass spectrum for opposite sign muon pairs is shown in Fig 2 for masses above 2. GeV/c^2. The processes contributing to this spectrum are Drell-Yan , J/ψ , ψ' , $D\bar{D}$ pairs and the background from mesons (π and K) decay. The π and K mesons decay leads also to like sign pairs which are used to estimate the background using the relation:

$$N^{Bckg} = 2 \times \sqrt{N^{++} \times N^{--}} \ ,$$

where N^{++} (N^{--}) is the number of pairs made of two positive (negative) charged muons. The signal to background ratio is larger than 10 at the J/ψ peak. There are roughly 5% non-J/ψ signal events under the J/ψ peak, mainly originating from Drell-Yan events since the $D\bar{D}$ contribution is expected to be unsignificant at a mass of 3. GeV/c^2.

The number of J/ψ , ψ' , and Drell-Yan events is obtained from a fit to the invariant mass spectrum. This fit is performed for masses above 3.05 GeV/c^2, using the maximum likelihood method with a superposition of four contributions as described hereafter:

$$\frac{dN^{+-}}{dM_{\mu^+\mu^-}} = A_{J/\psi}\frac{dN^{J/\psi}}{dM_{\mu^+\mu^-}} + A_{\psi'}\frac{dN^{\psi'}}{dM_{\mu^+\mu^-}} + A_{DY}\frac{dN^{DY}}{dM_{\mu^+\mu^-}} + \frac{dN^{Bckg}}{dM_{\mu^+\mu^-}} \ .$$

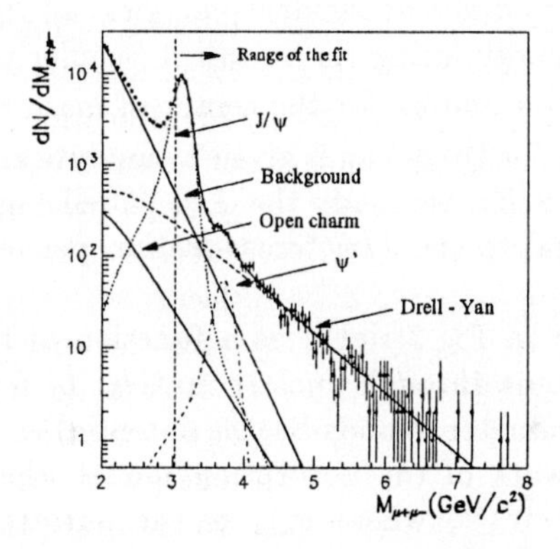

Fig. 2. Opposite sign muon pairs mass distribution in Pb-Pb collisions from the 1995 data collection.

The shape of the J/ψ , ψ' and Drell-Yan contributions are determined by Monte-Carlo simulations. Analytical functions are used to smooth the reconstructed simulated events as can be seen in Fig 2. The fit is performed with five free parameters including the three parameters $A_{J/\psi}$, $A_{\psi'}$, A_{DY} and the mass and width of the gaussian-like distribution for J/ψ . The corresponding parameters (mass, width) for the ψ' are related to those of the J/ψ according to the nominal mass difference and width ratio obtained from the Monte-Carlo simulation. Each contribution can be studied as a function of the centrality of the collision. For this purpose, we use the neutral transverse energy spectrum and divide it into equipopulated bins. The higher value bins account for the more central collisions.

Drell-Yan cross-section in A+B collisions is a superposition of primary independent nucleon-nucleon Drell-Yan components, expressed as $\sigma_{DY}^{AB} =$

$\sigma_{DY}^{NN} \times (A \times B)^{\alpha}$. All the measurements in NA38 as well as in Pb-Pb collisions experiment agree with an α value of 1 when taking into account isospin effects[1]. Thus, Drell-Yan scales as $(A \times B)$ as expected. This behaviour is illustrated in Fig 3(left) where the ratio of the measured Drell-Yan cross-section to the lowest order theoretical Drell-Yan value is shown as a function of $(A \times B)$. This ratio is expressed through the so-called K factor which is found to be around 2.5 in good agreement with the well known Drell-Yan theoretical calculations.

The nuclear dependance of charmonia production in A+B collisions is also expressed through the expression $\sigma_{J/\psi}^{AB} = \sigma_{J/\psi}^{NN} \times (A \times B)^{\alpha}$. The α value is found to be around 0.91 for all NA38 data as shown in Fig 3(center) where the J/ψ cross-section per nucleon-nucleon collision is plotted as a function of $(A \times B)$. Relatively to the $(A \times B)^{0.91}$ dependence, the 1995 Pb J/ψ cross-section is lower by a factor 0.74 ± 0.06 and thus exhibits an unexpected abnormal behaviour [6]. In order to compare the J/ψ cross-sections from different systems at different incident energies, all the values have been rescaled to the same energy, using the so-called Schuler parametrization [9], $\sigma = \sigma_0 (1 - \frac{M_{\mu\mu}}{\sqrt{s}})^{12}$, $\sqrt{s}$ standing for the center of mass energy per nucleon. The energy dependence for Drell-Yan is given by analytical calculations. Since Drell-Yan scales as $(A \times B)$, we study the J/ψ /Drell-Yan cross-sections ratio which is proportional to the J/ψ "cross-section per nucleon-nucleon collision".

This ratio is shown in Fig 3(right) as a function of the mean length of the path of the charmonia through nuclear matter, $\bar{L}$. $\bar{L}$ is calculated using the standard three-parameter Woods-Saxon potential as explained in reference [8]. In the framework of the absorption model, the J/ψ cross-section is proportional to $e^{-\rho_0 \bar{L} \sigma_{abs}}$, where σ_{abs} is the nuclear absorption cross-section. As clearly seen in this figure, except for the Pb data, all the other data decrease exponentially with $\bar{L}$ leading to an absorption cross-section of 6.2 ± 0.7 mb. The last Pb bin lies a factor 0.62 ± 0.04 below the absorption line, off by 10 standard deviations from the absorption expectation. It has to be noticed that the Pb-Pb data and the proton data are rescaled to 200 GeV/c.

4 1996 Pb-Pb results

A new event selection has been applied in order to increase the statistics of peripheral collisions. Indeed, applying the usual event selection criteria which includes the identification of the primary vertex leads to a sample of

[1] As the Drell-Yan mechanism depends on the isospin of the nucleon, the measured Drell-Yan cross-section is corrected to account for the relative amount of interacting protons and neutrons . The Drell-Yan corrected cross-section is equivalent to the Drell-Yan cross-section we would have measured for the same two nuclei made only of protons and is therefore expected to scale exactly with the number of nucleon-nucleon collisions.

Recent Results from the NA49 Experiment on Pb-Pb Collisions at 158 GeV per Nucleon

Dieter Röhrich (roehrich@ikf.uni-frankfurt.de) for the NA49 Collaboration

University of Frankfurt, Germany

Abstract. Baryon stopping and transverse energy production in central Pb+Pb collisions at 158 GeV per nucleon lead to an estimate of the energy density of 2-3 GeV/fm^3. Multiplicities of strange particles as compared to negatively charged hadrons are higher in heavy ion collisions than in p+p and p+A reactions. Single particle momentum spectra and two particle correlations suggest a rather low temperature (120 MeV) at thermal freeze-out for central Pb+Pb collisions at 158 GeV per nucleon. The matter at freeze-out shows an almost boost-invariant strong longitudinal expansion. The $k_\perp$ dependence of the transverse radius parameter together with the inverse slopes of the transverse mass spectra indicate significant transverse flow.

1 Introduction

The ultimate goal of studying collisions of ultrarelativistic heavy nuclei is to observe the state of deconfined quarks and gluons. Recent lattice QCD calculations predict such a state for bulk hadronic matter at energy densities of about 2.5 GeV/fm^3 or even less. Such an energy density is produced in the interaction volume of central Pb+Pb collisions at the SPS (158 GeV per nucleon). Single particle momentum spectra and two particle correlations can provide information on the space-time extent and evolution of the system, particle ratios can determine conditions at the time of chemical freeze-out of hadron species. The transient existence of a quark-gluon-plasma in the collision is expected to modify this evolution in comparison to a scenario of confined hadronic matter. The aim of experiment NA49 is to measure the hadronic final state of a central Pb+Pb collision as completely as possible in order to enable a study of single particle spectra and multiplicities of various hadrons, of particle correlations and of event-by-event observables.

2 Experiment NA49

The NA49 experiment is a large acceptance hadron spectrometer comprised of four large volume Time Projection Chambers (TPCs). Two of the TPCs are situated in separate superconducting dipole magnets downstream of the target while two larger devices are situated outside the magnetic field on either side of the beam. These tracking chambers record the trajectories of a large fraction of all charged particles produced in the event. Multiple measurements of the specific energy loss of charged particles in the relativistic rise regime allow particle identification. Four highly segmented time-of-flight (TOF) walls augment the capability to identify particles. Neutral strange particles are detected by their decay topologies as recorded in the tracking chambers. Central events are selected by a cut in the forward energy distribution measured in the veto calorimeter (top 5% of the inelastic cross-section).

3 Stopping and Energy Density

The net proton yield $(p - \overline{p})$ can be determined to a good approximation by the charge excess distribution, which is the difference between spectra of positively and negatively charged hadrons. The $(p - \overline{p})$ spectra have been corrected for the inbalance between the K^+ and K^- yields and for protons from the decay of Λ-particles, a low p_T-cut - corrected by an extrapolation procedure - removes the excess π^--yield [1].

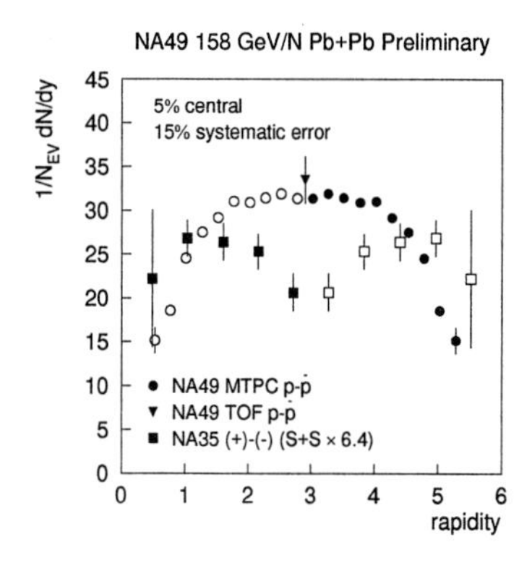

Fig. 1. Rapidity distribution of net protons $(p - \overline{p})$ in central Pb+Pb collisions and S+S interactions. For comparison the S+S data are scaled by the ratio of participant nucleons in both systems.

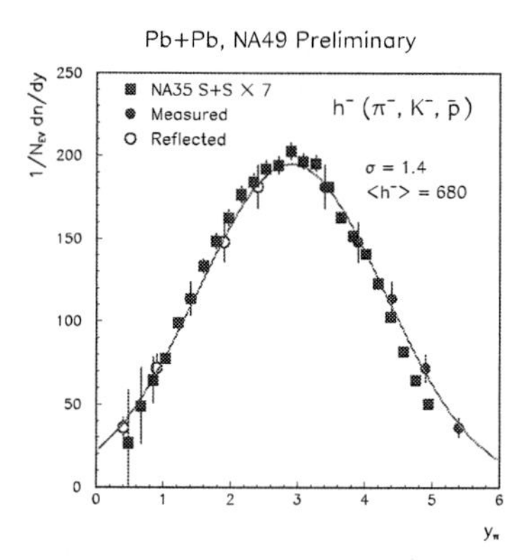

Fig. 2. Rapidity distribution of negatively charged hadrons produced in central Pb+Pb collisions and S+S interactions. For comparison the S+S data are scaled by the ratio of participant nucleons in both systems.

The rapidity distribution of net protons is shown in Fig. 1. Also shown is a data point of identified $(p-\overline{p})$, not corrected for Λ-decays. The net proton $(p-\overline{p})$ rapidity distribution is approximately flat over three units of rapidity and is somewhat narrower than the net proton yield in S+S collisions [2], scaled for comparison by the ratio of participant nucleons in both systems, suggesting a somewhat greater stopping in the heavier system. Fig. 2 shows the rapidity distribution of negatively charged hadrons produced in central Pb+Pb and S+S (scaled) collisions. The shape of the S+S distribution is found to closely resemble that of the heavier system. Consequently the multiplicity of h^- per participant nucleon is the same in both reactions. Based both on these measurements and those of the transverse energy production one can estimate the energy density to 2-3 GeV/fm^3 [3].

4 Strangeness Production

Strangeness production can shed light on the timescale of chemical freeze-out and therefore can carry information about the early stages of a heavy ion collision. NA49 measures strange mesons (K^0_S, K^+, K^-, Φ), strange baryons (Λ, $\overline{\Lambda}$) and double strange baryons (Ξ, $\overline{\Xi}$). The K/π ratio for various systems is shown in Fig. 3. The ratio is similar for heavy ion reactions, nucleon-nucleon and p+A data are lower. Strangeness enhancement is observed in S+S and Pb+Pb collisions. This strangeness enhancement is also reflected in the ratio $\overline{\Lambda}/\overline{p}$. Preliminary NA49 data (the correction of $\overline{p}$ stemming from $\overline{\Lambda}$-decays was estimated to be 50%) is shown in Fig. 4 together with ratios near midrapidity from p+p, p+A and S+A collisions [4]. The ratio $\overline{\Lambda}/\overline{p}$ in central Pb+Pb collisions is at least as large as the one in S+A reactions and therefore significantly larger than the values for p+p and p+A data.

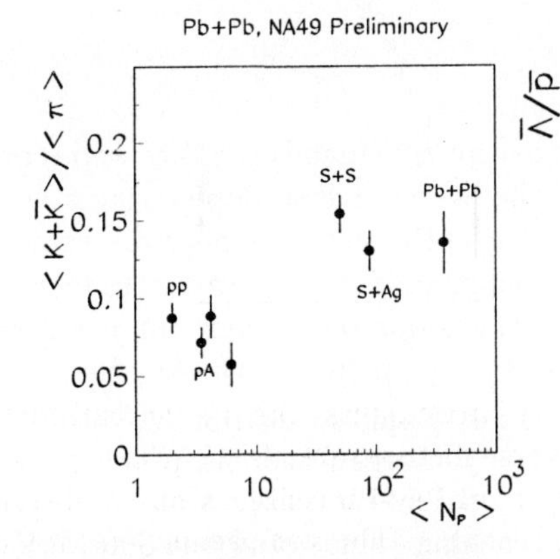

Fig. 3. Ratio of the multiplicities of kaons and pions for various target–projectile systems as a function of the number of participants.

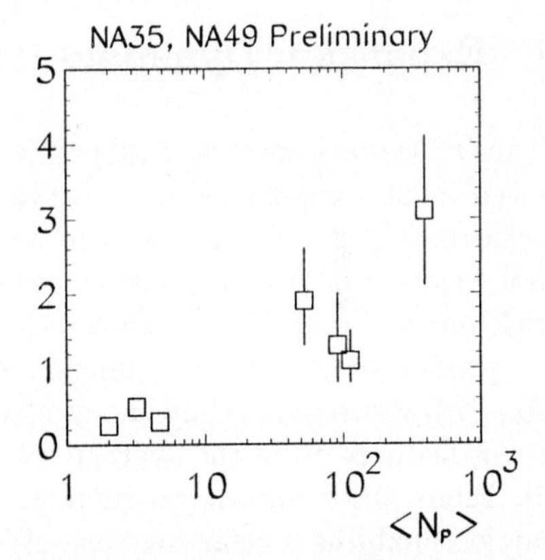

Fig. 4. Ratio of the rapidity densities of $\overline{\Lambda}$ and $\overline{p}$ near midrapidity for various target–projectile systems as a function of the number of participants.

The Φ-meson, a carrier of hidden strangeness, is of special interest in heavy ion collisions because its mass, width and branching ratios could be influenced by the surrounding medium. Fig. 5 shows the invariant mass spectrum of the charged kaons. A prominent Φ resonance peak is observed at a mass 1019.0 ± 0.3 MeV/c^2. The solid line shows a relativistic Breit-Wigner function convoluted with a Gaussian experimental mass resolution. The width of the resonance is consistent with the free Φ width. Fig. 6 shows the rapidity distribution of Φs, giving an average multiplicity of 5.4 ± 0.7. The Φ/h^- is more than a factor of two larger in central Pb+Pb collisions than in p+p interactions.

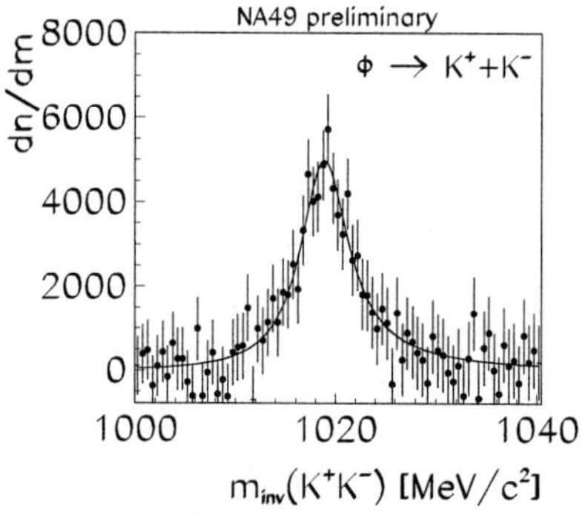

Fig. 5. K^+K^- invariant mass spectrum, background subtracted. The solid line shows a relativistic Breit-Wigner function convoluted with a Gaussian experimental mass resolution.

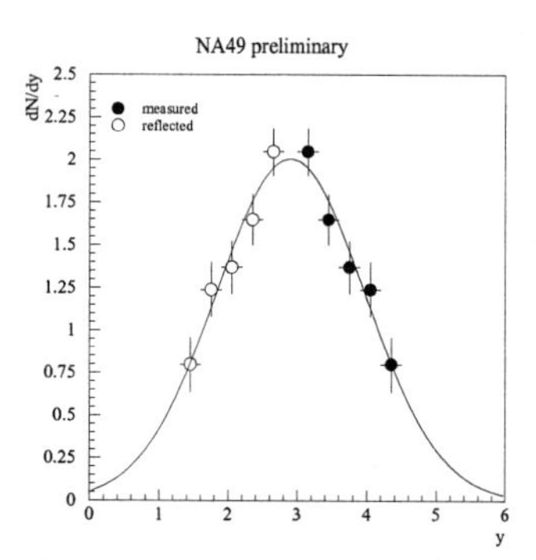

Fig. 6. Rapidity distribution of Φ. The solid circles are measured, the open circles are reflected at midrapidity ($y = 2.9$). The solid line is a Gaussian fit with $\sigma = 1.07$.

5 Expansion Dynamics

Transverse mass spectra of all particles except h^- (mainly π^-) are well reproduced by an exponential in m_T. Fig. 7 shows the inverse slope parameters T for various particles produced in central Pb+Pb collisions near midrapidity as a function of their respective mass. The inverse slope parameters increase with particle mass. Fig. 8 shows the inverse slope parameters for $p - \overline{p}$ and h^- produced in different collision systems (p+p, p+A, O+Au, S+A and Pb+Pb), characterized by the number of participants. For h^- the parameter T was deduced from the average transverse momentum: $T = (4/3\pi) \cdot < p_T >$ [5]. Pions show almost no increase with the system size, whereas heavier particles exhibit a clear increase of T, reaching values of about 300 MeV in central Pb+Pb collisions. Such high inverse slope parameters might be explained by a large transverse flow within the framework of a hydrodynamical model for the freeze-out stage. A fit to the m_T-spectra yields a temperature of the system at thermal freeze-out of about 120 MeV and an averaged transverse flow velocity of about $v_\perp = 0.43c$ [6]. Although temperature and $v_\perp$ are anticorrelated and usually many $(T, v_\perp)$-values fit the data equally well, the dependence of the two-pion Bose-Einstein correlation on the average transverse momentum of the pion pair can constrain the range of transverse flow velocities.

The correlations between the momenta of two identical bosons have been shown to provide detailed and largely model independent information on the

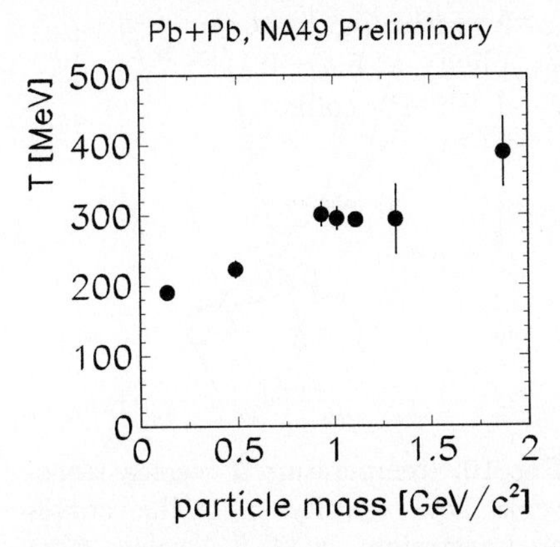

Fig. 7. Inverse slope parameters of various particles produced in central Pb+Pb collisions at SPS energy.

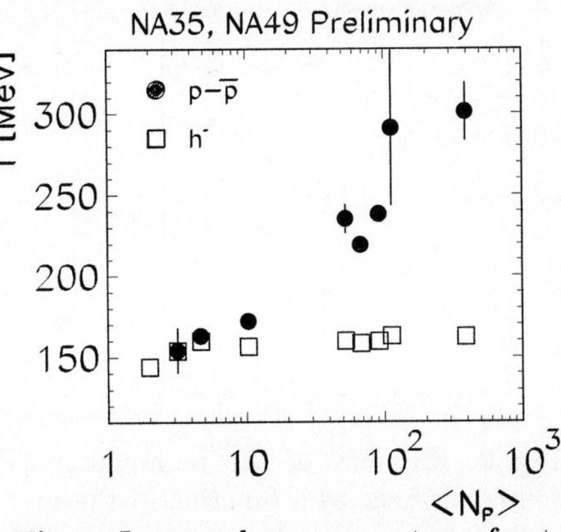

Fig. 8. Inverse slope parameters of net protons and negative hadrons (from $< p_T >$, see text) for various target–projectile systems as a function of the number of participants.

space-time properties of the fireball at thermal freeze-out from which the bosons are emitted. The Yano-Koonin-Podgoretski scheme for the decomposition of the correlation function allows the disentanglement of the spatial and temporal properties of the source [7]. It describes the source as consisting of homogeneous regions parametrized by their space-time extent ($R_{\|}(k_\perp, y_{\pi\pi})$, $R_\perp(k_\perp, y_{\pi\pi})$, $R_0(k_\perp, y_{\pi\pi})$) and their average longitudinal velocity $\beta(k_\perp, y_{\pi\pi})$, where $k_\perp$ and $y_{\pi\pi}$ are the average transverse momentum and rapidity of the pion pairs. The rapidity of the source element that emits pions as a function of the average rapidity of the emitted pion pairs is shown in Fig. 9 [8]. The source rapidity is nearly proportional to the pair rapidity, suggesting a boost-invariant longitudinal expansion of the source. The duration of the pion emission R_0 is $3 - 5$ fm/c, the lifetime of the system can be estimated to be $7 - 9$ fm/c [8, 9].

The k_T dependence of the radius parameters and the single particle momentum spectra can be described within the framework of a hydrodynamically motivated model for the freeze-out stage [10, 11] Fig. 10 shows the $(T, v_\perp)$-plane with constraints on possible $(T, v_\perp)$-values from m_T-spectra and correlation measurements. Temperatures of 120 MeV and average transverse flow velocities of $0.5 - 0.6c$ are consistent with both the transverse mass spectra and the k_T dependence of the $R_\perp$ parameter. One should, however, take into account that resonances have not yet been included in the model and that (systematic) errors have also not been estimated.

6 Summary and Conclusion

Central Pb+Pb collisions create energy densities of $2 - 3$ GeV/fm^3. Particle ratios show an enhanced production of strangeness. The Pb+Pb collision

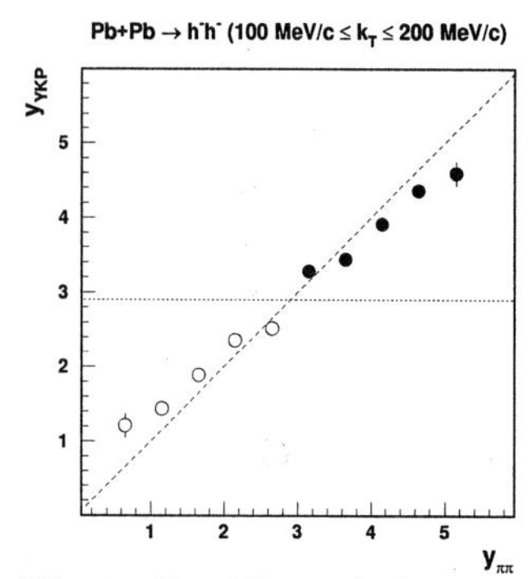

Fig. 9. Rapidity of the pion-emitting source element as a function of the rapidity of the emitted pion pairs in central Pb+Pb collisions.

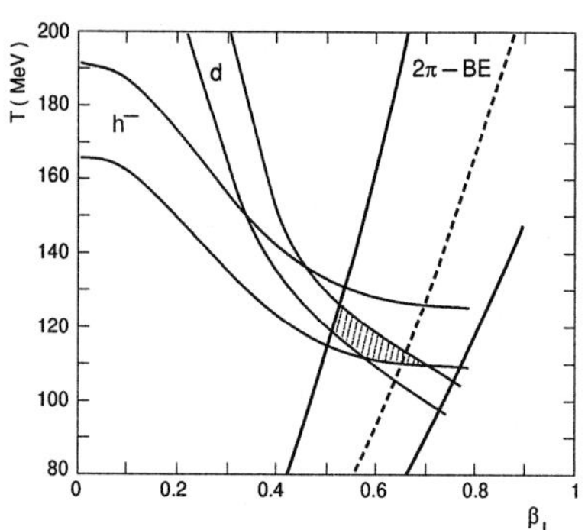

Fig. 10. Temperature T versus transverse flow velocity $\beta_\perp$. The curves are constraints on $(T, \beta_\perp)$ values from m_T-spectra and from two particle correlations.

system at the very end of its strong interaction phase (thermal freeze-out) expands almost boost-invariant longitudinally, shows a significant transverse flow and a low temperature of about 120 MeV. Single particle spectra, two particle correlations and particle ratios can map the evolution of the Pb+Pb system on its way towards thermal freeze-out and may provide the means for tracing the trajectory of the system back to a very early stage by dynamical models.

References

[1] P. Jones et al. (NA49 Collaboration), Nucl. Phys. A610 (1996) 188c.
[2] J. Bächler et al. (NA35 Collaboration), Phys. Rev. Lett. 72 (1994) 1419; D. Röhrich et al. (NA35 Collaboration), Nucl. Phys. A566 (1994) 35c.
[3] T. Alber et al. (NA49 Collaboration), Phys. Rev. Lett. 75 (1995) 3814.
[4] T. Alber et at. (NA35 Collaboration), Phys. Lett. B366 (1996) 56.
[5] R. Hagedorn, Ref. TH.3684-CERN (1983).
[6] B. Kämpfer, preprint FZR-149, hep–ph/9612336.
[7] F. Yano and S. Koonin, Phys. Lett. B78 (1978) 556; M. Podgoretski, Sov. J. Nucl. Phys. 37 (1983) 272.
[8] H. Appelshäuser, PhD thesis, Universität Frankfurt, 1996.
[9] S. Schönfelder, PhD thesis, Technische Universität München, 1997.
[10] J. Sollfrank et al., Z. Phys. C52 (1991) 593; E. Schnedermann et al., Phys. Rev. C48 (1993) 2462; U. Wiedemann et al., ncul–th/9611031 preprint; S. Chapman et al., Heavy Ion Physics I 1 (1995) 1; U. Heinz, Nucl. Phys. A610 (1996) 264c.
[11] S. Chapman et al., Phys. Rev. C52 (1995) 2694; U. Wiedemann et al., Phys. Rev. C53 (1996) 918.

PHOBOS Experiment at the RHIC Collider

Barbara Wosiek (barbara.wosiek@ifj.edu.pl)

Institute of Nuclear Physics
Kawiory 26a, 30-055 Krakow, Poland
for the PHOBOS Collaboration:
B.Back[a], M.D.Baker[f], D.Barton[b], R.Betts[a,i], A.Bialas[e], C.Britton[h], A.Budzanowski[d], W.Busza[f], A.Carroll[b], Y.-H.Chang[g], A.E.Chen[g], T.Coghen[d], W.Czyz[e], P.Decowski[f], K.Galuszka[d], R.Ganz[i], E.Garcia-Solis[j], K.H.Gulbrandsen[f], S.Gushue[b], C.Halliwell[i], P.Haridas[f], A.Hayes[k], R.Holynski[d], B.Holzman[i], U.Jagadish[h], E.Johnson[k], J.Kotula[d], P.Kulinich[f], M.Lemler[d], W.Lin[g], S.Manly[l], D.McLeod[i], J.Michalowski[d], A.Mignerey[j], M.Neal[f], A.Olszewski[d], R.Pak[k], H.Pernegger[f], M.Plesko[f], L.P.Remsberg[b], G.Roland[f], L.Rosenberg[f], A.Sanzgiri[l], P.Sarin[f], P.J.Stanskas[j], S.G.Steadman[f], P.Steinberg[f], G.S.F.Stephans[f], M.Stodulski[d], C.Taylor[c], A.Trzupek[d], G.Van Nieuwenhuizen[f], R.Verdier[f], B.Wadsworth[f], F.Wolfs[k], D.Woodruf[f], B.Wosiek[d], K.Wozniak[d], A.Wuosmaa[a], B.Wyslouch[f], K.Zalewski[e]
[a] Argonne National Laboratory, [b] Brookhaven National Laboratory, [c] Case-Western Reserve University, [d] Institute of Nuclear Physics, Krakow, [e] Jagellonian University, Krakow, [f] Massachusetts Institute of Technology, [g] National Central University, Taiwan, [h] Oak Ridge National Laboratory, [i] University of Illinois at Chicago, [j] University of Maryland, [k] University of Rochester, [l] Yale University

Abstract. PHOBOS is a compact silicon detector designed to study the physics of Au - Au collisions at a center-of-mass energy of 200 A GeV at the Relativistic Heavy Ion Collider. The main goal of the PHOBOS experiment is a comprehensive study of various hadronic observables that should reveal the onset of new phenomena expected to occur in central collisions of gold nuclei. In this paper the detector and the physics program of the PHOBOS experiment is presented.

1 Introduction

In heavy ion collisions at the energies of the Relativistic Heavy Ion Collider (RHIC) one expects to produce large volumes of high energy density. Under such conditions Quantum ChromoDynamics predicts [1] a formation of a new phase of matter: a deconfined, chirally symmetric plasma of quarks and gluons (QGP). The number of hadronic and leptonic observables have been proposed as signatures of the QGP formation. These predicted signatures are, however, based on a simplified description of the plasma which depends essentially on the assumption that local equilibrium is quickly attained. In view of this, the initial experimental program at the RHIC was planned to allow the study of all aspects of heavy ion collisions. This program will be carried out by four heavy ion experiments: the two large experiments (STAR and PHENIX) provide wide physics capabilities, the other two, smaller experiments (PHOBOS and BRAHMS) have unique features which complement the performance of the large experiments.

The distinguishing feature of the PHOBOS detector is its capability of measuring particles with very low transverse momenta (p_t) in the central rapidity region, the region of high energy density where the presence of new physics is most likely to show. Low p_t particles are directly related to large time and distance scales, not attainable in hadron-hadron collisions. Thus they provide a sensitive probe for revealing the existence of a QGP or any other collective phenomena.

2 PHOBOS detector

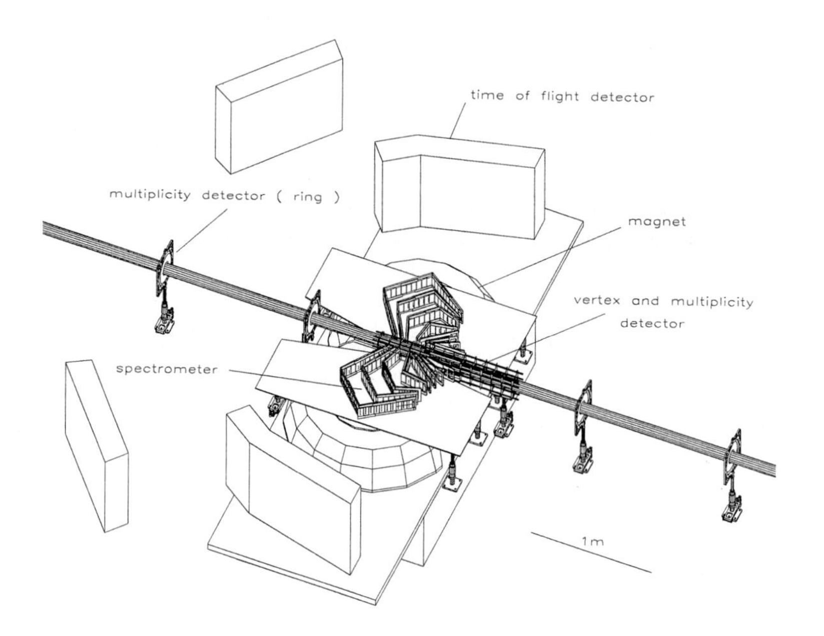

Fig. 1. Schematic view of the PHOBOS detector.

The PHOBOS detector [2] is almost exclusively constructed of silicon pad and strip detectors. The apparatus, shown in Fig.1, consists of two main subsystems, a multiplicity array and a two-arm multiparticle spectrometer. The multiplicity array includes an octagonal strip detector centered on the nominal interaction point and 6 rings of silicon pads placed along the beam pipe at distances up to $\sim$ 5 m down and up-stream from the interaction point. It covers more than 10 units in pseudorapidity ($|\eta| \leq 5.3$) and 2π in azimuth. The multiplicity array incorporates also a vertex detector. The two-arm multiparticle spectrometer is located on either side of the beam pipe in a 2 Tesla magnetic field. Each spectrometer arm contains 14 planes of Si sensors. The spectrometer has in total ~ 1 % coverage in solid angle and covers the midrapidity region. A plastic scintillator time-of-flight system augments particle identification and momentum acceptance of the spectrometer.

3 Physics capability of the multiplicity array

The multiplicity array provides event-by-event measurements of particle multiplicity and pseudorapidity and azimuthal angle distributions. It has an excellent granularity in both angles which is determined by the 14,000 electronics read out channels. The geometry and segmentation of the silicon sensors allows measurements of particle densities in very small phase space domains, down to 0.1 η units and 0.2 radians in azimuth. These measurements allow the analysis of particle density fluctuations which may be associated with the phase transition and the subsequent expansion of the QGP. Monte Carlo studies, based on a full GEANT simulations of the detector and followed by a detailed analysis of background corrections [3], showed that the instrumental uncertainty for measuring large particle multiplicities is smaller than the statistical noise. The event-by-event measurements of pseudorapidity distribution are shown in Fig.2. A good agreement between the measured and generated η distributions is observed in the entire range of pseudorapidity. The data on global event properties, obtained from the multiplicity array, can be used to select interesting events for further detailed study in the PHOBOS spectrometer.

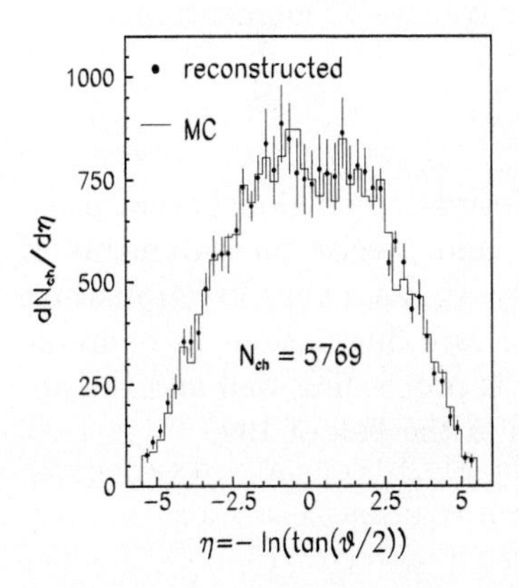

Fig. 2. Pseudorapidity distributions for a single central Au-Au collision.

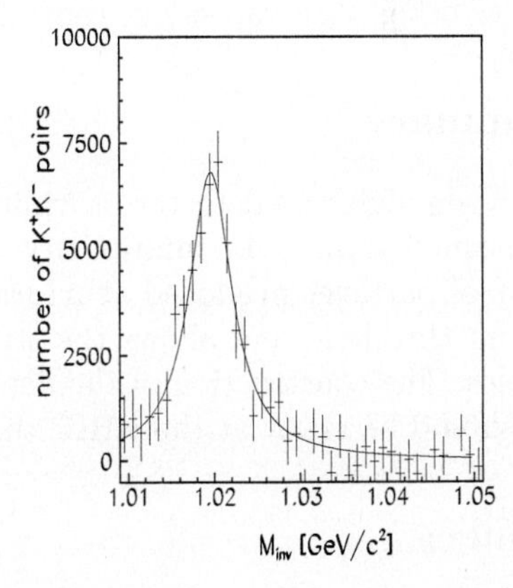

Fig. 3. The background subtracted distribution of M_{inv} with the Breit-Wigner fit.

4 Physics performance of the PHOBOS spectrometer

The high quality two-arm spectrometer covers the region of large energy density and will detect about 50 charged particles per central Au-Au collision. Its silicon sensors are coupled to about 135,000 electronics read out channels. The spectrometer provides precise measurements of particle momenta and particle identification in a wide kinematical range. The reconstruction efficiency of particle trajectories is better than 85 % and independent of the particle momentum. The momentum resolution is of the order of 5 MeV/c for pions and slightly larger for kaons and protons. The measurements of ionization energy loss in silicon sensors allow for separation of π/K up to about 0.650 GeV/c and protons and kaons up to 1.2 GeV/c. These momentum ranges can be extended by using the time of flight information [3]. The lowest transverse momenta accessible in the spectrometer are about 20 MeV/c for pions, 50 MeV/c for kaons and 70 MeV/c for protons and antiprotons.

The Monte Carlo study of the correlations between identical particles showed that we can reliably measure correlations down to very small relative momenta, and consequently derive sizes of emission sources as large as about 20 fm [3]. The ability of our detector to measure the parameters of Φ mesons decaying into charged kaons is shown in Fig.3, where the background subtracted invariant mass distribution of kaon pairs, obtained after one week of running at the nominal luminosity, is presented. The resulted resolution in the invariant mass is of the order of 1.6 MeV/c^2, and the uncertainty in the measured width is $\sim$ 1.0 MeV. The fact that we can achieve such good precision, with the small acceptance of the spectrometer, is due to our high data taking rate, the highest for any of RHIC experiments. For a full year of running we should be able to measure the mass and width of Φ decay as a function of other observables like centrality or transverse momentum.

5 Summary

PHOBOS is a high rate detector providing information on global event properties obtained from a 4π multiplicity array and precise measurements of about 1% of particles produced at midrapidity. It has a very low transverse momentum threshold and allows the study of rare fluctuations in hadronic observables. The construction of the detector is proceeding well and the apparatus should be ready at the RHIC turn on in the Fall of 1999.

References

[1] See Proc. of QM'96 Conference, Nuclear Physics A610 (1996)
[2] PHOBOS Conceptual Design Report, (1994)
[3] A. Trzupek and the PHOBOS Collaboration, Acta Physica Polonica B27 (1996),3103- 3111

Heavy Ion Physics at LHC with the CMS Detector

Ramaz Kvatadze (Ramazi.Kvatadze@cern.ch) ,

Institut de Physique Nucléaire de Lyon, France

Abstract. The CMS detector with its high quality central tracker, large acceptance muon system and fine granularity of calorimeters is well adapted for specific heavy ion studies. The possibility of the detection of dimuons and high transverse energy jets in heavy ion collisions are investigated.

1 Introduction

The main motivation for heavy ion experiments at LHC is to study strong interaction thermodynamics and phase transition from hadronic matter to the plasma of deconfined quarks and gluons (QGP). Among the proposed probes heavy quarkonium ($c\bar{c}$), ($b\bar{b}$) and high transverse momentum jet production are especially promising, because they are produced at early stages of the collision process and carry out information about the evolution of the produced system [1, 2].

2 General conditions for $Pb-Pb$ collisions and impact parameter measurement

At LHC in the case of Pb nuclei the energy per nucleon pair will be 5.5 TeV and expected luminosity for a single experiment is $L \approx 10^{27}$ cm^{-2} s^{-1}. The interaction cross-section for $Pb - Pb$ collisions is about 7.5 b.

For the study of total and transverse energy flow in nucleus-nucleus collisions the HIJING event generator is used [3]. It is shown that detected transverse energy strongly increases with decreasing impact parameter, thus allowing to measure the centrality of the collision and select central $Pb - Pb$ interaction events with the E_T trigger. The production cross-sections of hard processes in minimum bias nucleus-nucleus collisions are extrapolated from those in pp interactions according to the parametrization $\sigma_{AA} = A^{2\alpha} \times \sigma_{pp}$, with $\alpha = 0.9$ for (J/ψ, ψ'), $\alpha = 0.95$ for (Υ, Υ', Υ'') states, and $\alpha = 1.0$ for other hard processes. To estimate the influence of the soft secondary particles, hard process events are superimposed on the "thermal" background obtained from different parametrizations using HIJING, PYTHIA [4] and SHAKER [5] programs.

3 Dimuon detection

According to calculations [6] and to the Fermilab results [7], extrapolated to LHC energies and to $Pb-Pb$ collisions, we assume the following values of $\sigma \times BR(\rightarrow \mu^+\mu^-)$: for J/ψ state production 140 mb, while for $\Upsilon : \Upsilon' : \Upsilon'' = 410 : 120 : 41\mu$b. The resonances are accepted if both decay muons cross any of the muon stations. The dependence of Υ state acceptance on its transverse momentum is rather flat in both parts of the detector: barrel $|\eta| < 1.3$ and endcaps $1.3 \leq |\eta| \leq 2.4$. The acceptance for the J/ψ state is strongly suppressed in the barrel part for $P_T < 5$ GeV/c. The integrated acceptances for full CMS detector are 7% and 39% for J/ψ and Υ states respectively. The efficiency to reconstruct dimuons in the pseudorapidity range $|\eta| \leq 0.8$ from Υ decay is 66% for most central $Pb-Pb$ collisions (8000 charged particle per unit of rapidity) with the purity larger than 95% [8]. Fig. 1 shows dimuon mass distribution for J/ψ and Υ mass regions in minimum bias $Pb-Pb$ collisions. The main contribution to the background under the Υ peak comes from uncorrelated muon pairs from π/K decays (76% in the barrel part). Expected numbers of heavy quark states in $Pb-Pb$ collisions for two weeks of running time detected in different parts of the detector are presented in table 1 together with the signal/background ratios [9]. The S/B ratio for J/ψ state in full CMS detector is rather low 0.17. For Υ state signal/background ratio ranges from 0.8 in the full detector to 1.6 for barrel part only. For lighter ions S/B ratio for Υ is much larger, for example in $Ca-Ca$ collisions this ratio is about 14 for the full detector. The obtained mass resolution, about 40 MeV, is enough to make clean separation between Υ and Υ' states.

Detector	Full CMS			Barrel		
Resonance type	J/ψ	Υ	Υ'	J/ψ	Υ	Υ'
Statistics	580000	55000	20000	31000	23000	8300
S/B	0.17	0.8		1.8	1.6	
$S/\sqrt{S+B}$	290	150		140	120	

Table 1. Expected statistics and signal/background ratios for J/ψ and Υ resonances in $Pb-Pb$ collisions.

In the high invariant mass region $M(\mu^+\mu^-) \geq 20$ GeV main sources of dimuons are Drell-Yan and Z boson production and semileptonic decays of heavy flavours ($c\bar{c}$,$b\bar{b}$). Fig. 2 presents $\mu^+\mu^-$ pair invariant mass distribution for muons with $P_T > 5$ GeV/c. The expected number of Z decays in ± 10 GeV mass window in $Pb-Pb$ collisions for two weeks of running time is 11000 with low background ($\approx 4\%$). In the mass range 20 GeV$\leq M(\mu^+\mu^-) \leq 50$ GeV the dominant contribution comes from $b\bar{b}$ fragmentation (about 75%). The

Drell-Yan and $c\bar{c}$ processes are giving nearly equal number of $\mu^+\mu^-$ pairs in this range – 15% and 10% respectively.

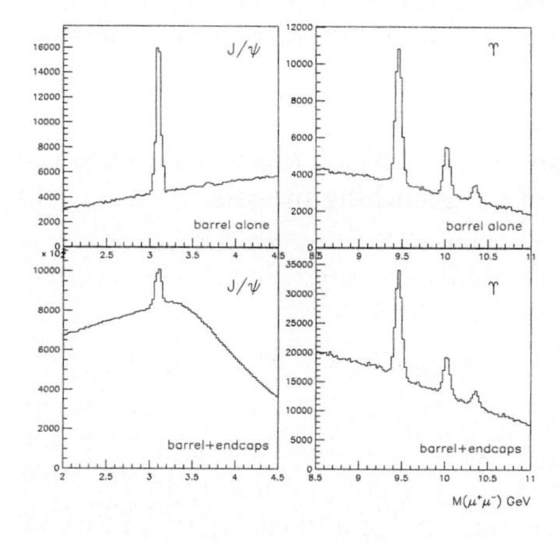

Fig. 1. Dimuon mass spectra for J/ψ and Υ mass regions in different parts of detector.

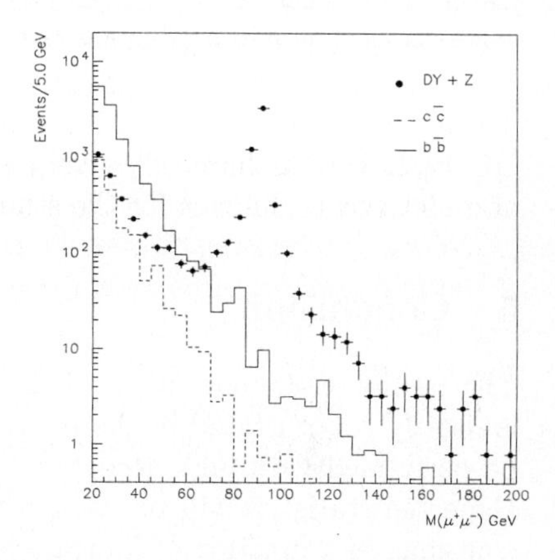

Fig. 2. Invariant mass spectra of $\mu^+\mu^-$ pairs.

4 Jet recognition and production rates in central *Pb* – *Pb* collisions

To study the jet quenching process three different channels have been considered: QCD jet pair [10], Z($\rightarrow \mu^+\mu^-$)+jet [11] and γ+jet production. The jet recognition efficiency and production rates are studied only in the barrel part of the calorimeters ($|\eta| \leq 1.5$). The modified UA1-type jet finding algorithm is used in $\eta - \phi$ space. Table 2 presents the ratios of reconstructed to simulated jet numbers, relative contribution of "false" jets, transverse energy resolution and QCD jet production rates in central $Pb - Pb$ collisions for two weeks of running time. The jet reconstruction efficiency is close to 100%. Contribution from "false" jets is 12± 3% for $E_T \geq 50$ GeV and becomes negligible for higher transverse energies. The resolution in jet transverse energy improves from 16.7% for $E_T \geq 50$ GeV to 11.6% for $E_T \geq 100$ GeV. For jets with transverse energy lower than 50 GeV background contribution increases rapidly and the energy resolution is worse.

Estimated statistics for Z($\rightarrow \mu^+\mu^-$)+jet and γ+jet production with transverse momentum of Z(γ) and jets greater than 50 GeV/c are 600 and 36500 events respectively in minimum bias $Pb-Pb$ collisions for the same period of running time. However, for the γ+jet events background contribution from

$E_T \geq$, GeV	ε_1	ε_2	$\sigma(E_T)/E_T$,%	Rate
50	0.94 ± 0.03	0.120 ± 0.030	16.7	$1.7 \cdot 10^7$
100	1.03 ± 0.02	0.010 ± 0.004	11.6	$1.2 \cdot 10^6$

Table 2. Reconstruction efficiency ε_1, contribution of "false" jets ε_2, transverse energy resolution and production rates, for jets with different transverse energies.

QCD jets with isolated π^0 is very high (about 100%), which makes the use of γ+jet events difficult for the study of jet quenching process.

5 Conclusion

The results of simulation shows that the acceptance for J/ψ state detection is mostly concentrated in the endcaps. Detection of $\Upsilon \rightarrow \mu^+\mu^-$ decays can be performed with high acceptance in large pseudorapidity range $|\eta| \leq 2.4$. Expected statistics will be enough to study Υ state production in $Pb - Pb$ collisions as a function of its transverse momentum and centrality of events. At the same time about 11000 events of Z$\rightarrow \mu^+\mu^-$ decays will be detected and can be used as a benchmark process for understanding nuclear effects.

The jet recognition efficiency for $E_T \geq 50$ GeV jets is found to be close to 100% with background 12$\pm$ 3%. Production rate of QCD jets will be sufficient to study high E_T jet production for the different impact parameter of the collisions. Expected number of Z($\rightarrow \mu^+\mu^-$)+jet events in $Pb - Pb$ collisions is rather low only about 600 events. However, for lighter ions this process looks very promising because of much higher luminosities.

References

[1] H. Satz, in ECFA LHC Workshop Proceedings, CERN 90-10, ECFA 90-133 (1990), Vol. I, 188.
[2] C. Singh, Phys. Rep. 236 (1993) 147.
[3] X.-N. Wang and M. Gyulassy, Phys. Rev. D 44 (1991) 3501.
[4] T. Sjostrand, Computer Phys. Comm. 82 (1994) 74.
[5] F. Antinori, ALICE/MC 93-09.
[6] R. Gavai et al., CERN-TH.7526/94; BI-TP 63/94.
[7] F. Abe et al., (CDF Collab.), Fermilab-Pub-95/271-E.
[8] O. Kodolova and M. Bedjidian, CMS Note 1997/095.
[9] M. Bedjidian, CMS Conference Report 1997/008.
[10] M. Bedjidian, R. Kvatadze and V. Kartvelishvili, CMS Note 1997/082.
[11] V. Kartvelishvili, R. Kvatadze and R. Shanidze, Phys. Lett. B 356 (1995) 589.

Charm Transverse Momentum as a Thermometer of the Quark–Gluon Plasma

Benjamin Svetitsky (bqs@julian.tau.ac.il) and Asher Uziel

School of Physics and Astronomy, Raymond and Beverly Sackler Faculty of Exact Sciences, Tel Aviv University, 69978 Tel Aviv, Israel

Abstract. A charmed quark experiences drag and diffusion in the quark-gluon plasma, as well as strong interaction with the plasma surface. Our simulations indicate that charmed quarks created in heavy ion collisions will be trapped in the mixed phase and will come to equilibrium in it. Their momentum distribution will thus reflect the temperature at the confinement phase transition.

1 Introduction

Consider charm created in a high-energy nuclear collision. Some 99% of the charmed quarks created in hadronic collisions are *not* to be found in $c\bar{c}$ bound states, but rather in the open charm continuum. Much like their bound counterparts, the unbound charmed quarks are created early, move through the interaction region slowly, and react strongly with their environment. They should contain as much information about the collision region as the J/ψ, although this information may be harder to extract.

Assuming invariance of the collision kinematics under longitudinal boosts [1], a space-time picture shows that a charmed quark, created near $t = z = 0$ (the initial nucleus–nucleus collision), moves according to $z = v_0 t$ which is exactly a longitudinal streamline of the fluid. Thus there is not much information to be gained from the quark's eventual longitudinal momentum; it is to the transverse momentum that we turn in order to learn about the quark's interaction with the surrounding matter.

The quark is created in the nascent plasma and, assuming it is not created too near the edge, sees this plasma cool to the mixed phase and beyond, to the hadron gas which dissociates soon after. The mixed phase lasts a long time, typically ten times as long as the pure plasma which precedes it. Our calculation thus concentrates on the charmed quark's interaction with the mixed phase [2]. We track the quark's diffusion through the initial plasma, its hadronization upon emerging into the hadron phase, its collisions (as a D meson) with pions, and its possible reabsorption by a plasma droplet. For a given initial transverse momentum, we calculate the distribution of the final $p_\perp$ of the D meson. (For details of the following see [3].)

2 Simulation

Our simulation of the mixed phase is based on the cascade hydrodynamics code of Bertsch, Gong, and McLerran [4]. At proper time $\tau = \tau_p$, the uniform plasma breaks up into droplets of radius $r_0 = 1.0$ or 1.5 fm (a parameter of which we have little knowledge). Each droplet has a mass equal to its volume times the Stefan-Boltzmann energy density at the transition temperature, $M = (\frac{4}{3}\pi r_0^3) \times (\gamma T^{*4})$, and its temperature is fixed at T^*. Each droplet then radiates pions as a black body. At each pion emission, the droplet shrinks (to conserve energy) and recoils (to conserve momentum). Similarly, a droplet absorbs any pion that comes close enough, with the concomitant growth and recoil. Pion–pion collisions are included in the simulation as well. The dominant motion of the droplet–pion mixture is a rarefaction due to the longitudinal expansion imposed by the initial conditions. This leads to the eventual evaporation of the droplets.

A charmed quark is created inside the plasma at $\tau_0 < \tau_p$ and is allowed to diffuse according to a Langevin process [5]. The plasma cools as $T = T_i(\tau_i/\tau)^{1/3}$ until it reaches T^*, whereupon it breaks into droplets. (We make sure the quark is *inside* a droplet.) The quark continues to diffuse. If the quark hits the surface of a droplet, it stretches a flux tube out into the vacuum. This flux tube has a tension $\sigma = 0.16$ GeV^2 and a fission rate (per unit length) $d\Gamma/d\ell = 0.5$ to 2.5 fm^{-2}. (This is the range of values used in string-based event generator programs.) If the flux tube breaks in time, then the c quark finds itself to be a D meson outside the droplet; otherwise it is reflected back into the droplet.

If the quark indeed hadronizes, it is tracked as a D meson in the pion gas. If this D meson hits a droplet, it is absorbed: The light quark gets stripped off and the c quark proceeds back into the plasma.

Very often, a c quark coming to the surface of its droplet has insufficient energy to emerge as a meson. (It needs about 350 MeV.) In that case the quark is *trapped.* Unless it gains energy through diffusion or through droplet recoil, it will remain trapped until the droplet evaporates away much later.

3 Results

We show in Fig. 1 the mean $p_\perp$ of a D meson as a function of the initial $p_\perp$ of its parent c quark. Note the very weak dependence on the initial momentum. This feature, together with the width of the final-momentum distribution, indicates that the D meson is *thermalized.* The mean thermal $p_\perp$ of a D meson at $T^* = 150$ MeV is 820 MeV, and we attribute the higher $\langle p_\perp \rangle$ shown in the figure to the flow of the droplet fluid.

In interpreting our results we distinguish between two populations of charmed quarks, those that emerge from flux-tube fission and then escape the system—*fragmentation mesons*—and those that are trapped within their

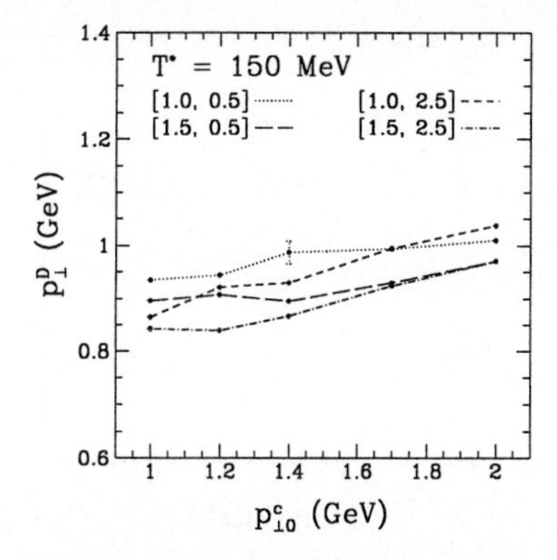

Fig. 1. RMS transverse momentum of D meson vs. initial transverse momentum of c quark, for transition temperature $T^* = 150$ MeV. The four sets of points correspond to different values of r_0 and $d\Gamma/d\ell$ (in fm and fm^{-2}, respectively) as shown. A typical statistical error bar is shown.

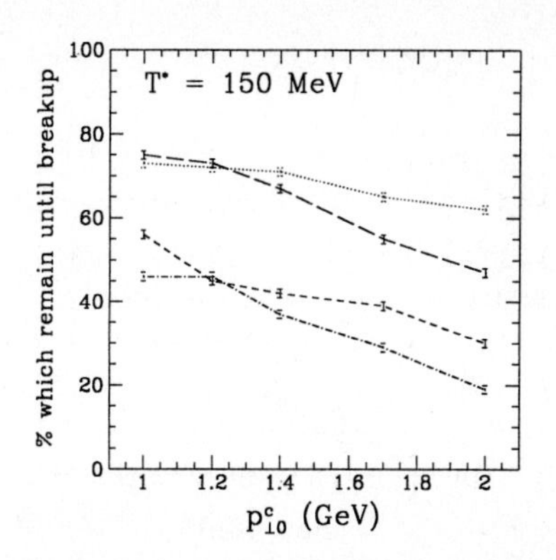

Fig. 2. The proportion of c quarks which are trapped inside their original droplets until their breakup. The four sets of points are for different values of r_0 and $d\Gamma/d\ell$ as in Fig. 1.

original droplets until the latter evaporate—*breakup mesons*. As can be seen in Fig. 2, between 40% and 80% of the quarks are trapped until breakup. Fig. 3 shows that the breakup mesons are well thermalized while the fragmentation mesons carry some memory of their quarks' initial momentum.

Finally, we compare results for two different transition temperatures. Fig. 4 shows a comparison of the predictions for T^*=150 MeV with those for T^*=200 MeV. The D transverse momentum shows a sensitivity to a temperature difference of 50 MeV which rises above the effects of the uncertainty in the two model parameters r_0 and $d\Gamma/d\ell$.

To summarize our qualitative conclusions:

1. The mean $p_\perp$ of a D meson is almost independent of the initial $p_\perp$ of the parent c quark.
2. Many c quarks (40% – 80%) are trapped until the droplets evaporate, and come to equilibrium at the transition temperature.
3. Other D mesons come close to equilibrium in the plasma and in the pion gas, and in any case decouple shortly after the transition is complete.
4. Recapture by the droplets (which we haven't discussed) is an important effect, affecting the equilibration of 10% – 40% of the particles.

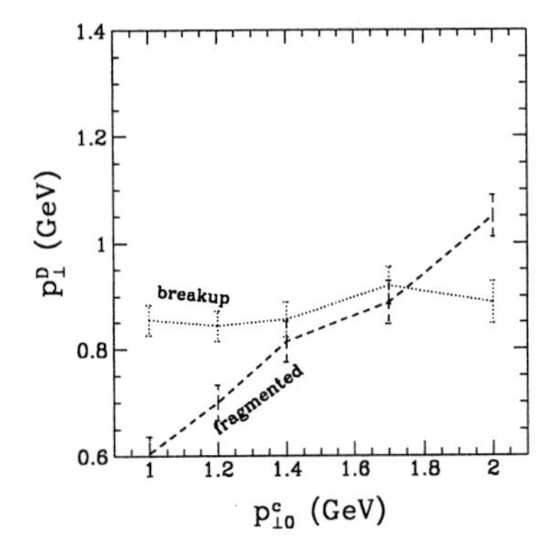

Fig. 3. Comparison of two populations of D mesons. The RMS $p_\perp$ of D mesons is plotted against the initial $p_\perp$ of the c quarks, for breakup mesons (dots) and for fragmentation mesons (dashes). Here $T^* = 150$ MeV, $r_0 = 1$ fm, $d\Gamma/d\ell = 2.5$ fm^{-2}.

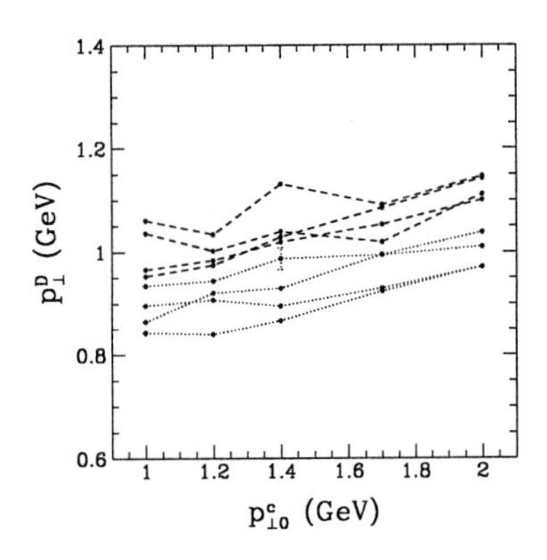

Fig. 4. Comparison of the mean D momentum for $T^* = 150$ MeV (dots, same data as in Fig. 1) with that for $T^* = 200$ MeV (dashes).

The distribution of D momenta is roughly thermal, but the mean momentum is above the thermal average because of transverse flow. Presumably the latter may be gauged by the momentum distributions of the lighter species. Everything considered, the $p_\perp$ of the D mesons does provide a rough thermometer of the phase transition.

Acknowledgments

This work was supported by a Wolfson Research Award administered by the Israel Academy of Sciences, and by the Basic Research Fund of Tel Aviv University.

References

[1] J. D. Bjorken, Phys. Rev. D **27**, 140 (1983).
[2] B. Svetitsky, Phys. Lett. **227B**, 450 (1989).
[3] B. Svetitsky and A. Uziel, Phys. Rev. D **55**, 2616 (1997).
[4] G. Bertsch, M. Gong, L. McLerran, V. Ruuskanen, and E. Sarkkinen, Phys. Rev. D **37**, 1202 (1988).
[5] B. Svetitsky, Phys. Rev. D **37**, 2484 (1988).

Strong Coupling Improved Equilibration in High Energy Nuclear Collisions

S.M.H. Wong (wong@theorie.physik.uni-wuppertal.de)

Fachbereich Physik, Universität Wuppertal, D-42097 Wuppertal, Germany

Abstract. In high energy nuclear collisions, most calculations on the early equilibration of the parton plasma showed that the system does not come close to full equilibrium, especially for the fermion components. However, since the system is constantly evolving, the cooling effect due to expansion will lead to a decrease of the average parton energies. So the interactions should be enhanced by a corresponding increase of the running coupling. We show that this leads to a faster and improved equilibration. This improvement is more important for the fermions compensating partially for their weaker interactions and slower equilibration.

1 Introduction

High energy nuclear collisions have become a subject of great interest in the last decade or so, bolstered by the experiments at Brookhaven AGS and CERN SPS and culminating in the constructions of the soon-to-be running larger higher energy colliders like RHIC and LHC. One attempts in these experiments to create and study deconfined matter. An obvious question is how to identify matter in the new phase. This is done by the various proposed particle signatures such as J/ψ suppression, strangeness enhancement and excess at low mass dileptons. A question of similar importance is the evolution of matter in the new state which influences directly these signatures produced during the deconfined phase and indirectly in modifying the initial conditions of the subsequent hydrodynamic expansion. Final hadron yields are closely related to the generated entropy which is mainly produced in the initial parton creation and subsequent parton interactions. This precisely depends on the time evolution. We study this in order to gain a better understanding of the evolution and the equilibration process.

The main questions of equilibration is whether thermalization is fast, if parton chemical equilibration can be completed before the start of the phase transition and the possible duration of the deconfined phase. Whereas approximate thermalization should be achieved in the first few fm/c as shown by models like parton cascade [1, 2], completion of chemical equilibration is impossible for all present existing models [3, 4, 5, 6]. As to the duration of the parton phase, it depends on various factors. Actually the answers to all three questions depend on the inherent uncertainties associated with these studies. They are the initial conditions, infrared cutoff parameter and the value of the coupling used. One can raise the question as to whether some of these can be exploited to determine better or improve the equilibration process such as parton chemical equilibration. To change the initial conditions to get better answers is too arbitrary and irrelevant since one can vary no more than the energy/nucleon and centrality of the collision in the experiments. Infrared cutoff, which is usual in perturbative QCD is actually not needed in a deconfined medium because of the screening effect of the latter. Finally the third possibility is the coupling. One usually uses a value of $\alpha_s = 0.3$ for an average momentum transfer of around 2.0 GeV. However, during the time evolution, the parton energies drop considerably due to the longitudinal expansion and particle creation. Therefore so will the average momentum transfer drops in time and the coupling, which is treated as a kind of average value here, cannot stay at a constant value. The exploitation of this effect will be the main subject of this talk.

2 Strong Coupling Improvements on Equilibration

Our method for this investigation is based on solving Boltzmann equation with the relaxation time approximation and explicit construction from perturbative QCD for the collision terms. All binary collisions and 2-to-3 gluon multiplication as well as the reverse process are included at the tree level [5, 6]. In accordance with our discussion in the introduction, we study the effect of the coupling on the equilibration. This is done by comparing results with various fixed values for the coupling and a time varying coupling obtained by using the one-loop running α_s formula and evaluating the coupling at the scale, at any time, of the average parton energy. The reason being that the average momentum transfer in parton scatterings should be of the order of the average parton energy. This latter approach allows the system to determine its interaction strength and removes an otherwise free parameter, whose value was usually chosen with perturbative QCD in mind. This will be an advantage as we will see presently because the results on equilibration do depend on the value of α_s used [7]. We use the initial conditions from HIJING for our time evolution after allowing for the resulting parton gas to free stream to an isotropic distribution [4].

To examine the state of the equilibration, we plotted in Fig. 1, the quark and gluon fugacities and their temperature estimates as a function of time and in Fig. 2 the pressure ratios, which is a check of kinetic equilibration. The various curves are for $\alpha_s = 0.3, 0.5, 0.8$ and for the time evolving coupling α_s^v. As can be seen in Fig. 1, larger coupling leads to faster chemical equilibration. The fugacities for both gluons and quarks at $\alpha_s = 0.8$ rise faster than those at $\alpha_s = 0.5$, which in turn, are faster than those at $\alpha_s = 0.3$. For gluon, the fugacities at around 4.0 fm/c at LHC and at around 6.0 fm/c at RHIC in all cases are already reasonably close to 1.0, whereas those for quarks vary widely with the value of α_s. Larger α_s gives much better results for quarks. The curves for α_s^v tend to move in time across those with fixed α_s because

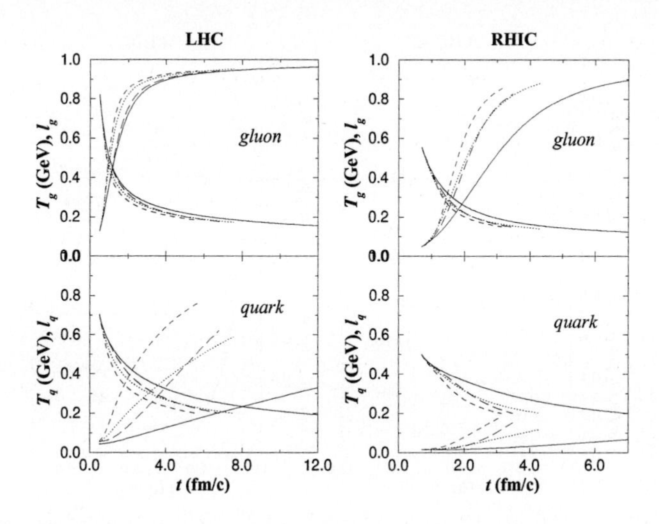

Fig. 1. Time evolution of parton fugacities and temperature. Solid, dotted, dashed and long dashed are for $\alpha_s = 0.3$, 0.5, 0.8 and α_s^v, respectively.

of the drop in the average parton energies due to the expansion and parton creation which lead to a decrease of the average momentum exchange and hence an increase of α_s^v. The resulting curves then tend to move from curves of smaller to larger α_s. Thus if we use α_s^v, which seems to be a more consistent choice, parton chemical equilibration will be faster and improved significantly in the case of quarks. For gluon, it is hard to get any obvious improvement when chemical equilibration is already very good near the end of the time evolution.

In the case of kinetic equilibration, something similar happens. In Fig. 2, the ratio of longitudinal to transverse pressure as well as one third of the energy density to transverse pressure are plotted. When the momentum distribution is isotropic as in a thermalized system, these ratios are 1.0. This is the case at the start of the evolution because of our momentarily thermalized initial conditions. As the evolution starts, the expansion pushes the plasma out of equilibrium and isotropy is lost. The ratios are seen to deviate from 1.0. At some point in time, net interactions become fast enough in response to the expansion and bring the plasma back towards equilibrium so the curves make a turn and rise again towards isotropy.

In the initial stage, there is not much difference between cases with different values of α_s because this is the expansion dominated phase. Larger α_s, however, causes the interactions to take over sooner and achieve at the end a better degree of kinetic equilibration both for quarks and for gluons. Like chemical equilibration, only quarks but not gluons show considerable improvement in the end degree of kinetic equilibration. Again, α_s^v is better equilibrated than that of the $\alpha_s = 0.3$ case due to the increasing α_s.

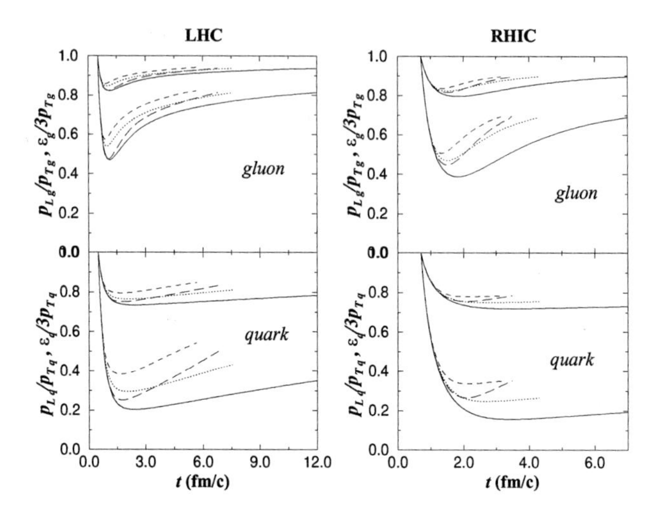

Fig. 2. Time evolution of the pressure ratios (bottom set of 4 curves in each plot) and energy to pressure ratios (upper set of 4 curves) . The assignment of the values of α_s to the curves are the same as in Fig. 1.

These improvements are, however, accompanied by a reduction of the duration of the parton phase resulting from more rapid cooling (Fig. 1). This comes about because of enhanced gluon multiplication and conversion into quark-antiquark pairs while the quarks have to expand against a stronger pressure. So a consistent $\alpha_s = \alpha_s^v$ speeds up and improves equilibration for the partons but only at the expense of the duration of the deconfined phase.

In summary, because different values of the coupling have rather large effects on equilibration, we argued that a time evolving averaged coupling, which reflects the decrease of the average parton energies of the system, should be used in studying the time evolution in high energy nuclear collisions. This more consistent approach enhances the equilibration process both in speed and in bringing the system closer to full equilibration.

References

[1] K. Geiger and B. Müller, Nucl. Phys. B 369 (1991) 600.
[2] K. Geiger, Phys. Rev. D 46 (1992) 4965, 4986.
[3] K. Geiger and J.I. Kapusta, Phys. Rev. D 47 (1993) 4905.
[4] T.S. Biró, E. van Doorn, B. Müller, M.H. Thoma and X.N. Wang, Phys. Rev. C 48 (1993) 1275.
[5] S.M.H. Wong, Nucl. Phys. A 607 (1996) 442.
[6] S.M.H. Wong, Phys. Rev. C 54 (1996) 2588.
[7] S.M.H. Wong, Phys. Rev. C 56 (1997) 1075.

Jupiter Effect

Peter Filip (filip@savba.sk)

Comenius University Bratislava, SK-84215, Slovakia

Abstract. Specific results of the computer simulation of dilepton production from expanding pion gas created in Pb+Pb 160 GeV/n collisions are presented. Azimuthal asymmetry of dilepton pairs in non-central collisions and interesting shape of the rapidity distribution of dilepton pairs are predicted. These results are explained on theoretical level as a consequence of momentum and space asymmetries in the initial state of pion gas without any assumption of thermalization. Implications on the production of dileptons in pre-hadronic phase of HIC are drawn.

1 Introduction

Goal of heavy ion collision (HIC) experiments is to reveal properties of the compressed hot nuclear matter created during the collision of heavy nuclei. Final momentum distributions of most abundant particles - hadrons are however influenced during the dilute and late freeze-out stage of heavy ion collision. Interesting information about the early stage of the collision is in this case hidden by subsequent collective effects of strongly interacting hadrons.

Fortunately this is not valid for leptons or photons. Distributions of these types of particles can provide us with more direct information about the early stages of the collision process.

In most of theoretical estimates for production of dileptons from the hadronic matter created in HIC experiments an assumption about the equilibrium - thermalized stage of the collision is used.

Our study of dilepton production from the expanding pion gas is not based on the assumption of thermalization. Phenomena described in next sections are generated also in the case of 1 collision per particle approximation what means that the mean free path of particles participating in mutual interactions is comparable with the size of the system.

In subsequent sections we report about two phenomena revealed by the computer simulation of the dilepton production from expanding pion gas created in Pb+Pb 160 GeV/n collisions: a) Azimuthal asymmetry of dilepton pairs in non-central collisions, b) Rapidity distribution of dilepton pairs.

2 Azimuthal Asymmetry of Dilepton Pairs

Azimuthal asymmetries of secondary produced particles have been clearly identified in relativistic non-central collisions [1]. This phenomenon is well understood as a consequence of collective behavior of nuclear matter [2] or explained by the absorption of secondary produced particles in spectator parts of nuclei [3]. Recently also azimuthal asymmetries in transverse momentum distributions of less abundant hadrons - K mesons and Λ baryons [4] have been studied in HIC experiments. However azimuthal asymmetries in transverse momentum distributions of dileptons have not been addressed experimentally or theoretically so far.

From theoretical point of view mechanisms generating azimuthal asymmetries in transverse momentum distributions of hadrons are not applicable for dileptons. Dileptons leave freely collision volume after being produced without final state interactions or absorption processes.

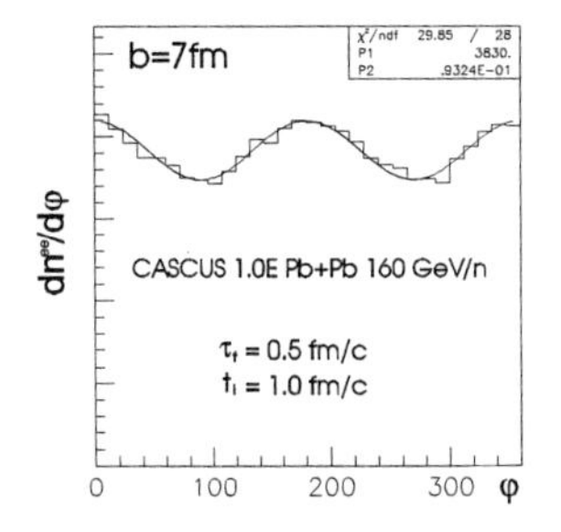

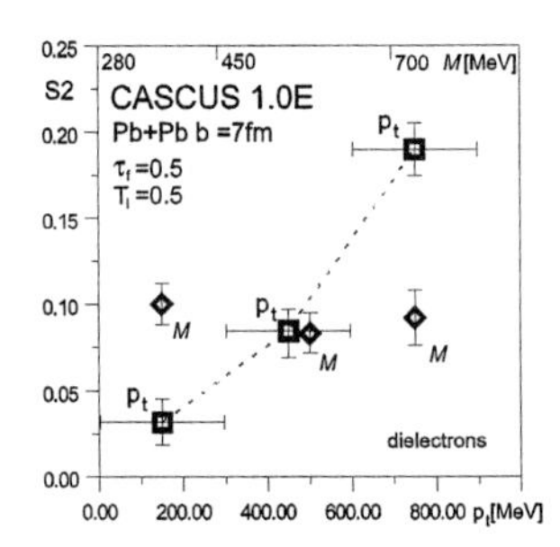

Fig. 1. Azimuthal asymmetry of dilepton pairs and its transverse momentum and invariant mass dependence for Pb+Pb 160 GeV/n $b = 7$fm events.

In spite of this computer simulation [5] predicts significant second-order asymmetry in transverse momentum distribution of dilepton pairs. Result shown in Fig.1 is obtained by the simulation of Pb+Pb 160 GeV/n non-central b=7fm events. The fit of the histogram shown in Fig.1 to the function:

$$R(\phi) = S_0[1 + S_2 \cdot \cos(2\phi)] \quad (1)$$

gives numerical value of the asymmetry coefficient $S_2 = 0.093 \pm 0.004$. Corresponding value of R_p parameter $R_p \approx \langle p_y^2 \rangle / \langle p_x^2 \rangle = 1.202$. Asymmetry of dilepton pairs is oriented *in* the reaction plane, it increases with p_t and does not depend on invariant mass region. Theoretical understanding of the origin of this asymmetry is sketched in Section 4 and studied more carefully in [6].

3 Rapidity Distribution of Dilepton Pairs

Rapidity distribution of dilepton pairs produced via $\pi^+\pi^-$ annihilation channel is determined by the rapidity distribution of momentum sum $p_{\pi^+} + p_{\pi^-}$ of pions annihilating. Result of simulation [5] is shown in Fig.2 where also rapidity distribution of pions participating in the rescattering process is shown.

For parameters of the simulation $\tau_f = 0.5$ fm, $T_i = 1.0$ fm number of collisions per pion is close to 2.5 and minimum in the rapidity distribution of dileptons is strong. For higher collision rates (smaller values of τ_f and T_i) the minimum becomes weaker and for $n_c = 10$ coll./π the minimum disappears.

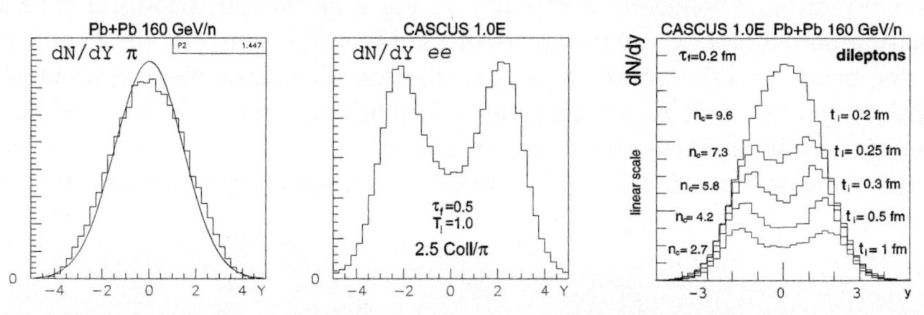

Fig. 2. Rapidity distribution of pions and dilepton pairs predicted by the simulation of Pb+Pb 160 GeV/n collisions for different values of τ_f, T_i parameters.

From experimental point of view data on rapidity distribution of dileptons produced in $p - A$ or $A - A$ collisions are rather rare. It seems that the only published rapidity distribution of dileptons which seem to originate from $\pi^+\pi^-$ annihilation was obtained in pioneering experiments of DLS collaboration at Bevalac accelerator in Berkeley [7].

4 Theoretical understanding of results

Azimuthal asymmetry of dilepton pairs and rapidity distribution of dileptons obtained in the simulation can be understood without the assumption of thermalization. Initial transverse momentum distribution of pions is azimuthally symmetrical - constant in the simulation. Therefore azimuthal asymmetry of dileptons is not generated by transverse momentum asymmetry of pions. It is a consequence of the initial spatial distribution of pions in transverse plane which is asymmetrical in the case of non-central collisions.

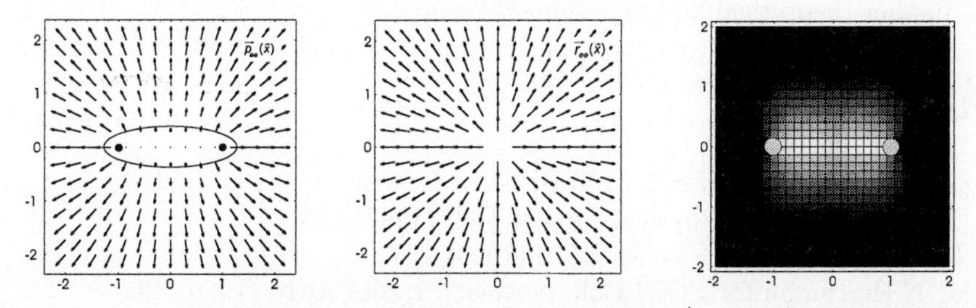

Fig. 3. Vector field $p_{ee}(x, y)$ of dilepton pairs created by $\pi^+\pi^-$ annihilation of pions emitted from two sources. The two-point source serves as a rough approximation of the asymmetrical shape of the initial distribution of pions in transverse plane. Density-plot of the difference $|p_{ee}(x, y) - r_{ee}(x, y)|$ ($r_{ee}(x, y)$ is radial field) shows that the asymmetry is generated between emission points - in the volume of source.

Rapidity distribution of dilepton pairs can be explained as a consequence of the asymmetry of pions in momentum space. Average transverse momentum of pions in Pb+Pb 160 GeV/n collisions is much smaller compared to average longitudinal momentum of pions. This strongly influences distribution of pion-pion collisions in rapidity. In Fig.4 we show distribution of annihilation points of $\pi^+\pi^-$ pairs emitted from two-point source oriented parallel to the beam direction. Momentum distribution of emitted pions is asymmetrical: $< |p_l| > \approx 2 < p_t > = 380$ MeV. Annihilation events in the middle of the sources are suppressed due to the shape of $\pi^+\pi^- \to e^+e^-$ cross section which is peaked at $M = 770$MeV (detailed description can be found in [6]).

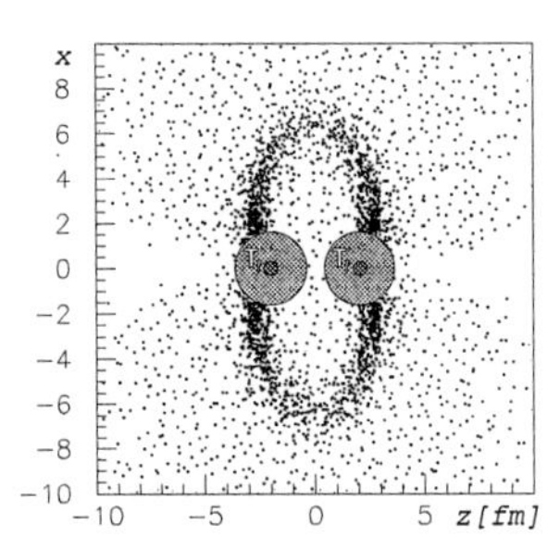

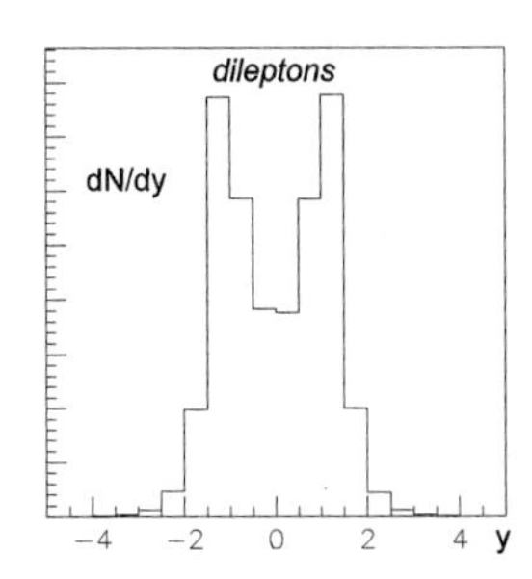

Fig. 4. Distribution of annihilation points of pions emitted from two-point source described in the text and rapidity distribution of the generated dilepton pairs. Formation time T_f [8] excludes annihilations close to emission points.

5 Conclusions

Phenomena described in preceding sections are generated in pre-equilibrium stage of the pion gas expansion. Similar effects might occur also in parton gas possibly created in heavy ion collision experiments. In this case the azimuthal asymmetry of dilepton pairs in the invariant mass region 2-5 GeV might be considered as a signature of secondary collisions [9] among partons. Gas-like behavior of the system of partons [10] created in HIC experiments is worth of further theoretical and experimental study.

References

[1] J.Barrette et al. E877 Coll. Physical Review C55 (1997) 1420
[2] J.-Y.Ollitrault, Physical Review D46 (1992) 229
[3] B.A.Li, Nuclear Physics A570 (1994) 797
[4] N.Herrmann for FOPI Coll. Nuclear Physics A610 (1996) 49c
[5] P.Filip, Acta Physica Slovaca 47 (1997) 53; nucl-th/9709075
[6] P.Filip, PhD Thesis, Comenius University Bratislava (1997)
[7] DLS Coll. Phys.Rev.Lett 61 (1988) 1069; Phys.Rev.Lett. 62 (1989) 2652
[8] J.Pišút et al. Zeit.f.Phys. C67 (1995) 467
[9] P.V.Ruuskannen, Quark-Gluon Plasma 1; World Scientific (1993) 519
[10] T.Wienold et al. Discussion in Jupiter Bar (1996) Berkeley

Multiplicity and Transverse Energy Distributions Associated to Rare Events in Nucleus–Nucleus Collisions

J. Dias de Deus[1], C. Pajares, C. A. Salgado

Depto. Física de Partículas, Universidade de Santiago de Compostela,
15706 SANTIAGO DE COMPOSTELA, SPAIN

Abstract. It is shown that in high energy nucleus-nucleus collisions the transverse energy or multiplicity distribution P_C, associated to the production of a rare, unabsorbed event C, is universally related to the standard or minimun bias distribution P by the equation $P_C(\nu) = \frac{\nu}{<\nu>}P(\nu)$ with $\nu \equiv E_T$ or n. Deviations from this formula are discussed, having in view the formation of the plasma of quarks and gluons. This possibility can be distinguished from absorption or interaction with comovers, looking at the curvature of the J/Ψ over Drell-Yan pairs as a function of E_T.

Particle production in hadron-hadron, hadron-nucleus and nucleus-nucleus collisions is generated by superposition of elementary, particle emitting, collisions. Assuming that the elementary collisions may be treated as equivalent, particle production fluctuations will contain a contribution from the fluctuation in the number of elementary collisions and from the fluctuations in particle production resulting from the elementary collisions. The fact that the width of the nucleus-nucleus particle distribution is very large and larger than the hadron-hadron distribution and this larger than e^+e^- distribution is an indication that multiparticle fluctuations in nucleus-nucleus collisions are dominated by fluctuations in the number ν of elementary collisions [1] (this is true also for hadron-hadron collision [2]). Therefore, we can write

$$<n>=<\nu>\bar{n} \; ; \qquad \frac{D^2}{<n>^2} = \frac{<\nu^2> - <\nu>^2}{<\nu>^2} \tag{1}$$

where $<n>$ and D are the average multiplicity and dispersion respectively. The same kind of approximation is valid for higher moments [1]. Therefore, is a good approximation to take

$$P(n) = P(\nu), \qquad n = \nu\bar{n} \tag{2}$$

As the transverse energy E_T is a good measure of the multiplicity, the equation 2 can be written as well for E_T distributions.

[1] Iberdrola visiting professor. On leave of absence from Instituto Superior Técnico, 1096 LISBOA.

Let us next consider the particle distribution associated to a rare event C [2] (Drell-Yan pairs, high p_T particle production at ISR energies, some weak process, J/Ψ and Ψ' production, Υ production; etc) which does not suffer strong absorption. These are events of type C in the classification of reference [4] and are events that are shadowed only by themselves. If α_C is the probability of event C to occur in an elementary collision and $N(\nu)$ is the number of events with ν elementary collisions, the number of events where event C occurs $N_C(\nu)$ is

$$N_C(\nu) = \sum_{i=1}^{\nu} \binom{\nu}{i} (1-\alpha_C)^{\nu-i} \alpha_C^i N(\nu) \tag{3}$$

If the event C is rare we can approximate 3 by the leading term in α_C (this is our definition of rare event)

$$N_C(\nu) = \alpha_C \nu N(\nu) \tag{4}$$

If N is the total number of events, $\sum_\nu N(\nu) = N$, $\sum_\nu \nu N(\nu) = <\nu> N$, we have

$$P_C(\nu) = \frac{N_C(\nu)}{\sum_\nu N_C(\nu)} = \frac{\alpha_C \nu N(\nu)}{\alpha_C <\nu> N} = \frac{\nu}{<\nu>} P(\nu) \tag{5}$$

within the approximation 2 we finally obtain

$$P_C(n) = \frac{n}{<n>} P(n) \quad or \quad P_C(E_T) = \frac{E_T}{<E_T>} P(E_T) \tag{6}$$

The relations of equation 6 are universal, independent of α_C. This universality is due to the smallness of α_C. If α_C is not small enought, we should use equation 3 instead of 4 and our final equation should be changed, losting the universality. The first comparison of equation 6 was done in reference [5], long time ago, for $\alpha\alpha \to high\ p_T + X$ at $\sqrt{s}$=31 GeV, showing a remarkable agreement. In Fig.1 the NA38 experimental associated multiplicity to Drell-Yan pairs in S-U collisions [6] is compared to $\frac{E_T}{<E_T>} P(E_T)$ where $P(E_T)$ is the experimental multiplicity distribution of S-U collisions. A very good agreement is also obtained for rare events in hadron-hadron collisions as can be seen in reference [2], where the CDF experimental associated multiplicity distribution for $W^\pm$ and Z^0 events produced in $\bar{p}p$ collisions at $\sqrt{s}$=1.8 TeV [7] is compared to $\frac{n}{<n>} P(n)$, being $P(n)$ the minimum bias distribution. Also a good agreement is obtained for the case of annihilation events in $\bar{p}p$ collisions [8].

The linearity of the dependence of the number of events C on ν will not stand up in the case of final state destructive absorption, as in J/Ψ production. The effective number of collisions where event C appears is smaller. This can be taking into account by doing the changes

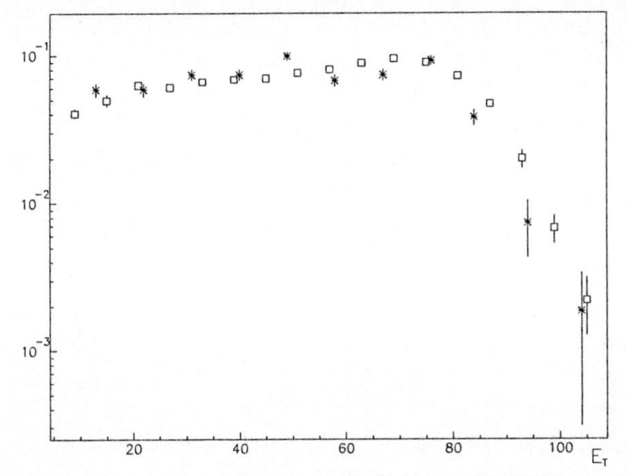

Fig. 1. NA38 experimental associated multiplicity to Drell-Yan pairs in S-U collisions (cross points) compared to $\frac{n}{<n>}P(n)$ where $P(n)$ is the experimental multiplicity distribution of S-U collisions (squared points).

$$\alpha_C \nu N(\nu) \longrightarrow \alpha_C \nu^\varepsilon N(\nu), \quad \varepsilon < 1 \quad and \quad P_C(n) = \frac{n^\varepsilon}{< n^\varepsilon >} P(n) \tag{7}$$

As $\varepsilon < 1$, it is clear that absorption makes the associated distribution closer to the minimum bias distribution. In Fig.2 the NA38 experimental associated E_T distribution [6] for J/Ψ production is compared to formula 7. In this case, to obtain agreement we need $\varepsilon = 0.7$ indicating that the absorption can not be neglected. According to our formulas

$$N_{J/\Psi}(E_T)/N_{DY}(E_T) \sim 1/E_T^\gamma, \quad \gamma \simeq 1 - \varepsilon > 0 \tag{8}$$

This means that this ration decreases with E_T (the first derivative is negative) and the curvature (the second derivative) is positive. In all calculations of J/Ψ absorption, including destruction by comovers the tendency for a large E_T saturation occurs, which implies positive curvature. In the case that the J/Ψ formation were prevented as it will happen if a transition to quark-gluon occurs, we would have

$$\begin{aligned} \alpha_{J/\Psi}(\nu) &= \alpha_{J/\Psi} \quad \nu \leq \nu^* \\ \alpha_{J/\Psi}(\nu) &= 0 \quad \nu > \nu^* \end{aligned} \tag{9}$$

or, making a more reasonable gaussian approximation to (9),

$$\alpha_{J/\Psi}(\nu) = \alpha_{J/\Psi} exp(-\nu^2/\nu^{*^2}) \tag{10}$$

and the J/Ψ to DY ratio becomes now:

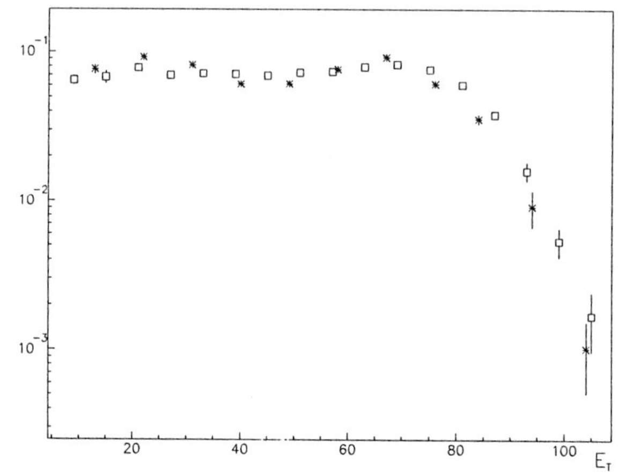

Fig. 2. Same as Fig.1 but now the cross points are the experimental associate multiplicity to J/Ψ production and comparison is made with $\frac{n^{\varepsilon}}{<n^{\varepsilon}>}P(n)$, and $\varepsilon = 0.7$.

$$N_{J/\Psi}(E_T)/N_{DY}(E_T) \sim exp(-E_T^2/E_T^{*^2}) \tag{11}$$

to be compared to 8. There is an essential difference: if the quark-gluon plasma is produced the curvature in the E_T dependence of the ratio changes, being negative for $E_T < E_T^*$ and positive for $E_T > E_T^*$. From the NA50 experimental data presented at this conference [9] the reader can form his own conclusion, about whether or not there is a change on the curvature of the ratio for Pb-Pb collisions.

References

[1] J. Dias de Deus, C. Pajares and C.A. Salgado Phys.Lett.**B407** (1997), 335.

[2] J. Dias de Deus, C. Pajares and C.A. Salgado Phys.Lett.**B408** (1997), 417.

[3] J. Dias de Deus, C. Pajares and C.A. Salgado Phys.Lett.**B409** (1997), 474.

[4] B. Blackenbecler, A. Capella, J. Tran Thanh Van, C. Pajares and A. V. Ramallo Phys. Lett. **107B** (1981), 106.

[5] M. Faessler, Phys.Rep. **115** (1984), 1.

[6] NA38 Collaboration, C. Charlot in 'High Energy Hadronic Interactions', Proceedings of the XXV Rencontres de Moriond, Ed. Frontières, 1990.

[7] CDF Collaboration, F. Rimondi Proc XXII Int. Symposium on Multiparticle Dynamics, Ed. C. Pajares, World Scientific 1992.

[8] J.G. Rushbrooke and B. R. Webber Phys. Rep. **44** (1978) 1.

[9] NA50 Collaboration, F. Ohlsson-Malek these proceedings.

Part VII

Precision Tests of the Standard Model

Precision Electroweak Results from N-N Scattering in CCFR/NuTeV

Jaehoon Yu, (for the CCFR/NuTeV Collaborations) (yu@fnal.gov)

Fermi National Accelerator Laboratory, Batavia, IL60510, USA

Abstract. We present a recent precision electroweak result from ν-N scattering experiment at CCFR. The latest measurement of the weak mixing angle from the CCFR experiment is $\sin^2\theta_W = 0.2236 \pm 0.0041$ which corresponds to on-shell mass of the W boson, $M_W = 80.35 \pm 0.21 GeV/c^2$. This result is the best measured weak mixing angle in all ν-N scattering experiments to date. We also present the improvements in the measurement from the new experiment, the NuTeV, whose expected M_W uncertainty is $110 MeV/c^2$. This expectation is compatible to the combined direct measurements from the Tevatron collider experiments.

1 Introduction

The weak mixing angle, $\sin^2\theta_W$, is one of the fundamental parameters in the electro-weak sector of the Standard Model (SM). Neutrino-Nucleon (νN) scattering experiment is an excellent probe of electro-weak physics, because neutrinos interact with quarks only via weak interactions. The measurement of the weak mixing angle in νN scattering experiment is also a complementary measurement of the mass of the W bosons (M_W), via the SM on-shell relationship, $\sin^2\theta_W = 1 - M_W^2/M_Z^2$. In addition, high precision comparisons among the distinct electro-weak processes in wide range of q^2 provide an excellent test of the theory and a window to searching for non-SM physics contributions [1].

In this report, we present the new extraction of $\sin^2\theta_W$ from the CCFR experiment which supersedes the previous result [2], and the expectation of the $\sin^2\theta_W$ measurement from the succeeding NuTeV experiment.

2 CCFR $\sin^2\theta_W$ result

The data used in this analysis were taken between 1984 and 1988 in FNAL experiments E744 and E770. The total number of candidate events used in this analysis was 8.1×10^5.

Since the cross section of charged current (CC) interactions of neutrinos is proportional to weak isospins ($I^{(3)}_{Weak}$) while that for neutral current (NC) interactions is proportional to ($I^{(3)}_{Weak} - Q_{EM}\sin^2\theta_W$), the ratio of the CC to NC cross sections is proportional to $\sin^2\theta_W$ via Llewellyn-Smith fomula [3]:

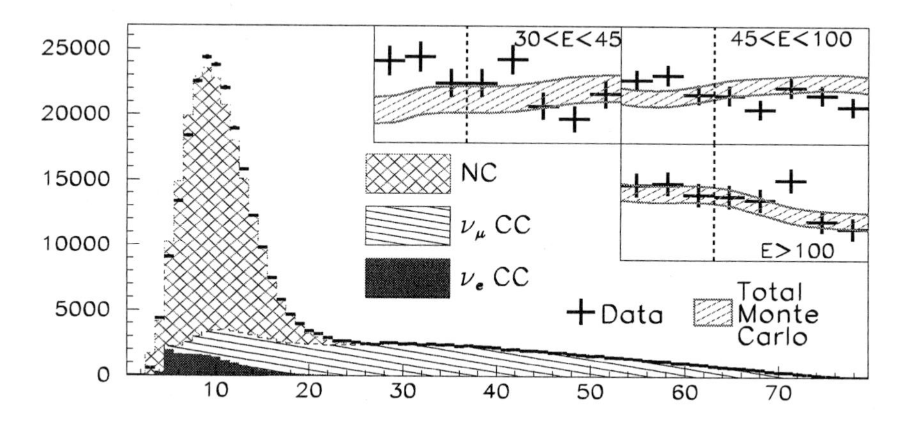

Fig. 1. Data and MC comparisons of the event length variable in CCFR. The hatched bands in the insets illustrate the MC uncertainties.

$$R^{\nu(\overline{\nu})} = \frac{\sigma_{NC}^{\nu(\overline{\nu})}}{\sigma_{CC}^{\nu(\overline{\nu})}} = \rho^2 \left(\frac{1}{2} - \sin^2\theta_W + \frac{5}{9}\sin^4\theta_W \left(1 + \frac{\sigma_{CC}^{\overline{\nu}(\nu)}}{\sigma_{CC}^{\nu(\overline{\nu})}} \right) \right). \quad (1)$$

The most important element in this measurement is separating NC from CC events. This separation is obtained statistically by using the event length variable. The event length is defined for each event to be the number of counters with energy deposition above that of a single muon. The NC events have short length due to the absence of the muons in the event, while CC events are long. Thus, the experimentally measured ratio $R_{meas} = N_{short}/N_{long}$, represent the NC to CC cross section ratio, $R^{\nu(\overline{\nu})}$. We vary only the NC couplings in a detailed Monte Carlo (MC), until $R^{\nu(\overline{\nu})}$ from the MC matches with R_{meas}.

The cross section model in the MC used a leading order (LO) corrected Quark-Parton-Model (QPM), using LO parton distribution functions from the CCFR structure function measurements. The second part of the Monte Carlo simulates detailed detector and beam effects, such as "short" CC events caused by ranging out low energy muons or by the muons exiting the detector fiducial volume, "long" NC events due to π/K decays and punch through, electron neutrino (ν_e) CC events which appear short in the detector, and other detector effects that affect the event length variable. The detailed MC also includes corrections, such as : electromagnetic and electroweak radiative corrections ; target isovector effects ($\sim$ 6% for iron target); higher order QCD effects due to longitudinal structure functions, R_L; effect due to heavy quark productions.

Figure 1 shows the comparisons of the "event length" variable between data and MC for the CCFR experiment. The NC events peak at shorter lengths, while ν_μ CC events distribute under the peak as a continuous band, stretching out to longer lengths. The CCFR data is indicated with crosses, while the insets compare the lengths in various hadronic energy bins with vertical dashed lines indicating the cut separating "short" and "long" events. The separation between "short" and "long" events is made at 20, 25, and 30 counters for the three different hadronic energy bins indicated in the insets.

There are two major sources of systematic uncertainties in the CCFR measurements. The first is the theoretical uncertainties due to mass threshold effect in the CC production of charm-quark which is modeled by LO slow-rescaling. The paremeters in the slow rescaling model were measured by the CCFR experiment, using events with two oppositely charged muons, where $m_c = 1.31 \pm 0.24 GeV/c^2$ and $\kappa = 0.37 \pm 0.05$ [4]. The uncertainties in the above two parameters results in $\delta \sin^2\theta_W = 0.0027$.

The second source is the lack of precise knowledge on ν_e flux. Approximately 80% of the total ν_e come from $K^{\pm}_{e3}$ decays. The ν_e from this source is well constrained by observed $K^{\pm}_{\mu 2}$ spectra. Remaining 16% of ν_e come from neutral K decays, K_{Le3}, whose production mechanism is known only to 20%. These two sources constitute 4.1% uncertainty to N_{ν_e}, based on a Monte Carlo study. A direct measurement, based on longitudinal shower development [5], gives 3.5% uncertainty. Averaging these two uncertainties give 2.9% uncertainty in N_{ν_e} which results in $\delta \sin^2\theta_W = 0.0015$.

Combining all the errors, final CCFR result is $\sin^2\theta_W = 0.2236 \pm 0.0041$ which corresponds to on-shell mass of W boson $M_W = 80.35 \pm 0.21 GeV/c^2$.

3 NuTeV Improvements

The NuTeV experiment is a successor of the CCFR experiment, which minimizes the major systematic uncertainties discussed in the previous section. The uncertainty due to the CC charm production effect can be minimized, using Paschos-Wolfenstein relationship [6]:

$$R^- = \frac{\sigma^{\nu}_{NC} - \sigma^{\overline{\nu}}_{NC}}{\sigma^{\nu}_{CC} - \sigma^{\overline{\nu}}_{CC}} = \rho^2 \left(\frac{1}{2} - \sin^2\theta_W \right) \quad (2)$$

By taking the ratios of differences in NC to CC cross sections for ν and $\overline{\nu}$ interactions, the CC charm production uncertainties largely cancel out. In order to use this relationship, one needs to be able to distinguish neutrino NC interactions from anti-neutrino interactions. To achieve this discrimination, the NuTeV modified the beam line to use a Sign-Selected-Quadrupole-Train (SSQT) to select either ν or $\overline{\nu}$ at a given running period.

The uncertainty caused by the ν_e flux is minimized by adding a slight upward angle to the incident proton beam relative to the horizontal axis which directs to the detector, causing the neutral secondaries (K_L) to be directly absorbed into the dump immediately behind the target.

As of this conference, the NuTeV has analyzed about $\sim$ 70% of expected full data set. The NuTeV has measured R^{ν} and $R^{\overline{\nu}}$ which results

Table 1. $\sin^2\theta_W$ uncertainties of CCFR and NuTeV experiments.

Source	CCFR $\delta\sin^2\theta_W$ (R^ν)	NuTeV $\delta\sin^2\theta_W(R^-)$
Statistical	0.0019	0.0018
ν_e flux	0.0015	0.0004
Other Experimental Systematics	0.0012	0.0009
Total experimental Systematics	0.0019	0.0010
Model	0.0030	0.0008
Total $\delta\sin^2\theta_W$	0.0041	0.0022
δM_W	0.21 (GeV/c^2)	0.11 (GeV/c^2)

R^- to exploit the CC charm production model independence of the Paschos-Wolfenstein relationship. Table 1 shows the comparison of the uncertainties between the CCFR and NuTeV measurements of $\sin^2\theta_W$. It is worth to point out that the expected final uncertainty from the NuTeV $\sin^2\theta_W$ measurement is equivalent to $\delta M_W = 0.11(GeV/c^2)$ which is compatible to the combined uncertainty of the two Fermilab hadron collider experiments, $\delta M_W = 0.09(GeV/c^2)$ [7].

4 Conclusions

The CCFR experiment has reduced systematic uncertainties in $\sin^2\theta_W$ compared to the previous measurement, resulting in $\sin^2\theta_W = 0.2236 \pm 0.0041$ which is equivalent to $\delta M_W = 0.21(GeV/c^2)$. The expected reduction in the $\sin^2\theta_W$ uncertainty from the NuTeV experiment is equivalent to $\delta M_W = 0.11(GeV/c^2)$. Thus this new measurement will contribute significantly to the world M_W measurement as well as constraining the SM Higgs mass.

References

[1] E. Gallas' talk (31305) in this conference.
[2] C.Arroyo, B.J.King *et. al.*, Phys. Rev. Lett. **72**, 3452 (1994).
[3] C.H.Llewellyn Smith, Nucl. Phys. **B228**, 205 (1983)
[4] S.A.Rabinowitz *et. al.*, Phys. Rev. Lett. **70**, 143 (1993).
[5] A.Romosan *et. al.*, Phys. Rev. Lett. **78**, 2912 (1997).
[6] E.A.Paschos and L. Wolfenstein, Phys. Rev. **D7**, 91 (1973)
[7] Q.Li, Talk given at FNAL user's meeting (1997).

Low Energy Tests of the Standard Model from *beta*-Decay and Muon Capture

Jan Govaerts (govaerts@fynu.ucl.ac.be)

Institute of Nuclear Physics
Catholic University of Louvain
2, Chemin du Cyclotron
B-1348 Louvain-la-Neuve
Belgium

Abstract. Two recent low energy precision experiments are considered, in order to illustrate how limits set by these measurements for couplings beyond the Standard Model are complementary to high energy constraints.

1 Introduction

Low energy precision experiments are complementary to high energy ones, both in the diversity of available experimental techniques as well as in the probed range of the parameter spaces of different model extensions of the Standard Model (SM).

It is the purpose of the present contribution to highlight such complementarity with the example of two recent precision measurements, one in nuclear β-decay, and the other in nuclear muon capture on ^{3}He. The latter measurement constrains different hadronic corrections to the charged electroweak current within the SM, while both measurements constrain a variety of model extensions of the SM. Some of these constraints appear to be new, or to be competitive with existing ones derived from high energy measurements [1].

2 Nuclear β-decay

The polarisation-asymmetry correlation in β-decay presents an increased sensitivity to any deviation from the $(V-A)$ structure of the charged electroweak interaction [2]. Such experiments consist in the measurement of the longitudinal polarisation of the β particle emitted in a direction either antiparallel or parallel to the polarisation of an oriented nucleus. The ratio of these two polarisations—a relative measurement less prone to systematic corrections—is of the form [2, 3] $R(J) = R_0(J)\,[1 - k(J)\,\Delta]$, J being the degree of nuclear polarisation, while $R_0(J)$ and $k(J)$ are known functions of the β energy and asymmetry, quantities which are experimentally accessible. Finally, Δ is a vanishing quantity in the SM, which may be expressed in terms of the underlying effective charged current interaction [3]. In particular, for J close to unity, the factor $k(J)$ can become appreciable, thus enhancing the sensitivity to a possible deviation $\Delta \neq 0$.

Two such experiments have been performed, one using polarised ^{12}N [4, 5], the other polarised ^{107}In [6]. The combined data result in the precision value [5],

$$\Delta = 0.0004 \pm 0.0026 \quad , \qquad (1)$$

thus in perfect agreement with the SM prediction. Prospects are to perform a similar measurement for ^{17}F at ISOLDE/CERN, as well as for μ^+ decay at PSI, with at least a 50-fold improvement in the precision of the Michel parameter ξ''.

3 Nuclear muon capture

The statistical muon capture rate on ^{3}He to the triton channel was measured in a recent experiment to a precision of 0.3%. The experimental result is [7],

$$\lambda_{\text{exp}} = 1496 \pm 4 \text{ s}^{-1} \quad , \qquad (2)$$

to be compared to the prediction [8] $\lambda_{\text{theor}} = 1497 \pm 12 \text{ s}^{-1}$. Prospects are to perform at PSI a similar measurement for muon capture on the proton to better than 1%, with as by-product a 2-fold improvement in the precision of Fermis' coupling constant G_F.

4 Tests within the SM

The result (2) allows for tests of QCD chiral symmetry predictions, namely tests of PCAC and of second-class currents[1]. The vector and axial current matrix elements are parametrised in terms of six nuclear form factors. These include the pseudoscalar one F_P whose value is related to that of the axial one F_A through PCAC, as well as the second-class ones, namely the scalar and tensor form factors F_S and F_T, which vanish in the limit of exact isospin and charge conjugation invariance. Using CVC and the ^{3}H β-decay rate, the values of the remaining form factors may be determined with sufficient reliability [8].

On that basis, ignoring first the contributions of F_S and F_T, the result (2) leads to a value for F_P which, when compared to the PCAC prediction, is in a ratio of [7] 1.004 ± 0.076. At the level of the nucleon, the same ratio is then [10] 1.05 ± 0.19. Consequently, the value (2) provides the most precise test of nuclear PCAC available, a situation to be contrasted with the recent result [11] from radiative muon capture on the proton which deviates by more than a factor of 1.5 from the PCAC prediction.

On the other hand, assuming the PCAC value for F_P, (2) may be used to set a value either for F_S or for F_T, ignoring in each case the contribution of the other second-class form factor. One then obtains[2] [9] $F_S = -0.062 \pm 1.18$ or $F_T = 0.075 \pm 1.43$, values which agree of course with expectations [12], and do improve on the existing situation [13].

5 Tests beyond the SM

Physics beyond the SM may be parametrised in terms of 4-fermion effective interactions at the quark-lepton level. For muon decay, a representation in the charge exchange form has

[1] For further details, see Ref.[9].

[2] The normalisation of these form factors is relative to $q^\mu/(2M)$, M being the average ^{3}He-^{3}H mass.

become standard in terms of effective couplings $g^{S,V,T}_{\eta_1\eta_2}$, where the lower indices indicate the chiralities of the electron and muon, respectively [14]. Similarly, β-decay is parametrised in terms of coupling coefficients $f^{S,V,T}_{\eta_1\eta_2}$, with the index η_2 being the chirality of the down quark, while muon capture is parametrised in terms of coefficients $h^{S,V,T}_{\eta_1\eta_2}$, η_1 (resp. η_2) being the muon (resp. d quark) chirality [9].

Assuming only vector and axial couplings $f^{V}_{\eta_1\eta_2}$, the result (1) implies $|f^V_{RR} - f^V_{RL}|^2 = 0.0004 \pm 0.0026$. The quantity Δ also involves scalar and tensor contributions, but the ensuing limits do not improve existing constraints [5].

Under different assumptions, the result (2) sets new constraints [9]. L-handed vector couplings only imply $|h^V_{LL}/f^V_{LL}|^2 = 0.9996 \pm 0.0083$, namely a universality test which at present does not improve the usual such test from pion decay[3]. For both L- and R-handed vector couplings, one finds $h^V_{LR} = 0.0005 \pm 0.0102$. Scalar (resp. pseudoscalar) couplings are such that $(h^S_{RR} + h^S_{RL})G_S = -0.0012 \pm 0.022$, (resp. $(h^S_{RR} - h^S_{RL})G_P = -0.078 \pm 1.49$), G_S, G_P being the nuclear matrix elements for the scalar and pseudoscalar quark densities. And for tensor couplings, one has $h^T_{RL}\, G_T/2 = -0.00008 \pm 0.00143$ (G_T being the tensor nuclear matrix element). The scalar constraint is quite stringent, but the tensor one is especially restrictive.

Within specific model extensions of the SM, these results translate into constraints on the parameters of such models. The involved observables being different from those accessible usually from high energy experiments, the probed regions of these parameter spaces are complementary to one another. Here, only a few such instances are indicated [5, 9].

Within left-right symmetric models not manifestly symmetric between their two chiral sectors, as a function of the heavier charged gauge boson mass M_2, the result (1) probes regions in the right-handed mixing matrix element V^R_{ud} or in the ratio g_R/g_L of gauge couplings constants which are inaccessible [5, 15] to the collider experiments [16]. In particular, the latter are totally insensitive to a mass M_2 close to the W mass provided for example the V^R_{ud} quark mixing is sufficiently small, while the ratio V^R_{ud}/V^L_{ud} is much constrained in that mass region by the result (1) [5]. Such a possibility is thus still worth exploring also at high energies. The result (2) only constrains the charged gauge boson mixing angle, to a level comparable to existing limits [9].

Contact interactions are analysed similarly, replacing the coupling coefficients by $\pm 4\pi/\Lambda^2_{\eta_1\eta_2}$. The result (1) translates into $\Lambda^V_{R\eta_2} > 2.5$ TeV (90% CL) for charged vector interactions within the first generation. Note that these limits as such are not directly accessible to unpolarised high energy measurements. Eq.(2) and h^V_{LR} imply $\Lambda^V_{LR} > 4.9$ TeV (90% CL) for charged vector interactions between the first quark generation and the second lepton generation. Eq.(2) also sets limits for such scales associated to scalar or tensor interactions. For $G_T = 1$, one has $\Lambda^T_{RL} > 9.3$ TeV (90% CL). These limits on contact interactions

[3] Precise to better than 0.4%.

for charged currents interactions within the first or the first two generations are certainly comparable to existing ones, if not better or altogether new in some cases. Most analyses of contact interactions based on the excess of large Q^2 events at HERA have concentrated on neutral current interactions, for which the limits are in the 2.5 – 3.0 TeV range [1].

When extending the by-now standard approach of Ref.[17] for leptoquarks with a right-handed neutrino for each generation, a new scalar and a new vector leptoquark is possible, with three new coupling coefficients for each. The result (1) sets a limit on the coupling of the S_0 or V_0 leptoquarks (in the notation of Ref.[18]) to $(\nu_e)_R$, namely $|\lambda^R_{S_0}\lambda^{\nu_R}_{S_0}/M^2(S_0)|$ and $2|\lambda^R_{V_0}\lambda^{\nu_R}_{V_0}/M^2(V_0)|$ each less than 4.1 TeV^{-2} (90% CL), limits which obviously are not available so far from high energy measurements. The result (2) sets stringent limits on couplings and masses for interactions between the first two generations, some of which improve existing limits [18]. This is especially true for the effective tensor interactions induced by the S_0 and $S_{1/2}(Q=-2/3)$ leptoquarks, leading to $|\lambda^L_{S_0}\lambda^R_{S_0}/M^2(S_0)|\,|G_T|$ and $|\lambda^L_{S_{1/2}}\lambda^R_{S_{1/2}}/M^2(S_{1/2})|\,|G_T|$ both being less than 0.29 TeV^{-2} (90% CL).

References

[1] See contributions to these Proceedings.

[2] P. Quin and T.A. Girard, *Phys. Lett.* **B229** (1989) 29.

[3] J. Govaerts, M. Kokkoris and J. Deutsch, *J. Phys.* **G21** (1995) 1675.

[4] M. Allet *et al.*, *Phys. Lett.* **B383** (1996) 139.

[5] E. Thomas, Doctoral Dissertation (July 1997, unpublished); E. Thomas *et al.*, in preparation.

[6] N. Severijns *et al.*, *Phys. Rev. Lett.* **70** (1993) 4047; (E) *ibid* **73** (1994) 611; J. Camps *et al.*, in preparation.

[7] P. Ackerbauer *et al.*, `hep-ph/9708487`, to appear in *Phys. Lett. B*.

[8] J. Congleton and H. Fearing, *Nucl. Phys.* **A552** (1993) 534.

[9] J. Govaerts, `hep-ph/9701385`, to appear in *Nucl. Instr. Meth.*; J. Govaerts and J.L. Lucio, in preparation.

[10] J. Congleton and E. Truhlík, *Phys. Rev.* **53** (1996) 956.

[11] G. Jonkmans *et al.*, *Phys. Rev. Lett.* **77** (1996) 4512.

[12] H. Shiomi, *Nucl. Phys.* **A603** (1996) 281.

[13] B.R. Holstein, *Phys. Rev.* **C29** (1984) 623.

[14] The Particle Data Group, *Phys. Rev.* **D54** (1996) 251.

[15] J. Govaerts, Internal Report UCL-IPN-95-R05 (July 1995, unpublished).

[16] F. Abe *et al.*, *Phys. Rev. Lett.* **74** (1995) 2900; S. Abachi *et al.*, *Phys. Rev. Lett.* **76** (1996) 3271.

[17] W. Buchmüller, R. Rückl and D. Wyler, *Phys. Lett.* **B191** (1987) 442.

[18] S. Davidson, D. Bailey and B.A. Campbell, *Z. Phys.* **C61** (1994) 613.

$\mathcal{O}(N_f\alpha^2)$ Radiative Corrections in Low-Energy Electroweak Processes

Robin G. Stuart (stuartr@umich.edu)

Randall Laboratory of Physics, 500 E. University, Ann Arbor, MI 48109-1120, USA

Abstract. Of the the three best-measured electroweak observables, α, G_μ and M_Z, the first two are extracted from low-energy processes. Both G_μ and M_Z are now known to an accuracy of about 2 parts in 10^5 and there is a proposal to improve the measurement of the muon lifetime by a factor of 10 in an experiment at Brookhaven. Yet calculations of electroweak radiative corrections currently do no better than a few parts in 10^3–10^4 and therefore cannot exploit to available experimental precision. We report on the calculation of the $\mathcal{O}(N_f\alpha^2)$ corrections to Thomson scattering and the muon lifetime from which α and G_μ, respectively, are obtained. The $\mathcal{O}(N_f\alpha^2)$ corrections are expected to be a dominant gauge-invariant subset of 2-loop corrections.

We have studied the $\mathcal{O}(N_f\alpha^2)$ corrections to Thomson scattering and muon decay [1, 2]. and present some of our results below. These. are corrections that come from 2-loop diagrams containing a massless fermion loop where N_f is the number of fermions. Because, N_f is quite large, the $\mathcal{O}(N_f\alpha^2)$ corrections are expected to be a dominant subset of 2-loop graphs contributing at about the 1.5×10^{-4} level. Moreover, since N_f provides a unique tag, the complete set of $\mathcal{O}(N_f\alpha^2)$ corrections contributing to a particular physical process will form a gauge-invariant set.

The vast majority of $\mathcal{O}(N_f\alpha^2)$ diagrams, including box diagrams, can be reduced to expressions involving a master integral of the form

$$\int \frac{d^n p}{i\pi^2}\frac{1}{[p^2]^j[p^2+M^2]^k}\int \frac{d^n q}{i\pi^2}\frac{1}{[q^2]^l[(q+p)^2]^m}$$
$$= \frac{\pi^{n-4}}{(M^2)^{k+j+l+m-n}}\Gamma\left(l+m-\frac{n}{2}\right)\Gamma\left(\frac{n}{2}-l\right)\Gamma\left(\frac{n}{2}-m\right)$$
$$\times\frac{\Gamma(n-j-l-m)\Gamma(k+j+l+m-n)}{\Gamma\left(\frac{n}{2}\right)\Gamma(k)\Gamma(l)\Gamma(m)\Gamma(n-l-m)}$$

using techniques described in [1].

There are a few diagrams in which the internal fermion mass cannot be set to zero initially but they can be treated via a mass expansion along the lines given in ref.[3].

Dimensional regularization is employed throughout and we find that simpler results are often obtained by carrying the full analytic dependence on

n, the number of space-time dimensions, rather than expanding up to finite terms in $\epsilon = 2 - n/2$. A case in point is the Z-γ mixing at zero momentum transfer that contributes to Thomson scattering and for which the diagrams are shown in Fig.1.

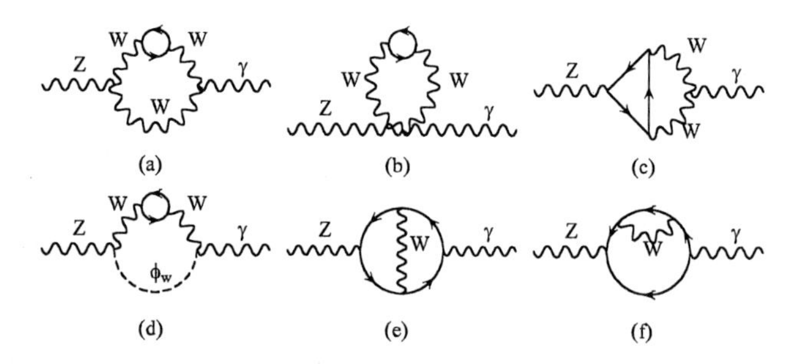

Fig. 1. $\mathcal{O}(N_f \alpha^2)$ corrections to the Z-γ mixing, $\Pi^{(2)}_{Z\gamma}(0)$.

The result is

$$\Pi^{(2)}_{Z\gamma}(0) = \left(\frac{g^2}{16\pi^2}\right)^2 8 s_\theta c_\theta M_Z^2 \frac{(\pi M_W^2)^{n-4}}{n} \Gamma(4-n)\Gamma\left(2-\frac{n}{2}\right)\Gamma\left(\frac{n}{2}\right)$$

where s_θ and c_θ are $\sin\theta_W$ and $\cos\theta_W$ respectively and we assume one generation of massless fermions.

The diagrams contributing at $\mathcal{O}(N_f \alpha^2)$ to the photon vacuum polarization are shown in Fig.2.

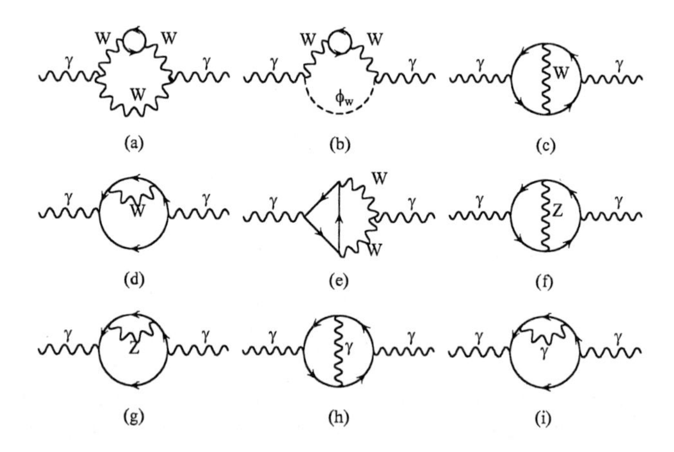

Fig. 2. $\mathcal{O}(N_f \alpha^2)$ corrections to the photon vacuum polarization, $\Pi'^{(2)}_{\gamma\gamma}(0)$.

The result of calculation of these diagrams yields

$$
\begin{aligned}
\Pi'^{(2)}_{\gamma\gamma}(0) &= \left(\frac{g^2 s_\theta}{16\pi^2}\right)^2 \frac{8}{3}\frac{(n+2)}{n}(\pi M_W^2)^{n-4}\Gamma(4-n)\Gamma\left(2-\frac{n}{2}\right)\Gamma\left(\frac{n}{2}-1\right) \\
&+ \left(\frac{g^2 s_\theta}{16\pi^2}\right)^2 \frac{4}{27c_\theta^2}(44s_\theta^4 - 27s_\theta^2 + 9) \\
&\quad \times \frac{(n-6)}{n}(\pi M_Z^2)^{n-4}\Gamma(5-n)\Gamma\left(2-\frac{n}{2}\right)\Gamma\left(\frac{n}{2}-1\right) \\
&- \left(\frac{g^2 s_\theta}{16\pi^2}\right)^2 \frac{4(n-2)}{3n}\pi^{n-4}(M_W^2)^{\frac{n}{2}-2}\Gamma\left(3-\frac{n}{2}\right)\Gamma\left(2-\frac{n}{2}\right) \\
&\quad \times \sum_f Q_f^2 (m_f^2)^{\frac{n}{2}-2} \\
&- \left(\frac{g^2 s_\theta}{16\pi^2}\right)^2 \frac{8}{3}\pi^{n-4}(M_W^2)^{\frac{n}{2}-2}\Gamma\left(2-\frac{n}{2}\right)^2 \sum_f Q_f t_{3f}(m_f^2)^{\frac{n}{2}-2} \\
&- \left(\frac{g^2 s_\theta}{16\pi^2}\right)^2 \frac{16(n-2)}{3n}\pi^{n-4}(M_Z^2)^{\frac{n}{2}-2}\Gamma\left(3-\frac{n}{2}\right)\Gamma\left(2-\frac{n}{2}\right) \\
&\quad \times \sum_f Q_f^2 \left(\frac{\beta_{Lf}^2 + \beta_{Rf}^2}{2}\right)(m_f^2)^{\frac{n}{2}-2} \\
&+ \left(\frac{g^2 s_\theta}{16\pi^2}\right)^2 \frac{16n}{3(n-2)}\pi^{n-4}(M_Z^2)^{\frac{n}{2}-2}\Gamma\left(3-\frac{n}{2}\right)\Gamma\left(2-\frac{n}{2}\right) \\
&\quad \times \sum_f Q_f^2 \beta_{Lf}\beta_{Rf}(m_f^2)^{\frac{n}{2}-2} \\
&+ \left(\frac{g^2 s_\theta}{16\pi^2}\right)^2 \frac{4s_\theta^2}{3}\frac{(5n^2-33n+34)}{n(n-5)}\pi^{n-4}\Gamma\left(3-\frac{n}{2}\right)\Gamma\left(2-\frac{n}{2}\right) \\
&\quad \times \sum_f Q_f^4 (m_f^2)^{n-4} \qquad (1)
\end{aligned}
$$

where β_{Lf} and β_{Rf} are the left- and right-handed couplings of the Z^0 to a fermion, of charge, Q_f, weak isospin, t_{3f}, and mass, m_f, given by

$$
\beta_{Lf} = \frac{t_{3f} - s_\theta^2 Q_f}{c_\theta}, \qquad \beta_{Rf} = -\frac{s_\theta^2 Q_f}{c_\theta}.
$$

Contributions that are suppressed by factors m_f^2/M_W^2 relative to the leading terms have been dropped. The first term on the right hand side of eq.(1) comes from diagrams Fig.2a–e that contain an internal W boson, the second comes from diagrams Fig.2f&g containing an internal Z^0. The last term in eq.(1) comes from diagrams Fig.2h&i and is pure QED in nature. Contributions for which the fermion mass can be safely set to zero without affecting the final result were obtained using the methods described in ref.[1]. The terms in which the fermion mass appears are obtained using the asymptotic expansion of ref.[3]. Note that setting $m_f = 0$ in Fig.2c–g does not immediately cause any obvious problems in the computation because the diagram still contains one non-vanishing scale. A certain amount of care is thus required to identify situations in which the fermion mass cannot be discarded.

Although the calculation of the photon vacuum polarization is somewhat lengthy one has the non-trivial check that longitudinal part vanishes.

Dependence on the fermion mass, m_f, is eliminated in all diagrams, with the exception of Fig.2h&i, by fermion mass counterterms. The diagrams of Fig.2h&i obviously become non-perturbative when the internal fermions are light quarks. In that case the hadronic contribution is treated in the usual manner by writing

$$\Pi'^{(f)}_{\gamma\gamma}(0) = \mathrm{Re}\,\Pi'^{(f)}_{\gamma\gamma}(\hat{q}^2) - [\mathrm{Re}\,\Pi'^{(f)}_{\gamma\gamma}(\hat{q}^2) - \Pi'^{(f)}_{\gamma\gamma}(0)]$$

with $\hat{q}^2$ being chosen to be sufficiently large that perturbative QCD can be used. The term in square brackets on the rhs can be obtained in the usual way using dispersion relations and we have calculated for $|\hat{q}^2| \gg m_f^2$

$$\begin{aligned}\Pi'^{(2\mathrm{QED})}_{\gamma\gamma}(\hat{q}^2) = &-\sum_f \left(\frac{g^2}{16\pi^2}\right)^2 Q_f^4 s_\theta^4 (\pi\hat{q}^2)^{n-4} \\ &\times \left\{ 8(n^2-7n+16)\frac{\Gamma\left(2-\frac{n}{2}\right)^2 \Gamma\left(\frac{n}{2}\right)^3 \Gamma\left(\frac{n}{2}-2\right)}{\Gamma(n)\Gamma(n-1)} \right. \\ &\left. + 24\frac{(n^2-4n+8)}{(n-1)(n-4)}\cdot\frac{\Gamma(4-n)\Gamma\left(\frac{n}{2}\right)^2\Gamma\left(\frac{n}{2}-2\right)}{\Gamma\left(\frac{3n}{2}-2\right)}\right\}. \end{aligned} \quad (2)$$

Despite appearances, the expression on the right hand side of eq.(2) has only a simple pole with a constant coefficient at $n = 4$ that can be canceled by local counterterms. The leading logarithmic expressions can be found in ref.[4, section 8-4-4] where the authors invite the "foolhardy reader" to check that the finite parts are transverse. Here we have gone further and demonstrated this property in the exact result. Eq.(2) could also be obtained by applying analytic continuation relations for the hypergeometric functions, ${}_2F_1$ and ${}_3F_2$, to formulas given by Broadhurst *et al.* [5].

References

[1] R. Akhoury, P. Malde and R. G. Stuart, hep-ph/9707520.
[2] P. Malde and R. G. Stuart, in preparation.
[3] A. I. Davydychev and J. B. Tausk, *Nucl. Phys.* **B 397** (1993) 123.
[4] C. Itzykson and J.-B. Zuber, *Quantum Field Theory*, McGraw-Hill (1980).
[5] D. J. Broadhurst, J. Fleischer and O. V. Tarasov, *Z. Phys.* **C 60** (1993) 287.

Measurements of the Tau Lifetime and Branching Ratios

Isidoro Ferrante (ferrante@pisa.pi.infn.it)

Università di Pisa and INFN sezione di Pisa, Italy.

Abstract. New results on τ lifetime and branching ratios obtained by DELPHI and OPAL are presented, together with a short review of older measurements of the same quantities. Lepton universality is tested using the most recent values of τ decay parameters.

1 Introduction

In this paper I will present the status of τ lepton lifetime and branching ratio measurement, showing new results recently published or presented at this conference, ending with some considerations about the tests on lepton universality. The new results come from DELPHI (on topological branching ratios, inclusive and exclusive hadronic decays, muonic branching ratio and lifetime) and from OPAL (muonic branching ratio) [1].

In the following, when possible, world averages have been calculated including also the preliminary results presented at the 1996 Tau workshop [2] and at the XXVIIIth International Conference on High Energy Physics: no increase of errors as suggested by the Particle Data Group has been applied; otherwise the PDG fit values as published in the 1996 Review of Particle Properties [4] have been used. When world averages are used for comparison with the new results, these are excluded from the average itself.

2 Topological branching ratios

DELPHI has measured the topological branching ratios into one or three charged particles. Their result takes into account the fraction of five prongs decays so as calculated by the Particle Data Group. The DELPHI analysis treats the K^0_S as a truly neutral particle, so that the decays $\tau^- \to h^- n K^0_S m \pi^0$ enter only the one prong measurement. The results are shown in table 1, where are compared with the preliminary ones obtained by ALEPH [5], which use the same definition for 1-prong topology, and with the current world averages [6]: to allow this last comparison, the K^0_S contribution ($0.51 \pm 0.05\%$) has been subtracted from the DELPHI value, to be consistent with the PDG definition of this decay. The world average has large contributions from older results which are in not good agreement among them, and with the newer ones: so the two sigma discrepancy with the DELPHI result is not worrying. It's interesting however to see that also this result, like the ALEPH one, seems to favour a lower value for the one prong branching ratio and an higher one for the three prong.

Table 1. Topological branching ratios

Topology	DELPHI prel. (%)	Other results	
1 prong (incl. $K_S^0 \to \pi^+\pi^-$)	$85.06 \pm 0.10 \pm 0.17$	ALEPH	$85.39 \pm 0.11 \pm 0.13$
1 prong (excl. $K_S^0 \to \pi^+\pi^-$)	$14.85 \pm 0.10 \pm 0.17$	ALEPH	$14.51 \pm 0.11 \pm 0.13$
3 prong (excl. $K_S^0 \to \pi^+\pi^-$)	$84.55 \pm 0.10 \pm 0.18$	W.A.	85.04 ± 0.13
3 prong (incl. $K_S^0 \to \pi^+\pi^-$)	$15.36 \pm 0.10 \pm 0.18$	W.A.	14.87 ± 0.11

3 Hadronic branching ratios

The branching ratios to the inclusive τ decay modes to one or three charged hadron plus zero, one or two π^0's have been measured by DELPHI identifying the various modes with standard cuts for low multiplicity π^0's and with a neural network technique for all topologies. The efficiency for the π^0 identification was found to be about 60%, with a misidentification probability of 20% for decays without any π^0 and of 5-10% for single photons from radiative muon pairs. The preliminary results for the one and three prong modes are shown in table 2. Since the branching ratio for the mode to a single charged hadron is relevant for the universality tests, all the measurements which contribute to the world average are shown in fig. 1.a: the final result is determined almost exclusively by the ALEPH, CLEO and DELPHI values. The agreement among these last ones is reasonable.

Table 2. Inclusive hadronic branching ratios.

Mode	DELPHI prel. (%)	Other results	
$\tau^- \to h^-\nu_\tau$	$11.98 \pm 0.13 \pm 0.11$	W.A.	11.66 ± 0.10
$\tau^- \to h^-\nu_\tau\pi^0$	$25.92 \pm 0.19 \pm 0.25$	W.A.	25.61 ± 0.15
$\tau^- \to h^-\nu_\tau 2\pi^0$	$9.23 \pm 0.34 \pm 0.17$	W.A.	9.21 ± 0.15
$\tau^- \to h^-\nu_\tau \geq 2\pi^0$	$10.25 \pm 0.19 \pm 0.14$	PDG	10.95 ± 0.16
$\tau^- \to h^-\nu_\tau \geq 3\pi^0$	$1.02 \pm 0.21 \pm 0.16$	PDG	1.46 ± 0.11
$\tau^- \to h^-h^+h^-\nu_\tau$	$9.30 \pm 0.07 \pm 0.11$	PDG	9.48 ± 0.10
$\tau^- \to h^-h^+h^-\nu_\tau \geq 1\pi^0$	$5.44 \pm 0.09 \pm 0.14$	PDG	4.88 ± 0.11

The a_1 problem. The most successful models in describing the decay $\tau^- \to \pi^-\pi^+\pi^-\nu_\tau$ (KS, Kühn and Santamaria, and IMR, Isgur, Morningstar and Reader) treat this decay as dominated by the decay into $a_1\nu_\tau$, where the a_1 subsequently decays to $\rho\pi$. However, ARGUS [8], OPAL [9], and now DELPHI have observed a not perfect agreement between these two models and the data for high values of the 3π invariant mass squared. DELPHI has tried to fit the data with a new model (Feindt) which can mimic both the KS or IMR models with an appropriate choice of parameters, but takes also into account the possibility of the existence of an intermediate state in which the τ decays into a radial excitement of the a_1 (a') with a mass of 1700 MeV and a width of $300 MeV$, which could decay in turn with equal probability into $\rho'\pi$ or $\rho\pi$. In this way, the agreement between data and the two models improves, and they find a preliminary result for the branching ratio $BR(\tau \to a'\nu_\tau \to 3\pi\nu_\tau) = (2.0 \pm 0.6) \cdot 10^{-3}$

Decay modes with K_S^0's. DELPHI has performed also a measurement of some inclusive and exclusive branching ratio of final states containing at least one K_S^0. The K_S^0 have been identified looking to neutral pion pairs forming a V_0 with great decay length (to suppress the ρ^0 background), rejecting electron pairs from photon conversion. The resulting K_S^0 selection efficiency is about 10%. The selected sample has been subjected also to a deeper analysis, looking for charged π which could come from the decay of a K^*, or for the presence or absence of a neutral electromagnetic or hadronic shower to identify the presence of a π^0 or a K_L^0 accompanying the decay. The preliminary results are shown in table 3:

Table 3. Branching ratios of modes containing K_S^0's.

Mode	DELPHI prel. (%)	Other results	
$\tau^- \to K_S^0 X^- \nu_\tau$	$0.97 \pm 0.10 \pm 0.05$	PDG	1.58 ± 0.10
$\tau^- \to K^* \nu_\tau (K_S^0 \pi)$	$1.50 \pm 0.33 \pm 0.06$	PDG	1.28 ± 0.08
$\tau^- \to h^- K^0 \nu_\tau$	$1.05 \pm 0.26 \pm 0.16$	PDG	0.92 ± 0.08
$\tau^- \to h^- K^0 \pi^0 \nu_\tau$	$0.51 \pm 0.31 \pm 0.23$	PDG	0.55 ± 0.05
$\tau^- \to h^- K_S^0 K_L^0 \nu_\tau$	$0.13 \pm 0.10 \pm 0.07$	ALEPH	$0.13 \pm 0.04 \pm 0.02$

The decay into $K_S^0 K_L^0$ is worth of more attention: taking the average with the ALEPH result [7], which is the only existing other direct measurement, one has $BR(\tau \to h^- K_S^0 K_L^0 \nu\tau) = 0.130 \pm 0.045$. In a similar way, taking the ratio between the two only direct measurements of the branching ratio into two K_S^0, one has: $BR(\tau \to h^- K_S^0 K_S^0 \nu\tau) = 0.0239 \pm 0.0057$. Taking the ratio, one finds: $BR(K_S^0 K_S^0)/BR(K_S^0 K_L^0) = 0.184 \pm 0.077$, a result which is substantially lower than the naive 0.5 expected, suggesting the existence of some unknown intermediate state.

4 Leptonic branching ratios

Two new measurements of the muonic branching ratio have been made available at this conference by OPAL and DELPHI. OPAL uses two complementary muon identification procedures to check efficiency and systematics on data alone: one is based on muon chambers, while the other uses information from calorimeters. The efficiency in muon reconstruction is about 90% with a background of about 2.5%. On 1994 data (100K τ pairs), they obtain a preliminary result of $B_{\mu,OPAL} = 17.48 \pm 0.12 \pm 0.08\%$. This value is strongly correlated with the previous one published by the same experiment. DELPHI uses an OR on muon chamber and calorimeter information, thus obtaining a better efficiency (97%) at the price of an higher background (4%). The analysis is performed on 1993-1994 data (56K pairs). When combining this result with the previous published ones, they obtain a new preliminary number of:

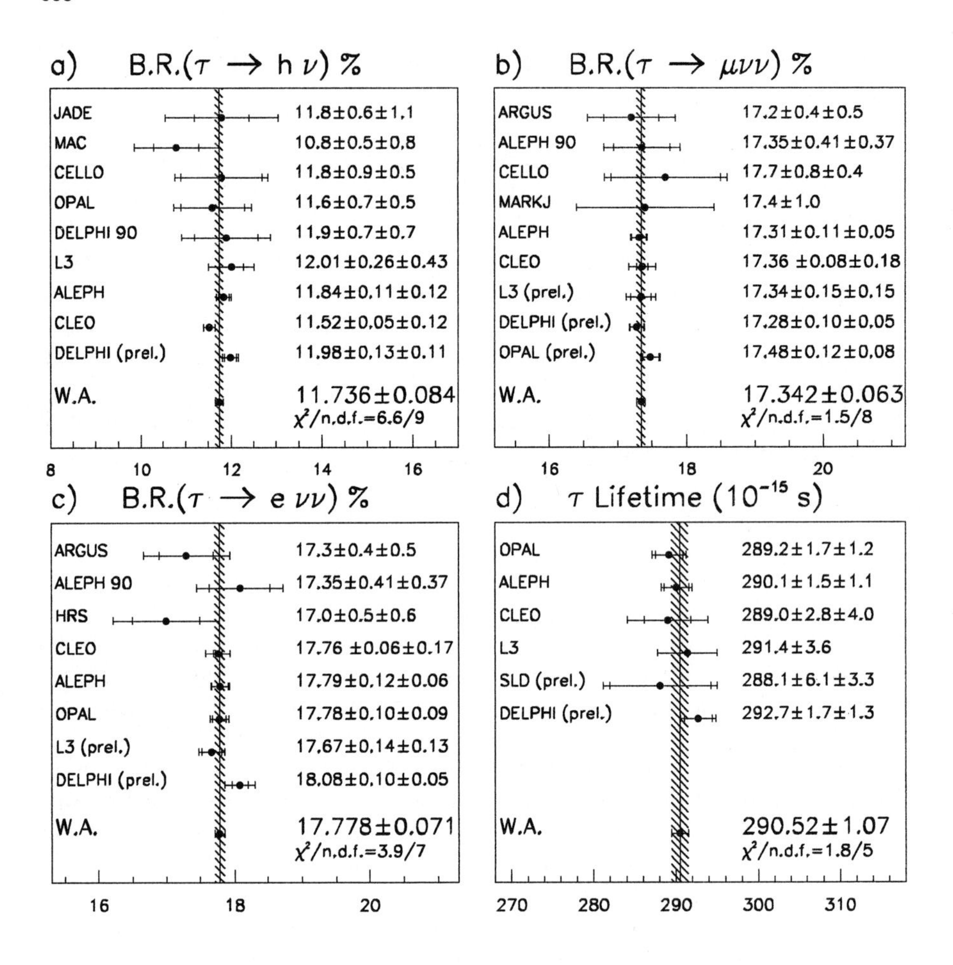

Fig. 1. Tau branching ratios and lifetime measurements: a) BR($\tau^- \to h^- \nu_\tau$) b) BR($\tau^- \to \mu^- \bar{\nu}_\mu \nu_\tau$) c) BR($\tau^- \to e^- \bar{\nu}_e \nu_\tau$) d) τ lifetime

$B_{\mu,DELPHI} = 17.28 \pm 0.10 \pm 0.08\%$ The experimental status of the muonic branching ratio is shown in fig. 1.b. The smallness of the χ^2 is mainly due to the presence of four old measurements with big error bars.

No new measurements of the electronic branching ratio are available: the actual status is the one shown in fig. 1.c.

5 Lifetime measurements

The situation of τ lifetime measurements is rapidly evolving toward a status which could remain stable for many years: almost all LEP I data have been analyzed, while no big improvements are expected neither from SLD or CLEO. With the exception of the OPAL one, all the measurements which contributed to form the average of the 1996 Review of Particle Properties have now been updated:

- ALEPH has analyzed data up to 1994, using also a new technique on hadronic decays which uses a full reconstruction of τ flight up to a twofold ambiguity not relevant to the final result [10]. Their result is the most precise so far[11].
- L3 has published an analysis based on the first data collected with their vertex detector [12].
- SLD has published a new preliminary result based on 4316 τ pair events: while this number is not big enough to compete with the other experiments, the smallness of their beam size allow them to reach an amazing low statistic and systematic error [14].
- CLEO has analyzed data corresponding to approximatively 3 millions produced τ pairs, using only the direct measurement of the secondary vertex displacement in three prong decays [13].

In addition to the above, DELPHI presents to this conference a new preliminary number based on a re-analysis of the three prong decays of the τ in data collected from 1991 to 1995 and in an extension to 1994 data of the two methods (impact parameters sum and difference) used in τ pair decays with 1-1 topology. Their result is $\tau_\tau = 292.7 \pm 1.7 \pm 1.3 fs$, with a precision of 0.7%, very close to the ones obtained by ALEPH and DELPHI.

The status of τ lifetime measurements is shown in fig. 1.d: as can be seen, there are now three measurements better than 1%. A fit to a common value gives the result: $\tau_\tau = 290.47 \pm 1.07 fs$, with a χ^2 of 1.8 for five degrees of freedom. The results appear thus to be very consistent: it should be noted that the method used by OPAL (classical impact parameter) in one prong events is rather different from the ones used by DELPHI and ALEPH. The probability of a common systematic error is thus very low.

6 Tests of lepton universality

The equality of the τ and muon coupling constants to the W can be checked by confronting their decay widths to electron [16, 15]. One has:

$$\frac{g_\tau^2}{g_\mu^2} = 0.9996 \frac{\tau_\mu}{\tau_\tau} \left(\frac{m_\mu}{m_\tau}\right)^5 BR(\tau \to e\nu_e\nu_\mu) \tag{1}$$

An alternative test can be performed by confronting the decay width of the tau to one single charged hadron (π or K) to the hadron's width to muon.

$$\frac{g_\tau^2}{g_\mu^2} = \frac{m_\mu^2}{\tau_\tau m_\tau^3} \frac{B_h}{H_\pi + H_k} = 14021 fs GeV^3 \frac{B_h}{\tau_\tau m_\tau^3} \tag{2}$$

where H_h is function of the τ and hadron masses, and of the hadron branching ratio to muon. These two test are somewhat complementary, since the former tests the equality of the coupling to the transverse component of the W boson, while the latter is sensitive to the longitudinal one.

The equality of electron and muon coupling constants to the transverse component of the W boson can be tested by confronting instead the τ decay widths to these two particles respectively, and correcting for the different phase space factor:

$$\frac{g_\mu^2}{g_e^2} = 1.0282\frac{B_\mu}{B_e} \tag{3}$$

Inserting into equation (1) the world average values, one finds: $g_\tau/g_\mu = 0.9996 \pm 0.0027$, where the error comes in roughly equal size from the lifetime and branching ratio measurements. The second test (2) gives instead: $g_\tau/g_\mu = 1.0049 \pm 0.0040$, where this time the biggest contribution to the error (0.35%) comes from the determination of the hadronic branching ratio. To check the last equality (3) it is not in principle possible to take the ratio of the world averages of electronic and muonic branching ratios, since there are correlations in the same experiment. It is better instead to take the values for the ratio g_e/g_μ as given by the four LEP experiments, plus CLEO, and then take the average. In this way, one has: $g_e/g_\mu = 0.9996 \pm 0.0027$. The error is only a factor two bigger than the result obtained comparing pion decay widths.

References

[1] Abstracts N. 133,147,315,316,317,318,319 presented to his conference.

[2] J.G.Smith and W.Toki (editors), Proceedings of the Fourth Workshop on τ Lepton Physics, Estes Park, Colorado, Nuclear Physics B (Proc. Suppl.) 55C (1997) MAY 1997.

[3] A. Ajduk and A.K. Wroblewski (editors), Proceedings of the XXVIIIth International Conference on High Energy Physics.

[4] R. M. Barnett *et al.* (Particle Data Group), Phys. Rev. D54 (1996) 1.

[5] D. Buskulic *et. al*)(ALEPH collaboration), pa07-011 in [3]

[6] H. G. Evans, in [2]

[7] P. Campana, in [2]

[8] H. Albrecht *et al.* (ARGUS Coll.), Phys. LEtt. B349 (1995) 576.

[9] R. Akers *et al.* (OPAL Coll.), Z. Phys. C67 (1995) 45.

[10] R. Barate *et al.* (ALEPH Collaboration), Z. Phys. C74 (1997) 387-398

[11] R. Barate *et al.* (ALEPH Collaboration), Updated measurement of the τ lepton lifetime, CERN preprint PPE-97-090, Subm. to Phys. Lett., B.

[12] M. Acciarri *et al.* (L3 Collaboration), Phys. Lett. B389 (1996) 187-196

[13] R. Balest *et al.*, Phys. Rev. Lett. 76 (1996) 6037-6053.

[14] SLD collaboration, XXVIIIth International conference on high energy physics, Warsaw, Poland, 1996, pa07-064.

[15] H. Videau, in [3]

[16] P. Weber, in [2]

Theoretical Aspects of $\tau \to K\pi\pi\nu_\tau$ Decays

J.H. Kühn, E. Mirkes and J. Willibald

Institut für Theoretische Teilchenphysik, Universität Karlsruhe,
D-76128 Karlsruhe, Germany

Abstract. Predictions based on the chirally normalized vector meson dominance model for decay rates and distributions of τ decays into $K\pi\pi\nu_\tau$ final states are discussed. Disagreements with experimental results can be traced back to the K_1 widths.

Hadronic τ decays into final states with kaons can provide detailed information about low energy hadron physics in the strange sector. Predictions for final states with 2 and 3 meson final states [1, 2] based on the "chirally normalized vector meson dominance model" are in good agreement with recent experimental results [3]. Problems in the axial vector part in the $K\pi\pi$ final states are discussed in this contribution. We believe that these can be traced back to the K_1 widths.

The matrix element $\mathcal{M}$ for the hadronic τ decay into $K\pi\pi$ final states $\tau(l,s) \to \nu(l',s') + K(q_1,m_1) + \pi(q_2,m_\pi) + \pi(q_2,m_\pi)$ can be expressed in terms of a leptonic and a hadronic current as $\mathcal{M} = G/\sqrt{2}\,\sin\theta_c\, M_\mu J^\mu$ with $M_\mu = \bar{u}(l',s')\gamma_\mu(1-\gamma_5)u(l,s)$. The most general ansatz for the matrix element of the hadronic current $J^\mu(q_1,q_2,q_3)$ is characterized by four form factors F_i, which are in general functions of $s_1 = (q_2+q_3)^2$, $s_2 = (q_1+q_3)^2$, $s_3 = (q_1+q_2)^2$ and Q^2 (chosen as an additional variable)

$$J^\mu = T^{\mu\nu}\,[\,(q_1-q_3)_\nu\,F_1 \,+\, (q_2-q_3)_\nu\,F_2\,] + \,i\,\epsilon^{\mu\alpha\beta\gamma}q_{1\,\alpha}q_{2\,\beta}q_{3\,\gamma}\,F_3 \tag{1}$$

In Eq. (1) $T_{\mu\nu} = g_{\mu\nu} - (Q_\mu Q_\nu)/Q^2$ denotes a transverse projector. A possible pseudo-scalar form factor F_4 is neglected in Eq. (1). The form factors F_1 and $F_2(F_3)$ originate from the $J^P = 1^+$ axial vector hadronic current ($J^P = 1^-$ vector current) and correspond to a hadronic system in a spin one state.

The resulting choice for the form factors F_i for the $\pi^0\pi^0K^-$, $K^-\pi^-\pi^+$, $\pi^-\overline{K^0}\pi^0$ decay modes is summarized by [2]

$$F_{1,2}^{(abc)}(Q^2,s_2,s_3) \quad = \frac{2\sqrt{2}A^{(abc)}\sin\theta_c}{3f_\pi}G_{1,2}^{(abc)}(Q^2,s_2,s_3) \tag{2}$$

$$F_3^{(abc)}(Q^2,s_1,s_2,s_3) = \frac{A_3^{(abc)}\sin\theta_c}{2\sqrt{2}\pi^2 f_\pi^3}G_3^{(abc)}(Q^2,s_1,s_2,s_3) \tag{3}$$

where the Breit-Wigner amplitudes $G_{1,2,3}$ are listed in table. 1. The normalization factors are $A^{(\pi^0\pi^0K^-,K^-\pi^-\pi^+,\pi^-\overline{K^0}\pi^0)} = 1/4, -1/2, 3/(2\sqrt{2})$ and $A_3^{(\pi^0\pi^0K^-,K^-\pi^-\pi^+,\pi^-\overline{K^0}\pi^0)} = 1, 1, \sqrt{2}$. The form factors F_1 and F_2 are gov-

Table 1. Parametrization of the form factors F_1 F_2 and F_3 in Eqs. (2,3) for $K\pi\pi$ decay modes.

channel (*abc*)	$G_1^{(abc)}(Q^2, s_2, s_3)$	$G_2^{(abc)}(Q^2, s_1, s_3)$
$\pi^0\pi^0K^-$	$T_{K_1}^{(a)}(Q^2)T_{K^\star}^{(2m)}(s_2)$	$T_{K_1}^{(a)}(Q^2)T_{K^\star}^{(2m)}(s_1)$
$K^-\pi^-\pi^+$	$T_{K_1}^{(a)}(Q^2)T_{K^\star}^{(2m)}(s_2)$	$T_{K_1}^{(b)}(Q^2)T_\rho^{(1)}(s_1)$
$\pi^-\overline{K^0}\pi^0$	$\frac{2}{3}T_{K_1}^{(b)}(Q^2)T_\rho^{(2m)}(s_2)$ $+\frac{1}{3}T_{K_1}^{(a)}(Q^2)T_{K^\star}^{(2m)}(s_3)$	$\frac{1}{3}T_{K_1}^{(a)}(Q^2)\times$ $\left[T_{K^\star}^{(2m)}(s_1) - T_{K^\star}^{(2m)}(s_3)\right]$
	$G_3^{(abc)}(Q^2, s_1, s_2, s_3)$	
$\pi^0\pi^0K^-$	$\frac{1}{4}T_{K^\star}^{(3m)}(Q^2)\left[T_{K^\star}^{(2m)}(s_1) - T_{K^\star}^{(2m)}(s_2)\right]$	
$K^-\pi^-\pi^+$	$\frac{1}{2}T_{K^\star}^{(3m)}(Q^2)\left[T_\rho^{(2m)}(s_1) + T_{K^\star}^{(2m)}(s_2)\right]$	
$\pi^-\overline{K^0}\pi^0$	$\frac{1}{4}T_{K^\star}^{(3m)}(Q^2)\left[2T_\rho^{(2m)}(s_2) + T_{K^\star}^{(2m)}(s_1) + T_{K^\star}^{(2m)}(s_3)\right]$	

erned by the $J^P = 1^+$ three particle resonances with strangeness

$$T_{K_1}^{(a)}(Q^2) = \frac{1}{1+\xi}\left[\mathrm{BW}_{K_1(1400)}(Q^2) + \xi \mathrm{BW}_{K_1(1270)}(Q^2)\right]$$

$$T_{K_1}^{(b)}(Q^2) = \mathrm{BW}_{K_1(1270)}(Q^2) \tag{4}$$

with $\xi = 0.33$ [2]. Here, BW denote normalized Breit-Wigner propagators

$$\mathrm{BW}_{K_1}[s] \equiv \frac{-m_{K_1}^2 + i m_{K_1}\Gamma_{K_1}}{[s - m_{K_1}^2 + i m_{K_1}\Gamma_{K_1}]} \tag{5}$$

with [4] (all numbers in GeV)

$$\begin{array}{ll} m_{K_1}(1400) = 1.402 & \Gamma_{K_1}(1400) = 0.174 \\ m_{K_1}(1270) = 1.270 & \Gamma_{K_1}(1270) = 0.090 \end{array} \tag{6}$$

The three meson vector resonance in the form factor F_3, denoted by $T_{K^\star}^{(3m)}$, and the two meson ρ and $K^\star$ resonances, denoted by $T_\rho^{(2m)}$ and $T_{K^\star}^{(2m)}$ in table 1, are discussed in detail in [2].

Our predictions for the branching ratios of the various $K\pi\pi$ final states based on the above parameterization are listed in the second column of table 2. The predictions are considerably larger than the world averages for the experimental results presented at the TAU96 conference (fourth column in

Table 2. Predictions for the branching ratios $\mathcal{B}(abc)$ in % for the $K\pi\pi$ decay modes. Results for K_1 parameters in Eq. (6) (second column, vector contribution in parentheses) and for $\Gamma_{K_1}(1400) = \Gamma_{K_1}(1270) = 0.250$ GeV (third column) are compared with the experimental world average (WA) as given at the TAU96 conference (fourth column).

channel (abc)	Γ_{K_1} [Eq. (6)]	$\Gamma_{K_1} = 0.250$GeV	WA (TAU96) [3]
$\pi^0\pi^0K^-$	0.14 (0.012)	0.095	0.098 ± 0.021
$K^-\pi^-\pi^+$	0.77 (0.077)	0.45	0.228 ± 0.047
$\pi^-\overline{K^0}\pi^0$	0.96 (0.010)	0.53	0.399 ± 0.048

table 2). The predictions in the second column in table 2 are based on the particle data group values for the widths of the two K_1 resonances (see Eq. (6)). We believe that these numbers are considerably too small (see below). The strong sensitivity of the branching ratios to the K_1 width is demonstrated by the numbers in the third column of table 2, where predictions based on $\Gamma_{K_1} = 0.250$ GeV are shown. The results are now much closer to the measured values. Our direct fit to recently measured differential decay distributions for the $\tau \to K^-\pi^-\pi^+\nu_\tau$ decay mode by the ALEPH [5] and DELPHI [6] collaborations shown in Fig. 1 yields for the K_1 widths (numbers in GeV):

$$\begin{array}{lll} \Gamma_{K_1}(1270) = 0.37 \pm 0.1 & \Gamma_{K_1}(1400) = 0.63 \pm 0.12 & \text{ALEPH} \\ \Gamma_{K_1}(1270) = 0.19 \pm 0.07 & \Gamma_{K_1}(1400) = 0.31 \pm 0.08 & \text{DELPHI} \end{array} \quad (7)$$

with $\chi^2 = 38/30$ and $\chi^2 = 15.8/12$, respectively. The predicted two meson resonance structure based on these values is shown in Fig. 2 and is in good agreement with the experimental data.

References

[1] J.H. Kühn, A. Santamaria, Z. Phys. C48 (1990) 445.

[2] M. Finkemeier, E. Mirkes, Z. Phys. C69 (1996) 243.

[3] M. Finkemeier, J.H. Kühn and E. Mirkes, Proceedings of the Fourth Workshop on Tau Lepton Physics (TAU 96).

[4] Rev. of Particle Physics, Particle Data Group, Phys. Rev. D 54 (1996) 1.

[5] M. Daview, Proceedings of the Fourth Workshop on Tau Lepton Physics (TAU 96).

[6] DELPHI coll., W. Hao. et al., DELPHI 96-76 Conf. 8, ICHEP'96.

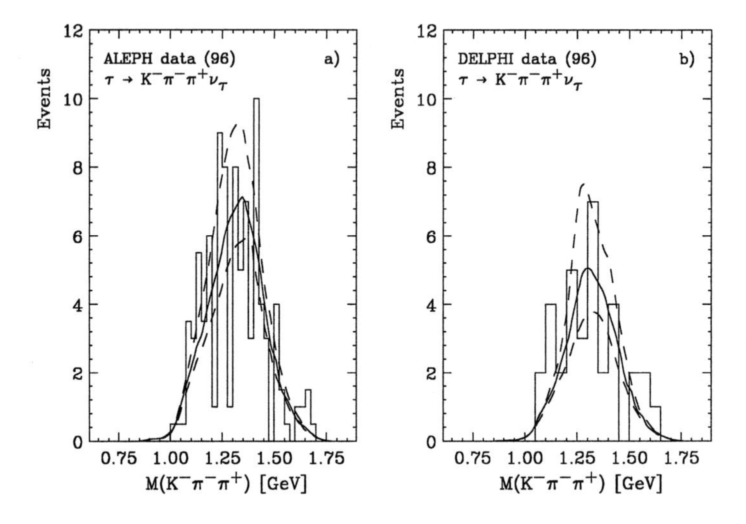

Fig. 1. Invariant mass $m(K^-\pi^-\pi^+)$ distributions for the $\tau \to K^-\pi^-\pi^+\nu_\tau$ decay mode. The histograms show recent data from (a) ALEPH [5] and (b) DELPHI [6]. The solid line shows the fit result to the K_1 widths parameters in Eq. (5) yielding the values in Eq. (7). The dashed lines represent the errors given in Eq. (7) for the K_1 widths. The experimental branching ratios are 0.23 ± 0.05 % (ALEPH) and 0.49 ± 0.08 % (DELPHI). The theoretical predictions for these branching ratios based on the values in Eq. (7) are in good agreement with these numbers.

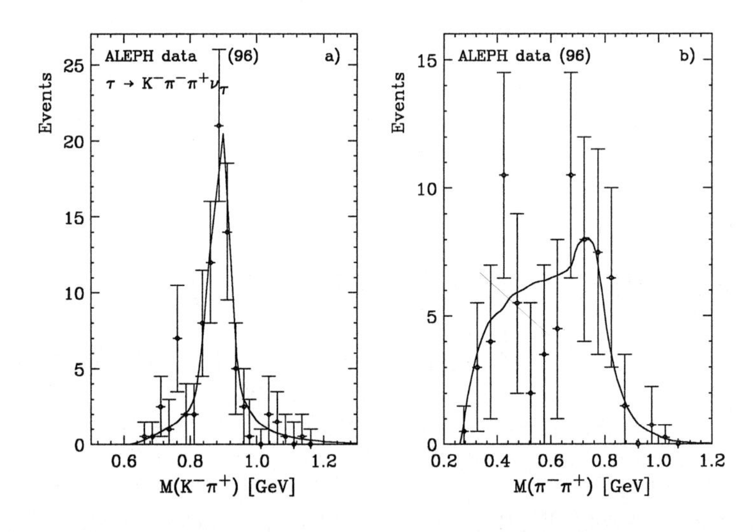

Fig. 2. $K^-\pi^+$ (a) and $\pi^+\pi^-$ (b) invariant mass distributions for the $\tau \to K^-\pi^-\pi^+\nu_\tau$ decay mode. Data are shown from ALEPH [5]. The solid line is the theoretical prediction based on the K_1 parameters in Eq. (7).

The Measurement of R_{uds}, R_c and R_b at LEP and SLD

Chiara Mariotti (Chiara.Mariotti@cern.ch)

CERN (and on leave of absence from INFN-Italy)

Abstract. The new measurements of the partial widths of the Z into quarks at LEP and SLD are reviewed.

1 Introduction

The precise measurement of the partial width of the Z into quarks $R_{q_i} = \Gamma_{Z\to q_i\bar{q}_i}/\Gamma_{Z\to had}$ is a good test of the standard model, because R_{q_i} has negligible dependence on α_s, on the top and Higgs masses. Only R_b is affected by electroweak corrections involving the top quark mass.

The crucial problem in these measurements is the identification of the quark flavour with high purity and efficiency. A very high efficiency is needed in order to minimize the statistical error; a very high purity and a good understanding of the background decrease the systematic error coming from the contamination of the other quark species. Any selection should aim to minimize the correlation between the 2 hemispheres and the corresponding systematic error. The high statistics collected during the LEP1 phase (about 3.5M hadronic Z decays per experiment) together with the analyses developed by the different LEP collaborations and SLD allow the measurement of the partial widths with very high precision.

2 The Measurement of R_u and R_{ds}

The measurement of R_{uds} has been done with a high purity (88%) and efficiency (25%) by ALEPH [1]. More difficult instead is to measure separately the three light quark flavour partial widths. A new measurement from OPAL [2] assumes flavour universality to measure $R_{d,s}$ and R_u separately. Identifying the high momentum stable particles (K, π, p, Λ) in each hemisphere as a tag of the primary quark, OPAL measures:
$R'_u = 0.258 \pm 0.031 \pm 0.032, \qquad R'_{d,s} = 0.371 \pm 0.016 \pm 0.016$
where $R'_q = R_q/(R_u + R_d + R_s)$, thus independent of the measurement of the heavy flavour fractions. The results are in good agreement with the standard model expectations of 0.282 and 0.359 respectively. The largest systematic contribution come from the particle identification.

3 The new measurements of R_c

A deep understanding of the uncertainties dominating the measurement of R_c has been reached recently, and the value of R_c has been measured to be in agreement with the standard model expectation. The measurement of R_c is dominated by the large contamination of the b quark: only 7-10% precision has been achieved so far by each experiment.

Two new measurements of R_c come from ALEPH [3], using the *charm counting method* and all LEP1 statistics, and from OPAL [4] using the double tag methods with exclusive and inclusive D^* reconstruction. Both methods are systematically limited. The main contributions to the systematic error come from fragmentation, vertexing and b-background.

A new method to measure R_c has been proposed by SLD [5]. They developed a pure and efficient c-tag reconstructing secondary vertices [6] and computing their masses and momenta. The region of high momentum and of mass between 0.6-2.0 GeV is enriched in charm and a purity of 68% with an efficiency of 14% is achieved. Since b is the main background, in each hemisphere both a b and a c tag are performed, comparing then the rate of single, double and mixed tag. The reached precision is very high: the statistical error is comparable to the one of LEP experiments with a better systematic error. The main systematic left is from the charm tag hemisphere correlations.

The results are summarized in figure 1.a.

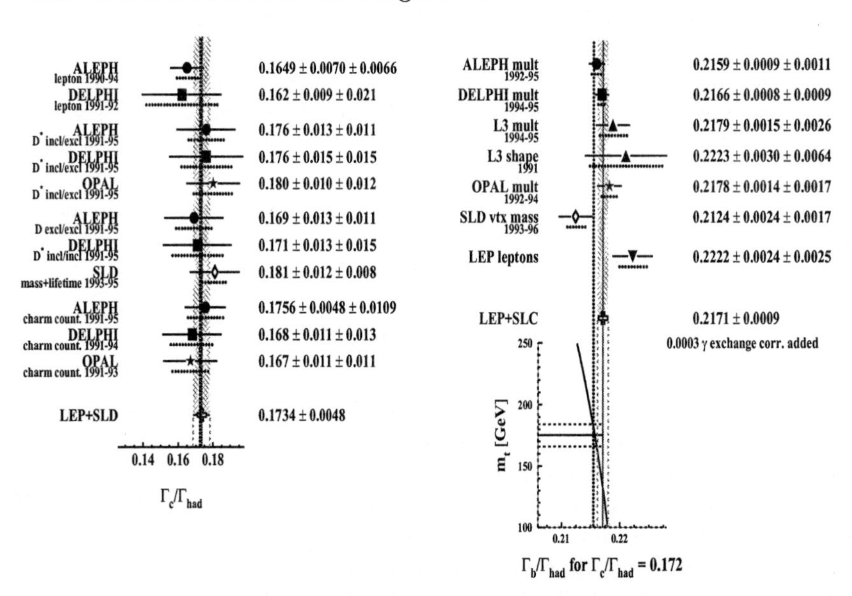

Fig. 1. a) The R_c results. b) The R_b results.

4 The new precise measurements of R_b

The measurement of the partial width of the Z into b quarks has reached very high precision (about 0.4%) and together with the discovery of the top allow good test of the standard model.

SLD and the LEP experiments are equipped with microvertex detectors (see table 1) allowing the identification of b-quarks with very high purity and efficiency via the reconstruction of secondary vertices or the measurement of the impact parameter distributions. The individual detector b-tag performance are shown in table 1.

Two of the largest contributions affecting the measurement of R_b, i.e. the charm background and the correlations between hemispheres, have been reduced considerably. The former by developing very pure b-tags and thus reducing considerably the other quark contaminations. The latter by fitting a primary vertex in each hemisphere, leaving the angular correlation and the gluon radiation as the only sources of correlation. The data, the simulation and the theoretical expectation [7] are in good agreement, resulting in a small systematic uncertainty on the correlation itself.

Three new measurement have been presented this year: the DELPHI result [8] is based on a new and very pure b-tag that combines impact parameter with reconstructed secondary vertices informations, reaching 98% purity at 32% efficiency (see table 1). The b-tag is used in a multivariate analysis that further decreases the statistical error reaching a precision of 0.55%. The systematic error coming from the detector resolution understanding is substantially smaller than for the other experiments, thanks to the 3 layers of the vertex detector that allows the pattern recognition and the tracking to be cleaner and less ambiguous. The result is very stable as a function of the cut and up to purity larger than 99.5% [8].

The measurement of L3 [9], a combined result from an impact parameter analysis and a lepton momentum spectra analysis, reaches a precision of 1.4%, due to the low purity of the working point.

The highest purity and efficiency is reached by the SLD detector that profits from the very small interaction region and thus to the small radius of the first vertex detector layer (see table 1). The b-tag is based on the mass of secondary vertices [6]. The result [10] suffers from the low statistics available.

The results from all the collaborations are listed in figure 1.b. The combined LEP and SLD result of $R_b = 0.2171 \pm 0.0009$ reaches the 0.4% precision and it is consistent with the standard model expectation of 0.2158 ± 0.0003. The systematic error contributions for all the experiment are listed in table 2. The most precise results suffer from the uncertainty on the gluon splitting into b quark rate. With the final results from all the experiment based on the full available statistic and with the forseeable improved measurements on various b and c related quantities the precision on R_b will likely decrease even more.

5 Summary

A very high precision has been reached in the measurement of the partial widths of the Z into quarks. The combined LEP and SLD results reached the 3% precision for the R_c measurement, while 0.4% precision has been achieved in the R_b measurement, thank to new powerful b identification techniques. All the measurements agree with the Standard Model expectations.

	ALEPH	DELPHI	L3	OPAL	SLD
Coordinates Used	$R\Phi, z$	$R\Phi, z$	$R\Phi, z$	$R\Phi$	$R\Phi, z$
Num. of layers	2	3	2	2	3
Radius of layers	6.5-11.3	6.3/11	6.4-7.9	6.1-7.5	2.9/4.1
$R\Phi$ I.P. resolution μm	25	20	30	18	13
z I.P. resolution μm	25	30	30		24
Primary Vtx resol. ($x \times y \times z$ μm)	$58 \times 10 \times 60$	$22 \times 10 \times 22$	$42 \times 10 \times 42$	40×10	$6.4 \times 6.4 \times 15$
b Purity %	97.8	98.0	86.4	90.5	97.6
b Efficiency %	22.7	32.4	23.7	23.1	47.9

Table 1. Vertex Detector Characteristics and b-Tagging performance

	ALEPH	DELPHI	L3	OPAL	SLD
detector resolution	0.0005	0.0002	0.0004	0.0004	0.0011
ϵ_b correlation	0.0003	0.0004	0.0007	0.0010	0.0004
udsc simulation	0.0005	0.0005	0.0018	0.0009	0.0005
gluon splitting	0.0006	0.0005	0.0002	0.0006	0.0006
MC statistic	0.0005	0.0005	0.0008	0.0003	0.0009
event selection	0.0001	0.0001		0.0003	0.0003
total systematic	0.0011	0.0009	0.0026	0.0017	0.0016
statistics	0.0009	0.0008	0.0015	0.0014	0.0034

Table 2. R_b Error Breakdown

References

[1] ALEPH Collab., Contribution ICHEP Warsaw 1996, Ref.Pa10-017
[2] OPAL Collab., K.Ackerstaff et al., CERN-PPE/97-063.
[3] ALEPH Collab., Contribution to this conference, (1997) Ref.623
[4] OPAL Collab., CERN-PPE/97-093 (1997)
[5] SLD Collab., SLAC-PUB-7594 (1997) and Ref. 120.
[6] D.Jackson, Nucl. Inst. Meth. **A 388** (1997) 247.
[7] P.Nason and C.Oleari, Phys. Lett. **B 387** (1996) 623.
[8] DELPHI Collab., Contribution to this conference, (1997) Ref.419.
[9] L3 Collab., Contribution to this conference, (1997) Ref.489
[10] SLD Collab., SLAC-PUB-7585, (1997) and Ref.118.

Measurements of Hadronic Asymmetries in e^+e^- Collisions at LEP and SLD

Erez Etzion (Erez@lep1.tau.ac.il)

School of Physics University of Wisconsin, Madison, WI 53706, U.S.A
and Stanford Linear Accelerator Center, Stanford University, CA94309, U.S.A
currently: School of Physics and Astronomy, Tel Aviv University, ISRAEL

Abstract. High precision experimental new electroweak results measured at the four LEP experiments and the SLD collaboration are discussed. Heavy quark ($b\bar{b}$ and $c\bar{c}$) forward-backward asymmetries measured at LEP are presented along with polarized forward-backward and left-right asymmetries measured at SLD. The results are compared, and the combined averages are used to evaluate the Standard Model parameters.

Introduction

Around the Z^0 peak the fermion-pair are produced mainly through the Z^0 channel, where the γ exchange contribution is very small. Asymmetry measurements, Forward-Backward (FB) and polarized asymmetries are sensitive to the right handed $Zf\bar{f}$ couplings complementary to the partial widths measurements which are more sensitive to the left handed couplings. For unpolarized beams (LEP) the FB asymmetry, $A_{FB}^f = \frac{3}{4}A_eA_f$, is sensitive to the (initial) electron and the outgoing fermion couplings to the Z^0.
Given the longitudinal polarization of the electron beams at SLD, one can use that knowledge to simply measure the difference between left and right handed cross-section, $A_{LR} = \frac{\sigma_L - \sigma_R}{\sigma_L + \sigma_R} = P_eA_e$, where P_e is the polarization of the incident e^- beam. One can also measure the FB polarized asymmetry,

$$A_{FB}^{pol(f)} = \frac{(\sigma_{L,F} - \sigma_{R,F}) - (\sigma_{L,B} - \sigma_{R,B})}{(\sigma_{L,F} - \sigma_{R,F}) + (\sigma_{L,B} - \sigma_{R,B})} = \frac{3}{4}P_eA_f.$$

While the Asymmetries expected from neutrinos, charged leptons, u-type quarks and d-type quarks are: 1, 0.15, 0.67 and 0.94 respectively, the sensitivity of these to the weak mixing angle, $\frac{\delta A_f}{\delta \sin\theta_W}$ are 0, -7.9, -3.5 and -0.6. For comparison all the LEP and SLD asymmetries are given in terms of the effective mixing angle which is defined as: $\sin^2\theta_W^{eff} \equiv 0.25(1 - v_e/a_e)$, where the v_e/a_e is extracted from the asymmetry measurements.
Complementary to the leptonic asymmetries which are sensitive to the oblique radiative corrections, the heavy flavour measurements ($R_{b/c}, A_{b/c}$) test the Standard Model (SM) [1] through vertex corrections. The c and the b quarks being the heaviest quarks accessible at the Z^0 peak, may provide a potential window to physics beyond the SM.

The measurements are based on about 4.4M events collected at each of the LEP experiments and the 200K events accumulated at SLD, produced with highly polarized electron beam. All the experiments have a good leptonic identification, DELPHI and SLD use Cerenkov ring imagine devices for particle identification, where the other 3 experiments use dE/dx for that purpose. The b and c event tag is based mainly on the experiments vertexing capability. All LEP experiments have installed double sided silicon micro-strip vertex detector, providing a typical high momentum track impact parameter resolution of $\sim 25\mu$m, where SLD uses silicon pixel detector which its upgraded version, installed before the 1996 run, gives a high momentum track impact parameter resolution of $\sim 13\mu$m.

Polarized Asymmetries

SLD has a new preliminary measurement of A_{LR} based on the data collected in 1996. The event sample, mostly consists of hadronic Z^0 decays, has 28,713 and 22,662 left- and right-handed electrons respectively. The resulting measured asymmetry is thus $A_m = (N_L - N_R)/(N_L + N_R) = 0.1178 \pm 0.0044(\text{stat})$. To obtain the left-right (LR) cross-section asymmetry at the SLC center-of-mass energy of 91.26 GeV, a very small correction $\delta = (0.240 \pm 0.055)\%(\text{syst})$ is applied which takes into account residual contamination in the event sample and slight beam asymmetries. As a result,

$$A_{LR}(91.26\ \text{GeV}) = \frac{A_m}{\langle P_e \rangle}(1+\delta) = 0.1541 \pm 0.0057(\text{stat}) \pm 0.0016(\text{syst})$$

where the systematic uncertainty is dominated by the systematic understanding of the beam polarization. Finally, this result is corrected for initial and final state radiation as well as for scaling the result to the Z^0 pole energy:

$$\begin{aligned} A^0_{LR} &= 0.1570 \pm 0.0057(\text{stat}) \pm 0.0017(\text{syst}) \\ \sin^2\theta^{eff}_W &= 0.23025 \pm 0.00073(\text{stat}) \pm 0.00021(\text{syst}). \end{aligned}$$

The 1996 measurement combined with the previous measurements [2] yield:

$$A^0_{LR} = 0.1550 \pm 0.0034 \ ; \quad \sin^2\theta^{eff}_W = 0.23051 \pm 0.00043,$$

which is the single most precise determination of weak mixing angle.
SLD has presented a direct measurement of the Z^0-lepton coupling asymmetry parameters based on a sample of 12K leptonic Z^0 decays collected in 1993-95 [3]. The couplings are extracted from the measurement of the double asymmetry formed by taking the difference in number of forward and backward events for left and right beam polarization data samples for each lepton species. This measurement has a statistical advantage of $(P_e/A_e)^2 \sim 25$ on the LEP FB asymmetry measurements. It is independent of the SLD A_{LR} using Z^0 decays to hadrons, and it is the only measurement which determines

A_μ not coupled to A_e. The results are: $A_e = 0.152 \pm 0.012(stat) \pm 0.001(sys)$, $A_\mu = 0.102 \pm 0.034 \pm 0.002$, and $A_\tau = 0.195 \pm 0.034 \pm 0.003$ or assuming universality $A_\ell = 0.151 \pm 0.011$. A new preliminary measurement based on the μ pairs collected at the 1996 run $A_\mu(1996) = 0.164 \pm 0.046$ is given, and combined with the 93-95 yield $A_\mu = 0.123 \pm 0.027$.
The SLD preliminary weak mixing angle value combining A_{LR}, Q_{LR} and leptons asymmetries measurements is: $\sin^2\theta_W^{eff} = 0.23055 \pm 0.00041$, which is more than 3σ below the LEP average.

Heavy Quark Asymmetries

The heavy quark final state asymmetries at the Z^0 peak are large and more sensitive to the EW parameters than the leptonic asymmetries. Since the LEP FB asymmetries are sensitive to both the electron and the out going fermion coupling to the Z^0 it can be interpreted in two ways:

1. Assume universal SM coupling of the out going quarks and derive $\sin\theta_W$, with a much higher sensitivity than the leptonic measurements gaining from the high asymmetry values of $A_c \sim 0.67$ and even better $A_b \sim 0.94$.
2. Use A_e value as given by the leptonic (LEP+SLD) measurements to determine the parity violation parameters, $A_{b,c}$.

The LEP FB asymmetries [4]- [7] were measured on the Z^0 peak and above and below the peak. Off peak measurements were corrected to the peak energy before combining to a pole asymmetries $A_{FB}^{0,f}$.
The SLD LR FB asymmetries [8] are independent of the electron coupling and therefore provide a direct measurement of $A_{b,c}$ again with that statistical advantage of $(P_e/A_e)^2 \sim 25$ compared to the LEP measurements.
Three different principal methods to select a sample of $b\bar{b}$ and measure the b asymmetry have been presented:

- Use the charge of the e or μ in b leptonic decays to sign the b quark direction, where the analyzing power is estimated from the momentum and transverse momentum distribution studied with simulation. (Used by ALEPH, DELPHI, L3, OPAL and SLD).
- Selection of b sample based on lifetime/mass distribution, and momentum weighted track charge to sign the b quark direction. This is a self calibrated technique where the analyzing power is derived directly from the data. (Done by DELPHI, L3, OPAL and SLD).
- A unique method introduced by SLD using mass tag to select the $b\bar{b}$ events and the Cerenkov Ring Imagine Detector (CRID) to identify kaons. The b quark charge is assigned on the basis of the charge of the kaons from the B decay chain.

The new b asymmetry analyses presented at 1997 are jet charge measurements by L3 (new), OPAL (final numbers - winter 97) and SLD (updated

at winter 97), lepton tag by L3 and SLD (both updated for winter 97) and the SLD k-tag analysis updated for summer 97. The ALEPH preliminary jet charge analysis introduced in summer 96 which gave the single most precise determination of A_{FB}^{b} has been withdrawn before the conference and a correction is expected. The LEP measurements of A_{FB}^{b} and A_{FB}^{c} extrapolated to the Z^0 peak are given in Fig. 1 (a) and (c). The LEP measurements translated to a b or c quark asymmetry using the LEP/SLD combined value for $A_e = 0.1505 \pm 0.0023$, are given in Fig. 1 (b) and (d) along with the SLD direct parity violating parameters $A_{b/c}$ measurements.
The measurement of the charm hadrons asymmetry is performed in a similar way, and again three principal methods were used for these analyses:

- Lepton tag similar to the b asymmetry analysis used by all five experiments. (SLD's number updated at the winter conferences).
- D^*, D^0 and D^+ exclusive or semi-exclusive reconstruction used by ALEPH, DELPHI, OPAL and SLD to select a sample enriched with c hadron events, and to determine the direction of the c quark direction.
- In a new method introduced by SLD, charm quarks were tagged by an inclusive mass tag based on topological vertexing, resulting in efficiency of 20% with a purity of 75%. Again the CRID was used to identify kaons and the charge separation is obtained from a combination of the vertex charge and sign of kaon from the D decay chain.

After calculating the overall averages, the following corrections are applied (using ZFITTER [9]) in order to derive the quark pole asymmetries $A_{FB}^{0,b}$ ($A_{FB}^{0,c}$): -0.0013 (-0.0034) for the energy shift from 91.26 GeV/c^2 to M_z, +0.0041 (+0.0104) QED correction and -0.0003 (-0.0008) for γ exchange and γZ interference. QCD corrections (see e.g. [10]) depend strongly on the experimental analyses. Therefore the numbers quoted for the experiments were already corrected for QCD effects as described in [11].
A fit to the LEP and SLD data [12] gives the following combined results for the electroweak parameters:

$$\begin{array}{ll} R_b^0 = 0.2170 \pm 0.0009 & R_c^0 = 0.1734 \pm 0.0048 \\ A_{FB}^{0,b} = 0.0984 \pm 0.0024 & A_{FB}^{0,c} = 0.0741 \pm 0.0048 \\ A_b = 0.900 \pm 0.050 & A_c = 0.650 \pm 0.058 \end{array}$$

with $\chi^2/d.o.f = 0.65$. The R_b-R_c correlation is -20%, the $A_{FB}^{0,b}$-$A_{FB}^{0,c}$ correlation is 13% and other elements in the correlation matrix are between 1 to 8%, where the parameters A_b and A_c have been treated as independent of the FB asymmetries.

The Hadronic Charge Asymmetry $\langle Q_{FB} \rangle$

The LEP experiments [13]- [16] have provided measurements of the average charge flow in the inclusive samples of hadronic decays which is related to

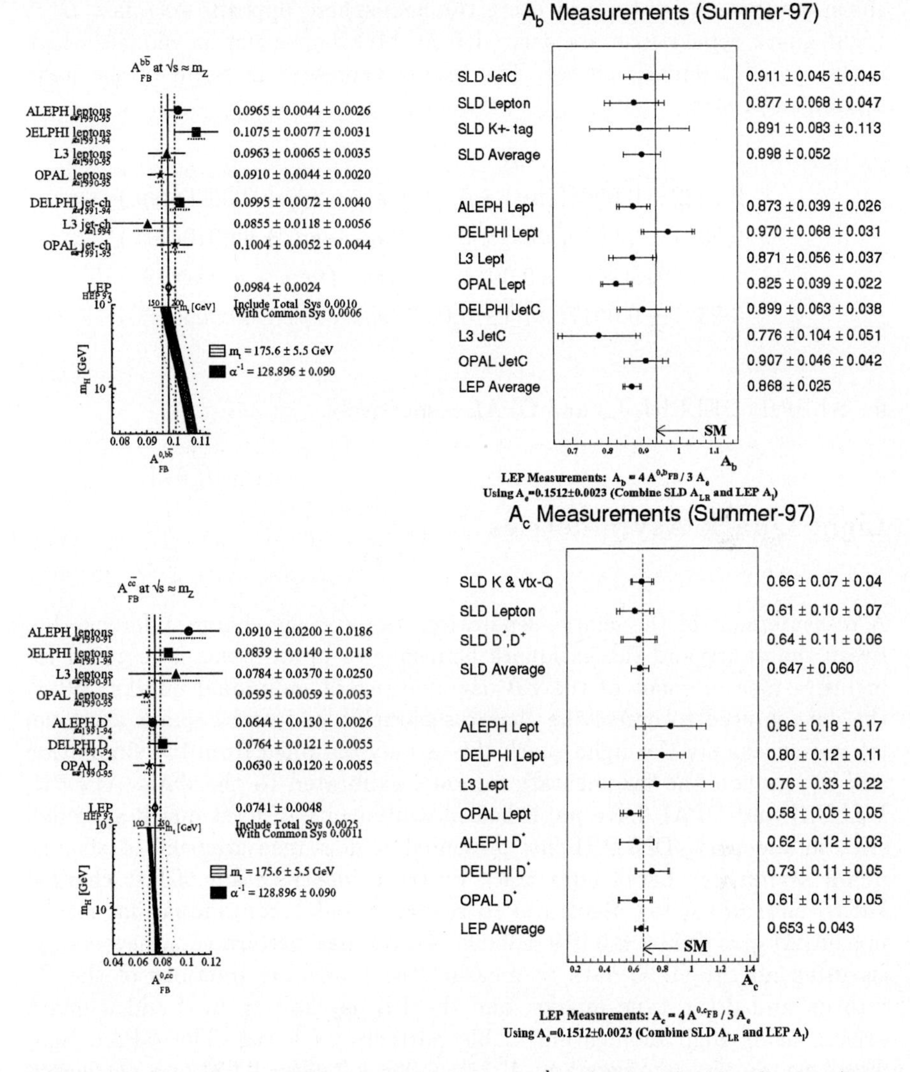

Fig. 1. On the left the LEP measurements of A^b_{FB} (top) and A^c_{FB} (bottom), on the right A_b and A_c asymmetries as measured by SLD and LEP.

FB of the individual quarks asymmetry as following:

$$< Q_{FB} >= \sum_{quark\ flavour} \delta_f A^f_{FB} \frac{\Gamma_f}{\Gamma_{had}}.$$

The charge separation, δ_f, is the average charge difference between quark and antiquark in an event. The b and c are extracted from the data, the δ_b as a by-product of the b asymmetry measurement (self calibration) where the charm separation is obtained using the hemisphere opposite to a fast $D^{*\pm}$. Light quark separations are derived from MC hadronization models which is the main systematic source. The results expressed in terms of the weak mixing angle are:

$$
\begin{aligned}
0.2322 \pm 0.0008(stat) \pm 0.0007(sys.\ exp) \pm 0.0008(sep.) \\
0.2311 \pm 0.0010(stat) \pm 0.0010(sys.\ exp) \pm 0.0010(sep.) \\
0.2336 \pm 0.0013(stat) \pm 0.0014(sys.\ exp) \ \ (new - winter\ \ 97) \\
0.2321 \pm 0.0017(stat) \pm 0.0027(sys.\ exp) \pm 0.0009(sep.)
\end{aligned}
$$

for ALEPH, DELPHI, L3 and OPAL respectively.

Light Quark Asymmetries

A measurement of the charge separation, the average charge difference between the quark and the antiquark hemispheres in an event, is required for inclusive measurement of the FB asymmetries of individual quarks. When the data is used to derive the charge separation in b and c quark asymmetry measurements, for light quark this is only obtained from the simulation and depend on the fragmentation model calibrated to the data. ALEPH, DELPHI and OPAL have published measurements of light quark asymmetries in the past. DELPHI has presented a new measurement of strange quark asymmetry based on a track by track identification of fast charged kaons, and anti b tag algorithm to reduce c and b contamination, resulting in: $A^{s}_{FB} = 0.114 \pm 0.019 \pm 0.005$. OPAL has performed a new analysis using all the 90-95 data to measure the branching fractions of the Z^0 into up and down type quarks, and the FB asymmetry in d and s quark events, using high momentum stable particles as a tag. The OPAL light quark asymmetry numbers are: $A^{d,s}_{FB} = 0.068 \pm 0.035 \pm 0.011$ or a correlated result $A^{u}_{FB} = 0.040 \pm 0.067 \pm 0.028$.

Weak Mixing Angle Measurement

The different values of $\sin\theta_W$ as obtained from the leptons and quark asymmetries, tau polarization and LR asymmetries are given in table 1. A χ^2/d.o.f of 12.6/6 is obtained where the largest disagreement at a level of 3 σ is between the most precise measurements A^{b}_{FB} at LEP and A_{LR} at SLD.

Observable	$\sin\theta_W$
Lepton FB	0.23102 ± 0.00056
τ polarization	0.23228 ± 0.00081
τ FB polarization	0.23243 ± 0.00093
b FB asymmetry	0.23237 ± 0.00043
c FB asymmetry	0.2315 ± 0.0011
Jet charge	0.2322 ± 0.0010
LEP average	0.23199 ± 0.00028
Left-Right (SLD)	0.23055 ± 0.00041
LEP+SLD	0.23152 ± 0.00023

Table 1: Summary of the LEP and SLD $\sin\theta_W$ measurements.

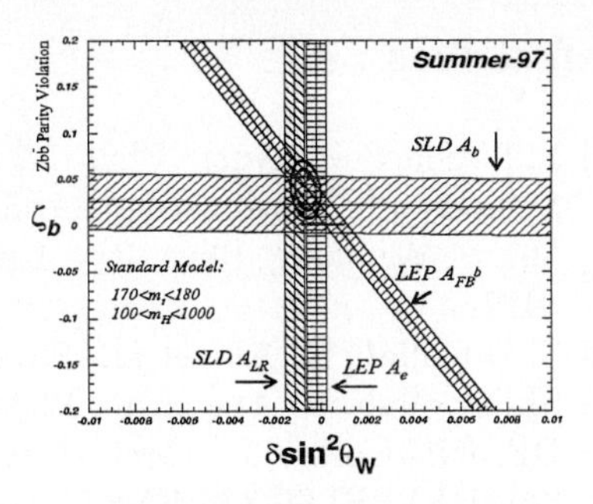

Fig. 2: The measured values of $\sin\theta_W$ as derived from SLD's A_{LR} and leptons asymmetries at LEP, LEP A^b_{FB} and SLD A_b measurements plotted on $\delta\sin\theta_W$ versus $Z_{b\bar{b}}$ parity violation plot (following Takeuchi, Grant and Rosner [17]).

Summary

Quark final state asymmetries provide experimmental powerful tools to study the SM. New measurements and some new techniques were preseted at the conference. The combination of many precise elctroweak results agree well with the theory predictions. The current LEP+SLD average value of A_b is 3 σ lower than predicted by the SM, or deriving the weak mixing angle from the LEP A_{FB} measurements makes the two most precise $\sin\theta_W$ measurements (A_{LR} and A^b_{FB}) disagree at a level of 3 σ. An interesting illustration of the current LEP and SLD $\sin\theta_W$ measurements plotted against the b parity asymmetry measurements is given in Fig. 2. The use of new tagging techniques which have been introduced in the R_b measurements [18] and analyzing the data which still has not being analyzed can further improve the precision of the b and c asymmetry measurements. SLD has started a new run and is expected to reach a precision on $\delta\sin\theta_W \sim 0.0002$.

Acknowledgment

I would like to thank the LEP and SLD collaborators and our colleagues from the LEP and SLC accelerators who brought the field to such an impressive level of precision. Many of the averages for this talk were derived by the detailed careful work of the LEP/SLD EW group [12]. It is a pleasure to thank the organizers for the interesting and successful conference.

References

[1] S. L. Glashow, Nucl. Phys **22** (1961)579. A. Salam, Proc. 8th Nobel Symposium, Aspengarden, Almqvist and Wiksell ed., Stokholm (1968) 367. S. Weinberg, Phys. Rev. Lett. **19** (1967)1264; Phys. Rev. **D5** (1972) 1412.

[2] SLD Collab. K. Abe et al., Phys. Rev. Lett. **78** (1997) 2075-2079.

[3] SLD Collab. K. Abe et al., Phys. Rev. Lett. **78** (1997) 804.

[4] DELPHI Collab, P. Abreu et al., Z. Phys. **C65** (1995) 569; **C66** (1995) 341; DELPHI 95-87 PHYS 522 contributed paper to EPS-HEP-95, Brussels eps0571.

[5] L3 Collab, L3 Note 2129 (1997).

[6] OPAL Collab, R. Akers et al., Z Phys. **C67** (1995) 365; K. Ackerstaff et al., Z. Phys. **C75** (1997) 385; G. Alexander et al., Z. Phys. **C73** (1996) 379.

[7] ALEPH Collab, D. Buskulic et al., Z Phys. **C62** (1994) 1; Contributed paper to EPS-HEP-95, Brussels, eps0634.

[8] SLD Collab, SLAC-PUB-7629, SLAC-PUB-7630, SLAC-PUB-7595, contributed papers to this conference: EPS-122, EPS-123, EPS-124 and EPS-126.

[9] D. Bardin et al., Z. Phys. **C44** (1989) 493; Comp. Phys. Comm. **59** (1990); Nucl. Phys. **B351** (1991)1; Phys. Lett. **B255** (1991) 290 and CERN-TH 6443/92 (1992).

[10] J. B. Stav and H. .A. Olsen, Phys. Rev. **D52** (1995) 1359; Phys. Rev. **D52** (1995) 1359.

[11] The LEP Heavy Flavour Group, LEPHF/97-01, http://www.cern.ch/LEPEWWG/heavy/.

[12] The LEP Electroweak Working Group & the SLD Heavy Flavor group, *LEPEWWG/97-02, SLD Physics Note 63*, “A Combination of Preliminary Electroweak Measurements and Constraints on the Standard Model”.

[13] ALEPH Collab, D. Camp et al., Phys. Lett. **B259** (1991) 377; ALEPH-Note 93-041 (1993); 93-042 (1993); 93-044 (1993); D. Buskulic et al., Z. Phys. **C71** (1996) 357.

[14] DELPHI Collab, P. Abreu et al., Phys. Lett. **B277** (1992) 371; DELPHI 96-19.

[15] L3 Collab, L3-note-2064, (1997).

[16] OPAL Colab, P. D. Acton et al., Phys. Lett. **B294** (1992) 396; OPAL-Physics note **PN195** (1995).

[17] T. Takeuchi, A. Grant, J. Rosner, FermiLab-Conf-94/279-T, talk presented at the DPF'94 Meeting, Albuquerque, NM, Aug/94.

[18] Chiara Mariotti, these proceedings.

Updated Z Parameters, and Standard Model Fits from Electroweak Precision Data

Günter Quast[1] (Gunter.Quast@cern.ch)

Johannes Gutenberg-Universität, Mainz

Abstract. Data taking around the Z resonance at LEP I ended in 1995, and determinations of the mass, width, hadronic pole cross section and leptonic couplings of the Z boson are approaching a final status. This is accompanied by an improved understanding of the beam energy during the high-statistics data taking approximately two GeV above and below the Z resonance in 1993 and 1995. Together with other precision measurements presented at this conference impressive tests of the consistency of the Standard Model can be made and an upper limit on the mass of the Higgs boson can be derived.

1 LEP energy calibration

The major outstanding problem in the determination of the energy of the beams in LEP was due to the energy rise during fills, which was first observed in 1995, when new nuclear magnetic resonance probes ("NMRs") were installed in LEP. Since precise calibrations of the beam energy using the technique of resonant depolarisation of the transversely polarised electron beam were performed at the end of fills, knowledge on the time behaviour of the magnetic field is crucial to a precise determination of the (luminosity weighted) average beam energy. Furthermore, the energy rise during fills was certainly also present during the 1993 energy scan, when no NMR probes in the LEP dipoles were available.

The temperature dependence of the magnetic dipole field was studied with a dedicated test setup consisting of a spare dipole magnet equipped with current bars, beam pipe, NMRs, cooling water system and temperature probes; this was complemented by measurements with 16 NMR probes in the bending dipoles during the 1996 running. Six fills calibrated at the beginning and end confirmed the NMR measurements. Parametrisations of the temperature behaviour in combination with the effects due to parasitic currents running on the LEP beam pipe could thus be tested and the uncertainties on the parameters of the energy model could be estimated.

The beam energy was determined as a function of time in 15 min time slices. Starting from the precise beam energy measurements by resonant depolarisation, several corrections had to be applied. The most important ones are:
- the expected energy rise parameterised as a function of time-of-day and time-into-fill;

- beam displacements from the central orbit position caused by tidal effects and hydro-geological deformations in the Geneva area;
- local effects at the interaction points due to the radio frequency acceleration system and due to local dispersion.

The Z mass, m_Z, is only affected by errors correlated between the energy points, and the width, Γ_Z, is affected by anti-correlated errors. The full correlation matrix of the errors at the seven scan points from 1993 to 1995 was therefore constructed and used in the fits to extract the mass and width of the Z. The results are summarised below.

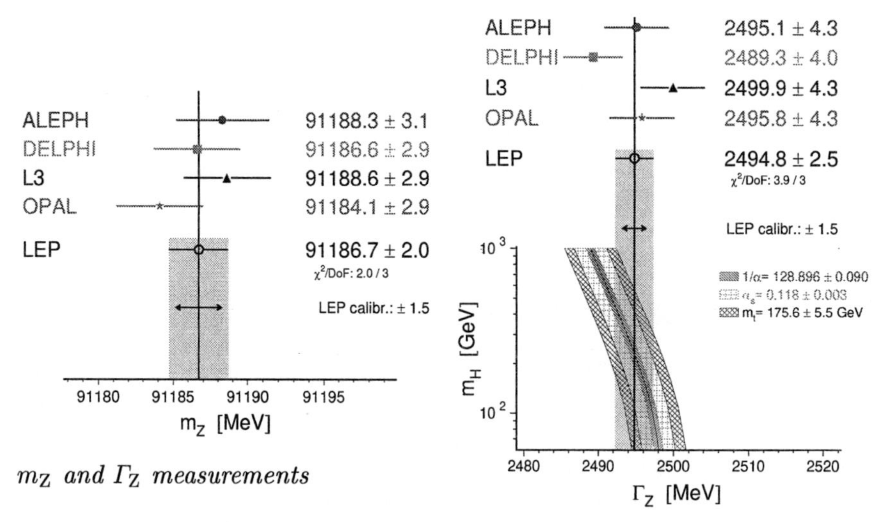

m_Z and Γ_Z measurements

The value of χ^2 per degree of freedom represents the diagonal χ^2 of the mass measurements, taking as errors the experimental ones without the common LEP error of 1.5 MeV in both cases. The ultimate energy error on Γ_Z will come down to 1.2 MeV once all experiments will have updated their analyses with the final energy error matrix. The graph on the lower part of the figure shows the dependence of the Z width on the Higgs mass as expected in the Standard Model ("SM").

2 Electroweak precision data

The electroweak precision data presented at this conference [1, 2] are summarised in the figure below. Since all averages include preliminary data, the numbers given here and in the next section are preliminary. The agreement of each observable with the SM is shown on the right-hand side in terms of the pull, defined as the difference between a measurement and the SM value divided by the error in the measurement; the parameters of the SM are those of the fit to all precision data in the first column of the table below. The χ^2 of this fit is 17 for 15 degrees of freedom, corresponding to a χ^2-probability of 30 %. The largest contribution to χ^2 arises from the two most precise determinations of the effective Weinberg angle, $\sin^2\theta_W^{eff}$, from the b-quark forward-backward asymmetry and from the SLC left-right asymmetry, which show a discrepancy of about three standard deviations. The average effective Weinberg angle from all asymmetry-type measurements is $\sin^2\theta_W^{eff} = 0.23152 \pm 0.00023$; the error is as large as the parametric uncertainty on the SM prediction arising from the error in the electromagnetic coupling constant, α.

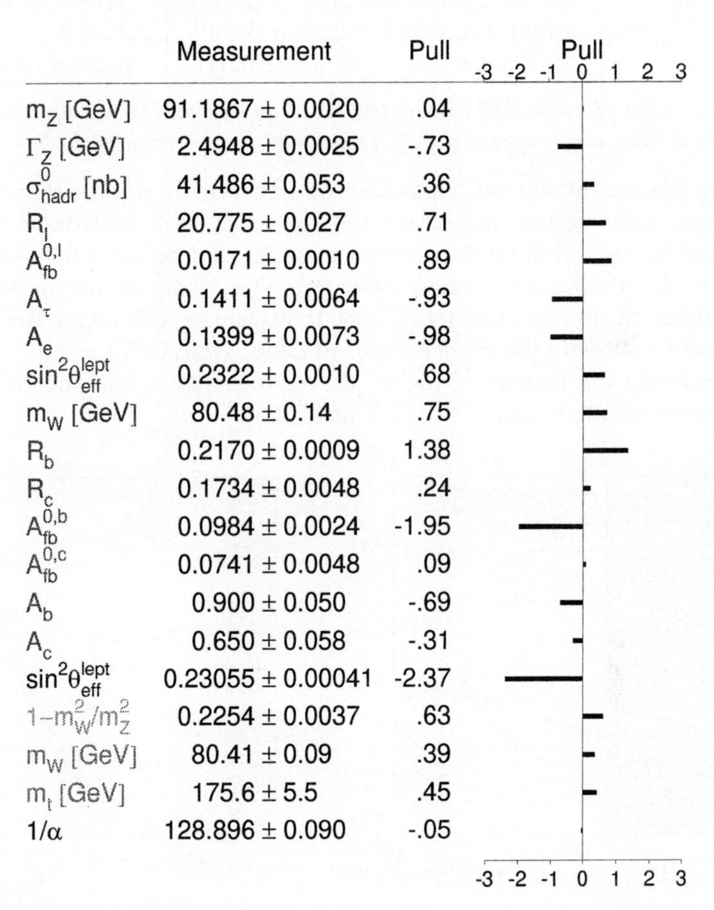

3 Standard Model fits

Fundamental SM parameters are extracted from the electroweak precision data using recent electroweak calculations [3]. The results of such fits for different sets of input variables are shown in the table below.
Allowing for an extra contribution from new physics to the Z decay width leads to $\Gamma^x_{inv} = -1.2 \pm 1.8$ MeV or $\Gamma^x_{inv} < 2.8$ MeV at 95% C.L.

	all data	all data w.o. $m_{\rm top}$ and $m_{\rm W}$	LEP I &LEP II $m_{\rm W}$
$m_{\rm top}$ [GeV]	173.1 ± 5.4	157^{+10}_{-9}	158^{+14}_{-11}
$m_{\rm H}$ [GeV]	115^{+116}_{-66}	41^{+64}_{-21}	83^{+168}_{-49}
α_s	0.120 ± 0.003	0.120 ± 0.003	0.121 ± 0.003
$\sin^2 \theta_W^{eff}$	0.23152 ± 0.00022	0.23153 ± 0.00023	0.23188 ± 0.00026
$1 - m_{\rm W}^2/m_{\rm Z}^2$	0.2231 ± 0.0006	0.2240 ± 0.0008	0.2246 ± 0.0008
$m_{\rm W}$	80.375 ± 0.030	80.329 ± 0.041	80.298 ± 0.043

Fits to data [2] with ZFITTER; the error quoted on the strong coupling constant does not contain any QCD contribution, estimated to be ~0.002

Contour lines in the $m_{\rm H} - \alpha_s$ plane and the χ^2 curves for $m_{\rm H}$ are shown in the last figure. Good agreement between the value of α_s from electroweak fits and the world average from QCD studies is seen. Small differences between three electroweak calculations are also indicated. The χ^2 curves on the last plot reflect different resummation techniques, treatment of scale dependencies and factorisation schemes [3]; these amount to $\Delta\log_{10}(m_{\rm H}/{\rm GeV}) \simeq \pm 0.1$. Taking conservatively the highest of the χ^2 curves to derive a limit on the Higgs boson mass results in $m_{\rm H} < 420$ GeV at 95 % C.L.

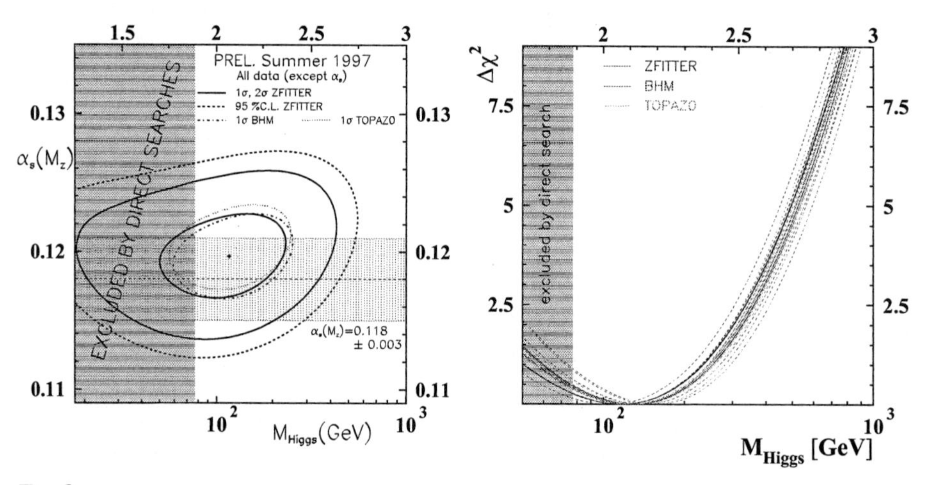

References

[1] Talks at this conference: 701, J. Yu; 707, C. Mariotti; 708 E. Etzion.

[2] The LEP Collaborations, the LEP Electroweak Working Group and the SLD Heavy Flavour Group, *A Combination of Preliminary Electroweak Measurements and Constraints on the Standard Model*, Contributed paper to this conference.

[3] Computer codes BHM, TOPAZ0 and ZFITTER, CERN Yellow Report 95-03, Geneva, 31 March 1995, eds. D. Bardin, W. Hollik and G. Passarino, and references therein.

Sensitivity of Two-Loop Corrections to Muon Decay to the Higgs-Boson Mass

Georg Weiglein (georg@itpaxp5.physik.uni-karlsruhe.de)

Institut für Theoretische Physik, Universität Karlsruhe, 76128 Karlsruhe, Germany

Abstract. The Higgs-mass dependence of the two-loop contributions to muon decay is analyzed at the two-loop level. Exact results are given for the Higgs-dependent two-loop corrections associated with the fermions, i.e. no expansion in the top-quark and the Higgs-boson mass is made. The remaining theoretical uncertainties in the Higgs-mass dependence of Δr are discussed.

1 Introduction

The experimental accuracy meanwhile reached for the electroweak precision observables allows to test the electroweak Standard Model (SM) at its quantum level, where all parameters of the model enter the theoretical predictions. In this way one is able to derive constraints on the mass of the Higgs boson, which is the last missing ingredient of the minimal SM. From the most recent global SM fits to all available data one obtains an upper bound for the Higgs-boson mass of 420 GeV at 95% C.L. [1]. This bound is considerably affected by the error in the theoretical predictions due to missing higher-order corrections, which gives rise to an uncertainty of the upper bound of about 100 GeV. The main uncertainty in this context comes from the electroweak two-loop corrections, for which the results obtained so far have been restricted to expansions for asymptotically large values of the top-quark mass, m_{t}, or the Higgs-boson mass, M_{H} [2, 3].

In order to improve this situation, an exact evaluation of electroweak two-loop contributions would be desirable, where no expansion in m_{t} or M_{H} is made. In this paper the Higgs-mass dependence of the two-loop contributions to Δr in the SM is studied [4]. Exact results for the corrections associated with the fermions are presented.

2 Higgs-mass dependence of Δr

The relation between the vector-boson masses in terms of the Fermi constant G_μ reads [5]

$$M_{\mathrm{W}}^2\left(1-\frac{M_{\mathrm{W}}^2}{M_{\mathrm{Z}}^2}\right)=\frac{\pi\alpha}{\sqrt{2}G_\mu}\left(1+\Delta r\right), \tag{1}$$

where the radiative corrections are contained in the quantity Δr. In the context of this paper we treat Δr without resummations, i.e. as being fully expanded up to two-loop order, $\Delta r = \Delta r_{(1)} + \Delta r_{(2)} + \ldots$. The theoretical predictions for Δr are obtained by calculating radiative corrections to muon decay. We study the variation of the two-loop contributions to Δr with the Higgs-boson mass by considering the subtracted quantity

$$\Delta r_{(2),\mathrm{subtr}}(M_\mathrm{H}) = \Delta r_{(2)}(M_\mathrm{H}) - \Delta r_{(2)}(M_\mathrm{H} = 65\,\mathrm{GeV}), \qquad (2)$$

where $\Delta r_{(2)}(M_\mathrm{H})$ denotes the two-loop contribution to Δr. Potentially large M_H-dependent contributions are the corrections associated with the top quark, since the Yukawa coupling of the Higgs boson to the top quark is proportional to m_t, and the contributions which are proportional to $\Delta\alpha$.

The methods used for the calculations discussed in this paper have been outlined in Ref. [6]. The generation of the diagrams and counterterm contributions is done with the help of the computer-algebra program *FeynArts* [7]. Making use of two-loop tensor-integral decompositions, the generated amplitudes are reduced to a minimal set of standard scalar integrals with the program *TwoCalc* [8]. The renormalization is performed within the complete on-shell scheme [9], i.e. physical parameters are used throughout. The two-loop scalar integrals are evaluated numerically with one-dimensional integral representations [10]. These allow a very fast calculation of the integrals with high precision without any approximation in the masses.

We first consider the contribution of the top/bottom doublet, which is denoted as $\Delta r^{\mathrm{top}}_{(2),\mathrm{subtr}}(M_\mathrm{H})$. From the one-particle irreducible diagrams obviously those graphs contribute to $\Delta r^{\mathrm{top}}_{(2),\mathrm{subtr}}$ that contain both the top and/or bottom quark and the Higgs boson. The technically most complicated contributions arise from the mass and mixing-angle renormalization. Since it is performed in the on-shell scheme, the evaluation of the W- and Z-boson self-energies is required at non-zero momentum transfer.

The contribution of the terms proportional to $\Delta\alpha$ has the simple form $\Delta r^{\Delta\alpha}_{(2),\mathrm{subtr}}(M_\mathrm{H}) = 2\Delta\alpha\,\Delta r_{(1),\mathrm{subtr}}(M_\mathrm{H})$ and can easily be obtained by a proper resummation of one-loop terms [11]. The remaining fermionic contribution, $\Delta r^{\mathrm{lf}}_{(2),\mathrm{subtr}}$, is the one of the light fermions, i.e. of the leptons and of the quark doublets of the first and second generation, which is not contained in $\Delta\alpha$. Its structure is analogous to $\Delta r^{\mathrm{top}}_{(2),\mathrm{subtr}}$, but because of the negligible coupling of the light fermions to the Higgs boson much less diagrams contribute.

The total result for the one-loop and fermionic two-loop contributions to Δr, subtracted at $M_\mathrm{H} = 65\,\mathrm{GeV}$, reads

$$\Delta r_{\mathrm{subtr}} \equiv \Delta r_{(1),\mathrm{subtr}} + \Delta r^{\mathrm{top}}_{(2),\mathrm{subtr}} + \Delta r^{\Delta\alpha}_{(2),\mathrm{subtr}} + \Delta r^{\mathrm{lf}}_{(2),\mathrm{subtr}}. \qquad (3)$$

It is shown in Fig. 1, where separately also the one-loop contribution $\Delta r_{(1),\mathrm{subtr}}$, as well as $\Delta r_{(1),\mathrm{subtr}} + \Delta r^{\mathrm{top}}_{(2),\mathrm{subtr}}$, and $\Delta r_{(1),\mathrm{subtr}} + \Delta r^{\mathrm{top}}_{(2),\mathrm{subtr}} + \Delta r^{\Delta\alpha}_{(2),\mathrm{subtr}}$ are shown for $m_\mathrm{t} = 175.6\,\mathrm{GeV}$. The two-loop contributions $\Delta r^{\mathrm{top}}_{(2),\mathrm{subtr}}(M_\mathrm{H})$ and $\Delta r^{\Delta\alpha}_{(2),\mathrm{subtr}}(M_\mathrm{H})$ turn out to be of similar size and to cancel each other to a large extent. In total, the inclusion of the higher-order contributions discussed here leads to a slight increase in the sensitivity to the Higgs-boson mass compared to the pure one-loop result.

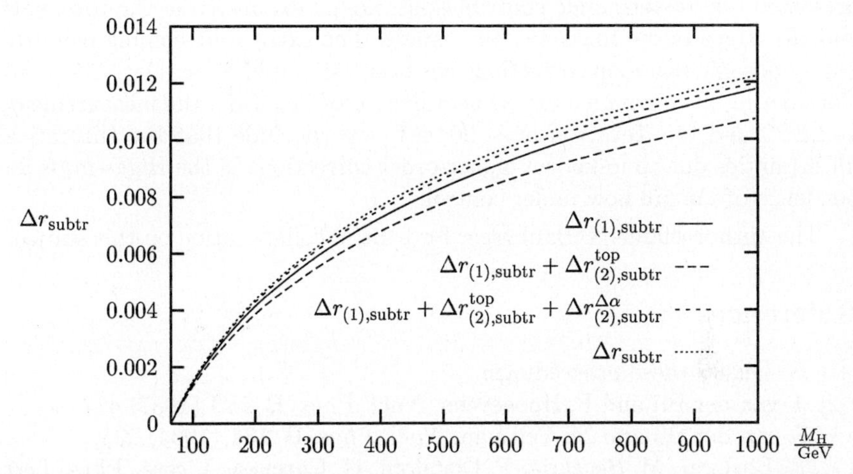

Fig. 1. One-loop and two-loop contributions to Δr subtracted at $M_{\rm H} = 65\,\rm GeV$. $\Delta r_{\rm subtr}$ is the result for the full one-loop and fermionic two-loop contributions to Δr, as defined in the text.

We have compared the result for $\Delta r^{\rm top}_{(2),\rm subtr}$ with the result obtained via an expansion in $m_{\rm t}$ up to next-to-leading order, i.e. $\mathcal{O}(G_\mu^2 m_{\rm t}^2 M_{\rm Z}^2)$ [3]. The results agree within about 30% of $\Delta r^{\rm top}_{(2),\rm subtr}(M_{\rm H})$, which amounts to a difference in $M_{\rm W}$ of up to about 4 MeV [4].

Regarding the remaining Higgs-mass dependence of Δr at the two-loop level, there are only purely bosonic corrections left, which contain no specific source of enhancement. They can be expected to yield a contribution to $\Delta r_{(2),\rm subtr}(M_{\rm H})$ of about the same size as $\left.\left(\Delta r^{\rm bos}_{(1)}(M_{\rm H})\right)^2\right|_{\rm subtr}$, where $\Delta r^{\rm bos}_{(1)}$ denotes the bosonic contribution to Δr at the one-loop level. The contribution of $\left.\left(\Delta r^{\rm bos}_{(1)}(M_{\rm H})\right)^2\right|_{\rm subtr}$ amounts to only about 10% of $\Delta r^{\rm top}_{(2),\rm subtr}(M_{\rm H})$ corresponding to a shift of about 2 MeV in the W-boson mass. This estimate agrees well with the values obtained for the Higgs-mass dependence from the formula in Ref. [12] for the leading term proportional to $M_{\rm H}^2$ in an asymptotic expansion for large Higgs-boson mass. The Higgs-mass dependence of the term proportional to $M_{\rm H}^2$ amounts to less than 15% of $\Delta r^{\rm top}_{(2),\rm subtr}(M_{\rm H})$ for reasonable values of $M_{\rm H}$.

3 Conclusions

We have analyzed the Higgs-mass dependence of the relation between the gauge-boson masses at the two-loop level by considering the subtracted quantity $\Delta r_{\rm subtr}(M_{\rm H}) = \Delta r(M_{\rm H}) - \Delta r(M_{\rm H} = 65\,{\rm GeV})$. Exact results have been presented for the fermionic contributions, i.e. no expansion in the top-quark and the Higgs-boson mass has been made. The extra shift coming from the purely bosonic two-loop corrections has been estimated to be relatively small. Considering the envisaged experimental error of $M_{\rm W}$ from the measurements at LEP2 and the Tevatron of $\sim 20\,{\rm MeV}$, we conclude that the theoretical uncertainties due to unknown higher-order corrections in the Higgs-mass dependence of Δr are now under control.

The author thanks S. Bauberger for fruitful collaboration on this subject.

References

[1] G. Quast, these proceedings.
[2] J. van der Bij and F. Hoogeveen, *Nucl. Phys.* **B 283** (1987) 477;
J. van der Bij and M. Veltman, *Nucl. Phys.* **B 231** (1984) 205;
R. Barbieri, M. Beccaria, P. Ciafaloni, G. Curci, A. Vicere, *Phys. Lett.* **B 288** (1992) 95, *err.:* **B 312** (1993) 511; *Nucl. Phys.* **B 409** (1993) 105;
J. Fleischer, O.V. Tarasov and F. Jegerlehner, *Phys. Lett.* **B 319** (1993) 249; *Phys. Rev.* **D 51** (1995) 3820.
[3] G. Degrassi, P. Gambino and A. Vicini, *Phys. Lett.* **B 383** (1996) 219;
G. Degrassi, P. Gambino and A. Sirlin, *Phys. Lett.* **B 394** (1997) 188.
[4] S. Bauberger and G. Weiglein, hep-ph/9707510, to app. in *Phys. Lett.* **B**.
[5] A. Sirlin, *Phys. Rev.* **D 22** (1980) 971;
W.J. Marciano and A. Sirlin, *Phys. Rev.* **D 22** (1980) 2695.
[6] S. Bauberger and G. Weiglein, *Nucl. Instr. Meth.* **A 389** (1997) 318.
[7] J. Küblbeck, M. Böhm, A. Denner, *Comp. Phys. Com.* **60** (1990) 165;
H. Eck, J. Küblbeck, *Guide to FeynArts1.0* (Univ. of Würzburg, 1992);
H. Eck, *Guide to FeynArts2.0* (Univ. of Würzburg, 1995).
[8] G. Weiglein, R. Scharf and M. Böhm, *Nucl. Phys.* **B 416** (1994) 606;
G. Weiglein, R. Mertig, R. Scharf and M. Böhm, in *New Computing Techniques in Physics Research 2*, ed. D. Perret-Gallix (World Scientific, Singapore, 1992), p. 617.
[9] see e.g. A. Denner, *Fortschr. Phys.* **41** (1993) 307, and references therein.
[10] S. Bauberger, F.A. Berends, M. Böhm and M. Buza, *Nucl. Phys.* **B 434** (1995) 383;
S. Bauberger, F.A. Berends, M. Böhm, M. Buza and G. Weiglein, *Nucl. Phys.* **B** *(Proc. Suppl.)* **37B** (1994) 95, hep-ph/9406404;
S. Bauberger and M. Böhm, *Nucl. Phys.* **B 445** (1995) 25.
[11] A. Sirlin, *Phys. Rev.* **D 29** (1984) 89.
[12] F. Halzen and B.A. Kniehl, *Nucl. Phys.* **B 353** (1991) 567.

Fermion-Pair Production Above the Z and Limits on New Physics

Pat Ward (cpw1@hep.phy.cam.ac.uk)

University of Cambridge, UK

Abstract. Measurements of cross-sections and asymmetries for hadron and lepton pair production in e^+e^- collisions at centre-of-mass energies of 130–183 GeV by the four LEP experiments are presented. Results are compared with the predictions of the Standard Model and used to search for new physics.

1 Cross-section and Asymmetry Measurements

During 1995-6 each LEP experiment accumulated about 25 pb^{-1} of data at e^+e^- centre-of-mass energies $\sqrt{s}$ of 130–172 GeV. Cross-sections for the production of hadrons and lepton pairs, and lepton-pair asymmetries have been measured [1, 2, 3, 4]. First results on flavour-separated hadronic samples at high energy have been obtained by two experiments [4, 5]. Some very preliminary cross-section and asymmetry results have also been obtained from about 6 pb^{-1} of data taken at 183 GeV in August 1997 [6, 7, 8].

Above the Z peak, a large part of the e^+e^- cross-section consists of radiative return to the Z pole. Non-radiative events are separated from the radiative returns by cutting on the value of s', the square of the effective mass of the e^+e^- system after initial-state radiation. The value of s' is calculated from the particles observed in the detector. Cross-sections and asymmetries are thus measured for inclusive samples (typically $s'/s > 0.01$) and 'non-radiative' samples (typically $s'/s > 0.7 - 0.8$). The definition of s' is ambiguous because of interference between initial- and final-state radiation, but the OPAL experiment has developed a well-defined procedure to correct measurements to 'no interference' before comparison with Standard Model predictions; the corrections are small for inclusive events, but 1–2% for non-radiative events [4].

All measured cross-sections and asymmetries are in good agreement with Standard Model predictions calculated using ZFITTER [9], as shown in table 1 and figures 1, 2 and 3.

In electroweak fits to data at the Z peak, γ-Z interference is not well constrained by the data, and fixed to the Standard Model expectation. Cross-section and asymmetry measurements above the Z peak can be used to constrain the interference term. Fits are performed by each experiment using the S-matrix formalism [10], in which the free parameter $j_{\mathrm{had}}^{\mathrm{tot}}$ describes the γ-Z interference in the hadronic cross-section; results from the four experiments

$\sqrt{s}$ /GeV	130.26	136.24	161.32	172.10	182.7
$\sigma(q\bar{q})$ /pb	77.6±3.1	65.1±2.8	36.5±1.1	29.8±1.0	23.9±1.2
SM	82.7	66.6	35.3	28.9	24.3
$\sigma(\mu^+\mu^-)$ /pb	9.4±1.0	8.7±1.0	5.05±0.41	3.92±0.35	3.32±0.42
$\sigma(\tau^+\tau^-)$ /pb	10.4±1.3	7.1±1.1	5.82±0.55	3.87±0.44	4.25±0.85*
SM	8.4	7.3	4.62	3.96	3.45

Table 1. LEP combined measured non-radiative cross-sections, corrected to $\sqrt{s'}/\sqrt{s} > 0.85$, compared with Standard Model predictions. The value marked with an asterisk is from L3 and OPAL only.

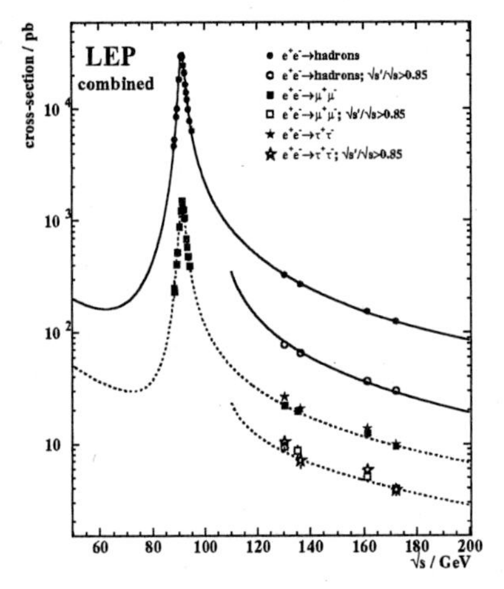

Fig. 1. LEP combined measured cross-sections for inclusive and non-radiative events compared with Standard Model predictions.

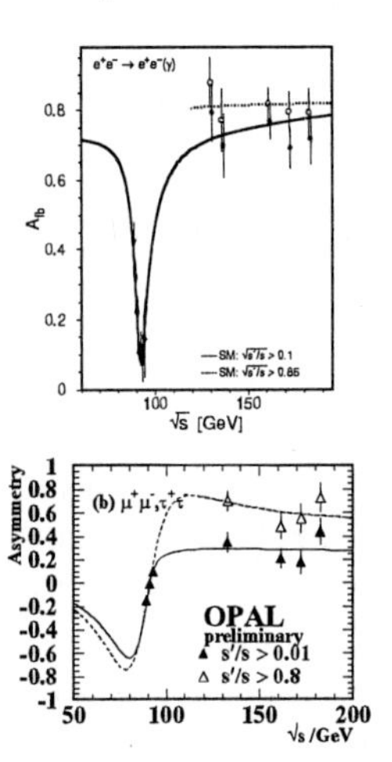

Fig. 2. Measured asymmetries for electron pairs from the L3 experiment (top), muon and tau pairs from the OPAL experiment (bottom).

are then averaged [11]. Adding 130-172 GeV data to those taken at the Z peak reduces the uncertainty on j^{tot}_{had} by about 60%, as shown in figure 4. The OPAL experiment has used non-radiative cross-section and asymmetry measurements to measure the value of the electromagnetic coupling constant at high energies [4], obtaining a value of $1/\alpha_{em}(157.42\ \text{GeV}) = 119.9^{+5.1}_{-4.1}$.

2 Search for New Physics

Measurements of two-fermion processes at high energies have been used to place limits on many potential new physics processes.

The L3 experiment has included 130-172 GeV data in a search for Z′ boson [12]; whereas data taken on the Z peak constrain the Z-Z′ mixing angle,

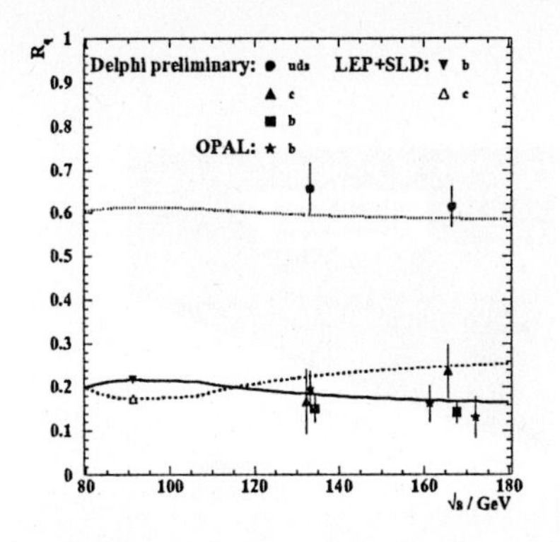

Fig. 3. Measured values of R_b, R_c and R_{uds} compared with Standard Model predictions.

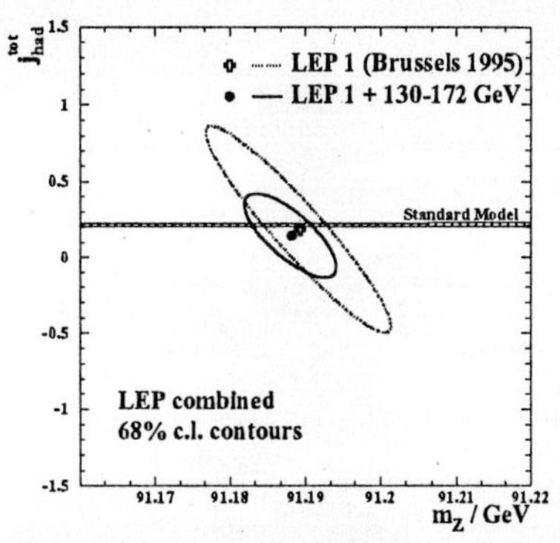

Fig. 4. Central values and two standard deviation contours (68% probability content) in the j_{had}^{tot} vs. m_Z plane resulting from model-independent fits to cross-section and asymmetry data.

high energy data constrain the Z′ mass if mixing is small. For various E6 and LR models, a lower mass limit of about 250 GeV is obtained at 95% confidence level. For a sequential Z′ boson with Standard Model couplings and no mixing, the lower mass limit is 550 GeV at 95% confidence level, similar but still lower than the limit of 690 GeV from the CDF experiment. DELPHI have also obtained similar results [13].

The most general framework for searching for new physics is to consider a four-fermion contact interaction with an energy scale Λ. Sensitivity to such a contact interaction increases with centre-of-mass energy. ALEPH [1, 6], DELPHI [5, 14] and OPAL [4] have all placed limits on the energy scale of contact interactions. For example, the limits obtained by OPAL are shown in figure 5.

The t-channel exchange of a new heavy particle, for example a leptoquark or a squark in supersymmetric theories with R-parity violation, may alter the hadronic cross-section. ALEPH [1, 6], DELPHI [5] and OPAL [4] have searched for such effects, placing limits on the mass and couplings of such particles. For example, figure 6 shows the limits obtained by ALEPH for one particular case.

The L3 [15] and DELPHI [14] experiments have used lepton pair data to put limits on sneutrinos in R-parity violating SUSY models, while OPAL [4] has used hadronic cross-section measurements to place limits on light gluinos.

I would like to thank members of the LEP experiments for providing the results for this talk, and in particular members of the S-Matrix subgroup of the LEP Electroweak Working Group for averaging the cross-section measurements and performing the S-Matrix fits.

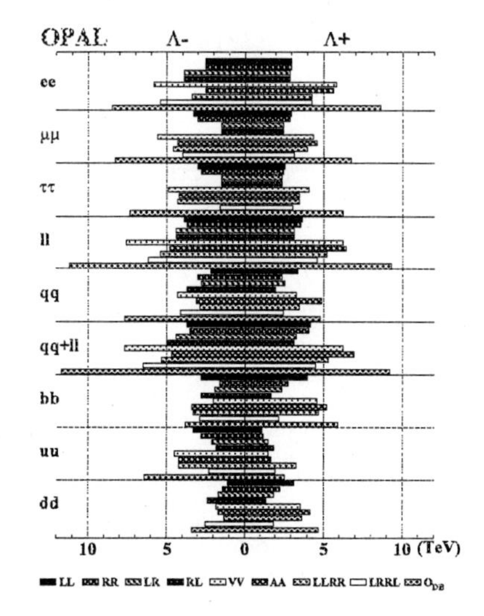

Fig. 5. 95% confidence level limits on the energy scale Λ resulting from the contact interaction fits to OPAL data at 130–172 GeV. For each channel, the bars from top to bottom indicate the results for models LL to $\overline{\mathcal{O}}_{\rm DB}$ in the order given in the key. Limits are given for the usual convention $g^2 = 4\pi$.

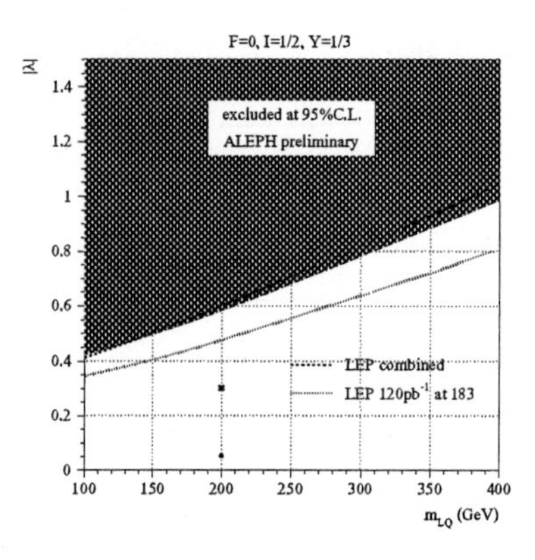

Fig. 6. 95% confidence level limit on the coupling λ as a function of leptoquark mass, $m_{\rm LQ}$, for a leptoquark with fermion number 0, isospin 1/2 and hypercharge 1/3, or equivalently a $\tilde{u}_L$ squark, obtained from ALEPH data including preliminary 183 GeV data. The points show the values of coupling and mass needed to explain the excess of high-Q^2 events at HERA.

References

[1] ALEPH Collab., contributed paper 602.
[2] DELPHI Collab., contributed paper 464.
[3] L3 Collab., M. Acciarri et al., CERN PPE/97-52, contributed paper 509.
[4] OPAL Collab., K. Ackerstaff et al., CERN-PPE/97-101, contr. paper 172.
[5] DELPHI Collab., contributed paper 466.
[6] ALEPHI Collab., contributed paper 856.
[7] DELPHI Collab., contributed paper 860.
[8] L3 Collab., contributed paper 857.
[9] D. Bardin et al., CERN-TH 6443/92 (May 1992); Phys. Lett. **B255** (1991) 290; Nucl. Phys. **B351** (1991) 1; Z. Phys. **C44** (1989) 493.
[10] A. Leike, T. Riemann, J. Rose, Phys. Lett. **B273** (1991) 513; T. Riemann, Phys. Lett. **B293** (1992) 451.
[11] S-Matrix Subgroup of the LEP Electroweak Working Group, LEPEWWG/LS/97-02, September 1997.
[12] L3 Collab., contributed paper 510.
[13] DELPHI Collab., contributed paper 469.
[14] DELPHI Collab., contributed paper 467.
[15] L3 Collab., M. Acciarri et al., CERN PPE/97-99.

Tests of QED and the SM with Photonic Final States at LEP

Vitaly Choutko (vitaly.choutko@cern.ch)

Institute of Theoretical and Experimental Physics, Moscow, Russia

Abstract. Experimental studies of the events with photonic final states at LEP are summarised. Emphasis is given to searches for the possible deviations from QED and SM, such as contact interactions, anomalous couplings and new resonances.

1 QED Tests

QED process $e^+e^- \to n\gamma$, where $n \geq 2$, is free of complications due to weak and/or strong interactions. The overall weak corrections at the LEP energy range are below 1%.

The well known QED cross section [1] has the form

$$\sigma = \frac{2\pi\alpha^2}{s}\left(\log\frac{1+\cos\theta_m}{1-\cos\theta_m} - \cos\theta_m\right) \times \left(1 + O(\alpha\log\frac{s}{m_e})\right) \tag{1}$$

The deviations from the QED can be parameterised in terms of effective energy scale form factors. The following general forms are usually considered[2]:

$$\frac{d\sigma}{d\Omega} = \left(\frac{d\sigma}{d\Omega}\right)_{QED}\left(1 + \frac{s^2}{\alpha}\frac{1}{\Lambda^4}\sin^2\theta\right) \tag{2}$$

$$\frac{d\sigma}{d\Omega} = \left(\frac{d\sigma}{d\Omega}\right)_{QED}\left(1 \pm \frac{s^2}{2}\frac{1}{\Lambda_\pm^4}\sin^2\theta\right) \tag{3}$$

$$\frac{d\sigma}{d\Omega} = \left(\frac{d\sigma}{d\Omega}\right)_{QED}\left(1 + \frac{s^3}{32\pi\alpha^2}\frac{1}{\Lambda'^6}\frac{\sin^2\theta}{1+\cos^2\theta}\right) \tag{4}$$

1.1 Event Selection

The following generic cuts were used to select the $e^+e^- \to n\gamma$ events:

- Two or more photon candidates with energy greater than $O(1\text{ GeV})$ in some angular interval $\theta_{min} < \theta_\gamma < \pi - \theta_{min}$;
- Small transverse missing energy or total event energy greater than $O(\frac{\sqrt{s}}{2})$;
- Energy deposition in hadron calorimeter less than $O(1\text{ GeV})$;
- No tracks/hits in vertex detectors.

After the selection background is on the level of 0.1-0.5% of the signal.

1.2 Cross Section Measurements

Table 1 shows the numbers of the multiphoton events found by LEP experiments in 161-172 GeV runs, together with the predicted by QED (shown in brackets).

Experiment	2γ	3γ	4γ or more
L3	240(239±1)	9(13.5±0.3)	2(0.5)
OPAL	169(170±1)	5(4.1±0.2)	0

Table 1. The number of multiphoton events.

The measured by the L3 experiment cross section at various centre-of-mass energy is shown at Fig. 1.

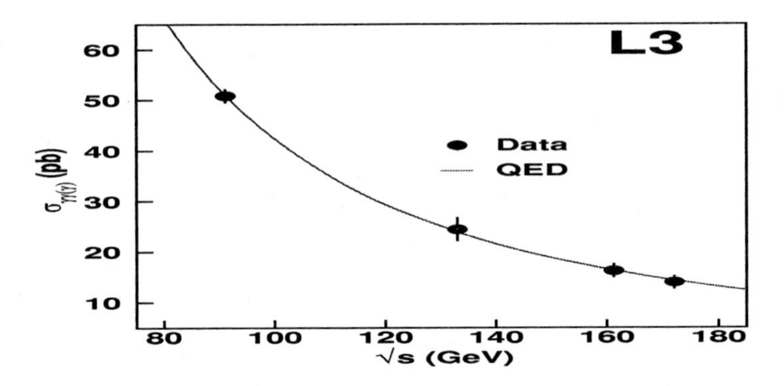

Figure 1. Measured cross section as function of centre-of-mass energy.

Good agreement between measured and predicted values is clearly seen.

1.3 Search for QED Deviations

To extract the limits on the QED scale parameters, binned likelihood fit to $\left(\frac{d\sigma}{d\Omega}\right)_{\mathrm{Born}}$ is usually used.

Table 2 shows the 95% CL limits on the energy scale form factors and mass of the excited electron, obtained by OPAL and L3 experiments.

Experiment	Λ_+	Λ_-	Λ	Λ'	m_{e^*}
L3	207	205	844	507	210
OPAL	194	211	789	479	192

Table 2. Limits on QED deviation parameters.

2 SM Tests

Events with photons and missing p_t can probe the variety of the physics processes beyond SM, such as production of neutralino and/or excited neutrino as well as presence of anomalous $WW\gamma$ and $ZZ\gamma$ couplings. The SM source of the above events at LEP is the reaction $e^+e^- \to \nu\bar{\nu}\gamma(\gamma)$.

The cross section of this process is about 7 pb at 170 GeV. The theoretical accuracy is 1-2% [3].

2.1 Event Selection

The following generic cuts were used to select the $e^+e^- \to \nu\bar{\nu}\gamma$ events:

1. One or more γ candidates in $O(10°) < \theta_\gamma < 180 - O(10°)$;
2. Energy deposition in hadron calorimeter and/or very forward calorimeters less than $O(1$ GeV$)$;
3. No activity in muon detector ;
4. No tracks/hits in vertex detector(s);
5. γ candidate(s) energy and/or p_t more than $O(5$ GeV$)$;
6. Acoplanarity more than $O(5°)$ if two or more photons are present.

Cuts 1-4 reject cosmic ray and beam gas events. Cut 5 rejects the $ee\gamma$ events with two invisible electrons. Last cut rejects $ee \to \gamma\gamma(\gamma)$ events.

2.2 Cross Section Measurements

Fig. 2 shows the ratios of the measured $\sigma_{\nu\nu\gamma(\gamma)}$ to the Standard Model one, obtained by LEP experiments in 161-172 GeV runs.

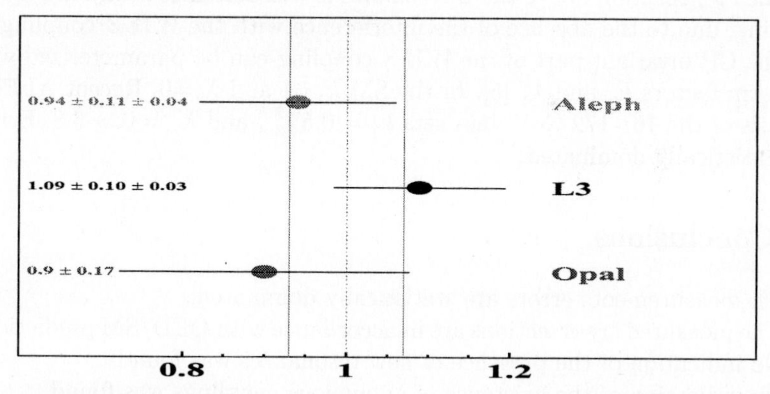

Figure 2. Ratios of the measured $\sigma_{\nu\nu\gamma(\gamma)}$ to the Standard Model one.

The number of observed and predicted events for the $e^+e^- \to \nu\bar{\nu}\gamma\gamma(\gamma)$ reaction is shown in Table 3.

Experiment	$\cos\theta_m$	N^{Obs}	N^{SM}
ALEPH	0.95	3	5
L3	0.97	6	7.8
OPAL	0.97	8	11

Table 3. Number of observed and predicted twophoton events.

All measured cross sections are in accordance with the SM expectations.

2.3 Search for Anomalous Couplings

The $\nu\bar{\nu}\gamma$ reaction at the Z resonance is well suited to study the $ZZ\gamma$ coupling.

The presence of such coupling would manifest itself in the enhancement of the hard photon tail in the $e^+e^- \to \nu\nu\gamma(\gamma)$ reaction.

Fig. 3 shows the upper limits on the $ZZ\gamma$ coupling parameters h^Z_{30} and h^Z_{40} [4] obtained by L3 for Λ_Z=500 GeV.

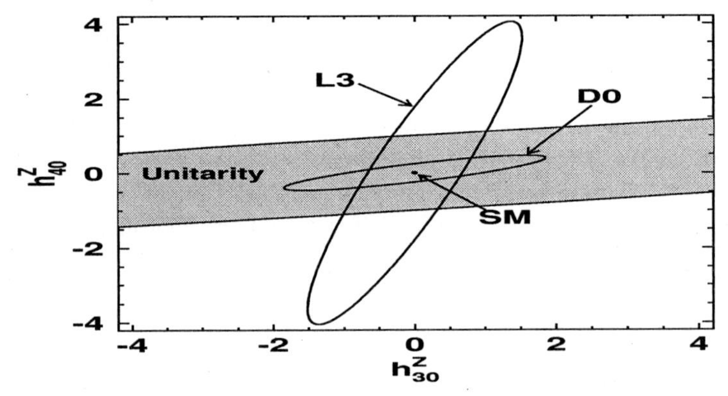

Figure 3. Upper limits on the $ZZ\gamma$ coupling parameters.

The $\nu\bar{\nu}\gamma$ reaction above the Z resonance is well suited to study the $WW\gamma$ coupling, due to the absence of the interference with the WWZ coupling.

The CP invariant part of the $WW\gamma$ coupling can be parameterized with two form factors k_γ and λ_γ [5]. In the SM k_γ=1 and λ_γ=0. Recent ALEPH analysis of the 161-172 GeV data sets $k_\gamma = 0.5^{-2.8}_{+3.3}$ and $\lambda_\gamma = 0. \pm 5.8$. Errors are statistically dominated.

3 Conclusions

- All measurements errors are statistically dominated;
- The measured cross sections are in accordance with QED/SM predictions;
- No indication of the presence of new resonances was found.
- No indication of the existence of anomalous couplings was found.
- New limits on the QED scale parameters are established ($\Lambda_\pm \approx$ 200 GeV).

References

[1] F.A. Berends and R. Gastmans, Nucl. Phys. B61 (1973) 414.
[2] O.J.P. Eboli, A.A Natale and S.F. Novaes Phys. Lett. B271 (1991) 274.
[3] G. Montagna et al., Nucl. Phys. B452 (1996) 161.
[4] U. Baur and E.L. Berger, Phys. Rev. D47 (1993) 4889.
[5] K. Hagiwara, R.D Peccei and D. Zeppenfeld, DESY 86-058.

Exact 1-Loop QED Corrections to $e^+e^- \to 4f$

Alessandro Vicini (vicini@phoenix.ifh.de)

DESY-Zeuthen, Zeuthen (Germany)

Abstract. The status of the calculation of the exact 1-loop radiative corrections to $e^+e^- \to 4f$ and some technical progresses are presented.

The phase 2 of LEP, running above the W pair production threshold, is providing a very important set of informations concerning the gauge structure of the Standard Model (SM). It is possible that in the future a new e^+e^- collider with a center of mass energy of about 1 TeV will improve further our knowledge of this sector of elementary particle physics. The measurement of the mass of the W boson, the nature of the tri- and quadrilinear gauge boson vertices, the investigation of the mechanism of symmetry breaking and the study of the properties of the top quark are some of the main issues under investigation. Thanks to the clean environment characteristic of e^+e^- collisions, it should be possible to perform precision studies of all these topics. In order to obtain a sensitive comparison between experiments and theory, it is mandatory that the theoretical predictions include in the most complete and accurate way the effects of the SM 1-loop radiative corrections, which are known to be potentially sizeable.

The full electroweak 1-loop corrections to on-shell W pair production have been calculated in the past [1], but this is not sufficient to analyze in detail the processes at energies above the production threshold. We can distinguish, in the process $e^+e^- \to W^+W^- \to 4f$, two kinds of radiative corrections: those affecting either the W pair production or the W's decay, and those that connect these subprocesses by means of the exchange of a virtual particle. The former are usually called *factorizable*, while the latter are *non-factorizable*. At the moment only some classes of universal leading QED and QCD corrections are available for the process $e^+e^- \to W^+W^- \to 4f$. Only recently a first estimate of *non-factorizable* QED corrections has been presented, in the double pole approximation [2]. It is well known that the *non-factorizable* corrections yield terms which are potentially large, growing like $\log \frac{s}{m_W^2}$ and thus will be important at the energies of a Next-Linear-Collider. An estimate [3] of the effect of the missing 1-loop corrections in $e^+e^- \to 4f$ shows that they could modify the presently used approximations by 1.5% at LEP2 energies, up to 22% at $\sqrt{s} = 2$ TeV. It is therefore mandatory to take into account also these *non-factorizable* corrections.

One serious problem we face in performing the calculation of the virtual corrections to the amplitude with six external fermions is the presence of 5 and 6 points integrals, with tensorial structure involving up to 3 powers of the integration momentum in the numerator. We want to reduce all the loop integrals to a combination of basic scalar integrals. This reduction can be achieved by means of a covariant decomposition, but these standard techniques are not satisfactory for two main reasons. The large number of tensorial structures which can be formed using the external momenta and the metric tensor yields complicated Gram determinants. The zeroes of the latter introduce spurious kinematical singularities, which should be avoided as much as possible, in order to prevent the appearance of numerical instabilities. Furthermore the final number of scalar integrals is definetely large and therefore the CPU time to evaluate them will increase correspondingly.

The point is that, as long as the integration momentum appears inside one fermion line, it is not possible to simplify it by any means, in order to reduce the rank of the integral. On the other hand the Dirac algebra is very simple for massless fermions in the helicity representation and can be exploited [4] to extract out of the fermion lines some scalar products which contain some powers of the integration momentum. These scalar products can be expressed in terms of the propagators which appear in the denominator of the expression and therefore can be easily simplified. In this way we are able to reduce at the same time the rank and the number of points of the tensor integrals. We implemented this observation [4] in two Mathematica packages, developing several optimization criteria. The two main goals are to reduce as much as possible the rank of 5 and 6 point integrals and to keep the resulting number of terms small. Compared to the standard covariant decomposition, we obtain a reduction by a factor 10 to 50, depending on the example we choose, in the number of final terms which must be numerically evaluated. Clearly the reduction of the rank simplifies also the form of the Gram determinants. This packages can be used to simplify any arbitrary amplitude which has external fermions and produce an output which is expressed in terms of scalar and lower rank tensor integrals. In the calculation of the 1-loop corrections to $e^+e^- \to 4f$ the most involved integrals which remain after the simplification procedure are rank 1, 6 point and rank 2, 5 point integrals. We have found a stable numerical procedure to express these integrals in terms of scalar functions (1,2,3 and 4 point integrals).

This work has been done in collaboration with S.Herrlich and F.Jegerlehner.

References

[1] J.Fleischer, F.Jegerlehner and K.Kolodziej, Phys.Rev.D47 (1993) 380 and reference therein;

[2] W.Beenakker, F.A.Berends and A.P. Chapovsky, hep-ph/9706339 and hep-ph/9707326;

[3] A.Denner and S.Dittmaier, hep-ph/9706388;

[4] R.Pittau, Comput.Phys.Commun.104 (1997), 23-36.

Power Corrections from Short Distances

Georges Grunberg (grunberg@pth.polytechnique.fr)

CERN, Theory Division, CH-1211 Geneva 23, Switzerland, and Centre de Physique Theorique de l'Ecole Polytechnique, 91128 Palaiseau Cedex, France

Abstract. It is argued that power contributions of short distance origin naturally arise in the infrared finite coupling approach. A phenomenology of $1/Q^2$ power corrections is sketched.

1 Introduction

Power-behaved contributions to hard processes not amenable to operator product expansion (OPE) have been derived [1] in recent years through various techniques (renormalons, finite gluon mass, dispersive approach), which all share the assumption that these contributions are of essentially *infrared* (IR) origin. In this talk, I point out that the IR finite coupling approach [2, 3] naturally suggests the existence of additionnal non-standard contributions of *ultraviolet* (UV) origin, hence not related to renormalons (but which may be connected [4, 5] to the removal of the Landau pole from the perturbative coupling).

2 Power corrections and IR regular coupling

Consider the contribution to an Euclidean (quark dominated) observable arising from dressed virtual single gluon exchange, which takes the generic form (after subtraction of the Born term):

$$D(Q^2)=\int_0^\infty \frac{dk^2}{k^2}\,\bar{\alpha}_s(k^2)\,\varphi\left(\frac{k^2}{Q^2}\right) \tag{1}$$

The "physical" coupling $\bar{\alpha}_s(k^2)$ is assumed to be IR regular, and thus must differ from the perturbative coupling $\bar{\alpha}_s^{PT}(k^2)$ (which is assumed to contain a Landau pole) by a non-perturbative piece $\delta\bar{\alpha}_s(k^2)$:

$$\bar{\alpha}_s=\bar{\alpha}_s^{PT}+\delta\bar{\alpha}_s \tag{2}$$

To determine the various types of power contributions, it is appropriate [2, 6] to disentangle long from short distances "a la SVZ" with an IR cutoff Λ_I:

$$D(Q^2)=\int_0^{\Lambda_I^2} \frac{dk^2}{k^2}\,\bar{\alpha}_s(k^2)\,\varphi\left(\frac{k^2}{Q^2}\right)+\int_{\Lambda_I^2}^\infty \frac{dk^2}{k^2}\,\bar{\alpha}_s^{PT}(k^2)\,\varphi\left(\frac{k^2}{Q^2}\right)+\int_{\Lambda_I^2}^\infty \frac{dk^2}{k^2}\,\delta\bar{\alpha}_s(k^2)\,\varphi\left(\frac{k^2}{Q^2}\right) \tag{3}$$

The first integral yields, for large Q^2, "long distance " power contributions which correspond to the standard OPE "condensates". If the Feynman diagram kernel $\varphi\left(\frac{k^2}{Q^2}\right)$ is $\mathcal{O}\left((k^2/Q^2)^n\right)$ at small k^2, this piece contributes an $\mathcal{O}\left((\Lambda^2/Q^2)^n\right)$ term from a dimension n condensate, with the normalization given by a low energy average of the IR regular coupling $\bar{\alpha}_s$. The integral over the perturbative coupling in the short distance part represents a form of "regularized perturbation theory " (choosing the IR cut-off Λ_I above the Landau pole). The last integral in eq.(3) usually yields (unless $\delta\bar{\alpha}_s(k^2)$ is exponentially supressed) new power contributions at large Q^2 of short distance origin , unrelated to the OPE. Assume for instance a power law decrease:

$$\delta\bar{\alpha}_s(k^2) \simeq c\left(\frac{\Lambda^2}{k^2}\right)^p \tag{4}$$

The short distance integral will then contribute a piece:

$$\int_{Q^2}^{\infty} \frac{dk^2}{k^2}\, \delta\bar{\alpha}_s(k^2)\, \varphi\left(\frac{k^2}{Q^2}\right) \simeq A\ c\left(\frac{\Lambda^2}{Q^2}\right)^p \tag{5}$$

where $A \equiv \int_{Q^2}^{\infty} \frac{dk^2}{k^2}\left(\frac{Q^2}{k^2}\right)^p\ \varphi\left(\frac{k^2}{Q^2}\right)$ is a number. If one assumes moreover that $p < n$, the lower integration limit in eq.(5) and in A can actually be set to zero [7], and one gets a *parametrically leading* $\mathcal{O}\left((\Lambda^2/Q^2)^p\right)$ power contribution of UV origin, unrelated to the OPE.

3 Power contributions to the running coupling

A power law decrease of $\delta\bar{\alpha}_s(k^2)$ is a natural expectation for a coupling which is assumed to be defined at the non-perturbative level, and could eventually be derived from the OPE itself as the following QED analogy shows. In QED, the coupling $\bar{\alpha}_s(k^2)$ should be identified, in the present dressed single gluon exchange context, to the Gell-Mann-Low effective charge $\bar{\alpha}$, related to the photon vacuum polarisation $\Pi(k^2)$ by:

$$\bar{\alpha}(k^2) = \frac{\alpha}{1+\alpha\ \Pi(k^2/\mu^2,\alpha)} \tag{6}$$

One expects $\Pi(k^2)$, hence $\bar{\alpha}(k^2)$, to receive power contributions from the OPE. Of course, this cannot happen in QED itself, which is an IR trivial theory, but might occur in the "large β_0", $N_f = -\infty$ limit of QCD . Instead of $\Pi(k^2)$, it is convenient to introduce the related (properly normalized) renormalisation group invariant "Adler function" (with the Born term removed):

$$R(k^2) = \frac{1}{\beta_0}\left(\frac{d\Pi}{d\log k^2} - \frac{d\Pi}{d\log k^2}|\alpha = 0\right) \tag{7}$$

which contributes the higher order terms in the renormalisation group equation:

$$\frac{d\bar{\alpha}_s}{d\ln k^2} = -\beta_0(\bar{\alpha}_s)^2\,(1+R) \tag{8}$$

where β_0 is (minus) the one loop beta function coefficient. Consider now the $N_f = -\infty$ limit in QCD. Then $R(k^2)$ is expected to be purely non-perturbative, since in this limit the perturbative part of $\bar{\alpha}_s$ is just the one-loop coupling $\bar{\alpha}_s^{PT}(k^2) = 1/\beta_0 \ln(k^2/\Lambda^2)$. Indeed, OPE-renormalons type arguments suggest the general structure [8] at large k^2:

$$R(k^2) = \sum_{p=1}^{\infty}\left(a_p \log\frac{k^2}{\Lambda^2} + b_p\right)\left(\frac{\Lambda^2}{k^2}\right)^p \tag{9}$$

where the log enhanced power corrections reflect the presence of double IR renormalons poles [9]. Eq.(8) with R as in eq.(9) can be easily integrated to give:

$$\bar{\alpha}_s(k^2) = \bar{\alpha}_s^{PT}(k^2) + \frac{\Lambda^2}{k^2}\left[a_1\,\bar{\alpha}_s^{PT}(k^2) + \beta_0\,(a_1+b_1)\left(\bar{\alpha}_s^{PT}(k^2)\right)^2\right] + \mathcal{O}\left((\Lambda^4/k^4)\right) \tag{10}$$

One actually expects : $a_1 = b_1 = 0$ (corresponding to the absence of $d = 2$ gauge invariant operator), and $a_2 = 0$ (corresponding to the gluon condensate which yields only a single renormalon pole). It is amusing to note that keeping only the $p = 2$ (gluon condensate) contribution in eq.(9) with $a_2 = 0$, eq.(8) yields :

$$\bar{\alpha}_s(k^2) = \frac{1}{\beta_0\left(\ln\frac{k^2}{\Lambda^2} + \frac{b_2}{2}\frac{\Lambda^4}{k^4}\right)} \tag{11}$$

which coincides with a previously suggested ansatz [10] based on different arguments.

4 $1/Q^2$ power corrections

The previous QED - inspired model (with $a_1 = b_1 = 0$) remains of academic interest, since it is clear that the short distance power corrections induced in QCD observables by the OPE-generated corrections in $\bar{\alpha}_s$ are then parametrically consistent with those expected from the OPE, and are actually probably numerically small compared to those originating directly from the long distance piece in eq.(3) (although it is still an interesting question whether such short distance contributions will not mismatch the expected OPE result for the coefficient functions). The situation changes if one assumes [4, 5] the existence of $1/k^2$ power corrections of non-OPE origin in $\bar{\alpha}_s$. Evidence for such corrections has been found in a lattice calculation [11] of the gluon condensate, and physical arguments have been given [5, 12] for their actual occurence. For instance, consider the case where $n = 2$ in the low energy behavior of the kernel $\varphi\left(\frac{k^2}{Q^2}\right)$, i.e. w here the leading OPE contribution has dimension 4 (the gluon condensate). Then, setting $p = 1$ in eq.(4), the parametrically leading power contribution will be a $1/Q^2$ correction of short distance origin given by the right-hand side eq.(5) with:

$$A \equiv \int_0^\infty \frac{dk^2}{k^2} \frac{Q^2}{k^2} \varphi\left(\frac{k^2}{Q^2}\right) \tag{12}$$

Assuming further the physical coupling $\bar{\alpha}_s$ is *universal* [2, 3], so is the *non-perturbative* parameter c in eq.(4). Then the process dependance of the strength of the $1/Q^2$ correction is entirely contained in the *computable* parameter A. In particular, it is interesting to check [13] whether A in the pseudoscalar channel is substantially larger then in the vector channel, which could help resolve [14] a long standing QCD sum rule puzzle [15], in addition to provide further evidence for $1/Q^2$ corrections. Note that the proposed mechanism is different from the (in essence purely perturbative) one based on UV renormalons [14, 5]. The latter yields [16] an enhancement factor of 18 already in the single renormalon chain approximation (consistent with the present dressed single gluon exchange picture), but is subject to unknown arbitrarily large corrections from multiple renormalons chains, at the difference of the present argument.

I am grateful to M. Beneke, Yu.L. Dokshitzer, G. Marchesini and V.I. Zakharov for useful discussions.

References

1 For recent reviews, see R. Akhoury and V.I. Zakharov, Nuclear Physics Proc. Suppl. 54A (1997) 217; M. Beneke, hep-ph/9706457
2 Yu.L. Dokshitzer and B.R. Webber, Physics Letters B352 (1995) 451
3 Yu.L. Dokshitzer, G. Marchesini and B.R. Webber, Nuclear Physics B469 (1996) 93
4 G. Grunberg, hep-ph/9705290, hep-ph/9705460
5 R. Akhoury and V.I. Zakharov, hep-ph/9705318, hep-ph/9710257
6 M. Neubert, Physical Review D51 (1995) 5924
7 I am indepted to M. Beneke for a related observation
8 G. Grunberg, Physics Letters B349 (1995) 469
9 M. Beneke, Nuclear Physics B405 (1993) 424
10 V.N. Gribov (private communication from Yu.L. Dokshitzer)
11 G. Burgio, F. di Renzo, G. Marchesini and E. Onofri, hep-ph/9706209
12 R. Akhoury and V.I. Zakharov, hep-ph/9710487
13 G. Grunberg, in preparation
14 K. Yamawaki and V.I. Zakharov, hep-ph/9406373
15 V.A. Novikov, M.A. Shifman, A.I. Vainshtein and V.I. Zakharov, Nuclear Physics B191 (1981) 301
16 S. Peris and E. de Rafael, Nuclear Physics B500 (1997) 325

Part VIII

W-Boson Physics

Four-Fermion Production at LEP2 and NLC

Roberto Pittau (Roberto.Pittau@cern.ch)

Theoretical Physics Division, CERN CH-1211 Geneva 23

Abstract. The present knowledge on four-fermion production in electron-positron collisions is reviewed, with emphasis on W boson physics. Different methods to extract M_W from the data are presented and the role of QCD loop corrections discussed.

1 Introduction

Four-fermion processes represent the experimentally measured signal for W boson physics at LEP2 and NLC. In fact, the produced W's always decay, giving four-fermion final states [1]. One is primarily interested in measuring the W mass [2], but also measurements of the trilinear gauge boson couplings [3] and the four-fermion cross sections [4] provide useful information. An accurate determination of M_W can be combined with the precision data coming from LEP1 to further constrain the Standard Model of the Electroweak Interactions. In fact, a global fit to the LEP1 data [5] predicts the W mass with an error of the same order as the expected final LEP2 error (35 MeV), while NLC will presumably reduce that error down to 15 MeV [6]. A strong discrepancy between predicted and measured M_W would be a signal for new physics. Alternatively, an improvement on the measurement of M_W can significantly tighten the present bounds on the Higgs mass through the relation

$$G_\mu = \frac{\alpha\pi}{\sqrt{2}M_W^2(1-M_W^2/M_Z^2)} \cdot \frac{1}{1-\Delta_r(m_t,m_H)}\,, \tag{1}$$

where Δ_r is a calculable contribution coming from radiative corrections.

Several tree-level four-fermion codes are available [4]. Electroweak radiative corrections are usually included at the leading log (LL) level, but very recently, non-factorizable QED corrections have been computed as well [7]. While tree level programs + LL corrections seem in general to be adequate to deal with LEP2 physics, further refinements, especially in the sector of the loop radiative corrections, are needed in view of the NLC precision physics.

In this contribution, I focus my attention on two particular aspects of four-fermion W physics, namely M_W measurement and QCD corrections.

In the next section, I describe two techniques to measure the W mass, and present a new method to extract M_W from the best measured variables.

When quarks are present in the final state, QCD loop contributions have to be included as well. Those corrections are discussed in the last section of the paper.

2 M_W measurement

Two methods are mainly used to extract the W mass: the threshold method and the direct reconstruction technique [2]. In the first case the total W^+W^- cross section is measured near threshold (161 GeV), where the sensitivity to M_W is stronger, and plotted as a function of the W mass. At LEP2, that gives $M_W = 80.40 \pm 0.22$ GeV [5].

The direct reconstruction method is applied at higher energy, where the statistics increases. It requires three steps:

1. From the experimental data the invariant mass distribution $\frac{d\sigma}{dM}$ is reconstructed. To improve the mass resolution, a constrained fit is usually performed event by event, assuming no Initial State Radiation (ISR) and equality between the invariant masses coming from different W's.
2. A theoretical distribution is taken for $\frac{d\sigma}{dM}$ (usually a convolution of a Breit-Wigner with a Gaussian) and a mass M'_W fitted.
3. The value M'_W is then corrected by Monte Carlo, from the bias introduced by the constrained fit, to get the reconstructed W mass M_R with an error ΔM_R .

Such a measurement gave $M_W = 80.37 \pm 0.19$ GeV [5], in the LEP2 run at 172 GeV.

Recently, a new method has been proposed [8] (direct fit method), in which only the best measured quantities are used to extract the W mass. The idea is simple. Given a set of well measured quantities $\{\Phi\}$ one computes, event by event, the theoretical probability P_i of getting the observed set of values $\{\Phi_i\}$ for $\{\Phi\}$. This is a function of M_W and is given by the ratio of the differential cross section in those variables, divided by the total cross section in the experimental fiducial volume

$$P_i(M_W) = \frac{\frac{d\sigma}{d\Phi_i}}{\sigma} . \tag{2}$$

Given N observed events, the logarithm of the likelihood function L

$$\log L(M_W) \equiv \log \prod_{i=1}^{N} P_i(M_W) = \sum_{i=1}^{N} \log \frac{d\sigma}{d\Phi_i}(M_W) - N \log \sigma(M_W) \tag{3}$$

is distributed, for large N, as a quadratic function of M_W. The previous equation is then computed for different values of M_W and a parabola fitted, from which the reconstructed W mass M_R is obtained with an error ΔM_R.

In order to construct a tool for the evaluation of $P_i(M_W)$ one has to choose the set $\{\phi\}$ of accurately measured variables. Although one can always consider more sets $\{\phi\}$, the following choices seem reasonable in practice [9] for different four-fermion final states:

1. Semileptonic case: $q_1 q_2 \ell \nu$
   1a $\{\phi\} = \{E_\ell, \Omega_\ell, \Omega_{q_1}, \Omega_{q_2}\}$
   1b $\{\phi\} = \{E_\ell, \Omega_\ell, \Omega_{q_1}, \Omega_{q_2}, E_h\}$, where E_h is the total energy of the jets.
2. Purely hadronic case: $q_1 q_2 q_3 q_4$
   $\{\phi\} = \{\Omega_{q_1} \Omega_{q_2} \Omega_{q_3} \Omega_{q_4}\}$.
3. Purely leptonic case: $\ell_1 \nu_1 \ell_2 \nu_2$
   $\{\phi\} = \{E_{\ell_1}, \Omega_{\ell_1}, E_{\ell_2}, \Omega_{\ell_2}\}$

Since eight variables determine an event when no ISR is present, sets 1a and 3 would require one or two integrations, cases 1b and 2 none. Including ISR adds two integrations. When jets cannot be assigned to specific quarks a folding over the various possibilities should be included.

As an example of the direct fit method, I show, in table 1, the reconstructed masses obtained by fitting a sample of 1600 EXCALIBUR [1] CC3 events [4] including ISR, generated with an input mass of $M_W = 80.23$ GeV, at $\sqrt{s} = 190$ GeV. In all four cases, the W mass is correctly reconstructed.

1a	1b	2	3
80.238 ± 0.049	80.238 ± 0.032	80.255 ± 0.036	80.209 ± 0.077

Table 1. Reconstructed W mass (GeV) with four choices of the set $\{\Phi\}$ (see text).

Cases 1b and 2 give better errors, because less information is integrated out. Conversely, in the leptonic case 3, the error is worse, since a large part of the kinematical information is actually missing.

In the direct reconstruction method, one wants to keep as much information as possible, also preferably photon momenta, in order to reconstruct the kinematics event by event. On the contrary, in the direct fit method, one has to integrate over all information that is not well determined, in particular ISR. Because of the fact that the integral over the p_T distribution of ISR photons is theoretically better known than the distribution itself, one expects the details of the radiation to matter less in the direct fit method. That can be viewed as an advantage with respect to the direct reconstruction technique.

However, a last remark is in order. All numbers presented in table 1 refer to the partonic level, without inclusion of hadronization effects and detector resolution. Therefore one still has to prove that the fitting procedure survives those effects. This question is currently under investigation [10].

It is also clear that the direct fit method is not only applicable to measure M_W, but can be used, in principle, to extract any parameter - as Γ_W or a set of anomalous trilinear gauge couplings (TGCs) - from the data sample. Not surprisingly, the whole strategy for the direct fit method has been first discussed in the final report of the Workshop on Physics at LEP2, in the context of TGCs determination [3].

3 QCD corrections

QCD loop corrections to four-fermion production in e^+e^- collisions can be divided in two classes, namely QCD corrections to $\mathcal{O}(\alpha^2\alpha_s^2)$ and $\mathcal{O}(\alpha^4)$ processes, respectively.

The first corrections appear as $\mathcal{O}(\alpha_s^2)$ contributions to four-jet production via QCD [11], while the second ones are relevant for studying W boson physics, and, more in general, semileptonic four-fermion processes and fully hadronic final states mediated by electroweak bosons.

I shall concentrate here on the latter contributions, which can all be obtained by defining suitable combinations of loop diagrams plus real gluon radiation, as shown in ref. [12]. The calculation is simplified a lot by using the reduction procedure presented in ref. [13].

In table 2, cross sections computed with the program in ref. [12] are presented, for the semileptonic process $e^+e^- \to \mu^-\bar{\nu}_\mu u\,\bar{d}$.

$\sqrt{s}$	Born	NLO	nQCD
161 GeV	.24962 ± .00002	.24760 ± .00002	.24790 ± .00002
175 GeV	.96006 ± .00007	.94519 ± .00007	.94613 ± .00007
190 GeV	1.184003 ± .00009	1.16681 ± .00009	1.16766 ± .00008
500 GeV	.46970 ± .00006	.47109 ± .00007	.46131 ± .00006

Table 2. Cross sections in pb for $e^+e^- \to \mu^-\bar{\nu}_\mu u\,\bar{d}$ with canonical cuts [4] .

The exact calculation (NLO) is compared with a "naive" approach to strong radiative corrections (nQCD), where the QCD contributions are simply included through the substitutions

$$\Gamma_W \to \Gamma_W \left(1 + \frac{2}{3}\frac{\alpha_s}{\pi}\right), \qquad \sigma \to \sigma\left(1 + \frac{\alpha_s}{\pi}\right). \tag{4}$$

For the direct reconstruction of the W mass, the quantity $\langle \Delta M \rangle = \frac{1}{2\sigma}\int (\sqrt{s_+} + \sqrt{s_-} - 2\,M_W)\, d\sigma$ is relevant [4], where M_W is the input mass in the program. One gets, with canonical cuts at $\sqrt{s} = 175$ GeV:

$$\begin{aligned}\langle \Delta M \rangle_{\rm NLO} &= -\ 0.5585 \pm 0.0002 \text{ GeV}\\ \langle \Delta M \rangle_{\rm nQCD} &= -\ 0.5583 \pm 0.0002 \text{ GeV}.\end{aligned} \tag{5}$$

Also, one can show that the angular distributions are distorted, in the NLO calculation, with respect to the nQCD prediction [12].

From the previous results it is clear that nQCD is adequate at LEP2 for semileptonic processes, but exact calculations are important at NLC and for anomalous couplings studies, where the angular distributions matter to constrain the anomalous contributions.

In table 3, I show results for the fully hadronic process $e^+e^- \to u\bar{d}s\bar{c}$ at $\sqrt{s} = 175$ GeV. The numbers are obtained by using the program described in ref. [14]. In case (a) only canonical cuts are applied. In (b) two reconstructed masses M_{R1} and M_{R2} are determined by minimizing the quantity $\Delta'_M = (M_{R1} - M_W)^2 + (M_{R2} - M_W)^2$, and a cut $|M_{Ri} - M_W| < 10$ GeV is imposed. In (c) a smearing with a Gaussian with a 2 GeV width is introduced in addition, to mimic the experimental resolution.

σ(pb)	(a)	(b)	(c)
NLO	1.1493(4)	0.7895(5)	0.7758(9)
nQCD	1.1069(3)	1.0545(3)	1.0479(3)

Table 3. Cross sections for $e^+e^- \to u\bar{d}s\bar{c}$ at $\sqrt{s} = 175$ GeV.

For the process at hand one gets, for case (b),

$$\begin{aligned}\langle \Delta M \rangle_{\rm NLO} &= -\ 0.2290 \pm 0.0010 \text{ GeV}\\ \langle \Delta M \rangle_{\rm nQCD} &= -\ 0.0635 \pm 0.0004 \text{ GeV},\end{aligned} \tag{6}$$

where now $\langle \Delta M \rangle = \frac{1}{2\sigma}\int (M_{R1} + M_{R2} - 2\,M_W)\, d\sigma$.

We are easily convinced, by the above results, that the naive QCD implementation fails in describing hadronic four-fermion final states at LEP2. In particular, the reduction in cross section (compare cases (a) and (b) in table 3) shows that many soft gluons are exchanged between decay products of different W's that are not taken into account using nQCD. A failure of nQCD can also be proved at NLC energies [14].

4 Conclusions

Precise measurements of M_W are performed at LEP2, using the threshold method and the direct reconstruction of $\frac{d\sigma}{dM_W}$.

An alternative technique is also available in which only well measured quantities are directly used to extract M_W (and TGCs) from the data.

QCD loop corrections to semileptonic $\mathcal{O}(\alpha^4)$ processes are well approximated at LEP2, by nQCD, except for angular distributions, while the naive approach fails in describing hadronic four-fermion final states.

All existing calculations have to be refined and radiative corrections better understood in view of the NLC precision measurements.

References

[1] F.A. Berends, R. Pittau and R. Kleiss, Nucl. Phys. B424 (1994) 308 and B426 (1994) 344, Comput. Phys. Commun. 85 (1995) 437.
[2] Z. Kunszt et al., in Physics at LEP2, CERN 96-101 (1996), eds. G. Altarelli, T. Sjöstrand and F. Zwirner, Vol. 1, p. 141, hep-ph/9602352.
[3] G. Gounaris et al., ibidem, Vol. 1, p. 525, hep-ph/9601233.
[4] D. Bardin et al., ibidem, Vol. 2, p. 3, hep-ph/9709270.
[5] LEP Electroweak Working Group in http://www.cern.ch/LEPEWWG/.
[6] E. Accomando et al., DESY 97-100, hep-ph/9705442.
[7] W. Beenakker, A. P. Chapovskii and F. A. Berends, CERN-TH-97-114, hep-ph/9706339,CERN-TH-97-158, hep-ph/9707326;
A. Denner, S. Dittmaier and M. Roth, CERN-TH-97-258, hep-ph/9710521;
see also S. Dittmaier in these proceedings.
[8] F. A. Berends, C. G. Papadopoulos and R. Pittau, CERN-TH-97-224, hep-ph/9709257, to appear in Phys. Lett. B.
[9] F. A. Berends et al., hep-ph/9709413.
[10] F. A. Berends, C. G. Papadopoulos and R. Pittau, in preparation.
[11] J. M. Campbell, E. W. N. Glover and D. J. Miller, hep-ph/9706297;
Z. Bern, L. Dixon and D. A. Kosower, hep-ph/9708239;
A. Signer and L. Dixon, Phys. Rev. D56 (1997) 4031;
Z. Nagy and Z. Trócsányi, Phys. Rev. Lett. 79 (1997) 3604.
[12] E. Maina, R. Pittau and M. Pizzio, Phys. Lett B393 (1997) 445.
[13] R. Pittau, Comput. Phys. Commun. 104 (1997) 23.
[14] E. Maina, R. Pittau and M. Pizzio, hep-ph/9709454 and 9710375.

Theoretical Aspects of W Physics

Stefan Dittmaier (Stefan.Dittmaier@cern.ch)

CERN, Theory Division, Switzerland

Abstract. High-precision predictions for W-production processes are complicated by the instability of the W bosons, requirements of gauge invariance, and the necessity to include radiative corrections. Salient features and recent progress concerning these issues are discussed for the process ee $\to$ WW $\to 4f$.

1 Introduction

The investigation of the W boson and its properties at LEP2 [1] and possible future linear e^+e^- colliders [2] is very promising. Together with the Fermi constant and the LEP1 observables, an improvement of the empirical value of the W-boson mass M_W will put better indirect constraints on the mass of the Standard-Model Higgs boson and on new-physics parameters. The W-boson mass can be obtained by inspecting the total W-pair production cross-section near threshold, where it is most sensitive to M_W, or by reconstructing the invariant masses of the W decay products. W-boson production in ee-, eγ-, and $\gamma\gamma$-collisions also yields direct information on the vector-boson self-couplings, which are governed by the gauge symmetry. For low and intermediate centre-of-mass (CM) energies, useful information can be obtained by investigating the distributions over the W-production angles. For higher energies also the total cross-sections become very sensitive to anomalous couplings.

The described experimental aims require the knowledge of the Standard-Model predictions for the mentioned observables to a high precision, e.g. for the cross-section of W-pair production at LEP2 to $\sim$0.5% [3]. The instability of the W bosons, the issue of gauge invariance, and the relevance of radiative corrections render this task highly non-trivial. In this short presentation these sources of complications and their consequences for actual calculations are discussed, and special emphasis is laid on recent developments. For definiteness, we consider the process ee $\to$ WW $\to 4f$, which is the most important one for W physics at LEP2.

2 Gauge invariance and finite-width effects

At and beyond a per-cent accuracy, gauge-boson resonances cannot be treated as on-shell states in lowest-order calculations, since the impact of a finite decay width Γ_V for a gauge boson V of mass M_V can be roughly estimated to Γ_V/M_V, which is, for instance, $\sim$3% for the W boson. Therefore, the full set of tree-level diagrams for a given fermionic final state has to be taken into account. For ee $\to$ WW $\to 4f$ this includes graphs with two resonant W-boson lines ("signal diagrams") and graphs with one or no W resonance ("background diagrams"), leading to the following structure of the amplitude [3–5]:

$$\mathcal{M} = \underbrace{\frac{R_{+-}(k_+^2, k_-^2)}{(k_+^2 - M_W^2)(k_-^2 - M_W^2)}}_{\text{doubly-resonant}} + \underbrace{\frac{R_+(k_+^2, k_-^2)}{k_+^2 - M_W^2} + \frac{R_-(k_+^2, k_-^2)}{k_-^2 - M_W^2}}_{\text{singly-resonant}} + \underbrace{N(k_+^2, k_-^2)}_{\text{non-resonant}} . \tag{1}$$

Gauge invariance implies that $\mathcal{M}$ is independent of the gauge fixing used for calculating Feynman graphs (gauge-parameter independence), and that gauge cancellations between different contributions to $\mathcal{M}$ take place. These gauge cancellations are ruled by Ward identities. For a physical description of the W resonances, the finite W decay width has to be introduced in the resonance poles. However, since only the sum in (1), but not the single contributions to $\mathcal{M}$, possesses the gauge-invariance properties, the simple replacement

$$\left[k_\pm^2 - M_W^2\right]^{-1} \quad \to \quad \left[k_\pm^2 - M_W^2 + iM_W \Gamma_W(k_\pm^2)\right]^{-1} \tag{2}$$

in general violates gauge invariance.

Although such gauge-invariance-breaking terms are formally suppressed by a factor Γ_V/M_V [6], they can completely destroy the consistency of predictions if they disturb gauge cancellations [7–9]. Gauge cancellations can occur if a current $\bar{u}(p_1)\gamma^\mu u(p_2)$ that is associated to a pair of external fermions becomes proportional to the momentum k of the attached gauge boson:

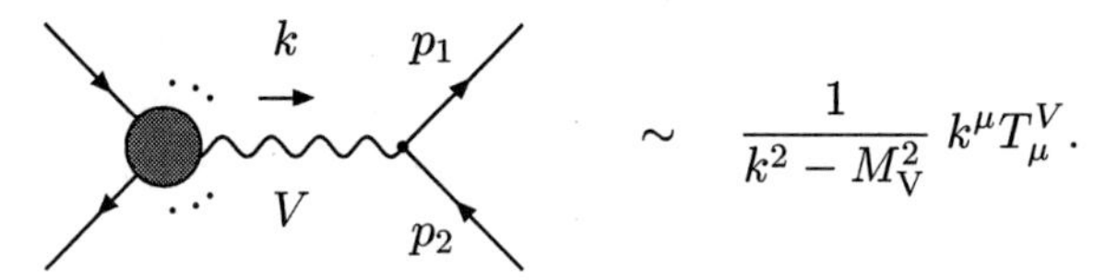

T_μ^V represents the set of subgraphs hidden in the blob. The cancellations in $k^\mu T_\mu^V$ are governed by the Ward identities

$$k^\mu T_\mu^\gamma = 0, \qquad k^\mu T_\mu^Z = iM_Z T^\chi, \qquad k^\mu T_\mu^{W^\pm} = \pm M_W T^{\phi^\pm}. \tag{3}$$

The first one expresses electromagnetic current conservation and is relevant, e.g., for forward scattering of $e^\pm$ ($k \to 0$). The others imply the Goldstone-boson equivalence theorem, which relates the amplitudes for high-energetic longitudinal W and Z bosons ($k^0 \gg M_V$) to the ones for their respective would-be Goldstone bosons ϕ and χ.

Among the proposed methods (see Refs. [3,9] and references therein) to introduce finite widths for W and Z bosons in tree-level amplitudes, the field-theoretically most convincing one is provided by the "fermion-loop scheme". This scheme goes beyond a pure tree-level calculation by including and consistently Dyson-summing all closed fermion loops in $\mathcal{O}(\alpha)$. This procedure introduces the running tree-level width in gauge-boson propagators via the imaginary parts of the fermion loops. The Ward identities (3) are not violated, since the fermion-loop (as well as the tree-level) contributions to vertex functions obey the simple linear (also called "naive") Ward identities that are related to the original gauge invariance rather than to the more involved BRS invariance of the quantized theory. Owing to the linearity of the crucial Ward identities for the vertex functions, the fermion-loop scheme works both with the full fermion loops and with the restriction to their imaginary parts. The full fermion-loop scheme has been worked out for ee $\to$ WW $\to 4f$ in Ref. [9], where applications are discussed as well. Simplified versions of the scheme have been introduced in Ref. [8].

The fermion-loop scheme is not applicable in the presence of resonant particles that do not exclusively decay into fermions. For such particles, parts of the decay width are contained in bosonic corrections. The Dyson summation of fermionic *and* bosonic $\mathcal{O}(\alpha)$ corrections leads to inconsistencies in the usual field-theoretical approach, i.e. the Ward identities (3) are broken in general. This is due to the fact that the bosonic $\mathcal{O}(\alpha)$ contributions to vertex functions do not obey the "naive Ward identities". The problem is circumvented by employing the background-field formalism [10], in which these naive identities are valid. This implies [11] that a consistent Dyson summation of fermionic and bosonic corrections to any order in α does not disturb the Ward identities (3). Therefore, the background-field approach provides a natural generalization of the fermion-loop scheme. We recall that any resummation formalism goes beyond a strict order-by-order calculation and necessarily involves ambiguities in relative order α^n if not all n-loop diagrams are included. This kind of scheme dependence, which in particular concerns gauge dependences, is only resolved by successively calculating the missing orders.

Note that the consistent resummation of all $\mathcal{O}(\alpha)$ loop corrections does not automatically lead to $\mathcal{O}(\alpha)$ precision in the predictions if resonances are involved. The imaginary parts of one-loop self-energies generate only tree-level decay widths so that directly on resonance one order in α is lost. To obtain also full $\mathcal{O}(\alpha)$ precision in these cases, the imaginary parts of the two-loop self-energies are required. However, how and whether this two-loop contribution can be included in a practical way without violating the Ward identities (3) is still an open problem. Taking the imaginary parts of all two-loop contributions solves the problem in principle at least for the background-field approach, but this is certainly impractical.

Fortunately, the full off-shell calculation for the process ee $\to$ WW $\to 4f$ in $\mathcal{O}(\alpha)$ is not needed for most applications. Sufficiently above the W-pair threshold a good approximation should be obtained by taking into account only the doubly-resonant part of the amplitude (1), leading to an error of the order of $\alpha\Gamma_{\mathrm{W}}/(\pi M_{\mathrm{W}}) \lesssim 0.1\%$. In such a "pole scheme" calculation [4,12] the numerator $R_{+-}(k_+^2, k_-^2)$ has to be replaced by the gauge-independent residue $R_{+-}(M_{\mathrm{W}}^2, M_{\mathrm{W}}^2)$. The structure of this approach, which is in fact non-trivial and has not been completely carried out so far, is described below.

3 Electroweak radiative corrections

Present-day Monte Carlo generators for off-shell W-pair production (see e.g. Ref. [13]) include only universal electroweak $\mathcal{O}(\alpha)$ corrections[1] such as the running of the electromagnetic coupling, $\alpha(q^2)$, leading corrections entering via the ρ-parameter, the Coulomb singularity [15], which is important near threshold, and mass-singular logarithms $\alpha \ln(m_e^2/Q^2)$ from initial-state radiation. In leading order, the scale Q^2 is not determined and has to be set to a typical scale for the process; for the following we take $Q^2 = s$. Since the full $\mathcal{O}(\alpha)$ correction is not known for off-shell W pairs, the size of the neglected $\mathcal{O}(\alpha)$ contributions is estimated by inspecting on-shell W-pair production, for which the exact $\mathcal{O}(\alpha)$ correction and the leading contributions were presented in Refs. [16] and [17], respectively. These $\mathcal{O}(\alpha)$ corrections have already been implemented in an event generator for on-shell W pairs [18]. The following table shows the difference between an "improved Born approximation" δ_{IBA}, which is based on the above-mentioned universal corrections, and the corresponding full $\mathcal{O}(\alpha)$ correction δ to the Born cross-section integrated over the W-production angle θ for some CM energies $\sqrt{s}$.

θ range	$\sqrt{s}$/ GeV	161	175	200	500	1000	2000
$0^\circ < \theta < 180^\circ$	$(\delta_{\mathrm{IBA}} - \delta)/\%$	1.5	1.3	1.5	3.7	6.0	9.3
$10^\circ < \theta < 170^\circ$		1.5	1.3	1.5	4.7	11	22

Here the corrections δ_{IBA} and δ include only soft-photon emission. For more details and results we refer to Refs. [3, 19]. The quantity $\delta_{\mathrm{IBA}} - \delta$ corresponds to the neglected non-leading corrections and amounts to ~1–2% for LEP2 energies, but to ~10–20% in the TeV range. Thus, in view of the aimed 0.5% level of accuracy for LEP2 and all the more for energies of future linear colliders, the inclusion of non-leading corrections is indispensable. The large contributions in $\delta_{\mathrm{IBA}} - \delta$ at high energies are due to terms such as $\alpha \ln^2(s/M_W^2)$, which arise from vertex and box corrections and can be read off from the high-energy expansion [20] of the virtual and soft-photonic $\mathcal{O}(\alpha)$ corrections.

As explained above, a reasonable starting point for incorporating $\mathcal{O}(\alpha)$ corrections beyond universal effects is provided by a double-pole approximation. Doubly-resonant corrections to ee $\to$ WW $\to 4f$ can be classified into two types: factorizable and non-factorizable corrections [3–5]. The former are those that correspond either to W-pair production [16] or to W decay [21]. Since these corrections were extensively discussed in the literature, we focus on the non-factorizable corrections. They are furnished by diagrams in which the production subprocess and/or the decay subprocesses are not independent. Among such corrections, doubly-resonant contributions only arise if a "soft" photon of energy $E_\gamma \lesssim \Gamma_W$ is exchanged between the subprocesses.

In Ref. [22] it was shown that the non-factorizable corrections vanish if the invariant masses of both W bosons are integrated over. Thus, these cor-

[1] The QCD corrections for hadronic final states are discussed in Ref. [14].

rections do not influence pure angular distributions, which are of particular importance for the analysis of gauge-boson couplings. For exclusive quantities the non-factorizable corrections are non-vanishing and have been calculated in Refs. [23–25][2]. It turns out [25] that the correction factor to the differential Born cross-section is non-universal in the sense that it depends on the parametrization of phase space. The calculations [23–25] have been carried out using the invariant masses $M_\pm$ of the $W^\pm$ bosons, which are identified with the invariant masses of the respective final-state fermion pairs, as independent variables. Since all effects from the initial e^+e^- state cancel, the resulting correction factor does not depend on the W-production angle and is also applicable to processes such as $\gamma\gamma \to WW \to 4f$. Figure 1 shows that non-factorizable corrections to a single invariant-mass distribution are of the order of ~1% for LEP2 energies, shifting the maximum of the distribution by an amount of 1–2 MeV, which is small with respect to the experimental uncertainty at LEP2 [26]. For higher energies the non-factorizable correction is more and more suppressed.

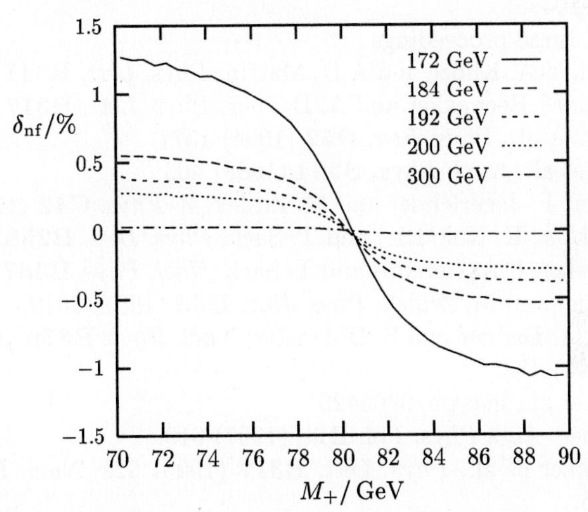

Fig. 1. Relative non-factorizable corrections to the invariant-mass distribution $d\sigma/dM_+$ in $e^+e^- \to \mu^+\nu_\mu\tau^-\bar{\nu}_\tau$ for some CM energies (plot taken from Ref. [25]).

References

[1] Report CERN 96-01, *Physics at LEP2*, eds. G. Altarelli, T. Sjöstrand and F. Zwirner.

[2] E. Accomando et al., hep-ph/9705442, to appear in *Phys. Rep.*

[3] W. Beenakker et al., in Ref. [1], hep-ph/9602351.

[2] The recent evaluations [24, 25] are in complete mutual agreement, but confirm the analytical results of Ref. [23] only for the special final state $f\bar{f}f'\bar{f}'$.

[4] A. Aeppli, G.J. van Oldenborgh and D. Wyler, *Nucl. Phys.* **B428** (1994) 126.
[5] W. Beenakker and A. Denner, *Int. J. Mod. Phys.* **A9** (1994) 4837.
[6] A. Aeppli, F. Cuypers and G.J. van Oldenborgh, *Phys. Lett.* **B314** (1993) 413.
[7] Y. Kurihara, D. Perret-Gallix and Y. Shimizu, *Phys. Lett.* **B349** (1995) 367.
[8] U. Baur and D. Zeppenfeld, *Phys. Rev. Lett.* **75** (1995) 1002;
C. G. Papadopoulos, *Phys. Lett.* **B352** (1995) 144;
E.N. Argyres et al., *Phys. Lett.* **B358** (1995) 339.
[9] W. Beenakker et al., *Nucl. Phys.* **B500** (1997) 255.
[10] A. Denner, S. Dittmaier and G. Weiglein, *Nucl. Phys.* **B440** (1995) 95;
X. Li and Y. Liao, *Phys. Lett.* **B356** (1995) 68.
[11] A. Denner and S. Dittmaier, *Phys. Rev.* **D54** (1996) 4499.
[12] R.G. Stuart, *Phys. Lett.* **B262** (1991) 113.
[13] D. Bardin et al., in Ref. [1], hep-ph/9709270, and references therein.
[14] E. Maina, R. Pittau and M. Pizzio, *Phys. Lett.* **B393** (1997) 445 and hep-ph/9710375;
R. Pittau, these proceedings.
[15] V.S. Fadin, V.A. Khoze and A.D. Martin, *Phys. Lett.* **B311** (1993) 311;
D. Bardin, W. Beenakker and A. Denner, *Phys. Lett.* **B317** (1993) 213;
V.S. Fadin et al., *Phys. Rev.* **D52** (1995) 1377.
[16] M. Böhm et al., *Nucl. Phys.* **B304** (1988) 463;
J. Fleischer, F. Jegerlehner and M. Zralek, *Z. Phys.* **C42** (1989) 409;
W. Beenakker, K. Kolodziej and T. Sack, *Phys. Lett.* **B258** (1991) 469;
W. Beenakker, F.A. Berends and T. Sack, *Nucl. Phys.* **B367** (1991) 287;
K. Kolodziej and M. Zralek, *Phys. Rev.* **D43** (1991) 3619.
[17] M. Böhm, A. Denner and S. Dittmaier, *Nucl. Phys.* **B376** (1992) 29; E: **B391** (1993) 483.
[18] S. Jadach et al., hep-ph/9705429.
[19] S. Dittmaier, *Acta Phys. Pol.* **B28** (1997) 619.
[20] W. Beenakker et al., *Phys. Lett.* **B317** (1993) 622; *Nucl. Phys.* **B410** (1993) 245.
[21] D.Yu. Bardin, S. Riemann and T. Riemann, *Z. Phys.* **C32** (1986) 121;
F. Jegerlehner, *Z. Phys.* **C32** (1986) 425;
A. Denner and T. Sack, *Z. Phys.* **C46** (1990) 653.
[22] V.S. Fadin, V.A. Khoze and A.D. Martin, *Phys. Rev.* **D49** (1994) 2247;
K. Melnikov and O. Yakovlev, *Phys. Lett.* **B324** (1994) 217.
[23] K. Melnikov and O. Yakovlev, *Nucl. Phys.* **B471** (1996) 90.
[24] W. Beenakker, F.A. Berends and A.P. Chapovsky, hep-ph/9706339 and hep-ph/9707326.
[25] A. Denner, S. Dittmaier and M. Roth, hep-ph/9710521.
[26] Z. Kunszt et al., in Ref. [1], hep-ph/9602352.

Review of W Results from D

Ashutosh V. Kotwal (kotwal@fnal.gov) for the DØ Collaboration

Columbia University, New York City, U.S.A.

Abstract. The DØ experiment collected ~15 pb^{-1} of data in 1992-93 and ~89 pb^{-1} in 1994-95 at the Fermilab Tevatron Collider using $p\bar{p}$ collisions at $\sqrt{s} = 1.8$ TeV. From $W \to e\nu, \mu\nu$ and $Z \to ee, \mu\mu$ decays, the W and Z production cross sections and the W width are determined. Events with $W \to \tau\nu$ decays are used to determine the ratio of W couplings. Using $W \to e\nu$ and $Z \to ee$ decays, the W boson mass is measured. Limits on the anomalous trilinear gauge boson couplings are obtained at the 95% confidence level from four diboson processes: $W\gamma$, WW, WZ and $Z\gamma$.

1 *W*, *Z* Cross Sections, *W* Width and Ratio of *W* Couplings

Events with decays to electrons and muons are selected by requiring a high p_t e or μ and large missing E_T ($\not{E}_t$) for W's and two high p_t e's or μ's for Z's. Electrons were restricted to a region $|\eta| < 1.1$ and $1.5 < |\eta| < 2.5$ and muons to the region $|\eta| < 1.0$. The hadronic decay of the τ is used to select the $W \to \tau\nu$ events.

The published 1992-93 results [1] and the preliminary 1994-95 results of $\sigma \cdot B$ are shown in figure 1 along with the results from CDF, and the Standard Model (SM) predictions obtained using the CTEQ2M parton distribution functions (pdf).

Many common systematic errors cancel in the ratio

$$R = \frac{\sigma_W \cdot B(W \to l\nu)}{\sigma_Z \cdot B(Z \to ll)} = \frac{\sigma_W \cdot \Gamma(W \to l\nu) \cdot \Gamma_Z}{\sigma_Z \cdot \Gamma(Z \to ll) \cdot \Gamma_W} \tag{1}$$

Combining the electron and muon measurements, we obtain the preliminary 1994-95 result $R = 10.32 \pm 0.43$. Using the measured value of R, the branching ratio $B(Z \to ll)$ from LEP measurements and σ_W/σ_Z and $\Gamma(W \to l\nu)$ from the SM, the preliminary 1994-95 result for Γ_W is

$$\Gamma_W = 2.159 \pm 0.092 \text{ GeV} \tag{2}$$

Comparison of the published world average $\Gamma_W = 2.062 \pm 0.059$ GeV [1] (not including the preliminary 1994-95 result) with the SM prediction $\Gamma_W = 2.077 \pm 0.014$ gives a 95% confidence level (CL) upper limit of $\Delta\Gamma_W < 109$ MeV on non-SM decays of the W.

The hadronic decays of the τ are identified by the presence of a narrow, isolated jet with width $rms_{jet} < 0.25$, E_T(jet)>25 GeV, $|\eta_{jet}| < 0.9$, $\not{E}_t$ >25

GeV, and the absence of an opposite side jet. QCD background is suppressed by using the jet E_T profile. The final $W \to \tau\nu$ sample consists of 1202 events with an estimated 18.5% background which is dominated by QCD and noise events. The preliminary result of $\sigma \cdot B(W \to \tau\nu) = 2.38 \pm 0.09$ (stat) $\pm$ 0.10 (syst) nb is obtained, where the luminosity uncertainty has not been included. The ratio of the W couplings is determined to be $g_\tau^W / g_e^W = 1.004 \pm$ 0.019 (stat) $\pm$0.026 (syst). This measurement supports the hypothesis of $e-\tau$ universality.

2 *W* Mass

A precise measurement of M_W tests the SM and in conjunction with the top mass constrains the Higgs mass. The 1992-93 published result [2] is $M_W =$ 80.35 $\pm$ 0.27 GeV. We present a preliminary measurement of the W mass from the 1994-95 data sample using the $W \to e\nu$ events. The events were selected by requiring an isolated electron with $E_T^e > 25$ GeV and $| \eta |< 1.0$, $p_t^\nu \equiv \not{E}_t > 25$ GeV and $p_t^W < 15$ GeV. The distribution of the transverse mass, defined as $M_T^W = \sqrt{2p_t^e p_t^\nu - 2\mathbf{p}_t^e \cdot \mathbf{p}_t^\nu}$ of the resulting 28323 events is used to extract the W mass by performing a maximum likelihood fit using Monte Carlo distributions generated for different M_W.

The Monte Carlo simulation uses a theoretical model of the W production and decay and a parametrized model of the detector response. The detector model, including the electromagnetic and hadronic energy scales and resolutions, is tuned on collider data, mainly $Z \to ee$ decays, and also $J/\psi \to ee$ and converted $\pi^0 \to \gamma\gamma \to 4e$. The production model is constrained using the p_t^Z spectrum measured using the electrons from $Z \to ee$ decays.

The preliminary result from the 1994-95 data is $M_W = 80.45 \pm 0.070$ (stat)$\pm$ 0.065 (em scale) $\pm$ 0.070 (syst) GeV. The fit to the p_t^e spectrum gives $M_W =$ 80.49 $\pm$ 0.14 GeV. The fit to the p_t^ν spectrum gives $M_W = 80.42 \pm 0.18$ GeV. Combining the 1992-93 and 1994-95 results from the M_T^W fit, we obtain $M_W = 80.44 \pm 0.11$ GeV, which is currently the most accurate direct measurement. This result, with the direct measurement of the top quark mass [3], is shown with the SM predictions in figure 2.

3 $W\gamma$, WW and WZ Production

The $W(l\nu)\gamma$ candidates were selected by requiring an isolated, high E_T photon, large $\not{E}_t$, and an isolated high p_t electron (and high $e\nu$ transverse mass) or muon. The selection criteria yielded 57 $W(e\nu)\gamma$ and 70 $W(\mu\nu)\gamma$ candidates. The measured $W\gamma$ $\sigma \cdot B(W \to l\nu)$ is consistent with the SM prediction. A U(1)-only $W\gamma$ coupling is excluded at 96% CL.

The W boson pair candidates were selected by requiring two isolated leptons ($ee, e\mu$ or $\mu\mu$) with high E_T and large $\not{E}_t$. The major sources of

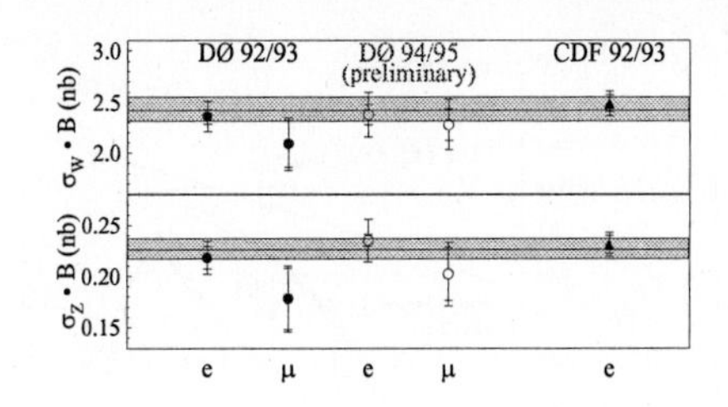

.0

Fig. 1. W and Z cross section measurements compared with the prediction using CTEQ2M pdf (band indicates theoretical uncertainty).

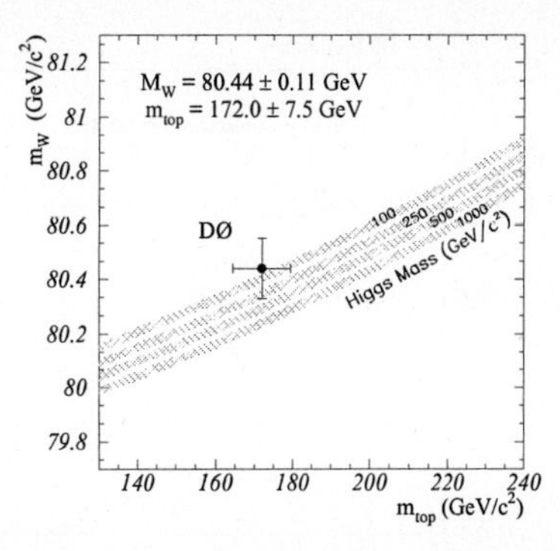

Fig. 2. Direct measurement of W and top mass compared with SM predictions for various assumed Higgs masses.

background are Drell-Yan, $t\bar{t}$, $W\gamma$, $Z \to \tau\tau \to$ dileptons and $W+$ jets production. For the three channels combined, five events are observed, with the SM expectation of 2.10±0.15 and expected background of 3.3±0.4 events.

The WW/WZ candidates were obtained by requiring an isolated high E_T electron, large $\not{E}_t$ and M_T, and two high E_T jets with the dijet invariant mass consistent with W or Z decay. 483 data events are observed, consistent with the expected background of 463±40 events, mainly from QCD multijet and W + jets processes. The SM predicts 20.7±3.1 events. The U(1) point is excluded at the 99% CL.

A simultaneous fit was performed to the photon E_T spectrum of $W\gamma$ events, the E_T of the two leptons in the $WW \to$ dileptons candidates and the $p_t^{e\nu}$ spectrum in the $WW/WZ \to e\nu$ + dijet candidates, from the combined 1992-95 data. Correlated uncertainties in integrated luminosities and predicted cross sections are taken into account. The fit yields significantly better limits than the analyses of the individual production modes [4]. Quoting the preliminary 95% CL limits using the assumption that the $WW\gamma$ and WWZ couplings are equal:

$$-0.33 < \Delta\kappa < 0.45\ (\lambda = 0);\ -0.20 < \lambda < 0.20\ (\Delta\kappa = 0) \qquad (3)$$

The limits shown in figure 3 represent significant progress in constraining $WW\gamma/WWZ$ couplings in the past several years and are competitive with expectations from a completed LEP2 program.

4 $Z\gamma$ Production

In the charged lepton channels, we search for two high p_t, isolated electrons or muons and an isolated, high p_t photon. In the combined 1992-95 data,

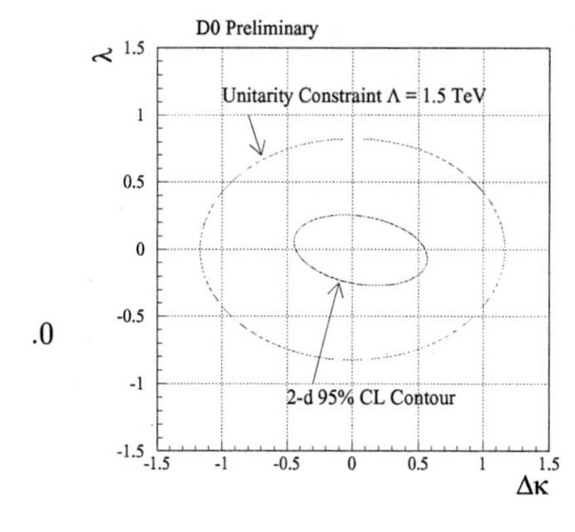

Fig. 3. Limits on anomalous couplings from a combined fit to $W\gamma$, $WW \to$ dileptons and $WW/WZ \to e\nu jj$ data.

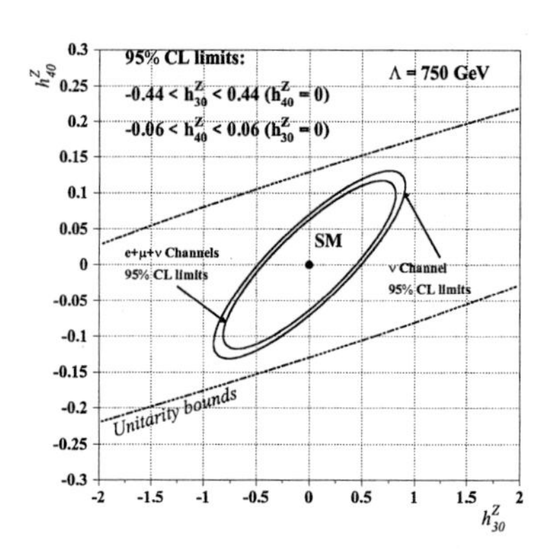

Fig. 4. Limits on $ZZ\gamma/Z\gamma\gamma$ anomalous couplings.

we find 18 $Z(ee)\gamma$ and 17 $Z(\mu\mu)\gamma$ events. The expected background from $Z+$ jets, multijet and direct photon production in both channels is 5.9±1.1 events.

An analysis of the $Z\gamma \to \nu\nu\gamma$ process, requiring the photon to have E_T >40 GeV and $\not{E}_t > 40$ GeV yields four events, with an expected background of 5.8±1.0 events from $W \to e\nu$ and cosmic or beam halo muon bremsstrahlung, and the SM expection of 1.8±0.2 events.

Limits are extracted using a maximum likelihood fit to the photon E_T. Figure 4 shows the limits on anomalous $Z\gamma$ couplings from the 1992-93 data. The 1994-95 preliminary 95% CL limits from the charged lepton decay modes, assuming $\Lambda = 0.5$ TeV, are:

$$\begin{aligned} -1.3 < h^Z_{30}(h^Z_{10}) < 1.3 \; (h^Z_{40}(h^Z_{20}) = 0) \\ -0.26 < h^Z_{40}(h^Z_{20}) < 0.26 \; (h^Z_{30}(h^Z_{10}) = 0) \end{aligned} \quad (4)$$

References

[1] S. Abachi *et al.*, (DØ Collaboration), Physical Review Letters 75 (1995), 1456 and references therein.

[2] S. Abachi *et al.*, (DØ Collaboration), Physical Review Letters 77 (1996) 3309.

[3] S. Abachi *et al.*, (DØ Collaboration), Phys. Rev. Lett. 79 (1997) 1197; S. Abachi *et al.*, (DØ Collaboration), FERMILAB-PUB-97-172-E, submitted to Phys. Rev. Lett.

[4] S. Abachi *et al.*, (DØ Collaboration), submitted to Phys. Rev. D, Fermilab-Pub-97/088-E.

The Cross Sections for WW and Four-Fermion Production at LEP2

Christoph Burgard (christoph.burgard@cern.ch)

CERN, PPE, 1211 Geneva 23, Switzerland

Abstract. The first year of LEP2 running gave a number of results on the production of four-fermion final states in e^+e^- collisions above the Z^0 pole. The results of four-fermion production through neutral current, single W and WW pair events are presented and discussed.

1 Introduction

After six years of running on the Z^0 pole, LEP gradually increased its center-of-mass energy in steps to 130-136 GeV, 161 GeV, 172 GeV in 1995 and 1996. At these energies the four LEP experiments each recorded about 6 pb^{-1}, 11 pb^{-1} and 10 pb^{-1} respectively. At 161 GeV the WW-pair production threshold was passed. In 1997 183 GeV was reached. At the time of this conference 7 pb^{-1} were recorded by each of the experiments at this energy [1].

In this talk I will present results on the production of four-fermion final states, as measured by the four LEP experiments. Depending on the specific final state and phase space region under consideration, diagrams involving only neutral vector bosons, a single W boson or WW boson pairs dominate, with sensitivities to different aspects of the Standard Model (SM) or extensions thereof.

2 Neutral Current four-fermion production

In the Standard Model the neutral current diagrams for four-fermion production include two neutral vector boson propagators. Depending on the bosonic decay modes, events can be classified into four classes and be selected using their distinct topologies [1]. The "low multiplicity" events from $\ell\ell\ell'\ell'$ final states are characterised by four charged tracks and large visible energy. Lepton identification can be applied but is not mandatory due to the clear topology. When one of the lepton pairs is a tau pair, the decays of the tau leptons can increase the number of tracks and typically reduces the visible energy. The "high multiplicity" topology from $\ell\ell q\bar{q}$ final states can be selected by requiring two leptons, two jets and large visible energy. The $\nu\bar{\nu}f\bar{f}$ final states give a "missing momentum" topology. Requiring that the missing

[1] At the end of 1997 this increased to about 60 pb^{-1} per experiment.

momentum does not to point along the beam pipe significantly reduces the background from two-photon processes. The "four-jet" topology from $q\bar{q}q'\bar{q}'$ final states has a large background from QCD which can only be reduced for diagrams with at least one on-shell massive vector boson. For neutral current events below the ZZ threshold only events with a radiative return to the Z^0 pole exhibit a resonant structure. Such events are not of foremost interest from the four-fermion production point of view and are not considered here further. However, in terms of searches it was hoped to find a new resonance in this channel [2].

The number of selected neutral current four-fermion candidates and the expectation from SM processes are listed in table 1[2]. Some excess is seen in the OPAL high multiplicity sample for the 130-136 GeV run coming from $q\bar{q}\mu\bar{\mu}$ final states (with 5 events observed over 0.6 expected in this specific final state). The other experiments do not see such an excess, suggesting it to be a statistical fluctuation. Otherwise, good agreement is found.

		130-136 GeV			
		ALEPH	DELPHI	L3	OPAL
$q\bar{q}\ell\ell$	exp. signal	4.41 ± 0.23	$2.38 \pm 0.14^*$	2.10 ± 0.04	$2.46 \pm 0.21^*$
$+$	exp. backg.	$0.08^{+0.17}_{-0.04}$	$0.32 \pm 0.15^*$	0.80 ± 0.30	$0.31^{+0.23*}_{-0.15}$
$\ell\bar{\ell}\ell'\bar{\ell}'$	observed	2	2*	2	7*
$ff\nu\bar{\nu}$	exp. signal	2.26 ± 0.12	$0.27 \pm 0.03^*$		
	exp. backg.	$0.05^{+0.08}_{-0.02}$	$0.27 \pm 0.17^*$		
	observed	3	0*		

		161 GeV			172 GeV	
		DELPHI	L3	OPAL	DELPHI	L3
$q\bar{q}\ell\ell$	exp. signal	$2.30 \pm 0.11^*$	2.70 ± 0.10	$3.44 \pm 0.20^*$	$2.00 \pm 0.10^*$	2.56 ± 0.09
$+$	exp. backg.	$0.91 \pm 0.17^*$	0.71 ± 0.09	$0.80 \pm 0.19^*$	$0.95 \pm 0.15^*$	0.80 ± 0.20
$\ell\bar{\ell}\ell'\bar{\ell}'$	observed	3*	3	7*	5*	4
$ff\nu\bar{\nu}$	exp. signal	$0.27 \pm 0.02^*$			$0.22 \pm 0.03^*$	
	exp. backg.	$0.80 \pm 0.15^*$			$1.44 \pm 0.17^*$	
	observed	1*			0*	

Table 1. Number of expected and observed events for neutral current production of different four-fermion final states. The DELPHI $ff\nu\bar{\nu}$ numbers are for $q\bar{q}\nu\bar{\nu}$ only.

3 Single W production

At tree level single W production [3] is described by the three diagrams depicted in figure 1. Diagram (a) is the most interesting one due to the presence

[2] Throughout this paper preliminary numbers are indicated by an asterix. Errors quoted include statistical and systematic errors.

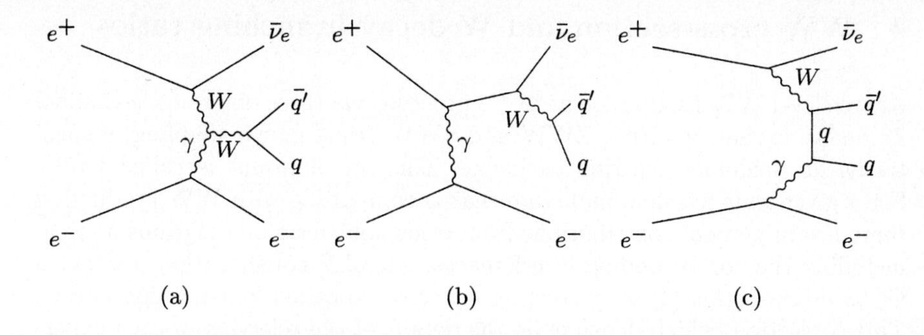

Fig. 1. Lowest order Feyman diagrams for single W production.

of a triple-gauge-coupling (TGC). These couplings while an important ingredient of the SM, are not very well tested experimentally. The q^2 of the neutral boson is small due to the t-channel nature of this diagram and the TGC is almost exclusively photonic.

The four-fermion final state from single W production is $\mathrm{e}^-\bar{\nu}_\mathrm{e}\mathrm{f}\bar{\mathrm{f}}'$. When the W decays hadronically, the event has two acoplanar jets and large transverse missing momentum, while for leptonically decaying W bosons one finds one isolated energetic lepton in the detector, nothing else. For both topologies, the backgrounds from two-photon and 2f(γ) processes can be significantly reduced by requiring no activity in the forward detectors. The number of expected and observed events, as well as the signal efficiencies for the measurements performed by L3 and OPAL, are listed in table 2. From these, the following limits on anomalous photonic TGC can be derived. L3: $-3.6 < \lambda_\gamma < 3.6$, $-3.6 < \Delta\kappa_\gamma < 1.5$; OPAL: $-3.1^* < \lambda_\gamma < 3.1^*$, $-3.6^* < \Delta\kappa_\gamma < 1.6^*$.

	$\mathrm{e}^-\bar{\nu}_\mathrm{e}\mathrm{q}\bar{\mathrm{q}}$		$\mathrm{e}^-\bar{\nu}_\mathrm{e}\ell^+\bar{\nu}_\ell$	
comb.	L3	OPAL	L3	OPAL
observed	10	4*	2	2*
exp. signal	2	1.3*	1	0.8*
exp. backg.	6.2	1.1*	0.5	1.2*
ϵ^{sig}	41%	37%*	53%	50%*

Table 2. Number of observed and expected events and signal efficiencies for single W production.

4 WW cross-section and W decay branching ratios

At tree level WW production at LEP2 proceeds via three diagrams. s-channel Z^0 and γ exchange with a ZWW and γWW triple gauge coupling, respectively, or t-channel neutrino exchange. This set diagrams is called CC03. For a given four-fermion final state that is compatible with WW production there are in general contributions from other four-fermion diagrams as well, including the corresponding interferences. All LEP collaborations extract a CC03 cross-section [4] by correcting for these so called four-fermion effects. This correction, which depends on the details of the selection of each experiment, is obtained by comparing the results from a Monte Carlo with only the CC03 diagrams switched on to those obtained when the complete set of four-fermion diagrams is included.

There are three distinct topologies which are characterized as follows. If both W decay leptonically (11%), the visible part of the event consists of two acoplanar, energetic leptons of opposite charge. If only one of the W decays hadronically (44%), two well-separated jets and one energetic, isolated lepton are found in general. The rest of the events (46%), consist of four jets of similar energies coming from hadronic decays of both W bosons. Such a four-jet topology has a sizeable QCD background from $Z/\gamma \to q\bar{q}$ events. To extract the cross-section of this channel all LEP experiments apply an event-weight technique rather than just counting the number of events that pass a certain set of cuts. From the number of events observed in the different channels (or the sum of the event weights in the fully hadronic channel), the estimated size of the background and the measured integrated luminosity, the WW production cross-section and the W branching ratios can be extracted. The corresponding results are listed in table 3. It can be seen that within the

	ALEPH	DELPHI	L3	OPAL	LEP
$W \to e\bar{\nu}_e$ [%]	9.7 ± 2.2	$10.2 \pm 3.8^*$	16.5 ± 3.7	9.8 ± 2.1	$10.8 \pm 1.3^*$
$W \to \mu\bar{\nu}_\mu$ [%]	11.2 ± 2.2	$10.7 \pm 3.2^*$	8.4 ± 2.8	7.3 ± 1.8	$9.2 \pm 1.1^*$
$W \to \tau\bar{\nu}_\tau$ [%]	11.3 ± 2.9	$13.4 \pm 5.0^*$	10.9 ± 4.2	14.0 ± 2.9	$12.7 \pm 1.7^*$
$W \to q\bar{q}$ [%]	67.7 ± 3.2	$66.0^{+3.7*}_{-3.8}$	$64.2^{+3.7}_{-3.8}$	69.8 ± 3.2	$67.2 \pm 1.7^*$
σ_{WW}(161GeV)[pb]	4.23 ± 0.75	$3.67^{+0.98}_{-0.87}$	$2.89^{+0.82}_{-0.71}$	$3.62^{+0.94}_{-0.84}$	3.69 ± 0.45
σ_{WW}(172GeV)[pb]	11.71 ± 1.26	$11.58^{+1.48*}_{-1.39}$	$12.27^{+1.43}_{-1.34}$	12.3 ± 1.3	$11.96 \pm 0.70^*$

Table 3. Results on W decay branching ratios and WW production cross sections including the LEP combined numbers [5].

present statistics lepton universality is valid in W decays.

The hadronic branching ratio can be used to test the unitarity of the three known quark families. Mixing with a heavy, fourth family would spoil this unitarity. From

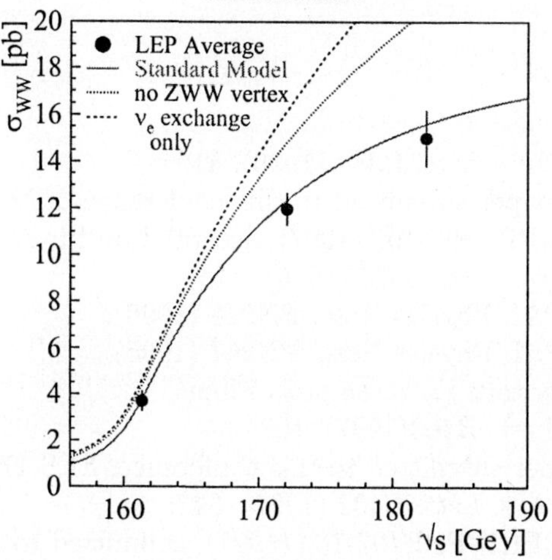

Fig. 2. WW production cross section as function of the center-of-mass energy.

$$\frac{Br(\mathrm{W} \to \mathrm{q\overline{q}})}{1 - Br(\mathrm{W} \to \mathrm{q\overline{q}})} = (1 + \frac{\alpha_s}{\pi}) \sum_{i=\mathrm{u,c};\, j=\mathrm{d,s,b}} |\mathrm{V_{ij}}|^2$$

and the LEP combined hadronic branching ratio one finds that

$$\sum_{i=\mathrm{u,c};\, j=\mathrm{d,s,b}} |\mathrm{V_{ij}}|^2 = 1.97 \pm 0.15^*$$

which is compatible with the value of 2 demanded by unitarity. From this, it is also possible to extract a value for $|\mathrm{V_{cs}}|$, the least well measured CKM matrix element, by subtracting the measured values of all other matrix elements. This gives a value of $|\mathrm{V_{cs}}| = 0.96 \pm 0.08^*$, which can be compared to $|\mathrm{V_{cs}}| = 1.01 \pm 0.18$ as obtained from D decays [7].

In figure 2 the WW production cross-section is shown, including some preliminary results on the first 7 pb^{-1} of the 183 GeV run [6]. The cross-section exhibits a prominent rise once the WW threshold is crossed. From the measured energy dependence it is obvious that all three CC03 graphs contribute, i.e. both the photonic and ZWW triple gauge couplings must be present. Depending on the mass of the W boson the theoretically expected cross-section curve moves right or left. Thus, the cross-section measurement at 161 GeV is sensitive to M_W and a value of $M_W = 80.40 \pm 0.22$ GeV can be extracted from the LEP combined number.

References

[1] ALEPH Coll., Phys. Lett. B388 (1996), 419;
DELPHI Coll., paper submitted to this conference, EPS 322 (1997);
L3 Coll., CERN-PPE/97-106 (1997), accepted by Phys. Lett. B;
OPAL Coll., Phys. Lett. B376 (1996), 315;
OPAL Coll., OPAL Physics Note, PN228 (1996);
OPAL Coll., OPAL Physics Note, PN267 (1996).
[2] P. Janot, plenary talk 17, these proceedings.
[3] L3 Coll., Phys. Lett. B403(1997), 168;
OPAL Coll., paper submitted to this conference, EPS 171 (1997).
[4] ALEPH Coll., Phys. Lett. B401 (1997), 347;
ALEPH Coll., CERN-PPE/97-102 (1997), submitted to Phys. Lett. B;
DELPHI Coll., Phys. Lett. B397 (1997), 158;
DELPHI Coll., paper submitted to this conference, EPS 347 (1997);
L3 Coll., Phys. Lett. B398 (1997), 223;
L3 Coll., Phys. Lett. B407 (1997), 419;
OPAL Coll., Phys. Lett. B389 (1996), 416;
OPAL Coll., CERN-PPE/97-116, accepted by Z. Phys. C.
[5] The LEP Electroweak Working Group, LEPEWWG/97-02.
[6] ALEPH Coll., paper submitted to this conference, EPS 856 (1997);
DELPHI Coll., paper submitted to this conference, EPS 865 (1997);
L3 Coll., paper submitted to this conference, EPS 857 (1997);
OPAL Coll., OPAL physics note, PN313.
[7] Particle Data Group, Phys. Rev. D54 (1996),1.

Results on the Mass of the W Boson from LEP2

Dimitris Fassouliotis (fassoul@vxcern.cern.ch)

N.C.S.R. Demokritos, Athens

Abstract. This paper reports on the measurement of the mass of the W boson at LEP. In 1996, an integrated luminosity of approximately 20 pb^{-1} per experiment was delivered at centre-of-mass energies 161 and 172 GeV. The mass of the W boson was derived from the W pair threshold cross section and through the study of the invariant mass distribution of the W decay products. Combining the results of the two methods from the four LEP collaborations yields M_W=80.48+-0.14 GeV.

1 Introduction

One of the principal goals of the LEP2 programme is the measurement of the mass of the W boson M_W. The comparison of this direct measurement and the value of M_W determined from precise electroweak analyses can provide an important test of the Standard Model (SM) of the electroweak interactions. Further more, through the determination of the effect of the radiative corrections on M_W, constraints on the mass of the Higgs can be obtained.

Two methods have been used at LEP to measure M_W. The first is based on the strong dependence of the W pair production cross section near threshold from M_W (threshold cross section measurement method) and the second, on the study of the invariant mass distribution of the W decay products (direct reconstruction method). The two methods have comparable statistical power, for a given integrated luminosity, while they are complementary, having rather different concepts and systematic uncertainties. This paper is organized as follows. In section 2, the threshold cross section method is briefly described. In section 3 the invariant mass method is described followed by a short discussion on the sytematic uncertainties. Finally a combined value for M_W and the outlook are presented in section 4.

2 Threshold cross section measurement

The cross section for the process $e^+e^- \to W^+W^-$ depends strongly on M_W for centre-of-mass energies close to the threshold. It is therefore possible to extract M_W, in the context of the SM, by comparing the measured cross section with the theoretical expectation. As shown in reference [1], the centre-of-mass energy with the optimal sensitivity on M_W is $\sqrt{s}^{opt} \simeq 2M_W + 0.5\, GeV$. It has to be noted that the sensitivity as a function of the centre-of-mass energy remains essentially flat around the optimal value, in the region that previous measurements constrain M_W allowing an *a priori* choice of $\sqrt{s}^{opt}$.

The measured cross sections of the four experiments [4], were corrected to correspond to W pair production through the three doubly resonant tree-level diagrams and have been combined assuming the smallest quoted systematic error to be 100% correlated. Gentle [3] program was used to calculate the cross section as a function of the W mass.

The expected cross section as a function of W mass, together with the LEP average cross section measurement are shown in figure 1, from which the W mass is determined to be:

$$M_W = 80.40^{+0.22}_{-0.21}\ GeV$$

Approximately 70 MeV of the error is due to common systematics. In addition to the ones reported in reference [4] for the cross section measurement, there are two sources of systematic uncertainties in the W mass evaluation. One comes from theoretical uncertainties in the cross section prediction which amount to about 2% and the other from the LEP energy determination uncertainty (50 MeV).

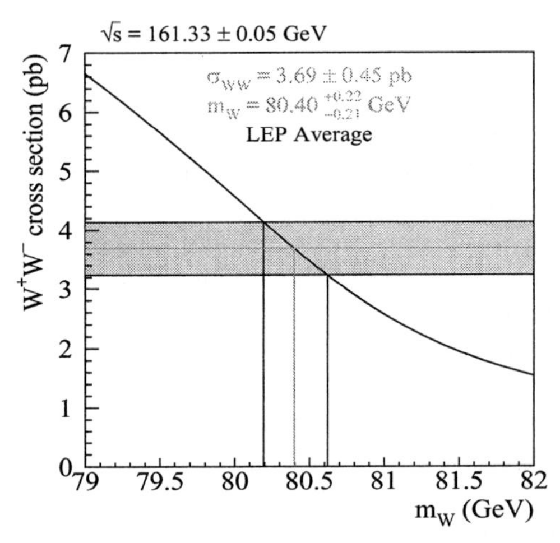

Fig. 1. The cross section of W pair production as a function of M_W. The curve shows the SM prediction and the horizontal band the cross section measurement with its error.

3 Direct reconstruction

Within the direct recosntruction method the fully hadronic ($WW \to q\bar{q}q\bar{q}$) and the semileptonic ($WW \to q\bar{q}l\nu_l$) decay channels were used. The event selections used were similar to the ones of reference [4] for the cross section measurements. Emphasis, however, was made on applying only criteria that leave the mass measurement unbiased.

3.1 Event kinematic reconstruction

In order to improve the mass resolution, in all analyses, results of which are reported in this paper, a kinematic fit is applied to the W decay products (leptons, jets or neutrinos). This fit consists of imposing constrains of energy and momentum conservation and in most of the cases, equal masses for the two W bosons. Variations to the kinematic quantities (angles and energies) of the visible W decay products are applied, according to the experimental resolutions and a χ^2 constructed from these is subsequently minimized. This procedure leads to the determination of the average of the two W masses, on single events, with a precision at the level of the width of the W. However, a direct consequence of this kind of treatment is the induction of systematic uncertainties coming from the Initial State Radiation (ISR) and the LEP energy determination.

3.2 Fitting methods

To extract M_W from the observed mass distribution the maximum likelihood method is used. In three of the four analyses the expected mass spectrum as a fumction of M_W is evaluated through Monte Carlo reweighting techniques. Subsequently the likelihood is constructed either by binning the distributions [5],[6] or on single event basis using the "box" technique to account for detector effects [7]. In the fourth analysis [8] the likelihood of observing a single event is extracted by the convolution of a Breit-Wigner $\times$ Phase Space function, denoting the expected distribution of the average of the two W masses, with a Gaussian resolution function based on event-by-event error estimations. In all cases the behaviour of the accepted background is evaluated through Monte Carlo integration.

In the fully hadronic channel, there are additional complications coming from the possibility of multiple pairings of the jets and the treatment of hard gluon radiation, which introduce further differences in the above discussed approaches. In general it is argued that accepting more than one possible pairings per event increases the sensitivity, while in reference [8], the probability of observing a 5-jet event due to gluon radiation is explicitly introduced to the likelihood estimator.

3.3 Result and systematic uncertainties

Extesnive consistency and stability checks have been made and several sources of systematic uncertaities have been studied. It has to be emphasized that the size of most of them (ISR, background contamination, fragmentation parameters, production diagrams used in reference samples, detector callibration) depends mainly on the Monte Carlo statistics and/or the size of reference real data samples used for their evaluation. Therefore, they can "easily" be reduced if they arrive to have significant contribution to the final error of the measurement. Two systematic errors have been treated as being 100% correlated between the experiments. First, the LEP energy uncertainty which can also be reduced, provided the number of depolarization measurements at different beam energies, which are used for calibration, is increased. Another common systematic error of 100 MeV is assigned in the case of the $q\bar{q}q\bar{q}$ channel due to potential effects of interconnection[1] of the hadronic decay

[1] Colour Reconnection and Bose-Einstein correlations

products. From the theoretical point of view their effect on M_W measurement appears to be model dependent. Recent studies [9], though, imply that their effect in a realistic measurement of M_W might have been overestimated. However, to obtain conclusive results in this sector, experimental studies [10] are highly demanded to justify the theoretical predictions.

At the present level of statistics the M_W determined from the semileptonic channel and that from the fully hadronic channel are consistent. The combined result of the measurements reads:

$$M_W = 80.53 \pm 0.18\ GeV$$

Approximately 60 MeV of the error is due to common systematics.

4 Combined measurement

Combining the results of the two methods discussed above, yields as the current LEP average:

$$M_W = 80.48 \pm 0.14\ GeV$$

with a common systematic error of 40 MeV.

As discussed above, an increase in size of the available Monte Carlo samples and the accumulation of more experimental data will allow better understanding of the systematic uncertainties. Concequently, it can realisticly be expected that the sensitivity in the determination of M_W at the end of LEP2 phase will be of the order of 30-40 MeV.

References

[1] Z. Kunst, W.J. Stirling et al, *Determination of the mass of the W boson*, Physics at LEP2, eds. G. Altareli, T. Sjöstrand and F. Zwirner, CERN 96-01 (1996) Vol 1, 141

[2] ALEPH Collaboration, R. Barate, et al., Phys. Lett.**B401** (1997) 347.
DELPHI collaboration, P. Abreu, et al., Phys. Lett.**B397** (1997) 158.
L3 collaboration, M. Acciarri, et al., Phys. Lett.**B398** (1997) 223.
OPAL Collaboration, K. Ackerstaff et al.,Phys. Lett.**B389** (1996) 416.

[3] D. Bardin et al, *GENTLE/4fan v. 2.0*, DESY 96-233, hep-ph/9612409.

[4] C. Burgard, contributed paper 805 to EPS-HEP-97, Jerusalem.

[5] ALEPH Collaboration, R. Barate, et al., CERN-PPE/97-102, contributed paper 600 to EPS-HEP-97, Jerusalem.

[6] OPAL Collaboration, K. Ackerstaff et al., CERN-PPE/97-116, contributed paper 168 to EPS-HEP-97, Jerusalem.

[7] L3 Collaboration, M. Acciarri et al., CERN-PPE/97-98, contributed paper 512 to EPS-HEP-97, Jerusalem.

[8] DELPHI Collaboration, P. Buschmann et al., DELPHI 97-108 CONF 90, contributed paper 347 to EPS-HEP-97, Jerusalem.

[9] V. Kartvelishvili ,R. Kvatadze and R. Moller, contributed paper 233 to EPS-HEP-97, Jerusalem, hep-ph/9704424.

[10] P. Perez, contributed paper 808 to EPS-HEP-97, Jerusalem.

Results on Triple Gauge-Boson Couplings from LEP2

Alfons Weber (Alfons.Weber@cern.ch)

I.Phys.Inst., RWTH-Aachen, Germany

Abstract. This report presents a combination of published and preliminary measurements of triple-gauge-boson coupling parameters from the four LEP experiments. The measurements result from the analysis of the data recorded at LEP2 during the 1996 run, which yielded approximately 10 pb^{-1} at 161 GeV and $10\mathrm{pb}^{-1}$ at 172 GeV centre of mass energy, per experiment.

1 Introduction

During the initial year of operation of LEP2 (1996), centre-of-mass energies have been attained which allow, for the first time, the production of $\mathrm{W}^+\mathrm{W}^-$ boson pairs in $\mathrm{e}^+\mathrm{e}^-$ collisions. A total integrated luminosity of approximately 10 pb^{-1} was recorded at a centre-of-mass energy of 161 GeV and $10\mathrm{pb}^{-1}$ at 172 GeV, per experiment.

The $\mathrm{W}^+\mathrm{W}^-$ production process involves the triple gauge boson vertices between the $\mathrm{W}^+\mathrm{W}^-$ and the Z^0 or photon. The measurement of these triple gauge boson couplings (TGCs) and the search for possible anomalous values is one of the principal physics goals at LEP2.

The parameterisation of anomalous TGCs is described in [1] to [6]. The most general Lorenz invariant Lagrangian which describes the triple gauge boson interaction has fourteen independent terms, seven describing the $\mathrm{WW}\gamma$ vertex and seven describing the WWZ vertex. Assuming electromagnetic gauge invariance and C and P conservation the number of parameters reduces to five. One common set is $\{g_1^{\mathrm{z}}, \kappa_{\mathrm{z}}, \kappa_\gamma, \lambda_{\mathrm{z}}, \lambda_\gamma\}$ where $g_1^{\mathrm{z}} = \kappa_{\mathrm{z}} = \kappa_\gamma = 1$ and $\lambda_{\mathrm{z}} = \lambda_\gamma = 0$ in the Standard Model. Another such set is $\{\delta_{\mathrm{z}}, \Delta\kappa_{\mathrm{z}}, \Delta\kappa_\gamma, y_{\mathrm{z}}, y_\gamma\}$ which are all zero in the Standard Model.

Different sets of parameters have also been proposed which are motivated by $\mathrm{SU(2)}\times\mathrm{U(1)}$ gauge invariance and constraints arising from precise measurements at LEP1. One such set [6] is:

$$\begin{aligned}
\alpha_{W\phi} &\equiv \Delta g_1^{\mathrm{z}} \cos^2\theta_W \\
\alpha_W &\equiv \lambda_\gamma \\
\alpha_{B\phi} &\equiv \Delta\kappa_\gamma - \Delta g_1^{\mathrm{z}} \cos^2\theta_W
\end{aligned}$$

with the constraints that $\Delta\kappa_{\mathrm{z}} = -\Delta\kappa_\gamma \tan^2\theta_W + \Delta g_1^{\mathrm{z}}$ and $\lambda_{\mathrm{z}} = \lambda_\gamma$. The Δ indicates the deviation of the respective quantity from its Standard Model value and θ_W is the weak mixing angle. Each of the α parameters has the value zero in the Standard Model. A similar approach, HISZ [2], reduces this set to two parameters with the extra constraint $\alpha_{B\phi} = \alpha_{W\phi}$. This extra constraint is equivalent to $\Delta g_1^z = \Delta\kappa_\gamma/(2\cos^2\theta_W)$, and $\Delta\kappa_\gamma$ and $\lambda_\gamma \equiv \alpha_W$ are normally used as the variable parameters.

The four LEP experiments have recently presented measurements of anomalous coupling parameters [7, 8, 9, 10] using the 1996 data set. Each of the experiments has measured several of the many possible TGC parameters and the references should be consulted for the complete set of measurements in each case.[1] In this note I present the results of a combination to obtain LEP averages of $\{\alpha_{W\phi}, \alpha_W, \alpha_{B\phi}\}$, which have been measured by all four experiments[11], and of other parameters that have been measured by L3 and Opal. Some of these parameters are already constrained by measurements performed at TEVATRON[12].

2 Methods and Results

Each of the experiments has analysed the 1996 data set, which corresponds to approximately 25 and 100 events per experiment at 161 GeV and 172 GeV respectively. Anomalous TGCs can affect both the total production cross-section and the shape of the differential cross-section as a function of the W^- production angle. The relative contributions of each helicity state of the W bosons are also changed, which in turn affects the distributions of their decay products. The analyses presented by each experiment make use of different combinations of each of these quantities, and of different decay modes of the WW system (see references for details). In general, however, all analyses use at least the expected variations of the total production cross-section and the W^- production angle.

Each of the experiments have provided the full negative log likelihood curve, $\log\mathcal{L}$, as a function of each of the measured TGC parameters. The $\log\mathcal{L}$ curves from each experiment include both statistical and systematic effects. It is necessary to use the $\log\mathcal{L}$ curves directly for the combination as they are not parabolic, and it is therefore not possible to combine the results correctly by taking simple averages of one standard deviation measurements.

In principle there are some common systematic effects which should be included in a correlated way. These include the uncertainties of the LEP beam energy and the W mass, and some effects estimated by varying Monte Carlo generators. However, in this first combination common errors have been neglected as they have a negligible effect compared to the uncorrelated systematic and statistical errors.

[1] Some experiments have since published results, which were still preliminary at the time of the conference [10]. These updates have not been considered in this note.

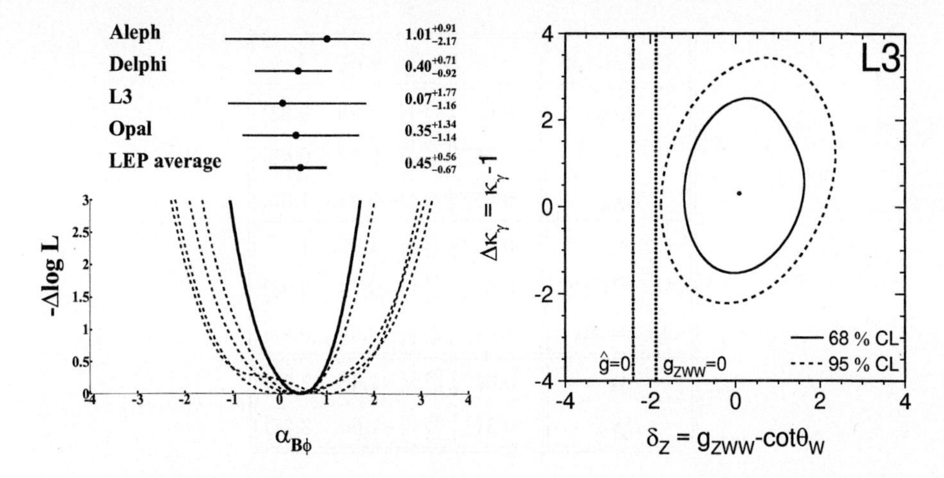

Fig. 1. Left: The LEP combined $\log \mathcal{L}$ cruve as a function of the parameter $\alpha_{B\phi}$. This is obtained from the combination of the curves from the individual LEP experiments, which are schown as dotted lines. The measurement and one standard diviation errors are also listed. **Right:** Two parameter fit to δ_z and $\Delta\kappa_\gamma$ illustrating the the existance of the ZWW vertex, independent of the W boson magnetic moment $\Delta\kappa_\gamma$. The expectation due to vanishing ZWW and weak couplings, $g_{ZWW} = 0$ and $\hat{g} = 0$, are indicated by the dotted and dashed-dotted lines.

The individual $\log \mathcal{L}$ curves for each parameter are added together. As an example, the result of this procedure is shown in Fig. 1(left), where each curve is plotted relative to its minimum value.

The one standard deviation limits (S.D.) are directly obtained from the curves by taking the values of each TGC parameter where $\Delta \log \mathcal{L} = +0.5$ from the minimum. The 95% confidence level (C.L.) limit is given by the values of each TGC parameter where $\Delta \log \mathcal{L} = +1.92$. The results obtained are given in table 1. The value of each TGC parameter in the table is consistent with the expectation of the Standard Model, which is 0 in each case.

As an indication for further prospects, two parameter fits have been performed by L3 and Opal. Fig. 1(right) shows, that the ZWW vertex exists at more than 95% C.L. even when allowing for an anomalous magnetic moment $\Delta\kappa_\gamma$ of the W boson.

Acknowledgements

I would like to thank the LEP experiments for providing the data used in this note and the LEP TGC combination group for performing some of the combinations quoted here[11].

Parameter	1 S.D.	95% C.L.
$\alpha_{W\phi}$	$0.02^{+0.16}_{-0.15}$	[−0.28, 0.33]
α_W	$0.15^{+0.27}_{-0.27}$	[−0.37, 0.68]
$\alpha_{B\phi}$	$0.45^{+0.56}_{-0.67}$	[−0.81, 1.50]
δ_z	$-0.01^{+0.62}_{-0.57}$	[−1.06, 1.19]
$\Delta\kappa_\gamma$ (HISZ)	$0.03^{+0.46}_{-0.41}$	[−0.74, 0.98]
$\Delta\kappa_\gamma = \Delta\kappa_z$	$0.02^{+0.42}_{-0.38}$	[−0.69, 0.87]
δ_z	$0.09^{+1.05}_{-0.94}$	[−1.48, 2.01]
$\Delta\kappa_\gamma$	$0.31^{+1.81}_{-1.12}$	[−1.60, 3.27]

Table 1. The combined one standard deviation and 95% confidence intervals obtained after combination of the results from up to four LEP experiments. The anomalous TGC parameters are those listed in the introduction. Both statistical and systematic errors are included. The results for δ_z, $\Delta\kappa_\gamma$(HISZ) and $\Delta\kappa_\gamma = \Delta\kappa_z$ are based on L3 and Opal data only. The two last lines are the the Result of the L3 two parameter fit.

References

[1] K.Gaemers and G.Gounaris, Z.Phys. **C1** (1979) 259
[2] K.Hagiwara et al., Nucl.Phys. **B282** (1987) 253
[3] M.Bilenky et al., Nucl.Phys. **B409** (1993) 22
M.Bilenky et al., Nucl.Phys. **B419** (1994) 240
[4] I.Kuss and D.Schildknecht,Phys.Lett. **B383** (1996) 470
[5] G.Gounaris and C.G.Papadopoulos, HEP-PH/9612378
[6] Physics at LEP2, Ed. G.Altarelli et al.,CERN 96-01 (1996) Vol.1
[7] ALEPH Collab., Contrib. paper to EPS-HEP97 Jerusalem, EPS/97-601
[8] DELPHI Collab., P.Abreu et al.,Phus.Lett. **B397** (1997) 158
DELPHI Collab., Contrib. paper to EPS-HEP97 Jerusalem, EPS/97-295
[9] L3 Collab., M.Acciarri et al., Phys.Lett. **B398** (1997) 223
L3 Collab., M.Acciarri et al., CERN-PPE/97-98, accepted by Phys.Lett.
[10] OPAL Collab., K.Ackerstaff et al., Phys.Lett. **B397** (1997) 147
OPAL Collab., CERN-PPE/97-125, submitted to Z.Phys. **C**
[11] LEP TGC combination Group; A Combination of Preliminary Measurements of Triple Gauge Boson Coupling Parameters Measured by the LEP Experiments, LEPEWWG/TGC/97-01
[12] in this proceedings:
A.Ward, PL15; P.de Barbaro, PA803; A.KOTWAL, PA804

Properties of WW Events and W Decays at LEP2

Patrice Perez (perez@hep.saclay.cea.fr)

C.E.A. Saclay, France

Abstract. A summary is presented of results from the four experiments at LEP2: the direct measurement of the Vcs element of the CKM matrix from W decays into charm, and a first look at interconnection effects like Bose-Einstein correlations and colour reconnection in W pair production with their consequences on the measurement of the W mass.

1 Charm tagging and measurement of Vcs

Two studies [1] have been contributed to this conference by ALEPH with an integrated luminosity $\mathcal{L}$ of 10.65 pb^{-1} at 172 GeV c.m energy, and DELPHI which combines 161 and 172 GeV data with $\mathcal{L} = 9.9$ and 10 pb^{-1}.

Both qqqq and $\ell\nu$qq W decay channels are used. The flavour of hadronic jets is tagged with a lifetime probability determined from the impact parameters measured by the vertex detectors, lepton identification and leading particle energy. DELPHI also uses the RICH to identify Kaons and pions, as well as the correlation between the jet direction and the flavour of the primary quark.

The distributions of the output of a neural net or of the probability of a dijet to be $c\bar{s}$ are then fitted with a maximum likelihood method with one unknown: Vcs or equivalently $R_{cs} = \Gamma(W^+ \to c\bar{s})\ /\ \Gamma(W^+ \to hadrons)$ in DELPHI, $R_{cW} = \Gamma(W^+ \to cX)\ /\ \Gamma(W^+ \to hadrons)$ in ALEPH. Results are given in table 2 together with indirect measurements at LEP2 from the hadronic width of the W and the semileptonic D decay.

Systematic errors arise from the choice of a jet algorithm and jet pairing, charm fragmentation modelling and decay properties, lifetime tag. However present errors are governed by statistics.

2 Interconnection effects in WW events

Two effects have been studied which are interesting in themselves but may also distort the reconstructed W mass distributions in the qqqq channel hence shifting the measured mass value.

Color reconnection between two partons originating from different W's in a W pair event may lead to a sizeable difference in the charged multiplicity from fully hadronic W pair events w.r.t semi-leptonic events [2]. Results from

L3 [3], OPAL [4] and DELPHI [5] are summarized in table 2. DELPHI has two compatible analyses, only one is quoted in the table which sees a 2 sigma "excess" in the yield of low momentum charged tracks for fully hadronic events w.r.t semi-leptonic events. However, this is not confirmed by OPAL.

At the present level of statistics, there is no evidence for colour reconnection effects.

Bose-Einstein correlations arise when identical particles are emitted close in phase space. Such correlations have already been seen at lower e^+e^- collider energies, and are looked for in the distribution of the 2-particle probability density $P(Q)$ where $Q^2 = M^2(\pi\pi) - 4m_\pi^2$. The Bose-Einstein correlations are evident when this distribution is compared to a reference distribution of pairs with no expected correlation $R(Q) = P(Q)/P_{ref}(Q)$. To obtain such reference distributions, pairs of opposite sign pions are used, but with the caveat that they contain resonances not present in $\pi^\pm\pi^\pm$. Also one mixes particles from different events with similar topologies, but where kinematical correlations are lost. Thus analyses are made using double ratios like: $R'(Q) = \frac{R^{data}(Q)}{R^{MC}(Q)}$, or even

$$\left(\frac{N\pi^{++,--}_{WW\to q\bar{q}q\bar{q}} - 2N\pi^{++,--}_{WW\to \ell\nu q\bar{q}}}{N\pi^{+-}_{WW\to q\bar{q}q\bar{q}} - 2N\pi^{+-}_{WW\to \ell\nu q\bar{q}}}\right)^{data} \Big/ \left(\frac{N\pi^{++,--}_{WW\to q\bar{q}q\bar{q}} - 2N\pi^{++,--}_{WW\to \ell\nu q\bar{q}}}{N\pi^{+-}_{WW\to q\bar{q}q\bar{q}} - 2N\pi^{+-}_{WW\to \ell\nu q\bar{q}}}\right)^{MC}$$

These distributions are fitted with the form $\kappa(1+\epsilon Q)(1+\lambda \exp{(-\sigma^2\ Q^2)})$ when the particle source is assumed to have a spherical space-time distribution with a radius σ for a gaussian model of its density.

New analyses at the Z [6] from L3, and ALEPH yield compatible results (see table 3).

In semi-leptonic W pair events, with only one W hadronic decay, ALEPH [6] and DELPHI [5] show an indication for an enhancement at low Q over simulations which do not contain B.E. correlations.

The data favour absence of correlations between hadronic decay products of different W's as already stated in [7] with $\lambda = -0.20 \pm 0.22 \pm 0.08$ for DELPHI and $\lambda = -0.64 \pm 1.14$ for ALEPH (see for instance figure 1)

Effects on the W mass measurements have recently been studied with event weighting techniques to simulate the B.E. effect in a quantum mechanical approach [8]. After taking into account the experimental procedures used to extract the W mass (mainly that a fit to the mass distribution is less sensitive to tails than the average mass), the possible mass shift is estimated to be less than 20 MeV/c^2.

The above mentioned measurements are all limited by statistics. The 1997 run yielded already a factor 5 improvement in luminosity over these data, and we expect a factor 10 more by the end of LEP2. There is already an indication of Bose-Einstein correlations in WW $\to \ell\nu$qq events. Interconnection between the decay products of different W's is small.

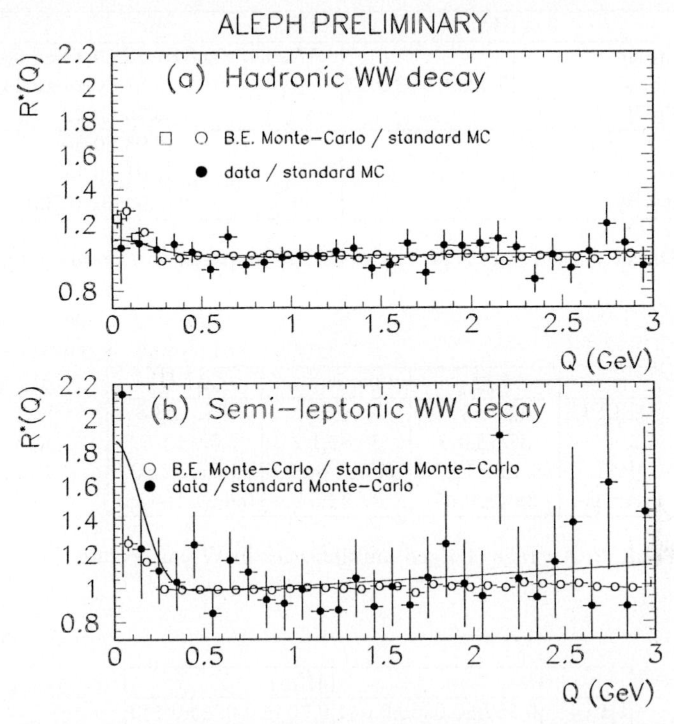

Fig. 1. Example of Data/MC ratios for semileptonic and fully hadronic W pair events. An excess at low Q is a sign for Bose-Einstein correlations.

References

[1] ALEPH Coll., EPS-HEP/97-630 , DELPHI Coll., EPS-HEP/97-535

[2] J. Ellis and K. Geiger, Physics Letters B404 (1997) 230-237

[3] L3 Collaboration, EPS-HEP/97-814

[4] OPAL Collaboration, EPS-HEP/97-169

[5] DELPHI Collaboration, EPS-HEP/97-307, EPS-HEP/97-667

[6] L3 Coll., EPS-HEP/97-505 , ALEPH Coll., EPS-HEP/97-590

[7] DELPHI Collaboration, Physics Letters B401 (1997) 181

[8] Jadach and Zalewski, Acta Physica Polonica B28 (1997) 1363-1380, Kartvelishvili et al., Physics Letters B408 (1997) 331-339

	Rcs	RcX	Vcs
ALEPH direct		0.57±0.18±0.04	1.13±0.43±0.08
DELPHI direct	0.42±0.13±0.06		0.87±0.24±0.11
average direct			0.95±0.20
LEP $\Gamma(W \to hadrons)$			0.95±0.08
$D \to K\ell\nu$			1.01±0.18
CKM unitarity			0.9493±0.0008

Table 1. Direct measurements of Vcs at LEP2 compared with indirect ones

	N^{ch}_{qqqq}	$2 \times N^{ch}_{\ell\nu qq}$	ΔN^{ch}
DELPHI	41.7±1.8	36.4±2.2	+5.3±2.8
L3	37.9±0.9	38.8±1.2	-0.9±1.5
OPAL	38.3±1.1±0.6	36.8±1.8±0.8	+1.5±2.2±0.8
average	38.5±0.65	37.9±0.9	0.6±1.1(stat)

Table 2. Average charged multiplicities in W pair events

	λ	σ(fm)
L3	0.252±0.006±0.044	0.764±0.024±0.153
ALEPH	0.3±0.02	0.84±0.03

Table 3. Bose-Einstein correlation fits at the Z

	$qq\ell\nu$	$qqqq$
DELPHI	$\lambda = 0.41 \pm 0.13$	$\lambda = 0.09 \pm 0.06$
	$\sigma = 0.5$ fm fixed	$\sigma = 0.5$ fm fixed
ALEPH	$\lambda = 0.96 \pm 0.84$	$\lambda = 0.12 \pm 0.12$
	$\sigma = 0.99 \pm 0.41$ fm	$\sigma = 0.64 \pm 0.38$ fm

Table 4. Bose-Einstein correlation fits in W pair events

Part IX

Quark Masses and Mixing, *CP* Violation and Rare Phenomena

Penguins, Charmless B Decays and the Hunt for CP Violation

Atwood, David[1] (atwood@iastate.edu)
Amarjit Soni[2] (soni@bnl.gov)

[1]Dept. of Physics and Astronomy, Iowa State Univ., Ames, IA 50010
[2]Brookhaven National Laboratory, Upton, NY 11973; (presenter)

1 Introduction

By a multitude of very important observations involving the QCD penguin, CLEO [1] has declared 1997 to be the year of the Strong Penguin. The observed size of the various modes suggest that the penguin is rather robust. The experimental observations have fueled intense theoretical activity some of which I will briefly review here. For convenience, I have divided the theoretical activity into four categories (See Table 1): Exclusive modes, inclusive η' modes, the charm deficit and implications for CP violation.

2 The challenge of pure hadronic (exclusive) modes

There are renewed attempts to understand B-decays to two-body hadronic modes involving the penguins. Most of the works are centered around modifying factorization. The starting point in these calculations is the next-to-leading-order (NLO) short-distance (SD) Hamiltonian which at the quark level is clearly rather precise with a minimal uncertainty due to scale. However, it is quite unclear as to the advantage of using the NLO apparatus as the calculations of the hadronic matrix elements is highly uncertain. See below.

3 Charming-Penguins

Martinelli *et al.* [2] have made the interesting observation that graphs containing $c\bar{c}$ loops (the so-called "charming penguins") could be important for final states involving the light states (such as $K\pi$, $\pi\pi\cdots$) and therefore should not be ignored as they are commonly done in calculation of matrix elements. The idea of charming penguin is closely related to the eye-graphs which are believed to be important for the emergence of the observed $\Delta I = 1/2$ rule in K-decays. Of course, their role in B-decays is not known.

Table 1. Sample of Recent Theoretical Works

Who	What	Comments
Ciuchini *et al.* [2]	2 Body Modes	Charming Penguins
Ali & Greub [3]	2 Body Modes	Improve Factorization, $\eta_c \leftrightarrow \eta' \cdots$
Cheng & Tseng [4]	$\eta'(\eta) + K(K^*, \rho, \pi)$	Improve Factorization, $\eta_c \leftrightarrow \eta' \cdots$
Kagan & Petrov [5]	$\eta' K$	SM Factorization OK$\cdots$
Datta *et al.* [6]	$\eta' K(K^*)$	Factorization and Anomaly
Shuryak & Zhitnitsky [7]	$\eta' K$	$\eta_c \leftrightarrow \eta' \cdots$
Halperin & Zhitnitsky [8]	$\eta' K$	$\eta_c \leftrightarrow \eta' \cdots$
Halperin & Zhitnitsky [8]	$\eta' X_s$	$\eta_c \leftrightarrow \eta' \cdots$
Hou & Tseng [9]	$\eta' X_s$	New Physics
Kagan & Petrov [5]	$\eta' X_s$	New Physics
Yuan & Chao [10]	$\eta' X_s$	$(\bar{c}c)_8 \to \eta' X_s$
Datta *et al.* [6]	$\eta' X_s$	Factorization + Anomaly
Atwood & S [11]	$\eta' X_s, \eta X_s \cdots$	QCD Anomaly
Dunietz *et al.* [12]	Charm Deficit	
Lenz *et al.* [13]	Charm Deficit	
Fleischer & Mannel [14]	$K\pi$	γ
London & S [15]	$\eta' K + \cdots$	$\beta_{penguin}$
Dighe *et al.* [16]	$\eta'(\eta) + K(\pi)$	Direct CP
Hou & Tseng [9]	$\eta' X_s$	Non-Std-CP
Kagan & Petrov [5]	$\eta' X_s$	Non-Std-CP
Atwood & S [17]	$\eta' X_s \cdots$	Non-Std-CP

4 η_c-η' mixing [18, 3, 4, 7, 8, 10]

Since the decay $b \to c\bar{c}s$ is accompanied by a hefty CKM factor, $\eta_c \leftrightarrow \eta'$ mixing could possibly become important in $B \to \eta'$ decays. However, it is quite tricky to isolate this mixing as such. For one thing, some of the glue in the $b \to sg^*$ originates from $c\bar{c}$ annihilation via $b \to c\bar{c}s$. So one may equally well think of the η' originating from this glue. More specifically mixing with η_c and with 2-glue are interrelated. Quite understandably there is an enormous variation in the estimated $\eta_c \leftrightarrow \eta'$ mixing.

Table 2 shows a comparison of some of the *recent* calculations of exclusive 2-body modes. Notice that $\eta' K^*$ is an excellent discriminator amongst various models. The $\pi^0\pi^0$ mode which is especially important for α-extraction appears most difficult to pin down; almost all the models seem to find it below 10^{-6} but the predictions are not at all reliable.

All of these theoretical calculations are extremely uncertain and have a huge range. In large part this is a reflection of the fact that scores of assump-

Table 2. Recent studies of exclusive modes; Br's in units of 10^{-5}

Mode	Sample Studies				CLEO [1]
	AG [3]	CT [10]	KP [5]	Romans [2]	
$K^+\pi^-$	1–3				$1.5^{+.5}_{-.4} \pm .1 \pm .1$
$K^0\pi^+$			1–2		$2.3^{+1.1}_{-}1.0 \pm .2 \pm .2$
$\pi^+\pi^-$	.4–2.4				< 1.5
$\pi^0\pi^0$	.02–.08			$(.05\text{–}.1)^*$	
$\eta' K^\pm$	5–6	6–7	1–12		$7.8^{+2.7}_{-2.2} \pm 1.0$
$\eta' K^{*\pm}$	.07–.16	1–2		6–9	< 29
$\eta K^\pm$	.01–.04	.2–.5	.1–.5	.01–.5	< 0.8
$\eta K^{*\pm}$	.1–.2	.3–.8		.1–4	< 24

tions and approximations have been made to arrive at these numbers. It is perhaps useful to list some of the assumptions and approximations typically used in these calculations:

1. Effective Wilson coefficients are same for $B \to D\pi$ and $B \to K\pi$.
2. Color suppression works as well for $(\pi\pi,\ K\pi\cdots)$ as for $(D\pi,\ D_s K\cdots)$.
3. Color suppression works as well for matrix elements of penguin operators as well as it does for matrix elements of tree operators.
4. Factorization works just as well for penguin operators as for tree ones.
5. The eye-graph with the c-loop (i.e. the charming penguin) does not contribute to light final states (such as $K\pi$, $\pi\pi\cdots$) even though lattice studies over the years have suggested that such graphs are important for the emergence of the $\Delta I = 1/2$ rule in $K \to \pi\pi$ decays.
6. Annihilation or exchange contributions are neglected even though these are intimately related to "factorizable contributions" via FSI.
7. Penguin matrix elements are extremely sensitive to the numerical value of current quark mass, m_s.

Due to this very long list of assumptions and approximations the calculations for hadronic decays are highly unreliable. Thus deviations of the experimental numbers for the absolute rates from the crude theoretical estimates that are available cannot be used as a reliable hint for the presence of new physics. Search for CP violating asymmetries in some of the modes can be a much more reliable test of new physics.

5 A possible faint silver lining

Despite the morass of dealing with pure hadronic modes there is perhaps sign of a silver lining. The point is that B decays have a multitude of light-light (2-body) final states. The theoretical information that goes into these calculations is highly correlated. So despite the plethora of assumptions, theoretical models can easily run into trouble. For example, it is difficult to get a hierarchy $Br(B^+ \to K^0\pi^+) > Br(B^0 \to K^+\pi^-) > Br(B^0 \to \pi^+\pi^-)$. This is especially true for $K^0\pi^+$ versus $K^+\pi^-$. So if improved experiments confirm the present trend then it would provide useful constraint on the models. The modes wK, $\eta' K^*$, ηK, ηK^* can also be very useful tests of models.

6 $B \to \eta' + X_s$

Some of the important issues are: SM vs. new physics, the form factor for $g^* \to \eta' g$, the $c\bar{c}$ content of the η', direct CP violation etc.

We have made a specific proposal that a large fraction of the inclusive signal $B \to \eta' + X_s$ originates from the penguin graph through the fragmentation $g^* \to g\eta'$ via the QCD anomaly [11]. The form factor $[H(q_1^2, q_1^2, m_{\eta'}^2)]$ at the anomalous vertex was estimated by using the measured rate for $\psi \to \gamma\eta'$[11]. Explicit calculations indicate that $\psi \to \gamma\eta'$ is dominated by near "on-shell" gluons, i.e. $q_1^2 \sim q_2^2 \sim 0$. Since the gluon in $b \to sg^*$ is typically with $q_1^2 \sim 5$–10 GeV^2, q_1^2 dependence of H becomes quite important.

While we are not aware of any theoretical study on the dependence of H on q_1^2 for $g^* \to \eta' g$, the corresponding case of the QED anomaly for $\pi^0 \to \gamma^* + \gamma$ has received some theoretical attention [19]. Although the details vary there is broad agreement as to how the form factor for $\pi^0 \to \gamma^*\gamma$ due to the quark loop scales. If one assumes that $g^* \to \eta' g$ form factor is essentially the same as $\gamma^* \to \pi^0\gamma$ then the anomaly contribution to $B \to \eta' X_s$ would become negligible. However, there are at least two reasons to think that $g^* \to \eta' - g$ effective form factor is quite different from $\gamma^* \to \pi^0 - \gamma$.

Interactions of η' with a gluon are likely to be significantly different from that of the π^0 with γ. Specifically, for the former case the interaction does not have to proceed through a $\bar{q}q$ loop. Given that the η' owes its existence to the gluons, the η' wave function should contain $G\cdot\tilde{G}$. The important dimensionful parameter for the g^*-η'-g form factor may not be f_π (or $f_{\eta'} \sim f_\pi$) but rather the effective gluon mass, m_g^{eff}. Lattice calculations as well as phenomenological arguments suggest $m_g^{eff} \sim 500$–700 MeV. Since $(m_g^{eff}/f_\pi)^2 \sim 16$–25 the anomaly contribution to $B \to \eta' + X_s$ will be significantly more than estimated by the use of the pionic form factor.

Furthermore, the g^*-η'-g form factor may also be greatly influenced by the presence of nearby gluonia or states rich in gluonic content. The gluon from the penguin can combine with a soft glue to make such a state which subsequently decays to η' + light hadrons. These states can be searched by the resonant structure in $\eta' + \pi\pi$, $\eta' + K\bar{K}\cdots$ amongst the $\eta' + X_s$ events.

6.1 Enhanced $b \to sg$ via new physics

Hou and Tseng and Kagan and Petrov suggest that the large $B \to \eta' X_s$ signal is due to a very large branching ratio for $b \to sg$ due to new physics:

i.e. about 100 times the SM value. It is difficult to see how such a major perturbation from new physics would not affect $B \to X_s\gamma$; after all gluon emission enters this calculation in important ways [20].

Also Ref.[5] argues that the exclusive $B \to \eta' K$ signal is OK for the SM and the inclusive is problematic. Since the inclusive/exclusive ratio is about 8 ± 2 and is in the same ball-park as for $(B \to \gamma X_s)/(B \to \gamma K^*) \sim$ it is difficult to appreciate their concern.

6.2 Enhanced $b \to sg^*$ in the SM

An important point to note is that $b \to sg^*$ may be significantly enhanced over the expectation of perturbation theory due especially to the possibility of enhanced FSI in the dominant decay, $b \to c\bar{c}s$.

7 Correspondence with ψ decays

Examination of ψ decays reveals several interesting final states which presumably result from fusion of two gluons. In general one expects these states to have $J^{PC} = 0^{++}, 0^{-+}, 2^{++} \cdots$ etc. Many of these states appear in radiative ψ decays with branching ratios comparable to $\psi \to \gamma\eta'$. In particular, there appears to be a close correspondence between $G \cdot \tilde{G}$ (i.e. 0^{-+}) and $G \cdot G$ (i.e. 0^{++}) [17]. So we should expect f_0, with appreciable BR's as well.

8 Non-Standard CP

As is well known in $b \to s$ transitions, CP violation effect due to the SM are expected to be suppressed as the relevant CKM phase $\sim 0(\eta\lambda^2)$. Due to their big rates they can be poweful in searching for non-standard CP phase.

For simplicity we can assume that $b \to s$ penguin has no SM CP-odd phase, but due to the $u\bar{u}$, $c\bar{c}$ threshold the SM contribution possesses a CP-conserving strong phase. Non-SM interactions possessing a CP-odd phase may contribute another amplitude. Interference between the two leads to direct CP which can cause partial rate asymmetry in e.g. $B \to \eta' X_s$.

Three recent studies have examined such asymmetries. In the HT [9] and KP [5] works, the non-standard phase is assumed to be in the chromomagnetic form factor for $b \to sg$. Also, as mentioned before, they assume that the rate for $b \to sg$ is enhanced over the SM by about two orders of magnitude. In our work [17] we allow the non-standard CP-odd phase to reside in the chromo-electric or chromo-magnetic form factor. In fact the chromo-electric form factor is found to lead to larger asymmetries. Our study also showed that appreciable asymmetry (8–18%) can arise even if Non-Standard-Physics contributes only 10% to the production rate. This is especially significant as comparison of the rate between experiment and theory is extremely unlikely to show the presence of such a source of new physics and yet CP search can prove to be a viable thermometer. We also emphasize that there are many final states with similar asymmetries. A specially interesting mode is $B \to K\bar{K}X_s$ i.e. with three kaons [17].

9 Summary

CLEO has seen signals of a rather robust QCD penguin in several exclusive channels providing new impetus for improved theoretical understanding. The observed modes so far do not seem to require (a large) intervention by new physics. Many of them are useful to search for a non-standard CP phase.

10 Acknowledgements

We thank Ahmed Ali, Karl Berkelman, Tom Browder, Hai-Yang Cheng, Robert Fleischer, Christopher Greub, Michael Gronau, Laura Reina and Jim Smith. This research was supported under DOE contracts DE-AC02-76CH00016 (BNL) and DE-FG02-94ER40817 (ISU).

References

[1] See D. Miller [CLEO] talk in these proceedings.
[2] M. Ciuchini *et al.*, hep-ph/9703353; hep-ph/9708222.
[3] A. Ali and C. Greub, hep-ph/9707251.
[4] H.-Y. Cheng and B. Tseng, hep-ph/9707316.
[5] A. Kagan and A. Petra, hep-ph/9707354.
[6] A. Datta *et al.*, hep-ph/9707259.
[7] E. Shuryak and A.R. Zhitnitsky, hep-ph/9706316.
[8] I. Halperin and A. Zhitnitsky, hep-ph/9704412; hep-ph/9705251.
[9] W.-S. Hou and B. Tseng, hep-ph/9705304.
[10] F. Yuan and K.T. Chao, hep-ph/9706294.
[11] D. Atwood and A. Soni, Phy. Lett. B**405**, 150 (1997).
[12] I. Dunietz *et al.*, hep-ph/9612421.
[13] A. Lenz *et al.*, hep-ph/9706501.
[14] R. Fleischer and T. Mannel, hep-ph/9704423.
[15] D. London and A. Soni, Phys. Lett. B**407**, 61 (1997).
[16] A. Dighe *et al.*, hep-ph/9707521.
[17] D. Atwood and A. Soni, hep-ph/9706512 to appear in PRL (Dec. 1997).
[18] K. Berkelman, private communication.
[19] S.J. Brodsky and P. Lepage, PRD**24**, 1808 (1980); A. Anselm *et al.*, hep-ph/9603444; J.M. Gerard and T. Lahana, PLB**356**, 381 (1995).
[20] A. Ali and C. Greub, PLB**361**, 146 (1995); ZPC**49**, 431 (1991).

CP Violation in Selected B Decays*

F. Krüger** and L.M. Sehgal

Institut für Theoretische Physik (E), RWTH Aachen,
D-52056 Aachen, Germany

Abstract. We summarize the results of two papers in which we have studied CP violation in inclusive and exclusive decays $b \to d\, e^+e^-$. Two CP-violating effects are calculated: the partial rate asymmetry between b and $\bar{b}$ decay, and the asymmetry between e^- and e^+ spectra for an untagged $B/\bar{B}$ mixture. These asymmetries, combined with the branching ratio, can potentially determine the parameters (ρ, η) of the unitarity triangle. We also summarize a paper by Browder *et al.* on a possible CP-violating asymmetry in the inclusive reaction $B \to K^-(K^{*-})X$.

1 Decay $b \to d\, e^+e^-$

The decay $b \to d\, e^+e^-$ invites attention as a testing ground for CP for the following reason [1, 2]. The effective Hamiltonian for $b \to q\, e^+e^-$ $(q = d, s)$ calculated on the basis of electroweak box and penguin diagrams (see, for example, Ref. [3]) has the structure

$$H_{\text{eff}} = \frac{G_F \alpha}{\sqrt{2}\pi} V_{tb} V_{tq}^* \Big\{ c_9 (\bar{q}\gamma_\mu P_L b)\bar{l}\gamma^\mu l + c_{10}(\bar{q}\gamma_\mu P_L b)\bar{l}\gamma^\mu\gamma^5 l - 2\, c_7^{\text{eff}}\, \bar{q} i\sigma_{\mu\nu}\frac{q^\nu}{q^2}(m_b P_R + m_q P_L) b\, \bar{l}\gamma^\mu l \Big\}, \qquad (1)$$

where the coefficients have numerical values $c_7^{\text{eff}} = -0.315$, $c_9 = 4.227$, and $c_{10} = -4.642$, and $P_{L,R} = (1 \mp \gamma_5)/2$. There is, however, a correction to c_9 associated with $u\bar{u}$ and $c\bar{c}$ loop contributions generated by the nonleptonic interaction $b \to q\, u\bar{u}$ and $b \to q\, c\bar{c}$:

$$c_9^{\text{eff}} \approx c_9 + (3c_1 + c_2)\Big\{ g(m_c, s) + \lambda_u \left[g(m_c, s) - g(m_u, s) \right] \Big\}\,, \qquad (2)$$

where the loop functions $g(m_c, s)$ and $g(m_u, s)$ have absorptive parts for $s > 4m_c^2$ and $s > 4m_u^2$, and the coefficient $(3c_1 + c_2) \simeq 0.36$. The complex coefficient $\lambda_u = (V_{ub}V_{uq}^*)/(V_{tb}V_{tq}^*)$ is of order λ^2 in the case of $b \to s\, e^+e^-$ $(q = s)$, but of order unity in the case of $b \to d\, e^+e^-$ $(q = d)$. Notice that $\arg[(V_{ub}V_{ud}^*)/(V_{tb}V_{td}^*)] = \beta + \gamma$, where β, γ are the base angles of the unitarity triangle. For this latter reaction, therefore, the Hamiltonian possesses both

---

* Presented by L.M. Sehgal.

** Supported by the Deutsche Forschungsgemeinschaft (DFG) through Grant No. Se 502/4-3.

the weak (CKM) and dynamical (unitarity) phases that are mandatory for observable CP violation.

The inclusive aspects of the decay $B \to X_d e^+e^-$ can be calculated in the parton model approximation. The differential cross section $\mathrm{d}\Gamma/\mathrm{d}s$ is a quadratic function of the couplings c_7^{eff}, c_9^{eff} and c_{10}. The spectrum of the antiparticle reaction $\bar{b} \to \bar{d} e^+e^-$ is obtained by replacing λ_u by λ_u^*, giving rise to a partial rate asymmetry

$$A_{CP}(s) = \frac{\mathrm{d}\Gamma/\mathrm{d}s - \mathrm{d}\bar{\Gamma}/\mathrm{d}s}{\mathrm{d}\Gamma/\mathrm{d}s + \mathrm{d}\bar{\Gamma}/\mathrm{d}s} = \frac{\eta}{(1-\rho)^2+\eta^2} \times (\text{kinematical factor}) \,. \qquad (3)$$

A second spectral feature is the angular distribution of the e^- in the e^-e^+ centre-of-mass system, $\mathrm{d}\Gamma/\mathrm{d}\cos\theta$, which contains a term linear in $\cos\theta$, producing a forward-backward asymmetry [4]

$$A_{\mathrm{FB}}(s) = c_{10}\left[\hat{s}\,\mathrm{Re}\,c_9^{\mathrm{eff}} + 2c_7^{\mathrm{eff}}\left(1+\hat{m}_d^2\right)\right] \times (\text{kinematical factor}) \qquad (4)$$

with the notation $\hat{s} = s/m_b^2$ and $\hat{m}_d = m_d/m_b$. The corresponding asymmetry of the e^+ in $\bar{b} \to \bar{d} e^+e^-$ is obtained by replacing λ_u by λ_u^* in c_9^{eff} [cf. Eq. (2)]. The difference of A_{FB} and $\bar{A}_{\mathrm{FB}}$ is a CP-violating effect

$$\delta_{\mathrm{FB}} = A_{\mathrm{FB}} - \bar{A}_{\mathrm{FB}} = c_{10}\frac{\eta}{(1-\rho)^2+\eta^2} \times (\text{kinematical factor}) \,. \qquad (5)$$

Notice that δ_{FB} is twice the forward-backward asymmetry of the *electron* produced by an equal mixture of B and $\bar{B}$ mesons. Equivalently, δ_{FB} is twice the energy asymmetry of e^+ and e^- produced by a $B/\bar{B}$ mixture:

$$A_{\mathrm{E}} \equiv \frac{\Gamma(E_+ > E_-) - \Gamma(E_+ < E_-)}{\Gamma(E_+ > E_-) + \Gamma(E_+ < E_-)} = \frac{1}{2}\delta_{\mathrm{FB}} \,, \qquad (6)$$

where $E_\pm$ denote the lepton energies in the B rest frame. The branching ratio of $b \to d e^+e^-$ is clearly proportional to $|V_{td}|^2 \sim (1-\rho)^2 + \eta^2$. A measurement of the branching ratio, together with the asymmetry A_{CP} or δ_{FB} can potentially provide a self-contained determination of the parameters (ρ, η) of the unitarity triangle, as illustrated in Fig. 1 (for related discussions, see [5]). In [2], we have investigated the exclusive channels $B \to \pi\, e^+e^-$ and $B \to \rho\, e^+e^-$, employing different models for the form factors [6, 7]. Some representative results are given in Table 1.

2 *CP* Violation in Inclusive $B \to K^{(*)}X$

In a paper by Browder *et al.* [8], attention is focussed on the decays $B^- \to K^{(*)-}X$ and $\bar{B}^0 \to K^{(*)-}X$ with a highly energetic K (K^*), the system X containing u and d quarks only. By requiring $E_{K^{(*)}} > 2\,\mathrm{GeV}$, the background from $b \to c \to s$ is effectively suppressed. An attempt is made to relate such "quasi-inclusive" decays to the elementary process $b \to sg^* \to su\bar{u}$

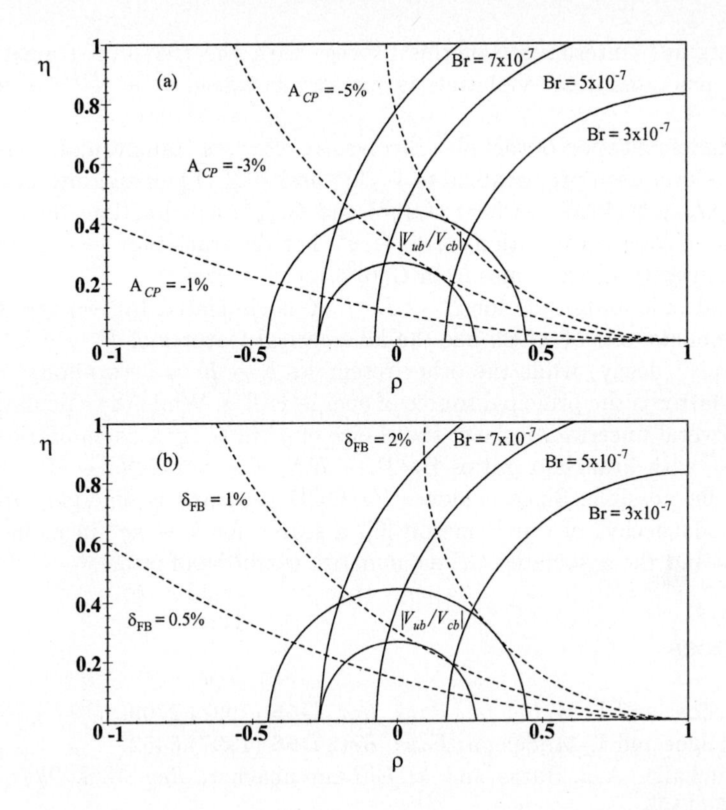

Fig. 1. Constraints on (ρ, η) imposed by measurement of branching ratio Br for $B \to X_d\, e^+e^-$, combined with the CP-violating asymmeties A_{CP} (a) or $\delta_{\rm FB}$ (b).

Table 1. Branching ratios, forward-backward asymmetries, and the CP-violating asymmetries A_{CP} and $\delta_{\rm FB}$ in inclusive and exclusive $b \to d\, e^+e^-$ reactions ($\rho = -0.07$, $\eta = 0.34$).

	$B \to X_d\, e^+e^-$	$B \to \pi\, e^+e^-$ [b]	$B \to \rho\, e^+e^-$ [b]
Br	5.5×10^{-7}	3.1×10^{-8}	5.0×10^{-8}
$\langle A_{\rm FB}\rangle$	-9%	$\equiv 0$	-17%
$\langle A_{CP}\rangle$ [a]	-2.7%	-3.1%	-2.8%
$\delta_{\rm FB}$	$+\,1\%$	$\equiv 0$	$+2\ \%$

[a] for $1\,{\rm GeV} < \sqrt{s} < 3\,{\rm GeV}$.

[b] using form factors of Melikhov and Nikitin [7].

(QCD penguin). Interference of this process with the tree-level transition $b \to su\bar{u}$ generates a CP-violating asymmetry between $B \to K^{(*)-}X$ and $\bar{B} \to K^{(*)+}X$.

The implementation of this idea involves an effective Hamiltonian containing a tree-level term proportional to $V_{ub}V_{us}^*$ and a QCD penguin interaction $\sim (V_{ub}V_{us}^*G_u + V_{cb}V_{cs}^*G_c)$, where $G_u(q^2)$ and $G_c(q^2)$ are functions denoting the $u\bar{u}$ and $c\bar{c}$ loop contributions to $b \to sg^*$. For the transition $b \to su\bar{u}$, the essential unitarity phase comes from $G_c(q^2)$ for $q^2 > 4m_c^2$.

The hadronic amplitude for $B \to K^{(*)-}X$ is simulated by two types of matrix elements, one of which has the kinematical features of $B^- \to K^-u\bar{u}$ ("three-body" decay) while the other resembles $b \to K^-u$ ("two-body" decay). The latter is the principal source of energetic K's. While the calculation involves several uncertainties, e.g. the choice of q^2 in $G_c(q^2)$, asymmetries of order 10%, with branching ratios $\mathrm{Br}(B \to KX,\, E_K > 2\,\mathrm{GeV}) \sim 10^{-4}$ are shown to be possible. Since evidence for QCD penguins is emerging from the two-body decays of the B meson [9], a search for $b \to sg^*$ in inclusive $B \to KX$, and the associated CP asymmetry, would be of interest.

References

[1] F. Krüger and L. M. Sehgal, *Phys. Rev.* **D55** (1997) 2799.
[2] F. Krüger and L. M. Sehgal, *Phys. Rev.* **D56** (1997) 5452.
[3] G. Buchalla, A. J. Buras and M. E. Lautenbacher, *Rev. Mod. Phys.* **68** (1996) 1125.
[4] A. Ali, T. Mannel and T. Morozumi, *Phys. Lett.* **B273** (1991) 505.
[5] L. T. Handoko, *preprint* hep-ph/9707222; C. S. Kim, T. Morozumi and A. I. Sanda, *Phys. Rev.* **D56** (1997) 7240; S. Rai Choudhury *Phys. Rev.* **D56** (1997) 6028.
[6] P. Colangelo, F. De Fazio, P. Santorelli and E. Scrimieri, *Phys. Rev.* **D53** (1996) 3672.
[7] D. Melikhov and N. Nikitin, *preprint* hep-ph/9609503.
[8] T. E. Browder, A. Datta, X. G. He and S. Pakvasa, *preprint* hep-ph/9705320.
[9] M. Neubert, these proceedings.

CP Violation from Charged Higgs Exchange in Hadronic Tau Decays with Unpolarized Beams

J.H. Kühn and E. Mirkes

Institut für Theoretische Teilchenphysik, Universität Karlsruhe,
D-76128 Karlsruhe, Germany

Abstract. CP violating signals in semileptonic τ decays induced by an exotic scalar exchange are studied in a completely model-independent way. These can be observed in decays of unpolarized single τ's even if their rest frame cannot be reconstructed. No beam polarization is required. The importance of the two-meson channel, in particular the $K\pi$ final state is emphasized.

1 Introduction

CP violation has been experimentally observed only in the K meson system. The effect can be explained by a nontrivial complex phase in the CKM flavour mixing matrix [1]. However, the fundamental origin of this CP violation is still unknown. In particular the CP properties of the third fermion family are largely unexplored. Production and decay of τ leptons might offer a particularly clean laboratory to study these effects. In this contribution, we investigate the effects of CP violation [2] to be observed in semileptonic τ decays which could arise in a framework outside the mechanism prosposed by Kobayashi and Maskawa. We show that the structure function formalism of Ref. [3] allows for a systematic analysis of possible CP violation effects in the two and three meson channels. Special emphasis is put on the $\Delta S = 1$ transition $\tau \to K\pi\nu_\tau$ where possible CP violating signals from multi Higgs boson models [4] would be signaled by a nonvanishing difference between the structure functions $W_{SF}[\tau^- \to (K\pi)^-\nu_\tau]$ and $W_{SF}[\tau^+ \to (K\pi)^+\nu_\tau]$. Such a measurement is possible for unpolarized single τ's without reconstruction of the τ rest frame and without polarized incident e^+e^- beams. It is shown that this CP violation requires both nonvanishing hadronic phases and CP violating phases in the Hamiltonian, where the hadronic phases arise from the interference of complex Breit-Wigner propagators, whereas the CP violating phases could arise from an exotic charged Higgs boson. An additional independent test of CP violation in the two meson case is possible, but would require the knowledge of the full kinematics and τ polarization.

2 CP Violating Signals in the $\tau \to K\pi\nu_\tau$ Decay Mode

Transitions from the vacuum to two pseudoscalar mesons h_1 and h_2 are induced through vector and scalar currents only. Expanding these hadronic matrix elements along the set of independent momenta $(q_1 - q_2)_\beta$ and $Q_\beta = (q_1 + q_2)_\beta$ we define $(T_{\alpha\beta} = g_{\alpha\beta} - (Q_\alpha Q_\beta)/Q^2)$

$$\langle h_1(q_1)h_2(q_2)|\bar{u}\gamma_\beta d|0\rangle = (q_1 - q_2)^\alpha T_{\alpha\beta}\, F(Q^2) + Q_\beta\, F_S(Q^2) \quad (1)$$

$$\langle h_1(q_1)h_2(q_2)|\bar{u}d|0\rangle = F_H(Q^2). \quad (2)$$

The representation of the hadronic amplitude $\langle h_1 h_2|\bar{u}\gamma_\beta d|0\rangle$ corresponds to a decomposition into spin one and spin zero contributions, *e.g.* the vector form factor $F(Q^2)$ corresponds to the $J^P = 1^-$ component of the weak charged current, and the scalar form factor $F_S(Q^2)$ to the $J^P = 0^+$ component. A fictitious scalar Higgs exchange contribution is proportional to $\eta_S F_H$. The general amplitude for the $\Delta S = 1$ decay (where $h_1 \equiv K, h_2 \equiv \pi$)

$$\tau^-(l,s) \to \nu(l',s') + h_1(q_1,m_1) + h_2(q_2,m_2)\,, \quad (3)$$

can thus be written as

$$\mathcal{M} = \sin\theta_c \frac{G}{\sqrt{2}} \bar{u}(l',s')\,\gamma_\alpha(1-\gamma_5)\,u(l,s)\left[(q_1-q_2)_\beta\, T^{\alpha\beta}\, F + Q^\alpha\, \tilde{F}_S\right] \quad (4)$$

$$\text{with} \qquad \tilde{F}_S = F_S + \frac{\eta_S}{m_\tau}\, F_H \quad (5)$$

In Eq. (4) s denotes the polarization 4-vector of the τ lepton. The complex parameter η_S in Eq. (5) transforms like

$$\eta_S \xrightarrow{\mathrm{CP}} \eta_S^* \quad (6)$$

and thus allows for the parametrization of possible CP violation. Up to the small isospin breaking terms, induced for example by the small quark mass difference, CVC implies the vanishing of F_S for the two pion ($h_1 \equiv \pi^-, h_2 \equiv \pi^0$) case. For the transition $\tau \to K\pi\nu$ the $J = 1$ form factor F is dominated by the $K^*(892)$ vector resonance contribution. The scalar form factor F_S is expected to receive a sizable resonance contribution ($\sim 5\%$ to the decay rate) from the $K_0^*(1430)$ with $J^P = 0^+$ [5]. The corresponding τ^+ decay is obtained from Eq. (4) through the substitutions

$$(1-\gamma_5) \to (1+\gamma_5), \qquad \eta_S \to \eta_S^*. \quad (7)$$

Reaction (3) is most easily analyzed in the hadronic rest frame $\mathbf{q}_1 + \mathbf{q}_2 = 0$ [3]. After integration over the unobserved neutrino direction, the differential decay rate in the rest frame of $h_1 + h_2$ is given by [5, 3]

$$d\Gamma(\tau^- \to 2h\nu_\tau) = \{\bar{L}_B W_B + \bar{L}_{SA} W_{SA} + \bar{L}_{SF} W_{SF} + \bar{L}_{SG} W_{SG}\}$$

$$\frac{G^2}{2m_\tau} \sin^2\theta_c \frac{1}{(4\pi)^3} \frac{(m_\tau^2 - Q^2)^2}{m_\tau^2} |\mathbf{q}_1| \frac{dQ^2}{\sqrt{Q^2}} \frac{d\cos\theta}{2} \frac{d\alpha}{2\pi} \frac{d\cos\beta}{2} . \quad (8)$$

For the definition and discussion of the angles and leptonic coefficients $\bar{L}_X$ in Eq. (8) we refer the reader to Ref. [2]. The coefficients $\bar{L}_X$ contain all α, β and θ angular and τ-polarization dependence. The hadronic structure functions W_X, $X \in \{B, SA, SF, SG\}$, depend only on Q^2 and the form factors F and $\tilde{F}_S$ of the hadronic current. One has [5, 3, 2]:

$$\begin{aligned} W_B[\tau^-] &= 4(\mathbf{q}_1)^2\,|F|^2 & W_{SF}[\tau^-] &= 4\sqrt{Q^2}|\mathbf{q}_1|\,\mathrm{Re}\left[F\tilde{F}_S^*\right] \\ W_{SA}[\tau^-] &= Q^2\,|\tilde{F}_S|^2 & W_{SG}[\tau^-] &= -4\sqrt{Q^2}|\mathbf{q}_1|\,\mathrm{Im}\left[F\tilde{F}_S^*\right] \end{aligned} \quad (9)$$

The hadronic structure functions $W_X[\tau^+]$ are obtained by the replacement $\eta_S \to \eta_S^*$ in $\tilde{F}_S$ in Eqs. (9,5). CP conservation implies that all four structure functions are identical for τ^+ and τ^-. With the ansatz for the form factors formulated in Eq. (4) CP violation can be present in W_{SF} and W_{SG} only and requires complex η_S.

As demonstrated in Ref. [2] W_{SF} can be measured in e^+e^- annihilation experiments in the study of single unpolarized τ decays even if the τ rest frame cannot be reconstructed. In this respect the result differ from earlier studies of the two meson modes where either polarized beams and reconstruction of the full kinematics [6] or correlated fully reconstructed τ^- and τ^+ decays were required [7]. The determination of W_{SG}, however, requires the knowledge of the full τ kinematics and τ polarization [2] which is possible with the help of vertex detectors. The corresponding distributions in this second case are equivalent to the correlations proposed in Refs. [6, 7].

The crucial observation made in Ref. [2] is that one can measure the following CP-violating differences

$$\Delta W_{SF} = \frac{1}{2}\left(W_{SF}[\tau^-] - W_{SF}[\tau^+]\right) \quad \Delta W_{SG} = \frac{1}{2}\left(W_{SG}[\tau^-] - W_{SG}[\tau^+]\right) \quad (10)$$

under the above mentioned conditions. The hadronic structure functions $W_X[\tau^-]$ and $W_X[\tau^+]$ differ only in the phase os η_S and one obtains

$$\Delta W_{SF} \;=\; 4\sqrt{Q^2}|\mathbf{q}_1|\,\frac{1}{m_\tau}\,\mathrm{Im}\,(FF_H^*)\;\;\mathrm{Im}\,(\eta_S)\;. \quad (11)$$

In essence the measurement on ΔW_{SF} analyses the difference in the correlated energy distribution of the mesons h_1 and h_2 from τ^+ and τ^- decay in the laboratory. As already mentioned, ΔW_{SF} is observable for single τ^+ and τ^- decays without knowledge of the τ rest frame. Any nonvanishing experimental result for ΔW_{SF} would be a clear signal of CP violation. Note that a nonvanishing ΔW_{SF} requires nontrivial hadronic phases (in addition to the

CP violating phases η_S) in the form factors F and F_H. Such hadronic phases in F (F_H) originate in the $K\pi\nu_\tau$ decay mode from complex Breit Wigner propagators for the K^* (K_0^*) resonance. Sizable effects of these hadronic phases are expected in this decay mode [5].

Once the τ rest frame is known and a preferred direction of polarization exists one may also determine ΔW_{SG} which is theoretically given by

$$\Delta W_{SG} = 4\sqrt{Q^2}|\mathbf{q}_1| \frac{1}{m_\tau} \mathrm{Re}\,(F F_H^*) \; \mathrm{Im}\,(\eta_S) \; . \tag{12}$$

Any observed nonzero value of $\Delta W_{SF}, \Delta W_{SG}$ would signal a true CP violation. Eqs.(11) and (12) show that the sensitivity to CP violating effects in ΔW_{SF} and ΔW_{SG} can be fairly different depending on the hadronic phases. Whereas ΔW_{SF} requires nontrivial hadronic phases ΔW_{SG} is maximal for fixed η_S in the absence of hadronic phases.

3 Three Meson Decays

The structure function formalism [3] allow also for a systematic analysis of possible CP violation effects in the three meson case. Some of these effects have already been briefly discussed in Ref. [8]. The $K\pi\pi$ and $KK\pi$ decay modes with nonvanishing vector <u>and</u> axial vector current are of particular importance for the detection of possible CP violation originating from exotic intermediate vector bosons. This would be signalled by a nonvanishing difference between the structure functions $W_X(\tau^-)$ and $W_X(\tau^+)$ with $X \in \{F, G, H, I\}$. A difference in the structure functions with $X \in \{SB, SC, SD, SE, SF, SG\}$ can again be induced through a CP violating scalar exchange. CP violation in the three pion channel has been also discussed in Ref. [9] and in the $K\pi\pi$ and $KK\pi$ channels in Ref. [10], where the latter analysis is based on the "$T-$odd" correlations in Ref. [3] and the vector meson dominance parametrizations in Ref. [11].

References

[1] M. Kobayashi and T. Maskawa, Prog. Theor. Phys. **49**, 652 (1973).
[2] J.H. Kühn and E. Mirkes, Phys. Lett. B**398**, 407 (1997).
[3] J.H. Kühn and E. Mirkes, Z. Phys. C**56**, 661 (1992); erratum *ibid.* **C67**, 364 (1995); Phys. Lett. B **286**, 281 (1992).
[4] Y. Grossman, Nucl. Phys. B**426**, 355 (1994), and references therein.
[5] M. Finkemeier and E. Mirkes, Z. Phys. C**72**, 619 (1996).
[6] Y.S. Tsai, Phys. Rev. D**51**, 3172 (1995).
[7] C.A. Nelson *et al.*, Phys. Rev. D**50**, 4544 (1994), and references therein.
[8] M. Finkemeier and E. Mirkes, Proceedings of the Workshop on the Tau/Charm Factory, Argonne, IL, (1995), [hep-ph/9508312].
[9] S. Choi, K. Hagiwara and M. Tanabashi, Phys. Rev. D**52**, 1614 (1995).
[10] U. Kilian, J. Körner, K. Schilcher and Y. Wu, Z. Phys. C**62**, 413 (1994).
[11] R. Decker, E. Mirkes, R. Sauer and Z. Was, Z. Phys. C**58** 445 (1993).

Invariant Formulation of CP Violation

F. del Aguila[1] (faguila@goliat.ugr.es),
J. A. Aguilar-Saavedra[1] (aguilarj@goliat.ugr.es);
G. C. Branco[2] (d2003@beta.ist.utl.pt)

[1]University of Granada, Spain; [2]Instituto Superior Técnico, Lisboa, Portugal

Abstract. CP violation can be described with expressions invariant under redefinitions of the weak fermion basis. This formulation is explicitly known for the Standard Model extensions with one sequential or vector-like quark and for its left-right version. The invariant formulation is very useful in model building and is specially suitable to the syudy of physical limits as the chiral limit $m_{u,d,s} = 0$, where CP violation can only arise from Physics Beyond the Standard Model.

In the Standard Model (SM) there is only one physical phase parametrising CP violation in the Cabibbo-Kobayashi-Maskawa (CKM) matrix [1]. It is also known that CP conservation can be characterised by the vanishing of [2]

$$\begin{aligned} I &\equiv \det\,[M_u M_u^\dagger, M_d M_d^\dagger] \\ &= -2i(m_t^2 - m_c^2)(m_t^2 - m_u^2)(m_c^2 - m_u^2)(m_b^2 - m_s^2) \\ &\quad \times (m_b^2 - m_d^2)(m_s^2 - m_d^2)\,\mathrm{Im}\,V_{ud}V_{cd}^* V_{cs} V_{us}^* \,, \end{aligned} \tag{1}$$

where $M_{u,d}$ are the up and down quark mass matrices, m_i the mass of the quark i and V_{ij} the elements of the CKM matrix, with $\mathrm{Im}\,V_{ud}V_{cd}^* V_{cs} V_{us}^* = c_{12}s_{12}c_{13}^2 s_{13}c_{23}s_{23}\sin\delta_3$ and δ_3 the CP violating phase. It has been shown [3] that by considering the most general CP transformation of the quark fields which leave invariant the kinetic energy terms and the gauge interactions written in the weak basis, one can readily derive not only Eq. (1) but also invariant necessary conditions for CP invariance applicable to extensions of the SM. In models beyond the SM, the search for a complete set of invariants characterizing the CP properties is quite involved. The vanishing of the weak basis invariants are necessary and sufficient conditions for CP conservation. They are necessary by construction, for they involve the determinant or the trace of products of mass matrices which are invariant under generalised weak basis refefinitions. In the SM I is invariant under the transformations $M_u \to U_L^\dagger M_u U_R^u$, $M_d \to U_L^\dagger M_d U_R^d$, with unitary matrices U_L, $U_R^{u,d}$ [3]. To prove that the set of invariant conditions is sufficient requires to solve a complicated set of algebraic equations. The number of equations, and hence of invariants, grows faster than the number of CP violating phases. Thus in the simplest extensions of the SM with one extra family or an extra singlet quark the number of invariants required to characterise CP violation is more than twice the number of CP violating phases (see Table 1). The explicit

	Masses	Mixing angles	Phases	Invariants
3 families	6	3	1	1
4 families	8	6	3	8
3 families + 1 singlet	7	6	3	7

Table 1. Physical parameters and CP invariants for the SM and simple extensions

expressions of these invariants can be found in Refs. [4, 5]. In general to find the complete set of invariants one needs an efficient symbolic program [6]. Examples involving the lepton sector can be found there.

Once the complete set is known, however, the invariant formalism is more suited for answering model dependent questions, for the invariants can be calculated in any weak basis. In particular one can ask for definite mass limits as the chiral limit $m_{u,d,s} = 0$. For the two cases above the number of independent invariants reduces in the chiral limit to 3 (and the number of CP violating phases to 2). These, up to non-zero mass factors, are functions of $B_1 \equiv \mathrm{Im}\, V_{cb} V_{4b}^* V_{4b'} V_{cb'}^*$, $B_2 \equiv \mathrm{Im}\, V_{tb} V_{4b}^* V_{4b'} V_{tb'}^*$ and $B_3 \equiv \mathrm{Im}\, V_{cb} V_{tb}^* V_{tb'} V_{cb'}^*$. Then the size of CP violation in these cases is bounded by $|B_i| \leq 10^{-2}$ for an extra sequential family and $|B_i| \leq 10^{-4}$ for an extra quark singlet. Similarly, in the left-right extension of the SM CP violation is described by 2 CP violating phases and 2 invariant expressions in the chiral limit, as it will be shown elsewhere

References

[1] N. Cabibbo, Phys. Rev. Lett. **10**, 531 (1963); M. Kobayashi and T. Maskawa, Prog. Theor. Phys. **49**, 652 (1973); Particle Data Group, R. M. Barnett *et al.*, Phys. Rev. D **54**, 1 (1996)

[2] C. Jarlskog, Phys. Rev. Lett. **55**, 1039 (1985); Z. Phys. **C29**, 491 (1985)

[3] J. Bernabéu, G. C. Branco and M. Gronau, Phys. Lett. **169B**, 243 (1986)

[4] F. del Aguila and J. A. Aguilar–Saavedra, Phys. Lett. **B386**, 241 (1996); M. Gronau, A. Kfir and R. Loewy, Phys. Rev. Lett. **56**, 1538 (1986)

[5] F. del Aguila, J. A. Aguilar–Saavedra and G. C. Branco, hep-ph/9703410, to appear in Nucl. Phys. **B**; see also G. C. Branco and L. Lavoura, Nucl. Phys. **B278**, 738 (1986)

[6] F. del Aguila, J. A. Aguilar–Saavedra and M. Zrałek, Comp. Phys. Comm. **100**, 231 (1997); F. del Aguila and M. Zrałek, Nucl. Phys. **B447**, 211 (1995)

A New Symmetry of Quark-Yukawa Couplings

H. González[1,3], S.R. Juárez W.[2], P. Kielanowski[1,4] and G. López Castro[1]

[1]Dept. de Fís., CINVESTAV del IPN, México D.F., Mexico., [2]ESFM–IPN, U.P.-A.L.M., México D.F., Mexico, (rebeca@esfm.ipn.mx), [3]Prog. de Mat. y Fís., U. Surcolombiana, Neiva, Colombia, [4]Dept. of Physics, Warsaw University, Poland.

Abstract. We discuss a new kind of symmetry that relates the quark Yukawa couplings (QYC) of the up and down quarks. In this scheme we naturally include the hierarchy of quark masses and reproduce the quark mixing matrix with CP violation phase.

The structure of the QYC before symmetry breaking cannot be uniquely recovered from the measured values of the quark masses and mixings. Nevertheless the hierarchy observed in the low energy values of the masses and mixings [1] can be used as a guide to search for unifying relations between QYC. Here our approach to the problem of the quark masses and mixings consists in the postulation of a relation between the QYC of up and down quarks valid at the scale of Grand Unification (GU). Then, the evolution of these couplings to the scale of top quark mass furnishes a quark mixing matrix in very good agreement with present experimental data. Based on the hierarchy of quark masses at low energies let us propose [2]

$$Y_u = C Y_d \cdot Y_d, \quad C \sim \frac{v\, m_t}{m_b^2 \sqrt{2}},$$

($Y_{u,d}$ are the QYC of up and down quarks), as an exact property valid at the energy scales of GU, where v is the *vev* of the Higgs field.

We will assume a hierarchy for Y_d in terms of the Cabibbo angle, which coincides with one of the two possible forms derived in [3]. The form of Y_d was chosen because it is also compatible with the parameterization of the Cabibbo-Kobayashi-Maskawa (CKM) matrix given in Ref. [4]. As is well known, the CKM matrix can be defined as $V_{ckm} = V_u V_d^\dagger$, where V_u (V_d) is a unitary matrix that diagonalizes the up (down) quark mass matrix and it depends on 4 parameters (3 angles and a phase). There are different parameterizations [5] of the CKM matrix. We will consider here the parameterization in terms of the Eigenvalues and Eigenvectors (EE) of the CKM matrix [4]. In this parameterization V_{ckm} satisfies the properties:

$$V_{ckm}^3 = I, \quad V_{ckm} = \hat{A} D \hat{A}^\dagger,$$

$$D = \mathrm{Diag}\left(e^{-2\pi i/3}, \quad e^{2\pi i/3}, \quad 1\right).$$

The unitary matrix $\hat{A}$ can be interpreted as a universal matrix that diagonalizes the mass matrices of up and down quarks. At the GU scale we choose the matrix $\hat{A}$ to be orthogonal and it can be written in the following form in terms of new hierarchical parameters ($\lambda \approx 0.22$, A) which are similar to the Wolfenstein's ones

(λ_W, A_W) as $\hat{A} =$

$$\begin{pmatrix} \frac{1}{k_1} & \frac{-\lambda}{\sqrt{3}k_2}(1+\frac{A^2\lambda^4}{3}) & 0 \\ \frac{\lambda}{\sqrt{3}}\frac{1}{k_1} & \frac{1}{k_2} & \frac{-A\lambda^2}{k_3\sqrt{3}} \\ \frac{A\lambda^3}{3}\frac{1}{k_1} & \frac{A\lambda^2}{\sqrt{3}}\frac{1}{k_2} & \frac{1}{k_3} \end{pmatrix},$$

where $k_i = k_i(\lambda, A), i = 1,2,3$.

To obtain the evolution of the QYC down to low energies in the EE scheme we consider the Renormalization Group Equations (RGE). This evolution of the QYC is obtained in the case of the standard model (SM) and its two-Higgs doublet (DHM) and minimal supersymmetric (MSSM) extensions. The RGE for the QYC [6], become modified by a quark field rephasing transformation which is compatible with our parameterization (see [2] for further details).

Upon this transformation only the weak charged current and the Yukawa couplings are affected. The charged current matrix becomes D which may therefore be interpreted as the zeroth order CKM matrix, *i.e.* $V_{ckm} = \hat{A}D\hat{A}^\dagger$ after moving to mass eigenstates (V_{ckm} is symmetric at the GU scale). As a consequence of the RG evolution from the GU to the m_t scale, the values of A and λ do not change for Y_u and they become complex for Y_d. Then

$$V_{ckm}(t) = \hat{A}(A,\lambda)D\hat{A}_d^+(\tilde{A},\tilde{\lambda}),$$

where $\hat{A}_d =$ has a similar structure as $\hat{A}$ in terms of $\tilde{\lambda} \equiv \lambda(m_t)$ and $\tilde{A} \equiv A(m_t)$. The explicit expressions of $\tilde{\lambda}$ and $\tilde{A}$ can be found in [2] and their specific values depend on the model considered. The results of the fit to the experimental values for the CKM matrix for the SM and MSSM models are

Data	*MSSM*	*SM*
$\lvert V_{ud}\rvert$	.9751	.9751
$\lvert V_{us}\rvert$	.2217	.2216
$\lvert V_{cd}\rvert$	.2216	.2215
$\frac{\lvert V_{ub}\rvert}{\lvert V_{cb}\rvert}$	.0892	.0898
$\lvert V_{cb}\rvert$	.04076	.04083
$\lvert V_{ub}\rvert$	0.00364	0.00366
$\lvert V_{td}\rvert$	0.00730	0.00808
χ^2	2.92	2.966
$A \pm \Delta A$	$1.48 \pm .01$	$2.11 \pm .04$
$\lambda \pm \Delta\lambda$	$0.16 \pm .01$	$0.16 \pm .01$

In conclusion, our results for the CKM matrix are compatible with the experimental data. While the CKM matrix is symmetric at GUT scales, it exhibits the correct asymmetries at low energies. These facts strongly supports the validity of our ansatz at the GU scale.

S. R. J. W. gratefully acknowledges partial support by COFAA del IPN.

References

[1] Particle Data Group: Phys. Rev. D54 (1996), 1.

[2] H. González et al, hep-ph/9703292.

[3] P. Fishbane et al, Phys. Rev. D45 (1992), 293; Y. Koide et al,, Phys. Rev. D46 (1992), R4813.

[4] P. Kielanowski, Phys. Rev. Lett. 63 (1989), 2189.

[5] N. Cabbibo, Phys. Rev. Lett. 10 (1963), 531; M. Kobayashi and et al,, Prog. Theor. Phys. 49 (1973), 652; L. Wolfenstein, Phys. Rev. Lett. 51 (1983), 1945; M. Gronau et al, Phys. Rev. Lett. 54 (1985), 385.

[6] B. Grzadkowski et al, Phys. Lett.B198 (1987), 64.

Charmless and Rare B Decays. Recent and Inclusive Results and the Charm Puzzle

Peter M. Kluit (kluit@vxcern.cern.ch)

NIKHEF, Amsterdam

1 Motivation

An outstanding problem in B physics is the so-called charm puzzle [1]. The problem is that the measured semi-leptonic B branching ratio is lower than the predicted value [2]. There are three solutions to this problem proposed, that all increase the hadronic b decay width and consequently lower the predicted semi-leptonic B branching ratio.

The Standard solution is to enlarge the branching ratio for the process $Br(b \to c\bar{c}s)$ by 0.10-0.15. This implies a larger number of charmed particles per B decay of $N_C = 1.15 - 1.3$ [2],[3]. A second solution is to increase the branching ratio $Br(b \to (c\bar{c})s)$, where the $(c\bar{c})$ system annihilates [4]. A third solution implies new physics in the form of a large charmless B decay branching ratio $Br(b \to no\,charm)$ of 0.10-0.15, e.g. in the form of additional $b \to s\,g$ decays [5]. Up to now no experimental data exists to constrain the last two solutions.

The present measurements of the number of charmed particles per B decay N_C are derived from branching ratio measurements into final states with charmed particles. The CLEO result is $N_C = 1.10 \pm 0.05$ [6] and the present LEP average is 1.17 ± 0.07 [2]. Both measurements suffer from large common systematic errors. The experimental values are rather low with respect to the theoretical expectation.

2 Experimental results

The experimental results discussed below are based on conference contributions from the ALEPH [7], DELPHI [8] and CLEO [9] experiments [1]. All results are preliminary [2] and the author should be held responsible for the combination of the results.

---

[1] The preliminary SLD study on the $Br(b \to s\,g)$ is discussed by M. Daoudi in these proceeedings.

[2] Some results are being finalized and published. Here only the results submitted to HEP97 are discussed.

2.1 Measurements of $Br(b \to s\gamma)$

A new measurement of the $Br(b \to s\gamma)$ has been performed by the ALEPH experiment [7]. The analysis used tagged photons in the calorimeter and reconstructed the B direction and energy after applying a b-tag. A fit to the photon energy $E^{\star}_{\gamma}$ in the B rest frame gave the result quoted in Table 1. The measured branching ratio is in agreement with the results and upper limits from other experiments. It is compatible with the Standard Model prediction of $Br(b \to s\gamma) = (3.28 \pm 0.33)\ 10^{-4}$.

expt	$Br(b \to s\gamma)$	ref.
ALEPH	$(3.38 \pm 0.74 \pm 0.85)\ 10^{-4}$	[7]
CLEO	$(2.32 \pm 0.57 \pm 0.35)\ 10^{-4}$	[10]
DELPHI	$< 5.4\ 10^{-4}$ (90% CL)	[11]
L3	$< 1.2\ 10^{-3}$ (90% CL)	[12]

Table 1.

2.2 Measurement of $Br(b \to$ *no open charm*)

The DELPHI experiment has measured the inclusive charmless B decay branching ratio [8]. The experimental technique is based on the vertex topology: (i) B decays without open charm (Br_{0C}) will give one B hadron vertex; (ii) B decays into one charmed particle (Br_{1C}) will have one B hadron and one charm decay vertex, while (iii) B decays into two open charmed particles (Br_{2C}) will give one B hadron and two charm decay vertices. The symbol C refers to all charmed particles: D mesons and charmed baryons. The first category includes charmless B decays ($b \to s\ \gamma$, $b \to s\ g$, $b \to u\bar{u}d$ etc.), B decays into hidden charm (J/ψ and excited states) and possible B decays with annihilating charm (solution three).

The experimental technique is based on the b-tagging probability, defined as the probability for the hypothesis that a given set of tracks comes from the primary vertex. The b-tagging probability is precisely studied and used for the measurement of R_b. The events are divided into one tagging hemisphere and one measurement hemisphere.

The distribution of P_H^+ in the measurement hemisphere was fitted using the shapes for the $0C$, $1C$ and $2C$ distributions from the simulation. P_H^+ is the combined b-tagging probability per hemisphere for all tracks with a positive lifetime sign. The shapes are quite different and allow a simultanuous fit of the branching ratios Br_{0C} and Br_{2C} ($Br_{1C} = 1 - Br_{0C} - Br_{2C}$). An extensive study of the systematic errors has been made. Two main sources have to be distinguished: firstly, the modelling of B decays and secondly detector effects, uncorrelated from year to year. For δBr_{0C} they amount to resp. 0.009 and 0.012 and for δBr_{2C} to 0.033 and 0.035. A detailed breakdown of the systematic errors for 17 sources can be found in ref. [8].

The preliminary results for the branching ratios were:

$$Br_{0C} = 0.044 \pm 0.025$$
$$Br_{2C} = 0.163 \pm 0.046.$$

After subtraction of the measured (and extrapolated) hidden charm branching ratio of 0.026 ± 0.004 [4],[6], one obtains the charmless B branching ratio of

$$Br(b \to no\, charm) = 0.018 \pm 0.025 \pm 0.004.$$

This is compatible with the Standard Model expectation of 0.026 ± 0.01 [4]. It is possible to put an upper limit on new physics in charmless B decays by subtracting the Standard Model contribution:

$$Br(b \to no\, charm)^{NEW} < 0.045\ (95\%\ CL).$$

2.3 Measurements of $Br(b \to c\bar{c}s)$

DELPHI has also measured the branching ratio Br_{2C} using partially reconstructed D meson with an identified kaon, exploiting the charge correlation between the D meson and the kaon. A fit to the transverse momentum distribution of the kaons gave the preliminary result:

$$Br_{2C} = 0.170 \pm 0.047.$$

The CLEO experiment has measured the Br_{2C} branching ratio, reconstructing charmed particles (D^0,D^+,$\bar{D}^0$,D^-) from a B or $\bar{B}$ meson, where the B or $\bar{B}$ is tagged with a lepton [9].

¿From the measured ratio $\Gamma(B \to DX)/\Gamma(B \to \bar{D}X) = 0.100 \pm 0.026 \pm 0.016$ and $Br(B \to D_s X)$ the following preliminary result was obtained:

$$Br_{2C} = 0.189 \pm 0.037.$$

The three results from the DELPHI and CLEO experiments are compatible and can be combined:

$$Br_{2C} = 0.178 \pm 0.026.$$

2.4 An upper limit on $Br(b \to s\, gluon)$

DELPHI has put an upper limit [8] on the process $b \to sg$ using as a signature a high p_t kaon (above 1 GeV/c). The kaon was identified in RICHes or in the TPC using the measured ionization loss. The $b \to sg$ decays were simulated by creating a $s - g$ string, giving spectator quark the fermi momentum, and then the system was fragmented using the JETSET program [5]. The following preliminary upper limit was obtained:

$$Br(b \to s\, g) < 0.05\ (95\%\ CL),$$

where in the upper limit the total systematic error is included.

3 Summary

A new measurement of the radiative B decay branching ratio has been presented. The charmless B branching ratio has been measured to be $Br(b \to no\,charm) = 0.018 \pm 0.025 \pm 0.004$, compatible with the Standard Model expectation. The combined result for the B branching ratio into double charm is $Br_{2C} = 0.178 \pm 0.026$.

Theories that predict a large annihilating charm or charmless branching ratio are constrained by two measurements: $Br(b \to no\,charm)^{NEW} < 0.045\,(95\%\,CL)$ and $Br(b \to s\,g) < 0.05\,(95\%\,CL)$.

The number of charmed particles per B decay can be obtained from the double charm branching ratio, the measured B branching ratio into hidden charm, and the SM value for the charmless B branching ratio. It gives

$$N_C = 1.178 \pm 0.026 \pm 0.01.$$

The result is more precise than previous measurements. It is compatible with the Standard explanation for the semi-leptonic B branching ratio that predicts $N_c = 1.15 - 1.3$.

References

[1] G. Altarelli and S. Petrarca, Phys. Lett. B261 (1991) 303.

[2] M. Neubert, 'Heavy-Quark Efffective Theory and Weak Matrix Elements', plenary talk at HEP97, see these Proceedings;
M. Feindt, 'B Physics', plenary talk at HEP97, see these Proceedings.

[3] M. Neubert and C.T. Sachrajda, Nucl. Phys. B482 (1997) 339.

[4] I. Dunietz et al., FERMILAB-PUB-96/421-T, hep-ph/9612421.

[5] A.L. Kagan, 'Hints for enhanced $b \to s\,g$ and inclusive searches for $B \to K\,X$', see these Proceedings;
G.Hou, 'Enhanced $b \to s\,g$ Decay from New Physics', see these Proceedings.

[6] L. Gibbons et al. (CLEO Collaboration), CLNS 96/1454, hep-ex/9703006.

[7] ALEPH Collaboration, 'An Inclusive Measurement of $Br(b \to s\gamma)$', submitted to HEP97, ALEPH 97-023 PHYSIC 97-018.

[8] DELPHI Collaboration, 'First Measurement of the Inclusive Charmless B Decay Branching Ratio and Determination of the $b \to c\bar{c}s$ Rate', submitted to HEP97-448, DELPHI 97-80 CONF 66.

[9] CLEO Collaboration, 'Flavor-Specific Inclusive B decays to Charm', submitted to HEP97-383, CLEO-CONF 97-27.

[10] M.S. Alam et al. (CLEO Collaboration), Phys. Rev. Lett. 74 (1995) 2885.

[11] W. Adam et al. (DELPHI Collaboration), Z. Phys. C72 (1996) 207.

[12] O. Adriani et al. (L3 Collaboration), Phys. Lett. B317 (1993) 637.

Measurements of the B_d^0 Oscillation Frequency with the L3 Experiment

Bruna Bertucci (Bruna.Bertucci@cern.ch)

I.N.F.N. and University of Perugia, Italy

Abstract. The time-dependent B_d^0-$\bar{B}_d^0$mixing has been studied using about two million hadronic Z decays registered by L3 in 1994 and 1995. The measured mass difference Δm_d between the B_d^0mass eigenstates is:

$$\Delta m_d = 0.445 \pm 0.032 \pm 0.027 \text{ ps}^{-1}$$

1 Introduction

In the Standard Model, oscillations between particle-antiparticle states of the neutral B mesons are due to second order weak interactions through box diagrams. The probability density for a pure B_d^0 state to decay as a $\bar{B}_d^0$ as a function of time depends on the lifetime of the b meson and on the mass difference between the mass eigenstates. A measurement of the oscillation frequency thus gives a direct measurement of the mass difference Δm_d. This measurement requires the identification of the initial and final flavour of the B^o (particle-antiparticle) as well as the reconstruction of its decay time. The selected sample in which these informations are available is analysed using an unbinned maximum likelihood method to extract the Δm_d value. In the next section the methods to tag the B hadron states will be presented as well as the proper time reconstruction technique. Finally the results for the single analysis and the combination of the two will be given. Details on the analysis can be found in [1, 2].

2 Event selection

Hadronic events are selected in the L3 detector according to the cuts described in [1] and divided in two hemispheres by the plane orthogonal to the thrust axis. Events containing b hadrons are then selected by requiring at least one high energy lepton, as in [1]. The sign of the lepton charge tags the state of the parent B hadron: when both B hadrons decay semileptonically like-sign leptons are a signature for mixing, and a likelihood is assigned to each dilepton event according to the probability density to find it at the measured decay time. If only one B decays semileptonically in the *same side* hemisphere, its state at the production time is determined using the jet charges in both hemispheres. The jet charge is the weighted sum of the track charges in

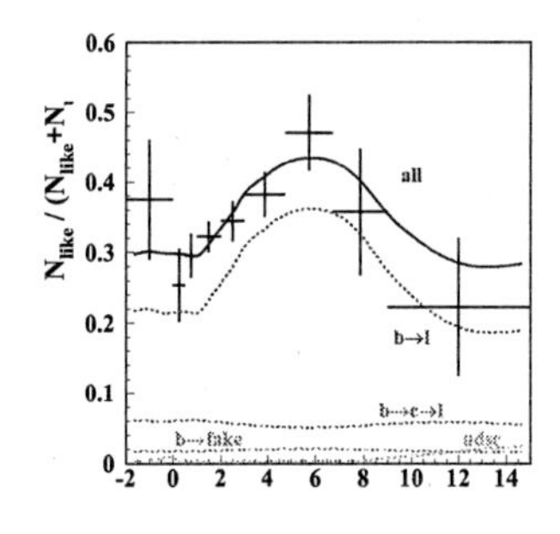

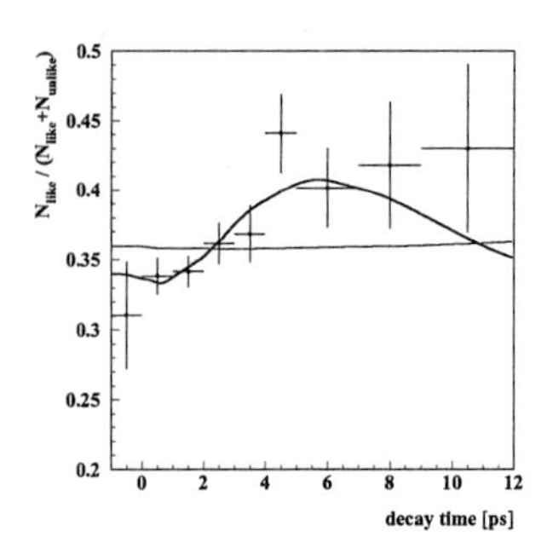

a given hemisphere, the weight for each track being proportional to the *k-th* power function of the longitudinal component of the tracks's momentum with respect the thrust axis. The power k is decided in Monte Carlo studies by maximizing the probability of correct reconstruction of the jet charge. In the same way the optimum variable to reconstruct the primordial quark flavour is found to be a linear combination of the jet charges in the hemispheres with different k factors. The chosen interval of this variable corresponds to an overall reliability of 71% in the flavour tag with an efficiency of 73%.
The proper time is related to the decay lenght in the laboratory system by $t = lm_B/p_B$ where m_B and p_B are the mass and the momentum respectively of the b-hadron.m_B is taken to be 5.3 GeV. The decay lenght is the distance between primary and secondary vertex reconstructed in the plane orthogonal to the beam axis, and converted to a 3D distance using the polar angle of the jet associated to the lepton. p_B is assumed a constant fraction of the beam energy, a value of 0.85 for this fraction optimizes the overall proper time resolution, better than 1 ps in 90% of the measurements.

3 Results

The unbinned likelihood fit performed on 1490 dilepton events gives as result:

$$\Delta m_d = 0.458 \pm 0.046 \pm 0.032 \text{ ps}^{-1}$$

From the 8707 events of the jet charge + lepton analysis we get:

$$\Delta m_d = 0.437 \pm 0.043 \pm 0.044 \text{ ps}^{-1}$$

In the figures the like-sign fraction of events as a function of the decay time is shown, for the dilepton analysis (left) and the jet-charge + lepton. The fit result is superimposed. The combined result is:

$$\Delta m_d = 0.445 \pm 0.032 \pm 0.027 \text{ ps}^{-1}$$

References

[1] L3 Collaboration Phys. Lett. B383 (1996), 487-498
[2] Contributed paper to this conference n.491

B_d^0 and B_s^0 Oscillation with DELPHI

Ivano Lippi (lippi@pd.infn.it) for the DELPHI collaboration

I.N.F.N. Sezione di Padova, Italy

Abstract. The $\mathrm{B_d^0}$ oscillation frequency Δm_d and the lower bound of the $\mathrm{B_s^0}$ oscillation frequency Δm_s are measured by DELPHI.

Introduction At LEP the back-to-back production of $\mathrm{b\bar{b}}$ quark pairs allows to study the $\mathrm{B^0}$ oscillation, tagging the produced and decaying b flavours in the opposite hemispheres. The large boost of the b-quarks and the high precision vertex devices allow to analyse the time evolution of the $\mathrm{B^0}$ oscillation and hence to calculate its frequency Δm, evaluating the B meson proper time $t_\mathrm{B} = m_\mathrm{B}(d_\mathrm{B}/p_\mathrm{B})$, where d_B is the B decay length and p_B its momentum.

The b-quark content of the sample is enhanced selecting the events with a high total and transverse momentum lepton in one hemisphere. The tracks associated to the D meson from the semi-leptonic B decay chain are identified through an efficient (94%) original inclusive vertex finding algorithm developed in Delphi. d_B and p_B are calculated from the $\mathrm{D} - \ell$ system.

The maximum likelihood method is used to fit the proper time of all the sample components and to extract the oscillation signal.

Δm_d analysis Four channels are used in Delphi to measure Δm_d. The signal significance is dominated by the tagging purity factors.

$[\,Q_{hem} - \ell\,]$ ($\Delta m_\mathrm{d} = 0.493 \pm 0.042 \pm 0.027\,\mathrm{ps}^{-1}$) : The lepton charge is used to tag the decaying b quark, and the jet charge Q_{hem} in the opposite hemisphere to tag the produced b quark. The produced tagging purity is fitted together with Δm_d ($\Pi_p = 67\%$). The statistics is $\sim$ 60000 events.

$[\,\ell - \ell\,]$ ($\Delta m_\mathrm{d} = 0.480 \pm 0.040 \pm 0.051\,\mathrm{ps}^{-1}$) : A higher sample ($\sim$ 94%) and tagging ($\Pi_p > 80\%$) purities are obtained requiring a high momentum lepton in both hemispheres. The charge correlation of the two leptons is used. Even if $\sim$ 4800 events are selected, a similar significance is obtained.

$[Q_{hem} - (\pi^* - \ell)]$ ($\Delta m_\mathrm{d} = 0.499 \pm 0.053 \pm 0.015\mathrm{ps}^{-1}$) : From the $Q_{hem} - \ell$ sample a sub-sample highly enriched in $\mathrm{B_d^0}$ content is extracted through an inclusive selection of $\mathrm{D^*}$ using the pion produced in $\mathrm{D^*} \to \mathrm{D^0}\pi$. The signal contains a large $\mathrm{B_d^0}$ fraction ($> 80\%$). The $\pi^* - \ell$ charge correlation allows to carefully study the background and to obtain a high decay tagging purity ($> 90\%$). This channel provides the worldwide lowest systematic error.

$[\,Q_{hem} - \mathrm{D^*}\,]$ ($\Delta m_\mathrm{d} = 0.523 \pm 0.072 \pm 0.043\,\mathrm{ps}^{-1}$) : Instead of the lepton a fully reconstructed $\mathrm{D^*}$ is used to select the b-quark events. The decay chain $\mathrm{D^*} \to \mathrm{D^0}\pi$ with $\mathrm{D^0} \to \pi, \pi\pi^0, \pi\pi\pi$ is used, relying on the small $\mathrm{D^*} - \mathrm{D^0}$ mass difference. This channel suffers from a low sample purity ($\sim$ 20%).

The Delphi average is $\Delta m_\mathrm{d} = 0.497 \pm 0.026(stat.) \pm 0.023(syst.)\,\mathrm{ps}^{-1}$. The correlations between the samples are taken into account.

Δm_s analysis The B_s^0 oscillation amplitude dumping effect due to the proper time resolution, enhanced by the larger Δm_s value, and the lower statistics are such that only a lower bound of Δm_s can be given at LEP.

Δm_s is calculated using the so called amplitude method : for a fixed Δm_s the oscillation amplitude $\mathcal{A}$, given by $(1 \pm \mathcal{A} \cdot cos(\Delta m_s \cdot t))$, is fitted from the likelihood, resulting in a value $\mathcal{A} \pm \sigma_{\mathcal{A}}$. If $\mathcal{A} = 1$ is excluded at 95% confidence level, i.e. $\mathcal{A} + 1.65\sigma_{\mathcal{A}} < 1$, the corresponding Δm_s is also excluded.

For the inclusive channels $Q_{hem} - \ell$ and $\ell - \ell$, already described for Δm_d, the same sample and the same likelihood are used, with the Δm_d and tagging purity values fixed to the ones found in the Δm_d analysis.

The exclusive channel $\overline{B_s^0} \rightarrow D_s^+ \ell^- \overline{\nu} X$, where a D_s is fully reconstructed in the lepton hemisphere, provides a better performance because of an improved time resolution and a better the sample purity ($\sim$ 90%). A larger statistics, but a lower sample purity ($\sim$ 40%), are obtained from the channels $D_s^+ h^- X$, where, instead of a lepton a large impact, large momentum hadron with the proper charge is required to accompany the D_s, and $\phi \ell^- \overline{\nu} X$, where instead of a D_s, only a ϕ is fully reconstructed. Moreover, they suffers respectively from the ambiguity in the hadron selection and from a worse time resolution.

A further improvement is obtained using a new method to tag the produced b-quark, based on 9 discriminating variables defined using the charge structure of both hemispheres. The $b/\overline{b}$ tagging probability, extracted from the simulation, is applied on an event by event basis in the likelihood. This new method, respect to the Q_{hem} cut, provides a higher efficiency (100% w.r.t. 70%) and a higher tagging purity (78% w.r.t. 69%).

The upper picture on the side shows the Δm_s values excluded at 95% confidence level for the studied channels. The lower plot shows the combined result. The data points are the oscillation amplitudes $\mathcal{A}$ fitted at different fixed Δm_s values, the curve giving the amplitude values excluded at 95% confidence level. The dashed curve is the 1.65 $\sigma_{\mathcal{A}}$ error band with $\mathcal{A} = 0$. It describes the analysis sensitivity, i.e. the values of the amplitude that the error on its measurement would allow to exclude at 95% confidence level.

The Delphi limit $\Delta m_s > 8.5 \text{ ps}^{-1}$ is the Δm_s value for which $\mathcal{A} = 1$ is excluded at 95% confidence level.

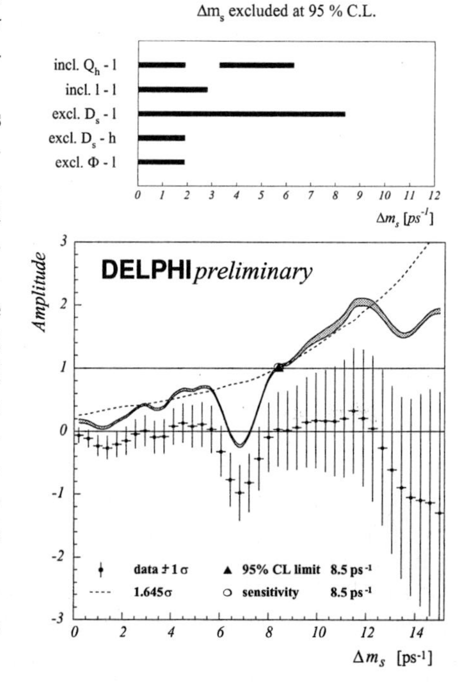

B^0_{ds} Mixing Results from ALEPH

Paschal Coyle[1] (COYLE@CPPM.IN2P3.FR)

Centre de Physique des Particules de Marseille (CPPM)

Abstract. A new Δm_d analysis using jet charge in both hemispheres is reported. Combining with previous ALEPH analyses yields $\Delta m_d = 0.446 \pm 0.027\ \mathrm{ps}^{-1}$. A new Δm_s limit from an improved inclusive lepton analysis is also presented. Combining with previous ALEPH analyses gives $\Delta m_s > 10.4\ \mathrm{ps}^{-1}$ at 95 % CL. All results are preliminary.

1 Introduction

Within the standard model $B^0_{d,s}$ mixing is controlled by box diagrams involving the top quark. Theoretical uncertainties in the extraction of CKM matrix elements from an oscillation frequency measurement, are significantly reduced if the ratio $\Delta m_s/\Delta m_d$ is considered. Two new ALEPH analyses optimised for measurement of Δm_d and Δm_s are now discussed.

2 B^0_d Mixing

The new Δm_d measurement [1] uses a 3D grid search method to assign tracks to the primary vertex or a secondary vertex in each thrust hemisphere. The B^0_d fraction in the sample is enhanced by selecting events containing well displaced vertices. The secondary vertex fit having the best χ^2/dof is choosen as the B^0_d candidate, the other is not used. The proper time of the B^0_d candidate is estimated from the decay length and momentum of both the charged and neutral tracks assigned to its vertex. The charge of the b quark at production (decay) is estimated from the momentum weighted hemisphere charge in the opposite (same) hemisphere to the B^0_d candidate. About 420K events are selected from the total LEP I sample. The mistag is $\approx 44\%$ with an average proper time resolution of ≈ 0.9 ps.

A binned likelihood fit to the time dependence of the product of the hemisphere charges in both hemispheres yields $\Delta m_d = 0.441 \pm 0.026 \pm 0.029\ \mathrm{ps}^{-1}$. Combining this result with other previously published ALEPH Δm_d measurements [3] yields $\Delta m_d = 0.446 \pm 0.027\ \mathrm{ps}^{-1}$.

3 B^0_s Mixing

For the new Δm_s analysis [5], events containing high P_t leptons are selected and vertexed with the other charged tracks in the lepton hemisphere assigned to the charm candidate. Compared to the previously published ALEPH inclusive lepton analysis [2] the improvements are (a) the "optimal tagging" procedure developed for the D_s based analyses [4] is applied to obtain a better mistag rate, (b) subsamples of the events are assigned an enhanced or reduced B_s^0 fraction based on properties of the event such as, vertex charge, presence of kaons etc. and finally (c) stringent selection requirements reduce the fraction of events for which the event-by-event proper time resolution is poorly estimated.

A total of 33K events are selected from the LEP I data. The average effective mistag is $\approx 29\%$ and the average proper time resolution is ≈ 0.3 ps. An amplitude fit to the data yields a limit (sensitivity) for this analysis of 10.2(10.6) ps^{-1} at 95 % CL. The dominant systematic uncertainty is from f_s and is small. This is the single most sensitive Δm_s analysis to date. Fig. 3 shows the amplitude spectrum resulting from a combination of this analysis with the ALEPH D_s^- based analyses [4]. The combined ALEPH limit (sensitivity) is 10.4 (11.7) ps^{-1}at 95 % CL.

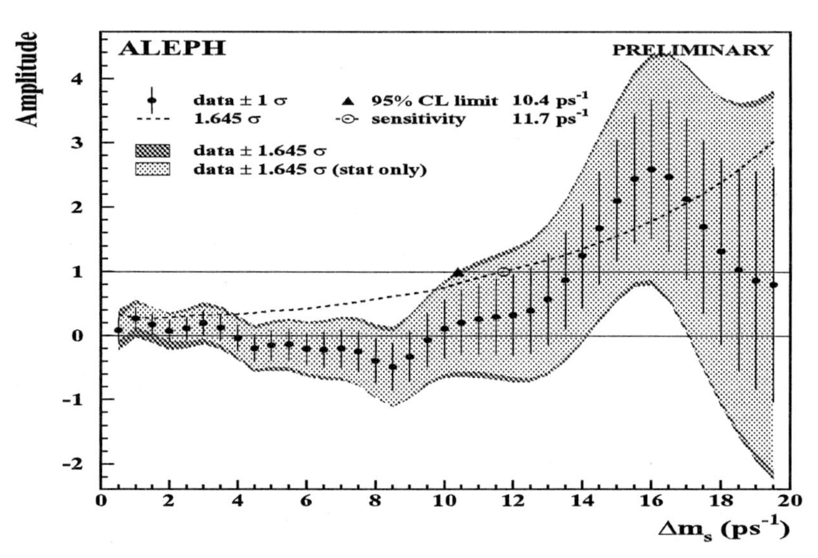

Fig. 1. Amplitude spectrum for the combination of the ALEPH Δm_s analyses.

References

[1] ALEPH Collaboration, "Inclusive Lifetime and Mixing Using Topological Vertexing", EPS-596, contribution to the EPS-HEP Jerusalem, 1997.
[2] ALEPH Collaboration, Physical Review Letters B356 (1995), 409.
[3] ALEPH Collaboration, Zietschrift fur Physik C75 (1997), 397.
[4] ALEPH Collaboration, "Search for B_s^0 oscillations using inclusive D_s^-events", EPS-611, contribution to the EPS-HEP Jerusalem, 1997.
[5] ALEPH Collaboration, "Search for B_s^0 oscillations using inclusive lepton events", EPS-612, contribution to the EPS-HEP Jerusalem, 1997.

OPAL Measurements of Δ_{m_d} and Search for CP and CPT Violation in the B_d System

Martin Jimack (martin.jimack@cern.ch)

Birmingham University, Edgbaston, Birmingham B15 2TT, UK and CERN

Abstract. Four measurements of $\Delta m_{\rm d}$ from OPAL are presented, giving

$$\Delta m_{\rm d} = 0.467 \pm 0.022^{+0.017}_{-0.015}\ {\rm ps}^{-1}$$

Using a time-dependent analysis, results on the CP and CPT violating parameters ${\rm Re}\,\epsilon_{\rm B}$ and ${\rm Im}\,\delta_{\rm B}$ were also obtained, and the parameters were found to be consistent with zero.

1 Introduction

Second-order weak interactions, involving virtual top quarks, allow $\rm B^0 \leftrightarrow \bar{B}^0$ transitions. These weak eigenstates are then not equal to the mass eigenstates, $\rm B_1$ and $\rm B_2$:

$$|{\rm B_1}\rangle = \frac{(1+\epsilon+\delta)|{\rm B^0}\rangle + (1-\epsilon-\delta)|{\rm \bar{B}^0}\rangle}{\sqrt{2(1+|\epsilon+\delta|^2)}}$$
$$|{\rm B_2}\rangle = \frac{(1+\epsilon-\delta)|{\rm B^0}\rangle - (1-\epsilon+\delta)|{\rm \bar{B}^0}\rangle}{\sqrt{2(1+|\epsilon-\delta|^2)}}, \quad (1)$$

where ϵ and δ are complex, and parametrise indirect CP and CPT violation [1]. Note that ϵ allows for CP and T violation while respecting CPT symmetry, and δ allows for CP and CPT violation, but respects T symmetry. The mass difference, $\Delta m_{\rm d}$, between the $\rm B_1$ and $\rm B_2$ depends on the top mass and the CKM element $V_{\rm td}$ as well as decay and bag constants, obtained from lattice calculations. The uncertainty on these two factors currently dominates the uncertainty on the derivation of $|V_{\rm td}|$ from $\Delta m_{\rm d}$.

The time evolution of a $\rm B^0$ state depends on both the initial and final b-flavours ($\rm B^0$ or $\rm \bar{B}^0$) [2, 3], which need to be tagged experimentally. Here, I report on four studies of $\rm B^0\bar{B}^0$ oscillations by OPAL, using $\rm D^{*\pm}\ell$ combinations together with jet charge, $\rm D^{*\pm}$ opposite leptons, dilepton events, and inclusive lepton events with jet charge.

2 Results

A total of 1200 $\rm D^{*\pm}\ell$ candidates were selected with a good jet charge production flavour tagging [4]. The $\rm D^0$ from the $\rm D^{*\pm}$ decay was selected in the

$K^-\pi^+$ and $K^-\pi^+\pi^0$ (satellite) modes, but, in the latter case, the π^0 was not reconstructed. By fitting the fraction of 'mixed' events (where the lepton and jet charges disagree) as a function of time, the oscillation frequency was measured: $\Delta m_d = 0.539 \pm 0.060 \pm 0.024\,\mathrm{ps}^{-1}$. A total of 348 $D^{*\pm}$ candidates opposite leptons [4] were selected with $D^0 \to K^-\pi^+$. The fitted result was $\Delta m_d = 0.567 \pm 0.089^{+0.029}_{-0.023}\,\mathrm{ps}^{-1}$.

For the dilepton analysis, a total of 5357 events were selected with leptons in different jets and at least one proper time reconstructed [5]. The fitted result was $\Delta m_d = 0.430 \pm 0.043^{+0.028}_{-0.030}\,\mathrm{ps}^{-1}$.

A total of 94 843 inclusive lepton events [3] were selected with a proper decay time reconstructed. The production flavour was determined using jet charge calculated from both the opposite jet and the lepton jet. The fitted result was $\Delta m_d = 0.444 \pm 0.029^{+0.020}_{-0.017}\,\mathrm{ps}^{-1}$, the single most precise measurement of this quantity. The systematic errors are controlled by allowing key parameters to vary in the fit.

Combining all four Δm_d results [5] gives $\Delta m_d = 0.467 \pm 0.022^{+0.017}_{-0.015}\,\mathrm{ps}^{-1}$, very competitive with other measurements.

CP violation would cause the rate of positive leptons to differ from the that of negative leptons in a time dependent way. Therefore, in the inclusive lepton analysis, the fraction of leptons with negative charge was studied as a function of time. Fitting $\mathrm{Re}\,\epsilon_B$ and $\mathrm{Im}\,\delta_B$ simultaneously gave the result: $\mathrm{Re}\,\epsilon_B = -0.006 \pm 0.010 \pm 0.006$ and $\mathrm{Im}\,\delta_B = -0.020 \pm 0.016 \pm 0.006$. If CPT is assumed to be a good symmetry ($\mathrm{Im}\,\delta_B = 0$), then a somewhat more precise result is obtained for $\mathrm{Re}\,\epsilon_B$: $\mathrm{Re}\,\epsilon_B = 0.002 \pm 0.007 \pm 0.003$.

The result for $\mathrm{Re}\,\epsilon_B$ is the first time-dependent CP-violation test in the B system, and is considerably more precise than previous measurements of this quantity [6]. The result is consistent with the Standard Model expectation ($\leq 10^{-3}$) and with SuperWeak models ($\leq 10^{-2}$), but is starting to restrict the range for the latter. The $\mathrm{Im}\,\delta_B$ result is the first test of CPT symmetry in the B system. The level of this test, $\mathrm{Im}\,\delta_B \cdot \Delta m_d / M_{B^0}$, indicates that this is the most precise test of CPT violation outside the K system.

References

[1] V.A. Kostelecky and R. Potting, Phys. Rev. **D 51** (1995) 3923.
[2] V.A. Kostelecky and R. Van Kooten, Phys. Rev. **D 54** (1996) 5585.
[3] OPAL Collaboration, K. Ackerstaff *et al.*, CERN-PPE/97-036, Accepted by Z. Phys. C.
[4] OPAL Collaboration, G. Alexander *et al.*, Z. Phys. **C 72** (1996) 377.
[5] OPAL Collaboration, K. Ackerstaff *et al.*, CERN-PPE/97-064, Accepted by Z. Phys. C.
[6] CLEO Collaboration, J. Bartelt *et al.*, Phys. Rev. Lett. **71** (1993) 1680; CDF Collaboration, F. Abe *et al.*, Phys. Rev. **D 55** (1997) 2546.

Measurement of Time-Dependent $\overline{B^0} - \overline{B^{10}}$ Flavor Oscillation at CDF

Alberto Ribon (ribon@fnal.gov)

Fermilab (USA), and Padova University (Italy)

Abstract. The time dependence of $B^0 - \overline{B}^0$ oscillations has been studied using several techniques by the CDF experiment at Fermilab. The data comprise $110\,pb^{-1}$ of $p\bar{p}$ collisions at $\sqrt{s} = 1.8\,TeV$. Preliminary measurements of the $B^0 - \overline{B}^0$ oscillation frequency (Δm_d) will be presented.

Large samples of B hadron decays are selected using the following triggers:

1. inclusive single lepton (e, μ). The transverse momentum threshold is about $8\,GeV/c$.
2. dilepton ($e\,\mu$; $\mu\,\mu$). The transverse momentum threshold is $2\,GeV/c$ for the dimuon trigger; for the other trigger it is $5\,GeV/c$ for the electron and $2.5\,GeV/c$ for the muon.

Four analyses are presented here for the $B^0 - \overline{B^0}$ mixing: two of them use the single lepton triggers, and the other two use dilepton triggers.

Jet Charge inclusive lepton analysis : the sign of the trigger lepton tags the flavor of the B at decay time; the flavor at production is obtained either from the sign of the other lepton, if another lepton is found, or from the the sign of jet charge:

$$Q_{jet} = \frac{\sum_i q_i(\mathbf{p}_i \cdot \mathbf{a})}{\sum_i(\mathbf{p}_i \cdot \mathbf{a})}$$

where **a** is the jet axis, and the sum is over the tracks in the jet. The result obtained is: $\mathbf{\Delta m = 0.47 \pm 0.06\,(stat) \pm 0.04\,(syst)\ ps^{-1}}$.

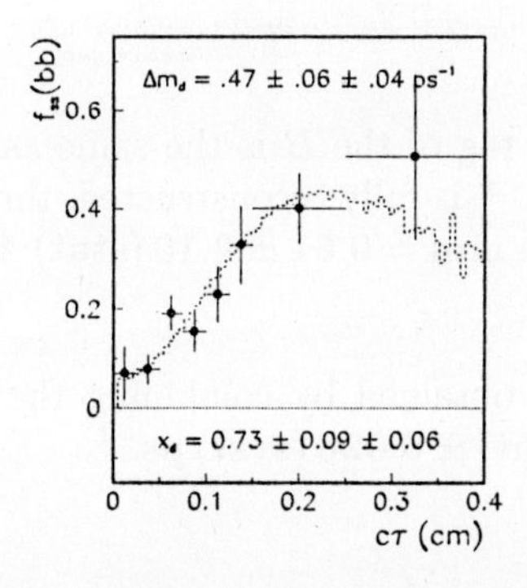

l + D$^{(*)}$ with Same Side Tagging : the lepton trigger tags the flavor of the B at decay time; the flavor at production is obtained from the sign of the track, selected on the same side, which has the minimum transverse momentum with respect to the $l^- D^*$ system, and is within a 0.7 cone in $\eta - \phi$. In fact, $\overline{B}^0_d$ (B^0_d) mesons are expected to be produced in association with π^- (π^+), as also would happen if they were produced from B^{**} resonances. Several channels are used to reconstruct D^*'s.
The result obtained is: $\boldsymbol{\Delta m = 0.45 \pm 0.06\,(stat) \pm 0.03\,(syst)\,ps^{-1}}$.

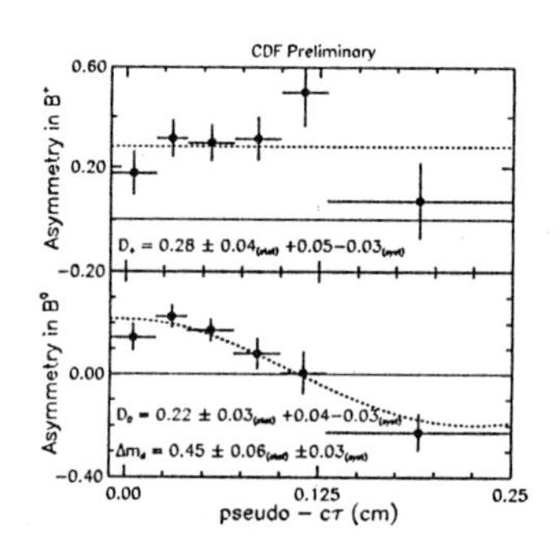

e − μ analysis : the sign of the lepton on the side where the secondary vertex is found tags the flavor of the B at decay. The sign of the other lepton is used to tag the flavor at production.
The result obtained is: $\boldsymbol{\Delta m = 0.45 \pm 0.05\,(stat) \pm 0.05\,(syst)\,ps^{-1}}$.

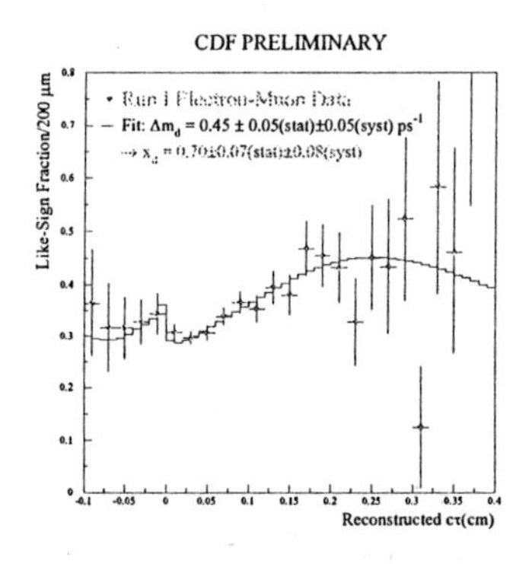

l vs l + D$^{(*)}$: the flavor tag of the B is the same as in the $e - \mu$ analysis, but in this case the $D^{*\pm}$ is fully reconstructed, through several channels. The result obtained is $\boldsymbol{\Delta m = 0.51 \pm 0.10\,(stat) \pm 0.04\,(syst)\,ps^{-1}}$.

The CDF average value, obtained by combining the above four results, is $\boldsymbol{\Delta m = 0.464 \pm 0.030\,(stat) \pm 0.026\,(syst)\,ps^{-1}}$.

Time-Dependent B Mixing at SLD

David Jackson (djackson@slac.stanford.edu)

Rutherford Appleton Laboratory, Chilton, Didcot, Oxon, OX11 0QX, England

Abstract. Measurements of B_d mixing using 150,000 Z^0 decays collected during 1993-5 are reviewed. Preliminary studies for B_s mixing are also discussed.

1 B_d mixing measurements

In order to measure time dependent B^0 mixing the analysis must determine, for each decay, the B decay length (or proper time t) and the particle-antiparticle nature of the B at production (the initial state at $t = 0$) and at decay (final state). The initial state tag combines the polarized forward-backward asymmetry and opposite hemisphere jet charge to yield an overall average correct tag probability of $\sim 80\%$ with 100% efficiency.

Four final state tags have been employed. Two of the analyses are based on the inclusive vertex reconstruction described in Ref. [1] to tag the final state. The first uses the charge sum of kaons identified by the CRID and attached to the secondary vertex. The tag exploits the fact that most B decays occur via the dominant quark transition $b \to c \to s$. The simulation shows that the probability for correctly tagging the final state is 77% for B_d decays [2]. The second inclusive analysis relies on a novel method which exploits the $B \to D$ cascade charge structure. The charge dipole δq of the reconstructed vertex is defined as the relative displacement between the weighted mean location of the positive tracks and of the negative tracks along the vertex axis [2]. The correct final state tag probability is a function of $|\delta q|$.

The other B_d mixing analyses are based on semileptonic reconstruction of the B decay. In the first method the intersection of the lepton trajectory and a topologically reconstructed D meson identifies the B decay vertex [3]. In the second semileptonic analysis the B decay location is estimated as the weighted average position along the lepton path of the points of closest approach of all tracks in the jet containing the high-p_t lepton. This analysis also estimates the relativistic boost of the B hadron by assuming energy and momentum conservation to determine the total energy of the jet containing the lepton [4]. In both semileptonic analyses, the charge of the lepton tags the final state with a correct sign probability of $\sim 85\%$.

Upon averaging the results of the four analyses, taking into account correlated statistical and systematic errors, we find the following SLD average of $\Delta m_d = 0.525 \pm 0.043(stat) \pm 0.037(syst)$ ps^{-1}.

2 B_s mixing studies

The dipole and lepton final state $b/\overline{b}$ tags have been studied in the context of B_s mixing. Due to the spectator s quark, positively and negatively charged kaons are nearly equally produced in B_s decays and cannot be used as a final state tag. Initial studies indicate that the excellent decay length resolution achieved with the upgrade vertex detector, VXD3, allows high oscillation frequencies to be probed with a sensitivity around three times greater than the old vertex detector. During the 1996 run 50,000 Z^0 events were collected with VXD3. Fig 1 shows the charge dipole δq for 1996 data and Monte Carlo. The MC B_s and $\overline{B}_s$ are shown amplified by a factor of five to allow the separation of the final states to be seen.

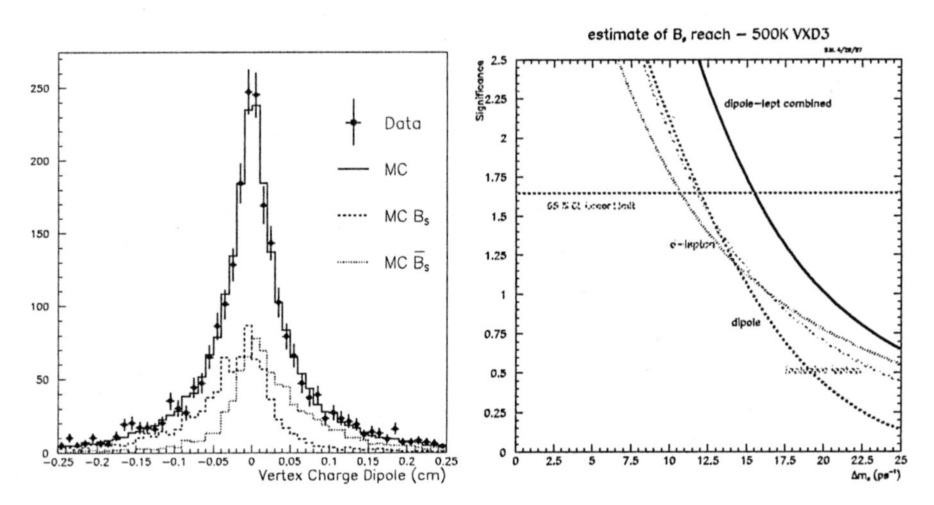

Fig. 1. The charge dipole distribution for 1996 data and Monte Carlo is shown on the left. The plot on the right shows the sensitivity to Δm_s with future data.

The second plot in fig 1 shows the sensitivity to B_s oscillations as a function of Δm_s for the various final state tags individually and combined. This plot indicates that with 500,000 Z^0 decays with VXD3 the region $\Delta m_s < 16.0$ ps^{-1} could be excluded at the 95% confidence level.

References

[1] D. J. Jackson, Nucl. Inst. and Meth. **A388** (1997) 247.
[2] K. Abe *et al.*, SLAC-PUB-7230, July 1996.
[3] K. Abe *et al.*, SLAC-PUB-7229, July 1996.
[4] K. Abe *et al.*, SLAC-PUB-7228, July 1996.

CP Tests and the Weak Dipole Moment in τ-Pair Production at LEP

Markus Schumacher, Physikalisches Institut der Universität Bonn, Nussallee 12, 53115 Bonn, Germany

Tests of $\mathcal{CP}$ invariance in the reaction $e^+e^- \rightarrow \tau^+\tau^-$ on the Z^0 peak are reviewed. Averaging the present LEP measurements, upper limits on the real and imaginary parts of the weak dipole moment of the τ lepton of $|Re(d_\tau^w(M_Z^2))| < 3.3 \cdot 10^{-18}$ ecm and $|Im(d_\tau^w(M_Z^2))| < 9.8 \cdot 10^{-18}$ ecm with 95% c.l. are obtained.

1. Introduction

The violation of the $\mathcal{CP}$ symmetry is one of the necessary ingredients to explain the apparent baryon asymmetry in the universe [1]. The magnitude however may not be sufficiently explained by the strength of $\mathcal{CP}$ violation offered by the Standard Model (SM). This motivates the search for new $\mathcal{CP}$ phenomena studying reactions where the SM does not predict any measurable effect. An observation of $\mathcal{CP}$ violation in τ pair production at the Z^0 peak would indicate physics beyond the Standard Model.

New $\mathcal{CP}$ violating physics can be described by extending the SM langrangian by an effective $\mathcal{CP}$ violating lagrangian in a model independent way [2]. The strength of the new interaction is parametrized by a complex form factor $d_\tau^w(M_Z^2)$, termed the weak dipole moment of the τ lepton. The total cross section for τ pair production is then given by: $\sigma_{tot} \propto M_{SM} + |d_\tau^w|^2 \cdot M_{\mathcal{CP}} + Re(d_\tau^w) \cdot M_{\mathcal{CP}}^{Re} + Im(d_\tau^w) \cdot M_{\mathcal{CP}}^{IM}$. Only the inteference term which is linear in d_τ^w is $\mathcal{CP}$ odd and gives rise to $\mathcal{CP}$ violation. The $\mathcal{CP}$-even squared amplitude $M_{\mathcal{CP}}$ contributes to the partial width $Z \rightarrow \tau^+\tau^-$ [2].

Many $\mathcal{CP}$ violating extensions of the SM predict mass dependences of fermion dipole moments which favour the search in the τ sector. In models with extended Higgs sector [3] one finds $d_f \propto m_f^3$, while leptoquark models [4] predict the ratio for lepton dipole moments to be $d_\tau : d_\mu : d_e = m_\tau m_t^2 : m_\mu m_c^2 : m_e m_u^2$.

2. Method

Observables $\mathcal{O}$ which change sign under $\mathcal{CP}$ transformation are used to test $\mathcal{CP}$ invariance. In the most recent analyses [5–7] optimal observables [8], have then been employed, which are defined as

$$\mathcal{O}^{Re} = \frac{M_{\mathcal{CP}}^{Re}}{M_{SM}} \quad ; \quad \mathcal{O}^{Im} = \frac{M_{\mathcal{CP}}^{Im}}{M_{SM}} \quad . \qquad (1)$$

They allow to measure real and imaginary parts of $d_\tau^w(M_Z^2)$ separately. The calculation of the observables requires the reconstruction of the τ flight and spin directions [9], [5]. L3 used a different (asymmetry) method [10].

The experimental signature of $\mathcal{CP}$ violation are non vanishing expectation values for the $\mathcal{CP}$ odd observables, which are proportional to d_τ^w as

$$\langle \mathcal{O}^{Re} \rangle \propto \mathrm{Re}(d_\tau^w) \quad ; \quad \langle \mathcal{O}^{Im} \rangle \propto \mathrm{Im}(d_\tau^w) \quad . \qquad (2)$$

The proportionality constants are called sensitivities. Their magnitude depends on the spin analyzing power and the quality of the reconstruction of the τ flight direction for the considered decay modes. The ideal sensitivity are reduced by acceptance cuts, the detector resolution and background contamination in the different selected event classes.

The $\mathcal{CP}$ symmetry of the detectors is vital for this measurement. Detector asymmetries e.g. arising from misalignment of detector components can fake $\mathcal{CP}$ violation or hide a true $\mathcal{CP}$ effect. In order to assess the level of $\mathcal{CP}$ symmetry of the LEP detectors, data events are studied for which non-vanishing expectation values of $\mathcal{O}$ arising from true $\mathcal{CP}$ violation can be excluded: e.g. $Z \rightarrow \mu\mu$ events [6] and mixed τ events for which the spin correlations are purposely destroyed [5, 7]. The LEP experiments have verified that their detectors are $\mathcal{CP}$ symmetric to a level which is one order of magnitude smaller that the statistical precision of the $\mathcal{CP}$ test.

3. Results and Conclusion

Figure 1 summarizes the measurements obtained by the LEP experiments on the weak dipole moments. No evidence for $\mathcal{CP}$ violation was observed. Combining the results on d_τ^w the 95% c.l. limits are

$$\begin{aligned} |Re(d_\tau^w)| &< 3.3 \cdot 10^{-18}\,\text{ecm} \\ |Im(d_\tau^w)| &< 9.8 \cdot 10^{-18}\,\text{ecm} \\ |d_\tau^w| &< 10.3 \cdot 10^{-18}\,\text{ecm} \end{aligned}$$

In order to assess the relative importance of the results one may consider models with a $\mathcal{CP}$ violating Higgs sector [3] or with lepto-quarks [4]. Assuming that weak and electric dipole moments can be directly compared and are expected in these models to have the same order of magnitude, the above result on $|d_\tau^w|$ has been scaled in table 3 with the ratio of the fermion masses[1] as described in section 1.

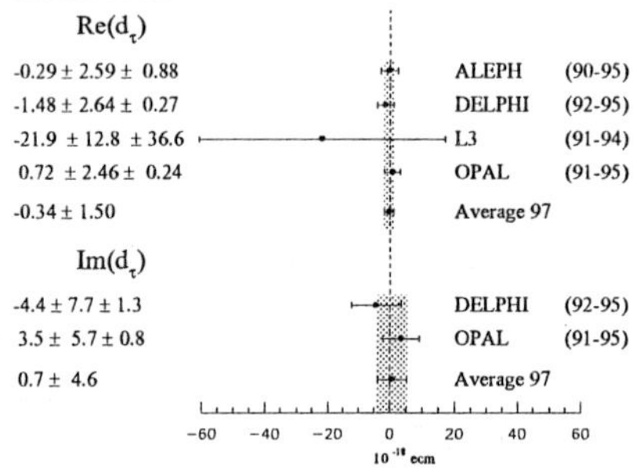

Figure 1. *Summary of measurements of $Re(d_\tau^w)$ and $Im(d_\tau^w)$ obtained by the LEP experiments.*

Other models also exist which predict a weaker dependence on the mass of the fermion. Nevertheless, table 1 indicates that in terms of most model expectations the present limit on d_τ^w is more restrictive than existing measurements on the electric dipole moment of electron and muon.

Another useful comparison is obtained by defining a $\mathcal{CP}$ parameter ϵ_τ as

$$\epsilon_\tau = \frac{\Delta\Gamma_{Z\to\tau\tau}}{\Gamma_{\tau\tau}} = \frac{0.237 \times 10^{31}\,|d_\tau(ecm)|^2}{\Gamma_{\tau\tau}\,(GeV)} \qquad (3)$$

[1] For the quark masses the values $m_u = 5$ MeV, $m_c = 1.3$ GeV, $m_t = 180$ GeV have been used.

where $\Delta\Gamma_{Z\to\tau\tau}$ is the change in the τ partial width due to the new $\mathcal{CP}$ violating interaction.

Using the limit for $|d_\tau^w|$ and for $|Re(d_\tau^w)|$ (i.e. assuming $Im(d_\tau^w) = 0$), respectively, and $\Gamma_{\tau\tau} = (83.88 \pm 0.39)$ MeV [11] one obtains

$$\epsilon_\tau^{|d_\tau^w|} < 3.0 \cdot 10^{-3} \quad \text{and} \quad \epsilon_\tau^{Re(d_\tau^w)} < 3.0 \cdot 10^{-4}.$$

The limits on ϵ_τ indicate that the precision of test of $\mathcal{CP}$ invariance in $Z \to \tau^+\tau^-$ has reached a level of one in a thousand.

	meas. limit on d_f^{el} [12]	Multi Higgs $d_f \propto m_f^3$	Leptoquark $d_f \propto m_f m_q^2$
e	4×10^{-27}	2×10^{-28}	2×10^{-30}
μ	1.0×10^{-18}	2×10^{-21}	3×10^{-23}

Table 1
Comparison of limits on electric dipole moments from direct measurements with scaled limits from d_τ^w. All units in (ecm).

REFERENCES

1. A. D. Sakharov, *JETP Lett.* **5** (1967) 24.
2. W. Bernreuther, O. Nachtmann, *Phys. Rev. Lett.* **63** (1989) 2787 and references therein.
3. W. Bernreuther, T. Schröder, T.N. Pham, *Phys. Lett.* **B279** (1992) 389.
4. W. Bernreuther, A. Brandenburg and P. Overmann *Phys. Lett.* **B391** (1997) 413.
5. OPAL Collaboration, K. Ackerstaff et al., *Z. Phys.* **C 74** (1997) 403.
6. ALEPH Collaboration, D. Buskulic et al., *ICHEP'96*, Ref. PA08-030 (1996)
7. DELPHI Collaboration, M.C. Chen et al., *HEP'97*, Ref. PA07-321 (1997).
8. D. Atwood, A. Soni, *Phys.Rev.* **D 45** (1992) 2405; P. Overmann, *Univ. of Dortmund preprint DO-TH* **93-24**, (1993); M. Diehl, O. Nachtmann, *Z.Phys.* **C 62** (1994) 397.
9. N. Wermes, *Nuclear Physics B (Proc. Suppl.)* **55C** (1997) 313 and references therein.
10. E. Sanchez Alvaro, *Nuclear Physics B (Proc. Suppl.)* **55C** (1997) 305.
11. Particle Data Group, R. M. Barnett et al., *Phys. Rev.* **D 54** (1996) 1.
12. E.D. Commins et al., *Phys. Rev.* **A 50** (1994) 2960.

Search for CP-Violation and $b \longrightarrow sg$ in Inclusive B Decays

Mourad Daoudi (daoudi@SLAC.Stanford.EDU)

Stanford Linear Accelerator Center, Stanford, CA, USA

Abstract. We present preliminary results on two analyses performed by the SLD Collaboration using inclusive B decays: a search for CP violation and a search for the $b \longrightarrow sg$ transition.

1 Search for CP violation in inclusive B decays

Because they involve large branching fractions and sizable CP asymmetries, (semi-) inclusive B decays have been proposed[1] as a means of searching for CP violation, and extracting measurements of CKM parameters. The totally inclusive asymmetry provides a measurement of the CP observable $a = \mathcal{I}m \frac{\Gamma_{12}}{M_{12}}$. It is the focus of this analysis. Its time dependence is[1]:

$$\mathcal{A}(t) = \frac{\Gamma(B^0(t) \rightarrow all) - \Gamma(\bar{B}^0(t) \rightarrow all)}{\Gamma(B^0(t) \rightarrow all) + \Gamma(\bar{B}^0(t) \rightarrow all)} = a \left(\frac{\Delta m \ \tau_B}{2} \sin \Delta m \ t - \sin^2 \frac{\Delta m \ t}{2} \right).$$

A non-zero value of a implies CP violation. Due to the large value of Δm_s, this analysis is only sensitive to asymmetries in B_d decays for which a_d is expected to be $\approx 10^{-3}$ in the Standard Model.

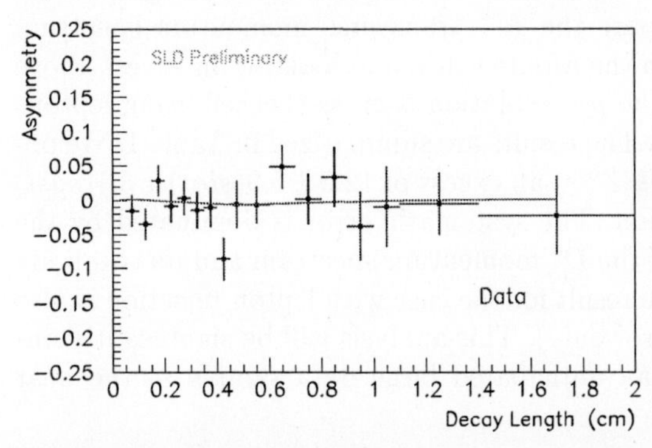

Fig.1. Asymmetry as a function of decay length.

B-decay vertices are reconstructed using a topological technique[2]. From the 1993-96 data sample ($200k$ Z^0's), about $11k$ neutral and $19k$ charged vertices are selected, with a B_d content of 50% and 35%, respectively. The crucial part of the analysis is tagging the flavor of the b quark at production. This is done mainly using the left-right forward-backward asymmetry (given by the electron beam longitudinal polarization and the thrust axis polar angle) and the opposite hemisphere momentum-weighted jet charge. These two tags which have an efficiency of 100%, are complemented by additional information from the opposite hemisphere when it is available (vertex charge, sign of a high-p_T lepton, charge sum for kaons from a B-decay). The overall b-flavor tag purity is estimated to be 84%. The measured asymmetry is shown in Fig. 1 as a function of decay length. A binned χ^2 fit is performed and a value of $a_d = -0.04 \pm 0.12(stat) \pm 0.05(sys)$ is obtained. This gives a 95% C.L. limit of $-0.29 < a_d < 0.22$.

2 Search for enhanced $b \to sg$ in inclusive B decays

It was suggested recently that a branching ratio $\sim$ 10% for the $b \to sg$ transition (0.2% in the Standard Model) could resolve a variety of B decay puzzles (e.g., b semileptonic branching ratio, number of c quarks produced per B decay, etc)[3]. The search strategy consists of looking for an excess in kaon production at high p_T, where the signal-to-background ratio is expected to be of the order of 1:1 (for a 10% branching ratio).

Table 1. Number of kaons with $p_T > 1.8$ GeV/c

	1-Vertex		2-Vertex	
	All	No Lepton	All	No Lepton
Data	35.0	30.0	30.0	23.0
M.C.	22.1	14.1	27.5	20.3
Diff.	12.9 ± 5.9	15.9 ± 5.5	2.5 ± 5.5	2.7 ± 5.5

At SLD, we select B vertices that contain an identified $K^{\pm}$ using the Čerenkov Ring Imaging Detector. We measure the kaon transverse momentum w.r.t. the direction defined by the small SLC interaction point and the well-reconstructed B decay vertex. The signal is enhanced by: *i*) separating the data into a one-vertex sample (signal) and a two-vertex sample (control) according to the probability for all B-decay tracks to originate from a single point, *ii*) rejecting decays that contain an identified lepton. The efficiency for isolating true one-vertex decays (e.g., charmonium) is estimated at 80%, whereas only 45% of standard $b \to c$ transitions satisfy the one-vertex requirement. We compare the $K^{\pm}$ transverse momentum spectrum observed in the data to that in the Monte Carlo and look for an excess above 1.8 GeV/c. Our modeling of the p_T resolution is cross-checked using leptons whose spectrum is well known. The results are summarized in Table 1. We observe in the 1993-95 data (150k Z^0's) an excess of $12.9 \pm 5.9(stat) \pm 3.1(syst)$ decays, without lepton rejection. The systematic error is dominated by the uncertainty in the modeling of the D^0 momentum spectrum and its two-body decay branching fractions. The result for the case with lepton rejection is also given in Table 1 (statistical error only). This analysis will be significantly improved with the addition of an anticipated large data sample in the near future.

References

[1] M. Beneke, G. Buchalla, and I. Dunietz, Phys. Lett. B393 (1997) 132.
[2] D.J. Jackson, Nucl. Inst. and Meth. A388 (1997) 247.
[3] A. Kagan and J. Rathsman, HEP-PH/9701300.

Direct Measurement of $|V_{tb}|$ at CDF

G. F. Tartarelli (tarta@mail.cern.ch), for the CDF Collaboration

CERN, Geneva (Switzerland)

Abstract. We present a first direct measurement of the ratio of branching fractions $R = B(t \to Wb)/B(t \to Wq)$, obtained by the CDF Collaboration. We measure $R = 0.99 \pm 0.29$(stat+syst) in agreement with the Standard Model predictions. We use this result to measure the element $|V_{tb}|$ of the Cabibbo-Kobayashi-Maskawa quark mixing matrix. We obtain $|V_{tb}| = 0.99 \pm 0.15$(stat+syst) which translates in the 90% confidence limit $|V_{tb}| \geq 0.8$.

In the Standard Model, assuming three-generation unitarity, a global fit to the elements of the CKM matrix yields the indirect allowed range: $|V_{tb}| = 0.9989 \div 0.9993$ (at 90% CL). This implies that R, the ratio of branching fractions $B(t \to Wb)/B(t \to Wq)$ (where q is a down-type quark), is close to unity. If we release the three generations hypothesis, $|V_{tb}|$ is essentially unconstrained: $|V_{tb}| = 0. \div 0.9993$ (at 90% CL) [1]. Neither R nor $|V_{tb}|$ have been measured directly up to now.

We present results obtained using $109 \pm 7\ \text{pb}^{-1}$ from $p\bar{p}$ collisions at $\sqrt{s} = 1.8$ TeV recorded at the Collider Detector at Fermilab (CDF) during the 1992-1995 run of the Fermilab Tevatron.

SVX	SLT	W4J	DIL
none	none	126	6
none	one	14	n/a
one	n/a	18	3
two	n/a	5	0
TOTAL:		163	9

Table 1. Number of events in the W4J and DIL samples, according to SVX and SLT tag results (no tag, one tag, two tags, not applied).

At the Tevatron, the top quark has been observed only when produced in pairs. Assuming top decays always involve a real W, $t\bar{t}$ samples are categorized according to the W's decay modes. We use the W4J sample (one W decays to e or μ and the other to two jets) which has a final state characterized by a high P_T lepton, $\not{E}_T$ and four jets, and the DIL sample (both W's decay to e or μ), characterized by a final state with $\not{E}_T$, two high P_T leptons and two jets. By construction, the W4J and DIL samples have no overlap. The presence of the decay $t \to Wb$ can be successively desumed by searching the jets in these events for displaced vertices (SVX tagging) or for low-P_T e and μ from b hadrons decays (SLT tagging) [2, 3].

The quantity R is measured by comparing the observed number of tags in the data (see table 1) with expectations based on acceptances, tagging efficiencies and backgrounds. Acceptances and efficiencies are obtained using a top Montecarlo ($M_{top} = 175\ \text{GeV/c}^2$) followed by a detailed simulation of the detector. The SVX (SLT) algorithm is characterized by an efficiency to tag a b jet in a top event of $30.5 \pm 3.0\%$ ($10.2 \pm 1.0\%$) and by a fake rate at the 0.5% (2%) level. The acceptance for having one (two) b-jets in the events is normalized to the acceptance to have no b-jets (so that trigger and lepton identification efficiencies cancel out in the ratio) and is measured to be 12.4 ± 1.2 (40.5 ± 4.0) for the SVX sample and 14.5 ± 1.4 (58.5 ± 5.8) for the DIL one. Backgrounds are calculated using a combination of data and Montecarlo information [2, 3].

The determination of R is obtained by mean of a likelihood fit. As the W4J and DIL sample are independent, the likelihood is the product of two individual likelihoods of the form: $\mathcal{L} = \prod_i P(N_i; \bar{N}_i) \prod_j G(x_j; \bar{x}_j, \sigma_j)$, where $P(N_i; \bar{N}_i)$ is the Poisson probability of observing N_i events in bin i with expected mean $\bar{N}_i$ (see table 1). The function $G(x_j; \bar{x}_j, \sigma_j)$ is a Gaussian in x_j, with mean $\bar{x}_j$ and width σ_j. Each $\bar{x}_j$ is, in turn, one of the tagging efficiencies, backgrounds and acceptances (with σ_j being its extimated error). The total number of t quarks in the samples is a free fit parameter (so that the result does not depend upon the top production cross section).

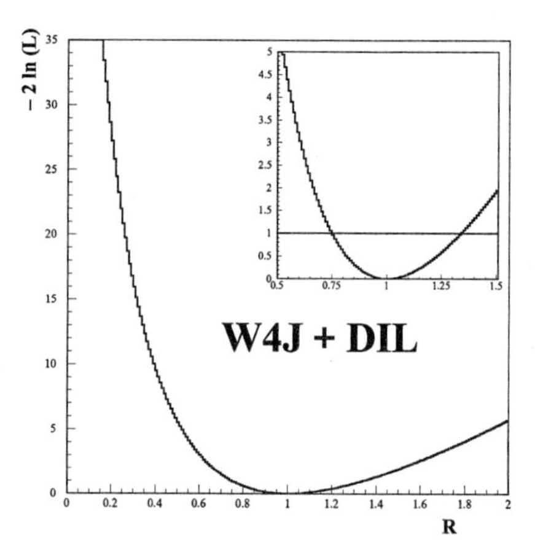

Fig. 1. Variation of $\mathcal{L}$ as a function of R (in two different scales).

The fit yields $R = 0.99 \pm 0.29$, where the error includes both statistical and systematical uncertainties, with the first one being the dominant contribution. The behaviour of $\mathcal{L}$ as a function of R is shown in fig. 1. The 90(95)% CL, obtained by numerical integration, is $R > 0.64(0.58)$.

The CKM element $|V_{tb}|$ is directly related to R (although with the model-dependent assumption that top decays to non-W final states are negligible) by the following relation: $R = |V_{tb}|^2/(|V_{ts}|^2 + |V_{td}|^2 + |V_{tb}|^2)$. If we assume three generation unitarity, the denominator is unity and we measure $|V_{tb}| = 0.99 \pm 0.15$ or $|V_{tb}| > 0.80(0.76)$ at the 90(95)% CL. Releasing this hypothesis, a lower bound on $|V_{tb}|$ is obtained by using for $|V_{ts}|$ and $|V_{td}|$ a mean value deduced from their 90 % CL intervals [1]. If we set $|V_{ts}| = 0.009$ and $|V_{td}| = 0.04$, $|V_{tb}| > 0.055(0.048)$ at the 90(95)% CL.

References

[1] R.M. Barnett *et al.*, Phys. Rev. **D54**, 1 (1996).
[2] F. Abe *et al.*, Fermilab-Pub-97/284-E (submitted to Phys. Rev. Lett.).
[3] F. Abe *et al.*, Fermilab-Pub-97/286-E (submitted to Phys. Rev. Lett.).

Search for FCNC in B Decays

Gilberto Alemanni[1] (gilberto.alemanni@ipn.unil.ch)

University of Lausanne, Switzerland

Abstract. Here are reported some results about searches for Flavour Changing Neutral Current in B mesons decays and also a result about a search performed at LEP I by the L3 collaboration, which establishes the first preliminary limit on flavour-changing decay of a real Z^0 boson.

1 Introduction

Within Standard Model, FCNC processes are forbidden at tree level by the GIM mechanism. At higher order, SM predicts typical branching fractions from 10^{-6} to 10^{-10}. However, some new physics beyond the SM can increase these rates significantly.
The most general effective neutral current, between quarks of flavour i and j, can be written as [1]:

$$J_\mu = \bar{q}_i(p')\left(\gamma_\mu A_{ij} + i\sigma_{\mu\nu}\frac{k_Z^\nu}{M_Z}B_{ij}\right)q_j(p) \tag{1}$$

The term B_{ij} (anomalous term) arises only from loop effects, while the A_{ij} arises either from loop or at tree level; the model dependence is parametrized by these terms.
By studying FCNC processes, one can investigate the off-diagonals terms ($i \neq j$) and therefore test the validity of physics models.
In low-energy processes, like B meson decays, the Z^0 is highly virtual; therefore, because of the factor $\frac{k_Z^\nu}{M_Z}$, the B_{ij} term contribution should be negligible, and only the A_{ij} can be probed. On the other hand, at LEP I, where the Z^0 is real, the B_{ij} term could contribute significantly. That is why low phenomenological limits on FCNC B decays rates don't preclude the observation of rare Z^0 decays at LEP [2].

2 FCNC in B decays

Table 1 summarizes results on FCNC search in B meson decays [3]. So far, no signal has been seen. The experimental limits are still one to three order of magnitude larger than the SM predictions.

Experiment	Process	Rate $\times 10^{-5}$
L3	$B_d^0 \to e^+e^-$	< 1.4
L3	$B_s^0 \to e^+e^-$	< 5.4
L3	$B_d^0 \to \mu^+\mu^-$	< 1.0
L3	$B_s^0 \to \mu^+\mu^-$	< 3.8
L3	$B_d^0 \to e^\pm\mu^\pm$	< 1.6
L3	$B_s^0 \to e^\pm\mu^\pm$	< 4.1
CLEO	$b \to s\mu^+\mu^-$	< 5.6
CLEO	$b \to see$	< 5.7
CLEO	$b \to se^\pm\mu^\mp$	< 2.2
CLEO/CDF	$B \to l^+l^-K^+$	< 1.0
CLEO	$B \to l^+l^-K^*$	< 1.6
CDF	$B \to l^+l^-K^*$	< 2.5
CDF	$B_d^0 \to \mu^+\mu^-$	< 0.026
CDF	$B_s^0 \to \mu^+\mu^-$	< 0.077
CDF	$B_s^0 \to e^\pm\mu^\pm$	< 2.3
CDF	$B_d^0 \to e^\pm\mu^\pm$	< 0.44

Table 1. FCNC in B mesons decays

3 FCNC at LEP I

At LEP I it is possible to search for flavour-changing decays of a real Z^0, e.g. $Z^0 \to b\bar{s}$ (or $b\bar{d}$) [1]. This kind of analysis gives an independent test of FCNC and, as mentioned before, it can probe the anomalous term in the expression (1).
The idea is, in two-jet events, to tag a "b jet" and a "light jet". For this, one needs powerful tagging tools.

A Neural Network b-tag has been developed, which combines two standard tagging techniques used in L3: the Impact Parameter and the Decay Length tag. The improvement by this method is shownn in figure 1.a

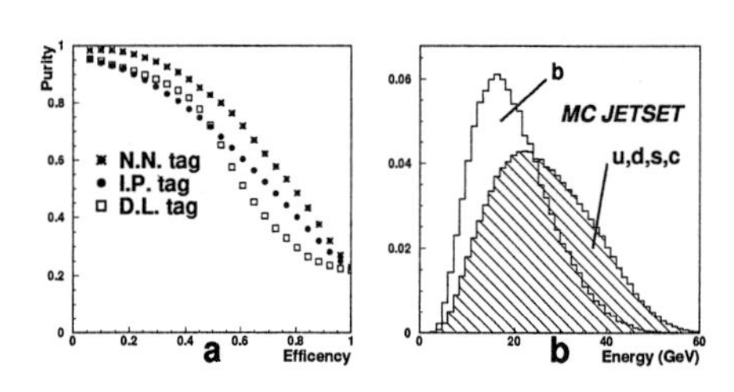

Fig. 1. (a) b jet tagging. (b) Light jet tagging.

The "light jet" selection makes use of NN information combined with the energy of the "Most Energetic Calorimetric Cluster" associated to a jet. Figure 1.b shows the tagging power of the latter.

In this analysis one measures the *flavour-violating branching ratio*:

$$R_{bl} = \sum_{q=s,d} \frac{\Gamma(Z \to bq)}{\Gamma(Z \to hadrons)} \tag{2}$$

which can be written in terms of the selection efficiencies ϵ^{data}, ϵ^{signal} and ϵ^{bkg}:

$$R_{bl} = \frac{\epsilon^{data} - \epsilon^{bkg}}{\epsilon^{signal} - \epsilon^{bkg}} \tag{3}$$

The Monte Carlo signal sample has been produced using a modified version of JETSET 7.4 [4].

The result has been obtained by analysing 27.6 pb^{-1} events collected by L3 in 1994:

$$R_{bl} = (\, 8 \pm 19\,(stat) \pm 17\,(sys)\,) \times 10^{-4}$$

This result is compatible with $R_{bl} = 0$, and permits to fix the first preliminary limit on Z^0 flavour-changing decay rate:

$$\sum_{q=d,s} B_r(Z^0 \to bq) < 5.6 \times 10^{-3} \text{ at } 95\% \text{ C.L.}$$

Table 2 summarizes the contributions to the systematic error.

Effect	$\times 10^{-4}$
ΔR_{bl}^{sys} (MC statistics)	11.9
ΔR_{bl}^{sys} (R_b)	0.9
ΔR_{bl}^{sys} (R_c)	3.5
ΔR_{bl}^{sys} ($g \to cc$)	0.5
ΔR_{bl}^{sys} ($g \to bb$)	0.8
ΔR_{bl}^{sys} (fragmentation model)	4.9
ΔR_{bl}^{sys} (energy resolution)	8.7
ΔR_{bl}^{sys} (track resolution)	6.2
ΔR_{bl}^{sys} (all)	17.2

Table 2. Systematic errors on R_{bl}

4 Conclusions

A brief overwiew on searches for FCNC in B decays has been presented. No signal has been seen. Future machines and detectors, dedicated to B physics, should improve present limits and to get a signal which could reveal an underlying new physics.
A search for flavour-changing Z^0 decay at LEP has been also presented. This search permits to fix the limit:

$$\sum_{q=d,s} B_r(Z^0 \to bq) \ < 5.6 \times 10^{-3}$$

This is a preliminary result, which will be completed by analysing the full L3 data statistics (49.7 pb^{-1}) and by improving the light quark jet tagging.

References

[1] D. Cocolicchio and M. Dittmar, *"The radiative and hadronic flavour changing decay of the Z"*, **CERN-TH 5753/90** (1990)

[2] M.J. Duncan, *"Flavour changing decays of the Z^0 at LEP"*, **CERN-TH 5429/89** (1989)

[3] M. Acciarri *et al.* (L3 Collaboration) Phys. Lett. **B391**, 474 (1997)
J.P. Alexander *et al.* (CLEO Collaboration), **CLEO CONF 97-15**, submission to IECHEP97 conference, Jerusalem (1997)
F. Abe *et al.* (CDF Collaboration) Phys. Rev. Lett. **76**, 4675 (1996)

[4] L. Cuénoud, *"Générateurs de Flavour Changing Neutral Currents"*, Travail de diplôme, Université de Lausanne IPN (1996)

Rare B Decays and V_{ub} from CLEO

Ahren Sadoff[1] (ajs@lns62.lns.cornell.)

Ithaca College and Cornell University, Ithaca,N.Y.

Abstract. CLEO has measured the CKM matrix element V_{ub} using data from inclusive and exclusive semileptonic $b \to u$ decays. In both cases, the measurements are consistent with a value of about 3×10^{-3}. CLEO has also reported on searches for many rare B decay modes. Results for the particular decays: $B \to \pi\pi, K\pi, KK$ will be given here. In addition, the status of $b \to s\gamma$ will be discussed.

1 Introduction

In the Standard Model, weak decays can be understood through the Cabibbo-Kobayashi-Maskawa (CKM) matrix which relates the physical quarks to their weak eigenstates. In the Wolfenstein representation, the elements V_{ub} and V_{td} depend on real and imaginary parameters, ρ and η. It is the existence of η that allows for the possibility of CP violation in weak decays.

Rare B meson decays are dominated by either $b \to u$ decays or by single loop penguin decays as shown in Fig. 1. These are of particular interest

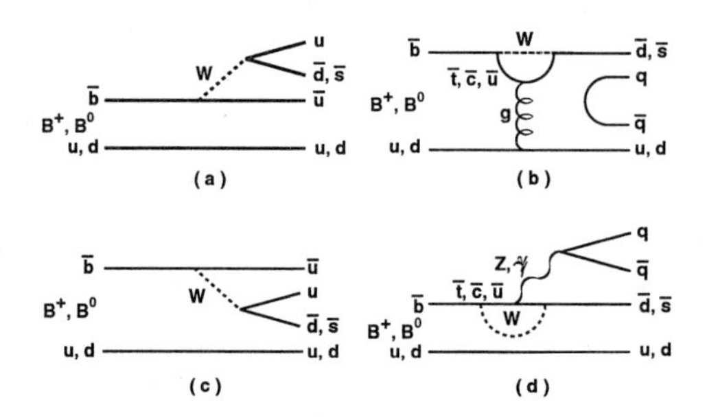

Fig. 1. The dominant decay proecesses are expected to be a) external W emission, b) gluonic penguin, c) internal W emission, d) external electroweak penguin

since it is expected that CP violating effects via these diagrams will be a good deal larger in B decays than in K decays. For example, interference between the penguin and spectator diagrams in $B^0 \to K^+\pi^-$ can give rise to direct CP violation which would manifest itself as a rate asymmetry in B^0 and $\bar{B}^0$ decays. This could lead to a determination of the angle γ, the phase of V_{ub}. In addition, it has recently been pointed out [1] that the ratio, $R = \mathcal{B}(B^0 \to K^+\pi^-)/\mathcal{B}(B^+ \to K^0\pi^+)$, could also lead to a constaint on γ.

The decay $B^0 \to \pi^+\pi^-$ can be used to measure indirect CP violation due to $B^0 - \bar{B}^0$ mixing. This decay is thought to be dominated by the spectator diagram shown in Fig. 1(a). But, if there is a substantial penguin contribution (penguin pollution), this complicates the interpretation of any measured asymmetry. In order to disentangle the penguin and spectator effects, a variety of neutral and charged B decay modes must be measured. CLEO has previously reported upper limits on a large number of these rare B decay modes [2].

In addition, rare B decays could provide a window for physics beyond the Standard Model. Due to the small predicted branching fractions, these modes could be sensitive to new physics.

2 Measurement of V_{ub}

As indicated above, if V_{ub} were to be zero, no CP violation would be predicted by the Standard Model. CLEO has measured V_{ub} using both inclusive and exclusive semileptonic $b \to u$ decays [3], [4]. For inclusive decays, one takes advantage of the fact that $b \to ul\nu$ transitions produce leptons with higher momentum than can be produced from $b \to cl\nu$. From the excess of leptons in the momentum range 2.3-2.6 Gev/c, CLEO extracted a value for V_{ub} of $(3 \pm 1) \times 10^{-3}$. The error is mainly systematic due to the large model dependence in the fraction of leptons in the high momentum range compared to the entire spectrum for the $b \to ul\nu$ transition.

In measuring exclusive decays, CLEO studied B decays to $\pi l\nu$, $\rho l\nu$, and $\omega l\nu$ by reconstructing the missing neutrino. This was then used for both background suppression and for reconstruction of the candidate B mass. Fig. 2 shows the relevant distributions. From the measured branching fractions and averaging over the different modes, CLEO found $V_{ub} = (3.3 \pm 0.8) \times 10^{-3}$, consistent with the value obtained in the inclusuve measurement. Here, again, the dominant error is systematic due to model dependence.

3 $B \to K\pi, \pi\pi, KK$

The most recently published CLEO results [2] on charmless hadronic B decays were based on a sample of 2.6 million $B\bar{B}$ pairs. The results presented here are based on a sample of 3.3 million $B\bar{B}$ pairs and a reoptimized analysis which leads to a 20% increased efficiency.

The decays of intertest are characterized by two high momentum (~ 2.6 GeV/c) hadronic tracks. Two kinematic variables are used to define the signal. The energy difference, $\Delta E = E_1 + E_2 - E_{beam}$, where E_1 and E_2 are the energies of the daughters of the B candidate, is centered around 0 for signal events. The resolution in ΔE, $\sigma_{\Delta E} = 25 \pm 2$ MeV. At momenta near 2.6 GeV/c, $K\pi$ and $\pi\pi$ events differ in energy by 42 MeV, or $1.6\sigma_{\Delta E}$. The second variable is the beam-constained mass, $M_B = \sqrt{E_{beam}^2 - \mathbf{p}_B^2}$, where

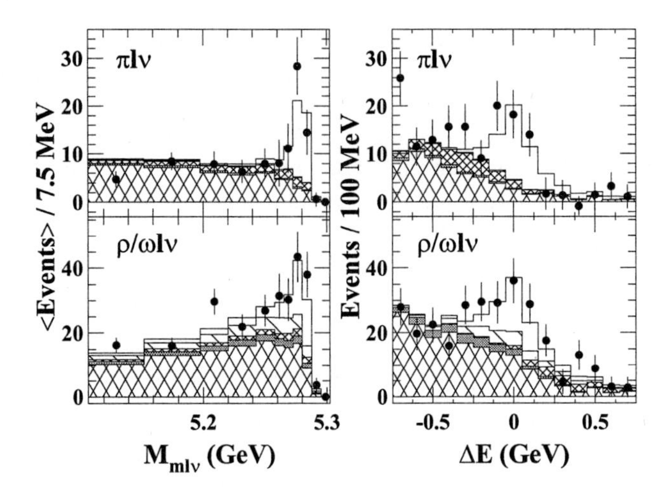

Fig. 2. M_B, ΔE in the ΔE and M_B signal bands, respectively, for the combined π(top) and vector (bottom) modes. The data (points) are continuum and fake subtracted. The course crosshatch, grey, and unshaded components are $b \to cX$, $B \to H_u l\nu$, and signal, respectively.

$\mathbf{p}_B$ is the B candidate momentum. The resolution in M_B ranges from 2.5 to 3.0 MeV/c^2 where the larger resolution corresponds to decay modes with high momentum π^0's.

Backgrounds are entirely due to $e^+e^- \to q\bar{q}$ from the continuum. Backgrounds from $b \to c$ are kinematically excluded and $b \to u$ backgrounds are negligible. Continuum suppression is achieved by taking advantage of the fact that the signal and continuum decay distributions have very different shapes. Since B mesons are produced approximately at rest, their decay products tend to have a spherical structure while hadrons from the continuum tend to have a two-jet structure. This distinction is exploited in two ways.

In the first case, the angle θ_T between the thrust axis of the B candidate and the sphericity axis of the rest of the event is calculated. The distribution of $cos\theta_T$ is strongly peaked near ± 1 for qq events and is nearly flat for $B\bar{B}$ events. A cut of $|cos\theta_T| < 0.8$ eliminates 83% of the background.

Secondly, a Fisher discriminant, $\mathcal{F}$, is used as described in detail in Ref. [2]. $\mathcal{F}$ is a linear combination of , in our case, 11 variables where the coefficients are chosen to maximize the difference between Monte Carlo signal and background using continuum data. There is about a 1σ separation between the $\mathcal{F}$ distributions for signal and background.

Charged particles are identified by using the specific ionization (dE/dx) measured in the 51 layers of the central drift chamber. A study of pions and kaons with momenta p $\sim$ 2.6 GeV/c using D^{*+}-tagged $D^0 \to K^+\pi^-$ events indicates a dE/dx separation of $(1.7 \pm 0.1)\sigma$.

For all modes except $B^0 \to K^0\bar{K}^0$, an unbinned maximum likelihood fit was performend using ΔE, M_B, $\mathcal{F}$, dE/dx (where applicable), and $|cos\theta_B|$ (θ_B is the angle between the B meson momentum and the beam axis) as the input variables. Fits were performed for five different modes: h^+h^-, $h^+\pi^0$, $\pi^0\pi^0$, $h^+K^0_S$, and $K^0_S\pi^0$ where h^{+-} indicates a charged kaon or pion. Fig. 3 shows likelihood contours for the three modes indicated. Fig. 4 shows the M_B projection for events with $|\Delta E| < 2\sigma_{\Delta E}$ and a cut on $\mathcal{F}$. The results

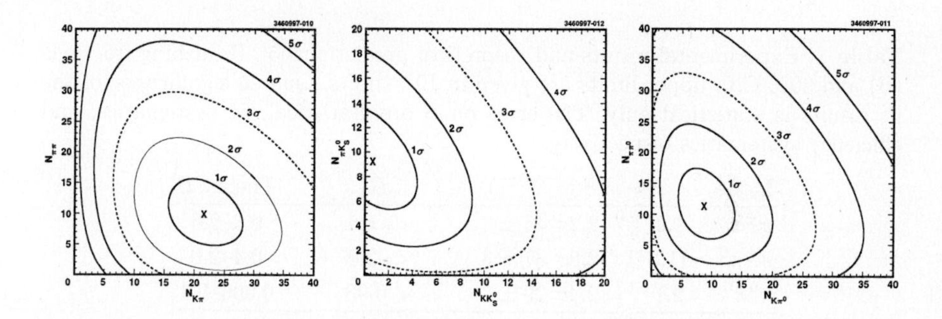

Fig. 3. Contours of the likelihood fit to (a) $N_{K\pi}$ and $N_{\pi\pi}$ for $B^0 \to K^+\pi^-, \pi^+\pi^-$; (b)$N_{K^0_S K}$ and $N_{K^0_S\pi}$ for $B^+ \to K^+\bar{K}^0$, $K^0\pi^+$; (c)$N_{K\pi^0}$ and $N_{\pi\pi^0}$ for $B^+ \to K^+\pi^0$, $\pi^+\pi^0$.

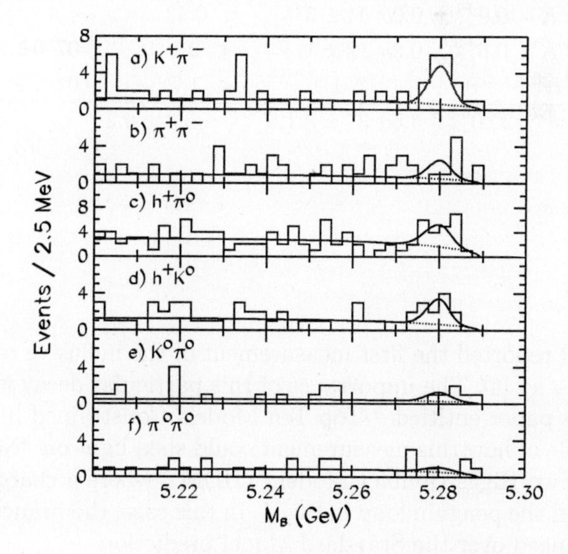

Fig. 4. M_B distributions for the modes indicated. The projection of the likelihood fit (solid curve) and the continuum background component (dotted curve) are included.

for all modes are shown in Table 1. Statistically significant signals are found for the decays $B^0 \to K^+\pi^-$ and $B^+ \to K^0\pi^+$ and for $B^+ \to h^+\pi^0$. In these cases, branching fractions are quoted with their statistical and systematic errors. If no significant signal is observed, then 90% C.L. upper limits are quoted. For the $B^+ \to K^0\pi^+$ mode, this is the first observation of a pure gluonic penguin decay. In addition, while there may be a $b \to u$ spectator contribution, the $B^0 \to K^+\pi-$ mode is expected to be dominated by the penguin decay amplitude.

Table 1. Experimental results and theoretical predictions [5]. Branching fractions ($\mathcal{B}$) and 90% C.L. upper limits are given in 10^{-5} units. Quoted significance of the fit results is statistical only. The erros on $\mathcal{B}$ are statistical, fit systematics, and efficiency systematics respectively.

Mode	N_S	Sig.	$\mathcal{E}(\%)$	$\mathcal{B}$	Theory $\mathcal{B}$
$\pi^+\pi^-$	$9.9^{+6.0}_{-5.1}$	2.2σ	44 ± 3	< 1.5	0.8–2.6
$\pi^+\pi^0$	$11.3^{+6.3}_{-5.2}$	2.8σ	37 ± 3	< 2.0	0.4–2.0
$\pi^0\pi^0$	$2.7^{+2.7}_{-1.7}$	2.4σ	29 ± 3	< 0.93	0.006–0.1
$K^+\pi^-$	$21.6^{+6.8}_{-6.0}$	5.6σ	44 ± 3	$1.5^{+0.5}_{-0.4} \pm 0.1 \pm 0.1$	0.7–2.4
$K^+\pi^0$	$8.7^{+5.3}_{-4.2}$	2.7σ	37 ± 3	< 1.6	0.3–1.3
$K^0\pi^+$	$9.2^{+4.3}_{-3.8}$	3.2σ	12 ± 1	$2.3^{+1.1}_{-1.0} \pm 0.3 \pm 0.2$	0.8–1.5
$K^0\pi^0$	$4.1^{+3.1}_{-2.4}$	2.2σ	8 ± 1	< 4.1	0.3–0.8
K^+K^-	$0.0^{+1.3}_{-0.0}$	0.0σ	44 ± 3	< 0.43	–
$K^+\bar{K}^0$	$0.6^{+3.8}_{-0.6}$	0.2σ	12 ± 1	< 2.1	0.07–0.13
$K^0\bar{K}^0$	0	–	5 ± 1	< 1.7	0.07–0.12
$h^+\pi^0$	$20.0^{+6.8}_{-5.9}$	5.5σ	37 ± 3	$1.6^{+0.6}_{-0.5} \pm 0.3 \pm 0.2$	–

4 $b \to s\gamma$

In 1995 CLEO reported the first measurement of the inclusive radiative penguin decay $b \to s\gamma$ [6]. The importance of this particular decay is outlined by J. Hewitt in a paper entitled " Top Ten Models Constrained by $b \to s\gamma$ [7]. As one example of how this measurement could shed light on "New Physics", consider the Two-Higgs Doublet Model (THDM), where a charged Higgs replaces the W in the penguin loop in Fig. 1. In this case, the branching fraction could be enhanced over the Standard Model prediction.

Experimentally, even though CLEO has a factor of two more data, there has been no update on their original results of $\mathcal{B} = (2.32 \pm 0.57 \pm 0.35) \times 10^{-4}$. On the other hand, there has been new theoretical calculations by

Chetyrkin, Misiak, and Münz where the full next-to-leading order calculations have been completed. Their value is $\mathcal{B} = (3.28 \pm 0.33) \times 10^{-4}$. This increases the central value from $2.8 \rightarrow 3.28$, and, more importantly, decreases the theoretical uncertainty from $0.8 \rightarrow 0.33$. At this point, this uncertainty is less than the experimental error. Figure 5 shows that for THDM Model II these new results put a lower limit of about 650 GeV/c^2 on the charged Higgs mass.

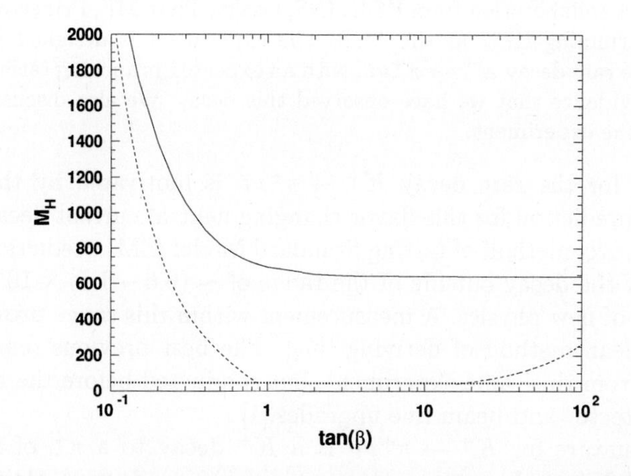

Fig. 5. Limits on Model II charged Higgs mass. The solid line shows the limit due to the $b \rightarrow s\gamma$ measurement. The dashed curve is due to the $b \rightarrow s\mu^+\mu^-$ measurement from D0 and the dotted line near 0 is due to the LEP I direct search.

References

[1] R. Fleischer and T. Mannel, Univ. of Karlsruhe preprint TTP 97-17

[2] D.M. Asner *et al.*, Phys. Rev. D **53**, 1039 (1996)

[3] J. Bartelt *et al*, .Phys. Rev. Lett. **71**, 4111 (1993)

[4] J.P. Alexander *et al.*, Phys. Rev. Lett. **77**, 5000 (1996)

[5] N. G. Deshpande and J. Trampetic, Phys. Rev. D **41**, 895 (1990); L. L. Chau *et al*, Phys. Rev. D **43**, 2176 (1991); A. Deandrea *et al*, Phys. Lett. B **318**, 549 (1993); A. Deandrea *et al*, Phys. Lett. B **320**, 170 (1994); G. Kramer and W. F. Palmer, Phys. Rev. D **52**, 6411 (1995); D. Ebert, R. N. Faustov, and V. O. Galkin, Phys. Rev. D **56**, 312 (1997); D. Du and L. Guo, Zeit. Phys. C **75**, 9 (1997)

[6] M. S. Alam *et al*, Phys. Rev. Lett. **74**, 2885 (1995)

[7] J. L. Hewitt, SLAC-PUB-6521 (May 1994)

Evidence for $K^+ \to \pi^+\nu\overline{\nu}$ from E787

James S. Frank, for the E787 collaboration (frank@bnlku5.phy.bnl.gov)

Brookhaven National Laboratory, Upton, NY 11973 USA

Abstract. A collaboration from KEK, INS, Osaka, TRIUMF, Princeton, and BNL is currently running E787 at the AGS. The experiment is designed primarily to search for the rare decay $K^+ \to \pi^+\nu\overline{\nu}$, with an expected branching ratio of $\sim 10^{-10}$. We report evidence that we have observed this decay. We also discuss the future outlook for the experiment.

The search for the rare decay $K^+ \to \pi^+\nu\overline{\nu}$ is motivated by the theoretically clean prediction for this flavor changing neutral current decay and thus provides a novel method of testing Standard Model (SM) predictions. A measurement of the decay outside of the range of $\sim (0.6 - 1.5) \times 10^{-10}$ may be a signature of new physics. A measurement within this range provides a theoretically clean method of deriving $|V_{td}|$. The best previous search for this decay was from this same experiment, but conducted before the most recent series of detector and beam line upgrades.[1]

The signature for $K^+ \to \pi^+\nu\overline{\nu}$ is a K^+ decay to a π^+ of momentum $P < 227$ MeV/c and no other observable decay product. A claim of observation of this decay requires suppression of all backgrounds to well below the sensitivity for the signal. In addition, one must have reliable estimates of residual background levels, which come primarily from the two-body decays $K^+ \to \mu^+\nu_\mu$ ($K_{\mu 2}$) and $K^+ \to \pi^+\pi^0$ ($K_{\pi 2}$). The range (R), energy (E), and momentum (P) of the decay products from K^+ decays at rest were measured in a drift chamber and a range stack (RS) of 2 cm scintillation counters in order to provide adequate background suppression. The single charged-particle track was required by kinematics to be a π^+ with P, R and E between the $K_{\pi 2}$ and $K_{\mu 2}$ peaks. Pions were distinguished from muons by kinematics and by observing the $\pi \to \mu \to e$ decay sequence in the RS using 500-MHz flash-ADC transient digitizers (TD). At the rates at which we took data, the TD system provided a suppression factor 10^{-5} for muons. The inefficiency for detecting events with π^0s was 10^{-6} for a photon energy threshold of about 1 MeV. A description of the detector may be found elsewhere.[2]

The data were analyzed with the goal of reducing the total expected background to ~ 0.1 event in the final sample. We took advantage of redundant independent constraints available on each source of background to establish two independent sets of cuts. One set of cuts could be relaxed or inverted to enhance the background so that the rejection of the other group could be evaluated. For example, the background from $K_{\pi 2}$ was evaluated by separately measuring the rejections of the photon detection system and kinematic cuts. Small correlations in the separate groups of cuts were investigated for each background source and corrected for if necessary.

After cuts, we determined that there were 0.08 ± 0.03 background events expected in the signal region. We were able to check this by simultaneously relaxing all cuts in order to deliberately increase the background by several orders of magnitude. The observed backgrounds scaled as expected. This tested the independence of the two sets of cuts and the validity of the background estimates.

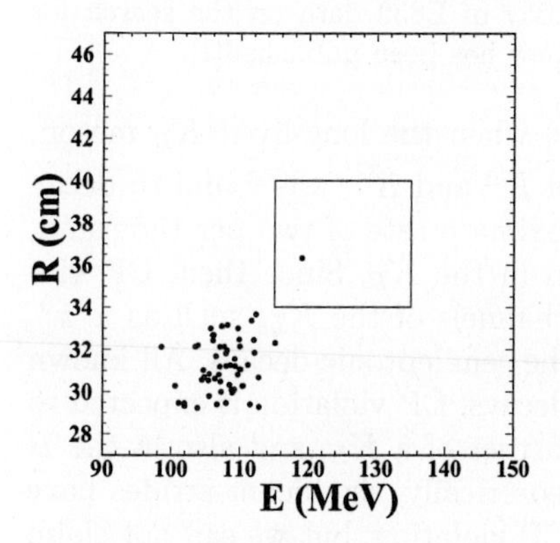

Fig. 1.

Range (R) vs. energy (E) distribution for the $K^+ \to \pi^+ \nu\bar{\nu}$ data set with the final cuts applied. The box enclosing the signal region contains a single candidate event.

Figure 1 shows R vs. E for the events surviving all other analysis cuts. Only events with measured momentum in the accepted region $211 \leq P \leq 230$ MeV/c are plotted. The box containing one event shows the signal region which encloses the upper 16.2% of the $K^+ \to \pi^+ \nu\bar{\nu}$ phase space. The cluster of events below the signal region centered at $E = 108$ MeV is consistent with $K_{\pi 2}$ decays for which both photons had been missed.

The event also satisfied the most demanding criteria designed in advance for candidate evaluation. This put it in a region with an additional background rejection factor of 10. Since the explanation of the observed event as background is highly improbable, we conclude that we have likely observed a $K^+ \to \pi^+ \nu\bar{\nu}$ decay.

Based upon the acceptance for $K^+ \to \pi^+ \nu\bar{\nu}$ of $0.0016 \pm 0.0001^{stat} \pm 0.0002^{syst}$, and the total exposure of 1.49×10^{12} kaons entering the target, the the branching ratio is $B(K^+ \to \pi^+ \nu\bar{\nu}) = 4.2^{+9.7}_{-3.5} \times 10^{-10}$. The observation of an event with the signature of $K^+ \to \pi^+ \nu\bar{\nu}$ is consistent with the expectations of the SM and $|V_{td}|$ lies in the range $0.006 < |V_{td}| < 0.06$. This work has recently been published[3].

E787 has recently collected a factor of > 2 additional data. Additional detector and data acquisition upgrades have also been made for future data collection. With some modest additional upgrades and long, efficient run cycles at the AGS, we expect to improve acceptance while maintaining sufficient background rejection in order to reach a single event sensitivity of $\sim 10^{-11}$ over the next several years.

This work was supported by the U. S. Department of Energy under Contract No. DE-AC02-76 CH00016.

References

[1] S. Adler *et al.*, Phys. Rev. Lett. **76**, 1421 (1996).

[2] M.S. Atiya *et al.*, Nucl. Instr. Meth. **A321**, 129 (1992).

[3] S. Adler *et al.*, Phys. Rev. Lett. **79**, 2204 (1997).

New Results from KTEV : 1

Sunil V. Somalwar (somalwar@physics.rutgers.edu)(KTeV Collaboration)

Rutgers University, Piscataway, New Jersey 08855, USA

Abstract. I summarize the status of the Fermilab KTeV-E832 experiment. The principle goal of this experiment is to probe the origin of CP violation by measuring Real(ϵ'/ϵ) to a precision of 1×10^{-4}. A successful run during 1996-97 has resulted in the collection of approximately 4-5 million of the statistics-limiting $K_L \to 2\pi^0$ decays, which corresponds to a measurement of Real(ϵ'/ϵ) with statistical precision in the vicinity of 1.5×10^{-4}. While the Real(ϵ'/ϵ) measurement is in progress, an unrelated result from approximately one day of E832 data on the search for supersymmetric particles containing light gluinos has been published[1].

CP violation was discovered in 1964[2] when the long-lived K_L meson, thought to be the pure CP-odd mixture of K^0 and $\overline{K}^0$, was found to decay into the CP-even $\pi^+\pi^-$ state at the approximate rate of two per thousand, indicating a small CP-even contamination in the K_L. Since then, CP violation has been observed in other decay channels of the K_L, such as $\pi^0\pi^0$, $\pi^+\pi^-\gamma$, and in the charge asymmetry in the semileptonic decays. All known manifestations of CP violation are in K_L decays. CP violation is expected to occur in the K_S meson, the short-lived partner of a K_L, and also in the B meson, the heavier version of a kaon. Theoretically, significant strides have been made towards the understanding of CP violation, but we can not claim to understand the origins of CP violation. The standard model offers the best explanation as yet through the imaginary phase of the Cabibbo-Kobayashi-Maskawa (CKM) three-generation mixing matrix. However, the predictions of this hypothesis have so far eluded vigorous experimental scrutiny. Adequate CP violation is a necessary ingredient in baryogenesis, and thus a fundamental understanding of CP violation is of cosmological significance as well.

Real(ϵ'/ϵ) is a measure of the difference of CP violation in the $K_L \to \pi^+\pi^-$ and $K_L \to \pi^0\pi^0$ decays, given by

$$\mathrm{Real}(\epsilon'/\epsilon) \approx 1/6 \left(\left| \frac{A(K_L \to \pi^+\pi^-)/A(K_S \to \pi^+\pi^-)}{A(K_L \to \pi^0\pi^0)/A(K_S \to \pi^0\pi^0)} \right|^2 - 1 \right). \quad (1)$$

A consequence of accomodating CP violation in the CKM matrix is that the CP violation in these two channels is expected to be different by a very small amount, characterized by Real(ϵ'/ϵ) in the range of $1\text{-}10 \times 10^{-4}$[3]. The measurement from CERN-NA31[4] with Real(ϵ'/ϵ) $= (23 \pm 7) \times 10^{-4}$ indicates this difference, whereas Fermilab E731's measurement[5] of $(7.4 \pm 5.9) \times 10^{-4}$ is inconclusive and compatible with the superweak hypothesis[6].

The KTeV experiment(figure 1) is designed for precision measurements of the $\pi^+\pi^-$ and $2\pi^0$ final states. It has two neutral double beams, with an alternating regenerator to produce the requisite K_S component in one of the two beams. The charged decay products are measured with a precision spectrometer consisting of four drift chambers and a large aperture magnet. The photons from the $2\pi^0$ decays are measured with a pure CsI calorimeter.

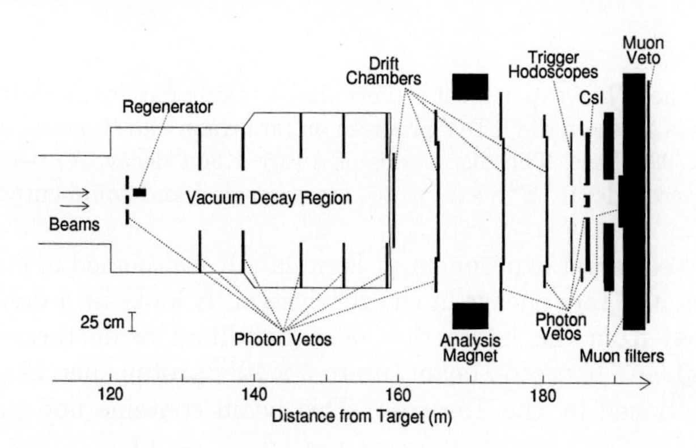

Fig. 1. Plan view of the KTeV detector in E832 configuration

The analysis of a partial dataset is in progress. The kaon mass resolutions in all four modes of interest for the Real(ϵ'/ϵ) measurement are below 2 MeV's. Compared to the predecessor experiment E731, the backgrounds from the incoherently regenerated kaons are significantly smaller due to the deployment of an active regenerator instead of a passive one. The background from $K_L \to 3\pi^0$ decays is smaller because of improved photon vetos and due to better calorimeter resolution. Overall accidental rates are also reduced due to a cleaner beam. Given these improvements, it is expected that the systematic error will be considerably less than that in E731.

This work is supported by the NSF, DOE, and the US-Japan Foundation. We thank the Fermilab staff for their dedication.

References

[1] S.V. Somalwar, Report No. 1403 in this volume; J. Adams *et al.*, Phys. Rev. Lett. **79**, 4083 (1997).

[2] J.H. Christenson, J.W. Cronin, V.L. Fitch, and R. Turlay, Phys. Rev. Lett. **13**,138 (1964).

[3] A.J. Buras *et al.*, Phys. Lett. **B389**, 749 (1996).

[4] G.D. Barr *et al.*, Phys. Lett. **B317**, 233 (1993).

[5] L.K. Gibbons *et al.*, Phys. Rev. D **55** 6625 (1997).

[6] L. Wolfenstein, Phys. Rev. Lett. **13**, 562 (1964).

New Results from KTEV : 2

Emmanuel Monnier (For the KTeV Collaboration) (monnier@hep.uchicago.edu)

Enrico Fermi Institute, The University of Chicago, Chicago, Illinois 60637, USA

Abstract. The KTeV experiment at Fermilab is taking data since October 1996. A status report is given on the KTeV program on rare Kaon and Hyperon decays called the KTeV/E799 phase. The discoveries of a rare Kaon decay, $K_L \to \pi^+\pi^- e^+ e^-$, and of a Hyperon decay, $\Xi^0 \to \Sigma^+ e^- \overline{\nu_e}$, are presented and commented upon.

KTeV is a fixed target experiment at Fermilab. It is designed to study mainly CP violation and rare decays in the K^0 system. It looks at a double neutral beam created from the interaction on a beryllium oxide target of a high intensity 800 GeV/c proton beam (up to 5×10^{12} protons per 19 s spill every minute), produced by the Tevatron. This beam contains not only K^0 and $\overline{K}^0$ (as well as neutrons and photons), but also a good fraction of $\Lambda^0, \overline{\Lambda^0}, \Xi^0$, and $\overline{\Xi}^0$. Thus, Hyperon decays and particularly rare Ξ^0 decays can also be studied. The KTeV/E799 phase of the experiment, has taken six weeks of data in winter 1997 and is currently taking more data since the end of July.

The KTeV detector showed and described in the previous talk report (# 920 of this proceeding) is based on a precise tracking system and a high resolution Cesium Iodide electromagnetic calorimeter. The tracking system which provides the momentum of the charged particles and identifies the charged vertex, is made of four drift chambers with a precision dipole magnet inbetween that gives roughly a 205 MeV/c kick along the horizontal axis to the charged particles. For the E799 phase, the regenerator has been removed and a nine chamber transition radiation detector (TRD) has been added, in front of the calorimeter to improve the pion/electron identification. A pion rejection of 276 ± 38 to one at 90% electron efficiency has been measured for this detector. In addition to these various elements, a set of sub-detectors, mainly scintillators, are used to define the fiducial volume. All the informations provided by the various sub-detectors of the KTeV experiment are used by the different levels of the trigger system to select the best events to keep on tapes. For the Hyperon studies, a specific trigger component, aimed to identify high momentum charged particles traveling down the beam holes, has been added to the trigger logic.

The KTeV/E799 rare Kaon decay program will allow a single event sensitivity (SES) close to 10^{-11} for decay modes with 10% acceptance (most of the two track modes). When the detailed analyses that are on going will unfold their results, many rare Kaon decay modes such as the CP violating ones will be measured for the first time and may give some new information on the CP violation. In addition, KTeV/E799 data makes possible a rich assortment of Hyperon physics. But, even though the data taking has not yet been completed, two new decay modes have already been discovered.

Within a typical one day of data taken during the winter KTeV/E799 campaign, the decay $K_L \to \pi^+\pi^-e^+e^-$ has been looked for in four charged track events. The right part of figure 1 represents, for the selected events, the distribution of the invariant $\pi^+\pi^-e^+e^-$ mass that exhibit a clear peak at the Kaon mass value. The background peak to the left of the signal peak is due to $K_L \to \pi^+\pi^-\pi^0$ with $\pi^0 \to e^+e^-\gamma$ events where the γ is lost. From the 36 candidates found on top of a calculated background of 8, a branching ratio of $(2.6 \pm 0.5_{stat+syst}) \times 10^{-7}$ has been extracted. A lot of other analysis are in progress to study rare Kaon decays and soon will bring new physics results.

In all the data recorded during the winter 97 campaign, we looked for the decay of Ξ^0 into a Σ^+, an electron and a $\overline{\nu_e}$ with the Σ^+ decaying into a proton and a π^0 A final sample of 200 events has been selected with 153 $\Xi^0 \to \Sigma^+e^-\overline{\nu_e}$ on top of a background of 6 events. The left part of figure 1 shows, for these events, the distribution of the proton-π^0 invariant mass which exhibits a clear peak centered on the Σ^+ mass value. In combination with 30715 events selected as $\Xi^0 \to \Lambda^0\pi^0$, the main decay mode, a branching ratio of $(2.6 \pm 0.2_{stat} \pm 0.3_{syst}) \times 10^{-4}$ has been extracted. The theoretical calculation from the Cabbibo theory of 2.61×10^{-4} agrees well with this measurement. In the future, with all the data taken in 1997, the analysis of this decay mode will allow us to measure the Ξ^0 form factors. Many other Hyperon decay modes, such as the Ξ^0 radiative decays, are being studied and results will be presented in the near future.

In a mid term future, a new KTeV campaign is in preparation for 1999. It will improve the rare Kaon decay physics program by a factor of three and will increase the Hyperon decays statistics by a factor of four allowing better precision in the Ξ^0 form factor measurement.

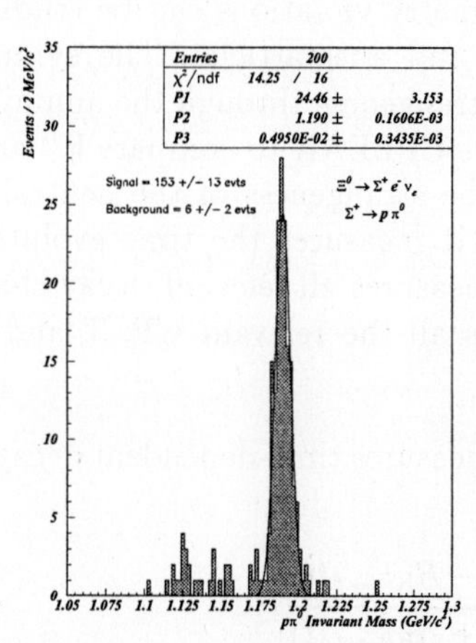

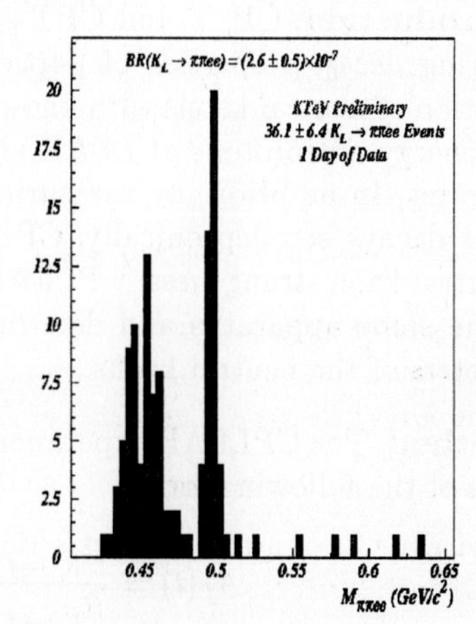

Fig. 1. Left: Reconstructed $p\pi^0$ mass for $\Xi^0 \to \Sigma^+e^-\overline{\nu_e}$ candidates. Right: Reconstructed $\pi^+\pi^-e^+e^-$ mass for $K_L \to \pi^+\pi^-e^+e^-$ candidates.

CPLEAR Experiment at CERN: CP, T, and CPT Measurements in the Neutral Kaon System

Danilo Zavrtanik (danilo.zavrtanik@ses-ng.si)

J. Stefan Institute and University of Ljubljana, SI-1000 Ljubljana, Slovenia
for the CPLEAR Collaboration
University of Athens, University of Basle, Boston University, CERN, LIP and University of Coimbra, Delft University of Technology, University of Fribourg, University of Ioannina, University of Liverpool, J. Stefan Inst. and Phys. Dep., University of Ljubljana, CPPM, IN2P3-CNRS et Université d'Aix-Marseille II, CSNSM, IN2P3-CNRS, Orsay, Paul-Scherrer-Institut(PSI), CEA, DSM/DAPNIA, CE-Saclay, Royal Institute of Technology, Stockholm, University of Thessaloniki, ETH-IPP Zürich.

Abstract. Using strangeness tagging at production time, CPLEAR measures time-dependent asymmetries between K^0 and $\overline{K}^0$ decay rates for pionic and semileptonic decays. In the latter, strangeness tagging at decay time with the lepton charge is also performed. A complete set of parameters describing CP, T and CPT violation in neutral kaon mixing and decay can be determined. This includes the first direct experimental observation of time reversal violation. The mass equality of K^0 and $\overline{K}^0$ is tested to the level of 10^{-19} GeV. In the framework of quantum mechanics for open systems, CPLEAR results allow physics to be probed on a scale approaching the Planck mass.

Introduction. CP, T and CPT symmetry violations can be studied by comparing decay properties of particles and antiparticles. The symmetric production of neutral kaons with known strangeness through the annihilation of low energy antiprotons at LEAR allows CPLEAR to compare K^0 and $\overline{K}^0$ decay rates. In addition, by measuring the strangeness of the neutral kaon when it decays semileptonically, CPLEAR measures the time evolution of the neutral kaon strangeness. CPLEAR measures all relevant decay channels with the same apparatus and determines all the relevant CP, T and CPT parameters of the neutral kaons.

Method. The CPLEAR experiment measures time-dependent decay asymmetries of the following form:

$$A_f(t) = \frac{R_{\overline{K}^0 \to \bar{f}}(t) - R_{K^0 \to f}(t)}{R_{\overline{K}^0 \to \bar{f}}(t) + R_{K^0 \to f}(t)} \tag{1}$$

where $R_{\overline{K}^0 \to \bar{f}}(t)$ and $R_{K^0 \to f}(t)$ are the decay rates of neutral kaons that were produced as $\overline{K}^0$ and K^0, and are decaying into the charge conjugate final states $\bar{f}$ and f. The neutral kaons are produced by $\mathrm{p\bar{p}}$ annihilation at rest, $\mathrm{p\bar{p}} \to \mathrm{K}^{\pm}\pi^{\mp}\mathrm{K}^0(\overline{\mathrm{K}^0})$. The strangeness of the neutral kaon is tagged by the accompanying charged kaon. In the case of semileptonic decays, the strangeness of the neutral kaons may also be tagged at the decay time ($\Delta S = \Delta Q$ rule). The CPLEAR detector allows different final states to be selected. In the case of hadronic decays, CPLEAR measures the rates of $\pi^+\pi^-$, $\pi^0\pi^0$, $\pi^+\pi^-\pi^0$ and $\pi^0\pi^0\pi^0$ decays, where $f = \bar{f}$. In the case of semileptonic decays, CPLEAR measures $\pi^{\pm}\mathrm{e}^{\mp}\nu(\bar{\nu})$ decay rates, where K^0 or $\overline{\mathrm{K}^0}$ are the final states ($\Delta S = \Delta Q$ rule). The use of the asymmetries defined in Eq. (1) minimizes systematic errors as most of the acceptances cancel.

The description of the CPLEAR detector as well as the experimental details are given elsewhere [1]. In total, $\approx 10^8$ $\overline{\mathrm{K}^0}$ and K^0 decays were reconstructed, i.e. $7 \cdot 10^7$ $\pi^+\pi^-$ decays with $\tau > 1\tau_S$, $1.8 \cdot 10^6$ $\pi e \nu_{\mathrm{e}}$ decays, $0.5 \cdot 10^6$ $\pi^+\pi^-\pi^0$ decays, $2 \cdot 10^6$ $\pi^0\pi^0$ and $1.3 \cdot 10^4$ $\pi^0\pi^0\pi^0$ decays.

Corrections. Although $\overline{\mathrm{K}^0}$ and K^0 are symmetrically produced, corrections have to be applied to construct the asymmetries from the measured decay rates. This is done on an event by event basis, in the same way for all the decay channels.

The tagging efficiencies of $\overline{\mathrm{K}^0}$ and K^0 differ due to different interactions of charged kaons and pions with the detector material. This leads to a relative normalization which depends on the kinematics and topology of the primary $\mathrm{K}^{\pm}\pi^{\mp}$ pair.

Corrections due to the regeneration of neutral kaons in the detector material have to be applied. Because of the lack of experimental data on regeneration amplitudes, the systematic error of some CP violating parameters (for example ϕ_{+-}) would be dominated by regeneration effects. A dedicated run was then devoted to the measurement of the regeneration amplitudes. Details and results are given elsewhere [2].

Background contributions and decay time resolution are obtained from the Monte Carlo simulation and the asymmetries are modified accordingly.

An additional normalization has to be applied to semileptonic decays and is due to the difference in relative detection efficiency of the $\pi^+\mathrm{e}^-$ and $\pi^-\mathrm{e}^+$ pairs. The normalization factor is determined from pionic annihilation and photon conversion data.

Hadronic decays. From the rate asymmetries as defined in Eq. (1), with $f = 2\pi, 3\pi$, we can determine the CP violating parameters η_{+-}, η_{+-0}, η_{00} and η_{000}. Details of the analysis and the sources of systematic errors are described in Ref. [3] for the $\pi^+\pi^-$ decays, in Ref. [4] for the $\pi^0\pi^0$ decays and in Ref. [5] for the $\pi^+\pi^-\pi^0$ decays. These measurements provide the most precise determination of ϕ_{+-} and the best limits on η_{+-0} and η_{000}.

Semileptonic decays. The CPLEAR experiment can simultaneously measure the four decay rates:

Parameter	Symmetry	CPLEAR analysis
$\|\eta_{+-}\|$	CP	$(2.316 \pm 0.039) \times 10^{-3}$
ϕ_{+-}	CP	$43.6^0 \pm 0.6^0$
$\|\eta_{00}\|$	CP	$(2.5 \pm 0.4) \times 10^{-3}$
ϕ_{00}	CP	$42.0^0 \pm 5.9^0$
$Re(\eta_{+-0})$	CP	$(-2 \pm 8) \times 10^{-3}$
$Im(\eta_{+-0})$	CP	$(-2 \pm 9) \times 10^{-3}$
$Re(\eta_{000})$	CP	0.15 ± 0.30
$Im(\eta_{000})$	CP	0.29 ± 0.40
A_{T}	T, CPT	$(6.3 \pm 2.8) \times 10^{-3}$
$Re(\varepsilon)$	T	$(1.63 \pm 0.09) \times 10^{-3}$
$Re(\delta)$	CPT	$(2.0 \pm 2.7) \times 10^{-4}$
Δm		$(530.1 \pm 1.1) \times 10^7 \hbar s^{-1}$
$Re(x)$		$(8.5 \pm 10.2) \times 10^{-3}$
$Im(x + \delta)$		$(0.5 \pm 2.5) \times 10^{-3}$
$Im(\delta)$	CPT	$(-0.4 \pm 1.9) \times 10^{-5}$
$\|M_{22} - M_{11}\|$	CPT	$(0.56 \pm 2.65) \times 10^{-19}$ GeV
α	CPT	$(-0.5 \pm 2.8) \times 10^{-17}$ GeV
β	CPT	$(2.5 \pm 2.3) \times 10^{-19}$ GeV
γ	CPT	$(1.1 \pm 2.5) \times 10^{-21}$ GeV

Table 1. Main results of the CPLEAR analysis. The value of $|M_{22}-M_{11}|$ is obtained assuming $\Gamma_{22} = \Gamma_{11}$.

$$R_{\overline{\mathrm{K}}^0 \to \pi^+ \ell^- \bar{\nu}}\,, \quad R_{\overline{\mathrm{K}}^0 \to \pi^- \ell^+ \nu}\,, \quad R_{\mathrm{K}^0 \to \pi^+ \ell^- \bar{\nu}} \text{ and } R_{\mathrm{K}^0 \to \pi^- \ell^+ \nu}\,.$$

With $f = \overline{\mathrm{K}}^0$ and $\bar{f} = \mathrm{K}^0$ in Eq. (1), forming the asymmetry A_{T}, we directly compare the rate of a neutral kaon produced as a $\overline{\mathrm{K}}^0$ and decaying as a K^0 with the rate of a kaon produced as a K^0 and decaying as a $\overline{\mathrm{K}}^0$. Any asymmetry is evidence of T violation. Similarly, the asymmetry A_{CPT}, which we get from Eq. (1) with $f = \mathrm{K}^0$ and $\bar{f} = \overline{\mathrm{K}}^0$, is zero if CPT is conserved. The fit of A_{T} and A_{CPT} represent the first direct measurements of T violation and CPT invariance. Our results prove experimentally that CP violation is indeed dominated by T violation and compatible with CPT invariance.

The asymmetry

$$A_{\Delta m}(t) = \frac{R_{\Delta S=0}(t) - R_{\Delta S=2}(t)}{R_{\Delta S=0}(t) + R_{\Delta S=2}(t)} \tag{2}$$

where $R_{\Delta S=2}(t) = R_{\overline{\mathrm{K}}^0 \to \pi^- \ell^+ \nu} + R_{\mathrm{K}^0 \to \pi^+ \ell^- \bar{\nu}}$ and $R_{\Delta S=0}(t) = R_{\overline{\mathrm{K}}^0 \to \pi^+ \ell^- \bar{\nu}} + R_{\mathrm{K}^0 \to \pi^- \ell^+ \nu}$, is sensitive to Δm and $Re(x)$ while the asymmetry

$$A_x(t) = \frac{R_{\overline{\mathrm{K}}^0 \to \pi \ell \nu}(t) - R_{\mathrm{K}^0 \to \pi \ell \nu}(t)}{R_{\overline{\mathrm{K}}^0 \to \pi \ell \nu}(t) + R_{\mathrm{K}^0 \to \pi \ell \nu}(t)} \tag{3}$$

is sensitive to $Im(x + \delta)$. The asymmetry (2) yields the independent determination of Δm [6], free of correlations with ϕ_{+-}. The asymmetry (3) gives a limit on the validity of $\Delta S = \Delta Q$ rule which is more than a factor of 10 better than the best value quoted by the PDG [7].

CPLEAR has made a global analysis [8] of this set of measurements, referring to all the main kaon decay modes, also in conjunction with the Bell-Steinberger equation. The results of this analysis are reported in Table (1). We note the most precise values of ϕ_{+-}, $Im(\delta)$ and $Re(\delta)$. The value of $Im(\delta)$ leads to limits on the K^0 and $\overline{K}^0$ mass and width differences.

Test of CPT symmetry and QM. The relation of CPT symmetry to the existence of the 'time arrow' through causality and gravity is fundamental. According to Hawking [9], quantum gravity suggests that quantum field theory should be modified in such a way that pure quantum states evolve into mixed states. This necessarily entails a violation of CPT. Such modification of the quantum-mechanics description of the neutral-kaon system induces three new parameters α, β and γ (see Ref. [10]) describing the loss of quantum coherence in the observed system. By fitting the CPLEAR data on $\pi^+\pi^-$ and $\pi e \nu$ decays [11], we constrain the parameters to values approaching the Planck mass (see Table 1).

Summary. The CPLEAR experiment obtained the most precise values for ϕ_{+-}, Δm, η_{+-0} and η_{000}. In addition, it provides the first direct test of T violation and confirms CPT invariance with high precision. The CPLEAR experiment has also demonstrated the impressive sensitivity of the neutral kaon system to the physics beyond the Standard model.

References

[1] R. Adler et al., CPLEAR Collaboration, Nucl. Instr. Methods A379 (1996) 76.
[2] A. Angelopoulos, CPLEAR Collaboration, Phys. Lett. B 413 (1997) 422.
[3] R. Adler et al., CPLEAR Collaboration, Phys. Lett. B363 (1995) 243.
[4] R. Adler et al., CPLEAR Collaboration, Z. Phys. C70 (1996) 211.
[5] R. Adler et al., CPLEAR Collaboration, Phys. Lett. B407 (1997) 193.
[6] R. Adler et al., CPLEAR Collaboration, Phys. Lett. B363 (1995) 237.
[7] Review of Particle Properties, Phys. Rev. D 54 (1996) 1.
[8] P. Pavlopoulos, CPLEAR Collaboration, Proc. Workshop on K physics, ed. L. Iconomidou-Fayard, Edition Frontières, Gif-sur-Yvette, France (1997) 307.
[9] S. Hawking, Comm. Math. Phys. 87 (1982) 395.
[10] J. Ellis et al., Nucl. Phys. B241 (1984) 381.
[11] R. Adler et al., CPLEAR Collaboration, Phys. Lett. B364 (1995) 239.

Part X

Neutrino Physics

Determination of the Number of Light Neutrinos Species with the OPAL Detector at LEP

Günter Duckeck (Guenter.Duckeck@cern.ch)

Ludwig-Maximilians-Universität, Munich, Germany

Abstract. The determination of the number of light ν species from the Z^0 lineshape with the OPAL detector at LEP is presented. The experimental challenges and the improvements made to fully exploit the high statistics data from the LEP1 running in 1989–1995 are discussed. Combining all four LEP experiments results in $N_\nu = 2.993 \pm 0.011$, compatible with three light ν generations.

1 Introduction

The measurement of the Z^0-lineshape allows a high precision determination of the Z^0-mass and its couplings to leptons and quarks. From this measurement one can derive the number of light neutrino species, N_ν:

$$N_\nu \cdot \Gamma_\nu^{SM} \equiv \Gamma_{\mathrm{inv}} = \Gamma_Z - \Gamma_h - 3 \cdot \Gamma_l$$

where Γ_Z is the total Z^0 width, Γ_h, Γ_l and Γ_{inv} are the measured partial Z^0 decay widths into hadrons, charged lepton pairs and the invisible width, respectively. Γ_ν^{SM} is the expected partial decay width into neutrinos as predicted by the Standard Model. In order to minimize theoretical uncertainties it is favorable to determine the ratio $\Gamma_{\mathrm{inv}}/\Gamma_l$, which is rather insensitive to the values assumed for the mass of the Higgs boson and the top quark.

2 Experimental Challenges

Using typical numerical values one derives the following error propagation formula in terms of the experimental observables:

$$\Delta\left(\frac{\Gamma_{\mathrm{inv}}}{\Gamma_l}\right) \approx \underline{6}\,\frac{\Delta N_l}{N_l} \oplus \underline{21}\,\frac{\Delta N_h}{N_h} \oplus \underline{15}\,\frac{\Delta L}{L}$$

Uncertainties in the selection of hadronic events and the luminosity determination, L, are most important for the measurement. Statistically it is limited by the number of hadrons with $\Delta N_h/N_h = 0.06\,\%$. To fully exploit the data one has to match the systematic uncertainty for the hadron selection and luminosity to this accuracy and reach about $0.2\,\%$ for the leptons.

Much effort went into optimizing the event selections. The dependence on Monte Carlo generators and detector simulation could be largely reduced by using the redundancy of different sub-detectors to identify the final states. A thorough calibration and careful cross-checks maintained the stability of the detector response throughout the years.

For the luminosity the need for such a high accuracy was not anticipated in the original design. The measurement is based on low angle Bhabha–scattering as reference process and depends crucially on the knowledge of the angular region the apparatus is sensitive to. To reach 0.1 % accuracy the inner edge of the detector must be determined to 25 μm. Such a precision could only be reached with a new generation of detectors based on calorimetry with Si-layers. They were proposed, designed and installed in 1991–1993. With this new device the experimental luminosity uncertainty could be reduced below 0.05 %, after elaborate metrology and numerous test-beam measurements. However, there is still a 0.11 % theoretical uncertainty in the luminosity determination. Currently this limits the precision of the measurement.

3 Results

Using all $4 \cdot 10^6$ Z^0 events recorded in 1989–1995, results in $\Gamma_{\rm inv}/\Gamma_l = 5.955 \pm 0.033$. With the Standard Model value for $\Gamma_\nu/\Gamma_l = 1.991 \pm 0.001\,(M_{\rm t}, M_{\rm H})$ it can be translated in a measurement of the number of light neutrino species:

$$N_\nu = 2.991 \pm 0.017\,(exp) \pm 0.002\,(M_{\rm t}, M_{\rm H})$$

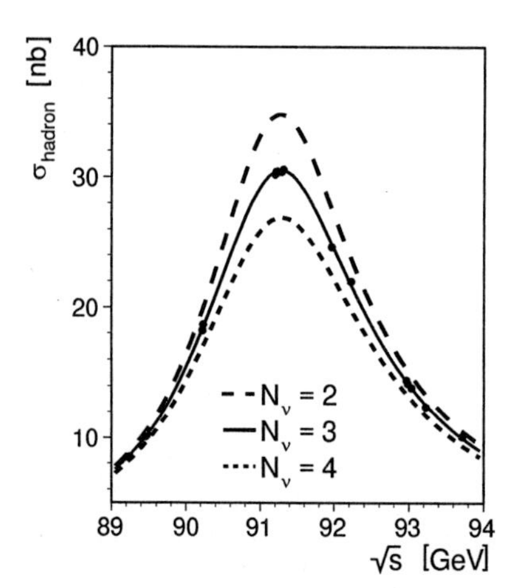

The figure shows the hadronic cross-sections as measured with the OPAL detector. Also shown are three curves corresponding to the Standard Model prediction for $N_\nu = 2$, 3 and 4.

Combining all four LEP experiments [1] yields $\Gamma_{\rm inv}/\Gamma_l = 5.960 \pm 0.022$ and $N_\nu = 2.993 \pm 0.011\,(exp) \pm 0.002\,(M_{\rm t}, M_{\rm H})$.

A precise determination of $\Gamma_{\rm inv}$ also sets significant constraints for extensions of the Standard Model. For example, in a scenario with a Higgsino-like lightest supersymmetric particle [2] one gets a mass limit for the Higgsino of 37 GeV@95 % $C.L.$ for $\tan\beta = 1.5$.

I would like to thank M.Mannelli, D.Strom and C.Paus for their help in preparing this contribution.

References

[1] G. Quast, EPS presentation 709, these proceedings.
[2] G.L. Kane, J.D. Wells, Phys. Rev. Lett. 76 (1996) 4458.

A Limit on the τ Neutrino Mass. DELPHI Results in the Channel $\tau \to 3\pi^{\pm}\nu_\tau$

Lars Bugge (lars.bugge@fys.uio.no),
representing the DELPHI collaboration.

University of Oslo, Norway

Abstract. Fits to the invariant mass squared spectrum of the three pion system in $\tau \to 3\pi^{\pm}\nu_\tau$ decays are used to constrain the tau neutrino mass. It is shown that the results are dependent on the models used to describe the spectra theoretically. The data favour the inclusion of a resonance at 1700 MeV with a width of 300 MeV in the description of the tau decay.

An observation of non-zero neutrino masses could have far-reaching consequences for fundamental cosmological problems (dark matter). It would also imply an extension of the minimal standard model with righthanded neutrinos and a CKM mixing matrix in the lepton sector. In the present work parameters of theories describing the tau decay to three charged pions plus a neutrino are fitted to data. Fits with the tau neutrino mass as a free parameter are used to set upper limits on the tau neutrino mass.

The DELPHI detector at LEP is described in detail in ref. [1]. Tracks with all the barrel tracking detectors contributing have a resolution $\sigma(1/p_t) = 8 \times 10^{-4}(\text{GeV})^{-1}$, leading to a mass resolution of the 3 charged pion system of $\sigma(m_{3\pi})$ ~30 MeV. The detector resolution was parametrised by three Gaussians. This parametrisation was checked on real data reconstructing the D^0 mass in the channel $D^* \to D^0\pi \to K\pi\pi$ with p_{D^0} >30 GeV. The data, collected from 1992 through 1995, corresponded to 4 million hadronic Z^0 decays, or about 200k $\tau^+\tau^-$ pairs. Tau pairs were selected in the 1 vs 3 and 3 vs 3 topologies. For details of the analysis, see ref. [2]. A total of 7180 event candidates survived all cuts, corresponding to an efficiency inside $|\cos\theta| <0.73$ of 36% with a purity of 80%. The background was dominated by $\tau \to 3\pi^{\pm}\pi^0\nu_\tau$ (15%) and was predominantly at $s <2$ GeV^2, where s is the invariant mass squared of the 3 charged pion system.

Three models were used in the study of the hadronic structure in $\tau \to 3\pi^{\pm}\nu_\tau$ decays: the Kühn and Santamaria model (KS), the Isgur, Morningstar, and Reader model (IMR), and the Feindt model (MF). References and discussions of these models can be found in ref. [2]. As a first step the KS and IMR models were fitted to the data with a fixed zero neutrino mass leaving the a_1 mass and width as free parameters. The models did not describe the data well for squared invariant masses $s >2.3$ GeV^2. This was particularly evident in the Dalitz plot projection onto $\sqrt{s_1}$, the higher mass of the two oppositely charged pion combinations. The fits to both models gave a_1 widths of the order 500 MeV, consistent with earlier measurements using tau decays [4]. The inadequacy of the models to describe the high s data might be interpreted as evidence for the existence of a higher mass resonance which the tau can decay through if this in turn decays via a state at $\geq$1250 MeV. An excellent candidate for this is the proposed [3] 1^{++} state a_1', possibly observed by the VES collaboration [5]. The MF model was used in fits based on the KS and IMR models, but now including an a_1' state with a mass of 1700 MeV and a width of 300 MeV, allowed to decay with equal probabilities through $\rho'\pi$ (S-wave) and $\rho\pi$ (D-wave) as suggested by the VES observation. The a_1 parameters as well as the relative a_1' contribution were left as free parameters. The fitted a_1 width turned out significantly smaller, less than 400 MeV for both models. The models described the Dalitz projection better, and gave consistent fits for the relative a_1' contribution of $\sim$2.3%.

Fits to the KS and IMR models as described above, but with the tau neutrino mass as an additional free parameter, were performed. The results were compatible with zero, with upper limits at 95% CL of 33 and 31 MeV for the KS and IMR models, respectively, systematics included. A similar fit was made to the MF model with the a_1' contribution and the tau neutrino mass as free parameters. The result was m_{ν_τ} =37$\pm$12 MeV (systematic errors included). The corresponding 95% CL upper limit was 62 MeV. We regard this as a conservative estimate of the upper limit. Details on the fits can be found in ref. [6].

Conclusions. We have demonstrated that neutrino mass fits in the $\tau \rightarrow 3\pi^{\pm}\nu_\tau$ channel are model dependent. The KS and IMR models fail to describe the data for $s >$2.3 GeV^2. The inclusion of an a_1' of 1700 MeV mass and 300 MeV width in agreement with the VES observations [5], results in a better agreement with the data, gives smaller fitted a_1 widths, in better agreement with hadronic production experiments [4], and leads to a conservative 95% CL upper limit on m_{ν_τ} of 62 MeV.

References

[1] P. Aarnio et al., DELPHI Coll., Nucl.Instr.Meth.**A303**(1991), 233-276
P. Abreu et al., DELPHI Coll., Nucl.Instr.Meth.**A378**(1996), 57-100

[2] R. McNulty and A. Galloni, Paper HEP'97 #319, subm.to this conf.

[3] R. Kokoski and N. Isgur, Phys.Rev.**D35**(1987), 907-933

[4] R. M. Barnett et al. (Particle Data Group), Phys.Rev.**D54**(1996), 345

[5] D. V. Amelin et al., VES collaboration, Phys.Lett.**B356**(1995), 595-600

[6] A. Galloni and R. McNulty, Paper HEP'97 #320, subm.to this conf.

A Limit on the τ Neutrino Mass from the ALEPH Experiment

Maria Girone (Maria.Girone@cern.ch)

Imperial College, London, United Kingdom

Abstract. A bound on the tau neutrino mass is established using the data collected from 1991 to 1995 at $\sqrt{s} \simeq m_Z$ with the ALEPH detector. An upper limit of 18.2 MeV/c^2 at 95% confidence level is derived by fitting the distribution of visible energy vs invariant mass in $\tau^- \to 2\pi^-\ \pi^+\ \nu_\tau$ and $\tau^- \to 3\pi^-\ 2\pi^+\ (\pi^0)\ \nu_\tau$ decays.

1 The method

In ALEPH, a bound on the neutrino mass is obtained from the study of the $\tau^- \to 2\pi^-\ \pi^+\ \nu_\tau$, $\tau^- \to 3\pi^-\ 2\pi^+\ \nu_\tau$ and $\tau^- \to 3\pi^-\ 2\pi^+\ \pi^0\ \nu_\tau$ decay modes[1] of the tau lepton, using a technique based on a two dimensional likelihood fit in the variables invariant mass, m_h, and energy, E_h, of the hadronic systems [1]. The limit on the mass is derived from a maximum likelihood function giving the probability density of obtaining the observed distribution in the plane (m_h, E_h). For a given event i, the probability density function $P_i(m_\nu)$ takes the form:

$$\mathcal{P}(m_\nu) = \frac{1}{\Gamma} \cdot \frac{d^2\Gamma}{dE_h dm_h} \otimes \mathcal{G}(E_{beam}, E_\tau) \otimes \mathcal{R}(m_h, E_h, \rho, \sigma_{m_h}, \sigma_{E_h}, \ldots) \otimes \varepsilon(m_h, E_h)$$

where $\frac{1}{\Gamma} \cdot \frac{d^2\Gamma}{dE_h dm_h}$ is the theoretical distribution of the given decay mode, $\mathcal{G}(E_{beam}, E_\tau)$ is the radiation kernel and $\mathcal{R}(m_h, E_h, \sigma_{m_h}, \sigma_{E_h}, \ldots)$ and $\varepsilon(m_h, E_h)$ are the detector resolution and the selection efficiency of each mode, respectively.

The exact functional form of the spectral functions entering the expression $\frac{1}{\Gamma} \cdot \frac{d^2\Gamma}{dE_h dm_h}$ is not predicted by theory. Nevertheless, since the spectral functions are expected to vary slowly with q^2 in the region close to the kinematic boundary, the uncertainty in their form plays only a minor role in the determination of the bound on m_ν. The decay $\tau^- \to 2\pi^-\ \pi^+\ \nu_\tau$ is described using the model of Kühn and Santamaria (KS) [2], inspired by the asymptotic limit of chiral theory. For the $\tau^- \to 3\pi^-\ 2\pi^+\ (\pi^0)\ \nu_\tau$ mode, there are very few studies of the spectral functions, mainly because the number of observed candidates is very small. Experimentally, it is seen that the invariant hadronic mass spectrum peaks at high values of q^2 and seems unlikely to be dominated by a single resonance.

---

[1] The inclusion of charge conjugate modes is always implied throughout this paper.

2 Data selection and background

The data selection aims at introducing the smallest possible bias towards lower values for the determination of the upper limit. Since at LEP the separation of $\tau^+\tau^-$ events from other processes is relatively easy, the main concern is the rejection of background from misidentified tau decays. The topology of the background which lowers the neutrino mass limit is the one with a true final state multiplicity lower than the observed one, because in this case the reconstructed values of the hadronic mass and energy are systematically higher than the true ones. The event selection has been designed to reduce such contamination to a negligible level, while it tolerates a moderate background from tau decays with multiplicities higher than the observed one.

The analyses presented here are based on the data collected by ALEPH from 1991 to 1995 in the proximity of the Z resonance. The ALEPH detector and its performance are described in detail in [3].

The final selection efficiencies for the $\tau^- \to 2\pi^- \pi^+ \nu_\tau$, $\tau^- \to 3\pi^- 2\pi^+ \nu_\tau$ and $\tau^- \to 3\pi^- 2\pi^+ \pi^0 \nu_\tau$ channels are 49%, 24.7% and 7.0%, respectively. The lower efficiency of the last mode is caused by stringent cuts on π^0 reconstruction which are needed to suppress the cross-channel contamination from $\tau^- \to 3\pi^- 2\pi^+ \nu_\tau$ decays.

The contamination of this selection from $q\bar{q}$ events amounts to 0.3% for the $\tau^- \to 2\pi^- \pi^+ \nu_\tau$, and 0.1% for the $\tau^- \to 3\pi^- 2\pi^+ (\pi^0) \nu_\tau$ decay modes. The background from tau decays amounts to 6.7%, 7.6% and 0.6% for the three channels, respectively.

A total of 2939 $\tau^- \to 2\pi^- \pi^+ \nu_\tau$, and 52 (3) $\tau^- \to 3\pi^- 2\pi^+ (\pi^0) \nu_\tau$ candidates are selected in the data, in good agreement with the expectations. The distributions in the upper part of the (E_h, m_h) plane are shown in Fig. 1. Due to the large number of candidates, the selection and the fit in the $\tau^- \to 2\pi^- \pi^+ \nu_\tau$ channel are restricted to the region of the (E_h, m_h) plane $0.89 < E_h/E_{beam} < 1.07$ and $0.76 < m_h < 1.83$ GeV$/c^2$. The fitted region is shown in Fig. 1. The size of region has been chosen large enough to make the limit on the tau neutrino mass insensitive to variations of the region boundaries.

3 Results and systematic effects

The fits to the 2939 $\tau^- \to 2\pi^- \pi^+ \nu_\tau$ and to the 55 $\tau^- \to 3\pi^- 2\pi^+ (\pi^0) \nu_\tau$ events give 95% CL upper limits on the tau neutrino mass of 22.3 MeV$/c^2$ and 21.5 MeV$/c^2$, respectively. The 95% confidence level is taken as the point where the logarithm of the likelihood is 1.92 lower than its maximum.

Several sources of systematic errors have been considered. For each source a new fit was performed, having changed in the likelihood the appropriate quantity by one standard deviation. The difference between the value of the 95% CL upper limit on m_{ν_τ} obtained from the original fit and the one with the modified likelihood has been taken as the systematic error due to that

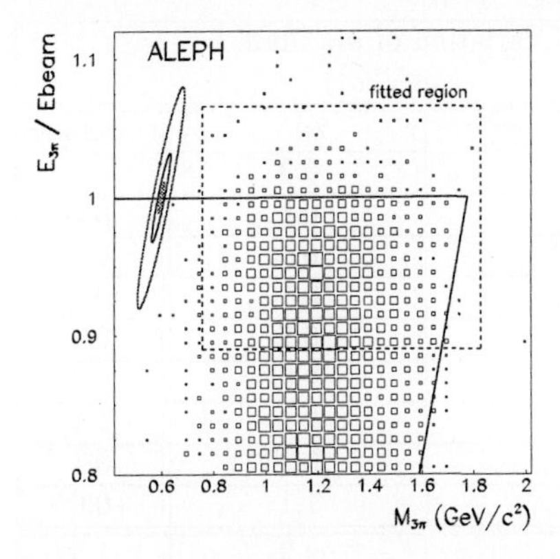

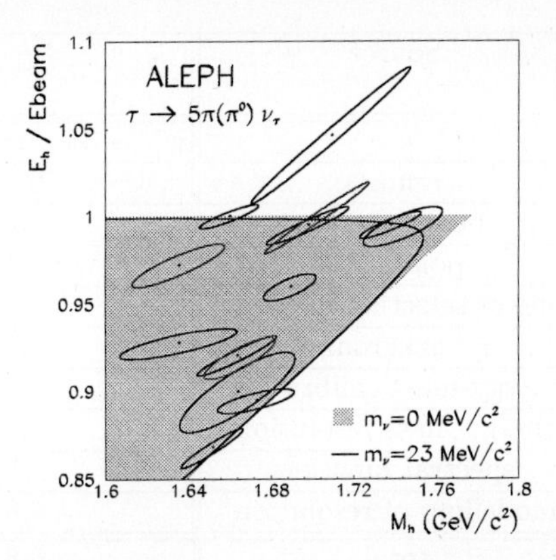

Fig. 1. Left: Distribution in the the (m_h, E_h) plane for $\tau^- \to 2\pi^-\ \pi^+\ \nu_\tau$ candidates in the data (the three ellipses show the typical size of the resolution. Right: Distribution in the (m_h, E_h) plane for $\tau^- \to 3\pi^-\ 2\pi^+\ (\pi^0)\ \nu_\tau$ candidates in the data.

source. All the variations were then summed in quadrature to give the global systematic error, which was added linearly to the result of the original fit.

The sources of systematics considered belong to four major categories: tau properties, such as tau mass, energy and polarisation; detector effects, such as absolute momentum calibration and resolution; selection efficiency and background contamination; and tau decay modelling. The corresponding variations of the neutrino mass limit are reported in Table 1. The variations for both three- and five-prong final states are separately summed in quadrature to obtain the two total systematic errors of 4.2 and 0.8 MeV/c^2, respectively. These errors are summed linearly to the measured mass limits to obtain 95% CL upper limits of 25.7 MeV/c^2 and 23.1 MeV/c^2 for the three-prong and five-prong modes, respectively. Interestingly, the $\tau^- \to 2\pi^-\ \pi^+\ \nu_\tau$ mode is competitive with the $\tau^- \to 3\pi^-\ 2\pi^+\ (\pi^0)\ \nu_\tau$ mode thanks to the larger number of candidates, which compensate for the less favourable distribution in the (E_h, m_h) plane. The two limits are complementary since the limit derived from the $\tau^- \to 2\pi^-\ \pi^+\ \nu_\tau$ mode is more sensitive to the energy distribution and the others to the mass distribution of the hadronic system.

The combined upper limit has been determined from a new likelihood $\mathcal{L}^{comb}$, constructed as the product of the individual $\tau^- \to 2\pi^-\ \pi^+\ \nu_\tau$ and $\tau^- \to 3\pi^-\ 2\pi^+\ (\pi^0)\ \nu_\tau$ likelihoods $\mathcal{L}^{3\pi}$ and $\mathcal{L}^{5(6)\pi}$. This likelihood limits m_{ν_τ} below 16.6 MeV/c^2 at 95% CL. Table 1 summarises the variation of the two limits and the variation of the combined limit, for each source of error. In this way, a total systematic error of 1.6 MeV/c^2 and a final 95% CL limit of 18.2 MeV/c^2 were obtained.

Recently the DELPHI Collaboration has suggested the existence of a hitherto unseen decay mode of the tau lepton in a radial excitation of the a_1 [5].

Source	Variation of m_ν limit (MeV/c^2)		
	$\tau^- \to 3\pi^-\ 2\pi^+\ (\pi^0)\ \nu_\tau$	$\tau^- \to 2\pi^-\ \pi^+\ \nu_\tau$	combined
τ mass	0.2	0.3	0.2
beam energy	< 0.1	0.1	0.2
τ polarisation	< 0.1	0.1	0.1
slope of selection efficiency	< 0.1	0.1	0.1
τ background	0.3	0.1	0.2
energy-mass calibration	0.3	2.6	0.9
energy-mass resolution	0.2	3.1	1.1
spectral function	< 0.1	0.3	0.1
modelling of resolution	0.6	1.1	0.6
total	0.8	4.2	1.6

Table 1. Systematic variation of the 95% CL upper limit on m_ν (in MeV/c^2) for the individual and combined $\tau^- \to 3\pi^-\ 2\pi^+\ (\pi^0)\ \nu_\tau$ and $\tau^- \to 2\pi^-\ \pi^+\ \nu_\tau$ likelihoods.

In that analysis this a' resonance is assigned a mass of 1700 MeV/c^2 and a width of 300 MeV. Its contribution is fitted to be (2.3 ± 0.6) %. If 2.5 % of this resonance is introduced in the fit of the $\tau^- \to 2\pi^-\ \pi^+\ \nu_\tau$ mode, the agreement between the model and the data deteriorates giving a $\chi^2/n.d.f.$ of 1077/999 with respect to the value of 1059/999 obtained with the KS spectrum alone. If this resonance were considered in the fit, the limit from the $\tau^- \to 2\pi^-\ \pi^+\ \nu_\tau$ sample would increase by 6 MeV/c^2, and the combined limit would increase from 18.2 to 19.2 MeV/c^2.

4 Conclusions

ALEPH has used the modes $\tau^- \to 2\pi^-\ \pi^+\ \nu_\tau$ and $\tau^- \to 3\pi^-\ 2\pi^+\ (\pi^0)\ \nu_\tau$ to bound the tau neutrino mass by fitting the distribution of events in the (m_h, E_h) plane. An upper limit of 18.2 MeV/c^2 on the tau neutrino mass is obtained at 95% confidence level.

References

[1] ALEPH Collaboration, Phys. Lett. B 349 (1995) 585; ALEPH Collaboration, CERN-PPE/97-138.
[2] J. H. Kühn *et al*, Z. Phys. C 48 (1990) 445.
[3] ALEPH Collaboration, Nucl. Instr. Meth. A 294 (1990) 121; 303 (1991) 393; ALEPH Collaboration, Nucl. Instr. Meth. A 360 (1995) 481.
[4] ALEPH Collaboration, Z. Phys. C 62 (1994) 539.
[5] DELPHI Collaboration, contributed papers 319(PA 7 PL 6) and 320(PA 7, 10 PL 6, 11) to this conference.

Status of the Mainz Neutrino Mass Experiment

H. Barth[a], A. Bleile[a], J. Bonn[a], L. Bornschein[a], B. Degen[a], L. Fleischmann[a], O. Kazachenko[b], A. Kovalik[c], E.W. Otten[a], M. Przyrembel[a], Ch. Weinheimer[a]

[a] Institute for Physics, Johannes Gutenberg University Mainz, Germany
[b] Institute for Nuclear Research, Russian Academy of Sciences, Troitsk/Russia
[c] Joint Institute for Nuclear Research, Dubna/Russia

presented by Jochen Bonn (Jochen.Bonn@uni-mainz.de)

Abstract. The present status of the Mainz tritium β decay experiment is given. Recent improvements of the setup and first tritium data are presented.

1 Introduction

Whether neutrinos have a non-vanishing mass or not is still one of the most interesting questions of particle physics and cosmology. In contrast to neutrino oscillations or neutrinoless double β decay the determination of the electron antineutrino mass from the tritium β decay spectrum does not depend on assumptions on neutrino properties. Its sensitivity is reaching a few eV, which is most relevant for a possible dark matter contribution and in context with the LSND experiment, which claims the observation of neutrino oscillations, for which at least one of the mass eigenstates should be as heavy as 1 eV.

2 Neutrino mass measurement from tritium β decay

A non-zero neutrino mass changes the phase space of the 3-body decay such, that a bend down below the endpoint occurs in the electron energy spectrum. This signature of a non-zero mass does not vanish further below the endpoint but turns into a constant offset fading away relative to the quadratically rising β spectrum. Additionally systematic effects make the spectrum further below the endpoint more complex. To reduce the influence of systematic uncertainties it is clearly the task for a tritium β decay experiment to determine the electron neutrino mass from the shape of the β spectrum at its very end requiring a high signal to background ratio and a resolution of a few eV.

3 The Mainz Neutrino Mass experiment

At Mainz the so-called "solenoid retarding spectrometer" was set up to investigate the tritium β spectrum close to its endpoint following the requirements mentioned above. [1]. The tritium source is a film of T_2 molecules quench-condensed on a graphite substrate at temperatures between 2.8 and 4.2 K.

With this setup the tritium β spectrum was investigated in 1994 [2]. The very high signal to background ratio combined with the high energy resolution allowed to determine the neutrino mass (more exactly m_ν^2) from a reasonably small interval of the last 140 eV below the endpoint resulting in:

$$m_\nu^2 = -22 \pm 17_{\text{stat}} \pm 14_{\text{sys}} \ (eV/c^2)^2 \ , \tag{1}$$

for which a limit of $m_\nu < 5.6 \ eV/c^2$ (95% C.L.) can be calculated [2].

Looking to the experimental spectrum further below this interval a clear excess was observed compared to the extrapolation of the fit over the last 140 eV. If this would be common it could be the reason for negative values of m_ν^2 reported earlier by other experiments. An input to all T_2 decay experiments is the final states distribution of the known only from theory. After being discussed in this respect the final states distribution was carefully checked confirming the old calculations.

The Mainz group has carried out investigations of sources of systematic uncertainties of the experiment like backscattering and energy loss. No significant deviation was found to explain the excess count rate further below the endpoint except for one open question: Has the T_2 film undergone a roughening transition resulting in a harder energy loss spectrum, which could have produced the observed spectrum?

Detailed investigations using conversion electron spectroscopy and scattered light techniques yielded the following results: The roughening transition cannot be avoided but its speed can drastically be slowed down by using lower temperatures. Extrapolating from the stable hydrogen isotopes a T_2 film at 1.7 K should have a time constant of thousands of years if the β decay does not matter. A first test with a T_2 film at 1.7 K resulted in no observation of any roughening transition within 4 days.

This gives a clear recipe for the future, but unfortunately no clear statement for the past.

4 The upgraded Mainz setup

To look for a non-zero neutrino mass in the range of a few eV is getting more and more important for reasons mentioned above. Additionally the excess count rate further below the endpoint observed at Mainz and moreover the monoenergetic anomaly just below the endpoint reported by the Troitsk experiment[3] clearly need further independent checks.

To fulfil these expectations the following problems had to be solved:

- to avoid the T_2 film roughening transition
- to increase the signal to background ratio
- to make long term runs feasible

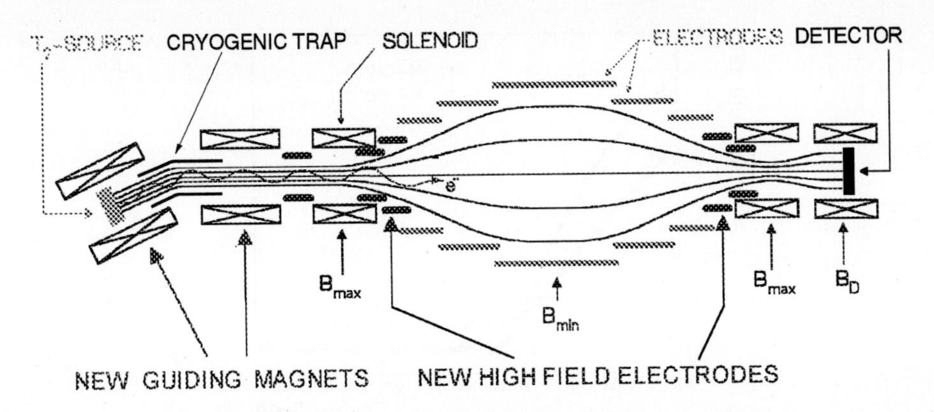

Fig. 1. The improved and enhanced Mainz setup

For these reasons the Mainz experiment has improved its setup substantially (see figure 1) by the following items:

- A new, automatically controlled source cryostat was installed to slow down the T_2 film roughening transition to a negligible speed by working at temperatures down to 1.7 K.
- A new doublet of superconducting solenoids, rotated by 20° to each other, was installed. The source is placed in the left most solenoid. Electrons are guided into the spectrometer, whereas tritium from the source is prohibited from contaminating the spectrometer and residual gas molecules of the spectrometer from condensing onto the T_2 film.
- The electrodes were redesigned to lower spectrometer background.
- A control system was set up to run the experiment fully automatically.

5 First measurement with the improved Mainz setup

Up to this conference calibration measurements, a first test run and 4 weeks of data taking were carried out with the improved Mainz setup.

- The full setup, especially the new cryostat running at a temperature of 2.5 K, was working stable over 4 weeks. This behaviour allowed to do the whole run fully automatically, human intervention was needed only for filling of LHe and LN_2.
- The thickness of the T_2 film of about 280 monolayer was measured before and after the runs with ellipsometry. Within the precision of a few monolayers no signal of condensing residual molecules were found, demonstrating the performance of the cryotrap.
- Although the T_2 film was about 6 times thicker the background was even lower than in 1994 (see figure 2) resulting in an about a factor 10 higher signal to background ratio.

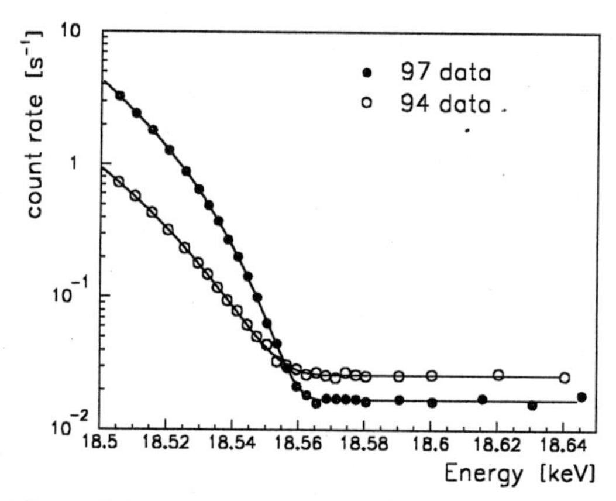

Fig. 2. Mainz tritium β spectra close to the endpoint

- Due to a better alignment of the whole system the spectrometer could run at a higher energy resolution of 4.4 eV compared to 6.3 eV in 94.
- A first fit to the raw data gave a statistical uncertainty on $m_\nu^2 c^4$ of ± 6 eV^2 (94: ± 35 eV^2) and of ± 8 eV^2 (94: ± 63 eV^2) for the last 75 eV or 50 eV of the spectrum respectively. More data taking is planed for this autumn.

6 Conclusions

The Mainz experiment has improved its setup to run with very high signal to background ratio over longer periods fully automatically. The success of this upgrade was demonstrated by a first 4 week run, showing a much better statistical accuracy than in all previous runs. The ongoing data analysis will very likely result in a better sensitivity on the neutrino mass and in statements on the checks of the two anomalies reported by the Troitsk or by the Mainz experiment earlier.

Acknowledgement

This work was sponsored by the Deutsche Forschungsgemeinschaft. A.K additionally thanks the Grant Agency of the Czech Republic for financial support (contr. no. 202/96/0552).

References

[1] A. Picard *et al.*, Nucl. Inst. Meth. B63 (1992) 345
[2] J. Bonn, Proc. Neutrino 96, World Scientific/Singapure
[3] A.I.Beleshev *et al.*,Physics Letters B350 (1995) 263

An Explanation of Anomalies in Electron Energy Spectrum from Tritium Decay

Jacek Ciborowski[1] (cib@fuw.edu.pl)
Jakub Rembieliński[2](jaremb@mvii.uni.lodz.pl)

[1]Department of Physics, University of Warsaw, Poland
[2]Department of Physics, University of Łódź, Poland

Abstract. Anomalies observed in the electron energy spectrum from tritium decay can be explained if electron (anti)neutrino is a tachyon. This hypothesis offers also a natural explanation of the V–A structure of the weak leptonic current.

In recent tritium decay experiments [1]-[3] the fitted values of neutrino mass squared come out negative. This is caused by excess of counts near the end point which has the form of a 'bump' in a linearized electron energy spectrum. There is no explanation of this effect on the grounds of conventional physics [4]. In order to explain this we investigate consequences of a hypothesis that neutrino is a tachyon. A unitary (causal) theory of tachyons, recently proposed by Rembieliński [5], is the basis for calculations presented in this paper. Parameter κ stands for the *tachyonic mass* in the relation: $E^2 - \overline{p}^2 = -\kappa^2$.

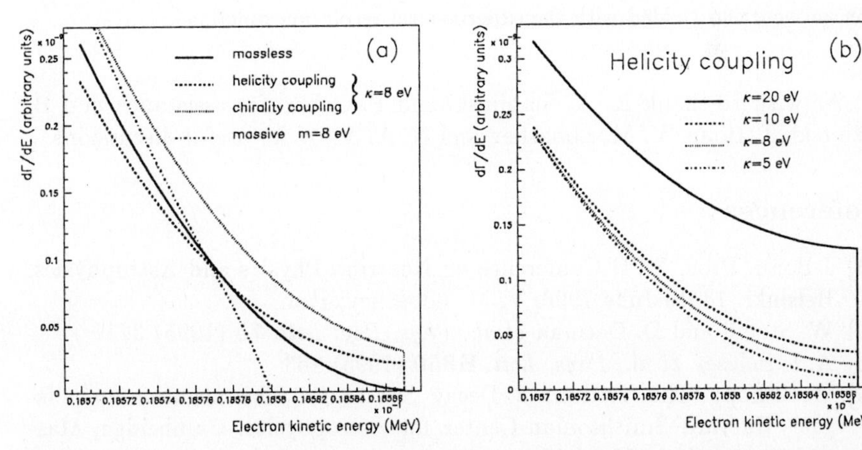

Fig. 1. (a) Differential electron energy spectra in the vicinity of the end point, for tritium decay with: a tachyonic antineutrino of mass $\kappa = 8\ eV$, massless neutrino and massive neutrino of mass $m = 8\ eV$; (b) as above for the helicity coupling and a range of masses, κ.

In order to calculate the decay amplitude we choose the leptonic current in the form (helicity coupling): $J_l^\mu \sim \bar{u}_e \gamma^\mu w$ where the tachyonic neutrino field w is an eigenvector of the helicity operator (chirality coupling differs by an additional term $\frac{1}{2}(1-\gamma^5)$). Therefore *helicity coupling offers a natural explanation of the V–A structure of the weak leptonic current.* Differential electron energy spectra are shown in fig. 1. In both tachyonic cases the almost step-like termination at E_{max} leads to the observed excess of events close to the end point. Comparing with the measurement of Troitsk (fig. 2) we observe a striking similarity of the predicted behaviour with the bump-like structure in the data. For more details see ref. [6] and references therein.

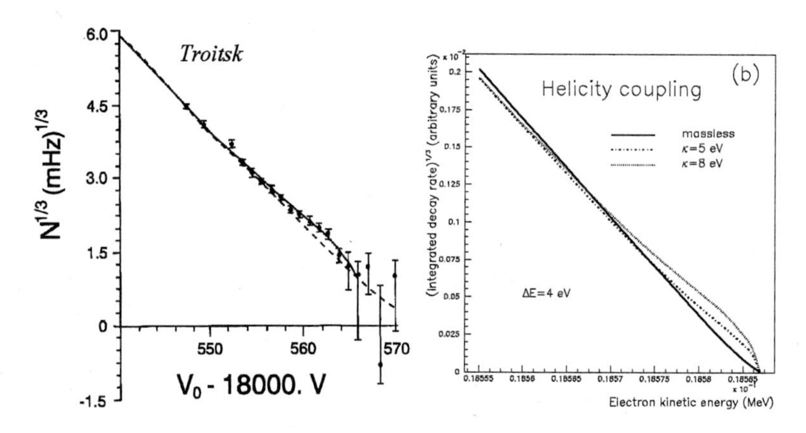

Fig. 2. (a) Integral linearized (cube root) electron energy spectrum obtained in Troitsk; (b) corresponding theoretical prediction for a decay with a tachyonic electron antineutrino (folded with the experimental resolution function).

We wish to thank K. A. Smoliński and P. Caban for assistance and B. Jeziorski, J. Bonn, V. M. Lobashev and N. A. Titov for useful discussions.

References

[1] J.Bonn, Proc. XVII Conference on Neutrino Physics and Astrophysics, Helsinki, 13–19 June 1996; V. M. Lobashev, *ibid.*
[2] W. Stoeffl and D. Decman *et al.*, *Phys. Rev. Lett.* **75** (1995) 3237
[3] A. I. Belesev *et al.*, *Phys. Lett.* **B350** (1995) 263
[4] Workshop: 'The Tritium β Decay Spectrum: The negative $m^2_{\nu_e}$ Issue', Harvard–Smithsonian Center for Astrophysics, Cambridge, Massachusetts, USA, 20–21 May 1996
[5] J. Rembieliński, *Int. J. Mod. Phys.* **A12** (1997) 1677
[6] J. Ciborowski and J. Rembieliński, this conference, paper #744; to be published

Neutrino Masses and Mixing from Neutrino Oscillation Experiments

S.M. Bilenky[1], C. Giunti[2] and W. Grimus[3]

[1] Joint Institute for Nuclear Research, Dubna, Russia, and Institute for Advanced Study, Princeton, N.J. 08540

[2] INFN, Sezione di Torino, and Dipartimento di Fisica Teorica, Università di Torino, Via P. Giuria 1, I–10125 Torino, Italy

[3] Institute for Theoretical Physics, University of Vienna, Boltzmanngasse 5, A–1090 Vienna, Austria

Abstract. We discuss which information on neutrino masses and mixing can be obtained from the results of all neutrino oscillation experiments in the cases of three and four massive neutrinos. We show that in the three-neutrino case the neutrino oscillation data are not compatible with a hierarchy of couplings. In the case of four neutrinos, a hierarchy of masses is disfavored by the data and only two schemes with two pairs of neutrinos with close masses separated by a gap of the order of 1 eV can accommodate the results of all experiments.

The search for neutrino oscillations is one of the most active branches of today's high-energy physics. From the LEP measurements of the invisible width of the Z-boson we know that there are three light active flavor neutrinos: ν_e, ν_μ and ν_τ. In general, flavor neutrinos are not mass eigenstates and the left-handed flavor neutrino fields $\nu_{\alpha L}$ are superpositions of the left-handed components ν_{kL} of the fields of neutrinos with a definite mass ($k = 1, 2, 3, \ldots, n$): $\nu_{\alpha L} = \sum_{k=1}^{n} U_{\alpha k}\, \nu_{kL}$, where U is a unitary mixing matrix. The number n of massive neutrinos can be three or more, without any experimental upper limit. If $n > 3$, there are $n-3$ sterile flavor neutrino fields, i.e., fields of neutrinos which do not take part in standard weak interactions; in this case $\nu_\alpha = \nu_e, \nu_\mu, \nu_\tau, \nu_{s_1}, \nu_{s_2}, \ldots, \nu_{s_{n-3}}$. Neutrino oscillations is a direct consequence of neutrino mixing; the probability of $\nu_\alpha \to \nu_\beta$ transitions is given by (see [1])

$$P_{\nu_\alpha \to \nu_\beta} = \left| \sum_{k=1}^{n} U_{\beta k} U_{\alpha k}^{*} \exp\left(-i \frac{\Delta m_{k1}^2 L}{2E} \right) \right|^2 , \tag{1}$$

where L is the distance between the neutrino source and detector, E is the neutrino energy and $\Delta m_{kj}^2 \equiv m_k^2 - m_j^2$.

In this report we discuss which information on the neutrino mass spectrum and mixing parameters can be obtained from the results of neutrino oscillation experiments. Many short-baseline (SBL) neutrino oscillation experiments with reactor and accelerator neutrinos did not find any evidence of neutrino oscillations. Their results can be used in order to constrain the allowed values of the neutrino masses and of the elements of the mixing matrix. In our analysis we use the most stringent exclusion plots obtained in the $\bar{\nu}_e \to \bar{\nu}_e$ channel by the Bugey experiment [2], in the $\nu_\mu \to \nu_\mu$ channel by the CDHS and CCFR experiments [3], and in the $\overset{(-)}{\nu}_\mu \to \overset{(-)}{\nu}_e$ channel by the BNL E734, BNL E776 and CCFR [4] experiments.

There are three experimental indications in favor of neutrino oscillations that come from the anomalies observed by the solar neutrino experiments [5], the atmospheric neutrino experiments [6] and the LSND experiment [7]. The solar neutrino deficit can be explained with oscillations of solar ν_e's into other states and indicates a mass-squared difference of the order of $10^{-5}\,\mathrm{eV}^2$ in the case of resonant MSW transitions or $10^{-10}\,\mathrm{eV}^2$ in the case of vacuum oscillations. The atmospheric neutrino anomaly can be explained by $\overset{(-)}{\nu}_\mu \to \overset{(-)}{\nu}_x$ oscillations ($x \neq \mu$) with a mass-squared difference of the order of $10^{-2}\,\mathrm{eV}^2$. Finally, the LSND experiment found indications in favor of $\bar{\nu}_\mu \to \bar{\nu}_e$ oscillations with a mass-squared difference of the order of $1\,\mathrm{eV}^2$.

Hence, three different scales of mass-squared difference are needed in order to explain the three indications in favor of neutrino oscillations. This means that the number of massive neutrinos must be bigger than three. In the following we consider the simplest possibility, i.e. the existence of four massive neutrinos ($n = 4$). In this case, besides the three light flavor neutrinos ν_e, ν_μ, ν_τ, there is a light sterile neutrino ν_s.

However, before considering the case of four neutrinos, we discuss the minimal possibility of the existence of only three massive neutrinos ($n = 3$). In this case one of the experimental anomalies mentioned above cannot be explained with neutrino oscillations (we choose to disregard the atmospheric neutrino anomaly).

In both cases of three and four massive neutrinos the oscillations in the LSND experiment imply that the largest mass-squared difference $\Delta m^2_{n1} \equiv m^2_n - m^2_1$ is relevant for SBL oscillations, whereas the other mass-squared differences are much smaller. Hence, the mass spectrum must be composed of two groups of massive neutrinos with close masses ($\nu_1, \ldots, \nu_{r-1}$ and $\nu_r, \ldots, \nu_n$) separated by a mass difference in the eV range ($m_1 < \ldots < m_{r-1} \ll m_r < \ldots < m_n$) and in SBL experiments we have $\frac{\Delta m^2_{n1} L}{2E} \gtrsim 1$, $\frac{\Delta m^2_{k1} L}{2E} \ll 1$ for $k < r$, $\frac{\Delta m^2_{nk} L}{2E} \ll 1$ for $k \geq r$. The formula (1) written as

$$P_{\nu_\alpha \to \nu_\beta} = \left| \sum_{k=1}^{r-1} U_{\beta k} U^*_{\alpha k} e^{-i \frac{\Delta m^2_{k1} L}{2E}} + e^{-i \frac{\Delta m^2_{n1} L}{2E}} \sum_{k=r}^{n} U_{\beta k} U^*_{\alpha k} e^{i \frac{\Delta m^2_{nk} L}{2E}} \right|^2 \tag{2}$$

leads to the following expression for the transition ($\beta \neq \alpha$) and survival ($\beta = \alpha$) probabilities of neutrinos and anti-neutrinos in SBL experiments:

$$P^{(\mathrm{SBL})}_{\overset{(-)}{\nu}_\alpha \to \overset{(-)}{\nu}_\beta} = A_{\alpha;\beta} \sin^2 \frac{\Delta m^2 L}{4E}, \qquad P^{(\mathrm{SBL})}_{\overset{(-)}{\nu}_\alpha \to \overset{(-)}{\nu}_\alpha} = 1 - B_{\alpha;\alpha} \sin^2 \frac{\Delta m^2 L}{4E}, \tag{3}$$

with $\Delta m^2 \equiv \Delta m^2_{n1}$ and the oscillation amplitudes

$$A_{\alpha;\beta} = 4\left|\sum_{k=r}^{n} U_{\beta k}\, U_{\alpha k}^{*}\right|^{2}, \quad B_{\alpha;\alpha} = 4\left(\sum_{k=r}^{n} |U_{\alpha k}|^{2}\right)\left(1-\sum_{k=r}^{n} |U_{\alpha k}|^{2}\right). \quad (4)$$

The formulas (3) have the same form of the standard expressions for the oscillation probabilities in the case of two neutrinos (see [1]) with which the data of all the SBL experiments have been analyzed by the experimental groups. Hence, the results of these analyses can be used in order to constrain the possible values of the oscillation amplitudes $A_{\alpha;\beta}$ and $B_{\alpha;\alpha}$.

First, we consider the scheme 3H of Tab.1, with three neutrinos and a mass hierarchy. This scheme (as all the schemes with three neutrinos) provides only two independent mass-squared differences, Δm^2_{21} and Δm^2_{31}, which we choose to be relevant for the solution of the solar neutrino problem and for neutrino oscillations in the LSND experiment, respectively.

Let us emphasize that the mass spectrum 3H with three neutrinos and a mass hierarchy is the simplest and most natural one, being analogous to the mass spectra of charged leptons, up and down quarks. Moreover, a scheme with three neutrinos and a mass hierarchy is predicted by the see-saw mechanism for the generation of neutrino masses, which can explain the smallness of the neutrino masses with respect to the masses of the corresponding charged leptons.

In the case of scheme 3H we have $n = r = 3$ and Eq.(4) implies that

$$A_{\alpha;\beta} = 4\,|U_{\alpha 3}|^{2}\,|U_{\beta 3}|^{2}, \qquad B_{\alpha;\alpha} = 4\,|U_{\alpha 3}|^{2}\left(1-|U_{\alpha 3}|^{2}\right). \quad (5)$$

Hence, neutrino oscillations in SBL experiments depend on three parameters: $\Delta m^2 \equiv \Delta m^2_{31}$, $|U_{e3}|^2$ and $|U_{\mu 3}|^2$ (the unitarity of U implies that $|U_{\tau 3}|^2 = 1 - |U_{e3}|^2 - |U_{\mu 3}|^2$). From the exclusion plots obtained in reactor $\bar{\nu}_e$ and accelerator ν_μ disappearance experiments it follows that at any fixed value of Δm^2, the oscillation amplitudes $B_{e;e}$ and $B_{\mu;\mu}$ are bounded by the upper values $B^0_{e;e}$ and $B^0_{\mu;\mu}$, respectively, which are small quantities for $0.3 \lesssim \Delta m^2 \lesssim 10^3\,\mathrm{eV}^2$. From Eq.(5) one can see that small upper bounds for $B_{e;e}$ and $B_{\mu;\mu}$ imply that the parameters $|U_{e3}|^2$ and $|U_{\mu 3}|^2$ can be either small or large (i.e., close to one):

$$|U_{\alpha 3}|^2 \leq a^0_\alpha \quad \text{or} \quad |U_{\alpha 3}|^2 \geq 1 - a^0_\alpha, \quad \text{with} \quad a^0_\alpha = \frac{1}{2}\left(1 - \sqrt{1 - B^0_{\alpha;\alpha}}\right), \quad (6)$$

for $\alpha = e, \mu$. Both a^0_e and a^0_μ are small ($a^0_e \lesssim 4\times 10^{-2}$ and $a^0_\mu \lesssim 2\times 10^{-1}$) for any value of Δm^2 in the range $0.3 \lesssim \Delta m^2 \lesssim 10^3\,\mathrm{eV}^2$ (see Fig.1 of Ref.[8]).

Since large values of both $|U_{e3}|^2$ and $|U_{\mu 3}|^2$ are excluded by the unitarity of the mixing matrix ($|U_{e3}|^2 + |U_{\mu 3}|^2 \leq 1$), at any fixed value of Δm^2 there are three regions in the $|U_{e3}|^2$–$|U_{\mu 3}|^2$ plane which are allowed by the exclusion plots of SBL disappearance experiments: Region I, with $|U_{e3}|^2 \leq a^0_e$ and $|U_{\mu 3}|^2 \leq a^0_\mu$; Region II, with $|U_{e3}|^2 \leq a^0_e$ and $|U_{\mu 3}|^2 \geq 1 - a^0_\mu$; Region III, with $|U_{e3}|^2 \geq 1 - a^0_e$ and $|U_{\mu 3}|^2 \leq a^0_\mu$.

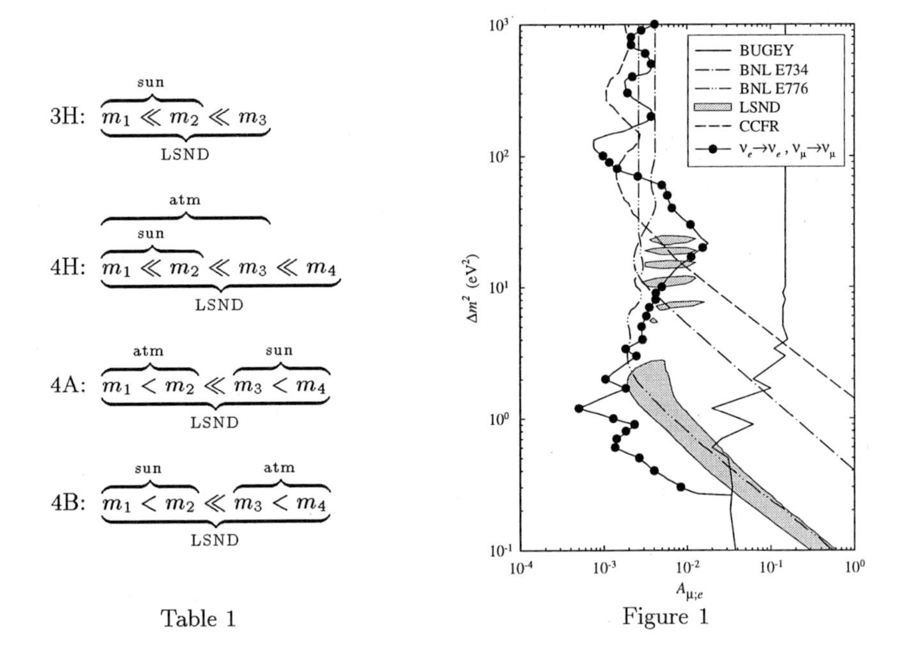
$$3\text{H}: \overbrace{m_1 \ll m_2}^{\text{sun}} \ll m_3 \quad (\underbrace{m_1 \ll m_2 \ll m_3}_{\text{LSND}})$$

$$4\text{H}: \overbrace{\overbrace{m_1 \ll m_2}^{\text{sun}} \ll m_3}^{\text{atm}} \ll m_4 \quad (\underbrace{m_1 \ll m_2 \ll m_3 \ll m_4}_{\text{LSND}})$$

$$4\text{A}: \overbrace{m_1 < m_2}^{\text{atm}} \ll \overbrace{m_3 < m_4}^{\text{sun}} \quad (\underbrace{m_1 < m_2 \ll m_3 < m_4}_{\text{LSND}})$$

$$4\text{B}: \overbrace{m_1 < m_2}^{\text{sun}} \ll \overbrace{m_3 < m_4}^{\text{atm}} \quad (\underbrace{m_1 < m_2 \ll m_3 < m_4}_{\text{LSND}})$$

Table 1

Figure 1

In region III $|U_{e3}|^2$ is large and ν_e has a small mixing with ν_1 and ν_2, which is insufficient for the explanation of the solar neutrino problem. Indeed, the survival probability of solar ν_e's is bounded by $P^{\text{sun}}_{\nu_e\to\nu_e} \geq |U_{e3}|^4$ (see [8]). If $|U_{e3}|^2 \geq 1 - a_e^0$, we have $P^{\text{sun}}_{\nu_e\to\nu_e} \gtrsim 0.92$ at all neutrino energies, which is a bound that is not compatible with the solar neutrino data. Hence, region III is excluded by solar neutrinos.

In region I $A_{\mu;e} \leq 4\, a_e^0\, a_\mu^0$, which means that the probability of $\overset{(-)}{\nu}_\mu \leftrightarrows \overset{(-)}{\nu}_e$ transitions in SBL experiments is strongly suppressed. The corresponding upper bound obtained from the 90% CL exclusion plots of the Bugey [2] $\bar{\nu}_e$ disappearance experiment and of the CDHS and CCFR [3] ν_μ disappearance experiments is represented in Fig.1 by the curve passing trough the circles. The shadowed regions in Fig.1 are allowed at 90% CL by the results of the LSND experiment. Also shown are the 90% CL exclusion curves found in the BNL E734, BNL E776 and CCFR [4] $\overset{(-)}{\nu}_\mu \to \overset{(-)}{\nu}_e$ appearance experiments and in the Bugey experiment. One can see from Fig.1 that in region I, the bounds obtained from the results of $\bar{\nu}_e \to \bar{\nu}_e$, $\nu_\mu \to \nu_\mu$ and $\overset{(-)}{\nu}_\mu \to \overset{(-)}{\nu}_e$ experiments are not compatible with the allowed regions of the LSND experiment [8]. Therefore, region I is disfavored by the results of SBL experiments. This is an important indication, because region I is the only one in which it is possible to have a hierarchy of the elements of the neutrino mixing matrix analogous to the one of the quark mixing matrix.

Having excluded the regions I and III of the scheme 3H, we are left only with region II, where ν_μ has a large mixing with ν_3, i.e., ν_μ (not ν_τ) is the "heaviest" neutrino.

Let us now consider the possible schemes with four neutrinos, which provide three independent mass-squared differences and allow to accommodate in a natural way all the three experimental indications in favor of neutrino oscillations. We consider first the scheme 4H of Tab.1 with four neutrinos and a mass hierarchy. The three independent mass-squared differences, Δm^2_{21}, Δm^2_{31} and Δm^2_{41}, are taken to be relevant for the oscillations of solar, atmospheric and LSND neutrinos, respectively. In the case of scheme 4H we have $n = r = 4$ and Eq.(4) implies that the oscillation amplitudes are given by

$$A_{\alpha;\beta} = 4\,|U_{\alpha 4}|^2\,|U_{\beta 4}|^2\,, \qquad B_{\alpha;\alpha} = 4\,|U_{\alpha 4}|^2\left(1 - |U_{\alpha 4}|^2\right)\,. \tag{7}$$

In this case, neutrino oscillations in SBL experiments depend on four parameters: $\Delta m^2 \equiv \Delta m^2_{31}$, $|U_{e3}|^2$, $|U_{\mu 3}|^2$ and $|U_{\tau 3}|^2$. From the similarity of the amplitudes (7) with the corresponding ones given in Eq.(5), it is clear that replacing $|U_{\alpha 3}|^2$ with $|U_{\alpha 4}|^2$ we can apply to the scheme 4H the same analysis presented for the scheme 3H. Hence, also the regions III and I of the scheme 4H are excluded, respectively, by the solar neutrino problem and by the results of SBL experiments. Furthermore, the purpose of considering the scheme 4H is to have the possibility to explain the atmospheric neutrino anomaly, but this is not possible if the neutrino mixing parameters lie in region II. Indeed, in region II $|U_{\mu 4}|^2$ is large and the muon neutrino has a small mixing with the light neutrinos ν_1, ν_2 and ν_3, which is insufficient for the explanation of the atmospheric neutrino anomaly [9].

Hence, the scheme 4H is disfavored by the results of neutrino oscillation experiments. For the same reasons, all possible schemes with four neutrinos and a mass spectrum in which three masses are clustered and one mass is separated from the others by a gap of about 1 eV (needed for the explanation of the LSND data) are disfavored by the results of neutrino oscillation experiments. Therefore, there are only two possible schemes with four neutrinos which are compatible with the results of all neutrino oscillation experiments: the schemes 4A and 4B of Tab.1. In these two schemes the four neutrino masses are divided in two pairs of close masses separated by a gap of about 1 eV. In scheme A, Δm^2_{21} is relevant for the explanation of the atmospheric neutrino anomaly and Δm^2_{43} is relevant for the suppression of solar ν_e's. In scheme B, the roles of Δm^2_{21} and Δm^2_{43} are reversed.

From Eq.(4) and using the unitarity of the mixing matrix, the oscillation amplitudes $B_{\alpha;\alpha}$ in the schemes 4A and 4B ($n = 4$, $r = 3$) are given by $B_{\alpha;\alpha} = 4\,c_\alpha\,(1 - c_\alpha)$, with $c_\alpha \equiv \sum_{k=1,2} |U_{\alpha k}|^2$ in the scheme 4A and $c_\alpha \equiv \sum_{k=3,4} |U_{\alpha k}|^2$ in the scheme 4B. This expression for $B_{\alpha;\alpha}$ has the same form as the one in Eq.(7), with $|U_{\alpha 4}|^2$ replaced by c_α. Therefore, we can apply the same analysis presented for the scheme 4H and we obtain four allowed regions in the c_e–c_μ plane (now the region with large c_e and c_μ is not excluded by the unitarity of the mixing matrix, which gives the constraint $c_e + c_\mu \leq 2$): Region I, with $c_e \leq a^0_e$ and $c_\mu \leq a^0_\mu$; Region II, with $c_e \leq a^0_e$ and $c_\mu \geq 1 - a^0_\mu$; Region III, with $c_e \geq 1 - a^0_e$ and $c_\mu \leq a^0_\mu$; Region IV, with $c_e \geq 1 - a^0_e$

and $c_\mu \geq 1 - a_\mu^0$. Following the same reasoning as in the case of scheme 4H, one can see that the regions III and IV are excluded by the solar neutrino data and the regions I and III are excluded by the results of the atmospheric neutrino experiments [9]. Hence, only region II is allowed by the results of all experiments.

If the neutrino mixing parameters lie in region II, in the scheme 4A (4B) the electron (muon) neutrino is "heavy", because it has a large mixing with ν_3 and ν_4, and the muon (electron) neutrino is "light". Thus, the schemes 4A and 4B give different predictions for the effective Majorana mass $\langle m \rangle = \sum_k U_{ek}^2 m_k$ in neutrinoless double-beta decay experiments: since $m_3 \simeq m_4 \gg m_1 \simeq m_2$, we have $|\langle m \rangle| \leq (1 - c_e) m_4 \simeq m_4$ in the scheme 4A and $|\langle m \rangle| \leq c_e m_4 \leq a_e^0 m_4 \ll m_4$ in the scheme 4B. Hence, if the scheme 4A is realized in nature, the experiments on the search for neutrinoless double-beta decay can reveal the effects of the heavy neutrino masses $m_3 \simeq m_4$. Furthermore, the smallness of c_e in both schemes 4A and 4B implies that the electron neutrino has a small mixing with the neutrinos whose mass-squared difference is responsible for the oscillations of atmospheric neutrinos (i.e., ν_1, ν_2 in scheme 4A and ν_3, ν_4 in scheme 4B). Hence, the transition probability of electron neutrinos and antineutrinos into other states in atmospheric and long-baseline experiments is suppressed [10][1].

References

[1] S.M. Bilenky and B. Pontecorvo, Phys. Rep. **41**, 225 (1978).

[2] B. Achkar *et al.*, Nucl. Phys. B **434**, 503 (1995).

[3] F. Dydak *et al.*, Phys. Lett. B **134**, 281 (1984); I.E. Stockdale *et al.*, Phys. Rev. Lett. **52**, 1384 (1984).

[4] L.A.Ahrens *et al.*, Phys. Rev. D **36**, 702 (1987); L.Borodovsky *et al.*, Phys. Rev. Lett. **68**, 274 (1992); A.Romosan *et al.*, *ibid* **78**, 2912 (1997).

[5] B.T. Cleveland *et al.*, Nucl. Phys. B (P.S.) **38**, 47 (1995); K.S. Hirata *et al.*, Phys. Rev. D **44**, 2241 (1991); GALLEX Coll., Phys. Lett. B **388**, 384 (1996); SAGE Coll., Phys. Rev. Lett. **77**, 4708 (1996).

[6] Y. Fukuda *et al.*, Phys. Lett. B **335**, 237 (1994); R. Becker-Szendy *et al.*, Nucl. Phys. B (P.S.) **38**, 331 (1995); W.W.M. Allison *et al.*, Phys. Lett. B **391**, 491 (1997).

[7] C. Athanassopoulos *et al.*, Phys. Rev. Lett. **77**, 3082 (1996).

[8] S.M. Bilenky *et al.*, Phys. Rev. D **54**, 1881 (1996).

[9] S.M. Bilenky, C. Giunti and W. Grimus, preprint hep-ph/9607372.

[10] S.M. Bilenky, C. Giunti and W. Grimus, preprint hep-ph/9710209.

[1] After we finished this paper the results of the first long-baseline reactor experiment CHOOZ appeared (M. Apollonio *et al.*, preprint hep-ex/9711002). The upper bound for the transition probability of electron antineutrinos into other states found in the CHOOZ experiment is in agreement with the limit obtained in Ref.[10].

KARMEN: Present Limits for Neutrino Oscillations and Perspectives after the Upgrade

Bernd Armbruster, KARMEN Collaboration (armbruster@ik1.fzk.de)

Forschungszentrum Karlsruhe, Institut für Kernphysik I
D-76021 Karlsruhe, Postfach 3640, Germany

Abstract. The neutrino experiment KARMEN is operated at the beam stop neutrino source ISIS. It provides ν_μ's, ν_e's and $\bar{\nu}_\mu$'s in identical intensities from the π^+–μ^+–decay at rest. The oscillation appearance channels $\nu_\mu \rightarrow \nu_e$ and $\bar{\nu}_\mu \rightarrow \bar{\nu}_e$ are investigated with a 56 t liquid scintillation calorimeter at a mean distance of 17.6 m from the ν–source. So far no evidence for oscillations has been found with KARMEN, resulting in 90% CL exclusion limits of $\sin^2(2\Theta) < 8.5 \cdot 10^{-3}$ ($\bar{\nu}_\mu \rightarrow \bar{\nu}_e$) and $\sin^2(2\Theta) < 4.0 \cdot 10^{-2}$ ($\nu_\mu \rightarrow \nu_e$) for $\Delta m^2 > 100\,\mathrm{eV}^2$. In 1996, the KARMEN neutrino experiment was upgraded by an additional veto system. First measurements after the upgrade show a substantial reduction of the cosmic background in the $\bar{\nu}_\mu \rightarrow \bar{\nu}_e$ channel meeting the expected factor of 40. This significantly enhances the experimental sensitivity towards $\sin^2(2\Theta) \approx 1 \cdot 10^{-3}$ for large Δm^2.

1 Neutrino Production and Detection at ISIS

The **K**arlsruhe **R**utherford **M**edium **E**nergy **N**eutrino experiment KARMEN is being performed at the neutron spallation facility ISIS of the Rutherford Appleton Laboratory. The neutrinos are produced by stopping 800 MeV protons in a beam dump Ta-D_2O-target. ν_μ, ν_e and $\bar{\nu}_\mu$ emerge with equal intensities from the decay sequence $\pi^+ \rightarrow \mu^+ + \nu_\mu$ and $\mu^+ \rightarrow e^+ + \nu_e + \bar{\nu}_\mu$. The ν_μ's from π^+–decay at rest are monoenergetic (E_ν=30 MeV) whereas the continuous energy distributions of ν_e and $\bar{\nu}_\mu$ up to 52.8 MeV can be calculated using the V–A theory. Two parabolic proton pulses of 100 ns base width and 225 ns apart are produced with a repetition frequency of 50 Hz. The different lifetimes of π^+ ($\tau = 26$ ns) and μ^+ ($\tau = 2.2\,\mu$s) allow a clear separation in time of the ν_μ–burst from the following ν_e's and $\bar{\nu}_\mu$'s and provide a duty cycle of 1×10^{-5} and 5×10^{-4} respectively.

The neutrinos are detected in a 56 t liquid scintillation calorimeter [1] with an active volume of 96% and an excellent energy resolution of $\sigma_E = 11.5\%/\sqrt{E[\mathrm{MeV}]}$. The detector is segmented by double acrylic walls with an air gap allowing efficient light transport via total internal reflection. Gd_2O_3 coated paper within the module walls provides efficient detection of thermal neutrons via Gd(n,γ) capture. A massive 7000 t iron blockhouse in combination with two layers of active anti counters provides shielding against beam correlated spallation neutron background and suppression of hadronic cosmic radiation as well as reduction of the flux of cosmic muons.

2 Oscillation limits on $\bar{\nu}_\mu \rightarrow \bar{\nu}_e$

In this contribution we only present results concerning the oscillation search in the $\bar{\nu}_\mu \rightarrow \bar{\nu}_e$ appearance channel obtained in the data taking period from 1990–1995. The investigation of the $\nu_\mu \rightarrow \nu_e$ channel and results for ν–nucleus interactions on ^{12}C are described elsewhere [2]. The intrinsic contamination of the neutrino source with $\bar{\nu}_e$'s is only $\bar{\nu}_e/\nu_e < 5 \cdot 10^{-4}$. Thus the detection of $\bar{\nu}_e$'s via $p(\bar{\nu}_e, e^+)n$ ($\langle\sigma\rangle = 0.934 \cdot 10^{-40}\text{cm}^2$) would strongly indicate oscillations $\bar{\nu}_\mu \rightarrow \bar{\nu}_e$ in the appearance channel. The signature for the detection of $\bar{\nu}_e$'s is a spatially correlated coincidence of positrons from $p(\bar{\nu}_e, e^+)n$ with energies up to $E_{e^+} = E_{\bar{\nu}_e} - Q = 52.8 - 1.8$ MeV $= 51$ MeV and delayed γ emissions of either of the two neutron capture processes $p(n,\gamma)d$ or $Gd(n,\gamma)Gd$ with γ energies of 2.2 MeV or up to 8 MeV, respectively. The positrons are expected 0.6 to 10.6 μs after beam-on-target whereas the neutrons are thermalized and captured typically with $\tau \approx 120\,\mu$s. The neutron detection efficiency for the analyzed data is 28.2%, but has been enhanced to 50.0% after the upgrade.

A prebeam analysis of cosmic ray induced sequences results in an accumulated background level of (8.6 ± 0.2) events per μs in the prompt 10 μs–window. The data set remaining after applying all cuts results in a rate of $(12.4 \pm 1.3)/\mu$s which corresponds to a beam excess of $2.4\,\sigma$ compared with the prebeam level including ν_e–interactions (9 events) and $\bar{\nu}_e$–contamination (1.2 events). Although the sequential events show the typical signature of thermal neutron capture, the prompt time and energy distribution does not follow the expectation from $\bar{\nu}_\mu \rightarrow \bar{\nu}_e$ oscillations.

To extract a possible small contribution of $\bar{\nu}_\mu \rightarrow \bar{\nu}_e$, the data set is scanned with a two-dimensional maximum likelihood analysis on time, position and energy distributions of the positrons as well as the sequential neutron captures. The spectroscopic measurement of the e^+ energy is highly sensitive to changes in the energy spectrum due to variations of the oscillation mass term Δm^2. The energies of the positrons used in the likelihood analysis therefore have been tested with Δm^2 in the range from 0.01 to 100 eV^2. For most of the investigated parameter range the likelihood analysis results in best fit values compatible with a zero signal within a 1σ error band. Only for a region at $\Delta m^2 = 6.2\,\text{eV}^2$ there is a positive signal 2.3σ above zero, which, however, is not considered as statistically significant.

On this basis of no evidence for oscillations, 90% CL upper limits for the oscillation parameters Δm^2 and $\sin^2(2\Theta)$ have been deduced. For large Δm^2 an upper limit of the mixing angle $\sin^2(2\Theta) < 8.5 \cdot 10^{-3}$ 90% CL is gained. Fig. 1 shows the KARMEN exclusion curves for the appearance channels $\nu_\mu \rightarrow \nu_e$ and $\bar{\nu}_\mu \rightarrow \bar{\nu}_e$ in comparison with results from other ν–oscillation

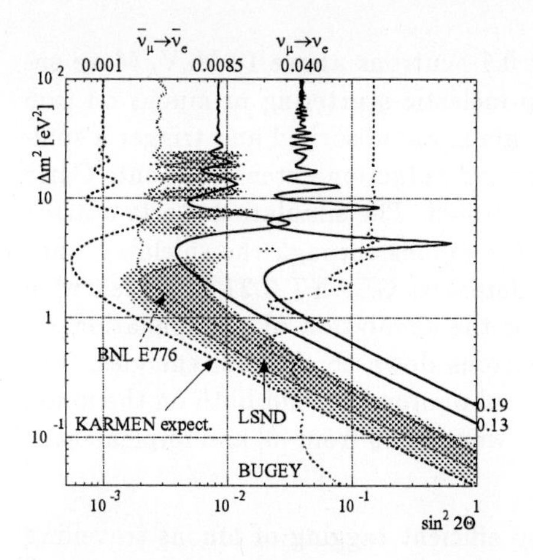

Fig. 1. 90% CL exclusion curves from KARMEN for $\nu_\mu \to \nu_e$ and $\bar{\nu}_\mu \to \bar{\nu}_e$ as well as the expected sensitivity for $\bar{\nu}_\mu \to \bar{\nu}_e$ after the upgrade; oscillation limits from BNL E776 [3] and Bugey [4]; LSND favoured regions [5] are shown as shaded areas.

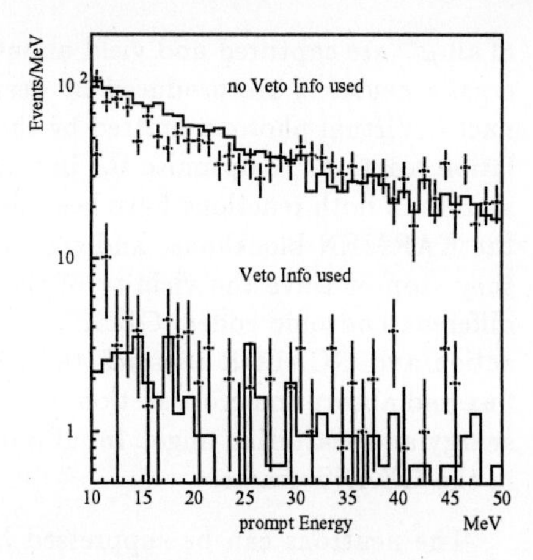

Fig. 2. Energy of cosmic induced neutrons measured after the KARMEN Upgrade (dots with errors) before and after the new Veto information was used are shown together with Monte Carlo predictions (solid lines).

experiments at accelerators [3] and reactors [4]. The KARMEN result is in contradiction with the positive result of LSND [5]. The sensitivity of the KARMEN experiment for $\bar{\nu}_\mu \to \bar{\nu}_e$ before the upgrade was only comparable to that of LSND and therefore not high enough to cover the entire parameter space favoured by LSND.

3 The KARMEN Upgrade

The sensitivity in the $\bar{\nu}_\mu \to \bar{\nu}_e$ channel is essentially limited by the cosmic induced background. Neutrons produced by cosmic muons in the massive 7000 t iron shielding surrounding the KARMEN detector can cross the two layer anti–system untagged and simulate a $\bar{\nu}_\mu \to \bar{\nu}_e$ signature by elastic n-p scattering followed by the capture of the thermalized neutron via Gd(n,γ) or p(n,γ). The sequential event is thus indistinguishable from the sequential event of a true $\bar{\nu}_\mu \to \bar{\nu}_e$ sequence. However, it is possible to separate an oscillation signal from the background using the time and energy information of the prompt event within the limits given by the statistical fluctuations of the well known cosmic background.

There are two reactions producing neutrons above 10 MeV in iron: captures of stopped μ^- produce neutrons with an energy spectrum $N_n(E_n) = N_0 \exp(-E_n/7.2\,\text{MeV})$ via $\mu^- + {}^{56}\text{Fe} \to {}^{55}\text{Mn} + \text{n} + \nu_\mu$. A fraction of 90.9%

of all μ^- are captured and yield about 0.1 neutrons above 10 MeV. More energetic neutrons are produced by deep inelastic scattering of muons on iron nuclei. Virtual photons emitted by the muon are absorbed and trigger a spallation reaction. To optimize the background reduction extensive Monte Carlo studies of both reactions have been performed. The simulation starts outside the KARMEN blockhouse and tracks the muons through the shielding until they stop or leave the vicinity of the detector. GEANT 3.21 was used with different hadronic codes: GHEISHA for the simulation of the spallation reaction and GCALOR to track the neutrons down to thermal energies. The flux and absorption cross section of virtual photons depend both on the muon energy and scattering angle. Input data were taken from [6] and implemented in the GEANT frame.

The neutrons can be suppressed by efficient tagging of muons travelling through the iron in the vicinity of the main detector. During 1996, a 300 m^2 additional veto system was installed in the 2 m thick walls and 3 m thick roof of the KARMEN blockhouse in a limited space of 14 cm width given by its slab structure. In total 136 modules of 65 cm wide, 5 cm thick and 3.15 to 4 m long BICRON BC412 plastic scintillator bars have been installed. A 180° light bending system guides scintillation light to four 2" Philips XP 2262 PMTs on each module end. The modules have been optimized for uniform lightoutput, high muon detection efficiency and excellent μ/γ separation.

A detection efficiency of 99.4% for cosmic muons entering the main detector has been achieved, although the area coverage of the veto is only 87.7%. The module positions within the walls are optimized with respect to minimum total area of the veto system and a minimum amount of iron of one metre between the veto sides and the main detector. This attenuates neutrons produced *outside* the veto volume by an untagged muon by two orders of magnitude. The simulations showed that a reduction of the cosmic background by a factor of 40 could be expected.

The upgraded KARMEN detector takes data since February 1997. Figure 2 shows the energy spectra of event sequences before and after using the information from the additional Veto compared with the Monte Carlo expectation for both cases. The background rate in the energy interval 20–50 MeV is reduced by a factor (41.3 ± 7.7) to (0.088 ± 0.014) mHz. This reduces the expected cosmic background in the beam window after two years of measuring time down to 1.3 events. The improved KARMEN sensitivity for $\bar{\nu}_\mu \rightarrow \bar{\nu}_e$ is expected to exclude the whole parameter region of evidence suggested by LSND if no oscillation signal will be found (fig. 1). In that case, mixing angles with $\sin^2(2\Theta) > 1 \cdot 10^{-3}$ will be excluded for large Δm^2. The upgrade increases also the signal to background ratio of single prong ν-induced events on ^{12}C and improves the investigation of the anomaly in this time distribution [7].

We acknowledge the financial support of the German Bundesministerium für Bildung, Wissenschaft, Forschung und Technologie.

References

[1] G. Drexlin et al. Nucl. Instr. and Meth. A289 (1990) 490
[2] B. A. Bodmann et al. Phys. Lett. B339 (1994) 215.
B. Zeitnitz et al. Prog. Part. Nucl. Phys. 32 (1994) 351
[3] L. Borodovsky et al. Phys. Rev. Letters 68 (1992) 274
[4] B. Achkar et al. Nucl. Phys. B434 (1995) 503
[5] C. Athanassopoulos et al. Phys. Rev. C54 (1996) 2685
[6] L. B. Bezrukov, É. V. Bugaev, Sov. J. Nucl. Phys. 33(5)635, 1981
[7] B. Armbruster et al. Phys. Lett. B348 (1995) 19

Preliminary Results from the NOMAD Experiment at CERN

Viatcheslav Valuev (Slava.Valouev@cern.ch)

LAPP Annecy, France/JINR Dubna, Russia

Abstract. NOMAD is a short baseline neutrino oscillation experiment searching for $\nu_\mu \to \nu_\tau$ and $\nu_\mu \to \nu_e$ oscillations in the CERN SPS wide band neutrino beam. A preliminary analysis of the 1995 data sample set 90% confidence limits of $\sin^2 2\theta_{\mu\tau} < 3.4 \times 10^{-3}$ for $\nu_\mu \to \nu_\tau$ and $\sin^2 2\theta_{\mu e} < 2 \times 10^{-3}$ for $\nu_\mu \to \nu_e$ oscillations at large Δm^2.

1 Introduction

The Neutrino Oscillation MAgnetic Detector (NOMAD) experiment [1] was designed to search for ν_τ appearing from $\nu_\mu \to \nu_\tau$ oscillations in the CERN SPS wide band neutrino beam, which consists primarily of ν_μ neutrinos ($\langle E_{\nu_\mu} \rangle \approx 24$ GeV) with a small ν_e component (less than 1%) and a negligible ($< 5 \times 10^{-6}$) contamination of prompt ν_τ [2]. If $\nu_\mu \to \nu_\tau$ oscillations occur, ν_τ's would be detected via their charged current (CC) interactions $\nu_\tau N \to \tau^- X$ in an active target using the kinematical characteristics of the subsequent τ^- decays. Located at a distance of 640 m from the average neutrino production point, NOMAD is sensitive to the cosmologically interesting ν_τ mass range $\Delta m^2 \sim 10 \div 100$ eV^2 and is expected to reach a sensitivity of $\sin^2 2\theta_{\mu\tau} < 3.8 \times 10^{-4}$ for $\Delta m^2 \geq 40$ eV^2.

The detector has been optimized to efficiently detect electrons, in particular ν_e CC interactions. Their analysis is relevant for the search for $\nu_\mu \to \nu_e$ oscillations, since an oscillation signal would manifest itself both as an excess of events in the ν_e CC sample and as a change in the shape of the ν_e CC energy spectrum. The interest for this kind of study has recently highly increased, following the LSND claim for evidence for $\nu_\mu \to \nu_e$ oscillations [3]. In case of $\nu_\mu \to \nu_e$ oscillations with $\Delta m^2 > 10$ eV^2 and with the probability of 3×10^{-3} observed by LSND, a signal should be seen in the NOMAD data.

2 The Detector

The NOMAD detector is described in detail elsewhere [4]. The crucial part of the apparatus are the 44 3×3 m^2 drift chambers used to reconstruct charged particle tracks and to measure their momenta in a 0.4 T magnetic field. The target mass (2.7 t) is given by the chamber structure. The active target is followed by the transition radiation detector which provides a

10^3 pion rejection factor for a 90% electron efficiency [5]. The preshower and the lead-glass electromagnetic calorimeter are used to improve the electron identification and to provide the measurement of electron energy ($\sigma(E)/E = 3.2\%/\sqrt{E(\mathrm{GeV})} \oplus 1\%$) together with the reconstruction of electromagnetic showers induced by photons. Large area drift chambers located outside the magnet are used to identify muons with $p_\mu > 2.5$ GeV. The detector is complemented by iron-scintillator hadronic and front calorimeters.

The upstream veto and the two trigger planes are used to select neutrino interactions in the detector. During the 1995 run the NOMAD experiment collected data for a total exposure of 0.85×10^{19} protons on target, out of which a sample of about 1.6×10^5 neutrino interactions in the fiducial volume was selected and analysed.

3 Search for $\nu_\mu \to \nu_e$ Oscillations

Since the search for $\nu_\mu \to \nu_e$ oscillations implies a direct comparison between the observed and the expected spectra, a precise knowledge of the neutrino fluxes and spectra is crucial. A Monte-Carlo package based on the GEANT and FLUKA libraries has been developed to describe the entire SPS neutrino beam line, simulated neutrino fluxes are further constrained by the recent NA56 results [6]. Another approach uses data accumulated by NOMAD to extract the flux of all ν_e sources (K^+, K^0_L, μ^+, etc.) and predict the ν_e spectrum in the absence of oscillations.

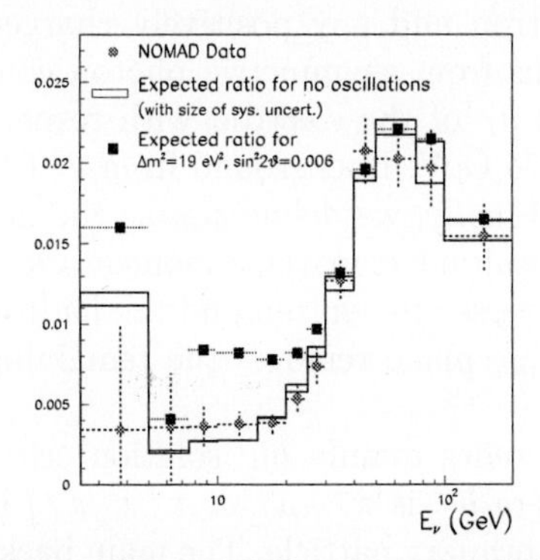

Fig. 1. Observed $\mathcal{R}_{e\mu}$, compared to expectations in the absence of oscillations and to oscillations with $\sin^2 2\theta = 0.006$, $\Delta m^2 = 19$ eV^2.

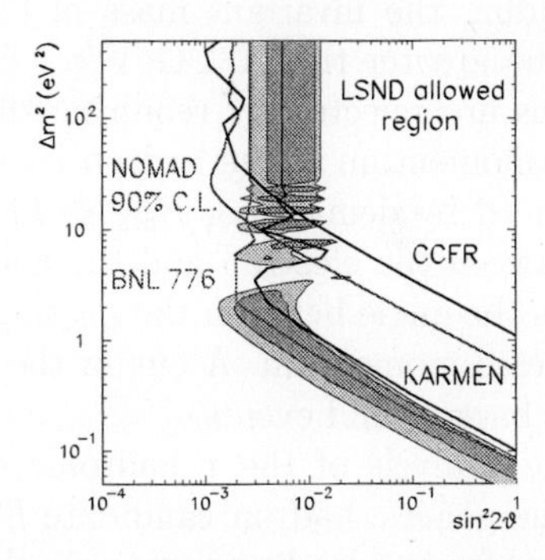

Fig. 2. Preliminary exclusion plot, obtained using NOMAD 1995 data. The limits set by other experiments and the LSND allowed region are also shown.

In order to reduce systematic uncertainties, it is preferable to study the ratio between the number of ν_e and ν_μ charged current interactions as a function of the neutrino energy E_ν:

$$\mathcal{R}_{e\mu}(E_\nu) = \left[\frac{\text{Nr. of } \nu_e\text{CC events}}{\text{Nr. of } \nu_\mu\text{CC events}}\right](E_\nu) \ . \tag{1}$$

In figure 1 the ratio measured with the 1995 data sample (grey circles) is superimposed on the one expected in the absence of oscillations, for which the size of systematic uncertainties is also given (area between histograms); error bars on data points include statistical errors only. For comparison, the expected $\mathcal{R}_{e\mu}$ for oscillations with $\sin^2 2\theta = 0.006$ and $\Delta m^2 = 19$ eV^2 is also shown (black squares). The measured ratio fits expectations well, indicating that there is no evidence for oscillations in NOMAD data.

The preliminary 90% confidence upper limit, which takes into account a 10% systematic uncertainty on the K/π ratio, is shown in figure 2. It corresponds to $\sin^2 2\theta_{\mu e} < 2 \times 10^{-3}$ for large values of Δm^2. This result excludes the $\Delta m^2 > 10$ eV^2 region of oscillation parameters suggested by LSND.

4 Search for $\nu_\mu \to \nu_\tau$ Oscillations

The selection of ν_τ CC interactions in NOMAD relies on kinematic criteria. A careful study of event kinematics, particle isolation and momentum balance in the transverse plane allows one to distinguish the ν_τ CC interactions from ν_μ and ν_e CC or neutral current background events. NOMAD is able to search for τ^- decays into two leptonic channels: $e^-\nu_\tau\bar{\nu}_e$, $\mu^-\nu_\tau\bar{\nu}_\mu$ and into three hadronic ones: $\pi^-\nu_\tau$, $\rho^-\nu_\tau$ and $\pi^-\pi^-\pi^+(n\pi^0)\nu_\tau$, i.e. about 88% of the τ decays.

The $\tau^- \to e^-\nu_\tau\bar{\nu}_e$ decay is particularly attractive because of a small background. Photon conversions and Dalitz decays of π^0's are removed by demanding the invariant mass of the electron and any positively charged track be greater than 0.1 GeV/c^2. Electrons from asymmetric photon conversions are rejected by requiring that the p_T of the electron with respect to the momentum of the hadron jet is > 0.75 GeV. Background from ν_e CC is reduced by demanding $E_{vis} < 40$ GeV. Finally, we define ϕ_{eh} as the angle between the electron and the hadron resultant transverse momenta and ϕ_{mh} as the angle between the missing transverse momentum and the hadron transverse momentum. A cut in the ϕ_{eh}–ϕ_{mh} plane removes the remaining ν_e CC background events.

The analysis of the τ hadronic decays relies mainly on isolation criteria. The generic hadron candidate h^- (where h^- is π^-, ρ^- or $\pi^+\pi^-\pi^-$) is searched for as a leading negatively charged primary particle. The main background comes from neutral current neutrino interactions and charged current interactions where the leading lepton is not identified. The Q_T, component of the momentum of the hadron system $\vec{p_{h^-}}$ perpendicular to the total visible momentum $\vec{p_{tot}}$, is computed for each event. Since hadrons in the background events are produced in the fragmentation of the hit quark and the nucleon remnants, they have a limited transverse momentum with respect to the total system, while the expected signal exhibits a long tail. After the requirement $Q_T > 1.7$ GeV there are no expected background events left and no survivors in the data.

Decay mode	Br, %	Efficiency, %	Observed	Background
$\tau^- \to e^- \nu_\tau \bar{\nu}_e$	17.8	4 ± 1	0	0.6
$\tau^- \to \mu^- \nu_\tau \bar{\nu}_\mu$	17.4	0.8 ± 0.2	0	0
$\tau^- \to h^- \nu_\tau$	12.0	1.6 ± 0.2	0	0
$\tau^- \to h^- \pi^0 \nu_\tau$	25.8	0.5 ± 0.2	0	0
$\tau^- \to \pi^- \pi^- \pi^+ (n\pi^0) \nu_\tau$	14.9	1.2 ± 0.2	0	0.4
TOTAL	87.9		0	1.0

Table 1. The summary of $\nu_\mu \to \nu_\tau$ oscillation search.

The summary of the $\nu_\mu \to \nu_\tau$ oscillation search for different τ^- decay channels is given in table 1. No candidate events were observed in the 1995 data sample, while the estimated total background amounts to 1 event. In order to reliably determine the expected backgrounds and the τ selection efficiencies, ν_μ CC data events are used as a simulator to cross-check Monte-Carlo estimations.

The preliminary limit on the probability of $\nu_\mu \to \nu_\tau$ oscillations is:

$$P(\nu_\mu \to \nu_\tau) < 2.41/1441 \approx 1.7 \times 10^{-3} \quad (90\% \text{ C.L.})$$

which corresponds to $\sin^2 2\theta_{\mu\tau} < 3.4 \times 10^{-3}$ for large Δm^2.

5 Conclusion

A preliminary analysis of the NOMAD 1995 data results in an upper limit of $\sin^2 2\theta_{\mu\tau} < 3.4 \times 10^{-3}$ for $\nu_\mu \to \nu_\tau$ and $\sin^2 2\theta_{\mu e} < 2 \times 10^{-3}$ for $\nu_\mu \to \nu_e$ oscillations at large Δm^2 (at 90% C.L.).

The experimental data from 1996 run (about 4.7×10^5 neutrino interactions) are currently being analysed. The NOMAD detector is continuing data taking in 1997 and 1998.

References

[1] P. Astier et al., CERN-SPSLC/91-21 (1991); Add. 1, CERN-SPSLC/91-48 (1991); Add. 2, CERN-SPSLC/91-53 (1991).

[2] M. C. Gonzalez-Garcia and J. J. Gomez-Cadenas, Physical Review D55 (1997), 1297; B. Van de Vyver, Nuclear Instruments and Methods A385 (1997), 91.

[3] C. Athanassopoulos et al., Physical Review Letters 77 (1996), 3082; Physical Review C54 (1996), 2685; nucl-ex/9709006.

[4] J. Altegoer et al., CERN-PPE/97-059, in press in Nuclear Instruments and Methods A.

[5] G. Bassompierre et al., LAPP-EXP-97/05, LAPP-EXP-97/06, in press in Nuclear Instruments and Methods A.

[6] M. Bonesini, contributed paper 1019, these proceedings.

$\nu_\mu \longrightarrow \nu_\tau$ Oscillations Search by CHORUS

Lucio Ludovici (lucio.ludovici@cern.ch)

CERN/PPE, 1211 Genève 23, Switzerland

Abstract. CHORUS is searching for $\nu_\mu \to \nu_\tau$ oscillation in the CERN Wide Band Neutrino Beam. A fraction of the events collected in 1994-95 has been located in the emulsion target and a search for τ produced in charged current interactions of ν_τ has been performed in the $\tau^- \to \mu^- \bar{\nu}_\mu \nu_\tau$ and $\tau^- \to \mathrm{h}^- (n\pi^\circ)\nu_\tau$ decay channels. No candidates have been found. A limit on the mixing angle of $sin^2 2\theta_{\mu\tau} < 2.3 \cdot 10^{-3}$ at 90% C.L. for large $\Delta \mathrm{m}^2_{\mu\tau}$ can be set, improving the previous best result.

1 Introduction

The CHORUS experiment has been designed to search for $\nu_\mu \to \nu_\tau$ oscillation in the parameter space region of small $\theta_{\mu\tau}$ mixing angle and $\Delta \mathrm{m}^2_{\mu\tau} \sim 10\,\mathrm{eV}^2$. A neutrino with this mass would solve the puzzle of the missing hot dark matter in the Universe [1].

The experiment searches for $\nu_\tau \mathrm{N} \to \tau^- + \mathrm{X}$ charged current interactions in the CERN Wide Band Neutrino Beam (WBB), which is an almost pure ν_μ beam, with a ν_τ contamination well below the reachable sensitivity. Nuclear emulsion was chosen as neutrino target because of its exceptional spatial resolution (better than one micrometer) and hit density (300 grains/mm along a track). It is the only detector which allows a direct observation of a three-fold experimental signature: i) the primary vertex where the ν_τ interaction takes place; ii) the τ flight path; iii) the τ decay vertex topology.

2 The experimental setup

The CERN WBB consists mainly of muon neutrinos from π^+ and K^+ decay, with an average energy of 27 GeV. The $\bar{\nu}_\mu$ contamination is 5% and the ν_e,$\bar{\nu}_\mathrm{e}$ contamination is at the level of 1%. The ν_τ background in the beam has been evaluated to be of the order of $3.3 \cdot 10^{-6}$ charged current interactions per ν_μ charged current interaction [2], and then can be neglected.

The CHORUS experiment is described in details elsewhere [3]. It is a *hybrid* setup, composed of an emulsion target and electronic detectors.

The emulsion target has a mass of 770 kg and a surface area of $1.42 \times 1.44\,\mathrm{m}^2$. It is segmented longitudinally in four stacks of 36 plates each. Each plate consists of two 350 μm thick layers of nuclear emulsion on both sides of a supporting 90 μm thick plastic base. Downstream of each stack, three emulsion sheets are used as interface to the electronic detectors (see fig.1). They have a thicker plastic base (800 μm) to provide a good angle measurement out of a single interface emulsion sheet.

A set of scintillating fiber trackers is used to locate the trajectories of the charged particles produced in the neutrino interaction. The tracker resolution is 150 μm on the lateral position and 2 mrad on the angle.

In the target region, downstream the emulsion target, an air-core magnet of hexagonal shape and three additional planes of scintillating fibers allow the reconstruction of charge and momentum of charged particles. Downstream the target region the lead scintillating fiber calorimeter detects neutral particles and measures the energy and direction of hadronic showers. Finally, the muon spectrometer measures the charge and momentum of muons.

3 The data collection

The detector has been taking data in the CERN WBB from 1994 to 1997. After a first two years run (1994-95), the target emulsions were replaced and developed. In its four years data taking, CHORUS has been collecting neutrino interactions corresponding to about 5.1×10^{19} protons on target, out of which 2.01×10^{19} in 1994-95. This corresponds to about $320,000$ ν_μ charged current interactions which are expected to have occured in the emulsion target in 1994-95 and $520,000$ in 1996-97.

The results presented here refer to data collected in 1994-95. $250,932$ events have been reconstructed in the electronic detectors with an identified negative muon from a neutrino interaction vertex in emulsion.

4 Event location in emulsion

Negative muons with $P_\mu < 30\,\mathrm{GeV/c}$ and negative hadrons, in events with no muons, with $1 < P_h < 20\,\mathrm{GeV/c}$, are considered for emulsion scanning.

The emulsion scanning procedure is fully automated, using computer controlled microscopes equipped with CCD cameras and fast processors. The size of one CCD view corresponds to 150 μm × 120 μm. The reconstructed tracks are first measured in the interface emulsion sheets and then the track found there is extrapolated to the downstream surface of the emulsion stack. Here, 16 CCD views taken changing the focus plane depth about every 6 μm, are processed online by the fast processor. The track found is followed upstream, plate by plate, until it disappears and then the interaction vertex is found. In this process, fiducial cuts are applied on the lateral position and direction of the tracks essentially to avoid the emulsion edges and the cone corresponding to the muon beam (used for calibration).

5 τ decay search

An event with a τ lepton is identified by the presence of a change of direction (*kink*) of the τ^- track due to its one-prong decay. The track measurements in the vertex emulsion plate are used to select events on the basis of the impact parameter of the τ daughter track or on the transverse momentum of the daughter track w.r.t. the candidate τ track. The retained events undergo a computer assisted *eye scan* of the vertex plate (and donwstream plates if necessary) in order to confirm the presence of a τ decay topology. Events are considered as ν_τ candidates if i) a kink is found along the selected (muon or hadron) track; ii) there are no other charged leptons at the primary vertex; iii) the transverse momentum P_t of the selected track, w.r.t the τ candidate direction, is larger than 250 MeV/c; iv) the kink is located within 5 plates downstream of the vertex plate. The cut ii) suppress the background from charm in anti-neutrino interaction, and the cut iii) eliminates K^- decays.

6 Preliminary results

Presently (end of 1997) the location in emulsion and the τ decay search has been completed on about 45% of the 1994-95 statistics for the muon events. The analysis of the muon-less events started later and the analysed sample correspond to only 64.4% of the muon events statistics.

In the muon sample, 31,279 events have been located and analysed at the vertex, searching for $\tau^- \to \mu^- \bar{\nu}_\mu \nu_\tau$ decays. No candidates have been found. The expected background is 0.1 events, dominated by the charm production from the anti-neutrino component of the beam, where the positive lepton at the primary vertex escapes detection (or it is mis-identified) and the leptonic decay of the negative charmed meson fakes the $\tau^- \to \mu^- \bar{\nu}_\mu \nu_\tau$ decay topology.

In the muon-less sample, 4,553 events have been located and analysed at the vertex, searching for $\tau^- \to h^-(n\pi^\circ)\nu_\tau$ decays. No candidates have been found. The expected background is about 0.3 events, dominated by the kinks due to the hadron scattering on nuclei without visible recoil (*white kinks*).

The following formula relates the oscillation probability $P(\nu_\mu \to \nu_\tau)$ with N_τ, the ν_τ observed in the $\tau^- \to \mu^- \bar{\nu}_\mu \nu_\tau$ and $\tau^- \to h^-(n\pi^\circ)\nu_\tau$ channels:

$$N_\tau = N_\mu \cdot < P(\nu_\mu \to \nu_\tau) > \cdot (\Sigma_i^{\mu,h} \epsilon_i^{kink} \cdot BR_i \cdot r_i^{acc} \cdot r_\sigma \cdot s_i) \qquad (1)$$

where

- $N_\mu = 31,279$ is the number of analysed muon events;
- $P(\nu_\mu \to \nu_\tau)$ is the oscillation probability ($P \simeq \frac{1}{2} \cdot sin^2 2\theta_{\mu\tau}$ for large $\Delta m^2_{\mu\tau}$)
- ϵ_i^{kink} are the kink detection efficiencies;
- BR_i are the branching ratios of $\tau^- \to \mu^- \bar{\nu}_\mu \nu_\tau$ and $\tau^- \to h^-(n\pi^\circ)\nu_\tau$;
- $r_i^{acc} = \frac{<A_i(\nu_\tau)>}{<A(\nu_\mu)>}$ are the cross section weighted acceptance ratios for ν_μ and ν_τ interactions;

- $r_\sigma = \frac{<\sigma(\nu_\tau)>}{<\sigma(\nu_\mu)>} = 0.53$ is the neutrino energy weighted cross section ratio;
- s_i account for the reduced statistics analysed in the muon-less sample.

Table 1. Values used for the oscillation limit calculation.

Decay mode	$\frac{<A_i(\nu_\tau)>}{<A(\nu_\mu)>}$	BR_i	$\epsilon_i(kink)$	s_i
$\tau^- \rightarrow \mu^- \bar{\nu}_\mu \nu_\tau$	1.05	0.1735	0.466	1.000
$\tau^- \rightarrow \mathrm{h}^-(n\pi^\circ)\nu_\tau$	0.43	0.4978	0.234	0.644

The numerical values are given in table 1. Equation 1 can be used to calculate the 90% C.L. upper limit $sin^2 2\theta_{\mu\tau} \leqslant 2.3 \times 10^{-3}$, for large $\Delta m^2_{\mu\tau}$. A graphical representation of the oscillation parameter region excluded by the present analysis is shown in fig.2.

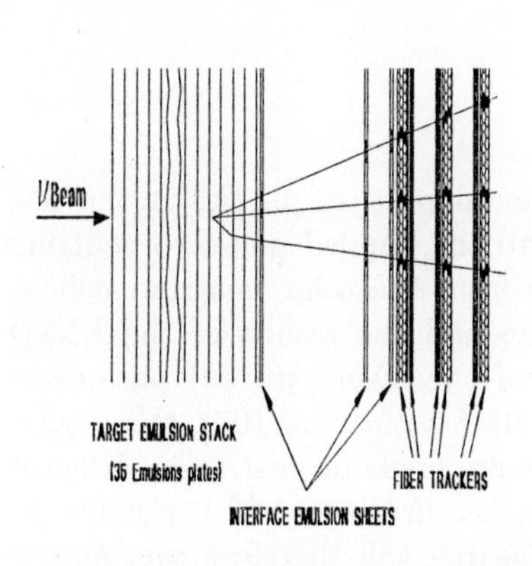

Fig. 1. An emulsion stack and the fiber tracker.

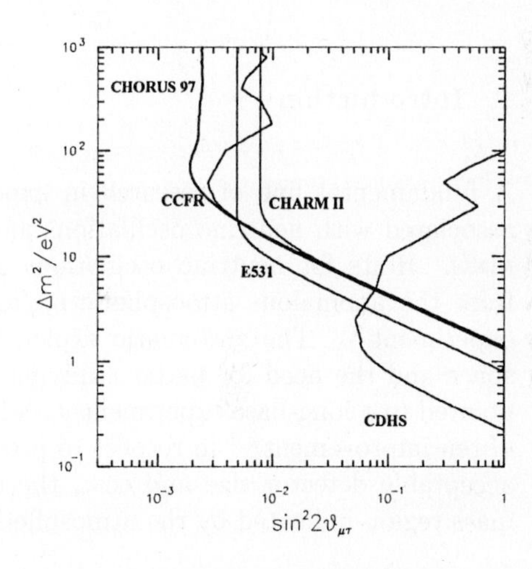

Fig. 2. Exclusion plot (90% C.L.).

References

[1] H. Harari, *Phys. Lett.* B**216** (1989), 413;
J. Ellis, J.L. Lopez, D.V. Nanopoulos, *Phys. Lett.* B**292** (1992), 189;
H. Fritzsch, D. Holtmannspotter, *Phys. Lett.* B**338** (1994), 290.

[2] B. Van de Vyver, *Nucl. Instrum. Methods* A**385** (1997), 91.
M.C. Gonzalez-Garcia, J.J. Gomez-Cadenas, *Phys. Rev.* D**55** (1997), 1297.

[3] E. Eskut et al., *Nucl. Instrum. Methods* A**401** (1997) 7.

A Long-Base Search for Neutrino Oscillations at a Reactor; the CHOOZ Experiment

C. BEMPORAD

INFN and Dipartimento di Fisica dell'Università, Pisa

CHOOZ, the first long-base reactor-neutrino vacuum-oscillation experiment, has been tuned in 1996 and is running since early 1997. The liquid scintillator calorimeter is located in a tunnel, at a distance of 1 Km from the neutrino source, and is protected by a 300 mwe rock cover, which strongly reduces the cosmic ray generated background. The experiment wants to probe Δm^2 down to 10^{-3} eV^2 and $sin^2 2\theta$ in the interval $1 \leftrightarrow 0.1$. After measuring the no-reactor background, the experiment collected data for several months, at an increasing and now at full, reactor power. Results of the neutrino oscillation search will be published in autumn 1997; information relative to the detector performance and to the data taking is presented.

1 Introduction

A fundamental line of research in experimental particle physics is the one associated with neutrino oscillations and with the implied non-zero neutrino mass. Hints for neutrino oscillations came from the solar neutrino deficit, from the anomalous atmospheric ν_μ/ν_e ratio and the results of the LNSD experiment [1]. The systematic exploration of the $(\Delta m^2, sin^2 2\theta)$ parameter space and the need for better studying the $10^{-2} < \Delta m^2 < 10^{-3}$ eV^2 region pointed to a long-base experiment with low-energy reactor neutrinos. A factor of ten improvement [2,3] in respect to previous measurements [4,5,6] is possible at acceptable detector size and cost; the new search will therefore overlap the mass region indicated by the atmospheric neutrino signal.

---

[a]

The Chooz Collaboration:
Drexel University:C.E.Lane,R.Steinberg,J.Steele,S.Tomshaw;*Kurchatov Institute*:A.Etenko,A.Martemyanov,M.Skorokhvatov,S.Sukhotin,V. Vyrodov:*INFN and University of Pisa*:A.Baldini,C.Bemporad,F.Cei,M. Grassi,D.Nicolò,R.Pazzi,G.Pieri;*INFN and University of Trieste*:M. Apollonio,E.Caffau,P.Cristaudo,G.Giannini;*LAPP Annecy*:Y.Déclais,M. Laiman,A.Oriboni;*LPC Collège de France*:H.de Kerret,D.Kryn, B.Lefièvre,M.Obolensky,D.Marchand,P.Salin,D.Veron;*University of California, Irvine*:S.Riley,H.Sobel;*University of New Mexico, Albuquerque*:B.Dieterle,J.George.

2 Description of the CHOOZ Experiment

CHOOZ is the name of the new power station built by Électricité de France close to the village of Chooz in the Ardennes region; the station has two pressurized water reactors for a total thermal power of $2 \times 4.25\ Gwth$; the first reactor reached full power in May 1997, the second reactor will reach full power this summer. The reactors are of a type of well known characteristics; the neutrino flux is $2\ 10^{20}\ \bar{\nu}_e\ Gw^{-1}s^{-1}$. The apparatus is located in an underground experimental hall at a distance of $1\ Km$ from the neutrino source; the $300\ mwe$ rock overburden reduces the cosmic ray μ-flux by a factor ≈ 300. While it is possibile to partially compensate the $\overline{\nu}_e$-flux reduction due to the neutrino source distance by increasing the detector sensitive mass, the only way to maintain a good signal/background ratio is to go underground, since the most dangerous background is indeed due to fast neutrons from μ-induced nuclear spallation events in the materials surrounding the detector; with respect to that, the CHOOZ site is unique. The apparatus is a low energy calorimeter using two paraffin based liquid scintillators; it is made of three concentric detectors: an outer 90 *tons* active veto shield; an intermediate 17 *tons* optically separated γ-radiation containment calorimeter; a central 5 *tons* target, a transparent plexiglass container filled with a 0.1 % gadolinium loaded scintillator. Neutrino events are detected by the positron from $\bar{\nu}_e + p \rightarrow n + e^+$ and by the neutron delayed capture by Gadolinium (30 μs mean capture time) which produces γ's with a total energy $E \approx 8\ MeV$. The scintillator light is measured by 240 PMTs; 192 PMTs observe the internal calorimeter (15 % surface coverage), 48 PMTs belong to the veto shield; PMT properties were fully studied before their installation [7]. The detector structure is simple and the detector can be easily calibrated and be reliably simulated by the Montecarlo method. We use a particularly simple primary trigger, which depends on two threshold levels, one set on the total number of photoelectrons measured by the PMT system, the other set on the number of fired photomultipliers. A second level trigger requires a delayed coincidence of two primary triggers within $\approx 100\mu s.$; they correspond to the positron signal and to the delayed neutron capture. The light level produced by neutrino events in the inner target is $\approx 0.5\, phe\, MeV^{-1}\, PMT^{-1}$. The primary trigger has a rate $\approx 130\ Hz$, while the secondary trigger rate is $\approx 0.15\ Hz$. In addition to the main trigger, a trigger requiring a delayed coincidence of two lower threshold primary triggers within $\approx 380\mu s$ is also present; it has a rate of $6\ Hz$ and it operates in connection with a neural-network-based event reconstruction fast processor (NNP) which reduces the rate to $\approx 0.35\ Hz$ [8]. The acquisition system uses a SUN computer and FASTBUS, VME and CAMAC circuitry. Each PMT is read into one ADC and TDC FASTBUS channel and by the multiplexed-by-8 CAMAC ADC of the NNP processor; in addition to that, the event history is recorded by multiplexed-by-8 fast ($150\ MHz, 206ns, 8pages$) and slow ($20\ MHz, 200\mu s$) waveform digitizers. The NNP is based on a CNAPS/VME parallel processor [9] on which the JETNET 3.0 [10] was implemented. NNP reconstructs the position and the energy of events in $\approx 180\mu s$; this makes possible the study of events corresponding to much lower energy thresholds than what is allowed by the main CHOOZ trigger; events corresponding to neutron capture on hydrogen or to low energy radioactivity become therefore accessible. Calibrations are performed, almost daily, by ^{60}Co and ^{252}Cf sources at the center of the detector. Other measurements, routinely performed, are the determination of PMT amplifications at one photoelectron ($\approx 25\ mV/phe$), the determination of the total number of photoelectrons for an energy loss of 1 MeV at the center of the detector ($\approx 130\ phes/MeV$), the energy calibration of the trigger thresholds, both at the center and at other positions in the detector, etc. Fig. 1 shows some data relative to the ^{252}Cf source at the detector center; the neutron capture lines, 2.2 MeV on hydrogen, 8 MeV on Gadolinium, are cleanly seen and the energy resolutions are $\sigma/E = 9\%$ and $\sigma/E = 6\%$ respectively; also shown are the global (neutron slowing-down and diffusion + gamma absorption) neutron position determination; $\sigma_x \approx 25\ cm$.

3 Data Analysis

The scintillator calorimeter is watched by 192 $PMTs$; the event reconstruction procedure is mainly based on the processing of their pulse height information. It can be shown that there is no loss of energy and position resolutions if one processes the signals from 24 *patches*, i.e.: groups of 8 $PMTs$.. The reconstruction procedure by a MINUIT fit takes a long time ($\approx 1\ s\, event^{-1}$ on a *alpha* 300 computer); things greatly improve ($\approx 1\ ms\, event^{-1}$) if one uses neural network techniques (NN) based on the package $JETNET\,3.0$ [10], trained by Monte Carlo method events. Since the reconstruction by a MINUIT fit is more precise than the one by the NN, we prefer first to reconstruct the events by the NN and then to perform a MINUIT fit, with the initial values supplied by the NN. The selection of neutrino events is based on simple cuts: 1) the requirement that the first (e^+) of the two correlated events be of energy $E < 10\ MeV$ and the second (n) be of energy $5 < E < 15$, 2) that both be at a 30 cm distance from the PMT supporting structure, 3) a distance between the two $< 100\ cm$, a time interval bewteen the two $< 100\mu s$. The positron and neutron "densities" are shown in fig. 2, as a function of R, the distance from the detector vertical axis (the events at position R are weighted by $1/V(R)$,

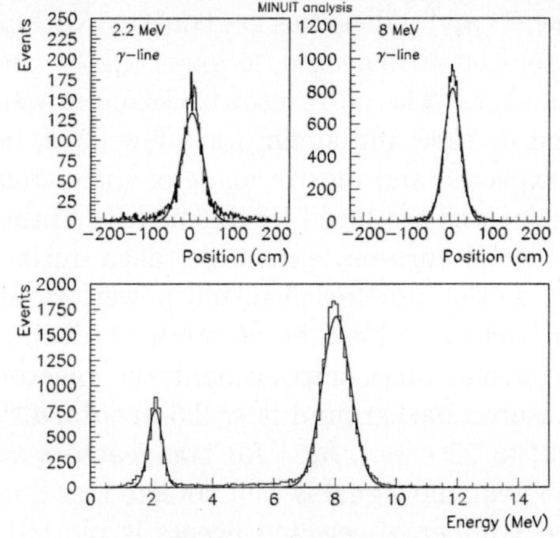

Fig.1 Capture lines for Gd source.

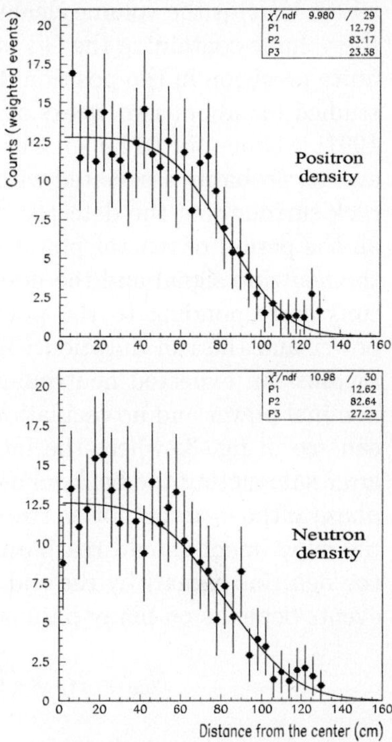

Fig.2 e+ and n densities in target.

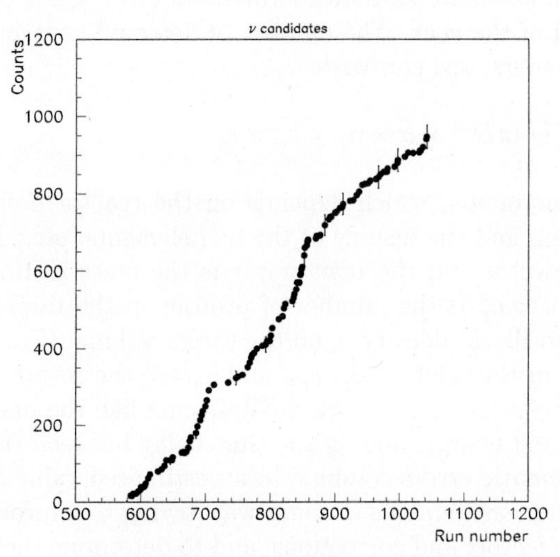

Fig.3 Neutrino candidates vs. time.

where $V(R)$ is the volume element. The e^+ and n densities are constant within the volume containing the Gd scintillator and are smeared, as expected, by the finite precision in the position determination. The no-reactor background was studied for about two weeks at the end of 1996 and again, for a few days, in 1997; it came out to be smaller than expected and mostly made of correlated events, probably associated with fast neutrons produced by cosmic rays in the rock surrounding the detector. Since reactor turn-on, data were taken during all the period of reactor power rise, up to the, now reached, full power. Both the neutrino signal and the no-reactor background can be determined, during runs corresponding to the power rise if one takes into account the reactor power and the run duration. The measured background is $\approx 2.5\ events/d^{-1}$ (against an expected neutrino signal of $\approx 25\ events/d^{-1}$ for two reactors at nominal power and no oscillations); the neutrino signal is well detected, as one can see in fig. 3, where the integrated number of selected events is plotted, after subtracting the background, as a function of the run number, during the phase ofthe reactor power rise. If the reactor program goes on as foreseen the experiment should reach an adequate statistics (statistical error $< 4\%$ on the neutrino signal) by the end of the year. The number of detected neutrino events depends on many parameters; one can write:

$$N_{ev} = N_\nu \times (1/4\pi D^2) \times \sigma \times n_p \times \epsilon_n \times \epsilon_{e+}$$

where N_ν is the number of reactor $\bar{\nu}_e$s, which depends on the reactor daily program, on the fuel composition and the history of the fuel elements, etc.; D is the distance between the detector and the reactors; σ is the cross section for the reaction $\bar{\nu}_e + p \to n + e^+$; n_p is the number of protons in the target, which depends on the liquid scintillator density ρ, on the target volume V and on the percentage of hydrogen in the scintillator; ϵ_{e+} and ϵ_n are the positron and neutron detection efficiencies,which depend on analysis cuts like the ones on the event position, on the event energy, and on the time delay between the e^+ and the n events. The systematic errors combine to an estimated value of $\approx 4\%$. A program of checks and measurements is under way, trying to improve on the knowledge of the various factors and corrections, and to determine their residual uncertainties.

4 Conclusions

Chooz has measured the no-reactor background at the end of 1996; it has successfully taken neutrino data for several months, and will probably reach a statistical error $< 3\%$ by the end of the year. The experiment should be in a position to publish results, in a written form (by collaboration agreement), by late autumn 1997.

References

1. for a review see M.Nakahata, Neutrino masses and oscillations, in these Proceedings.
2. The CHOOZ Collaboration, Proposal. Available on the internet at http://duphy4.physics.drexel.edu/chooz_pub/
3. The PALO VERDE collaboration, Proposal.
4. B. Achkar *et al.*, *Nucl. Phys.* B **434**, 503 (1995).
5. G. Zacek *et al.*, *Phys. Rev.* D **34**, 2621 (1986).
6. G.S. Vydiakin *et al.*, *JEPT Lett.* **59**, 364 (1994).
7. A. Baldini *et al.*, *Nucl. Instrum. Methods* A **372**, 207 (1996).
8. A. Baldini *et al.*, *Nucl. Instrum. Methods* A **389**, 1441 (1997).
9. Adaptive Solutions, 1400 N.W. Compton Dr., Suite 3340, Beaverton OR 97006.
10. C. Peterson et al., CERN-TH, 71335/94.

The Palo Verde Neutrino Oscillation Experiment

Yi Fang Wang (yfwang@hep.stanford.edu)

Stanford University, USA

Abstract. A new long baseline neutrino oscillation experiment[1] is being built at the Palo Verde Nuclear Generating Station near Phoenix, Arizona. Its sensitivity is $\Delta m^2 \sim 10^{-3} eV^2$ and $sin^2(2\theta) \sim 0.1$. The lab construction has been finished, the detector installation is almost completed and we are about to take data soon.

The Palo Verde Neutrino Oscillation Experiment(formally known as San Onofre experiment) [1] at the Palo Verde Nuclear Generating Station near Phoenix, Arizona, is designed to search for neutrino oscillations with a baseline of $\sim$ 750 m. The total thermal power of three reactors is 10.9 GW, giving 2.3×10^{31} electron anti-neutrinos per second. The energy spectrum of these neutrinos, peaking at 2 MeV, will be measured using inverse β decay $\bar{\nu}_{\mathrm{e}} + \mathrm{p} \longrightarrow \mathrm{n} + \mathrm{e}^{+}$. Any distortions of measured spectrum from the known neutrino spectrum [2] of the reactor would indicate $\bar{\nu}_e \rightarrow \bar{\nu}_X$ oscillations.

The detector consists of a 6×11 array of 9 m long cells filled with Gadolinium-loaded liquid scintillator. Positrons deposit their energies in the scintillator and annihilate, yielding two γ's. This process generates a triple coincidence signal in the detector. Neutrons thermalize and are captured within $100\mu s$ in the Gd, giving a γ-ray shower of 8 MeV total energy. The 12-ton central detector is surrounded by a 1 m thick water shield and by large cosmic ray veto counters as shown in Figure 1. The detector is located in a shallow underground laboratory at a depth of 25 m to be shielded from cosmic rays.

The backgrounds come from cosmic-muon-generated neutrons(correlated) and environmental radioactivity(uncorrelated). The correlated background from muon capture and muon spallation can be measured during the refueling time of reactors which is scheduled to be two periods of 40 days per year with one reactor down each time. It can also be measured by triggering on cosmic muons. This uncorrelated background can be measured with a great precision by varying the time correlation window between the prompt positron event and delayed neutron event. We have carefully selected construction materials

[1] The Palo Verde Collaboration: D. Lawrence, B. Ritchie, Physics Department, Arizona State University; F. Boehm, B. Cook, J. Hanson, H. Henrikson, K. Lou, N. Mascarenas, D. Michael, V.M. Novikov, A. Piepke, P. Vogel, S. Yang, Division of Physics and Astronomy, Caltech; S. Pittalwala, R. Wilferd, S. Young, Palo Verde Nuclear Generating Station; G. Gratta, L. Miller, D. Tracy, Y. F. Wang, Physics Department, Stanford University; J. Busenitz, A. Vital, J. Wolf, Department of Physics and Astronomy, University of Alabama.

for the lab and all detector components in order to minimize the uncorrelated background from environmental radioactivity. Extensive Monte Carlo Study shows that neutrino signals will be about 51 events per day while the correlated background is about 34 events/day and uncorrelated background is about 15 events/day. For one year of running, we expect to reach the sensitivity of $\Delta m^2 \sim 10^{-3} eV^2$ and $sin^2(2\theta) \sim 0.1$.

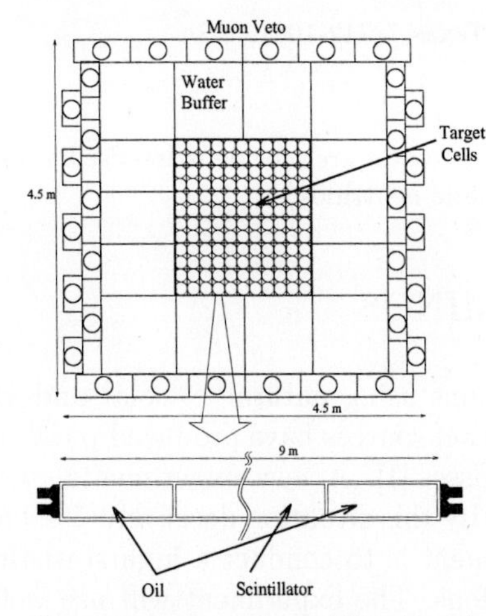

Figure 1. The front view of the Palo Verde Detector

The mineral-oil-based liquid scintillator is loaded with 0.1% Gd to reduce the neutron capture time, which is about 28 μs on Gd, allowing a smaller time correlation window, thus reducing dramatically the uncorrelated background. The light attenuation length of the scintillator has been achieved 7.8 m over three years without significant deterioration. Up to now (Sep. 1997) nine out of 66 cells have been filled with scintillator and installed. We have been taking calibration data using cosmic rays and radioactive sources.

The civil construction of the lab started in May 1996 and was completed in Dec. 1996. The mechanical installation of the detector was completed in March 1997. The muon veto system has been installed and is fully functional. The custom-made front-end electronics and the FPGA-based trigger system were designed, manufactured and installed this summer. The Fastbus-based DAQ system has been installed and is running. The whole experiment will be started at the end of this year.

References

[1] G. Gratta *et al.*, in Proceedings of the "XVII International Conference on Neutrino Physics and Astrophysics", Helsinki, June 96;
F. Bohem *et al.*, "Proposal for the San Onofre Oscillation Experiment", Caltech, Jan. 1994, Unpublished.

[2] B. R. Davis *et al.*, Phys. Rev. C19 (1979) 2259; P. Vogel *et al.*, Phys. Rev. C24 (1981) 1543; G. Zacek *et al.*, Phys. Rev. D34 (1986) 2621.

MINOS: Long Baseline Search for Neutrino Oscillations at Fermilab

Karol Lang[1] (lang@hep.utexas.edu)

The University of Texas at Austin, Austin, Texas 78712-1081, USA

Abstract. MINOS — a long baseline two-detector search for neutrino oscillations mixing until now studied only with atmospheric neutrinos.

1 Motivations and Goals of MINOS

Recent explorations of neutrino interactions using natural — solar and atmospheric — and accelerator-based neutrino sources have produced puzzling results which hint non-zero neutrino masses [1]. A new experimental program is being planned at Fermilab to clarify this situation decisively [2]. The main motivation for the MINOS experiment is to conduct a high statistics two-detector search for neutrino oscillations. The experiment will use well-controlled accelerator beams with the goal to cover a large range of Δm^2 and $\sin^2(2\theta)$ mixing parameters which would include the region previously covered only with atmospheric neutrinos. For $\nu_\mu \to \nu_e$ transitions MINOS will span Δm^2 below 10^{-3} eV^2 and $\sin^2(2\theta)$ close to 10^{-3} (at $\Delta m^2 \simeq 10^{-2}$ eV^2). The $\nu_\mu \to \nu_\tau$ search will have a somewhat poorer reach. Several oscillation signatures and optimized neutrino beams (both wide and narrow band) will provide a variety of control tools for understanding and minimizing measurement uncertainties and to determine the mixing parameters — if a positive signal is detected.

2 The Beam and Detectors

A new neutrino beamline, 750 m long and declining towards the Soudan mine in Minnesota at the 58 mrad angle, will accept up to 4×10^{13} protons with energy of 120 GeV delivered every 2 seconds from the Main Injector. A three-horn wide band ν_μ beam and a lithium lens narrow band ν_μ beam are being designed. Additional hadron focusing using a current carrying axial wire running along the decay pipe (a "hadron hose") is also under investigation. This and a number of target, transport, and focusing elements are being optimized to assure that the beam energy spectrum, divergence, and composition will be well understood and controlled. The smaller mass, 1 kton, "near" detector will be placed about 500 m downstream of the end of the decay pipe and about 100 m underground. It will assure precise determination of the beam characteristics such as the energy spectrum and ν_e contamination, expected at the 0.5% level. The entire detector will provide continuous beam monitoring while only the small region within 25 cm radius around the beam axis will be used for near-far neutrino oscillation analysis. The large mass, 8 kton, "far" detector will be placed 713 m underground in the Soudan mine 731 km from the near detector. The role of this detector will be to measure the rate of the ν_μ interactions and identify statistically a fraction of ν_e and ν_τ events. Both detectors will be as similar in structure as possible. The baseline design of the far detector is an 8 m diameter octagon composed of 730 magnetized 1 inch thick steel plates interspersed with 1 cm thick and 4.1 cm wide extruded polystyrene-based scintillator strips.

Each strip will be instrumented with a light-collecting 1 mm diameter wavelength shifting fiber imbedded in a surface groove. The fibers will be read out at both ends in a multiplexed fashion by grouping 8 fibers spaced by 1 m on an individual pixel of a multianode photomultiplier. By grouping differently the two sides of the detector an unambiguous pattern recognition is possible. The near detector will be a similar 6 m octagon instrumented strategically for optimized hadron shower energy and muon momentum measurements.

The main ν_μ oscillation signatures will be the ratio of the number of charged current ν_μ events (CC_μ) — expected in the far detector at the rate of about 30,000 per year — to the number of neutral current (NC) events in the two detectors, the rate of CC_μ events in the far detector as predicted by the CC_μ rate in the near detector, the shape of the energy spectra of the CC_μ events, and the explicit ν_e and ν_τ identification. In these comparisons some unavoidable differences between the two detectors will have to be accounted for. Principally, these differences stem from the relative proximity of the near detector to the neutrino source which results in higher event rate (by almost a factor of 10^6) and somewhat different energy spectrum.

3 Look Ahead

An R&D program is underway to further optimize and extend the low Δm^2 reach of the baseline MINOS design. This effort includes optimization of low energy beams, the engineering of the steel plates and improving the quality of extruded scintillator. In addition, possible use of hybrid photodetectors is investigated. The technical design report of MINOS will be prepared in the Spring of 1998. The plan will assume a funding profile aiming to start data collection in 2002.

References

[1] Super-Kamiokande's and M. Nakahata's contributions to this conference
[2] Fermilab Proposal P875, February, 1995

Results from Super-Kamiokande

Kai Martens *for the Super-Kamiokande Collaboration (kai@icrr.u-tokyo.ac.jp)*

Institute for Cosmic Ray Research, University of Tokyo, Japan

Abstract. 300 days of data taking with the Super-Kamiokande detector give preliminary new constraints on neutrino oscillation parameters. 4400 solar neutrinos were detected with energies between 6.5 MeV and 20 MeV. No evidence for a day/night effect or spectrum distortion is observed so far. The preliminary value for the observed flux is $2.44^{+0.06}_{-0.06}(stat.)^{+0.25}_{-0.09}(sys.) \times 10^6 \mathrm{cm}^{-2}\mathrm{s}^{-1}$. A preliminary $\nu_\mu \rightarrow \nu_\tau$ analysis of our atmospheric neutrino data indicates Δm^2 of order a few times 10^{-3} and a large mixing angle.

1 The Super-Kamiokande Experiment

Super-Kamiokande (SK) is a 50 kt water cherenkov detector built and run by a Japanese-American collaboration. Data taking 1000 m (2700 MWE) underground in the Kamioka Mining Company's Mozumi Mine 30 km south of the Japan Sea port city of Toyama started on April 1st 1996. Apart from its neutrino related physics program SK sets new limits on the proton lifetime.

The cylindrical detector volume is divided into an inner and an outer detector (ID and OD), separated by a light barrier. The 32.5 kt ID is 36.2 m high, has a diameter of 33.8 m and is viewed by 11146 20 inch photomultiplier tubes (PMT). The OD completely surrounds the ID with > 2 m of water viewed by 1885 8 inch PMTs. Together with 55 cm of dead volume inside the light barrier this amounts to a water shield of > 2.6 m all around the ID.

Vital for SK is its water purification system (WS), which maximizes water transparency and minimizes natural radioactivity. The equivalent of one detector volume is recirculated through it in about one month. By the end of May 1996 changes in the attenuation length for cherenkov photons had dropped to the 1% level. Special precautions are taken to prevent admission of radon into the tank water, since the β-decay of ^{214}Bi in its decay chain is a serious source of background at low energies. The current level of radon activity is < 5 mBq - a factor 100 lower than in Kamiokande.

PMT signals from the ID are processed in custom made ATM boards that digitize time and charge of PMT hits. The boards are self-triggering with a threshold of 1/4 photoelectron (PE) equivalent. A readout of the experiment is initiated when a certain number of PMTs receive hits within 200 ns. To avoid dependency of the detector trigger threshold on changing PMT noise levels the corresponding signal is AC-coupled into a discriminator. Current trigger thresholds for the ID correspond to 29 and 31 hit PMTs, the lower one implying an energy threshold of 5.5 MeV. The corresponding trigger rates are

10 – 12 Hz and 4.3 Hz. OD-triggers are generated independently at a rate of 2.7 Hz. A 25 hit threshold is just becoming operational in an effort to lower the analysis threshold for solar neutrinos. Data are collected at a rate of 10 Gbytes/day. Calibration efforts limit the livetime of the experiment to 85% over the whole period of operation since April 1st 1996.

2 ID Calibration Procedures

Calibration of the ATM electronics boards is done with a standard calibrated charge/time generator. The timing range spans 1.2 μs with 0.3 ns resolution and charge is recorded up to 550 pC with 0.2 pC resolution. Charge measurement reaches overflow at the level of 250 PE.

Relative gains of ID PMTs are balanced using light from a scintillator ball driven by a Xe flashlamp. After adjustment the spread in gain amounts to 7%. For timing calibration the short pulse of a nitrogen laser is diffused from the center of the ID. The average timing resolution of the inner detector PMTs is found to be 3 ns at the 1 PE level, reaching a limit of 0.6 ns at about 100 PE. Energy calibration depends on the energy range of interest and is discussed in the corresponding analysis section.

3 Solar Neutrinos

Two of the reactions in the solar fuel cycle produce neutrinos with energies above 2 MeV: β-decay of ^{8}B and proton capture on ^{3}He. Predictions for their rates are inferred from solar modelling (SM). The rate of ^{3}He-neutrinos is expected to be $< 10^{-3}$ that for ^{8}B neutrinos, which itself constitutes only $\sim 10^{-4}$ of the total flux of ν_{sol}. Thus only the ^{8}B neutrinos are accessible to an experiment like SK with a threshold of $\sim$ 5 MeV. Their spectrum ends at 14.1 MeV, its shape being independend of solar physics[1].

The data discussed here were taken between May 31st 1996 and June 23rd 1997. The accumulated livetime is 306.3 days. The fiducial volume for solar neutrino (ν_{sol}) and atmospheric neutrino analysis contains 22.5 kt of water. It is defined by a minimum distance of 2 m from the ID PMT surfaces.

Background to the ν_{sol} analysis has three main sources. Cosmic ray muons may produce radioactive spallation products. Decays of spallation products are identified using their space and time correlation with preceeding muons. Deadtime introduced amounts to 20%, while the background is cut by 43%. γ-rays originating in the surrounding rock or PMT glass produce an excess of events near the confines of the detector. Events for which their back projected track intersects with the rock within 5 m are removed from the sample at the cost of 7.8% deadtime, yielding a further reduction of background by 30%. Radon decay events can not be isolated.

The analysis discussed in this contribution is the first ν_{sol} analysis with its energy scale based on SK calibration with a small linear accelerator for electrons (LINAC). Single electrons are injected into the tank at various radii and depth. At each position data are taken at discrete electron energies ranging from 5 to 16 MeV. Energy definition by the LINAC system is better than the 0.5% resolution of the Germanium detector used to calibrate it. Only two LINAC positions in the ID were completed in time for the analysis discussed here. After careful evaluation the energy scale for ν_{sol} events was shifted by $\sim$ 2% from what had previously been inferred from data taken with a γ-ray source. Subsequent LINAC measurements at other points are consistent with the initial two and validate this shift. Yet the systematic error on the ν_{sol}-flux is currently dominated by energy scale (+9.9%) and resolution (-3.1%). These errors have less impact on relative measurements like day/night (D/N) and seasonal variations. Vertex and direction fitting contribute -1.3% and +1.7% respectively. All other systematics are estimated to be less than 0.7%, bringing the total to $^{+10.1\%}_{-3.5\%}$ for the flux measurement.

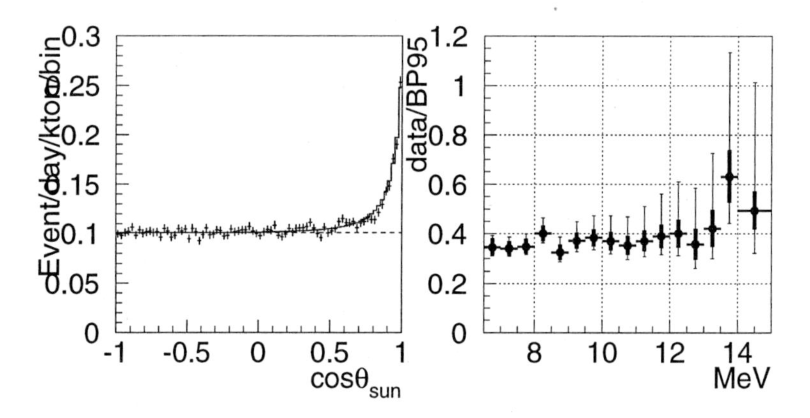

Fig. 1. 4400 ν_{sol} with energies 6.5 – 20 MeV from 306 d of SK data and their spectrum, normalized to the prediction from BP95

The left side of figure 1 shows the distribution of directions relative to the direction to the sun in the final sample. Of its almost 60000 events with energies from 6.5-20 MeV 4400 are in the peak towards the sun. The flux value extracted from the fit (histogram) to the data is $2.44^{+0.06}_{-0.06}(stat.)^{+0.25}_{-0.09}(sys.) \times 10^6 \mathrm{cm}^{-2}\mathrm{s}^{-1}$. Comparison to the solar model of Bahcall and Pinsonneault[2] (BP95) yields $0.368^{+0.010}_{-0.009}(stat.)^{+0.037}_{-0.013}(sys.) \times$ BP95. The measured D/N asymetry $(\mathrm{D}-\mathrm{N})/(\mathrm{D}+\mathrm{N})$ is $-0.017 \pm 0.026(stat.) \pm 0.017(sys.)$. The right side of figure 1 shows the measured energy spectrum normalized to the ^{8}B neutrino spectrum with BP95 normalization. The inner part of the error bars represents the statistical error, the outer part the systematic error estimate. No deviation from uniform suppression can be inferred.

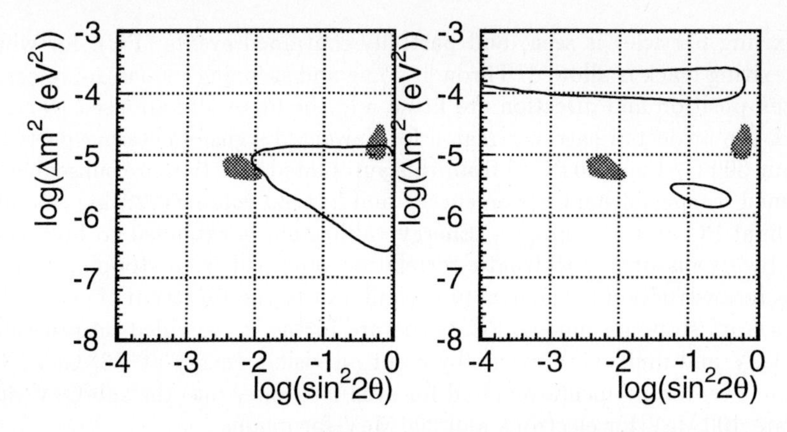

Fig. 2. Exclusion plots based on 300 d of SK D/N (left) and spectral (right) data

Based on these results 90% CL exclusion plots for neutrino mixing parameters in the MSW region are generated employing a χ^2-method. In figure 2 these are shown together with the 95% CL allowed regions obtained by Hata and Langacker[3] from a combined fit to older solar neutrino data from various experiments (shaded areas). The exclusion plot on the right is derived purely from the spectral information without constaint on the overall flux. For the D/N plot on the left the nighttime data were further subdivided into five $\cos(\theta_{zenith})$ bins with θ_{zenith} being the zenith angle of the momentary position of the sun. Vacuum oscillation parameters too are further constrained.

4 Atmospheric Neutrinos

Primary cosmic rays impinging on the atmosphere produce hadronic showers. Atmospheric neutrinos (ν_{atm}) originate from the decays of secondaries in these showers. For neutrino energies below 1 GeV almost all ν_{atm} stem from pion decay. That leads to the prediction $r = (\nu_\mu + \overline{\nu}_\mu)/(\nu_e + \overline{\nu}_e) \approx 2$. Interest in ν_{atm} was sparked by the observation that in all high statistics experiments this ratio differed from expectation. Since predictions for the absolute rate of ν_{atm} are still rather uncertain it remains unclear whether this anomaly reflects a lack of ν_μ or an excess of ν_e. Yet for the ratio r large parts of these uncertainties cancel out and predictions by various authors agree within 5%. Thus the double ratio $R = r_{Data}/r_{MC}$ has become the focus of attention.

Like proton decay, neutrino interactions produce contained events, i.e. no particle enters through the OD. With muons losing about 200 MeV/m in water high energy muons from a ν_μ interaction may exit from the detector. Thus two event samples are collected from the data stream for the ν_{atm} analysis: Fully contained events (FC), for which no evidence of associated entering or exiting particles is seen, and partially contained events (PC), for which one exiting track is allowed. Through-going and stopping muons are rejected. Vertex position and direction are fitted and the fit results are used to cross-check the projected entrance region for unwanted signs of entering particles. About 30 FC/d and 20 PC/d from this automated selection are subsequently scanned by specialists. Cuts on energy and fiducial volume (22.5 kt) establish the final PC and FC samples. Energy calibration is extended to high energies by means of a tracklength correlation obtained from stopping muons, mass reconstruction for neutral pions and a fit to the spectrum of decay electrons from stopping muons. FC events are further subdivided into so-called sub-GeV and multi-GeV events by a cut on visible energy at 1.33 GeV. The minimal lepton momenta required for events to enter into the sub-GeV sample are 100 MeV for electrons and 200 MeV for muons.

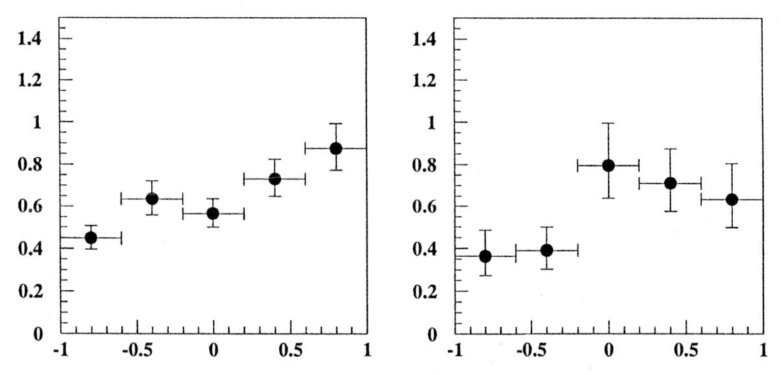

Fig. 3. Zenith angle dependence of R for 326 d of SK sub-GeV (left) and multi-GeV+PC (right) data. Horizontal axis is $\cos(\theta_{zenith})$, vertical is R

In charged current (CC) reactions on the nucleons in the detector water the outgoing lepton preserves the lepton flavor of the interacting neutrino. To identify the lepton flavor events are classified as showering or non-showering[4], introducing systematic errors of 2% on the sub-GeV sample and 4% on the multi-GeV sample. PC events are a 97% pure ν_μ sample. Pion production in the interaction process gives rise to multi ring events. To obtain the cleanest possible sample of CC quasi-elastic scattering events only single ring events are used, giving rise to a 4% error from ring counting. Systematic errors arising from uncertainties in the CC and NC interaction cross sections amount to 3.6% and 2-3% respectively, bringing the estimated total systematic errors to 8.2% on the sub-GeV data and 10.7% on the combined multi-GeV+PC samples.

The preliminary result for R from 325.8 d of ν_{atm} observation with SK is $0.604^{+0.065}_{-0.058} \pm 0.018 \pm 0.065$. The errors given are the statistical error of the data sample, MC sample, and the systematic error.

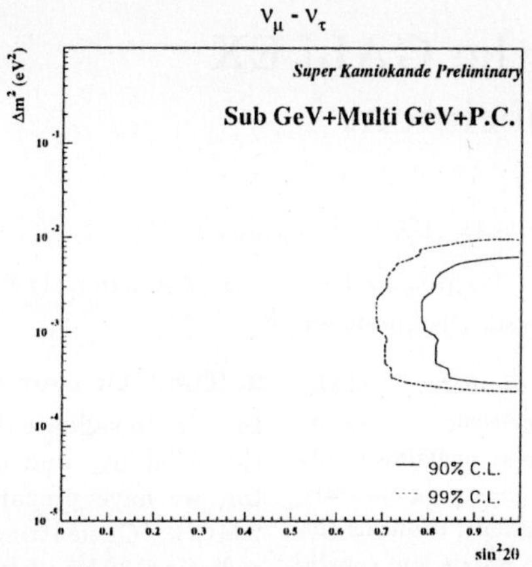

Fig. 4. 90 and 99% CL allowed regions for $\nu_\mu \to \nu_\tau$ oscillation from 326 d of SK atmospheric neutrino data

Figure 3 shows the dependence of R on the zenith angle for the sub-GeV and multi-GeV+PC samples. $\cos\theta_{zenth} = 1$ corresponds to downward going events. The zenith angle dependence of the sub-GeV data forces Δm^2 down to a few times 10^{-3} in a $\nu_\mu \to \nu_\tau$ oscillation analysis as shown in figure 4.

5 Conclusion

Solar neutrino observation with SK entered a new era with the advent of the first LINAC data. Oscillation analysis so far remains inconclusive. Extending the spectal range of this measurement to lower neutrino energies is a clear priority for SK.

Atmospheric neutrinos show a zenith angle dependence when comparing the observed to expected μ-like/e-like event ratio. In the context of $\nu_\mu \to \nu_\tau$ oscillations our measurement suggests a large mixing angle and Δm^2 of order a few times 10^{-3}. Suitable long baseline experiments will soon explore large parts of this parameter space.

References

[1] J.N. Bahcall,Phys. Rev. D **44**, 1644 (1991)
[2] J.N. Bahcall, M.H. Pinsonneault, Rev. Mod. Phys. **67**, 781 (1995)
[3] N. Hata, P. Langacker, Preprint No. HEP-PH/9705339
[4] S. Kasuga et al., Phys. Let. B **374**, 238 (1996)

Results from the GALLEX Solar ν– Experiment

Michael Altmann (for the GALLEX Collaboration)

Physik Department E15, Technische Universität München, D-85747 Garching, Germany; altmann@e15.physik.tu-muenchen.de

Abstract. After six years of operation GALLEX has completed its large-scale experimental program. The overall solar neutrino result of 65 exposures, 76 ± 8 SNU, being significantly below all solar model predictions, confirms the long-standing solar neutrino puzzle and constitutes an indication for non-standard neutrino properties. This conclusions is validated by the results of ^{51}Cr neutrino source experiments and ^{71}As doping tests.

1. Solar neutrino observations

The radiochemical GALLEX detector, using 30 tons of gallium in the form of 101 t $GaCl_3$ solution, has been measuring the integral solar neutrino flux exploiting the capture reaction $^{71}\text{Ga} + \nu_e \rightarrow {}^{71}\text{Ge} + e^-$. Referring to [4,5] for a detailed description of the detector setup and experimental procedure we concisely summarize the GALLEX experimental program in tab.1.

Table 1. GALLEX experimental program. In column 2 's' and 'b' denote 'solar run' and 'blank run', respectively. For Gallex-IV pulse shape information is used only for K-signals.

period	no. of runs	result
Gallex I	15 s + 5 b	$81 \pm 17 \pm 9$ SNU
Gallex II	24 s + 22 b	$75 \pm 10 {}^{+4}_{-5}$ SNU
Source I (^{51}Cr)	11 source	$R = 1.01 {}^{+0.11}_{-0.10}$
Gallex III	14 s + 4 b	$54 \pm 11 \pm 3$ SNU
Source II (^{51}Cr)	7 source	$R = 0.84 {}^{+0.12}_{-0.11}$
Gallex IV	12 s + 5 b	$118 \pm 19 \pm 8$ SNU
^{71}As-test	4 arsenic	$Y = 1.00 \pm 0.03$

The combined result of all 65 solar runs is $76.4 \pm 6.3 {}^{+4.5}_{-4.9}$ SNU. This result is only about 60% of the predictions from solar model calculations, which constitutes an indication for non-standard neutrino properties, even without considering the results from other solar neutrino experiments.

2. The ^{51}Cr source experiments

In order to validate this conclusion and check the reliability and efficiency of the detector, we have prepared an intense (almost 2 MCi) ^{51}Cr neutrino source by neutron irradiating 36 kg of isotopically enriched Cr [3]. ^{51}Cr decays by EC and emits neutrinos of energy 750 keV (decay to g.s., 90%) and 430 keV (decay to 320 keV excited level, 10%), nicely accommodating the neutrino energies of the solar ^{7}Be branch. The source has been inserted in the central tube of the GALLEX target tank. Altogether 18 extractions have been performed with the source in place, divided into two series of measurements with the source being re-activated in between. Fig.1 shows the individual run results.

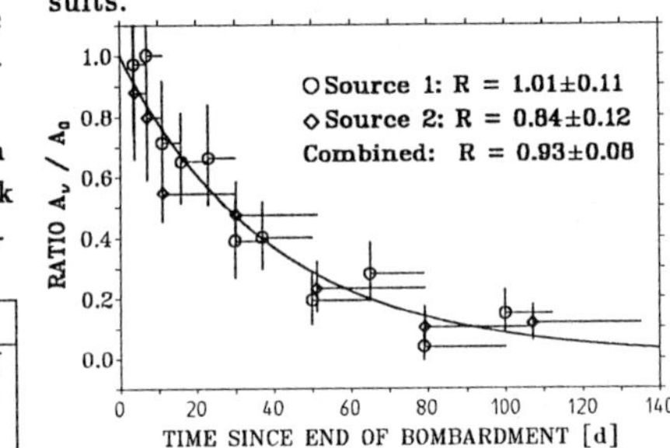

Figure 1. Individual run results, normalized to the known source activity, as a function of time after insertion of the source.

An analysis of Source I and Source II leads to $R = 1.01 {}^{+0.11}_{-0.10}$ and $R = 0.84 {}^{+0.12}_{-0.11}$, respectively, where R is the source activity as deduced from the neutrino measurement normalized to the known true source activity. The combined analysis of both series results in $R = 0.93 \pm 0.08$, clearly demonstrating the absence of large systematic errors which could account for the observed 40% solar neutrino deficit.

3. ^{71}As experiments

However, though the ^{51}Cr source did outperform the sun by more than a factor 15 after insertion in the target tank, the experiments still are low statistics, involving only several dozens of neutrino produced ^{71}Ge atoms. Therefore we have at the very end of GALLEX performed a large-scale test of potential effects of hot chemistry, which might lead to a different chemical behaviour of ^{71}Ge produced in a nuclear reaction compared to the stable Ge carrier isotope: The in-situ production of ^{71}Ge by $\beta-$decay of ^{71}As. A known quantity of ^{71}As (O(10^5) atoms) has been added to the tank (t-sample), where it decayed with $T_{1/2} = 2.9$ d to ^{71}Ge. Four runs have been made under different operating conditions (mixing, carrier addition, standing time), cf. tab.2. For every spike a reference sample (e-sample) was kept aside, making possible to calculate the ratio of t- to e-sample which does not suffer from most of the systematics associated with ^{71}Ge-counting.

Table 2. Experimental conditions and results (ratio t-sample/e-sample) of the ^{71}As runs.

run	mixing cond. [$h \times m^3/h$]	Ge-carr. addition	stand. time	result (t/e)
A1	22×5.5 $+0.17 \times 60$	with As	19.9 d	1.01 ± 0.03
A2	6×5.5	no	19.9 d	1.00 ± 0.03
B3-1	24×5.5	after As	2.0 d	1.01 ± 0.03
B3-2	—	after As	22.0 d	1.00 ± 0.03

In all cases we got a quantitative recovery of 100%. This demonstrates on a 3%-level the absence of withholding effects even under unfavourable conditions like carrier-free operation.

4. Gallium Neutrino Observatory

With the ^{71}As-tests GALLEX has completed its large-scale experimental program. However, solar neutrino measurements with a gallium target at Gran Sasso will be recommenced in spring 1998 in the frame of the Gallium Neutrino Observatory (GNO) [2], which is designed for long-term operation covering at least one solar cycle. In its first phase GNO will use the 30 t target of GALLEX. However, for the second phase we plan to increase the target mass to 60 t and later to 100 t. In addition, as it is equally important to decrease the systematic uncertainty, effort is made to improve ^{71}Ge counting, both by improving the presently used proportional counters, and by investigating novel techniques like semiconductor devices and cryogenic detectors [1].

5. Conclusion

The GALLEX result at the end of the experimental program is $76.4 \pm 6.3 \ ^{+4.5}_{-4.9}$ SNU, only about 60% of the solar model predictions. The observed deficit has been validated by ^{51}Cr neutrino source experiments and ^{71}As spiking tests which both gave results perfectly consistent with expectation. This serves to exclude experimental artifacts or an improperly determined capture cross section to be the reason for the solar neutrino deficit. Solar neutrino observations with a gallium target will continue at Gran Sasso for many years in the frame of the GNO experiment which is going to commence operation early in 1998.

Acknowledgement This work has been supported by the german BMBF.

References

[1] M.Altmann et al., Development of cryogenic detectors for GNO, Proc. 4th Int. Solar Neutrino Conf., ed.: W. Hampel, Heidelberg, Germany, 1997.

[2] E.Bellotti et al., Proposal for a permanent gallium neutrino observatory at Gran Sasso, 1996.

[3] M.Cribier et al., Nucl. Inst. Meth. A 378 (1996) 233.

[4] GALLEX Collab., Phys.Lett. B 285 (1992) 376.

[5] E.Henrich et al., Angew. Chem. Int. Ed. Engl. 31 (1992) 1283.

HELLAZ - the Third Generation Solar Neutrino Experiment

Thomas Patzak (Patzak@cdf.in2p3.fr)

Laboratoire de Physique Corpusculaire et Cosmologie, Collège de France, IN2P3 - CNRS, 11, Place Marcelin Berthelot, F-75231 Paris, France

Abstract. A new generation solar neutrino experiment, HELLAZ, has been proposed and is currently under development. The experiment is dedicated to measure the spectrum of neutrinos originating in the sun. In a first running period we will concentrate on neutrinos from pp and ^{7}Be weak interactions. Hellaz is a real-time experiment measuring the neutrino energy in elastic neutrino-electron scattering reactions. The target consists of cooled helium contained in a 2000 m^3 TPC. The expected rate of neutrino events is about 12 per day.

1 Physics with HELLAZ

Data from all running solar neutrino experiments indicate a large difference between the flux of neutrinos calculated in the standard solar model, [1], and the measured neutrino flux.
To provide an essential contribution to solve the enigma of solar neutrinos a new generation experiment, HELLAZ (**HEL**ium at **L**iquid **AZ**ote - nitrogen - temperature), has been proposed [2]. The most interesting solar neutrinos are formed in the reaction $p + p \longrightarrow {}^2_1D + e^+ + \nu_e$ with a continuous spectrum and an end-point energy of 420 keV (pp - neutrinos). Furthermore, the mono-energetic ^{7}Be neutrinos produced in the reaction ${}^7_4Be + e^- \longrightarrow {}^7_3Li + \nu_e$ are highly interesting because a best fit to all data from experiments is compatible with zero flux for ^{7}Be neutrinos.
Our main objective with the HELLAZ experiment is to measure the spectrum of pp - neutrinos and ^{7}Be neutrinos. Predictions from various solar models for the flux of ^{7}Be and ^{8}B - neutrinos show large differences since the involved nuclear cross sections are not well determined in this low energy region. All existing model predictions for the flux of pp - neutrinos are in very good agreement. This comes from the fact that the flux of pp - neutrinos is strongly constrained by the solar luminosity and measurements from helio seismology. Therefore, any observed difference in spectral shape, flavour or flux from model predictions has to be interpreted in terms of new physics (flavour oscillations) or in drastic changes of the solar model, which is very difficult because of the above mentioned constraints.

2 The HELLAZ experiment

The concept of the HELLAZ experiment is based on the measurement of the solar neutrino energy, E_ν, by measuring the kinetic energy, T_e, and the scattering angle, θ, of recoil electrons from elastic neutrino - electron scattering. The kinetic energy of recoil electrons is measured by counting the individual electrons in an ionisation cloud generated by the energy loss of the recoil electron due to ionisation in the helium gas. The scattering angle is measured by determining the relative position in space of these ionisation electrons. The coordinates are measured using a x,y detector covering the surface of each end-cap of the TPC. The relative z coordinate is obtained by measuring the arrival time of the individual electrons using TDC's.
An angular resolution of about 35 mrad will be achieved. Therefore, the energy of the incoming neutrino can be measured with a precision better then 10 %. This neutrino-energy resolution is good enough to detect a possible deformation of the pp spectrum as it is predicted for the vacuum oscillation solution [3].

To develop the HELLAZ detector a large R & D program is currently undertaken. We are focusing on tests of different detector options, micro-gap chamber (MGC) in combination with gas electron multipliers(GEM), micro-gap wire chamber (MGWC) and Micromegas. These solutions are chosen because of their rapidity.
In parallel a test facility to study the radiation of different materials has been constructed. The setup is situated in the underground laboratory Modane in France and consists of an ultra-pure Ge-detector shielded with archaeological lead.

Acknowledgements

I would like to thank Professor S.T. Petcov for numerous pleasant and inspiring discussions. I would like to thank my friends and colleagues from the HELLAZ collaboration for their great help and support.

References

[1] J.N. Bahcall, M.H. Pinsonneault, rev. Mod. Phys. 67 (1995) 781.
[2] F. Arzarello et al., CERN-LAA/94-19.
[3] P.I. Krastev and S.T. Petcov, Phys. Rev. D53 (1996) 1665.

Double-β Decay of Various Nuclei with the NEMO Experiment

NEMO Collaboration, presented by F. Laplanche (laplanch@lal.in2p3.fr)

LAL, IN2P3-CNRS et Université Paris-Sud, 91405 Orsay, France

Abstract. The NEMO-2 detector, built for background measurements and located in the Fréjus Underground Laboratory has provided results for three $\beta\beta$ decay sources (^{100}Mo, ^{116}Cd and ^{82}Se) with data accumulated over a total of 23000 h. The NEMO-3 detector which will be able to accomodate sources of 10 kg is under construction and the starting of the experiment is foreseen in 1999.

1 Introduction

Two prototype detectors, NEMO-1 [1] and NEMO-2 [2] have been constructed as research and development tools. The NEMO-2 detector, designed for $\beta\beta$ and background studies, was operating in the Fréjus Underground Laboratory (4800 w.m.e.). From 1992 to 1995 $\beta\beta2\nu$ decays of ^{100}Mo [3] and ^{116}Cd [4] were studied. These already published results are recalled. Enriched and natural sources of selenium were installed in the NEMO-2 detector in autumn 1995, results of measurements with ^{82}Se are presented here.

The NEMO Collaboration started to build the tracking detector NEMO-3 [5] for $\beta\beta$ decay experiments which will be capable of studying $\beta\beta0\nu$ decays of ^{100}Mo and other nuclei up to half-lives $\sim 10^{25}$ y, which corresponds to a Majorana neutrino mass $\sim$ 0.1 eV. The sensitivity to $\beta\beta0\nu\chi^0$ and $\beta\beta2\nu$ decays will be $\sim 10^{23}$ y and $\sim 10^{22}$ y, respectively.

2 NEMO-2 detector

NEMO-2 (Fig. 1) consists of a 1m^3 tracking volume filled with helium gas and 4% ethyl alcohol. Vertically bisecting the detector is the plane of the source foil (1m $\times$ 1m). Tracking is accomplished with long open Geiger cells with an octagonal cross section defined by 100 μm nickel wires. On each side of the source foil there are 10 planes of 32 cells which alternate between vertical and horizontal orientations. A calorimeter made of scintillators covers two vertical opposing sides of the tracking volume. In the configuration used with molybdenum sources, the calorimeter consisted of two planes of 64 scintillators associated with "standard" PMTs. After this first experiment the calorimater was modified, in each plane 25 scintillators were associated with low radioactivity PMTs. The tracking volume and scintillators are surrounded with a lead (5 cm) and iron (20 cm) shield. The performances and parameters of NEMO-2 are described elsewhere [2].

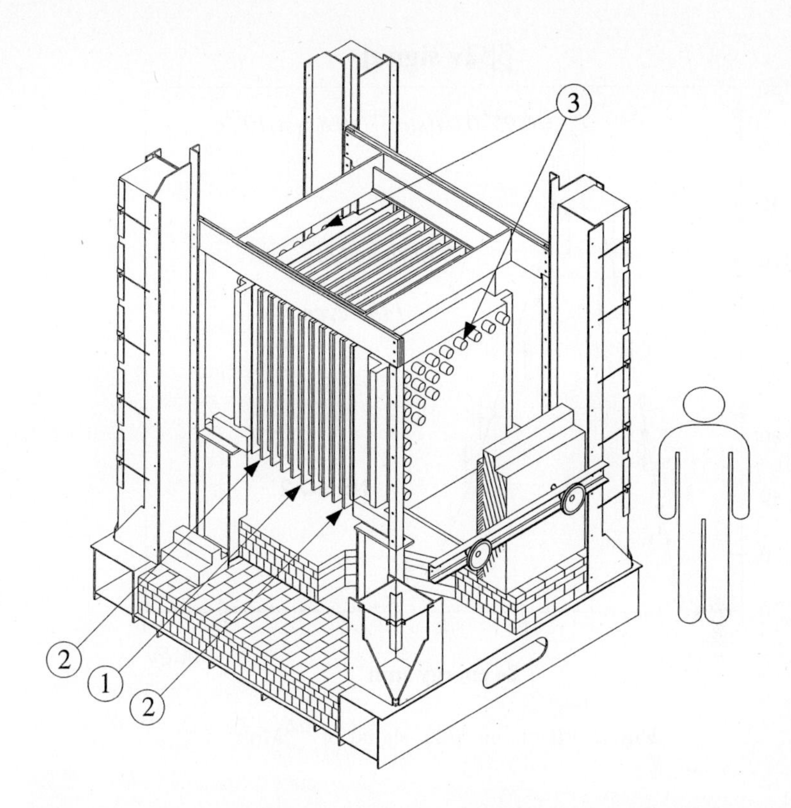

Fig. 1. The NEMO-2 detector without shielding. (1) Central frame with the source plane. (2) Tracking device of 10 frames of Geiger cells. (3) Two scintillator arrays.

3 Molybdenum data

In this experiment [3] data were accumulated during 6140 h (1.18 mol·y of ^{100}Mo.) The 1 m^2 source foil was divided into two parts, the first one was 172 g (40μm thick) of enriched molybdenum (98.4% in ^{100}Mo, Q = 3.03 MeV) and the second one was 163 g (44μm thick) of natural molybdenum (9.6% in ^{100}Mo). Only limits on the activity of ^{214}Bi, ^{208}Tl and ^{234m}Pa (most troublesome isotopes), were obtained from Germanium detectors and NEMO-2 itself. The background due to internal radioactivity was a few percent of the signal. The "external" background induced by the photon flux inside the shield was subtracted by using the natural foil events. The $\beta\beta 2\nu$ energy spectrum is shown in Fig. 2 (background subtracted). A one parameter fit of this spectrum leads to the half-life,

$$T^{2\nu}_{1/2} = [0.95 \pm 0.04(st) \pm 0.09(sy)] \cdot 10^{19} y.$$

Half-life limits on $\beta\beta 0\nu$ and $\beta\beta 0\nu\chi^0$ are given in Table 1.

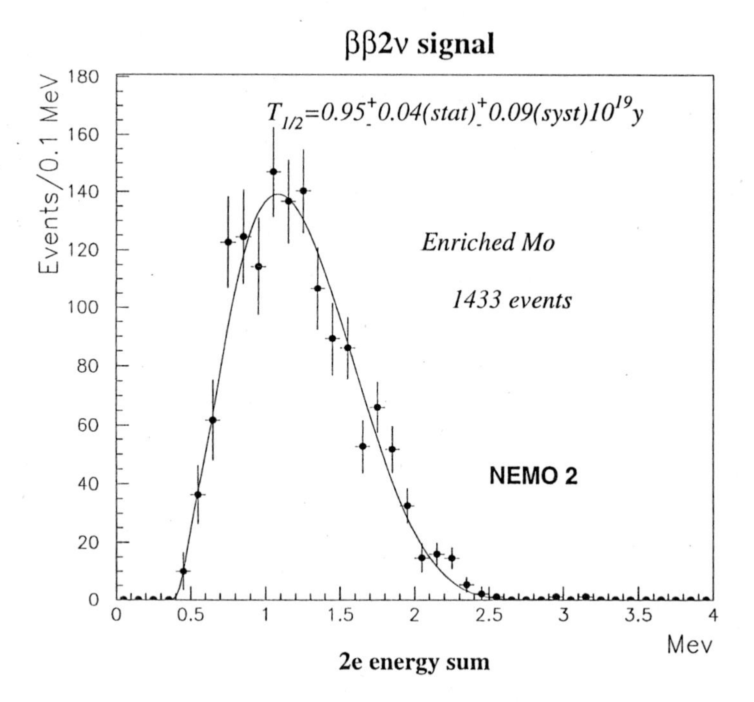

Fig. 2. Fit of the $\beta\beta 2\nu$ signal of ^{100}Mo.

4 Cadmium data

The source plane was divided into two halves, one was 152 g ($\sim 50\mu$m thick) of enriched cadmium
(93.2% in ^{116}Cd, Q = 2.802 MeV), the other was 143 g ($\sim 43\mu$m thick) of natural cadmium (7.49% in ^{116}Cd). During a useful running time of 6588 h, 0.92 mol·y were accumulated. Few additional 2e events due to neutron flux were detected in natural cadmium because of the huge capture cross-section in ^{113}Cd (isotopic abundance of 12.2%). A cut on the angle between the two electrons ($cos(\alpha) < 0.6$) was applied to lower the external background. The summed electron energy spectrum after background subtraction is shown in Fig. 3 and one gets a half-life,

$$T_{1/2}^{2\nu} = [3.75 \pm 0.35(st) \pm 0.21(sy)] \cdot 10^{19} y.$$

The thickness of the scintillators and the use of low background PMTs have reduced the systematic error by almost a factor two with respect to the molybdenum experiment. In Table.1 are given the 0ν mode limits.

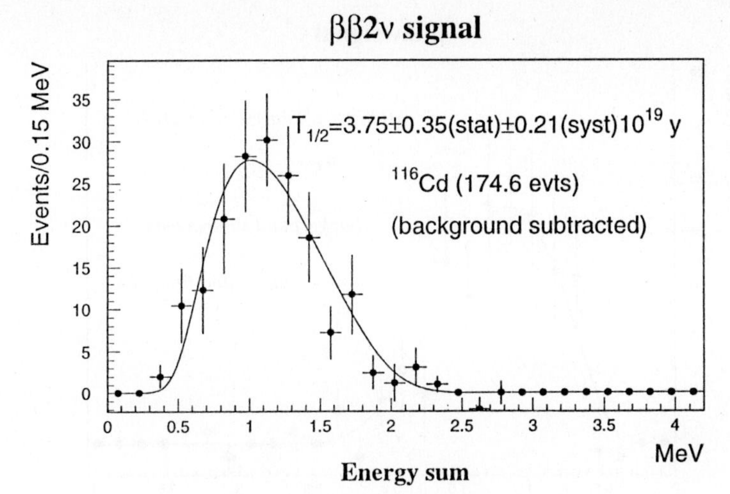

Fig. 3. Fit of $\beta\beta2\nu$ signal of ^{116}Cd.

5 Selenium data

The source plane is divided into two halves, the first one is a 156.6 g enriched selenium foil (97.02% in ^{82}Se, Q = 2.995 MeV), the second half is a 133.7 g foil of natural selenium of which 8.73% is ^{82}Se, during a data taking of 10357 h 2.17 mol·y were accumulated. Some activity from ^{40}K was found in the foils, 200 mBq/kg in enriched and 117 mBq/kg in natural selenium. The 2e events are selected by time-of-flight cuts. The background in the enriched foil is calculated by using the 2e events in the natural foil and the measured contaminations. In natural selenium 8.4±1.2 two-electron events are induced by ^{40}K activity (17.7 events in enriched selenium) and the contribution of the $\beta\beta2\nu$ decay is found to be 12.4±1.4. An external background of 64.2±7.8 events is calculated in the enriched foil. The $\beta\beta2\nu$ energy spectrum in enriched selenium (Fig. 4) is obtained by subtracting from the raw data spectrum (231 events) the background spectrum (81.9 events). This spectrum is fit with the simulated spectrum and one gets,

$$T^{2\nu}_{1/2} = [0.83 \pm 0.09(st) \pm 0.07(sy)] \cdot 10^{20} y.$$

The $\beta\beta2\nu$ half-life can be compared with previous geochemical [6] experiments, the published values are ranging from 1. to 1.3 10^{20} y. Another direct experiment [7] has found a half-life of $1.08^{+0.26}_{-0.06} \cdot 10^{20} y$.

6 Limits on 0ν modes

Half-life limits for $\beta\beta0\nu$ decays to the ground state, with and without Majoron emission, and for decays to 2^+ excited states are given in Table 1. The

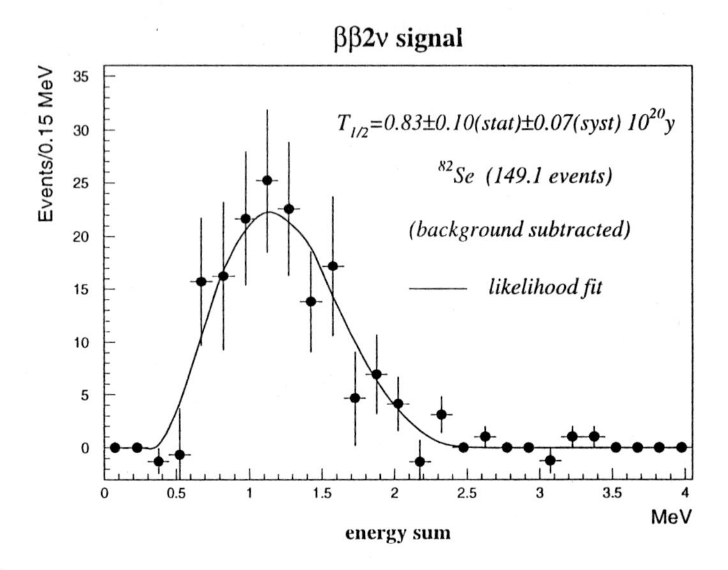

Fig. 4. Fit of the $\beta\beta2\nu$ signal of ^{82}Se.

Table 1. Half-life limits (90% C.L.) on 0ν modes for the three $\beta\beta$ emitters investigated with NEMO-2.

$T_{1/2}(10^{21}$y)	$0^+_{g.s.}$	Majoron	2^+
^{100}Mo	>6.4	>0.5	>0.8
^{116}Cd	>5.0	>1.2	>0.6
^{82}Se	>9.5	>2.4	>2.8

half-life limits on 0ν modes listed in Table 1 are of the same order as those from other direct experiments (see the discussions in the published papers on molybdenum [3] and cadmium [4]). For selenium the limits given here are comparable to those given in the direct experiment [7] already mentionned.

7 NEMO-3 detector

A general view of the detector in the Fréjus Underground Laboratory is shown in Fig. 5. The sources foils divide this volume into two concentric cylinders defined by an inner and an outer scintillator array, two end caps of scintillators close the volume.

The tracking detector consists of 6180 Geiger cells 2.7 m long and parallel to the cylinder axis.

The calorimeter is equipped with 1960 low radioactivity PMTs and the scintillator weight will reach 7 tons. A 30 Gauss magnetic field produced by

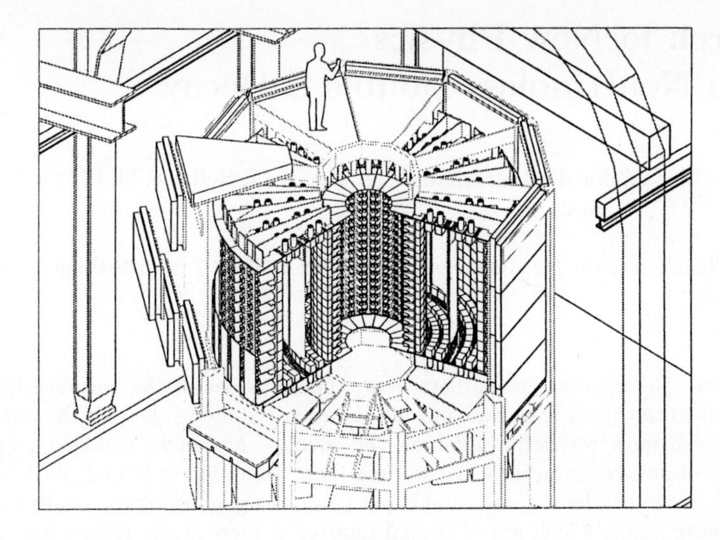

Fig. 5. Layout of the NEMO-3 detector installed in the Fréjus Underground Laboratory.

a solenoid will improve the rejection of pair creation events and incoming electrons. A 20 cm iron shield will protect the detector from external radioactivity, and a paraffin neutron shield will be added.

8 Conclusion

The selenium data are the last ones taken with the NEMO-2 detector which is now dismounted. This prototype has provided experience for the successful operation of NEMO-3.

The mounting of NEMO-3 in the Fréjus Underground Laboratory will start in 1998 and data collection is foreseen to begin about one year later.

References

[1] D. Dassié et al.: Nucl. Instr. Meth. A309 (1991) 465
[2] R. Arnold et al.: Nucl. Instr. Meth. A354 (1995) 338
[3] D. Dassié et al.: Phys. Rev. D51 (1995) 2090
[4] R. Arnold et al.: Z.Phys. C72 (1996) 239
[5] NEMO-3 Proposal. LAL preprint 94-29 (1994)
[6] O.K. Manuel: J.Phys. G17 (Supplement) (1991) 221
[7] S.R. Elliot et al.: Phys. Rev. C46 (1992) 1535

Search forNew Physics with Neutrinoless Double β Decay

1018: H.V. Klapdor–Kleingrothaus **, L. Baudis, J. Hellmig, M. Hirsch, S. Kolb, H. Päs ***, Y. Ramachers

Max–Planck–Institut für Kernphysik, P.O. Box 103980, D–69029 Heidelberg, Germany

Abstract. Neutrinoless double beta decay ($0\nu\beta\beta$) is one of the most sensitive approaches to test particle physics beyond the standard model. During the last years, besides the most restrictive limit on the effective Majorana neutrino mass, the analysis of new contributions by the Heidelberg group led to bounds on left-right-symmetric models, leptoquarks and R-parity violating models competitive to recent accelerator limits, which are of special interest in view of the HERA anomaly at large Q^2 and x. These new results deduced from the Heidelberg-Moscow double beta decay experiment are reviewed. Also an outlook on the future of double beta decay, the GENIUS proposal, is given.

1 Introduction

Double beta decay [1, 2] corresponds to two single beta decays occuring in one nucleus and converts a nucleus (Z,A) into a nucleus (Z+2,A). While even the standard model allowed process emitting two antineutrinos

$$^A_Z X \rightarrow ^A_{Z+2} X + 2e^- + 2\overline{\nu}_e \tag{1}$$

is one of the rarest processes in nature with half lives in the region of 10^{21-24} years, more interesting is the search for the neutrinoless mode,

$$^A_Z X \rightarrow ^A_{Z+2} X + 2e^- \tag{2}$$

which violates lepton number by two units and thus implies physics beyond the standard model.

2 The Heidelberg–Moscow Double Beta Decay Experiment

The Heidelberg–Moscow experiment [3, 4, 1] is a second generation experiment searching for the $0\nu\beta\beta$ decay of ^{76}Ge. It has been described recently in

* Talk presented by Heinrich Päs
** Spokesman of the Collaboration
*** present address: Laboratori Nazionali del Gran Sasso (INFN), 67010 Assergi (AQ), Italy; E-mail: Heinrich.Paes@mpi-hd.mpg.de

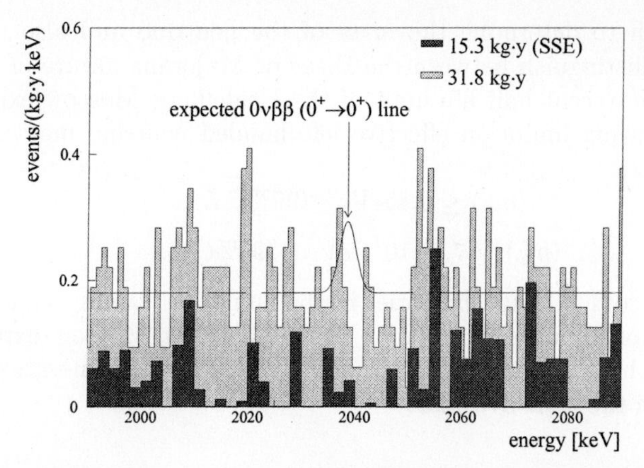

Fig. 1. Combined spectrum in the $0\nu\beta\beta$ peak region: Complete data with 31.8 kg ·y, with pulse shape analysis detected single site events with 15.3 kg · y and 90% C.L. excluded peak.

detail in [3, 4, 1]. Fig. 1 shows the results after 31.8 kg y measuring time for all data corresponding to a half life limit of

$$T_{1/2}^{0\nu\beta\beta} > 1.2 \cdot 10^{25} y. \tag{3}$$

The limits from the pulse shape [5, 4] data with 15.3 kg y (filled histogram in Fig. 1), corresponding to $T_{1/2}^{0\nu\beta\beta} > 1.1 \cdot 10^{25} y$, are not yet competitive to the large data set without application of pulse shape analysis. However the background improvement will allow to test the half life region up to $6 \cdot 10^{25}$ y, corresponding to a neutrino mass limit of 0.1–0.2 eV, during the next five years.

The standard model allowed $2\nu\beta\beta$ decay is measured with the highest statistics ever reached, containing 21115 events in the energy region of 500-2040 keV, and yields a half life of [3]

$$T_{1/2}^{2\nu\beta\beta} = (1.77^{+0.01}_{-0.01}(stat.)^{+0.13}_{-0.11}(syst.)) \cdot 10^{21} y. \tag{4}$$

This result, confirming the theoretical predictions of [6] with an accuracy of a factor $\sim \sqrt{2}$, provides a consistency check of nuclear matrix element calculations. It also for the first time opens up the possibility to search for deviations of the $2\nu\beta\beta$ spectrum such as those due to emission of exotic scalars [7].

3 Double Beta Decay and Physics Beyond the Standard Model

3.1 Neutrino Mass

The search for $0\nu\beta\beta$ decay exchanging a massive left–handed Majorana neutrino between two standard model vertices at present provides the most sensitive approach to determine the mass of the neutrino and also a unique possibility to distinguish between the Dirac or Majorana nature of the neutrino. With the recent half life limit of the Heidelberg–Moscow experiment [4, 1] the following limits on effective left–handed neutrino masses can be deduced:

$$\langle m_\nu \rangle \leq 0.45 eV \quad (90\% C.L.) \tag{5}$$

$$\langle m_\nu \rangle \geq 7.5 \cdot 10^7 GeV \quad (90\% C.L.) \tag{6}$$

Taking into account the uncertainties in the numerical values of nuclear matrix elements of about a factor of 2, the Heidelberg–Moscow experiment, improving its half life limit up to $6 \cdot 10^{25} y$, will test degenerate neutrino scenarios [8] in the next five years.

3.2 Left–Right–Symmetric Models

In left–right symmetric models the left–handedness of weak interactions is explained as due to the effect of different symmetry breaking scales in the left– and in the right–handed sector. $0\nu\beta\beta$ decay proceeds through exchange of the heavy right–handed partner of the ordinary neutrino between right-handed W vertices, leading to a limit of

$$m_{WR} \geq 1.2 \Big(\frac{m_N}{1TeV} \Big)^{-(1/4)} TeV. \tag{7}$$

Including a theoretical limit obtained from considerations of vacuum stability [9] one can deduce an absolute lower limit on the right–handed W mass of [10]

$$m_{W_R} \geq 1.2 TeV. \tag{8}$$

3.3 Supersymmetry

Supersymmetry (SUSY), providing a symmetry between fermions and bosons and thus doubling the particle spectrum of the SM, belongs to the most prominent extensions of the standard model. While in the minimal supersymmetric extension (MSSM) R–parity is assumed to be conserved, there are no theoretical reasons for R_P conservation and several GUT [11] and Superstring [12] models require R_P violation in the low energy regime. Also recent reports concerning an anomaly at HERA in the inelastic e^+p scattering at high Q^2 and x [13] renewed the interest in $\not{R}_P$ SUSY (see for example [14]). In this case $0\nu\beta\beta$ decay can occur through Feynman graphs involving the exchange of superpartners as well as $\not{R}_P$–couplings λ' [15, 16, 17, 18]. The half–life limit of the Heidelberg–Moscow experiment leads to bounds in a multidimensional parameter space [15, 17]

$$\lambda'_{111} \leq 3.2 \times 10^{-4} \Big(\frac{m_{\tilde{q}}}{100GeV} \Big)^2 \Big(\frac{m_{\tilde{g}}}{100GeV} \Big)^{1/2} \tag{9}$$

(for $m_{\tilde{d}_R} = m_{\tilde{u}_L}$), which are the sharpest limits on $R\!\!\!/_P$SUSY. This limit also excludes the first generation squark interpretation of the HERA events [19]. In the case of R_P conserving SUSY, based on a theorem proven in [21], the $0\nu\beta\beta$ mass limits can be converted in sneutrino Majorana mass limits being more restrictive than what could be obtained in inverse neutrinoless double beta decay and single sneutrino production at future linear colliders (NLC) [21].

3.4 Leptoquarks

Leptoquarks are scalar or vector particles coupling both to leptons and quarks, which appear naturally in GUT, extended Technicolor or Compositeness models containing leptons and quarks in the same multiplet. Also production of a scalar leptoquark with mass of $m_{LQ} \simeq 200GeV$ has been considered as a solution to the HERA anomaly (see for example [14]). However, TEVATRON searches have set stringent limits of [22] $m_{LQ} > 240GeV$ for scalar leptoquarks decaying with branching ratio 1 into electrons and quarks. One possibility to keep the leptoquark interpretation interesting is therefore to reduce the branching ratio due to the mixing of different multiplets leading to a significant weakening of the CDF/D0 limits [23]. This kind of mixing can be obtained by introducing a leptoquark–Higgs coupling – which would lead to a contribution to $0\nu\beta\beta$ decay [24]. Combined with the half–life limit of the Heidelberg–Moscow experiment bounds on effective couplings can be derived [25]. Assuming only one lepton number violating $\Delta L = 2$ LQ–Higgs coupling unequal to zero and the leptoquark masses not too different, one can derive from this limit either a bound on the LQ–Higgs coupling

$$Y_{LQ-Higgs} = (few) \cdot 10^{-6} \qquad (10)$$

or a limit implying that HERA does not see Leptoquarks with masses of $\mathcal{O}(200GeV)$. This excludes most of the possibilities to relax the TEVATRON bounds by introducing LQ–Higgs couplings to reduce the branching ratio [19].

4 Outlook on the Future of Double Beta Decay: The Heidelberg Project GENIUS

To render possible a further breakthrough in search for neutrino masses and physics beyond the standard model, recently GENIUS, an experiment operating a large amount of naked Ge–detectors in a liquid nitrogen shielding, has been proposed ([1] and studied in detail in [26]). The possibility to operate Ge detectors inside liquid nitrogen has already been demonstrated by the Heidelberg group. Operating 1 ton of enriched ^{76}Ge improves the sensitivity to neutrino masses down to 0.01 eV, which allows to confirm or exclude Majorana neutrinos as hot dark matter, test the atmospheric neutrino problem, as well as SUSY models, leptoquarks and right–handed W–masses comparable to the LHC [1, 26]. A ten ton version would probe neutrino masses even down to 10^{-3} eV, testing the MSW large angle solution of the solar neutrino problem. As direct dark matter detection experiment it would allow to test almost the entire MSSM parameter space already in a first step using only 100 kg of enriched or even natural Ge [1, 26, 27].

5 Conclusions

Neutrinoless double beta decay has a broad potential to test physics beyond the standard model. The possibilities to constrain neutrino masses, left–right–symmetric models, SUSY and leptoquark scenarios have been reviewed. Experimental limits on $0\nu\beta\beta$ decay are not only complementary to accelerator experiments but at least in some cases competitive or superior to the best existing direct search limits. The Heidelberg–Moscow experiment has reached the leading position among double beta decay experiments and as the first of them now yields results in the sub–eV range for the neutrino mass. A further breakthrough will be possible realizing the GENIUS proposal. For the application of double beta technology in WIMP dark matter search we refer to [27].

References

[1] H.V. Klapdor–Kleingrothaus, in [20]

[2] W.C. Haxton, G.J. Stephenson, Progr. Part. Nucl. Phys. 12 (1984) 409; M. Moe, P. Vogel, Ann. Rev. of Nucl. Part. Science 44 (1994) 247; M. Doi, T. Kotani, E. Takasugi, Progr. Theor. Phys. Suppl. 83 (1985) 1

[3] HEIDELBERG–MOSCOW collab., Phys. Rev. D 55 (1997) 54

[4] HEIDELBERG–MOSCOW collab., Phys. Lett. B 407 (1997) 219

[5] J. Hellmig, PhD thesis, University of Heidelberg, 1996 J. Hellmig, H.V. Klapdor–Kleingrothaus, to be published

[6] A. Staudt, K.Muto, H.V. Klapdor–Kleingrothaus, Europhys. Lett. 13 (1990) 31

[7] M. Hirsch, H. V. Klapdor–Kleingrothaus, S. G. Kovalenko, H. Päs, Phys. Lett. B 372 (1996) 8; H. Päs *et al.*, in: Proc. *Int. Workshop on Double Beta Decay and Related Topics*, Trento, 24.4.–5.5.95, World Scientific Singapore; HEIDELBERG–MOSCOW collab., Phys. Rev. D 54 (1996) 3641

[8] A.Y. Smirnov, in Proc. of the Internation Conference on High Energy Physics (ICHEP), Warsaw, 1996; D.G. Lee, R.N. Mohapatra, Phys. Lett. B 329 (1994) 463; S.T. Petcov, A.Y. Smirnov, Phys. Lett. B 322 (1994) 109; A. Ioanissyan, J. Valle, Phys. Lett. B 332 (1994) 93; H. Fritzsch, Z.Z. Xing, Phys. Lett. B 372 (1996) 265 R.N. Mohapatra, S. Nussinov Phys. Lett. B 346 (1995) 75

[9] R.N. Mohapatra, Phys. Rev D 34 (1986) 3457
[10] M. Hirsch, H.V. Klapdor–Kleingrothaus, O. Panella, Phys. Lett. B 374 (1996) 7
[11] L. Hall, M. Suzuki, Nucl. Phys. B 231 (1984) 419; D. Brahm, L. Hall, Phys. Rev D 40 (1989) 2449; K. Tamvakis Phys. Lett. B 382 (1996) 251; G.F. Guidice, R. Rattazzi, hep-ph/9604339; R. Barbieri, A. Strumia, Z. Berezhiani, hep-ph/9704275; K. Tamvakis, Phys. Lett B 383 (1996) 307; R. Hempfling, Nucl. Phys. B 478 (1996) 3; A. Y. Smirnov, F. Vissani, Nucl. Phys. B 460 (1996) 37
[12] M.C. Bento, L. Hall, G.G. Ross, Nucl. Phys. B 292 (1987) 400; N. Ganoulis, G. Lazarides, Q. Shafi, Nucl. Phys. B 323 (1989) 374; A. Faraggi, Phys. Lett B 398 (1997) 95; A. Faraggi, in: [20]
[13] C. Adloff *et al.* (H1 collab.), Z.Phys. C 74 (1997) 191; J. Breitweg *et al.* (ZEUS collab.) Z.Phys. C 74 (1997) 207
[14] J. Kalinowski, R. Rueckl, H. Spiesberger, P. M. Zerwas, Z.Phys. C 74 (1997) 595
[15] M. Hirsch, H.V. Klapdor–Kleingrothaus, S.G. Kovalenko, Phys. Rev. Lett. 75 (1995) 17
[16] M. Hirsch, H.V. Klapdor–Kleingrothaus, S. Kovalenko, Phys. Lett. B 352 (1995) 1
[17] M. Hirsch, H.V. Klapdor–Kleingrothaus, S. Kovalenko, Phys. Rev. D 53 (1996) 1329
[18] M. Hirsch, H.V. Klapdor–Kleingrothaus, S.G. Kovalenko, Phys. Lett. B 372 (1996) 181
[19] M. Hirsch *et al.*, in [20]
[20] H.V. Klapdor–Kleingrothaus, H. Päs (Eds.): *Beyond the Desert – Accelerator and Non-Accelerator Approaches*, Proc. Int. Workshop on Particle Physics beyond the Standard Model, Castle Ringberg, June 8-14, 1997, IOP Publ., Bristol, Philadelphia
[21] M. Hirsch, H.V. Klapdor–Kleingrothaus, S.G. Kovalenko, Phys. Lett. B 398 (1997) 311 and 403 (1997) 291; M. Hirsch, H.V. Klapdor–Kleingrothaus, S.G. Kovalenko, Phys. Rev. D in press, 1997; S. Kolb. M. Hirsch, H.V. Klapdor–Kleingrothaus, S. Kovalenko, in [20]
[22] J. Conway, in [20]; S. Eno, in [20]; M. Kraemer, T. Plehn, M. Spira, P. M. Zerwas Phys. Rev. Lett. 79 (1997) 341; M. Kraemer, in [20]
[23] K. S. Babu, C. Kolda, J. March-Russell, hep-ph/9705414
[24] M. Hirsch, H.V. Klapdor–Kleingrothaus, S.G. Kovalenko, Phys. Lett. B 378 (1996) 17
[25] M. Hirsch, H.V. Klapdor–Kleingrothaus, S.G. Kovalenko Phys. Rev. D 54 (1996) R4207
[26] J. Hellmig, H.V. Klapdor–Kleingrothaus, Z. Phys. A, in press 1997; H.V. Klapdor–Kleingrothaus, M. Hirsch, Z. Phys. A, in press 1997;
[27] H.V. Klapdor–Kleingrothaus and Y. Ramachers, in [20]

The NA56/SPY Experiment at CERN

M. Bonesini, on behalf of the NA56/SPY collaboration (bonesini@mi.infn.it)

Sezione INFN, via Celoria 16, 20133 Milano, Italy

1 Introduction

The NA56/SPY experiment [1] has measured the production rates of charged pions and kaons and their ratios below 60 GeV/c from 450 GeV/c protons hitting beryllium targets of different thickness and shape. These data are of great importance to predict energy spectra and compositions of neutrino beams for present and future neutrino oscillation experiments at accelerators. Measurements of the $\pi^{\pm}$ and $K^{\pm}$ production rate by 400 GeV/c protons on Be targets were performed by Atherton et al. [2] for secondary momenta above 60 GeV/c and up to 500 MeV/c of transverse momentum. At lower momenta extrapolations of the existing data [3] or Montecarlo calculations [4, 5] had to be used.

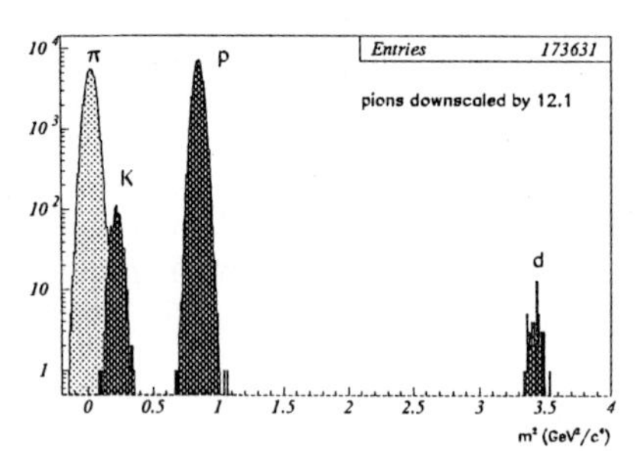

Fig. 1. Mass recontruction with TOFs at 15 GeV/c: K/π separation is performed with the $\check{C}0$ and $\check{C}1$ Cerenkov counters.

The SPY experiment was performed with the NA52 spectrometer [6] in the H6 SPS beamline in the North Area at CERN. Data were collected in the secondary momentum range of 7-135 GeV/c with angular scans at fixed momenta (15 and 40 GeV/c). Redundant particle identification was provided by a set of time of flight (TOF) detectors, threshold $\check{C}0 \div \check{C}2$ and differential (CEDAR) Čerenkov counters along the beamline and a hadron calorimeter for the electron, muon and hadron separation. A set of proportional chambers was used for particle tracking along the spectrometer. An example of the attainable particle identification capabilities is shown in fig. 1.

2 Experimental results

To extract the $K^{\pm}/\pi^{\pm}$ production ratios, data were corrected for particle decays in flight, particle dependent losses along the beamline and interactions of the primary beam with the material around the target. The contribution to the pion flux from strange particle decays ($K_s^0 \mapsto \pi^+\pi^-, \Lambda \mapsto p\pi^-, ...$) outside the target has been subtracted.

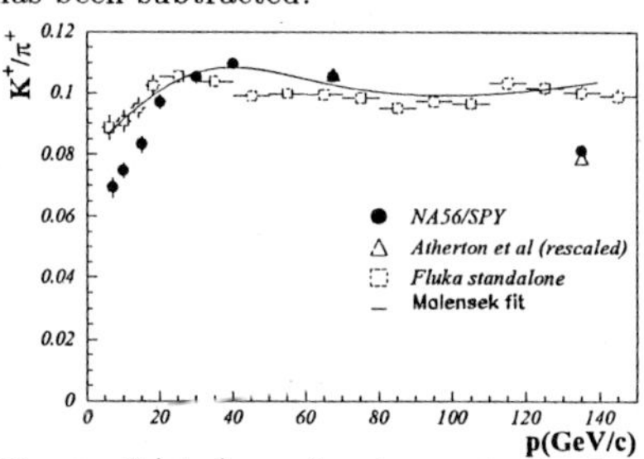

Fig. 2. K^+/π^+ production ratio in the forward direction.

The measured K^+/π^+ production ratio for the 100 mm slat beryllium target is reported in fig. 2 and 3 respectively. Error bars account for systematic and statistical errors added in quadrature (the total uncertainty is $\sim 3\%$). In the angular scans (fig. 3a) the common systematics ($\sim 1\%$) is not included.

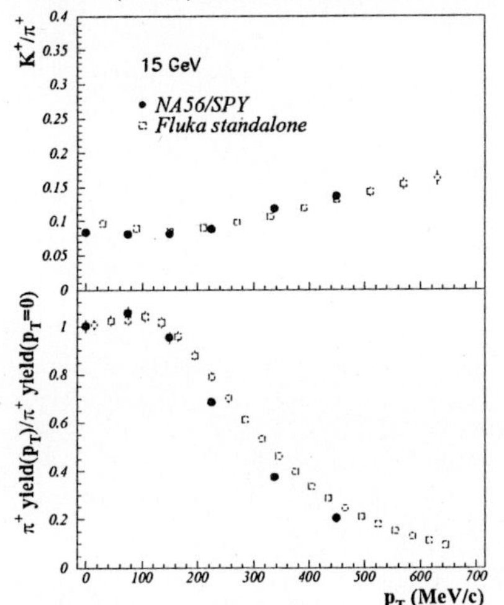

Fig. 3. Angular dependence of the K^+/π^+ ratio (a) and the shape of the π^+ yield (normalized to p_T=0 MeV/c) (b), at 15 GeV/c.

In the same plots the measurement of Atherton et al. [2] (rescaled for the different beam momentum), the parameterization of Malensek[3] and the results of the FLUKA Montecarlo simulation [5] are reported.

To extract π and K production rates, data were corrected for trigger and data acquisition efficiences, particle decays in flight, strange particle decays outside the target and contributions from interactions of the primary beam with material around the target area. Moreover, the knowledge of the primary proton beam intensity and of the H6 spectrometer acceptance was required.

The uncertainty on the π,K yields evaluation is dominated by the systematic error on the calculation of the H6 beamline acceptance: 5-10 % depending on beam momentum. Being a common systematics, this error is not included in the angular scan for π^+, reported in fig. 3b. Results in the forward direction for the 100 mm slat beryllium target are shown in fig. 4.

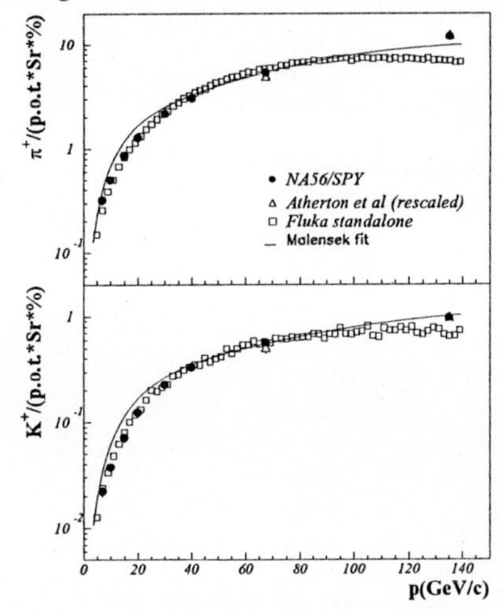

Fig. 4. K^+ and π^+ production yields in the forward direction.

3 Conclusions

The NA56/SPY measurements show that an experimental accuracy around 3% on the K/π production ratios and better than 10% on the π, K yields in p-Be interactions at 450 GeV/c has been achieved. These results will reduce the uncertainty on the estimation of the ν_e component of neutrino beams and provide accurate flux estimations for present and future neutrino experiments.

References

[1] G. Ambrosini et al., CERN-SPSLC/96-01.

[2] H.W. Atherton et al., CERN 80-07, 1980.

[3] A. J. Malensek, Fermilab FN - 341 (1981).

[4] R. Brun et al., "GEANT", W5013.

[5] A. Fassó et al., CERN TIS-RP 97-05.

[6] K. Pretzl et al., Conference Proceedings of "International Symposium on Strangness and Quark Matter", 1994, Krete (Greece).

Part XI

Physics Beyond the Standard Model

Gauge-Mediated Supersymmetry Breaking at an Intermediate Scale

Stuart Raby (raby@mps.ohio-state.edu)

The Ohio State University, Department of Physics, 174 W. 18th Avenue, Columbus, OH 43210 U.S.A.

Abstract. In gauge-mediated SUSY breaking models, messengers transmit SUSY breaking from a partially hidden sector to the standard model sector via common standard model gauge interactions. The minimal set of messengers has quantum numbers of a $5 + \bar{5}$ of SU(5); identical to the quantum numbers of the minimal Higgs sector of an SU(5) GUT. We show in a simple model with messenger masses of order the GUT scale that Higgs - messenger mixing quite naturally leads to a low energy MSSM with gluinos as the lightest supersymmetric particles [LSP]. We study the phenomenological consequences of such a model.

1 Gauge-mediated SUSY Breaking

Gauge-mediated SUSY breaking [GMSB] models [1, 2] solve the problem of flavor changing neutral currents inherent in the MSSM [3, 4]. Consider for the purposes of this talk, flavor changing processes of charged leptons. Supersymmetric charged lepton mass terms are of the form

$$\bar{e}\; \mathbf{m}_e\; e$$

where e ($\bar{e}$) represents 3 families of left-handed (right-handed) fermions and their scalar partners and $\mathbf{m}_e$ is a complex 3 x 3 mass matrix. In addition, scalars necessarily have soft SUSY breaking mass terms given by

$$\tilde{e}^*\; \mathbf{m}^2_{\tilde{e}}\; \tilde{e} \;+\; \tilde{\bar{e}}^*\; \mathbf{m}^2_{\tilde{\bar{e}}}\; \tilde{\bar{e}}$$

where $\tilde{e}$ ($\tilde{\bar{e}}$) represents the left-handed (right-handed) sleptons and $\mathbf{m}^2_{\tilde{e}}$ ($\mathbf{m}^2_{\tilde{\bar{e}}}$) is an hermitian 3 x 3 mass squared matrix. One may always diagonalize the supersymmetric mass term $\mathbf{m}_e$ by a simultaneous rotation of the charged lepton and slepton fields. This rotation however will not, in general, diagonalize $\mathbf{m}^2_{\tilde{e}}$, $\mathbf{m}^2_{\tilde{\bar{e}}}$, unless they are proportional to the identity matrix. Note, off diagonal slepton masses lead to flavor violating processes such as $\mu \to e\gamma$, $\mu \to 3e$, $\mu \to e$ conversion, etc.

In GMSB, SUSY breaking occurs in an almost hidden sector of the theory due to the expectation value F_X, the F component of a superfield X. Moreover, standard model [SM] squarks, sleptons and gauginos do not couple directly to X. Hence they do not obtain SUSY breaking masses at tree level. The states which couple directly to X are the messengers of SUSY breaking.

They carry SM gauge interactions, but otherwise do not couple to squarks and sleptons directly. Thus SUSY breaking enters the SM sector at one loop to gauginos and at two loops to squarks and sleptons. These SUSY breaking effects are dimensionally of order $\Lambda \equiv F_X/M$ where M is the messenger mass. Moreover, they are determined by gauge quantum numbers; thus, for example, the matrices $\mathbf{m}^2_{\tilde{e}}$, $\mathbf{m}^2_{\bar{\tilde{e}}}$ are proportional to the identity matrix at M. As a result, flavor violating effects due to physics at the GUT scale, M_G, are suppressed by factors of $(M/M_G)^2$.

2 The Minimal Messenger Sector

The messenger states must carry both color and electroweak quantum numbers. In addition, the messengers should be in complete SU(5) representations, to preserve GUT predictions for gauge couplings. The minimal set of states satisfying these criteria transform as a $5 + \bar{5}$ with the color triplet (weak doublets) denoted as follows t, $\bar{t}$ (d, $\bar{d}$). In the minimal models, all messengers have a common mass M. The resulting soft breaking masses are as follows.

Gauginos obtain mass at one loop given by

$$m_{\lambda_i} = \frac{\alpha_i(M)}{4\pi}\Lambda \text{ (for } i = 1, 2, 3). \tag{1}$$

The scalar masses squared arise at two-loops

$$\tilde{m}^2 = 2\Lambda^2 \left[\sum_{i=1}^{3} C_i \left(\frac{\alpha_i(M)}{4\pi}\right)^2\right] \tag{2}$$

where $C_3 = \frac{4}{3}$ for color triplets and zero for singlets, $C_2 = \frac{3}{4}$ for weak doublets and zero for singlets, and $C_1 = \frac{3}{5}\left(\frac{Y}{2}\right)^2$, with the ordinary hypercharge Y normalized as $Q = T_3 + \frac{1}{2}Y$ and α_1, GUT normalized.

In the limit $M << M_G$, we have $\alpha_3(M) >> \alpha_2(M) > \alpha_1(M)$. Thus right-handed sleptons are expected to be the lightest SUSY partners of SM fermions and binos are the lightest gauginos.

3 SUSY GUT and Higgs - Messenger Mixing

In the minimal SU(5) SUSY GUT, Higgs doublets are contained in a $5_H + \bar{5}_H$. In order to avoid large baryon number violating nucleon decay rates, the color triplet Higgs t_H, $\bar{t}_H$ must have mass of order M_G, while the Higgs doublets d_H, $\bar{d}_H$ remain massless at the GUT scale. The latter are responsible for electroweak symmetry breaking at M_Z.

Our main observation[5] is that the Higgs in a SUSY GUT and the messengers of GMSB have identical quantum numbers. Thus, for messengers with mass at an intermediate scale, Higgs-Messenger mixing is natural. Moreover as a result of doublet-triplet splitting in the Higgs sector, the doublet and triplet messengers will also be split. This can have significant consequences for SUSY breaking masses.

As a simple example, consider the natural doublet-triplet splitting mechanism in SO(10) [6]. The 10 of SO(10) decomposes into a $5 + \bar{5}$ of SU(5) and the adjoint 45 can be represented by an anti-symmetric 10×10 matrix. The Higgs sector superspace potential is given by

$$W_{Higgs} = 10_H \; 45 \; 10 \; + \; X \; 10^2 \tag{3}$$

where 10_H contains $5_H + \bar{5}_H$, 10 is an auxiliary $5 + \bar{5}$ introduced for doublet-triplet splitting and X is a singlet. Assuming $< 45 >= M_G(B - L)$, i.e. 45 obtains an SO(10) breaking vacuum expectation value in the $B - L$ direction and $< X >= M$, we obtain the triplet (doublet) mass terms given by

$$\begin{aligned} W_{Higgs} = t_H \; M_G \; \bar{t} \; + \; t \; M_G \; \bar{t}_H \; + \; M \; t \; \bar{t} \\ + \; M \; d \; \bar{d} \end{aligned} \tag{4}$$

Note, the triplets naturally have mass of order the GUT scale, while the auxiliary doublets have mass M and the Higgs doublets are massless.[1] The doublet mass is necessarily smaller than the triplets in order to suppress baryon number violating interactions [7]. Specifically, if only 10_H couples to quarks and leptons, then the effective color triplet Higgs mass $\tilde{M}_t$ which enters baryon decay amplitudes is given by $\tilde{M}_t = M_G^2/M$. Hence $\tilde{M}_t > M_G$ implies $M/M_G < 1$.

The theory we propose, with Higgs-messenger mixing, is quite simple. Assume the auxiliary 10 is the messenger of SUSY breaking, i.e. assume that X gets both a SUSY conserving vev M and SUSY breaking vev F_X –

$$\begin{aligned} < X > &= M \; + \; F_X \, \theta^2; \\ \Lambda \; &= \; \frac{F_X}{M} \quad \sim 10^5 \text{ GeV}; \\ A \; &\equiv \; \frac{M}{M_G} \quad \sim 0.1 \end{aligned} \tag{5}$$

Since the triplet messengers (mass $O(M_G)$) are heavier than the doublets (mass $O(M)$), *SUSY breaking effects mediated by color triplets are suppressed.* This has significant consequences for gluinos which only receive SUSY violating mass corrections through colored messengers.

Gauginos obtain mass at one loop given by

$$m_{\lambda_i} = D_i \tfrac{\alpha_i(M)}{4\pi} \Lambda + \tfrac{\alpha_i(M)}{2\pi} \Lambda B^2 \quad (\text{for } \; i = 1, 2, 3) \tag{6}$$

where $D_1 = \frac{3}{5}$, $D_2 = 1$, $D_3 = 0$.

---

[1] There are several different ways that a μ term for the Higgs doublets can be generated once SUSY is broken. We will not discuss this issue further here.

• In order to generate SUSY violating gaugino masses, both SUSY and R symmetry must be broken. In this theory, F_X breaks SUSY and the scalar vev M breaks the R symmetry which survives GUT symmetry breaking. Thus both are necessary to generate the SUSY violating effective mass operator given by

$$\frac{1}{\mathcal{M}^2} \int d^4\theta \, X^\dagger X \, W_i^\alpha \, W_{\alpha\, i} \quad \text{for } i = 1, 2, 3 \tag{7}$$

where $\mathcal{M}$ is determined by the heaviest messenger entering the loop.
• Note the terms proportional to B^2. Without them the gluino mass vanishes at one loop due to an accidental cancellation.[2] In order to compensate for this one loop cancellation we include additional messengers with a common mass of order M_G, and an R symmetry breaking mass M. This sector thus contributes a common mass correction proportional to $B \sim M/M_G$.

The scalar masses squared arise at two-loops. We obtain

$$\begin{aligned} \tilde{m}^2 = 2\Lambda^2 & \left[C_3 \left(\frac{\alpha_3(M)}{4\pi}\right)^2 (A^2 + 2\, B^2) + C_2 \left(\frac{\alpha_2(M)}{4\pi}\right)^2 (1 + 2\, B^2) \right] \\ & + 2\Lambda^2 \left[C_1 \left(\frac{\alpha_1(M)}{4\pi}\right)^2 (\tfrac{3}{5} + \tfrac{2}{5}\, A^2 + 2\, B^2) \right] \end{aligned} \tag{8}$$

where $C_i, \;\; i \; = \; 1, 2, 3$ are defined after equation 2.

4 Low Energy Spectrum

The heaviest SUSY particles are electroweak doublet squarks and sleptons and weak triplet winos, while right-handed squarks, sleptons and binos are lighter. Finally gluinos are expected to be the lightest SUSY particle [LSP]. We have the approximate mass relations[3], after renormalization group running to M_Z,

$$\begin{aligned} M_2 &= \frac{\alpha_2(M_Z)}{4\pi}\, \Lambda & &\approx 3 \times 10^{-3}\, \Lambda \\ M_3 &= \frac{\alpha_3(M_Z)}{2\pi}\, B^2\, \Lambda & &\approx 9 \times 10^{-5}\, (B/0.1)^2\, \Lambda. \end{aligned} \tag{9}$$

In addition, the gravitino mass (which sets the scale for supergravity mediated soft SUSY breaking effects) is given by

$$m_{3/2} = \left(\frac{F_X}{\sqrt{3} M_{pl}}\right). \tag{10}$$

Hence

$$m_{3/2} = \left(\frac{M_G}{\sqrt{3} M_{pl}}\right) B\, \Lambda \approx 6 \times 10^{-4}\, (B/0.1)\, \Lambda. \tag{11}$$

[2] I thank Kazuhiro Tobe for pointing this out to me. Note, eqns. (6, 8) are corrections for similar equations in ref. [5].
[3] neglecting terms of order A^2 or B^2, when possible

The gluino mass M_3 depends on the arbitrary parameter B, the ratio of the R symmetry breaking scale M to the typical messenger mass of order M_G. The gravitino mass also depends parametrically on B when expressed in terms of the SUSY breaking scale Λ.

With $\Lambda = 10^5$ GeV, we obtain

$$\begin{aligned} M_2 &\approx 300 \text{ GeV} \\ m_{3/2} &\approx 60\ (B/0.1) \text{ GeV} \\ M_3 &\approx 9\ (B/0.1)^2 \text{ GeV} \end{aligned} \tag{12}$$

5 Signatures of SUSY with a Gluino LSP

Gluinos are stable.[4] They form color singlet hadrons, with the lightest of them[8] given by

$$\begin{aligned} R_0 &= \tilde{g}\, g \\ \tilde{\rho} &= \tilde{g}\, q\, \bar{q} \quad \text{with } q = u,\ d \\ S_0 &= \tilde{g}\, u\, d\, s \end{aligned} \tag{13}$$

where R_0 is an iso-scalar fermion [glueballino]; $\tilde{\rho}$ is an iso-vector fermion and S_0 is an iso-scalar boson with baryon number 1.

It is unclear which one is the stable color singlet LSP. For this talk, we assume R_0 is lighter and that both $\tilde{\rho}$, S_0 are unstable, decaying via the processes $\tilde{\rho} \to R_0 + \pi$ and $S_0 \to R_0 + n$.

Consider the consequences of a gluino LSP. First, the missing energy signal for SUSY is seriously diluted. An energetic gluino, produced in a high energy collision, will fragment and form an hadronic jet containing an R_0. The R_0 will deposit energy in the hadronic calorimeter. Thus collider limits on squark and gluino masses must be re-evaluated. Gluinos with mass as large as 50 GeV may have escaped detection.[5] A lower bound on the gluino mass of 6.3 GeV has been obtained using LEP data on the running of α_s from m_τ to M_Z.[9] Thus glueballinos may be expected in the range from 6 – 50 GeV.

At LEP a 4 jet signal is expected above the squark threshold, since squarks decay into a quark plus gluino.

Now consider possible constraints from exotic heavy isotope searches. Stringent limits exist on heavy isotopes of hydrogen. However, an R_0 must be in a bound state with a proton in order for these searches to be relevant. Such a bound state is unlikely due to the short range nature of the interaction of

[4] Assuming R parity is conserved.

[5] This is a rough estimate using CDF limits on squarks and gluinos based on missing energy. Clearly if the gluino is heavy enough it will leave a missing energy signal.

R_0 with hadrons; predominantly due to the exchange of a glueball (the lightest of which is 10 times heavier than a pion) or multiple pions. Strong limits on heavy isotopes of oxygen also exist. An R_0 can certainly be trapped in the potential well of a heavy nucleus. However, for these searches to be restrictive, the expected abundance of R_0-nucleus bound states must be above the experimental bounds. The dominant process for forming such bound states is for R_0s, produced by cosmic ray collisions in the earth's atmosphere, to be captured into nuclei. A back of the envelope estimate gives an expected abundance bordering on the observable limit. A more detailed calculation is therefore needed to say more.

Finally, what about the cosmological abundance of R_0s. Since R_0s annihilate via strong processes $R_0 + R_0 \to 2\,\pi$, the cosmological abundance is quite suppressed. A rough estimate gives

$$n_{R_0} = 10^{-10} \left(\frac{m_{R_0}}{m_\pi}\right) n_B \tag{14}$$

where n_B is the cosmological baryon density. As a result, R_0s are NOT dark matter candidates.

Acknowledgements This short talk greatly benefited from extensive discussions with M. Carena, C.E.M. Wagner, G.G. Ross, J. Gunion, H. Haber, C. Quigg, X. Tata, H. Baer, U. Sarid, K. Tobe, A. Mafi, S. Katsanevas, E.W. Kolb, A. Riotto, L. Roszkowski, G. Steigman, R. Boyd, A. Heckler; the hospitality of the Aspen Center for Physics and the organizers of this conference.

References

[1] M. Dine, W. Fischler, and M. Srednicki, Nuclear Physics B189 (1981), 575; S. Dimopoulos and S. Raby, Nuclear Physics B192 (1981), 353; M. Dine and W. Fischler, Physics Letters B110 (1982), 227; M. Dine and M. Srednicki, Nuclear Physics B202 (1982), 238; L. Alvarez-Gaumé, M. Claudson, and M. Wise, Nuclear Physics B207 (1982), 96; C. Nappi and B. Ovrut, Physics Letters B113 (1982), 175.

[2] M. Dine, A.E. Nelson and Y. Shirman, Physics Review D51 (1995), 1362; M. Dine, A.E. Nelson, Y. Nir and Y. Shirman, Physics Review D53 (1996), 2658.

[3] S. Dimopoulos and H. Georgi, Nuclear Physics B193 (1981), 150.

[4] L.J. Hall, V.A. Kostelecky and S. Raby, Nuclear Physics B267 (1986), 415; H. Georgi, Physics Letters 169B (1986), 231.

[5] S. Raby, Physical Review D56 (1997), 2852-2860.

[6] S. Dimopoulos and F. Wilczek, Proceedings Erice Summer School, Ed. A. Zichichi (1981).

[7] K.S. Babu and S.M. Barr, Physical Review D48 (1993) 5354-5364.

[8] G.R. Farrar and P. Fayet, Physics Letters 76B (1978), 442; ibid. 575.

[9] F. Csikor and Z. Fodor , Physical Review Letters 78 (1997) 4335-4338.

Constraints on Finite Soft Supersymmetry Breaking Terms

T. Kobayashi[1], J. Kubo[2], M. Mondragón[3] and G. Zoupanos[4,5]

[1]Inst. of Particle and Nuclear Studies, Tanashi, Tokyo 188, Japan
[2]Dept. of Physics, Kanazawa Univ. , Kanazawa 920-1192, Japan
[3]Inst. de Física, UNAM, Apdo. Postal 20-364, México 01000 D.F., México
[4]Physics Dept., Nat. Technical University, GR-157 80 Zografou, Athens, Greece
[5]Insitüt f. Physik, Humboldt-Universität, D10115 Berlin, Germany

Abstract. A new solution to the requirement of two-loop finiteness of the soft supersymmetry breaking terms (SSB) parameters is found in Finite-Gauge-Yukawa unified theories. The new solution has the form of a sum rule for the relevant scalar masses, relaxing the universality required by the previously known solution, which leads to models with unpleasant phenomenological consequences. Using the sum rule we investigate two Finite-Gauge-Yukawa unified models and we determine their spectrum in terms of few parameters. Some characteristic features of the models are that a) the old agreement of the top quark mass prediction with the measured value remains unchanged, b) the lightest Higgs mass is predicted around 120 GeV, c) the s-spectrum starts above 200 GeV.

1 Introduction

Finite Unified Theories (FUTs) have fundamental conceptual importance in the search of the final theory describing Elementary Particle Physics.

In our attempts here, following the usual minimality assumption, we are restricted in unifying only the known gauge interactions, and it is interesting to point out that *finiteness does not require gravity.* Finiteness is based on the fact that it is possible to find renormalization group invariant (RGI) relations among couplings that keep finiteness in perturbation theory, even to all orders [4].

2 Finiteness and Reduction of Couplings in $N = 1$ SUSY Gauge Theories

Let us then consider a chiral, anomaly free, $N = 1$ globally supersymmetric gauge theory based on a group G with gauge coupling constant g. The superpotential of the theory is given by

* Supported partly by the mexican projects CONACYT 3275-PE, PAPIIT IN110296, by the EU projects FMBI-CT96-1212 and ERBFMRXCT960090, and by the Greek project PENED95/1170;1981.

$$W = \frac{1}{2} m^{ij} \Phi_i \Phi_j + \frac{1}{6} C^{ijk} \Phi_i \Phi_j \Phi_k \ , \tag{1}$$

where m^{ij} and C^{ijk} are gauge invariant tensors and the matter field Φ_i transforms according to the irreducible representation R_i of the gauge group G.

The one-loop β-function of the gauge coupling g is given by

$$\beta_g^{(1)} = \frac{dg}{dt} = \frac{g^3}{16\pi^2} [\sum_i l(R_i) - 3\, C_2(G)] \ , \tag{2}$$

where $l(R_i)$ is the Dynkin index of R_i and $C_2(G)$ is the quadratic Casimir of the adjoint representation of the gauge group G. The β-functions of C^{ijk}, by virtue of the non-renormalization theorem, are related to the anomalous dimension matrix γ_i^j of the matter fields Φ_i as:

$$\beta_C^{ijk} = \frac{d}{dt} C^{ijk} = C^{ijp} \sum_{n=1} \frac{1}{(16\pi^2)^n} \gamma_p^{k(n)} + (k \leftrightarrow i) + (k \leftrightarrow j) \ . \tag{3}$$

At one-loop level γ_i^j is

$$\gamma_i^{j(1)} = \frac{1}{2} C_{ipq}\, C^{jpq} - 2\, g^2\, C_2(R_i) \delta_i^j \ , \tag{4}$$

where $C_2(R_i)$ is the quadratic Casimir of the representation R_i, and $C^{ijk} = C^*_{ijk}$.

As one can see from Eqs. (2) and (4) all the one-loop β-functions of the theory vanish if $\beta_g^{(1)}$ and $\gamma_i^{j(1)}$ vanish, i.e.

$$\sum_i \ell(R_i) = 3C_2(G) \ , \qquad \frac{1}{2} C_{ipq} C^{jpq} = 2\delta_i^j g^2 C_2(R_i) \ . \tag{5}$$

A very interesting result is that the conditions (5) are necessary and sufficient for finiteness at the two-loop level.

The one- and two-loop finiteness conditions (5) restrict considerably the possible choices of the irreps. R_i for a given group G as well as the Yukawa couplings in the superpotential (1). Note in particular that the finiteness conditions cannot be applied to the supersymmetric standard model (SSM), since the presence of a $U(1)$ gauge group is incompatible with the condition (5), due to $C_2[U(1)] = 0$. This naturally leads to the expectation that finiteness should be attained at the grand unified level only, the SSM being just the corresponding, low-energy, effective theory.

A natural question to ask is what happens at higher loop orders. The finiteness conditions (5) impose relations between gauge and Yukawa couplings. We would like to guarantee that such relations leading to a reduction of the couplings hold at any renormalization point. The necessary, but also sufficient, condition for this to happen is to require that such relations are solutions to the reduction equations (REs)

$$\beta_g \frac{dC^{ijk}}{dg} = \beta^{ijk} \tag{6}$$

and hold at all orders. Remarkably the existence of all-order power series solutions to (6) can be decided at the one-loop level.

There exists an all-order finiteness theorem [2] based on: (a) the structure of the supercurrent in $N=1$ SYM and on (b) the non-renormalization properties of $N=1$ chiral anomalies [2]. Details on the proof can be found in ref. [2].

One-loop finiteness implies that the Yukawa couplings C^{ijk} must be functions of the gauge coupling g. To find a similar condition to all orders it is necessary and sufficient for the Yukawa couplings to be a formal power series in g, which is solution of the REs (6).

3 One and two-loop finite supersymmetry breaking terms

Here we would like to use the two-loop RG functionsm to re-investigate their two-loop finiteness and derive the two-loop soft scalar-mass sum rule.

Consider the superpotential given by (1) along with the Lagrangian for SSB terms,

$$-\mathcal{L}_{\rm SB} = \frac{1}{6}\, h^{ijk}\, \phi_i\phi_j\phi_k + \frac{1}{2}\, b^{ij}\, \phi_i\phi_j + \frac{1}{2}\, (m^2)^j_i\, \phi^{*\,i}\phi_j + \frac{1}{2}\, M\, \lambda\lambda + \text{H.c.} \tag{7}$$

where the ϕ_i are the scalar parts of the chiral superfields Φ_i , λ are the gauginos and M their unified mass. Since we would like to consider only finite theories here, we assume that the gauge group is a simple group and the one-loop β function of the gauge coupling g vanishes. We also assume that the reduction equations (6) admit power series solutions of the form

$$C^{ijk} = g \sum_{n=0} \rho^{ijk}_{(n)} g^{2n} \ , \tag{8}$$

According to the finiteness theorem of ref. [2], the theory is then finite to all orders in perturbation theory, if, among others, the one-loop anomalous dimensions $\gamma_i^{j\,(1)}$ vanish. The one- and two-loop finiteness for h^{ijk} can be achieved by

$$h^{ijk} = -MC^{ijk} + \ldots = -M\rho^{ijk}_{(0)}\, g + O(g^5) \ . \tag{9}$$

Now, to obtain the two-loop sum rule for soft scalar masses, we assume that the lowest order coefficients $\rho^{ijk}_{(0)}$ and also $(m^2)^i_j$ satisfy the diagonality relations

$$\rho_{ipq(0)}\rho^{jpq}_{(0)} \propto \delta^j_i \text{ for all } p \text{ and } q \ \text{ and } \ (m^2)^i_j = m^2_j \delta^i_j \ , \tag{10}$$

respectively. Then we find the following soft scalar-mass sum rule

$$(m_i^2 + m_j^2 + m_k^2)/MM^\dagger = 1 + \frac{g^2}{16\pi^2}\,\Delta^{(1)} + O(g^4) \tag{11}$$

for i, j, k with $\rho_{(0)}^{ijk} \neq 0$, where $\Delta^{(1)}$ is the two-loop correction, which vanishes for the universal choice in accordance with the previous findings of ref. [3].

4 Finite Unified Theories

A predictive Gauge-Yukawa unified SU(5) model which is finite to all orders, in addition to the requirements mentioned already, should also have the following properties:

1. One-loop anomalous dimensions are diagonal, i.e., $\gamma_i^{(1)\,j} \propto \delta_i^j$, according to the assumption (10).
2. Three fermion generations, $\overline{\mathbf{5}}_i$ $(i = 1,2,3)$, obviously should not couple to **24**. This can be achieved for instance by imposing $B-L$ conservation.
3. The two Higgs doublets of the MSSM should mostly be made out of a pair of Higgs quintet and anti-quintet, which couple to the third generation.

In the following we discuss two versions of the all-order finite model.

A: The model of ref. [1].
B: A slight variation of the model **A**.

The superpotential which describe the two models takes the form [1, 5]

$$W = \sum_{i=1}^{3} [\, \frac{1}{2} g_i^u\, \mathbf{10}_i \mathbf{10}_i H_i + g_i^d\, \mathbf{10}_i \overline{\mathbf{5}}_i\, \overline{H}_i \,] + g_{23}^u\, \mathbf{10}_2 \mathbf{10}_3 H_4 \tag{12}$$

$$+ g_{23}^d\, \mathbf{10}_2 \overline{\mathbf{5}}_3\, \overline{H}_4 + g_{32}^d\, \mathbf{10}_3 \overline{\mathbf{5}}_2\, \overline{H}_4 + \sum_{a=1}^{4} g_a^f\, H_a\, \mathbf{24}\, \overline{H}_a + \frac{g^\lambda}{3}\, (\mathbf{24})^3 \;,$$

where H_a and $\overline{H}_a$ $(a = 1,\ldots,4)$ stand for the Higgs quintets and anti-quintets.

The non-degenerate and isolated solutions to $\gamma_i^{(1)} = 0$ for the models $\{\mathbf{A}\,,\,\mathbf{B}\}$ are:

$$(g_1^u)^2 = \{\frac{8}{5},\frac{8}{5}\}g^2\;,\;(g_1^d)^2 = \{\frac{6}{5},\frac{6}{5}\}g^2\;,\;(g_2^u)^2 = (g_3^u)^2 = \{\frac{8}{5},\frac{4}{5}\}g^2\;, \tag{13}$$

$$(g_2^d)^2 = (g_3^d)^2 = \{\frac{6}{5},\frac{3}{5}\}g^2\;,\;(g_{23}^u)^2 = \{0,\frac{4}{5}\}g^2\;,\;(g_{23}^d)^2 = (g_{32}^d)^2 = \{0,\frac{3}{5}\}g^2\;,$$

$$(g^\lambda)^2 = \frac{15}{7}g^2\;,\;(g_2^f)^2 = (g_3^f)^2 = \{0,\frac{1}{2}\}g^2\;,\;(g_1^f)^2 = 0\;,\;(g_4^f)^2 = \{1,0\}g^2\;.$$

According to the theorem of ref. [2] these models are finite to all orders. After the reduction of couplings the symmetry of W is enhanced [5].

The main difference of the models **A** and **B** is that three pairs of Higgs quintets and anti-quintets couple to the **24** for **B** so that it is not necessary to mix them with H_4 and $\overline{H}_4$ in order to achieve the triplet-doublet splitting after the symmetry breaking of $SU(5)$.

5 Predictions of Low Energy Parameters

Since the gauge symmetry is spontaneously broken below $M_{\rm GUT}$, the finiteness conditions do not restrict the renormalization property at low energies, and all it remains are boundary conditions on the gauge and Yukawa couplings (13) and the $h = -MC$ relation (9) and the soft scalar-mass sum rule (11) at $M_{\rm GUT}$. So we examine the evolution of these parameters according to their renormalization group equations at two-loop for dimensionless parameters and at one-loop for dimensional ones with these boundary conditions. Below $M_{\rm GUT}$ their evolution is assumed to be governed by the MSSM. We further assume a unique supersymmetry breaking scale M_s so that below M_s the SM is the correct effective theory.

The predictions for the top quark mass M_t are $\sim$ 183 and $\sim$ 174 GeV in models **A** and **B** respectively. Comparing these predictions with the most recent experimental value $M_t = (175.6 \pm 5.5)$ GeV, and recalling that the theoretical values for M_t may suffer from a correction of less than $\sim 4\%$ [4], we see that they are consistent with the experimental data.

Turning now to the SSB sector we look for the parameter space in which the lighter s-tau mass squared $m^2_{\tilde{\tau}}$ is larger than the lightest neutralino mass squared m^2_{χ} (which is the LSP). For the case where all the soft scalar masses are universal at the unfication scale, there is no region of $M_s = M$ below O(few) TeV in which $m^2_{\tilde{\tau}} > m^2_{\chi}$ is satisfied. But once the universality condition is relaxed this problem can be solved naturally (provided the sum rule). More specifically, using the sum rule (11) and imposing the conditions a) successful radiative electroweak symmetry breaking b) $m_{\tilde{\tau}^2} > 0$ and c) $m_{\tilde{\tau}^2} > m_{\chi^2}$, we find a comfortable parameter space for both models (although model **B** requires large $M \sim 1$ TeV). The particle spectrum of models **A** and **B** in turn is calculated in terms of 3 and 2 parameters respectively.

In the figure we present the $m_{\mathbf{10}}$ dependence of m_h for for $M = 0.8$ (dashed) 1.0 (solid) TeV for the finite Model **B**, which shows that the value of m_h is stable.

6 Conclusions

The search for realistic Finite Unified theories started a few years ago [1, 4] with the successful prediction of the top quark mass, and it has now been complemented with a new important ingredient concerning the finiteness of the SSB sector of the theory. Specifically, a sum rule for the soft scalar masses has been obtained which quarantees the finiteness of the SSB parameters up

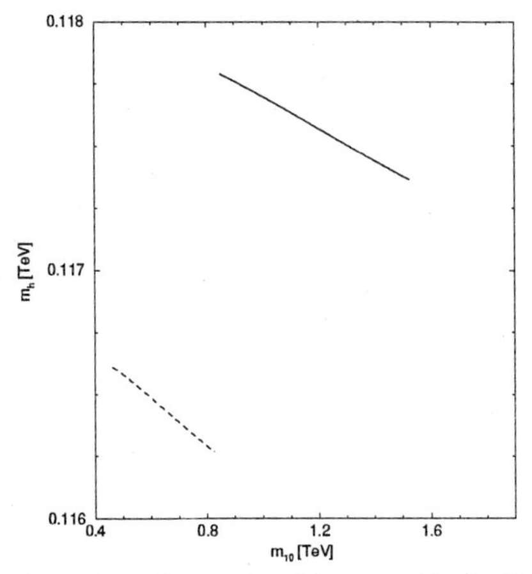

Fig. 1. m_h as function of m_{10} for $M = 0.8$ (dashed) 1.0 (solid) TeV.

to two-loops [5], avoiding at the same time serious phenomenological problems related to the previously known "universal" solution. It is found that this sum rule coincides with that of a certain class of string models in which the massive string modes are organized into $N = 4$ supermultiplets. Using the sum rule we can now determine the spectrum of realistic models in terms of just a few parameters. In addition to the successful prediction of the top quark mass the characteristic features of the spectrum are that 1) the lightest Higgs mass is predicted $\sim$ 120 GeV and 2) the s-spectrum starts above 200 GeV. Therefore, the next important test of these Finite Unified theories will be given with the measurement of the Higgs mass, for which the models show an appreciable stability in their prediction. In the figure we show the dependence of the lightest Higgs mass m_h in terms of the two free parameters of the model **B**.

References

[1] D. Kapetanakis, M. Mondragón and G. Zoupanos, *Zeit. f. Phys.* **C60** (1993) 181; M. Mondragón and G. Zoupanos, *Nucl. Phys.* **B** (Proc. Suppl) **37C** (1995) 98.

[2] C. Lucchesi, O. Piguet and K. Sibold, *Helv. Phys. Acta* **61** (1988) 321; *Phys. Lett.* **B201** (1988) 241.

[3] I. Jack and D.R.T. Jones, Phys.Lett. **B333** (1994) 372.

[4] For an extended discussion and a complete list of references see: J. Kubo, M. Mondragón and G. Zoupanos, Acta Phys. Polon. **B27** (1997) 3911.

[5] T. Kobayashi, J. Kubo, M. Mondragón and G. Zoupanos, *Constraints on Finite Soft SUSY Breaking Terms* to be published in Nucl. Phys. B.

Closing the Light Gluino Window

F. Csikor[1] (csikor@hal9000.elte.hu) Z. Fodor **[2] (fodor@theory4.kek.jp)
[1] Inst. Theor. Phys., Eötvös University
[2] KEK, Theory Group, 1-1 Oho, Tsukuba 305, Japan

Abstract. The running of the strong coupling constant, $R_{e^+e^-}$, R_Z and R_τ is studied on the three-loop level. Based on experimental data of $R_{e^+e^-}$, R_Z and R_τ and the LEP multijet analysis, the light gluino scenario is excluded to 99.97% CL (window I) and 99.89% CL (window III).

1 Introduction and motivation

Asymptotic freedom is one of the most interesting predictions of QCD. One can study this by measuring the running coupling constant at different energies and comparing the results. Fig. 1 shows α_s running as obtained from different experiments together with QCD predictions. Of course QCD is in very good shape. Nevertheless one may quantify this statement. This is not the subject of the present talk, for details see [1]. One may also speculate on slower running than required by QCD. A popular possibility is the light gluino extension of QCD. In this paper we discuss that this scenario can be excluded using three-loop perturbative results and the existing experimental data [1].

The experimentally excluded regions of light gluinos (1996 status) – mass v. lifetime – are shown in Fig. 2. The moral is that window I., i.e. lighter than 1.5 GeV and window III. i.e. masses between 3 GeV and 5 GeV are allowed by these results.

Last year has brought important technical development of the subject, namely 3-loop results have been calculated [2]–[4]. It has been

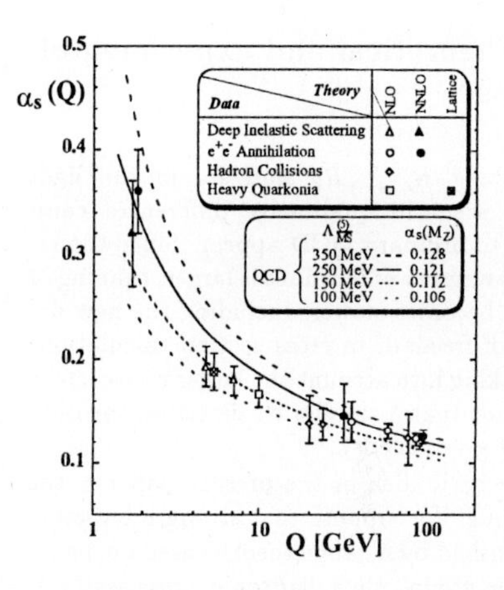

Fig. 1. α_s running and QCD predictions

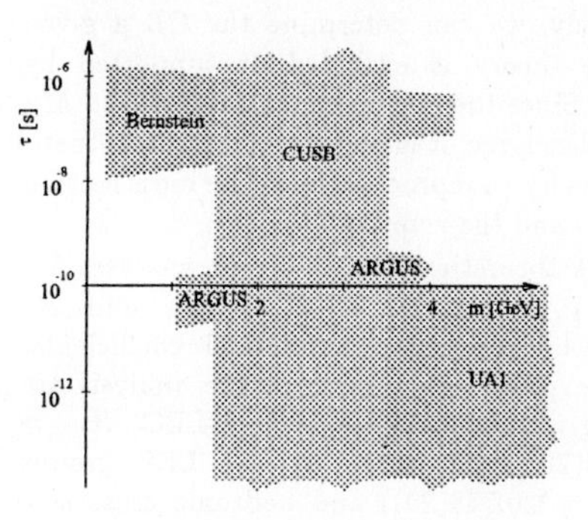

Fig. 2. Light gluino exclusion limits

* Presented by F.Cs.

** On leave from Inst. Theor. Phys., Eötvös University

emphasized that light gluinos are obtained naturally from gauge mediated SUSY breaking [5, 6]. Also some new papers both experimental [7]–[9] and theoretical [10, 11] presenting results on light gluino exclusion have appeared. [12]–[14] critisizes the exclusion results.

2 Theoretical and experimental inputs

Describing $R_{e^+e^-}, R_Z$ and R_τ in the light gluino scenario qualitative differences compared to ordinary QCD appear. Since the effective number of fermions is larger, running of $\alpha_s(Q)$ becomes slower. Including the new degrees of freedom in cross section calculations and taking into account the larger phase-space one finds that a smaller α_s describes the data at any given energy.

The basic idea of the present paper is the following. We suppose that strong interaction is described by a gauge theory based on a simple Lie-group, thus hadronic cross-sections, widths, β-function are given by C_F, C_A, T_F and the active number of fermions. Then by looking for $C_F, C_A, T_F, n_{\tilde{g}}$ to describe data accurately, we can determine the CL a given gauge theory is excluded or supported by data. Since the energy range from M_τ to M_Z is rather large, it is nontrivial for a hypothetical theory to reproduce both the cross section values and the running of $\alpha_s(Q)$.

The theoretical inputs of our analyses are $\mathcal{O}(\alpha_s^3)$ calculations of $R_{e^+e^-}, R_Z$ and R_τ for arbitrary group theorctical coefficients. The experimental inputs of the analysis are R_τ given by ALEPH and CLEO R_τ = 3.616(20)), R_Z given by the LEP groups (R_Z = 20.778(29)) and hadronic cross sections at energies 5 GeV–M_Z (all existing published data (and some unpublished)). The total number of data points included is 182.

3 Treatment of different sources of errors

Besides the data the errors are also very important for a fit. Being too optomistic destroys reliability of the results. When realized that data imply gluino exclusion, we have choosen to use very conservative error estimates.

We treat uncertainties in a unified manner, add the systematic errors linearly and total systematic and statistical errors quadratically. The accuracy of $R_{e^+e^-}$ and R_Z is limited by experiments, while for R_τ theoretical uncertainty dominates. For higher order perturbative corrections we suppose that the error is the last computed term (asymptotic series). For R_τ this gives significantly larger error than usually assumed. For R_τ mass and nonperturbative corrections are taken into account following [15].

Since experimental errors are correlated, we minimize

$$\chi^2 = \Delta^T V^{-1} \Delta \ . \tag{1}$$

Δ is an n-vector of the residuals of $R_i - R_{fit}$, n is the number of individual results, V is $n \times n$ error matrix.

4 Light gluino results from running of $\alpha_s(Q)$

First we present results from fiting $R_{e^+e^-}$, R_Z, R_τ, i.e. α_s running. An example of a two parameter ($x = C_A/C_F, y = T_f/C_F$) fit is shown in Fig. 3 for the case of window III. More quantitatively: we exclude window III light gluinos with a small mass dependence to 93(91)% CL for $m_{\tilde{g}}$=3(5) GeV. For window I we do not use R_τ in the fit and get only 71 % exclusion CL.

Following [10] we fixed the underlying group to SU(3) and determined the number of gluinos. This corresponds to a one parameter fit for $n_{\tilde{g}}$. For window III we get

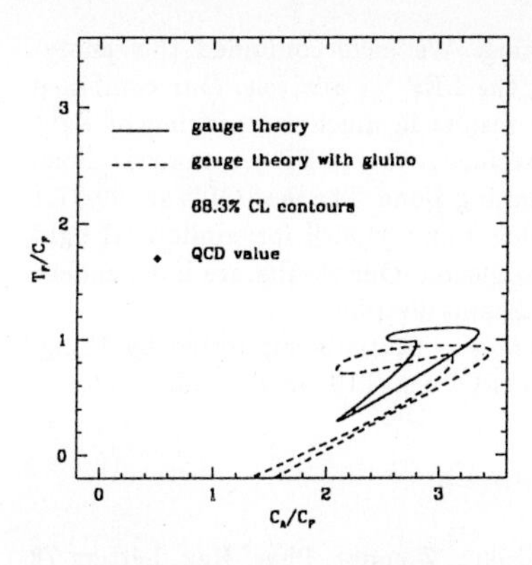

Fig. 3. Exclusion from α_s running for $m_{\tilde{g}}$=3 GeV (window III)

$n_{\tilde{g}}$=0.0078±0.52, 88% CL exclusion and for window I $n_{\tilde{g}}$=-0.070±0.7, 80% CL exclusion (without R_τ).

Farrar [12] claims that increasing our R_τ error estimate by a factor 2 leads to 68% exlusion for window III case. Performing the calculation we get 87(85)% and 81% for window III 2 and 1 parameter fits, respectively.

5 Combination with jet analysis at LEP

LEP groups have determined the group theoretical factors x,y comparing experimental results with leading order 4 jet and higher order 2,3 jet predictions. A summary of the earlier results was given in [16] and shown in Fig. 4. To illustrate what happens when our analysis based on α_s running is combined with a LEP jet analysis, Fig. 5 shows the 1σ excluded regions for the OPAL jet analysis alone [17] (which actually does not exclude light gluinos) and ours (copied from Fig. 3). Note that the overlap of the two regions is rather small, *resulting in stronger exclusion than any of the*

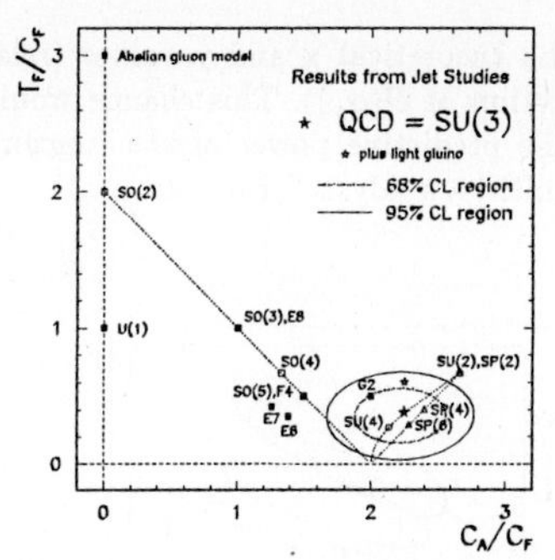

Fig. 4. LEP 4 jet analyses exclusion results

uncombined analyses. It is also clear that increasing the jet analysis ellipsis or shifting it does not change the overlap region too much, i.e. the exclusion changes only very little.

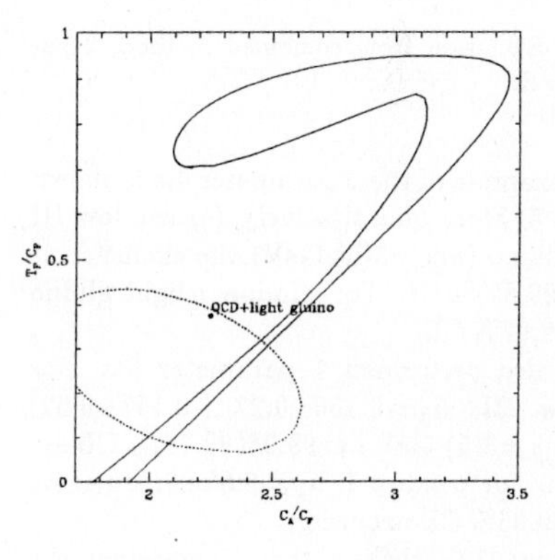

Fig. 5. Comparison of excluded regions from α_s running and OPAL 4 jet analysis

In the actual calculation we have included the older LEP jet analyses and the new ALEPH result [8]. To take into account higher order corrections of 4 jet QCD predictions we have increased the axes of the error ellipse by

12% of the theoretical x and y values (relative correction of $\mathcal{O}(\alpha_s)$). This change would destroy the predictive power of the (uncombined) multi-jet analysis for old data.

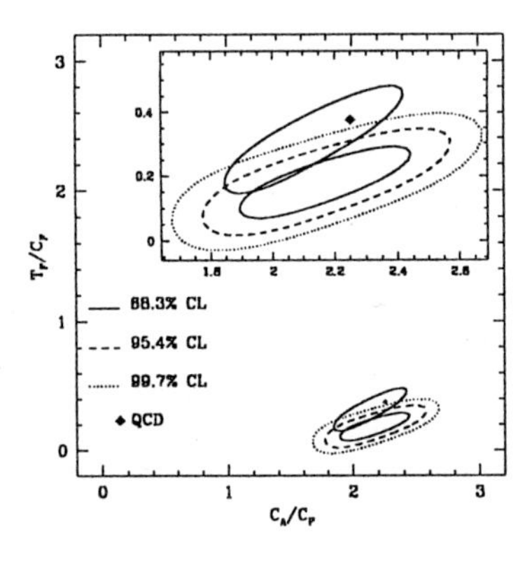

Fig. 6. Exclusion from combined method, 2 parameter fit

An example of the 2 parameter fits is shown in Fig. 6. More quantitatively, for window III light gluino ($m_{\tilde{g}}$ =3(5) GeV) the exclusion is 99.99(99.89)% CL. For window I light gluino it is: 99.97% CL.

We also performed 1 parameter fits. For window III: $n_{\tilde{g}}$=-0.156±0.27 (-0.197±0.32) with $m_{\tilde{g}}$ =3(5) GeV i.e. 99.96(99.76)% CL exclusion. For window I: $n_{\tilde{g}}$=-0.35±0.33 i.e. we have 99.96% CL exclusion.

Farrar [12] claims that increasing the ALEPH errors by a factor 3 would lead to 68% gluino exclusion. Performing the actual calculation we get at least 95% exclusion for all cases of the combined analysis.

6 Summary, conclusion

We have obtained light gluino exclusion from fits of $R_{e^+e^-}$, R_Z and R_τ, i.e. essentally from α_s running. We have combined this analysis with the LEP jet analysis. Our combined analysis results in much more stringent light gluino exclusion than LEP jet analysis alone or α_s running alone. The best CL's are 99.97% for window I and 99.89% for window III light gluino exclusion. Our results are independent of light gluino lifetime.

This work was partially supported by Hung. Sci. grants OTKA-T16248,T22929.

References

[1] F.Csikor, Z.Fodor, Phys. Rev. Letters 78 (1997) 4335 and to be published
[2] K.G. Chetyrkin, Phys. Lett. B391 (1997) 402
[3] L.J. Clavelli, L.R. Surguladze, Phys. Rev. Lett. 78 (1997) 1632
[4] L.J. Clavelli, P.W. Coulter, L.R. Surguladze, Phys. Rev. D55 (1997) 4268
[5] S. Raby, Phys. Rev. D56 (1997) 2852 and this proceedings
[6] R. N. Mohapatra, S. Nandi, Phys. Rev. Letters 79 (1997) 181
[7] F. Abe et al. (CDF), Phys. Rev. D56 (1997) 1357
[8] R. Barate et al., Zeit. Phys. C76 (1997) 191
[9] J. Adams et al. (KTeV), hep-ex/9709028
[10] A. de Gouvêa, H. Murayama, Phys. Lett. B400 (1997) 117
[11] Z. Nagy, Z. Trócsányi, hep-ph/9708343
[12] G. Farrar, hep-ph/9707467
[13] G. Farrar, hep-ph/9710277
[14] G. Farrar, hep-ph/9710395
[15] G. Altarelli, P. Nason, G. Ridolfi, Z. Phys. C68 (1995) 257
[16] M. Schmelling, , in Proc. of 15th Int. Conf. Physics in Collision, Ed. Frontieres, Gif-Sur-Yvette, p. 287 (1995)
[17] R. Akers et al. (OPAL), Z. Physik C65 (1995) 367

QCD Corrections to Electroweak Precision Observables in SUSY Theories

Georg Weiglein (georg@itpaxp5.physik.uni-karlsruhe.de)

Institut für Theoretische Physik, Universität Karlsruhe, 76128 Karlsruhe, Germany

Abstract. The two-loop QCD corrections to the ρ parameter are derived in the Minimal Supersymmetric Standard Model. They turn out to be sizable and modify the one-loop result by up to 30%. Furthermore exact results for the gluonic corrections to Δr are presented and compared with the leading contribution entering via the ρ parameter.

1 Introduction

The Minimal Supersymmetric Standard Model (MSSM) provides the most predictive framework beyond the Standard Model (SM). While the direct search for supersymmetric particles has not been successful yet, the precision tests of the theory provide the possibility for constraining the parameter space of the model and could eventually allow to distinguish between the SM and Supersymmetry via their respective virtual effects. While the SM predictions for Δr and the Z-boson observables include leading terms at two-loop and three-loop order, the corresponding predictions within the MSSM have been restricted so far to one-loop order [1]. In order to treat the MSSM at the same level of accuracy as the SM, higher-order contributions should be incorporated. In this paper results for the QCD corrections to the ρ parameter in the MSSM are presented [2, 3]. In addition, the result for the gluonic contribution to Δr is derived and compared with the approximation based on the contribution entering via the ρ parameter.

2 QCD corrections to the ρ parameter in the MSSM

In the MSSM, the leading contributions of scalar quarks to Δr and the leptonic Z-boson observables enter via the ρ parameter. The contribution of squark loops to the ρ parameter can be written in terms of the transverse parts of the W- and Z-boson self-energies at zero momentum-transfer,

$$\Delta\rho = \frac{\Sigma^{\rm Z}(0)}{M_{\rm Z}^2} - \frac{\Sigma^{\rm W}(0)}{M_{\rm W}^2}. \quad (1)$$

The one-loop result for the stop/sbottom doublet in the MSSM reads [4]

$$\Delta\rho_0^{\rm SUSY} = \frac{3G_\mu}{8\sqrt{2}\pi^2}\left[-scF_0(m_{\tilde{t}_1}^2, m_{\tilde{t}_2}^2) + cF_0(m_{\tilde{t}_1}^2, m_{\tilde{b}_L}^2) + sF_0(m_{\tilde{t}_2}^2, m_{\tilde{b}_L}^2)\right],$$

where $s = \sin^2\theta_{\tilde{t}}$, $c = \cos^2\theta_{\tilde{t}}$, $\theta_{\tilde{t}}$ is the stop mixing angle, and mixing in the sbottom sector has been neglected. The function $F_0(x,y)$ has the form $F_0(x,y) = x+y-\frac{2xy}{x-y}\log\frac{x}{y}$. It vanishes if the squarks are degenerate in mass. In the limit of a large mass splitting between the squarks it is proportional to the heavy squark mass squared. This is in analogy to the case of the top/bottom doublet in the SM [5], $\Delta\rho_0^{\rm SM} = \frac{3G_\mu}{8\sqrt{2}\pi^2}F_0(m_{\rm t}^2, m_{\rm b}^2) \approx \frac{3G_\mu m_{\rm t}^2}{8\sqrt{2}\pi^2}$.

Since the contribution of a squark doublet vanishes if all masses are degenerate, in most SUSY scenarios only the third generation contributes. In the third generation the top-quark mass enters the mass matrix of the scalar partners of the top quark and can give rise to a large mixing in the stop sector and to a large splitting between the stop and sbottom masses.

The two-loop Feynman diagrams of the squark loop contributions to $\Delta\rho$ at $\mathcal{O}(\alpha\alpha_s)$ can be divided into diagrams in which a gluon is exchanged, into diagrams with gluino exchange, and into pure scalar diagrams. After the inclusion of the corresponding counterterms the contribution of the pure scalar diagrams vanishes and the other two sets are separately ultraviolet finite and gauge-invariant (see Ref. [3]).

The result for the gluon-exchange contribution is given by a simple expression resembling the one-loop result

$$\Delta\rho_{1,\rm gluon}^{\rm SUSY} = \frac{G_\mu \alpha_s}{4\sqrt{2}\pi^3}\left[-scF_1(m_{\tilde{t}_1}^2, m_{\tilde{t}_2}^2) + cF_1(m_{\tilde{t}_1}^2, m_{\tilde{b}_L}^2) + sF_1(m_{\tilde{t}_2}^2, m_{\tilde{b}_L}^2)\right].$$

The two-loop function $F_1(x,y)$ is given in terms of dilogarithms by

$$\begin{aligned} F_1(x,y) = {} & x + y - 2\frac{xy}{x-y}\log\frac{x}{y}\left[2 + \frac{x}{y}\log\frac{x}{y}\right] \\ & + \frac{(x+y)x^2}{(x-y)^2}\log^2\frac{x}{y} - 2(x-y)\mathrm{Li}_2\left(1 - \frac{x}{y}\right). \end{aligned} \tag{2}$$

It is symmetric in the interchange of x and y and vanishes for degenerate masses, $F_1(m^2, m^2) = 0$, while in the case of large mass splitting it increases with the heavy scalar quark mass squared: $F_1(m^2, 0) = m^2(1 + \pi^2/3)$.

The gluon-exchange contribution is of the order of 10–15% of the one-loop result [2, 3]. It is remarkable that contrary to the Standard Model case [6], $\Delta\rho_1^{\rm SM} = -\Delta\rho_0^{\rm SM}\,\frac{2}{3}\frac{\alpha_s}{\pi}(1 + \frac{\pi^2}{3})$, where the QCD corrections are negative and screen the one-loop result, $\Delta\rho_{1,\rm gluon}^{\rm SUSY}$ enters with the same sign as the one-loop contribution. It therefore enhances the sensitivity in the search for the virtual effects of scalar quarks in high-precision electroweak measurements.

The analytical expression for the gluino-exchange contribution is much more complicated than for gluon-exchange. In general the gluino-exchange diagrams give smaller contributions compared to gluon exchange. Only for gluino and squark masses close to the experimental lower bounds they compete with the gluon-exchange contributions. In this case, the gluon and gluino contributions add up to about 30% of the one-loop value for maximal mixing [2]. For higher values of $m_{\tilde{g}}$, the contribution decreases rapidly since the gluino decouples in the large-mass limit.

3 Gluonic corrections to Δr

The leading contribution to Δr in the MSSM can be approximated by the contribution to the ρ parameter according to $\Delta r \approx -c_{\mathrm{W}}^2/s_{\mathrm{W}}^2 \Delta\rho$, where $c_{\mathrm{W}}^2 = 1 - s_{\mathrm{W}}^2 = M_{\mathrm{W}}^2/M_{\mathrm{Z}}^2$. In order to test the accuracy of this approximation, we have derived the exact result for the gluon-exchange correction to the contribution of a squark doublet to Δr. It is given by

$$\Delta r_{\mathrm{gluon}}^{\mathrm{SUSY}} = \Pi^{\gamma}(0) - \frac{c_{\mathrm{W}}^2}{s_{\mathrm{W}}^2}\left(\frac{\delta M_{\mathrm{Z}}^2}{M_{\mathrm{Z}}^2} - \frac{\delta M_{\mathrm{W}}^2}{M_{\mathrm{W}}^2}\right) + \frac{\Sigma^{\mathrm{W}}(0) - \delta M_{\mathrm{W}}^2}{M_{\mathrm{W}}^2}, \tag{3}$$

where $\delta M_{\mathrm{W}}^2 = \mathrm{Re}\,\Sigma^{\mathrm{W}}(M_{\mathrm{W}}^2)$, $\delta M_{\mathrm{Z}}^2 = \mathrm{Re}\,\Sigma^{\mathrm{Z}}(M_{\mathrm{Z}}^2)$, and Π^{γ}, Σ^{W}, and Σ^{Z} denote the transverse parts of the two-loop gluon-exchange contributions to the photon vacuum polarization and the W-boson and Z-boson self-energies, respectively, which all are understood to contain the subloop renormalization.

The gluon-exchange correction to the contribution of the stop/sbottom loops to Δr is shown in Fig. 1 together with the $\Delta\rho$ approximation, $\Delta r \approx -c_{\mathrm{W}}^2/s_{\mathrm{W}}^2 \Delta\rho$, as a function of the common scalar mass parameter $m_{\tilde{q}} = M_{\tilde{\mathrm{t}}_L} = M_{\tilde{\mathrm{t}}_R} = M_{\tilde{\mathrm{b}}_L} = M_{\tilde{\mathrm{b}}_R}$, where the $M_{\tilde{q}_i}$ are the soft SUSY breaking parameters appearing in the stop and sbottom mass matrices as specified in Ref. [3]. In this scenario, the scalar top mixing angle is either very small, $\theta_{\tilde{\mathrm{t}}} \sim 0$, or almost maximal, $\theta_{\tilde{\mathrm{t}}} \sim -\pi/4$, in most of the MSSM parameter space. The plots are shown for the two cases $M_{\mathrm{t}}^{LR} = 0$ (no mixing) and $M_{\mathrm{t}}^{LR} = 200$ GeV (maximal mixing) for $\tan\beta = 1.6$.

The two-loop contribution $\Delta r_{\mathrm{gluon}}^{\mathrm{SUSY}}$ is of the order of 10–15% of the one-loop result. It yields a shift in the W-boson mass of up to 20 MeV for low values of $m_{\tilde{q}}$ in the no-mixing case. If the parameter M_{t}^{LR} is made large or the assumption of a common scalar mass parameter is relaxed, much bigger effects are possible [3]. As can be seen in Fig. 1, the $\Delta\rho$ contribution approximates the full result rather well. The two results agree within 10–15%.

4 Conclusions

The two-loop $\mathcal{O}(\alpha_s)$ corrections to the ρ parameter has been derived in the MSSM. The gluonic corrections are of $\mathcal{O}(10\%)$: they are positive and increase the sensitivity in the search for scalar quarks through their virtual effects in high-precision electroweak observables. The gluino contributions are in general smaller except for relatively light gluinos and scalar quarks; the contribution vanishes for large gluino masses. The exact result for the gluon-exchange correction to the contribution of squark loops to Δr has also

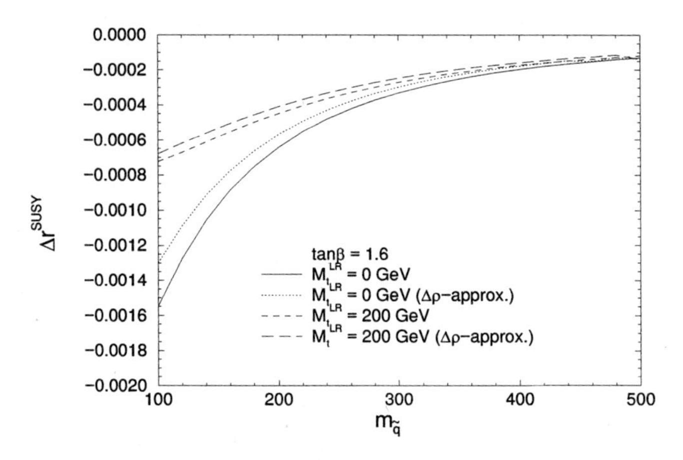

Fig. 1. Contribution of the gluon-exchange diagrams to Δr^{SUSY}. The exact result is compared with the $\Delta\rho$ approximation.

been presented. It gives rise to a shift in the W-boson mass of up to 20 MeV. The result has been compared with the leading contribution entering via the ρ parameter, and good agreement has been found.

The author thanks A. Djouadi, P. Gambino, S. Heinemeyer, W. Hollik and C. Jünger for collaboration on this subject.

References

[1] P. Chankowski, A. Dabelstein, W. Hollik, W. Mösle, S. Pokorski and J. Rosiek, *Nucl. Phys.* **B 417** (1994) 101;
D. Garcia and J. Solà, *Mod. Phys. Lett.* **A 9** (1994) 211;
D. Garcia, R. Jiménez, J. Solà, *Phys. Lett.* **B 347** (1995) 309 and 321;
D. Garcia and J. Solà, *Phys. Lett.* **B 357** (1995) 349;
P. Chankowski and S. Pokorski, *Nucl. Phys.* **B 475** (1996) 3;
W. de Boer, A. Dabelstein, W. Hollik, W. Mösle and U. Schwickerath, *Z. Phys.* **C 75** (1997) 627.

[2] A. Djouadi, P. Gambino, S. Heinemeyer, W. Hollik, C. Jünger and G. Weiglein, *Phys. Rev. Lett.* **78** (1997) 3626.

[3] A. Djouadi, P. Gambino, S. Heinemeyer, W. Hollik, C. Jünger and G. Weiglein, KA-TP-8-1997, hep-ph/9710438.

[4] R. Barbieri and L. Maiani, *Nucl. Phys.* **B 224** (1983) 32;
M. Drees and K. Hagiwara, *Phys. Rev.* **D 42** (1990) 1709.

[5] M. Veltman, *Nucl. Phys.* **B 123** (1977) 89.

[6] A. Djouadi and C. Verzegnassi, *Phys. Lett.* **B 195** (1987) 265;
A. Djouadi, *Nuovo Cimento* **A 100** (1988) 357.

Spontaneous CP Violation and Higgs Spectrum in a Next-to-Minimal SUSY Model

A.T. Davies, C.D. Froggatt and A. Usai (a.davies@physics.gla.ac.uk)

University of Glasgow, Glasgow G12 8QQ, Scotland

Abstract. We explore the possibility of spontaneous CP violation within the Next to Minimal Supersymmetric Standard Model. In the most general form of the model, without a discrete Z_3 symmetry, we find that even at tree level spontaneous CP violation can occur, while also permitting Higgs masses consistent with experiment.

1 NMSSM

CP violation, the Higgs spectrum and supersymmetry are at the forefront of experimental investigation and theoretical interest. We focus on spontaneous CP violation in the Next to Minimal Supersymmetric Standard Model (NMSSM) which contains a singlet N in addition to the two doublets H_1 and H_2 of the MSSM. Spontaneous CP violation is achievable in both the MSSM and NMSSM with the usual Z_3 discrete symmetry, but only by invoking radiative corrections to raise a negative $(\text{mass})^2$ mode to a small, experimentally unacceptable, real mass [1, 2, 3]

Our main result is that spontaneous CP violation is possible for the general NMSSM potential *even at tree level*, and that Higgs masses need not be small.

Above some SUSY-breaking scale, M_S, unknown but hopefully not far beyond experimental reach, the most general superpotential for these fields is [4, 5]

$$W = \lambda N H_1 H_2 - \frac{k}{3} N^3 - rN + \mu H_1 H_2 + W_{Fermion} \tag{1}$$

μ has dimension of mass and r of $(\text{mass})^2$. At lower energy a general quartic form is adopted, with 8 couplings λ_i, which at M_S may be expressed in terms of the gauge couplings and the superpotential coupling constants, and may be determined at the electroweak scale M_{Weak} using renormalization group (RG) equations, if $M_S > M_{Weak}$. New SUSY-breaking soft cubic and quadratic terms are added. In much of the work on the NMSSM, RG equations are used to run down the soft couplings from a hypothesized universal form at the scale M_{GUT}, but we do not assume this, and regard $m_i, i = 1 \ldots 7$, below as arbitrary parameters.

The effective potential is thus [4, 5]

$$V_0 = \frac{1}{2}\lambda_1(H_1^\dagger H_1)^2 + \frac{1}{2}\lambda_2(H_2^\dagger H_2)^2$$
$$+(\lambda_3+\lambda_4)(H_1^\dagger H_1)(H_2^\dagger H_2) - \lambda_4 \left|H_1^\dagger H_2\right|^2$$
$$+(\lambda_5 H_1^\dagger H_1 + \lambda_6 H_2^\dagger H_2)N^\star N + (\lambda_7 H_1 H_2 N^{\star 2} + h.c.)$$
$$+\lambda_8(N^\star N)^2 + \lambda\mu(N + h.c.)(H_1^\dagger H_1 + H_2^\dagger H_2)$$
$$+m_1^2 H_1^\dagger H_1 + m_2^2 H_2^\dagger H_2 + m_3^2 N^\star N - m_4(H_1 H_2 N + h.c.)$$
$$-\frac{1}{3}m_5(N^3 + h.c.) + \frac{1}{2}m_6^2(H_1 H_2 + h.c.) + m_7^2(N^2 + h.c.) \qquad (2)$$

A restricted version of this has been advocated to explain why μ is of electroweak scale, the 'μ-problem' of the MSSM. The terms in the superpotential involving dimensionful couplings and the soft terms in m_6 and m_7 are dropped, leaving a Z_3 discrete symmetry which protects the hierarchy. The VEV $< N >$ replaces μ, thereby introducing a domain wall problem [6, 7]. This restricted model provides only a limited improvement on the MSSM:

(a) Surveys of the parameter space tend to favour a low energy Higgs spectrum very similar to the MSSM [8, 9]

(b) As in the MSSM, spontaneous CP violation is possible, but only as a result of radiative corrections [10, 11, 12]. The lightest neutral Higgs has a mass less than 45 GeV, and typically tan$\beta \leq 1$.

We do not impose the Z_3 symmetry, so do not have the above domain wall problem, but are left with the μ-problem. We find that spontaneous CP violation is more easily achieved in this model than in the MSSM or the NMSSM with Z_3.

We consider real coupling constants, so that the tree level potential is CP conserving, but admit complex VEVs for the neutral fields, giving

$$V_0 = \frac{1}{2}(\lambda_1 v_1^4 + \lambda_2 v_2^4) + (\lambda_3+\lambda_4)v_1^2 v_2^2 + (\lambda_5 v_1^2 + \lambda_6 v_2^2)v_3^2$$
$$+2\lambda_7 v_1 v_2 v_3^2 cos(\theta_1+\theta_2-2\theta_3) + \lambda_8 v_3^4 + 2\lambda\mu(v_1^2+v_2^2)v_3 cos(\theta_3)$$
$$+m_1^2 v_1^2 + m_2^2 v_2^2 + m_3^2 v_3^2 - 2m_4 v_1 v_2 v_3 cos(\theta_1+\theta_2+\theta_3)$$
$$-\frac{2}{3}m_5 v_3^3 cos(3\theta_3) + 2m_6^2 v_1 v_2 cos(\theta_1+\theta_2) + 2m_7^2 v_3^2 cos(2\theta_3) \qquad (3)$$

$$\langle H_i^0\rangle = v_i e^{i\theta_i} (i=1,2), \qquad \langle N\rangle = v_3 e^{i\theta_3}. \qquad (4)$$

where, without loss of generality, $\theta_2 = 0$, and $0 \leq \theta_3 < \pi$.

This potential has 3 extra terms as compared with a Z_3 invariant potential: a cubic term arising from the μ and λ terms in the superpotential, and two quadratic terms with coefficients m_6^2 and m_7^2. If $\mu = 0$ the effective potential loses this cubic term, and the quadratic terms alone account for the difference between our results and previous ones in the literature.

The scalar mass matrix gives rise to 1 charged and 5 neutral particles. If the angles θ_1 and θ_3 are non-zero, the neutral matrix does not decouple into sectors with CP = +1 and -1.

2 Searches and results

We search the parameter space of this potential to see if spontaneous CP violation is compatible with experimental bounds on the Higgs spectrum.

There are 11 parameters, but we impose some restrictions.

(a) 4 superpotential parameters $\mu, \lambda, k,$ and h_t, the top Yukawa coupling: Running (λ, k) up from the electroweak scale using RG equations [13] requires them to be small to avoid blow-up at high energy. In the examples below we fix $\lambda = 0.5, k = 0.5$. We take the fixed point value $h_t = 1.05$ at the SUSY scale, bearing in mind that at the electroweak scale the running top mass is $h_t v_0 \sin\beta$.

(b) Soft breaking mass parameters $m_i, i = \ldots 7$:
Five of these can be traded for VEV magnitudes and phases $v_1, v_2, v_3, \theta_1, \theta_3$. We eliminate one by the condition $v_0 = \sqrt{v_1^2 + v_2^2} = 174$ GeV and replace 2 others by the conventional $\tan\beta \equiv v_2/v_1$ and $R \equiv v_3/v_0$. A sixth mass, m_4 say, can be exchanged for the mass of the charged Higgs, M_{H^+}, which $b \to s + \gamma$ experiment suggests to be greater than 250 GeV [14]. We can obtain an analytic form for this:

$$M_{H^+}^2 = -\lambda_4 v_0^2 - \frac{2(\lambda_7 v_3^2 sin(3\theta_3) + m_6^2 sin\theta_3)}{sin(2\beta) sin(\theta_1 + \theta_2 + \theta_3)} \tag{5}$$

This shows how the parameter m_6^2, not necessarily positive, introduces extra freedom to raise the charged Higgs mass. This leaves one parameter m_5, with no particular interpretation.

In the case $\mu = 0$, λ and k cannot be too small. In the simplest case, where the Higgsino-gaugino mixing is small, the (unbroken SUSY) charged Higgsino mass is $|\mu + \lambda v_3 \cos\theta_3|$. The experimental lower bound, conservatively $\frac{1}{2}M_Z$, then disallows small $R(\equiv v_3/v_0)$. There is no upper bound on R, and indeed the cubic and quartic terms in the field N can in some cases provide deep global minima of the potential for large R, thus excluding some otherwise acceptable values of the other parameters.

Our *modus operandi* was to scan over a grid of parameters chosen to make the first derivatives of the potential vanish at prescribed VEVs. Numerical minimization was performed to ensure that these were minima, giving positive $(\text{mass})^2$, and to reject local minima. In many apparently reasonable cases the spontaneous CP violating minimum was metastable, with a lower electroweak and CP conserving vacuum lying at $v_1 = v_2 = 0$ and v_3 the order of TeV. We present two indicative examples from preliminary searches.

Parameters giving spontaneous CP violation							
CASE	$\tan\beta$	R $(= v_3/v_0)$	θ_1	θ_3	M_{H^+}	m_5	μ
A	2.0	2.0	1.20π	0.65π	250 GeV	60 GeV	-20 GeV
B	2.0	2.0	0.60π	0.35π	250 GeV	-100 GeV	0

Case (A) is for a SUSY scale of 1 TeV, with quartic couplings radiatively corrected using RG equations, assuming that all squarks and gauginos lie at the SUSY scale. This has neutral Higgs massses from 89 to 318 GeV, a lightest neutralino of 48 GeV and a chargino of 99 GeV. Larger neutralino masses can be obtained by increasing R.

Case (B) is for $M_S = M_{Weak}$ i.e. a SUSY quartic potential. It gives neutral Higgs masses of 81 to 372 GeV, a lightest neutralino of 47 GeV, and a chargino of 79 GeV, ignoring gaugino mixing which should be included if indeed the SUSY scale is as low as the electroweak scale.

This case is presented as a go-theorem, a counter example to no-go examples [10]. As $\mu = 0$ here, the quartic potential has the standard Z_3 invariant form and avoids the conclusion of Romao [10] due to the two Z_3 violating soft terms. Nor do the conditions of the Georgi-Pais [15] no-go theorem apply. It relates to the situation where CP is conserved at tree level and is broken only by perturbative radiative corrections. The soft terms here are not in this category.

We are encouraged to explore further the phenomenology of such models.

References

[1] N. Maekawa, Phys. Lett. B282 (1992), 387;
Nucl. Phys.B (Proc. Suppl) 37A (1994), 191

[2] A. Pomerol, Phys. Lett. B287 (1992), 331

[3] K.S. Babu and S.M. Barr, Phys. Rev. D49 (1994), R2156

[4] J. Gunion, H.E. Haber, G.L. Kane and S.Dawson, 'The Higgs Hunter's Guide' (Addison-Wesley, Reading MA, 1990).

[5] J. Gunion and H.E. Haber, Nucl. Phys. B272 (1986), 1

[6] J. Ellis, J. Gunion, H.E. Haber, L. Roszkowski and P. Zwirner, Phys. Rev. D39 (1989), 844

[7] S. A. Abel, S. Sarker and P. L. White, Nucl. Phys. B454 (1995), 663;
S.A. Abel, Nucl. Phys. B480 (1996), 55

[8] S.F. King and P.L.White, Phys. Rev. D52 (1995), 4183

[9] C. Ellwanger, M. Rausch de Traubenberg, C.A. Savoy, Nucl. Phys. B492 (1997), 21

[10] J.C. Romao, Phys. Lett. B173 (1986), 309

[11] M.M. Asatrian and G.K. Eguian, Mod. Phys. Lett. A10 (1995), 2943

[12] N. Haba, M. Matsuda and M. Tanimoto, Phys. Rev. D54 (1996), 6928

[13] T. Elliott, S. F. King and P. L. White, Phys. Rev. D49 (1994), 2435

[14] CLEO collaboration, M.S. Alam et al, Phys. Rev. Lett 74 (1995), 2885

[15] H. Georgi and A. Pais, Phys. Rev. D10 (1974), 1246

R-Parity Breaking in Minimal Supergravity

Marco Aurelio Díaz (mad@flamenco.ific.uv.es)

Universidad de Valencia, Spain

Abstract. We consider the Minimal Supergravity Model with universality of scalar and gaugino masses plus an extra bilinear term in the superpotential which breaks R–Parity and lepton number. We explicitly check the consistency of this model with the radiative breaking of the electroweak symmetry. A neutrino mass is radiatively induced, and large Higgs–Lepton mixings are compatible with its experimental bound. We also study briefly the lightest Higgs mass. This one–parameter extension of SUGRA–MSSM is the simplest way of introducing R–parity violation.

The Minimal Supersymmetric Standard Model (MSSM) [1] contains a large number of soft supersymmetry breaking mass parameters introduced explicitly in order to break supersymmetry without introducing quadratic divergencies. When the MSSM is embedded into a supergravity inspired model (MSSM–SUGRA), the number of unknown parameters can be greatly reduced with the assumption of universality of soft parameters at the unification scale. In addition, in MSSM–SUGRA the breaking of the electroweak symmetry can be achieved radiatively due to the large value of the top quark Yukawa coupling.

The most general extension of the MSSM which violates R–parity [2] contains almost 50 new parameters, all of them arbitrary although constrained by, for example, proton stability . The large amount of free parameters makes R–parity violating scenarios less attractive. Nevertheless, models of spontaneous R–parity breaking do not include trilinear R–parity violating couplings, and these models only generate bilinear R–parity violating terms [3].

Motivated by the spontaneous breaking of R–parity, we consider here a model where a bilinear R–parity violating term of the form $\epsilon_i \widehat{L}_i^a \widehat{H}_2^b$ is introduced explicitly in the superpotential [4]. We demonstrate that this "ϵ–model" can be successfully embedded into supergravity, i.e., it is compatible with universality of soft mass parameters at the unification scale and with the radiative breaking of the electroweak group [5].

For simplicity we consider that only the third generation of leptons couples to the Higgs. Therefore, our superpotential is

$$W = \varepsilon_{ab}\left[h_t \widehat{Q}_3^a \widehat{U}_3 \widehat{H}_2^b + h_b \widehat{Q}_3^b \widehat{D}_3 \widehat{H}_1^a + h_\tau \widehat{L}_3^b \widehat{R}_3 \widehat{H}_1^a - \mu \widehat{H}_1^a \widehat{H}_2^b + \epsilon_3 \widehat{L}_3^a \widehat{H}_2^b\right] \quad (1)$$

where the last term is the only one not present in the MSSM. This term induces a non–zero vacuum expectation value of the tau sneutrino, which we denote by $\langle \tilde{\nu}_\tau \rangle = v_3/\sqrt{2}$.

The ϵ_3–term cannot be rotated away by the redefinition of the fields

$$\widehat{H}_1' = \frac{\mu\widehat{H}_1 - \epsilon_3\widehat{L}_3}{\sqrt{\mu^2+\epsilon_3^2}}, \qquad \widehat{L}_3' = \frac{\epsilon_3\widehat{H}_1 + \mu\widehat{L}_3}{\sqrt{\mu^2+\epsilon_3^2}}, \tag{2}$$

and in this sense the ϵ_3–term is physical. If the previous rotation is performed, the bilinear R–Parity violating term disappear from the superpotential. Nevertheless, a trilinear R–Parity violating term is reintroduced in the Yukawa sector and it is proportional to the bottom quark Yukawa coupling. In addition, bilinear terms which induce a non–zero vacuum expectation value of the tau sneutrino reappear in the soft terms, and therefore, the vacuum expectation value of the tau sneutrino is also non–zero in the new basis: $\langle\tilde{\nu}_\tau'\rangle = v_3' \neq 0$. These terms are

$$V_{soft} = (B_2 - B)\frac{\epsilon_3\mu}{\mu'}\widetilde{L}_3' H_2 + (m_{H_1}^2 - M_{L_3}^2)\frac{\epsilon_3\mu}{\mu'^2}\widetilde{L}_3' H_1' + h.c. + ... \tag{3}$$

where $\mu'^2 = \mu^2 + \epsilon_3^2$, B and B_2 are the bilinear soft breaking terms associated to the next-to-last and last terms in eq. (1), and m_{H_1} and M_{L_3} are the soft mass terms associated to H_1 and $\widetilde{L}_3$.

The presence of the ϵ_3 term and of a non–zero vev of the tau sneutrino induce a mixing between the neutralinos and the tau neutrino. As a consequence, a tau neutrino mass is generated which satisfy $m_{\nu_\tau} \sim (\epsilon_3 v_1 + \mu v_3)^2$. The quantity inside the brackets is proportional to v_3', thus a non–zero vev of the tau sneutrino in the rotated basis is crucial for the generation of a mass for the tau neutrino.

We assume at the unification scale universality of soft scalar masses, gaugino masses, soft bilinear mass parameters, and soft trilinear mass parameters. Using the RGE's given in [5] we impose the correct electroweak symmetry breaking. In order to do that, we impose that the one–loop tadpole equations are zero, and find the three vacuum expectation values. This tadpole method is equivalent to use the one–loop effective potential [6]. The solutions we find are displayed as scatter plots. In Fig. 1 we have the induced tau neutrino mass m_{ν_τ} as a function of the combination $\xi \equiv (\epsilon_3 v_1 + \mu v_3)^2$, which is related to the v.e.v. of the tau sneutrino in the rotated basis through $\xi = (\mu' v_3')^2$.

In Fig. 1 we see that we find plenty of solutions with values of the tau neutrino mass compatible with experimental bounds. The reason is that in models with universality of soft supersymmetry breaking parameters it is natural to find small values of the v.e.v. $v_3' \sim (\epsilon_3 v_1 + \mu v_3)$. This can be understood if we look at the tree level tadpole corresponding to the tau sneutrino in the rotated basis. The relevant linear term is $V_{linear} = t_3'\tilde{\nu}_\tau'^R + ...$, with $\tilde{\nu}_\tau'^R = \sqrt{2}Re(\tilde{\nu}_\tau') - v_3'$, and the tree level tadpole equation is

$$\begin{aligned} t_3' = (m_{H_1}^2 - M_{L_3}^2)\frac{\epsilon_3\mu}{\mu'^2}v_1' + (B_2 - B)\frac{\epsilon_3\mu}{\mu'}v_2' + \frac{m_{H_1}^2\epsilon_3^2 + M_{L_3}^2\mu^2}{\mu'^2}v_3' \\ + \tfrac{1}{8}(g^2 + g'^2)v_3'(v_1'^2 - v_2^2 + v_3'^2) = 0 \end{aligned} \tag{4}$$

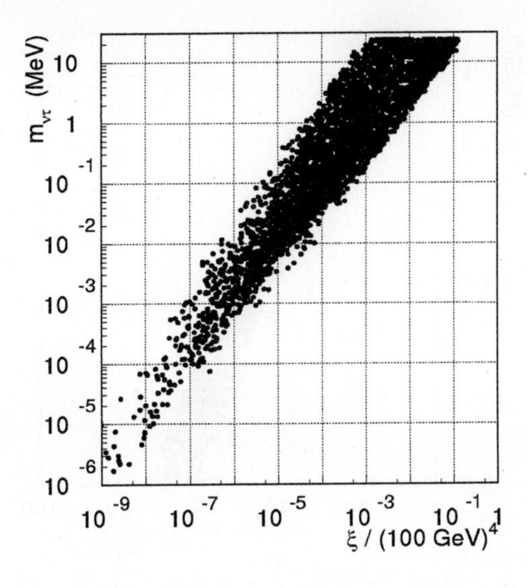

Fig. 1. Tau neutrino mass as a function of $\xi \equiv (\epsilon_3 v_1 + \mu v_3)^2$, which is related to the v.e.v. of the tau sneutrino in the rotated basis through $\xi = (\mu' v_3')^2$.

It is clear that the first two terms are generated radiatively, because at the unification scale we have $m_{H_1} = M_{L_3}$ and $B_2 = B$. The RGE's of these parameters are such that at the weak scale we have non–zero differences $(m_{H_1}^2 - M_{L_3}^2)$ and $(B_2 - B)$ generated at one–loop and proportional to $h_b^2/(16\pi^2)$, where h_b is the bottom quark Yukawa coupling. If for a moment we neglect these radiative corrections, the first two terms in eq. (4) are zero and as a consequence $v_3' = 0$, implying that the induced tau neutrino mass is zero. In reality this is not the case, and the tau neutrino mass is radiatively generated.

In this model, the CP–even Higgs bosons of the MSSM mix with the real part of the tau sneutrino [7]. For this reason, the neutral CP–even scalar sector contains three fields and the mass of the lightest scalar is different compared with the lightest CP–even Higgs of the MSSM. In Fig. 2 we plot the ratio between the mass of the lightest CP–even neutral scalar in the ϵ–model and the lightest CP–even Higgs of the MSSM, as a function of the v.e.v. of the tau sneutrino in the unrotated basis. In the radiative corrections to these masses we have included the most important contribution which is proportional to m_t^4. As it should, the ratio approaches to unity as v_3 goes to zero. Most of the time the effect of v_3 is to reduce the scalar mass, but there are a few points where the opposite happens.

In summary, we have proved that a bilinear R–Parity violating term can be successfully embedded into supergravity, with universality of soft mass terms at the unification scale, and with radiative breaking of the electroweak

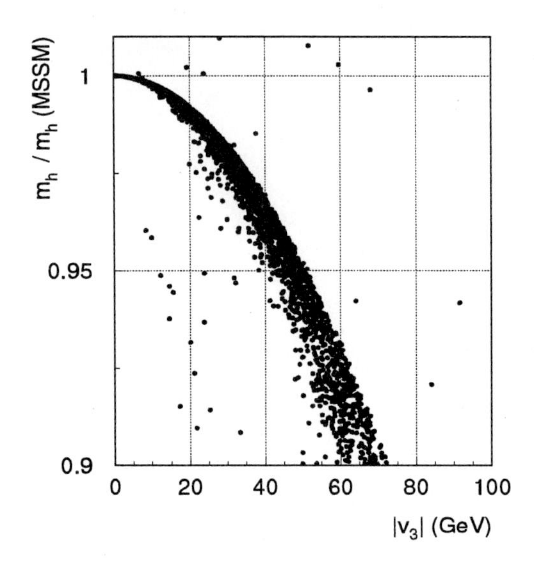

Fig. 2. Ratio between the lightest CP-even neutral scalar mass in the ϵ–model and the lightest CP–even Higgs mass in the MSSM, as a function of the tau sneutrino vacuum expectation value v_3.

symmetry. In addition, the induced neutrino mass is generated radiatively at one–loop, and therefore it is naturally small. This is a one parameter (ϵ_3) extension of MSSM-SUGRA, and therefore the simplest way to study systematically R–Parity violating phenomena.

References

[1] H.P. Nilles, *Phys. Rep.* **110**, 1 (1984); H.E. Haber and G.L. Kane, *Phys. Rep.* **117**, 75 (1985); R. Barbieri, *Riv. Nuovo Cimento* **11**, 1 (1988).
[2] L. Hall and M. Suzuki, *Nucl.Phys.* **B231**, 419 (1984).
[3] A. Masiero and J.W.F. Valle, *Phys. Lett.* **B251**, 273 (1990); J.C. Romão, A. Ioannissyan and J.W.F. Valle, *Phys. Rev. D***55**, 427 (1997).
[4] A.S. Joshipura and M. Nowakowski, *Phys. Rev.* **D51**, 2421 (1995); R. Hempfling, *Nucl. Phys.* **B478**, 3 (1996); F. Vissani and A.Yu. Smirnov, *Nucl. Phys.* **B460**, 37 (1996); H.P. Nilles and N. Polonsky, *Nucl. Phys.* **B484**, 33 (1997); B. de Carlos, P.L. White, *Phys. Rev.* **D55**, 4222 (1997); S. Roy and B. Mukhopadhyaya, *Phys. Rev.* **D55**, 7020 (1997).
[5] M.A. Díaz, J.C. Romao, and J.W.F. Valle, hep-ph/9706315.
[6] M.A. Díaz and H.E. Haber, *Phys. Rev. D* **46**, 3086 (1992).
[7] F. de Campos, M.A. García-Jareño, A.S. Joshipura, J. Rosiek, and J.W.F. Valle, *Nucl. Phys.* **B451**, 3 (1995); A. Akeroyd, M.A. Díaz, J. Ferrandis, M.A. Garcia–Jareño, and J.W.F. Valle, hep-ph/9707395.

On the Expectations for Leptoquarks in the Mass Range of 0(200 GeV)

Johannes Blümlein (blumlein@ifh.de)

DESY-Zeuthen, Platanenallee 6, D-15738 Zeuthen, Germany

Abstract. We investigate to which extent an interpretation of the recently observed excess of events in the high Q^2 range at HERA in terms of single leptoquark production is compatible with bounds from other experiments.

1 Introduction

Recently both the H1 [1] and ZEUS [2] experiments reported an excess of deep inelastic neutral current events in the range $Q^2 \gtrsim 15000\,\text{GeV}^2$. There is also a slight indication of an excess for charged current scattering in the H1 and ZEUS data, cf. [1, 2]. Various phenomenological and theoretical analyses were performed shortly after this finding to seek an interpretation [3], though the experimental signals have first to stabilize adding in more data in the current runs. One possible interpretation for these events might be single leptoquark production in the e^+q or $e^+\overline{q}$ channels, respectively. In this note we summarize the status of the phenomenological studies which were performed during the last months. [1] Other interpretations, such as a signal for substructure, effects due to R–parity violating supersymmetry, as well as implications for experiments at e^+e^- colliders, were also discussed, cf. e.g. [6].

2 $p\overline{p}$ Scattering

Leptoquarks may be searched for at hadron colliders both studying single and pair production processes. In the case of the single production processes [7] the reaction cross sections are $\propto \lambda^2$ and amount to $\sigma_{\text{sing}} \sim 0.4\ fb...1.3\ fb$ at Tevatron energies for $(\lambda/e)\sqrt{Br} \approx 0.075$, eq. (10), only, [3]. They are therefore too small to be detected currently.

On the other hand, given the small fermionic couplings, the pair production processes depend on the leptoquark–gluon couplings only. In the case of scalar leptoquarks the production cross section is completely predicted, whereas it depends on anomalous couplings, such as κ_G and λ_G, in the case of vector leptoquarks. As was shown in ref. [8], however, there exists a global minimum $\mathsf{min}_{\kappa_G,\lambda_G}\left[\sigma(\Phi_V\overline{\Phi}_V)\right] > 0$ allowing for a model-independent analysis.

The production cross sections for scalar leptoquarks in the partonic subsystem read, cf. [8],

[1] For surveys of the earlier literature on leptoquarks see e.g. refs. [5].

$$\sigma^{q\overline{q}}_{\Phi_S\overline{\Phi}_S} = \frac{2\pi\alpha_s^2}{27\hat{s}}\beta^3 \tag{1}$$

$$\sigma^{gg}_{\Phi_S\overline{\Phi}_S} = \frac{\pi\alpha_s^2}{96\hat{s}}\left[\beta(41-31\beta^2)-(17-18\beta^2+\beta^4)\ln\left|\frac{1+\beta}{1-\beta}\right|\right], \tag{2}$$

with $\beta = \sqrt{1-4M_\Phi^2/\hat{s}}$. The $O(\alpha_s)$ correction to the production cross section was calculated in ref. [9] and amounts to a K-factor of 1.12 only for the choice $\mu = M_\Phi$ of the factorization scale. The cross sections in the case of vector leptoquark pair production are more complicated, cf. [8], due to the presence of the anomalous couplings κ_G and λ_G and have the general structure

$$\sigma^{q\overline{q}}_{\Phi_V\overline{\Phi}_V} = \frac{4\pi\alpha_s^2}{9M_V}\sum_{i=0}^{5}\chi_i^q(\kappa_G,\lambda_G)\widetilde{G}_i(\hat{s},\beta) \tag{3}$$

$$\sigma^{gg}_{\Phi_V\overline{\Phi}_V} = \frac{\pi\alpha_s^2}{96M_V}\sum_{i=0}^{14}\chi_i^g(\kappa_G,\lambda_G)\widetilde{F}_i(\hat{s},\beta)\ . \tag{4}$$

The functions $\chi^{q,g}, \widetilde{G}_i$ and $\widetilde{F}_i$ are given in ref. [8]. For $\kappa_G = \lambda_G = 0$ one obtains

$$\sigma^{q\overline{q}}_{\Phi_V\overline{\Phi}_V} = \frac{\pi\alpha_s^2}{54M_V}\left[\frac{\hat{s}}{M_V^2}+23-3\beta^2\right], \tag{5}$$

$$\sigma^{gg}_{\Phi_V\overline{\Phi}_V} = \frac{\pi\alpha_s^2}{96M_V}\left\{\beta A(\beta)-B(\beta)\ln\left|\frac{1+\beta}{1-\beta}\right|\right\},$$

$$A(\beta) = \frac{523}{4}-90\beta^2+\frac{93}{4}\beta^4, \qquad B(\beta) = \frac{3}{4}\left[65-83\beta^2+19\beta^4-\beta^6\right]\ . \tag{6}$$

Choosing the factorization and renormalization scales by $\mu = M_\Phi$ the pair production cross sections for scalar and vector leptoquarks (minimizing for κ_G and λ_G) at Born level and using the parametrization [10] for the parton densities are [3] :

$$\sigma_S(M_\Phi = 200\,\text{GeV}) = 0.16\,pb \qquad \sigma_V(M_\Phi = 200\,\text{GeV}) = 0.29\,pb. \tag{7}$$

3 Excluded Mass Ranges

The most stringent limits on leptoquark masses, which are independent of the fermionic couplings, were derived by the Tevatron experiments, see [11, 12], searching for leptoquark pair production. The following mass ranges are excluded for scalar leptoquarks associated to the first fermion generation :

$$\begin{array}{llll} M < & 213\ \text{GeV} & \text{CDF}, & Br(eq)=1 \\ M < & 225\ \text{GeV}(204\ \text{GeV}) & \text{D0}, & Br(eq)=1\ (1/2) \end{array} \tag{8}$$

at 95% CL. A combined exclusion limit of $M < 240$ GeV, $Br(eq) = 1$, or $M < 200$ GeV, $Br(eq) = 1/2$, can be derived, cf. [13]. The mass bounds for vector leptoquarks are correspondingly higher because of the production cross section being larger by at least a factor of two, eq. (7), but have not yet been derived by the Tevatron experiments for the general case [8].

4 The HERA Events

If the observed high-Q^2 excess is interpreted in terms of single leptoquark production [14] constraints on the fermionic coupling λ of the leptoquarks Φ, which may be either scalars or vectors, can be derived. Due to the location of the excess found by H1 in the range $M = \sqrt{xS} \sim 200$ GeV we assume this scale in the estimates given below. In the narrow width approximation the production cross section reads

$$\sigma = \frac{\pi^2}{2}\alpha\left(\frac{\lambda}{e}\right)^2 q(x, \langle Q^2 \rangle)\left\{\begin{matrix} 2 & : & V \\ 1 & : & S \end{matrix}\right. \times Br(\Phi \rightarrow eq) \,. \tag{9}$$

For the observed excess in the $e^+ + jet$ channel at H1

$$\frac{\lambda_S}{e}\sqrt{Br} \sim 0.075 \;\; (0.15) \quad u \;\; (d), \quad \Phi = S \tag{10}$$

is derived [3] using the parametrization [10] of the quark densities, while $\lambda_V = \lambda_S/\sqrt{2}$ and $\lambda_{\rm ZEUS} = 0.55\lambda_{\rm H1}$. These couplings are well compatible with the limits derived from low energy data [15]. For the production cross section of scalar leptoquarks the QCD corrections were calculated in [16]. They amount to +23%. Both for the measurement of the total cross section and differential distributions, such as the mass distribution $M = \sqrt{xS}$ or the y distribution, a precise treatment of the QED radiative corrections is of importance, because these corrections can be very large depending on the way in which the kinematic variables are measured, see ref. [17] for a general discussion. The universal QED corrections, also accounting for higher orders, can be calculated using the code HECTOR [17].

An information on the spin of the produced state can be derived from the y distribution of the events. The statistics is yet to low to allow for such an analysis. The average value $\langle y \rangle_{\rm H1} = 0.59 \pm 0.02$ is still compatible with both the expectation for a scalar $\langle y \rangle_S = 0.65$ or a vector $\langle y \rangle_V = 0.55$, cf. [3]. One also may consider the scattering process $eq \rightarrow \Phi g$ both for scalar and vector leptoquarks Φ [14c,18] to get further constraints on the spin of the state produced.

A severe constraint on the leptoquark states which may be produced in e^+q scattering is imposed by the $SU(2)_L \times U(1)_Y$ quantum numbers[2]. If besides the e^+q final states the indication of also νq final states becomes manifest, scalar leptoquarks are **not** allowed since low energy constraints demand either $\lambda_L \ll \lambda_R$ or $\lambda_R \ll \lambda_L$. For the vector states $U_{3\mu}^0$ or $U_{1\mu}$, which may be produced in the e^+d channel, on the other hand, the branching ratios are $Br(e^+d) = Br(\nu u) = 1/2$.

The combination of the bounds in eq. (8) exclude both a scalar and a vector leptoquark in the mass range currently explored at HERA at 95% CL

[2] For a classification in the case of family-diagonal, baryon- and lepton number conserving, non-derivative couplings, see [14b].

if $Br(eq) = 1$. For $Br(eq) = 1/2$ this applies as well for vector leptoquarks. These bounds are widely model independent. If the excess of events persists in the data at higher statistics a measurement of the $\nu jet/e^+ jet$ ratio of the effect will be possible. In the future high-luminosity runs at HERA the search potential for leptoquarks reaches beyond the mass range being excluded by the Tevatron measurements at present. It will therefore be interesting to see whether there are signals above the presently excluded ranges.

Acknowledgement. For discussions I would like to thank Y. Sirois.

References

[1] C. Adloff et al., H1 collaboration, *Z. Phys.* **C74** 191 (1997); H1 collaboration, these proceedings.
[2] J. Breitweg et al., ZEUS collaboration, *Z. Phys.* **C74** 207 (1997); ZEUS collaboration, these proceedings.
[3] For a detailed list of references see e.g. J. Blümlein, *Z. Phys.* **C74** (1997) 605 and ref. [4].
[4] J. Blümlein, in : Proc. of the 5th Int. Conf. on Deep Inelastic Scattering, Chicago, April 1997, DIS '97, DESY 97-105, `hep-ph/9706205`.
[5] J. Blümlein and E. Boos, *Nucl. Phys.* **B** *(Proc. Suppl.)* **37B** (1994) 181 and ref. [3].
[6] J. Ellis, these proceedings.
[7] See e.g.: A. Djouadi et al., SLAC–PUB–95–6772 and references in ref. [4].
[8] J. Blümlein, E. Boos, and A. Kryukov, *Z. Phys.* **C76** (1997) 137.
[9] M. Krämer et al., *Phys. Rev. Lett.* **79** (1997) 341.
[10] H. Lai et al., *Phys. Rev.* **D51** (1995) 4763.
[11] F. Abe et al., CDF collaboration, `hep-ex/9708017`; X. Wu, these proceedings.
[12] B. Abbott et al., D0 collaboration, `hep-ex/9710032`; B. Klima, these proceedings, `hep-ex/97110019`.
[13] M. Krämer, `hep-ph/9707422`.
[14] J. Wudka, *Phys. Lett.* **B167** (1986) 337; W. Buchmüller, R. Rückl, and D. Wyler, *Phys. Lett.* **B191** (1987) 442; A. Dobado, M.J. Herrero, and C. Muñoz, *Phys. Lett.* **B191** (1987) 449.
[15] M. Leurer, *Phys. Rev. Lett.* **71** (1993) 1324; *Phys. Rev.* **D49** (1994) 333; *Phys. Rev.* **D50** (1994) 536; S. Davidson et al., *Z. Phys.* **C61** (1994) 613.
[16] Z. Kunszt and W. Stirling, *Z. Phys.* **C75** (1997) 453; T. Plehn et al., *Z. Phys.* **C74** (1997) 611.
[17] A. Arbuzov et al., `HECTOR 1.00`, `hep-ph/9510410`, *Comp. Phys. Commun.* **94** (1996) 128.
[18] J. Blümlein and A. Kryukov, DESY 97–067.

Part XII

Phenomenology of GUTs, Supergravity and Superstrings

Updated Combined Fit of Low-Energy Constraints to Minimal Supersymmetry

W. de Boer*, R. Ehret*, A.V. Gladyshev†, D.I. Kazakov†

* Inst. für Experimentelle Kernphysik, Univ. of Karlsruhe, Germany
†Bogoliubov Lab. of Theor. Physics, JINR Dubna, Russia

Abstract. The new precise LEP measurements of α_S and $\sin^2\theta_W$ as well as the new LEP II mass limits for supersymmetric particles and new calculations for the radiative (penguin) decay of the b-quark into $s\gamma$ allow a further restriction in the parameter space of the Constrained Minimal Supersymmetric Standard Model (CMSSM).

1 Introduction

In this paper we repeat our previous fit of the supergravity parameters[1] with the new input data from LEP concerning the coupling constants and new higher order calculations for the $b \to s\gamma$ decay rate branching ratio [2]. The latter indicate that next to leading log (QCD) corrections increase the SM value by about 10%. This can be simulated in the lowest level calculation by choosing a renormalization scale $\mu = 0.57m_b$, which is done in the following. In addition, a new preliminary measurement of the $b \to s\gamma$ rate has been given by the ALEPH Coll.[3]: $BR = (3.38 \pm 0.74 \pm 0.85) * 10^{-4}$, which is slightly higher than the previously measured value by the CLEO Coll.[4]: $BR = (2.32 \pm 0.57 \pm 0.35) * 10^{-4}$. Both are consistent with the SM expectation $BR = (3.52 \pm 0.33) * 10^{-4}$. This value was calculated for $\alpha_s = 0.122$, which is the best fit value to the electroweak data from LEP, thus excluding the SLC value of $\sin^2\Theta_{eff}^{lept}$, which is inconsistent with the present Higgs limit of 79 GeV[5], as shown in fig. 1. It remains to be seen, if the SLC value is a statistical fluctuation or due to a systematic uncertainty. For the moment we have taken the LEP values for the coupling constants ($\alpha_s = 0.122 \pm 0.003$ and $\sin^2\Theta_{\overline{MS}} = 0.2319 \pm 0.00029$), which are slightly higher than the fit values from the combined LEP and SLC data ($\alpha_s = 0.120 \pm 0.003$ and $\sin^2\Theta_{\overline{MS}} = 0.2315 \pm 0.0004$) and lead to different unification conditions, as shown in fig. 1on the right hand side. In addition to the statistical errors on $\sin^2\Theta_{\overline{MS}}$ we added in quadrature the error from the uncertainty in the QED coupling constant at M_Z, which is 0.0026 and arises mainly from the uncertainty of the hadronic vacuum polarization. We did not use the world average of $\alpha_s = 0.118$, but used only the LEP value, for which the 3rd order QCD corrections have been calculated and found to be small, so the uncertainties from the higher order corrections are small.

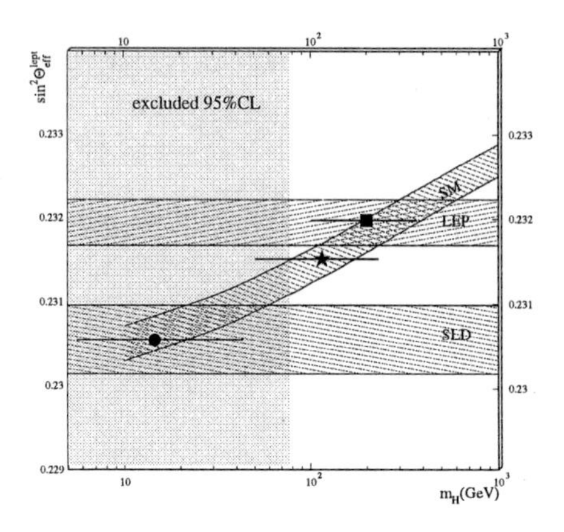

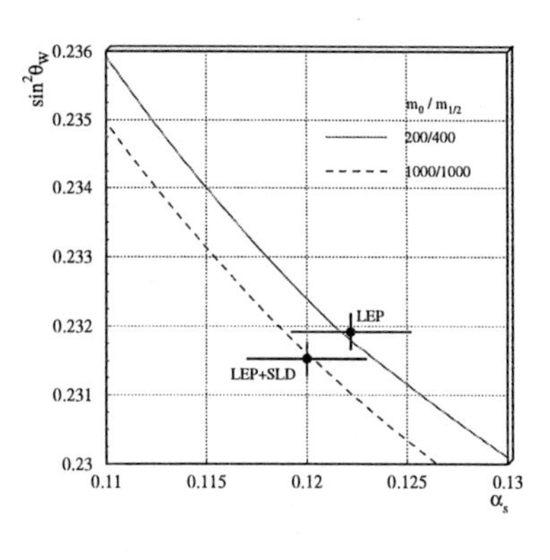

Fig. 1. Left: The electroweak mixing angle versus Higgs mass. The diagonal band corresponds to the SM prediction; the width of the band is determined by the uncertainty in the top mass. The point at $m_h = 14.6^{+28.6}_{-14.6}$ GeV corresponds to the SLC value of the mixing angle, while the point labeled LEP corresponds to the LEP value and the intermediate point corresponds to the combined data.
Right:The values of α_s and $\sin^2\Theta$ yielding unification for $m_0 = 200$, $m_{1/2}$=400 and $m_0 = 1000, m_{1/2} = 1000$ GeV. The upper curve fits better the LEP couplings, while the lower curve fits better the combined data of LEP and SLC.

In addition LEP has provided a new chargino mass limit around 90 GeV in case of a gaugino-like chargino and a heavy sneutrino [6], which is exactly the MSSM case considered here: ($\mu > m_{1/2}$ and $m_{\tilde{l}} > 100$ GeV). The Higgs mass limit is now 79 GeV at 95% C.L. for the SM Higgs [5] , which corresponds to the CMSSM case too, since all the other Higgses are heavy and decouple in the CMSSM limit considered here.

2 Results

The fitted supergravity parameters are mainly sensitive to the following input data: The GUT scale $M_{\rm GUT}$ and coupling constant $\alpha_{\rm GUT}$ are determined from gauge coupling unification, the Yukawa couplings Y_t, Y_b, Y_τ at the GUT scale from the masses of the third generation, μ from electroweak symmetry breaking and $\tan\beta$ from $b\tau$-unification. Of course, in a χ^2-fit all parameters are determined simultaneously, thus taking all correlations into account.

Since m_0 and $m_{1/2}$ enter in all observables, and are strongly correlated, we perform the fit for all combinations of m_0 and $m_{1/2}$ between 100 GeV and 1 TeV in steps of 100 GeV.

The most restrictive constraints are the gauge coupling constant unification and the requirement that the unification scale has to be above 10^{15}

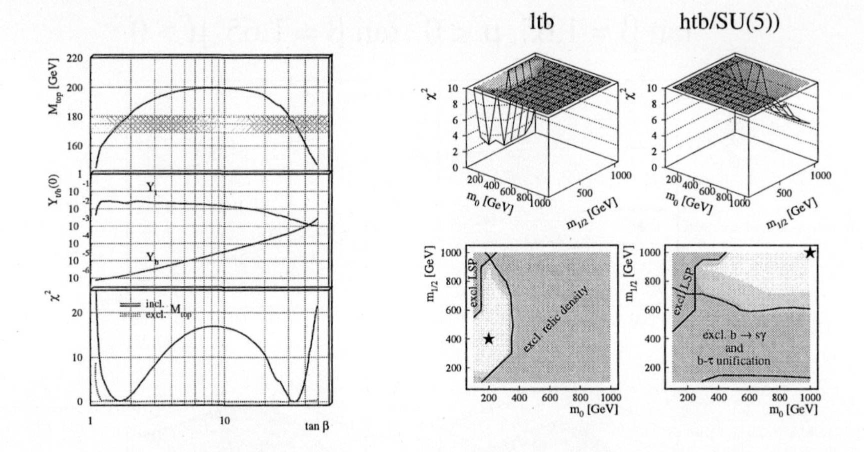

Fig. 2. Left:The top quark mass as function of $\tan\beta$ (top) for values of $m_0, m_{1/2} \approx 1$ TeV. The curve is hardly changed for lower SUSY masses. The middle part shows the corresponding values of the Yukawa coupling at the GUT scale and the lower part the χ^2 values. If the top constraint ($m_t = 175 \pm 6$, horizontal band) is not applied, all values of $\tan\beta$ between 1.2 and 50 are allowed (thin dotted lines at the bottom), but if the top mass is constrained to the experimental value, only the regions $1 \leq \tan\beta \leq 3$ and $20 \leq \tan\beta \leq 40$ are allowed.
Right:The χ^2-distribution for the low and high $\tan\beta$ solutions. The different shades in the projections indicate steps of $\Delta\chi^2 = 1$, so basically only the light shaded region is allowed. The stars indicate the optimum solution. Contours enclose domains excluded by the particular constraints used in the analysis.

GeV from the proton lifetime limits, assuming decay via s-channel exchange of heavy gauge bosons. They exclude the SM as well as many other models.

The requirement of bottom-tau Yukawa coupling unification strongly restricts the possible solutions in the m_t versus $\tan\beta$ plane. For $m_t = 175.6$ ± 5.5 GeV only two regions of $\tan\beta$ give an acceptable χ^2 fit, as shown in fig. 2. Y_t is left free independent of $Y_b = Y_\tau$.

In fig. 2 the total χ^2 distribution is shown as a function of m_0 and $m_{1/2}$ for the two values of $\tan\beta$ determined above. One observes minima at m_0, $m_{1/2}$ around (200,400) and (1000,1000), as indicated by the stars. The contours in the lower part show the regions excluded by the different constraints used in the analysis.

The requirement that the LSP is neutral excludes the regions with small m_0 and relatively large $m_{1/2}$, since in this case one of the scalar staus becomes the LSP after mixing via the off-diagonal elements in the mass matrix. The LSP constraint is especially effective at the high $\tan\beta$ region, since the off-diagonal element in the stau mass matrix is proportional to $A_t m_0 - \mu \tan\beta$.

The $b \to s\gamma$ rate is too large in most of the parameter region for large $\tan\beta$, at least if one requires $b-\tau$ unification. Both m_b and $b \to s\gamma$ have loop

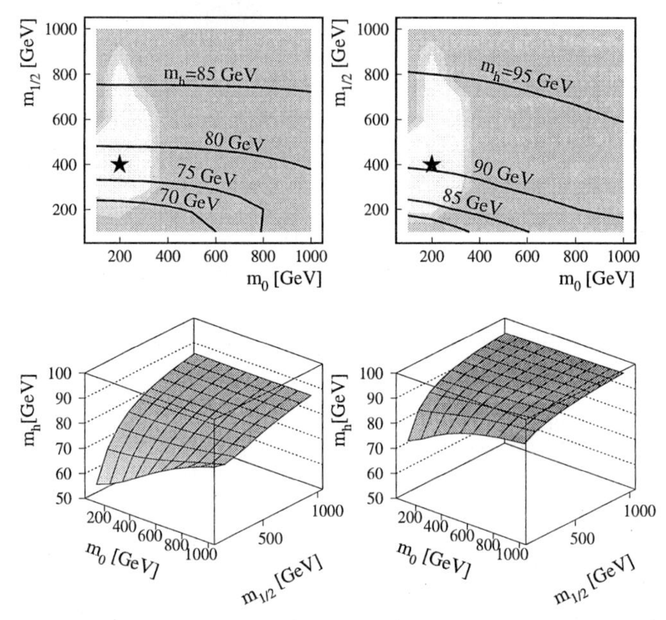

Fig. 3. Contours of the Higgs mass (solid lines) in the $m_0, m_{1/2}$ plane (above) and the Higgs masses (below) for both signs of μ for the low $\tan\beta$ solution $\tan\beta = 1.65$.

corrections proportional to $\mu \tan\beta$ and it is difficult to get both constraints fulfilled simultaneously unless μ can be choosen to be small, which can only be obtained for non-unified Higgs masses at the GUT scale.

The long lifetime of the universe requires a mass density below the critical density, else the overclosed universe would have collapsed long ago. This requires that the selectron is sufficiently light for a fast annihilation through t-channel selectron exchange. For large $\tan\beta$ the Higgsino admixture becomes larger, which leads to an enhancement of $\tilde{\chi}^0 - \tilde{\chi}^0$ annihilation via the s-channel Z boson exchange, thus reducing the relic density. As a result, in the large $\tan\beta$ case the constraint $\Omega h_0^2 < 1$ is almost always satisfied unlike in the case of low $\tan\beta$.

3 Summary

It is shown that the CMSSM can fit simultaneously the constraints from

- gauge couplings unification;
- $b - \tau$ Yukawa couplings unification;
- radiative EWSB;
- life time of the universe.

For the high $\tan\beta$ *scenario* the present limits of LEP [6] on the chargino mass and the more precise NLO calculations of $b \to s\gamma$ decay rate together with the measurements from ALEPH and CLEO leave only the small upper right corner region in parameter space ($m_0 > 500, m_{1/2} > 700$ GeV) (see fig. 2), where all squark masses are above 1 TeV.

For the low $\tan\beta$ *scenario* the Higgs mass has to be below 97 ± 6 GeV. The present run at 183 GeV did not observe a positive signal and limits above 83 GeV were quoted[7], which practically excludes the $\mu < 0$ solution and requires $m_{1/2} > 200$ GeV for the $\mu > 0$ solution (see fig. 3). In addition the relic density constraint requires $m_0 < 350$ GeV (see fig. 2), thus leaving the light shaded area on the right hand side of fig. 3 as the preferred parameter space of the CMSSM.

References

[1] W. de Boer et al., *Z. Phys.* **C71** *(1996) 415.*

[2] K.Chetyrkin et al, hep-ph/9612313
A. Buras et al., hep-ph/9707482;
C. Greub and T.Hurth, hep-ph/9708214.

[3] ALEPH Coll., contributed paper to HEP 97.

[4] R. Ammar et al.,CLEO-coll., *Phys. Rev. Lett.* **74** *(1995) 2885.*

[5] P. Janot, contr. to these proceedings.

[6] Contributed papers from LEP Coll. to HEP 97.

[7] Presentations by the LEP Coll. at the LEPC meeting, CERN, Geneva, Nov. 11, 1997.

Low-Energy Gravitino Interactions

Ferruccio Feruglio (feruglio@padova.infn.it)

Theory Division, CERN, Switzerland
on leave from University of Padova, Italy

Abstract. I discuss the low-energy limit of several processes involving only ordinary particles and gravitinos. Astrophysical and laboratory applications are briefly addressed.

It is well known that the solution of the hierarchy problem in supersymmetric extensions of the Standard Model (SM) requires a mass splitting $\Delta m \sim 1$ TeV between ordinary particles and their superpartners. This requirement, however, leaves largely undetermined the supersymmetry-breaking scale $\sqrt{F}$, or, equivalenty, the gravitino mass $m_{3/2} = F/(\sqrt{3}M_P)$, M_P being the Planck mass. The ratio $\Delta m^2/F$ is given by the coupling of the goldstino to the matter sector under consideration. If this coupling is gravitational, of order $m_{3/2}/M_P$, then Δm is of order $m_{3/2}$ and supersymmetry breaking takes place at the intermediate scale $\sqrt{F} \sim 10^{10}$ GeV. On the other hand, if the goldstino coupling to matter is of order 1, then $\sqrt{F}$ is comparable to the mass splitting Δm, and the gravitino becomes superlight, with a mass of about 10^{-5} eV. In the absence of a theory of supersymmetry breaking, F should be treated as a free parameter.

If the gravitino is superlight, then one expects a substantially different phenomenology from that characterizing the Minimal Supersymmetric Standard Model (MSSM). In this case, only the $\pm 3/2$ gravitino helicity states can be safely omitted from the low-energy effective theory, when gravitational interactions are neglected. The $\pm 1/2$ helicity states, essentially described by the goldstino field, should instead be accounted for at low energy, because of their non-negligible coupling to matter. The lightest supersymmetric particle is the gravitino and peculiar experimental signatures can arise from the decay of the next-to-lightest supersymmetric particle into its ordinary partner plus a gravitino [1].

Moreover, even when all superymmetric particles of the MSSM are above the production threshold, interesting signals could come from those processes where only ordinary particles and gravitinos occur. As soon as the typical energy of the process is larger than $m_{3/2}$, a condition always fulfilled in the applications discussed below, one can approximate the physical amplitudes by replacing external gravitinos with goldstinos, as specified by the equivalence theorem [2]. If the masses of the ordinary particles involved are negligible with respect to the energy of interest, these processes are controlled by just one dimensionful parameter, the supersymmetry-breaking scale $\sqrt{F}$, entering the amplitudes in the combination $(\tilde{G}/\sqrt{2}F)$, $\tilde{G}$ denoting the goldstino wave function.

This class of processes includes $\gamma\gamma \to \tilde{G}\tilde{G}$, $e^+e^- \to \tilde{G}\tilde{G}$, which may influence primordial nucleosynthesis, stellar cooling and supernovae explosion [3]. Of direct interest for LEP2 and for the future linear colliders is the reaction $e^+e^- \to \tilde{G}\tilde{G}\gamma$. Partonic reactions such as $q\bar{q} \to \tilde{G}\tilde{G}\gamma$, $q\bar{q} \to \tilde{G}\tilde{G}g$ and $qg \to \tilde{G}\tilde{G}q$ can be indirectly probed at the Tevatron collider or in future hadron facilities. In the absence of experimental signals, one can use these processes to set absolute limits on the gravitino mass. At variance with other bounds on $m_{3/2}$ discussed in the literature [4], these limits have the advantage of not depending on detailed assumptions about the spectrum of supersymmetric particles. Finally, the study of these processes can reveal unexpected features of the low-energy theory, which were overlooked in the standard approach to goldstino low-energy interactions.

The natural tools to analyse the above processes are the so-called low-energy theorems [5]. According to these, the low-energy amplitude for the scattering of a goldstino on a given target is controlled by the energy–momentum tensor $T_{\mu\nu}$ of the target. To evaluate the physical amplitudes, it is more practical to make use of an effective Lagrangian, containing the goldstino field and the matter fields involved in the reactions, and providing a non-linear realization of the supersymmetry algebra. For instance, in the non-linear construction of [6], the goldstino field $\tilde{G}$ and the generic matter field φ are incorporated into the following superfields:

$$\Lambda_\alpha \equiv \exp(\theta Q+\overline{\theta Q})\,\tilde{G}_\alpha = \frac{\tilde{G}_\alpha}{\sqrt{2}F}+\theta_\alpha+\frac{i}{\sqrt{2}F}(\tilde{G}\sigma^\mu\overline{\theta}-\theta\sigma^\mu\overline{\tilde{G}})\partial_\mu\frac{\tilde{G}_\alpha}{\sqrt{2}F}+\dots, \quad (1)$$

$$\Phi \equiv \exp(\theta Q + \overline{\theta Q})\,\varphi = \varphi + \frac{i}{\sqrt{2}F}(\tilde{G}\sigma^\mu\overline{\theta} - \theta\sigma^\mu\overline{\tilde{G}})\partial_\mu\varphi + \dots . \quad (2)$$

The goldstino–matter system is described by the supersymmetric Lagrangian:

$$\int d^2\theta d^2\bar{\theta}\Lambda^2\bar{\Lambda}^2\left[2F^2 + \mathcal{L}(\Phi, \partial\Phi)\right] \quad , \quad (3)$$

where $\mathcal{L}(\varphi, \partial\varphi)$ is the ordinary Lagrangian for the matter system. This non-linear realization automatically reproduces the results of the low-energy theorems, in particular the expected goldstino coupling to the energy–momentum tensor $T_{\mu\nu}$ associated to φ.

An alternative approach consists in constructing a low-energy Lagrangian, starting from a general supersymmetric theory defined, up to terms with more than two derivatives, in terms of a Kähler potential, a superpotential and a set of gauge kinetic functions. The effective theory can be obtained by integrating out, in the low-energy limit, the heavy superpartners [3].

When applied to the process $\gamma\gamma \to \tilde{G}\tilde{G}$, the two procedures yield the same result. The only independent, non-vanishing, helicity amplitude for the process is:

$$a(1,-1,1/2,-1/2) = 8\sin\theta\cos^2\frac{\theta}{2}\frac{E^4}{F^2} \quad , \tag{4}$$

where $(1,-1)$ and $(1/2,-1/2)$ are the helicities of the incoming and outgoing particles, respectively; E and θ are the goldstino energy and scattering angle in the centre-of-mass frame. The total cross section is $s^3/(640\pi F^4)$.

In earlier cosmological and astrophysical applications, with a typical energy range from about 1 keV to 100 MeV, a cross-section scaling as $\Delta m^2 s^2/F^4$ was assumed, giving rise to a lower bound on $m_{3/2}$ close to 10^{-6} eV. When the correct energy dependence is taken into account, this bound is reduced by at least a factor 10, and becomes uninteresting compared to those obtainable at colliders.

When considering $e^+e^- \to \tilde{G}\tilde{G}$, in the limit of massless electron, one has to face an unexpected result [7]. On the one hand, by integrating out the heavy selectron fields, one finds the following helicity amplitude:

$$a(1/2,-1/2,1/2,-1/2) = \frac{4(1+\cos\theta)^2 E^4}{F^2} \quad , \tag{5}$$

all other non-vanishing amplitudes being related to this one. On the other hand, by using the non-linear realization of [6], one obtains:

$$a(1/2,-1/2,1/2,-1/2) = \frac{4\sin^2\theta E^4}{F^2} \quad . \tag{6}$$

The amplitudes of eqs. (5) and (6) scale in the same way with the energy, but have a different angular dependence. We should conclude that the low-energy theorems do not apply to the case of a massless fermion. A particularly disturbing aspect is that the non-linear realization of eq. (3) is supposed to provide the most general parametrization of the amplitude in question, independently of any considerations about the low-energy theorems. In the case at hand, the Lagrangian of eq. (3) reads:

$$\mathcal{L}_e = \int d^2\theta d^2\bar{\theta}\Lambda^2\bar{\Lambda}^2\left[2F^2 + iE\sigma^\mu\partial_\mu\bar{E} + iE^c\sigma^\mu\partial_\mu\bar{E}^c\right] \quad , \tag{7}$$

where E and E^c are the superfields associated to the two Weyl spinors e and e^c describing the electron, according to eq. (2). The solution to this puzzle [7] is provided by the existence of an independent supersymmetric invariant that has been neglected up to now in the literature:

$$\delta\mathcal{L}_e = \int d^2\theta d^2\bar{\theta}(\Lambda E\bar{\Lambda}\bar{E} + \Lambda E^c\bar{\Lambda}\bar{E}^c) \quad . \tag{8}$$

The amplitudes of eq. (5) are reproduced by the combination $\mathcal{L}_e + 8\,\delta\mathcal{L}_e$.

On the other hand, there is no reason to prefer either the result of eq. (5) or that of eq. (6). The process $e^+e^- \to \tilde{G}\tilde{G}$ does not have a universal low-energy behaviour. In the framework of non-linear realizations, this freedom can be described by the invariant Lagrangian $\mathcal{L}_e + \alpha\, \delta\mathcal{L}_e$, where α is a free parameter of the low-energy theory.

The process $e^+e^- \to \tilde{G}\tilde{G}\gamma$ suffers from a similar ambiguity. Indeed the soft and collinear part of the cross-section, which is the dominant one, is associated to the initial-state radiation and hence is determined by the cross-section for $e^+e^- \to \tilde{G}\tilde{G}$. The total cross-section, with appropriate cuts on the photon energy and scattering angle, scales as $\alpha_{em} s^3/F^4$. The photon energy and angular distributions are not universal, as for the case of the goldstino angular distribution in $e^+e^- \to \tilde{G}\tilde{G}$. They could be completely determined only through a computation performed in the fundamental theory. From the non-observation of single-photon events above the SM background at LEP2, one can roughly estimate a lower bound on $\sqrt{F}$ of the order of the machine energy. The precise value of this limit would require the analysis of the relevant background, as well as the inclusion of the above mentioned theoretical ambiguity. However, in view of the quite strong power dependence of the cross-section on the supersymmetry-breaking scale, one expects only a small correction to the limit obtained by a rough dimensional estimate.

The partonic processes $q\bar{q} \to \tilde{G}\tilde{G}\gamma$, $q\bar{q} \to \tilde{G}\tilde{G}g$ are also expected to have cross-sections scaling as $\alpha_{em} s^3/F^4$ and $\alpha_s s^3/F^4$ respectively. The agreement between data and SM expectations in $p\bar{p} \to \gamma + X$ and $p\bar{p} \to \text{jet} + X$ at the Tevatron collider could then be used to infer a lower limit on $\sqrt{F}$. This is expected to be around the typical total energy of the partonic subprocess, about 600 GeV.

Acknowledgements

I would like to thank Andrea Brignole and Fabio Zwirner for the pleasant collaboration on which this talk is based.

References

1 S. Ambrosanio et al., Phys. Rev. D54 (1996) 5395 and refs. therein.
2 P. Fayet, Phys. Lett. B70 (1977) 461 and B86 (1979) 272; R. Casalbuoni et al., Phys. Lett. B215 (1988) 313 and Phys. Rev. D39 (1989) 2281.
3 A. Brignole et al., preprint hep-ph/9703286 and refs. therein.
4 See ref. [25] in [3].
5 B. de Wit and D.Z. Freedman, Phys. Rev. D12 (1975) 2286 and Phys. Rev. Lett. 35 (1975) 827.
6 S. Samuel and J. Wess, Nucl. Phys. B221 (1983) 153; J. Wess, in *Quantum Theory of Particles and Fields* (B. Jancewicz and J. Lukierski eds.), World Scientific, 1983, p. 223.
7 A. Brignole et al., preprint hep-th/9709111 and refs. therein.

Baryogenesis, Inflation and Superstrings

J.Alberto Casas (casas@cc.csic.es)

Instituto de Estructura de la Materia, CSIC, Spain

Abstract. We study the conditions for successful Affleck–Dine baryogenesis in generic inflation and supergravity scenarios, finding powerful restrictions on them. String-based SUGRA models are especially interesting since they are surprisingly suitable for the implementation of AD baryogenesis and also inflation itself, presenting a nice solution to the ubiquitous η-problem.

1 Introduction

In the absence of better alternatives, the Affleck and Dine (AD) mechanism [1, 2] is a very attractive method for baryogenesis. It takes place if in the early universe some scalar "AD fields", ϕ, carrying baryon or lepton number (B or L), get large initial vacuum expectation values (VEV's), ϕ_{in}. Then the equations of motion of ϕ (together with the presence of some baryon-violating operator involving the ϕ fields) lead to a net final baryon number [1]. The initial large VEV's can hardly be originated by quantum fluctuations along flat directions during inflation. The reason is that during inflation SUSY is necessarily spontaneously broken by the VEV of the scalar potential $\langle V \rangle = 3H^2M_P^2$, triggering effective SUSY soft breaking terms (in particular soft masses) of the order of the Hubble constant H. Consequently, the initial VEV's, ϕ_{in}, will occur if those effective masses squared are *negative*, something that is possible in a generic SUGRA theory [2]. Whether this happens or not depends crucially on the type of inflation.

2 D–Inflation

The possibility of D–inflation [3, 4, 5] (i.e. inflation triggerred by a non-vanishing D–term) in SUGRA scenarios is attractive since it can overcome the ubiquitous η-problem (see end of sect.4).

In order to be concrete (with no loss of generality), let us suppose that inflation is mainly triggered by one "anomalous" $U(1)$ D–term [6, 7]

$$V_D = \frac{1}{2}D^2 = \frac{1}{2}g^2 \ (\xi + \sum_j q_j |z_j|^2 K_{j\bar{j}})^2 \ , \tag{1}$$

where g is the corresponding gauge coupling, q_j are the charges of all the chiral fields, z_j, under the anomalous $U(1)$ and $K_{j\bar{j}}$ is the Kähler metric (we adopt a z_j-basis where the latter is diagonal). Finally, $\xi = g^2M_P^2(\sum_j q_j/192\pi^2)$ is the apparent anomaly. At low energy the D–term is cancelled by the VEV's of some of the scalars entering Eq.(1), but initially $\langle D \rangle$ may be different from zero, thus triggering inflation. The scenario is quite insensitive to the details of the Kähler potential K (note that $(K_{j\bar{j}})^{1/2} z_j$ are simply the canonically normalized fields).

Concerning AD baryogenesis, there are two main scenarios to consider, depending on $q_\phi \neq 0$ or $q_\phi = 0$. In the first case, it is clear that the D–term induces an effective mass term for ϕ, which is *negative* provided sign(ξ) sign$(q_\phi) = -1$. Then, we expect $\langle \phi \rangle_{in} \neq 0$. Nevertheless, this cannot be the whole story, since in the absence of additional ϕ–dependent terms in V, $\langle \phi \rangle_{in}$ would adjust itself to cancel the D–term, thus disabling the inflationary process and breaking B or L at low energy. Thus, we need extra contributions yielding $\langle D \rangle_{in} \neq 0$, $\langle \phi \rangle_f = 0$. These may come from *(a)* low-energy soft breaking terms, *(b)* F–terms, *(c)* extra D–terms. However, no one of these three possibilities really works. The reason is that either the extra contributions are too small (case *(a)*) or they lead to a potential which is lifted by renormalizable terms, which does not work for baryogenesis purposes [2] (for further details see ref.[8]).

This leaves us just with the second scenario, namely, $q_\phi = 0$. Then, from (1), $m_\phi^2 = 0$ during inflation. So, there is a truly flat direction along ϕ and the AD mechanism can be implemented in an old-fashioned way (through quantum fluctuations). This argument is only exact at tree level. Strictly speaking, there are small contributions to m_ϕ coming from higher loop corrections [9] and the expected O(TeV) low-energy supersymmetry breaking effects. In any case, ϕ will acquire a large VEV during inflation due to quantum fluctuations if the correlation length for de Sitter fluctuations, $l_{coh} \simeq H^{-1} \exp(3H^2/2m_\phi^2)$, is large compared to the horizon size. This translates into the condition $H^2/m_\phi^2 \gtrsim 40$ [2], which is easily fulfilled in this context.

3 F–Inflation

F–Inflation occurs when inflation is triggerred by a non-vanishing F–term of the appropriate size. Concerning the implementation of the AD mechanism in F–inflationary scenarios, the main question (see Sect. 1) is whether it is possible to get an effective mass squared $m_\phi^2 < 0$ or $m_\phi^2 \simeq 0$ for the AD field, ϕ, during inflation. To answer this question, we need to examine the F–part of the effective potential V. In a standard notation

$$V = e^G \left(G_{\bar{j}} K^{\bar{j}l} G_l - 3 \right) = F^{\bar{l}} K_{j\bar{l}} F^j - 3e^G \; . \tag{2}$$

Here $G = K + \log |W|^2$ where W is the superpotential, $K^{\bar{l}j}$ is the inverse of the Kähler metric and $F^j = e^{G/2} K^{j\bar{k}} G_{\bar{k}}$ are the corresponding auxiliary fields. During inflation $\langle V \rangle_{in} = V_0 \simeq H^2 M_P^2$, which implies that some F fields are different form zero, thus breaking SUSY. The effective gravitino mass is given by $m_{3/2}^2 = e^G = e^K|W|^2$ in M_P units. The SUSY breakdown induces soft terms for all the scalars, in particular for ϕ. More precisely, the value of the effective mass squared, m_ϕ^2, is intimately related to the form of K. It is convenient to parametrize K as

$$K = K_0(I) + K_{\phi\bar{\phi}}|\phi|^2 + \cdots , \tag{3}$$

where I represents generically the inflaton or inflatons. Plugging (3) in (2) it is straightforward to see [8] that if there is no mixing between ϕ and I in the quadratic term of K, i.e. if $K_{\phi\bar{\phi}} \neq K_{\phi\bar{\phi}}(I)$, then the effective mass squared for the canonically normalized field, $(K_{\phi\bar{\phi}})^{1/2}\phi$, is $m_\phi^2 = m_{3/2}^2 + V_0/M_P^2$. Hence m_ϕ^2 is of $O(H^2)$ and *positive* and, therefore, the AD mechanism cannot be implemented. This result is remarkable, excluding for instance minimal SUGRA. A successful implementation of the AD mechanism thus requires and AD fields ϕ, i.e. $K_{\phi\bar{\phi}} = K_{\phi\bar{\phi}}(I)$ in Eq.(3). This mixing should be remarkably strong and even so the possibility of a negative effective mass term is not guaranteed (see ref.[8]). Hence, the mixing is a necessary but not sufficient condition for $m_\phi^2 \leq 0$ Finally, let us note that the possibility of a very small mass $m_\phi^2 \sim 0$ (also welcome for a successful inflation itself, as we will discuss shortly) does not seem natural at first sight, since it would imply some conspiracy between the various contributions to m_ϕ^2 coming from (2). However, string-derived SUGRA models provide beautiful surprises in this sense.

4 String Scenarios

The best motivated SUGRA scenarios are those coming from string theories. The corresponding Kähler potential, K, is greatly constrained and, therefore, the implementation of AD baryogenesis for the F-inflation framework is not trivial at all. In order to be concrete we will consider orbifold constructions, where the (tree–level) Kähler potential is given by

$$K = -\log(S + \bar{S}) - 3\log(T + \bar{T}) + \sum_j (T + \bar{T})^{n_j}|z_j|^2 . \tag{4}$$

Here S is the dilaton and T denotes generically the moduli fields, z_j are the chiral fields and n_j the corresponding modular weights. The latter depend on the type of orbifold considered and the twisted sector to which the field belongs. The possible values of n_j are $n_j = -1, -2, -3, -4, -5$. The discrete character of n_j will play a relevant role later on.

Since a strong mixing between the inflaton and the AD field ϕ in the quadratic term ($\propto |\phi|^2$) of K is required (see Sect. 3), our first conclusion is that T is the natural inflaton candidate in string theories. S–dominated inflation cannot work, and this is a completely general result since the (tree-level) S–dependence of K is universal in string theories. We should recall, however, that a strong mixing in K is a necessary but not sufficient condition for $m_\phi^2 \leq 0$. We must then examine the precise value of m_ϕ^2 in the presence of a non-vanishing cosmological constant $\langle V\rangle_{in} = V_0 > 0$. Restricting ourselves to the moduli-dominated case, m_ϕ^2 is given by [10]

$$m_\phi^2 = m_{3/2}^2 \left\{(3+n_\phi)C^2 - 2\right\} , \tag{5}$$

where $m_{3/2}^2 = e^K|W|^2$, $C^2 = 1 + [V_0/(3M_P^2 m_{3/2}^2)]$ and n_ϕ is the modular weight of the AD field ϕ. Since $C^2 > 1$, it is clear that $n_\phi \leq -3$ is a sufficient condition to get $m_\phi^2 \leq 0$. Actually, if $C^2 \leq 2$, then $m_\phi^2 \leq 0$ whenever $n_\phi \leq -2$, which is a very common case.

There is a particularly interesting limit of eq.(5) that could well be realized in practice. Namely, since $V_0 = K_{T\bar{T}}|F^T|^2 - 3m_{3/2}^2 = 3H^2M_P^2$, it may perfectly happen that $K_{T\bar{T}}|F^T|^2 \gg m_{3/2}^2$, and thus $C^2 \gg 1$. Then, from eq.(5)

$$m_\phi^2 \simeq H^2(3+n_\phi). \tag{6}$$

Hence for $n_\phi = -3$ we get $m_\phi^2 \simeq 0$, as in the D–inflation case. There is no fine-tuning here, since n_ϕ is a discrete number which can only take the values $n_\phi = -1, -2, -3, -4, -5$.

What is even more important: this is also good news for F–inflation itself. As has been pointed out in the literature, F–inflation has the problem that if the inflaton mass is $O(H)$, as expected at first sight during inflation, then the necessary slow rollover is disabled (this is the so-called η–problem). We see here, however, that a hybrid–inflation scenario [11] in which T is the field responsible for the large V_0 and a second field (any one with $n = -3$) is responsible for the slow rollover is perfectly viable. This is a nice surprise since F-inflation is very difficult to implement in generic SUGRA theories, even with fine-tuning!

References

[1] I. Affleck and M. Dine, *Nucl. Phys.* **B249** (1985) 361.

[2] M. Dine, L. Randall and S. Thomas, *Nucl. Phys.* **B458** (1996) 291.

[3] J.A. Casas and C. Muñoz, *Phys. Lett.* **B216** (1989) 37; J.A. Casas, J.M. Moreno, C. Muñoz and M. Quirós, *Nucl. Phys.* **B328** (1989) 272.

[4] E.D. Stewart, *Phys. Rev.* **D51** (1995) 6847

[5] P. Binetruy and G. Dvali, *Phys. Lett.* **B388** (1996) 241; E. Halyo, *Phys. Lett.* **B387** (1996) 43; T. Matsuda, *hep-ph/9705448*.

[6] M. Dine, N. Seiberg and E. Witten, *Nucl. Phys.* **B289** (1987) 589.

[7] J.A. Casas, E. Katehou and C. Muñoz, *Nucl. Phys.* **B317** (1989) 171.

[8] J.A. Casas and G.B. Gelmini, *Phys. Lett.* **B410** (1997) 36

[9] G. Dvali, *Phys. Lett.* **B355** (1995) 78; *hep-ph/9605445*.

[10] A. Brignole, L.E. Ibáñez and C. Muñoz, *Nucl. Phys.* **B422** (1994) 125.

[11] A. Linde, *Phys. Lett.* **B259** (1991) 38; *Phys. Rev.* **D49** (1994) 748.

Phenomenology of Antigravity in N=2,8 Supergravity

Stefano Bellucci (bellucci@lnf.infn.it)

INFN-Laboratori Nazionali di Frascati, Frascati, Italy

Abstract. $N = 2, 8$ supergravity predicts antigravity (gravivector and graviscalar) fields in the graviton supermultiplet. Data on the binary pulsar PSR 1913+16, tests of the equivalence principle and searches for a fifth force yield an upper bound of order 1 meter (respectively, 100 meters) on the range of the gravivector (respectively, graviscalar) interaction. Hence these fields are not important in non-relativistic astrophysics (for the weak-field limit of $N = 2, 8$ supergravity) but can play a role near black holes and for primordial structures in the early universe of a size comparable to their Compton wavelengths.

The quest for a unified description of elementary particle and gravity theories led to local supersymmetry [1]. The large symmetry content of supergravity yields, in spite of its lack of renormalizability, powerful constraints on physical observables, e.g. anomalous magnetic moments [2].

It has been shown that a clear case for antigravity theories arises, when considering $N > 1$ supergravity theories [3, 4]. Combining laboratory data together with geophysical and astronomical observations has provided us restrictions on the antigravity features of some extended supergravity theories [5, 6]. This can have important consequences for high precision experiments measuring the difference in the gravitational acceleration of the proton and the antiproton [7]. A review of earlier ideas about antigravity is found in [8].

The $N = 2, 8$ supergravity multiplets contain, in addition to the graviton, a vector field A^l_μ [9], [10, 11]. This field, which we refer to as the gravivector, carries antigravity, because it couples to quarks and leptons with a positive sign and to antiquark and antileptons with a negative one. The coupling is proportional to the mass of the matter fields and vanishes for self-conjugated particles. The other antigravity field is the scalar σ entering the $N = 8$ supergravity multiplet [3, 4]. We refer to it as the graviscalar.

We are bound, in force of the result of the Eötvös experiment, to take a nonvanishing mass for the field A^l_μ [3, 4].

$$m_l = \frac{1}{R_l} = \sqrt{4\pi G_N}\, m_\phi \langle\phi\rangle \, , m_\phi = \langle\phi\rangle \, , \qquad (1)$$

where the Higgs mechanism has been invoked.

The presence of the gravivector in the theory introduces a violation of the equivalence principle on a range of distances of order the Compton wavelength R_l. At present, the equivalence principle is verified with a precision $|\delta\gamma/\gamma| <$

3×10^{-12} [12].[1] This number was used in [6], in order to constrain the *v.e.v.* of the scalar field ϕ, and therefore its mass $m_\phi > 15$ (31) GeV, N=2 (8). The above constraint on the field that gives to the gravivector its mass corresponds to an upper bound of order 1 m for R_l [6].

It is worth to remind the reader that there are interesting connections between antigravity in $N = 2, 8$ supergravities and CP violation experiments, via the consideration of the K^0–$\overline{K}^0$ system in the terrestrial gravitational field [3].[2] However, the present experiments on CP violations yield bounds on the range of the gravivector field which are less stringent than those obtained from the tests of the equivalence principle [5].

The null results of the search for possible deviations from Newton's law reported in [15] forbid values in the following ranges [6]:

$$82\ \text{GeV} < m_\phi < 376\ \text{GeV} \qquad (N = 2)\ , \tag{2}$$

$$46\ \text{GeV} < m_\phi < 461\ \text{GeV} \qquad (N = 8)\ . \tag{3}$$

A high precision test of the equivalence principle in the field of the Earth is currently under planning in Moscow [16]. The precision expected to be achieved in this experiment is $|\delta\gamma/\gamma| < 3 \times 10^{-15}$. In the case that the new experiment verifies the equivalence principle with the expected accuracy, the limits on m_ϕ would be pushed to $m_\phi > 0.5$ (1) TeV, $N = 2$ (8). [3]

If supersymmetry is unbroken, the violation of the equivalence principle due to the graviscalar of $N = 8$ supergravity takes the form of a universal spatial dependence in the effective (gravitational) Newton's coupling, $G_N = G_N(r)$ [4]. However there are corrections due to the breaking of supersymmetry - which we hope to account for in a forthcoming publication - and those depend on the composition of the material. If we neglect them for the time being, then there is no effect of σ in Eötvös-like experiments, where the acceleration difference between two bodies of different composition is measured. In this case it is still possible to constrain the effective range of the σ-mediated interaction, analyzing data from the binary pulsar PSR 1913+16 [17], constraints from the observations searching for a fifth force [18], and some of the experiments aimed at testing Newton's inverse square law [15]. In this way we got the following bounds on the Compton wavelength of the graviscalar: $R_\sigma < 0.15$ cm, 70 m$< R_\sigma < 100$ m [6].

[1] The equivalence principle is advocated in a recent proposal of a consistent quantum gravity [13].

[2] For a low-energy theorem in gravity coupled to scalar matter, see e.g. [14].

[3] A caveat concerning our results on the gravivector is that the presence of a $U(1)$ symmetry for the $D = 4$ extended supergravity theory obtained by dimensional reduction from a higher dimension implies a mass for this field of order the Planck mass [10]. In this particular instance it is unlikely that experimental limits on the gravivector have any physical application (aside perhaps from applications to inflationary models, if it were possible to use a vector field instead of a scalar). We plan to come back to this issue in a further study.

There have been many papers on the effects of non-Newtonian gravity in astrophysics, in particular those due to a fifth force like the one obtainable from $N = 2, 8$ supergravity in the weak field limit (see references in [8]).[4] However, the upper bounds of order 1 m (100 m) on the Compton wavelength R_l (R_σ) of the gravivector (graviscalar) field found in [6] imply that antigravity effects induced by the extended supergravity theories do not play any role in nonrelativistic astrophysics, since the length scales involved in stellar,[5] galactic and supergalactic structures dominated by gravity are much larger than R_l and (R_σ).

This supergravity-induced antigravity could affect, in principle, processes that take place in the strong gravity regime, where smaller distance scales are involved. Examples of these situations are processes occurring near black hole horizons or in the early universe, when the size of the universe is smaller than, or of the order of, R_l and (R_σ). The relevance of the supergravity-induced antigravity in such situations will be studied in future publications.

Our final remark concerns a point that apparently went unnoticed in the literature on supergravity: the detection of gravitational waves expected in a not too far future will shed light on the correctness of supergravity theories. In fact, after the dimensional reduction is performed, the action of the theory contains scalar and vector fields as well as the usual metric tensor associated to spin 2 gravitons [4]. These fields are responsible for the presence of polarization modes in gravitational waves, whose effect differs from that of the spin 2 modes familiar from general relativity. Therefore, extended supergravities and general relativity occupy different classes in the $E(2)$ classification of gravity theories [23]. The extra polarization states are detectable, in principle, in a gravitational wave experiment employing a suitable array of detectors [23]. However, it must be noted that a detailed study of gravitational wave generation taking into account the antigravity phenomenon is not available at present. Such a work would undoubtedly have to face the remarkable difficulties well known from the studies of gravitational wave generation in the context of general relativity.

We are grateful to G.A. Lobov for drawing our attention to the ITEP experiment. We acknowledge useful comments by C. Kounnas and G. Veneziano.

References

[1] P. van Nieuwenhuizen, Phys. Rep. C68 (1981) 189.

[4] The effect of the cosmological constant is shadowed by Newtonian-gravity effects [19, 20]. Large-scale topology-changing configurations can suppress its effect also for 1-loop scalar-QED, see e.g. [21].

[5] The conclusion that the stellar structure is unaffected by antigravity might change if the non-Newtonian force alters the equation of state of the matter composing the star [22].

[2] S. Ferrara and E. Remiddi, Phys. Lett. B53 (1974) 347; F. del Águila, A. Mendez and F.X. Orteu, Phys. Lett. B145 (1984) 70; S. Bellucci, H.Y. Cheng and S. Deser, Nucl. Phys. B252 (1985) 389; F. del Águila et al., preprint UG-FT-71/96, hep-ph/9702342.

[3] J. Scherk, Phys. Lett. B 88 (1979) 265.

[4] J. Scherk, in Supergravity, Proc. 1979 Supergravity Workshop at Stony Brook, eds. P. van Nieuwenhuizen and D.Z. Freedman (North-Holland, Amsterdam, 1979) p. 43.

[5] S. Bellucci and V. Faraoni, Phys. Rev. D 49 (1994) 2922.

[6] S. Bellucci and V. Faraoni, Phys. Lett. B 377 (1996) 55.

[7] N. Beverini et al., CERN report CERN/PSCC/86–2 (1986); N. Jarmie, Nucl. Instrum. Methods Phys. Res. B 24/25 (1987) 437; P. Dyer et al., Nucl. Instrum. Methods Phys. Res. B 40/41 (1989) 485; R.E. Brown, J.B. Camp and T.W. Darling, Nucl. Instrum. Methods Phys. Res. B 56/57 (1991) 480.

[8] M.M. Nieto and T. Goldman, Phys. Rep. 205 (1991) 221; erratum 216 (1992) 343.

[9] C.K. Zachos, Phys. Lett. B 76 (1978) 329.

[10] J. Scherk and J.H. Schwarz, Phys. Lett. B 82 (1979) 60; ibid., Nucl. Phys. B 153 (1979) 61.

[11] E. Cremmer, J. Scherk, and J.H. Schwarz, Phys. Lett. B 84 (1979) 83.

[12] Y. Su et al., Phys. Rev. D 50 (1994) 3614.

[13] S. Bellucci and A. Shiekh, preprint LNF-97/003 (P), gr-qc/9701065; ibid., LNF-97/012 (P), gr-qc/9703085.

[14] A. Akhundov, S. Bellucci and A. Shiekh, Phys. Lett. B 395 (1997) 16, gr-qc/9611018.

[15] R. Spero et al., Phys. Rev. Lett. 44 (1980) 1645;
J.K. Hoskins et al., Phys. Rev. D 32 (1985) 3084; M.U. Sagitov et al., Dokl. Akad. Nauk. SSSR 245 (1979) 567; V.K. Milyukov, Sov. Phys. JETP 61 (1985) 187.

[16] S.M. Kalebin, Preprint ITEP 93-26 (in Russian).

[17] R.A. Hulse, Rev. Mod. Phys. 66 (1994) 699; J.H. Taylor,Jr., ibid., 711.

[18] C. Talmadge and E. Fischbach, in: 5th Force-Neutrino Physics, Proc. XXIII Rencontre de Moriond, eds. O. Fackler and J. Tran Thanh Van (Editions Frontières, Gif-sur-Yvette, 1988).

[19] E. Elizalde and S.D. Odintsov, Phys. Lett. B333 (1994) 331.

[20] S. Bellucci, preprint LNF-97/016 (P), hep-th/9704039, to appear in Phys. Rev. D.

[21] S. Bellucci and D. O'Reilly, Nucl. Phys. B364 (1991) 495.

[22] M.M. Nieto et al., Phys. Rev. D 36 (1987) 3684; 3688; 3694; P. De Leon, Can. J. Phys. 67 (1989) 845.

[23] D.M. Eardley et al., Phys. Rev. Lett. 30 (1973) 884; Phys. Rev. D 8 (1973) 3308; see also C. M. Will, Theory and Experiment in Gravitational Physics (Cambridge University Press, Cambridge, 1981).

Supersymmetry Breaking in M-Theory

Mariano Quirós (quiros@pinar1.csic.es)

Instituto de Estructura de la Materia, Serrano 123, 28006-Madrid, Spain

Abstract. We describe the breaking of supersymmetry in M-theory by coordinate dependent (Scherk-Schwarz) compactification of the eleventh dimension. Supersymmetry is spontaneously broken in the gravitational and moduli sector and communicated to the observable sector, living at the end-points of the semicircle, by radiative gravitational interactions.

1 Introduction

The strong coupling limit of the $E_8 \times E_8$ heterotic superstring compactified on a CY manifold is believed to be described by the eleven-dimensional M-theory compactified on $CY \times S^1/Z_2$, where the semicircle has a radius ρ [1]. The value of the unification scale M_G becomes $\sim M_{11}$, the M-theory scale, which is lower than the string scale M_H, while the radius ρ of the semicircle is at an intermediate scale $\rho^{-1} \sim 10^{13}$ GeV and, for isotropic CY, the compactification scale $V^{-1/6}$ is of the order of M_{11}[2]. Fortunately, this is inside the region of validity of M-theory, $\rho M_{11} \gg 1$, $(2\pi)^6 V \kappa_{11}^{-4/3} \gg 1$. As a result, the effective theory above the intermediate scale behaves as 5-dimensional, but only in the gravitational and moduli sector; the gauge sectors coming from $E_8 \times E_8$ live at the 4D boundaries of the semicircle.

2 Supersymmetry breaking by Scherk-Schwarz on the eleventh dimension

The $N = 1$ supersymmetry transformations in the 5D theory are [3, 4]:

$$\delta e_M^m = -\frac{i}{2}\overline{\mathcal{E}}\Gamma^m \Psi_M; \qquad \delta\Psi_M = D_M \mathcal{E} + \cdots \tag{1}$$

where e_M^m is the fünfbein, $\Gamma^m = (\gamma^\mu, i\gamma_5)$ are the Dirac matrices, Ψ_M is the gravitino field, $\mathcal{E}$ the spinorial parameter of the transformation, and the dots stand for non-linear terms. Similar transformations hold for the components of vector multiplets and hypermultiplets for which our subsequent analysis can be generalized in a straightforward way.

All 5D fermions can be represented as doublets under the $SU(2)$ R-symmetry whose components are subject to a (generalized) Majorana condition [5]. It is convenient to decompose the spinors in terms of $SU(2)_R$ doublets, and define the Γ_5-chirality and $\mathcal{R}$-chirality as,

$$\Gamma_5 \Psi_{L,R} = \pm \Psi_{L,R}; \quad \Gamma_5 = \begin{pmatrix} \gamma_5 & 0 \\ 0 & -\gamma_5 \end{pmatrix}, \tag{2}$$

$$\mathcal{R} \Psi_{L,R}(x^\mu, y) = \pm \eta \Psi_{L,R}(x^\mu, -y) . \tag{3}$$

with $\eta = 1$ for $\Psi = \Psi_\mu$ and $\eta = -1$ for $\Psi = \Psi_5$. The $\mathbf{Z}_2$ projection is defined by keeping the states that are even under $\mathcal{R}$. It follows that the remaining massless fermions are the left-handed components of the 4D gravitino $\Psi_{\mu L}$, as well as the right-handed components of Ψ_{5R}. Taking into account the $\mathbf{Z}_2$ action in the bosonic sector, which projects away the off-diagonal components of the fünfbein (e_μ^5), the above massless spectrum is consistent with the residual $N = 1$ supersymmetry transformations at $D = 4$ given by eq. (1) with a fermionic parameter $\mathcal{E}$ reduced to its left-handed component $\mathcal{E}_L$.

In order to spontaneously break supersymmetry, we apply the Scherk-Schwarz mechanism on the fifth coordinate y [6]. For this purpose, we need an R-symmetry, which transforms the gravitino non-trivially, and impose boundary conditions, around S^1, which are periodic up to a symmetry transformation:

$$\Psi_M(x^\mu, y + 2\pi\rho) = e^{2i\pi\omega Q} \Psi_M(x^\mu, y) , \tag{4}$$

where Q is the R-symmetry generator and ω the transformation parameter. The continuous symmetry is in general broken by the compactification to some discrete subgroup, leading to quantized values of ω. For instance, in the case of $\mathbf{Z}_N$ one has $\omega = 1/N$ and $Q = 0, \ldots, N-1$. For generic values of ω, eq. (4) implies that the zero mode of the gravitino acquires an explicit y-dependence:

$$\Psi_M(x^\mu, y) = U(y) \Psi_M^{(0)}(x^\mu) + \cdots ; \; U = e^{i \frac{\omega}{\rho} y Q} \tag{5}$$

where the dots stand for Kaluza-Klein (KK) modes.

Consistency of the theory requires that the matrix U commutes with the reflection $\mathcal{R}$, which defines the $N = 1$ projection [7, 8]. A solution is given by [9] (for the general solution see [5]):

$$Q = \sigma_2; \quad U = \cos \frac{\omega y}{\rho} + i \sigma_2 \sin \frac{\omega y}{\rho} , \tag{6}$$

where σ_i are the Pauli matrices representing $SU(2)_R$ generators. Another solution is found in the particular case $\omega = 1/2$: $\exp\{i\pi Q\} = (-)^{2s}$ [10].

Inspection of the supersymmetry transformations (1), together with the requirement that the fünfbein zero mode does not depend on y, $e_M^m = e_M^m(x^\mu)$, shows that the y-dependence of the supersymmetry parameter is the same as that of the gravitino zero-mode [6], i.e. $\mathcal{E}(x^\mu, y) = U(y) \mathcal{E}^{(0)}(x^\mu)$. Supersymmetry in the 4D theory is then spontaneously broken, with the goldstino being identified with the fifth component of the 5D gravitino, $\Psi_5^{(0)}$. Indeed, for global supersymmetry parameter, $D_\mu \mathcal{E}^{(0)} = 0$, its variation is:

$$\delta \Psi_5^{(0)} = i \sigma_2 \frac{\omega}{\rho} \mathcal{E}^{(0)} + \cdots \tag{7}$$

while no other fermions can acquire finite constant shifts in their transformations.

The above arguments are also valid in the $N = 1$ theory, obtained by applying the $\mathbf{Z}_2$ projection defined through the $\mathcal{R}$-reflection (3). The y-dependence of the remaining zero modes is always given by eq. (5). Supersymmetry is spontaneously broken and the goldstino is identified as the right-handed component $\psi_{5R}^{(0)}$, which, from eq. (7), transforms as: $\delta\psi_{5R}^{(0)} = \frac{\omega}{\rho}\varepsilon_R^{(0)*} + \cdots$ The surviving gravitino is $\Psi_{\mu L}^{(0)}$. Its mass is given by $m_{3/2} = \omega/\rho$. In the limit $\rho \to \infty$, supersymmetry is locally restored: $m_{3/2} \to 0$, $\delta\psi_{5R}^{(0)} \to 0$.

Note, however, that the above analysis in the $N = 1$ case is valid, strictly speaking, for values of y inside the semicircle, obtained from the interval $[-\pi\rho, \pi\rho]$ through the identification $y \leftrightarrow -y$. This leads to a discontinuity in the transformation parameter $\mathcal{E}$ around the end-point $y = \pm\pi\rho$, since $U(-\pi\rho) = U^{-1}(\pi\rho) \Rightarrow \mathcal{E}(-\pi\rho) \neq \mathcal{E}(\pi\rho)$. This discontinuity survives even in the large-radius limit $\rho \to \infty$ where the gravitino mass vanishes and supersymmetry is restored locally. This phenomenon is reminiscent of the one found in ref. [11], where the discontinuity at the weakly coupled end $y = \pi\rho$ is due to the gaugino condensate of the hidden E_8 formed at the strongly coupled end $y = 0$. In fact the two results become identical for the transformation parameter $\mathcal{E}$ in the neighbourhood $y \sim \pi\rho$, in the limit $\rho \to \infty$:

$$\varepsilon_L(y) \sim \cos\pi\omega\,\varepsilon_L^{(0)} + \epsilon(y)\sin\pi\omega\,\varepsilon_R^{(0)*}\,. \tag{8}$$

On the other hand, it is easy to see that the goldstino variation vanishes in this limit, since the discontinuity in $\partial_y\varepsilon(y)_L$ is proportional to $\delta(y)\sin(y\omega/\rho)$. The transformation parameter $\varepsilon_L(y)$ is thus identified with the spinor η' of ref. [11], which solves the unbroken supersymmetry condition $\delta\psi_{5R} = 0$.

3 Supersymmetry breaking in the observable sector

At the lowest order, supersymmetry is broken only in the five-dimensional bulk (gravitational and moduli sector), while it remains unbroken in the observable sector. The communication of supersymmetry breaking is then expected to arise radiatively, by gravitational interactions [10]. At the one-loop level, the diagrams that contribute to the scalar masses in the observable sector were studied in ref. [10]. After adding the contribution of diagrams related by supersymmetry, we obtain (for the above mentioned case of $\omega = 1/2$) the following expressions for the scalar and gaugino masses:

$$m_0^2 \sim f_0(\mathcal{G})\frac{m_{3/2}^4}{M_p^2}\mathcal{J}_0; \qquad m_{1/2} \sim f_{\frac{1}{2}}(\mathcal{G})\frac{m_{3/2}^3}{M_p^2}\mathcal{J}_{\frac{1}{2}} \tag{9}$$

$$\mathcal{J}_s = \int_0^\infty \frac{dx}{x^{3-2s}}\left(\frac{\pi}{x}\right)^{1/2}\left[\theta_3\left(\frac{i\pi}{x}\right) - \theta_4\left(\frac{i\pi}{x}\right)\right]; \quad (s = 0, 1/2) \tag{10}$$

where f_0 and $f_{\frac{1}{2}}$ are functions of the Kähler potential $\mathcal{G}$, θ_i are the Jacobi theta-functions and we have used the Poisson resummation formula. The fact that all scalar squared mass splittings are of order $m_{3/2}^4/M_p^2$ is a consequence of the absence of quadratic divergences in the effective supergravity. Inspection of eq. (10) shows that cancellation of quadratic divergences arises non-trivially. Indeed, any single excitation n of the sum gives a contribution to the integral, which is quadratically divergent, so that after introducing an ultraviolet cutoff $\propto 1/M_p^2$ one would get a contribution of order $m_{3/2}$ to the mass. However, after summing over all modes and performing the Poisson resummation, one finds that the integrand has an exponentially suppressed (non-analytic) ultraviolet behaviour.

The phenomenology provided by the masses (9) depends a lot on the functions f_0 and $f_{\frac{1}{2}}$. If the Kähler potential is of no-scale type, as suggested by explicit calculations [9], then $f_0 = 0$ $(m_0 = 0)$ and supersymmetry breaking in the observable sector is triggered by gauginos ($m_{3/2} \sim 10^{14}$ GeV, $m_{1/2} \sim m_{3/2}^3/M_P^2 \sim 1$ TeV) and transmitted (gauge mediated) to the scalar sector by (universal) one-loop gauge interactions. The scalars would then acquire masses $m_0 \sim m_{1/2}$ [12]. If the Kähler potential turns out to be not of no-scale type, then the scalar masses in (9) are not zero ($m_{3/2} \sim 10^{12}$ GeV, $m_0 \sim m_{3/2}^2/M_P \sim 1$ TeV) but the gaugino masses are negligible and require to be generated by (supersymmetric) massive fields transforming non-trivially under the gauge group [10].

References

[1] P. Hořava and E. Witten, *Nucl. Phys.* **B460** (1996) 506; *Nucl. Phys.* **B475** (1996) 94; E. Witten, *Nucl. Phys.* **B471** (1996) 135.
[2] I. Antoniadis and M. Quirós, *Phys. Lett.* **B392** (1997) 61.
[3] A.C. Cadavid et al., *Phys. Lett.* **B357** (1995) 76; I. Antoniadis, S. Ferrara and T.R. Taylor, *Nucl. Phys.* **B460** (1996) 489.
[4] M. Günaydin et al., *Nucl. Phys.* **B242** (1984) 244.
[5] I. Antoniadis and M. Quirós, *Phys. Lett.* **B** to appear [hep-th/9707208].
[6] J. Scherk and J.H. Schwarz, *Nucl. Phys.* **B153** (1979) 61 and *Phys. Lett.* **B82** (1979) 60.
[7] C. Kounnas and M. Porrati, *Nucl. Phys.* **B310** (1988) 355; S. Ferrara, C. Kounnas, M. Porrati and F. Zwirner, *Nucl. Phys.* **B318** (1989) 75; C. Kounnas and B. Rostand, *Nucl. Phys.* **B341** (1990) 641.
[8] I. Antoniadis, *Phys. Lett.* **B246** (1990) 377; Proc. PASCOS-91 Symposium, Boston 1991 (World Scientific, Singapore, 1991), p. 718; I. Antoniadis, C. Muñoz and M. Quirós, *Nucl. Phys.* **B397** (1993) 515.
[9] E. Dudas and C. Grojean, e-print [hep-th/9704177].
[10] I. Antoniadis and M. Quirós, *Nucl. Phys.* **B** to appear [hep-th/9705037].
[11] P. Hořava, *Phys. Rev.* **D54** (1996) 7561.
[12] I. Antoniadis, E. Dudas and M. Quirós, in preparation.

Spontaneous SUSY Breaking in Super Yang–Mills Theories and Their Chiral Structure

Peter Minkowski (mink@itp.unibe.ch)

University of Bern, Switzerland

Abstract. It is shown that spontaneous susy and chiral symmetry breakings are linked in pure super Yang-Mills theories.

1. Outline

We discuss spontaneous parameters (potential and real) for N = 1 pure super Yang Mills theories based on a simple gauge group. An adequate minimal set of generating functionals and derived (thermodynamic) potentials arise naturally in the thermodynamic limit, i.e. the limit of conditioning the quantum mechanical system within finite volumes tending to infinity. The $U1$ transformation associated with the (overall) chiral fermionic current [1] exhibits — in this limit — a restoring effect, due to the relaxation of potential θ angles. The following results are derived :

1. Equilibrium conditions pertinent to the thermodynamic limit involve a set of external parameters, in particular arbitrary (complex) gauge invariant fermion mass terms.

2. Spontaneous breaking of chiral charge is linked to the spontaneous breaking of susy in the limit of vanishing fermion masses. Respective order parameters are the gaugino and gauge boson condensates.

3. In the susy limit the fermionic condensate cannot exist without the bosonic one and vice versa.

2. The thermodynamic limit

We consider a class $\mathcal{K} = \{O_{1,\dots,n}(x)\}$ of local operators within the given field theory. The thermodynamic limit assigns to each operator its zero momentum limit

$$O_{1,\dots,n}(x) \longrightarrow o_{1,\dots,n} = \lim_{\Upsilon \to \infty} \frac{1}{\Upsilon} \int_{\Upsilon} d^4y \, O_{1,\dots,n}(y) \tag{1}$$

Υ stands for the four dimensional volume. The above limit involves further nonlocal operators which derive in the functional description of field variables from the generating action functional. Thus we associate external sources and classical field variables with the operators in $\mathcal{K}$

$$O_{1,\ldots,n}(x) \longleftrightarrow J_{1,\ldots,n}(x) \longleftrightarrow O^{cl}_{1,\ldots,n}(x) \tag{2}$$

The functions $J_k(x)$, $O^{cl}_k(x)$ in eq. (3) perturbatively satisfy boundary conditions for $x \to \infty$ which ensure that their respective four dimensional volume averages vanish. Then the J_k generate the connected Green functions, which are represented by functional integrals over the basic gauge boson and gaugino fields, i.e. right and left chiral (or circular) components

$$\begin{array}{l} F^a_{\mu\nu} = \partial_\nu V^a_\mu - \partial_\mu V^a_\nu - f_{abc} V^b_\nu V^c_\mu \;\; ; \;\; \lambda^a_\alpha \;\; , \;\; \lambda^{*\,a}_{\dot\alpha} \\ \left(F_{\{\alpha\delta\}} = 2\,\Sigma(k)_{\{\alpha\delta\}} \left[E_k + iB_k \right] \right)^a \; ; \; \left(\Sigma(k) \; = \; -\sigma(k)\,\varepsilon \right)_{\{\alpha\delta\}} \end{array} \tag{3}$$

In eq. (4) $\Sigma(k) = \Sigma_R(k) \;\; = \;\; - \; \sigma_k \varepsilon$ denote the symmetric base representation matrices for right circular spinor components, reversed upon complex conjugation. The label a refers to a basis in the Lie algebra of the simple gauge group considered. The connection curvature form is (Lie algebra representation) matrix valued, $\tau^a(\mathcal{D})$ denoting the (general) Lie algebra representation, whereas susy Yang Mills theory only needs the adjoint representation

$$\begin{array}{l} \left(F_{\{\alpha\delta\}} \;\; , \;\; \lambda_\alpha \right)^a \; = \; U^a \;\; \to \;\; U(\mathcal{D}) = U^a \left[i\tau^a \right] \\ \mathcal{D} \;\; \to \;\; \tau^a(\mathcal{D}) \;\; ; \;\; \left[\tau^a , \tau^b \right] = i f_{abc}\, \tau^c \;\; ; \;\; \mathcal{D} = \mathrm{ad} \;\; \to \;\; \left(i\tau^a \right)_{st} = f_{ast} \end{array} \tag{4}$$

Manifest supersymmetry is enforced if coordinates together with the gauge group are extended analytically to their complex stature. The curvature forms in eqs. (4) merge into the fermionic susy connections

$$W_\alpha = \exp(-V)\, D_\alpha \, \exp V \;\; ; \;\; \widetilde{W}_{\dot\gamma} = \exp V \; D^*_{\dot\gamma} \, \exp(-V) \tag{5}$$

obtained from the hermitian susy vector field

$$V = V(C, X, M, \lambda, v, D) = V^* \;\; ; \;\; v_\mu = \frac{1}{i} \mathcal{W}_\mu \tag{6}$$

Details of BRS gauge fixing were discussed in [2]. Here we use the derived extended functional measure. Full susy covariance is lost in the process, reducing to propagating components in the Wess-Zumino gauge, in contrast to full susy-chiral gauge fixing [3]. We turn to the covariant susy field strengths

$$w_\alpha = \bar{d}^2\, W_\alpha \quad ; \quad \bar{d}^2 = \frac{1}{2} D^{*\dot{\gamma}} D^*_{\dot{\gamma}}$$

$$w_\alpha = \left\{ \begin{array}{l} \vartheta^2\, i D_{\alpha\dot{\beta}} \lambda^{*\dot{\beta}} \\ + \vartheta^\gamma f_{\alpha\gamma} + \lambda_\alpha \end{array} \right\} \; ; \; f_{\alpha\gamma} = \frac{i}{2} F_{\alpha\gamma} \; ; \; f_\alpha^\gamma = i \boldsymbol{\sigma}_{\alpha\gamma} \left(\mathbf{E} + i \mathbf{B} \right) \tag{7}$$

The quadratic invariant chiral superfield derived from w_α in eq. (8) $\Phi = \frac{1}{2}\,\mathrm{tr}\, w^\alpha w_\alpha$ is associated with the external source superfield $\mathbf{J}$:

$$\Phi = \left\{ \begin{array}{l} \vartheta^2 F \\ + \vartheta^\alpha \psi_\alpha + \frac{1}{2} z \end{array} \right\} \quad \leftrightarrow \quad \mathbf{J} = \left\{ \begin{array}{l} \vartheta^2 (-\mathbf{m}) \\ + \vartheta^\alpha \boldsymbol{\eta}_\alpha + \frac{1}{2} \mathbf{j} \end{array} \right\} \tag{8}$$

In eq. (9) fields and their associated sources $F \leftrightarrow \mathbf{j}$ and $z \leftrightarrow -\mathbf{m}$ are

$$\begin{array}{l} F = \Lambda^* \sigma^\mu \frac{1}{2} i \overleftrightarrow{D}_\mu \Lambda - i \frac{1}{2} \partial_\mu \Lambda^* \sigma^\mu \Lambda - \frac{1}{4} F^a_{\mu\nu} F^{a\,\mu\nu} + i \frac{1}{4} F^a_{\mu\nu} \widetilde{F}^{a\,\mu\nu} \\ \psi_\alpha = \frac{1}{\sqrt{2}} \left(f_\alpha^{a\,\gamma} \Lambda^a_\gamma \right) \quad ; \quad z = \Lambda^{a\,\alpha} \Lambda^a_\alpha \quad ; \quad \mathbf{j}(x) = \frac{1}{g^2(x)} - i \frac{\theta(x)}{8\pi^2} \end{array} \tag{9}$$

In eq. (10) the gaugino fields are rescaled : $\lambda^a_\alpha = \sqrt{2}\, \Lambda^a_\alpha$. In the following we omit external fermion fields $\boldsymbol{\eta}_\alpha$, focusing on the bosonic degrees of freedom. Chiral and antichiral external field bilinears and the external field Lagrangean (density) $\mathcal{L}_E$ become

$$\begin{array}{l} E(\Phi, \mathbf{J}) = \Phi \mathbf{J} = \left\{ \vartheta^2 \left[\frac{1}{2} (\mathbf{j} F - \mathbf{m} z) \right] + \vartheta^\alpha \frac{1}{2} \left(\mathbf{j} \psi_\alpha \right) + \frac{1}{4} \mathbf{j} z \right\} \quad ; \quad E \leftrightarrow \overline{E} \\ \mathcal{L}_E = \left[\frac{1}{2} (\mathbf{j} F - \mathbf{m} z) + h.c. \right] \end{array} \tag{10}$$

The local restriction of the external superfield $E, \overline{E}$ in eq. (11) extends to nontrivial boundary conditions, determining the thermodynamic limit

$$\lim_{x \to \infty} \mathbf{j}(x) = \mathbf{j}(.) = \frac{1}{g^2(.)} - i \frac{\theta(.)}{8\pi^2} \quad ; \quad \lim_{x \to \infty} g^2(x)\, \mathbf{m}(x) = \mathcal{M}(.) \tag{11}$$

In eq. (12) the argument of limiting components $E(.)$ stands for the attribute: subject to renormalization. The complex quantity $\mathcal{M}(.) \to \mathcal{M}$

is a chiral external mass parameter, an explicit breaking of susy and of chiral symmetry, quite similar to the situation in QCD [4]. Every spontaneous phenomenon can be traced through an external field extension. The actual occurrence of spontaneous phenomena becomes manifest upon relaxation of external symmetry breaking fields.

3. Chiral and trace anomaly in pure super Yang Mills theories

We give here the form of the two central anomalies relating renormalization group invariant and accordingly rescaled operators. For details we refer to ref. [2] , [5] , [6] .

$$j^{\mu(5)} = \frac{1}{g^2} j_\mu^{(\Lambda)} \quad ; \quad F^a_{\mu\nu} = g\, G^a_{\mu\nu} \quad ; \quad \kappa = \left(\frac{g}{4\pi}\right)^2$$

$$\begin{aligned} \left(\partial_\mu j^{\mu(5)}\right)_{an} &= 2\, C_2(G) \frac{1}{8\pi^2} \left[\frac{1}{4} F^a_{\mu\nu} \widetilde{F}^{a\mu\nu}\right] = 2\, C_2(G)\, ch_2(F) \qquad (12) \\ \left(\vartheta^\varrho_\varrho\right)_{an} &= -3\, C_2(G) \frac{1}{8\pi^2} \left[\frac{1}{4} F^a_{\mu\nu} F^{a\mu\nu}\right] \end{aligned}$$

The two field strength bilinears in eq. (13), related by susy, are identically renormalized to two loop order [7] , [8].

4. The thermodynamic limit of N = 1 pure super Yang Mills theories

We consider the generating functional for external fields $\mathbf{J}$ and associated quantized fields Φ defined in eqs. (9), (11) and (12), with initially vanishing boundary values

$$\Phi = \left\{ \vartheta^2 F + \vartheta^\alpha \psi_\alpha + \frac{1}{2} z \right\} \leftrightarrow \mathbf{J} = \left\{ \vartheta^2 (-\mathbf{m}) + \frac{1}{2} \mathbf{j} \right\} ; \; \lim_{x \to \infty} \mathbf{J}(x) = \mathbf{J}_\infty = 0 \qquad (13)$$

The quantum mechanical Lagrangian $\mathcal{L}_E$ and the source field are given by eq. (11). The (external field) generating functional takes the form

$$\begin{aligned} \mathcal{D}\mu(.) &= \Pi_y \left(\mathcal{D}V_y\right) \left(\mathcal{D}c_y^*\right) \left(\mathcal{D}c_y\right) \exp i \left(S_{g.f.}\right)_{\text{extr.}} \\ \exp i\Gamma(\mathbf{J}) &= Z_0^{-1} \int \mathcal{D}\mu(.)\, T \exp i \big(S_{\text{qu}} + S(\mathbf{J})\big) \qquad (14) \\ S(\mathbf{J}) &= S\big(E\,(\mathbf{J})\big) \quad ; \quad E(\Phi_{\text{qu}}, \mathbf{J}) = \Phi_{\text{qu}}\, \mathbf{J} \end{aligned}$$

The gaugino mass term $\sim \mathbf{m}$ is the auxiliary field of the external multiplet $\mathbf{J}$. Thus a specific off shell susy prevails, which we call extrinsic, encompassing on shell or intrinsic susy. The external field functional $\Gamma(\mathbf{J})$ exhibits the relations

$$\mathcal{D}\Gamma(\mathbf{J}) = \tfrac{1}{2}\langle \Phi_{qu}(x,\vartheta)\rangle\, \delta\mathbf{J}(x,\vartheta) + c.c. \quad ; \quad \langle \Phi_{qu}(x,\vartheta)\rangle = \Phi_{cl}(x,\vartheta)$$

$$\frac{\delta\Gamma(\mathbf{J})}{\delta\mathbf{j}(x)} = \tfrac{1}{2} f(x) \quad , \quad \frac{\delta\Gamma(\mathbf{J})}{\delta\mathcal{M}(x)} = -\tfrac{1}{2} Z(x)$$

$$f(x;\mathbf{J}) = \left\langle - \left(\tfrac{1}{4} F^a_{\mu\nu} F^{a\,\mu\nu} \; - i\tfrac{1}{4} F^a_{\mu\nu} \widetilde{F}^{a\,\mu\nu} \right)(x) \right\rangle \; ; \; Z(x;\mathbf{J}) = \langle \ell^{a\,\alpha} \ell^a_\alpha(x) \rangle \tag{15}$$

The functions $f(x)$ and $Z(x)$ in eq. (16) are the classical field variables. The gaugino fields ℓ^a_α are normalized to unit of kinetic energy ($\Lambda = g_R\, \ell$) . We turn to the Legendre transform $\mathcal{V}$ of the external field functional Γ, the effective or more precisely internal potential

$$\begin{aligned} \mathcal{V}(f, Z; \overline{f}, \overline{Z}) &= \mathrm{Re} \int d^4 x \,(f(x)\, \mathbf{j}(x) - Z(x)\, \mathcal{M}(x)) - \Gamma \\ f(x) = 2\, \frac{\delta\Gamma(\mathbf{J})}{\delta\mathbf{j}(\mathbf{x})} &\; ; \; Z(x) = -2\, \frac{\delta\Gamma(\mathbf{J})}{\delta\mathcal{M}(x)} \end{aligned} \tag{16}$$

The internal potential $\mathcal{V}$ determines the external fields by the conjugate relations to eq. (16)

$$\begin{aligned} \frac{\delta\mathcal{V}(f, Z; \overline{f}, \overline{Z})}{\delta f(x)} &= \frac{1}{2}\, \mathbf{j}(x) & \frac{\delta\mathcal{V}(f, Z; \overline{f}, \overline{Z})}{\delta Z(x)} &= -\frac{1}{2}\, \mathcal{M}(x) \\ \frac{\delta\Gamma(\mathbf{J})}{\delta\, \mathbf{j}(\mathbf{x})} &= \frac{1}{2}\, f(x) & \frac{\delta\Gamma(\mathbf{J})}{\delta\mathcal{M}(\mathbf{x})} &= -\frac{1}{2}\, Z(x) \end{aligned} \tag{17}$$

The thermodynamic limit corresponds to large wave lengths characterizing the fields $f \leftrightarrow \mathbf{j}$ and $Z \leftrightarrow \mathcal{M}$. It characterizes a (space time) subvolume $\Upsilon_{\prec} \subset \Upsilon$ such that $\mathbf{j}$ and $\mathcal{M}$ take constant values in $\Upsilon_{\prec}$. At this stage we relax the trivial boundary conditions in eq. (16). Thus eq. (18) represents equilibrium conditions for constant values $\mathbf{j}$ and $\mathcal{M}$ inside $\Upsilon_{\prec}$ and vanishing in the complement $\Upsilon / \Upsilon_{\prec}$ in all conceivable infinite volume limits $\Upsilon_{\prec} < \Upsilon \to \infty$.

$$\lim_{x \to \infty} \mathbf{J}(x) = \mathbf{J}_\infty = \begin{cases} \dfrac{1}{g^2_\infty} - i\, \dfrac{1}{8\pi^2}\, \theta_\infty & \text{for } \mathbf{j}, \\ \mathcal{M}_\infty & \text{for } \mathcal{M} = g^2 \mathbf{m} \end{cases} \tag{18}$$

In comparing with eq. (14) we see that the boundary values g^2_∞ and $\mathcal{M}_\infty$ can be absorbed into a redefinition of $\mathbf{J}_0$ as defined there

$$\mathbf{J}_0 \to \left\{ \begin{array}{l} \vartheta^2 \left(- \dfrac{1}{g_\infty^2} \, \mathcal{M}_\infty(.) \right) \\ + \dfrac{1}{16\pi^2} \left(\dfrac{1}{2\,\kappa(.)} + \dfrac{1}{2\,\kappa(\infty)} - i\,(\theta(.) + \theta_\infty) \right) \end{array} \right\} \tag{19}$$

$$\kappa_\infty = \frac{g_\infty^2}{16\pi^2}$$

Herewith we reach the thermodynamic limit.

5. The free Gibbs potential

The transfer of boundary values of the fields $\mathbf{J}$ to the quantum mechanical Lagrangian allows to construct the free Gibbs potential from the quantum mechanical Lagrangean with constant renormalized parameters $g = g_R$, $\vartheta = \vartheta_R$ and $\mathcal{M} = \mathcal{M}_R$. This means one type of boundary condition within the respective volumes $\Upsilon_\prec \subset \Upsilon$. The equilibrium conditions then establish equality of the so chosen separate boundary conditions. These conditions turn into minimum conditions for $\mathcal{V}_{free}$, which is of the form

$$\mathcal{V}_{\text{free}}(\,f, Z\,,\, \overline{f}, \overline{Z}\,;\, \mathcal{M}, \overline{\mathcal{M}}\,,\, \vartheta)\,_{\vartheta'} \;\; ; \;\; Z = \exp\left(-i\mathcal{V}_{\text{free}}\right) \tag{20}$$

6. The chiral anomaly and the free Gibbs potential

The chiral anomaly in eq. (13) implies the relation for the generating functional Z in eq. (21)

$$2\left(\nu\frac{\partial}{\partial\theta} + \frac{\partial}{\partial A}\right) Z = 0 \quad \to \quad Z = Z\,(\theta - \nu A) \;\; ; \;\; A = \arg\mathcal{M} \tag{21}$$

The functional dependence of Z in eq. (22) indicates a discrete symmetry

$$Z_\nu \,:\, \theta \to \theta + 2\pi\frac{r}{\nu} \;\; ; \;\; r = 0, 1, \ldots, \nu - 1 \tag{22}$$

involving the variables

$$\begin{array}{l} \{\,\chi, \overline{\chi}\,\} \;\; ; \;\; \chi = \exp\left(-8\pi^2\,\mathbf{j}_0\right) = |\chi| e^{i\theta} \;\; ; \;\; |\chi| = \exp\left(-\dfrac{8\pi^2}{g^2}\right) \\ \chi_{1/\nu} = \chi^{1/\nu} = |\chi_{1/\nu}| e^{i\theta/\nu} \quad \text{mod } Z_\nu \;\; ; \;\; |\chi_{1/\nu}| = \exp\left(-\dfrac{8\pi^2}{\nu g^2}\right) \end{array} \tag{23}$$

It follows from eq. (24) that all phase dependence of $\mathcal{V}_{\text{free}}$ is through the products

$$\overline{\chi}_{1/\nu}\,\mathcal{M} \;\; ; \;\; \chi_{1/\nu}\,\overline{\mathcal{M}} \;\; ; \;\; \chi_{1/\nu}\,Z \;\; ; \;\; \overline{\chi}_{1/\nu}\,\overline{Z} \tag{24}$$

The discrete symmetry Z_ν in eq. (23) is related to fixed time large gauge transformations and the associated Chern-Simons 3-form in representations differing from the adjoint one. We therefore do not allow Z_ν to be spontaneously broken. The equilibrium conditions in eq. (22) with respect to the imaginary parts of $\mathbf{j}$, $\mathbf{j}'$ - as discussed in [4] - are tantamount to determine the minimum of $\mathcal{V}_{\text{free}}$ with respect to the boundary values θ, θ'. As far as θ, θ' relaxation - to which we turn now - is concerned the situation in QCD is identical. The minimum condition relaxes θ, θ' to the value(s)

$$\theta - \theta' - \nu \arg \mathcal{M} \;\equiv\; 0 \tag{25}$$

for generic $\mathcal{V}_{\text{free}}$, guaranteed by CPT invariance. This allows us to eliminate the free variables dual to $\mathbf{j}$, $\mathbf{j}'$ first and restrict the discussion to the remaining variables related to $\mathcal{M}$. Thereby CP invariance is restored, when all equilibrium conditions are met. The relaxation phenomenon has not been accounted for by Veneziano and Yankelovich [9] in their treatment of super Yang-Mills systems. Relaxation of θ(s) does not only restore CP invariance but also chiral symmetry in the thermodynamic limit, overriding the anomaluos Ward identity. It does not follow that to a spontaneously broken *restored* symmetry there exists a massless Goldstone mode. The discussion of these modes is beyond the scope of this paper.

7. Linking spontaneous breaking of susy and restored chiral symmetry

As a consequence of the last section the remaining equilibrium variables are, in the notation of eq. (14)

$$\Phi = \left\{ \vartheta^2 f + \frac{1}{2} Z \right\} \quad \leftrightarrow \quad \mathbf{J} = 0 \tag{26}$$

The natural variables relative to the free Gibbs potential $\mathcal{V}_{\text{free}}$ are f, Z, forming a constant chiral superfield. The factors $\chi_{1/\nu}$ and $\overline{\chi}_{1/\nu}$ in eq. 25 are now subsummed in the definition of the variables Z and $\overline{Z}$. The relations in eq. (18) become

$$\begin{aligned} &\Phi \rightarrow (f, Z; \overline{f}, \overline{Z}) \quad , \quad \mathbf{J}' \rightarrow (\mathbf{j}' = 0, -\mathcal{M}' = 0) \\ &\frac{\partial}{\partial f} \mathcal{V}_{\text{free}}(\Phi; \mathbf{J}'') = \frac{1}{2} \mathbf{j}'(x) \rightarrow \mathbf{0} \end{aligned} \tag{27}$$

The relation eq. (28) is the appropriate one for f, $\overline{f}$ being auxiliary fields of respective chiral multiplets and the associated potential $\mathcal{V}_{\text{free}}$ being minimized in the process of eliminating them. Extrinsic susy implies the form of the free Gibbs potential

$$\begin{aligned} &\mathcal{V}_{\text{free}}(f, Z, \overline{f}, \overline{Z}; \mathcal{M}, \overline{\mathcal{M}}) = -\overline{f} K_{Z,\overline{Z}} f + (f W_Z + c.c.) \\ &K = K(Z, \overline{Z}; \mathcal{M}, \overline{\mathcal{M}}) \quad ; \quad W = W(Z; \mathcal{M}, \overline{\mathcal{M}}) \\ &K_{Z,\overline{Z}} = \frac{\partial^2 K}{\partial Z \, \partial \overline{Z}} \quad ; \quad W_Z = \frac{\partial W}{\partial Z} \end{aligned} \tag{28}$$

The functions K, W in eq. (29) denote the Kähler and super potential respectively. The free Gibbs potential becomes the effective action, for prescribed constant masses $\mathcal{M}$ in $\mathcal{L}_{\text{qu}}$, and for prescribed constant values of $Z = \langle \varrho^{a\,\alpha} \varrho^{a\,\alpha} \rangle$. The equilibrium condition becomes minimum condition with respect to all possible trial values of the variables f, $\overline{f}$; Z, $\overline{Z}$ for given external masses $\mathcal{M}$. First we determine the auxiliary fields f, $\overline{f}$

$$\begin{aligned} &\overline{f} = \frac{W_Z}{K_{Z,\overline{Z}}} \; ; \; \mathcal{V}_{\text{free}} = \frac{|W_Z|^2}{K_{Z,\overline{Z}}} \; ; \; K = K(Z, \overline{Z}; \mathcal{M}, \overline{\mathcal{M}}) \; ; \; W = W(Z; \mathcal{M}, \overline{\mathcal{M}}) \\ &\mathcal{V}_{\text{free}} \rightarrow \mathcal{V} = \mathcal{V}(Z, \overline{Z}; \mathcal{M}, \overline{\mathcal{M}}) \; \rightarrow \; \frac{\partial \mathcal{V}}{\partial Z} = 0 \end{aligned} \tag{29}$$

The induced Kähler metric $K_{Z,\overline{Z}}$ must be positive to ensure stability of large volume (normal) fluctuations, so we find the familiar positivity property of the Gibbs potential, implied by susy. Moreover the relaxation of $\theta(\text{s})$ discussed at the end of section 6 implies the restored chiral invariance under the phase rotations

$$Z \rightarrow e^{-i\xi} Z = Z_\xi \quad ; \quad \mathcal{M} \rightarrow e^{+i\xi} \mathcal{M} = \mathcal{M}_\xi \quad \text{and} \quad c.c. \tag{30}$$

The phase invariance in eq. (31) — assigns chiral charges 1, −1 to Z, $\mathcal{M}$ respectively. It is only valid in the thermodynamic limit and provided the generic θ phase relaxation proceeds as stated in the last section. Hence the derivative of the superpotential W_Z must have a well defined chiral charge, whereas the Kähler potential must be neutral. Finally we relax the gaugino masses. They may continue to serve well to explore a minimal set of hysteresis lines.

$$\mathcal{M} \to 0 \;\leftrightarrow\; \mathcal{V}\left(|Z|^2 ; Z\mathcal{M}, \overline{Z\mathcal{M}}, |\mathcal{M}|^2 \right) \longrightarrow \mathcal{V}\left(|Z|^2 ; 0, 0, 0 \right) \quad (31)$$

It follows from eq. (32) that the potential $\mathcal{V}$ attains its minimum along a circle in the complex Z plane. However the only circle where $\overline{f} \propto W_Z$ can vanish is at the origin $Z = 0$. Therefore the only consistent solution with no spontaneous breaking of susy does not admit *any* condensates. Then neither gauginos nor gauge bosons can be confined, since a 'wall' to reverse their chirality upon reflection is absent. Our conclusions are at variance with the treatment of boundary conditions by Witten [10] . The remaining alternative of nontrivial condensate formation — *short of spontaneous breaking of gauge invariance* — links the gaugino condensates $Z, \overline{Z}$ with their gauge boson counterparts $f, \bar{f}$. The latter spontaneously break susy at a positive value of $\mathcal{V}$. Relaxation of the θ parameter(s) implies the absence of spontaneous CP violation *also* for arbitrary gaugino mass. While we can not prove this last situation to prevail from first principles, the lessons from hadron dynamics and QCD teach us that it does.

Acknowledgments
We acknowledge interesting discussions with C. Kounnas, G. Zoupanos and G. Veneziano.

References

[1] R. J. Crewther, *Riv. Nuovo Cim.* **2** (1979) 63,
G. 't Hooft *Phys. Rev.* **D14** (1976) 3432, (E) *Phys. Rev.* **D18** (1978) 2199.

[2] M. Leibundgut and P. Minkowski, Bern University preprint, BUTP-97/22, hep-th/9708061 (1997).

[3] M. T. Grisaru, W. Siegel and M. Roček, *Nucl. Phys* **B159** (1979) 429.

[4] P. Minkowski, *Phys. Lett.* **B237** (1990) 531.

[5] J. Kodeira, *Nucl. Phys.* **B165** (1980) 129.

[6] P. Minkowski, Bern University preprint (1976) ,
N. K. Nielsen, *Nucl. Phys.* **B120** (1977) 212 ,
S.L. Adler, J.C. Collins and A. Duncan, *Phys. Rev.* **D15** (1977) 1712,

J.C. Collins, A. Duncan and S.D. Joglekar, *Phys. Rev.* **D16** (1977) 438.

[7] V.A. Novikov, M.A. Shifman, A.I. Vainshtain and V.I. Zakharov, *Phys. Lett.* **B157** (1985) 169.

[8] M.T. Grisaru, B. Milewski and D. Zanon, *Phys. Lett.* **B157** (1985) 174.

[9] G. Veneziano and S. Yankielowicz, *Phys. Lett.* **113B** (1982) 231.

[10] E. Witten, *Nucl. Phys.* **B202** (1982) 253.

Part XIII

Searches for New Particles at Present Colliders

New States of the Charmed Strange Baryon, Ξ_c

Richard D. Ehrlich[1] (rde4@cornell.edu)

Cornell University, Ithaca, New York 14853

Abstract. Using almost $5fb^{-1}$ of e^+e^- data accumulated at the CESR collider in the energy region around 10.5 GeV , the CLEO II collaboration has found evidence for several narrow excitations of the Ξ_c. Two of these are seen in the modes $\Xi_c^+\gamma$ and $\Xi_c^0\gamma$, and are identified with Ξ_c', the $J^p = \frac{1}{2}^+$ c{sq} state, which is **symmetric** under exchange of light quark quantum numbers. The third state is seen in its decay to $\Xi_c^+\pi^+\pi^-$ via an intermediate Ξ_c^*, and is consistent with the Ξ_{c1}^+, a $J^p = \frac{3}{2}^-$ p-wave state analogous to the $\Lambda_{c1}^+(2625)$. The mass differences relative to the ground state are: $M(\Xi_c^{+\prime}) - M(\Xi_c) = 107.8 \pm 1.7 \pm 2.5 MeV/c^2$, $M(\Xi_c^{0\prime}) - M(\Xi_c) = 107.0 \pm 1.5 \pm 2.5 MeV/c^2$, and $M(\Xi_{c1}^+) - M(\Xi_c^+) = 349.4 \pm 0.7 \pm 1.0 MeV/c^2$

1 Introduction to Ξ_c Spectroscopy

The baryons formed from a (heavy) charmed quark and two light quarks (s, u, or d) can be conveniently classified[1] by the considering a) the relative angular momentum, L, betwen the light diquark and the charmed quark, or the angular momentum , ℓ, within the diquark, b) the combination of L or ℓ with the light spins to give a J^p_{light}, and c) the coupling of this light quark "spin" to the charmed quark spin to give the final J^p. The states fall into sextets (**6**'s) and anti-triplets (**3***'s) under the SU(3) of light quark flavor. The **3***'s are antisymmetric under interchange of light quark quantum numbers; the **6**'s are antisymmetric. When L=ℓ=0, the lowest lying **3*** has $J^p = \frac{1}{2}^+$ and contains the Λ_c , and the Ξ_c. The lowest lying **6** contains the Σ_c, Ω_c, as well as another symmetric $J^p = \frac{1}{2}^+$ state , the Ξ_c', which is expected to decay to $\Xi_c\gamma$, and has been heretofore undetected, but will be reported in what follows. The other s-wave multiplet is a **6** with $J^p = \frac{3}{2}^+$, which contains the recently reported Ξ_c^*[2]. All of the states with $l = 0$ have been found,except for the $\frac{1}{2}^+$ Σ_c^{*+} and Ω_c^{*0}, which are experimentally challenging, requiring detection of a soft π^0.

The lowest-lying states with one unit of angular momentum are expected to have L=1, $\ell = 0$, with little configuration mixing. We thus expect to find a pair of antisymmetric **3***'s with $J^p = \frac{1}{2}^-$ and $J^p = \frac{3}{2}^-$, and five **6**'s: one with $\frac{5}{2}^-$, two $\frac{3}{2}^-$, and two $\frac{1}{2}^-$. Earlier experiments have found a likely candidate for a member of the $\frac{3}{2}^-$ **3***, the $\Lambda_c^+(2625)$[3].

2 CESR, CLEO II, and Charmed Baryons

The results presented herein were obtained with the CLEO II detector[4], operating at the Cornell Electron Storage Ring, CESR. The data derive from nearly $5fb^{-1}$ of integrated luminosity, acquired at the $\Upsilon(4S)$ and on the continuum just below the resonance. In this analysis, however, we have no contributions from B-decay, since we place a lower limit on the parent charmed baryon's scaled momentum, $x_p = p/p_{max} > 0.6$ to 0.7, where $p_{max}^2 = E^2 - M_{baryon}^2$, in order to optimize the signal-to-noise.

The detector employs a cylindrical drift chamber system and a 1.5T solenoidal field to measure charged particles, and 7800 CsI(Tl) crystals provide excellent energy and spatial resolution for photons, electrons, and π^0's. The reconstruction of charmed-strange states begins with collection of a sample of Λ 's (antiparticles, too, are always implied) which do not come from the primary interaction point. These are combined with pions to form Ξ^0 and $\Xi^{\pm}$, or with charged kaons to form Ω^-. The Ξ or Ω^- must be consistent with the primary vertex, and must have decayed closer to the origin than did the daughter Λ. Charged tracks are required to be consistent with the particle identification information provided by the drift chamber's dE/dX system and (when available) by the time-of flight system. The "raw" Λ, Ω, and Ξ, and π^0 masses must be within ~ 3 standard deviations of their known values. They are later kinematically re-fit for improved resolution in the charmed baryon mass and momentum. We have almost 10,000 Ξ^-, 6000 Ξ^0, and 400 Ω^- in the data sample. The Ξ_c^0 are reconstructed in the modes $\Xi^-\pi^+$, $\Xi^-\pi^+\pi^0$, and Ω^-K^+; Ξ_c^+ in the modes $\Xi^-\pi^+\pi^-$ (dominant) , $\Xi^0\pi^+$, and $\Xi^0\pi^+\pi^0$.

3 Analysis of $\Xi_c\gamma$ State

3.1 Ξ_c , Ξ_c' reconstruction

Ξ_c^+ are reconstructed in the modes $\Xi^-\pi^+\pi^+$ and $\Xi^0\pi^+\pi^0$. Ξ_c^0 modes are $\Xi^1\pi^+$, $\Xi^-\pi^+\pi^0$, and Ω^-K^+. Events which are within 2σ of the published CLEO II Ξ_c^+ and Ξ_c^0 masses for each channel are combined with well-measured γ's to form Ξ_c' candidates. The γ must have an energy 100 MeV, and must not form π^0's with other γ's in the detector. A decay-channel dependent x_p cut of 0.5 to 0.6 is applied. Figure 1 shows the resultant distribution of mass-difference between the $\Xi_c\gamma$ and the Ξ_c candidates, with evident peaks containing 28 ± 7 and 25 ± 6 events in the neutral and charged channels, respectively. Each peak has a statistical significance of more than 4σ, and is robust when the π^0 veto is removed, and is fit with a width determined by Monte Carlo. The mass differences relative to the ground state are: $M(\Xi_c^{+\prime}) - M(\Xi_c) = 107.8 \pm 1.7 \pm 2.5 MeV/c^2$, and $M(\Xi_c^{0\prime}) - M(\Xi_c) = 107.0 \pm 1.5 \pm 2.5 MeV/c^2$.

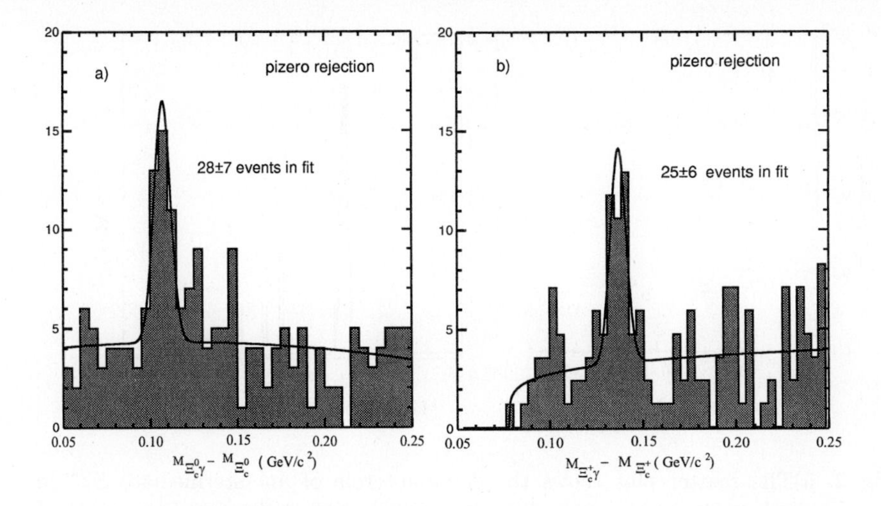

Fig. 1. Distributions of mass-difference between $\Xi_c\gamma$ and Ξ_c candidates: a) neutral channel, b) charged channel.

3.2 $\Xi_c^+\pi^+\pi^-$ analysis

We start with a sample of more than 200 Ξ_c^+ events and combine these with remaining π^- tracks to form Ξ_c^{*0} candidates. A fit (not shown) finds $30.1 \pm 6.6\Xi_c^{*0}$, consistent in mass and width with the Reference 2 result. We combine events within 8 MeV of the Ξ_c^{0*} mass with remaining π^+ tracks, requiring the finalx_p to exceed 0.7, as was done in our earlier Λ_c^* analysis. We plot the mass-difference distribution in Figure 2(b), where a clean enhancement of 11 events can be seen above an estimated 0.3 background events. Its width is fit with the Monte Carlo expectation of 2.5 MeV, consistent with the detector resolution. To avoid uncertainties from the imperfectly known Ξ_c^{0*} width, we quote the mass difference relative to the Ξ_c^+. The result is: $M(\Xi_{c1}^+) - M(\Xi_c^+) = 349.4 \pm 0.7 \pm 1.0 MeV/c^2$. In Figure 2(a), we present the data in scatter plot form, where we see clustering in the $M(\Xi_c^+\pi$ variable, rather than a vertical band. From the absence of decays outside the Ξ_c^{0*} region, we can estimate that the new state decays via an intermediate Ξ_c^{0*} more than 76% of the time, at the 90% confidence level.

4 Conclusions

Even though no spin-parity analysis has been performed, the features of the two $\Xi_c\gamma$ decays support the identification of these narrow peaks with the $\Xi_c^{'}$, the charmed-strange members of the **6** which contains the Σ_c triplet and Ω_c. Its electromagnetic decay is allowed by the quantum numbers and expected by theory to give a mass difference of about 110 MeV, in accord with this

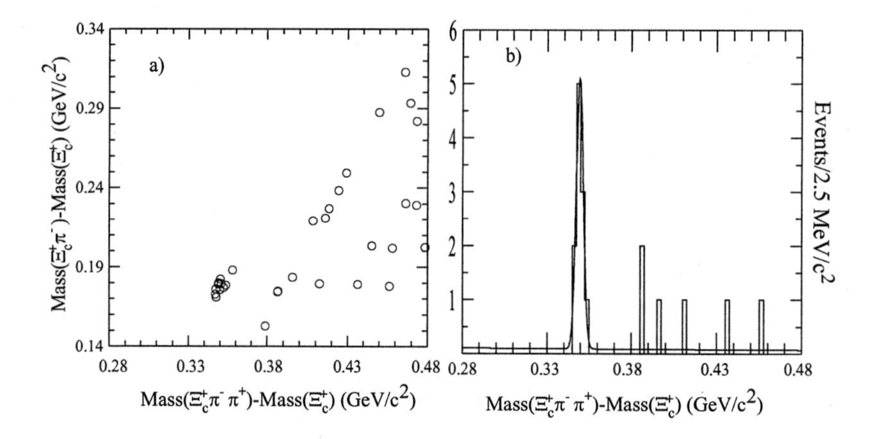

Fig. 2. a)The scatter plot shows the dominant role of an intermediate Ξ_c^{0*} in the $\Xi_c^+\pi^-\pi^+$ peak, b) the mass-difference distribution with an enhancement of 11 events on a (linear) background of 0.3 events

experiment. The $\Xi_c*(3/2^+)$ states, having been already filled, are excluded as an option.

The $\Xi_c^+\pi^+\pi^-$ state, which decays similarly to the Λ_{c1}, is identified as a probable $3/2^-$ entity, one member of an isospin doublet. CLEO is still investigating the partner state. The measured mass difference, relative to the Ξ_c accords well with expectations, and with the analogous number for the Λ_c. Although a D-wave decay to $\Xi_c\pi$ is allowed by total angular momentum and parity , HQET forbids it because of light-quark quantum numbers. We expect that the other L=1 doublet $(1/2^-)$ will decay to $\Xi_c^{'\pi}$, rather than to $\Xi_c^*\pi$.

This work depended on the hard work and skill of thc CESR staff, and was supported by the US National Science Foundation.

References

[1] J.G. Koerner, M. Kramer, and D. Pirjol, Prog. in Part. Nucl. Phys. 33, (1994), 787-868 , D. Pirjol and T.-M. Yan, Phys. Rev. D56, (1997), 5483-5510

[2] (CLEO II) L. Gibbons, et al., Phys. Rev. Lett. 77, (1996), 810-813

[3] H,. Albrecht, et al., Phys. Lett. B317 , (1993), 227-232, (CLEO II) P. Avery, et al., Phys. Rev. Lett. 74, (1995), 4364-4368, P. Frabetti, et al., Phys. Lett. B364, (1996), 461-469

[4] (CLEO II) Y. Kubota, et al., Nucl. Instrum. Meth., A320, (1992), 66-113

Search for Anomalous Photon Events at LEP

Michel Chemarin (chemarin@cern.ch)

Institut de Physique Nucléaire de Lyon, IN2P3-CNRS
127 Université Claude Bernard, F-69622 Villeurbanne Cedex, France

Abstract. At e^+e^- high energy colliders, events with photons and missing energy can offer the first hint of the emergence of new physics. We report on the results obtained by the four LEP experiments concerning such anomalous events, in the 1996 data at $\sqrt{s} = 161 \div 172$ GeV.

1 Introduction

Events with photons, missing energy and nothing else in the detector can offer the first hint of the emergence of new physics. Near the Z^0 energy, the measurement of the cross section for the single photon process has long been recognised as the direct method for the determination of the number of light neutrino generations and all experimental groups at LEP1 reported studies of the single photon production [1]. But new physic processes, implying weakly interacting particles, could also arise through the same channel and this can be particularly true at LEP2 for supersymmetric particles.

Two photons + missing energy search is more recent and is mainly accountable to the 1995 CDF event $ee\gamma\gamma + \not{E}$, where two high energy electrons and two high energy photons with large missing transverse energy have been observed [2]. In the hypothesis where supersymmetry would be relevant for the explanation of the topology of the event, photonic states with 1 or 2 photons and missing energy, would be likely to appear at LEP energies.

In this paper, we present the results obtained in 1996 by the 4 LEP experiments. During this period, the machine delivered about $11pb^{-1}$ at centre-of-mass energies of both 161 and 172 GeV. All the results given use both $\sqrt{s}$ and most are preliminary. Limits are at 95 % confidence level.

2 Results, interpretations and limits

No deviation from the Standard Model(SM) expectations have been observed in the single photon data. For illustration, one can see in Fig.1 and Fig.2 the results from L3. The recoil mass spectrum has photons with energy $E_\gamma > 10$ GeV and the energy spectrum is due to a second analysis for the selection of low energy photons. The main contribution to the samples comes from the $\nu\bar{\nu}\gamma$ reaction, mainly due to a radiative return to the Z^0 energy, followed by its decay to 2 neutrinos. Generators agree at the few % level. Others contributions are negligible, except for radiative low angle Bhabhas in the case of the low energy sample.

Fig.3 collects all the 2γ events from the 4 experiments together with the cumulative expected contribution of the $\nu\bar{\nu}\gamma\gamma$ process. It has been produced by the LEP Susy Working Group (LSWG), a group setup to combine and optimise the results of LEP experiments. A few events are seen with a recoil mass $M_R > 100$ GeV/c^2. A correct background estimation is essential if one would like to address the significance of such events with large missing energy. There were some concerns about the Monte Carlo predictions. New results [3] in the LSWG and the collaborations show better agreement. The low statistics taken into account, nothing really puzzling is seen in these data.

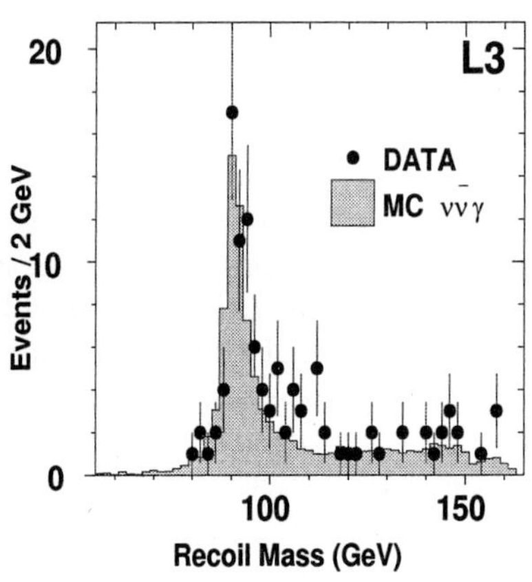

Figure 1: L3 recoil mass spectrum for high energy 1γ data.

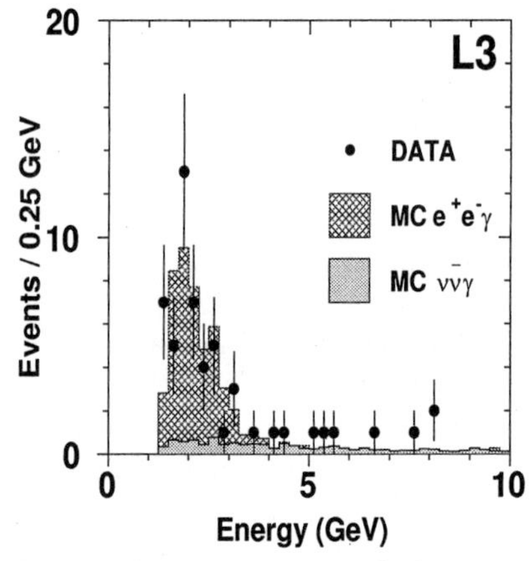

Figure 2: L3 energy spectrum for low energy 1γ data.

Absence of any obvious anomaly in $\gamma \not{E}$ or $\gamma\gamma \not{E}$ spectra is interpreted by LEP collaborations in the framework of supersymmetric models with R-parity conservation and thus with the existence of a stable lightest supersymmetric particle(LSP).

In a first class of studies, the neutralino $\tilde{\chi}_1^0$ is the LSP and the CDF event is interpreted in the framework of the Minimal Supersymmetric extension of the Standard Model(MSSM), in a particular mass scenario, proposed in [4] which is consistent with all existing data. At LEP, the best possibility to verify this scheme, is the production of $\tilde{\chi}_2^0\tilde{\chi}_2^0$ which leads to $\gamma\gamma + \not{E}$. The channel has been studied by ALEPH which has produced cross section limits and an exclusion plot of the CDF allowed region(Fig.4).

Numerous studies have been done using a second class of models where a very light gravitino $\tilde{G}$, is the LSP. This happens in Gauge Mediated Supersymmetry Breaking models (GMSB) where supersymmetry breaking is realized at low energy scale [5] and in "no-scale" supergravity models, such as the LNZ model [6] where gravity is still responsible of supersymmetry breaking. They imply, always for LNZ, most of the time for GMSB models, the neutralino as the Next Lightest Supersymmetric particle (NLSP), opening channels with very attractive signatures such as $1\gamma + \not{E}$ or acoplanar $2\gamma + \not{E}$. All LEP experiments have presented interpretations in this frame.

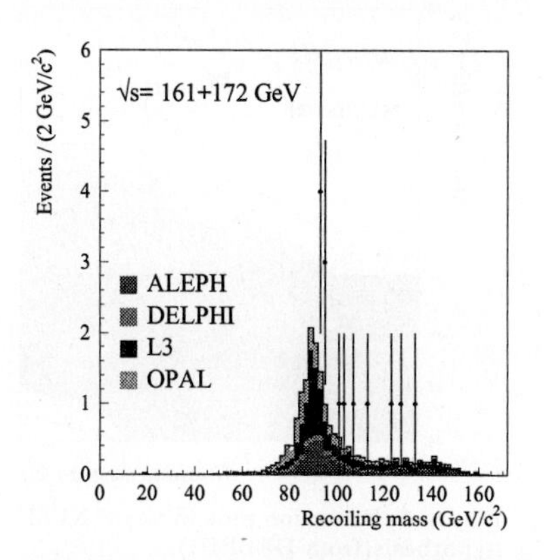

Figure 3: 2γ data. Recoil mass spectrum using the combined data of the 4 LEP experiments, compared with the KORALZ generator results [7].

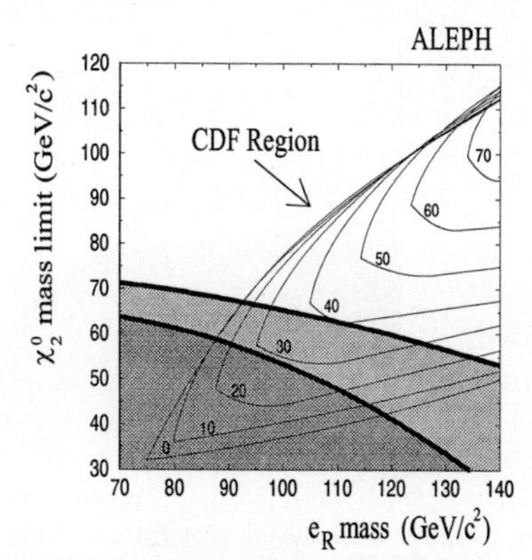

Figure 4: Exclusion plot from 2γ data in the $\tilde{\chi}_1^0$ LSP hypothesis. The shaded area is for $m_{\tilde{e}_L} = m_{\tilde{e}_R}$ and the darker area for $m_{\tilde{e}_L} \gg m_{\tilde{e}_R}$. Curves are labelled with $\tilde{\chi}_1^0$ masses.

The first example, with the gravitino as LSP, is in the "single photon" channel. Excess above the SM contribution can be interpreted as signal for the direct production of a gravitino with an accompanying neutralino which decays radiatively to gravitino. The 4 experiments present results and cross section limits. Using the LNZ model, it is possible to constrain the $\tilde{\chi}_1^0$ mass versus the gravitino mass. For $m_{\tilde{G}} \sim 10^{-5}$ eV$/c^2$, a neutralino mass of 100 GeV$/c^2$ is excluded.

In the gravitino LSP scenario, acoplanar photon pairs come from $\tilde{\chi}_1^0$ pair production followed by the decay to photons and invisible gravitinos. The 4 experiments have set limits on the production cross section and have given lower limit on the neutralino mass in the frame of the LNZ or GMSB models. The LSWG has combined their results and put limit on the neutralino mass, $m_{\tilde{\chi}_1^0} > 76$ GeV$/c^2$. The interpretation of the CDF event in the light gravitino scenario leads to a specific area in the $m_{\tilde{\chi}_1^0}, m_{\tilde{e}_R}$ plane seen in Fig.5. A significant fraction of the allowed region is already excluded.

There exists another possibility, always with the gravitino as LSP. For part of the parameter space, an $\tilde{l}$ and in particular a $\tilde{\tau}$ can be the NLSP. The decay length is function of $m_{\tilde{\tau}}$ and $m_{\tilde{G}}$ and, depending of its value, experiments must develop different strategies : search for leptons + $\not{E}$, large impact parameter tracks, secondary vertices or stable heavy ionising particles (HIP). ALEPH, DELPHI and OPAL have searched for such HIPs and combined their results inside the LSWG, to give an experimental limit on the mass of the $\tilde{\tau}$ in this scenario, $m_{\tilde{\tau}_R} > 75.6$ GeV$/c^2$ or $m_{\tilde{\tau}_L} > 76.5$ GeV$/c^2$. DELPHI has specific analyses for all the different signatures of $\tilde{\tau}$ decay and has combined all of them to give cross section limits and an exclusion area in the plane $m_{\tilde{\tau}}, m_{\tilde{G}}$ (Fig.6). The $\tilde{\tau}$ must have a mass higher than 55 GeV/c^2 for $m_{\tilde{G}} > 1$ eV/c^2 or higher than 65 GeV/c^2 for $m_{\tilde{G}} > 40$ eV/c^2 [8].

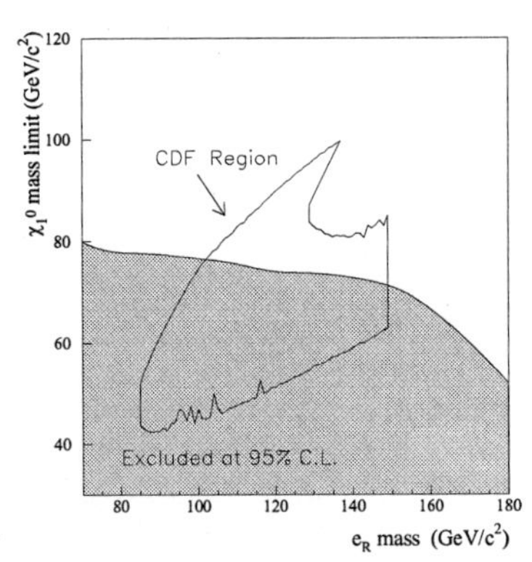

Figure 5: Exclusion plot from 2γ data in the $\tilde{G}$ LSP hypothesis(from LSWG).

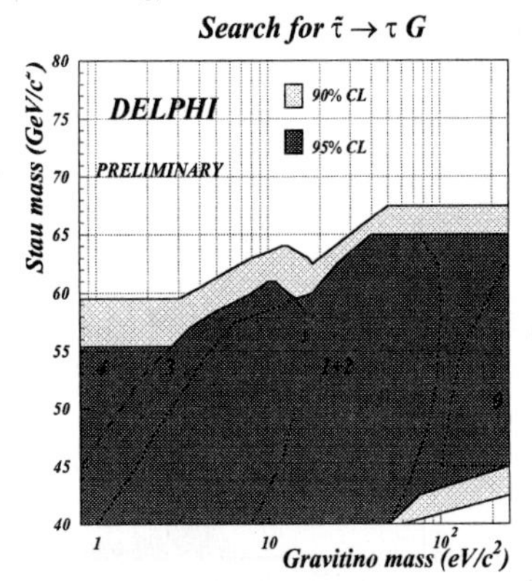

Figure 6: Exclusion plot in the $\tilde{\tau}$ NLSP hypothesis(from DELPHI) .

3 Conclusions

No anomaly with respect to standard model predictions has been reported by any experiment. In both the neutralino NLSP and the $\tilde{\tau}$ NLSP option, constraints on the mass of the intervening particles have been set and a significant part of the region deduced from the CDF event in the gravitino scenario has been excluded. Final results can be seen in [9].

I want to thank all the people from the 4 LEP experiments and from the LSWG who helped me in providing all the informations.

References

[1] ALEPH Collab., D. Buskulic *et al.*, Phys. Lett. **B 313** (1993) 520; DELPHI Collab., P. Abreu *et al.*, Z. Phys. **C 74** (1997) 577; L3 Collab., O. Adriani *et al.*, Phys. Lett. **B 292** (1992) 463; OPAL Collab., R. Akers *et al.*, Z. Phys. **C 65** (1995) 47.

[2] S.Park, in 10th Topical Workshop on Proton-Antiproton Collider Physics, Fermilab, ed. R.Raja and J.Yod, (AIP, New York, 1995).

[3] P.Bain and R.Pain, Preprint Delphi note 97-89, (1997).

[4] G.L.Kane and G.Mahlon, (1997), UM-TH-97-08 hep-ph/9704450.

[5] P.Fayet, Phys. Lett. **117B** (1982) 460.

[6] J.L.Lopez, D.V.Nanopoulos and A.Zichichi, Phys. Rev. Lett. **77** (1996) 5168.

[7] S.Jadach, B.F.L.Ward and Z.Was, Comp. Phys. Commun. **79** (1994) 503.

[8] C.Lacasta *et al.*, Preprint Delphi note 97-97, (1997).

[9] ALEPH, Preprint CERN-PPE/97-122, CERN, (1997), sub. Phys.Lett.B; DELPHI, Preprint CERN-PPE/97-107, CERN, (1997), sub. Zeit. für Phys.; L3, Preprint CERN-PPE/97-76, CERN, (1997), sub. Phys.Lett.B; OPAL, Preprint CERN-PPE/97-132, CERN, (1997), sub. Zeit. für Phys..

Searches for New Particles in Photon Final States at the Tevatron

John Womersley (womersley@fnal.gov)

Fermi National Accelerator Laboratory, Batavia, IL 60510, U.S.A.

Abstract. Searches for new physics at the Tevatron collider experiments, using signatures involving photons, are described.

1 Photon Signatures

Photon final states can provide a clean and interesting channel for new physics at hadron colliders:

- electromagnetic objects can be triggered on with good efficiency;
- there is better missing E_T resolution in events with photons than in multijet signatures, which may be important for supersymmetry;
- the background cross section for real and fake photons is of order 10^{-3} times that for jets, so one can hope to access rare processes by selecting photon-rich final states.

In both CDF and DØ, photons are identified as isolated electromagnetic clusters in the calorimeter, with a shower shape consistent with expectations and no associated track. The results described here are all based on the full Run I dataset of approximately 100 pb^{-1} per detector accumulated in 1992–95.

2 Supersymmetry

Much interest has been generated by the observation of a single anomalous event at CDF[1], containing two photons, two electrons and missing transverse energy($\not\!\!E_T$) (Fig. 1). The standard model probability to have observed such an event is extremely small, and the presence of $\not\!\!E_T$ has led many to interpret this as a supersymmetry (SUSY) candidate (see references in [1]).

There are two ways in which photon-rich final states can result from supersymmetric particle production:

- "light neutralino" models in which the second lightest neutralino decays to the lightest plus a photon, $\chi_2^0 \to \chi_1^0\gamma$. This can occur in a small part of the MSSM parameter space.

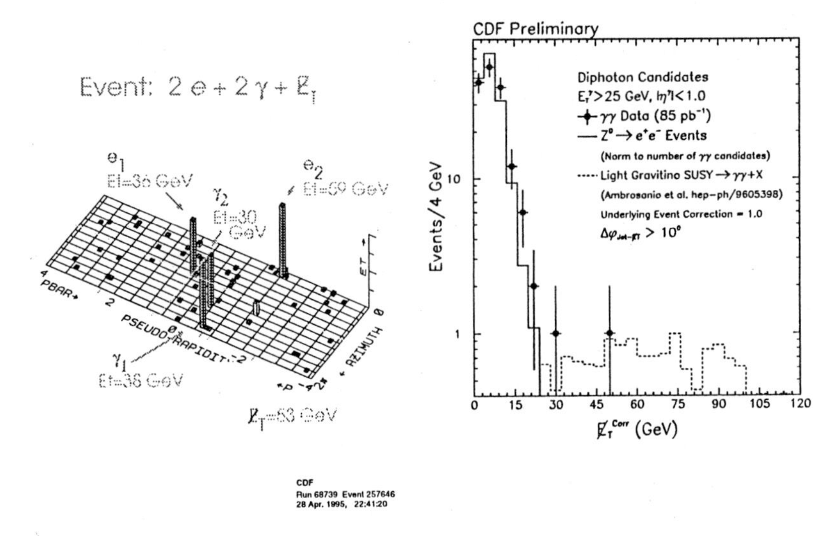

Fig. 1. (left) CDF $ee\gamma\gamma$ $\not\!\!E_T$ event; (right) Spectrum of $\not\!\!E_T$ for events with two photons in CDF.

– "light gravitino" models in which the lightest neutralino itself can decay to a gravitino plus a photon, $\chi_1^0 \to \tilde{G}\gamma$. This occurs in models with gauge-mediated SUSY breaking, where supersymmetry is broken at scales as low as 100 TeV, and also in "no-scale" supergravity-inspired models.

2.1 Two photons + $\not\!\!E_T + X$

Both CDF and DØ have carried out searches inspired by these possibilities. The first is a generic search for the pair production of SUSY particles with decays to photons:

$$p\overline{p} \to \gamma\gamma \not\!\!E_T + X.$$

CDF required two photons with $E_T > 25$ GeV and $|\eta| < 1.0$; no excess is seen at large $\not\!\!E_T$. (see Fig. 1).

DØ has also carried out two analyses in this channel[2, 3]. The most recent requires two photons with $E_T > 20$ and 12 GeV respectively, and $|\eta| < 2.0$. For $\not\!\!E_T > 25$ GeV, two events are seen where 2.3 ± 0.9 are expected. In the "light gravitino" scenario, this can be translated into limits in the (μ, M_2) plane as shown in Fig. 2) which are sufficient to rule out the "chargino" interpretation of the CDF event quoted in [4]. They also exclude lightest charginos $\chi_1^{\pm}$ and neutralinos χ_1^0 with masses less than 150 GeV/c^2 and 75 GeV/c^2 respectively (95% C.L.).

DØ has also obtained limits on production cross sections for selectrons, sneutrinos, and neutralinos in the "light neutralino" model. They range from 400 fb to 1 pb for $m_{\chi_2^0} - m_{\chi_1^0} > 20$ GeV/c^2. A general limit on the cross section

$$\sigma(p\overline{p} \to \gamma\gamma \not\!\!E_T + X) < 185 \text{ fb (95\% C.L.)}$$

for $E_T^\gamma > 12$ GeV, $|\eta| < 1.1$, and $\not\!\!E_T > 25$ GeV, is also obtained.

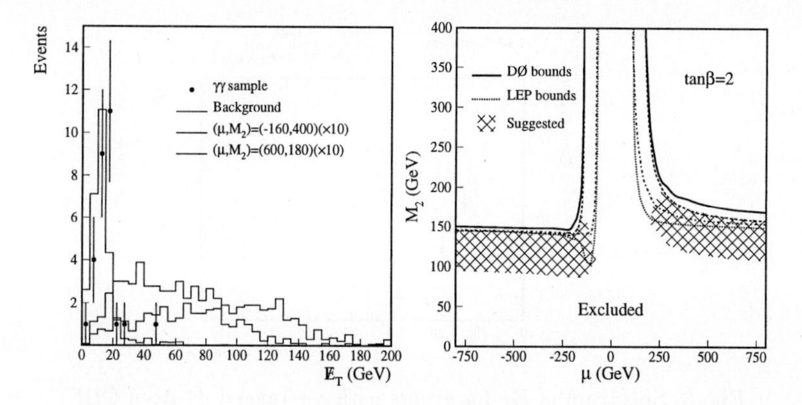

Fig. 2. (left) Spectrum of $\not{E}_T$ for events with two photons in DØ; (right) Exclusion region in the (μ, M_2) plane from this analysis. The hatched region is that proposed in [4] to explain the CDF $ee\gamma\gamma$ $\not{E}_T$ event. These limits are only very weakly dependent on $\tan\beta$.

2.2 Photon + Tagged Jet + $\not{E}_T$

CDF has also searched for single photons produced together with b-jets and $\not{E}_T$. This could occur from the light stop squark which is expected in the light neutralino scenario.

The analysis requires one photon with $E_T > 25$ GeV, one jet with $E_T >$ 15 GeV, and an SVX b-tag. The $\not{E}_T$ distribution for these events is shown in Fig. 3. Two events are observed with $\not{E}_T > 40$ GeV. Without attempting to perform a background subtraction, the upper limit is then 6.4 events, which marginally excludes (6.7 events expected) a baseline model of Ambrosanio *et al.*[5] with a 40 GeV χ_1^0, a 70 GeV χ_2^0, 60 GeV stop squark, 225 GeV gluinos and 250 GeV squarks. If the lighter sparticle masses are held constant, limits may be set on the squark and gluino masses: $m_{\tilde{q}}, m_{\tilde{g}} < 200$ GeV/c^2 is excluded, and for $m_{\tilde{q}} = m_{\tilde{g}}$ the limit is 225 GeV/c^2.

3 Search for High-Mass Photon Pairs

DØ has carried out a search for massive objects decaying into two photons which might be produced in association with a vector boson:

$$p\bar{p} \rightarrow X(W/Z) \rightarrow \gamma\gamma \text{ jet jet}$$

Here X might be technipion or a Higgs with nonstandard couplings. In particular the case where X is a so-called "bosonic Higgs" was investigated: this is a Higgs with SM couplings to vector bosons but zero couplings to fermions[6]. Two photons were required with $E_T > 20$ and 15 GeV and $|\eta| < 2.0$ and 2.25 respectively; also two jets satisfying the same kinematic requirements,

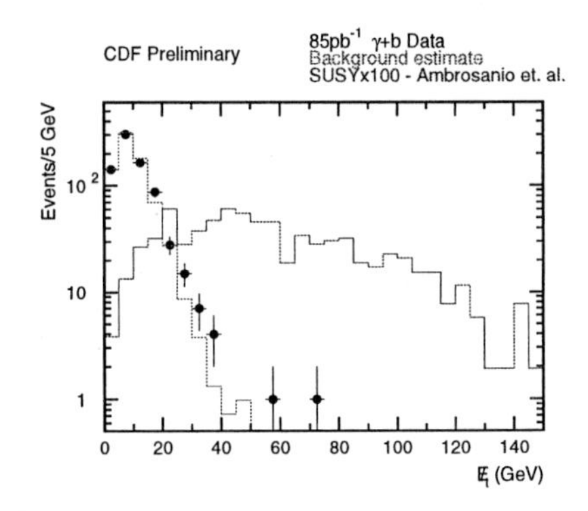

Fig. 3. Spectrum of $\not{E}_T$ for events with $\gamma+$ tagged jet from CDF.

and with a dijet mass between 40 and 150 GeV/c^2 (consistent with W or Z decay). Additionally it was required that both the diphoton system and the dijet system have $p_T > 10$ GeV/c, to select events where the diphoton and dijet recoil against each other. Seven events are observed, with 10.5 ± 4.0 expected. No indication of new physics is seen, and the case of a "bosonic Higgs" can be excluded for $m_H < 81(86)$ GeV/c^2 at 90(95)% C.L.

4 Conclusions

Photons offer productive search channels for new physics. While the CDF $ee\gamma\gamma$ event is still interesting, there is no sign of a confirming observation in any other channel. Twenty times the present dataset will be accumulated in Run II, starting in 1999; if new physics is indeed just around the corner, it will be hard to miss. For example, we might have 20 "$ee\gamma\gamma$" events in CDF and DØ by 2002.

References

[1] R. Culbertson, FERMILAB-CONF-97/277-E. to be published in the proceedings of the 5th International Conference on Supersymmetries in Physics (SUSY 97), Philadelphia, PA, May 27-31, 1997.

[2] S. Abachi *et al.* (DØ Collaboration), Phys. Rev. Lett. **78** (1997) 2070.

[3] B. Abbott *et al.* (DØ Collaboration), hep-ph/9708005, accepted for publication in Phys. Rev. Lett.

[4] J. Ellis, J. Lopez, and D. Nanopoulos, Phys. Lett. **B 394** (1997) 354.

[5] S. Ambrosanio *et al.*, Phys. Rev. Lett. **76** (1996) 3498; **77** (1996) 3502.

[6] A. Stange, W. Marciano and S. Willenbrock, FERMILAB-Pub-93/142-T; A.G. Akeroyd, Phys.Lett. **B368** (1996) 89.

SUSY Searches at the Tevatron

John Conway[1] (conway@physics.rutgers.edu)

Rutgers University, Piscataway, New Jersey, USA

Abstract. Searches continue for evidence of supersymmetric particles in both the fixed-target and collider experiments at the Tevatron at Fermilab. The results of recent searches place new limits on the masses of squarks, gluinos, charginos, neutralinos, and the charged Higgs.

1 Tevatron Runs

The CDF and D0 experiments at Fermilab amassed a very large sample of data from proton-antiproton collisions at $\sqrt{s} = 1.8$ TeV in the years 1992-1996. These data have led to many new physics results, including the discovery of the top quark [1], precision measurements of the W mass, and searches for new phenomena. This talk covers some of the recent results in the searches for supersymmetric particles at CDF and DØ, and also a new result from the KTeV fixed-target experiment which collected data in the years 1996-1997.

2 KTeV Light Gluino Search

If supersymmetry gives rise to new fermion and boson partners in nature, and R-parity is conserved, there exists the possibility that the gluino (the fermionic partner of the gluon) could form a bound state with a gluon in a relatively stable neutral bound state. [2] This state, denoted R^0, would be pair-produced strongly, and could have a lifetime of up to milliseconds.

The KTeV detector comprises a kaon decay vacuum region with an active regenerator, a charged particle magnetic spectrometer, CsI electromagnetic calorimeter, and muon shield and scintillator hodoscopes. If R^0 particles are produced in the primary target, then if their lifetime is short enough to decay before reaching the spectrometer and not so short that the decay products miss the spectrometer, their decay to (for example) ρ plus photino will give two charged pions in the final state.

Figure 1 shows the region of R^0 mass and photino mass excluded by KTeV, using one day's worth of data.[3] The plot also shows the regions excluded on cosmological grounds. With more data and refined cuts, KTeV plans to extend this search to lower R^0 masses and other decay modes.

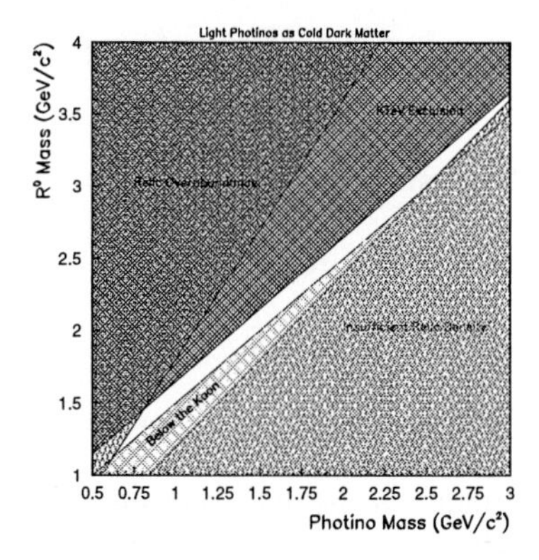

Fig. 1. KTeV and cosmological limits at 95% CL in the space of R^0 mass and photino mass.

3 Search for Squarks/Gluinos

At hadron colliders the production of squarks and gluinos proceeds via their strong coupling. The classic method is to look in final states with hadronic jets and missing transverse energy carried away by the lightest supersymmetric particle to which the squarks and gluinos ultimately decay. But there can be final states with leptons in addition, with relatively small backgrounds, which complement the search in the jets plus missing E_T final state.

The D0 collaboration has performed updated searches for squark-gluino production in the final states either having two electrons, two jets, and missing E_T, or having multiple high-E_T jets with large missing E_T. The number of events observed agrees well with the expected background from standard model processes. One can set limits in either the squark mass-gluino mass plane or in the plane of the assumed SUGRA unification masses m_0 and $m_{1/2}$, as shown in figure 2.

4 Chargino/Neutralino Search

Chargino-neutralino production in $\bar{p}p$ annihilation can lead to distinctive final states with three leptons and missing transverse energy. One lepton comes from the decay of the chargino, and two others come from the neutralino. The only standard model background which can contribute significantly to this channel is WZ production, which can be rejected with a mass requirement on same-flavor oppositely charged leptons.

CDF has updated the search for this process; the selection requirements retain no events from the full 107-pb^{-1} data sample. The remaining background processes would be expected to contribute 1.2±0.3 events. Figure 3

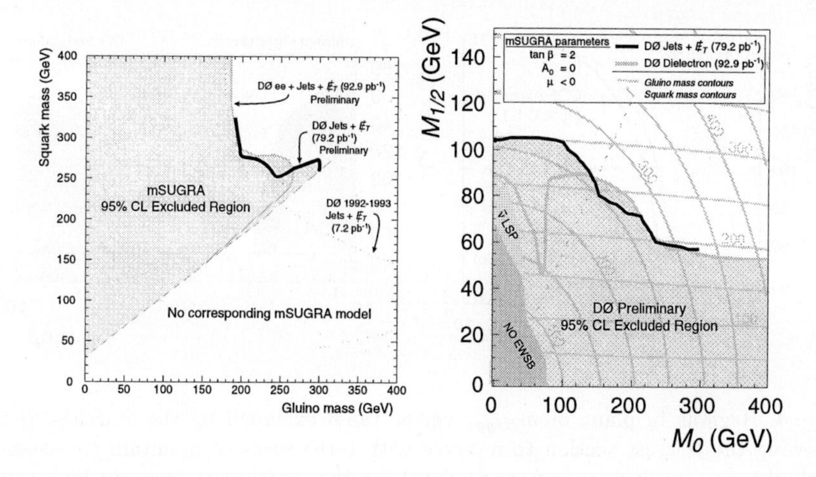

Fig. 2. Limits at 95% CL on (left) the masses of the squark and gluinos, and (right) on the parameters $m_{1/2}$ and m_0.

shows the resulting 95% CL limits on this process as a function of chargino mass, assuming the SUGRA unification mass relationships and branching ratios. The limit on chargino mass lies in the range 70-80 GeV/c^2, depending on the parameters μ and $\tan\beta$ of the model.

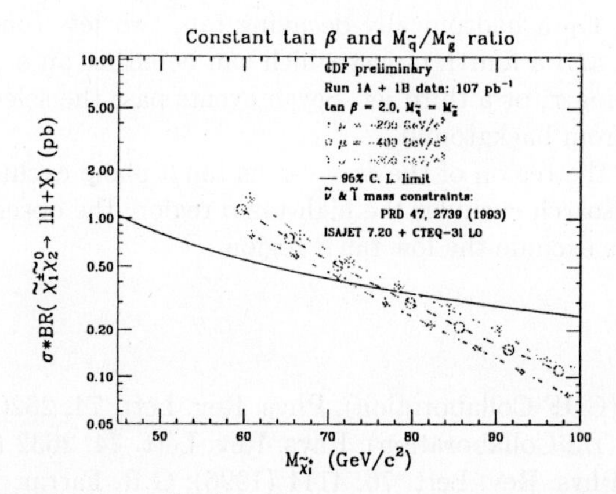

Fig. 3. CDF limits at 95% CL on chargino/neutralino production as a function of the chargino mass.

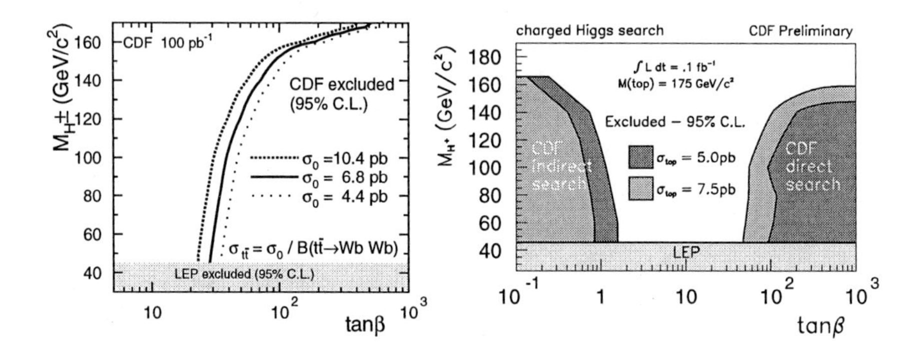

Fig. 4. Regions in plane of m_{Higgs} versus $\tan\beta$ excluded by the searches, (left) allowing the $t\bar{t}$ cross section to increase with $\tan\beta$ so as to maintain consistency with the top quark discovery, and (right) for the searches at low and high $\tan\beta$ with constant top cross sections.

5 Charged Higgs

Charged Higgs bosons arise in extensions to the standard model (such as supersymmetry) with two Higgs doublets. If the charged Higgs is less massive than the top quark, and if the parameter $\tan\beta$ (the ratio of the vacuum expectation values of the two Higgs doublets) is large, then the top can decay predominantly to the charged Higgs and a b quark. In this regime, the charged Higgs decays almost exclusively to $\tau\nu$.

CDF has performed a search for such events[4] by selecting those events with large missing E_T a hadronically decaying tau, two jets (one of which must be b-tagged), and a fourth object which can be either an e, μ, another hadronically decaying τ, or a third jet. Seven events pass the selection, with 7.4±2.0 expected from background.

Figure 4 shows the region of the m_H versus $\tan\beta$ plane excluded by the search. The direct search excludes the high-$\tan\beta$ region; the observed $t\bar{t}$ lepton plus jets events exclude the low $\tan\beta$ region.

References

[1] F. Abe *et al.* (CDF Collaboration), Phys. Rev. Lett. 74, 2626 (1995); S. Abachi *et al.* (D0 Collaboration), Phys. Rev. Lett. 74, 2632 (1995).
[2] G.R. Farrar, Phys. Rev. Lett. 76, 4111 (1996); G.R. Farrar, Phys. Rev. D 51, 3904 (1995).
[3] J. Adams, *et al* (KTeV Collaboration), Phys. Rev. Lett. 79, 4083 (1997). (See also paper 1403 in these Proceedings.)
[4] F. Abe *et al.* (CDF Collaboration), Phys. Rev. Lett. 79, 357 (1997).

Searches for Sfermions at LEP

Shoji Asai (asai@hpoputsv.cern.ch)

University of Tokyo, Japan

Abstract. The LEP combined results of searches for the scalar fermions at centre-of-mass energies of 130-172 GeV are summarised.

Since November 1995, the LEP e^+e^- collider was operated at centre-of-mass energies, $\sqrt{s}$, of 130-136, 161 and 172 GeV. Each of the four LEP experiments, ALEPH, DELPHI, L3 and OPAL collected integrated luminosities of about 5, 10 and 10 pb^{-1} at 130-136, 161 and 172 GeV, respectively. Using these data above the Z^0 peak, searches for various new particles have been performed. In this note, the LEP combined results on charged scalar leptons (sleptons) and the third generation scalar quarks (stop and sbottom) are summarised.

The supersymmetric (SUSY) standard models are most promising extensions of the standard model, because the SUSY can naturally deal with the problem of the quadratic Higgs mass divergence. In these theories, each elementary particle has a superpartner whose spin differs by 1/2 from that of the particle. The Minimal Supersymmetric Standard Model (MSSM) is a most simple and predictable SUSY model. After imposing GUTs conditions, there are only five parameters: $\tan\beta(\equiv v_2/v_1)$, M_2(SU(2) gaugino mass), μ(superpotential Higgs mass), m_0(universal scalar mass), and A(scalar trilinear coupling). This model is used to guide the analysis but more general cases are also studied.

R-parity is defined to be even for the ordinary particles and odd for the superpartners($R \equiv (-1)^{2S+3B+L}$). If R-parity is conserved, (1) SUSY particles should be produced in pairs, (2) the lightest SUSY particle (LSP) should be stable, and (3) other SUSY particles should decay finally into the LSP and ordinary particles. The candidate for the LSP is the lightest neutralino ($\tilde{\chi}^0_1$) or the scalar neutrino ($\tilde{\nu}$). Throughout this note the R-parity conservation is assumed.

1 Charged scalar leptons

Each lepton has two scalar partners, the right- and left-handed scalar leptons ($\tilde{\ell}_R$ and $\tilde{\ell}_L$), according to the helicity states of their non-SUSY partners. They could be pair-produced through γ or Z^0 exchange processes (s-channel). Scalar electrons ($\tilde{e}$) could also be produced through t-channel neutralino exchange enhancing the production cross-section compared to those for the other sleptons. Then the cross-section of $\tilde{e}$ depends on the SUSY parameters.

In general the cross-sections for right-handed sleptons are smaller than for left-handed, so only the right-handed sleptons are considered here.

Slepton decay dominantly into a lepton and the $\tilde{\chi}_1^0$, appearing in the detector as a pair of acoplanar leptons. Searches were performed for this topology. If the second lightest neutralino, $\tilde{\chi}_2^0$, is lighter than sleptons, cascade decay is also possible. But this decay is allowed in relatively small parameter regions, and the branching fraction is expected to be small for $\tilde{\ell}_R$.

No evidence of the $\tilde{\ell}$ production was observed at all four experiments [1]. Figure 1 shows the LEP combined results on the excluded regions in $(M_{\tilde{\ell}}, M_{\tilde{\chi}_1^0})$ for $\tilde{e}_R$ and $\tilde{\mu}_R$. The cross-section of $\tilde{e}_R$ and branching fraction of the cascade decay are calculated with the MSSM model, in which $\tan\beta$=1.5 and μ=-200 GeV are chosen. The branching fraction for the cascade decay is not large in this choice and the detection efficiencies for this were set to be zero conservatively. The 95% C.L. levels were calculated using the PDG method [2], and the systematic uncertainties on the signal efficiencies have been taken into account following Ref. [3]. The lower-limits on $M_{\tilde{\ell}}$ are 75 GeV for $\tilde{e}_R$ and 73 GeV for $\tilde{\mu}_R$, if $M_{\tilde{\chi}_1^0}$ is larger than 10 GeV and smaller than $M_{\tilde{\ell}} - 10$ GeV.

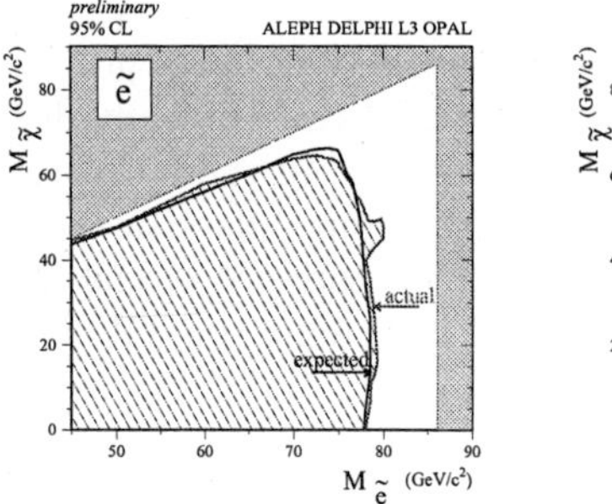

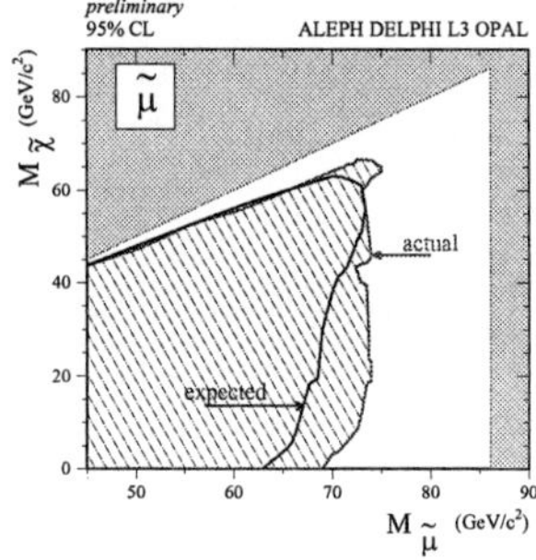

Fig. 1. Excluded regions in $(M_{\tilde{\ell}_R}, M_{\tilde{\chi}_1^0})$ plane, for $\tan\beta$=1.5 and μ=200 GeV. The hatched regions are excluded by the LEP combined results. The 'expected' show the limits calculated from the sensitivities. (These results are preliminary results.)

If the mass difference between $\tilde{\ell}$ and $\tilde{\chi}_1^0$ is smaller than 100 MeV, $\tilde{\ell}$ can not decay in the detectors. Such long-lived $\tilde{\ell}$'s are also searched for using the energy-loss measurements in the central tracking system (dE/dx) and RICH particle ID informations. No evidence was observed [4] in ALEPH, DELPHI and OPAL, and the combined lower-limits on mass of the long-lived $\tilde{\mu}_R$ is 75.6 GeV.

2 Scalar top and bottom quark

The scalar top quark ($\tilde{t}$), the bosonic partner of the top quark, can be the lightest charged SUSY particle for two reasons. Firstly, one loop radiative corrections to the $\tilde{t}$ mass through Higgsino-quark loops and Higgs-squark loops are always negative. The correction is large for a large top quark mass of about 175 GeV. Secondly, the supersymmetric partners of the left-handed and right-handed top quarks ($\tilde{t}_L$ and $\tilde{t}_R$) mix, and the resultant two mass eigenstates ($\tilde{t}_1$ and $\tilde{t}_2$) have a large mass splitting. The lighter mass eigenstate ($\tilde{t}_1 = \tilde{t}_L \cos(\theta_{\tilde{t}}) + \tilde{t}_R \sin(\theta_{\tilde{t}})$) can be lighter than any other charged SUSY particle, and lighter than the top quark itself. The $\theta_{\tilde{t}}$ can be determined by the top quark mass and the SUSY parameters and all SUSY parameters are hidden in $\theta_{\tilde{t}}$. In the case with the large $\tan\beta$, the scalar bottom quark ($\tilde{b}$) can be light for the similar reasons to $\tilde{t}$. The lighter mass eigenstate($\tilde{b}_1$) of $\tilde{b}$ is also the mixing state of $\tilde{b}_L$ and $\tilde{b}_R$, *i.e.* $\tilde{b}_1 = \tilde{b}_L \cos(\theta_{\tilde{b}}) + \tilde{b}_R \sin(\theta_{\tilde{b}})$.

The $\tilde{t}_1(\tilde{b}_1)$ is pair-produced through γ or Z^0 exchange processes. If the $\theta_{\tilde{t}}$=0.98 rad ($\theta_{\tilde{b}}$=1.17 rad), the $\tilde{t}_1$ ($\tilde{b}_1$) decouples from Z^0, and the production cross-section becomes small.

In the case that the $\tilde{\nu}$ is heavier than the $\tilde{t}_1$, the dominant decay mode of the $\tilde{t}_1$ is restricted to be the simple two body decay: $\tilde{t}_1 \to c\tilde{\chi}^0_1$ via one-loop processes. In the opposite case, the dominant decay mode is the three-body decay: $\tilde{t}_1 \to b\ell^+\tilde{\nu}$. Because the lifetime of the $\tilde{t}_1$ is much longer than the typical time scale of the hadronisation in the both cases, the $\tilde{t}_1$ would hadronise to form a $\tilde{t}_1$-hadron before it decays. The experimental signature for $\tilde{t}_1\bar{\tilde{t}}_1$ events is an acoplanar sharp two-jet topology with large transverse momentum or an acoplanar two-jet plus two leptons with large missing momentum. The $\tilde{b}_1$ decays dominantly into b-quark and the $\tilde{\chi}^0_1$, appearing in the detector as an acoplanar two-jet. The cascade decay into $b\tilde{\chi}^0_2$ is ignored in this analysis, since this decay is allowed in relatively small parameter regions.

No evidence of $\tilde{t}_1$ and $\tilde{b}_1$ was observed [5], and the combined results on the excluded regions are shown in Fig. 2. The lower-limit on $\tilde{t}_1$-mass is 75 GeV even for $\theta_{\tilde{t}}$=0.98 rad, if the $\tilde{t}_1$ decays into $c\tilde{\chi}^0_1$, and the mass difference between $\tilde{t}_1$ and $\tilde{\chi}^0_1$ is larger than 10 GeV. The $\tilde{b}_1$-mass limit is 71 GeV, if the $\theta_{\tilde{b}}$=0.0 and the mass difference between the $\tilde{b}_1$ and $\tilde{\chi}^0_1$ is larger than 7 GeV.

I would like to thank members of the LEP SUSY working group, especially, L.Pape, M.Schmitt, F.Cerutti and P.Kluit.

References

[1] ALEPH Collab., Phys. Lett. B407 (1997) 377 and Phys. Lett. B373 (1996) 246; DELPHI Collab., Phys. Lett. B387 (1996) 651; L3 Collab., CERN-PPE-97-130 (1997), submitted to Phys. Lett. B and Phys. Lett. B377 (1996) 289; OPAL Collab., Phys. Lett. B396 (1997) 301 and Z. Phys. C73 (1997) 201.

[2] Particle Data Group, Phys. Rev. D54 (1996) 159.

[3] R.D.Cousins and V.L. Highland, Nucl. Instr. Meth. A320 (1992) 331.

[4] ALEPH Collab., Phys. Lett. B405 (1997) 379; DELPHI Collab., Phys. Lett. B396 (1997) 315.

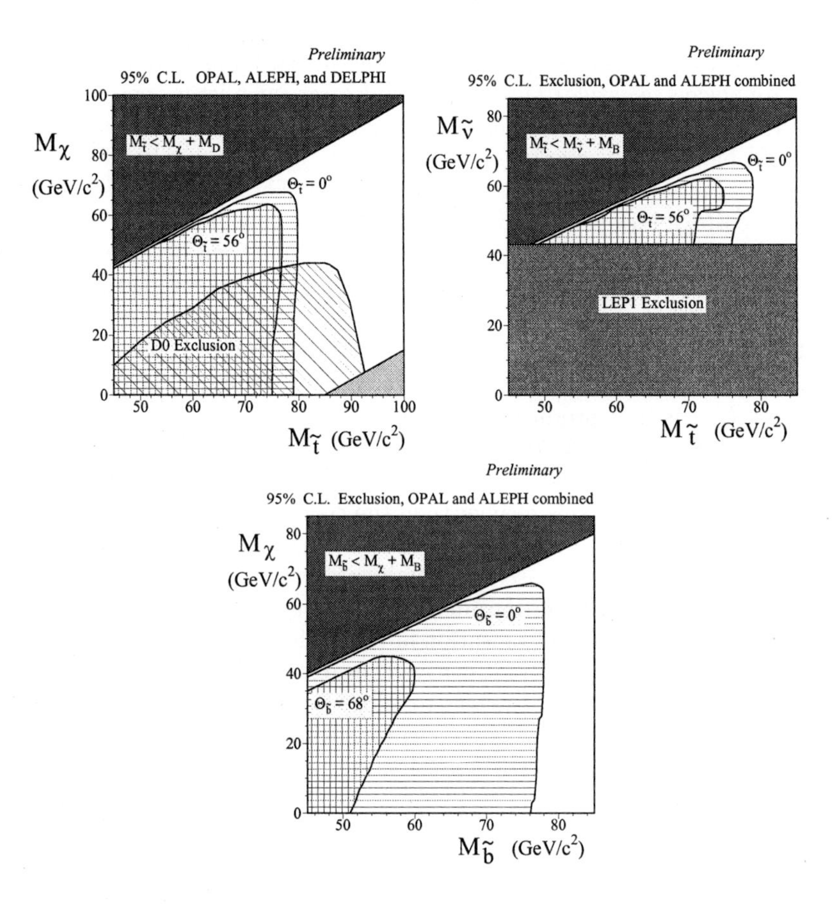

Fig. 2. (Top-left) Excluded regions in $(M_{\tilde{t}_1}, M_{\tilde{\chi}_1^0})$ plane for $\theta_{\tilde{t}}$=0.0 and 0.98 rad, in the case of $\tilde{t}_1 \to c\tilde{\chi}_1^0$. The region excluded by the D0 collaboration [6] is also shown. (Top-right) Excluded regions in $(M_{\tilde{t}_1}, M_{\tilde{\nu}})$ plane for $\theta_{\tilde{t}}$=0.0 and 0.98 rad in the light $\tilde{\nu}$ case. (Down) Excluded regions in $(M_{\tilde{b}_1}, M_{\tilde{\chi}_1^0})$ plane for $\theta_{\tilde{t}}$=0.0 and 1.17 rad. (These results are preliminary.)

[5] ALEPH Collab., CERN-PPE 97-084 (1997), submitted to Phys. Lett. B and Phys. Lett. B373 (1996) 246; DELPHI Collab., Phys. Lett.B387 (1997) 651; L3 Collab., Phys. Lett. B377 (1996) 289; OPAL Collab. Z. Phys. C75 (1997) 409, Phys. Lett. B389 (1996) 197 and Z. Phys. C73 (1997) 201.
[6] D0 Collab., Phys. Rev. Lett. 76 (1996) 2222.

Search for SUSY Signatures at LEP

Marta Felcini (Marta.Felcini@cern.ch)

Institute for Particle Physics, ETH, Zürich

Abstract. The searches for signals of SUperSYmmetry at LEP are reviewed. The results of the searches in the data collected at center-of-mass energies between 130 GeV and 183 GeV are reported and interpreted in terms of production and decay of charginos and neutralinos.

The LEP collider has started operation at energies well above the Z peak since the Autumn of 1995. At that time a test run at center-of-mass energies ($\sqrt{s}$) between 130 and 140 GeV resulted into about 6 pb^{-1} collected by each of the four LEP experiments, ALEPH, DELPHI, L3 and OPAL. In 1996 about 11 pb^{-1} and 10 pb^{-1} where collected by each experiment at $\sqrt{s}$=161 GeV and 172 GeV respectively. In 1997 the energy was raised up to 183 GeV and about 60 pb^{-1} have been collected by each experiment. The center-of-mass energy of the machine will be increased to 189 GeV in 1998, it will reach up to 200 GeV in 1999 and it will continue to run at this energy in the year 2000. The luminosity expected to be delivered to each experiment per year is $\sim$ 150 pb^{-1}. Thus LEP is at present the most powerful tool for the direct search of SUperSYmmetric (SUSY) particles, particularly the weakly interacting charginos, neutralinos and sleptons, in a mass region never explored before.

At LEP, SUSY particles, such as charginos, $\chi^{\pm}$, neutralinos χ^0 and scalar fermions, $\tilde{f} = \tilde{e}, \tilde{\mu}, \tilde{\tau}, \tilde{\nu}, \tilde{q}$, are expected to be produced in pairs through the reactions: $e^+e^- \rightarrow \chi_i^+\chi_j^-, \chi_i^\circ\chi_j^\circ, \tilde{f}\bar{\tilde{f}}$ (the indices i, j indicate the different mass states: within SUSY models with minimal particle content $i = 1, 2$ for charginos and $i = 1$ to 4 for neutralinos, the lightest state corresponding to $i = 1$, the next to lightest to $i = 2$ and so on). They should decay into known Standard Model (SM) particles plus the lightest SUSY particle (LSP). If the simplest scenario is assumed (as hereafter), namely that R-parity is conserved, the LSP is stable and neutral, so that, as a heavy neutrino, it escapes detection. Thus the "classical" signature of SUSY particle production and decay is an event with large missing energy. The experimental SUSY signatures and their backgrounds, as shown in Fig. 1, depend on the mass difference $\Delta M = M_{SP} - M_{LSP}$, where M_{SP} and M_{LSP} are the mass of the decaying SUSY particle and the mass of the lightest, stable, SUSY particle. In a model parameter independent approach all the ΔM range, from $\sim$0 GeV up to the maximum kinematically allowed, should be investigated. However, if one restricts oneself to the most popular Minimal Supersymmetric Standard Model (MSSM) and assumes the GUT relation $M_1 = 5/3 \tan\theta_W M_2$ (M_1 and M_2 being the U(1) and SU(2) gaugino mass parameters at the electroweak scale), the minimum ΔM favoured by this model is $\gtrsim 3$ GeV. If the GUT relation assumption is removed then any ΔM is allowed.

Different search strategies have been devised in different ΔM ranges. In

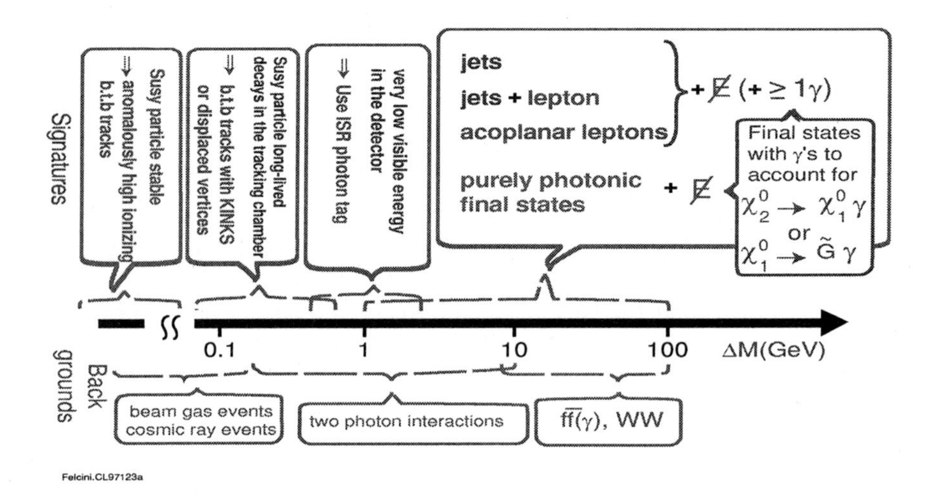

Fig. 1. Experimental SUSY signatures and dominating backgrounds as function of the mass difference $\Delta M = M_{SP} - M_{LSP}$.

the range $\Delta M \lesssim 0.1$ GeV, the SUSY particle is long-lived. The search for stable heavy charged particles is based on the identification of events with only two back-to-back tracks, with anomalously large ionization loss in the central tracking chamber [1]. Signal efficiencies between 60% and 80% have been obtained for charged particle masses between 50 and 90 GeV. No anomalous events have been found. Using all the data collected at $\sqrt{s}$ between 130 and 183 GeV, stable charged particle cross section upper limits between 50 and 200 fb have been set in the mass range between 45 and 90 GeV, excluding the existence of MSSM stable charginos within this mass range, in most of the allowed parameter space [2].

In the range 0.1 GeV $\lesssim \Delta M \lesssim$ 0.5 GeV a SUSY particle is very likely to decay inside the central tracking detector. If it is charged, both the primary and the secondary tracks are reconstructed, with an angle, 'kink', among them. The search [3] is based on the identification, in the central tracking detector, of events with two back-to-back tracks with kinks. The signal efficiency depends on the chargino decay length, which in turn depends on ΔM (larger ΔM, smaller the decay length). Efficiencies between 20% and 40% (depending on the chargino mass) are obtained in the central region of the tracking detector, reducing to few % in the inner and outer regions of this detector. No excess was observed over the expected background in the data collected at $\sqrt{s} \leq 172$ GeV. Thus, in a MSSM scenario without GUT relation, chargino mass lower limit as large as 85 GeV are set for $\Delta M \lesssim 0.2$ GeV, reducing to 45 GeV for ΔM approaching 0.5 GeV.

The range 0.5 GeV $\lesssim \Delta M \lesssim 3$ GeV is the most difficult to investigate, since the visible energy in the signal events is very small ($< 5\%\sqrt{s}$) and the background from two-photon interactions is very high. To improve trigger efficiency and suppress the large background, an initial state radiation photon with large energy and transverse momentum is required in the event [3]. This requirement reduces the signal efficiency to few %, but also reduces dramatically the background: in the data collected at $130 \leq \sqrt{s} \leq 172$ GeV, 1 event is observed, while ~ 2 are expected, mainly from two-photon physics. This result allows to set a lower limit on the chargino mass of 53.4 (54.4) GeV for $\Delta M = 0.5$ (3) GeV in the case of maximum expected chargino cross-section (chargino with large gaugino component and assuming large sneutrino mass).

In the range $\Delta M \gtrsim 3$ GeV the main signatures can be purely hadronic or purely leptonic final states, or final states with jets and an isolated lepton. They can be accompanied by one or more photons, if , for instance, the lightest neutralino is not the LSP and it decays into a gravitino plus a photon ($\chi_1^0 \rightarrow \tilde{G}\gamma$). In this case purely photonic final states (single or multi photon final states with missing energy) are also possible SUSY signatures: they are discussed in an other contribution [4] to these proceedings. The event selections are optimized in several ΔM sub-ranges, using discriminating quantities, such as acoplanarity, transverse momentum, missing mass and several others [5]. The signal detection efficiencies range from few percents in the very low ΔM (high two-photon background) and very high ΔM (high WW background) regions to 60–70% in the intermediate region 20 GeV $\lesssim \Delta M \lesssim 60$ GeV. After selections, less than 50 events (the exact number depends on the specific analysis of each experiment) are observed in total by each experiment, always in agreement (unfortunately for new physics searcher) with the expectations for Standard Model processes. This result is then translated into a 95% CL upper limit on the minimum observable cross section as function of M_{SP} and M_{LSP}. With the data collected at 183 GeV [2], chargino (neutralino) production cross sections as low as $\sim 0.5 - 1$ pb are excluded for chargino (neutralino) masses very close to the kinematic limit and for $\Delta M \sim$ 5(10) GeV. For larger values of ΔM, smaller values of the cross sections are excluded.

The upper limits on the cross sections are used to set bounds on the parameters of specific SUSY models, the most popular being the MSSM with GUT relation. Within this framework, the masses and couplings of charginos and neutralinos depend mainly on three parameters: M_2, the SU(2) gaugino mass parameter, μ, the Higgs mass parameter, and $\tan\beta$, the ratio of the vacuum expectation values of the two Higgs doublets. The cross sections of charginos and neutralinos can depend also on sneutrino and selectron masses which, in turn, depend on an additional free parameter, m_0, the common sfermion mass at the GUT scale. In particular, the chargino production cross

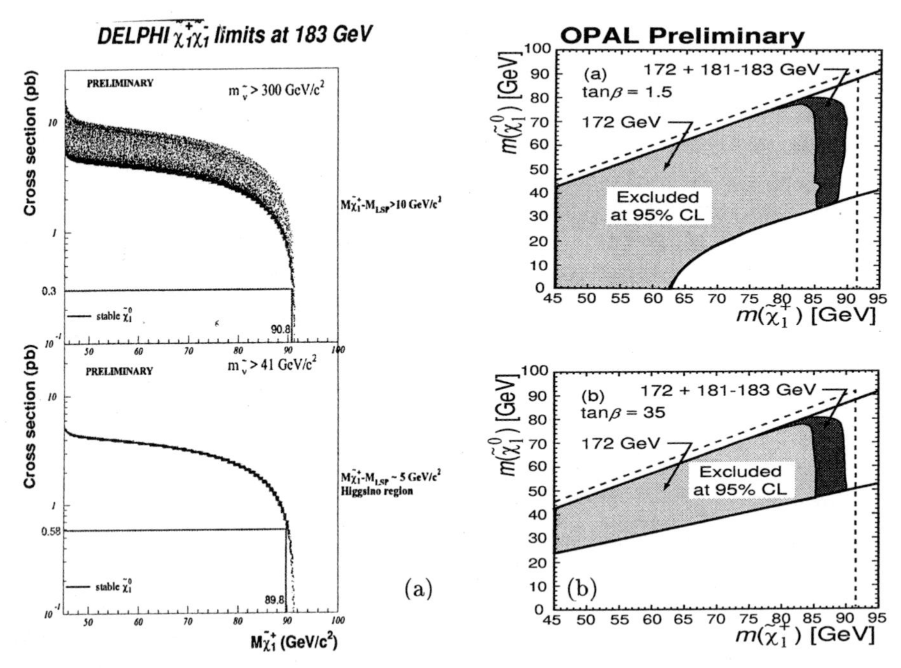

Fig. 2. (a) Expected $\chi_1^+\chi_1^-$ production cross sections (black dots) as function of the chargino mass, for different values of the MSSM parameters, and cross section upper limits set by the DELPHI experiment using the data at $\sqrt{s}$ = 183 GeV and (b) regions of the plane $M_{\chi_1^+} - M_{\chi_1^0}$ which, within the MSSM framework, for $\tan\beta$=1.5 (upper plot), or $\tan\beta$=35 (lower plot), and large sneutrino mass ($m_0 = 1$ TeV), are excluded by the OPAL experiment with the data at $\sqrt{s} \leq 183$ GeV.

section at a given energy can be greatly reduced when the sneutrino mass is below 300 GeV, if the chargino has a large gaugino component, that is in the region $|\mu| \gg M_2$ (gaugino region). In this region, for a given chargino mass, the chargino cross section is maximum and indepenpendent of sneutrino mass, for $m_{\tilde{\nu}} > 300$ GeV. Smaller cross sections, but independent of $m_{\tilde{\nu}}$, for any $m_{\tilde{\nu}}$ value, are obtained in the so called Higgsino region $M_2 \gg |\mu|$. In Fig. 2(a) we see the MSSM chargino cross sections (black dots) expected at $\sqrt{s} = 183$ GeV [2], as a function of the chargino mass, for different sets of the parameters μ, M_2 and $\tan\beta$. Also shown are the minimum values of cross section upper limits (horizontal lines) set by the DELPHI experiment with the data collected at 183 GeV, for $\Delta M > 10$ GeV and $m_{\tilde{\nu}} > 300$ GeV (upper plot), resulting into a lower limit $M_{\chi^+} > 90.8$ GeV, as well as for $\Delta M \sim 5$ GeV and $m_{\tilde{\nu}} > 41$ GeV (lower plot), resulting into a lower limit $M_{\chi^+} > 89.8$ GeV. In Fig.2(b) the regions in the plane $M_{\chi_1^+} - M_{\chi_1^0}$ excluded by the OPAL experiment with the data collected at $\sqrt{s} \leq 183$ GeV [2] is shown for $m_0 = 1$ TeV (large sneutrino masses) and for $\tan\beta = 1.5$ (upper plot) or $\tan\beta = 35$

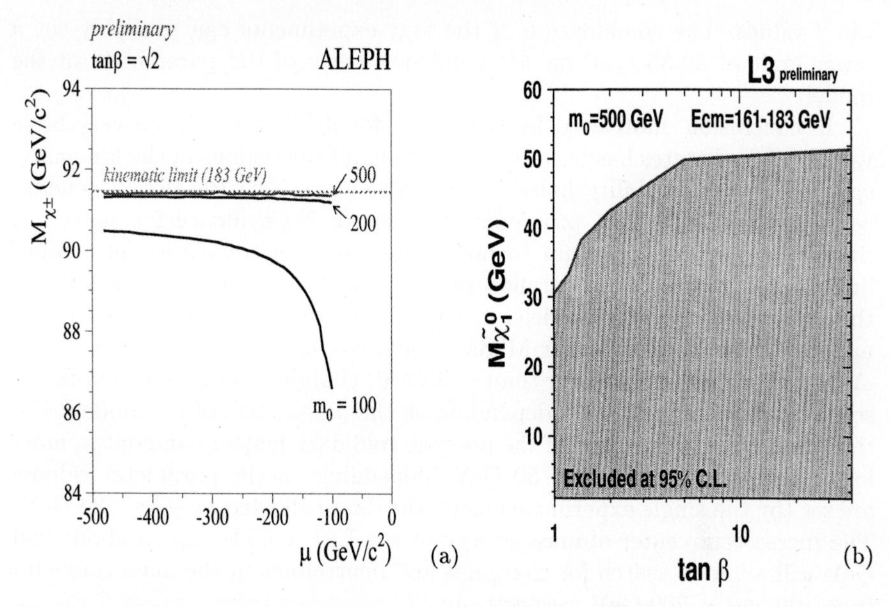

Fig. 3. (a) Chargino mass lower limit dependence on the MSSM parameters μ and m_0 in the gaugino region ($|\mu| > M_2$) by the ALEPH experiment with the data at $\sqrt{s}$ = 183 GeV and (b) neutralino mass lower limit as function of $\tan\beta$ and for large sneutrino mass (m_0 = 500 GeV), set by the L3 experiment using the data collected at $161 \leq \sqrt{s} \leq 183$ GeV.

(lower plot). The light grey and dark grey areas show the regions excluded by the data at $\sqrt{s} \leq 172$ GeV and ≤ 183 GeV, respectively. The dashed lines show the kinematic limits for chargino production and decay. The thick solid lines in these plots delimit the regions (in white) theoretically unaccessible in the MSSM. We see that for neutralino mass $\sim$ 80 GeV the chargino is excluded up to 85 GeV while for lower neutralino masses charginos up to $\sim$ 90 GeV are excluded. Fig. 3(a), by the ALEPH experiment using the 183 GeV data [2], shows the dependence on m_0 of the chargino mass lower limit in the gaugino region. We see that for large m_0 values this lower limit is very close to the kinematic limit of 91.5 GeV, independently of μ. For smaller m_0 values (that is small sneutrino masses) the limit is weaker (for m_0 = 100 GeV μ = -100 GeV it is reduced to $\sim$ 86 GeV). A lower limit on the lightest neutralino mass can be derived, combining all the results for chargino, neutralino and slepton searches. Fig.3(b) shows the $M_{\chi_1^0}$ lower limit as function of $\tan\beta$, for m_0 = 500 GeV, derived by the L3 experiment using the data collected at 161 GeV $\leq \sqrt{s} \leq$ 183 GeV [2]. The $M_{\chi_1^0}$ lower limit independent of $\tan\beta$ is 30 GeV, while, if $\tan\beta > 5$, it reaches 50 GeV. For small m_0 values ($m_0 \lesssim 100$ GeV), this limit is weaker but still is expected to be (analyses are still in progress) around 20-25 GeV (for the single experiment), for any m_0 and

$\tan\beta$ values. The combination of the four experiments can probably set a lower limit of 30-35 GeV on $M_{\chi_1^0}$, independently of the parameters of the model.

In conclusion, motivated by the search for SUSY particles, a very large variety of final states has been investigated in all the regions of the kinematic space. The "detectability holes", where SUSY, or New Physics in general, could escape, have been progressively reduced. No evidence for signals of SUSY has been detected up to now. This result is translated into upper limits on chargino and neutralino production cross sections, as well as, in the framework of SUSY models, in lower limits on chargino and neutralino masses. Thus, in the popular MSSM scenario with GUT relation, in which ΔM is expected to be larger than ~ 3 GeV, chargino mass lower limits are set between 85 and 91 GeV (depending on the parameters of the model). For the lightest neutralino, one of the possible cold dark matter candidates, mass lower limits between 20 and 50 GeV (depending on the parameter values) are set (by the single experiment) using the data collected at $\sqrt{s} \leq 183$ GeV. The increase in center-of-mass energy of the LEP collider up to about 200 GeV will allow to search for charginos and neutralinos in the mass range up to ~ 100 and ~ 200 GeV, respectively.

References

[1] Details of the analyses can be found in the publications by the experiments: ALEPH Coll., Phys. Lett. B405 (1997) 379; DELPHI Coll., Phys. Lett. B396 (1997) 315.

[2] The results concerning the 183 GeV data are preliminary. They were presented by ALEPH, DELPHI, L3 and OPAL Collaborations, in the Open presentations at the LEPC meeting, 11 Nov. 1997.

[3] DELPHI Coll., "Search for charginos mass degenerate with the LSP", Paper # 427 submitted to this Conference.

[4] M. Chemarin, "Search for anomalous photonic events at LEP", these proceedings.

[5] Details of the analyses can be found in the publications by the experiments: ALEPH Coll., "Searches for Charginos and Neutralinos in e^+e^- Collisions at $\sqrt{s} = 161$ and 172 GeV", CERN PPE/97-128, submitted to Eur. Phys. Jour. C; DELPHI Coll., "Search for charginos, neutralinos and gravitinos at LEP" CERN PPE/97-107, submitted to Eur. Phys. Jour. C; L3 Coll., "Search for Scalar Leptons, Charginos and Neutralinos in e^+e^- collisions at $\sqrt{s} =$ 161-172 GeV", CERN-PPE/97-130, submitted to Eur. Phys. Jour. C; OPAL Coll., "Search for Chargino and Neutralino Production at $\sqrt{(s)}=$ 170 and 172 GeV at LEP". CERN-PPE/97-083, submitted to Eur. Phys. Jour. C; OPAL Coll., Phys. Lett. B389 (1996) 616.

Search for the Standard Model Higgs Boson at LEP

William Murray[1] (w.murray@rl.ac.uk)

Rutherford Appleton Laboratory, Chilton, Didcot, Oxon, OX11 0QX, UK

Abstract. The results of the searches for the Higgs boson made by the 4 LEP experiments in data between 161 and 172GeV are presented. Also the preliminary results of combining these measurements which gives an improved mass limit of $77GeV/c^2$. Finally, there are is a taste of the 1997 data.

1 Standard Model Higgs at LEP 2

The main production process is Higgs-strahlung, as at LEP 1, but WW and ZZ fusion can contribute up to 10%. There is therefore a significant cross-section for Higgs masses up to nearly $\sqrt{(s)} - M_z$, which is around $80GeV/c^2$ at 172GeV. The typical cross-sections are of order 0.5pb when 10GeV below this cut-off.

If the Higgs mass is around $70GeV/c^2$, the branching ratio into $b\bar{b}$ is around 85%, with decays to $\tau^+\tau^-$ being 8% and the remainder nearly all other hadronic modes. The experiments search for a Z^0 in ll, $\nu\nu$, $q\bar{q}$ or $\tau\tau$ with a hadronically decaying Higgs, and Z^0 to $q\bar{q}$ with $\tau\tau$ from the Higgs. 64.6% of the rate is in the four jet channel, and the Z^0 to $\nu\nu$ at 20% gives the next largest. The leptonic channels have low rates, but are very clean, and can have better mass resolution.

2 Individual results at 161 and 172GeV

All four experiments took around $10pb^{-1}$ at each of 161 and $172GeV$ in 1996.

2.1 OPAL results[1]

These include one particularly interesting candidate in the four jet channel, taken at 172GeV. The expected background is 0.88 events in this channel.

This event contains two jets compatible with a Z, making a mass of $96.4GeV/c^2$ in a 4C fit. The Higgs candidate jets, selected as giving the best kinematic fit probability, have a mass of $75.6GeV/c^2$.

Figure 1 shows the limit in numbers of events versus Higgs mass, and the predicted numbers for different data sets. The OPAL limit is $69.4GeV/c^2$

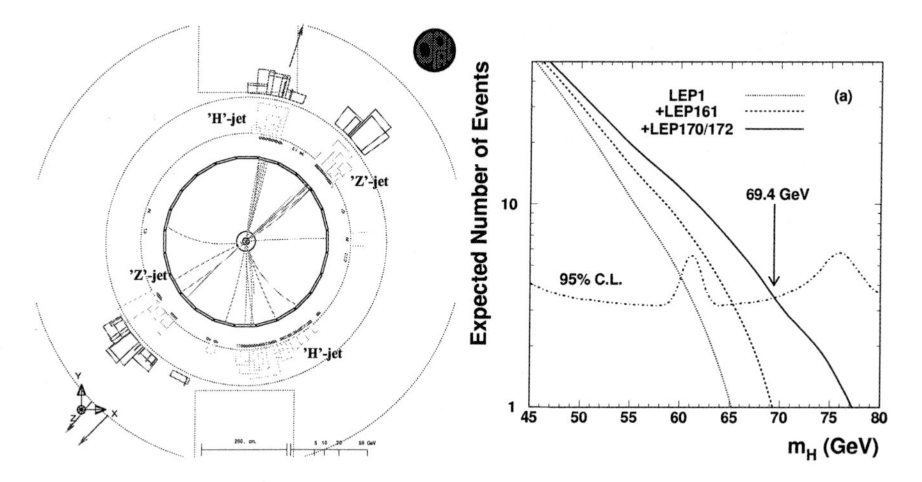

Fig. 1. *Left: OPAL candidate at 170-172 GeV energies. Right: OPAL limits with various standard Model event rates superimposed.*

2.2 ALEPH results[3]

No candidates were observed at 161 or 172GeV. The efficiency for the Hll is very good, around 75%, The four jet efficiency is not, but the background is very low. This was automatically selected to optimise the expected limit.

The ALEPH result is $69.5GeV/c^2$ using LEP 2 data alone, and $70.7GeV/c^2$ when combined with LEP 1 results. This is the best single limit achieved.

2.3 DELPHI results[2]

DELPHI found 2 candidates, one at 161 GeV with mass $64.6GeV/c^2$ in H$\nu\nu$ and one at 172GeV, mass $58.7GeV/c^2$ in four jets, in agreement with background.

The limit is $66.2GeV/c^2$ at 95% C.L.; this is significantly reduced by the candidate at the limit.

2.4 L3 results[4]

33 candidates are selected, but each has a weight depending upon how Higgs-like it appears. This weight is calculated from the b-tag output, kinematic information and the consistency of the measured mass with the Higgs mass hypothesis. The combined weight is used in the likelihood ratio to extract the Higgs cross-section limit.

L3 deduce a limit is $69.5GeV/c^2$, using all data.

2.5 Performance summary

	ALEPH	DELPHI	OPAL	L3
Efficiency	29%	30%	33%	55%
Signal expected	2.34	2.29	2.43	4.03
Background	0.48	2.57	2.04	17.8
LEP 1&2 limit	70.7	N.A.	69.5	69.4
LEP 2 only limit	69.5	65.8	69.3	N.A.
Expected limit (using DELPHI method)	68.8	65.2	66.1	65.7

The efficiencies and backgrounds correspond to a $70GeV/c^2$ Higgs at 172GeV. All experiments have found better than average limits.

3 The Combined Limit

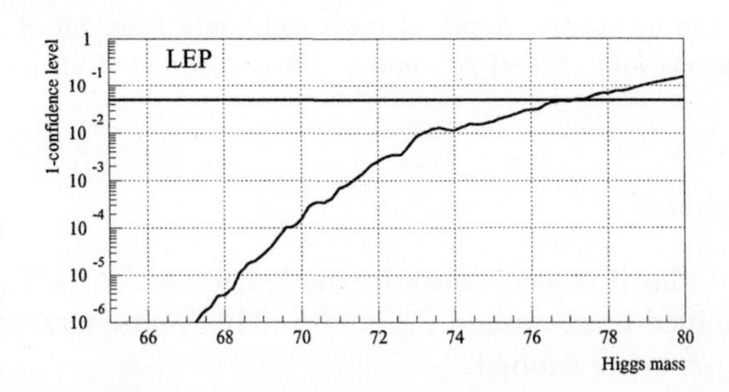

Fig. 2. *Confidence level to which background hypothesis is excluded, versus Higgs mass, for the four experiments.*

The LEP Higgs W.G.[5] has combined these results using 4 techniques, all giving broadly comparable results. This limit is shown in figure 2. The compatibility of the background observed with that predicted was checked. The backgrounds are generally lower than expected, but not unreasonable. The 1996 data gives a (preliminary) limit of $77GeV/c^2$, at 95%CL

4 First results from 1997

LEP is running at around 183 GeV. We hope for 50pb^{-1}, which could give a $90GeV/c^2$ Higgs limit, as shown in figure 3[6].

Around 5pb^{-1} were available at the time of the conference, and ALEPH, DELPHI and L3 quoted improved limits of 74.0, 73 and $72GeV/c^2$, respectively. [1] One candidate, a $qq\nu\nu$ event from L3, is shown in figure 3.

[1] At the LEPC meeting on 11th November 1997, preliminary limits were reported from around $50pb^{-1}$ by the four experiments. The highest limit was $88.6GeV/c^2$, from ALEPH.

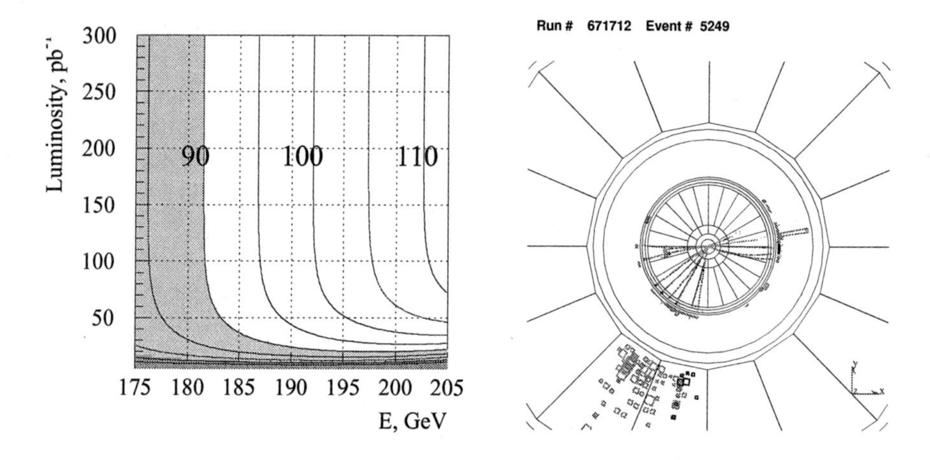

Fig. 3. *Left: The luminosity required to exclude a particular Higgs boson mass at LEP, as a function of energy. Right: A $qq\nu\nu$ candidate from L3 at 183 GeV beam energy. The two jets have $81 GeV/c^2$ mass, which suggests $We\nu$ as an alternative hypothesis.*

5 Conclusions

The experiments have placed limits in the range 66.2 - $70.7 GeV/c^2$. These will give a combined limit around $77 GeV/c^2$. This years data will see masses around $90 GeV/c^2$ being probed.

LEP will improve these results in future years. If run at 200GeV in 1999 and 2000, the sensitivity should go to over $100 GeV/c^2$

References

[1] Ackerstaff et al., CERN-PPE-97-115, Submitted to Z. Phys., C

[2] Abreu et al., "Search for neutral and charged Higgs bosons in e^+e^- collisions at $\sqrt{s} = 161 GeV$ and $172 GeV$", CERN-PPE-97-085, Subm. to Z. Phys., C

[3] Barate et al., ALEPH collaboration. "Search for the Standard Model Higgs Boson in e^+e^- Collisions at $\sqrt{s}$=161, 170 and 172 GeV. " CERN-PPE-97-070, Subm. to Phys. Lett., B.

[4] Acciarri et al. L3 Collaboration. "Search for the Standard Model Higgs Boson in e^+e^- Interactions at $161 \leq \sqrt{s} \leq 172$ GeV." CERN-PPE-97-081, Subm. to Phys. Lett., B .

[5] "Lower bound for the SM Higgs boson mass: combined result from the four experiments." CERN/LEPC 97-11, LEPC/M 115

[6] Allanach et al. "Report of the 1997 LEP2 Phenomenology Working Group on "Searches"." RAL-TR-97-037; CERN-TH-97-234. (hep-ph/9708250), Subm. to J. Phys., G.

Search for Higgs Bosons Beyond the Standard Model at LEP

Yibin Pan (Yibin.Pan@cern.ch)

Department of Physics, University of Wisconsin
1150 University Ave. Madison, WI 53706, USA

Abstract. Recent results of searches for Higgs bosons in MSSM, charged Higgs and Higgs bosons with rare decays are summarized. These results are based on data and analyses from the four LEP experiments: ALEPH, DELPHI, L3, and OPAL. The data were collected during the summer and fall of 1996 with center-of-mass energies of 161 and 172 GeV.

1 Introduction

One of the major objectives of the LEP program is the search for new phenomena beyond the standard model. In the summer of 1996, LEP was upgraded (LEP2), and, for the first time, a center-of-mass energy above the threshold for the production of W pairs was achieved. The higher center-of-mass energy provides a great opportunity to search for new phenomena.

In 1996, two sets of data were collected by each of the four LEP experiments: one at a center-of-mass energy of 161 GeV and the other at 172 GeV. An integrated luminosity of about 10 to 11 pb^{-1} per experiment was collected at each of these two energy points. The results presented in this review are based on these 1996 LEP2 data.

This review summarizes recent results of searches performed by the four LEP experiments; it includes searches for neutral Higgs bosons in MSSM, charged Higgs bosons and Higgs bosons with rare decays.

2 Neutral Higgs Bosons

In minimal extensions of the Standard Model, two Higgs doublets are introduced in order to give masses to up-type quarks and down-type quarks separately. In these models, the Higgs sector therefore consists of five physical states, namely three neutral bosons – two CP-even h and H, and one CP-odd A – and a pair of charged bosons $\mathrm{H}^{\pm}$. Six independent parameters are required: four Higgs boson masses, the ratio $v_1/v_2 \equiv \tan\beta$ of the vacuum expectation values of the two Higgs doublets, and α, the mixing angle in the CP-even sector.

Predictions can therefore only be made in specific models, of which the most popular is the Minimal Supersymmetric extension of the Standard

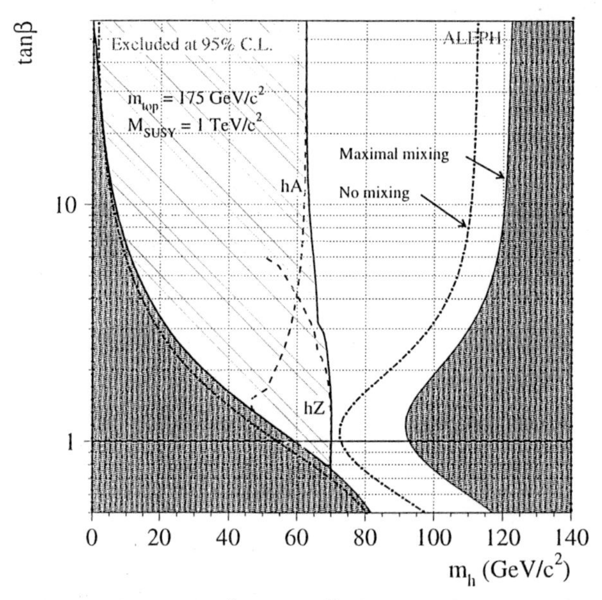

Fig. 1. ALEPH results on the $[m_h, \tan\beta]$ plane in the maximal mixing configuration. The dark area are theoretical disallowed. The hatched area is excluded at the 95% confidence level by the combined search for $e^+e^- \longrightarrow hA$ and $e^+e^- \longrightarrow hZ$. The dot-dashed lines show the change in the theoretical region in the no mixing configuration.

Model (MSSM). In this model, both H and $H^\pm$ are predicted to be too heavy to be discovered at LEP2. The analysis presented here is consequently restricted to the search for the lighter Higgs bosons, h and A, which can be produced by two complementary processes, the Higgs-strahlung process $e^+e^- \longrightarrow hZ$ with a cross section proportional to $\sin^2(\beta-\alpha)$ and the associated pair-production $e^+e^- \longrightarrow hA$ with cross section proportional to $\cos^2(\beta-\alpha)$.

At tree-level, only two parameters are needed to dertermine all the other relevant quantities (masses, couplings, and therefore cross sections). Here, these are chosen to be $\tan\beta$ and the mass m_h. For $m_h = 60$ GeV/c^2 and $\tan\beta = 10$ the branching fraction of h and A to $b\bar{b}$ is 92% and to $\tau^+\tau^-$ is 8%, giving $b\bar{b}b\bar{b}$ final states in 84% of the events and $\tau^+\tau^- b\bar{b}$ in 14%. Analyses of both these channels are performed.

Given the fact that neutral Higgs bosons decay predominantly to two b jets, all four experiments have used effective b-tagging algorithms in the analyses. b jets are identified mainly by exploiting the longer lifetime and higher mass of b hadrons compared to other hadrons, but also by the presence in the jets of high p_T leptons from semileptonic decays. To use the lifetime information in a given jet, track impact parameters and secondary decay vertices are reconstructed relative to an event-by-event interaction point. As an example, in ALEPH three variables which discriminate between b jets and light quark jets are combined using neural networks to tag b quark jets. The first two variables are lifetime-based; the third is based on the transverse momentum of identified leptons.

	AELPH	DELPHI	L3	OPAL
m_h >	62.5	59.5	58.4	56.1
$m_{H^\pm}$ >	52.0	54.5		52.0

Table 1. Lower mass limits (in GeV/c^2) at 95% C.L. for neutral Higgs bosons in MSSM and charged Higgs bosons

No significant excess of events from the search of the process of $e^+e^- \longrightarrow hA$ were observed. The results are combined with the results of the search for the Higgs-strahlung process $e^+e^- \longrightarrow hZ$. Table 1 gives the limits on m_h for the four LEP experiments [1, 2, 3, 4]. The highest individual limit was at 62.5 GeV/c^2. Fig.1 gives the 95% C.L. excluded region in the $[m_h, \tan\beta]$ plane for ALEPH.

For the more general models containing two Higgs doublets, the MSSM constraints do not exist. Figs.2 and 3 show the DELPHI results [5] of the excluded regions for CP-conserving and non CP-conserving scenarios respectively.

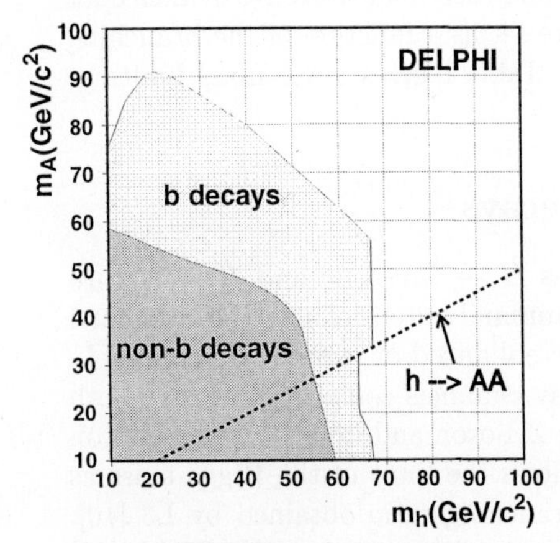

Fig. 2. DELPHI results of excluded regions at the 95% CL in the CP-conserving model with dominant b-decays (light grey) and non-b decays (dark grey).

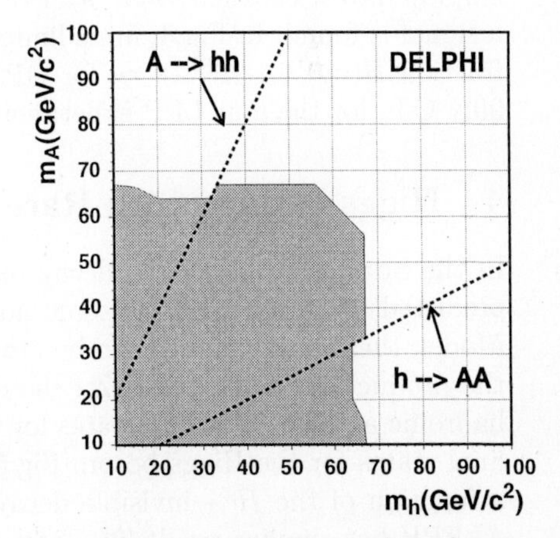

Fig. 3. DELPHI results of excluded regions at the 95% CL in the non CP-conserving model.

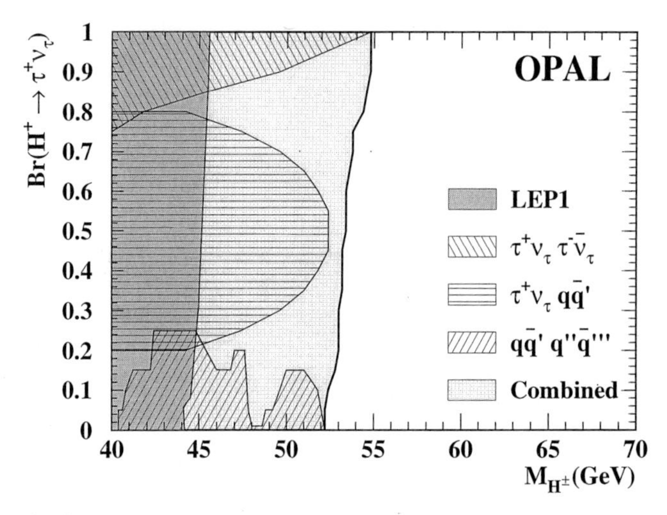

Fig. 4. OPAL limitsat 95% CL on the mass of charged Higgs bosons as a function of $B_r(\tau\nu)$.

3 Charged Higgs Bosons

Charged Higgs pair production has also been searched for at LEP [6, 7, 8]. The analyses are performed in a model independent way. Only two decay modes $\tau\bar{\nu}$, $c\bar{s}$ of $\mathrm{H}^\pm$ are assumed. As a result, in each experiment three analyses are employed to select the $\tau^+\nu_\tau\tau^-\bar{\nu_\tau}$, $c\bar{s}\tau^-\bar{\nu_\tau}$and $c\bar{s}\bar{c}s$ final staes. No evidence for a signal is found. In Fig.4, mass limits are set as a function of the branching fraction $B(\tau\nu)$ for $\mathrm{H}^\pm \rightarrow \tau\nu$ for OPAL. Table 1 gives lower mass limits at 95% C.L. for the four LEP experiments.

4 Higgs Bosons with Rare Decays

In the Standard Model, the decay modes $H \rightarrow \textit{invisible}$ and $H \rightarrow \gamma\gamma$ are predicted to be rare. But in some non-minimal extensions to the Standard Model, these decay modes can be greatly enhanced or even be dominant. In the production of $\mathrm{e}^+\mathrm{e}^- \longrightarrow \mathrm{hZ}$, the decay channels considered involve both hadronic and leptonic final states for the Z boson and invisible or di-photon final states for the Higgs boson. Fig.5 shows the limit of the Higgs mass as a function of the $H \rightarrow$invisible decay branching ratio obtained by L3 [10]. ALEPH has similiar result [9]. Fig.6 gives the limit obtained by OPAL [11] on $B_r(H^0 \rightarrow \gamma\gamma)$.

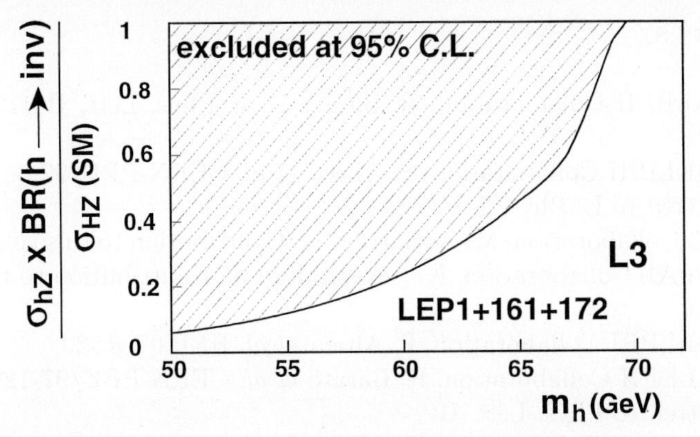

Fig. 5. Upper limits on the rate of invisible Higgs decays, relative to the standard Model Higgs production rate by L3.

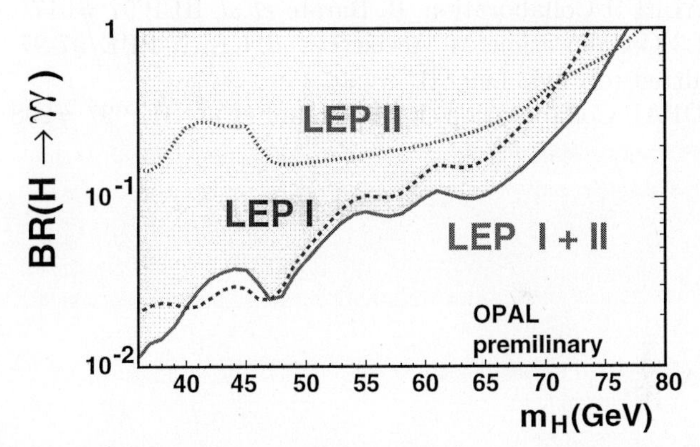

Fig. 6. OPAL 95% C.L. upper limit on $B(H^0 \to \gamma\gamma)$ for Standard Model Higgs boson production using data from LEP1 (dashed line), LEP2 (dotted line), and all data combined (solid line).

References

[1] The ALEPH Collaboration, R. Barate *et al,* Phys. Lett. B 412 (1997) 173-188
[2] The DELPHI Collaboration, P. Abreu *et al,* CERN-PPE/97-85 Submitted to E. Phys. J. C
[3] The L3 Collaboration, M. Acciarri *et al,* Contribution to this conference.
[4] The OPAL Collaboration, K. Ackerstaff *et al,* Contribution to this conference.
[5] The DELPHI Collaboration, P. Abreu *et al,* HEP'97 #529
[6] The ALEPH Collaboration, R. Barate *et al,* CERN-PPE/97-129 Submitted to Phys. Lett. B
[7] The DELPHI Collaboration, P. Abreu *et al,* CERN-PPE/97-145 Submitted to Phys. Lett. B
[8] The OPAL Collaboration, K. Ackerstaff *et al,* CERN-PPE/97-168 Submitted to Phys. Lett. B
[9] The ALEPH Collaboration, R. Barate *et al,* HEP'97 #617
[10] The L3 Collaboration, M. Acciarri *et al,* CERN-PPE/97-97 Submitted to Phys. Lett. B
[11] The OPAL Collaboration, K. Ackerstaff *et al,* HEP'97 #208

Part XIV

Astroparticle Physics and Cosmology

Improved Constraints on the Neutralino Cold Dark Matter Candidate from the Search for Chargino and Neutralino Production Using the OPAL Detector at 170 ÷ 172 Gev at LEP

Homer Neal,Jr. (hneal@hpopcw.cern.ch)

CERN, Switzerland

Abstract. New limits on the cold dark matter candidate, the lightest neutralino ($\tilde{\chi}_1^0$), are obtained from a search for charginos and neutralinos using a data sample of 10.3 pb^{-1} at the center-of-mass energies of $\sqrt{s}$ =170 and 172 GeV with the OPAL detector at LEP.

The observed luminous matter accounts for <1% of the critical closure density of the universe [1]. A flat universe (currently favored) requires a matter density equivalent to the closure density. The problem can be solved by the existence of dark matter. Nucleosynthesis calculations show that most of the dark matter must be of non-baryonic, massive, stable, weakly interacting particles. The particle called the lightest neutralino which appears in super-symmetric theories (SUSY) is an excellent candidate for this cold dark matter [2]. Neutralinos are the mass eigenstates, $\tilde{\chi}_i^0$, formed by the mixing of the field of the fermionic partners of the γ and the Z boson. In this analysis [3], new limits on the neutralino cold dark matter are obtained from searches for the production of neutralinos and charginos in e^+e^- collisions at the OPAL [4] experiment at LEP. Data are used from the October-November 1996 run at centre-of-mass energies ($\sqrt{s}$) of 170 and 172 GeV, to obtain more stringent exclusion mass and cross-section limits as compared to the previous results from the analysis of data near the Z peak (LEP1), at $\sqrt{s}$ = 130 GeV and 136 GeV (LEP1.5) and at $\sqrt{s}$ = 161 GeV by the OPAL and the other LEP collaborations [5].

R-parity conservation is assumed. Therefore, the lightest supersymmetric particle (LSP) is stable. This is a necessary conditions for the neutralino to be a source of dark matter. It escapes detection due to its weakly interacting nature. Due to the energy and momentum carried away by the LSP (and possible neutrinos), the experimental signature for the signal events is large missing energy and large missing momentum transverse to the beam axis.

The results are presented in the context of the Constrained Minimal Supersymmetric Standard Model (CMSSM). Within this model, a common gaugino (supersymmetric partners of the gauge bosons) mass, $m_{1/2}$, and a common sfermion mass, m_0, at the GUT scale is assumed. The mass spectra and couplings of charginos and neutralinos are mainly determined by the ratio of the vacuum expectation values of the two Higgs doublets ($\tan\beta$), the

U(1) and SU(2) gaugino mass parameters at the weak scale (M_1 and M_2), and the mixing parameter of the two Higgs doublet fields (μ). Furthermore, M_2 and $m_{1/2}$ are related by $M_2 = 0.82\ m_{1/2}$.

Neutralino pairs ($\tilde{\chi}_1^0\tilde{\chi}_2^0$) could be produced through an s-channel virtual Z or γ, or by t-channel scalar electron (selectron, $\tilde{e}$) exchange. The $\tilde{\chi}_2^0$ would decay into the final states $\tilde{\chi}_1^0\nu\bar{\nu}$, $\tilde{\chi}_1^0\ell^+\ell^-$ or $\tilde{\chi}_1^0 q\bar{q}$.

Charginos could be pair-produced in e^+e^- collisions through a γ or Z in the s-channel or through scalar electron neutrino (electron sneutrino, $\tilde{\nu}_e$) exchange in the t-channel. The lightest chargino $\tilde{\chi}_1^+$ could decay into $\tilde{\chi}_1^0\ell^+\nu$, or $\tilde{\chi}_1^0 q\bar{q}'$, via a virtual W, scalar lepton (slepton, $\tilde{\ell}$), sneutrino ($\tilde{\nu}$) or scalar quark (squark, $\tilde{q}$). In much of the MSSM parameter space $\tilde{\chi}_1^+$ decays via a virtual W are dominant.

Limits are presented for $\tan\beta$ from 1.0 to 35. For $\tan\beta$ close to 1.0, the ordinary analysis based on large missing momentum is insensitive in the region $M_2 \approx \mu \approx 0$ of the (M_2,μ) plane. However, at $\sqrt{s}$ = 172 GeV, the sum of the cross-sections for the four chargino pair production processes ($e^+e^- \rightarrow \tilde{\chi}_1^+\tilde{\chi}_1^-$, $\tilde{\chi}_2^+\tilde{\chi}_2^-$ and $\tilde{\chi}_1^\pm\tilde{\chi}_2^\mp$) near $M_2 = \mu = 0$ is as large as 6 pb, which is comparable to the W^+W^- production cross-section which is about 13 pb. Since, in the (M_2,μ) region considered here, the event shapes of the chargino pair events ($\tilde{\chi}_1^+\tilde{\chi}_1^-$, $\tilde{\chi}_2^+\tilde{\chi}_2^-$ or $\tilde{\chi}_1^\pm\tilde{\chi}_2^\mp$) are similar to those of ordinary W^+W^- events, a search for an excess of W^+W^--like events with respect to the Standard Model expectation (mainly pairs of W bosons) is performed.

1 Analyses and Limits

The overall efficiencies for $\tilde{\chi}_1^+\tilde{\chi}_1^-$ events are 29–57% if the mass difference between $\tilde{\chi}_1^\pm$ and $\tilde{\chi}_1^0$ is $\geq$ 10 GeV and $\leq m_{\tilde{\chi}_1^\pm}/2$. No events were observed to pass all the cuts in either category. This is consistent with the total background expected for 10.3 pb^{-1} which is 0.87 events.

The overall efficiencies for $\tilde{\chi}_2^0\tilde{\chi}_1^0$ events are 29–43% if $m_{\tilde{\chi}_2^0} - m_{\tilde{\chi}_1^0} \geq 20$ GeV and $\leq$ 70 GeV. No events were observed to pass all the cuts. This is consistent with total background expected for this search (0.96 events).

For the analysis of W^+W^--like events, the overall efficiencies for $\tilde{\chi}_1^+\tilde{\chi}_1^-$ events are 35–45%. The total number of expected background events is 17.4, while 15 events are observed.

A model-independent interpretation is formed by calculating the 95% confidence level (C.L.) upper limits on the production cross-sections for $\tilde{\chi}_1^+\tilde{\chi}_1^-$ and $\tilde{\chi}_1^0\tilde{\chi}_2^0$ assuming the specific decay modes $\tilde{\chi}_1^\pm \rightarrow \tilde{\chi}_1^0 W^{*\pm}$ and $\tilde{\chi}_2^0 \rightarrow \tilde{\chi}_1^0 Z^*$. From the observation of no events at $\sqrt{s} = 170 - 172$ GeV in the analyses, excluding the special analysis of W^+W^--like events, and using the signal detection efficiencies and their uncertainties, exclusion regions are determined using the procedure outlined in [6], and incorporating systematic errors following the method given in [7] by numerical integration, assuming Gaussian errors. To compute the 95% C.L. upper limits, the previous results obtained

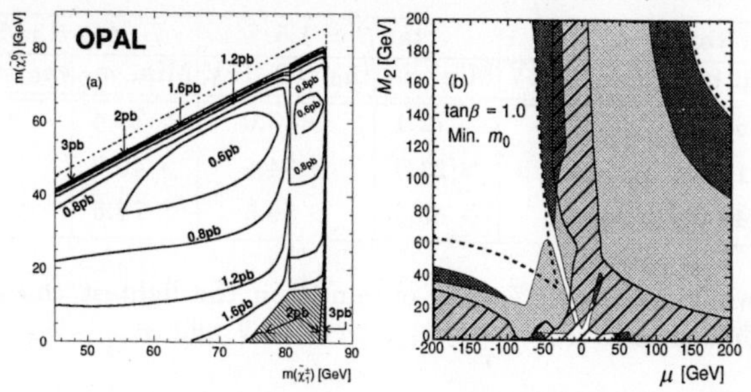

Fig. 1. (a) Contours of the 95% C.L. upper limits for the $e^+e^- \to \tilde{\chi}_1^+\tilde{\chi}_1^-$ production cross-sections at $\sqrt{s} = 172$ GeV are shown. The hatched area indicates where the analysis of W^+W^--like events contributes. (b) Exclusion regions at 95% C.L. in the (M_2,μ) plane for the the minimum m_0 case are shown. The light shaded areas show the LEP1 excluded regions. Using data from $\sqrt{s} = 161 - 172$ GeV, the dark shaded areas show the additional excluded region the analysis of W^+W^--like events, and the hatched area shows what is excluded by the analysis of W^+W^--like events. The dashed curves show kinematical boundary for $\tilde{\chi}_1^+\tilde{\chi}_1^-$ production.

at $\sqrt{s} = 161$ GeV have been combined with the present results. Contours of the upper limits for the $\tilde{\chi}_1^+\tilde{\chi}_1^-$ cross-sections are shown in Fig. 1a assuming $\tilde{\chi}_1^\pm \to \tilde{\chi}_1^0 W^{*\pm}$ with 100% branching fraction. The Standard Model branching fractions are used for the W^* and Z^* decays, including the invisible decay mode $Z^* \to \nu\bar{\nu}$ and taking into account phase-space effects for decays into heavy particles.

Within the constrained MSSM (CMSSM), a light m_0 results in low values of the masses of the $\tilde{\nu}$ and $\tilde{\ell}$, thereby enhancing the contribution of the t-channel exchange diagrams that may have destructive interference with s-channel diagrams, thus reducing the cross-section for chargino pair production. Small values of m_0 also tend to enhance the leptonic branching ratio of charginos, often leading to smaller detection efficiencies. The results are therefore presented for two cases: $m_0 = 1$ TeV and the smallest value of m_0 that is compatible with current limits on the $\tilde{\nu}$ mass ($m_{\tilde{\nu}_L} > 43$ GeV [6]), and OPAL limits on the $\tilde{\ell}$ mass [8]. This latter "minimum m_0" case gives the lowest $\tilde{\chi}_1^+\tilde{\chi}_1^-$ production cross-section for $\tan\beta = 1.0$ but not necessarily for larger $\tan\beta$ values. The problem associated with $\tan\beta$ approaching 1.0 is clear from Fig. 1b where the region near $M_2 = \mu = 0$ is only excluded if the analysis of W^+W^--like events is included.

The restrictions on the CMSSM parameter space presented can be transformed into exclusion regions in ($m_{\tilde{\chi}_1^0}$,$m_{\tilde{\chi}_1^\pm}$) mass space. A given mass pair is considered excluded only if *all* CMSSM parameters in the scan which lead to the same values of mass pairs being considered are excluded at the 95% C.L. This procedure is followed for a range of $\tan\beta$ to find the lower limit on the mass of the $\tilde{\chi}_1^0$ as a function of $\tan\beta$ for $m_0 = 1$ TeV and minimum m_0 with the result shown in Fig. 2(a).

Mass GeV	$\tan\beta = 1.0$ Min. m_0	$\tan\beta = 1.0$ $m_0 = 1$ TeV	$\tan\beta = 1.5$ Min. m_0	$\tan\beta = 1.5$ $m_0 = 1$ TeV	$\tan\beta = 35$ Min. m_0	$\tan\beta = 35$ $m_0 = 1$ TeV
$m_{\tilde{\chi}_1^\pm}$	> 65.7	> 84.5	> 72.1	> 85.0	> 74.4	> 85.1
$m_{\tilde{\chi}_1^0}$	> 13.3	> 24.7	> 23.9	> 34.6	> 40.9	> 43.8
$m_{\tilde{\chi}_2^0}$	> 46.9	> 46.9	> 45.3	> 56.5	> 74.6	> 85.5

Table 1. Lower limits at 95% C.L. obtained on the lightest chargino and two lightest neutralino masses for $m_{\tilde{\chi}_1^+} - m_{\tilde{\chi}_1^0} \geq 10$ GeV and $m_{\tilde{\chi}_2^0} - m_{\tilde{\chi}_1^0} \geq 10$ GeV.

2 Summary

Using a data sample corresponding to an integrated luminosity of 10.3 pb^{-1} at $\sqrt{s}$ =170 and 172 GeV collected with the OPAL detector new and more stringent limits on the neutralino cold dark matter candidate have been established. The new lower limits for the neutralino and chargino masses are listed in Table 1. The lower limit on the lightest neutralino mass ($m_{\tilde{\chi}_1^0}$) at 95% C.L. for $\tan\beta \geq 1.0$ is 24.7 GeV for $m_0 = 1$ TeV and 13.3 GeV for the minimum m_0 prescription.

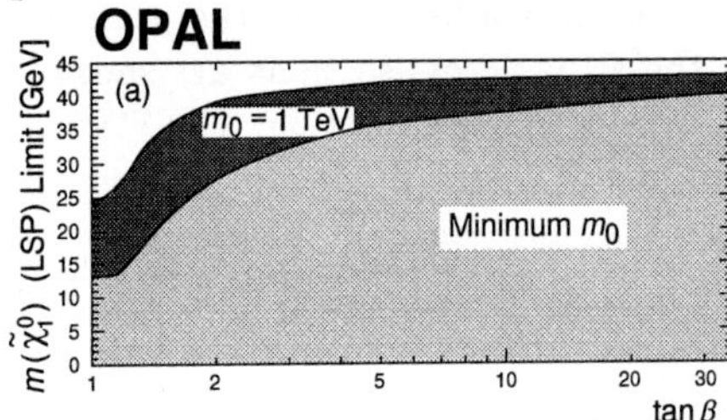

Fig. 2. The lower mass limit on the mass of the $\tilde{\chi}_1^0$ as a function of $\tan\beta$ for $m_0 = 1$ TeV and minimum m_0.

References

[1] K. Olive, astro-ph/9707212 (1997).
[2] M. Drees, hep-ph/9402211 (1994).
[3] OPAL Collab., K. Ackerstaff *et al.*, CERN preprint CERN-PPE/97-083.
[4] OPAL Collab., K. Ahmet *et al.*, Nucl. Instr. Meth. **A305** (1991) 275; P.P. Allport *et al.*, Nucl. Instr. Meth. **A346** (1994) 476; B.E. Anderson *et al.*, IEEE Trans. on Nucl. Science **41** (1994) 845.
[5] ALEPH Collab., D. Buskulic *et al.*, Phys. Lett. **B373** (1996) 246; DELPHI Collab., P. Abreu *et al.*, CERN preprint CERN-PPE/96-75; L3 Collab., M. Acciarri *et al.*, Phys. Lett. **B377** (1996) 289.
[6] Particle Data Group, Phys. Rev. **D50** (1994) 1173.
[7] R.D. Cousins and V.L. Highland, Nucl. Instr. Meth. **A320** (1992) 331.
[8] OPAL Collab., K. Ackerstaff *et al.*, Phys. Lett. **B396** (1997) 301.

Implications of ALEPH SUSY Searches for the MSSM

Laurent Duflot[1], on behalf of ALEPH collaboration (Laurent.Duflot@cern.ch)

Laboratoire de l'Accélérateur Linéaire, Orsay, France

Abstract. A lower limit on the lightest neutralino mass is derived in the MSSM from the ALEPH data. Constraints from Higgs searches are explored.

1 Introduction

The Lightest Supersymmetric Particle is a good cold dark matter candidate if R-parity is conserved. Limits on the mass of the neutralino LSP (more generally of an invisible lightest neutralino) of the Minimal Supersymmetric Standard Model are derived from the searches for supersymmetry using data collected by the ALEPH detector at $\sqrt{s} = 161$ and 172 GeV [1, 2], similarly to the analysis [3] based on the LEP1.5 ALEPH data.

The neutralino masses in the Minimal Supersymmetric Standard Model depend on the $U(1)$ and $SU(2)$ gaugino mass parameters M_1 and M_2, on μ, the supersymmetric Higgs mass parameter, and on $\tan\beta$, the ratio of the two Higgs doublet vacuum expectation values. Chargino masses depend on M_2, μ and $\tan\beta$, so neutralino and chargino searches can be combined without further assumptions. In the following, the gaugino mass unification condition $M_1 = \frac{5}{3}\tan^2\theta_W M_2$ is assumed. (If no relation between M_1 and M_2 is assumed, the limit $M_1 \to 0$ and large M_2 leads to no visible SUSY process accessible at LEP.)

The slepton masses influence the production and decay of neutralinos and charginos. The simplest case of heavy sleptons is considered in section 2 and the case of light sleptons in section 3. The neutral Higgs boson search results [4] are then used to constrain the SUSY parameter space, strengthening the M_χ limit for low m_0 and $\tan\beta$. In the following, limits are derived for a 95% confidence level and all results should be considered as preliminary. Additional information can be found in [5].

2 Heavy sneutrinos and sleptons

In the case of heavy sneutrinos and sleptons, neutralinos (charginos) are produced mainly via a Z (γ/Z) exchange and decay via a Z (W) exchange. To retain sensitivity for small neuralino masses, specific chargino analyses have been developed. The searches for charginos and neutralinos at LEP2 [1] and at LEP1 [6] constrain the MSSM parameter space, and limits on the neutralino mass are derived. For large $\tan\beta$, the neutralino mass limit is found at large $|\mu|$ while it is found at small negative μ for low $\tan\beta$. The M_χ limit is shown as a function of $\tan\beta$ in figure 1(a), leading to:

$$M_\chi > 25\ \mathrm{GeV}/c^2 \qquad (\text{for } M_{\tilde{\nu}} \gtrsim 200\ \mathrm{GeV}/c^2). \tag{1}$$

3 Light sneutrinos and sleptons

The chargino cross-section decreases when the sneutrinos become light. When sneutrinos and/or sleptons are light, the leptonic branching ratios of charginos increase, leading to an acoplanar lepton topology which suffers from the irreducible WW background. As a consequence, charginos searches are less constraining. Even more, when sneutrinos are lighter than the chargino, two body decays to lepton sneutrino often dominate. The final state lepton is very soft when the sneutrino-chargino mass difference becomes small, giving rise to a "corridor" of unexcluded $(M_{\tilde{\nu}}, M_{\chi^\pm})$ values, down to the LEP1 limit $M_{\chi^\pm} > 45$ GeV. In the regions of parameter space of interest in this case, the neutralinos and sneutrinos decay mostly to invisible final states, hence only indirect constraints, with additional theoretical hypotheses, can improve the neutralino mass limit.

A universal sneutrino and slepton mass m_0 at the GUT scale is assumed and sneutrino and slepton masses are derived from m_0, M_2 and $\tan\beta$ using renormalisation group equations. Direct searches for sleptons [2] exclude domains of the (μ,M_2) plane for given values of m_0 and $\tan\beta$. The combination of the excluded domains from neutralino/chargino and slepton searches gives a lower limit on the neutralino mass shown in figure 1(b). The effect of the weakening of the chargino limits is clearly visible, partially recovered by sleptons for low m_0. The lower limit on the neutralino mass, independent of $\tan\beta$ and m_0 is reached for $m_0 = 68.4\ \mathrm{GeV}/c^2$ and $\tan\beta = 2.4$:

$$M_\chi > 14\ \mathrm{GeV}/c^2 \qquad (\text{for any } M_{\tilde{\nu}}). \tag{2}$$

4 Neutral Higgs boson constraints

The search for supersymmetric Higgs bosons [4] with the data collected by ALEPH in 1995 and 1996 allows to constrain the lightest scalar Higgs mass m_{h}, depending on its coupling to the Z governed by $\sin^2(\alpha-\beta)$ ($m_{\mathrm{h}} >$ 70.7 GeV for SM couplings). At tree level, the m_{h} upper limit is given by $m_{\mathrm{Z}}|\cos 2\beta|$ and a limit on m_{h} would imply a lower limit on $\tan\beta$. Neglecting mixing, the main one-loop contribution to m_{h} is from the stop mass, hence a limit on m_{h} translates into a lower limit on the stop mass as a function of $\tan\beta$. The situation is more complex when stop mixing is taken into account. Large mixing gives large corrections to m_{h} but also a light stop state. Searches for stops [7] can therefore be used to reduce the allowed parameter space.

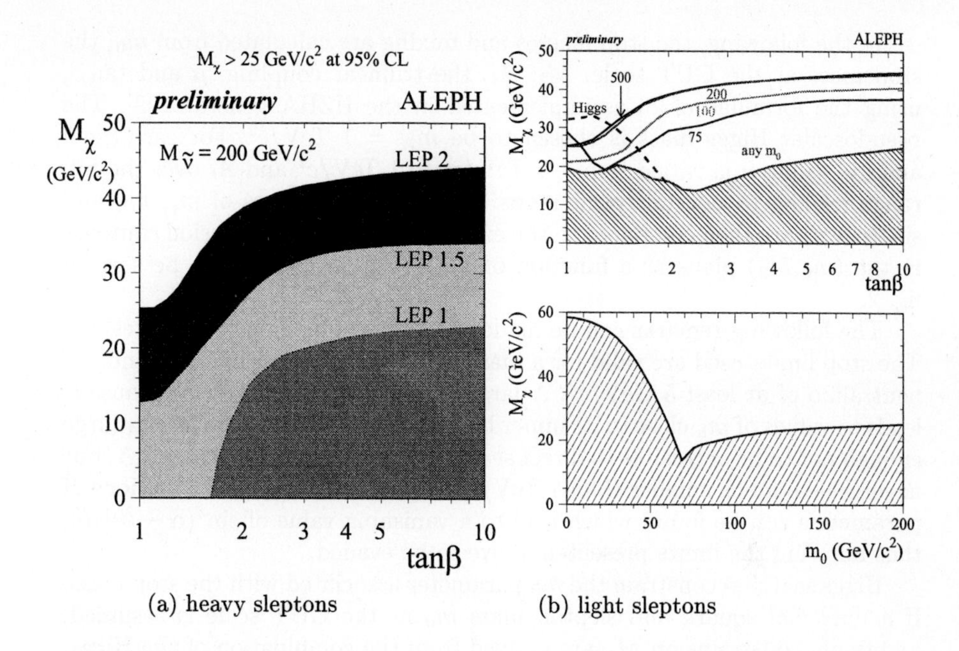

Fig. 1. Lower limit on the mass of the lightest neutralino as a function of $\tan\beta$, for heavy sleptons (a) and light sleptons for a series of m_0 values (b) (top). The curve labelled 'any m_0' shows the result obtained allowing m_0 to be free. The dashed curve shows the result coming from the Higgs constraints, for $m_0 = 75$ GeV/c^2. The lower plot shows the M_χ limit as a function of m_0, independent of $\tan\beta$.

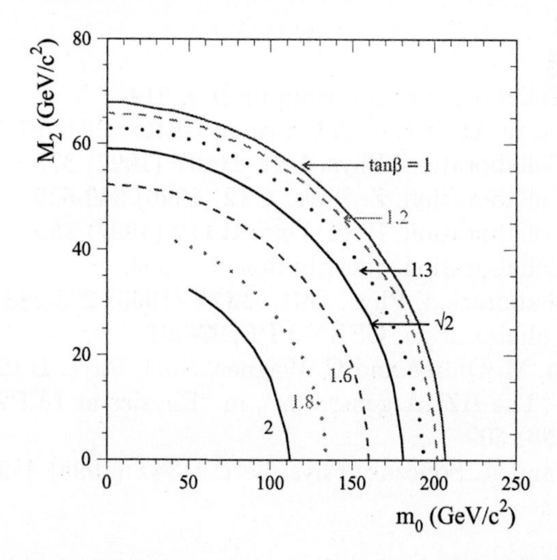

Fig. 2. In the (m_0, M_2) plane, limits from the Higgs searches, for various values of $\tan\beta$. The curves for $\tan\beta = 2$ and 1.8 are interrupted where sneutrinos become tachyonic.

In the following, the stop masses and mixing are calculated from m_0, the stop mass at the GUT scale, M_2, A_t, the trilinear coupling, μ and $\tan\beta$, using the formulae of [8] as implemented in the HZHA generator [9]. The pseudoscalar Higgs mass is chosen to be $m_A = 1$ TeV/c^2. For each $\tan\beta$ and m_0 value, μ is varied from -1 TeV/c^2 to 1 TeV/c^2 and A_t over the full range leading to non-tachyonic stops. The predicted values of $m_{\tilde{t}}$, m_h and $\sin^2(\alpha-\beta)$ are then compared to the experimental limits. Exclusion contours in the (m_0,M_2) plane as a function of $\tan\beta$ are derived, as can be seen in fig. 2.

The following remarks can be made as to the robustness of these results. The stop limits used are valid for a mass difference between the stop and the neutralino of at least 5 GeV/c^2. A large value of m_A was chosen because it leads to values of m_h close to its upper bound. However, for values of m_A large enough not to be accessible to direct searches for the process $e^+e^- \to hA$, but nevertheless much smaller than 1 TeV/c^2, highly fine tuned combinations of parameters can be found which lead to a vanishing value of $\sin^2(\alpha-\beta)$[10], thus allowing the limits presented above to be evaded.

Higgs searches constrain the m_0 parameter associated with the stop mass. If a universal squark and slepton mass m_0 at the GUT scale is assumed, additional constrains on M_χ are derived from the combination of the Higgs, stop, sleptons and gauginos searches as can be seen in the curve labelled "Higgs", derived for $m_0 = 75$ GeV, in figure 1(b). However, the minimum of the neutralino mass is found for a $\tan\beta$ value too large to be constrained by Higgs searches.

References

[1] ALEPH Collaboration, contribution Ref. 614.
final results in ALEPH Collaboration, CERN PPE/97-128

[2] ALEPH Collaboration, Phys. Lett. **B407** (1997) 377

[3] ALEPH Collaboration, Z. Phys. **C72** (1996) 549-559

[4] ALEPH Collaboration, Phys. Lett. **B412** (1997) 155

[5] ALEPH Collaboration, contribution Ref. 594.

[6] OPAL Collaboration, Phys. Lett. **B377** (1996) 273-288

[7] ALEPH Collaboration, CERN-PPE/97-084

[8] M. Carena, M. Quiros and C. Wagner, Nucl. Phys. **B461** (1996) 405

[9] P. Janot, "The HZHA generator", in "Physics at LEP2", CERN 96-01 Vol. 2 (1996) 309

[10] J. Rosiek and A. Sopczak, Phys. Lett. **B341** (1995) 419

Direct Search for Light Gluinos

Sunil V. Somalwar (somalwar@physics.rutgers.edu)(KTeV Collaboration)

Rutgers University, Piscataway, New Jersey 08855, USA

Abstract. I summarize the search for the appearance of $\pi^+\pi^-$ pairs with invariant mass $\geq$648 MeV/c^2 in the KTeV neutral beam. Our null result severely constrains the existence of an R^0 hadron, which is the lightest bound state of a gluon and a light gluino ($g\tilde{g}$). Depending on the photino mass, we exclude the R^0 in the mass and lifetime ranges of 1.2 – 4.6 GeV/c^2 and 2×10^{-10}–7×10^{-4} s, respectively.

This search [1] is motivated by recent discussions [2] of the possible existence of long-lived hadrons that contain light gluinos. The gluons (g), gluinos, and quarks can form bound states, the lightest of which is a spin-1/2 $g\tilde{g}$ combination called the R^0. The approximate degeneracy of R^0 with the 0^{++} glu-onia [2] suggests that the R^0 mass could be of the order of 1.3–2.2 GeV/c^2. The stable photino in this theory is a cold dark matter candidate [3]. Estimates from particle physics [2] and cosmology [3] place the R^0 lifetime in the 10^{-10}–10^{-5} s range. The R^0 decays into a $\tilde{\gamma}$ and hadrons. Because of the approximate C invariance of supersymmetric QCD, the R^0 decay into $\rho\tilde{\gamma}$ is expected to be the dominant decay mode [2]; depending on the extent of C violation, the R^0 may also decay into $\pi^0\tilde{\gamma}$ or $\eta\tilde{\gamma}$.

The R^0's can be produced in pN collisions by processes such as quark-antiquark annihilation, gluon fusion, etc. Since squarks need not be involved, the R^0 production is in the realm of conventional QCD. Dawson, Eichten, and Quigg[4] have calculated cross sections for gluino production in tree approximation. Their calculations suggest an order-of-magnitude cross section estimate[5] of $\sim$ 10 μb per nucleon for the production of an R^0 with 2 GeV/c^2 mass in 800 GeV/c pN collisions.

In the KTeV experiment [6], protons are incident on a 30 cm long beryllium oxide target at a vertical angle of 4.8 mr with respect to the neutral beam channel. The interaction products are filtered through a 50.8 cm beryllium absorber and a 7.6 cm lead absorber. Two neutral beams, each 0.25 μstr in solid angle, emerge after collimation and sweeping. One of the beams passes through a regenerator, but the decays from this beam are not used in this analysis. The beam transport and decays took place in an evacuated region with a vacuum of 0.5–1.0 $\times$ 10^{-4} torr. The results presented here are from the delivery of 1.9×10^{15} protons.

The most crucial detector element for this analysis, the charged spectrometer, consists of four planar drift chambers, two on either side of an analyzing magnet. Each chamber measures positions in two orthogonal views. Each view consists of two planes of wires, with cells arranged in a hexagonal geometry. Each chamber has approximately 100 μm single-hit position resolution per plane. The spectrometer magnet has a 2 m by 1.7 m fiducial region over which a transverse momentum impulse of 411 MeV/c is imparted to the charged particles. A set of helium bags integrated into the spectrometer system minimizes multiple scattering. The invariant mass resolution for the decay $K_L \to \pi^+\pi^-$ is better than 2 MeV/c^2. A pure CsI electromagnetic calorimeter of dimension 1.9 m by 1.9 m is used to reconstruct photon and electron energies to better than 1% precision. The calorimeter is used to match the orthogonal track views and to reject background from K_{e3} decays. A set of 12 photon vetos provides hermetic photon coverage up to angles of 100 mr. A counter bank (muon veto) located at the downstream end of the detector is used to reject $K_{\mu 3}$ decays. The event trigger is initiated by signals from two scintillator hodoscopes located downstream of the spectrometer. The primary trigger requires two hits in these counters consistent with two oppositely charged tracks, at least one hit in each of the two upstream drift chambers, and lack of hits in the muon veto bank.

A combination of online and offline cuts required two oppositely charged tracks matched to clusters in the calorimeter. Individual track momenta were required to be more than 8 GeV/c, and the scalar sum of two track momenta was required to be between 30 GeV/c and 160 GeV/c. A longitudinal vertex position between 126 m and 155 m downstream of the target defined the fiducial region. Electrons were rejected by requiring the energy deposited in the calorimeter to be less than 80% of the particle momentum. The signals in various veto devices were required to be no more than those due to accidental activity. The momentum ratio of the two tracks was required to be between 0.2 and 5 to reduce various backgrounds. Figure 1 shows the

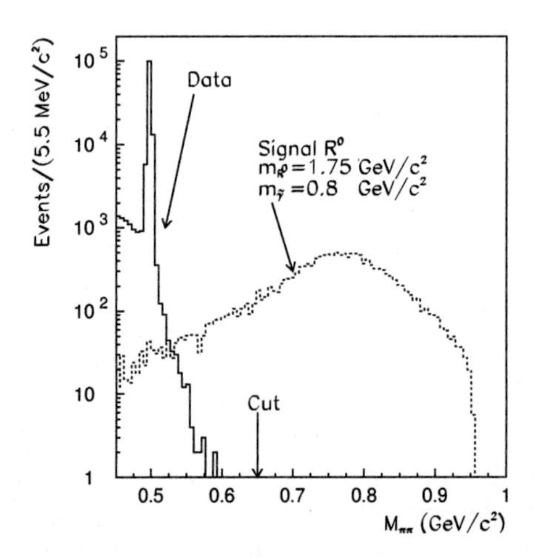

Fig. 1. $m_{\pi^+\pi^-}$ distribution for the data (solid) and R^0 signal Monte Carlo (dashed, arbitrary scale). The peak at 500 MeV/c^2 corresponds to $K_L \to \pi^+\pi^-$ decays.

$m_{\pi^+\pi^-}$ distribution for the events surviving all cuts. Note the peak corresponding to the $K_L \to \pi^+\pi^-$ candidates and the rapidly falling background to the right of the kaon peak. The figure also shows the $m_{\pi^+\pi^-}$ distribution for an R^0 with mass m_{R^0}=1.75 GeV/c^2 and photino mass $m_{\tilde{\gamma}}$=0.8 GeV/c^2. There are no R^0 candidates in the signal window between 648 MeV/c^2 (i.e. 150 MeV/c^2 above the kaon mass) and 1.0 GeV/c^2. For the simulated R^0

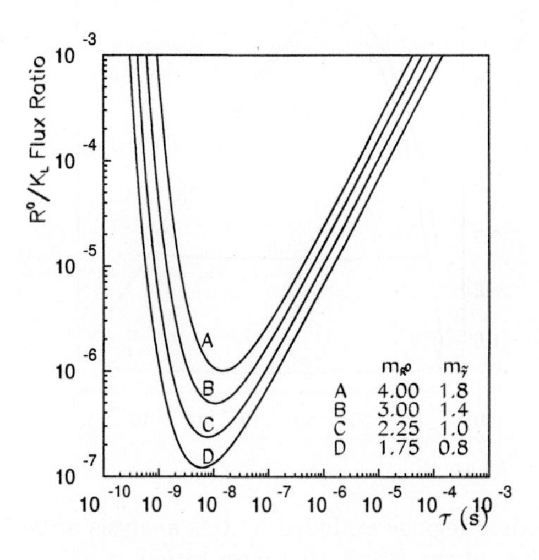

Fig. 2. Upper limits with 90% confidence level on the R^0/K_L flux ratio as a function of R^0 lifetime. m_{R^0} and $m_{\tilde{\gamma}}$ are in GeV/c^2.

shown, 91% of the decays are contained in the signal window. We define the R^0/K_L flux ratio to be the ratio of the number of R^0's to the number of K_L's exiting the beam absorbers, calculated with the assumptions that the $R^0 \to \tilde{\gamma}\rho$, $\rho \to \pi^+\pi^-$branching ratio is 100% and that the photino does not interact significantly in the detector material. The K_L flux was determined using a total of 116,552 $K_L \to \pi^+\pi^-$ candidates with two-body transverse momentum squared (p_t^2) less than 250 MeV^2/c^2 and $m_{\pi^+\pi^-}$ within 10 MeV/c^2 of the K_L mass. We calculate a total of 1.33×10^{10} K_L's of all energies exiting the absorbers. Figure 2 shows the upper limits on R^0/K_L flux ratio as a function of R^0 lifetime for various R^0 and $\tilde{\gamma}$ masses, but the same R^0 to photino mass ratio r(=2.2). The upper limits on the R^0/K_L flux ratio constrain the light gluino scenario assuming a specific model for the R^0 production. The perturbative QCD calculations[4, 5] suggest $\sigma(R^0) \simeq 860e^{-2.2m_{R^0}}$ μb in 800 GeV/c pN collisions, where m_{R^0} is in the units of GeV/c^2. This estimate is approximately consistent with the heavy flavor production cross sections. For our experiment, this $\sigma(R^0)$ implies that the R^0/K_L flux ratio for an R^0 of mass m_{R^0} is expected to be $9.2 \times 10^{-2}e^{-2.2m_{R^0}}$. We have taken into account a factor of 10.2 for the R^0 absorption in the target and the beam absorbers, assuming the R^0N cross section to be the same as NN cross section. The mass-lifetime region for which this R^0/K_L flux ratio expectation exceeds the measured upper limits is taken to be ruled out. Figure 3 shows the excluded regions for two different values of the mass ratio r. Apart from the low mass

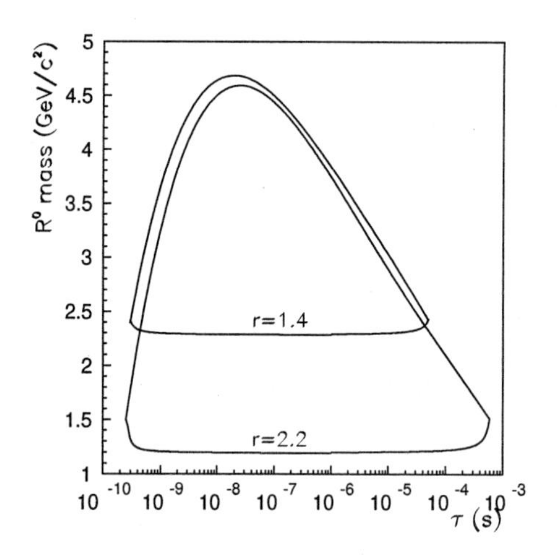

Fig. 3. R^0 mass-lifetime regions excluded by this analysis at 90% confidence level for two different values of the R^0-photino mass ratio r.

cutoff for $m_{R^0} \lesssim 0.648r/(r-1)$ GeV/c^2, the contour shapes in figure 3 are fairly insensitive to the value of mass ratio r. The cosmologically interesting region corresponds to $r \lesssim 1.8$ [3].

In summary, our search places upper limits on the R^0/K_L flux ratio approaching 1×10^{-7}. These stringent limits rule out the production of R^0 in a wide range of masses (~1.2 – 4.6 GeV/c^2) and lifetimes ($\sim 3 \times 10^{-10}$ – 7×10^{-4} s).

This work is supported by the NSF, DOE, and the US-Japan Foundation.

References

[1] J. Adams *et al.*, Phys. Rev. Lett. **79**, 4083 (1997).
[2] G.R. Farrar, Phys. Rev. Lett. **76**, 4111 (1996); G.R. Farrar, Phys. Rev. D **51**, 3904 (1995).
[3] D.J.H. Chung, G.R. Farrar, and E.W. Kolb, Report No. astro-ph/9703145; G.R. Farrar and E.W. Kolb, Phys. Rev. D **53**, 2990 (1996).
[4] S. Dawson, E. Eichten, and C. Quigg, Phys. Rev. D **31**, 1581 (1985).
[5] C. Quigg (private communication).
[6] S.V. Somalwar, Report No. 920 in this volume.

Status Report and Recent Results from the UK Dark Matter Collaboration

G.J. Davies on behalf of the UKDMC (g.j.davies@ic.ac.uk)

Imperial College, UK

Abstract. The UK Dark Matter Collaboration (UKDMC) has an ongoing programme of detector development and operation with the eventual aim of reaching dark matter rates below 0.1/day/kg. Limits for the interaction rate have been set by this group and others using pulse shape discrimination in sodium iodide scintillation detectors. Methods for improving these limits through increased light output will be discussed, including avalanche photodiode R&D . To reach the lowest interaction rates we are collaborating with members of ICARUS on liquid xenon systems and this work is briefly reported.

1 Introduction

Dark matter (DM) in the form of weakly interacting massive particles (WIMPs) can be detected via the very small (<50keV) energy deposit of a nuclear recoil resulting from the elastic scatter of the incoming WIMP. Such collisions will be rare, 1-0.01 c/kg/day [1]. DM detectors are in general located underground to exclude cosmic rays. Though the ambient radioactivity can be reduced by high purity shielding, the irreducible radioactive contamination in the target/detector itself ultimately limits the sensitivity. Thus new detectors must be able to distinguish the nuclear recoil signal events from the background electron recoils resulting from gammas and beta decay. This discrimination has been studied by several groups in both room temperature and cooled sodium iodide (NaI) and liquid xenon (LXe) [2][3][4][5]. Neutron sources can be used to produce similar nuclear recoils as would dark matter particles. We discuss results from a 6kg room temperature NaI(Tl) target which exploits the difference in the pulse shape between nuclear recoils and background. For full details see [6]. We have also operated pure NaI detectors underground and are developing LXe detectors.

2 Experimental Assembly

The UKDMC's experiments are situated 1100m underground (2900m water equivalent) in the Boulby salt mine. We employ two types of shielding, either a 6m by 6m cylindrical high purity water tank, which also removes neutrons from radioactivity in the rock, or low activity electroformed copper and lead which have been stored underground.

The 6kg device was operated in the water tank and read out by two PMTs in coincidence.

3 Data sample

The integrated pulse from each incident pulse was recorded on disk after digitisation with a LeCroy 9430 oscilloscope. The sensitivity was 1.6 photoelectrons (pe)/keV and the threshold set at 2.4 p.e./PMT. The data reported here are for 173 days running. The temperature was measured but not stabilised during this period and drifted slowly in the range $32 \pm 1^\circ$. A correction was made for this slow deviation [1]. Energy calibrations were carried out using a ^{57}Co source; gamma pulse shape calibration was achieved both with a ^{60}Co source and by raising the detector to expose it to the gamma background from the cavern; neutron pulse shape calibration was achieved by lowering a Cf source to 1m from the detector.

4 Data analysis

The data set in the range 4-25keV was grouped into 8 energy bins: 4-5keV, 7-10keV and thereafter in 3keV intervals. Discrimination occurs in NaI(Tl) at ambient temperature through a variation in a single time constant. Normalised calibration distributions for the 6kg crystal are shown in fig.1 (left) for the 7-10keV energy range, together with 6 months data from the fully-shielded underground detector.

Time constants (τ) were obtained by fitting to a single exponential and the distribution in τ was well described by a Gaussian in $\ln\tau$. At energies below 10keV this distribution overlaps a distribution of faster PMT noise pulses. The noise-free portion of the spectrum was used to give an accurate estimate of the true number of events in these energy bins. See curves (a) and (b) in fig. 1 (left).

Any dark matter signal would be expected to follow the shape of the neutron calibration spectrum. Thus a combination of dark matter signal (or neutrons) and radioactive background would result in a spectrum lying somewhere between the neutron and gamma calibration spectra. In fact the data can be expressed as a linear combination of these. The most likely fraction of nuclear recoils and gammas can then be found. We see from fig. 1 that the data are consistent with being entirely radioactive background. A 90% confidence level (c.l.) upper limit on the nuclear recoil signal can be obtained by running Monte Carlo simulations of the data [7]. This process was carried separately for each energy bin.

[1] Temperature stabilisation is now in operation for all systems and in fact the detectors are run below ambient to enhance the discrimination.

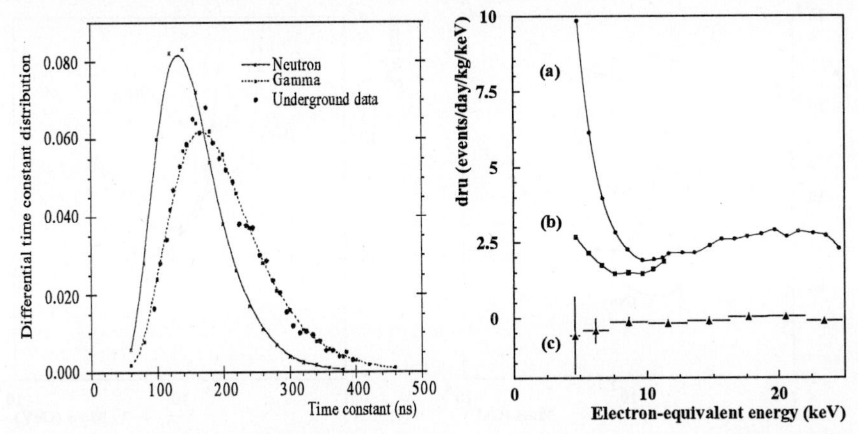

Fig. 1. Time constant distribution for neutrons and gammas in the 7-10keV range for the 6kg target (left). Background differential spectra (right): (a) observed rate; (b) after subtraction of fast noise pulses; (c) 2σ limits on nuclear recoil events for energy spans shown by horizontal bars.

We conclude, at the 90% c.l., that we have discrimination gain factors of 0.025-0.18 depending on which bin is considered. This enables us to reduce the 90% c.l. upper limit on the dark matter counting rate for each bin. See curve (c) in fig. 1 (left).

5 Results

We are thus able to set dark matter limits for each of the 8 energy bins, labeled by the index k (k=1 to 8). Following [8], the differential nuclear recoil rate, $S_{nk} = [\Delta R_n/\Delta E]_k$ where R is the nuclear recoil rate for dark matter of mass, m_D, and energy, E_o, with a target nucleus of mass, m_T,E_r is the recoil energy and form factor, F, is given by: $S_{nk}\Delta E = (R_0/r)(c_1/E_0)$x $\exp(-c_2 E_r/E_0 r)F^2 dE_r$ where R_0 is the total rate unit in c/kg/day and r is $(4m_T m_D)(m_T + m_D)^2$.

Thus energy span k yields a limit curve of rate (or cross-section) versus mass, with the minimum $(R_0/r)_k$ at higher m_D for higher k. Since each limit is statistically independent the best result is obtained by forming a single weighted (R_0/r). The 90% c.l. in R_0/r is shown in fig.2 for the spin dependent and coherent ($\propto A^2$) interactions.

6 Developments towards lower event rates

Further gains can be achieved by improved light collection via alternative light detectors, such as avalanche photodiodes; indeed we have an programme with Silicon Sensors in Germany to develop suitable drift devices.

The group, in collaboration with UCLA and members of ICARUS at CERN, is also investigating the use of LXe, which appears to offer far greater

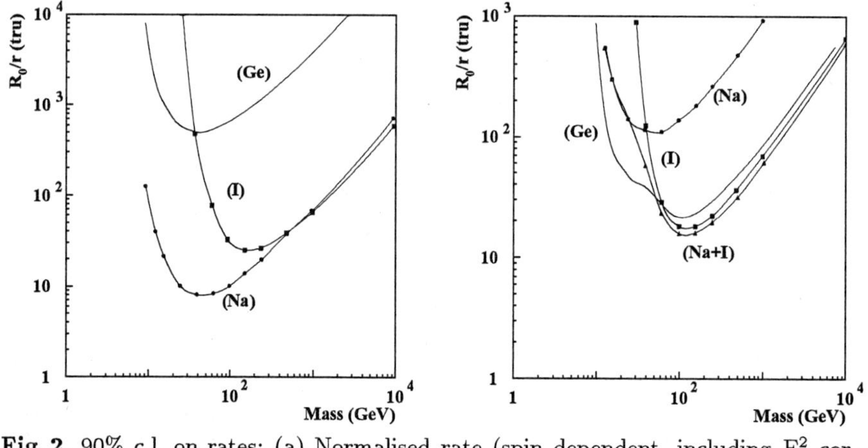

Fig. 2. 90% c.l. on rates: (a) Normalised rate (spin dependent, including F^2 correction); (b) Normalsed rate (spin-independent, including F^2 correction and normalised to Ge using I = $(A\text{-}Z)^2$.

discrimination [9][10]. By extrapolating higher energy data for electrons and alphas (which approximate to a nuclear recoil) simulations show that it is possible to reduce the background by factors of ~100, even at low energies, by timing measurements on the scintillation light [4]. Alternatively, the LXe can be operated in the proportional scintillation mode. This method has been used to separate 5.5MeV alpha particles and 122keV gammas [5]. Preliminary results with neutron sources indicate that very significant discrimination is possible using this technique [10]. We suggest that such a detector could reach event rates below 0.1/day/kg.

References

[1] J. Ellis and R. A. Flores, Physics Letters B263 (1991), 259.

[2] P. F. Smith, Review talk, *Proc. 1st Int. Symp. on Sources of Dark Matter in the Universe*, World Scientific (1995) pp. 227, Ed. D.B. Cline; C. Bacci et al., Physics Letters B293 (1992), 330.

[3] N. J. C. Spooner and P. F. Smith, Physics Letters B314 (1993), 430.

[4] G. J. Davies et al., Physics Letter B320 (1994), 395.

[5] P. Benetti et al., Nucl. Instr. and Meth. A327 (1993), 203.

[6] P. F. Smith et al., Physics Letters B379 (1996), 299.

[7] N.J.T. Smith et al., Nucl. Phys. B48 (1996), 67.

[8] J. D. Lewin, P. F. Smith, Astroparticle Physics 6 (1996) 87.

[9] W.G. Jones et al.,Proceedings of the 1st International Workshop on the Identification of dark matter, World Scientific, 1997 428.

[10] H. Wang et al., Proceedings of the 1st International Workshop on the Identification of dark matter, World Scientific, 1997 427.

EROS-II Status Report

Olivier Perdereau (perdereau@lalcls.in2p3.fr)

Laboratoire de L'Accélérateur Linéaire, F-91405 Orsay

Abstract. EROS-II is a second generation microlensing experiment. The experimental setup, in operation at the European Southern Observatory (ESO) at La Silla (Chile) since mid-1996 is briefly described together with its scientific objectives. Preliminary results concerning our microlensing searches towards the Small Magellanic Cloud (SMC) and the Galactic plane are presented. We conclude by overviews of the semi-automated supernovae search and the cepheid studies being pursued in parallel.

1 Introduction

The microlensing effect was proposed ten years ago by B. Paczynski[1] as a unique experimental signature of MACHOs. Dark compact baryonic objects (MACHOs) are plausible components of the galactic dark matter. The flux of an observed star is gravitationnally deflected if one of these objects passes close to the line of sight. If the image distorsion is undetectable one is left with a transcient amplification of the total flux.

Few years later, several occurences of the microlensing amplification of stars were observed in two different directions, the LMC (Large Magellanic Cloud)[2] [3] and the Galactic Bulge [4]. This field is now entering a more quantitative era. EROS-I has isolated 2 candidates over 3 years of running[5]. The Macho collaboration has taken data from 1993 until now and has a handfull microlensing candidates towards LMC[6]. These observations indicate a total halo MACHO mass fraction within a factor of two from the total required to explain the Galactic rotation curve. In addition the time scales associated with these events indicate large MACHO masses, which is difficult to accomodate with known stellar populations.

To address these questions, EROS started as early as 1993 to build a new apparatus, which started observations in June 1996. It is outlined in this paper and first results of some of its programs are also presented.

2 The instrument

The EROS-II[1] instrument is a 1m diameter telescope, the MarLy, which has been specially refurbished and automated to enable a reliable microlensing survey. It is in operation at the European Southern Observatory at La Silla (Chile). The optics include a dichroïc beam splitter allowing simultaneous

[1] participating institutes : CEA-Saclay, IN2P3-CNRS, INSU-CNRS

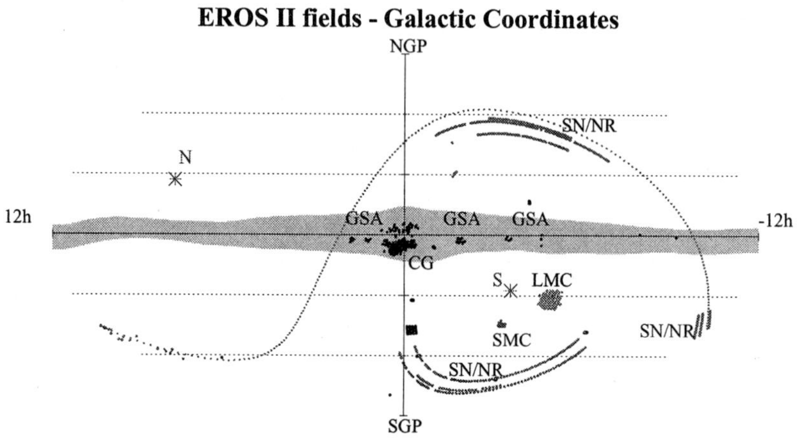

EROS II fields - Galactic Coordinates

Fig. 1. All-sky view of our fields (equatorial coordinates). The shaded areas represents our Galaxy. We indicated the celestial equator and poles (N and S). fields from specific programs are labelled respectively CG (Galactic Center), LMC and SMC (Magellanic Clouds), SN/NR (SN or red dwarves search).

observations in two wide pass-bands (a blue and a red one). The field of the instrument is observed in each band by a mosaic of 2×4 Loral $2k \times 2k$ pixels thick CCDs. The total usable field is $.7 \times 1.4\,\mathrm{deg}^2$. The median seeing (FWHM of the signal from a point-like source) is about 2. arcsec. CCDs from each mosaic are readout in parallel by DSPs. The total readout time is 50s. The data acquisition in controlled by two VME crates (one per camera). Images are analyzed in the CCIN2P3 in Lyons. We are also developping an alert capability by monitoring on site, the day following the observation, a sample of stable stars from the microlensing fields.

3 The EROS-II programs

EROS-II is primarily aimed at giving a better understanding of the matter distribution in the Galaxy using microlensing. In order to achieve this, a number of direction are currently studied : the Magellanic Clouds (60 fields for the Large, 10 for the Small), the Galactic Center (80 fields) and 4 areas within the Galactic plane ($\approx$ 6 fields each). We are currently giving the highest priority to the microlensing search in new line of sights (the Small Magellanic Cloud and the Galactic plane fields), from which we present some preliminary results.We also address other cosmologicaly important programs, such as a systematic study of Cepheids, a search for high proper motion stars and a semi-automated supernovæ search. These programs use images from different regions in the sky, as shown on figure 1. The schedule of each night optimizes the observation conditions for each program.

4 SMC data analysis

The analysis of the SMC data from our first running year is now completed ; more details can be found elsewhere[7]. The SMC was observed for July 1996 to February 1997 and then after July 1997 (our analysis includes data up to August 1997). A hundred images of each field are usable for subsequent analysis. In total more than 5.10^6 light-curves could be analyzed. They were searched for microlensing events. This selection is described in [7]. The cuts are designed to select light curves with a unique and achromatic amplification, a sufficient S/N ratio and no known variable star contamination. The global selection efficiency is about 15%. In our data 10 light-curves passed all cuts and where checked visually, one being finally selected : it is shown in figure 2. This event has also been seen by the Macho Collaboration[8]. Both groups

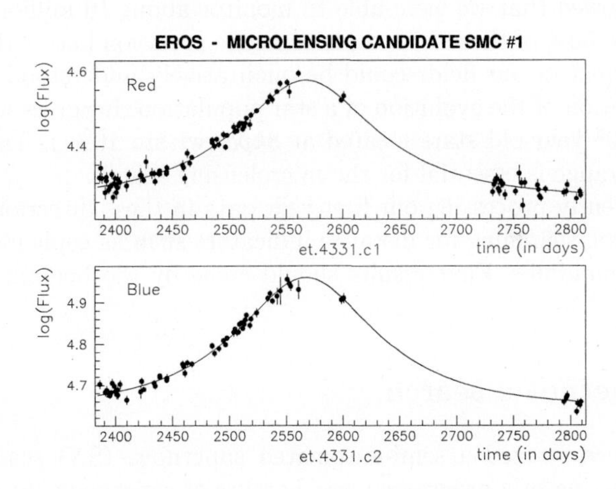

Fig. 2. The SMC-1 candidate light-curve with a standard microlensing fit superimposed (no blending assumed). Our time origin is Jan 0,1990.

signal that the amplified star is a superposition of two stars with close line of sight with flux ratios of 70% and 30% . Only the brightest of these two stars is amplified. EROS also reports a periodic modulation of the combined flux with a period of 5.128 days and an amplitude of about 3% of the brightest flux.

Using a simulation to compute our efficiency we also reported a first measurement of the Galactic halo optical depth τ (instantaneous probability that a given star be magnified by more than a factor 1.34) towards SMC:

$\tau \simeq 2.2\ 10^{-7}$. This value is similar to that measured towards the LMC. Comparisons with several halo models show that this sole event contributes by at least 40% of the optical depth due to the halo of our Galaxy. However, the absence of parallax effect (caused by the modulation of the Earth velocity) in this long duration event tends to imply a heavy lens (a few $M_\odot$) or a deflector near the source. This event may in fact also be interpreted with both deflector and source lying within the SMC, with a lens mass of order $.1M_\odot$. In that case, we derive $\tau_{SMC-SMC} \simeq 1.3\ 10^{-7}$ which does not conflict with the SMC structure[7].

5 The Galactic Plane

Measuring optical depths in various directions in the Galactic plane would help constraining the different Galactic components to the Bulge and LMC or SMC optical depths. We chose several directions to look at, grouped respectively near galactic longitudes of 25° (5 fields), 30° (6 fields), 310° (6 fields) and 320^0 (12 fields). A feasibility study of a microlensing search in these directions[9] showed that we were able to monitor about 10 million stars in these directions, 50% of these with a photometric precision better than 10%. The stellar content of our fields could be qualitatively understood with the help of a simulation of the evolution of a star population. It agrees with what we expect of 10^8 year old stars located at $8kpc$ (within 10%). This rather small distance range is essential for the microlensing search.

We are currently processing our first year data in these directions with a twofold objective, searching for distance indicators such as cepheids and for microlensing candidates. First results should come by the begining of next year.

6 The supernovæ search

EROS-II has also started a semi-automated supernova (SN) search. It is aimed at discovering in a programed way batches of supernovæ. Spectrography and photometry observations may thus be planned in order to classify and study them accurately. Supernovæ are rare phenomena (≈ 1 per century and per galaxy) and study a large (≈ 100) number of them offers interesting cosmological perspectives, for example measuring their rate, or study type Ia SN. Systematic surveys for nearby SNe have takled their usability as distance indicators (with ≈ 20 SNe)[10]. This could be well studied in a intermediate redshift search such as that of EROS. Our detection threshold is estimated to be about a 22d magnitude, corresponding to a redshift of 0.2. Such systematic study are also essential in extracting cosmological informations from the distant SNe searches[11][12].

Our program goes as follows. Using the EROS-II instrument, we take images near two new moons. The more recent (current) image is compared to

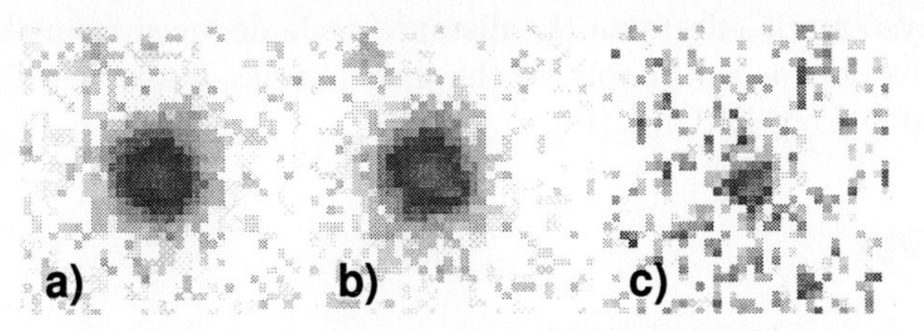

Fig. 3. Pictorial view of the discovery of sn1997bl. The image (a) is our reference image, taken on 03/02/97, (b) is the discovery image taken on 07/03/97, showing the SN superimposed onto the host galaxy. (c) is the difference between (b) and (a), on which the supernova was detected.

the older one (reference) during daytime to search for SN, as shown on figure 3. This program was tested in spring 1997. Three supernovae were discovered, in good agreement with our estimated discovery rate of .1 $SN.\deg^{-2}$. A more intensive campaign has started this autumn. We obtained follow-up time on the ESO 1.5m Danish telescope, and have also few spectrospic nights on the ARC 3.5m telescope (New-Mexico, USA). In October we discovered 5 new SN (within 55 $\deg^2$) and 4 up to November 28th. We do also follow these SN on our own instrument.

7 Magellanic cepheids

EROS-I reported a possible metallicity effect on the period-luminosity relation for SMC and LMC cepheids[13]. We pursued last year a specific program in order to study more accurately this effect. Two fields in LMC and two in SMC were taken each night. These data were then searched for cepheids. Our new sample contains about 300 cepheids in LMC and 600 in SMC.Beeing observed with the same instrument and in similar conditions will also reduce systematics. Results from this analysis will come by the end of the year.

8 Conclusion

EROS-II started taking data more than one year ago. During this period, the detector has run quite well. We present a first measurement of the galactic halo microlensing optical depth towards SMC based on one event. This event could however as well be interpreted as due to a lens lying within SMC, and presents some intriguing properties. We demonstrated the feasibility of a microlensing search towards the galactic plane, and expect first candidates to be isolated soon. Our SNe search has been quite intense this autumn, leading to the discovery of $\approx$ 10 SNe that are being followed on our instrument and others. We are also analysing a large LMC and SMC cepheids database to check for systematic effects on the distance scale deduced from them. These subjects are only a subsample of the rich physics outcome that is to be expected soon from EROS-II[14].

References

[1] B. Paczynski ApJ 304 (1986),1
[2] C. Alcock et al (MACHO Coll.), Nature 365 (1993), 621
[3] E. Aubourg et al (EROS coll.), Nature 365 (1993), 623
[4] A. Udalski et al, (OGLE Coll.) Acta Astronomica 43 (1993), 289
[5] R. Ansari et al (EROS Coll.) A&A 314 (1996), 94
[6] C. Alcock et al (Macho Coll.) ApJ 486 (1997), 697
[7] N. Palanque-Delabrouille et al, astro-ph/9710194 (submitted to A&A)
[8] C. Alcock et al (MACHO Coll.), ApJ Letters 491 (1996), 11
[9] B. Mansoux PhD thesis (1997), LAL 97-19
[10] Hamuy et al ApJ 112 (1996), 2391
[11] Kim et al ApJ 483 (1997), 565
[12] B. Leibundgut et al ApJ Letters 466 (1996), 21
[13] D. Sasselov et al, A&A 324 (1997), 374
[14] see our WWW site at URL http://www.lal.in2p3.fr/EROS/

Graceful Exit in String Cosmology

Ram Brustein and Richard Madden (ramyb, madden @bgumail.bgu.ac.il)

Department of Physics, Ben-Gurion University, 84105 Beer-Sheva, Israel

Abstract. The graceful exit transition from a dilaton-driven inflationary phase to a decelerated Friedmann–Robertson–Walker era requires certain classical and quantum corrections to the string effective action. Classical corrections can stabilize a high curvature string phase while the evolution is still in the weakly coupled regime, and quantum corrections can induce violation of the null energy condition, allowing evolution towards a decelerated phase.

1 Introduction

String theory predicts gravitation, but the gravitation it predicts is not that of standard general relativity. In addition to the metric fields, string gravity also contains a scalar dilaton, that controls the strength of coupling parameters. An inflationary scenario [1], is based on the fact that cosmological solutions to string dilaton-gravity come in duality-related pairs, an inflationary branch in which the Hubble parameter increases with time and a decelerated branch that can be connected smoothly to a standard Friedmann–Robertson–Walker (FRW) expansion of the Universe with constant dilaton. The scenario (the so called "pre-big-bang") is that evolution of the Universe starts from a state of very small curvature and coupling and then undergoes a long phase of dilaton-driven kinetic inflation and at some later time joins smoothly standard radiation dominated cosmological evolution, thus giving rise to a singularity free inflationary cosmology.

However, in the lowest order effective action these two branches are separated by a singularity. Additional fields or correction terms need to be added to make this "graceful exit" transition possible. It has been studied intensely in the last few years, but all attempts failed to induce graceful exit. In [2, 3] it was shown that the transition is forbidden for a large class of fields and potentials. In [4] we proposed to use an effective description in terms of sources that represent arbitrary corrections to the lowest order equations and were able to formulate a set of necessary conditions for graceful exit and to relate them to energy conditions appearing in singularity theorems of Einstein's general relativity [7]. In particular, we showed that a successful exit requires violations of the null energy condition (NEC) and that this violation is associated with the change from a contracting to an expanding universe (bounce) in the "Einstein frame", defined by a conformal change of variables. Since most classical sources obey NEC this conclusion hints that quantum effects, known to violate NEC in some cases, may be the correct sources to look at.

Because the Universe evolves towards higher curvatures and stronger coupling, there will be some time when the lowest order effective action can no longer reliably describe the dynamics and it must be corrected. Corrections to the lowest order effective action come from two sources. The first are classical corrections, due to the finite size of strings, arising when the fields are varying over the string length scale $\lambda_s = \sqrt{\alpha'}$. These terms are important in the regime of large curvature. The second are quantum loop corrections. The loop expansion is parameterized by powers of the string coupling parameter $e^{\phi} = g^2_{string}$, which in the models that we consider is time dependent. So quantum corrections will become important when the dilaton ϕ becomes large, the regime we refer to as strong coupling.

In [5] we were able to find an explicit model that satisfies all the necessary conditions and to produce the first example of a complete exit transition. The specific model we present here makes use of both classical and quantum corrections. We allowed ourselves the freedom to choose the coefficients of correction terms which generically appear in string effective actions. Our reasoning for allowing this stems in part from a lack of any real string calculations and in part by our desire to verify, by constructing explicit examples, the general arguments of [4].

2 A Specific Exit Model

String theory effective action in four dimensions takes the following form in the string frame ($\underline{S|}$),

$$S_{eff}^{\underline{S|}} = \int d^4x \left\{ \sqrt{-g} \left[\frac{e^{-\phi}}{16\pi\alpha'} \left(R + \partial_\mu \phi \partial^\mu \phi \right) \right] + \tfrac{1}{2}\sqrt{-g}\mathcal{L}_c \right\}, \tag{1}$$

where $g_{\mu\nu}$ is the 4-d metric and ϕ is the dilaton. The Einstein frame ($\underline{E|}$) is defined by the change of variables $g_{\mu\nu} \to e^{\phi/2} g_{\mu\nu}$, diagonalizing the metric and dilaton kinetic terms and resulting in equations of motion similar to those of standard general relativity.

We are interested in solutions to the equations of motion derived from the action (1) of the FRW type with vanishing spatial curvature $ds^2 = -dt_S^2 + a_S^2(t) dx_i dx^i$ and $\phi = \phi(t)$. The contribution of $\mathcal{L}_c$ is contained in the correction energy-momentum tensor $T_{\mu\nu} = \frac{1}{\sqrt{-g}} \frac{\delta \sqrt{-g}\mathcal{L}_c}{\delta g^{\mu\nu}}$, which will have the form $T^{\mu}_{\ \nu} = diag(\rho, -p, -p, -p)$. In addition we have another source term arising from the ϕ equation, $\Delta_\phi \mathcal{L}_c = \frac{1}{2} \frac{1}{\sqrt{-g}} \frac{\delta \sqrt{-g}\mathcal{L}_c}{\delta \phi}$.

The 00 equation of motion is quadratic and may be conveniently written,

$$\dot{\phi} = 3H_S \pm \sqrt{3H_S^2 + e^{\phi}\rho}, \tag{2}$$

where $H_S = \dot{a}_S / a_S$, and we have fixed our units such that $16\pi\alpha' = 1$. The choice of sign here corresponds to our designation of (+) and (−) branches.

For the corrections, we propose classical, one and two loop terms of a particularly simple form. The first term in (3) is the form of α' corrections examined in [6], the second and third are plausible forms for the one and two loops corrections respectively. The large coefficients account for the expected large number of degrees of freedom contributing to the loop. The signs of these terms are deliberately chosen to force the exit.

$$\tfrac{1}{2}\mathcal{L}_c = e^{-\phi}\left(\frac{R_{GB}^2}{4} - \frac{(\nabla\phi)^4}{4}\right) - 1000(\nabla\phi)^4 + 1000e^{\phi}(\nabla\phi)^4. \tag{3}$$

We have checked that qualitatively similar evolution is obtained for a range of coefficients, of which (3) is a representative.

We set up initial conditions in weak coupling near the (+) branch vacuum and a numerical integration yields the evolution shown in Fig. 1 in the $\dot{\phi}$,H_S phase space. We have also plotted lines marking important landmarks in the evolution, the (+) and (−) vacuum ($\rho = 0$ in (2)), the line of branch change (+) → (−) (square root vanishing in (2)) and the position of the $\underline{E|}$ bounce ($H_E = 0$, $\dot{\phi} = 2H_S$). We see the that the evolution falls into three distinct phases which we will discuss in turn. **Phase (i)** The solution begins

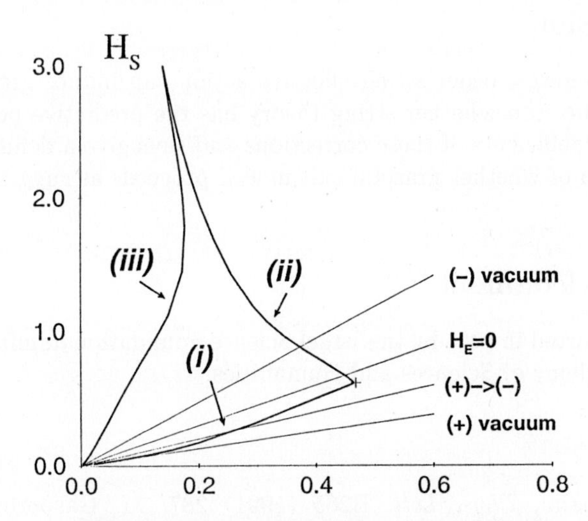

Fig. 1. The correction induced graceful exit

with a long evolution near the (+) vacuum, this is the inflationary phase. As curvature becomes large we see deviation induced by the α' corrections in (3) and without influence from other corrections the solution would settle into the fixed point noted in [6], marked with a '+'. The solution does cross the line of branch change, but does not execute the $\underline{E|}$ bounce required by a complete exit, corresponding to the fact it does not violate NEC in the $\underline{E|}$. **Phase (ii)** While the Universe sits near the fixed point the dilaton is still increasing

linearly, so eventually the loop corrections in (3) will become important. The first to do so is the one loop correction. Since we require further NEC violation to complete the $\underline{E|}$ bounce, we have chosen the sign of the one loop correction to provide this and in this phase corrections are dominated by this term. As a result a bounce occurs and the evolution proceeds into the $\rho > 0$ region. We checked that other forms of loop correction will have the same effect if they are introduced with a coefficient allowing NEC violation. But without further corrections this solution would continue to grow into regions of larger curvature and stronger coupling. We refer to this era as "correction dominated" and we also find there are obstacles to stabilizing the dilaton with standard mechanisms like capture in a potential or radiation production.
Phase (iii) To offset the destabilizing NEC violation we have introduced the two loop correction with the opposite sign, allowing it to overturn the NEC violation when it becomes dominant as ϕ continues to grow. Indeed during this phase we see the expansion decelerating, dilaton growth stabilizing, and the corrections vanishing. We have also checked that in this phase the dilaton can be captured into a potential minimum or halted by radiation production. This phase can be smoothly joined to standard cosmologies.

3 Conclusion

Corrections to lowest order string effective action can induce graceful exit. It remains to be seen whether string theory has the predictive power to fix the form and coefficients of these corrections and thus give a definite answer to the question of whether graceful exit indeed proceeds as suggested in our model.

4 Acknowledgment

Research supported in part by the Israel Science Foundation administered by the Israel Academy of Sciences and Humanities.

References

[1] G. Veneziano, *Phys. Lett.* B265 (1991) 287; M. Gasperini and G. Veneziano, *Astropart. Phys.* 1 (1993) 317.
[2] R. Brustein and G. Veneziano, *Phys. Lett.* B329 (1994) 429.
[3] N. Kaloper, R. Madden and K.A. Olive, *Nucl. Phys.* B452 (1995) 677.
[4] R. Brustein and R. Madden, hep-th/9702043.
[5] R. Brustein and R. Madden, hep-th/9708046.
[6] M. Gasperini, M. Maggiore and G. Veneziano, *Nucl. Phys.* B494 (1997) 315.
[7] S. W. Hawking and G. F. R. Ellis, The large scale structure of space-time, Cambridge University Press, Cambridge, England, 1973

Status of the RICE Experiment

Dave Besson[1] (dzb@kuhep4.phsx.ukans.edu)

University of Kansas, USA

Abstract. The Radio Ice Cerenkov Experiment (RICE) is designed to detect ultrahigh energy (>100 TeV) neutrinos from astrophysical sources, such as AGN. RICE will consist of an array of compact radio receivers (100 to 500 MHz) buried in ice at the South Pole. The objective is to construct an array of greater than one cubic kilometer effective volume, in order to be complementary to TeV energy optical neutrino telescopes currently under construction. The effective volume using the radio technique increases faster with energy than the optical technique, making the method more efficient at ultrahigh ($\geq$ 100 TeV) energies. During the 1995-96 and 1996-97 austral summers, several receivers and transmitters were deployed in bore holes drilled for the AMANDA project, at depths of 120 m to 250 m. This was the first *in situ* test of radio receivers in deep ice for neutrino astronomy. (For more information, see http://kuhep4.phsx.ukans.edu/~iceman/index.html.)

1 Introduction

The detection of ultrahigh energy ($\geq$ 100 TeV) neutrinos represents a unique opportunity in astrophysics. Photons of such high energy are attenuated on the cosmic microwave background, while protons, being charged, are deflected in intergalactic magnetic fields and do not point back to their source. The possibility therefore exists to observe objects not seen by any other method. The exciting potential of high energy neutrino astrophysics has stimulated much theoretical and experimental work in the past few years[1]. Several potential production sites of high energy neutrinos have been theorized, including massive black holes at the center of the Milky Way or Active Galactic Nuclei (AGN), supconducting cosmic strings, the source of the highest energy cosmic rays (which may be related to one of the previous two), gamma ray bursts, weakly interacting massive particles trapped in the sun or earth, young supernova remnants, and X-ray binary systems[2]. In proton-acceleration models of AGN, protons are accelerated to extremely high energy at shocks in the accretion disk around the central black hole or at the base of the jets observed on such objects. The protons collide with photons, photoproducing pions which decay to gamma rays and neutrinos. While most of the protons and gamma rays will not escape the AGN, the neutrinos do. Escaping photons may be responsible for the multi-TeV gamma rays observed from two nearby AGNs. Fluxes from all of these potential sources are small enough that detectors with active volumes on the order of one cubic kilometer are needed in order to observe them.

Several projects (AMANDA,NESTOR,Baikal,ANTARES[3]) are currently underway to attempt to observe these high energy neutrinos. These projects all use photomultiplier tubes to observe visible and UV Cerenkov light emitted by muons created in charged current interactions by muon neutrinos, and all are optimized for detection at TeV energies.

2 RICE Description

RICE is a new experimental effort optimized to detect electron neutrinos at ≈PeV energies, making it complementary to current optical Cerenkov efforts. Neutrinos are detected through their interactions with ice molecules in the Antarctic icecap, based on the principle of "radio coherence". An UHE electron neutrino that undergoes a charged current interaction in the icecap will transfer most of its energy to the resulting electron, which will initiate an electromagnetic shower. A charge imbalance will develop in the shower as positrons are annihilated and atomic electrons are scattered in. Detailed Monte Carlo calculations[4] find that the net charge is about 20 percent of the total number of electrons. This moving pancake of net negative charge will produce coherent Cerenkov radiation at wavelengths smaller than its own spatial extent, which is tens of cm, corresponding to radio frequencies ($f \leq 1$ GHz). This radiation is then observed with an array of radio receivers buried in the ice cap or on the surface at the South Pole. The location, direction, and size of the shower are determined from the timing and signal size information from several receivers in the array.

The radio power radiated scales like the primary energy squared, and the volume of ice sensed by a single detector grows as E^3. This growth with energy is greater than that in the optical regime, making the radio method more efficient at higher energies but much less efficient at TeV energies.

Several features of Antarctic ice make it an attractive target medium. Cold ice has extremely long attenuation lengths at radio frequencies, of order 1 km for 100 MHz to 1 GHz frequencies[5]. Antarctic ice is a very abundant, cheap, high purity material. Also, the AMANDA project has already drilled and installed instruments in several deep bore holes at the South Pole, demonstrating that deployment in deep ice is feasible and providing a ready-made infrastructure to support operations.

3 Pilot Experiment

RICE pilot experiments have been undertaken at the South Pole during the last two Antarctic field seasons (12/95 to 2/96, and 12/96 to 2/97). In both cases, radio receivers and transmitters were deployed on the surface and in bore holes drilled by the AMANDA project. In the 1995-96 season, one transmitter was placed on the snow surface and two receivers were deployed in the ice, at depths of 140 m and 250 m. Both consist of dipole antennas with center frequencies of 100 MHz, and a pair of 36dB HEMT amplifiers in series enclosed in a cylindrical pressure vessel. In the 1996-97 season, four more receivers and one transmitter were deployed in the ice. The receiver depths vary from 152 m to 213 m, while the transmitter is at a depth of 201 m. The 1996-97 receivers consist of dipole antennas with a center frequency of 250 MHz, and a single 36dB HEMT amplifier in a pressure vessel. The signals are sent to the surface via LMR-500 low-loss coaxial cable. Another 36dB HEMT amplifier magnifies the signal at the surface. The buried transmitter consists of a dipole antenna identical to the 1996-97 receiver antennas, connected to the surface by LMR-500 cable. It is driven from the surface by an HP8133A signal generator.

The data acquisition system consists of an HP-54542A digital oscilloscope that is read out by a Macintosh computer programmed with LabView software. The oscilloscope can operate in two trigger modes: noise mode, where random 10 μsec samples of data are taken, and glitch mode, where data samples are recorded when a particular channel exceeds a given threshold.

4 Results and Conclusions

4.1 Objectives of current RICE

There are several objectives of the current RICE array:

1. Engineering/feasibility test. This is the first attempted deployment of radio receivers in the icecap. Clearly, numerous engineering issues must be addressed in order to deploy antennas and receivers (including active amplifiers) under ice at high pressure in order to have the system work at all.
2. Measure man-made local background noise. While Antarctica is relatively free of radio noise compared to most sites in the world today, there are still numerous background sources due to the presence of the manned research base at the Pole. These must be measured in order to determine how to prevent them from producing false neutrino events.
3. Measure noise temperature of ice. Background thermal noise from the ice determines the minimum detection threshold for the array.
4. Demonstrate reconstruction of transmitter location in pulse mode. In order to determine the energy of an event, the shower location must be determined from the timing information in several channels. This can be tested by pulsing the buried transmitter and attempting to reconstruct its known location.
5. Measure ice transparency at radio frequencies.
6. See if presence of AMANDA cable distorts beam pattern of receivers. So far all RICE modules have been deployed in bore holes containing AMANDA strings. This has an obvious advantage: RICE can be deployed along with AMANDA at no additional drilling cost. However, the

presence of AMANDA electrical cables in close proximity to the RICE receivers may affect the antenna beam pattern, and this must be checked before proceeding.

4.2 Results

1) All four receivers deployed in 1996-97 appear to be functioning. However, only one of three deployed transmitters appears to function. Continuous waves at frequencies of 200 to 300 MHz that are sent to the working transmitter can be seen in all four receivers. However, no signals are seen from the other transmitters. 2) The effective noise temperature has been measured to be of order 200 K, consistent with naive expectation. This confirms that the South Polar site is, indeed, a low noise environment. 3) The beam pattern and frequency response are affected by AMANDA cable.

5 Future Plans

5.1 1997-98

Aims for the upcoming 1997-98 campaign are: a) Improve transmitter reliability, b) Upgrade DAQ system to a crate-based system, c) Add 10 more modules at shallow depths, d) Improve Monte Carlo simulations. e) Prototype wide-band biconical antennas, using optical fiber signal transmission.

6 Acknowledgements

This experimental effort has been made possible only through the cooperation, assistance, and support of the AMANDA collaboration, and the financial support of the National Science Foundation's Office of Polar Programs and the K*STAR/NSF EPSCoR Program, the State of Kansas, and the Cottrell Research Corporation.

References

[1] Barwick, S.W., et al., Proc. 25th Int. Cosmic Ray Conf., 7, 1, (Durban, 1997).

[2] Bezrukov, L.B. et al., presented at 2nd Workshop on the Dark Side of the Universe: Experimental Efforts and Theoretical Framework, Rome, Italy (1995).

[3] P. Billoir, presented at this conference.

[4] Halzen, F., Zas, E., Stanev, T., Phys. Lett. **B257**, 432 (1991).

[5] Bogorodsky, V.V., and Gavrilo, V.P., Ice: Physical Properties, Modern Methods of Glaciology (Leningrand, 1980). Salamantin, A.N., et al., in Antarctic Committee Reports, 24, 94 (Moscow, 1985).

New Results of the HEGRA Air Shower Detector Complex

Axel Lindner (axel.lindner@desy.de), for the HEGRA Collaboration[1]

University of Hamburg, Germany

Abstract. Recent highlight results obtained with the air shower detector complex HEGRA at the Canary island La Palma are presented. Observations and searches for γ sources above an energy of 500 GeV are covered, where special emphasis is put on the recent flaring activities of the Blazar Mkn 501. A second main research topic of the experiment is the measurement of the energy spectrum and the elemental composition of the charged cosmic rays. Data covering the energy region of the "knee" around 3 PeV are presented.

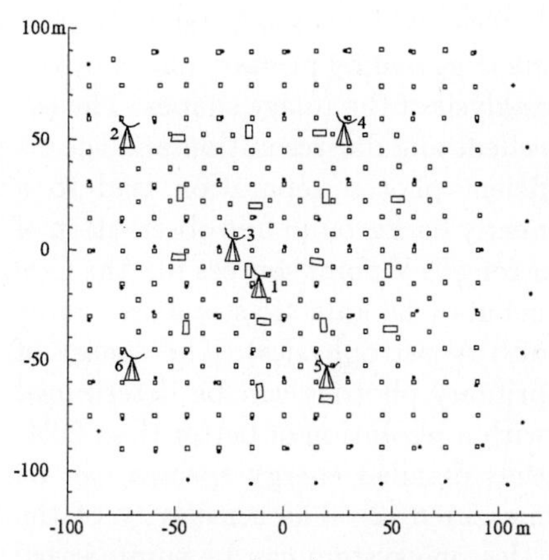

Fig. 1. *The layout of the HEGRA experiment at La Palma*

1 The Setup

The HEGRA (High Energy Gamma Ray Astronomy) [1] experiment is located at the astronomical observatory of the Canary island La Palma at a height of 2200 m. It is operated by a collaboration of the German Max-Planck-Institutes für Kernphysik (Heidelberg) and für Physik (München), the univers. of Hamburg, Kiel, Wuppertal (Germany) and Madrid (Spain) and the Yerevan Physics Institute (Armenia). The experimental setup covers a square area of $40000\,\mathrm{m}^2$ (Figure 1).

The HEGRA experiment measures extensive air showers (EAS) initiated by high energetic cosmic photons and nuclei. It can roughly be divided into two parts: imaging air Cherenkov telescopes (IACTs) observe selected targets with an energy threshold of 500 GeV, while three arrays of detectors monitor the whole sky above the installation in the energy range beyond 15 TeV.

1.1 The HEGRA IACTs

Since the end of 1996 six IACTs are operational. Four identical telescopes and an upgraded prototype can operate as a system. This unique installation observes the air Cherenkov light images of EAS in a stereo mode. The segmented mirror area of each system telescope is $8.5\,\mathrm{m}^2$ large. Each camera consist of 271 pixels (photomultipliers each viewing $0.25°$ of the sky) covering a field of view of $4.3°$.

[1] supp. by the BMBF, DFG, CICYT

For photons above the energy threshold of 500 GeV the direction of the primary photon is deduced from the orientation of the air Cherenkov light image in the camera, where the stereo observations allow for an angular resolution of better than 0.1°. Photon induced EAS are separated from showers triggered by primary nuclei by the analysis of the image shapes. The excellent angular resolution and the efficient photon separation lead to a nearly background free observation of strong TeV point sources like the Crab nebula (the galactic standard candle of TeV astrophysics). The energy of primary photons can be determined with a resolution of better than 20%, thus detailed energy spectra can be measured [2]. The sensitivity of the telescope system can be summarized such, that within 100 h observation a detection on the 5 σ level is possible for a point source with a flux of only 3% of the Crab.

1.2 The HEGRA Arrays

The HEGRA arrays sample the EAS showerfront. 243 scintillator huts on a grid with 15 m spacing and a more dense part around the center register charged particles. The so called AIROBICC array of 77 stations (consisting of 20 cm diameter open photomultipliers attached to a Winston cone) provides a non imaging measurement of the air Cherenkov light. A Geiger tower array of 17 stations is located in the central part of the array. They allow for the measurement of the electromagnetic energy of an air shower at detector level and the identification of muons. The energy threshold for vertical photon induced showers was around 20 TeV for scintillator triggered events, which has been lowered to 5 TeV after a DAQ upgrade in spring 1997. Air showers above 11 TeV trigger AIROBICC. The angular resolution for photon induced showers at threshold amounts to 0.3° for AIROBICC triggered events and 0.9° for scintillator data (above 20 TeV).

2 TeV Photon Sources

The most exciting result in 1997 were the measurements with the imaging air Cherenkov telescopes (IACT) of the emission of TeV photons with a very high and variable flux from the extragalactic source Mkn 501 [3, 4]. This object in a distance of 500 million lightyears belongs to the Blazar class of active galactic nuclei (AGN) and is thought to contain probably a black hole of 10^8 solar masses. Blazars are AGNs which emit relativistic jets pointed towards the observer. Under certain conditions they can also be detected in the TeV energy regime. TeV γ's from Mkn 501 were first measured in 1995 [5] with a flux corresponding to 8% of the galactic Crab nebula (the strongest known TeV emitter before). The activity rose to 0.3 Crab in 1996 [6]. In 1997 the mean flux lies around 2.0 Crab, but flares with intensities of up to 10 times the Crab flux have been registered approx. once a month (Fig. 2). The energy spectrum in comparison to the spectrum of the Crab nebula is shown in Figure 3. The Mkn 501 spectrum follows a power law up to 10 TeV. At higher energies the study of systematic detector effects is not finished yet so that the data points are not included in the plot.
Due to the absorption of high energy

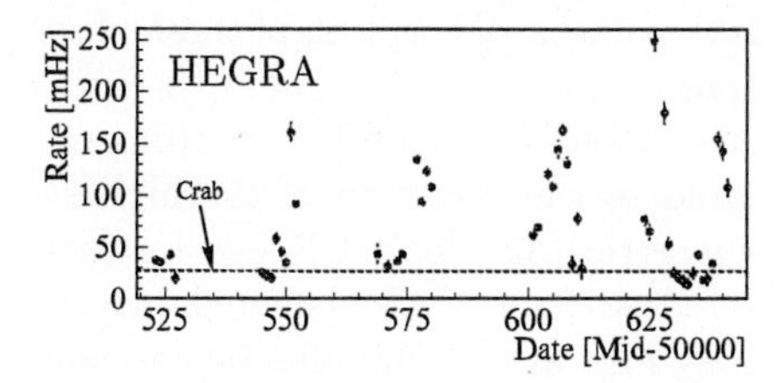

Fig. 2. *The HEGRA IACT system trigger rate of photons from Mkn 501 in a part of 1997. April 1st corresponds to 539 on the horizontal axis, July 1st to 630.*

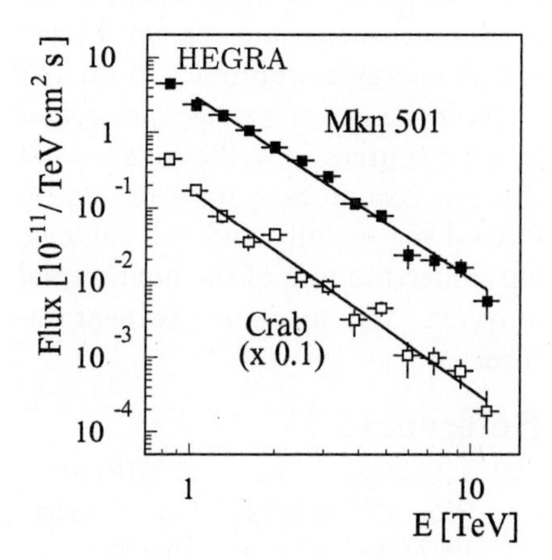

Fig. 3. *The energy spectra of Mkn 501 (averaged over all measurements in 1997) and of the Crab nebula. Data beyond 10 TeV are not shown.*

photons via interaction with infrared photons resulting in pair production these data will provide an important tool to measure indirectly the intergalactic infrared background light intensity, which is tightly connected to the history of star formation [7]. Multiwavelength observations of Mkn 501 [8] indicate that the measurements may be explained by synchrotron emission of $e^{\pm}$ and inverse Compton scattering. Other sources observed with the Cherenkov telescopes are Mkn 421, an object very similar to Mkn 501 at nearly the same distance, but not as spectacular as Mkn 501 in 1997, and the Crab nebula. No evidence for TeV photon emission could be found from other AGNs. Also no TeV photons could be detected from the supernova remnants (SNR) IC443 and G78+2. The upper limits provide significant constraints on models which try to explain the acceleration of the bulk of charged cosmic rays in the shells of SNRs.
Source searches with the HEGRA arrays (with a higher energy threshold and less flux sensitivity compared to the IACTs) have not been successful up to now. Only small excesses have been found from the Crab nebula, Hercules X-1 and a sample of near Blazars, but all need further confirmation.

3 Charged Cosmic Rays

The energy spectrum of charged cosmic rays (CR) follows a power law up to roughly 3 PeV, where the CR spectrum becomes steeper and follows another power law. The understanding of this so called "knee" feature may provide a clue to understand the origin of CR. HEGRA developed new methods to measure the energy spectrum and the coarse composition in the energy region around the knee. One method is described in more detail: the air Cherenkov light measurements of AIROBICC allow to reconstruct the distance between the detector and the air shower maximum. Together with the registered number of charged particles at detector level or the total amount of Cherenkov light the energy contained in the electromagnetic shower component can be determined. This is combined with the energy per nucleon of the primary nucleus estimated from the depth of the shower maximum to calculate the primary energy and the nucleon number. While the primary energy can be determined for cosmic ray nuclei with an accuracy of 30%, the determination of the nucleon number is possible only in a coarse manner due to natural fluctuations in the shower developments. Figure 4 shows the reconstructed (preliminary) fraction of light nuclei (H and He) in the CR. The results below 1 PeV are in good

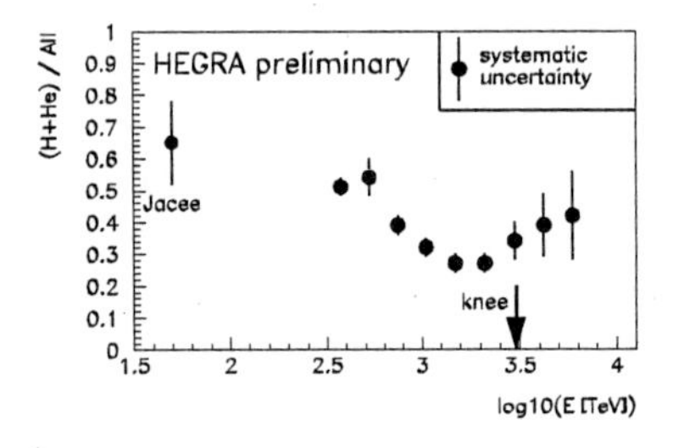

Fig. 4. *The fraction of light nuclei in cosmic rays. "JACEE" marks results obtained by direct balloon measurements.*

agreement to data obtained by balloon and satellite measurements. The decrease near the knee is in agreement with models describing the CR acceleration in supernova remnants, but also a constant composition is not excluded.

4 Other HEGRA Results

No Gamma Ray Burst (GRB) counterparts have been found for energies above 5 TeV, neither coincident with the GRBs or for long time measurements. This is in accordance with an extragalactic origin of GRBs, where no TeV emission is expected due to intergalactic absorption of these photons.

A small evidence for TeV photons related to the direction of the highest energetic (320 EeV) CR event registered so far [9] was detected, which may hint at an intergalactic cascade initiated by extremely energetic particles. Currently IACT measurements in this region are under way.

5 Summary

The last years have established air shower measurements as a new branch of high energy astrophysics. Both TeV γ observations of extragalactic and galactic sources as well as analyses of charged cosmic rays have been performed successfully and will enhance our understanding of the nonthermal universe even more in the near future.

References

[1] A.Lindner et al. (HEGRA), Proc. 25th ICRC **5**, 113 (Durban 1997) and references therein
[2] W.Hofmann, astro-ph/9710297 (1997)
[3] F.Aharonian et al. (HEGRA), astro-ph/9706019 (1997), acc. by Ast. & Astrophys. Lett.
[4] R.J.Protheroe et al., astro-ph/9710118 (1997)
[5] J.Quinn et al. (WHIPPLE), ApJL. 456, L83 (1996)
[6] S.M.Bradbury et al. (HEGRA), Ast. & Astrophys. **320**, L 5 (97)
[7] D.MacMinn, J.R.Primack, Space Sci. Rev **75**, 413 (1996) F.W.Stecker, O.C.De Jager, astro-ph/9710145 (1997)
[8] E.Pian et al. (BEPPO-SAX), astro-ph/9710331 (1997)
[9] D.J.Bird et al. (FLY'S EYE), Astrophys. J. **441** 1, 144 (1995)

Ultra High Energy Cosmic Rays: Physics Issues and Potentialities of the Auger Hybrid Detector

Pierre Billoir[1], on behalf of the AUGER collaboration (billoir@in2p3.fr)

LPNHE, Universités Paris VI-VII, Paris, France

Abstract. The Pierre Auger Observatory Project is designed to collect a large number of cosmic rays of highest energy (above $10^{19}\,eV$), whose origin is yet unknown. The atmospheric showers are observed through the combination of fluorescence detectors and a large ground array of Čerenkov water tanks. Neutrino interactions above $10^{17}\,eV$ could also be extracted from background at large zenith angles.

1 The cosmic rays of highest energy

1.1 Observations and possible interpretations [1]

Sixty years ago Pierre Auger observed extended air showers at ground level; from their size he concluded that the spectrum of primary cosmic rays should extend up to $10^{15}\,eV$. In the last 30 years, several detectors using various techniques observed energies up to $3.10^{20}\,eV$, with a power-law spectrum (roughly E^{-3}). At these energies the interaction with the cosmic microwave background radiation (photoproduction of pions and photodisintegration for nuclei) gives rise to the GZK cutoff : the source should be relatively close (typically: less than $100\,Mpc$ for protons at $5.10^{19}\,eV$, less than $20\,Mpc$ for nuclei).

At present, there is no satisfactory explanation of their origin. The models can be classified into:
- "*bottom-up*" processes: the most probable is a stochastic acceleration of protons or nuclei through shocks fronts (Fermi mechanism), in an accretion disk or in lobes of radio-galaxies, or in colliding galaxies. However the parameters of such objects have to be pushed to their limits to explain the highest observed energies. Such sources would be pointlike, and the intergalactic and galactic magnetic fields are not expected to scatter the particles above $10^{20}\,eV$ by more than a few degrees.
- "*top-down*" phenomena: for example supermassive GUT particles could be radiated by *topological defects* formed in the early universe. They could be characteristized by their angular dispersion and the presence of "exotic" components, such as photons or neutrinos above $10^{18}\,eV$.

1.2 Atmospheric shower development

The shower is a cascade of interactions and decays; for convenience, two components are distinguished:
- An *electromagnetic* cascade ($\gamma, e^{\pm}$), induced directly by an electromagnetic primary particle, or by π^0's if the primary is hadronic. It contains most of the energy and it is practically the only source of fluorescence.
- A *hadronic* cascade with nucleons and charged mesons, decaying into muons when their interaction length (determined by the cross section and the density of matter) is larger than their decay length (depending on their Lorentz factor); if the zenith angle is small, the muons are created at intermediate altitudes ($\simeq 10\,km$) with a few GeV, so that most of them reach the ground.

Although the first steps of the cascade correspond to c.m. energies (up to hundreds of TeV) beyond the present experimental results from colliders, the main features of the shower development do not depend strongly on the extrapolated interaction models, and the observations in present detectors are in good agreement with the simulations.

2 The Pierre Auger Observatory : an hybrid detector

2.1 General design

The basic idea of the project is to combine *atmospheric fluorescence* "fly's eye" detectors (FD), which gives the longitudinal development of the shower core, with a *ground array* (GA) of water tanks, which samples its lateral distribution through the Čerenkov radiation produced by charged relativistic particles. The tanks are cylinders (radius $1.8\,m$, depth $1.2\,m$) spaced of $1.5\,km$, so that a shower of $10^{19} eV$ or more triggers at least 5 ground stations; an array of 1600 stations, covering $3000\,km^2$, would then observe 3000 events per year above this energy. A full-scale water tank prototype was installed in the AGASA area in AKENO (Japan); the signals recorded in coincidence with this array are in good agreement with the predictions. Three fly's eyes give a full coverage of the array, with some overlap regions: this allows a precise stereoscopic reconstruction, with possible cross-calibration of FD and GA. The duty cycle of the FD is 10% (15% if some protection against moonlight can be achieved). Wireless communication between the tanks and the central station is used to build the global trigger and the data acquisition. All stations are synchronized with GPS receivers.

In order to cover the full sky and to evaluate unbiased anisotropies, two sites have been chosen, in Northern and Southern hemisphere (Utah and Argentina).

2.2 Primary Identification

As a first approximation, a nucleus of mass A, energy E interacts as A nucleons of energy E/A: the shower develops earlier than with a single nucleon of the same energy, and the fraction of muons in ground particles is larger (less interaction steps implies less electromagnetic conversion through π^0's). If the primary is a photon, the hadronic shower and then the muon rate are reduced. The muonic fraction is evaluated in GA from the rise time (the muons arrive earlier than $\gamma, e^{\pm}$ in average) and the lateral distribution (their density decreases less rapidly with the distance from the shower axis).

Two independent discriminating variables are then defined : the position of the maximum seen by the FD, and the (μ:e.m.) ratio at ground. When combining these informations, the separation between γ and p, or p and Fe, represents roughly twice the fluctuations, allowing a good statistical evaluation of the composition.

2.3 Energy and angle precision

The energy is estimated independently from the fluorescence profile and the GA signals. A full simulation of the shower and the detector gives a precision going from 30% to 15% between 10^{19} and $10^{20}\,eV$, with the GA alone (20% to 10% with a hybrid reconstruction).The direction is computed from the arrival times on GA stations (precision 2 to 1 deg); the geometrical constraint of the FD improves greatly the reconstruction (0.25 to 0.2 deg).

3 Horizontal atmospheric showers

3.1 Muonic tail of "normal" showers

Strongly interacting particles entering the atmosphere with a large zenithal angle (above 70 deg) produce a shower with specific features : the development of the hadronic cascade is achieved at high altitude, with a low air density. Then the charged pions decay early and give muons at high energy (typically 50 to 100 GeV), well collimated to the shower axis; these muons travel over tens of kilometers and keep enough density to trigger GA stations. On the contrary the electromagnetic shower is completely extinguished. As a result, the shower front becomes very thin and flat, giving in the stations very short signals, well aligned in time. Another interesting feature is the distortion due to the geomagnetic field: the shape of the muon spot on the ground depends on the azimuth angle. An example is shown in Fig.1 (left)

3.2 "Deep" showers produced by weakly interacting objects

At $10^{17}\,eV$ and above, neutrinos (or any object with similar cross section on matter) have a small, but not negligible probability to interact in the atmosphere (especially at large zenith angles), proportionnally to the density, hence mainly at low altitude. A charged current interaction of a ν_e gives rise to a shower with its full energy ; the acceptance to these "deep" showers (with zenith angle larger than $70\,deg$) was evaluated at 10^{17}, 10^{18}, $10^{19}\,eV$ to be respectively 2, 15 and $30\,km^3sr$ of equivalent water (with zenith angle larger than $70\,deg$). Then AUGER would see around 1 neutrino event per year (extrapolation of known astrophysical flux) or more in "top-down" models. These rare events could be discriminated from the large background of "normal" showers, using the time spread of the GA signals and the curvature of the front. Fig.1 (right) gives an example of a shower induced $3\,km$ above the ground.

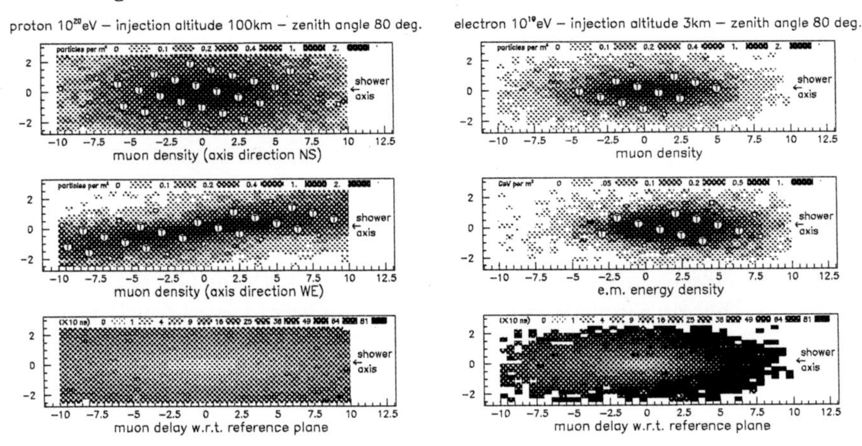

Fig. 1. Ground detection of horizontal showers. Left : muonic tail of a "normal" shower, with geomagnetic effects, and deviations of the front from a plane. Right : "deep" neutrino-like shower, with muon and electromagnetic densities, and deviations from a plane

4 Conclusions

The Auger Observatory Project is aimed to solve the puzzle of the origin of extremely high energy cosmic rays. It is an optimal combination of well proven techniques, with a good precision on the energy and on the direction (allowing the localization of sources and possibly the analysis of cosmic magnetic fields), with an unique capability of cross checking independent measurements from fluorescence and ground array detectors. Statistical evaluation of the chemical composition is possible. Neutrinos or exotic weakly interacting objects could be detected in almost horizontal showers.

References

[1] Details on the project and references may be found in the WWW sites of the Collaboration, for example :
http://www-lpnhep.in2p3.fr/auger/
http://www-td-auger.fnal.gov:82/

Part XV

Lattice Calculations and Non-pertubative Methods

Consequences of Magnetic Stability in Yang–Mills Theory at High Temperature

A.C. Kalloniatis (ack@theorie3.physik.uni-erlangen.de)

University of Erlangen-Nuremberg, Germany

Abstract. SU(2) Yang-Mills theory at finite temperature is probed by a homogeneous chromomagnetic field. We use a recent axial gauge formulation which has the feature of respecting Z_2-symmetry in perturbation theory. In the Z_2-symmetric phase a temperature-dependent mass term and antiperiodic boundary conditions arise, providing stabilisation against magnetic field formation at sufficiently high temperature. In the broken phase the combined constraints of thermodynamic and magnetic stability lead to a slow approach to the Stefan-Boltzmann limit, as seen in lattice SU(2) gauge theory.

Monte Carlo lattice simulations of pure Yang-Mills theory and QCD have unfolded a consistent picture for hadronic physics at high temperature. In pure Yang-Mills theory a deconfinement phase transition occurs for static quarks signalled by a change in the Polyakov loop corresponding to a change from unbroken center symmetry to a spontaneously broken phase. Also, there is now evidence [1] of nontrivial gluonic structure above T_c which persists over a significant temperature range. The approach to the high temperature limit of a free gluon gas is slow, and even at very high temperatures standard perturbation theory has only met with limited success. In this report, we show that this behaviour can be qualitatively recovered in pure SU(2) gauge theory by simultaneously demanding thermodynamic and magnetic stability in the broken phase. We exploit certain nonperturbative features which are analytically accessible in the static temporal gauge, in particular the generation of changes in boundary conditions and mass terms as observed recently in [2].

At finite temperature, the Euclidean time direction is compactified ($x_0 \in (0, 1/T)$). We use the static temporal gauge, $\partial_0 A_0 = 0$ followed by a diagonalisation of the surviving time-independent zero mode so that we are left with $a_0^3(\mathbf{x})\tau^3/2$ in the temporal component. We relabel now $a_0^3 \to a_0$. This object appears in the eigenphases of the field $\mathcal{L}(\mathbf{x}) = \mathrm{P}\exp(ig\int_0^{1/T} dx_0 A_0(x))$, the trace of which can be identified with the Polyakov loop, $L(\mathbf{x}) = \frac{1}{2}\mathrm{Tr}\,\mathcal{L}(\mathbf{x})$. The Polyakov loop is then given by the expression $L(\mathbf{x}) = \cos[ga_0(\mathbf{x})/2T]$ which shows that in the static temporal gauge this important variable appears as an elementary rather than composite field. There is a residual gauge freedom with respect to Abelian transformations of the neutral fields which can be easily fixed with no associated Faddeev-Popov ghosts. The details of this are not important for what follows. In the generating functional one must include a nontrivial Jacobian $\prod_{\mathbf{y}} \sin^2(ga_0(\mathbf{y})/2T)$. As a consequence the corresponding part of the Haar measure enters with a finite range of integration precluding any ordinary perturbative treatment. In QED the same procedure yields the standard flat measure for $a_0(\mathbf{x})$.

As shown in [2], we can nonetheless integrate out a_0 in Yang-Mills theory prior to any perturbative expansion keeping all other fields external. First, we regulate the functional integral by a spatial lattice (lattice constant ℓ). Subsequently, the variable a_0 is shifted to $a_0' = a_0 - \pi T/g$. As a consequence, the $\sin^2$ factor in the functional measure is replaced by $\cos^2$, and the integration limits become $[-\pi T/g, \pi T/g]$. It is advantageous to make the redefinition $A_i^1 \mp iA_i^2 \to \exp(\pm i\pi T x_0)(A_i^1 \mp iA_i^2)$. This does not change the physics, but yields charged gluon fields which obey *antiperiodic* boundary conditions. With the above spatial discretisation, the functional integral over a_0 can be explicitly computed while keeping all other fields external. The correlator of two a_0 fields can be computed order by order in the dimensionless variable $\ell T/g^2$. Ultralocality of the Jacobian leads to the leading order term being a delta function: a_0 is non-propagating. Higher order terms give 'hopping', but in the continuum limit $\ell \to 0$ these are always suppressed. No ultraviolet divergent expressions occur in this subsector of the theory, so that the lattice regulator ℓ can be safely taken to zero after completion of the a_0 integration. One is left with an effective action for the remaining degrees of freedom with $A_0 = 0$ and a term $\frac{1}{2}M(T)^2 \sum_{a=1,2} \int d^4x A_i^a(x) A^{a,i}(x)$; the Polyakov loop variable has left its signature in a "geometrical" mass term of the charged gluons $M(T)^2 = \left(\pi^2/3 - 2\right) T^2$ and in a change to antiperiodic boundary conditions $A_i^{1,2}(1/T, \mathbf{x}) = e^{i\chi} A_i^{1,2}(0, \mathbf{x})$, with, here, $\chi = \pi$. The neutral gluons A_i^3 remain massless and periodic. As discussed in [2], the expectation value of the Polyakov loop $\langle L(\mathbf{x})\rangle$ vanishes in zeroeth order perturbation theory, indicating that this formulation is particularly well suited for describing the Z_2-symmetric phase.

In this formalism, we proceed now to study magnetic stability by probing the above effective action with a background magnetic field, in the first instance keeping the variables χ and $M(T)$ unspecified for reasons which will become clear. We choose a homogeneous external chromomagnetic field in the colour 3-direction, with the potential of the form $A_\mu^a|_{\text{bg}} = -\delta^{a3} g_{\mu 1} x_2 H$. Although this looks identical to Savvidy's choice [3] there is a difference: unlike the background-gauge case, the 3-direction has *already* been specified by diagonalizing the Polyakov loop variables. We thus investigate the stability of the vacuum against quite specific homogeneous magnetic field fluctuations which point in the same direction in colour space as the Polyakov loop.

The one-loop energy density in the presence of the external field is the sum over the single particle energies, determined by solving the the Landau level problem in the temporal gauge [4]. At one loop we need only consider the charged gluon contributions. In this scheme, we evaluate the following expression for the one loop correction to the vacuum energy density,

$$\varepsilon_1 = \frac{gHT\mu^{2\epsilon}}{2\pi} \sum_{s=\pm 1} \sum_{n=0}^{\infty} \sum_{k=-\infty}^{\infty} [2gH(s+n+1/2) + (2\pi k + \chi)^2 T^2 + M(T)^2]^{\frac{1}{2}-\epsilon} \tag{1}$$

to which should be added the classical contribution $\varepsilon_0 = H^2/2$. The parameter ϵ regularises the ultraviolet behaviour in the sense of the zeta-function method [5]. The arbitrary scale parameter μ^2 has been introduced to keep dimensions correct.

The Casimir contribution from the charged gluons will emerge from $\varepsilon = \varepsilon_0 + \varepsilon_1$ in the $H \to 0$ limit and will correspond to the standard Casimir energy for massive charged scalars with quasiperiodic boundary conditions. From this the pressure can be computed. We do not give the result here but two aspects can be noted: the temperature dependent mass leads to logarithmic terms which thwart the approach to the Stefan-Boltzmann limit at high temperature. Secondly, even in the absence of the mass the antiperiodic boundary conditions mean the system is thermodynamically unstable. In the confinement phase where the true vacuum is highly nontrivial these are natural features in the above context. However we will see below that these parameters must vary if one is to go to the deconfined phase.

With the external field non-zero the renormalisation proceeds in two steps: first a coupling constant renormalisation after identification of the renormalisation group invariant $B = gH$ is performed. This leads to the correct SU(2) Yang-Mills beta function to one loop order, a confirmation that the non-perturbative aspects of the present formulation do not spoil asymptotic freedom. The arbitrary scale μ^2 is fixed by choosing $(11/48\pi^2)\ln(B/\Lambda_{\mathrm{MS}}^2) \equiv 1/(2g_{\mathrm{R}}^2(\mu)) + (11/48\pi^2)\ln(B/\mu^2)$ where Λ_{MS} is the scale parameter for SU(2) in the minimal subtraction scheme. There remain poles independent of B, but dependent on M, T. These are eliminated by performing a subtraction at $B = 0$.

In [4] we have numerically computed the thus renormalised energy density for arbitrary magnetic field and checked that the only stable minimum appears at zero field $B = 0$ for a range of temperatures. Namely, because of the mass and boundary conditions there is no imaginary part (at values of B large compared to the mass we recover the unstable Savvidy result [3] with imaginary subleading terms). In the neighborhood of $B = 0$, we obtain for the renormalised total energy density to one loop

$$\varepsilon_{\mathrm{R}}(B \ll M(T)^2) \approx \frac{11B^2}{48\pi^2}[\ln\left(\frac{M(T)^2}{\Lambda_{\mathrm{MS}}^2}\right) + 1 + \psi(-1/2) - 4\sum_{n=1}^{\infty} \cos(n\chi) K_0(nM(T)/T)] \tag{2}$$

Note that due to the existence of the additional mass scale, we do not have the magnetic field appearing in the logarithm. Now it is simple to determine the value of T, T_0, above which the sign of B^2 in Eq. (2) is positive for $\chi = \pi$

and $M(T)^2 = (\pi^2/3 - 2)T^2$:

$$1/T_0 = \frac{M(T)}{\Lambda_{\rm MS} T} \exp[(1 + \psi(-1/2) - 4\sum_{n=1}^{\infty}(-1)^n K_0(nM(T)/T))/2] \approx \frac{3.36}{\Lambda_{\rm MS}}. \tag{3}$$

For $T \geq T_0$ the system can be said to be preferring the "empty" vacuum ($B = 0$), even though the calculation does not yet contain enough dynamics to make statements about the true high temperature behaviour. Given that lattice measurements give a critical temperature for deconfinement $T_c \approx \Lambda_{\rm MS}$, then the center-symmetric phase is stable above roughly $T_c/3$. Thus, it is plausible to assume that in the broken phase this stability will persist.

We thus appear to have arrived at a partly satisfactory description of the response of the Z_2 symmetric Yang-Mills phase to a homogeneous magnetic field at some intermediate temperature. Increasing the temperature will improve the stability with respect to magnetic field fluctuations. Above the critical temperature however the thermodynamic instability becomes distinctly unnatural (at sufficiently high temperature, if the Stefan-Boltzmann law is approximately satisfied, then *positive pressure* suffices to characterise thermodynamic stability). To allow for variations in the parameters $M(T)$ and χ is thus clearly necessary for a correct description of the high temperature phase: to recover the Stefan-Boltzmann law and the correct dimensional reduction. Neither this variation, nor even the phase transition itself, can be generated in a one loop calculation. We thus make the minimal assumption that the essential nonperturbative dynamics persisting above the phase transition remain encoded in these – now varying – quantities: $M(T)$ and χ. We consider the scenario that at the critical point the charged gluons become periodic, reflecting the rapid change in the Polyakov loop, but that their mass falls off gradually to zero. We shall use the simultaneous demand of thermodynamic and magnetic stability to determine bounds on the rate of this fall off. So, with $\chi = 0$ the condition for the critical mass, $M_0(T)$, above which the system is magnetically stable reads (see Eq. (2))

$$\ln\frac{\Lambda_{\rm MS}}{4\pi T} = \frac{1}{2} - \gamma + \frac{\psi(-1/2)}{2} - \frac{\pi T}{M_0(T)} - \pi\sum_{n=1}^{\infty}\left(\frac{2}{\sqrt{\frac{M_0(T)^2}{T^2} + (2\pi n)^2}} - \frac{1}{\pi n}\right). \tag{4}$$

From this we get that stability is only possible for values of $M(T)$ larger than a certain limiting value, $M = \Lambda_{\rm MS}\exp\left(-\frac{1}{2}(1 + \psi(-1/2))\right) \approx 0.60\Lambda_{\rm MS}$. More significantly the high T behaviour of $M_0(T)$ can be derived analytically from Eq. (4),

$$M_0(T \to \infty) \approx \frac{2\pi T}{(3(1-\gamma) + 2\ln(2\pi T/\Lambda_{\rm MS}))}. \tag{5}$$

Inserting the standard running coupling constant this corresponds to the lower bound

$$M_0(T) \geq \frac{11}{12\pi} T g^2(T) \approx 0.29\, T g^2(T) \qquad (T \to \infty) \tag{6}$$

at high temperatures. This result guarantees that we will asymptotically recover the Stefan-Boltzmann behaviour of an ideal gas. It is remarkable that this coincides with both theoretical expectations and lattice results for the temperature dependence of the gluon magnetic mass; thus for instance, a recent determination in the Landau gauge in a wide temperature range [6] has been fitted with the formula $M_{\mathrm{m}}(T) = (0.46 \pm 0.01) T g^2(T)$, surprisingly close to (but larger than) our lower bound from magnetic stability, Eq. (6). We emphasise that we have obtained Eq. (6) within a perturbative calculation in a region where perturbation theory becomes increasingly more reliable, but with the single assumption that the nonperturbative charged gluon mass obtained in the static temporal gauge does not immediately vanish above the critical temperature. This is not inconsistent with the understanding that the 'magnetic gluon mass' is a fundamentally nonperturbative quantity.

We now estimate the energy density and pressure by inserting this critical gluon mass into the well known expressions for the Casimir effect. We need one subtraction to get a finite result (which reduces somewhat the predictive power of this approach): we adjust the energy density at one reference temperature $\tilde{T}$. After this subtraction, we obtain the Casimir energy density for charged gluons

$$\varepsilon(T) = M^4 \left[\frac{\tilde{\varepsilon}}{\tilde{M}^4} + \frac{2}{16\pi^2} \ln \frac{M}{\tilde{M}} - \frac{2}{\pi^2} \sum_{n=1}^{\infty} \frac{1}{n^2} \left(\frac{T^2}{M^2} K_2\left(\frac{nM}{T}\right) - \frac{\tilde{T}^2}{\tilde{M}^2} K_2\left(\frac{n\tilde{M}}{\tilde{T}}\right) \right) \right] \tag{7}$$

where $M = M_0(T)$, $\tilde{M} = M_0(\tilde{T})$ and $\tilde{\varepsilon} = \varepsilon(\tilde{T})$. As a sample calculation, we set $\tilde{\varepsilon}$ equal to a certain fraction of the Stefan-Boltzmann value (0.8) at the highest temperature; this value has been selected since it is consistent with the reduction of 30% at $2T_c$ reported in the SU(2) lattice calculation of [7]. The pressure was evaluated by numerical differentiation. The result is shown in Fig. 1 where the massless neutral gluon contribution has been added, since our formalism gives no hint of modifications in the neutral sector. Note that the low temperature region should not be taken seriously as we cannot in this calculation describe the phase transition. We observe the slow (logarithmic) approach to the Stefan-Boltzmann limit as $T \to \infty$, highly reminiscent of the lattice results of [1].

We have also studied the insertion of the mass bound Eq. (6) into a computation of the interaction measure and the electric screening mass. In the latter case, we succeed in reproducing the linear $T-$ rather than $gT-$ dependence for the electric screening mass seen by [6]. These results, together with more detailed explanations of the above, are reported in [4].

This work was supported by the grants 06ER747 and 06ER809 from the BMBF.

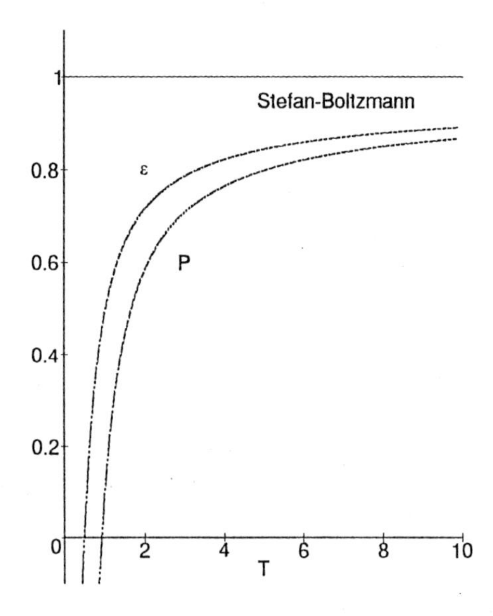

Fig. 1. Energy density ϵ and pressure P, normalised to Stefan-Boltzmann values, vs. temperature in units of Λ_{MS}. The curves correspond to periodic gluons with massless neutral components and charged components of mass as in Eq. (6), and should be regarded as upper bounds at high T.

References

[1] J. Engels, et al., Z. Phys. C. **42** (1989) 341; G. Boyd, et al., Nucl. Phys. **B469** (1996) 419

[2] F. Lenz and M. Thies, hep-ph/9703398

[3] G.K. Savvidy, Phys.Lett. **71B** (1977) 133

[4] V. Eletsky, A.C. Kalloniatis, F. Lenz, M. Thies, hep-ph/9711230

[5] E. Elizalde, et al., "Zeta Regularization Techniques with Applications", World Scientific, Singapore, 1994

[6] U.M. Heller, F. Karsch, J. Rank, Phys. Lett. **B355** (1995) 511

[7] J. Engels, F. Karsch, K. Redlich, Nucl. Phys. **B435** (1995) 295

[8] U.M. Heller, F. Karsch and J. Rank, hep-lat/9710033

Topological and Chiral Structure in Lattice QCD at Finite Temperature

Stefan Thurner, Markus Feurstein and Harald Markum
(thurner@kph.tuwien.ac.at)

Institut für Kernphysik, TU Wien, A-1040 Vienna, Austria

Abstract. We perform an analysis of the topological and chiral vacuum structure of four-dimensional QCD on the lattice at finite temperature. It is demonstrated that at the places where instantons are present, amplified production of monopole currents and quark condensate takes place. We observe such nontrivial correlation functions also in the deconfinement phase and for single gauge field configurations. Values of the topological susceptibility across the temperature phase transition are reported.

As a matter of fact topology of gauge fields is a characteristic feature of nonperturbative QCD. Instantons on one hand, are known to be relevant in the following aspects: First, topological excitations are needed to theoretically understand particle masses, as in the case of the η'. Here the topological susceptibility χ contributes to the mass according to the Witten-Veneziano formula. Second, the topological charge Q of gauge fields is related to the zero modes of the fermionic matrix via the Atiyah-Singer index theorem. This gives hope that topology can be used to clarify the chiral symmetry breaking mechanism of QCD. Third, one believes that the instantons as carriers of the topological charge might play a crucial role in understanding the confinement mechanism, if one assumes that they form a so-called instanton liquid [1]. Color magnetic monopoles on the other hand can explain the confinement mechanism if one assumes a dual Meissner effect, where monopoles condense and force the color electric flux into tubes, leading to a linear confinement potential. There is evidence that monopoles actually condense in the confinement phase of QCD.

To study nonperturbative QCD from first principles, the lattice approach is an appropriate method. Recently, it was demonstrated with lattice calculations that monopole currents appear preferably in the regions of nonvanishing topological charge density [2, 3]. It has been conjectured that both instantons and also monopoles are related to chiral symmetry breaking [1, 4]. In this contribution we further support these ideas, by a direct investigation of the local correlations of the quark condensate and the topological charge and monopole density. We demonstrate a nontrivial coexistence also on single gauge field configurations, by looking at correlation functions of individual fields and by direct visualization. At the end we present the topological susceptibility obtained from a Wilson renormalization group approach [5, 6].

There exist several definitions of the topological charge on a Euclidian lattice. The field theoretic prescriptions are a straightforward discretization of the continuum expression, $q(x) = \frac{g^2}{32\pi^2}\epsilon^{\mu\nu\rho\sigma}\,\mathrm{Tr}\left(F_{\mu\nu}(x)F_{\rho\sigma}(x)\right)$ [7]. To investigate monopole currents we project $SU(N)$ onto its abelian degrees of freedom, such that an abelian $U(1)^{N-1}$ theory remains. We employ the so-called maximum abelian gauge [8, 9]. From the monopole currents $m_i(x,\mu)$ we define the local monopole density as $\rho_m(x) = \frac{1}{4NV_4}\sum_{\mu,i}|m_i(x,\mu)|$. Mathemati-

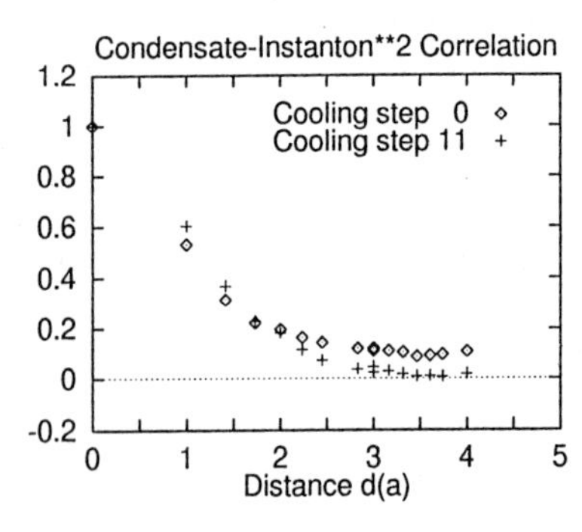

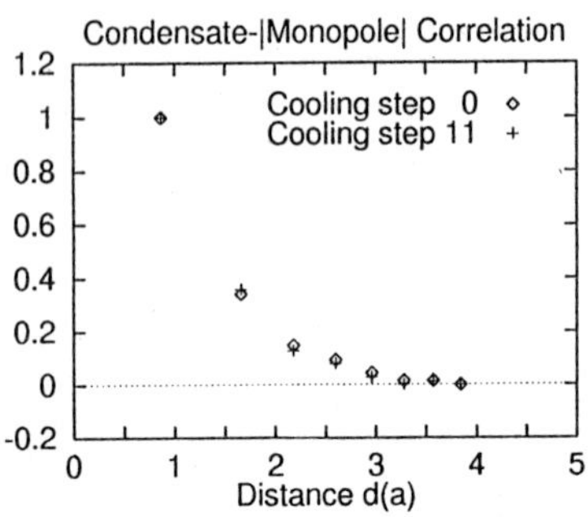

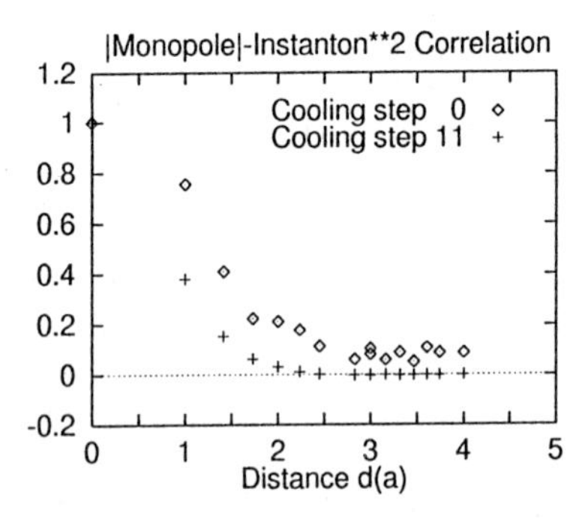

Fig. 1. Correlation functions in the presence of dynamical quarks in the confinement phase. All correlations extend over two lattice spacings and indicate local correlations of the chiral condensate and topological objects.

cally and numerically the local quark condensate $\bar{\psi}\psi(x)$ is a diagonal element of the inverse of the fermionic matrix of the QCD action.

In order to study low energy excitations as instantons and monopoles, one has to smooth quantum fluctuations and get rid of renormalization constants. For most of our studies so far we employed the Cabbibo-Marinari cooling method. However this method drives gauge field configurations into classical configurations in a theoretically uncontrolled way. It might destroy relevant physical information, like instanton sizes or the color orientation of instantons with respect to each other and leads to changes of instanton positions and to annihilations of instantons with antiinstantons. We also present results relying on an alternative method to cooling, which smoothens gauge fields by interpolating from a coarse to a fine lattice. This method is a realization of Wilson's original concept of renormalization group transformations on the lattice.

The following simulations were performed for full $SU(3)$ QCD on an $8^3 \times 4$ lattice with periodic boundary conditions in the confinement phase at $\beta = 5.2$. Dynamical quarks in Kogut-Susskind discretization with 3 flavors of degenerate mass $m = 0.1$ were taken into account using the pseudofermionic method. We compute correlation functions between two observables $\mathcal{O}_1(x)$ and $\mathcal{O}_2(y)$, $g(y-x) = \langle \mathcal{O}_1(x)\mathcal{O}_2(y)\rangle - \langle \mathcal{O}_1\rangle\langle \mathcal{O}_2\rangle$, and normalize them at the smallest lattice distance. Since topological objects with opposite sign are equally distributed, we correlate the quark and monopole density with the square or the absolute value of the topological charge density.

Figure 1 shows results for the correlation functions with $\mathcal{O}_{1,2} = \bar{\psi}\psi(x), \rho_m(x)$ and $q^2(x)$. All correlation functions exhibit an extension of more than two lattice spacings, indicating nontrivial correlations. The correlation of the local quark condensate and the topo-

5 Cooling steps

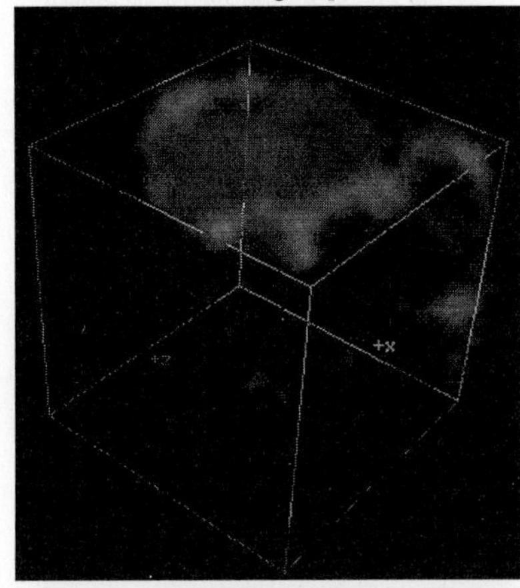

10 Cooling steps

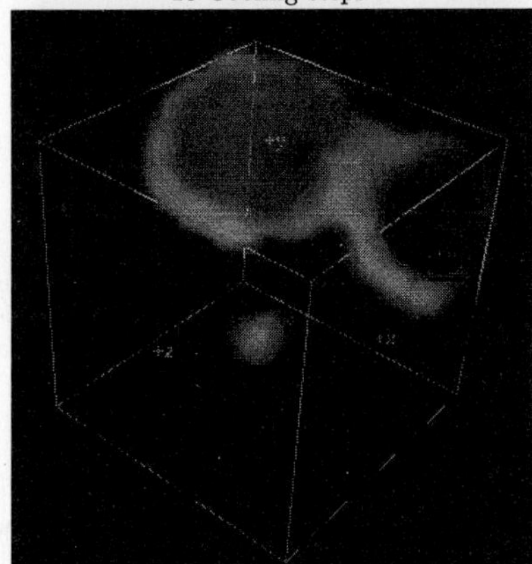

Fig. 2. Cooling history for a time slice of a single gauge field configuration of $SU(3)$ theory with dynamical quarks, after 5 and 10 cooling steps. The dark and medium grey shades represent the positive and negative topological charge density respectively; the surrounding light grey tone the density of the quark condensate. It turns out that the quark condensate takes a non-vanishing value at the positions of the instantons.

logical charge density is cooling dependent, whereas the correlation between the condensate and the monopole density is not. The monopole-instanton correlation is similar as in pure $SU(3)$ theory.

It is assumed that the size ρ of a t'Hooft instanton $q_\rho(x) \sim \rho^4 (x^2 + \rho^2)^{-4}$ centered around the origin enters also into the associated distribution of the chiral condensate $\bar{\psi}\psi_\rho(x) \sim \rho^2 (x^2 + \rho^2)^{-3}$ [10]. To estimate ρ we fitted a convolution of the functional form $f(x) = \int \bar{\psi}\psi_\rho(t) q_\rho^2(x-t)\, dt$ to the data points. This was evaluated after 11 cooling steps where the configurations are reasonably dilute. Our fit yields $\rho(\bar{\psi}\psi q^2) = 1.8$ in lattice spacings corresponding to about 0.4 fm.

We now visualize densities of the quark condensate and topological quantities from individual gauge fields. In Fig. 2 a time slice of a typical configuration from $SU(3)$ theory with dynamical quarks in the confinement phase is shown. We display the positive instanton density by dark grey shades and the negative density by a medium grey-tone if the absolute value $|q(x)| > 0.003$. The quark-antiquark density is indicated by a light grey shade whenever a threshold for $\bar{\psi}\psi(x) > 0.066$ is exceeded. By analyzing dozens of gluon and quark field configurations we found the following results. The topological charge is covered by quantum fluctuations and becomes visible by cooling of the gauge fields. For 0 cooling steps no structure can be seen in $q(x)$, $\bar{\psi}\psi(x)$ or the monopole currents, which does not mean the absence of correlations between them. After 5 cooling steps clusters of nonzero topological charge density and quark condensate are resolved. This particular configuration possesses a cluster with a positive and a negative topological charge corresponding to an instanton and an antiinstanton, respectively. For more than 10 cooling steps both topological charge and chiral condensate begin to die out and eventually vanish. Combin-

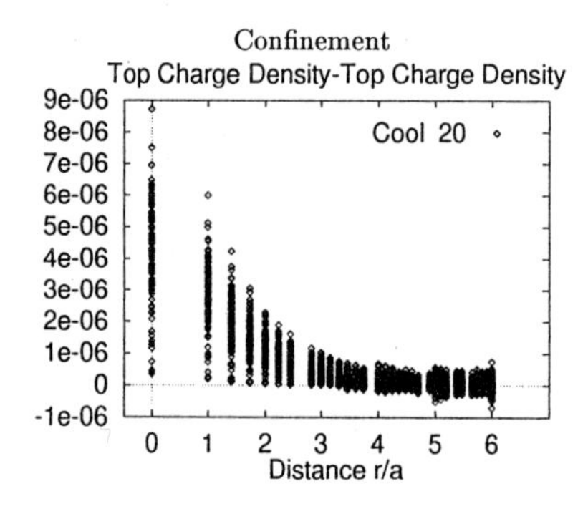

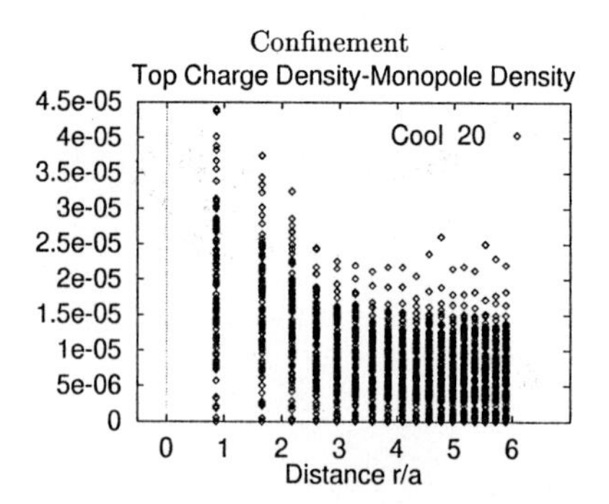

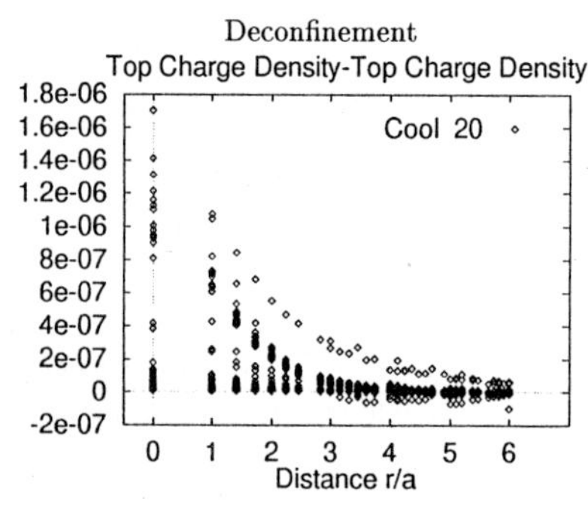

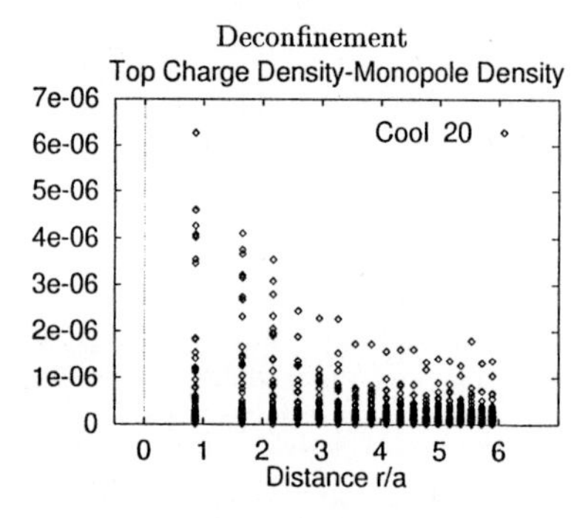

Fig. 3. Auto-correlation functions of the topological charge density after 20 cooling steps for 100 independent configurations in both phases of pure SU(2).

Fig. 4. The $\rho_m|q|$-correlations are displayed in both phases. All configurations with nonvanishing qq-auto-correlation give rise to a nontrivial $\rho_m|q|$-correlation.

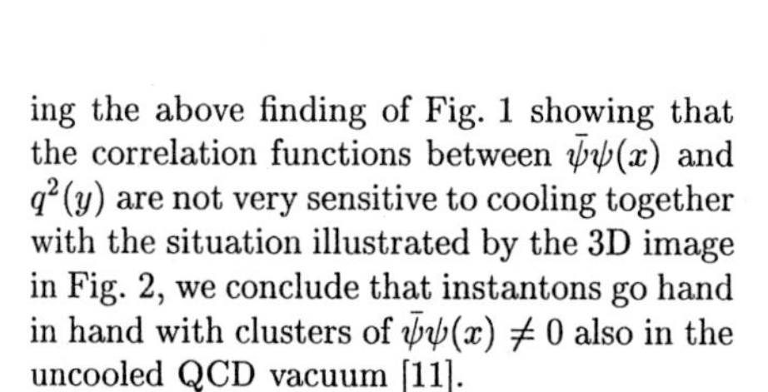

ing the above finding of Fig. 1 showing that the correlation functions between $\bar{\psi}\psi(x)$ and $q^2(y)$ are not very sensitive to cooling together with the situation illustrated by the 3D image in Fig. 2, we conclude that instantons go hand in hand with clusters of $\bar{\psi}\psi(x) \neq 0$ also in the uncooled QCD vacuum [11].

To trace back to the origin of the nontrivial correlation between monopoles and instantons, we analyzed pure SU(2) theory on a $12^3 \times 4$ lattice both in the confinement and deconfinement phase at $\beta = 2.25$ and 2.4, respectively. Fig. 3 presents the auto-correlations of the topological charge and Fig. 4 the $\rho_m|q|$-correlations after 20 cooling steps for 100 independent configurations. In the confinement phase the auto-correlation functions have many different amplitudes reflecting a large variety of topologically nontrivial configurations. Also the corresponding monopole-instanton correlations show many different amplitudes. In the deconfinement phase only about 15% of the auto-correlation functions are nontrivial. All of these configurations give rise to a nontrivial $\rho_m|q|$-correlation. This indicates that the relation between monopoles

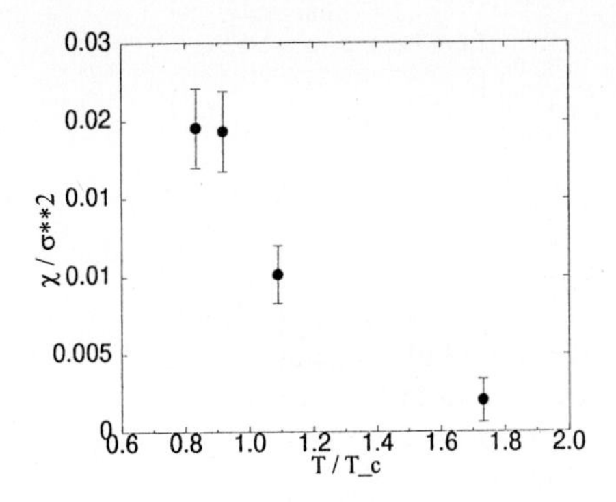

Fig. 5. Topological susceptibility across the temperature phase transition for an SU(2) fixed point action. The drop is due to diminished topological activity in the deconfinement phase.

and topological charge found on gauge average also holds for single configurations.

In Fig. 5 we show the expected drop of the topological susceptibility χ (divided by the string tension squared σ^2) over the phase transition. This study was performed with an SU(2) fixed point action using a renormalization group method instead of cooling [6]. This method allows for a more realistic vacuum where no cooling artifacts are present.

To summarize, we have demonstrated the local coexistence of instantons not only with maximum abelian monopoles, but also with the distribution of the "chiral condensate" in QCD with dynamical fermions. We have evidenced that this relation holds already on single gauge fields and in the deconfinement phase. A fixed point action with a renormalization group concept was shown to work and might help to decide if the instanton liquid model is realized in nature and if instanton-antiinstanton pairs appear in the plasma phase.

References

[1] E.V. Shuryak, Nucl. Phys. B302 (1988) 559.

[2] S. Thurner, M. Feurstein, H. Markum, and W. Sakuler, Phys. Rev. D 54 (1996) 3457; S. Thurner, M. Feurstein, and H. Markum, Phys. Rev. D 56 (1997) 4039; M. Feurstein, H. Markum, and S. Thurner, Phys. Lett. B 396 (1997) 203.

[3] M.N. Chernodub and F.V. Gubarev, JETP Lett. 62 (1995) 100;
M. Fukushima, S. Sasaki, H. Suganuma, A. Tanaka, H. Toki, and D. Diakonov, Phys. Lett. B 399 (1997) 141; R.C. Brower, K.N. Orginos, and Chung-I Tan, Phys. Rev. D 55 (1997) 6313.

[4] O. Miyamura, Nucl. Phys. B (Proc. Suppl.) 42 (1995) 538.

[5] T. DeGrand, A. Hasenfratz, P. Hasenfratz, and F. Niedermayer, Nucl. Phys. B454 (1995) 578; Nucl. Phys. B454 (1995) 615.

[6] M. Feurstein, E.-M. Ilgenfritz, M. Müller-Preußker, and S. Thurner, Nucl. Phys. B, in print; hep-lat/9611024.

[7] P. Di Vecchia, K. Fabricius, G.C. Rossi, and G. Veneziano, Nucl. Phys. B192 (1981) 392; Phys. Lett. B 108 (1982) 323; Phys. Lett. B 249 (1990) 490.

[8] G. 't Hooft, Nucl. Phys. B190 (1981) 455.

[9] A.S. Kronfeld, G. Schierholz, and U.-J. Wiese, Nucl. Phys. B293 (1987) 461.

[10] T. Schäfer and E.V. Shuryak, to appear in Rev. Mod. Phys., hep-ph/9610451.

[11] S.J. Hands and M. Teper, Nucl. Phys. B347 (1990) 819.

Topology in QCD

G. Boyd[a], (boyd@rccp.tsukuba.ac.jp)
B. Allés[b], (alles@sunmite.mi.infn.it)
M. D'Elia[c d], (delia@dirac.ns.ucy.ac.cy)
A. Di Giacomo[d], (digiaco@mailbox.difi.unipi.it)

[a]Centre for Computational Physics, University of Tsukuba, Japan
[b]Dipartimento di Fisica, Sezione Teorica, Università di Milano, Italy
[c]Department of Natural Sciences, University of Cyprus, Cyprus
[d]Dipartimento di Fisica, Università di Pisa, Italy

Abstract. Topology on the lattice is reviewed. In quenched QCD topological susceptibility χ is fully understood. The Witten-Veneziano mechanism for the η' mass is confirmed. The topological susceptibility drops to zero at the deconfining phase transition. Preliminary results are also presented for χ and χ' in full QCD, and for the spin content of the proton. The only problem there is the difficulty of the usual Hybrid Monte Carlo algorithm to bring topology to equilibrium.

1 Introduction

Topology plays a fundamental role in QCD. The key equation is the $U_A(1)$ anomaly

$$\partial_\mu J^5_\mu = -2N_f Q(x) \tag{1}$$

where $J^5_\mu = \sum_{i=1}^{N_f} \overline{\psi}_i \gamma^5 \gamma_\mu \psi_i$ is the singlet axial current, N_f the number of light flavours and

$$Q(x) = \frac{g^2}{64\pi^2} F^a_{\mu\nu} F^a_{\rho\sigma} \epsilon_{\mu\nu\rho\sigma} \tag{2}$$

is the topological charge density. $Q(x)$ is related to the Chern current $K_\mu(x)$

$$\partial_\mu K_\mu(x) = Q(x) \tag{3}$$

with

$$K_\mu = \frac{g^2}{16\pi^2} \epsilon_{\mu\nu\rho\sigma} A^a_\nu \left(\partial_\rho A^a_\sigma - \frac{1}{3} g f^{abc} A^b_\rho A^c_\sigma \right) \tag{4}$$

and as a consequence $Q = \int Q(x)$ is an integer on smooth classical configurations with finite action. Eqs. (1-3) yield

$$\partial_\mu \left(J^5_\mu(x) + 2N_f K_\mu(x) \right) = 0 \tag{5}$$

whence Ward identities can be derived. At the leading order in the $1/N_c$ expansion, ($N_c \longrightarrow \infty$, with $g^2 N_f$ fixed), $Q(x)$ is zero, $U_A(1)$ is a symmetry and η' is its Goldstone particle, $m_{\eta'} = 0$. The anomaly acts as a perturbation and shifts the position of the pole from zero to the actual η' mass. From Eq.(5) by inserting the quark mass terms, the relation follows [1, 2]

$$\chi = \frac{f_\pi^2}{2N_f}\left(m_\eta^2 + m_{\eta'}^2 - 2m_K^2\right) \tag{6}$$

where χ is the topological susceptibility of the unperturbed vacuum ($1/N_c = 0$) defined as

$$\chi = \int \mathrm{d}^4(x-y)\partial_\mu^x \partial_\nu^y \langle 0|T(K_\mu(x)K_\nu(y))|0\rangle_{\text{quenched}}. \tag{7}$$

The subscript "quenched" indicates that the matrix element has to be computed on the ground state of the $1/N_c = 0$ theory. In particular this implies that fermion loops, which are $O(g^2 N_f)$ are put to zero. Eq.(6) gives

$$\chi = (180\ \text{MeV})^4 \tag{8}$$

which is expected to be valid within an order $O(1/N_c)$ of accuracy. Eq.(6) is a peculiar equation relating physical quantities (masses, f_π, N_f) to χ, which exists in an artificial $1/N_c = 0$ world. Its verification, however, is a check of the validity of the expansion, which is a fundamental issue. Lattice is an ideal tool to produce this artificial world, and in particular the absence of fermions in it simplifies the numerical work. Eq.(7) uniquely fixes the prescription for the singularity in the product of operators $K_\mu(x)K_\nu(y)$ as $x \longrightarrow y$: δ-like singularities disappear after integration and this uniquely determines χ. Eq.(7) is a specific prescription for the notation

$$\chi = \int \mathrm{d}^4x \langle 0|T(Q(x)Q(0))|0\rangle_{\text{quenched}}. \tag{9}$$

When computing χ by any regularization scheme, like lattice, an appropriate subtraction must be performed to satisfy the prescription Eq.(7).

The behaviour of χ at finite temperature, and more specifically, at deconfinement, is an important key to understand the structure of QCD vacuum [3]

A regularized version of the operator $Q(x)$, $Q_L(x)$, can be defined on the lattice. There is a large arbitrariness in this definition, by terms of higher order in the lattice-spacing which go to zero in the continuum limit. In general $Q_L(x)$ will not be an exact divergence, so that a multiplicative renormalization Z with respect to continuum can exist [4]. If a is the lattice-spacing,

$$Q_L(x) = a^4 Z Q(x) + O(a^6). \tag{10}$$

A lattice susceptibility χ_L is defined as

$$\chi_L = \sum_x \langle Q_L(x)Q_L(0)\rangle = \frac{\langle Q_L^2\rangle}{V}. \tag{11}$$

In general the definition (11) will not satisfy the prescription Eq. (7). It will be [5]

$$\chi_L = a^4 Z^2 \chi + M + O(a^6) \tag{12}$$

where M is a mixing with the continuum operators having dimension ≤ 4 ($\overline{\beta}$ is the beta function)

$$M(\beta) = A(\beta)\langle \frac{\overline{\beta}(g)}{g} F^a_{\mu\nu} F^a_{\mu\nu} \rangle a^4 + P(\beta) \cdot 1. \tag{13}$$

From Eq.(12)

$$\chi = \frac{\chi_L - M}{Z^2 a^4}. \tag{14}$$

χ_L is measured on the lattice numerically. a^4 is determined as usual by comparison to a physical quantity (ρ mass, string tension); M and Z can be determined non-perturbatively by a procedure known as heating [6]. The idea is that classical configurations with known topological charge, can be dressed by quantum fluctuations without modifying the topological content since topological charge is difficult to change by the usual local Monte Carlo algorithms. Z is determined by measuring the total charge Q_L on a configuration with an instanton, where $Q = 1$, Eq.(10). M is determined from Eq.(12) by measuring χ_L on the sector $Q = 0$, where as a consequence of Eq.(7), $\chi = 0$.

In Eq.(14) χ_L, M and Z strongly depend on the choice of Q_L, on the action and on the coupling constant $\beta = 2N_c/g^2$. a depends on the action and on β. χ must be independent of all these parameters. Fig. 1 shows χ for $SU(3)$ [7], determined for several β's and by use of different operators: as visible in the figure $\chi = (175 \pm 5\ \text{MeV})^4$. For $SU(2)$ (Fig. 2) [8] χ is somewhat larger: $\chi = (198 \pm 6\ \text{MeV})^4$.

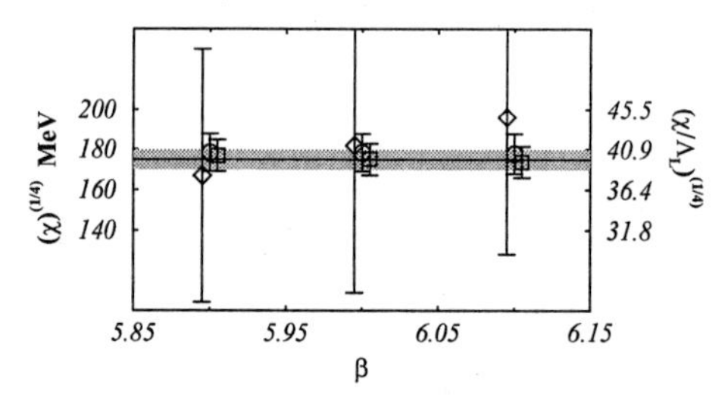

Fig. 1. χ for $SU(3)$. Diamonds, circles and squares correspond to the 0, 1 and 2-smeared operators.

The determination by using the so-called geometrical method, if accompained by the appropriate subtraction, agrees with the other choices of Q_L.

Fig. 3 shows the behaviour of χ across deconfinement. The drop is stronger for $SU(3)$ than for $SU(2)$ [8].

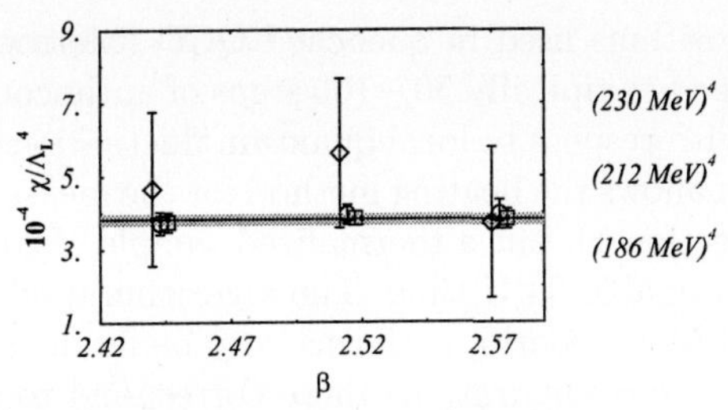

Fig. 2. χ for $SU(2)$. Diamonds, circles and squares correspond to the 0, 1 and 2-smeared operators.

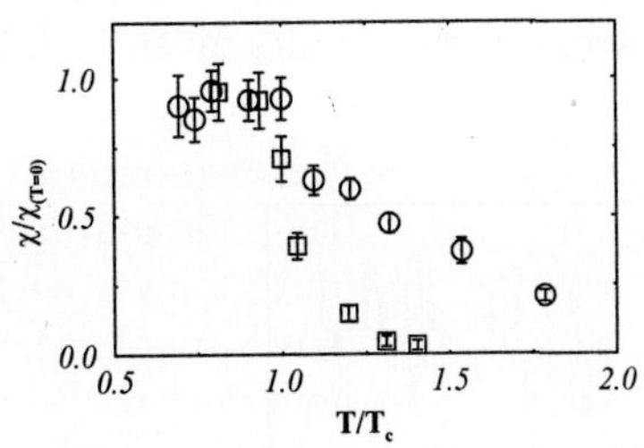

Fig. 3. χ across the deconfinement transition for $SU(2)$ (circles) and $SU(3)$ (squares).

2 Full QCD

By use of the same procedure described in the previous section, χ can be determined in full QCD. The expectation from the Ward identities is that

$$\chi \approx \langle \sum_{i=1}^{N_f} m_i \overline{\psi}_i \psi_i \rangle + O(m_q^2). \tag{15}$$

Our preliminary result, extracted from simulating at $\beta = 5.35$ where $a = 0.11(1)$ fm with 4 staggered fermions at $am = 0.01$ is

$$\chi = (110 \pm 8 \text{ MeV})^4 \tag{16}$$

to be compared to the predicted value

$$\chi = \frac{m_q}{N_f} \langle \overline{\psi}\psi \rangle_{m=0} \sim (109 \text{ MeV})^4. \tag{17}$$

On the same sample of configurations we obtain for χ' the preliminary value

$$\chi' = 258 \pm 100 \text{ MeV}^2 \quad \text{or} \quad \sqrt{\chi'} = 19 \pm 4 \text{ MeV} \tag{18}$$

which is compatible with the value expected from sum rules [9] $\sqrt{\chi'} = 25 \pm 3$ MeV. However both these determinations are preliminary because of the effect shown in Fig. 4 where we display the history of the topological charge Q along the Monte Carlo updating which produces the configurations [10].

In the updating algorithms used in quenched QCD (Metropolis, heat-bath) the topological charge has tipically $50-100$ steps of authocorrelation time. It thermalizes slowly with respect to local quantum fluctuations (and this is the basic property which allows the heating method for the measurement of Z and M as explained in section 1), but a thermalized sample of configurations can be prepared in a reasonable CPU time. The algorithm used with dynamical fermions, the hybrid Monte Carlo, performs very badly in that respect, as is visible from Fig. 4. The configurations there correspond to about 700 CPU hours of an APE Quadrics with 25 GFlop which is a huge time. Our sample has thus a much smaller number of independent configurations than shown in Fig. 4, and therefore the errors in the results given in Eqs. (16) and (18) are underestimated.

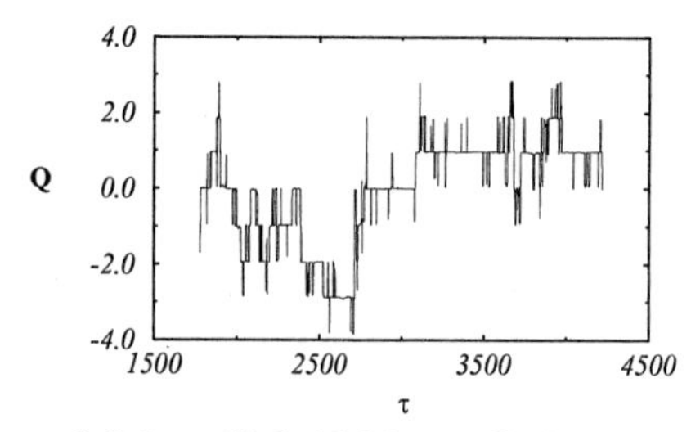

Fig. 4. History of Q in a Hybrid Monte Carlo run as a function of the molecular dynamic time τ.

The same uncertainty affects our determination, on the same sample, of the spin content of the proton. The matrix element of J^5_μ between proton states can be parametrized as

$$\langle \mathbf{p}', s'|J^5_\mu|\mathbf{p}, s\rangle = \overline{u}(\mathbf{p}', s')\left(G_1(k^2)\gamma_\mu\gamma^5 - G_2(k^2)\gamma^5 k_\mu\right) u(\mathbf{p}, s) \tag{19}$$

with $k = p - p'$. The form factor G_1 is related to the so-called spin content of the proton $\Delta\Sigma$, $G_1(0) = \Delta\Sigma$ where $\Delta\Sigma \equiv \Delta u + \Delta d + \Delta s$ is the contribution of the different quarks species to the spin of the proton. The naïve expectation would be $\Delta\Sigma \sim 0.7$. The value determined from the moments of the spin dependent structure functions of inelastic scattering of leptons on nucleons is much lower: $\Delta\Sigma = 0.2(1)$. The lattice allows a determination of $\Delta\Sigma$ from first principles. One possible technique consists in the direct measurement of the matrix element (19). An alternative is to use the anomaly equation, which after taking the divergence of both sides of Eq.(19),

$$\langle \mathbf{p}', s'|Q|\mathbf{p}, s\rangle = \frac{m_N}{N_f} A(k^2)\overline{u}(\mathbf{p}', s')i\gamma^5 u(\mathbf{p}, s) \tag{20}$$

$$A(k^2) = G_1(k^2) + \frac{k^2}{m_N} G_2(k^2). \tag{21}$$

As $k \longrightarrow 0$ Eq.(21) determines $G_1(0)$, unless $G_2(k^2)$ has a pole at $k^2 = 0$ and this is the case in the quenched approximation but not in full QCD. Eq. (20) gives thus $\Delta\Sigma$ in terms of the matrix element $\langle \mathbf{p}', s'|Q|\mathbf{p}, s\rangle$, which can be measured on the lattice. In principle the lattice operator Q_L would mix with $\partial_\mu J^5_\mu(x)$ and $\overline{\psi}\gamma^5\psi$, but this mixing, as well as the small anomalous dimension of Q can be neglected [11]. Our preliminary value is $\Delta\Sigma = 0.04(4)$. Here again the error could be larger and in any case the value is preliminary, due to the bad sampling of topology in our ensemble of configurations.

3 Conclusions

Measurement of the topological susceptibility χ on the lattice is fully under control. For quenched $SU(3)$ the value is in good agreement with the prediction of [1, 2]. χ drops to zero at the deconfining transition. Preliminary determinations of χ' in full QCD agree with sum rules. The spin content of the proton is at hand. The practical problem is the thermalization of topology on the lattice. Our huge sample of configurations is not thermalized with respect to it. This creates in principle a problem for the lattice determination of any quantity: a priori, indeed, it is not known how it could depend on the topological sector and therefore if the ensemble is biased with respect to topology, this could affect the result in an impredictable way. Solutions of this numerical problem are currently under study.

References

[1] E. Witten, Nucl. Phys. B156 (1979) 269.
[2] G. Veneziano, Nucl. Phys. B159 (1979) 213.
[3] E. Shuryak, Comments Nucl. Part. Phys. 21 (1994) 235.
[4] M. Campostrini, A. Di Giacomo, H. Panagopoulos, Phys. Lett. B212 (1988) 206.
[5] M. Campostrini, A. Di Giacomo, H. Panagopoulos, E. Vicari, Nucl. Phys. B329 (1990) 683.
[6] A. Di Giacomo, E. Vicari, Phys. Lett. B275 (1992) 429.
[7] B. Allés, M. D'Elia, A. Di Giacomo, Nucl. Phys. B494 (1997) 281.
[8] B. Allés, M. D'Elia, A. Di Giacomo, Phys. Lett. B412 (1997) 119.
[9] S. Narison, G. M. Shore, G. Veneziano, Nucl. Phys. B433 (1995) 209.
[10] B. Allés, G. Boyd, M. D'Elia, A. Di Giacomo, E. Vicari, Phys. Lett. B389 (1996) 107.
[11] B. Allés, A. Di Giacomo, H. Panagopoulos, E. Vicari, Phys. Lett. B350 (1995) 70.

Update on Lattice QCD with Domain Wall Quarks

Tom Blum[1] (tblum@bnl.gov)
Amarjit Soni[1] (soni@bnl.gov); Presenter

Brookhaven National Laboratory, Upton, NY 11973

Abstract. Using domain wall fermions, we estimate $B_K(\mu \approx 2\,\mathrm{GeV}) = 0.602(38)$ in quenched QCD which is consistent with previous calculations. We also find ratios of decay constants that are consistent with experiment, within our statistical errors. Our initial results indicate good scaling behavior and support expectations that $O(a)$ errors are exponentially suppressed in low energy ($E \ll a^{-1}$) observables. It is also shown that the axial current numerically satisfies the lattice analog of the usual continuum axial Ward identity and that the matrix element of the four quark operator needed for B_K exhibits excellent chiral behavior.

1 Introduction

We recently reported[1, 2] on calculations using a new discretization for simulations of QCD, domain wall fermions (DWF) [3, 4], which preserve chiral symmetry on the lattice in the limit of an infinite extra 5th dimension. There it was demonstrated that DWF exhibit remarkable chiral behavior[1] even at relatively large lattice spacing and modest extent of the fifth dimension.

In addition to retaining chiral symmetry, DWF are also "improved" in another important way. In the limit that the number of sites in the extra dimension, N_s, goes to infinity, the leading discretization error in the effective four dimensional action for the light degrees of freedom goes like $O(a^2)$. This theoretical dependence is deduced from the fact that the only operators available to cancel $O(a)$ errors in the effective action are not chirally symmetric [5, 2]. For finite N_s, $O(a)$ corrections are expected to be exponentially suppressed with the size of the extra fifth dimension. Our calculations for B_K show a weak dependence on a that is easily fit to an a^2 ansatz. Preliminary results for the ratios f_π/m_ρ and f_k/f_π indicate good scaling behavior as well.

We use the boundary fermion variant of DWF developed by Shamir. For details, consult Kaplan[3] and Shamir[4]. See Ref. [6] for a discussion of the $4d$ chiral Ward identities (CWI) satisfied by DWF. Our simulation parameters are summarized in Table 1.

2 Results

We begin with the numerical investigation of the lattice PCAC relation. The CWI are satisfied exactly on any configuration since they are derived from the

Table 1. Summary of simulation parameters. M is the five dimensional Dirac fermion mass, and m is the coupling between layers $s = 0$ and $N_s - 1$.

$6/g^2$	size	M	m(# conf)
5.85	$16^3 \times 32 \times 14$	1.7	0.075(34) 0.05(24)
6.0	$16^3 \times 32 \times 10$	1.7	0.075(36) 0.05(39) 0.025(34)
6.3	$24^3 \times 60 \times 10$	1.5	0.075(11) 0.05(15) 0.025 (22)

corresponding operator identity. We checked this explicitly in our simulations. In the asymptotic large time limit, we find for the usual PCAC relation

$$2\sinh\left(\frac{am_\pi}{2}\right)\frac{\langle A_\mu|\pi\rangle}{\langle J_5|\pi\rangle} = 2m + 2\frac{\langle J_{5q}|\pi\rangle}{\langle J_5|\pi\rangle}, \qquad (1)$$

which goes over to the continuum relation for $am_\pi \ll 1$ and $N_s \to \infty$ (see Ref. [6] for operator definitions). The second term on the r.h.s. is anomalous and vanishes as $N_s \to \infty$. It is a measure of explicit chiral symmetry breaking induced by the finite 5th dimension. At $6/g^2 = 6.0$ and $N_s = 10$ we find the l.h.s. of Eq.1 to be 0.1578(2) and 0.1083(3) for $m = 0.075$ and 0.05, respectively. The anomalous contributions for these two masses are $2\times$(0.00385(5) and 0.00408(12)), which appear to be roughly constant with m. Increasing N_s to 14 at $m = 0.05$, the anomalous contribution falls to ($2\times$) 0.00152(8) while the l.h.s. is 0.1026(6), which shows that increasing N_s really does take us towards the chiral limit.

Next we investigate the matrix element of the four quark operator O_{LL} which defines B_K. $\langle K|O_{LL}|\bar{K}\rangle$ vanishes linearly with m in the chiral limit in excellent agreement with chiral perturbation theory (Fig. 1).

In Fig. 2 we show the kaon B parameter at each value of $6/g^2$ versus af_π which is used to set the lattice spacing. The results for B_K depend weakly on $6/g^2$, and are well fit to a pure quadratic in a. We find $B_K(\mu = a^{-1}) = 0.602(38)$ in the continuum limit. This value is already consistent with previous results[7] though it does not include the perturbative running of B_K to a common energy scale. This requires a perturbative calculation to determine the scale dependence of O_{LL}, which has not yet been done[8]. From Table 2, the energy scale at $6/g^2 = 6.0$ is roughly 2 GeV.

At $6/g^2 = 6.0$, we have also calculated B_K using the partially conserved axial current $A^a_\mu(x)$ (and the analogous vector current). This point split conserved current requires explicit factors of the gauge links to be gauge invariant. Alternatively a gauge non-invariant operator may be defined by omitting the links; the two definitions become equivalent in the continuum limit. Results for the gauge non-invariant operators agree within small statistical errors with those obtained with naive currents, Fig. 2(see Ref.[6, 1] for operator definitions). The results for the gauge invariant operators are somewhat

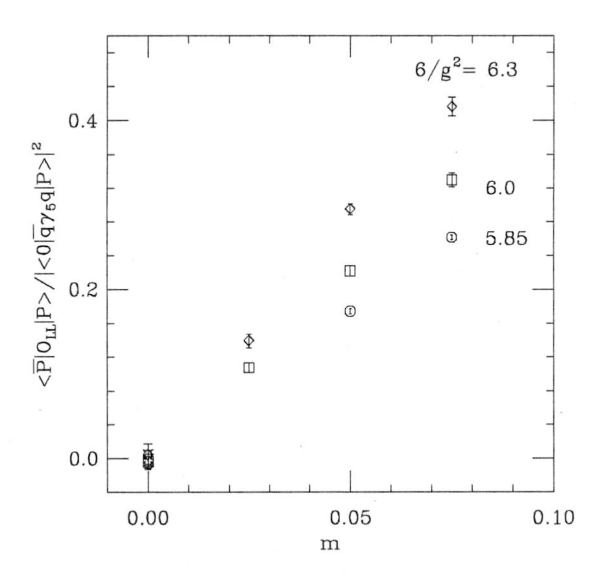

Fig. 1. The matrix element $\langle \bar{P}|O_{LL}|P \rangle$ *vs.* m. m is proportional to the quark mass in lattice units.

larger: $B_K^{inv}(\mu = a^{-1}) = 0.857(20)$ and $0.946(28)$ at $m = 0.05$ and 0.075, respectively. A similar situation holds in the Kogut-Susskind case where it was shown that the gauge invariant operators receive appreciable perturbative corrections which bring the two results into agreement [9].

Table 2. Lattice spacing and decay constant summary.

$6/g^2$	$a^{-1}(m_\rho)$	$a^{-1}(f_\pi)$	f_π/m_ρ	f_K/f_π	N_s
5.85	1.49(29)	1.62(27)	0.154(56)	1.206(15)	14
6.0	1.89(14)	2.06(15)	0.155(23)	1.205(15)	10
6.3	2.96(25)	3.24(27)	0.155(26)	1.14(14)	10

Using Eq. 1, neglecting the anomalous contribution, and using the definition of the decay constant, we can determine the pseudoscalar decay constant from the measurement of $\langle 0|J_5^a|P\rangle$. The results are summarized in Table 2. As with B_K, the estimates of the physical ratios f_π/m_ρ and f_K/f_π indicate good scaling behavior. They are also consistent with experiment, within rather large statistical errors. The errors in Table 2 are crude estimates derived using the statistical uncertainties only; the small sample size precludes

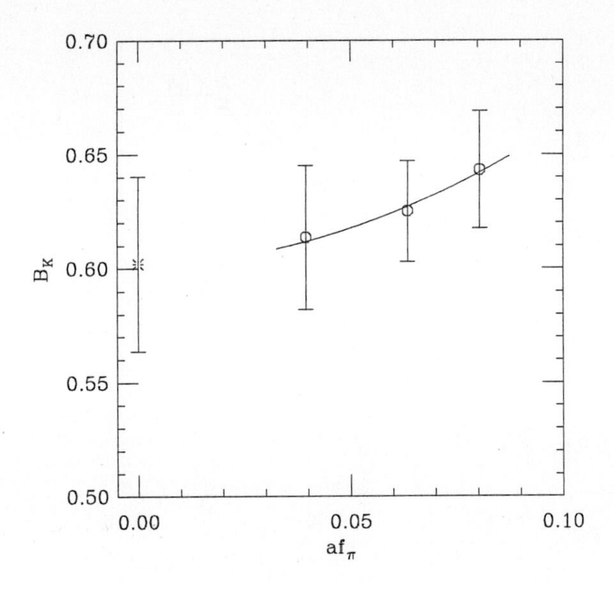

Fig. 2. The kaon B parameter.

us from properly accounting for the correlations in the data. All of the results are so called effective values, calculated from the two-point correlators without χ^2 fits and averaged over suitable plateaus. In the last column at the two larger couplings, an attempt has been made to carry out a more sophisticated jackknife error analysis all the way through to the final ratio.

While all of the above results indicate good scaling, they must be checked further with improved statistics and a fully covariant fitting procedure. Also systematic effects like finite volume still need to be investigated. Only then can the continuum limit can be reliably taken. We note that a recent precise calculation using quenched Wilson quarks by the CP-PACS collaboration yields values of f_π and f_K in the continuum limit that are inconsistent with experiment [10].

In Fig. 3 we show the pion mass squared as a function of m. For $N_s =$ 10 the data at $6/g^2 = 6.0$ and 6.3 are consistent with chiral perturbation theory. However, at $6/g^2 = 5.85$, m_π^2 extrapolates to a positive non-zero value in the chiral limit for $N_s = 10$ and 14. There is a large downward shift in the line as N_s goes from 10 to 14, but it still does not pass through the origin. However, at $N_s = 18$, m_π^2 extrapolates to -0.004(19) at $m = 0$. The anomalous contribution on the r.h.s. of Eq. 1 drops from 0.0098(5) to 0.0038(2) as N_s varies from 10 to 18. It is interesting to note that at $N_s = 10$ the anomalous piece is more than double the value at $6/g^2 = 6.0$ whereas the value at $N_s = 18$ is roughly the same. When the anomalous term is sufficiently small compared to the bare parameter m, the chiral symmetry is effectively

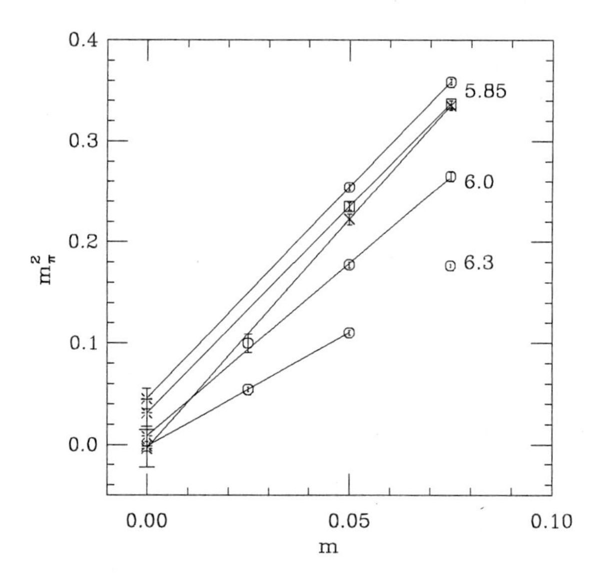

Fig. 3. The pion mass squared.

restored. From the above, it seems we must have $\langle J_{5q}|\pi\rangle \,/\, \langle J_5|\pi\rangle \lesssim 0.1\, m$. Of course, as $m \to 0$, one must also take N_s larger, which is analogous to the situation with the ordinary spatial volume. Fewer sites in the extra dimension may be sufficient at 5.85 if M is increased still further.

Research supported by US DOE grant DE-AC0276CH0016. The numerical computations were carried out on the NERSC T3E.

References

[1] T. Blum and A. Soni, PR **D56**, 174 (1997).
[2] T. Blum and A. Soni, hep-lat/9706023, PRL **79**, 3595 (1997).
[3] D. Kaplan, PLB **288**, 342 (1992).
[4] Y. Shamir, NPB **406**, 90 (1993).
[5] Y. Kikukawa, R. Narayanan, and H. Neuberger, PLB 399, 105, (1997); H. Neuberger hep-lat/9710089.
[6] Y. Shamir and V. Furman, NPB **439**, 54 (1995).
[7] S. Aoki, *et al.* , NPB (Proc.Suppl. **53**), 341 (1997); S. Sharpe, NPB (Proc.Suppl. **34**) , 403 (1994); S. Aoki, *et al.* , hep-lat/9705035; G. Kilcup, PRL **71**, 1677 (1993); W. Lee and M. Klomfass, hep-lat/9608089.
[8] However, see S. Aoki and Y. Taniguchi, hep-lat/9711004.
[9] N. Ishizuka, *et al.* , PRL **71**, 24 (1993).
[10] Plenary talk by T. Yoshié, at LAT'97, hep-lat/9710056.

The Λ-Parameter and m_s of Quenched QCD

Rainer Sommer (sommer@ifh.de) for the ALPHA collaboration

DESY-IfH, Zeuthen, Germany

Abstract. We explain how scale dependent renormalized quantities can be computed using lattice QCD. Two examples are used: the running coupling and quark masses. A reliable computation of the Λ-parameter in the quenched approximation is presented.

1 Introduction

In a perturbative treatment, one takes as parameters of QCD the running coupling and running quark masses in a specific renormalization scheme. If we adopt a mass-independent scheme, the coupling and quark masses satisfy rather simple renormalization group equations. Their asymptotic high energy behavior is given by the Λ-parameter and the renormalization group invariant masses $\{M_f\}$ (f labeling the flavors). These parameters have the advantage that M_f are independent of the renormalization scheme and the Λ-parameters of different schemes can be related exactly by 1-loop calculations.

Once one considers QCD on the non-perturbative level, it is natural to take $N_{\rm f}+1$ hadronic observables to be the parameters that fix the theory instead of Λ and $\{M_f\}$. We call this a hadronic scheme (HS). For QCD with u and d taken as mass-degenerate and an additional s-quark, one may take for example the proton mass, $m_{\rm p}$, the pion mass, m_π, and the Kaon mass, $m_{\rm K}$, as the basic parameters. Of course, the two sets of parameters are related once a non-perturbative solution of QCD is available. It is a challenge for lattice QCD to provide this relation with completely controlled errors. In this brief report we describe a method that allows to solve this problem. The numerical results (sect. 4) are for quenched QCD, i.e. the fermion determinant is omitted in the path integral; one may think of this as a version of QCD with $N_{\rm f}=0$ flavors of sea-quarks.

The problem described above is the connection of the short distance and long distance dynamics of QCD. As such it appears at first sight to be intractable by numerical simulations, since it requires that various scales are treated on one and the same lattice. This is impossible, if discretization errors are supposed to be under control! However, one may define a running coupling and running quark masses in QCD in a finite volume of linear size L with suitably chosen boundary conditions (cf. sect. 3). Then the masses and coupling run with an energy scale $q=1/L$ and their *evolution can be computed recursively.* The only requirement to keep discretization errors small in each step of the recursion is $L \gg a$ which is easy to satisfy (a denotes the lattice spacing)[1, 2].

2 Strategy

We start by giving an overview of the strategy to compute short distance parameters in figure 1. One first renormalizes QCD replacing the bare pa-

$$
\begin{array}{lcccc}
L_{\rm max} = {\rm O}(\tfrac{1}{2}{\rm fm}): & & {\rm HS} & \longrightarrow & {\rm SF}(q = 1/L_{\rm max}) \\
 & & & & \downarrow \\
 & & & & {\rm SF}(q = 2/L_{\rm max}) \\
 & & & & \downarrow \\
 & & & & \bullet \\
 & & & & \bullet \\
 & & & & \bullet \\
 & & & & \downarrow \\
 & & & & {\rm SF}(q = 2^n/L_{\rm max}) \\
 & & & & {\rm PT:}\ \downarrow \\
 & {\rm jet-physics} & \stackrel{\rm PT}{\longleftarrow} & & \Lambda_{\rm QCD}, M
\end{array}
$$

Fig. 1. The strategy for a non-perturbative computation of short distance parameters.

rameters by hadronic observables. This defines the HS introduced above. It can be related to the finite volume scheme (denoted by SF) at a low energy scale $q = 1/L_{\rm max}$, where $L_{\rm max}$ is of the order of 1 fm. Within this scheme one then computes the scale evolution up to a desired energy $q = 2^n/L_{\rm max}$. It is no problem to choose the number of steps n large enough to be sure that one has entered the perturbative regime. There perturbation theory (PT) is used to evolve further to infinite energy and compute Λ and $\{M_f\}$. Inserted into perturbative expressions, these parameters provide predictions for jet cross sections or other high energy observables. In figure 1, all arrows correspond to relations in continuum QCD; the whole strategy is designed such that lattice calculations for these relations can be extrapolated to the continuum limit.

For the practical success of the approach, the finite volume coupling and quark masses must satisfy a number of criteria.

- They should have an easy perturbative expansion, such that the β-function (and τ-function, which describes the evolution of the running masses) can be computed to sufficient order in the coupling.
- They should be easy to calculate in MC (small variance!).

– Lattice effects must be small to allow for safe extrapolations $a/L \to 0$.

Careful consideration of the above points led to the introduction of the running coupling, $\bar{g}$, and quark mass, $\overline{m}$, through the Schrödinger functional (SF) of QCD [3, 2, 4, 5].

3 Schrödinger functional scheme

We describe the definition of $\bar{g}$ and $\overline{m}$ in a formal continuum formulation. The SF is the partition function of the Euclidean path integral on a hyper-cylinder with Dirichlet boundary conditions in time, $(P_\pm = \frac{1}{2}(1 \pm \gamma_0))$

$$\begin{array}{llll} A_k(x) = C_k(\mathbf{x}), & P_+\psi(x) = \rho(\mathbf{x}), & \overline{\psi}(x)P_- = \bar{\rho}(\mathbf{x}) & \text{at } x_0 = 0 \,, \\ A_k(x) = C'_k(\mathbf{x}), & P_-\psi(x) = \rho'(\mathbf{x}), & \overline{\psi}(x)P_+ = \bar{\rho}'(\mathbf{x}) & \text{at } x_0 = L \,. \end{array} \quad (1)$$

Here, $A_k, \ldots, \bar{\rho}'$ are classical prescribed boundary fields. In space, the gauge fields are taken periodic under shifts $x \to x + L\hat{k}$ and the fermion fields are periodic up to a phase θ.

The renormalized coupling is defined through the response of the SF to an infinitesimal change of the boundary gauge fields C_k, C'_k. Since the boundary fields are taken with a strength proportional to $1/L$, there is no scale apart from L and the coupling $\bar{g}(L)$ runs with L. Details have been discussed in several publications [3, 2].

A natural starting point for the definition of a renormalized quark mass is the PCAC relation,

$$\partial_\mu A^a_\mu(x) = 2mP^a(x) \,, \quad (2)$$

which expresses the proportionality of the divergence of the axial current, A^a_μ, to the pseudo-scalar density P^a, where

$$P^a(x) = \overline{\psi}(x)\gamma_5 \tfrac{1}{2}\tau^a \psi(x), \quad A^a_\mu = \overline{\psi}(x)\gamma_\mu\gamma_5 \tfrac{1}{2}\tau^a \psi(x) \,, \quad (3)$$

and τ^a are the Pauli-matrices acting in flavor space (we take $N_\mathrm{f} = 2$ degenerate flavors from now on). Renormalizing the operators in equation 2,

$$(A_\mathrm{R})^a_\mu = Z_\mathrm{A} A^a_\mu \,, \quad P^a_\mathrm{R} = Z_\mathrm{P} P^a \,, \quad (4)$$

we define a renormalized quark mass by

$$\overline{m} = \frac{Z_\mathrm{A}}{Z_\mathrm{P}} m \,. \quad (5)$$

Here, m is to be taken from equation 2 inserted into an arbitrary correlation function. It does not depend on the chosen correlation function, since equation 2 is an operator identity. The scale independent renormalization Z_A can be fixed through current algebra relations (also in the lattice regularization [6, 7]) and we are left to give a normalization condition for the pseudo-scalar density,

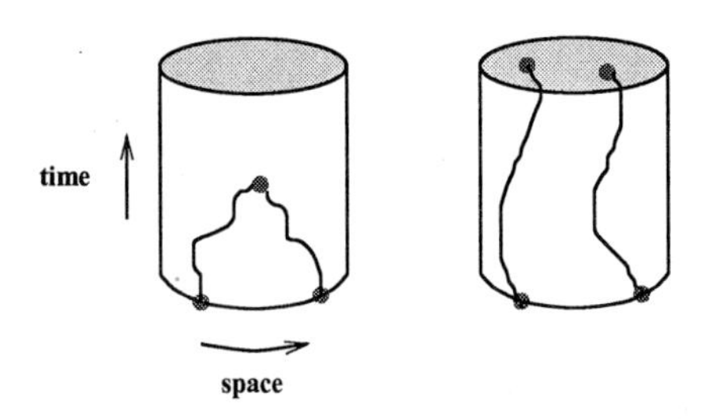

Fig. 2. $f_{\rm P}$ (left) and f_1 (right) in terms of quark propagators.

$P_{\rm R}$. The running mass, $\overline{m}$, then inherits its scheme- and scale-dependence from the corresponding dependence of $P_{\rm R}$.

We define $Z_{\rm P}$ in terms of correlation functions in the SF. To start, we construct (isovector) pseudo-scalar fields at the boundary of the SF [1],

$$\mathcal{O}^a = \int \mathrm{d}^3\mathbf{u} \int \mathrm{d}^3\mathbf{v}\ \bar{\zeta}(\mathbf{u})\gamma_5 \tfrac{1}{2}\tau^a \zeta(\mathbf{v}) \ , \tag{6}$$

from the "boundary quark fields", $\zeta, \bar{\zeta}$. Their precise definition [8] involves a functional derivative, e.g. $\zeta(\mathbf{x}) \equiv \frac{\delta}{\delta \bar{\rho}(\mathbf{x})}$. After taking this functional derivative one sets the boundary values, ρ... ,to zero. Analogously one defines a field $\mathcal{O}'^a$, which resides at $x_0 = L$. These fields are used in the correlation functions

$$f_{\rm P}(x_0) = -\tfrac{1}{3}\langle P^a(x)\mathcal{O}^a\rangle \ , \quad f_1 = \langle \mathcal{O}'^a \mathcal{O}^a\rangle \ , \tag{7}$$

(see figure 2). In the ratio

$$Z_{\rm P} = \mathrm{const.}\sqrt{f_1}/f_{\rm P}(x)|_{x_0=L/2} \ , \tag{8}$$

the renormalization of the boundary quark fields [4] cancels out; it can therefore be taken as the definition of the renormalization constant of P. The proportionality constant is to be chosen such that $Z_{\rm P} = 1$ at tree level. Further details of the definition are $C = C' = 0$ and a specific choice for θ. The renormalization constant is to be evaluated for zero quark mass, m. This defines a mass-independent scheme.

By construction, the SF scheme is non-perturbative and independent of a specific regularization. For a concrete non-perturbative computation, we do, however, need to evaluate the expectation values by a MC-simulation of the corresponding lattice theory. For the details of the lattice formulation we refer to [8] but mention that it is essential to use an O(a)-improved formulation in order to keep lattice spacing effects small [9]. We now explain the computation of the scale dependence, omitting the matching to the HS scheme [2, 5] for lack of space.

---

[1] At this point, the SF provides us with an important advantage compared to other boundary conditions, namely we can project the boundary quark fields onto zero momentum. As a result, the correlation functions vary slowly with x_0, leading to both small statistical and small discretization errors.

4 Scale evolution, Λ and RGI masses

Each step in the recursive computation of the scale evolution consists of:

1. Choose a lattice with L/a points in each direction.
2. Tune the bare coupling, g_0, such that the renormalized coupling $\bar{g}^2(L)$ has the value u and tune the bare mass, m_0, such that the mass (equation 2) vanishes; compute $Z_{\rm P}$.
3. At the same values of g_0, m_0, simulate a lattice with twice the linear size; compute $u' = \bar{g}^2(2L)$ and $Z'_{\rm P}$. This determines the lattice step scaling functions for the coupling, $\Sigma(u, a/L)$=u', and for the mass, $\Sigma_{\rm P}(u, a/L) = Z_{\rm P}/Z'_{\rm P}$.
4. Repeat steps 1.–3. with different resolutions L/a and extrapolate $\sigma(u) = \lim_{a/L\to 0} \Sigma(u, a/L)$, $\sigma_{\rm P}(u) = \lim_{a/L\to 0} \Sigma_{\rm P}(u, a/L)$. This extrapolation can be done by allowing for small linear a/L-discretization errors [2].

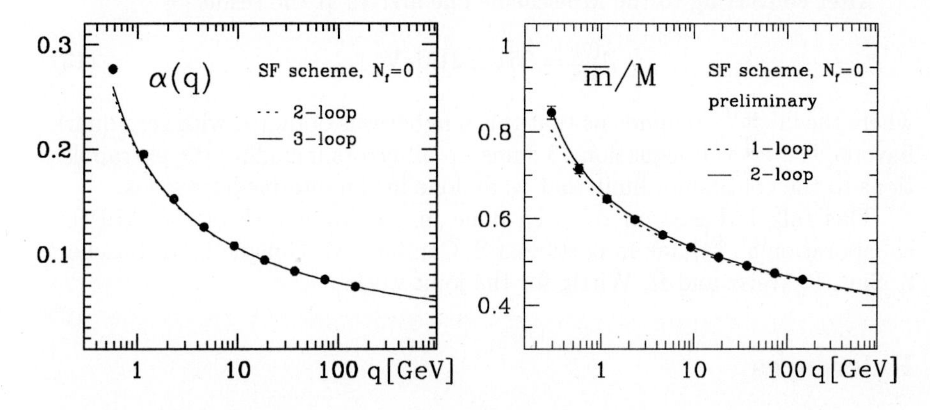

Fig. 3. Non-perturbative running coupling and quark mass as a function of $q \equiv 1/L$. Uncertainties are smaller than the size of the symbols.

In this way, $\sigma(u)$, $1.1 \le u \le 2.4$ and $\sigma_{\rm P}(u)$, $1.1 \le u \le 3.48$ have been computed in the quenched approximation. The scale evolution is monotonous in this range: $\sigma(u) > u$, $\sigma_{\rm P}(u) > 1$. σ and $\sigma_{\rm P}$ can therefore be used to follow figure 1 downwards (up in the energy scale).

Starting from the box size $L_{\rm max}$, defined by $\bar{g}^2(L_{\rm max}) = 3.48$, we use

$$\bar{g}^2(L_{k-1}) = \sigma(\bar{g}^2(L_k)),\;\; \overline{m}(L_{k-1}) = \sigma_{\rm P}(\bar{g}^2(L_k))\,\overline{m}(L_k),\;\; L_k = 2^{-k}L_{\rm max} \quad (9)$$

to compute the sequences [5]

$$\bar{g}^2(L_k),\quad \overline{m}(L_k)/\overline{m}(L_{\rm max}),\quad k = 0, 1, \ldots 8 \;. \quad (10)$$

In figure 3, these are compared to the perturbative evolutions (starting at the weakest coupling). It is surprising that the perturbative evolution is quite precise down to rather low energy scales. This property may of course not be generalized to other schemes but has to be checked in each scheme separately.

On the other hand, the figure clearly demonstrates that the SF-scheme is perturbative in the large energy regime, say $q > 50\mathrm{GeV}$. We may therefore start from the exact relations

$$\Lambda = q \left(b_0 \bar{g}^2\right)^{-b1/(2b_0^2)} \mathrm{e}^{-1/(2b_0\bar{g}^2)} \exp\left\{-\int_0^{\bar{g}} \mathrm{d}x \left[\frac{1}{\beta(x)} + \frac{1}{b_0 x^3} - \frac{b_1}{b_0^2 x}\right]\right\} \quad (11)$$

$$M = \overline{m}\,(2b_0\bar{g}^2)^{-d_0/2b_0^2} \exp\left\{-\int_0^{\bar{g}} \mathrm{d}g\left[\frac{\tau(g)}{\beta(g)} - \frac{d_0}{b_0 g}\right]\right\} \ , \quad (12)$$

insert for the β-function (τ-function) its 3-loop (2-loop) approximation and set q to the maximal value reached, to compute Λ and $\overline{m}/M$. One can easily convince oneself that the left over perturbative uncertainty is negligible compared to the accumulated statistical errors.

After converting to the $\overline{\mathrm{MS}}$-scheme one arrives at the result [5]

$$\Lambda^{(0)}_{\overline{\mathrm{MS}}} = 251 \pm 21\mathrm{MeV} \ , \quad (13)$$

where the label $^{(0)}$ reminds us that this number was obtained with zero quark flavors. The error in equation 13 sums up all errors including the extrapolations to the continuum limit that were done in the intermediate steps.

This talk is based on research done in the framework of the ALPHA collaboration[5]. I thank in particular S. Capitani, M. Guagnelli, M. Lüscher, S. Sint, P. Weisz and H. Wittig for the joint work.

References

[1] M. Lüscher, P. Weisz and U. Wolff, Nucl. Phys. B359 (1991) 221
[2] M. Lüscher, R. Sommer, P. Weisz and U. Wolff, Nucl. Phys. B389 (1993) 247, Nucl. Phys. B413 (1994) 481
[3] M. Lüscher, R. Narayanan, P. Weisz and U. Wolff, Nucl. Phys. B384 (1992) 168
[4] S. Sint, Nucl. Phys. B421 (1994) 135, Nucl. Phys. B451 (1995) 416
[5] S. Capitani et al., hep-lat/9709125
[6] M. Bochicchio, L. Maiani, G. Martinelli, G.C. Rossi and M. Testa, Nucl. Phys. B262 (1985) 331
[7] M. Lüscher, S. Sint, R. Sommer and H. Wittig, Nucl. Phys. B491 (1997) 344
[8] M. Lüscher, S. Sint, R. Sommer and P. Weisz, Nucl. Phys. B478 (1996) 365
[9] R. Sommer, O(a)-improved lattice QCD, hep-lat/9705026

Part XVI

New Developments in Field Theory and String Theories

Finding the Vacuum Wave Function in Supersymmetric Matrix Theories

Charles M. Sommerfield (charles.sommerfield@yale.edu)

Yale University, New Haven, CT, USA

Abstract. Several examples of solvable cases of extended supersymmetric matrix theories are discussed with a special emphasis on the ground state.

1 Introduction

Recently there has been a resurgence of interest in supersymmetric matrix models following the suggestion that the basic interactions of nature might be manifestations of a master formulation known as M theory [1]. In particular one of the realizations of M theory in the infinite momentum frame can be expressed in terms of a matrix model with $N = 16$ extended supersymmetry and SU(∞) internal symmetry.

An important question associated with this particular model is the nature of the vacuum or ground state. Attempts to find the wave function of the ground state go back to 1985 [2]. The analogous $N = 2$ and $N = 4$ models were also considered at that time. Acceptable solutions were found for the $N = 2$ vacuum and excited states for $SU(2)$ internal symmetry. Very recently solutions were also given for $U(n)$ and $SU(n)$ internal symmetry [3].

In this paper I investigate the $N = 4$ and $N = 16$ theories for SU(n) internal symmetry. The SU(∞) case can be obtained from these by replacing commutators with Moyal brackets as suggested in Ref. [4].

One can motivate the techniques used in the more complicted models by considering a generalized Landau model of a charged particle with spin in a constant magnetic field in multidimensional space. This model exhibits the supersymmetry in a relatively simple manner. Due to space limitations, I cannot present it here.

The organization of the paper is as follows. Sec. 2 contains a general introduction to the notation of supersymmetric matrix models. In Sec. 3, using techniques analogous to those used for the Landau model, I will exhibit a ground-state solution for the $N = 4$ matrix model. Although appearing unacceptable at first glance, this solution can be interpreted in a path-integral formalism by imposing a legitimate restriction of the bosonic integral to half of the normal bosonic configuration space. As pointed out in Sec. 4, the techniques that work for the Landau model and for $N = 4$ do not work for $N = 16$. In principle, there is a method that involves diagonalizing 2^{8d_n} matrices, hardly an encouraging prospect. Here $d_n = n^2 - 1$, the rank of the internal gauge group.

2 Supersymmetric matrix models

The hamiltonian for the supersymmetric matrix models I consider is

$$H = \tfrac{1}{2}(\pi^a_m)^2 + \tfrac{1}{4}g^2(f^{abc}\phi^b_m\phi^c_n)^2 - i\tfrac{1}{2}g\psi^a_\alpha f^{abc}(\Gamma_m)_{\alpha\beta}\phi^b_m\psi^c_\beta. \qquad (1)$$

The matrices Γ_m are real and symmetric and satisfy $\{\Gamma_m, \Gamma_n\} = 2\delta_{mn}$. The structure constants f^{abc} of the SU(n) internal symmetry group are normalized so that $f^{abc}f^{cda} = -\delta^{bd}$. The bosonic dynamical variables are ϕ^m_a whose canonical conjugates are π^m_a. The Fermi spinors, denoted by ψ^a_α, are hermitian and satisfy $\{\psi^a_\alpha, \psi^b_\beta\} = \delta^{ab}\delta_{\alpha\beta}$.

The two situations I will consider are $N = 4$ and $N = 16$, which indicate the number of supersymmetries and also the dimensionality of the spinor. The hamiltonian in (1) is derived by dimensional reduction from $N = 1$ supersymmetric theories in higher dimensions [5]. The vector indices $m, n, ...$ take on values from 1 to s, where s, the number of spatial dimensions in the higher-dimensional space, is 3 for $N = 4$ and 9 for $N = 16$. The supersymmetry transformations are generated by the N hermitian operators

$$Q_\alpha = [(\Gamma_m)_{\alpha\beta}\pi^a_m + \tfrac{1}{2}gf^{abc}(\Gamma_m\Gamma_n)_{\alpha\beta}\phi^b_m\phi^c_n]\psi^a_\beta.$$

Since $\{Q_\alpha, Q_\beta\} = H\delta_{\alpha\beta}$, I can try to find the wave function F of a supersymmetric zero-energy ground state by solving $Q_\alpha F = 0$.

The higher-dimensional theories used to derive (1) are gauge theories. They give rise to first-class constraints that must be satisfied by the physical wave functions. In the dimensionally reduced theory the generators of these constraints are the generators of the global SU(n) symmetry, namely

$$J^a = f^{abc}[\pi^b_m\phi^c_m + \tfrac{1}{2}i\psi^b_\alpha\psi^c_\alpha]$$

and the constraints read $J^aF = 0$. As a practical requirement this means that any physical wave function should saturate the SU(n) indices.

3 N=4 matrix model

A simplifying feature of the $N = 4$ theory is that the product of any two of the 4×4 Γ_m matrices is directly expressed in terms of the third. I introduce the hermitian, antisymmetric matrix $\Gamma = -i\Gamma_1\Gamma_2\Gamma_3$ which commutes with each Γ_m, so that $\Gamma_m\Gamma_n = i\epsilon_{mnp}\Gamma\Gamma_p = i\epsilon_{mnp}\Sigma_p$, $(m \neq n)$, where ϵ_{mnp} is the completely antisymmetric symbol in three dimensions. The 4×4 matrices Σ_m are hermitian and antisymmetric. They obey the SU(2) commutation relations $[\Sigma_m, \Sigma_n] = 2i\epsilon_{mnp}\Sigma_p$. I then find that Q_α may be written

$$\begin{aligned} Q_\alpha &= [\pi^a_m\Gamma\Sigma_m + \tfrac{1}{2}igf^{abc}\epsilon_{mnp}\Sigma_p\phi^b_m\phi^c_n]_{\alpha\beta}\psi^a_\beta \\ &= [\Gamma e^{g\Gamma\Delta}\pi^a_m\Sigma_m e^{-g\Gamma\Delta}]_{\alpha\beta}\psi^a_\beta = [\Gamma e^{g\Gamma\Delta}]_{\alpha\beta}\tilde{Q}_\beta \end{aligned} \qquad (2)$$

where

$$\Delta = \tfrac{1}{6} f^{abc} \epsilon_{pmn} \phi^a_p \phi^b_m \phi^c_n = f^{abc} \phi^a_1 \phi^b_2 \phi^c_3.$$

Solutions F to $Q_\alpha F = 0$ are therefore the same as solutions to $\tilde{Q}_\alpha F = 0$.

The spinor components of ψ^a may be expressed in terms of non-hermitian $2d_n$-component spinors by expansion in orthonormal eigenspinors $(u_\mu)_\alpha$ and $(u^*_\mu)_\alpha$ of Γ:

$$\psi^a_\alpha = \hat{\psi}^{\dagger a}_\mu (u_\mu)_\alpha + \hat{\psi}^a_\mu (u^*_\mu)_\alpha.$$

The expression for $\tilde{Q}_\alpha$ is linear in $\hat{\psi}$ and $\hat{\psi}^\dagger$. The term in the hamiltonian that contains fermions is a linear combination of products of $\hat{\psi}$ and $\hat{\psi}^\dagger$ and therefore conserves "fermion number." Thus I look for solutions $F(\phi, \hat{\psi})$ of $\tilde{Q}_\alpha F = 0$ in the $\hat{\psi}$ representation that are homogeneous in $\hat{\psi}$. The simplest cases are when the degree of homogeneity is 0 or the maximum possible, (namely, $2d_n$, where all of the $\hat{\psi}$'s are multiplied together). I will therefore consider solutions of the form $F_0(\phi)$ and $F_\omega(\phi)\Pi_{\mu,a}\hat{\psi}^a_\mu$. With $\hat{\psi}^{\dagger a}_\mu$ represented by $\partial/\partial\hat{\psi}^a_\mu$ and π^a_m by $-i\partial/\partial\phi^a_m$, it follows from $\tilde{Q}_\alpha F = 0$ that

$$\frac{\partial}{\partial \phi^a_m}[e^{g\Delta} F_0(\phi)] = 0 \quad \text{and} \quad \frac{\partial}{\partial \phi^a_m}[e^{-g\Delta} F_\omega(\phi)] = 0$$

so that two possible solutions are $f_0 = e^{-g\Delta}$ and $f_\omega = e^{g\Delta}\Pi_{\mu,a}\hat{\psi}^a_\mu$.

Since Δ is odd in ϕ, both $e^{\pm g\Delta}$ blow up unacceptably in some regions of configuration space, as has long been recognized [2]. Due to singularities arising from the discontinuity in the step functions $\theta(\pm\Delta)$, an expression of the form $F = \theta(-g\Delta)f_0(\phi, \hat{\psi}) + \theta(g\Delta)f_\omega(\phi, \hat{\psi})$, which immediately suggests itself, would not satisfy $HF = 0$. Presumably quantities similar to F obtained from other fermion-number sectors suffer from the same problems.

My proposal to deal with this is to restrict the configuration space of ϕ so that $g\Delta$ has a definite sign, say positive. In support of this I point out that from a Feynman path-integral point of view the quantity that governs the theory is

$$\int [D\phi][D\hat{\psi}][D\hat{\psi}^\dagger] \exp i \left[\int dt\; L(\phi, \hat{\psi}, \hat{\psi}^\dagger)\right]$$

where L is the lagrangian. In fact this path integral involves double-counting since H and L and the measure in the integral are all unchanged if $\phi \to -\phi$ and $\hat{\psi}^a_\mu \to C_{\mu\nu}\hat{\psi}^{\dagger a}_\nu$ where C is the charge-conjugation matrix. Hence it is legitimate to insert the positivity constraint $\theta(g\Delta)$ into the path integral. This will change the Hilbert-space interpretation of the theory, however. For example, the unitary operator that would represent a finite translation in a given ϕ^a_m, if it would normally result in a change of sign of Δ, would now have to involve all of the other components of ϕ, as well as $\hat{\psi}$ and $\hat{\psi}^\dagger$. It is clear that further investigation of this proposal is necessary.

4 N = 16 matrix model

For $N = 16$ there is no matrix that plays the role that Γ does in $N = 4$. The product of all nine Γ_m is in fact ± 1 depending on the representation. If any of these matrices is diagonalized and $\hat{\psi}$'s and $\hat{\psi}^\dagger$'s introduced as above the resulting expression for the fermion part of H does not conserve the artificially created fermion number. This, in turn, means that a solution F of the equations $Q_\alpha F = 0$ in the $\hat{\psi}$ representation, would involve many terms, perhaps up to the maximum of 2^{8d_n}.

Nevertheless, there are some simplifying features in the $N = 16$ model that may be of use in helping eventually to find a solution. There is an analogue of (2) in that I can express

$$\pi_m^a \Gamma_m + \tfrac{1}{2} g f^{abc} \Gamma_m \Gamma_n \phi_m^b \phi_n^c = \Gamma_m \left(\pi_m^a + g[\pi_m^a, V]\right)$$

where $V = \frac{1}{42} f^{abc} \Gamma_m \Gamma_n \Gamma_p \phi_m^a \phi_n^b \phi_p^c$. This was also noticed in Ref. [6]. The requirement $QF = 0$ can then be written as $(\Gamma_m U)_{\alpha\beta} (\partial/\partial\phi_m^a)(U_{\beta\gamma}^{-1} \psi_\gamma^a F) = 0$ with $(\partial/\partial\phi_m^a) U = -g[(\partial/\partial\phi_m^a) V] U$. At the moment U is known only perturbatively.

Given that the solution to the Landau model uses an equivalence between finite dimensional matrices and fermion operators (Jordan-Wigner transformation), one might try to introduce the appropriate matrices in the $N = 4$ and $N = 16$ matrix models. But the dimensionality $2^{Nd_n/2}$ of these matrices is rather intimidating. It would appear that yet a different method will be required for a complete analysis.

5 Acknowledgments

Much of this work was done in collaboration with Alan Chodos, to whom I am indebted for many critical recommendations, although I take the blame for any misguided or mistaken speculation. Financial support for this work has come, in part, from the U.S. Department of Energy, grant DE-FG02-92ER40704.

References

[1] T. Banks, W. Fischler, S. Shenker and L. Susskind, Physical Review D55 (1997) 5112
[2] M. Claudson and M. B. Halpern, Nuclear Physics B250 (1985), 689
[3] S. Samuel, "Solutions of extended supersymmetric matrix models for arbitrary gauge groups," e-Print Archive: hep-th/9705167
[4] D. Fairlie and C. Zachos, Physics Letters B224 (1989), 101
[5] L. Brink, J. H. Schwarz and J. Scherk, Nuclear Physics B121 (1977), 72
[6] E. Keski-Venturi and A. Niemi, Physics Letters B231 (1989) 75

Self-dual Perturbiner in Yang–Mills Theory

Konstantin Selivanov (selivano@heron.itep.ru)

ITEP, Moscow, Russia

Abstract. The *perturbiner* approach to the multi-gluonic amplitudes in Yang-Mills theory is reviewed.

1 Definition and motivation

Perturbiner is a solution of field equations which can be defined in any field theory [1]. *Definition.* (I use for a moment the scalar field theory) Consider linear part of the equations (that is, with no interactions), and take its solution of the type of $\phi^{(1)} = \sum_j^N \mathcal{E}_j$, where $\mathcal{E}_j = a_j e^{ik^j x}$, k's are on-shell momenta, $k^2 = m^2$ and a_j is assumed to be nilpotent, $a_j^2 = 0$. The perturbiner is a (complex) solution of the nonlinear field equations which is: i)polynomial in $\mathcal{E}_j$ with constant coefficients and ii)whose first order in $\mathcal{E}_j$ part is precisely $\phi^{(1)}$.

Generalization for higher spins is obvious: one should put a polarization factor in front of $\mathcal{E}_j$ in $\phi^{(1)}$.

Motivation. The perturbiner so defined is a generating function for the tree form-factors. The set of the plane waves entering the perturbiner is essentially the set of asymptotic states in the amplitudes which the perturbiner is the generating function for. The nilpotency condition $a_j^2 = 0$ means that one considers only amplitudes with no multiple particles in the same state.

Notice: i) due to this definition one actually works with a finite-dimensional space of polynomials in N nilpotent variables instead of an infinite-dimensional function space ; ii)this definition is different from the one traditionally considered in the stationary phase approach to the S-matrix [2].

In general case, one cannot proceed in a different from the usual perturbation theory way. However in some theories and/or for some sets of the asymptotic states included one can use other powerful methods. In the Yang-Mills (YM) theory one can consider only the same helicity states which leads to considering only the self-duality (SD) equations instead of the full YM ones. This type of SD solutions has been discussed in refs.[4], [5], [6]. In ref.[4] and independently in ref.[5], the tree like-helicity amplitudes were related to solutions of the SD equations. In [4] it was basically shown that the SD equations reproduce the recursion relations for the tree form-factors (also called one-gluonic currents) obtained originally in ref.[9] from the Feynman diagrams; the corresponding solution of SD equations was obtained in terms of the solution of ref.[9] of the recursion relations for the "currents". In ref.[5] an example of SD perturbiner was obtained in the SU(2) case by a 'tHooft anzatz upon further restriction on the asymptotic states included. The consideration of ref.[6] is based on solving recursion relations analogous to refs. [9]. In [10] the YM SD perturbiner was constructed by the twistor methods [7] which also allowed us to obtain perturbiner with one opposite-helicity gluon and thus to obtain a generating function for the so-called maximally helicity violating Parke-Taylor amplitudes [8], [9]. In [10] we also constructed the SD perturbiner in the background of an arbitrary instanton solution.

Briefly, the twistor approach to the SD equations goes as follows. One introduces an auxiliary twistor variable $p^\alpha, \alpha = 1, 2$, which can be viewed on as a pair of complex numbers, and form objects $A_{\dot\alpha} = p^\alpha A_{\alpha\dot\alpha}, \bar\partial_{\dot\alpha} = p^\alpha \frac{\partial}{\partial x^{\alpha\dot\alpha}}, \bar\nabla_{\dot\alpha} = \bar\partial_{\dot\alpha} + A_{\dot\alpha}$. In terms of $\bar\nabla_{\dot\alpha}$, the SD equations turn to a zero-curvature condition, $[\bar\nabla_{\dot\alpha}, \bar\nabla_{\dot\beta}] = 0$, at any $p^\alpha, \alpha = 1, 2$, which can be solved for $A_{\dot\alpha} = p^\alpha A_{\alpha\dot\alpha}$ as

$$A_{\dot\alpha} = g^{-1}\bar\partial_{\dot\alpha} g \tag{1}$$

where g is a function of $x^{\alpha\dot\alpha}$ and p^α with values in the complexification of the gauge group. g must depend on p^α in such a way that the resulting $A_{\dot\alpha}$ is a linear function of p^α, $A_{\dot\alpha} = p^\alpha A_{\alpha\dot\alpha}$. Actually, g is sought for as a homogeneous of degree zero rational function of p^α. Such function necessary has singularities in the p^α-space and it is subject to condition of regularity of $A_{\dot\alpha}$. Then by construction, $A_{\dot\alpha}$ is a homogeneous of degree one regular rational function of two complex variables $p^\alpha, \alpha = 1, 2$. As such, it is necessary just linear in p^α.

An essential moment is that in the case of perturbiner g^{ptb} can only be a polynomial in the variables $\mathcal{E}_j$. First order in $\mathcal{E}_j$ term in g^{ptb} is fixed by the plane wave solution of the free equation (that is by the set of asymptotic states included), while the demand of regularity of $A_{\dot\alpha}$ fixes g^{ptb} up to the gauge freedom.

2 The plane wave solution of the free equation

A solution of the free (i.e. linearized) SD equation consisting of N plane waves looks as follows

$$A^{(1)N}_{\alpha\dot\alpha} = \sum_j^N \epsilon^{+j}_{\alpha\dot\alpha} \hat{\mathcal{E}}_j \tag{2}$$

where the sum runs over gluons, N is the number of gluons, $\epsilon^{+j}_{\alpha\dot\alpha}$ is a four-vector defining a polarization of the j-th gluon, $\hat{\mathcal{E}}^j = t_j \mathcal{E}^j = t_j a_j e^{ik^j x}$, t_j is a matrix defining color orientation of the j-th gluon. $k^j_{\alpha\dot\alpha}$, as a light-like four-vector, decomposes into a product of two spinors $k^j_{\alpha\dot\alpha} = æ^j_\alpha \lambda^j_{\dot\alpha}$.[1] The polarization $\epsilon^{+j}_{\alpha\dot\alpha}$, as a consequence of the linearized SD equations, also

[1] the reality of the four momentum in Minkowski space assumes that $\lambda_{\dot\alpha} = \bar{æ}_\alpha$

decomposes into a product of spinors, such that the dotted spinor is the same as in the decomposition of momentum k, $\epsilon^{+j}_{\alpha\dot{\alpha}} = \frac{q^j_\alpha \lambda^j_{\dot{\alpha}}}{(æ^j, q^j)}$ where normalization factor is defined with use of a convolution $(æ^j, q^j) = \varepsilon^{\gamma\delta} æ^j_\gamma q^j_\delta = æ^{j\,\delta} q^j{}_\delta$. Indexes are raised and lowered with the ε-tensors. The free anti-SD equation would give rise to a polarization $\epsilon^-_{\alpha\dot{\alpha}} = \frac{æ_\alpha \bar{q}_{\dot{\alpha}}}{(\lambda, \bar{q})}$ The auxiliary spinors q_α and $\bar{q}_{\dot{\alpha}}$ form together a four-vector $q_{\alpha\dot{\alpha}} = q_\alpha \bar{q}_{\dot{\alpha}}$ usually called a reference momentum. The normalization was chosen so that $\epsilon^+ \cdot \epsilon^- = \varepsilon^{\alpha\beta} \varepsilon^{\dot{\alpha}\dot{\beta}} \epsilon^+_{\alpha\dot{\alpha}} \epsilon^-_{\beta\dot{\beta}} = -1$

3 The solution for g^{ptb} and $A^{ptb}_{\dot{\alpha}}$

First order in $\mathcal{E}$ terms in g^{ptb} are easily found from equation (1), first order version of which reads $A^{ptb(1)}_{\dot{\alpha}} = \bar{\partial}_{\dot{\alpha}} g^{ptb(1)}$ and hence, with use of (2),

$$g^{ptb(1)} = 1 + \sum_j \frac{(p, q^j)}{(p, æ^j)} \frac{\hat{\mathcal{E}}^j}{(æ^j, q^j)} \tag{3}$$

Thus $g^{ptb(1)}$ has simple poles on the auxiliary space at the points $p_\alpha = æ^j_\alpha$. Actually, one can see that the condition of regularity of $A^{ptb}_{\dot{\alpha}}$ dictates that the full g^{ptb} has only simple poles at the same points as $g^{ptb(1)}$. Moreover, it also fixes residues of g^{ptb} to all orders in $\mathcal{E}$ in terms of the residues of $g^{ptb(1)}$ Eq.(3).

The known singularities of g^{ptb} fix it up to an independent of the auxiliary variables $p^\alpha, \alpha = 1, 2$ matrix, i.e., up to a gauge freedom. The problem of reconstructing g^{ptb} from its singularities essentially simplifies if one considers color ordered highest degree monomials in g^{ptb}, for which one obtaines ([10])

$$g^{ptb}_{N(N,\dots,1)} = \frac{(p, q^N)(æ^N, q^{N-1}) \dots (æ^2, q^1)}{(p, æ^N)(æ^N, æ^{N-1}) \dots (æ^2, æ^1)} \frac{\hat{\mathcal{E}}^N}{(æ^N, q^N)} \cdots \frac{\hat{\mathcal{E}}^1}{(æ^1, q^1)} \tag{4}$$

This is, essentially, a solution of the problem. Substituting g^{ptb} (4) into equation (1) determines the perturbiner $A^{ptb}_{\dot{\alpha}}$ (see ref.[10]).

4 The Parke-Taylor amplitudes

The SD perturbiner can be used as a base point for a perturbation procedure of adding one-by-one gluons of the opposite helicity, or other particles, say, fermions, interacting with gluons.. The explicit expression for g^{ptb} (4) is very useful in this procedure. The SD perturbiner itself describes the tree form-factors - objects including an arbitrary number of on-shell SD gluons and one arbitrary off-shell gluon. To obtain the Parke-Taylor amplitudes, those with two gluons of the opposite helicity, one essentially needs to construct the perturbiner including one on-shell gluon of the opposite helicity, that is, to solve the linearized YM equation in the background of SD perturbiner. Details of this solution can be found in ref.[10]. The resulting generating function for the Parke-Taylor amplitudes reads

$$M(k'',k',\{a_j\}) = -i(æ'',æ')^2 \int d^4x\, tr$$

$$\hat{\mathcal{E}}''(g^{ptb}|_{(p=æ',q=æ'')})^{-1}\hat{\mathcal{E}}' g^{ptb}|_{(p=æ',q=æ'')} \tag{5}$$

Considering cyclic ordered terms in this expression with g^{ptb} (4) one easily reproduces the Parke-Taylor maximally helicity violating amplitudes [8], [9].

5 The SD perturbiner in a topologically nontrivial sector

The concept of perturbiner can be generalized to a topologically nontrivial sector (see [10]). In the latter case it provides a framework for the instanton mediated multi-particle amplitudes. All what we need to know about the instanton, $A_{\dot{\alpha}}^{inst}$, that it can be represented in the twistor-spirit form $A_{\dot{\alpha}}^{inst} = g_{inst}^{-1}\bar{\partial}_{\dot{\alpha}}g_{inst}$ g_{inst} is assumed to be a rational function of the auxiliary variables p^{α}, such that $A_{\dot{\alpha}}^{inst}$ is a linear homogeneous function of p^{α}. Then the SD topologically nontrivial perturbiner $A_{\dot{\alpha}}^{iptb}$ is represented in the form $A_{\dot{\alpha}}^{iptb} = (g^{iptb})^{-1}\bar{\partial}_{\dot{\alpha}}g^{iptb}$ and the corresponding g^{iptb} is found to be

$$g^{iptb}(\hat{\mathcal{E}}^1,\hat{\mathcal{E}}^2,\ldots) = g^{inst}g^{ptb}(\hat{\mathcal{E}}^1_g,\hat{\mathcal{E}}^2_g,\ldots) \tag{6}$$

where $\hat{\mathcal{E}}^j_g$ stand for twisted harmonics $\hat{\mathcal{E}}^j_g = (g_{inst}|_{(p=æ^j)})^{-1}\hat{\mathcal{E}}^j g_{inst}|_{(p=æ^j)}$ and g^{ptb} as in Eq. (4). This is the sought for SD perturbiner in an arbitrary instanton background.

References

[1] A.Rosly, K.Selivanov, preprint ITEP-TH-43-96, hep-th/9610070

[2] A.A.Slavnov, L.D.Faddeev, Introduction to the Theory of Quantum Qauge Fields,Nauka, Moscow, 1978

[3] C.Itzykson, J.-B.Zuber, Quantum Field Theory, McGrow-hill, NY, 1980

[4] W.Bardeen, Prog.Theor.Phys.Suppl, 123 (1996) 1

[5] K.Selivanov, preprint ITEP-21-96, hep-ph/9604206

[6] V.Korepin, T.Oota, J.Phys.A 29 (1996) 625, hep-th/9608064

[7] R.S.Ward, Phys.Lett.61A (1977) 81

[8] S.Parke, T.Taylor, Phys.Rev.Lett. 56 (1986) 2459

[9] F.Berends, W.Giele, Nucl.Phys. B306 (1988) 759

[10] A.Rosly, K.Selivanov, Phys.Lett.B 399 (1997) 135

[11] M.Mangano, S.Parke, Phys.Rep. 200 (1991) 301

‘Soft’ Phenomenology

Carlos Muñoz

Departamento de Física Teórica, Universidad Autónoma de Madrid, Cantoblanco, 28049 Madrid, Spain
and
Department of Physics, Korea Advanced Institute of Science and Technology, Taejon 305–701, Korea

Abstract. A review of the soft SUSY–breaking parameters and the μ term arising in SUGRA and superstrings is performed paying special attention to their phenomenological implications. In particular, the violation of the scalar mass universality which may lead to dangerous FCNC phenomena and the existence of CCB minima are discussed.

1 Introduction and Summary

Superstring theory is by now the only theory which can unify all the known interactions including gravity. Although this is a remarkable theoretical success, experimental evidence is still lacking. Since, unlike philosophy, physics must be subject to experimental test, if we do not want to regard superstring theory in the future as a philosophical theory some experimental evidence must be obtained. Unfortunately, whereas the natural scale of superstring models is $\mathcal{O}(M_{Planck})$, the ‘natural’ scale of particle accelerators is $\mathcal{O}(1TeV)$, and therefore to obtain any experimental proof seems very difficult. Nevertheless, if Nature is supersymmetric (SUSY) at the weak scale, as many particle physicists believe, eventually the spectrum of SUSY particles will be measured providing us with a possible connection with the superhigh–energy world of superstring theory. The main point is that the SUSY spectrum is determined by the soft SUSY–breaking parameters and these in its turn can be computed in the context of superstring models. To compare then the superstring predictions about soft terms with the experimentally–observed SUSY spectrum would allow us to test the goodness of this theory. This is the path, followed in the last years by several groups, that I will try to review briefly in sect. 2 paying special attention to the constraints deriving from flavor changing neutral current (FCNC) phenomena. In sect. 3 other phenomenological constraints on soft parameters are studied. In particular, those deriving from charge and color breaking (CCB) minima. Finally, in sect. 4 I will discuss the effects of pure supergravity (SUGRA) loop corrections on soft parameters and the μ term and apply them to superstring models. New sources of the μ term and FCNC will appear.

2 Soft Parameters from Superstrings and FCNC

As is well known the soft parameters can be computed in generic hidden sector SUGRA models [1]. They depend on the three functions, K, W and f which determine the full N=1 SUGRA Lagrangian. Expanding in powers of the observable fields C^α these are given by

$$\begin{aligned}
K &= \hat{K} + \tilde{K}_{\alpha\overline{\beta}} C^\alpha \bar{C}^{\overline{\beta}} + \frac{1}{4}\tilde{K}_{\alpha\overline{\beta}\gamma\overline{\delta}} C^\alpha \bar{C}^{\overline{\beta}} C^\gamma \bar{C}^{\overline{\delta}} \\
&+ \left[\frac{1}{2} Z_{\alpha\beta} C^\alpha C^\beta + \frac{1}{2} Z_{\alpha\beta\overline{\gamma}} C^\alpha C^\beta \bar{C}^{\overline{\gamma}} \right. \\
&\left. + \frac{1}{6} Z_{\alpha\beta\gamma\overline{\delta}} C^\alpha C^\beta C^\gamma \bar{C}^{\overline{\delta}} + \ h.c.\right] + \ldots \\
W &= \hat{W} + \frac{1}{2}\tilde{\mu}_{\alpha\beta} C^\alpha C^\beta + \frac{1}{6}\tilde{Y}_{\alpha\beta\gamma} C^\alpha C^\beta C^\gamma + \ldots
\end{aligned}$$

$$f_{ab} = \hat{f}_{ab} + \frac{1}{2}\tilde{f}_{ab\alpha\beta}C^\alpha C^\beta + ... \quad (1)$$

where the coefficient functions depend upon hidden sector fields and the ellipsis denote the terms which are irrelevant for the present calculation. Although at tree level the cubic and quartic terms in K and the quadratic terms in f_{ab} are also irrelevant, we will see in sect. 4 that when loop effects are included they may be important for the μ term and FCNC phenomena.

For example, assuming vanishing cosmological constant, the *un-normalized* scalar masses arising from the expansion of the (F part of the) tree-level SUGRA scalar potential are given in this context by

$$m^2_{\alpha\overline{\beta}} = m^2_{3/2}\tilde{K}_{\alpha\overline{\beta}} - F^m R_{m\overline{n}\alpha\overline{\beta}}\overline{F}^{\overline{n}} \quad (2)$$

$$R_{m\overline{n}\alpha\overline{\beta}} = \partial_m\partial_{\overline{n}}\tilde{K}_{\alpha\overline{\beta}} - \tilde{K}^{\gamma\overline{\delta}}\partial_m\tilde{K}_{\alpha\overline{\delta}}\partial_{\overline{n}}\tilde{K}_{\gamma\overline{\beta}} \quad (3)$$

where F^m denote the hidden field auxiliary components. Notice that, after normalizing the fields to get canonical kinetic terms, the first piece in (2) will lead to universal diagonal soft masses but the second piece will generically induce *off-diagonal* contributions. Actually, universality is a desirable property for phenomenological reasons, particularly to avoid FCNC. We will discuss below how string models may get interesting *constraints* from FCNC bounds.

On the other hand, from the fermionic part of the tree-level SUGRA Lagrangian, the *un-normalized* Higgsino masses are given by

$$\mu_{\alpha\beta} = e^{\hat{K}/2}\frac{\hat{W}^*}{|\hat{W}|}\tilde{\mu}_{\alpha\beta} + m_{3/2}Z_{\alpha\beta} - \overline{F}^{\overline{n}}\partial_{\overline{n}}Z_{\alpha\beta} \quad (4)$$

As is well known the presence of this μ term is crucial in order to have correct electroweak symmetry breaking [1].

The arbitrariness of SUGRA theory, one can think of many possible SUGRA models (with different K, W and f) leading to *different* results for the soft terms, can be ameliorated in SUGRA models deriving from superstring theory, where K, f, and the hidden sector are more constrained. The heterotic string models have a natural hidden sector built-in: the dilaton field S and the moduli fields T_i, U_i. Without specifying the supersymmetry-breaking mechanism, just assuming that the auxiliary fields of those multiplets are the seed of supersymmetry breaking, interesting predictions for this simple class of models are obtained [1].

Let us focus first on the very interesting limit where the dilaton S is the source of all the SUSY breaking. At superstring tree level the dilaton couples in a universal manner to all particles and therefore, this limit is compactification *independent*. In particular, since the VEVs of the moduli auxiliary fields F^i are vanishing, and $\tilde{K}_{\alpha\overline{\beta}}$ is independent on S, the second piece in (2) is vanishing and therefore, after normalization, the soft scalar masses turn out to be universal, $m_\alpha = m_{3/2}$. Because of the simplicity of this scenario, the predictions about the low-energy spectrum are quite precise [1].

Let us finally remark that although the soft parameters are universal in this dilaton-dominated SUSY-breaking limit, this result may be spoiled due to superstring loop effects [2]. In particular, $\tilde{K}_{\alpha\overline{\beta}}$ receive an S-dependent contribution and therefore the second piece in (2) will be non-vanishing. In sect. 4 we will find another source of non-universality due to generic SUGRA loop effects.

In general the moduli fields, T_i, U_i, may also contribute to SUSY breaking, i.e. $F^i \neq 0$, and in that case their effects on soft parameters must also be included (see e.g. (2)). Since different compactification schemes give rise to different expressions for the moduli-dependent part of K (1), the computation of the soft parameters will be model *dependent*.

Let us concentrate first on the simple situation of diagonal moduli and matter metrics. This is the case of most $(0,2)$ symmetric Abelian orbifolds which, at superstring tree level, have $\tilde{K}_{\alpha\bar{\beta}} = \delta_{\alpha\beta}\Pi_i(T_i + \overline{T}_i)^{n^i_\alpha}$ where n^i_α are the modular weights of the matter fields C^α. Plugging this result in (2) we get the following scalar masses after normalizing the observable fields:

$$m_\alpha^2 = m_{3/2}^2 + \sum_i \frac{n^i_\alpha}{(T_i + \overline{T}_i)^2}|F^i|^2 \qquad (5)$$

With this information one can analyze the structure of soft parameters and in particular the low–energy spectrum [1]. If we assume that also continuous Wilson–line moduli contribute to SUSY breaking, the analysis turn out to be more involved since off–diagonal moduli metrics arise due to the mixing between T_i, U_i and Wilson lines. However, the final formulas for scalar masses are similar to (5) (with some extra contributions due to the Wilson–line auxiliary fields), and therefore the low–energy spectrum is also similar [3].

Notice that the scalar masses (5) show in general a *lack of universality* due to the modular weight dependence [4]. So, even with diagonal matter metrics, FCNC effects may appear. However, we recall that the low–energy running of the scalar masses has to be taken into account. In particular, in the squark case, for gluino masses heavier than (or of the same order as) the scalar masses at the boundary scale, there are large flavour–independent gluino loop contributions which are the dominant source of scalar masses. This situation is very common in orbifold models. The above effect can therefore help in fulfilling the FCNC constraints [5].

Although diagonal metrics is the generic case in most orbifolds, there are a few cases, Z_3, Z_4 and Z_6', where off–diagonal metrics are present in the untwisted sector. In particular, at superstring tree level, the moduli–dependent part of K (1) is given by: $K = -\ln\det(T_{i\bar{j}} + \overline{T}_{i\bar{j}} - C^i\bar{C}^{\bar{j}})$. This clearly implies that the scalar mass eigenvalues will be in general non-degenerate, which in turn may induce FCNC [6]. The same potential problem is present in Calabi–Yau compactifications where off–diagonal metrics is the generic situation: $K = \hat{K}^T + \hat{K}^U + \tilde{K}^T_{i\bar{j}}\mathbf{27}^i\mathbf{27}^{*\bar{j}} + \tilde{K}^U_{k\bar{l}}\overline{\mathbf{27}}^k\overline{\mathbf{27}}^{*\bar{l}}$ where $\tilde{K}^T_{i\bar{j}} = (\partial^2\hat{K}^T/\partial T_i\partial\overline{T}_j)e^{(\hat{K}^U-\hat{K}^T)/3}$, $\tilde{K}^U_{k\bar{l}} = (\partial^2\hat{K}^U/\partial U_k\partial\overline{U}_l)e^{(\hat{K}^T-\hat{K}^U)/3}$, and $\hat{K}^T \approx -\ln k_{ijk}(T_i + \overline{T}_i)(T_j + \overline{T}_j)(T_k + \overline{T}_k)$, Obviously, using (2), the mass eigenvalues are typically non–degenerate [7].

3 CCB Constraints on Soft Parameters

We already noticed above that the high–energy constraints on the soft parameter space obtained from studying superstring models can be combined with low–energy phenomenological constraints as e.g. FCNC phenomena. But, in fact, we can go further and impose the (theoretical) constraint of demanding the no existence of low–energy CCB minima deeper than the standard vacuum [8]. In the particular case of the dilaton–dominated scenario, the restrictions coming from the CCB minima are very strong and the whole parameter space $(m_{3/2}, B, \mu)$ turns out to be excluded on these grounds [9]. Given these dramatic conclusions, a way–out must be searched. The simplest possibility is to assume that also the moduli fields T_i contribute to SUSY breaking. Since then the soft terms are modified and new free parameters beyond $m_{3/2}$ and B appear, possibly some regions in the parameter space will be allowed. Although now the situation is clearly more model dependent, a good and simple starting point might be to study the case of orbifolds with diagonal moduli and matter metrics. Assuming that SUSY breaking is equally shared among T_i's, i.e. the 'overall modulus' T scenario, basically only one more free parameter must be added [10].

4 Supergravity Radiative Effects on Soft Parameters and the μ Term

The soft parameters and μ are usually computed at the tree level of SUGRA interactions. However, there can be a significant modification in this procedure due to quadratically divergent SUGRA one–loop effects [11]. For example, in the case of the μ term these quantum corrections do not only modify the already known contributions from $\tilde{\mu}_{\alpha\beta}$ and $Z_{\alpha\beta}$ in the tree–level matching condition (4) but also provide *new* sources of the μ term, naturally of order the weak scale. These sources depend upon the coefficients $Z_{\alpha\beta\overline{\gamma}}$ and $Z_{\alpha\beta\gamma\overline{\delta}}$ in K and also the coefficients $\tilde{f}_{ab\alpha\beta}$ in f_{ab} (1).

Another interesting feature of the SUGRA corrections is the *lack of universality*. If the Riemann tensor (3) can be factorized as $R_{m\overline{n}\alpha\overline{\beta}} = c_{m\overline{n}}\tilde{K}_{\alpha\overline{\beta}}$, as it happens e.g. in no–scale SUGRA models or in the dilaton–dominated scenario explained in sect. 2, the tree–level matching conditions (2) would give a universal soft scalar mass for the *normalized* observable fields. But now it is possible to see that due to the SUGRA radiative corrections, particularly due to the contributions of the pieces depending upon $Z_{\alpha\beta\overline{\gamma}}$ and $\tilde{K}_{\alpha\overline{\beta}\gamma\overline{\delta}}$ in K (1), the scalar masses will have a generic matrix structure with non–degenerate eigenvalues [11].

Let us apply these general results to the case of superstring effective SUGRA models reviewed in sect. 2. We will concentrate here on symmetric Abelian orbifolds. Although not computed explicitly yet, one can imagine the following modular–invariant form of $\tilde{K}_{\alpha\overline{\gamma}\beta\overline{\delta}}$: $\tilde{K}_{\alpha\overline{\gamma}\beta\overline{\delta}} = \delta_{\alpha\gamma}\delta_{\beta\delta}X_{\alpha\beta}\Pi_i(T_i + \overline{T}_i)^{n^i_\alpha}\Pi_j(T_j + \overline{T}_j)^{n^j_\beta}$, where $X_{\alpha\beta}$ are constant coefficients of order one. Using now the SUGRA corrections computed in [11], the scalar masses (5) get an extra contribution proportional to $m^2_{3/2}\sum_\gamma X_{\alpha\gamma}$. One can now see more explicitly the interesting feature of the one–loop SUGRA corrections which were discussed briefly above: even when the tree–level matching condition (5) leads to a universal soft mass, which would be the case if all C^α have the same modular weight or if all $F^i = 0$ (this corresponds to the case of dilaton-dominated SUSY breaking), the SUGRA corrections depending upon $X_{\alpha\gamma}$ are *no longer universal.*

References

[1] For a recent review, see: A. Brignole, L.E. Ibáñez and C. Muñoz, *hep-ph/9707209*, and references therein

[2] J. Louis and Y. Nir, *Nucl. Phys.* **B447** (1995) 18

[3] H.B. Kim and C. Muñoz, *Mod. Phys. Lett.* **A12** (1997) 315

[4] L.E. Ibáñez and D. Lüst, *Nucl. Phys.* **B382** (1992) 305

[5] A. Brignole, L.E. Ibáñez and C. Muñoz, *Nucl. Phys.* **B422** (1994) 125; P. Brax and M. Chemtob, *Phys. Rev.* **D51** (1995) 6550

[6] A. Brignole, L.E. Ibáñez, C. Muñoz and C. Scheich, *Z. Phys.* **C74** (1997) 157

[7] H.B. Kim and C. Muñoz, *Z. Phys.* **C75** (1997) 367

[8] For a recent review, see: C. Muñoz, *hep-ph/9709329*, and references therein

[9] J.A. Casas, A. Lleyda and C. Muñoz *Phys. Lett.* **B380** (1996) 59

[10] A. Ibarra, J.A. Casas and C. Muñoz, in preparation

[11] K. Choi, J.S. Lee and C. Muñoz, *hep-ph/9709250*

Generalized 2d-Dilaton Models, the True Black Hole and Quantum Integrability

M.O. Katanaev[1,2], W. Kummer[1], H. Liebl[1] and D.V. Vassilevich[1,3]

[1]Institut für Theoretische Physik, Technische Universität Wien
Wiedner Hauptstr. 8–10, A-1040 Wien, Austria

[2]Steklov Mathematical Institute,
Vavilov St. 42, 117966 Moscow, Russia

[3]Dept. of Theoretical Physics, St. Petersburg University
198904 St. Petersburg, Russia

Abstract. $1+1$ dimensional dipheomorphism-invariant models can be viewed in a unified manner. At the quantum level for *all* generalized dilaton theories — in the absence of matter — the local quantum effects are shown to disappear. This enables us to compute e.g. the second loop order correction to the Polyakov term.

It is important for any application of the dilaton black hole (DBH) or its generalizations [1, 2, 3, 4] to compare the respective singularity structure with the ones encountered in General Relativity (GR), e.g. for the (uncharged) spherically symmetric case. However, careful studies of the singularity structure in such theories seem to be scarce. During our recent work [4] we noted that for the ordinary dilaton black hole of [1] null extremals are complete at the singularity. Of course, non–null extremals are incomplete, and so at least that property holds for the DBH, but, from a physical point of view, it seems a strange situation that massive test bodies fall into that singularity at a finite proper time whereas it needs an infinite value of the affine parameter of the null extremal (describing the influx e.g. of massless particles) to arrive.

In order to pave the way for a more realistic modelling of the Schwarzschild black hole we considered a two parameter family of generalized dilaton theories which interpolates between the DBH and other models, several of which have been suggested already in the literature [5] . The Eddington–Finkelstein (EF) form of the line element, appearing naturally in 2d models [6] is very helpful in this context. We indeed find large ranges of parameters for which possibly more satisfactory BH models in $d=2$ may be obtained.

Here we concentrate on generic dilaton gravity

$$\mathcal{L}_{(1)} = \sqrt{-g}\left(-X\frac{R}{2} - \frac{U(X)}{2}(\nabla X)^2 + V(X)\right) \tag{1}$$

which can be shown to be equivalent to the first order Lagrangian

$$\mathcal{L}_{(2)} = X^+ De^- + X^- De^+ + X d\omega + \epsilon(V(X) + X^+X^-U(X)), \tag{2}$$

where $De^a = de^a + (\omega \wedge e)^a$ is the torsion two form, the scalar curvature R is related to the spin connection ω by $-\frac{R}{2} = *d\omega$ and ϵ denotes the volume two form $\epsilon = \frac{1}{2}\varepsilon_{ab}e^a \wedge e^b = d^2x \det e_a^\mu = d^2x\,(e)$. The auxiliary fields $X^\pm$ and X form (2) clearly can be interpreted as momenta. The dynamics is encoded in the factor of ϵ. We quantize following the canonical BVF [7] approach and work in a 'temporal' gauge :

$$e_0^+ = \omega_0 = 0 \quad , \quad e_0^- = 1 \tag{3}$$

After computing the extended Hamiltonian by the introduction of the usual two types of ghosts for a stage 1 Hamiltonian, we finally arrive at the generating functional for the Green functions. Integrating out the ghosts the generating functional becomes

$$W = \int (\mathcal{D}X)(\mathcal{D}X^+)(\mathcal{D}X^-)(\mathcal{D}e_1^+)(\mathcal{D}e_1^-)(\mathcal{D}\omega_1)F \exp\left[i\int_x \mathcal{L}_{(2)} + \mathcal{L}_s\right], \tag{4}$$

where F denotes the determinant

$$F = \det(\delta_i^k \partial_0 + C_{i2}^k) = (\det \partial_0)^2 \det(\partial_0 + X^+ U(X)). \tag{5}$$

$\mathcal{L}_{(2)}$ is the gauge fixed part of the Lagrangian (2) and $\mathcal{L}_s$ denotes the contribution of the sources:

$$\mathcal{L}_s = j^+ e_1^- + j^- e_1^+ + j\omega_1 + J^+ X^- + J^- X^+ + JX \tag{6}$$

Contrary to the standard approach for the path integral in phase space we now integrate *first* the 'coordinates' $e_1^\pm, \omega_1$ and use the resulting δ-functions to perform the X-integrations yielding the generating functional of connected Green functions

$$Z = -i \ln W = \int JX + J^- \frac{1}{\partial_0} j^+ + J^+ \frac{1}{\partial_0 + U(X)\frac{1}{\partial_0} j^+} \left(j^- - V(X)\right), \tag{7}$$

where X has to be replaced by (with suitable definitions for ∂_0^{-1}, ∂_0^{-2} [8])

$$X = \frac{1}{\partial_0^2} j^+ + \frac{1}{\partial_0} j \quad . \tag{8}$$

Eq. (7) gives the exact non-perturbative generating functional for connected Green functions. It clearly describes tree–graphs only. Hence no quantum effects remain.

In the presence of matter we consider perturbation theory in the *matter field*, treating the geometrical part still exactly by a *nonperturbative* path integral [9]. Our approach thus differs fundamentally from the conventional 'semiclassical' one [1] in which (mostly only one loop) effects of matter are added and the resulting effective action subsequently is solved classicaly.

Our matter contribution is a minimally coupled scalar field whose Lagrangian

$$\mathcal{L}^m = \frac{1}{2}\sqrt{-g}g^{\mu\nu}\partial_\mu S \partial_\nu S = -\frac{1}{2}\frac{\varepsilon^{\alpha\mu}\varepsilon^{\beta\nu}}{e}\eta_{ab}e^a_\mu e^b_\nu \partial_\alpha S \partial_\beta S \quad . \tag{9}$$

is to be added to (2). For technical reasons, we restrict ourselves to a subclass of 2d models with $U(X) = 0$. Also a source for S is added to $\mathcal{L}$.

The generating functional of Green functions with proper nontrivial measure for S is obtained after integrating the ghosts and using (3):

$$W = \int (\mathcal{D}\sqrt{e_1^+}S)(\mathcal{D}X)(\mathcal{D}X^+)(\mathcal{D}X^-)(\mathcal{D}e_1^+)(\mathcal{D}e_1^-)(\mathcal{D}\omega_1)F \exp\left[\frac{i}{\hbar}\int_x \mathcal{L}_{gf}\right] . \tag{10}$$

In this case we *first* integrate the gauge fixed Lagrangian $\mathcal{L}_{gf}$ over e_1^-, X^- and ω_1 and use the resulting delta functions to integrate out the remaining variables X^+, e_1^+ and X.

Finally the S integration has to be performed. According to our approach we now expand the effective L , obtained so far, in terms of S^2 where the one loop term $\mathcal{O}(S^2)$ is integrated (Gaussian integral) providing a Polyakov term, corrected by the dilaton field which is present in our case. The next order S^4 is treated perturbatively using the 'full' propagator obtained also from the S^2 - integral.

It is now rather straightforward [9] to show that the two loop contribution γ is *independent* of the fields and in the generating functional for connected Green functions the $\hbar^2$ term expresses a renormalization of the 'potential' V.

As a final remark we may add that in the effective action the Polyakov term does not acquire 2-loop corrections.

Acknowledgement

This work has been supported by Fonds zur Förderung der wissenschaftlichen Forschung (FWF) Project No. P 10221-PHY. One of the authors (D.V.) thanks the Russian Foundation for Fundamental Research, grant 97-01-01186, for financial support. Another author (M.O. K.) is grateful to the Erwin Schrödinger International Institute, The International Science Foundation (grant NFR 000) The Russian Fund for Fundamental Investigations (grant RFFI-96-010-0312) and the Austrian Academy of Sciences.

References

[1] E. Witten, Phys. Rev. **D44**, (1991) 314; C. G. Callan, S. B. Giddings, J. A. Harvey, and A. Strominger, Phys. Rev. **D45**, (1992) 1005

[2] S. Elitzur, A. Forge and E. Rabinovici, Nucl. Phys. **B359**, (1991) 581; D. Banks and M. O'Loughlin, Nucl. Phys. **B362** (1991) 649; D. Louis–Martinez, J. Gegenberg and G. Kunstatter, Phys. Lett. **B321** (1994) 193; D. Louis–Martinez and G. Kunstatter, Phys. Rev. **D49** (1994) 5227;

[3] M.O. Katanaev and I. V. Volovich; Phys. Lett. **175B**, (1986) 413; Ann. Phys., **197** (1990) 1; W. Kummer and D.J. Schwarz, Phys. Rev. , **D45**, (1992) 3628; T. Strobl, Int. J. Mod. Phys., **A8** (1993) 1383; F. Haider and W. Kummer, Int. J. Mod. Phys. **9** (1994) 207; M. O. Katanaev, Nucl. Phys. **B416** (1994) 563.

[4] M.O. Katanaev, W. Kummer and H. Liebl, Phys. Rev. **D53**, (1996) 5609.

[5] J.S. Lemos and P.M. Sa, Phys. Rev. **D49** (1994) 2897; S. Mignemi, Phys. Rev. **D50**, (1994), R4733; A. Fabbri and J.G. Russo, Phys. Rev. **D53**, (1996), 6995.

[6] T. Strobl, Thesis, Tech. Univ. Vienna 1994; T. Klösch and T. Strobl, Class. Quantum Grav. **13** (1996), 965 and 2395; An early version of this method can be found in N. Walker, J. Math. Phys. **11** (1970) 2280.

[7] I.A. Batalin and G.A. Vilkovisky, Phys. Lett. **B69**, (1977), 309.

[8] W. Kummer, H. Liebl and D. Vassilevich, Nucl. Phys. **B493**, (1997), 491.

[9] W. Kummer, H. Liebl and D. Vassilevich, *Nonperturbative path integral of 2d dilaton gravity and two-loop effects from scalar matter*, hep-th/9707115.

Part XVII

New Detectors and Experimental Techniques

Recent Advances in Vertex Detection

John Conway[1] (conway@physics.rutgers.edu)

Rutgers University, Piscataway, New Jersey, USA

Abstract. This talk reviews the recent advances in vertex detectors for high-energy colliders, focussing on CVD diamond as a radiation-hard alternative to silicon, the SLD 307-million-pixel tracking system, and the SVX-II detector in CDF.

1 CVD Diamond Detectors

In hadron colliders, the intense radiation field near the interaction region presents a major challenge to the survivability of particle detectors. The presently favored tracking detector technology, silicon microstrip detectors, will not be able to survive these hadron fluxes indefinitely. The silicon microstrip detectors uunder construction for CDF and D0 will undergo type inversion and then require ever-higher bias voltage to keep the detector bulk fully depleted. Eventually leakage current noise and voltage breakdown will necessitate replacement of the devices.

Chemical vapor deposition (CVD) diamond represents a radiation-hard alternative to silicon. The large crystal cohesive energy and small transmutation cross section of carbon in diamond form give it its radiation hardness, and its other physical properties, particularly its high resistivity, low mass, and large thermal conductivity, make diamond a nearly ideal material for charged particle detectors. Such detectors can be fabricated by depositing metal electrodes (Cr-Au) on the surfaces; no p-n junction is needed due to the large resistivity.

Since 1994 the RD42 collaboration [1] has developed charged particle microstrip detectors based on CVD diamond obtained from collaboration with industrial sources. This relationship has improved the signal size of these detectors by orders of magnitude, as indicated in Figure 1. The collection distance, the average distance by which electrons and holes separate in the bulk, has improved from a few μm to about 200 μm. Since minimum-ionizing particles deposit 3600 e-h pairs per 100 μm material traversed, this corresponds to over 7200 e^- signal.

Figure 2 shows the pulse height distribution from one of the recent diamond detectors, measured in a 100-GeV pion beam using a low-noise VA-2 readout chip. The mean pulse height is 7450 electrons, and the most probable is about 4000 electrons. The distribution does not follow a Landau shape, and in particular the distribution shows a minimum pulse height relatively near zero, at about 500 electrons in this case. Equipped with 25-nsec electronics, the SCT-32A chip developed at CERN for the ATLAS silicon detectors, this diamond gives a mean signal-to-noise ratio of over 7.

Eight planes of silicon strip detectors allow determination of the position resolution of the detector. Figure 2 shows the distribution of the difference between hit position and track position; the resolution of the detector is 16 μm.

The radiation hardness of diamond has been demonstrated by exposure of samples to gamma rays, pions, protons, and neutrons at fluences either exceeding or comparable to those anticipated after several years of operation near the interaction region of the LHC. Neither exposure to 10^{15} pions/cm^2 nor to 10^{14} protons/cm^2 shows any degradation of performance. Exposure to nearly 10^{15} neutrons/cm^2, however, results in a roughly 20% loss of signal, as shown in figure 3; a silicon detector exposed simultaneously with the diamond shows a large increase in leakage current during the exposure, whereas the current in the diamond follows the neutron flux.

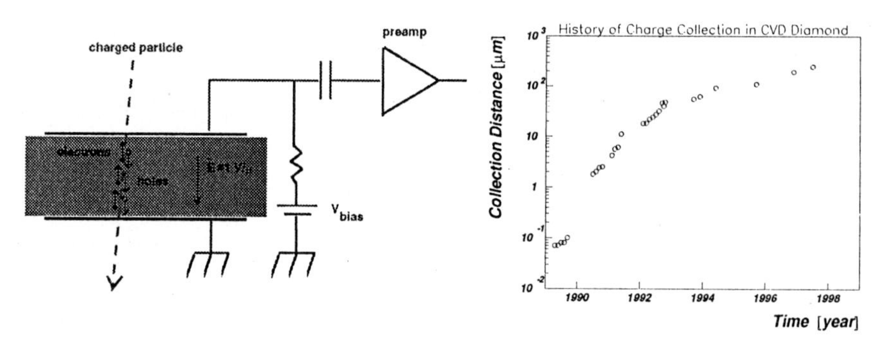

Fig. 1. Left: principle of operation of diamond particle detectors. Right: History of improvement in diamond charge collection distance.

2 SLD Pixel Detector

In early 1996 the SLD experiment at the SLC at SLAC began operating a 307-million-channel silicon pixel vertex detector called the VXD3, the successor to the VXD2. The three-layer device gives three-dimensional space points along charged particle trajectories. The basic sensitive units are the approximately 20-μm-cube sensing elements of CCD arrays, in which a minimum-ionizing particle deposits typically 1200 electrons. Figure 4 shows the construction of the basic detector module, and also shows the improvement in spatial resolution obtained with the VXD3 relative to that of the VXD2. For high-transverse-momentum tracks the impact parameter resolution is about 12

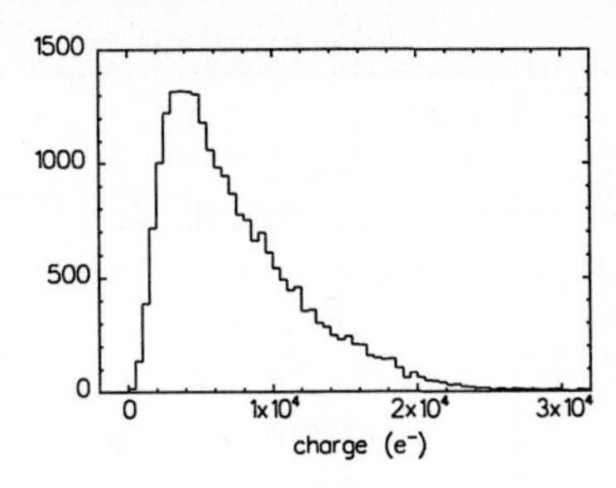

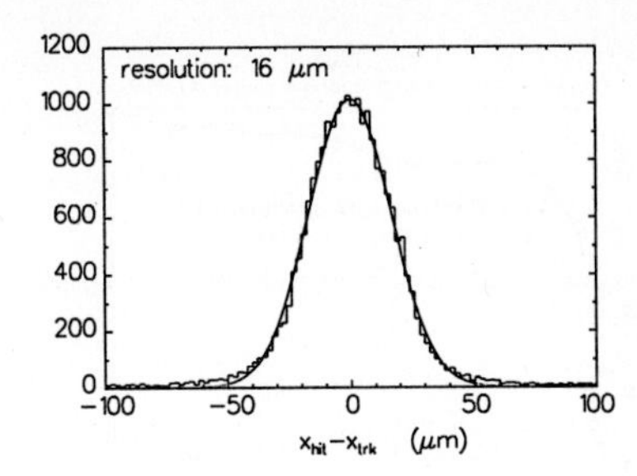

Fig. 2. Left: Distribution of collected charge in 50-μm-pitch CVD diamond detector. Right: Distribution of difference between track and hit positions in a 50-μm-pitch CVD diamond detector; position resolution is 16 μm.

μm in the $r\phi$ view, and about 18 μm in the rz view. This greatly aids, for example, the efficiency for reconstruction of b-quark jets; they obtain 98.5% purity at an efficiency of 50%.

3 SVX-II at CDF

The CDF Collaboration at the Tevatron at Fermilab has embarked on a major upgrade, which includes a 1-meter-long five-layer system of double-sided silicon microstrip detectors, the SVX-II, as the innermost charged particle tracking device. The SVX-II will significantly extend the physics capabilities of CDF by providing better secondary vertex information and better z information than the old tracking system.

The inner layer of the SVX-II must be radiation hard to doses exceeding 1 Mrad to survive Run 2 at the Tevatron. Prototype detectors have been exposed to radiation (protons) in beams at KEK and Fermilab, and tested in beam telescopes at Fermilab. Though not employing the final version of the SVX readout chip, these tests have indicated that the signal size of detectors irradiated to 1.3 Mrad does not decrease, (see figure 5) but the noise increases by about a factor of 1.5.

The third and fifth layers of the SVX-II have low-angle (1.2°) stereo strips, connected to the readout by a double-metal technique. Prototypes of these detectors perform well. Figure 5 shows the distribution of the difference between hit position and track position; the intrinsic position resolution is about 12-13 μm.

References

[1] The RD42 Collaboration, LHCC Status Report/RD42, CERN LHCC 97-3 (Dec. 96).

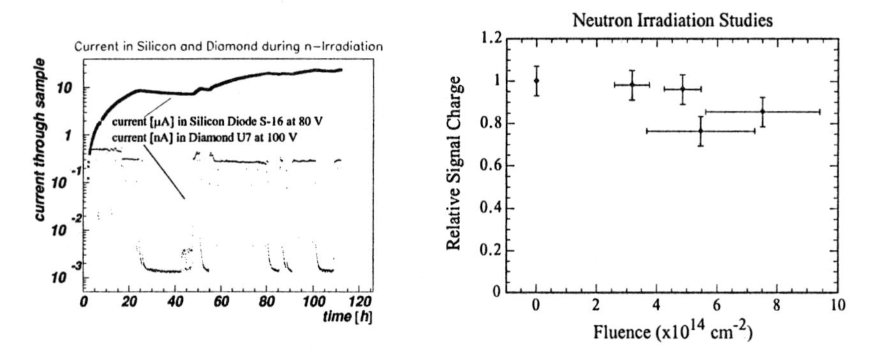

Fig. 3. Left: Current in diamond and silicon detectors during neutron irradiation. Right: Collection distance versus received neutron dose.

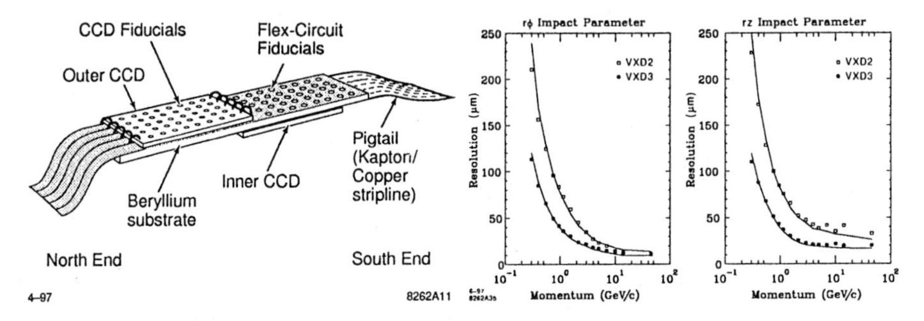

Fig. 4. Left, construction of an SLD pixel detector module. Right, resolution of the VXD-2 and VXD-3 detectors.

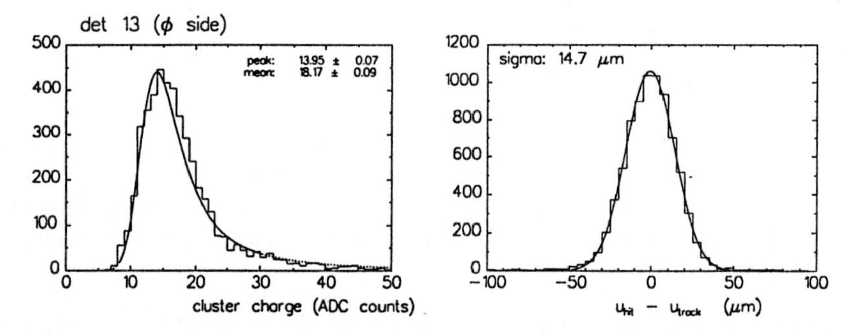

Fig. 5. Left: Signal distribution in SVX-II detector irradiated in a proton beam to 1.3 Mrad. Right: Distribution of hit-track position difference in the stereo side of a 60-μm-pitch SVX-II detector; the intrinsic detector resolution is about 13 μm.

A Radiation-Hard Vertex Detector Based on Liquid-Core Fibres

Kerstin Hoepfner[1] (kerstin.hoepfner@cern.ch)

Technion - Israel Institute of Technology, Haifa

For future detection of short-lived particles, such as tau or beauty, vertex detectors with a high spatial resolution are mandatory. Heavy flavour experiments have to cope with a high radiation environment and an interaction rate which requires a fast readout (<100 ns) or a very selective trigger. Not many detector technologies fulfill these stringent requirements. Silicon detectors are the sufficient solution for a low radiation environment.

An alternative is the technology of liquid-core fibres. After great improvements during the last years mainly by the Pilot Target Group [1] and its successor the RD46 collaboration [2], their performance [3, 4, 5, 6] allows a large-scale appplication. Liquid scintillators are proven radiation hard till doses of 180 Mrad [4], 20 times as high as silicon. Thin layers, of millimetre thickness, could be assembled from capillaries of 20-50 μm and filled with liquid scintillator to form a vertex detector. With a high-precision readout a resolution of 6-20 μm is reachable [5, 7].

To study the tracking capabilities and vertex reconstruction for high-multiplicity events a prototype based on 20 μm capillaries is taking data in the Cern neutrino beam and has recorded a first sample of 150 events. After reconstruction using an algorithm based on a circular projection of pixel pulse heights and an overall vertex fit [6] a transverse vertex resolution of 30 μm was found. Applications of the capillary technology as an active target for neutrino physics are under study, in particular to detect tau-neutrino interactions by the observation of the short-lived tau lepton [6, 8].

Another considered application, which takes advantage of the radiation hardness of capillaries, is an upgrade of the vertex detector for the Hera-B experiment [9]. Due to the fixed target mode of operation in the proton beam the radiation level is very high (10^7 cm^{-2}s^{-1} at 1 cm [9]). By decreasing the distance from the beam, angular acceptance and resolution improve [10]. The drawback is an increase of the radiation level (r^2-dependance). If thin capillary layers of similar geometry and detection properties as silicon layers are used, a frequent replacement (as for silicon) is not necessary. The influence of multiple scattering is reduced due to the larger radiation length of capillary layers (X_{cap}=38 mm, X_{si}=22 mm). The densities of recorded signals are similar for the two devices.

---

[1] now at DESY Hamburg, Germany

Several stations should be placed along the beam direction. Each station contains 1 mm thick layers in x and x',y and y' projection. The readout needs to be able to detect single photons and fast enough to handle the 40 MHz interaction rate. The slow readout (~ms) of a high-resolution CCD readout is not adapted to such high frequencies. An alternative would be the technique of electron-bombarding [11] provided a resolution comparable to the capillary diameter can be reached.

References

[1] Pilot target group: NIKHEF Amsterdam the Netherlands, Humboldt University Berlin Germany, IIHE(ULB–VUB) Bruxelles Belgium, JINR Dubna Russia, CERN Switzerland, Universitè Catholique de Louvain Belgium, Universitá "Federico II" and INFN Napoli Italy, IHEP Protvino Russia, Universitá "La Sapienza" and INFN Rome Italy and SCHOTT Fibre Optics Inc. USA, GEOSPHAERA Research Center Inc. Russia

[2] RD46 collaboration: NIKHEF Amsterdam the Netherlands, Humboldt University Berlin Germany, IIHE(ULB–VUB) Bruxelles Belgium, JINR Dubna Russia, CERN Switzerland, Technion Haifa Israel, Wilhems-University Münster Germany, Universitá "Federico II" and INFN Napoli Italy, IHEP Protvino Russia, Universitá "La Sapienza" and INFN Rome Italy

[3] RD46 coll.: R. van Dantzig et al., CERN/LHCC 95-7, Feb-27-1995, R&D proposal

[4] S.V. Golovkin et al., Nucl. Instr. and Meth. A362(1995) 283.

[5] RD46 coll.: S. Buontempo et al., Nucl. Instr. and Meth. A360(1995) 7.

[6] K.Hoepfner, PhD thesis, Humboldt University Berlin 1996

[7] C. Cianfarani et al., Nucl. Instr. and Meth. A339 (1994) 449.

[8] K. Winter, Acta Physica Hungaria 68 (1990) 135.
R. Nahnhauer, V. Zacek, CERN-PPE/90-138
K. Winter, Proc. of the General Meeting on LHC Physics & Detectors, Evian-les-Bains 1992, p.449

[9] Hera-B coll., T.Lohse et al., HERA-B proposal and technical design report

[10] J.Thom, Diploma thesis, Hamburg University 1996

[11] The technology of electron bombarding has been developed by the CERN-LAA project together with Delft Instruments (The Netherlands) and Canberra Semiconductor hv (Belgium).

Recent Developments in Gas Tracking Detectors for High Luminosities

Archana Sharma*

GSI, Darmstadt, Germany

Abstract: With the high luminosities of upcoming colliders, the frontiers of present generation gas tracking detectors have been pushed, resulting in a large number of innovations in their design, technology and readout systems of which I will discuss some examples. The ATLAS muon spectrometer despite comprising 'simple' proportional tubes, has been extensively studied to better understand and optimize its performance inclusive of the front-end electronics. The MicroStrip Gas Chambers have come long way since their introduction, their spin-offs being Micro-Gap Chambers and the MICRODOT chambers which are described. The introduction of a novel device called the Gas Electron Multiplier GEM, seems promising, and is being considered by HERA-B for improved reliability.

1 Introduction

The demanding challenges of LHC and high luminosity B-factories, have brought about impressive new developments in tracking detectors. In this paper, some of the progress will be reviewed. The Monitored Drift Tube (MDT) system of the muon spectrometer of the ATLAS detector consists of a large number (~ 7.10^4) of proportional counters which have been exhaustively investigated for detector performance and large system aspects. The developments on MicroStrip gas chambers (MSGCs) in preparation of the CMS tracker for example, will be described. The MICRO-MEsh GAseous Structure (MICROMEGAS) has the advantages of being a robust, highly granular and a highly rate detector. Recent developments of the novel Gas Electron Multiplier GEM, which can improve detector reliability for operation in a high rate environment when coupled with a readout element either the MSGC, or simply a printed circuit, are discussed. Detectors for UV-photons, imaging and RICH applications have been developed as MICRO-DOT detectors on silicon wafers, and some recent results will be described.

2 Muon Detection

For the ATLAS Muon spectrometer, aluminum proportional tubes with 3 cm diameter, and lengths ranging from 70 cm to 630 cm, have been designed to operate at high rates with acceptable ageing properties. Assembly includes several layers with measurement accuracies commensurate with the physics requirements giving momentum resolutions $\partial p/p \sim 10^{-4}$ p/GeV. Details of large system aspects may be found in refs. [1, 2]. The operating gas mixture has been optimized for a linear space-time relation, and a large effort has been invested in understanding the detector physics and response of the front-end electronics. Large progress has also been made in the field of simulating detector performance. For the resolution and full chain studies, some of the following software packages have been used.

MAGBOLTZ [3, 4] is a program optimized to compute gas transport parameters, and has been incorporated into GARFIELD [5], which is by now a popular drift chamber simulation program including HEED [6] for energy loss, gas gain and gain variation with a Polya distribution for cylindrical geometry. The limitation here is the simulation of odd shaped electrodes with wires, and no possibility to include insulators, which as shown in the next section, are necessary for several applications. This is made feasible by commercial programs like MAXWELL [7] and MAFIA [8]. The front end electronics may be simulated with PSPICE [9].

Fig. 1 shows the linear space-time correlation of the drift tube, the maximum drift time is 500 ns, the gas mixture being ($Ar/N_2/CH_4$ 91:4:5). The positive ions have a drift time of 4 to 5 μs which

* Address for correspondence Archana.Sharma@cern.ch

constrains the operation at high rates in some regions [2]. Pulse-shapes recorded with a digital scope were compared to GARFIELD+SPICE simulated signals. The simulation of resolution was corroborated by test-beam results. After getting confidence in the measurements, a scan over front-end parameters like preamplifier peaking time, filter constants and threshold settings was performed in order to optimize resolution, hit rate, trailing edge resolution and dead time. Resolution improvement by performing a time slewing correction with a double threshold discriminator or a leading edge charge measurement was studied. Fig. 2 from [10] shows the resolution, measured and simulated, as a function of drift distance for a single threshold, leading edge charge corrected and a double threshold corrected signal with a peaking time of 15 ns. Resolution is good for low threshold, but threshold is limited by noise; one could increase the gain but that is a sensitive parameter leading to ageing. Resolution has also been found to be better for fast (peaking times of 10 ns) preamps, which dictates the choice of CMOS technology [1].

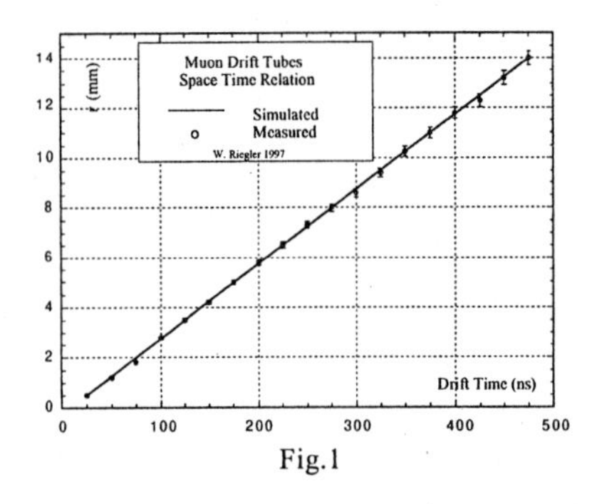

Fig.1

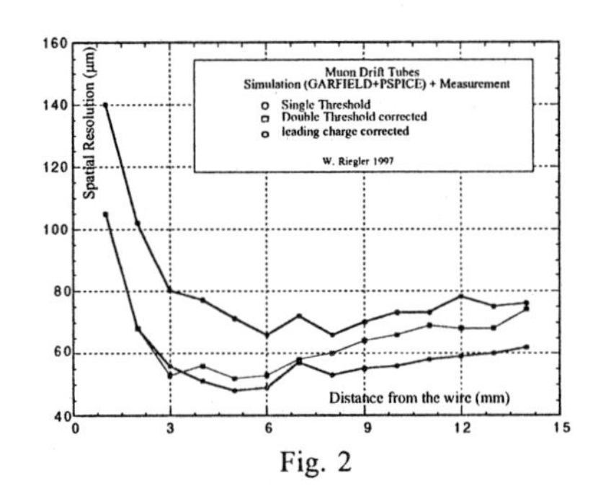

Fig. 2

3 CENTRAL TRACKING DETECTORS

MicroStrip gas chambers have been introduced in 1988 by Oed [11], comprising glass or plastic substrates few hundred µm thick with metal anode and cathode strips patterned on its surface using micro-photolithography. A drift plane placed a few mm away from this surface defines the active area and delimits the gas volume. By applying appropriate potentials on the electrodes this detector has been demonstrated to be capable of high rates (~10^6 Hz/mm^2), with position resolutions of <40 µm and fully efficient for minimum ionizing particles [12, 13]. Long term stable operation has been demonstrated in clean conditions by several groups [14-17] while severe ageing was observed in other cases with different metalizations and operating conditions [18]. In view of the sizes required for the tracking volumes, large areas (27cm x25 cm) have been manufactured [19]. Performances in magnetic fields up to 2.3 T have been verified with good spatial accuracies [20, 21].

Due to the deposition of ions from the avalanche on the substrate between the anodes and the cathodes, a rate dependent gain drop was observed. This was overcome by using either an electron-conducting substrate [22-26] or a resistive over- or under-coat on the substrate [27, 28].The diamond-like coating [29] completely eliminates the gain loss at high rates as shown in fig. 3 from ref. [28]. Several other kinds of coatings have also been investigated [30, 31], the electron-conducting glass being difficult to produce technologically. Fig. 4 shows a comparison of the behaviour of MSGCs with some such coatings with surface resistivities of the order of 10^{14} -10^{15} Ω/□ from ref. [30]. Comparing the electric field in the case of an insulating substrate and a substrate with a thin resistive coating, it was seen that this field is higher along the surface of the insulator in the latter case. Another observation from studying the electric field is that avalanches originating from the region close to the edges of the cathodes have a gain ten times higher than those coming from the rest of the active volume [32, 33]. This has led to the hypothesis that in the presence of a heavily ionizing particle (few 100 keV of

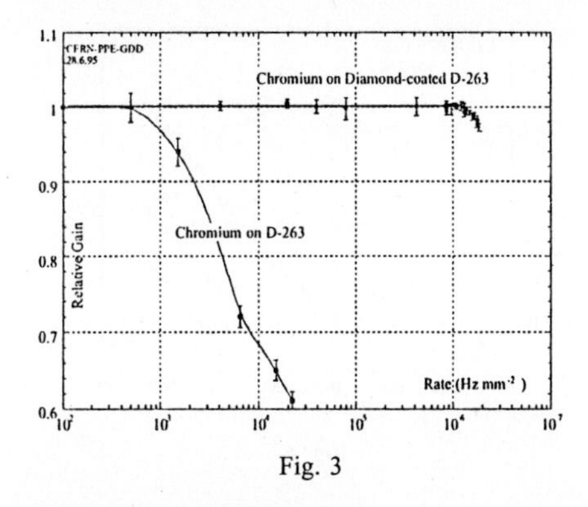

Fig. 3

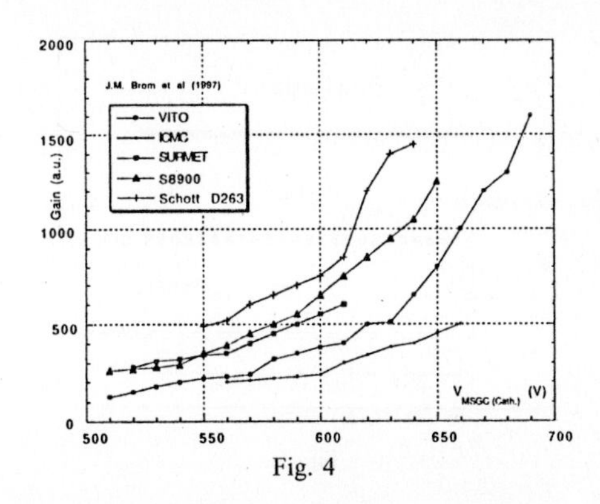

Fig. 4

energy loss compared to few keV of minimum ionizing particles) avalanches that originate close to the edges of cathodes transform into streamers, which due to the higher field along the surface lead to gliding discharges between anodes and cathodes [34] damaging their edges, resulting in yet higher field and thus larger gains from neighboring channels. A large effort has also gone in the understanding of these effects [25, 34-36] and several techniques have been proposed to overcome this problem; some authors [37] have proposed to passivate the edges of the cathode strips by an insulating polyimide, which would prevent the onset of Malter effect at the cathode edges. This concept has been extended in ref. [38, 39] and [40] passivating all along the edges of the cathodes in the inter-electrode space. This technique permits the increase of the efficient working potentials on the MSGC, and reduces the probability of discharges as reported in [41, 42].

A novel technique for operation at high rates is afforded by the MICROMEGAS detector [43]. It consists of a fine mesh kept at ~ 100 μm from the read-out plane by polyimide spacer pillars and separated from the drift (conversion) plane by a few mm; held at a high potential, it provides a high multiplication field. With a large ratio of amplification to conversion field this mesh is fully transparent and the detector operates with large gains ~ 10^4. Long term stable operation, good spatial resolutions and operation in magnetic field have still to be demonstrated.

The novel self supporting Gas Electron Multiplier (GEM) was proposed in ref. [44]; this detector element has the specific feature that it subdivides the gain of a critically operating detector into several sub-critical parts, thus easing the operation [45]. It consists of a polyimide (Kapton) foil metallized on both sides, with through holes of 80-100 μm in diameter and 140-200 μm pitch. With the drift plane a few mm away and appropriate voltages applied, coupled with an MSGC as shown in fig. 5 [45], the GEM provides pre-amplification (PA) factors upto ~ 100. Operating the GEM say at a PA factor of 30, and the MSGC at a gain of 100 (see fig. 6), implies a total gain of 3000, and the operation of the MSGC at a much lower voltage, thus eliminating the problem of discharges [46]. The combined detector works at full efficiency for minimum ionizing particles with a much longer plateau. Spatial resolution has been measured in a cosmic ray hodoscope as shown by fig. 7 [47]] to be better than 30 μm. In another test with minimum ionizing particles full efficiency, timing, and spatial resolution have been measured with a GEM and an MSGC [48]. Higher PA factors are demonstrated with thicker GEMs and operation in a magnetic field of ~ 1 T has been demonstrated without any effect on the gain. Several large sizes of GEMs (27 x 25 cm^2) are built for the high rate HERA-B tracker and are operational at PA factors of ~ 30 [49].

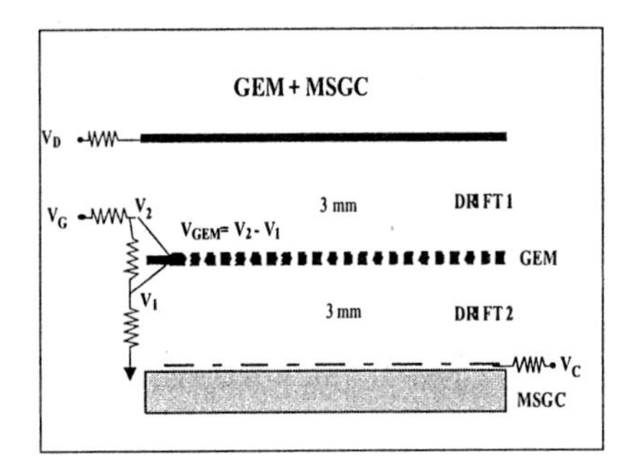

Fig. 5

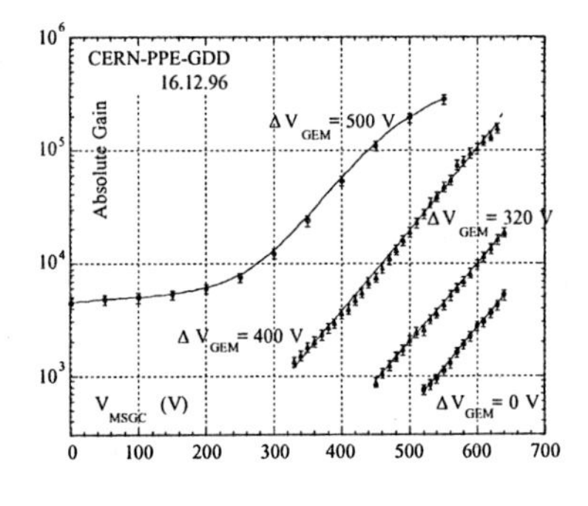

Fig. 6

Employing two GEMs in cascade coupled with a Micro-Gap Chamber [50] Seguinot and Ypsilantis have shown a high pressure operation with detector gains ~ 10^5, suited for single photoelectron detection; see fig. 8 [51]. On the other hand Breskin et al [52] have demonstrated (see fig. 9) the high gain operation of the GEM at low pressure with sensitive photocathodes for UV and visible light detection. The photocathodes are protected since ions no longer recede towards it, due to the special GEM geometry, thus increasing their longevity.

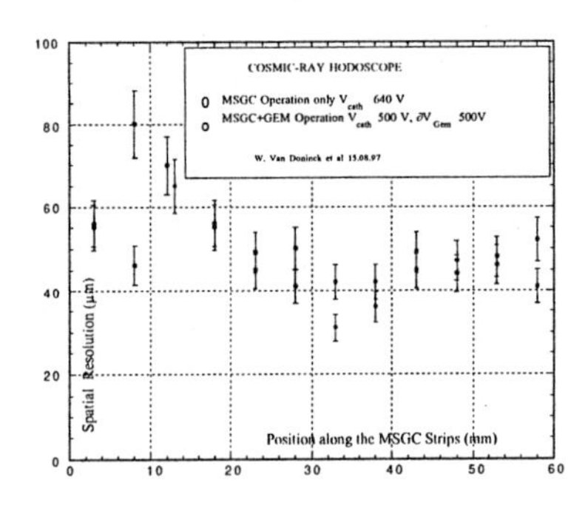

Fig. 7

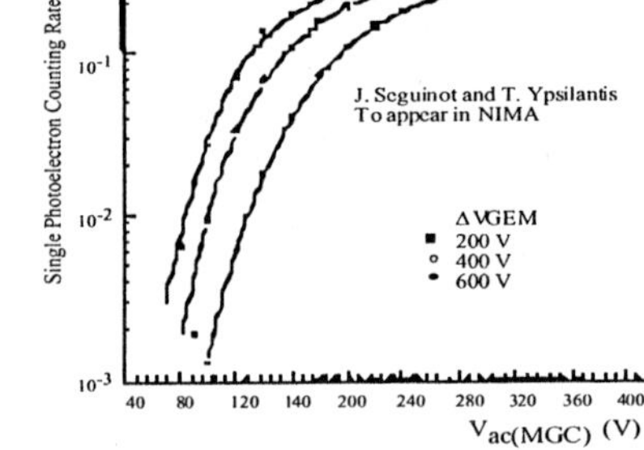

Fig. 8

4 Gas Pixel Detectors

As a spin off from the MSGC design and technology, the Micro-Dot Gas Avalanche Detector introduced by Biagi [53] consists of anode dots surrounded by cathode rings on a silicon wafer on which a layer of oxide has been deposited to maintain a 'safe' distance from the cathodes. With a gas delimiting drift plane and appropriate voltages applied, this detector is conceptually a true pixel device. Small sizes have been operational. A floating cathode 'ring' was introduced in order to defocus the electric field from the buried anode bus. Large gains ~ 10^4 have been reported [54] while charging up and ageing studies are still in progress. Coating the drift electrode with a photosensitive layer, one can operate this device both in a parallel plate mode or a gas avalanche mode with preamplification in the drift region and multiplication at the MICRODOT [55] giving very high gains 10^5 - 10^7 as shown in fig. 11 at low pressure. Very fast single electron pulses are obtained with rise times of ~ 3 ns with propane as filling gas at low pressure. One can avoid the floating rings by coating the surface with a low resistance overlayer as shown in a variation called the MICROSQUARE detector [56]. The advantage of the MICRO-DOT detectors is that the field close to the cathodes is not as high as encountered in MSGCs; if one 'dot' discharges the effective working area lost is one pixel. Nevertheless large areas and single pixel read-out are still difficult to be obtained by this technology.

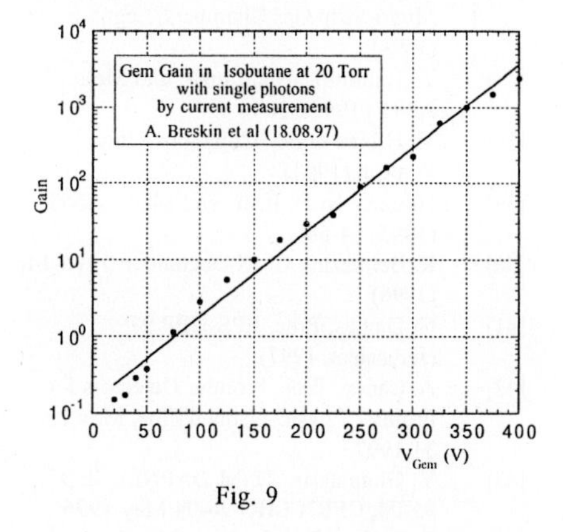

Fig. 9

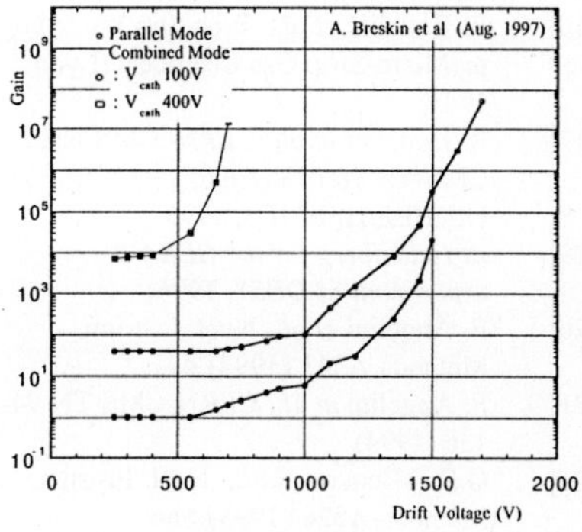

Fig. 10

5 SUMMARY AND OUTLOOK

For the ATLAS Muon detectors, the MDTs are well understood in terms of space-time relationship, efficiency and simulations of resolution corroborated by measurements. The choice of gas mixture is pending for ongoing studies on ageing, since larger gain and hence better signal/noise work in the opposite direction. The problems relating to the performance of MSGCs in high rate environments in the presence of highly ionizing particles have led to several new developments namely the MICROMEGAS and the GEM detectors. Stable operation, large areas, full efficiencies, high rate capabilities, operation in magnetic field without gain loss and good spatial resolutions have been reported for the GEM. Long term operation is still an open issue. In conjuction with UV and visible photocathodes MICRODOT detectors have been shown to operate stably with high gains at low pressure, with very fast signals. Addition of the GEM protects the sensitive photocathodes from ion/photon feedback increasing their lifetime. Despite impressive progress, this exciting field has still some work to be done for successful charged particle detection at high luminosities.

References

[1] ATLAS MUON Collaboration, CERN/LHCC 97-22 (1977)

[2] Y. Hasegawa, Proc. International Conference on High Energy Physics, Jerusalem, Israel (1997)

[3] S.F. Biagi, Nucl. Instrum. Methods **A** **283** (1989) 716

[4] S.F. Biagi, *MAGBOLTZ*, 1994

[5] R.Veenhof,*GARFIELD*,1996,CERN, also *http://wwwcn.cern.ch/writeup/gas/examples/trans2000.html*

[6] I. Smirnov, *http://wwwcn.cern.ch/writeup/heed*

[7] Simulation Program, *MAXWELL 2D and 3D Parameter Extractor*, 1994, Ansoft Corporation: Pittsburgh, PA

[8] Simulation Program, *MAFIA*, CST GmbH, Darmstadt, Germany

[9] Simulation Program, *PSPICE*, MicroSim Corporation

[10] W. Riegler, CERN-ATLAS MUON-NO137 and 173 (1997)

[11] A. Oed, Nucl. Instrum. Methods **A263** (1988) 351

[12] Proc. International Workshop on Micro-Strip Gas Chambers, Legnaro, Italy, (1994)

[13] Proc. Int. Workshop on Micro-Strip Gas Chambers, Lyon (1995)

[14] F. Angelini *et al.*, Nucl. Instr. and Meth. **A 382** (1996) 461-469
[15] R. Bouclier *et al.*, Proc. Int. Conf. on Micro-Strip Gas Chambers (Legnaro, 1994), 48
[16] R. Bouclier *et al.*, Proc. Int. Workshop on Micro-Strip Gas Chambers (Lyon, 1995)
[17] R. Bouclier *et al*, *CERN*- CMS Note CMS-TN 96-018 (1996)
[18] NIKHEF/HERMES, 1996
[19] *At Heidelberg for the HERA-B experiment at DESY*, 1996
[20] F. Angelini *et al*, Nucl. Instrum. Methods **A343** (1994) 441
[21] F. Angelini *et al.*, CERN-CMS TN 94-136 (1994)
[22] G.D. Minakov *et al.*, Nucl. Instrum. Methods **A326** (1993) 566
[23] N. Lumb *et al.*, Proc. Int. Workshop on Micro-Strip Gas Chambers (Lyon, 1996) 291
[24] V. Grishkevich *et al.*, ATLAS Int. Note INDET-NO-064 (1994)
[25] J.P. Duerdoth *et al.*, Proc. Int. Workshop on Micro-Strip Gas Chambers (Lyon, 1995)
[26] L. Alunni *et al.*, Nucl. Instrum. Methods **A348** (1994) 344
[27] R. Bouclier *et al.*, Nucl. Instrum. Methods **367** (1995) 168
[28] R. Bouclier *et al.*, Nucl. Instrum. Methods **A 369** (1995) 328
[29] *Diamond-like coating done by* SURMET Corp. Burlington MA, USA
[30] Private Com. J.M. Brom *et al.* (1997)
[31] S. Brons *et al.*, Nucl. Instrum. Methods **A342** (1994) 411
[32] J.J. Florent *et al.*, Nucl. Instrum. Methods **A329** (1993) 125
[33] R. Bouclier *et al.*, Nucl. Instrum. Methods **A367** (1995) 163
[34] V. Peskov *et al.*, (Proceedings, Manchester, 1996)
[35] B. Boimska *et al.*, CERN-PPE/96-201, see also Proceedings of the 5th International Conference on Advanced Technology and Particle Physics, Villa Olmo, Como (1996)
[36] M.R. Bishai *et al.*, Proc. Int. Conf. on Micro-Strip Gas Chambers, Legnaro, (1994)
[37] T. Tanimori, Nucl. Instr. and Meth. **A381** (1992) 280
[38] V. Hlinka *et al.*, Comenius Univ. Preprint (1993)
[39] Hlinka, Nucl. Instr. and Meth. **A365** (1995) 54-58
[40] R. Bellazzini, CMS Document 1996-146 (1996)
[41] G. Davies, Proc. EPS-HEP 97 (Jerusalem, 1997)
[42] A. Caner, Proc. Frontier Detectors for Frontier Physics, Isola d'Elba, May 25-31,1997
[43] Y. Giomataris, DSM DAPNIA/SED 95-04, CERN/LHC/96-04 May 1996
[44] F. Sauli, Nucl. Instr. and Meth. **A 386** (1996) 531-534
[45] R. Bouclier, CERN-PPE 97-32, 97-73 (1997)
[46] F. Sauli, Proc. Frontier Detectors for Frontier Physics (Isola d'Elba, 1997)
[47] W.V. Doninck et al, *GEM and MSGC, Results obtained in the cosmics hodoscope* (1997), Private Com.
[48] F. Sauli, *Beam Tests with GEMs*, Private Com. Aug-1997
[49] F. Sauli and B. Schmidt for HERA-B Collaboration at DESY Private Com., July 1997
[50] F. Angelini *et al.*, Nucl. Instrum. Methods **A335** (1993) 69
[51] J. Seguinot and T. Ypsilantis, Nucl. Instr. and Meth. To Appear (1997)
[52] A. Breskin, *Low Pressure Operation of GEM*, 1997, Private Com.
[53] S.F. Biagi *et al.*, Nucl. Instrum. Methods **A361** (1995) 72
[54] S.F. Biagi, (Manchester, 1996)
[55] A. Breskin, WIS-96/45/Nov.-PH (1996)
[56] W.S. Hong, Proc. IEEE-96 (1996).

Recent Developments in Particle Identification

Mehdi Benkebil (benkebil@lal.in2p3.fr)

Laboratoire de l'accélérateur linéaire,
IN2P3-CNRS et Université Paris XI, F-91405 Orsay(France).

Abstract. Two types of particle identification detectors will be described in the present article, the RICH and the DIRC. They will be installed in CLEO III at Cornell and in BaBar at PEP II, respectively.

1 Introduction

The construction of B meson factories led increased interest in particle identification detectors. Indeed, a new developments on the Ring Imaging Cerenkov counters (RICH) has been done during these last years. On the other hand, a new technique based on the total reflection of Cerenkov light in a radiator has been adopted for one of the B-factories detector, BaBar[2]. This type of device is called the DIRC, acronym for Detection of Internally Reflected light. In the RICH technique, used by CLEO III detector[1], the detected lights emerge from the radiator and expand in a cone. However, in the DIRC detector the emitted photons propagate in the radiator by total reflection and conserving the Cerenkov angle until the detector plane.
The physics requirement for both experiments CLEO III and BaBar are similar in the principle. They need to achieve a good K/π separation. In case of CLEO III, the separation should be performed in the momentum region up to $2.8 GeV/c$. Due to the boost, the particle identification in BaBar must achieve this separation in a larger momentum range (1.5 to 4 GeV/c). Also, a kaon identification is required for the B mesons tagging in a 1 GeV/c momentum range. Finally, the CLEO III and BaBar should be capable to detect low energy photons. This requires that the detector in front of the electromagnetic calorimeter should be thin in the physical dimensions and radiation length.

2 The RICH of CLEO III experiment

The RICH of CLEO is aimed to achieve, with the dE/dx, a 4σ separation between kaons and pions. It also represents 13% of a radiation length for a track of normal incidence. Its design is based on the work done by the College de France and Strasbourg group [3]. It contains some modifications from the initial design: analog readout, a sawtooth shape for the radiator and some modification of the wire chamber parameters.

The detector (fig. 1) is composed of a LiF radiator medium where the photons are refracted. Then, an expansion zone filled with Nitrogen due to the fact that the emitted Cerenkov light is in the ultra-violet region. Finally, a photon detector which consists of a multi-wire chamber filled with a mixing of 93% CH_4 gas and 7% Triethylamine (TEA) vapor. The 230.000 cathode pads in the chamber will detect the induced charge from the avalanche.

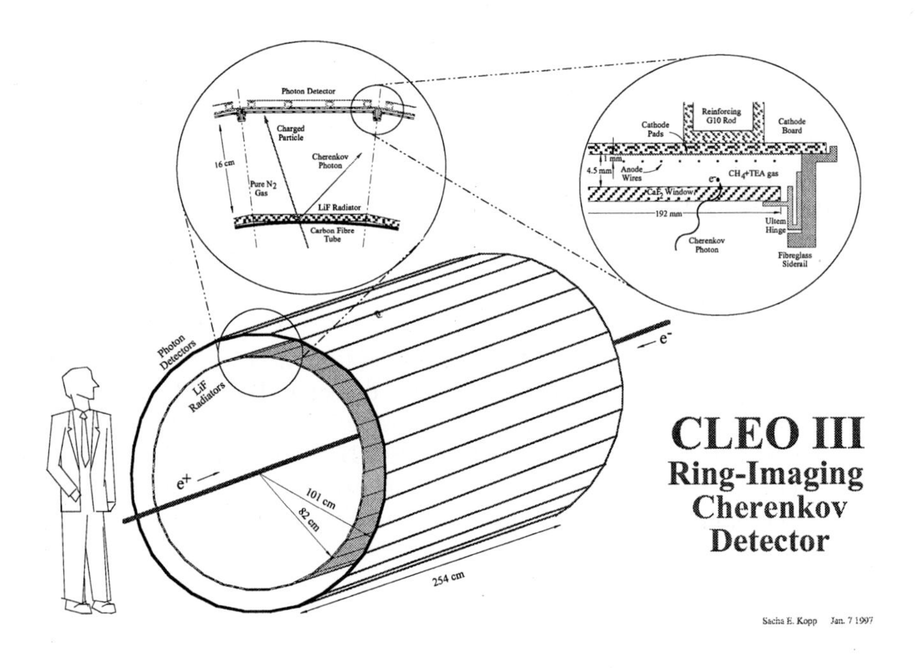

Fig. 1. The mechanical elements of the RICH detector

2.1 The performance of the prototype in the beam test

A prototype composed of a $15 \times 80\ cm^2$ chamber with its final gas mixture has been exposed to cosmic rays at Syracuse University. This test's goals are to study the mechanical feasibility such as corrosion of the materials to the TEA, the behavior of the wire chamber and the resolutions. For the chamber, no loss in the gain has been observed. The plateau has been reached with a gain of 2×10^4. At that gain, the number of reconstructed Cerenkov photons is 12 with a resolution per photon of $15.5 \pm 0.5\ mrad$. By applying correction due to the limited acceptance, the resolution per Cerenkov photons will be 14 $mrad$ with 13 photoelectrons detected per track of $\beta = 1$. These results are consistent with what it is expected from the Monte Carlo study. Thus,

the CLEO III RICH could reach a 3.3σ separation between kaons and pions at 2.8 GeV/c.
For the prototype with the final geometry of the radiator, the beam test is underway at Fermilab. The initial results show that the chambers are stable and the lost of read-out channels is only 1%.

3 The DIRC detector for BaBar experiment

It is aimed to achieve more than 3σ separation up to 4.0 GeV/c with 20 to 50 detected photoelectrons. It is fast and tolerant to the background. Its radiation length and physical dimensions allow to the electromagnetic calorimeter to detect low energy photons. The device 2 is composed of 144 synthetic quartz bars in groups of 12 with the dimension $1.7\times\ 3.5\times 490\ cm^3$. Each bar is obtained by gluing 4 bars of 122.5 cm length which is the longest bars of good quality commercially available. At the end of the bars, a mirror is placed in order to recover the forward going photons. When the Cerenkov photons leave the bar, they pass a wedge made of quartz which is used to reflect the photons coming from the upper side of the quartz. Then, these photons reach the expansion zone, called Standoff box, filled with a volume $\approx 6\ m^3$ of purified, de-ionized water. Finally, the detector plane is populated of 11000 photomultipliers placed on a toroidal surface.

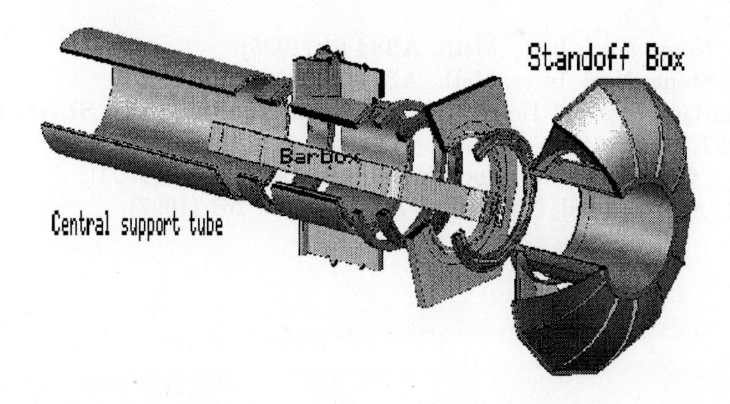

Fig. 2. The mechanical elements of the DIRC detector

3.1 The performance of the prototype II in the beam test

The DIRC prototype II has a simple geometry compared to the final design. The standoff box has triangular cross section, and 500 PMTs were placed on

a quartz windows in order to facilitate rearranging the coverage of Cerenkov ring. For the radiator, two quartz bars have been glued together to form a 2.4 m long bar.
The angular resolution per photon obtained during the beam test at CERN [4] is 10.2 $mrad$ which is in agreement with the Monte Carlo simulation. The measured separation between kaons and pions for 2.7 GeV/c momentum tracks is 3.6σ. This results corresponds to the detection of about 20 Cerenkov photons at track dip angle equal to 20° which is the worst case in BaBar. Since the mean travel length of the photons in BaBar is about 5 m, the attenuation of the light in the quartz bar is an important parameter and has been found to be only $4.1\%/m$. Other results could be explored in the ref. [4]. Other tests on the final components of the DIRC have been successfully performed such as: the operation of PMTs in water, light catchers, background study.

4 Conclusion

We have reviewed two novel techniques of particle identification: RICH and DIRC. The detectors have the same goals but not at the same momentum range. The beam tests on the devices showed that both can meet the physics requirement needed for B physics at CLEO III and BaBar experiments.

References

[1] S. Kopp, Nucl. Instr. Meth. **A384** 61 (1996).
S. Stone, Nucl. Instr. Meth. **A368** 68 (1995).
[2] BaBar Technical Design Report, BaBar Collaboration, **SLAC-R-95-457**, March 1995.
[3] J.L Guyonnet et al., Nucl. Instr. Meth. **A350**, 430 (1994).
[4] R. Aleksan et al, Nucl. Instr. Meth. **A397**, 261 (1997).

Towards Gaseous Imaging Photomultiplier for Visible Light

E.Shefer, A.Breskin, A.Buzulutskov[1], R.Chechik and M.Prager[2]
(fnefrat@wis-weizmann.ac.il)

Weizmann Institute of Science, Rehovot, Israel
[1]currently at BINP Novosibirsk, Russia
[2]ELAM Ltd., Jerusalem, Israel

Abstract. We discuss the possibility of detecting visible photons using a solid photoconverter coupled to a gaseous electron multiplier. The sensitive alkali-antimonide photocathodes are protected with thin dielectric films, to prevent their deterioration under gas avalanche operation. We report on the results of coating K-Cs-Sb photocathodes with 300Å thick CsBr films. The coated photocathodes have a quantum efficiency superior to 5% at 300-350 nm spectral range and can withstand exposure to considerable doses of oxygen and dry air. Their behavior under intense photon flux is also presented.

1 Introduction

Photon detectors based on the conversion of photons in a thin solid photocathode, followed by the multiplication in gas of the emitted photoelectrons, were proposed some years ago [1, 2, 3, 4]. Such devices can have large gain, fast response, and efficiently localize single photoelectrons [5]. Unlike vacuum photomultipliers, they can be made very large and should be less sensitive to magnetic fields. UV detectors based on CsI photocathodes were extensively investigated [6] and are successfully implemented in Ring Imaging Cherenkov (RICH) detectors in high energy physics experiments [7].

In the visible range, alkali- and bialkali-antimonide photocathodes are highly efficient and widely used in vacuum photomultipliers. These materials are extremely sensitive to even minute amounts of oxygen and humidity, commonly present in gases. For this reason, coupling these materials to gaseous electron multipliers encountered serious difficulties [1, 3].

A solution to this problem is coating the photocathode with a thin protective dielectric layer [8], allowing for the transport of photoelectrons through the film to the gas, while preventing contact between the gas impurities and the photocathode. Finding a coating material, with a good protection capability and the smallest reduction of the quantum efficiency (QE), is a difficult task. The photoelectron transport to the gas depends on the photocathode and the coating material [9], requiring an optimization for each photocathode.

We have investigated the coating of Cs_3Sb and K-Cs-Sb photocathodes by numerous materials: CsBr, NaI, CsI, CsF, NaF, MgF_2, SiO_2 and calciun stearate [10, 11, 12]. Our best results in terms of quantum efficiency and stability were achieved with K-Cs-Sb photocathodes coated with 300Å of CsBr, and are summarized here.

2 Results

We constructed an ultra-high vacuum system, allowing us to produce bialkali-antimonide photocathodes, evaporate protective film on top of them, measure the absolute QE and study the effects of exposure to gases in situ. A detailed description of the experimental setup can be found elsewhere [13].

Fig. 1 shows typical QE plots of a K-Cs-Sb and K-Cs-Sb/300Å CsBr reflective photocathodes. The quality of the bare photocathodes manufactured in our system is comparable to that of commercial photomultipliers, and they are stable in vacuum and in pure CH_4. Coating the photocathode with 300Å CsBr, results in a decrease of the QE by a factor of about 6-7 at 312 nm. Nevertheless, the composite photocathode has more than 5% quantum efficiency in the wavelength range of 300-350 nm. As observed for other coated photocathodes [9], the decrease in QE has two components. The first 50Å layer creates a new potential barrier, resulting in a sharp decrease by a factor of 1.5-7 for 250-550 nm respectively. For layers thicker than 50Å there is a moderate decay in QE, exponential with thickness.

The results of exposure of bare and protected photocathodes to O_2 is shown in fig. 2. One can see that the CsBr-protected K-Cs-Sb photocathode is not affected by an exposure to up to 150 Torr of O_2, which corresponds to its partial pressure in air. After exposing the photocathode to 150 Torr of O_2 for 25 min, it still had 3.5% QE at 350nm [12]. The stability of Cs_3Sb photocathodes under exposure is shown in fig. 2 for comparison. Such photocathodes coated with 250Å CsBr, 250Å CsI and 150Å NaI have lower QE. It is also apparent that CsI and NaI have lower protection capability. The exposure of bare K-Cs-Sb and Cs_3Sb photocathodes for five minutes to 10^{-5} Torr of O_2 results in their total loss as shown in fig. 2. Exposing K-Cs-Sb/300Å CsBr photocathodes for 5 minutes to ambient air, had no effect on the composite photocathodes up to 0.1 Torr. However, exposing them to higher pressures resulted in a fast decay in QE, probably due to the moisture.

We investigated the effects of high photon flux on the QE. An upcharging of CsBr-coated K-Cs-Sb photocathodes was measured in 1 atmosphere of CH_4 for photon flux between 2 and 30 GHz/mm^2. For the latter, which is unrealistically large, a decrease of 25% in QE was observed. The photocathode QE recovered within 15 minutes after stopping the illumination. For photon flux lower by an order of magnitude, practically no upcharging was observed. Bare K-Cs-Sb photocathodes displayed no upcharging, even at the highest photon flux.

Aging of the photocathodes by photon impact was measured in the same conditions, using 5 eV photons at a flux of of 30GHz/mm^2. A decay of the

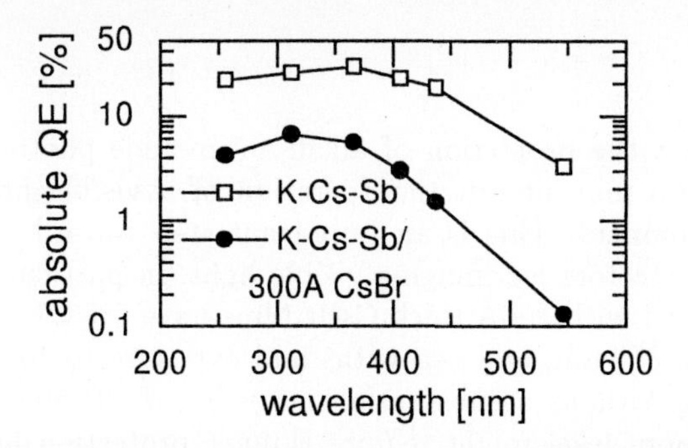

Fig. 1. Quantum efficiency spectra of a reflective K-Cs-Sb photocathode, uncoated and coated with 300Å thick CsBr film.

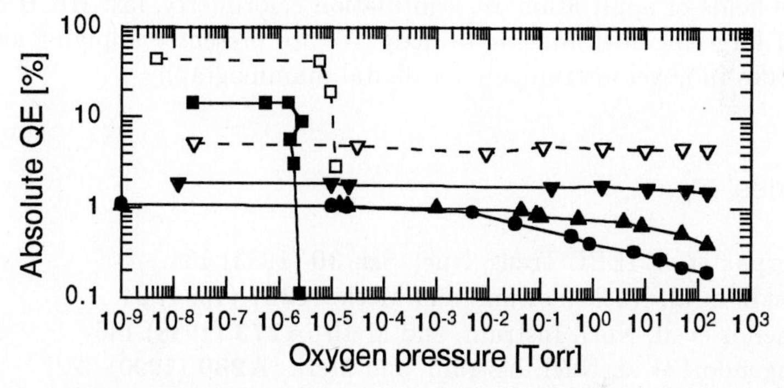

Fig. 2. Photocathodes exposure to O_2. Shown is the QE, measured at 312nm, of K-Cs-Sb photocathodes: uncoated (open squares) and coated with 300Å CsBr (open triangles down), and of Cs_3Sb photocathodes: uncoated (full squares) and coated with 150Å NaI (triangles up), 250Å CsI (circles) and 250Å CsBr (full triangles down), as a function of the residual oxygen pressure. Exposure time at each data point is 5 min. followed by measurement in vacuum.

QE by 50% was observed after an accumulated charge of $10\mu C/mm^2$. The QE of unprotected photocathode remained unchanged after accumulating $25\mu C/mm^2$. The mechanism responsible for the aging process is being investigated. It may be related to photolysis of CsBr or to ion displacement in the coating layer due to electron accumulation in the interface between the photocathode and the coating film and in the coating film.

Our current experimental setup does not allow for in situ multiplication of electrons emitted from the CsBr-coated visible photocathodes. However, multiplication of photoelectrons from a CsBr photocathode has been investigated in a gas counter, revealing stable operation under high photon flux and in avalanche gain of 10^4.

3 Summary

We have shown that the protection of alkali-antimonide photocathodes by a thin dielectric film may permit the operation of a visible photocathode in a gaseous environment. This is an important step towards the conception of large area detectors for imaging visible light. In particular, K-Cs-Sb photocathodes coated with 300Å thick CsBr films have 5% QE in the wavelength range of 300-350 nm, and can withstand exposure to 150 Torr of O_2 for several minutes without a substantial decay in QE. Protection against impurities at the ppm level might require thinner protection layers, hence yielding photocathodes with higher QE.

The observed protection level, permits photocathode handling in a glove box for a few hours. This will allow simpler handling and detector assembly.

Possible fields of application are scintillation calorimetry, fast RICH detectors and large medical imaging devices. We are presently applying such photocathodes in novel instruments for digital mammography.

References

[1] G.Charpak et al, IEEE Trans. Nucl. Sci. **30** (1983) 134.
[2] A.Breskin et al, Nucl. Instrum. and Meth. **A227** (1984) 24.
[3] J.Edmends et al, Nucl. Instrum. and Meth. **A273** (1988) 145.
[4] V.Dangendorf et al, Nucl. Instrum. and Meth. **A289** (1990) 322.
[5] V.Dangendorf et al, Nucl. Instrum. and Meth. **A308** (1991) 519.
[6] A.Breskin Nucl. Instrum. and Meth. **A371** (1996) 116 and references therein.
[7] S.Esumi et al, Fizika **B4** (1995) 205, and A.Braem et al, CERN-PPE/97-120 August 1997.
[8] V.Peskov et al, Nucl. Instrum. and Meth. **A348** (1994) 269, and Nucl. Instrum. and Meth. **A353** (1994) 184.
[9] A.Buzulutskov et al, J. Appl. Phys. **81** (1997) 466.
[10] A.Breskin et al, Appl. Phys. Lett. **69** (1996) 1008.
[11] A.Buzulutskov et al, Nucl. Instrum. and Meth. **A387** (1997) 176.
[12] A.Buzulutskov et al, The protection of K-Cs-Sb photocathodes with CsBr films, Nucl. Instrum. and Meth. A, in press.
[13] A.Breskin et al, SLAC-PUB-7175,(1996) 29; http://www.slac.stanford.edu/pubs/icfa/

BABAR, an Ambitious Detector for *CP* Violation Measurements at PEP II Beauty Factory

Guy Wormser , LAL Orsay (wormser@lal.in2p3.fr)

IN2P3-CNRS et Université Paris-XI, Orsay, 91405-France

Abstract. The stringent requirements on detector performances needed to complete an exhaustive study of CP violation in the B meson system are described. The BABAR detector designed to best meet these goals is presented with special emphasis on the innovative technologies used in the different subsystems. Recent results obtained with large scale prototypes have shown that design performances are quite likely to be achieved.

1 CP violation measurements and detector requirements

The detailed description of the CP violation study in the B meson system is outside the scope of this paper and can be found in these proceedings [1] and in [2]. Three conditions have to be fulfilled on the same event for its use in a CP violation measurementand lead to the following requirements:

-Requirements for B spectroscopy

Most final states used to study CP violation present a combination of low branching fraction and potentially high background (e.g., $\pi^+\pi^-$, D^+D^-). The detector must then have :

-An excellent momentum and energy resolution

-A good acceptance in the forward region. This is made necessary by the large track density in this region induced by the machine boost.

-Track reconstruction down to p_t below 50 MeV/c. Many useful modes involve D* which produce a low p_t pion in its decay.

-Powerful particle identification up to 4 GeV. The mode $B_d^0 \to K\pi$ forms the most dangerous background to the mode $B_d^0 \to \pi^+\pi^-$ and can have in addition a different CP asymmetry. The highest degree of separation at high momemtum is thus required.

-'No' material in front of the calorimeter. Many modes including low momentum π^0 and photons will be of great importance, for instance $B_d^0 \to \pi^+\pi^- \pi^0$.

- K_L^0 detection capability. It will be very interesting to compare CP asymmetries obtained in the modes $J/\psi\ K_S^0$ and $J/\psi\ K_L^0$.

Requirements for B tagging

The most common methods used so far for B tagging are based on the lepton tagging and on the kaon tagging.

-Lepton identification down to 300 MeV/c. Recent studies [3] have shown that it is possible to separate soft secondary leptons coming from charm semileptonic decays from the primary leptons, produced in B semileptonic decays.

-Kaon Identification between 0.5 GeV/c and 2 GeV/c. This represents the momentum range for kaons coming from the b$\rightarrow$ c $\rightarrow$ s chain. In these conditions , kaon and lepton tagging contribute equally to the overall tagging efficiency.

Requirement from Δz

-$\sigma_{\Delta z} << 250\mu m$. The resolution must be a small fraction of the average distance between the two B vertices induced by the machine boost.

-Very transparent vertex detector. Since most particles have very low momentum, multiple scattering in the vertex detector plays a key role in the resolution.

2 The PEP-II Beauty Factory

The PEP-II beauty factory [4] is an asymmetric e^+e^- collider with a 9 GeV electron beam colliding with a 3 GeV positron beam. The design luminosity is 3 10^{33} cm^{-2}s^{-1}, leading to 30 fb^{-1} per year. The very high currents in the machine (1 and 2 A in the high and low energy rings respectively) lead to the following issues:

Background control

-The commissioning detector. A special purpose background detector for PEP-II will enable a detailed understanding of PEP-II background during its commissioning and while BABAR is in construction. This initiative enhances significantly the probability of a fast BABAR turn-on in spring 1999.

-Pileup in the calorimeter. The background will induce about 1 MeV per μs in each calorimeter CsI crystal, making the reconstruction of low momenta photons more difficult.

-Flexible and robust trigger Photoproduced hadrons from machine background will dominate the trigger rate. Multiplicity and p_t cuts are required in order to eliminate this source.

Radiation damage isssues

Two subdetectors elements have to pay a special attention to radiation hardness problems:

-Radiation hard Silicon Electronics. These chips, mounted very close to the beam pipe, will receive 250 krad/year in the worst regions.

-Synthetic quartz for DIRC The very good optical transmission required for these quartz bars has been shown to be lost after exposure to a few kilorads for natural quartz. Synthetic quartz has therefore to be used.

The machine schedule

In June 1997, the first beams were stored in the PEP-II high energy ring and impressive results were already achieved : beam lifetime in excess of 6 hours, beam currents of 0.1 A, 1000 bunches in circulation. In October 1997, 0.3 A was obtained. The low energy ring will start its operation in April 1998 and the first collisions are expected in July 1998. A six month period is thus available for tuning the machine for high luminosity and low background before BABAR rolls on early 1999, with a good chance to start its operation with a high initial luminosity.

3 BABAR, an ambitious detector

The design of the BABAR detector has been based on the various requirements exposed above. The overall view of the BABAR detector can be seen on Fig. 1, with its five subsystems : Si vertex tracker, drift chamber, Detector of Internally Reflected Cerenkov light (DIRC), CsI calorimeter and Instrumented Flux Return (IFR).

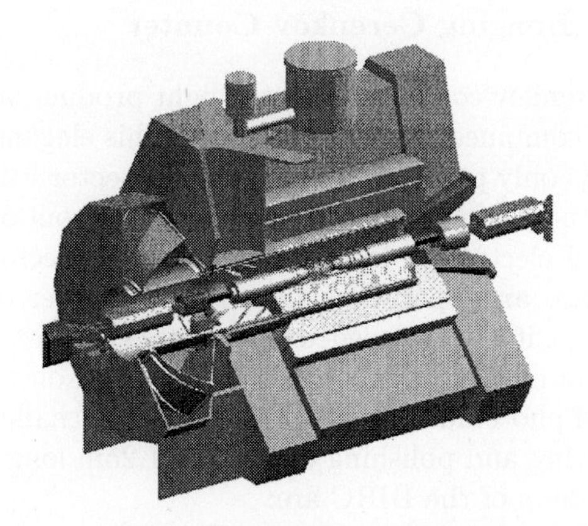

Fig. 1. The BABAR detector

3.1 The Si Vertex Tracker (SVT)

The SVT is a 5 layer double sided Si vertex detector with 50 and 100 μm pitch read out for both coordinates z and ϕ. The read out scheme is time over threshold, achieved with a custom-made rad hard low noise mixed analog/digital chip. The resolution is given by:

$\sigma_z = 15\mu m + 50\mu/p_t; \sigma_\theta = 1.6mrad/p_t$

The requirements for good acceptance in the forward region and for low p_t tracks led to the 5 layer design allowing an autonomous pattern recognition and tracking for p_t as low as 50 MeV/c and to the innovative arch design which allows to put all passive materials (support, electronics, cooling, cables) outside of the tracking acceptance (down to 350 mrad). An arch layer and a barrel layer have been exposed to test beam at CERN PS in August 1997 to measure their efficiency and resolution.

3.2 The Drift Chamber

The key feature of this chamber of this chamber is low multiple scattering thru low-mass gas (80 %He-20%C_2H_6) and wires (Al). This allows a factor 2 improvment over previous similar chambers (e.g. CLEO-II) [1]. It will be segmented in 40 layers, both axial and stereo with a resolution given by :

$$\sigma_{p_t}/p_t = 0.21\% + 0.14\% p_t$$

A full length prototype has been built at SLAC and operated with cosmic rays. A 100 μm cell resolution is achieved in the center of the cell, in agreement with expectation. A small chamber with similar cell structure has been run in a test beam at the CERN PS to study the dE/dx properties. Preliminary results indicate a 6.5% resolution, as expected.

3.3 The DIRC Imaging Cerenkov Counter

The DIRC is a Cerenkov counter where the light production and light transport functions are combined in the quartz bars. This elegant concept leads to several advantages : only passive material in the detector volume, small radial occupancy, small and uniform amount of material in front of the calorimeter, photodetectors and electronics located outside the detector volume leading to easy maintenance and operation, increase of number of photons in the forward region where it is most needed since tracks are more energetic in this domain. Although a novel concept, the DIRC is based on reliable techniques : quartz, water and photomultipliers tubes. The main challenge resides in the precise manufacturing and polishing of the 576 1.25m long quartz bars. The expected performances of the DIRC are:

-N_γ *between* 20 (*at* $\theta = 20^0$) *and* 50 (*at* $\theta = 60^0$)

-$\sigma_{\theta_c} = 4$ *mrad* (*at* $\theta = 20^0$) *and* 2.5 *mrad* (*at* $\theta = 60^0$)

-π -K separation: $> 4\ \sigma$ from 0.3 GeV/c (veto mode) up to 4 GeV/c.

The powerful technique of Cerenkov ring imaging will bring also bring the following benefits:

(i) A useful $\mu - \pi$ separation exists below 700 MeV/c, quite complementary to the one provided by the Instrumented Flux Return, which is fully efficient above 800 MeV/c (see below).

(ii) The possibility to reconstruct photons conversions in the latter part of the chamber and in the DIRC itself by seeing the rings generated by the electrons.

---

[1] CLEO-III chamber will be very similar to BABAR's

(iii) The possibility to tag K decays in flight by observing rings for which the reconstructed θ_c will not lie on the normal (p,θ_c) curves.

Extensive tests of a large scale DIRC prototype have been carried out at the CERN PS in 1995 and 1996. They are described in detail in [5] in these proceedings [6]. The number of detected photons, the angular resolution and the light attenuation in the quartz bars have been found to be in good agreement with expectations. The π-proton separation at 5.4 GeV/c, equivalent to the π-K separation at 2.7 GeV/c - the momentum that would have pions produced in the decay $B_d^0 \rightarrow \pi^+\pi^-$ at 20^0 (the worst case) - is 3.6 σ as can be seen in Fig.2.

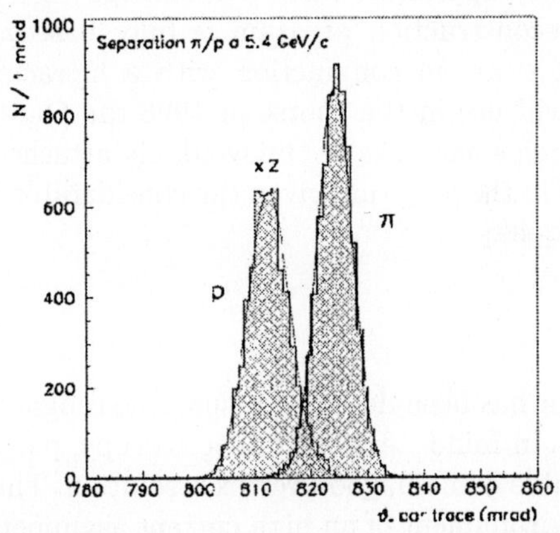

Fig. 2. π-p separation in the DIRC prototype at 5.4 GeV/c

3.4 The CsI calorimeter

The calorimeter consists of 7000 4.8x4.8 cm^2 CsI crystals covering the barrel and the forward region. The design energy resolution is :

$$\sigma_E/E = 1\%/E^{1/4} + 1.2\%$$

which matches well the charged track momentum resolution and represents a significant improvment over CLEO-II CsI calorimeter, obtained through increased light output and lower noise electronics This will allow a photon detection threshold as low as 20 MeV. The nominal resolution has been measured in a test beam at PSI [7] using a 5x5 crystal matrix.

3.5 The Instrumented Flux Return (IFR)

The Instrumented Flux Return consists of 19 layers of Resistive Plate Chambers (RPCs) inserted in the magnet return yoke. The iron thickness varies from 2 cm to 5 cm, allowing a muon detection threshold as low as 600 MeV/c, reaching full efficiency at 800 MeV/c. This low threshold is important for lepton tagging as explained above. All barrel RPCs have been installed in the

BABAR magnet iron representing more than 1000 m^2. The noise rate is quite low, 0.1 Hz/cm^2.

The IFR fine granularity will also allow to tag K_L^0 and to measure their direction (not their energy)and thus to study the decay $B_d^0 \to J/\psi\, K_L^0$.

3.6 Software Issues

Given the very large data volumes to be collected by BABAR, it is very important to use state-of-the-art techniques in this area also to perform BABAR physics program in a satisfactory manner. BABAR has chosen to use a fully Object Oriented (OO) approach, with OO langage (C++), on which the presently running reconstruction program is fully based, OO database as condition and event store, in conjunction with a hierachical mass storage system (HPSS). It will use in the course of 1998 the OO GEANT4 simulation package. Although some risks are unavoidably attached to this strategy, it will clearly pay off in the long run, given the considerable potential brought by these new techniques.

4 Conclusion

The BABAR detector has been designed having the toughest requirements of CP violation physics in mind : excellent spectroscopy, particle identification over the largest possible domain, good vertex resolution. This, in conjunction, with the unusual environnment of an high current asymmetric e^+e^- collider, lead to many innovative features which have been described in some details. All subsystems have exposed large scale prototypes to test beams, showing that the needed performances are quite likely to be met in BABAR.

BABAR and PEP-II schedules allow a 6-month period to tune the colliding beams in PEP-II with a dedicated commissioning detector while finishing BABAR construction. This important feature in addition to the very impressive start of PEP-II high energy ring gives a lot of confidence for a fast startup of BABAR in spring 1999.

References

[1] K. Schubert, Talk PA18-20, these proceedings.
[2] BABAR Technical Design Report, SLAC-R-95-457, March 1997.
[3] BABAR Technical Design Report, p. 72 and S. Plaszczynski, M-H. Schune and G. Wormser, BABAR Note 209, (1995)
[4] PEP-II Conceptual Design Report,SLAC-PUB-418 (1993)
[5] Test of a large scale prototype of the DIRC, R. Aleksan et al., NIMA397(1997)261.
[6] M. Benkebil, Talk PA17-03, these proceedings and LAL 97-92.
[7] R.J. Barlow et al., BABAR Note 367

The KLOE Drift Chamber and DAQ System

M. Primavera on behalf of the KLOE Tracking and DAQ Group (listed in the Appendix) (primavera@le.infn.it)

INFN, Lecce, via Arnesano, 73100 Lecce, Italy

Abstract. In order to fulfil the demanding tasks of the KLOE experiment in the reconstruction of the charged decays of the K mesons, the drift chamber design exhibits unconventional solutions in terms of dimensions, layer and cell geometry and gas mixture. A full size prototype was built and successfully tested with a 50 GeV/c pion beam at CERN. A description of the chamber is presented, together with a discussion of the prototype performances, and, finally, a general overview of the KLOE DAQ system is given.

1 Introduction

The main aim of the KLOE experiment [1] at the Frascati Φ-factory (DAΦNE) is to investigate the CP violation in the K system, by measuring $Re(\varepsilon'/\varepsilon)$ with an accuracy of the order of 10^{-4}. The KLOE detector [2],[3], designed to reach this goal, consists of, in order of increasing radius, a cylindrical drift chamber, a hermetic electromagnetic calorimeter, made of lead and scintillating fibers, and a solenoidal superconducting coil, providing a 0.6 T magnetic field. Due to the peculiar topology of the K_L decays (long decay path and isotropic angular distribution of the charged products) and the momentum range of the charged particles at KLOE, the chamber must assure high and uniform tracking efficiency in the active volume, accurate determination ($\sim$ 1 mm) of the decay vertices, good momentum resolution ($\sim$ 0.5%) and, finally, high transparency in order not to impair low energy photon detection in the calorimeter.

2 The KLOE drift chamber

The requirements of uniformity and transparency cited above have determined the choice of the whole design of the chamber (whose construction is now completed at LNF in Frascati): cell topology, mechanical structure and materials employed, basic gas of the operating mixtures and value of the magnetic field. The detector is a large cylinder of $\sim$ 2 m radius and $\sim$ 3.3 m length, in which $\sim$ 12600 single sense wire square cells, arranged in 58 circular and coaxial stereo layers, provide the most uniform sampling of the active volume. Consecutives layers have stereo angles of opposite signs and, in order to minimize the distortion of the cell in the radial direction, naturally introduced by such a geometry, the inward radial displacement at the wire center (stereo waist) has been kept constant, 1.5 cm, for all layers, implying an increase of the stereo angle along the chamber radius from ± 60 to ± 150 mrad. The cell size, which realizes a compromise between a good granularity and a reasonable number of wires, is $3 \times \pi$ cm^2 for the outer 46 layers and $2 \times 2\pi/3$ cm^2 for the 12 inner ones, to improve the efficiency in K_S charged decays reconstruction. The choice of the field to sense wires ratio (3:1) seems appropriate to provide a good electrostatic field definition inside the cell and to prevent ageing problems, but keeping the chamber opacity under control. For what concerns the gas mixture, helium has been chosen as basic component, in order to minimize multiple scattering, photon conversion before the calorimeter and K_S regeneration effects. The low drift velocity and the small Lorentz angle, together with a limited diffusion, allow for a good spatial resolution. From the resolution and efficiency point of view, a 90%-10% He-iC_4H_{10} looks a quite satisfactory mixture, allowing also for stable operating conditions. The mechanical structure has been designed in order to maximize the trasparency and to reduce to acceptable limits (<0.5 mm) the deformations of the end plates under the total load ($\sim$ 3.5 tons) of the wires. The two end plates are spherical, with $\sim$ 10 m radius of curvature and made of quasi-isotropic carbon fiber-epoxy composites. They are kept apart by twelve rods of unidirectional carbon fiber. Two high Young module stiffening rings, attached to the end plates, reduce the plates deformations to less than 0.5 mm. The inner cylinder (0.7 mm thick carbon fiber) and the 12 panels (sandwich of hex-cell and carbon fiber), which make out the outer cylinder, carry no load. The expected performance of the detector on the invariant K mass resolution is: $\Delta M_{\pi^+\pi^-} = (1.10 \pm 0.01)\ MeV/c^2$.

3 The chamber full scale prototype

In order to check the design performances and all the materials and techniques, later used in the KLOE chamber, a full size prototype representing a cylindrical sector of the chamber, $\sim$ 1 m in radius and $\sim$ 3 m long, was built [4]. It consists of 30 coaxial layers, with radii between 28 cm and 104 cm, of squared single sense wire cells, $3\times\pi$ cm^2 in size in the twenty outermost layers and $1.5\times\pi/2$ cm^2 in the remaining ones. The layers are strung at alternating sign stereo angles in the range of $50\div120$ mrad between conical end plates, 6.2 mm thick, made of carbon fiber. The detector was operated with a gas mixture 90%-10% He-iC_4H_10, at a gas gain of 10^5. A data sample of 27×10^6 tracks from 50GeV/c pions has been collected at the SPS T1-X7 area, allowing to study the cell response as a function of the relevant parameters: the angles a) ϕ between the track and the radial direction, b) β_u and β_l describing the cell shape deformations due to the stereo geometry. Tracking has required the adoption of a "dedicated "algorithm [5]. The distance of closest approach $r(t_d)$ of the track to the hit sense wire has been parameterized as a function of the drift time t_d by joining three 4th order polynomials. While the first branch polynomial is stable as ϕ, β_l and β_u change, the anisotropy and the cell deformations affect $r(t_d)$ with increasing the impact parameter (larger than 1 cm for $(3 \times \pi)$ cm^2 cells) by amounts much larger than the resolution. A complete description of the cell response has been obtained in terms of the $r(t_d)$ parameterizations, coupled to the corresponding resolution function $\sigma(t_d)$ [6]. The space resolution, of the order of 120μm for impact parameters between 0.2 and 1.0cm, is well below the KLOE goal. As an example, a set of $\sigma(t_d)$ curves is shown in fig.1, for different ranges of the track angle ϕ in the cell reference frame. The sharp increase of $\sigma(t_d)$ at $t_d > 700$ ns is well understood and due to the influence of the field wire at the edge of the cell. As a result of the space resolution, angular resolutions of the order of 0.2 mrad in the transverse plane and of 2 mrad in the longitudinal plane have been obtained. The hardware efficiency of the cell, averaged over the cell area, exceeded 99.7% when a discrimination threshold of 4mV, corresponding to a noise frequency per event of 0.002 hits/cell, was set in. Finally, charge measurements have exhibited good dE/dx performances of the gas mixture. A 27% resolution is observed on single cell measurements, while a 15% truncated mean, over the average number of hits for tracks in the $K \rightarrow \pi^+\pi^-$ process, allows to pull down the dE/dx resolution to 3.5%.

4 The KLOE DAQ system

The trigger rate in KLOE at full luminosity is estimated to be 10 KHz, corresponding to a total throughput of 50 Mbytes/s. These requirement imposes the design of a DAQ system with fast electronics, fast data trasmission and efficient on-line software [7]. The system is organized in 10 readout chains, VME based, including both commercial and custom designed modules and buses, connected to 10 farms of processors (event builders) via a FDDI Gigaswitch. The data pass through a two levels concentration procedure. The first level, in which sub-events are processed, is performed in a single crate to a hardwired readout controller (ROCK), located in the crate itself. In the second level, in which strings of sub-events are prepared for transmission to the farms, each crate chain is connected to a readout controller manager (ROCKM). A central processor (Data Flow Control) in the VME system is responsible for the dynamic management of data flow, optimizing the distribution of the traffic on the different switch ports and the load of the farms. Since the whole architecture is distributed in a multi-platform (UNIX based) environment, the network functionalities are crucial for the system operation. Standard TCP/IP protocol is used for data transmission, while the network management is performed by the KLOE message system, based on the Simple Network Management Protocol (SNMP), which sees all commands issued by the Run Control process and the requests by monitoring and debugging tasks in the form of SNMP messages and implements a command acknowledge mechanism. A scaled-down prototype of the KLOE DAQ system was tested using cosmic rays as input, getting results quite comfortable about the final system performances. On this prototype, the monitoring tasks are performed with a new package, the So package [8], based on the O-packages of Barrand [9], which realize the real time on-line and off-line visualization combining the O-packages functionalities with the KLOE process communication system. This package can be regarded as a template for the KLOE monitor

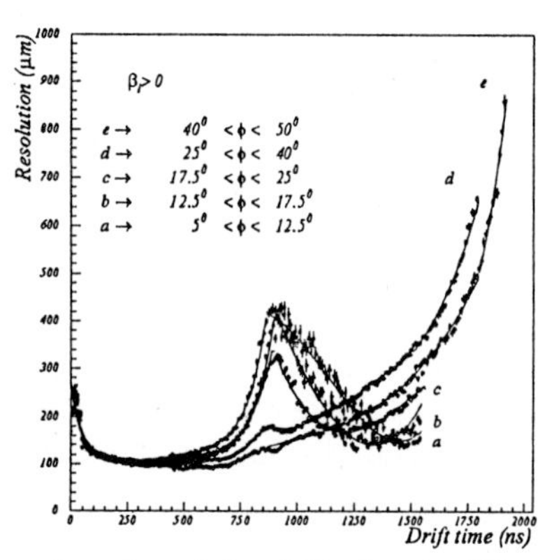

Fig. 1. Resolutions as a function of the drift time in different ranges of the ϕ angle (see text).

References

[1] The KLOE Collaboration, KLOE, a general purpose detector for DAΦNE, LNF-92/019 (IR) (1992).

[2] The KLOE Collaboration, The KLOE detector, Technical Proposal ,LNF-93/002 (IR) (1993).

[3] The KLOE Collaboration, The KLOE Central Drift Chamber, addendum to the KLOE technical proposal, LNF-94/028 (IR) (1994).

[4] F. Grancagnolo, Nucl. Instr. Meth. A367 (1995)108; S. Spagnolo, Nucl.Phys.B 54B (1997)70; A. Andryakov et al., to appear in Nucl. Instr. and Meth. NIMA11692 (1997).

[5] M. Primavera and S. Spagnolo, KLOE note 153, Dec. 1995; G. Cataldi et al., Nucl. Instr. Meth. A388 (1997)127;

[6] M. Primavera and S. Spagnolo, KLOE Memo N. 117

[7] The KLOE Collaboration, The KLOE data Acquisition system, addendum to the KLOE technical proposal, LNF-95/014 (IR) (1995).

[8] internet site : http://www.lnf.infn.it/kloe/private/online/

[9] Internet site : http://www.lal.in2p3.fr/opacs

The DØ Detector Upgrade at Fermilab

Michael G. Strauss *for the DØ Collaboration* (strauss@mail.nhn.ou.edu)

University of Oklahoma, USA

Abstract. The DØ detector at the Fermilab Tevatron is undergoing a major upgrade to prepare for data taking in the Main Injector era with luminosities reaching 2×10^{32} cm^{-2} s^{-1}. The upgrade includes a new central tracking array, new muon detector components, and electronic upgrades to many subsystems. The DØ upgraded detector will be operational for Run II in early 2000.

1 Overview of the Upgrade

The DØ upgrade at the Fermilab Tevatron is designed to operate in the high luminosity environment of the upgraded Tevatron, where the instantaneous luminosity will reach 2×10^{32} cm^{-2} s^{-1} and the minimum time between bunch crossings will be reduced initially from the present 3.5 μs to 396 ns, and then to 132 ns. The upgrade builds on the strengths of DØ, including nearly complete solid angle calorimetric and muon detection, while enhancing tracking and triggering capabilities. [1] A major element of the upgrade is the replacement of the inner tracking system with a silicon vertex detector and a scintillating fiber tracker inside a two Tesla superconducting solenoid. The new tracking system will improve secondary vertex finding and allow momentum measurements for charged hadrons up to pseudorapitidity ≈ 3. Between the solenoid and the inner radius of the central calorimeter cryostat is a preshower detector with wavelength shifter readout. In the forward region, a preshower detector is installed on the faces of the end calorimeter cryostats. The preshower detectors and the fiber tracker use a visible light photon counter readout. The improved muon system includes scintillator trigger detectors over the full pseudorapidity region to handle the higher event rates, and replaces the forward proportional drift tubes with mini-drift tubes. Electronic upgrades are needed because of the smaller bunch spacing and to provide pipelining of the various signals from the new tracking, calorimeter, and muon front-end electronics. More extensive triggering, together with a new trigger control system, is planned to help contain the raw event rates. The DØ upgrade detector is shown in Fig. 1, and details of the new tracking system are shown in Fig. 2.

2 The Silicon Vertex Detector

The 840,000 channel silicon vertex detector [2] will provide high resolution position measurements of charged particle tracks within a radius of 2.5 to 10 cm from the beam line and up to $\eta \approx 3$. Many of the design features of the

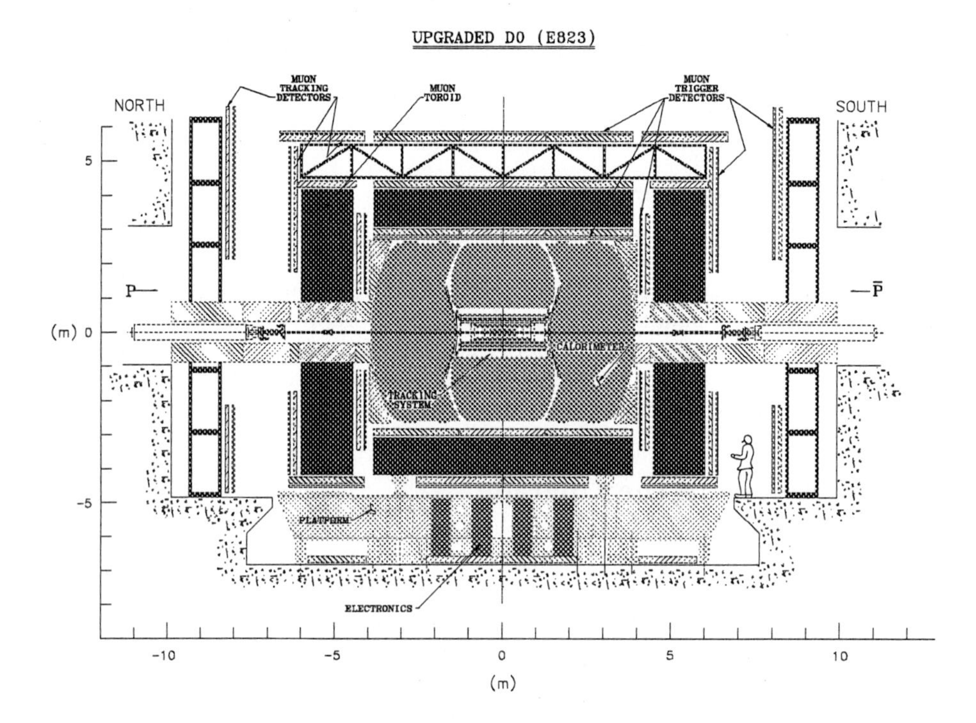

Fig. 1. The DØ Upgrade Detector.

vertex detector are influenced by the expected performance of the upgraded Tevatron. For instance, the increased luminosity requires that the silicon operate after receiving a total radiation dose of about 1 Mrad, corresponding to an integrated luminosity of about 2 fb^{-1}. The radiation hardness of the silicon has been tested, and we have demonstrated that this lifetime can be achieved at an operating temperature of about 10° C[3].

The Tevatron interaction region will have $\sigma_z \sim 25$ cm which makes it difficult to deploy detectors such that the tracks are generally perpendicular to the detector surfaces for all η. This has motivated a barrel/disk hybrid geometric design. The silicon detector is constructed from six modules, each having four layers of silicon oriented parallel to the beam axis (barrels) and one layer of silicon oriented perpendicular to the beam axis (disks). Each barrel/disk module is 12 cm in length so that the six modules extend to $|z| \sim 36$ cm. Barrels 1 through 4 are located at nominal radii of 2.77, 4.61, 6.83, and 9.16 cm, respectively. Barrels 2 and 4 are constructed from double sided AC-coupled silicon devices with 2° stereo angle strips. Barrels 1 and 3 are constructed from double sided AC-coupled devices with 90° stereo strips on the inner four modules and single sided AC-coupled axial strips on the outer two modules. At the end of each module, the disks extend from a radius of 2.57 to 9.96 cm and consist of double sided AC-coupled silicon with $\pm 15°$ stereo strips. At each end of the barrel/disk array a series of 3 more identical disks are added to enhance the tracking in the forward region. Finally, two larger disks are located at $z = \pm 110$ cm and $z = \pm 120$ cm with a radius between 9.5 and 26 cm to allow tracking in the large η region. These disks are composed of two single sided AC-coupled silicon devices, placed back to back with strips set at an angle of $\pm 7.5°$. The single point resolution for the silicon detectors has been shown to be $< 10\,\mu$m.

Electronic readout will be done using the 128 channel SVX II readout chip developed by Fermilab and Lawrence Berkeley Laboratory. Each channel contains a charge sensitive preamp, 32 stages of analog pipeline delay, an 8 bit analog to digital converter, and sparse data readout. The SVX II chip has been tested is fully functional.

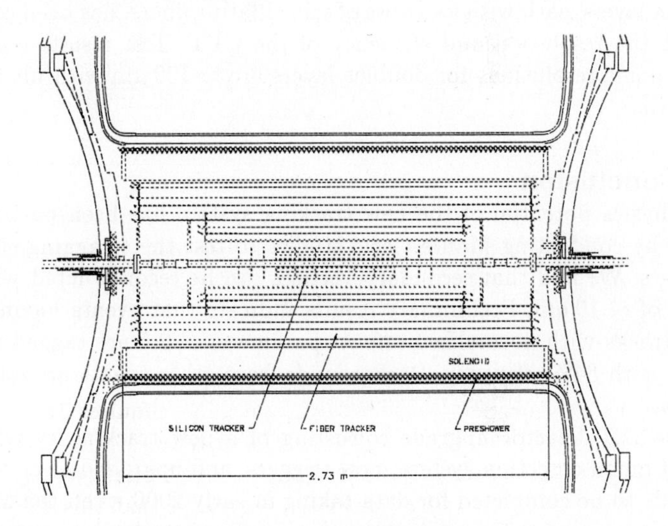

Fig. 2. The DØ tracking system.

3 The Scintillating Fiber Tracker

The central fiber tracker (CFT) surrounds the silicon vertex detector and provides a momentum measurement, a first level trigger, and full track reconstruction for a pseudorapidity range up to $|\eta| = 1.7$. The detector consists of 77,000 scintillating fibers mounted on eight concentric cylinders at radii from 20 to 52 cm. Each of the eight cylinders has two layers of 835 μm diameter multiclad fibers oriented axially, and two layers oriented at small angle stereo, either $\pm 2°$. The inner (outer) two (six) cylinders have fiber length of 1.6 (2.5)

m. The fibers are mated to 8-11 m multiclad clear fiber waveguides which conduct the light to photodetectors situated under the central calorimeter. Each fiber will be precisely located to $< 40\,\mu$m to allow for efficient tracking and triggering.

The photodetectors must be capable of detecting single photons with high efficiency at high rates and with large gain. We will make use, for the first time in a high energy physics experiment, of a large number (77,000) of visible light photon counters (VLPC's). VLPC's are impurity band conduction devices derived from solid state photomultipliers. Test results indicate they can detect single photons with quantum efficiencies up to 80% and gains $> 10^4$. The VLPC's operate at a temperature of 6-12° K with a bias voltage of 6-10 V. They can operate at rates of at least 10 MHz, and $> 40 \times 10^6$ photoelectrons/sec.

A cosmic ray test stand consisting of 3000 fiber channels arranged in doublet layers, each with two rows of scinitillating fibers, has been constructed to test the resolution and efficiency of the CFT. Test results confirm that single point resolutions for doublet layers are $\approx 100\,\mu$m with hit efficiencies of 99.5%.

4 Conclusion

The physics potential of the new tracking system has been partially determined by conducting Monte Carlo studies to test the b tagging efficiency in $t\bar{t}$ events. We find that secondary vertices can be reconstructed with a resolution of $\sim 10\,\mu$m. Preliminary results indicate that events having three or more tracks with normalized impact parameter > 2.5 are tagged as b quark events with 64% efficiency. Backgrounds from other events are estimated to be $< 5\%$.

The DØ detector upgrade consisting of a new tracking system, an improved muon detection system, new triggers, and upgraded electronics, is on schedule to be completed for data taking in early 2000 when the Main Injector is operational at Fermilab. These improvements will allow us to operate in the high luminosity environment of the upgraded Tevatron and will enhance the already excellent physics potential of the DØ detector.

References

[1] "The DØ Upgrade: The Detector and Its Physics," DØ Collaboration, Fermilab Pub-96/357-E, (1996), Also available on the WWW at http://higgs.physics/lsa/umich.edu/dzero/d0doc96/d0doc.html.

[2] "DØ Silicon Tracker Technical Design Report," DØ Collaboration, DØ Note 2169, July 1, 1994.

[3] "Summary of Results on Prototype Single-sided Silicon Microstrips for the D0 Barrel Tracker," DØ Note 2812, unpublished, 1995, "Effects of Radiation on the D0 Silicon Tracker," DØ Note 2679, unpublished, 1995.

DAQ/Trigger Simulations of CDF II

Henry Kasha and Michael Schmidt

Physics Department, Yale University
New Haven, Connecticut 06520-8121, USA

Introduction

Fermilab's Tevatron $\overline{p}p$ collider is currently being upgraded both in energy (from $\sqrt{s}$ = 1.8 TeV/c^2 to 2.0 TeV/c^2) and luminosity (from $2\cdot10^{31}$ cm^{-2}s^{-1} to $2\cdot10^{32}$ cm^{-2}s^{-1}). The order-of-magnitude increase in the luminosity will be achieved essentially by reducing the time between bunch collisions from 3.5 μs to 132 ns. The data acquisition (DAQ) and trigger electronics had to be redesigned to meet the new conditions while minimizing the dead time of the detector. We describe here the Monte Carlo simulations which we carried out to guide this undertaking to achieve a deadtime of $\approx$5% at the canonical maximum rates of 45 kHz for events accepted by the Level 1 trigger and 300 Hz for events accepted by Level 2.

CDF II DAQ/Trigger System

Three trigger levels are used to accept events. In each channel of the front-end electronics the digitized data from the drift chamber TDC's and calorimeter ADC's enter 42-step synchronous pipelines clocked every 132 ns. The purpose of this pipelining is to provide the Level 1 trigger with $\approx$ 5.5ms time to reach a decision without loss of data due to deadtime. Similarly, analog data from the SVX II silicon vertex detector and the ISL intermediate silicon tracker are kept in Ring Buffers awaiting the Level 1 decision. The Level 1 trigger is designed to reduce the event rate from 7.6 MHz to 50 kHz.

Events accepted by Level 1 are stored in Level 2 FIFO buffers distributed among all readout crates to await acceptance or rejection by the Level 2 trigger processor. This processor is pipelined: its loading stage (10μs latency) and processing stage $((10+exp < 1.75 >)\mu$s latency) can simultaneously work on two different events. The Level 2 trigger will reduce the event rate to 300 Hz.

Upon acceptance by Level 2, transfer begins of the event data (about 250 kb) towards the Level 3 trigger. The data in the Level 2 buffers of each detector element is tranferred to "Scan Buffers" serially over optical links at 256 Mb/s (1 Gb/s for SVX II). and then the event fragments are trnsferred from the Scan Buffers to an event builder at the input of the Level 3 processor farm by a commercial ATM 16 x 16 switch capable of speeds in excess of 10 Mb/s per link. Each event will undergo a full analysis by a processor of the farm, which will decide its final acceptance or rejection.

The MODSIM model of the DAQ/Trigger System

We used the MODSIM simulation language[1] to create and study a behavioral model of the CDF II DAQ/trigger system. A unique feature of the language is the built-in concept of time flow. Different processes can be modelled to run simultaneously and can

[1]CACI Products Co.,La Jolla, CA 92037

be easily synchronized. Another powerful feature of the object-oriented language is its library of easily adaptable queuing constructs (objects).

The model of the CDF II DAQ/trigger system developped by us consists of several processes evolving independently in time:

- CollisionClock trigger process generates every 132 ns a collision if the proton and antiproton beam buckets are both filled. (Various beam patterns were studied.) Level 1 threshold is set to simulate a specific acceptance rate. If an event is accepted and if there is a free Level 2 buffer, the event is added to SvxQueue and (Level 2) DecisionQueue. If no Level 2 buffer is available, deadtime is incurred.
- SvxQueue is checked for events awaiting processing to the rhythm of SvxClock. Events found in the queue are processed by the SvxProc model.
- Level 2 DecisionQueue is checked for events awaiting processing to the rhythm of L2Clock. If the event is accepted it is added to ScanQueue if a Scan Buffer is available. If no Scan Buffer is free, BUSY flag is asserted and no further Level 2 accepts can be issued. As soon as all of the Level 2 buffers are filled with events accepted by Level 1, the system begins to incur deadtime. The BUSY line is deasserted when, in all detector elements, a Scan Buffer becomes free again.
- ScanQueue is polled by the Switch for events to be transferred to Level 3. The frequency of this is governed by the SwitchClock cycle.

A run simulating one second of real time takes 40 min using an executable compiled by MODSIM III in conjunction MS VisualC/C++ Ver.5.0, running on a 166 MHz Pentium under Windows 95.

Results

Simulations of the CDF II DAQ/Trigger system were instrumental in establishing the DAQ protocol according to which each data acquisition device has locally four-deep FIFO buffers for events awaiting the Level 2 trigger decision. In terms of queuing theory, the ratio K of the mean request rate to the mean service rate is not very dissimilar for events waiting to be processed by the Level 2 trigger and for the events accepted by Level 2 and waiting to be switched over to Level 3. Confirmed by the simulations, this led to the incorporation of a four-deep FIFO "Scan Buffer" for event fragments awaiting switching to Level 3 trigger processor farm. The simulations helped to evaluate the constraints and effects on DAQ deadtime of various architectures and latencies of the Level 2 decision making, of the speed of the various data transfer processes such as the transfer of data from the Level 2 Buffers to the Scan Buffers and of the speed of the ATM, and to assess the effect of different beam bunch distributions in the Tevatron ring on the effective luminosity and the detector deadtime.

This research was supported in part by DOE Grant DE-AC02-76ER03075.

Operations in Several RPC Conditions

Sergio P. Ratti (for the CMS-Bari-Pavia Collaboration*) (ratti@pv.infn.it)

University of Pavia and I.N.F.N. - Sezione di Pavia

In this paper we report on the performances of Resistive Plate Counters (RPC)[1] operated with different electrode surface treatments, different gas mixtures and different background conditions. RPCs are gas detectors in which high voltage (HV) is applied to resistive electrodes (bakelite plates) 2-3 mm apart. Signal readouts are obtained with proper capacitive coupled aluminum strips [2]. RPCs can be operated either in streamer mode (high gas gain) or in avalanche mode (low gas gain)[3]c,d,[4]. Traditional bakelite electrodes of the RPC's were treated with linseed oil thus smoothing the surface.

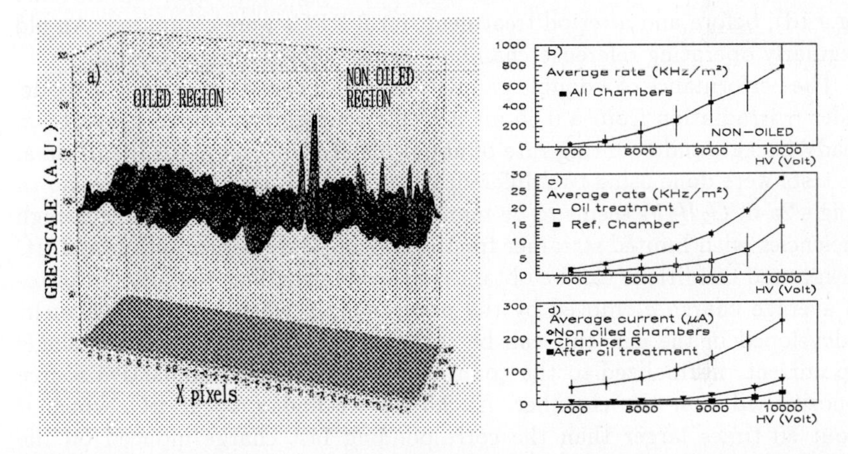

Fig. 1. a: A $0.25\mu m x 0.25\mu m$ pixel greyscale readout of a micropicture (arbitrary scale, $100\mu m x 12\mu m$ area) of oiled (lefthand part) and non-oiled (righthand part) "old" bakelite; b,c: Average single rate vs HV at a fixed discriminating threshold; b- non oiled gap (threshold $-80\,mV$); c- oiled gap (threshold $-50\,mV$); reference chamber (full squares); d: Current vs HV averaged on eight non oiled (open cross) and oiled gaps (full square). Reference chamber (full triangles).

Numerical values of the measured surface "roughness" are given in ref. [3]a. In fig. 1a the greyscale readout of a microphoto[3]b of two adjacent surface regions (oiled and non oiled) is shown. The numerical values of the roughness vary from $R_a \simeq 0.55\mu$m for the non oiled region to $R_a \simeq 0.2\ \mu$m for the oiled region [3]b. Bakelite of new production can reach values as low as $R_a \simeq 0.2\mu$ m. The linseed oil provides a relatively large absorption band

---

* co-authors are: M.Abbrescia, A.Colaleo, G.Iaselli, F.Loddo, M.Maggi, B.Marangelli, S.Natali, S.Nuzzo, G.Pugliese, A.Ranieri, F.Romano Bari; V.Arena, G.Bonomi, G.Gianini, M.Merlo, S.P.Ratti, C.Riccardi, L.Viola, P.Vitulo, Pavia.

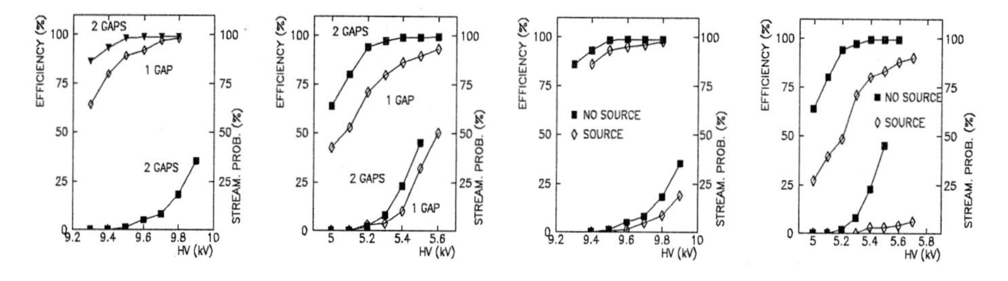

Fig. 2. a,b: efficiency and streamer probability without source: a- in $C_2H_2F_4$; b- in argon; c,d: comparison of efficiency and streamer probability with and without source: c- in $C_2H_2F_4$; d- in argon.

in the ultraviolet region that might quench secondary electron photoemission from the electrode surface. The electrode conditions affect also noise single rate and dark current as shown in fig.s 1b,c and d, while efficiencies remain substantially unaffected[3]f. Average single rates (fig.s 1b,c) and dark currents (fig.s 1d), before and after oil treatment, for 8 RPCs were compared. An old "regularly operating reference chamber" is also shown for comparison.

The performance of a double gap RPC was also tested, in avalanche mode, under γ irradiation from a 0.65 mCu ^{137}Cs source. The source generates a steady background counting rate of about $500Hz/cm^2$ on a $10x10cm^2$ area; the tests were done using two different gas mixtures[1] and the data are shown in fig.s 2a-d. $C_2H_2F_4$ seems to be more suitable than argon[3]e to reach high efficiencies with limited streamer fractions. The power consumption at LHC background conditions can be obtained by evaluating first the amount of total average charge q_s moved by the power supply when a single avalanche is developed in the gas; this can be computed as the increase of the single gap current, normalized to the counting rate, when the radioactive source is positioned near the chamber. At 9.5 kV we find q_s = 41 pC, which is about 50 times larger than the corresponding fast charge induced on the pickup electrode. The power consumption, for a 1 m^2 chamber, uniformly irradiated at an equivalent counting rate of 100 Hz/cm^2, can then be estimated as $(41pC) \times (100Hz/cm^2) \times (10^4cm^2) \times (9500V) = 400mW$; i.e., a 1 m^2 double gap RPC operating in avalanche mode with $C_2H_2F_4$ in typical LHC background conditions would dissipate less than a 1 W/m^2, acceptable for LHC running. For "equal efficiency", the argon mixture shows a power consumption which is about a factor 5 smaller.

To understand whether the operation conditions can be kept under control, we developed a Monte Carlo simulation[5] and compared the results to the experimental charge distributions. The signal formation mechanism is modelled (described in details in [4]f) based upon physical parameters such

[1] "argon mixture": 90% $C_2H_2F_4$,10%i-C_4H_10; "$C_2H_2F_4$ mixture": 70% Ar,5%i-C_4H_10,10%CO_2,15% $C_2H_2F_4$.

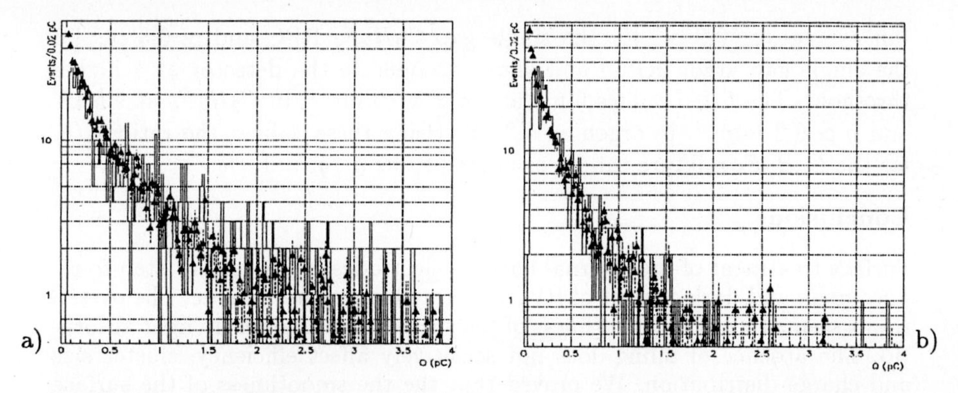

Fig. 3. Montecarlo charge spectrum distributions compared to experimental data:a- $C_2H_2F_4$ mixture; b- argon mixture.

as the Townsend coefficient α (number of ionizing encounters per unit length undergone by one electron), the attachment coefficient β and the effective ionization coefficient $\eta = \alpha - \beta$. The induced fast charge q_e and the total charge q_s are related by the simple formula $q_e/q_s = k/\alpha d$ where k depends upon geometrical and electrical factors such as ϵ_r of the bakelite, the gas gap width d and the electrodes thickness s. That is: $k = (\frac{\epsilon_r d}{s})/(\frac{\epsilon_r d}{s} + 2)$. The simulation includes some effects in the avalanche multiplication inside an RPC, such as the fluctuations on the initial position and on the size and number of primary clusters and the fluctuations on the total number of final electrons. The crossing of the gas by an ionizing particle is simulated by generating primary clusters along its path according to Poisson statistics. The number of electrons in any primary cluster is either taken from experiments, or from distributions available in the literature. The fast charge q_e due to the avalanche originated by the j-th primary cluster at position x_j is computed according to the formula $q_e = [k/\eta d]Q_j g[e^{\eta(d-x_j)} - 1]$ where Q_j=$q_{el}n_j$ (q_{el} is the electron charge) and g is a factor which accounts for the fluctuation in the avalanche gain. [determined by randomly extracting a number from an exponential distribution with average value equal to $n_j e^{\eta(d-x_j)}$]. Finally, the total fast charge q_e is obtained by summing the contributions of all clusters. Predictions for the charge distribution can be obtained by inserting in the simulation the values of the cluster density λ of the gas mixture and η given above. By fitting the results of the simulation to the available experimental charge distributions, λ and η can be obtained. The comparison of experimental and simulated charge spectra are shown in fig.s 3, for a 2 mm single RPC gap filled with $C_2H_2F_4$,ixture (fig. 3a) and filled with argon mixture (fig. 3b), both operated at their knee of the efficiency plateau. The agreement is fairly good. The Monte Carlo charge distributions in fig.s 3 do not take into account the experimental threshold ($\simeq$ 100 fC) for signal detection; therefore it extends to very low charge values where no signal can be detected experimentally. The use of heavy gas produces more charge on the pickup electrode (on average, $\simeq 0.9$ pC for the single gap $C_2H_2F_4$ RPC and $\simeq 0.5$ pC for the single gap argon RPC) and allows to operate the detector at a higher threshold. The fitted values for η are: $\eta \simeq 9.2$ mm^{-1}, in $C_2H_2F_4$ at 9.5 kV and $\eta \simeq 9.2$ mm^{-1}, in argon at 5.2 kV. Using these values, the ratio q_e/q_s for the $C_2H_2F_4$ mixture can be evaluated to be $q_e/q_s = 1.9\%$.

Conclusions

Surface treatment of the internal bakelite electrodes is a necessary step to reduce noise and dark current of RPCs operated in streamer mode. The overall result seems to be the smoothness of the surface and a larger uv light absorptio. The absence of oiling does not sensitively affect efficiency, cluster size and charge distribution. We proved that the the smoothness of the surface can be obtained within the industrial process of the bakelite foil production. Chambers made by smoother surface bakelite and linseed oiled bakelite show similar performances. In no way oil treatment degrades the performance of an RPC at least when operated in streamer mode. The study of double gap RPCs -operated in avalanche mode- under irradiation of a ^{137}Cs source at an equivalent rate of 500 Hz/cm^2 shows that environmental friendly $C_2H_2F_4$ mixtures seem to be more suitable than argon mixtures to reach high efficiency with a limited streamer fraction. The power consumption, which could be an issue being $C_2H_2F_4$ electronegative, is limited and acceptable for LHC background running conditions. Experimental charge spectra are well explained by a proper Monte Carlo simulation. The model is a tool suitable to explain the physical process taking place in the RPC gaps during the development of the avalanche. The advantage of $C_2H_2F_4$ mixtures (larger average fast charge), which enables to reach high efficiency with relatively higher thresholds, seems to be evident. Comparable efficiencies in argon mixtures would require much lower thresholds which could hardly be used.

References

[1] a- Proceedings of the 2nd and 3rd International Workshops on RPCs: Scientifica Acta **8**(1993) b- ibidem **11**(1996);

[2] R.Santonico and R.Cardarelli: N.I.M. in Phys. Res. **187**(1981)377;

[3] a- CMS Technical Design Report (to be published); b- P.Vitulo, priv. comm.; c- I.Duerdoth et al.: N.I.M. in Phys. Res. **A348**(1994)303; d- I.Crotty et al.: ibidem **A337**(1994)370; e- M.Abbrescia et al.: ibidem **A392**(1997)155; f- ibidem **A394**(1997)13;

[4] a- M. Abbrescia et al.: Scientifica Acta **11**(1996)217; b- A. DiCiaccio: Proc. EPS-ICHEP (Ed. J. Lemonne, C. Vander Velde, F. Verbeure. World Sci. 1996) p. 605; c- A. DiCiaccio: Scientifica Acta **11**(1996)263; d- E. Cerron Zeballos et al.: N. I. M. **A367**(1995)388; e- C. Bacci et al.: N.I.M. in Phys. Res., **A352**(1995)552; f- M. Abbrescia et al.: *Properties of $C_2H_2F_4$ based mixtures for avalanche mode operation of RPCs.* CMS Note 1997/004, (submitted to N.I.M. in Phys. Res.);

[5] a- M.Abbrescia: *Resistive Plate Chambers in Avalanche Mode: a Comparison between Model Predictions and Experimental Results* (to be published);

The ATLAS Transition Radiation Tracker and Muon Chamber System

The ATLAS Collaboration
represented by
Yoji Hasegawa (Yoji.Hasegawa@cern.ch)

ICEPP, University of Tokyo, Japan

Abstract. The design of the ATLAS detector is in the final phase towards construction. In this talk, recent results from the development of the ATLAS transition radiation tracker and muon chamber system are presented.

1 ATLAS Detector

The ATLAS detector is a general purpose detector at the CERN's Large Hadron Collider starting in 2005. The detector consists of the inner detector, calorimeter and muon spectrometer. The LHC will provide very high luminosity ($10^{34}\mathrm{cm}^{-2}\mathrm{s}^{-1}$), therefore, the studies on rate capability and ageing effects of each detector are essential. In order to achieve the design performance, quality control and precise alignment of the detectors in such a very large system are also important. In the following sections, the results of the development of the ATLAS transition radiation tracker, the monitored drift tube chambers (MDT) and thin gap chambers (TGC) are presented.

2 ATLAS Transition Radiation Tracker[1]

The inner detector consists of discrete (semiconductor trackers) and continuous (transition radiation tracker, TRT) trackers, the combination gives excellent tracking performance. In addition, the TRT gives powerful particle identification. The ATLAS TRT is 680.2cm long and its outer and inner diameter are 206cm and 96(barrel)-128cm(endcaps), respectively. The TRT is made of 372000 kapton straw tubes. The straw's inner diameter is 4mm and wire diameter is 30 μm. It is surrounded by radiator material. The straw is operated with 70% Xe+20%CF_4+10%CO_2 and the gas gain of 2.5-3.5$\times10^4$. The track position is determined from drift time measurement.

The rate capability and ageing effects of the straw tubes have been tested. Fig. 1 shows the rate capability of the straw. At 20MHz counting rate, the efficiency of the drift time measurement is around 60%, and about 170μm of the accuracy of the track position is achieved. In this case, a track has at least 36layers$\times$0.60 = 22 high precision points which give sufficient resolution.

The ageing effects have been investigated in various conditions: gas flow rate of 0.01-10cm^3/min(nominal 0.1cm^3/min), gas gain of 2.5-6$\times 10^4$ (nominal 2.5$\times 10^4$) and dose rate of 0.14-5μA/cm(expected 0.14μA/cm). As for the anode wires, no ageing effect was observed up to 8C/cm corresponding to 10 years LHC. The straw cathode behaviour has been examined up to 18C/cm(more than 20 years LHC) and no change of the cathode properties was observed.

A full scale prototype of an endcap module of the TRT has been constructed and examined. The outer and inner diameters of the prototype (TRT wheel) are 200cm and 100cm, respectively, and the wheel contains 9600 straws which make up 16 layers. In order to examine the accuracy of the alignment of the straws in the wheel, the wire offsets are measured in a X-ray calibration stand with an accuracy of better than 20 μm. The average of the offsets at the inner radius of the wheel is about 26.9μm which is much better than the required accuracy of 50 μm. We confirmed that we can construct the whole system with good accuracy.

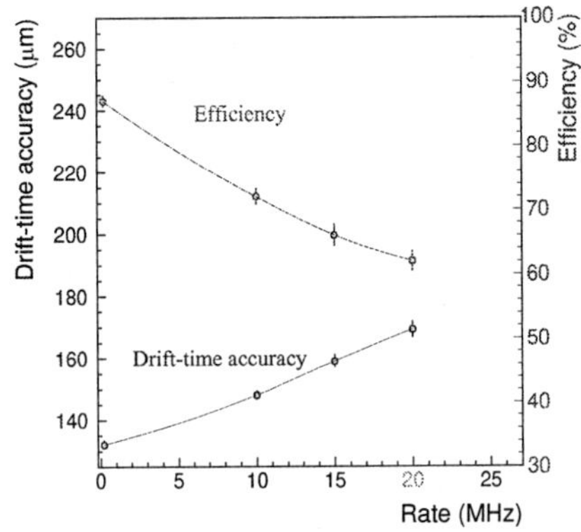

Fig. 1. The rate capability of the straw tube is shown.

3 ATLAS Muon Chamber System

The ATLAS muon spectrometer consists of chambers for precise 1D- position measurement, chambers dedicating to triggering and 2nd coordinate measurement and air-core superconducting toroids generating magnetic field of 0.5-2T. Fig. 2 shows a schematic r-z view of the spectrometer. The muon spectrometer has the capability of standalone measurement of muon momentum. The technologies used for the precision chambers are monitored drift tubes chambers(MDT) and cathode strip chambers(CSC). For the trigger chambers, two kinds of technologies are used: resistive plate chambers (RPC) in the barrel and thin gap chambers (TGC) in the endcaps.

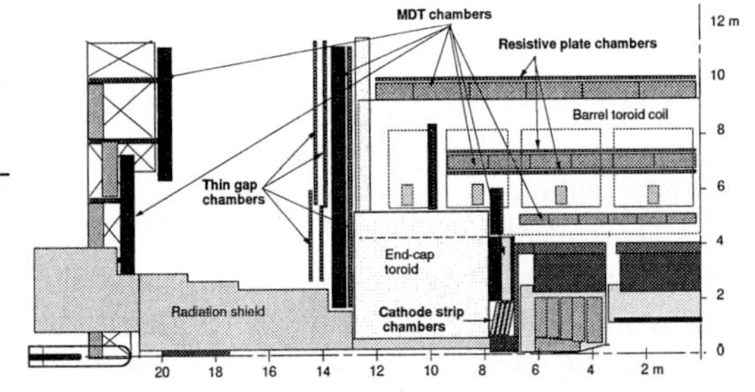

Fig. 2. A schematic RZ view of the muon spectrometer is shown.

3.1 Monitored Drift Tube Chamber[2]

The precise track position in the muon system is determined by drift time measurement using monitored drift tube chambers (MDT). Each drift tube is 0.7-6.3m long and its outer diameter is 30mm. The anode wire diameter is 50μm. The drift tube is operated with a gas mixture of 91%Ar+4%N_2+5%CH_4 which is pressurized to 3 bar. The gas gain is about 2×10^4. Three or four layers of drift tubes are assembled to form a multilayer. A MDT has two multilayers separated by about 20-30cm and in-plane alignment monitoring device. 'Monitored' is named after the concept of the monitoring. Total number of MDT's is 1194 containing 37000 of drift tubes.

The muon spectrometer performance is affected by the following factors;

Drift tube resolution The resolution of a drift tube depends on the gas gain, the peaking time of the front-end electronics and the threshold. Higher gas gain, fast electronics and lower threshold give better resolution. However, higher gas gain increases ageing effects, fast electronics is more expensive and lower thresholds make the chambers more sensitive to noise. As the results of detailed studies, our baseline was set as follows: 2×10^4 of gas gain, 11ns of peaking time and 20 electrons of threshold. The position resolution is expected to be 80μm.

Mechanical precision The mechanical precision is measured by X-ray tomograph which is similar to the technology used for the TRT. As the results of the measurements, the mechanical precision of the drift tubes is controlled at the level of 20μm within a single layer and 25μm within a multilayer.

Alignment precision There are thee kinds of alignment systems in the muon spectrometer: in-plane, projective and axial alignment. With the in-plane alignment system, the internal chamber deformations are measured. The projective alignment system measures relative position of the chambers in three stations. The axial alignment system links the chambers in the same plane and helps to limit the number of rays for the projective alignment. The precision of the projective and axial alignment is about 30μm.

Material and non-uniformity of the toroid field The influence of the material and non-uniformity of the toroid field on the muon system performance are investigated by Monte-Carlo simulation.

Fig. 3 shows the simulated momentum resolution of the muon spectrometer. The momentum resolution of the muon system is 2-5% for muons with p_T of 100 GeV/c and 10-20% of muons with 1TeV/c.

A system test facility, called DATCHA, has been constructed. DATCHA gives us operating experience with a complete sector of the barrel muon spectrometer. DATCHA consists of MDT's, RPC's, alignment, data acquisition and detector control systems. Using DATCHA, we can verify the validity of the alignment concept.

3.2 Thin Gap Chamber[2]

The thin gap chambers (TGC) as the trigger chambers in the endcaps of the ATLAS detector is a sort of multi-wire proportional chamber, but it differs

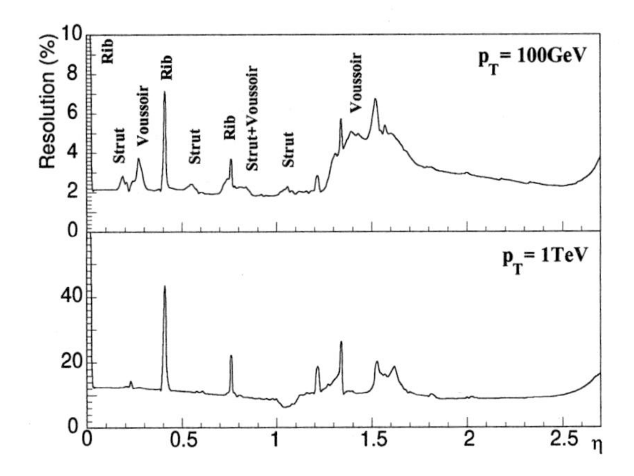

Fig. 3. Simulated momentum resolution of the muon spectrometer is shown. The resolution is averaged over ϕ.

from MWPC in regard of its cell structure: the anode-to-anode distance(1.8mm) is larger than anode-to-cathode distance(1.4mm). The chamber is operated with a highly quenching gas mixture of 55% CO_2 + 45% n-pentane in saturated mode. The operation results in no streamers and small sensitivity to mechanical deformations and variations in primary ionization.

We investigated the rate capability of the TGCs and ageing effects on its properties. There is no significant rate dependence up to 20kHz/cm^2 which corresponds to factor 20 above the expected environment. For ageing effects, small chambers (10×10cm^2) were tested with various irradiation rates and gas flow rates. There was no signal degradation up to an accumulated charge of 0.1C/cm corresponding to 18 years LHC.

3.3 Conclusion

The straw tube design and working point for the ATLAS TRT have been fully specified and a prototype wheel gave valuable experience for the final TRT wheel design. Detailed studies on the basic properties of the drift tubes of the muon spectrometer have been done, and system tests are in progress. The TGCs show good rate capability and no ageing effect up to 18 years LHC.

The Technical Design Reports of the inner detector [1] and the muon spectrometer [2] have been submitted to the LHC committee and are expected to be approved this fall. Construction of module 0 of all chambers will start in 1998 and series production is expected in 1999-2003.

References

[1] ATLAS Inner Detector Technical Design Report, CERN/LHCC/97-16,17, 30 April 1997.

[2] ATLAS Muon Spectrometer Technical Design Report, CERN/LHCC/97-22, 31 May 1997.

Recent Results from $PbWO_4$ Crystals and MSGC's in CMS

G.J. Davies on behalf of the CMS Collaboration (g.j.davies@ic.ac.uk)

Imperial College, London

Abstract. CMS is one of the two general purpose detectors for the LHC. The ECAL is fully active, consisting of ~100,000 lead tungstate crystals. Progress on crystal R&D, including radiation hardness and energy resolution is discussed. Inside the ECAL is the tracker which consists of several silicon pixel layers followed by silicon microstrips and finally at radii between 60-110cm MSGCs. The successful operation of CMS Performance Prototype MSGCs in the high rate T10 testbeam is reported.

1 Introduction - ECAL

The discovery of the Higgs boson is one of the main goals of the LHC. In the range $80<m_H<120$GeV the benchmark decay is H-→$\gamma\gamma$. The need to detect this two photon decay in the harsh LHC environment requires the ECAL to be both precise and radiation hard.

The effect of irradiation on lead tungstate is to reduce its optical transmission but leave the scintillation mechanism unaffected. Hence we can quantify any effect in terms of the induced absorption coefficient. By the end of 1994 we had shown that even for doses in excess of 10Mrad the attenuation length remains >0.5m and hence the contribution to the constant term in the energy resolution from the longitudinal response is kept below the design goal of 0.3%. However in the testbeam runs of mid 1996 a decrease in light yield of 5% was seen at low doses (~650rad). It is the significant progress made in resolving this issue that will be discussed, in terms of recent testbeam results and crystal R&D.

2 Results -ECAL -testbeam

Beam tests on a 7x7 crystal (closely resembling those to be used at η=0) matrix were carried out in H4 during 1996 and early 1997. The test set-up was similar to the one previously used [1]. Optical fibres fed normalised light pulses to the front of the crystals. Red and green laser light, a pulse Xe system and a red LED were used. As radiation damage only affects the attenuation length the longitudinal uniformity and hence the energy resolution should not be significantly altered. Fig 1 shows the energy distribution in a sum of 9 crystals for 120GeV electrons on crystal 1283 before and after 650rad.

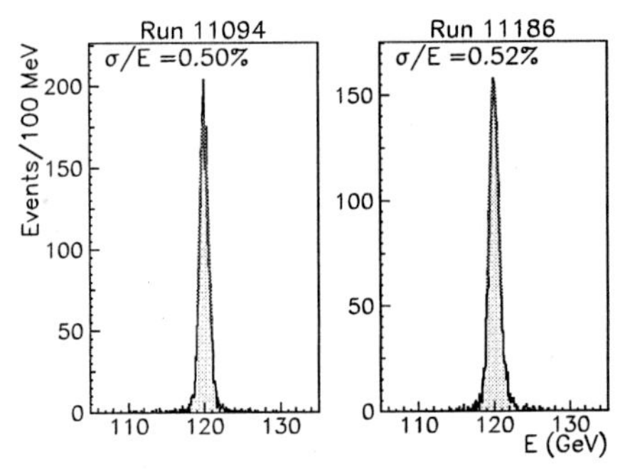

Fig. 1. Energy distribution seen in sum of 9 crystals for 120 GeV electrons incident in a 4x4mm^2 area of the central crystal before the irradiation.

To track the radiation damage it is necessary to determine the constant of proportionality between the change observed by the monitoring system and the change in the beam signal -fig. 2 (left).

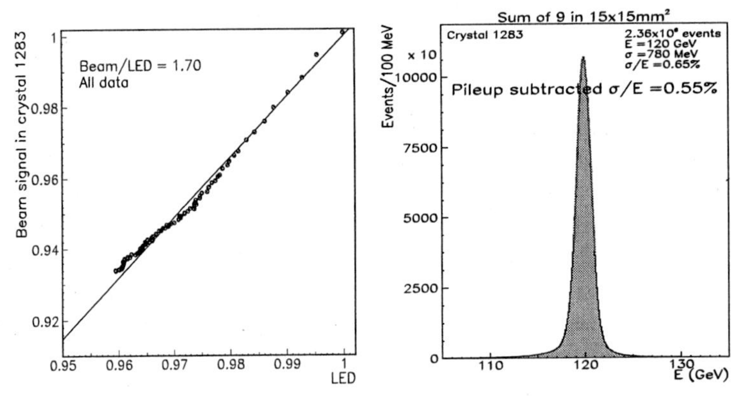

Fig. 2. Correspondence between monitoring LED signals and beam signals during irradiation to 650 rads (left). Energy distribution in a sum of 9 crystals obtained using LED monitoring to follow the calibration, for all electrons incident in a 15x15mm^2 area centred on crystal 1283 taken throughout the irradiation (right).

This constant will not be 1 as the light from the monitoring system follows a different path to the scintillation light and because they have different wavelengths. Fig. 2 (right) shows the corrected energy distribution for all events, ie before/during and after the irradiation. When the pileup noise (due to the high rate during the irradiation) has been subtracted the energy resolution is 0.55%, compared to 0.53% before irradiation.

In April 1997 9 crystals, deliberately chosen to have different radiation tolerances, were irradiated and the drop in beam signal compared to that seen by the green laser. A good correlation, with the same constant of proportionality was found for all nine crystals.

The good energy resolutions achieved in 1996 were partly due to a better understanding of how to effectively uniformise the longitudinal response from the crystals. This work continued in 1997 and resulted in us reaching our design goal of <0.6% at 120GeV with a large number of crystals (15).

3 Results -ECAL -crystal progress

CMS ECAL is developing harder crystals whilst optimising how to monitor out residual calibration fluctuation. To this end an extensive R&Dprogramme is underway. Its main goals are to the control purity of the raw material, specific doping, optimised stoechiometry and annealing.

Figure 3 illustrates the variation in the induced absorption with the amount of PbO and WO_3 in the crystal melt, after 50krad.

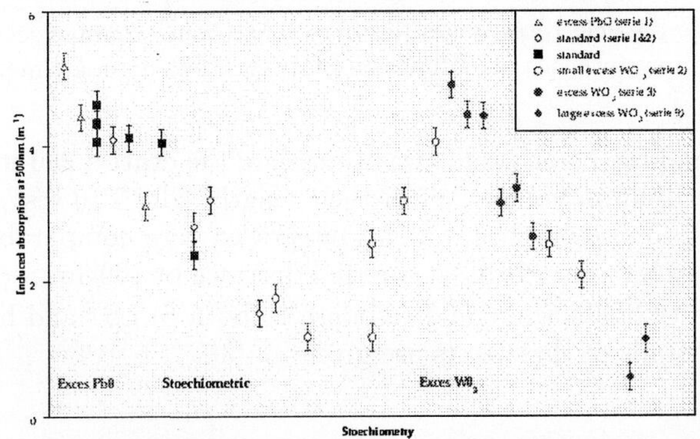

Fig. 3. Induced absorption at 500nm for 29 full size Russian crystals

Doping with Nb, Lu and La has also been shown to greatly reduce the induced absorption [2]. Finally crystals with optimum doping and stoechiometry will be produced. Significant progress has also been made as a result of varying the atmosphere in which the crystals are annealed (oxygen being favoured).

4 Introduction -MSGCs

A set of CMS performance prototype MSGCs were exposed, for the first time, to a high intensity 3GeV/c pion beam (in T10 at the CERN SPS). Not only was the m.i.p. rate comparable to that expected at LHC but inelastic interactions generate approximately the same number of h.i.p.s as expected at LHC. These devices had both the "standard passivation" - the substrate edges perpendicular to the strips were passivated for 2mm with a polymide film - and the "advanced passivation" - individual cathode edges were coated by a 3μm thick polymide film, 4μm wide in each dimension across the strip.

A signal to noise (S/N) ratio of 20 allows for a hit reconstruction efficiency $> 98.5\%$. Thus as the S/N ratio at CMS is expected to be 2.2 worse than that in T10 the chambers were operated at a S/N ratio of 44. The corresponding voltage settings are V_{Drift} =3500V and $V_{Cathode}$ =520V. The duty cycle in

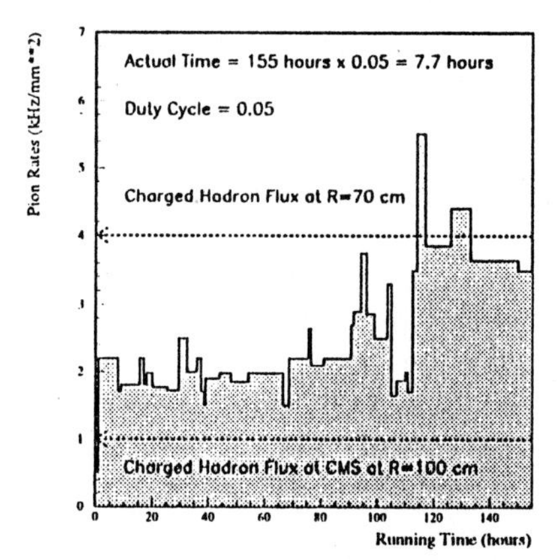

Fig. 4. Average particle rates measured by a 0.5x0.5cm^2 counter set close to the region of maximum intensity as a function of integrated running time.

T10 was 5% and the beam profile was Gaussian; the central 250 strips of each detector were illuminated for half of their length. The T10 test amounts to ~7.7 hours of continuous beam at an integrated rate comparable or larger than the one expected at LHC at a radial distance of 100cm. See fig 4.

The stability of efficiency and response to both m.i.p.s and h.i.p.s is discussed. Full details of these tests are given in [3].

5 Results -MSGCs -stability

The detectors performances have been analysed over the whole data taking period and a summary of the study is reported in fig. 5 (left) for a typical chamber.

As can be seen, the charge, noise, S/N, cluster size and hit-multiplicity have been extremely stable over a period of 225hrs.

6 Results -MSGCs -no sparking

The condition of the detectors was monitored frequently to check for the onset of a possible streamer regime. Fig. 5 (right) shows the beam profile recorded by one of the detectors during the full run period. The binning in the plots is equal to the strip pitch and so a dead or shortened strip would be detected as an inefficiency. No damage was seen in any of the chambers, nor in the visual inspection carried out afterwards.

Furthermore high luminosity runs were selected and transient charging of the substrate looked for — no evidence was found. Additionally the cathode voltage on several chambers was raised, up to 600V and no deviation from the usual observed.

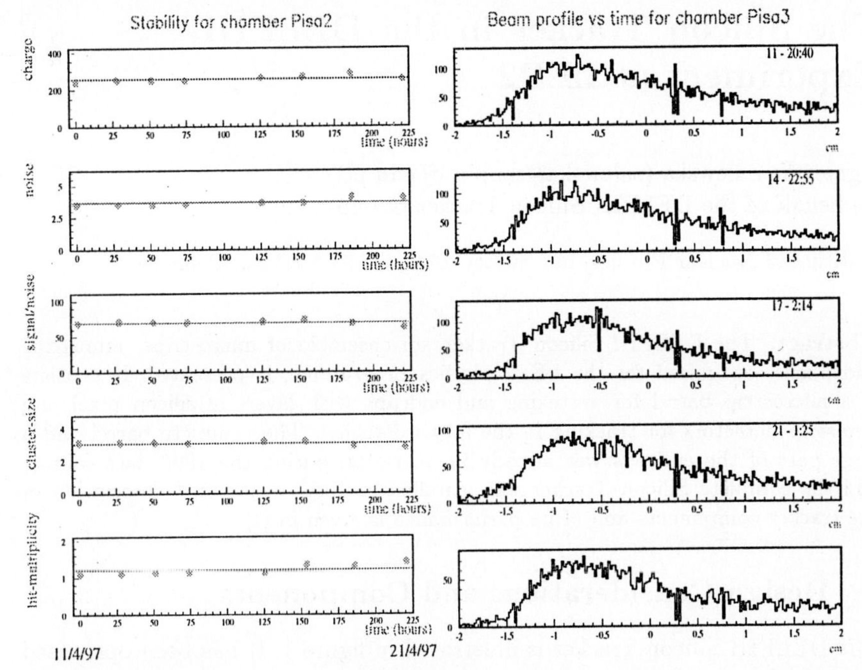

Fig. 5. Stability study for an MSGC chamber. All the variables are plotted along the 225hrs of running time, for low intensity runs.

7 Conclusions -ECAL and MSGCs

The ECAL community has demonstrated good progress on energy resolution and crystal radiation hardness. Many crystals met the design goal of <0.6% at 120GeV and crystals that only loose a few% of light after $\sim$ 650rad have been produced. 1997 will continue as a full R&D year.

The tracker community successfully operated performance prototype MS-GCs in a high rate environment. Further tests are planned in September at PSI, Villigen.

8 Acknowledgments

It is a great pleasure to thank all the members of the ECAL and tracker community who made this work, and indeed this report, possible.

References

[1] G. Alexeev et al., NIM A385 (1997), 425.
[2] A. N. Annenkov et al., CMS Note 1997/055.
[3] D. Abbaneo et al., Proceedings of 7th Pisa meeting on Advanced Detectors, Elba, May 1997, to be published in NIM (1997).

The Silicon Tracker in the DELPHI Experiment at LEP2

Agnieszka Zalewska (zalewska@chopin.ifj.edu.pl)
on behalf of the DELPHI Silicon Tracker Group

Institute of Nuclear Physics, ul.Kawiory 26A, 30-055 Kraków, Poland

Abstract. The DELPHI Silicon Tracker, an ensemble of microstrips, ministrips and pixels, optimised for the LEP2 physics programme, is presented. It consists of a microstrip barrel for vertexing and endcaps with layers of silicon pixel and ministrip detectors for tracking in the forward region. The complete barrel and a large part of the endcaps was already in operation during the 1996 data taking, and the complete Silicon Tracker was installed in 1997. A detailed description of the tracker components and of its performance is given in [1].

1 Design Considerations and Components

The DELPHI Silicon Tracker is illustrated in figure 1. It has been optimised to cope with the requirements posed by the physics programme at LEP2, i.e. the measurements of the four fermion processes, $e^+e^- \to q\bar{q}\gamma$ or $e^+e^- \to \gamma\gamma$ and the search for the Higgs boson and for supersymmetric particles.

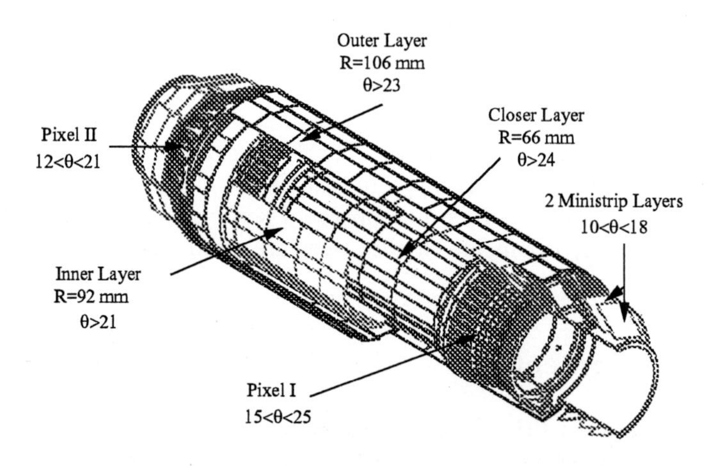

Fig. 1. Layout of the DELPHI Silicon Tracker

The Silicon Tracker contains 888 detecting elements having a total active surface of about 1.6 m^2 of silicon, has 1399808 readout channels and covers polar angles between 11° and 169°. It consists of the Barrel, extending from 21° to 159° and playing the role of vertex detector, and the Very Forward Tracker (VFT) in the form of two silicon endcaps, providing standalone pattern recognition and increasing the track reconstruction efficiency between 11° and 25°. Throughout the detector there is great emphasis on the overlap of sensitive silicon. This provides redundancy and makes possible a self alignment procedure, but places great constraints on the assembly, with silicon plaquettes from different layers often lying less than 1 mm apart.

The Barrel contains 640 AC coupled microstrip silicon detectors, arranged in three concentric layers at average radii of 6.6 cm, 9.2 cm and 10.6 cm. The 149504 electronics channels read signals collected on the strips which give $R\phi$ measurements with a readout pitch of 50 μm and Rz measurements with pitches varying between 42 μm and 176 μm. The material in the sensitive region is kept to a minimum by the use of double-sided detectors, double-metal readout and light mechanics. The radiation lengths of material at a polar angle $\theta = 90°$ are 0.4%, 1.1% and 3.5% for points after the beam pipe, the first measured point and the whole Silicon Tracker respectively.

Each of the two VFT endcaps contains two layers of silicon pixel detectors and two layers of ministrip detectors. The pixels have dimensions of 330×330 μm^2 and there are 1225728 in total. They are connected to the readout electronics channels using an industrial bump bonding method and their readout is performed by a sparse data scan circuit. The AC coupled ministrip detectors (96 in total, corresponding to 24576 electronics channels) have a strip pitch of 100 μm and a readout pitch of 200 μm. To help the pattern recognition the ministrips are mounted at a small stereo angle.

2 Tracker Performance

In the Barrel, three dimensional b tagging information is available down to a polar angle of 25°. In the $R\phi$ plane the resolution of the microstrip detectors is around 8 μm, and in the Rz plane it varies between 10 μm and 25 μm for tracks of different inclination. Impact parameter resolutions have been measured of $28\ \mu\text{m} \oplus 71/(p \sin^{\frac{3}{2}}\theta)\ \mu\text{m}$ in $R\phi$ and $34\ \mu\text{m} \oplus (69/p)\ \mu\text{m}$ in Rz, where p is the track momentum in GeV/c.

The forward part of the detector shows average efficiencies of more than 96%, has signal-to-noise ratios of up to 40 in the ministrips, and noise levels at the level of less than one part per million in the pixels. Measurements of space points with low backgrounds are provided, leading to a vastly improved tracking efficiency for the region with polar angle less than 25°.

References

[1] P.Chochula et al., The Silicon Tracker in the DELPHI Experiment at LEP2, paper 306 submitted to HEP97, DELPHI 97-121 CONF 103.

The LXe Calorimeter for $\pi \rightarrow \mu\nu\gamma$ Decay

A.T.Meneguzzo (a) (meneguzzo@pd.infn.it), G. Carugno (a), E. Conti (a), G. Bressi (b) , S. Cerdonio (c) and D. Zanello (c)

(a)I.N.F.N. Sez. di Padova, (b) I.N.F.N. Sez. di Pavia-Italy and Università di Padova-Italy ,(c) I.N.F.N. Sez. di Roma I, Università di Roma "La Sapienza"-Italy

Abstract. A large Liquid Xenon ionization chamber was operated for the detection and measure of the low-energy gamma-rays from the decay $\pi \rightarrow \mu\nu\gamma$. The scintillation light is successfully used for the trigger and the ionization charge for the energy measurement. The detector was calibrated with radioactive sources and its performances were compared to a Monte Carlo simulation.

Introduction - The spectrum of the γ from the rare decay $\pi^+ \rightarrow \mu^+\nu\gamma(1)$ behaves approximately as $1/E_\gamma$ with an end-point at 29.8 MeV. With a Lxe Time Projection chamber(TPC) the RAPID experiment [1] aimed to detect the γ exploiting the scintillation light and to measure the spectrum by means of the ionization signal with good efficiency ,fine resolution and an energy cutoff as low as possible (of the order of a few hundreds keV). The detector was exposed to γ-ray radioactive sources for calibration during the period of the data acquisition at the Paul Scherrer Instutute.

The TPC (64 l of LXe kept at a temperature of 180 K) is housed in a safety tank. The sensitive volume is a cylinder segmented into 6 drift region, 6.3 cm deep [3]. Each anode for the charge collection is segmented in two concentric pads. The operating electric drift field is 500 V/cm. The 175 nm scintillation light is detected by 12 UVphotomultipliers (UVPM) coupled to the LXe by UV-transparent quartz windows symmetrically placed 3 by 3 at 120^0 angle in 4 planes along the TPC axis.

Big effort was devoted on the material cleaning and on the procedure of TPC filling in order to reach and maintain a high liquid purity (to avoid degradation of the ionization charge signal due to the 33 μs of drifting time) and to get long term time stability . We measured an electron lifitime $\tau \geq 2.7$ ms (95% C.L.) and stable along the 2 month of data taking [2] [3].

Calorimeter perfomances - Before and after data taking for decay (1) we calibrated the calorimeter by means of γ-ray sources placed in the position where the γ-ray of reaction (1) were generated.

The following radioactive sources were exploited: ^{22}Na, ^{60}Co, ^{88}Y and Am-Be. The ^{22}Na emits simultaneously a 1275 keV γ-ray and a β^+ with a 90% probability.The β^+ promptly yields two 511 keV back-to-back rays with isotropic distribution. With the help of a NaI(Tl) detector placed in front of TPC and set in order to select γ of 1275 or 511 keV, the TPC was checked with two almost monochromatic and externally triggered γs. These data compared with the results of a MC simulation of the detector and trigger allowed to know the UVPMs gain and threshold from fits and to compute the total trigger acceptance $\alpha = \epsilon(E_\gamma) \cdot d\Omega$ [4]. α takes into ac-

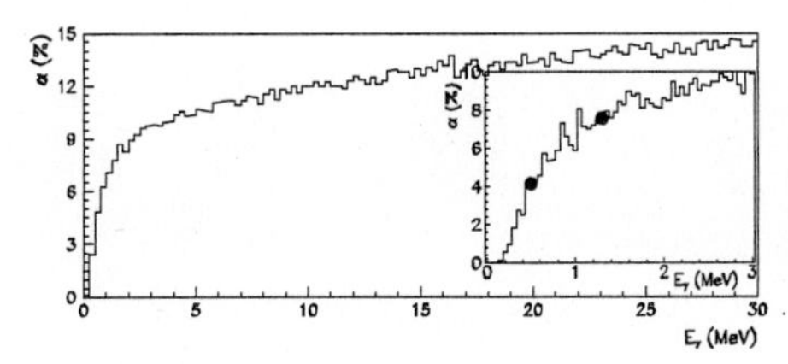

count the PM trigger efficiency $\epsilon(E_\gamma)$ and the geometrical acceptance Ω of the calorimeter. The passive materials are the major responsible of the dependence of ϵ on the γ-ray energy. The curve of $\alpha(E_\gamma)$ in the whole range of decay (1) was determined simulating a gamma source with a flat spectrum from 0 to 30 MeV. The result is shown in the figure : the full dots in the insert are the experimental points (they lie in the region where $\alpha(E_\gamma)$ shows the highest derivative) . The minimum triggerable energy is 170keV.

We reconstruct [4] the charge spectrum for each anode and compare it to the MC results. With a minimisation procedure the electronic noise, the ADC-energy conversion factor and the minimum measurable energy is extracted. The experimental spectra together with the MC best fits are shown in the fig(1). The intrinsic TPC energy resolution is $7\%/\sqrt{E_\gamma(Mev)}$ as several authors report [5]. The electronic noise turns out to be about 140 keV almost for all anodes and the minimum measurable energy around 230 keV .

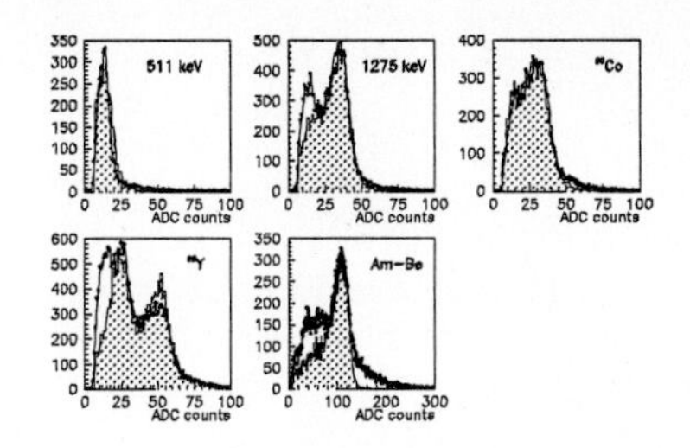

Fig. 1. Total spectra (full dots) and MC (histograms) for different γ-sources.

Conclusions - A liquid xenon TPC calorimeter was exposed to γ-rays with energies up to 5 MeV and its response studied and compared to MC simulation results. The detector trigger efficiency agrees with the MC calculations. The minimum triggerable energy turns out to be 170 keV. Regarding the charge signal, the calorimeter behaviour is reproducible by the MC with good accuracy. The TPC intrinsic resolution is 7% at 1 MeV. The electronic noise is around 140 keV and the minimum measurable energy 230 keV with an electric field of 0.5 kV/cm. The overall apparatus acceptance reaches 10% at 2 MeV and above is almost independent on the γ-ray energy.

References

[1] A.Meneguzzo et al., Hep97 proceeding 214; G.Bressi et al. ,in press in Nucl.Phys B.

[2] G. Carugno et al., Nucl. Instr. and Meth. A 292 (1990), 580

[3] G. Carugno et al., Nucl. Instr. and Meth. A 376 (1996), 149

[4] G. Bressi et al., Nucl. Instr. and Meth. A 396 (1997), 67

[5] E. Aprile et al., Nucl. Instr. and meth. A 302 (1991), 177

Part XVIII

Physics at Future Machines

The Next 20 Years: Machines for High Energy Physics

Peter Mättig (peter.mattig@cern.ch)

CERN, Geneva, Switzerland and Weizmann Institute, Rehovot, Israel

Abstract. The upgrades of existing and the new accelerators for High Energy physics are summarised.

1 Introduction

During the last 25 years High Energy Physics accelerators and experiments have pushed the explored mass regions from a few GeV to about hundred GeV. This increase in resolution power lead to the triumph of what is today deemed the 'Standard' Model. It rests on three cornerstones: fermions, gauge bosons and the Higgs boson. All fermions and also the gauge bosons have been experimentally studied, in most cases, with very high accuracy. Less precisely tested are the properties of the top quark and of the W boson like its mass and its interactions with the γ and Z^0, the triple gauge interactions. Another outstanding aspect is CP violation which has only been seen in the strange system. The only experimentally missing part of the Standard Model is the Higgs boson, on which only recently some first, albeit indirect, experimental constraints have been obtained. It is one principle aim of future high energy machines to close these still open chapters of the Standard Model. But even if all these measurements would only confirm the Standard Model, there is a general consensus that it cannot be the final theory. The repetition of fermion structures, the common scheme of all interactions and their converging couplings around 10^{14} GeV, suggests some more general underlying principle. There is a strong belief that new physics exists below 1 TeV. This new energy domain and the search for signs of new physics is the second focal point of the future machines.

The tools experimentalists have at hand are quark-quark, quark-lepton and lepton-lepton colliders, realised by pp($\bar{\text{p}}$), ep, and e^+e^- collisions. These machines provide collisions between fermions, between gauge bosons and fermion - gauge boson interactions. For example, the accessible mass scale $\sqrt{s'}$ for fermion - fermion scattering depends on the total c.m.energy $\sqrt{s}$, but, because of the proton substructure, for pp and ep collisions also on the luminosity $\mathcal{L}$. The typical dependence is indicated in table 1.

Table 1. Typical dependence of accessible fermion - fermion mass scale $\sqrt{s'}$ on the c.m. energy $\sqrt{s}$ and luminosity $\mathcal{L}$ for different kinds of colliders. Rows 3 and 4 give their required increase for a doubling of $\sqrt{s'}$.

	$pp(\bar{p})$	ep	e^+e^-
effective mass scale $\sqrt{s'} \sim$	$\sqrt{s}^{\,0.6}\mathcal{L}^{\,0.2}$	$\sqrt{s}^{\,0.8}\mathcal{L}^{\,0.25}$	$\sqrt{s}$
Required luminosity increase	25	16	
Required $\sqrt{s}$ increase	3.2	2.4	2

To study new mass ranges a simple increase in $\sqrt{s}$ is not enough. Rather it should be accompanied by an increase in luminosity to compensate for the $1/s'$ drop of the pointlike cross section. Only with a higher luminosity the event yield can be maintained at a constant level. In contrast, the last two decades saw a significant increase in $\sqrt{s}$ but a rather constant luminosity (Fig 1,2). For example, for the several e^+e^- colliders, covering an energy range between 1 and 180 GeV, the obtained luminosities [1] are in general between 1 and $5{\cdot}10^{31}$. A notable exception is the e^+e^- collider CESR that has reached by now a maximum luminosity in excess of 10^{32}. This machine is in operation for almost 20 years and, during this time, has experienced continuous improvements and major upgrades. Such a long time for machines to operate in collider mode is unusual.

The future challenges of low energy precision measurements in the beauty system and the exploration of the TeV scale demand significantly higher luminosities. As also shown in Fig 1,2, a common feature of all new projects is a projected quantum jump in luminosity.

2 Major Upgrades and New Machines

2.1 pp colliders

The machine which has reached the highest mass range is the Tevatron at Fermilab. It has started operation ten years ago and provided until '96 interactions at $\sqrt{s}$ = 1.8 TeV with a maximum luminositiy of $\mathcal{L} \sim 1.6{\cdot}10^{31}$. In total some 120 pb^{-1} were delivered to the two experiments CDF and D0. Currently it undergoes a major upgrade, mainly the construction of a new main injector which will allow to increase the current, and a recycler which will improve the $\bar{p}$ accumulation. Between '99 and '02 it is planned to run at

[1] The instantaneous luminosities will be given in the usual dimension $cm^{-2}sec^{-1}$ throughout this report.

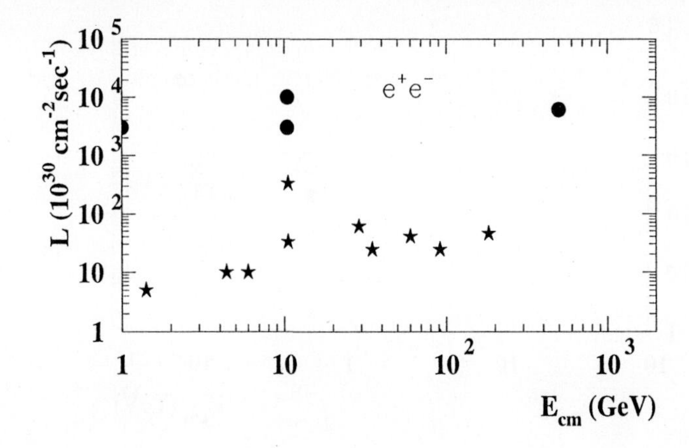

Fig. 1. $\mathcal{L}$ for various e^+e^- colliders as a function of $\sqrt{s}$. Stars indicate luminosities achieved in operation, the dots those projected for new machines.

$\sqrt{s}$ = 2 TeV and $\mathcal{L}$ of 10^{32} such that a total of $\sim$ 2 fb^{-1} will be delivered to the experiments. After 2002 it is envisaged to continously upgrade the Tevatron, for example allowing for more bunches, such that eventually a luminosity of $2{\cdot}10^{33}$ will be reached and a total of about 20 fb^{-1} will be delivered. In particular the boost in luminosity by two orders of magnitude will extend the physics reach significantly. Beauty mixing and CP violation can be explored, and the masses of the top and W-boson can be determined to 2-3 GeV and 20-40 MeV precision, respectively. New particles with masses of up to 700 GeV can be detected.

An even higher energy and luminosity is projected for the CERN LHC which will be installed in the LEP tunnel. The LHC was approved in 1994 and will start data taking in 2005 at $\sqrt{s}$ = 14 TeV. The luminosity should reach 10^{33} in the initial phase and increase to 10^{34} after operational experience. The main technological challenges are the superconducting magnets to provide a field of 8.3 Tesla and the handling of the 2385 bunches, corresponding to a stored energy equivalent to 75 kg of TNT. These high currents and luminosities produce a very straining environment also for the detectors which have to cope with high radiation levels, backgrounds, and a huge amount of data. The main physics goal of the LHC is to be sensitive to Higgs boson masses from 100 GeV to 1 TeV, thus providing the final word on a Standard Model Higgs. In addition, the several 10 millions of bottom and top quarks, and also the huge number of W - bosons will allow rather

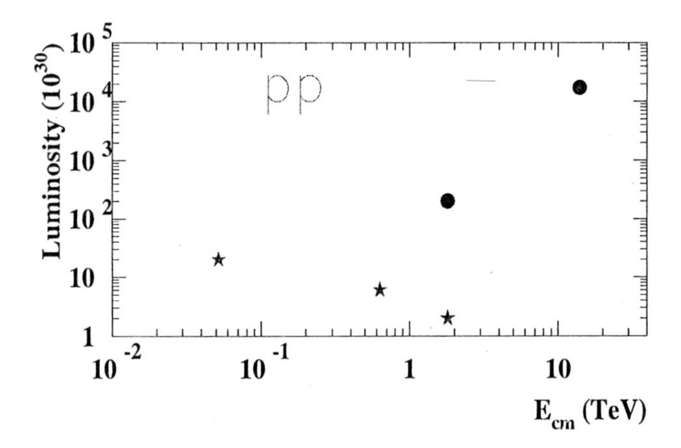

Fig. 2. $\mathcal{L}$ for various pp($\bar{\text{p}}$) colliders as a function of $\sqrt{s}$. Stars indicate luminosities achieved in operation, the dots those projected for new machines.

precise measurements on their properties. New particles can be detected for masses of up to 3000 GeV.

2.2 ep colliders

HERA is the world's only ep collider. It operates since '91 at DESY and has up to now reached luminosities of $1.4{\cdot}10^{31}$. In total HERA has delivered some 75 pb^{-1} with electrons or positrons of 27.5 GeV and protons of 820 GeV yielding $\sqrt{s}$ = 300 GeV. A major upgrade is planned in 1999/2000, when the interaction regions are restructered allowing for a reduced beam size at the collision point. HERA will then be able to reach luminosities of $7.5{\cdot}10^{31}$ and to deliver some 200 pb^{-1} per year to the H1 and ZEUS experiments. This large increase in luminosity will allow electroweak phenomena, particularly the charged current, to be studied and new particles with masses of 300 GeV and small production rates to be probed.

In the far future ep - collisions may also be possible at CERN if the protons from the LHC are scattered with the electrons from LEP. With such collisions c.m. energies of $\sqrt{s}$ $\sim$ 1.3 TeV will be reached. The option is left open. But it seems unlikely that it will be realised before 2015.

2.3 e^+e^- colliders

In its first stage of operation '89 - '95 LEP at CERN has produced some 20 million Z^0's on resonance. In the current LEP2 period its c.m.energy is increased well above the threshold of W-pair production. The crucial technological issue are the super conducting cavities which operate stable at an acceleration gradient of 6MV/m and beam currents of 5mA. LEP2 has up to now reached a maximum $\sqrt{s}$ of 184 GeV and delivered a total luminosity of some 85 pb^{-1}. It will operate another two, probably three years and may reach a maximum energy of 200 GeV with additional cavities and higher gradients. A total of 500 pb^{-1} should be collected by each experiment. With such an amount of data the W- mass can be determined with a precision of $\sim$ 40 MeV. LEP will be able to detect the Higgs boson and other new particles for masses of up to $\sim$ 100 GeV. If no new phenomena are found, significant constraints could be put on various extensions of the Standard Model. For example, the MSSM parameter space of $\tan\beta \leq 2$ is excluded.

Beyond LEP2, collisions of e^+e^- require a linear collider based on new technologies. A vigorous international R&D program is under way and various options are considered. The principal difference among these schemes is the choice of the RF. Its value largely constrains the machine parameters. The two main lines of thought are to use conventional cavities working in the S, C, or X band (3.5, 7 or 11.4 GHz) or super conducting cavities in the L - band (1.3 GHz). The technology for the former option exists, however, the high frequency leads to strong wakefields that may spoil the performance. As a result alignment of machine components is very demanding and the beam size at the interaction point must not exceed a few nm. The super conducting solution, on the other hand, would have a good stability and less stringent other constraints, but it requires a substantial development program to reduce the cost for the cavities. Some machine parameters based on these options are listed in table 2.

Table 2. Some machine parameters for the two options of NC (NLC/JLC) and SC (TESLA) cavities.

	NLC/JLC	TESLA
RF	S, C, or X	L
Length (km)	14-28	32
Beam size at I.P. (nm)	3-6	19
Beam power (MW)	3-5	8

The main contenders for a linear collider have produced conceptual design reports in '96/'97. The technical design reports and the governmental approval are not expected before 2000. Once approved it requires some six years to build the machine. There is a consensus that the final goal are energies significantly above 1 TeV, but that this will only be reached in steps. The c.m. energy reached in the first step could be 250 GeV, aiming at a light Higgs particle, 380 GeV to act as a top factory, or 500 GeV which allows a wide range of studies.

2.4 CP violation in the Beauty System

In addition to the high energy frontier there are some 2000 physicists trying to find experimental clues to the outstanding problem of CP violation. The strange sector is part of both the CERN and Fermilab fixed target program. It will also be studied with the e^+e^- collider DAΦNE, the Frascati ϕ factory. The new frontier in $\not{C}\!P$ is the beauty sector which is approached both by dedicated e^+e^- machines: CESR, KEKB and PEPII and dedicated detectors at proton machines: HERAB, LHCB and possibly BTEV. In addition it is a significant aspect of the physics program of the general purpose detectors at the Tevatron and LHC. Whereas the gold plated process to measure $\not{C}\!P$ is $(B_d^0, \bar{B}_d^0) \to J/\psi K_s^0$, a full understanding requires measurements in additional decays like $B^0 \to \pi\pi$, $\rho\pi$, $B_s^0 \to K_s^0\rho^0$, D_sK etc. This profits from the complementary strengths of the various approaches.

The new e^+e^- machines use the resonance enhancement of the $\Upsilon(4S)$ and work with asymmetric collisions of around 9 on 3 GeV. This will gain sensitivity compared to symmetric machines like CESR. A 50 times higher luminosity than typically achieved in previous e^+e^- colliders is required, which is reached by a high number of bunches made possible by two separate beam pipes. The proton experiments take advantage of the very high cross section for beauty production but have to fight a substantial background of S/B $\sim 10^{-3}$ (LHC) - 10^{-6} (HERAB). This poses strong requirements on the trigger and data reduction. The important characteristics of some planned projects is given in table 3. The race is on to see the first significant signal of CP violation in the beauty system!

Table 3. Overview over the next round of dedicated b-physics experiments and colliders.

	e^+e^-			pp		
	CESR	PEPII	KEKB	HERA-B	CDF/D0	LHCB
Prod.	symm	asymm	asymm	fixed target	coll.	coll
Start	ongoing	'99	'99	>'98	'00	'05
$\mathcal{L}$	$\geq 10^{33}$	$\geq 3 \cdot 10^{33}$	$\geq 10^{33}$	$\sim 10^{33}$	$\geq 10^{32}$	$\geq 10^{33}$

3 Plans and Reality

As discussed before, quantum jumps in luminosity are perceived and required for future accelerators. Past experience suggests that experimenters should be careful in assuming a certain event yield. They frequently use the design luminosity $\mathcal{L}_{des}$ to expect an annual luminosity of

$$\int_{\text{year}} \mathcal{L} \;=\; \mathcal{L}_{des} \cdot 100 \text{ days} \;\sim \mathcal{L}_{des} \cdot 10^7 \text{ sec}$$

Comparing this expected luminosity to the ones actually collected at previous machines, one finds, for typical machines, an evolution of luminosities as shown in Fig 3. According to this experience, a more realistic estimate seems to be

$$\text{for first (next) 4 years}: \int_{\text{year}} \mathcal{L} \;\sim\; \mathcal{L}_{des} \cdot 10 \text{ (40) days}$$

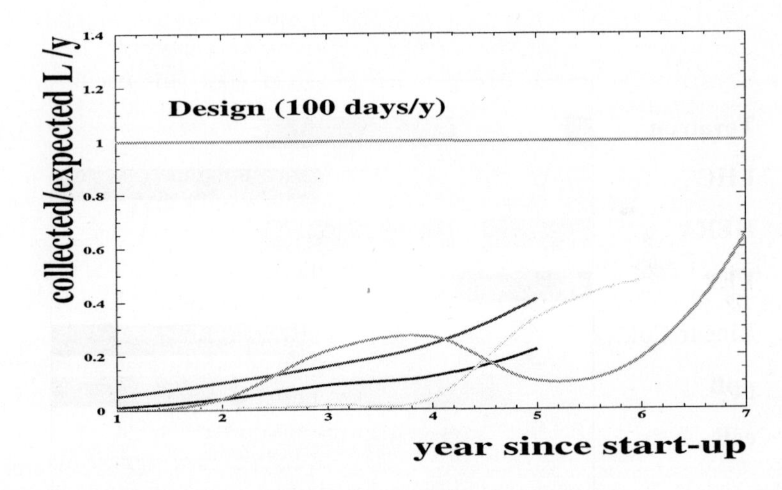

Fig. 3. The achieved $\mathcal{L}$ compared to the projected one for various types colliders as a function of the time of operation.

However, this is not the whole story. Once the data are at hand, and the detectors are understood, new methods and ideas come up and the projected precisions and sensitivities may well be superseeded by more refined methods and a thorough understanding of the data. Some prominent examples are the mass and the width of the Z^0. Before LEP turn-on they were assumed to be measurable to a precision of 25 and 50 MeV, respectively. After LEP1 they are known to 2 and 2.7 MeV.

4 Summary

In the next 20 years high energy physics is based on a broad and comprehensive program covering energy ranges from 1 GeV to 10 TeV. The time table for the various upgrades and new projects is summarised in Fig 4. Almost all of these are already approved and under construction. The exception is the linear e^+e^- collider but there is hope that some time before 2010 first collisions can be observed.

These new machines will allow us to indeed probe the physics at the TeV scale and to obtain a comprehensive picture of rare processes like CP violation. Based on the projected precisions that can be reached, our knowledge may evolve as listed in table 4. With today's prejudice such reach will lead us into mass scales where the Standard Model looses its validity. But, of course, in 15 years, something completely new and unexpected may have turned up!

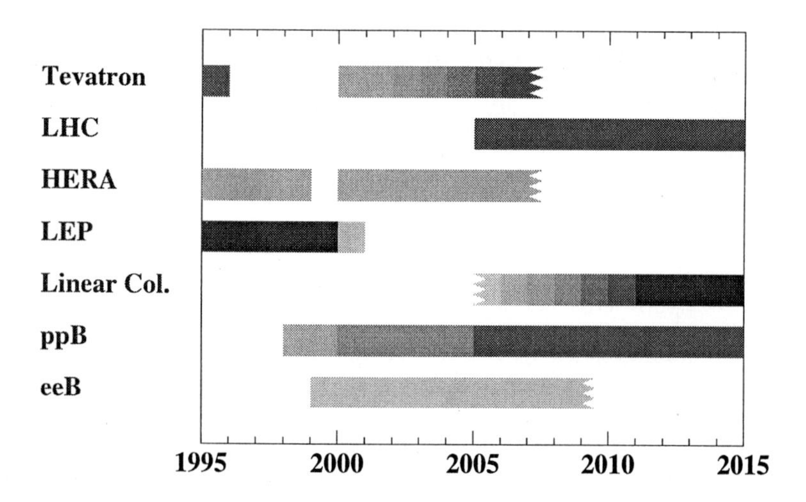

Fig. 4. Existing, upgraded and projected machines until 2015. ppB denotes beauty physics at proton machines, eeB at e^+e^- machines. The shading indicates either upgrades (Tevatron/HERA/LEP), additional facilities to become operational (ppB) or some degree of likelihood of operation (Linear e^+e^- Collider).

Apart from the increase in energy, an outstanding issue is the projected boost in luminosity. This will not only be challenging for the construction and

Table 4. Projected sensitivities for some outstanding physics goals as a function of time

	today	2000	2005	2015
δM_W (MeV)	100	40	25	15
δM_{top} (GeV)	6	6.	2.	0.2
TGC's	0.5	0.1	0.07	0.001
Higgs range (GeV)	75	100	130	1000
Contact Interaction (TeV)	3	10		50
$\not{C}P(\delta(\sin 2\beta))$		0.1	0.02	<0.01
MSSM range ($\tan\beta$)		< 2		complete
Leptoquarks (GeV)	~220		360	1500
Excited leptons (GeV)	~150	200		500
Z' (GeV)	~500		1300	> 7000

operation of these machines themselves, they will also require the experiments to handle the consequences of high beam currents and of high data rates.

In addition to these forseeable projects, during the next two decades the ground has to be laid for the next-to-next generation of High Energy machines. New projects which are considered and for which an R&D program has started are, for example, a very large hadron collider of 100 TeV (VLHC), a $\mu^+\mu^-$ collider of $\sim$ 4 TeV and new methods of acceleration.

5 Acknowledgments

I gratefully acknowledge discussions with R.Heuer, S.Komamiya, R.Schmidt and W.Zeuner.

6 References

Apart from numerous machine design studies, dedicated reports on the performance of machines and physics prospects, I profited from the recent reviews on proton and electron machines:

E.Keil, Future proton colliders, invited talk at the 18th Symposium on Lepton - Photon interactions, Hamburg, July 26 - August 1, 1997 to be published in the proceedings, LHC-Project Report 138

A.Mosnier, Future lepton colliders, invited talk at the 18th Symposium on Lepton - Photon interactions, Hamburg, July 26 - August 1, 1997 to be published in the proceedings

B.Wiik, New Electron Collider Outlook, invited talk at the International Conference on High Energy Physics, Jerusalem, August 19-26, 1997, to be published in these proceedings,

Standard Model Physics at TeV2000

John M. Butler (*jmbutler@bu.edu*)

Department of Physics, Boston University, Boston MA 02215 USA

Abstract. *tev_2000* describes the collider physics program at the Fermilab Tevatron for the next decade.

1 Introduction to *tev_2000*

tev_2000 is the term used to describe the Fermilab Tevatron collider physics program beyond the 1992-96 run ("Run I"). The program is rich in its array of physics topics, including precision top quark studies, intermediate vector boson (IVB) physics, and searches for the Higgs boson, supersymmetry (SUSY), and other exotica. These topics are best addressed at the Tevatron since it is the highest energy accelerator in the world today and therefore has the farthest physics reach. For example, the top quark is produced and studied exclusively at the Tevatron. In addition, there will be detailed studies of QCD at increasingly higher jet E_T and a raft of b physics studies, including searches for CP violation, rare decays, and mixing. *tev_2000* is remarkable not only for its variety but for its incisiveness: the program outlined here addresses many of the fundamental questions in HEP today – the mechanism for electroweak symmetry breaking (EWSB), the origin of mass, CP violation, and what may lie beyond the minimal Standard Model (SM).

The *tev_2000* study group consisted of experimenters from CDF and DØ, accelerator physicists, and theorists. The group produced an extensive report [1], further details on the material presented here can be found in Ref. [1] and references therein. The results presented in Ref. [1] have considerable credibility, being based primarily on Run I data and experience. Existing detector simulations of the Run I CDF and DØ detectors were used along with algorithms from published analyses. The effects of high luminosity conditions were studied using Run I data. Finally, the improvements from the upgrades of CDF and DØ for Run II were, conservatively, not included. It is expected that the experiments' physics reach will easily exceed what is presented here.

Two factors are crucial to *tev_2000*: the upgrade of the Tevatron and the upgrades of the detectors. Improvements in magnet cooling will allow the Tevatron to increase $\sqrt{s}$ from $1.8 \rightarrow 2.0$ TeV. While the 10% increase appears modest, for high mass objects, like top quarks, the production cross section increases by 30%. The future Tevatron runs are currently envisaged to take place in two stages, designated Run II and Run III. Run II will take place during 1999–2002, have a peak luminosity of $2 \times 10^{32}\text{cm}^{-2}\text{s}^{-1}$, and $\int \mathcal{L}dt \approx 4\,\text{fb}^{-1}$. After a shutdown in 2003, Run III will follow during 2004–2007. While the accelerator will be capable of a peak luminosity of $1 \times 10^{33}\text{cm}^{-2}\text{s}^{-1}$, the Tevatron will run in "luminosity leveled" mode where the average number of interactions per crossing will be maintained at ~ 5. That translates into running at a constant $\mathcal{L} = 5 \times 10^{32}\text{cm}^{-2}\text{s}^{-1}$. An integrated luminosity of 20 fb^{-1} is assumed for Run III. Upgrades of the CDF and DØ detectors were described in other talks at this conference [2, 3].

2 Physics Reach

2.1 Top Quark Physics

The Tevatron is the world's only "top factory" and is thus the only place to study this most interesting elementary particle. It is possible that top could be a window onto new physics. The high mass of the top quark means that the coupling of top to exotica is potentially large. $M_{top} \approx$ scale of EWSB, is there a connection? In addition, the large M_{top} implies top decays before hadronization, allowing, for the first time, the study of a bare quark. The dozens of top events collected in Run I were sufficient to claim discovery and make first measurements of M_{top} and the production cross section $\sigma_{t\bar{t}}$. With dramatically larger data sets in Run II and beyond, see Table 1, precise determinations of many top quark properties will be made.

Mode	Run II 4 fb^{-1}	Run III 20 fb^{-1}
Dilepton: $ee, \mu\mu, e\mu$	330	1640
$W + 4j$ single tag	2070	10340
$W + 4j$ double tag	1030	5200

Table 1. Top event yields per experiment. Offline selection efficiencies have been included.

With the data from Run II only, the following is a sample of what to expect on top quark properties:

- The statistical uncertainty on M_{top} will be ≤ 1 GeV and the largest systematics are determined by the data itself, thus scaling like $1/\sqrt{N}$. It is expected that M_{top} will be measured to ≈ 2 GeV in the lepton + jets mode.
- $\sigma_{t\bar{t}}$ will be determined to 7%, providing a non-trivial test of QCD predictions. The precision will also limit possible anomalous top production. For example, a $t\bar{t}$ resonance from a topcolor Z' can be ruled out for $M_{Z'} \leq 1$ TeV.

- Detailed study of top decays: limits on $|V_{tb}|$, $t \to H^+ b$, rare decays, and FCNC.
- Single top production provides the only practical method for measuring the top width. The Tevatron, being a $q\bar{q}$ machine, has a significant S/N advantage for single top over the LHC.

2.2 Physics with IVB's

The Tevatron is a copious source of events with W, Z, and γ. For example, in the electron final state alone, each experiment will detect, including selection efficiencies, $\approx 3 \times 10^6 W \to e\nu$ and $\approx 3 \times 10^5 Z \to ee$ in Run II. With the data from Run II only, the following measurements of the properties of IVB's and their mutual coupling will be made:

- The statistical uncertainty on M_W will be $\leq$ 20 MeV and the largest systematics are determined by the data itself, thus scaling like $1/\sqrt{N}$. It is expected that M_W will be measured to $\approx$ 40 MeV, comparable to LEP II expectations. Unless a new systematic "brick wall" exists, $\delta M_W \approx 20$ MeV is possible after Run III.
- Γ_W is sensitive to non-standard coupling and will be determined to $\approx$ 15 MeV.
- Asymmetries: measurement of the W charge asymmetry constrains the viable set of pdf's, crucial in reducing the error on M_W.
- There will be a host of limits on anomalous couplings WWV and $Z\gamma V$ where $V = \gamma, Z$. Vertices will be determined with a precision of 10% for WWV and $10^{-2} - 10^{-3}$ for $Z\gamma V$. The measurements of WWV are comparable and complimentary to those made at LEP II while the Tevatron does a superior job on $Z\gamma V$.
- The search for the radiation zero in $q\bar{q} \to W\gamma$ production, this is best done at a $q\bar{q}$ machine like the Tevatron.

2.3 Light Higgs

In the SM, M_W, M_{top} and the Higgs mass are related so that knowledge of M_W and M_{top} provides an indirect measurement of M_{Higgs}. This relationship is shown in Fig. 1 along with the current measurements. After Run II, the combined CDF and DØ results will yield $\delta M_{Higgs} \approx 40\% M_{Higgs}$.

In addition to indirect measurements of M_{Higgs}, the Tevatron experiments have sensitivity for discovery of a Higgs with mass 60–125 GeV. This range exceeds the reach of LEP II and covers a difficult window in M_{Higgs} for the LHC. It is also intriguing that the current central value for M_{Higgs}, predicted by precision electroweak measurements, lies in this range [4]. The best channel for this search is $q\bar{q} \to WH$ where $H \to b\bar{b}$ (the dominant branching fraction in this mass range). The analysis requires both b jets are tagged and looks for a bump in the two jet mass spectrum, see Fig. 2. With 20 fb^{-1}, a $\geq 5\sigma$ excess

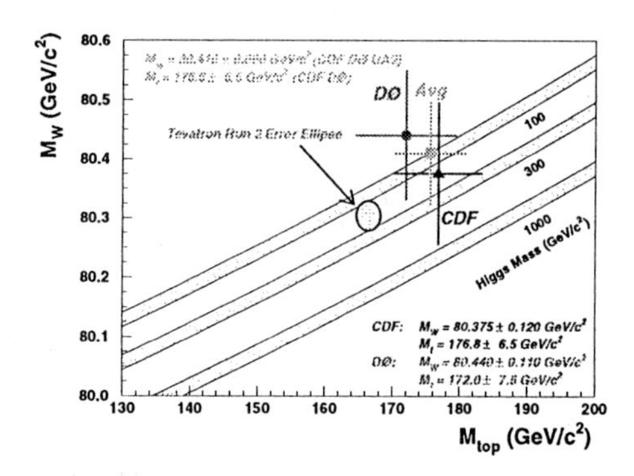

Fig. 1. M_W vs. M_{top} where the bands correspond to different Higgs masses. The CDF and DØ measurements are shown along with the expected error ellipse after Run II.

can be observed for $M_{Higgs} \leq 125$ GeV. Other modes, such as $q\bar{q} \to ZH$ and $H \to \tau\tau$ decays, can be used to confirm the signal.

2.4 SUSY Searches

SUSY is arguably the leading candidate for physics beyond the SM. At the Tevatron, the most promising SUSY signals include the search for charginos in a trilepton final state and gluinos in the missing $E_T + jets$ final state [5]. With the data from Run II alone, the mass reach will be up to $M_{\tilde{\chi}^\pm} \sim 220$ GeV and up to $M_{gluino} \sim 400$ GeV. With 20-30 fb^{-1} in hand after Run III, CDF and DØ have an excellent chance for discovery of SUSY. If SUSY is not observed then Tevatron measurements will significantly constrain the available parameter space. On the other hand, if no sparticle is found, the Tevatron results themselves can not exclude the existence of SUSY.

2.5 Searches for Exotica

Many models for new physics beyond the SM exist and are testable at the Tevatron. A list of new phenomena to search for includes W' and Z', leptoquarks, technicolor, contact interactions, and excited quarks. The benefits of Runs II& III, higher Tevatron energy and large integrated luminosities, are mitigated somewhat by the generally steeply falling production spectra of new phenomena with mass/scale. The Run II mass reach will be extended by a factor of ≈ 1.5 over that achieved in current Run I analyses. Nevertheless, until a higher energy machine is operational, the Tevatron will be the best place to search for exotica.

3 Conclusions

The high energy frontier will remain at the Tevatron for roughly the next decade. The *tev_2000* group has outlined the rich program of physics avail-

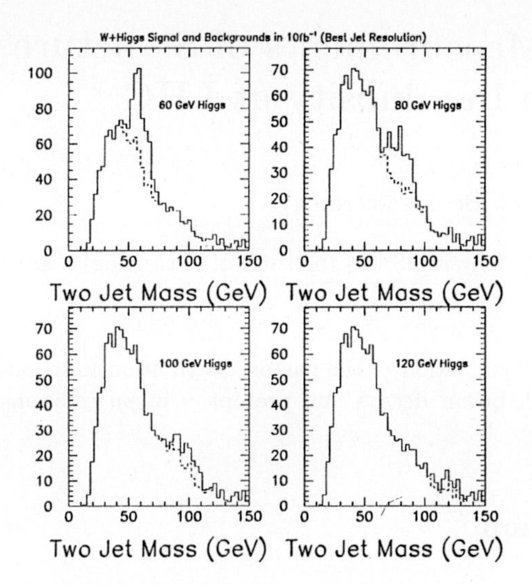

Fig. 2. The expected Run II two jet mass distribution for the $q\bar{q} \to WH$ process using 4 assumed Higgs masses. The solid line is signal+background, the dashed line is the sum of all backgrounds [1].

able in Runs II & III, of which only a small sample was presented here. The opportunities in *tev_2000* will be exploited fully by the upgraded CDF and DØ detectors. Many precision measurements will provide stringent tests of the SM and possibly reveal chinks in the SM armor. There exist very good chances to either discover new physics or, in its absence, to severely constrain extensions to the SM.

References

[1] D. Amidei and R. Brock, eds., Fermilab-Pub-96/082.
[2] D. Bisello, these proceedings.
[3] M. Strauss, these proceedings.
[4] D. Ward, these proceedings.
[5] S. Mrenna *et al.*, Phys. Rev. D53, 1168 (1996).

Precision Measurements of Structure Function and Parton Luminosity at LHC

Frank Behner (Frank.Behner@cern.ch)

Institute for Particle Physics, Swiss Institute of Technology, Zurich

Abstract. The determination of the parton distribution functions and luminosity using leptonic weak boson decays and prompt γ in pp collisions at the LHC is discussed.

1 Introduction

So far it is assumed that the proton luminosity can be measured to a $\approx 5\%$ accuracy [1] [2]. Thus a large uncertainty in the cross sections measured is introduced. In the general used ansatz to factorise the parton distribution functions and the cross sections on the parton level one can write the expected rate for two given processes $pp \to X$ and $pp \to Y$ as:

$$N_{pp\to X} = L_p \; f(x_1, Q^2) \; f(x_2, Q^2) \; \sigma_{p_1 p_2 \to X} \; , \tag{1}$$

$$N_{pp\to Y} = L_p \; f(x_1', Q'^2) \; f(x_2', Q'^2) \; \sigma_{p_1 p_2 \to Y} \; , \tag{2}$$

where L_p is the proton luminosity, f the parton distribution function for parton 1 and 2 respectively and σ the cross section on the parton level for the two reactions in question. Thus taking one as a precise measured reference reaction, where rate and parton distribution are determined, the proton luminosity uncertainty cancels. One obtains a parton luminosity for the process to be measured.

In this article single W, Z production with leptonic decays and γ-jet final states are discussed as such a reference processes. They are well reconstructible, since they contain high transverse momentum, p_t , isolated leptons or photons. Also this events have a well defined kinematic, since these are two body decay processes. The initial parton momentum fractions can be determined using two observables,

$$M^2 = x_1 x_2 s \; , \tag{3}$$

$$y = \frac{1}{2} \log \frac{x_1}{x_2} \; , \tag{4}$$

the mass M of the system and the rapidity y, assuming negligible masses of the partons. Thus, the event rates provide the overall normalisation M and y determine the shape of the relevant parton distribution function.

2 Quark and anti-quark structure function

Single weak boson production and their leptonic decays can act as reference to measure the quark, anti-quark structure function. Here the mass of the system or Q^2 is given by the mass of the weak boson. With the rapidity distribution of the W^+ a measurement of the $f_u \cdot f_{\bar{d}}$ distribution is obtained. With those of the W^- and the Z a measurement of $f_d \cdot f_{\bar{u}}$ and the sum $f_u \cdot f_{\bar{u}} + f_d \cdot f_{\bar{d}}$ is made. The rapidity distribution of the weak W boson can be determined using Monte Carlo by measuring the pseudorapidity distribution of the charged lepton from the W decay, as shown with the correlation of the rapidity of the W and the pseudorapidity of the lepton in figure 1. The kinematic of the Z is fully reconstructible, since both decay leptons are seen. Figure 1a) shows also

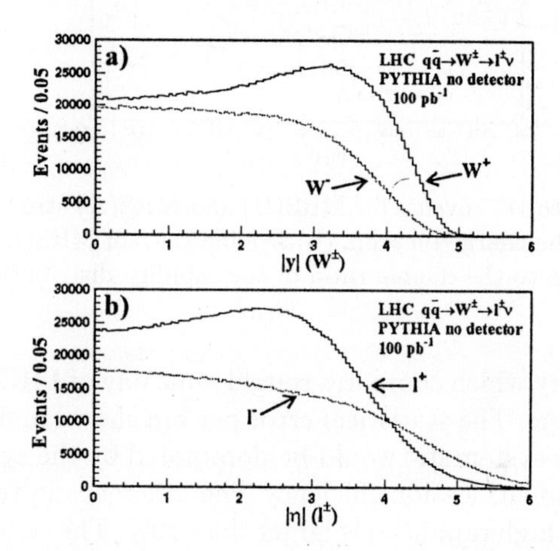

Fig. 1. Rapidity (y) of the W boson (a) compared with the pseudorapidity (η) of the decay charged lepton (b)

an enhancement in the W^+ distribution for large rapidities. This is explained by the harder x in the valence-sea quark scattering of the two u quarks in the proton. In contrast to proton anti-proton collision, in proton proton the anti-quark has to come from the sea. Consequently, at low rapidities sea-sea quark scattering can be measured. We have studied this method using a PYTHIA simulation of 14 TeV LHC events, where W are selected using the following criteria. A charged lepton with transverse momentum $p_t > 30$ GeV is accepted with a pseudorapidity $|\eta| < 2.4$. The lepton has to be isolated, which means the sum of the p_t of all particles with a $p_t > 0.5$ GeV and $|\eta| < 3$ in cone of $R = \sqrt{\Delta\phi^2 + \Delta\eta^2} < 0.6$ is smaller then 5 GeV. Furthermore, there is no jet in the event with a $p_t > 20$ GeV. For the Z selection the mass of the dilepton system must be in a range of 2 GeV around the Z mass. Using these selections different structure functions are compared in detail [3]. Figure 2 shows for example differences of very similar structure functions MRS(H) and MRS(A). The fluctuations in this figure correspond to about 100 pb^{-1} of

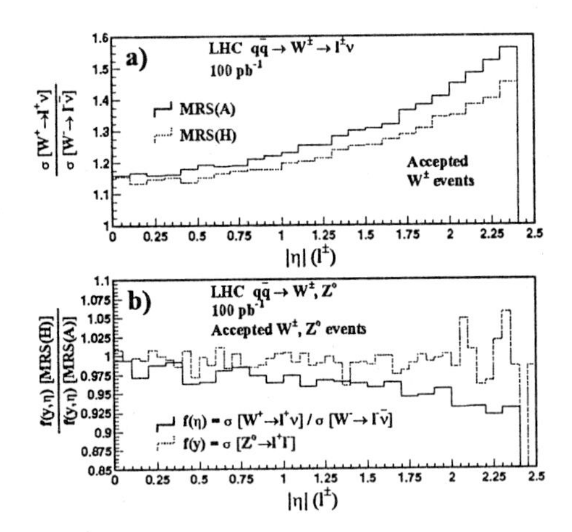

Fig. 2. Ratio of W^+ to W^- events for MRS(H) and MRS(A) structure function (a) for different $|eta|$ of the charged lepton. (b) Double ratio of MRS(H) to MRS(A) for W^+ to W^- compared to the double ratio of the rapidity distributions of Z events.

integrated luminosity which compares roughly one day of LHC running in the low luminosity regime. The statistical error per bin shown in figure 2 is of the order of 1% and the systematic would be dominated by the systematic of the lepton and photon identification efficiency The difference in the distributions shown in figure 2 at high rapidities is larger than 10%. The structure functions chosen here differ only in isospin breaking, resulting in a difference between the $\bar{u}$ and $\bar{d}$ component of the sea in the. Therefore in figure 2b) the double ratio for the Z decays, which is sensitive to the sum $f_u \cdot f_{\bar{u}} + f_d \cdot f_{\bar{d}}$, is flat, while the distribution falls in the double ratio sensitive to $(f_u \cdot f_{\bar{d}})/(f_d \cdot f_{\bar{u}})$.

3 Gluon structure function

A similar approach can be chosen to study the gluon structure functions. The processes used here are $qg \to \gamma q$ and $qg \to Zq$ with leptonic Z decays, where the first process is expected to give high rates while the second one is cleaner. First estimates of the background rate of $qg \to \gamma q$ yield to 30% contamination. The γ-jet events were selected requesting a photon with high transverse momentum $p_t > 40\ GeV$ and in the isolation cone the sum of the p_t of all other particles is less than 7% of the photon p_t. Jet and photon have to be back to back, which means $\Delta\phi > 174°$. The jet has to have a minimal p_t of 30 GeV and only one jet in the event is allowed. For the Z-jet events the criteria are similar. Since in this channel the Q^2 has to determined in contrast to the resonant production processes, the rapidity distribution can be studied in different p_t bins of the photon or Z to obtain also the Q^2 evolution. In our study we have compared structure functions GRV and MRS(G) with a almost identical quark sector. This could be measure with the single weak boson production as shown. Figure 3a) shows that differences from more than 10% for low rapidities can be observed, while also in 3b) the identical behaviour for the quark sector is observed.

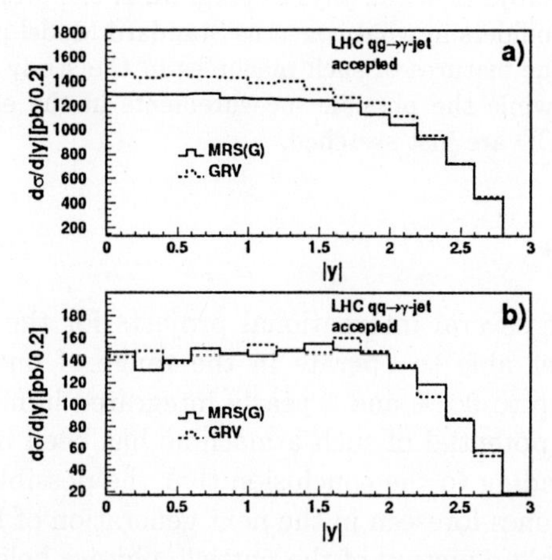

Fig. 3. Rapidity distribution of the photon jet system for $qg \rightarrow \gamma q$ (a) and the background process $qq \rightarrow \gamma g$ (b).

4 Conclusion

A new ansatz to measure the luminosity at the LHC has been introduced. It is based on simple kinematics using leptonic weak boson decays and prompt photons. The statistical errors can be neglected and the systematic of this method based on lepton and photon identification is expected to be low. The method should constrain the parton distribution function on the % level, especially in the range $Q^2 \approx 10^4 \ GeV^2$, important for H^0 production and its relevant backgrounds.

References

[1] Baytatian et al., CMS Technical Proposal, CERN/LHCC 94-38
[2] Armstrong et al.,ATLAS Technical Proposal, CERN/LHCC 94-43
[3] Dittmar et al.,Phys.Rev. D56 (1997) 7284-7290.

Standard Model Physics at Future Electron–Positron Linear Colliders

Manel Martinez (martinez@ifae.es)

IFAE Barcelona, Spain

Abstract. The main subjects of the physics program of the proposed next generation of e^+e^- Linear Colliders in what concerns Standard Model physics are briefly summarized. The unique features of such machines in the study of the top quark physics are discussed while the possible measurements in the electroweak gauge boson sector and in QCD are just sketched.

1 Introduction

At present, there are several international projects for the construction of a linear e^+e^- collider able to operate in the range of one TeV, reaching beam polarizations up to 90 % and a yearly integrated luminosity of 10-100 fb^{-1}[1]. The physics potential of such a machine has been investigated over the last few years, leading to the conclusion that the possible measurements will complement the ones foreseen in the next generation of Hadron colliders helping to get a complete picture of the particle physics below one TeV [2].

One important subject for such a machine is the study of Standard Model physics, since the present e^+e^- colliders have proven to be the most powerful machines for the precision tests of the Standard Model. In this article we give a brief summary of which measurements can be expected from such a machine in what concerns Standard Model physics, concentrating our scope in three different subjects: top quark physics, gauge boson physics and QCD.

2 Top quark physics.

One of the most important topics of the physics programme of a linear e^+e^- collider with centre-of-mass energies in the range 350-1000 GeV is the detailed study of the top quark properties. The dominant top production channel goes through the e^+e^- annihilation to a virtual photon or Z, which subsequently will decay to a $t\bar{t}$ pair. The $t\bar{t}$ production cross section is about 650 fb at $\sqrt{s}$ = 500 GeV. At the foreseen luminosities of $10^{33} - 10^{34}\ \text{cm}^{-2}\ \text{s}^{-1}$ (or 10–100 fb^{-1} per year) the event sample is sufficient for detailed studies.

In the Minimal Standard Model (MSM) the top quark decays essentially 100% of the times into a bottom quark and a W boson. Depending on the decays of the two W's present in the event, the $t\bar{t}$ event will consist of six jets (45.5%); four jets, a high-momentum lepton and a neutrino (43.9%); or two jets, two high-momentum leptons and two neutrinos (10.6%). In all cases there will be two b jets. It has been shown [3] that no demanding detector performances are required to efficiently study these events.

The fact that the total top quark width is about $\Gamma_t \sim 1.4$ GeV $>> \Lambda_{QCD}$, acts as an effective energy cut-off for soft non-perturbative and infrared perturbative QCD effects. This has two kind of consequences[2]:

- On the one hand, it implies that tops decay before the toponia bound states have had time to form and even before top-flavored hadrons can be effectively produced. Therefore, unlike for the other heavy flavors, no top-spectroscopy is expected.
- On the other hand, the impact of non-perturbative effects on production and decay can be neglected to a high level of accuracy, so that the top quark sector can then be safely analyzed within perturbative QCD.

This second feature, implies that the top resonance behaviour should be reliably described in perturbative QCD, allowing hence precision studies, and also that, at the continuum, the top initial helicity should be conserved and transmitted to final state without depolarization (behaving in this sense somehow like tau leptons).

The study of the top threshold resonance is one of the physics subjects which can only be attacked in an e^+e^- collider. An energy scan around the $t\bar{t}$ threshold can be used to do a precise measurement of the top quark mass. Three observables, for which precise theoretical predictions exist [4], have been studied: the total production cross section, $\sigma_{t\bar{t}}(s)$, which is particularly sensitive to m_t and also quite sensitive to the strength of the $t\bar{t}$ binding potential and therefore to α_s, the top momentum distribution, $d\sigma/dp(s)$, which gives mainly information on the top mass and finally the forward-backward asymmetry, $A_{FB}(s)$, coming to the overlap of S- and P-wave amplitudes due to the large top width.

The shape of the threshold is severely distorted by initial state radiation effects (ISR) as well as by beam energy smearing due to the accelerator (BS) as can be seen in figure 1. Both effects decrease substantially the sensitivity of the cross section measurement to the top mass. In order to analyze the cross section in the resonance region, the measurement of the luminosity, the machine energy and the luminosity spectrum plays an important role in this study and is discussed in detail in ref. [5].

The aforementioned observables have been used in a χ^2 fit assuming a nine-point scan around $\sqrt{s} = 350$ GeV with an integrated luminosity of 5 fb^{-1} in each point. Another 5 fb^{-1} was assumed to be taken below the threshold in order to measure the background. From that simulation [6] the overall precision expected in the top quark mass is better than 110 MeV and in the strong coupling constant at M_Z^2 obtained from the binding potential is of about 0.003. In the continuum, the sample of tops produced can be used to measure the production and decay top formfactors. The tops are produced

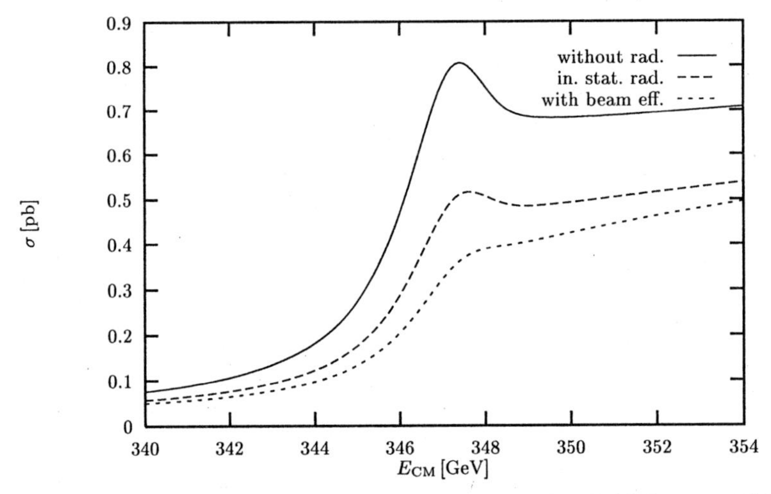

Fig. 1. Total cross section for top production near threshold for $m_t = 175$ GeV/c^2. The theoretical cross-section is given by the solid curve to which radiative effects have been successively added: initial state radiation (dashed curve) and beam effects (dotted curve).

with high longitudinal polarization and due to the undistorted transmission of the helicity to their decay products, the top helicities can be determined from the distributions of jets and leptons in $t \to bW+ \to bf\bar{f}'$. The top is the only quark in which this can be done. The concept is similar to the one so successfully used at LEP-SLC for detailed studies of the tau couplings. Detailed simulations [7] show that this approach allows the precise determination of the top production anomalous moments as well as the determination of its decay vertex structure.

3 Electroweak gauge bosons and QCD.

The properties of the electroweak gauge bosons can be determined with high accuracy at a high energy e^+e^- collider following basically identical techniques to the ones presently applied at LEP and SLC.

In what concerns the W boson couplings, the main difference with respect to LEP-2 is that, since the deviations from the unitarity cancellation grow up as $(\beta\gamma)^2$, the sensitivity to the anomalous couplings increases drastically with energy, a part from having the possibility of disentangling the different contributions by using the beam polarization. If the goal at LEP-2 is reaching accuracies of $O(10^{-1})$, a linear collider at 500 GeV should provide accuracies of $O(10^{-3})$ and up to $O(10^{-4})$ if it would reach an energy of 1.5 TeV [2, 3]. A determination of the W mass at $E_{cm} = 500$ GeV using the $WW \to l\nu q\bar{q}'$ channel, which is not affected by theoretical uncertainties such as the ones related to colour-reconnection effects and Bose-Einstein correlations among the Ws, would lead to an uncertainty of $\Delta M_W \sim 15$ MeV [2, 3].

If the linear collider would be operated back to the present LEP energies, as the machine proposals claim [3], it could benefit from having simultaneously a very high luminosity and high beam polarization for repeating some of the statistically dominated LEP-SLC studies. If, for instance, such a machine would be run back to the WW threshold with a luminosity of 50 pb^{-1} per year, it would produce about 100 times the collected statistics of LEP-2 at threshold, leading to a determination of the W mass, independent from the previously mentioned, with an uncertainty of about 15 MeV (assuming $\Delta E_{cm} \sim 10$ MeV) [2]. In the same way, if such a machine would be operated back to the Z peak energy, it would be able to produce about 10^7 highly polarized Zs per month, allowing an extremely accurate measurement of the left-right asymmetry A_{LR}. Some educated extrapolation of the known systematics affecting this measurement indicate that a precision in the effective weak mixing angle of $\Delta \sin^2 \theta^{lept}_{eff} < 0.0001$ could be obtained [2].

Similarly, the possibility of running the machine in a broad range of energies with the same detector, has been suggested as a powerful tool for studying the energy dependence of some relevant observables, avoiding the systematic uncertainties coming from combining different experiments at different energies. One of the subjects that could benefit more from this possibility would be the precision test of predictions from QCD like the running of the strong coupling constant or the existence of asymptotic freedom. Detailed studies using the fraction of events with $2, 3, 4, \ldots$ jets show that unprecedented precision could be obtained for both tests [2].

References

[1] P. Mattig, contribution to this proceedings.

[2] "Physics with e^+e^- linear colliders", DESY 97-100, P. M. Zerwas et al., and references therein.

[3] "Conceptual Design of a 500 GeV e^+e^- linear collider with an Integrated X-ray Laser Facility", DESY 48-97, vol I, R.Brinkmann et al.

[4] K. Fujii, T. Matsui and Y. Sumino, *Phys. Rev.* **D50** (1994) 4341; R. Harlander, M. Jeżabek, J. H. Kühn and T. Teubner, *Phys. Lett.* **B346** (1995) 137, and references therein.

[5] D. Miller and Y. Kurihara in DESY 123E-97.

[6] A. Juste, M. Martinez and D. Schulte in DESY 123E-97.

[7] M. Schmitt in DESY 123D-96.

Higgs Boson Production and Decay at Future Machines

Michael Spira (Michael.Spira@cern.ch)

CERN, Theory Division, CH-1211 Geneva 23, Switzerland

Abstract. Higgs boson production at future colliders within the Standard Model and its minimal supersymmetric extension is reviewed. The predictions for decay rates and production cross sections are presented including all relevant higher-order corrections.

1 Introduction

Standard Model [SM]. The SM contains one isospin doublet of Higgs fields, which leads to the existence of one elementary neutral CP-even Higgs boson H [1]. Only its mass is unknown. The direct search for the Higgs boson at the LEP experiments excludes Higgs masses below ~ 77 GeV [2]. Unitarity of scattering amplitudes requires the introduction of a cut-off Λ [3], which imposes an upper bound on the Higgs mass. For the minimal value $\Lambda \sim 1$ TeV the upper bound on the Higgs mass is $M_H \lesssim 700$ GeV. For $\Lambda \sim M_{GUT} \sim 10^{15}$ GeV, the Higgs mass has to be smaller than ~ 200 GeV.

Minimal Supersymmetric Extension [MSSM]. Supersymmetry provides a solution to the hierarchy problem of the SM, which arises for small Higgs masses. The MSSM requires the introduction of two Higgs doublets, which leads to five elementary Higgs particles: two neutral CP-even (h, H), one neutral CP-odd (A) and two charged $(H^\pm)$ Higgs bosons. In the Higgs sector only two parameters have to be introduced, which are usually chosen as $\text{tg}\beta = v_2/v_1$, the ratio of the two vacuum expectation values of the CP-even Higgs fields, and the pseudoscalar Higgs mass M_A. Radiative corrections to the MSSM Higgs sector are large, since the leading part grows as the fourth power of the top quark mass. They increase the upper limit on the light scalar Higgs mass to $M_h \lesssim 130$ GeV [4].

2 Standard Model

2.1 Decay Modes

$H \to f\bar{f}$. For $M_H \lesssim 140$ GeV the branching ratios of $H \to b\bar{b}\,(\tau^+\tau^-)$ reach values of $\sim 90\%$ ($\sim 10\%$). Above the $t\bar{t}$ threshold the branching ratio of $H \to t\bar{t}$ amounts to $\lesssim 20\%$. QCD corrections to the $b\bar{b}, c\bar{c}$ decays are large, owing to large logarithmic contributions, which can be absorbed in the running quark mass $\overline{m}_Q(M_H)$. Far above the $Q\bar{Q}$ threshold they are known up to three loops [5, 6]. The NLO corrections have been evaluated including the full quark mass dependence [5]. They are moderate in the threshold region $M_H \gtrsim 2m_Q$.

$H \to W^+W^-, ZZ$. The $H \to WW, ZZ$ decay modes dominate for $M_H \gtrsim 140$ GeV. Electroweak corrections are small in the intermediate Higgs mass range, while they enhance the partial widths by about 20% for $M_H \sim 1$ TeV due to the self-interaction of the Higgs particle [7]. For $M_H < 2M_{W,Z}$ off-shell decays $H \to W^*W^*, Z^*Z^*$ are important. They lead to WW (ZZ) branching ratios of about 1% for $M_H \sim 100$ (110) GeV.

$H \to \gamma\gamma$. The decay $H \to \gamma\gamma$ is mediated by fermion and W-boson loops, which interfere destructively. The W-boson contribution dominates [8]. The photonic branching ratio reaches a level of $\gtrsim 10^{-3}$ for Higgs masses $M_H \lesssim 150$ GeV. This decay mode plays a significant rôle for the Higgs search at the LHC for $M_H \lesssim 140$ GeV. QCD corrections are small in the intermediate mass range [9].

Branching ratios and decay width. The branching ratios of the Higgs boson are presented in Fig. 1. For $M_H \lesssim 140$ GeV, where the $b\bar{b}, \tau^+\tau^-, c\bar{c}$ and gg decay modes are dominant, the total decay width is very small, $\Gamma \lesssim 10$ MeV. Above this mass value the WW, ZZ decay modes become dominant. For $M_H \lesssim 2M_Z$ the decay width amounts to $\Gamma \lesssim 1$ GeV, while it reaches $\sim$ 600 GeV for $M_H \sim 1$ TeV. Thus the intermediate Higgs boson is a narrow resonance.

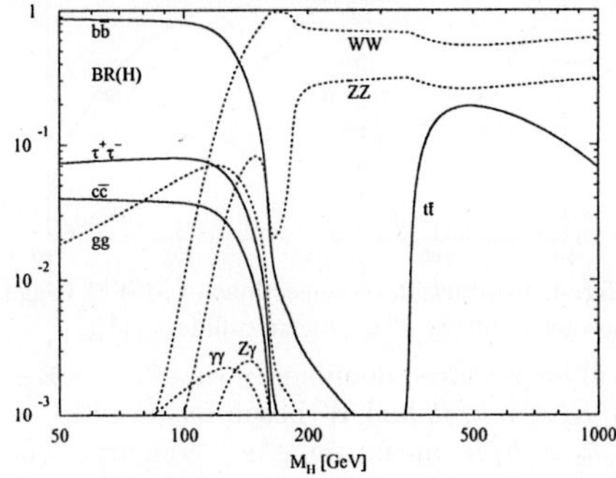

Fig. 1. Branching ratios of the SM Higgs boson as a function of its mass [10].

2.2 Higgs Boson Production.

<u>e^+e^- *colliders.*</u> At lower energies Higgs bosons

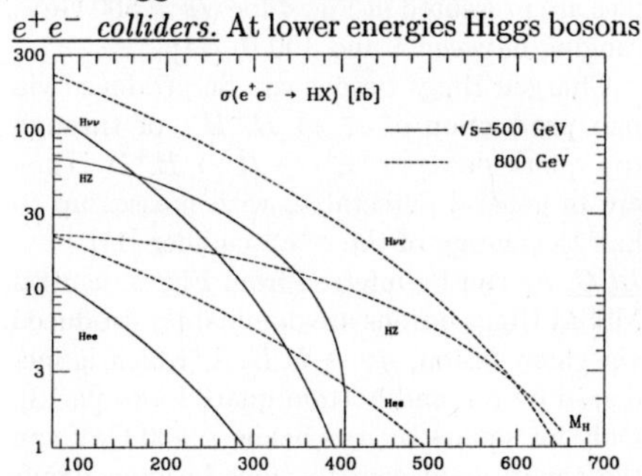

Fig. 2. Production cross sections of SM Higgs bosons at future e^+e^- linear colliders [11].

are dominantly produced via Higgs-strahlung off Z bosons, $e^+e^- \to ZH$, while at high energies the W fusion process $e^+e^- \to \nu_e\bar{\nu}_e H$ dominates [11]. Electroweak corrections to the Higgs-strahlung process are moderate [12]. The production cross sections at future e^+e^- colliders are shown in Fig. 2 for c.m. energies of 500 and 800 GeV. They range between 1 and 300 fb in the relevant mass range and provide clean signatures for the Higgs boson. The angular distribution of Higgs-strahlung is sensitive to the spin and parity of the Higgs particle [11].

<u>*LHC.*</u> Higgs boson production at the LHC

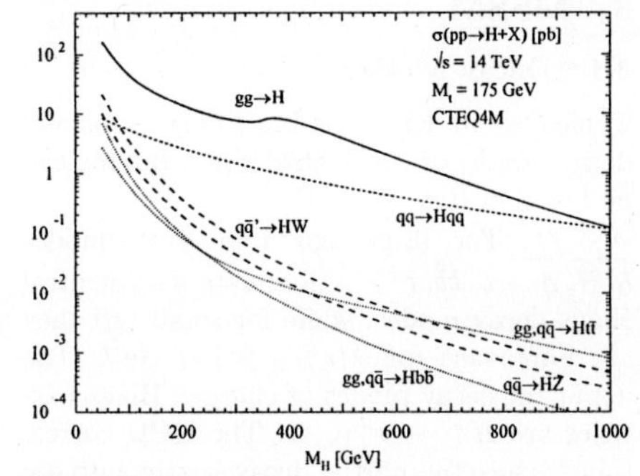

Fig. 3. Higgs production cross sections at the LHC for the various mechanisms as a function of the Higgs mass [10].

is dominated by the gluon fusion mechanism $gg \to H$, which is mediated by top and bottom triangle loops [9, 13]. This can be inferred from Fig. 3, which presents all relevant Higgs production cross sections as a function of the Higgs mass. The two-loop QCD corrections increase the production cross section by 60–90%, so that they can no longer be neglected [9]. [The QCD corrections to most of the background processes at the LHC are also known.] In spite of the large size of the QCD corrections the scale dependence is reduced significantly, thus rendering the NLO result reliable.

For large Higgs boson masses the vector boson fusion mechanism $WW, ZZ \to H$ becomes competitive, while for intermediate Higgs masses it is about an order of magnitude smaller than gluon fusion [14]. The QCD corrections are small and thus negligible [15].

Higgs-strahlung $W^*/Z^* \to HW^*/Z^*$ plays a rôle only for $M_H \lesssim 100$ GeV. The QCD corrections are moderate, so that this process is calculated with reliable accuracy [16].

Higgs bremsstrahlung off top quarks, $gg, q\bar{q} \to Ht\bar{t}$, is sizeable for $M_H \lesssim 100$ GeV [17]. The QCD corrections to this process are unknown, so that the cross section is uncertain within a factor of ~ 2.

3 MSSM

3.1 Decay Modes

Typical examples of the branching ratios and decay widths of the MSSM Higgs bosons can be found in Refs. [10, 11].

$\phi \to f\bar{f}$. For large tgβ the decay modes $h, H, A \to b\bar{b}, \tau^+\tau^-$ dominate the neutral Higgs decay modes, while for small tgβ they are important for $M_{h,H,A} \lesssim 150$ GeV. The dominant decay modes of charged Higgs particles are $H^+ \to \tau^+\nu_\tau, t\bar{b}$. The QCD corrections reduce the partial decay widths into b, c quarks by 50%–75% as a result of the running quark masses, while they are moderate for decays into top quarks [5, 6].

Decays into Higgs and gauge bosons. The decay modes $H \to hh, AA, ZA$ and $A \to Zh$ are important for small tgβ below the $t\bar{t}$ threshold. Similarly the decays $H^+ \to WA, Wh$ are sizeable for small tgβ and $M_{H^+} < m_t + m_b$. The dominant higher-order corrections can be absorbed into the couplings and masses of the Higgs sector. Below the corresponding thresholds decays into off-shell Higgs and gauge bosons turn out to be important especially for small tgβ [18]. The decays $h, H \to WW, ZZ$ are suppressed by kinematics and, in general, by the SUSY couplings and are thus less important in the MSSM. The decay $h \to \gamma\gamma$ is only relevant for the LHC in the decoupling limit, where the light scalar Higgs boson h has similar properties to those of the SM Higgs particle.

Decays into SUSY particles. Higgs decays into charginos, neutralinos and third-generation sfermions can become important, once they are kinematically allowed [19]. Thus they could complicate the Higgs search at the LHC, since the decay into the LSP will be invisible.

3.2 Higgs Boson Production

e^+e^- colliders. Neutral MSSM Higgs bosons

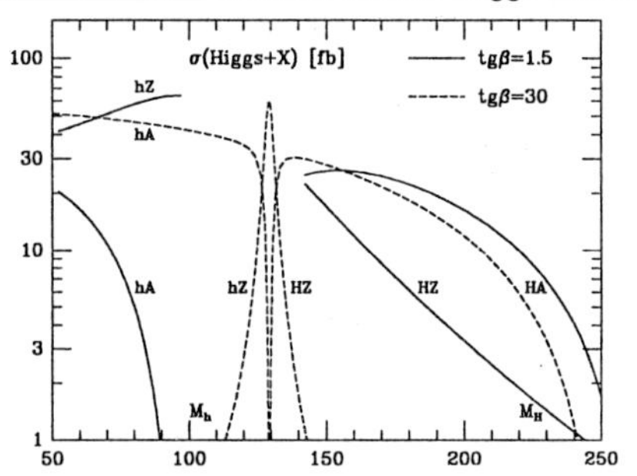

Fig. 4. Production cross sections of MSSM Higgs bosons at future e^+e^- linear colliders [11].

will be produced dominantly via $e^+e^- \to Z + h/H, A + h/H$ and W boson fusion $e^+e^- \to \nu_e\bar{\nu}_e + h/H$ at future e^+e^- colliders. The size of the individual cross sections depends strongly on tgβ, but the sum is always of the order of the SM cross section. Typical examples are presented in Fig. 4 for $\sqrt{s} = 500$ GeV, ranging between 1 and 100 fb [11].

Charged Higgs bosons can be produced via pair production $e^+e^- \to H^+H^-$ or through top quark decays $e^+e^- \to t\bar{t} \to H^+b\bar{t}$. They are in general detectable, with masses up to half the energy of the e^+e^- collider [11].

LHC. As can be inferred from Fig. 5, neutral MSSM Higgs bosons are dominantly produced via gluon fusion, $gg \to h, H, A$, which is mediated by top and bottom quark loops [9, 13]. Only for squark masses below ~ 400 GeV can the squark loop contributions become significant [20]. The two-loop QCD corrections increase the production cross sections by 10%–100% and thus cannot be neglected [9].

For large tgβ Higgs bremsstrahlung off b quarks, $gg, q\bar{q} \to b\bar{b} + h/H/A$, dominates in a large part of the relevant Higgs mass ranges [17, 21]. The QCD corrections are unknown.

Vector boson fusion $WW/ZZ \to h/H$ and Higgs-strahlung, $W^*/Z^* \to W/Z + h/H$, are suppressed in most of the parameter space by

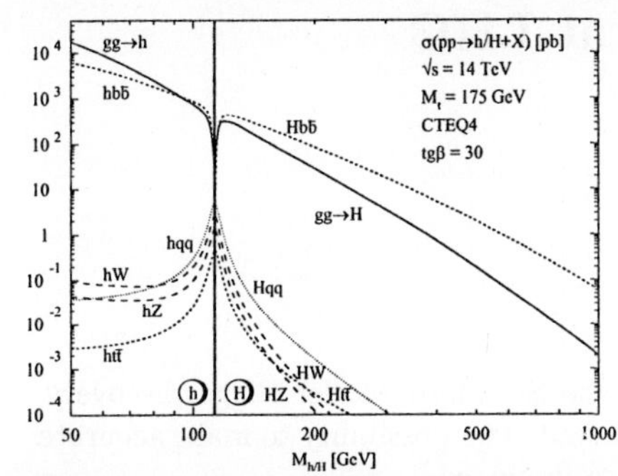

Fig. 5. Scalar Higgs production cross sections at the LHC for the various mechanisms as a function of the Higgs mass for $tg\beta = 30$ [10].

SUSY couplings compared with the SM and thus less important.

References

[1] P.Higgs, Phys. Rev. Lett. **12** (1964) 132 and Phys. Rev. **145** (1966) 1156; F.Englert and R.Brout, Phys. Rev. Lett. **13** (1964) 321; G.Guralnik et al., Phys. Rev. Lett. **13** (1964) 585.

[2] P. Janot, these proceedings.

[3] N.Cabibbo et al., Nucl. Phys. **B158** (1979) 295; M.Lindner, Z. Phys. **C31** (1986) 295; A.Hasenfratz et al., Phys. Lett. **B199** (1987) 531; J.Kuti et al., Phys. Rev. Lett. **61** (1988) 678; M.Lüscher and P.Weisz, Nucl. Phys. **B318** (1989) 705. M.Sher, Phys. Rep. **179** (1989) 273; G.Altarelli and G.Isidori, Phys. Lett. **B337** (1994) 141; J.Casas et al., Phys. Lett. **B342** (1995) 171;

[4] Y.Okada et al., Prog. Theor. Phys. **85** (1991) 1; H.Haber and R.Hempfling, Phys. Rev. Lett. **66** (1991) 1815; J.Ellis et al., Phys. Lett. **B257** (1991) 83; M. Carena et al., Phys. Lett. **B355** (1995) 209; H.Haber et al., Z. Phys. **C75** (1997) 539.

[5] E.Braaten and J.Leveille, Phys. Rev. **D22** (1980) 715; M.Drees and K.Hikasa, Phys. Lett. **B240** (1990) 455 and (E) **B262** (1991) 497.

[6] S.Gorishny et al., Mod. Phys. Lett. **A5** (1990) 2703; A.Kataev and V.Kim, Mod. Phys. Lett. **A9** (1994) 1309; K. Chetyrkin, Phys. Lett. **B390** (1997) 309.

[7] J.Fleischer and F.Jegerlehner, Phys. Rev. **D23** (1981) 2001; B.Kniehl, Nucl. Phys. **B352** (1991) 1 and **B357** (1991) 357; D. Bardin et al., JINR-P2-91-140; A. Ghinculov, Nucl. Phys. **B455** (1995) 21; A. Frink et al., Phys. Rev. **D54** (1996) 4548.

[8] J.Ellis et al., Nucl. Phys. **B106** (1976) 292; A.Vainshtein et al., Sov. J. Nucl. Phys. **30** (1979) 711.

[9] M.Spira et al., Nucl. Phys. **B453** (1995) 17 and references therein.

[10] M.Spira, CERN-TH/97-68 (to appear in Fort. Phys.).

[11] E.Accomando et al., hep-ph/9705442.

[12] J.Fleischer and F.Jegerlehner, Nucl. Phys. **B216** (1983) 469; A.Denner et al., Z. Phys. **C56** (1992) 261; B.Kniehl, Z. Phys. **C55** (1992) 605.

[13] H.Georgi et al., Phys. Rev. Lett. **40** (1978) 692.

[14] R.Cahn and S.Dawson, Phys. Lett. **B136** (1984) 196; K.Hikasa, Phys. Lett. **B164** (1985) 341; G.Altarelli et al., Nucl. Phys. **B287** (1987) 205.

[15] T.Han et al., Phys. Rev. Lett. **69** (1992) 3274.

[16] T.Han et al., Phys. Lett. **B273** (1991) 167.

[17] Z.Kunszt, Nucl. Phys. **B247** (1984) 339; J.Gunion, Phys. Lett. **B253** (1991) 269; W.Marciano and F.Paige, Phys. Rev. Lett. **66** (1991) 2433.

[18] A.Djouadi et al., Z. Phys. **C70** (1996) 437; S.Moretti and W.Stirling, Phys. Lett. **B347** (1995) 291, (E) **B366** (1996) 451.

[19] A.Djouadi et al., Z. Phys. **C74** (1997) 93.

[20] S.Dawson et al., Phys. Rev. Lett. **77** (1996) 16.

[21] D.Dicus et al., Phys. Rev. **D39** (1989) 751.

Standard Model Physics at LHC

Luc Poggioli (Luc.Poggioli@cern.ch)

LPNHE, University of Paris 6 & 7, France

Abstract. The potential of the LHC for the Standard Model Higgs discovery, using the ATLAS and CMS detectors, is reviewed. The possibility to make accurate Standard Model physics measurements is also discussed.

1 Standard Model Higgs

The search of the Standard Model Higgs can be performed at LHC from a mass of 80 GeV (below the LEPII limit) up to the TeV region. One can distinguish three mass regions (low, intermediate, and high mass region) where dedicated strategies are worked out to discover the Standard Model Higgs. In each region, the strategy is a trade-off between lepton-photon final state (low signal rate, low QCD background) and hadronic final state (higher signal yield, but large QCD backgrounds).

1.1 Low mass region ($m_H < 120$ GeV)

Direct and associated production $t\bar{t}H$, WH, $H \to \gamma\gamma$ The Higgs mass resolution is here essential to be able to observe a narrow peak above a very large background. The contribution comes from the energy resolution itself - $3\%(5\%)/\sqrt{E}$ for CMS at low (high) luminosity, $10\%/\sqrt{E}$ for ATLAS - the angular (polar) resolution, and the potential degradation of performance due to the presence of upstream material. CMS has developed algorithms based on the tracker information and the good calorimeter granularity to recover 87% of the converted photons, without loss in performance [1], as shown in Fig. 1, yielding an overall efficiency of 74% per photon for a mass resolution of 690 MeV at 100 GeV. ATLAS has a overall photon efficiency of 77%, and a mass resolution of 1.3 GeV for 100 GeV Higgs mass at high luminosity. The tails are at the 5% level [2], see Fig. 2.
For the direct production, the irreducible background is about 50 times the signal. The reducible QCD background, made by jets faking photons, can be reduced to 30% of the continuum by using the calorimeter information [3]. A further reduction of about 3 is obtained in ATLAS by using the fine η-segmented first sampling, and in CMS by using the fine lateral granularity of the calorimeter (and a preshower in the end-cap region) [4].
The associated production yields low signal rate, typically 15 signal events and 15 background events for 100 GeV and one year at high luminosity.

Associated production WH, $H \to b\bar{b}$ This channel gives about 10 times more $\sigma \times BR$ than the $H \to \gamma\gamma$ mode, with $BR(H \to b\bar{b}) \sim 100\%$. The first issue is the Higgs mass reconstruction, which is worse for $H \to b\bar{b}$ ($\sim$ 15 GeV) than for $W \to jj$ ($\sim$ 7 GeV), due to final state radiation and hadronization, [2], as shown in Fig. 3.
An other critical issue is the ability to tag b-jets to reject QCD backgrounds. Using pixel layers at small radii allows to achieve efficiencies around 70% (50%) at low (high) luminosity, for rejection of about 100 versus u-jets [5]. As shown in Fig. 4, the backgrounds, dominated by $Wb\bar{b}$ and Wjj where a jet is misidentified as a b-jet are large, and the signals are broad. This channel should be considered as complementary to the $\gamma\gamma$ mode, and can be complemented by the $t\bar{t}H$ channel.
Figure 5 summarizes the discovery potential of the LHC in the low Higgs mass region, showing that after 3 years at low luminosity a convincing signal can be observed, by combining the $\gamma\gamma$ and the $WH, H \to b\bar{b}$ channels.

1.2 Intermediate mass region (120 GeV $< m_H <$ 2 m_Z)

This region is accessible via the gold-plated mode $H \to ZZ^* \to l^+l^-l^+l^-$. Here again a good mass resolution, both for electrons and muons is essential. For the 4-electrons mode, both internal and external bremsstrahlung spoil the response, yielding mass resolution for m_H = 130 GeV of 1 GeV (1.5) with 30% (20%) tails outside $\pm 2\sigma$ for CMS (ATLAS) [4, 2]. For the 4-muon mode, the mass resolution is 1 GeV at 150 GeV for CMS, where the resolution is dominated by the central tracker one [4]. In ATLAS, at 130 GeV, a stand-alone measurement using the muon system information alone gives 2 GeV [6], whereas combining with the tracker will yield 1.4 GeV.
The reducible backgrounds, $Zb\bar{b}$ and $t\bar{t}$, can be safely reduced by using isolation and vertexing [3]. One is left with the ZZ^* continuum, implying low signal yields, 10-30 events for 3 years at low luminosity, and $S/B \sim 10$. This channel allows to observe a Higgs on the whole mass range, with a difficult region near $2m_W$.

1.3 High mass region ($m_H > 2m_Z$)

$H \to ZZ \to l^+l^-l^+l^-$ For $m_H >$ 200 GeV, the Higgs becomes broad and the detector performance is less crucial. The dominant background is the ZZ continuum, easily rejected with loose cuts on p_T^Z. The observation is easy up to 600 GeV, with large signal yields and significance [7].

$H \to ZZ \to l^+l^-\nu\bar{\nu}$ This mode delivers 6 times more rate than the 4-lepton mode, and the dominant background is the ZZ continuum, for $E_t^{miss} >$ 200 GeV. The Z+jets background where mismeasured jets may create fake E_t^{miss} is under control, provided the calorimeter covers down to 5 in $|\eta|$ [3]. The signal peaks are nevertheless broad, and the large uncertainty on the WZ and ZZ rates make the observation difficult.

$H \to WW \to l\nu jj$, $H \to ZZ \to l^+l^-jj$ This mode delivers 150 (20 in ZZ mode) times more rate than the 4-lepton mode, and may be used for Higgs search up to $\sim$ 1 TeV. The final state features are a central $W \to jj$ with jets close-by in space ($\Delta R \sim 0.4$), and 2 outgoing jets at large rapidity ($|\eta| > 2$) coming from the WW fusion process. The W reconstruction needs dedicated algorithms which allow to maintain a good reconstruction efficiency and mass resolution, see Fig. 6. After central jet veto, the dominant background is W/Z + jets, where jets can fake a real W. The forward jet tagging, see Fig. 7, allows to reduce this background, and yields large significance. The observation of a broad signal over a large background is thus possible, with some careful analysis [8].

1.4 $H \to WW \to l\nu l\nu$

This channel is extremely interesting near the $2m_W$ mass threshold, where the 4-lepton mode has a small sensitivity. By applying kinematical cuts on the outgoing leptons, based on different kinematics between the Higgs and the WW processes, one can observe a broad excess of events (in p_T) above the dominant WW background, see Fig. 8, with $S/B \sim 1$ [9].
This study can be extended over the full Higgs mass range. As shown in Fig. 9, below the ZZ threshold, the significance are the same as for the 4-lepton mode but with $S/B \sim 0.2$ (it is around 10 in the 4-lepton mode). Above the ZZ threshold, the significances are definitely smaller than for the 4-lepton mode. This channel is nevertheless extremely precious over the full mass range as a cross-check.
In summary, Fig. 10 shows the expected sensitivity for the Standard Model Higgs discovery, for ATLAS as an example. There is a good sensitivity over the full mass range, where in most cases 2 decay modes are available. In addition, the $H \to WW \to l\nu l\nu$ gives an extra safety in the $2m_W$ region.

1.5 Properties

As discussed in [10], the LHC will allow to make accurate measurements on the Higgs spin, CP, width and couplings by looking at various cross-sections and branching ratios. For $m_H > 2m_Z$ one has access to Γ_{tot}, and measuring the rate of $WW \to H \to WW$ will allow to measure the WWH coupling.

2 Other Standard Model channels

2.1 Top

At LHC, and for a moderate luminosity of 10^{32} cm^{-2}s^{-1}, one gets 100 reconstructed $t\bar{t} \to (l\nu b)(jjb)$ ans 10 isolated $e\mu$ pair per day. The total cross-section can be measured with an accuracy of 5-10%, limited by the integrated luminosity uncertainty. The measurement of the top mass in the jjb mode and by using leptons (combining 2 leptons from the same top, which is the least sensitive to p_T^{top}) should reach an accuracy of ± 2 GeV (including the systematics coming from the b-fragmentation and the polarization effects). The top decays can be also investigated: $BR(t \to Wb)$ can be evaluated to $\pm$ 5%, and a sensitivity of 5.10^{-5} can be achieved on rare decays like $BR(t \to Zc)$ [3].

2.2 Gauge bosons pair production

This channel provides a test of the 3 vector-boson couplings. The non-standard physics are described by 2 parameters κ and λ (in SM, $\kappa = 1$ and $\lambda = 0$). By looking at $W\gamma \to l\nu\gamma$ and $WZ \to l\nu ll$, a deviation to the SM is characterized by an excess of events in the p_T^{γ} ot the p_T^{Z} distribution [3], see Fig. 11.

2.3 W mass

At LHC, for $10^4 \mathrm{pb}^{-1}$, one expects about 2.10^7 $W \to l\nu$ events with $65 < m_T < 100$ GeV. The measurement of the $(l\nu)$ transverse mass, used in UA1, UA2 and at the Tevatron, is less sensitive to p_T^W, but sensitive to the pile-up. Extrapolating the Tevatron results at the LHC yields $\delta m_W < 15$ MeV for $10^4 \mathrm{pb}^{-1}$. Others methods, like comparing m_T^W to m_T^Z, or inferring p_T(lepton) versus p_T^W (not well known) by using p_T(lepton) versus p_T^Z (known), should yield $\delta m_W \sim 10$ MeV [11]. Combined with $\delta(m_{top}) \sim 2$ GeV, this will constrain m_H to 20%.

3 Conclusions

LHC can discover a SM Higgs over the full mass range, from the LEPII limit up to the TeV region. In most cases, at least 2 decay modes are accessible. Most of the studies are based on full simulation results, assessed by test beam results obtained on full scale prototypes. If a Standard Model Higgs is discovered, the LHC will allow accurate measurements on the Higgs parameters. In addition, a wide range of SM processes are available, yielding precise measurements (top, m_W).

References

[1] K. Lassila, 6th International Calorimetry Conference, Frascati, 1996.
[2] ATLAS Calorimeter Performance TDR, 1996.
[3] ATLAS, Technical Proposal, 1994.
[4] CMS, Technical Proposal, 1994.
[5] ATLAS Inner Tracker TDR, 1997.
[6] ATLAS Muon TDR, 1997.
[7] E. Richter-Was et al., ATLAS Internal Note PHYS-048, 1996.
[8] S. Zmushko et al., ATLAS Internal Note PHYS-103, 1997.
[9] M. Dittmar and H. Dreiner, Phys. Rev. D55, 167-172,1997.
[10] J. Gunion et al., Snowmass proceedings, 1996.
[11] U. Baur et al., Snowmass proceedings, 1996.

Search for the Higgs Bosons at the LHC

Jorma Tuominiemi (Jorma.Tuominiemi@helsinki.fi)

Helsinki Institute of Physics, Helsinki, Finland

Abstract. Discovery limits for the MSSM Higgs bosons in the ATLAS and CMS experiments at LHC are reviewed. The most interesting decay channels are discussed and comments on future improvements in the analysis are given.

1 Introduction

There are strong theoretical arguments suggesting that the theory of elementary particles should obey supersymmetry. It is therefore necessary to investigate the consequences supersymmetry would have for the Higgs sector of the theory and to see how it would influence the search for the Higgs bosons at LHC. The most interesting choice for the SUSY model is the one with the simplest Higgs sector, the Minimal Supersymmetric Standard Model, MSSM. In this model there are two Higgs doublet fields, the minimum number needed for the masses of up and down type quarks. In addition, the relative evolution of the gauge couplings is exactly what is desired in grand unification schemes. The Higgs sector contains two CP-even (h, H), one CP odd (A^0) neutral states and one charged $(H^{\pm})$ state. At tree level there are two free parameters determining the masses and couplings of h,H, A^0 and $H^{\pm}$. Usually m_A and the ratio of the expectation values of the Higgs doublets, $tan\beta$, are chosen. When loop corrections are included, other parameters enter, e.g. there is a strong dependence on the top and stop masses as well as on the mixing of the stop quarks.

Extensive simulation work has been done to study the possibilities to search for the MSSM Higgs bosons with the CMS and ATLAS detectors. We present here an update that includes the latest calculations on the loop corrections, including two-loop, RGE-improved radiative corrections [1],[2]. With the top mass fixed at 175 GeV/c^2, the sparticle masses at 1 TeV/c^2 and assuming no stop mixing, the mass limits for the lightest neutral Higgs bosons become $m_h \leq$ 113 GeV and $m_H \geq$ 113 GeV. With maximal stop quark mixing the limit moves up to 124 GeV. The widths of the neutral bosons remain smaller than 1 GeV for all reasonable values of $tan\beta$, say $tan\beta$ <50.

In generating the Higgs production and decay processes ATLAS and CMS have used the event generator PYTHIA, version 5.7 [3], completed with the improved radiative corrections in [1],[2]. For detector simulation the GEANT program has been used. Generated events were fully simulated and reconstructed in the detectors at particle level. Where adequate, parametrizations of detector response and resolutions were used. Descriptions of the CMS and ATLAS detectors can be found in refs. [4] and [5], respectively.

The results presented here are discussed in detail in ref. [6] for CMS and in ref. [7] for ATLAS.

2 Decay channels of interest

If all sparticle masses are large enough that they can be ignored in the decay of the MSSM Higgs bosons the same decay channels as in the Standard Model appear, together with a few new channels. The couplings are modified and hence the branching ratios and production cross sections can be very different from those of the Standard Model.

In MSSM the $\sigma\times$branching ratio of $h, H \to \gamma\gamma$ is smaller than in SM but the associated production channels Wh and $t\bar{t}h$ play an important role, albeit at high luminosity. The small width of h allows to benefit from high precision calorimetry to improve the possibility to see the $\gamma\gamma$ decay channel despite its small signal-to-background ratio. The estimated 5σ discovery limits for this channel are shown in fig. 1 for CMS for an integrated luminosity of $10^{-5}pb^{-1}$, in both inclusive $\gamma\gamma$ search and in $\gamma\gamma$ -lepton associated production modes.

The couplings of h, H to W and Z are at most as large as in SM. Only a small region in the parameter space with $tan\beta$ <2 is available for the observation of $H \to 4$ leptons. The light Higgs boson h becomes visible in the four lepton channel only if the stop mixing is large. As the expected numbers of events in these channels are small, the mass resolution is all the more important. The obtained discovery limits in CMS are shown in fig. 1 for maximal stop mixing.

The couplings of the neutral bosons to $\tau^+\tau^-$ are larger than in SM when $tan\beta$ is large. The most promising way to observe this decay channel is to trigger with a lepton from the leptonic decay of one τ and then search for a hadronic decay of the second τ. The single hadron decay of the τ can be recognized requiring an isolated pion in the electromagnetic and in the hadron calorimeter. Detailed GEANT simulations have shown that a rejection factor of the order of 10^{-3} against the QCD jets can be achieved, with the τ selection efficiency at the level of 30 %. The discovery limits for this channel are shown in fig. 1 for an integrated luminosity of $3 \times 10^{-4}pb^{-1}$. The mass of the Higgs boson can be reconstructed in an approximate way for a part of the events, using the E_t^{miss} to reconstruct the energies of the neutrinos. This is rather sensitive to the precision of the E_t^{miss} measurement. The obtained mass peak is, however, small compared to the background from irreducible $Z \to \tau\tau$, from $t\bar{t}$ production and from W+jet production. However, through identification of b jets with impact parameter measurements, the associated $b\bar{b}H$ production channels can be used to suppress the background and to improve the signal-to-background ratio of the Higgs mass peak, although with a loss of statistics. This is still under study in CMS.

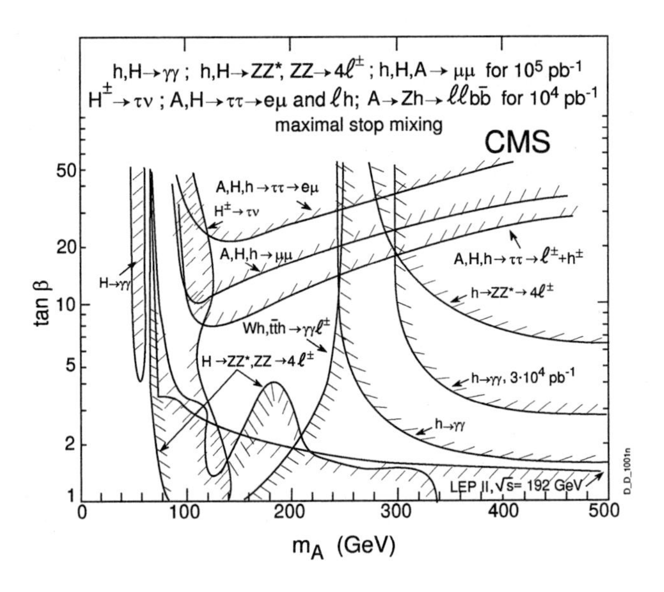

Fig. 1. Estimated discovery limits for MSSM Higgs bosons in CMS with maximal stop mixing.

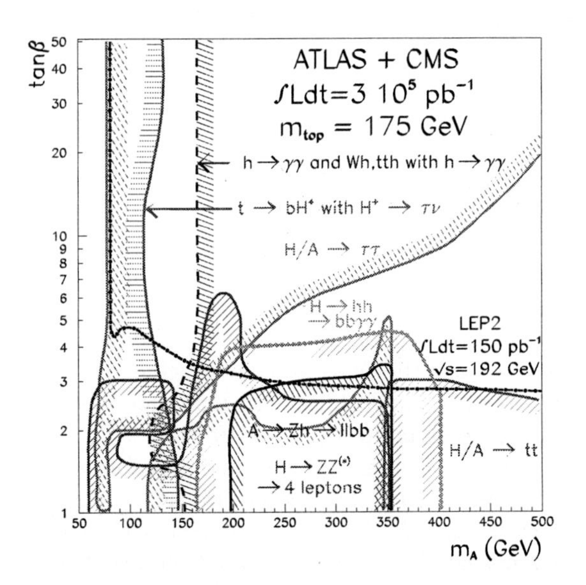

Fig. 2. Estimated discovery limits for MSSM Higgs bosons CMS and ATLAS combined for an integrated luminosity of $3 \times 10^{-5} pb^{-1}$, no stop mixing.

In the channel $H^{\pm} \to \tau\nu$ b-tagging is again necessary to suppress the large background of events with $W + jet$. The obtained discovery limit is shown in fig. 1 for $10^{-4}pb^{-1}$.

The excellent muon momentum resolution in CMS makes it possible to study the decay channels $h, H, A \to \mu^{+}\mu^{-}$, which suffer from small branching ratios and large backgrounds. The yields are largest at large $tan\beta$ as can be seen from the discovery limits in fig. 1. At m_A near the Z mass the signal is masked by the $Z \to \mu^{+}\mu^{-}$ decay. Here again good b tagging helps and the associated production channel $Hb\bar{b}$ offers a possibility to investigate this mass region.

3 Conclusions

As is seen from fig.1 already a large part of the MSSM parameter space can be covered at least with one decay mode during the first years, with an integrated luminosity of $3 \times 10^{-4}pb^{-1}$ in CMS. With additional running to get $10^{-5}pb^{-1}$ only the region $130 < m_A < 250$ GeV, $2 < tan\beta < 10$ remains unexplorable. Similar results are obtained in ATLAS. If LEPII will run at 200 GeV c.m. energy, the LEP discovery limit for Zh production channel can be pushed up to $tan\beta \sim 5$. Finally fig. 2 shows that the full MSSM parameter space can be explored with an integrated luminosity of $3 \times 10^{-5}pb^{-1}$ and combining the results of ATLAS and CMS.

Further improvements in the analysis are still under way. The use of dynamical algorithms to reconstruct photons and gammas in the presence of conversions and bremstrahlung in the tracker, as well as development of sophisticated τ reconstruction algorithms can still improve the resolution of the H mass measurement and hence lead to higher observability. Secondly, the potential of b tagging has to be studied fully, this being an important tool both to suppress the backgrounds to the associated production channels and possibly to exploit the decay channels $H, h \to b\bar{b}$ [8].

References

[1] J.F.Gunion, A.Stange and S.Willenbrock, UCD-95-28, ILL-(TH)-95-28.
[2] M.Carena, J.R.Espinosa, M.Quiros, and C.E.M.Wagner, Phys.Lett. B355 (1995) 209.
[3] T.Sjöstrand, CERN TH/7112/93 (1993), revised 1995.
[4] CMS Technical proposal, CERN/LHCC 94-38 (1994).
[5] ATLAS Technical proposal, CERN/LHCC 94-43 (1994).
[6] Ritva Kinnunen, D.Denegri, CMS Note 1997/057 (1997).
[7] Elzbieta Richter-Was, D.Froidevaux, Fabiola Gianotti, L.Poggioli, Donatella Cavalli, Silvia Resconi, ATLAS Note, PHYS-No-074 (1996).
[8] S.Abdullin, D.Denegri, CMS Note 1997/070 (1997).

Higgs Physics
at 300 - 500 GeV e^+e^- Linear Collider

H. Jürgen Schreiber (schreibe@ifh.de)

DESY-Zeuthen

Abstract. We summarize discovery potentials for the Standard Model Higgs boson produced in e^+e^- collisions and measurements of its detailed properties. Prospects for Higgs boson detection within the Minimal Supersymmetric Standard Model are also discussed.

1 Introduction

Electroweak breaking due to the Higgs mechanism [1] implies the existence of at least one new particle, the Higgs boson. Its discovery is the most important missing link for the formulation of electroweak interactions. LEP has established a lower bound of the Higgs mass $M_H \gtrsim 77$ GeV [2]. High precision data, interpreted within the Standard Model (SM), favour a Higgs boson with a mass somewhere between 100 and 180 GeV.

Future e^+e^- linear colliders are the ideal machines for a straightforward Higgs boson discovery, its verification and precision measurements of the Higgs sector properties.

Recent simulation studies [3] involve, thanks to the effort of several groups, i) the full matrix elements for 4-fermion final states, ii) all Higgs decay modes (with SM decay fractions $\gtrsim 1\%$), iii) initial state QED and beamstrahlung (TESLA design), iv) a detector response [4] and v) all important background expected to contribute.

2 SM Higgs discovery potentials

The Higgs boson can be produced by the Higgsstrahlung process $e^+e^- \rightarrow Z^* \rightarrow ZH^o$ (1), or by the fusion of WW and ZZ bosons, $e^+e^- \rightarrow \nu\bar{\nu}H^o$ (2) and $e^+e^- \rightarrow e^+e^-H^0$ (3), respectively, or by radiation off top quarks, $e^+e^- \rightarrow t\bar{t}H^o$, with however a very small cross section. At e.g. $\sqrt{s} = 360$ GeV and $M_H = 140$ GeV, the Higgsstrahlung process is about four times more important than the fusion reactions. Reaction (1) admits two strategies for the Higgs search: i) calculation of the recoil mass against the $Z \rightarrow e^+e^-/\mu^+\mu^-$, $M_{rec}^2 = s - 2\sqrt{s}(E_{l^+} + E_{l^-}) + M_Z^2$, which is independent of assumptions about Higgs decay modes, and ii) the direct reconstruction of the invariant mass of the Higgs decay products.

In order to achieve the best experimental resolution, energy-momentum as well as $M(l^+l^-) = M_Z$ constraints have been imposed when appropriate, and to make a signal-to-background analysis as meaningful as possible a consistent evaluation of the signal and all expected background rates has been made [3].

The leptonic channel, $e^+e^- \rightarrow ZH^o \rightarrow (e^+e^-/\mu^+\mu^-)(b\bar{b})$, allows Higgs detection either in the mass recoiling against the Z or in the hadronic two b-quark jet mass. Typically, an integrated luminosity of $\sim 10fb^{-1}$ is needed to observe the Higgs boson with a significance $S/\sqrt{B} > 5$ after application of appropriate selection procedures [3].

The tauon channel, $e^+e^- \rightarrow ZH^o \rightarrow (q\bar{q})\ (\tau^+\tau^-)$, requires some more refined selection procedures [3, 5] due to missing neutrinos from τ decays. Hence, the accumulated luminosity to observe a clear H^o signal in the recoil resp. $\tau^+\tau^-$ mass should be between 50 to 80 fb^{-1}. Energy-momentum constraints are required to determine the original τ and jet energies by a fit; without that no H^o signal would be visible.

The missing energy channels, $e^+e^- \rightarrow ZH^o \rightarrow (\nu\bar{\nu})(b\bar{b})$ and $e^+e^- \rightarrow \nu\bar{\nu}H^o \rightarrow \nu\bar{\nu}(b\bar{b})$ are very important due to their large discovery potential for the Higgs boson. It has been demonstrated [3, 6] that with $\sim 1fb^{-1}$ of integrated luminosity a convincing signal over some small remaining background in the doubly-tagged b-jet mass should be obtained (Fig.1). Thus, few days of running a 300 GeV e^+e^- linear collider at its nominal luminosity would suffice to discover the SM Higgs in the intermediate mass range.

With increasing integrated luminosity other H^o decay modes like $WW^{(*)}$ and $c\bar{c}$ + gg would be measurable, allowing to verify the Higgs interpretation; 50 to 100 fb^{-1} are needed for significant signals. As an example, Fig.2 shows the signal from selected $H^o \rightarrow c\bar{c}$ + gg decays in presence of a huge background in the 2-jet missing energy event topology.

Observation of the $H^o \rightarrow \gamma\gamma$ decay requires large statistics ($> 200fb^{-1}$) and a fine-grained electromagnetic calorimeter with very good energy resolution [7].

3 Higgs properties

If a Higgs candidate is discovered, it is imperative to understand its nature. In the following we assume an integrated luminosity of $100fb^{-1}$.

Measurements of the Higgs boson mass are best performed by searching for $Z \rightarrow l^+l^-(l = e/\mu)$ decay products and reconstructing the recoil mass peak. The mass resolution depends on $\sqrt{s}$ and the detector performance, in particular on the momentum resolution of the tracking system. For $M_H = 140$ GeV and $\sqrt{s} = 360$ GeV, we expect an error for the Higgs mass of ~ 50 MeV for the detector design of ref.[4].

When the Higgsstrahlung ZH^o rate is significant, the CP-even component of the H^o dominates. It is possible to cross check this by studying the

Higgs production and the Z decay angular distributions. Thus, observation of angular distributions as expected for a pure CP-even state implies that the Higgs has spin-0 and that it is not primarily CP-odd [8]. Further, studying decay angular correlations between the decay products of the tauons in the reaction $e^+e^- \to ZH^o \to Z\tau^+\tau^-$ provides a democratic probe of the Higgs CP-even and CP-odd components [8, 9]. If the $H^o \to \gamma\gamma$ decay is visible (or the Higgs is produced in $\gamma\gamma$ collisions) the Higgs must be a scalar and has a CP-even component. Whether a CP-odd component also exists can be studied by comparing Higgs production rates in $\gamma\gamma$ collisions with different photon polarizations [10].

The determination of the branching fraction of the Higgs into a final state X requires to compute $BF(H \to X) = [\sigma(ZH) \cdot BF(H \to X)]/\sigma(ZH)$, where $\sigma(ZH)$ is the inclusive Higgs cross section, see Fig. 3. Its error is expected to be $\pm 5\%$ [3], while the numerator $\sigma(ZH) \cdot BF(H \to X)$ and its precision can be obtained from the X invariant mass distribution.

Typical statistical branching fraction uncertainties expected for a 140 GeV SM Higgs boson are [3]

$BF(H \to b\bar{b})$	$BF(H \to \tau^+\tau^-)$	$BF(H \to WW^*)$	$BF(H \to c\bar{c} + gg)$
$\pm 6.1\%$	$\pm 22\%$	$\pm 22\%$	$\pm 28\%$

As can be seen, many of the Higgs decay modes are accessible to experiment and the measurements allow a discrimination between the SM-like and e.g. SUSY-like Higgs bosons. However, more fundamental than branching fractions are the Higgs partical widths. If $M_H \gtrsim 140$ GeV, $\Gamma(H \to WW)$ is measurable from the WW fusion cross section, whereas for $M_H \lesssim 130$ GeV a $\gamma\gamma$ collider allows to measure $\Gamma(H \to \gamma\gamma)$. Combined with the corresponding branching fraction (from LHC or NLC), Γ_{tot}^{Higgs} is calculable, which in turn allows the determination of the remaining partial widths. In addition, the $H^o \to ZZ$ coupling-squared is obtained from $\sigma(ZH)$ with an error comparable to that of the inclusive cross section.

4 MSSM Higgs bosons: $h^o, H^o, A^o, H^\pm$

In the minimal supersymmetric scenario five Higgs particles are expected, two CP-even states h^o, H^o (with $M_{h^o} < M_{H^o}$), one CP-odd state A^o and two charged states $H^\pm$. An upper bound on the mass of the h^o, including radiative corrections [11], is predicted to be $M_{h^o} \lesssim 130$ GeV. If M_{h^o} exceeds 130 GeV, the MSSM is ruled out.

In the limit of large $M_{A^o} (\gg M_Z)$ the properties of the h^o are very close to those of the SM Higgs and a discrimination between the two states is only possible if precise measurements show $h^o \neq H^o_{SM}$ or further Higgs states are discovered.

Besides Higgsstrahlung and fusion production mechanisms for h^o and H^o, Higgs pair production $e^+e^- \to h^oA^o$, H^oA^o, H^+H^- is possible, with the relations $\sigma(e^+e^- \rightarrow h^oZ, h^o\nu\bar{\nu}, H^oA^o) \propto sin^2(\beta-\alpha)\,(I)$ and $\sigma(e^+e^- \rightarrow H^oZ, H^o\nu\bar{\nu}, h^oA^o) \propto cos^2(\beta-\alpha)\,(II)$ (α, β are mixing angles). One process in each line (I), (II) always has a substantial rate and the sum of the h^o or H^o cross sections is constant and equals $\sigma(H^o_{SM})$ over the whole $(M_A, \tan\beta)$-plane. Therefore, a h^o or H^o cannot escape detection if it is kinematically accessible.

Concerning the branching fractions it is usually assumed that the masses of the supersymmetric particles are so heavy that SUSY-Higgses cannot decay into s-particles; only decays into standard particles are possible. Whatever the masses of the Higgs particles and tanβ values are, we expect i) b-quark decays of h^o, H^o and A^o to occur in large regions of the parameter space, underlining the importance of excellent b-tagging, ii) vector boson decays are never significant for the A^o while strong restrictions exist for the H^o and iii) if $M_{H^\pm} \gtrsim 180$ GeV, the $H^\pm \rightarrow tb$ decay rate is large leading in most cases via $e^+e^- \rightarrow H^+H^- \rightarrow W\,b\bar{b}\,W\,b\bar{b}$ to 8-jet final states involving four b-quark jets, a very challenging task to recognize and analyze these events. An example for MSSM Higgs boson detection is shown in Fig.4 for the process $e^+e^- \rightarrow H^oA^o \rightarrow b\bar{b}b\bar{b}$ at $\sqrt{s} = 800$ GeV [4].

If however s-particle decays of Higgses are possible, the decays end (in many SUSY models) with the lightest 'neutrino-like' supersymmetrie particle. Hence, event signatures are characterized by missing $p_\perp/E_\perp$ + leptons + jets which warrant further experimental simulations.

5 Conclusions

An e^+e^- collider focusing on $e^+e^- \rightarrow ZH^o$ production in combination with a $\gamma\gamma$ collider allows to fully explore the properties of an intermediate SM Higgs boson in the shortest time.

Within the MSSM, one light CP-even Higgs boson must be found or the MSSM is ruled out. A 500 GeV collider offers very good prospects to discover all five MSSM Higgs particles, $h^o, H^o, A^o, H^\pm$, if $M_A \lesssim 220$ GeV. If these particles are discoverd large statistics experiments are needed to measure their parameters.

6 Figures

References

[1] P.W.Higgs, Phys.Rev.Lett. 12 (1964) 132 and Phys. Rev. 145 (1966) 1156.

[2] W.Murray, Search for the SM Higgs, presented at the Int. Europhysics Conf. on High Energy Physics, 16-26 August 1997, Jerusalem, Israel.

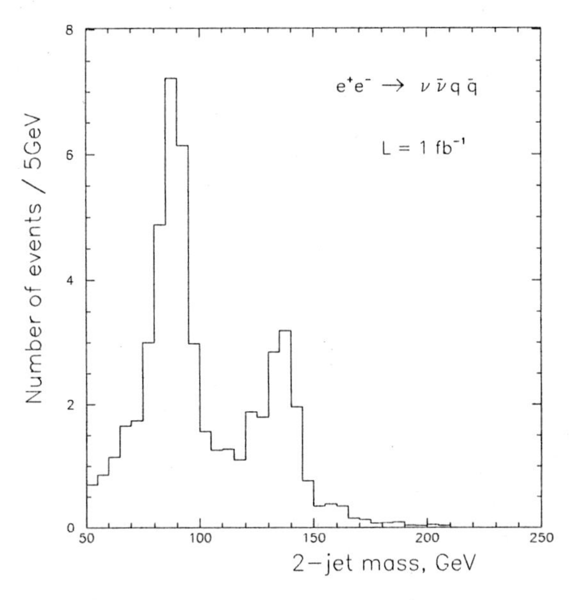

Fig. 1. *2-jet mass distributions of the reaction* $e^+e^- \to \nu\bar{\nu}q\bar{q}$, *for an integrated luminosity of* $1fb^{-1}$ *at* $\sqrt{s} = 360$ *GeV. Background contributions are included.*

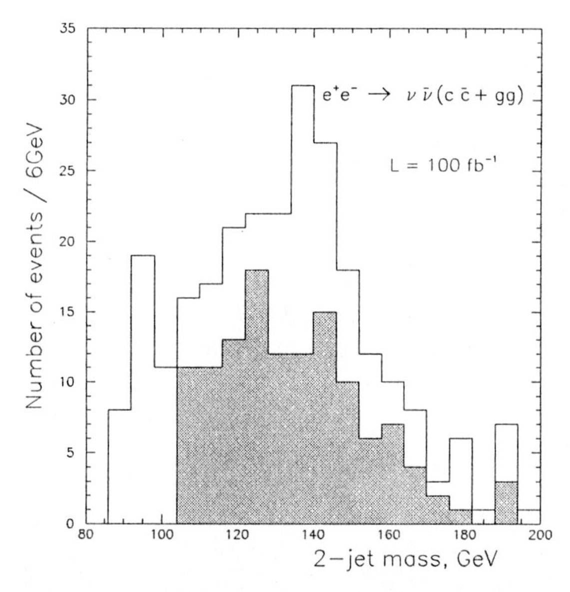

Fig. 2. *2-jet mass distribution of the reaction* $e^+e^- \to \nu\bar{\nu}$ $(c\bar{c} + gg)$ *at* $\sqrt{s} = 360$ *GeV. The shaded histogram represents the reducible background expected.*

Towards a Complete Calculation of $\gamma\gamma \rightarrow 4f$

E. Boos[1,2] (boos@ifh.de), T. Ohl[2] (Thorsten.Ohl@Physik.TH-Darmstadt.de)

[1]Moscow State University, Russia; [2]Darmstadt University of Technology, Germany

Abstract. We present a general classification of all four fermion final states in $\gamma\gamma$ collisions and a calculation of cross sections below the W^+W^--threshold.

The planned e^+e^- Linear Collider will provide the opportunity to study $\gamma\gamma$-collisions with energies of several hundred GeV in the center of mass system. Of immediate interest will be gauge and Higgs boson production. However, neither the Higgs, nor the W or the Z will be directly observed and we typically have to study four fermion production processes, for which the resonant diagrams $\gamma\gamma \rightarrow VV \rightarrow 4f$ do *not* form gauge invariant subsets and calculations of $\gamma\gamma \rightarrow 4f$ beyond on-shell gauge boson production are required for precision studies.

As a step towards the construction of a complete event generator for $\gamma\gamma \rightarrow 4f$, we have introduced [1] a classification of all four fermion final states in $\gamma\gamma$-collisions and the corresponding Feynman diagrams. Fig. 1 shows all

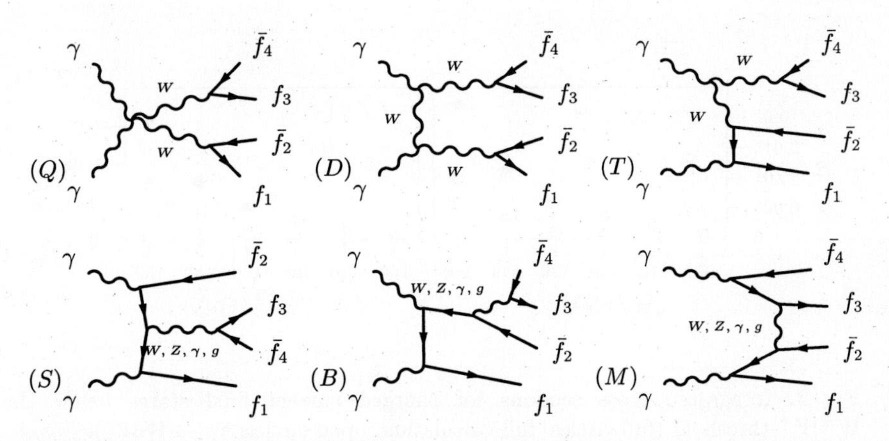

Fig. 1. The six topologies for $\gamma\gamma \rightarrow 4f$ in unitary gauge (see [1] for R_ξ-gauges).

six topologies appearing in the Feynman diagrams for $\gamma\gamma \rightarrow 4f$. The flavors of the final state fermions determine which of W, Z, γ, or g contribute for the lines labelled such. In a three generation standard model, there are 98 possible four fermion final states that fall into only eight different classes. They are listed in table 1 with the number of diagrams for each topology. The charged

Class		Q	T	D	S	B	M	$\sum$	$\mathcal{O}(\alpha_S)$	e. g.
CC13	l	1	4	2	0	4	2	13		$e^-\bar{\nu}_e\mu^+\nu_\mu$
CC21	sl	1	6	2	2	6	4	21		$e^-\bar{\nu}_e u\bar{d}$
CC31	h	1	8	2	4	8	8	31		$u\bar{d}\bar{c}s$
NC06	l/sl	0	0	0	2	4	0	6		$\nu_e\bar{\nu}_e\mu^-\mu^+$
NC20	l/h	0	0	0	4	8	8	20	+10	$e^-e^+e^-e^+$
NC40	l/sl/h	0	0	0	8	16	16	40	+20	$e^-e^+\mu^-\mu^+$
mix19	l	1	4	2	2	8	2	19		$e^-e^+\nu_e\bar{\nu}_e$
mix71	h	1	8	2	12	24	24	71	+20	$u\bar{u}d\bar{d}$

Table 1. The eight classes of diagrams in $\gamma\gamma \to 4f$.

current classes (*CCn*) are gauge invariant completions of W-pair production and, unlike $e^+e^- \to 4f$, there are *no* final states without multi-peripheral diagrams. The neutral current classes (*NCn*) do not contain any cubic or quartic gauge couplings. Finally, the mixed classes (*mixn*) correspond to final states that are invariant under the transformation from "charge exchange" to "charge retention" form and vice versa. For each class, one typical final state is given (see [1] for complete lists). The column $\mathcal{O}(\alpha_S)$ shows the number of one gluon exchange diagrams to be added in the neutral current classes for purely hadronic final states.

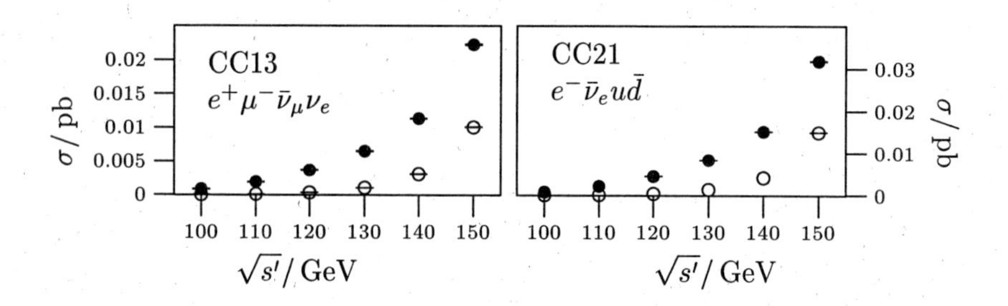

Fig. 2. Integrated cross sections for charged current final states below the W^+W^--threshold (full circles: full calculation, open circles: $\gamma\gamma \to WW^*$).

Supported by Deutsche Forschungsgemeinschaft. Presented by E. B., supported by the HEP97 Organizing Committee and Russian Ministry of Science.

References

[1] E. Boos and T. Ohl, Phys. Lett. **B407** (1997) 161.

Testing the $Z\gamma H$ Vertex at Future Linear Colliders for Intermediate Higgs Masses

E. Gabrielli (egabriel@wave.phys.nd.edu), Univ. of Notre Dame, IN, USA
V.A. Ilyin (ilyin@theory.npi.msu.su), Moscow St. Univ., Russia
B. Mele (mele@roma1.infn.it), INFN, Univ. "La Sapienza", Italy

Abstract. Higgs production in $e\gamma$ collisions, through the one-loop reaction $e\gamma \to eH$ at large p_T, can provide a precise determination of the $Z\gamma H$ vertex.

Among other couplings, the interactions of the Higgs scalar with γ and Z are particularly interesting, since they depend on the relation between the spontaneous symmetry breaking mechanism and the electroweak mixing of the two gauge groups $SU(2)$ and $U(1)$. In this respect, three vertices can be studied: ZZH, $\gamma\gamma H$ and $Z\gamma H$. While in the SM the ZZH vertex stands at the tree level, the other two contribute only at one-loop. This means that the $\gamma\gamma H$ and $Z\gamma H$ couplings can be sensitive to the contributions of new particles circulating in the loop. For the Higgs masses discussed here, $m_H \lesssim 140$ GeV, a measurement of the $\gamma\gamma H$ coupling should be possible by the determination of the BR for the decay $H \to \gamma\gamma$, e.g. in the LHC Higgs discovery channel, $gg \to H \to \gamma\gamma$, or in $\gamma\gamma \to H$ at future $\gamma\gamma$ linear colliders. A chance of measuring the $Z\gamma H$ vertex is given by collision processes, e.g. in $e^+e^- \to \gamma H\,, ZH$. However, in the ZH channel the $Z\gamma H$ vertex contributes to the one-loop corrections, thus implying a large tree-level background. The reaction $e^+e^- \to \gamma H$ has been extensively studied [1]. Unfortunately, it suffers from small rates, $\approx 0.05 \div 0.001$ fb at $\sqrt{s} \sim 500 \div 1500$ GeV, and (as we estimated) the main background $e^+e^- \to \gamma b\bar{b}$ process has large cross sections: $\approx 4 \div 0.8$ fb for $m_{b\bar{b}} = 100 \div 140$ GeV, at $\sqrt{s} \sim 500 \div 1500$GeV, assuming reasonable kinematical cuts. Recently, the one-loop process $e\gamma \to eH$ was analysed in details [2]. The total rate for this reaction is rather high, > 1 fb for $m_H <$ 400 GeV. The main strategy to enhance the $Z\gamma H$ vertex effects consists in requiring a final electron tagged at large angle. E.g., for $p_T^e > 100$GeV, $Z\gamma H$ is about 60% of the generally dominant $\gamma\gamma H$ contribution. The main irreducible background comes from $e\gamma \to eb\bar{b}$. A further background is the charm production through $e\gamma \to ec\bar{c}$, when the c quarks are misidentified into b's. At $\sqrt{s} = 500$ GeV the cut $\theta_{b(c)} > 18°$ (between each $b(c)$ quark and the beams) makes the background comparable to the signal [2]. Resolved $e\gamma(g) \to eb\bar{b}(ec\bar{c})$ production, where the photon interacts via its gluonic content, could also contribute but, as we found, it is quite small. We studied polarization effects and found they are rather strong. E.g., for right handed electrons there is a strong destructive interference between the terms $\gamma\gamma H$ and $Z\gamma H$.

Now we discuss the prospects of the $e\gamma \to eH$ reaction in setting experimental bounds on a possible anomalous $Z\gamma H$ coupling. We assume, that the anomalous $\gamma\gamma H$ contributions have been well tested in some other experiment (e.g., through $\gamma\gamma \to H$). Then, one would like to get limitations just on the anomalous $Z\gamma H$ contributions. Anomalous CP-even and CP-odd operators contributing to $e\gamma \to eH$ [3] are:

$$\mathcal{O}_{UW;UB} = \left(\frac{|\Phi|^2}{v^2} - \frac{1}{2}\right)\{WW;BB\} , \qquad \bar{\mathcal{O}}_{UW;UB} = \frac{|\Phi|^2}{v^2}\left\{W\tilde{W};B\tilde{B}\right\} ,$$

where $\mathcal{L}^{eff} = d \cdot \mathcal{O}_{UW} + d_B \cdot \mathcal{O}_{UB} + \bar{d} \cdot \bar{\mathcal{O}}_{UW} + \bar{d}_B \cdot \bar{\mathcal{O}}_{UB}$. The corresponding $Z\gamma H$ anomalous terms in the helicity amplitudes of $e\gamma \to eH$ are

$$\frac{4\pi\alpha}{M_Z(M_Z^2 - t)}\sqrt{-\frac{t}{2}}\,\{d_{\gamma Z}[(u-s) - \sigma\lambda(u+s)] - i\bar{d}_{\gamma Z}[\lambda(u-s) + \sigma(u+s)]\} ,$$

where s, t and u are the Mandelstam kinematical variables, $\sigma/2$ and λ are the electron and photon helicities, and $d_{\gamma Z} = d - d_B$, $\bar{d}_{\gamma Z} = \bar{d} - \bar{d}_B$.

At $\sqrt{s} = 500$ GeV, for $m_H = 120$ GeV, one can then constrain the CP-even coupling in the following way: $-0.0025 < d_{\gamma Z} < 0.004$ in the unpolarized case, $|d_{\gamma Z}| < 0.0015$ for left-handed and $-0.007 < d_{\gamma Z} < 0.004$ for right-handed electrons. The corresponding bounds on the CP-odd coupling depends only slightly on the electron polarization, and are $|\bar{d}_{\gamma Z}| \lesssim 0.006$. Here we have taken into account the contributions for background from $e\gamma \to eb\bar{b}(ec\bar{c})$, assuming 10% of the c/b misidentifying, and from the resolved photons. The cuts $\theta_{b(c)} > 18°$, $p_T^e > 100$ GeV and $|m_{b\bar{b}(c\bar{c})} - m_H| < 3$ GeV are applied. The bounds presented have been computed by using the requirement that no deviation from the SM cross section is observed at the 95% CL, with an integrated luminosity 100 fb^{-1}. If the anomalous terms appear as contributions of new particles in the $Z\gamma H$ loop with the mass M_{new}, then one gets $d_{\gamma Z}, \bar{d}_{\gamma Z} \sim (v/M_{new})^2$. By using this relation, one obtains the bounds $M_{new} \gtrsim 6.2$ TeV in the CP-even case and $M_{new} \gtrsim 3.5$ TeV in the CP-odd case. All the results presented here were obtained with the help of the CompHEP package [4].

References

[1] A. Barroso et al,, Nuclear Physics B267 (1985), 509; B272 (1986), 693
A. Abbasabadi et al, Physical Review D52 (1995), 3919
A. Djouadi et al, Nuclear Physics B491 (1997), 68

[2] E. Gabrielli, V.A. Ilyin and B. Mele, Physical Review D56 (1997), 5945

[3] W. Buchmüller and D. Wyler, Nuclear Physics B268 (1986), 621
K. Hagiwara et al, Physical Review D48 (1993), 2182
G.J. Gounaris et al, Nuclear Physics B459 (1996) 51

[4] P.A. Baikov et al, in: Proc. X Int. Workshop QFTHEP'95 (Zvenigorod, Sept. 1995), ed. B. Levtchenko and V. Savrin, MSU, Moscow, 1996, p.101, hep-ph/9701412;
E.E Boos et al, SNUTP-94-116, 1994, hep-ph/9503280

Search for SUSY at LHC: Discovery and Inclusive Studies

Salavat Abdullin (abdullin@mail.cern.ch) ,

Institute for Theoretical and Experimental Physics, Moscow, Russia

Abstract. We discuss the expected discovery potential of the LHC in sparticle searches. The study is done within the framework of the mSUGRA-MSSM model. The domain of parameter space where SUSY can be discovered in squark/gluino, $h \to b\bar{b}$, slepton, chargino-neutralino searches is investigated. The results show that LHC will be capable to detect sparticles in the domain of parameter space where SUSY would be relevant at the electro-weak scale.

1 Introduction

Supersymmetry (SUSY) is a possible scenario for new physics beyond the Standard Model (SM), allowing to relieve difficulties with quadratic divergence of SM Higgs mass [1]. The minimal extension of the SM, MSSM [2] implies the existence of supersymmetric partners to all the ordinary particles, with masses of the order of electro-weak scale. In MSSM some "soft-breaking" terms are introduced by hand to parameterize the effects of supersymmetry breaking in a more fundamental theory. A general parameterization of SUSY breakings in MSSM introduces about 100 free parameters [1]. This hampers a phenomenological analysis. The studies presented here are carried out in the framework of a minimal Supergravity model (mSUGRA) [3], which implies the simplest symmetry breaking at a GUT scale and leads to only 5 extra parameters : a common gaugino mass ($m_{1/2}$), a common scalar mass (m_0), a common trilinear interaction amongst the scalars (A_0), the ratio of the vacuum expectation values of the Higgs fields at M_Z (tabβ), and a Higgsino mixing parameter sign ($sign(\mu)$). Models with R-parity conservation imply pair production of sparticles and the existence of Lightest Supersymmetric Particle (LSP), which is $\tilde{\chi}_1^0$ in mSUGRA and appears at the end of the decay chain of each sparticle. It escapes detection and reveals itself through missing E_T. Since masses of sparticles cannot be reconstructed, the detection of these sparticles relies on the observation of an excess of events over SM expectations, characterized by a signal significance expressed as $S = N_S/\sqrt{N_S + N_B}$, where N_S is a number of particular SUSY signal events, and N_B is the number of background events. Background can be a mixture of both SM and SUSY events.

The purpose of this study is to evaluate the LHC discovery potential in sparticle searches, in particular to determine the domain of mSUGRA parameter space where SUSY reveals itself in squark/gluino, scalar Higgs, sleptons, chargino-neutralino searches. The simulation of mSUGRA signal events is performed with ISAJET [4] and of SM backgrounds with PYTHIA [5]. The CMS detector performance is simulated with a fast Monte-Carlo program CMSJET [6], which is based on detailed simulation parameterizations.

2 Gluinos and squarks at LHC

Since the total SUSY production is dominated by strongly interacting sparticles, a typical signal contains squarks and gluinos decaying through a number of steps to quarks, gluons, charginos, neutralinos, W, Z, Higgs bosons and ultimately a stable $\tilde{\chi}_1^0$. The final state has thus a number of jets, missing energy (2 LSP + neutrinos) and a variable number of leptons, depending on the decay chain. Among the final states discussed here are : one lepton (1l), two leptons of opposite sign (2l OS), two leptons of same sign (2l SS), three leptons (3l), four leptons (4l) and five leptons (5l). The SM background considered are : $t\bar{t}$, W+jets, Z+jets, WW, ZZ, ZW, QCD ($2 \to 2$, including $b\bar{b}$). The study of Baer *et al.* [7] carried out with a generic LHC detector description and with fixed kinematical cuts shows that a significant signal can be obtained in a wide range of mSUGRA parameters. Here we summarize

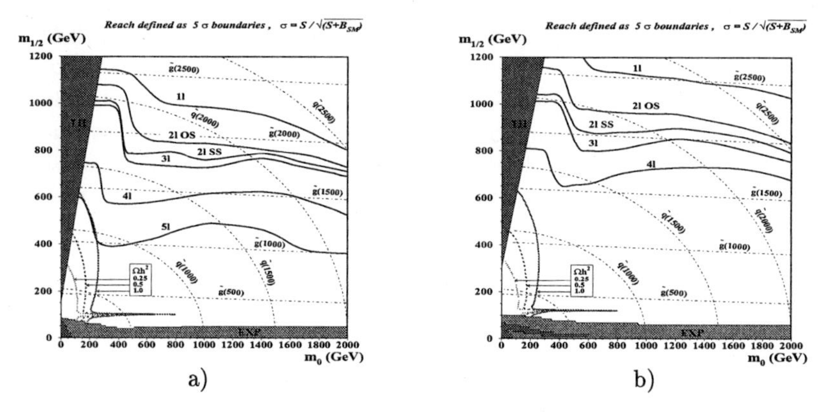

Fig. 1. Explorable domain of (m_0, $m_{1/2}$) parameter space in for 10^5 pb^{-1} integrated luminosity, in various final states with leptons + E_T^{miss} + $\geq$2 jets : a) A_0=0, tanβ=2, μ <0 and b) A_0=0, tanβ=2, μ >0.

the results of a detailed study [8]. The following kinematical variables are most useful ones for the SM backgrounds suppression : E_T^{miss}, number of jets (N_j), jet E_T^j, scalar transverse energy sum $E_T^{sum} = E_T^{miss} + \Sigma E_T^j + \Sigma E_T^l$. In some cases cuts on lepton p_T^l, *Circularity* (C) and transverse angle between hardest lepton and E_T^{miss} ($\delta\varphi$) are also used. The cuts are optimized for each mSUGRA domain (kinematics of the signal) and final state topology and are typically : $E_T^{miss} > 100 \div 600$ GeV, $N_j \geq 2 \div 6$, $E_T^j > 40 \div 300$ GeV, $E_T^{sum} > 500 \div 1200$ GeV, $p_T^l > 10 \div 50$ GeV, C > 0.1, $\delta\varphi > 10 \div 20$ degrees. Electrons are always isolated to be unambiguously identified. It is shown [8] that taking into account non-isolated muons (which come from decays of abundantly produced b-jets) allows one to improve significantly the reach in some cases. Figs. 1a and 1b show the 5σ discovery contours in various final states in $\tilde{q}$, $\tilde{g}$ searches for two sets of mSUGRA parameters. The regions of parameter space space excluded either by theory (TH) or experimental data up to now (EX) are indicated by shaded regions. The upper limit of neutralino relic density [9] corresponding to the $\Omega h = 1$ contour is fully contained in the explorable region.

3 Possibility to observe $h \to b\bar{b}$ in $\tilde{q}$, $\tilde{g}$ decays

The most general way to detect the lightest CP-even Higgs of MSSM is to search for the inclusive channel $h \to \gamma\gamma$, which, with 10^5 pb^{-1}, would allow to explore a domain approximately given by m$_A \geq 250$ GeV, tan$\beta \geq 3$ [10]. We can expect a signal on a top of a large irreducible background with S/B $\leq 1/20$ Other production or decay modes of h have also some disadvantages or limitations. On the other hand, it is well known that the MSSM h can be abundantly produced in decays of charginos and neutralinos (primarily $\tilde{\chi}_2^0$). In turn, the $\tilde{\chi}_2^0$ is a typical decay product of squarks and gluinos, which are produced with large (strong-interaction) cross sections. The idea is thus to search the h in the $b\bar{b}$ mode in cascade decays of sparticles [11]. The SM backgrounds considered are : $t\bar{t}$, Wtb, QCD ($2 \to 2$ including $b\bar{b}$). We find that there is a significant part of the mSUGRA parameter space, where the $h \to b\bar{b}$ peak can be observed with S/B ~ 1, taking into account nominal CMS performance [12]. The outcome of this study is shown in Fig.2 for two sets of mSUGRA parameters.

4 Slepton production

Slepton pairs are produced in a Drell-Yan processes and decay leptonically into final states characterized by two hard, same flavour, opposite sign isolated leptons, E$_T^{miss}$ and little jet activity. The sum of signal lepton transverse momenta and E$_T^{miss}$ are mainly back-to-back. Lepton isolation and jet vetoing capability of the detector are essential issues to control the large backgrounds. Typical cuts applied are : 2 isolated leptons with p$_T^l > 30$ GeV; veto on jets with E$_T^j > 30$ GeV in $\mid \eta \mid < 4$; E$_T^{miss} > 80$ GeV; relative azimuthal angle between leptons and E$_T^{miss}$ $> 160°$. Cuts are optimized depending on the domain of mSUGRA parameter space.The dominant backgrounds are : reducible $t\bar{t}$ and irreducible WW, $\tilde{\chi}_1^{\pm}\tilde{\chi}_1^{\mp}$. Fig.3 shows the results of study [13]. The slepton mass reach is $\sim$ 400 GeV.

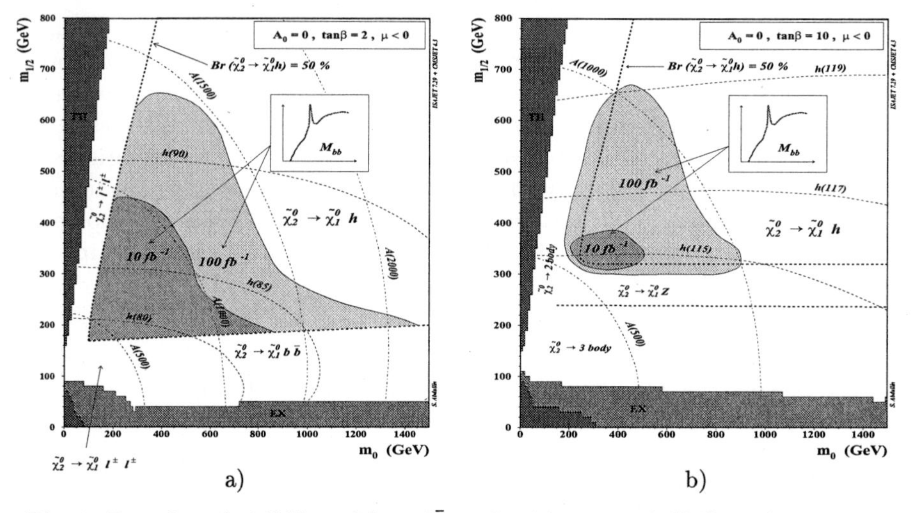

Fig. 2. Domains of visibility of $h \to b\bar{b}$ peak with nominal CMS performance in mSUGRA-SUSY for two sets of the model parameters : a) A_0=0,tanβ=2, μ <0 and b) A_0=0, tanβ=10, μ <0.

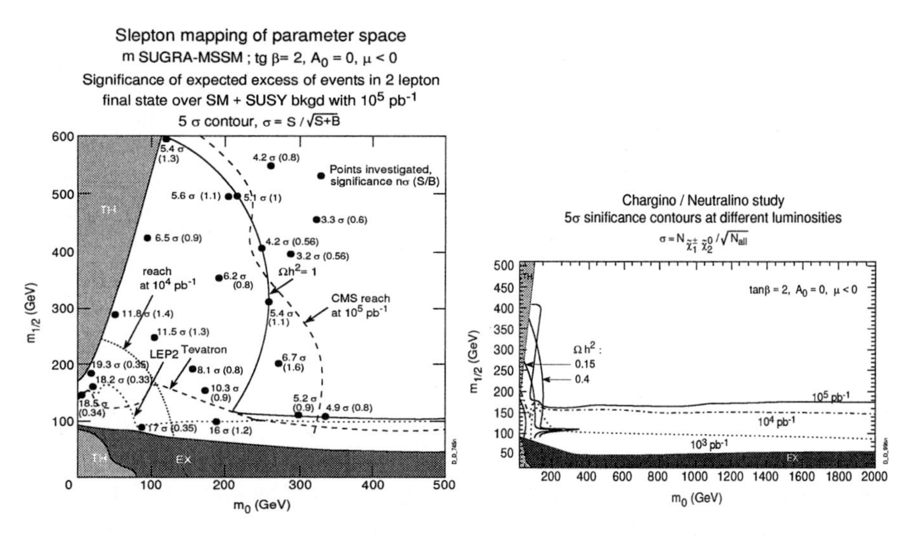

Fig. 3. Slepton search probed points, with expected signal significance and S/B ratio. The 5σ contour is shown by dashed line, the cosmological $\Omega h^2 = 1$ boundary is shown by solid line.

Fig. 4. $\tilde{\chi}_i^{\pm}\ \tilde{\chi}_j^0 \to 3$ leptons + E_T^{miss} + no jets final states 5σ significance contours for various luminosities; relic neutralino matter density contours are indicated.

5 Chargino-neutralino pair production

There are 21 reactions for chargino-neutralino pair production through a Drell-Yan mechanism. The $\tilde{\chi}_1^{\pm}\tilde{\chi}_2^0$ production has the largest cross section. The most promising experimental signature to detect the signal is $3l$ + *no jets* + E_T^{miss} [14]. The SM backgrounds considered are : WZ, ZZ, $t\bar{t}$, Wtb, $Zb\bar{b}$, $b\bar{b}$. The SUSY background is dominated by strong production, but the jet veto efficiently reduces $\tilde{q}$, $\tilde{g}$ multijet final state contribution. The typical set of cuts imposed is : 3 isolated leptons with p_T^l > 15 GeV; veto on jets with E_T^j > 25 GeV in $\mid \eta \mid$ < 3.5; a Z mass window cut of ± 10 GeV. The signal observability contours at various integrated luminosities are shown in Fig.4 [15].

6 Characteristic shape of 2l OS SF effective mass as an evidence for SUSY

It has been known for some time that there is a "structure" in the l^+l^- invariant mass spectrum in exclusive $\tilde{\chi}_1^{\pm}\tilde{\chi}_2^0$ production [14]. The question is how general is this observation. The edge in the invariant mass distribution originates from 3- or 2-body decays of $\tilde{\chi}_2^0$. In first case the maximum of the

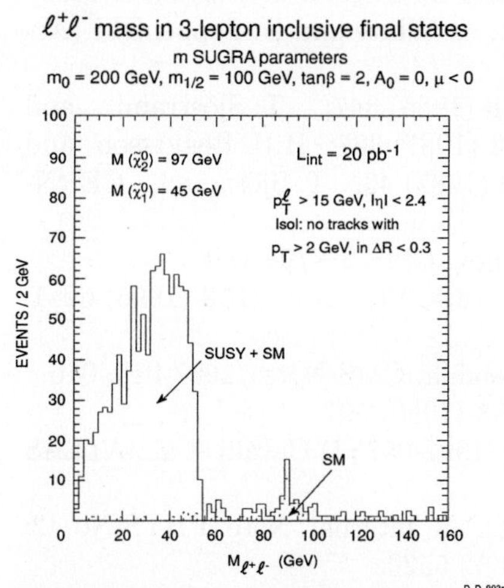

Fig. 5. Dilepton mass spectrum in an inclusive 3-lepton final state.

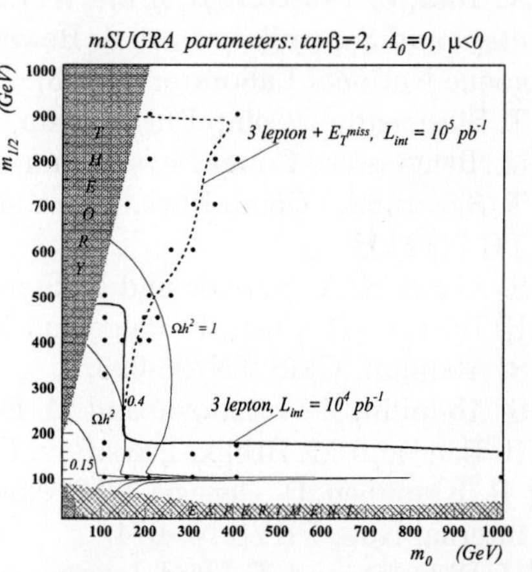

Fig. 6. Explorable region of mSUGRA parameter space, where dilepton mass characteristic edge (Fig.5) is observed in 3 lepton (+ E_T^{miss}) final states.

decay spectrum is at $M_{l^+l^-}^{max}$ = m($\tilde{\chi}_2^0$) - m($\tilde{\chi}_1^0$), in the latter case at $M_{l^+l^-}^{max}$ = m($\tilde{\chi}_1^0$) · $\sqrt{\{[1 - m^2(\tilde{l})]/m^2(\tilde{\chi}_2^0)\} \cdot \{[1 - m^2(\tilde{\chi}_1^0))/m^2(\tilde{l})\}}$. The studies [16] show that the "edge" can be visible in a significant part of parameter space in inclusive $\tilde{\chi}_2^0$ production. The SM background can be easily suppressed by requiring a third lepton or/and E_T^{miss}, as seen in Fig.5. An example of the explorable domain of the mSUGRA parameter space is shown in Fig.6.

7 Conclusion

At LHC supersymmetry will reveal itself by an excess production over SM expectations. LHC experiments will be able to test SUSY at electro-weak scale in a decisive way. The plausible part of mSUGRA-MSSM parameter space will be explored in a number of characteristic signatures. Expected mass reach is : for $\tilde{q}$, $\tilde{g} \sim 2$ TeV, for $\tilde{\chi}_1^0$, $\tilde{l} \sim 400$ GeV. The $h \to b\bar{b}$ may be much easier to observe in $\tilde{q}$, $\tilde{g}$ decays than $h \to \gamma\gamma$ in significant part of parameter space.

References

[1] M. Drees, APCTP-5, KEK-TH-501.
[2] Z. Kunszt and F. Zwirner, Nucl.Phys. **B385** (1992) 3.
[3] H. Baer et al., FSU-HEP-950401, CERN-PPE/95-45.
[4] F. Paige and S. Protopopescu, in *Supercollider Physics*, p. 41, ed. D. Soper (World Scientific, 1986); H. Baer, F. Paige, S. Protopopescu and X. Tata, in *Proceedings of the Workshop on Physics at Current Accelerators and Superolliders*, ed. J. Hewett, A. White and D. Zeppenfeld (Argonne National Laboratory, 1993).
[5] T. Sjöstrand, Comp.Phys.Comm. **39** (1986) 347; T. Sjöstrand and M. Bengtsson, Comp.Phys.Comm. **43** (1987) 367; H.U. Bengtsson and T. Sjöstrand, Comp.Phys.Comm. **46** (1987) 43; T. Sjöstrand, CERN-TH.7112/93.
[6] S. Abdullin, A. Khanov and N. Stepanov, CMS TN/94-180.
[7] H. Baer, C.-H. Chen, F. Paige and X. Tata, Phys.Rev. **D53** (1996) 6241.
[8] S. Abdullin, CMS TN/96-095;
S. Abdullin, Ž. Antunović and M. Dželalija, CMS Notes/1997-015, 016.
[9] H. Baer and M. Brhlik, Phys.Rev. **D53** (1996) 597
[10] R. Kinnunen, D .Denegri, CMS Note/1997-057 ; D.Cavalli et al., ATLAS Internal Note PHYS-No-074.
[11] L.Didenko and B.Lund-Jensen, ATLAS Internal Note PHYS-No-42; I.Hinchliffe et al., Phys.Rev. **D55** (1997) 5520.
[12] S. Abdullin, D. Denegri, CMS Note/1997-070.
[13] D. Denegri, L. Rurua and N. Stepanov, CMS TN/96-059.
[14] R. Barbieri et al., Nucl.Phys. **B367** (1993) 28; H.Baer, C.-H. Chen, F. Paige and X. Tata, Phys.Rew. **D50** (1994) 4508.
[15] I. Iashvili, A. Kharchilava and K. Mazumdar, CMS Note/1997-007.
[16] I. Iashvili, A. Kharchilava, CMS Note/1997-065; L.Rurua et al., CMS Note in preparation.

Search for SUSY at LHC: Precision Measurements

Frank E. Paige (paige@bnl.gov)

Brookhaven National Laboratory, Upton, NY 11973 USA

Abstract. Methods to make precision measurements of SUSY masses and parameters at the CERN Large Hadron Collider are described.

1 Introduction

It is quite easy to find signals for SUSY at the LHC.[1, 2] But every SUSY event contains two missing $\tilde{\chi}_1^0$'s, so it is not possible to reconstruct masses directly. A strategy developed recently[3, 4] is to start at the bottom of the SUSY decay chain and work up it, partially reconstructing specific final states and using kinematic endpoints to determine combinations of masses. These are then fit to a model to determine the SUSY parameters. This paper is limited to discussion of this approach; search limits and inclusive measurements are discussed by Abdullin.[5]

The LHCC (LHC Program Committee) selected five points in the minimal SUGRA model[2] for detailed study. The parameters of this model are m_0, the common scalar mass; $m_{1/2}$, the common gaugino mass; A_0, the common trilinear coupling; $\tan\beta = v_2/v_1$, the ratio of Higgs vacuum expectation values; and $\mathrm{sgn}\,\mu$, the sign of the Higgsino mass. These parameters are listed in Table 1, and representative masses are listed in Table 2. Point 3 is the

Table 1. Parameters for the five LHC SUGRA points.

	m_0 (GeV)	$m_{1/2}$ (GeV)	A_0 (GeV)	$\tan\beta$	$\mathrm{sgn}\,\mu$
1	400	400	0	2.0	+
2	400	400	0	10.0	+
3	200	100	0	2.0	−
4	800	200	0	10.0	+
5	100	300	300	2.1	+

Table 2. Representative masses for the five LHC SUGRA points in Table 1.

	$M_{\tilde{g}}$ (GeV)	$M_{\tilde{u}_R}$ (GeV)	$M_{\tilde{W}_1}$ (GeV)	$M_{\tilde{e}_R}$ (GeV)	M_h (GeV)
1	1004	925	325	430	111
2	1008	933	321	431	125
3	298	313	96	207	68
4	582	910	147	805	117
5	767	664	232	157	104

"comparison" point; LEP would have already found the light Higgs at this point. Point 5 is constructed to give the right cold dark matter. Points 1 and 2 have heavy masses, while Point 4 has heavy squarks.

2 Specific Final States

This section describes only a few of the final states that have been studied. For all of these studies, signal and background events were gener-

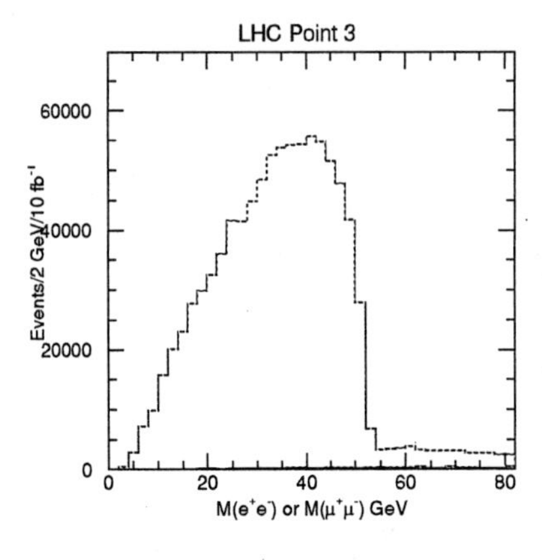

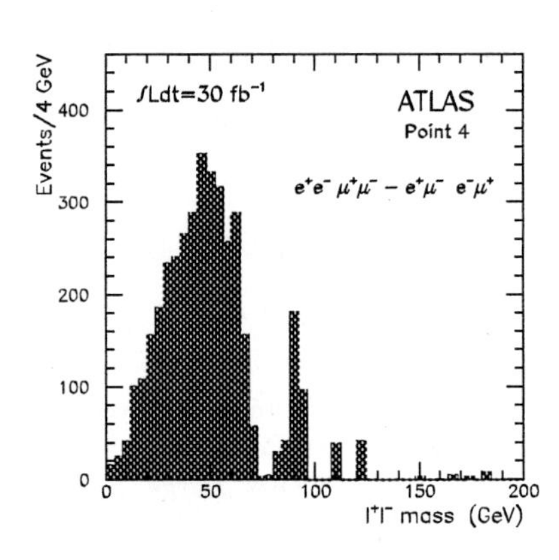

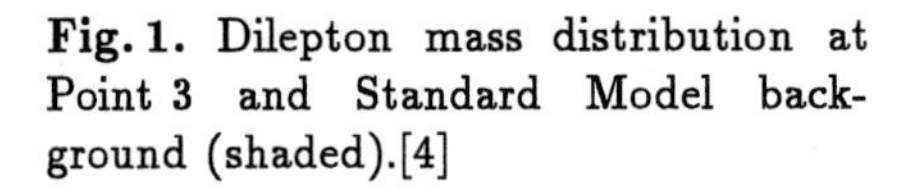

Fig. 1. Dilepton mass distribution at Point 3 and Standard Model background (shaded).[4]

Fig. 2. Dilepton mass distribution for Point 4.[8]

ated using ISAJET[6] or PYTHIA[7], the response of the detector was simulated, and an analysis was done to select the signal from the background.

$M(\tilde{\chi}_2^0) - M(\tilde{\chi}_1^0)$: The prototype of precision measurements[3] is based on the decay $\tilde{\chi}_2^0 \to \tilde{\chi}_1^0 \ell^+ \ell^-$ at Point 3. Point 3 has unusual branching ratios:

$$B(\tilde{g} \to \tilde{b}_1 \bar{b} + \text{h.c.}) = 89\%$$
$$B(\tilde{b}_1 \to \tilde{\chi}_2^0 b) = 86\%$$
$$B(\tilde{\chi}_2^0 \to \tilde{\chi}_1^0 \ell^+ \ell^-) = 2 \times 17\%$$

Events were selected with an $\ell^+\ell^-$ pair with $p_{T,\ell} > 10\,\mathrm{GeV}$ and $\eta < 2.5$ and at least two jets tagged as b's with $p_T > 15\,\mathrm{GeV}$ and $\eta < 2$. Efficiencies of 60% for tagging b's and 90% for lepton identification were included. No $\not{E}_T$ cut was used. The resulting dilepton mass distribution, Figure 1, has a spectacular edge at the $M(\tilde{\chi}_2^0) - M(\tilde{\chi}_1^0)$ endpoint with almost no Standard Model background. Determining the position of the edge is much easier than measuring M_W at the Tevatron, and the statistics are huge. The estimated error for $10\,\mathrm{fb}^{-1}$ is $\Delta(M(\tilde{\chi}_2^0) - M(\tilde{\chi}_1^0)) = 50\,\mathrm{MeV}$.

The low masses and unusual branching ratios make Point 3 particularly easy. But there is a similar edge at Point 4 plus a Z peak coming from decays of the heavier gauginos, as can be seen in Figure 2.[8] In this case the estimated error is $\Delta\left(M(\tilde{\chi}_2^0) - M(\tilde{\chi}_1^0)\right) = \pm 1\,\mathrm{GeV}$. A scan of the SUGRA parameter space[10] finds an observable signal for $m_{1/2} \lesssim 200\,\mathrm{GeV}$ and for a region of small m_0 in which the sleptons are light.

$\tilde{g}$ and $\tilde{b}_1$: The next step at Point 3 is to combine an $\ell^+\ell^-$ pair near edge with jets. Events are selected as before. If the $\ell^+\ell^-$ pair has a mass near the endpoint, then the $\tilde{\chi}_1^0$ must be soft in $\tilde{\chi}_2^0$ rest frame, so

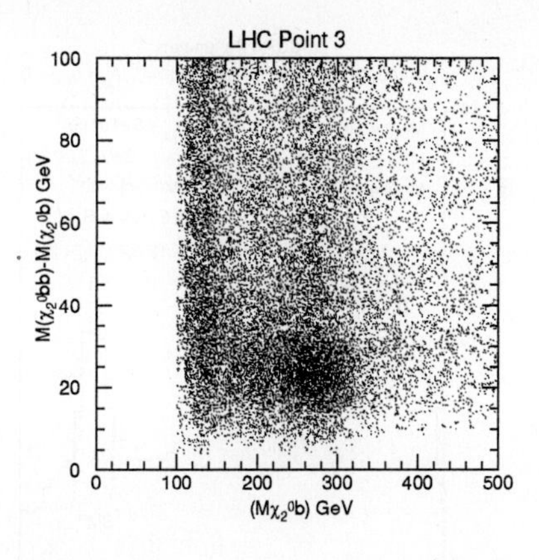

Fig. 3. Scatter plot of $M(\tilde{g}) - M(\tilde{b})$ vs. $M(\tilde{b})$. [4]

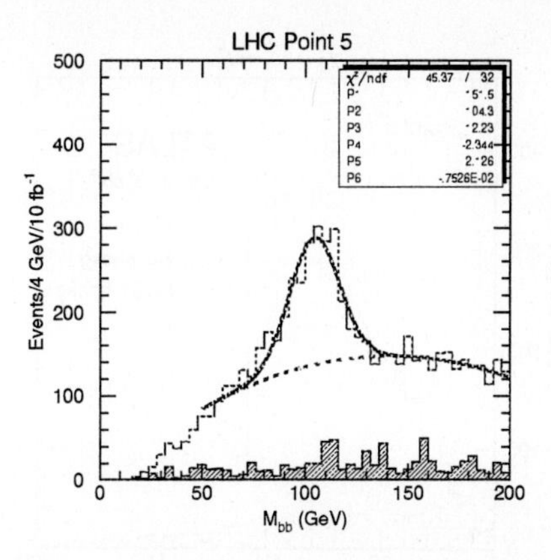

Fig. 4. $M(bb)$ at Point 5 and Standard Model background (shaded).[4]

$$\mathbf{p}(\tilde{\chi}_2^0) \approx \left(1 + \frac{M(\tilde{\chi}_1^0)}{M(\ell\ell)}\right) \mathbf{p}(\ell\ell)$$

where $M(\tilde{\chi}_1^0)$ must be determined. Lepton pairs were selected with masses within 10 GeV of the endpoint and were combined with one b to make $M(\tilde{b}_1)$ and then with a second b to make $M(\tilde{g})$. Figure 3 shows a scatter plot of all combinations. Since the $\bar{b}$ jet from $\tilde{g} \to \tilde{b}\bar{b}$ is soft, there is good resolution on the $M(\tilde{g}) - M(\tilde{b}_1)$ mass difference — c.f. $D^* \to D\pi$. Varying the assumed $\tilde{\chi}_1^0$ mass gives $\Delta M(\tilde{b}_1) = \pm 1.5 \Delta M(\tilde{\chi}_1^0) \pm 3\,\mathrm{GeV}$ and $\Delta(M(\tilde{g}) - M(\tilde{b}_1)) = \pm 2\,\mathrm{GeV}$.

$h \to b\bar{b}$: For Point 5, $\tilde{\chi}_2^0 \to \tilde{\chi}_1^0 h$ is kinematically allowed. Events are selected with at least four jets with $p_T > 50\,\mathrm{GeV}$, $p_{T,1} > 100\,\mathrm{GeV}$, transverse sphericity $S_T > 0.2$, $M_{\mathrm{eff}} = \not{E}_T + \sum_{i=1}^4 p_{T,i} > 800\,\mathrm{GeV}$, and $\not{E}_T > \max(100\,\mathrm{GeV}, 0.2 M_{\mathrm{eff}})$. Then M_{bb} is plotted for jets tagged as b's with $p_{T,b} > 25\,\mathrm{GeV}$ and $\eta_b < 2$. There is a clear peak with a substantial SUSY background and small Standard Model background.

The two jets from $h \to b\bar{b}$ can be combined with one of the two hardest jets in the event to determine the squark mass: the smaller of the two $b\bar{b}q$ masses must be less than a function of the squark mass and the other masses in the decay $\tilde{q} \to \tilde{\chi}_2^0 q \to \tilde{\chi}_1^0 hq$.

$\ell^+\ell^-$ Again: For Point 5 after standard cuts one finds an edge in Figure 5[9] for $> M_Z$. Since the two-body decay $\tilde{\chi}_2^0 \to \tilde{\chi}_1^0 h$ has been reconstructed at this point, this edge cannot come from the three-body decay $\tilde{\chi}_2^0 \to \tilde{\chi}_1^0 \ell^+\ell^-$, since the phase space is much smaller. It must come instead from $\tilde{\chi}_2^0 \to \tilde{\ell}^\pm \ell^\mp \to \tilde{\chi}_1^0 \ell^\pm \ell^\mp$. Thus the edge determines

$$M_{\tilde{\chi}_2^0} \sqrt{1 - \frac{M_{\tilde{\ell}}^2}{M_{\tilde{\chi}_2^0}^2}} \sqrt{1 - \frac{M_{\tilde{\chi}_1^0}^2}{M_{\tilde{\ell}}^2}}$$

with an error of $\pm 1\,\mathrm{GeV}$.

It is possible to have both $\tilde{\chi}_2^0 \to \tilde{\ell}_R \ell$ and $\tilde{\chi}_2^0 \to \tilde{\chi}_1^0 \ell\ell$ edges for some

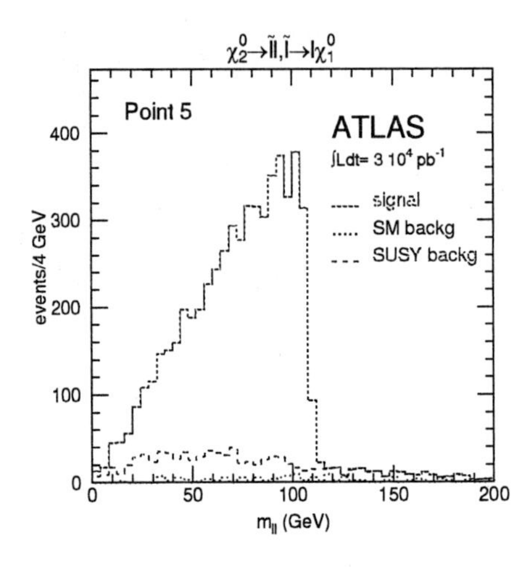

Fig. 5. $\ell^+\ell^-$ mass distribution for Point 5.[9]

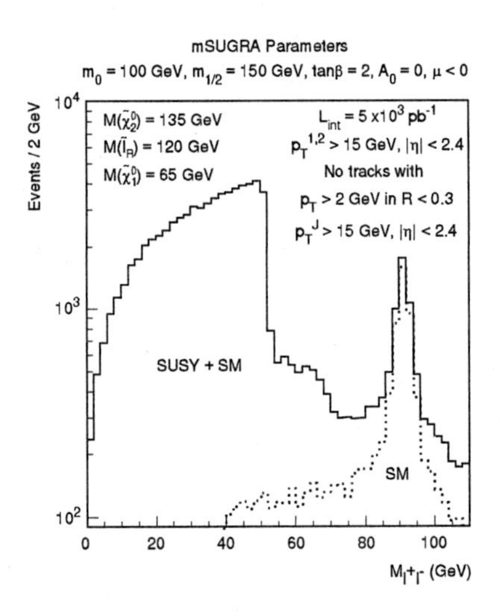

Fig. 6. SUGRA $\ell^+\ell^-$ distribution showing edges from both $\tilde{\chi}_2^0 \to \tilde{\ell}\ell$ and direct $\tilde{\chi}_2^0 \to \tilde{\chi}_1^0\ell\ell$ decays.[10]

choices of the SUGRA parameters. An example is shown in Figure 6.[10]

It should in principle be possible to extract the $\tilde{\chi}_2^0$, $\tilde{\ell}$, and $\tilde{\chi}_1^0$ masses from a fit to all the dilepton data. This has not been studied, but as a first step the distribution for the ratio $p_{T,2}/p_{T,1}$ of lepton p_T's has been examined for $m_0 = 100, 120\,\mathrm{GeV}$. This distribution is clearly exhibits sensitivity to the slepton mass. The same distribution can also be used to distinguish two-body and three-body decays.

$M(\tilde{g}) - M(\tilde{\chi}_2^0), M(\tilde{\chi}_1^\pm)$: Gluino production dominates at Point 4. Previously, an $\ell^+\ell^-$ edge was found at this point, determining $M(\tilde{\chi}_2^0) - M(\tilde{\chi}_1^0)$. The strategy for this analysis is to select

$$\tilde{g} + \tilde{g} \to \tilde{\chi}_2^0 q\bar{q} + \tilde{\chi}_1^\pm q\bar{q}$$

using leptonic decays to identify $\tilde{\chi}_2^0$ and $\tilde{\chi}_1^\pm$ and so to reduce the combinatorial background. Then the jet-jet mass should have a common endpoint since $M(\tilde{\chi}_2^0) \approx M(\tilde{\chi}_1^\pm)$.

The analysis[8] requires three isolated leptons with $p_T > 20$, 10, 10 GeV and $|\eta| < 2.5$, one opposite-sign, same-flavor pair with $M_{\ell\ell} < 72\,\mathrm{GeV}$, four jets with $p_T > 150$, 120, 70, 40 GeV, $|\eta| < 3.2$, and no additional jets with $p_T > 40\,\mathrm{GeV}$ and $|\eta| < 5$ to minimize combinatorics. There are three pairings per event. The pairing of the two highest and the two lowest p_T jets is unlikely and is discarded. The distribution for the remaining pairings, Figure 7, shows an edge at about the right endpoint.

3 Fitting SUGRA Parameters

Points were generated in SUGRA parameter space, and the masses were calculated and compared with the

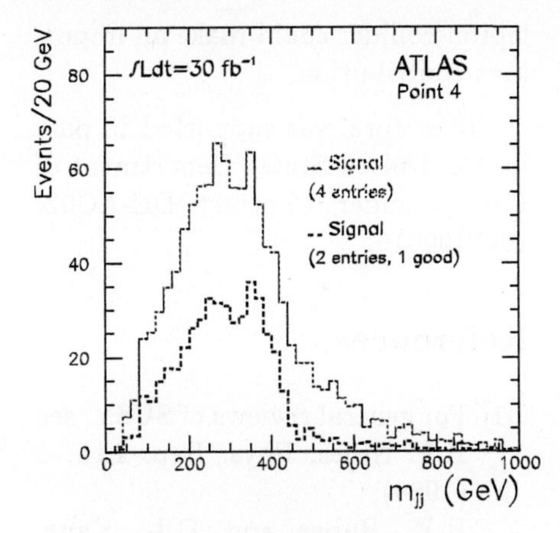

Fig. 7. Jet-jet mass distribution for Point 4 after cuts described in the text and corresponding distribution for correct pairing (dotted).[8]

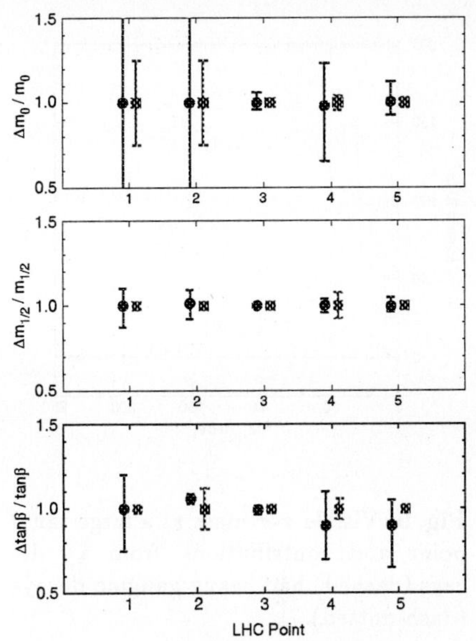

Fig. 8. Results for Fit I (circles) and Ultimate Fit II (squares).

combinations of masses determined by precision measurements. Fit I[4] uses a smaller set of such measurements, assumes that the Higgs mass can be related to the SUGRA parameters with an error of 3 GeV, and uses an integrated luminosity of $10\,\mathrm{fb}^{-1}$. Fit II[11] uses a larger set of precision measurements plus a few other measurements, e.g., from changing squark mass and seeing the effect on the highest p_T jet, assumes a negligible theoretical error on the Higgs mass, and uses an integrated luminosity of $300\,\mathrm{fb}^{-1}$.

For both fits the SUGRA parameter space was scanned to determine the 68% confidence interval for each parameter. The results are summarized in Figure 8. Clearly the parameters are quite well determined. No disconnected regions of parameter space were found. In particular, $\operatorname{sgn}\mu$ could always be determined. The gluino and squark masses are insensitive to m_0 at Points 1 and 2, so Fit I gives large m_0 errors. Finally, A_0 is poorly constrained in all cases. It is possible to determine the weak scale parameters A_t and A_b, but these are insensitive to A_0.

4 τ Modes at Large $\tan\beta$

For large $\tan\beta$ the $\tilde{\tau}_1$ can be relatively light. At the SUGRA point $m_0 = m_{1/2} = 200\,\mathrm{GeV}$, $A_0 = 0$, $\tan\beta = 45$, $\mu < 0$, the decays $\tilde{\chi}_2^0 \to \tilde{\tau}_1^\pm \tau^\mp$ and $\tilde{\chi}_1^\pm \to \tilde{\tau}_1^\pm \nu_\tau$ are dominant. Discovery is still straightforward, but all the analyses discussed in Section 2 do not apply. One possible approach is to select 3-prong τ decays to enhance the visible τ-τ mass. This is shown in Figure 9; it has a clear endpoint at $M(\tilde{\chi}_2^0) - M(\tilde{\chi}_1^0)$ plus a continuum from heavier gauginos. This

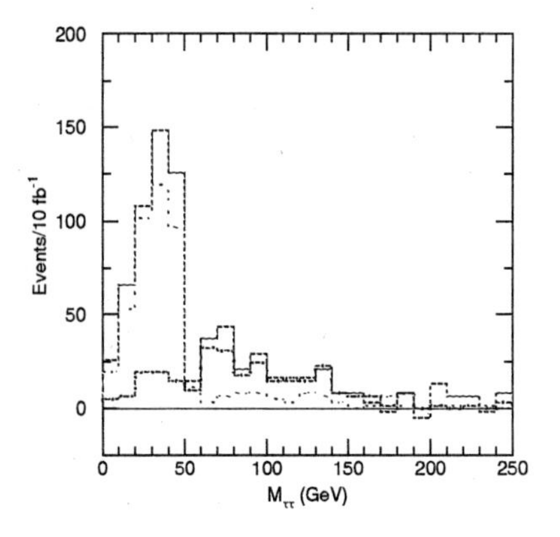

Fig. 9. Visible τ-τ mass at a large $\tan\beta$ point and contributions from $\tilde{\chi}_2^0$ decays (dashed) had heavy gaugino decays (dash-dotted).

example shows that the five LHCC points do not exhaust the possibilities even of the minimal SUGRA model.

5 Summary

If SUSY exists at electroweak scale, it should be easy to find signals for it at the LHC. The new result described here is that it is possible in many cases to make precision measurements of combinations of SUSY masses, and these measurements can at least in favorable cases determine the underlying SUSY parameters. While these results are quite encouraging, it seems likely that some SUSY particles — including heavy gauginos, sleptons unless $\tilde{\chi}_2^0 \to \tilde{\ell}\ell$ or $M(\tilde{\ell}) \lesssim 200\,\mathrm{GeV}$ to allow substantial Drell-Yan production, and heavy Higgs bosons — will be hard to study at the LHC, so a future lepton-lepton collider could make an important contribution.

This work was supported in part by the United States Department of Energy under Contract DE-AC02-76CH00016.

References

[1] For general reviews of SUSY, see H.P. Nilles, Phys. Rep. **111**, 1 (1984);
H.E. Haber and G.L. Kane, Phys. Rep. **117**, 75 (1985).

[2] For a review of SUSY phenomenology see H. Baer, et al., FSU-HEP-950401 (1995).

[3] A. Bartl, et al., in *New Directions for High Energy Physics* (Snowmass, 1996) p. 693.

[4] I. Hinchliffe, et al., Phys. Rev. D55, 5520 (1997).

[5] S. Abdullin, these Proceedings.

[6] H. Baer, F.E. Paige, S.D. Protopopescu, and X. Tata, *Physics at Current Accelerators and Supercolliders*, ed. J. Hewett, A. White and D. Zeppenfeld, (Argonne National Laboratory, 1993).

[7] T. Sjostrand, LU-TP-95-20 (1995);
S. Mrenna, Comput. Phys. Comm. **101**, 232 (1997).

[8] F. Gianotti, ATLAS Phys-No-110 (1997).

[9] G. Polesello, L. Poggioli, E. Richter-Was, and J. Soderqvist, ATLAS Phys-No-111 (1997).

[10] A. Kharchilava, CMS CR 1997/012 (1997).

[11] D. Froidevaux, ATLAS Phys-No-112 (1997).

Search for SUSY Particles at Future e^+e^- Colliders

Hans-Ulrich Martyn (martyn@mail.desy.de)

I. Physikalisches Institut, RWTH Aachen, Germany

Abstract. Experimental aspects of searches and studies of SUSY particles, scalar leptons, charginos and the stop quark, at future e^+e^- colliders are discussed.

1 Introduction

The minimal supersymmetric extension of the Standard Model assigns to each particle a SUSY partner differing by spin 1/2. In GUT scenarios the particle spectrum is completely specified by five parameters (m_0, $m_{1/2}$, A_0, $\tan\beta$, $sign\,\mu$). Generally the lightest supersymmetric particle is the neutralino χ_1^0.

A high energy e^+e^- linear collider is an ideal place to discover SUSY particles [1, 2] up to the kinematic limit already with moderate luminosities of $\mathcal{L} \sim 10$ fb^{-1}. Here, experimental aspects to study their properties with high precision will be discussed [3]. The requirements on the detector performance are modest: good particle identification over a large solid angle $|\cos\theta| < 0.98$, a momentum resolution of $\delta p_\perp/p_\perp = 1.5 \cdot 10^{-4}\, p_\perp$ [GeV] and a calorimetric resolution of $\delta E_h/E_h = 0.50/\sqrt{E_h\,\text{[GeV]}} \oplus 0.04$. It is important that for precision mass measurements the beamstrahlung effects of the $e+e-$ collider have to be kept small.

2 Scalar Leptons

Mass Determination of $m_{\tilde{\mu}_R}$ and $m_{\chi_1^0}$ Scalar leptons $\tilde{\ell}^\pm$ are pair-produced via s channel γ/Z exchange and t channel $\tilde{\nu}_e$ exchange ($\tilde{e}$). The production of scalar muons $\tilde{\mu}_R$ has been studied at $\sqrt{s} = 800$ GeV [3]

$$e^+e^- \to \tilde{\mu}_R^+ \tilde{\mu}_R^- ,$$
$$\tilde{\mu}_R \to \mu\,\chi_1^0 \qquad \mathcal{BR} = 99.5\,\% ,$$

assuming masses of $m_{\tilde{\mu}_R} = 275.1$ GeV and $m_{\chi_1^0} = 88.1$ GeV. The neutralino χ_1^0 escapes detection, thus the experimental signature is an acoplanar $\mu^+\mu^-$ pair plus missing energy. The cross section is $\mathcal{O}(20$ fb) and simple selection criteria give an efficiency of $\epsilon \simeq 0.5$. The most severe background comes from $e^+e^- \to W\,W$ with $W \to \mu\,\nu$. This background can be reduced by an order of magnitude by running with right-handed polarised electrons of $P_{e_R^-} = 90\,\%$. Simultaneously, the signal can be enhanced by a factor of two.

The masses of the scalar muon $\tilde{\mu}_R$ and the neutralino χ_1^0 can be accessed through the energy spectrum of the final state leptons. The scalar muon decays isotropically and the two-body kinematics results in a flat muon energy distribution with endpoint energies

$$\frac{m_{\tilde{\mu}_R}}{2}\left(1-\frac{m^2_{\chi_1^0}}{m^2_{\tilde{\mu}_R}}\right)\gamma\,(1-\beta)\ \le E_\mu \le\ \frac{m_{\tilde{\mu}_R}}{2}\left(1-\frac{m^2_{\chi_1^0}}{m^2_{\tilde{\mu}_R}}\right)\gamma\,(1+\beta)\ .$$

The lepton energy spectra for unpolarised and right-handed polarised electrons are shown in figure 2. The sharp rise at E_μ^{min} is most sensitive to

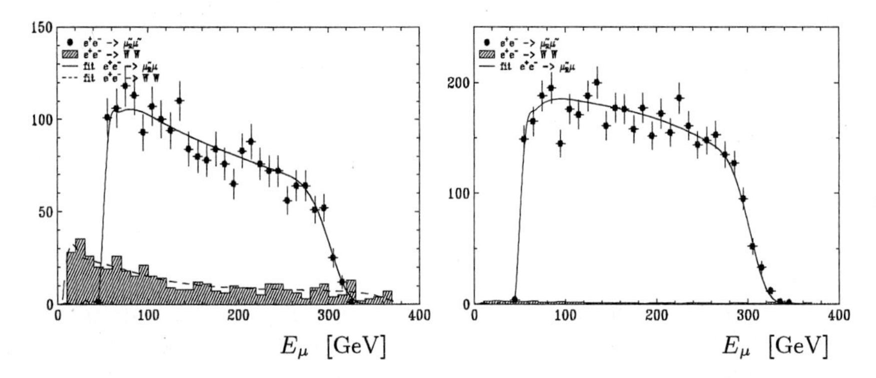

Fig. 1. Left: Energy spectrum of the final state $\mu^\pm$ for the reaction $e^+\,e^- \to \tilde{\mu}_R\,\tilde{\mu}_R$ (• and —) and the background $e^+\,e^- \to W\,W$ (histogram), assuming $\mathcal{L} = 125$ fb^{-1} at $\sqrt{s} = 800$ GeV. **Right:** Same for $e^+\,e^-_R \to \tilde{\mu}_R\,\tilde{\mu}_R$, polarisation $P_{e^-_R} = 90\,\%$

$m_{\tilde{\mu}_R}$, while E_μ^{max} is more sensitive to $m_{\chi_1^0}$. A two-parameter mass fit yields $m_{\tilde{\mu}_R} = 274.1\,^{+1.7}_{-1.9}$ GeV and $m_{\chi_1^0} = 87.2\,^{+2.1}_{-2.4}$ GeV for unpolarised beams, respectively $m_{\tilde{\mu}_R} = 275.3\,^{+0.9}_{-0.7}$ GeV and $m_{\chi_1^0} = 88.2\,^{+1.8}_{-1.6}$ GeV for right-handed polarised electrons.

Threshold Scan A more precise $\tilde{\mu}_R$ mass determination can be obtained by a scan around the production threshold. The signature are two almost monoenergetic leptons, not spoiled by cascade decays. A fit to the excitation curve, shown in figure 2, gives $m_{\tilde{\mu}_R} = 275.1\,^{+0.35}_{-0.30}$ GeV. At the same time a cross section measurement, $\sigma \propto \beta^3$, allows to determine the spin 0 of the scalar muon.

3 Charginos

Mass determination of $m_{\chi_1^\pm}$ and $m_{\chi_1^0}$ Chargino production $\chi^\pm$ occurs via vector bosons γ, Z in the s channel and via $\tilde{\nu}_e$ exchange in the t channel. A study of the lightest chargino production $\chi_1^\pm$ and its decay into $W^*\,\chi_1^0$ at $\sqrt{s} = 500$ GeV has been performed [3]

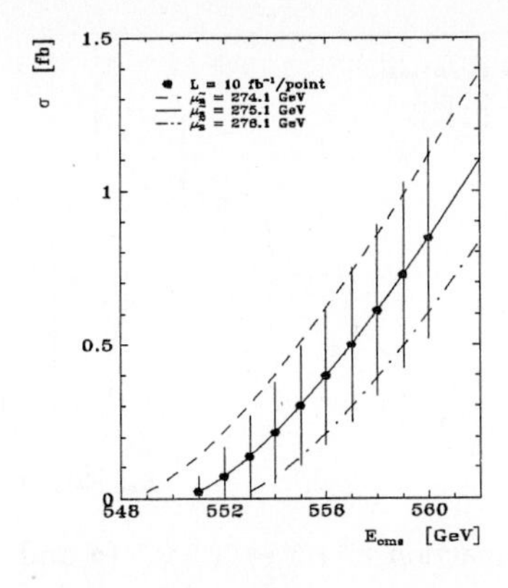

Fig. 2. Cross section for $e^+ e^-_R \to \tilde{\mu}_R \tilde{\mu}_R$ around the kinematic threshold, assuming $\mathcal{L} = 10\ \text{fb}^{-1}$ per measurement (•). The curves indicate scalar muon masses of $m_{\tilde{\mu}_R} = 275.1 \pm 1.0$ GeV

$$
\begin{aligned}
e^+ e^- &\to \chi_1^+ \chi_1^- \,, \\
\chi_1^\pm &\to \ell^\pm \nu \chi_1^0 \qquad \mathcal{BR} = 34.5 \,, \quad \ell = e,\ \mu,\ \tau \,, \\
&\to q \bar{q}' \chi_1^0 \qquad \mathcal{BR} = 65.5\,\% \,,
\end{aligned}
$$

assuming masses of $m_{\chi_1^\pm} = 168.2$ GeV, $m_{\chi_1^0} = 88.1$ GeV and $m_{\tilde{\nu}_e} = 304$ GeV. The experimental signature is an isolated lepton e or μ and two hadronic jets plus missing energy. The cross section reveives large t channel contributions, the χ_1^- being preferentially emitted into the forward direction. This behaviour is quite different from the dominant $e^+e^- \to W\,W$ background. The event rates are comfortably high ($\sigma_{\chi_1^+ \chi_1^-} = 245$ fb, $\epsilon \simeq 0.12$ mainly due to decay branching ratios) and the background amounts to $\sim 10\,\%$.

The masses of the chargino $m_{\chi_1^\pm}$ and the neutralino $m_{\chi_1^0}$ can be accessed through a study of the continuous energy spectrum of the di-jet system, shown in figure 3. A two-parameter fit to the spectrum yields a statistical accuracy of $m_{\chi_1^\pm} = 167.6 \pm 1.1$ GeV and $m_{\chi_1^0} = 88.2 \pm 0.6$ GeV.

Threshold Energy Scan An alternative method to determine the chargino mass with high precision is to measure the $\chi_1^+ \chi_1^-$ production cross section at the kinematic threshold, shown in figure 3. The chargino mass resolution, $m_{\chi_1^\pm} = 168.2 \pm 0.10$ GeV, can be improved by an order of magnitude with a moderate integrated luminosity. A measurement of the excitation curve, $\sigma \propto \beta\,(3 - \beta^2)/2$, allows to determine the spin 1/2 of the chargino $\chi_1^\pm$.

Polarised Beams Polarisation does not improve the signal to bacckground ratio of chargino production. However, polarised beams offer the possibility to disentangle the s and $\tilde{\nu}_e$ t channel diagrams and to determine the gaugino and higgsino components of the chargino: $\chi_1^\pm = \alpha\,\tilde{W}^\pm + \beta\,\tilde{H}^\pm$. At high energy γ and Z decouple to B^0/W^3 and the wino couples only to left-handed electrons, while the higgsino couples to both left-handed and right-handed electrons (s channel). Thus, precise measurements of σ_L and σ_R allow to determine the chargino mixing parameters. With a luminosity of $\mathcal{L} = 10\ \text{fb}^{-1}$ one expects statistical accuracies of $\delta\sigma_L \simeq 4.5\,\%$ and $\delta\sigma_R \simeq 15\,\%$.

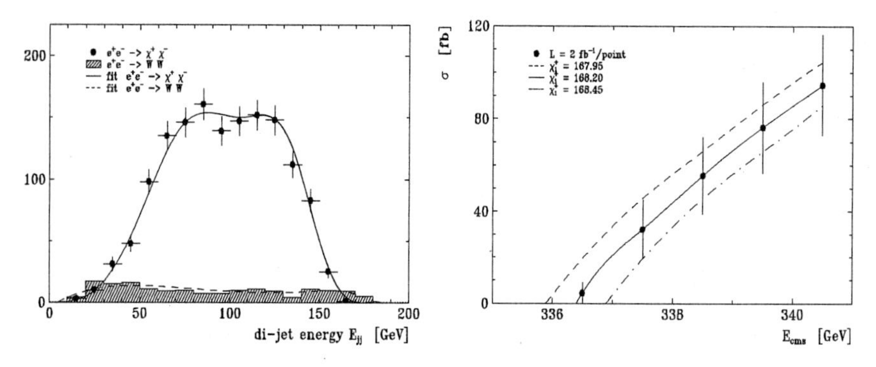

Fig. 3. Left: Di-jet energy spectrum E_{jj} of the reaction $e^+ e^- \to \chi_1^+ \chi_1^-$ (• and —) and the background reaction $e^+ e^- \to W W$ (histogram and $--$), assuming $\mathcal{L} = 50\ \text{fb}^{-1}$ at $\sqrt{s} = 500$ GeV. **Right:** Cross section for $e^+ e^- \to \chi_1^+ \chi_1^-$ around the kinematic threshold assuming $\mathcal{L} = 2\ \text{fb}^{-1}$ per measurement (•). The curves indicate chargino masses of $m_{\chi_1^\pm} = 168.2 \pm 0.25$ GeV

4 Scalar Top Quark

The scalar top quark $\tilde{t}_1$ may be the lightest squark. Due to large Yukawa terms its mass may evolve to lower values than for the first generations and the off-diagonal elements of the mass matrix can be large leading to strong $\tilde{t}_R - \tilde{t}_L$ mixing in the mass eigenstates

$$\begin{aligned} \tilde{t}_1 &= \quad \cos\theta_{\tilde{t}}\, \tilde{t}_R + \sin\theta_{\tilde{t}}\, \tilde{t}_L \ , \\ \tilde{t}_2 &= -\sin\theta_{\tilde{t}}\, \tilde{t}_R + \cos\theta_{\tilde{t}}\, \tilde{t}_L \ . \end{aligned}$$

Scalar top production has been studied at $\sqrt{s} = 500$ GeV assuming a mixing angle of $|\cos\theta_{\tilde{t}}| = 0.57$, *i.e.* implying low cross sections, and the following masses and decay modes [3]

$$\begin{aligned} e^+ e^- &\to \tilde{t}_1 \tilde{t}_1 & m_{\tilde{t}_1} &= 180\ \text{GeV}\ , \\ \tilde{t}_1 &\to b\, \chi_1^\pm & m_{\chi_1^\pm} &= 150\ \text{GeV}\ , \\ &\to c\, \chi_1^0 & m_{\chi_1^0} &= 100\ \text{GeV}\ . \end{aligned}$$

Since the final state contains many invisible particles, a direct mass reconstruction is impossible. However, the event topology can be clearly identified and polarised electron cross section measurements can be used to determine the scalar top mass and mixing angle. For a polarisation of $P_{e^-} = \pm 90\,\%$ and $\mathcal{L} = 10\ \text{fb}^{-1}$ one expects a precision of $\sigma_L(\tilde{t}_1 \tilde{t}_1) = 48.6 \pm 6.0$ fb and $\sigma_R(\tilde{t}_1 \tilde{t}_1) = 46.1 \pm 4.9$ fb. The results are displayed in figure 4, leading to a measurement of the scalar top mass of $m_{\tilde{t}_1} = 180 \pm 7$ GeV and a mixing angle of $|\cos\theta_{\tilde{t}}| = 0.57 \pm 0.06$.

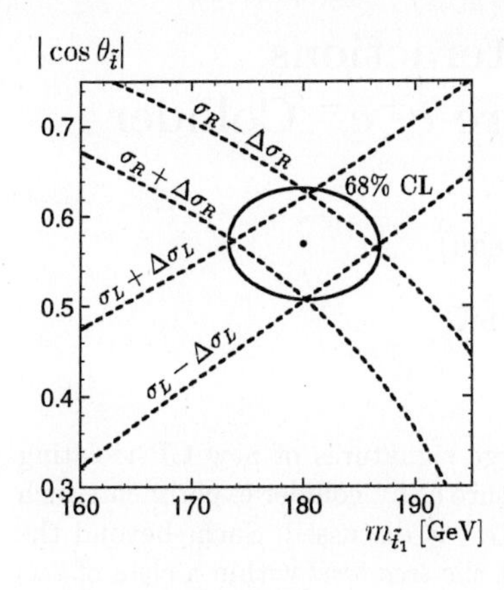

Fig. 4. Measurement of the stop mass $m_{\tilde{t}_1}$ and the mixing angle $\theta_{\tilde{t}}$ in the reactions $e^+ e^-_L \to \tilde{t}_1 \bar{\tilde{t}}_1$ and $e^+ e^-_R \to \tilde{t}_1 \bar{\tilde{t}}_1$ assuming $\mathcal{L} = 10\ \text{fb}^{-1}$ at $\sqrt{s} = 500$ GeV

5 Conclusions

The physics programme of a future e^+e^- linear collider extends the SUSY particle searches at LEP and is in many aspects complementary to LHC. The great advantage of e^+e^- colliders is the high precision of experimental analyses – masses, cross sections and couplings – covering the kinematically accessible domain in a conclusive manner. An overconstrained set of observables allows to check the basic SUSY–GUT parameters, *e.g.* the gaugino mass relations $M_i = \alpha_i/\alpha_{\text{GUT}}\, m_{1/2}$ and slepton mass differences, to a high precision. At a future e^+e^- collider with an energy upgrade 500 GeV $\to$ 1 TeV $\to$ 2 TeV and highly polarised beams the complete SUSY particle spectrum may be explored.

References

[1] Proceedings *'$e^+ e^-$ Collisions at 500 GeV: The Physics Potential'*, ed. P.M. Zerwas, DESY 92-123 A+B, DESY 93-123 C and DESY 96-123 D.
[2] T. Tsukamoto *et al.*, Phys. Rev. D 51 (1995) 3153.
[3] *'Conceptual Design Report of a 500 GeV $e^+ e^-$ Linear Collider'*, DESY 1997-048 and http://www.desy.de/HEP/desy-hep.html.

Probing New Higgs-Top Interactions at the Tree-Level in a Future e^+e^- Collider

Shaouly Bar-Shalom (shaouly@phyun0.ucr.edu)

University of California, Riverside CA 92521, USA

Abstract. [1,2] The possibility of observing large signatures of new CP-violating and flavor-changing Higgs-Top couplings in a future e^+e^- collider experiments such as $e^+e^- \to t\bar{t}h$, $t\bar{t}Z$ and $e^+e^- \to t\bar{c}\nu_e\bar{\nu}_e$, $t\bar{c}e^+e^-$ is discussed. Such, beyond the Standard Model, couplings can occur already at the *tree-level* within a class of two Higgs doublets models.

1 Introductory Remarks

The simplest extension of the SM with an additional scalar doublet, can give rise to rich new phenomena beyond the SM associated with top-Higgs systems, e.g., tree-level CP-violation and tree-level flavor-changing-scalar (FCS) transitions, in interaction of neutral scalars with the top quark.

In this talk we present four distinct reactions which are very powerful probes of the tth and $t\bar{c}h$ Yukawa couplings [1]. The first two reactions, $e^+e^- \to t\bar{t}h$ and $e^+e^- \to t\bar{t}Z$, exhibit large CP-violating asymmetries, at the order of tens of percent, already at the *tree graph level.* These CP-nonconserving effects are driven by the CP-phase in the $\mathcal{H}t\bar{t}$ interaction lagrangian piece of a general two Higgs doublets model (2HDM) ($\mathcal{H}$ is a neutral Higgs particle):

$$\mathcal{L}_{\mathcal{H}tt} = -\frac{g_W}{\sqrt{2}}\frac{m_t}{m_W}\mathcal{H}\bar{t}\left(a_t^{\mathcal{H}} + i b_t^{\mathcal{H}}\gamma_5\right)t\ , \tag{1}$$

Where the couplings $a_t^{\mathcal{H}}$ and $b_t^{\mathcal{H}}$ depend on the ratio between the two vacuum expectation values, $\tan\beta \equiv v_u/v_d$, and on the three Euler angles, $\alpha_{1,2,3}$ which parameterize the neutral Higgs mixing matrix [1].

The other two reactions are *tree-level* $t\bar{c}$ production through the W^+W^- and ZZ fusion processes, $e^+e^- \to W^+W^-\nu_e\bar{\nu}_e \to t\bar{c}\nu_e\bar{\nu}_e$ and $e^+e^- \to ZZe^+e^- \to t\bar{c}e^+e^-$, respectively, which appear to be extremely sensitive to a $t\bar{c}h$ FCS interaction. The FCS interactions in these reactions arise from

[1] For an expanded version of this talk see: S. Bar-Shalom, hep-ph/9710355. This talk is based on works done in collaboration with D. Atwood, G. Eilam, A. Soni and J. Wudka.

[2] In what follows we refer only to the expanded version of this talk [1]. Additional references relevant for this discussion are listed there.

the $tc\mathcal{H}$ lagrangian piece of a 2HDM in which no discrete symmetries are imposed:

$$\mathcal{L}_{\mathcal{H}tc} = -\frac{g_W}{\sqrt{2}} \frac{\sqrt{m_t m_c}}{m_W} f^{\mathcal{H}} \mathcal{H} \bar{t} (\lambda_R + i\lambda_I \gamma_5) c \,, \tag{2}$$

where $f^{\mathcal{H}}$ depends on the mixing angle $\tilde{\alpha}$ which is determined by the parameters of the Higgs potential. Also, we note that $\lambda_R, \lambda_I \sim \mathcal{O}(1)$ is not ruled out by existing experimental data [1].

For reasons explained in [1], only two out of the three neutral Higgs particles of these 2HDMs are relevant for both the flavor changing effect and CP-violating effect presented in this talk. We denote them by h and H, corresponding to the light and heavy neutral Higgs-boson, respectively.

2 Tree-Level CP-Violation in $e^+e^- \to t\bar{t}h$, $t\bar{t}Z$

A novel feature of these reactions is that the effect arises at tree graph level. Basically, for the $tth(ttZ)$ final states, Higgs(Z) emission off the $t, \bar{t}$ interferes with the Higgs(Z) emission off the s-channel Z-boson (see Fig. 1). The CP-odd asymmetries are therefore very large and may reach $10 - 30\%$.

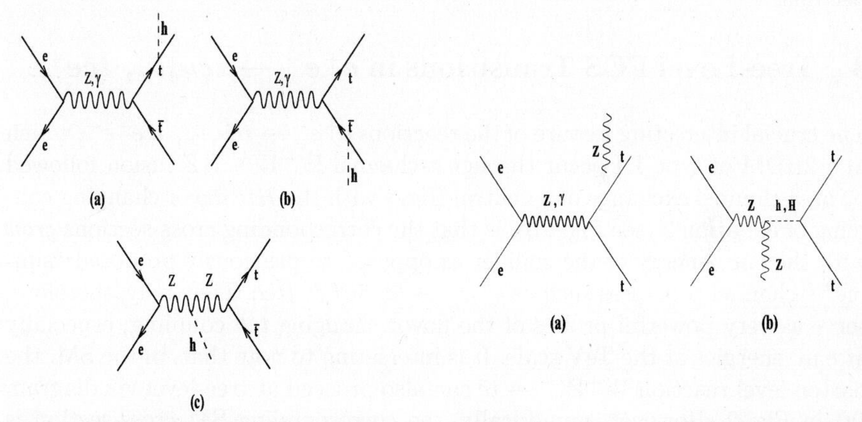

Fig. 1. Tree-level Feynman diagrams contributing to $e^+e^- \to t\bar{t}h$ (left hand side) and $e^+e^- \to t\bar{t}Z$ (right hand side) in the unitary gauge, in a two Higgs doublet model. For $e^+e^- \to t\bar{t}Z$, Diagram (a) on the right hand side represents 8 diagrams in which either Z or γ are exchanged in the s-channel and the outgoing Z is emitted from e^+, e^-, t or $\bar{t}$.

In Table 1 we give the statistical significance, N_{SD}, with which these CP-asymmetries may be detected in a future high energy e^+e^- linear collider. The CP-asymmetries were calculated within a 2HDM of type II. The numbers in the Table are given for $\{\alpha_1, \alpha_2, \alpha_3\} = \{\pi/2, \pi/4, 0\}$ and we have used $\tan\beta = 0.5(0.3)$ for the $t\bar{t}h(t\bar{t}Z)$ case. The heavy Higgs mass is set to be

$m_H = 1$ TeV. Also, a yearly integrated luminosity of $\mathcal{L} = 200$ and 500 $[\text{fb}]^{-1}$ was assumed for $\sqrt{s} = 1$ and 1.5 TeV, respectively, and an efficiency reconstruction factor of $\epsilon = 0.5$ for both energies.. We see that in the best cases a 3-4 sigma CP-violating signal will be possible.

		N_{SD} for $e^+e^- \to t\bar{t}h(t\bar{t}Z)$		
$\sqrt{s}$	j	O_{opt}		
(TeV) ⇓	(GeV) ⇒	$m_h = 100$	$m_h = 160$	$m_h = 360$
1	-1	2.2(1.7)	2.0(1.8)	1.1(2.2)
	unpol	2.0(1.6)	1.9(1.6)	1.0(2.0)
	1	1.8(1.5)	1.7(1.5)	0.9(1.8)
1.5	-1	4.0(2.9)	3.9(3.0)	3.2(3.3)
	unpol	3.6(2.6)	3.5(2.7)	2.9(3.0)
	1	3.2(2.3)	3.1(2.3)	2.6(2.6)

Table 1. The statistical significance, N_{SD}, in which the CP-nonconserving effects in $e^+e^- \to t\bar{t}h$ and $e^+e^- \to t\bar{t}Z$ (in parenthesis) can be detected in one year of running of a future high energy e^+e^- linear collider with either unpolarized or polarized incoming electron beam. $j = 1(-1)$ stands for right(left) polarized electrons.

3 Tree-Level FCS Transitions in $e^+e^- \to t\bar{c}\nu_e\bar{\nu}_e$, $t\bar{c}e^+e^-$

The crucial interesting feature of the reactions $e^+e^- \to t\bar{c}\nu_e\bar{\nu}_e$, $t\bar{c}e^+e^-$, which in a 2HDM of type III occur through t-channel W^+W^-, ZZ fusion followed by an $\hat{s}$-channel exchange of a neutral Higgs with the $\mathcal{H}t\bar{c}$ flavor changing coupling of equation 2 (see Fig. 2b), is that the corresponding cross-sections *grow* with the c.m. energy of the collider as opposed to previously proposed "simple" s-channel processes such as $e^+e^- \to t\bar{c}$; $t\bar{c}f\bar{f}$; $t\bar{t}c\bar{c}$. They may, therefore, serve as very powerful probes of the flavor changing tch coupling, especially at c.m. energies at the TeV scale. It is interseting to note that, in the SM, the parton level reaction $W^+W^- \to t\bar{c}$ can also proceed at tree-level via diagram (a) in Fig. 2. However, numerically, the corresponding SM cross-section is found to be vanishingly small, of no experimental relevance, due to a severe CKM suppression [1]. The ZZ fusion process is even smaller in the SM as, apart from being GIM supressed, it occurs at one-loop.

In Fig. 3 we show the dependence of the scaled W^+W^- fusion cross-section, $\sigma^{\nu\nu tc}/\lambda^2$ (we have set $\lambda_I = 0$ and $\lambda_R \equiv \lambda$ in equation 2) on the mass of the light Higgs m_h for four values of s and for $\tilde{\alpha} = \pi/4$, i.e., $f^h = f^H$ in equation 2. Evidently, the cross-section is of the order of ~few fb with a Higgs mass at the order of the EW scale. We also find that the ZZ fusion process exhibits the same behaviour, however, it is typically one order of magnitude smaller than the W^+W^- one, basically due to the difference between the W and the Z structure functions. Therefore, in a high energy e^+e^- linear

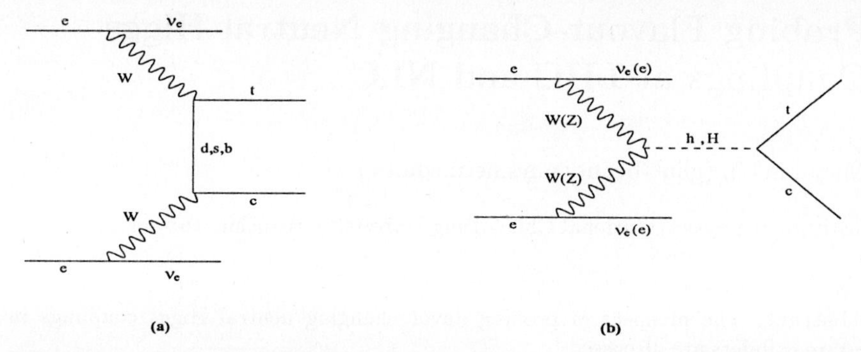

Fig. 2. (a) The Standard Model diagram for $e^+e^- \to t\bar{c}\nu_e\bar{\nu}_e$; (b) Diagrams for $e^+e^- \to t\bar{c}\nu_e\bar{\nu}_e(t\bar{c}e^+e^-)$ in a 2HDM of type III.

collider, running at energies of $\sqrt{s} = 1 - 2$ TeV and with a yearly integrated luminosity of the order of $\mathcal{L} \gtrsim 10^2$ [fb]$^{-1}$, a 2HDM of type III (with $\lambda = 1$) predicts hundreds and up to thousands of $t\bar{c}\nu_e\bar{\nu}_e$ events and several tens to hundreds of $t\bar{c}e^+e^-$ events.

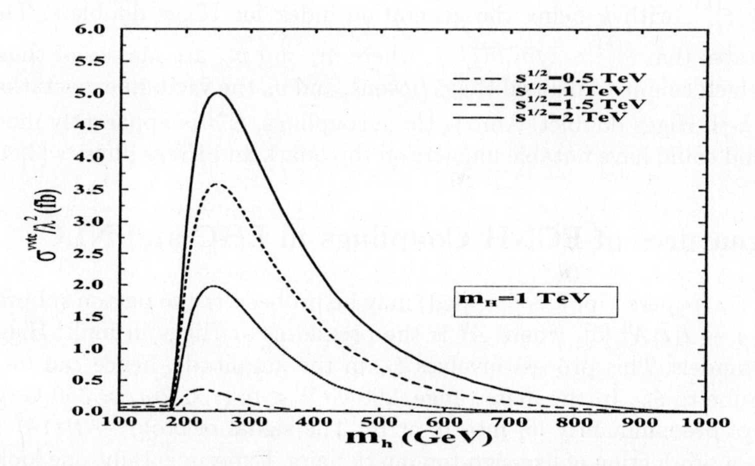

Fig. 3. The cross-section $\sigma^{\nu\nu tc} \equiv \sigma(e^+e^- \to t\bar{c}\nu_e\bar{\nu}_e + \bar{t}c\nu_e\bar{\nu}_e)$ in units of λ^2 as a function of m_h for $\sqrt{s} = 0.5, 1, 1.5$ and 2 TeV. $\tilde{\alpha} = \pi/4$ and $m_H = 1$ TeV.

References

[1] S. Bar-Shalom, hep-ph/9710355.

Probing Flavour-Changing Neutral Higgs Couplings at LHC and NLC

Guey-Lin Lin (glin@beauty.phys.nctu.edu.tw)

Institute of Physics, National Chiao-Tung University, Hsinchu, Taiwan

Abstract. The prospect of probing flavor changing neutral Higgs couplings in future colliders are discussed.

1 Flavor changing neutral Higgs couplings

In the electroweak theory, the quark masses and CKM matrix elements display an interesting hierarchy pattern. Inspired by this pattern, Cheng and Sher suggested [1] that low energy FCNC can be naturally suppressed without invoking discrete symmetries [2]. For multi-Higgs-doublet models, they proposed an ansatz for flavor changing neutral Higgs (FCNH) couplings denoted by $\xi_{ij}^{(k)}$, with k being the generation index for Higgs doublets. The ansatz states that $\xi_{ij}^{(k)} \sim \sqrt{m_i m_j}/v_k$, where m_i and m_j are masses of those quarks which couple to neutral Higgs bosons, and v_k the vacuum expectation value of k-th Higgs doublet. Among these couplings, $\xi_{tc}^{(k)}$ is apparently most sizable and could have notable impacts on top quark and Higgs physics [3, 4].

2 Signatures of FCNH Couplings in LHC and NLC

In LHC, ξ_{tc} (superscript k is dropped) may be probed via the parton subprocess $c(\bar{c})g \to t(\bar{t})A^0$ [5], where A^0 is the pseudoscalar Higgs in multi-Higgs doublet models. This process involves ξ_{tc} in the amplitude, hence can be a direct probe to ξ_{tc}. In the mass range, 200 GeV $< m_{A^0} < 2m_t \approx 350$ GeV, A^0 decays predominantly [6] into $t\bar{c}$ or $\bar{t}c$. The signal of $c(\bar{c})g \to t(\bar{t})A^0$ is therefore a production of like-sign top quark pairs. Experimentally, one looks for like-sign dilepton events, accompanied by two b-jets, large missing energy, plus one additional jet,

$$cg \to tA^0 \to \ell_1^+ \ell_2^+ \nu\nu + bb + \bar{c}, \tag{1}$$

and similarly for $\bar{c}g \to \bar{t}A^0 \to \ell_1^- \ell_2^- \bar{\nu}\bar{\nu} + \bar{b}\bar{b} + c$. Taking the Cheng-Sher ansatz and choosing $m_{A^0} = 250$ GeV, we found $\sigma(pp \to t(\bar{t})A^0 + X) =$ 37 fb in LHC. With an integrated luminosity of 100 fb^{-1} and 50% double b-tagging efficiency, we expect for each $\ell^+\ell^+$ and $\ell^-\ell^-$ modes 40 events per year. The event rate for other m_{A^0} values can be read off from Fig. 2 of Ref. [5] and Fig. 1 of Ref. [6], which respectively show the m_{A^0} dependences of $\sigma(pp \to t(\bar{t})A^0 + X)$ and BR$(A^0 \to t\bar{c},\ \bar{t}c)$. The major background is $q\bar{q}' \to W^+(W^-)t\bar{t}$ [8] with both top quark decaying semi-leptonically. The resulting final state also have like-sign dileptons accompanied by two b-jets. It however has *two* addtional jets, while the signal has only one. The background rate is found to be few times greater than the signal. However we expect a naïve jet counting can sufficiently suppress it [5].

What is the prospect of probing FCNH couplings at NLC? Previously it was suggested that ξ_{tc} may be probed at NLC via $e^+e^- \to Z^* \to t\bar{c}$ [7] or $e^+e^- \to Z^* \to h(H)A \to tt\bar{c}\bar{c}$ [6]. However the former process is loop suppressed, producing less than 0.1 $t\bar{c}$ event annually for a 500 GeV NLC with 50 fb^{-1} integrated luminosity. The latter process gives rise to intriguing like-sign top pair final states. Unfortunately its event rate is also low, only a few like-sign dilepton events produced per year. As both processes are s-channel type, increasing CM energy of NLC does not improve their event rates. Recently, a t-channel process $e^+e^- \to W^+W^-\bar{\nu}_e\nu_e \to t\bar{c}(\bar{t}c)\bar{\nu}_e\nu_e$ was suggested as a probe to ξ_{tc} [9]. There the $t\bar{c}(\bar{t}c)$ pairs are produced via W-fusion with neutral Higgs bosons H^0 and h^0 as intermediate states. The authors of Ref. [9] focus on the mass range where either H^0 or h^0 is as heavy as 1 TeV. We however focused on the mass range 200 GeV $< m_{H^0},\ m_{h^0} < 2m_t \approx 350$ GeV, and obtained a greater $t\bar{c}(\bar{t}c)$ production cross section. For $\sqrt{s} \geq 1$ TeV, the total cross section of $t\bar{c}$ and $\bar{t}c$ production is always greater than 1 fb for the above range of Higgs masses. This amounts to an annual production of 50 $t\bar{c}$ or $\bar{t}c$ pairs. With CM energy raised to 2 TeV, the cross section can almost reach 10 fb for optimal choice of parameters. This mode is clearly very promising for probing ξ_{tc} in NLC.

This work is supported in part by National Science Council of R.O.C. under the grant number NSC 87-2112-M-009-038.

References

[1] T. P. Cheng and M. Sher, Phys. Rev. D35 (1987), 3484.

[2] S. L. Glashow and S. Weinberg, Phys. Rev. D15 (1977), 1958.

[3] W. S. Hou, Phys. Lett. B296 (1992), 179.

[4] L. J. Hall and S. Weinberg, Phys. Rev. D48 (1993), R979.

[5] W.-S. Hou, G.-L. Lin, C.-Y. Ma and C. P. Yuan, Phys. Lett. B409 (1997), 344.

[6] W.-S. Hou and G.-L. Lin, Phys. Lett. B379 (1996), 261.

[7] D. Atwood, L. Reina and A. Soni, Phys. Rev. D53 (1996), 1199.

[8] Z. Kunszt, Nucl. Phys. B247 (1984), 339; V. Barger *et al.*, Phys. Rev. D42 (1990), 3052.

[9] S. Bar-Shalom, G. Eilam, A. Soni and J. Wudka, Phys. Rev. Lett. 79 (1997), 1217.

[10] W.-S. Hou, G.-L. Lin and C.-Y. Ma, hep-ph/9708228, to appear in Phys. Rev. D.

The HERA-B Experiment

Peter Križan (peter.krizan@ijs.si), on behalf of the HERA-B collaboration

University of Ljubljana and J. Stefan Institute, Ljubljana, Slovenia

Abstract. The HERA-B experiment, under construction at the DESY laboratory, aims at measuring CP violation in the B system. In the report, the status of the experiment is reviewed together with its expected physics reach.

Motivation and reach

The HERA-B experiment aims at measuring: CP violation in $B^0 \to J/\psi K^0_S$ and $B^0 \to \pi^+\pi^-$ decays, $B^0_s\bar{B}^0_s$ mixing and $B\bar{B}$ decays with two leptons in the final state. The B mesons will be produced in fixed target collisions of protons with momenta up to 1000 GeV/c. The target will consist of 8 wires in the halo of the proton beam in order not to disturb experiments measuring ep collisions.

With the HERA-B spectrometer [1], currently under construction at DESY, Hamburg, the CP reach can be summarized by quoting the expected error in the asymmetry measurement $\Delta sin2\beta = 0.13$ after one year of running. In the present contribution, we shall briefly present the spectrometer and discuss some recent modifications in the design. Results of the tests with a partially equipped spectrometer will be discussed together with the plan for the 1998/99 running.

The spectrometer

The HERA-B apparatus is a spectrometer with a dipole magnet of 2 Tm field integral. The set-up (Fig. 1) reflects the fixed target nature of the experiment, with finer detector granularity for small production angles, where higher occupancies and high momenta tracks are expected.

The tracking of charged particles is performed by a Si-strip vertex detector, followed by an outer tracking system (honeycomb drift chambers [2]) at distances from the proton beam larger than 20 cm, and an inner tracking system, Si-strip detectors and microstrip gas chambers closer to the beam.

Kaon identification is performed by a RICH counter, which uses a gas (C_4F_{10}) as the radiator, and a photon detector built of multianode PMTs. Electron identification is provided by an electromagnetic calorimeter of the shashlik type. At small production angles it is supplemented by a TRD counter using a fibre radiator and straw tubes as X-ray detector. Muons are identified in the muon system, composed of three chamber types, gas tubes, pad and pixel chambers, organized in four chamber systems, the first three interleaved with absorber.

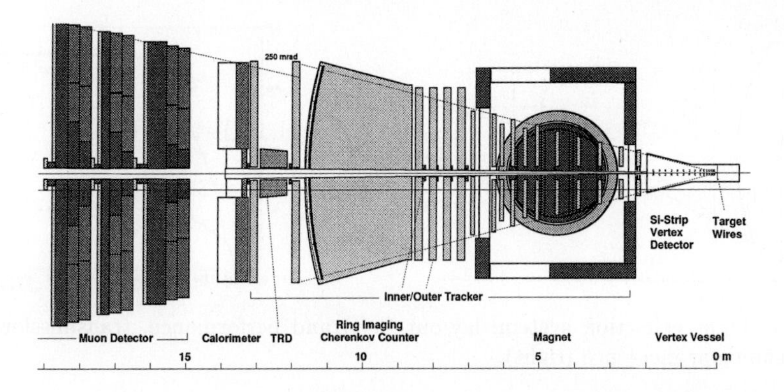

Fig. 1. The HERA-B spectrometer, top view.

The huge background of inelastic proton-nucleon scattering over the signal (10^{11}) requires a highly selective and efficient trigger [3]. In particular the first level is of high importance with a latency of 12 μs and a rate reduction factor of 200. It searches for e^+e^- and $\mu^+\mu^-$ track pairs with a high invariant mass. In the subsequent stages the reconstructed tracks are fitted, and a search for a common vertex is performed. The last stage is the complete event reconstruction [4]. An additional high p_T trigger looking for pions from the decay $B^0 \rightarrow \pi^+\pi^-$ will be provided by a dedicated system of gas pad and pixel chambers.

Two major modifications have been carried out in the design of the spectrometer since the proposal in 1995. They concern the photon detector of the RICH counter, as well as the main component of the inner tracking system, the microstrip gas chambers.

RICH: detection of photons The initially foreseen detectors based on wire chambers were abandoned; the TMAE detector showed a prohibitive decrease of avalanche gain due to aging effects, while the CsI photocathode could not be routinely produced and maintained with sufficiently high quantum efficiency in addition to problems with rates in excess of a few kHz per pixel [5]. The final photon detector thus consists of Hamamatsu R5900 M16 and M4 photomultipliers for which it was proven that they can be used for single photon counting with a high efficiency and very low cross-talk [6]. Due to the smaller photocathode surface compared to the photomultiplier cross section, a two lens demagnification system (2:1) was designed (Fig. 2) in order to adjust the required pixel size to the PMT pad size. The granularity of the photon detector (either 9 x 9 mm^2 or 18 x 18 mm^2) has been chosen on the basis of expected occupancy and the required position resolution.

From beam tests with 3 GeV electrons, and from the measurements of the light collection efficiency and mirror reflectivity, the expected number of detected Čerenkov photons in the final detector can be estimated to 27. The results of a Monte Carlo calculation indicate that under these conditions

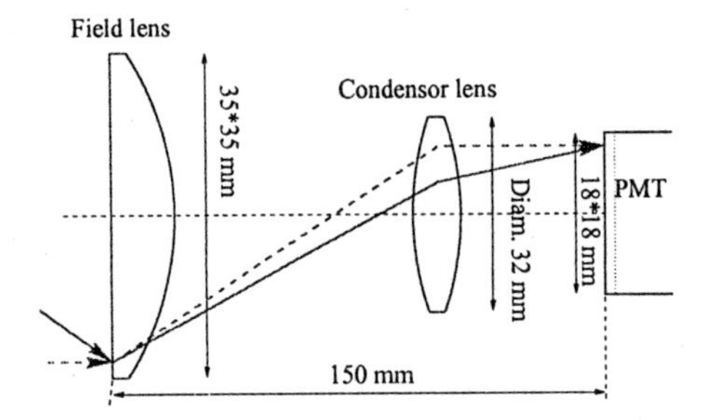

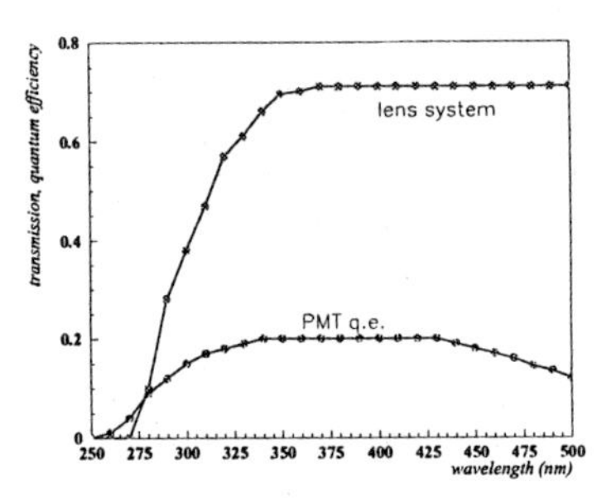

Fig. 2. Light collection system: lay-out (left) and performance, transmission for perpendicular incidence (right).

a total kaon identification efficiency of 75% with 5% pion misidentification probability could be achieved.

Inner tracking system The microstrip gas chamber (MSGC), the originally foreseen detector, was tested for ageing. While the tests with X rays showed satisfactory behaviour, it was discovered in hadron beam tests that heavily ionizing particles deposit locally enough charge to start discharges, which then destroy the gold anodes. To remedy this problem, other anode materials were tried instead of gold, e.g. Ro, Cr. No cure was found, since only anodes made of Cr were not destroyed, most probably because of their somewhat smaller conductivity, which, in turn, prohibitively reduces the signals.

In a second approach, the MSGC was combined with a preamplifier stage GEM (gas electron multiplier), a concept that was introduced by F. Sauli el al. [7], and extensively tested in the last year [8]. The results of tests look very promising and a system of 16 chambers will be prepared for the beginning of the 1998 data taking.

Results of the 96/97 measurements

The results of tests with a partially equipped spectrometer [9] can be summarized as follows. The design of the 1996/97 wire target set-up was chosen as close to the final one as possible, with two stations of four wires, with various target materials (C, Al, Cu). With these tests its was shown that automatic steering of wires at fixed rate works reliably, and rates around 40 MHz could be kept over several hours. A signal to noise ratio of around 15 was measured on the three double sided silicon detectors employed in the present set-up. By using recorded data it was possible to reconstruct tracks and to associate them with individual target wires.

One of the main questions to be answered by the test measurements was whether the measured rates in various detector components agree with the predictions of Monte Carlo simulations. This, in turn, is critically connected to the question whether in the first trigger level a fast and efficient reconstruction of lepton pairs is possible. As can be seen from Fig. 3, the agreement is

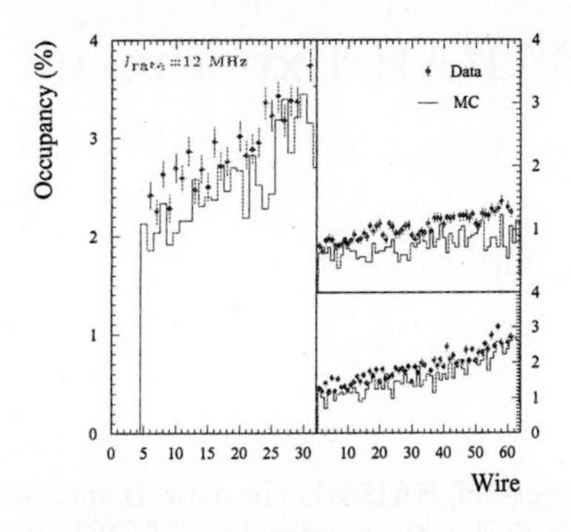

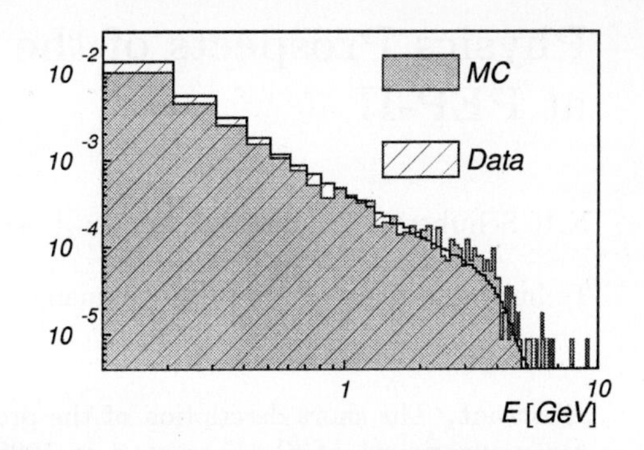

Fig. 3. Comparison of measured and predicted rates: occupancy in three different sections of the tracking system (left), spectrum of energies deposited in the calorimeter (right).

satisfactory, both in case of the occupancy in various tracking system sections, as well as for the spectrum of the deposited energy in the calorimeter.

Outlook: Time table

During the five months shutdown in winter/spring 1997/98 major parts of the spectrometer will be installed. The main purposes of the 1998 running period are the commissioning of the detector, as well as a measurement of the $b\bar{b}$ production cross section. In the next shutdown in winter 1998/99, the rest of the spectrometer will be mounted.

To summarize we note that the cohabitation of four HERA experiments was established at 40 MHz HERA-B interaction rate, that MC and data are in good agreement, and that long term high rate operation of all subdetectors was studied.

References

[1] T. Lohse et al. Proposal for HERA-B, DESY PRC-94/02, May 1994.
[2] C. Stegmann, Nucl. Instr. Meth. **A384**(1996) 196-200
[3] D. Ressing, Nucl. Instr. Meth. **A384**(1996) 131-135
[4] R. Mankel, Nucl. Instr. Meth. **A384**(1996) 201-206
[5] P. Križan et al., Nucl. Instr. and Meth. A 387 (1997) 146-149
[6] P. Križan et al., Nucl. Instr. and Meth. A 394 (1997) 27-34
[7] R. Bouclier et al., IEEE Trans. Nucl. Sci. 44 (1997) 646-650.
[8] A. Sharma, these proceedings.
[9] J. Spengler, Nucl. Instr. Meth. **A384**(1996) 106-112

Physics Prospects of the BABAR Experiment at PEP-II

K.R. Schubert (Schubert@physik.tu-dresden.de)

Technische Universität Dresden, Germany

Abstract. The short description of the prospects of BABAR, the new B meson decay experiment at SLAC to start in 1999, includes the parameters of PEP-II, possibilities for going beyond its design luminosity, and the main physics goals of BABAR. These include precision measurements of the sides and angles of the CKM Unitarity Triangle. With an integrated luminosity of 30/fb, corresponding to design luminosity and one year of operation, V_{cb} can be measured to ±5%, $|V_{\mathrm{ub}}|$ to ±10%, $\sin 2\beta$ to ±0.06, and $\sin 2\alpha$ to ±0.09.

1 Introduction

The main properties of the BABAR detector [1, 2] have been discussed at this Conference by Guy Wormser [3]. I will, therefore, limit the writeup of my contribution to the the physics programme and its potential with BABAR and PEP-II. The storage ring will be completed end of 1998, the detector beginning of 1999. Data taking will start mid 1999.

2 The PEP-II Storage Ring

The luminosity for an $\mathrm{e^+e^-}$ storage ring can be expressed as

$$\dot{\mathcal{L}} = 2.2 \cdot 10^{34}\ \mathrm{cm^{-2}s^{-1}} \cdot \frac{\Delta Q \cdot E \cdot I \cdot (1+r)}{\beta_y^*} \, \frac{\mathrm{cm}}{\mathrm{GeV \cdot A}} , \qquad (1)$$

where ΔQ is the tune shift, E the beam energy, I the beam current, r the ratio of vertical to horizontal beam dimensions, and β_y^* the vertical β function in the interaction point. For asymmetric beam energies in PEP-II with $E(\mathrm{e^-}) =$ 9.0 GeV and $E(\mathrm{e^+}) = m^2(\Upsilon 4\mathrm{S})/4E(\mathrm{e^-}) = 3.1$ GeV, the parameters ΔQ, I, r, and β^* of the two beams are adjusted for nearly equal values of $\Delta Q \cdot E \cdot I \cdot (1+r)/\beta_y^*$. With the initial design parameters as given in Table 1,

$$\dot{\mathcal{L}}(\mathrm{design}) = 3 \cdot 10^{33}/\mathrm{cm^2/s} \ . \qquad (2)$$

Reaching this ambitious goal will require long experience with the operating machine, and we expect that it can be obtained after one year of operation. The design luminosity is equivalent to two times the present CESR value scaled by the number of bunches per ring length.

Table 1: Initial design parameters for PEP-II. HER and LER = High and Low Energy Ring, IP = Interaction Point.

Parameter	HER	LER	Unit
Beam Energy	9.000	3.109	GeV
Center-of-Mass Energy	10.580		GeV
Circumference	2219.3		m
Number of Bunches	1658		
Bunch Spacing	1.26		m
Particles per Bunch	$2.7 \cdot 10^{10}$	$5.9 \cdot 10^{10}$	
Beam Current	0.99	2.14	A
Horizontal Emittance ϵ_x	48	64	nm · rad
Horizontal β Function β_x^* in the IP	50	37.5	cm
Horizontal Beam Size σ_x^* in the IP	155		μm
Vertical Emittance ϵ_y	1.9	2.6	nm · rad
Vertical β Function β_y^* in the IP	2.0	1.5	cm
Vertical Beam Size σ_y^* in the IP	6.2		μm
Tune Shift $\Delta Q_x = \Delta Q_y$	0.03		

Since the competing project in Japan, KEK-B, has a more daring design (bunch distance 0.6 m instead of 1.26 m, crossing angle 11 mrad instead of zero, superconducting RF cavities instead of normalconducting) with a potential for $\dot{\mathcal{L}} = 10^{34}/\text{cm}^2/\text{s}$, it is important to know the possibilities for later improvements of PEP-II. The number of bunches in each ring is limited by the zero crossing angle. In the case of good experience with the 11 mrad of KEK, the PEP-II interaction region could be rebuilt with non-zero crossing without major changes to the BABAR detector and to the rest of the storage ring. In addition to this increase in the number of bunches, i. e. in the beam currents, there could be room to change ΔQ from 0.03 to 0.06, β_x^* from 2 cm to 1 cm, and $1 + r$ from 1.04 to 2, i. e. from flat to round beams. All these changes do not factorize, but $\dot{\mathcal{L}} = 10^{34}/\text{cm}^2/\text{s}$ is not excluded for a later stage of PEP-II.

3 Main Physics Goals

The scientific goals of BABAR can be formulated in two simple questions:

- Is the Weak Interaction of the present Standard Model fully responsible for the observed CP violation in the K^0 meson system?
- Will the CP asymmetries in B meson decays be consistent with the Standard Model or will they show New Physics?

CP violation by the Weak Interaction in the Standard Model requires non-trivial imaginary parts in the elements of the Cabbibbo-Kobayashi-Maskawa matrix [4]. For its notation we use Wolfenstein's [5] parametrisation with λ, A, ρ, and η. This is not an approximation since the equalities

$$\sin\theta_{12} \equiv \lambda \ , \quad \sin\theta_{23} \equiv A\lambda^2 \ , \quad \sin\theta_{13}\mathrm{e}^{\mathrm{i}\delta} \equiv A\lambda^3(\rho + \mathrm{i}\eta) \ , \qquad (3)$$

with the Particle Data Group parameters θ_{12}, θ_{23}, θ_{13}, and δ fulfill strict unitarity of the matrix [6, 7]. Within a precision of 10^{-3} for ρ and η, i. e. sufficient for the lifetime of BABAR, we have

$$V_{\mathrm{td}} = A\lambda^3(1 - \overline{\rho} - \mathrm{i}\overline{\eta}) \ \text{with} \ \overline{\rho} = \rho(1 - \lambda^2/2) \ , \ \overline{\eta} = \eta(1 - \lambda^2/2) \ , \qquad (4)$$

and the unitarity triangle becomes

$$(1 - \lambda^2/2)V_{ub}^* - A\lambda^3 + V_{\mathrm{td}} = 0 \qquad (5)$$

in the $\overline{\rho}, \overline{\eta}$ plane as shown in Fig. 1. The bands in this figure represent the measured values and error bars [8] of $|V_{\mathrm{ub}}/V_{\mathrm{cb}}|$, $\Delta m(\mathrm{B}^0)$, and $\epsilon(\mathrm{K}^0)$. All measurements are in agreement with each other, i. e. with their Standard Model description. The shaded area is the allowed region for $\overline{\rho}$ and $\overline{\eta}$, it gives our present errors on these two Standard Model parameters.

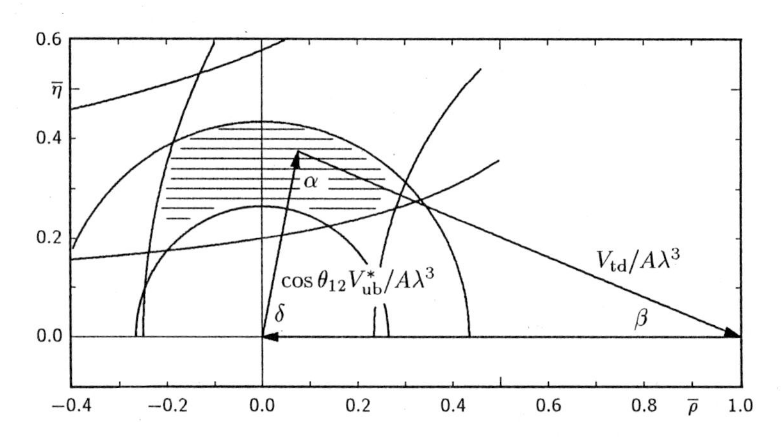

Fig. 1. The unitarity triangle of the CKM matrix in the $\overline{\rho}, \overline{\eta}$ plane as defined in equation 4.

The two goals of BABAR are precision measurements for the sides and angles of the Unitarity triangle in Fig. 1 in order

– to check if all Standard Model descriptions of these mesurements are consistent with each other, and if so
– to reduce the shaded area in Fig. 1.

4 Expected Contributions

The discussion of BABAR's potential will be limited here to the main program points obtained from running PEP-II on the $\Upsilon(4S)$ resonance,

- the reduction of $\sigma(V_{cb})$ from presently 10% to about 5%,
- the reduction of $\sigma(|V_{ub}|)$ from presently 25% to about 10%,
- the determination of $\sin 2\beta$ with $\sigma = 0.06/\sqrt{t_{run}/1y}$,
- the determination of $\sin 2\alpha$ with $\sigma = 0.09/\sqrt{t_{run}/1y}$, and
- a short list of additional physics with $\sqrt{s}(e^+e^-) = m(\Upsilon 4S)$.

4.1 Inclusive Determination of V_{cb}

The present discrepancy between the results for $\mathcal{B}(B \to \ell\nu X)$ from the $\Upsilon(4S)$ and from the Z^0 [9] requires precise remeasurement. BABAR plans to reuse the ARGUS method [10] to tag inclusive electrons of $p^* > 0.5$ GeV/c with leptons of $p^* > 1.4$ GeV/c from the second B meson. The precision will rely on the quality of electron identification which is expected from the quality of the CsI calorimeter, but which also requires large data samples of pure electrons (from radiative Bhabha events) and pure hadrons (from K^0 and D decays).

4.2 Exclusive Determination of V_{cb}

Theoretical uncertainties in the extraction of V_{cb} from the value of the Isgur-Wise function [11] at minimum $v \cdot v'$ have presently decreased to the 5% level. The experimental aim, to be reached within the first few years, is the same statistical error level in both $B \to D^*\ell\nu$ and $B \to D\ell\nu$ decays.

4.3 Exclusive Determination of $|V_{ub}|$

As already studied in detail for the B-Meson Factory at CERN [12] and successfully measured at CLEO with a small number of events [13], the decays $B \to \pi\ell\nu$, $B \to \rho\ell\nu$, and $B \to \omega\ell\nu$ can be reconstructed with low background over a wide range of the Dalitz plot by tagging them with events where the second B is "fully contained" in the detector, i. e. where the four-momenta of all decay particles, with the exception of one neutrino in the first B, are properly reconstructed. This requires BABAR's good charged particle identification and good photon measurement over a wide angular range. With a tagging efficiency between 5 and 10%, and with a reconstruction efficiency around 50% for the semileptonic decay, BABAR expects about 600 events in $\dot{\mathcal{L}}\, t_{run} = 30/\text{fb}$ and signal to background ratios better than 3. Given the measured two-dimensional decay fractions $d\mathcal{B}/dp_\ell dq^2$, theoretical uncertainties on $|V_{ub}|$ should then be on the 10% level.

4.4 Inclusive Determination of $|V_{\mathrm{ub}}|$

There will be no improvement over the present CLEO result [14] from $\mathrm{B} \to \ell\nu\mathrm{X}$ with $p_\ell^* > 2.3$ GeV/c if only this part of the lepton spectrum beyond the kinematical limit from decays with charm is used again. Extension to lower lepton momenta requires exclusive reconstruction of the second B, with efficiencies between 0.5% and 1%, and inclusive mass determination of the hadronic system in the semileptonic decay.

4.5 Determination of sin 2β in $\mathrm{B} \to \mathrm{J}/\Psi\ \mathrm{K}^0_\mathrm{S}$ Decays

Observation of the CP asymmetry $A_{\Psi\mathrm{K}} = [N(\mathrm{J}/\Psi\ \mathrm{K}^0_\mathrm{S}, \overline{\mathrm{T}}) - N(\mathrm{J}/\Psi\ \mathrm{K}^0_\mathrm{S}, \mathrm{T})] / [N(\mathrm{J}/\Psi\ \mathrm{K}^0_\mathrm{S}, \overline{\mathrm{T}}) + N(\mathrm{J}/\Psi\ \mathrm{K}^0_\mathrm{S}, \mathrm{T})]$, where the tag T comes from a B^0 decay and $\overline{\mathrm{T}}$ from a $\overline{\mathrm{B}}^0$, requires measurement of both times t of the $\mathrm{J}/\Psi\ \mathrm{K}^0_\mathrm{S}$ decay and t_T of the tag decay since the time-integrated asymmetry is zero;

$$A(t, t_\mathrm{T}) = \frac{N(\mathrm{J}/\Psi\mathrm{K}^0_\mathrm{S}, t; \overline{\mathrm{T}}, t_\mathrm{T}) - N(\mathrm{J}/\Psi\mathrm{K}^0_\mathrm{S}, t; \mathrm{T}, t_\mathrm{T})}{N(\mathrm{J}/\Psi\mathrm{K}^0_\mathrm{S}, t; \overline{\mathrm{T}}, t_\mathrm{T}) + N(\mathrm{J}/\Psi\mathrm{K}^0_\mathrm{S}, t; \mathrm{T}, t_\mathrm{T})} = -\sin 2\beta \cdot \sin(t - t_\mathrm{T}) \ . \tag{6}$$

The statistical error on $\sin 2\beta$ is estimated as

$$\sigma(\sin 2\beta) = \sqrt{\frac{1 + B/S}{I \cdot \tilde{N}}} \ , \tag{7}$$

where B/S is the background to signal ratio for the $\mathrm{J}/\Psi\ \mathrm{K}^0_\mathrm{S}$ reconstruction, I is the (Fisher) information per event, and $\tilde{N}$ is the total number of tagged events (both tags including mistags). This number is

$$\tilde{N} = \dot{\mathcal{L}}\ \sigma_{\Upsilon 4\mathrm{S}}\ t_\mathrm{run} \cdot 2\mathcal{B}_1\mathcal{B}_2\mathcal{B}_3 \cdot \eta_\mathrm{rec} \cdot \eta_\mathrm{tag}(1 - 2w)^2 \ , \tag{8}$$

where $\sigma_{\Upsilon 4\mathrm{S}}$ is the $\Upsilon(4\mathrm{S})$ formation cross section, t_run the running time, $\mathcal{B}_1$ =0.50 the decay fraction of $\Upsilon(4\mathrm{S}) \to \mathrm{B}^0\overline{\mathrm{B}}^0$, $\mathcal{B}_2$ that of $av(\mathrm{B}^0, \overline{\mathrm{B}}^0) \to \mathrm{J}/\Psi\mathrm{K}^0_\mathrm{S}$, $\mathcal{B}_3$ that of $J/\Psi \to \ell^+\ell^-$, η_rec the reconstruction efficiency for $\mathrm{J}/\Psi\mathrm{K}^0_\mathrm{S}$, η_tag the tag efficiency (including mistags), and w the ratio of wrong-sign tags to all tags. The information per event is estimated as

$$I = I_0 \mathrm{e}^{-\sigma_t^2/2\tau_\mathrm{B}^2} \ / \ [1 - (1 - 2w)^2 \sin^2 2\beta] \ , \tag{9}$$

where $I_0 \approx 0.31$ [12], σ_t the time resolution, and τ_B the B meson mean life.

Table 2 gives the efficiencies for the reconstruction of $\mathrm{B} \to \mathrm{J}/\Psi\mathrm{K}^0_\mathrm{S}$ with K^0_S into $\pi^+\pi^-$ and $\pi^0\pi^0$. The results from the estimate of BELLE [15] at KEK are given for comparison. The background to signal ratio B/S is negligeable in both estimates. With $\dot{\mathcal{L}}\ t_\mathrm{run} = 30$/fb, $\sigma_{\Upsilon 4\mathrm{S}} = 1.20$ nb, and $\mathcal{B}_2 = 5 \cdot 10^{-4}$, BABAR estimates $\tilde{N} = 375$ and $\sigma(\sin 2\beta) = 0.10$, whereas BELLE estimates $\tilde{N} = 190$ and $\sigma(\sin 2\beta) = 0.13$ for the same integrated luminosity but using $\sigma_{\Upsilon 4\mathrm{S}} = 1.15$ nb and $\mathcal{B}_2 = 4 \cdot 10^{-4}$. There are no non-understood

Table 2: Efficiencies for the determination of $\sin 2\beta$ from $J/\Psi K^0_S$ decays.

	$\pi^+\pi^-$	$\pi^0\pi^0$
η_{rec} for J/Ψ, geometry, lepton ID, mass	0.82	0.82
η_{rec} for K^0_S, geometry, γ cuts, mass	0.73	0.45
B^0 mass and momentum cuts	0.99	0.95
η_{rec} for $B \to J/\Psi K^0_S$ in BABAR	0.59	0.35
η_{rec} for $B \to J/\Psi K^0_S$ in BELLE	0.47	0.18
$\eta_{tag}(1-2w)^2$ in BABAR	0.34	0.34
$\eta_{tag}(1-2w)^2$ in BELLE	0.30	0.30

discrepancies between BABAR and BELLE; the two estimates are based on different assumptions. BABAR has continued its studies of reconstruction and tagging efficiencies with more realistic Monte Carlo simulations. The tagging efficiency will probably be larger than $\eta_{tag}(1-2w)^2 = 0.34$ by weighted combinations of primary and secondary leptons, $K^\pm$, and high-momentum $\pi^\pm$. On the other hand, the final reconstruction efficiency will be slightly lower than the $\pi^+\pi^-, \pi^0\pi^0$ average of Table 2, $\eta_{rec} = 0.51$. The 1997 data of CLEO [16] tell us, using isospin symmetry for B^0 and B^+, that $\mathcal{B}_2 = (4.7 \pm 0.4) \cdot 10^{-4}$. My private best guess for $\sigma(\sin 2\beta)$ with 30/fb uses $\mathcal{B}_2$ as above, $\sigma_{\Upsilon 4S} = (1.15 \pm 0.05)$ nb, $I = 0.28 \pm 0.02$, $\eta_{rec} = 0.43 \pm 0.06$, $\eta_{tag}(1-2w)^2 = 0.36 \pm 0.02$, $1-2w = 0.78 \pm 0.03$, $\sin 2\beta = 0.55 \pm 0.20$, and gives $\tilde{N} = 301 \pm 45$, $\sigma(\sin 2\beta) = 0.098 \pm 0.011$ (which would be 0.109 ± 0.009 without the "binomial" factor $1/[1-(1-2w)^2 \sin^2 2\beta]$ in equation 9).

4.6 Other Channels for $\sin 2\beta$

All esimates for $\sigma(\sin 2\beta)$ will be given for $\dot{\mathcal{L}}\, t_{run} = 30$/fb and are taken from the Technical Design Report of BABAR [2].

$B^0 \to J/\Psi K^0_L$. Here, the expected asymmetry is opposite in sign compared to that in equation 6, i. e. $A(t, t_T) = +\sin 2\beta \cdot \sin(t - t_T)$. The Instrumented Flux Return of BABAR, primarily designed for muon identification, is also a K^0_L detector with good efficiency, good angular information, but poor energy resolution. The reconstruction efficency is 0.33, yielding $\tilde{N} = 242$. Because of the poor K^0_L energy resolution, there is some background from $B^0 \to J/\Psi\, K^{0*}, K^{0*} \to K^0_L \pi^0$, giving $\sqrt{1 + B/S} = 1.18$ and $\sigma(\sin 2\beta) = 0.16$. The background has a CP asymmetry of opposite sign requiring samples of different background levels and extrapolation to $B = 0$.

$B^0 \to \Psi(2S) K^0_S$. $\Psi(2S) \to J/\Psi \pi^+\pi^-$, $\eta_{rec} = 0.32$, $\tilde{N} = 80$, $\sigma(\sin 2\beta) = 0.19$.

$B^0 \to J/\Psi K^{0*}$. $K^{0*} \to K^0_S \pi^0$, $\mathcal{B} = (13 \pm 2) \cdot 10^{-4}$, one CP value dominates, $\eta_{rec} = 0.39$, $\tilde{N} = 104$, $B/S \approx 0$, $\sigma(\sin 2\beta) = 0.19$.

$B^0 \to D^+D^-$. This final state is not yet observed, but the expectation $\mathcal{B} =$

$\sin^2\theta_{12}\cdot\mathcal{B}(\mathrm{B}\to\mathrm{DD_s})$ is a good estimate. With reconstruction of 6 D channels, but only 25 channel combinations excluding those with more than one π^0, the BABAR simulation finds $\eta_{\mathrm{rec}} = 1.2\%$, $\tilde{N} = 90$, $B/S \approx 1$, $\sigma(\sin 2\beta) = 0.24$.

$\mathrm{B^0 \to D^{*+}D^{*-}}$. With the assumption $\mathcal{B} = \sin^2\theta_{12}\cdot\mathcal{B}(\mathrm{B}\to\mathrm{D^*D_s^*})$ the simulation gives an estimate of $\sigma(\sin 2\beta) = 0.15$ if one CP state dominates.

$\mathrm{B^0 \to D^{\pm}D^{*\mp}}$. This final state is not a CP eigenstate. But there is a time-dependent asymmetry in the sum of the two states $\mathrm{D^+D^{*-}}$ and $\mathrm{D^-D^{*+}}$ giving $\sin 2\beta$ and $\cos 2\beta$. Full simulation studies are still underway; an estimate yields $\sigma(\sin 2\beta) = 0.15$.

Combining the seven channals including $\mathrm{J}/\Psi\ \mathrm{K_s^0}$ gives
$\sigma(\sin 2\beta) = 0.059$ for $\dot{\mathcal{L}}\, t_{\mathrm{run}} = 30/\mathrm{fb}$.

4.7 $\sin 2\alpha$ from $\mathrm{B^0} \to \pi^+\pi^-$, $\rho^{\pm}\pi^{\mp}$, and $\mathrm{a_1^{\pm}}\pi^{\mp}$

Here we have two major difficulties. The decay fractions are unknown (CLEO [16] has only limits), and the unknown penguin to tree ratios influence the relations between $\sin 2\alpha$ and the measured time-dependent asymmetries A.

$\mathrm{B^0} \to \pi^+\pi^-$. The known CLEO limit is $\mathcal{B} < 1.5\cdot 10^{-5}$, the BABAR simulation assumed $\mathcal{B} = 1.2\cdot 10^{-5}$. DIRC, the BABAR particle ID system, is a powerful tool for discrimination against $\mathrm{K}^{\pm}\pi^{\mp}$, but there is large background from continuum $\mathrm{e^+e^-} \to \mathrm{q\bar{q}}$ events. A cut with $\cos\alpha < 0.7$, where α is the angle between the reconstructed B direction and the sphericity axis of the event, gives 54% reconstruction efficiency, $S/B = 1/2$, and $\sigma(A) = 0.20$ for $\dot{\mathcal{L}}\, t_{\mathrm{run}} = 30/\mathrm{fb}$. The $\mathrm{q\bar{q}}$ background is likely to be CP symmetric, but it might be helpful to control this by using lepton tags only. The translation between A and $\sin 2\alpha$ needs theoretical assumptions and measurements of $\mathcal{B}(\mathrm{B^0}\to\pi^0\pi^0)$ and $\mathcal{B}(\mathrm{B^+}\to\pi^+\pi^0)$. Both channels have the same high background problem from $\mathrm{q\bar{q}}$ events as $\mathrm{B^0}\to\pi^+\pi^-$.

$\mathrm{B^0}\to\pi^0\pi^0$. Here we need only the decay fraction $\mathcal{B}$ and not the CP asymmetry. The reconstruction efficiency is 55% with $B/S < 4$ if $\mathcal{B} = 5\cdot 10^{-6}$. The CLEO limit is $\mathcal{B} < 9\cdot 10^{-6}$.

$\mathrm{B^+}\to\pi^+\pi^0$. $\eta_{\mathrm{rec}} = 0.53$, $B/S < 3$ if $\mathcal{B} = 1.2\cdot 10^{-5}$. $\mathcal{B}(\mathrm{CLEO}) = (9\pm 4)10^{-6}$.

$\mathrm{B^0}\to\rho^{\pm}\pi^{\mp}$. This requires reconstruction of two different channels with two different tags each. The sensitivity on $\sin 2\alpha$ varies with $R = \mathcal{B}(\mathrm{B^0}\to\rho^-\pi^+)/\mathcal{B}(\mathrm{B^0}\to\rho^+\pi^-)$. For the decay fraction of $\mathrm{B^0}\to\rho^-\pi^+$ we assume $(f_\rho/f_\pi)^2\cdot\mathcal{B}(\mathrm{B^0}\to\pi^-\pi^+)$. The simulation yields $\eta_{\mathrm{rec}} = 0.65$, $B/S < 1.5$, and $\sigma(\sin 2\alpha) = 0.11$ if $R = 1$.

$\mathrm{B^0}\to\mathrm{a_1^{\pm}}\pi^{\mp}$. An estimate without full simulation gives $\sigma(\sin 2\alpha) = 0.24$.

Combining the three channals gives $\sigma(\sin 2\alpha) = 0.085$ for $\dot{\mathcal{L}}\, t_{\mathrm{run}} = 30/\mathrm{fb}$.

4.8 Other Physics on the $\Upsilon(4S)$ Resonance

BABAR will mainly operate on the $\Upsilon(4S)$ resonance during the first years. There will be some running in the continuum at an energy slightly below the $\Upsilon(4S)$ and may be some on the $\Upsilon(3S)$ for calibration checks and possibly for a search of the states $h_{b1}(1P)$ and $\eta_b(1S)$. Nothing is planned on the $\Upsilon(5S)$ since the main point there, a determination of $\Delta m(B_s)$, cannot be reached with the sensitivity of LEP. Continuum running is necessary for determining B meson backgrounds, not for continuum physics. This physics is a rich programme point at the $\Upsilon(4S)$ energy where $\sigma(\text{continuum}) \approx 3$ nb.

The most exciting items of the "continuum" programme are charmed meson and baryon spectroscopy, charmed baryon lifetimes, rare charm decays, search for $D^0\overline{D}^0$ oscillations, rare τ decays, precise separation of $\pi\pi\pi$, $K\pi\pi$, and $K\overline{K}\pi$ decays of the τ, Michel parameters of leptonic τ decays, the τ-neutrino mass, where a sensitivity around 5 MeV can be reached with 30/fb, and light meson spectroscopy in fragmentation and two photon reactions.

There are also important B meson decay items which contribute to the potential of BABAR but were not mentioned in this paper. A short list includes $B \to \tau\nu$ measuring $f_B \cdot |V_{ub}|$, $b \to s\gamma$, $b \to sg$, other rare B decays, search for CP violation in B^+/B^- decays, and a time-dependent precision measurement of $\Delta m(B^0)$.

The optimum detector design for measuring $\sin 2\beta$ ensures the high potential of the detector for measuring all the rest of the wide BABAR programme. Both ARGUS and CLEO have shown us how rich this programme is.

References

[1] Boutigny D. et al, BABAR Letter of Intent, SLAC-443 (1994)
[2] Boutigny D. et al, BABAR Tech. Design Report, SLAC-R-95-457 (1995)
[3] Wormser G. (BABAR), Contribution 1705 to this Conference.
[4] Kobayashi M. and Maskawa T., Progr. Theor. Physics 49 (1973) 652
[5] Wolfenstein L., Physical Review Letters 51 (1983) 1945
[6] Schmidtler M. and Schubert K.R., Zeitschr. f. Physik C 53 (1992) 347
[7] Buras A. et al, Physical Review D 50 (1994) 3433
[8] Buras A., Proc. 28th Int. Conf. High En. Phys., Warsaw 1996; updated
[9] Feindt M., Plenary Talk at this Conference
[10] Albrecht H. et al (ARGUS), Physics Letters B 318 (1993) 397
[11] Isgur N. and Wise M.B., Physics Letters B 232 (1989) 113
[12] Feasibility Study for a B-Factory in the ISR Tunnel, CERN 90-02 (1990)
[13] Gibbons L. (CLEO), Proc. 7th Int. Symp. on Heavy Flavor Physics, Santa Barbara, July 1997
[14] Bartelt F. et al (CLEO), Phys. Rev. Letters 71 (1993) 4111
[15] BELLE Collaboration, Letter of Intent, KEK 94-2 (1994)
[16] Browder T. (CLEO), private communication, July 1997

B Physics at the Tevatron: Now, Soon, and Later

Joel N. Butler[1] (butler@fnal.gov)

Fermi National Accelerator Lab,
Batavia, IL, US 60510

Abstract. We discuss recent B physics results obtained at the Fermilab Tevatron by CDF, prospects for observing CP violation in the next run, and a proposal for a Dedicated B Collider Experiment, BTeV, at the Tevatron after the next collider run.

The main points of this paper are:

- The next round of experiments will certainly establish evidence for CP violation in B decays. The Tevatron Collider experiments can play a key role in this effort;
- Because of the limited statistics of this first round of CP experiments and with the growing appreciation of the theoretical difficulties of interpreting the results, this round will open up the study of CP violation in B decays but is not likely to close it out; and
- Hadron collider B experiments will play an even larger role in the second generation of "precision" CP violation and mixing studies because of the large number of B's they produce. Experiments with forward rapidity coverage may have the best chance of achieving the high efficiency required for this effort.

Below, we discuss what has to be measured and with what accuracy; the reach of the next round of experiments – BaBar, Belle, CLEO, and HERA-B; CDF's prospects for their next run; and a new proposed experiment for Fermilab, BTeV.

1 Brief Review of CP Violation

The weak interaction can transform quarks of one flavor into another. This 'quark mixing' is expressed by the Cabibbo-Kobayashi-Maskawa (CKM) matrix[1], which is unitary and, for three generations of quarks, has four independent parameters. The CKM Matrix is shown here in the Wolfenstein representation [2], to order λ^3 ($\lambda = 0.22$, the sine of the Cabibbo angle):

$$\left\{ \begin{array}{ccc} 1-\lambda^2/2 & \lambda & A\lambda^3(\rho - i\eta) \\ -\lambda & 1-\lambda^2/2 & A\lambda^2 \\ A\lambda^3(1-\rho-i\eta) & -A\lambda^2 & 1 \end{array} \right\}$$

CP violation produces a difference in the rates of the decays of a particle into some final state and its antiparticle into the CP conjugate state. These asymmetries are due to interference effects. It is the non-zero value of η (or equivalently, the complex phase) which is responsible for CP violation in the Standard Model (SM). The interference can occur by 3 mechanisms [3]: 'indirect CP violation'; 'direct CP violation'; or 'mixing-induced CP violation'. To observe mixing-induced CP violation, which is expected to be large in B decays, one must 'tag' the initial flavor of the produced B. CP asymmetries can be written as functions of ρ and η directly or as angles, α, β, and γ of the famous unitarity triangle.

The precision for measuring a CP asymmetry is given as:

$$\sigma_{asym} = \frac{\sqrt{1 - asym^2}}{D \times \sqrt{N \times \epsilon \times BR}}$$

where N is the effective number of produced B's of the parent species of interest; BR is the branching fraction into the final state of interest; ϵ is the overall efficiency including the tag; and D is the so-called 'dilution factor'. To achieve high precision, one needs a large number of produced B's and high efficiency for reconstructing the signal particle and for detecting and correctly assigning the tagging particle(s). All predictions of CP reach need to be evaluated in light of this expression.

2 CP Reach of Various Machines and Experiments

Table 1 shows the expected luminosity (peak and integrated over a 'Snowmass' year), the B pair cross section, and the total number of B pairs produced per year at various facilities. HERA-B (not shown) will run at an interaction rate of about 30 Mhz and the cross section is between 6 and 20 nb.

BaBar's projected sensitivity [4] for $\sin 2\beta$ and $\sin 2\alpha$ which are based on extensive simulation and evaluation of the resulting reconstruction and tagging efficiencies, is shown in table 2.

3 CP Violation: The Next Generation

The first observation of CP violation in B decays is within reach of the e^+e^- machines if they achieve their design luminosities. However, a full program of B physics investigations requires one to measure the three angles and the three sides of the CKM triangle in order to check for consistency with the Standard Model. Moreover, one needs to make redundant measurements

[1] estimated $B_s\bar{B}_s$

[2] Main Injector Design. Upgrades have been proposed.

[3] LHC-B reference design.

Table 1. Luminosity goals, cross sections, and rates of produced B's

facility	luminosity	$B-\bar{B}$ cross section	luminosity/ year fb^{-1}	$B-\bar{B}$ pairs per year
CESR II $\Upsilon(4S)$	4×10^{32}	$1.15nb$	4.0	4.6×10^{6}
LEP	1.6×10^{31}	$7.0nb$	0.16	1.1×10^{6}
FNAL Run I	1×10^{31}	$100\mu b$	0.1	1.0×10^{10}
e^+e^- $\Upsilon(4S)$	3×10^{33}	$1.15nb$	30	3.0×10^{7}
e^+e^- $\Upsilon(5S)$	3×10^{33}	$0.1nb$ [1]	30	3.0×10^{6}
FNAL [2] Run II	2×10^{32}	$100\mu b$	2.0	2.0×10^{11}
LHC [3]	1.5×10^{32}	$500\mu b$	1.5	7.5×10^{11}

Table 2. Estimated Sensitivity for $\sin 2\beta$ and $\sin 2\alpha$ for an integrated luminosity 30 fb^{-1} for BaBar.

Final State	BR	$\sin\phi$	σ
$J/\psi K_s^o$	0.5×10^{-3}		0.10
$J/\psi K_L^o$	0.5×10^{-3}		0.16
$J/\psi K^{*o}$	1.6×10^{-3}		0.19
D^+D^-	6×10^{-4}		0.21
$D^{*+}D^{*-}$	7×10^{-4}		0.15
$D^{*\pm}D^{\mp}$	8×10^{-4}		0.15
Combined		$\sin 2\beta$	0.06
$\pi^+\pi^-$	1.2×10^{-5}		0.20
$\rho\pi$	5.8×10^{-5}		0.11
$a_1\pi$	6×10^{-5}		0.24
Combined		$\sin 2\alpha$	0.09

because nearly every measurement, except $\sin 2\beta$, has problems associated with its interpretation. The measurement of α via the decay $B^o \to \pi^+\pi^-$ is complicated by the presence of Penguin diagrams which can compete with mixing induced CP violation. The measurement of γ presents very difficult experimental and theoretical problems. We need to carry out CP studies of B_s decays and make a good measurement of B_s mixing to complete the picture. We also need to check that a variety of rare decays have the rate expected from the SM and to do precision searches for decays which are explicitly forbidden in the SM. We hope to find deviations from the SM that will signal new physics.

It is clear that we face a long and difficult program which will challenge the ingenuity of experimenters and theorists for many years to come.

4 B Physics at Hadron Colliders

The Tevatron, at a luminosity of 10^{32}, produces 10^{11} B-pairs per year. It produces all species of B's: B_d, B_u, B_s, B-baryons of all sorts, and B_c states. It is thus a **Broadband, High Luminosity B Factory**. Cross sections at the LHC are approximately 5 times larger. At both machines, B experiments will be able to have all the luminosity they can handle.

With these advantages come some serious challenges. The B events are accompanied by a very high rate of background events. Only one event in a thousand at the Tevatron is a B pair event, The B's are also produced over a very large range of momenta and angles. Even in the B events of interest there is a complicated underlying event so one does not have the stringent constraints that one has at an e^+e^- machine.

These lead to questions about the triggering, tagging, and reconstruction efficiency and the background rejection that can be achieved at a hadron collider. On the other hand, with an almost **four order of magnitude advantage in B production**, it is likely that hadron collider B experiments will play a significant role in future high precision phase of CP violation studies.

4.1 B Physics at CDF

There is already a positive indication that the challenges of doing B physics at a hadron collider can be overcome. CDF has used the data set from its last run to do excellent B physics [5] and in the process has developed many of the tools and techniques needed to look for CP violation.

To observe mixing-induced CP violation, in particular to measure $\sin 2\beta$, one needs three capabilities:

- to reconstruct the final state, in this case ψK_s with good efficiency and low background;
- to measure the proper time distribution of the decays; and
- to tag the decays with high efficiency and with a well understood dilution factor.

CDF has achieved each of these capabilities. Figure 1 shows the signal for B^o decay into the so-called gold-plated mode ψK_s. CDF has a precision silicon detector, called the SVX, which enables it to resolve the secondary B decay vertex from the production vertex. Using this device to measure the proper time of the decays, CDF has made some of the best lifetime measurements of the B^o, B^+, B_s, and Λ_b. Finally, Fig. 2 shows a tagged time-dependent study of B^o decays showing clear evidence for mixing. This study, done with a signal of a D meson and a lepton making a single detached vertex, uses for a tag the same sign pion charge correlation. In addition to measuring the mixing parameter, x_d, this study also measures the dilution factor for this tagging method.

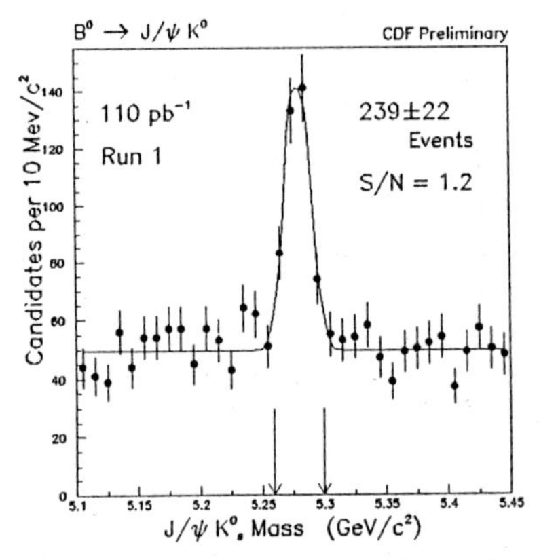

Fig. 1. Signal for B^o decaying to $\psi(\mu^+\mu^-)K_s(\pi^+\pi^-)$ from CDF.

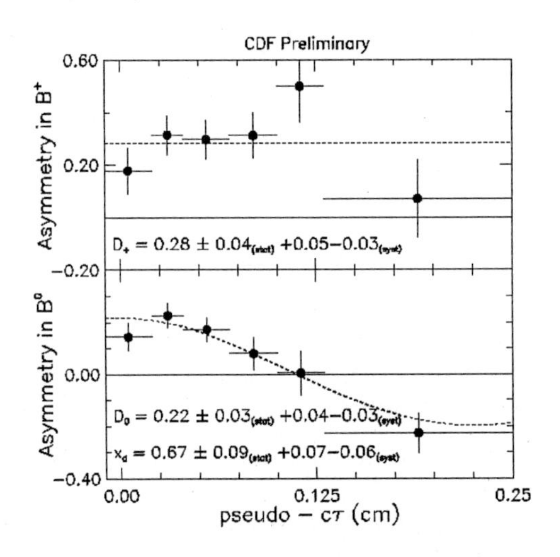

Fig. 2. Tagged pseudo proper time distribution from CDF for a) $B^{\pm}$ showing no mixing and b) B^o showing clear oscillation from mixing.

Using the information from the last run, CDF can make good projections of the of CP sensitivity they can expect for the next run which is expected to have 20 times the integrated luminosity or 2 fb^{-1}. Their conservative projection gives an error on $\sin 2\beta$ of 0.07.

4.2 BTeV: A Dedicated B Collider Experiment

While CDF has demonstrated that many of the problems associated with doing B physics at a hadron collider are tractable, it is not optimized for doing the studies required of a second generation CP violation experiment. It covers a kinematic region where the B's have very low momentum so its vertex resolution and proper time resolution are badly degraded by multiple scattering. This limits their reach in x_s. Moreover, it lacks good charged hadron identification which is viewed as essential for many B physics measurements. Finally, its trigger and data acquisition system are not optimized for B physics studies.

BTeV is a dedicated B Collider experiment being designed to run at the Tevatron in a new interaction region, C0, whose construction has just been approved. The experiment is optimized for high precision CP violation and mixing studies. The experiment can also study rare charm decays and D-meson mixing. The BTeV proposal is still under development. An extensive Expression of Interest has been submitted to Fermilab [6]. The experiment's goal is to start in the 2003-2004 timeframe.

The key design features of BTeV which give it high efficiency for a wide variety of B decays are:

- A dipole located on the IR which gives BTeV effectively two spectrometers – one covering the forward rapidity region and one covering the backward rapidity region;
- A precision vertex detector based on planar pixel arrays;
- A vertex/impact parameter trigger at Level I which makes BTeV especially efficient for states with no leptons in them.
- Strong particle identification based on a Ring Imaging Cerenkov counter. Many states that will be of interest in this phase of B studies will only be separable from other B states if this capability exists. It also permits one to use charged kaons for tagging.

The vertex tracker consists of planar silicon detectors shown schematically in figure 3.

The trigger is based on a heavily pipelined processing architecture based on high speed switches and inexpensive processsing nodes. Hits from the pixel planes are organized into subunits based on azimuthal sectors. In stage 1, station hits are formed from each triplet of planes and these are formed into minivectors. In stage 2, the minivectors are passed to a processor farm which does track finding. Finally, in stage 3, the tracks are passed to a farm of vertex processors. Detailed hit-level simulations of the trigger have been carried out. The trigger efficiency for the decay $B^o \to \pi^+\pi^-$ and the rejection against minimum bias events is shown in figure 4. Here N is the number of

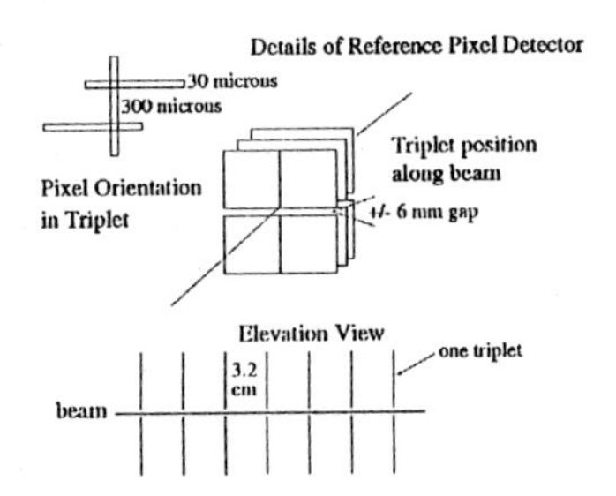

Fig. 3. Schematic of BTeV pixel vertex detector

tracks exceeding some P_t cut which are required not to point to the primary vertex based on a normalized impact parameter cut, $\frac{b}{\sigma_b}$, whose value is M. For N=2, and M = 3.5, the trigger efficiency is 40% and the probability to trigger on a minimum bias event is only 5×10^{-3}.

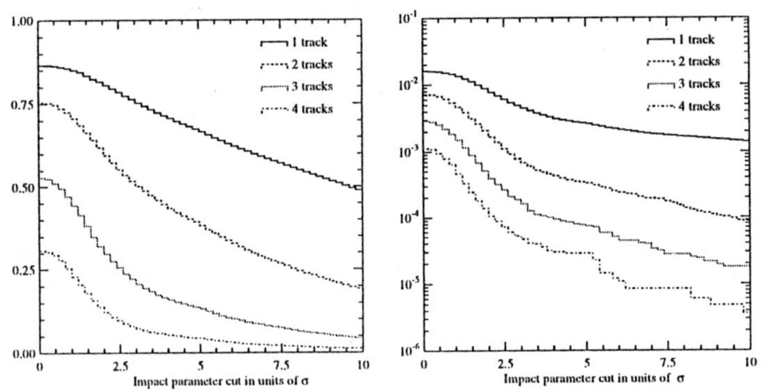

Fig. 4. BTeV vertex trigger efficiency for (left) $B^o \to \pi^+\pi^-$ and (right) minimum bias.

Particle identification must operate from about 3 GeV/c, for efficient away-side kaon tagging, to 70 GeV/c, for two body decays such as $B^o \to \pi^+\pi^-$. To cover this range, a gas Ring Imaging Cerenkov (RICH) Counter, is employed. The π, K, and p thresholds are 2.5 (2.4), 9.0 (8.4), and 17.1 (15.9) GeV/c for C_4F_{10} (C_5F_{12}).

Sensitivities of BTeV for one year of running at 5×10^{31} are:

parameter	uncertainty
$\sin 2\beta$	0.04
$\sin 2\alpha$	0.10
x_s reach	<50
$BR(B^o \to K\mu^+\mu^-)$	10%

The sensitivity for γ is still being evaluated. The sensitivity to x_s is shown in Fig. 5 for two decay modes. The experiment expects to operate at a higher luminosity of 2×10^{32} after its first year or two of running.

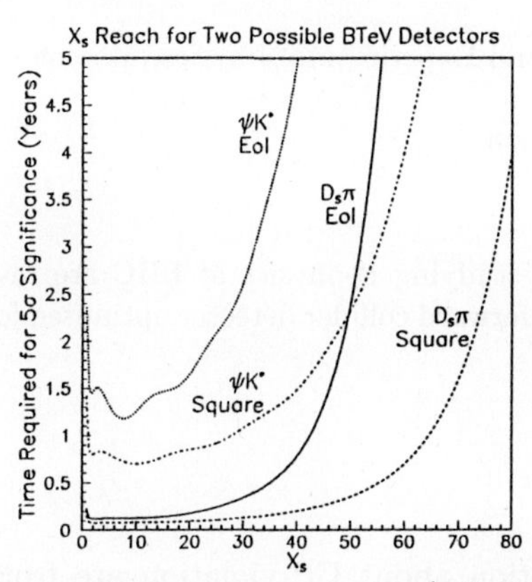

Fig. 5. The x_s reach of BTeV using $B_s \rightarrow D_s^{\pm}\pi^{\mp}$. The solid curve indicates, for a given value of x_s, the number of years of running the EOI detector at $5 \times 10^{31}\mathrm{cm}^{-2}\mathrm{s}^{-1}$ to obtain a 5σ effect. The dotted curve gives the same information for $\psi\overline{K}^{*0}$. The other two lines show the improvement obtained by using a modified design.

References

[1] M. Kobayashi and T. Maskawa, Prog. Theor. Phys. 49, 652 (1973)
[2] L. Wolfenstein, Phys. Rev. Lett. 51, 1945 (1983)
[3] For a concise explanation, see H. Quinn, *$B^o - \bar{B}^o$ Mixing and CP Violation in B Decay*, Review of Particle Properties, Phys. Rev. D 50, 1632 (1994).
[4] F. Bianchi, *Status of BaBar at SLAC*, in *Beauty '96*, Elsevier Science B.V. 1996 p 67.
[5] The CDF II Detector, Technical Design Report, November 1996.
[6] Available at http://fnsimu1.fnal.gov/btev.html.

[7] This work was supported by the Fermi National Accelerator Laboratory, which is operated by Universities Research Association, Inc., under contract DE-AC02-76CH03000 with the U.S. Department of Energy.

B Physics at the LHC

S.Semenov (ssemenov@iris1.itep.ru)

Institute for Theoretical and Experimental Physics, Moscow

For the LHCB Collaboration

Abstract. Prospects for studying B-physics at LHC are reviewed. In particular LHCB, an open-geometry forward collider detector optimised for studying B-physics is described in detail.

1 Introduction

Up to now all information about CP-violation are from the kaon system studies. Decays of B-mesons offer a wide range of possibilities to expand our knowledge of CP-violation and provide a stringent test for the CKM-origin of CP-asymmetry. The determination of the angles in the unitarity triangle relies on the measurement of CP-asymmetry in B-decays.

B $\rightarrow J/\psi K^0_S$ decay channel is proposed to use for measurement of $\sin 2\beta$ [1]. This channel is very attractive both from experimental (clean dilepton signature) and theoretical (less than 1% uncertainties) point of view. In general, time-dependent CP-violating rate asymmetry is defined as:

$$A_{CP}(t) = \frac{N(t) - \overline{N}(t)}{N(t) + \overline{N}(t)} = a \cos \Delta m t + b \sin \Delta m t$$

where a is due to CP-violation in the decay amplitude and b is due to the interplay between decays and oscillations. For B $\rightarrow J/\psi K^0_S$, we have

$$A_{CP}(t) = -\sin 2\beta \sin \Delta m t.$$

The situation is more complicated for α angle measurements. For extracting this angle the decay $B_d \rightarrow \pi^+\pi^-$ is proposed. In this case the penguin diagrams contribution is not negligible and estimated to be 10%-20% and we have

$$a = 2\frac{|P|}{|T|} \sin(\delta_P - \delta_T) \sin \alpha; \quad b = -\sin 2\alpha - 2\frac{|P|}{|T|} \cos(\delta_P - \delta_T) \sin \alpha \cos 2\alpha$$

where P and T represent the magnitudes of the penguin and tree diagrams contributions. Therefore to determine the ratio, $\frac{|P|}{|T|}$ and extract $\sin 2\alpha$ measurements of other modes such as $B_d \rightarrow \pi^0\pi^0$, $B_u^\pm \rightarrow \pi^\pm\pi^0$ are necessary

[2]. If it cannot be carried out, the error induced on $\sin 2\alpha$ is of the order of $\frac{|P|}{|T|}$.

A very clean way to extract the angle γ is to use time dependent decay rates $B_s \to D_s^{\pm} K^{\mp}$ [3]. One needs to study four time-dependent decay rates since the final state is not a CP-eigenstate. The proper time distribution depends on mixing parameter in B_s system, final state interaction strong phase and angle γ .

Another approach is to measure the exclusive decay rates in the following three channels $B^0 \to D_1^0 K^{*0}$, $B^0 \to \overline{D^0} K^{*0}$, $B^0 \to D^0 K^{*0}$ and their CP-conjugates, where D_1^0 is a CP-eigenstate, $D_1^0 = (D^0 + \overline{D^0})/\sqrt{2}$ [4].

There are a few experiments under construction which may observe CP-violation in B-decays in the nearest future. The experimental conditions for these experiments are listed in Table 1. They have a good chance to get an evidence of CP-asymmetry in B-decay if CP-violation is mainly due to the Standard Model. However, detailed CP-studies with these detectors are restricted either by the B-production rate (B-factories and HERA-B) or by the detectors properties which are not optimised for B-physics (D0, CDF at Tevatron).

The LHC will provide a unique opportunity for an extensive B-physics study due to the enormous amount of B's produced and rather moderate background conditions (see Table 1). In this article, a short review on B-physics at LHC is presented.

Table 1. Comparison of different B-meson production facilities.

Accelerator	HERA	B-factory	Tevatron	LHC
Reactions	p Cu	e^+e^-	$p\bar{p}$	pp
$\sigma_{b\bar{b}}$	~ 12 nb	~ 1 nb	$\sim 100\ \mu$b	$\sim 500\ \mu$b
$\sigma_{b\bar{b}}/\sigma_{hadronic}$	$\sim 10^{-6}$	$\sim 2\times10^{-1}$	$\sim 2\times10^{-3}$	$\sim 5\times10^{-3}$
N_{bb}/y	$\sim 10^8$	$\sim 10^8$	$\sim 2\times10^{11}$	$\sim 5\times10^{12}$

2 CP-violation studies with ATLAS and CMS

ATLAS [5] and CMS [6] are two general-purpose detectors which are designed to work at LHC. Both detectors are mostly optimised for Higgs boson study, as well as for searches and studies of the supersymmetric particles. However, during the beginning of the LHC operation when the machine is expected to run with low Luminosity ($L \approx 10^{33} cm^{-2} s^{-1}$) they plan to perform B-physics study. Using very good lepton identification they expect to reach excellent results in measuring $\sin 2\beta$ already in one year of operation.

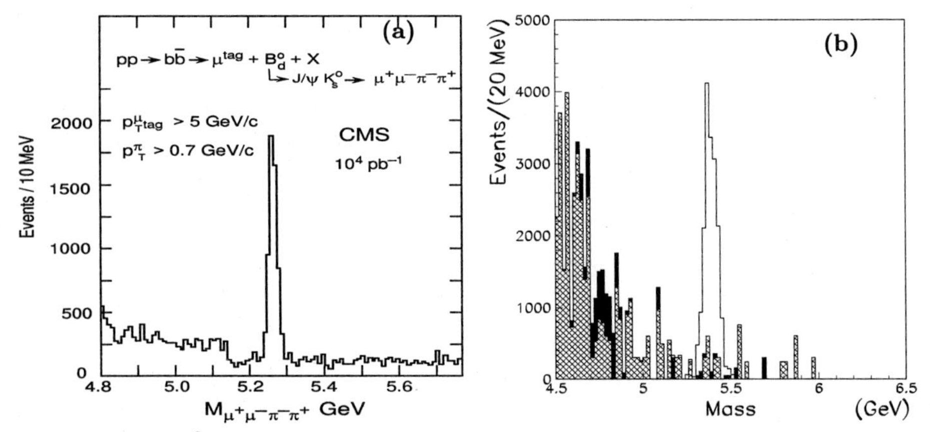

Fig. 1. $J/\psi K_S^0$ invariant mass distribution expected in one year of operation for CMS (a) and ATLAS (b)

In Figure 1, $J/\psi K_S^0$ invariant mass distributions expected after one year of operation for ATLAS and CMS detectors are shown. There are very clean peaks with very good signal to background ratio. For measuring $\sin 2\alpha$ from $B_d \to \pi^+\pi^-$ decay the capability to separate kaons from pions is essential. Both experiments are unable to do it. Although they expect to obtain rather good statistics for $B_d \to \pi^+\pi^-$ decay but arising systematic errors could be large. Without K/π separation it is practically impossible for both experiments to extract the γ angle, since the signal decay channel $B_s \to D_s K$ is considerably weaker than the background channel $B_s \to D_s \pi$.

One can conclude that ATLAS and CMS are limited in study B physics. Therefore, a dedicated experiment optimised for B-physics study is necessary to exploit the rich B-physics potential at LHC.

3 LHCB, a dedicated experiment to study B-physics

LHCB [7] is a dedicated detector optimised for B-physics studies at LHC. A schematic view of the detector is shown in Figure 2. The detector is a forward single-dipole spectrometer, covering a polar angle interval of 30 - 300 mrad.

A forward spectrometer offers several important advantages:

- The produced b and $\bar{b}$ are typically correlated in one unit of rapidity. Therefore the geometric efficiency to detect all B decay tracks plus a tagging particle from the accompanying $\overline{\mathrm{B}}$ is large.
- The large Lorentz boost of accepted B-mesons (corresponding to about 7 mm mean decay distance) allows proper-time measurements with a good accuracy.
- Forward planar detector systems are less expensive, easier to maintain and can be optimised for the best resolution.

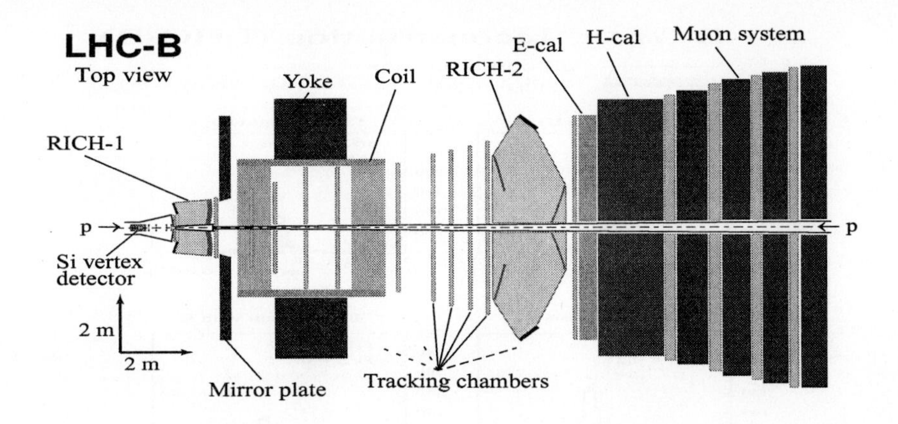

Fig. 2. The scheme of LHCB detector.

The detector consists of silicon microvertex detector, a tracking system, RICH counters, electromagnetic and hadronic calorimeters and a muon system.

Microvertex detector is installed inside the vacuum tank at the interaction region. Independent inner and outer tracking systems are proposed. For outer tracker Honeycomb strip chambers are considered as a candidate. For inner tracker possible options are micro-strip Si counters, micro-cathode strip chambers and micro-strip-gas counters.

As shown in Figure 2 the spectrometer utilises two RICH counters, one before and one after the magnet. The momentum range of the decay products for accepted B-meson extends from 1 GeV to 150 GeV. Proposed RICH system provides 3σ kaon-pion separation over this momentum interval. In Figure 3, reconstructed B-mesons from $B \to \pi\pi$ decays are shown for different requirements on K/π separation. One can see that a clean sample of $B \to \pi\pi$ decays could be obtained with a reasonable efficiency. Particle identification is very important for extracting both the angle α and the angle γ.

Electromagnetic calorimeter based on Shashlik design provides electron-pion separation and high-p_t electron trigger. The hadron calorimeter is necessary for the high-p_t hadron trigger.

LHCB muon detection system serves for muon identification and for high-p_t muon trigger.

A summary of LHCB performance in the CP-violation study are presented in Table 3. The results are based on Letter of Intent (LOI) [7]. Since the LOI, a significant change has been made in the trigger scheme which may result in an increase of a factor of two to three in yield of reconstructed B-mesons [8].

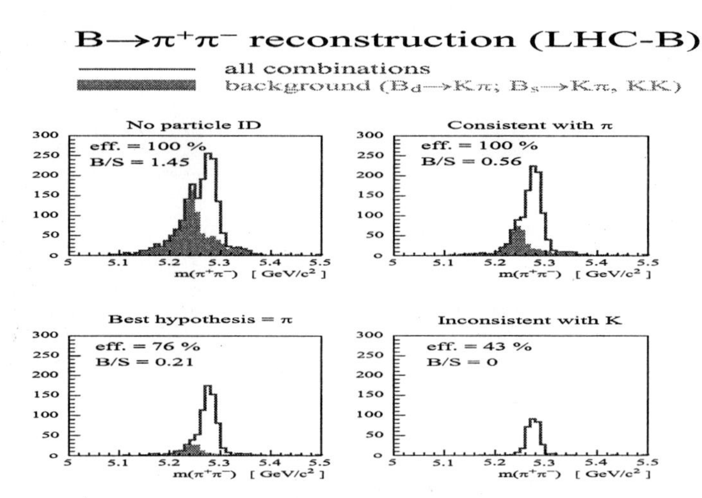

Fig. 3. $\mathrm{B} \to \pi^+\pi^-$ reconstruction for different requirement on purity of pions

Table 2. LHCB reach for one year of operation with Luminosity $1.5 \times 10^{32} cm^{-2}s^{-1}$.

	Decay mode	Number of events	Uncertainty in one year
$\sin 2\beta$	$\mathrm{B}_d \to J/\psi K_S$	55 k	0.02
$\sin 2\alpha$	$\mathrm{B}_d \to \pi^+\pi^-$	14 k	0.04*
γ'	$\mathrm{B}_s \to \mathrm{D}_s K$	3.3 k	$6^o - 16^o$
γ	$\mathrm{B}_d \to \mathrm{D}K^\star$	0.9 k	$7^o - 12^o$
x_s	$\mathrm{B}_s \to \mathrm{D}_s\pi$	35 k	$x_s \leq 55$

⋆ - assuming small penguin pollution

4 Other B-physics studies at LHC

B-physics studies at LHC will not be restricted to measurements of the CP-asymmetry in B-decays. In this section other topics are briefly reviewed.

Enormous amount of B-mesons produced at LHC provides an opportunity to study Flavour Changing Neutral Current (FCNC) B decays. For instance, branching ratios for leptonic B-meson decays are predicted to be very small in the Standard Model (SM): $BR(\mathrm{B}_s \to \mu^+\mu^-) = 3.5 \times 10^{-9}$, $BR(\mathrm{B}_d \to \mu^+\mu^-) = 1.5 \times 10^{-10}$ [9] and any deviations from these values would be a signal of new physics (beyond the SM).

Measurements of exclusive FCNC processes $\mathrm{B}_d \to \rho^0\mu^+\mu^-$, $\mathrm{B}_d \to K^\star\mu^+\mu^-$ provide us with additional information on the CKM matrix elements.

CP-violation in radiative penguin decays is also a good place to search for new physics. CP-asymmetries within the SM are reliably estimated to be about 1%. In different extensions of the SM, CP-asymmetries are predicted to be significantly larger [10] and therefore could be measured at the LHC.

The very large data sample containing heavy flavour particles may be used for precise spectroscopy of beauty hadrons, as well as for B_c studies [11].

5 Conclusion

LHC opens a unique opportunity to study CP-violation in B-meson decays with very high statistics. LHCB [7], a forward collider detector optimised for studying B-physics, will be able to exploit this potential. LHCB will provide detailed study of CP-violating decay modes measuring all angles of the unitarity triangle (α, β, γ) as well as B_s oscillation up to $x_s \sim 55$.

LHC detectors contribute to study rare B-decays with $BR < 10^{-8}$, B_c study, beauty hadron spectroscopy and many other topics.

Aknowledgements

I wish to thank the organisers of this Conference for financial support. I would like to thank all members of the LHCB collaboration. I am grateful to I. Belyaev, R. Kagan, T. Nakada and F. Ratnikov for their assistance in preparing this talk.

References

[1] I. I Bigi and A. I. Sanda , Nucl. Phys. **B193**, 85 (1981).

[2] M. Gronau, D. London , Phys. Rev. Lett. **65**, 3381 (1990);
M. Gronau, Phys. Lett. **B300**, 163 (1993).

[3] R. Aleksan, I Dunietz, B. Kayser, Z.Phys. **C54** (1992) 653.

[4] M. Gronau, D. Wyler, Phys. Lett. **B265** (1991) 172.

[5] W.W. Armstrong et al . CERN/LHCC/94-43, LHCC/P2 (1994).

[6] G.L. Bayatian et al. CERN/LHCC/94-38, LHCC/P1 (1994).

[7] CERN/LHCC/95-5 (1995).

[8] LHCB, Technical Proposal, in preparation.

[9] A.Ali In Proceedings of the 4th KEK Topical Conference on Flavour Physics,Japan,29-31, October,1996.

[10] L.Wolfenstein, Yu.L.Wu, Phys. Rev. Lett. **73** (1994) 2809;
R.N.Mohaptra, J.C.Pati, Phys.Rev. **D11** (1975) 566.

[11] A.V.Bereznoy,V.V.Kiselev,A.K.Lihoded,A.I.Onischenko, Preprint IHEP 97-2 (hep-ph/9703341).

Part XIX

High Q^2 Events at HERA

New Results on Neutral and Charged Current Scattering at High Q^2 and a Search for Events with a Lepton and $\not{P}_t$

Darin Acosta[1] (acosta@phys.ufl.edu) on behalf of the ZEUS Collaboration

The Ohio State University, Physics Department, Columbus, Ohio 43210, USA

Abstract. Using the ZEUS detector at HERA, we have studied the deep inelastic scattering reactions $e^+p \to e^+X$ and $e^+p \to \overline{\nu}X$ for $Q^2 > 1000$ GeV^2 with 33 pb^{-1} of data collected during the years 1994 to 1997. Preliminary results indicate that the data lie above the Standard Model predictions at high x and Q^2. In addition, we have searched for events containing an isolated high P_t electron or muon and missing transverse momentum.

1 Introduction

The HERA ep collider provides a unique opportunity to study ep collisions at high center-of-mass energy ($\sqrt{s} = 300$ GeV) and high Q^2 (negative squared four-momentum transfer). Recently, the two HERA experiments H1 [1] and ZEUS [2] have reported an excess of neutral current (NC) deep inelastic scattering (DIS) events at high Q^2 in their 1994 to 1996 data samples, for which the integrated luminosities were 14.2 and 20.1 pb^{-1} respectively. This paper presents a preliminary update of the NC DIS reaction $e^+p \to e^+X$, which proceeds via the exchange of a photon or Z^0 boson, using an additional 13.4 pb^{-1} of data collected by ZEUS during 1997 (33.5 pb^{-1} total). Preliminary results are presented also for the charged current (CC) DIS reaction $e^+p \to \overline{\nu}X$, which is mediated by the W boson, using 33 pb^{-1} of data. During this time HERA provided collisions between 27.5 GeV positrons and 820 GeV protons. Both results are compared to Standard Model (SM) predictions. In addition, we present a preliminary search for events containing an isolated high P_t electron or muon and missing transverse momentum.

2 Kinematic Reconstruction

In NC scattering, four quantities are measured: the energy and polar angle of the scattered electron (E'_e, θ), and the energy and effective polar angle of the hadronic system (E_h, γ). Only two measurements are necessary at the Born level to specify the kinematics. ZEUS uses the double-angle (DA) method for NC DIS events, whereby

[1] Now at the University of Florida, Gainesville, Florida USA

$$\cos\gamma = [\,(P_t)_h^2 - [E_h - (P_z)_h]^2\,] \,/\, [\,(P_t)_h^2 + [E_h - (P_z)_h]^2\,] \quad (1)$$
$$Q^2_{DA} = 4E_e^2\,[\,\sin\gamma(1+\cos\theta)\,] \,/\, [\,\sin\gamma + \sin\theta - \sin(\theta+\gamma)\,] \quad (2)$$
$$y_{DA} = [\,\sin\theta(1-\cos\gamma)\,] \,/\, [\,\sin\gamma + \sin\theta - \sin(\theta+\gamma)\,] \quad (3)$$

where E_e is the positron beam energy, and all hadronic quantities are measured from the uranium-scintillator calorimeter. In CC DIS, only the hadron (h) measurements are available, so

$$y_h = (E - P_z)_h \,/\, (2E_e) \quad (4)$$
$$Q^2_h = (P_t)_h^2 \,/\, (1 - y_h) \quad (5)$$

In either case, x is determined from the relation $x = Q^2/sy$.

3 Neutral Current Results

The NC DIS selection procedure is identical to that used in the published 1994–1996 results [2]. Briefly, an isolated electromagnetic cluster must be found in the calorimeter with an energy greater than 20 GeV and an associated track measured by the central tracking detector if $\theta_e > 17.2°$. If $\theta_e < 17.2°$, then the transverse energy of the cluster must be larger than 30 GeV and the track match condition is dropped. A vertex consistent with the nominal interaction region is required, and the quantity $E - P_z$ (nominally twice the positron beam energy) must lie between 40 GeV (44 GeV if $\theta_e < 17.2°$) and 70 GeV. Events must satisfy $Q^2_{DA} > 5000$ GeV^2. The overall acceptance is 81%, and residual background from photoproduction is estimated to be less than 1% in the region $x_{DA} > 0.45$ and $y_{DA} > 0.25$.

Figure 1 shows the high-Q^2 event sample as a function of y_{DA} and x_{DA}. Table 1 compares the number of observed events with SM expectations (using the MRSA parton density parameterizations) for several regions in Q^2_{DA} and x_{DA}, and also lists the Poisson probability to observe at least as many events as seen in data. Specifically, in the region $x_{DA} > 0.55$ and $y_{DA} > 0.25$, defined as the region of excess in the 1994–1996 data sample, ZEUS observes one additional event in 1997 for a total of 5 events where 1.51 are expected from the SM. In contrast, in a 25 GeV window centered at an eq mass of $M_{eq} = \sqrt{sx} = 200$ GeV, where H1 observed an excess in [1], the three events observed by ZEUS are in good agreement with the expected 2.92. The estimated errors on the SM predictions take into account both experimental and theoretical uncertainties (parton densities account for ±6.5%), but are not included in the Poisson probability calculation. Finally, Fig. 2 shows both the x_{DA} distribution and the Q^2_{DA} distribution of the selected NC DIS events compared to SM expectations.

The cross sections for Q^2 above various minimum values have been obtained using the relation $\sigma = N_{obs}/\mathcal{A}\mathcal{L}$, where $\mathcal{L}$ is the integrated luminosity and $\mathcal{A}$ is the ratio of the number of events with $Q^2_{DA} > Q^2_{min}$ to the number

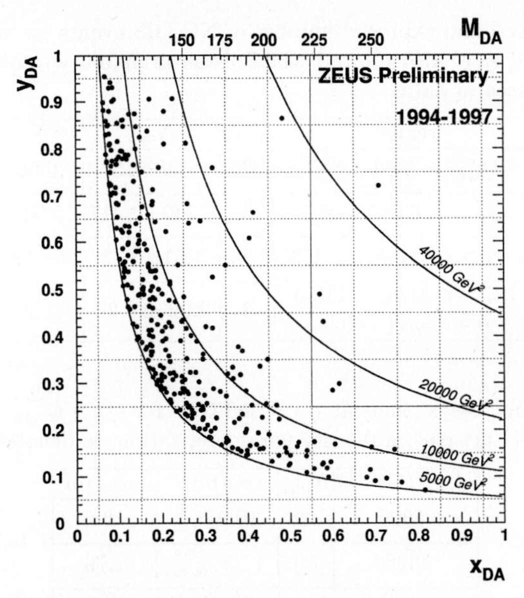

Fig. 1. The distribution of the selected NC DIS events in $y_{\rm DA}$ and $x_{\rm DA}$ (equivalent $M_{\rm DA}$ values are shown along the upper axis). The curves indicate constant values of $Q^2_{\rm DA} = x_{\rm DA} y_{\rm DA} s$ for $Q^2_{\rm DA} = 5000$, 10000, 20000, and 40000 GeV^2.

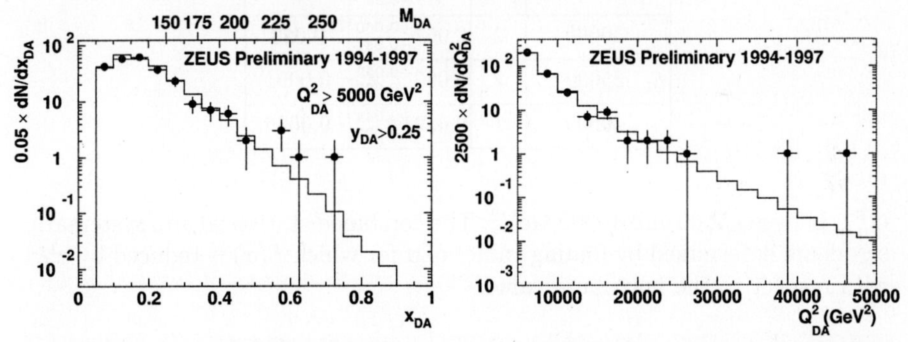

Fig. 2. The $x_{\rm DA}$ distribution (left) and the $Q^2_{\rm DA}$ distribution (right) of the selected NC DIS events with the cuts shown (full dots), compared to the Standard Model e^+p NC expectation (histogram). Errors shown are statistical only.

of events with a propagator Q^2 exceeding $Q^2_{\rm min}$ (as obtained from a Monte Carlo event sample which includes electroweak radiative corrections). The cross sections, listed in Table 2, thus are not corrected to the Born level. To estimate the errors in the cross section, we define the likelihood:

$$L(\sigma) = \frac{1}{\sqrt{2\pi}\delta} \int_0^\infty d\eta' \exp\left[-\frac{1}{2}\left(\frac{\eta' - \eta}{\delta}\right)^2 - \eta'\sigma \right] \frac{(\eta'\sigma)^{N_{\rm obs}}}{N_{\rm obs}!} \qquad (6)$$

where $\eta = \mathcal{AL}$ is the effective luminosity and δ is its corresponding systematic uncertainty. This experimental uncertainty ranges from 4.5% to 6.5% for

Table 1. Observed and expected number of NC DIS events for various selections in $Q^2_{\rm DA}$, $x_{\rm DA}$, and $y_{\rm DA}$. Also shown is the Poisson probability to observe at least as many events as seen in data.

region	$N_{\rm obs}$	$N_{\rm exp}$	$\mathcal{P}(N_{\rm obs})$
$Q^2_{\rm DA} > 5000$ GeV^2	326	328±15	0.55
$Q^2_{\rm DA} > 35000$ GeV^2	2	0.242±0.017	0.025
$x_{\rm DA} > 0.55$, $y_{\rm DA} > 0.25$	5	1.51±0.13	0.019
$187.5 < M_{\rm DA} < 212.5$ $0.4 < y_{\rm DA} < 0.9$	3	2.92 ± 0.24	0.56

Table 2. NC DIS cross sections at $\sqrt{s} = 300$ GeV for $Q^2 > Q^2_{\rm min}$. The SM prediction using the CTEQ4 parton density parameterizations is denoted by $\sigma_{\rm SM}$.

$Q^2_{\rm min}$ (GeV^2)	$N_{\rm obs}$	σ (pb)	$\sigma_{\rm SM}$ (pb)
5000	326	$10.9^{+0.8}_{-0.8}$	10.6
10000	50	$1.73^{+0.28}_{-0.25}$	1.79
15000	18	$0.60^{+0.16}_{-0.14}$	0.49
20000	7	$0.24^{+0.11}_{-0.08}$	0.161
25000	3	$0.10^{+0.07}_{-0.05}$	0.059
30000	2	$0.067^{+0.060}_{-0.037}$	0.023
35000	2	$0.060^{+0.059}_{-0.037}$	0.0091
40000	1	$0.032^{+0.044}_{-0.023}$	0.0036

$Q^2_{\rm min}$ between 5000 and 40000 GeV^2. The combined statistical and systematic errors are determined by finding values of σ for which $L(\sigma)$ is reduced by $e^{1/2}$ with respect to the maximum value.

4 Charged Current Results

The CC DIS selection requires a clean tracking vertex within the interaction region. The missing transverse momentum, $\not{P}_{\rm t}$, as measured by the calorimeter, must be greater than 15 GeV; and $\not{P}_{\rm t}/E_{\rm t}$ must exceed 0.4 where $E_{\rm t} = \sum_i P_{{\rm t},i}$ is the scalar transverse energy sum over all calorimeter cells. Topological cuts are applied to reject cosmic and halo muons as well as NC DIS events, and $Q^2_{\rm h} > 1000$ GeV^2 is required. Visual scanning removes 17 of 472 events which are obvious backgrounds not from ep collisions. The overall acceptance is 73%, and backgrounds are expected to be $< 1\%$ of the total.

Figure 3 shows the distribution of the 455 events which satisfy the event selection as a function of $y_{\rm h}$ and $x_{\rm h}$. Figure 4 shows both the $x_{\rm h}$ and $Q^2_{\rm h}$

distributions compared to the SM expectations. The estimated errors in the SM predictions take into account both experimental (energy scale) and theoretical (parton densities) uncertainties. These comparisons are summarized in Table 3, which presents the observed and expected number of events above various minimum values of $x_{\rm h}$ and $Q^2_{\rm h}$. There is a tendency for the data to lie above the SM predictions for $x_{\rm h} > 0.3$ and $Q^2_{\rm h} > 10000\ \mathrm{GeV}^2$.

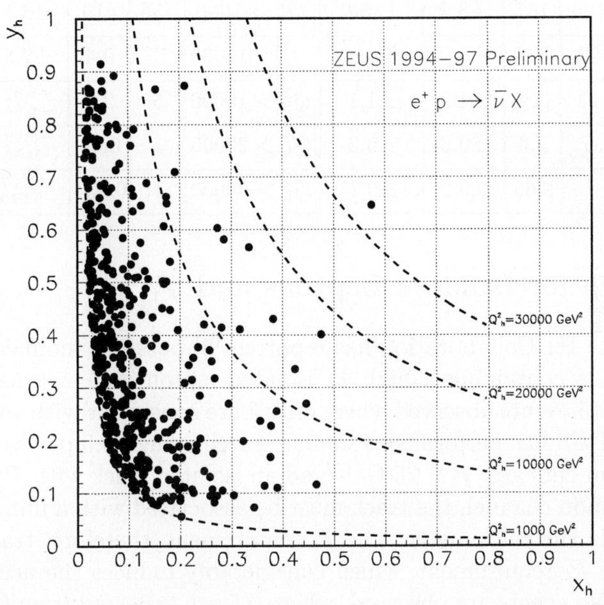

Fig. 3. The distribution of the selected CC DIS events in $y_{\rm h}$ and $x_{\rm h}$. The lines indicate constant values of $Q^2_{\rm h} = x_{\rm h} y_{\rm h} s$ for $Q^2_{\rm h} = 1000$, 10000, 20000, and 30000 GeV^2.

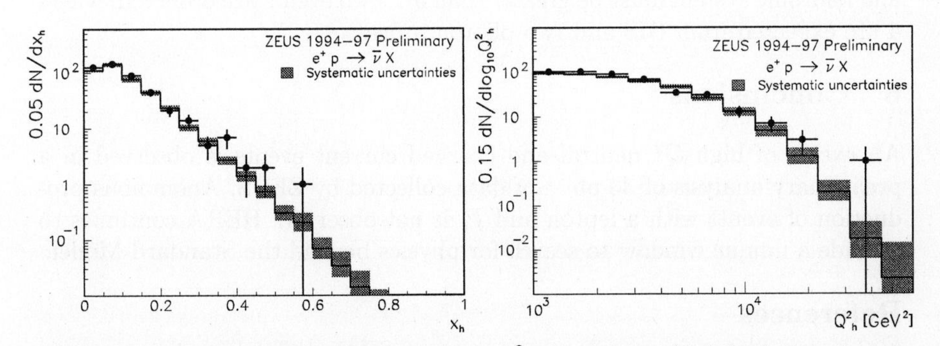

Fig. 4. The $x_{\rm h}$ distribution (left) and the $Q^2_{\rm h}$ distribution (right) of the selected CC DIS events (full dots), compared to the SM e^+p CC expectation (histograms). The systematic errors in the SM predictions are indicated as a shaded band.

Table 3. Observed and expected number of CC DIS events for various selections in x_h ($Q_h^2 > 1000$ GeV^2) and in Q_h^2. Also shown are the experimental and theoretical uncertainties in the expected number of events from the calorimeter energy scale and the parton densities, respectively.

x_h cut	N_{obs}	N_{exp}	N_{exp} uncertainty		Q_h^2 cut (GeV^2)	N_{obs}	N_{exp}	N_{exp} uncertainty	
			E scale	PDF				E scale	PDF
$x_h > 0.1$	186	167	± 8.5	± 16.0	$Q_h^2 > 1000$	455	419	± 13	± 33
$x_h > 0.2$	52	45.5	± 5.0	± 4.6	$Q_h^2 > 10000$	15	9.4	± 2.5	± 1.6
$x_h > 0.3$	17	11.5	± 1.9	± 1.1	$Q_h^2 > 15000$	5	2.0	$^{+0.81}_{-0.54}$	± 0.4
$x_h > 0.4$	5	2.8	± 0.6	± 0.3	$Q_h^2 > 20000$	1	0.46	$^{+0.28}_{-0.16}$	± 0.10
$x_h > 0.5$	1	0.65	± 0.2	± 0.1	$Q_h^2 > 30000$	1	0.034	$^{+0.037}_{-0.016}$	± 0.008

5 Search for Isolated Leptons and $\not{P}_t$

Recently, the H1 Collaboration has reported [3] possible anomalous production of events containing a high P_t isolated lepton and missing transverse momentum (5 events observed where only 2 are consistent with single W production). ZEUS has performed a similar search using 33.5 pb^{-1} of e^+p data. The selection cuts are: $\not{P}_t > 25$ GeV and an isolated track with $P_t > 10$ GeV.

In the muon channel, the track must be associated with a minimum ionizing cluster in the calorimeter. The $\not{P}_t$ requirement, therefore, translates into an hadronic P_t requirement, which considerably reduces the acceptance for W bosons. No events are observed, where 0.5 are expected from CC DIS and W production.

In the electron channel, the track must be associated with an electromagnetic cluster with $E_t > 10$ GeV. Also, the acoplanarity between the electron and hadronic system must be greater than 5°. Two events are observed, where 4 are expected from DIS and two-photon processes.

6 Conclusions

An excess of high Q^2 neutral and charged current events is observed in a preliminary analysis of 33 pb^{-1} of data collected by ZEUS. Anomalous production of events with a lepton and $\not{P}_t$ is not observed. HERA continues to provide a unique window to search for physics beyond the Standard Model.

References

[1] H1 Collaboration, C. Adloff et al., Z. Phys **C74** (1997) 191.
[2] ZEUS Collaboration, J. Breitweg et al., Z. Phys **C74** (1997) 207.
[3] U. Bassler (H1), these proceedings.

New H1 Results at High Q^2

Ursula Bassler[1] (bassler@mail.desy.de)

LPNHE - Paris, France

Abstract. From the 1994-1996 data samples,the H1 and ZEUS collaborations at HERA reported an excess of events at high Q^2 [1, 2]. We present here the update of the H1 publication using in addition the first half of the data taken in 1997, which corresponds to a total luminosity of 23.7 pb^{-1}. These data allow also for an improved measurement of the double differential neutral current cross-section up to $Q^2 = 5000$ GeV^2 and x=0.32. Within these data events with missing p_t and isolated leptons were observed, which are compared to expectations from W production.

1 Introduction

At the HERA ep collider, the interaction of 27.5 GeV positrons with 820 GeV protons allows for precise measurements of the proton structure functions via t-channel exchange of virtual photons or weak bosons [3]. New particles can be produced in the s-channel, such as leptoquarks, leptogluons or R-parity violating squarks with masses up to 300 GeV. Above these masses, contact term interactions may be found in the t or u channel.

The H1 detector is a multipurpose, nearly 4π detector. Its highly segmented liquid Argon calorimeter has a resolution of $12\%/\sqrt{E}$ for the electromagnetic section and $50\%/\sqrt{E}$ for the hadronic section. The energy scale of the scattered electron is known within 3% and its polar angle with a precision of 2-5 mrad. The energy scale error on the inclusive measurement of the hadronic final state is about 4%. The redundancy in the measurement of the scattered lepton and the hadronic final state allows for a variety of methods to reconstruct the kinematics, which can also be used for detector calibration, studies of initial QED radition and a good control of different systematic errors.

The ω method was developped, in order to compare the high Q^2 events of the H1 and ZEUS collaboration. The explanation of the excess observed by the two experiments [1, 2] as due to a single narrow resonnance was rendered unlikely by the comparison of the two datasets [4], as shown in figure 1.

Since then, the data taken until the end of june 1997, has been analysed, increasing the total luminosity from positron-proton scattering accumulated since 1994 to 23.7 pb^{-1}. Besides an update of the observation of events at very high Q^2, we present a measurement of the Neutral Current (NC) cross-section at high Q^2. We could also identify five events in the 94-97 data sample with a high missing p_t in the liquid Argon calorimeter and a clearly identified isolated lepton. These events are compared to expectations from W production.

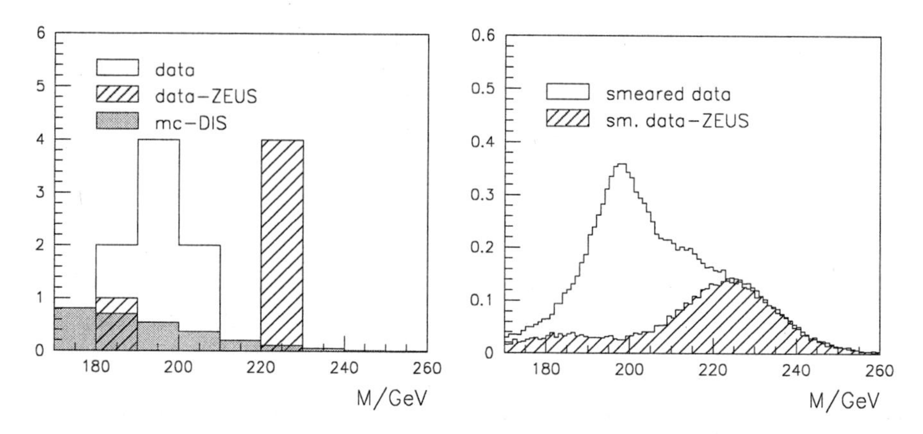

Fig. 1. Comparison of H1 and ZEUS high Q^2 events using the ω method.

2 F_2 Measurement at High Q^2

In deep inelastic scattering (DIS) the double differential NC cross-section can be written as

$$\frac{d^2\sigma}{dx\,dQ^2} = \frac{2\pi\alpha^2}{xQ^4}\left[Y_+F_2(x,Q^2) - y^2F_L(x,Q^2) - Y_-xF_3(x,Q^2)\right] \qquad (1)$$

with the helicity term $Y_\pm(y) = 1 \pm (1-y)^2$. In this notation, F_L is the longitudinal structure function, F_2 is the generalized structure functions which contains a contribution from pure photon exchange F_2^{em}, a weak contribution from the Z^0 exchange supressed by M_Z^4, and a contribution from the interference of both terms suppressed by M_Z^2, while xF_3 is due to the Z^0 exchange only. In the following we are presenting $F_2 = F_2^{em}$. The correction to F_2 arising from the weak contributions to the cross-sections is below 1% at $Q^2 < 1500$ GeV^2 and about 10% at $Q^2 = 5000$ GeV^2 and $x = 0.08$. The correction from the longitudinal cross-section is about 5% at $y = 0.9$ and negligible at $y < 0.5$.

The figure 2(a) shows the measurement of F_2 from $Q^2 = 150$ GeV^2 to $Q^2 = 5000$ GeV^2 compared to NLO-QCD fit performed on low Q^2 H1, BCDMS and NMC data ($Q^2 < 120$ GeV^2) [5]. The DGLAP extrapolation of this fit into the measured kinematic region shows good agreement and verifies the validity of perturbative QCD (pQCD) within this kinematic domain.

In figure 2(b) the measurement of the reduced cross-section

$$\sigma(e^+p) = \frac{xQ^4}{2\pi\alpha^2}\frac{1}{Y_+}\frac{\mathrm{d}^2\sigma}{\mathrm{d}x\mathrm{d}Q^2} \qquad (2)$$

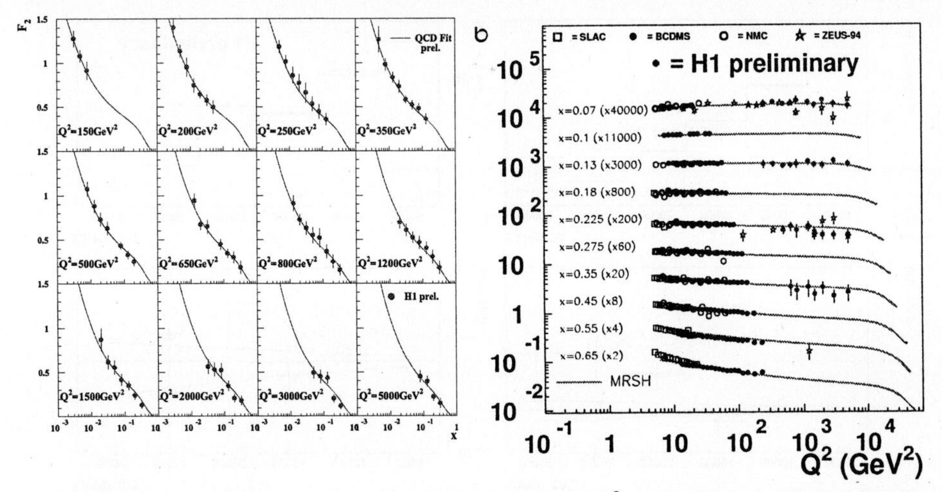

Fig. 2. (a) Preliminary F_2 measurement at high Q^2 compared to NLO QCD-fit performed on moderate Q^2 data ($Q^2 < 120$ GeV^2). (b) Reduced cross-section as function of Q^2 for fixed values of x, compared to fixed targed date and the MRSH parton density parametrisation.

is presented as a function of Q^2. The data are compared to the MRSH parton density parametrisation [6], which is used in the Standard Model expectation at very high Q^2. The new measurement shows no significant deviation from this parametrisation and supports a 6% uncertainty of the parton densities in the very high Q^2 region. The cross-section obtained from the QCD evolution is well constrained in this kinematic region by the low Q^2 fixed target data. It has essentially a flat behaviour in Q^2, except at $Q^2 > 10000$ GeV^2, where the suppression of the cross-section by the weak contribution becomes visible.

3 Events at Very High Q^2

The event selection applied for the very high Q^2 event sample can be found in [1]. For the NC event sample, it is essentially based on an identified scattered electron with $Q^2 > 2500$ GeV^2 within a y range between 0.1 to 0.9. The geometric acceptance is limited to an electron-scattering angle of 10° [1]. The background events, as well as events with strong initial state QED radiation are rejected using mainly kinematic criterias from energy-momentum conservation. The remaining background contribution was estimated to be less than 0.1%, the dominating source being jet-jet events from photoproduction.

Figure 3(a,c) is showing the Q^2 distribution of these events compared to the expectation from DIS using the MRSH parton densities parametrisation. We see a good agreement between the data and the expectation up

[1] the polar angle is measured at HERA w.r.t. the proton beam direction

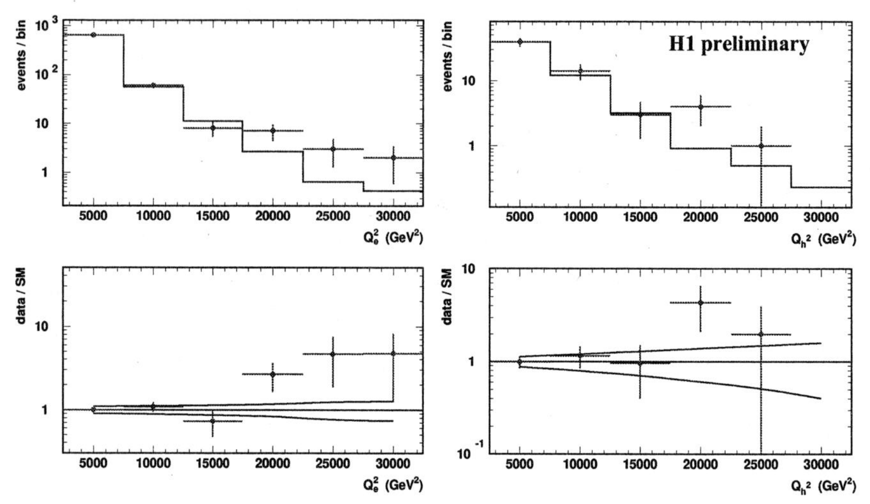

Fig. 3. Q^2 distribution of 94-97 data compared to expectation (MRSH) for Neutral Current (a) and Charged Current (b).

to Q^2 values of 15000 GeV^2. Above this Q^2 value the data are lying above the expectation: for $Q^2 > 2500$ GeV^2 724 events are observed compared to 714 ± 69 expected, whereas at $Q^2 > 15000$ GeV^2 18 events are observed compared to 8 ± 1.2 expected. This corresponds to a Poisson probability $\mathcal{P}(\mathcal{N} \geq \mathcal{N}obs) = 3{\times}10^{-3}$. The significance of the excess has slightly decreased compared to the observation on the 94-96 data only, where for $Q^2 > 15000$ GeV^2 12 events were observed againts 4.71 ± 0.76 expected corresponding to a Poisson probablity $\mathcal{P}(\mathcal{N} \geq \mathcal{N}obs) = 6 \times 10^{-3}$. The number of charged current events (figure 3 b,d) as function of Q^2 shows also good agreement up to Q^2 values of 15000 GeV^2. At $Q^2 > 15000$ GeV^2 6 events are observed against 2.9 expected.

In the mass bin of 200 GeV $\pm$ 12.5 GeV with y> 0.4, where in 94-96 data 7 events accumulated against 0.95 expected, only one new event was added during the first half of the 1997 data taking (fig.4).Now 8 events are observed within this mass window compared to 1.53 ± 0.29 expected. This corresponds to a Poisson probability $\mathcal{P}(\mathcal{N} \geq \mathcal{N}obs) = 3.3 \times 10^{-4}$.

4 Events with Isolated Leptons

A first event showing an isolated μ and large missing p_t has been found during the 1994 data taking [7]. Based on the charged current event selection, i.e. on events with missing $p_t > 25$ GeV in the liquid Argon calorimeter, four more events were found with isolated leptons carrying more than 10 GeV of p_t since then. The isolation crieria are based both on track and calorimetric isolation in rapidity and azimuth. The isolated leptons could be identified as

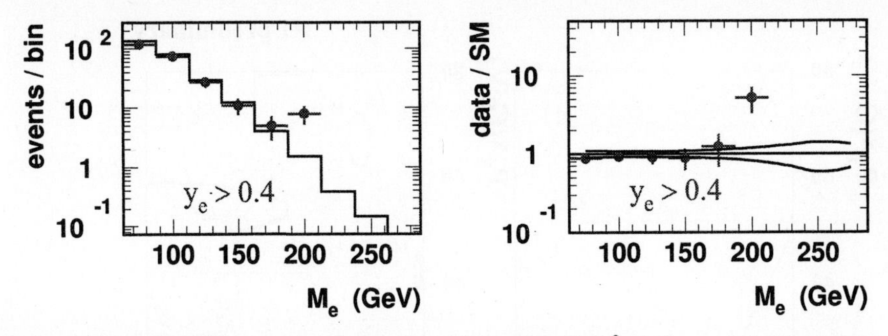

Fig. 4. Mass distribution of events with $Q^>$ 15000 GeV^2 and y> 0.4. One more event was observed in 97 in the mass bin around 200 GeV.

being μ^+ in two events (μ-1, μ-2), μ^- in two events (μ-3, μ-4) and an electron in one event. The event properties are summarized in table 2 of [8].

These events can be compared to expectations from W production with subsequent leptonic decay (fig.4). The expected kinematic signature for W-production would be a small hadronic p_t and a Jacobian peak in $M_{T(l\nu)}$ close to the W-mass. Figure 6 compares the observed events to W-production events simulated in the H1 detector with 500 times the luminosity of the data sample. The observed events are indicated by the lines which are giving the errors on the recontruction of p_t and the transverse mass. In the case of W-production with an $e\nu$ decay, the observed event can be compared to 1.34 events expected and its kinematics is lying in the bulk part of events from W-production. For W-production with a $\mu\nu$ decay we expect 0.41 events, and 4 events are observed with muons in the final state. The μ-1 and μ-2 events are at very low mass and high p_t, whereas μ-3 and μ-4 have masses close to the W mass. The estimated background for these events is below 0.05 events, mainly events from heavy quark production in γp and $\gamma\gamma \to \tau^+\tau-$.

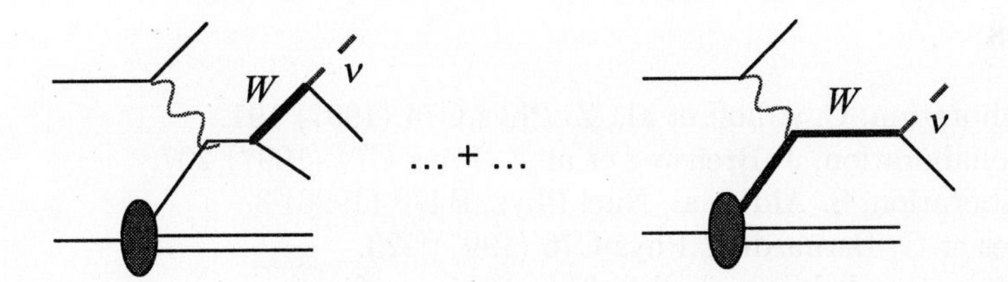

Fig. 5. The W production with leptonic decay.In the H1 detector the scattered electron would escape undected in the beam-pipe and large missing p_t would be observed due to the neutrinos.

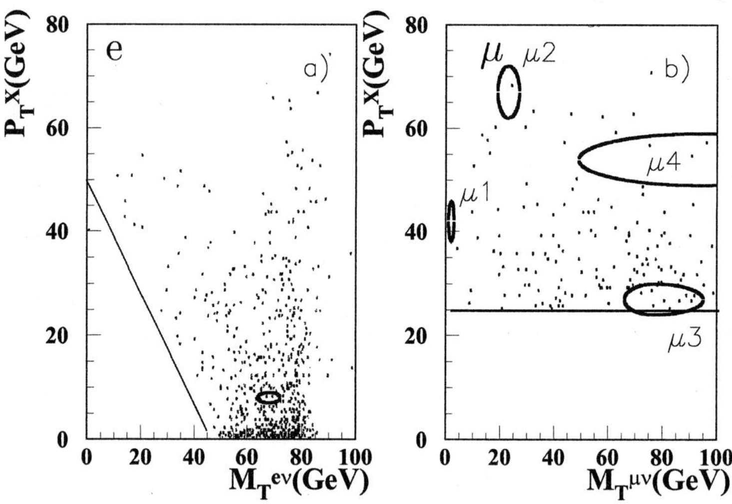

Fig. 6. Transverse mass distribution of the event with an isolted electron in the hadronic final state (a) and of the 4 events with an isolated muon compared to expectations from W-production. The lines indicate the acceptance due to the selection cuts.

5 Summary

The 1997 data taking has been very succesfull, already with the data accumulated up to june 97, the accuracy of the proton structure function measurement in the kinematic region of Q^2 below 5000 GeV^2 has been substantially improved. The cross-section in this kinematic region is in good agreement with the expectation from the pQCD. At very high Q^2, above 15000 GeV^2, 18 Neutral Current events are observed in this data sample againts 8 events expected, decreasing slightly the significance of the observation on the 94-96 data sample only. Four events with isolated muons and missing p_t were observed and compared to W production.

References

[1] H1 Collaboration, C. Adloff et al., Z. Phys C74 (1997) 191.
[2] ZEUS Collaboration, J. Breitweg et al.,Z. Phys C74 (1997) 207.
[3] H1 Collaboration, S. Aid et al. Nucl.Phys. B470 (1996) 3.
[4] U. Bassler et G. Bernardi, Z.Phys C76 (1997) 223.
[5] V. Lemaitre, Parallel Session Talk 314 at this conference.
[6] A.D. Martin, R.G. Roberts and W.J. Stirling, Phys.Lett B387 (1996) 419.
[7] H1 Collaboration, T. Ahmed et al., DESY 94-248 (1994).
[8] Eckhard Elsen, Plenary Talk 26 at this conference.

Leptoquarks and R-Parity Violating Supersymmetry Search at CDF

Xin Wu (xin.wu@cern.ch) (for the CDF collaboration

University of Geneva, Geneva, Switzerland

Abstract. We present results from direct searches for leptoquarks and R-parity violating SUSY particles by the CDF experiment. These results strongly constrain the interpretation of the recently reported high Q^2 event excess at HERA as leptoquark or R-parity violating SUSY production.

1 Search for Leptoquarks

Leptoquarks (LQ) refer to a general class of particles that couple both to leptons and quarks. They are bosons possessing both lepton numbers and color. These particles arise in many extensions to the SM where a new lepton-quark symmetry is assumed, for example GUT, technicolor, compositeness and SUSY models. The specific *qlLQ* couplings are model dependent but in order to be compatible with many low energy measurements these couplings are severely constrained. To avoid Flavor Changing Neutral Currents which are not observed experimentally, LQ's at the 100-1000 GeV/c^2 range are assumed to couple only within one generation [1].

There is recent strong interest in leptoquarks with masses around 200 GeV/c^2 due to the observation of an excess of events at high Q^2 at HERA [2] which can be speculated as the production of a first generation leptoquark [3].

At hadron colliders, gluon fusion and $q\bar{q}$ annihilation can produce pairs of leptoquarks. This is a strong process, thus its cross section can be calculated with perturbative QCD, independent of the *qlLQ* coupling. All three generations of leptoquarks can be produced at hadron colliders. In general leptoquark decays are assumed to conserve both lepton and baryon numbers in order to avoid rapid proton decay. Therefore there are two possible decay channels for a leptoquark: either to a quark and a charged lepton, or to a quark and a neutrino. Defining the charged lepton branching ratio as β, for the leptoquark pair produced the branching ratio of two charged lepton final state is β^2, while that of the one charged lepton plus neutrino state is $2\beta(1-\beta)$.

CDF has searched for all three generation leptoquarks in its data. Search results are summarized in Table 1. In this note we present the details of the analysis of the first two generation leptoquarks. The third generation leptoquark search has been published in [4].

generation	first	first	second	third
channel	$eejj$	$e\nu jj$	$\mu\mu jj$	$\tau\tau jj$
95% CL limit (GeV/c^2)	213 (β=1)	180 (β=0.5)	195 (β=1)	99 (β=1)
status	published [5]	preliminary	preliminary	published [4]

Table 1. Summary of CDF scalar leptoquark search results.

1.1 Search for first generation leptoquarks with $eejj$

The analysis begins with selecting events with two high energy electrons ($E_T > 25$ Gev) and 2 jets ($E_T > 30$ and $E_T > 15$ GeV). Z^0 events are removed by a cut on M_{ee}: $M_{ee} < 76$ or $M_{ee} > 106$ GeV/c^2 . To reduce the background from Drell-Yan dilepton continuum, the sum E_T of two electrons and the sum E_T of two jets are required to be greater than 70 GeV. The electrons are then paired with jets to reconstruct two leptoquarks. Two parings are possible. The paring that gives closer ej masses are chosen. To select candidate events that are consistent of leptoquark pair production of a mass M_{LQ}, the mass difference of two ej masses, ΔM_{ej}, is required to be less than $0.2M_{LQ}$, and the average of the two ej masses within 3σ of M_{LQ}.

With these requirements, only one event passed the selection for $140 < M_{LQ} < 170$ GeV/c^2 in a data sample of 110pb^{-1}, which is consistent of the background expectation. Acceptance for the signal is evaluated by the PYTHIA program and a full detector simulation. The result of this analysis is interpreted as a 95% CL limit on leptoquark pair production cross section derived from Poisson statistics, taking into account systematic uncertainties (13.4% in total) on the acceptance, but without background subtraction. Figure 1 shows this limit, as function of M_{LQ}, compared to the NLO QCD calculation [6]. From this comparison, a first generation leptoquark with $\beta = 1$ is excluded for $M_{LQ} < 213$ GeV/c^2 at 95%CL.

1.2 Search for first generation leptoquarks with $e\nu jj$

If $\beta < 1$, leptoquark pairs can decay to $e\nu jj$ final states. The maximum braching ratio occurs at $\beta = 0.5$.

The analysis selects events with one high energy electron, ($E_T > 20$ GeV), large missing energy ($\not{E}_T > 35\ GeV$) and 2 jets ($E_T > 30$ and $E_T > 15$ GeV). The main background in this case is W plus multi-jets and they are rejected by the following cuts: the sum E_T of two jets must be greater than 80 GeV and the transverse mass of electron and neutrino must be greater than 120 GeV/c^2 .

Using the paring that gives closer M_{ej} and $M^T_{\nu j}$, a 3σ cut around a nominal M_{LQ} on the M_{ej} vs $M^T_{\nu j}$ plane is applied to select candidate events that are consistent with leptoquark pair production. One candidate with $100 < M_{LQ} < 130$ GeV/c^2 is selected from 110 pb^{-1} data.

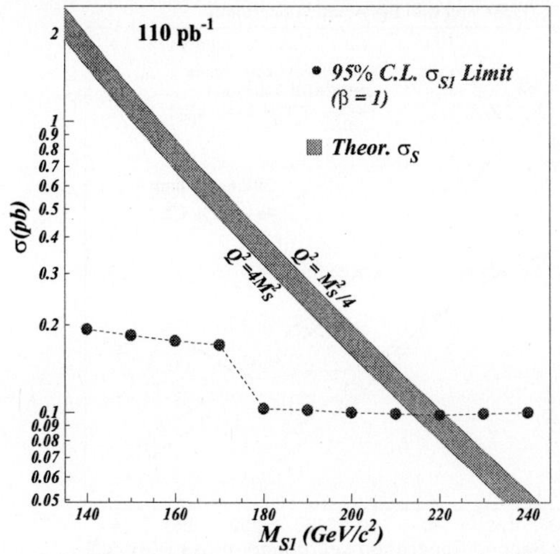

Fig. 1. 95 % C.L. cross section limit for the first generation LQ.

Acceptance for the signal is evaluated by the PYTHIA program and a full detector simulation. The total systematics uncertainty is 15%. Using a Poisson probability distribution without background subtraction, a conservative 95%CL limit on first generation leptoquark production cross section with $\beta = 0.5$ is derived.This gives a lower mass limit of 180 GeV/c^2 comparing to the NLO QCD cross section calculation.

1.3 Search for second generation leptoquarks with $\mu\mu jj$

This analysis is very similar to that of the dielectron channel. Events with two high energy muons ($P_T > 30$ and 20 GeV/c) and 2 jets ($E_T > 30$ and $E_T > 15$ GeV) are selected. Z^0 and J/ψ events are removed by a cut on $M_{\mu\mu}$: $M_{\mu\mu} < 76$ or $M_{\mu\mu} > 106$ GeV/c^2 , and $M_{\mu\mu} > 11$ GeV/c^2 . Using the paring that gives closer μj masses, a 3σ cut around a nominal M_{LQ} on the $M_{LQ}(1)$ vs $M_{LQ}(2)$ plane is used to select candidate events that are consistent of leptoquark pair production. Two candidates with $100 < M_{LQ} < 140$ GeV/c^2 and one candidate with $M_{LQ} > 220$ GeV/c^2 are selected from 110 pb^{-1} data, with 3.8 and 0.09 background expected, respectively.

Acceptance for the signal is evaluated by the PYTHIA program and a full detector simulation. The total systematics uncertainty is 20%. Using a Poisson probability distribution without background subtraction, a conservative 95%CL limit on second generation leptoquark production cross section is derived, as shown in Figure 2 in comparison with the NLO QCD calculation. We exclude a second generation leptoquark with $\beta = 1$ below 195 GeV/c^2 .

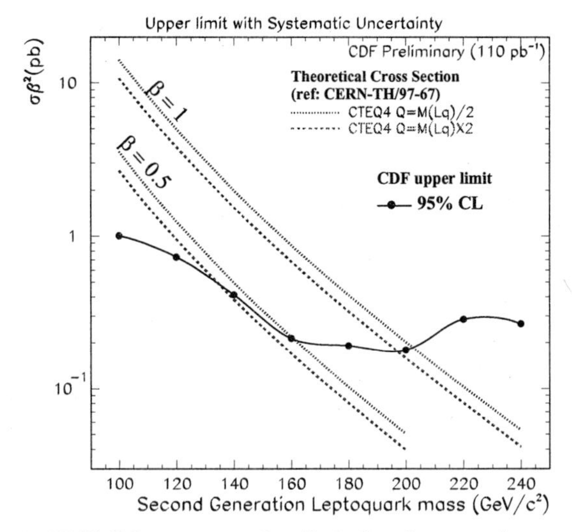

Fig. 2. 95 % C.L. cross section limit for the second generation LQ.

2 Search for R-parity violating SUSY particles

In the Minimal Supersymmetric Standard Model, R parity is assumed to be conserved to avoid rapid proton decay. However this requirement can be relaxed: either baryon number or lepton number is conserved, but R parity can be violated. This gives rise to new decay channels for SUSY particles. In fact, one of the interpretations for the HERA high Q^2 events is the R parity violating production and decay of a squark: $e^+d \rightarrow \tilde{c}_L, \tilde{t}_L \rightarrow e^+d$.

We have studied two complementary R-parity violating SUSY processes:

$$p\bar{p} \longrightarrow \tilde{g}\tilde{g} \longrightarrow (\bar{c}\tilde{c}_L)(\bar{c}\tilde{c}_L) \stackrel{\not{R}_p}{\longrightarrow} \bar{c}(e^{\pm}d)\bar{c}(e^{\pm}d) \tag{1}$$

$$p\bar{p} \longrightarrow \tilde{c}_L\bar{\tilde{c}}_L \longrightarrow (c\tilde{\chi}_1^0)(\bar{c}\tilde{\chi}_1^0) \stackrel{\not{R}_p}{\longrightarrow} c(q\bar{q}'e^{\pm})\bar{c}(q\bar{q}'e^{\pm}) \tag{2}$$

Process 1 is important when $M_{\tilde{\chi}_1^0} > M_{\tilde{c}_L}$ and $\mathrm{Br}(\tilde{c}_L \rightarrow e^{\pm}d)$ is large, while process 2 is important when $M_{\tilde{\chi}_1^0} < M_{\tilde{c}_L}$ and $\mathrm{Br}(\tilde{c}_L \rightarrow c\tilde{\chi}_1^0)$ is large. In the second case, the $\tilde{\chi}_1^0$ decay is R-parity violating.

These final states have the distinct feature of two lepton of same charge (like sign) and multiple jets, with little missing energy. Like sign dilepton signature have the benefit of low SM background.

CDF has conducted a search in ~105 pb^{-1} of data. Events are required to have two electrons of the same charge, both with $E_T > 15$ Gev, and two jets with $E_T > 15$ GeV. A cut on the missing energy significance, $S \equiv \not{E}_T/\sqrt{\Sigma E_T} < 5$, is used to remove events with large missing energy. No event passed these cuts. This null result is then used to derive limits on SUSY parameters for the above two R parity violating processes.

2.1 Limit on $\tilde{g}\tilde{g} \longrightarrow (\bar{c}\tilde{c}_L)(\bar{c}\tilde{c}_L) \longrightarrow \bar{c}(e^{\pm}d)\bar{c}(e^{\pm}d)$

This process is studied with $M_{\tilde{c}_L}$=200 GeV/c^2 and $M_{\tilde{q}\neq\tilde{c}_L}$=400 GeV/$c^2$ ($M_{\tilde{s}_L}$ is fixed by the mass relation in MSSM). The acceptance is then evaluated with ISAJET for $M_{\tilde{g}}$ of 210-400 GeV/c^2 . The dependance on $M_{\tilde{g}}$ is not strong so the average acceptance of 15.8%±2.4% is used. This gives a result of $\sigma Br(\tilde{g}\tilde{g} \longrightarrow e^{\pm}e^{\pm}X) < 0.19$ pb at 95% CL. The NLO calculation [8] of $\sigma(p\bar{p} \to \tilde{g}\tilde{g})$, with the above fixed $M_{\tilde{c}_L}$ and $M_{\tilde{q}\neq\tilde{c}_L}$ values, is a function of $M_{\tilde{g}}$. Therefore for a given $Br(\tilde{g}\tilde{g} \longrightarrow e^{\pm}e^{\pm}X)$, the above cross section limit can be used to derive a limit on $M_{\tilde{g}}$, as presented in Figure 3. The result is that for $Br(\tilde{g}\tilde{g} \longrightarrow e^{\pm}e^{\pm}X) = 0.5$, $M_{\tilde{g}}$ must be heavier than 270 GeV at 95% CL.

Limits are also set on one particular scenario [7] of this process proposed to explain the HERA events.

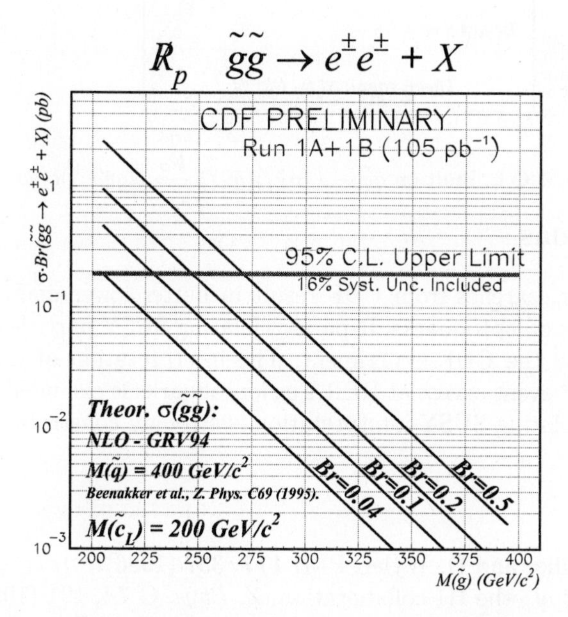

Fig. 3. 95 % C.L. limit on $M_{\tilde{g}}$ for $\tilde{g}\tilde{g} \xrightarrow{\not{R}_p} e^{\pm}e^{\pm}X$

2.2 Limit on $\tilde{c}_L\bar{\tilde{c}}_L \longrightarrow (c\tilde{\chi}_1^0)(\bar{c}\tilde{\chi}_1^0) \longrightarrow c(q\bar{q}'l^{\pm})\bar{c}(q\bar{q}'l^{\pm})$

The acceptance for this process is a strong function of $M_{\tilde{\chi}_1^0}$ and $M_{\tilde{c}_L}$ so the null search result can be used to derive a limit on $\sigma(p\bar{p} \to \tilde{c}_L\bar{\tilde{c}}_L)Br(\tilde{c}_L\bar{\tilde{c}}_L \to e^{\pm}e^{\pm} + X)$ as function of $M_{\tilde{\chi}_1^0}$, $M_{\tilde{c}_L}$.

Assuming $Br(\tilde{c}_L \to c\tilde{\chi}_1^0) = 1$ and $Br(\tilde{\chi}_1^0\tilde{\chi}_1^0 \to LSee) = 1/8$, one can exclude regions of $M_{\tilde{\chi}_1^0}$-$M_{\tilde{c}_L}$ by comparing to $\sigma(p\bar{p} \to \tilde{c}_L\bar{\tilde{c}}_L)$ calculation.

One can also consider the case where $M_{\tilde{q}}$ is degenerate except for $\tilde{t}$, and then compare the experimental limit with NLO calculation of $\sigma(p\bar{p} \to \tilde{q}\bar{\tilde{q}})$ to set exclusion areas in $M_{\tilde{\chi}_1^0}$, $M_{\tilde{q}}$ and $M_{\tilde{g}}$, since the NLO $\sigma(p\bar{p} \to \tilde{q}\bar{\tilde{q}})$ is a function of $M_{\tilde{g}}$. The result for $M_{\tilde{q}}/2 < M_{\tilde{\chi}_1^0} < M_{\tilde{q}} - Mq$ is shown in Figure 4.

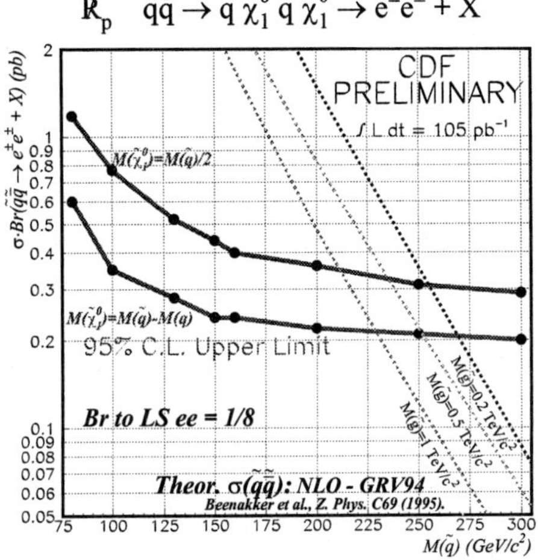

Fig. 4. 95 % C.L. limit on $\tilde{q}\bar{\tilde{q}} \longrightarrow (q\tilde{\chi}_1^0)(\bar{q}\tilde{\chi}_1^0) \xrightarrow{R\!\!\!/_p} c(q\bar{q}'e^\pm)\bar{c}(q\bar{q}'e^\pm)$

3 Conclusions

The latest direct searches from CDF result in better lower mass limits for three generations of leptoquarks. R-parity violating SUSY particles have also been searched by the CDF experiment. The interpretation of the recently reported high Q^2 event excess at HERA as leptoquarks is excluded for $\beta = 1$. The R-parity violating SUSY interpretation is strongly constrained.

References

[1] W. Buchmüller and D. Wyler, PLB **117**, 337 (1986).
[2] C. Adloff *et al.*, the H1 collaboration, *Z. Phys.* C **74**, 191 (1997).
J. Breitweg *et al.*, The ZEUS collaboration, *Z. Phys.* C **74**, 207 (1997).
[3] G. Altarelli *et al.*, hep-ph/9703276; J. Blümlein, *Z. Phys.* C **74**, 605 (1997); K.S. Babu *et al.*, PLB **402**, 367 (1997); J. L. Hewett and T. Rizzo, *Phys. Rev.* D **56**, 5709 (1997).
[4] F. Abe *et al.*, The CDF Collaboration, *Phys. Rev. Lett.* **78**, 2906 (1997).
[5] F. Abe *et al.*, The CDF Collaboration, FERMILAB-PUB-97-280-E, Submitted to Phys.Rev.Lett.
[6] M. Krämer *et al.*, *Phys. Rev. Lett.* **79**, 341 (1997).
[7] D. Choudhury and S. Raychaudhuri, *Phys. Rev.* D **56**, 1788 (1997).
[8] W. Beenakker *et al.*, *Z. Phys.* C **69**, 163 (1995).

Searches for Pair Production of First Generation Leptoquark at DØ

Boaz Klima (klima@fnal.gov) (for the DØ Collaboration)

Fermilab, Batavia, Illinois, U.S.A.

Abstract. We have searched for the pair production of first generation scalar leptoquarks using the full data set (123 pb^{-1}) collected with the DØ detector at the Fermilab Tevatron during 1992–1996. We observe no candidates, consistent with the expected background. We combine these new results from the $ee+$jets and $e\nu+$jets channels with the published $\nu\nu+$jets result to obtain a 95% CL upper limits on the LQ pair production cross section as a function of mass and β, the branching ratio to a charged lepton and a quark. Comparing to the NLO theory predictions, we set 95% CL lower limits on the LQ mass of 225, 204, and 79 GeV/c^2 for $\beta = 1$, $\frac{1}{2}$, and 0, respectively. The results of this analysis rule out an interpretation of the excess of high Q^2 events at HERA as leptoquarks with LQ mass below 200 GeV/c^2 for values of $\beta > 0.4$.

1 Introduction

Leptoquarks (LQ) are hypothesized exotic color-triplet bosons which couple to both quarks and leptons. They appear in extended gauge theories and composite models and have attributes of both quarks and leptons such as color, fractional electric charge, and lepton and baryon quantum numbers [1]. Leptoquarks with universal couplings to all flavors would give rise to flavor-changing neutral currents and are severely constrained by studies of low energy phenomena [2]. Therefore, only leptoquarks which couple within a single generation are considered. The H1 and ZEUS experiments at HERA have reported an excess of events at high Q^2 in e^+p collisions [3] [4]. One possible interpretation of these events is production of first generation leptoquarks at a mass near 200 GeV/c^2 [5] .

Leptoquarks would be dominantly pair-produced via strong interactions in $\overline{p}p$ collisions, independently of the unknown LQ–l–q Yukawa coupling. Each leptoquark would subsequently decay to a lepton and a quark. For first generation leptoquarks, this leads to three possible final states: ee+jets, $e\nu$+jets and $\nu\nu$+jets, with rates proportional to β^2, $2\beta(1-\beta)$ and $(1-\beta)^2$, respectively, where β denotes the branching fraction of a leptoquark to an electron and a quark (jet).

Lower limits on the mass of a first generation leptoquark were published by the LEP experiments [6], by CDF and DØ [7], and by H1 [8].

This report describes a search for the pair production of first generation scalar leptoquarks in the $ee+$ jets and $e\nu+$jets final states using 123 ± 7 pb^{-1} of data collected by DØ at the Fermilab Tevatron with $\sqrt{s}=$ 1.8 TeV during 1992–1996. A similar search was conducted by the CDF collaboration [9]. The DØ results reported in this article are described in Refs. [10] and [11]. The DØ detector and data acquisition system are described in detail in Ref. [12].

2 The ee + jets Channel

A base data sample of 101 events with two electrons and two or more jets was selected. The electrons ($E_T^e >$ 20 GeV) were required to be separated from jets ($E_T^j >$ 15 GeV). Events whose ee invariant mass lies within the Z boson mass region were rejected. The efficiency of the trigger used to collect the base data sample exceeded 99% for the leptoquark mass range addressed by this analysis.

Monte Carlo (MC) signal samples were generated for leptoquark masses between 120 and 260 GeV/c^2 using the ISAJET event generator and a detector simulation based on the GEANT program. Leptoquark production cross sections were taken from the recently available next-to-leading order (NLO) calculations [13]. The primary backgrounds to the ee + jets decay mode are Drell-Yan+2 jets production (DY), $t\bar{t}$ production, and misidentified multijet events. The ISAJET DY cross section normalization was fixed by comparing it with Z+2 jets data. The DØ measured $t\bar{t}$ production cross section of 5.5 ± 1.8 pb [14] was used for top quark MC HERWIG events. The multijet background was estimated directly from data [10] .

To search for leptoquarks, a random grid search method was used to optimize cuts on the data and MC samples. Consistent results were obtained using a neural network [15]. The limit setting criterion of a maximum number of signal events for a fixed number of background events was adopted. The background level chosen was 0.4 events, corresponding to a 67% probability that no such events would be observed.

The set of cuts which optimally separates signal from background was determined by a systematic search using distributions of signal MC events. Many sets of selection criteria were explored. A cut on a single, relatively simple variable, $S_T \equiv H_T^e + H_T^j$, where $H_T^e \equiv E_T^{e1} + E_T^{e2}$ and $H_T^j \equiv \sum_{\text{jets}} E_T^j$, satisified the limit setting criterion. Approximately 0.4 background events are expected for $S_T >$ 350 GeV. No events remain in the base data sample after this S_T cut is applied. The highest value of S_T seen in the data is 312 GeV. The background which is roughly equally distributed among the three main sources is estimated to be 0.44 ± 0.06 events.

To investigate the background further, constrained mass fits were performed on the events in the base data sample, on background samples, and on the 200 GeV/c^2 leptoquark signal MC sample. The 3C mass fit was based on the SQUAW kinematic mass fitting program and required the two ej masses to be identical. Figures 1(a–c) show S_T as a function of the fit mass for the estimated background, 200 GeV/c^2 leptoquark events, and the base data sample. The distribution from the data agrees with that of the expected background. Figure 1(d) shows the one dimensional mass distributions for the same samples. Inset in Fig. 1(d) are the same distributions after a cut on $S_T >$ 250 GeV.

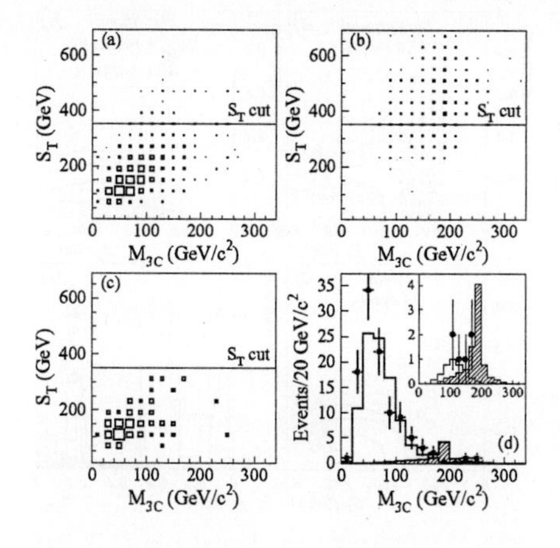

Fig. 1. S_T *vs.* **3C fit mass distributions for (a) background, (b) 200 GeV/c^2 leptoquarks, and (c) the base data sample. (d) Mass distribution of the data events (solid circles), expected background (solid line histogram), and 200 GeV/c^2 leptoquarks (hatched histogram). The inset plot shows these distributions for $S_T > 250$ GeV.**

As can be seen, the data are consistent with the background prediction.

The overall signal detection efficiency is 9–37% for leptoquark masses of 120–250 GeV/c^2. A 95% confidence level (CL) upper limit on the cross section σ was set using a Bayesian approach with a flat prior distribution for the signal cross section. The statistical and systematic uncertainties on the efficiency, the integrated luminosity, and the background estimation were included in the limit calculation with Gaussian prior distributions. The resulting upper limit on the cross section is shown in Fig. 2 together with the NLO calculation of Ref. [13]. The intersection of our limit curve with the lower edge of the theory band ($\mu = 2M_{LQ}$) is at $\sigma = 0.068$ pb, leading to a lower limit on the mass of a first generation scalar leptoquark of 225 GeV/c^2.

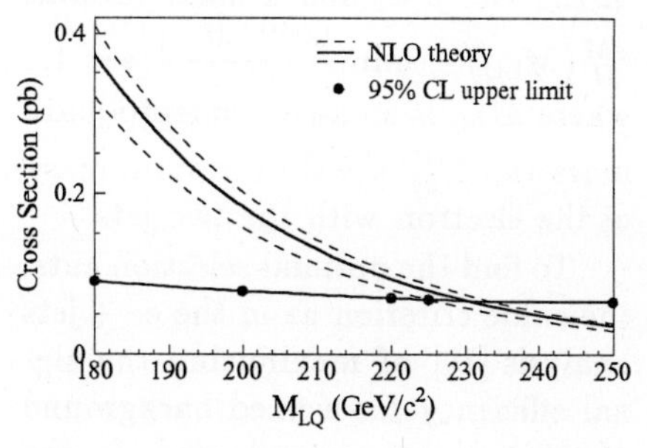

Fig. 2. **Upper limit on the leptoquark pair production cross section for 100% decay to** ***eq*****. Also shown is the NLO calculation [13] where the central solid line corresponds to $\mu = M_{LQ}$, and the lower and upper dashed lines to $\mu = 2M_{LQ}$ and $\mu = M_{LQ}/2$, respectively.**

3 The $e\nu$ + jets Channel

A base data sample of 14 events with one electron, two or more jets, missing E_T ($\not\!\!E_T > 30$ GeV), and $M_T^{e\nu} > 110$ GeV/c^2 was selected. The estimated background is 17.8±2.1 events, of which 11.7±1.8, 4.1±0.9, and 2.0±0.7 events are from W + jets, QCD multijets, and $\bar{t}t$ production, respectively.

Monte Carlo samples similar to those in the ee+jets were used in this analysis. The W + jets background which is dominant prior to requiring $M_T^{e\nu} > 110$ GeV/c^2 was simulated using the VECBOS program.

Two variables that provide significant discrimination between signal and remaining background were identified. They are the energy variable $S_T \equiv E_T^e + E_T^{j1} + E_T^{j2} + \not{E}_T$, where E_T^{j1} and E_T^{j2} are the transverse energies of the two jets, and a mass variable $\frac{dM}{M}(M_{\rm LQ}) \equiv \min(\frac{|M_{ej}^{(i)} - M_{\rm LQ}|}{M_{\rm LQ}}; i = 1, 2)$, where M_{LQ} is an assumed leptoquark mass and $M_{ej}^{(i)}$ are the invariant masses of the electron with the two jets.

To find the optimal selection cuts, the same criterion as in the ee + jets analysis [10] of maximizing the signal efficiency for a fixed background of $\approx$0.4 events was adopted. In the low mass range ($M_{LQ} \leq$120 GeV/c^2), where the LQ production rates are high, requiring $S_T > 400$ GeV is sufficient. For $M_{LQ} > 120$ GeV/c^2 , neural networks (NN) were used since they provide higher efficiency than an S_T cut alone. At each mass where MC events were generated, a three layer feed-forward neural network [15] was used with two inputs (S_T and $\frac{dM}{M}(M_{\rm LQ})$), five hidden nodes, and one output node. Each NN was trained using simulated LQ events as the signal (desired output $\mathcal{D}_{NN} = 1$) and a mixture of $t\bar{t}$, W + jets, and multijet events as background (with desired output $\mathcal{D}_{NN} = 0$). Cuts on $\mathcal{D}_{NN}$ that yield background estimates in closest proximity to the desired background were obtained by varying $\mathcal{D}_{NN}$ in steps of 0.05. The expected background after the cut ranges between 0.29 ± 0.25 and 0.61 ± 0.27. No data events pass the cuts.

Figures 3 (a)-(c) show the 2-dim. distributions of $\frac{dM}{M}(180)$ *vs.* S_T for simulated LQ signal events with M_{LQ} = 180 GeV/c^2, the combined background, and data. The contours corresponding to constant values of $\mathcal{D}_{NN}$ demonstrate the separation achieved between signal and background. The distribution of $\mathcal{D}_{NN}$ for data is compared with the predicted distributions for background and signal in Fig. 3 (d). The data are described well by background alone. The highest $\mathcal{D}_{NN}$ observed in the final data sample is 0.79.

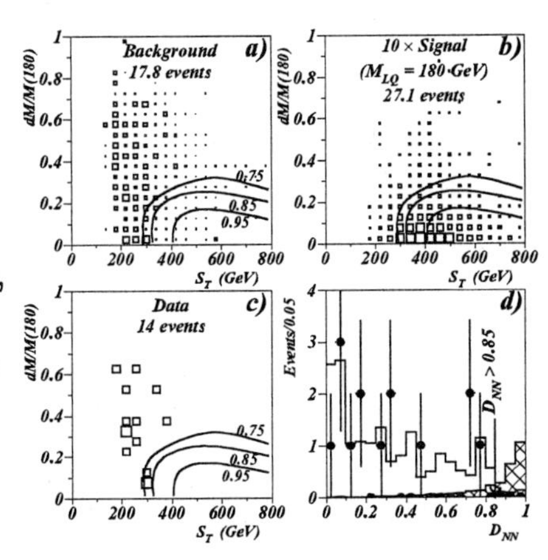

Fig. 3. $\frac{dM}{M}(180)$ *vs.* S_T distributions for (a) predicted background, (b) simulated LQ events (M_{LQ} = 180 GeV/c^2), and (c) data, after all cuts except that on $\mathcal{D}_{NN}(180)$. The contours correspond to $\mathcal{D}_{NN}$ = 0.75, 0.85, and 0.95. The area of a box is proportional to the number of events in the bin, with the total number of events normalized to 115 pb^{-1}. (d) Distributions of $\mathcal{D}_{NN}$ for data (solid circles), background (solid hist.) and expected LQ signal for M_{LQ} = 180 GeV/c^2 (hatched hist.).

Using Bayesian statistics, a 95% CL upper limit on the leptoquark pair production cross section was obtained for $\beta = \frac{1}{2}$ as a function of leptoquark mass. The statistical and systematic uncertainties in the efficiency, the integrated luminosity, and background estimation were included in the limit calculation with Gaussian prior probabilities. The measured 95% CL cross section upper limits for various LQ masses are also plotted in Fig. 4 together with the NLO calculations [13] for $\beta = \frac{1}{2}$. The intersection of the limit curve with the lower edge of the theory band ($\mu = 2M_{LQ}$) is at 0.19 pb, leading to a 95% CL lower limit on the LQ mass of 175 GeV/c^2.

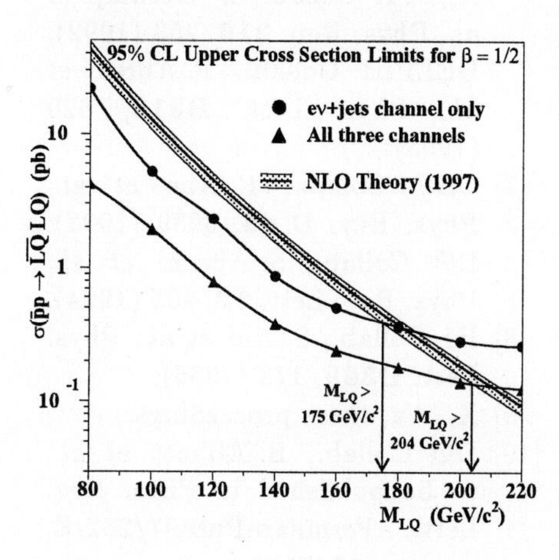

Fig. 4. Measured 95% CL upper limits on the leptoquark pair production cross section in the $e\nu+$jets channel (circles) and all three channels combined (triangles) for $\beta = \frac{1}{2}$. Also shown are the NLO calculations of Ref. [13] where the central line corresponds to $\mu = M_{LQ}$, and the lower and upper lines to $\mu = 2M_{LQ}$ and $\mu = M_{LQ}/2$, respectively.

4 Combined Limits

An analysis of the $\nu\nu+$jets channel is accomplished by making use of our published search ($\int Ldt \approx 7.4$ pb^{-1}) for the supersymmetric partner of the top quark [16]. Three events survive the selection criteria ($\not{E}_T > 40$ GeV and 2 jets with $E_T > 30$ GeV) consistent with the estimated background of 3.5±1.2 events. The signal efficiency for $M_{LQ} = 80$ GeV/c^2 is calculated to be 2.2%. This analysis yields the limit $M_{LQ} > 79$ GeV/c^2 at 95% CL for $\beta = 0$.

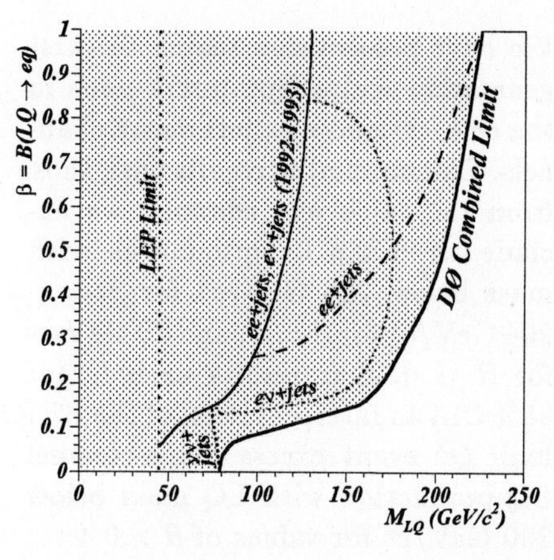

Fig. 5. Lower limits on the first generation scalar leptoquark mass as a function of β, based on searches in all three possible decay channels for leptoquark pairs. Limits from LEP experiments and from our previous analysis [7] of 1992-93 data are also shown. The shaded area is excluded at 95% CL.

Combining the $ee+$jets, $e\nu+$jets, and $\nu\nu+$jets channels, 95% CL upper limits on the LQ pair production cross section were calculated as a function of LQ mass for various values of β. These cross section limits for $\beta = \frac{1}{2}$ (shown in Fig. 4), when compared with NLO theory ($\mu = 2M_{LQ}$), yield a 95% CL lower limit on the LQ mass of 204 GeV/c^2. The lower limits on the LQ mass derived as a function of β, from all three channels combined as well as from the individual channels are shown in Fig. 5. These results can also be used to set limits on pair production of any heavy scalar particle decaying into a lepton and a quark, in a variety of models.

5 Conclusion

We have presented a search for first generation scalar leptoquark pairs in the ee+jets and $e\nu$+jets decay channels. Combining the results with those from the $\nu\nu$ + jets channel, we exclude at 95% CL leptoquarks with mass below 225 GeV/c^2 for $\beta = 1$, 204 GeV/c^2 for $\beta = \frac{1}{2}$, and 79 GeV/c^2 for $\beta = 0$. Our results exclude (at 95% CL) an interpretation of the HERA high Q^2 event excess via s-channel LQ production with LQ mass below 200 GeV/c^2 for values of $\beta > 0.4$.

6 Acknowledgements

I would like to express my appreciation to the organizers of this excellent conference. It is also my pleasure to thank my colleagues from DØ who helped me in preparing this report.

References

[1] J.L. Hewett and T.G. Rizzo, Phys. Rep. **183**, 193 (1989), and references therein.

[2] M. Leurer, Phys. Rev. D **49**, 333 (1994).

[3] H1 Collab., C. Adloff *et al.*, Z. Phys. **C74**, 191 (1997); ZEUS Collab., J. Breitweg *et al.*, Z. Phys. **C74**, 207 (1997).

[4] D. Acosta (ZEUS) and U. Bassler (H1), these proceedings.

[5] J.L. Hewett and T.G. Rizzo, to be published in Phys. Rev. D, SLAC-PUB-7430, hep-ph/9703337.

[6] OPAL Collab., G. Alexander *et al.*, Phys. Lett. **B263**, 123 (1991); L3 Collab., O. Adriani *et al.*, Phys. Rep. **236**, 1 (1991); ALEPH Collab., D. Decamp *et al.*, Phys. Rep. **216**, 253 (1992); DELPHI Collab., P. Abreu *et al.*, Phys. Lett. **B316**, 620 (1993).

[7] CDF Collab., F. Abe *et al.*, Phys. Rev. D **48**, 3939 (1993); DØ Collab., S. Abachi *et al.*, Phys. Rev. Lett. **72**, 965 (1994).

[8] H1 Collab., S. Aid *et al.*, Phys. Lett. **B369**, 173 (1996).

[9] X. Wu, these proceedings.

[10] DØ Collab., B. Abbott *et al.*, to be published in Phys. Rev. Lett., Fermilab-Pub-97/252-E, hep-ex/9707033.

[11] DØ Collab., B. Abbott *et al.*, to be submitted for publication in Phys. Rev. Lett., Fermilab-Pub-97/344-E.

[12] DØ Collab., S. Abachi *et al.*, Nucl. Instrum. Methods A **338**, 185 (1994).

[13] M. Krämer, T. Plehn, M. Spira, and P.M. Zerwas, Phys. Rev. Lett., **79**, 341 (1997).

[14] DØ Collab., S. Abachi *et al.*, Phys. Rev. Lett. **79**, 1203 (1997).

[15] L. Lönnblad *et al.*, Comput. Phys. Commun. **81**, 185 (1994).

[16] DØ Collab., S. Abachi *et al.*, Phys. Rev. Lett. **76**, 2222 (1996).

Tevatron Searches for Compositeness

Elizabeth Gallas (eggs@fnal.gov)

University of Texas at Arlington, USA

Abstract. Tevatron experiments have recently set or improved limits on quark and lepton compositeness. Included here are results from $\sqrt{s} = 1.8$ TeV $p\bar{p}$ collider experiments DØ and CDF and from fixed target neutrino experiment CCFR, who report limits on quark-quark, quark-lepton, and quark-neutrino compositeness, respectively.

1 Introduction

Compositeness theories hypothesize that any or all Standard Model (SM) fundamental particles may have substructure. Recent interest in compositeness is due to the CDF observation of an excess in the inclusive jet E_T spectrum at high E_T and the excess of events with high x and Q^2 in deep inelastic scattering at HERA.

In general, composite models define an energy scale Λ_C (a *compositeness scale*) at which preons (quark or lepton subparticles) are bound within the quark or lepton states. In the SM, $\Lambda_C = \infty$, and consequently quarks and leptons are pointlike. But if Λ_C is finite, then substructure would be observed at energy scales above Λ_C. At lower energies, other effects may be observed as described in the searches herein.

Three Tevatron experiments have recently set or improved limits on quark and lepton compositeness. These searches assumed that the center-of-mass energy of the parton-parton system is much smaller than the compositeness scale ($\sqrt{\hat{s}} << \Lambda_C$). Under these conditions, quarks and leptons appear nearly pointlike and effective new interactions due to compositeness can be described by additional four fermion contact interactions. They also assumed that the gauge bosons are elementary, and that in the quark-lepton compositeness searches, the compositeness scale of quarks and leptons is the same. The datasets and event topologies in each search are distinct and therefore are described in separate sections below.

2 DØ Compositeness - Dijet Angular Distribution

One manifestation of quark compositeness would be an increase in the cross section for the production of jets in $p\bar{p}$ collisions. This increase is expected to be most apparent at high transverse energy (E_T), which is the source of the acute interest in the excess of events observed by CDF in the high end of the E_T spectrum (see E.Buckley-Geer talk 413 in these proceedings). Interest turned to controversy when DØ reported no such excess (see R.McCarthy talk 412 in these proceedings). A more sensitive measure of compositeness in jet production is found by studying the dijet angular distribution. This measurement is not subject to the large uncertainties of the jet cross section prediction due to the choice of parton distribution functions (pdf).

DØ exploits it's high resolution, hermetic calorimetry [2] to search for compositeness in QCD dijet events using their inclusive jet data sample collected during the 1994-1995 run. The angle θ^* is defined as the parton-parton center of mass scattering angle. By transforming to the variable $\chi = (1+\cos\theta^*)/(1-\cos\theta^*)$, the expected next-to-leading order (NLO) QCD distribution $1/NdN/d\chi$ is relatively flat while a contact term features a steep rise at low χ, enhancing the signal to background in this region. The data are separated into four bins in dijet invariant mass, the highest of which is expected to be most sensitive to a compositeness signal.

The data points of Figure 1 show the dijet invariant mass distribution in the variable χ for the highest mass bin ($M_{jj} > 635$GeV). Statistical errors are indicated on the data points while systematic uncertainties are shown as a band below. The NLO predictions (generated using JETRAD [3]) are indicated by the solid curve. Dashed and dotted curves indicate the change in the prediction due to the addition of contact interaction terms. For this and the other mass bins, the NLO prediction is in good agreement with the data and is found to be insensitive to the pdf choice. The result changes somewhat with the choice of renormalization/factorization scale. Using a Bayesian technique and an assumption that the prior probability distribution is flat in $1/\Lambda_C^2$, limits on compositeness at the 95% CL were found. Assuming all quarks to be composite, the resulting lower limits on the quark compositeness scale range from 2.1 to 2.4 TeV. Results vary with the choice of renormalization scale and the sign of the interference term in the contact interaction Lagrangian, as detailed in Reference [1].

3 CDF Compositeness - Drell-Yan Spectrum

If the HERA excess of events (see D.Acosta/U.Bassler talks 31301/31302 in these proceedings) at high Q^2 is indeed due to quark and lepton compositeness, it would also appear in the dilepton invariant mass spectra in $p\overline{p}$ collisions. CDF exploits its fine tracking resolution [4] to study dielectron and dimuon production and search for quark-lepton compositeness.

The model predicts that four-fermion contact interaction partonic cross sections are proportional to the mass squared. Therefore, substructure would be observable in the l^+l^- pair mass spectra at high masses, atop of the steeply falling spectrum of pairs produced via the Drell-Yan process. This search channel is ideal because the Drell-Yan topology is very distinctive, containing two well separated, oppositely charged leptons in the final state which tend to be isolated from jets or other activity.

Figure 2 shows the dielectron and dimuon invariant mass spectra obtained from an analysis including 110 pb^{-1} of $p\overline{p}$ collisions observed in the CDF detector. The measured e^+e^- and $\mu^+\mu^-$ invariant mass spectra are normalized to the SM cross section over $50 < M_{l^+l^-} < 150$ GeV. The SM prediction (solid curve) is obtained from a next-to-leading-logarithmic (NLL) QCD calculation using MRS A′ pdf's, the results of which are clearly consistent with the data. Additional dotted and dashed curves indicate the change in the theoretical prediction due to a 2 TeV scale contact interaction.

The compositeness model assumed a general Lagrangian with interacting fermions having either chiral, vector, scalar, or axial current interactions. Dielectron, dimuon and combined datasets were used to obtain quark-electron, quark-muon, and quark-lepton (assuming lepton universality) limits on compositeness scales, a complete list of which is available in Reference [5]. To briefly summarize, lower limits on chiral qe and qμ compositeness scales range from 2.5 to 4.2 TeV.

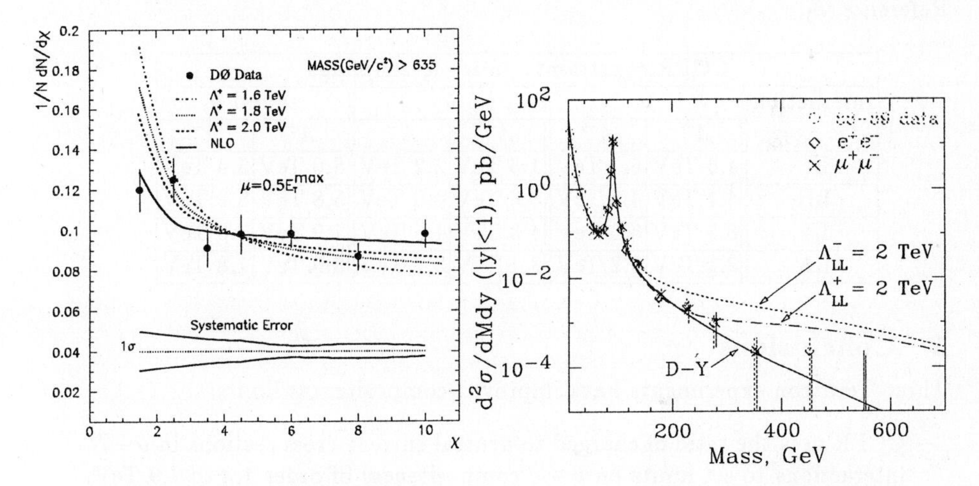

Fig. 1. Datapoints indicate the measured DØ dijet angular distribution in χ compared to the NLO QCD prediction (solid) and the change in the theory with additional contact terms.

Fig. 2. CDF e^+e^- (diamonds) and $\mu^+\mu^-$ (crosses) invariant mass distribution compared to a SM NLL calculation (solid) and SM + compositeness prediction (dashed/dotted curves).

4 CCFR - Ratio of Charged to Neutral Current Events

The CCFR experiment is a Tevatron fixed target neutrino experiment which utilizes a beam containing ν and $\bar{\nu}$ from decay secondaries produced by 800 GeV protons on a BeO target. Neutrino (or $\bar{\nu}$) induced events are observed in a 690 ton neutrino target calorimeter. In the SM, the interaction of a ν with nuclear matter proceeds via heavy charged and neutral vector boson exchange commonly referred to as charged (CC) and neutral current (NC) interactions. The ratio (R) of CC to NC cross sections is predicted by the SM. A measure of this ratio yields a precision measurement of a SM constant,

$\sin^2\theta_W$. For the extraction of $\sin^2\theta_W$, SM couplings $u_{L,R}$ and $d_{L,R}$ (the left and right-handed couplings of up and down quarks to the Z^0) are fixed to their SM expectations while allowing $\sin^2\theta_W$ to vary. Results are consistent with SM expectations (see J.Yu talk 701 in these proceedings).

A measure of the ratio R can also be used to constrain the couplings $u_{L,R}$ and $d_{L,R}$. The measurement is used to constrain linear combinations of deviations from standard model expectations in the variables $g^2_{L,R}$ and $\delta^2_{L,R}$, the sum and difference of the squares of the q-Z^0 couplings. To set limits on compositeness, a general contact interaction Lagrangian is assumed which allows chiral, vector or axial vector couplings between ν and/or $\bar{\nu}$ and up and down quarks. Resulting lower limits on $\nu - q$, $\bar{\nu} - q$, and combined-q contact interactions are shown in the table below. Further details are available in Reference [6].

CCFR constraints on $\nu - q$ couplings						
Interaction	ν		$\bar{\nu}$		ν and $\bar{\nu}$	
	Λ^+	Λ^-	Λ^+	Λ^-	Λ^+	Λ^-
LL	4.6 TeV	5.1 TeV	1.3 TeV	2.2 TeV	5.0 TeV	5.4 TeV
LR	4.1 TeV	4.3 TeV	3.8 TeV	4.0 TeV	5.8 TeV	5.8 TeV
LV	6.5 TeV	6.5 TeV	4.3 TeV	4.5 TeV	7.9 TeV	7.8 TeV
LA	2.1 TeV	3.2 TeV	3.1 TeV	3.7 TeV	3.0 TeV	1.8 TeV

5 Conclusions

Three Tevatron experiments have improved compositeness limits:

- CCFR uses the ratio of charged to neutral current cross sections in $\nu - N$ interactions to set limits on $\nu - q$ compositeness of order 1.3 to 7.9 TeV.
- CDF sets improved limits on q-l compositeness in the range from 2.5 to 4.2 TeV by studying the high mass dilepton production spectrum in $p\bar{p}$ collisions.
- DØ, another collider detector, sets improved limits on q-q compositeness ranging from 2.1 to 2.4 TeV by looking for deviations from NLO predictions in the dijet angular distribution.

References

[1] DØ collaboration, submitted to Phys. Rev. Letters, hep-ex/9707016.
[2] DØ collaboration, S.Abachi et al, NIM A 338 (1994), 185.
[3] W.T.Giele,E.W.N.Glover and D.A.Kosower, Nucl.Phys.B403(1993)633.
[4] CDF Collaboration, F.Abe et al, NIM A 271 (1988) 387.
[5] CDF Collaboration, F.Abe et al, Phys. Rev. Lett 79 (1997), 2198.
[6] CCFR collaboration, K.S.McFarland et al, submitted to Phys Rev Letters, Fermilab-Pub-97/001-E.

Search for R-Parity Violation and Leptoquark Signature at LEP

Ivor Fleck (Ivor.Fleck@cern.ch)

CERN, PPE–OPAL, Geneva, Switzerland

Abstract. At LEP many searches for physics beyond the Standard Model are performed. In this talk searches for SUSY particles decaying via R–parity violating couplings and for leptoquarks are presented for data taken at LEP2. No evidence for any such particle has been observed.

1 R–Parity Violation

In SUSY theories a new discrete multiplicative symmetry, called R-parity is introduced, $R_p = (-1)^{3B+L+2S}$, which is 1 for SM particles and -1 for SUSY particles. If R-parity is conserved, the LSP does not decay and the experimental signature is large missing momentum. If R-parity is violated, the LSP does decay and the experimental signature shows **no** missing momentum. Therefore exclusion limits for SUSY particles made under the assumption of R-parity conservation are not valid under R-parity violation.

The R-parity violating super–potential W is given by [1]:

$$W = \lambda_{ijk} L^i_L L^j_L \overline{E}^k_R + \lambda^{'}_{ijk} L^i_L Q^j_L \overline{D}^k_R + \lambda^{''}_{ijk} \overline{U}^i_R \overline{D}^j_R \overline{D}^k_R$$

$L^i(Q^i)$ are the lepton (quark) $SU(2)_L$ doublet superfields, $\overline{E}^i(\overline{U}^i, \overline{D}^i)$ are the singlet superfields; $\lambda, \lambda^{'}$, and $\lambda^{''}$ are Yukawa couplings and i, j, and k are generation indices. The first two terms break L and the third term breaks B.

From non LEP experiments limits on the size of the Yukawa couplings $\lambda, \lambda^{'}$, and $\lambda^{''}$ exist and are, for a sparticle mass of 100 GeV, of $\mathcal{O}$ (10^{-2}) for most of the couplings [1]. Much more stringent limits exist on the product of two of these couplings. Therefore it is assumed, that only one of the couplings is significantly different from zero. At LEP these couplings can be probed down to values of 10^{-7} for sparticle masses up to the beam energy. Sneutrinos and squarks can also be produced singly and a direct search for these particles up to the centre-of-mass energy is possible. These particles contribute also to two fermion final state production via the t-channel.

Under R-parity violation any sparticle can decay directly via the R-parity violating potential. This kind of decay will be called direct decay. On the other hand decays like under R-parity conservation are allowed with only the LSP decaying via the R-parity violating potential. These decays are called indirect decays. The event topologies depend on which coupling is assumed to be dominant. Consequently the analyses are separated into topologies involving either a $\lambda, \lambda^{'}$, or $\lambda^{''}$ coupling.

1.1 Gauginos

λ coupling: $\chi_1^+\chi_1^-$ production is excluded for most of the phase space up to the kinematic limit assuming the indirect decay mode, and even above, when including $\chi_2^0\chi_1^0$ production [2, 3, 4]. A typical exclusion plot is shown in figure 1. χ_1^0 are excluded at 95 % C.L. with a mass below 23 GeV [2] for any phase space point, as shown in figure 2. The limits for $\chi_1^+\chi_1^-$ production in the case of direct decay do not reach the kinematic limit for most phase space points, but masses below 73 GeV are excluded [2].

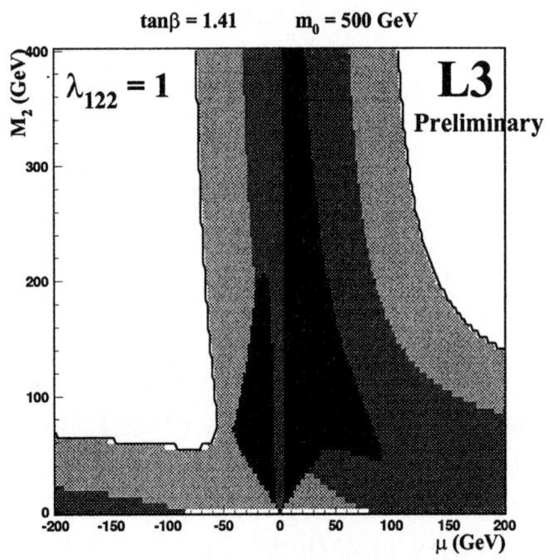

Fig. 1. *Excluded area in the M_2, μ plane for $\tan\beta = \sqrt{2}$ and $m_0 = 500$ GeV. The black area is excluded from $\chi_2^0\chi_1^0$ production, the dark grey one from LEP1 line shape analyses and the light grey one from $\chi_1^+\chi_1^-$ production.*

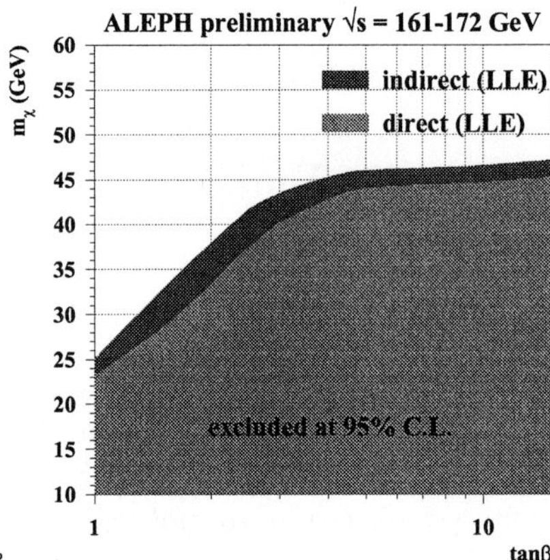

Fig. 2. *χ_1^0 mass limit as a function of $\tan\beta$. The limits are valid for all couplings λ_{ijk} and shown for both, direct and indirect decays.*

λ' coupling: For the indirect decay mass limits of m($\chi_1^\pm$) of 83, 85 GeV are given for $\tan\beta = \sqrt{2}, 35$ and $m_0 = 500$ GeV [2]. For the direct decay production cross sections greater than 1.4 pb are excluded for $\chi_1^0\chi_1^0$ and $\chi_1^+\chi_1^-$ production for masses between 65 and 85 GeV [5] for final states with electrons or muons.

λ'' coupling: Cross sections greater than 1.4 pb for $\chi_1^0\chi_1^0$ and $\chi_1^+\chi_1^-$ production are excluded for masses between 45 and 60 GeV for the direct decay[3].

1.2 Sleptons and Sneutrinos

λ coupling: The final states consist of 2, 4, or 6 charged leptons. The exclusion limits in case of the direct decay are given in table 1.

λ' coupling: So far only analyses for the direct decay have been performed [2, 3, 5]. The final state consists of 4 jets, similar to that of W-pair production. Current limits are of $\mathcal{O}$(70 GeV) for the 1. generation and of $\mathcal{O}$(50 GeV) for the other generations.

limits	$\tilde{e}_R$	$\tilde{\mu}_R$	$\tilde{\tau}_R$	$\tilde{\nu}_e$	$\tilde{\nu}_\mu, \tilde{\nu}_\tau$
ALEPH	60 GeV	49 GeV	49 GeV	76 GeV	59 GeV
DELPHI	70 GeV	58 GeV	52 GeV	62 GeV(λ_{122})	
OPAL	56 GeV	–	–	79 GeV ($\lambda_{121}, \lambda_{122}$)	49 GeV ($\lambda_{1j1}, \lambda_{1j2}, j = 2, 3$)

Table 1. *Mass limits for sleptons and sneutrinos in case of direct decay [2, 3, 5]. The limits hold for all of the λ couplings, unless otherwise stated. The mass limits for the 1st generation sparticles should not be directly compared, as they have been calculated for different points in the MSSM parameter space.*

s-channel resonance production: Considering the indirect decay of a $\tilde{\nu}$ generated in the s-channel, a mass of 130 GeV$< m_{\tilde{\nu}} <$ 170 GeV can be excluded for a decay via λ_{121}, with $\lambda_{121} > 5 \times 10^{-3}$ [3].

s- and t-channel sneutrino exchange: These processes contribute to the two electron final state. From cross section and forward backward asymmetry measurements a mass of 130 GeV$< m_{\tilde{\nu}} <$ 170 GeV can be excluded for a decay via λ_{131}, with $\lambda_{131} > 5 \times 10^{-3}$ [6, 7].

1.3 Squarks

For the scalar t-quark the mixing of the left- and right handed state could produce a rather light observable $\tilde{t}_1$. The production cross section is maximal for a mixing angle of 0 rad and minimal for 0.98 rad.

λ' **coupling:** Exclusion limits for the direct decay are given below [3, 5]. The final state consists of two leptons of the same flavour and two jets.

LIMITS		$\tilde{t}_1 \to e + q$	$\tilde{t}_1 \to \mu + q$	$\tilde{t}_1 \to \tau + q$
$\theta_{\tilde{t}} = 0$ rad	DELPHI	66.9 GeV	69.5 GeV	59.3 GeV
$\theta_{\tilde{t}} = 0$ rad	OPAL	72.0 GeV	67.0 GeV	–
$\theta_{\tilde{t}} = 0.98$ rad	DELPHI	50.0 GeV	58.7 GeV	–
$\theta_{\tilde{t}} = 0.98$ rad	OPAL	61.0 GeV	47.0 GeV	–

λ'' **coupling:** For the direct decay the final state consists of four jets and the mass limits achieved are 61 GeV for $\theta_{\tilde{t}} = 0$ rad [3, 5], and 47 GeV for $\theta_{\tilde{t}} = 0.98$ rad [5].

single production: Masses up to the centre of mass energy can be probed. The final state consists of a lepton and a quark.

t-channel exchange: The SM quark pair production cross section gets a contribution from t-channel s-quark exchange. This method has a sensitivity for masses well above the centre of mass energy [3, 8].

Exclusion limits for the different production modes are shown in figure 3 for the case of $\tilde{t}_L$ [3].

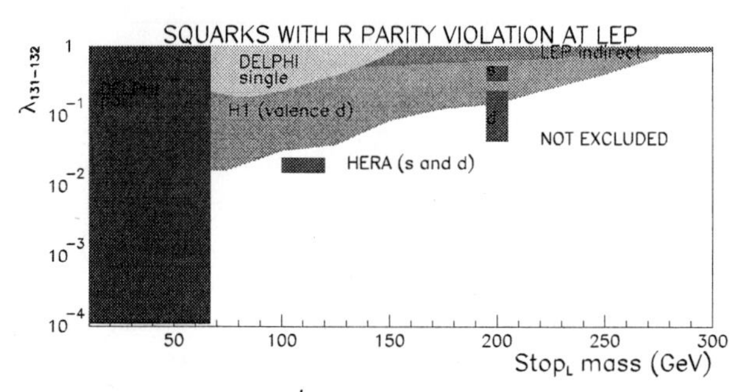

Fig. 3. *Excluded regions in the* $\lambda^{'}$–$m_{\tilde{t}_L}$ *plane from pair and single production and from t-channel exchange (labelled 'LEP indirect'). Also shown is the region excluded by HERA. Possible solutions to the excess of high* Q^2 *events observed at HERA are shown as the rectangular boxes.*

2 Leptoquarks

pair production: For final states with 2 quarks and 2 neutrinos mass limits achievable at LEP2 for scalar leptoquarks, assuming multiplets are not mass degenerate, could become better than existing limits. The best limit achieved so far is 73 GeV [9].

single production: The mass limits of around 140 GeV (130 GeV) for $\beta = 1.0(0.5)$ [10, 11] for an assumed coupling of $\sqrt{4\pi\alpha_{em}}$ are much smaller than those from the TEVATRON.

t-channel exchange: The cross section for $q\bar{q}$ final states would be modified by the t-channel exchange of leptoquarks. Limits well above the centre of mass energy (and above those set by TEVATRON) can be set for large values ($\mathcal{O}(0.5)$) of the coupling constant [8, 12]. Using b-tagging in the final states, these limits can be improved even further [8].

References

[1] G. Bhattacharyya, Nucl Phys B (Proc. Suppl.) 52A (1997), 83-88
H. Dreiner, hep-ph/9707435
[2] ALEPH Collab., contributed paper 621
[3] DELPHI Collab., contributed paper 589
[4] L3 Collab., private communication
[5] OPAL Collab., contributed paper 213
[6] DELPHI Collab., contributed paper 467
[7] L3 Collab., CERN-PPE/97-99
[8] OPAL Collab., CERN-PPE/97-101
[9] OPAL Collab., contributed paper 217
[10] DELPHI Collab., contributed paper 352
[11] OPAL Collab., contributed paper 205
[12] ALEPH Collab., contributed paper 602

Search for New Physics at LEP: Excited Fermions, Compositeness and Z'

Marc Besancon (besanco@dphrsb.saclay.cea.fr)

CEA-Saclay/DSM/DAPNIA/SPP

Abstract. We review the results of the search for new physics from the four LEP experiments where new physics stands for excited fermions, compositeness and new gauge bosons Z'. These searches have been performed with the data collected from the large e^+e^- collider LEP running in its phase two i.e. LEP2, in 1995 and 1996, except for the search for new Z′ where data from LEP1 runs at $\sqrt{s} = M_Z$ have been included. No evidence for new effects have been found in the data.

1 Introduction

Among all the possibilities leading to physics beyond the standard model, two main options are usually considered. The first option envisages a deeper level of elementariness in which the known particles of the standard model are bound states of more fundamental constituants i.e compositeness. The second option extends the symmetries of the standard model. This latter extension can be either of space-time origin i.e. supersymmetry, or of gauge origin i.e. grand unification, or a combination of both. The present experimental report tackles these two main options since we review the results on compositeness from the four LEP experiments, ALEPH, DELPHI, L3 and OPAL, with data taken during 1995 and 1996 at LEP2 running between $\sqrt{s} = 130$ GeV and $\sqrt{s} = 172$ GeV, which amount to about 26 pb^{-1}, as well as results on new gauge boson Z′ including LEP1 and LEP2 data.

2 Compositeness

2.1 Excited fermions

Assuming that fermions of the standard model are composed of bound states of new elementary particles, the first excited level called excited fermions, with mass determined by the compositeness scale Λ, can be directly searched for at LEP2. Excited fermions are assumed to have spin and weak isospin 1/2 and to be heavier than the ordinary fermions. They are assumed to decay promptly since for masses above 20 GeV their mean lifetime is predicted to be less than 10^{-15} seconds. The effective lagrangian describing excited fermions interactions can be found in [1]. The independent parameters are the compositeness scale Λ and the effective changes from the standard model couplings f,f' and f_s.

Assuming $| f |=| f' |=| f_s |$, one is left with only one parameter which is $f/\Lambda = \sqrt{2}\lambda/m_{l^\star}$ where λ is the excited lepton coupling.

In e^+e^- collisions, single excited quarks and leptons of the second and third family are produced through s-channel γ and Z exchanges, while in the first family there is an additional contribution due to t-channel exchanges (W exchange in the case of $\nu_e^\star$, and γ and Z for $e^\star$). The double production of charged excited fermions proceeds via the s-channel γ and Z exchanges, while for excited neutrinos only the s-channel Z exchange contributes. Although t-channel contributions are also possible, they correspond to double de-excitation, and give a negligible contribution to the overall production cross-section.

Excited leptons can decay by radiating a γ, a Z or a W. The decay branching ratios are functions of the f and f' couplings of the model and depend on the mass of the excited leptons. Excited quarks are assumed to decay by radiating either a γ or a gluon. All these possible decays give rise to a large amount of possible signatures with photonic, leptonic and hadronic final states. No evidence for excited fermions have been found in the LEP2 data. 95 % Confidence Level (CL) exclusions on excited fermions masses are shown in table 1 for double excited lepton production. Exclusion region in the

limits	ALEPH	DELPHI		L3		OPAL	
in GeV/c^2		$f = f'$	$f = -f'$			$f = f'$	$f = -f'$
$e^\star$	84.7	84.6	74.6	85.1		85.1	80.6
$\mu^\star$	84.8	84.6	74.6	85.0		85.1	80.6
$\tau^\star$	84.5	84.6	74.6	84.1		84.1	80.6
					$\nu_e^\star$	84.0	85.1
$\nu^\star$	80.5	67.9	80.1	78.9	$\nu_\mu^\star$	83.9	85.1
					$\nu_\tau^\star$	79.5	85.1

Table 1: Excited leptons mass limits from double production searches

plane ($\lambda/m_{l^\star}$,$m_{l^\star}$) from the DELPHI collaboration on single excited leptons production are shown in figures 1 where the 95 % C.L. excluded regions are located above the 3 different lines corresponding to the 3 different lepton flavors. Excited electron can be search for indirectly from the $2\gamma(\gamma)$ topology in addition to the direct search thus extending the sensitivity to higher excited electron masses. Again no evidence for such topologies have been found in the data. The OPAL collaboration, fixing $\Lambda_{e^\star}$ to $m_{e^\star}$, can exclude the mass domain $m_{e^\star} < 192 GeV/c^2$ at 95 % confidence level [3].
Finally singly produced excited quarks searches have been pioneered at LEP2 by the DELPHI collaboration assuming the excited quark decay into either a quark and a γ or a quark and gluon. There were again no evidence for these

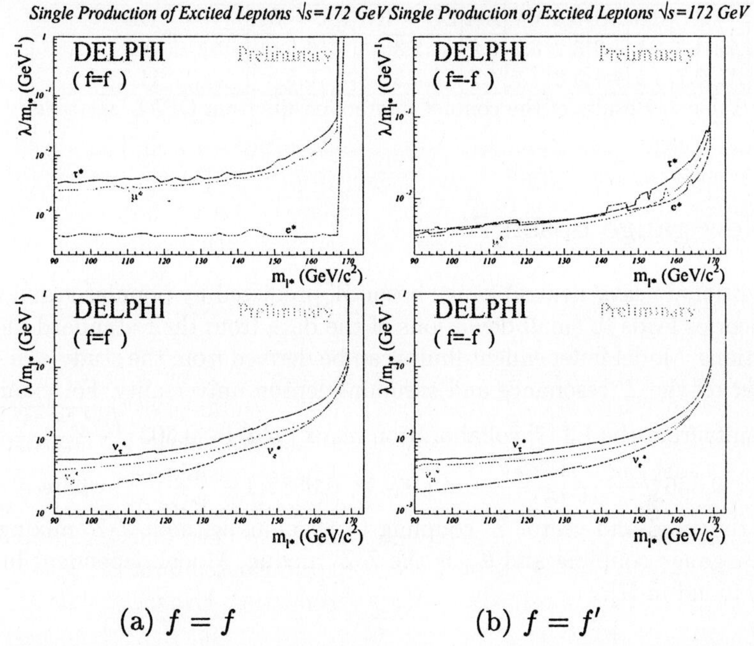

(a) $f = f$ (b) $f = f'$

Fig. 1. Exclusion region in the plane $(\lambda/m_{l^*}, m_{l^*})$ on single l^* production

types of topologies in the data and exclusion region in the plane $(\lambda/m_{l^*}, m_{l^*})$ can be found in [2].

2.2 Contact interactions

A complementary way to search for compositeness lies on the search for the remnant interactions of the binding force between the subconstituants which are then considered as new effective four fermions interactions, namely the contact interactions. The contact interactions can be described by an effective lagrangian [4] where Λ is the energy scale of the contact interactions. The chiral structure of the contact interactions (i.e. left left (LL), right right (RR) etc) can be parametrized by η_{ij} and, for example, in table 2, DB stands for ($\eta_{LL} = \pm 1$, $\eta_{RR} = \pm 4$, $\eta_{LR} = \eta_{RL} = \pm 2$) [5]. Deviation from the expectation of the standard model on the angular distributions for the non radiative ($e^+e^- \to l^+l^-$) and ($e^+e^- \to q\bar{q}$) are expected so that fits to the data are performed. The results are shown in table 2 for the OPAL collaboration. The fitted value is $\epsilon = 1/\Lambda^2$. $\Lambda_\pm$ are the 95 % C.L. level limits in TeV deduced from ϵ. Only the combined results i.e. ($e^+e^- \to f\bar{f}$), on $\Lambda_\pm$ are given. For a more exhaustive table see [6].

	LL	RR	LR	RL	VV	AA	LL + RR	LR + RL	DB
Λ_+	4.2	4.1	3.3	3.2	6.2	6.9	5.3	4.5	9.2
Λ_-	3.7	3.5	4.4	5.0	7.7	4.7	5.3	6.5	11.7

Table 2: Results of the contact interaction fits from OPAL, see text

3 New gauge boson Z′

An additional heavy neutral gauge boson Z′ predicted by popular grand unified theories leads to small deviations of the data from their standard model predictions. Model independent limits can be derived from the study $e^+e^- \rightarrow l^+l^-$ far off the Z′ resonance and assuming lepton universality. For example the results from the L3 [7] collaboration reads $\mid g'a'_l \mid < 0.303\sqrt{\frac{m^2_{Z'} - 156^2 GeV}{156^2 GeV}}$, $\mid g'v'_l \mid < 0.336\sqrt{\frac{m^2_{Z'} - 156^2 GeV}{156^2 GeV}}$ and $\mid \theta_M \mid < 0.0582(1 - \frac{m^2_Z}{m^2_{Z'}})^{1/2}$, where a' and v' are the axial and vector Z′ coupling to fermions without Z-Z′ mixing, g' the new gauge coupling and θ_m is the Z-Z′ mixing. Model dependent limits can be found in [7].

References

[1] F.Boudjema, A.Djouadi and J.L.Kneur, Zeitschrift fur Physik **C57** (1993), 425 and K.Hagiwara, S.Komamiya and D.Zeppenfeld, Zeitschrift fur Physik **C29** (1985), 115.

[2] DELPHI Collaboration, DELPHI note 97-111 Jerus.CONF 93 and CERN preprint PPE in preparation. See also ALEPH Collaboration, conference paper ref. 615, OPAL Collaboration, CERN-PPE/97-123, accepted by Zeit.f.Physik C and and L3 Collaboration, L3 note 2074 and CERN-PPE/97-12.

[3] OPAL Collaboration, CERN-PPE/97-109, accepted by Zeit.f.Physik C. See also L3 Collaboration, CERN-PPE/97-77.

[4] E.Eichten, K.Lane, M.Peskin, Phys. Rev. Lett. **50** (1983)811.

[5] G.J.Gounaris, D.T.Papadamou, F.M.Renard, hep-ph 9703281, submitted to Phys. Rev. D.

[6] OPAL collaboration, CERN-PPE/97-101. See also ALEPH collaboration, conference paper ref. 602, and DELPHI collaboration, note 97-116 Jerus.CONF 98.

[7] L3 Collaboration, L3 note 2126. See also, DELPHI Collaboration, note 97-133 Jerus.CONF 111.

Search for New Particles at ZEUS

Aleksander Filip Żarnecki (zarnecki@vxdesy.desy.de)
for the ZEUS Collaboration

DESY – F1, Notkestraße 85, 22607 Hamburg, Germany

Abstract. Using the ZEUS detector at HERA, we have searched for heavy new particles in e^+p collisions at a center-of-mass energy of 300 GeV. With an integrated luminosity of 9.4 pb^{-1} (1994-95 data), no evidence was found for production of excited states of electrons, neutrinos, or quarks. In the framework of R-parity conserving supersymmetric extensions of the Standard Model, using 20.0 pb^{-1} of ZEUS data (1994-96), no evidence was found for the production of a selectron and a squark, with direct decay into the lightest neutralino. Limits on the production cross section times branching ratio are derived for different particle masses.

1 Introduction

This paper presents results from e^+p running with the ZEUS detector during the years 1994 to 1996, at proton and positron beam energies of $E_p = 820$ GeV and $E_e = 27.5$ GeV. With the integrated luminosity of up to 20 pb^{-1} collected in this period, we search for production of excited fermion states as well as selectron–squark pair production.

For details on the analysis presented here, the interested reader is referred to [1, 2]. A description of the ZEUS detector can be found in references [3, 4].

2 Search for excited fermions

At the HERA electron-proton collider, single excited electrons (e^*) and quarks (q^*) could be produced by t-channel γ/Z^0 boson exchange, and excited neutrinos (ν^*) could be produced by t-channel W boson exchange. We report on a search for resonances with masses above 30 GeV, decaying to a light fermion through electroweak mechanisms. The data sample, collected by ZEUS during the years 1994–1995, corresponds to an integrated luminosity of 9.4 pb^{-1}. This represents a 17-fold statistical increase over our previously published [5] limits from e^-p collisions. A search based on 2.75 pb^{-1} of e^+p data has been reported recently by the H1 Collaboration [6].

We choose a model [7] which couples left-handed fermions to right-handed excited states, and in which the excited fermions form both left- and right-handed weak isodoublets. Using the Lagrangian [8, 9], the excited fermion interactions are described by the scale parameter Λ and the coefficients f, f', and f_s, multiplying the $SU(2)_L$, $U(1)_Y$, and $SU(3)_C$ couplings respectively.

For specific assumptions relating f, f', and f_s, the branching fractions are known and the cross sections can be described by a single parameter (e.g., f/Λ) with dimension, GeV^{-1}.

2.1 Event Selection

The processes of interest for production of the various f^* are

$$\begin{aligned} e + p &\rightarrow e^* + X \\ e + p &\rightarrow \nu^* + X \end{aligned} \qquad (1)$$

for excited leptons, and

$$e + p \rightarrow e + q^* + X \qquad (2)$$

for excited quarks. Here X represents the proton remnant (or proton in the case of elastic e^* production). We have considered 11 different excited fermion decay channels, resulting in 8 possible final state topologies. Event selection results are summarized in Tab. 1. For each f^* decay mode decay signature, typical acceptance, number of observed events, and number of expected events from background sources are shown.

2.2 Results

We have no positive evidence for excited leptons or quarks in any of the decay chains described above. The strongest upper limits on the production cross section times branching ratio are obtained for the decay channels with photons in the final state, and are typically less than 1 pb.

In order to extract the limits on coupling strengths we choose $f = f'$ for e^* and q^*, and $f = -f'$ for ν^*. For excited quarks, we consider only electroweak couplings and set $f_s = 0$. The upper limits on f/Λ as a function of the excited electron mass are shown in Fig. 1. These limits are derived under the narrow width approximation. The searches encompass multiple decay modes; the combined limit is shown as dashed lines.

It is possible to set mass limits on excited fermions if one makes the further assumption that $f/\Lambda = 1/M_{f^*}$. For this case, excited electrons are ruled out at the 95% confidence level in the mass interval 30–200 GeV using the combined limit from all three decay modes. Excited neutrinos are excluded over the range 40–96 GeV. Excited quarks with only electroweak coupling are excluded over the range 40–169 GeV using the combined limit from $q^* \rightarrow q\,\gamma$ and $q^* \rightarrow q\,W$.

3 Search for selectron–squark pair production

In the supersymmetric extensions of the Standard Model (SM), each SM fermion has a scalar partner. In models where the R-parity is conserved,

Channel	Signature	A	$N_{\rm obs}$	$N_{\rm exp}$
$e^* \to e\,\gamma$	2 EM clusters	75%	103	97.5 ± 4.8 (COMPTON: 81.5) (NC DIS: 16.0)
$e^* \to e\,Z^0 \to e\,q\,\bar{q}$ $\nu_{\rm e}^* \to e\,W \to e\,q\,\bar{q}$	EM cluster large $E_{\rm t,had}$ and $M_{\rm had}$	55% 70%	21	15.3 ± 1.5 (NC DIS)
$e^* \to \nu_{\rm e}\,W \to \nu_{\rm e}\,q\,\bar{q}$ $\nu_{\rm e}^* \to \nu_{\rm e}\,Z^0 \to \nu_{\rm e}\,q\,\bar{q}$	$\not{P}_{\rm t}$ large $E_{\rm t,had}$ and $M_{\rm had}$	65% 75%	13	8.4 ± 1.5 (CC DIS: 4.8) (NC DIS: 1.0) (PHP: 2.6)
$\nu_{\rm e}^* \to \nu_{\rm e}\,\gamma$	$\not{P}_{\rm t}$ EM cluster (no track)	50%	0	0.4 ± 0.1 (CC DIS)
$e^* \to e\,Z^0 \to e\,\nu\,\bar{\nu}$ $e^* \to \nu_{\rm e}\,W \to e\,\nu_{\rm e}\,\bar{\nu}_{\rm e}$	$\not{P}_{\rm t}$ EM cluster (track)	70% 60%	1	1.0 ± 0.2 (CC DIS: 0.1) (NC DIS: 0.45) (W PROD'N: 0.45)
$\nu_{\rm e}^* \to e\,W \to e\,\bar{e}\,\nu_{\rm e}$	$\not{P}_{\rm t}$ 2 EM clusters	60%	0	≈ 0
$q^* \to q\,\gamma$	EM cluster (no track) large $E_{\rm t,had}$	60%	18	23.5 ± 2.5 (PROMPT γ: 14.4) (PHP: 5.1) (NC DIS: 4.0)
$q^* \to q\,W \to q\,\bar{e}\,\nu_{\rm e}$	$\not{P}_{\rm t}$ EM cluster large $E_{\rm t,had}$	50%	0	1.5 ± 0.3 (CC DIS: 0.5) (NC DIS: 0.75) (W PROD'N: 0.2)

Table 1. The f^* decay signature, typical acceptance (A), number of observed events ($N_{\rm obs}$), and number of expected background events ($N_{\rm exp}$) for the excited fermion decay topologies investigated.

supersymmetric particles can be produced only in pairs and the lightest supersymmetric particle is stable. In these models, the production of a slepton and a squark is the lowest order process in which supersymmetric particles are produced at HERA [13]. The selectron (squark) can decay directly to the lightest neutralino χ_1^0 and an electron (quark): $\tilde{e} \to e\chi_1^0$ ($\tilde{q} \to q\chi_1^0$). If the lightest neutralino is the lightest supersymmetric particle (LSP), it is stable and, being neutral, escapes undetected from the experimental apparatus. Therefore, the signature for the production of a selectron and a squark is one electron from selectron decay, high p_t hadronic system from squark decay and missing momentum due to the escaping neutralinos [14].

To reduce the number of free parameters, the right and left-handed selectrons are assumed to have the same mass $m_{\tilde{e}}$ and all the squarks (except stop, which is not considered in this analysis) to have the same mass $m_{\tilde{q}}$. Furthermore, we assume no mixing between the L-R scalar fermions. The calculation of the cross section and the simulation of the signal have been done with a

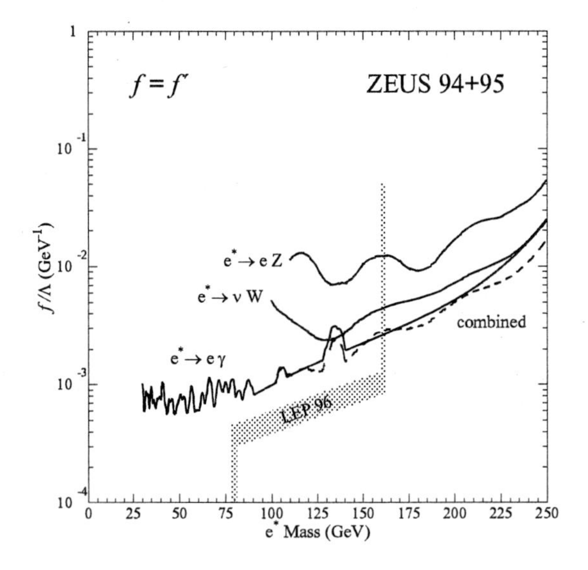

Fig. 1. Upper limits at the 95% confidence level on the coupling f/Λ as a function of the excited electron masses. The dashed line is the combined limit from indicated decay modes. The shaded line is the limit reported recently by the LEP experiments [11, 12].

Monte Carlo (MC) generator based on the differential cross sections calculated in [13] and including also the effect of the initial state radiation from the incoming lepton. The branching ratios for the direct selectron and squark decay in the first neutralino are calculated using ISASUSY code [15].

3.1 Event Selection

The data sample used in this analysis consists of e^+p runs taken by ZEUS in the 1994-96 running period. The total integrated luminosity used is $\mathcal{L} = 20.0\ \text{pb}^{-1}$. To select selectron and squark events we require a well identified electron with polar angle $\theta < 2.5$ rad[1] and transverse momentum $p_{te} > 4$ GeV and an hadronic system with $p_{th} > 4$ GeV. To reduce the background from NC DIS, we require that the hadronic system and the electron are not back to back in azimuth by cutting on the acoplanarity $A = |\phi_e - \phi_h| - \pi > 0.25$ rad. The final sample is selected with a cut in the plane defined by the net transverse momentum P_t and the variable $E - P_z$, that, for contained events, should be close to zero and two times the electron beam energy respectively. We accept events with $P_t > 12$ and $E - P_z < 50$ and $E - P_z < 2P_t$.

Two events survive the selection, in agreement with the 2.74 ± 0.47 events expected from the SM background.

[1] In the ZEUS frame the positive Z axis ($\theta = 0$) is in the direction of the proton beam

3.2 Results

Since no excess was found, we set limits on supersymmetric models. The efficiency on the signal can be conveniently parametrized as a a function of a linear combination of the selectron, squark and neutralino mass: $M_{vis} = m_{\tilde{q}} + m_{\tilde{e}} - 2m_{\chi_1^0}$. MC simulation shows that the selection efficiency is above 50% for $M_{vis} > 50$ GeV. We derive the 95% CL upper limit on the cross section times the branching ratio for the direct decays to the lightest neutralino. The limit is of the order of 1 pb for $M_{vis} \sim 25$ GeV and goes down to about 0.4 pb at high M_{vis}.

The upper limit on the cross section is used to exclude regions of the MSSM parameter space. Figure 2 shows the excluded region in the plane defined by the lightest neutralino mass and by the selectron or squark mass, assuming that selectron and squark have similar mass, for different values of μ and $\tan\beta$. In the region where the first neutralino is mostly a gaugino $(\tilde{\gamma}, \tilde{Z})$, i.e. for $|\mu| >> M_1, M_2$, the process has a sizeable cross section and we can exclude selectron and squark mass, up to 70 GeV. As the cross section and the selection efficiency depend essentially on the sum of the masses of squark and selectron, the limit on $m_{\tilde{e}} + m_{\tilde{q}}$ also applies (neglecting small differences coming from branching ratios) for different squark and selectron masses.

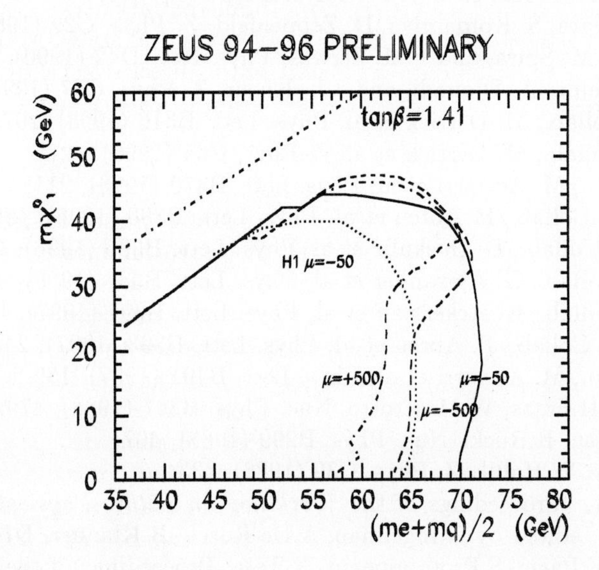

Fig. 2. Regions excluded at the 95 % CL in the plane defined by the lightest neutralino mass and the mass of squark and selectron, assumed to be equal, for different μ values, as indicated on the plot. The H1 95 limit [16] is also shown for $\mu = -50$.

4 Summary

We have searched for heavy excited states of electrons, neutrinos, and quarks using 9.4 pb^{-1} of e^+p collisions at a center-of-mass energy of 300 GeV recorded with the ZEUS detector at HERA. No evidence of a signal was found. We exclude excited electrons with mass between 30 and 200 GeV, excited electron neutrinos with mass between 40 and 96 GeV, and excited quarks coupled electroweakly with mass between 40 and 169 GeV. A search for the production of a selectron and a squark, with direct decay into the lightest neutralino has been performed on 20 pb^{-1} of ZEUS data. No evidence for the production of supersymmetric particles has been found.

References

[1] ZEUS Collab., J.Breitweg et al, DESY 97-112 (June 1997), accepted by Zeitschrift f. Physik - MS 739.

[2] ZEUS Collab., paper contributed to this conference, N-689.

[3] ZEUS Collab., M.Derrick et al, Phys. Lett. B293 (1992), 465; ZEUS Collab., M.Derrick et al, Z. Phys. C63 (1994), 391.

[4] ZEUS Collab., The ZEUS Detector, Status Report 1993 (DESY 1993).

[5] ZEUS Collab., M. Derrick et al, Phys. Lett. B316 (1993), 207; ZEUS Collab., M. Derrick et al, Z. Phys. C65 (1995), 627.

[6] H1 Collab., S. Aid et al, Nucl. Phys. B483 (1997), 44.

[7] K. Hagiwara, S. Komamiya, D. Zeppenfeld, Z. Phys. C29 (1985), 115.

[8] U. Baur, M. Spira, and P.M. Zerwas, Phys. Rev. D42 (1990), 815.

[9] F. Boudjema, A. Djouadi, and J.L. Kneur, Z. Phys. C57 (1993), 425.

[10] ZEUS Collab., M. Derrick et al, Phys. Lett. B316 (1993), 207; ZEUS Collab., M. Derrick et al, Z. Phys. C65 (1995), 627.

[11] L3 Collab., M. Acciarri et al, Phys. Lett. B370 (1996), 211; DELPHI Collab., P. Abreu et al, Phys. Lett. B380 (1996), 480; ALEPH Collab., D. Buskulic et al, Phys. Lett. B385 (1996), 445; OPAL Collab., G. Alexander et al, Phys. Lett. B386 (1996), 463.

[12] OPAL Collab., K. Ackerstaff et al, Phys. Lett. B391 (1997), 197; DELPHI Collab., P. Abreu et al, Phys. Lett. B393 (1997), 245; L3 Collab., M. Acciarri et al, Phys. Lett. B401 (1997), 139.

[13] A.Bartl, H.Fraas, W.Majerotto, Nuc. Phys. B297 (1988), 479; H.Komatsu, R.Rückl, Nuc. Phys. B299 (1988), 407; T.Bartels, W.Hollik, Z. Phys. C39 (1988), 433;

[14] P.Shleper, "Proceedings of the *Workshop on Future Physics at HERA* 1995/96", edited by G.Ingelman, A.De Roeck, R.Klanner, DESY 1996.

[15] H.Baer, F.Paige, S.Protopopescu, X.Tata, Proceedings of the workshop on "Physics at Current Accelerators and Supercolliders", edited by J.Hewett, A.White, D. Zeppenfeld, ANL 1993.

[16] H1 Collab., S.Aid et al, Phys. Lett. B380 (1996), 461.

Searches for Heavy Exotic States at the Tevatron

Kara Hoffman (kara@fnal.gov)

Purdue University, West Lafayette, IN, U.S.A.

Abstract. We present the results of searches for Standard Model Higgs, charged stable massive particles, dijet mass resonances, and heavy neutral gauge bosons in Tevatron $p\bar{p}$ collisions at $\sqrt{s} = 1.8$ TeV using the CDF and DØ detectors.

While the Standard Model has enjoyed many phenomenological successes, the mechanism of electroweak symmetry breaking remains at large. This suggests that the Higgs boson may be heavy enough to remain elusive, or, perhaps, there is additional physics beyond the Standard Model. Until the LHC comes online, the Tevatron will remain at the energy frontier and thus plays a unique role in the search for Higgs and non-Standard Model physics. Here we report on the results of searches within the Tevatron's exclusive energy reach.

1 Standard Model Higgs

The CDF collaboration has searched for the Standard Model Higgs decay signature WH^0 with $W \to \ell\nu$ and $H^0 \to b\bar{b}$ [1]. Events with an isolated high p_T lepton ($p_T > 20$ GeV/c) and large missing energy ($\not{E}_T > 20$ GeV) are selected if they are not consistent with a Z^0 or a top dilepton decay. The events are additionally required to have at least one jet associated with a b hadron decay using reconstructed secondary vertices from CDF's silicon vertex detector. The event is considered double tagged if an additional b-jet is tagged by the same algorithm or an algorithm that searches for soft leptons consistent with a semileptonic b decay. The events are classified by jet multiplicity with the signal assumed to be in the W + 2 jet bin. The other bins are compared to the simulated background from QCD + top.

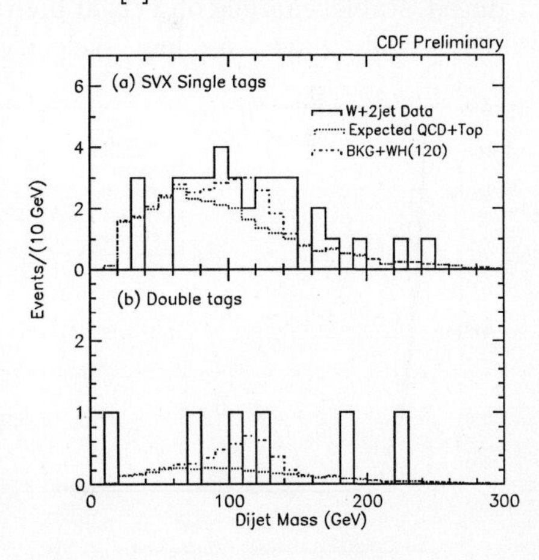

Fig. 1. $b\bar{b}$ dijet mass spectrum for (a) single tags, and (b) double tags.

An excess of double tagged events is observed in the W + 2 jet bin, however, the $b\bar{b}$ mass spectrum is consistent with expected background as shown in Figure 1.

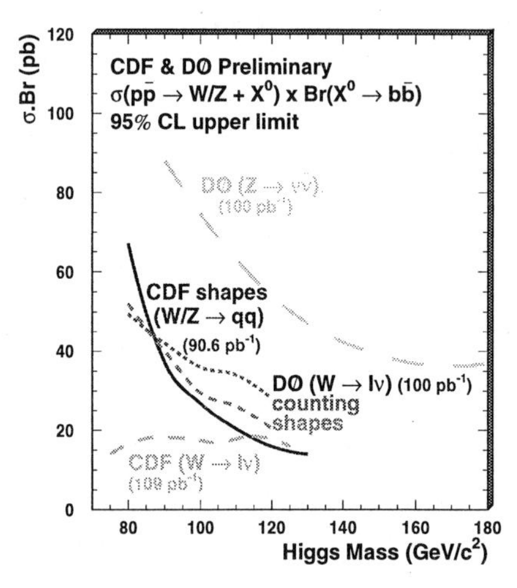

Fig. 2. Summary of limits on Standard Model Higgs from the Tevatron.

The DØ collaboration has also searched for heavy neutral scalars[2]. They use the decay signature Z^0X with $Z \to \nu\nu$ and $X \to b\bar{b}$ where the b jets are identified by their decays to soft muons. The muon is required to have $p_T > 3.5$ GeV$/c$ and to be central in pseudorapidity, $|\eta| < 1.0$. Events consistent with $W \to \ell\nu$ decays are removed. To identify the presence of neutrinos, a cut on missing transverse energy of $\not{E}_T \geq 35$ GeV is chosen because it is efficient for a simulated 90 GeV Higgs signal. After all cuts are applied, there are 2 events left which is consistent the expected background from top, W, and Z^0 of 2.6±0.7 events.

Figure 2 summarizes the limits on the Higgs cross section based on these and other analyses at the Tevatron.

2 Charged stable massive particles

CDF has exploited it's central tracking system in a search for strongly produced, stable, charged objects at high mass. Because such objects are massive, they will have a low velocity and therefore will suffer large ionization losses in the tracking chambers: $\frac{dE}{dx} \approx \frac{1}{\beta^2}$. At low values of $\beta\gamma$ ($\beta\gamma < 0.85$) the mass can be uniquely determined from dE/dx when combined with a momentum measurement as illustrated in Figure 3. If these objects are stable ($\gamma\tau > 10^{-8}s$), they will penetrate the detector and trigger as high p_T muons.

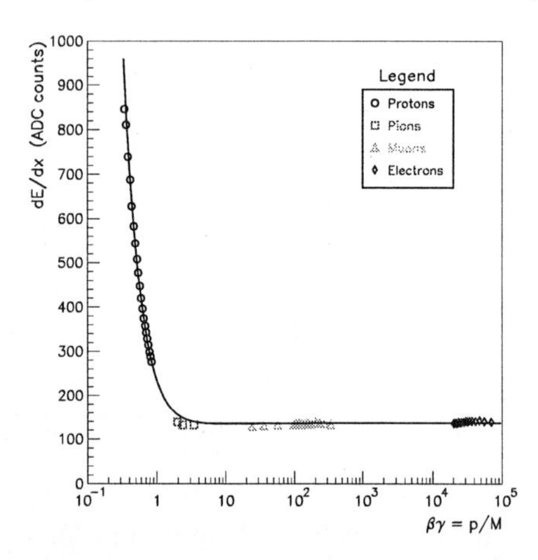

Fig. 3. dE/dx in CDF's silicon vertex detector.

Events with good track quality are selected for this analysis from high p_T muon samples which do not have a minimum ionizing requirement. The events are further required to have $|\mathbf{p}| > 35$ GeV, and the dE/dx cut is tightened at lower momentum to reduce background. After all kinematic cuts are applied, no events remain above 100 GeV. Using color triplet quarks as a reference model, lower mass limits of 195 GeV and 220 GeV are obtained at the 95% confidence level for a charge 1/3 and a charge 2/3 quark respectively.

3 Dijet mass resonances

Using the dijet mass spectrum at CDF and DØ to search for resonances, many particles may be sought with a single spectrum. The large background in such a search is compensated by large cross sections for dijet production. *B*-tagging may be added for background suppression or to detect particles decaying preferentially to the third generation, such as those predicted in Topcolor models.

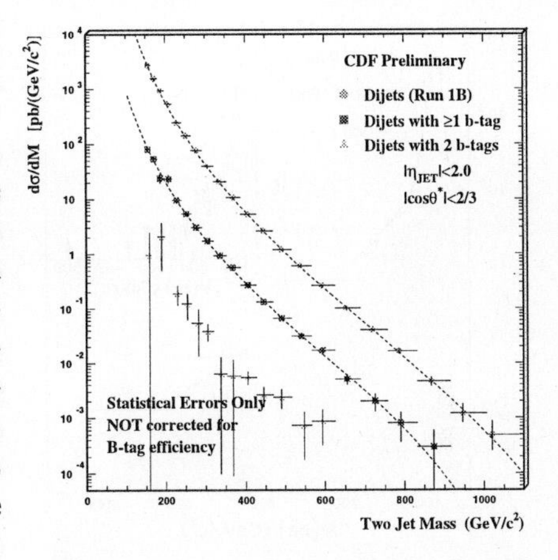

Fig. 4. The untagged, *b*-tagged, and double tagged dijet mass mass spectra from CDF.

The CDF dijet mass spectrum is fit to a smooth parameterization of the form:

$$\frac{d\sigma}{dm} = \frac{A(1 - m/\sqrt{s} - cm^2)^N}{m^p} \quad (1)$$

as shown in Figure 4. Limits are set by fitting the spectrum to the above parameterization plus a signal resonance and minimizing the likelihood.

Axigluons and colorons are excluded between 200 and 980 GeV, excited quarks between 80 and 570 GeV and 580 and 760 GeV, color octet technirhos between 260 and 480 GeV, W' between 300 and 420 GeV, and E_6 diquarks are excluded between 290 and 420 GeV. A 2.6 σ fluctuation in the 550 GeV bin weakens some limits [3].

For the DØ dijet mass search, the data is normalized to a detector smeared JETRAD simulation of the QCD background. The simulated background, along with a comparison of the data to the normalization are shown in Figure 5. Minimizing the background plus a signal, they exclude excited quarks below 725 GeV, Z' between 365 GeV and 615 GeV, and W' between 340 and 680 GeV [4].

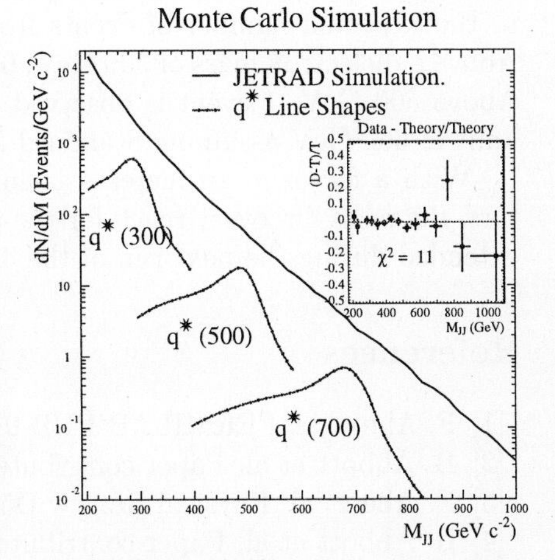

Fig. 5. Simulations of the DØ dijet spectrum and excited quark shapes after detector smearing.

4 Heavy neutral gauge bosons

Many extensions of the Standard Model require the existence of additional neutral gauge bosons. CDF has searched for both ee and $\mu\mu$ decays of such particles by fitting the ee and $\mu\mu$ mass spectra to the expected backgrounds. After lepton identification cuts, a sample of 7234 ee candidates and 2566 $\mu\mu$ candidates remain. The expected background in the $Z' \to \mu\mu$ channel comes from Z^0 and Drell-Yan production. The $\mu\mu$ spectrum is fit to the predicted background distributions normalized to the height of the Z^0 peak, and is found to be consistent with Standard Model processes. Misidentified dijet events contribute an additional background in the $Z' \to ee$ channel. Rather than subtract these events, they are fit to the parametric form in Equation 1 and included the background. A comparison of the ee mass spectrum to the predicted background is shown in Figure 6, with the raw data shown in the inset. Combining these two analyses using a binned maximum likelihood method, a lower mass limit of 690 GeV is found for a Z' with Standard Model couplings [5].

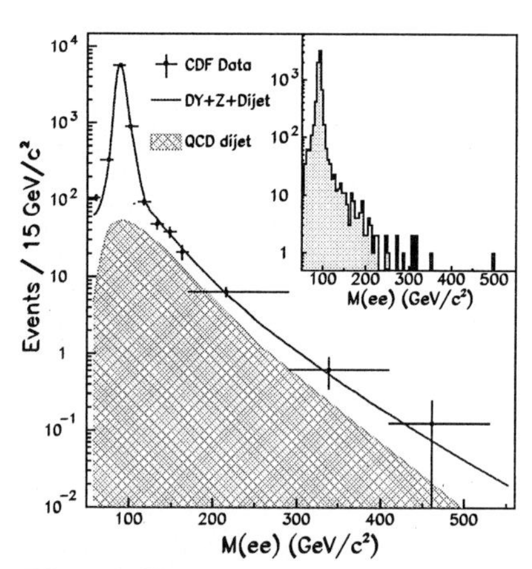

Fig. 6. CDF ee mass spectrum fit to the Drell-Yan + Z + dijet continuum.

DØ has searched for $Z' \to ee$ by counting the number of observed events with a mass window of $M_{Z'} \pm 4\Gamma_{Z'}$ for each Z' value tested, and comparing to the expected number of events from Z^0 and Drell Yan in that window. Above a dielectron mass of 300 GeV, 6 events are observed with 5.8 expected. Above 500 GeV, 1 event is observed with 0.3 expected. This yields a mass limit of 660 GeV assuming Standard Model couplings [6].

With a factor of 20 increase in luminosity and upgrades of both detectors, the mass discovery reach for the particles presented here will be greatly extended during the next run of the Tevatron.

References

[1] F. Abe et al, FERMILAB-PUB-97-280-E, submitted to Phys. Rev. Lett.
[2] B. Abbott et al. Paper contributed to HEP97 #808 (1997)
[3] F. Abe et al, Physical Review D55 (1997), 5263-5268
[4] B. Abbott et al, Paper contributed to HEP97 #101 (1997)
[5] F. Abe et al, Physical Review Letters 79 (1997), 2198-2203
[6] B. Abbott et al, Paper contributed to HEP97 #806 (1997)

Hunting Down Interpretations of the HERA Large Q^2 Data

John Ellis (John.Ellis@cern.ch)

Theoretical Physics Division, CERN, CH 1211 Geneva 23

Abstract. Possible interpretations of the HERA large-Q^2 data are reviewed briefly. The possibility of statistical fluctuations cannot be ruled out, and it seems premature to argue that the H1 and ZEUS anomalies are incompatible. The data cannot be explained away by modifications of parton distributions, nor do contact interactions help. A leptoquark interpretation would need a large τq branching ratio. Several R-violating squark interpretations are still viable despite all the constraints, and offer interesting experimental signatures, but please do not hold your breath.

1 Introduction

The large-Q^2, large-x HERA data reported by the H1 [1] and ZEUS [2] collaborations early in 1997 excited considerable interest, being subject to many theoretical interpretations and stimulating many related experimental searches. The new instalment of HERA data and other experimental data reported here cast new light on many of the suggested interpretations. Now is an opportune moment to take stock of the situation, assess which interpretations remain viable, and review how those remaining might be elucidated by future analyses. Space restrictions do not permit me to do justice to the many papers on this subject, and what follows is a personal selection of material: for other recent reviews, see [3].

2 Statistical Fluctuations?

H1 and ZEUS reported here the current status of their 1997 neutral-current data analysis [4], in addition to the data published previously [1, 2]. H1 has by now found 7+1=8 events vs. 0.95+0.58 = 1.53 ($\pm$ 0.29) background in the window $M_e = 200\pm12.5$ GeV, $y_e > 0.4$ [5], and ZEUS has 4+1=5 events vs. 0.91 + 0.61 = 1.51 ($\pm$ 0.13) background in the window $x > 0.55, y_{DA} > 0.25$ [6], and they quote respectively probabilities of 1% (1.9%) for such a fluctuation in any window (in the quoted window). These probabilities are not negligible, but one should be impressed if one could be convinced that the two experiments were seeing compatible anomalies – more on this shortly. It is important to note that, although the 1997 data do not repeat the magnitudes of the anomalies reported in the earlier data, this year's data are not below the rates expected in the Standard Model.

In addition, the HERA collaborations have reported apparent excesses in their charged-current data: H1 has found 14 events vs. $Q^2 > 10^4$ GeV^2, compared with 8.33 ± 3.10 expected [5], and ZEUS has 15 events vs 9.4 ± 2.5 ± 1.6 expected [6]. These excesses are intriguing, but not as significant as those reported in the neutral-current samples.

For the moment, we conclude that statistical fluctuations cannot be excluded, even if they are formally unlikely.

3 Are the H1 and ZEUS Anomalies Compatible?

The reported excesses are apparently in different kinematic regions: H1 discusses the region $M_e = 200 \pm 12.5$ GeV (corresponding naïvely to $x_e = 0.45 \pm 0.06$) and $y_e > 0.4$ [5], whereas ZEUS discusses $x_{D_A} > 0.55$ and $y_{D_A} > 0.25$ [5]. However, the two collaborations use different measurement methods – H1 using electron measurements and ZEUS the double-angle method – with different statistical and systematic errors. In particular, the methods are affected differently by initial-state photon radiation [7, 8, 9], the possible effects of gluon radiation have not been evaluated completely, and detector effects such as the energy calibrations cannot be dismissed entirely.

The most complete analysis, using all the available measured quantities, indicates that the experiments' apparent mass difference $\Delta m = 26 \pm 10$ GeV should be reduced to 17 ± 7 GeV, on the basis of which it seems that the decay of a single narrow resonance is unlikely [10, 5].

In my view, it is premature to exclude a resonance interpretation. I personally would like to see a more complete treatment of gluon radiation using the full perturbative-QCD matrix element now available [11], the Standard Model backgrounds are non-negligible, and one should perform a global fit to all the large-x data of both collaborations. Nevertheless stocks in models with production of a single leptoquark or $\not{R}$-violating squark decaying into eq final states are surely going down.

4 Modified Parton Distributions?

Forgetting for the moment about possible structures in the x distributions, could the global excess at large x and Q^2 be accommodated by modifications of the conventional parton distributions used to predict Standard Model rates? One suggestion was a possible "lump" in the valence quark distribution at $x \sim 1$ at low Q^2, but evolution of this "lump" does not have a large effect in the HERA range of x and Q^2 [12]. An alternative suggestion was intrinsic charm [13], but this is tightly constrained by EMC data on $\mu N \to \mu \bar{c} c X$, and unable to modify significantly the Standard Model predictions if $c(x) = \bar{c}(x)$ [14]. However, even if one makes $\bar{c}(x)$ harder than $c(x)$, the effect on the HERA neutral-current rate is $\lesssim 10\%$ in the range $M = 200 \pm 10$ GeV, though effects up to 50% might be possible in the charged-current rate, and this type of model might provide an alternative interpretation of the CDF high-E_T jet excess [15].

My conclusion is that the proposed modifications of the parton distributions cannot make a significant contribution to explaining the neutral-current data, but they might make a more important contribution to resolving the possible charged-current excess. This is in any case less significant, and I would recommend waiting to see how it develops.

5 New Contact Interactions?

Those relevant to the HERA data may be written in the general form

$$\frac{\eta}{m^2}(\bar{e}\gamma_\mu e)_{L,R}(\bar{q}\gamma^\mu q)_{L,R} \; : \; \eta = \pm 1, \; q = u, d \tag{1}$$

giving 16 possibilities, which are tightly constrained by experiments on parity violation in atoms (which, however, allow parity-conserving combinations: $\eta^{eq}_{LR} = \eta^{eq}_{RL}$ and other, more general, possibilities), by LEP2 measurements of $e^+e^- \to \bar{q}q$, and by CDF measurements of the Drell-Yan cross section. A recent global analysis [16] of atomic-physics parity-violation experiments, lower-energy deep-inelastic eN experiments, LEP, Drell-Yan and the Q^2 distribution of the first instalment of large-Q^2 HERA data found fits with $\chi^2 = 176.4$ for 164 degrees of freedom with no additional contact interactions: $\chi^2 = 187.1$ for 163 d.o.f. with $\eta^{eu}_{LR} = \eta^{eu}_{RL}$, which is *worse*, and $\chi^2 = 167.2$ for 156 d.o.f. when all possible contact interactions are allowed, which "represents no real improvement over the Standard Model". The real killer for the contact-interaction fits is the constraint imposed by the CDF Drell-Yan data. Needless to say, no contact interaction would give a resonance peak in the eq invariant mass! Moreover, the constraints of weak universality make it difficult to arrange a large charged-current signal.

Therefore, I personally conclude that there is no motivation currently to pursue further the possibility of contact interactions.

6 Leptoquark?

The tree-level cross section for producing a leptoquark or R-violating squark in ep collisions is $\sigma = (\pi/4s)\ \lambda^2\ F_J$, where $F_J = 1(2)$ for scalar(vector) leptoquarks. One-loop QCD corrections to this cross section have been calculated, as well as a number of final-state kinematic distributions [11]. Using the known parton distributions at $x \sim 0.5$, it was found that, in order to explain the first batch of HERA data, one needed $\lambda \simeq 0.04/\sqrt{B(e^+q)}$ if production was via e^+d collisions, and $\lambda \gtrsim 0.3/\sqrt{B(e^+q)}$ for production off some q or $\bar{q}$ in the sea, where $B(e^+q)$ is the branching ratio for decay into the observed e^+-jet final states [7].

The dominant production mechanism at the FNAL Tevatron collider is pair production via QCD, which is independent of λ, and for which the one-loop QCD corrections have also been calculated [17]. Using these, CDF [18] and D0 [19] together imply $m_{LQ} \gtrsim 240$ GeV for $B(e^+q) = 1$ for a scalar leptoquark, and a far stronger (though somewhat model-dependent) limit for a vector leptoquark, which option we shall not discuss further. Could the leptoquark have $B(e^+q) < 1$? A large $B(\bar{\nu}q)$ is difficult to reconcile with charged-current universality [20, 21]. Also, if $B(e^+q) > B(\bar{\nu}q)$ as suggested by the possible relative magnitudes of the neutral- and charged-current anomalies (19 events with 8 background vs 7 events with 3 background for $M > 175$ GeV, $Q^2 > 15{,}000$ GeV^2) [5, 6], the FNAL data together imply $m_{2Q} > 220$ GeV, beyond the range suggested by the HERA data [22]. Moreover, a significant $B(\mu q)$ is excluded by upper limits on anomalous muon capture on nuclei: $\mu N \rightarrow eN$ [7]. The only remaining unexcluded decay mode into Standard Model particles is τq, and it would be worthwhile looking for this decay mode at FNAL and/or HERA, so as to pin down or exclude finally the leptoquark interpretation. Failing τq, one needs some decay mode involving particles beyond the Standard Model, and these are provided by the final interpretation we discuss.

7 Squarks with *R* Violation?

The supermultiplet content and symmetries of the minimal supersymmetric extension of the Standard Model (MSSM) admit extra couplings beyond those responsible for the quark and lepton masses: $\lambda_{ijk} L_i L_j E^c_k + \lambda'_{ijk} L_i Q_j D^c_k + \lambda''_{ijk} U^c_i D^c_j D^c_k$. Any one of these couplings would violate $R = (-1)^{3B+L+2S}$. They would provide dilepton, leptoquark and dijet signatures, respectively, and a combination of the λ' and λ'' couplings would cause baryon decay. The apparent HERA excess could be due to a λ' coupling: $\lambda'_{ijk} L_i Q_j D^c_k \Rightarrow \lambda'_{ijk} e^+_R d_{R_k} \bar{\tilde{u}}_{L_j}$, $\lambda'_{ijk} e^+_R \bar{u}_L \tilde{d}_{R_k}$ via the specific mechanisms $e^+ d \rightarrow \tilde{u}_L, \tilde{c}_L, \tilde{t}$ with $\lambda' \sim 0.04$, or $e^+ s(b) \rightarrow \tilde{u}_L, \tilde{c}_L, \tilde{t}$ and conceivably $e^+ \bar{u}(\bar{c}) \rightarrow \tilde{d}, \tilde{s}, \tilde{b}$ with $\lambda' \sim 0.3$ [23, 7, 24, 25, 26]. The absence of neutrinoless $\beta\beta$ decay [27] imposes

$$|\lambda'_{111}| < 7 \times 10^{-3} \left(\frac{m_{\tilde{q}}}{200 \text{ GeV}}\right)^2 \left(\frac{m_{\tilde{g}}}{1 \text{ TeV}}\right)^2 \tag{2}$$

which rules out $e^+ d \rightarrow \tilde{u}_L$ (and $e^+ \bar{u} \rightarrow \tilde{d}$), but allows $e^+ d \rightarrow \tilde{c}_L, \tilde{t}$. The upper limit on $K^+ \rightarrow \pi^+ \bar{\nu}\nu$ decay [28] imposes

$$|\lambda'_{ijk}| \lesssim 0.02 \left(\frac{m_{\tilde{d}_{R_k}}}{200 \text{ GeV}}\right) \tag{3}$$

up to mixing and cancellation factors, which barely allows the $e^+ d \rightarrow \tilde{c}_L$ mechanism. Many constraints on large λ' couplings, such as charged-current universality, atomic-physics parity violation and neutrino masses exclude almost all sea production mechanisms. The only mechanisms that survive this initial selection are the $e^+d \to \tilde{c}_L$ (down-scharm), $e^+d \to \tilde{t}$ (down-stop) and $e^+s \to \tilde{t}$ (strange-stop) interpretations [23, 7, 24, 25, 26].

The strange-stop interpretation is quite severely constrained by precision measurements at LEP1, notably the ρ parameter, which requires non-trivial $\tilde{t}_L$–$\tilde{t}_R$ mixing, and limits on the violation of universality in $Z^0 \to e^+e^-, \mu^+\mu^-$ decay, which impose $|\lambda'_{13j}| \lesssim 0.6$, so that $B(\tilde{t} \to e^+q)$ cannot be very small [29]. On the other hand, we recall that the FNAL upper limits on leptoquark production require $B(e^+q) \gtrsim 0.5$ for $m \sim 200$ GeV [22].

Competitive branching ratios for the R-violating decay $\tilde{c}_L \to e^+d_R$ and the R-conserving decay $\tilde{c} \to c_L\chi$ (followed by R-violating decay of the lightest neutralino χ) are possible in generic domains of parameter space, as a result of a cancellation in the R-conserving coupling of the lightest neutralino [7]. This is much more difficult to arrange in the down-stop interpretation, where $\tilde{t} \to e^+d$ either dominates (if $m_{\tilde{t}} < m_t + m_\chi$) or is dominated by $\tilde{t} \to t\chi$ (if $m_{\tilde{t}} > m_t + m_\chi$). The strange-stop scenario is intermediate: after taking into account the LEP constraints, there is very little parameter space where $B(\tilde{t} \to e^+s) \sim B(\tilde{t} \to t\chi)$ if $m_{\tilde{t}} = 200$ GeV, but considerably more if $m_{\tilde{t}} \sim 220$ GeV [29].

At the moment none of the three supersymmetric interpretations is very healthy, but none can yet be ruled out.

8 Tests to Discriminate Between Models

More statistics should soon clarify the Q^2 and x distributions, whether there is really an excess, telling us whether it is peaked in any particular mass bin, and whether there is any unusual pattern of gluon radiation.

We should also know soon whether there is really an excess of charged-current events, which would be difficult to understand in most scenarios invoking new contact interactions, leptoquarks, or even R-violating squarks (except in some interesting corners of parameter space, where cascade squark decays imitate the simple charged-current topology [21, 30]). In 1998, it is planned to run HERA in e^-p mode: most scenarios predict the absence of a signal, though one could appear in the $e^+s \to \tilde{t}$ scenario.

Turning to other signatures, in the $e^+d \to \tilde{c}_L$ scenario one predicts observable $\tilde{c} \to c\chi$ decays followed by $\chi \to \bar{q}qL, \bar{q}q\nu$. There should be equal numbers of $\ell^\pm$ + jets final states, and there could be $\mu^\pm$ and/or $\tau^\pm$ + jets as well as $e^\pm$ + jets. These decay modes should also be observable at the FNAL Tevatron collider [7]. A first search for them has been made [18], but it does not yet have the required sensitivity.

Effects could show up in $e^+e^- \to \bar{q}q$ at LEP2, either via contact interactions or via virtual leptoquark or squark exchange [31, 32], which might be detectable in models invoking production from the sea. Identification of the final-state quark flavours would improve the sensitivity, e.g., to $\tilde{t}$ exchange in $e^+e^- \to \bar{s}s$.

One exciting possibility is the production of a direct-channel $\tilde{\nu}$ resonance if there is a $\lambda_{ijk}L_iL_jE^c_k$ coupling [33]. The LEP experiments have already provided some limits on $|\lambda|$ as a function of $m_{\tilde{\nu}}$ [34]. Unfortunately, sensitivity is rapidly lost if $E_{cm} < m_{\tilde{\nu}}$. It would be a shame if LEP missed out on discovering a squark or a sneutrino because it was not pushed to the maximum possible centre-of-mass energy, so let us all hope that a way can be found to operate LEP at its design energy of 200 GeV, with an integrated luminosity sufficient to exploit fully its capabilities.

References

[1] H1 collaboration, C. Adloff et al., Zeitschrift für Physik C74 (1997), 191-206.
[2] ZEUS collaboration, J. Breitweg et al., Zeitschrift für Physik C74 (1997), 207-220.
[3] J. Blumlein, Parallel Session Talk 1109 at this conference, Zeitschrift für Physik C74 (1997), 605-609 and hep-ph/9706362;
G. Altarelli, hep-ph/9708437, hep-ph/9710434;
R. Rückl and H. Spiesberger, hep-ph/9710327, hep-ph/9711352.
[4] E. Elsen, Plenary Talk 26 at this conference.
[5] U. Bassler, Parallel Session Talk 31302 at this conference.
[6] D. Acosta, Parallel Session Talk 31301 at this conference.
[7] G. Altarelli, J. Ellis, G.F. Giudice, S. Lola and M. Mangano, Nuclear Physics B506 (1997),3-28.
[8] G. Wolf, hep-ex/9704006.
[9] M. Drees, Physics Letters B403 (1997), 353-356.
[10] U. Bassler and G. Bernardi, Zeitschrift für Physik C76 (1997), 223-230.
[11] M. Heyssler and W.J. Stirling, Physics Letters B407 (1997), 259-267;
T. Plehn, H. Spiesberger, M. Spira and P. Zerwas, Zeitschrift für Physik C74 (1997), 611-614.
[12] S. Kuhlmann, H.L. Lai and W.K. Tung, Physics Letters B409 (1997), 271-276.
[13] J.F. Gunion and R. Vogt, hep-ph/9706252.
[14] K. Babu, C. Kolda and J. March-Russell, Physics Letters B408 (1997), 268-274.
[15] W. Melnitchouk and A. Thomas, hep-ph/9707387.
[16] V. Barger, K. Cheung, K. Hagiwara and D. Zeppenfeld, hep-ph/9707412, and references therein.
[17] M. Kramer, T. Plehn, M. Spira and P. Zerwas, Physical Review Letters 79 (1997), 341-344.
[18] X. Wu, Parallel Session Talk 31303 at this conference.
[19] B. Klima, Parallel Session Talk 31304 at this conference.

[20] K. Babu, C. Kolda and J. March-Russell, Physics Letters B408 (1997),261-267.
[21] G. Altarelli, G.F. Giuduice and M. Mangano, Nuclear Physics B506 (1997), 29-47.
[22] P. Janot, Plenary Talk 17 at this conference.
[23] D. Choudhury and S. Raychaudhuri, Physics Letters B401 (1997), 54-61.
[24] K. Babu, C. Kolda, J. March-Russell and F. Wilczek, Physics Letters B402 (1997), 367-373.
[25] H. Dreiner and P. Morawitz, Nuclear Physics B503 (1997), 55-78, B428 (1994), 31-60.
[26] T. Kon and T. Kobayashi, Physics Letters B409 (1997), 265-270;
T. Kon, T. Kobayashi and S. Kitamura, Physics Letters B333 (1994), 263-270.
[27] M. Hirsch, H.V. Klapdor-Kleingrothaus and S. Kovalenko, Physical review Letters 75 (1995), 17-20.
[28] S. Adler et al., Physical Review Letters 79 (1997), 2204-2207.
[29] J. Ellis, S. Lola and K. Sridhar, Physics Letters B408 (1997), 252-260.
[30] M. Carena, D. Choudhury, S. Raychaudhuri and C. Wagner, hep-ph/9707458.
[31] J. Kalinowski, R. Rückl, H. Spiesberger and P. Zerwas, Zeitschrift für Physik C74 (1997), 595-603.
[32] I. Fleck, Parallel Session Talk 31306 at this conference;
C.P. Ward, Parallel Session Talk 712 at this conference.
[33] J. Kalinowski, R. Rückl, H. Spiesberger and P. Zerwas, Physics Letters B406 (1997), 314-320, and references therein.
[34] L3 collaboration, M. Acciarri et al., CERN preprint PPE/97-099 (1997).

Druck:	Strauss Offsetdruck, Mörlenbach
Verarbeitung:	Schäffer, Grünstadt